ISBN 978-1-5277-9493-1
PIBN 10898360

1 MONTH OF
FREE
READING

at
www.ForgottenBooks.com

By purchasing this book you are
eligible for one month membership to
ForgottenBooks.com, giving you
unlimited access to our entire
collection of over 1,000,000 titles via
our web site and mobile apps.

To claim your free month visit:
www.forgottenbooks.com/free898360

35 CENTS

RADIO BROADCAST

F. J. Edgars

HOW TO ELECTRIFY YOUR PHONOGRAPH

A Survey of the Preferences of Listeners—Facts About Modern Receiving Tubes—
Problems of Series Filament Connection—The March of Radio—An Improved Socket
Power Device—Facts About A-Potential Supplies—A New Short-Wave Receiver

Doubleday, Page & Company, Garden City, New York

Q · R · S

TRADE MARK
REGISTERED

Gaseous
Rectifier Tubes

The Sensation of Radio
Development for your
Radio Battery Eliminator

When your B Eliminator needs a new tube ask for a Q · R · S Rectifier Tube with a guarantee of a full year's service or a refund for every month less than a year it fails to function.

In conjunction with a filter circuit for which we will furnish diagram and instructions free, the Q · R · S 400 Milliampere Tube will eliminate all A, B, and C Batteries and run your set from your house current supply.

65 Milliampere and 85 Milliampere type for B Battery Eliminators,

Price, $5⁰⁰ each

400 Milliampere type for A, B, and C Battery Eliminators

$7⁰⁰

For over a quarter of a century we have been manufacturing Quality merchandise—Quality always costs a little more but that's why

Q · R · S Radio Tubes
Are Better

For full details ask your dealer or if he is unable to serve you yet, address

The Q · R · S Music Company

Mfrs. of Q · R · S Player Rolls used in over a Million Homes

New York	Chicago	San Francisco
135th St. & Walnut Ave.	306 S. Wabash Ave.	306 7th St.

RADIO BROADCAST. May, 1927. Published monthly. Vol. XI. No. 1. Published at Garden City, N. Y. Subscription price $4.00 a year. Entered at the post office at

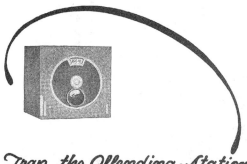

Trap the Offending Station!

The MADISON-MOORE

WAVE TRAP

enables the user of a loop set in practically every locality to get within 10 to 20 kilocycles nearer the offending station. If, with your present receiver, you have trouble in eliminating powerful near-by stations, the Madison-Moore Wave Trap will materally assist the selectivity.

With this latest achievement in radio engineering, even small sets give satisfactory performance, making them more like loop sets.

The Madison-Moore Wave Trap is made of the finest materials, precision tested, in our laboratory. A model adapted for use with any set. It is encased in a No. 12 aluminum solid box; it tunes so sharply that a Vernier dial must be used. Simple to operate—comes complete with Vernier dial and instructions for connecting and operation. Priced within the reach of everyone.

The MADISON-MOORE WAVE TRAP is made by the
manufacturers of the famous
ONE●SPOT TRANSFORMER

〖 *Try your dealer first. If he can't supply you, write us.* 〗

MADISON-MOORE RADIO CORPORATION
2524 K Federal Boulevard
Denver, Colorado · U. S. A.

MADISON~MOORE
The Finest RADIO APPARATUS *in the World!*

ALUMINUM BOX SHIELDS

Made of "Alcoa Aluminum"

Consist of:
Top, Bottom, Sides
4 Extruded Corner Posts
8 Aluminum Screws

MEETING the highest radio standards—shipped to you in the most convenient knocked-down form for easy assembly. These Box Shields are made of heavy aluminum (.080"—No. 12 B. & S.) and are supplied 5" x 9" x 6", which will cover most requirements. If the size does not meet your exact needs, change it—Aluminum is easy to work.

Manufacturers can obtain these shields made to their exact specifications or they can secure the necessary corner-post moulding and sheet to manufacture under their own supervision.

Those who use Aluminum have ample proof of its advantages. Insist on "Alcoa Aluminum," ask your dealer or write us.

"ALCOA ALUMINUM" *is furnished to manufacturers in the following forms:*

> Sheet: for shields, chassis, variable condensers, cabinets.
> Panels finished in walnut and mahogany.
> Die and Sand Castings.
> Screw Machine Products.
> Foil for fixed condensers.
> High Purity Rods for rectifiers.
> Stamping, rod, wire, rivets.

ALUMINUM IN EVERY COMMERCIAL FORM

ALUMINUM COMPANY of AMERICA
2320 OLIVER BUILDING, PITTSBURGH, PA.

Output of Battery Eliminators

THE ordinary type of low resistance voltmeter cannot be used to measure the output of battery eliminators because most eliminators cannot deliver sufficient current to operate the voltmeter and still maintain their voltage. For this service Weston now offers a special voltmeter with an exceptionally high self-contained resistance of 1,000 ohms per volt—Model 489 Battery Eliminator Voltmeter. This new instrument, because of its high resistance, requires a current of only one milliampere to produce full scale deflection. It is made in two double range combinations of 200/8 and 250/50 volts. The latter range can be supplied with an external multiplier to increase it to 500 volts. Ask your dealer or address—

WESTON ELECTRICAL
INSTRUMENT CORPORATION
179 Weston Avenue, Newark, N. J.

RADIO BROADCAST

Willis K. Wing, Editor

MAY, 1927

Keith Henney
Director of the Laboratory

John B. Brennan
Technical Editor

Edgar H. Felix, Contributing Editor

Vol. XI, No. 1

CONTENTS

AMONG OTHER THINGS. . .

WITH this issue, RADIO BROADCAST starts its sixth year of publication. The five years just concluded have seen many changes in the radio industry and among radio experimenters and fans. American radio magazines of to-day are quite different from those of 1922; RADIO BROADCAST has changed considerably, both in physical appearance and in type of contents, since those days. We believe the changes are improvements, for certainly our files are filled with countless letters of approval. The aim of Doubleday, Page & Company has been to publish the highest class radio magazine possible. This we have attempted to do, first, in setting the standard of physical appearance of the magazine where it now is, and secondly, to spare no pains or expense in making our technical material as completely accurate and helpful as possible. Two large laboratories are maintained in Garden City where practical and theoretical experiments are constantly under way; and our advertising pages are carefully supervised.

MANY editorial features in RADIO BROADCAST have given the magazine a unique position in its field. The "March of Radio" provides an editorial comment and suggestion about all branches of radio. "The Listeners' Point of View" stands alone as a national review of broadcasting, particular and general. "As the Broadcaster Sees It" has turned out to be a unique department (prepared, incidentally, by one of the ablest broadcast engineers in the country) where engineer, program director, listener, and general reader alike, may meet. The review of current radio periodicals, the Laboratory Data Sheets, the listing of informative manufacturers' booklets, all furnish valuable information for our readers. Our constructional articles are chosen carefully for accuracy and greatest help and interest for the reader. We are at work on an editorial schedule now which holds much for everyone interested in radio in all its branches. Unfortunately there is insufficient space here to outline that schedule, but we prefer to let each issue of RADIO BROADCAST speak for itself in that connection.

PRINTERS' INK, in its tabulation of advertising lineage for March magazines, shows that RADIO BROADCAST led the field with a total of 20,621 lines, followed by Radio News with 18,930 lines, Popular Radio with 14,872, Radio with 12,770, and Radio Age with 4395.

IN THE June RADIO BROADCAST, a fine story for the home experimenter is scheduled, describing the construction and use of a modulated oscillator. Other articles deal with short waves, the problems of series filament connection for 201-A type tubes, how to use new apparatus, technical problems for broadcast operators and others, and many other features of unusual interest.

Willis K. Wing.

Doubleday, Page & Co.
MAGAZINES

COUNTRY LIFE
WORLD'S WORK
GARDEN & HOME BUILDER
RADIO BROADCAST
SHORT STORIES
EDUCATIONAL REVIEW
LE PETIT JOURNAL
EL ECO
FRONTIER STORIES
WEST

Doubleday, Page & Co.
BOOK SHOPS
(Books of all Publishers)

NEW YORK
{ LORD & TAYLOR
  PENNSYLVANIA TERMINAL (2 Shops)
  GRAND CENTRAL TERMINAL
  48 WALL ST. and 526 LEXINGTON AVE.
  848 MADISON AVE. and 166 WEST 32ND ST.
ST LOUIS: 223 N. 8TH ST. and 4914 MARYLAND AVE.
KANSAS CITY: 920 GRAND AVE. and 206 W. 47TH ST.
CLEVELAND: HIGBEE CO
SPRINGFIELD, MASS.: MEEKINS, PACKARD & WHEAT

Doubleday, Page & Co.
OFFICES

GARDEN CITY, N. Y.
NEW YORK: 285 MADISON AVE.
BOSTON: PARK SQUARE BUILDING
CHICAGO: PEOPLES GAS BUILDING
SANTA BARBARA, CAL.
LONDON: WM. HEINEMANN LTD.
TORONTO: OXFORD UNIVERSITY PRESS

Doubleday, Page & Co.
OFFICERS

F. N. DOUBLEDAY, President
NELSON DOUBLEDAY, Vice-President
S. A. EVERITT, Vice-President
RUSSELL DOUBLEDAY, Secretary
JOHN J. HESSIAN, Treasurer
L. A. COMSTOCK, Asst. Secretary
L. J. McNAUGHTON, Asst. Treasurer

DOUBLEDAY, PAGE & COMPANY, Garden City, New York

Copyright, 1927, in the United States, Newfoundland, Great Britain, Canada, and other countries by Doubleday, Page & Company. All rights reserved.
TERMS: $4.00 a year; single copies 35 cents.

THE TOWERS OF STATION WOW, AT OMAHA, NEBRASKA

Formerly known as WOAW, this station has enjoyed a wide national popularity. In April, this station celebrates its fourth anniversary. The illustration shows the hoisting of a large section of plate glass, which now forms the front of a unique studio atop this 19-story building. The studio is arranged as a stage, properly insulated against sound inside, but with a front panel of glass to allow those seated in a small auditorium to watch and to hear the broadcasting as it occurs

RADIO BROADCAST

VOLUME XI   NUMBER 1

MAY, 1927

With MacMillan to the Arctic

Tales from the Pen of the Sachem's Radio Operator—Not so Stupid, These Eskimo Flappers—Some Notes on the Aurora, Mirages, and Radio—Abe Puts One Over!

By AUSTIN G. COOLEY

TAKING up the threads of our story from where we dropped them last month, we on the *Sachem* and our friends on the *Bowdoin* find ourselves at Godhaven, Disko Island, the northernmost port of call of the expedition.

Let us deviate a little, though, to tell of the inhabitants of these arctic regions, for an account of an expedition into the land of the Eskimos is not complete without some mention of these most interesting people. Although the full-blooded Eskimos are few and far between, their habits and methods of living are preserved by those who possess considerable Danish blood. They are all wards of the Danish Government and are well cared for, and every effort is made to preserve their native customs and practices.

The Eskimo flappers are not so stupid; in fact, they have many of the same instincts that are common among the American beauties. In the matter of short skirts, they are years ahead of the best of our stenographers. The complete absence of skirts is due in a large measure to the Danish rulings that prevent them from changing from their native custom of sealskin pants and boots. They take great pride in making their boots the most attractive in the settlement. Generally they are dyed with a brilliant white or bright red. The girls all make large boat collars for themselves of their own design, with beads imported from Denmark. Ordinarily it is very difficult to obtain these collars but this year a few were purchased for thirty kroner (Danish money). Many of the girls, if they have considerable Danish blood in them, are quite pretty; otherwise they look too much like Eskimos.

A few of the Eskimos live in small frame houses or shacks while the others have their homes built of sod. These houses, or igloos, are built very scientifically with a selected sod that is a good heat insulator. The walls are built very thick and the doors all have an angle to the door. Each igloo has a bed consisting of a raised platform large enough to accommodate the entire family. The Greenland Eskimos, and the igloos that we saw, were very clean.

Many of the expedition found it great sport trading with the Eskimos for souvenirs, such as harpoons, bird spears, model kyaks, etc. In Sukkertoppen, Mrs. Metcalf wanted to get a model of a woman's skin boat. She asked some of the Eskimos in their language if they had any "coumiaks."

CAPTAIN "MAC" AND ABE

The latter was the Eskimo interpreter of the Expedition. He returned with the others to America when the *Sachem* and *Bowdoin* came back and, among many other things, saw a train for the first time in his life

Considerable laughter among the Eskimos immediately resulted. Abe Bromfield, Commander MacMillan's interpreter, had to come to her assistance. It was an "umiak" she wanted. "Coumiak" is the Eskimo word for "lice"!

For exchange, the Eskimos generally wanted tobacco and calico, although some asked for kroner which could be used at the stores run in the large settlements by the Danish government. The Danish money used in Greenland bears pictures of animals, such as polar bears, whales, seals, and ducks, so that the value may be recognized by the Eskimos.

During certain seasons, the Eskimos work in fisheries operated by the Government, and at other times they are busy sealing and hunting. The womens' work includes the chewing of seal skins so as to soften them, and the making of boots and clothing.

Anyone espied walking ashore with a camera was sure to have a large following, for the Eskimo girls are as anxious to have their pictures taken as are some American girls to appear in the movies. When preparing for a shore trip with the cameras, I always considered a supply of gum and cigarettes as important as the films. After taking pictures, these would be passed around to the eager natives, who were especially anxious for cigarettes; even the babies in their mother's arms made manifest a desire to smoke. They seemed to enjoy smoking as much as their grandmothers.

The Eskimo's Santa Claus is more of a fact than a myth. His phase angle displacement is 180 degrees from ours! He comes from the south in the summer time in a white schooner, and brings toys and candy for the bad little boys and girls as

well as the good ones. He is known to them
as 'Captain Mac." Captain Mac's own
generosity is always backed up by various
candy manufacturers who see that the expedition does not leave American shores
without a plentiful supply of all kinds of
candy for the natives. The toys distributed
by Captain Mac are generally of the nickel
and dime variety, but some of the most
deserving Eskimo men receive presents
of excellent jack knives.

Some of the unfortunate Eskimos receive presents of lemons and a pleasant
visit by the expedition's doctor—Doctor
Thomas of Chicago. Doctor Thomas treated a number of cases among the Labrador
Eskimos and in one instance performed an
operation.

At practically all of the settlements at
which we stopped, Captain Mac put on
movie shows for the natives. The attendance was always one hundred per cent.
"Robbie," the mate of the *Bowdoin*, was
skilled in packing whole settlements into
a small room so as to give each Eskimo a
good View of the screen.

WE START OUR TRIP HOME

COMMANDER MacMillan planned
on running south from Godhaven,
Greenland, then across to Holsteinborg, then across
to Baffin Land, but the problem of re-
plenishing our fuel supply required some
changes in the schedule. We went south to
a whaling station at Edgesminde where
ten barrels of fuel oil were obtained, but
this was not enough to take us home.
Using the engines seemed more desirable
than the sails, so we went further on, to
Sukkertoppen, where a larger supply could
be had.

This southbound trip included a few
days' stop at Holsteinborg, formerly a center for American halibut fisheries. In the
cemetery were found a number of graves of
Gloucester fishermen. All foreign fishermen
are now barred from the coast of Greenland
and the Danish Government is operating
the halibut fishing industry at Holsteinborg with great success. During our stay
at this port, we enjoyed many meals of halibut. It was delivered to the ship at a price
corresponding to one cent a pound.

Except for the fog at sea, the weather we
experienced in Greenland was of the best.
During the day, the temperature generally
ran around 70 degrees. This weather, on
the middle western Greenland coast, is
due to a warm ocean current which prevents the winters from being any more
severe than they are in Maine.

Our call at Sukkertoppen for fuel lasted
only a few hours. Before leaving, plans were
made to set a course for Cape Murchison,
Bevoort Island. If the ice was not too
heavy, we planned to run up Robinson
Sound, between Bevoort Island and Baffin
Island, and anchor in a harbor that was
dotted in on the chart. The dots meant:
"Harbor about here."

The radio conditions during our last
few days in Greenland were very poor.
Even the powerful commercial stations

faded out completely at times. While going
across to Baffin Land it was hardly possible
to get through any amount of traffic to
the States because of complete fading.
Radio 1 AAY (Kenneth M. Gold, Holyoke,
Massachusetts) was worked a couple of
times but it was only possible to get one
complete message of about fifty words
from him. Other stations were worked but
communications only lasted a few minutes
before they faded out completely.

Davis Straits had received considerable
publicity as being a "blind spot" because
it so happened that short-wave radio communication could not be established from
this location on the previous year. On the
recent outward trip, Paul j. McGee, the
operator at WNP, the *Bowdoin*, succeeded
in transmitting to the States a 200-word
message when going north through the
Straits. At the same time I observed that
the signals from the States were coming
through with good intensity. Why the conditions on our return trip were so poor
seemed to have some connection with the
aurora because strong displays of the
northern lights were visible as we left
Sukkertoppen at 2 A.M. on Friday, August
the thirteenth.

On the return trip, both McGee and I
seemed to have considerable trouble with
swinging waves. When we reached the ice
fields off the Labrador coast we were
hardly able to get through any communications although, after considerable effort, we exchanged position messages. The
ice was fairly heavy but we considered it
not too bad to prevent us from making
Cape Murchison. About 8 P.M. on the
fourteenth, we sighted the cape through
the fog about a half mile off.

The weather kept getting thicker and
colder and a strong gale came up that increased the danger of navigating these un-
charted waters. The harbor where we
were to meet the *Bowdoin* was full of ice.
The storm became more severe and the
night was very dark, so it was considered
too much of a risk to try to anchor the
ship even though we could find a harbor.
A broken reverse gear added to the hazards.
The entire crew stood watch that night.
We all wore all the clothes we had
aboard under our oilskins, together with
hip boots so as to offer maximum resistance to the cold rain, sleet, and snow that
was driven hard by the gale.

Radio communication was established
with radio 1 AAV, and as I was asking him
to give our position to the *Bowdoin* if he
could, all signals faded out completely.
At frequent intervals I called WNP, the
Bowdoin, but I received no answer.

As the day started to break, two har-
bors were located. One was well sheltered
but we could not get in because the entrance was blocked by a few large icebergs.
The other was full of rocks and bergs but,
with the skillful handling of the ship, a
reasonably safe place was located and then
the hook dropped. Dropping the hook was
not so easy. The gale was still strong and
the water twelve fathoms deep, and we had

to let out about forty fathoms of ancho-
chain. The last twenty fathoms came out
of the chain locker in installments of one
and two feet at a time while it was being
untangled and untwisted.

Communication was established with
WNP that afternoon on a schedule. His
wave was swinging badly but I was able
to learn his position and that he was
headed for Hall Island, near the southern
end of Baffin Island. Apparently the *Bowdoin* got the worst of the storm as they
stayed at sea and rode it out, for Commander MacMillan considered the ice too
heavy to attempt to run into the Sound.
The danger of the ice was that, if the wind
changed, it might pack in on us and push
the vessel on to the beach.

The following morning we pulled in our
40 fathoms of anchor chain and headed for
Hall Island.

While passing through the ice, we saw
many seals, and after clearing it, walruses
were plentiful. After quite an exciting time
with artillery, harpoons, and killing irons,
we managed to get one of the latter aboard
that weighed about 1500 pounds.

Upon arriving at Hall Island that evening, a radio message was received from the
Commander advising us that he had gone
on and that we should meet him at Saglek
Bay, Labrador. At four o'clock the following morning, we started out on this long
trip across the Hudson Straits. Fair weather
permitted us to make Saglek Bay by six
P.M. the following day.

OBSERVATIONS ON RADIO FADING

OUR activities at Saglek Bay deserve
considerable mention. The *Bowdoin*
had gone up far into one arm of the bay so
that the scientists could make a collection
of some kind of field mice that exist there.
Some of the other boys found sport hunting caribou but they did not enjoy the
success the scientists had.

McGee planned on setting up a receiving set ashore that night so that Kennett
Rawson, a member of the *Bowdoin* crew,
could hear his father who was scheduled to
broadcast from WJAZ, Chicago. An elaborate
installation was made with the help of
"Ken," who was so strong he stretched the
antenna wire enough to break it. The heavy
equipment had been carried about two
miles up into the hills to an "ideal" location. All this work did not even result in
a squeal from WJAZ. This was most disappointing as reception in Greenland from
this station had been quite successful at
times.

During the time that McGee was trying
to receive WJAZ, I was trying to receive anyone I could. At times signals seemed fairly
good, and then they would fade out completely. Someone informed me that the
aurora was quite active, so I went on deck to
investigate. Never before had I ever seen
such a violent and brilliant display of the
northern lights.

At times we were completely surrounded
by the bright blue bands of aurora, and
streamers from all around the horizon

would shoot up to the zenith, making a complete umbrella of aurora. In places, the bands were fringed with a dark red.

Until making this trip, I felt that no relation existed between the aurora and radio conditions, as I had made rough observations during the part of one winter in Alaska and was able to detect no connections between the two on wavelengths of six hundred meters or higher (500 kc. or less). No one seemed to really know whether any relation existed although some were quite certain that it did and others were more positive that it did not.

An excellent opportunity was offered me to gather some data that might throw a little light on the subject, so I started work, but in a rather crude fashion. My notes consisted of brief descriptions of the aurora together with a log of the stations that were coming through and remarks as to intensity. For three hours I kept running up and down the companionway taking notes on the aurora and then listening to the radio. I was thoroughly convinced after these notes had been gone over that a definite relation did exist between the aurora and radio conditions.

The following night we were at sea, bound for Nain. The aurora appeared in thin bands sweeping across the sky from the northeast. I took data as on the night before. Mr. Warren was at the wheel at the time and noticed compass variations of

five or six degrees as the bands of aurora passed over us.

We were following the *Bowdoin* along the half-charted coast of northern Labrador when our engines stopped. The chief engineer found the fuel pipe lines full of water. A little over an hour was required to take down all the lines and drain about a barrel of water out of the fuel tank that had been filled at Sukkertoppen. The *Bowdoin* was a mile or two ahead of us when we stopped and her lights soon dropped out of sight altogether. There was a light breeze up so we were able to keep clear of the rocks with the use of the sails. Nain was no easy place to find as it was hidden away somewhere among a great conglomeration of islands and bays

not appearing on the charts with which we were supplied.

The following afternoon we ran into an entrance that appeared to be the correct one for Nain. After hours of sailing around islands and into bays that all looked alike, we came to anchor. Commodore Metcalf, the Captain, and the Chief immediately set to chasing a large bear we sighted along the shore.

One time that afternoon, while we were sailing up a wide clear channel at full speed, we hit a rock. The ship lurched a little to one side as we slid over it and went right on. The hull was well built and especially constructed to stand any shock of hitting rocks or running on to ice, so we were none the worse for the mishap.

Locating Nain was no problem for Commander MacMillan on the *Bowdoin* as he knew the Labrador coast equally as well as the streets of his home town, but while originally acquiring his familiarity of the country he tells how he spent two weeks locating this settlement.

The Moravian missionaries here had a motive in locating their station in a place so difficult to find. Before the fishermen began intruding upon this coast, the Eskimos were a strong healthy lot of people making a prosperous living from fishing and hunting. Fighting took place between the Eskimos and fishermen at their first appearance. To protect the Eskimos and

SOME PICTURES TAKEN IN ARCTIC REGIONS

The upper circle shows an Eskimo boy in his kayak at Sukkertoppen, Greenland. The top left "snap" was taken at the northernmost port of call of the Expedition—Godhaven. The Eskimo girls are interested in the camera and are not a bit bashful. The pile of sod in front of them is being dried for use in the construction of an igloo. This material is also used for fuel. The *Bowdoin* and *Sachem* are shown at anchor at the foot of the towering rocky coastline at Godhaven in the top right-hand picture. The inner circle depicts Maude Fisher at the wheel of the *Sachem*, cheerful despite a painfully swollen face caused by Labrador mosquitoes. The *Sachem* is at anchor in Battle Harbor, Labrador, in the lower left-hand picture. The group of three to the right includes: Captain Crowell, skipper of Commodore Metcalf's *Sachem*; Marion Smith; and Maude Fisher

prevent their annihilation, Moravian missionaries from Germany risked and sacrificed their lives for the cause. They have been able to keep the race alive for over a hundred and forty years by their persistent, self sacrificing work. Protection from the diseases of the fisherman has been one of the most important problems they faced, and their station at Nain was hidden away so as to make its approach by the white man very difficult.

In hunting for Nain, Commodore Metcalf had given way to the opinions of others as to where we should head the *Sachem*. His good sense of direction finally won the courage of his convictions when he informed us that we would hoist anchor at five A. M. and arrive at Nain at eight. At 8:20 we dropped the hook alongside the *Bowdoin* at Nain.

In going south from Nain to Battle Harbor, we stopped nearly every night. We had two days at Jack Lane's Bay where Abe Bromfield lived. Arrangements were made to take him to the States for the winter. We were all as much interested in his trip to civilization as Abe himself. Abe had never seen a railroad train, an automobile, a horse, or many of the other things that are so common to us. Just how he would be impressed by all these things was a matter of much speculation on our part.

Some of our stops on the Labrador coast represented agony because of the flies and mosquitoes. At one place I arranged to set up some apparatus for measuring earth currents that exist during aurora displays. I soaked myself with Flit and went ashore to locate a suitable place. The effect of the Flit did not last long. I was compelled to retreat to the ship with all the speed I could put behind the dory I was in. The swellings and itching from the fly bites lasted for days.

In comparison with what the three women who were with us on the *Sachem* suffered from the flies, I had no reason to complain. They went ashore, protected by head nets, but the flies easily worked their way through. Miss Fisher's face was swollen up like a toy balloon while Miss Smith suffered with hundreds of bites but was not affected by such swelling.

I was able to get additional data on the relation between the aurora and radio every night. In addition to the aurora data, notes were taken on the barometer readings and on the mirages that were so common. At times we were able to get photographs of the ice mirages. After reporting to RADIO BROADCAST about the mirages, the editor sent me a message quoting a press report that read as follows: "Captain Rose, of Steamer *President Adams*, at 8 P.M. July 15, in Mediterranean Sea, bound for Port Said, states they saw 'large field of floating ice cakes suspended above horizon and presently a number of small pieces drifted into view followed by a large one. The latter was so clear we could see blue and green veins in the ice.'"

Early in the trip, Commander MacMillan said that if conditions were right, it would be possible to see mirages of images half way around the earth. Considering light and radio waves to be the same thing, except for difference in frequency, it seemed that there might be some connection between the mirages and radio phenomena.

In the data collected, there appeared to be a rather definite relation between barometric pressure, mirages, aurora, magnetic storms, and radio fading. The data is by no means complete but the observations substantiate the following statements:

1.   Mirages and aurora only occur with heavy air pressure.
2.   The relation between the aurora and radio fading depends upon the following:
     (a).   Formation of aurora and its location in respect to the approaching radio wave and the receiving station.
     (b.)   Frequency of the radio signal.

In accounting for the fading, the temperature is an important factor in the formulas already worked out for mirages. A more detailed account of the data and the conclusions drawn will be taken up in a later article.

It might be well to mention here that communication between the *Bowdoin* and *Sachem* was never hampered due to "skip distance" effect. The distance between the two ships varied between three feet and three hundred miles during the trip and we always communicated on waves between thirty-two and forty-two meters (9370 and 7140 kc.).

We found Battle Harbor filled up tight with eighteen fishing vessels, and we were compelled to anchor in what is known as "Outer Harbor." We had prepared to wait there until the mail boat arrived on the following day hoping that a new reverse gear would be aboard for us. According to calculations, the chances seemed very poor and the next boat was not due for two weeks.

The story about the reverse gear, however, displays the efficiency of modern means of communication. The extent of the damage done to the gear which was caused when we ran aground on the Arctic Circle, was not determined until an examination was made at Holsteinborg. A radiogram was sent through on the night of the examination to the offices of John G. Alden, the ship's designers, in Boston. This message was received by them when the offices were opened at nine in the morning. A wire was sent from there to the makers of the gear but not knowing the type used, they wired the builders of the engine in Columbus, Ohio, for this information. By noon a wire was received at the Alden offices from the makers of the gear that a new one had been shipped express, special handling. Only by a margin of a few hours, the gear reached the mail steamer before sailing, and we received it the morning after arriving at Battle Harbor.

It was not until we reached Battle Harbor that I put any broadcast music on the loud speaker for the entertainment of the crew. I held back on the broadcast music because we had an excellent Sonora portable phonograph with a large number of records and because the reception up to that time did not have much entertainment value. It was possible to hear broadcast stations faintly as they faded in and out but to try to get them regularly seemed only a waste of time.

I noticed that many of the missionaries in Labrador and governors in Greenland have receivers that had been given them the previous year by Commander E. F. MacDonald. With their large antennas, they are able to hear stations in the States reasonably well in the summer. Winter reception is reported as excellent.

The 63-meter (4760-kc.) signals from KDKA came in well during most of the summer but were subject to bad fading. The 32½-meter (9225-kc.) signals from WGY appeared very steady and fairly strong while we were in Greenland.

BACK TO CIVILIZATION

IF YOU want to find out how great a fine big juicy beef steak can taste, just take a three months' trip into the north and live on canned goods as we did. When we loaded up with fresh supplies on our arrival at Sydney, Nova Scotia, we all ate like wild men. With the crew fed up on red meat, the captain was afraid to let the men handle the lines for fear of breaking all we had.

On the last leg of the voyage before reaching the States, we ran into some very heavy weather off the Nova Scotia coast. Most of us on the *Bowdoin* and *Sachem* experienced seasickness but it did not last long on the *Sachem* as Commodore Metcalf put aboard a supply of "seasick pills" at Sydney which served their purpose nicely. I believe their trade name is "Sea Oxyl." This is not a free advertisement—it is sympathetic advice.

Our arrival at Wiscasset on September the eleventh does not end the story. Abe Bromfield has to be accounted for. At Sydney, Abe appeared to be most interested in the railroad whistle and a long train of cars. At Wiscasset, he could not understand why they had small light houses in the middle of the streets.

Commander MacMillan was engaged to speak at the New York Radio Show. Abe went with him. In a day and a half Commander E. F. MacDonald showed Abe more of New York than I had seen in a year and a half. Abe remembers every detail of this trip, for his memory is remarkable. Reporters all shot the stock question at him: "What do you think of American flappers?" Abe's confidential report is that the Wiscasset girls are much prettier than the ones in New York.

Abe is a great one for shaking hands. I am sure he would enjoy exchanging places with President Coolidge at times. On one occasion, Abe was following Commander MacDonald into the New York Radio Show when the ticket taker put out his hand. Abe shooks hands with him. This first attempt at crashing gates was a grand success as the ticket taker almost passed out while Abe walked on in.

THE MARCH OF RADIO

News and Interpretation of Current Radio Events

What Does the Listener Want? Let Him Speak

THE Radio Act of 1927 provides that "the licensing authority shall make such a distribution of licenses, bands of frequency or wavelengths, periods of time for operation, and of power among the different states and communities as to give fair, efficient, and equitable service to each of the same." Licenses shall be issued and renewed only "if public interest, convenience, or necessity are served thereby."

These are the sole bases in the new radio law upon which broadcasting stations shall be distributed and licensed. The criterion, then, by which a station's right to broadcast is to be judged is its service value to the listening public in its own service area and the effect of its operation upon broadcasting in other areas.

What a trail of broken hearts must follow in the wake of that formula! Even now we can hear the groans and lamentations of the disappointed, who have pleaded in vain before the relentless Commission. Here is a reformer, who spent the hard-earned money of generous contributors, to broadcast messages of uplift to an immense aud-

The illustration forming the heading shows a general view of the first international radio telegraph station at Alfragide, nine miles west of Lisbon, Portugal. On December 5, 1926, direct high-speed service to London opened. The station will also communicate with East and West Africa, and with South America. The short-wave "beam" is used.

ience, an enthusiastic audience which bought so many of his expensive little pamphlets in the past; we see his bowed form as he staggers dizzily out of the offices of the Radio Commission, realizing that his work of saving the public soul can no longer be continued behind the comfort of the microphone. There is a philanthropic business man who, out of the goodness of his great heart, has spent thousands and thousands of his excess profits to spread gladness and cheer and—advertising, to a receptive public. And here is a little fellow, so he says, who has erected a modest little station, but a nice little station nevertheless, just to make folks happy but, he sobs, the accursed monopoly has bought the souls of the Radio Commission; they get their place in the ether without a struggle; he must put his tubes and condensers back in stock now and sell them to his unsuspecting customers, not without a little free monopoly publicity in the home town paper though, to help him on his way.

Gentlemen of the Radio Commission, let but one voice rule you! The voice of the broadcast listener! Give him fair, efficient, and equitable service! Remember, not one of those who seek to broadcast has anything but a selfish purpose, however disguised, in seeking a place in the ether. Big and little, alike, have something to sell, whether it be

a cause depending upon contributions for revenue, a commodity feeding its sales through goodwill, or a community encouraging capital and population to its territorial limits.

The most important evidence to guide the commissioners in determining what is fair, efficient, and equitable service to the different states and communities will be the expressions of the listening public of their desires. The broadcast listener must become articulate if this wise provision in the law is to have the opportunity to mean what it says.

Already there has been recognition of the necessity for listener organization. We noted, for instance, in a recent issue of the *Northwest Radio News*, the official organ of the Northwest Radio Trade Association, a strong plea that the trade support radio listener organizations. It urges such listener groups to consider questions of receiver operation and repair, to conduct set building contests, to give opportunity for unbiased demonstration of commercial sets and accessories, to criticize, condemn, or commend sales methods and radio advertising, to encourage new uses for radio, to improve radio reception conditions by exposing broadcasting station interference, discouraging radiating receivers, and locating power line leaks, and, most important of all, to voice constructive criticism and crystalized opinion on broadcasting conditions.

Listener clubs are successful in a few isolated instances. In Scranton, the Lackawanna Radio Club, two years ago, had forty-seven members. It now has 680 and is growing so rapidly that it expects soon to reach a total of 1500. In British Columbia, the Victoria Radio Club is an active organization and has eliminated serious power line interference with the aid of apparatus and engineering talent paid for out of the club's treasury.

Although listener clubs are likely to attract only the most enthusiastic, if but five per cent. of the listening audience became articulate through such organizations, the problems of the Radio Commission in determining listener sentiment would be immeasurably reduced. Frankly, we are a little pessimistic about the probabilities of powerful listener organizations because radio is a pastime which does not lend itself easily to group enjoyment. Radio listening is a personal, or a family affair. The membership of camera clubs and automobile clubs, facing similar conditions, is restricted only to the most enthusiastic. But only by extensive listener organization is it possible to ascertain with certainty what the listener wants.

It requires no power of divination to decide that all heterodyning and overlapping of stations must be eliminated but, beyond this, the Radio Commission has little upon which to base its policy. It may designate broadcasting stations so that the maximum number are crowded in the ether bands without interference or it may decide upon the minimum number of stations for each area which give sufficient variety of service, which leave the widest possible gaps for distant stations, and offer the listener the choice of the greatest possible number of program sources.

RADIO BROADCAST seeks to aid the Commission by securing a definite expression of the station or stations desired by listeners of various areas. We urge our readers to fill in the questionnaire which appears on this page. The questionnaires returned to us will be carefully tabulated and the result will be presented to the Radio Commission for its information.

The expressions of the listening audience, so far demonstrated in a wholly unorganized and spasmodic way, have already shown their effectiveness in securing desired action. When Senators Pittman, Howell, Copeland, and Heflin endangered the passage of radio legislation by their continued wrangling over the bill, listener

Which Stations Shall Broadcast?

THE answers to the few questions below, if answered and returned to RADIO BROADCAST at once, will form the basis for a presentation to the Federal Radio Commission. The Radio Act of 1927 provides that the Commission shall assign licenses to broadcast stations in the light of "public interest, convenience, and necessity." Unless the Commission has some means of determining the feeling of radio listeners in each of the five new radio districts, that task is going to be nearly impossible. The importance of terse, complete answers to the questions below is apparent. If our readers desire to make their wishes articulate, we believe that an answer to these questions will enable us to help them and to aid the Radio Commission.

Please typewrite your answers wherever possible. If you do not desire to cut this page from the magazine, please answer the questions in order on another sheet and mail it to us. Many readers complained when answering the recent questionnaire in "The Listeners' Point of View" that they were loath to cut up their copy since each was carefully preserved. Mechanical difficulties make it impossible to insert this questionnaire in any other part of the magazine. Please fill in the answers at once and mail to:

The Editor,
RADIO BROADCAST,
Garden City, New York.

Please Answer These Questions

1. List local stations (within 100 miles) you wish retained on the air in order of your choice:

5. List the stations you wish eliminated, the most unpopular ones first. (Please list present call letters only).

2. List favorite stations which are now excluded to you by radio interference:

6. What kind of receiver have you? (Please be brief: neutrodyne, tuned r.f., etc.)

7. Name _____

3. List favorite out of town stations you wish retained:

Town _____

State _____

8. Additional remarks and suggestions should be made on a type-written sheet attached to this questionnaire.

4. Do you prefer a maximum number of local (within 100 miles) stations to the exclusion of the greatest number of out-of-town stations?
(A.) Yes. (B.) No. (C.) Maximum number of "locals" _____

City listeners can answer completely all of the questions in the above list. Those listeners who have no local stations can indicate their opinion satisfactorily by answering all questions except No. 1; it is particularly valuable, however, to have their opinions on No. 4.

protests came in such numbers that these very vocal Senators were forced to capitulate. But there was no truly organized listener opinion represented. Indeed, Senators clutched upon the feeblest ray of evidence to cite listener opinion in support of their various arguments. For example, references appear in the *Congressional Record* to a protest against the Radio Bill made by one C. Wood Arthur, speaking for a so-called Radio League of America. Letters from this magazine addressed to Mr. Arthur have received no acknowledgment or reply. Nor have we been able to learn from any other source who Mr. Arthur represents beside himself. In the absence of a recognized listener organization, anyone, however unknown and unrepresentative, may take it upon himself to influence the course of legislation and the future action of the Commissioners. The lack of organization of broadcast listeners is a menace to their interests, a condition which would be quickly alleviated by the formation of a truly nationwide listener organization.

Objections to the Radio Law

WHILE the Radio Act was the subject of debate in the House and Senate, certain objections were brought to its provisions which, were they valid, would undoubtedly be serious reflections upon the soundness of the Act. It is worth while to consider some of these points lest confidence in the Radio Law be undermined.

The radio listener has every reason to be congratulated on the fine piece of legislation which Senator Dill so ably marshalled through the Senate. We regret, of course, that the original White Bill, which did not provide the cumbersome and expensive radio commission, was not passed, but the ingenious compromise worked out is the best law which could be hoped for. The commission, we fear, is a permanent expense. It will probably remain in session practically all the time even after the first year, since anyone questioning the Department of Commerce's rulings automatically brings his case before the Commission.

Some of the objectors to the bill apparently failed to read it carefully or they have no adequate understanding of the English language. One provision of the Act states that broadcasters cannot acquire or construct wire lines of their own for the purpose of securing unfair advantage over rival broadcasters. Senator Copeland (D. New York) spoke harshly against this provision, stating that the telegraph and telephone companies (against whom this provision is directed) make so much money now out of telegrams sent in response to radio programs that they fear broadcasters might take away this business by means of a communication system of their own!

He also stated that light, heat, and power would "soon be transmitted by radio" and

that, therefore, licenses should not be granted for three years. The Act specifically provides that a station's license may be revoked if it is used for a service differing in any way from that for which the license was granted. This same provision is adequate protection against the transmission of "scrambled" programs for which the listener would be compelled to use a special

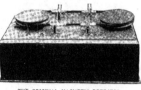

THE ORIGINAL MAGNETIC DETECTOR USED BY MARCONI

patented receiver. The Commission has full power to limit and prescribe the kind of equipment to be used in transmission. Fortunately, no specific barrier to charging for programs was incorporated in the law because it is quite conceivable that there may, some day, be demand for high-grade programs for which the listener might be asked to pay. On the other hand, such a private service could not be undertaken, under the Act, if it did not meet the requirement that it was fair and equitable service to the listener. One Representative created quite

a splutter by insisting that, as soon as the "interests" were licensed under the new Act, they might begin transmitting scrambled programs, so that the listener would have to pay for them or else junk his set. Such transmission could not be undertaken without express permission of the Commission, which would be derelict in its duty if it granted that permission without public demand for it.

The Question of Vested Rights

THE only serious criticism of the Act in the Senate centered upon the point that no declaration was made in the law to the effect that the ether is owned by the people. The law ignores the matter of ether ownership but asserts unequivocally the full power of the government to regulate its use. Ether ownership is a subject of considerable legal complexity, since we cannot accurately define what the ether is. The "ether" is international and hence cannot be owned by any one nation or any individual. If the ether were considered as "property," divestment of a station of its frequency would automatically make it a matter of property confiscation. As a matter of fact, had broadcasting stations, by any legal interpretation, possessed vested rights prior to the passing of the Radio Act, the mere statement by Congress that they did not possess such rights would not automatically dissolve them. The inclusion of a declaration of ownership of the ether by the people would have in-

TWO WELL-KNOWN RADIO EXECUTIVES

Powel Crosley, Jr., President of the Crosley Radio Corporation, and his newly appointed assistant, R. H. Langley. For the last six years, Mr. Langley has been in charge of receiving set development for the General Electric Company and is widely known in radio engineering circles. The first American airplane transmitter is credited to Mr. Langley

volved the law in precarious legal entanglements. Confinement of the Act's scope to regulation of any and every device using ether waves in no way jeopardizes or weakens its effectiveness. If you have entire control of the use of a thing, its technical ownership becomes of minor importance. If the government can prescribe when, where, and for what purpose you may use your motor car, it, in effect, can exert all of the advantages of complete ownership.

Legal experts state that a vested right could be secured to a wavelength, provided unrestricted ownership had been exercised for a period of many years. In historic covered wagon days, a prospector had only to stake out his claim and to occupy his property to obtain a good title to it. His occupation and improvement thereof eventually gave him legal possession. Complete possession and utilization, without existence of any rights conflicting with it, might in time bring about the establishment of a vested right to a radio frequency. But no such condition obtains with respect to broadcasting frequencies. The right of the government to regulate interstate radio communication is clearly given in the Constitution and the government asserted and exercised that right long before the first broadcasting station went on the air. No broadcasting station at any time has operated without first securing a government license and, in so doing, it has recognized the government's power to regulate the use of the ether.

Undoubtedly there will be efforts on the part of disgruntled stations to prove that their investment in radio transmitting equipment is, in effect, confiscated if they are not allowed on the air. They will therefore demand compensation as is guaranteed under the Constitution when private property is confiscated. But they will be wasting their time and money in seeking compensation because the government's power to revoke a license for the general good, even if it involves depreciation or destruction of private property, has been abundantly established by endless court decisions. One need only recall the actions brought many years ago when local option closed saloons. Although a license to sell intoxicating liquor was property so tangible that it could be sold, mortgaged, and subjected to execution, the courts decided that a license to sell liquor is not a vested right and the state could, in the exercise of its police power, revoke it. Revoking broadcasting licenses is not different. Although revocation of a license may involve making a radio transmitter valueless, the license is nevertheless subject to revocation for the good of the people as a whole. It is extremely unlikely, particularly since broadcasting stations were officially warned that the ether lanes were overcrowded, that any of the existing stations whose licenses must be revoked will be able to obtain compensation on the grounds of confiscation.

We are indebted to the Hon. S. H. Rollinson, an eminent New Jersey lawyer, who has made a thorough study of the legal

B. L. SHINN
New York City

Associate Director, National Better Business Bureau. Especially written for RADIO BROADCAST:

"Many radio firms as well as radio dealers and manufacturers have written to the National Better Business Bureau to express appreciation of its work in promoting the accurate retailing, advertising, and selling of radio storage batteries. This appreciation is most welcome, and in acknowledging it, I take the occasion to point out that this is but one of a number of instances in which the Bureau has worked out, with the engineering and advertising leaders in the radio industry, methods of selling, which have assisted the public to obtain accurate and dependable facts regarding radio products. Even the woods appearing on exposed surfaces of radio cabinets and consoles are now accurately described by most manufacturers, instead of by only a few, because of the activities of this Bureau.

"The Better Business Bureau's service to industry and the public, however, is by no means limited to radio. There is scarcely an important item of merchandise in your home in whose advertising, branding, and marketing the influence of the Bureau's recommendations are not felt. If you have saved money for investment, the National Better Business Bureau and forty-three local Bureaus stand ready to supply reliable and disinterested information upon the security which you are considering."

aspects of the broadcasting stituation, for other examples of a governmental police power, the exercise of which has involved what amounts to, but is not, confiscation of property. The State of New Jersey passed a law prohibiting the use of repeating firearms capable of firing more than two shots. This rendered useless repeating rifles possessed by many sportsmen and, in the legal action which followed as a result, brought by sportsmen and manufacturers of firearms, the state proved its complete right to regulate the matter of firearms under the law. Likewise, the government may or

may not grant charters to national banks, as its chooses. If it does not consider it in the public interest to license an additional national bank in any locality, it need not do so. It is empowered to revoke the charter of a national bank in the public interest. A reckless automobile driver may have his license revoked in the maintenance of law and order, even though, in effect, it makes a valuable and expensive piece of property useless to him. The regulation of the ether is a police power which cannot be overthrown upon any grounds of vested rights and cannot be hampered by demands for compensation. We are convinced that the actions brought by broadcasting organizations for compensation or broadcasting rights will be unsuccessful, unless and only if they can show that failure to license them is denying the public fair, efficient, and equitable radio service.

The case of the disgruntled broadcasting station must be overwhelming proof that closing it down is contrary to the public interest. The personal desire of those maintaining the station is of absolutely no value under the law. The Radio Act of 1927 has wiped the broadcasting slate clean for the Radio Commission. At this writing, no broadcasting station is licensed for more than sixty days. Let the Commission have courage! The vast majority will support it to the limit in its difficult task of reducing the number of broadcasting stations by at least sixty per cent. A noisy and selfish minority will always oppose and criticize it. Let the clamoring be outweighed by an organized and powerful listener sentiment and there will be no difficulty in deciding what is meant by fair, efficient, and equitable radio service.

"Christian" Mud Throwing

WE WERE pleased to receive a complete disclaimer from the Christian Science Mother Church in Boston, stating that the destructive propaganda sent out by WHAP in New York is in no way sanctioned by and does not in any way represent the views of that organization. Station WHAP has used its broadcasting station to disseminate large quantities of mud, trained largely against Catholicism. The action of WHAP has disgusted listeners of every shade of religious belief. Broadcasting, fortunately, has been very largely free of intolerance, every kind of religious belief having free access to the microphone to spread its thought constructively. Broadcaster WHAP, disregarding its obligation to the diversified radio audience, has chosen a course of intolerance and villification. Its attacks on Catholicism can not be condoned, and tend to undermine the faith of impressionable persons in any religion. For the good of radio, let us hope that it will no longer be used as a means of breaking down anyone's belief, be it Catholicism, Protestantism, Buddhism, Christian Science, or atheism. A cardinal virtue of Christianity is tolerance. No doubt we will be suspected of intolerant WHAP of being subsidized by the

Pope, but this item is written by a non-Catholic who resents, as do all liberals, the besmirching of any religious belief, his own or some one else's. Let us have no more of WHAP.

A New Term for "A. C. Supply" Units

THE National Better Business Bureau has endorsed the use of the term "socket power unit" to describe devices for the purpose of furnishing A or B power for radio sets. The term "A eliminator" or "B eliminator" is declared as obsolete as the term "horseless carriage," which was at first applied to the motor car. "Socket power" may be applied to describe devices employing combined storage batteries and chargers, thus powering the set indirectly rather than directly from the light socket. It thereby covers the numerous trickle charging storage battery combination units now being so widely sold. We suggest the general acceptance of a term such as "electric set" in order that one may differentiate between receivers using trickle charger—storage battery devices and those powered directly from the light socket through filtering devices without the aid of a secondary battery.

Our attention has been called to advertising, having wide publication, describing A battery devices which "banish the storage battery forever," "eliminate A battery troubles," and similar claims. Investigation has proved that these devices frequently comprise storage batteries combined with trickle chargers. The implication of these phrases in advertising is that the device eliminates the storage battery. Combination trickle chargers and A batteries reduce maintenance attention to a very desirable minimum but we believe the declaration that they completely eliminate it is both exaggerated and misleading.

There Are No Radio Engineers

FROM D. A. Johnston of New Britain, Connecticut, we receive a comment on our item in a previous issue, urging that the education of radio engineers be better balanced with respect to economic and commercial phases of their work:

"Insofar as radio engineers are concerned, I am not quite convinced that there is any such thing. Engineering is very nearly an exact science and an engineer should be able to tell on paper what his product will do so that another engineer can tell exactly whether his product is better or not. Did you ever see any firm producing radio apparatus who would give you information comparing to that which you would expect in buying an electric motor? For example, how many engineers working on sets know what the curves of their particular set look like? I doubt if many of those producing simple articles like battery chargers do know what actual efficiency is. The few who do know these things are not sufficiently satisfied with the product to be willing to say much about them."

This comment is often made by engineers in fields better established than radio. The

EDWARD E. SHUMAKER
New York
President of the Victor Talking Machine Company:

"The question as to who is to pay for radio broadcasting appears to have been temporarily solved. The bills are being met by those who benefit directly from it. While I do not believe that the broadcasting of radio entertainment can be made to take the place of other established forms of advertising, it is an additional medium for creating demands for some products, and a good-will builder when properly used. We have found that the broadcasting of Victor recording artists results in an immediate and traceable demand for their records. We are convinced, also, that anything we may do to raise the standards of radio programs will be reflected in a healthier condition in our business and in other branches of the music industry.

"Radio and the talking machine may at times appear to overlap somewhat. In actual practice they do not overlap. Each has its own place as an instrument for home entertainment. This is borne out by the experience of more than 6000 Victor dealers in the United States. It is also a fact that thousands of new talking machines which are not equipped with radio receiving sets are being sold annually in homes which also contain radio sets.

"In 1924, and the early part of 1925, when the talking machine industry was at a low ebb due to its failure to improve its products, the general impression was that recorded music was being replaced by radio broadcasting. Subsequent developments have demonstrated clearly that such was not the case."

fact is that we have no standardized method of rating the efficiency or selectivity of receiving sets, which nets down to a figure of merit of standardized valuation.

A Survey of Radio Conditions

RADIO RETAILING has issued a report on the broadcasting situation based upon telegraphic summaries from 21 cities, scattered throughout the United States and Canada. Washington, District of Columbia, and Portland, Oregon, were the only two districts re-

porting satisfactory conditions, but it was noted that almost every center of population has one or two high-grade stations which are not being interfered with seriously. The most enterprising leadership in handling the situation was demonstrated on the Pacific Coast, where the Pacific Radio Trade Association not only secured pledges from broadcasting stations that they would abide by the district radio inspector's decisions as to changes in wavelengths, but exerted strong influence in having them observed. Twenty-five per cent. of the midwest stations conflict with local wavelengths in San Francisco on sets of average sensitiveness. From this report and other sources, we learn that among the important stations seriously heterodyned are KOA, WCCO, WOR, WEAF, WTAM, WHN, WEEI, WNAC, CKCL, and KFKX. *Radio Retailing* is to be congratulated upon the excellence and thoroughness of its survey.

The Month In Radio

THE sales of combination radio and talking machine instruments made by the Victor Company during 1926 had a retail value of something over seventeen million dollars and amounted to one sixth of the total business of the company.

FROM Mr. C. R. Cuchins, Vice President of the First National Bank of Bessemer, Alabama, we learn that the Birmingham *News* decided, upon suggestion from broadcast listeners who forwarded copies of RADIO BROADCAST's editorial on the subject, to resume running radio programs in its columns in a manner which makes them intelligible to the listener.

PORTLAND, Oregon, has passed an ordinance making it unlawful to operate without a permit any apparatus generating high-frequency oscillations which interfere with broadcast reception. Violet ray machines, quenched spark devices, and X-ray machines must be licensed and may not be used, except in emergencies, between the hours of seven and eleven. Power interference, being a local matter, appears to be suited to local regulation.

A NEW transmitter for WEAF will be erected at Bellmore, Long Island. Bellmore is on the south shore of Long Island, the nearest towns being Freeport, Hempstead, and Farmingdale. An advantage of this location is the relatively small population which suffers from proximity to the station and the fact that the new station will impress the strongest signal where ship interference is most likely to mar its programs. Plans for the new station have been drawn by Dr. Alfred N. Goldsmith, Chief Research Engineer of the Radio Corporation of America, Dr. E. F. W. Alexanderson, of the General Electric Company, and Frank Conrad, Assistant Chief Engineer of the Westinghouse Company. This station may be in operation by October, 1927.

RADIO beacons have been installed at McCook Field, Dayton, and at the Detroit Ford airport for the guidance of the Stout-Ford commercial airplane between those two points.

The Electrical Phonograph

A Non-Technical Explanation of the Principles Involved in Electrical Recording and Reproduction—The New "Panatrope" and "Electrola"—Data for the Home-Constructor Wishing to Build His Own Electrical Phonograph

By
JAMES MILLEN

THE "PANATROPE"

An entirely electrically operated phonograph. Provision is made so that the amplifier system, including the baffleboard loud speaker, may be used for radio purposes after the detector in any radio circuit. This photograph was taken in the RADIO BROADCAST Laboratory

ALMOST fascinating experience for one interested in radio—especially one who has long been connected with its development—is to spend an evening with the early issues of some of the older radio magazines.

A study of the advertisement section not only recalls the queer contraptions that were looked upon in their day as the acme of engineering perfection, but also throws light upon the founding and first products of small companies, regarded now as leaders in the radio industry. Not only do large things often develop from a humble start, but also large concerns of one decade often pass into oblivion by the next.

Aside from this, we may also read what the 'prophets" of but just a few years ago outlined for the future of the radio industry. For instance, in the November, 1922, issue of this magazine, appeared an article entitled: "Will Radio Replace the Phonograph?"

Apparently there existed some doubt in the mind of the public as to whether the new novelty, radio, could ever reach the "perfection" of the phonograph as regards tone quality, service, and reliability. Now, on the other hand, there appears to be some doubt in the mind of the public regarding the same question, but from a different angle: "Can today's phonograph compete with the radio in tone quality, service, and reliability?"

But, why not, from comparisons of the two, draw one's own conclusions? It is not at all a difficult or costly task to construct a truly fine electric phonograph. Before going into the construction of such devices, however, let us first find out just what this new instrument, known commercially as the "Panatrope" and "Electrola," really is, and just how it works.

The grooves in a phonograph record are so cut as to cause the needle to vibrate from side to side. In the old phonograph these transverse vibrations of the needle were conveyed mechanically by a system of multiplying levers to a mica diaphragm located at the small end of a horn. The vibration of the needle caused vibration of the diaphragm which set the air column in the horn moving, and thus produced sound.

Such a system, while low in manufacturing cost and reliable in operation, resulted in considerable distortion. The horn was resonant at certain frequencies and the diaphragm at others. Thus some notes were greatly over-emphasized while others were entirely missing.

The new electric system depends for its operation upon the vibrations of the needle to produce a constantly varying electric voltage. The minute electric voltage generated by the movement of the needle is amplified by a high-quality audio-frequency amplifier, such as is a component part of the better radio sets, and then converted into sound by a loud speaker.

While the new phonograph will play the old records, the results are not the same as when the new Orthophonic (correct-tone) type records, made especially for the purpose, are used. In former years, records were made mechanically by a machine much resembling the old-style phonograph, into the horn of which orchestras played while closely and uncomfortably huddled together.

Now, however, studios much the same as those of modern broadcast stations, are employed for record making. Standard radio microphones, as many as needed, are so placed as to properly blend all the instruments of even a large symphony orchestra. No longer must the player in front play softer than natural in order not to "drown out" those in back. Everyone plays as if giving a regular recital, and the various microphones are so placed as to produce the proper results.

In fact, the output of the electric phonograph, when one gets right down to the matter, is but a standard broadcast program which, instead of being sent over the air, is recorded and delivered to the consumer without picking up static, heterodyne whistles, and other disturbances en route.

The acoustical difference between the same piece played by the same orchestra over a high-grade broadcast station and played on an electric phonograph, is nil, assuming, of course, that the same quality audio-frequency amplifier and loud speaker are used in both instances.

With these improvements, and one other—the elimination of the record scratch—the modern phonograph becomes a highly desirable companion to the modern radio receiver. Static and SOS signals no longer need spoil an evening's entertainment. Favorite selections, beautifully rendered and reproduced, are available at a moment's notice when the radio program is not tempting.

Even the inconvenience of constantly changing records seems soon to be done away with according to recent announcements, of some of the leading phonograph concerns.

But do not understand this article to be an argument in favor of the phonograph over the radio. The phonograph can never take the place of the radio. First, the radio brings news and entertainment into the home as it is actually occurring—banquets, speeches, spotting events and many others; and second, radio supplies its own program; its *repertoire* is not limited by the number of records in the album.

But, as a companion to modern radio, there is a distinct service to be performed by the electrically operated phonograph.

THE NEW PHONOGRAPH

BEFORE considering the re-vamping of the old-style phonograph, perhaps a brief semi-technical description of the "Panatrope" or the "Electrola" may not be amiss. In order to make such a description more clear and to better "tie it up" with what is to follow, the complete device will be divided into its various sub-units.

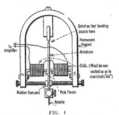

FIG. 1

Details of a home-made pick-up. It consists essentially of a permanent horse-shoe magnet, an armature, and two small electromagnets. The armature is reduced in cross section at its upper end by grinding. The purpose of this is to provide a hinge-like action at this point

First there is the turntable upon which the record is placed. As all of the present models are designed solely for a. c. operation, the motor employed to drive the turntable is generally of the induction type, which, due to lack of the sparking brushes and commutator of the motor, will not cause any pseudo static disturbances in the loud speaker.

Then there is the pick-up, which converts the mechanical vibrations imparted by the record to the needle into electrical impulses.

As will be seen from Figs. 1 and 2, the pick-up consists of a permanent magnet, to which is bolted two pole pieces, two small electromagnets, and a movable armature. Movement of the armature, at one -extremity of which is located the needle, results in a variation of the electromagnetic flux passing through the cores of the electromagnets. This variation in flux induces a varying or alternating voltage in the coils of the electromagnets. The home constructed pick-up illustrated in Fig. 1 clearly indicates the form of construction employed in several high-grade units. Some commercial pick-ups employ variations of the design, such as the use of but one coil, that coil being wound on the armature rather than the pole pieces of the permanent magnet.

Although but few readers will have the facilities for making their own pick-ups, a brief description of an electromagnetic unit should prove interesting to even those who intend purchasing a unit.

A pick-up, as will be seen from Fig. 1, consists of a special shape and size horse-shoe type permanent magnet to the ends of which are attached pole pieces carrying coils of fine wire. Between the pole pieces is the reed, or armature, which carries the needle.

The permanent magnet should be made from tool steel, cut, drilled, and formed, before being hardened. The larger the cross section of the magnet, the more volume, within limits, that will be obtained from the pick-up. An excessively large magnet will press too heavily upon the record. To magnetize, when a magnetizing machine is not available, wind about a hundred turns of heavy cotton-covered wire around the steel horse-shoe and connect the ends of the wire across a six-volt storage-battery for a few minutes. Fewer turns of wire than specified will draw an excessive current from the battery.

The pole pieces and coils may be taken from a radio headset. The headset should be of fairly high impedance and good make.

The armature should also be made of hardened steel, and may either be pivoted or spring mounted. In order that there may be no sustained res-

RADIO BROADCAST Photograph
THE "MEROLA" PICK-UP
An adapter makes possible the use of this device in combination with any good quality amplifier. Merely plug the adapter into the detector tube's socket in the receiver and connect a second lead to the B battery 45-volt post

onance effect from the armature or reed, it is necessary that it should be mechanically damped. The greater the mass and stiffness of the reed, the greater must be the damping if transient free vibrations are to be quickly checked. The elasticity can be made very small by pivoting the reed, but if the mass is too much reduced, the sensitivity is lowered by the consequent reduction of the iron circuit. Rubber may be used for damping.

Proper adjustment of the completed pick-up is important. The faces of the two pole pieces and the armature must be parallel. The armature must be centered between the pole pieces, and the pole pieces must be as close together as possible without danger of the armature hitting. Damping should not be greater than necessary.

ELIMINATING NEEDLE SCRATCH

ASIDE from tone quality, one of the outstanding achievements of the new phonograph is the elimination of surface noises and needle "scratch." The use of a new material for the manufacture of records has done much to eliminate this annoyance of the past but the final and complete elimination is accomplished by means of an electrical filter circuit so tuned as to suppress frequencies in the neighborhood of the scratch frequency. Such an electrical filter is quite simple,

A COMBINATION OF RADIO AND PHONOGRAPH
Representing the ultimate in luxury so far as home entertainment is concerned. This " Electrola " retails for a sum not considerably below a thousand dollars. The reproduction from the records is by electrical pick-up, electrical amplification, and a cone loud speaker

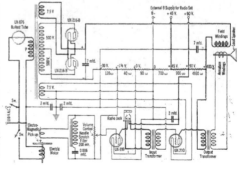

FIG. 2
The circuit diagram of the "Panatrope." The purpose of the five tubes employed is as follows: The two UX-216-B's are wired to form a full-wave rectifier system. The UX-876 is a ballast tube connected in the a, c. line to compensate line voltage variations. The UX-199 is the first stage audio amplifier while the UX-210 is a power output tube—the second audio stage

THE GRIMES PICK-UP
An excellent example of the
electro-magnetic type of pick-up

as will be seen from Fig. 3, and is connected
between the pick-up and the amplifier. The
filter is so located before, rather than after the
amplifier, in order to prevent unnecessary overloading of the amplifier.

The volume is controlled by a wire-wound potentiometer which permits the audio-frequency
voltage supplied to the amplifier to be made any
desired fraction of the pick-up output.

The amplifiers employed in the present commercial models of the electric phonograph operate entirely from a. c. The circuit, as will be
seen from Fig. 2, comprises a stage of voltage
amplification, using a UX-199 tube with filament
lighted from rectified a, c., and a stage of power
amplification using a UX-210 tube with filament
lighted from raw a. c. This power tube is supplied with a potential of close to 400 volts, and
the requisite negative grid bias. Transformer
coupling is employed between the two stages of
amplification as well as between the amplifier
output and the loud speaker.

The power supply, which is the same as that
employed in the RCA-104 loud speaker, embodies several unique features.

First, a line ballast tube is employed to compensate variations in line voltage so as to prevent changes in filament current of the 199 tube,
and to adapt the complete phonograph to power
lines of various voltages located in different sections of the country.

Second, the field winding of the baffleboard

type of Rice-Kellogg cone loud speaker is used
as the filter choke for the B supply.

Third, the necessity of a ground connection
is eliminated by the use of a 2-mfd. condenser
between one side of the 110-volt line and the
center-tap of the power tube filament winding.
In inserting the 110-volt plug in a socket or other
outlet, the plug should first be tried one way
and then the other as best results are generally
secured when the grounded side of the line is the
side to which the condenser is connected.

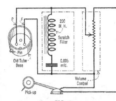

FIG. 3

The electrical connections of the pick-up, scratch
filter, volume control, and adapter. The scratch
filter as shown is tuned to approximately 4600
cycles

A special input jack makes the amplifier—
loud speaker unit available for operation directly from the detector output of any radio set,
or better yet, a small two- or three-tube set, such
as the two-tube R. B. "Lab." receiver, described in the January RADIO BROADCAST by
John B. Brennan, may be built just for the purpose.

Terminals are also provided on the phonograph for supplying B power to the radio receiver.

The loud speaker, as already mentioned, is
of the cone type and is electrically "tied into"
the amplifier due to the use of its field winding
as a filter inductance.

THE HANSCOM "SUPERUNIT".
This pick-up is constructed from
a Baldwin phone unit

PICK-UPS

THERE are at least four different types of
pick-ups. The electro-magnetic and the
crystal types function by generating voltages,
while the condenser and the carbon forms operate by variations of impedance.

Of these four, the only one of commercial
importance, and, it would seem, the most practical, is the electromagnetic form. To this class
belongs those used in the "Electrola," "Panatrope," and the "Vitaphone." This latter is in
reality a special electric phonograph designed
for synchronization with a motion picture projector.

One of the large corporations interested in the
development of radio is experimenting with a
pick-up of the crystal type but as yet this unique
device is still in the experimental stage.

Pick-ups of the carbon type have been in existence for a number of years but only recently
has one become capable of high-quality reproduction
been perfected. This is the "Bristophone,"
shown in a photograph on page 23.

This pick-up, which depends for its operation
upon the change of its electrical resistance with
vibrations of the needle, is in a somewhat different class than the others as it is designed to operate a loud speaker without the use of an amplifier.

Some pick-ups are made of the same parts as
used in the construction of an ordinary radio
type head telephone. That such should be the
case is quite reasonable, for the function of the
pick-up is exactly the reverse of the loud speaker
unit or telephone receiver.

Generally, however, such devices tend to over-
emphasize certain notes and, unless used in
connection with a "trick" amplifier, will sound
"flat." An exception is the unit from the Western Electric 540 AW cone loud speaker. This
unit can be converted into a very fine, but needlessly expensive, pick-up. One of the tricks to
such a conversion is the counter balancing of the
tone-arm so that the unit, which is quite heavy,
does not press too hard on the record.

As a rule, however, for best results, a unit designed from the "ground-up" as a high grade
pick-up should be used. Several units of this
type are now being manufactured.

USING A PICK-UP WITH ANY SET

BY REMOVING the base from an old 201-A
tube (this may readily be accomplished
by heating over a gas flame to melt the cement
in the base and the solder on the prongs), and
connecting the leads from a pick-up to the plate
and filament prongs, as shown in Fig. 3, the
home-made pick-up shown in Fig. 1 may be
used with the amplifier in any radio set. Merely
remove the detector tube and in its place insert
the "adapter." Fig. 3 also shows the inclusion
of a scratch filter and volume control.

The audio amplifier channels of many radio
sets in use to-day are so poor as to make such
a system of using the pick-up most undesirable.
There is little good in purchasing or constructing
a pick-up for obtaining the same quality from a
phonograph as from a radio set if the audio quality of the radio set is poor.

WITHIN THE "PANATROPE"
The rear of the baffleboard loud speaker is seen in the center of this picture. A metal funnel is placed over the UV-876 tube to deflect the heat

1. Use of high-grade coupling mediums. The use of the new large-size audio transformers, resistance coupling, or impedance coupling, is to be recommended.

2. Use of a power tube in last audio stage with proper B and C voltages to prevent overloading.

3. Use of an output device. Inductance-capacity units are, in the author's opinion, preferable to the transformer units.

4. Use of a high-grade loud speaker.

RADIO BROADCAST Photograph

A HOME-CONSTRUCTED PHONOGRAPH–RADIO COMBINATION
Built by the author. A more detailed description of this particular instrument is scheduled to appear in RADIO BROADCAST for next month

SCRATCH FILTERS

WHILE an electrical filter circuit of the type indicated in Fig. 3 will remove all objectionable scratch from the music issuing from the loud speaker, it will not prevent one hearing the un-amplified scratch noise directly from the record. For this reason the lid of the turntable compartment should be kept closed while records are playing.

A high-resistance potentiometer, such as the Centralab or the type L Electrad Royalty unit, makes a very excellent volume control, when wired as indicated in Fig. 3. As a rule, it will generally be found rather convenient to mount this volume control on a small square or disc of bakelite in the turntable compartment of the phonograph. This same panel may also well contain the amplifier control switch.

To get the full benefit from a home-constructed electric phonograph, it is essential to use the new electrically cut records and a good loud speaker of the cone type. Of course the pick-up and amplifier must be good, but generally, after devoting much attention to them, the final results are spoiled by failure to use a good loud speaker.

In the June RADIO BROADCAST, details of a console phonograph cabinet in which a two-tube R. B. "Lab" receiver, baffleboard loud speaker, and lamp-socket powered amplifier with record pick-up, will be described. Data on the construction of a special amplifier, entirely lamp-socket operated and designed primarily for use with a pick-up device, will also be given.

A desirable thing to do is to construct a new high-quality lamp socket operated audio channel of the type described by the writer in recent issues of RADIO BROADCAST so as to secure well nigh perfect phonograph and radio quality.

Where the amplifier in the radio set is to be used with a pick-up, the following essentials of good quality should be kept in mind:

RADIO BROADCAST Photograph

THE "BRISTOPHONE"
A pick-up of the carbon type which operates a loud speaker without using an amplifier

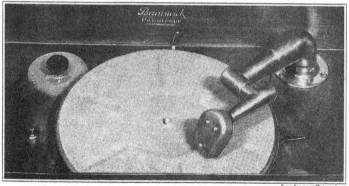

RADIO BROADCAST Photograph

A VIEW OF THE TURNTABLE COMPARTMENT OF THE BRUNSWICK "PANATROPE"
This picture shows the magnetic pick-up arm to the right, the "start-stop" switch at the left, and the volume control above the switch. An induction type of motor turns the turntable, making the operation of the "Panatrope" completely electrical

A Balanced Short-Wave Receiver

A Description, by the Designer of the Best Receiver Submitted in Radio Broadcast's Recent Contest, of a Short-Wave Receiver That Won't Radiate

By FRANK C. JONES

Amateur Station, 6 AJF

THE short-wave receiver described by the author in the September, 1926, issue of RADIO BROADCAST was an effort toward the design of a short-wave receiver which would not radiate and interfere with other near-by receivers. Although it had many desirable features, it was not as sensitive as a standard short-wave receiver. The receiver described in this article is the result of further experiments and calculations carried along the same lines. If made carefully, it does not radiate at all and is, in the author's opinion, just as sensitive as the ordinary carefully built short-wave receiver. The improvement over the first receiver described is considerable; in fact, there is no comparison between the two, both as to non-radiating qualities and sensitivity.

The receiver, as shown in the accompanying pictures, was a model built up after the circuit was conceived and not in its final form, since a thorough job of shielding had not been done when the photographs were taken. Imperfect though the photographed receiver was, it was possible to copy New Zealand amateurs on a second receiver while this incompleted one was connected to the same antenna and ground system and permitted to oscillate on the same frequency as the distant station. No interference whatsoever was apparent from the oscillating set during the reception of the New Zealand stations. The writer has never seen any other short-wave receiver which could approach that mark for non-radiating qualities When listening to the same New Zealand station on the balanced receiver described here, the standard receiver could not be tuned to the same frequency without absolutely swamping everything with its whistling. In fact, it would ruin any attempt to receive anything except a local station when it was oscillating on the same frequency as the received station. Capacity coupling to the same antenna was used by both receivers, one being the balanced and the other a standard Reinartz receiver.

The circuit of the balanced receiver is shown in Fig. 1, and, as can be readily seen, is a form of Wheatstone bridge. It is absolutely necessary to use some form of a bridge with the antenna and ground across zero potential points of the oscillating circuit or circuits. In Fig. 1, the midget coupling condensers, C_4 and C_5, form the two capacitive arms of the bridge and $L_4C_3C_6$ and $L_3C_6C_7$ form the essential arms on the other side of the bridge. Both of these latter arms are tuned simultaneously to the same frequency, *i.e.*, that of the incoming frequency and plus or minus say 1000 cycles for heterodyning purposes, and C_4 and C_5 are left set in a certain relation to each other. C_6 is a resultant capacity from the combination of C_3 and the grid-filament capacity, C_F. The condensers C_3 and C_6 are on the same shaft and should be exactly similar so that the two circuits will be tuned to the same frequency or wavelength at any point of the tuning scale. The tickler coils, L_3 and L_4, are coupled inductively to their respective coils, L_1 and L_2, in the proper phase relation to cause the oscillatory currents in the bridge to just neutralize each other, or balance out. L_4 is coupled to L_2 so that the detector will oscillate, and L_3 is reversed with respect to direction of winding to L_4. C_7 is a very small neutralizing type condenser which is set to the same capacity as C_6. This latter setting is easy to determine in practice. C_6 is simply a "throttle" condenser to control the amount of regeneration or oscillation in the detector circuit, such as is used in practically all modern short-wave receivers. Thus the balanced receiver has one tuning control and one feed-back control, the latter being adjusted only once or twice throughout the whole tuning range of the receiver. Simplicity and ease of tuning have been accomplished in this receiver.

Now for some simple theory as to why the receiver does not radiate when properly balanced. The energy of the incoming signal at any one instant can be represented by the dotted arrows. This energy splits and part of it goes through each coupling condenser, C_4 and C_5. The tuning circuits associated with L_4 and L_3 are tuned to the frequency of the desired incoming signal energy and so offer an extremely high impedance to this energy. This means that most of the signal energy is impressed equally across the grid-filament capacity of the detector tube and the small capacitance C_3. The action thus far is the same as for any receiver. The energy component in the plate circuit of the detector through the tickler coils L_4 and L_3 induce energy in the coils L_1 and L_2 respectively, that in L_1 in a direction as indicated by the single-headed solid arrow, and in L_2 in a direction as shown by the double-headed solid arrow. This is obtained by having the direction of the windings of the coils in reverse directions. The energy induced in L_1 adds to the incoming signal energy of the same frequency and this continuous feed-back causes the detector to oscillate. Unfortunately, part of this feed-back energy, as shown by the single solid arrows, splits at the top of the bridge, part of it going across the grid and the rest through C_1 to the antenna, which can, for our purposes, be represented as an inductance, resistance, and capacitance across the points A and G, as shown in Fig. 1. The energy induced in L_2, however, is in such a direction as to be opposite to that induced in L_1 in its effect across the points A and G. The net effect in the antenna circuit is zero output from the receiver. Thus we have a one way circuit, the antenna gives energy to the detector but does not take any away.

Since the circuit $C_3L_3C_7$ is tuned to the desired frequency, as is also $L_4C_6C_6$, these two arms of the bridge offer high impedance to the feed-back energy from the opposite tickler coils respectively, and most of the energy finds its way across the points A and G. The reactance or impedance of the antenna circuit is comparatively low so that the main components of energy are as shown by the arrows. The minute quantities, in comparison with those shown by the arrows, can be automatically eliminated or compensated

SHOWING THE DISPOSITION OF PARTS WITHIN THE CABINET

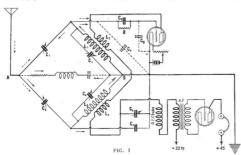

FIG. 1

is nothing new or different in the audio amplifier, though a peaked transformer could be used to advantage here, say with a resonant peak somewhere between 600 and 1000 cycles. The coupling condensers, C_4 and C_5 are midget variable condensers of very low capacity, of which there are numerous types on the market. The grid condenser, C_5, should be about 0.0001 mfd., and the grid leak, R, of about 8 or 10 megohms in value. The detector socket should be well cushioned, a piece of sponge rubber being used in the receiver described.

In balancing the receiver, a separate short-wave receiver should be set up, preferably using separate batteries, but coupled to the same antenna and ground. By listening-in on the regular short-wave receiver, a loud squeal will probably be heard when the two sets are tuned to the same wavelengths. It is a good plan to use about 45 volts on the plate of the detector in the balanced receiver in order to make sure that the oscillations will be quite strong. C_4 and C_5, together with C_7, are then varied until the bridge is balanced, which condition will be indicated by there being no interference in the regular receiver when the two are tuned to the same frequency. If the same relative setting of C_4 and C_5 holds true for the whole tuning range of the receivers, then the receiver is nicely balanced and the tickler coils are coupled to coils L_3 and L_4 correctly, and the small capacity C_7 is correctly adjusted. If the condensers C_4 and C_5 have to be changed, then try readjusting C_7, and also the tickler coil couplings, until a correct balance is obtained.

In order to improve the receiver shown, the coupling condensers C_4 and C_5 should be shielded from the rest of the set so that there is no capacitive coupling to the antenna on the antenna side of the condensers. In this set, the antenna lead was brought into the set in a bunch of battery leads, which actually shield it to some extent until it reaches the condensers C_4 and C_5. A good arrangement would be to have the common connection point to the antenna as near as possible to the place where the down lead comes through the shielded box. Another better arrangement would be to have the twin condenser placed in the center of the front panel and the two circuits placed symmetrically on each side of it instead of in the arrangement as shown in the photograph. The set should of course be completely shielded, in order that no radiation will take place from the coils, etc., of the set itself.

in adjusting the values of C_4 and C_5 and the relative positions of the tickler coils L_3 and L_4 with respect to their associated coils, L_1 and L_2.

In order to simplify as much as possible the explanation of this receiver circuit, values of energy were spoken of instead of induced currents and effective voltages. In speaking of induced energy, it was meant that portion of energy which was available at the points of the tuned circuits which would cause a radiation in the antenna system. In tracing through the circuit using currents and voltages, all of the phase differences must be taken into account, which would make it a very complicated explanation of the functioning of this circuit.

CONSTANTS OF THE CIRCUIT

A DIAGRAM which may be somewhat easier to follow in wiring up such a receiver is shown in Fig. 2. The coils L_1 and L_2 should be exactly similar, with preferably spaced winding on a form about 2½ inches in diameter. For covering the 40-meter (7500-kc.) band, 9 turns for use with the 199 type of tubes and 8 turns if 201-A type tubes are used, is about right. In the original receiver, an old 17-plate condenser was rebuilt so that there were two separate condensers in the one unit, each with four plates. This set tunes from about 30 up to 50 meters (about 10,000 to 6000 kc.) when using UX-199 type tubes. The tickler coils, L_3 and L_4, are similar and are wound with about No. 26 wire on a 2-inch form with 7 turns apiece. Celluloid dissolved in acetone was used in holding these coils in shape and makes a minimum amount of dielectric in the field of the coils, L_3 and L_4, where the losses should be kept as low as possible. The coils L_1 and L_2 were space wound with No. 18 wire on a 2½-inch cardboard tube. Four narrow strips of celluloid were laid at equal distances around the cardboard tube and the wire wound over them. Where the wire touches the strips, it was painted with the acetone celluloid solution and allowed to dry, after which the cardboard could be torn out leaving the four strips of cemented celluloid to support the coil turns. The two coil mountings for the four coils were made from strips of hard rubber acting as clamps over the coils, and the whole unit was screwed down to the baseboard of the receiver. This arrangement makes it possible to mount the

The Facts About this Receiver

Name of Receiver	Balanced Short-Wave Receiver (Non-Radiating).
Type of Circuit	The circuit is of the autodyne type in which the detector acts as an oscillator. The local oscillations are prevented from going out on the antenna by a special Wheatstone Bridge arrangement.
Number and Kind of Tubes	Two tubes are used, one as an oscillator and detector, and the other in a stage of audio frequency. Either 199 or 201-A type tubes can be used, a slight change in the number of turns in the tuning coils being required for the different tubes, as explained in the text.
Wavelength Range	30 to 50 meters (9994 to 5996 kc.).

This receiver is the result of further work on the part of Mr. Jones in his efforts to develop a truly non-radiating short-wave receiver which may be used by the average amateur. His first attempt was described in RADIO BROADCAST in September, 1926. The present receiver is more sensitive and radiates far less than the first because of refinements in the bridge circuit.

tickler coils about an inch from the filament ends of the coils L_1 and L_2, and to vary the coupling to L_3 and L_4, which is necessary in balancing the receiver. Incidentally, plug-in coils could be used in order to cover the other amateur bands, providing exactly similar coils with their associated ticklers were obtainable with plug-in mountings. The feedback control condenser, C_6, can be of any type of some value near 0.00025 mfd. maximum, and could as well be controlled by a small knob as a large dial since it has practically no effect at all on the tuning of this receiver. The radio-frequency choke, in series with the primary of the audio-frequency transformer, is in this case a midget honeycomb coil of about 400 or 500 turns of fine wire. Any kind of a small r.f. choke coil can be used here, though one with a very small external field should be used. There

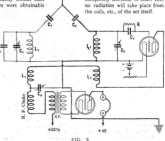

FIG. 2

What About the A Battery?

The Importance of Balance in Trickle Charging—Those Misleading
Statements About A Battery "Elimination"—Comparing the Methods
of Charging and Their Costs—The Best Rate and When to Charge

By EDGAR H. FELIX

SOCKET Power is a term used to describe devices which enable a receiving set to draw its power directly from the electric light mains. To the uninformed enthusiast, it usually implies a complete elimination of all the elements of the power supply system which require care and attention.

Socket power devices now on the market fall slightly short of accomplishing this desirable result. They do, however, reduce to the very minimum the care and attention required to keep the receiver's power supply at full efficiency. The storage battery and rectifier, which these devices invariably combine in a single unit, do nevertheless require a little intelligent care, and, given that care, are both reliable and convenient.

The most advanced types of socket power equipment comprise a charging device and storage battery so designed that the battery is left on charge at a slow or trickle rate during the entire time the set is not in use. Therefore the storage battery is either discharging through the tubes of the receiver or being charged by the trickle charger. Theoretically, the only attention which must be given such socket power devices is the occasional addition of distilled water to the storage battery and the trickle charger. Even when turning on and off the set, it is not necessary to think about the trickle charger connections because, by the use of an automatic device, the receiver's "on-off" switch disconnects the charger when the switch is turned on, and connects it when in the "off" position. With most devices provision is also made for connecting the power line to a B supply device when the control switch is in the "on" position.

A minimum of attention is indeed attained, *provided* that the trickle charge is of the correct rate to offset the current drawn when the set is used. But, alas, this condition of perfect balance of input to output is not always happily secured. Therefore the filament current require ments of the receiver and the number of hours a day it is used must be determined fairly accurately, and the charging rate by the trickle charger adjusted to meet them. For this reason, A power combinations are equipped with means of adjusting the trickle charge rate to suit the power require-ments of different receiving sets.

The importance of balancing charge and discharge becomes apparent when we consider the effects of undercharge and over-charge upon the life and con-dition of the storage battery. If the trickle charge rate is too low, the battery voltage will fall gradually until the set at last

fails to function because of a discharged A battery. In this case, unless the trickle charger has a high charging adjustment which permits bringing back the battery from its discharged state to full charge quickly, it is difficult to restore it to good working order without the assistance of a high-rate charger. When a user experiences a run-down A battery, he naturally increases the trickle charging rate to the maximum, which should bring the battery back to full efficiency in several days. But often, he considers such an experience as evidence that the charging rate should permanently be kept at maximum. For a long time, perhaps even two or three months, this rate gives satisfaction because the battery is always kept fully charged, but, sooner or later, the battery breaks down completely from prolonged and continued overcharging. Active material starts to fall out of the plates and, also, the speci-fic gravity of the electrolyte may rise to a high value and the distilled water of the solution requires renewal every few days. Usually the manufacturer is condemned for these misfortunes, while the cause is simply incorrect setting of the trickle charging rate.

Different socket power devices use different combinations of charging rate. Philco, for ex-ample, has three adjustments, low, medium, and high, giving 0.2, 0.33, and 0.6-ampere charging rates respectively. Willard uses a low and a high adjustment, giving 0.5 and 2.0 amperes, while Gould offers no less than five adjustments, 0.2 to 0.25 ampere, 0.275 to 0.320 ampere, 0.350 to 0.425 ampere, 0.450 to 0.750 ampere, and 0.9 to 1.5 amperes, according to line voltage.

RADIO BROADCAST Photograph

A THREE-IN-ONE SOCKET POWER DEVICE

Combining trickle charger, storage battery, and B power supply in one unit. Al-though there are many manufacturers combining the first two in one unit, there are comparatively few who manufacture such a device supplying B current also

With charging rates as high as two amperes available, the danger of grave overcharge, resulting in quick destruction of the battery, is constantly present. On the other hand, a few hours a week with a high charging rate, and a low trickle charge for the balance of the time, will keep the battery in the best of condition.

There are always occasions when a set is used a great deal for a few days for special reasons, and in this case a higher charging rate is temporarily necessary to bring the battery back to full charge. The table on page 27 shows the charging adjustments recommended by Philco. This table gives the average number of hours a day a set may be used at various current drains and charging rates. For example, with a five-tube set used 3.02 hours a day, each tube drawing a quarter of an ampere, the trickle charger at the 0.2 ampere adjustment will keep the battery up to full charge. If the set is used but two hours a day, the trickle charger should be shut off by removing the attachment plug occasionally so that constant overcharging is avoided.

On the other hand, such a five-tube receiver may be used for twelve hours within two days because of some special broadcasting event, and this would draw a total of fifteen ampere hours from the battery. For example, the set may be in use from eight o'clock until one a. m. on one evening and from six o'clock to one a. m. on the next, drawing a total of fifteen ampere hours from the storage battery. Between these two periods of listening, and between the second period and a third period which we shall sup-pose commences at eight o'clock on the third evening, the battery is recharged for thirty-six hours at the low rate, which gives it back 7.2 ampere hours. Now, if the cur-rent drain, after this experience of twelve hours' use in two days, is continued at its normal average for about three hours daily, and the normal charging rate of 0.2 ampere is maintained, the battery may never be re-stored to full charge.

Understanding these condi-tions, it is not difficult to under-stand the correct charging condi-tions. Estimate the average needs of your set in ampere drain and hours of use and adjust the trickle charging rate recommended by the manufacturer for that load. If you depart from this average use by extended use on special oc-casions, use a higher charging rate for a few days to counter-balance it. On the other hand, if the receiver is subject to long periods of disuse, disconnect the attachment plug for a few days when you estimate the battery to be in a fully charged condition. These are operations of the ut-most simplicity and, by observing them, you get the true convenience of socket power.

The only other attention required to keep a socket power device in working order is to keep it constantly filled with fresh distilled water. The Gould "Unipower" has an ingenious method of reminding you of this water matter because the rectifier cell is so designed that the rectifier cuts off just before the storage battery needs renewal of water. Hence the battery will go dead before its plates have been unduly exposed because of lack of water. Philco supplies a convenient hydrometer with certain of its larger storage batteries, fastened in the top of the cell case. If the water is at the proper level, squeezing the hydrometer bulb draws electrolyte into the charge indicator. If it fails to do so, you are plainly warned that the water must be renewed. Philco and others also use built-in charge indicators which permit one to see the electrolyte level through the glass walls of the cell, simple markings indicating the danger point.

MISLEADING STATEMENTS

WE HAVE observed the advertising and literature of several concerns purporting to describe A-battery eliminators or insinuating that their devices eliminate A-battery troubles, storage battery attention, etc. A number of these devices which we have examined, are simply storage batteries with trickle chargers. To imply "elimination of storage battery troubles" with such devices is plain deception. If manufacturers are not above using such deception in their advertising, dealers can hardly be blamed for extending this misrepresentation to the consumer. Readers are therefore warned to examine so-called A-battery eliminators carefully before purchase, lest they prove to be only the conventional trickle charger-storage battery combination. They then require the simple attentions herein recommended and will fail in service if they are not given it.

The condition of a storage battery, whether used in connection with a trickle charging device or otherwise, is not difficult to check so long as it receives fairly normal treatment. Voltmeter readings, however, should not be relied upon unless a very accurate, high-grade instrument is used.

A hydrometer is very useful for the purpose, because it is inexpensive and somewhat more accurate in its indications than a voltmeter. However, it too is subject to certain slight discrepancies which, when understood, are not difficult to account for.

When a battery is fully charged, the specific gravity is at its maximum. Constant overcharge produces boiling, reducing the water in the solution and accordingly increasing the specific gravity. Extended and continued overcharging (such as that resulting from an excessive trickle charging rate) tends to force the sulphuric acid to the top of the cell so that we sometimes get a high specific gravity reading although the battery goes mysteriously dead after a few minutes of use.

Another misleading state is manifest when an unusually strong sulphuric acid solution is used. This subterfuge is sometimes employed to make a storage battery appear better than it really is. In automobile service, where extremely heavy current is drawn for a few seconds, it is of advantage to use a strong solution because an electric starter sometimes draws as much or more

than ninety amperes from the battery but, with a radio battery, no such heavy loads are drawn. The only advantage gained by the use of a strong solution is a little higher terminal voltage and better specific gravity reading; on the other hand, it causes more rapid deterioration of the plates. The specific gravity reading for a radio storage battery at full charge should be between 1225 and 1300, depending upon the recommendations of the maker. The majority of radio

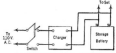

THE CHARGER ARRANGEMENT

The double-pole single-throw switch is for disconnecting the a. c. power source when the set is in use. The charger is left connected to the storage battery while the set is operating, but unless you are sure that the grounded side of the power line tallies in sign with that of the A battery, the latter should be disconnected from the set during charging hours, unless otherwise specified by the manufacturer of the charger

batteries use an electrolyte showing a reading between 1265 and 1285 at full charge. Recharging should be begun when the specific gravity falls to 1200. Complete discharge is between 1100 and 1150, according to the strength of the electrolyte.

The actual value of a storage battery to a user is determined by:

1. Its voltage output.
2. Its capacity, as determined by its service requirements.
3. Its life.

The voltage is simply a matter of having the correct number of cells and, for radio purposes, the six-volt standard is established by the requirements of storage battery tubes. Four-volt storage batteries are sometimes used for large dry cell sets employing the UX-199 type tubes.

DISCHARGE REQUIREMENTS

THE rated capacity of a storage battery is usually fixed at the maximum amperage which the battery will deliver in a complete discharge of 100 hours. Thus, a one-hundred ampere-hour battery will deliver twelve amperes for a little over eight hours. At higher discharge rates, it

gives a lower output in ampere hours, while at a lower discharge rate, somewhat higher capacity.

The effect of drawing an exceedingly high amperage from a storage battery is to concentrate the chemical action of discharge upon the surface of the plates. One of the by-products of discharge is the gradual formation of a high resistance coating of lead sulphate on the plates, and this increases the internal resistance, thus causing battery voltage to fall. A slow discharge rate permits the withdrawal of energy from the very heart of the plate itself and, in so doing, causes the formation of lead sulphate to some depth in the plates. Consequently, the small drains and low charging rates of radio service are harder on storage batteries than the heavy periodical drains of the automobile storage battery.

For this reason, a much heavier and better built form of plate must be used to withstand the ravages of radio service. This is particularly the case in trickle chargers, which are constantly subjected to chemical action by charging or discharging electric potentials. The service life of a storage battery is largely determined by the thickness and quality of the plates. For example, a plate with active material a quarter of an inch thick is good for about 600 to 800 cycles of complete charge and discharge, while one of half that thickness is good for only about 300 cycles of complete charge and discharge. Hence thickness of plates is one of the hidden qualities which determine the value of a storage battery as an investment. The value of purchasing a battery made by a manufacturer of established reputation is apparent when we realize that the true value of a storage battery is completely concealed by its case.

Although a storage battery should not be subjected to continued and prolonged overcharge, it should, on the other hand, be given at least two to four hours overcharge about once a month. This tends to reduce the deleterious effect of sulphation and maintains the battery at its highest efficiency. By overcharging is meant continuation of the charging process after the battery has reached the full charge point. During this part of the charging process, gassing takes place freely and, for this reason, the vents of the battery should be removed.

Only distilled water should be used in replacing that lost by evaporation. It should be added before charging the battery and not afterward in order that it may be thoroughly mixed with the electrolyte during charging. The use of other than distilled water introduces traces of metals and other impurities into the cell, resulting in parasitical electro-chemical actions within the cell, which reduce its output. Rain water, collected in clean containers by direct precipitation, without first passing down roofs and through leader pipes, may be used, as well as clean snow, gathered in clean bottles immediately after a snowfall, if there is no dust or smoke in the air. Distilled water, however, is quite inexpensive and there is little reason for trying to save on this item, thus possibly risking the serviceability of a storage battery worth fifteen or twenty dollars.

The excessive evaporation of distilled water to below the level of the tops of the plates means that a part of the plates is not used in charge and discharge, reducing the capacity of the battery, also that the electrolyte is more highly concentrated, both of these being undesirable results.

FILAMENT CURRENT, AMPERES	RATE AND HOURS SERVICE		
	0.2 AMP.	0.33 AMP.	0.6 AMP.
0.25	10.0	13.1	16.4
0.50	6.35	9.00	12.4
0.75	4.65	6.86	10.0
1.00	3.66	5.54	8.40
1.25	3.02	4.64	7.24
1.50	2.57	4.00	6.35
1.75	2.24	3.50	5.56
2.00	2.00	3.13	5.10

A RATE-OF-CHARGE TABLE

Which shows at a glance how many hours a day you may use your set and keep the storage battery fully charged by leaving the trickle charger on for the other hours of the day. For example, suppose you had a five-tube receiver, using four 201-A type tubes and a UX-171 type power output tube. This combination would draw 1.5 ampere hours from the battery per hour. The set could then be used for four hours daily providing the trickle charger was left on for the remaining twenty hours at an 0.33-amp. charging rate

The cases of storage batteries are now so well designed that they are usually quite sanitary. Hydrometers are equipped with non-drip points so that keeping the battery clean is reduced to a small labor. Another convenient form of hydrometer, as stated previously is designed as part of the vent caps, to be screwed permanently into the top of the cell cases. This type eliminates the risk of dripping electrolyte, which damages carpets and furniture. If acid sometimes gets on the outside of the battery, it can be taken up with blotting paper or with a cloth which has first been soaked in ammonia. Local chemical action forms a coating on the terminals of the battery as a result of moisture or spray from overcharging. This should be carefully removed by means of soda water applied with an old tooth-brush, followed by final cleaning with a file or scraper. This precaution is apt to be slighted and is often the cause of receiving set noises. After cleaning the terminals and connecting clips, cover them with vaseline. It will keep them uncorroded for a period of several months.

When a trickle charger or A-battery socket power unit is not employed, other types of chargers are available. Those of the vibrator type are the least expensive. They have the disadvantage of causing noise and, if the vibrator contacts stick, feeding alternating current into the storage battery. New contacts are easily installed at low cost and a fine file can be used to keep the contacts smooth and fairly silent.

Electrolytic chargers of various kinds can be purchased separately or in combination with small storage batteries, forming a socket power unit. These require the addition of distilled water about as frequently as the storage battery itself needs it and, generally after a year or two of use, renewal of the active electrode. Occasionally new acid is needed for the rectifier and that is best renewed at a battery service station, usually at the cost of a dollar or two.

Bulb rectifiers are generally of the two-ampere or five-ampere type. A bulb is usually good for two thousand or three thousand hours of charging. For example, a set which requires a twenty-five ampere-hour storage battery supply a week can be recharged and maintained by five hours charging a week from a five-ampere charger. This is the average A power requirement of a four-tube storage battery set used about three hours a day. A charger bulb used in this modest service would last for many years. In trickle charger service, however, the bulb would be in use about 140 hours a week instead of but five and would therefore last only five or six months. The bulb depreciates as rapidly when furnishing half an ampere as when furnishing two amperes.

Some trickle charger outfits are fitted with bulb chargers. Only when the current drain from a storage battery is so large, by the use of a multi-tube set for many hours a day, is a trickle charger of such large capacity necessary. Otherwise the overcharge danger, to which we have already referred at length, occurs, and the life of the charger bulb is reduced to a matter of months, unless the trickle charger is disconnected from the line a good part of the time. For example, a half-ampere rate, supplied forty hours a week, instead of the entire time the set is not in use, may tenfold the life of the rectifier tube, cut the electric power required to one tenth, and greatly increase the life of the storage battery, by avoiding continued overcharging.

Since the crux of the whole matter of storage battery life and economy is a matter of correct charging rate so as to assure ample current in the battery when you need it but not at the cost of continued and damaging overcharging, it may be of advantage to suggest a simple charging policy which meets the requirements of the average individual.

With bulb type chargers (whether combined in trickle charger units or not) it is preferable to use the two-ampere rate on the required number of overnight charges per week necessary to keep the storage battery in condition. Experience will determine whether one night a week or two nights a week is sufficient to keep the battery in good condition, showing full charge at the end of the charging period and still giving reasonable service voltage before charging is begun again. Five-ampere chargers bring the battery back into condition quickly, but they offer a rate which is generally a little high for radio receiving requirements. A high charging rate means more intensive chemical action with consequently intensified deterioration of the plates.

With chemical types, low charging rates are generally available. Perhaps the easiest policy to pursue is to use a low charging rate during the week, which seems to keep the set going satisfactorily under ordinary conditions, without permitting the storage battery to reach a nearly discharged condition at any time. It may be below full charge a good part of the time without danger so long as it is not showing a specific gravity below 1150 at any time. Then, over the week-end, give it a high charging rate, if that is necessary to bring it to the full charge condition. If the set is used for prolonged periods, one or two nights during the week, so that the balanced condition is upset, restore the battery to its normal condition by a few hours at a high charging rate.

There may be occasions when a low trickle charge rate keeps the battery constantly at full charge and the renewal of distilled water is found necessary more frequently than once a month. This indicates continued overcharging and should be avoided by disconnecting the trickle charger from the set one or two days a week. When some such formula, which gives the balanced condition, is established, storage battery maintenance becomes a simple and easy matter.

CONCLUSION ABOUT CHARGING

IN THOSE few areas served by direct current, rectifiers are not suited to storage battery charging. The only thing required is some form of resistance so as to reduce the current supplied to the battery during charge to the value which it requires for the purpose. A convenient form of resistance is an electric light bulb connected in series with the battery, of a wattage determined by the battery's charging requirements. A fifty-watt lamp, connected with a six-volt storage bat-

tery, supplies about half an ampere; four such lamps in parallel will deliver two amperes. The circuit diagram on this page shows the correct circuit connections.

Two precautions should invariably be observed when charging storage batteries. First, be certain that the positive or plus, charger lead is connected with the red, plus, or positive A-battery terminal and that the negative charger lead is connected with the minus A-battery terminal. Second, disconnect the radio set from the A battery terminals, unless you are absolutely certain that the power line is not short-circuited through the charger.

To determine the polarity of the power line leads when charging through a bank of lamps plunge the terminals to be connected with the battery into a glass of salt water solution. Bubbles will rise from the negative terminal. Rectifiers of the electrolytic and bulb type have plain markings, usually employing a red wire for the positive lead.

Many misleading statements have been made as to the cost of battery maintenance. Service stations charge from fifty cents to one dollar for charging a radio battery and, if this service is required twice a month, the cost comes to one or two dollars a month, plus inconvenience and "time out" because the set is not in service.

Chargers of various types are generally advertised as drawing negligible current from the line but this is not always true. The most efficient way to charge a battery is with a standard charger capable of supplying several amperes—say two. Such chargers draw 25 or 30 watts from the line so that, if used five hours a week, only about one seventh of a kilowatt, costing from two to three cents depending upon the locality, will be used. Trickle chargers draw from 5 to 20 watts from the line, the average being about 10 watts. Assuming you use a set three hours a day, this means about 630 hours of non-use a month which, at 10 watts, means 6.3 kilowatt hours. The average rate is ten cents a kilowatt hour so that the upkeep cost for electric power is about sixty three cents a month. The chemical type of charger is somewhat more efficient than the bulb type when they are used for trickle charging and, for this reason, from the standpoint of economy, the electrolytic type is preferable although this advantage is offset by the fact that the chemical type requires more care in its upkeep than does the bulb type.

Since it is advantageous to disconnect the trickle charger if it subjects the battery to overcharge, we find in power economy an added argument for disconnecting the trickle charger two or three days a week. In this way current cost, with practically any type of charger, may be reduced to less than a dollar a month.

We therefore find that A power satisfaction, with due consideration to convenience, economy, and reliability, is most readily attained by observing the following:

1. Use a charging system, whether trickle or high-rate, which balances the load from the battery.
2. Keep the electrolyte level of rectifier and storage batteries correct by the addition of distilled water.

The first rule may require occasional disconnection of the trickle charger or the use of a higher charging rate for limited periods. The reward for obeying these most simple precautions is unfailing and reliable A battery service at low cost.

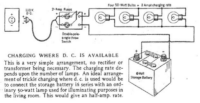

CHARGING WHERE D. C. IS AVAILABLE

This is a very simple arrangement, no rectifier or transformer being necessary. The charging rate depends upon the number of lamps. An ideal arrangement of trickle charging where d. c. is used would be to connect the storage battery in series with an ordinary 50-watt lamp used for illuminating purposes in the living room. This would give an half-amp. rate.

THE LISTENERS' POINT OF VIEW

Conducted by John Wallace

How Long Will Radio Broadcasting Prosper?

DOPING out what is going to happen in the future is always a pleasurable, if hardly ever a profitable, pastime. Most dopesters of radio's future are extravagant in their prophecies of ever-widening influence and ever-increasing prosperity. We are inclined to agree with them, providing they do not attempt to push their claims too far into the future. For it is our conjecture that radio will arrive at its full maturity in a very few years. From then on its course will be no more sensational than is' that of the talking machine at the present time. And in a decade or so will come some new contraption that will relegate radio to as exciting and varied career as that of a telephone operator.

But this inevitable cycle could, conceivably, be nipped in the bud (astounding mixed metaphor) by the failure of radio to keep up with the exactions of a public which has already become inured to its novelty. Concerning this possibility we have an interesting letter from a reader, J. R. Coolidge 3rd, of Brookline, Massachusetts, which presents a good analysis of the factors which will determine radio's longevity. Mr. Coolidge says in part:

"Answering your questionnaire has prompted me to go further into the fundamental questions which affect the future of radio in this country, questions which radio enthusiasts discuss eagerly whenever they meet and for some strange reason rarely put on paper.

"Without having any statistics at my disposal I will assume that there is one radio receiver in use for 50 per cent. of the families in America, and that B. C. L.'s may be divided into three important classes:

"1. Average normal healthy citizens who buy standard sets in good looking cabinets as an investment in entertainment and as a piece of furniture for the living room and because their neighbors have one. This class spends from $50.00 to $300.00 on an outfit once in about five years and then about $25.00 per year in upkeep.

"For them, this investment cannot be worth while in the long run if most of the 'entertainment' is dance music, jazz, cheap popular songs, etc., because the same amount of money would permit them to hear the original performances at frequent intervals and to go to the dance halls, etc., where this class of music is current. If this is to be the prevailing type of program, the majority will cease to use their receivers after the novelty wears off. To hold the interest of this large class permanently it will be necessary to provide music of very high quality indeed, not necessarily heavy classics only but any good music played by well trained orchestras, organ recitals by the most skilled players, good instrumental solos, etc., in the field of music. In other fields, such as politics, education, religion, athletics, and radio plays, the same standards must be maintained if broadcasting is to be permanent. Give the B. C. L. something at home which would be expensive or impossible to obtain outside, and his receiver will be in use every night.

"2. Invalids, shut-ins, convalescents, aged people, those who are hard of hearing, and those who have poor sight. This class is not nearly so large numerically as the first but it is very important because nearly every family has at least one member. To this class, the radio is more than mere entertainment; it is their constant companion. Instead of mere 'news flashes' a good reader could read them in detail extracts from the morning papers every morning, including the best editorial comment. The same could be pursued with regard to magazines and popular novels (a chapter a day). For this class a certain amount of popular music, etc., could be worked in at appropriate intervals before the healthy members of the family returned from their daily occupations. This class does not buy radio receivers, but uses the family receiver hard and constantly. Often it is because of them that the receiver is purchased.

"3. The smallest class, the enthusiasts and experimenters, who buy radio parts and build and rebuild their own receivers, constantly striving to improve the quality, the selectivity and sensitivity of reception, always spending more than they can afford, whether it be in tens, hundreds, or even thousands of dollars, on their equipment. Most of the readers of RADIO BROADCAST must belong to this group, because they alone, among B. C. L.'s, are interested in technical questions and details of construction.

"This group has special requirements —It wants high grade programs just as much as the first group, but in addition it must have DX because DX is a definite measure of comparison between receiving sets in the matter of sensitivity."

The writer's classification of radio listeners seems to us an accurate one, as does also his outline of their respective demands. The first group, he says, must be guaranteed first class entertainment or they will gradually lose interest and finally abandon radio altogether. There need be no grave fears on this score however, for this fact is realized by a sufficient number of radio station operators. Of course it is not realized by all. But that is their hard luck. The

AN INFANT PRODIGY AT WRVA
Conrad Rianhard, aged nine years, who is said to have mastered, to an incredible degree, many difficult piano compositions

NO STRANGERS TO THE MICROPHONE

The first is Eva Gruninger Atkinson, contralto, who sings from KPO; next is Keith McLeod, studio manager and pianist at WJZ; the third is Miss Honarine La Pee, a "syncopating pianist" at KMOX. Miss La Pee recently won a contest as the most popular artist on the staff of this St. Louis station. The fourth photograph shows Miss Josephine Holub, violinist on the "Pilgrim" program, heard from KGO on Tuesday nights

listeners, instead of abandoning their receiving sets, will abandon the inferior stations, whereupon the inferior stations will go out of business.

The service which radio can perform for Mr. Coolidge's second group is undeniable. However, it is a *service*. That is, it is a department which can bring the broadcaster no demonstrable financial return. Evidently, in this day of cut throat operation, no station manager can afford to devote much time and thought to a group which is numerically small. But when time and the operation of the 1927 Radio Law has winnowed the existing number of stations down to a comparative few, these stations will find it not only possible, but advisable, to provide features for minority groups.

Why a Good Program Was Good

THE inaugural program of the National Broadcasting Company, which took place last winter, has come in for quite a lot of praise. Perhaps the most authoritative testimonial to it was that of Samuel Chotzinoff, music critic of the New York *World* in his column: "Concert Pitch." Commenting on the arrangement of the program, Mr. Chotzinoff said.

If Mr. Aylesworth constructed this masterpiece of program making he proves himself a better psychologist than all the orchestra conductors we have ever listened to. Mr. Aylesworth knows that the untutored radio ear cannot be much different from the sophisticated concert ear. Four Beethoven sonatas on the same program are as boring to the trained musician as they are to the public at large, even more

boring, in fact, because the musician knows what he is in for.

Mr. Aylesworth might have followed the ethereal "Lohengrin" prelude with the equally ethereal prelude to "Parsifal" and by doing so alienate the radio fans from Wagner forever. Instead,

JAMES PEARSON

The "Newsboy" at KFNF, Shenandoah, Iowa. Mr. Pearson gives a digest of the news at 7:45 P. M. "The Newsboy" is also heard in a health talk at 7:50 A. M. On Sundays, Mr. Pearson gives religious talks

he isolates the "Lohengrin" piece by following its last pianissimo with the lustful, mundane chorus from "Tannhauser." Who but Mr. Aylesworth, ever thought of presenting only the first movement of Schumann's concerto? This movement contains the best of Schumann's inspiration—the other two are anti-climax. During the intermezzo of this concerto, people have a tendency to scan the ads. in the program. Mr. Aylesworth must have noticed this and decided not to take any chances. Even the first movement demands the listeners' closest attention, so Mr. Aylesworth gives his invisible audience a chance to relax by bringing on Mary Garden at once. Mary chooses her own songs, but her personality is intriguing, whether she decides to sing Bach or Irving Berlin. Bear in mind that it is Mr. Aylesworth's purpose to keep his twenty millions from straying out of earshot of the loud speaker.

Those who feel that the Chorale from "Die Meistersinger" is a little too steep will stay to hear the "Lost Chord," which follows it. Will Rogers follows the "Lost Chord." etc.

Compare this amazing program with, let us say, that of the Cleveland Orchestra, which paid us a visit last week. The Clevelanders played a Mozart symphony, the Stravinsky "Fire Bird," and three new American compositions. With the exception of the symphony, for which Mr. Aylesworth can match the Schumann concerto and the Wagner numbers, the highbrow orchestra gave us nothing near as good as any number on the radio program.

As I see it, it isn't the radio that needs encouragement. Mr. Aylesworth should be consoled. The so-called legitimate musical events get the great volume of critical comment because they need it. The radio seems to be getting along beautifully without it.

AT STATION CKNC, TORONTO

Frank Blachford, violinist of the Toronto Conservatory of Music Instrumental Trio; R. H. Combs, general manager of CKNC, which is owned by the Canadian National Carbon Company; Arthur Blight, baritone, frequently heard from this Toronto station

Thumb Nail Reviews

WEBH (and others)—The last act of "Il Trovatore" by the Chicago Civic Opera Company from the Auditorium stage. This was the second trial at broadcasting the Chicago Opera and it proved an incredible improvement on the first job. The voices, which in the first attempt were blurred and echo-y, came out clear and undistorted. The orchestra, as before, was well picked up. In fact the whole broadcast made us reconsider our statement that we had little faith in theater broadcasts. It was in every way a success, and to our taste vastly more pleasant than watching the same opera. An unforeseen delay brought this act on almost an hour late. The attempt of the broadcasters to fill in this wait without specially prepared "filler" was indeed a sorry one.

WMAQ—The Woman's Symphony Orchestra of Chicago playing the Caesar Franck symphony, which is heard all too infrequently by radio. The first movement of this symphony, full as it is of luscious tunes, could be made quite as popular as the hackneyed William Tell overture if radio orchestras would give it sufficient airings. And as real music goes, it would constitute a large improvement on the Tell piece.

WBAP—Jazz, pre-war jazz of a vintage we thought entirely exhausted. But evidently it is making a last stand in the hinterlands. Jiggly-jazz with stops, panting jazz with a hurry-up tempo, noisy jazz with neither rhyme nor reason.

WHT—Al and Pat, the ultra-lowbrow Hello-folkers-ers of a super-lowbrow station, supplying us with one of the very few stomach laughs that have ever got out of radio. With Al at the organ, Pat commenced reeling off the weather reports in delightful burlesque of our old friend the Pianologue. While Pat improvised a melodramatic and quavering accompaniment on the organ, Al recited, in the manner of one depicting little 'Liza's flight across the ice: "For Iowa increasing cloudiness and rising temperature Saturday (sob). Rain Saturday night (tremolo). Sunday cloudy to p-p-p-partly c-c-c-cloudy (blubber), preceded by rain of snow (tears and a complete breakdown)."

WLS—Haymaker's Minstrels. Oh how sad! Minstrel Shows just naturally don't get over by radio. Even the best of minstrel jokes need the reinforcement of a clowning End Man and the coöperation of a lenient For-the-Benefit-of-Charity audience. As for the worst of minstrel jokes—well those were the type essayed by the Haymakers. F'rinstance: A long discussion between two End Men as to the definition of a new moon, capped after many minutes of futile introduction with the side splitting climax "If a man by the name of Moon had a son he'd be a new Moon!"

KOA—Monologue by a Mrs. (or was it Miss?) Harrison. First rate, and in excellent style for radio delivery. Mrs. (Miss?) Harrison is acquainted with one of the principal secrets of success in humorous broadcasting, to wit—that of not pausing after each wise crack for a laugh. Nobody laughs out loud at a radio joke anyway so such pauses are simply flat. This monologist rushed through with her lines at top speed and without underscoring her jokes by a changed inflection of the voice. The ludicrous laugh which punctuated her remarks lost nothing of its mirth in the broadcasting. The sketch was original and had to do with a club woman's busy day.

WJZ and Blue Network—A new radio team, Vernon Dahlhart, Carson Robinson, and Maruy Kellner, specializing in light popular selections,

comedy numbers, and songs of the South. Dahlhart, the singer and spokesman, has a non-obnoxious Southern accent, striking a happy mean between the orthodox "number" and the Southern "numbah." The accompanying instruments are a guitar and a violin. Rather good.

WHT—Al Barnes and Pat Carney again; this time in the "Your Hour," presenting "A Trip Through the Dials." Highly comical in a boisterous, infectious way. Why, we ask ourself, have we been missing these perfectly elegant low-brows all these years? To those of you who become wearied at the lofty pomposity of Doctor Damrosch, we heartily recommend the artfully artless Al Barnes as an antidote.

WJZ and associates—The First National To-Be-Weds issuing propaganda for various picture shows in the guise of a controversy on the merits of a movie just witnessed. Terrible!

WEAF, WJZ, WEEI, WBZ, WTIC, WJAR, WTAG, WCSH, WGR, WGY, WLIT, WRC, WCAE, KDKA, WTAM, WSAI, WLW, WWJ, WGN, WMAQ, KYW, KSD, WOC, WCCO, WDAF, WFAA, WSM, WMC, WSB, KFKX, KVOO, KOA, KSL, KPO, KGO, KFI, KGW, KOMO, KFOA, KHQ, WJAX, WHAS—A speech by one Calvin Coolidge on the occasion of Washington's Birth-

ON THE AIR AT WBBM, CHICAGO

Howard Osburn's International Radio Orchestra. Left to right: C. Mason, P. Beckler, H. Osburn (leader), G. Moorehead, and N. Sherr

day. But of course you heard it, so write your own review. This is simply for purposes of record.

WAIU—Celebrating the installation of a new 5000-watt transmitter by reading a lot of telegrams and sending the best regards of the chief engineer to some personal friend of his in Florida.

WAAT—Nut Club at 2 A. M. Oh well, it was our own fault for staying up that late.

KOIL—A radio play "The Scoop." Good in that it was easy to follow and had a quick moving plot. But the lines were amateurishly written and read in a none too convincing manner.

WFAK—(or WFAD or WFAJ or WPCX!?)—Clearwater, Florida. A garrulous announcer who took up at least as much time between numbers as the numbers themselves, and in spite of all his wordiness never succeeded in pronouncing his call letters so they could be deciphered by any one other than a magician. In striking contrast to—

KFI—whose announcers, realizing, no doubt, that they represent the most sought after DX station, call their letters with a pause between each one making the signal intelligible through even the worst static disturbances.

KMA—Henry Powell doing some excellent old time fiddling. Followed by a trio playing conventional, uninspired jazz.

Microphone Miscellany

WE ARE eternally deluged with printed matter telling us how this feature drew three thousand telegrams of commendation and how that artist receives 'em at the rate of three a minute while his concert is in progress. While we frequently scoff, there must be some truth in the matter. If so, it is only reasonable to surmise that the advent of radio has brought the telegraph companies an enormous increase in revenue. Then if this be so, we suggest, with no very valid argument to back up the suggestion, that the telegraph companies return some of this gold to the people by sponsoring a weekly program. If the reports of large telegraphic returns made by radio press agents are veritable, which the telegraph companies will probably now deny, we can see no reason why they shouldn't spend several grand a week dishing out a first rate program. And if their programs were good enough they might receive enough of telegrams to pay the entire cost of said programs —which is something like the worm devouring himself.

THE KFI-KPO-KGO network has been broadcasting a series of concerts by the Los Angeles Philharmonic Orchestra. The final concert will be on April 23 from 9:10 to 10:45 P. M. Walter Henry Rothwell is conductor.

Communications

Sir:

The National broadcasts form 95 per cent. of what's worth while on the air to-day and they are making, or rather saving, radio, but unless one goes up and down the dials on the hour and the half-hour he's liable to miss much that is really good and worth while, for nowhere are these programs all listed—the chaff winnowed from the wheat. Many metropolitan newspapers own or are affiliated with some one broadcasting station, and labor under the short-sighted policy that they should make no mention in their daily radio programs of any big event that is going out through stations other than their own. On New Year's Day, for instance, the Chicago *Tribune* featured the fact that it would put on the air that afternoon, through its station, WGN, a telegraphic report of the Leland Stanford—Alabama football game in Pasadena, "with the WGN Quartet singing college songs and furnishing the local color." Thousands of fans in this part of the country listened to Quin Ryan's "kiss-proof-lip-stick" voice broadcast this big game from the ticker, not knowing that KYW and the WJZ Chain was putting it on the air direct from the Rose Bowl!

And one other thing in this connection: If I were a national advertiser, spending from $1000 to $10,000 a week in creating good-will by giving the radio public the splendid programs that such advertisers are giving, I'd see to it that I had real coöperation from certain participating stations and the newspapers which own them. Station WGN never misses an opportunity of cutting off the New York announcer; and the Chicago *Tribune* does all it can to create the impression that it, The World's Greatest, at great expense, out of pure love, is furnishing this wonderful program to its Dear Readers. It hides

altogether, or keeps well hidden in the background, even the name of the advertiser.

These great, national broadcasts can't continue unless they bring their sponsers at least a fair return, and they are not going to bring such return unless they are tied up with a certain amount of definite publicity for the advertiser in question, both in the news and over the air.

R. H. J.
MILWAUKEE, Wisconsin.

SIR:

Your fourth question on the enclosed questionnaire has been answered in a general manner because I think American people tend to judge the quality of their cigars by the price. ("What are the six best broadcasts you have heard?" was the fourth question in the recent RADIO BROADCAST questionnaire). As a simple matter of fact I believe it can be demonstrated that there are hundreds of voices as good as the best advertised ones, dozens of comparatively unknown orchestras that would be ranked near the top, and so on.

What I desire in radio is entertainment. If I wish education in a fresher, more interesting, and more permanent form than by radio, I should like to hear a few great men talk over the radio just to discover how human they are—but the others can't ride on my electrons. Neither can

the flatted and fluttery vocalists, or the jazzbos who keep time with a pick handle. I am interested in DX because if the set will bring in KDKA in daylight, or 4 QG at night, then I know that the machine is keeping step in a halfhearted manner, and will bring home entertainment if the weather will let it.

Even in this stationless part of the country nearly every wave carries two stations and a pack of coyotes. It is to weep.

ROBERT T. POUND.
LAVINA, Montana.

SIR:

Of all the "technical" journal's articles I like your facetious articles best of all. Your self-admitted ignorance is refreshing, as is your style of writing.

P. L.
SAN FRANCISCO.

SIR:

I think there is too much constructive criticism of the destructive type in your dept. "As the Broadcaster Sees It" is my preference.

F. H. S.
PITTSBURGH, Pennsylvania.

There seems to be some difference of opinion here!

WHAT THE LISTENER LIKES AND HOW HE LIKES IT

IN THE January and February RADIO BROADCAST a full page questionnaire appeared in this department which was answered in great detail by readers of the magazine. About 1000 answers were received and the results are tabulated below.

It is difficult to interpret the results fairly, because we had answers from city dwellers in congested radio districts and from listeners in remote points, many miles from the nearest broadcasting station. Many answers were not definite but were interestingly qualified. It will be seen that the replies are classified according to the district from which they came. Metropolitan centers, such as New York, Chicago, and San Francisco, were separated from the others in the tabulation.

Among the many important conclusions to be drawn from this survey, perhaps the most impor-

tant is that, under present radio conditions, the city listener, especially in the large city, relies on his local stations for the most part, while those living some distance from the so-called "key stations" rely on DX. The comparatively few listeners who answered our specific question: Is the DX listener disappearing? gave conclusive reply that the DX listener is here now and for evermore. Another interesting conclusion is that "instrumental music" is favored by above all other classifications by more than 60 per cent. Serious music, so called, was most popular, although not overwhelmingly so.

The fourth question about the six most popular broadcasts received a variety of answers, which fell into three classes, as the tabulation shows. Our purpose was to discover what six broadcasts had attained great popularity. The

tabulation of this question shows that the regular feature or "hour" of greatest popularity was always the one which was broadcast over the largest number of stations. But the comparative popularity of broadcasts such as that of the Happiness Boys and Sam 'n' Henry is remarkable, for each feature is broadcast over but one station. In special events, sports broadcasts top the list. Among individual artists, the name of John McCormack appears most often, although that of Walter Damrosch is a close second.

After this tabulation was completed a large number of questionnaires were received which arrived too late to be tabulated and we regret that these could not be included. And to all those who sent in letters of comment and appreciation and the filled-out questionnaire, we offer our hearty thanks.—THE EDITOR.

The Results

1. Do you listen to your radio evenings as you would to a regular show, or do you simply turn it on and use it as a background to other activities?

	New York	Chicago	San Francisco	National
As a background	10	3	4	91
As a show	45	10	18	370
As a background except for features	80	17	40	448

2. Do you regularly tune-in on distant stations or do you regularly rely on your local stations?

Locals only	61	13	19	314
Regularly rely on distance	25	8	27	433
Distance occasionally	49	3	14	187

(They tell us that the DX hound is a fast-disappearing breed. Is he?)

Yes	1	0	6	36
No	3	3	3	92

3. If you had a hundred minutes to listen to all, or any part of the following broadcasts, how would you apportion your time?

Instrumental Music:				
Serious	26.3%	30.1%	28.5%	23.3%
Light	18.	18.2	14.3	18.4
Popular	16.6	14.8	16.1	17.4
	60.9	63.1	58.9	59.1

Vocal Music	12.	13.6	13.2	12.6
Radio Play	5.3	1.7	8.0	4.7
Speech	4.7	4.3	3.0	3.8
Educational Lecture	7.6	8.2	4.6	7.0
Miscellaneous Novelties	9.5	9.2	12.3	13.8

4. What are the six best broadcasts you have heard?

Hours or Regular Events

Atwater Kent	228
Victor Hour	128
N. Y. Symphony	93
Eveready	89
Boston Symphony	70
Maxwell Hour	53
A. & P. Gypsies	49
N. Y. Philharmonic	48
Whitall Anglo-Persians	47
Clicquot Club	44
Capitol Theatre	44
Ipana	40
Goldman's Band	38
San Francisco Symphony	33
Happiness Boys	31
Roxy	30
Royal Hour	28
George Olsen	28
Sam 'n' Henry	26
Zippers	26
KDKA Symphony	25
Balkite	22
Cook's Travelogue	19
Silvertown	19
Goldy and Dusty	18

Little Jack Little	18
Marine Band	18
Record Boys	16
Davis Saxophone Octette	15

Special Events

Dempsey-Tunney Fight	153
Inaugural N. B. C.	131
World's Series	124
Army-Navy Game	83
Coolidge's Inaugural	39
Radio Industries Banquet	38
Democratic Convention	38
Pershing's Farewell	31
Alabama-Stanford Game	28
Dempsey-Firpo Fight	20
Election 1924	13

Individuals

John McCormack	41
Walter Damrosch	39
Schumann-Heinck	33
Calvin Coolidge	30
Vincent Lopez	28
Joseph Hoffman	19
Will Rogers	18
Rev. Cadman	18
Reinald Warrenrath	17
Madame Homer	16
Marian Talley	16
Godfrey Ludlow	14
Mary Lewis	12
Mary Garden	12
Rev. Fosdick	10
Wendell Hall	10

Filament Lighting from the A. C. Mains

A Discussion of a Practicable Method for Batteryless Receiver Design—The Advantages of Series Connection of Filaments

By ROLAND F. BEERS

A NATURAL query when discussing the subject of series filaments is to inquire why such an arrangement is either necessary or desirable. This inquiry ably illustrates that growth of tradition which radio is rapidly accumulating. With a few exceptions, modern radio receivers up to the present time have employed the parallel filament connection. The filaments of the radio tubes have been operated from a constant 6-volt source, and variations of individual tube filaments have been made by changing the applied voltage.

It is not difficult to trace the reason for the prevalence of the parallel filaments scheme. In the beginning, audion filaments often required one or two amperes for each tube. A multi-tube set would therefore require 3 to 6 amperes. What better source of power was there, then, than the familiar 6-volt storage battery—a low-voltage, high current capacity affair? Its availability through the regular channels of distribution was assured, and everyone was acquainted with its operation and maintenance. As the first uses of the vacuum tube were determined by the amateurs, so were its tendencies bound to follow along their pioneer activities. It did not take long, therefore, for the 6-volt parallel filament scheme to become firmly entrenched in the minds of those who were to become radio set designers and builders of later periods.

In answer to the query concerning the system wherein radio tube filaments are arranged consecutively in series, it is first necessary to show why A batteries have not yet been completely eliminated by the use of the parallel filament scheme. Let us first consider the general principles of power supply, illustrated by the modern B-power unit. Here we have the customary transformer, rectifier, and filter circuit, as shown in Fig. 2. The current and voltage capacity of such a device is of the order of 85 mA. and 200 volts d. c. The degree of filtering of high-grade units of this type is such that a variation or ripple of but 0.1 per cent. is attained in the current output.

If high-quality reproduction is desired, with freedom from hum and "motor-boating," it is absolutely necessary that the variation in current supply be of this order. In order to achieve this degree of smoothing in plate supply devices, a filter structure of such design as is shown in Fig. 2 is used. The retail cost of a structure of this type is approximately $30.00.

If we are to adhere to the same standard of quality with regard to the A power-supply source, it is also necessary that the current ripple shall not exceed 0.1 per cent. This statement has been theoretically demonstrated and experimentally verified, using an average radio receiver with no potentiometers which may be used to balance out any hum. With such a receiver the storage battery might be replaced with this theoretical A power-supply unit. Basing our judgment upon the design of the filter circuit shown in Fig 2, we may in-

crease the values of the constants required until we obtain the same degree of filtering for 2½ amperes at 6 volts. Our structure will then look like Fig 1. The weight of such a device would be approximately 300 pounds, and its cost, on the same basis as previously assumed, would average $1500.00! For the reason of its weight and cost, it is therefore not feasible to effect A-battery elimination at 2½ amperes.

There are two main reasons why the series filament connection is so much more desirable from the power standpoint. The first reason is that the total current to be filtered is very much less than with the parallel system. With any number of tubes in series, the maximum current required is only that taken by one filament, and as the size and cost of filter chokes increases rapidly with the increase in the amount of current they are required to pass, we realize the economy possible by a reduction of current.

We are somewhat assisted by the fact that the

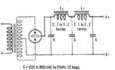

C = 2000 to 8000 mfd. for 5 Volts, 2.5 Amps.

FIG. 1

filtering efficiency of a choke coil increases with the current through it; therefore, we should be able to design a more efficient filter system for A supply than for B supply, as far as the chokes are concerned. The amount of energy that can be stored in a given choke coil is equal to one half the inductance times the square of the current. In spite of this fact, however, there are other considerations which work to our disadvantage in the design of a high-current choke coil. For example, on account of the direct-current saturation of the steel core it is necessary to include large air gaps in high-current chokes. These air gaps reduce the effective permeability of the core to low values, making it necessary to increase the amount of copper and iron to large quantities, in order to attain sufficient inductance at high direct-current saturations. We are also limited in this consideration by the amount of copper

that can be used in a choke coil, on account of the d. c. resistance. If the choke coils have high d. c. resistance, the voltage output of a power unit intended for parallel filament supply would be much too large on a 3-tube set, for example, if the unit were designed to supply sufficient voltage for a 6- or 8-tube set. A power unit having sufficiently low regulation for parallel filament operation would therefore require monstrous inductances.

A second reason why the series filament connection is desirable is that higher voltages are available for filtering. For example, a 5-tube receiver with 199 tubes in series requires 15 volts for the filament supply, and an additional 15 volts may be advantageously employed for grid bias. Therefore, the total voltage required is 30 volts instead of 3 volts (grid bias is obtained from external batteries) for the parallel filament connection. Additional voltage is also available at the filter circuit by virtue of the fact that the A current is obtained through a series rheostat, the voltage drop through which may range from 100 to 150 volts (Regulation in this circuit has not the same importance as in the parallel filament scheme, because it is a constant-current system and not a constant-voltage system). The total voltage applied to the filter circuit is equal to the sum of these values or approximately 200 volts.

Now it is a matter of common knowledge that condenser efficiency in a filter circuit is much greater at high voltages than at low voltages. The amount of energy stored in a condenser is equal to one-half the capacity times the square of the voltage applied. Therefore, from this standpoint alone, a great saving is gained. For example, if a total capacity of 12 mfd. is required for a given degree of smoothing at 200 volts, the capacity required for the same degree of smoothing at 3 volts would be in excess of 50,000 mfd. The saving in inductance and capacity affected by the use of the series filament scheme is therefore enormous.

Fig 3 shows a typical filter circuit for use in connection with 199 type tubes in series. This arrangement actually gives the same degree of filtering found in Figs. 1 and 2. Its cost is within reasonable limits, in comparison with other methods of power supply, while its size is not beyond the scope of average radio cabinets.

One obvious way in which to attack the problem of series filament connections is to place the vacuum tubes consecutively in sequence, in the same order in which they normally occur. That is, we may begin with the radio-frequency stages and run through the detector and audio stages. While this method may appear to be the most straightforward, it is open to several criticisms. In the first place, it has been demonstrated that less difficulty with hum will be experienced if the detector is placed nearest the B-minus or ground connection. Several different theories have been advanced to explain this situation, but the important fact to

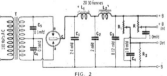

FIG. 2

the average set constructor is that a real advantage is gained by this position of the detector tube.

A second difficulty, frequently encountered in this method of series connection is that the proper values of negative grid bias are not always available. In certain commercial receivers, notably the Western Electric 4-B super-heterodyne, it is common practice to connect the grid return lead of those tubes requiring negative grid bias to some preceding filament in the series. This connection takes advantage of the voltage drop

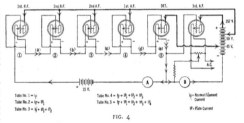

FIG. 3

in the filament circuit, and employs it as a grid bias. It can be readily demonstrated that while some of the tubes in the series will be benefited by such a connection, there will always be at least one of the tubes for which the required negative grid bias is not available. The third objection to the straight series connection is that all of the 199 tube filaments, with the exception of the first, must have some current overload in them, if the applied plate voltage exceeds 40 volts. For an analysis of this situation, see Fig. 4.

In this diagram, with no potential applied to any of the plates, the filament current ammeter, A, could be adjusted to read 60 mA., and the correct current would then be passing through each tube filament. However, when B voltage is applied to the plates, current flows in all the plate circuits, as indicated by the plate milliammeter, B. This current must return to the negative B. If we check the current in the filament for tube No. 3 at the point (b) we would find that it was carrying the regular filament current plus the plate current of tubes Nos. 1 and 2, because, as stated above, these plate currents must return to the negative B, and the most direct path is through the filaments. Consequently, the filament of tube No. 3 would be overloaded by an amount depending upon the plate current of tubes Nos. 1 and 2.

The data printed on the diagram, Fig. 4, give the amount of current in the filaments of each of the first five tubes. The amount of overload depends on the plate current of the preceding tubes. Thus, in the case of tube No. 3 (the 2nd audio stage), the overload is equal to the sum of the plate currents of tubes Nos. 1 and 2.

Examination of the other filament circuits would indicate similar overloading.

On account of these various difficulties involved, it was determined to make a study of the errors which existed in the straight series connection and to make such alterations as were required to bring the radio receiver to its normal operating characteristics.

The first step in this direction was to place the normal grid bias on each tube, with 90 volts on the plates of the amplifier tube and 45 volts on detector plates. It was frequently found that, if the various filament voltage drops were relied upon as the only source of grid bias, three difficulties arose. First, because of the need of a tube in the series ahead of each one requiring grid

bias, the number of available biases was insufficient. A receiver wired in this manner very frequently blocked completely in one or more stages on account of improper grid bias. When C batteries were inserted in these troublesome stages, amplification was again at normal values, but as this receiver was intended to be a batteryless set, other means of grid biasing were sought.

The second difficulty was encountered in the r.f. stages, where regeneration often occurred. In this situation the operator is unable to control his receiver on account of the incessant squeals and whistles of heterodyned carrier waves. A similar situation may also exist in a.f. stages. In this case it may cause an audio howl that will completely prevent reception.

The means finally adopted in obtaining correct grid biases was to place a 60- to 75-ohm adjustable resistance in series with each filament, for 199 type tubes, as shown in Fig. 5. This method is entirely satisfactory and has been used to some extent in Western Electric receivers. If adjustable resistances are used, the plate current of each tube can be set at the right value when final tests are made on the receiver. Fixed resistances may be used, however, if the constructor does not want to make this refined adjustment. The plate current of each 199 type amplifier tube can be reduced to about 2.2 mA., with 90 volts on the plates.

When the 60-ohm resistances were placed in position in the filament circuit, it was found that there was no tendency to regenerate, as outlined above. From a study of Fig. 5 it may be seen that r.f. currents in tube No. 1 must traverse the filament of tube No. 2; likewise, those currents in tube No. 2 must pass through the filament of tube No. 5. The question might arise whether or not this would be a source of regeneration. It was found, however, that the inclusion of the 60-ohm resistances in the grid circuit of the respective tubes was of the proper value to eliminate the tendency to oscillation. This phenomenon is a well known fact, and in many modern radio receivers a resistance is deliberately included in the grid circuit for the purpose of stabilizing the tendency to oscillate. The principle involved here is that, if the resistance of an oscillatory circuit is of sufficiently large value, sustained oscillations

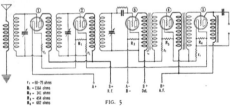

Tube No. 1 = I_F
Tube No. 2 = $I_F + I_P_1$
Tube No. 3 = $I_F + I_P_1 + I_P_2$

Tube No. 4 = $I_F + I_P_1 + I_P_2 + I_P_3$
Tube No. 5 = $I_F + I_P_1 + I_P_2 + I_P_3 + I_P_4$

I_F = Normal Filament Current

I_P = Plate Current

FIG. 4

r_1 = 60-75 ohms
R_1 = 1364 ohms
R_2 = 341 ohms
R_3 = 454 ohms
R_4 = 682 ohms

FIG. 5

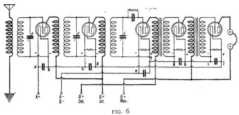

FIG. 6

cannot be produced therein. Such a circuit is said to have a high decrement.

When the series filament receiver had been provided with proper grid biasing, as described, it was found that its performance was at least equal to that of a similar set with parallel filaments. This particular receiver had two stages of tuned r.f., detector, and three stages of resistance-coupled a.f. amplification. Five UX-199 tubes and one UX-112 power tube were used. A filament battery of 36 volts was employed feeding 60 mA. to the set. A plate battery of 157 volts delivered a total current of 22 mA.

If the results of Fig. 4 are now applied to the set in consideration, it is readily seen that the filaments of all the 199 tubes in series, with the exception of tube No. 1, will carry a current which is in excess of 60 mA. The excessive current in the last tube in the series may rise to a value which is 30 per cent. above the normal value. The actual current overloads may be computed by reference to the table on this page. In this table, the overload figures are obtained by deducting 60 from the filament-current figures given in the second column of figures. These data are based on the assumption that all tubes have proper grid bias, that amplifier tubes are provided with 90 volts on the plates, and that detector tubes have 45 volts on the plates. This arrangement will give a plate current in each amplifier tube of approximately 2.2 mA., while that of the detector tubes will average 1.0 mA.

A very simple and satisfactory method of reducing the filament current overload is to shunt a bypass resistance across each filament, as shown in Fig. 5. These resistances do not all have the same values, and their proper size, for 199 tubes, is determined as follows:-

$$R = \frac{3}{\text{Excess Filament Current, Amperes.}}$$

For example, the last tube in a 6-tube series was measured, and its filament current was found to be 71 mA. The excess filament current is therefore 11 mA., or 0.011 amperes. The correct value of the bypass resistance is then equal to $\frac{3}{0.011}$, or 273 ohms.

The table shows the correct values of bypass resistance to be used in various combinations of 199 tubes.

When the receiver has been provided with the correct biasing resistances and filament shunts, it is now ready to be connected up for use. If the receiver is to be used in connection with an electric power supply device, it may be necessary to take additional precautions to prevent "motor-boating."

For the present discussion, the following rules summarize the procedure in laying out a series filament receiver using 199 tubes:

1. Wire all filaments in series, including a

60-ohm resistor ahead of each amplifier tube for grid bias. (see Fig. 5).

2. Choose the proper order of wiring the filaments, so that the detector is next to the B minus or ground connection.

3. Place shunt resistances across each filament, as shown in Fig. 5.

The foregoing discussion is based upon a plate voltage of 90 volts on all amplifier tubes, so that large signal voltages may be applied without distortion. Under this condition it is necessary that a negative grid bias be obtained in some manner for these tubes, such as by the use of the 60-ohm

resistances in series with each amplifier filament. It is entirely possible, however, to build a satisfactory series filament radio receiver without the use of these 60-ohm resistances, for use on moderate volumes. In this case it is essential that the amplifier plate voltage shall not exceed 50 volts, so that the average plate current will lie between 2.0 and 2.5 mA. This situation frequently prevails in radio-frequency stages where the signal voltage is low enough not to warrant the use of a high negative grid bias, and also in resistance-coupled a.f. stages where the actual voltage on the plates is frequently of the order of 40 volts.

In such cases as these a very simple procedure may be followed—that of connecting the grid return leads to the negative leg of the respective filaments, as shown in Fig. 6. It is, of course, apparent that the use of shunt resistors across each filament is always necessary, in order to prevent current overload. It will also be advisable to provide some means of stabilizing the r.f. amplifier, thus preventing the tendency to oscillation. The inclusion of the 60-ohm biasing resistances reduces this tendency to a very low degree, but without their use, difficulties are at hand. A satisfactory means of overcoming this situation is to bypass these r.f. currents through a 0.1–1.0 mfd. condenser shunted across each filament, as shown in Fig. 6. The use of these bypass condensers insures complete control over the receiver at all times and affords a greater possibility of high-quality reproduction.

Type Of Set	Order Of Tubes	Plate Current, Ma.	Filament Current, No Shunt. Ma.	Shunt Resistance.
3-Tube Regenerative 3-199's	2nd a. f. 1st a. f. Det.	2.2 2.2 1.0	60.0 62.2 64.4	None 1364 682
Browning-Drake 3-199's; 1 UX-112 on A. C.	1st r. f. 1st a. f. Det.	2.2 2.2 1.0	60.0 62.2 64.4	None 1364 682
Browning-Drake 4-199's; UX-112 on A. C.	1st r. f. 2nd a. f. 1st a. f. Det.	2.2 2.2 2.2 1.0	60.0 62.2 64.4 66.6	None 1364 682 454
T. R. F. 4-199's; 1 UX-112 on A. C.	1st r. f. 2nd r. f. 1st a. f. Det.	2.2 2.2 2.2 1.0	60.0 62.2 64.4 66.6	None 1364 682 454
T. R. F. 5-199's	1st r. f. 2nd r. f. 2nd a. f. 1st a. f. Det.	2.2 2.2 2.2 2.2 1.0	60.0 62.2 64.4 66.6 68.8	None 1364 682 454 341
T. R. F. 5-199's; 1 UX-112 on A. C.	1st r. f. 2nd r. f. 3rd r. f. 1st a. f. Det.	2.2 2.2 2.2 2.2 1.0	60.0 62.2 64.4 66.6 68.8	None 1364 682 454 341
T. R. F. 6-199's	1st r. f. 2nd r. f. 3rd r. f. 2nd a. f. 1st a. f. Det.	2.2 2.2 2.2 2.2 2.2 1.0	60.0 62.2 64.4 66.6 68.8 71.0	None 1364 682 454 341 273
Super-Heterodyne 8-199's	1st i. f. 2nd i. f. 3rd i. f. 1st Det. Osc. 2nd a. f. 1st a. f. 2nd Det.	2.2 2.2 2.2 1.0 1.0 2.2 2.2 1.0	60.0 62.2 64.4 66.6 67.6 68.6 70.8 73.0	None 1364 682 454 400 348 277 231

AS THE BROADCASTER SEES IT

By CARL DREHER

Drawings by Franklyn F. Stratford

One Explanation for the Plethora of Broadcasting Stations

FOR some time, as the number of broadcasting stations in the United States mounts toward the thousand mark, I have been wondering where they all came from, and what process caused them to multiply. Did they reproduce by simple cell division, like the amoeba? Did they cast spores over the countryside, which, falling on fertile soil, took root and became new transmitters? Or was it necessary to mate a pair of them in order to produce a third? Whatever the mechanism, I have found one of the sources of their nourishment, which heretofore I had overlooked. In small towns, it may well be the principal urge leading to expression in kilocycles. It is the booster spirit, the "Bigger and Better Blattville" pressure which all but blows off the safety valve in the would-be metropoli of the inland states. Among other activities, it builds broadcasting stations.

The rivalry between country towns, while usually good-natured, is exceedingly intense. When it goes so far as to lead to the posting of signs about the town, urging the citizens to buy round-trip tickets at the local railroad station, in order that the total for the year may be higher than the ticket sales of some rival settlement down the line, it is apt to extend to all the other activities of community life. One of these is broadcast reception—and transmission. Some village of 2000 out on the plains possesses a broadcasting station, perhaps by accident. A manufacturer of babies' diapers, say, has erected it to advertise his product. The next village, with 4000 population and a natural feeling of superiority, feels an irresistible impulse to have a broadcasting station bigger than the diaper broadcasting station. The local manufacturer of varnished pretzels thinks he might take a whack at it. His primary object is, of course, to advertise his varnished pretzels. But he also wants to shine at his luncheon club among his fellow business men. He wants to be slapped on the back by the President of the Chamber of Commerce. He wants to be pointed out as a benefactor of the town. These considerations weigh with him as much as his itch for gold. No doubt they are the determining factor in many cases where a station stays on the air although it is actually losing in

dollars and cents. The boosters must broadcast. Some of them are even willing to hold the bag for a while, as a matter of civic pride.

In some cases the stations actually receive community support through the local Chamber of Commerce or some other semi-official or official agency. Then, of course, there are municipally owned and operated transmitters, such as those of Atlantic City, New Jersey (WPG), and New York City (WNYC). Whether municipally maintained or not, broadcasting stations tend to function as the mouthpieces of their respective communities—"The Voice of Whatever-It-Is" being a common slogan derived from this rôle.

Of course there are very definite limits, economic and electrical, to this process of community-magnification by radio. The towns may compete all they like in fire engines, railroad ticket sales, and fraternal orders; such matters are their own business. But radio knows no county lines, and there is no such thing as one's own business when it comes to letting loose ether waves of frequencies between 10 and 50,000 kilocycles per second. Furthermore, to plant a broadcasting station in every township is about as rational a procedure, economically, as maintaining a dozen telephone systems. In wire telephone systems, for reasons too elementary to require statement, unity is the aim. And right there we have the answer to the broadcasting problem as it affects the civic ambitions of the smaller towns. Instead of Podunk having

"HE ALSO WANTS TO SHINE AT HIS LUNCHEON CLUB"

its own precious twenty-watt squeaker manipulated by one of the local apprentice electricians, and Peadunk, fifteen miles away, competing desperately with a twenty-five watter, let Podunk and Peadunk and a few others, if they *must* broadcast, contribute toward a common transmitter, and maintain studios connected to it by wire lines. There is plenty of No. 12 copper running on poles over the countryside, and, at any rate, one transmitter with a half-dozen studios is better than six transmitters with one studio apiece.

Broadcasting and Social Upheavals

ABOUT a year ago the British general strike flared up, ran its course, and came to an end. Maybe it was the last general strike of this industrialized world, but it takes a confirmed optimist to believe that. If it should come to pass again, whether in the British Isles or on some other portion of the globe, what part will be played by radio broadcasting, the latest and most rapidly developing of all the means of mass communication?

The British conflict afforded some indications of what may be expected to happen in the future. Once the battle was on, the regular newspapers either ceased publication or were reduced to little more than handbills. The government published a newspaper of its own, the *British Gazette*, in opposition to the *British Worker*, representing the Labor side. Judged by normal newspaper standards, neither paper was a shining star. But they were about all that remained of British journalism. Piling transportation difficulties on to the confusion in the newspaper plants themselves, the strike raised hurdles too high to surmount. Had it not been for the chain of the British Broadcasting Company, with its principal stations at Aberdeen, Belfast, Birmingham, Bournemouth, Cardiff, Glasgow, London, Manchester, Newcastle, and Daventry, the public would have been practically uninformed during the early days of the strike. At such times the tide toward panic runs swiftly. In the absence of reliable information, sensational rumors spring up and circulate rapidly, gaining horrors, by a well-known process of accretion, as they go along. An agency which helps to keep people's feet on the ground, at such a time, is performing no small service. This is substantially what radio broadcasting did in England during the crisis of May, 1926.

Let it be emphasized that our discussion is concerned with the function of broadcasting during times of grave social disturbance in industrial communities. During normal periods, broadcasting occupies a field quite distinct from that of journalism. Since fairly normal periods, until the world gets a good deal worse, may be expected to cover 0.999 of the total time, the newspapers are in a secure position. Their facilities for news-gathering and catering to the interests of great masses of people are in a class by themselves. Broadcasting, in quiet times, interferes with the newspapers about as much as the theatres do; that is to say, not at all. If a man intends to go to a show in the evening, or to listen to his radio, he reads his morning and evening papers just the same.

But when there is an acute industrial crisis, the tables may be turned temporarily, as the British strike showed. The reason lies in the contrasting conditions of news dissemination by radio telephony and by printing. Publishing a modern newspaper of large circulation is a formidable project. We do not realize what a huge undertaking it is merely because we are accustomed to it. The thing has been organized and built up on such a scale that we feel it must come and every morning, like the sun. That is a palpable mistake. The newspaper is produced by the concerted action of hundreds or thousands of men. If the men quit, there is no newspaper.

Even if the newspaper is produced, it means nothing unless the distribution system remains intact. Modern newspapers are bulky. One copy does not weigh much, but try lifting fifty and then visualize the motor trucks and mail cars required to transport fifty thousand. Reduce the size, and you have ameliorated the difficulty, but you cannot remove it. Paper is gross matter, subject to the physical limitations of physical things.

Contrast the radio telephone station. Instead of hundreds of workers, it requires only a handful of men. A station of 1000 watts output is considered fairly large; its night program coverage in an urban district is, in fact, comparable with that of a good sized newspaper's circulation—say 100,000 listeners. Plenty of 500- and 1000-watt stations with the studios, control room, and power plant in close proximity can be, and are, run by one technician. One of the largest broadcasting plants in the world, with forty or fifty field points, and the studios and radio power plant separated by thirty-five miles, is operated by a technical staff of sixteen men. In a pinch, with the field work tossed overboard, the two engineers in charge, whose functions in normal times are mainly administrative, could run the whole plant alone. They might need a wire chief for the lines connecting the studio and radio station, but a telephone engineer could substitute for the wire chiefs if the latter all went out on strike. In short, three or four professional men, who are likely to be on the "White" side in a serious industrial conflict, can operate the largest broadcasting stations, and one or two men each can take care of the rest. They might not turn out a one hundred per cent. transmission job, but that is beside the point. The station would radiate and say what the proprietors wanted it to say.

The second factor, that of transportation, presents an even more striking contrast. A broad-

casting station generates its own "carrier," as the high frequency wave is aptly termed. Its content is not printed on a few ounces of paper or other tangible medium. It is of the nature of radiation—weightless, impalpable, invisible, and, once released, it penetrates to every point within range of the station without the aid of a single man or vehicle. In distribution, even more than in production, the radio is free where the newspaper is shackled, when the men walk out.

Not very many powerful radio stations are required to cover a country of moderate area. In Great Britain a single high-power station, Daventry, of twenty-five kilowatts rating, can cover the entire island kingdom. Countries like France and Germany are similarly protected. Even in the United States a single fifty-kilowatt station located in the North-East can provide usable service, in daylight, for South-Eastern Canada, New England, New York, Pennsylvania, New Jersey, Maryland, Delaware, the District of Columbia, and the Virginias, with a possibility of service to regions beyond. The daylight range of such a transmitter is about 400 miles. The population of the area of a circle of this radius, in this part of the United States, is of the order of thirty millions. Three or four such stations strategically placed over the country, getting

"OUR LISTENING WILL BE DONE AT THE ORIFICES OF LOUD SPEAKERS"

news over telephone or telegraph lines remaining in service, or by airplane if the worst comes to the worst, could solve the problem even in the United States. Of course, not every family, even in such industrialized countries as the United States and Great Britain, owns a radio receiver. But millions of them do, and each set is potentially a focus of information when information is hard to get by other means.

The weaknesses of radio distribution of news fall under two heads: First, physical vulnerability analogous to, but less serious than that of the newspaper; second, the limitations of the spoken word, as such. The first division may in turn be considered under two subheads: Power supply and wire connections. A broadcasting station requires electric power, normally obtained from central stations. During a general strike this power might not be available. But the amount required is not excessive, being in the ratio of five or six times the energy output of the transmitter. Ten horsepower would be an ample sup-

ply for a 500-watt station. A gasoline driven alternator of two hundred and fifty kilowatts capacity would supply the largest broadcasting station in existence at the present time. Such a machine is readily obtainable, and might be included as an integral part of broadcasting plants whose continuance in operation is vital. As for the telephone lines connecting studio and power plant, the former being located in the city and the latter rurally, a detachment of infantry with a few motor cyclists, could safeguard twenty miles of aërial cable without special difficulty. And in many instances the stretch of wire is short. In New York City, for example, there are two five-kilowatt stations with studios practically adjacent to the power rooms. The assailability of radio stations, even if overt force is employed, is not greater than that of water works and similar utilities, and far less than that of newspapers. As for the telegraph circuits on which the radio stations would have to depend, in the main, for news—armies and navies usually maintain very effective radio telegraph systems. In the United States there would be no insurmountable difficulty on that score.

But, it may be objected, when the radio speaks, its words go in at one ear and out at the other; it lacks the relative permanency of the printed phrase. This defect is of only moderate consequence. For the general public the newspaper certainly contains no element of permanency; it lasts a morning or evening and goes into the fire. The readers remember principally the headlines of the articles which interest them. These salient points are impressed just as well by oral communication, and, by frequent repetition, or by some coördination of printing with radio, the defect may be overcome entirely. For example, in New York City, several hundred police booths and precinct station houses are being fitted out with receiving apparatus capable of responding selectively to the municipal radio broadcasting station. During periods of civil disturbance, these official receiving posts could be utilized as secondary distribution points for news, with no more additional equipment than simple lettering materials for printing bulletins. Such ideas have their ramifications, which we need not trace in this sketch; the developments will follow when the necessity for them arises.

Against the limitations of radio broadcasting, even admitting them to be more serious than they actually are, we must balance the directness and speed of this form of communication. When the radio audience receives blow-by-blow descriptions of prize fights, the impulse of pain has scarcely passed along the nerve paths of the man struck before the radio listeners know as much about it as he does. This quality of immediate contact, as opposed to the tedious mechanical interventions of printing, is of special importance at times when event follows event and conditions change from hour to hour.

The next general strike, wherever it bobs up, will provide food for further reflection. It is a plausible guess that we shall hear more than we shall see, and that our listening will be done at the orifices of loud speakers.

Glad Tidings from the West

A CLIPPING kindly contributed by Mr. Zeh Bouck to this department reveals astounding leaps forward in the progress of the radio art, as set down for posterity in the Santa Cruz *Morning Sentinel*. It is entitled: "Of

Interest to All DX Fans." Only one paragraph need be lifted:

The inventor's claims regarding the application of polarized harmonics was (sic) a little too deep for the writer but to demonstrate what he meant he tuned in KYW where Mr. Meehan was singing an Indian song and by manipulation of the loop he entirely eliminated the accompaniment, also reversing the situation by bringing the accompaniment up so loud that it interfered with the singer. Another demonstration on an orchestra was the elimination or making of the string instruments predominate at will. On a mixed duet the soprano could be almost completely tuned out leaving the tenor singer predominating and vice versa. Altogether, the demonstration was very remarkable.

The only comment I can think of is: "You bet!"

The Radio Club of America

FOR one evening each month, in the large lecture room of Havemeyer Hall, at Columbia University, the portrait of the venerable James Renwick, LL. D., Professor of Natural Philosophy and Chemistry, 1820–1854, looks down on a group of young men, and some older ones, gathered to discuss a subject which did not exist during his life. There are a hundred of these young men, more or less. They talk about power packs, loud speakers, shortwave transmitters, tendencies in modern radio receivers. The portrait of Professor Renwick, who lived while Clerk Maxwell was evolving the electro-magnetic theory, seems to bear a slightly puzzled frown. But the members of the Radio Club of America never look up at him. He is too far back for them. Their thoughts sometimes regress to the early days of radio, to the year 1909, say, when the Club was founded. In radio that is a long time. Radio men think mainly in the present; they find plenty there to occupy them.

The Club has two principal objects—one concerned with the engineering aspects of the radio art, the other with the perpetuation of the amateur tradition. The Year Book for 1926 states: "The Club now has among its members many prominent scientists, inventors, and engineers, as a glance at the membership list will show. However, it is always anxious to embrace amateurs of the present day, in order that its membership shall never lack the renewed life given by embryo scientists." It is primarily an amateur association, as the Institute of Radio Engineers is fundamentally a professional body. Amateurs and professionals belong to both organizations, but the Radio Club members never forget that they were or are radio amateurs, while the members of the I. R. E., carrying the burden of radio scholarship, seldom forget that they are professionals.

The Club now has some 400 members, of whom 108 are of Fellow grade. Fellows qualify by five years of membership in the Club or by contributions to the radio art, at the discretion of the Board of Directors, the governing body. It consists of a President, Vice-President, Treasurer, Corresponding Secretary, Recording Secretary, and thirteen Directors. The officers and seven Directors are elected annually by the membership; the remaining six Directors are elected by a majority vote of the newly constituted Board of Direction at its first meeting. This year, the President of the Club is Ernest V. Amy; C. R. Runyon, Jr. is Vice-President. Past Presidents are W. E. D. Stokes, Jr., Frank King; George J. Eltz, Jr., Edwin H. Armstrong (1916–1920), and George E. Burghard (1921–1925). Mr. Amy was re-elected in 1926 for the present year.

Among the authors who have presented papers before the Radio Club of America are included Armstrong, Farrand, Van Dyck, Weagant, Hazeltine, W. C. White, Godley, Conrad, Heising, Aceves, Clement, Morecroft, Grebe, John Stone Stone, Lowenstein, Dubilier, Goldsmith, Marriott, Logwood, Pacent, and Hogan. The complete list is a formidable one, and the names above represent it only partially. As

AT A RADIO CLUB BANQUET
Seated at the table, from left to right: George H. Clark, E. E. Bucher, Gano Dunn, Edwin H. Armstrong, Michael I. Pupin, E. V. Amy, George J. Eltz, Jr., David Sarnoff, George Burghard, John L. Hogan, Paul F. Godley

early as 1912 the members of the Club were talking about square-law variable condensers and directional radio transmission. The value of the papers presented, to the radio art as a whole, has been inestimable. They form a contribution to the engineering literature second only to the collected Proceedings of the Institute of Radio Engineers. Incidentally, these papers are printed in RADIO BROADCAST.

The discussions following the presentation of papers are lively and informal. Nobody tries to display his knowledge and no one is afraid to show his ignorance. The attitude is simply that of a group of men vastly interested in radio, whether or not they are making any money out of it, who meet once a month to talk about the subject which happens to entertain them most.

And now, what course is to be charted for the future? Will the Club remain a sort of junior engineering society, or will it become a club literally? One plausible guess is that the organization will become a club in fact as well as in name, within the next few years, but without sacrificing its historic status. The two rôles are in no way incongruous. The formation of some sort of radio club in New York City, with permanent quarters, is obviously within reach.

The Radio Club of America has the name. It is a going concern. It has a membership of 400 at this writing. An inspection of the 1926 Year Book reveals that 78 per cent. of the members therein listed reside in New York City, or its suburbs. But it is not a club, as yet. It maintains an office at 55 West 42nd Street, New York City, but it has no lounging rooms where the members can guzzle bottles of cool ginger ale, boast about their distance records of fifteen years ago, recall fondly the days when Doc. Hudson called FNK with a service on 600 meters to remind John V. L. Hogan that he had left his pipe at the Doctor's house the night before, and consummate million dollar deals. If half the present membership wanted a club, it could be started on a modest scale, this year or next. Such a development, of course, would not interfere with the monthly technical sessions, and technical membership on the present order, with dues of $3 and $5 a year, could be retained for those not interested in the proposed aspect of the Club's activities. No one who has attended meetings of the Club, or been entertained at one of its annual banquets, can doubt that the organization has the vitality and energy necessary to take the lead in this enterprise. Steps have already been taken in this direction. It is interesting to note that the Club has recently appointed a special "House Committee" for the purpose of investigating the desirability of such an understanding and to submit recommendations for organization and financing. A modest Club House, with comfortable lounging room or rooms, is planned. The radio men in and around New York will await developments with much interest, now that the wheels have actually begun to turn.

Technical Operation of Broadcasting Stations

15.   Volume Indicators

THE sight of a milliammeter in the plate circuit of an overloaded tube, its needle fluctuating violently, is familiar to every broadcast operator and engineer. When the tube is in the speech circuits of the transmitter the sight, for obvious reasons, is a deplorable one, but the principle, or something on its order, is valuable for preventing the very overloading which the fluctuating milliammeter shows. This is when it is embodied in the instrument called the "volume indicator," much used in broadcasting as a visual guide in setting energy levels.

Before going into the description of one type of volume indicator I should like to make it clear that it is an indicator, as the name implies, rather than a measuring instrument. Basically, the instrument is an alternating current voltmeter, and in certain forms it may be calibrated as an accurate low-reading instrument of this class, extremely useful in telephone and radio work for measuring audio potentials. Again, it may be designed and calibrated to measure telephone levels in TU's. The form shown in Fig. 1, however, is intended primarily as an indicator.

The input transformer shown may be an audio amplifying transformer, preferably with a high-step-up ratio (1:10 or 1:5). The primary impedance must be sufficiently high so that, even at low frequencies, the instrument will not affect circuits of about 500 ohms impedance, across which it is bridged. This means that at 100 cycles the primary of the transformer should have an impedance of 20,000 ohms, which requires 40 henrys inductance, approximately. The secondary of this step-up transformer feeds a vacuum tube, which may be of the UX201-A type, or its equivalent. The grid of this tube is biased negatively to about 10 volts, so that the plate current, measured on a milliammeter, MA., of 0-10 range, is only about 0.5 milliampere with 90 volts plate potential. A filament voltmeter is provided to keep the voltage across the filament constant.

The theory of operation is as follows: Through the negative biasing of the grid, the tube is being worked so far down on the curve that when the transformer secondary contributes a further negative potential to the grid the plate current is only insignificantly reduced. See the characteristic curve of Fig. 3. On the other hand, positive potentials carry the current up on to the steep portion of the curve once more, resulting in a flick upward of the milliammeter pointer. The amplitude of this movement indicates the magni-

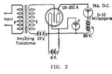

FIG. 2

tude of the alternating potential impressed on the grid, and hence, divided by the transformer voltage ratio, the audio potential on the line across which the volume indicator has been bridged. The device is merely a special form of rectifier. Fig. 4 is a table, in which columns 1 and 2 are taken from the characteristic curve of Fig. 3, showing the variation of plate current with grid potential. Column 3 shows the plus rise in voltage calculated from the base point, which is the negative grid bias potential, 10 volts in this case. Column 4 is this figure divided by the transformer ratio, assumed as 10; this gives us a rough measure of the audio voltage across the input of the instrument. By tapping various potentials along the grid bias battery one may calibrate a volume indicator, for comparative purposes, in this way. The results should not be taken as absolute values, because the response characteristic of the meter, the terminal conditions in the circuit being measured, and other factors, all affect the readings. However, these unknowns may be disregarded when only indications are required. For example, with a tube of normal emission in the volume indicator, and proper filament and plate voltages, it may have been ascertained that, when the volume indicator is bridged across the line leading to the radio station, with given settings there, the maximum allowable deflection of the meter,

short of transmitter overloading, is 5.0. Then it is the control operator's business not to allow his peaks on the volume indicator meter to hit more than 5.0, and, even if he does not know the absolute value of the r. m. s. audio voltage he

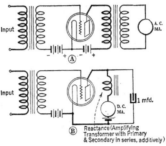

FIG. 1

FIG: 1 — Reactance (Amplifying Transformer with Primary & Secondary in series, additively)

STATIC CHARACTERISTIC
UX-201-A
90 V. Plate

FIG. 3

is putting out on the line, the volume indicator gives him something to go by, and he is far better off than if he were trying to regulate his output only by ear.

Just as loud speakers may be coupled to the plate of the last tube through an output transformer, so an a.c. milliammeter or galvanometer in a volume indicator may be connected in the same way. The instrument then reads only the audio variations in the plate circuit of the tube, the pointer being at zero in the intervals, while when a d.c. millammeter is directly connected in the plate circuit there is always a residual reading of a fraction of a milliampere even when there is no modulation on the circuits. Fig. 1A shows the connection with an output transformer, 1B with a choke-condenser connection, which has proved effective in practice.

For increased range and sensitiveness, volume indicators are sometimes built with two tubes, the first stage being an audio amplifier to bring up low voltages to a point where they can register

sufficiently on the second tube, which is the volume indicator proper. Both in the one- and two-tube outfits, various combinations of transformers, tubes, and indicating instruments may be utilized. An ordinary amplifying transformer of 1:3 or 1:5 ratio, a high-mu tube, and a galvanometer of about 1.5 milliamperes full scale reading, is a good single-tube combination. The voltages may be about the same as with the high ratio input transformer, UX-201-A tube, and 0-10 milliammeter described below. Mr. W. K. Aughenbaugh sends in a sketch of a volume indicator using two UX-112's, 45 volts plate, no grid bias, ordinary amplifying transformers, and a 0-10 d.c. voltmeter as an indicator. A swing from 2 to 6 on the meter gives him the proper modulation on the WFBG transmitter, which he runs. This is not really an orthodox volume indicator, and it would be improved with the negative bias feature, but it serves the purpose. Since the instrument is used only as a level indicator, the proper limits of operation having been established by listening checks on quality and overload observations in the amplifiers and transmitter, the operators at each

GRID VOLTS	PLATE MILLIAMPS.	POSITIVE GRID SWING VOLTS	INPUT VOLTS *
–12	0.3		
–10	0.5	0	0
–8	1.0	2	0.2
–6	2.25	4	0.4
–4	3.75	6	0.6
–2	5.6	8	0.8

* With 1:10 Input Transformer Ratio

FIG. 4

station may utilize whatever apparatus is handy and indulge their fancy for experimentation at the same time.

Something About Gain Control

FOR the edification of any members of the operating fraternity who may not have thought about the matter of proper gain regulation, I would point out a fundamental difference in methods. The right way is to bring up the gain moderately on pianissimo passages, reducing it once more to a safe level before or at the beginning of the next crescendo. That is, a skilled gain operator works with a certain base level, set for the particular soloist or ensemble being broadcast as being within the overload limits of the amplifier and transmitter. He brings up the gain cautiously on low portions of the music. He does not have to cut down in a panicky fashion on peaks, because his base level is set to take care of peaks. This ideal procedure requires, in practice, familiarity with the artist's volume range and the score, which can generally be acquired only through a rehearsal.

It also requires a feeling for music. To entrust the gain control to a man without this feeling is like committing a $1200 Sèvres vase to a drunken chambermaid.

Some Facts About Coil Design

The Oscillating Circuit as a Form of Voltage Amplifier—The "Gain" in the Broadcast Spectrum of Some Typical Coils—The Essentials for a Well-Designed Coil

By ROSS GUNN

THE oscillating circuit, as applied to radio reception, has been so thoroughly discussed in the past that it might appear the matter has been completely exhausted. A certain point of view is always most fruitful in the consideration of oscillating circuits, especially in giving one a thorough understanding of the physical phenomena that take place. In the first part of the present paper an approach to the problem is made from a particular angle, and it will be shown that an oscillating circuit is not only a device to select a given frequency but may also be considered as a type of amplifier. In the second part of the paper, the radio-frequency inductance and its relation to selectivity and gain in a tuned amplifier, considering the oscillating circuit, is discussed. Finally, the prime factors that should be considered to make up the ideal coil are given.

The resistance or "loss" in an oscillating circuit is made up of several components, and it is convenient to divide the circuit resistance into two parts, namely, the part due to coil resistance, and the part due to condenser resistance. Modern condensers are relatively efficient since they have resistances of only a fraction of an ohm at broadcast frequencies, while the ordinary coils have resistances of from 5 to 50 ohms. It is then obvious that, even if a manufacturer of good condensers would succeed in cutting the resistance of his condenser in half, the total circuit resistance would be but slightly affected since the condenser resistance is, in general, but a small percentage of the total circuit resistance. On the other hand, if the coil resistance were cut in half, the circuit resistance would be cut almost in half, and the signal impressed on the grid of the amplifier will be double what it would be with the original poor coil. An improvement of this magnitude would be well worth while, and may be accomplished by giving sufficient attention to the design of the coil used in the oscillating circuit. Attention is directed to the coil particularly, because the good modern condensers are so much better than the best coils that there is no particular gain made by the use of the highest grade condensers. The coil, then, should be the object of considerable attention and study until its losses are reduced to a minimum.

In the following discussion it is shown that an oscillating circuit may be considered as a type of voltage amplifier, and a physical interpretation of the relation of the circuit constants to the gain or desirability of any given circuit will be given. The physical interpretation that this method leads to, simplifies to a great extent the perplexing questions that arise, when one tries to state definitely just what is desired in an oscillating circuit.

Every electrical oscillating circuit consists of an inductance, capacity, and a resistance (with usually an oversupply of the latter). The inductance serves to store the energy in the form

of a magnetic field, the capacity serves to store the energy in the form of an electrostatic field, and the resistance serves to transform this stored energy into heat. The physical picture of what happens as the energy shifts from the condenser to the coil, and back again, is undoubtedly familiar to the reader, and will not be given at this time. In case the reader is unfamiliar with

ROSS GUNN, the author of this article, which is the first of two on the subject of receiving set inductors, was formerly a radio research engineer for the United States Air Service. Mr. Gunn is now attached to the Sloane Laboratory at Yale University and has devoted considerable attention to the problems involved in the design and use of coils in high-frequency oscillating circuits. The present article does not pretend to be revolutionary, for the basic theory involved is at least 50 years old. The author, instead of discussing power factor and decrement, which have little meaning to many readers, has chosen to use a factor which has a simple physical interpretation. The idea that there are two potentials in an oscillating circuit and that their ratio is the power factor, is of course well known, but the use of the factor "G" to explain the best ratio of inductance to capacity, has rarely been pointed out. The casual reader will find this story an excellent discussion of circuit problems which are by no means abstract, and those who are better informed should be interested in the work Mr. Gunn and his assistants have done, and in the way in which it is presented.—THE EDITOR.

this process, he is referred to almost any good elementary textbook on physics or electricity.

When the theory of oscillating circuits is examined, it is found that, in any series resonant circuit, there may be considered to be two high-frequency potentials which are approximately 90 degrees out of phase with each other. That is,

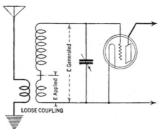

LOOSE COUPLING

FIG. 1

when one potential reaches its instantaneous maximum, the other potential is going through its instantaneous zero value. These two potentials have a certain interesting relation in terms of the circuit constants. One of these potentials will be referred to as the *applied* potential and the other as the *generated* potential. The applied potential is the potential introduced into the circuit from some outside influence, such as an oscillator or an antenna. This applied potential is in phase with the high-frequency current flowing in the oscillating circuit. The other potential, or the generated potential, is produced within the circuit, by the action of the oscillating current, and is 90 degrees out of phase with the current and the applied potential.

The current flowing in any alternating current circuit is determined by the applied potential and the circuit constants, in the manner given by the well known equation:

$$I = \frac{Ea}{\sqrt{R^2 + \left(2\pi fL - \frac{1}{2\pi fC}\right)^2}} \quad (1)$$

where I is the current, Ea is the applied potential, L is the inductance, C the capacity, R the resistance, f the frequency, and π is a numerical constant of value 3.1416. If the circuit is resonated to a definite frequency, the reactances are balanced and equation (1) reduces to:

$$I = \frac{Ea}{R} \quad (2)$$

where I, Ea, and R are the same as previously mentioned. The greatest interest, however, is not in the current flowing, but in the signal produced by the applied potential. Considering now the application of the oscillating circuit to a radio set, it is assumed to be connected to a vacuum tube as shown in Fig. 1, and so neutralized that there is no appreciable regeneration. Since the vacuum tube is a device controlled solely by potential, the greater the potential we are able to apply to the grid, the greater will be the output signal from the tube.

THE "GAIN" OF A COIL

THE diagram of connections in Fig. 1 shows that the condenser and the inductance are in parallel across the grid and filament of the amplifying tube. Therefore, any potential that appears across the inductance or condenser will be impressed on the grid of the tube. It is well known that, when an alternating current flows through an inductance, there is generated a potential which is determined by the following equation:

$$Eg = 2\pi fLI \quad (3)$$

where Eg is the potential generated across the coil, I is the current in the coil, L the inductance, and f the frequency. Similarly, the potential generated across the condenser is given by:

$$Eg = \frac{I}{2\pi fC} \quad (4)$$

where the symbols are as before.

If we eliminate the unknown current from the equations (3) and (4) by substituting the relation expressed in (2), we obtain the following results:

$$\frac{Eg}{Ea} = \frac{2\pi f_L}{R} = \frac{1}{2\pi f CR} = \frac{1}{R}\sqrt{\frac{L}{C}} = G \quad (5)$$

or rewriting in the more convenient form, we have, due to inductance:

$$Eg = \frac{(2\pi f_L)}{R} Ea \quad (6)$$

or:

$$Eg = \left(\frac{1}{R}\sqrt{\frac{L}{C}}\right) Ea \quad (7)$$

or simply:

$$Eg = G\, Ea \quad (8)$$

where G is either of the factors given above and always has a value greater than unity. That is, the potential applied to the grid of the amplifier tube is always greater than the applied potential by a factor "G". The factor "G" is therefore of the nature of an amplifying factor, and the writer has usually referred to it as the "circuit voltage amplifying factor."

It has been suggested that this factor be called the "gain" and since this is a much more convenient term, we shall now call "G" the gain due to the oscillating circuit. Obviously then, the factor "G," or the gain, should be made as large as possible for any given circuit. To illustrate this idea, suppose we had two oscillating circuits, one with a value for "G" of 50 and the other a value of 200. Suppose further, that there was such an antenna current and such coupling in both cases that the applied potential was $\frac{1}{50}$ of a volt. Then since $Eg = G \times Ea$, the potential applied to the grid of the tube (that is the generated potential) would be $50 \times \frac{1}{50} = 1$ volt in the first case and $200 \times \frac{1}{50}$ (4 volts) in the second case. Obviously then, the coil with a high value of "G" would be the best. It may then be said, that in an oscillating circuit, the potential impressed on the grid of an associated tube depends on the potential *applied* to the oscillating circuit, and depends *equally* on a gain factor "G," which in turn is dependent for its value on the circuit constants only.

Before considering the application of the above results to the determination of the proper coil, it will be valuable to point out the intimate relation between this factor "G" and other characteristics of the oscillating circuit. Consider first, the relation between the factor "G" and the selectivity of the circuit. The selectivity, or the "sharpness of resonance," of an oscillating circuit is directly proportional to the factor "G," and if we define selectivity or sharpness of resonance as the ratio of the natural frequency of the tuned circuit at resonance to the difference of the natural frequencies of the circuit when on each side of resonance, such that the oscillating energy on each side or resonance is just one half the energy at resonance, then the selectivity so defined is exactly equal to the factor "G" or:

$$\text{Selectivity} = \frac{f_r}{f_2 - f_1} = G \quad (9)$$

where f_r is the natural frequency of the oscillating circuit when in resonance with the incoming wave, f_2 is the natural frequency of the tuned circuit at a point *above* resonance, such that the oscillating current is 70.7 per cent of the current at resonance, and f_1 is the natural frequency of the tuned circuit at a point *below* resonance, such that the oscillating current is 70.7 per cent of the current at resonance.

This definition is equivalent in every way to

the usual one. The fact that the selectivity of a circuit and the gain go hand in hand is indeed a very pleasant one since, if one increases, the other automatically increases also.

THE EFFICIENCY

THE expression for the efficiency of an oscillating circuit may be easily obtained by defining the efficiency in the following manner:

$$\text{Efficiency} = \frac{\text{ENERGY OBTAINED PER HALF CYCLE}}{\text{ENERGY SUPPLIED PER HALF CYCLE}}$$

This for convenience may be written as:

$$\text{Efficiency} = \frac{\text{ENERGY SUPPLIED} - \text{ENERGY LOST}}{\text{ENERGY SUPPLIED}}$$

Now, the energy supplied per half cycle is the maximum energy stored in the inductance which is given by the well known equation:

$$\text{Energy} = \frac{L\, I^2 \max.}{2}$$

where L is the inductance and I max. is the maximum current. Moreover, the energy lost *per half cycle* is:

$$\frac{R\, I^2}{2\, f}$$

where R is the equivalent series resistance, I is the root mean square current, and f is the frequency. Making use of the relation between root mean square current and maximum current, namely:

$$I^2 = \frac{I^2 \max.}{2}$$

the efficiency may be written as follows:

$$\text{Efficiency} = \frac{\dfrac{L\, I^2 \max.}{2} - \dfrac{R\, I^2 \max.}{4\, f}}{\dfrac{L\, I^2 \max.}{2}} = \frac{L - \dfrac{R}{2\, f}}{L}$$

and canceling out the common factors, the efficiency reduces to:

$$\text{Efficiency} = 1 - \frac{R}{2\, f\, L} = 1 - \frac{\pi}{G} \text{ since } G = \frac{2\pi\, f\, L}{R}$$

or

$$\text{Efficiency} = \frac{G - \pi}{G} = \frac{G - 3.1416}{G}$$

or for per cent. efficiency:

$$= \left(\frac{G - 3.1416}{G}\right) 100\%$$

Thus, if the gain is known for any given circuit, the efficiency may readily be computed.

This equation gives the efficiency of the circuit in the strict engineering sense and should not be construed loosely as a measure of the general performance of a coil. The poorest coils and condensers yield efficiencies of 90 per cent. and better while the best coils show efficiencies of 99 per cent., an apparent increase of but 9 per cent.. The increase in performance would, however, amount to approximately 1300 per cent. for the good coil in comparison to the poor one.

Other things must be considered in radio telephony besides the intensity of the received signal. The quality of the received signal is of great importance and if we should increase "G" indefinitely we would soon have a circuit that would be so selective that the only frequency that would be amplified would be the carrier frequency, and the side bands, which give rise to the speech or music, would be cut off or, at the least, badly mutilated. There is an upper limit to the value of the "gain" if the quality of the speech is to be maintained. In the case of telegraphy, the upper limit is much higher than for telephony.

Calculations which would be tiresome and out of place in a discussion of this type, show that if all voice frequencies are to be faithfully reproduced up to 4000 cycles at a mean wavelength of 450 meters (666 kc.) the value of the factor "G" should not materially exceed a value of 250. This puts a definite upper limit to the gain or selectivity that can be used in the ordinary radio set.

Up to this point the discussion relates to the non-regenerative oscillating circuit as a whole. Now if the condenser used with the inductance is a good one, having a low resistance, then the losses of the condenser may, in comparison with the losses of the coil, be neglected. With this assumption, which is perfectly legitimate in the average case, we may then talk about the "gain" due to the coil, and understand by this, that this is the gain that would be obtained if the coil were used with any condenser of negligible resistance. These conditions have been met with in all experimental data given in this article since a precision condenser was used which was insulated with amber and which had a resistance of not over 0.2 ohm at 1000 cycles. Thus, at 300 meters (1000 kc.), the resistance would be entirely negligible.

Considering how the application of the principles just discussed are applied to an inductance, we may say that the inductance for any given circuit should be so chosen that the factor "G," which, as we have shown, is a measure of the selectivity and the efficiency, should be as large as possible, provided it does not greatly exceed the limiting value for good quality.

Returning now to the factors that determine how large the "gain" will be in any case, equation (5) shows how this factor varies with the inductance, associated capacity, resistance, and the frequency. It is impossible to examine the equation and say off hand that the ratio of the inductance to the capacity should be very large, as equation (7) would seem to indicate, because, as the inductance of the coil is increased, the resistance increases, perhaps very rapidly, depending on the physical structure of the coil. On the other hand, if the capacity of the condenser is increased the circuit resistance will decrease, but at a relatively slow rate. It would then appear that in order to determine just what combination of inductance and capacity to use or to determine what types of coils are the best, it will be necessary to make measurements of this factor "G" and determine experimentally its values under different circumstances.

The factor for the "gain" may be determined in several different ways. The ratio of the generated potential to the applied potential may be measured directly by means of a vacuumtube voltmeter, and since by definition this ratio is the "gain," we have a simple method of getting its value for different coils. The factor may also be determined by use of the definition for selectivity as given in equation (9) which can be transformed so that known condenser settings may be used instead of frequencies. The values for "G" in this paper were computed from the known values of the resistance, inductance, and frequency, since these data were already available in some cases from previous work.

SOME COIL EXPERIMENTS

A GREAT many measurements on different types of coils have been made during the last year in an attempt to obtain sufficient information to point the way to the design of still better types of coils. Curves showing the "gain" of various representative coils are included in this paper. In every case but one, the coils were

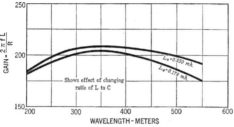

FIG. 2

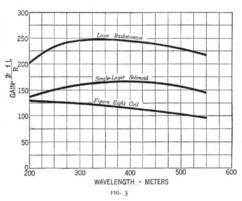

FIG. 3

those points considered the most important, or necessary, are tabulated below. The requirements in their order of importance follow:

(a) The "gain" of the coil as previously defined, and hence also, its selectivity, should in general, be made as large as possible but should not exceed the limit of 250 for good quality.

This item in itself says a great deal, implicitly, since in order for the gain factor to be even as large as 200, say, all known methods to keep the coil losses down must be employed. This would mean, in general, that the coil would be self supporting, it would be wound with relatively small wire since small wire serves to keep the eddy current losses down, and it would be so wound that individual turns were reasonably well separated. It seems to have been pretty definitely established that the best coil is obtained when the wire is spaced by an amount about equal to the wire diameter. It would mean a great many other smaller things which can all be grouped under the statement given in the preceding paragraph.

(b) The exterior field of the coil should be zero or, certainly, very small. This is essential in order that there shall be as little stray energy exchange within the various circuits of the radio receiver as possible. Putting an ordinary coil having an exterior field in a metal can is to be discouraged, since it will greatly reduce the value for the "gain." For the same reason, the value of an ordinary coil may be greatly reduced by mounting the coil in the vicinity of metal or poor dielectric.

(c) The distributed capacity of the coil should be very low, and the high- and low-potential ends of the coil should be well separated. The terminals of the coil should be well separated so that connecting wires to the coil will not introduce an excessive amount of distributed capacity into the circuit.

(d) Mechanically the coil should be strong and able to withstand a reasonable amount of abuse.

(e) Its physical structure should not be excessively large taking into account item (a).

(f) A commercial coil should have some marking which would give its "gain" over the broadcast band.

---

purchased on the open market, and may be taken as reasonably representative of the commercial coils now available.

One of the first things investigated, in the light of the previous discussion, was the effect of changing the ratio of inductance to capacity in any circuit using coils of nearly identical construction. A single-layer solenoid, consisting of 43 turns of No. 24 s. c. c. wire was wound on a cardboard form of 3¼¼ inches in diameter. The values for the gain of this coil over the broadcast range, using a condenser with maximum capacity of 350 mmfd., were plotted as shown in Fig. 2, labeled $L_0 = 0.250$ mh. Wire was then taken off this coil until the coil resonated to 550 meters (545 kc.) with a maximum capacity of 485 mmfd. The values for the gain were again determined over the broadcast band and plotted as shown in the curve Fig. 2, labeled $L_0 = 0.178$ mh. These curves, with others, seem to show that the gain in any given oscillating circuit does not change greatly with moderate changes in the ratio of inductance to capacity. There does, however, seem to be some *slight* advantage in using the higher values of inductance.

Fig. 3 shows the value for the gain for a certain commercial "figure eight" coil. It was wound in the conventional manner, the winding beginning at one end and progressing in a series of "figure eights" to the other end. The coil was made up of 80 complete loops of No. 26 s. s. c. wire and had an inductance of 0.250 mh approximately. The coil was 1⅜ inches long and 1⅜ by 4 inches in the other directions. The gain for this type of coil was surprisingly low, as it varied from 100 to 130, as shown by the curve.

The curve for a typical commercial type coil is shown in Fig. 3 (b). This coil was a single-layer open solenoid of 0.245 mh. It was made up of 58 turns of No. 23 s. s. c. wire closely wound so it could be made self supporting with the aid of a binder. It was 3¼" in diameter and 1⅞" long. The curve shows that the coil is not what one would call a bad coil.

Recent work of the Bureau of Standards has shown that in general the best type of solenoid construction is the type known as the "loose basketweave." The curve, Fig. 3 (c), shows the gain for such a coil. It was wound with 53 turns of No. 24 s. c. c. wire, and was 3½ inches in diameter. This coil is a very good coil and would give the best of results if it is not placed too close to other apparatus.

Much better coils can, of course, be built by

the use of litzendraht wire, but the writer has only considered the solid wire-wound coil so far. A graduate student getting his Master's degree at Yale in 1926 succeeded in building a solenoid out of "litz" with which he obtained an average value for the gain of over 450 for the entire broadcast band. Such coils are, of course, exceptional and are more or less laboratory curiosities because of their size and large external field. They cannot be used in the vicinity of other apparatus without increasing the losses and hence decreasing the value for the "gain." This is because their field sets up either eddy currents in the metal parts, or causes dielectric losses in insulators. These losses naturally decrease the value of the gain.

THE IDEAL COIL

THIS brings up consideration of the ideal coil or high-frequency inductance. Considerable thought has been given to this question and

Perfecting the B Socket Power Device

How the Oscillograph Is Utilized to Obtain Visual Indication of the Voltage Characteristics at Different Points of the Circuit—The Construction of a Better B Device with Uniform Output Properties

By HOWARD E. RHODES

B SOCKET power units are now in their second or third year of popularity, and during this time many excellent units have been designed. Some of the first models were described in this magazine as far back as September, 1924, but it was not until the present gaseous and thermionic type rectifiers became available that B power from the light socket became really practical for home use. It is surprising that since these rectifiers first became available there have been no important advances in design over the first models.

Power supply units are simple devices, employing principles that have been known many years, although their special application to radio is comparatively recent. Many construction articles have appeared in the various magazines but in these articles there has been little or no information as to how the devices function. It will be worth while to devote some space to a brief explanation of the fundamentals underlying the operation of any ordinary power unit.

For the purposes of our description we will use the power unit illustrated and described in this article. Sufficient information will be given to enable anyone to construct the unit and it is hoped that many will do so because this particular device gives very good results and has several design features, explained in this article, to particularly commend it.

B socket power units are devices intended to take power from an ordinary 110-volt 60-cycle a. c. house lighting system and convert it into high-voltage direct current suitable for the operation of a radio receiver. In order to accomplish this it is first of all necessary to transform the 110-volt a. c. power into energy of the same type but of higher voltage. This is accomplished by means of a transformer, which merely consists of two coils of wire wound on an iron core. If these two windings have a turn ratio of 1 to 2, placing 110 volts across the smaller, or primary winding, would give 220 volts across the other, or secondary; with a 1 to 3 ratio transformer the voltage across the secondary would be 330, and so on.

What does this voltage look like? We cannot see it directly, but with the aid of an instrument called the oscillograph, we are able to obtain a visual picture of it. The oscillograph consists of a thin wire, or vibrator, strung between two strong permanent magnets. A light is thrown on this wire and the shadow of the wire, by means of revolving mirrors, is thrown on a screen. If a current is passed through the wire, the magnetic lines of force set up by the magnets will react with the current variations through the wire, causing the latter to vibrate in

accordance with these current variations, and the way in which it vibrates will therefore indicate the character of the current flowing through it. It is possible to connect an oscillograph at various points in a circuit and in this way determine the nature of the current—whether it is alternating or direct, or whether it is a combination of the two.

In analyzing the a. c. line voltage at the input of the power device, the oscillograph, indicated by the circle with an "X" in it in Fig. 1, was connected across the primary of the transformer of the B power unit, as shown at "A," and at "B" is a copy of the picture that was seen. The line marked "zero" indicates the position of the vibrator when there is no voltage applied. The voltage starts at zero, rises to a maximum in one direction, decreases to zero, and rises to a maximum in the opposite direction, decreases to zero, and then starts the same cycle over again. The voltage across the secondary of the transformer has the same form but it is larger.

Across the secondary of the transformer we have a high a. c. voltage and the next thing to do is to change it to d. c. In the particular unit we are working with, this is accomplished with a cx-313 double-wave rectifier tube. This rectifier has two plates and two paralleled filaments so that both the halves of the wave—above and below the zero line in "B"—can be utilized. The cx-313 tube was arranged as shown at "C" in Fig. 1, only one of the plates being connected, and the oscillograph was connected as indicated. We now get a wave as shown in Fig. 1 at "D." Notice that we get a current through the circuit for each positive half of the a. c. voltage but nothing during the negative half, because the rectifier will only conduct current when the plate is positive, while during the half cycle when the plate is negative, no conduction occurs. If we connect

the other plate, as shown in "E," we will be able to take advantage of each half wave because, during the half cycle when terminal No. 5 of the transformer is negative, terminal No. 7 is positive and thus, during the half cycle when no conduction is taking place from terminal No. 5, there will be current flowing at terminal No. 7. Consequently, we get a wave form in the oscillograph as shown in diagram "F" of Fig. 1. Now, for each half cycle of the original a. c. voltage shown in "B," we get a corresponding impulse in the oscillograph. But because of the rectifying action of the tube, which will conduct current only one way, all the impulses shown in "F" are in the same direction. We now have a circuit in which the current flows in only one direction, and this is the essential characteristic of a direct current. But, although the current is unidirectional, it is pulsating; that is, it goes from zero to its maximum value and then drops to zero again, and then repeats the same thing over again. Such direct current is not suitable for application to radio purposes for it must be perfectly constant for this use.

THE FILTER CIRCUIT

THIS brings us to the third part of a B power unit—the filter. In Fig. 1, at "G," we have added a filter to the output of the rectifier. The filter consists of choke coils L_1 and L_2 and condensers C_1, C_2, and C_3. The function of the choke coils is to block any sudden changes in current while the function of the condensers is to act as storage tanks or reservoirs. Energy is delivered to the filter in the form shown in "F" and the filter serves to smooth out these impulses and to allow non fluctuating direct current to be drawn from its output. The oscillograph was first connected at X_1, and the curve obtained is shown at "H." Here we notice a great change.

The ripple is not nearly as great as in the preceding picture and it never goes down to the zero line. Through the ripple we show a dotted line indicating the average value of the ripple current. What the curve means is that we now have a direct current, I_{dc}, with a small a. c. ripple in it, I_{ac}. The ripple must be gotten rid of so we pass the output through another filter section. The oscillograph drawing "I" was obtained by connecting the instrument in the circuit at the point marked X_2 in "G". Notice how much smaller the ripple is. The filtration is almost complete. Connecting the oscillograph on the other side of the last filter condenser, C_3, at the point marked X_3, we obtain the drawing shown at "J," and the ripple is entirely absent. This means that the voltage at the end of the filter is satisfactory to apply to a receiver.

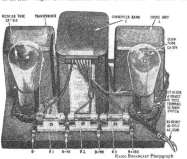

RADIO BROADCAST Photograph

THE B DEVICE DESCRIBED IN THE TEXT

The filter is one of the most important parts of a B socket power device, and it is essential that it be carefully designed if it is to give best operation. The best filter will, of course, be that one which gives the least hum in the output consistent with reasonable cost. It is preferable to accomplish the filtration with a filter as small as is feasible because, in this way, it is possible to keep the resistance of the filter circuit at a low value. In ' F" the filter is supplied with energy at the rate of 120 cycles per second. It therefore seems logical to design the first section of the filter to eliminate this 120-cycle ripple and to then use one additional filter stage to eliminate any other harmonic frequencies or any residual ripple that gets through the first section. A selective filter of this type has been patented by Kendall Clough and is used in the Silver-Marshall 331 Unichoke and also is incorporated in this unit because, with such a circuit, excellent filtering action can be obtained and, at the same time, it is possible to keep the filter resistance down to a fairly low value.

At the output of the filter we obtain the maximum voltage delivered by the device. A receiver, however, requires several different potentials for its operation. In most cases a high-voltage tap, say about 180 for a 1,₁' power tube, a 90-volt tap for the first audio amplifier and the r. f. stages, and a 45-volt tap for the detector, are required.

To obtain the various voltages it is necessary to equip the device with an output potentiometer, which merely consists of a resistance, or several resistances, connected in series across the output, as will be seen in drawing "G," Fig. 1. This circuit diagram is similar to many power units in use to-day. It has one important disadvantage, which is that the voltage obtained from the various terminals depends upon the load. If one tube is supplied from the tap marked 90-volts it might really receive 110 volts, two tubes 90 volts, and three tubes only 70 volts, and so on. This is one important respect in which a B power unit differs from a dry battery. From a battery we obtain the same voltage at all reasonable loads, but from an ordinary power unit we obtain voltages that depend on the milliampere drain.

"MOTOR BOATING" CAUSES

THAT the output of most B power units varies with load is important because it is very likely that one of the causes of "motorboating" is to be found in this fact.

"Motor-boating" is seldom experienced (although it is not an impossibility) when an amplifier is operated from good new B batteries. Ordinary B batteries must be discarded as the voltage runs down, the actual reason for this being that the internal resistance of a battery goes up as the voltage goes down, with the result that the high resistance, being common to the circuits of all the tubes in a receiver, may cause oscillation, and most certainly distortion. Considering a B power supply from this angle, we find a comparatively high internal resistance unavoidable with standard rectifying devices, and we further find a comparatively high resistance is necessary for the mechanical construction of satisfactory filter chokes. The net result is poor voltage regulation; that is, as the current drawn from the B supply increases, the voltage does not remain constant as with a battery but, instead, falls off at a fairly rapid rate. Curve A, Fig. 3, indicates the regulation that may be expected from a standard B supply with an extremely low-resistance filter. The higher the filter choke resistance, the poorer the voltage regulation; and, likewise, the higher the internal

resistance of the rectifier, the poorer the voltage regulation. In the case of batteries in good condition, the internal resistance being very low, the voltage remains substantially constant under normal current loads and a regulation curve for a battery is substantially a straight line. In itself, a falling characteristic may not, at first, appear to be a disadvantage so long as the curve indicates that the device is capable of supplying sufficient voltage at the particular load at which it is to be used, but, a poor regulation curve, indicating a high internal resistance, is doubtlessly an important contributory cause of "motor-boating" in a receiver.

"Motor-boating" can also be produced by overloading, which causes a change in the average plate current drawn by the receiver. If the average current drawn by the receiver varies,

the B power unit will be called upon to supply a varying load. If its regulation curve is not flat, any variation in the milliampere load will cause a change in the voltage supplied to the receiver and this will produce another surge in the receiver again affecting the average plate current and repeating the process. "Motor-boating" due to this cause will be prevented if the regulation curve of the power unit be made substantially flat, and this has been accomplished in the particular unit illustrated on page 43 by connecting into the output circuit a glow tube, as shown in Fig. 2. The problem of eliminating "motorboating" is complicated and for this reason RADIO BROADCAST Laboratory expects to make a series of tests in an endeavor to determine the exact causes of the trouble. These data will be published in a later article.

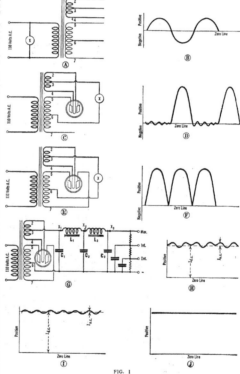

FIG. 1

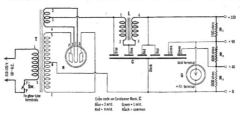

FIG. 2

The use of a glow tube to improve the regulation of a socket power device is a comparatively recent development. The effectiveness of this tube can be readily seen by referring to the curves B, C, and D in Fig. 3, and noticing that the voltage at the 90-volt tap varies only 5 volts between a load of 0 milliamperes to 45 milliamperes. The tube also affects the regulation of the other voltage taps on the device so that all of the taps show a comparatively small variation voltage with change in load. The voltage regulation at the 90-volt tap is most important since this tap generally supplies a majority of the tubes in a receiver. The use of a glow tube with the concomitant improvement in the regulation curve will eliminate a factor (poor regulation) which certainly has a very definite tendency to cause "motor-boating," and it is suggested that, where possible, the tube be used in conjunction with any B power unit that is causing a great deal of trouble.

As was mentioned at a previous point in this article, an important feature of the socket power unit illustrated herein is that it may be connected to any standard receiver drawing normal current with the certainty that the voltages will fall within the practically required operating limits, by virtue of the fact that a glow tube is employed. In an ordinary B unit, variable high resistances are used to control the output voltage, and it is seldom that the user has any definite idea what voltages are being applied to his receiver.

The parts necessary to build a B socket power employing the principles outlined in this article, are listed below:

LIST OF PARTS

R— Cunningham CX-313 Rectifier Tube	$ 5.00
G— Cunningham CX-374 Glow Tube	5.50
T— S-M 329 Power Transformer, with 5-volt balanced filament winding, two 200-volt secondaries, and electrostatic shield	9.00
L— S-M 331 Unichoke	8.00
C— Dubilier Type PL 381 Condenser Bank	14.00
Two S-M 511 Tube Sockets	1.00
R₁—Amsco 1500-Ohm Resistor with Mounts—50MA. Type 125	2.00
R₂—Amsco 4000-Ohm Resistor—10MA. Capacity	.90
R₃—Amsco 6000-Ohm Resistor—10MA. Capacity	.90
Two Amsco Resistor Mounts	.30
Four Fahnestock Clips	.08
7″ x 10″ Wood Base with 17 No. 6, ½″, R. H., wood screws, and 15 feet of Kellogg fabric insulated hook-up wire	.50
TOTAL	$47.18

The parts mentioned in the above list were used in the model constructed in the Laboratory and are known to give good results. There is no reason why the parts of other manufacturers might not be substituted provided care is taken to make certain that the electrical characteristics of the substituted parts are equivalent.

The Amsco type 125 resistance was not used in the model made up and illustrated in the photograph on page 43. It will be found that the type 125 unit is one inch longer than the resistance shown in the photograph.

Satisfactory block condensers are also made by Silver-Marshall, Tobe Deutschmann, Aerovox, Sangamo, Muter, Faradon, and the resistors that are used may be the products of any reputable manufacturer provided they are capable of carrying the currents that are specified. The S-M 653 resistor (Ward Leonard S 11500), costing $2.50, can be used in place of the Amsco resistors and mounts.

The construction of the unit is very simple. Upon the 7″ by 10″ wood base, beginning at the left-hand rear, the power transformer is screwed down with the Dubilier condenser bank along side of it, and the Unichoke at the right-hand end. The socket for the type 313 tube is placed directly in front of the transformer, while that for the 374 glow tube is placed directly in front of the Unichoke. The three resistance mounts are lined up in front of the sockets, and then, in front of the resistance mounts, are placed four Fahnestock clips. The left-hand clip is negative, and progressing to the right, we have the plus 45-, plus 90-, and plus 180-volt taps. Wiring to most of the units can be done without soldering, if desired, by simply fastening the scraped ends of the connecting wire under the terminal screws of those units which are so equipped. The circuit diagram given in Fig. 2 is marked with terminal numbers which correspond to those on the parts that were used in this particular model, and this will aid in making it a simple matter to correctly wire the

entire unit. No particular care is necessary in the way in which the wires are run since there is little danger of feed-back. Just run the wires in the most convenient way to the various terminals. By reference to the circuit diagram it will be noted that a switch, marked Sw., is indicated in the primary of the power transformer. This switch is part of the glow tube base. The terminals in the glow tube corresponding to the plate and minus A (minus A as indicated on the Silver-Marshall socket) are short-circuited inside of the tube base wiring. Therefore, the lead from the unit that connects to the 110 volts a. c. is cut and one end connected to the plate terminal of the glow tube socket (right-hand socket) and the other end connected to the negative filament terminal of this same socket. With this connection, the power is thereby automatically cut off if the glow tube is pulled out of its socket. This is necessary because, with the glow tube removed, there will be practically 180 volts at the 90-volt tap.

In operation, there are practically no precautions to be observed; the unit is foolproof. The four B-battery leads from the receiver are simply connected to the similarly marked posts on the B unit, the receiver and B unit turned on, and reception obtained by tuning the receiver in the usual fashion. The 175 volts at 20 milliamperes, obtainable from the 180-volt tap, will operate a UX-171 type tube to perfection. The 90-volt tap will supply up to 45 mils. at this voltage; the 45-volt tap up to 10 mils., as shown by Fig. 3. There is no danger of damaging tubes and condensers in the receiver due to high-voltage surges from the B unit, since the voltage of the high tap can never rise above 190 volts.

In use, the two lighted filaments of the CX-313 rectifier tube will cause the tube to get rather warm; at no time should the plates ever become red due to heating. The glow tube will glow with a bluish or pinkish light, and this may possibly flicker when a very strong signal is being received, due to the reciprocating action of the tube. If too great current—more than 45 milliamperes—is drawn from the 90-volt tap of the unit, the glow tube may cease to glow. Turning off the receiver will immediately result in this tube re-lighting, after which the set may again be turned on. If more than 45 milliamperes is drawn by the 90-volt receiver circuit, it should be examined for trouble. The brilliancy of the glow in the CX-374 regulator tube will vary with different loads.

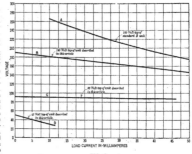

FIG. 3

Methods of Measuring Tube Characteristics

A Paper Delivered Before the Radio Club of America Discussing the Several Bridge and D. C. Systems for Use in Obtaining Tube Characteristics

By KEITH HENNEY

Director, Radio Broadcast Laboratory

IT IS neither the desire nor purpose of the writer to burden this article with an eulogy on the vacuum tube. Nearly every article that has appeared on the tube in the popular radio press has done that, pointing out that it is a most wonderful device, the modern Aladdin's Lamp, and a number of other superlatives that fill up space. There can be no doubt that the tube is important. Witness our present broadcasting structure, our long-distance telephone service, our communication by telephone across the Atlantic, and by high frequencies and exceedingly low powers to all parts of the world. All of these things depend upon the tube.

No study of the tube can be complete without a knowledge of its varied services. For example, it is possible with a tube to convert direct current from a set of batteries to alternating current of all frequencies from as near zero as one likes up to 60,000 or more kilocycles. It is then possible, with another tube exactly similar to the first, to convert these extremely high frequencies back to direct current. It is also possible to amplify both direct and alternating currents, and hence amplify power. It is also possible to separate what is placed on the input of the tube into direct and alternating currents of practically any ratio desired. All of these varied functions are carried out without any moving parts, without noise, with practically no loss of power, and with so little fuss that tubes now exist that are capable of giving service for 20,000 hours, a life greater than that of the circuit in which they are used. There can be no doubt about the importance of the tube in the field of electrical engineering. In addition, tubes and their associated circuits are now being used to measure the rate of growth of plants, to measure extremely small differences of thickness, the strength and rapidity of a man's pulse, and from this latter, to determine whether he is a lover, a thief, a liar. The tube has been harnessed and trained to do a vast number of interesting tricks.

Now this little assembly of glass and metal performs its multitudinous functions with the aid of three elements. The first and most important of these elements is the filament. This filament has undergone rather remarkable changes since tubes first came into existence. The first ones were made of tungsten which operated with a high temperature, then going to low temperature oxide-coated filaments manufactured by a complicated and difficult process, thence to our most recent filament, the thoriated wire. The measure of efficiency of the filament is its emission per watt expended in heating it, and the newest thoriated wire is exceedingly efficient. Pure tungsten filaments operate at a very high temperature. Oxide filaments consume considerable current at low voltage and at a much lower temperature. Thoriated filaments are somewhere between.

The other two elements are the grid and plate, and because these elements can be changed in size and relative position, tubes differ in characteristics. There must, then, be some means by which engineers can compare tubes just as they rate generators and motors or other electrical apparatus.

TUBE CONSTANTS

IN TUBE engineering, there are two very important factors which, when known, define the tube in exactly the same manner that we used to say in school that the United States is bounded on the north by Canada, on the east by the Atlantic Ocean, etc. The two constants—which really are not constants at all—are the amplification factor and the plate impedance,

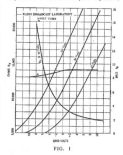

FIG. 1

and every function that the tube performs and its efficiency in doing so may be discovered by a knowledge of these factors and the constants of the circuits into which the tube works.

Another factor is the mutual conductance which, contrary to popular opinion, is not so important as it may seem. The term is somewhat difficult to picture physically. It has the dimensions of a conductance, *i.e.*, a current divided by a voltage, but the current exists in one circuit and the voltage in another, with the tube as the connecting link. It is due to Professor Hazeltine.

These constants, or factors, are variable within rather wide limits. For example, the amplification factor may range from 3 to 30, while the impedance varies, as someone has said, from Hell to Peru. The amplification factor is pretty well determined when the tube is sealed and pumped; that is, it depends to within very narrow limits upon the geometry of the tube. The mesh of the grid and its spacing with respect to the other elements are the governing factors. At low grid voltages the amplification factor falls off somewhat, rising to a maximum at zero grid voltage, and remaining constant thereafter, or falling gently in some tubes.

The plate impedance depends upon a lot of things, the filament efficiency, the amplification constant, and the grid and plate voltages. No one curve or graph can show how it varies. To properly represent it would require a three-dimensional model, such as has been constructed by Doctor Chaffee and others. Some photographs of very beautiful models of this nature may be seen in the *Proceedings of the I. R. E.*

After the tube is sealed and placed in operation the impedance changes with each change in instantaneous plate or grid voltage—all of which makes the theory of the tube more or less complicated.

These three elements, the filament, the grid, and the plate, cause any current flowing in the plate circuit to change, making it go through very wide fluctuations. The plate current is defined by the equation:

$$I_p = f(E_p + \mu E_g)$$

By maintaining constant any one of the three variables in this equation and varying the other two, we arrive at the relation between the plate current and the voltage on the grid or plate that we usually know as characteristic curves, and it is by means of these curves that the important tube factors are defined. For example, both grid and plate potential have some effect on the grid voltage, but the grid is relatively more important than the plate. In Fig. 1 it may be seen that at zero grid bias, changing the plate voltage from 90 to 135 changes the plate current by 5.2 milliamperes, while changing the grid bias by 5 volts will do the same thing.

The amplification factor is then defined as the ratio of the change in plate potential to the change in grid potential which produces the same effect in plate current. In this case the amplification factor is 9:

$$\mu = \frac{\Delta E_p}{\Delta E_g} = \frac{135-90}{5-0} = \frac{45}{5} = 9$$

The other factor of importance, the plate impedance, is the ratio between the change in plate voltage to the resultant change in plate current. In this case it is 45 volts divided by 0.0052 amperes, or roughly 8700 ohms:

$$R_p = \frac{\Delta E_p}{\Delta I_p} = \frac{135-90}{0.0116-0.0064} = \frac{45}{0.0052} = 8700$$

Now, as has been indicated by the Greek letter Δ in the definitions, these factors are defined by changes, and for accuracy the changes must be small.

The mutual conductance, defined as the ratio between a change in plate current and the change in grid voltage that produced it, is also the ratio between the amplification factor and the plate impedance, as can be seen from the mathematics below. For comparing tubes under exactly the same conditions, this factor is somewhat import-

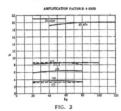

FIG. 2

ant, but as will be shown later, it serves little purpose in telling an engineer how well such a tube will function in the circuit:

$$G_m = \frac{\Delta I_p}{\Delta E_g} = \frac{\mu}{R_p} = \frac{\Delta E_p}{\Delta E_g} \cdot \frac{\Delta E_p}{\Delta I_p}$$

Figs. 2, 3, and 4 show how the tube factors change. The fact that the plate impedance is the reciprocal of the slope of the E_p-I_p curve is shown in Fig. 4. When the plate-current curve straightens out, the R_p curve is parallel to the E_p axis, but finally rises again as the saturation point is reached. These data were taken on a rather poor 199 tube.

It is a simple matter to get the tube factors from a set of characteristic curves which may show the effect upon the plate current of the grid or plate voltage. It is somewhat tedious, however, to take a mass of data and to plot it and then to pick off points on the resulting curves to determine the tube's factors. In actual practice it is simpler to go through a little routine, say of measuring the plate current under certain conditions of plate and grid voltage and then to get a new current by changing the grid voltage. This gives the mutual ..conductance. Then the plate voltage can be changed to get the plate impedance, and to multiply these factors together to get the amplification factor. At the risk of too much repetition the writer wishes to emphasize here that changes in grid and plate voltage must be small if the resultant determinations of amplification factor and plate impedance are to be representative of the tube's characteristics.

MEASURING THE TUBE CONSTANTS

THE various bridge methods of measuring tubes were developed to provide quick and simple means of measuring tubes.

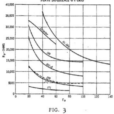

FIG. 3

One of the first methods of making quick measurements was due to J. M. Miller, and is shown in Fig. 5. In practice, the resistances, r_1 and r_2, are varied until closing the switch causes no change in plate current. Under these conditions the amplification factor is given by the ratio of r_2 and r_1. This follows from a consideration of the law governing the plate current as a function of grid and plate voltages given above.

If, with the switch closed, the voltage across plate and filament is increased, the corresponding grid-filament .voltage is decreased. If no change in plate current takes place, however, the following relation holds: :

whence $(\Delta E_p + \mu \Delta E_g) = 0$

and $I_n + \mu I r_1 = 0$

$$\mu = \frac{r_2}{r_1}$$

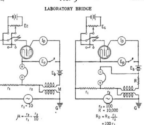

LABORATORY BRIDGE

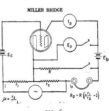

$$r_2 = 100$$
$$R = 10,000$$
$$\mu = \frac{r_2}{r_1} = \frac{r_2}{10}$$
$$R_p = R \cdot \frac{r_1}{r_2} = 100 \, r_1$$

FIG. 7

The first improvement on this simple bridge was to substitute a.c. voltages and to use a pair of telephones in place of the plate ammeter. Under these conditions the amplification factor is found in exactly the same manner. When the bridge is balanced, indicated by silence in the receivers, the amplification factor is the ratio indicated above.

Miller described also the simple addition to this scheme which permits measurements of the plate impedance to be made as shown in Fig. 6. It is not difficult to prove that the bridge balance indicates that the following relation holds:

$$R_p = R\left(\mu \frac{r_1}{r_2} - 1\right)$$

There is one disadvantage in this system. It is necessary to measure the amplification factor before the other factor may be obtained. Since

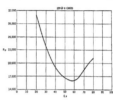

FIG. 4

the amplification factor is a constant over the ordinary ranges of grid.and plate voltages, one determination will suffice for a given tube. It must be remembered, however, that any error in measuring mu will cause an error in Rp.

In the RADIO BROADCAST Laboratory a bridge has been in use for several years which will measure either the mu or plate impedance, independently of each other, and has the additional advantage that the desired factors may be read directly. This bridge in its two forms, is shown in Fig. 7. The method of obtaining the amplification factor is exactly that of the Miller bridge while the other constant is measured in another arrangement of the same parts used by Miller. .

In all of these bridges it is necessary to use some amplification to indicate a balance unless the work is done in a quiet room. The amplifier should preferably use batteries separate from those used for the bridge. The source of tone may be a buzzer, a hummer, or an oscillator. As a matter of fact, a radio receiver might easily be used, since there is practically no change in tube characteristics at audio frequencies. The inclusion of a small variometer in the plate circuit to balance out the quadrature of plate voltage due to the grid-plate capacity, is also useful. Care 'must be taken with regard to the way it is connected into the circuit so that it will not be necessary to take into account its reactance in the final calculation of tube factors. When connected correctly the reactance, which is never greater than 10 or 15 ohms, is in series with the plate impedance. In the other connection this reactance is in series with the balancing resistance and will give absurd results.

MILLER D.C. BRIDGE

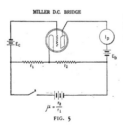

$$\mu = \frac{r_2}{r_1}$$

FIG. 5

MILLER BRIDGE

FIG. 6

$$\mu = \frac{r_2}{L}$$ $$R_p = R\left(\frac{r_1}{r_2} - 1\right)$$

MUTUAL CONDUCTANCE

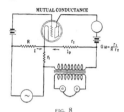

FIG. 8

INPUT CONDUCTANCE

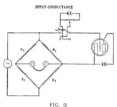

FIG. 9

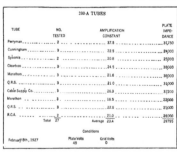

FIG. 10

There are two sources of error in the bridges shown so far. One is in the drop in plate voltage due to the plate current flowing through the balancing resistance. The other is the resistance drop of the indicating device, whether this be a pair of receivers or the primary of a transformer. Trouble from this source may be avoided by making certain that the actual plate-filament voltage is of the value desired after the balance is obtained, and by the use of a low-resistance transformer, such as a modulation transformer, to connect the bridge with the associated audio amplifier. On the Laboratory bridge, a push button connects the plate voltmeter when its reading is desired. Otherwise it is out of the circuit.

In practice, the normal filament, grid, and plate potentials are applied to the tube, and a 1000-cycle current of about one to two milliamperes flows through the bridge arms. To obtain a balance in measuring mu, the resistance in the grid circuit is set at 10 ohms and the plate resistance and variometer varied for balance. Thus, a tube whose mu is 8 requires 80 ohms in the plate side. In a quiet room, and with leads from the oscillator well separated from the amplifier mu can be read to the second decimal place. When tubes of high amplification factor, say 20 to 30, are measured, it is well to reduce the input grid resistance so that a smaller resistance is placed in the plate circuit.

To measure plate impedance with the Miller bridge, the switch is closed and a new balance obtained, when the factor desired is obtained as a function of mu. In both this bridge and in the Laboratory bridge, the resistance, R, against which the tube impedance is compared, may be fixed exactly at 10,000 ohms, although the value is not important as long as it is definitely known. On the Laboratory bridge the plate resistance is set at 100 ohms and the grid resistance varied for balance. If the impedance is 12,000 ohms, R_g will be 120.

MUTUAL CONDUCTANCE BRIDGE

THERE are a number of other bridge schemes for measuring the various factors in which engineers are interested. Ballantine has described several in the *Proceedings of the I. R. E.* One of these is a method of measuring the mutual conductance directly, and for comparing a great many tubes of the same sort under the same

conditions, it provides a useful instrument for the laboratory or tube manufacturer, or even dealer. This is shown in Fig 8. It will be seen that it differs but little from the other arrangements.

Another set-up of apparatus will measure the input characteristics of the tube at low frequencies. It is shown in Fig. 9. At balance, the grid conductance (the inverse of the input impedance) is given by:

$$K_g = \frac{R_1}{R_4 \, R_2}$$

and in practice, with normal tubes, all balancing is done with R_1, since R_2 is held constant at 100 ohms, and R_4 at 10,000 ohms. With soft tubes the conductance may change sign at high plate potentials, and it is then necessary to give the input a positive conductance by actually connecting across the grid and filament a high resistance, say of 50,000 ohms. The tube conductance may be obtained by subtracting from this known value, that determined by the bridge. Data given in Fig. 10 on a soft tube show the effect of ionization in giving the input circuit a negative conductance.

TABLE NO. 1

Type	Ep	Ex	A.C. Rp	D.C. Res.	Factor	Where d.c. =a.c.
201–A	90	—4.5	11000	30000	2.7	Ep 90, Eg—2
199	90	—4.5	18500	37900	2.0	Ep 90, Eg—3
112	135	—9.0	5500	21400	3.9	Ep 135, Eg—0.2

COMMERCIAL TUBE TESTERS

THERE are a number of tube testers on the market some of which are very expensive. As far as the writer knows, there is but one which measures tubes according to the bridge schemes outlined above. This is made by the General Radio Company and uses the Miller connections. Others are made by the Hickok Company of Cleveland, by Jewell, Hoyt, and others. The Hickok uses a 60-cycle voltage obtained from a step-down transformer. The others vary the grid and plate voltages by fixed—and it is to be hoped, small—amounts, and obtain the various factors either directly or by simple calculations. They are quite reliable and valuable instruments although not capable of the precision that a true bridge circuit can attain.

It is also possible to obtain tube constants from a knowledge of the d.c. resistance. For a given type of tube, the plate impedance at certain conditions of plate and grid voltage may be obtained by multiplying the d.c. resistance under those conditions by a constant factor; at some other value of plate and grid voltage it will be exactly equal to the d.c. resistance. For example, dividing the d.c. resistance of a 201–A type tube at 90 volts on the plate and a negative 4.5 bias on the grid by 2.7 will give an approximate idea of the a.c. impedance.

Table No. 1 is representative of what may be expected from such methods of estimating tube impedance.

A tube tester with a d.c. plate current meter calibrated with several scales will read the plate impedance with an accuracy that may be all that is desired by dealers and others who do not need to use the values in circuit calculations.

The factors of tubes commonly used to-day are shown in the accompanying tables, Figs. 11, 12, 13, and 14. It is a fact to be thankful for that tubes are now so uniform, since a tube with odd constants placed into a well-engineered receiver is often enough to change conditions from good to very bad. A year ago such standardization had not been reached, as data on file in the Laboratory show.

Now, having shown how various tube factors may be calculated from curves, or measured with bridges or d.c. instruments, it remains to be shown how useful such factors are, and in some measure to justify

200-A TUBES

TUBE	NO. TESTED	AMPLIFICATION CONSTANT	PLATE IMPEDANCE
Perryman	2	37.5	31750
Cunningham	3	22.5	24000
Sylvania	2	20.8	25000
Clearbon	3	24.5	38000
Marathon	3	21.6	38000
Q.R.S.	3	21.0	31000
Cable Supply Co.	3	26.3	37200
Marathon	3	16.5	22000
Q.R.S.	3	22.8	25000
R.C.A.	2	21.0	26000
	Total 27	Average 23.4	29795

	Conditions	
February 8th, 1927	Plate Volts 45	Grid Volts 0

FIG. 11

the statement that mutual conductance is not the determining factor in a tube's goodness or unfitness for particular tasks.

In receivers, as we have them to-day, the first tube generally acts as a radio-frequency amplifier with inductance in both plate and grid circuits. It is necessary that the plate-grid capacity be small and that a given make and type of tube will be uniform. It is also necessary that the input impedance be high and the output impedance be low, for maximum gain. For example, the maximum possible gain from an amplifier is given by the well known expression:

$$K = \frac{\mu}{\sqrt{R_p}} \times \sqrt{\frac{K_g}{2}}$$

and a little mathematics will show that when the proper load is inserted into the plate circuit of a high-frequency amplifier that the maximum amplification will be given by the formula:

$$K_m = \frac{\mu}{\sqrt{R_p}} \cdot \frac{L\omega}{\sqrt{R}}$$

where L is the inductance of the secondary coil and R the effective resistance of the circuit. Tube

CeCo Type K				
No.	μ	Rp	Gm	μ²/Rp
1	11.1	11,700	940	10.5
2	12.1	13,000	930	11.3
3	12.5	15,700	795	10.0
4	11.0	12,200	900	9.9
5	13.1	14,000	880	11.5
6	11.3	13,100	855	9.6
		Rp = 90	Eg = —3	

TABLE NO. 2

were measured. These data should be on the carton of every tube sold.

Table No. 2 is illustrative of the fact that the mutual conductance of a tube may be lower than another and still have a higher "gain" factor; e.g., compare No. 3 and No. 6 of this table.

DETECTOR IMPEDANCE

THERE has been much speculation about the output impedance of a detector tube. This is important in order that amplifier engineers know exactly under what conditions their products will work. For example, it is well known that a transformer-coupled amplifier will have one characteristic working with a tube of 10,000 ohms and another out of 30,000 ohms. Just what is the average impedance of a detector?

It is somewhat difficult to picture what happens when we measure this impedance in one of our bridges. The detector is a distorting device and a pure thousand-cycle note used for balancing the bridge will no longer be a pure note in the output. Furthermore, a detector tube has both high- and low-frequency voltages in both input and output. What effect has this combination of frequencies upon its impedance, if any? Is the impedance of a C-battery detector the same as that of a grid leak and condenser with grid slightly positive? It seems reasonable that the C battery demodulator will have a much higher impedance which may cause us to sit and think when such a device is recommended because of the superior quality of reproduction possible by its use.

There are other problems. For example, shall we place the voltage directly on the grid; shall

112 TUBES

TUBE	NO. TESTED	AMPLIFICATION CONSTANT	PLATE IMPEDANCE	MUTUAL CONDUCTANCE
Cunningham	3	8.3	4660	1785
Daven Mu 6	1	5.3	5150	1060
Diatron	1	6.5	4670	1400
Hercultron	1	9.0	8700	1035
Q.R.S.	1	6.5	5700	1140
Regal	1	9.0	7400	1215
Zetka	4	8.1	7850/9200	1114/1250
Total 12		Average 7.5		

Conditions

Plate Volts 135   Grid Volts —9

February 8th, 1927

FIG. 12

171 TUBES

TUBE	NO. TESTED	AMPLIFICATION CONSTANT	PLATE IMPEDANCE	MUTUAL CONDUCTANCE
Cunningham	2	2.9	2100	1380
Cleartron	3	2.6	2610	1000
Perryman	2	3.15	2650	1190
DeForest	2	2.5	2500	1000
Ureco Special	1	2.8	2000	1250
R.C.A.	5	2.85	2200	1300
Hercultron	1	3.25	3800	1160
Sylvania	2	2.7	2100	1290
Marathon	7	3.1	3600	1190
Total 25		Average 2.84	2395	1195

Conditions

Filament Volts 5   Plate Volts 135   Grid Volts 27

February 8th, 1927

FIG. 13

constants enter into other circuit calculations as shown below:

$$\text{Voltage Amplification} = \tfrac{1}{2}\sqrt{R_i} \times \frac{\mu}{\sqrt{R_p}}$$

$$\text{Power Amplification} = \frac{R_i}{4} \times \frac{\mu^2}{R_p}$$

$$\text{Power Output} = \frac{E_g}{8} \times \frac{\mu^2}{R_p}$$

where Ri is the input resistance and Eg is the input volts, peak.

In every case it will be seen that the tube enters in some ratio of its amplification constant squared, divided by its plate impedance. Knowing this factor, it is only necessary to insert it into circuit equations and calculate the result at once.

There has been much talk among tube manufacturers regarding standardization, and a universal desire is evidenced for a single term by which tubes could be rated. Unfortunately, no such term has been provided simply because no mathematics has been invented that will make such a thing possible. The important factors are the plate impedance, the amplification factor, and the figures for grid and plate voltage under which the values

201-A TUBES

TUBE	NO. TESTED	AMPLIFICATION CONSTANT	PLATE IMPEDANCE	MUTUAL CONDUCTANCE
Apco	5	6.3	8950	755
Armor	5	7.25	8360	850
Boehm	9	6.98	9030	780
Cable Supply Co.	2	7.4	9810	754
Ceco	6	8.15	11150	736
Champion	5	8.54	13320	643
Cleartron	17	7.1	10420	680
Cunningham	9	8.30	10825	750
DeForest DL 5	4	9.65	11100	885
DeForest DL 2	4	7.15	8625	830
DeForest DL 4	1	8.3	11200	740
Empiretron	3	7.1	10230	694
Fultone	4	8.85	11800	750
Gormac	5	7.15	12600	578
Hytron	3	9.43	14833	635
Ken-Rad	20	8.55	12450	695
Magnatron	2	8.0	12250	652
Marathon	9	8.2	11540	713
Perryman	14	7.6	10350	740
Q.R.S.	24	8.7	11440	715
Schickerling RS 10	8	7.93	10167	790
Sky Sweeper	8	7.84	13100	600
Sonatron	4	8.43	13350	632
Strongson	6	8.2	9300	885
Supertron	3	9.5	14600	660
Sylvania	11	8.36	11300	757
Televocal	3	7.1	8000	900
Ureco	15	7.2	9820	800
Van Horne	12	8.63	12800	677
Voltron	19	5.64	7000	845
Zetka	8	8.1	13300	605
Total 225		Average 7.9	10000	795

Conditions

Plate Volts 90   Grid Volts —4.5

February 8th, 1927

FIG. 14

the plate circuit have a load other than the resistance; shall the radio-frequency voltages in the circuit?

Several schemes have been suggested to determine whether bridge measurements on distorting tubes mean anything. The one described here is due to Mr. Howard Rhodes of the staff of RADIO BROADCAST Laboratory. It follows from the succeeding consideration. In Fig. 15 is the symbolic representation of a simple circuit in which Rp is the usual tube impedance, and Ro is some other resistance inserted into the circuit and whose value is variable and known. It is simple enough to measure the voltage across this resistance.

Let us suppose the tube impedance, Rp, is 5000 ohms and that we measure and plot the voltage across Ro as the latter is varied. When the two resistances are equal, Eo will be ½ μ Eg and when Ro is 3Rp, Eo will be ¾ μ Eg, or 1.5 times as much as when Ro and Rp are equal. We shall then get a curve similar to that of Fig. 15. From these data a triangle may be formed.

whose base is fixed at three units and whose vertical leg is, when $R_p = R_0$, equal to 1.5. It is then only necessary to plot the voltages developed across known resistances in the plate circuit of the detector tube and to form the above triangle on this curve. Some data on detector impedance measured by several methods will be available later. Table No. 3 is the result of bridge methods.

POWER OUTPUT

THE final measurement in which we shall be interested at present is that of undistorted power output. With the advent of tubes of the 112, the 171, and the 210 class, honest-to-goodness amplifiers have been possible, and many strange misconceptions have arisen from a none too clear understanding of their nature. Some people think that a great increase in volume will result from the substitution of a 171 for a 201-A. Of course such a result is impossible. As a matter of fact the 201-A, with its larger mu, will produce twice as much voltage amplification as a 171, provided the proper impedances are used, but it is certain that more power, with less

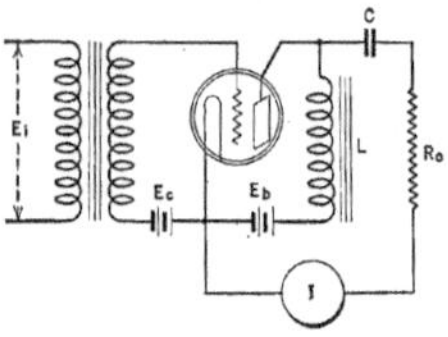

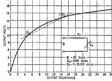

FIG. 15

Tube	E_g	E_p	R_p	R, megs.	C
201-A	—4.5	45	30,000		
301-A	G.L.	45	9500	1.5	0.00025
112	—4.5	45	14,000		
112	G.L.	45	6800	1.5	0.00025

TABLE NO. 3

where E_g is a peak voltage. If $R_0 = 2R_p$ this value becomes:

$$\frac{3 \mu^2 E_g^2}{9 R_p}$$

It is well known that the maximum power will be delivered to the loud speaker when the latter's impedance equals that of the tube, and recent data published in this country and in England indicate that the greatest amount of undistorted power will be delivered when the loud speaker impedance is twice that of the output tube. Fig. 17 shows how the power and the voltage gain of a tube vary with input voltage. When the lower bend of the characteristic curve is traversed, considerable rectification takes place, with corresponding change in d.c. plate current. When the plate current has changed roughly

10 per cent. the voltage amplification falls off, and the power output-voltage input curve flattens out.

Fig. 18 differs from Fig. 17 in several respects. Here the grid circuit is fed through a transformer. When the grid goes positive, the characteristic curve flattens out, duplicating roughly the curve at the lower bend. This results in smaller change in average d.c. plate current but a greater loss in amplification. In either case the change in plate current is a fair means of indicating distortion due to positive grid or to rectification at the lower bend.

NEW TUBES

TWO new tubes have been announced recently. One is a 300-milliampere rectifier which will make it possible to run 201-A type tubes in series from rectified a.c., while the other follows a suggestion of Mr. B. F. Meissner, whose paper, delivered before the Radio Club of America, on lighting filaments from a.c., was printed in the February and March issues of RADIO BROADCAST. This new tube requires two amperes at

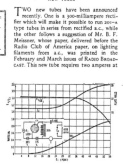

FIG. 17

0.6 volts. Naturally the voltage drop across such tubes to r.f. currents is remarkably low. Its thermal inertia is increased vastly over that of even 112 type tubes. The problem seems to be one of making a filament that will have a life comparable to that of other tubes now procurable.

It must be admitted that there are a great many tube measurements that have not been discussed in this brief paper. For example, there is much to be done with detectors; with the possibility of using amplifiers in which the grid takes considerable current; on the effect of the amplitude of input a.c. voltages upon tube factors; and a host of other interesting and important measurements. There is, at the present time, too much taking the tube for granted, not only by the hundreds of thousands of users, but by manufacturers and engineers as well. A tube is not merely a thing to shove into a radio receiver socket; it is a sensitive and delicate device with a patient and willing nature.

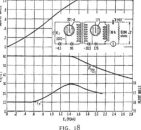

$W_0 = I^2 R_0$

FIG. 16

distortion, will be delivered to the loud speaker when a true-power tube is used.

In the first place it may be said that measurement of undistorted power output from present-day tubes seems impossible, for the simple fact that there is no such thing. The question is one of allowable distortion, which involves not only matters of opinion but the particular amplifier and loud speaker used.

It is simple enough to measure the power from a tube. It is only necessary to measure the current through a known resistance, and if the tube constants are known to within ten per cent., and if the grid does not take over 10 microamperes in 350,000 ohms, approximately, the measured power will check the mathematical value to within 10 per cent.

The power developed in the load resistance in Fig. 16 is:

$$\frac{\mu^2 E_g^2 R_0}{(R_0 + R_p)^2}$$

And when $R_0 = R_p$ this simplifies to:

$$\frac{\mu^2 E_g^2}{8 R_p}$$

FIG. 18

RADIO BROADCAST is the official publication of the Radio Club of America, through whose courtesy, the foregoing paper has been printed here. RADIO BROADCAST does not, of course, assume responsibility for controversial statements made by authors of these papers. Other Radio Club papers will appear in subsequent numbers of this magazine

FUTURE PERFECTION OF RADIO RECEPTION DEMANDS RADIO TUBES DESIGNED FOR EACH RADIO FUNCTION.

deForest

Like the Benvenuto Cellini cup from the Altman collection of the Metropolitan Museum in New York, the De Forest Audions are fine examples of craftsmanship and painstaking skill.

DE FOREST engineers have recognized certain characteristics in the functioning of tubes in all radio units. Our laboratories have labored long to advance these characteristics that so improve radio reception, and now, these highly desirable elements have been developed in De Forest Audions for specific operations in the various radio reception departments.

Fans who are keen to bring their radios up to the highest degree of efficiency will eagerly adopt these Specialist Audions, the idea of which has long been appreciated in England.

These new De Forest Specialist Audions are now available for detector work, radio frequency amplification and use in all audio stages in types taking up to 500 volts on the plate.

To demonstrate the advantages of this idea and the improvement possible in your radio's performance tune in a weak and distant station or turn down the volume of a local until you can just barely hear it in the loud speaker. Substitute De Forest DL-4 Specialist radio frequency Audions in place of the RF amplifiers you have been using. Note the remarkable increase in volume—how much louder the distant station and how the music of a local is raised to room filling proportion.

Radio amateurs will appreciate the characteristics of these efficient tubes. We must remember that regardless of RF circuits, tubes for best results must be uniform. The rigid limits, both electrical and mechanical, to which De Forest Audions are held assure a high standard of uniformity. With a very constant grid-plate capacity and high mutual conductance the volume these Audions obtain from distant reception is both amazing and satisfying.

These DL-4's are recommended for trial before you change *all* the tubes in your set.

De Forest Audions have been standard since 1906. The same genius who has made the broadcasting of voice and music possible is still hard at work for greater perfection and greater achievement in radio reception.

Of course, De Forest has designed a general purpose Audion. It is a good one and where price is a consideration the DOI-A Audion is an unequalled value at $1.65. This tube is built to the same high standards of quality that mark all De Forest Audions.

De Forest dealers are pretty much everywhere. Look for displays of the brilliant black and orange Audion containers in shop windows. (Metal boxes in which De Forest Specialist Audions are packed insure their safety and dependability.)

If dealer is not available write for booklet which describes characteristics of each Audion and for chart indicating proper replacements for all standard makes of radio.

DE FOREST RADIO COMPANY

Powell Crosley, Jr.   President

JERSEY CITY, N. J.

The *Radio Broadcast*
LABORATORY INFORMATION
SHEETS

INQUIRIES sent to the Questions and Answers department of RADIO BROADCAST were at one time answered either by letter or in "The Grid." The latter department has been discontinued, and all questions addressed to our technical service department are now answered by mail. In place of "The Grid," appears this series of Laboratory Information Sheets. These sheets contain much the same type of information as formerly appeared in "The Grid," but we believe that the change in the method of presentation and the wider scope of the information in the sheets, will make this section of RADIO BROADCAST *of much greater interest to our readers.*

The Laboratory Information Sheets cover a wide range of information of value to the experimenter, and they are so arranged that they may be cut from the magazine and preserved for constant reference. We suggest that the series of Sheets appearing in each issue be cut out with a razor blade and pasted on 4" by 6" filing cards, or in a notebook. The cards should be arranged in numerical order. Several times during the year an index to all sheets previously printed will appear in this department. The first index appeared in November.

Those who wish to avail themselves of the service formerly supplied by "The Grid," are requested to send their questions to the Technical Information Service of the Laboratory, using the coupon which appears on page 60 of this issue. Some of the former issues of RADIO BROADCAST, *in which appeared the first sets of Laboratory Sheets, may still be obtained from the Subscription Department of Doubleday, Page & Company at Garden City, New York.*

No. 89   RADIO BROADCAST Laboratory Information Sheet   **May, 1927**

Short-Wave Coils

SOME DATA ON THEIR RESISTANCE

THERE are, at present, a great many excellent coils on the market for use in short-wave receivers. They are generally of the "plug-in" type so that different coils are used to obtain the various ranges required.

These coils should have as low a radio-frequency resistance as is possible, consistent with a construction sufficiently rugged to prevent their being damaged if they are handled somewhat roughly. It would be preferable if the coils could be wound on some solid form but the question then arises whether or not a form can be used without increasing the resistance of the coil to a considerable extent.

The General Radio Company has conducted some experiments along this line to determine just how much the form used affects the coil's resistance and also to determine what size wire is best to use. Tests were made using a standard bakelite form having a diameter of 2¼". The curve given on this Sheet indicates how the radio-frequency resistance of the coil varies with the size of the wire used. Evidently, from the curve, the wire size is not especially critical but best results are obtained with a wire size of about No. 12 or 14 gauge.

It was found that the use of good binders to hold the turns in place has no appreciable effect upon the resistance. A coil was wound in such a manner that a form could be slipped in and out of it without disturbing the wire. Measurements on the coil with and without the form indicated that the difference in efficiency was negligible.

Tests were also made with regard to shielding and it was found that the shielding could be placed very near the coil and have no appreciable effect. The result of the tests may be summed up as follows:

When designing a coil for use on the 40-meter

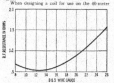

(7500-kc.) short-wave band (all these tests were made at this frequency), it is well to (1.) use about No. 12 to 14 wire; (2.) use a coil form if desired; (3.) use any good dope as a binder; (4.) use any reasonable amount of shielding where advantageous; (5.) keep the form factor (diameter divided by length) around 1 to 2.5.

These data are taken from the February, 1927, issue of the General Radio *Experimenter.*

No. 90   RADIO BROADCAST Laboratory Information Sheet   **May, 1927**

Loop Antennas

SOME OF THEIR ADVANTAGES

THE operation of a transformer is usually explained by saying that the current flowing in the primary sets up an alternating magnetic field which in turn causes a current to flow in the secondary. This is also the simplest way to explain the operation of a loop antenna, the only difference being that the alternating magnetic field that causes the current to flow in the loop is in the form of radio waves.

The number of volts induced in a loop by the passage of radio waves is:

$$2 \pi f_n A H x 10^{-8}$$

where H is the amplitude of the wave, f the frequency, n the number of turns in the loop, and A the area of the loop. The voltage calculated from this formula is only correct when the plane of the loop is vertical and perpendicular to the direction of the magnetic field. That is, the loop must be pointing toward the transmitting station. If rotated about a vertical axis only a quarter of a turn, no voltage will be induced.

This feature is the most important advantage of a loop, for two stations using exactly the same wavelength may often satisfactorily be separated (provided they do not lie in the same or exactly opposite directions) by simply turning the loop at right

angles to the interfering station. Loops are coming into greater use as transmitting stations become more powerful, and they will probably ultimately be used almost exclusively on account of the small space required, ease of installation, portability, lack of necessity to safeguard against lightning, and the improvement of the ratio of signal strength to interfering noises, due to their directional properties.

If a loop is compared in size to an antenna of the ordinary type it would appear that the amount of energy intercepted by the loop would be exceedingly small indeed. The fact is, however, that a good loop antenna, tuned with a condenser having low insulation losses, will pick up signals much better than might be expected from a comparison of its size to that of an outdoor antenna. This is due to the fact that the loop has a very much lower resistance than an elevated antenna.

The loop type antenna has been used most frequently in conjunction with super-heterodynes because, with this type of receiver, it is easy to obtain a large amount of radio-frequency amplification. During the last year, however, several receivers of the neutrodyne type have been placed on the market designed for use with a loop. These receivers are generally completely shielded so as to prevent interaction between the loop and the coils in the receiver.

Model AC-9
7-Tube, 2-Dial, Batteryless

This two-dial control set is designed especially for AC power, for use with the Amrad A B & C Power Unit. Easily operated, marvelous selectivity. Furnished with Power Unit but without tubes.

$142

Console Model $192

$192.

Model S-733
7-Tube, 2-Dial, Battery Type

Owners of this Amrad Neutrodyne report complete satisfaction. High ratio vernier controls simplify tuning. Volume is controlled by a single adjustment. Beautifully designed cabinet finished in two-toned mahogany. Without accessories.

$77

Console Model $127

AMRAD

7 tube Neutrodynes of Quality and Precision

A MRAD Neutrodynes are built with the greatest skill and precision. Each set must pass certain high standard tests before it leaves the factory.

The great skill and engineering feats of the Amrad Laboratories are manifest in the circuit as well as in the beautifully designed cabinets.

Produced under mass production methods influenced by Powel Crosley, Jr., combined with Amrad's engineering skill, these genuine neutrodynes are the greatest values on the market.

The console model AC-9-C is an unusual value. It is a 7-tube set with two-dial control. All the necessary power is furnished by the Amrad A B & C Power Unit, an efficient power supply tested under actual home conditions

for more than a year and operating from AC current, 100-120 volts, 60 cycle. No trickle charger is concealed in this unit. No more power supply troubles. Just snap the switch and set is in full operation. The cabinet is of beautiful two-toned mahogany finish, with the genuine Crosley musicone built in. This is a wonderful value at $192, with the power unit, but without the tubes.

Write Dept. 2E7 for descriptive literature and information.

AMRAD CORPORATION
Medford Hillside, Mass.

"B" ELIMINATOR
Will furnish B current voltages 22½, 30 or 45, 60, 90, 135 or 180. Maximum volts, 180 at 50 mils. Unit is housed in a metal cabinet and finished in black enamel.
$35

Efficient 5 tube genuine Neutrodynes, unsurpassed in the radio market anywhere at this price /

$60

Model S-522
5-Tube, 3-Dial, Battery Type

Amrad quality is again exemplified in this beautifully made and proportioned set. The simple, yet elegant lines of this set are pleasing to the eye. Actual reports of performance are remarkable. Simple to tune and easy to operate. Also made in console model at $110.

$125

Model AC-5
5-Tube, 3-Dial, Batteryless

A compact, efficient set delivering the utmost in radio enjoyment at the lowest possible cost. No batteries to fuss with. Operates direct from light current. Unusual selectivity, volume, and tone make this the greatest neutrodyne value on the market. Console Model S175.

No. 91   RADIO BROADCAST Laboratory Information Sheet   **May, 1927**

A Simple Tube Tester

HOW TO GET CHARACTERISTICS OF TUBES

CONTRARY to the opinion of many experimenters, a set-up of instruments to measure the characteristics of vacuum tubes is not excessively costly nor is it complicated. The diagram of connections of a tester is shown on Laboratory Sheet No. 92; this Laboratory Sheet will explain how to measure tube characteristics using the tester. The procedure can be explained most easily by taking an actual example.

Suppose we desire to measure the characteristics of a 201-A tube. We would first place the tube in the socket and then, with switch No. 2 in position B and switch No. 3 in position A, the rheostat would be adjusted until the filament voltage, as read on the voltmeter, is correct. In this case the correct voltage would be 5. Then, with switch No. 3 in position B the plate voltage is adjusted to 90 volts. The grid bias is next adjusted to 4.5 volts by throwing switches Nos. 1 and 2 to the A position and adjusting the potentiometer P. The milliammeter will now read about 0.002 amperes (2 m A.). Note down the plate voltage, the grid voltage, and the resulting plate current.

Now adjust the potentiometer until the grid bias is, say 3.5, and read the plate current. It should read about 0.003 amperes (3 mA.). Leaving the grid bias at 3.5, next adjust the switches to read the plate voltage. Reduce the plate voltage so as to make the milliammeter read exactly the same as

before (2 mA.). The new reading of plate voltage may be 82. We now have all the necessary data to calculate the constants of the tube.

The amplification constant will be equal to the difference of the two plate voltages, 90-82, or 8, divided by the difference of the two grid voltages, 4.5—3.5 = 1. The amplification constant is therefore 8. The plate impedance is equal to the difference of the plate voltages divided by the difference in the plate currents, or 8 divided by 0.001. The quotient is 8000, which is the plate impedance. The mutual conductance is the plate current difference divided by the grid voltage difference, or 0.001 divided by 1 = 0.001 mhos or 1000 micromhos.

In measuring tube factors with this apparatus, care must be taken that the actual changes in voltages—and the corresponding current changes—are small. If the plate-current meter, and the grid-voltage meter, can be read with sufficient accuracy, very small changes should be made—say a plate voltage change of 5 volts. This, however, would make it necessary to read grid bias changes of less than a volt. The investigator, then, is between two fires in his endeavor to measure his tubes accurately. If he takes plate current readings resulting from large voltage changes, he gets a factor which represents working the tube over a large part of its characteristic curve. On the other hand, if he uses small voltage changes, the accuracy depends upon the accuracy of his meters and his ability to read them.

No. 92   RADIO BROADCAST Laboratory Information Sheet   **May, 1927**

Circuit Diagram of Tube Tester

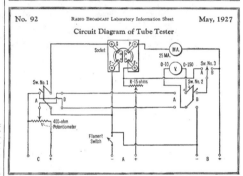

No. 93   RADIO BROADCAST Laboratory Information Sheet   **May, 1927**

Audio Amplifying Systems

DUAL-IMPEDANCE COUPLED AMPLIFIERS

ON THIS Sheet we give some facts regarding dual-impedance coupled amplifiers. A circuit diagram of such an amplifier will be found on Laboratory Sheet No. 87 (Apr., 1927).

Double-impedance amplifiers are capable of giving excellent results if care is taken in the selection of the apparatus and in the layout of the parts. The plate impedances should have an inductance around 100 henries; if the inductance is much less, the low frequencies will be lost. Well-made 0.1-mfd. blocking condensers are essential to prevent leakage.

The amplification of each stage is generally equal to about nine tenths of the amplification constant of the tube. If we lose one tenth on each stage, then the total amplification in three stages will be equal to 0.9 x 0.9 x 0.9 = 0.73 times the product of the amplification constants of the three tubes concerned. Suppose two 201-A's, each with an amplification of eight, and one 171 with an amplification of three, are used. Then the total amplification will be equal to 8 x 8 x 3 x 0.73 = 140.16. This value is rather too low for best results, and for this reason high-mu tubes, having an amplification constant of anything up to about thirty, are generally used in this type of amplifier.

From some tests made in the Laboratory, it ap-

pears possible to overload the power stage of an impedance amplifier to a considerable extent, without introducing very objectionable distortion. This comes about in the following way.

In a transformer-coupled amplifier the maximum signal that can be placed on the grid is limited by the fact that, if the signal voltage is too large, grid current will flow in the grid circuit of the power tube. This current flowing through the secondary of the transformer saturates the core and prevents the transformer from properly amplifying the signal. In an impedance amplifier there are no transformers, and the grid current only has the effect of slightly lowering the inductance of the impedance unit in the power tube's grid circuit. Slight overloading is therefore less noticeable in an impedance amplifier than in a transformer-coupled one.

As stated above, the amplification obtained at low frequencies depends upon the use of high-inductance impedances in the plate and grid circuits. There has however been a recent development in the design of double-impedance amplifiers by which it is possible to obtain very good low-note amplification without using very large coils. This design feature consists in so determining the inductance of the plate and grid coils and the capacity of the coupling condenser, that the entire combination tunes or resonates at about 30 cycles, with the result that the amplification of these low frequencies is unusually good.

No. 94   RADIO BROADCAST Laboratory Information Sheet   May, 1927

The Principle of Reflexing

AN EXPLANATION OF THE ACTION

WHEN a tube capable of amplifying a fairly strong signal is used to amplify a very weak one, it is evident that its power amplifying ability is not being made use of to the fullest possible extent.

"Reflexing" is a system for getting more out of a tube by making it amplify two things—the incoming signal, which is a radio-frequency current, and the detected signal, which is an audio-frequency current. The accompanying diagram indicates a simple receiver using one stage of reflexed amplification.

In this receiver, the radio-frequency current enters the receiver via the antenna and is impressed on the tube by the tuned circuit, A. It then passes through the tubes and into the tuned transformer, B, the output of which is impressed on a crystal detector. The audio-frequency currents resulting from the detecting action of the crystal pass through the primary of the audio transformer, T. The voltage induced in the secondary of this transformer is impressed on the grid of the tube and is amplified. A pair of phones is used in the plate circuit of the tube for receiving the signal.

So long as the variations of potential due to these two different signals do not cause the tube to overload, neither interferes with the other. Some circuits use a reflex principle consisting of several stages of radio-frequency amplification and several stages of audio-frequency amplification. In such sets it is advantageous to use the system due to David Grimes and known as the Inverse Duplex

system. In this system, the tube handling the smallest amount of radio-frequency energy is made to handle the largest amount of audio-frequency energy and, vice versa, the tube handling the greatest

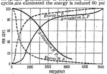

amount of radio energy handles the smallest amount of audio energy. In this way the point of overloading is not reached as quickly, and it is possible to obtain high efficiency from such a receiver.

No. 95   RADIO BROADCAST Laboratory Information Sheet   May, 1927

Storage Batteries

NECESSARY CARE

THE storage battery has been developed to a remarkable degree of perfection so that it will function over a long period of time with only a small amount of attention. Such attention consists more than anything else in keeping the battery properly filled with pure distilled water and correctly charged at all times. The efficiency and the life of the battery will decrease considerably if these two points are not carefully watched. The charging rate should be as close as possible to that recommended by the manufacturer, this information generally being given on the name plate of the battery. Although the state of charge of a battery can be measured with some accuracy by means of a voltmeter if the proper precautions are taken, the readings made in this way are not generally to be relied upon. A better method for use in testing a storage battery is to determine the state of charge by means of a hydrometer. The specific gravity, which is what the hydrometer measures, will be found to increase the reading of the hydrometer as the battery is charged, up to a certain point. The specific gravity reading for full charge is not the same for all batteries. For this reason, an endeavor should be

made to obtain from the manufacturer of the battery information regarding the hydrometer reading which should be obtained using his battery when it is fully charged and when it is fully discharged. Frequently, but not always, these same data will be found on the name plate. In the event that this information cannot be obtained, it is a safe rule to charge the battery until the hydrometer reading does not change during a period of one hour. When this condition holds true, the battery has absorbed all the charge possible. It will generally be found also that, when this condition of constant specific gravity reading throughout an hour is reached, the electrolyte will also begin to gas or bubble.

Care should be taken in charging the battery to make certain that its positive terminal is connected to the positive terminal of the source being used for charging purposes. If the battery is charged in the opposite direction the plates will be reversed in chemical character, and if the charging is continued for any great length of time, the battery will be destroyed. If a battery has only been charged in the wrong direction for a short length of time it can generally be brought back to normal by charging in the right direction for a very long time at a low charging rate.

No. 96   RADIO BROADCAST Laboratory Information Sheet   May, 1927

Analysis of Voice Frequencies

RELATIVE IMPORTANCE OF LOW AND HIGH FREQUENCIES

MANY investigations have been made to determine the relative importance of the various frequencies that are found in the human voice. For these investigations a high-quality audio amplifying system must be employed over which it is possible to hear equally as well as by direct transmission through the air.

Tests have been made by the Western Electric Company using such an amplifying system to determine the relative importance of different frequencies in the voice frequency range, and the results of these tests are shown in the curves on this sheet. These curves were obtained by inserting in the circuit low-pass (L.P.) filters, which will only pass low frequencies, and high-pass (H. P.) filters designed to pass no frequencies below a certain point. First of all let us consider the curve marked "Articulation H. P." The curve shows that the articulation was 40 per cent. when a high-pass filter was used that eliminated all frequencies below 2000 cycles. The articulation rises to 70 per cent. when a high-pass filter was used to cut off all frequencies below 1400 cycles. The curve marked "Energy L.P." shows that 60 per cent. of the total energy in the voice remained when a low-pass filter was used to cut off frequencies above 500 cycles.

These curves indicate, then, that the lower frequencies furnish most of the energy in the voice and that the higher frequencies are most important for proper articulation. If frequencies below 500 cycles are eliminated the energy is reduced 60 per

cent., and the articulation is only reduced 2 per cent. Eliminating all frequencies above 400 cycles leaves remaining 60 per cent. of the total energy but the articulation is only about 5 per cent.

The curves on this sheet were traced from an excellent book by K. S. Johnson entitled: *Transmission Circuits for Telephone Communication*.

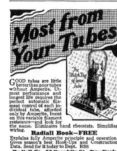

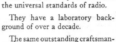

Equipment for the Home‑ Constructor

How to Use Some of the New Equipment Tested and Approved by the "Radio Broadcast" Laboratory

By THE LABORATORY STAFF

INTERFERENCE ELIMINATORS

MAN-MADE, as well as the natural sort of "static," will hinder the reception of radio signals. Fortunately, however, the former can be eliminated in almost all cases. Any kind of electrical spark will set up high-frequency oscillations similar to those used for broadcasting, the difference between the two being that the electrical spark from a motor or other household appliance is of an intermittent and varying intensity, which results only in noise in the receiver. There are many classes of electrical apparatus which can cause interference, such as oil burners, battery chargers, violet ray apparatus, etc. The commutator type of motor commonly used in connection with household appliances is one of the most common offenders. The spark takes place at the commutator, but the high-frequency oscillations may be carried along wires for a block or more and blanket reception over a large area.

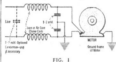

FIG. 1

A combination of fixed condensers and choke coils, placed in the electric line near the offending apparatus, will usually eliminate, or at least greatly reduce, the interference from this source. The common connections of such a filter apparatus are shown in Fig. 1. A choke coil is placed in each leg of the line supplying the motor or other apparatus. These coils have a high impedance to high-frequency oscillations, and are usually of approximately 200 milli-henries inductance. They should be wound with wire heavy enough to carry the current required without undue heating. Chokes of this character may be constructed by winding 175 turns of No. 16 d.c.c. on a porcelain tube 1 inch in diameter and 12 inches long. The condensers act as a by-path to ground for the high-frequency current. The value of these condensers may be from 1 to 2 microfarads, depending upon the nature of the interference. Sometimes another condenser, shown by the dotted lines, is placed across the line side of the filter. These condensers should have a high enough working voltage to take care of any voltage which may be impressed on them.

There are several interference filters on the market, two of which are shown in the photograph. While the circuit diagram given may not be exactly that used in the commercial type of filter, these latter are all based on the same general principle. In each device there are five wires leading out of the case, two of which go to the electric line, two to the motor or other apparatus, and the fifth to the grounded frame of the motor. The Tobe Interference Filter No. 1 is made by the Tobe-Deutschmann Company, of Cambridge, Massachusetts, and sells for $15.00.

This filter is designed for household appliances and will work effectively on motors up to ½ horse-power, and may be used on d.c. or a.c. lines. It is understood that these filters may be obtained, by special order, for installations of as high as 500-kilowatts and 1000 volts potential. The Day-Fan Electric Company, of Dayton, Ohio, also make a filter, known as the "Quietus," which is obtainable in two models, one, the No. 6001, for general use in the home, which sells for $10.00, and another, the No. 6003, for use with a household lighting plant, such as the Delco-Light, which is priced at $8.00.

RESISTORS FOR SOCKET-POWER DEVICES

RESISTORS designed for use in socket-power devices may be of the wire-wound type or of the metallic filament type, but in every case, they must be designed to dissipate the heat generated fast enough so that they will not burn out. Some manufacturers rate their resistors according to how much current may be carried with safety, while others rate them in watts. To choose the correct resistance when the rating is given in current, the amount of current which is to flow in the circuit must be known. This may be measured or it may be calculated by Ohm's Law. For instance, if the resistance is 1000 ohms and the voltage 100, the voltage divided by resistance will give the amount of current which will flow, which in this case is 0.10 amperes (100 milliamperes). The current-carrying capacity of the resistor should therefore be rated at 100 or more mils. In the case of the resistor rated in watts, the procedure is slightly different. The wattage is found by multiplying the voltage by the current, or in the case of the above resistor, the voltage (100) times the current (0.10) gives us 10 watts. That is, a 1000-ohm resistor having a rating of 10 watts would carry the 100 milliamperes with-

Radio Broadcast Photograph

THE "QUIETUS" AND TOBE-DEUTSCHMANN INTER- FERENCE FILTERS .

out overheating. Another way in which the carrying capacity may be figured is by the formula in which watts are equal to the current squared times the résistance. Transposing the formula we get:

$$I = \sqrt{W/R}$$

where I is the current, w is the watts rating, and R is the resistance. Substituting the actual values where R = 1000 ohms and w = 10, in the above formula, we have 1 equal to the square root of 10 ÷ 1000. Taking the square root, we obtain the safe carrying capacity in current as 0.1 amperes, or 100 milliamperes.

The resistances of various units as measured in the Laboratory by means of the Wheatstone bridge were found to differ by about 3 per cent. from the rated value, which is close enough for all ordinary purposes. Resistors of this type are made by Amsco Products, Incorporated, New York City; Ward-Leonard Electric Company, Mount Vernon, New York; the Tobe-Deutsch-mann Company (Veritas), Cambridge, Massachusetts; Arthur H. Lynch, Inc., New York, and the C. E. Mountford Company (Kroblak), New York City.

SHIELDS

ARGUMENTS still continue among engineers, experimenters, and home-constructors, regarding the shielding of radio broadcast receivers. Some say that coils, condensers, etc., if properly designed and properly spaced in the receiver, will need no shielding, while others claim that shielding is an absolute necessity.

Until quite recently the home-constructor was hampered in the use of shielding, because he had no tools for working the material; besides, the latter was hard to get. Several types of aluminum box shields can now be had, and they may easily be incorporated in almost any receiver requiring a shield. These shields usually come in knock-

Radio Broadcast Photograph

INDIVIDUAL STAGE SHIELDS

down form, and may easily be assembled by means of ingeniously designed corner strips. Their size is sufficient to allow a tube, radio-frequency transformer, and variable condenser to be mounted within them without coming too close to the sides. If the size is not exactly right they may be cut down before assembly, as they are in flat pieces. An accompanying photograph shows one of these shield boxes, made by the Aluminum Company of America, both in knocked-down and assembled form. The Aluminum Company's box is 12 gauge, 6 inches high, 5 inches wide, and 9 inches deep, and is sold for $3.00. The Hammarlund Manufacturing Company's box shield is 22 gauge, 5⅜ inches high, 6 inches wide, and 9 inches deep, and sells for $2.00. The Silver-Marshall Company makes a shield of this type which, however, is not knocked down. The box is 21 gauge, 5 inches high, 3⅞ inches wide, and 7½ inches deep. The price is $2.00.

SOCKET-POWER UNITS

THE Balkite socket-power B units are all of the electrolytic type, having eight cells in series which rectify the a.c. current from the line.

Radio Broadcast Photograph

THE BALKITE "KX" UNIT

These cells are of the acid type having one electrode of lead and the other of a special material. As a rule, this type of rectifier, combined with the proper transformers, chokes, and condensers gives very quiet operation, and the only attention required is the addition of pure water at intervals of several months to take care of the evaporation which naturally takes place.

In the instructions accompanying the apparatus, it is stated that the formation of a brown or white sludge in the bottom of the cell is a normal condition and no attention should be paid to it. The solution sometimes turns pink, and this also may be regarded as natural. It should never be necessary to replace the electrolyte except in case of accident when the solution is spilled. A new solution should be obtained from the manufacturer or from the dealer who sold the unit. If the electrolyte gets spilled in the metal case, the whole unit should be shipped back to the factory immediately for a thorough cleaning, otherwise the acid may cause considerable damage to the electrical mechanism in the device and cause it to cease functioning.

Sometimes when the electrolytic type of rectifier is left for some time without being operated, a thin film of sulphate will form on the surface of the electrodes which will materially lower the output voltage. There are two methods of bringing the unit back to normal. For ordinary cases, the unit should be disconnected from the receiver and a short circuiting wire connected across the output terminals marked "High" and "Negative." The line current is then turned on for half an hour, after which the jumper wire is disconnected and the unit is again hooked up to the receiver, and normal operation should be obtained. Another method which may be used is to put a jumper across the output as before and then short out half of the rectifier cells at a time for a period of about thirty seconds. In the Laboratory the breaking down of the sulphate film was tried out, the voltage, in one case, being raised from 125 to 140 volts. On one of the smaller units, the voltage was raised from 105 to 120 volts. Both units had been standing for some time. One model, the type KX, which includes a trickle charger in the same case, is shown in the accompanying photograph on this page.

A table of the price list and capacities is appended at the bottom of this page. All of these Balkite units are for use with 110 to 120 volts a.c. Models are sold for either 50 or 60 cycles. They are manufactured by Fansteel Products, Incorporated, of North Chicago, Illinois.

Model	Watts Consumption	Terminals	No. of Tubes	Capacity	Price
B —W	7	Det., Amp., B neg.	5 or less	67–90 volts, 20 mils.	$27.50
B —X	12	Det., Low. Mod., High, B neg.	5 to 8	135 volts, 30 mils.	42.00
B —Y	17	Det., Low, Med., High, Power, Gnd., B neg.			
KX*	14	Det., Low, Med., High, A plus, A minus, B neg.	Any receiver	150 volts, 40 mils.	69.00
			5 to 8	135 volts, 30 mils.	59.50

*Includes a trickle charger for the A battery.

PRICE LIST AND CAPACITIES OF BALKITE UNITS

A KEY TO RECENT RADIO ARTICLES

By E. G. SHALKHAUSER

THIS is the nineteenth installment of references to articles which have appeared recently in various radio periodicals. Each separate reference should be cut out and pasted on 4" x 6" cards for filing, or pasted in a scrap book either alphabetically or numerically. An outline of the Dewey Decimal System (employed here) appeared last in the January RADIO BROADCAST.

R550. BROADCASTING. BROADCASTING.
Popular Radio. Jan., 1927. Pp. 11-ff. *Who Pays for,*
"Who Pays the Broadcaster?" O. E. Dunlap, Jr.
The problem of supporting radio broadcast programs, either through actual contributions by the listening public, or through indirect channels, such as advertising, and thus creating good will and deriving publicity, is discussed. A list of sponsors of various program features, and the stations broadcasting, is tabulated, showing what is being done to-day in furnishing entertainment. A chart is also appended stating the toll charges of the various stations for given periods of time on the air.

R141.2. RESONANCE. RESONANCE.
Popular Radio. Jan., 1927. Pp. 31-ff. *Fading.*
"What Happens When You Tune Your Set," Sir Oliver Lodge.
A sample treatise on the conversion of radio waves into electrical impulses in the tuning circuits of the radio set and its analogy to mechanical or sound resonance is given in this second of a series of articles. The function of the detector tube is clearly outlined.

R620.08. INSTALLATIONS. INSTALLATIONS.
Bureau of Standards Handbook No. 9. *Safety Rules.*
"Safety Rules for Radio Installations."
The material contained herein is a reprint of Part 5 of the National Electrical Safety Code, dealing with proper radio installations as approved by the American Engineering Standards Committee.

R000. HISTORY. HISTORY.
Radio Service Bulletin. Dec. 31, 1926., Pp. 24-33.
"Important Events in Radio—Peaks in the Waves of Wireless Progress."
A brief outline of the important happenings in the history of radio, opposite the Year in which these events occurred, is given. Starting with the year 1877, when the principle of magnetism was first discovered, and leading down through the year 1926, the progress made in radio is presented.

R550. BROADCASTING. BROADCASTING.
Radio Service Bulletin. Dec., 31, 1926. *List of Stations.*
Pp. 9-21.
"Broadcasting Stations, Alphabetically by Call Signals."
A list of broadcasting stations, giving call signals, location of station, owner of station, power, wavelength, and frequency, is presented. The list is complete up to December 31, 1926. This publication may be procured from the Superintendent of Documents, Government Printing Office, Washington, District of Columbia, at five cents per copy.

R330. ELECTRON TUBES. ELECTRON TUBES,
Radio. Jan., 1927. Pp. 20-ff. *4 Element.*
"The Shielded Grid Vacuum Tube," E. E. Turner, Jr.
By placing a second grid between the plate and the control grid of a vacuum tube, Doctor Hull, of the General Electric Company, has succeeded in reducing inter-electrode capacity between the elements of the tube. This second grid is said to act as an electrostatic shield when positively charged. It is composed of a number of flat circular discs, or rings, 3 mm. wide and spaced 3 mm. apart. The characteristics of the tube are discussed.

R344.5. ALTERNATING CURRENT SUPPLY. SOCKET-POWER
Radio. Jan., 1927. Pp. 23-24. DEVICE, *A Battery.*
"A Home-Built A Battery Eliminator," G. M. Best.
Constructional data concerning a tungar A battery eliminator employing a transformer, chokes, two tungar 2-ampere bulbs, and electrolytic condensers or rheostats, are presented.

R382. INDUCTORS. INDUCTORS,
Radio. Jan., 1927. Pp. 26-27. *Chart.*
"A Reversible Inductance Chart," A. C. Kulmann.
A very comprehensible and useful inductance chart for single-layer solenoids, giving either the inductance of a coil or the coil size required for a given inductance, is shown. The range in inductance is from 4 to 20,000 microhenries. The use of the chart is illustrated.

R134.8. REPLEX ACTION. REFLEX
Radio. Jan., 1927. Pp. 29-ff. ACTION.
"More About the New Inverse Duplex," D. Grimes.
In this second of a series of articles on the new Inverse Duplex System, complete data on the construction of the separate coils, data on wiring and assembly, and elimination of trouble that might occur, are given.

R343. ELECTRON-TUBE RECEIVING SETS. RECEIVER,
Radio. Jan., 1927. Pp. 30-ff. *Infradyne.*
"How and Why the Infradyne Works," R. B. Thorpe.
The writer discusses theoretically the operation of the new infradyne receiving circuit, analyzing the action taking place within the various parts of the circuit. Various diagrams and graphs are shown, in order to explain clearly the arrangement employed.

R344.3. TRANSMITTING SETS. TRANSMITTERS,
Proc. I. R. E. Jan., 1927. Pp. 9-36. *Crystal-Controlled.*
"Piezo-Electric Crystal-Controlled Transmitters," A. Crossley.
A discussion of piezo-electric crystals and the early history of the development of the art, are given.
The development of crystal-controlled vacuum-tube oscillators by the Naval Research Laboratory is outlined, and various means of amplifying the output of a crystal-controlled oscillator are cited. The best method being described. This method consists of balancing or neutralizing the various stages of amplification and also observing proper precautions for reducing grid circuit losses by using high values of biasing voltage.
A complete high-power low-frequency crystal-controlled transmitter is described, and a schematic wiring diagram of the circuits employed in this transmitter is shown. A schematic wiring diagram, and illustrations of one type of low-power high-frequency transmitter, complete the subject matter covered in this paper.

R344. ELECTRON-TUBE GENERATORS. ELECTRON-TUBE
Proc. I. R. E. Jan., 1927. Pp. 37-39. GENERATOR.
"Simultaneous Production of a Fundamental and a Harmonic in a Tube Generator," H. J. Walls.
A method whereby a single generator tube may be used to transmit two frequencies simultaneously, thus eliminating the necessity of using two separate tubes, is described. In addition to the fundamental, some harmonic is amplified in a separate oscillating circuit. A broadcasting station may thus send out the same modulated energy on two frequencies using but one transmitter.

R113.1. FADING. FADING.
Proc. I. R. E. Jan., 1927. Pp. 41-47. *Recorder for.*
"An Automatic Fading Recorder," T. A. Smith and G. Redwin.
A device for automatically recording signal intensities is described, with the method employed to amplify the signal sufficiently to operate a commercial type graphic meter. Sample fading records of various transmissions are also presented.

R331. CONSTRUCTION; EVACUATION ELECTRON TUBES,
OF TUBES. *Gas.*
Proc. I. R. E. Jan., 1927. Pp. 49-55. *Alkali Vapor.*
"Behavior of Alkali Vapor Detector Tubes," H. A. Brown and C. T. Knipp.
The comparative efficiency of gas-filled tubes now on the market, and certain potassium sodium alloy tubes, is described. It is shown how the operation of gas-filled tubes depends on the temperature of the gas and the tube walls. It is also shown how these K Na tubes compare with the 201-A and the 200 type tubes, and that the K Na tubes are ideal for durability, true tone reproduction, and non-critical adjustment of plate and filament voltage.

International Short-Wave Test

FROM April 18th to 30th, short-wave tests on 7000-kc. (43 meters) will be conducted from WAQ-2 XAI, the Westinghouse experimental station at Newark, New Jersey. The schedule is: 8 to 8:30 P. M. (Eastern Standard Time) "ABC de 2 XAI" sent automatically on a crystal-controlled transmitter. From 8:30 to 9, amateurs will be worked and other tests made. The manager of the station, E. Gundrum, Westinghouse Electric & Manufacturing Company, Plane and Orange Streets, Newark, New Jersey, welcomes reports on audibility, fading, and keying, from listeners throughout the world. Reports should be forwarded either to RADIO BROADCAST or directly to the station.

Manufacturers' Booklets Available

A Varied List of Books Pertaining to Radio and Allied Subjects Which May Be Obtained Free by Using the Accompanying Coupon

AS AN additional service to RADIO BROAD-CAST readers, we print below a list of booklets on radio subjects issued by various manufacturers. The publications listed below cover a wide range of subjects, and offer interesting reading to the radio enthusiast. The manufacturers issuing these publications have made great effort to collect interesting and accurate information. RADIO BROADCAST hopes, by listing these publications regularly, to keep its readers in touch with what the manufacturers are doing. Every publication listed below is supplied free. In ordering, the coupon printed on page 62 must be used. Order by number only.—THE EDITOR.

1. FILAMENT CONTROL—Problems of filament supply, voltage, regulation, and effect on various circuits. RADIALL COMPANY.

2. HARD RUBBER PANELS—Characteristics and properties of hard rubber as used in radio, with suggestions on how to "work" it. B. F. GOODRICH RUBBER COMPANY.

3. TRANSFORMERS—A booklet giving data on input and output transformers. PACENT ELECTRIC COMPANY.

4. RESISTANCE-COUPLED AMPLIFIERS—A general discussion of resistance coupling with curves and circuit diagrams. COLE RADIO MANUFACTURING COMPANY.

5. CARBORUNDUM IN RADIO—A book giving pertinent data on the crystal as used for detection, with hook-ups, and a section giving information on the use of resistors. THE CARBORUNDUM COMPANY.

6. B-ELIMINATOR CONSTRUCTION—Constructional data on how to build. AMERICAN ELECTRICAL COMPANY.

7. TRANSFORMER AND CHOKE-COUPLED AMPLIFICATION—Circuit diagrams and discussion. ALL-AMERICAN RADIO CORPORATION.

8. RESISTANCE UNITS—A data sheet of resistance units and their application. WARD-LEONARD ELECTRIC COMPANY.

9. VOLUME CONTROL—A leaflet showing circuits for distortionless control of volume. CENTRAL RADIO LABORATORIES.

10. VARIABLE RESISTANCE—As used in various circuits. CENTRAL RADIO LABORATORIES.

11. RESISTANCE COUPLING—Resistors and their application to audio amplification, with circuit diagrams. DEJUR PRODUCTS COMPANY.

12. DISTORTION AND WHAT CAUSES IT—Hook-ups of resistance-coupled amplifiers with standard circuits. ALLEN-BRADLEY COMPANY.

13. B-ELIMINATOR AND POWER AMPLIFIER—Instructions for assembly and operation using Raytheon tube. GENERAL RADIO COMPANY.

13a. B-ELIMINATOR AND POWER AMPLIFIER—Instructions for assembly and operation using an R. C. A. rectifier. GENERAL RADIO COMPANY.

16. VARIABLE CONDENSERS—A description of the functions and characteristics of variable condensers with curves and specifications for their application to complete receivers. ALLEN D. CARDWELL MANUFACTURING COMPANY.

17. BAKELITE—A description of various uses of bakelite in radio, its manufacture, and its properties. BAKELITE CORPORATION.

19. POWER SUPPLY—A pocket-size booklet with particular reference to lamp-socket operation. Theory and constructional data for building power supply devices. ACME APPARATUS COMPANY.

20. AUDIO AMPLIFICATION—A booklet containing data on audio amplification together with hints for the constructor. ALL-AMERICAN RADIO CORPORATION.

21. HIGH-FREQUENCY DRIVER AND SHORT-WAVE WAVEMETER—Constructional data and application. BURGESS BATTERY COMPANY.

46. AUDIO-FREQUENCY CHOKES—A pamphlet showing positions in the circuit where audio-frequency chokes may be used. SAMSON ELECTRIC COMPANY.

47. RADIO-FREQUENCY CHOKES—Circuit diagrams illustrating the use of chokes to keep out radio-frequency currents from definite points. SAMSON ELECTRIC COMPANY.

48. TRANSFORMER AND IMPEDANCE—Tables giving the mechanical and electrical characteristics of transformers and impedances, together with a short description of their use in the circuit. SAMSON ELECTRIC COMPANY.

49. BYPASS CONDENSERS—A description of the manufacture of bypass and filter condensers. LESLIE F. MUTER COMPANY.

50. AUDIO MANUAL—Fifty questions which are often asked regarding audio amplification, and their answers. AMERTRAN SALES COMPANY, INCORPORATED.

51. SHORT-WAVE RECEIVER—Constructional data on a receiver which, by the substitution of various coils, may be made to tune from a frequency of 16,660 kc. (18 meters) to 1900 kc. (150 meters). SILVER-MARSHALL, INCORPORATED.

54. AUDIO QUALITY—A booklet dealing with audio-frequency amplification of various kinds and the application to well-known circuits. SILVER-MARSHALL, INCORPORATED.

56. VARIABLE CONDENSERS—A bulletin giving an analysis of various condensers together with their characteristics. GENERAL RADIO COMPANY.

57. FILTER DATA—Facts about the filtering of direct current supplied by means of motor-generator outfits used with transmitters. ELECTRIC SPECIALTY COMPANY.

59. RESISTANCE COUPLING—A booklet giving some general information on the subject of radio and the application of resistors to a circuit. DAVEN RADIO CORPORATION.

60. RESISTORS—A pamphlet giving some technical data on resistors which are capable of dissipating considerable energy; also data on the ordinary resistors used in resistance-coupled amplification. THE CRESCENT RADIO SUPPLY COMPANY.

62. RADIO-FREQUENCY AMPLIFICATION—Constructional details of a five-tube receiver using a special design of radio frequency transformer. CAMFIELD RADIO MANUFACTURING COMPANY.

63. FIVE-TUBE RECEIVER—Constructional data on building a receiver. AERO PRODUCTS, INCORPORATED.

64. AMPLIFICATION WITHOUT DISTORTION—Data and curves illustrating the use of various methods of amplification. ACME APPARATUS COMPANY.

65. RADIO HANDBOOK—A helpful booklet on the functions, selection, and use of radio apparatus for better reception. BENJAMIN ELECTRIC MANUFACTURING COMPANY.

66. SUPER-HETERODYNE—Constructional details of a seven-tube set. G. C. EVANS COMPANY.

70. IMPROVING THE AUDIO AMPLIFIER—Data on the characteristics of audio transformers, with a circuit diagram showing where chokes, resistors, and condensers can be used. AMERICAN TRANSFORMER COMPANY.

71. DISTORTIONLESS AMPLIFICATION—A discussion of the resistance-coupled amplifier used in conjunction with a transformer, impedance, or resistance input stage. Amplifier circuit diagrams and constants are given in detail for the constructor. AMSCO PRODUCTS INCORPORATED.

72. PLATE SUPPLY SYSTEM—A wiring diagram and layout plan for a plate supply system to be used with a power amplifier. Complete directions for wiring are given. AMERTRAN SALES COMPANY.

80. FIVE-TUBE RECEIVER—Data are given for the construction of a five-tube tuned radio-frequency receiver. Complete instructions, list of parts, circuit diagram, and template are given. ALL-AMERICAN RADIO CORPORATION.

81. BETTER TUNING—A booklet giving much general information on the subject of radio reception with specific illustrations. Primarily for the non-technical home constructor. BREMER-TULLY MANUFACTURING COMPANY.

82. SIX-TUBE RECEIVER—A booklet containing photographs, instructions, and diagrams for building a six-tube shielded receiver. SILVER-MARSHALL, INCORPORATED.

83. SOCKET POWER DEVICE—A list of parts, diagrams, and templates for the construction and assembly of socket power devices. JEFFERSON ELECTRIC MANUFACTURING COMPANY.

84. FIVE-TUBE EQUAMATIC—Panel layout, circuit diagrams, and instructions for building a five-tube receiver, together with data on the operation of tuned radio-frequency transformers of special design. KARAS ELECTRIC COMPANY.

85. FILTER—Data on a high-capacity electrolytic condenser used in filter circuits in connection with a power supply units, are given in a pamphlet. THE ABOX COMPANY.

Accessories

22. A PRIMER OF ELECTRICITY—Fundamentals of electricity with special reference to the application of dry cells to radio and other uses. Constructional data on buzzers, automatic switches, alarms, etc. NATIONAL CARBON COMPANY.

23. AUTOMATIC RELAY CONNECTIONS—A data sheet showing how a relay may be used to control A and B circuits. YAXLEY MANUFACTURING COMPANY.

31. ELECTROLYTIC RECTIFIER—Technical data on a new type of rectifier with operating curves. KODEL RADIO CORPORATION.

36. DRY CELLS FOR TRANSMITTERS—Actual tests given, well illustrated with curves showing exactly what may be expected of this type of B power. BURGESS BATTERY COMPANY.

37. DRY-CELL BATTERY CAPACITIES FOR RADIO TRANSMITTERS—Characteristic curves and data on discharge tests. BURGESS BATTERY COMPANY.

38. B BATTERY LIFE—Battery life curves with general curves on tube characteristics. BURGESS BATTERY COMPANY.

39. HOW TO MAKE YOUR SET WORK BETTER—A nontechnical discussion of general radio subjects with hints on how reception may be bettered by using the right tubes. UNITED RADIO AND ELECTRIC CORPORATION.

30. TUBE CHARACTERISTICS—A data sheet giving constants of tubes. C. E. MANUFACTURING COMPANY.

31. FUNCTIONS OF THE LOUD SPEAKER—A short, nontechnical general article on loud speakers. AMPLION CORPORATION OF AMERICA.

32. METERS FOR RADIO—A catalogue of meters used in radio, with connecting diagrams. BURTON-ROGERS COMPANY.

33. SWITCHBOARD AND PORTABLE METERS—A booklet giving dimensions, specifications, and shunts used with various meters. BURTON-ROGERS COMPANY.

34. COST OF B BATTERIES—An interesting discussion of the relative merits of various sources of B supply. HARTFORD BATTERY MANUFACTURING COMPANY.

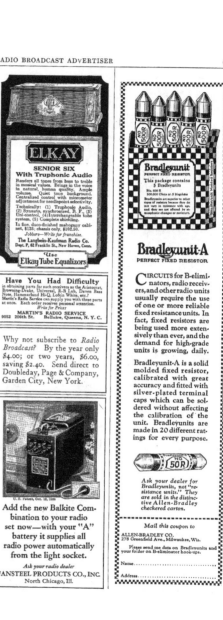
Why not subscribe to *Radio Broadcast?* By the year only $4.00; or two years, $6.00, saving $2.40. Send direct to Doubleday, Page & Company, Garden City, New York.

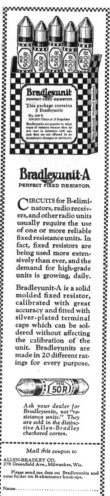

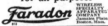

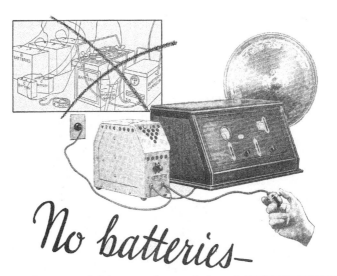

No batteries—

Radio's most revolutionary development! Run this radio direct from house current outlet. Ordinary 110 volt 60 cycle domestic electricity transformed mechanically into smooth, quiet radio A, B and C power as you use it. Radio power supply annoyances ended for all time. A snap of the switch is the only demand radio will make upon you from NOW ON.

No more batteries to fuss with.
No more trickle chargers to watch.
No more keeping something filled with water.
No batteries to renew or recharge.
No upsetting the home to have the radio serviced.

See these wonderful sets at any Crosley dealers, or write Dept. 20 for descriptive literature.

The Crosley Radio Corporation

Powel Crosley, Jr.
President

CINCINNATI
OHIO

Radio Energy Unit

Crosley Radio Energy Unit weighs only 13 lbs., is only half the size of an ordinary A storage battery—operates without interfering hum and with the certainty of an electric motor.

$50

6-tube AC Receivers
for use with Crosley Radio Energy Unit

Crosley radios designed for use with this marvelous power supply are the AC-7, a 6-tube table model at $79, and the AC7-C, a 6-tube console at $95.

Crosley sets are licensed under Armstrong U. S. Patent No. 1,113,149, or under patent applications of Radio Frequency Laboratories, Inc., and other patents issued and pending

Prices slightly higher west of the Rocky Mts

CROSLEY
RADIO

CROSLEY
ULTRA
MUSICONE
$9.75

Radio is better with *battery* power

Eveready Layerbilt "B" Battery No. 486, the Heavy-Duty battery that should be specified for all loud-speaker sets. Price $5.00.

The Layerbilt patented construction revealed. Each layer is an electrical cell, making automatic contact with its neighbors, and filling all available space inside the battery case.

Here is *battery* power at

LIKE every other good battery, the Eveready Layerbilt provides only pure DC (Direct Current), steady, noiseless, the only current that can give you the best results of which your set is capable. Radio is better with *Battery* Power, always, for batteries alone produce pure DC, and are entirely reliable, convenient, available anywhere, always ready to work. For best results and satisfaction, use batteries, and for greatest economy, get the Eveready Layerbilt "B" Battery No. 486.

For years you have known Eveready Radio Batteries as "the kind that last longer"—
and now you are finding that

lasts longest of all. So long does it last in proportion to its price that hundreds of thousands of people have found it to be the most economical battery they ever used. It is not only that, but on the basis of exceedingly careful scientific tests it is by far the most economical, dependable and satisfactory source of "B" power on the market today. These tests, unerringly revealing what each type of "B" power will do, have proved the superiority of the Layerbilt, and have shown why more and more people are adopting it.

It is a wonderful battery—
no wonder when you look at its
construction! It is built in

the "Lay-
patented
utilizes
inside the
waste spa
soldered
more act
the batte
more effi
longer-las

NATIONA
New York
Unit of Union

Tuesday nigh
9 P. M.,

WEAF–*New Y*
WJAR–*Provide*
WEEI–*Bost*on
WTAG–*Worcest*
WFI–*Philadelp*
WGR–*Buffalo*
WCAE–*Pittsbur*

Cunningham RADIO TUBES

LAST AUDIO STAGE ONLY

LAST AUDIO ST

CUNNINGHAM
CX-371

CUNNINGHAM
CX-220

CUNNINGHAM
CX-112

Since 1915 Standard for all Sets-

At the Reproducer

AT the reproducer—where quality counts most—there Cunningham Power Tubes prove their indispensability to finished, well-rounded tone.

Just as CX-371, CX-112 and CX-220 are leaders in the crusade for more natural reproduction, so other Cunningham types are leaders in their various fields.

Consult your dealer. He knows the right combination of radio tubes for your receiver.

Sixteen Types all in the Orange and Blue Carton.

E. T. CUNNINGHAM, INC.

New York   Chicago   San Francisco

DeForest, always the pioneer, leads the advance to better radio reception with the creation of special tubes for specific radio functions

The first Radio Tube in the World, 1906, from which has sprung the present gigantic radio industry. The device through which natural sounds audible to the human ear were transmitted by radio for the first time.

Skilled hands that fashioned so exquisite an object as the Nuremburg covered cup of 1850, now carefully guarded in the Metropolitan Museum, must be as deft today in the delicate precision required in making DeForest Audions.

CERTAIN tube characteristics that make for improved reception in the various functions of a radio hook-up have been carefully developed by DeForest engineers. Those invisible factors specifically performing in their recognized spheres are making radio reception more and more enjoyable and dependable every day.

Take no one's word but your own. Try the new DeForest Specialist DL-4 Audion in your radio frequency stages. The decided improvements you will get are an indication of the superiority of all genuine Audions. Weak signals hardly heard before become loud and clear. Distant stations move up close like locals. Better performance because these Audions are especially designed to do a radio amplification job.

The rigid limits, both electrical and mechanical, to which DeForest Specialist Audions are strictly held assure a high standard of uniformity. Radio amateurs appreciate such efficiency. Constant grid-plate capacity and high mutual conductance provide a quality-volume from distant reception which is heartily satisfying to the critical radio fan.

You are earnestly urged to test the features of these tubes by replacing in your RF stages with these Specialist DL-4 Audions. Such a trial will show you their superiority definitely. Expense is slight. DeForest Audions perform amazingly.

A new audion—going any price appeal one better—is the general purpose Audion —the D-01A. It is an unmatchable value at $1.65. It offers the same standard of quality that has made DeForest Audions the recognized perfection in radio tube manufacturing.

DeForest dealers display the distinct black, green and orange Audion container. Write us direct, mentioning name of your Dealer for booklet giving full characteristics of each Audion together with the chart indicating proper DeForest Audion in all standard makes of radio.

Write Dept. 13 for descriptive literature.

THE DeFOREST RADIO CO.
Powel Crosley, Jr., Pres.   Jersey City, N. J.

Assure
Set Performance
with the

WES TON

Convertible "Pin-Jack"
Voltmeter

A VOID the pit-falls to set per-formance and satisfac-tion, avoid unnecessary replacement expense for tubes and batteries and at the same time get better re-ception. ¶This "Pin-Jack" Voltmeter plugs directly into the filament jacks of Radiola, Victor, Brunswick and Bosch sets. It can also be used on any other make of set by simply installing the pin-jacks provided with each instru-ment. ¶To prevent overloading tubes just plug the voltmeter into the fil-ament jacks and turn the control rheo-stat until the meter indicates the correct filament voltage. ¶To know the condition of batteries it is only necessary to plug the voltmeter into its High Range Stand and then make the simple test using the thirty inch cables attached to the instrument.

WESTON ELECTRICAL
INSTRUMENT CORPORATION
179 Weston Avenue, Newark, N. J.

STANDARD THE WORLD OVER
WESTON
Pioneers since 1888

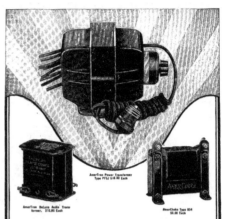

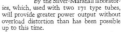

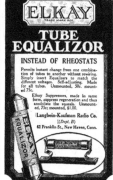

Put a Power tube
(¼ amp. ZP 201-A)
in *Every* socket—

ZETKA PROCESS ZP-201-A

without re-wiring your set

Radio's sensational development has seen nothing more startling than the new ZP 201 A, oxide filament, ¼ amp. Power Tube—the Zetka Process power tube for every stage.

Now for the first time, can power tubes be used in every socket of your set. Now you can enjoy *real* power. And *without changing a single wire!*

Replace your old 201 A's with a set of these new ZP 201 A's and enjoy the most startling achievement in tube history.

The current drain is small in both A and B—it is not necessary to hook up additional C's—but—what an improvement in tone, in volume, in sensitivity.

Only $2.50 each at your nearest dealer puts a complete Power tube set well within your radio budget. And then, and then only you'll enjoy real radio satisfaction —with power tubes of longer life, that increase in efficiency as you use them.

Ask your Zetka dealer to show you the Zetka Weston meter test. You'll want nothing but Zetka thereafter.

YOUR SET DESERVES THIS FINER EQUIPMENT—
PRICES NO HIGHER

ZETKA
The *Clear Glass* Tube

ZETKA LABORATORIES INC.
73 Winthrop Street   Newark, New Jersey

RADIO BROADCAST

WILLIS KINGSLEY WING, Editor

JUNE, 1927

KEITH HENNEY
Director of the Laboratory

JOHN B. BRENNAN
Technical Editor

Vol. XI, No 2

EDGAR H. FELIX, Contributing Editor

AMONG OTHER THINGS. . .

AS ANNOUNCED on page 78, the July number of RADIO BROADCAST will appear with a new cover design. For the last six months, we have been examining sample designs and searching the field for an artist who could supply us with a design which properly reflected the character of this magazine. Finally, Harvey Hopkins Dunn, of Philadelphia, presented a design which drew unqualified approval. Mr. Dunn is internationally known as a typographical designer and his most recent cover design has been used by International Studio. Our July cover will appear in an unusually attractive shade of yellow, green, and black, and is distinguished by its simplicity and effectiveness.

THIS number of RADIO BROADCAST contains an exceptionally wide range of editorial features, starting with R. W. King's review of present knowledge of how radio waves are propagated. Then comes a short article on what one should know about shielding, of distinct help to those who sell and those who buy shielded receivers. James Millen's second article on radio and the electric phonograph gives a wealth of practical information to those who are planning to make their present phonograph equipment work with their high-quality radio receiver. The spring and summer months offer an excellent time to make these not very difficult changes. We call especial attention to the "Strays' from the Laboratory"—a new feature of this magazine, which will, from month to month, provide a place where our readers may find comment and news of great interest. Another new feature in the popular department "As the Broadcaster Sees It" is found in the presentation of a technical problem, and its appended solution. These problems will cover a wide range and will prove of value not only to the technical broadcaster but to the general reader as well. Homer Davis' "How to Design a Loop Antenna" has long been awaited by our readers and is of distinctly practical nature. The second of Roland F. Beers' articles on the problems of series filament connections in receiving sets appears, presented in a very clear manner. The Radio Club of America paper this month casts more light on the design of power amplifiers and should attract wide interest because of the increasing popularity of this accessory.

PRINTERS' INK, in its tabulation of advertising lineage for April magazines shows that RADIO BROADCAST printed 15,315 lines, being exceeded only by Radio News with 15,454. Radio printed 11,129 lines, Popular Radio 10,510 and Radio Age 3,728.

ANSWERS from our readers to the question posed in our May number: "Which Stations Shall Broadcast?" are coming in great numbers, and are rapidly being tabulated for the Federal Radio Commission. These reflections of opinion from our readers provide an unusually reliable cross-section of opinion and important information about local conditions from the entire country and from Canada.

THE July magazine, among many other important features, will contain an exclusive article on the new Raytheon "A" tube, showing its use and value in A-battery charging circuits; two complete constructional articles will tell how to use the QRS 400-mA. rectifier tube and the Raytheon 350-mA. tube.
—WILLIS KINGSLEY WING.

Doubleday, Page & Co. MAGAZINES	Doubleday, Page & Co. BOOK SHOPS (Books of all Publishers)	Doubleday, Page & Co. OFFICES	OFFICERS
COUNTRY LIFE WORLD'S WORK GARDEN & HOME BUILDER RADIO BROADCAST SHORT STORIES EDUCATIONAL REVIEW LE PETIT JOURNAL EL ECO FRONTIER STORIES WEST	{ LORD & TAYLOR BOOK SHOP NEW YORK: { PENNSYLVANIA TERMINAL (2 Shops) { GRAND CENTRAL TERMINAL { 38 WALL ST. and 526 LEXINGTON AVE. { 848 MADISON AVE. and 166 WEST 32ND ST. ST. LOUIS: 223 N. 8TH ST. and 4914 MARYLAND AVE. KANSAS CITY: 920 GRAND AVE. and 206 W. 47TH ST. CLEVELAND: HIGBEE CO. SPRINGFIELD, MASS.: MEEKINS, PACKARD & WHEAT	GARDEN CITY, N. Y. NEW YORK: 285 MADISON AVENUE BOSTON: PARK SQUARE BUILDING CHICAGO: PEOPLES GAS BUILDING SANTA BARBARA, CAL. LONDON: WM. HEINEMANN LTD. TORONTO: OXFORD UNIVERSITY PRESS	F. N. DOUBLEDAY, President NELSON DOUBLEDAY, Vice-President S. A. EVERITT, Vice-President RUSSELL DOUBLEDAY, Secretary JOHN J. HESSIAN, Treasurer L. A. COMSTOCK, Asst. Secretary L. J. McNAUGHTON, Asst. Treasurer

DOUBLEDAY, PAGE & COMPANY, Garden City, New York

Copyright, 1927, in the United States, Newfoundland, Great Britain, Canada, and other countries by Doubleday, Page & Company. All rights reserved.
TERMS: $4.00 a year; single copies 35 cents.

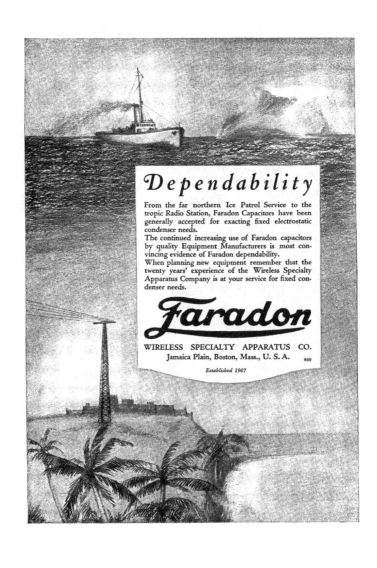

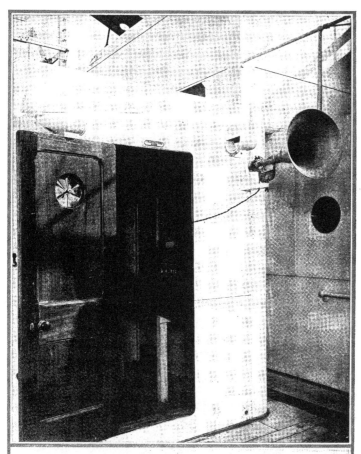

RADIO PROGRAMS AT SEA

One of a number of loud speakers of the speech equipment installed aboard the motor vessel Asturias, one of the newest passenger ships. The various loud speakers may be connected to an electrical attachment to reproduce phonograph music, to a local microphone, or to the output of a broadcast receiver. The transmitting room of the ship's radio installation is in the foreground, while at the top left extreme can be seen the antenna frame of the radio direction finder

RADIO BROADCAST

VOLUME XI NUMBER 2

JUNE, 1927

What Do We Know About Radio Waves in Transit?

How Radio Waves Behave After Leaving the Transmitter— The Queer Influence of Various Phenomena Upon Them

By R. W. KING

American Telephone and Telegraph Company

THE ancient Greek god with the winged sandals has been replaced by invisible electric waves. These waves are the conveyors for all of our many forms of electrical communication. Speeding at well-nigh incredible velocities, they bear alike the code message of the familiar Morse telegraph, the voice that flows over the wires of the household telephone, and the music which leaves the antenna of the radio broadcasting station.

In the case of the wire telephone and telegraph, the waves are closely harnessed and obediently follow a given pair of wires or a single wire with "ground return." As radiated from the broadcasting antenna, we commonly think of the waves as being free and spreading out in all directions. Yet they cannot be entirely free and unconstrained. Apparently the atmosphere exerts upon them a sort of guiding influence sufficient to prevent their being lost in space. A fundamental fact of radio transmission as we observe it on the earth is its following not straight but curved paths. The earth is a ball and radio waves, instead of traveling out along a tangent plane, curve their course sufficiently to conform to the rotundity of the earth,

Just how great is the curvature involved we can visualize more readily by constructing a small-scale model. If we have a single medium propagating waves without absorption or dispersion, a principle of optics states that all wave paths remain similar when the scales of time and space are reduced in the same ratio. Application of this principle leads to the result that if, on the one hand, we have waves 100 meters long traveling over an earth 8000 miles in diameter, and on the other hand, waves of red light (rather less than one millionth of a meter long) and a sphere about two

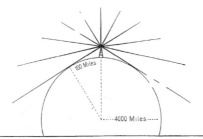

FIG. 1

If the earth's atmosphere had no effect on the transmission of radio waves, their propagation would be along straight lines as shown in this drawing. An antenna, one mile high, would be capable of transmitting signals to a distance of about 100 miles before the shadow cast by the earth's curvature would cut them off. If the receiving antenna were also a mile high, the distance of transmission would be increased to about 200 miles

inches in diameter, then the geometry of the two cases will be identical. Experience tells us, however, that red light would not creep to any appreciable degree around a sphere the size of a billiard ball. The sphere would cause a very apparent shadow if held so as to intercept light falling upon a screen, and if radio waves crept around the earth to no greater extent than this light would around the darkened portion of the ball, long-distance signaling by radio would be quite impossible.

Among those skilled in the science of optics no small amount of surprise was occasioned, therefore, by Marconi's announcement about twenty-five years ago that radio signals had been successfully transmitted across the Atlantic Ocean. This at once gave rise to speculation as to how the beam of waves could bend itself around the protuberance of the curved earth, it having naturally been taken for granted that to such waves the atmosphere would be merely a uniform and transparent medium.

Besides constraining radio waves to travel around our spherical earth—a very fortunate fact—the atmosphere causes in them variations of strength with time of day and year, erratic fluctuations known as "fading," and sometimes irregularities of such a sort as to result in serious loss of quality to the messages it carries. These latter effects of the atmosphere are not particularly fortunate so far as the radio engineer is concerned. However, all effects, fortunate and unfortunate, may be so related in cause that if the atmosphere were to yield the secret of why it guides waves to follow the surface of the earth, it might also yield secrets helping to master the troublesome phenomena of fading and loss of quality. The near future is likely to see us in possession of the reason why the atmosphere acts as a guide to waves and if, in acquiring this knowledge, we gain an insight into other facts, the radio engineer will have occasion to be thankful that he lives on an earth that is round.

Unfortunately it is only in thought that we can pass from considering 100-meter waves traveling around the earth to waves of red light traveling around a billiard ball. If we could but experiment with the latter with its proper diminutive atmosphere surrounding it, we might hope to vary one factor at a time and thus more readily discover the cause or combination of causes for radio waves traveling in curved beams. But it appears that we are limited to experiment on the full-scale earth itself. And here not even Joshua of old, commanding the sun to stand still, could help us very much. By stopping our celestial luminary, he could, to be sure, supply uninterrupted daylight and night conditions, and these would undoubtedly help in understanding the differences they present to radio transmission, but it seems likely that the presence or absence of daylight is only one of many factors entering the problem. To mention just a few others, there are lightning flashes attended by the possible liberation of very high speed electrons, the "cosmic rays" which are known to increase in intensity as we ascend to high altitudes, probable electric discharges from the sun (these are now commonly supposed to be the principal cause of the aurora borealis), and, in turn, the pressure, temperature, and compositions of the upper strata of the atmosphere.

Here is a complexity of influences almost sufficient to satisfy the mathematician who aspired to develop his technique to the point that he could predict the orbit of a house fly.

Returning to the original problem, an explanation of · why radio waves bend around the earth has been sought in many directions. One of the first attempts was to call upon the fact that the earth is

an electrical conductor. Waves started from a grounded antenna would, of course, be in contact at the outset with the earth, and it was thought possible that the interaction between the waves and the currents they would induce in the surface of the earth would be such as to cause them to remain closely attached to the earth as they spread out. This tendency actually does exist but calculations soon showed that it was far too minute to account for the known strength of signals received at a great distance.

The way was prepared, however, for the suggestion of the so-called Heaviside layer, a conducting region in the upper atmosphere which would imprison radio waves between itself and the conducting earth below, and constrain them to move only in the annular region between. Whether such a conducting layer actually exists, it is as yet impossible to say. So far as the electrical properties of gases are known, it could re-

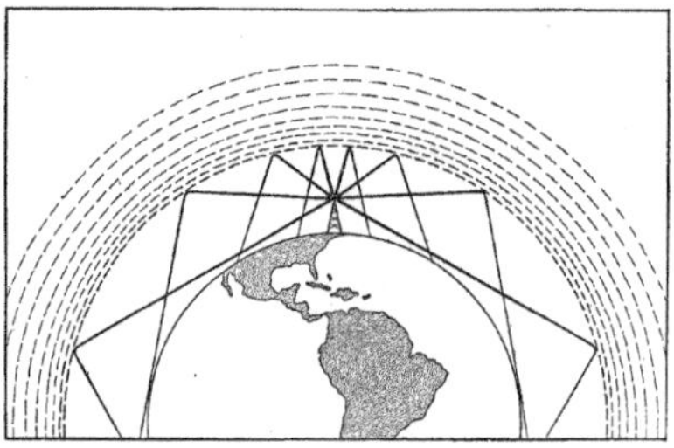

FIG. 2

This drawing illustrates the general manner of radio propagation assuming a rather sharply defined conducting (reflecting) layer in the upper atmosphere. The Heaviside layer is ordinarily assumed to possess this character. By means of the conducting stratum of air above, and the conducting earth below, radio waves are guided essentially as waves are guided along a pair of wires in wire communication

sult only from rather sharply defined ionization in the upper atmosphere. When we speak of a gas as ionized, we mean that certain of its molecules have been robbed of one or more of their normal electrons. These ionized molecules then constitute freely moving positive charges. The electrons which they have lost may continue to circulate as individual negative charges, or they may attach themselves to neutral molecules of gas, still retaining their negative charges but losing considerably in mobility because of the relatively heavy molecules with which they are associated.

The air at the surface of the earth is always somewhat ionized—on an average to the extent of perhaps 1000 negative ions and 1000 positive ions for the 30 quintillion molecules actually present per cubic centimeter of gas. A small amount, indeed, but very important perhaps. This ionization may originate partly through radioactive materials in the earth and in the air, and to a smaller extent through the agency of the so-called cosmic rays. Ultra-violet light

from the sun may also cause appreciable ionization in the upper strata where it is largely absorbed, but this doubtless does not persist through the night, and therefore is merely one of the agencies causing radio waves to behave as they do.

It should be borne in mind, however, that the current explanation of the aurora is based upon ions which are hypothecated as streaming from the sun and bending around into the darkened hemisphere under the influence of the earth's magnetic field. It may be such ions as these that account for the bulk of both day time and night time conductivity in the upper atmosphere. See Fig. 5.

Another factor, recently pointed out by Professor C. T. R. Wilson, which may be by no means the least important in determining the electrical state of the upper atmosphere, is thunder storms. From a statistical study of the distribution of thunder storms, the English Meteorological Office concludes that about 1800 thunder storms are, on the average, in progress at a given moment, producing about 100 lightning flashes per second. The quantity of electricity discharged in a flash is of the order of 20 coulombs, and the potential difference which causes the discharge may rise to as high, it is estimated, as one billion volts. Thus Professor Wilson suggests that the power expended in producing lightning by thunder clouds the world over may be as great as one ten-thousandth part of the total power received by the earth from the sun. Now, as a celestial receiving set, the earth does a very creditable job, picking up, from the power which the sun broadcasts, about one hundred trillion horsepower. It is indeed noteworthy, therefore, that if 1–10,000th part, or even a much smaller fraction, of this power is turned loose in the form of lightning—and therefore in a form which generates static—it offers to our man-built radio stations competition of no insignificant order of magnitude.

Not only is the electric power expended by thunder clouds large, but both the current and the voltage involved are of the order of magnitude which suggests the possibility of important effects on the electrical state of the upper atmosphere. Thus, the voltage of a thunder cloud is such that it may act as a source of extremely high speed electrons and also X-rays.

It is conceivable that high-speed electrons or beta rays from lightning flashes may pass upward through the outer atmosphere, and, due to the earth's magnetic field, reënter the atmosphere in widely scattered regions contributing, perhaps, to auroral phenomena and to the·

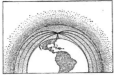

FIG. 3

This drawing illustrates the bending effect of ionization permeating the atmosphere. Downward refraction toward the earth would be due to an ionization that becomes greater with increased altitude

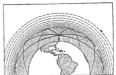

FIG. 4

If the effects shown in Figs. 2 and 3 were combined, they would result in a type of transmission shown in this drawing

penetrating "cosmic" radiation as well as to such atmospheric ionization as is important in radio transmission. In this connection, it is interesting to recall that the green spectrum line which is so characteristic of the light of the aurora has been found by Lord Rayleigh in the radiation from the normal night sky.

Attempts are being made to associate radio reception conditions with the weather (see articles by J. C. Jensen and Eugene van Cleef in RADIO BROADCAST, and a recent paper, by G. W. Pickard read before the Institute of Radio Engineers) and, in accordance with this theory of C. T. R. Wilson, it may not be impossible to find a connection between reception and static and thunder storms, both local and remote.

Another theory, and one of the latest to be proposed (Sir J. Larmor, *Phil. Mag.*, Vol. 48, page 1025, 1924), is based upon ionization in the atmosphere but does not require that this exist in the form of a layer whose lower surface is sharply defined. Larmor shows that if the amount of ionization increases with altitude, the more elevated portions of a wave train will travel faster than the portion near the earth and the train as a whole will be deflected downward. This downward deflection may easily be sufficient to cause the wave train to conform to the curved surface of the earth or even to dive into the earth at a slight angle. On this theory the bending of radio waves downward is rather analogous to the bending upward of light waves to produce a mirage. It will be recalled that a mirage is seen across a flat, heated, landscape, such as desert sand, the layers of air closest to the earth being most highly heated and, therefore, of lowest density, transmitting light with a slightly greater velocity than the cooler overlying layers

Which of these two theories is better, or whether there is some theory, or

combination of theories, as yet unproposed, which will ultimately win out, it is, of course, scarcely worth our while to speculate at this juncture. As matters now stand, some may prefer to entertain the idea of a more or less sharply defined and reflecting layer of ionization, located perhaps ten miles, perhaps fifty miles, above the surface of the earth, while others may prefer to think in terms of a widely diffused ionization gradually increasing with increasing altitude, which acts prismatically to bend radio waves earthward. A paper given at the 1926 midwinter Convention of the A. I. E. E. by Baker and Rice, attempts a quantitative statement of the distribution of ions.

WORK FOR THE AMATEUR

IN THIS work the radio amateur occupies a very strategic position. Curiously enough, it is short-wave transmission that is bringing to light facts so striking that they cannot fail to be crucial to any theory of transmission. One of the outstanding pieces of work in this new field is that of Dr. A. H. Taylor who has correlated many results obtained by amateurs with the short-wave experience of the Navy. Among the striking phenomena now commonly recognized may be mentioned "skip distance." No one was very much surprised when it became known that short wavelengths die away very rapidly as one recedes from the transmitting antenna. But when these short waves, which presumably had disappeared entirely at a distance of, say, 200 miles, were found to reappear at a distance of 600 or 1000 miles, it became evident that a new phenomenon had to be reckoned with. Data recently published (for example, by Heising, Schelleng, and Southworth, of the Bell Telephone System) indicate, furthermore, that the "skip distance" lengthens as the wavelength decreases. This fact is not out of harmony with the theory of a refracting ionization distributed throughout the atmosphere, but neither can it be said as yet to prove that such ionization exists.

The "skip distance" displayed by short-wave transmission seems to indicate that, of the waves which leave the transmitting antenna, those which travel near the surface of the earth are rapidly absorbed, while those starting upward, perhaps at a slight angle to the earth's surface, soon reach a rarified region through which they travel with little absorption, but which gradually bends them downward, either by reflection or refraction, until they again reach the earth, hundreds or even thousands of miles away.

If this view is correct, it means that a beam of short waves constitutes a messenger which we can send up perhaps to the outermost confines of the atmosphere and have return to us again. There is no doubt but what scientists will quickly devise means for determining what alterations these returning waves have undergone, and thus interpret the message which they bring back regarding the constitution of the upper atmosphere.

In studying electric waves as messengers returning from the upper strata of the atmosphere, it will be increasingly important to take into account the effect of the earth's magnetic field in such ways as were pointed out by Nichols and Schelleng in the Bell System *Technical Journal*, April, 1925. In enlarging the theory of Larmor to include the earth's magnetic field, they found that under some circumstances it is possible for a ray to follow the bend of the earth, even though the number of ions decreases with altitude. They also concluded that, for low frequencies (long wavelengths), the magnetic field prevents the electrons from moving in as large orbits as they otherwise would describe, which results, in turn, in smaller absorption of energy and therefore in reduced attenuation. A magnetic field permeating the ionized atmosphere may also divide a beam into differently polarized components, thus giving additional data on the nature of the transmission path supplied by the upper atmosphere.

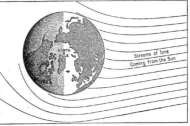

FIG. 5

This drawing illustrates how streams of negative electrons or ions emitted by the sun would be swerved by the earth's magnetic field. Some of these rapidly moving ions would reach the upper atmosphere over the darkened hemisphere of the earth and would therefore be available to influence night-time transmission. A stream of positive ions coming from the sun would be deflected downward and not upward

THE MARCH OF RADIO

News and Interpretation of Current Radio Events

Why Radio Home Construction Continues to be Fascinating

WHEN the moguls of the radio industry gather nowadays, they no longer discuss heatedly the relative prospects of the home-built versus the manufactured receiver. The normal level of the parts business is now founded upon that percentage of listeners which finds pleasure and satisfaction in its own receiver; economy no longer acts as the chief stimulant to augment the ranks of the army of set builders. It is easier and in many cases, cheaper to buy a good receiver of all around efficiency than to build one. The man who builds his own does so because he likes to work with his hands or because he prefers exceptional performance in one or more particular qualities, like selectivity, sensitivity, efficiency, or economy. Other hobbies and pastimes in a similar position have thrived for years with no firmer foundation for their prosperity.

The small tool business, for example, owes half its volume to the man who putters with tools for the fun he gets out of them. There is no economic reason for building your own chair, desk, or bookshelf. With less trouble and expenditure, you can buy

an article. But no one worries especially about the future of the small tool business.

Our contact with thousands of readers through questionnaires and letters shows that the set builder recognizes his enthusiasm as a pastime and not altogether as a money making proposition. A very significant point, brought out by a recent study,

WATCH FOR OUR NEW COVER

STARTING with July, RADIO BROADCAST will *make its appearance with a new cover, designed by Harvey Hopkins Dunn of Philadelphia. Mr. Dunn, who is internationally known as a decorative designer and typographer, in creating this new cover, has provided one of unusual and effective design, which we feel will add great distinction to the magazine, and be a just reflection of the quality of contents which we strive always to maintain.*

—THE EDITOR

is the fact that many set builders have made ten and fifteen sets in the last few years and eagerly await new and superior designs so that they may make more. Set building is a habit; the turnover in set builders is small, while their ranks continue to grow steadily. These same home constructors stated that they recommend manufactured sets to their friends, sets pos-

sessing characteristics frequently different from those of the delicate and sensitive receivers which they delight particularly in making. Predominant in their recommendations for the many friends who consult them as radio experts are such well recognized manufactured types as the superheterodyne, the neutrodyne, and tuned radio-frequency receivers, a decided contrast to the sensitive, super-efficient, and often more difficult to tune, radio frequency-regenerative detector sets which are the favorites of the home constructor.

This stabilized relationship between the parts and the complete set business removes to a large degree any uncertainty prevailing as to the future of the parts business. So long as parts engineers continue to show ingenuity in developing really improved and novel designs for the home builder, the position of the parts business remains impregnable. True, the number of complete set owners grows more rapidly than that of set builders, but the fruitfulness of the latter field is ample enough to warrant a hopeful and promising outlook.

One important recommendation which we would make to the complete set manufacturer as a result of our investigation is that advertising in the radio magazines be prepared with better appreciation of what

---

The photograph forming the heading shows the antenna system of station 3 XN, the experimental station of the Bell Telephone Laboratories at Whippany, New Jersey. Important radio television tests were made between 3 XN, to New York. Transmitters of from 5 to 50 kw. are installed here.

kind of information their readers want. All of them are better informed on radio subjects than those who respond to general advertising; in fact, one third of our readers are professional radio men—dealers, manufacturers, and engineers. Many have asked us to give them technical and quantitative analyses of the performance of manufactured products in order that they might be better qualified to advise those who seek them out as experts. The advertiser may well take heed of this demand by substituting for his general claims about tone quality, selectivity, and sensitiveness, in his radio magazine advertising specific facts about the mechanical and electrical construction of his receiver and its efficiency and performance as indicated by gain, selectivity, and frequency curves. The reader of radio magazines is entitled to this special attention because he is the local radio oracle, whose advice influences the purchase of from three to twenty individual radio sets a year, while his professional activities control huge quantity purchases.

Patent Licensing Points to a Bright Radio Future

THE recent licensing by the Radio Corporation of the Radio Receptor, Zenith, All-American and Splitdorf companies is the first step toward widespread inter-licensing somewhat along the lines suggested by the articles by French Strother in this magazine, for October, November, and December, 1926, which aroused widespread and favorable comment in the industry. Upon the heels of these encouraging signs comes notice of the Hazeltine suit against Zenith over their Latour patents. The impending battle between these power-·ful interests, Hazeltine and Radio Corporation, already preceded by many preliminary skirmishes, promises at last to enter into a final and decisive stage which will make possible a comprehensive and satisfying appraisal of the value of their respective rights. It is a question of weighing the work of Latour and Hazeltine against that of Alexanderson Langmuir, Rice, Hartley, and others. We regard the peace which will follow this engagement as perhaps the last important step in patent stabilization which will establish definitely the position not only of the interested parties but of all the independents as well. We are certain, also, that liberal licensing on an equitable basis will end the patent bootlegging now current and establish the industry finally on a sound business basis.

Stabilizing the Broadcasting Situation

BRIGHT and hopeful as our two preceding items are, even more significant are the encouraging aspects of the broadcasting situation. We write just as· we return from the public hearings before the Federal Radio Commission in Washington. At these hearings, there were practic-

A WAVEMETER FOR BROADCASTING STATIONS

A Bureau of Standards employee with a portable type wavemeter for use in measuring the emitted frequency of a broadcast station

ally no representatives from the listening public. Only the interests of the broadcasting stations themselves were actually represented. Nevertheless, out of this group, came the exposition of sound principles and promising recommendations which we hope will be translated into action before these lines appear in print. RADIO BROADCAST presented a plea for the broadcast listener along lines familiar to our readers through these columns. It found before it four commissioners disposed to weigh and not to jump hurriedly to conclusions. They repre-

sent fearlessness and stability, if ever goverment commission represented those qualities. Until their course is finally set, these men are conscientious and open minded. They pleaded for a free expression of views and were rewarded by recommendations which, one by one, deprived them of means of accommodating all the 732 broadcasters who seek to continue to impress themselves upon the listener. Broadening the band was disposed of with a finality which leaves little hope for the revival of that pernicious proposition; division of time was frowned upon as uneconomical; narrowing the width of the ten-kilocycle channel and simultaneous broadcasting of interconnected stations on the same channel, were dismissed as technically unfeasible at the moment; power reduction by large stations was demonstrated as a restriction of the broadcast listeners' opportunity, so that, toward the end of the sessions, the commissioners were convinced that less stations was the only answer. Although hundreds of stations were represented, no one was heroic enough to offer to close down voluntarily, although, doubtless, many realized that their doom had been spelled by the recommendations of the conference.

Two plans for the administration of broadcasting station allocation were presented. One suggested that two thirds of the broadcast band be given over to ten per cent. of the existing stations, a glorious and courageous conception, altogether too easily criticised as the forerunner of a monopoly. Rome was not built in a day and we rather doubt that radio heaven can be created by any means so drastic. A second plan provides a somewhat more liberal reprieve to the less able stations by confining them to fifty-watt power on thirty channels set aside for that purpose and

© Barratt's

THE TRANSMITTING STATION OF A FAMOUS BRITISH AMATEUR

The name of Gerald Marcuse, G 2 NM, is well known among the amateur operating fraternity throughout the world. The illustration shows his 1000-watt transmitter with a silica Mullard tube in the foreground

permitting all the well established key stations to continue on their present powers. Eventually, it is likely that the fifty-watt station will prove uneconomic and, in the execution of that plan, lies hope that we will gradually find less than three hundred medium and high-power stations remaining, giving equitable service to the listener in every part of the country.

One could not help but be impressed by the way in which this commission went about its business. It hinted at no convictions or opinions; it was there to absorb. We regret that the fifth commissioner and chairman, Rear Admiral W. H. G. Bullard, could not be present. We heard occasional whisperings, from those having a sensitive ear to monopoly scandal, that the Admiral might have monopolistic leanings. These silly, back-stair comments arise out of the fact that he, almost alone, is responsible for the existence of an American owned system of worldwide communication, the creation of which he fostered as a measure of national defense, by encouraging the electrical interests to acquire the British owned Marconi Company and to form the Radio Corporation of America. For this accomplishment, he deserves the homage of his fellow citizens. Now that he has been entrusted with the equally important task of assuring an equitable service to the broadcast listener, we are sure he will be just as fearless in the attainment of that end. In that task, his eye will be ever vigilant to prevent the usurpation of the ether channels by any domestic monopoly just as he once so ably released American radio communication from foreign monopoly.

Pleas for a Blue Radio Sunday

THE church folk, as usual were present at the conference, seeking special privileges; in effect, they requested exclusion of all competition for their Sunday programs. At the risk of being called irreligious, we will take the liberty of trying to show that such a monopoly would certainly curtail the effectiveness of the Sunday religious radio offerings. We wish we could use the same subtlety and skill in presenting that argument as did Mr. George Furness before the conference. He answered the churchmen so effectively that we heard one of them remark that his reply was a "blamed good job."

The broadcast listener now has choice between two kinds of Sunday afternoon programs—second-rate jazz and unadulterated moral food. There are exceptions, of course, to so broad a generality. Purely from the broadcasting standpoint, neither of these program types are capable of attracting large Sunday radio audiences. Indeed it is fair to say that the one day of the week on which the largest radio audiences could be obtained is, from the program attractiveness standpoint, the weakest of the seven. Singing congregations if for no better reason than the acoustic conditions under which they appear before the microphone, naturally make poor broadcasting, while the spirited admonitory messages which predominate from the pulpit cause many sinful listeners to shut down their radio sets for relief. A monopoly of Sunday programs would serve only to eliminate those who like dance music from the listening audience; certainly it would not augment the "religious" listener group.

Now, supposing that Sunday programs were highly developed with a good balance of classical and chamber music, of educational material and well-presented religious material. The first effect would be to tenfold Sunday afternoon audiences. Better radio showmanship on the part of the churches would enhance the opportunity to win a following to the magnificence of church music. The religious element need not fear that its radio appeal will suffer

seriously from good competition. There are no finer soloists than those of the great churches nor more impressive music than that of the church organ. With a larger Sunday audience, the church can find no greater opportunity than radio offers to augment its following.

The Wisely Chosen Radio Commissioners

BY THE generous coöperation of various governmental departments in the matter of space, furniture, and personnel, the Radio Commission is able to function despite Senator Howell of Nebraska who, at one time, was the only Senator to object to the passage of the appropriation bill which would give the commission funds with which to work. The Senate had time to confirm three of the commissioners but awaited the opportunity to better acquaint itself with two members of the commission. Confirmation was further complicated by the fact that one or two Senators had private secretaries whom they wanted to have appointed. The personnel of the commission has been so well selected, from the standpoint of ability and judgment, that there can be little doubt of their ultimate confirmation.

What About the Canadian Demand for More Wavelengths?

ONE of the first problems which focussed attention on the Radio Commission was the demand of Canada for additional broadcasting channels. Prior to radio chaos (we can now use that expression safely), Canada had six exclusive channels and shared twelve others. Inasmuch as the twelve shared channels were so allocated in the United States as to permit of the use of 500-watt Canadian stations on them, and considering that the populated

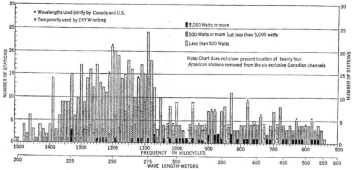

FREQUENCY DISTRIBUTION OF BROADCASTING STATIONS OF THE UNITED STATES

area of Canada is roughly only one-fifth that of the United States and that its total population is less than one tenth that of the United States, eighteen channels for Canadian use seemed a fairly liberal proportion of the 95 available. This allocation permits the simultaneous operation of twelve stations of 500 watts power and six more of practically unlimited power. The latest Canadian demand is for fifteen exclusive channels in addition to the twelve shared channels, the granting of which would involve giving Canada substantially more channels and broadcasting facilities, in ratio to population and inhabited area, than we ourselves enjoy. The negotiations are being handled by the Department of State and they are seriously complicated by the antagonism which our ruthless wave jumpers so deliberately aroused by the appropriation of Canadian channels. Again we renew our plea that any station which took upon itself the endangering of our friendly relations with neighboring countries by injuring their broadcasting service, be promptly and forever disbarred from the ether.

Conditions for Long-Distance Receiving Do Vary

AN EXCELLENT indicator of the changes in the responsiveness of the ether medium to the transmission of radio signals is given by the varying experience of British listeners in hearing American broadcasting. During the 1923-1924 season, WGY was heard much more frequently than other American stations, although KDKA, WOR, and WJZ were occasionally reported. During the following season all of these except WGY disappeared from the picture, but WBZ and WPG took their places. The 1925-1926 season was notable for its complete absence of American signals, checking with our own season of mediocre long-distance reception. This season is remarkably good, many listeners having reported as many as five American stations on the loud speaker in a single evening. The improvement is in part due to the use of increased power, but also to better transmission characteristics of the ether.

Direct Advertising Over the Air from Corsets to Calliopes

FROM the Middle-West come many complaints regarding certain undesirable broadcasting stations which do nothing but inflict atrocious direct advertising upon the listening audiences. Some of these stations are of considerable power and we have been able to hear them for short periods. The surprising thing about these disgraces is that the bargain corsets and harnesses which they offer are purchased by numerous dullwitted listeners who are thereby filling with numerous shekels the coffers of these miserable ether polluters. Easterners, accustomed to a comparatively pure ether

from the advertising standpoint (and this includes only that large proportion which concentrate on the bigger and better stations), sometimes become irritated when WEAF ventures to tell with excessive detail how many gas, electric, and traction companies certain of its clients control. These listeners would have a surprising disillusionment were they to hear the bargain ballyhoo to which Central Western listeners must submit. Fortunately most of these stations are pirates of the worst order and will probably be among the first to go at the behest of our courageous radio commission.

Latest Television Developments

DURING the first week of April, a remarkable German film was showing in New York. The picture dealt with the mechanical advance of civilization. Television—the local reproduction of distant scenes; played a leading part. On April 7th, the American Telephone & Telegraph Company fulfilled the prophecy and demonstrated an operative television system which brought the voice and face of speakers in Washington over telephone wires to New York in an astoundingly

clear fashion. After the wire television method was shown to be workable, the immensely more difficult television by radio was undertaken between the Bell Laboratories in New York and experimental station 3 XN at Whippany, New Jersey, about 30 miles away. Visual and audible reproduction was excellent in this medium as well.

With the attainment of television, the Bell Telephone Laboratories have, in the space of but a few months, contributed to the perfection of three extraordinary scientific tools of modern life. First came the transatlantic radio telephone, then the Vitaphone, and now workable television. It is of course true that, in each of these accomplishments, the telephone engineers borrowed freely from all that had gone before, but all honor to the Telephone Company for actually accomplishing what others have attempted. Much notice has been attracted by the work of John L. Baird with television in England, and our readers will recall a descriptive article on his accomplishments in these pages recently. Since there has been no demonstration of the Baird system in this country, it is impossible to compare the two systems.

TELEVISION APPARATUS AT THE BELL TELEPHONE LABORATORIES

The large illustration shows President Walter S. Gifford of the A. T. & T. Company talking into an ordinary subscriber's set placed in front of the television screen which shows a moving and illuminated image about the size of a vest pocket size snapshot. Dr. Herbert Ives, the man who is chiefly responsible for the development of the system, is standing at the extreme right. In the circle is Dr. Frank Gray, standing next to the large glass screen which is used when the distant scene is made visible to a large group. A loud speaker is in the lower section of the standard. At present, the small screen for individual use is much more successful than the large screen, although the results with the large screen, considering the problems involved, are remarkable

Baird claims the use of a mysterious "secret cell" but, by the lack of full disclosure, aids the doubters. The Telephone Company state that they have done nothing essentially new and shroud their accomplishment with no mystery. It became necessary to use a large photoelectric cell; forthwith Dr. Herbert Ives developed the largest ever built (it is nearly as large as a 250-watter). An operative method· of synchronism was required; H. M. Stoller and E. R. Morton developed a motor control synchronizing cleverly with 18 and 2000 cycles, and the requirement is met; a score of specialist-engineers work unceasingly on the wire and radio problems involved. Under the synthesizing guidance of Doctor Ives, television is accomplished. The story is simple and not dramatic, but the world of practical science owes much to these Bell Laboratories men. They are content to let the other fellow do the talking while they iron out the trouble.

Speculation is wide and commentators both serious and humorous are hard at work pointing out what can and will be done with television. Application of the system is certainly not to be expected very soon, and time, not prophesy, will tell what use the world will make of this distance-conquering eye.

RAY H. MANSON
—— Rochester ——

Chief Engineer, Stromberg-Carlson Telephone Manufacturing Company, Rochester, New York. Especially written for RADIO BROADCAST:

"*Radio is, in my estimation, rapidly advancing into a new era. Much research has been made along electrical and mechanical lines in the laboratories of many radio manufacturers, but in the minds of many engineers this has not been sufficient and, accordingly, there has been established in a very few manufacturing plants what is coming to be known as the electro-acoustical laboratory. We of the Stromberg-Carlson Company, being thoroughly convinced of the necessity of this phase of the work, feel fortunate in having established at our plant one of the few such laboratories in the country.*

"*The electro-acoustical laboratory deals with the experiment and measurement of audio frequencies. The work, which, by the way, is just beginning, has made rapid strides and has furnished much data for the design of radio receivers, but the main work centers around loud speaker design.*

"*It is my expectation that through the work of the electro-acoustical laboratories of the country, a new conception of radio entertainment, and a new naturalness and realism of reproduction, will be effected.*"

The Month In Radio

IN our March issue, we suggested the elimination of a great number of stations in the New York area and received in response a well-worded protest from WAAM which clearly demonstrated priority of service and adherence to its assigned wavelength, despite encroachments from wavejumpers. These considerations entitle that station to better treatment than that suggested for some of those mentioned in our proposal. We are pleased to accord WAAM our assurance of continued support of the principle of priority, based upon length of service on the frequency actually in use when the Radio Act went into effect.

RUMORS of significant radio mergers are becoming more frequent. Many of these surround the already powerful Crosley Radio Corporation which has already, in effect, absorbed DeForest, Amrad, and other radio concerns. Now that patent and broadcasting stabilization is in sight, the next trend in the radio industry will be the merger of small concerns, which find their limited opportunities a handicap. The accomplishment of similar mergers in the automobile business has had an important influence in bringing that industry to its healthy and enviable position. Radio mergers will make for better and cheaper receivers and for more efficient coördination in the encouragement of good broadcasting.

WESTINGHOUSE engineers announce a new rectifier suited to use in A, B, and C battery elimination, consisting of units of copper and copper oxide which introduce some sort of electrolytic action. By the assembly of a sufficient number of units, currents of an order of magnitude adequate for radio reception purposes are secured. It must be kept in mind that this development is only in the laboratory stage. It is one of great promise because the life of the elements appears to be practically indefinite.....

THE National Electrical Manufacturers Association took gratifying cognizance (or at least we flatter ourselves that they did) of our series of articles by French Strother on the patent situation and also our recommendations regarding the limitations of the total number of broadcasting stations. Resolutions adopted at its recent Briarcliff Convention, urged cross-licensing and the reduction of broadcasting stations to 200.

ONE of the recommendations of the American Engineering Council to the Radio Commission provides the adoption of a unit of ten millivolts per meter as the criterion which should determine the service range of a broadcasting station. A signal of that strength, even with our huge layout of broadcasting stations to-day, is available in an area less than one per cent. of the total of the United States and therefore represents an ideal which is so far from immediate attainment that it is of little practical value. The practical service range of a broad-

casting station is far in excess of that indicated by this ideal unit. It would be most unfortunate if· this value were adopted by the commission on the basis that it represented the limit of satisfactory reception, because it would involve the acceptance of heterodyne interference on ranges beyond this ideal service range as being of no account. Thus, if a listener twenty miles from a 500-watt broadcasting station complained of heterodyne interference from a distant station, he would be disregarded on the ground that he is beyond the service range of the station in question. Admittedly, we would all be very happy to have a signal of this strength, but while that privilege is now available to so small a proportion of the country, we recommend·the adoption of a substantially smaller unit, permitting of unheterodyned reception of 500-watt stations for at least 200 miles until better coverage of the United States is an accomplished fact.

WE ALMOST suffered collapse when we received the extraordinary and gratifying announcement that the Radio Manufacturers' Association· and the National Association of Broadcasters had combined in the establishment of a single office to serve both organizations in New York, Washington, and Chicago. L. S. Baker has been chosen to direct this work. Such working together in the industry is an unheard of precedent, the result of coöperation during the difficult days of the halting progress of the radio bill through Congress. This new 'tendency toward coöperation·and the disappearance of mutual jealousies will perform miracles for the rejuvenated industry. Now we are prepared for the shock of an announcement that the R. M. A. and the N. E. M. A. standards committees have joined forces, but perhaps that is too much to hope for at this stage.

STATION KFRC of San Francisco, has the distinction of being the first to rebroadcast a Japanese program. This feat was accomplished on the morning of March 16 when the listeners of the Frisco station enjoyed an hour's reception of JOAK, stated by the publicity men to be with "perfect volume and clarity." Considering that the distance is about six thousand miles, these Pacific coast geniuses must therefore be at least twice as good as the Radio Corporation·engineers, whose·transatlantic rebroadcasting was hardly of high standard.

THE following patents are now involved in litigation: 1,018,502, General Electric vs. Crown Electric Co. Inc. (bill dismissed without prejudice); 1,050,441; 1,050,728, and 1,113,149, Westinghouse Electric & Manufacturing Co. vs. W. Egert; 1,180,264, Westinghouse Lamp Co. vs. C. E. Manufacturing Co. Inc.; 1,195,632, 1,231,764, 1,251,377. Radio Corporation of America vs. Radio Receptor C. (presumably this is the case which resulted in licensing of the latter by the former); 1,231,764, 1,251,377 Radio Corporation of America vs. Epom Corp.; 1,018,502 and others, General Electric Co. vs. H. & D. Radio Co.; 1,271,527 and others, Lektophone Corp. vs. Brandes Products Co. (decree, dismissing bill, January 27, 1927); 1,377,405, DeForest Radio Co. vs. North Ward Radio Co.; 1,571,501, Dubilier C. & R. Corp. vs. N. Y. Coil Co. (decree of D. C. affirmed); design 68,770, The Popley Co. vs. Blue Bird Furniture Manufacturing Co.; 1,271,527, Lektophone Corp. vs. Western Electric Co. (claims 2930, held not infringed); 1,271,529, as above (claims 1 to 4 and 8 held not infringed).

THE number of radio beacons on the coast of France will soon be increased from four to ten, according to the Department of Commerce.

Why Shielding?

What to Look for and What to Look Out for in Judging a Set's Shielding

By EDGAR H. FELIX

THE superior performance of completely and carefully shielded receivers has led to the widespread adoption of the phrase "fully shielded" in connection with any set having a stray piece of sheet metal in its anatomy. Shielding is generally considered as a sort of electrical mudguard which prevents the spattering of undesired electrons upon neighboring circuits. So indeed it is, but the significant influence of shielding upon the performance of a receiver is hardly indicated by this limited conception. The confinement of the energy in every element of the receiver strictly to the performance of useful service, accomplished by effective shielding, tremendously enhances selectivity, sensitiveness, and permissible amplification of the instrument.

Specifically, the principal results of complete shielding are: (1) Compactness, permitting the embodiment of many stages of radio and audio amplification in a receiver of small proportions without destructive interactions; (2) greater permissible amplification because relatively large radio- and audio-frequency currents can be conducted through circuits without consequent couplings to neighboring stages; (3) stable neutralization throughout the wavelength range, because all unwanted inductive and capacitative coupling is eliminated; (4) increased selectivity resulting from the use of more stages of radio-frequency amplification with consequently greater filter action; (5) uniform amplification throughout the wavelength range without increased tendency toward self-oscillation at the higher frequencies; (6) elimination of electromagnetic pick-up (except that coupling purposely introduced through the primary of each stage's transformer) from the antenna and preceding stages and, with it, the resultant broadened tuning; (7) reduced influence of static and power line induction because pick-up is limited to the antenna circuit itself; (8) greater mechanical rigidity attained by supporting effect of substantial shielding and chassis construction; (9) foolproof wiring, largely concealed in enclosed cans; and (10) reduced losses due to dust and dirt on condenser plates and other exposed parts.

The term "shielding" describes any metallic conductor installed in the radio set for the purpose of eliminating undesired electrostatic or electromagnetic reactions, whether it be a modest strip of tin-foil pasted to the back of the panel near the dials to reduce hand capacity, or the complete sealing cases for each stage of amplification, embodied in the latest receivers. Shielding has been fully discussed in various technical papers, references to which are given in the bibliography at the end of this article. Our concern here, however, is not a technical discussion but a simple exposition on how to judge, by inspection and demonstration, the effectiveness of shielding in a receiver. This is a relatively difficult task, however, because a receiver can be built to appear perfectly shielded while the only thing actually accomplished by improperly applied shielding may be the introduction of high-resistance losses and consequent reduction in amplification and efficiency.

The theory of shielding is quite simple. Any

Radio Broadcast Photograph

INDIVIDUAL SHIELD CANS

Several manufacturers now supply them in knock-down form

circuit carrying a radio-frequency current is constantly surrounded by electromagnetic and electrostatic fields. The extent of these fields is proportional to the energy in the circuits. The greater the amplification, the greater the need for shielding between stages. With small amplification, no shielding is essential, although it may serve usefully even in a receiver consisting of but one stage of radio amplification combined with a detector. A receiver with four efficient stages of radio-frequency amplification, however, approaches the limit of practical amplification and

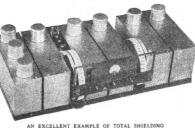

AN EXCELLENT EXAMPLE OF TOTAL SHIELDING

The five sections to the right all enclose tuned stages while the three audio stages are to the left

also the upper limit of energy which practical shielding can confine.

Experimenters who have tried to use shielding without a thorough understanding of its proper application have been known to argue against it because they have obtained poorer results by adding shields in their receivers. But to argue against any and all shielding because of an experience with mis-applied shielding is like condemning a twelve-room house because it cannot be built on a twenty-five foot lot. Unless a receiver is specially designed for it, shielding is as likely to decrease efficiency as it is to improve it.

The most fundamental and simple principles of electricity explain the functioning of shielding. You probably remember the experiment of rubbing a piece of glass with fur and using it to pick up small bits of paper. The effect of rubbing the glass rod is to charge it. The effect of the charge on neighboring objects is to induce an opposite charge on them, an influence sufficiently strong to actually lift a bit of paper to the glass rod. Any part of a circuit in a radio set, the potential of which is raised or lowered with respect to the potential in neighboring conductors, exerts its influence by drawing or repelling electrons in neighboring conductors. A condenser is a device especially designed to accentuate such electrostatic effects. Obviously, any change of potential in a conductor in a radio set will cause potential changes in all near-by conductors. Warding off the influence of the electrostatic field is simple and the most elementary application of shielding will accomplish it. The electrostatic influence is restricted by placing a grounded conductor between any point where potentials rise or fall, and neighboring objects which are likely to be influenced by the electrostatic effects resulting therefrom. If a good conducting path is provided to the ground, the influence of the electrostatic field does not penetrate beyond the shield.

In a circuit in which there are changes in current taking place, a varying magnetic field surrounds the conductors. The extent of the magnetic field is determined by the intensity of the current and the form of windings used. It is not difficult to cause response in a detector circuit following two stages of radio-frequency amplification with a coupling of two feet between the second r. f. stage and the detector grid coil, provided the incoming signal is strong. The interactions taking place in radio receivers are surprisingly extensive and complex, every tuned circuit actually influencing every other tuned circuit. Such effects are minimized, but by no means eliminated, by skillful placing of parts so that these influences are neutralized.

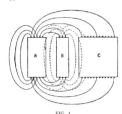

FIG. 1

With alternating currents of low frequency and through the audio ranges, the magnetic fields may be deflected by soft iron. The reason iron or steel reduces the extent of the magnet's field is because magnetic lines of force seek the easiest way and concentrate through the iron path so provided. The fields surrounding conductors carrying radio-frequency currents, however, are practically uninfluenced by the presence of iron or steel. Shielding accomplishes the constraint of magnetic fields, not by deflection or insulation, but by setting up equal and opposite magnetic fields; in other words, by magnetic neutralization. Recognition of this principle makes it easier to identify effective and ineffective shielding.

An elementary electrical theory calls to mind that a conductor within the influence of a changing magnetic field has an electric potential induced in it as a result. Suppose we have three coils, A, B, and C, as in Fig. 1, A being the primary and C the secondary of a radio-frequency transformer. A radio-frequency current in A induces a similar current in coil B, lagging half a cycle behind that flowing in A. The fields resulting from the current in B induce a potential in C, lagging half a cycle behind that in B. Now suppose these coils are so placed that the influence of coil A on coil C is exactly equal to that of coil B on coil C. The result would be two similar and equal magnetic fields, half a cycle apart, influencing coil C. Being equal and opposite to each other, each would neutralize the other's influence, so that their total effect would be nil.

A thick metal shield performs the same function as coil B in our illustration by having induced in it currents by any magnetic field playing upon it. These eddy currents, in turn, set up magnetic fields but, if the shielded conductor is sufficiently thick and of low resistance, the eddy currents are so diffused that the magnetic fields set up by them do not influence surrounding circuits. Even so, with five stages of radio-frequency amplification, the eddy currents are of such magnitude that the receiver is unstable.

It is clear that the effectiveness of shielding bears a close relation to the energy in the circuit. Since eddy currents are built up in the shield by radio-frequency energy withdrawn from useful purposes, the resistance of the shield should not be too high, or the tuning of the circuits will be broadened. Too thin shielding, or shielding of high resistance, broadens tuning and fails to eliminate magnetic interaction between circuits.

THICKNESS OF SHIELDING MATERIAL

PROFESSOR John H. Morecroft, in his comprehensive paper appearing in the August, 1925, issue of the *Proceedings of the Institute of Radio Engineers*, shows by a series of curves the influence of thickness of shielding upon its effectiveness. The general conclusion

of authorities is that, for the shielding of broadcast frequencies, copper at least $\frac{1}{32}$ of an inch thick should be used in order that resistance losses are not introduced into the tuned circuits. Aluminum, having lower specific conductivity than copper, must be $\frac{2}{64}$ of an inch in thickness, and brass $\frac{1}{8}$ of an inch thick to secure the same result. Copper is widely used for effective shielding because it is easily nickel plated to give good appearance, and is easily soldered. Aluminum has the advantage of light weight and good appearance, but is difficult to solder. Sheet aluminum of the necessary thickness, though half again as thick as the equivalent copper sheet, is the lightest of the three materials; copper $\frac{1}{32}$ of an inch thick is half again as heavy; brass $\frac{1}{8}$ of an inch thick is six times as heavy as aluminum sheet.

The good conductivity of shielding must not be limited to the shielding plates themselves, but also to the connections through which they are grounded. This becomes of increasing importance in succeeding radio-frequency amplifier stages. The detector stage of the Freed-Eisemann shielded set, following four stages of radio frequency, has special bonding between the detector and and that of the third r. f. stage to insure it good ground. Referring again to Morecroft's paper, he states: "An ordinary piece of copper mesh was used as a shield for different frequencies and the results showed very little shielding effect. A border of solder 0.5 cm. wide was put around the edge of this shield so as to make good contact between the ends of the wires, and the shielding was increased approximately seventy-five per cent."

You need, therefore, not only a shield of adequate thickness and conductivity, but also it must have good electric contact with the ground and with neighboring shields along all its edges. To quote Professor Morecroft again: "Any imperfect joint in the shield, which tends to constrain the eddy currents to restricted paths, will seriously interfere with the shielding obtainable."

Furthermore, shielding can be reduced in ef-

FIG. 2

fectiveness by horizontal slits or cuts at points where eddy currents are likely to be induced. Morecroft states in his paper that the effect of a slit diametrically across a shield of copper 0.021 cm. thick reduced its effectiveness as a shield by four per cent. and a second diametric slit, at right angles to the first, subtracted another four per cent. from the effectiveness of the shield. While Mr. Morecroft's measurements were made at audio and low radio frequencies, it must be borne in mind that shielding seriously cut up with slits to accommodate wiring is not as effective as solid sheet. For example, in Fig 2, a long slit in the base of a shield will confine eddy currents to the two legs at A and B, thus building up relatively strong fields at those points, influencing the tuned circuits of the adjacent stage.

A receiver having each stage confined in a can of suitable thickness and conductivity, may yet fail to accomplish the desirable results of shielding if the power supply leads, both of the A and B sources, carry radio-frequency energy from one circuit to the next. Such shielding is about as useful as a pail with a hole in the bottom.

The Stromberg-Carlson receiver, the pioneer

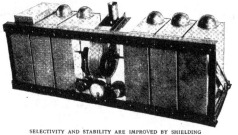

SELECTIVITY AND STABILITY ARE IMPROVED BY SHIELDING
Many modern commercial receivers are now totally enclosed in metal

of fully shielded sets, has a 1-mfd. bypass condenser between the plus A and minus A filament terminals of each tube, and another condenser of the same size for carrying the radio-frequency currents to ground, thus confining the radio-frequency energy entirely within the cans of the radio-frequency amplifier. As a further precaution, a 200-henry choke is used in the detector circuit to keep all audio-frequency currents out of the B battery, while radio-frequency chokes are used in the r. f. plate power leads of each stage.

Speaking of these and other bypass and filtering precautions, Dreyer and Manson, in their valuable paper on the subject, delivered before the Institute of Radio Engineers, state that all of these are not necessary in a three-stage receiver but: ". . . . they become of more and more importance the higher the total amplification, and all of them appear to be necessary in a four-stage receiver. Even with precautions of this type, there seems to be an upper limit to the amplification that may be obtained at the broadcasting frequency. This limit is not reached with a three-stage receiver but may be with a four-stage one. It appears to be due to radio-frequency potentials which build up upon the shielding of the last stage. These potentials are capable of feeding back energy to the antenna through the inherent capacity between these shields and the antenna."

An added advantage of bypassing and filtering of audio- and radio-frequency currents from the A and B power sources is that it minimizes the effect of resistance coupling in the B power source. Increased B battery resistance of old batteries, or that resistance inherent in B socket power devices, with receivers having well filtered and bypassed power leads, does not contribute any tendency to whistle.

NEUTRALIZATION SIMPLIFIED

BY THE observance of these elaborate precautions, applied to expensive receivers, the useful energy of receiving circuits is limited and confined to where it belongs. There being no magnetic or electrostatic interstage influences to consider, except that unavoidably introduced by the capacity between the grid and plate of each tube, the problem of neutralization is tremendously simplified. There is only one capacity coupling to neutralize and that practically independent of frequency. As a consequence, not only is perfect neutralization obtainable, but the influence of neutralization is more uniform throughout the frequency range. The receiver, consequently, delivers more even amplification over the entire wavelength band. Furthermore, the highest possible amplification per stage, by the use of r. f. transformers giving the largest gain, is made possible by shielding. The tendency toward oscillation at the high frequencies no longer limits the amplification of middle and lower frequencies. Thus we gain much greater sensitiveness by the aid of a correctly designed shielded receiver. Transformer design is largely a matter of compromise among selectivity, sensitiveness, efficiency, and tendency to oscillate. By elimination of the last consideration through effective shielding, the necessity for sacrificing the first three factors in favor of the last is practically done away with.

L. A. Hazeltine, commenting on Manson and Dreyer's paper, shows that we may select an r. f. transformer ratio which gives, within a few per cent., maximum amplification at resonance and, at the same time, gives twenty per cent. lower amplification at a frequency only ten per cent. off resonance. This applied successively through three or four stages gives a considerable advantage to a weak signal at resonance over a powerful signal ten per cent. off resonance. This accounts for the vastly superior selectivity of well-designed shielded receivers.

One of the rules of successful application of shielding is not to place any inductance too close to the metal shield itself. Look for at least an inch of separation between the shield, high potential wiring, and tuning inductances. The end of the coil is preferably so placed that it does not come up against a shielded plate, but at right angles to it. Variable condensers of higher maximum capacity are sometimes necessary in connection with completely shielded receivers. It is well to confirm when testing a shielded receiver, that tuning condensers of adequate size are used, by tuning

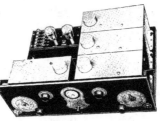

WHERE SHIELDING HELPS STABILITY

There are three stages of tuned r.f. in this receiver, a condition which calls for the utmost care in design. Shielding of the tuned stages greatly improves results

in to stations at both ends of the frequency scale.

As we have seen, the electromagnetic and electrostatic fields surrounding a radio-frequency circuit are proportionate to the strengths of the radio-frequency currents flowing in it. Receivers employing but one stage of radio-frequency amplification and a regenerative detector have most of their energy confined in the detector circuit with comparatively small currents flowing in the radio-frequency amplifier. A single sheet of shielding between the end or side of the radio-frequency coil next the detector coil is often sufficient to eliminate a great deal of the interaction occurring. With two stages of tuned neutralized radio-frequency amplification, adequately spaced, it is possible to eliminate any serious influence of interaction, if the coils are properly placed. A small shield between the radio-frequency transformers contributes to stability. But any attempt at three stages of r. f. without heavy, highly conductive, shielding, completely enclosing each stage, is bound to result in so much interaction that stability, selectivity, and high amplification per stage is quite impossible. With four stages of radio fre-

quency, we must not only meet the foregoing requirements, but also be certain that there is adequate bypassing of all r. f. and a. f. currents and, in addition, that filters are used in the power leads to further suppress any of these currents which tend to stray from their appointed paths.

The attainment of highest possible gain per stage is not so easily determined as the physical qualities of shielding which mark the well-designed receiver. However, any very marked indication of lower stability at the higher frequency end of the receiver than at the lower is one sign that shielding is not complete or that some element of design has not been successfully carried out. In testing a receiver, increase the filament brilliancy to the maximum and tune to the shortest wavelength of which the receiver is capable. If there is a tendency toward oscillation at that end of the spectrum, which disappears at the upper end, it is tell tale evidence that shielding has not been perfectly applied, or that the receiver is not properly neutralized.

Another significant indicator is the sharpness with which a near-by, high-power, local station is tuned out. With three or more well shielded stages, extraordinary selectivity without sacrifice of quality is attainable. This should manifest itself conclusively when tuning, both at the high and particularly at the low end of the frequency scale. The customary broadness of tuning experienced with average receivers at the upper wavelengths is due to reduced regenerative amplification and also to induction from the antenna and the first and second radio-frequency transformers upon the detector circuit. Good shielding should correct both of these undesirable effects.

Recapitulating, the obvious mechanical qualities of shielding are adequately thick shielding material of high conductivity, low resistance grounding and firm corner connections at all edges, absence of any extensive slits which tend to force eddy currents to a restricted path, and suitable placement of instruments in each stage at a sufficient distance from the shielding. Electrically, tuning condensers of sufficient capacity to cover the wavelength range and, for multi-stage sets, adequate bypassing and filtration of A and B power leads are necessary. In performance, watch for equal stability and sensitiveness throughout the frequency range and, with multi-stage radio-frequency receivers, for great selectivity even with high-power locals. There are an encouraging number of high-grade shielded receivers on the market which fulfill these exacting requirements.

Bibliography

John H. Morecroft and Alva Turner, "The Shielding of Electric and Magnetic Fields," *Proc. I. R. E.,* August, 1925.

John F. Dreyer, Jr. and Ray H. Manson, "The Shielded Neutrodyne Receiver," *Proc. I. R. E.,* April, 1926.

L. A. Hazeltine, Discussion of Above, *Proc. I. R. E.,* June, 1926.

SOME ADVANTAGES OF SHIELDING

SHIELDING Having the Following Properties:

1. Thickness adequate to assure dissipation of all eddy currents.
2. Conductivity of a high order to prevent resistance losses.
3. Bonding and grounding of a character avoiding resistance losses and concentration of eddy currents.
4. Solidity of material, by absence of long cuts or slits for wires.
5. Completeness, by confinement, through efficient bypassing and choking of power leads, of each circuit's magnetic and electrostatic fields within its respective shield.
6. Spacing between shields and coils sufficient to prevent excessive electrostatic capacity at minimum tuning adjustment and to minimize eddy currents in the shield.

—PERMITS the Design of a Receiver with the Following Properties:

1. Compactness, allowing the embodiment of many stages of amplification in a relatively small cabinet.
2. Higher permissible amplification because relatively larger radio- and audio-frequency currents can be conducted through circuits without interaction.
3. Stable neutralization throughout the wavelength range with more equal amplification of both low and high radio frequencies.
4. High selectivity, resulting from filter effect of added radio-frequency amplifier stages.
5. Elimination of external pick-up except that introduced by the antenna, reducing power line influence, static, and interference from near-by high-power stations.
6. Greater mechanical strength and reduced resistance losses from dust and corrosion.

Building an Electrical Phonograph

Suggestions for a High-Quality Combined Radio and Electrical Phonograph—Making a Baffleboard Loud Speaker —An A. C. Amplifier for the Electrical Phonograph

By
JAMES MILLEN

RADIO BROADCAST Photograph

THE AUTHOR'S RADIO-PHONOGRAPH COMBINATION

The circuit arrangement of this particular layout is shown in Fig. 2, the combination illustration on page 89. The cabinet originally housed only the Pathé phonograph

THERE are at least two distinct angles from which the design of an electrically operated phonograph may be approached. The first considers the phonograph as a companion instrument and component physical part of a high-grade radio receiver, while the second views it as an instrument entirely separate from the radio.

From the first point of view, it is hard to find as a basis for the design of the radio and electric phonograph a more satisfactory combination than that described in the January issue of RADIO BROADCAST by John B. Brennan and the writer. A two-tube "Lab" circuit receiver and a separate high-quality lamp socket powered amplifier with a console cabinet, record turntable, pick-up, needle scratch filter, and the other essential components of the combined radio-phonograph, may be gathered.

The writer's own machine, as shown in the accompanying illustrations, consists of a revamped Pathé console arranged to accommodate the receiver proper as well as the automatic relay switch and A power-supply unit in the left-hand record compartment. In the rear of the central compartment, formerly taken up by the horn, is placed the amplifier, while immediately behind the grille work is mounted a baffleboard type loud speaker. The electromagnetic pick-up is mounted in place of the mechanical one in the turntable compartment and the leads brought down through the hollow base and the opening to which was formerly fitted the neck of the horn. Such an arrangement is very satisfactory and if the reader decides to use this set-up, he should proceed as outlined in this article.

Before starting the cabinet work associated with the re-vamping of a console, remove the shelf carrying the turntable and motor. In almost every phonograph, this shelf may be removed by first removing the crank, and then unfastening the screws that hold the shelf in place. The entire back of the cabinet may generally be removed by taking out a few more screws. It is very seldom

that the back is glued in place. However, should the back panel be glued, or prove difficult to remove, then leave it in place, as the primary reason for taking it off is to facilitate the removal of the horn.

The horn may, with a little juggling, be unfastened and removed, when necessary, through the opening made by the removal of the motor shelf. The horns are usually fastened by means of wood screws but without glue.

To remove the partitions in the record compartments is not always quite as simple as removing the horn. In the case of the Pathé console shown in the illustration, it was necessary to slightly raise the entire left-hand section of the top, which is held in place with wood dowel pins. A small piece of wood was placed along the under edge of the top and gently tapped with a hammer until the glue holding the dowel pins became loose.

After the removal of the partitions, the panel for the radio set may be cut and fitted into place, keeping in mind that it must be set back sufficiently far to permit the closing of the compartment door without hitting any of the knobs. In many instances it will be found that one of the veneer partitions removed from the former record compartment will make a very fine panel for the radio set.

Should some readers have the parts for, or prefer to use some other type of set, such as the "Universal," the Browning-Drake, or any of the many other receivers, then such a receiver may of course be used in place of the RADIO BROADCAST "Lab" receiver. It is essential, however, that it be equipped with a good audio amplifier with an output power tube, or, if a separate amplifier is used, the amplifier may well be any of those combination power supply and high-quality audio channel units described by the writer in the last few issues of RADIO BROADCAST.

The amplifier unit shown in the photographs, and described in the January RADIO BROADCAST, consists of a stage of impedance and two of resistance coupling, with a phase shifting grid choke in the last or power stage. The power supply uses a Raytheon BH tube. Filament current for the power tube is obtained from a special filament winding on the power transformer while that for the first two amplifier tubes, as well as the tubes in the receiver proper, is obtained from an A power unit, such as the Westinghouse "Autopower." This latter consists of a battery and trickle charger combination. The receiver should be so constructed that the output of the detector tube may be fed directly into the amplifier unit.

The detector tube socket should be of the spring suspended type, such as the Benjamin, and a lead or heavy rubber cap or weight placed on top of the detector tube in order to prevent microphonic howling when the loud speaker is mounted in the same cabinet.

Some constructors may wish to place a lamp on top of the cabinet after the construction work has been completed. It will generally be found that when this is done some hum will be induced in the receiver due to pick-up from the lamp cord. Placing the lamp cord in a different position may eliminate the trouble but in those cases where this is not effective it will be necessary to use a shielded receiver. Also, when a lamp is placed on top of the console, it tends to place the tuning control in a shadow and for this reason any constructor who expects to top off the job by using a lamp on the console will do well to use illuminated dials in the receiver.

THE LOUD SPEAKER

GOOD loud speakers are of two types; the large cones, such as the Western Electric, and the small baffle cones such as those due to Messrs. Rice and Kellogg and used in the "Electrola" and "Panatrope." This latter type of loud speaker is particularly well suited, because of its small size and method of mounting, for use in a home-constructed electrical phonograph. Baffle cone loud speakers are made by several different loud speaker companies, such as RCA, Rola, Magnavox, and Peerless.

If an RCA No. 100 loud speaker is available it should be removed from its case and fastened directly to a baffleboard which should be constructed as follows:

It should be about ½ inch thick and of white pine or other soft wood. A round hole, approximately 8 inches in diameter, should be cut in the center of the board and so beveled that the outside diameter is slightly larger than the inside diameter. The cone frame-work is then bolted in place so that the cone is centered behind the hole.

Very thin veneer should not be used as a baffleboard as it may prove to be resonant at some frequency within the audible range.

The baffleboard may be mounted so as to close the opening of the phonograph made by the removal of the horn. In mounting the baffleboard, it should be set back far enough to permit replacement of the silk-covered grille work if desired. By making the baffleboard slightly larger than the opening, it may be mounted from the rear in much the same manner as a glass is inserted in a picture frame. The filter circuit contained in the small box in the bottom of the RCA No. 100 loud speaker case should be disconnected from the loud speaker unit and discarded.

A large opening, which may well be covered with a light dust-proof silk curtain, should be provided in the rear of the compartment in which the cone is placed. Such an opening will be found to improve materially the acoustical performance of the loud speaker.

In the case of the Pathé console, it was found necessary to raise the shelf on which is mounted the motor and turntable, a quarter of an inch, in order to place the amplifier under the motor. The only difficulty encountered in this process was the necessity of enlarging the hole through which the crank passed.

While it is desirable to so place the amplifier that the controls are readily accessible, such a location is not of prime importance, as once the controls have been properly set, they will require no further attention.

When the mechanical pick-up is removed, a large hole is exposed through which the neck of the horn formerly passed. By mounting the electric pick-up in the same place as the former one, this hole is covered. Furthermore, the cord from the pick-up to the control switch may be brought down through a hole made in the base of the pick-up stand directly over the hole in the turntable shelf, thus exposing no wires. In mounting the pick-up, fasten its base in such a position that the needle, when swung to the center of the record, will rest in the exact center of the turntable shaft. Most pick-ups have a volume control located on the base of the stand. If not, then a 25,000-ohm resistance, such as a Royalty or the Centralab Radiohm, may either be mounted in the turntable compartment or on the panel of the receiver. The pick-up selected must be capable of high-quality reproduction or the best results cannot be expected. The Pacent, the Grimes "Gradeon," the Crosley "Merola," the Baldwin, and the Bosch "Recreator," have been found satisfactory.

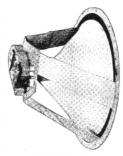

THE RCA 100 CONE

Few will recognize this popular loud speaker as shown here. It is stripped of its surrounding bronze metal case and gauze front and back. An excellent baffleboard loud speaker can be made with this unit, as explained in the text

THE SCRATCH FILTER

THE scratch filter, as described in detail in the May article, is for the purpose of electrically removing from the output of the loud speaker the "hiss" due to the contact of the needle on the record.

While the connection of a 0.006-mfd. fixed condenser across the output of the pick-up or input to the amplifier, will remove this noise, such an arrangement will also remove many of the higher audio frequencies and thus lower the quality of reproduction. For this reason an

electrical filter circuit tuned to stop the passage of only those currents in the neighborhood of the scratch frequencies is used.

The difficulty in completely eliminating the scratch lies in the fact that the scratch frequency is not any one frequency, but quite a wide frequency band. If, however, the filter circuit is tuned to approximately 4500 cycles, the greater part of the scratch noise is removed without the sacrifice of tone quality. The residual hiss, when a scratch filter is employed, is practically unnoticeable and cannot be detected except for the first few seconds or so before the music starts.

Such a device may either be purchased as a complete unit (one is made by the National Co.), or may be home constructed from a choke coil and condenser so selected as to be most effective at about 4500 cycles. This frequency peak should be somewhat "broadened" by the use of a very small quantity of iron in the construction of the inductance. A scratch filter can be assembled by employing a 1500-turn honeycomb coil with a 0.008-mfd. fixed condenser. The circuit for the scratch filter is shown to the extreme left of Fig. 1.

A Yaxley No. 63 triple-pole double-throw jack switch is mounted at some place convenient and so connected, as shown in Fig. 2, that when in the central position, both the phonograph and the radio are shut off, while on one side the radio is turned on and the other, the pick-up system is thrown into use.

One pole controls the input of the amplifier, either radio or phonograph, while the other two control the A power either to both the amplifier and the radio set or just to the amplifier alone. A Yaxley automatic relay switch is used to control the 110-volt power, to both the amplifier and the "Autopower." The switch on the front of the "Autopower" must at all times be kept in the "off" position and connections of the two A leads soldered directly to the battery terminals and not fastened to the terminals provided on the "Autopower." This arrangement is necessary in

Radio Broadcast Photograph

A REAR VIEW OF THE AUTHOR'S OUTFIT

This picture gives some idea of the simplicity of the equipment necessary for the phonograph-radio combination. This is a rear view of the outfit shown in the photograph at the top of the first page of this article. The two-tube "Lab" receiver is hidden in front of the Westinghouse Autopower. Fig. 2 is the circuit diagram for this particular arrangement

order to use the relay switch for automatic control rather than the manual switch on the front of the A power unit; which is generally rather difficult to get at.

And now, after the above changes have been made, we will have a truly modern instrument as far as performance is concerned. The only "antique" device in the system being the spring motor of the phonograph. Though the preferred arrangement for an a. c. operated phonograph is the use of an induction motor to operate the turntable, such motors of the proper size are rather difficult to obtain. Most phonograph dealers, however, stock excellent motors of the universal type, which, as they are only run when the radio receiver is not in operation and will therefore cause no interference, are entirely satisfactory. Such motors cost from $15.00 to $30.00, depending upon the type, and generally an allowance is made for the spring motor if it is turned in at the same time.

If the spring motor is retained in the phonograph then the amplifier should be so placed that the Raytheon and power tubes, which become quite warm in operation, are not directly under the motor, for such placement might result in the graphite in which the spring is packed being melted.

If the spring motor is replaced with an electric drive, then, of course, no thought need be given to the location of the amplifier.

ELECTRIC PHONOGRAPH WITHOUT RADIO

THERE are doubtless some who wish to modernize their old phonograph and make it electrically operated but entirely independent of the radio installation. Such a person will do well to construct a compact power amplifier, preferably lamp-socket operated, or to make up an ordinary amplifier and use it in conjunction with a B power unit capable of supplying sufficient voltage for a power tube.

Several manufacturers are at present working on completely a. c. operated amplifiers which

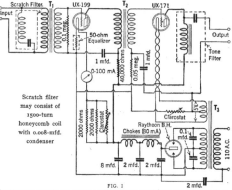

Scratch filter
may consist of
1500-turn
honeycomb coil
with 0.008-mfd.
condenser

FIG. 1

For those who wish to build an amplifier solely for the purpose of using it with the phonograph, the diagram above should be followed. Both of the amplifier tubes obtain their A, B, and C potentials from the mains, no batteries whatsoever being required. A list of parts is given in the text—on this page and a photograph is shown below

will be satisfactory for use in conjunction with a phonograph pick-up. At the present time, however, there are no completely a.c. operated units on the market. It is, of course, possible for anyone with the necessary apparatus to home construct an a. c. operated power amplifier. The author has constructed such an amplifier and it gives very satisfactory results. The circuit diagram is given in Fig. 1. This amplifier consists of an input transformer, T_1, with a scratch filter connected in its primary, a first-stage amplifier tube of the 199 type supplied with filament current from the filter system, a second transformer, T_2, and a power tube feeding the output, to which should be connected a high-quality cone type loud speaker. The list of parts used in constructing this amplifier is given below:

National Power Transformer	. .	16.50
National Filter Choke, Type 80	.	10.00
National Filter Condenser Block	.	17.50
National Tone Filter	.	8.00
National Scratch Filter . . .	.	5.50
1 Pair AmerTran DeLuxe Transformers		20.00
2 General Radio Sockets . . .	.	1.00
2 Clarostats	.	4.50
2 1-Mfd. Tobe Condensers . .	.	2.50
1 2-Mfd. Tobe Condenser . .	.	1.75
2 2000-Ohm Wire-Wound Heavy Duty		
Resistors	.	2.50
1 0.1-Meg. Metalized Resistor .	.	.75
1 0.05-Meg. Metalized Resistor .	.	.75
3 Mounts	.	1.05
1 50-Ohm Fixed Filament Resistor	.	1.00
1 40,000-Ohm Resistor .		1.10
1 Weston 0-100 Milliameter . .	.	8.00
1 Terminal Strip	.	.75
1 Baseboard	.	.50
1 Set Tobe Buffer Condensers . .	.	1.40
1 Ceco Type B (ux-199) . . .	.	2.00
1 CeCo Type J-71 (ux-171) . .	.	4.50
1 Raytheon BH	.	6.00

TOTAL $117.55

In this amplifier two transformer stages are used so as to keep low the number of stages of amplification required. It is not essential that the parts specified be used, and other high-grade apparatus manufactured by other companies will be quite satisfactory provided their electrical characteristics are satisfactory.

It is essential that the chokes employed in the filter system be capable of giving satisfactory results with 80 milliamperes flowing through

THE A. C. AMPLIFIER SHOWN IN FIG. 1

With this amplifier, a pick-up device, and good cone loud speaker, we have all the essentials for converting a phonograph of the mechanical reproduction type to one of the electrical reproduction form. No batteries are required—not even the usual C batteries. The two center binding posts are for connecting the milliameter used to adjust the filament current of the 199 tube. The others are the input and output connections

them. Two 2000-ohm resistors are employed in parallel rather than a single 1000-ohm unit, first because 2000-ohm units are more readily obtainable, and secondly because two 2000-ohm units in operation will heat up much less than a single 1000-ohm resistance.

C voltage for the 199 tube is obtained by taking the voltage drop across a 50-ohm resistance in the negative filament lead of this tube, while C voltage for the power tube is obtained by utilizing the voltage drop across a resistance in the lead to the center tap of the power tube's filament transformer. With this arrrangement there is no common interstage coupling in the C bias resistances and this is distinctly advantageous in preventing distortion. Care will be necessary in adjusting an amplifier of the type illustrated in Fig. 1, to make certain that the filament current in the 199 tube is exactly 60 mils. Unfortunately it will be found that using the resistance units that are available the current through the 199 will increase as the resistances heat up. It is therefore best to properly adjust the 199 filament after the amplifier has been in operation for about ten minutes. An occasional check on the filament current through the 199 should be made to make

certain that it is not exceeding the rated 60 milliamperes. The two 2000-ohm resistors are shunted around the Clarostat controlling the filament current of the 199 in order to reduce the current that must be passed by this latter resistor. Also, the wire-wound resistor has a positive temperature coefficient whereas the Clarostat has a negative temperature coefficient and the combination of these two tends to neutralize each other and therefore keep at a minimum the variation current that occurs as the resistors heat up.

It is not at all essential that the amplifier be included in the phonograph proper, and the installation will be made much simpler if the old phonograph is not disturbed and the amplifier placed instead in a small external cabinet of its own. The pick-up can be mounted so as not to disturb the mechanical pick-up ordinarily used with the phonograph. Just how the installation is arranged will depend entirely upon the individual taste and desire of the constructor. Electrically, one arrangement is just as satisfactory as the other.

A great deal of information prepared by the author regarding the construction of satisfactory

amplifiers can be obtained by glancing through several recent issues of RADIO BROADCAST (the January, February, March, and April issues) in which many different types of units have been illustrated. In the March issue of RADIO BROADCAST will be found a description by John B. Brennan of a simple two-stage transformer-coupled amplifier that might be used in conjunction with a separate B power unit to operate the loud speaker.

GROUND CONNECTIONS

A ground connection to the negative side of the filter circuit is absolutely essential. In most instances this connection may be obtained by connecting through a 2-mfd. condenser to one side of the 110-volt line as shown in Fig. 1. The 110-volt line connections may have to be reversed to see which way is best. This is readily accomplished by merely plugging into the lamp socket or base outlet first one way and then the other.

In conclusion, the writer wishes to emphasize the importance of using the new electrically cut records for obtaining the best of results with the electric phonograph.

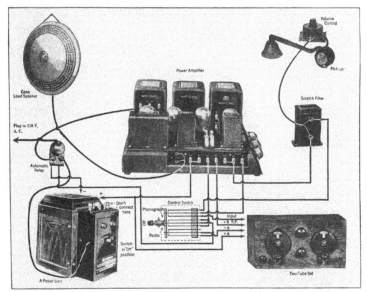

FIG. 2

This is the layout of apparatus recommended by the author for a combination electrical reproduction phonograph and radio, and employed by him in the phonograph illustrated on pages 86 and 87 of this article. The power amplifier was described in RADIO BROADCAST for January of this year, as was also the two-tube "Lab" receiver. By a simple turn of the triple-pole double-throw Yaxley jack switch (marked "Control Switch" in this diagram) either the phonograph or radio receiver is connected to the power amplifier, while a central position of the switch disconnects both. The automatic relay switches either the trickle charger or B socket-power device to the electric light mains, depending on whether the outfit is in use or not. Although this illustration shows a cone loud speaker (which may certainly be used) the author explains the construction of a baffleboard loud speaker in the text

"Strays" from the Laboratory

Growth of the Laboratory

RADIO BROADCAST Laboratory began its existence in its present form about the beginning of 1925. There were several reasons why a well-equipped and well-staffed laboratory was necessary. In the first place, nearly everyone who manufactured any kind of electrical apparatus, or automobile accessories, or did a seasonable business of any kind, had in some way or other got into the radio game. These manufacturers found it necessary to advertise their products in nationally known magazines with the result that the advertising staff of RADIO BROADCAST and the fellows that set type downstairs did a flourishing business. Since very few people knew much about radio—even in 1925—the sky was the limit to what a manufacturer could say about his product with the moral certainty that no one could catch him up on it. As a matter of fact the manufacturer himself probably believed what his advertising stated because he, too, was in a new field with little or no background of past experience.

About the middle of the 1925 season, however, a change was taking place. Already there were manufacturers in the field who were more conscientious and readers were becoming more critical—and a bit wary.

To protect the readers, not only of RADIO BROADCAST, but of all Doubleday-Page publications which ran radio advertising, and to protect those manufacturers whose products were chaff, as well as to aid the latter in case they desired technical advice, the Laboratory became a necessity. Soon the entire "Quality Group," consisting of World's Work, Harper's Magazine, The Atlantic Monthly, The Golden Book, Scribners, and Review of Reviews, availed themselves of the Laboratory's service so that their advertising pages too could be protected.

It was also true that radio circuits were coming out at that time in all radio publications so thick and fast that only a few readers could keep up. Here again it had become necessary to scrutinize carefully, to test receivers before they were described, and to develop circuits with the certainty that they were technically correct. Here was another crying need for a well-equipped laboratory.

Due to the far-sighted policy of the publishers of RADIO BROADCAST, the Laboratory was well equipped from the start— and this equipment has steadily increased. It is now possible to make practically any radio- or audio-frequency measurement with the aid of instruments that are always on hand in the Laboratory.

The very first task of the Laboratory is representative of what the staff has done from the time of its inception. Several pages of lurid advertising were sent in from a manufacturer of tubes for insertion in RADIO BROADCAST. These several pages contained some very exaggerated statements which, when boiled down, came to these claims:

These tubes require half the battery current and deliver twice as much volume. Distance doubled. Will last indefinitely.

A simple test in the Laboratory did indeed reveal the fact that the plate current of the tubes was about half that of a normal 201-A tube under the same conditions of filament voltage and current. So far, so good. In a receiver, however, the volume was "way down," as was to be expected. The filament wire was poor, having low emission, which gave the tubes a high impedance.

When the advertising was rewritten to conform with more reasonable claims, the advertising manager of the manufacturer said "nothing doing" in emphatic tones, and the magazine lost the advertising.

Not long after, the largest retail store in New York refused to sell these particular tubes, and in a few months the manufacturer was out of business. During this time various other publications ran the advertising without any hesita-

*E*VER *since its establishment, the* RADIO BROADCAST *Laboratory has served as a means of contact between our readers, the work of the manufacturer, and indeed, the entire technical field of radio. It is the purpose of these pages to extend that contact by presenting an opportunity for the Director of the Laboratory and his associates to discuss various subjects of technical interest in a way which should be of greatest value to our readers. "'Strays' from the Laboratory" will treat of a wide range of subjects —suggestions for productive fields of experiment, reference to interesting developments of manufacturers, brief abstracts of important technical articles, and suggestions from readers.*
—THE EDITOR.

tion although they claimed to have tested and passed all apparatus before accepting advertising.

Since that time the Laboratory has tested tubes of every manufacturer known to be building tubes in the United States. Not only have tubes come from the manufacturer for test but nationally known manufacturers of receiving sets have relied upon the Laboratory for information regarding tubes that could be recommended to their dealers and jobbers. Some tube manufacturers have sent representative lots of tubes as often as once a month for two years.

As a clearing house for unbiased technical data and information, the Laboratory soon became well known. It was possible to obtain data here that were untainted by manufacturing jealousy or secrecy. Representative apparatus from every manufacturer of note was examined in the Laboratory and data kept on file. It was possible for dealers and manufacturers to have various units tested in the Laboratory without the expense of building a laboratory of their own.

For example, a builder of an "extraordinary" new loud speaker came to Garden City bringing with him his inventor and one of the strangest contraptions we had seen. It looked like a butter bowl and was guaranteed to be more sensitive than any existing loud speaker. The inventor, who had little to say, was touted as having designed the Western Electric 540-AW cone which, we decided, was to be the basis of comparison.

Our "beat" oscillator, whose frequency is continuously variable from zero to 20,000 cycles, was wound up and applied to the two loud speakers in turn, and everyone noted when, in his estimation, sound could· be heard from the 540-AW, and when the "butter bowl" took interest in the proceedings. The same process was repeated at 1000 cycles but with varying input voltages. The Western Electric cone could be heard long after the "butter bowl" was perfectly dead, not only with respect to :frequency but to input voltage as well—all of which did not speak well for the inventor who, as he claimed, was responsible for the 540-AW.

"What do we care," said the manufacturer, "it's price that counts. People won't know the difference. I'm going to sell a million of these."

That loud speaker never came on the market.

In the two years and a half that the Laboratory has existed, its activities have gradually changed. Organized to protect the advertising pages of RADIO BROADCAST and other magazines, it was only a question of time until the chaff manufacturers were weeded out; and those who made wheat remained. There are few radio junk manufacturers who advertise nationally. Some newspapers still carry considerable wildcat claims, especially in their Saturday radio sections, and some magazines dig up advertising of this nature, but the Quality Group and RADIO BROADCAST no longer worry.

In those two years and a half the magazine has lost enough advertising to make the salesmen weep, but in nearly every instance the manufacturer went out of business before the accounting department could have collected, had the magazine run the copy.

At the present time the duties of the Laboratory staff are many and varied. It still exercises a strict censorship over advertising; it examines and tests all manner of radio equipment—every piece of apparatus that is described in RADIO BROADCAST for home constructors is put through its paces. It has complete laboratory tests on two thousand tubes from nearly one hundred manufacturers; data on socket power devices, many of which never saw the daylight of a dealer's shelf; life tests on batteries and tubes; tests on condensers, resistances, and audio transformers. The staff scans radio and scientific periodicals of this and many foreign countries in search of material for its clientèle—readers of the magazine.

In such foreign papers the staff finds material that delights the heart of any one who is interested in radio, and an attempt will be made to bring the gist of these articles to the readers of RADIO BROADCAST through this department every month as well as news that floats in on all manner of carrier waves, news from manufacturers, from technical laboratories, from amateur stations, and all other sources of radio information and trade gossip. It is the staff's fervent hope that these pages will bring the Laboratory into closer touch with those who are seriously interested in radio engineering and experimenting.

THE New York public, and

The Facts About that of other cities, has been
the A. C. Tube stangely upset by an announcement, more or less premature, about the RCA's new tube that uses raw a. c. and, as the papers say, "eliminates all batteries." It seems that at every opportunity the papers use that phrase as though they had a grudge against batteries. Of course the tube is reported as " revolutionizing radio," another pet phrase.

The statements of Mr. Elmer Bucher, general sales manager of the Radio Corporation, and Herbert H. Frost, general sales manager of E. T. Cunningham, Incorporated, the following day, are significant, and will allay the fears of all who see in this news story the utter destruction of present plans for the coming year. The unipotential cathode tube has been known for years and engineers are excited periodically by reports of tubes with amplification factors of 20 and plate impedances of 2000 ohms, for such seems possible with tubes of this type. Many efforts have been made to bring it to a point of reality. The nearest approach is the McCullough tube. Mr. Frost says in the New York *Times* of March 26th:

Considerable publicity has been given this week to a so-called new a. c. tube. This alleged development has been designated as "revolutionary." The following statements and claims for this tube appeared in the news article in one of New York's leading morning newspapers on Wednesday, March 24th:

"Batteries and current supply devices will be dispensed with in broadcast receivers by a new alternating current tube . . ."
". . . a revolutionary development . . ."
"The tube seen here yesterday was marked UX-225."
". . . an alternating current detector and amplifier tube which could be used in direct connection with the 110-volt house lighting sockets in much the same fashion as an incandescent lamp, thereby dispensing with all B batteries, trickle chargers, storage batteries, dry cells, and current-supply devices, such as A and B eliminators."
". . . 1927's greatest contribution to the revolutionary developments in radio."
". . . a set which will not require batteries or current supply devices."
"The necessity for batteries and battery eliminators is obviated. . ."

Mr. Frost continues:

Type CX-325, our equivalent of UX-225, referred to above, has been in an experimental and developmental stage for nearly two years. Its output and capabilities are similar to those of our well-known type CX-301-A. In CX-325 we are attempting to replace the filament with a cathode heated directly by house a. c. supplied through a step-down transformer. When and if successful,

PRESENT AMERICAN PLAN	REVISED PLAN	DESCRIPTION
UX-199	3V1o6	Dull Emitter—General Purpose.
UX-120	3V1V12	Dull Emitter—Power Amplifier.
UX-201-A	5¹⁄₂X25	Five Volt—General Purpose.
UX-112	5¹⁄₂X50	Five-Volt Power—Rₚ=6000.
UX-171	5¹⁄₂V50	Five-Volt Power—Rₚ=2500.
UX-210	7¹⁄₂X125	Seven-Volt—Rₚ=6000.
UX-216-B	7.5R125	Half-Wave Rectifier.
UX-213	{ 5R200 } { 5R200 }	Full-Wave Rectifier.
UX-876	50B176	Ballast Tube.
UX-174	90G50	Protective Tube.
UX-200-A	5D25	Detector (Vapor-Filled).
UX-877	{ 5P20 } { 3P90 }	Protective Tube.

this tube would eliminate the A battery, substituting raw a. c. It will not eliminate the B and C batteries, or B eliminators. This tube has not yet reached a commercial stage. It is difficult to manufacture and would have to sell at a price from $6.00 to $9.00 each. It is our opinion that the practical difficulties connected with the manufacture of CX-325 will prevent it ever being commercialized. If perfected, it could not by any stretch of the imagination be called a revolutionary development. The CX-325 tube could not be used in present equipment without substantial wiring changes, and then would not improve reception but merely eliminate the A battery.

Crystal-   ONE OF a number of ex-
Control on   cellent papers recently de-
Short Waves   livered before the Radio Club
of America was the work of an old time amateur, Mr. C. R. Runyon, Jr., and described the equipment of his more or less "high-powered" amateur short-wave station. As is the case with many of our best stations, Mr. Runyon's 2 AG is crystal controlled, which brings up a point that many amateurs evidently debate among themselves: Is crystal control worth while?

A quartz-controlled transmitter is simply a master oscillator system in which a small tube whose frequency is determined by the crystal drives a large tube acting as an amplifier. Owing to the fact that the quartz plate oscillates only in an extremely small band of frequencies, the transmitting wavelength for all practical purposes is fixed. Changes in tube constants, antenna height, even adjustments of inductance and capacity have no effect upon the emitted frequency. It is also true that the note of the station will be dis-

tinctive, and that even with raw a. c. on the plates of the tubes, the note will be good.

All of this means that a crystal-controlled station will have a good note on a definite frequency.

For stations that carry on schedule communications, or point-to-point service, or transmit traffic for considerable periods, nothing could be better than absolute steadiness of frequency. For amateur work the case is somewhat different.

For example, at 2 GY, the experimental station of the Laboratory, considerable effort was put forth to make a 500-watt crystal-controlled station. The apparatus was as extensive as Mr. Runyon's. There were two 7.5-watt oscillators, two 7.5-watt amplifiers operating at double the oscillator frequency, two 50 watters amplifying the output of the two 7.5-watt tubes, and at the same frequency, and finally two 250-watt tubes which fed into the antenna. It was possible to deliver considerable power to the antenna on the 40-meter (7500-kc.) band and the note was beautiful, but no sooner had the station gone on the air than FW, in France, settled on the same frequency, or so near it that interference was unavoidable. The result was that distant amateurs could not hear 2 GY when the St. Assise station was in operation, which was from about 4:30 p. m., E. S. T. until much later in the morning than 2 GY could be kept open.

On amateur bands, and with purely amateur traffic, there is need occasionally to change the frequency slightly to avoid interference. Crystal-controlled stations are sadly handicapped here, unless several quartz slabs are handy and unless the operator can go through a somewhat complicated rigmarole in re-tuning. Quartz-controlled oscillators using a good source of plate supply (batteries or good d. c.), turn out a note that is extremely tiring to the ears; and on the higher frequency bands a series of tubes are necessary with more or less critical adjustments.

It is true that the Army stations which use raw a. c. for filament, grid, and plate voltage, have notes that are penetrating and fairly easy on the ear, and when night after night one hears them pounding away somewhat below the 40-meter amateur band, it is easy to get the idea that crystal transmitters are greatly to be desired. But few amateurs have the funds, few care to bother adjusting many circuits, and few seem to realize that a 250-watt tube which will handle 500 watts as an oscillator will only take care of 250 when used as an amplifier.

New   ONE OF the most encourag-
Nomenclature   ing signs in the radio industry
for our Tubes?   is the manifest desire for standardization on the part of the better manufacturers. Under the untiring efforts of Mr. George Lewis, chairman of the vacuum-tube committee of the Radio Manufacturers' Association, tubes have come in for their share of attention. At the Chicago convention and trade show of this association, June 13-17, an important question will be discussed, *viz.*, nomenclature. Criticizing our present system of naming tubes by nondescript numbers, *viz.*, 199, 213, 171, Mr. Lewis suggests the following system and will welcome suggestions. Tubes will be known by a number giving first the filament voltage in Arabic, then a Roman numeral denoting the amplification factor, and finally the filament current in Arabic. Special tubes will be designated by a letter indicative of the service performed. The table on this page gives Mr. Lewis' suggested nomenclature.

An Instrument for the Home Laboratory

How to Construct a Useful Device—a Modulated Oscillator—That Will Measure Coils, Calibrate Receivers, and Help Locate Trouble

By

KEITH HENNEY

Director of the Laboratory

THE MODULATED OSCILLATOR

With the several tuning coils necessary to cover four slightly overlapping wave bands, is illustrated here. A specially made Karas dial, shown in front, was satisfactorily substituted for the aluminum one in the Laboratory

EVERY laboratory, regardless of its size, must possess certain equipment without which the investigator is handicapped. The home radio laboratory, housed in a corner of the den, or the attic, or the basement, and costing probably no more than a high-grade receiver, differs in this respect not a whit from the huge industrial or college laboratory with an endowment of many thousands of dollars within the walls of which may work hundreds of highly trained and skilled investigators.

One of these essential pieces of equipment was described nearly two years ago, September, 1925, in RADIO BROADCAST. It was called a modulated oscillator, and so many uses for the instrument have been found in the Laboratory and by the many readers who followed the series of articles written around that small generator of radio waves, that a new description at this time seems more than worth while.

The present unit differs but little from that described before. It consists essentially of two oscillators coupled together. As described, one covers the frequency range from 500 to 6000 kilocycles (50 to 600 meters), by means of four plug-in coils, and will be useful either as a source of radio-frequency waves for various laboratory experiments, or as an accurate and sensitive frequency meter or wavemeter. The other oscillator is a fixed frequency generator, oscillating at approximately 1000 cycles and providing a source of audible tone for certain other laboratory work. When desired, the radio-frequency tube may be operated alone, emitting a pure unmodulated carrier wave, or it may be modulated with the 1000 cycles of the second tube, and be useful in another set of experiments.

USES OF THE OSCILLATOR

THE most obvious use of the complete instrument is as a source of modulated high-frequency signals. By their aid, a receiver can be calibrated in kilocycles or wavelengths without the bother of waiting until broadcasting stations come on the air. The frequency of either the audio or radio oscillator can be varied, the entire output being under the control of the experimenter. The method of calibration is simple. A receiver is set up and the oscillator turned on. When the receiver is tuned to the same frequency as the oscillator the latter's tone will be heard in the loud speaker. Of course it is possible to set a receiver to a known frequency by this same process if, for example, the experimenter desires to listen for some definite station the frequency of which is known. It is also possible to measure the frequency of incoming signals. It is only necessary to tune the oscillator until its signals are heard at the same receiver condenser setting as the distant station.

The grid meter of the radio oscillator makes this unit a very sensitive and accurate resonance indicator. For example, suppose we have a coil and condenser combination and we desire its range in kilocycles or meters. Or suppose we have a choke coil for a short-wave transmitter and we want to know its natural wavelength so that it will not absorb energy on any wavelength to which the transmitter may be tuned.

The coil, or choke, is brought near the oscillator inductance, and when the latter is tuned to resonance, a sharp dip in the grid current reading will be noted. Extremely loose coupling can be used, materially increasing the sensitivity of the instrument. It is also possible to show the effect of coupling two circuits too closely together. As the tuning condenser is varied, there will be two dips, indicating that the two circuits are so closely coupled that neither has a chance to oscillate at its own frequency, or that the two circuits are not closely enough tuned to the same frequency to give a single response.

With the radio oscillator as a source of unmodulated energy, it is possible to measure the overall gain of radio-frequency amplifiers. The same is true of audio amplifiers, by using the low-frequency tube.

The tone alone, obtained from the low-frequency tube, can be used for testing circuits (where a dry cell and buzzer could not be used) or as a source of tone for bridge work in measuring inductance, resistance, or capacity.

The circuit diagram of the complete modulated oscillator is shown in Fig. 1, and is not difficult to follow; nor is the set-up difficult to get working. The tubes can be any of the general purpose tubes now used, *viz.*, 199, 12, or 201-A type. In the Laboratory WD-12 type tubes were used, since it was possible to place both A and B batteries for the outfit in the same cabinet which housed the tubes, transformer, and condenser, thereby making the combination frequency meter and oscillator a truly portable affair that can be "lugged" out of the Laboratory into the field for experiments on antennas, etc.

The layout of this useful instrument is due to Mr. John Brennan. The only important change in the circuit from that first described is the inclusion of the grid-current meter in the radio oscillator, which makes it an extremely sensitive frequency meter; other minor variations in the circuit resulted from the natural desire to make the completed equipment more useful.

Both oscillators follow the Hartley circuit, presenting several points of especial interest. One of these points relates to the inclusion of the grid-current meter in the radio oscillator, as stated above. The second point to note is the radio-frequency choke through which the B-battery voltage is fed to the second oscillator. This is necessary to prevent the radio-frequency energy from going into the center tap of the audio transformer, which is at ground potential with respect to r. f. voltages. It will be noted, too, that the tuning condenser is across half the coil only.

The grid meter increases the

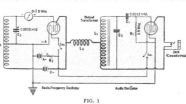

FIG. 1

The circuit diagram of the modulated oscillator

cost of the instrument somewhat, and is not vitally necessary for the operation of the modulated oscillator. It does, however, make the radio oscillator into a sensitive wavemeter, extending the uses to which the complete equipment may be put. Its inclusion is strongly urged.

The low-frequency oscillator consists of an output transformer connected as shown in Fig. 2. A push-pull output transformer may be used if desired, as also shown in Fig. 2. The frequency of the audio oscillator is controlled by the inductance of the transformer and the capacity across

Push-pull Output Transformer   1-1 Output Transformer

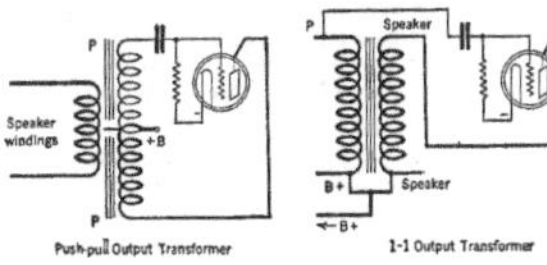

FIG. 2

Either push-pull or output transformer may be employed in the audio-frequency oscillator

it, as well as by the plate and filament voltage and grid leak values. In such a simple piece of apparatus it is neither necessary to know the exact audio frequency or to go to great labor to maintain it constant. The radio oscillator varies in frequency too, but not sufficient to be measured on an ordinary wavemeter.

A Jefferson push-pull output transformer, with 0.015 mfd. C_2 across the grid and filament of the tube, oscillated in the Laboratory at 900 cycles, and a Pacent output transformer 27B tuned according to the table below when a WD-12 tube was used and connected as shown in Fig 1:

C_2 Capacity	Frequency
None	3000 cycles
0.000125 mfd.	2000 "
0.0008 "	1000 "
0.00125 "	650 "
0.00275 "	550 "
0.012 "	240 "
0.03 "	160 "
0.04 "	130 "

If it is desired to use more than one tone from the oscillator, a pair of binding posts may be brought out on the panel and a small capacity unit, consisting of several condensers connected to a switch, may be attached to the low-frequency tube. The exact frequencies obtained may differ from those shown in the table. For this reason, it is not necessary or useful to get condensers of exactly the same capacity as specified. They may be obtained by connecting several condensers together so that the resultant capacities are similar to those given.

With the tuning condenser across the grid-filament half of the coil, the tuning ranges are approximately as shown below when WD-12 tubes are used as oscillators. If the dial is divided into 100 degrees, and has a straight frequency-line condenser, each degree will cover about the number of kilocycles shown in the table:

Coil	Wavelength Range	Frequency Range	Kilocycles Per Degree
15 turns	45-120	2500-6660	31.60
30 turns	80-210	1450-3750	23.30
60 turns	165-400	750-1820	10.70
90 turns	265-620	485-1130	6.50

The dial should be large and provided with a fine adjustment. That shown in the photograph

was made for the Laboratory from aluminium by Mr. Harold Benner of Cruft Laboratory, Harvard University, and has the dimensions shown in Fig. 3. Very accurate adjustments are possible with this dial, and its accompanying vernier makes it possible to read to within one kilocycle in the broadcast frequency band, two kilocycles in the intermediate band, and three in the higher band. Any good vernier or slow-motion dial can be used.

An even more accurate instrument could be made by reducing the overlap so that each coil covered only the frequency band desired. In this way the smallest coil would cover exactly 3000 kilocycles, the intermediate ranges would be 2000 and 1000, and the largest coil would cover 500 kilocycles. Fewer turns will be necessary and the experimenter will be forced to rely upon his own ingenuity in winding them.

COILS

THE coils used in the Laboratory oscillator were General Radio type 277 and have the following dimensions:

Coil	Turns	Size Wire	Diameter	Length of Winding	Inductance
277-A	15	21	2½″	1⅜″	0.014mh.
277-B	30	21	"	1⅜″	0.055 "
277-C	60	21	"	1⅜″	0.217 "
277-E	90	27	"	1⅜″	0.495 "

CALIBRATION

IT IS a simple matter to calibrate such an oscillator. It is best to do this without the low-frequency tube oscillating. The accessory apparatus is not extensive, only a two-tube blooper, or any other receiver that can be made to oscillate, being necessary.

In addition to the oscillator and blooper, a single accurately known frequency is required.

FIG. 3

The dimensions of the vernier dial

This is provided by any well-known broadcasting station, preferably near the lower edge of the frequency band, say 400 or 500 meters. If such a station cannot be heard in the daytime, or when the calibration is to take place, the station should be picked up wherever possible on the blooper (which should be set up and running for several minutes before the final tuning is done) and no changes made, except to turn off the tube. The next day, or whenever the calibration takes place, the tube should be lit for several minutes to permit it and the batteries to settle down to a steady state. The detector tube is made to oscillate and the blooper is set accurately on the broadcaster's frequency by tuning to zero beat with him.

This must be done with a vernier condenser or with a single-plate condenser shunted across the large tuning condenser. As the blooper's frequency nears that of the broadcasting station, a lowering in tone of the beat note will be heard. It will soon become inaudible, to rise again on the other side of exact resonance, as indicated in Fig. 4. Halfway between the limits of this inaudible area is the exact frequency desired. With care, the experimenter can set his blooper to within 100 cycles of a station operating at 100 kilocycles.

The blooper is now oscillating at exactly, say 750 kilocycles (400 meters), and should not be touched or approached by the hand throughout the following procedure. At least one stage of audio should be used on the blooper and a pair of phones in the output of this amplifier should be used to indicate resonance between harmonics of the blooper and harmonics of the oscillator.

The blooper and oscillator have in their output not only the wavelength to which the coil and condenser are tuned but harmonics of this wavelength as well. That is, if the wavelength is 400 meters, in the output will also be one half,

Radio Broadcast Photograph

THE SIDE VIEW

Showing the interior of the modulated-oscillator and indicating plainly the mode of construction employed. Ample room is to be found under the sub-panel to locate two dry cells and two small 22½-volt B batteries

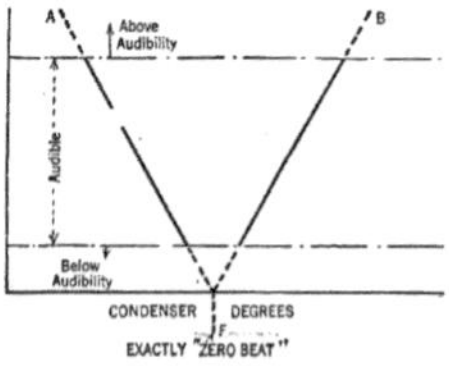

FIG. 4

one third, one fourth, etc., of this figure, or 200, 133, 100, etc., meters. Whenever the oscillator is tuned to one of these wavelengths a beat note between the oscillator fundamental and the blooper's harmonic will be heard. A beat will also take place when the second harmonic of the blooper beats with the third of the oscillator, and so on down the line. The number of wavelengths that can be secured in this manner are shown by Table No. 1. which has 400 meters as the basis of reference. The figures at the top represent the oscillator's harmonics, while those along the left-hand vertical line represent the harmonics of the blooper. For example, if the oscillator is set at 600 meters, its third harmonic will beat with the second of the blooper. The other wavelengths in the table may be found by multiplying 400 by $\frac{1}{2}$, $\frac{1}{3}$, $\frac{2}{3}$, $\frac{4}{5}$, $\frac{7}{8}$, etc. In calibrating the Laboratory oscillator with the coupling indicated in Fig. 5, twenty-two of these wavelengths were found and, with closer coupling all of those given in Table No. 1 and others should be found. The ones in this table to be expected are denoted by heavy type. In our note book we set down our data as shown in Table No. 2, noting the condenser setting for each beat note, and marking with an asterisk the strongest. These will correspond to exact harmonics of the blooper, that is, the second, third, fourth, etc., or small fractional harmonics, that is $\frac{3}{2}$, $\frac{2}{3}$, $\frac{4}{5}$, $\frac{3}{4}$, etc.

Now, all of this probably sounds complicated and difficult to the uninitiated. There is no reason, however, for getting stage fright at this point. Suppose we have set up the blooper and that it is tuned accurately to 400 meters. We are listening in the plate circuit of a stage of audio amplification behind this oscillating detector. The oscillator, too, is ready to operate, the filament having been lighted for several minutes. We place the 90-turn coil in the oscillator, arrange the coupling about as shown in Fig. 5, and turn the oscillator dial. We get a strong whistle at 34 degrees. We make mental record of how loud the beat note is and are certain that we have hit 400 meters, and turn the dial again. At about 92, 12, and 4 we get other squeals. Now we know that the 90-turn coil will go to about 600 meters and we note from our table of those wavelengths at which beat notes will be heard that, when the oscillator is tuned to 600 meters, its third harmonic (200 meters) will beat with the second of the blooper (also 200 meters), to produce an audible note. By the same reasoning, when the oscillator is tuned to 300 meters, its third harmonic, or 100 meters, will beat with the fourth of the blooper, also 100 meters. The loud beat at 4 degrees corresponds to 266 meters.

In the calibration process so far we have been certain of but one point, 400 meters at 34. If we listen closely, or use somewhat closer coupling, we shall also hear beat notes at 69, 59, 16, and 9 degrees on the oscillator dial. Using our table of wavelengths we guess again, and assume that the wavelengths are 533, 500, 320, and 280 meters. Everything but the 400-meter point has been guessed at, but if we plot a curve of wavelengths against condenser degrees it will tell at once if we have erred. If a good curve results, our guesses have been correct. If one or more points refuse to fall in line we have guessed wrong and we must listen for beat notes again.

Blooper Harmonics	Oscillator Harmonics									
	1	2	3	4	5	6	7	8	9	10
1	**400**	800	1200	—	—	—	—	—	—	—
2	**200**	—	**600**	—	—	—	—	—	—	—
3	**133**	**266**	—	**533**	—	—	—	—	—	—
4	**100**	—	**300**	—	**500**	—	—	—	—	—
5	**80**	**160**	**240**	**320**	—	480	—	—	—	—
6	**66.6**	—	—	—	**333**	—	466.6	—	—	—
7	**57.1**	**114.3**	—	**228**	286	**343**	—	457	514	571
8	**50**	—	150	—	250	—	350	—	450	—
9	**44.5**	89.0	**134**	178	222	—	311	356	—	445
10	40	—	120	—	—	—	**280**	—	360	—

TABLE NO. 1

On the 60-turn coil, we get a very loud note at 98, weaker notes at 50, 37, and 15, and still weaker notes at 69, 65, 59, 45, 28, 24, and 6 degrees. By our process of guessing — assuming that the 98 degree point is 400 meters — we put down the wavelengths as given in Table No. 2. The same process is carried out in getting points on the 30- and 15-turn coils.

Blooper Set at 400 Meters							
90-Turn Coil		60-Turn Coil		30-Turn Coil		15-Turn Coil	
Condenser Degrees	Meters	Condenser Degrees	Meters	Condenser Degrees	Meters	Condenser Degrees	Meters
92*	600	98**	400	96*	200	77*	100
69	533	69	343	68	171		
		65	333				
59	500	59	320	58	160		
		50*	300				
34**	400	45	280			46*	80
		37*	266	36.5*	133		
16	320	28	240			28.5	66.6
		24	230	23	114.3		
12*	300	15*	200	15*	100	18	57
						11	50
9	280						
		6	171				
4*	266					3	44.4

TABLE NO. 2

If the coils are well made, the smaller inductances having half the number of turns of the preceding coil, beats will be heard at approximately the same condenser settings as Table No. 2 shows. If a condenser with a straight wavelength calibration is used, the problem is simplified, since we can guess fairly accurately where the next beat note is to be found. The same applies with a straight frequency calibration except that we must think in terms of frequencies and not wavelengths. With a straight capacity-line condenser, the wavelengths can be estimated by remembering that the wavelengths corresponding to two condenser settings are proportional to the square roots of the capacities involved. If the condenser has one of the modified calibrations, the experimenter is up against it. He had better use a condenser whose calibration is known. Practically all well-known condenser manufacturers have units that will have a straight frequency-line. An excellent condenser is the National Equicycle, which uses a 270-degree arc instead of 180 degrees. This will spread out the frequency band considerably.

For general work around the laboratory, the modulated signal may be used, but for careful measurements the pure carrier wave is better. Coupling to a receiver using one stage of audio need be no closer than the opposite sides of a rather large room. When the high-frequency oscillator is modulated it is possible to pick up the tone in two places very near the carrier which will be comparatively quiet. These two signals represent the side bands which are one thousand cycles on either side of the carrier in the case of a thousand-cycle signal.

In a later article a bridge for measuring vacuum-tube characteristics will be described. The tone will be very useful in conjunction with this instrument. Among other articles in this series for the home laboratory will be methods of measuring inductance and capacity, and here again the tone will be necessary. Still another article has been prepared on the vacuum-tube

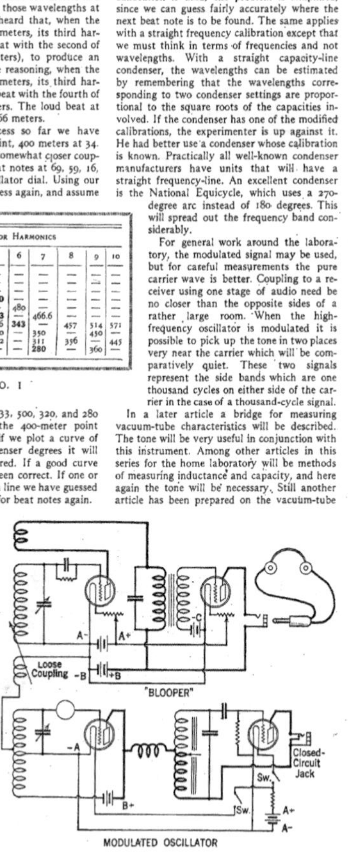

FIG. 5

The circuits employed in calibrating the modulated oscillator

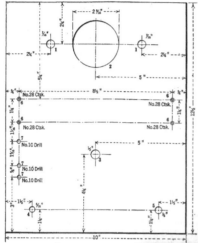

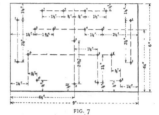

FIG. 7

The sub-panel layout, as shown here, supports the following pieces of apparatus, as numbered: 1, brackets; 2, sockets; 3, transformer; 4, grid leak; 5, radio-frequency choke coil; 6, fixed condenser; 7, binding posts or machine screws

USES OF THE OSCILLATOR

1. Radio Oscillator
  (a) Heterodyne wavemeter.
  (b) Radio-frequency driver.
  (c) Measuring gain and frequency range of amplifier.
  (d) Tuning range of coil — condenser combination.

2. Audio Oscillator
  (a) Source of tone for circuit testing.
  (b) Source of tone for measuring capacity, inductance, or resistance.
  (c) Source of tone for measuring gain of audio amplifiers.

3. Modulated Oscillator
  (a) Measure unknown frequencies.
  (b) Measuring tuning range of receivers.
  (c) Setting receiver to known frequency.

FIG. 6 (*Above*)

In this panel layout the various parts are mounted in the holes which are numbered, as follows: 1, filament switches; 2, meter; 3, tuning condenser; 4, rheostat; 5, output jack; 6, brackets; 7 binding posts

voltmeter, and for this instrument both the high and low frequencies will be needed.

A jack is provided in the plate circuit of the low-frequency oscillator so that the tone may be utilized at any desired point.

The necessary dimensions and layout for the various pieces of apparatus are shown in the accompanying diagrams, and the following list of parts covers the apparatus actually used in the Laboratory oscillator:

LIST OF PARTS

L_4—1 Pacent 27B Output Transformer	$	7.50
L_4—4 General Radio Coils		
277-A		1.25
277-B		1.25
277-C		1.25
277-E		1.50
L_5—1 Samson Radio-Frequency Choke		
No. 85.		1.50
C_1—1 Sangamo Condenser 0.0012 Mfd.		.50
C_2—1 Small Fixed Condenser (see p. 93)		.50
C_3—1 Karas Straight Frequency-Line		
Condenser, 0.0005 Mfd.		7.00
R_1—1 Carborundum Grid Leak, 0.5		
Megs.		.35
R_2—1 Frost Rheostat, 10 Ohms		.50
2 Carter "Imp" Battery Switches		1.30

(*Continued on page 96*).

TO THE RIGHT

This picture shows clearly the disposition of the parts employed in the construction of the modulated oscillator

RADIO BROADCAST Photograph

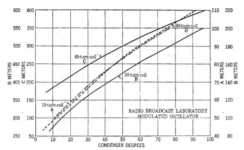

THE CALIBRATION CHART

For the four coils employed in the radio-frequency end of the modulated oscillator, will closely resemble the one shown here, when a straight capacity-line condenser, such as the General Radio 247 type is used

1 Carter Closed-Circuit Short Jack		.80
1 Weston Milliammeter—0-1.5		
Mils.		12.00
2 Benjamin Brackets		1.40
3 General Radio Binding Posts		.45
2 Benjamin Sockets		1.50
1 Main Panel 10″ x 12½″ x $\frac{3}{16}$″		2.50
1 Sub-Panel 6″ x 9″ x $\frac{3}{16}$″		1.08
Machine Screws, Wire, Solder,		
Etc.		.50
TOTAL		**$44.63**

Other parts than those specified in this list may be used provided they are of equal quality. The main requirements are that the condenser be rigid in its bearings, that the dial be easily and accurately read, and that the device be recalibrated when tubes are changed. It is a good idea to lock the tubes and batteries into the cabinet.

Following articles will discuss in detail the varied uses of this modulated oscillator. The Laboratory will be glad to hear from readers at any time regarding the uses which they have found for the apparatus described here.

The following construction hints are from Mr. Brennan who is responsible for the construction of the unit used in the Laboratory.

The panel layouts shown are based on the use of the material listed here and should be altered accordingly if substitutions are made

The first step in the construction of the oscillator is the preparation of the main panel, Fig. 6, for drilling. The various hole centers should be spotted with the aid of a hammer and centerpunch, and then drilled with a small drill, say No. 28. Then those holes which require enlarging may be so enlarged by the use of Stevens tapered reamers, which are indispensable for such work. The hole for the meter may be made by drilling a number of small holes around the periphery of the circle and then filing it clean. Better yet, a circle drill may be used to make a clean-cut hole just slightly larger than the diameter of the body of the meter.

The sub-panel is next prepared, and after all the holes have been drilled in accordance with Fig. 7, the two Benjamin brackets are mounted in place on the main panel in the holes numbered "6." Round-head machine screws, $\frac{1}{4}$″ x $\frac{3}{16}$″,

pass through the holes numbered "1" on the sub-panel to fasten it to the top of the brackets. After this operation is completed, the audio transformer, sockets, fixed condenser, r. f. choke,

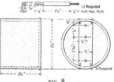

FIG. 8
Dimensions for the coil forms and connector pins

and grid leak are mounted on top of the sub-panel.

Prop up the rear edge of the sub-panel to maintain the assembly in an upright position and then mount the main panel instruments thereon.

The wiring is next. Point to point connections are made which make for short direct leads. This system of wiring is more to be desired over the system where right-angle turns add to the neatness of wiring but otherwise do not materially improve matters.

The specifications for the coil forms are given in Fig. 8. If General Radio plug pins are used, a different arrangement for connecting them to the unit will have to be devised. The pins shown in Fig. 8 allow the coils to be mounted in the binding posts and to extend somewhat over the edge of the panel.

FOR CALIBRATION

SIGNALS from any of these stations may be used in calibrating wavemeters. They are maintained very closely to the frequency indicated here.

STANDARD FREQUENCY STATIONS

Station	Location	Frequency
WEAF	New York, New York	610
WRC	Washington, District of Columbia	640
WJZ	Bound Brook, New Jersey	660
WGY	Schenectady, New York	790
WBZ	Springfield, Massachusetts	900
KDKA	East Pittsburgh, Pennsylvania	970

CONSTANT FREQUENCY STATIONS

Station	Location	Frequency
WHO	Des Moines, Iowa	570
KFRU	Columbia, Missouri	600
WOC	Davenport, Iowa	620
WTIC	Hartford, Connecticut	630
WMAQ	Chicago, Illinois	670
KLDS	Independence, Missouri	680
KPO	San Francisco, California	700
WLW	Harrison, Ohio	710
WCCO	St. Paul-Minneapolis, Minnesota	720
WTAM	Cleveland, Ohio	770
KTHS	Hot Springs, Arkansas	800
WJJD	Mooseheart, Illinois	810
KGO	Oakland, California	830
WJAD	Waco, Texas	850
WWJ	Detroit, Michigan	850
WLS	Crete, Illinois	870

RIGHT
The layout of parts on the sub-panel

RADIO BROADCAST
Photograph

"Radio Pests" and How to Be One

WHEN we proposed to the readers of this department the question: "Do you listen to your radio in the evenings as you would to a regular show or do you simply turn it on and use it as a background to other activities?" it was with malice aforethought. We were gathering ammunition for a little tirade against the radio pest who turns on his radio at about six A. M. and leaves it in that distressing condition until midnight.

The questionnaire showed that one out of every ten listeners employs his radio almost exclusively as a background. This does not mean that one out of every ten radio receiver operators is a pest; if that were true, it would put us in the pest class, for there are many occasions when we utilize the radio as a background. But the flagrant offenders are included within that number; just how large a part of it they make up you can decide for yourself.

At any rate you know the beast. When you open the door of his house for a friendly little call of an evening you are immediately greeted by a resounding blare from a loud speaker in the next room. And whether the order of the evening be bridge or poker, ping pong or conversation, that infernal loud speaker continues to vibrate viciously the evening through. It vibrates from string quartet to brass band, from election speech to bed time story, from wheezy soprano to musical saw player, from stock market report to bee keeping instruction, and back again—and all over again.

All the while nobody is paying it the slightest bit of attention. To add to the misery, some fluctuation in factors radio and electrical has thrown it out of tune and interference has been set up by some powerful local bird store.

When we say nobody is paying attention to it we refer particularly to its owner and the members of his family. They have, through long exposure, developed an immunity to it. But you, however (we assume you to be a sensitive being whose ears are normally free of impedimenta), are woefully aware that it is "going." Gradually its cumulative effect is to grind down your nerves to rather small shreds. Finally, at the end of a couple of hours, the one nerve left you unground tautens itself in the superhuman paroxysm and you arise, devour the pack of cards, kick over the receiving set, shoot your host, and rush out into the night.

To radio receiver owners of this ilk may be laid the blame for the not at all small army of scoffers who steadfastly refuse to consider radio seriously. The scoffer is a scoffer, nine times out of ten, because his first introduction to radio was in the home of a Radio Pest.

Naturally he considers the device an instrument of the devil; for an instrument of the devil it is when so operated.

Any radio owner, no matter who he be, and including ourself, can profit by an occasional

THE SMITH BROTHERS

"Trade" and "Mark" who in their non-microphone life are "Scrappy" Lambert and Billy Hillpot. You may remember hearing them with Ben Bernie and his orchestra. They are the Smith Brothers for a half hour beginning at ten p.m. over the WEAF network on Wednesday nights

REAL HAWAIIANS AT KHJ

The Moana Hawaiian Entertainers, reported to have come "direct from the Islands," played a week's engagement recently at KHJ, Los Angeles

recalling of the old saw: "Familiarity breeds contempt." To get the maximum enjoyment out of a set it must be treated with a certain amount of respect. The greatest advantage of radio is that it requires no more than the flip of a switch to bring music and other entertainment right into your own home. But, paradoxical though it may seem, this is also radio's greatest drawback. By the time one has struggled into his dinner clothes, taxied a half hour or more to the theater, and laid out eight-eighty for tickets he is in a frame of mind to enjoy the performance or die in the attempt.

The fact that it takes some effort to get to the source of entertainment undeniably adds something to the pleasure to be derived. On this score radio is at a decided disadvantage. Asking little it receives even less (that is, understand us correctly, in the home of the Radio Pest).

We hesitate to adopt such a smug and paternalistic course as to set forth a set of rules for the proper operation of a receiver, especially because we suspect that the great majority of our readers knows a lot more about it than we do. But perhaps you number a Radio Pest among your friends and would like to have the table of laws to clip and mail to him anonymously. Hence we advance these suggestions on;

HOW TO OPERATE A RADIO RECEIVER FOR MAXIMUM PLEASURE

First: Keep the fool thing in proper mechanical shape. A receiver with a defective tube or a weak battery or some other wheeze-provoking ailment should be immediately retired from service and kept retired until its ills are attended to. A distorting set is ninety-nine times worse than no set at all.

Second: As a general rule, don't use the radio simply as a background, at least while some sensitive soul like ourself is parked in your parlor. Like all rules this one is largely invalid because certain types of broadcasts are genuinely fitting as background. Almost any program of instrumental music (excepting jazz) may serve as a pleasant obbligato to bridge or conversation, or even reading, providing it is rheostated down to a sufficiently *sotto* volume. But speeches, songs, and static are a suitable background for nothing we know of outside of a tea party in a mad house.

Third: The most genuine enjoyment from a radio program is to be secured by preparing for it in advance and listening to it with reasonable reverence. Consult some advance program and find out the exact hour that some feature you want to hear is scheduled. Let the receiver remain deathly silent until two minutes before that hour. Then provide

ETHEL AND JANICE
Who are known to the listening world who tune-in
on KMOX at St. Louis as the "music mixers,"
whatever that means

yourself with an easy chair (formal clothes not
imperative) and tune-in the desired program right
on the instant of the first word of the first announcement. If you can provide yourself in advance with an itemized program of the feature
it will contribute to the illusion of being at a
concert. Try it some time if you haven't already
(selecting, of course, a program that's worth the
ritual) and you'll be surprised how much this
slight mark of respect for the program repays
you in enjoyment of it.

Making Up the "Sponsored" Program

AN INTERESTING insight into the advertising man's ideas concerning radio
advertising was afforded in an article by
Uriel Davis in *Printers' Ink*. He suggests methods of procedure for the prospective air advertiser and says in part:
"The product or company should be human-
ized, dramatized. The narrator's voice, without

which the dramatic element will not succeed,
should be studied and analyzed for timbre, intensity, and pitch, and its quality so finely determined that a suitable musical accompaniment,
in proper register, may be provided by a single
instrument or an ensemble.
"The name of the company, or its product,
should be introduced in as subtle a manner as
possible. The announcement of the name or
product could recur from time to time during the
actual performance instead of between musical
or other selections, which is the custom to-day.
Under no circumstances should the advertisement appear too obvious. Where there is no good
reason, from the standpoint of entertainment, to
mention the name of the product or the company,
it should be omitted.
"There can be no advertising loss in a plan of
this kind. As a matter of fact, the value of subtle
advertising of this kind should be far greater than
that obtained from 'program interference,' as the
present-day announcements may be termed.
"An hour or any period of radio entertainment
must have continuity. Continuity is essential
to hold the listener's attention over even a short
period. He is constantly looking for something
he likes to hear. When he attends a theatrical
performance, his mind is not on the show appearing at another theatre. Why could not his mind
be put in just as receptive a mood when he listens
in on a radio hour? To obtain such continuity it
is necessary to build programs upon a basis
similar to that employed in the preparation of
moving picture scenarios.
"Since music is essential in successful radio
performance, there should be a continuous tie-up
between the spoken word and the orchestra or
whatever group of musical instruments that
may be used. The music, when not an actual part
of the program, may be employed as a background or setting for the voices used for descriptive purposes as well as for song.
"Pauses between selections as well as between
the voices and music may be eliminated with ease
if the scenario, so called, will provide for absolute
synchronization whereby the music will always
shortly begin before the voice ceases speaking,
and *vice versa*.

"Continuity of performance or program, in
other words, the use of a scenario, should enable
advertisers to provide not only a high grade
of entertainment, but a program sufficiently
interesting to hold the listener's attention
throughout its entire length. It should arrest
the attention of the casual tuner-in to such extent
that once the program is brought in on a set it
will not be dismissed until completion.
"Successfully to carry out the suggestions
I have made would require not only careful
preparations of the scenario and attendent
musical program, but a careful study of the
advertiser's product as well."
We are glad if the advertisers are to be convinced that "where there is no good reason, from
the standpoint of entertainment, to mention
the name of the product or company, it should
be omitted."
But we are filled with grief at the other doctrine Mr. Davis preaches: "An hour or any
period of radio entertainment must have continuity."
Heaven deliver this listener from the type of
"continuity" employed in the Don Amaizo
hour. Such continuity smacks of the earliest
days of the movie subtitle. It is not our business
as a radio reviewer to tell the advertising men
how to run theirs. So we will acquiesce to their
superior wisdom and concede that there may be
some necessity for continuity to hold the attention of some of the listeners. But it gives us the
willies. If the artists on a given hour are good
we will stay with them regardless of the omission or inclusion of such stuff as:
". . . . and now our Coral Throated Baritone
and Sarah Szbisco, the Soprano with a Soul, bid
tearful farewell to the sturdy Volga boatmen and
boarding their Peerless Yacht Company Auxiliary barge they drift slowly down the Danube,
past the lovely rock of the Loselei, where, amid
the twitter of the birds and the bees, we hear
emerging from the forest, as from the mellifluous
dulcimer, the tender strains of 'Down on the
Mississip'."
It is notable that several of the most popular
"hours" omit continuity entirely or use it very
sparingly. But if we are to have it with us permanently, as seems inevitable, how's for the
advertising agencies who handle these programs
applying a little ingenuity to the writing of better
continuity? It could be made non-obnoxious.
In fact it could be made right entertaining. But
it would take a lot more labor than it is now
accorded. If one sixth of the time in an advertising hour is to be turned over to an announcer
we do not see why he should not be supplied with
text written by a first rate copy writer and at a
fee at least equal to one sixth of the cost of the
hired performers.
As a schooling for the prospective composer
of continuity we can think of nothing better
than a thorough study of the evolution of movie
subtitles. As we said a while back, radio continuity is in the same stage of development as
was the movie subtitle in the days of "Came
The Dawn."
The "Came The Dawn" subtitles have
largely disappeared and very excellently and
succinctly composed ones have supplanted them.
However, in the movies, it took fifteen years or
more. Let the radio advertisers profit by this
example and see if they can't cut down this fif-
teen-year wait by a month or two.
And now to present the broadcasting station's
point of view of the advertising program, which,
by coincidence, comes to our desk as we are writing this article:
Program features originating with the National Broadcasting Company fall into two main
classifications—'sponsored" features put on the

A POPULAR FEATURE AT WTAM, CLEVELAND
Guy Lombardo's Royal Canadians are among the most popular of the Middle-West radio dance
orchestras. Where the other instruments in the band are, we cannot say. The players at least, have
reported for duty

air under the auspices of commercial concerns for the purpose of building institutional good-will, and "sustaining" features, including broadcasts by the various National Broadcasting Company "stock" companies, educational and religious and musical programs of all kinds from hotels, night clubs, and prominent motion picture theatres.

The life of a sponsored feature really begins, so far as the whole personnel of WEAF and WJZ is concerned, when a contract has been made between the Company's Commerical Department and a commercial concern for the use of time on the air. Immediately, the machinery of the Program Department of the station involved starts to function.

The contract itself may specify what entertainers are to broadcast during the time allotted to the new feature, and in this case, the work of the Program Department is lessened. Usually, however, the Commercial Department, the new client, the Station Manager, and the Program Department will combine to decide upon the artists, leaving the working out of the details to the Program Department.

The period of planning may involve almost any amount of work. The elusive idea must be pursued and captured and a definite scheme of entertainment mapped out. In some cases, three or four complete plans are made. Conferences are held between the Program Department, the Commercial Department, and the sponsor and the type of entertainment is decided upon. The time at which the feature will appear on the air must also be decided, a process which involves many considerations. The station management must balance its entire program for the evening and make sure that every feature attains as much prominence as possible. In other words, a whole evening's program must be varied if it is to be effective. Two periods of the same sort of entertainment should not follow each other, or both of them will lose in effectiveness because of the fact.

When a plan has been approved, work is begun on detailed programs. Artists are engaged, a process which may require auditions attended by representatives of the various departments and by the sponsor. A continuity is prepared for the opening program and an announcer is chosen for the feature. The artists are given the detailed program in order that they may start rehearsing. In short, a sample program is prepared for presentation.

In preparing the continuity, care is taken that the program shall merely create good-will rather than describe the sponsor's products. The spoken portion of any sponsored feature should relate to the musical selections, if the entire program is to accomplish its object. The detailed program is submitted to the Department of Musical and Literary Research in order that all copyrights on the various selections may be investigated. In some cases, numbers are changed to comply with copyright restrictions.

When the sample program has been prepared, it is assembled as a unit for rehearsal at the studio. This rehearsal is attended by a Commercial Department representative, a member of the Program Department, and the sponsor. In instances which involve unusual pick-up problems, a member of the Operations and Engineering Department is also present to work out proper microphone placement and insure the best possible pick-up.

In the meantime, three other activities have been begun, looking forward to the time when the feature will first be heard on the air. The Traffic Department has communicated with the various stations through which the sponsor desires his program to be heard and has arranged for telephone facilities to carry the program to these stations.

The clerical force of the Program Department has prepared program material on the feature and forwarded it to the Publicity Department, so that proper announcement of the coming feature may be made.

The rehearsal at the studio is criticized by those who attend it and any desired changes are made in the program. Other rehearsals will take place before the initial broadcast of the series goes on the air, and rough spots in the presentation will be smoothed off. Shortly after the first rehearsal, however, the various departments which have helped to get the first program ready start to work on the second and third appearances of the feature. Detailed programs are made up and given to the artists three weeks in advance so that every detail of each presentation may be carefully worked out. The final step in the presentation of the first program takes place when it goes on the air. The broadcast is listened to by the Station Manager or his representatives, for he is really the stage manager of the station. In every case where contact occurs between various departments, printed forms are used to make sure that information is transmitted accurately. No details are left to memory or to oral agreement. Once the first program has been broadcast, a regular rehearsal schedule is maintained for further features in the series. Every broadcast must be rehearsed in the studio twice before it goes on the air, necessitating an elaborate schedule.

An Evening of Chain Programs

OCCASIONALLY there come to our ears certain distant mumblings and mutterings of rage against the "Big Radio Chain Monopoly." It is characteristic of the American to growl gutturally when confronted by anything exuding the faintest odor of monopoly. It is also characteristic of him to slough off his prejudices and form one himself the minute he gets a chance.

It may be that some of the epithets, such as "Radio Trust," "Un-American," "Capitalistic," etc., hurled at the big chain are not without justification. However, we do not propose to examine the facts of the case. For, frankly, we do not care. Our duty as radio reviewer requires us to make use, not of our vague recollections of Economics 51, nor of our theories of business ethics, but simply of our ears. And our

LOUIS KATZMAN
Leader of the Anglo-Persians. The Whittall Anglo-Persian Orchestra has been a regular and well-liked feature of the WEAF network pgograms for many months

ears find it good. As a "listener" we are little concerned with what goes on at the other end of this wireless transaction. Our concern is with the things that come out of the loud speaker and our special concern is with the things that are good.

So heark ye to the very pleasant evening we were enabled to put in on Friday, March 18, relying exclusively on chain broadcasts from three of the local (Chicago) stations:

WLIB—7 P. M. About the best brass band available to the radio audience, that of Edwin Franko Goldman, playing, with nice regard for the exactions of the microphone, such acceptable pieces as Tschaikowsky's "1812" Overture, the "Peer Gynt" suite, and the Procession of the Knights of the Holy Grail from Wagner's "Parsifal."

THE HERMANN TRIO AT WLW
This trio, composed of Emil Hermann, concert master of the Cincinnati Symphony Orchestra; Thomie Prewitt Williams, and Walter Hermann, is heard each Wednesday evening at 10, Eastern Standard Time. Indications are that this group of artists is one of the most popular heard from WLW

KYW—7:30 P. M. The Royal Hour featuring Helen Clark, the "Royal Heroine" and Charles Harrison the "Hero" in a series of solos and duets alternated with orchestral selections. This program was made up of popular songs of Arabian and Egyptian coloring, which same coloring, unauthentic though it may be, is responsible for one of the most ingratiating modes of popular composition. Such titles as "Sand Dunes," "Song of Araby," "There's Egypt in Your Dreamy Eyes," "Lady of the Nile" and "Africa."

WGN—8 P. M. The National Grand Opera Quartet composed of Zielinska, soprano, Nadworney, contralto, di Benedetto, tenor, and Ruisi, basso—all voices specially selected for their adaptability to broadcasting. They were able to sing the Rigoletto quartet so that it "came over" in assimilable form, without all the voices being blurred together like the kernels of over-cooked rice—a common enough fault with most radio renditions of this exacting and amazing composition.

KYW—8 P. M. Two Metropolitan stars, Mario Chamlee, tenor, and Florence Easton, soprano, and the violinist, Max Rosen, collaborating in a Brunswick Concert. To be sure, the program was made up of somewhat over-familiar compositions such as the Volga Boatmen's song, "Connais Tu Le Pays?" "Songs My Mother Taught Me," and so forth, but such is the practice of radio, and besides they were so well presented as to give new zest to the hearing. Florence Easton's is a good radio soprano voice.

WGN—9 P. M. Louis Katzman and his very good orchestra, the Whittall Anglo-Persians. Here was music well played, though again we might have been better pleased had the program been of less orthodox radio makeup. It included Thomas' "Mignon Overture," Sibelius' "Valse Triste," Gautier's "Le Secret," Dvorak's "Humoresque," and Jessel's "Parade of the Wooden Soldiers."

But all in all, we point out a most enjoyable and varied evening of music, and one which would require much jumping about in taxi-cabs, and the outlay of a pretty penny, to obtain first hand at the concert halls.

THUMB NAIL REVIEWS

WJZ—Announcing that, of the next three numbers to be played from the Commodore Hotel, the third was to be a brand new number from "Lucky," then only twenty-fours hours old on Broadway,"—a number which we particularly desired to hear. A miscalculation in time made necessary a return to the studio right in the middle of the second selection. Station WJZ

preacher and the other a female evangelist. If I could be drawn into a debate over the 'mike' of a neutral station I don't think the other stations in town would have 25 listeners apiece that night. The above mentioned each have powerful broadcasting stations. Draw your own conclusions." The which we print in hopes that our correspondent's suggestion may reach the "above mentioned" and start the fun.

WHAT CENSORS CENSOR

ON THE subject of the freedom of the air we note, since last writing, that complaint was registered by Representative Celler of New York that WEAF censored his address on George Washington of all statements that the father of his country liked his toddy, gambled, and in one of his Virginia campaigns, supplied rum and rum punch to the voters. Mr. Celler protested that "the radio management should permit discussion of the foibles of our great men if it adheres to the truth and is not phrased in a tone of disrespect. The things I proposed to say about Washington were not prepared with a view to belittling him. They are human qualities that detract none from his stature as a statesman or a man."

As a counterbalance to this censorship we have as exhibit B the extraordinary temerity evinced by KOA in allowing Judge Ben B. Lindsey to talk from its studio on companionate marriage, "the revolutionary theory that has scandalized society, brought down the wrath of the church, and set the whole world talking." We regret that interference kept us from hearing this decidedly controversial broadcast but we did "get" KOA the night before and heard advance information from director Freeman H. Talbot to the effect that: "Our station, like a magazine, will not assume responsibility for the principles advocated. But because of Judge Lindsey's prominence, because of his work in behalf of women and children, not only in the United States but in foreign lands, we believe he is entitled to an unbiased hearing."

Microphone Miscellany

BETWEEN the hours of eight and nine February 11, KFI, and ten other Pacific Coast stations presented what they termed an Interference Hour. The stations were paired off and so changed their wavelengths as to interfere seriously with one another.

After an hour of squeals, howls, indistinguishable announcements, and distorted music, the stipulated wave-lengths were resumed, following which pleas were made from each of the stations in support of the radio bill then before the senate.

18; MacDowell, 16; Puccini, 15; Strauss, 12; Saint-Saens, 9; Weber, 7; Rimsky-Korsakoff, 7; Gluck, 4; Moussorgsky, 4; Massenet, 3; Rossini, 2; Berlioz, 2; Mascagni, 1; and Leoncavallo, 0.1

URIEL Davis, who controls the destinies of more than a thousand musicians, organized in a hundred combinations, made some observations in The Billboard on the selection of musical numbers for different kinds of audiences, which should interest every ambitious program director. He points out that the numbers selected for resort orchestras, for provincial towns, and for principal cities, vary greatly. The best index found to audience tastes of any locality is through the kind of phonograph records which sell the best. The resort audiences want popular music but numbers which have had an opportunity over a period of months to become well known; the small town and country audiences, time-worn numbers, known to the trade as semistandard music; the big cities will not tolerate a number if the ink has had a chance to dry on the music. This may give a new angle to some program directors who have difficulty in accounting for the popularity of certain programs which nobody with whom they come in contact seems to like.

A DEPARTMENT of Commerce report states that the Tokio broadcasting station is soon to increase power to ten kilowatts. It states further that there are already 326,000 subscribers to the present one-kilowatt station. Eighty per cent. use crystal sets, although there is a strong demand now for much better vacuum tube receivers.

Correspondence

Los Angeles, Calif.
SIR:

Right here and now I want to protest your statement in the March RADIO BROADCAST. In the "Answers to questionnaires" story, you refer to the lack of feminine radio fans.

First, I am feminine,—and the most rabid kind of a fan,—tone, DX, quality programs, home-built, and factory-made sets. In fact, I have been completely "broke" buying radio parts ever since your magazine caused my downfall some eighteen months ago with a circuit. You see, the trouble was that the durn thing worked, otherwise I should have become discouraged and been saved! To date, I have built twenty-three sets, and lost hours of sleep trying them out.

But I believe that there is a reason; I have had no intolerant males to push me aside and let them show me how it is done. My trials and triumphs have been my own, and there have been considerably more trials than triumphs.

I'll make a wager with you: Let every woman

Problems of A. C. Filament Supply

Solving Some of the Difficulties Encountered in A. C. Operation of 201-A Type Tubes

By ROLAND F. BEERS

THE principles of series filament connections outlined in the first article of this group, published in the May RADIO BROADCAST, related the facts which pertained to 199 type radio tubes. It will be remembered that several typical circuits were discussed in detail, and two methods were outlined whereby the radio constructor could build up a receiver of this type.

In this, the second article of the series, the discussion will be extended to include 201-A type tubes, and data will be given regarding the theory of the power supply apparatus necessary to supply this type of tube.

The problem of supplying filament and plate current to 201-A tubes in series finds difficulties in the design and construction of the power unit rather than in the actual connection of the series filaments. There are three main problems which confront the home constructor of this type of unit. They relate to the transformer, rectifier, and filter circuit. In general, it is not possible to employ apparatus which is not specially designed for the circuit. The reasons for this statement will appear later.

In order to have an adequate understanding of the design of the power transformer, we must first consider the type of rectifier with which it is to be used. The requirements of the circuit demand a device of reasonably long life (say a minimum of 1000 hours), with uniform and stable characteristics throughout this period. The rectifier should also show good efficiency, particularly at full load. This quality is exceedingly desirable in order that the size and cost of the power transformer may be kept as small as possible. Early attempts of the writer to build a ½-ampere power supply unit led to the design of a transformer which weighed nearly 25 pounds because it was called upon to supply a great deal of power which was lost by the inefficient rectifiers used at the time. It was not until efficient high-current rectifiers were available that an economically satisfactory transformer could

be built. The efficiency of the rectifier also has considerable bearing on the design of the filter circuit, for the ease of filtering a given amount of power is dependent directly upon the voltage rectifying efficiency of the rectifier.

A third quality which must be considered in the design of a high-current rectifier is the voltage regulation which it contributes to the entire circuit. By this we mean the change in output voltage for either a change in load or a change in line voltage. What would be most desirable, of course, is a constant voltage output for all loads

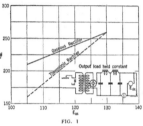

FIG. 1

and for any line voltage. This is not entirely possible, on account of losses in the filter circuit which cannot be completely offset, but it is possible to achieve a great improvement in regulation over that of many existing power units. This improvement is to be found in the design of the rectifier, as will be shown from the following discussion.

Let us first consider the thermionic rectifier. The performance of a device of this type is so well known that only those features will be considered that are important when using a rectifier

to obtain the high currents necessary to supply a 201-A filament. In the first place, a thermionic rectifier has a point of temperature saturation. Temperature saturation is that state which occurs when the plate draws the entire electron emission from the filament and under such conditions a further increase in applied plate voltage produces no increase in plate current. The effect of this characteristic is to limit the amount of current that can be taken from a power unit, and it is essential that in ordinary operation no attempt be made to exceed this limit.

Line voltage variations assume relatively major importance with respect to the regulation of a power supply unit using a thermionic rectifier, by virtue of this temperature saturation effect. For example, if the rectifier were to be entirely independent of line voltage for its filament supply, a 10 per cent. decrease in line voltage would produce a certain decrease in d. c. output voltage, but if the rectifier filament were entirely dependent on the line voltage, it too would drop 10 per cent. and there would be an additional falling off in d. c. output. The total loss in output would then be the sum of those two separate effects. The curve in Fig. 1, marked "Thermionic Rectifier," shows how the output voltage of such a unit varies with different input voltages across the primary of the power transformer.

A gaseous rectifier contains no filament and therefore this double effect upon lowering the input voltage is not noticed. As a result, the regulation for the gaseous rectifier is much better than for the thermionic rectifier.

There is yet another effect inherent in a thermionic rectifier which also affects the regulation of a power unit in which it may be incorporated, i. e., the voltage lost in a rectifier increases with increases in load, as shown by curve A, in Fig. 2. The voltage loss constantly increases with increasing current, and at the same time those losses in the filter circuit increase in the same manner, and the overall effect of the tube and the filter is shown by curve A in Fig. 3.

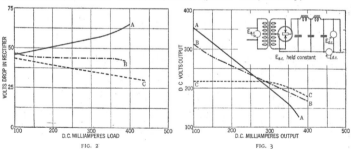

FIG. 2

FIG. 3

It is possible to design a gaseous rectifier which will have a nearly constant voltage drop throughout the entire range of its load. Such a device will have a characteristic as shown in "B," Fig. 2. This rectifier would neither add to nor subtract from the overall regulation of a power unit, and the curve of this type of unit is shown at "B," Fig. 3. It can be seen that the improvement gained is considerable. However, a still greater benefit is available from the rectifier design. It is possible to construct a device of this type which will have a negative regulation characteristic such as that shown by "C," Fig. 2, in which the voltage loss in the tube decreases with increases in load. The advantages of a device of this type are at once apparent. If the negative slope of this line can be made equal in magnitude, and opposite in sign to the positive slope of the filter circuit, the overall regulation of the power unit will be nearly a horizontal straight line such as "C," Fig. 3. The curves in Figs. 3 and 4 were taken using various rectifiers constructed by the author.

The question might be asked most logically: Why is regulation so desirable in a series filament receiver? The answer is tied up in several fine points, no one of which seems important by itself, but their accumulated effect is great. In the first place, consider the result of removing a tube from its socket in a series filament receiver. The current load on the power unit drops from—say,

FIG. 4

A FILTER FOR 350 MA.

THE design of a filter circuit for 350 ma. presented a large number of problems at the outset. The main question was how to obtain satisfactory and reasonably proportioned filtering in a circuit. The first attempts along this line resulted in filter circuits weighing nearly 50 pounds, and while a device of this size might be acceptable to some set builders, it certainly would not meet universal approval. A study was therefore made to determine the smallest weight of filter circuit which would give satisfactory quality.

The first problem of design was that of the filter chokes. Very little data were available on the performances of inductances under high d. c. loads. It was known in a general way that the inductance of a given choke coil decreased as the d. c. through it was increased. The starting point of the design was therefore to determine the actual relations which obtain in high-current choke coils. Several different models were made up having different sized cores, a different number of turns, and adjustable air gaps. Fig. 4 shows the performance of various models constructed by the writer in the development of a high-current choke coil. The measurements of inductance were taken at the Massachusetts Institute of Technology. By the variation of different factors, a set of curves was finally evolved whereby a design for a given value of inductance at a given d. c. load could be determined. The curves of Fig. 5 show why it is not possible to employ ordinary filter circuits for 350 ma d. c. load. As the direct current through the windings of a choke coil is increased, the core becomes more saturated, until the effective permeability becomes very low. At extreme values of current saturation, the permeability approaches that of air. This portion of the saturation curve is shown by the dropping off of the inductance curves. Curves 1 and 2 show how ordinary B power-supply chokes lose their inductance with increasing d. c. load.

The second reason why ordinary filter circuits cannot be used at high 350 mA. load is that there is not sufficient condenser capacity in which to store up energy. The rate at which energy is being taken from the filter circuit is proportional to the current output. It is therefore apparent that

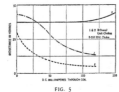

FIG. 5

350 milliamperes to less than 20 milliamperes. In a unit such as that represented by "A," Fig. 3, there is immediately an enormous increase in the output voltage. This high voltage is extremely hazardous to the filter condensers and may puncture any of them. Therefore, in order to build a reliable power unit of this type a very costly condenser must be employed. If our power unit has the characteristics shown by "C," Fig. 4, it is readily seen that the no-load voltage on the condensers is not excessive. There is also much less danger to associated apparatus. The cost of construction is therefore lowered.

Further considerations for good regulation lie in the design of the filter circuit. If the rectifier has a characteristic such as that shown by "C," Fig. 2, it is possible to match this curve by the positive slope of the filter circuit regulation. There is the possibility of building a fairly small filter choke whose resistance would ordinarily be too great but which would give satisfactory regulation in connection with the improved type of rectifier. The fact that the rectifier has less voltage loss at high current than at low current contributes a considerable gain to the ease of filtering at high loads. All of these factors greatly assist in reducing the ultimate cost of filtering.

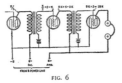

FIG. 6

more capacity will be required for high-current filter circuits than for low-current ones. The result of using insufficient capacity is an increase in the hum.

The magnitude of this hum depends very largely upon the amplification characteristics of the radio receiver. Poor quality sets which give very little amplification at low frequencies can be operated from a power unit which has a large ripple in the output, without giving objectionable hum. As the amplification of low frequencies is increased, however, it is necessary that the ripple in the power unit be reduced to not more than 0.1 per cent.

One very bad effect which will be produced by the lack of sufficient capacity in the filter circuit is the tendency to audio regeneration in the receiver. From the standpoint of the power supply unit, this may come from two separate causes. The first cause is that the terminal capacity on the filter circuit is too small. If this is the case, it is possible that the effective impedance of the filter in the plate circuit of the amplifier tubes will be very large. It is, of course, well known that the impedance of a condenser, such as the last one in the filter circuit, increases as the frequency decreases. At ordinary audio frequencies this impedance may be a very few ohms, and in this case its effect on reproduction is negligible. If the amplifier is slightly unstable, it is possible that an incipient oscillation of from 2 to 10 cycles per second might arise in one stage, and being coupled to the other stages through the filter condenser, it might cause trouble. At this low frequency, the effective impedance of the filter condenser would be very large, and would offer sufficient coupling between the amplifier stages to induce sustained oscillation. This effect is commonly known as "motor-boating," and its remedy is frequently found in an increase in the terminal capacity of the filter circuit.

A second cause for this type of regeneration is found in the presence of a large a. c. ripple in the output of the power unit. If this variation exceeds 0.1 per cent. it can readily be shown from Fig. 6 that the amplification of the hum from the detector plate to the power amplifier stage would result in an enormous a. c. grid voltage.

Assume the plate voltage on the amplifier tubes is 100 volts and that the plate voltage on the detector is 50 volts. Also assume that the ripple in the power unit has a frequency of 120 cycles and that the transformer amplification at this frequency is 2. The amplification in the tube would be 8. If K equals the percentage ripple in the B supply source then:

$$\text{A. C. voltage} = K \times \text{D. C. volts}$$
$$\text{Detector} = \frac{K}{100} \times 50 = \frac{K}{2} \text{ volts}$$
$$\text{Amplifiers} = \frac{K}{100} \times 100 = K \text{ volts}$$

If we then calculate the a. c. voltage on the grid of the power tube at various percentages of ripple in the power supply unit we obtain the following values:—

K%	A. C. hum voltage on power amplifier grid
.05	0.9
.10	1.8
.20	3.6
.30	5.4
.40	7.2
.50	9.0
.60	10.8
.70	12.6
.80	14.4
.90	16.2
1.00	18.0
2.00	36.0

If the a. c. hum voltage is greater than the normal C bias on the power amplifier grid at any part of the cycle, there would result a great increase in the plate current of this tube. This increase would occur periodically, and frequently would attain such a magnitude that the output condenser of the filter would be almost completely discharged. It would result in a failure of voltage at the terminals of the power-supply unit, until the filter circuit could fill up again with energy.

The remedy for this situation is to improve the quality of filtering or to increase the negative bias on the power amplifier tube. The latter remedy sacrifices some amplification, and for best results it is therefore recommended to add capacity to the filter circuit until the difficulty is stopped. The subject of "motor-boating" is given such consideration here because it gives much more trouble in series filament receivers than is customary, The reason for this is that the failure of voltage at the power supply output caused by "motor-boating" results not only in decreased plate potential, but also in a dropping of the filament temperature. The addition of these two separate effects frequently stops the operation of the receiver entirely, because they are both accumulative in the same direction. One general method of overcoming the difficulty just mentioned is to isolate troublesome stages from the common power-supply source. This can be very well done by the use of choke coils and bypass condensers, which are placed in the plate circuit as shown in Fig. 7. These units effectively restrain the alternating currents to their proper paths and prevent mutual coupling in the impedance of the power supply unit.

When we actually connect up a radio receiver with 201-A tubes in series there are not more than two or three principles which must be remembered. The first of these is the manner in which grid bias is to be obtained, and the second is the order in which the tubes shall be arranged in sequence. As a matter of secondary importance there is a possibility that it will be necessary to place shunt resistances around certain of the filaments in series, in order to limit the current through them to safe amounts. .

A method of obtaining grid bias is to place in the filaments, series resistances of the proper value, whose voltage drop will give the required grid bias. The value of resistance depends upon the amount of bias required, and is equal to the required voltage multiplied by 4. For example—if 4.5 volts grid bias is desired, this is obtained

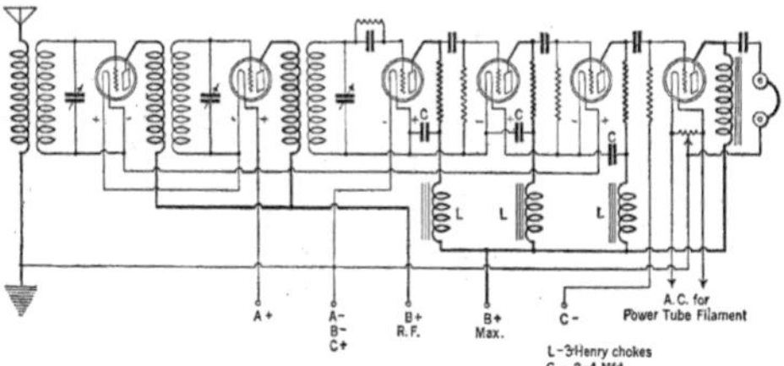

by a resistance of 4.5 x 4, or 18 ohms. This is placed in the circuit as shown in Fig. 8. and the grid return connection is made to point A. These biasing resistors carry the entire filament current, or 0.25 ampere. One convenient form in which they are obtainable is the customary 20-ohm rheostat, which can be inserted in the filament line as shown, and then adjusted to any required value. It is advisable to incorporate bypass condensers in this circuit.

The order in which the tubes are arranged has been discussed to some extent in the first article of this series. The main point to be remembered, of course, is the location of the detector filament nearest the B-minus or ground point. After that may come in order the first a. f. and second a. f. followed by the radio-frequency stages. Aside from the actual effect of a. c. fields on the performance of the receiver, another point to be remembered is the amount of effective plate voltage required in the various stages. It is apparent at once that those radio tubes which are nearest the A-plus terminal have a lower effective plate voltage than those at the negative end of the series. Very frequently increased amplification can be obtained by choosing the location of certain stages, so that they will receive the optimum value of plate potential.

As a matter of general procedure, it is not ordinarily necessary to provide filament shunt resistances for 201-A tubes in series. The writer has used as many as six of these tubes in series without bypass shunts, as a protection from plate-current overload. It was necessary, of course, to limit the plate current in the amplifier

stages to moderate values by the use of proper C-bias voltages. If more than six 201-A tubes are connected in series, or if the total plate current consumed by a receiver of this type is more than, say 35 mA., it will be advisable to use the method described in the preceding article for protecting the last two tubes in the series, which consists of shunting the filaments with resistances of such a value that the current through the filaments is reduced to normal.

The proper value of protective resistance is found by dividing the normal filament voltage by the total plate current of the preceding tubes. For example, if the total plate current at the 6th tube in a series of 8 is found to be 40 mA., it is apparent that the total current in the 7th filament will equal 250 plus 40, or 290 mA. This represents an overload of 40 mA., which must then be bypassed through a resistor across the 7th filament. The value of this resistor will be equal to 5 divided by 0.04, or 125 ohms.

In the next article of this series the constructor will be shown how to build up an A, B, C power unit incorporating the principles previously discussed. It will also include a description of the new Raytheon BA 350-milliampere rectifier, which is a full-wave gaseous rectifier operating without a filament.

The characteristics of this rectifier have made possible the development and construction of an efficient and highly satisfactory power unit for 201-A tubes in series. This unit supplies complete radio power for any type of modern receiver, and several popular circuits will be discussed at some length.

FIG. 7

L – 3-Henry chokes
C – 2-4 Mfd.

A. C. for
Power Tube Filament

A+   A–   B+   B+   C–
     B–   R.F. Max.
     C+

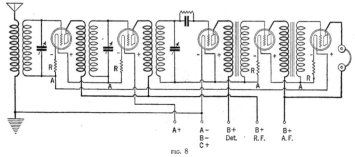

A+   A–   B+   B+   B+
     B–   Det. R.F. A.F.
     C+

FIG. 8

How to Design a Loop Antenna

Details for Using the Accompanying Chart Which Makes Difficult Mathematical Calculations Unnecessary—Properties Governing Loop Efficiency

By HOMER S. DAVIS

THE loop antenna has been in use for many years and of course needs no introduction. For the broadcast listener it has a number of very desirable characteristics. Being compact and self-contained, it permits the placing of the receiver in any desirable part of the home, whereas the usual antenna and ground connections often make this location fall in a most unseemly place. The inherent selectivity of the loop is contributed to by its directional properties; that is, it receives best from directions parallel to its plane, and very poorly, or not at all, from directions at right angles to it. An interfering station, such as a powerful local, or one on a neighboring wavelength to the desired station, may often be thus eliminated by turning the loop at right angles to it. The loop is also generally regarded to be less susceptible to static and other atmospherics.

The outstanding disadvantage of the loop is that its effectiveness as an antenna is, comparatively speaking, small, usually only a very low percentage of that of the average outdoor type. Its use thus requires a very sensitive multi-tube receiver, such as the super-heterodyne. But this objection seems about to be overcome with the development of high-gain tuned radio-frequency amplifiers, and of many battery eliminators which are making the multi-tube receiver

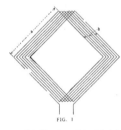

FIG. 1

economical. Thus the popularity of the loop antenna is steadily growing.

The design of a good loop is largely a matter of compromise between a number of variable and often conflicting factors. It has been found that, with all other conditions remaining the same, the received current in a loop is greater. (1)the larger the number of turns of wire used, (2) the larger the area enclosed by the loop, and (3) the greater its inductance. The required inductance of the loop is determined by the capacity of its tuning condenser; the area or physical size is limited by available space, or good appearance in the home. The necessary number of turns of wire to obtain the required inductance depends upon the spacing of the turns.

As in the design of the inductance coils within the receiver, the distributed capacity of the loop

should be kept as low as possible. This capacity increases with the number of turns, and is at a maximum when the turns are close together, but decreases rapidly as the wires are separated. Spacing the turns not only reduces the distributed capacity but enables a larger number of them to be used for a given inductance, thus adding to the received current. Distributed capacity of a loop is also increased by two factors seldom of importance with small inductance coils. One of these applies to the lead wires to the set, which should be kept short and separated. The other factor is that a loop is an arrangement of wires above the ground, and therefore forms a condenser with the ground, and also with various parts of the receiver. This tends to limit the highest frequency (lowest wavelength) to which the loop will respond, and there is little that the builder can do about it.

Two forms of loops are in common use, the "box," or single-layer square type, Fig. 1, and the "spiral," or flat square type, Fig. 2. The spiral loop is the simplest and cheapest to construct, but is less desirable since the inner turns rapidly become less useful as the area diminishes. The first step in designing a loop antenna is to decide upon its physical proportions, keeping in mind the factors mentioned in the discussion above. A good idea of present practice may be had in a visit to the retail shops. Having chosen the most desirable size of the loop, and the size of the variable condenser for tuning, the next question is the necessary number of turns of wire to wind upon it. On page 264 of Bureau of Standards Circular No. 74 is given the formula:

$$L = 0.008\, an^2 \left[2.303 \log_{10} \frac{a}{b} + 0.726 + 0.2231 \frac{b}{a}\right] - 0.008\, an\,[A + B]$$

Even for one having the mathematical ability to use this formula, it is not possible to solve directly for the number of turns. But for use in designing loops for broadcast receivers, a simplified approximate formula has been developed, and the accompanying alignment chart constructed from it.

USING THE CHART

THE drawing of two straight lines with a pencil and ruler is all that is required to use this chart. It consists of four numbered scales. Scale "a" represents the effective side of the square; in the case of the box loop, this is the actual length of a side, as indicated in Fig. 1; for the spiral type, the average length should be used, as in Fig. 2, since the inner turns are shorter. The effective breadth of the winding, scale "b," is a little greater than the actual breadth indicated as "b" in Figs. 1 and 2, overlapping each side by half the space between wires. Scale "c" represents the capacity of the tuning condenser, while scale "n" is the number of turns of wire.

The procedure is best illustrated by working out one or two examples. Suppose a box loop is to be built with 18" sides, for use with a 500-mmfd. (0.0005-mfd.) tuning condenser, and a 6" effective breadth of winding is decided upon. The key at the bottom of the chart indicates

which scales are to be connected together. Draw a line, therefore, from 500 on the "c" scale to 18 on the "a" scale. Then, through the intersection of the first line with the Index Line, draw a second line from 6 on the "b" scale across to the "n" scale, and you will read 16 as the number of turns of wire required. To take another typical example, it may be necessary to know the size of tuning condenser to use with a certain spiral loop of 18 turns, with 20" sides at the rim and an effective breadth of 4". The average length of a side is then 16". According to the key, connect 18 on the "n" scale with 4 on the "b" scale, then draw another line from 16 on the "a" scale through the intersection until it crosses the "c" scale, reading 350 mmfd. (0.00035 mfd.) as the required capacity of the tuning condenser.

Care should be taken always to connect the proper scales as shown by the key. The chart is an approximation of the complicated formula reproduced elsewhere on this page and its accuracy is 5 per cent. or better, when used with a loop whose side is not more than about five times greater than its breadth.

In choosing the size of tuning condenser to use with a loop, it should be borne in mind that the

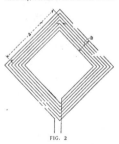

FIG. 2

distributed capacity of the latter will usually be found to be comparatively large, and a large condenser is desirable since it has a greater ratio of maximum to minimum capacity. With the smaller sizes, difficulty is likely to be experienced in reaching the higher frequencies (lower wavelengths). A capacity of 500 mmfd. (0.0005 mfd.) is usually regarded as most satisfactory, and has been adopted as standard by most loop manufacturers.

Rectangular loops of artistic design have recently appeared, but the formula for this type is nearly impossible to chart. If the reader desires to build a rectangular loop, the chart for square loops may be utilized as a guide, first using it for the shorter side, then for the longer. The number of turns actually required will lie somewhere between these limits.

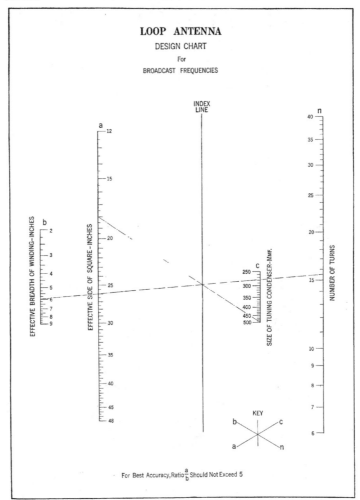

LOOP ANTENNA

DESIGN CHART

For

BROADCAST FREQUENCIES

AS THE BROADCASTER SEES IT
by CARL DREHER

Drawings by Franklyn F. Stratford

The Difficult Business of Running a Broadcasting Station

EVERYBODY wants to run a radio broadcasting station. The general public, not having access to one, must do without this luxury, but among the fortunate beings who happen to be employed in the broadcasting business, from the owners of stations down to the functionaries who have dedicated their lives to folding and unfolding the camp-chairs in the studios—too many want to be broadcast admirals. There are considerable numbers of badly managed stations at which a tug-of-war, more or less violent, is constantly in progress, with the proprietors, program people, and technicians all trying to make final decisions and dominate the works. In general, this condition obtains at the smaller stations.

As a general rule, no one of the above groups is competent to run a broadcast station unaided. If they were, considerable money could be saved. The effects of undue domination by any of the functionaries will now be set down, and the audience is politely requested not to throw pop-bottles.

The owner of a broadcasting station might be thought a safe person to entrust with its fortunes, because of his monetary interest in the establishment, but it does not necessarily work out that way. I could offer in evidence an actual station, member of the United States Supreme Court or of Mr. Ziegfeld's chorus who will apply in person. My duties led me, recently, to a point within such a short distance of this transmitter that I was able to judge the technical quality of its emissions accurately. It was terrible. The difficulty seemed to be a harrowing loss of the high frequencies, the cut-off, apparently, being lower than that of a commercial telephone circuit. Speech was barely intelligible, and a jazz orchestra sounded like an organ. I thought the set had been built by the local tinsmith, but my informants told me that it was a 1 kw. product of a nationally known manufacturer, practically new, and secured at a cost of something like $30,000. After a while I got the story. The proprietor of the outfit, it appears, had objected to the microphone hiss —the "blow," as they call it in that neighborhood. The operator thereupon instituted experiments to eliminate this disturbance. His labors culminated in the connection of a 0.5-mfd. condenser *across the line between the output of the 5-watt amplifier and the input of the 50-watt*

stage, these being 500-ohm circuits. The microphone hiss disappeared as by magic, and so did all other frequencies above 1800 or so. (The reactance of a 0.5-mfd. condenser is about 340 ohms at 1000 cycles; 170 at 2000; 85 at 4000, etc.) The operator had some vague idea of what he was doing, but, as he had got rid of the hiss, he let the owner listen to the results. The owner was pleased. He declared that the music sounded "mellow." So it does; if "mellow" means absence of high frequencies, then this is the mellowest station in the world.

Possibly the proprietor has an ear unusually rich in subjective harmonics, or his receiving set may be so high-pitched that the combination of drummy broadcasting station and tinny receiver is fairly flat. At any rate, the station continues to run with the 0.5-mfd. bypass. The operator, by now, would take it off, but the owner likes it, so it stays on. As a crusader for good broadcasting, I should be pleased to dip owner and operator in a strong saline solution, applying 1000 volts to the operator and 1500 to the owner while they are still wet. The laws forbid this punishment, so between the two of them they continue to ruin their station.

The case cited has no direct bearing on the question of how to arrange the internal relations of a broadcast station so that the best results may be secured on the air. All that it shows is that, if the owner-manager and technician are both nitwits, as far as broadcasting goes, the station will be a peanut-roaster, whether the two work together or cut each other's throats. But, even if there is a considerable amount of brains

THEY ALL WANT TO BE BROADCAST ADMIRALS

in one or the other division, maximum efficiency cannot be attained unless there is a sound division of labor and power between the groups concerned.

In broadcast stations, as in other technical fields, there is frequently more or less of a gap between the technical and non-technical forces. This condition is only partly remediable. The fact is that most technical enterprises are run—and probably it is for the best—by laymen. The specialists and managers start with a totally different training and background. At some stations, they hardly seem to realize that the aim they have in common is to make the business run. It is admittedly difficult for the layman to be patient with the technician in certain situations. Technical development, for example, is tedious, costly, and apt to interfere with operation. Only the man who knows all the obstacles and how they must be overcome, one by one, can judge how well or how badly the job is being done. At a given stage in the evolution of an art only a partial solution may be possible. In 1921, the production of an acoustically excellent broadcast receiver was impossible, because several of the essential elements were still lacking. These components had to be worked out, in several special fields, before they could be combined by still another group of workers. Not only technical facilities were lacking, but also knowledge. It takes time to secure both the material and the immaterial elements. The non-technical manager often travels with the idea that all technology is built up on exact ideas, and that if he could only find the man who had the knowledge required, all would be well. Sometimes this is true, but as a matter of fact every branch of industry is constantly pushing out against a circumference of dubious knowledge, or downright ignorance; this is particularly true of radio communication and the arts of visual and acoustic reproduction. Again, while the modern engineer is forced to specialize, his work frequently carries him into fields where a broader training would stand him in good stead. He may be working out a problem in auditorium acoustics one day and wrestling with hydraulic equations the next. Some motion is lost with each change, but there is no help for that as long as we want to retain our present economic and technological structure. All these things make co-operation between technical and lay workers somewhat difficult, and where they are forced into intimate relations, as in a broadcast station, the fur is apt to fly. The managerial or program group presses for results, often unintelligently; the mechanical forces feel oppressed and inarticulate in the face of people who have power over them and do not speak their language. The solution lies in intelligent executive direction. The successful executive gives the engineers a free hand in their operations, within the economic limits of the enterprise, and yet exerts a constant, gentle pressure for results. The greatest weakness of technical men is their penchant for getting lost in the mazes of technical endeavor; they lose sight of the goal which they are paid to work for. At such times they must be firmly grasped by the bridle and led back to the path. But they do not require a check-rein every day of the year, and, as far as broadcasting specifica ly is concerned, one never sees a station functioning in healthy fashion unless the technical group is given latitude and power in their end of the business. If the special knowledge of the mechanical experts is ignored, the station is always pretty sick.

The program people suffer from similar disabilities. Program managers, like telephone operators and elevator runners, rarely receive praise for their efforts. Everybody thinks he could put on a better show, given the same money to spend. It is sometimes embarrassing when the owner of the establishment thinks so. The owner of a broadcasting station, also, takes a much greater interest in it than he might in some other business. The proprietor of a great merchandising enterprise, for example, will sometimes spend most of his time on his broadcasting station, even though it is a small proposition compared to other divisions of his business. This disproportion arises from the fact that broadcasting is more interesting, and richer in publicity, than other branches of commerce. It is a relatively romantic field of endeavor, and hard-boiled business men have a streak of romance in them, whether they admit it or not. Thus the employees of a broadcasting station, even when the boss is a man of great means and wide interests, generally work very much under his eye. This has advantages and disadvantages. They are subject to all sorts of caprice and unpredictable reaction. The technicians can always find some sort of refuge in their mechanical jargon, but the program people are pitiably exposed. Yet there is some utility in the situation. The object of the station, after all, is to please the public, and the owner is apt to be a pretty good representative of public taste. He may have more money than the average, but his general psychology, prejudices, emotional interest, and fears, are usually basically the same as those of the bulk of the audience. Broadcast entertainment falls within the wide latitude of middle class life, and a successful business man is generally well within the borders of middle class culture. If he owns a broadcasting station, and the programs make his high blood pressure worse, he is probably right in canning the impresario responsible. When the programs suit the owner, they will probably suit the Smiths, Joneses, and Robinsons likewise, and that is the purpose for which the amperes chase each other up and down the antenna.

The fundamental difficulty with broadcasting is that it is complicated, very complicated. It is one of the most intricate sectors of the whole tangle of industrial life. Telephone and radio

engineers, musicians, program arrangers, advertising experts, salesmen, publicity men, executives, are thrown together helter-skelter and expected to coöperate in a common endeavor. It is a wonder that they work together as well as they do, and that the results are as good as they are.

Man and Modulation

MODULATION, in broadcasting, is a variation of electric carrier amplitude in accordance with the vibrations of sound. The result is the "intelligence-bearing side band"—one of the most important scientific creations of the twentieth century. But in its broad sense, "modulation" was known long before the days of radio telephony. Before radio phrases got into the vocabulary of the man in the street, modulation was any variation or inflection in tone, as in speech or music. The Greeks knew as much about modulation of sounds, or rather, they could modulate sounds as skilfully, as modern men. But they did not know how to generate or modulate electric waves, the advantage of which is that they can reach out to enormously greater distances than the unaided organs of speech.

"HARD-BOILED BUSINESS MEN HAVE A STREAK OF ROMANCE IN THEM"

Yet, while the purposeful generation of electric oscillations is a very recent invention, periodicity and radiation are among the fundamental processes of the universe. The cosmos is full of waves, short and long. The shortest and fastest of terrestrial origin which we know of now—if no faster ones are discovered in the interval between the writing and printing of this discussion, are the gamma rays of radium, which vibrate no less than 150 quintillion (150×10^{18}) times a second. The wavelength, therefore, must be of the order of 2×10^{-8} meters or, expressed in the Angstrom units preferred by learned scientists (an Angstrom unit is 10^{-10} meter)—about 200 Angstrom units. Relatively speaking—and we can't speak in any other way—this is quite short. It seems still shorter when compared to the length of the radiation from a 60-cycle circuit, which emits waves 5,000,000 meters, or

3,100 miles in length. (Inasmuch as radiation varies as the fourth power of the frequency, a 60-cycle circuit emits only feeble waves, but it does radiate. V. Bush has calculated that an antenna which radiates 100 kw. at a radio communication frequency of 30,000 cycles per second, will radiate only 1.6 microwatts if excited to the same potential at 60 cycles.) The ratio between the long wave which we have selected for purposes of illustration (60 cycles—5×10^6 meters) and the short gamma ray (150×10^{18} cycles—2×10^{-8} meters) is 2.5×10^{14}, a figure which only an astronomer could gaze at unmoved. And there are waves of all lengths between the shortest and the longest, many existing in nature and others producible by man, and therefore also existing in nature.

Man lives not by bread alone, but by radiation as well. Where would he be without the light and heat of the sun? He cannot even live without his share of ultra-violet rays. And recently the great spiral nebulae have been discovered to be sources of extremely short and penetrating X-rays, which, it is speculated, may have something to do with the origin and maintenance of life on the earth. "The primary radiation of the universe," according to Prof. J. H. Jeans, "is not visible light, but short-wave radiation of a hardness which would have seemed incredible at the beginning of the present century."

But, as found in nature, these electric waves carry only their own vibratory energy. It was left for man to generate sustained waves which he could modify intelligently, so that they would bear through space the murmurs of his speech and music, so transient and yet so important to this strange creature endowed with consciousness. The energy of these waves is insignificant. Were it abstracted from the sum total of energy in the universe, or even on this little earth, its absence, as energy, would not be noticeable. It is important only because it is a carrier of human emotion, which, ultimately, in endlessly varying forms, is what man lives for. It would avail him little if the blind energy of the universe, streaming through all living creatures, kept him alive as if he were a plant or a simple, satisfied animal. He would be bored to death. The skill he has developed in the art of modulation is one of his ways of seeking refuge from that fate.

Technical Problem. No. 1

EACH issue hereafter will contain a practical broadcast problem, involving some calculation, on which professional broadcast technicians can test their skill. The question will be stated, and the same issue will contain a full solution, confirming the answer reached by those readers who handled the problem successfully, and showing the others how they should have gone about it. The problem for this month (See Fig. 1) is the following:

A carbon transmitter M_1, whose output is 30 TU's down, is used in the studio of a broadcasting station, feeding the first stage of a series of amplifiers leading to the modulators. At a point 100 miles distant another transmitter, M_2, 20 TU's down, feeds a remote control amplifier which has a power amplification of 10,000 times. The line is largely open wire and has a loss of 12 TU's in all. What size of artificial line should be used at the control room termination of the 100-mile circuit to bring its level with the output of

FIG. 1

the studio microphone, assuming all impedances to be matched so that the line and studio microphone feed the first control room stage of amplification with equal efficiency?

(Solution on page 109)

Design and Operation of Broadcasting Stations

16. Studio Design

A LARGE part of the material in this discussion is taken from Bureau of Standards Circular No. 300, *Architectural Acoustics*, obtainable from the Superintendent of Documents, Government Printing Office, Washington, D. C., at five cents a copy. This pamphlet in turn leans heavily, as do all works on the subject, on the investigations of the late Wallace Clement Sabine, now being carried on by P. E. Sabine. These studies are concerned with the acoustics of auditoriums, but the results are adaptable to studios of broadcasting stations on the basis that the best reverberation time for electrical reproduction is about half that of an auditorium of the same volume where the performance is intended only for the audience physically present.

The principal acoustic characteristic of a room is its reverberation time, which, as originally defined and measured by Sabine, was the time required for a sound a million times audibility to die down to just audibility, both expressed in power units. In other words, you take a sound of a certain intensity, and let it die down 60 TU (in telephone terminology) in an enclosure, and the time you have to wait is the reverberation time of that enclosure. Actually, reverberation times are generally calculated, and their principal use is comparative and empirical. For example, you have a certain studio whose characteristics have been found to be satisfactory. Then, if you have occasion to design another studio, it is wise to calculate the reverberation time of the first room and duplicate it in the second, or to make such modifications as are indicated by past performance.

Sound may be reflected by the walls, floor, and ceiling of a room, or absorbed, depending on the materials of which the surfaces are composed, the frequency, etc. Generally, there is partial reflection and partial absorption. Open space is taken as the perfect absorbing material; thus, in a room, open windows are perfect absorbing areas. If a square foot of open window is taken as the unit of perfect absorption, various coefficients may be applied to square feet of wall materials to indicate their power of absorption. A few such absorption coefficients are given below:

Brick wall, 18 inches thick	0.032
Carpets and rugs	.20
Cork tile	.03
Linoleum	.03
Hair felt, 1 inch thick, unpainted	.55
Hair felt, 2 inches thick, unpainted	.70
Plaster on tile	.025
Marble	.01
Varnished wood	.03
Human beings	.90
Chenille curtains	.23
Curtains, in heavy folds	0.5 to 1.00

In many cases these coefficients vary with frequency; theory indicates, in fact, that a flat characteristic for a homogeneous absorbing material is impossible. The coefficients given are supposed to be correct for 512 cycles per second, which is a good mean musical frequency, as 800 cycles per second is a good mean speech frequency for commercial telephone calculations. As would be expected, hard materials like marble, plaster, etc., absorb sound very little, whereas

curtains, felt, human beings, etc. have high absorbing power. This is borne out by practical observation; everyone has noticed the persistence of sound in empty rooms and auditoriums, especially where the walls are hard and unbroken, while when curtains, furniture, and persons are present, reverberation is reduced. As the volume of the enclosure increases, the reverberation time also increases. Given the volume of a room, and the areas and absorption coefficients of the materials used in its construction, the reverberation time may be approximately calculated by the empirical formula:

$$t = \frac{0.05 \, V}{A} \tag{1}$$

where t is the calculated reverberation time in seconds; V the volume of the room in cubic feet, and A the total absorption. We find A by taking the area of the various materials in the room and multiplying each by its coefficient of absorption, then adding the quantities so obtained. Obviously this assumes that the absorbing power of a surface depends only on its area and acoustic characteristics, and is independent of its position in the room. This is substantially true.

The application of the formula is most readily learned by the actual solution of a problem:

Given a room 20 by 15 by 10 feet, with plaster walls and ceiling, and a varnished hardwood floor, calculate the reverberation time, and apply acoustic treatment to reduce this period to an allowable value for broadcast purposes.

V equals 20 times 15 by 10, or 3000 cubic feet.

Let W represent the wall area, C the ceiling area, and F the floor area. Then we find in the above case that $W=700$ square feet, $C=300$ square feet, and $F=300$ square feet. A may now be calculated for the bare room:

Wall plus ceiling area (1000 sq. ft.) multiplied by the coefficient for plaster (0.025)	25
Floor (300 sq. ft.) multiplied by the coefficient for hardwood (0.03)	9
Total absorption (A)	34

Substituting in Formula (1) above, we have:

$$t = \frac{0.05 \, (3000)}{34} = 4.4 \text{ seconds}$$

This is much too high for a small room. Sabine found that good musical taste required a reverberation time of slightly over one second, for piano music, in a room of moderate size. Musicians check each other with fair accuracy in such determinations. Even a large concert hall will have an optimum reverberation time of between two and three seconds only. (The allowable reverberation time increases approximately as the first power of the linear dimensions of the room, for the same absorbing materials, since V in the numerator of formula (1) is a cube function of the linear dimensions, while A, in the denominator, varies as the square of the linear dimensions.) Hence it is necessary to apply additional absorption to the room we are considering, in order to reduce the reverberation time to about 0.5 second. This is on the theory that the optimum period would be about 1.0 second, for a room of this size, considering results in the room only, and that this quantity should be halved for our purpose of electrical reproduction at a distance.

Suppose, now, that the ceiling of our 20 by 15 by 10 room be covered with some acoustic board material, or hair felt under muslin, with a coefficient of, say, 0.5, and that the floor is carpeted, the walls remaining plaster. The total absorption, A, then becomes:

Ceiling area (300 sq. ft.) multiplied by 0.5	150.
Floor area (300 sq. ft.) multiplied by 0.2	60.
Wall area (700 sq. ft.) multiplied by 0.025	17.5
	227.5

Substituting in (1) above, we have:

$$t = \frac{0.05 \, (3000)}{227.5} = 0.66 \text{ seconds}$$

This is nearer a satisfactory value, and may give good results on the air. If, now, one of the 15 by 10 walls of the room, preferably that nearest the contemplated position of the microphone, be covered with curtains having an absorption coefficient of 0.5, the total absorption of the studio becomes 298.7, a value which the reader may readily verify by going through the simple computation himself. The reverberation time is then reduced to 0.5 second. In order to provide margin against reflecting surfaces of the piano, tables, etc., and to allow variation of the period to secure different effects, it might be well to drape two of the walls of the room with curtain material suspended from rods and movable at will, and to have the rug only partly covering the floor. This will permit variation of the reverberation period between, say, 0.3 and 1.2 seconds, which gives sufficient latitude for experiments. Various other combinations are of course possible, and each designer may work out schemes to suit his own requirements or fancies.

In selecting patented sound proofing materials it is well to try to secure a curve showing variation of the absorption coefficient with frequency. The nearer this characteristic approaches the ideal flatness between 100 and 5000 cycles, the more favorably it should be considered for general use. It is better to choose material with a moderate absorption coefficient which remains fairly constant over the audio range, than a highly selective surface which has high absorption at one or two pitches and reflects badly at other points.

Formula (1) above is not as accurate as a later equation worked out by P. E. Sabine, given by Crandall in his *Theory of Vibrating Systems and Sound* (D. Van Nostrand Co.):

$$t = \frac{0.0083 \, V \, (9.1 - \log_{10} A)}{A} \tag{2}$$

Formula (2) gives results for t slightly lower than (1), so that for electrical reproduction it is just as well to use the simpler expression (1), which gives additional margin against excessive reverberation.

Where lively surface materials are used in studio or auditorium design, a protection against echo and objectionable reverberation is to break up flat surfaces by recessing, coffering, etc., thus diffusing the reflected energy.

The circular of the Bureau of Standards referred to at the beginning of this article gives a table of allowable number of instruments in a room of given volume, for best artistic performance. This data should interest musical directors and broadcast technicians who habitually overcrowd their studios:

VOLUME OF ROOM	NUMBER OF INSTRUMENTS
50,000	10
100,000	20
200,000	30
500,000	60

While allowing 5000 cubic feet and over per instrument may be excessive, this table is probably nearer a sensible compromise than the comical broadcast practice of jamming thirty instruments into a room of 6000 cubic feet, allowing about 200 cubic feet per instrument. The greater absorption of broadcast studios certainly does not justify such acoustic atrocities.

Shrapnel from Cleveland

FROM Mr. J. D. Disbrow and Mr. Ross J. Plaisted, of the Engineering Department of WTAM, come separate roars of protest, addressed to the Editor of this journal, regarding my remarks "Concerning B. C. Operators" in the January issue. Mr. Disbrow issues a warning: "Your yearly subscriptions will certainly drop if you continue to print such 'poison' as Mr. Dreher has in his columns this month." He adds: "This article may be correct for some of the operators but it certainly doesn't say very much for the fellows who have spent their life at radio and worked like slaves putting radio where it is to-day. . . . The men behind the scenes get little enough praise as it is, so why rub it in with an unjust article like this one." Mr. Plaisted adopts the *argumentum ad hominem.* "It is entirely possible," he writes, "that the class of operators of Mr. Dreher's station is that of which he speaks, but I must say that the majority of those I have met in the middle-west take their business seriously and spare no effort in learning everything they can about it."

The position of the gentlemen from Cleveland would be tenable, it seems to me, only if *all* broadcast operators read the literature and took their work seriously. Then the article in question would be uncalled for. But both critics admit that some technical broadcasters are more or less delinquent in these respects. What, then, makes it immoral for me to remonstrate with this minority, on occasion? There was nothing in "Concerning B. C. Operators" which could be taken as an assertion that all, or a majority of technical broadcasters, neglect the intellectual foundation of their profession. Therefore there was no offense. Those whom the shoe does not fit are certainly not forced to put it on. Those whom it does fit might be benefited by wearing it for a time.

As for my general attitude toward my fellow workers, I do not think that anyone who has read even a small portion of my material can accuse me of any desire to disparage the men of my profession. It is true that I am not a habitual dispenser of perfumed vaseline; that rôle I leave to those who have talent for it. I could point to numerous passages in praise of people who deserve recognition, and as for the technical men, I have consistently maintained that their work is the foundation of the art, supporting the whole superstructure of program service and commercial returns. I need not stress this point; let the articles speak for themselves. On the other hand, I do not feel, as apparently Messrs. Disbrow and Plaisted do, that radio men constitute a fraternal organization, existing for mutual praise and support through thick and thin. Radio is a business, with its good features and bad, personnel competent and incompetent, its failures, successes, and hopes. These things I discuss from month to month, as entertainingly as I am able, and instructively when my knowledge permits. I try to learn from what other radio men say and write, and I value criticism insofar as it is based on logical considerations and a desire to get at the facts. That sort of criticism we need badly in radio; we have none too much of it. When it comes to emotional bias we have an ample supply, and it does not help us in any

of our problems, whether the question is a major one like the allocation of frequency bands, or a minor one like what should be printed in "As the Broadcaster Sees It." I invite Messrs. Disbrow and Plaisted to abandon what seems to me a singularly juvenile attitude. But on the basis I have suggested above, their opinions, as that of all other broadcasters and readers, are invariably welcome.

Memoirs of a Radio Engineer: *XVIII*

WE HAVE had to interrupt the publication of these yarns for several issues, but we shall now take up the narrative where we left off—with the entrance of the United States into the War, in April, 1917. The ink had scarcely dried on the official signatures to the Congressional resolution declaring war between the United States and Germany when the U. S. Navy, with a yell, clasped all domestic radio activities to its bosom and held them there for two years.

" . . . . I MADE NUMEROUS CALCULATIONS "

Also with a yell, my classmates and I, who were receiving the benefits of instruction at the College of the City of New York, escaped from the academic groves, which, in fact, never saw us more. After some negotiation, the Dean of the College offered the five of us in Doctor Goldsmith's radio engineering course credit for the balance of the term, and our degrees, if we would engage in some radio enterprise helpful to the U. S. Navy. This was in May, 1917, and in June we would have graduated anyway—we were Upper Seniors, by now—but there was the opportunity to avoid the final examinations, and, anyway, we wanted to get out. Doctor Goldsmith took charge of the preliminary arrangements, saw us placed, gave us his benediction, and the class scattered. Freed went to the Washington Navy Yard, where, about then, were also gathered Lester Jones (likewise a C. C. N. Y. man), Hazeltine, and Priess, none as yet famous. Buchbinder occupied himself at the Philadelphia Navy Yard, where Forbes, Ballantine, and Commander Lowell were working. Kayser stayed nearest home; the Brooklyn yard was his berth. Marsten and I were allocated to the Aldene factory of the Marconi Wireless Telegraph Company of America, staggering under the weight of government contracts and groaning—so we thought—for our aid.

A comical incident of this period shows the

atmosphere of suspicion, which assumed pathological intensity during the war, although in this particular case it was not unreasonable. The story begins in the radio laboratory of the College, before the United States entered the conflict. One of the requirements of the course was the writing of a thesis, which included the complete design of a radio station for a specific service. In my case I had the job of designing a transmitter, receiver, power plant, antenna system, and all auxiliaries for a vessel of the size and type of the SS. *Leviathan*, one of the requirements being that communication with land should be maintained throughout a transatlantic voyage. I made numerous calculations, most of them as mythical as the vessel, and got together a design after some months of work—it took this long because all the component parts had to be individually specified, the high tension transformer, for example, being detailed as to size of core, number of primary and secondary turns, gauge of wire, etc. I addressed a number of inquiries to manufacturers, including one to an antenna insulator factory in Brooklyn.

I do not recollect whether they answered or not. I completed my thesis, and left for the Aldene front. A few months later I received the following letter:

NAVY DEPARTMENT
United States Naval Communication
Service.

Office of District Communication Superintendent Third Naval District
Navy Yard, New York, N. Y.
July 26, 1917.

SIR:

1. This office has information that you desire to purchase Radio Antenna Insulators.

2. As all Radio Apparatus has been ordered dismantled by the Executive Order of the President, you are requested to forward to this office reasons why, and for what purpose, you desire insulators.

Respectfully,

J. C. LATHAM,
Lieutenant (JG), U. S. Navy,
District Communication Superintendent

I replied with equivalent grandeur, and I presume gave the D. C. S. a laugh.

Solution of Technical Problem No. 1

The amplification of the remote control amplifier fed by the microphone M_2 is given as 10,000 times in power units. The gain in TU given by the formula:

$$TU = 10 \log_{10} \frac{P_1}{P_2}$$

is found to be 40. As M_2's output is given as minus 20 TU, the output of the amplifier is +20 TU, which, incidentally, is quite a whale of a level to put on a telephone line. This being done, however, and the attenuation along the line being given in the problem as 12 TU, the level reaching the control room is +8 TU. This is to be reduced to the output level of the microphone transmitter M_1, which is —30 TU. The artificial line required for this purpose will have to drop 8+30 TU, or 38 TU. A 38-TU pad is therefore required.

As an aid in the solution of this problem the reader may consult the following past publications in this department:
Technical Operation of Broadcasting Stations, No. 10. "Calculation of 'Gain.'" September, 1926. Abstract of Technical Article: *The Transmission Unit and Telephone Transmission Reference Systems,* by W. H. Martin, *Journal of the American Institute of Electrical Engineers,* Vol. XLIII, No. 6, June, 1924; October, 1926.

Book Reviews

A Scholarly Contribution to the Art of Sound Reproduction—A Comprehensive Radio Dictionary

Not for the Kindergarten

THEORY OF VIBRATING SYSTEMS AND SOUND: *By Irving B. Crandall, Ph.D. Published by D. Van Nostrand Co., New York City. 272 pages. Price, $5.00.*

ALTHOUGH the reviewer considers this book an extremely valuable contribution to the art of sound reproduction, of which radio broadcasting is a part, it cannot be recommended as indispensable to every owner of a receiving set. Even the most sapient listeners, no matter how many sets they have built, and the most inveterate readers of the semi-technical press, no matter how many articles on flat amplifiers they have digested, will get into difficulties early in the volume. Let them but gaze at the list of symbols on Page 9, and they will recoil in fright from Lagrange's determinant of coefficients of motion, the kinematic viscosity, and Laplace's operator, the last denoted by a triangle standing on its apex, and squared. Doctor Crandall does not deceive them; he begins his preface with the statement: "This treatment of the theory of sound is intended for the student of physics who has given a certain amount of attention to analytical mechanics . . ." The material was first presented in one of the "out-of-hour" courses at the Bell Telephone Laboratories, where, since 1913, the author has specialized in the field of speech and various other phases of sound phenomena; a portion, also, was given as a course at the Massachusetts Institute of Technology during the spring of 1926. In other words, "The Theory of Vibrating Systems and Sound" is a serious work for the scholarly engineer and physicist, who has taken the trouble to arm himself with the requisite tools of analysis, and who realizes that the distinction between "theory" and "practice" is merely a convenient fiction for those who are unwilling or unable to think exactly in physical fundamentals. This book is theoretical and highly practical at the same time. It is not a handbook full of convenient tables which will enable one immediately to design a broadcast studio or a cone loud speaker, but there are problems at the ends of the chapters by which the student may test his progress in acoustic analysis, and one of the appendices consists of an invaluable summary of "Recent Developments in Applied Acoustics," containing references mainly to articles printed since 1922.

The book begins with a review of some of the fundamentals of mechanical theory applied to acoustics, such as the equations of motion of a simple vibrating system, free and forced oscillations, and the vibration of circular membranes in terms of Bessel's functions. Here the going becomes slightly heavy, and the reader who has not kept up with modern mathematical analysis will sink somewhat in his own esteem. He will find little relief in an analysis of air damping, in which the active portion of a condenser transmitter diaphragm is considered as an imaginary "equivalent piston," an assumption valid below the first natural frequency. This chapter is a good preparation for the experimental study of actual telephone diaphragms.

Systems of two degrees of freedom, in natural oscillation and the steady state, are described in considerable detail, references to the classical work of Rayleigh, Lamb, Heaviside, and others, being introduced, as throughout the book, to avoid tedious re-statement. Resonators are treated alone, and then coupled to a diaphragm. There follows the interesting problem of a loaded string—a stretched fibre weighted at equal intervals with equal masses, which is tied in with the highly important technical field of electrical networks.

The properties of the medium and the equations of wave motion carry the treatment into three dimensions, after a view of transmission in tubes and pipes, and resonance in such containers. Point and spherical sources, and the reaction of a viscous medium on a vibrating string, as in the Einthoven galvanometer, are some of the subjects.

In Chapter IV (Radiation and Transmission Problems) there is a brief non-mathematical interlude in the form of an outline of the underwater signalling experiments of Professors Pupin, Wills, and Morecroft, in 1918. It is not generally known that frequencies of the order of 50,000 cycles per second (which would correspond to a wavelength of 6000 meters in electromagnetic radiation) are employed in high-frequency submarine signalling. The analysis proceeds with other forms of radiation from a piston, into conical and exponential horns. Crandall cites Webster, Hanna, Slepian, and others in this section, but disagrees with one of the conclusions of Goldsmith and Minton relative to a comparison between exponential and conical horns. The analysis of the finite exponential horn is not sketchy; it covers eleven pages and a chart.

To a broadcaster, Chapter V, largely on "Architectural Acoustics," is naturally of greatest interest. The work of Wallace Clement Sabine is reviewed, with P. E. Sabine's additions. As it happened, while the reviewer was reading Crandall's book for this outline, he was treated to a swell luncheon by an experienced acoustic engineer. His practical ideas were sound, but when one of my colleagues inquired why some absorbing materials showed marked frequency selection, our host explained that it was because of the resonant action of the equal interstices between the fibres of the structure. I suspected at the time that this theory was not kosher, so to speak, but was too prudent to fight over the sweetbread patty. If you want the real explanation, take a day off and steep yourself in Section 51 and Appendix A of Crandall's book. There you will be led through the conception of the absorbing material as a honeycomb of narrow conduits in which sound waves are dissipated by frictional resistance, the impedance of the mouths of the conduits, the velocity of flow of air particles in the tubes, the resistance coefficients, the esoteric law of Poiseuille, kinetic viscosity, and calculations resulting in values for a certain absorbing wall of the absorption coefficient, showing how it varies with frequency. "If the absorption coefficient is plotted against frequency," writes Crandall, "a very good resonance curve is apparently obtained. This resemblance is evidently accidental, as no resonance

phenomenon has been implied in the problem we are considering." Several other facile assumptions are unobtrusively knocked into cocked hats in Doctor Crandall's text. Space limitations preclude extended comment on the general treatment of reverberation in three dimensions and reaction of the enclosure on the source of sound.

The book is written throughout in elegant and forceful scientific English. The object—to train the properly prepared student in the handling of the mathematical and physical principles involved—is never lost sight of. The reasoning is pellucid from cover to cover. Doctor Crandall has made a worthy contribution to an understanding of the advanced classical literature in his field, and brought the treatment in various subdivisions up to date with distinguished competence.

CARL DREHER.

A Radio Dictionary

THE RADIO ENCYCLOPEDIA: *By S. Gernsback. Published by Sidney Gernsback, New York City. 168 Pages. Price $2.00.*

THE "Radio Encyclopedia" is a comprehensive radio dictionary, clearly printed and, with the exception of the biographies, well illustrated. Messrs. Squier, Goldsmith, and Hogan we know to be much handsomer men than it would appear from their pictures in the encyclopedia, and probably there are others suffering similar injustices at the hands of the photographer, engraver, and printer.

The work will doubtless be useful to experimenters, particularly those who need occasional aid in the interpretation of technical terms used in newspaper and magazine articles on radio subjects. It is not sufficiently detailed for the broadcast or radio engineer and indeed, were it so, the work could hardly be simple enough for the broadcast enthusiast.

Such terms as "gain," "split announcing," "scrambling," "mixing panel," "vernier,"—the kind of thing a broadcast station employee might look up—are not defined, although the popular circuits, usual instruments used in the receiving set, electrical terms applying to broadcasting, etc., are fully and adequately defined. The cross reference notes with each definition are commendably complete and the work is obviously the product of meticulous labor. Since the book is primarily designed for enthusiasts, we regret the absence of definitions for a few words which occurred to us, such as "sound," "tone," "quality," "reproduction," "single-control," "blooper," "radiating receiver," "mike," and "condenser microphone."

A peculiarity of the indexing is that television is described under "television," but "telephotography" is listed neither under that term nor under "photographic transmission" nor "photographs, radio transmission of," but under "transmission of photographs by radio." Perhaps this is an outstanding exception to correct indexing and certainly one of minor importance. This the reviewer cannot judge because a reading of forty pages of material was considered sufficient to determine the general value of the work.

Terms, used in their correct and limited sense, will be found quite complete. The names of almost every conceivable receiving instrument and part, the more popular circuits, such as Cockaday, neutrodyne, Armstrong (Grimes and Roberts are not found under their names), and a scattering number of contemporary and historic biographies are included as the subject matter of Gernsback's dictionary.

EDGAR H. FELIX.

Analyzing the Power Amplifier

How Oscillograms Help to Indicate the Transient Voltages and Currents in the Circuits—A Radio Club of America Paper

By D. E. HARNETT

Engineering Dept., Pacent Electric Company

THE commercial need for high-quality reproduction has lately focused the attention, not only of engineers, but of the whole radio world, on the audio-frequency circuits of radio receivers. Selectivity and sensitivity for a number of years received much attention, but during that time, those engineers who realized how much better broadcast receiver reproduction could be, were trying to persuade the public to recognize good quality and to prefer it. Their efforts have finally met with success. The modern radio set has a much better amplifier than was the case a year and a half ago.

One of the lessons learned from the extensive experimenting on audio circuits is that it is necessary to have the loud speaker driven by a tube which can develop considerably more power than is required from the other tubes in the set. This means that the B-battery supply must furnish much more power to the output tube than was previously thought necessary. As a result, the B socket-power device, which operates from the a. c. house mains, has become very popular. The superiority of this type of socket-power device over the type which operates from some other source, such as a d. c. house-lighting circuit, is that is is possible to maintain the voltage by a suitable transformer at practically any desired value. The limiting factors are the safe operating voltage of the rectifier and the insulation in the device, particularly the insulation in the condensers. These factors are, of course, economic, for it is possible to build rectifying tubes for practically any voltages and to insulate properly for these voltages.

The device under discussion is the "Power-former." It is a combination power amplifier and B socket-power device. The device was developed by a group of engineers under the direction of Mr. L. G. Pacent.

It is desirable to have all that wiring which connects with a high-voltage circuit (such as the plate circuit of the power amplifier tube) confined within a metal case, in order to eliminate danger of injury to the broadcast listener. The rectifier tube is capable of rectifying considerably more current than is required for the power tube. This excess current is supplied to a potentiometer from which the plate potentials for the different tubes in the receiver are drawn. These different plate potentials—of the order of 150 volts, or less—are brought out to binding posts, which connect to the plate circuit of the radio receiver. The audio circuit connects through the jacks. The schematic diagram is shown in Fig. 1.

The operation of the separate parts of this circuit has been studied by a good many engineers and reported before the Radio Club of America and before the Institute of Radio Engineers. When, however, these different units are combined into one circuit, certain complications arise. It has been found that the actual

functioning of the different parts of this circuit is not generally understood.

The principal object of this device is to enable the broadcast listener to have good quality reproduction, with a fair amount of volume. In order to obtain this result, an amplifier tube must be employed which is capable of delivering about one watt without distortion. The audio transformers must be properly designed so that they will not distort the signal. This means that they must have a high primary inductance, a

FIG. 1

high mutual impedance, and a ratio which is somewhat lower than was thought best three or four years ago. With the use of ordinary magnetic materials, this necessitates a large core and a winding employing from two to three times the copper used in the average transformer of the past.

Consider first the rectifier circuit—here a 60-

60-cycle Transformer Normal Operation

Primary Current

Rectifier Wdg. Current

Primary Voltage

FIG. 2

cycle transformer is operated with the primary on a 110-volt lighting circuit. There are three other windings besides the primary. One of these operates the filament of the rectifier tube; the second, the filament of the amplifier tube; the third supplies the rectifier with high voltage, that is, it supplies the power for operating the plate circuits of the amplifier tube and those of the other tubes of the broadcast receiver. The current and voltages in the transformer are shown in Fig. 2. In Figs. 2 and 3, the arrow indicates the direction of travel in the film. In other words, the head of the arrow indicates an earlier time than the stem of the arrow. In the later photographs, the arrow indicates the direction of time; in other words, the meaning of the arrow is reversed. It will be noticed that the device is supplying the rectified current during approximately one third of the cycle; that is, instead of rectifying during all of the half cycle, which has the proper polarity for rectification, it is rectifying only during the portion of the cycle during which the transformer voltage exceeds the voltage across the first condenser in the filter circuit. Abrupt changes in the impedance of the transformer load introduce a large number of harmonics in the high-voltage winding which are reflected in the primary, giving the wave shape shown at the top of the photograph. Naturally, the direct-current component of this secondary current cannot be reflected through the transformer. This is shown by the fact that the area of the top half of the primary wave is equal to the area of the lower half. The power factor of this device is approximately 70 per cent.

$$\text{Power Factor} = \frac{\text{Power}}{EI}$$

The power is the product of primary voltage, the 60-cycle component of primary current, and the cosine of the angle between them. E is the primary voltage, and I is the r. m. s. value of primary current; that is, r. m. s. value of 60-cycle primary current and all of the harmonic currents. Thus there are two effects tending to decrease the power-factor—the phase difference between the primary voltage and current, and the presence of the harmonics in the primary. The current in the other two secondary windings is a sine wave working into a resistive load, that is, into the tube filaments. Approximately two thirds of the power drawn from the transformer is supplied to the rectifying circuit; the remaining one third is used to light the tube filaments. Fig. 3 shows the effect of the rectifier winding on the primary current much more clearly than does Fig. 2. In Fig. 2, the presence of the harmonics appearing in the primary current wave is partially obscured by the presence of the power current which supplies the power to the tube filaments. In Fig. 3 the transformer was operated with the tube filament windings open-circuited, that is, the tube filaments were supplied by a separate transformer. Consequently, the primary current is the sum of the magnetiz-

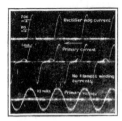

FIG. 3

ing current and the rectifier winding current reflected through the transformer.

THE FILTERING CIRCUIT

THE operation of the filtering circuit is shown in Fig. 4. The lower curve shows the current through the first condenser, and the upper curve shows the current through the first of the two chokes. It has been observed that the ratio of the r. m. s. value of the current of this first condenser to the direct-current rectified output is approximately constant, and lies between 1.8 and 2.2. for normal loads. It is practically independent of the capacity of this first condenser. The current through the second of the two chokes has very little ripple in it.

In a circuit such as this filter circuit, it might seem that the presence of the chokes and condensers would cause rather large transient voltages when the device is connected to or disconnected from the line. Actually, there are probably no transient increases and certainly none which exceed about ten per cent. of the normal voltage. The transient occurring when the switch was shut off is shown in Fig. 5. In this particular case, the switch was shut off during the time interval when the condenser was being charged. This is indicated by the fact that the last charging current peak is lower than the others. Oscillograms which were taken when the switch was not rectifying, do not show this decrease in the last of the charging current peaks. It will be noticed that the current through the condenser drops to zero within approximately one fiftieth of a second after the switch is turned off. During the same time interval, the current through the first choke drops to zero. At the time that the current through this first choke has dropped to zero, there is still a charge on this first condenser. The choke current drops to zero, however, since the charge on the second condenser is still appreciable, and is sufficiently greater than the charge on the first condenser to stop the current. The current through the second choke starts to drop off at about the time when the current through the first choke is reduced to one third of its normal value. This current then drops to approximately zero at the time when the voltage on the third filter condenser is sufficiently greater than the voltage on the second to

stop the current. By this time, the voltage on the second condenser is so low that the remaining charge on the first condenser proceeds to discharge through the choke, giving the slight current through the first choke that is shown in the illustration. A close inspection of the original film will show a slight current through the condenser at this instant. The reason it does not show up in the photograph is due to the fact that the current scale for the choke current is approximately four times that for the condenser current. A very short time after this discharge in current, the charge on the final condenser has been dissipated through the resistance and consequently the remaining charge on the second condenser discharges through the second choke, giving the slight rise shown. These steps in the discharge transient do not, of course, completely discharge the condensers, and very small current can be seen on the original film which repeats the first two slight rises in the choke currents that are plainly distinguished on this film. Several oscillograms of this transient were taken and all of them verify the results shown in this photograph. The right-hand portion of the middle curve on this line shows the amount of ripple in the second choke current before the switch is turned off. Practically all of this ripple is bypassed through the third of the filter condensers. Fig. 6 is another oscillogram taken under the same conditions. The condenser current is not faithfully recorded in this photograph because the oscillograph element was loose. The other two oscillograph elements were working properly,

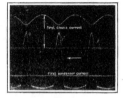

FIG. 4

shown. This decay is shown even more clearly in Fig. 6.

In order that any considerable voltage will be built up across one of the condensers due to the switching transient, it would be necessary that the direction of the flow of energy in the steady state would be, for the moment, reversed. Since none of the photographs give any indication of the reversed current through the chokes, it follows that there can be no high voltage across any of the condensers during the discharging interval.

A moment's consideration of the energy stored in the filter circuit will show that it is very unlikely that a high-voltage transient across one of the condensers will occur. In this filter circuit we have three fifty-henry chokes in series with the high-voltage lead, with three 2 - mfd. condensers connected across the line. See Fig. 7. Under normal conditions, the voltages across the condensers are 500, 375, and 350 volts respectively, 50 milliamperes of direct current flowing through each inductance. The energy stored in the first condenser will be $\frac{1}{2}$ CE^2, which is 0.25 joules. The other condensers store 0.14 and 0.112 joules. The energy stored in the chokes will be considerably less than this energy stored in the condensers. It is $\frac{1}{2}$.LI^2, or 0.0625 joules for each choke. The total stored energy in the circuit is then 0.637 joules. Thus we see that, even though the entire amount of this energy should by chance all pour into one of the condensers at once, the voltage across this condenser would be less than double the normal voltage across the first condenser. The solution is as follows:

$$\tfrac{1}{2} CE^2 = \tfrac{1}{2} \times 2 \times 10^{-6} \times E^2 = 0.637$$

Therefore, $E = 800$ volts. But since the oscillogram shows that there is no reverse in the flow of energy, even this cannot occur, and there will be no increase in the voltage across any one of the condensers when the switch is turned off.

The starting transient is shown in Fig. 8. When the transformer is switched on to the line there may be a high secondary voltage transient if it is connected during a particular part of the cycle. In the device under consideration, this will have no effect except on the transformer itself. The reason is that the rectifier filament is cold, so the rectifier tube acts as an open circuit during the first few cycles, and therefore it separates the transformer from the filter. As the tube filament heats up, the tube rectifies and gradually builds up the necessary voltage to maintain the steady state. From the appearance of the

FIG. 5

so that the choke currents are faithfully recorded.

In order to understand why the current drops to zero in pulses, consider the network as two separate circuits. The first is the circuit of the inductances, resistance, transformer, and rectifier in series. Assume the rectifier to act as a resistance. This is a reasonable assumption, since the direction of current flow is correct and the filament will remain hot for several cycles. The current in this circuit will decay logarithmically. The oscillatory discharge of the second circuit, inductances and condensers, will be superimposed on this discharge, giving the characteristics

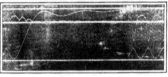

FIG. 6

oscillogram, it is evident that this voltage is built up gradually. If there were a transient voltage exceeding the normal, some of the charging current peaks would be slightly higher than the steady charging peaks. It will be noted that the area of the first few of these charging peaks is greater than the areas under the following discharging alternations of the condenser current. This is the interval during which the first condenser is accumulating its steady charge.

The drop in B voltage supplied by the "Powerformer" is directly proportional to the current drawn. The regulation curve for the high tap slopes down in a straight line from 175 volts, zero current, to zero volts, 37 milliamperes. The socket-power supply will give 20 milliamperes at 90 volts. The characteristic falls rapidly because it is impossible to use more than half the voltage at the output of the filter. The plate-circuit requirements of the power amplifier tube necessitate a high voltage at the filter output. The excess must be absorbed in a resistance. The scheme gives much poorer regulation than can be obtained from a good socket-power device where the voltage output of the filter is correct for the set. The power supply is ample for the usual two-step r. f. amplifier, detector, and one step audio amplifier. The second audio step is in the "Powerformer."

With the natural high impedance of socket-power devices, considerable trouble has been caused by the coupling between the audio plate circuits through the impedance of the socket device circuit. This trouble is entirely eliminated with this scheme. The detector plate circuit is completed through the 2-mfd. bypass condenser. The 30,000-ohm series resistance eliminates the coupling effects between the detector circuit and the two audio circuits.

The 7500-ohm sections of the potentiometer separate the two audio circuits so that each audio current goes through its proper bypassing condensers. Even at as low a frequency as 60 cycles, the condenser path for the audio component of the power amplifier plate current is only one thirtieth that of the potentiometer path.

THE AMPLIFIER

WE NOW consider the amplifier portion of the circuit. If we are to get fair reproduction, it is necessary that the current through the secondary of the output transformer, when connected to a normal load, will have the same wave shape as the signal voltage impressed across the primary of the input transformer. In other

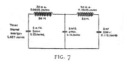

FIG. 7

words, the condition illustrated in Fig. 9 should obtain. The wave forms of the input voltage plate current and output transformer current are shown. The signal voltage is 4.22 volts r. m. s., corresponding to 6 volts peak signal or 18 on the grid through the 3:1 step-up transformer. The bottom curve represents the plate current of the tube. The zero of this current is above the curve

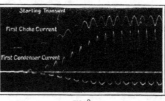

FIG. 8

shown. The center curve shows the current through the secondary of the output transformer working into a 5000-ohm resistive load used to simulate a loud speaker load. It will be noticed that the output current has about the same wave shape as the input signal voltage. The plate voltage was 390 volts; the grid bias 27 volts; the direct-current component of the plate current— 34 milliamperes. This, and the succeeding oscillograms, were taken with a 60-cycle signal. The amplifier stage was connected to a more flexible battery supply than the rectifier and filter circuit we have been discussing. We used the separate battery supply so that it could be easily adjusted to illustrate the effect of having improper B and C voltages on the tubes. Fig. 10 shows the effect of using too small a C battery in this circuit. The signal voltage is still approximately 4½ volts. The plate voltage is 400 volts. The C bias has been reduced to 10 volts, giving a plate current of 77 milliamperes. This condition obtains as the grid signal swings positive enough to draw current. The IZ drop of the current flowing through the high impedance of the transformer secondary will absorb practically all of the voltage induced in the secondary winding; consequently, the grid potential will remain approxim-

ately constant during the positive half of the grid signal, as shown in the photograph. When there is no alternating current in the plate circuit, the current through the output transformer secondary will, of course, fall off, just as any current decays when the potential across an inductance is removed and the inductance is discharged through a resistance. This gives the wave form shown in the center curve of the film. It will be noticed that the signal voltage shown at the top of the curve is no longer a sine wave. This is due to the fact that this particular curve was taken at 12:35 A. M. at Columbia University, using the United Service for the alternating current supply. It was noticed during that evening, as well as on other evenings, that the wave shape of the supply was much poorer after midnight than it was earlier in the evening. Fig. 11 is another curve representing approximately the same condition. The difference in this case is that another output transformer was substituted for the standard one. This output transformer has a large step-down ratio of the type which is adaptable to certain of the moving coil type loud speakers. In this case the signal voltage was considerably lower, that is, ¾ volt instead of 4½ volts, accounting for the change in the plate current and the output transformer current. Since the output transformer secondary was much lower in inductance, the current decays more rapidly than in the other case.

The next oscillogram, Fig. 12, shows the amplifier tube operating with too large a C battery. The signal voltage in this case was 4½ volts; B voltage—400 volts; the C voltage—44 volts; 14 milliamperes flowing in the plate circuit. This signal carried the grid too far negative that the curvature on the lower end of the tube characteristic distorted the signal. This distortion which appears both in the plate current curve (the top curve in the photograph), and the output transformer curve (the middle curve), is very similar to the distortion introduced when the grid swings too far positive, where the poor regulation of the high impedance grid circuit causes the peaks of the wave to be cut off. Fig. 13 illustrates the case where the signal voltage is too high, or the B voltage is too low. In this case, the signal voltage was 4½ volts; the plate voltage—175 volts; the C bias—15 volts. This C bias was found to be about the best adjustment for 175 volts B voltage. The plate current was 14 milliamperes. It will be seen that this signal carried the grid down near the cut off point, giving distortion at the lower end, and also carried it sufficiently positive to cause

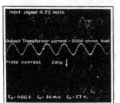

FIG. 9

FIG. 10

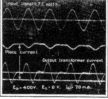

FIG. 11

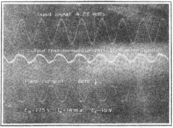

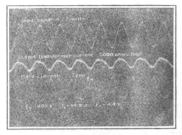

FIG. 12   FIG. 13

similar distortion at the positive end of the cycle. This distortion appears in both the plate current curve and in the output transformer current curve.

Observations were made in several homes to find out how much signal voltage the average broadcast listener desires for what he would call normal volume. It was found that 6 volts peak audio-frequency voltage on the primary of the input transformer to the power stage was about the proper value. This corresponds to approximately 150 to 185 volts on the loud speaker. This value of 6 volts would of course be changed were a different transformer ratio to be used, or were a tube having a different amplification factor to be substituted. This power amplifier will take about 8 volts peak signal before the signal will suffer any distortion. It will take up to 12 volts on occasional peaks with such a small amount of distortion that the average broadcast listener will not notice it. If the plate voltage is reduced much below 300 volts, it will be found that the average broadcast listener will tune-in the signal so that there will be very objectionable tube overloading.

There is one other possible source of distortion, namely, the effect of the direct current flowing through the primary of the input transformer in saturating the core and so causing distortion. This form of distortion will not be encountered in the output transformer, since a small air gap is

provided to prevent core saturation. There are two forms of distortion which might be expected from core saturation in the input transformer. One would be that form where the iron is operated so high up on the magnetization curve that the effect of the part of the signal cycle which would tend to carry it a little further up, would not be as effective in changing the flux as the part which tends to reduce the flux in the iron. This would give an unbalanced effect, destroying the symmetry between the two halves of the output voltage. This effect does not exist in the type of amplifier we are considering, for the reason that the signal voltage causes such an extremely small change in the flux density in the iron that the difference in the slope of the magnetization curve in different parts of the cycle is inappreciable. Fig. 14 shows this effect. The smaller of the two curves shows the signal current flowing through the secondary of the output transformer into a 5000-ohm resistive load when a sine wave is impressed on the primary of the first transformer. For the larger of the two curves, the same conditions were repeated with the change that 31 milliamperes of direct current were flowing through the primary of the input transformer. The change in magnitude of the wave is of no significance since it is due to a change in magnitude of the impressed signal voltage. It will be noted that this direct current, 31 milliamperes, is far in excess of any current that would flow through the primary of this transformer in a normal circuit. It will be noticed that the two wave forms are similar, showing that even this large direct current had no effect as far as introducing this particular type of distortion is concerned.

There is still another type of distortion to be considered. When the direct current is flowing through the primary, that is, when the iron is worked higher up on the magnetization curve, the impedance of the transformer primary will be reduced. This acts to reduce the amplification of low frequencies when the signal is impressed through a high series resistance, such as the plate circuit of the tube. Fig. 15 illustrates an exaggerated example of this effect. In the larger of the two curves, the 60-cycle voltage was impressed on the primary of the transformer through a 27,000 ohms. Under these circumstances, the transformer is operating under more

unfavorable conditions than is usually the case. The departure from the sine wave is due to the better amplification of the harmonics of the signal than of the fundamental. The fluctuation in the height of the peaks of the same wave was due to the fluctuation in the B-potential supply. The smaller of the two curves shows the effect of 20 milliamperes flowing through the primary while the signal is impressed across the primary through a 27,000-ohm series resistance. It will be noticed that the magnitude of the output signal has decreased, due to the reduction in the primary impedance. The harmonics are also more pronounced than they are in the upper curve. The conclusion to be drawn from these two photographs is that the effect of the direct current flowing through the primary of the transformer is to tend to cut off the lower frequencies. From experimental data which were taken in addition to these two oscillograms, it seems that a current of 5 milliamperes has very little effect in reducing the effectiveness of the transformer at the lower frequencies.

In conclusion, I wish to thank Dr. S. L. Quimby for his assistance in taking the oscillograms, and for his helpful suggestions. I also wish to thank Mr. Goudy, Mr. Brown, Mr. Corbett, and Mr. Lundahl, of the Pacent Electric Company's engineering department, for making this paper possible.

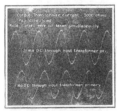

FIG. 14

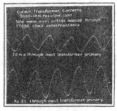

FIG. 15

RADIO BROADCAST is the official publication of the Radio Club of America, through whose courtesy, the foregoing paper has been printed here. RADIO BROADCAST does not, of course, assume responsibility for controversial statements made by authors of these papers. Other Radio Club papers will appear in subsequent numbers of this magazine

Raytheon *Leads*

**Raytheon BA
350 m. a.**
(Half size)

Raytheon's leadership in the rectifier field is demonstrated again by the announcement of these sensational new rectifiers, worthy to take their places alongside Raytheon Types B and BH.

**Raytheon A
2½ amps.**
(Actual size)

Light Socket A-B-C Power
With One Rectifier

By this amazing rectifier, Raytheon BA-350 m. a., the final great step in the development of radio power is accomplished. Compact, built-in A-B-C power, without accessories, becomes a fact.

Soon a specially selected group of leading radio manufacturers will announce their newest receivers, using Raytheon BA 350 m. a. Standard 201-A tubes are used, and all batteries, chargers, accessories, and outside power equipment are eliminated.

Raytheon BA-350 m. a., the crowning accomplishment of the Raytheon Research Laboratories, has at last given the radio world a practical, proven solution to the problem of simple light socket receiver operation.

A Revolutionary Scientific Achievement

In High Current, Low Voltage Rectification

Raytheon A-2½ amps. is revolutionary in principle, in construction, in performance. It is an unbreakable metal cartridge, compact and simple, without liquids or filaments.

Above all, it is The Efficient Rectifier. Its operating cost over the period of a year, compared with that of other types of rectifiers designed for similar uses, will show a cash saving of many dollars.

Raython A-2½ amps. was invented by Monsieur André of La Radio technique in Paris, and developed by the Raytheon Research Laboratories with his cooperation.

Battery chargers and A power units using this remarkable rectifier will bear the same seal of approval that distinguishes all Raytheon-equipped power devices.

RAYTHEON MANUFACTURING COMPANY
Cambridge, Massachusetts

Only manufacturers whose units have been tested and approved by Raytheon may use this seal on their products.

New Prices

Type B	. .	$4.50
Type BH	. .	6.00
Type BA	. .	7.50
Type A	. .	4.50

The Heart of Reliable Radio Power

The *Radio Broadcast*
LABORATORY INFORMATION SHEETS

INQUIRIES sent to the Questions and Answers department of RADIO BROADCAST were at one time answered either by letter or in "The Grid." The latter department has been discontinued, and all questions addressed to our technical service department are now answered by mail. In place of "The Grid," appears this series of Laboratory Information Sheets. These sheets contain much the same type of information as formerly appeared in "The Grid," but we believe that the change in the method of presentation and the wider scope of the information in the sheets, will make this section of RADIO BROADCAST of much greater interest to our readers.

The Laboratory Information Sheets cover a wide range of information of value to the experimenter, and they are so arranged that they may be cut from the magazine and preserved for constant reference. We suggest that the series of Sheets appearing in each issue be cut out with a razor blade and pasted on 4" by 6" filing cards, or in a notebook. The cards should be arranged in numerical order. Several times during the year an index to all sheets previously printed will appear in this department. The first index appeared in November.

Those who wish to avail themselves of the service formerly supplied by "The Grid," are requested to send their questions to the Technical Information Service of the Laboratory, using the coupon which appears on page 127 of this issue. Some of the former issues of RADIO BROADCAST, in which appeared the first sets of Laboratory Sheets, may still be obtained from the Subscription Department of Doubleday, Page & Company at Garden City, New York.

No. 97   RADIO BROADCAST Laboratory Information Sheet   June, 1927

Methods of Generating High-Frequency Energy

THE ARC

BEFORE the invention of the three-electrode tube, and its subsequent use as a source of large amounts of high-frequency energy, the arc was a common type of continuous-wave generator.

In the drawing on this Sheet is given the circuit diagram of a simple arc. The ordinary arc light used for street lighting might be used, but much more efficient operation is obtained from an especially designed arc. The elementary theory of the arc is given below.

The drawing indicates the simplest arrangement of the apparatus. "G" is a direct current generator. "r" is a resistance to control the current, L_1 and L_2 are two choke coils to keep the r. f. energy out of the generator and to keep the current practically constant, "K" is the arc, and "C," "L," and "R" are respectively, the capacity, inductance, and resistance of the oscillating circuit.

The arc, which consists of two electrodes, is different from ordinary electrical conductors in one important respect, which is that its resistance is not a constant quantity but a variable one, depending on the current flowing through it. At high current values the resistance is low and at low current the resistance is high. Consequently, an increase in current will produce a decrease in resistance.

Now, when the switch is closed, certain currents flow and the condenser begins to charge, and, therefore, part of the current is diverted from the arc. Since the current through the arc is decreased

by this action, the voltage across the arc must rise and it continues to rise as long as the condenser continues to charge. As soon as the condenser becomes fully charged, the arc voltage stops rising and the condenser begins to discharge itself through the arc.

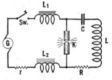

When the discharge is complete, the cycle of charge and discharge repeats itself with a frequency determined by the constants of the inductance L and the capacity C. By carefully choosing these values, large amounts of high-frequency energy can be obtained.

No. 98   RADIO BROADCAST Laboratory Information Sheet   June, 1927

Audio Amplifying Systems

RESISTANCE-COUPLED AMPLIFIERS

A VERY satisfactory method of audio amplification is that employing resistance coupling. The usual resistance-coupled amplifier requires three stages of amplification in order to obtain sufficient overall gain to satisfactorily operate a loud speaker. The introduction, however, of a new tube with a very high amplification constant, makes it possible in some cases to obtain sufficient amplification using only two stages. This new tube is known as the type 240 and data on it will appear on Laboratory Sheet No. 106 (July, 1927).

Several factors must be given attention if satisfactory results are to be obtained from a resistance-coupled amplifier. The mere fact that it is resistance coupled will not insure good quality. A poorly designed resistance-coupled amplifier is capable of creating as much distortion as can be obtained from a poorly designed amplifier of any other type. Some data regarding the constants of a resistance-coupled amplifier were given on Laboratory Sheet No. 74 (March, 1927). The constants given were for an ordinary tube for use in the resistance-coupled amplifier with an amplification constant of about 20. For the new type high-mu tube, however, with an amplification factor of about 30, it is necessary to use somewhat different values of resistance. See Laboratory Sheet No. 106.

The coupling condenser is a very important factor, and it is essential that this condenser have a very high insulation resistance, otherwise some of the B voltage will leak through the condenser to the grid circuit, and the amplifier will no longer function satisfactorily. In building up a resistance-coupled amplifier the best condensers should be used.

It is essential that high-quality plate and grid resistances be used to prevent noise in the amplifier. Also, the plate resistor should be capable of carrying the plate current of the tube without overheating.

Another important point is the amount of plate voltage used. It should be realized that most of the plate voltage supplied to the amplifier is lost in the resistance in series with the plate circuit of the tube. For this reason, it is necessary that fairly high voltages be available in order that there will be sufficient voltage left at the plate of the tube to obtain satisfactory operation. At least 135 volts should be used, and it should preferably be 180. The C-battery voltages should be kept as low as possible. It will generally be found that in an ordinary resistance-coupled amplifier a C-battery voltage of about 3 volts will be necessary on the grid of the tube preceding the last tube, if the latter is of the 171 type. The C voltage on the first tube of the amplifier need not be more than one volt.

No. 99   RADIO BROADCAST Laboratory Information Sheet

Data on the "Universal" Receiver

PARTS REQUIRED

ON LABORATORY Sheet No. 100J is given the circuit diagram of the new "Universal" receiver which was described in the December, 1926, issue of RADIO BROADCAST. In constructing this receiver, the following parts are necessary:

L_1—Antenna coil consisting of 13 turns of No. 26 d. s. c. wire wound at one end of a 2¼-inch tube.
L_2—Secondary coil consisting of 50 turns of No. 26 d. s. c. wire wound on the same tube as L_1. The separation between L_1 and L_2 should be ¼ inch.
L_3—Primary of interstage coil constructed in same manner as L_2 and tapped at the exact center.
L_4—Secondary winding constructed in same manner as L_2 and tapped at point No. 9, the 15th turn from that end of L_4 which is nearest to L_3
C_1, C_2—Two 0.0005-mfd. variable condensers.
C_3—Neutralizing condenser, variable, 0.000015 mfd.
C_7—Regeneration condenser, 0.00005 mfd.
L_5—R. F. choke coil, made by winding 400 turns of No. 38 wire on an ordinary spool.
T_1, T_2—Two audio-frequency transformers.

J—Interstage doub
J_1—Single-circuit fil
R_1—30-ohm rheosta
R_2—Fixed filament
201-A tubes.
R_3—Fixed filament power tube. One 0.
3-megohm grid leak.
Four Sockets.
Eleven Binding post

In operation, conde the tuning, and C_3 regeneration. Various tried on the plate of voltage used which g Frequently 22½ volts Make certain that exce used on the grid of the tion obtained will be such conditions. If th regeneration or howl stages, reverse the con transformer, T_2.

No. 100   RADIO BROADCAST Laboratory Information Sheet

Circuit Diagram of the "Universal" R

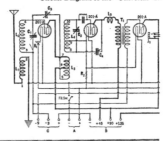

No. 101   RADIO BROADCAST Laboratory Information Shee

Lightning Arresters

HOW THEY WORK

AN ESSENTIAL part of any radio installation is the lightning arrester, which should be connected in the circuit as indicated in the diagram.

The arrester should preferably be located outside of the building at that point where the antenna lead-in enters the building. One terminal of the lightning arrester connects to the antenna and the other terminal connects to a good ground. A lightning arrester is a very simple device and actually consists of two metal electrodes which are spaced to within about five thousandths of an inch of each other. A radio-frequency current is too weak and too low in voltage to jump across these points which form the gap in the arrester and hence there is no path for the signal except that through the antenna to the receiver and thence to ground. The receiver is therefore actuated by this radio-frequency current and a signal is produced in the telephones, or the loud speaker, as the case may be. Suppose, however, that a high-potential atmospheric electrical discharge takes place near the antenna. Such discharges are always erratic in character and of high frequency. The antenna coil of the set therefore exerts a powerful choking action upon them even though the coil is quite small. For this reason, and also due to the very high voltage of the lightning

discharge, it jumps arrester and passes to more effect on the set will possibly drown o Also, during electri

Lightning Arrester
Gap

approaching, there static electricity pre tends to accumulate such time as the vol across the small gap voltage is generally a

No. 102 — Radio Broadcast Laboratory Information Sheet — June, 1927

Efficiency of Amplifying Systems

ARTICULATION

IN AUDIO-amplifying systems, efficiency must be judged from two standards. One of the standards is, in the terminology of telephone engineers, the "volume efficiency" of the system, which tells us how much increase in loudness of sound is produced by the system. The other standard is known as "articulation efficiency." The "articulation efficiency" of any system is a measure of its effectiveness in the transmission of detached speech sounds. In these tests, sounds are grouped into meaningless monosyllables and the efficiency is measured by the percentage of sounds which are correctly received.

In actual tests on a system the monosyllables are spoken into the input of the system and listeners at the output record what sounds they think were spoken. In very high quality systems it is possible to obtain an articulation efficiency of almost 100 per cent.

The articulation efficiency, depends upon the frequency distortion in the system, the amount of noise, and the volume efficiency. On this Sheet is an interesting curve the data for which were taken from a paper by Mr. R. L. Jones in the April, 1924, issue of the *A. I. E. E. Journal*, which shows how articulation varies with variations in intensity of sound. At the zero point the intensity of the received speech is equal to the intensity of the speech as it leaves the mouth and the articulation is about 91

per cent. With an intensity 100 times greater (10^2), the articulation falls to 87 per cent. If the intensity is decreased to a million times less than when it leaves the mouth, (10^6) the articulation is still very

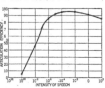

good, being 85 per cent. These tests were made under quiet conditions and, of course, under noisy conditions the results would have been somewhat different.

No. 103 — Radio Broadcast Laboratory Information Sheet — June, 1927

Audio Transformers

HIGH-IMPEDANCE PRIMARY NEEDED

THIS Sheet will explain why better quality is obtained from transformers with high-impedance primaries.

Drawing A shows how a transformer is connected in the plate circuit of a tube. Now, a voltage, Eg, on the grid of a tube is equivalent to a voltage of μEg (amplification constant times Eg) in the plate circuit of the tube. Also, the plate circuit of a tube acts like a resistance equal in value to the plate impedance of the tube (12,000 ohms for a 201-A type tube). These two facts were used in drawing the equivalent circuit diagram, B. In this diagram μEg indicates the voltage acting in the plate circuit and Rp represents the plate impedance of the tube. It is evident that the total voltage, μEg, available in the circuit, must divide itself between Rp and T, the transformer, and therefore the percentage of the total voltage across the transformer, increases with increased impedance in the transformer. Now, the impedance varies with the frequency, becoming greater as the frequency rises and decreasing as the frequency becomes lower. It is evident, then, that

the percentage of the total voltage across the transformer will also vary with frequency, and if this variation is very great it will be a source of distortion. Practically, the result will be that, at the low frequencies where the transformer impedance is low, very little of the total voltage will be across the transformer, most of it being across the tube. As a result, the low frequencies will not receive as much amplification as do the moderate and high frequencies. The problem then is to so design the transformer that this variation of amplification with frequency is as small as possible consistent with economy of manufacture. The problem evidently comes down to one of designing a transformer to have as large an impedance as possible at the low frequencies for, since the impedance increases with frequency, there is no difficulty in obtaining high impedance at other than the low frequencies.

The impedance at low frequencies depends upon the inductance. The larger the inductance the greater the impedance. In order to get a large inductance, a large number of primary turns are required. It is also essential that the core of the transformer be very efficient so that the turns will have as much inductance as possible.

No. 104 — Radio Broadcast Laboratory Information Sheet — June, 1927

Socket Power Units

VOLTAGE OUTPUT CURVES

AT THE present time, it is common practice in rating B power units, to specify the voltages at various current drains which the unit will deliver from the high-voltage tap. These data are obtained by connecting a variable resistance, in series with a milliammeter, between the negative B and the terminal giving the highest voltage, and then measuring the output voltage with different values of current through the resistance. The data may be collected in the form of a table or a curve may be plotted. It is best to plot a curve for, from it, we can determine the voltage at any current drain. Also, the slope of the curve gives us visually an idea of how constant the voltage is.

If full benefit is to be obtained from such curves, it is essential that we thoroughly understand what they signify. We must first determine the total plate-current drain of our receiver. This information can be obtained by connecting a milliammeter in series with the negative B lead, where it will measure the total plate current. Suppose the reading to be 35 milliamperes. This value of current is now located on the curve and we find that the corresponding plate voltage is (in this particular case) 135 volts. This is the maximum voltage that the socket-power unit will supply at 35 milliamperes. If you require a maximum of 135 volts for your receiver, then the unit is satisfactory. If you cannot use as much as 135 volts and there is no adjustment on the device to lower this voltage, then the unit is not satisfactory; or you might want to use a 171 type tube

with 180 volts and in this case also the unit will be unsatisfactory for it can only deliver 135 volts at the requisite current drain.

The curve tells us nothing concerning the voltages supplied by the other terminals on the power

unit. These other voltages are generally controlled by variable resistances so that any voltages from zero to maximum can be obtained.

Equipment for the Home-Constructor

How to Use Some of the New and Interesting Radio Equipment Which the Market Offers

By THE LABORATORY STAFF

MIDGET CONDENSERS

SMALL midget condensers of a capacity variable between about 15 and 130 mmfd. find many uses in the modern radio receiver. In Fig. 1 are shown several circuit positions where they might quite successfully be applied. For example, one may be connected in the antenna circuit to reduce the electrical length of the antenna without actually removing any wire, as shown at "A." "B" indicates the connection to be used to obtain vernier tuning—almost equivalent to a slow-motion dial—with an ordinary variable condenser. The midget is in this case shunted across its larger brother, and generally should have a maximum capacity of about 50 mmfd.

It should not be overlooked that the presence of a condenser in this latter position will upset somewhat the readings of the main dial so, if the receiver is to be logged accurately, the midget vernier condenser must always be set at approximately the same position before tuning is proceeded with. Midget condensers are frequently used with gang condensers in order to compensate any inequalities between the various stages.

denser is set at the proper value and left in that position.

The last drawing in Fig. 1, "F," shows one of several methods of regeneration control by means of a midget condenser. In this instance, the tickler coil is fixed in relation to the secondary, the amount of radio-frequency current going through it being controlled by the midget condenser.

Of the many possible circuit connections for these small variable condensers, the foregoing paragraphs list only a few of the most common.

The following are some of the manufacturers making midget condensers of capacities varying between 0.00001 and 0.00015 mfd: Hammarlund Manufacturing Company, New York City; Allen D. Cardwell Manufacturing Company, Brooklyn, New York; Silver-Marshall, Incorporated, Chicago, Illinois; Precise Manufacturing Company, Rochester, New York.

PICK-UP DEVICE

REALIZATION that the new phonographs employing electrical reproduction are far ahead of the old type ones using mechanical reproduction, has created a fertile market for

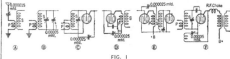

FIG. 1

In a circuit in which the tube capacity must be neutralized a small variable condenser connected in the proper way can be used for stabilization. Either a midget condenser of the type shown in Figs. 2 and 3, or a neutralizing condenser such as shown in Fig. 4, may be used. In the first two types of condenser, the capacity is readily variable by means of a knob, and may be, if desired, mounted on the panel. The latter type is usually mounted at the rear of the sub-panel, and capacity adjustments are made by means of a screwdriver. If the neutralizing condenser is of the midget type it may be used either as an oscillation control or to neutralize the tube capacity, at will. In the first case, the set may be made to regenerate over the full band of frequencies by simply adjusting the capacity so that more or less feedback is obtained. There are many different circuits making use of a small condenser to prevent oscillation (obtain neutralization). In Fig. 1, "C" shows the Rice system of neutralization, "D" the Roberts system, and "E" the Hazeltine method. For neutralization, the con-

equipment which makes possible the conversion of an old phonograph to one of the new type. Generally speaking, such devices consist of an electromagnetic pick-up, volume control, scratch filter, and often an adapter for plugging into the detector tube socket of a radio receiver; thus the audio amplifier of the receiver affords the necessary amplification of the electrical vibrations originating in the pick-up. These vibrations vary in electrical character depending upon the motion of the needle, which, in turn, relies for its minute movements upon the grooves in the rotating phonograph record.

The newest in the pick-up field is the Bosch "Recreator." Judging by the very satisfactory results obtained in our tests on this instrument, its somewhat belated appearance is due to a desire on the part of its manufacturers to produce a really first-rate piece of equipment. Fig. 5 is a diagram showing how the Bosch "Recreator" is used. In this diagram, "A" is the electromagnetic pick-up device; "B" is the swinging arm pivoted to the base "E" by a swivel arrange-

FIG. 2
Precise

FIG. 3
Silver-Marshall

RADIO BROADCAST Photograph
FIG. 4
Hammarlund

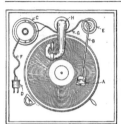

FIG. 5

ment; "G" is a flexible lead with pin jacks which are plugged into the volume control, "C"; "F" is another flexible lead between the volume control and the adapter "D," this latter, of course, being plugged into the detector tube socket of the receiver; "H" in the diagram is the regular tone arm and mechanical pick-up of the phonograph.

It is perhaps irrelevant to emphasize the necessity of employing a high-quality amplifier and loud speaker with this device. The advantages that accrue from the use of electrical reproduction will be forfeited if an amplifier system of satisfactory characteristics is not employed.

The volume obtainable from an outfit employing a Bosch pick-up necessarily depends upon the amplifier utilized. Although commercial electrical phonographs can deliver sufficient volume to make necessary the use of a 210 type tube, a two-stage transformer-coupled amplifier, or its equivalent, employing a 171 type output tube, will give satisfactory volume for average purposes.

Features of this device which are particularly commendable are as follows: (1) The arm between the pick-up and its base is adjustable in length; (2) the base, "D," is small yet sufficiently heavy for satisfactory balance; (3) the flexible leads, "G" and "F," are long, thus making it unnecessary to place the volume control close to the turntable, where generally there is inadequate room for it; (4) the Volume control functions as a volume control should, i. e., it satisfactorily and gradually reduces the music to a whisper, or vice versa, as desired; (5) the pick-up is very sensitive; (6) a clamp is supplied so that the electromagnetic pick-up may be permanently attached to the existing tone arm of the phonograph if desired, thus doing away with the arm and base supplied.

Manufactured by the American Bosch Magneto Corporation, Springfield, Massachusetts. Price $30.00, attractively boxed.

TRANSMITTING TUBE

A NEW transmitting tube, particularly adaptable to short-wave work, has recently made its appearance. This is the new UX-852, offered by the Radio Corporation of America.

The UX-852 is of the round bulb design, with three arms or extensions of the glass envelope. The largest or filament arm of the tube is provided with a standard UX base, suitable for use in either the UX or the UV type socket. The socket must be mounted so that the filament of the tube will be in a vertical position. Contrariwise, the horizontal grid and plate arms are not based. Instead, two heavy stranded leads, arranged in parallel, are brought from each of the widely separated stems for connection with grid and plate, respectively.—Double grid and double plate leads serve greatly to increase the current-carrying capacity at exceptionally high frequencies, and both leads for each element should be employed at all times so as to carry safely the large circulating currents which flow at very high frequencies.

The fact that inter-electrode capacitance has been reduced to a minimum, permits of operating

the UX-852 at wavelengths below 100 meters (3000 kc.). The tube provides excellent results on the popular 80-, 40-, and 20-meter channels (3750, 7500, and 15,000 kc.), and it has been successfully operated on wavelengths below 5 meters (60000 kc.) and even down to 77 centimeters (0.77 meters), or a frequency of 390,000 kilocycles, which is by no means the limit for the amateur who desires to explore the lowest wavelengths of short-wave radio.

The wiring of the ultra-short-wave transmitter is materially simplified when employing the UX-852 tube, since, with the grid and plate leads coming out of the bulb at different points, all connections do not have to be concentrated at the base, and the wiring can be made proportionately shorter and with wider spacing. The base, while of the UX type with four contact prongs, makes use of only two of the filament connections. Mounted upright in the usual UX push type or UV type socket, the tube has ample air circulation and operates much cooler than the 50-watt UX-203-A and other tubes. The ample cooling capacity is due in large part to the large area of the glass envelope. The new tube will handle plate voltages of 2000 normally, and even up to 3000 with proper precautions, without internal breakdown.

Alternating current should be used to operate the filament when possible. A center tap on the secondary of the filament transformer should be used for the grid and plate circuit returns. Rheostat control should be provided on the power supply side of the transformer. When it is necessary to use direct current to light the filament, the plate and grid return leads should be connected together and to the positive lead. Filament voltmeter leads should always be connected as closely as possible to the socket terminals.

The characteristics of the UX-852 tube are as follows:

Filament Voltage................10
Filament Current................3.25 amp.
Filament Power.................32.5 watts
Filament Type...........Thoriated Tungsten
Plate Voltage..................2000 to 3000
Plate Current (Oscillating)........0.075 amp.
Input Power....................150 watts
Maximum Safe Power
  Dissipation...................100 watts
Amplification Constant..........16
Nominal Output ....75 watts
Plate Impedance (Eg = o) ....8000 ohms.

THE UX-852
Transmitting Tube

BROADCAST RECEIVER

THE Ferguson Model 14 broadcast receiver is a ten-tube set employing a loop antenna. Provision is made for connecting an outside antenna, but under ordinary circumstances its use should be unnecessary. The set has six stages of radio frequency, a detector, and three audio stages: The 201-A type tubes are used in all

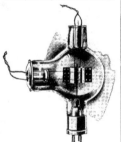

stages except the output, which is designed to take either the 112 or 171 type tube. C-battery connections are provided for use in the power stage. This receiver should not be operated without a power tube otherwise considerable distortion will result, due to overloading .

Only one dial is used for tuning, the condensers being arranged in gang formation. The dial is of the popular window type, and is illuminated from the rear by a small light operated by the battery supply. The dial is marked off directly in wavelengths and is graduated from 200 to 550 meters. During a test in the Laboratory the scale readings were found to check very closely with the wavelength of the stations logged. A volume control is placed directly beneath the tuning knob and consists of a small handle which regulates a rheostat in the filament circuit of the first, third, and fifth radio-frequency tubes. Throwing the lever as far as it will go in a counter clockwise direction turns off all the tubes. The dial is centered on a wooden panel 9 inches high and 20 inches long. The whole receiver is contained in a mahogany finished cabinet 12 inches high, 24 inches wide, and 16 inches deep. A battery cable runs through the cabinet to the rear and affords connections for A, .B, and C

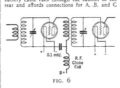

FIG. 6

batteries. The loud speaker jack is also in the rear, as are also the antenna and ground connections.

In spite of the number of tubes employed in the receiver there is a drain of only 30 milliamperes from the B supply. Either heavy-duty dry cells or a socket-power device may be used to supply the B voltage. The A supply is obtained from a storage battery. A photograph of the receiver appears on this page. Manufactured by J. B. Ferguson, Incorporated, Long Island City New York. Price $235.00 without tubes or accessories.

PLATE POWER-SUPPLY RESISTOR

THIS resistor is used as standard equipment with the Amertran Power Supply Kit to provide the various voltages needed for the receiver. It can easily be applied to any type of plate supply device. The total resistance is 41,000 ohms, taps being taken at 32,000 ohms, 21,000 ohms, 16,500 ohms, 12,500 ohms, and 9000 ohms, as indicated in the accompanying diagram, Fig. 7. No definite values of voltage can be given for the different taps as this is dependent on the amount of current taken and the voltage of the rectifier element. The voltages across the resistances are not hard to calculate, however, and full instructions are given on Laboratory

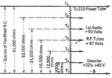

FIG. 7

Information Sheet No. 63 (page 302, January, 1927, RADIO BROADCAST). In the diagram, the taps used in the Amertran device are indicated, together with the tubes they supply. A 210 type is used in the last stage with 400 volts on the plate. The resistors are connected in series and the resistance values are accurate within approximately two per cent. The current-carrying capacity of the unit is 30 milliamperes continuously, which is ample to take care of most of the conditions to which the resistor will be subject in actual service. The resistor is wire-wound and slightly inductive. There are no doubt many uses for this resistor. For instance, it may be used as a potentiometer or as a grid leak at low-power transmitting stations. Manufactured especially for the Amer-Tran Sales Company, Newark, New Jersey. Resistor type 400. Price $7.50.

RADIO-FREQUENCY CHOKE COIL

THE radio-frequency choke coil is a fairly recent addition to the family of parts which go to make up the refinements of modern receivers. There are many choke coils on the market and most are designed for a specific use, though they may be used by the experimenter in many different positions in the circuit. The inductance of the Bremer-Tully choke coil was found in the Laboratory to be 12.75 millihenries, and the direct current resistance, 83.7 ohms. This means that the voltage drop (d. c.) with a current of 2 mils. would be only about 0.0167 volts, or a figure which can be neglected. The impedance to a radio-frequency current, however, of 1500 kc. (200 meters) would be approximately 122,000 ohms, and at 545 kc. (550 meters), would be approximately 45,000 ohms. When such a choke is placed in the radio-frequency battery leads as shown in the diagram, Fig. 6, the loss of d.c. voltage is of no practical importance, while the impedance offered to radio-frequency currents is very high and effectually keeps these currents from the battery leads. By keeping the radio currents from the batteries, there is less chance for coupling between stages and the receiver may be more easily neutralized. In all cases, a bypass condenser must be provided, as shown in the diagram, to provide a path for the radio-frequency currents, as it is not the object to stop the r.f. currents entirely but to sidetrack them in such a way that. they will not cause trouble in the receiver. Manufactured by Bremer-Tully Manufacturing Company, Chicago, Illinois. Price $0.90.

THE FERGUSON MODEL 14 RECEIVER

Manufacturers' Booklets

A Varied List of Books Pertaining to Radio and Allied Subjects Obtainable Free With the Accompanying Coupon

*A*S *AN additional service to* RADIO BROAD-
CAST *readers, we print below a list of booklets
on radio subjects issued by various manufacturers. The publications listed below cover a wide
range of subjects, and offer interesting reading to
the radio enthusiast. The manufacturers issuing
these publications have made great effort to collect
interesting and accurate information.* RADIO
BROADCAST *hopes, by listing these publications
regularly, to keep its readers in touch with what the
manufacturers are doing. Every publication listed
below is supplied free. In ordering, the coupon
printed on page 124 must be used. Order by
number only.*—THE EDITOR.

1. FILAMENT CONTROL—Problems of filament supply, voltage, regulation, and effect on various circuits. RADIALL COMPANY.
2. HARD RUBBER PANELS—Characteristics and properties of hard rubber as used in radio, with suggestions on how to "work" it. B. F. GOODRICH RUBBER COMPANY.
3. TRANSFORMER—A booklet giving data on input and output transformers. PACENT ELECTRIC COMPANY.
4. RESISTANCE-COUPLED AMPLIFIERS—A general discussion of resistance coupling with curves and circuit diagrams. COLE RADIO MANUFACTURING COMPANY.
5. CARBORUNDUM IN RADIO—A book giving pertinent data on the crystal as used for detection, with hook-ups, and a section giving information on the use of resistors. THE CARBORUNDUM COMPANY.
6. B-ELIMINATOR CONSTRUCTION—Constructional data on how to build. AMERICAN ELECTRIC COMPANY.
7. TRANSFORMER AND CHOKE-COUPLED AMPLIFICATION—Circuit diagrams and discussion. ALL-AMERICAN RADIO CORPORATION.
8. RESISTANCE UNITS—A data sheet of resistance units and their application. WARD-LEONARD ELECTRIC COMPANY.
9. VOLUME CONTROL—A leaflet showing circuits for distortionless control of volume. CENTRAL RADIO LABORATORIES.
10. VARIABLE RESISTANCE—As used in various circuits. CENTRAL RADIO LABORATORIES.
11. RESISTANCE COUPLING—Resistors and their application to audio amplification, with circuit diagrams. DEJUR PRODUCTS COMPANY.
12. DISTORTION AND WHAT CAUSES IT—Hook-ups of resistance-coupled amplifiers with standard circuits. ALLEN-BRADLEY COMPANY.
15. B-ELIMINATOR AND POWER AMPLIFIER—Instructions for assembly and operation using Raytheon tube. GENERAL RADIO COMPANY.
15A. B-ELIMINATOR AND POWER AMPLIFIER—Instructions for assembly and operation using an R. C. A. rectifier. GENERAL RADIO COMPANY.
16. VARIABLE CONDENSERS—A description of the functions and characteristics of variable condensers with curves and specifications for their application to complete receivers. ALLEN D. CARDWELL MANUFACTURING COMPANY.
17. BAKELITE—A description of various uses of bakelite in radio, its manufacture, and its properties. BAKELITE CORPORATION.
19. POWER SUPPLY—A discussion on power supply with particular reference to lamp-socket operation. Theory and constructional data for building power supply devices. ACME APPARATUS COMPANY.
20. AUDIO AMPLIFICATION—A booklet containing data on audio amplification together with hints for the constructor. ALL AMERICAN RADIO CORPORATION.
21. HIGH-FREQUENCY DRIVER AND SHORT-WAVE WAVEMETER—Constructional data and application. BURGESS BATTERY COMPANY.
46. AUDIO-FREQUENCY CHOKES—A pamphlet showing positions in the circuit where audio-frequency chokes may be used. SAMSON ELECTRIC COMPANY.
47. AUDIO-FREQUENCY CHOKES—Circuit diagrams illustrating the use of chokes to keep out radio-frequency currents from definite points. SAMSON ELECTRIC COMPANY.
48. TRANSFORMER AND IMPEDANCE DATA—Tables giving the mechanical and electrical characteristics of transformers and impedances, together with a short description of their use in the circuit. SAMSON ELECTRIC COMPANY.
49. BYPASS CONDENSERS—A description of the manufacture of bypass and filter condensers. LESLIE F. MUTER COMPANY.
50. AUDIO MANUAL—Fifty questions which are often asked regarding audio amplification, and their answers. AMERTRAN SALES COMPANY, INCORPORATED.
51. SHORT-WAVE RECEIVER—Constructional data on a receiver which, by the substitution of various coils, may be made to tune from a frequency of 16,660 kc. (18 meters) to 1999 kc. (150 meters). SILVER-MARSHALL, INCORPORATED.
52. AUDIO QUALITY—A booklet dealing with audio-frequency amplification of various kinds and the application to well-known circuits. SILVER-MARSHALL, INCORPORATED.
56. VARIABLE CONDENSERS—A bulletin giving an analysis of various condensers together with their characteristics. GENERAL RADIO COMPANY.
57. FILTER DATA—Facts about the filtering of direct current supplied by means of motor-generator outfits used with transmitters. ELECTRIC SPECIALTY COMPANY.
59. RESISTANCE COUPLING—A booklet giving some general information on the subject of radio and the application of resistors to a circuit. DAVEN RADIO CORPORATION.
60. RESISTORS—A pamphlet giving some technical data on resistors which are capable of dissipating considerable energy; also data on the ordinary resistors used in resistance-coupled amplification. THE CRESCENT RADIO SUPPLY COMPANY.

63. RADIO-FREQUENCY AMPLIFICATION—Constructional details of a five-tube receiver using a special design of radio-frequency transformer. CAMFIELD RADIO MANUFACTURING COMPANY.
64. FIVE-TUBE RECEIVER—Constructional data on building a receiver. AERO PRODUCTS, INCORPORATED.
64A. AMPLIFICATION WITHOUT DISTORTION—Data and curves illustrating the use of various methods of amplification. ACME APPARATUS COMPANY.
65. RADIO HANDBOOK—A helpful booklet on the functions, selection, and use of radio apparatus for better reception. BENJAMIN ELECTRIC MANUFACTURING COMPANY.
66. SUPER-HETRODYNE—Constructional details of a seven-tube set. G. C. EVANS COMPANY.
70. IMPROVING THE AUDIO AMPLIFIER—Data on the characteristics of audio transformers, with a circuit diagram showing where chokes, resistors, and condensers can be used. AMERICAN TRANSFORMER COMPANY.
71. DISTORTIONLESS AMPLIFICATION—A discussion of the resistance-coupled amplifier used in conjunction with a transformer, impedance, or resistance input stage. Amplifier circuit diagrams and constants are given in detail for the constructor. AMSCO PRODUCTS INCORPORATED.
72. PLATE SUPPLY SYSTEM—A wiring diagram and layout plan for a plate supply system to be used with a power amplifier. Complete directions for wiring are given. AMERTRAN SALES COMPANY.
80. FIVE-TUBE RECEIVER—Data are given for the construction of a five-tube tuned radio-frequency receiver. Complete instructions, list of parts, circuit diagram, and template are given. ALL-AMERICAN RADIO CORPORATION.
81. BETTER TUNING—A booklet giving much general information on the subject of radio reception with specific illustrations. Primarily for the non-technical home constructor. BREMER-TULLY MANUFACTURING COMPANY.
83. SIX-TUBE RECEIVER—A booklet containing photographs, instructions, and diagrams for building a six-tube shielded receiver. SILVER-MARSHALL, INCORPORATED.
84. SOCKET POWER DEVICE—A list of parts, diagrams, and templates for the construction and assembly of socket power devices. JEFFERSON ELECTRIC MANUFACTURING COMPANY.
85. FIVE-TUBE EQUAMATIC—Panel layout, circuit diagrams, and instructions for building a five-tube receiver, together with data on the operation of tuned radio-frequency transformers of special design. KARAS ELECTRIC COMPANY.
85. FILTER—Data on a high-capacity electrolytic condenser used in filter circuits in connection with A socket power supply units, are given in a pamphlet. THE ABOX COMPANY.
86. SHORT-WAVE RECEIVER—A booklet containing data on a short-wave receiver as constructed for experimental purposes. THE ALLEN D. CARDWELL MANUFACTURING COMPANY.
88. SUPER-HETERODYNE CONSTRUCTION—A booklet giving full instructions, together with a blue print and necessary data, for building an eight-tube receiver. THE GEORGE W. WALKER COMPANY.
89. SHORT-WAVE TRANSMITTER—Data and blue prints are given on the construction of a short-wave transmitter, together with operating instructions, methods of keying, and other pertinent data. RADIO ENGINEERING LABORATORIES.
90. IMPEDANCE AMPLIFICATION—The theory and practice of a special type of dual-impedance audio amplification are given. ALDEN MANUFACTURING COMPANY.
93. B-SOCKET POWER—A booklet giving constructional details of a socket-power device using either the BH or 313 type rectifier. NATIONAL COMPANY, INCORPORATED.
94. POWER AMPLIFIER—Constructional data and wiring diagrams of a power amplifier combined with a B-supply unit are given. NATIONAL COMPANY, INCORPORATED.

ACCESSORIES

22. A PRIMER OF ELECTRICITY—Fundamentals of electricity with special reference to the application of dry cells to radio and other uses. Constructional data on buzzers, automatic switches, alarms, etc. NATIONAL CARBON COMPANY.
23. AUTOMATIC RELAY CONNECTIONS—A data sheet showing how a relay may be used to control A and B circuits. YAXLEY MANUFACTURING COMPANY.
25. ELECTROLYTIC RECTIFIER—Technical data on a new type of rectifier with operating curves. KODEL RADIO CORPORATION.
26. DRY CELLS FOR TRANSMITTERS—Actual tests showing actual life curves showing exactly what may be expected of this type of B power. BURGESS BATTERY COMPANY.
27. DRY-CELL BATTERY CAPACITIES FOR RADIO TRANSMITTERS—Characteristic curves and data on discharge tests. BURGESS BATTERY COMPANY.
28. B BATTERY LIFE—Battery life curves with general curves on tube characteristics. BURGESS BATTERY COMPANY.
29. HOW TO MAKE YOUR SET WORK BETTER—A non-technical discussion of general radio subjects with hints on how reception may be bettered by using the right tubes. UNITED RADIO AND ELECTRIC CORPORATION.
30. TUBE CHARACTERISTICS—A data sheet giving constants of tubes. C. E. MANUFACTURING COMPANY.
31. FUNCTIONS OF THE LOUD SPEAKER—A short, non-technical general article on loud speakers. AMPLION CORPORATION OF AMERICA.
32. METERS FOR RADIO—A catalogue of meters used in radio, with connecting diagrams. BURTON-ROGERS COMPANY.
33. SWITCHBOARD AND PORTABLE METERS—A booklet giving dimensions, specifications, and shunts used with various meters. BURTON-ROGERS COMPANY.
34. CLIP OF B BATTERIES—An interesting discussion of the relative merits of various sources of B supply. HARTFORD BATTERY MANUFACTURING COMPANY.

NATIONAL
TUNING UNITS
New Type

**Designed and Officially Approved
by Glenn H. Browning**

These new NATIONAL TUNING-UNITS comprise the OFFICIAL BROWNING-DRAKE Coil and R. F. Transformers—mounted on the New NATIONAL EQUI-TUNE Variable Condensers with their light, rigid *Girder-Frames.*

Each TUNING-UNIT is

fitted with a NATIONAL ILLUMINATED VEL-VET-VERNIER DIAL—Type G, with variable ratio and smoothest action. Each tuning unit is packed mounted as shown so that it may be used without change for experimental work or easily installed on a panel.

Price BD-1E, with Genuine
BD Antenna Coil and .0005
Condenser $10.75

Price BD-2E, with Genuine
B-D Transformer and
.00025 Condenser $14.25

Deduct 50c each for dials without illumination

NATIONAL Tuning
Units are standard for
good Radio sets. See
our NATIONAL Impedaformers for
quality audio. NAT-
IONAL True-Filters,
for power tube output
connection.

NATIONAL CO. makes
heavy-duty B-Supply
Units and B-stage Power
Amplifiers. National
Company, Inc., Cambridge, Mass. Send for
Bulletin 116-B 6.

handle cash remittances for parts, but when the coupon on page 126 is filled out, all the information requested will be forwarded.

201. SC FOUR-TUBE RECEIVER—Single control. One stage of tuned radio frequency, regenerative detector, and two stages of transformer-coupled audio amplification. Regeneration control is accomplished by means of a variable resistor across the tickler coil. Standard parts; cost approximately $58.85.

202. SC-II FIVE-TUBE RECEIVER—Two stages of tuned radio frequency, detector, and two stages of transformer-coupled audio. Two tuning controls. Volume control consists of potentiometer grid bias on r.f. tubes. Standard parts cost approximately $60.35.

203. "HFD" KIT—A five-tube tuned radio-frequency set having two radio stages, a detector, and two transformer-coupled audio stages. A special method of coupling in the r.f. stages tends to make the amplification more nearly equal over the entire band. Price $63.01 without cabinet.

204. R. G. S. KIT—A four-tube inverse reflex circuit, having the equivalent of two tuned radio-frequency stages, detector, and three audio stages. Two controls. Price $69.70 without cabinet.

205. PIERCE AIRO KIT—A six-tube single-dial receiver; two stages of radio-frequency amplification, detector, and three stages of resistance-coupled audio. Volume control accomplished by variation of filament brilliancy of r.f. tubes or by adjusting compensating condensers. Complete chassis assembled but not wired costs $43.50.

206. H & H-T. R. F. ASSEMBLY—A five-tube set; three tuning dials, two stages of radio frequency, detector, and a transformer-coupled audio stages. Complete except for baseboard, panel, screws, wires, and accessories. Price $35.00.

207. PREMIER FIVE-TUBE ENSEMBLE—Two stages of tuned radio frequency, detector, and two steps of transformer-coupled audio. Three dials. Parts assembled but not wired. Price complete, except for cabinet, $35.00.

208. "QUADRAFORMER VI"—A six-tube set with two tuning controls. Two stages of tuned radio frequency using specially designed shielded coils, a detector, one stage of transformer-coupled audio, and one stage of resistance-coupled audio. Gain control by means of tapped primaries on the r.f. transformers. Essential kit consists of three shielded double-range "Quadraformer" coils, a selectivity control, and an "Ampitrol," price $71.50. Complete parts $70.15.

209. GEN-RAL FIVE-TUBE SET—Two stages of tuned radio frequency, detector, and two transformer-coupled audio stages. Volume is controlled by a resistor in the plate circuit of the r.f. tubes. Uses a special r.f. coil ("Duo-Former") with figure eight winding. Parts mounted but not wired, price $37.50.

210. BREMER-TULLY POWER-SIX—A six-tube, dual-control set; three stages of neutralized tuned radio frequency, detector, and two transformer-coupled audio stages. Resistances in the grid circuit together with a phase shifting arrangement are used to prevent oscillation. Volume control accomplished by variation of B potential on r.f. tube. Essential kit consists of four r.f. transformers, two dual condensers, three small condensers, three choke coils, one 500,000-ohm resistor, three 1500-ohm resistors, and a set of color charts and diagrams. Price $41.50.

211. BRUNO DRUM CONTROL RECEIVER—How to apply a drum tuning unit to such circuits as the three-tube regenerative receiver, four-tube Browning-Drake, five-tube Diamond-of-the-Air, and the "Grand" 6.

212. INFRADYNE AMPLIFIER—A three-tube intermediate-frequency amplifier for the super-heterodyne and other special receivers, tuned to 3400 kc. (86 meters). Price $25.00.

213. RADIO BROADCAST "LAB" RECEIVER—A four-tube dual-control receiver with one stage of Rice neutralized tuned-radio frequency, regenerative detector (capacity controlled), and two stages of transformer-coupled audio. Approximate price, $78.15.

214. LC-27—A five-tube set with two stages of tuned-radio frequency, a detector, and two stages of transformer-coupled audio. Special coils and special means of neutralizing are employed. Output device. Price $85.20 without cabinet.

215. LOFTIN-WHITE—A five-tube set with two stages of radio frequency, especially designed to give equal amplification at all frequencies, a detector, and two stages of transformer-coupled audio. Two controls. Output device. Price $81.10.

216. K.H.-27—A six-tube receiver with two stages of neutralized tuned radio frequency, a detector, three stages of choke-coupled audio, and an output device. Two controls. Price $86.00 without cabinet.

217. AERO SHORT-WAVE KIT—Three plug-in coils designed to operate with a regenerative detector circuit and having a frequency range of from 10,900 to 2306 kc. (15 to 130 meters). Coils and plug only, price $12.50.

218. DIAMOND-OF-THE-AIR—A five-tube set having one stage of tuned-radio frequency, a regenerative detector, one stage of transformer-coupled audio, and two stages of resistance-coupled audio. Volume control through regeneration. Two tuning dials.

219. NORDEN-HAUCK SUPER 10—Ten tubes; five stages of tuned radio frequency, detector, and four stages of choke- and transformer-coupled audio. Two controls. Price $391.40.

220. BROWNING-DRAKE—Five tubes; one stage tuned radio frequency (Rice neutralization), regenerative detector (tickler control), three stages of audio (special combination of resistance- and impedance-coupled audio). Two controls.

221. LR4 ULTRADYNE—Nine-tube super-heterodyne; one stage of tuned radio frequency, one modulator, one oscillator, three intermediate-frequency stages, detector, and two transformer-coupled audio stages.

222. GREIFF MULTIPLEX—Four tubes (equivalent to six tubes); one stage of tuned radio frequency, one stage of transformer-coupled radio frequency, crystal detector, two stages of transformer-coupled audio, and one stage of impedance-coupled audio. Two controls. Price: complete parts, $50.00.

223. PHONOGRAPH AMPLIFIER—A five-tube amplifier device having an oscillator, a detector, one stage of transformer-coupled audio, and two stages of impedance-coupled audio. The phonograph signal is made to modulate the oscillator in much the same manner as an incoming signal from an antenna.

This is a good time to subscribe for

RADIO BROADCAST
Through your dealer or direct by the year, only $4.00

DOUBLEDAY, PAGE & CO.   GARDEN CITY, NEW YORK

A KEY TO RECENT RADIO ARTICLES

By E. G. SHALKHAUSER

THIS is the twentieth installment of references to articles which have appeared recently in various radio periodicals. Each separate reference should be cut out and pasted on 4" x 6" cards for filing, or pasted in a scrap book either alphabetically or numerically. An outline of the Dewey Decimal System (employed here) appeared last in the January RADIO BROADCAST.

R182. TRANSMISSION OF PHOTOGRAPHS. PHOTOGRAPH
Radio, Feb., 1927. Pp. 18-ff. TRANSMISSION.
"Radio Photography and Television," Dr. E. F. W. Alexanderson.
The writer gives a brief survey of the problems of telephotography and television, the developments made to date, the research being carried on in the laboratory, and the difficulties that will have to be overcome in order to make television practical.

R342.6. RADIO-FREQUENCY AMPLIFIERS. AMPLIFIER,
Radio, Feb., 1927. Pp. 22-ff. *Infradyne.*
"Infradyning R. F. Receivers," E. M. Sargent.
Wiring diagrams and data are presented showing how the infradyne amplifying principle may be adapted to four well-known circuits, namely, the Browning-Drake, the Hammarlund "Hi-Q," the Bremer-Tully Counterphase, and the Silver-Six.

R334. FOUR-ELECTRODE TUBES. VACUUM TUBES,
Radio, Feb., 1927. Pp. 23-24. Four-Electrode.
"A Four-Electrode Tube and Circuit," H. de A. Donisthorpe.
Several uses to which four-electrode tubes may be put are outlined. Circuits are given for rectification, for combined radio-frequency amplification, for detection, and for audio-frequency amplification. A circuit is also given where rectification and radio-frequency amplification are accomplished with the use of only one tube.

R007.1. UNITED STATES LAWS AND REGULATIONS. LAWS
AND REGULATIONS.
Public No. 632-69th Congress. H. R. 9971, Feb. 23, 1927.
"An Act for the Regulation of Radio Communication, and for Other Purposes."
The new radio law of 1927, enacted by the Senate and the House of Representatives, and signed by the President on the 23nd of February, 1927, is printed in this copy, which may be obtained from the Superintendent of Documents, Government Printing Office, at 5 cents per copy. Other acts and resolutions, previously in effect, are repealed by this act placing all control under the new law.

R182. TRANSMISSION OF PHOTOGRAPHS. PHOTOGRAPH
RADIO BROADCAST. March, 1927. Pp. 459-462. TRANS-
"Television: Europe or America First?" MISSION.
E. H. Felix.
An account is given of the theoretical and experimental work carried on by Dr. E. F. W. Alexanderson of the General Electric Company, in the field of television and radio photography. The apparatus consists of a source of light, a lens, and a revolving drum carrying a number of reflecting mirrors. It is stated that, in order to make television a success by the method outlined, "it will be necessary to transmit something like 300,000 pictures per second, a feat very difficult, if not impossible, to accomplish, unless other major difficulties are first overcome.

R343.7. ALTERNATING CURRENT SUPPLY. ELIMINATORS,
RADIO BROADCAST. March, 1927. Pp. 477-479. B-Battery.
"What You Should Know About B Power-Supply Devices," E. H. Felix.
A general discussion of B battery eliminators is given. The essential elements in such a unit consist of a transformer which steps up the line voltage to an amount determined by the requirements, a rectifier element, a system of inductances and filters to smooth out the pulsating output, and a potentiometer output device to obtain various voltages.
Tests made on a dozen different power-supply devices are shown in graph form. The operation, maintenance, and the causes of trouble are outlined in detail.

R343. Electron-Tube Receiving Sets. RECEIVER,
RADIO BROADCAST. March, 1927. Pp. 480-482. Reflex.
"Building the R. G. S. Inverse-Duplex Receiver. Part 3," D. Grimes.
This is the third of a series of articles describing the new Inverse-Duplex system of reception, and presents the constructional details for the adaptation of the previously outlined developments to the R. G. S. receiver. Wiring diagrams, a list of parts required, and data on winding the special coils used, are given.

R343.5. POWER AMPLIFIER. AMPLIFIER,
RADIO BROADCAST. March, 1927. Pp. 489-492. Power
"Constructing an Amplifier-Power Supply Device," Supply.
James Millen.
Detailed information is given on the construction of a three-stage resistance-coupled power amplifier operated directly from an a.c. source. The data include the winding of the power transformer, choke coils, and output impedance.

R343. Electron-Tube Receiving Set. RECEIVER,
RADIO BROADCAST. March, 1927. Model EA Garod.
Pp. 493-495.
"A C. as a Filament-Supply Source," B. F. Miessner.
Data are given on the operation of a receiver deriving all of its power from the lighting mains. Three 112 type tubes, one 199 type tube for detector, and one 210 type tube for the last stage of power amplification are used in this set. A description of the power conversion unit is given.

$63.05 LESS CABINET

Yes you CAN Build this Set!

Thousands of Fans Guarantee Success!

RADIO engineers—professional builders—dealers —*everyone in radio*—admits that the New Hi-Q Shielded Receiver is the most successful home-built set! Thousands have built it at home—every one a perfect success.

This background of nationwide success guarantees that you, too, can build this remarkable receiver and save $50 to $100 over any manufactured set of similar quality.

Send 25c for complete, illustrated instruction book or get copy from your dealer.

Approved parts from your dealer

ASSOCIATE MANUFACTURERS
Benjamin Elec. Mfg. Co.   Hammarlund Mfg. Co.
Carter Radio Co.   Radiall Co. (Amperite)
International Resistance Co.   Martin-Copeland Co.
(Durham Resistors)   Samson Electric Co.
Eby Mfg. Company   Sangamo Electric Co.
            Westinghouse Micarta

Hammarlund ROBERTS HiQ

HAMMARLUND-ROBERTS, Inc.
1182 Broadway, New York City

MU-RAD

Announces a new Super-Six

All Electric Receiver

Striking innovations and remarkable reception feature this new set

A NEW and even better Mu-Rad, beautifully encased in cabinets of two toned walnut decorated with burled panel effects, is now at your disposal. A Mu-Rad that you can operate with or without batteries, with *only one* tuning control, an indoor antenna if desired and every improvement that modern radio engineering can devise. You will be amazed at the extreme selectivity, the faithful tonal qualities, the volume range and the remarkable distance attainable from this new and unusual radio. Hear the Mu-Rad at your earliest opportunity. You will enjoy radio reception that has out-performed all competition in every radio test, since the birth of the industry.

No A. B. C. batteries.
No electrolyte.
One Tuning control.
One volume control.
Indoor antenna operation.
Extreme selectivity.
Ultimate in tone quality.
Volume to meet any requirement.
A go-getter on distance.
Unusually easy operation.

MU-RAD RADIO CORPORATION
DEPT. B, ASBURY PARK   **NEW JERSEY**

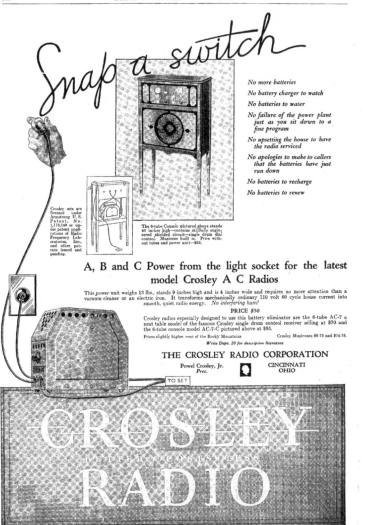

Snap a switch

No more batteries

No battery charger to watch

No batteries to water

No failure of the power plant just as you sit down to a fine program

No upsetting the house to have the radio serviced

No apologies to make to callers that the batteries have just run down

No batteries to recharge

No batteries to renew

Crosley sets are licensed under Armstrong U. S. Patent, No. 1,113,149 or under patent applications of Radio Frequency Laboratories, Inc., and other patents issued and pending.

The 6-tube Console pictured above stands 40 inches high—contains skillfully engineered shielded circuit—single drum dial control. Musicone built in. Price without tubes and power unit—$95.

A, B and C Power from the light socket for the latest model Crosley A C Radios

This power unit weighs 13 lbs., stands 9 inches high and is 4 inches wide and requires no more attention than a vacuum cleaner or an electric iron. It transforms mechanically ordinary 110 volt 60 cycle house current into smooth, quiet radio energy. *No interfering hum!*

PRICE $50

Crosley radios especially designed to use this battery eliminator are the 6-tube AC-7 a neat table model of the famous Crosley single drum control receiver selling at $70 and the 6-tube console model AC-7-C pictured above at $95.

Prices slightly higher west of the Rocky Mountains   Crosley Musicones $9.75 and $14.75.

Write Dept. 20 for descriptive literature

THE CROSLEY RADIO CORPORATION

Powel Crosley, Jr.   CINCINNATI
Pres.   OHIO

TO SET

CROSLEY

BETTER · COSTS · LESS

RADIO

New!

You don't have to be water boy to this battery charger

The Thordarson Battery Charger R-175 employs the Raytheon Rectifying Cartridge guaranteed as above.

THORDARSON
BATTERY CHARGER
R-175

Radically new—sound in principle—proven in performance.

The Thordarson Battery Charger makes its bow as a welcome relief to the army of butlers to thirsty battery chargers.

Dry—As dry as they make 'em. In fact the rectifying element is contained in a moisture-proof cartridge.

Silent—No vibrating parts. Current is rectified through a patented electro-chemical process.

Safe—There is no hazard to rugs or woodwork for there is no acid to spill. The tubes of the set are safe even if turned on when charger is in operation.

Compact—Fits into battery compartment easily. Only 2¾" wide, 5⅝" long and 4¼" high overall.

Efficient—This charger is always ready for service. No overhauling required. Rectifying element can be replaced in thirty seconds.

Guaranteed—The rectifying unit is guaranted for 1,000 hours full load operation, or approximately one year's normal service. The Transformer will last indefinitely.

Charging Rate—2 Amperes

For Sale at Good Dealers Everywhere or Direct from Factory

PRICE COMPLETE $12.50

THORDARSON ELECTRIC MANUFACTURING CO.
Transformer Specialists Since 1895
WORLDS OLDEST AND LARGEST EXCLUSIVE TRANSFORMER MAKERS
Huron and Kingsbury Streets — Chicago, Ill. U.S.A.

RADIO
BROADCAST

CONTENTS

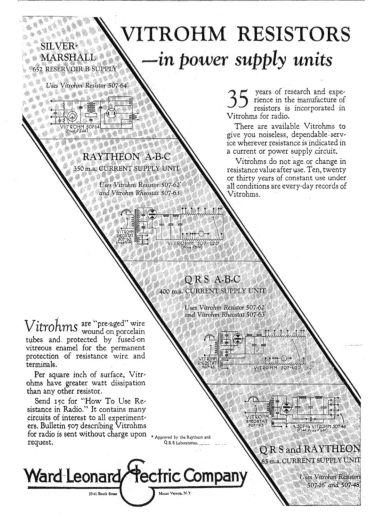

ZP-201-A power tubes

in every socket— Without Re-wiring!

—as many power tubes as there are sockets in your set

THINK of it—without adding or altering a single wire you can now use as many power tubes as there are sockets in your set . . . the greatest advance in power, volume and tone since the days of crystal receivers.

The new oxide filament, $\frac{1}{4}$ amp. ZP 201 A power tubes will give you everything you want in radio—power, extreme selectivity, realism of tone. *And more!*

They are economical in "A" and "B" current which materially lengthens the hours of battery service.

Each tube is serial-numbered and *guaranteed*. Every Zetka tube will give you the added satisfaction of increased efficiency plus longer life. You cannot realize what a difference these tubes will make in your set until you try them and convince yourself.

Go to your nearest Zetka dealer today and ask for a demonstration. Only $2.50 each.

ZP 201 A is one of a complete line of clear glass tubes—each one meeting a definite radio demand.

[Zetka's sensational new 6 volt $\frac{1}{10}$ amp. tube for electric sets is ready]

ZETKA LABORATORIES INC., 73 Winthrop St., NEWARK, N. J.

ZETKA
The *Clear Glass* Tube

RADIO BROADCAST. July, 1927. Published monthly. Vol. XI. No. 3. Published at Garden City, N. Y. Subscription price $4.00 a year. Entered at the post office at Garden City, N. Y., as second class mail matter. Doubleday, Page & Company, Garden City, N. Y.

You Can't Carry a Load of Hay on a Wheelbarrow

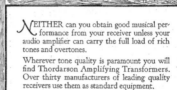

NEITHER can you obtain good musical performance from your receiver unless your audio amplifier can carry the full load of rich tones and overtones.

Wherever tone quality is paramount you will find Thordarson Amplifying Transformers. Over thirty manufacturers of leading quality receivers use them as standard equipment.

Follow the lead of the leaders. Whether buying or building a receiver—if you enjoy music —insist on Thordarson Amplification.

You don't have to be water boy to the

THORDARSON BATTERY CHARGER

DRY—As dry as they make them. In fact the rectifying element is contained in a moisture-proof cartridge.

SILENT—No vibrating parts. Current is rectified through a patented electro-chemical process.

SAFE—There is no hazard to rugs or woodwork for there is no acid to spill. The tubes of the set are safe, even if turned on when the charger is in operation.

COMPACT—Fits into battery compartment easily. Only 2¾" wide, 5¾" long and 4¾" high, overall.

EFFICIENT—This charger is always ready for service. No overhauling required. Rectifying element is held in spring clips and can be replaced in thirty seconds.

LONG LIFE—The Raytheon rectifying unit used in this charger is guaranteed by its manufacturer for 750 hours of full load operation, or approximately one year's service. The transformer will last indefinitely.

CHARGING RATE—2 amperes.

Price $12.50

THORDARSON
Power Compact

The only complete foundation unit for home-built power amplifiers. Contains power supply transformer, 2 30-henry choke coils, 2 buffer condensers and center tapped filament supply—all in one compound filled case. Supplied in two types: R-171 for Raytheon rectifier and UX-171 power tube, and R-210 for UX-216 B rectifier and UX-210 power tube.

Type R-171, $15.00
Type R-210, 20.00

THORDARSON R-200
Amplifying Transformer

A transformer designed for the musical epicure. The large core of finest silicon steel and the high inductance primary winding combine to give this instrument the most perfect transformer reproduction obtainable. Has a remarkably wide range of amplification. Ideal for use with cone type speakers. Designed for both first and second stage amplification. Weight, 2 lbs.

Price $8.00

THORDARSON

RADIO TRANSFORMERS

Supreme in Musical Performance!

THORDARSON ELECTRIC MANUFACTURING CO.
Transformer specialists since 1895
WORLD'S OLDEST AND LARGEST EXCLUSIVE TRANSFORMER MAKERS
Chicago, U.S.A.

RADIO BROADCAST

Willis Kingsley Wing, Editor

JULY, 1927

Keith Henney
Director of the Laboratory

Edgar H. Felix
Contributing Editor

Vol. XI, No. 3

CONTENTS

AMONG OTHER THINGS. . .

THIS issue of Radio Broadcast is almost an A.C. current supply number, for there is a wealth of information on the new methods for running vacuum tube filaments and supplying B and C potentials too. Although the four articles concerned do not total a great number of pages, they are the result of a great deal of laboratory work, and they contain plenty of information to aid the home constructor. Both the Raytheon and the Q. R. S. high-current rectifier units have been carefully tested for many weeks in the Radio Broadcast Laboratory under Howard Rhodes's direction and applied to receivers of various sorts. Much interest will also attach to James Millen's descriptions of the remarkable A battery charger tube developed in the Raytheon laboratory.

THOSE of a theoretical turn of mind will find Austin Cooley's story dealing with the aurora and radio fading, as observed on the last MacMillan Arctic expedition, of considerable interest. Mr. Cooley's deductions are ingenious and perhaps will cast considerable light on some of the problems of radio transmission. Neither Radio Broadcast, nor indeed, Mr. Cooley, regard the theory as more than an entering wedge. We hope other investigators may be encouraged to make actual long-time measurements which will give us a better basis for conclusion.

RADIO men, traditionally, are argumentative souls and those who have turned their attention to the economics of the current supply of radio receivers will find some interesting figures on page 146 of this number. Our position is neutral, but discussion is interesting.

OUR correspondents are constantly asking us for lists of radio reading matter, text book and periodical. A very helpful list of recent works is found on page 168 of this issue. There are, of course, many other standard books of value, many of which have in the past been reviewed in this magazine.

RADIO BROADCAST for September will contain the first of an usually complete series of articles on the elimination of interference. These articles are practical and definite and should help many a puzzled citizen to improve local reception. A shielded neutrodyne which can be built from standard parts will be described in the same issue. It was designed by H. G. Reich, a member of the physics department at Cornell University. A beautiful 80-meter code and phone transmitter will be described, also, and it will delight the heart of all of us who are interested in a compact efficient transmitter for this wave-band.

MANY features of Radio Broadcast, as our readers have discovered and generously appreciated, are designed to supply regularly concise information which is not to be found elsewhere. The Laboratory Data Sheets were the first of these features of this nature, then "The Best in Current Radio Periodicals," the "Manufacturers' Booklets Available," and, with the June issue, we began a listing of kits for building receivers, together with a brief technical summary of each. Information about kits can be secured through the Service Department of Radio Broadcast by exactly the same procedure that so many of our readers have followed with the manufacturers' booklets.

—Willis Kingsley Wing.

Doubleday, Page & Co.
MAGAZINES

Country Life
World's Work
Garden & Home Builder
Radio Broadcast
Short Stories
Educational Review
Le Petit Journal
El Eco
Frontier Stories
West

Doubleday, Page & Co.
BOOK SHOPS
(Books of all Publishers)

New York: {Lord & Taylor Book Shop
Pennsylvania Terminal (2 Shops)
Grand Central Terminal
38 Wall St. and 526 Lexington Ave.
848 Madison Ave. and 166 West 32nd St.
St. Louis: 223 N. 8th St. and 4914 Maryland Ave.
Kansas City: 920 Grand Ave. and 206 W. 47th St.
Cleveland: Higbee Co.
Springfield, Mass.: Meekins, Packard & Wheat

Doubleday, Page & Co.
OFFICES

Garden City, N. Y.
New York: 285 Madison Avenue
Boston: Park Square Building
Chicago: Peoples Gas Building
Santa Barbara, Cal.
London: Wm. Heinemann Ltd.
Toronto: Oxford University Press

Doubleday, Page & Co.
OFFICERS

F. N. Doubleday, President
Nelson Doubleday, Vice-President
S. A. Everitt, Vice-President
Russell Doubleday, Secretary
John J. Hessian, Treasurer
L. A. Comstock, Asst. Secretary
L. J. McNaughton, Asst. Treasurer

DOUBLEDAY, PAGE & COMPANY, Garden City, New York

Q·R·S

(Trade Mark Registered)

RADIO TUBES ARE BETTER

Q·R·S	Q·R·S	Q·R·S
201A Type	Power tube	
	$4.50	Super Detector
$1.75	Audio Amplifier Last Stage Only	
Detector Amplifier		$4.00

Q·R·S SPECIAL TUBES

Q·R·S	Q·R·S	Q·R·S
Voltage Regulator	Full Wave	Full Wave
Glow Tube	Gaseous Rectifier	Gaseous Rectifier
	175 Volts D. C.	200 Volts D. C.
$5.50	60 Mills	85 Mills
	$4.50	$5.00

Q·R·S

Rectifying Slug

2½ Amp.

$4.50

300 Volts D. C.
400 Mills D. C.

**Q·R·S
High Voltage
Full Wave Gaseous
Rectifier Tube with
ionizer
600 Volts D. C.
100 Mills
$6.00**

The Pioneer 400 Milliampere Q·R·S Full Wave Gaseous Rectifier. tube with ionizer.

The tube that makes possible the elimination of A, B & C batteries on standard 201A type receivers— Price $7.00.

We will furnish without charge— charts and diagram for building five, six and seven tube sets using 201A type tubes and Power Tube— operated direct from the house current supply.

THE Q·R·S MUSIC COMPANY, *Manufacturers*
Executive Offices, 306 S. Wabash Ave., Chicago
Factories—Chicago, New York, San Francisco, Toronto

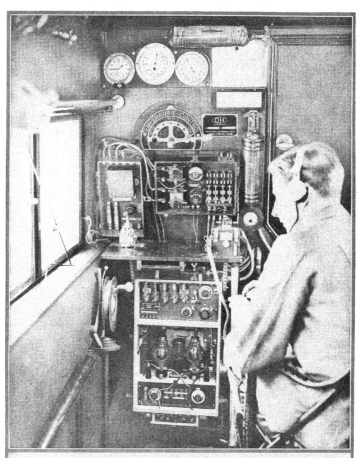

IN THE RADIO CABIN OF AN ENGLAND-EGYPT-INDIA AIRPLANE

Under the new international air regulations, a radio operator and mechanic is carried aboard passenger air liners. Previously the wireless apparatus was operated by the pilot and fitted in the cockpit. The 150-watt apparatus shown is aboard a De Haviland 66 Hercules multiple-engined machine on the England-Egypt-India air route. The reel for lowering the antenna is at the left

JUN 23'27

©C1B 747105

RADIO BROADCAST

VOLUME XI

NUMBER 3

JULY, 1927

The Aurora and Fading

A Report of Some Observations on the Relation Between Radio Signals and the Aurora Made During the Last MacMillan Arctic Expedition to Greenland and Labrador

By AUSTIN G. COOLEY

THE mystery of fading and freak radio conditions has been one of the most prominent problems in the science of radio since its earliest days. It has been observed in a general way that some relation exists between radio fading and other phenomena such as air pressure, temperature, humidity, aurora, etc., but no generally accepted hypothesis or evidence has been advanced for consideration that definitely ties these phenomena together.

Besides the studies made during the 1926 Rawson MacMillan Sub-Arctic expedition for the Field Museum of Chicago, were the efforts made by the writer to find a relation between the aurora and fading. As will be remembered, the *Sachem* (one of the two boats on the Expedition) was outfitted with radio equipment designed in the Laboratory of RADIO BROADCAST. The writer was the operator chosen to make the trip, and was thus able to make the studies which are outlined in this article.

Every effort was made to collect information that would prove valuable in determining what relation, if any, existed between aurora and radio fading. Owing to the lack of time and the limited amount of apparatus available, it was possible to collect only a fraction of the information that might be useful in the study, but sufficient was observed to point out a possible solution to one of the mysteries of radio fading.

During an early part of the Expedition, while on the coast of Labrador, mirages of ice and land caused considerable attention and comment. Commander MacMillan spent an evening telling of his experiences and observations on mirages to the members of the crew of the *Sachem*. He said that, if conditions were right, it would be possible to see mirages of objects half way around the world. This suggested the possible relation of mirages to short-wave radio transmission because radio and light waves are the same except for length. The shorter radio waves, which have a closer re-

lation to light, seem to behave in very much the same way as light in that they can be reflected or refracted so as to reach far distant parts of the globe.

A radio communication from the Expedition to RADIO BROADCAST commenting on this possible relationship, and telling of the mirages on the Labrador Coast, brought the following reply:

Greatly interested in your reflection idea. New York *Times* July 15 quotes Captain Rose of Steamer *President Adams* at 8 P. M. July 15 in Mediterranean Sea bound for Port Said quote Saw large field of floating ice cakes suspended above horizon and presently a number of small pieces drifted into view followed by a large one. The latter was so clear that we could see blue and green veins in the ice unquote Nearest ice field was 8000 miles away.

This news gave one member of the Expedition courage enough to tell a story he had heard about people in an Alaskan town seeing a mirage of an European city.

The mirages seen off the Labrador Coast formed very slowly. As a rough guess, the time averaged around five to fifteen minutes. They generally remained in position for a number of

hours. Because of this slow action, we did not suspect that there might be a connection with radio fading, but observations taken about a month later indicated that the occurence of mirages was an important factor to be considered.

Except for the last few days, no unusual radio conditions prevailed while the Expedition was off the Greenland coast. Reception on short waves seemed to be excellent regardless of position with respect to surrounding land or mountains. During the last three or four nights, however, it was very difficult to establish communication with any of the amateur stations in the States. For long periods, practically no amateur signals came through. The weather conditions at first were such that it could not be determined whether aurora existed at the time or not, but the weather cleared up enough the night of sailing west from Sukkertoppen, Greenland, to allow us good vision of the skies. Strong displays of aurora were observed but no time was then available to make any studies. While crossing Davis Straits to Baffin Land, fading on short waves made communication very difficult as the signals generally faded in and out in cycles varying in length from five to fifteen minutes. Again, thick weather prevented thorough investigation.

It was not until the Expedition arrived at Saglek Bay, Labrador, that anything definite could be determined regarding the aurora. Fortunately the displays at Saglek Bay were unusually strong and the weather perfectly clear. Here it was soon found that some association existed between the aurora and fading yet there were times when the signals were absolutely unaffected during strong displays. The problem resulted into one of determining just when and how the signals were affected in relation to the aurora. Having no information that would suggest what to look for, it was difficult to determine just what observations should be made. After watching the aurora and signals for a little over an hour, there appeared to be some

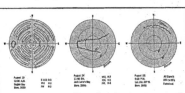

HOW THE AURORA WAS LOGGED.

The observer is supposed to be situated at the center of the circle and if some aurora is noted in the northern sky at an angle of somewhat less than 30° with the horizon, then some lines to indicate it are roughly drawn on the sketch between the horizon and the 30° circle. Such conditions existed when the right-hand sketch was made, the effect on radio signals being noted below the drawing. The center sketch shows aurora of a somewhat different form while the left-hand sketch indicates that the sky was, at this particular time, completely covered with bands of aurora, while a small cluster was also seen in the west

relation between the location of the aurora and strength of signals but it was all very puzzling. At times the sky would be completely covered with aurora yet the signals came through nicely, while at other times, the signals faded out completely when only a very small amount of aurora was visible.

While at sea the night after this display at Saglek Bay, further observations were taken but under quite different conditions. The aurora formation was entirely different in that it formed in long bands on the northern horizon then moved across the skies to the zenith—at times the bands traveled on to the south until they reached within thirty degrees of the southern horizon before they faded out. The relation of the aurora bands to radio fading appeared very definite. When the bands were between ten and thirty degrees above the horizon, the signals coming from the opposite direction to the aurora faded out completely. When the bands were more than forty-five degrees above the horizon, the signals were not sufficiently affected that any difference could be determined between reception then and under normal conditions. The actual intensity of the aurora seemed to have little or no effect on the signals.

The magnetic storm accompanying the aurora display was observed by taking bearings on islands and then watching the compass. As the streamers passed overhead, the compass was observed to swing about eight degrees, but when the streamers were close to the horizon, no effect on the compass was observed. This information appears to be interesting in that the radio fading did not occur during those periods when the magnetic swing was the strongest.

The district in northern Labrador where these studies were made is known as the zone of maximum amount of aurora. Consequently, the opportunities for making investigations on terrestrial magnetism and radio fading are great in these regions. For good work, a complete line of apparatus is necessary, the most essential thing being some device for scaring away the mosquitoes. Labrador is also the zone of maximum amount of flies and mosquitoes! Preparations were made to measure intensity and direction of earth currents during an aurora display near Nain, Labrador, but it was not possible to stay ashore for longer than ten minutes at a time on account of the flies, despite the fact that head nets and oils were used for protection.

The magnetic storms continued and the weather remained clear for a number of days, so it was possible to observe the aurora at leisure. All the information we gathered checked nicely with that obtained during the first two nights.

Fading was watched on two wave bands; one from twenty to two hundred meters (15,000 to 3000 kc.) and the other from two thousand to eighteen thousand meters (150 to 16 kc.). Because of familiarity with reception between thirty-two and forty-eight meters, this particular band was given the most attention. While at Saglek Bay, a broadcast receiver was set up ashore for the purpose of receiving a special program being broadcast from WJAZ, Chicago, but it was not possible to hear any of the broadcast stations. Under normal conditions, the broadcast stations could be heard reasonably well, even in Greenland.

Before trying to account for the fading during certain phases of the aurora displays, it probably would be interesting to consider a few of the definite cases. It was found that the European commercial short-wave stations were the last to fade out, their direction being a little south of east. Generally, when they were received weakly, it was found that they were sending test signals or sending traffic slowly and each word twice, this indicating that they were also experiencing the fading effects. Station KEL, near San Francisco, generally came through even when all the other stations on short waves had faded out completely. This station was operating on a wave of about thirty meters and its direction bore about thirty degrees south of west. At no time was it noticed that KEL was having difficulty getting his signals through to his receiving station which was in a westerly direction.

The schooner *Morrissey* of the Putnam Expedition, which was about a thousand miles to the north of us, operated without difficulty in communicating with a station near Chicago. The *Morrissey's* signals were received with normal intensity on the Labrador Coast when the aurora was very strong in his direction but his signals faded out when the aurora was confined chiefly to the southern horizon.

No fading effects were noticed on wavelengths above two thousand meters. No static noises were noticed at all during any of the aurora displays and all was quiet except for noises caused locally aboard the ship.

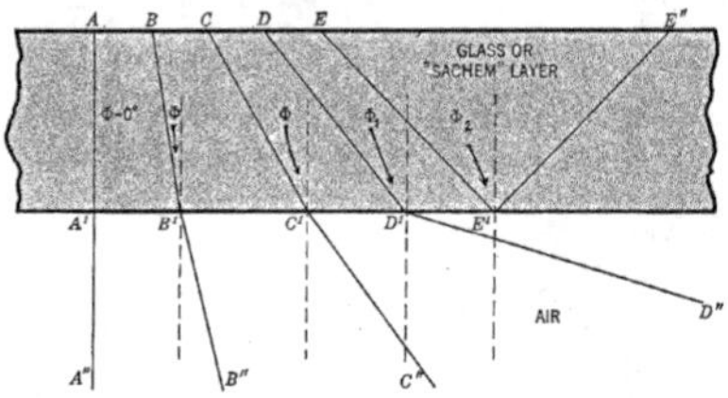

THEORIES INVOLVED

T HE physics of light seems to play an important part in the phenomena of radio wave propagation. In trying to account for radio fading it is best to consider first a few of the simple laws of reflection and refraction of light beams. When a light beam passes through one medium, e.g., glass, into another, e.g., air, the beam is bent or refracted. The amount of bending is a factor of the difference in speed with which the light travels through the two different mediums. The relative speeds of light in mediums are designated by their "index of refraction." The index of refraction of a vacuum is taken as unity. Glass has an index of refraction of about 1.6, and diamonds about 2.4. The "index of refraction" also varies with the wavelength. In the light and heat spectrum, the index decreases with the increase of wavelength. This has also been found to be true in the radio wave spectrum.

Fig. 1 shows how a light beam is refracted when passing through glass into air. The letters A, B, C, D, E, represent the sources of light beams. As the angle of incidence, φ, increases, a point will be reached where the ray will not pass into the air but will be totally reflected. The smallest angle at which total reflection occurs

is known as the "critical angle," and in Fig. 1 is some value between that of φ₁ and φ₂.

Considering that radio waves are the same as light waves except for the difference in length, we may expect them to behave similarly. With mirrors it can be definitely shown that short radio waves may be reflected in the same manner as light waves. It has also been demonstrated that radio waves may be refracted as light waves are when passed through a prism or lens. The amount of refraction for radio and light waves is almost the same in some prisms.

A radio wave is known to follow two paths, one along the surface of the earth and another into the higher atmospheres. Some of the waves taking the latter course appear to be reflected back to earth after they reach a certain height. The medium which causes this reflection is known as the Kennelly-Heaviside Layer and its height has been determined through the efforts of Dr. A. Hoyt Taylor and Doctor Hulburt to vary from 100 to 500 miles, depending upon the time of day and the season. The theory of reflection from the Kennelly-Heaviside Layer is generally accepted, so there is no need of considering it here in detail.

If a medium is placed over a receiving station so that the radio waves coming down from the Kennelly-Heaviside Layer are reflected, the receiving station will be completely shielded except for that portion of the radio waves which travels over the surface of the earth. Assuming that such a shielding medium does exist under certain conditions, let us consider just what effects may be realized. Referring again to Fig. 1, assume that the ray E-E¹ is a radio wave and the medium, instead of being glass, is the lower shielding layer or reflecting medium just referred to as existing between the station and the Heaviside Layer. This reflecting layer may for the sake of convenience, be called the "Sachem" Layer. If the radio wave strikes so that φ is more than or equal to the critical angle, the wave will be totally reflected. In Fig. 2 it is seen that, if the lower surface of the layer is cut at an angle, the same wave will pass on through to the receiving station.

If the lower surface is cut so as to incline the opposite way, Fig. 3, the waves C and D will strike at an angle more than the critical angle and will not pass on to the receiver.

If such a layer as suggested above does exist, it is most likely that its surface will be irregular and probably appears as waves that continually move due to slight disturbances. Any such layer then would cause radio fading which would be somewhat in proportion to the index of refraction of this layer, since the critical angle is dependent upon the refractive index.

Before referring any more to the observations made on the Expedition, it is best to consider the theoretical possibilities of any such layer. The requirements are that the index of refraction of the layer be greater than that of the atmosphere below if we are to obtain the reflections shown in the diagrams. It is also necessary to prove that the Heaviside Layer is above the Sachem Layer and that the two do not coincide and are not the same thing. There is not enough data available really to prove either of these two points; the best we can do is to attempt to draw reasonable conclusions from what information we have to work with.

In regard to relative heights of the two layers,

FIG. 1

the writer is guided by the work of Doctor Taylor and Doctor Hulburt and the studies made on the height of the aurora. The work of these two scientists appears to indicate that the Heaviside Layer at night on the Labrador Coast would be considerably over a hundred miles above the earth's surface. As will be mentioned later, it appears that the aurora occurs on the surface of the Sachem Layer. The aurora has been found to average about sixty miles in altitude.

The changes of refractive index at different altitudes cause light waves to be reflected and bent so objects at a distance may be seen. The images are known as mirages. The causes for the changes in refractive indices are variations in air pressure and temperature, but it stands to reason that the air pressure decreases with increase of height so we cannot expect to find in the upper atmospheres a layer having a higher refractive index than the air below unless some other factor than pressure has a control. From studies made with light it is found that the refractive index increases as the air gets colder. If there is a sudden drop in temperature in the upper atmosphere, we can plainly see that we will have the proper conditions for a reflecting medium that would reflect all waves striking at less than the critical angle.

At the irregular boundary line between the cold and warm air, the light waves are bent, or refracted, so as to give the wave appearance. With the air pressure as high as it is at the earth's surface, only a small change in temperature is necessary to produce these visible waves but, where the air pressure is less, a considerable change in temperature would be necessary to produce an appreciable effect. It was noticed that the fading effects that might be attributed to the Sachem Layer, only occurred when the air pressure was high. An increase of pressure at the line where the temperature drops would increase the possibilities of the existence of the Sachem Layer.

It is also to be considered that the surface of such a layer would be sharp since large temperature changes can occur in a very short distance. One often notices the sudden temperature changes in the air when driving in an automobile.

CHECKING UP THE HYPOTHESIS

THE investigations made on the Expedition appear to connect closely with the above hypothesis of the Sachem Layer in remarkable detail. Mirages and aurora appeared only as associated phenomena and not the direct cause of fading. From studies made, it appears that the aurora followed along the crest or high part of the Sachem waves. Because the waves were marked by the aurora, it was possible to tell just when signals would come through and when they would suffer reflection.

When the waves were very flat, the aurora appeared to fall from their crests into the troughs. This resulted in the entire sky being covered with aurora but, because the waves were rather flat, the radio waves were not subject to as much total reflection. Fig. 1 represents the case of a flat wave.

Since the inertia of the body of the Sachem Layer is very little, it is possible for the waves to travel at a very high speed. Judging from observations made in Labrador, it appeared that the waves traveled from five to thirty miles per minute. The major waves would undoubtedly be surfaced with smaller waves, as is the case with

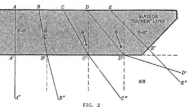

FIG. 2

ocean waves. The smaller, or minor Sachem waves, can cause additional fading that would appear more rapid.

It is to be noted that this Sachem wave hypothesis of fading takes into account the relation of temperature and humidity to the fading. When the temperature is high in the earth's atmosphere, a greater and sharper index of refraction will occur at the Sachem Layer. This will cause a larger portion of radio signals to be reflected so they will not reach the earth's surface. If the temperature is low, the change of index of refraction at the Sachem Layer may not be enough to cause any reflection even though the pressure may be high.

The humidity increases the index of refraction so fading would occur in a higher degree when the humidity is high. The effects of humidity on radio fading were given considerable study by Frank Conrad (8 xx, Pittsburgh) and James C. Ramsey (1 xa, Boston) when carrying on the first short-wave radiophone tests in 1922 on the sixty-meter band. They were thoroughly convinced that humidity affected fading.

It is also interesting to note that radio communication is chiefly affected in the north and south directions during aurora displays while land wire and cable communications are affected in the east and west directions, due to potentials generated in the earth probably by the swinging magnetic fields.

Vaudeville artists have been unable to determine whether the hen or the egg came first. With our present information it is difficult to say whether the aurora causes the Sachem Layer and waves or whether the Sachem Layer and waves cause the aurora. Probably neither is correct, but it seems certain that they occur at the same time and are dependent upon each other. The heavy air pressure on the surface of the earth has a direct relation to the Sachem Layer and may be

the cause of it. The Sachem Layer may produce the proper conditions so that, by means of the magnetic field revolving with the earth, high enough potentials may be produced to cause an electrical discharge along the Layer in places where the proper pressure or "critical pressure" exists. Such an electrical discharge would tend to put a load on the earth's magnetic field and distort it. This may be the answer to the swinging of the compass as the aurora bands pass overhead.

It appears that only a slight change of index of refraction is enough to refract a radio wave considerably. This may be accounted for by reflection of polarized waves but, to avoid confusion, a discussion on this point will be omitted.

As mentioned before, a change of pressure or temperature will cause the index change. The index also varies with different gases. A possibility exists that the change at the Sachem Layer may be due to a gas change. If this should be true, it is quite likely the gas responsible is nitrous oxide.

In 1912, Mount Katmai in Alaska erupted. A large amount of sulphur dioxide gas escaped and was so strong at Cordova, Alaska, about 400 miles away, that enough dissolved in the rain to bleach cloth. Radio communication was impossible for a number of days, even over a distance of a few miles. Apparently even the earth waves suffered such reflection that they were unable to travel any distance from the transmitter.

Another case of the apparent effect of gases was noticed by the writer when operating two receiving sets, one from a loop and the other from an antenna subjected to smoke from a chimney. Considerable fading from a local station was noticed on the set operating from the antenna while the loop set operated without any noticeable change in signal strength.

It is hoped that the information and hypothesis advanced in this article will encourage other investigators to work along lines to prove or disprove all that is claimed here, so that, by a process of substitution and elimination, more of the factors influencing radio fading will soon be known.

The writer wishes to thank Dr. J. H. C. Martens, geologist of the expedition, for his help in obtaining information on the atmospheric conditions; Commander MacMillan for his helpful suggestions; and Commodore Rowe B. Metcalf who built and financed the schooner *Sachem*.

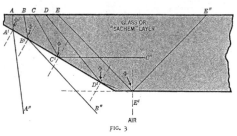

FIG. 3

What the Radio Commission Is Accomplishing

THE Federal Radio Commission has proceeded to its labors with caution and moderation. Its first step was to clear of American pirates, the six usurped Canadian channels. Next it cleared the various channels of those stations which elected to occupy two broadcasting channels rather than one, by restricting allocations to channels exactly ten kc. apart. It definitely decided against broadening of the broadcast band. It opened up the lower frequency end of the amateur band for experimental purposes. It adopted a policy of temporary licensing of broadcasting stations, reserving the option to change the assigned frequency or power of any station at any time or to revoke its license at will. This permits the Commission to arrive at the ultimate lineup of stations by a careful process of evolution rather than by a dangerous upheaval.

The Commission has made a start in cleaning up the New York and Chicago situations by assigning all stations to channels at least twenty kc. apart, thereby minimizing heterodyning and cross talk between stations in the same area. Some incomprehensible broadcasters had actually been using channels only five, three, and two kc. apart from other stations in the same area, thereby successfully eliminating all listeners from both channels. The ultimate plan is to require at least a fifty-kc. separation between stations, allowing the simultaneous operation of twenty stations in the same area.

Fifty-kc. separation, although a vast improvement, still means a nearly complete air blanket by local stations in New York and Chicago for all but the most expensive radio sets. It is far from the ideal broadcasting situation. The next step should be to eliminate a few more excess ether pollutors, of which there are many in New York and Chicago, whose demise will be loudly applauded by all listeners. For example, there is a group of anaemic New York stations, comprising the last word in program mediocrity,

which are linked up in a chain, thus inflicting their hopeless hours upon the listener at three places on the dial instead of one, There is no excuse for a chain of stations serving the same area except when the combination is a temporary one, linked only for program events of transcendant importance, such as presidential addresses or national election returns.

A further reduction in the number of channels required by New York and Chicago can be effected by requiring the splitting of time on the part of stations so limited in their ability to serve the public that they can afford to broadcast only during the prize hours of the evening. A station broadcasting only in the evening is not entitled to an exclusive channel. It should be limited to perhaps two evening programs a week. That would help to cut down the unpopular small fry.

Radio Freedom of Speech Impossible

IF WE omit those who make a profitable profession of agitating for freedom of speech, the radio audience is little exercised about this subject. We have examined several hundred replies to the questionnaire published in the May issue which, at this writing, are still being received from our readers in great numbers daily. Those who answered the questionnaire—and they are from all over the country—were emphatic and practically unanimous in their disapproval of the religious stations. In the Chicago district, another marked trend is indicated by the many requests for the elimination of WCFL and WJAZ. The feeling against WJAZ is particularly strong. Apparently undermining the Department of Commerce's legal status has caused distinct reaction. Other parts of the central west take particular delight in urging the immediate extinction of KFNF, KMA, and KTNT because of their blatant direct advertising. We regret that a few of our readers have even suggested diabolical forms of punishment for the owners of these stations, apparently a subconscious influence of previous incarnations, many being reminiscent of the Spanish Inquisition. A host of lesser stations come in for their share of blackballing, but the expressions toward them are not nearly as consistent or as emphatic as those directed against religious, direct advertising, and propaganda stations.

The old established stations, both of the chain and, we are glad to say, the better independents, rank highest among those stations which listeners wish retained. In our question, "What distant stations do you wish retained?" the percentage within range of KDKA putting that station first, is striking. The regularity with which KDKA, WEAF, WGY, and WJZ appear in that order is also remarkably consistent, while KYW, WGN, KPO, and KFI are strong in the approval of listeners within their respective ranges.

For making this statement of fact, we shall, no doubt, receive one or two letters from rabid demagogues, inquiring how much we were paid by the R. C. A. or the N. B. C. for mentioning the obvious superiority of some of their affiliated stations.

"BEAM" WIRELESS SERVICE DIRECT FROM ENGLAND TO AUSTRALIA

Messages are now sent at an average rate of 100 words per minute between London and Melbourne. The English receiver at Skegness uses the Marconi UG3 high-speed automatic syphon recorder shown in the illustration. An automatic tape puller is at the left

THE GOVERNORS OF THE UNITED STATES ETHER
The Radio Act of 1927 became law on February 24th. Its provisions did not go into force until two months later. From the time the Commission was appointed, it has been very active. In March, public hearings were held in Washington, giving a reliable cross section of opinion and suggestion for dealing with the broadcasting problem. The Commission has cleared the Canadian channels, reduced power of stations in residential districts to 500 watts, reënforced the 10-kc. separation between stations, and already reduced the total number of broadcasters from 732 to approximately 660.
From left to right: H. A. Bellows, E. O. Sykes, W. H. G. Bullard, J. F. Dillon, O. H. Caldwell

In response, we regret to advise that, to date, we have not been rewarded in any way. We would like particularly to publish in full a scurrilous letter along these lines, sent us by a certain Mr. O'Hara of Nebraska, because it is representative of the type of letter which sometimes follows favorable mention of the pioneer broadcasting stations. The ratio of letters, endorsing our stand for a drastic cut in the number of stations, to these rabid letters, accusing us of being bribed by the monopoly, is far better than two hundred to one. Mr. O'Hara's occasional profanity and excessive wordiness make it impossible to quote him in full, but these few words indicate his trend of thought:

> You want to see 500 stations eliminated and these 500 are to be the ones you New Yorkers and the chain don't own fine business for you, then you and your pals can get $4500.00 per for the use of your stations for advertising and you will have no competition so you can put your price up and make 'em pay it because there won't be any other stations. Well old dear lets see you put that over and get away with it. The half has never been told in this radio game yet and the dear public are going to be told a few things during the next few months that may make some of you fellows think this is a cruel old world. You know that is what Donahue and Sinclair both think this miserable U. S. Govt. has just treated them terrible. You fellows are not kidding the dear publick as much as you think you are. You and your gang want to hog the radio broadcasting, not because you are so interested in giving the dear people high class programmes and by the way just what do you mean by high class programmes? Some cigarette smoking female Dago or Russian warbling in upper C till they drive all the dogs in the

neighborhood crazy. If that is your idea of a high class programme and judging from the programmes we hear over WEAF it is just keep them in the cultured and protected east will you.

We inflict this twaddle on our readers to make certain that few, if any, agree with our correspondent. The suggestion that these columns are subsidized is silly;

what we endeavor to do is to promote the good of radio as a whole. We are using what influence we have to bring about the conditions which the majority of listeners want and should have. Only by following such a course can we hope to continue our stready growth. We welcome expressions of opinion of our editorial stand and hereby take the opportunity to thank publicly the hundreds of readers who have written us favorably during the last few weeks.

We continue unhesitatingly to urge further reduction in the number of stations until the total falls to 225 or 250 powerful and capable broadcasters in simultaneous operation. We hope for a spirited competition in program attractiveness among independent, well organized, and well financed broadcasting stations. We are in favor of eliminating a great number of small stations so that we may have more great stations. This is our stand because it is sound common sense and because it is the expressed wish of a majority of our readers.

What Stations Shall Be Eliminated?

THERE is no excuse for the existence of a station which serves only a special and limited interest —to the exclusion of general educational and entertainment services to which the broadcasting band should be devoted. Since broadcasting facilities are limited because there is room for only so many stations on the air, some form of selection must be applied. That may be censorship; call it what you will. "Freedom" of the air is impossible because more wish to broadcast at the same time than the ether will accommodate. That means restriction and restriction is not freedom.

The first stations which must go are the

THE CROSLEY SHORT-WAVE BROADCASTING UNIT AT CINCINNATI
All the WLW programs radiated on the standard broadcast frequency of 710 kc. (422.3 meters) are now also simultaneously broadcast on 5760 kc. (52.02 meters). The power is 250 watts. On the left is Russell Blair, the Crosley engineer, who built the set. On his right is Joe Whitehouse, chief engineer of WLW

direct advertisers because the public vigorously resents their existence. A broadcaster serving the private purposes of a harness maker or a seed salesman is not serving the public.

The next group which must go consists of stations representing perfectly legitimate special interests, but using the entire time of a private broadcasting channel for the benefit of only a fraction of the audience and for only a small part of the total hours of the day. Labor groups, sectarian religious appeals, socialism, Mormonism, atheism, vegetarianism, and spiritualism may require the use of microphones, but they do not require exclusive channels. If goodwill advertisers were as narrow minded as the religious and "ism" interests, they would demand separate and private channels also. Within a few evenings, we recently heard the La-Salle, the Willys Overland, and the Studebaker concerns broadcast through the same high-grade station. But suggest to a religious or educational group that they use a single station coöperatively with others of a different creed or line of thought and they wail that you discriminate against their interests.

There is need for a more generous spirit on the part of the smaller broadcasters, leading toward a healthy and desirable combination of stations with consequent increase of audiences, improvement of reception, reduced maintenance cost, and better program standards. There is no shortage of microphones to present any worthy cause. Broadcasters, accused of the most sordid commercialism, do not refuse the gentlemen of the clergy and educators free use of their broadcasting facilities. If the Radio Commission will let solely the interest of the listeners, rather than the private desires of the broadcasting station operators, determine their ultimate solution of the broadcasting problem, there will be a continued decrease in the number of stations on the air until they are less than 300.

It is unfair to condemn small broadcasters as a whole because here and there are one or two fine exceptions which have shown a farsighted spirit of coöperation. We spoke to the owner of a small station in Newark which did not increase power or change wavelength through the radio dark ages, in spite of the fact that two pirates completely hedged in its programs. After such maltreatment, however, the station's owner expressed a willingness that all small broadcasters were to be treated in the same way. Another example for small broadcasters is that set by G. S. Corpe, who has been in radio since 1909, and who wrote us recently as follows:

The writer owned KUY, one of the first three

or four stations on this Coast. It was a popular station, too, back in 1922; but two weeks after the Los Angeles *Times*—KHJ—installed a real Western Electric set, we discontinued KUY. Why? Because we felt that with inferior home-made equipment and a limited amount of talent in a small town, we were hurting radio by continuing rather than helping it. Apparently, there are some hundreds of broadcasters in the last year who have not been suffering from any such modest and retiring attitude.

But the majority of small broadcasters feel that self-elimination or combination is unthinkable. In the last analysis, there is but one choice. Broadcasting may remain stationary in comparative mediocrity and the radio industry will be thereby practically paralyzed, or the radio listener may be

A RADIO BEACON FOR SHIPS

Operating directly off the ship's a. c. mains. It has two 7½-watt tubes in parallel tuned to 352.7 kc. (850 meters). The signal—i. c. w.— is a series of one-second dashes for thirty seconds, followed by a silent period of thirty seconds. A single-wire antenna about 50 feet long is used, independent of the ship's antenna system. The beacon on shipboard enables ships using a radio compass to exchange radio bearings over a distance up to twelve miles. Naval radio compass stations on both coasts and on the Great Lakes now provide bearings to ships at sea on 374.8 kc. (800 meters)

served by substantially reducing the number of stations, encouraging capital to spend huge sums on improved broadcasting by reason of the greater audiences of the remaining stations. The axe must be applied either to the broadcasting stations or to the radio industry, and the Radio Commission is choosing between the two. Happily, it is bending its efforts toward station reduction, a saving grace for the radio industry; our only fear is that it will not go nearly far enough in the right direction.

Unnecessary Duplication of Radio Standards

EXPENSIVE trouble is being stirred up in the radio industry by the toleration of two separate organizations busily engaged in setting up standards. Standardization is vital but the establishment of two sets of standards is figurative suicide. We cannot be concerned with the respective claims or purposes of the organizations involved. Both groups are meritorious and well intentioned. In a previous issue, we stated that the radio industry was shortsighted in tolerating two sets of standards and were taken to task by one of these organizations for so doing. Its spokesman claimed to speak for the only representative organization. Whether his statement is true or not, the fact remains that two sets of standards are being drawn. We repeat that this is folly which in the end will be very costly. The rival organizations may differ entirely in purpose, membership, policies and what-not, but in the preparation of standards at least they should function as one organization.

Why Important Events Can't Always Be Broadcast

AN EDITORIAL in the New York *Evening Post*, entitled, "Radio and the Press," points to a serious evil of the broadcasting system, unavoidable under its present economic structure. The Butler-Borah debate, it pointed out, although of great public interest, could be heard only through two small New England stations, both of which were surrounded by interference from stations who had jumped their wavelength. None of the big broadcasting stations, because of commercial commitments, were in a position to offer their audiences this desirable event. The newspapers, on the other hand, were in a position to give the report of the debate plenty of space, hampered by no limitations as to cost of printing and paper, reportorial representation, and wire transmission; nor did the publication of this news require the sponsorship and capitalization of a commercial organization.

While the commercial broadcasting rate of even the best stations is necessarily limited at this stage of the game, a large proportion of their time must be given to commercial features which cannot, for very vital reasons of revenue, be set aside whenever an interesting broadcasting event heaves into sight. If the individual audience of stations be tenfolded, however, the increased revenue thereby gained would make it possible to set aside revenue producing features more freely in favor of such desirable news features.

Radio Reception Is No Longer Seasonal

A STATEMENT by the National Broadcasting Company to the effect that there is no summer seasonal slump in broadcasting program quality is supported by a list of regular weekly features which continue throughout the summer. The program fare of the listener is remarkably well sustained throughout this period and it is a pity that the tremendous prejudice against summer reception has been built up. This prejudice is a heritage of the days of headphone reception when even slight static impressed directly upon the ear drums of the listener caused acute discomfort. With modern pick-ups by loops or very small antennas, and up to date sound radiation through power loud speakers, reception from near-by and local stations is entirely satisfactory twelve months of the year. The seasonal usefulness of a good radio set is no less than that of a closed automobile which is out of service only for blizzard or hurricane for a few hours each year. Radio must make way only for thunderstorms. The only static eliminator required is the encouragement of the use of small pick-up devices, either loops or small antennas, rather than the all-too-long antenna frequently employed.

Radio Control Unsettled in Australia

A USTRALIA is in the midst of an argument about the regulation of broadcasting. The Naval Board desires to shift the broadcast band from its present 300 to 400 meters area (1000 to 750 kc. area) up to the 800-meter (375-kc.) region. Any attempt to regulate broadcasting from any constricted point of view always works hardship on the listening audience. The Australian Naval Board's proposition means that special receiving sets of higher cost will have to be made and these sets will not be able to pick up foreign broadcasting which may some day be within their range.

Where Are the Listeners' Organizations?

O UR comments in a recent issue regarding the general ineffectiveness of most broadcast listener societies caused a number of irate "executive" secretaries to uncover the bushel which hid their feeble light. Some of these accused us most petulantly of going out of our way to disregard their frantic efforts to form a broadcasting listener organization. We are not, however, guilty of any prejudice or oversight, since none of them have, through public statements or by public attention gained through constructive activities, come to our notice previously. We hope that the spirit of service to the

© Bell Telephone Laboratories

THE LATE IRVING BARDSHAR CRANDALL

Irving Bardshar Crandall, a member of the technical staff of the Bell Telephone Laboratories and an authority on the telephonic transmission of speech and methods of recording it, died on April 22, 1927, at the age of 36, in New York. He was graduated from the University of Wisconsin in 1909 with the degree of Bachelor of Arts and received the degree of Master of Arts from Princeton. In 1916, three years after he had become associated with the Bell Telephone Laboratories, Princeton made him a Doctor of Philosophy. At the time of his death, Doctor Crandall was engaged on important experiments. He recently published a book, *Vibrating Systems and Sound*, and he had previously written many monographs on the scientific aspects of speech, analyses of its mechanisms, and methods for its transmission and recording. He was a Fellow of the American Physical Society and the American Institute of Graphic Arts

listener, which should animate the proponents of any listener organization, will be sufficiently obvious to cause large numbers of radio folk to flock to their standards so that at least one of them will ultimately become a powerful and influential organization. The policies of bona fide societies should be actually shaped by the membership itself and not by the dictation of their handful of founders. Too often, however, are new organizations the figment of the imagination of a professional executive secretary who sees in them a great opportunity for profit. Whether this is the case is easily determined by examining the method by which the officers are elected and by determining to what degree the individual member has a real voice in shaping the policies.

The most promising organization from which we heard was the United States Radio Society, claiming 5000 members in and about Cincinnati, Ohio. Fred G. Gruen, President of the Gruen Watch Guild, is the principal source of encouragement and funds for this organization, and he is putting much effort into building up something which is to give form to the listeners' desires in an unmistakable and influential way: We have not had opportunity to examine the constitution of the organization but do know, in the person of Mr. Gruen, that it has intelligent and unselfish leadership to commend it.

The Frequent Misuse of the Word "Broadcasting"

T HE word "broadcasting" is frequently used in newspaper reports in referring to the dissemination of information through the regular radio telegraph channels. For example, it was recently announced that the Navy Department and the Weather Bureau in coöperation are broadcasting special weather information to aviators at 10:30 A. M. and 10:30 P. M. daily. The frequency employed is 8030 and 12045 kilocycles (37.24 and 34.89 meters). Eleven naval short-wave stations, well located, are used for the purpose. Obviously this is broadcasting in the sense that it is sent out in all directions by radio. However, the use of the term "broadcasting" should be confined to programs radiated in the broadcasting band. By so limiting the scope of the word "broadcasting," much confusion in the meaning of newspaper stories will be avoided.

PROBABLE PROGRESS OF TELEVISION

T HE stimulation of inventions due indirectly to the growth of broadcasting appears to be rapid. The first important contribution of scientific progress engendered by broadcasting is the application of vacuum-tube amplification to phonograph recording and reproduction. Now comes high-grade loud speaker entertainment to accompany motion pictures, a decidedly important contribution, particularly for small communities which have had to content themselves with second rate mechanical accompaniment or quite incompetent musical groups. The educational possibilities of the combination of loud speaker and motion picture are perhaps even more important although less spectacular than the entertainment aspects. The next step is to link the motion picture with radio transmission to give us television, a process already hatched in the laboratory incubator. Most people imagine television as a means of enabling them to see the radio artists perform in the studio, a privilege, we are frank to say, often of doubtful value. More than likely, the visual program broadcast for general reception will lean heavily upon the motion picture studio for its source. The improvement of the phonograph and the loud speaker at the moving picture theatre is accomplished by the use of radio developments with but slight adaptations. Television, on the other hand, will need much research before it is within range of home use. Considering the speed with which our laboratories work, stupendous as the problems yet to be overcome are, it may be a matter of only two or three years before you will be buying television receivers as a part of your radio equipment.

TWO ANNOUNCEMENTS BY THE WESTINGHOUSE COMPANY

T HE transmission of power by radio has frequently been made the subject of public demonstration during the last few years. The most recent announcement comes from Dr. Philips Thomas of the Westinghouse Company. He utilizes a high-frequency transmitter, a straight Hertzian antenna, and a metal screen output to obtain the beam effect. The radio-frequency output is approximately thirty-five watts and, by its means, lamps are lighted from a distance of ten to fifteen feet. It seems that the transmission of power must await some important new dis-

L. S. BAKER
New York

Executive Vice President Radio Manufacturers' Association. A special statement for RADIO BROADCAST:
"The question of whether or not the listener keeps his set in operation, tuned to his favorite station, is the keystone of the entire radio industry. Without trustworthy sets, good broadcasting is of no avail; without good broadcasting, the best set made is nothing more than junk.
"In recognition of this phase of public opinion which joins the two gigantic divisions of the radio industry and makes what are otherwise totally independent divisions entirely interdependent one upon the other, the National Association of Broadcasters and the Radio Manufacturers Association have joined hands by consolidating personnel, in an effort to keep all phases of the industry in constant liaison with each other."

coveries before we get beyond the technique of the average amateur "glow" wavemeter.

At about the same time, another Westinghouse engineer, D. D. Knowles, demonstrated his light-sensitive vacuum tube relay which does with light and shadow what our radio sets do with radio-frequency currents. One billionth of a watt of input light energy is sufficient to start a current as high as twenty-five milliamperes flowing through the tube which can, in turn, be magnified sufficiently to control an electric power system of any size. Thus it is quite possible by the use of this relay for a passing shadow to turn on the lights of a city, start or stop a railroad train, or move a battleship. The automatic turning on and off of street lamps at sunset and during dark storms is a valuable service which, accomplished automatically, may result in the saving of millions of horsepower annually.

THE PATENT SITUATION

THE patent examiner rejected claims of an interference under design patent 72,261, issued to M. C. Hopkins. Hopkins contended that the Atlas loud speaker was an infringement of his design patent. The patent examiner stated that merely to place a cone unit in the base "would not require an exercise of the inventive faculties. It would be regarded as a mere assembling of the elements which are old in the art of making radio loud speakers."

The Patent Office reports the following suits and decisions: Patent No. 1,589,308, J. A. Victoreen vs the Radio Art Company, decree pro

confesso and injunction granted; 1,616,207, J. W. Wardell vs. Utah Radio Products, suit filed February 28.

The Latour Corporation has begun suit against the Charles Freshman Company of New York and the Zenith Radio Corporation of Chicago. The latter company has been licensed by the Radio Corporation and consequently the suit represents a test to determine the relative strength of the R. C. A., Hazeltine, and Latour patent groups.

The Splitdorf Bethlehem Electric Company, the American Transformer Company, and the Crosley Radio Corporation, have joined the growing group which is licensed under the Radio Corporation patents.

A recent decision in the Exchequer Court of Canada respecting the Alexanderson radio-frequency tuning patent and the neutrodyne system, was favorable to the former. The contenders were the Canadian General Electric Co. and the Fada Radio Corp. Ltd. of Canada.

The Month In Radio

THE Hudson River Navigation Corporation is replacing its orchestras aboard the night boats, plying between New York, Albany, and Troy, with the highest grade of power radio equipment. The principal problem has been to obtain receiving sets sufficiently sensitive to overcome the dead spots created by ore deposits in the Catskills. This apparently has been achieved by three existing radio equipments, while an additional factor of safety will be provided by the fact that the amplifier system may be actuated by a high grade phonograph equipment to alternate with the radio programs.

A COMPANY has been formed in Africa to take over the bankrupt South African stations under the protection of a government monopoly for a period of five years. Stock will be offered to the public. The interests backing the plan are in control of the most important South African theatres and they promise better programs which should discourage the extensive evasion of license payments, the reef on which the original broadcasting plan was wrecked.

DR. A. HOYT TAYLOR, of the Naval Research Laboratory, states that station 2xs of the Radio Corporation of America, which theoretically should be inaudible at distances between two hundred and six hundred miles because of the skip distance effect, is nevertheless clearly heard in this so called area. Oddly, the much talked of skip distance influence does not, in this instance, follow any of the prohibitions laid upon it by radio observers.

THE Radio Corporation of America announces an improvement in its photographic reproduction system which accomplishes a ninefold enlargement of the picture received by radio through a new and ingenious method of printing the radio picture. Instead of putting a light beam on the surface of the sensitive paper, a special photographic paper is used upon which heat rays have the same influence that light waves have on the ordinary photographic paper. A jet of hot air is blown on the newly developed paper, making a black mark. A second jet of cold air, controlled by the radio signal, intercepts the hot air wave. The result is a more rapidly produced, clear print, nine times as large as one

C. P. EDWARDS
Ottawa

Director of Radio, Dominion of Canada:
"Broadcasting in Canada is indirectly paid for by the broadcast listener himself through the Canadian licensing system, whereby the owner of every radio receiving set must take out a license, for which he pays an annual fee of $1. For the fiscal year ended March 31, 1926, the proceeds from broadcasting license fees of all classes amounted to $139,742.40.
"In only one case is any of these funds used to assist in the support of a broadcasting station. Station CKY, Winnipeg, owned and operated by the Provincial Government of Manitoba, has entered into an arrangement with the Dominion Government whereby it enjoys a virtual monopoly of broadcasting in that Province, and fifty cents out of each dollar license fee collected from residents of the Province is paid to the Provincial Government to assist in maintaining the station."

made by light rays. It is such ingenuity that encourages us to predict practical television somewhat sooner than the scientists, steeped in the details of the problem, are willing to grant as a possibility. Incidentally, the transmission of photographs by wire has recently been supplemented by three-color service.

OBSERVERS at the recent demonstration of television at the Bell System Laboratory in New York stated that they sometimes observed two or three images reproduced in the background which appeared much as do the ghostly figures in spirit photographs. The engineers state that these "ghosts," which have only a fraction of the intensity of the main picture, are caused by signals which take a longer path through the atmosphere than the main incoming signal. In the reception of music, we have similar lag which expresses itself to the senses as indefinable distortion. Television may pay its debt to radio development by the spectacular way in which it will reveal the hidden causes of distortion in radio reception.

A DEPARTMENT of Commerce report states that there are 49.5 radio sets per thousand inhabitants in England. Sweden is the second European Country with 40.1, Austria, third, 37.8; Denmark 25.3, Germany 22, Norway 15.5, Czechoslovakia 12.9, Switzerland 12.8, Netherlands 7, Belgium 3.14, and Finland 3.1. No figures for France and Italy were reported.

A Low-Cost Battery Charger

Something About a New Raytheon Rectifier Tube Which Is Remarkably Efficient—A Home-Made and Compact 2½-Ampere Charger

By JAMES MILLEN

COMPACT AND RUGGED

This is the full-wave charger unit manufactured by the National Company, Cambridge, Massachusetts, and which employs two of the new "A" tubes described in this article. The circuit diagram of this charger is shown in Fig. 6

STORAGE batteries, the source of direct current for lighting the filaments of the tubes in a radio receiver, are just what their name implies; they have the property of storing up electrical energy in a chemical form within their cells, and when called upon to do work, discharge this energy by virtue of a chemical action which takes place inside the battery. As the current is drawn out of the battery, the nature of the plates and electrolyte which go to make up the cells of the battery, change their chemical character, and finally reach a point where it is not wise to make further current demands upon them. In this state, the battery is said to be discharged and it becomes necessary then to reverse the chemical action so that the plates are restored to their original state of usefulness.

This requires that a direct current be passed back through the storage cells, and our most convenient source of supply for this work is the energy to be derived from the house lighting circuit. However, in most cases, this energy is in the nature of alternating current and as such is unsuited for *immediate* use. The reason for this is that the current flows first in one direction and then in another direction, these changes occurring, in the case of a 60-cycle current, 120 times per second.

The job, therefore, is to employ a device which will make use of these alternations and provide a series of pulsations delivering current all in one direction. This device is termed a rectifier, and may be obtained in many different forms. Some, such as the electrolytic, depend for their operation upon chemical action. Others, such as the vibrator or magnetic type, depend upon a set of mechanical contacts which shift every time the alternating current reverses. Still others, such as the Tungar, depend upon the unidirectional conduction of the stream of electrons emitted by an incandescent filament in vacuo. All have their advantages and disadvantages.

Perhaps the ideal rectifier would consist of a piece of low-resistance wire with the property of conduction in one direction only. The recent perfection of a rectifier which approaches somewhat closely this ideal is what brings us to the main topic of this article.

Working under great handicap in his small laboratory just outside of Paris, a French physicist, M. Henri André, developed the forerunner of a new rectifier "tube," one of which is shown in a photograph on this page. But what could he do with his device? He did not have the facilities or financial backing so essential in order to complete his work and carry his idea to a finished and commercially

THE NEW 2½-AMP. "A" TUBE

practical state. Then radio stepped in, in the form of the research department of the Raytheon Company, specializing in the development and manufacture of rectifying devices for radio use. M. André came to America where with most complete laboratory facilities at his command and with several physicists and engineers to aid him, it took but a year to reach the long sought goal. A highly efficient, inexpensive, rugged, compact, and long-lived rectifier element, encased in a small steel tube hardly three inches

EMPLOYS THE NEW "A" TUBE

A commercially made charger employing the new tube described in this article. It is a product of the Acme Apparatus Company, Cambridge, Massachusetts

long and less than an inch in diameter, was the fruit of his long labors.

As this new rectifier cartridge is now on the market, it is the purpose of this article to give a brief description of its theory of operation, together with some data on the design and use of a battery charger in which the tube is employed as the rectifying medium.

A rectifier as we understand it, is a device which offers great resistance, or opposition, to the flow of current in one direction and little or no opposition to the flow of current in an opposite direction. The more completely the rectifying device prevents current from passing in the one direction, and the more easily it permits current to flow in the opposite direction, the more desirable is its use as a rectifier. This new rectifier very admirably fits in with these requirements, for it involves a metallic conducting path with an oriented junction at one point which exerts very little effect when current flows one way, but which effectively opens the circuit when current attempts to pass in the opposite direction.

There is much yet to be learned in regard to metallic conduction, and the behavior of electrons in solids is not at all clearly understood. Hence a clear explanation of the exact nature of this oriented condition is indeed difficult. However, we may quote from a report by Dr. V. Bush, of the Massachusetts Institute of Technology, as follows: "All materials contain electrons distributed in orbits about the nucleli of atoms. When conditions are such that electrons may with ease pass from an orbit about one nucleus to an orbit about an adjacent nucleus a motion of electrons through the material is readily produced and we have an electrical conductor. Metals have this property in large degree and are hence good conductors. When two metals are in contact a similar interchange of electrons ordinarily takes place between the adjacent atoms of the two metals, and conduction readily occurs in the two directions. A proper choice of metals in the presence of a suitable agent, however, may set up a condition in which this property is oriented or unilateral. Briefly this occurs when electron excursions of one metal are much extended in the presence of the agent, while the excursions of the other are inhibited. In this condition the far extending electrons readily pass to the opposed metal and conduction occurs, while for a potential in the other direction there is no overlap of orbits, and the device insulates."

In the new "A" rectifier, as this new development is called there are two metals—an anode of pure silver connected to the casing, and a cathode of a porous alloy connected to the central projection, brought into contact on the inside. The porous cathode contains in its interstices a non-conducting agent which has free access to the junction between the metals. The presence of this agent preserves the junction in an oriented condition, but the actual conduction is through ·the metals themselves. It not only creates the oriented condition, but preserves this function despite much abuse in the form of rough handling and usage.

Due to the fact that the conduction is metallic, the internal electrical resistance of the rectifier cell, and the power, or I^2R losses in the cell are exceedingly low, and thus its efficiency is quite high. As will be seen from Fig. 1, the efficiency of a charger employing the new tube is in the neighborhood of 60 per cent. When compared with existing types of chargers, this, as charger efficiencies go, is unusually high. Aside from the saving in power consumed, which may amount to as much as from $6.00 to $10.00 a year, the higher efficiency of a charger of this type permits it to be constructed from exceedingly compact, and in this case, less expensive parts.

Only the tube, fuse, and transformer, as indicated in Fig. 2, are required in the charger circuit. As no energy worth mentioning is wasted in the tube, it may be made quite small itself, and as the transformer does not have to supply a great deal of useless energy, its core need not be any larger than those of some of the new high-quality audio transformers.

RATE OF CHARGE

THE new "A" tube lends itself to either full-wave or half-wave rectification. When used as a half-wave rectifier, the maximum charging rate consistent with long life is 2½ amperes. By means of suitable ballast resistors of ½ ohm each, several "A" tubes may be operated in parallel for higher currents. In full-wave rectification, the charging rate may be double that for one "A" tube. While the tube

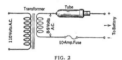

FIG. 2

will have a longer life when used as a trickle charger than when used as a comparatively high-rate charger, its use as a trickle·charger is not recommended, due to the necessity for a tremendously long life for rectifiers suitable for such use.

But is trickle-charging so desirable after all? As an answer to the demand made by the public for an A power unit and the elimination of the A battery and charger, many manufacturers brought out trickle chargers and circulated much information regarding the advantages of such a method of charging.

Both systems have their merits. The outstanding advantage of trickle charging is convenience.

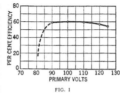

FIG. 1

There is no need to remember to turn on the charger for an overnight run each week, and if the charging rate is properly adjusted,

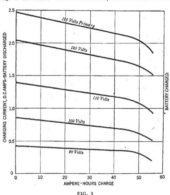

ANOTHER COMMERCIAL UNIT
This one is manufactured by the Mayolian Company, of New York, It is a half-wave charger

there is never the inconvenience of a run-down battery.

But· against the advantage of convenience must be placed ` lower electrical efficiency (which means greater operating costs), the difficulty of determining and then obtaining the optimum charging rate, and, finally, shorter battery life.

The storage battery manufacturers tell us that, as far as the life of the battery itself is concerned, the ideal charging system would consist of a high-rate initial charge, to remove any sulphate formation on the plates, and greatly reduce the time required for the complete charge, followed by a gradually decreasing rate of charge in order to prevent excessive gassing and thus slow disintegration of the positive plates as the charge nears completion.

In good battery service stations, charging

batteries at a high rate which gradually tapers off is accomplished by manually regulating the charging rate, as the state of the battery changes, by means of field rheostats on the motor-generators employed for charging.

In some types of chargers, where the secondary voltage is from 20 to 30 volts, the rise in back-voltage as the battery reaches its fully charged condition is only a small percentage of the total impressed or secondary voltage. Since the current flow is governed by the difference between· the impressed voltage and the back voltage of the battery on charge, no great change in current flow will take place and there is not the advantage to be gained of a tapered charge.

Because of the high efficiency of the charger discussed here which permits of·low secondary voltage, the variation in battery back voltage, as it approaches its fully charged condition, is a large percentage of the total effective voltage of the circuit, thus resulting in a very decided decrease in current flow, and thereby automatically producing a condition of tapered charge which is so beneficial in battery charging.

The curves given in Fig. 3 show this phenomenon clearly at various charging rates.

Fig. 4 shows. the wave form of· the output of the Raytheon "A" tube as indicated by an oscillograph. An oscillograph gives a visual indication of the variation of voltage or current, as the case may be, in an electrical circuit over a period of time. Thus, from Fig. 4, it will be seen that the current increases in a positive direction with time until a maximum is reached and then falls off to zero. Instead of then continuing to build up in a negative direction, as in the case of an alternating current, Fig. 5, the rectifier tube effectively opens the electrical circuit and the current remains zero until such time as the alternating line voltage has passed through another half cycle ($\frac{1}{120}$ of a second in the case of a 60-cycle supply) at which time the rectifier tube closes the circuit and permits the current to build

FIG. 3

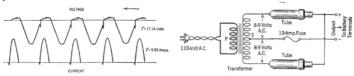

FIG. 4

FIG. 6

up again in the same direction as the first half cycle.

One of the purposes of making oscillograms of the operation of the rectifier tube is to ascertain just how completely it prevents current flow in the wrong direction during the half cycle when it should prevent current flow. A perfect rectifier would let no current through in a negative direction.

"A" TUBE CHARGER DESIGN

A NUMBER of prominent manufacturers, such as National, Thordarson, and Mayolian offer for sale complete chargers using this new tube. For those who wish to construct a charger at home using the new rectifier tube, details for the transformer are given in Fig. 7, and a photograph of the equipment appears on this page.

Consider the circuit diagram for the single-wave unit shown in Fig. 2.

Any well made transformer of about 20 watts capacity and with a low-resistance secondary having an open circuit voltage of between 8 and 9 volts may be used. It is preferable to mount the rectifier tube with the small end up. The fuse clips should be used, one making contact with the body of the tube and the other with the small cylinder projecting from the top.

The small cylinder (cathode) should be connected to the positive output circuit while the body of the rectifier (anode) should be connected through the transformer to the negative.

A fuse of not over 10 amperes capacity must be connected in the charging circuit to prevent damage should the output of the charger become short-circuited or the battery be connected in the reverse manner. Small automobile cartridge fuses are excellent for this purpose.

Perhaps it may occur to some readers that a charger with variable rate may be readily constructed from a transformer with a higher secondary voltage than that described, by inserting a rheostat in series with the tube. Such is not the case. The maximum back voltage that the tube will withstand without injury is 22 volts. As there is no current flowing during the half cycle in which the battery is not charging, IR drops become zero and the back voltage becomes equal to the peak a. c. secondary voltage plus the battery voltage. Thus, for long tube life, transformer voltages must be limited to 8

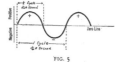

FIG. 5

FIG. 7

Here are the specifications for the construction of a transformer suitable for use in a home-made charger using the new "A" tube. The transformer should be of the customary shell type with a core having dimensions equivalent to those given here. The primary of the transformer consists of 750 turns of No. 24 enameled copper wire, while the secondary consists of 53 turns of No. 14 single cotton-covered enameled copper wire. The transformer coil should be layer wound with the secondary coil next to the core. The usual insulation between layers, between core and secondary, and between primary and secondary, should of course also be provided.

or 9 volts at no load. (Peak A. C. voltage = RMS × $\sqrt{2}$).

The National charger shown in the photograph on page 143 and diagrammatically in Fig. 6 is so designed as to be used either as a half- or full-wave charger. As a half-wave charger, only one rectifier tube is plugged in, while as a full-wave outfit both tubes are employed. The charging rate with one tube is approximately 2½ amperes and with both tubes 5 amperes. The manufacturer's price is $10.00 without tubes; the tubes are $4.50 each.

The Mayolian Charger, which is shown in a photograph on page 144 lists for $10.00, and the Thordarson charger lists for $12.50 complete.

The new rectifier unit is already being applied to A elimination and at least two manufacturers have complete A-units ready for the market at present.

There are many tricks to the successful design of such devices however, and the development of suitable filter circuits has been exceedingly difficult. The present commercial units employ several chokes with quite low inductance and exceedingly low d.c. resistance. Instead of ordinary condensers, special dry cells, offering a very high d.c., and at the same time extremely low a.c. path are employed.

Some of these units or cells are based on the principle of the electrolytic condenser, but use a paste electrolyte rather than a liquid.

THE HOME-MADE UNIT

Fig. 6 is the schematic diagram that should be followed in building this full-wave 5-amp. charger

"Strays" from the Laboratory

Measuring Audio Amplifiers

AS HAS already been mentioned in these pages, progress of standardization in the radio industry is encouragingly sponsored and aided by two organizations—the Radio Manufacturers Association and the radio section of the National Electrical Manufacturers Association. Considerable effort has been made by both organizations to standardize not only mechanical and electrical constants of radio apparatus but methods of measurement as well.

In the booklet giving the NEMA radio standards there can be found a method of testing audio-frequency coupling devices—transformers, resistance-condenser, and choke-condenser couplers. The problem of output devices has not, as yet, been included. The method of measuring transformers is not new, nor, in the opinion of many engineers, is it correct, in that it does not give a true picture of what the transformer will do under actual working conditions.

The circuit diagram of the test equipment is shown in Fig. 1. Briefly it consists of an oscillator whose frequency range is 30 to 7000 cycles and whose "percentage of harmonics present in the output shall be not more than 5 per cent." Current from this oscillator is passed through a potential divider and is read on a thermo-couple-milliammeter. A portion of the output voltage is impressed on the primary of the coupling device under test in series with a fixed resistance to simulate the plate impedance of the tube out of which, in normal practice, the coupling unit works. Arrangements are made for direct current to flow through the primary.

A given deflection on a vacuum-tube voltmeter is obtained by placing its terminals across the secondary or output side of the coupling device. Then the same deflection is obtained by impressing the oscillator output voltage directly on the tube voltmeter by means of a slider on the potential divider. Since the resistance through which the coupling device is fed is fixed, the ratio of these resistances gives the voltage gain of the coupling device.

Criticism leveled at this method arises from the fact that the input characteristics of the vacuum-tube voltmeter differ from those of a tube with an inductive load in the plate circuit—such as would normally work out of the coupling device. The frequency characteristic of the transformer depends upon what is shunted across its secondary or output terminals. In the Laboratory, and elsewhere, it has been found that the curve obtained in this way may have very little in common with that obtained from a complete amplifier, or even from a single coupling unit plus one amplifier tube.

The question naturally arises as to what is the best method of measuring a coupling device or an amplifier. Shall the curve obtained by the NEMA method be taken as standard; shall the overall characteristic of two or more coupling devices with their associated tubes be preferred; and if so, how shall the output be measured? That is, with what is the amplifier to be terminated; if a resistance, of what value? Are we interested in the amplifier as a power amplifying device; shall the output power be measured, or is it sufficient to measure the voltage appearing across, say, the primary of an output transformer and to divide it by the input voltage to the amplifier to obtain the overall voltage gain?

These questions have occurred in letters from laboratory experimenters who would use standard test methods if such could be agreed upon. It is possible that they have occurred to members of the standardization committee of the RMA and the NEMA too, and it is probable that readers who have suggestions will find willing ears.

A Fine British Magazine

IT IS disappointing to note that one of the best British technical journals, *Experimental Wireless and the Wireless Engineer*, published in London by Iliffe and Sons, has been compelled to raise its subscription price to two shillings and sixpence per copy on account of "lack of support by English advertisers." Without a doubt if Iliffe and Sons are compelled to pass the hat to keep this magazine going, they will receive strong support from American engineers, for this paper is without a peer as an organ for serious engineers and experimenters. In England a research is undertaken, it seems, with the love of pure science in the investigator's heart and not with one hand tied by production department threats. The English may pursue roundabout methods, but the end point is final and the answer is complete. No American deeply interested in radio science can afford to be without *Experimental Wireless*, even were the price raised to five or six "bob" per month.

Batteryless Receivers

THE Laboratory has witnessed the demonstration of two radically different systems of battery elimination in receivers within a month, one employing the high-voltage high-current rectifier tubes, the uses of which are outlined elsewhere in this issue of RADIO BROADCAST, and the other a method made possible by the so-called Meissner tube.

The Raytheon, QRS, and similar rectifier tubes, are designed to rectify enough current so that 201-A tubes can be wired in series and 250 milliamperes passed through them. The rectifier system must furnish five volts for each tube filament and also the high plate and grid bias voltages of the power output tube whose filament is heated from raw a. c. A five-tube set using four 201-A's and a 171 will require 20 volts for the filaments, 40 volts C grid bias, and 180 volts for the 171 plate; since the filament voltage is also available for the plate voltage of the last tube, the total will be 220 volts.

The Meissner tube is the high-current low-voltage valve predicted in the February and March RADIO BROADCAST and mentioned again in May of this year. The thermal inertia of the tube is so high, about sixteen times that of 112, and the voltage drop across the tube so low, that it is possible to place raw a. c. on the filament without introducing appreciable a. c. hum into the output. In fact, the ripple in the output, produced by incomplete filtering, can be made to serve a useful purpose. Since the plate current produces a ripple in the C bias which, is opposite in phase with that in the plate circuit, the a. c. hum which would otherwise emanate from the loud speaker is automatically reduced.

FIG. 1

And now that it is possible to throw away one's batteries, is it worth while? What are the advantages either way?

For the Raytheon and QRS outfits it must be conceded that the power problem indeed seems simplified. There are no acids or liquids, and it is possible to supply the high plate voltage of the final audio tube economically and continuously. There are no chargers constantly and sometimes inefficiently dissipating power; no batteries to fill with water, and few replacements.

On the side of the battery the advantages are also manifold. Battery voltages are pure unadulterated d. c. This means that a high-quality battery-operated amplifier and loud speaker will be quiet until a signal comes in and it will be possible to use headphones after a two-stage amplifier; this is frequently impossible with distance reception on account of the a.c hum

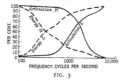

FIG. 3

which is more annoying in headphones than in a loud speaker. There is no a. c. hum whatsoever in the case of battery-operated sets, and there are no voltages higher than that necessary for the last tube. B batteries have a practically perfect regulation curve, i.e., they are of very low resistance when new. There is no heat to be dissipated. Either a. c. or d. c. may be used to charge the A battery, and the charging process may be made almost automatic.

The honors seem about even. Now let us look at the economics of the problem. Suppose the first cost of a battery and a charger is $30 while it costs $60 to equip one's set with a QRS or Raytheon A, B, C, supply. Further, suppose it costs $30 a year to run the power operated receiver and that batteries cost, for a year, about $50. This latter figure includes B batteries and charging the A battery.

At the end of the first year the power set has cost $90 compared to $80 for the battery receiver. At the end of the second year the respective total costs are $120 against $130. Thus, from the standpoint of economics, at the end of two years, there is little gained or lost by one method or the other. Figures on the Miessner system are not yet available.

Articulation Curves THERE are few independent laboratories that have the equipment to measure the effectiveness of a receiver from input radio-frequency voltage to output sound pressure from the loud speaker. For this reason, some data presented at a meeting of the Radio Manufacturers Association, held in New York on March 23, 1927, by Dr. John P. Minton and his associate, I. G. Maloff, are more than interesting.

Fig. 2 shows the frequency response of three loud speakers, curve A being a diaphragm horn speaker, curve B an average cone speaker, and curve C "one of the best cone speakers

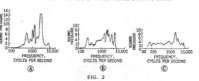

FIG. 2

that has been produced." It may be seen that the horn has an effective range of from 450 to about 3000 cycles, the average commercial cone covers from 200 to about 3600 with several severe "ups and downs," while the third curve is good from about 60 to 6000 cycles.

These curves of Dr. Minton's by themselves are nothing more than interesting, but when looked at with other data collected by such authorities as Dr. Harvey Fletcher and Dr. Minton himself, they become illuminating. For example, the curves shown in Fig. 3 give an idea of what happens when frequencies above any desired frequency are not reproduced, "L", and conversely "H" shows when only the high frequencies are passed through the receiver and loud speaker. From the standpoint of intelligibility or articulation the low frequencies are not so important, but if no frequencies below 500 cycles are reproduced, there will be but 40 per cent. of the speech energy present in the loud speaker output. On the other hand, the high frequencies carry little energy but are important from the standpoint of intelligibility. If all over 1000 cycles are cut off, 80 per cent. of the energy remains but only 40 per cent. of the articulation or intelligibility. From this consideration and others it seems certain that frequencies between about 75 and 5000 should be reproduced equally for faithful reproduction of music.

Fig 4 gives the comparison of sound pressures at various frequencies occurring in organ music, which explains why a poor loud speaker or poor amplifier makes organ music sound like anything but organ music.

Fig. 5 gives the overall radio amplification against wavelength of several common types of receivers. Curve S is for a super-heterodyne;

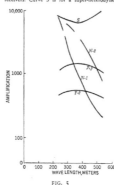

FIG. 5

curve N-2 is for a dual-control neutralized receiver of commercial make, while curve N-1 represents a single-control receiver of the same type; T-2 and T-3 are representative two- and three-stage resistance neutralized receivers.

Mr. Maloff makes several interesting and significant statements, of which the following are of special interest:

"For some unknown reason the importance of adequate audio amplifiers was overlooked by many manufacturers. Many sets on the market have very good r. f. circuits but very few of them have audio circuits of the same grade."

"In a certain commercial set the transformer form of coupling is used, the transformers being of high ratio and insufficient primary impedance. This set is very quiet even with the worst kind of battery eliminator, but responds only to the frequencies in the middle range, which means

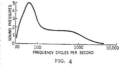

FIG. 4

lack of depth and lack of brilliancy of the reproduced sound."

New Equipment TWO instruments that deserve more than ordinary attention have recently been received by the Laboratory. One of these is the Taylor resonance tester, which consists of a small radio-frequency oscillator and a vacuum-tube voltmeter, all in one cabinet. It is for use in "lining up" two or more radio-frequency stages that are to be controlled by one dial. The second instrument is the Weston set tester, comprising a milliammeter with several shunt and series resistances and a multiple switch so that the meter can be used to measure plate current, and filament, grid, and plate voltages, as well as to test for opens, etc. Further interesting equipment received is as follows:

The Elkon dry rectifier, recently described by R. S. Kruse in *QST*; the Sprague tone control, for use between the amplifier and loud speaker; grid leaks and resistors made from Carborundum; electron relay tubes which will open or close a circuit through the influence of a distant signal; the Crosley AC-7 receiver, which operates without batteries; another automatic radio power relay, known as the Liberty 712S; a grand assortment of Fahnestock clips; tapped heavy-duty resistors from Mountford; a little vest pocket receiver by Flash, that actually receives signals from Manhattan, twenty miles away; experimental four-element tubes from Clearton and Van Horne; condensers for various uses from National and Wireless Specialty; output devices by Centralab, Muter, and Silver-Marshall; the new high-quality General Radio audio transformer, type 285 N; Carter's radio kit No. 400, comprising the necessary resistors for use in an A, B, C, device employing the new QRS 400-mil. tube; resistors from Cresradio; the Davy Vertrex Autocharger; new Paragon double-impedance units; Browning-Drake and Loftin-White receiver kits.

*Complete Construc-
tional Data for Con-
verting a Popular
19-Inch Cone Loud
Speaker Into a Three-
Foot One—Cone Kits
Available*

By
WARREN
T. MITHOFF

A THREE-FOOT CONE
It is a fairly simple matter for the
fan interested to make himself a
three-foot cone, and obtain superior reproduction of music. The
one illustrated to the left is described by the author and makes
use of a Western Electric 540-AW
unit

How to Build a 36-inch Cone

THE true dyed-in-the-wool radio fan is never content with his equipment, however excellent it may be. The constant urge is for improvement, advancement, and change. Hence, manufacturers are able to market with considerable success such items as super-sensitive detector tubes, $10 transformers, and improved loud speakers.

Thanks, in large measure, to the inquisitiveness of the aforementioned fan, radio has advanced more rapidly than most other sciences of a similarly complex nature. Take loud speakers, for instance. The moderate-sized cone of to-day is vastly superior to the tinny-sounding horn of five years ago; yet there looms on the horizon a much enlarged edition of the cone, costing nearly three times as much, and giving noticeably better quality. In theory, and in practice as well, perfect reproduction requires a large diaphragm—within reasonable limits, the larger the better. Therefore, the three-foot cone sets itself up as a contender for highest honors in the struggle for better radio music.

Many fans have doubtless wished for one of these giant cones, but for various reason have foregone the pleasure and pride of possession. For the owner of a Western Electric 540-AW loud speaker, however, there is a short cut to ownership without having to pay the penalty of excessive transportation and handling charges which attaches to the manufactured three-foot cone. He can build the cone himself, and install in it the excellent 540-AW driving unit, or any one of a number of other good units now obtainable, thereby gaining every advantage of the large cone at a minimum of expense. Where the 540-AW unit is removed from the cone for this purpose it does not necessarily follow that the smaller cone is permanently useless. The actuating unit is retained intact,

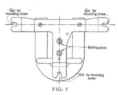

FIG. 1

so that on occasion it can be put back in the smaller cone should a more portable loud speaker be desired.

MOUNTING THE UNIT

IT IS the purpose of this article to describe the process necessary in building a giant cone and the 540-AW unit will be employed for the purpose

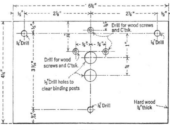

FIG. 2

although practically the same procedure will have to be followed where other units are used. First, we must gain access to this 540-AW unit. The five screws holding the perforated bronze screen are removed, and the screen taken out. The rear of the unit proper is then disclosed; in appearance, it is somewhat like Fig. 1. Now, the set screw at the tip of the cone is loosened, and the three mounting screws holding the unit to the circular frame are removed. The unit is then drawn out, care being exercised to see that the driving pin is not bent. For the large cone it is desirable first to mount the unit on a flat surface. A piece of hard wood, such as oak or maple, serves admirably, and it is easily worked. This should be at least ⅜ inch thick, and should be a heavy close-grained wood. Weight is needed back of the unit to preclude the possibility of the unit vibrating instead of the paper cone.

This mounting base is planed smooth, cut to size, and drilled as indicated in Fig. 2. The ⅛-inch holes near the center are simply to clear the unit, which project somewhat beyond the plane of the three slotted mounting feet.

The next step is the preparation of the supporting arm, Fig. 3. This may be of soft wood, and should, at the start, be made about 4½ inches long. The exact distance from tip to back of the three-foot cone may vary a little in individual cases, and it is easier to cut off than to add on.

The mounting base is now fastened to the supporting arm by means of three flat-head wood screws, fairly heavy, and about 1⅛ inches long. The arm is placed at the top of the base, in the exact center, and the three screws are passed through the countersunk holes and tightened.

Fig. 4 shows the details of the cross piece which supports the entire assembly inside the cone. It may be made of soft wood, drilled as indicated,

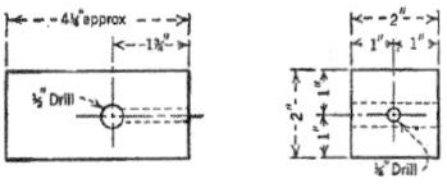

FIG. 3

and stained with walnut wood stain. A No. 14–20 machine screw, 3 inches long, is now obtained, and washers slipped over it in this order: Next to the head a lock washer, then a ¾″ washer with a ¼″ hole, then a 1″ washer with a larger hole. The screw is passed through the ½″ hole in the cross piece, through the ¼″ hole in the supporting arm, and the nut, either hex or square, is placed in the ½″ hole, so as to thread onto the screw, which is then tightened up. The reason for the apparently oversized ½-inch hole in the cross piece will be evident later on in the construction.

The unit is mounted on the baseboard by means of round-head machine screws 1 inch long. These may be of any size that will fit through the slots in the feet of the unit, and are passed through the slots, and through the ¼-inch holes in the base and drawn up tight. Lock washers are

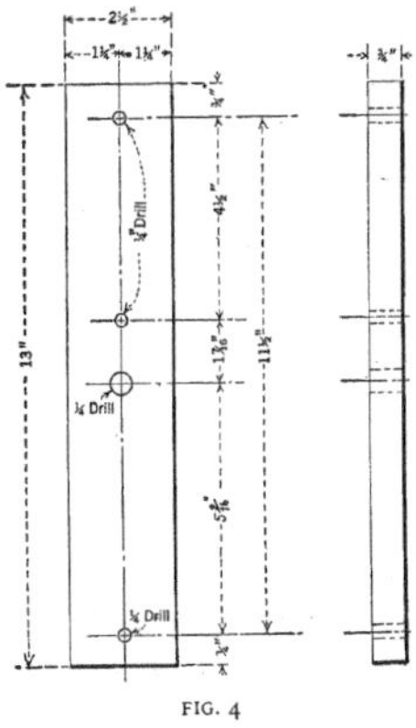

FIG. 4

advisable at this point, also, together with flat washers against the surface of the wood.

THE DIAPHRAGM

THE mechanism may be laid aside now while the cone itself is made. The first consideration is the choice of a paper for the purpose. Brown Alhambra Fonotex is probably best for quality of tone, ease of working, and appearance. It can be had in sheets 38 inches square at many radio stores, or from the manufacturers' agents. Another possibility is lamp shade parchment, which comes 40 inches wide. While this is not theoretically so good, being more compact and harder-surfaced, in actual practice it works very well, although it is a little more difficult to handle in the making. A thin coat of walnut wood stain gives it a rich brown color.

A few pointers regarding decoration may not be amiss at this point. If the Alhambra paper is used, it is a simple matter to decorate the face of the cone with water colors, and a very pleasing effect can be obtained by the use of simple bands of dark brown around the outer edge, as shown in the photograph. The procedure is as follows: A 38-inch sheet of the paper is laid, rough side up, on the floor or on a flat table and the exact center determined by means of crossed diagonals. A thumb tack is driven in at the center, and a piece of wire is used as a compass for drawing the circles. The largest circle should just fit on the sheet, and will have a diameter of 38 inches, or nearly that. The wire is then shortened in steps, and four more circles drawn to locate the bands. The outer margin is ⅞ inch; then a 2-inch band; then a ½-inch space; and then a ⅜-inch band. If the constructor has access to a draftsman's ruling pen, the decorating is simplified. A good water color to use is burnt umber, or Van Dyke brown, in tube form. A little is squeezed out in a dish and thinned with water. It may then be taken up on a brush (a No. 5 flat lettering brush is excellent) and applied to the ruling pen, which is used with the wire compass arm, and a circle drawn over each of the pencil lines. This makes it much easier to get a smooth edge on the bands when filling in with the brush.

The actual construction of the cone is accomplished by cutting out a 5-inch segment as shown in Fig. 5, *with the grain* of the paper. The direction of the grain is indicated on the wrapper. The circular form is then cut out and the sheet turned over, face down, and the two edges of the segment drawn close together and weighted down. A strip of paper about 1½ inches wide is cemented over the two edges, and a ruler laid over it and heavily weighted while the glue dries. A very good adhesive to use is Ambroid cement, a celluloid base mixture which is waterproof and will not buckle the paper.

While the face of the cone is drying, the back can be made, noting that the segment is to be 5½ inches at the outer edge, as shown in Fig. 5. This cone is cemented the same as the first one, with a strip of paper, and weighted down to dry.

MOUNTING THE CONE

NEXT, the wood ring, Fig 6, is cut, drilled, sandpapered, and stained with the walnut stain. When dry, the ring is laid flat on the table and the back cone is mounted on it so that the inner circle centers, and the seam in the paper is opposite one of the projections on the ring. Care must be exercised to see that the cone is not pulled out of shape while fastening it in place. It is best to drive six tacks, evenly spaced, through the edge of the paper to hold it temporarily and then check up the placing of the cone by putting the front cone in position and noting whether or not the edges meet evenly all around. If not, the tacks should be pulled out and the paper shaped so that it will. After the back cone has been adjusted properly, and tacked down, it is firmly secured by a coating of sealing wax applied quite liberally while very hot. The wax should cover the entire circumference thoroughly, and be allowed to harden.

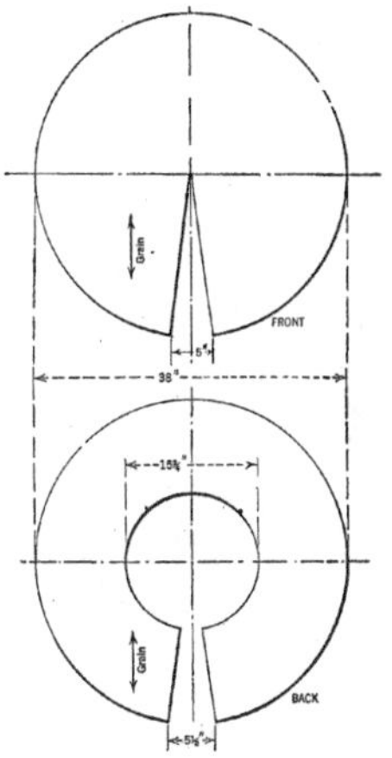

FIG. 5

The front cone is now placed, point down, in a round dish pan, which serves as a support, and the back cone fitted onto it, edge to edge, with the seams meeting. The two are joined together with the Ambroid cement, or with sealing wax. The front cone, having a smaller segment removed, will be a trifle larger than the other, and the cement or wax is applied along this slight extension. The cement or wax is applied freely, but not allowed to run over onto the face of the cone.

The next step is the making of the tip, Fig. 7.

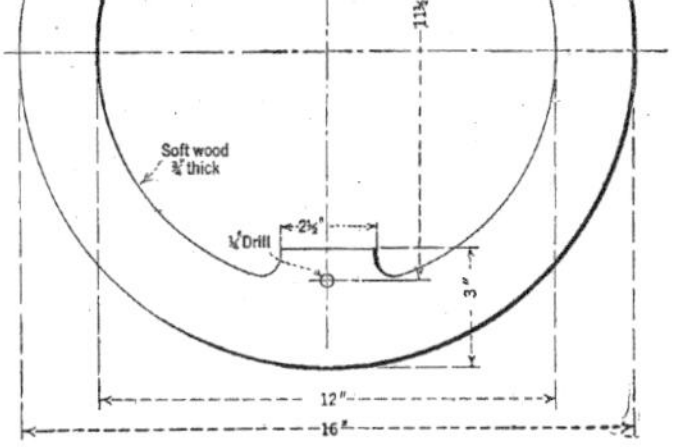

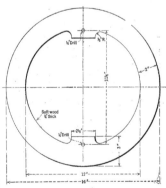

FIG. 6

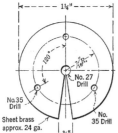

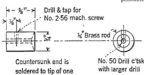

Countersunk end is
soldered to tip of one
of small brass cones

No. 50 Drill c'tsk
with larger drill

FIG. 7

The two circles are cut from thin sheet brass,
drilled, and segments cut out. These should then
be shaped into cones and the seams soldered
lightly. The tip is cut from ⅛-inch round brass
rod, drilled, and tapped as shown. This tip is
then soldered carefully onto the apex of one of
the brass cones, and excess solder removed with
a file.

The two brass cones are given a light coat
of the Ambroid cement, and secured to the point
of the paper cone by three No. 2–56 machine
screws passed through the holes drilled for them.
(See assembly, Fig. 8). The nuts are tightened on
the inside, and the cone is ready to receive the
actuating mechanism.

At this point the constructor will find it convenient to make a temporary stand to support
the cone while work progresses. This may consist
simply of a 12-inch board about 3 feet long, laid
flat on the floor, with uprights nailed to the edges
and braced. The uprights are fastened to the
wood ring on the back of the cone with wood
screws, and thus both hands are left free for adjusting the unit. Or, if the experimenter's workshop is in the basement, he may nail two narrow
strips to an overhead beam and hang the cone in
that fashion.

ASSEMBLING THE LOUD SPEAKER

THE cross piece, supporting the driving unit,
is now put in place, the driving pin being
pushed carefully through the metal tip. Should
the driving pin extend too far, or should the supporting arm be too long to permit the cross piece
to rest against the wood ring, the arm must be
removed and cut down accordingly. No. 14–20
machine screws are inserted in the holes in the
cross piece and wood ring, with flat washers and
lock washers on the inside, and flat washers on
the outside. The nuts are tightened on the inside,
and the loud speaker cord brought out through
the ½-inch hole in the cross piece. A No. 2–56

machine screw is inserted in the tapped hole in
the tip, and tightened, and the instrument is
ready to connect for a test.

The Western Electric unit is a very substantial one, and trouble is extremely unlikely,
though it may be necessary to make a slight adjustment to make sure that the driving pin is
centered exactly at the tip of the cone and not
bent or forced out of place. By loosening the
screw at "A," Fig. 8, slightly, the entire mechanism may be shifted from side to side, or up and
down, while the loud speaker is in operation.
Sometimes the quality of tone can be improved
materially by this process, and it should be tried,
even though it may not seem necessary. When
the best point is found, the screw is tightened
securely, and a final adjustment of the set screw
is made. The screw on the tip which holds the
driving pin to the cone is loosened to relieve any
possible strain and then tightened again.

For the final disposition of the cone in the
home, two methods suggest themselves. The experimenter who is handy with carpenter's tools
may wish to build a three-legged
stand similar to the one shown in
the photograph on page 148 A
simpler method is to hang the loud
speaker on the wall. Two small
rubber-tipped cast-iron door stops
are procured, such as are used to
prevent doors from banging into
the wall. These are screwed into
the lower part of the wood ring to
hold it away from the wall, and a
screw eye at the top will accommodate a cord. The cone may then
be hung from a hook on the picture
molding.

For the benefit of the experi-
menter who does not possess a Western Electric
540-aw loud speaker, there are on the market
several excellent units which can be handled in
a manner similar to that described here. Dimensions, of course, will differ,
but the general procedure is
the same. The fan can readily
adapt the individual unit to
the plan given here, and provided the unit chosen is a good
one, excellent results may be
obtained at a very nominal
cost.

NOTE

MR. MITHOFF'S instruc-
tions fully cover the
construction of a 3-foot cone
loud speaker where the Western Electric 540-aw cone loud
speaker is taken apart to
supply the driving unit. There
are, however, as Mr. Mithoff
explains, several other units
which may be obtained separately and installed to good
advantage in the 3-foot
cone.

Several companies have for
sale complete kits for the
home assembly of such cones,
comprising the driving unit,
mountings, and cone paper.
The following is a brief description of these kits:

Fenco Cone Speaker—This kit comprises two
pieces of cone paper, a driving unit, wooden
back ring, handle, cord, glue, and cement. The
two sheets of paper are cut in circular form, both
are marked for the cutting of the segment to
form the seam, and one is decorated with a border

stencil. Supplied by Fenco Cone Com-
Price $12.00.

Ensco Cone Speaker—The parts suppli
clude one sheet of cone paper, uncut, one c
mount block with four arms, one driving
and one loud speaker cord. Supplied by th
gineer's Service Company at a price of $1

Penn Cone Speaker—Consisting of one d
unit, two sheets of cone paper, uncut, one
back rings, one unit mounting, and one c
cement. Supplied by the Penn Radio Sales
pany, for $14.15.

Instruction booklets explaining the cons
tion of each of the cones listed above are sup
by the manufacturers and deal specifically
the assembly details involved.

Balsa Wood Reproducer—While this
speaker kit is not of the cone type it is of suff
size to be classed in that category. Inste
being circular or conical in shape it is rectan
and much like a picture frame. It is obtai
in three sizes: 24″ x 13″, 36″ x 21″, and 45″ x
The kit consists of frame material, three
slats of special balsa wood, a number of p
of narrow ribs of balsa wood, glue, brads, sc
chuck, and wood mounting for chuck.

The three slats are assembled inside th
tangular frame and the ribs are glued to the
of the slats in a radial fashion. At the c
where the ribs join, the chuck and mounti
placed to take any good loud speaker d
unit. Furnished by the Balsa Wood Repro
Corporation, price $5.50, $8.00, or $10, de
ing upon size.

—THE EDIT

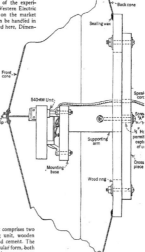

FIG. 8

Something About Single Control

Some of the Advantages and Disadvantages of Unified Control—Analyzing the Single-Control Receiver—Why Synchronism of Circuits Is Complicated

By EDGAR H. FELIX

SIMPLIFIED tuning is only one of the important advantages gained by the simultaneous adjustment of circuits by means of a single control. It is a most important advantage to the inexperienced, but even the most skilled dial twirler soon finds desirable qualities in a real one-control receiver.

For example, the attainment of true single control makes it possible to sample all the available programs throughout the frequency range in a brief time, thereby increasing the entertainment value of the receiver. Furthermore, appearance, in the opinion of many, is improved by the reduced number of controls.

One conspicuous knob plus several small ones, however, does not necessarily make a true single-control set. There are many pseudo-single-control sets, parading as one-dial receivers, which may possess several, but not all, of the advantages of unified control. A true single-control receiver should have only one tuning dial, and this should maintain each tuned radio-frequency stage and the detector circuit in perfect resonance without requiring the use of supplementary vernier adjustments. A compensating antenna circuit adjustment, which requires setting only once when the receiver is installed, does not disqualify a set from the single-control classification. If, however, such an adjustment must be used each time the listener diverts his attention from the lower to the higher frequencies, or vice versa, the receiver in which it is incorporated is not a genuine single-control receiver but a two-control receiver.

All single-control receivers employing more than one tuned circuit so far devised, use gang condensers mechanically coupled, a system thoroughly covered and controlled under the Hogan patents. The license fees charged under these patents are so moderate, however, that evasion has been practiced only in a few instances, and the development of single-control sets has not been hampered by costly patent litigation.

When seeking to judge the desirability of a single-control receiver, the first discrimination to be made is as to whether the set under consideration is really a true single-control set. Auxiliary controls are often concealed or camouflaged in order to give the impression that the receiver is tuned by only one manipulation while, in practice, each of these extra controls may require careful adjustment to tune-in a desired station. Even so, a receiver of such design may possess important advantages over the usual two- or three-dial receiver, provided there is only *one* adjustment of the main tuning dial which brings in each station.

In examining a receiver, therefore, observe every control upon it, no matter how it is labeled. A true single-control set has but two adjustable

SINGLE CONTROL BY MEANS OF ADJACENT KNURLED KNOBS

dials, *i.e.*, the tuning control and the volume control. In addition, there may be an "on-off" switch in the filament circuit. Any additional verniers, "fine tuning" controls, or compensators take the receiver out of the single-control classification.

The use of vernier controls to correct deviations from exact synchronism among several circuits greatly reduces the cost of manufacture, because close accuracy in making inductances and capacities is thereby dispensed with. If simplicity of control is the objective in purchasing a receiver, too great dependence upon vernier

adjustments may defeat the buyer's purpose. On the other hand, a well-designed receiver, in spite of verniers, can be a convenience, although their use should be reflected in lower cost of the receiver.

The writer has seen receivers upon which stations may be tuned-in in any position over a span of ten degrees of the main dial by correct manipulation of the verniers, thereby eliminating ease of adjustment, the most desirable quality of the single-control set. On the other hand, other receivers, although requiring vernier adjustments, can be properly tuned to a station at only one certain position of the main tuning control. Such receivers are frequently more convenient to tune than two- or three-dial sets.

Having determined whether the receiver is a true one-control set, or one equipped with verniers but so designed that only one adjustment of the main dial brings in any desired station, the efficiency of the mechanical coupling between the tuning elements should be tested. If there is back-lash, slip, or play in gears, the user can never be certain that his tuning circuits are in complete resonance.

To test the efficiency of the mechanical coupling of tuning elements, select a fairly weak station and tune to it accurately. Note the exact setting of the main dial. Turn the control to the opposite end of the scale and then restore it to precisely its original position. If the weak station is again heard to full volume, the mechanical construction is probably satisfactory. On the other hand, if the station is now found two or three degrees above or below the original setting, play and back-lash are likely to introduce tuning complications.

Some receivers depend upon the friction of adjacent knurled knobs, which may be adjusted separately or operated in unison by one hand at will thereby attaining the single-control ideal. With such a device, the user has choice of complete control over each individual circuit as well as unified control over all of them. The conveni-

SIMULTANEOUS TUNING OF CIRCUITS BY MEANS OF A CHAIN AND PULLEY SYSTEM

ence of this type depends chiefly upon how closely the circuits are synchronized throughout the scale, once the correct inter-relation between the adjacent knurled knobs has been determined.

The degree of synchronism attained throughout the dial scale with the knurled knob arrangement is easily ascertained. Tune-in a station at the low end of the scale, making fine adjustment of each of the knurled knobs separately. Next tune-in a station preferably of moderate volume at the high end of the scale, by turning the knurled knobs as a group, being careful as you do so not to change the relation between the knurled controls. Having thus tuned-in a station as well as possible, note the volume carefully. Then adjust each circuit separately, noting whether adjustment of the individual circuits is required to secure accurate resonance. If the same relationship between the several knurled knobs which establishes resonance at one end of the dial scale holds for the other, the set is, in effect, a single-control one. On the other hand, if a new relationship between the various knurled knobs is required at different points along the dials, the set may be considered as having as many controls as there are knobs. Then the only gain by the knurled knob over the ordinary dial is in appearance, and in the ease with which it is adjusted by one hand. There are several makes of receivers employing adjacent knurled knobs for control which, when correctly adjusted with respect to each other, can thereafter be operated as a single control. Such receivers represent the last word in convenience and attain simplicity of control without sacrifice in efficiency.

COMPLICATIONS OF SIN-GLE CONTROL

THE relative cost of obtaining exact synchronism of circuits manipulated by a single control depends on the sharpness of tuning attained for each individual circuit. With radio-frequency amplifiers of high gain and good sharpness of tuning, true single control is rather difficult to secure inexpensively without the aid of verniers. It is a question of precision manufacture, engineering tests, and painstaking design, all of which are naturally reflected in the cost of the receiver. If, at any point along the dial scale, one circuit falls slightly out of step, quality is seriously affected unless the set is fairly broad in its tuning.

There are a number of possible compromises involved in the design of a single-control receiver which account for the wide range in their prices. If quality of reproduction is sacrificed, great sensitiveness and selectivity may be secured at relatively low cost. If on the other hand, sharpness of tuning is sacrificed, a high standard of tonal quality without excessive cost is attainable. If high efficiency, high gain, perfect synchronism, and good quality of reproduction are combined without appreciable sacrifice of any of these qualities, such accuracy in design and manufacture requires that the cost of the set be necessarily high.

Eliminating the most expensive types, which have true single control, sensitiveness, selectivity, and good quality, how may we judge and balance all these diverse qualities in order to make the wisest purchase? To repeat: sensitivity, selectivity, or quality must be sacrificed to secure low cost.

Obviously the most desirable sacrifice is not quality of reproduction, but sensitiveness or selectivity. By choosing a receiver, admittedly a little broad in tuning, you may have single-control simplicity, good quality, and satisfactory volume from local stations at reasonable cost. If troubled by local interference, changes in the antenna installation often minimize the difficulty. A long antenna brings in considerable energy from near-by stations and therefore broadens tuning. By shortening the antenna, selectivity may be increased to a point where interference troubles are minimized. Shortening the antenna reduces sensitiveness as the penalty for improving selectivity.

For the long-distance enthusiast who desires single-control simplicity, sensitiveness, and low cost, there are a number of receivers which, in a measure, attain all of these qualities by introducing regeneration in the radio-frequency circuits. Such receivers usually have a volume control, which, when turned toward maximum, sets the receiver into oscillation throughout the tuning scale, or perhaps does so only at the higher frequencies. These are often radiating receivers of the most pernicious type, feeding oscillations directly into the antenna circuit. Such receivers may be recognized not only by the fact that they flop into vigorous oscillation at the high frequencies, and give a piercing whistle as the dial setting for a station is passed, but also by the fact that stations are often heard at two or three closely adjacent points on the dials. A

SIMPLIFIED ADJUSTMENT OF SEVERAL TAPPED COILS BY ONE KNOB

circuit closely approaching the regeneration point must be tuned with the utmost accuracy, a degree of accuracy almost unattained by any but the most precise of single-control receivers. Therefore, with single-control near-regenerative sets, stations are frequently heard at each point that the individual stages are precisely in tune with the incoming signal.

Such single-control near-regenerative receivers receive from long distances on the shorter waves, annoy the neighbors for miles around, and give good quality only with high-power local stations.

The better single-control receivers are absolutely non-regenerative. It must be realized that the absence of a tickler or regenerative adjustment is in no wise an indication that the receiver is not regenerative, nor need there be a plate coil feedback inductance or capacity or any other physical evidence of a regenerative circuit. The mere presence of a vacuum tube with elements having electrostatic capacity offers the foundations for a regenerative system. Whether this is avoided by neutralization or by the introduction of losses through counter-couplings or weak couplings is not obvious from any external inspection. Hence the real proof is an actual test of the receiver, particularly at the high frequencies where regenerative effects are most likely to be strongly manifest.

The high-grade, single-control receiver, which is well engineered through effective shielding and correct neutralization of regenerative effects, tunes sharply at both ends of the scale. The stations fall in and out of resonance like a parade as you go up and down the dial scale, with volume proportionate to their incoming signal strength. Such sets represent the last word in efficiency and simplicity of control—the product of precision manufacture and sound engineering. Many such receivers are equipped with three and four stages of radio-frequency amplification, associated with a directional loop.

An effective compromise in the attainment of convenient control without excessive cost is by the use of two controls, designed to operate as one, once their relative settings are fixed. Concentric or adjacent knurled knobs coupled by friction are used for this purpose. These are turned naturally as one, unless the user particularly desires to change their relative adjustment. Usually two controls are used in this way with receivers employing a separately tuned antenna circuit and unified control for the remaining stages of the radio-frequency amplifier. In this case, the input stage is made one of high gain, giving the receiver considerably greater sensitiveness. Different antennas have varying inductance, capacity, and resistance, and these factors affect the correct adjustment of the input circuit, making dual tuning necessary.

The difficulties of varying antenna constants may be obviated by using a loop designed as a part of the set, thus making dual control unnecessary. Other sets are designed for use with a very small antenna, or a large antenna in series with a small capacity, and the various circuits accordingly matched to meet such conditions.

Another method of attaining true single control, is to employ the input stage merely as a collector of antenna energy, without causing it to contribute any great amplification. The input stage is then untuned, a resistance or choke being used across the tube, through which the incoming signals are impressed upon its filament and grid. The remaining circuits, all being fed from the plate circuit of a preceding tube, are easily synchronized without any great manufacturing difficulties. Although contributing little or no amplification, the first tube introduces its share of tube noises and does not materially improve selectivity, a serious disadvantage if there is an excessively strong near-by signal. Considering the low cost of tubes, an inefficient stage is not a great disadvantage from an economy standpoint particularly as it makes possible single-control simplicity. On the other hand, it brings out the fallacy of rating a receiver's power by the number of tubes. The real criterion, by number of tubes, is dependent upon the number of stages of high-gain tuned radio-frequency amplification with which the receiver is equipped.

By this time, the reader will appreciate that judging a single-control receiver lends itself to no simple diagnosis. It is better to thrust aside all considerations of design, unless a most detailed study is made, in favor of a few simple observations and performance tests.

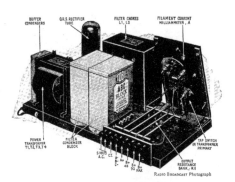

BUFFER CONDENSERS — Q.R.S. RECTIFIER TUBE — FILTER CHOKES L1, L2 — FILAMENT CURRENT MILLIAMMETER, A

POWER TRANSFORMER T1, T2, T3, T4 — FILTER CONDENSER BLOCK — 5 VOLTS A.C. C1 — C— A— B+ B+ A— 45 90 MAX. — TAP SWITCH IN TRANSFORMER PRIMARY — OUTPUT RESISTANCE BANK, R1

RADIO BROADCAST Photograph

A COMPACT A, B, C POWER-SUPPLY UNIT

This entire unit is constructed on a baseboard measuring 17" x 9". The following parts were used: Dongan power transformer and filter chokes, Tobe filter condensers, Carter output resistance, Yaxley tapped switch, Eby binding posts, Jewell milliammeter. Other parts that might be substituted are indicated in the table given in this article

A Lamp Socket A, B, C Device

The Design of a Power-Supply Unit Using the New Q. R. S. High-Current Gaseous Rectifier Tube

By GILBERT EDGAR

IT WOULD be difficult to estimate the amount of engineering research work that in the past has been devoted to the problem of complete a. c. operation of receivers. Probably the first really satisfactory all a. c. receivers were those using 199 tubes in series, with their filaments supplied with rectified a. c. from a B socket-power unit, and with a power tube in the last stage with its filament supplied by raw a. c. directly from a transformer. Such receivers as this were described in the October, 1926, RADIO BROADCAST, but evidently the development was somewhat premature, for not many such sets were built. Since then the question of a. c. operation has become a common subject of discussion and there is evidently considerable interest in this topic.

Two months ago, in the advertising pages of this magazine, there was announced a new rectifier tube, manufactured by the Q. R. S. Music Company, which could be satisfactorily used to supply complete power to a receiver from the power mains. One advantage that accrues from the use of this new tube is that 201-A type tubes may be used in the receiver, their filaments being wired in series. To wire a receiver with the filaments in series is not any more difficult than wiring them in parallel, and it is also an easy matter to rewire an existing receiver for series filament operation. The energy required by the new rectifier to supply A, B, and C voltage to a receiver is sufficiently low so that a. c. operation is not only convenient but also economically practicable. The audible hum heard in the output of a receiver operating from an A, B, C power unit is no greater than that experienced when the same receiver is operated with a storage-battery supply for the filaments and a B socket-power unit to supply the plate voltage. For normal operation the device must supply ¼ ampere (250 milliamperes) to light the filaments of the tubes and it must also supply plate current, so that the total load is generally about

280 to 300 milliamperes. This value of current is much greater than is found in a power unit designed to supply only plate voltage to a receiver and, therefore, the various chokes, condensers, and resistors used in such a unit cannot be used with this new tube. It is necessary that the various parts used have characteristics adapted to the use to which the device is to be put. The main power transformer has four windings, as shown in Figs. 1 and 2. T_1 is a primary preferably tapped for line voltages between 100 and 120; T_3 is a 5-volt winding capable of supplying ¾ ampere for a 171 type power tube; T_3 a high-voltage secondary winding capable of delivering 375 volts each side of the center tap, and T_4 is a 4-volt winding to supply 5 amperes to a small ionizer designed to lower the internal resistance of the tube. The two filter coils, L_4 and L_6, must carry all the filament and plate current for the receiver and, at this very high current drain, they should have an inductance of between 3 and 5 henries. The condensers in Fig. 1 should have the following characteristics:

C_1—0.1-mfd. Buffer Condenser, ⎫ 600 Volts.
C_2—0.2-mfd. Buffer Condenser, ⎬ D. C. Working Voltage.
C_3, C_4—0.1-mfd. Condenser, ⎭
C_6—5-mfd. Condenser, 400 Volts D. C. Working Voltage
C_7—10-mfd. Condenser, 400 Volts D. C. Working Voltage
C_8—1-mfd. Condenser, 200 Volts D. C. Working Voltage
C_9—1-mfd. Condenser, 200 Volts D. C. Working Voltage
C_5—1-mfd. Condenser, 150 Volts D. C. Working Voltage

The resistances used in the output system are required to carry large amounts of current and special units are therefore necessary which are capable of dissipating considerable power. The resistance R_1 should be capable of carrying about 330 milliamperes and should have a total resistance of about 800 ohms. It is preferable that this resistance should be semi-variable because its value will depend upon the number of tubes in the receiver. various manufacturers have designed transformers, chokes, condensers, and resistors especially for use with the QRS tube and a complete list of those companies making them is given in the table on page 154.

To determine the characteristics of the device, a complete power unit was constructed in accordance

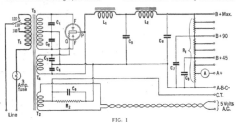

FIG. 1

with Fig. 1. Data were then taken on the unit, a photograph of which can be seen on page 153, to determine the input power and output voltage at various loads. In Fig. 3 is given a curve showing the power required by the device at various current drains. The total current drain from the device is equal to 250 milliamperes (that required to operate the series filaments of the tubes in the receiver) plus the total plate current required by all the tubes. Therefore if a receiver required 30 milliamperes of plate current, the total current drain from the device would be 250 milliamperes for the filaments plus 30 milliamperes for the plates, or a total of 280 milliamperes. The input power at this current drain is about 127 watts, which is approximately the power which the device requires when it is used to operate an ordinary five-tube receiver using a 171 type power tube. In Fig. 4, curve A, are plotted the data taken to determine the voltage delivered in normal operation. It is evident from the curve that, with a total plate-current load of 30 milliamperes (average load of a five-tube receiver), the unit will deliver about 200 volts, which is sufficient to satisfactorily supply plate and grid voltages for a 171 type tube. Curve B in Fig. 4 was made with transformer winding T_4 disconnected so that the ionizer filament was not functioning and, under such conditions, the device only delivers 145 volts with a plate load of 30 milliamperes. It is evident, therefore, that this tube will function without the ionizer but that under such conditions the output voltage is lowered.

FIG. 2

FIG. 3

OUTPUT SYSTEM

IT IS necessary to use at the output end of the unit a voltage divider, or potentiometer arrangement, to obtain the various voltages required by the receiver. The simplest, and in many cases an entirely satisfactory arrangement, is that shown in Figs. 1 and 2, in which the A and B voltages are taken from various taps along a tapped resistance, R_1.

The B voltages obtained by using such an arrangement are constant because the current flowing through the resistance due to the filament load is very much greater than the current drawn by the plate circuits from the various taps and, therefore, the plate-current is of a negligible quantity compared to the total current flowing through the resistors. With some receivers, however, especially those using audio amplification other than transformer coupling, some difficulty will frequently be experienced due to "motor boating," in which case it is necessary to use a different output arrangement. Complete information regarding the various output arrangements that should be used in such circumstances will be found in the article starting on page 157 of this issue.

C bias for the 171 tube is best obtained by connecting a resistance, R_2, in series with the center tap of the filament transformer supplying the filament of the 171 (See Fig. 1), while C bias for the other tubes is obtained by connecting resistances in series with their filaments at the correct points, as shown in the circuit diagrams accompanying the article beginning on page 157. A lead is brought out from the center tap of the filament winding T_2 to a terminal marked CT, and this terminal should connect to one side of the loud speaker jack in the receiver. In this way, the signal energy flowing through the loud speaker is brought back directly to the filament of the 171 tube, and this tends to make the operation of the receiver more stable.

FIG. 4

PARTS FOR THE A, B, C UNIT

SEVERAL manufacturers are now putting out parts capable of meeting the rigid requirements of high-current power-supply devices. Parts made by the following manufacturers, and suitable for such devices as that one described here, are approved by RADIO BROADCAST Laboratory.

TRANSFORMERS AND FILTER CHOKES: Dongan, Thordarson.
CONDENSERS: Aerovox, Dubilier, Fast, Muter, Wilson Electrical Laboratories, Tobe, Potter.
RESISTORS: Aerovox, Allen-Bradley, Amsco, Carter, Centralab, Clarostat, Electro-Motive Engineering Corporation, Lynch, Ward-Leonard.

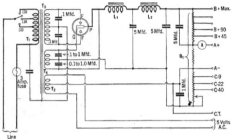

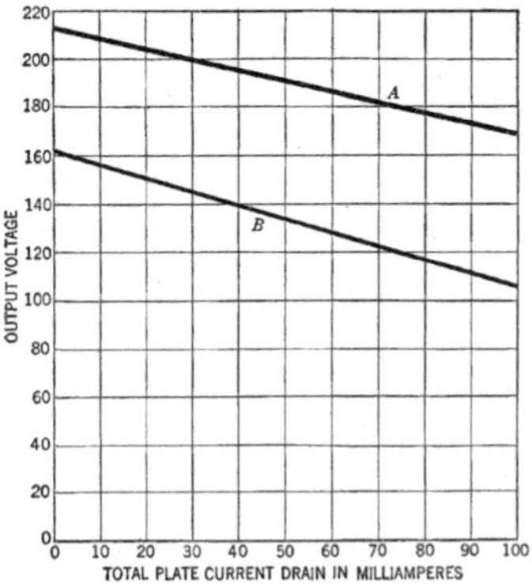

A complete A, B, C device is as simple to construct as an ordinary B power unit and is no more difficult to get operating properly. The entire device can be constructed on an ordinary baseboard. It will be found advisable to mount the baseboard on small rubber feet so that most of the wiring can be done under the baseboard. Use well insulated wire in the construction because the transformer supplies rather high voltages. The two leads from transformer winding T_2 supplying 5 volts a. c. to the filament of the 171 should be twisted together. In constructing a unit do not so arrange the apparatus that the filter chokes and transformer cores are near and in line with each other but endeavor to leave a space of several inches between them and place them out of line or at right angles to each other. The power unit shown schematically in Fig. 1 has been on test for some time, and has given very good results. The circuit diagram in Fig. 2 was supplied by the Q.R.S. Company. Very excellent results have been reported from the use of this latter circuit with all receivers.

For complete details regarding the correct manner in which to wire a receiver for series filament operation the reader is referred to the article starting on page 157 of this issue in which the subject is carefully explained. Pertinent data on this subject also appeared in articles in the two preceding issues of RADIO BROADCAST.

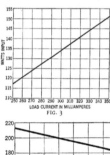

An A, B, C Power-Supply Unit for 201-A Type Tubes

A Power Device Which Provides for the Complete Elimination of All Batteries in a Radio Receiver—Some Design Hints and Precautionary Measures

By ROLAND F. BEERS

THE purpose of this article is to describe a satisfactory unit for supplying A, B, and C power to any radio receiver employing 201-A type radio tubes. It is the writer's belief that this article describes the first really satisfactory method, for the home-constructor, whereby complete radio power for 201-A tubes can be obtained from one unit. It is true that there have been separate A and B units in use for some time, and that their performance has been above reproach. For one who desires simplicity, however, the new method has many advantages to offer. Here is complete radio power from the light socket, without batteries or liquids, and the unit when completed requires absolutely no attention.

To use the power unit described necessitates that the filaments of the tubes in the receiver be wired in series. The writer has explained this method of connection in the two preceding articles of this series, and many diagrams and other data were given showing how to wire a receiver for series filament operation and how to obtain C bias for the various tubes; precautions necessary in using such an arrangement were also outlined. If the reader is not thoroughly familiar with the subject, the details thereof may be found on pages 33 to 35 of the May RADIO BROADCAST, and on pages 101 to 103 of the June issue. The writer has used series filament connection in practically all of the fundamental radio circuits and has found no reason why it cannot be used with any type of existing receiver if proper precautions are taken. An article starting on page 157 of this issue explains how the supply unit described here may be adapted to various well-known circuits.

The entire A, B, C supply unit is built around the new Raytheon Type BA-350 mA. rectifier tube. This new development is a full-wave device, operating upon the principles of gaseous conduction. It has no filament. The principles involved in the design of this new unit are an extension of the fundamentals of all Raytheon gaseous rectifiers, and reveal improvements which increase the efficiency of these devices

to a remarkable degree. These improvements were really a necessity, in order that a rectifier of reasonable proportions might be constructed. The loss in a device of this type takes the form

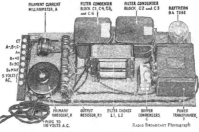

FIG. 1

Three output voltage curves for three different experimental designs of the Raytheon BA rectifier, curve C being the design that was finally adopted. Note that the output voltage is constant from 0 to 300 milliamperes. Curve A is a very unsatisfactory one for if a tube is suddenly removed from a receiver and the load on the power unit tube reduced to practically zero, the voltage will rise to a comparatively high value and endanger the filter condensers

of heat, which must be radiated by the glass bulb of the rectifier, and if the power losses were excessive, it would be essential to build a glass bulb several inches in diameter in order to keep the operating temperature at a safe value. By the improvement in the efficiency, it has been possible to keep the size of the BA rectifier down to moderate proportions.

The full load rating of the tube has been placed at 350 milliamperes and 200 volts d.c. output at the terminals of the filter circuit. This rating, which includes the watts which are bypassed through the filter circuit, is sufficient to operate any receiver having up to ten tubes and a power amplifier. A minimum life of 1000 hours may be expected from these new rectifiers provided they are not overloaded or abused. Fig. 1, "C," shows the characteristic of a typical BA rectifier used in a complete power unit. It should be noted that the voltage is practically constant at all loads and, therefore, if a tube is suddenly removed from the circuit and the load as a result drops from something around 300 milliamperes to about 20, the voltage will not rise excessively and endanger the filter condensers.

In other words, the voltage re mains practically constant independent of the load and, therefore, there is almost an unlimited current capacity in the device This large reserve of power is very desirable as long as it does not exceed the useful power required by the radio receiver, and means must evidently be taken to limit the output of the power unit to this useful value. It is therefore necessary to use special methods of design in the power transformer and filter circuit, and it is consequently not feasible to recommend that the radio constructor build his own power transformer or filter. Their characteristics are very unusual, and it is doubtful if the average constructor could meet the requirements of the design in home-built units. Manufacturers have coöperated with the designers of the BA rectifier, and have affected the design which is best suited for the purpose.

The power transformer is rated at approximately 175

A COMPLETE A, B, C POWER SUPPLY

Using a Raytheon BA tube. This particular unit uses a Thordarson transformer T, Thordarson choke coils L_4 and L_5, Dubilier condensers type BA2 for C_1 and C_5, type BA3 for C_2, C_4, C_6 and C_6 and Type BA1 for the two buffer condensers, C. The primary rheostat R, and the tapped output resistance R_1, are made by Ward-Leonard, while the 2000-ohm C bias resistance R_3 (this resistance cannot be seen in the photograph) is a 2000-ohm Tobe Veritas resistance. The filament milliammeter, A, is a 300-milliampere Weston model 506 meter. Eight X-L binding posts are used and the unit is wired with Kellogg Switchboard Company hook-up wire. It can be built at a cost of about $90.00

watts; it has a secondary winding delivering 320 volts per side for the BA rectifier, and a filament winding of 5 volts at 0.5 amperes for a 171 type power amplifier tube. The choke coils, L_1 and L_2, are of special design, and have a minimum inductance requirement of 10 henrys at 350 mA. direct current. Their d.c. resistance lies between 150 and 200 ohms each. Ordinary choke coils cannot be used in this circuit on account of the large amount of direct current which

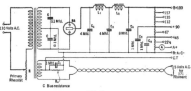

FIG. 2

must flow through the windings. The transformer and chokes are offered by the Acme Apparatus Company, Cambridge, Massachusetts; the Dongan Electric Manufacturing Company, Detroit, Michigan; and the Thordarson Electric Manufacturing Company, Chicago, Illinois.

The normal operating voltage of the filter condenser block should be at least 400 volts d.c. The two buffer condensers in series across the secondary of the power transformer have a capacity of 0.1 mfd., and an operating voltage rating of 1000 volts d.c. (Various manufacturers have arranged to supply condensers that meet these recommendations. At the present time they are being made by Dubilier, Tobe, Aerovox, Muter, Mayolian, Potter, and Fast, and very likely by the time this article appears in print other manufacturers will also be making them.—*Editor*.)

The resistance units used in the device are also special for they must have a wattage rating much in excess of that necessary for a resistance used in an ordinary B socket power unit. Excellent resistances have been designed, however, and are being made by Clarostat, Carter, Ward-Leonard, Centralab, Aerovox, Amsco, Lynch, and the Electro-Motive Engineering Corporation.

HOW THE UNIT FUNCTIONS

A BRIEF explanation of how the A, B, C unit functions will be interesting. The alternating current from the power transformer passes through the BA rectifier, where it is converted into pulsating direct current. This pulsating current may be considered as consisting of a pure direct current and an a.c. component. It is the function of the filter circuit to absorb or bypass the alternating current and to pass on only pure d.c. Approximately 60 mA. of a.c. are bypassed through the filter circuit for a d.c. output of 350 mA. Having obtained a uniform and smooth direct current at the output of the filter circuit, it is now our purpose to dispose of it with reference to the radio receiver, in the proper voltages and currents.

This is effectively accomplished in one of two ways. The first is to use a tapped fixed resistance across the output of the filter circuit as shown in Fig. 2. This unit has a resistance of about 470 ohms, and is supplied with a plurality of taps so that a great range of voltages is available. By the use of this single resistance all the plate and filament voltages are obtained with little difficulty, and since the points at which to take taps can be easily determined in designing the resistance, the method has the advantage that the voltages available at each tap are fairly definitely known. All the plate voltage taps which are used should be bypassed to A minus with a 1-mfd. condenser to prevent undesirable coupling effects.

Fig. 3 shows an alternative means of obtaining the proper plate and filament voltages, and in this arrangement practically any voltage from zero to maximum can be obtained from each voltage tap. This latter system is therefore quite flexible, but it has the disadvantage that the voltages are not known unless a voltmeter is available to measure them. In Fig. 3 the voltages are obtained through variable resistance units, R_2 and R_3, such as Clarostats, and filament control is obtained through a fixed or semi-variable resistance, R_1. In such an arrangement, R_1 must be capable of carrying $\frac{3}{4}$ ampere.

The most important control resistance is that shown in the primary of the power transformer and marked "R," in Figs. 2 and 3. The purpose of this resistance is to offer a degree of control over the primary line voltage and it will be found most valuable in obtaining satisfactory operation from the entire unit. The maximum value of this resistance should be about 15 ohms, and its minimum value about 4 ohms. In the case of the Acme transformer, a 4-ohm resistor is mounted on it so that the additional resistance may be obtained by the use of a 10- or 12-ohm rheostat. The rheostat must be capable of carrying $1\frac{1}{2}$ amperes continuously without undue heating. Rheostats for

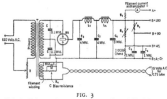

FIG. 3

This circuit diagram shows an arrangement using variable resistances in the output so that various voltages can be obtained whereas, in Fig. 2, the various voltages were obtained from taps in a fixed resistance. The primary rheostat, "R," should be adjusted so that the milliammeter, "A," in the filament circuit, reads between 240 and 250 milliamperes. The BA tube when it is first turned on will glow brilliantly but this glow should disappear in a few seconds

All the plate-, filament-, and C-battery potentials required by a receiver can be obtained from an A, B, C socket-power unit constructed in accordance with the circuit diagram given in this diagram. The various A and B voltages are obtained across a tapped fixed resistance in the output. A 171 power tube is supplied with 5 volts a.c. for its filament, while the C bias for this same tube is obtained by means of a voltage drop across resistance R_1. "R" is the primary rheostat which compensates variations in the line voltage

this purpose are made by Clarostat and Ward-Leonard.

If the experimenter desires, he may build up a power supply unit of this type as an accessory to his series filament radio receiver, or he may construct it as an integral part of the receiver. The performance of the unit will be the same in either case and it is simply a matter of choice which type of power supply the home constructor will build. The photograph shows the manner in which the various parts are assembled. Particular attention should be paid to the arrangement of the power transformer and the filter chokes, so that the effects of good filtering will not be offset by the cores of the filter chokes becoming saturated with leakage flux from the power transformer. Place the iron cores at right angles or out of line with each other.

In building a case or housing for the power supply unit, it should be borne in mind that there is a considerable amount of heat to be dissipated, and that sufficient ventilating means must be employed for this purpose. It is therefore not advisable to build a completely enclosed cabinet which will prevent the circulation of cool air throughout the interior. The rectifier bulb must not be allowed to become over-heated, neither must the transformer nor control resistances be prevented from radiating the normal amount of heat.

The operation of a series filament receiver supplied with the Raytheon power unit just described is not greatly different from that of the ordinary type of receiver. Variations in filament current of the order of 10 per cent. have very little effect upon the output of the radio receiver. In normal operation the Raytheon BA rectifier will be found to run fairly warm, as its efficiency is dependent upon a certain amount of heat within the rectifier, but if the bulb of the rectifier is hot enough to sizzle when it is touched with a moistened finger, it is evident that there is some overload on the rectifier, and the cause of this overload must be determined. It may be caused by a blown condenser in the filter circuit, or by a short-circuit of the output. Such a condition as this would of course be otherwise manifest by faulty reproduction, or total failure of the output voltage. Any such overload on the rectifier should be immediately removed in order not seriously to shorten its life. If a red-hot spot is seen, within the rectifier at any time, it is direct proof that there is a short-circuit in the output of the power unit, which should be corrected at once. The power transformer will run slightly warm under normal full load and the filter chokes should show a very slight rise in temperature. The output resistance units will all run approximately 190° F. If a higher temperature than this is attained it is not safe for inflammable material near by, but if the resistance is housed in a metal container it may be considered reasonable if these units do not exceed 300° F. It is of course a matter of common sense that the resistor units should be mounted in such a position that the hot surfaces do not come in contact with inflammable materials.

During the normal operation of the power unit it may be noticed that there are slight variations of the filament milliammeter which will be caused by fluctuations in the a.c. line voltage. It will be found convenient to maintain average filament currents at a figure between 240 and 250 mA., so that a 10 per cent. change will neither overload the filaments nor cause the amplification to go below normal values.

APPARATUS FOR A, B, C POWER UNITS

The equipment shown in this photograph, in addition to that specified for the units in the two preceding articles, has been especially designed for A, B, C devices. (1) Potter filter condensers; (2) Clarostat variable resistances, including the new power Clarostat for controlling the primary voltage; (3) Raytheon BA tube; (4) John E. Fast filter condensers; (5) Aerovox filter condensers; (6) Q. R. S. 400-mil. tube; (7) Acme filter chokes; (8) Amsco A, B, C resistor kit; (9) Lynch A, B, C resistor; (10) Acme power transformer; (11) Centralab A, B, C resistor kit

Receiver Design for A. C. Operation

Rewiring the Filament Circuits of Some Popular Receivers to Conform With the Requirements for A, B, C Socket Power Supply—New Circuits for the "Lab," "Universal," Browning-Drake, and Neutrodyne Receivers

By HOWARD E. RHODES

IN THE two preceding articles there has been given information regarding the construction of A, B, C power units, using two different rectifier tubes. In this article we will explain how to apply these power units to various receivers. The circuit diagrams of these two power units differ in some details but fortunately the output connections on both are exactly the same and, consequently, there is no need to differentiate between them in applying them to receivers. A receiver designed for use with the unit using the Raytheon tube will work equally well if supplied from the Q.R.S. power unit. Therefore, in the various diagrams in this article we will not include the power unit, and the reader should understand that in all cases the various terminals on the receiver are to be connected to the corresponding terminals on the power unit.

Tests have been conducted in the Laboratory using a five-tube receiver consisting of a stage of r.f. amplification, regenerative detector, and a three-stage double-impedance amplifier. A regenerative receiver using this type of amplification is frequently difficult to get working properly with a power-supply unit because of a marked tendency to "motor-boat," and it was felt that, if this receiver could be made to function satisfactorily with an A,B,C power unit, any other receiver would surely work satisfactorily. It was found that the receiver "motor boated" badly using the output circuit shown in Fig. 1, A, although when a two-stage transformer-coupled amplifier was substituted for the double-impedance affair, results were excellent. To prevent "motor boating" with the impedance amplifier, the output potentiometer circuits shown at B, C, D, or E, Fig. 1, had to be used.

The output arrangement indicated at "B"

differs from the output arrangement shown at "A" in that the filament and plate circuits have been separated. In this arrangement R_3 should have a value of about 700 ohms and be capable of carrying the filament current, and R_4 may be a fixed resistance of about 4000 ohms capable of carrying 50 milliamperes. The total resistance of R_4, R_4, and R_4 should be about 10,000 ohms with taps at various points for different plate voltages. The three latter resistances can be obtained from Amsco, Ward-Leonard, Lynch, or other reputable manufacturers.

In the arrangement shown at "C" the filament current and plate voltages for all the tubes except the detector are obtained from a common resistor. The detector plate voltage, however, is obtained from a separate resistor, R_1, which should have a value of 40,000 ohms. A variable resistance, such as the high-range Clarostat, may

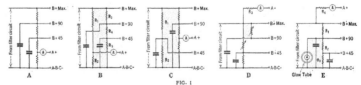

FIG. 1

also be used here. R_2 is a discharge resistor with a value of 10,000 ohms.

At "D" is shown an arrangement using variable resistances to obtain amplifier and detector voltages. Due to the use of two separate resistances for this purpose there is little or no common coupling and, therefore, freedom from "motor boating" is obtained. The filaments are here supplied from a separate resistance, R_4, which should be capable of carrying the filament current. All of these resistances may be obtained from the American Mechanical Laboratories in the Clarostat "A" kit, or they may be obtained as supplied by Centralab.

At "E" we have an arrangement using a glow tube which has also been found very effective in preventing "motor boating." In this circuit, R_1 should have a value of 1500 ohms, R_2 a value of 4000 ohms, and R_3 a value of 6000 ohms, and R_4 must be capable of carrying the filament current. A circuit somewhat similar to this has been worked out by Amsco and a drawing showing it is being supplied with the resistors which they have designed for the job.

ADAPTING POPULAR CIRCUITS

IN FIG. 2 is the circuit diagram of the four-tube "Lab" receiver, described in the November, 1926, RADIO BROADCAST, revised for series filament operation. The order of the tubes from the A plus terminal is r. f. tube first, first a.f. tube

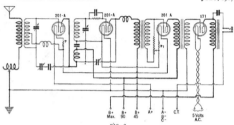

FIG. 2

brought down to two a.c. filament terminals, and it is absolutely necessary that the output circuit shown in the plate circuit of the 171 be used and that one side of the loud speaker jack connect to the center tap, CT, as indicated. The negative A should be grounded. No trouble should be experienced with this receiver due to "motor boating" and, for this reason, any output arrangement indicated in Fig. 1 may be used. However, in the event that the "Lab" receiver

In Fig. 3 we have a diagram of the new "Universal" receiver, described in RADIO BROADCAST for December, 1926, wired for a.c. operation. The order of filaments is the same as in the case of the "Lab" receiver, and the two resistances, r and r_1, should each also have a value of 18 ohms. This receiver will operate satisfactorily using any of the output arrangements indicated in Fig. 1.

In Fig. 4 is given the circuit diagram of an all-

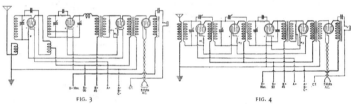

FIG. 3   FIG. 4

second, and detector tube third. The output tube is, of course, supplied with raw a. c. Resistances r and r_1 in the filament circuit are used to obtain grid bias for the r. f. and first audio tubes respectively. They should both have a value of 18 ohms (such as Carter type H-18), which will give each tube a grid bias of $4\frac{1}{2}$ volts. The Carter Radio Company makes a complete line of resistors for use in this connection and they are very satisfactory. The two leads from the 171 tube filament should be twisted together and

has been constructed using a different type of audio amplifier, such as an impedance- or resistance-coupled circuit, it will very likely be necessary to use a special output arrangement in the power unit to prevent "motor boating." Separating the detector plate voltage supply from the rest of the A and B circuits, as shown in B, Fig. 1, will generally eliminate the trouble, but if "motor boating" is very persistent it will be necessary to use the glow tube arrangement shown at "E" in Fig. 1.

a.c. neutrodyne receiver. The resistances r_1, r_2, and r_3 should each have a value of 18 ohms, which will give the two r. f. tubes and the first a. f. tube a grid bias of 4.5 volts. The receiver will function satisfactorily with any of the output arrangements indicated in Fig. 1.

Fig. 5 is the circuit diagram of the impedance-coupled Browning-Drake receiver which was described in the September, 1926, RADIO BROADCAST, now connected so as to be satisfactory for use in conjunction with A, B, C power units. The order of the filaments from the A plus terminal is r.f. tube first, second a.f. tube secondly, first a.f. tube third, and detector finally. The resistance r_1 should have a value of 12 ohms; r_2 should have a value of 4 ohms, and r_3, a value of 12 ohms. The negative A should be grounded and should also connect through a 0.5-mfd. fixed bypass condenser to the lower side of the coil in the antenna circuit. It is unlikely that this receiver will give satisfactory results with the output circuit arrangements of the power unit shown at "A," "B," or "C," Fig. 1. For this set it will be preferable to use the circuit arrangement indicated at "D" or "E."

The 15,000-ohm resistance, r, in the plate circuit of the r.f. tube, is used to cut down the voltage to about 65 from 90 to prevent any possibility of trouble due to oscillations in the r.f. amplifier. A. C. operation can also be applied to other receivers by following the suggestions given in this article.

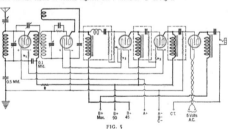

FIG. 5

THE LISTENERS' POINT OF VIEW

✎ Conducted by John Wallace ✎

Sunday Broadcasting Is Five Years Behind the Times

ALAS, what a sad affair is Sunday daytime broadcasting! In respect to its other hours of duty, radio has progressed remarkably since its earliest days, but Sunday has lagged behind and is as bad as it ever was. Sunday of all days!—the very day when radio should be neatly polished as to boot, washed as to ear, and comporting itself at its very best— the one day of the week when the man of the family who foots its bills is on hand to listen to it.

Before commencing to write this article we led ourself to a darkened room, sat ourself down in a not too comfortable chair, and forced ourself with the most hurculean efforts of will to ponder solemnly for one half hour as to the reason why. The conclusion of that thirty minutes of terrific concentration found us still entirely in the dark—in both senses. It is still utterly inexplicable why radio stations, the country over, and without exception, should put forth their very worst at the one time of the week when it would be most advantageous for them to put forth their very best. We pass the question on to you: Why?

That they blare forth either their worst—or nothing at all—is a matter of record, which record we shall presently submit to you. We are concerned in this present discussion simply with the program offerings of Sunday mornings and Sunday afternoons, that is, all those up to 6:00 P. M. The day time broadcasts may be divided into two kinds: First, those which are holy; and second, those which are not holy. In both departments a very low level of quality exists.

But to consider them one at a time, the non-holy offerings first. The complaint on this score can be briefly put—*where are they?* The few offerings aren't so bad, a couple of them are really good, but they are so lamentably few. The good soul who gets himself up early of a Sunday morning to go to church or to till the garden often finds himself with a couple of idle hours on his hands while he awaits dinner. Is there anything entertaining for him to listen to? There is not—unless he be interested in a radio reading of the funnies. It is practically impossible to get any respectable dinner music, unmixed with vocal solos, around Sunday noon, though on any given evening of the week a dozen dinner orchestras are available. In the afternoon he is little better off. In summer, few people are hanging around their parlors—one look at the roads proves that point. But in winter there are millions of potential listeners during the afternoon hours. Nevertheless the winter Sunday afternoon programs are no better than the summer ones.

We do our listening near Chicago and here is what is available to us on an average Sunday: Nothing at all before 1:15 when WGN offers a forty-five minute program from Lyon and Healy's; a string quartette from WMAQ at 1:45; a couple of mediocre orchestras from unimportant stations at 2:00 o'clock; at 3:00 (until recently) a very good concert by the Chicago Philharmonic Orchestra through WGN (but unfortunately its

A RELIGIOUS SERVICE AT WMBI, CHICAGO

season ended in April); at 4:00 o'clock a fair program from WBBM, and at 5:00 another fair one from WEBH.

We do not argue that these offerings, and some by other stations, are not likely to be good. But

SIGNING THE "BOOK OF FRIENDSHIP" AT WRVA

Following a program of Indian songs, Chief Wa-hun-sun-a-Cook, Chief of Chiefs and Great Sachem of the Pamunkey tribe, spoke from WRVA. He is shown signing up Studio Director Elmer G. Hoelzle in the "Book of Friendship." Assisting in the honors are Pocahontas, the Chief's daughter, and Minnehaha, his wife

we protest that our choice is so limited. The 2:00 to 6:00 hours on Sunday should discover us choosing from a plethora of good things; instead it more often finds us giving up a fruitless search in disgust and switching off the receiver.

Since daytime distance reception doesn't exist we can't be sure whether listeners in the East and in the West are as bad off as we are, but a reading of the Sunday program listings in those sections leads us to suppose so. A careful perusal of the New York programs for a recent Sunday in April suggested only five programs that we would consider tuning-in: WJZ at 2:00, Roxy and his gang; WGL at 3:00, Ortho-phonic Musicale; WPG at 3:15, Organ Recital; WOR at 4:30, Stúdio Guild Program; and WRNY at 4:30, Clarinet Quartette. Even then it would be a gamble whether all or any of them would be worth while.

In a west coast program magazine we could find, of the twenty-seven stations listed, only four programs that seemed to promise any interest whatsoever: KFWI at 1:00, Vocal Selections; KFWB at 2:00, Organ Recital; and KGW at 1:00 and 4:00, Orchestra.

Well, so much for the non-holy Sunday offerings—the only trouble with them is that they ain't!

As to the religious broadcasting, we hesitate to stick our foot into a so highly controversial subject, but perhaps we can maintain a perch on the fence and prod about in the material with a long and disinterested stick.

No one can deny that there is plenty of religious broadcasting. Turn to any part of the dial at any part of the day and hear rantings of all kinds and descriptions. As to whether all this talk accomplishes a very tremendous or very negligible good, it is decidedly out of our province to opine. As a radio program reviewer our concern is solely that the listeners, as a body, be given their money's worth in programs. Unquestionably there are great numbers of individuals interested in furnishing religious broadcasts. It is likewise unquestionable, if not so easily demonstrable, that there are great numbers of persons interested in hearing religious broadcasts, even such (to us) irritating bombast as that ladled out by the Reverend Bradley (WMAQ).

Religious broadcasting, as now done, should be continued. We would be the first one to protest if the Ultra-Advanced-Thinkers should attempt to eliminate the religious matter now being broadcast. However, still in our disinterested rôle of program critic, we think it should be clearly realized that there exists a great number of radio receiving set owners who can not conceivably be interested in this type of program. It is a matter of simple equity that they also be shown some consideration.

Zeh Bouck, whose comments on radio and things in general appear in the *Sun* (New York) writes us on this topic. He evidently finds himself in this unchurchly section of the radio audience. His arguments are convincing, if caustic. He says in part:

"I should like to see something done in the way of reasonable religious broadcasting. There is entirely too much hymn singing, damning of lost souls, and evangelizing in our Sunday ether. In other words, we have entirely too much theology and not enough religion. Is there no broadcasting station with sufficient courage to devote an hour each week to a philosophical discourse on right living—considering morality from a scientific, not theological, point of view?

"Aside from assisting Elmer Gantry in the preparation of sermons, Bob Ingersoll performed other useful functions during his life, one of which is summarized in his remark that, 'What we need is religion that will teach us how to live, not how to die.'

"While there are people who demand theology (the religion of golden streets, pearly gates, and seraphim), I see no really good reason why it should not be given to them. At the same time there are hundreds of thousands who are unable to reconcile Christian theology with their sense of criticism and scientific education.

"The type of religious broadcast that I suggest would be most welcome and helpful to this legion, which increases daily as the absurdities of fundamentalism—and even modernism—become more apparent in the light of a growing tendency to think clearly and independently."

Maintaining our position on the fence, we refuse to take sides either with the modernists, the fundamentalists, or with the skeptical group to which Mr. Bouck belongs. Each of these three groups should be given adequate attention, and service, in proportion to the numbers. Certainly under present-conditions the fundamentalists are getting a disproportionately large share of the Sunday broadcasting time. This is well and good for the fundamentalists. For the others who chance to listen-in on Sundays it is no good at all. Why? Because they don't listen to it.

It seems absurd to bring out this point, which is so plainly evident. But it would seem to be a fact that is not clearly realized by the religious broadcasters. Or perhaps they realize it and ignore it. If so, we do not see how they can conscientiously ignore it—and they are conscientious folk. The worthiness of the final aims of an evangelical fundamentalist preacher might be readily agreed to by both the other factions in the radio audience. Yet neither of these factions would lend an ear to this same preacher's radio sermons. The reason is plain enough—they can not understand the language he is speaking. You

might protest until blue in the face that they *should* understand the language he is speaking, but that will in no way alter the practical fact that they don't.

So it seems to us that the radio station which books its Sunday solidly with sermons of this type is discriminating unfairly, and deliberately closing the gates of salvation, or denying the formula for right living, or—call it what you will, to a vast section of its regular listeners.

If a radio station owner sincerely and genuinely assumes a certain satisfaction for his part

JEROME DAMONTE, AT KGO

This artist is heard regularly on the luncheon concert program over this station on Mondays, Wednesdays, and Fridays. He is a member of the Novelty Trio

in conveying words of wisdom to thousands of souls, he must, if he be truly sincere and genuine, also assume the burden of his crime of omission in denying these same messages to half his clientele.

If he feels duty bound to turn over his facilities to such orthodox preachers as ask his service, he is by that same sense of duty, bound, as a purveyor of public service, to turn over his facilities to such speakers as could conceivably reach the ignored half of his audience.

This does not mean that he must act in nonaccordance with his principles. It is his own radio station and surely he should not be asked to use it for the dissemination of doctrines which he sincerely believes to be pernicious. But it is perfectly possible for him to be of service to the unbelieving section of his listeners without violating any of his own pet beliefs. There are no end of contemporary writers and scientists, sociologists and philosophers who, without recourse to religious dogma at all, come to precisely the

same fundamental conclusions concerning right and wrong human conduct as have the venerable and understanding doctors of the churches.

Let the radio station manager select from this group of thinkers the men who reach, in their own quaint fashion, conclusions compatible with his own. Let him invite, urge, coax, or even pay these gentlemen to give Sunday talks from his radio station. In other words, let him appeal to the non-conformist group of his listeners in a language which they can understand.

To put our point concretely: Suppose some contemporary psychologist, through long years of scientific research and observation of his fellow humans, has come to the conclusion, moral aspects disregarded, that it is unscientific, unreasonable, unprogressive, and otherwise subversive of happiness, for a man to tread the primrose path. . . . Given a present day radio preacher holding forth on the same theme; he approaches it in an entirely different, if equally valid, method. But his method is so antagonistic to certain listeners, such as our correspondent, that they either refuse to listen to him or discount his conclusion. These same individuals might have been easily reached by a talk by the aforementioned psychologist.

The way is quite clear; it remains only for some enterprising radio station to take it up.

At any rate something should be done, and shortly, to improve Sunday broadcasting. In such a low state is it at the present time that we will hazard the guess that half the receiving sets in the nation lie idle the whole day through.

How It Feels to Face a Microphone For the First Time

MANY articles have been written on this subject, but none has afforded us more amusement, or seemed to more accurately portray the situation, than that written by Francis Hackett, the Irish novelist, for the *Radio Times* (London). A few paragraphs culled from the midst of others equally droll:

". . . In the studio there is perfect silence. You must begin. And for two instants you are struck by a dumb futility. How do you know that anyone is listening? This audience is a blank. It is inanimate. It cannot clap or boo or say 'Hear, hear.' For all you know, everyone has gone away to dinner and you are about to chatter to the void. This thick suspicion is so unbearable that you brace yourself to believe in something totally outside your experience.

"It is like a dive. In the way that a diver must say good-bye to his springboard and launch his body into the air, so must you pass from the sure footing of silence and launch into speech. With a rushing and breathless celerity you give your words to space, and what you are saying flicks by you unrecognized, like telegraph poles

MUSICIANS ON THE STAFF OF WSM, NASHVILLE

Edward Stockman, baritone; Mildred King, pianist; Vito Pellettieri, violinist and director of Vito's Radio Seven; Lillian Watt, soprano

from a train window. This is a strange confusion. You know you have actually begun to speak, but what exactly you are expressing, what the words are conveying, is not in your grasp. In the first moments you have more sensations than you can deal with. This plunge is headlong, dizzying, and obliterating. You have broken with the habit of a lifetime, have lost the earth. Whenever before you have spoken in public you have had your victims before you. They looked at you, you looked at them; they coughed if you bored them, and when they fell asleep you could enjoy their peaceful expression.

"After the first five minutes, what you want to say really takes possession of your mind, and you definitely want to communicate to these invisible listeners exactly what you have felt. As this conviction mounts, the act of speaking becomes more natural and more amusing. You are not courageous enough to look at the clock, which is glaring at you from the right, and you dare not glance away from the microphone lest it should turn its back on you.

"The dive is over; you are no longer gulping the water and gasping; you begin to time your strokes, to find a rhythm, to swim. And as you do this, the futility of your own ideas gradually becomes less apparent; you actually convince yourself that what you are saying is not so idiotic.

"Then the pleasure of speaking to invisible listeners begins to gain on you. Can they escape from you? You don't believe it. The disease which attacks all speakers seizes on you—verbal elephantiasis. Your words begin to swell. You feel you have a great deal more to say, and you turn away so that the ugly, sour-faced clock can no longer see you.

"Several athletic young men loom up at this point and make formidable gestures. You plead. They threaten. They drag you away."

British Listeners Want Lighter Broadcast Fare

THE business of operating a broadcasting monopoly, it would seem to us from our occasional reading of English periodicals, is no more of a sinecure than the operating of a competitive station in America. Complaints are continually visible in the British press concerning the manner in which the B. B. C. presents its programs. Most of the complaints seem to be to the effect that the B.B.C. is too highbrow and is taking advantage of its monopolistic position to "high hat" the common peepul.

The London *Daily Mail* conducted a ballot of the preferences of the listening public and received the astonishing number of 1,285,083 replies. A vote of this size, it seems to us, can be taken as a very adequate expression of the general opinion in Great Britain, and this is how it resulted:

SUBJECT	VOTES CAST
Variety and Concert Parties	238,489
Light Orchestral Music	179,153
Military Bands	164,613
Dance Music	134,027
Talk: Topical, Sport, and News	114,571
Symphony Concerts	78,781
Solos: Vocal and Instrumental	72,658
Opera and Oratorio	60,983
Outside Broadcasts	51,755
Short Plays and Sketches	49,857
Talk: Scientific and Informative	30,919
Glees, Choruses, Sea Chanties	30,445
Chamber Music	27,467
Revues	27,059
Long Plays	17,576
Readings and Recitations	2,717
Free votes not recorded	4,013
	1,285,083

The tenor of the vote was, as can be seen: (1.) A vote for more fun. (2.) A vote for fewer features which need sustained attention. The British conclusions rather parallel those gained from the questionnaire run in this department. Our readers have an overwhelming preference for instrumental music and a comparative indifference to plays, scientific talks, readings, and so forth. If a comparison to our questionnaire, which resulted in only one-hundredth as many replies, were fair, we might argue that the American public has a more sophisticated taste in music, since serious music, as of symphony orchestras, topped the list in our readers' vote, but is relegated to sixth place in the British vote.

THUMB NAIL REVIEWS

KYW and the blue network—The Philco Hour making its initial bow. We were playing bridge at the time and so couldn't give it very close attention, but we doubt if, conditions being otherwise, we would have. It struck us as an awful hodge podge of every sort and variety of entertainment that could be jammed into sixty minutes.

WMAQ and the red network—Another new advertising hour, this time an orchestra spons-

ored by the Cadillac-La Salle automobile manufacturers. The orchestra was all right but oh the drivel that was plentifully interlarded! Long spiels such as "and now the beckoning roads and the sunny skies call us to the great outdoors and the next number will be in the spirit of the springtime and of the motor car we are selling, the La Salle. Grieg's 'To Spring'."

WEAF (and network)—We listened to an Eveready Hour devoted to "musical hits of preradio days," a program we had looked forward to with greedy anticipation. Sadly, though, but few of the tunes we heard were other than those one might listen to on any dinner music program, from any broadcasting station, on any night. However, a poor Eveready hour these days is sufficiently rare to merit notice.

KDKA—"The Prisoner's Song"!!!

WABC—We accidentally happened upon WABC a while ago just as "An Evening at Tony Pastor's" had begun to unroll itself. "Tony Pastor's," so we gathered from a rather brief announcement, was a one-time music hall on 14th Street, New York City, and the radio audience of WABC was asked to imagine itself seated before the stage of the Hall in the year 1895. So efficacious were the efforts of those responsible for the staging of the program in the studio that we did

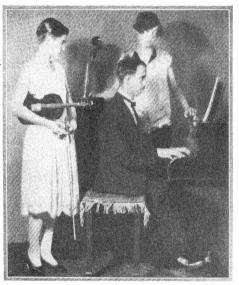

THE WHITNEY TRIO

A capable organization frequently heard through WMAQ, Chicago, of whose staff they are members. From left to right they are Noreen, Robert, and Grace Whitney, brother and sisters

not find it in the least difficult to imagine that we were present at "Tony's," actually enjoying the show on the stage. We got a great kick out of hearing such old timers as "Take Back Your Gold," "Everybody Works but Father," and "My Mother Was a Lady." And when the foreign gentleman got into difficulties trying to explain his act, in which his three lions were to leap through burning "hopps" (recognize "hoops?"), we thought we'd die laughing. The only thing that was a bit overdone, we thought, was the incidental conversation of the onlookers, which too obviously was spoken right into the microphone, and was rather inane anyhow.

Microphone Miscellany

CAMOUFLAGE FOR RADIO ARTISTS

THE new studios of the National Broadcasting Company in its Fifth Avenue home now under construction will make use of every device of color and decoration as a psychological means of egging the performers on to their best efforts. Operatic and stage stars, for example, will face the microphone in a large studio, with a spotlight playing upon them. The rest of the room will be dimly lighted, with the microphone placed in shadow where the artist cannot discern it, and the vista which will open before the performer will present the effect of a large auditorium, with a silent audience waiting to applaud the broadcaster's efforts.

One studio has been designed to appeal to prominent men. The suggested effect is that of the Roman Forum! Columns appear in the background, and a scheme of Pompeian decoration will be produced by hidden lights. Another studio is designed to stimulate minds to which the mystic carries a great appeal. Here the impression will be that of a Gothic church, with alternate light and dark sections suggesting the arch and aisles of such an edifice. From a concealed point near the ceiling the pattern of a church window will be thrown on the floor in light. The performer whose cosmic urge is titillated to activity by the proximity of pinchbeck, will be ushered into a studio of the style Louis XIV. Gilt and pastel colors will be much in evidence and through a window the effect of looking into an elaborate garden will be produced. One of the smaller studios will be decorated to stimulate jazz performers. In this room the decoration scheme will be wildly futuristic with plenty of color in bizarre designs. Another studio, designed to appeal to serious minds will pass itself off as a library. And, lo and behold, the advance report from the NBC goes on to state that two small studios will be left unadorned for the use of experienced broadcasters who react strongly to the mere presence of the microphone and the knowledge that millions of radio listeners are hearing them, although the audience is an invisible one.

We might suggest that a swimming pool be provided for Channel conquerors to make their speeches from; that a bed room with a yawping brat be supplied for the Uncle Charlies; that a saw-dust covered bar room floor be installed for singers of sentimental ballads; that a street scene in Madrid be improvised for temperamental Spanish instructors; that a—oh well, and so forth! But we refrain from this obvious pose. Adverse as our reaction is to such frumpery, we suspect that the powers of the National Broadcasting Company like it no more than we do and ordered the gew-gaws with their tongue in their cheek. Further, we extend to them our sympathy, for full well we know that they will become aw-

fully sick of spending their working days amid psuedo Gothic cathedrals and Rococo drawing rooms. Besides, if we were pressed, we would admit that the psychology behind all this sham work seems to us sound enough. The interpretative, or recreative, artist is notoriously devoid of good taste and his reactions to, and demands of, his surroundings continue to old age to be very childlike. This is only natural since his art is not self contained but comes from the outside. Your composer can sit him down in the most barrenly furnished, drab, little attic room and emerge hours later with a sublimely beautiful musical composition tucked under his arm. Set an average performer to playing that same tune in that same unattractive chamber and he will very likely protest that the surroundings are too depressing for him to do his best work. To rekindle in himself the emotions which charged the composer, he needs the warming lights, the dim auditorium, the silent, waiting audience, and all the other inspiriting adjuncts of the theater. Four walls and a microphone cannot be an adequate substitute. So we think he is justly entitled to all the illusory trappings the National Broadcasting Company is preparing for him. And we are thankful that radio is still non-visual!

UNIFORM WAVELENGTH FOR CHAIN FEATURES

KNOWING nothing ourself of the mechanical problems involved, and having heard none of the probable several valid objections to the plan, we shall not comment on the contention of C. B. Smith of wbbm in his open letter to the Federal Radio Commission that the broadcasting of all chain stations on a uniform wavelength would be more convenient for the listener. He says:

The chain stations since the first of the year have done much to raise the standard of air entertainment in the United States. They should be highly commended. Nevertheless I do not believe this is sufficient reason to allow them to take up several air channels for these programs. My thought in this case, therefore, is that you might well give to each chain now in existence or which may be organized, a certain air channel. When a broadcast is given by the chain over several stations, each station should be compelled to change its wavelength to the wavelength of the chain's key station.

Assuming that the key station of a chain was located in Chicago and that its channel was 400 meters, when there is to be a broadcast over this chain every station involved should change its wavelength to 400 meters. As it is now, the same program from a chain station may be heard at many points on the dial of a receiving set.

This would in no way interfere with the chain broadcasts, but it would permit of more programs. It would in fact simplify matters for the listeners. For instance if the channel of the XYZ Chain was 430 meters, the listeners would always know just where to pick up the programs from that chain.

Communications

Sir:

I am a bcl of five years experience. I long ago got over the notion that I owe anything to the broadcasting stations, but once in a while when the announcer says they have made some change in the station and desire reports on reception I still write a report. But I think I am through. You send a report, date, time, set, audibility, R3., etc., and in due course receive a circular letter beginning, "We are glad you enjoy our program." The chances are that you were listening on the phones, late at night when you

should be asleep, and that the program came in bits mixed up with other stations, static, etc. The rest of the letter tells you that their garden seeds, night club, etc., is as good as their station and will you send them an order or come in and spend something.

The more enterprising put you on their mailing list and there is that much more mail to put in the waste basket. I have reached a state of mind where I am inclined to regard requests for reports on reception—as distinguished from direct requests for applesauce—as just another way of asking for the same thing.

Of course if weaf asks for reports when they open up at Bellmore I'll send one, but the average station, even if they do boast a thousand watts, can get its reports elsewhere.

Yours for more sleep and less stamps,
                BEECHER OGDEN,
           PLEASANTVILLE, New York.

Sir:

I admit that I think enough of RADIO BROADCAST to read it carefully every month, but if I ever secretly entertained an ambition to become an announcer, your editorial in the January issue ought to be an antidote. If everybody is as hard on them as you are, I do not envy them the job. Even at that, I bet you were just sore because you could not grab a gun and go after the ducks that woc gave the whereabouts of. And if wsbt asked everybody who "could do something" to drop in at their studio, is that such a serious offence? Did you ever try to get up a program without Atwater Kent's money to do it with?

"Next we will play for you—" Well, who are they playing for anyway? "You" takes in quite a few, you know. And then comes the fellow who wants the announcer to say: "Sit by!" It never occurred to me to stand up when the announcer says: "Stand by." I just turn the dials.

The only time an announcer almost frightened me was when he boldly stated that he was going to "Sign off the air." I had a sort of choking sensation as I pictured him shutting off the oxygen, but we lived through, and found the air clearer after he was "off," because he took with him some of the most miserable jazz I ever heard.

Just let them go ahead and read wires and dedicate, and "play for you," just so they play. But, please shed a tear for us out here who are expected to listen for hours at a fiend, who is doing his "durndest" to sell automobile tires, plants, batteries, shrubbery, etc., and, as I write, on a stolen wavelength, too.
                JAMES NEWBURGH,
             SIOUX CITY, Iowa.

Sir:

Broadcasters are rather neglecting their country audience, as well as many of the worker class of their city audiences, by listing all their best programs at too late an hour. Even 9 P. M. is too late an hour for many of their most interested listeners, such as farmers, mechanics, the aged, etc. Of course these people do not write letters very freely, although they are most appreciative of good programs.
                H. T. DEMAREST,
             WARWICK, New York.

Sir:

Too much is said against Henry Field and Earl May stations, kfnf and kma. Both are clean and satisfy our Iowa farm homes. They are a temperance people and we need more like them.
                MRS. W. W. W.,
             ESTHERVILLE, Iowa.

A Combined Push-Pull Power Amplifier and Socket B Device

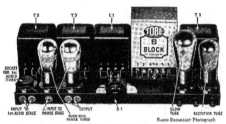

Glow Tube Used to Maintain Constant Voltage and Prevent "Motor Boating"

Provision Is Made to Include Extra Audio Stage for Electrical Phonograph Pick-up

WITH THE CASE REMOVED
The parts are lettered to coincide with Fig. 2
RADIO BROADCAST Photograph

By THE LABORATORY STAFF

IN THE May RADIO BROADCAST was described by Howard E. Rhodes an interesting power supply outfit designed by McMurdo Silver. This unit, shown here in Fig. 1, employed a full-wave rectifier, such as the UX-213 (CX-313) with the Clough filtering system, and delivered sufficient voltage for the operation of a 171 type tube in the final power stage in a receiver. Mr. Silver's unit, however, differed from others described recently in that it used the UX-874 (CX-374) glow tube to maintain the 90-volt tap at a constant voltage even though the current drain from that tap—and from the detector or 45-volt tap—varied through a wide range. The immediate effect of the glow tube is to decrease the apparent terminal impedance of the plate supply device so that "motor boating" trouble, experienced with resistance or impedance amplifiers when operated from socket power devices, is eliminated.

The photographs and Fig. 2 of the present article represent basically this same B supply device but with the addition of the necessary transformers to make up a power amplifier. Mr. Silver in this combination amplifier-B supply device has used a push-pull amplifier in which may be used either 112 or 171 type tubes.

If the broadcast listener who constructs this combination unit (all he really has to do is to wire it up) is near local high-quality broadcast stations of average power, he will use 171 tubes. If, on the other hand, he is some distance from powerful stations, 112 type tubes will produce more volume owing to their greater amplification factor, although their handling capacity is more limited.

The complete equipment is housed in a metal case that not only protects the tubes from damage

but lends a finished appearance to the unit. Space is provided for an extra socket so that, with a resistance coupling unit which may be installed below the baseboard, a two-stage amplifier results, which can be worked out of a detector. It will be necessary to use batteries for the filament and C bias of this first-stage audio tube, and since it is only to be used as a voltage amplifier, a 199 can be utilized with results almost comparable to a storage battery tube. Three dry cells may be used, two for filament and one for C bias; they will last long enough to provide, for all practical purposes, an extremely economical amplifier.

The development of the combination B supply device and power amplifier started from two separate angles, i.e., the demand for high-quality audio reproduction which necessitated the use of a power tube, and the need for a powerful B supply to furnish power for this last stage audio tube. At the time the combination unit was developed, many existing receiver installations were unable to supply the necessary voltage for the operation of a power tube inserted in the last audio stage, while the audio coupling devices found in many receivers of that period were poor, to say the least. The combined power amplifier stage and B supply was an important development then, not only from the standpoint of qual-

ity reproduction, but also from the standpoint of simplicity.

The power-supply device described here is capable of furnishing well rectified and filtered power to the receiver proper and to the power amplifier. This power is sufficient to take care of the demands of a good amplifier, that is, one which amplifies the lowest notes which are now being transmitted, and this power is furnished by a device with as low a terminal impedance as possible to prevent "motor boating" and kindred effects. This feature is accomplished by the use of the glow tube (UX-874 or CX-374). The amplifier has a good frequency characteristic, and secondly, it is capable of handling considerable input voltages. It is also efficient; that is, with a given input voltage, it will deliver to the loud speaker as great an undistorted power as possible.

PUSH-PULL AMPLIFICATION

THE push-pull form of amplification is employed in this power amplifier, and it has several advantages. In the General Radio *Experimenter* for May, 1927, Mr. C. T. Burke claims for the push-pull circuit "greater undistorted output than is possible with two tubes in parallel or a single tube. Even harmonics are eliminated. As most of the harmonics introduced by tube overloading are even, this permits operation of the tubes at heavier loads than is possible with the usual system. Another advantage is the elimination of d.c. magnetization of the output core as the direct current flows in opposite directions from the two tubes."

Considering several amplifiers each worked into its own impedance the power output may be found by multiplying the

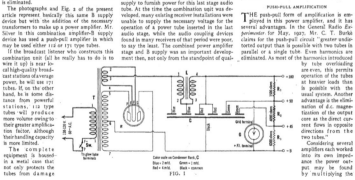

Color code on Condenser Bank, C
Blue = 2 mfd.   Green = 1 mfd.
Red = 4 mfd.   Black = common

FIG. 1

factors below by the square of the input r.m.s. volts:

SINGLE	SINGLE	SINGLE	PARALLEL OR PUSH-PULL:
171	112	210	171 112
1.125	3.34	3.04	2.25 6.7

The maximum power in milliwatts obtainable, however, under the condition of maximum allowable plate and grid voltages, is as follows:

SINGLE	SINGLE	SINGLE	PARALLEL OR PUSH-PULL:
171	112	210	171 112
930	184	1860	1840 368

There is another consideration. It has been demonstrated mathematically that the greatest undistorted power output will be delivered when the load impedance is double that of the amplifier output. It must be remembered too that the above figures are for a resistance load and that a single 171 will deliver its greatest power to a loud speaker whose impedance at some frequency is equal to about 2000 ohms. The parallel arrangement will do the same at a frequency where the impedance is 1000 ohms, while the push-pull amplifier "matches" at approximately 4000 ohms. The Silver output transformer is designed to match the average cone type loud speaker to the amplifier at approximately 30 cycles. As Mr. Burke points out, greater input voltages can be placed on a push-pull amplifier without distortion due to overloading becoming evident or objectionable. Thus this type of amplification which has been neglected since the advent of high-quality transformers, once more is made available for the home constructor, this time in an attractive form and with excellent electrical characteristics. The input transformer is similar in frequency characteristics to the s-m 220.

Some trouble may be had with the amplifier singing when 112 tubes are used—a difficulty that push-pull amplifiers of good construction frequently get into. The remedy is simple: Place a 0.0001-to 0.0004-mfd. fixed condenser across one-half of the input push-pull secondary windings, which will unbalance the amplifier enough at high frequencies to prevent singing.

The circuit diagram of the complete power amplifier and plate power supply is shown in an accompanying diagram, Fig. 2. In this case a gaseous rectifier such as the Raytheon BH or the

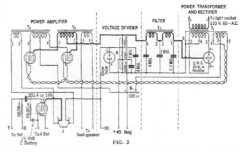

FIG. 2

QRS 85-milliampere tube is used instead of the thermionic rectifier that was used in the original device described in May. The filament winding on the s-m 329 transformer is used to light the filaments of the push-pull amplifier tubes. Connections are also shown for the resistance input

to the extra tube which may be used or not as the constructor desires. If this tube is a 199 it is possible to light its filament from the combined rectified current taken by the power tubes and the glow tube, although this makes the connections, adjustment, and regulation somewhat more complicated. It is much simpler to use dry cells. If a 201-A tube is used in the extra socket, a pair of leads must be brought out to the storage

battery, and some arrangement made whereby turning on the set turns on this tube. For example, its filament can be placed in parallel with that of any of the receiver's tubes, and with the proper ballast or rheostat so that both tubes get the proper current, automatic control over all filaments is secured.

The additional socket provided in the assembly makes it possible to use this combination amplifier-B supply device with a phonograph pick-up. Greater signal strength will be obtained by connecting the pick-up to the first tube by means of a good audio transformer. In the Laboratory several of the well-known pick-up devices were used with success.

The following is a list of parts used in the amplifier-B supply unit described here:

T₁—S-M No. 329 Power Transformer.	$ 6.00
T₂—S-M No. 230 Push-Pull Input Transformer	10.00
T₃—S-M No. 231 Push-Pull Output Transformer	10.00
L₁—S-M No. 331 Unichoke. . . .	6.00
Tobe No. 660 Condenser Block (Containing Two 4-Mfd., Two 1-Mfd., and Two 2-Mfd. Condensers) .	12.00
R—Ward-Leonard S-11,350 Tapped Resistance or S-M No. 655 .	2.50
R₁—Frost No. 834. 1000-Ohm Potentiometer	1.00
Four S-M No. 511 Tube Sockets .	2.00
Four Frost No. 253 Tip Jacks . . .	.60
Van Doorn No. 661 Steel Chassis and Cabinet with Hardware . . .	6.00
Three Eby Binding Posts (B Minus, Plus 45, Plus 90)	.45
Q.R.S. 85-Mil. Rectifier Tube . . .	5.00
cx-374 or ux-874 Voltage Regulator Tube	5.50
Two 112 or 171 Type Amplifier Tubes	9.00
TOTAL	**$76.05**

ADDITIONAL PARTS FOR FIRST AUDIO STAGE

S-M No. 511 Tube Socket	.50
Lynch Double Resistor Mount . .	.50
Lynch 1⁄10-Megohm Resistor . . .	.75
Lynch 1⁄4-Megohm Resistor .	.50
Tinytobe 0.01-Mfd. Fixed Condenser .	.55
Frost No. 951 Four-Contact, Double-Circuit Jack	.50
TOTAL	**$ 3.30**

UNDERNEATH THE UNIT

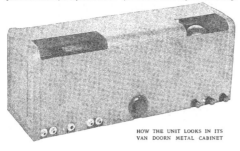

HOW THE UNIT LOOKS IN ITS
VAN DOORN METAL CABINET

A Portable Long-Wave Receiver

A Description of the Receiver Built by the Laboratory for the American Geographical Society —Some of the Signals That May be Heard

By KEITH HENNEY

Director of the Laboratory

THE FINISHED RECEIVER

This illustration clearly indicates how the aluminum sheets are arranged to divide the metal cabinet into compartments for batteries and phones, etc.

AMONG other noteworthy services due to radio is the increasing ease and accuracy with which navigators and explorers can determine their distance east or west from the Greenwich meridian. Time signals transmitted by wireless, which are used for such calculations, can be heard in practically all parts of the world with apparatus simple enough to be built by unskilled constructors, and light enough in weight to be carried, complete with batteries, on a man's back. The apparatus described and illustrated in this article is the result of several receivers constructed by Radio Broadcast Laboratory for explorers and for the American Geographical Society.

The receiver consists, in radio language, of a single-circuit long-wave set using honeycomb coils, and having two stages of audio amplification. The set first picks up and detects the signals, after which the signals are amplified sufficiently to be audible in a pair of headphones. Three dry-cell tubes (199 type), three A batteries, and two small B batteries (one is a spare) of 22.5 volts, are included in the metal case, together with antenna and ground wires, extra tubes, headphones, and simple tools.

The diagram of connections is shown in Fig. 1 and any one who has ever built or torn down a radio set will have no difficulty in constructing this simple receiver. Although the tubes are delicate, experience has shown that their life is quite long even when they must withstand severe shocks encountered in the field.

Several receivers of this general type have been built and placed in the hands of explorers in Brazil, Guatemala, and Venezuela. The first was placed in a Signal Corps telephone box having the approximate dimensions of four by eight by ten inches. The second was housed in a stout wooden box specially made and sufficiently large to accommodate the entire equipment. The third was placed in a metal tool box made by the Kennedy Manufacturing Company, with a tight fitting cover, and is perhaps the most satisfactory design. It is shown in the accompanying photographs. Complete with batteries, wire, and tools, it weighs about 22.5 pounds.

THE ANTENNA

THE antenna is very simple, consisting of a single wire from 50 to 100 feet long, and may be of any kind of wire, insulated or not. A simple manner of solving the antenna problem is to use a spool of rather fine wire so that many hundred feet may be included without adding much weight. Each antenna may be abandoned in this case after signals have been received, since the spool will contain enough wire to last an expedition several weeks listening-in once each day. A one-pound spool of No. 28 bare copper wire is about 2000 feet in length. Such a spool therefore, would provide sufficient wire for twenty antennas of 100 feet in length.

Some success has been had using loops in place of external antennas. They are useful in thick country where a long external wire would be difficult to erect. Signals, however, are not so loud with a loop although the directional effect is useful in avoiding interference and increasing the signal-to-static ratio.

A wire about 50 feet long, attached to a metal stake or plate and thrown into a creek or driven into moist earth, constitutes the ground connection. In dry or rocky territory the wire may be laid on the ground with practically the same results:

The accompanying diagrams and photographs give in sufficient detail the actual construction of this receiver. The Kennedy metal cabinet is divided into three compartments by means of two aluminum sheets 8" x 7" x 1' 16", and these are held in place by means of ¾" angle brass. The three 1½-volt A batteries are placed in the right-hand end metal compartment. An extra set of A batteries is usually carried with the exploring party, ready to slip into the receiver box. In tropical climates, and with reasonable care in handling the tubes, it should be possible to receive time signals daily for a period of about six months without requiring new batteries or tubes. All connections should be rigidly made of well insulated wire, and should be well soldered.

The receiver is placed into operation by opening the cover, making the ground connection, throwing a length of wire over a tree limb, or over any other elevation, plugging in the ear phones, and turning on the current to the tubes. At once signals should be received, since there are many high-powered long-wavelength stations pounding away at all hours of the day or night with sufficient intensity to be heard anywhere in the world.

In the United States the largest station sending time signals is Annapolis, a Navy station, whose call is NSS, wavelength 17,130 meters (17.6 kc.). With a 1500-turn honeycomb coil and a 0.001-mfd. tuning condenser, signals from this station will be heard from about 40 to 50 degrees on the condenser dial. About the time the signals are to be transmitted, this station may be heard emitting a continuous note while others will be transmitting in irregular dots and dashes. Time signals from this station—and from all other American stations—consist of a series of dots, and no difficulty should be experienced in distinguishing them from the more or less irregular sending of the other stations which will be heard at the same time. Adjusting the tuning condenser will enable the operator to find a point where NSS, or whatever station is being heard, is less bothered by interfering stations.

Time signals from NSS and from all United States stations consist of dots at every second from 11h. 55m. 00s. to 11h. 59m. 49s. and from 21h. 55m. 00s. to 21h. 59m. 49s., except at the 29th second of each minute and the last 5 seconds of each minute. The beginning of a dash at noon and at 22h. 00m. 00s. Standard Time, indicates the exact time. The lag constant for Annapolis time signals has been determined as 0.08 seconds.

Time signals transmitted from other stations are given in Table No. 1, and complete details of the transmission may be obtained from the stations themselves or from *Radio Aids*

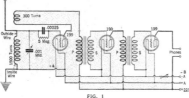

FIG. 1

to Navigation, published by the Hydrographic Office of the U. S. Navy, at 90 cents.

For those interested in learning the code, the receiver described here will be found to be very helpful. At all times of day or night, in any part of the world, signals may be heard, some very fast, others slow enough for the novice to copy. Some of it is in secret code, long words with absolutely no meaning or context, admittedly the best material for code practice. Other signals are in readable English and often the words are repeated twice. In the Laboratory, press has been received through LY in France and GSL in England.

Technically, the method of receiving used here is very inefficient. The receiver employs the beat note system of reception. That is, the detector tube oscillates, and the signals actually heard in the headphones are the beat notes caused by heterodyning of the incoming signals with those generated in the receiver. In other words, the receiver is actually detuned from the incoming frequency. For example, suppose we are listening to a station transmitting on 20,000 meters (15,000 cycles). Our ears and our headphones are most sensitive to notes of the order of 1000 cycles, so we detune our detector circuit to, say, 16,000 cycles, so that the desirable 1000-cycle beat note will result (the frequency of that beat note is equal to the difference of the two heterodyning frequencies, 16,000-15,000 kc. in this case. At the same time suppose a station to be transmitting on 16,000 meters (18,750 cycles), this will produce a second beat note in our headphones of 2750 cycles, so this latter station too may be heard.

When the detector is tuned to the exact frequency of the incoming signals, we shall hear nothing, for there is no beat note being produced.

Code listeners in the United States should be able to hear the stations listed in Table No. 2 and, under good conditions, many others in foreign countries. It is interesting to note that all of the stations in table No. 2 are operating in a frequency band only 8.2 kilocycles wide, a condition that seems appalling when one considers the wide bands available for broadcast or amateur work. In commercial receiving stations signals are picked up on special antennas which have considerable directive effect, after which they are filtered through circuits which pass a band only 200 cycles wide. They are then amplified until strong enough to operate relays which print the dots and dashes on tape. The receiving operator can copy either by sound or by watching the tape, or both.

An approximate calibration of a receiver similar to that shown in the photographs is given in Fig. 2. Owing to the broad tuning, a given

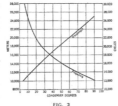

FIG. 2

station may be heard over rather a wide condenser variation, and several stations may be heard at the same time, as described above.

APPARATUS NECESSARY

THE list of apparatus gives the equipment that went into the metal encased receiver built for the American Geographical Society. The problem is one of limiting the space and

STATION	CALL LETTERS	WAVE-LENGTH	FRE-QUENCY
Marion, Massachusetts	WSO	11620	25800
Marion, Massachusetts	WRQ	13505	22205
Bolinas, California	KDU	13100	22890
Rocky Point, New York	WQK	16465	18250
Rocky Point, New York	WQL	17500	17130
Rocky Point, New York	WSS	16120	18600
Annapolis, Maryland	NSS	17030	17600
Tuckerton, New Jersey	WCI	16700	17950
Tuckerton, New Jersey	WGC	15900	18860

TABLE NO. 2

weight requirements. Unmounted transformers made by Modern, of Toledo, and the hedgehog type transformer of the Premier Electric Company, of Chicago, have been successfully used. The Kennedy tool box is made of sheet iron and

the coils must be mounted so that they will not be close or parallel to the metal wall. The tickler coil should be placed between the secondary coil and the iron, and if the detector does not at first oscillate, the tickler coil connections should be reversed. Some transformers may require a bypass condenser across the primary to insure good detector oscillations. The hedgehog did not require such a condenser but in case it is needed it should be of about 0.006 mfd

3 Benjamin Sockets	$1.80
2 Hedgehog Transformers . . .	7.00
1 0.001 Mfd. General Radio Condenser and Dial	7.50
1 1500-Turn Honeycomb Coil . . .	3.20
1 300-Turn Honeycomb Coil . . .	1.05
1 Baseboard 8″ x 7″ x ⅞″	.10
1 Kennedy Metal Cabinet, 16″x8″x9″	3.25
1 22½-Volt B Battery	1.75
1 1½-Volt Dry Cells	1.20
1 Pair Phones	5.00
1 20-Ohm Carter Combined Rheostat and Switch	1.00
4 XL Binding Posts	.60
1 8″ x 6″ Panel	1.00
3 UX-199 Tubes	5.25
1 Grid Condenser	.50
1 Grid Leak	.50
2 Pieces ¾″ Angle Brass 7″ Long . .	.30
2 Pieces Aluminum 8″ x 7″x₁¼″ . .	.25
Total	$41.25

The Kennedy metal cabinet referred to in the above list of parts is manufactured by the Kennedy Manufacturing Company, Van Wert, Ohio.

Using the two coils specified in the above list of parts, the receiver described will have a wavelength range of from about 10,000 to 25,000 meters (30 to 12 kilocycles). To cover other wavelengths, coils with a different number of turns than those specified should be used. The following list gives the various standard coils which should be used in the antenna circuit with a 0.001-mfd. tuning condenser together with the wavelengths they cover: 25 turns, 120-355 meters; 35 turns, 160 to 480 meters; 50 turns, 220-690 meters; 75 turns, 340-1020 meters; 100 turns, 430-1330 meters; 150 turns, 680-2060 meters; 200 turns, 900-2700 meters; 250, 1100-3410 meters; 300 1400-4120 meters; 400 turns, 1800-5500 meters; 500 turns, 2300-7000 meters; 600 turns, 2800-8200 meters; 750 turns, 3500-10400 meters; 1000 turns, 4700-13800 meters; 1250 turns, 6000-18000 meters.

The tickler coil should have from a third to half as many turns as the coil to which it is coupled.

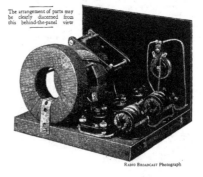

The arrangement of the parts may be clearly discerned from this behind-the-panel view

COUNTRY	STATION	CALL	G. C. T.			WAVE-LENGTH (Meters)	DIAL SETTING (Approx.)
			h.	m.	s.		
United States	Annapolis	NSS	17	00	00	17130	42
			3	00	00	17130	42
	San Diego	NPL	20	00	00	9801	0
Panama	Colon	NBA	10	00	00	6663	—
			18	00	00	6663	—
Hawaiian Islands	Pearl Harbor	NPM	00	00	00	11490	15
France	Bordeaux	LY	8	00	00	23400	80
	Lyons	YN	9	00	00	15500	35
Germany	Nauen	POZ	11	58	10	18050	50
			23	58	10	18050	50
Russia	Petrograd	RST	19	00	00	7100	—
	Moscow	RAI	21	00	00	7480	—
Eritrea	Massawa	ICX	10	00	00	11150	—
Java	Malabar	PKX	1	00	00	7700	—

Note: G. C. T. is an abbreviation for Greenwich Common Time. It is five hours ahead of Eastern Standard (75th meridian) Time. Thus 12 noon in New York is 17h. 00m. 00s. in G. C. T. which starts at midnight with 00h. 00m. 00s. and runs to 24h. 00m. 00 s. Thus 8h. 00m. 00s G. C. T. when LY sends time signals, is 3 a. m. E. S. T.

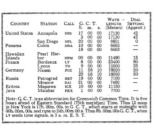

TABLE NO. 1

AS THE BROADCASTER SEES IT

&. CARL DREHER

Drawings by Franklyn F. Stratford

The Place of Television in the Progress of Science

ONE of the New York newspapers, commenting editorially on the recent television demonstration by the American Telephone and Telegraph Company, points out that there is no immediate commercial application for this latest marvel of technology, in view of the elaborate equipment and skilled attendance required. But, the writer adds, other inventions, such as the electric light, the phonograph, the airplane, and radio, have appeared with the same limitations, and he predicts that the televisor will undergo the same process of simplification and adaptation for everyday use. "Meanwhile," he speculates, "the invention may function in small but important fields. It would seem to make the identification of a kidnapped child or a murderer, found in some distant city, an easy matter."

The example is poorly chosen. For such purposes of identification the already more or less perfected and commercialized transmission of photographs would seem a much superior means. It takes only a fraction of an hour to take the photograph and reproduce it at a distance by wire or radio, and the results, at this stage of the game, are apt to be better. The expense should also be less. Television is inherently more complicated than telephotography, just as taking moving pictures is necessarily more difficult than snapping a still photograph, although with sufficient development both processes may be brought within the layman's reach. The editorial, however, suggests the need of thought on the subject of what rôles in the drama of modern life may best be played in the drama of modern life by such scientific applications as television, telephony, the phonograph, talking movies, aural broadcasting, and allied inventions.

In Table No. 1 below, the characteristics of the principal sense- and intelligence-reproducing inventions in this group are given:

TABLE No. I

Sound	Nature of Utility	Light
Phonograph	Permanent record	Photograph (static)
		Motion Picture (kinetic)
Telephone	Rapid reproduction at a distance	Television

Sight and hearing are the two principal senses of the higher animals. The other senses are quite limited in range and contribute less to the picture of the universe which man, especially, must try to construct for the purposes of his life. Accordingly, inventions in communication specialize in these two senses. We find, on this basis, two inventions—the phonograph and the motion picture—which permit the recording of sounds and sights, respectively, and their reproduction after a lapse of time. It should be noted that the

"THE WAX DISCS MAY BE CARRIED TO DISTANT POINTS"

motion picture, not the still photograph, is the logical counterpart of the phonograph. The motion picture gives a kinetic visual reproduction, corresponding to the kinetic aural reproduction of the phonograph. In music or speech one sound follows another, just as moving bodies are seen in one position following another in our visual perception of the external world. The motion picture is made possible, however, only through the physiological lag of the eye—the persistence of vision phenomenon—which enables us to merge a rapid succession of still pictures into apparently continuous motion.

Both the motion picture and the phonograph may be regarded as part of man's efforts to overcome the transitoriness of life. As Heraclitus pointed out some 3300 years ago, the most noticeable characteristic of the universe is that it exists in a state of constant flux. This means that many interesting or beautiful things will happen while some people are not present and must be reproduced artificially for those persons if they are to enjoy them. Even those who were present at the original occurrence in order to re-experience their sensations must have recourse to such machines as the motion picture camera-projector and the phonograph recorder-reproducer. Being machines, such devices are capable of mass reproduction of prototypes. In this way, John Barrymore and Caruso, alive and dead, are spread over the earth. In other words, by means of motion pictures and talking machines we try not only to protect ourselves against the fleeting nature of desirable events but also to multiply those events artificially by making them take place elsewhere than at the original location. Caruso and Barrymore cannot be transported to all the places where their presence is desired, but the wax discs and rolls of celluloid which are capable of reproducing their remarkable qualities may be carried to distant points very readily, and not less readily when the artists, having died, no longer emit beautiful tones nor present a pleasing appearance to the eye. Essentially, therefore, motion pictures, phonographs, and their synthesis, the talking motion picture, are means of, first, resisting the passage of time, and, secondly, overcoming the spatial and energic limitations of certain special human beings whose performances are of great interest to their fellows. By the refinement of machinery these aims are being accomplished with a constantly closer approach to perfection as regards sight and hearing, the two essential senses in the particular relationships involved.

As the phonograph and motion picture apparatus are basically systems to overcome the passage of time, so the telephone and televisor have the aim of overcoming the obstacle of intervening distance in the fields of sound and light respectively. By means of the telephone, sounds are transmitted practically instantaneously over distances which would otherwise render them inaudible, and now the television apparatus performs the same service for the sense of sight. The American Telephone and Telegraph Company's demonstration was really a combination of two inventions—the telephone and the televisor, in the same way that the talking motion picture combines the phonograph and the motion picture, but everyone is so used to the telephone that this aspect of the situation has been overlooked. Another explanation for this lies in the fact that voice and appearance are automatically linked in the television-telephone subjects, without the necessity for synchronization of sound and light vibrations which we must effect in the picture-phonograph combination. But, leaving this point to return to the main thread of our analysis, we note that by means of the telephone and the televisor we project ourselves, sensorially, through space; with the phonograph and motion picture we project ourselves backward through time.

The telephone and the televisor, like all the inventions of the "tele" group, utilize electric waves. This is because such waves, traversing space at a speed of 186,000 miles per second, cover terrestrial distances instantaneously as far as most activities of human beings are concerned. Even when, as in the telephoto process, the complete transmission takes an appreciable fraction of an hour, this is merely because the breaking up and re-integration of the picture, optically, takes time. In Table No. II the principal inventions of the "tele" group are summarized, with the dates, not of initial invention, which are controversial and difficult to determine in some cases, but of practical demonstration, when it became evident that the problem was well on the way to complete solution:

TABLE No. II

Date	Invention	Nature of Utility
1835	Telegraph	Transmission of symbols
1876	Telephone	" " sounds
1922	Telephoto	" " sights (static)
1927	Television	" " (kinetic)

The inventions of the "permanent record" group all use the device of impressing a performance which, being functional, passes with time, on some material substance which, to a degree, is independent of time and may also be multiplied indefinitely, each multiplication adding a large number of possible reproductions of the original event. The invention of printing is one of the early applications of this principle. A man has ideas, which are functional in their nature. By printing them he transmutes the ideas into the material form of symbols on paper, which may be read and reproduced as ideas by another man reading the symbols perhaps centuries later. The author projects himself functionally into the future, the reader into the past, by this physical device. The recent inventions of this group, which have to do with sight and hearing, are summarized below:

TABLE No. III

Date	Invention	Nature of Utility
1839	Photograph	Recording and later reproduction of sights (static)
1877	Phonograph	Recording and later reproduction of sights (static)
1893	Motion picture	Recording and later reproduction of sights (kinetic)
1926	Talking motion pictures	Recording and later reproduction of sights and sounds

From Tables No. II and III some of the relationships between these inventions may be traced. The telegraph, invented in 1835, utilized the crudest possible form of modulation of electric waves—simply starting and stopping a constant amplitude current according to a code. A little later the chemical fixation of the images in the *camera obscura* was accomplished. The telephone, following the telegraph in 1876 after an interval of 41 years, required a much more subtle modulation of the electric currents, and this has been further complicated in telephotography and television. The phonograph was invented almost at the same time as the telephone. The motion picture, following the still photograph after 54 years, is merely an ingenious elaboration of the latter. Talking movies, as has been pointed out before, are produced by combining the phonograph and cinematograph. In the same way the telephoto process may be considered a synthesis of photography and the principles of the telegraph and telephone. Television is derived from the two latter, the telephoto systems, and the motion picture art. The fundamental inventions, in this sense, are the telegraph, the camera, and the phonograph. The others are elaborations and cross-breedings. Taking them all in all, in a space of 100 years (1835–1935), allowing 8 more years for the development of television, the effective rapid transmission of symbols, sounds, and sights will have been accomplished; and the art of recording and reproducing at a later time sounds and sights, separately and in combination, will have been, for all practical purposes, perfected.

It will be noted that nothing has been said about broadcasting and wireless telegraphy in this discussion. That is because, coming down to fundamentals, radio is only a part of telegraphy and telephony. In wire telephony we modulate a direct current at audio frequency and send it directively along a wire. In broadcasting we superimpose the same variations on a radio-frequency current, and transmit them from an antenna in all directions through space. In each case our object is to reproduce sounds at a distance. One form happens to be suited for point-to-point communication between individuals, while the other is suited for one individual addressing an audience, or for the distribution of a single performance to a large number of individuals separated in space, but the difference is a secondary one. As in all telephony, the sound-emitters and sound-receivers are separated in space, but not in time. This will result in profound differences in the social application of the arts in question, but the metaphysics remain identical.

All these inventions are, in the last analysis, means by which human beings secure agreeable or necessary sensations, in the absence or because of the unavailability of the original sources of those sensations, owing to the movement of time and the non-movement of space. When agreeable sensations are involved we are dealing with entertainment; when the sensations are necessary, rather than merely pleasant, we speak of utility. There is no sharp dividing line. Broadly, one sustains life; the other helps make it worth while. Let us hope that television will do both. For the present we shall be satisfied with this outline of the hundred-year era which it closes, and, seeing it against this background, we shall be less likely to go astray in the hazardous business of prophecy.

Concerning the specific applications of television among the other arts of communication we shall have more to say hereafter.

What the Broadcast Technician Should Read

OVER two years ago (in the April, 1925, RADIO BROADCAST) we printed a short bibliography for broadcast operators and engineers, consisting mainly of references to articles in the *Proceedings of the Institute of Radio En-* gineers and the *Journal of the American Institute of Electrical Engineers*. Since then a considerable number of valuable contributions have been added to the literature, so that it now appears advisable to reprint the original list with the additions. This is also done in response to requests which we receive at intervals from readers interested in the technique of broadcasting or speech reproduction, and in need of help in selecting their reading matter. While this department is always ready to advise broadcasters with regard to the literature in this field, and to give individual attention to special problems, the present summary of available books and papers should serve the requirements of a majority of readers. The list is not confined to books on broadcasting as such, since to attempt to master the technique of broadcasting, without preparation in the general principles of radio communication and acoustics, is like trying to fence before one has learned how to hold a foil:

GOVERNMENT PUBLICATIONS

Obtainable from Superintendent of Documents, Government Printing Office, Washington, D. C.

Principles Underlying Radio Communication—Second Edition. (Radio Communication Pamphlet No. 40; Signal Corps, U. S. Army). $1.00
Radio Instruments and Measurements (Bureau of Standards Circular No. 74) 60
Telephone Service (Bureau of Standards Circular No. 112) 65
Sources of Elementary Radio Information (Bureau of Standards Circular No. 122) 05
Architectural Acoustics (Bureau of Standards Circular No. 300) 05

BOOKS

Morecroft: *Principles of Radio Communication.* John Wiley & Sons, Inc.
Van Der Bijl: *The Thermionic Vacuum Tube.* McGraw Hill Book Co.
Moullin: *Radio Frequency Measurements*, J. B. Lippincott Co.
Johnson: *Transmission Circuits for Telephone Communication.* D. Van Nostrand Co.
Miller: *Science of Musical Sounds.* Second Edition. Macmillan.
Sabine: *Collected Papers on Acoustics.* Harvard University Press.

JOURNAL OF THE AMERICAN INSTITUTE OF ELECTRICAL ENGINEERS

Issues obtainable from American Institute of Electrical Engineers, 33 West 39th Street, New York.

Martin and Clark: "Use of Public Address Systems with Telephone Lines." April, 1923.
Green and Maxfield: "Public Address System." April, 1923.
Arnold and Espenschied: "Transatlantic Radio Telephony." August, 1923.
Osborne: "Telephone Transmission over Long Distances." October, 1923.
Hitchcock: "Applications of Long-Distance Telephony on the Pacific Coast." December, 1923.
Jones: "The Nature of Language." April, 1924.
Martin and Fletcher: "High-Quality Transmission and Reproduction of Speech and Music." March, 1924.
Casper: "Telephone Transformers." March, 1924.
Martin: "The Transmission Unit." June, 1924.
Harden: "Practice in Telephone Transmission—Maintenance." December, 1924.
Ferris and McCurdy: "Telephony Circuit Imbalances." December, 1924.
Kellogg: "Design of Non-Distorting Power Amplifiers." May, 1925.
Discussion on above. June, 1925.
Maxfield and Harrison: "Methods of High-Quality Recording and Reproducing of Music and Speech Based on Telephone Research." March, 1926.
Nance and Jacobs: "Transmission Features of Transcontinental Telephony." November, 1926.
Espenschied: "Radio Broadcast Coverage of City Areas." January, 1927.
Discussion on above. April, 1927.

PROCEEDINGS OF THE INSTITUTE OF RADIO ENGINEERS

Obtainable from The Institute of Radio Engineers, 37 West 39th St., New York.

Espenschied: "Applications to Radio of Wire Transmission Engineering." October, 1922.
Nichols and Espenschied: "Radio Extension of the Telephone System to Ships at Sea." June, 1923.
Baker: "Description of the General Electric Company Broadcasting Station at Schenectady, N. Y." August, 1923.
Baker: "Commercial Radio Tube Transmitters." December, 1923.
Little: "KDKA, the Radio Telephone Broadcasting Station of the Westinghouse Electric and Manufacturing Company, East Pittsburgh, Pa." June, 1924.
Nelson: "Transmitting Equipment for Radio Telephone Broadcasting." October, 1924.
Weinberger: "Broadcast Transmitting Stations of the Radio Corporation of America." December, 1924.

Cummings: "Recent Developments in Vacuum-Tube Transmitters." February, 1925.
Heising: "Production of Single Side-Band for Transatlantic Radio." June, 1925.
Oswald and Schelling: "Power Amplifiers in Transatlantic Radio Telephony." June, 1925.
Espenschied, Anderson, and Bailey: "Transatlantic Radio Telephone Transmission." February, 1926.
Bown, Martin, and Potter: "Some Studies in Radio Broadcast Transmission." February, 1926.
Jensen: "Portable Receiving Sets for Measuring Field Strengths at Broadcasting Frequencies." June, 1926.
Little and Davis: "KDKA." August, 1926.
Goldsmith: "Reduction of Interference in Broadcast Reception." October, 1926.
Crossley: "Piezo-Electric Crystal-Controlled Transmitters." January, 1927.

In addition to the above engineering articles, the reader interested in the technique of practical broadcasting may be referred to the twenty short papers on various aspects of broadcast operation which have appeared in this department since September, 1925. These discussions have been on such subjects as "Microphone Placing," "Wire Lines," "Multiple Pick-Up," "Equalization," "Calculation of Gain," "Modulator Plate Current Variation," "The Condenser Transmitter," "Studio Design," etc. The object of the treatment in each case has been to aid broadcasters looking for information on their immediate problems, rather than to attain originality or high engineering calibre.

Technical Problems for Broadcasters. No. 2

A RECEIVING set is equipped with an output tube capable of delivering energy to a loud speaker, without distortion, at a level of not over +10 TU. It is normally tuned to a given broadcasting station and so adjusted that the audio output is +8 TU, allowing an overload margin of 2 TU. All other factors remaining constant, the power of the broadcasting station is doubled. By how many TU will the output tube of the receiving set be overloaded? Give all the steps in the solution. Answer on page 170.

Technical Problems for Broadcasters. No. 3.

THE ratio of energy between fortissimo and pianissimo passages of an orchestra is 1,000,000:1. The orchestra is reproduced without distortion by a public address system in another room. The operator brings up the pianissimo passages so that they are over the room noise level at a certain point near the projector, always returning to a base gain level for louder parts of the performance. The faintest pianissimo passages produce in the air at this point an R.M.S. pressure of 0.0001 dynes per square centimeter. The loudest passages produce a corresponding pressure of 1 dyne per square centimeter. By how many TU did the P.A. operator bring up the gain during pianissimo intervals? Solution on page 170.

Thirteen Years Ago

THE Institute of Radio Engineers was founded, as all good and even middling radio men know, in 1912, by the amalgamation of the "Society of Wireless Telegraph Engineers" and the "Wireless Institute." At the time of its formation the Institute boasted of less than 50 paid-up members. I do not know just how many members are on the rolls now, but Mr. Robert H. Marriott, the first President, re-

marked to me the other day that at the present rate of increase the membership is headed for the ten thousand mark within a few years. The recently compiled index to the *Proceedings*, which was issued quarterly in 1913 and now goes out to the members monthly, contains references to every conceivable radio subject, from acoustic tuning to wired wireless, and all the authors, great and small, from Alexanderson to Zenneck, are represented. I wrote an article some time ago, pointing out the advantages of membership to all serious-minded radio men, but it appears that there are still some strange creatures in the profession who fail to dig down for $6.00 a year, to their own advantage. The Institute needs them, but they need the Institute a lot more, if they only knew it. The Secretary will be pleased to mail application blanks from 37 West 39th Street, New York.

Important as this is, it is not precisely what I started to write about. The fact is that I thought it might amuse some of the faithful customers to peruse a few personal excerpts from Volume 2

"THERE ARE STILL SOME WHO FAIL TO DIG DOWN FOR $6 A YEAR"

of the *Proceedings*, issued in 1914. Unfortunately, I lack the first, or 1913, volume, and cannot get a copy for love or money. My only hope is that one of the reverend elders of the society will remember me in his will. At any rate, around that time a Year Book was issued (which I also lack), and Supplementary Lists of Members were printed in some of the issues. In these lists some of the great names of radio in our day appear. In some cases they were already great, but in others they had only begun the climb. I abstract a few below.

Henry E. Hallborg, who is now one of the Radio Corporation's chief short-wave experts and who, with Messrs. Hansell and Briggs, wrote a highly informing paper on that subject which should appear in the *Proceedings* around this time, was elevated to the grade of Member in 1914. He was then in charge of the newly finished or going-to-be-finished Marconi transatlantic stations at New Brunswick and Belmar. Lawrence M. Cockaday, the Technical Editor of *Popular Radio*, became an Associate that year; he was then General Secretary of the Cathedral Choir School; whether that helped him run his four-inch spark coil, he will have to tell us. Next among my victims I spy the name of the illustrious Paul F. Godley, he who first picked up

American amateur signals on the other side of the Atlantic (From 1 BCG, in 1921—listen to the Radio Club of America contingent yell!)—he was Radio Inspector for the Brazilian Government, and lived, then as now, in New Jersey. Presumably the Brazilians had placed an order for some radio equipment in the United States, and Mr. Godley was there to see that they did not get any celluloid spark gaps in the shipment. DeLoss K. Martin had not yet issued from the Polytechnic College at Oakland, California; to-day he is known as a colleague of Dr. Ralph Bown, the President of the Institute, and Mr. R. K. Potter, in the work which resulted in the paper on "Some Studies in Radio Broadcast Transmission," printed in the *Proceedings* for February, 1926, and recognized as one of the most brilliant pieces of research in all radio history. Elmo Neale Pickerill was exercising his perfect fist as an instructor at the Marconi Wireless Telegraph Company's school. Nowadays, as Chief on the *Leviathan*, he probably seldom touches a key. I should like to hear the fast but unhurried dots and dashes rippling once more from under his hand—I who was brought up among such artists of the brass lever, and now must earn my living among artists of another kind.

But to continue. There stands the name of Harry Sadenwater, now the engineer in charge of the General Electric Company's chain of broadcast stations, from Schenectady to Oakland. In 1914 Mr. Sadenwater was the radio instructor at the East Side Y. M. C. A. in New York City. If you didn't hear him in those days you missed another copperplate hand, and I do not refer to calligraphy. Thomas M. Stevens, Superintendent of the Eastern Division of the Marconi Company then, runs the whole Marine Department of the Radio Corporation now. Behold also Ellery W. Stone, a student at the University of California in 1914, and President of the Federal Telegraph Company to-day. Bowden Washington, an engineer of Cutting and Washington and Colonial Radio prominence these latter days, was testing for Clapp-Eastham thirteen years ago. Joseph O. Mauborgne was a First Lieutenant in the United States Army; First Lieutenants become Colonels. Allen D. Cardwell, one of the manufacturers who, with the advent of broadcasting, continued to take pride in turning out precise and decent apparatus when junk sold just as well, may have learned his craft as Chief Engineer of the American Telegraph Typewriter Company. Lewis Mason Clement, Chief Engineer of Fada, was a Shift Engineer in the Marconi station at Kahuku, in the Hawaiian Islands. Edwin H. Colpitts, in 1914, was a Research Engineer of the Western Electric Company; one does not become an Assistant Vice-President of the American Telephone and Telegraph Company overnight. Of William H. Howard, now in charge of the tube division in the Radio Corporation's Technical and Test Department, I shall have more to say hereafter, in the "Memoirs." He was a Laboratory Assistant of the Marconi Company in the ancient days to which we are applying the pick and spade. And here appears the name of Arthur H. Lynch, a Radio Operator of the same Marconi Company, a radio manufacturer now, and once Editor of this magazine. Also Lee L. Manley, Assistant Service Manager of R. C. A., came from about the same cradle, save that his title was "Radio Electrician." Did that mean that he got $2.00 a month more, or less, in the old Marconi days? Never mind, none of us got much. But here is a genuine pioneer of not only the radio, but the telegraph and cable business, Edward Butler Pillsbury, in 1914 Assistant Traffic Manager of the same Marconi Company, and now Vice-President and General Manager of the Radio Real Estate Corporation of America. With Mr. Pillsbury's name we may well close the present account. All these were members of the Institute of Radio Engineers in 1914, and all are members now.

Radio Revolutionized Again

ONCE again a revolutionary radio invention is heralded. A good-sized headline in one of the conservative journals, followed by 112 lines of type, informs us that "Crystal Device Turns Phone into Complete Radio System."

How? There are two pins, which connect the crystal detector or vacuum tubes—for, marvellous to relate, either may be used—to the telephone line and ground. The telephone line then acts as an antenna. The device may be clamped to the telephone, the only apparatus required for this job being a telephone and a clamp. The inventor is a "tea specialist" employed by the Department of Agriculture. "As paradoxical as it may seem," states the article, "this invention was conceived in a dearth rather than a wealth of radio knowledge. The inventor is recognized the country over as an authority on teas, but he boldly admits that he is not versed in the elemental principles of radio."

Patent specifications have been drawn up. "Patents are pending," as a sweetly familiar phrase has it. Alas, in the present instance, they will never cease to pend. Fifteen years ago some obscure amateur, whose name had been forgotten and whose body may be dust, first connected power and telephone lines, capacitively and even

conductively, to his primitive receiver, and heard MCC out on the Cape sending press in code to the Atlantic. If someone is cruel enough and has the time, let him dig up the precise number of *Modern Electrics* in which the discovery was chronicled. And let the telephone companies whose lines are unbalanced by this recrudescence of the marvel deal gently with the tea expert of Washington, D. C. He is, after all, a tea expert.

Solution to Technical Problem No. 2

THE power in the transmitting antenna, expressed in terms of antenna current and resistance, is:

$$Pt = It\ Rt \qquad (1)$$
$$\text{Whence} \quad It \propto \sqrt{Pt} \qquad (2)$$

From the Austin-Cohen formula and the theory of radio propagation, we know that the current and voltage in a receiving antenna vary directly as the current in the transmitting antenna. That is:

$$Ir \propto It \qquad (3)$$
$$\text{and} \quad Er \propto It \qquad (4)$$
$$\text{Therefore} \quad Er \propto \sqrt{Pt} \qquad (5)$$

That is, the radio-frequency voltage impressed on the receiver varies as the square root of the power in the transmitting antenna. This holds through the r. f. amplifier of the receiver, so that the voltage impressed on the grid of the detector (the second detector in a super-heterodyne receiver) varies as the square root of the power in the transmitting antenna. But, as the usual vacuum-tube rectifier follows a square law, the audio plate current or voltage varying as the square of the grid potential, we may write:

$$Ia \propto Er \propto Pt \qquad (6)$$

where Ia is the audio or post-detection current in the receiver.

But, impedance remaining constant, the TU level in any circuit is given by:

$$TU = 20\ \log_{10} \frac{I_1}{i_4} \qquad (7)$$

as explained in previous technical articles, based on engineering literature, in this department.

Hence, in the output of the receiver, a change in level is given by:

$$G = 20\ \log_{10} \frac{Ia_1}{Ia_2} \qquad (8)$$

Referring to (6) above, we may rewrite (8) in the final form:

$$G = 20\ \log_{10} \frac{Pt_1}{Pt_2} \qquad (9)$$

Equation No. 9 is the expression for level changes in the output of the receiver, in terms of different transmitter powers. The equation having been derived, the problem practically solves itself. The gain, with the doubling of power at the transmitter, is:

$$G = 20\ \log 2 = 20\ (0.301) = 6.0\ TU$$

As, by the conditions of the problem, the overload margin of the output tube is only 2.0 TU, with the doubling of transmitter power the receiver will overload by 6 minus 2, or 4 TU.

Solution to Technical Problem No. 3

THIS problem is not as complicated as it looks. We are given that the sound pressure in air at a certain point varies between 1.0 and 0.0001 dynes per square centimeter. This pressure is expressed in force per unit of area. By definition, energy is expressed as a force acting through a given distance. If E represents energy, F force, and d a distance through which F operates, then

$$E = Fd$$
$$\text{Hence} \quad \frac{E_1}{E_2} = \frac{F_1}{F_2}$$

And since power is directly proportional to energy, we may also write for the power P that

$$\frac{P_1}{P_2} = \frac{F_1}{F_2}$$

By definition the telephonic gain variation in TU, as we have seen in previous articles and problems, is given by

$$TU = 10\ \log_{10} \frac{P_1}{P_2}$$

Or, in the P. A. system described above, the gain range is given by

$$Ge = 10\ \log_{10} \frac{F_1}{F_2} = 10\ \log_{10} \frac{1.0}{0.0001}$$
$$= 10\ \log_{10} 10000$$
$$= 40\ TU$$

The subscript "o" in the above expression stands for "out" or "output."

We are also given that the orchestra itself plays with a power ratio of 1,000,000: 1, so that the gain range *into* the P. A. system is

$$G^i = 10\ \log_{10} \frac{1000000}{1} = 60\ TU$$

Therefore the answer is that the operator narrowed the volume range by 60 minus 40, or 20 TU.

A RADIO LABORATORY OF A DECADE AGO
This picture was taken in 1917 in the college of the City of New York. Doctor Goldsmith is sitting with face to the camera and Mr. Julius Weinberger, now of the Radio Corporation, is at his left. Carl Dreher is to the right of the picture, adjusting the Marconi transmitter

Description of a Short-Wave Station

Details of the 500-Watt Crystal-Controlled Transmitter at 2 AG—The Four-Tube Short-Wave Receiver—A Radio Club of America Paper

By C. R. RUNYON, JR.

THE TRANSMITTER

THE short-wave crystal-controlled transmitter at radio station 2 AG employs seven transmitting tubes in the following combination: One UX-210 as a crystal oscillator tube; two UX-210's in the first intermediate push-pull amplifier; two UV-203-A's in the second intermediate push-pull amplifier; two UV-204-A's in the push-pull radio-frequency amplifier.

A schematic diagram of the crystal oscillator stage, together with its power supply, is shown in Fig. 1. The crystal is connected between the grid and filament terminals of the UX-210 crystal oscillator tube. The crystal used has a fundamental frequency of 3665 kc. (81.9 meters), so, in view of the fact that this station transmits on the 80-meter band, there are no frequency multiplier stages between the crystal oscillator and the antenna system, straight amplification being all that is required.

While we are discussing the grid circuit of the oscillator, it is logical that we consider the system of modulation that can be employed when it is desired, as modulation is effected in this circuit.

When the transmitter is in operation, the chopper disc (shown in Fig. 1) rotates at such a speed that it opens and closes the circuit through the primary winding of the modulation transformer (T_1) at an audio-frequency rate. The shunt across the chopper contacts is removed when the set is in operation, by virtue of the fact that relay No. 5 is energized at that time.

When the circuit is closed through the chopper contacts, a direct current flows through the primary winding of the modulation transformer (T_1), this current being supplied by a 6-volt storage battery, and being limited by the 50-ohm rheostat (R_4) and the resistance of the primary winding in question.

Every time the circuit through the chopper contacts is closed, we get a direct-current flow through the primary winding of (T_1), which sets up a field which cuts through the secondary winding of (T_1) and produces a potential "kick" across it. This voltage "kick" is passed on to the terminals of the crystal oscillator tube, and subsequently to the grid of the first amplifier tube. The effect of each one of these "kicks" is to change the frequency, right at the source; hence, the frequency of the transmitted signals is also changed.

The amount of this change, as near as the ear can tell, is between 500 and 1000 cycles. Thus, by this novel method of modulation, the frequency of the transmitted signals is changed through a band 500 to 1000 cycles wide, at an audio-frequency rate (the rate of make and break at the chopper contacts). . . .

Plate voltage is supplied to the crystal oscillator tube from a full-wave rectifier system which employs two UX-216-B rectifier tubes.

This plate supply lead is prevented from offering a radio-frequency shunt across the crystal oscillator tube output circuit by means of the radio-frequency choke coil RFC₂ which is inserted in series with it. A d.c. milliammeter (0-100 mils.) is connected in series with this plate lead to indicate the plate current drawn by this tube.

The blocking condenser C_2, which has a capacity of 0.002 mfd., bypasses radio-frequency energy, but prevents the coil L_4 from short-circuiting the d.c. plate supply.

The output circuit of the crystal oscillator tube is tuned to 80 meters by means of the 0.0005-mfd. variable condenser C_6 and the coil L_4. The "0 to 3" thermo-ammeter, A_6, indicates the radio-frequency current circulating in the output circuit of the crystal oscillator tube.

The filament of this tube is supplied with alternating current from one of the low-voltage windings on the Acme 200-watt power transformer, T_3. This filament current is controlled by means of the 2-ohm General Radio rheostat, R_6. C_4 and C_5 are 0.002-mfd. radio-frequency bypass condensers.

The filaments of the two UX-216-B rectifier tubes are heated by means of current from another low-voltage winding on transformer T_3. The plates of the rectifier tubes are supplied with high-voltage alternating current from a secondary winding of T_3 which has a potential of 550 volts (r. m. s.) between its extremities and its mid-tap. The filament current to the rectifier tubes is limited by the General Radio 2-ohm rheostat, R_5.

In the rectifier filter circuit there are two 2-mfd. condensers and one 10-mfd. condenser for smoothing, designated as C_1, C_3, and C_3

respectively. Two 30-henry chokes, X_1 and X_8, are also used in this filter circuit. The high-voltage direct-current output of this No. 1 rectifier can be switched either to the plate of the crystal amplifier tube, or the plates of the tubes in a receiver, by means of the switch S_1.

The output voltage of the rectifier can be controlled to a certain extent by means of the 30-ohm rheostat R_1. When the master control switch at the operator's desk is thrown to the "send" position, relay No. 3 closes, closing the circuit through the primary winding of the power transformer, T_3, and thus lighting the filament of the crystal oscillator tube and applying plate potential to the crystal oscillator tube. Relay No. 5 also closes when the master control switch is thrown to the "send" position, and the resultant action of this relay is to remove the shunt across the chopper contacts.

When the master control switch is returned to the "receive" position, the circuit to the chopper motor is opened and it starts to slow down. The main function of relay No. 5 is to place a shunt across the chopper contacts when the master control switch is in the "receive" position, so that the make and break of the contacts will not cause Q R M (interference) when the chopper motor is slowing down.

Energy is transferred from the crystal oscillator tube output circuit to the input circuit of the first intermediate amplifier (the latter being push-pull, using two UX-210's), by means of the inductive coupling between the two coils, L_4 and L_6, the former being in the output circuit of the crystal oscillator, and the latter being in the input circuit of the first intermediate push-pull amplifier.

The mid-tap on the coil L_6 is connected to ground through a 45-volt bias battery, which applies a negative bias to the grids of both of the UX-210 tubes in the first intermediate amplifier. There is a 0.002-mfd. radio-frequency bypass condenser across this bias battery.

FIRST INTERMEDIATE AMPLIFIER AND ITS POWER SUPPLY

A SCHEMATIC diagram of the first intermediate amplifier is shown in Fig. 2. The attendant rectifier system is also shown in Fig. 2.

Midget, five-plate, neutralizing condensers are connected from the grid of one tube to the plate of the other in this stage of amplification, to neutralize the feed-back effect due to the inter-electrode capacity of the amplifier tubes used. This is an application of the "bridge" method of neutralization. For instance, when C_1 is adjusted to a value of capacity equal to the plate-grid capacity of the UX-210 whose grid is connected to No. 1 terminal in Fig. 2, the grid of the tube in question is at ground potential as far as the radio-frequency energy in the output cir-

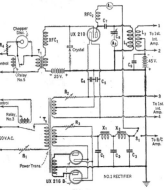

FIG. 1

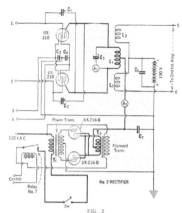

FIG. 2

cuit of this amplifier stage is concerned. Therefore, since the filament of this tube is metallically connected to ground, there can be no application of radio-frequency voltage to the grid of the tube in question, due to the radio-frequency energy in the output circuit of this amplifier stage.

The filaments of these two amplifier tubes are supplied with energy from the oscillator filament winding on the power transformer in the No. 1 rectifier assembly.

The output circuit of this stage of amplification is tuned by means of the coil L_1 and the 0.0005-mfd. variable condenser, C_4. The radio-frequency current flowing in this tuned circuit is indicated by the ' 0 to 5''' thermo-ammeter, A_1.

The plate potential is supplied to the midpoint of the coil L_1, and the plate current is indicated by the 0 to 150 mil. meter (A_4). The source of this high-voltage d. c. supply is the No. 2 rectifier system. This rectifier employs two UX-216-B rectifier tubes which receive their plate supply from the high-voltage secondary winding of a step-up transformer, T_1.

A separate filament transformer is used to supply filament heating energy to the two rectifier tubes used, due to the fact that keying is effected by opening and closing the circuit through the primary winding of the power transformer, T_1, which removes and applies, respectively, high-

voltage rectified a. c. to the plates of the two tubes in this amplifier stage.

If the filaments of the tubes in this No.2 rectifier were energized from a low-voltage winding on the transformer T_1, keying could not be satisfactorily effected, due to the time lag involved in bringing the rectifier tube filaments up to normal operating temperature, once the circuit through the primary winding of the transformer is closed.

When the master control switch is thrown to the "send" position, relay No. 7 closes, thus closing the circuit to the primary winding of the high-voltage transformer T_1, this circuit being under the control of the transmitting key.

Radio-frequency energy is induced into the input circuit of the second intermediate amplifier by means of the inductive coupling between L_1 in the output circuit of the first intermediate amplifier and L_4 and L_3 in the input circuit of the second intermediate amplifier.

The mid-point between the coils L_3 and L_4 is connected to ground through the 190-volt bias battery which maintains a negative bias on the grids of the two UV-203-A tubes in the second intermediate amplifier stage. This bias battery is bypassed by the radio-frequency bypass condenser, C_6.

SECOND INTERMEDIATE AMPLIFIER

THE schematic diagram of this stage of amplification is shown in Fig. 3. The "bridge" method of neutralization is used in this amplifier stage and is effected by means of the two neutralizing condensers C_1 and C_2. The output circuit of this push-pull amplifier is tuned by means of the coil L_4 and the condenser C_3. The circulating current in this tuned circuit is indicated by the "0 to 10" thermo-ammeter, A_1.

The filaments of these two 50-watt tubes are supplied with filament heating energy from a separate transformer T_1, this current being controlled by means of the rheostat R_4. The plate supply for these two tubes is obtained from a 2000-volt d.c. generator, a plate resistor, R_2, functioning to drop the plate voltage from 2000, at the generator source, to 1000 volts at the plates of the 50-watt tubes in this stage of amplification. There is a 0.002-mfd. bypass condenser from the low side of R_2 to ground.

When the master control switch is closed, it operates relay No. 4, which closes the circuit through the primary winding of the filament transformer T_1 for the two 50-watt tubes in this second intermediate amplifier stage.

RIGHT: IN A SPECIALLY BUILT CABINET

All the transmitting equipment, with the exception of the high-voltage generator, is included in the cabinet

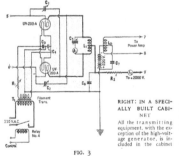

FIG. 3

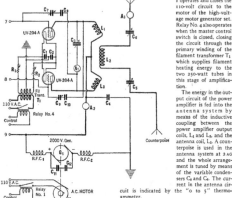

FIG. 4

The radio-frequency energy in the output circuit of this stage of amplification is induced into the input circuit of the succeeding stage, which is also push-pull, by means of the inductive coupling between the coil L_4 in the output circuit of the 50 watters, and the coil, L_5, in the input circuit of the power amplifier stage.

The mid point of the coil L_5 is connected to ground through a 220-volt bias battery, and, since the extremities of L_5 are connected to the grids of two UV-204-A 250-watt power tubes, the grids of the tubes in question are held 220 volts negative. There is a radio-frequency by-pass condenser, C_7, across this bias battery.

POWER AMPLIFIER AND ITS POWER SUPPLY

THE schematic diagram of the power ampli-
fier stage of amplification, its power supply, and the antenna system at station 2 AG, is shown in Fig. 4.

Filament heating energy for the two UV-204-A tubes in this stage of amplification is supplied from a separate step-down transformer, T_1. The filament current is controlled by means of two rheostats, R_1 and R_2.

Here again, the "bridge" method of neutralization is used, and is effected by neutralizing condensers, C_7 and C_8. The condensers C_4 and C_{10}, which are in series with the neutralizing condensers, are radio-frequency bypass condensers, and simply function to cut down the voltage drop across the neutralizing condensers and thus prevent the possibility of their arcing over.

The power amplifier output circuit is tuned by means of the coils L_4 and L_3 and the variable condenser C_3. The circulating current in this tuned circuit is indicated by the "0 to 20" thermo-ammeter. The mid-point between the two plate coils is connected to the positive 2000-volt terminal of the high-voltage d. c. generator through the "0 to 1000" milliammeter, A_2.

The high-voltage plate generator is driven by an a. c. motor which operates on 110 volts. When the master control switch is closed, relay No.

1 operates and closes the 110-volt circuit to the motor of the high-voltage motor generator set. Relay No. 4 also operates when the master control switch is closed, closing the circuit through the primary winding of the filament transformer T_1 which supplies filament heating energy to the two 250-watt tubes in this stage of amplification.

The energy in the output circuit of the power amplifier is fed into the antenna system by means of the inductive coupling between the power amplifier output coils, L_4 and L_3, and the antenna coil, L_5. A counterpoise is used in the antenna system at 2 AG and the whole arrangement is tuned by means of the variable condensers C_4 and C_5. The current in the antenna circuit is indicated by the "0 to 5" thermo-ammeter.

COMPLETE TRANSMITTER ASSEMBLY

A PHOTOGRAPH on page 172 shows the com-
plete transmitter assembly at station 2 AG. The entire equipment is included in a cabinet built for the purpose, with the exception of the high-voltage generator which is located in the basement.

The chopper unit is located in the lower right-hand corner of the cabinet. The crystal oscillator and first intermediate amplifier are located on the middle shelf behind a shield painted black. No. 2 rectifier is just to the left of this black box, and rectifier No. 1 is located on the top of the cabinet.

The second intermediate amplifier is located in the upper right corner, and the power amplifier is to the left of the latter, in the upper left-hand corner of the cabinet. The operators' desk is just to the left of this cabinet.

A schematic diagram of the transmitter is shown in Fig. 5. Note that the diagrams of the various stages have given the complete details concerning each stage, whereas the diagram of

the entire transmitter does not include the control relays or the power supply units.

CONTROL RELAY SYSTEM

WHEN the operator at station 2 AG closes
the master control switch to the "send" position, there are a great many actions that take place. This can best be explained by a study of Fig. 6, which shows the control relay system alone.

When the single-pole single-throw switches, S_1 and S_2, are closed, the control relays are under the control of the master switch. The pilot lamps are lighted when S_1 and S_2 are closed. The former is in an 8-volt circuit and the latter is in a 6-volt circuit.

With the master control switch thrown to the "transmit" position ("T" in the diagram), the following actions take place:

(A). Relay No. 2 closes. Relay No. 1 is thrown on the 110-volt a. c. line and it closes. When relay No. 1 closes, 110 volts a: c. is applied directly across the terminals of the a. c. motor which drives the high-voltage d. c. generator for the plates of the two 50 watters, and the plates of the two 250 watters.

(B). Relay No. 3 closes. The 110-volt a. c. circuit is closed through the primary winding of the power transformer in the No. 1 rectifier assembly.

(C). Relay No. 4 closes. An a. c. voltage of 110 is applied to the primary winding of the filament transformer for the two UX-216-B rectifier tubes in No. 2 rectifier.
A similar voltage is applied to the primary winding of the filament transformer for the two UV-203-A tubes in the second intermediate amplifier circuit.
A voltage of 110 is applied to the primary winding of the filament transformer for the two UV-204-A tubes in the power amplifier.
One hundred and ten volts a. c. is applied to the terminals of the motor that drives the chopper disc.

(D). Relay No. 5 opens. The shunt across the chopper contacts is removed.

(E). Relay No. 6 closes. The terminals of the headphones are connected to the output of the monitor receiver which allows the operator to hear the quality of his outgoing signals.

(F). Relay No. 7 closes. The circuit from the 110-volt a. c. supply, through the primary winding of the plate transformer for the No. 2 rectifier, is closed.

(G). Relay No. 8 closes. The A battery circuit to the filaments of the tubes in the monitor receiver is closed.

(H). The 8-volt control battery is connected in series with the modulation transformer and the chopper contacts.

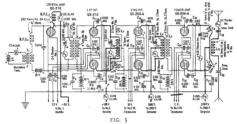

FIG. 5

A diagram of the complete 500-watt transmitter, 2 AG. The control relays and power supply equipment has been omitted in the diagram. This diagram combines Figs. 1, 2, 3, and 4 but the lettering of the parts is different

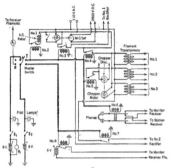

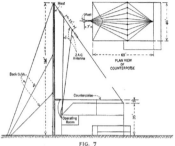

FIG. 7
The antenna system at 2 AG.

FIG. 6

The control relay system at 2 AG. Filament Transformers: No. 1 supplies filaments of rectifier tubes in No. 2 rectifier for the first intermediate amplifier; No. 2 for filaments of the second intermediate amplifiers (UV-203-A's); No. 3 for filaments of power amplifier tubes (UV-204-A's). The functioning of this control relay system is explained in the text on page 173

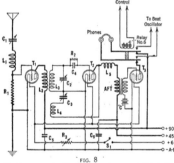

FIG. 8

The short-wave receiver. The following are the constants of the circuit: AFT, audio-frequency transformer (ratio 6-1); L_5, 50 turns No. 16 d. c. c. 3 inches diameter; L_4, 6 turns No. 16 bare wire, spaced wire width, 3 inches diameter; L_3, 23 turns No. 16 bare wire, spaced wire width, 3 inches diameter; L_2, 4 turns No. 16 bare wire, spaced wire width, 3 inches diameter; L_1, 175 turns No. 34 d. s. c. 1 inch diameter; C_1, 0.0005-mfd. variable condenser; C_5, 0.000075-mfd. variable condenser; C_2, 0.00025-mfd. variable condenser; C_4, 0.00025-mfd. fixed condenser; C_3, 0.1 mfd. fixed condenser; C_6, 0.1-mfd. fixed condenser; R_1, 50-ohm fixed resistance; R_2, 3-megohm grid leak; R_3, 6-ohm rheostat; S_1, Filament control switch (on master control switch); relay No. 6, 6-volt. d. p. d. t. The coils given above are for the 80-meter band. Those used for the 40-meter band are as follows: L_4, 10 turns No. 16 bare wire, spaced wire width, 3 inches diameter; L_3, 3 turns No. 6 bare wire, spaced wire width 3 inches diameter.

When the master control switch is thrown to the "receive" position, which is indicated by "R" in Fig. 6, all the control relays, with the exception of relay No. 5, open, and the contacts of this latter relay close, thus shunting the chopper contacts during the period that the chopper motor is coming to a stop. It is well to note that the filaments of the short-wave receiver are turned on by putting the master control switch in the "receive" position, and the headphones are disconnected from the output of the monitor receiver, and connected to the output of the short-wave receiver, this latter action being taken care of by relay No. 6.

THE ANTENNA SYSTEM

A DIAGRAM of the antenna system is shown in Fig. 7. The antenna consists of two verticals which are connected to the extremities of a very short flat-top, the latter being an active part of the antenna system, which is insulated from a guy wire which extends from the top of a 112-foot mast to

the top of the roof at the front of the house. The mast itself has three sets of back guys, this mast being about 18″ in diameter at the base and 6″ in diameter at the top.

The counterpoise is arranged on the top of the roof of the house, as shown in the plan view in the upper right corner of Fig. 7.

SHORT-WAVE RECEIVER

THE schematic diagram of the short-wave receiver used at 2 AG is shown in Fig. 8.

The first tube in this receiver, T_1, is simply a coupling tube. The antenna is connected to the grid of the receiver through the variable 0.0005-mfd. condenser C_1 and the coil L_1. There is a 50-ohm resistor, R_1, between the grid and filament of the coupling tube, and the filament is grounded.

The radio-frequency energy in the output circuit of the coupling tube is passed on to the input circuit of the detector tube through the medium of the inductive coupling between the coils L_2 and L_3, the latter being tuned to the incoming signals by means of the 0.000075-mfd. variable tuning condenser C_2.

Regeneration is accomplished by means of the inductive coupling between the feed-back coil L_4 and L_3, and the tuning is effected by the variable condenser C_5. The function of C_4 is to limit the amount of radio-frequency current flowing in the feed-back circuit, hence also limiting the amount of regeneration.

L_5 is a radio-frequency choke and AFT is the first audio-frequency interstage transformer. Only one stage of audio-frequency amplification is shown on the diagram, although in the actual receiver there are two stages.

C_6 is both a radio- and an audio-frequency bypass condenser, and C_3 is an audio-frequency bypass condenser.

MONITOR RECEIVER

THE monitor receiver is just an ordinary receiver which is tuned so that one of its harmonics beats with the fundamental frequency of the transmitter. In this way it is possible to monitor the outgoing signals by picking up a small amount of signal energy without danger of blocking the tubes in the receiver.

TOBE ~ EBY	**UNIPAC**	SILVER MARSHALL
VAN DOORN		Q.R.S. ~ LYNCH
FROST		WARD-LEONARD

The Unipac is Here!

L OOKING back a year to June of 1926, when Silver-Marshall cast a bomb-shell into the field of audio amplification with the now famous 220 and 221 audio transformers, it does not seem surprising that S-M engineering should still lead in A. F. amplification.

Now S-M offers the most powerful power pack yet devised, the amplifier stage of which can develop more undistorted power output than the average 210 power pack. And the Unipac amplifier has the same features of rising low note frequency gain and 5,000 cycle cut-off that have made 220's and 221's the largest selling high-grade audio transformers on the market—two features at first ridiculed by experts, then accepted and next season to be found in the most advanced high-class equipment.

The power supply of the Unipac, unlike average power supplies, gives practically constant output, and is substantially the Reservoir B unit so highly endorsed by Keith Henney of Radio Broadcast Laboratory. It furnishes B supply to any radio set and A, B and C power to the amplifier stage—power constant, unfluctuating and free from "motor-boating" and "putting."

A Unipac added to your set provides it with the finest quality of reproduction, handling capacity to spare, and replaces all B batteries, operating as it does directly from the 105 to 120 volt, 60 cycle, house lighting socket. Even though you may discard your set for a newer model, the Unipac will improve any receiver you ever buy or build—will remain the last word in distortionless power amplification and B power supply for years to come. And its applications are not limited—it may be used as a two stage amplifier, or to electrify any phonograph by means of a standard record pick-up, loud speaker and the Unipac.

The Unipac kit, with all parts including steel chassis and case, is available in two models. Type 660 contains the most powerful of all receiving amplifiers, a push-pull stage with 230 and 231 transformers, and is priced at $62.00. Type 660-B, with a slightly lower output level, includes a standard amplifier stage with 220 and 221 transformers, at $57.00.

The 440 Jewelers' Time Receiver consists of three R. F. amplifier stages and a detector, accurately tuned in the S-M laboratories to exactly .112 K.C., Arlington's wavelength, thus insuring reception of but one station at a time absolutely without interference.

The remarkable tone quality of the Unipac—its tremendous undistorted power output—is made possible only through the use of the S-M push-pull transformers—the new 230 input and 231 output models. You too can enjoy this tone quality by incorporating them in your audio amplifier or power pack. They are priced at $10.00 each.

SILVER-MARSHALL, Inc.

SM

838 West Jackson Blvd.   Chicago, U. S. A.

The *Radio Broadcast*
LABORATORY INFORMATION
SHEETS

THE RADIO BROADCAST Laboratory Information Sheets are a regular feature of this magazine and have appeared since our June, 1926, issue. They cover a wide range of information of value to the experimenter and to the technical radio man. It is not our purpose always to include new information but to present concise and accurate facts in the most convenient form. The sheets are arranged so that they may be cut from the magazine and preserved for constant reference, and we suggest that each sheet be cut out with a razor blade and pasted on 4″ x 6″ filing cards, or in a notebook. The cards should be arranged in numerical order. An index appears twice a year dealing with the sheets published during that year. The last index appeared on sheets Nos. 47 and 48, in November, 1926. This month an index to all sheets appearing since that time is printed.

The June, October, November, and December, 1926, issues are out of print. A complete set of Sheets, Nos. 1 to 88, can be secured from the Circulation Department, Doubleday, Page & Company, Garden City, New York, for $1.00. Some readers have asked what provision is made to rectify possible errors in these Sheets. In the unfortunate event that any such errors do appear, a new Laboratory Sheet with the old number will appear.

The Information Service of RADIO BROADCAST is conducted entirely by mail, the coupon on page 191 being used when application is made for technical information. It is the purpose of these Sheets to supply information of original value which often makes it possible for our readers to solve their own problems. —THE EDITOR.

No. 105   RADIO BROADCAST Laboratory Information Sheet   **July, 1927**

Measuring R. F. Resistance of a Coil

NECESSARY EQUIPMENT AND PROCEDURE

THE job of measuring radio-frequency resistance is not an especially difficult one, although it requires considerable apparatus. The circuit diagram of the test circuit is given on this Sheet. The apparatus used should have the following characteristics:

OSCILLATOR—This represents a source of radio-frequency energy which should be adjusted to the

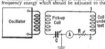

frequency at which the measurements are to be made. It should have plenty of power. In the Laboratory's a 210 tube with at least 300 volts on the plate is generally used, but it is doubtlessly possible to use a 201-A as an oscillator with about 100 volts on the plate. The important point is that adjustments in the test circuit should produce no change in the energy delivered by the oscillator.

A—This is a radio-frequency milliammeter with a range of about 200 milliamperes or preferably

somewhat less. It may be a hot-wire or thermocouple meter, or an ordinary crystal detector used with a low-range d. c. milliammeter.

C—The condenser should be a very carefully constructed one because it is essential that its resistance be low and constant. It should preferably be a laboratory type instrument although a well made receiving condenser can be used.

R—This resistance must be continuously variable and must be non-inductive. A decade resistance box is well suited for this purpose.

PICK-UP COIL—The pick-up coil functions to pick up energy from the oscillator and feed it into the test circuit. It may consist of just a few turns of wire coupled just close enough to the oscillator so as to give a good deflection on the meter, A.

The procedure in making a test is quite simple. Start with zero resistance at R and once the test has started make no changes at all in the oscillator or in the position of the pick-up coil. The oscillator should be turned on and the condenser varied until the circuit is in *exact* resonance, this condition being indicated by a maximum reading noted on meter A. Points 1 and 2 are now short circuited and the condenser readjusted so as to again bring the circuit into resonance. The reading of the meter will now be greater than before because the resistance of the coil under test is no longer in the circuit. Now add resistance to the circuit at R until the meter reading is decreased to the same value as was noted above, and under such conditions the resistance R is equal to the r. f. resistance of the coil under test.

No. 106   RADIO BROADCAST Laboratory Information Sheet   **July, 1927**

The UX-240 Type Tube

GENERAL CHARACTERISTICS

THE UX-240 type tube is designed for use in resistance-coupled amplifiers and under proper conditions will give an effective amplification of about 20 per stage. The plate resistor used with this tube should have a value of 250,000 ohms and the B and C voltages should be 180 volts or 135 volts and 3 or 1.5 volts respectively. The coupling condensers should have a value of 0.05 mfd. and the grid leak resistance should be of 2 megohms. These values are correct when the tube is used as an amplifier. It can also be used as a C-battery type detector in which case the C voltage should be 3 volts for a plate voltage of 135 or 4.5 volts for a plate voltage of 180. The plate resistor, coupling condenser, and grid leak should have the same values as given above.

The general characteristics of this tube are as follows:

Filament Voltage	. . . .	5.0 Volts
Filament Current	. . . .	0.25 Amperes
Maximum Plate Voltage	. .	180 Volts
Amplification Constant	. .	30
Plate Impedance	. . . .	150,000 Ohms
Plate Current	. . . .	0.2 Milliamperes

This tube can be used in any existing resistance-coupled amplifier provided the resistances used are of the proper value and the tubes are supplied with the proper A, B, and C voltages.

It is not possible to use this new tube in a transformer-coupled amplifier because its high plate

impedance will cause the transformer to have a rather sharp peak at some frequency. This fact, however, makes the tube very satisfactory as an amplifier for c. w. reception in short-wave receivers where we are interested in obtaining high amplification around 1000 cycles and very poor amplification at all other frequencies. The tube can also be used as a detector in a short-wave receiver.

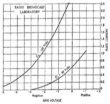

No. 107   RADIO BROADCAST Laboratory Information Sheet   July, 1927

Neutralization

EFFECTS OF MALADJUSTMENT

AT THE present time there is only one known way whereby a very high-gain high-frequency amplifier can be obtained, and that is by using several well-designed tuned radio-frequency amplifiers with each stage properly neutralized. Manufactured receivers are neutralized at the factory and consequently the problem of neutralizing a receiver or the effect of improper neutralization does not generally concern those who buy their receiver ready made. The home constructor, however, must neutralize his own receiver, and for this reason it is rather important that the effect of improper neutralization be known.

The first and most obvious manifestation of incorrect adjustment of the neutralizing device is oscillation in some or all of the radio-frequency circuits. These oscillations as a general rule become more severe as the frequency is increased, and a loud squeal or whistle will be heard as the tuning controls are adjusted to receive some station that is transmitting.

Such an effect will make it difficult for the user of the receiver to obtain satisfactory reception and the oscillations will be radiated from the antenna attached to the receiver and cause interference on other receivers located in the neighborhood. Such oscillations can be prevented by correct adjustment,

and it is essential that the proper setting be determined in order to make it possible to obtain best results from the receiver.

A second detrimental effect of maladjustment of the neutralizers is poor quality, which is generally due to the existence of too much regeneration. The quality under these conditions will generally sound drummy, indicating that the various frequencies in the carrier are being unequally amplified by the radio-frequency amplifiers. To preserve good quality, the radio-frequency amplifiers must amplify without distortion a band of frequencies extending about 5000 cycles above and 5000 cycles below the carrier frequency, and this condition does not exist unless proper neutralization is obtained.

Another effect of improper neutralization is to cause one or more of the tuned circuits in a single-control receiver to be thrown out of synchronism so that the set loses a great deal of its sensitivity, and as a result it is not possible to tune-in distant stations with satisfactory volume.

These three major effects of improper neutralization indicate how essential it is that neutralization be always carefully and completely accomplished. There are several satisfactory methods of neutralizing a receiver, and information regarding them can be found on Laboratory Sheet No. 38, published in the October, 1926, issue.

No. 108   RADIO BROADCAST Laboratory Information Sheet   July, 1927

High Voltage Supply for 210 Type Tube

THE DOUBLE TRANSFORMER METHOD

IF HIGH voltages up to 400 volts are required for operation of a 210 type power tube, it is generally best to use a B power unit incorporating a 216-B single-wave rectifier tube. This tube is capable of operating satisfactory at the high transformer voltages which must be used. It is possible, however, by using a somewhat complicated arrangement, to obtain the high voltage by using low-voltage rectifiers such as the Raytheon and Q. R. S.

An arrangement whereby 400 to 450 volts can be obtained using two gaseous rectifiers is shown in the drawing on this Sheet. Two power transformers, T_1 and T_2 are necessary, each supplying about 220 volts each side of the center tap. They are connected into the circuit as shown and supply two rectifiers which in turn feed a common filter system. The maximum permissible current drain is 20 milliamperes using Raytheon type B tubes and 35 milliamperes using type BH tubes. Condensers C_1, C_2, C_3, and C_4 each have a capacity of 0.1 mfd; C_5 and C_6 are of 2 mfd. capacity, and C_7, 5 mfd. All the condensers should have a working voltage of 750 volts d. c.

Filament current for the 210 tube should be obtained from a separate filament transformer

T_1 capable of supplying 1.25 amperes at 7.5 volts. The transformer should be tapped at the center as shown and a 1500-ohm resistance, R_1, connected between it and the negative B of the filter system. This resistance will supply C bias to the tube. Its bypass condenser should have a value of 2 mfd.

A 50,000- or 100,000-ohm resistance should be connected from B+ to B— if the unit is only to supply B potential to the 210, but if it is also to be used to supply B voltage to other tubes in a receiver the output should be shunted by several fixed resistors with taps at various points to obtain the desired voltages.

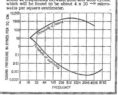

No. 109   RADIO BROADCAST Laboratory Information Sheet   July, 1927

The Threshold of Hearing and Feeling in the Ear

ENERGY REQUIRED FOR AUDIBILITY

A GREAT many important experiments in sound have been made in the various large laboratories. An interesting experiment is to determine how much energy is required by the ear in order to just hear tones of various frequencies between about 30 and 5000 cycles. Data of this sort can be plotted on a curve, a typical one being given on this Sheet. Such a curve is called a curve of "threshold audibility" because it indicates the amount of sound energy required to just produce an audible sound.

At 32 cycles a sound pressure of somewhat more than one dyne per square centimeter is required to produce an audible response, while at 2000 cycles only about 0.0003 dynes per square centimeter are required to produce an audible sound. The sound pressure required to produce a sound of minimum intensity is fairly constant between about 500 and 5000 cycles. Good speech articulation can be obtained within a frequency range of 250 and 2500 cycles; this band can, in fact, be narrowed to exclude all frequencies below 500 cycles and good articulation will still be retained. In the reproduction of music, however, it is necessary to include a much wider band having an upper limit of 5000 or 6000 cycles and a lower limit of about 32 cycles.

There is also an upper limit of sound pressure at which there is produced a sensation of feeling in the ear and it serves as a practical limit to the range of auditory sensation. At low frequencies the two curves of feeling and hearing meet each other, which indicates that these frequencies give a sensation

of feeling which is difficult to distinguish from a sensation of hearing.

The power in microwatts in each square centimeter of the sound wave under average conditions is related to the effective value of the pressure in dynes as follows:

$$\text{Power} = \left(\frac{\text{Pressure in Dynes}}{20.5}\right)^2$$

Using this formula we can calculate the average power required to produce a minimum audible sound at frequencies between 2000 and 4000 cycles, which will be found to be about 4×10^{-10} microwatts per square centimeter.

IN **R.F** by-pass circuits

SANGAMO
MICA CONDENSERS

A HIGH *self* inductance in condensers used in R. F. by-pass circuits means a *loss* in *capacity* at the lower wave lengths.

In many by-pass condensers the inductive reactance below 300 meters is appreciable. They become choke evils!

Use the larger capacities of Sangamo Mica Condensers in all R. F. circuits. Self inductance is negligible and direct current resistance more than 35,000-megohm! Sangamo Mica Condensers are all capacity.

SANGAMO
Accurate
Radio Parts

SANGAMO ELECTRIC COMPANY
6116-2   SPRINGFIELD, ILLINOIS

Condensers **TOBE** Resistors

At Booth 145 R. M. A. Show Chicago
June 13-18

RADIO FANS, a one-year's subscription to Radio Broadcast will cost you four dollars, two years six dollars. Consider this expenditure as being a necessary investment on your part for the future development of your own knowledge of Radio.

Convert your radio set into a light socket receiver with Balkite "B" and the Balkite Trickle and High-Rate Charger

Ask your radio dealer
FANSTEEL PRODUCTS CO., INC
North Chicago, Ill.

AmerTran DeLuxe audio transformers are guaranteed to amplify at 80% of their peak at 40 cycles, and their peak is above 10,000 cycles. Made for first and second stages, either type $10.00.

The AmerTran power transformer type PF52, $18.00, and the AmerChoke type 854, $6.00, (illustrated) are designed for use in the construction of power amplifiers to operate with UX-216B and UX-210 tubes at their correct voltages, supplying A, B and C to the last audio, and B and C to the other tubes.

AMERTRAN RADIO PRODUCTS CARRY THIS GUARANTEE

"AmerTran audio transformers, regardless of type, are fully guaranteed against defects for a period of one year from the date of purchase, and will be replaced free of charge either through your authorized AmerTran dealer or direct, if defective for any cause other than misuse. The individual parts are each carefully tested and inspected before assembly and the complete transformer receives a most rigid inspection and test before being packed for shipment."

This is the way the American Transformer Company has won confidence and wide use for its products. AmerTran De-Luxe audio transformers are recognized as reliable, efficient units for improving the tone quality and tone range of present sets and as the indispensable choice for new sets. Other AmerTran products have been adopted for power supply apparatus that on performance stand in the front rank of modern development.

Send for booklet "Improving the Audio Amplifier," and other useful data, free.

The American Transformer Co.
178 EMMET STREET   NEWARK, N. J.
"Transformer Builders for Over 26 Years"

No. 110   RADIO BROADCAST Laboratory Information Sheet   July, 1927

Dry-Cell Tubes

BEST FILAMENT VOLTAGE

ALTHOUGH dry-cell tubes are generally operated with 3 volts on the filament, somewhat better results can be obtained if 3.3 volts is used instead.

The two solid curves on the accompanying diagram are obtained by measuring the plate current at various values of negative grid bias with 3.0 and then 3.3 volts across the filament. If the tube is functioning properly this curve will be a straight line over most of its length. The 3.0-volt curve slopes off at low values of grid bias and this indicates that the filament emission is too low and a signal would be distorted. The 3.3-volt curve, however, is straight over a large portion of its length and therefore this same tube with somewhat higher filament voltage is capable of amplifying without distortion.

The two dotted curves show the plate impedance of the tube first with 3.0 volts and then with 3.3 volts on the filament. With 3.0 volts, and therefore a low filament emission, we obtain an erratic plate impedance curve, which rises to values as high as 80,000 ohms at zero grid voltage. The plate impedance curve taken with 3.3 volts again indicates the value of using this voltage, for it shows the plate impedance to be comparatively constant and low over a greater part of its length, and this is as it should be.

This recommendation that 3.3 volts be used on the filament is the result of many tests made in the Laboratory, and the Cunningham Tube Company has also recommended that this voltage be used.

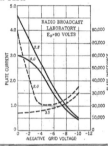

No. 111   RADIO BROADCAST Laboratory Information Sheet   July, 1927

Index
December, 1926, to July, 1927

No. 112   RADIO BROADCAST Laboratory Information Sheet   July, 1927

Index
December, 1926, to July, 1927

Equipment for the Home-Constructor

How to Use Some of the New and Interesting Radio Equipment Which the Market Offers

By THE LABORATORY STAFF

TAYLOR RESONANCE INDICATOR

THE design of a receiver in which several of the tuning condensers are actuated by a single control is a somewhat difficult job because it is essential that each of the tuned circuits be in exact resonance with each other to prevent loss in selectivity and sensitivity.

The coils and condensers used in such a receiver must be as electrically similar as possible, and it is generally advisable to check each coil-condenser combination separately to make certain that it tunes, throughout the entire frequency range of the receiver, exactly the same as the other tuned circuits. A simple and satisfactory method of making a test of this sort is possible with the use of the new Taylor resonance indicator, shown in a photograph on this page. The circuit diagram is given in Fig. 1. Two 201-A type tubes are necessary to operate the

THE TAYLOR RESONANCE INDICATOR

device. One of them is placed in socket A and acts as an oscillator, the oscillatory circuit consisting of C_1, L_1, and L_2; the frequency range of the oscillator is 500 to 1500 kc. The other tube is placed in socket B, and it acts as a rectifier of the current flowing through the galvanometer, "G." In the following paragraphs we will describe the procedure in testing a single-control receiver for synchronism.

Suppose that we have such a receiver consisting of three tuned circuits A, B, and C, all of them operated from a single dial. If the receiver is to give satisfaction these three circuits must tune to exactly the same frequency at exactly the same point, and the problem is to determine if such is the case. In making the test no batteries at all should be connected to the receiver. The resonance indicator itself, however, requires for its operation a 45-volt B battery and a 6-volt storage battery for filament supply.

There are two leads attached to the resonance indicator, one red, the other green, and these leads should be connected, respectively, to the stator and rotor of the first variable condenser in the receiver. Now set the dial on condenser C_1 at some medium frequency, say 1000 kilocycles,

which corresponds to 17 on the dial (See Fig. 2), and then turn the instrument on. The oscillations in tube A are fed through the small coupling condenser, C_5, to the tuned circuit of the receiver to which the leads are connected, and as a result, a voltage will be developed across the tuned circuit. This voltage will be a maximum when the tuned circuit in the receiver is in exact resonance with

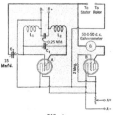

FIG. 1

the oscillations produced by tube A. This voltage across the tuned circuit, will cause a current to flow through the galvanometer and the tube B connected in series with it, and the current flowing through the galvanometer will be proportional to the voltage. It is evident that maximum galvanometer deflection means maximum voltage and, therefore, that the circuit is in resonance. The dial reading of the receiver at resonance should be noted down as accurately as possible.

Without in any way changing any of the settings on the resonance indicator, the two connections are moved over and connected to circuit B. The condenser in this circuit is then adjusted for maximum deflection as was done with the condenser

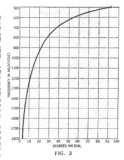

FIG. 2

Goodrich

Radio Panels

Black, Mahogany and Walnut.
Standard thickness 3-16" and ¼".
High softening point prevents
warping.
Low free sulphur content prevents
discoloration.
Machine easier and are better electrically than any other material in
common use.
Your radio dealer has them or will
get them for you—if not, write to
us and we will see that you are
supplied.
Also hard rubber for coils, spaghetti
tubing and miscellaneous rubber
items for radio manufacturers.

Radio Sales Division

The B. F. Goodrich Rubber Company
Established 1870   Akron, Ohio

SILVER-MARSHALL UNIPAC $62.
Wholesale Headquarters
FREE New 1927 Catalog FREE
Shows photographs and hook-ups of all latest kits, complete
line of cabinets and consoles, accessories and parts. We are
headquarters for all nationally advertised lines. Dealers and
professional set builders write on your letterhead today for
your copy of this big FREE CATALOG.
SHURE RADIO CO.,339 B-335 W. Madison St., CHICAGO, ILL.

Why not subscribe to *Radio
Broadcast?* By the year
only $4.00; or two years,
$6.00, saving $2.40. Send
direct to Doubleday, Page
& Company, Garden City,
New York.

GOOD tubes are little
better than poor tubes
without Amperite. Ut-
most performance and
longest life requires the
perfect automatic fila-
ment control of each in-
dividual tube, afforded
only by Amperite. Insist
on this variable filament
resistance—and look for
the name. Eliminates hand rheostats. Simplifies
wiring.
Radiall Book—FREE
Explains fully Amperite principle and operation.
Gives season's best Hook-Ups and Construction
Data. Send for it today to Dept. RB5
Radiall Co., 50 Franklin St., New York

Price $1.10
mounted (U.S.A.)
Sold Everywhere

AMPERITE
The "SELF-ADJUSTING" *Rheostat*

The NEW
and wonderfully improved

*Perfectly
Matched*

*Adaptable to all
Standard Tubes*

The supersensitive AERO coil has been improved! Always renowned for
its selectivity, power, and sensitivity, the AERO coil is destined to win
even greater favor in its new form. This new coil is the very last word
in inductance coil construction. It contains a host of new and exclusive
features. It possesses amazing adaptability and can be used in *all* R. F.
circuits—both bridge and loss balanced. What's more it is easily adaptable to 5, 6, or 7 tube sets!

Here are other sensational features of this new and improved AERO coil:

1. **Rugged**—Will keep appearance and original
electrical characteristics indefinitely.

2. **High electrical efficiency**—Unusually high
ratio of inductance to radio frequency resistance.

3. **Shape ratio**—Ratio of coil length to coil
diameter is such that magnetic coupling
between coils is at a minimum, thus eliminating necessity for shielding in many receivers.

4. **Good mounting facilities**—Terminals at
lower end of coil permit short connection to
tubes, sockets, etc.

5. **Adaptability**—Carefully designed primary
windings of proper impedance for any type
commercial tube immediately available.

6. **Ease of connection**—Screw type terminals
permit optional choice of connections
without soldering iron, or with soldering
iron.

The new AERO coils will be shown for the first time at our exhibition in Booth 12,
at the R. M. A. Trade Show. They will be available at your dealers after July 1.
You can get them in 3 coil kits, with refinement, at $12.00. 4 coil kit, with refinement,
$16.00. These new coils will also be used in the AERO R. F. R. Kit and the AERO
3 Circuit Tuner.

AERO PRODUCTS, Inc.
Dept. 109   1772 Wilson Ave., Chicago, Ill.

TO RADIO DEALERS!

The R. B. Laboratory Information Sheets have been appearing in RADIO
BROADCAST since June, 1926. They are a regular feature in each issue and
they cover a wide range of information of value to the radio experimenter
and set builder. We have just reprinted Lab. Sheets Nos. 1-88 from the
June, 1926, to April, 1927, issues of RADIO BROADCAST. They are arranged
in numerical order and are bound with a suitable cover. They sell at retail for one dollar a set. Write for dealers' prices. Address your letter to

Circulation Dept., RADIO BROADCAST, Garden City, N. Y.

in circuit A, and the dial reading noted. The same test is then made with the leads connected across the condenser in tuned circuit C. If it is found that the dial on the receiver must be tuned to exactly the same point in all three cases, we have a good indication that the three circuits are in synchronism. If, on the other hand, one of the condensers gives a different reading than the other two it must be readjusted so as to give exactly the same reading.

Usually only one test for alignment is necessary at some medium frequency. The set may be checked at three or more points if it is deemed necessary, by simply placing the condenser dial of the tester at a different setting and repeating the test outlined above. A precaution which must be observed is to keep the leads from the oscillator to the receiver well separated and well away from metallic parts, otherwise untrue results will be experienced. This is particularly true with short wavelength (high-frequency) readings.

This tester may also be used as a source of high-frequency oscillations for many different tests by simply attaching a short piece of wire to the "stator" binding post for an antenna and connecting the "rotor" binding post to ground. Manufactured by the Taylor Electric Company, Madison, Wisconsin. Price $32.50.

RADIO SET TESTER

WHEN a service man or a radio experimenter has to diagnose trouble in a radio receiver, he needs certain instruments and meters for the purpose. If these instruments and meters are individual pieces of apparatus, he will need a number of them. They are bulky and a good deal of time is wasted in making connections. The new Weston No. 519 set tester gives a combination of instruments in a compact form, which permits of easy operation and a considerable saving of time. All routine tests can be made on sets, tubes, and other accessories, with very little effort and a minimum number of connections.

Only one meter is used in the test kit, as shown in the accompanying photograph. This one meter, however, by means of an ingenious switch and a combination of meter scales, can be made to read A, B, and C voltages, and plate current, besides being able to check open or closed circuits. For testing tubes a socket is provided, and connections are made directly to the set by means of a cable and a plug arrangement which, by means of adapters, can be plugged into any type socket in the receiving set.

By simply placing the plug in one of the receiver sockets and manipulating the switch, all of the A, B, and C voltages may be read. This switch is of the double-contact type; that is, for each connection, both sides of the line are opened or closed, as the case may be. This prevents interconnections which might possibly cause some trouble. The switch is marked in the following manner:

(1) "OPEN." The meter is entirely disconnected from the cable or binding posts and is in the starting position.

(2) "VM. B.P." There are three binding posts at the right-hand side. With the switch in this position the meter is not connected with the cable but may be used by placing leads on the binding posts. By connecting between the minus binding post and the center one of the three, voltages up to 8 may be measured, while, by connecting between the minus post and the top one, voltages up to 200 may be read, thus giving the operator a simple double-range voltmeter.

(3) "C A-REV." This point on the switch gives the C-battery voltage (with plug in receiver) on a reversed socket. Some receivers have the filament connections on the socket reversed. While this condition does not affect the operation of the receiver it would make the d.c. meter read backward unless provision was made for it. The voltage as read on the 8-volt scale should be multiplied by 10.

(4) "C." This point on the switch gives the C-voltage reading when the socket is connected in the standard manner.

(5) "B." This point gives the voltage of the B battery at the socket and is read on the 200-volt scale.

(6) "PLATE MA." With a tube inserted in the socket of the tester and the cable connected to the receiver, the meter will read the plate current taken by the tube on the 200-volt scale. The reading on this scale should be divided by 10. That is, if the meter read 80, it would really indicate 8 mA.

(7) "A." This point on the switch connects the meter to the filament connections on the socket and gives A voltage on the 8-volt scale.

(8) "A-REV." Gives the A voltage if the socket happens to be reversed as mentioned above.

(9) "OPEN." This is the same as the "open" position in No. 1. The switch may be placed in either No. 1 or No. 9 position thus making it immaterial in which direction the switch is turned in starting a test.

In making tests, the plug is placed in one of the sockets of the receiver, the rest of the tubes remaining in their respective places. The batteries are left connected. The switch is then manipulated and the A, B, and C battery voltages read. If no reading is obtained, an open circuit is indicated. The tube may then be placed in the tester socket and its plate current determined.

A simple test is provided for tubes. A small button at the center bottom of the test panel is pressed. This places a zero grid bias on the tube. The difference in plate current reading with the button up or down indicates the worth of the tube. The difference can be compared directly with a table accompanying the tester.

All in all, any test on tubes, batteries, or receivers can be made quickly and definitely without loss of time with this instrument and it should recommend itself strongly to those who have to handle radio receivers in trouble. Manufactured by the Weston Electrical Instrument Corporation, of Newark, New Jersey. Price $75.00

RADIO BROADCAST *Photograph*

WESTON RADIO SET TESTER

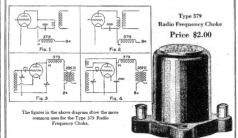

Manufacturers' Booklets

A Varied List of Books Pertaining to Radio and Allied Subjects Obtainable Free With the Accompanying Coupon

READERS may obtain any of the booklets listed below by using the coupon printed on page 188. Order by number only.

1. FILAMENT CONTROL—Problems of filament supply, voltage regulation, and effect on various circuits. RADIALL COMPANY.
2. HARD RUBBER PANELS—Characteristics and properties of hard rubber as used in radio, with suggestions on how to "work" it. B. F. GOODRICH RUBBER COMPANY.
3. TRANSFORMERS—A booklet giving data on input and output transformers. PACENT ELECTRIC COMPANY.
4. RESISTANCE-COUPLED AMPLIFIERS—A general discussion of resistance coupling with curves and circuit diagrams. COLE RADIO MANUFACTURING COMPANY.
5. CARBORUNDUM IN RADIO—A book giving pertinent data on the crystal as used for detection, with hook-ups, and a section giving information on the use of resistors. THE CARBORUNDUM COMPANY.
6. B-ELIMINATOR CONSTRUCTION—Constructional data on how to build. AMERICAN ELECTRIC COMPANY.
7. TRANSFORMER AND CHOKE-COUPLED AMPLIFICATION—Circuit diagrams and discussion. ALL-AMERICAN RADIO CORPORATION.
8. RESISTANCE UNITS—A data sheet of resistance units and their application. WARD-LEONARD ELECTRIC COMPANY.
9. VOLUME CONTROL—A leaflet showing circuits for distortionless control of volume. CENTRAL RADIO LABORATORIES.
10. VARIABLE RESISTANCE—As used in various circuits. CENTRAL RADIO LABORATORIES.
11. RESISTANCE COUPLING—Resistors and their application to audio amplification, with circuit diagrams. DEJUR PRODUCTS COMPANY.
12. DISTORTION AND WHAT CAUSES IT—Hook-ups of resistance-coupled amplifiers with standard circuits. ALLEN-BRADLEY COMPANY.
15. B-ELIMINATOR AND POWER AMPLIFIER—Instructions for assembly and operation using RayTheon tube. GENERAL RADIO COMPANY.
15A. B-ELIMINATOR AND POWER AMPLIFIER—Instructions for assembly and operation using an R. C. A. rectifier. GENERAL RADIO COMPANY.
16. VARIABLE CONDENSERS—A bulletin giving an analysis of the functions and characteristics of variable condensers with curves and specifications for their application to complete receivers. ALLEN D. CARDWELL MANUFACTURING COMPANY.
17. BAKELITE—A description of various uses of bakelite in radio, its manufacture, and its properties. BAKELITE CORPORATION.
18. POWER SUPPLY—A discussion on power supply with particular reference to lamp-socket operation. Theory and constructional data for building power supply devices. ACME APPARATUS COMPANY.
20. AUDIO AMPLIFICATION—A booklet containing data on audio amplification together with hints for the constructor. ALL AMERICAN RADIO CORPORATION.
21. HIGH-FREQUENCY DRIVER AND SHORT-WAVE WAVE-METER—Constructional data and application. BURGESS BATTERY COMPANY.
46. AUDIO-FREQUENCY CHOKES—A pamphlet showing positions in the circuit where audio-frequency chokes may be used. SAMSON ELECTRIC COMPANY.
47. RADIO-FREQUENCY CHOKES—Circuit diagrams illustrating the use of chokes to keep out radio-frequency currents from definite points. SAMSON ELECTRIC COMPANY.
48. TRANSFORMER AND IMPEDANCE DATA—Tables giving the mechanical and electrical characteristics of transformers and impedances, together with a short description of their use in the circuit. SAMSON ELECTRIC COMPANY.
49. BYPASS CONDENSERS—A description of the manufacture of bypass and filter condensers. LESLIE F. MUTER COMPANY.
50. AUDIO MANUAL—Fifty questions which are often asked regarding audio amplification, and their answers. AMERTRAN SALES COMPANY, INCORPORATED.
51. SHORT-WAVE RECEIVER—Constructional data on a receiver which, by the substitution of various coils, may be made to tune from a frequency of 16,660 kc. (18 meters) to 1990 kc. (150 meters). SILVER-MARSHALL, INCORPORATED.
52. AUDIO QUALITY—A booklet dealing with audio-frequency amplification of various kinds and the application to well-known circuits. SILVER-MARSHALL, INCORPORATED.
56. VARIABLE CONDENSERS—A bulletin giving an analysis of various condensers together with their characteristics. GENERAL RADIO COMPANY.
57. FILTER DATA—Facts about the filtering of direct current supplied by means of motor-generator outfits used with transmitters. ELECTRIC SPECIALTY COMPANY.
59. RESISTANCE COUPLING—A booklet giving some general information on the subject of radio and the application of resistors to a circuit. DAVEN RADIO CORPORATION.
60. RESISTORS—A pamphlet giving some technical data on resistors which are capable of dissipating considerable energy; also data on the ordinary resistors used in resistance-coupled amplification. THE CRESCENT RADIO SUPPLY COMPANY.
62. RADIO-FREQUENCY AMPLIFICATION—Constructional details of a five-tube receiver using a special design of radio-frequency transformer. CAMFIELD RADIO MANUFACTURING COMPANY.
63. FIVE-TUBE RECEIVER—Constructional data on building a receiver. AERO PRODUCTS, INCORPORATED.
64. AMPLIFICATION WITHOUT DISTORTION—Data and curves illustrating the use of distortionless amplification. ACME APPARATUS COMPANY.
65. RADIO HANDBOOK—A helpful booklet on the function, selection, and use of radio apparatus for better reception. BENJAMIN ELECTRIC MANUFACTURING COMPANY.
66. SUPER-HETERODYNE—Constructional details of a seven-tube set. G. C. EVANS COMPANY.
70. IMPROVING THE AUDIO AMPLIFIER—Data on the characteristics of audio transformers, with a circuit diagram showing where chokes, resistors, and condensers can be used. AMERICAN TRANSFORMER COMPANY.

71. DISTORTIONLESS AMPLIFICATION—A discussion of the resistance-coupled amplifier used in conjunction with a transformer, impedance, or resistance input stage. Amplifier circuit diagrams and constants are given in detail for the constructor. AMSCO PRODUCTS INCORPORATED.
72. PLATE SUPPLY SYSTEM—A wiring diagram and layout plan for a plate supply system to be used with a power amplifier. Complete directions for wiring are given. AMERTRAN SALES COMPANY.
80. FIVE-TUBE RECEIVER—Data are given for the construction of a five-tube tuned radio-frequency receiver. Complete instructions, list of parts, circuit diagram, and template are given. ALL-AMERICAN RADIO CORPORATION.
81. BETTER TUNING—A booklet giving much general information on the subject of radio reception with specific illustrations. Primarily for the non-technical home constructor. BREMER-TULLY MANUFACTURING COMPANY.
82. SIX-TUBE RECEIVER—A booklet containing photographs, instructions, and diagrams for building a six-tube shielded receiver. SILVER-MARSHALL, INCORPORATED.
83. SOCKET POWER DEVICE—A list of parts, diagrams, and templates for the construction and assembly of socket power devices. JEFFERSON ELECTRIC MANUFACTURING COMPANY.
84. FIVE-TUBE EQUAMATIC—Panel layout, circuit diagrams, and instructions for building a five-tube receiver, together with data on the operation of tuned radio-frequency transformers of special design. KARAS ELECTRIC COMPANY.
85. FILTER—Data on a high-capacity electrolytic condenser used in filter circuits in connection with A socket power supply units, are given in a pamphlet. THE ABOX COMPANY.
86. SHORT-WAVE RECEIVER—A booklet containing data on a short-wave receiver as constructed for experimental purposes. THE ALLEN D. CARDWELL MANUFACTURING CORPORATION.
88. SUPER-HETERODYNE CONSTRUCTION—A booklet giving full instructions, together with a blue print and necessary data, for building an eight-tube receiver. THE GEORGE W. WALKER COMPANY.
89. SHORT-WAVE TRANSMITTERS—Data and blue prints are given on the construction of a short-wave transmitter, together with operating instructions, methods of keying, and other pertinent data. RADIO ENGINEERING LABORATORIES.
90. IMPEDANCE AMPLIFICATION—The theory and practice of a special type of dual-impedance audio amplification are given. ALDEN MANUFACTURING COMPANY.
93. B-SOCKET POWER—A booklet giving constructional details of a socket-power device using either the BH or 313 type rectifier. NATIONAL COMPANY, INCORPORATED.
94. POWER AMPLIFIER—Constructional data and wiring diagrams of a power amplifier combined with a B-supply unit are given. NATIONAL COMPANY, INCORPORATED.

ACCESSORIES

22. A PRIMER OF ELECTRICITY—Fundamentals of electricity with special reference to the application of dry cells to radio and other uses. Constructional data on buzzers, automatic switches, alarms, etc. NATIONAL CARBON COMPANY.
23. AUTOMATIC RELAY CONNECTIONS—A data sheet showing how a relay may be used to control A and B circuits. YAXLEY MANUFACTURING COMPANY.
25. ELECTROLYTIC RECTIFIER—Technical data on a new type of rectifier with operating curves. KOBEL RADIO CORPORATION.
26. DRY CELLS FOR TRANSMITTERS—Actual tests given, well illustrated with curves showing exactly what may be expected of this type of B power. BURGESS BATTERY COMPANY.
27. DRY-CELL BATTERY CAPACITIES FOR RADIO TRANSMITTERS—Characteristic curves and data on discharge tests. BURGESS BATTERY COMPANY.
28. B BATTERY LIFE—Battery life curves with general curves on tube characteristics. BURGESS BATTERY COMPANY.
29. HOW TO MAKE YOUR SET WORK BETTER—A non technical discussion of general radio subjects with hints on how reception may be bettered by using the right tubes. UNITED RADIO AND ELECTRIC CORPORATION.
30. TUBE CHARACTERISTICS—A data sheet giving constants of tubes. C. E. MANUFACTURING COMPANY.
31. FUNCTIONS OF THE LOUD SPEAKER—A short, non technical general article on loud speakers. AMPLION CORPORATION OF AMERICA.
32. METERS FOR RADIO—A catalogue of meters used in radio, with connecting diagrams. BURTON-ROGERS COMPANY.
33. SWITCHBOARD AND PORTABLE METERS—A booklet giving dimensions, specifications, and shunts used with various meters. BURTON-ROGERS COMPANY.
34. COST OF B BATTERIES—An interesting discussion of the relative merits of various sources of B supply. HART FORD BATTERY MANUFACTURING COMPANY.
35. STORAGE BATTERY OPERATION—An illustrated booklet on the care and operation of the storage battery. GENERAL LEAD BATTERIES COMPANY.
36. CHARGING A AND B BATTERIES—Various ways of connecting up batteries for charging purposes. WESTINGHOUSE UNION BATTERY COMPANY.
37. CHOOSING THE RIGHT RADIO BATTERY—Advice on what dry cell battery to use: their application to radio with wiring diagrams. NATIONAL CARBON COMPANY.
53. TUBE REACTIVATOR—Information on the care of vacuum tubes, with notes on how and when they should be reactivated. THE STERLING MANUFACTURING COMPANY.
54. ARRESTERS—Mechanical details and principles of the vacuum type of arrester. NATIONAL ELECTRIC SPECIALTY COMPANY.
55. CAPACITY CONNECTOR—Description of a new device for connecting up the various parts of a receiving set, at the same time providing bypass condensers between the leads. KURZ-KASCH COMPANY.

68. CHEMICAL RECTIFIER—Details of assembly, with wiring diagrams, showing how to use a chemical rectifier for charging batteries. CLEVELAND ENGINEERING LABORATORIES COMPANY.

69. VACUUM TUBES—A booklet giving the characteristics of the various tube types with a short description of where they may be used in the circuit. RADIO CORPORATION OF AMERICA.

77. TUBES—A booklet for the beginner who is interested in vacuum tubes. A non-technical consideration of the various elements in the tube as well as their position in the receiver. CLEARTRON VACUUM TUBE COMPANY.

87. TUBE TESTER—A complete description of how to build and how to operate a tube tester. BURTON-ROGERS COMPANY.

94. VACUUM TUBES—A booklet giving the characteristics and uses of various types of tubes. This booklet may be obtained in English, Spanish, or Portuguese. DEFOREST RADIO COMPANY.

92. RESISTORS FOR A. C. OPERATED RECEIVERS—A booklet giving circuit suggestions for building a.c. operated receivers, together with a diagram of the circuit, used with the new 400-milampere rectifier tube. CARTER RADIO COMPANY.

97. HIGH-RESISTANCE VOLTMETERS—A folder giving information on how to use a high-resistance voltmeter, special consideration being given the voltage measurement of socket-power devices. WESTINGHOUSE ELECTRIC & MANUFACTURING COMPANY.

MISCELLANEOUS

38. LOG SHEET—A list of broadcasting stations with columns for marking down dial settings. U. S. L. RADIO, INCORPORATED.

41. BABY RADIO TRANSMITTER OF QXH-QKR—Description and circuit diagrams of dry-cell operated transmitter. BURGESS BATTERY COMPANY.

42. ARCTIC RADIO EQUIPMENT—Description and circuit details of short-wave receiver and transmitter used in Arctic exploration. BURGESS BATTERY COMPANY.

43. SHORT-WAVE RECEIVER OF QXM-QXR—Complete directions for assembly and operation of the receiver. BURGESS BATTERY COMPANY.

44. ALUMINUM FOR RADIO—A booklet containing much radio information with hook-ups of basic circuits, with inductance-capacity tables and other pertinent data. ALUMINUM COMPANY OF AMERICA.

45. SHIELDING—A discussion of the application of shielding in radio circuits with special data on aluminum shields. ALUMINUM COMPANY OF AMERICA.

58. HOW TO SELECT A RECEIVER—A commonsense booklet describing what a radio set is, and what you should expect from it, in language that any one can understand. DAY-FAN ELECTRIC COMPANY.

67. WEATHER FOR RADIO—A very interesting booklet on the relationship between weather and radio reception, with maps and data on forecasting the probable results. TAYLOR INSTRUMENT COMPANIES.

73. RADIO SIMPLIFIED—A non-technical booklet giving pertinent data on various radio subjects. Of special interest to the beginner and set owner. CROSLEY RADIO CORPORATION.

74. THE EXPERIMENTER—A monthly publication which gives technical facts, valuable tables, and pertinent information on various radio subjects. Interesting to the experimenter and to the technical radio man. GENERAL RADIO COMPANY.

75. FOR THE LISTENER—General suggestions for the selecting, and the care of radio receivers. VALLEY ELECTRIC COMPANY.

76. RADIO INSTRUMENTS—A description of various meters used in radio and electrical circuits together with a short discussion of their uses. JEWELL ELECTRICAL INSTRUMENT COMPANY.

78. ELECTRICAL TROUBLES—A pamphlet describing the use of electrical testing instruments in automotive work combined with a description of the cadmium-test for storage batteries. Of interest to the owner of storage batteries. BURTON-ROGERS COMPANY.

95. RESISTANCE DATA—Successive bulletins regarding the use of resistors in various parts of the radio circuit. INTERNATIONAL RESISTANCE COMPANY.

96. VACUUM TUBE TESTING—A booklet giving pertinent data on how to test vacuum tubes with special reference to a tube testing unit. JEWELL ELECTRICAL INSTRUMENT COMPANY.

98. COPPER SHIELDING—A booklet giving information on the use of shielding in radio receivers, with notes and diagrams showing how it may be applied practically. Of special interest to the home constructor. THE COPPER AND BRASS RESEARCH ASSOCIATION.

99. RADIO CONVENIENCE OUTLETS—A folder giving diagram and specifications for installing loud speakers in various locations at some distance from the receiving set. YAXLEY MANUFACTURING COMPANY.

USE THIS COUPON

RADIO BROADCAST SERVICE DEPARTMENT
RADIO BROADCAST, *Garden City, New York*
Please send me (at no expense) the following booklets indicat ed by numbers in the published list

....
....
....
....
....

Name...

Address..
    (Number)   (Street)

....
    (City)   (State)

ORDER BY NUMBER ONLY
This coupon must accompany every request.   RB 727

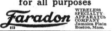

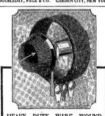

What Kit Shall I Buy?

*THE list of kits herewith is printed as an exten-
sion of the scope of the Service Department of
RADIO BROADCAST. It is our purpose to list here
the technical data about kits on which information
is available. In some cases, the kit can be purchased from your dealer complete; in others, the
descriptive booklet is supplied for a small charge
and the parts can be purchased as the buyer likes.
The Service Department will not undertake to
handle cash remittances for parts, but when the
coupon on page 191 is filled out, all the information requested will be forwarded.*

201. SC FOUR-TUBE RECEIVER—Single control. One
stage of tuned radio frequency, regenerative detector,
and two stages of transformer-coupled audio amplification.
Regeneration control is accomplished by means of a variable
resistor across the tickler coil. Standard parts; cost approximately $58.85.

202. SC-H FIVE-TUBE RECEIVER—Two stages of tuned
radio frequency, detector, and two stages of transformer-coupled audio. Two tuning controls. Volume control
consists of potentiometer grid bias on r.f. tubes. Standard
parts cost approximately $60.35.

203. "HI-Q" KIT—A five-tube tuned radio-frequency set
having two radio stages, a detector, and two transformer-
coupled audio stages. A special method of coupling in the
r.f. stages tends to make the amplification more nearly equal
over the entire band. Price $63.05 without cabinet.

204. R. G. S. KIT—A four-tube inverse reflex circuit,
having the equivalent of two tuned radio-frequency stages,
detector, and three audio stages. Two controls. Price $69.70
without cabinet.

205. PIERCE AIRO KIT—A six-tube single-dial receiver;
two stages of radio-frequency amplification, detector, and
three stages of resistance-coupled audio. Volume control
accomplished by variation of filament brilliancy of r.f.
tubes or by adjusting compensating condensers. Complete
chassis assembled but not wired costs $42.50.

206. H & H-T. & F. ASSEMBLY—A five-tube set; three
tuning dials, two steps of radio frequency, detector, and 2
transformer-coupled audio stages. Complete except for baseboard, panel, screws, wires, and accessories. Price $35.00.

207. PREMIER FIVE-TUBE ENSEMBLE—Two stages of
tuned radio frequency, detector, and two steps of transformer-coupled audio. Three dials. Parts assembled but
not wired. Price complete, except for cabinet, $35.00.

208. "QUADRAFORMER VI"—A six-tube set with two tuning controls. Two stages of tuned radio frequency using
specially designed shielded coils, a detector, one stage of
transformer-coupled audio, and two stages of resistance-
coupled audio. Gain control by means of tapped primaries
on the r.f. transformers. Essential kit consists of three
shielded double-range "Quadraformer" coils, a selectivity
control, and an "Amplitrol," price $17.50. Complete parts
$70.15.

209. GEN-RAL FIVE-TUBE SET—Two stages of tuned
radio frequency, detector, and two transformer-coupled
audio stages. Volume is controlled by a resistor in the plate
circuit of the r.f. tubes. Uses a special r.f. coil ("Duo-
Former") with figure eight winding. Parts mounted but
not wired, price $37.50.

210. BREMER-TULLY POWER-SIX—A six-tube, dual-
control set: three stages of neutralized tuned radio frequency,
detector, and two transformer-coupled audio stages. Resistances in the grid circuit together with a phase shifting
arrangement are used to prevent oscillation. Volume control
accomplished by variation of B potential on r.f. tube.
Essential kit consists of four r.f. transformers, two dual
condensers, three small condensers, three choke coils, one
500,000-ohm resistor, three 1500-ohm resistors, and a set
of color charts and diagrams. Price $41.50.

211. BRUNO DRUM CONTROL RECEIVERS—How to apply
a drum tuning unit to such circuits as the three-tube regenerative receiver, four-tube Browning-Drake, five-tube
Diamond-of-the-Air, and the "Grand" 6.

212. INFRADYNE AMPLIFIER—A three-tube intermediate-
frequency amplifier for the super-heterodyne and other
special receivers, tuned to 3400 kc. (86 meters). Price $25.00.

213. RADIO BROADCAST "LAB" RECEIVER—A four-tube
dual-control receiver with one stage of Rice neutralized
tuned-radio frequency, regenerative detector (capacity
controlled), and two stages of transformer-coupled audio.
Approximate price, $78.15.

214. LC-27—A five-tube set with two stages of tuned-
radio frequency, a detector, and two stages of transformer-
coupled audio. Special coils and special means of neutralizing
are employed. Output device. Price $85.20 without cabinet.

215. LOFTIN-WHITE—A five-tube set with two stages of
radio frequency, especially designed to give equal amplification at all frequencies, a detector, and two stages of transformer-coupled audio. Two controls. Output device. Price
$81.10.

216. K.H.-27—A six-tube receiver with two stages of
neutralized tuned radio frequency, a detector, three stages
of choke-coupled audio, and an output device. Two controls.
Price $86.00 without cabinet.

217. AERO SHORT-WAVE KIT—Three plug-in coils designed to operate with a regenerative detector circuit and
having a frequency range of from 19,990 to 2306 kc. (15 to 130
meters). Coils and plug only, price $12.50.

218. DIAMOND-OF-THE-AIR—A five-tube set having one
stage of tuned-radio frequency, a regenerative detector,
one stage of transformer-coupled audio, and two stages of
resistance-coupled audio. Volume control through regeneration. Two tuning dials.

219. NORDEN-HAUCK SUPER 10—Ten tubes; five stages of
tuned radio frequency, detector, and four stages of choke-
and transformer-coupled audio frequency. Two controls.
Price $495.00.

220. BROWNING-DRAKE—Five tubes; one stage tuned
radio frequency (Rice neutralization), regenerative detector
(tickler control), three stages of audio (special combination
of resistance- and impedance-coupled audio). Two controls.

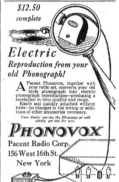

WHAT KIT SHALL I BUY (*Continued*)

221. LR4 ULTRADYNE—Nine-tube super-heterodyne: one
stage of tuned radio frequency, one modulator, one oscillator,
three intermediate-frequency stages, detector, and two
transformer-coupled audio stages.

222. GREIFF MULTIPLEX—Four tubes (equivalent to six
tubes): one stage of tuned radio frequency, one stage of
transformer-coupled radio frequency, crystal detector, two
stages of transformer-coupled audio, and one stage of
impedance-coupled audio. Two controls. Price complete
parts, $50.00.

223. PHONOGRAPH AMPLIFIER—A five-tube amplifier device having an oscillator, a detector, one stage of transformer-coupled audio, and two stages of imp dance-coupled
audip. The phonograph signal is made to modulate the
oscillator in much th: sam: manner as an incoming signal
from an antenna.

USE THIS COUPON

RADIO BROADCAST SERVICE DEPARTMENT
Garden City, New York.
Please send me information about the following kits in-
dicated by number:

....

....

....

....

Name..

Address ....
          (Number)   (Street)

....
          (City)   (State)
ORDER BY NUMBER ONLY. This coupon must
accompany each order.
                                        RB727

SET YOUR CLOCKS

THE correct time is now available nightly
from WEAF, WEEI, WJAR, WGY, WTAM, WWJ,
WGN, WCCO, WTIC, WFI, WSAI, WCAP, WOC, and
KSD. Through an arrangement with the Howard
Watch Company the time will be broadcast
simultaneously through these stations each week-
day night at 9:00 o'clock and on Sunday evenings at 9:15 Eastern Time. Station WJZ and network is likewise broadcasting the time, at 7:00
and 10:00 o'clock in the evening—or as close to
these hours as its program permits. It seems absurd to confine these valuable announcements
merely to one or two times in an evening.

KHJ COMPLETES 10,000 HOURS ON AIR

WITH the signing off of KHJ at Los Angeles on
the completion of its fifth anniversary por-
gram on April 13, the pioneer station of the west
coast completed 10,000 hours on the air during its
first five years of broadcasting. The present daily
schedule omits the afternoon hours but the station is on the air Tuesdays to Saturdays inclusive
from 6 to 10 P. M.; Sundays from 10:30 to noon
and from 7 to 10 P. M.; Monday is silent day.

TECHNICAL INFORMATION INQUIRY
BLANK

Technical Service,
RADIO BROADCAST LABORATORY,
Garden City, New York.

GENTLEMEN:
    Please give me fullest information on the
attached questions. I enclose a stamped
addressed envelope.

☐ I am a subscriber to RADIO BROADCAST,
and therefore will receive this information
free of charge.

☐ I am not a subscriber and enclose $1 to
cover cost of the answer.

NAME.................................

ADDRESS.........................R. B. Jy.

Drum notes not only heard —
but identified

THOUSANDS of radio listeners will now realize for the first time that radio orchestras have drums when they hook up this new, improved Crosley Musicone.

As originally produced the Musicone startled the radio world, eclipsing the old type horn and squeaky speaker.

Today, the new Musicone with its latest refinements and improvements correspondingly leads its host of imitators.

Prepare for a real surprise when you hear this amazing device with its beauty and fidelity of treble reproduction—clarity and breathless reality in middle tones—richness and resonance of bass. Today—infinitely bettered and superlatively developed, the Musicone is the world's finest loud speaker—and, at such extremely low prices, it's the world's greatest radio value.

The Crosley patented actuating unit (and *not* the cone) is the secret. There's nothing else like it.

Write Dept. 20 for descriptive literature.

The improved
~~CROSLEY~~
MUSICONE

| SUPER-MUSICONE 16 *inch Cone* $14.75 | THE CROSLEY RADIO CORPORATION POWEL CROSLEY, JR. *Pres.* CINCINNATI, OHIO *Prices slightly higher, west of the Rocky Mountains* | ULTRA-MUSICONE 12 *inch Cone* $9.75 |

BA

350 MILLIAMPERES

Complete Battery Elimination!

THIS one tube—the latest Raytheon achievement—provides noiseless, dependable ABC power from any light socket. The radio constructor can build a first class receiver using any of the standard radio circuits and incorporating this tube (with transformer, filter, etc.) to furnish *all* the power. Its 350 m.a. and 200 volts is more than adequate for the largest receiver, using power amplification.

Manufacturers of receivers and eliminators employing the BA

RAYTHEON A
2½ Amperes

tube will announce their new developments in the near future. If they bear the Raytheon seal of approval you are assured of design and construction that has successfully met Raytheon's rigid standards.

Complete technical information on design and operation of series filament receivers using this remarkable tube will be sent free by our Technical Service Bureau on request. Ask for Radio Power Bulletin SF-1. You may obtain the tube and approved parts at your dealer's. Tube Price—$7.50.

TYPE BH—85 m.a.

RAYTHEON BH—85 m.a. is designed for heavy duty applications of light socket power. Its introduction made possible the remarkable developments of power amplifiers and complete A. C. operated receivers using 199 tubes connected in series.

Rating: 85 m.a. output at 200 volts.

TYPE B—60 m.a.

RAYTHEON B—60 m.a. is the recognized standard for most of the B-power units now in use. Within the service for which it is rated, this rectifier is not surpassed by any rectifier made.

Rating: 60 m.a. output at 150 volts.

The Last Word in High Current, Low Voltage Rectification

A new type of rectifier, revolutionary in principle—appearance—performance. It is compact and simple, contains no liquids or filaments, and is enclosed in an unbreakable metal casing. The size and ability of Raytheon A makes possible the *smallest and most efficient* battery chargers and B power units. Leading radio manufacturers are designing apparatus to make full use of this new discovery.

RAYTHEON MANUFACTURING COMPANY
Cambridge, Massachusetts

THE HEART OF RELIABLE RADIO POWER

THE COUNTRY LIFE PRESS, GARDEN CITY, NEW YORK

Cunningham
RADIO TUBE

Round out the circle of vacation joys—

For real summer enjoyment you cannot do without Radio. To increase that pleasure, Cunningham Radio Tubes are most essential to your set.

No matter where you plan to be,—the mountains, the seashore or the back country—you can depend upon Cunningham Radio Tubes to bring you the world's best music with song, dance and laughter.

Your nearest Radio dealer will tell you the best combination of Cunningham Tubes for your particular set.

E. T. CUNNINGHAM, Inc.
NEW YORK   CHICAGO   SAN FRANCISCO

WHAT you want most in radio is smooth, well-rounded tones with just enough volume to make reproduction *real*.

The new ¼ amp. oxide filament ZP 201 A all-socket Power Tubes now contribute this missing quality to radio. ZP 201 A power tubes take just enough current from your battery to assure smoothness of tone. Differing from the majority of other tubes there is no excess—no "crowded" current active in ZP 201 A's to set up the blasting microphonic noises that distort true tone and give volume an artificial rasp. ZP. 201 A's in every socket—*installed without re-wiring*—give you music rich and clear, with as much power as you'll ever need.

The extremely conservative operating characteristics of ZP 201 A power tubes effect an appreciable economy in "A" power—reduces recharge bother and expense. And through elimination of magnesium coating they perform at topmost efficiency . . . actually improving with service.

A better, longer life power tube at only $2.50 each.
Ask for a demonstration at your nearest dealer.

ZETKA LABORATORIES, Inc., 73 Winthrop St., Newark, N. J.

ZETKA
The *Clear Glass* Tube

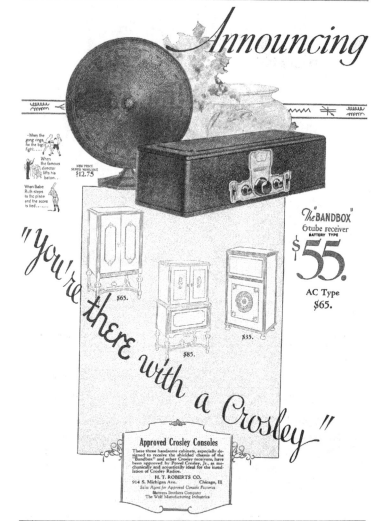

Announcing

NEW PRICE
SUPER MUSICONE
$12.75

- When the gong rings for the big fight.....

When the famous director lifts his baton...

When Babe Ruth steps to the plate and the score is tied.......

"you're there with a Crosley"

The "BANDBOX"
6 tube receiver
BATTERY TYPE

$55.

AC Type
$65.

$65.

$85.

$35.

Approved Crosley Consoles

These three handsome cabinets, especially designed to receive the shielded chassis of the "Bandbox" and other Crosley receivers, have been approved by Powel Crosley, Jr., as mechanically and acoustically ideal for the installation of Crosley Radios.

H. T. ROBERTS CO.
914 S. Michigan Ave.   Chicago, Ill.
Sales Agent for Approved Console Factories
Showers Brothers Company
The Wolf Manufacturing Industries

RADIO BROADCAST. August, 1927. Published monthly. Vol. XI, No. 4. Published at Garden City, N. Y. Subscription price $4.00 a year. Entered at the post office at Garden City, N. Y., as second class mail matter. Doubleday, Page & Company, Garden City, N. Y.

the ~~CROSLEY~~ "Bandbox"
and other new radio reception equipment
for the complete enjoyment of the 1927-28 radio season

Ever since Crosley entered the radio field their methods and developments have created a leading place for Crosley radio receivers.

Recent court decisions now greatly clarify radio patent situation

And now—completely available to Crosley—and amplifying Crosley supremacy in fullest measure, are the enormous resources, discoveries and ideas, embodied in patents of the Radio Corporation of America, The Westinghouse Co., The General Electric Co., and The American Telephone and Telegraph Co.—under which Crosley is now licensed to manufacture.

No wonder the new Crosley receivers are in the forefront, their amazing efficiency acknowledged and demanded by that section of the radio trade which insists on the latest and best at all times.

THE "BANDBOX"
It is a new 6-tube set of astonishing sensitiveness.

Many exceptional features commend the "Bandbox."

The metal outside case, 'tho keeping out strong local signals effectually enough, did not fully satisfy Crosley ideals of fine radio reception. Signals must be kept in order *inside* the set.

TILT-TABLE MUSICONE
$27.50

Although Musicones improve the reception of almost any radio set they are perfect affinities in finish, beauty and reproductive effectiveness for Crosley Radios. A new model built in the form of a Colonial Tilt-table and finished in brown mahogany stands 3 feet high.

12-inch Ultra Musicone $9.75   16-inch Super Musicone $13.75

Coils and condensers are like families living in a row of houses with no fences between. The children run around the yards; they meet, mix it up, quarrel and squabble. No harmony.

Magnetic and electric fields are the offspring of coils and condensers. With no fence between, they, too, run around the house, mix it up, quarrel and squabble. Howls and squeals result.

So, to keep each "family" or field of individual coils and condensers separated, metal fences are erected (copper fences for the coils) and the individual parts of the Bandbox are shielded as only found in the highest priced sets.

For fans who love to go cruising for faint, far-away signals the "Acuminators" intensify weak signals like powerful lens revealing distant scenes.

The "Bandbox" employs completely balanced or neutralized radio frequency stages, instead of the common or losser method of preventing oscillation. In presenting this important feature Crosley is exclusive in the field of moderate price radio.

Volume control is another big "Bandbox" feature. Signals from powerful local stations can *Volume for dancing* be cut from room filling volume to a whisper. Each "Bandbox" is fitted with a brown cable containing colored rubber covered leads for power and other connections.

The frosted brown crystaline finish harmonizes with the finest furniture and matches the frames of Musicones and the casing of the power unit. The bronze escutcheon creates an artistic control panel.

Withal, in the beautiful appearance and modest size of the "Bandbox" is the utmost in adaptability to requirements of interior arrangement or decoration. The outside case is easily and quickly removed *Soft and low thru volume control* for installation in console cabinets.

AC AND BATTERY OPERATION

The "Bandbox" is built both for battery and AC operation. The new R.C.A.—AC tubes make the operation of the set directly from house current both practical and efficient. In the AC set the radio stages and the first audio stage use the new R.C.A.—AC—UX-226 tubes. Filaments in these tubes are heated with raw AC current at proper voltage.

A Master Station Selector, with illuminated dial for shadowy corners, enables tuning for ordinary reception with a single tuning knob.

The UY-227, with indirectly heated emitter, is used with the detector. Power tube UX-171 at 180 volts plate.

There is no AC hum. The new R.C.A. Radiotrons do the work.

The power supply unit is a marvel of radio engineering ingenuity. Half the size of an ordinary "A" storage battery, it supplies A, B and C current direct from lamp socket to tubes.

Models for 25 and 60 cycles. Snap switch shuts down set and power unit completely.

~~CROSLEY~~
RADIO

Crosley Radio is licensed only for Radio Amateur, Experimental and Broadcast Reception.
RCA Radiotrons are supplied at standard prices, with each Crosley Receiver.
Prices slightly higher west of Rocky Mountains.

THE CROSLEY RADIO CORPORATION

Cincinnati, Ohio   Powel Crosley, Jr., Pres.

JUL 14 '27

RADIO BROADCAST

WILLIS KINGSLEY WING, Editor

AUGUST, 1927

KEITH HENNEY
Director of the Laboratory

EDGAR H. FELIX
Contributing Editor

Vol. XI, No. 4

CONTENTS

AMONG OTHER THINGS. . .

RADIO, or at least, its basic principles, be it known, is gradually influencing other fields, and one of the most interesting examples of this is told by James Millen in our leading article this month which describes the ingenious weightmeter developed by Albert Allen of Boston. While the theory of operation of this remarkable device is not complicated, many seemingly insurmountable problems were encountered and conquered before the instrument was practical enough for the strict requirements of commercial use.

A WORD about some of the authors in this issue may be of interest. Herbert G. Reich, whose shielded neutrodyne is described in these pages, is a member of the Department of Physics at Cornell University. A. V. Loughren, whose paper on vacuum tubes appears in this issue, is in the research laboratories of the General Electric Company at Schenectady.

THE Directory of Manufactured Receivers which appears in this number, beginning on page 250, should prove of wide interest. Tabulated data on receiving sets have appeared before, of course, but the information has been sketchy while our Directory is as complete in detail as it is possible to make it. From this listing, it is possible to determine much about the circuit of any receiver listed, how its volume is controlled, how many and what sort of tubes there are and how much current they take, what accessories are supplied and, important enough, the size of the cabinet. What is more, by using the Service Department coupon, more complete or additional information will be forwarded each inquirer with a minimum of trouble to him. The Directory will appear regularly and will be improved each month with additions and corrections. Naturally it has not been possible to include even nearly all the receivers on the market, but we believe that, with the monthly additions, the list will prove an adequate source of information to such as may find it necessary to refer to it.

WE SHALL soon publish a paper by B. F. Miessner, whose work with 2-amp. a.c. tubes has become so well known. There is wide interest in the characteristics and use of these tubes and Mr. Miessner has prepared a very interesting paper indeed. . . . The Laboratory has designed an inexpensive and extremely simple tube tester which for some time has been put to good use out here at Garden City. Complete constructional and operating information about this tester will appear soon.

—WILLIS KINGSLEY WING.

Doubleday, Page & Co. MAGAZINES	Doubleday, Page & Co. BOOK SHOPS	Doubleday, Page & Co.	Doubleday, Page & Co. OFFICERS
COUNTRY LIFE	(Books of all Publishers)	GARDEN CITY, N. Y.	F. N. DOUBLEDAY, President
WORLD'S WORK	LORD & TAYLOR BOOK SHOP	NEW YORK: 285 MADISON AVENUE	NELSON DOUBLEDAY, Vice-President
GARDEN & HOME BUILDER	PENNSYLVANIA TERMINAL (2 Shops)	BOSTON: PARK SQUARE BUILDING	S. A. EVERITT, Vice-President
RADIO BROADCAST	NEW YORK: GRAND CENTRAL TERMINAL	CHICAGO: PEOPLES GAS BUILDING	RUSSELL DOUBLEDAY, Secretary
SHORT STORIES	38 WALL ST. and 520 LEXINGTON AVE.		
EDUCATIONAL REVIEW	848 MADISON AVE. and 166 WEST 32ND ST.	SANTA BARBARA, CAL.	JOHN J. HESSIAN, Treasurer
LE PETIT JOURNAL	ST. LOUIS: 223 N. 8TH ST. and 4914 MARYLAND AVE.	LONDON: WM. HEINEMANN LTD.	L. A. COMSTOCK, Asst. Secretary
EL ECO	KANSAS CITY: 920 GRAND AVE. and 206 W. 47TH ST.	TORONTO: OXFORD UNIVERSITY PRESS	L. J. McNAUGHTON, Asst. Treasurer
FRONTIER STORIES	CLEVELAND: HIGBEE CO.		
WEST	SPRINGFIELD, MASS.: MEEKINS, PACKARD & WHEAT		

DOUBLEDAY, PAGE & COMPANY, Garden City, New York

Copyright, 1927, in the United States, Newfoundland, Great Britain, Canada, and other countries by Doubleday, Page & Company. All rights reserved.
TERMS: $4.00 a year; single copies 35 cents.

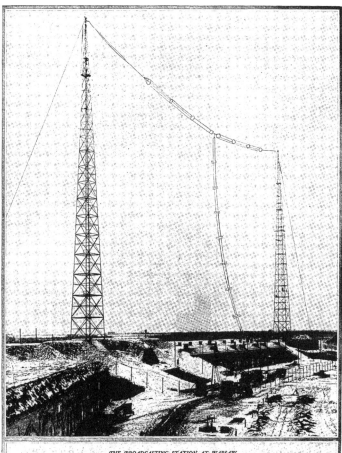

THE BROADCASTING STATION AT WARSAW

Located seven kilometers from Warsaw, this station is installed at Fort Mokotow. The Fort was built twenty-five years ago by the Russians, but has not been used for its original purpose for some years. The power is 8-kw. to the plates of the tubes. The antenna current is 30 amperes on a frequency of 269 kc. (1111.5 meters). The crystal range of the station is said to be 75 miles

RADIO BROADCAST

VOLUME XI NUMBER 4

AUGUST, 1927

Saving Paper!

Millions of Dollars are Saved Annually in the Paper and Rubber Industries by the Use of a Magic Device Operating With Radio Principles

By JAMES MILLEN

WHILE rapidly assuming a deserving place as a separate science in itself, radio is fundamentally a branch of electrical engineering. During the last few years, in which such great strides have been made toward the present-day ideas of perfection, many new scientific theories have been developed. Instead of these developments coming from the parent science, they have come from the engineers and other workers primarily interested in radio. At first, many electrical engineers were wont to consider radio as a mere toy or public plaything not worthy of their attention. Then, as time passed and startling progress in the new field became apparent, well-known electrical engineers began to devote their skill to the furtherance of development in the new field. Where "cut and try" ruled before, calculus and electrical engineering theory began to come to the front. Thus resulted the neutrodyne and many other genuine developments.

But, no sooner had these fatherly gentlemen begun to aid their offsprings in the perfection of a new "toy," than they began to realize that much of the theory involved was not only new and unique, but that it might be applied to their own established science as a panacea for long-endured ills.

Perhaps the first indication the public had of progress along such lines was the press reports of a year or two ago regarding the use made of radio by large electric light

and power companies to forewarn them, by means of static disturbances, of approaching thunder showers so that they might be prepared in advance for a sudden increased demand for power for lighting.

But while many such applications of radio principles to industrial use have been and are yet to be made, few indeed are as vitally important and yet seemingly impossible as that conceived some two or three years ago by a Maine man, then work-

THE "RADIO SCALES" IN OPERATION

Albert Allen, inventor of an instrument known as the weightmeter, shows a foreman in the paper mills of the Eastern Manufacturing Company at Bangor, Maine, how to weigh a strip of moving paper without actually touching it. Formerly it was necessary to tear a sample piece for weighing, with an ordinary pair of scales; from the edge of the moving paper

ing as a cadet engineer in a large paper mill. To-day his devices are in use in paper and rubber mills throughout the country, automatically regulating the uniformity of the mills' sheet products.

Several years ago, Albert Allen, like many other electrical engineers, realized that maybe there was something practical and worth while to this radio stuff after all.

He constructed a number of receiving sets and did much experimental work with them. One night, while his set was tuned to one of the local broadcasting stations, he took a piece of paper and inserted it between a pair of the plates in one of the tuning condensers that seemed on the point of short-circuiting. The result was that he immediately heard a different station from his loud speaker. Being a true electrical engineer, he realized that such was as might be expected. The paper, because of its different dielectric properties than air, had changed the electrical capacity of the condenser, and likewise the wavelength to which his set was tuned, bringing in a new station.

Later, when at his work in the mill, he stopped to watch the long sheet of new paper passing through numerous rollers and machines in an endless strip. Every little while, the foreman would tear a piece of paper from the strip, mark it, and send it off to the laboratory. Then a report would come back and he would readjust his machine. The purpose of this act was

TESTING THE EFFICACY OF THE WEIGHTMETER
The inventor of this remarkable device, Albert Allen, is to the right of this group in the laboratory. To the left is William Ready, of the National Company, while the author is in the center

the variation in weight of the stock being passed through it, but, by means of suitable relays, automatically controlled the paper machines so as to keep the stock at any desired weight.

HOW THE WEIGHTMETER WORKS

SOME may be interested in a more explicit explanation of just how the weightmeter functions—how a minute change in the capacity of an electrostatic condenser can result in the control of a several-horsepower motor operating the calenders in a rubber mill, for instance.

The fundamental principle by which a very minute variation in the electrostatic capacity of a condenser may result in a readily measurable change of current, is illustrated in Fig. 1. Here we have an oscillator, "A," coupled, by means of the coils "B" and "C," to a tuned circuit "D." Tuning the circuit "D" to resonance with the oscillator "A" results in a deflection of the thermal meter "E." By properly arranging the circuit constants, the sensitivity of the weightmeter, or in

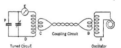

FIG. 1
The radio fan will readily recognize the elementary circuit of the weightmeter to consist of nothing more than an oscillator, wavemeter, and coupling circuit

to determine the weight per given area of the paper—whether it was up to the required standard, or overweight. If overweight, the company was losing money; if underweight, the customer might refuse to accept.

But just imagine how crude and costly such a process was? The piece torn out might not be representative. The strip was damaged, much time was wasted, and most important of all, all sorts of variations might take place between tests without anyone being any wiser.

Young Allen recalled his experience with his radio set and the piece of paper. Why not, thought he, apply the same principle to providing a continuous, instantaneous, check on the paper conditions. For two years, he devoted his entire time to the problem. At first he experimented alone, then, as success seemed certain, and as the president of the paper mill realized the tremendous value of his idea, he joined with

a staff of competent research engineers in the engineering laboratories at Harvard University.

Finally, finished working models were developed to the point where they were considered reliable and practical enough to be installed in the paper mills of the Eastern Manufacturing Company, at Bangor, Maine.

Then came necessary re-design as a result of the information gained under operating conditions. Trouble with dust accumulation, trouble due to variations in temperature causing greater variations in instrument readings than variations in paper weight, trouble due to the tremendous variations in the line voltage from which the instruments were operated—all had to be overcome. Troubles of all kinds were encountered, but even so, the new method was far better than the old.

But every cloud has its silver lining, and it was not long before the difficulties were overcome and the initial idea carried a step further, so that the "weightmeter," as the new instrument was called, not only recorded

this case, the response of the thermal meter, to minute capacity changes, can be any desired amount, from the condition illustrated by the peaked curve, "A," in Fig. 2, where even the very slightest capacity change results in a marked current change, to the other extreme condition illustrated by the low broad curve, "B," Fig. 2.

Now compare the elementary circuit diagram, Fig. 1, with the actual apparatus illustrated in the photograph on page 201. The oscillator "A," coupling circuit "B" and "C," and the tuned circuit "D," are located in the same case. The precision condenser, shown more clearly in a photograph on page 202, is used to tune the circuit

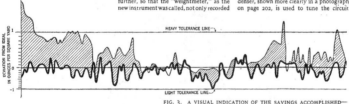

FIG. 3. A VISUAL INDICATION OF THE SAVINGS ACCOMPLISHED—
The straight center line shows the desired thickness of rubber, most economical production resulting from keeping the thickness as closely as possible—be tolerated. The heavy curve was made at a rubber mill when the weightmeter was in use, while the lighter one was made when no weightmeter—

"D" (Fig. 1) into a condition just off resonance. This condition is indicated by the
arrow on the side of the curve "A" in Fig.
2. In actual practice, either side of the resonance curve may be employed.

In parallel with the precision variable
condenser are two mechanically adjustable
plates forming a condenser. These are indicated by the lettering "P," in the photograph and Fig. 1. If the capacitance of
this condenser should change, due to a variation in the weight of the paper or other
material passing through it, the tuned circuit "D" Fig. 1, either comes more closely
into or further out of resonance, depending
upon which side of the resonance curve is
being used and whether the capacity change
is positive or negative, thus changing the
reading of the thermal ammeter.

By employing the recording type meter
shown alongside the weightmeter in the
photograph, a continuous curve may be
traced, showing variations in weight of the
material with time. A sample curve is shown

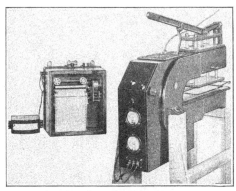

RECORDING METER AND WEIGHTMETER
The recording meter to the left is used to make curves similar to that of Fig. 3. The smaller
instrument to the extreme left is an indicating meter, calibrated to read weight in desired units.
The material to be weighed passes between the two horizontal invar plates marked "P," to the
right. The upper invar plate may be varied when desired by the long handle atop the instrument

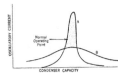

FIG. 2
The sensitivity of the weightmeter may be
varied by changing the shape of the resonance
curve. With a steep curve, a very minute change
in weight of material (and thus condenser
capacity) will cause a fairly large change in
meter reading

in Fig. 3. The straight center line indicates
the weight desired to be maintained. The
light curve shows a rather wide discrepancy between the weight desired and the
results obtained by hand feeding and the
old method of actual weighing. The heavy
curve indicates the results obtained when
using a weight meter control. The shaded
area between the two sets of curves illustrates the saving in material effected by the
new control method. In addition to the
actual saving in material, the product is
ever so much more uniform. It will be noticed that the "light" tolerance line is

nearer the "ideal" line than the "heavy"
tolerance line. This variation in the two
linear distances, however, does not represent a difference in upper and lower tolerance weights, but is the result of the changing slope of the resonance curve in either
direction.

The many irregularities in the two curves
in Fig. 3, are due to "feeding." Every time
more rubber is fed, the stock tends to gain
in weight, even though the calendar adjustments remain unchanged.

The absolute necessity for precision
built apparatus cannot be over emphasized.
Apparatus so substantial in design, and
expertly made, as to hold its calibration
under all kinds of abuse, in all kinds of
climate and conditions of service, is absolutely essential if any appreciable saving is
to accrue from its use.

Hence the massive cast iron frame for the
fixed condenser, and also the invar steel
rods and supports. If some other metal
were to be substituted for the expensive

invar, the least variation in temperature of
the condenser would change its capacity
far more than a little change in weight of
the stock being measured. In the earlier
weightmeters, compensating thermostatic
condensers were used to correct any changes
in temperature of the main condenser.
Such a system was found, in practice, to be
excessively complicated. The result was the
design of the present invar steel unit which
is not affected by temperature variations.

Another seemingly overwhelming difficulty encountered with the first commercial
installation was the tremendous variations
in the line voltage from which the oscillator
received its power. Fluctuations of as much
as 40 or 50 volts were not at all uncommon.

Ballast tubes, separate alternators, line
transformers, and many other complicated
and expensive remedies were tried. The
final solution arrived at, however, was not
only superior in performance to the other
methods, but required no additional equipment of any type.

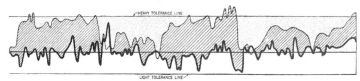

—BY THE USE OF THE WEIGHTMETER AT A RUBBER MILL
—to this value. The upper and lower horizontal lines show respectively the maximum and minimum deviation in ounces per square yard that can
—was used. The shaded portions represent material saved by the use of the weightmeter. This curve represents 5950 yards of rubber $\frac{1}{16}$" thick

As a power oscillator of the type used in the weightmeter is in reality a miniature radio transmitting station, one might imagine that radio reception in the vicinity of a mill in which the weightmeters are employed would not be all that could be desired. This apparatus, however, in no way interferes with radio reception. The oscillator is tuned to a wavelength slightly in excess of 600 meters so as to be outside of the broadcasting frequency band and, also, the oscillator is so completely shielded that an exceedingly sensitive super-heterodyne located but a hundred yards away is unable to pick up the carrier. The often long exposed leads to the indicating meters do not carry radio-frequency current as the thermal-junction, belonging to each meter, is located within the shielded oscillator cabinet.

Just what the saving amounts to in dollars and cents when the weightmeter is used depends greatly upon the quality and quantity of material being manufactured. Thus, in the production of a low grade of newsprint stock in a low-speed paper making machine, the saving effected per year per machine might be as low as $10,000 or so. On the other hand, where a highly priced stock is being manufactured in a rubber factory, such as tire tread for automobiles, the saving brought about as a result of automatic weightmeter control may be ten times as great.

In one automobile tire factory, the average last year saving per weightmeter employed was approximately $80,000. As this particular rubber company employed many machines, it will be seen that automatic control was of considerable financial benefit, aside from its fundamental purpose of aiding in the making of a uniform product.

The weightmeter is but one of many

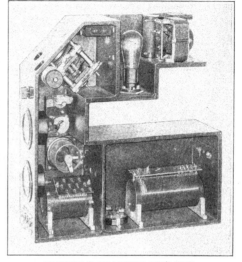

THE INTERIOR OF THE OSCILLATOR BEFORE WIRING
The case is of cast aluminum. The transformer at the upper right supplies A and B power for the UX-210 tube. The precision variable condenser is connected in parallel with the measuring plates and used to bring the tuned circuit to the desired point on the resonance curve

applications of radio principles to industrial purposes—principles which, before the advent and recognition of radio broadcasting as a major art, were, perhaps, little understood. That the development of this instrument is not an isolated instance of such progress, we have only to consult our journals for adequate and frequent proof.

The ideas obtained from the radio industry by Albert Allen and so successfully applied toward the elimination of a seemingly almost unsurmountable handicap encountered in the manufacture of paper and rubber—the difficulty of accurately weighing moving objects—is but a vivid example of such progress.

A DIFFERENT SET-UP OF APPARATUS
The equipment is basically the same as that shown in the photograph on the preceding page. The oscillator is the instrument to the right

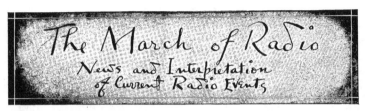

Revolutionary Methods Which Fail to Materialize

ILL-CONSIDERED publicity statements are the bane of the radio industry and the despair of newspaper and magazine publishers. Receiving, as we do, numbers of poorly prepared and inaccurate statements from radio manufacturers in every mail, we plead for mercy, not only to spare our weary eyes but to discourage their harmful influence on the well-being of radio.

Almost each week, a revolutionary invention is heralded in the press to disturb the confidence of the hesitant radio buyer. Many of these inventions are promises never performed; others announce hoary practices and inventions as new discoveries; all of them conspire to discourage the non-technically informed from purchasing his radio to-day in order that he may have the advantages of these doubtful but glorified inventions to-morrow.

In the pile of material before the writer are three announcements from one of the largest manufacturers in the radio and electrical industry. Considering the source of these statements, they would certainly be regarded as authoritative by any newspaper editor who did not possess an extensive background of technical knowledge and an intimate familiarity with the history and progress of the radio art. Since most newspaper men quite properly lack such detailed knowledge of radio, they naturally make much of these extravagant statements, although they belong only in the editor's waste basket.

The first of these statements announces the exponential horn as "a new invention," which received "its first public test" on May 12, 1927. Then follows a glamorous account of the magnificent improvements in tone quality which the "discovery" brings to the world. The statement totally overlooks the fact that the exponential horn, developed ten years ago, has been used in simple public address systems long before the first broadcasting station went on the air;

that the merits of the exponential horn have been discussed before technical societies time and again; that exponential horns are in use in thousands of homes as a part of the orthophonic phonograph and of certain power loud speakers. But how many potential loud speaker buyers decided, upon reading that statement in the press, to postpone buying a radio or a reproducer until they might have this new horn, said to "broadcast the natural human voice or tones of musical instruments without distortion for the greater part of a mile?"

The second statement deals with the transmission of radio-frequency energy, heralding a new era of wireless transmission of power. It discloses nothing, however,

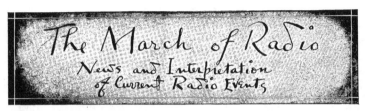

TELEPHONING TO SHIPS AT VANCOUVER, B. C.
The installation in the Vancouver Merchants' Exchange Building, with J. E. Harker, chief operator. More than forty British Columbian tug boats are equipped with 50-watt telephone sets, operating on 1507 kc. (199 meters). Any one aboard ship can operate the sets, the average range of which is from 60 to 150 miles, giving loud speaker volume at the receiving station. The sets are used to facilitate the delivery of lumber being towed down the waterways. Much of this is sold while enroute. It is the only service of its kind in the world and is operated by the radio branch, Department of Marine and Fisheries

which any amateur having a suitable wavemeter, cannot perform in his own home for the delectation of his friends.

The third effusion by this eminent manufacturer concerns itself with a new system of frequency modulation for broadcasting stations which enables one-half-kc. separation. Approximately 1900 broadcasters can operate simultaneously without sharing waves or splitting time. It is pointed out that this will solve the congestion problem which the Federal Radio Commission is now trying to overcome.

Extraordinary, absurd, and unwarranted claim! This invention entirely neglects the fact that this invention, before it can have any bearing on the work of the Radio Commission, must undergo the following tedious processes:

(1) Laboratory research and development; (2) standardization of transmitters and receivers utilizing the new principle and suited to quantity production; (3) remodeling of every existing broadcast transmitter to one using the new system and (4) the probable replacement or alteration of every existing receiving set for one of the new type receivers at a cost to the listening public of perhaps a billion dollars or so.

Each one of these problems has its share of hazards and pitfalls which those who issued the statement have reason to know as well as anyone. To state that the invention of frequency modulation has any bearing on the solution of present-day broadcasting problems is both absurd and well-nigh malicious, since it encourages small broadcasters, now gradually dying from neglect by the listener, to redouble their expiring efforts to remain on the air.

Some day, the radio industry will learn to restrain its blabbing tongue until it has the goods, ready to sell, on the dealers' shelves. The automobile industry does not announce a new type of car until every dealer all over the country has his demonstrator on hand. Thereby, motor car manufacturers capitalize public interest, aroused by their announcements, and convert that interest directly into cash sales. The radio industry, on the other hand, childishly chooses to make its announcements prematurely, thus bringing to an immediate stop any buying activity which may be manifest at the time. Then, months or years later, when it finally has its new product ready for sale, it cannot again stimulate any real public interest because the publicity men have already done their croaking.

The publicity disease has gone too far to be quickly remedied. Some encouragement might be given by a more conservative attitude on the part of newspapers. Instead of accepting as gospel truth the publicity statements offered them, they might first check with their own radio staffs or, in absence of a competent radio staff, take advantage of high-grade radio feature syndicate services which offer them the protection of expert and conservative censorship.

The Commission Regulates Broadcasting for the Broadcasters Not the Listener

THE Federal Radio Commission, still occupying the center of the radio stage, has made its first comprehensive re-allocation, effective June 15. In the congested areas, stations are now separated by at least fifty kc.; time-sharing is enforced upon the lesser stations and some interesting power reductions have been made. These power reductions are announced by the Commission as experimental. We hope that the experiment may soon be concluded and plenty of power be permitted to all our outstanding stations—which is not now altogether the case.

The enforced curtailment of the activities of the lesser stations has transferred much howling from the broadcast listener's receiver to the crowded doors of the Radio Commission. In deciding upon the case of individual stations, service to the public, as determined by program standards, quality of transmission, length of service, and orderly conduct through the late chaos, have been taken into consideration. Clean channels have been given to recognized old-timers and we are pleased to see the principle of priority of service, urged in these columns for more than a year, actually put into practice.

The New York allocation, for example, gives WNYC, WEAF, WJZ, WOR, WHN (with WQAO operating only on Sundays), WMCA (sharing with WEBJ which uses only six hours a week), and WABC (sharing with its own transmitter, WBOQ) practically undivided time. The most popular stations are therefore given a full opportunity to

serve their respective audiences which, combined, probably include 90 per cent. of all listeners in the area. In the questionnaire which so many of our readers filled out, considerably more than half the New York listeners expressed themselves identically as to their favorite stations, placing WEAF first, WJZ second, and WOR third, while more than 96 per cent. of the replies included at least one of these three stations. Only WNYC, which ranked eleventh in popularity, seems to have received curiously favored consideration on the part of the Commission in being included in the list of seven stations with virtually exclusive channels.

A point brought out with striking emphasis by our questionnaire, regarding the stations which have been compelled to divide time four or five ways on the less desirable channels, is that they are nobody's favorites. Some of our readers listed as many as ten "favorite" stations, yet 21 New York stations were not mentioned as favorites by a single listener. These same 21 insignificant stations are most prominent in the listings of stations which the listener wished eliminated, receiving 25 per cent. of the demands for a "permanent signing off." And, mind you, this 25 per cent. does not include such widely disliked stations as WHAP, WKBQ, and WLWL, which three stations alone polled 20 per cent. of the demands for elimination. Hence, the Commission's judgment, in giving the two National Broadcasting chain stations, WEAF and WJZ, and the five independents,

WNYC, WOR, WHN, WMCA, and WABC, virtually exclusive channels, is fully justified.

In the Chicago area, KYW is the only station favored with an exclusive channel. Our Chicago readers divided their first choices between KYW and WGN almost to the exclusion of other stations. Station WMAQ is the strongest second choice and WEBH the strongest third, while, strangely enough, neither of these latter two polled a substantial number of first choices. The superior standing of these four stations over all the others in the Chicago area is not, however, recognized by the Commission as in the case of the New York favorites. These four are treated no better than such wholly unpopular stations as WCFL, which is practically tied for first place with WGES in the demands for stations to be eliminated, and WJAZ, running a strong third in the unpopularity contest because of the prejudices aroused by its leadership in upsetting broadcasting conditions. It seems that the only discrimination made in the Chicago area, in deciding whether a station should divide time two, three, or more ways, is power and not service to the listener.

Considering the national situation as a whole, many assignments have been made which are impossible. Let us examine a few frequencies. The first on the list is 550 kc., shared by KSD and WMAK, the latter's 750-watt carrier certainly being sufficiently strong to cause serious interference with the former, at least in winter. The next channel, 560 kc., finds WNYC in the same bed with WHO, 5000 watts, which very likely will cause a steady whistle in New York. At 580 kc., we find an untenable assignment, WMC, Memphis, Tennessee, and WCAE, Pittsburgh, both 500-watt stations. The popular Philadelphia stations, WOO and WIP, must contend with a thousand-watt station, WOW, in Omaha, Nebraska, which probably limits the range of the former stations to their limited high-grade service area only. Two Baltimore stations, WCAO and WCBM, will not find WMBF's carrier, which pumps strongly into their locality, any improvement on their programs; nor will WCSH, the principal reliance of a large New England area, find its programs favorably affected by WSAI's powerful carrier. Unless WEW is sharing time with WOC, we cannot understand how either St. Louis, Missouri, or Davenport, Iowa, can be expected to enjoy its best local station with another powerful and nearby station sharing the 850-kc. channel. Nor can a listener enjoy WKRC's program in Cincinnati with WBZ, only six hundred miles to the east, sending its 15,000-watt carrier out on the same frequency. Station KMOX will

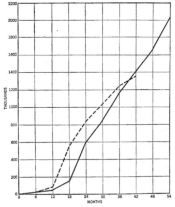

GROWTH OF LISTENERS IN ENGLAND AND GERMANY
The two curves, the plain one representing England and the dotted one Germany, show how licenses were issued for the reception of broadcast matter in those countries. A period of four and a half years is considered

A CORNER OF THE BUTTE, MONTANA, RADIO CLUB ROOMS

Among the complete equipment there is a broadcast receiver, a long-wave Reinatz receiver, a 50-watt short-wave transmitter using the call 7NT, and several short-wave receivers. The president of the Club is Judge W. E. Carroll; C. J. Trauerman is publicity director, M. E. Cooper, secretary, and J. R. Bartlett is vice president

probably sound like old times with WBAK in Harrisburg, Pennsylvania, on the air at the same time on the same frequency.

But, enough of this; all *local* heterodyning and cross-talk have been eliminated. Heterodyning carriers from distant stations will not cause serious annoyance during the summer, because of attenuation of the carrier. But these June 15 assignments leave us far from a solution of the broadcasting problem.

One might conclude that the Radio Commission has failed in its duty of regulating broadcasting in the interests of the listener because, on the face of things, it has fallen far short of eliminating the heterodyne squeal entirely. The Radio Commission has done the best possible job of trying to put a carload of apples into a bushel basket.

No matter how you juggle some 600 stations, many of considerable power, they cannot be accommodated on the broadcast band without heterodyning. We have preached the doctrine of station elimination consistently because it is the only practicable answer to the problem. The number of stations must be reduced to 200 or 225 so that the listener may have clear programs no matter where he is located. The only alternative is to cut the power of 400 to 450 stations to 50 watts and allocate only 75 or 100 stations to favored channels.

The Commission has elected to avoid the closing of any station, presumably because it does not care to have its prerogatives tested in the courts. It realizes that one successful injunction would bring chaos even worse than that of the last season and, with

it, the actual bankruptcy of the entire industry. The broadcasters have already shown that they act like small boys when the teacher is away—if the slightest fetters are placed upon the regulatory power. So the Commission is proceeding to do the impossible, namely, to accommodate the host of broadcasting stations in a frequency band which cannot hope to accommodate half of them properly; in effect, it is ruling broadcasting for the benefit of the broadcasting stations rather than for the broadcast listener.

If the Radio Commission must wear gloves, we wish they would choose boxing gloves and hand out a few knockout blows.

What Commissioner Bellows Thinks About DX Listening

COMMISSIONER Bellows is credited with saying that DX listening must be abandoned. The DX listener is frequently scored by the radio engineers, particularly those who have an appreciation of good tonal quality. They realize, as does everyone who secures really good quality, that the range limit of high-quality reception for a 500-watt station is some thirty miles and, for a 50,000-watt station, one hundred miles. Anything beyond this distance the engineer considers DX reception and frowns upon as unnecessary.

It is not generally realized, however, that these frowns shadow nearly 80 per cent. of the area of the United States! Only about 20 per cent. of our area lies within the high-quality reception range of any broadcasting station! If the radio business and the radio listeners of 80 per cent. of the United States are of no account, it is fair to set down DX listening as of no value.

Fortunately, the engineers do not rule the world. Some listeners are willing to listen to a station three or four hundred miles distant in spite of them. Indeed, a great many are compelled to listen to

A REPLICA OF MARCONI'S FIRST EXPERIMENTAL APPARATUS

With similar apparatus, Senatore Marconi demonstrated in 1895 the possibilities of "beam" transmission and reception, and later confirmed his results officially before representatives of the British Post-office and the military on Salisbury Plains, in September, 1896, when he communicated over a distance of 1¾ miles. The transmitter is at the left, and the receiver, at the right

stations at such "great" distances because the nearest high-grade broadcasting is at least that far away. In our recent questionnaire, we asked our readers to list the favorite out-of-town stations which they wish retained. More than 60 per cent. of New York listeners asked for the continuation of KDKA; well over 50 per cent. cast votes for WGY. Preferences continued in the following order: WBZ, WSB, WTAM, WLW, WOC, WLS, WPG, KYW, WMBF, WSAI, WEBH, WGN, and WBAL. The lowest of these were mentioned by more than 15 per cent. of our readers as "long-distance" favorites.

And still they say DX listening is of no importance!

Never Again Without Radio!

TWO continents waited breathlessly for news of the hero-aviator, Charles A. Lindbergh, as he braved the immense solitude of the transatlantic sky. But a few days before, that same sky had conquered Nungesser, that great fighting aviator, enveloping him in a blanket of complete mystery. There would have been neither suspense nor mystery had these intrepid fliers carried a radio transmitter.

Radio has repeatedly demonstrated its service to long-distance flying from the day, more than ten years ago, that Irwin's sos saved the crew of Vaniman's dirigible on the first transatlantic flight attempt. Lack of radio equipment nearly lost us the crew of the *Pn-9*. Radio made the whole world a by-stander as the *Norge* swept across the North Pole last summer. Where men venture into danger—whether it be transatlantic flight, polar expedition, or the conquest of equatorial jungle—radio has saved lives and removed isolation; without radio, they have vanished into heroic oblivion.

Why not profit from experience? We hope that never again will a transatlantic flight be undertaken without the protection which a radio transmitter affords. We cannot recall a single case in which a suitable radio transmitter has failed to bring prompt rescue to a plane which has made a safe forced landing on the sea.

Why We Need More Listeners

A NEWSPAPER report states that "one chain" of broadcasting stations will spend two million dollars for talent in the coming year. The cost for talent for a commercial hour ranges, on an average, between five hundred and two thousand dollars. Top-notch entertainers are said to receive from a thousand to two thousand dollars for a single studio appearance, while one jazz orchestra is booked for $1500 an hour. Still, one of the problems of radio is how the standard of programs may be improved. Commercial broadcasters cannot be expected to spend larger sums and to present better programs unless the numbers of the radio audience increase proportionately to their increased expenditures.

A most important problem of the radio industry is to make better known what it offers as an inducement to the purchase of a radio set. The sales barrage on the public has been concentrated upon selling the radio receiver as a perfected electrical instrument. The important work of making the big programs on the air better known has been more or less neglected. Broadcasting needs more listeners in order that programs may be improved and programs cannot improve unless there are more listeners. Advertising and sales effort, directed to the non-radio user, should stress the variety and quality of radio education and entertainment. On the other hand, the radio expert complains of the glittering generalities in the advertisements which are spread before him in the radio magazines. Why not recognize the distinction between radio users and non-radio users and plan advertising which appeals specifically to the radio user in the radio magazines and to non-radio users in the consumer magazines?

Unsnarling the Patent Tangle

THE National Electric Manufacturers' Association has appointed a committee to investigate the patent situation in the electric industry. Undoubtedly, this committee will deal with the radio situation, the branch of the industry now most seriously involved in the meshes of patent complication. Included in the membership of the committee are A. G. Davis of the General Electric Company, A. Atwater Kent of the Atwater Kent Manufacturing Company, and M. C. Rypinski of Federal-Brandes. It is unlikely that much progress can be made until certain important patent cases are adjudicated. Considering the tremendous cost of research and litigation and the long period before the various basic patents have become royalty producing, it is likely that there may be difficulty in restraining patent holders to sufficiently moderate royalties. In this respect, the committee can be very useful because the industry cannot suffer such burdens as 10 per cent. of the sales price of the receiver, which was once collected by one important patent holder. Considering that it would be difficult to make a radio set which does not infringe from twenty to thirty patents, the ultimate cost to the consumer may become prohibitive unless an agency, such as the N. E. M. A. committee, takes the situation in hand.

John V. L. Hogan has set an excellent example by charging only a modest royalty for the use of his single-control patent, which covers the combination of several interdependent resonant circuits, having an element which changes their respective electrical period equally and simultaneously. Thereby it covers every type of single-control receiver. In spite of its wide scope, evasions of royalty payment and legal costs have been reduced to a minimum because few, if any, manufacturers have at-

tempted to evade Mr. Hogan's modest license fees. One of the principal reasons why radio patents are litigated to the bitter end and every obstacle is raised to avoid payment of royalties is because licensing has been restricted wherever possible to a few manufacturers, and royalties have been exceedingly high. It is likely that, if a modest scale of royalties is established for all radio patents, few manufacturers will be unwilling to give the inventor his just due.

"Canned" Programs Good and Bad

JAMES C. PETRILLO, President of the Chicago Federation of Musicians, has started a crusade against Station WCRW of that city because it has been broadcasting dance music programs with the aid of a phonograph rather than with paid union musicians. A diet of phonograph records, so long as standard commercial records are used, is an imposition upon the listening audience. It would be unfortunate, however, if this crusade should result in a regulation against the use of any phonograph records in broadcasting. A suggestion made in Edgar H. Felix's new book, *Using Radio in Sales Promotion*, points out that special broadcasting programs may be rendered in recording studios and "assembled" and "cut" in the same way that motion picture films are perfected in the film laboratories. Every part of the program, by this recording process, may be rehearsed and fitted together to make the most perfect broadcasting performance. Numbers may be shortened or lengthened, re-rendered for more perfect reproduction, and worked over until the program is worked out to the utmost satisfaction. If such a carefully rehearsed and perfectly balanced program wins popular acceptance, it can be repeated at any time through any station or stations merely by running off the records again. The use of phonograph records as a program source in this manner is quite different, however, from broadcasting ordinary four- or six-minute commercial records. A station indulging in this practice is simply giving evidence of its incompetence in securing artists, and a way should be found to cancel its broadcasting license.

The Month In Radio

P. S. HILL, Vice-President of Herbert H. Frost, Inc., parts manufacturers, makes pertinent observations as to the value of the parts business, which indicate a larger total volume than is generally conceded to that branch of the industry. During 1926, the aggregate sale of complete receivers was $225,000,000 and of parts, $75,000,000. Accessories amounted to $230,000,000. Since the cost of parts of a home-built receiver, is, in general, less than that of a manufactured receiver, it is only reasonable to assume, says Mr. Hill, that half of the accessories, which include tubes, batteries, loud speakers, and power supply devices, were used in connection with home-built receivers. Making this division in accessory sales, the total business in manufactured sets, with the necessary auxiliaries, totalled $340,000,000 in 1926, while that of parts and their associated accessories was $190,000,000. In other words, reasons Mr. Hill, the sale of parts and their accessories was slightly more than 50 per cent. of the total sale of complete sets. This accounts for the prosperity of those dealers who learn to sell parts intelligently and concentrate upon that phase of the business.

THE appointment of distinguished music leaders to executive positions in broadcasting program direction, on both sides of the Atlantic almost simultaneously, is a striking recognition of radio as a force in the music world. In England, Sir Henry Wood, perhaps the foremost conductor of Great Britain, was retained by the British Broadcasting Company to develop its musical programs. On almost the same day, it was announced that Walter Damrosch had accepted the position of music counsel for the National Broadcasting Company. The two largest broadcasting organizations in the world called upon distinguished musical leaders and entrusted them with the guidance of the now most potent force in musical education.

LEGAL representatives of the American Radio Relay League successfully attacked the municipal regulation established by Portland, Oregon, which attempted to control federally licensed amateur radio stations. The municipality revised its ordinance so as to except amateur stations from its jurisdiction and the matter did not therefore become a test case in the courts.

ALTHOUGH New Zealand's two broadcasting stations are privately owned, the Government exercises close supervision over the broadcast listener. Before any make of radio set can be placed on the market, government inspectors test and pass upon its radiating qualities. If it fails in these tests, it cannot be sold without subjecting the dealer to penalties of the law.

THE Fourth Radio World's Fair will be held at the New Madison Square Garden in New York, September 19 to 24, under the direction of G. Clayton Irwin, Jr. In addition to an impressive collection of manufacturer's exhibits, many of the newest developments of the radio art will be featured, including, it is rumored, the first general public demonstration of telephotography. Last year, a quarter of a million people attended the exposition. Many favorite broadcasting artists will appear at the glass-enclosed studio, which will be the program source for the principal New York stations during show week.

THE opening of a 1000-watt broadcasting station in São Paulo, Brazil, has greatly stimulated the radio business in that country. The 55 per cent. ad valorem duty is a serious barrier to complete set sales, but the parts business is flourishing. While German and British headsets are selling in large quantities, other parts, such as rheostats, condensers, sockets, and binding posts, are largely imported from the United States.

A FEATURE of the Citizens' Military Training Camps, being conducted at Plattsburg, Fort Niagara, and Fort Ethan Allen, in the Second Corps Area this summer, will be a special evening course on advanced commercial radio

ALFRED E. WALLER
New York

Managing Director, National Electrical Manufacturers Association. A statement especially prepared for RADIO BROADCAST:

"Constant vigilance is essential if the radio industry is to keep its own house in order. As I see it, there are immediately required: (1.) A cleaning of the air; (2.) an organization of our manufacturing and distributing effort to level out buying and production; (3.) time, money, brains, and courage to provide for improved programs and their reception.

"The last-mentioned need is dependent upon the first two. The second point can be achieved most effectively through group action and constant, steady development of the product by the manufacturer, and by more competent selling and maintenance.

"The Federal Radio Commission has already worked marvels in clearing the air of what I have called 'electrostatic katy-dids.' The power company, the manufacturer, and the owner of noisy wave-generating equipment will be equally interested in a cleaning up process if we have sold him a good set, and provided him with continuously attractive programs."

practice, thus fitting those attending for the attainment of first-grade commercial radio licenses. Consequently, enrollment in the Signal Corps course at the Citizens' Military Training Camps will offer, in addition to the usual advantages, an opportunity to become a commercial radio operator.

ANNOUNCEMENT has at last been made of the much-rumored broadcasting chain, promoted by the Paramount-Famous-Lasky Corporation. It is understood that the network, which is to comprise about a dozen stations, will be known as the Keystone Chain. Most of the stations are to be in cities already served by the National Broadcasting chain. It is understood that KMOX, WHT, and WMAK are included in the chain, while several New York stations have been named as possible key stations.

Nothing would be better for radio than a real

rival to the National Broadcasting chain. The fact that the new organization is sponsored by a motion picture group, and probably by other companies, rather than by an independent company seeking to establish a commercial broadcasting business on a profitable basis, may cause it to concentrate so much on motion picture publicity that it will fail to offer real competition. But we hope for the best.

THE Circuit Court of Appeals reversed the decision of the District Court, unfavorable to the Radio Corporation, in its action to restrain the Twentieth Century Radio Company from further infringements of the Hartley and Rice patents. The decision affects the Hazeltine patents, the Twentieth Century Corporation being a distributor of Garod sets.

THE Indianapolis Broadcast Listeners' Association has been giving a series of twenty talks through WFBM on the general subject of radio interferences and how to identify, locate, prevent, and remedy them. A printed copy of one of these addresses reveals a most constructive attitude, urging the coöperation of listeners with power companies through the medium of listener organizations. Instead of wholesale abuse and fanatical antagonism to power companies and public pressure to bring about restricting municipal regulation of interference noises, the Indianapolis Association urges more effective measures, i.e., the employment of experts, remunerated by the contributions of broadcast listeners, to locate interference. Since radio listeners are their best customers, the power companies welcome their intelligent and helpful coöperation.

It has been frequently demonstrated that 90 per cent. of the noisy reception in any city can be eliminated or reduced to an unimportant minimum by an expenditure of less than $100 per square mile for an engineering investigation. In some instances, the broadcast listeners themselves, through a local radio club, have made this expenditure; in others, radio dealers have combined to do the work; and, in other instances, the power companies themselves have sought and cured the trouble. Any city troubled by notably poor reception, due to "man-made" interference, can, by coöperative means, eliminate most of it.

TEX RICKARD, who, not so long ago, was rated as an opponent of broadcasting, has become converted. "I have found," he says, "that the radio has done more to create new fans for sporting events than any amount of advertising could possibly do."

Accordingly, he has made arrangements to use the red and blue networks for important events, still maintaining WMSG for minor fight combats. A tribute to the popularity of fight broadcasts was found in many of our questionnaire returns. Practically all of those who favored the continuance of WMSG qualified their approval by the statement "for prize fight programs only."

THE National Electric Manufacturers' Association has issued the second Nema Handbook of Radio Standards. This covers such items as tests and necessary tolerances for cords, plugs, jacks, cable terminals, rheostats, variable condenser mountings, shaft extensions, standard tests for audio coupling devices, color codes for outside connections to set, battery dimensions and tests, socket power device markings, vacuum tube sockets, and constants for standard tubes. It establishes the frequency range of broadcast receivers from 550 to 1500 kilocycles, a most important requirement which many manufacturers are not yet regarding.

AS IT WAS IN THE BEGINNING!

RADIO BROADCAST Photograph

We would blush to introduce our friends to a receiver such as that depicted above in these days of low-loss apparatus. Yet this is a very typical example of pre-contemporary days. The Robert reflex, for it is none other, created quite a sensation at the time of its introduction a few years ago. The picture shows a two-tube model, a more popular one consisting of an identical circuit arrangement but with the addition of a push-pull audio stage

Modern Apparatus, Simple Changes, and Latest Tubes
Will Rejuvenate This One-Time Most Popular of Receivers

By JOHN B. BRENNAN

TO COUNT up the number of circuits which have remained in the public's eye over a lengthy period of time does not require many fingers. Such familiar names as Browning-Drake, Roberts, Haynes-Super, Knock-out Reflex, are all indelibly fixed in the minds of RADIO BROADCAST readers.

The most outstanding of these, at least to RADIO BROADCAST readers, is the Roberts circuit in its several forms. It was first brought to the attention of receiving set builders through a modest article by Dr. Walter Van B. Roberts in the April, 1924, issue of this magazine. Since then, the circuit has attained such prominence that those who have built it are numbered in the tens of thousands and, according to some estimations, approach the hundred-thousand mark. The most popular form of this circuit has probably been the Knockout four-tube Roberts reflex, a circuit diagram of which is shown in Fig. 1. Now, since April, 1924, considerable refinement has taken place in radio apparatus. Selectivity, sensitivity, volume, and ease of operation have all improved since that date and the thought has probably occurred to many owners of a Roberts receiver of the vintage of 1924, that some modern improvements might be made to bring the receiver

more or less up to date. The questions the owner should ask himself are these:

(1.) Is my Roberts receiver sensitive and selective enough?
(2.) Does it tune sharply enough?
(3.) How is the quality compared with receivers of to-day?
(4.) Do I secure sufficient volume from it?
(5.) Is it stable on the shorter wavelengths?
(6.) Is it difficult to operate compared to more modern receivers?

If the answers to these questions indicate to the reader and owner of the receiver that he is perfectly satisfied, there is no necessity for him to read further in the present article. On the other

hand, if his receiver seems to suffer somewhat compared to what he is led to expect is the best of present-day development, then the suggestions to follow may be useful. Boiled down to a simple, straightforward statement, this means that the owner of a Roberts receiver, if he desires, can bring it up to date in every respect so that it will be on a par with the latest receivers described, and incorporate all the niceties and refinements of present-day receiver design. It is possible that the answers to these questions indicate that the receiver in question needs a general overhauling. It is also possible that only one or more of the respective units need attention.

The sensitivity and selectivity of a receiver using the Roberts circuit go hand in hand; that is, a sensitive receiver will probably be selective and on the other hand, if it tunes broadly so that the signals desired have as a background signals from several other stations, it is probable that the receiver suffers on the side of sensitiveness as well. The old Roberts receiver used spiderweb coils, and these quite often were wound with wire having enamel insulation. Frequently the enamel wears off the wire at a point where it bends around the spiderweb form and if another adjacent turn is bared, a short-circuited turn results. As soon

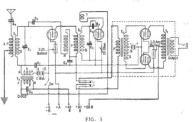

FIG. 1

as this happens, the set begins to tune broadly and signals are not as loud as they might be.

In the Roberts receiver there are two tuning controls, one on the antenna or radio-frequency stage, and the other on the detector. If either of the condensers seem to tune broadly, the coil should be looked at to see whether short-circuited turns are present on it. Sometimes it is impossible to determine definitely whether this is the cause of broad tuning or not and it is well, under such conditions, to remove the wire and either wind new coils or to substitute a more modern type of winding, such as a good solenoid inductance.

The antenna coil consists of 40 turns of No. 24 double cotton-covered wire with taps taken off about every 10 turns. The two secondary coils consist of 45 turns of the same size wire. The NP coil is wound with No. 26 double cotton-covered wire for 40 turns with a tap taken off at the 20th turn. This tap connects, via a jack, to the plate post of the primary of the input transformer of the push-pull unit and thence through that primary to the B battery. One end of the NP coil connects to the neutralizing condenser while the other connects to the plate of the radio-frequency tube. The tickler consists of 20 turns of No. 26 double cotton-covered wire. Incidentally, these same specifications will hold true whether the coils are to be wound in spiderweb, basketweave, or solenoid fashion. For the solenoid type of coil, the diameter of the form should be approximately 3 inches.

NEW COILS FOR THE ROBERTS

A MONG the manufactured coils which have been successfully used in the antenna stage of the Roberts receiver are those of Hammarlund, Sickles, Aero, General Radio, Bruno, and Silver-Marshall. Undoubtedly there are others which have the same characteristics and which may be employed in the antenna stage satisfactorily. The substitution of more up-to-date inductances

for the interstage spiderweb coil will also be a step toward modernization. The F. W. Sickles Company and the Hammarlund Manufacturing Company make coils intended specifically for the Roberts Circuit while the Silver-Marshall and Bruno coils consisting of primary, secondary, and tickler, may be used after the primary coil on the form has been removed and a new one wound according to the specifications given above. All of these coils have carefully been tried in a model of the receiver built in RADIO BROADCAST Laboratory according to the specifications contained in the September, 1924, issue of RADIO BROADCAST. No difficulty was experienced in substituting these coils for the original spiderwebs.

Broadness of tuning may be caused by poor connection between wires which are supposed to be making perfect contact. It is always advisable to solder such wires and even in soldering occasionally there exists high-resistance joints due to the failure of the constructor to clean properly the wires which are to be joined together. A good soldering flux, together with a soft wire solder, is a definite necessity in the wiring of a radio receiver.

It has been demonstrated since the early days of the Roberts receiver that no coil is more efficient than the solenoid of the proper dimensions. Efficiency with regard to coils means high inductance, low resistance, and low distributed capacity. These factors, in turn, mean high sensitivity and selective tuning circuits, and the owner of a Roberts receiver using old style spiderweb coils cannot make a better move than to substitute a highly efficient solenoid winding. This is especially true of the antenna stage where every bit of voltage amplification is useful. In the interstage coils, resistance is not so important and a coil of somewhat lower efficiency will not have the same importance as it would in the case of the radio-frequency amplifier. In other words,

the first coil changed is the antenna coil and if the set's sensitivity and selectivity are still to be improved, the detector coil should follow the antenna coil into the scrap basket and a new one be substituted.

Another remedy for broadness in tuning lies in the use of an antenna series-tuning condenser as shown in Fig. 1. This condenser should preferably be of the variable type and have a value of about 0.0001 mfd. However, if a variable type condenser of this value is not available, it will be found quite satisfactory to use a good make of fixed condenser of about the same value, such as those manufactured by Dubilier, Sangamo, Aerovox, Electrad, Tobe, etc.

There are other causes of broad tuning. As Fig. 1 shows, the input coil to the radio-frequency amplifier has between it and the tube a high-resistance winding—the secondary of the audio reflex transformer. This transformer secondary, unless properly bypassed, broadens the tuning of the antenna stage and it has been found in practice that it is advisable to shunt this secondary with a 0.0001-mfd. fixed condenser. It will be noticed that, when this shunt condenser is connected in the circuit as explained, the radio-frequency amplifier usually oscillates, particularly on the higher frequencies, and therefore it is more difficult to neutralize.

It will be found that the possibility of erratic action of the regeneration control is quite remote if coils such as those listed or manufactured specifically for use in the Roberts circuit are used. On the other hand, if the constructor is still employing his spiderweb type of coil and modern tubes, and cannot obtain smooth regeneration, he may experiment with the detector plate voltage and decrease the number of turns on the tickler coil one by one until the desired smoothness of regeneration is obtained. It was found in the receiver experimented upon in the Laboratory that 22½ volts was quite satisfac-

RADIO BROADCAST Photograph

THE REVISED ROBERTS REFLEX
The layout of parts for the three-tube arrangement. The circuit differs from Fig. 1 in that the push-pull audio arrangement has been omitted in favor of a straight audio stage

tory and about 6 turns had to be removed from the old 20-turn spiderweb tickler coil before the circuit was really satisfactory. Under these conditions the detector could be made to oscillate easily on 500 kilocycles, and be thrown out of oscillation on 1500 kilocycles. Experiments should be made with various sizes of grid leaks to obtain smooth control of regeneration. Grid leaks of a value of from 1 to 8 megs. may be used, smoother regeneration resulting from the use of higher valued leaks. The quality of reproduction, however, improves as lower grid leak values are used.

Where there is something wrong with the reflex part of the receiver it manifests itself in a number of ways. Perhaps the loud speaker will still continue to blare forth even though the detector tube is removed from its socket. Or perhaps reception is accompanied by a raucous squawk or howl when tuning-in to a station. Low volume is a good indication of trouble in the reflex stage and is often caused by the use of a cheap transformer in the circuit.

Besides having a good make of transformer in the reflex stage, it is essential that both the primary and secondary be bypassed intelligently. A 0.002-mfd. condenser across the primary and a 0.0001-mfd. one across the secondary will suffice, although other neighboring values should be tried and one selected which produces most satisfactory results, as it is always probable that no two similar circuits will require exactly the same condenser values to make them work properly.

THE AUDIO CHANNEL

IN THE few years since the appearance of the Roberts circuit, audio transformers have been developed which have much greater primary impedance than was necessary a few years ago. The reflex transformer may be one of the high grade units now on the market, such as Ferranti, Amertran Deluxe, Pacent 27-A, Samson, Thordarson, Silver-Marshall, or General Radio. The push-pull stage may be eliminated in favor of a

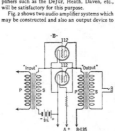

ORIGINAL CIRCUIT PUSH-PULL AMPLIFIER

OUTPUT IMPEDANCE

PARALLEL AUDIO AMPLIFIER

FIG. 2

In this figure are shown three arrangements of the power stage that can be satisfactorily used. At "A" is shown the original push-pull amplifier that was used in the old Roberts receiver. New push-pull transformers are now available and can be used with excellent results in this circuit. At "B" is an output arrangement consisting of two 112 tubes connected in parallel, both of them feeding into an output transformer. At "C" is an arrangement using a single 171 type power tube. In the plate circuit is a choke-condenser combination used to eliminate the direct current from the windings of the loud speaker.

straight stage employing one of the transformers used in the reflex stage, or a new push-pull amplifier may be used utilizing special Silver-Marshall or Samson transformers. Two 112 tubes in parallel, as shown in Fig. 2, will give somewhat greater output than a single tube

In a push-pull amplifier, the quality can be materially improved by substituting semi-power tubes of the 112 type for the two 201-A tubes. When this substitution is made, it is necessary that the plate voltage applied be 135 with 9 volts of C battery. The 171 type may be employed, but the volume will not be as great as with the 112 because the 171 has an amplification factor of 3 whereas the 112 has an amplification factor of 8. However, the 171 type tube has much greater handling capacity.

To determine whether the quality of the output from the reflex stage in your present Roberts receiver is everything that it should be, connect a pair of phones in series with the primary of the first push-pull transformer, or connect the phone tips in place of the terminals of the primary of this transformer. If the signal quality is not good here, then it is evident that the audio reflex transformer is at fault and the simplest way to improve the quality is to substitute another transformer of a better type.

If the tickler coil is adjusted up to the point where the detector is just about to spill over, it cannot be expected that the quality of the signal in the loud speaker will be everything that is desired and the tickler should always be backed off to prevent this condition.

Tubes offer a fertile field for the investigation of poor quality. A poor tube in the radio-frequency reflex stage may affect the tone quality and volume appreciably and the builder should switch around the tubes in the various sockets until the most satisfactory position for all tubes is found.

There are a number of unit amplifiers on the market which may be readily substituted for the second image audio originally employed. Such units as the Pacent "Powerformer," the Silver-Marshall "Unipack," those of Receptrad and General Radio, and several resistance amplifiers such as the DeJur, Heath, Daven, etc., will be satisfactory for this purpose.

Fig. 2 shows two audio amplifier systems which may be constructed and also an output device to

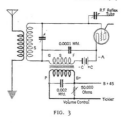

FIG. 3

couple an ordinary transformer stage to the loud speaker.

VOLUME CONTROL

THE volume of the receiver may be controlled by shunting the primary of the reflex audio transformer with a variable resistance of the value of 50,000 ohms, as shown in Fig. 3. The compression type of variable resistor, such as the Clarostat, may successfully be employed here.

In making the change to power tubes, it is not necessary to control the filaments with a rheostat since a fixed filament ballast will suffice. This, therefore, leaves one hole on the panel vacant by the removal of the rheostat which controls the tubes in the push-pull amplifier. In this hole may be mounted the volume control. Since no adjustment of the filament of the tube in the radio-frequency reflex stage is required, the rheostat originally provided may be dispensed with and a filament ballast substituted for it, thus making vacant another hole on the main panel. It is suggested that the rheostat for the detector tube be transferred to the hole formerly occupied by the rheostat for the radio-frequency reflex tube and in the hole made vacant by the rheostat on the detector tube, a filament switch be mounted. This switch should be connected in series with the minus A lead to the various tubes.

There are a number of neutralizing condensers now available which may take the place of the home-made tubular neutralizing condenser originally specified in the construction of the Roberts reflex receiver. These new condensers are of the variable plate or screw type and are much easier to adjust and to maintain adjusted than the tubular type. The primary requisite in neutralizing is to keep the connecting leads between the coil, neutralizing condenser, and tube, as short and direct as possible and not let these leads parallel any other leads. Satisfactory neutralizing condensers are those manufactured by Precise, Hammarlund, Cardwell, and XL Laboratories.

The changes as outlined here have centered mainly around the original popular Knockout four-tube reflex receiver and deal particularly with the substitution of new and improved parts for old ones, with one or two circuit changes. In its revamped form, it is a very fine receiver, worth the efforts of any home constructor.

It will give good quality and if the changes suggested are carefully made the receiver should equal, in sensitivity and selectivity, any of the more recent receivers which have been described. For best results the receiver should be used with some good cone-type loud speaker.

The
Balsa Wood
Loud
Speaker

Constructional Hints on a New Loud Speaker Which Was Found in the Laboratory to Give Very Excellent Quality

By THE
LABORATORY
STAFF

A NEW LOUD SPEAKER
The Balsa wood loud speaker is now obtainable in fully assembled form. The picture to the right shows a typical example

IT IS generally conceded that the loud speaker is the present weak link in broadcast receiving systems. It is possible to build excellent amplifiers both for transmitters and for receivers, and from the pick-up device in the radio broadcasting studio to the output of the final amplifier there is little to be desired as far as true reproduction of various frequencies is concerned. It is true that programs are still compressed somewhat as regards volume, partly because of the limited carrying capacity of the telephone lines connecting pick-up and transmitter, partly because of the limited power range of the transmitter, and partly because of the limited volume range of receivers; but the ear will stand for a great deal of distortion due to these causes without rebelling. The loud speaker, however, is far from perfect.

For this reason, when anything new in the way of loud speakers develops, the average radio engineer registers considerable interest. The latest idea in this realm is the Balsa wood loud speaker. The story goes that the Western Electric engineers, in whose laboratories have developed some of the best known translating devices of the present time, "played" for some time with Balsa wood and, like all good things the Western Electric engineers "play" with, sooner or later the idea got about. It is now possible to purchase in kit form the necessary Balsa wood, hardware, driving mechanism, and frame to put together a very excellent loud speaker.

Balsa is the lightest known wood, averaging about six or seven pounds per cubic foot, while cork averages about fourteen pounds. It grows in Panama and other tropical countries, and is used widely in the airplane industry. Because it is 92 per cent. air, it makes an excellent insulating medium against heat.

The Balsa wood slats in the kits for loud speaker construction come in three sizes ready to be assembled as diaphragms of one by two feet, two by three feet, or two by four feet, approximately. The middle sized one is the one most people build, and, in fact, is the best of the three. The smallest is too small; the largest is too awkward. Photographs of the medium sized one are shown here.

Heretofore, the best loud speaker that could be made was an elliptical

cone with about four feet as its longest dimension. This, driven with a Western Electric 540 AW unit, seemingly left little to be desired. It was, in many peoples' estimation, better than the 540 AW loud speaker purchased on the open market. The new Balsa loud speaker, if carefully made, is on a par with the elliptical cone.

Balsa wood is useful for loud speaker construction because of its extreme lightness and because it is possible to make the radiating part of the loud speaker extremely rigid. Lightness contributes toward the efficiency of the loud speaker,

which means that a greater percentage of the input power is radiated as sound energy and not wasted in friction. Rigidity is necessary in order that considerable energy can be fed into the device before anything begins to rattle or to fall apart. When it is considered that the best loud speaker now obtainable is less than five per cent. efficient, it may be seen that anything contributing toward greater efficiency is worth while.

The Balsa loud speaker consists of three boards or slats which are held together mechanically by smaller strips, as shown in the photographs. These strips should be stuck on with Ambroid cement. Glue is bad since it is hard, brittle, and heavy. In practice, the slats may be placed either with their edges flush or not. If they touch, they must be firmly cemented together. The strips may radiate from the center of the loud speaker, they may be placed at right angles to the long dimension of the speaker, or they may be curved as shown in one of the accompanying photographs. In the Laboratory, little difference was noted whichever way they were placed as long as everything was rigid and light.

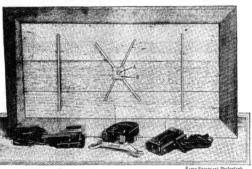

ONE OF THE BALSA LOUD SPEAKERS CONSTRUCTED IN THE LABORATORY
With several units in front of it. The Western Electric 540 AW unit was found to give most satisfaction. "A" in the photograph is the top of the bushing arrangement which grips the pin of the driving unit. The complete bushing unit is shown in Fig. 1. "B" is a piece of cedar shingle, afterwards replaced with a piece of Balsa wood. Since the Laboratory made tests on the Balsa loud speaker, a driving unit specially designed for it has been announced by the Balsa Corporation

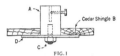

FIG. I

The bushing arrangement, or "gadget," which grips the pin of the driving unit employed. "A" is the shank into which the pin is placed and held by the set screw; "D" is a very thin piece of copper or aluminum sheet; "C" is a hex nut to hold "A" to "D." "B" should preferably be of Balsa wood and not cedar

Some Balsa loud speakers have been found to give better results than the 540 AW cone while others with the same unit and with the same care in construction are not so good, for some unknown reason. Reports have come to the Laboratory to the effect that not everyone is satisfied with the Balsa loud speaker after it is constructed. In our opinion, however, there is nothing better at the present time than a carefully constructed Balsa loud speaker, rigidly put together with Ambroid, and driven with a 540 AW unit. The following experience will reveal some of the difficulties.

A Balsa loud speaker of the familiar two-by-three-foot size was constructed of standard material consisting of slats, strips, and the "gadget," or bushing arrangement, which holds the drive rod of the unit, glue being used in the assembly. This "gadget" is shown diagrammatically in Fig. 1, and also at "A" in the photograph on page 211. In practice it varies somewhat in size and appearance. It was glued to the center of the diagonally radiating strips which were in turn glued to the slats, as shown in the photograph. A piece of cedar shingle, "B" in the same photograph, about two inches square, was also glued onto the strips with the shank and set screw of the "gadget" protruding through a hole in it. The purpose of the shingle is to hold the bushing arrangement rigidly to the diagonal strips.

The loud speaker was very fine until a certain power input, which was not very high, was exceeded, and then the thing rattled unmercifully. The first point of attack was the glued strips, which seemed to have a habit of breaking loose when certain frequencies were placed on the loud speaker. These glued strips were ripped off and re-cemented with Ambroid which not only stuck better than glue but was lighter. It formed a homogenous connection between slat and strip. Then the bushing which held the 540 AW unit drive rod to the Balsa wood was removed, the cedar shingle thrown away, and part "D" (see Fig. 1) of the bushing assembly cut down to its smallest possible dimensions, until it was about the size of a dime but thinner. Then a small piece of Balsa was placed over this dime dimensioned piece of thin copper sheet, to take the place of the cedar shingle, and the driving unit was replaced. This process seemed to release a great load from the unit for the high frequencies (which before had been about equal to the 540 AW 18-inch cone) were

now much better, and the low frequencies, which before had been much better than the cone, were not changed. It was possible to put the full output of an amplifier using a 210 type tube into it.

SOME SUGGESTIONS

THERE are several kinks that may be tried on this loud speaker. One is to drive the Balsa not at the center, but somewhere between the center and one end. Another idea is to place felt strips between the ends of the Balsa slats and the frame, insuring good contact and preventing any possible rattles.

In constructing a Balsa loud speaker from the standard kit, the procedure is as follows. The slats are first firmly fixed to the frame with Ambroid and then the moulding that is furnished is cemented in place. Next, the strips are cemented to the slats, again using Ambroid in sufficient quantities to insure perfect connection but not enough to run over the slats and increase the weight or spoil the appearance. At the center of the two diagonal strips, if they are arranged as shown in the photograph on page 211, and if the unit is to be placed there, a small depression will have to be cut in the strips in which the hexagonal nut, "C" in Fig. 1, is to fit; then the bushing assembly of Fig. 1 is cemented to the strips. Some means of mounting the driving unit must be provided, and since there are no two units that have the same dimensions, the constructor will have to exercise his own ingenuity at this point. The photograph on this page gives an idea of how it was done in one case in the Laboratory. The Balsa Corporation has an aluminum bracket that simplifies this problem considerably. Weights may be balanced on the strips, etc., while the Ambroid is drying to insure firm contact.

The loud speaker should not be tried until several hours after the final cementing has been completed. Then the weights, if these are used, should be removed from the strips, and the whole instrument looked over carefully for places where the strips might not be securely fixed to the slats.

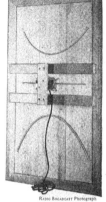

ANOTHER ASSEMBLY

The mounting of the unit is clearly shown in this illustration. This loud speaker differs in detail from that shown on the previous page

The matter of units is an important one. Several have been tried in the Laboratory, some of which are probably not representative of the best on the market. None so far tried can compare with the 540 AW, but as soon as the unit field has been covered by the Staff, the results of the investigation will be made known. Anyone who has a 540 AW loud speaker can experiment with the Balsa wood loud speaker for five dollars, which is the price of the Balsa wood. Unfortunately, the Western Electric 540 AW unit is not obtainable except in the complete cone loud speaker form. The Balsa material can be procured in New York from the Balsawood Reproducer Corporation, 331 Madison Avenue. Ambroid liquid cement is a product of the Ambroid Company, 1227 Miller Avenue, Brooklyn, and a 35-cent can is plenty.

In building a Balsa wood loud speaker, one should spend not less than three or four hours cementing slats to strips, and must bear in mind that the Balsa wood is easily broken or punctured. A moment's lack of care may cost a beautiful loud speaker. It should be remembered that it takes a very good power amplifier, and a very good program from a very good station to enable the average ear to tell the difference between the 18-inch 540 AW cone loud speaker and the Balsa wood loud speaker. Completely assembled Balsa loud speakers, attractively decorated and containing a specially constructed unit, are now on the market.

PREPARING TO REMOVE THE 540 AW UNIT

Operating Characteristics and Constructional Details of a 38- to 113- Meter Transmitter

Either Battery or A. C. Operation Possible—Transmits C. W. Signals or Phone at Option

RADIO BROADCAST Photograph

A FRONT VIEW

Of the short-wave transmitter described in this article. It can be used to transmit either c.w. or telephone signals, the changeover from one to the other being accomplished by merely throwing the switch in the center of the panel

A Flexible Short-Wave Transmitter

By HOWARD E. RHODES

THIS article describes the construction and operation of a low-power short-wave transmitter, designed for use on either c.w. or phone. The transmitter, which is illustrated in various photographs in this article, was designed and constructed in the RADIO BROADCAST Laboratory.

It uses the tuned-plate tuned-grid circuit, a type of oscillatory circuit which is very likely one of the easiest to adjust. The three dials on the front of the transmitter are used to tune the grid, plate, and antenna circuits, and properly setting these three condensers is all that is necessary in adjusting the transmitter. In this article considerable data will be given regarding the general characteristics of this type of circuit.

The transmitter is designed so that, by merely throwing a multiple-pole switch, it can be connected for use on either phone or c.w. On c.w. work this switch, located at the center of the panel, is thrown to the left, and then one of the 7½ watt tubes functions as an oscillator and the other two tubes are turned off. The set is keyed by connecting a pair of leads between a telegraph key and the two telephone tip jacks on the front of the panel (the photographs show an ordinary jack for this purpose but in practice it was found

Facts About This Transmitter

Circuit: Tuned-plate, tuned-grid
   The set can be used on either phone or c.w.
Tubes: One 210 oscillator and two 210 modulators
Wavelength Range: 19 meters (7900 kc.) to 113 meters (2700 kc.)
Power Input to Oscillator: 40 Watts

This transmitter can be operated from batteries or from a compact a.c. power unit also described in this article. The set is entirely controlled by a single master switch. The transmitter is not difficult to construct and is easily adjusted for efficient operation.

RADIO BROADCAST Photograph

THREE EXCELLENT MICROPHONES

To use on phone work. The Federal unit (price $7.00) is at the left. The Globe type E microphone (price $15.00) is the center one, and the microphone at the right hand is made by Kellog (price $8.90). The latter unit is equipped with a push button which is used to open and close the microphone circuit

that the voltages were high enough to break down the insulation on the jack, and therefore two pin jacks, one above the other, were substituted for the telephone jack).

When the control switch is thrown to the right, or" microphone," position, one tube functions as an oscillator and the other two tubes function as modulators in a Heising modulation system. The grids of the two modulator tubes are fed from the secondary of a General Radio modulation transformer, in the primary circuit of which is the microphone in series with two dry cells. The microphone is connected in the circuit through a telephone jack on the front of the panel. When the control switch is thrown to the "off" position it not only turns off the filaments of the tubes, but also opens the microphone circuit so that the microphone telephone plug can be left in the telephone jack without any danger of exhausting the two dry cells.

An accompanying photograph shows a group of microphones that are satisfactory for use with this transmitter. In all cases very good quality signals were transmitted using these microphones powered from two dry cells, although somewhat higher modulation can be obtained using three dry cells instead of two.

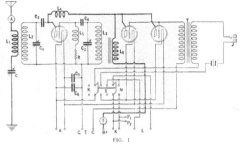

FIG. 1

When the transmitter was first constructed, the control switch was arranged so that in the "key" position all three tubes worked as oscillators. It was found, however, that with this arrangement, a considerable change in frequency took place on throwing the switch to microphone, and the circuit was then revised so as to have only one oscillator with the control switch in either position. With this latter circuit no change in frequency takes place on switching from microphone to key. It is therefore possible with this set to call an amateur on c.w. and then talk to him on phone without making him change his receiver adjustment except to take it out of oscillation, as modulated signals, unlike c.w. signals, cannot be satisfactorily received with an oscillating detector.

The transmitter is so wired that it can be operated from batteries or from a compact a.c. power unit, also described in this article, which

FIG. 2

will supply it with all the necessary filament, plate, and grid voltages. The circuit diagram of the transmitter is given in Fig. 1, and the connections for battery operation are given in Fig. 2. Fig. 3 is the circuit diagram of the power unit. No changes in the transmitter itself are necessary in changing from battery operation to a.c. operation; it is merely necessary to connect the various binding posts somewhat differently so as to obtain proper operation.

The transmitter can be operated on any wavelength between 38 and 113 meters (7900 kc. and 2650 kc.). The tuning chart shown in Fig. 4

was obtained by plotting wavelength against the settings of the grid (left-hand) tuning condenser dial. In adjusting the set, the grid-circuit tuning condenser is set at the desired reading, and the plate and antenna condensers are then adjusted for maximum efficiency as indicated by the plate milliammeter and antenna ammeter.

The transmitter has been carefully constructed so as to present a good appearance and its size and weight are such as to make it readily portable. It measures 18 inches long, 14 inches high, and 8 inches deep, and only weighs about 28 lbs. The a.c. power unit in its container measures 16 inches long, 8 inches high, and 8 inches deep, and also weighs about 28 lbs.

The entire transmitter can be duplicated at a cost of about $90.00 without tubes. If the set is to be operated from a.c., it will be necessary to build the power unit, which costs about $50.00.

Before the final model of the transmitter was

laid out and constructed, a temporary affair was put together on a large baseboard and a series of tests made to determine the characteristics of the circuit. Specifically, it was desired to determine: (a) The effect of varying the grid leak resistance; (b) the effect of varying the resistance of either the tuned grid or plate circuit; (c) the effect of varying the coupling between the plate and antenna coils; and (d) the effect of variations in plate voltage. Data were also obtained which showed how the plate current, antenna current, and the frequency, vary as the plate condenser is adjusted.

In all of the tests a single DeForest type DL9

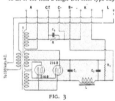

FIG. 3

tube was used, this tube being similar to a 210 type tube. The circuit used in making these tests was exactly the same as that given in Fig. 1 and all the data were taken at a frequency of 3750 kc. (80 meters). The plate potential was 385 volts, obtained from a bank of Exide storage batteries. A non-inductive resistance was used to represent the antenna. The coils and r.f. chokes used were those made by the Aero Products Company. The inductance of coils L_6 and L_4 is about 7.5 microhenrys each at 3750 kc. (80 meters). Fig. 5 shows how the inductance of either L_6 or L_4 varies with the wavelength.

RESISTANCE VARIATIONS OF THE GRID LEAK

WHEN a vacuum tube breaks into oscillation, it operates over its entire characteristic, and during part of each cycle the grid draws current. This current flowing through the grid leak resistance develops a voltage which impresses a negative bias on the grid of the tube. The group of curves given in Fig. 6 show the effect of using different values of grid leak resistance. Curve No. 1 shows how the average voltage across the grid-bias resistance varies

RADIO BROADCAST Photograph

THE SHORT-WAVE TRANSMITTER

As viewed from the rear. The upper and lower compartments into which the set has been divided makes possible a very compact arrangement. The Heising choke coil, modulation transformer, variable condensers, and microphone batteries, are all placed beneath the shelf

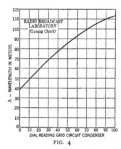

FIG. 4

with different values of resistance. At low resistance values the grid current is large but because of the small resistance only a low voltage is developed. With increasing resistance the voltage Eg also increases and reaches values as high as 124 volts when the grid leak resistance is 26,000 ohms. The plate-current, curve No. 2, is very large with low values of resistance but gradually decreases with increased resistance and reaches a minimum value with a grid leak resistance of about 16,000 ohms and then increases again. The current in the antenna circuit reaches a maximum with a grid leak resistance of about 9000 ohms although there is little change in antenna current between 4000 and 14,000 ohms.

The power input, Pi, is equal to the plate voltage times the plate current, and the power output is proportional to the square of the antenna current. The efficiency can therefore be expressed as:

$$\text{Efficiency} = \frac{\text{Power Output}}{\text{Power Input}} \times 100$$

$$= \frac{I_a^2}{Pi} \times 100 = \frac{I_a^2}{Eb \times Ip} \times 100$$

This expression is plotted in curve No. 4. Fig. 6, and shows a maximum with a grid leak resistance of between 12,000 and 14,000 ohms. It appears evident, then, that most efficient operation can be obtained with a grid leak resistance of about 13,000 ohms.

RESISTANCE IN THE GRID AND PLATE CIRCUITS

IN ORDER to obtain the most efficient operation from a tuned-plate, tuned-grid transmitter, it is essential that the coils used be carefully constructed so as to reduce to a minimum all losses. The curve in Fig. 7 shows the effect

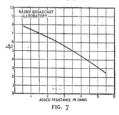

FIG. 7

of increasing the resistance of either the tuned-plate or tuned-grid circuit, and it indicates the quite rapid decrease in efficiency that occurs when resistance is added to either of these circuits. In a tuned-plate, tuned-grid transmitter, the tuned-grid circuit is coupled capacitively to the plate circuit through the inter-electrode capacity of the tube, the grid circuit acting more or less as a driver to keep the tube oscillating. It might appear, at first thought, that some loss in the grid circuit could be tolerated without any great decrease in efficiency, as is resistance in the plate circuit. It appears likely, from the tests made, that increasing the resistance in the grid circuit is practically as effective in reducing the efficiency, as is resistance in the plate circuit. The test was made by connecting small straight pieces of high-resistance manganin wire in series with the tuned circuit. The resistance of the small pieces of wire was measured on a d.c. bridge and their r.f. resistance was considered to be the same as their d.c. resistance. This procedure was followed because it is difficult to make accurate resistance measurements at very high radio frequencies.

VARYING ANTENNA AND PLATE COIL COUPLING

IN FIG. 8 are curves showing the effect of varying the coupling between the plate coil and the antenna-coupling coil. In the Aero coils the antenna coil is hinged at the bottom and in measur-

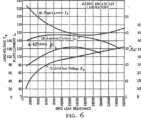

FIG. 6

ing the coupling merely the distance between the top of the plate coil and the top of the antenna coil was measured. The curves show that there is a considerable increase in efficiency when using very loose coupling. With very close coupling and with a coupling of 4½ inches the antenna current is practically the same but the difference

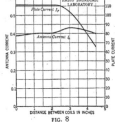

FIG. 8

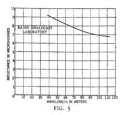

FIG. 5

in plate current between the two settings is of a ratio of 112 to 65, or practically 2 to 1. While 112 milliamperes at 385 volts is equivalent to an input power of 43 watts, with decreased coupling we are able to obtain practically the same antenna current with a reduction of input power to 25 watts. As the coupling between the two coils is decreased it will be found that the tuning of the antenna circuit becomes much sharper, but that, even with a coupling distance of 4½ inches between the coils, the coupling is still sufficiently close to produce two resonance peaks (a characteristic of coupled circuits). For example, in one of the tests, the grid condenser was set at the correct point for 80 meters, and with the antenna disconnected, the plate condenser was varied to a point where the tube oscillated most vigorously. With a coupling of 4½ inches between the antenna coil and the plate coil the antenna circuit was closed and the antenna condenser, starting at minimum capacity, was gradually increased. The antenna current gradually increased until it reached a peak and at this point the wavelength was measured and found to be 84 meters. Turning the antenna condenser only a fraction of a degree more, however, produced a very large increase in antenna current and on measuring the wavelength again it was found that it had decreased to 80 meters. This double effect will not be found if the tuning of the antenna circuit is done in the opposite direction, that is, starting with the antenna capacity at a maximum and bringing the circuit into resonance by decreasing the capacity of the antenna tuning condenser. When tuning is done in this way the antenna current will gradually increase until the maximum is reached and after the antenna tuning condenser has been adjusted for maximum current, a slight readjustment of

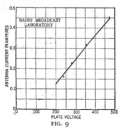

FIG. 9

the plate condenser is generally necessary to obtain maximum efficiency.

VARIATIONS IN PLATE VOLTAGE

THE power output from the transmitter of course increases with an increase in plate voltage, and Fig. 9 shows how the output power varies with plate voltages between about 200 and 500 volts. If the efficiency remained constant we would expect the power output to vary as the plate voltage squared. Doubling the plate voltage would then give four times as much *power* in the antenna circuit. The power in the antenna is proportional to the antenna current squared. Actually, however, doubling the plate voltage gives nine times as much power in the antenna, indicating that the efficiency also increases with increasing plate voltage and it is therefore advisable to use fairly high plate voltages in the operation of the transmitter. Every time the plate voltage is changed it is necessary to slightly readjust the condensers to obtain maximum power output.

In the operation of a tuned-plate tuned-grid transmitter it will be found that variations in the coupling between the antenna coil and the plate coil, and variations in the plate voltage, are the only practical means to control the amount of power going into the tube. Without any antenna load the input power to the tube is about 9½ watts. With the antenna connected, the input power can be made to go as high as 163 watts although most efficient operation was obtained with an input power of about 25 watts. The set operates very inefficiently with the high power and the plate of the oscillator tube gets very hot.

SOME GENERAL CHARACTERISTICS

IN FIG. 10 are given a group of curves which show in detail how the various quantities— plate current, antenna current, etc.,—vary as the plate condenser is tuned, the grid and antenna condensers having first been set at the correct point. With the plate condenser set at 100 (maximum capacity) the tube will not oscillate and the plate current is 85 milliamperes. The antenna current is, of course, zero, because the tube is not oscillating. If the tube is permitted to operate under these conditions for any length of time it will be ruined because the plate current is quite large and all of the power is

RADIO BROADCAST Photograph

A PHOTOGRAPH OF FIG. 3
The power unit is constructed on a baseboard with the power transformer on the left and the filter chokes and filter condensers at the right. The C-bias potentiometer can be seen at the rear center

dissipated on the plate, which becomes very hot. When the plate condenser dial is reduced to 50 the tube begins to oscillate and the plate current rapidly falls to 57 milliamperes. Further adjustments of the condenser cause the plate current to again increase. The plate current reaches a minimum when the dial reads 46 and the antenna

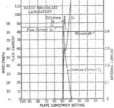

FIG. 10

current reaches a maximum with the dial set at 48.5 .The efficiency (antenna current divided by the plate current) as shown by the small curve is at a maximum with the dial set at 47.5 and this point does not correspond with the maximum of either the plate current or the antenna current. All this means that an accurate adjustment is necessary to obtain maximum efficiency. If the transmitter is operated from B batteries, the increase in efficiency resulting from accurate adjustment is well worth while. In this particular case it would mean a reduction in B battery current drain from 80 milliamperes to 62 milliamperes (18 mA.), and this means longer battery life.

In the course of the experiments, data were also taken on the effect of varying the size of the grid- and plate-coupling condensers. It was found that any size between 0.0001 mfd. and 0.001 mfd. would give equally efficient operation, values less than 0.0001 mfd. giving somewhat less efficient operation. These values of course vary with the frequency at which the transmitter is being worked. They hold for 80 meters.

It is likely that most efficient operation will be obtained at any given wavelength with some particular value of inductance and capacity in the tuned circuits. In a transmitter of the type illustrated in this article, which is designed to cover a wide band of wavelengths without changing the coils, the inductance-capacity ratio cannot be considered because the actual value of the inductance is that value which will tune with the various capacities in the circuit to the shortest wavelength on which it is desired to transmit. The variable condenser must be of such a size that it will tune with the coil to the longest wavelength to be used.

A decided advantage of the tuned-plate tuned-grid transmitter is that it is possible to calibrate it so that it is a simple matter to set the transmitter on any desired wavelength. The procedure in adjusting the transmitter is about as follows:

First, the grid tuning condenser is set at the correct point according to the tuning chart, Fig. 4, and then, with the antenna circuit detuned adjust the plate condenser to that point at which the plate current is at a minimum. With a coupling of about four inches between the plate coil and the antenna coil, the antenna tuning condenser is now adjusted for maximum antenna current. After the antenna circuit has been adjusted, a slight readjustment of the plate condenser is generally necessary to secure maximum efficiency. In making the preliminary adjustments on the set it is best to use somewhat low plate voltages so that the tube will not be damaged.

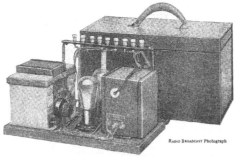

RADIO BROADCAST Photograph

A SIDE VIEW OF THE POWER UNIT
This photograph plainly shows the layout of apparatus. The center support of the binding post strip also acts as a holder for the 50,000-ohm fixed resistor. In the rear is the Kennedy tool case in which the power unit can be placed

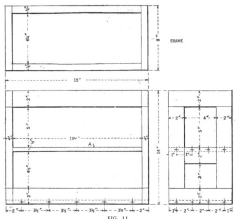

FIG. 11

PARTS USED IN THE TRANSMITTER

-l, L₂, L₃, L₄—Aero Coil Kit, Type 4080	$12.00
L₄—Amertran Choke Coil, Type 709 .	10.00
C, C₁, C₄—Three Cardwell 0.0005-Mfd. Variable Condensers, type 173-C .	15.00
C₂, C₅—Two Sangamo 0.00025-Mfd. Fixed Condensers	1.00
C₃, C₆—Two Tobe 1.0-Mfd. Bypass Condensers	1.80
T—General Radio Modulation Transformer	6.00
R—MountJord 12,000-Ohm Grid-Leak Resistor	1.50
S—Federal Four-Pole Double-Throw Switch	3.20
J—Yaxley Open-Circuit Jack . .	.50
P₁, P₂—Two Carter Telephone Pin Jacks	.60
A—General Radio 1-Amp. Antenna Ammeter Model 127-A . . .	7.75
I—Weston 200 MA. Milliammeter, Model 301	8.00
Westinghouse Micarta Panel, 14 Inches x 18 Inches	5.00
Three General Radio Sockets, Type 156	3.00
Three General Radio Dials, 2¾ Inches in Diameter with Indicators . .	1.80
Nine X-L Binding Posts . . .	1.35
Two Burgess Radio A Batteries . .	1.00
Federal Microphone	7.00
TOTAL	$86.50

The transmitter requires three 210 type tubes (DeForest type DL9) and the power supply employs two single-wave rectifiers of the 216-B type.

CONSTRUCTION

IN THE actual physical layout of the transmitter considerable care was taken to keep the size of the unit as small as possible so as to make it easily portable. The method of construction will be evident from the various photographs in this article. The transmitter was built around a wooden framework having the dimensions given in Fig. 11. The tube sockets, coils, fixed condensers, grid leak, radio-frequency choke coils, and filament bypass condensers, were mounted on the shelf "A," Fig. 11. On the floor of the framework are mounted the two dry cells for the microphone, the Heising choke coil, and the modulation transformer. Extending from the rear of the shelf is a terminal strip holding nine binding posts. At the end of the transmitter is another small strip on which are mounted two additional binding posts for the antenna and counterpoise connections.

The following procedure should be followed in constructing the transmitter:

(1) Assemble the wooden frame in accordance with the data given in Fig. 11. Do not fasten the shelf very tightly.

(2) Lay out and drill the main panel in accordance with Fig. 12.

(3) Mount the various instruments on the panel and then fasten it to the wooden frame.

(4) Remove the shelf from the frame and mount the various instruments as shown in the illustrations.

(5) On the floor of the frame mount the batteries, choke, and modulation transformer.

(6) Drill ₇⁄₆₄ " holes through the shelf at the points where connections are to be made to the tuning condensers.

(7) Run long bus bar connections from the terminals of the three tuning condensers to points tallying with the position of the holes previously made in the shelf.

(8) Mount the shelf in place, passing the wires from the condensers through the holes in the shelf.

(9) Complete the connection of these wires to the coils.

(10) Wire the rest of the transmitter using No. 18 lamp cord for the power and filament leads and bus bar for the "hot" connections—i.e., grid and plate leads.

THE POWER SUPPLY UNIT

IN THE first part of this article mention was made of the fact that a power unit had been designed to supply this transmitter. This power unit is illustrated in photographs on page 216. The construction is very simple. The power transformer, filter chokes, filter condensers, and tube sockets are all mounted on a baseboard measuring 7½" x 15½". The binding post strip is mounted on two brass supports so as to come even with the top of the Kennedy Tool Kit case in which the power unit is finally placed. The center support of the binding post strip is a piece of round brass, and this support is used as a mast for the 50,000-ohm discharge resistance. C bias for the grids of the two modulator tubes is obtained from a 1000-ohm potentiometer placed in the negative plate circuit, and with it, voltages up to about 100 volts are available for grid bias. The binding posts on the power unit are arranged in the same sequence as those on the transmitter so that in connecting the two together it is merely necessary to connect wires between the corresponding binding posts. The main switch on the panel has one set of contacts connected in series with the 110-volt lead so that, with the switch in the center position, the power unit is off, and with the switch thrown to either the right or left, the power is on.

POWER UNIT PARTS

T—Thordarson Power Transformer, Type T-2098 (containing two filament windings and a center-tapped high-voltage winding) . . .	$20.00
L—Thordarson Double Choke, Type T2099	14.00
C₁—Tobe 2-Mfd. High-Voltage Condenser	3.50
C₂—Tobe 4-Mfd. High-Voltage Condenser	6.00
C₃—Tobe 1-Mfd. Bypass Condenser .	.90
R—Yaxley 1000-Ohm Potentiometer .	2.00
R₁—Lynch 50,000-Ohm Fixed Resistance	3.00
Two Eby Sockets	1.00
Nine X-L Binding Posts . . .	1.35
TOTAL	$51.75

The parts given in the two lists were used in the two units designed by John B. Brennan and illustrated in this article, but the parts of any other reputable manufacturers may be used if they are electrically similar.

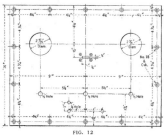

FIG. 12

THE LISTENERS' POINT OF VIEW

Conducted by John Wallace

Radio Is Doing a Good Job in Music

WITH all our skeptical remarks about radio education we have never yet expressed any mistrust of the important rôle radio can play, and is playing, in educating the populace in music appreciation. And when you get right down to it, even if the educational force of radio extends no further than this, there is small cause for complaint. Far more is gained by the initiating of ten people into the pleasures of music than by filling ten hundred with a miscellany of facts about economics or hog raising.

In the particular day and age in which we live there is a howling need for educating people in things esthetic and a very slight need for educating them in things practical. That is, there is very slight need for radio's exerting itself to educate them in practical things for a plethora of existing agencies is already busily engaged in bearing this burden.

Colleges and universities have increased their number and enlarged their enrollments to such a degree that it has come to be no longer a mark of distinction to attend one, but on the contrary rather classifies one as a member of the herd and the Babbittry. Technical schools, business colleges, night schools, and correspondence courses are likewise flourishing in a way never dreamed of a generation ago. The cities, through their health departments and citizenship bureaus, are daily educating thousands, and the bigger businesses and industries maintain after-hours schools for their employees. Even the newspapers devote columns to educational features and of course the current question book vogue is doing its bit toward equipping the man in the street with a knowledge of the date of the Battle of Waterloo and the specific gravity of lead.

In short, the man in the street is nowadays quite surrounded by persons who are both willing and anxious to educate him in any branch of practical knowledge, and who will do it for a song and at hours to suit his convenience. So even if radio does add its voice to those of the other practical educators it will augment the present clamor by only a small chirp.

Education in the useful arts is not to be railed at. In the nature of the present Western civilization it is as inevitable as tall buildings and prohibition. But it is not amiss to point out that for all the good it accomplishes, it in no wise assists in making more pleasurable man's journey through this terrestrial sphere. In the good old days before we became so highly civilized the average man found in his day's labor not only the means to support himself in the world, but also the means to escape said world. Almost everyone was a craftsman of some sort. If an individual's job happened to be the making of door hinges, it earned him his daily bread; but more, it enabled him to escape into the realms of fancy in inscribing on the hinge elegant bits of embellishment. A man in the hinge business to-day, if he be among the higher ups, is embroiled in an unvarying routine of paper transactions, or if he be lower down, passes his days before a whirling bit of machinery, pressing a specified lever at a specified time and thus drilling a specified hole in a specified hinge on an endless belt. So occupied, there is little chance for his spiritual half to go wandering about in the great unmowed spaces of the fancy.

Of course means of escape still exist, but such specialists has our civilization made of us that they seldom exist in our day's routine. The impractical, and hence wholly pleasurable things, have themselves succumbed to civilization and become departmentalized. The arts, as we find them to-day, have been isolated like mumps and typhoid germs, tucked away into tightly sealed compartments and, like a bottle of strychnine, have been plainly labeled "Music," "Drama," "Painting," "Poetry," and so forth.

In the middle ages, before man was thoroughly educated, he got his drama by putting on a costume and cutting up on the great public festival days. Now the drama has been segregated and is to be found only in certain buildings labeled "Theaters." In those same days if a

great and artistic cathedral were being built every man in the village would have a hand in it. Now all living art is cloistered away in "Latin Quarters" or placed under glass in gloomy and forbidding museums.

This is an unfortunate state of affairs, for if there was ever a time when mankind stood in need of the arts it is mankind of the Western civilization to-day. Being more highly civilized than his ancestor, man of the Twentieth Century is entitled to more highly civilized forms of recreation. The incongruous fact of the case is that his progress in pleasurable pursuits has not kept apace of his progress in practical pursuits. In his day's work he may be able to do things with a column of figures or with a lathe that would have completely mystified his prototype of the Thirteenth Century, yet when the five o'clock whistle blows he finds himself with no wit more knowledge of how properly to divert himself than did his ignorant forebear.

Logically, and by all rights, the arts should be enjoying their greatest usefulness to-day. They should rank high among man's various means of gaining pleasure and escaping dull routine. Evidently they are not so doing. Never in the history of the present culture have they been more exclusive, snobbish, and aloof.

But while these preceding remarks may hold true for the majority of the arts, they can no longer hold true for music. While the art of painting has steadily increased in complexity until it has reached a state where it is utterly beyond the grasp of the ordinary man, the art of music has intruded itself into his daily routine and once again become a part of his living.

There is no need to trot forth statistics, to quote questionnaires, or to tabulate radio station correspondence; the fact is beyond dispute that music is enjoyed and appreciated in the West to-day in a degree immeasurably greater than it was five years ago. Music is being heard to-day from all sorts of stations and on all sorts of programs that a few years ago would have been essayed only by the most highbrow station on its most highbrow hour.

We do not delude ourself that radio has been able to contribute a highly sophisticated taste in music to very many people. But (what is vastly more important as an opening wedge) it has given a taste for music to millions of people who previously hardly knew it existed.

However, great as the development has been, so far only the surface of music's possibilities has been scratched. The most radio has done so far is to acquaint the great mass of people with the fact that there is such a thing as music and that it is pleasant to listen to. Its next step is to cultivate the taste of its listeners to

ENTHUSIASM AT KPO

If this is the sort of thing you like, Hugh Barrett Dobbs, physical culture mentor, will be glad to put you through some pre-coffee exercises from KPO at some unearthly hour of the morning

the point where they may get the most—or at least more—out of it. For any given piece of music will yield a hundred times more enjoyment to the person who "knows what it's all about" than to the person who simply listens to it as a vaguely pleasing succession of sounds. Yet an understanding of music is never gained by accident but can be realized only as the result of definite instruction in music appreciation.

What with all the zeal there is to-day for the imparting and securing of "education" there seems to be no good reason why some of this zeal may not be turned to the interest of music appreciation education. Radio is the best fitted agent to do this work and can function to its own best advantage by forgetting all its silly aspirations to supplant the technical college and by devoting itself to this equally large and far more important task.

Not all the millions of people in the United States who own receiving sets have the intelligence to really get the low-down on what music is, but that does not controvert the fact that there are thousands upon thousands of people in the land who have got the mental equipment to enjoy music if they put themselves to it. It is amazing how many people who are apparently cultivated, well educated, and surrounded by opportunities, and who profess to enjoy music, can be discovered, by a couple of well-directed questions, not to have the remotest idea of what music really means.

There are many excellent books on the market and in the public libraries which offer primer courses in the understanding of music. A very acceptable one, devoting itself primarily to radio listeners, was recently reviewed in this magazine. The only objection to learning music from a book lies in the fact that the book can't play the music it is talking about. It can quote measures but if you are unable to read music this is of little use.

Herein lies the unique advantage of radio; it can offer explanations and at the same time illustrate them. There have been a number of music appreciation programs on the radio already but the saturation point has been far from reached. An impetus in this direction is furnished by the report, new at the time we write, that Walter Damrosch has accepted the post of Musical Counsel for the National Broadcasting Company and has already under way a comprehensive plan for promoting fine music through the medium of radio broadcasting. This plan provides for a series of concerts supplemented by talks which can reach the majority of the 25,000,000 students in American schools and colleges. At the time of the announcement Mr. Damrosch made a statement, which, in case you have not read it, we will here quote:

My experiences of the past winter in broadcasting orchestral concerts with the New York Symphony Orchestra and in giving Wagner lecture recitals at the piano have so amply borne out my belief in the extraordinary possibilities of radio broadcasting that I have accepted with the greatest interest the position offered me as Musical Counsel of the National Broadcasting Company.

The plan which I propose to follow is the outcome of over 25,000 letters which I received not only from the larger cities, but from the smallest country towns and Western farms and ranches. In many of these letters the wish is expressed that orchestral music by radio should be extended to our schools and colleges—that my concerts and explanatory comments could thus supplement the work done by local teachers in the schools. This suggestion appealed to me greatly. The possibility of playing and talking to an audience of 25,000,000 young people fascinates me to such

MICKEY MCKEE

An accomplished lady performer on the whistle —her own—and heard in the "Roxy and His Gang" broadcast on the Blue Network Monday evenings

an extent that I shall gladly carry out the plan if it proves acceptable to the school authorities.

I propose to give twenty-four orchestral concerts with explanatory comments on the works presented and on the instruments of a symphonic orchestra. These concerts shall be broadcast to every school and college in the country that chooses to accept them. There will be three series of eight concerts each, with carefully graded programs, one for the elementary schools, another for the high schools, and the third for colleges.

Previous to each concert I would send to every school that desires it a questionnaire on the music to be performed and on my explanatory comments, together with the proper answers. These answers would, of course, be intended only for the eyes of the music teachers. After each concert the pupils could be examined by them and rated accordingly. If the parents are interested as well, the questionnaires could be distributed to them also, either through the school authorities or the local newspapers. The papers could print the answers a few days later.

The "Continuity" Program

WEAF Network. The "Coca Cola Girl!" The opening program of a new series. Advance notices prepared us to expect great things of this new weekly serial. It would, we were told, "combine drama and music in a manner never before used in radio." At last, we told ourself, we are to hear the ideal continuity program. So much for our expectations—we are still looking for the perfect "continuity" hour. The soft drink damsel's program is not it.

The saddest part of it is that it could have been. The general idea of the new feature is not bad at all—in fact, it's very good. But its execution is—terrible.

Each broadcast in the series is to present an episode in the life of one "Vivian," a beautiful young Southern girl. The first one saw her at her "coming-out" party, an occasion which brought back to her home town her friend of childhood days, Dick Amberson, accompanied by his friend James. Jim promptly fell in love with her, and, it is to be presumed, will pursue her through the remaining episodes. During their conversations snatches of music are introduced illustrative of what they are saying. So you see, the idea had marvellous possibilities.

But neither the actors nor their lines are good; though acting and lines would seem to be two important elements in drama. The lines are amateurishly written and horribly stiff. Jim being presumably an average American youth would probably have said, upon viewing Vivian, "Pretty swell-looking bimbo!" or something equally inelegant. Instead he is made to say: "I suppose everyone that lays eyes upon her is captivated by her." And on one occasion he is made to ejaculate "Capital!" To the actor's credit, whenever he has a line that isn't too infernally impossible he puts quite a bit of conviction into it. But the Vivian, aside from the handicap of poor lines, is ill suited for the part. She may look young, and beautiful and Southern, but all we have to go by is her voice, which, though not displeasing, is certainly not young or Southern. It is decidedly matronly and decidedly eclectic in its accent.

THE MUNICIPAL BAND OF BALTIMORE

Heard over WBAL every Friday night at nine o'clock E. S. T.

We dwell thus long on a feature which certainly doesn't deserve the space simply because of the great possibilities which it has left unrealized and because it paves the way for some other advertiser to take up the same idea and handle it properly.

How is it to be done properly? Well, it would take a considerable outlay of funds. In the first place, it would take an expert to compose the lines. It seems to us that the best writer available would be none too good. To suggest a couple of names, either Mary Roberts Rhinehart or Booth Tarkington could do the job handily. Either one could, w₁ₕhou₁ much effort, compose a series of thirteen episodes that would be not only true to life and entertaining but which would provide an interesting commentary on the contemporary younger generation.

Next a couple of talented, well-seasoned actors would be necessary to acquit the lines convincingly. The program here reviewed lacked what is known to the stage as "click." That is, the speakers didn't follow, overlap, or interrupt each other in a way characteristic of natural conversation. It can be done, and the proof of it is "Sam 'n' Henry." Whether you like these two WGN entertainers or not you must admit they deliver their stuff in a most expert and realistic manner. These actors would not have to be the same ones who furnished the singing bits as it is an easy matter for radio to double a singer without anyone knowing the difference.

As to the interwoven theme of music, we think in this phase of the program, at least, the "Coca Cola Girl" did itself proud and any subsequent imitation of this program could study its method of handling the "score" to advantage.

We offer these suggestions for an ideal continuity program to whomsoever will take them. To engage the services of a "best-seller" author would, we grant, cost a lot of money. But we think it would be repaid in the prestige it lent the program. Moreover, it would intrigue many thousands of the more sophisticated radio set owners into listening without in any way diminishing the pleasure of the low-brow listeners who gobble up anything like this, good or bad.

Tolerating Jazz

VARIOUS readers of this department have belabored us for what they call our "leniency" toward jazz. But if "leniency" is not the attitude to adopt toward jazz we do not know what is. There is no use getting mad at it. It is no unnatural, monstrous thing to be shunned like a two-headed dog, but a perfectly normal manifestation of the present age. Jazz, as jazz, is no more reprehensible than is hot dog. Hence our failure to rant against it periodically. To be sure, we should not like to have a steady diet of jazz any more than we would like to have a steady diet of hot dogs. But there are times when the lowly weenie hits a spot that even a sirloin smothered in onions cannot reach.

There is still too much jazz on the radio, we grant; especially in the hinterlands. But fortunately most of it occurs at hours when good folk have gone to bed so it bothers us not. Those who like jazz are, we believe, entitled to it, and are not to be too much reviled for liking it. Those who understand real music realize, as no jazz addict ever can, that jazz is indeed a namby-pamby substitute for music; but, by way of vindicating them, list to what Paul Whiteman says in an article contributed to the New York Times Magazine:

I sincerely believe that jazz is the folk music of the machine age. There was every reason why this music sprang into being about 1915. The acceleration of the pace of living in this country,

the accumulation of social forces under pressure (and long before the war, too), mechanical inventions, methods of rapid communication, all had increased tremendously in the past 100 years—notably in the past quarter century. In this country especially the rhythm of machinery, the over-rapid expansion of a great country endowed with tremendous natural energies and wealth, have brought about a pace and scale of living unparalleled in history. Is it any wonder that the popular music of this land should reflect these modes of living? Every other art reflects them.

Like the folk songs of another age, jazz reflects and satisfies the undeveloped esthetic and emotional cravings of great masses of people. Such music in any age has not been entirely negligible. Jazz is a spirit, not a manner. Crude, unmusical perhaps, but as healthily vulgar and sincere as were the vulgarities of the Elizabethan age—the music of an uneducated, vigorous man struggling ungrammatically to express his response to the age in which he is living. Since when in music have these forms of music been pronounced dead and worth ignoring?

<hr>

THUMB NAIL REVIEWS

XXX—Due to a perhaps commendable reticence on the part of the broadcaster we were unable to discover what station was offering the program, but it was one of the best variations of the informational program we have yet heard. The two speakers were presumably touring in Alaska. The one asked the other various questions about the scenes and properties peculiar to that region with well-feigned curiosity. For instance: "What are those carved telegraph post things up in front of the houses?" Whereupon the other, who knew his Alaska, launched forth into a lucid and conversational explanation of the totem pole, and, led on by further questions, discussed their history, related how they were the personal insignia of the great families, just as are our family crests, explained why some are higher than others and included a bit of native gossip as to why one of them featured the bear among its carven decorations. By some hook or crook, music was worked into the feature to lighten it up a bit. We found ourself thoroughly interested, though ordinarily we loathe being "radicated" by radio, and picked up quite a number of miscellaneous facts about Alaska that we had never heard of before. Only pressing duties at other parts of the dial induced us to desert it before the concluding announcement.

WJZ and the Blue Network—George Olsen and his Stromberg-Carlson orchestra. Of all its departments radio is probably best represented in its "light" orchestral section. There are no end of first-rate dance orchestras, hotel orchestras, and advertiser sponsored orchestras that can be regularly relied upon to play what they play well. Of these the Olsen organization is among the very best. Its program of popular numbers on this occasion was made up of some juicy and not-too-often-played tunes, among them: A Shady Nook; An Olsen Tango; Pusata Marden Waltz; An American Fantasy; and Melancholy Melody.

WJZ and the Blue Network—One of a series of operatic concerts under the direction of Cesare Sodero, featuring as soloists, Astrid Fjelde, soprano; Elizabeth Lennox, contralto; Julian Oliver, tenor; and Frederick Baer, baritone. Another chain feature that can always be counted on to be exceedingly good. The ordinary operatic program is inevitably made up of only the most popular arias of the various operas and fails to consider the second most popular and third most popular arias, which are often quite as good.

The wjz operatic hour gives the second string tunes a chance. For instance, on this program "Samson and Delilah" was represented by "Printemps qui Commence" instead of by the customary "Mon Cœur à Ta Voix." The "Mon Cœur" aria is certainly a "wow" but it shouldn't be allowed eternally to displace the "Springtime" song, which is every bit as musicianly and even more interesting on repeated hearings.

<hr>

Broadcast Miscellany

ONE of the high spots in the history of KPI's broadcasting was the program contributed last spring by John Barrymore—a Shakespearian Hour. Mr. Barrymore is probably the foremost American Shakespearian actor. He included the soliloquies from Hamlet and King Richard the Third.

THERE ARE FEW BASS VIOL SOLOISTS

RADIO certainly permits you to hear musical curiosities which ordinarily wouldn't be stumbled across in a lifetime. Witness: the Edison Hour program from WRNY which featured Leon Ziporlin as soloist on the bass vɪoʟ! There are only a handful of players in the country to-day who have mastered this ponderous instrument to the point of virtuosity.

WHAT A WINDOW-DRESSER THINKS ABOUT

OCCASIONALLY our name gets on the wrong mailing list and we are privileged to get in on some of the trade talk of the radio beezness, with frequently laughable results. This contribution to genuine and heartfelt sentiment by a "Dealer Bulletin:"

Another opportunity for ————— dealers. Possibilities of unusual window displays and newspaper ads selling the idea of remembering Mother by giving her a ————— Radio. Mother's Day comes but once a year—make the best of this opportunity. Dress up those windows—suppose you use a few photos of typical mothers, flowers, etc.

MISCHA ELMAN TALKS A BIT OF NONSENSE

A LOT of red-hot hooey contributed by Mischa Elman to the Amplion Magazine (London):

Can an artist give his best over the wireless? Judging from some sets I have heard, the best modern receivers are capable of reproducing the tone of a violin—even that of my £10,000 Strad—almost perfectly, but I am afraid an artist's personality suffers loss when he broadcasts.

The question of personality and broadcasting has been discussed a great deal since the introduction of wireless. I believe that a speaker can develop to a very high degree what is generally called "wireless personality," but in the case of a musician, it is inevitable that a great deal must be lost in transmission through the ether. If you consider the question carefully you will find that in listening to a violinist in a concert hall you use all your senses—chiefly your eyes and your ears, but your other senses also to some extent. The combination of these feelings enables you to appreciate the music to the full—to "sense" the artist's personality to the full, and, after all, that is the object of going to hear him.

Sight is very important—not because it enables you to admire his virtuosity in a set of Paganini variations, but because it enables you to "take him in"—to sense his personality.

THE book review and literary period presented over wow every Saturday evening from 8 to 8:30 has proved one of the popular features of the station. The period is conducted by Eugene Konecky, of wow's staff. The latest books are reviewed by Mr. Konecky.

WALTER MALLORY AT WCCO, MINNEAP-OLIS-ST. PAUL

Mr. Mallory, a popular tenor, sings with the Buick Gold Seal Vagabonds from WCCO each Monday from 9 to 10 P. M. Central Time

GILBERT SELDES in the *New Republic:*

I have tried again and again to make myself a picture of the air at one of those moments when every tiny turn of the dial brings something new to the loud-speaker. It hardly seems possible that so many things could be of interest, that so many people would be trying to sell or persuade or exploit. Maxwell House Coffee presents old Southern melodies; Mrs. Augusta Stetson talks about God-de; *Collier's Weekly* transposes its forthcoming issue into music and drama; the political situation is summarized by Frederick William Wile; dinner music is broadcast direct from Janssen's Midtown Hofbrau House; Aimee MacPherson wishes that she could tell you how lovely Jesus has been to her; specialists speak on recondite subjects which suggest that they have collaborated with Robert Benchley; a lesson in Spanish from the municipality's own station; a plea for Jews to speak Hebrew; how to take care of an Airedale; Al Smith addresses newsboys and can't remember what year this is—waves, voices, personalities crowd each other, interfere with each other; a faint hum of jazz accompanies a Catholic priest; a prize-fight cuts into Bach; as you rapidly turn the dial from one end of the gauge to the other, you hear grunts and shrieks and the wild whistle of static. It is everything that America is interested in; it is America.

WARNINGS TO PURCHASERS

AN INCREASING number of local organizations of the National Better Business Bureau are taking up broadcasting as a means of disseminating their findings. The Bureau, as you probably know, busies itself at the ferreting out of dishonest practices in advertising or in selling, and its exposes are generally interesting and always valuable. The following stations are now used by Bureaus:

WNAC—Boston	KFJR—Portland,
WMAF—Chicago	Oregon
WSAI—Cincinnati	WJAR—Providence
WTAM—Cleveland	KFSD—San Diego
WAIU—Columbus	WIAN—Scranton
WWJ—Detroit	WSBP—St. Louis
WOWO—Fort Wayne	KWG—Stockton
KPRC—Houston	WFBL—Syracuse
WOQ—Kansas City	WIBX—Utica
KFON—Long Beach	WRC & WMAL—Washington, D. C.

A POPULAR WGN FEATURE

THE Salernos—Frank and Lawrence—are one of the best of WGN's several twenty-minute regular features. Their program opens

with Lawrence, the vocal member of the pair, singing an old Neapolitan street song, to his own guitar accompaniment, while Frank supplies an accordion accompaniment in the background. Then they go into old Italian folk-songs, Frank interjects several accordian solos, and Lawrence devotes his excellent baritone voice to classical and popular numbers.

Correspondence

NARRAGANSETT, RHODE ISLAND.

SIR:

It is primarily for the more technical articles that I read *R. B.*, but nevertheless your department is always interesting. Was pleased with your grilling at the "Radio-Pest," but your rules for operating a radio set, like too many of the suggestions and informative articles appearing in the radio magazines, are all very well for the city listener, but how about us, provincials? In my own case, the nearest broadcasting station with any power is 35 miles away with 500 watts. There is no high-powered station within 125 miles. Our first move in the evening is to find out whether the static is bad. If it is we go to the show, or maybe read. At any rate we have no local station to fall back on when a bad night comes along.

Suppose there is no static. Do we pick out a good program and wait for that time before tuning in on the set? We do not. We have learned better. Although I knew better than to try it, one Wednesday evening, recently, I decided to read until time for the Maxwell House program. There was little or no static, so that one should have no doubt about being able to enjoy the program. The concert started at 9:00 P. M., and all was fine until about 9:10 when a ship with a spark set started calling Tuckerton, WSC. The operator called almost steadily until 9:30. At that time another ship called and got WSC and QSG'd;

PARK V. HOGAN

Organist heard from WJZ during the Estey Organ Recital on Sundays at 7.00 o'clock

received instructions to QSR from the other ship, and did so. At 10:00 P. M., when the Maxwell House concert was presumably at an end, the tail end of the QSR'd message was just getting into Tuckerton.

The above is just one instance. Every evening we are bothered by spark sets. Since the S.S. *Boston* and *New York* started the season, conditions are worse than ever. WJK will call WEL for some time, then will start sending long dashes and the "three dot dash dot" combination. This is often kept up for the duration of a feature program, one which has been anticipated, maybe, for a week.

We don't mind the harmonics of the c.w. sets, but the sparks spoil the program on any wave down to 300 meters or below. This looks so much like a complaint that perhaps it should go to the Commission; but I will send it to you to show how we, in the outlying districts, need a little help.

H. C. Dow.

THE CROSLEY MOSCOW ART ORCHESTRA

A recent bi-weekly feature broadcast on Sunday afternoons over twenty-two stations of the Red Chain. This feature was on the air between 5:30 and 6:30, a time which had been practically open until the institution of the Crosley feature last winter. The violinist standing before the microphone is Arno Arriga, Director of the Orchestra. This group specialized in semi-classical selections, which were found to suit the popular taste best. They are scheduled to return in September

The "Myracycle"

AMONG the many suggestions made to the newly created Federal Radio Commission is that of a gentleman from the Middle West whose wishes, if they came true, would do away with the term "kilocycle" in favor of a new and perhaps more useful expression, the "myracycle." The myracycle would represent a unit of ten kilocycles, and since stations on the familiar broadcasting band are to be separated by ten-kilocycle—or one-"myracycle"—intervals, the suggestion should not be dismissed without a hearing.

Using the term myracycle would mean that a station operating on 660 kilocycles would be rated as 66 myracycles, and at the top of the broadcasting band a station now on 1500 kilocycles would be known as a 150-myracycle station. This term would do away with the final cipher in our present listing of stations.

It should be remembered, however, that it has taken a number of years to bring the term kilocycle into even the outer consciousness of the average radio listener who still prefers to think in terms of wavelengths, and to introduce another term might put the whole business of frequency designations back into the middle age method of designation by meters. There is already another frequency term, the "megacycle," which represents a thousand kilocycles (one million cycles). We first heard it used in Doctor Pickard's study after his summer spent in measuring the polarization of high-frequency signals, and although it sounded strange at first, it proved to be very useful in speaking of amateur frequencies. The relation between these several units is shown in the table below:

1 cycle = 10^0 cycles = 1 cycle
1 kilocycle = 10^3 cycles = 1000 cycles
1 myracycle = 10^4 cycles = 10,000 cycles = 10 kilocycles
1 megacycle = 10^6 cycles = 1,000,000 cycles = 1000 kilocycles

The Use of Exponents

WHICH brings up another interesting point—the use of exponents in the arithmetical calculations in which all radio engineers must indulge from time to time. Exponents are among the mathematician's most useful shorthand symbols, as the table below will indicate:

1 = 10^0 = Units
10 = 10^1 = Tens
100 = 10^2 = Hundreds
1000 = 10^3 = Thousands (Kilo.)
1,000,000 = 10^6 = Millions (Mega.)

1 = 10^0 = Units
.1 = 10^{-1} = Tenths
.01 = 10^{-2} = Hundredths
.001 = 10^{-3} = Thousands (Milli.)
.000001 = 10^{-6} = Millionths (Micro.)

Now the rules dealing with these complicated looking figures are simple, and when mastered, provide an exceptionally easy method of handling large numbers, or numbers in which the decimal point of the answer is in doubt. The rules are as follows:

When multiplying numbers, add exponents.
When dividing numbers, subtract exponents.
When squaring numbers, double exponents.
When getting square roots, halve exponents.
When transfering an exponent across the dividing line, change its sign.

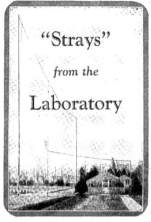

"Strays" from the Laboratory

For example the following problem may be simplified and solved:

$$\frac{1234 \times 0.02 \times 1000 \times 64}{2468 \times 800 \times 0.001 \times 100}$$

$$= \frac{1.234 \times 10^3 \times 2 \times 10^{-2} \times 10^3 \times 64}{2.468 \times 10^3 \times 8 \times 10^2 \times 10^{-3} \times 10^2}$$

$$= \frac{1.234 \times 2 \times 64 \times 10^3 \times 10^{-2} \times 10^3}{2.468 \times 8 \times 10^3 \times 10^2 \times 10^{-3} \times 10^2}$$

$$= \frac{1.234 \times 2 \times 64}{2.468 \times 8} = 8.0$$

As a somewhat more practical problem, let us consider the formula which states that the resonant frequency of a tuned circuit in cycles is as given below, when the inductance is in henries and the capacity in microfarads. If the coils and condensers in question are rated in millihenries and micro-microfarads, what will the constant above the line become?

$$f \text{ cycles} = \frac{159.2}{\sqrt{L \ b \ C} \ \text{mfd.}}$$

and since mh. = 10^{-3} hand mmfd. = 10^{-6} mfd., this formula becomes:

$$f \text{ cycles} = \frac{159.2}{\sqrt{L \times 10^{-3} \times C \times 10^{-6}}}$$

$$= \frac{159.2}{\sqrt{L \ C \times 10^{-9}}}$$

$$= \frac{159.2 \times 10^4}{\sqrt{L \ C} \sqrt{10^{-9}}}$$

$$= \frac{159.2 \times 10^3 \times 31.6}{\sqrt{L C}}$$

$$= \frac{5.033 \times 10^3}{\sqrt{L C}}$$

$$f \text{ kc.} = \frac{5.033}{\sqrt{L \ C}}$$

The Telephone Transmission Unit

THE use of exponents is the basis of our logarithms, as well as the foundation of the transmission unit, TU, which has been explained by Carl Dreher on several occasions in his department "As the Broadcaster Sees It." For example the exponent of 10^2 is 2, that of 10^3 is 3, and all numbers between .100 and 1000 have logarithms to the base 10 somewhere between 2 and 3. This amounts to saying that all numbers between 10^2 and 10^3 have exponents between 2 and 3.

The need for the transmission unit may be explained as follows. Let us suppose we have an electrical circuit—a telephone wire—connecting two points, and at the end of the first mile, the power has dropped to 0.9 of its original value. At the end of the second mile it has lost another 0.1 or is 0.9 of what it was at the end of the first mile, or 0.9 x 0.9, or 0.81, of its original value, and so on, to the end of the line. What is it at the end of eight miles? It is not only unwieldy to manage numbers of this sort but we must multiply them, which is less easy than addition.

Let us assign a unit to the ratio between the power at the end of the first mile to the original power, say A. This then represents the loss in that mile. Likewise, the ratio between the power at the end of the second mile to what appears at the end of the first is also A. In other words, at the end of the second mile we have lost twice A, or two units. If the line is eight miles long we shall have lost 8 units, and as Table No. 1 shows, the power will be 0.43 of its original value.

Reversing the direction of procedure along the line, as we approach the starting point we gain one unit for each mile of progress.

Again let us suppose that at the end of each mile as we go toward the end of the line we interpose an amplifier—a repeater as the telephone people call them—and that it boosts the power back to its original value. It is certainly much simpler to state that the amplifier has a power gain of A units than that it amplifies the power 1.111 times, for it must do such to raise 0.9 of the power back to its original value. The total gain of eight such repeaters will be 8 units.

The telephone engineer's foot rule, the transmission unit, is defined as ten times the logarithm to the base ten of the ratio between any two powers, or twenty times the logarithm of the ratios of voltages or currents into equal impedances. In mathematical language:

$$TU = 10 \log_{10} P_1/P_1 = 20 \log_{10} E_1/E_1 = 20 \log_{10} I_1/I_1$$

Thus an amplifier with a power gain of 100 has a gain of 20 TU since the exponent, or logarithm, of 100 is 2.0. Two of these amplifiers in series will have a gain of 40 TU or a power gain of 100 x 100, or 10,000. When a full orchestra plays fortissimo it is roughly 60 TU's more powerful than when playing pianissimo, a power ratio of 1,000,000. It is fortunate that the ear hears according to a logarithmic scale!

To become more familiar with the TU business the following facts may be useful. The average two-stage amplifier as used in broadcast receivers has a voltage gain of about 300, or 50 TU. The difference in power between a 500 watt station and one of 5000 watts is ten TU, and the latter enables a listener equidistant from the two to use 10 TU less amplification to get the same volume, which means that the receiver will be less susceptible to extraneous noises such as static. A variation of 10 TU at the two extremes of the audio-frequency spectrum, say at 100 or 5000 cycles, can be noted by the ear but will not make such an extraordinary difference in quality as some amplifier parts manufacturers would have us believe. That is, the volume at 100 cycles can be reduced by 10 TU, *i.e.*, the power can be reduced to $\frac{1}{10}$ or voltage to 0.316 of its former value before the ear notes it. A further reduction of 10 TU is appreciable.

Another example of the use of the TU, and one which merits very careful study, is shown in Fig. 1; it was taken from the *Bell System Technical Journal* for January, 1927, from an article by Lloyd Espenschied. The data for these curves which show the relative selectivity of several popular types of receivers was taken in the following manner. A laboratory oscillator was modulated at a fixed voice frequency, the receiver was tuned to the carrier, and the detected audio current measured as the oscillator was tuned in 10 kilocycles steps away from the original frequency.

The first significant point to note is that all receivers have a distinct cut-off within 10 kilocycles of resonance. Even at 5000 cycles the better grade receivers are 10 TU "down," which means that audio frequencies of this value will be down, and that a rising characteristic amplifier is probably a good idea. The curves show that the super-heterodyne or double-detector is considerably more selective than the others, and, as was to be expected, the simple single-circuit affairs have very little discrimination between wanted and unwanted signals.

Mr. Espenschied points out that undesired signals may not be bothersome when only 40 TU below the desired signals when the latter are strong, but when the program happens to call for a pianissimo passage the unwanted signals are disturbing. Reducing the level of the unwanted station to 60 TU eliminates this trouble.

The chart shows what may be expected in an area where stations are separated by 50 kilocycles and put an equal field strength about a given listening station. The sets with radio- or intermediate-frequency amplification give a 60-TU discrimination against unwanted signals with some to spare to take care of signals from more powerful stations. If the listener wants to "get out" he imposes a much greater task on his receiver. Suppose he receives 50,000 microvolts from a local station and 500 from a distant station (the example is Mr. Espenschied's). This represents a difference of 100 to 1, or 80 TU in favor of the local station. To reduce the local signals to the same level as the distant station, then, requires a selectivity of 80 TU and to take care of the added 60 TU necessary to reduce the local to the point where its signals will not bother during weak musical passes makes a total discrimination of 140 TU which the receiver must possess. This means a current reduction of the order of 10,000,000 to 1 —a high order of selectivity.

Colored Sockets and Bases ONE of the most interesting items of news from the "Nema" convention at Hot Springs, Virginia, is that regarding the new color arrangement for sockets and tube bases. According to this code, which was proposed by the Benjamin Electric Company, the bases of radio-frequency and first-

stage audio tubes, and the receiver sockets into which these tubes are used, will be colored maroon; the detector will be green, while the final tube, the power amplifier, and its socket, will be colored orange. The arrangement is, therefore, as follows:

Maroon: R. F. and 1st amplifier
$\left.\begin{array}{l} 201\text{-A} \\ 199 \\ 226 \end{array}\right\}$

Green: Detector only
$\left.\begin{array}{l} 200\text{-A} \\ 199 \\ 227 \end{array}\right\}$

Orange: Power tube only
$\left.\begin{array}{l} 112 \\ 171 \\ 210 \\ 120 \end{array}\right\}$

If we wanted to be facetious, we would suggest a crimson tube for Harvard and one for Lindbergh in whatever his favorite color may be.

We wonder what will be done with the other special-purpose tubes, such as the r. f. amplifier tubes with higher amplification factors and consequently higher impedances, rectifier tubes, high-mu tubes for resistance or impedance amplifiers, etc.?

The suggestion, admittedly, provides precaution against the danger of wrecking tubes or receiver through unfamiliarity with the inner workings of a complicated electrical machine. The average user of tubes has very little idea of the functions of the individual parts of his set. In receivers with an unusual arrangement of sockets, he can not know that the third tube from the end is not the detector unless he is warned before, either by reading a more or less dull and technical booklet, or by noting the color of the socket. While this color scheme has much to commend it, it seems somewhat inadequate in its present form.

New A. C. Tubes THE new R. C. A. tubes marked 226 and 227 operate from raw a. c.; their existence was vigorously denied by representatives of both the R. C. A. and Cunningham staffs only a short time ago. One day, seem-

ingly, the very thought of new tubes is abhorrent to these companies, while the next day complete operating data, photographs, etc., drop into the Laboratory, by special delivery, like a bolt from the blue. In the meantime—that is, between our statement last month and the time of going to press this month—a. c. tubes have been received and tested from the Armstrong Electric and Manufacturing Company, the Van Horne Company, the Sovereign Company, and the makers of the familiar Marathon tubes. Others proposed or available are the Arcturus, the Schickerling, the Quadrotron, the Zons, and probably others.

These tubes in general use are of two types, those using a low-voltage high-current filament, and those employing an extra heater which is not electrically connected to the receiving circuit. These latter tubes require several seconds to heat up and "get under way." The others are ready for reception as soon as the current is turned on. The R. C. A. has tubes—ready perhaps early in July—of both types; the R. C. A. high-current tube is for all positions except detector, and the heater type is for the detector socket. All of these tubes require different values of filament voltage and current, making the problem for the transformer manufacturer, or the home constructor, difficult, to say the least.

The new a. c. tubes can be used in place of d. c. tubes now in use, but we must admit that with a high-quality amplifier and a high-quality loud speaker, we have not heard a receiver using the a. c. tubes which was as "absolutely without hum" as the advertising would indicate. This, however, may be the fault of the receiver or plate-supply unit design. Time will tell.

Complete data on the various a. c. tubes will be prepared in time to appear in the September RADIO BROADCAST, we hope.

New Apparatus Received DURING the month of May, the following apparatus has been received in the Laboratory: Frost's new line of rheostats, jacks and sockets; resistances from American Resistor Company, Arthur H. Lynch Incorporated, De Jur, Electro-Motive Engineering Corporation, Aerovox, International Resistance Company, and Amsco; condensers from Aerovox, John E. Fast Company, Dubilier, and Globe Art Manufacturing Company; sockets and cables from Howard B. Jones; A. C. tubes from Marathon, Van Horne, Armstrong, Sovereigh Electric and Manufacturing Company; d. c. tubes from the Allan Manufacturing Company, Supertron, Magnavox, and Cable Supply Company; a pair of push-pull, high-quality transformers from Samson; a midget cone speaker from the Alden Manufacturing Company; a fine supply of Benjamin apparatus, including sockets, switches, condensers, etc.; a "Baritone", loud speaker unit; output filters from Federal, Erla, Muter, and Centralab; a new dry rectifier unit from Kodel; X-L binding posts; the Varion, a Raytheon A, B, C unit made by the Morrison Electric Supply Company, and recently described in *Popular Radio*; Marathon's new rectifier tubes; and as this is written, two very beautiful Westinghouse high-resistance volt meters.

The General Radio transformer type 285-N is for use as a coupling device between the element of the string oscillograph and a high-impedance bridge or tube circuit, and not as mentioned recently in these columns,

MILES	0	1	2	3	4	5	6	7	8
POWER	1	0.9	0.81	0.729	0.667	0.59	0.531	0.478	0.43
"UNITS LOSS"	0	1	2	3	4	5	6	7	8
POWER TO BASE 0.9	0.9^0	0.9^1	0.9^2	0.9^3	0.9^4	0.9^5	0.9^6	0.9^7	0.9^8

TABLE NO. I

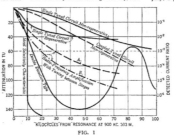

FIG. 1

Judging Tone Quality

*Training the Ear to Discriminate Between Good and Poor Quality—How Harmonics Cause
Different Instruments to Be Distinguishable—Practical Hints for the Potential Set Purchaser*

By EDGAR H. FELIX

APPRECIATION of good tonal reproduction is an art rather than a science. Scientific analysis of amplification curves are valuable to the engineer in determining possible causes of distortion, but the best and final test of reproduction quality is the ear of a trained and practiced observer. In working for good tone quality from a radio receiver, whether manufactured or home built, half the battle is understanding what good tone means.

In spite of the average radio enthusiast's opinion, not one in a thousand listeners is a discriminating judge of tone quality. Only most unusual exaggeration or the complete absence of low, high, or middle frequencies in the sound output of a receiver registers emphatically on the consciousness of the average observer. The most precise judgment which the self-appointed expert is able to pass upon a receiver's tonal capabilities is usually confined to such generalities as "good," "fair," and "poor." Yet the brain center of hearing is capable of making the most delicate discriminations. Only by learning to judge tonal defects with precision is one able to analyze the numerous possible causes of distortion in a radio receiver.

Much has been said and written regarding the causes of distortion. Oftentimes a particular make of transformer or a certain brand of sound reproducer is recommended as the panacea for deficient tone quality. But no one part of a radio set can, of itself, produce good tone. On the other hand, there is hardly a link in the radio chain which cannot exert a damaging influence on an otherwise perfect reproducing system.

In general, there are three kinds of distortion: (1.) Exaggeration of a particular frequency, causing a blasting effect; (2.) overloading, distorting all loud music of any pitch; and (3.) under or over amplification of a broad range of frequencies, such as the low, middle, or upper registers. Each of these is caused by specific faults in the design or operation of the receiver.

Simply that the causes of distortion may be viewed comprehensively at the outset, here is offered a partial list of the possible causes of distortion which may adversely affect the tonal output of a receiver, no matter what kind of audio and radio system is employed:

(1.) ANTENNA
   (a.) Too long, causing overloading of detector tube with sets lacking effective radio-frequency amplification control.
   (b.) Inadequate, requiring overworking of radio-frequency amplifier.
(2.) RADIO-FREQUENCY AMPLIFIER
   (a.) Regenerative or near-regenerative.
   (b.) Incorrect grid or plate voltages.
   (c.) Imperfectly synchronized circuits.
   (d.) Wrong type of tubes.
   (e.) Magnetic coupling with alternating current elements of power supply or with audio-transformers.
   (f.) Insufficient filtering or inadequate bypassing of power supply leads.
(3.) DETECTOR
   (a.) Overloaded:
     (1.) By radio-frequency amplifier.
     (2.) Because of incorrect grid voltage.
   (b.) Over-regeneration.
(4.) AUDIO AMPLIFIER

   (a.) Incorrect grid or plate voltages.
   (b.) Tube output impedance incorrect for load impedance.
   (c.) Audio-frequency regeneration due to:
     (1.) Magnetic coupling between transformers or coupling devices.
     (2.) Conductive coupling through imperfectly bypassed power leads.
     (3.) Acoustic or mechanical coupling of tube elements with resonant cabinet.
     (4.) Magnetic coupling with power circuits.
     (5.) Acoustic or mechanical coupling with reproducer.
(5.) REPRODUCER
   (a.) Saturation of magnetic element.
   (b.) Incorrect adjustment of vibrating element in relation to electro-magnet.
   (c.) Mechanical resonance in moving element.
   (d.) Discrimination against or favoring of certain frequencies in the electromagnetic unit.
   (e.) Acoustic resonance or filtering of certain frequencies in sound radiator.
   (f.) Limited frequency range of magnetic or acoustic elements.

This seemingly endless variety of possibilities may appear discouraging, so numerous are the hidden causes which may contribute to distortion. Nevertheless, modern amplification systems permit of relatively perfect reproduction, implying the conquest of each one of these possible causes of distortion. An optimist may be inclined to dismiss most of the list as unimportant and confine his attention to reproducer and transformers —favorite subjects for condemnation. But even the most superficial analysis will reveal that a few precautions in connection with the power supply, in the placing of the reproducer, and in the use of recommended tubes, will overcome most defects. Oftentimes, a slight readjustment will bring marked improvement in a receiver's tonal output. Other adjustments may make no perceptible contribution to improved tone quality, yet they may be, in fact, causing a really valuable improvement. Herein lies the most important obstacle to the attainment of true reproduction—the fact that the average listener is affected only *subconsciously* by most of the distortion to which he is subjected.

THE DELICATE SENSE OF HEARING

HOW delicate the discriminations which the ear can make, consciously or subconsciously, is indicated by the fact that there are 24,000 separate nerve endings in the ear membrane which may report impressions to the brain simultaneously. Each of these, habituated to the detection of conventional sounds of voice and musical instruments, is ready to report to the brain any deviation from the accustomed tone. Indeed, so complex and numerous are the impressions to which the sense of hearing may respond that we think definitely and specifically of only a very minute part of the actual impressions which the brain senses. If a loud speaker rattles harshly on a high note, you may notice it consciously and speak of it, but your trained ear has probably detected the existence of that rattle

or harshness a thousand times a minute before it was sufficiently acute to be perceived consciously! The millions of subconscious recognitions of distortion which your highly-developed hearing sense impresses upon your brain each minute of listening to a "fair" receiver are of great importance because they make radio music sound tiresome and unpleasant. They cause it to be mysteriously unreal and are the reason why you sometimes shut off your receiver without being able to offer a valid reason.

The sense of hearing is a creature of habit and, after a few hours or even a few minutes of listening, it can accustom itself to a persisting form of distortion. This is the cause of most of the misjudgments made with regard to tone quality. After listening habitually to a receiver having certain distorting characteristics, listening to a better set without that same distortion may be, at first, quite disagreeable. At the same time, a receiver having the opposite characteristics (one exaggerating the "lows," for example, after the listener has become habituated to one with marked absence of "lows") sounds peculiarly pleasing, although it is also a distorting receiver.

As evidence that habit accustoms one to distortion, the first telephone conversation in any individual's experience is generally quite difficult because the lower voice frequencies, which contribute mellowness, resonance, and sympathy in tone, are strangely absent. After a little practice, however, the ear accustoms itself to this form of distortion, so that almost any voice can be understood over the telephone without difficulty.

Obvious distortion in radio reproduction, such as rattles, whistles, and howls, etc., or the wholesale absence of low or high frequencies, can be appreciated by the most untrained ear, but the radio receiver that attracts admiration is the one designed not only to eliminate obvious causes of distortion but to bring out the many hidden fine points of tone quality which impress only the subconscious mind of the average listener. Once the earmarks of certain kinds of distortion are clearly understood, the trained observer becomes competent to distinguish forms of distortion otherwise felt only subconsciously. Appreciation of subtler kinds of distortion often enables you to find their cause in the radio-frequency amplifier, detector, audio-frequency amplifier, or loud speaker.

Mr. George Crom, Jr., of the Amertran engineering staff, has stated that the final test of a reproducing system is its effect upon the listener as measured over a period of several hours. If the loud speaker can be kept on continuously for many hours without wearying those within hearing or interfering with their regular occupations, it is likely to be reproducing with good quality.

Distortion due to resonance peaks and overloading is obviously distressing and most easily analyzed. But elimination of these obvious forms of distortion by correct adjustment of neutralization, elimination of undesirable interstage couplings, correction of overloading, and substitution of non-distorting coupling devices and reproducer, may yet fail in producing faithful reproduction.

Although there may be neither resonance peaks nor overloading, a receiver may still not give true music. In that case there is either insufficient or exaggerated reproduction of low, middle, or higher registers. Consequently, before proceeding with detailed analysis of causes, one must be able to recognize true reproduction—full, equal, and natural amplification of all of the essential range of frequencies.

Excellent reproduction is attainable in both transmitter and receiver, provided every element accomplishes its function properly. Perfect reproduction is not confined to any one system or amplification, such as resistance, impedance, or transformer, or to any one loud speaker. If this article appears to lay much stress, at first, on *what* good tone is, rather than *how* it is attained, it is because the understanding of this point is of great importance in analyzing the causes and capabilities of radio receivers and their audio-amplifiers and loud speakers.

INDEFINITE QUALITY TERMS

THERE is ample evidence in the advertising columns of any radio magazine that we speak of tone quality in most general and casual terms. Claims of instruments producing rich tones, mellow tones, and brilliant tones, and claims of good reproduction of low tones or upper harmonics, reduce frequently to admissions that such instruments distort. Yet tone is so little understood that a distorting quality may be claimed as a virtue. There is only one kind of reproduction which is to be sought as ideal and that is faithful reproduction, without any embellishments, exaggerations, or special qualities added to the original. Rich tone, for example, may be a pleasing contrast to thin tone, but it is a species of distortion which, in the end, is as tiring as thin tone.

True reproduction requires that the sound waves released from the loud speaker possess precisely the same characteristics as to frequency and amplitude as exist electrically in the audio-frequency modulation of the incoming carrier waves. This entails equal amplification of all the frequencies essential to good tonal reproduction, their transmission through all the various units of the reproducing system with equal facility, and their ultimate release as sound impulses, without any alteration. It means equal amplification of low, middle, and high registers, without marked resonance peaks. All these invisible qualifications of the perfect reproducer are manifest in its ultimate output—the music and speech impressed on the listener's consciousness.

Such faithful reproduction as described above is an attainable ideal but one quite beyond the reach of the average set owner. Satisfactory reproduction, however, requires fairly uniform amplification of the low, middle, and high registers rather than precisely equal amplification throughout the range, complete absence of sharp resonance peaks in the low and middle registers, and entire avoidance of overloading. But even granting these concessions to economy and simplicity from the ideal standard—true reproduction—it is surprising how far below the standard of satisfactory reproduction the average radio receiver falls.

Fundamentally, we are concerned with tone reproduction, with variations in air pressure, and with tiny air impulses mechanically set up by an electric machine. These impulses cause the diaphragm of the ear to vibrate, which, in turn, acts upon 24,000 sensitive nerve endings within the ear. Musical sounds are rhythmic variations in air pressure; noises or unmusical sounds are interrupted and irregular.

The most recognizable quality of tone is pitch. Pitch is the only characteristic of music which we

record in printed form as notes upon the musical staff. Melody is a pleasing succession of pitches. The clear, colorless, sound of the tuning fork is perfectly "pure" tone.

Pitch is our conscious recognition of the fundamental vibration of musical sound. The fundamental of the middle C on the piano, for example, is one of 256 impulses per second. So, also, is the fundamental set up by every instrument or musical sound sounding that particular pitch. The greatest amount of sound energy released by a musical instrument is concentrated on the fundamental.

The chart on page 226 shows the range of fundamental pitches set up by various instruments and voices. The lowest tone of a giant organ actually goes down to sixteen impulses per second. No commercial reproducing instrument goes so far down in frequency; in fact, only a few radio receivers respond to less than 100 cycles frequency, while most of them confine their attention only to frequencies higher than 150. Yet even these latter receivers reproduce a sound when the organ is broadcasting its lowest tone. That sound is due to the harmonics—energy released as double, triple, and higher multiples of the fundamental frequency.

The individual characteristics of different instruments, which enable a hearer to distinguish one from the other, even when sounding the same pitch, are embodied in these harmonics. When three instruments, for example, sound the same tone, not only is the fundamental alike in frequency, but so also are the harmonics. It is the *proportion* in which energy is distributed among these harmonics of identical frequency that gives instruments their individuality. The highest note of the French horn has a fundamental frequency of about 850. If the reproducer cut off at 900, you could recognize the highest pitch, but you could not distinguish what kind of an instrument was being broadcast. In the case mentioned, the French horn's individuality lies in the proportion of energy distributed among the fundamental frequency and higher harmonics. In a good reproducer, the French horn is distinguishable from the clarinet, for example, because the latter ranges the energy among the harmonics in a different way.

It is on account of the harmonics that it is important for a reproducing system to cover a wide frequency range, at least up to 6000 cycles. In judging the capabilities of an amplifier, the easiest method is to observe how clearly and easily distinction can be made among various instruments.

The following table, showing the relative energy distribution among fundamental and harmonics of the harp, piano, and violin, was calculated by Helmholtz:

HARMONICS AND THEIR RELATIVE INTENSITIES

Harmonic No.	1	2	3	4	5	6
Harp . . .	100.0	81.3	56.1	31.6	19.0	2.8
Piano . . .	100.0	99.7	8.9	2.3	1.2	0.01
Violin . . .	100.0	25.0	11.0	6.0	4.0	3.0

An amplifying system which is very weak in the region of the first harmonic of a certain note of the piano but which gravely exaggerates its third and fourth, might well make the piano sound like a harp. The existence of resonance peaks may so greatly alter the tonal quality of similar instruments, without however affecting the melody, that the instruments cannot be distinguished, and subtleties of composition and orchestration are thereby lost.

Perhaps the importance of unimpaired and unaltered reproduction of harmonics is best appreciated by considering the mechanism of our organs of speech. Through the so-called vocal

cords and diaphragm, we set up a fundamental tone or pitch and a wealth of harmonics. The different vowel sounds are produced by increasing or decreasing certain of these harmonics at will. In the mouth and throat are several resonance chambers, the size and shape of which we can alter with the aid of the tongue and jaws. "E," for example, is sounded by placing the tongue in such a position that one of the higher harmonics of the fundamental tone is stressed and the energy in intermediate harmonics reduced. It would be quite possible to make "ah" sound like "e" on a radio receiver for voices of a certain pitch by introducing a resonance peak at the proper point. Oftentimes the inability to understand speech from a radio set easily and without tiring is due to some such effect. Although we are never specifically conscious of a harmonic, the fundamental thing that makes a Stradivarius worth thirty thousand dollars and a cheap violin worth six dollars lies in minute percentage differences in the distribution of energy in their respective harmonics.

Certain instruments reproduce well on the poorest radio sets because their fundamental tones lie in the frequency range most easily reproduced and their harmonics are not important. A Hawaiian orchestra is particularly popular with the owner of an inferior radio set because distortion affects that particular musical combination the least. If an organ, concentrating as it does on the lower frequencies, sounds magnificent with a radio set, but sopranos are strangely colorless and without feeling, it is due usually to the absence of high frequencies in the reproduction.

The characteristic of radio sets prior to 1925 was the almost entire absence of low tones. The reaction which followed resulted in an era of exaggerated low tones, a much pleasanter form of distortion. We are about to enter upon an era of true reproduction. This involves curbing of the present trend to excessive booming mellowness and richness, when these are not present in the original music reproduced; and also the curbing of their predecessors, brilliance and sharpness of tone, the product of excessive amplification of high tones.

JUDGING QUALITY

THE expert sound critic, in testing a receiving system, listens to different kinds of music because each kind brings out the ability of the receiver with certain frequencies. The easiest way to learn to judge the deficiency of any reproducing system is to listen to distortion-free music and then to apply filters of known characteristics which take out various frequency bands. A little training will enable anyone of musical discrimination to analyze what is missing in music upon hearing it. An attempt will be made here to do this for you in words, describing familiar forms of distortion in the hope that you will be able to recognize them.

When a radio set is being demonstrated in a store, the most likely thing which you will hear is the male voice of an announcer. Fortunately, a male voice can reveal the principal characteristics of a sound reproducing system because it has a great wealth of harmonics. An absence of adequate amplification of the low tones, a characteristic of cheaper receivers incapable of great volume, is manifested by undue prominence of the consonants of speech. As we have seen, the lower frequencies give the pitch characteristic, the volume, and power, but it is the harmonics which give intelligibility and distinction. Hence, loss of the low tones does not detract materially from intelligibility. You could clearly understand what the announcer says if everything below a frequency of 500 cycles was cut off. If the announcer sounds as if he were talking over the telephone rather than speaking to you in person, it is likely that the "lows" are not adequately reproduced. The absence of "lows" detracts from the sympathy and resonance of the voice and makes it difficult to distinguish between individuals.

The other extreme is exaggeration of "lows" and loss of the "highs." If the male voice sounds throaty, booming, husky, muffled, and hashy, so that you think the speaker has asthma or needs a cough drop, you are likely to be listening to a receiver which exaggerates the low tones. Even with a moderate priced receiver, the male voice should be so clear that, when you turn from the loud speaker, you have difficulty in determining whether someone is speaking in person or whether you hear radio reproduction. The characteristic "radio speech" that makes you comment "it sounds like a radio," instead of "that sounds like a man speaking," is a recognition of distortion in one form or another.

A second form of radio reproduction likely to be heard at a demonstration is the piano. The piano has long been the most elusive instrument so far as reproduction is concerned. A majority of broadcasting stations are incapable of reproducing that instrument. Consequently, when listening to it critically through a radio receiver, you must be certain that you are tuned-in to a first class station. If that is the case, you will be able to detect inadequate amplification of low tones. If the piano sounds tingly, bell-like, sharp, and thin, and if the melody in the treble is prominent while the bass is weak, you can be sure that the low notes are not adequately reproduced. If you are fortunate enough to hear a fortissimo passage in the bass and find those low, crashing notes are only moderately reproduced and none of the grandeur of the piano is felt, you may rightly attribute these characteristics to imperfect reproduction of the lower frequencies.

If you are familiar with the tone of the clavichord, and piano reproduction reminds you of it, you may be certain that the low notes are not adequately reproduced. The essential improvement of the piano over its predecessor, the clavichord, lies in its possession of a large, resonant sounding board, which gives strength and body to the fundamental tones. If the lower frequencies are not reproduced by the radio receiver, obviously the piano will sound like the clavichord.

On the other hand, absence of the upper frequencies and exaggeration of the "lows" makes the fortissimo in the bass sound rich —almost organ-like. Reproduction in this case lacks the brilliance and brightness of the piano and possesses excessive richness and booming qualities. Fine trills in the treble or upper notes sound garbled and mixed, lacking entirely the penetrating brilliance characteristic of the piano.

A jazz orchestra also gives you opportunity for critical observation. The most prominent instruments, saxophone and cornets, use principally the middle registers. Reproduction in which the saxophone is permitted to dominate over piano, violin, and banjo, so that the latter form only a hazy background, exaggerates the middle registers. Usually, there is a set of instruments strumming time in the jazz orchestra, sometimes a piano, sometimes the banjos. It requires good reproduction of the higher harmonics to enable you to distinguish between piano and banjo instantly and decisively. If you have to stop and think about it before deciding whether piano or banjos are keeping time, the upper harmonics are not adequately reproduced.

When the piano contributes significantly to the melody by taking a lead in the treble, it should stand out prominently and clearly with its bright tingle. If it is submerged by the instruments in the low frequencies, the "highs" are being neglected. When the drums and the tympani come in, they should be emphatic and crashing, but a receiver which neglects the low frequencies makes the drums sound wooden and lacking in depth, instead of booming and rich. It is a good test of the power of the reproducing system with low tones.

Reproduction of the symphony orchestra should be sufficiently clear as to the upper harmonics so that there is not the slightest hesitation or difficulty in forming conscious distinction between the cellos and the violin. Given a good broadcasting station with a careful pick-up, the crashing of the tympani in the finale should be thoroughly imposing to the radio listener. When the violins take the lead and come forward busily, they should be sweet and strong and not scratchy and penetrating like an army of mosquitoes.

The last word in a test of reproduction of low tones is given by the organ. Few broadcasters are capable of sending out this instrument with anywhere near its true worth. Nevertheless, the organ should have majestic power reflected in its radio reproduction. Failure to reproduce the low notes makes it sound much like the flute or saxophone. An organ recital places the most severe test, not only upon the loud speaker and audio amplifier system, but also on the capacity of the plate potential supply to furnish the necessary current to accommodate the peaks of modulation present when the music of this instrument is broadcast.

It is not necessary to discuss in great detail that distortion due to sharp resonance points, because they are quite obvious. The listener notices an indefinable but disagreeable flatness in a certain pitch, usually a high note on the piano. That instrument, because of its sustained quality, brings out resonance points with particular emphasis. If that pitch is identified in the listener's consciousness, he will observe that a soprano singing that note sounds flat and slightly off key. Resonant points are most likely to be due to defective coupling devices in the audio-amplifier and mechanical resonance in the loud speaker and cabinet.

Overloading, also, is more easily considered as we go over the radio receiver part by part, to identify the causes of distortion. A receiver offering excellent tone with soft music but scratchy and blaring when it is turned on full, is likely to be suffering from overloading. Certain requirements with regard to tube output capacity, loud speaker volume capacity, and power supply, must be met to avoid distortion due to overloading. These points are fully discussed in an article on tone quality in the next RADIO BROADCAST.

For the present, if the reader has learned how to listen to a radio receiver, to appreciate its tonal quality, and to identify its weaknesses— whether they lie in the lower, middle, or upper registers—he will have taken an important step forward in judging the faithful reproduction qualities of a receiver.

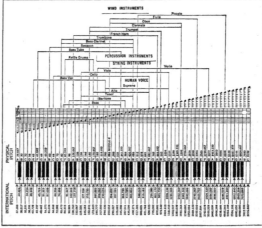

THE RELATION BETWEEN THE MUSICAL SCALE AND PIANO KEYBOARD
The organ range, not shown, is from 16 to 16,384 cycles. The chart is published through the courtesy of the Waverly Musical Products Company, Long Island City

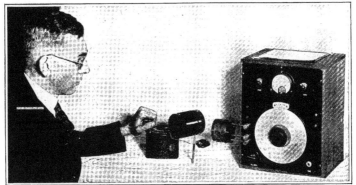

MEASURING THE CAPACITY OF A MICA CONDENSER BY MEANS OF THE MODULATED OSCILLATOR SHOWN IN FIG. 1

Condenser, Coil, Antenna Measurements

Some Simple and Practical Uses of the Modulated Oscillator— Comparing Its Accuracy with Other Systems of Measurement

By KEITH HENNEY

Director of the Laboratory

AMONG other uses to which the modulated oscillator described in the June RADIO BROADCAST and shown schematically here in Fig. 1 may be put, is that of comparing or measuring the capacities of small condensers. For example, it may be necessary to measure the maximum and minimum capacity of a variable condenser in terms of a standard, or it may be desirable to calibrate such a condenser. The modulated oscillator is very useful for this purpose, making it possible to substitute visible measurements for the audible indications of a bridge balance.

It is necessary only to have an inductance and a condenser whose capacity is definitely known. This latter should be a variable air condenser whose calibration is not likely to change with time, with as low a resistance as is practicable, and whose maximum capacity is from 500 to 1000 mmfd. Such a condenser is the General Radio Type 239 E or even the "can" condenser, type 247, made by the same company. The 247 E is inexpensive, has a dial calibrated in micromicrofarads, and can be read to within about 2 per cent. It has a maximum capacity of 500 mmfd., and, as an all round instrument it is to be highly recommended, although a better instrument is the 239 type. The latter is naturally more expensive.

Suppose, then, that we have a condenser whose calibration is known, and that around the laboratory we have an inductance, say about 60 turns on a three-inch diameter form, and another with about 15 turns on the same diameter form. These will enable us to tune to both low and high frequencies. Both windings may be on the same form.

The method of calibrating the unknown variable is essentially one of substitution. We connect the unknown across the inductance, couple the inductance to L_2 of the oscillator, and adjust C_2 until we get a reaction on the grid-current meter, when we know that resonance has occurred. Then the standard condenser is substituted and its capacity varied until resonance is again obtained, care being exercised to see that

the capacity of C_2 is not altered. Under these conditions we know that the settings of the two condensers—the standard and the unknown—when resonance is obtained, represent equal capacities.

When the unknown capacity is small it is simpler and more accurate, probably, to place both condensers in parallel, to resonate the circuit to some frequency so that about half of the unknown is used, then to remove the unknown and readjust for resonance, the difference in capacities of the standard under the two conditions representing the unknown capacity. If the unknown is large, the two condensers may be placed in series and resonance obtained. In this case the mathematics is not quite so simple, but the accuracy is the same and provides a good check on the parallel arrangement. In the series case the two should be connected and tuned first to some frequency so that the standard is near its top dial reading. Then the unknown is removed, and the standard lowered in value to a new resonance. The unknown value may then be found by using the formula and examples given below, where Cx represents the unknown capacity:

PARALLEL CASE

Standard alone, $88'' = 910$ mmfd.
Standard and Cx, $52'' = 415$ mmfd.
Cx $= 910 - 415 = 495$ mmfd.

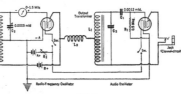

FIG. 1

SERIES CASE

Standard alone, $37° = 215$ mmfd.
Standard and Cx, $48° = 380$ mmfd.

$$Cx = \frac{Product}{Difference} = \frac{380 \times 215}{380 - 215} = 495 \text{ mmfd.}$$

Provided the dial on the condenser, and the calibration chart, could be read accurately, the percentage of accuracy of the unknown condenser, which was marked 0.0006 mfd., is as follows:

$$\frac{\text{Percentage}}{\text{Accuracy}} = \frac{\text{Measured Value}}{\text{Rated Value}} \times 100 = \frac{495 \text{ mmfd.}}{600 \text{ mmfd.}} = 82.5\%$$

and the percentage "off" is:

$$\frac{\text{Difference}}{\text{Rated Value}} \times 100 = \frac{105 \text{ mmfd.}}{600 \text{ mmfd.}} = 17.5\%$$

If it is impossible, however, to read the condenser dial closer than 5 mmfd. or to read the calibration closer than this figure, our error of reading may be 10 mmfd. and if the maximum capacity is 500 mmfd. the total error will be 2 per cent. With a large dial, and correspondingly large calibration chart, much greater accuracy may be obtained. If, in addition, the condenser is accurate to within 2 per cent., our total error of measurement may be 4 per cent.

MICA CONDENSERS

WHEN one comes to measure small fixed mica condensers by the method outlined above, curious results are obtained. At high frequencies, the distribution of charge on the condenser differs from that at 1000 cycles, and certain changes seem to occur in the dielectric. Manufacturers rate their condensers according to readings taken at 1000 cycles. It is also true that the equivalent series resistance of the condenser becomes appreciable at high frequencies. The result of all of these factors is that the apparent capacity of the condenser is smaller at high frequencies than at 1000 cycles. For example, the figures below give the values of several condensers measured at 1000 cycles and by the above method at 1000 kilocycles.

RATED	1000 CYCLES	1000 Kc.
250 mmfd.	230 mmfd.	220 mmfd.
250 "	200 "	190 "
500 "	405 "	350 "
500 "	500 "	480 "
600 "	530 "	500 "

TO MEASURE DISTRIBUTED CAPACITY OF COILS

ANOTHER interesting use for the oscillator is that of determining the natural wavelength, effective inductance, and distributed capacity of coils. The procedure is simple. The coil under test is resonated to various wavelengths by tuning it with known capacities, say with the variable standard. Then the added capacity is plotted against wavelength squared, as is illustrated in Fig. 2.

The data for this curve are given below:

TABLE NO. 1

ADDED CAPACITY	FREQUENCY	WAVELENGTH	(WAVELENGTH)²
500 mmfd.	562 kc.	534 meters	284,000
400 "	625 "	480 "	230,000
300 "	725 "	414 "	170,000
200 "	892 "	336 "	113,000
100 "	1290 "	243 "	59,000

The actual method of finding the inductance and distributed capacity is explained by the formula in Fig. 2. Where the line crosses the wavelength squared axis is the natural wavelength, squared, of the coil. The values for a certain coil, the dimensions of which are given below, as read from the graph, are given herewith, together with the figures obtained when different methods of measuring were employed:

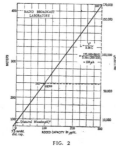

FIG. 2

Natural wavelength (by graph) .... 67.0 meters
Natural wavelength (by mod. osc.)........... 59.5 "
Distributed capacity .... 7.5 mmfd.
Inductance at 1000 cycles (bridge) .... 170 microhenries
Inductance at 1000 kc. (graph) .... 156 "
Coil diameter .... $2\frac{1}{4}$ inches
Length of winding ...... $1\frac{1}{2}$ "
Number of turns 60

TO MEASURE ANTENNA CAPACITY AND INDUCTANCE

THE oscillator can also be used to measure the capacity and inductance of antennas by the following methods. In the first method, two known inductances are inserted in the antenna system, and resonance obtained with the oscillator. If the inductance is in microhenries, and the capacity of the antenna is in microfarads, we shall have two equations:

$$\lambda = 1884 \sqrt{(L + La) \times Ca}$$
$$\lambda^1 = 1884 \sqrt{(L^1 + La) \times Ca}$$

where L and L^1 are the two known inductances, C_a is the antenna capacity, and L_a, the antenna inductance. C_a may be eliminated from the two equations, and after solving for L_a we have:

$$La = \frac{L^1 \lambda^2 - L \lambda^1{}_2}{\lambda^1{}_2 - \lambda^2}$$

After getting L_a, its value may be substituted in either of the two original wavelength equations and C_a determined. The second method, whose similarity with that mentioned above to determine the capacity and inductance of coils is apparent, consists in plotting wavelength squared against added inductance, as shown in Fig. 3. The intercept on the inductance axis represents the apparent inductance of the antenna; the intercept on the

wavelength squared axis represents the square of the natural wavelength. The capacity of course is the slope of the line as shown in the illustration.

The third method of determining the apparent capacity is to measure the wavelength, or frequency, of the antenna system with a certain, but not necessarily known, inductance, in series with it. Then the antenna and ground, or counterpoise, are removed and a calibrated condenser connected to the inductance. When resonance is obtained by tuning the condenser, its capacity is equal to the apparent capacity of the antenna.

All of these measurements at high frequencies will differ from those at audio frequencies, and will in general be true only at the frequency measured. They indicate an important class of experiments and measurements, however, that can be performed with the high-frequency part of the modulated oscillator. Care must always be taken that too close coupling to the oscillator does not vitiate the operation. Resonance should be indicated on the grid meter with as small a deflection as possible. The accuracy of measurements using these methods lies in the accuracy with which the oscillator is calibrated and read, and the accuracy of the standards.

Inductance standards are simple to construct. Their inductances at low frequencies may be calculated with sufficient accuracy for all home laboratory measurements. The interested experimenter is referred to the Bureau of Standards Bulletin 74 which gives formulas and descriptions of small standards of inductance. A subsequent article will describe a series of coils for use in the home laboratory as inductance standards.

The data for a short-wave antenna are given below and show how the values secured by the several methods compare:

ADDED INDUCTANCE	FREQUENCY	WAVELENGTH	(WAVELENGTH)²
16a microhenries	1612 kc.	186.6 meters	34,500
100 "	2030 "	147.5 "	21,800
18 "	3938 "	76.2 "	5,850

C_a = 55 mmfd. by method No. 1
" = 55 " " " No. 2
" = 54 " " " No. 3
L_a = 7.1 microhenries by method No. 1
= 12 microhenries by method No. 2
Natural λ = 47 meters by method No. 2

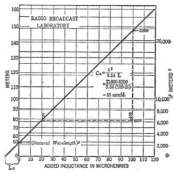

FIG. 3

A Super-Sensitive Five-Tube Receiver

Constructional Data for the Home Builder Who Desires a Selective Shielded Tuned Radio-Frequency Receiver with Controlled Regeneration in the Detector Circuit

By HERBERT J. REICH, M. S.

Physics Department, Cornell University

THE simplicity and selectivity of the various types of tuned radio-frequency receivers have for some time placed them in the foremost rank of popularity among the many receivers available to the radio enthusiast. A great deal of constructional data, on receivers of this type has been published and a large percentage of the tuned radio-frequency receivers now in use are probably home-built.

The chief difficulty that has to be overcome in the design and construction of a good tuned radio-frequency receiver is the tendency for the radio-frequency stages to oscillate. Oscillation in this type of set is obviously undesirable—for several reasons. In the first place oscillation ordinarily takes place at fairly low values of signal amplification, so that the overall efficiency of the receiver is low. Quality also suffers, and the tendency toward oscillation makes tuning critical, and when the set is in the hands of an inexperienced or thoughtless operator, it can cause serious annoyance to other receivers for miles around. If losses are introduced to prevent oscillation, the efficiency drops still further.

The use of some form of neutralization, and the careful placing of coils and other apparatus, partially overcomes this tendency toward oscillation. In spite of these precautions, however, there still remain capacitative and inductive feed-backs which careful placing of parts cannot prevent, and the obvious way to overcome these leakage feed-backs is to shield completely the individual stages. Many tests by the author and the recent appearance upon the market of several completely shielded neutrodynes indicates the value of properly designed shielding.

A receiver developed by the writer consisted of two stages of completely shielded and neutralized tuned radio-frequency amplification, a shielded regenerative detector, and two stages of high-quality audio-frequency amplification. From the standpoint of sensitivity and volume, this receiver is the equal of any tuned radio-frequency set and on a loop, or a ten-foot indoor antenna, it will give all the volume desired on local stations. Station WEAF is a rather difficult station to get with good volume at Ithaca, yet the writer has frequently had it with sufficient undistorted volume at 6:30 in the evening to be understood 400 feet from the cone loud speaker. Mexico City can often be heard with almost equal volume. Any time of day, from setting-up exercises to midnight dance orchestras, two or three stations at least may be had with as much loud-speaker volume as desired and with excellent quality on both local and DX. The limit of sensitivity is the general noise level, and the best evidence of a poor night is the number of whistles and yowls emitted by less fortunate neighbors. The receiver is simple to construct, extremely easy to operate, and comparatively inexpensive.

Fig. 1 shows the wiring diagram for the receiver. It will be seen that the circuit differs from the ordinary neutralized tuned radio-frequency circuit mainly by the addition of controlled

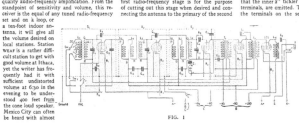

THE RECEIVER CONSTRUCTED BY THE AUTHOR

regeneration in the detector stage. The means of controlling regeneration, i. e., by means of a midget condenser, C_3, results in extreme simplicity and makes possible a wide variation in the amount of regeneration with only a very slight change of the tuning of the detector stage variable condenser. Regeneration also makes possible high amplification at the long wavelengths, where the gain would ordinarily tend to fall off.

The double-pole double-throw switch in the first radio-frequency stage is for the purpose of cutting out this stage when desired and connecting the antenna to the primary of the second

stage. This can frequently be done on good nights and somewhat simplifies tuning. The obvious objection is that selectivity is not so good.

Ordinarily, volume is controlled by means of the rheostat of the first r. f. stage, which is mounted on the panel. The regeneration control can usually be set at zero, but when additional volume is desired for weak stations or for daytime reception, additional amplification, equal to that of a third stage of radio-frequency, is instantly available with a turn of the midget condenser. The rheostat control makes possible the variation of volume from a mere whisper to the maximum the loud speaker can carry. When the first stage is cut out, volume is controlled by the midget condenser.

Separate terminals have been specified for the filaments of the second audio tube and the loud-speaker return, S, in order to make possible the operation of the power tube filament with a. c. from a transformer winding in a socket power device if this is desired. If the power tube is to be operated from the common A battery, these extra filament leads are connected directly to the main A terminals, and S, the loud speaker return, to A minus. The choke coil, L_{2b} may be put in any place in the B+ 180 volt lead if there is insufficient room in the set.

COIL CONSTRUCTION

THE coils are probably the most vital parts of the whole receiver, and particular care must be taken in their construction. Fig. 2 shows the design of the coil assemblies and the arrangements of the terminals, which are so placed as to give the most simple and direct wiring. The letters distinguishing the terminals correspond to those in the wiring diagram. At "A" and "B" is shown the detector stage transformer assembly, which contains three concentric tubes. The second stage radio-frequency transformer is identical with this except that the inner 2" tickler coil, and the M and T terminals, are omitted. The proper position for the terminals on the second and third stage transformers is determined by dividing the circumference of the outer tube into eight equal parts, as is indicated at "B." At "C" in Fig. 2 is given the arrangement for the first-stage transformer.

The secondaries are wound upon $2\frac{3}{4}''$ tubing and the primaries upon $2\frac{1}{2}''$ tubing placed within the secondaries and spaced by means of washers

FIG. 1

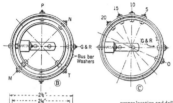

FIG. 2

made of bus bar placed over the terminal posts as they are inserted. The best length for the tubing is $2\frac{3}{4}''$, although $2\frac{1}{2}''$ can be used. The tubes should be assembled temporarily and the terminals located before the coils are wound. In order to insure the proper lining up of the holes in the different tubes, it is well to drill those in the outer tube first and locate the holes in the inner tubes by means of these, inserting the terminal posts and washers as each hole is completed. The tickler coil may be held by means of the screws which fasten the coils to the wooden upright and spaced with nuts or small pieces of panel. It will not matter if the tickler coil is not accurately centered. Round-head brass machine screws, 6–32 and $\frac{1}{2}''$ long, make excellent terminal posts.

The secondaries, L_2, L_4, and L_6, consist of 50 turns of No. 24 double cotton-covered wire. The upper ends of these windings should be started just far enough from the ends of the tubing to allow about a sixteenth of an inch clearance between the wire and the nut which holds the terminal post.

The primaries of the second and third stage transformers, L_3 and L_4, and L_5, consist of double windings of No. 28 or No. 30 double cotton-covered wire wound simultaneously, i. e., with the two wires side by side, turn for turn. Fig. 3 may help to make clear the proper way to make the primary connections to the terminals. One wire is fastened to the inside tubing at a point just above the "N" terminal, to which it should be connected after the coil has been wound; the other wire is fastened at a point just above the "B" terminal to which it will ultimately be connected, at the same distance from the lower end of the tubing as the lower end of the secondary is from the bottom of the outer tubing. The two wires are then wound simultaneously in the direction which carries them from "N" toward "B," the one connected to the "N" terminal above the other, until thirteen complete turns have been made with the upper wire (the one started at "N"). This will bring

the free ends of the wires above the "N" terminal. A small hole should then be drilled in the tubing just above the "P" terminal at such a height that the lower wire can pass into it and be fastened to the "P" terminal below. It is well to mark the proper point for the hole and unwind the upper turn of both wires while drilling the hole, or to determine the proper location and drill the hole before starting to wind. The upper wire should then be unwound a half turn, which will bring the free end above the "B" terminal. Another small hole should be drilled and this wire passed down to the "B" terminal, where it is finally fastened. If the terminals have been properly placed and the winding done as directed, there will be approximately $12\frac{1}{2}$ turns on each coil, and the lower ends of the coils will be at the same height as the lower ends of the secondaries.

The primary of the first-stage transformer, L_1, consists of a single winding of 20 turns of No. 24 d. c. c. tapped at 5, 10, and 15 turns. A neat way to connect the taps is to drill holes in the tubing at the proper points before starting the winding, and to bare the wire at these points when it is being wound, passing the tap wires through the holes and twisting them around the main wire. The joints can be soldered when the winding is completed.

The tickler winding, L_2, consists of 50 turns of No. 28 or No. 30 d. c. c. wire wound on the $2''$ tubing in the same direction as the primaries and secondaries. The lower end is connected to the "T" terminal and the upper end to the "M" terminal. The lower end of the winding is at the same level as the lower end of the secondary.

A good way to hold the ends of the wires to keep them from loosening up is to drill three holes close together and thread the wire in and out through them. A coat of collodion should be applied to the windings to hold them firmly in place. It is preferable to solder the ends of the windings directly to the heads of the machine screws rather than to terminal lugs, as this will save quite a few joints.

Wooden posts about $\frac{1}{2}''$ square with angle brackets at the bottom will serve to support the coils. The method of fastening the coils to the uprights is shown in Fig. 2, "A." The proper height is that which gives equal clearance from the shielding above and below the coils.

PARTS REQUIRED

THE choice of apparatus to be used in this receiver depends somewhat upon personal preference. In order to obtain the quality described in the discussion of performance, however, it is absolutely essential to use the very highest grade of audio transformers. For the faithful reproduction of voice and all forms of music can be expected only if each individual element of the receiver, including the audio transformers and loud speaker, is in itself capable of passing these frequencies. The use of old transformers salvaged from another receiver is to be discouraged.

The list below indicates what parts are necessary to construct the receiver; any standard makes may be used that have the proper characteristics:

C_2, C_3, C_7—Hammarlund Variable Condensers, 0.0005 Mfd.

L_1, L_2—
L_3, L_4, L_5—
L_6, L_7, L_8— } Home Constructed Coils. Specifications in Text
C_4, C_6—Hammarlund Neutralizing Condensers, 0.00015 Mfd. Approximately.
C_5—Sangamo 0.0001-Mfd. Fixed Condenser
C_7—Hammarlund Midget Variable Condenser, 0.00005 Mfd.
$C8$—Sangamo 4-Mfd. Output Condenser
C_9—Sangamo 0.00025-Mfd. Grid Condenser
C_{10}, C_{11}, C_{12}, C_{13}, C_{14}—Sangamo 1-Mfd. Bypass Condensers
C_{15}—Sangamo 0.001-Mfd. Fixed Condenser
R_1, R_2, R_3, R_4—Amperite Filament Resistors for 201-A Tubes, Type 1A.
R_5—Amperite Filament Resistor for Power Tube, Type 112 (not necessary with A. C.)
R_6—Yaxley 20-Ohm Rheostat
R_7—Yaxley 6-Ohm Rheostat
R_8—Tobe 2-Megohm Grid Leak
L_{10}—General Radio Output Choke Coil
J—Yaxley Single-Circuit Filament-Control Jack
J1—Yaxley Single-Circuit Jack
S_1, S_2, S_3—Copper Shields
SW_1—Yaxley Filament Switch
SW_2—Kresge Double-pole Double-Throw Switch
SW_3—Carter Antenna Tap Switch
T_1, T_2—Amertran Audio Transformers
Five Benjamin "CLE-RA-TONE" Sockets
Nine Binding Posts. Twelve if A. C. is used.
Three Dials
$7'' \times 24''$ Panel
Copper Panel shield, $6'' \times 18''$
Three Base Shields, $6'' \times 11''$
$13'' \times 23''$ Baseboard

The antenna coupling switch is best mounted within the shielding and should therefore be discerned to clear the tuning condenser and the side of the can.

SHIELDING

FIG. 4 shows the design of the shielding cans. Sixteen-ounce, or heavier, sheet copper should be used for all shielding. The cans are $10\frac{1}{2}''$ long, $5''$ wide, and $5\frac{1}{2}''$ high, and are open at the panel end. A $\frac{3}{8}''$ or $\frac{1}{2}''$ flange around the bottom affords the most satisfactory method of fastening the cans to the baseboard. A $3''$ hole in the top with its center $3\frac{1}{2}''$ from the back and right edges, makes it possible to insert and remove tubes without taking off the cans. The holes should be supplied with lids. Most builders will find it better to have a good tinsmith make the cans than to attempt to make them at home.

The panel shield (see photograph) is a single sheet $6''$ high and $18''$ long. It is a little better to use three separate $6'' \times 11''$ base shields for the three stages than one large one.

The general layout of apparatus may be discerned by references to the photograph. The condensers should be so placed as to allow as much clearance as possible on all sides of the stator plates and sufficient clearance for the rotor plates to insure freedom from short-circuits to the shielding cans. It is best to allow equal clearance on both sides of the coils and to allow at least $\frac{1}{4}''$ clearance between the coils and condensers.

The three shielded stages should be spaced just far enough apart so that the can flanges do not make contact with one another. As the frames of the first and second condensers are connected to C minus and that of the third to A plus, whereas the shielding is connected to A minus, it is necessary to cut holes in the panel shield around the condenser shafts and mountings. If metal dials are used, care should be taken to make certain that no short-circuit occurs between the dials and the screws which hold the shielding in place.

The screws which hold the sockets and coils will serve to fasten the base shields. The panel shield is held by four brass machine screws behind each condenser, placed so that they will be hidden by the dials. The base shields are soldered to the panel shield at two points in each stage.

If possible, the midget condenser should be placed within the shielding of the third stage, but it will probably be impossible to obtain sufficient clearance from the tuning condenser. The panel shield must be cut away where the antenna switch is mounted. The output condenser may be mounted on edge beside the second audio transformer, and the detector B plus bypass condenser, C_{15}, beside the first audio transformer. This will leave room for the r. f. amplifier B plus bypass condensers, C_{10}, and C_{11}, within the second and third stage cans, and for the C minus bypass condensers, C_{12} and C_{16}, within the first and second-stage cans.

In fastening the filament resistors it is best to separate the mountings from the shielding by means of strips of gasket rubber or similar insulation. Short-circuits are likely to occur if this precaution is not observed.

Little difficulty will be experienced in wiring the receiver, as most of the connections within the shielding are short and direct. Flexible wires tied into a cable will be found both neat and convenient for filament and battery leads leaving the cans, and for filament and battery leads going from the audio-frequency end of the board to the terminal strip at the rear of the set. Different colored insulations for the different leads will make it easier to check the wiring. Busbar should be used for all connections within the shielding, particularly for grid and condenser leads and for interstage plate and neutralizing leads. The rigidity of this type of wire will insure permanence of neutralization and of tuning condenser calibration.

There are two simple and equally good methods of leading wires out of the shielding. Probably the easier and neater way is to drill holes through the base shielding and board and cut short grooves in the underside of the baseboard under the edges of the cans to take the wires. The wires should be covered with spaghetti in the holes and grooves. After wiring, the grooves may be filled with tar or sealing wax taken from the tops of old dry cells. The two sets of interstage plate and neutralizing leads should be run in separate slots spaced about an inch or so apart.

The second method is to run the wires above the base shielding and cut notches in the bottom edges of the cans. The wires may be protected under the edges of the cans by means of bridges of copper soldered to the base shields. The notches in the bottoms of the cans should be as small as possible. If they are made too large, they form closed paths for eddy-currents and a slight amount of interstage coupling is likely to result. To minimize capacity to shielding, the neutralizing and plate leads should be kept a quarter to a half an inch above the base shielding except where they pass out of the cans. For the same reason, the copper bridges should be just long enough to afford protection to the wires.

The antenna, midget condenser, and detector plate leads may also be taken out of the shielding in either of the ways described, but as directly as possible. Each should be taken out individually through separate holes and grooves in order to prevent undesirable feed-back and capacity.

All leads within the shielding should be as direct as possible and close parallel wires should be avoided. The wires connecting the grid terminals with the secondaries and the condensers should be short and placed so as to keep them well away from the tops of the cans. This precaution will prevent the narrowing of the tuning range as the result of high minimum tuning capacity.

Obviously, since the shielding is connected to A minus, negative filament terminals, bypass condensers, midget condenser, etc., may be connected directly to the shielding, thus simplifying the wiring.

ADJUSTMENT AND OPERATION

BEFORE inserting the tubes into the sockets, it is well to check all connections by means of a battery and phones, buzzer, or voltmeter. The tubes should be inserted and lighted before the plate voltage is applied to the set. Another good test is to connect the A battery to the B-battery terminals and make sure that the tubes do not light. After this, all connections to the set may be made. If all connections are correct, the familiar microphonic ring will be heard when the detector tube is tapped. As the set has not yet been neutralized it will probably oscillate as soon as the three dials are brought into resonance, particularly at the shorter wavelengths, and tuning will be very critical.

Neutralization is a very simple process. The carrier wave of any fairly strong station at the short wavelength end of the dials will serve for this purpose. The regeneration control should be turned up far enough so that a strong heterodyne whistle is obtained when the first and second stages are detuned sufficiently to stop them from oscillating. The receiver should then be tuned so that the whistle is at maximum volume. The radio-frequency stages may be prevented from oscillating by turning down the rheostats. The first-stage shielding should be removed and the filament of the first tube disconnected by moving the filament resistor or turning off the volume control rheostat on the panel. It will be found that the whistle is still fairly loud and varies in intensity with the adjustment of the neutralizing condenser in the first stage. The neutralizing condenser should be adjusted until the whistle cannot be heard, or, at least, is at a minimum. It may be necessary to re-tune the first-stage tuning condenser slightly during this adjustment. The replacing of the can will probably upset the balance slightly. This difficulty may be avoided by adjusting the neutralizing condenser through a small hole in the can, but if this is done, the screw-driver must be thoroughly insulated by means of tape where it passes through the can. Failure to take this precaution will result in the short-circuiting of either the B or the C battery.

If neutralizing has been carefully done, it will be found that the receiver will not oscillate with any setting of the tuning condensers or antenna switch, providing detector regeneration is turned off. It should be possible to turn up the regeneration quite a bit without causing oscillation, and tuning should be very smooth. If the set oscillates when brought into resonance, there is some

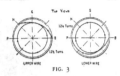

FIG. 3

error in the construction of the coils or in the wiring, or neutralization has not been perfectly accomplished.

The antenna coupling switch serves several purposes. Primarily it affords a means of adjusting the set to antennas of different lengths. It will, however, be found necessary to change the switch setting for stations of different wavelengths in order to obtain best results. The longer wavelength stations require a much greater amount of coupling for maximum volume than do those of short wavelengths. The antenna coupling tends to raise the wavelength of the first stage and broadens the tuning. This becomes very noticeable below about 300 meters (1000 kc.), and hence it is necessary to reduce the antenna coupling to ten or even five turns for the lower stations. With five turns, the tuning will be very sharp, down to the lower limit of the receiver. Decreasing the antenna coupling increases the selectivity; it also increases the volume of short wavelength stations but decreases that of long wavelength stations. This decrease of volume may be largely compensated by the use of regeneration, which also sharpens the tuning of the detector stage.

If the coils have been carefully constructed and the wiring carefully done, so that the three stages are as nearly alike as possible, the three dials should tune the same to within half or three-quarters of a degree over at least three-quarters of the broadcast band. At the lower end of the band, where the effect of the antenna circuit necessitates the change of antenna coupling, the first dial may read slightly different from the other two. For an average antenna the fifteen-turn tap will give the best coördination of dial settings. The use of regeneration will vary the detector dial setting by not more than half or three-quarters of a degree.

Fair results will be obtained if 199 tubes are used in the first three stages, but the volume will not be anywhere nearly as great as when 201-A tubes are used, and the quality will also suffer somewhat. In fact, much of the improvement that results from the use of shielding will be lost if small tubes are used, and it is therefore strongly recommended that 201-A tubes be employed. A 201-A tube must be used in the first audio stage, and either a 112 or 171, with proper plate and grid voltages, in the second audio stage. A 200-A detector tube will give slightly greater volume but is inclined to be quite noisy and microphonic, and broadens the detector tuning noticeably. A 2-megohm fixed grid leak of the best manufacture should be used with the 201-A tube. Excellent volume will be obtained if a loop is used in place of the secondary coil, L_4, of the first r. f. stage.

The set is not an experiment. It has been developed by a series of careful tests of various types of apparatus and circuits. Following the publication of data on an earlier model of the same receiver, the writer received a large number of letters from successful builders, some of whom have built a second and third set. It is partly in response to requests for data on modifications and improvements that the present article has been written.

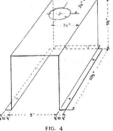

FIG. 4

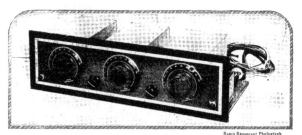

RADIO BROADCAST Photograph

Constructing a Five-Tube Neutrodyne

Selectivity, Sensitivity, Ease of Operation Embodied in this
Shielded Receiver—A Circuit for the New A. C. Tubes

By HOWARD E. RHODES

IN THE preceding article Mr. Herbert J. Reich described the construction of an efficient five-tube neutrodyne type tuned radio-frequency receiver. Mr. Reich's receiver, using home-made coils and shields, will appeal to experienced home-constructors, but there are also many who would prefer to construct the set using standard manufactured parts. Such a receiver has been constructed in the RADIO BROADCAST Laboratory and is illustrated herewith. In building this receiver some slight circuit changes were made which simplify the construction. For example, the circuit has been altered so that the variable condensers need not be insulated from the shields; insulating them was necessary in order to get a 4½-volt grid bias on the r. f. tubes. In the revised circuit shown in Fig. 1, the r. f. tubes have only one-volt bias on the grid and con-

sequently the total plate current drain is about 4 milliamperes greater than it would be with a 4½-volt bias. The change from 4½ volts to 1 volt for r. f. grid bias does not result in any loss of efficiency.

Some loss in efficiency possibly results through the use of coils not having the specifications recommended by Mr. Reich (which is the case with the receiver herein described), but even so, the completed model gave excellent results, it being possible during several tests at the Laboratory in Garden City to hear WGY, Schenectady, and WJAR, in Providence, R. I., during the daytime. Excellent local reception can also be had using a loop or a 12-foot piece of wire thrown across the floor.

The circuit diagram of the receiver given in Fig. 1, and the various photographs illustrate

very clearly how the receiver was assembled. The following parts were used in the construction:

L₁, L₄, L₇—
L₂, L₅, L₈— Three Silver-Marshall Coils, Type 116-A
L₃, L₆, L₉—
L₀—Samson R. F. Choke Coil, Type 85
C₁, C₂, C₃—Amsco 0.00035-Mfd. Variable Condensers
C₄, C₅—Hammarlund Neutralizing Condensers
C₆—Sangamo 0.00025-Mfd. Grid Condenser
C₇—Hammarlund 0.00005-Mfd. Midget Variable Condenser
C₈, C₉—Aerovox 1.0-Mfd. Bypass Condenser
C₁₀—Aerovox 4-Mfd. Condenser, Type 200
T₁, T₂—Thordarson R 200 Audio Transformers
T₃—Thordarson R 196 Choke Coil
S₁, S₂, S₃—Alcoa Aluminum Box Shields
R₁, R₂, R₃, R₄—Amperite Filament Resistors Type 1-A

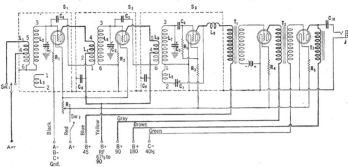

FIG. 1

R₄—Amperite Filament Resistor, Type 112
R₆—Lynch 4-Meg. Grid Leak with Mount
R₇—Carter Imp, 15-ohm Rheostat, Type IR-15
J—Yaxley No. 1 Single-Circuit Jack
SW₁—Yaxley No. 730 S. P. D. T.
SW₂—Yaxley No. 10 Filament Switch
Five Frost Sockets, Type 530
Three Karas Dials
Lignole Panel 7″ x 24″
Three Silver-Marshall Coil Sockets, Type 515
Yaxley No. 660 Seven-Wire Battery Cable and Plug
Kellogg Switchboard and Supply Co. Hook-up Wire
Kester Rosen Core Radio Solder
Hardwood Baseboard, 23″ × 12″.

In order to adapt the Silver-Marshall 116-A coils to this receiver, some slight changes in two of them are necessary. In drawing No. 1, Fig. 2, is shown the way in which the coils are connected when they are purchased, the numbers on the coil terminals corresponding to the numbers on the Silver-Marshall type 515 coil sockets. It should be noted that the lower end of the secondary coil "A" is connected to terminal No. 6 and that the lower end of coil "B" is also connected to the same terminal. The Silver-Marshall coil can be used without any changes whatsoever for the antenna stage and by referring to the circuit diagram, Fig. 1, it will be found that the various coil terminals are marked with figures indicating the terminals on the coil socket.

In sketch No. 2, Fig. 2, we show the coil revised for use ahead of the second r. f. tube of the receiver. The small winding connecting to terminals Nos. 1 and 2 should be removed entirely and then the three leads connecting the coil "B" terminals Nos. 4, 5, and 6 of the coil socket should be removed and the leads of this coil connected instead to terminals 1, 2, and 4, as indicated in drawing No. 2.

The detector coil is shown in sketch No. 3 and the 116-A coil should be so altered as to conform with this sketch. Coil L₄, connecting to terminals Nos. 1 and 2, is not the same coil as "C" in sketch No. 1 because this coil has not sufficient turns to produce regeneration. Coil "C" should be removed and a new coil wound in the same slot. It should consist of 40 turns of No. 30 wire and the winding should be started at terminal No. 1, which is also the terminal to which the lower end of the secondary is connected, and should be wound in the same direction as the secondary winding, until the required number of turns has been placed in the slot and then the winding should be completed at terminal No. 2. Those home-constructors desiring to make their own coils will find the

necessary data in the preceding article by Mr. Reich.

The various photographs given herewith will be very helpful in constructing the receiver since they show very clearly the layout of the apparatus. It will be found best to first lay out the apparatus in the three shielded stages and then drill holes in the aluminum bases of the shield cans at the correct points. The apparatus can best be fastened to the bases by means of 6-32 machine screws. The coil sockets should be mounted on 1¼″ 6-32 brass machine screws and in this way the coil socket can be supported so that it is about 1″ from the bottom of the can. The coil, when it is placed in the socket, will be more or less centered within the shield. In fastening the 6-32 screws that hold the coil sockets to the base it is a good idea to place

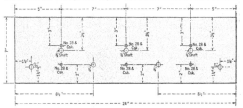

FIG. 2

no drilling template is given for the base of the aluminum cans. It will be found easier and in most cases more accurate to actually lay the apparatus on the base and mark with a center punch the various holes that must be drilled. The drilling dimensions for the front panel are given in Fig. 3. The panel is held to the receiver proper by means of six flat-head machine screws which

FIG. 3

a lug under each nut, which can later be used for making connections between the apparatus and the shield. The mountings for the Amperites should be prevented from touching the aluminum base by placing a couple of small washers between the mounting and the base when the former is fastened.

It is not at all essential that the apparatus be laid out in any exact fashion and for this reason

are passed through the six No. 28 holes indicated in Fig. 3 and then passed through six corresponding holes in the fronts of the three shields and finally held with a nut. In this way the fastening screws for the main panel will be concealed under the dials and the appearance of the front panel thereby improved.

In wiring the receiver, full advantage should be taken of the fact that the shields connect to the negative A. This makes it unnecessary, for example, to connect any wires between the rotor plates of the variable condensers and the lower end of the secondary windings. It is merely necessary to fasten the lower end of the secondary winding to the shield and then, since the variable condenser is also connected to the shield, the circuit will be complete. The leads connecting between the various stages can be run out of the shields by making small notches in the lower edge of the side of the shield and then passing the wire under these notches.

The C battery on the first audio stage is placed in the set itself, as can be seen in the accompanying photographs.

In connection with the building of this receiver, it is suggested that the reader carefully read Mr. Reich's manuscript because it contains many constructional hints that will be helpful. After the construction is completed it will be necessary to neutralize the receiver in order to obtain most efficient operation. The process of neutralizing the receiver should be the same as that explained by Mr. Reich except that with

COMPLETE EXCEPT FOR THE LIDS OF THE SHIELD CANS

A REAR ASPECT OF THE COMPLETED RECEIVER

this receiver the filament of the tube which is to be neutralized should not be turned off by removing the Amperite but, instead, should be disconnected from the circuit by removing from the socket the positive filament lead. This change in procedure is necessary because in Mr. Reich's receiver the Amperites are in the positive filaments and in this receiver they are in the negative filaments.

A. C. OPERATION

THERE have been announced several new tubes designed for operation on alternating current. Four types of these tubes have been received by the Laboratory and are illustrated in a photograph on this page. Many readers will doubtlessly be interested in the possibility of using these new tubes in conjunction with the shielded receiver described so as to make a complete a. c. receiver out of it. The Laboratory experimented with these tubes and all of them proved quite satisfactory when used in the receiver.

The Marathon and Sovereign tubes at the left are of the heater type, while the Van Horne and Armor tubes at the right are of the the a. c. filament type. In the case of the first two tubes, of the heater type, the alternating current flows through a filament which becomes hot and heats

up a special cathode which, when hot, emits electrons and acts similarly to an ordinary filament in a 201-A tube. In the case of the latter two tubes, the alternating current passes through a heavy filament which is itself the source of electrons necessary for the operation of the tube.

When using any of these a. c. tubes it was found best to change the detector circuit so as

SOME NEW A. C. TUBES

They are described in the text on this page

to make it a C-battery type detector because, with a grid leak and condenser for detection, there was a small amount of hum which practically disappeared when C-battery detection was used.

To use the Marathon tubes, no changes in the filament circuit of the receiver are necessary. These a. c. tubes are merely placed in their sockets and then extra filament wires are run from a filament transformer to the several tubes, which are connected in a parallel arrangement. These tubes have characteristics similar to the 201-A type tubes and are therefore suitable for use in the two r. f. stages, the detector, and first audio stage. An ordinary 171 should be used in the last stage with its filament operated on a. c.

The Sovereign tube can also be used without any filament circuit changes in the receiver except that the filament terminals on all the sockets, should be strapped together. The tubes are then energized by connecting the heater terminals in parallel and supplying them from a small transformer.

To use the Van Horne or Armor tubes it is necessary to slightly change the filament circuit of the receiver. The revised circuit diagram for these tubes is given in Fig. 4. It should be noted that both sides of each filament are connected across two terminals of a small transformer. A low-resistance potentiometer should be made from a 10-ohm rheostat and is shunted across the transformer, negative B, plus C, and ground connecting to the movable arm of the potentiometer. This circuit also indicates a C-battery type detector. The second winding on the filament transformer supplies 5 volts to the 171 power tube in the output stage. Transformers satisfactory for use in conjunction with these various tubes are made by Mayolian, Amertran, Dongan, Thordarson, Samson, and Enterprise.

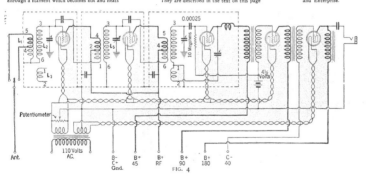

FIG. 4

AS THE BROADCASTER SEES IT

By CARL DREHER

Drawings by Franklyn F. Stratford

The Future of Television

IN THE July issue we discussed at considerable length the place of television among the communication arts, from what might be termed a metaphysical viewpoint. We tried to outline the fundamental relationships of the different branches of electrical, acoustic, and optical communication and reproduction in time and space. A moving picture showing Charlie Chaplin being hit in the nose by a custard pie, or a television apparatus reproducing for observers in New York the smile of a pretty telephone operator in Washington, may not seem very promising subjects for an analysis of this nature, but the content of the reproduction has nothing to do with the abstract functions of the machinery involved. Having disposed of these fundamentals to the best of our ability, we may proceed with an examination of the practical aspects of television. What part may it be expected to play in the industrial and social life of countries like the United States?

One thing should be noted before we embark on our self-appointed mission of prophecy. Practical television is not around the corner. Many moons will wax and wane before televisor screens in our homes show us distant events and people. Many sleepless nights are ahead of the engineers and scientists charged with the task of putting the new art on a production and commercial basis. If I had at this moment, in a lump sum, the money I expect to earn in very modest annual increments before television becomes a practical reality, I should depart forthwith for a pleasant winter in Algiers and live on the interest for the rest of my life. Anyone who hesitates to buy a radio receiver, for example, because he fears that one equipped with television features may be put on the market before he can realize on his investment, is taking a position almost as ludicrous as that of a man who decided not to buy a gasolene-driven automobile because some inventor might devise a vehicle which would run ten centuries on the intra-atomic energy of a pound of sodium bicarbonate. The everyday application of television is a remote possibility in five years, a fair possibility in ten, a probability in fifteen. Many good radio receivers, appealing to the ear only, will issue from the factories, play their melodies in millions of homes, and succumb to old age and new tastes, in that time. Or, if a slight mutilation of Horace is permissible: "To-morrow do thy worst, for I have listened to-day!"

But when the day arrives, and the engineers produce, for a reasonable expenditure, a television apparatus capable of reproducing distant events in a life-like manner on a sufficiently large screen, what then? Will the accomplishment be a blessing? Not always—take that from one who has seen many broadcast artists. For others, I wish we could have it to-morrow. Sometimes I have seen and heard some beautiful girl with an equally beautiful voice rehearsing her program in the afternoon, and, at the receiver in the evening, have had to be satisfied with hearing her only, and it has certainly been a deprivation. To be limited to one sense is more or less of a hardship in such cases. On the other hand, as I write this article I am listening to a capable group of musicians playing dinner music at one of the hotels. I hear the music and the announcements, although primarily my attention is focused on the typewriter. If my receiver had a television attachment now I should turn it off. In the first place I have no optic interest in the hotel musicians. They are probably just as homely as I am. In the second place, I don't want to look at anything; it would distract me. The ear is happily capable of carrying on its function soothingly and pleasantly, given the

"MANY SLEEPLESS NIGHTS ARE AHEAD OF THE ENGINEERS"

right material, without interfering with other activities. Many people rarely listen to a radio program with great concentration. Everyone does, at times, but, as frequently, radio music forms a background for conversation or reading. Often people talk during the music and pause to listen to the announcements. I generally read during the whole evening's program of a given station, but I hear the announcements and know pretty well what has happened afterwards. Obviously in such cases television is not a remarkable boon. It is even possible that when television has come to stay that some programs will be transmitted with the visual component and others without it.

The movie theatres, it would seem, can do without television. As we saw in the preceding article on television, the motion picture is primarily an art for transferring the past, optically, into the present. The moving picture audience finds no difficulty in attaining a feeling of being in contact with reality, even though the actors went through their parts months ago. There would be nothing gained by substituting television in such a case, and something would be lost. In the recording arts there is always the advantage of being able to try over and over until one gets the right results, and offering to the public only a finished product. Both the phonograph and the motion picture have this advantage. However, television in conjunction with telephony might be employed to reproduce distant spectacles, not only in the home, but in theatres. For example, a movie house might interrupt its program for a short speech, reproduced visually and audibly, by the President—an appeal for aid, for example, in time of national disaster. The theatres have not used broadcasting to any extent to serve their actual audiences; it has been merely a medium of advertising to them, a means of attracting audiences. If the visual component were added to broadcasting as we now know it, the motion picture theatres might find it a useful adjunct to their shows, probably in a form we do not foresee now. But it will not be a necessity; they can probably keep on doing a tidy business without it. Aural broadcasting has affected the motion picture industry only to a moderate extent. In the larger centers the hunger for amusement is so avid that every form of diversion is eagerly soaked

man (he was twenty-two years old) was instantly killed while working on a transmitter panel at 4 o'clock in the morning. The circumstances, as nearly as they can be ascertained, are worth recounting, in the hope that the analysis will lead some of the boys to take precautions which they might neglect otherwise.

The station in this case was not on the air. Repairs or alterations were being made during the night in order to avoid interference with program service. The man who was killed was working behind the panel, where there were two-thousand volt terminations fed from a generator in the basement of the building. This generator could not be heard running in the transmitter room. It is not known whether the victim of the fatal accident was working on live circuits with confidence that he could avoid getting across the high tension, or whether he imagined that the machine had been shut down. He took hold of the bare copper of one of the high-tension cables, while he was partially grounded through his shoes. This put him across the 2000-volts, but probably would not have killed him in itself. He exclaimed, "Wow, it's hot!" causing a fellow worker, who was in front of the panel, to run to his aid. In the meantime the man on the circuit,

"A YOUNG MAN CALLING UP HIS SWEETHEART WOULD FIND USE FOR THE TELEVISION SERVICE"

unable to let go of the wire, was thrown by the violence of his own reflex action into the panel itself, where he must have made better contact with live parts and grounded metal, so that the current through his body rose to a fatal level. His co-worker, rushing behind the panel, seized the cable by the insulation, apparently without shutting down the machine, and tore it out of the fallen man's hand. The wire was rubber-insulated against high tension, and the rescuer got away with his effort, which was both heroic and foolish. A third operator who was present attempted artificial respiration without success. The young man was dead.

What is the moral? There is only one: Be sure that the high tension is shut down before you do any work on potentially live parts of the transmitter. Two thousand volts is not terribly high; many of us have taken it and remained among the living. But if you make a good contact you are as good as dead on a much lower potential. Furthermore, once you are caught you may be thrown into any part of the machinery; you no longer have any control over your movements, and in a few seconds the tragedy is complete.

It is perfectly possible for a man to receive a shock on 110 volts which in itself would only be moderately painful, but if the high-tension portion of the equipment is running he may be flung onto a 10,000-volt circuit before anyone realizes what has happened.

But sometimes one thinks the juice is off when it is not off. Or someone may throw it on while a man is working on the circuits, through a mis-understanding. The remedy for that is simple; the cost is about five cents, and a little thought. Take a stick of dry wood one inch by one inch by, say, twenty inches. Screw on one end of this stick a piece of copper an inch wide, an eighth of an inch thick, and about six inches long, bent in the form of a hook. Let this copper hook be permanently grounded through a heavy, well-insulated flexible wire. When the set is running on program let the contraption lie under the frame or beside the panel. When anyone is going to work on the transmitter, take the copper hook by its insulating stick and hang it gently on the plate bus. If the bus is alive, by chance, there will be an explosion, but you won't get hurt. A new hook can be made in five minutes. If someone turns on the current while the plate bus is shorted, the hook will take the bulk of the flow, and a man's body, with its relatively high resist-ance, will not receive enough to kill him, perhaps not even enough to hurt him.

One more point is worth taking up. A great deal of testing is done at radio stations during the early morning hours, for obvious reasons. In fact, the regulations restrict broadcast stations to the hours between midnight and 11:30 A. M. for testing. During the early hours of the morn-ing, what with fighting sleep and fatigue, a man does not think as clearly as he might at other times. He is at a lower tension of consciousness, and not as distinctly aware of the outside world and its perils as he might be after eight hours of sleep. That is the more reason why a simple automatic device like the bus hook described above should be provided in every station and its use made mandatory, until it becomes so habitual that no one thinks of working without it. Even then, of course, there is no substitute for prudence and conscious direction of all one's movements when one is in the vicinity of a radio transmitter. The thing is dangerous, highly dangerous, and for innocently taking a chance with it one is apt to pay with one's life. And that chance need only be taken once.

Local Regulation of Broadcasting

DURING the partial interregnum between the Zenith decision and the passing of the Dill-White bill to provide for adequate Federal regulation of broadcasting, there was manifest a tendency on the part of some states and municipalities toward local handling of inter-ference problems and the like. Now that Federal administrative machinery is once more function-ing, this development will probably peter out by itself. It is worth discussing, however, because in at least one case—that of the City of Minneap-olis—an ordinance possessing broad regulatory powers was adopted (on February 11, 1927), and may still be in force.

The Minneapolis ordinance provides for an-nual licensing by the City Council of all broad-casting stations within the city, the fee being $50.00. Transmitters located within the city limits, or less than two miles outside, are limited to an antenna power of 500 watts. Between 500 and 1000 antenna watts the transmitter must keep at least four miles from the nearest city boundary line, or at least eight miles if the an-tenna power is 1000–5000 watts, and so on up to output power of the order of 50,000–100,000 watts, when the prescribed distance is increased to 25 miles. There are also provisions limiting the total number of hours of evening broadcasting allowed to any station in the city or within two miles thereof, to twelve per week; and the Building Inspector of the City is assigned power to regulate the hours of such stations. Another section is directed against radiating receiving equipment, violet ray machines, X-ray apparatus, and vibrating battery chargers of types capable of interfering with radio reception. Fines between $15.00 and $100.00, or imprisonment up to 90 days, are prescribed for violations. "Each and every day's continuance of any violation of this ordinance," reads the punitory section, "shall be deemed a separate offense." Another section forehandedly provides: " In case any section or any part of this ordinance is declared invalid, it shall not affect the validity of the remainder of this ordinance."

The manner of enforcement contemplated against transmitters outside of the city limits is of interest. The relevant section reads, " It shall be unlawful for any owner, lessee, or licensee of any metal wires, after written notice from the Building Inspector, to permit the use of such wires in the operation of any broadcasting station which is being operated by the owner or operator thereof in violation of the provisions of this ordinance." In other words, if the transmitter, located outside of the City of Minneapolis, does not conform in location, or otherwise, to the regulations of the municipality, the City will cut it off from electrical connection with any studio within the City, if the courts permit.

Obviously Federal regulation must intervene in all the major questions of broadcasting. The radiation of a broadcasting station does not proceed to the bounds of any city, state, or country and then drop dead. It is an interstate affair if there ever was one. Since there must be centralized regulation of the most complicated sort, it is not a very sapient procedure to attempt decentralized regulation at the same time. The two are bound to overlap. State and municipal control is probably unconstitutional. Professor Frank S. Rowley, of the University of Cincinnati College of Law, in a valuable article on " Problems in the Law of Radio Communication" (*University of Cincinnati Law Review*, January, 1927), writes as follows on the general principle involved:

. . . even though we concede that such stations (of low power and normal intrastate range) operate on a purely intrastate basis, there is excellent authority for sustaining Federal regulation of the same, by analogy to decisions governing regulation of other instrumentalities of intrastate traffic. The Commerce Clause of the Constitution necessarily excludes the states from direct control of subjects embraced within the clause which are of such a nature that, if regulated at all, their regulation should be prescribed by a single authority. That such a doctrine applies particularly in the case of regulation of radio communication is very apparent. A chaos of interference and confusion would result if powers of regulation were divided between federal and state authorities. To avoid such a possibility, a single authority must be given complete control.

Professor Rowley's discussion, although not written with the Minneapolis case in view, seems to me to say all that needs to be said about this or similar attempts. The Radio Commission, or the Secretary of Commerce, can take care of the radio needs of municipalities quite adequately. Such agencies possess much more effective means for compelling interference-causing transmitters to move out of congested districts than any municipality or state body can create or use. Whatever good might be accomplished locally would be overbalanced by possible evils which any

ROY A. WEAGANT
Said: "The salary will be $10.00 per week as a beginning"

layman can foresee. Local authorities had best keep their hands off, in my view. And I might add that I am a rabid states-rights exponent, a natural Jeffersonian, and a natural anti-Hamiltonian—in everything but radio. Radio is not a town-meeting affair, and can't be made one by the best of town legislators.

Memoirs of a Radio Engineer. XIX

ON A sunny afternoon in May my classmate, Jesse Marsten, and I, fresh from college, sat in an ante-room of the Marconi Company's factory at Aldene, waiting for Mr. Weagant, the Chief Engineer. The year was 1917. Did I say we sat? We stood, as a matter of fact. The office force, unimpressed by our appearance and demeanor, did not invite us within the enclosure wherein high-powered office managers, assisted by beautiful secretaries, guided the Marconi Company's destinies. Mr. Weagant was expected, so we waited. We had an appointment. We waited about an hour and a half. Mr. Weagant came in, accompanied by Adam Stein, Jr. He did not know us and passed into the office quickly. I had seen him at I. R. E. meetings and apprised Jesse. In a little while

"I WAS SUBJECT TO SAVAGE LAYING-OUT BY THE NEAREST FOREMAN"

Mr. Weagant came out again. He looked preoccupied; there was reason enough for it, as a matter of fact. In a small stone building, with a few bungalows scattered about it, the Marconi Company was trying to meet war-time demands for apparatus which would have overtaxed a plant six times as large. As Mr. Weagant passed the second time, going out, Marsten stopped him and reminded him of the appointment Doctor Goldsmith had made for us. Mr. Weagant looked dazed for an instant, then, recollecting, he apologized with his usual grave courtesy; and, dropping far more important business, took us to his office in the small wooden engineering building near the main plant. He questioned us and told about the engineering course which the Marconi Company had planned for such aspirants as ourselves. The idea was to move the student engineers about in the various departments of the plant, so that they could become acquainted with all the intricacies of design and manufacture of radio apparatus. We murmured our approval. The salary, continued Mr. Weagant, would be $10.00 per week as a beginning, with an increase of $1.60 per week every six months. If we had commercial operator's licenses, he added, six months' credit would be extended. I thereupon produced my first-class ticket, for which I had undergone the ordeal of examination at the Brooklyn Navy Yard some years before. Mr. Weagant seemed surprised; he examined the document carefully, and granted me the increase in salary which it rated.

Next, with the assistance of two of the engineers at the Marconi plant who were graduates of the City College, Messrs. Barth and McAusland, we secured board and lodging in the near-by village of Roselle, with a very charming middle-aged couple, a retired ship's purser and his wife. The lady was childless but maternal, so she lavished attention on the canary bird, Jesse, and me, and really made things unusually comfortable for us during our stay at the Aldene factory. The room and board amounted to $8.00 a week apiece. Each of us had a very nice room in a good neighborhood, with typical suburban streets shaded by old trees, and we were not badly off, except, of course, that we were not quite self-supporting on our respective salaries of $10.00 and $11.60 per week. This worried us considerably, and we felt that the world is very cruel to newly fledged college students. It is. Why shouldn't it be? But, as a matter of fact, we were able to knock down an average of fourteen dollars a week by working overtime. Some weeks I climbed to the dizzy height of sixteen dollars and fifty cents. A young man could live quite well on that sum in 1917.

We started to work on a Monday. Marsten, I believe, was relegated to the drafting room. I became a tester. I worked from seven in the morning to three in the afternoon, with a half-hour for lunch, punched a clock, and wore overalls. There were no halfway measures, either. If I came five minutes late in the morning my card showed it and I was docked fifteen minutes' pay. The overalls were dirty at the end of the first day I wore them. I was a factory hand like any other; subject to the same discipline, the same savage laying-out by the nearest foreman when I did anything wrong, the same difficulties of organization and manufacture which beset the other cogs in the machine. My B. S. was hidden under the overalls; nobody knew about it except my immediate superior, and he did not hold it against me after a few weeks. I forgot it. My life differed from that of my fellows only after 3 P. M., when, washed and arrayed in white flannels, I played tennis on the turf courts of Weequahic Park in Newark, and spent the evenings reading H. G. Wells.

Use of Tubes Having High Amplification

A Discussion of the Theory Underlying the Functioning of High-Mu Tubes—A Paper Delivered Before the Radio Club of America

By A. V. LOUGHREN

Research Laboratory, General Electric Company

THE amplifier tube has, within the last two years, undergone a notable change. In 1923 all amplifier tubes were of the general-purpose type and all were alike. To-day, on the other hand, we have, in addition to this general-purpose tube, two distinct groups of special-purpose amplifier tubes.

One of these groups—that having a low amplification factor and low plate-to-filament resistance as its chief characteristics—was brought into being to satisfy the demand for more speech power which arose when reception on the loud speaker became common. The performance of this group of tubes has already been discussed in some detail ("Output Characteristics of Amplifier Tubes," J. C. Warner and A. V. Loughren, *Proceedings Institute of Radio Engineers*, Vol. 14, No. 6, Dec., 1926).

The second group, consisting of tubes having amplification factors of 15 or more, will be treated in the present paper. It should be understood, of course, that tubes with high amplification factors are not at all new, as such tubes have been built since 1912. Some of their uses have not been completely investigated, however, until recently.

Any treatment of the uses of tubes having high amplification factors is of necessity primarily a discussion of their use in resistance-coupled amplifiers. In order to understand their application to this type of amplifier it is necessary to have its underlying theory clearly in mind.

Fig. 1 is a diagram of a two-stage amplifier with resistive-interstage coupling. The operations may be sketched roughly as follows:

When a signal is impressed on the input circuit, consider the phenomena at the instant when the grid potential has reached the highest value occurring during the cycle. In Fig. 2 this condition is represented by the point t_1. At this same instant the plate current reaches a maximum. Now, looking back at Fig. 1, we see that the voltage at the plate of tube No. 1 is equal to the battery voltage minus the IR drop in the resistance R_{p1}. The value of this drop is simply $I_p \times R_{p1}$. Therefore, it is directly proportional to the plate current. Thus, when the plate current is greater, as at t_1, the drop in the coupling resistance, R_{p1}, is greater, and hence the plate voltage on the tube is less. The curve of plate potential in Fig. 2 shows this exactly. Incidentally, it may be noted that the signal has been shifted 180° in phase by going through the first stage.

The alternating component of this plate po-

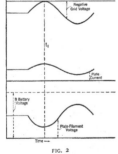

FIG. 1

tential is applied to the grid leak resistance, R_{g1}, of the second tube, while the direct component is kept back by the blocking condenser C_{b1}. The grid of tube No. 2 is so connected that the drop in the grid leak resistance is the alternating component of grid voltage. In this stage the action is exactly like that in the preceding one.

Now, having a slight familiarity with the circuit and its mode of operation, let us proceed to a more rigorous analysis of the performance. To do this we make use of a fundamental principle of physics which may be stated as follows: With circuit elements whose coefficients are constants, the circuit response to a complex emf. may be determined by evaluating the respective responses to each of the (simple) component emf's. and taking the summation, if desired. In the present case this principle permits us to analyze the performance of the amplifier circuit for the alternating quantities alone, neglecting entirely the direct components.

Fig. 3 is drawn in this way. It may be noted that only the quantities entering into the alternating-current phenomena are shown. The interstage coupling resistor, R_p, is shown connected from the plate back to the filament. In practice it is connected to the positive terminal of the B battery and through the latter to the filament, but here we do not need to show the

battery. The representation of a tube by a generator and the series resistance r_p is quite common in analyses of tube output characteristics.

By the use of Kirchoff's Laws for alternating current circuits an expression may be worked out for the relation between e_{g1}, the output voltage of the network, and μe_{g1}, the input voltage. The first voltage corresponds to the alternating voltage on the grid of the second tube while the second is proportional to that on the grid of the first tube. The steps of the mathematics are hardly worth showing in detail. The final expression is that given in Fig. 3. This expression contains a group of terms independent of frequency, and two terms containing the quantity P, where $P = 2\pi \times$ frequency, which are dependent on frequency. The first of these latter is directly proportional to frequency and thus may be expected to interfere with the amplifier performance more at higher frequencies; the second, on the contrary, is important at low frequencies since it is inversely proportional to frequency.

Before investigating the magnitudes of these two frequency effects we must see what the values of the various circuit coefficients are. Certain of them are quite familiar to radio workers, such as the internal plate circuit resistance r_p, the coupling resistance R_p, the grid leak R_g, and the blocking condenser C_b. The quantities C_p, C_g, and r_g have not been in common use, so a word about their magnitudes will hardly be amiss.

The plate-to-ground capacity, C_p, is very nearly the sum of the following:

(1.) Tube inter-electrode plate-filament capacity.
(2.) Tube inter-electrode plate-grid capacity.
(3.) Stray capacities in wiring.

For the Radiotron UX-240 the first two are 1.5 and 8.8 micro-microfarads, respectively.

The quantities r_g and C_g cannot be treated rigorously without going beyond the scope of the present paper. Their magnitudes are de-

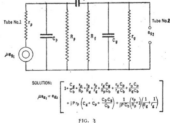

SOLUTION:

$$\mu e_{g_1} = e e_{g_2} \left[1 + \frac{C_g}{C_b} \cdot \frac{r_p}{R_p} + \frac{r_p}{R_g} + \frac{r_p}{r_g} + \frac{r_g C_g}{r_g C_b} + \frac{r_g C_p}{r_g C_b} + \frac{r_g C_p}{R_g C_b} + jP r_p \left(C_g + C_p + \frac{C_p C_g}{C_b} \right) + \frac{1}{jP C_b} \left(\frac{r_p}{R_p} + 1 \right) \left(\frac{1}{R_g} + \frac{1}{r_g} \right) \right]$$

FIG. 3

pendent on the constants of the plate circuit of the second tube. For our purpose it will be sufficiently accurate to consider r_g infinite as long as the second tube is biased sufficiently. C_g may be expressed as:

$$C_g = C_{gf} + C_{gp} (A_V + 1)$$

The reason for the appearance of the quantity $(A_V + 1)$ where A_V is the voltage amplification actually obtained between grid and plate of the second tube, is explained in the appendix on page 240. In practice C_{gf} is usually 3 to 4 mmfd. and C_{gp} 8 to 10 mmfd. while A_V may be between 2 and 25 or more. Accordingly C_g will vary over the range from 20 to perhaps 300 mmfd. The figures given, by the way, do not include the stray capacities in the wiring. It should further be noted that there is no marked difference in the values of these capacities for low and high amplification tubes; that is, the Radiotron UX-201-A for which $\mu = 8$ and the Radiotron UX-240 for which $\mu = 30$, have substantially the same inter-electrode capacities. The Radiotron UX-171 has nearly the same grid-to-plate capacity as the others, but as it has twice their

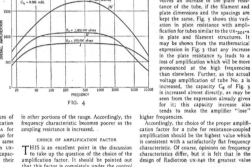

FIG. 4

in other portions of the range. Accordingly, the frequency characteristic becomes poorer as the coupling resistance is increased. .

CHOICE OF AMPLIFICATION FACTOR

THIS is an excellent point in the discussion to take up the question of the choice of the amplification factor. It should be pointed out that this factor is completely under the control of the tube designer so that there is little difficulty in building tubes having values anywhere

between 0.5 and 500. Any increase in the amplification factor, however, involves an increase in the plate resistance of the tube, if the filament and plate dimensions and the spacings are kept the same. Fig. 5 shows this variation in plate resistance with amplification for tubes similar to the UX-201-A in plate and filament structures. It may be shown from the mathematical expression in Fig. 3 that any increase in the plate resistance r_p leads to a loss of amplification which will be more pronounced at the high frequencies than elsewhere. Further, as the actual voltage amplification of tube No. 2 is increased, the capacity C_g of Fig. 3 is increased almost directly, as may be seen from the expression already given for it; this capacity increase also tends to make the amplifier "lose" higher frequencies.

Accordingly, the choice of the proper amplification factor for a tube for resistance-coupled amplification should be the highest value which is consistent with a satisfactorily flat frequency characteristic. Of course, opinions on frequency characteristics differ, but it is felt that in the design of Radiotron UX-240 the greatest value

FIG. 6

filament surface its grid-filament capacity is somewhat greater.

The frequency characteristic of each individual stage may be calculated by the relation shown in Fig. 3. The frequency characteristic of the complete amplifier is equal, of course, to the product of the frequency characteristics of the individual circuits. Fig. 4 shows certain of these overall frequency characteristics. It should be noted that the use of higher value coupling resistances increases the amplification, but that these increases are always less in magnitude at the upper end of the frequency characteristic than

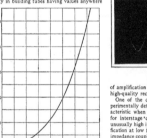

FIG. 5

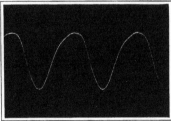

FIG. 7

of amplification factor that should be used in a high-quality receiver has been chosen.

One of the curves of Fig. 4 shows an experimentally determined overall frequency characteristic when iron-core inductances are used for interstage coupling. The units used have unusually high inductance, but the loss of amplification at low frequencies shows that, even so, impedance coupling was unsatisfactory.

So far we have assumed that the circuit coefficients are constants. Actually this may or may not be a valid assumption in a practical case. The two circuit coefficients which may

FIG. 8   FIG. 9

FIG. 10

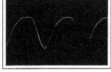

FIG. 11

FIG. 12

vary appreciably in practice are the plate resistance and the grid resistance.

Variation of the plate resistance becomes a matter of importance if the plate current is permitted to decrease to too low a value at any time during the cycle. The effect of this may be seen by comparing the oscillograms of Figs. 6 and 7. Fig. 6 shows the output of the UX-240 when operating normally, while in Fig. 7 the negative bias on the grid is excessive. The resulting distortion on the lower half of the plate-current wave is quite objectionable.

Variation of the grid resistance is negligible in magnitude when the grid potential is negative. When the grid becomes positive, however, its resistance falls quite rapidly, and it may, under some conditions, introduce appreciable distortion. If the source of the signal voltage has good regulation there is little likelihood of distortion occurring; the oscillogram of Fig. 8 demonstrates this. It shows the output of the UX-240 for the same signal voltage as in Fig. 6 and 7; that is, 1.06 volts effective, this time with no bias at all. The plate current is obviously undistorted. Fig. 9 shows the rather poor results obtained when the regulation of the signal source is unsatisfactory.

If the tube is operating with a blocking condenser and grid leak and receives sufficient signal to make the grid positive, the electrons which flow from the filament to the grid must continue through the leak and back to the filament. In doing so they will develop a voltage drop across the leak which will bias the grid negative; the trouble will thus be largely self-correcting. Fig. 10, which illustrates this point, was made under the same conditions as Fig. 9 except that a blocking condenser of 0.015 mfd. was interposed in the lead from the grid to the signal source and a grid leak of 1 megohm was connected from grid to filament. The improvement in output wave form over that of the preceding oscillogram is quite striking.

Fig. 11 shows the distortion that occurs with no bias and the same signal amplitude as before, when the signal is supplied from another tube through an interstage transformer. Here the grid current cannot bias the tube appreciably as it has a low-resistance return to the filament

FIG. 13

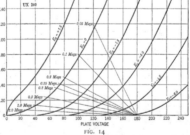

FIG. 14

and hence produces only a negligible IR drop. In the record of Fig. 12 this condition has been corrected by the introduction of a suitable bias.

Perhaps a word about the UX-240, the tube on which our work is based, would not be amiss. In external appearance this tube is identical with

the UX-201-A. The filament characteristics and ratings are also identical. Normal operating conditions for the tube are 1.5 to 3 volts grid bias, 135 to 180 volts plate supply voltage, with a 250,000 ohm resistance in series with the plate. Under these conditions the actual voltage amplification will be between 15 and 20, representing better than 50 per cent. utilization of the tube's inherent amplification factor of 30.

Fig. 13 shows the mutual characteristics of the tube and Fig. 14 shows a family of plate characteristics. These are the two forms in which static characteristics are conventionally shown and are reproduced for that reason.

APPENDIX

THE change in grid-to-ground capacity introduced by the plate and plate circuit of the tube may be treated as follows: In the case of a resistance-coupled circuit having substantially unity power factor, so that there is no appreciable phase shift, let us observe what happens when the grid is raised in potential 1 volt. The plate potential falls by an amount A_v volts, where A_v is the actual voltage amplification which the stage is furnishing. Now, across the direct grid-filament capacity we have introduced a net change in potential difference of 1 volt, and accordingly the quantity of electricity which will raise this potential difference 1 volt is added to that already providing the electro-static field between grid and filament. Across the direct grid-plate capacity we have introduced voltage changes of 1 volt at the grid side and of A_v volts at the plate side, the two changes being of the same sign insofar as their effect on the electrostatic phenomena is concerned. As a result, a quantity of electricity sufficient to change the grid-plate capacity to $1 + A_v$ volts must flow on the grid.

By combining these two terms we find that the effective capacity from grid to ground is;

$$C = C_{gf} + C_{gp} (1 + A_v)$$

It should be noted that a general treatment of this capacity effect is, ipso facto, a study of regeneration due to inter-electrode capacity.

RADIO BROADCAST is the official publication of the Radio Club of America, through whose courtesy the foregoing paper has been printed here. RADIO BROADCAST does not, of course, assume responsibility for controversial statements made by authors of these papers. Other Radio Club papers will appear in subsequent numbers of this magazine

The *Radio Broadcast*
LABORATORY INFORMATION
SHEETS

THE Radio Broadcast Laboratory Information Sheets are a regular feature of this magazine and have appeared since our June, 1926, issue. They cover a wide range of information of value to the experimenter and to the technical radio man. It is not our purpose always to include new information but to present concise and accurate facts in the most convenient form. The sheets are arranged so that they may be cut from the magazine and preserved for constant reference, and we suggest that each sheet be cut out with a razor blade and pasted on 4″ x 6″ filing cards, or in a notebook. The cards should be arranged in numerical order. An index appears twice a year dealing with the sheets published during that year. The first index appeared on sheets Nos. 47 and 48, in November, 1926. Last month an index to all sheets appearing since that time was printed.

The June, October, November, and December, 1926, issues are out of print. A complete set of Sheets, Nos. 1 to 88, can be secured from the Circulation Department, Doubleday, Page & Company, Garden City, New York, for $1.00. Some readers have asked what provision is made to rectify possible errors in these Sheets. In the unfortunate event that any such errors do appear, a new Laboratory Sheet with the old number will appear.

The Information Service of Radio Broadcast is conducted entirely by mail, the coupon on page 255 being used when application is made for technical information. It is the purpose of these Sheets to supply information of original value which often makes it possible for our readers to solve their own problems.　—The Editor.

No. 113　　　Radio Broadcast Laboratory Information Sheet　　　**August, 1927**

Output Circuits

THREE POSSIBLE CIRCUITS

IN THE sketch on this Sheet are shown three output arrangements that can be used to couple a loud speaker to a power tube in order to prevent the direct current in the plate circuit of the power tube from passing through the windings in the loud speaker and affecting its satisfactory operation. In sketches "A" and "B," the inductance of the choke coil, "L," should be at least 60 henrys and these coils should have a low resistance so as to prevent any great loss in voltage which would occur if the resistance was very high. A good unit should not cause a loss in voltage of more than 15 or 20 volts and this means that its d.c. resistance must not be greater than 750 ohms. The blocking condensers, "C," should have a capacity of from 2 to 4 mfd. The larger size theoretically gives somewhat better reproduction but practically little difference will be noticed with most loud speakers whichever size is used.

The arrangement shown at "A" has the advantage that, if the condenser breaks down, it will not result in any damage to the loud speaker because a breakdown in the condenser will merely cause the loud speaker to be shunted across the output choke "L," whereas, with arrangement "B," a breakdown of the condenser will cause the B battery to be short-circuited through the loud speaker and it is possible that the latter will be burned out. A disadvantage of arrangement "A" is that the a.c. current flowing through the loud speaker must

flow through the B supply in order to return to the negative filament, and a comparatively small amount of resistance in the B supply will frequently cause a howl in the amplifier. In the arrangement shown in "B," the a.c. currents in the loud speaker return directly to the negative filament and do not have to traverse the B power unit; consequently, with this latter arrangement, there is less danger of oscillation in the audio amplifier.

In one particular case, during experiments in the

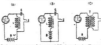

Laboratory, a resistance of 37 ohms in the B power unit, using circuit "A," produced continuous oscillations, whereas a resistance of 600 ohms was necessary in circuit "B" before oscillations were produced.

The arrangement at "C" shows an output transformer which is also a satisfactory method of coupling a loud speaker to a tube. It is essential, however, that the transformer be very carefully designed to prevent magnetic saturation because it must carry comparatively large direct current.

No. 114　　　Radio Broadcast Laboratory Information Sheet　　　**August, 1927**

The Transmission Unit

DEFINITION

ANY electrical system having anything to do with the transmission of electrical energy which is finally to be changed into sound energy should have its performance rated in some manner which bears a relation to the sensitivity of the ear. Two audio amplifiers might give power outputs of 800 milliwatts and 1000 milliwatts, and it appears from these figures as though the second amplifier would be capable of giving a considerable increase in volume over that obtained from the first amplifier, but actually this would not be so; the difference between the two amplifiers could hardly be detected by the ear. Evidently it would be of advantage to express the relation between the power outputs of the two amplifiers by some unit which would indicate their relative value as measured by the ear. The telephone companies have worked out such a unit, known as the transmission unit, abbreviated "TU." It is possible for the ear to just distinguish the difference between two powers that differ in intensity by 1 TU.

The two powers mentioned above, 800 and 1000 milliwatts, are in the ratio of 1.25 to 1. The TU difference between these two powers is equal to ten times the natural logarithm of the ratio of the two powers:

$$TU = 10 \log_e \frac{P_1}{P_2}$$

The ratio in this case is 1.25 and the natural logarithm is 0.097, which, multiplied by ten, gives 0.97 TU. The minimum perceptible change in loudness is 1 TU and therefore the difference between the two amplifiers would not be audible.

The equation given in the preceding paragraph gives the TU when two powers, or their ratio, are known. If instead of powers we deal with voltages, E_1 and E_2, then the formula is:

$$TU = 20 \log_e \frac{E_1}{E_2}$$

When using currents, I_1 and I_2, the equation is:

$$TU = 20 \log_e \frac{I_1}{I_2}$$

The logarithm of the ratio of two voltages differing by 12 per cent., is 0.05, and 20 times this gives 1 TU. Therefore, if two audio transformers differ in amplification by 12 per cent., they will give equally good results because a 1 TU change is not audible to the ear.

The natural logarithm of numbers can be found by using a slide rule or they can be determined from tables of logarithms which are frequently found in the appendix of text books.

Precision

T HIS special NATIONAL variable condenser is designed and built to read to 1/1000000 Mfd. and to do it just as accurately after years of use as when new. At the same time its construction is sufficiently rugged to withstand the day in, day out factory conditions under which it plays its essential part in the ATLANTIC WEIGHT METER.

All NATIONAL RADIO PRODUCTS are made with equal care and adherence to rigid standards.

NATIONAL
RADIO PRODUCTS

Send for Bulletin B-8

NATIONAL CO · INC · W · A · READY · PRES · MALDEN · MASS ·

No. 115   RADIO BROADCAST Laboratory Information Sheet   **August, 1927**

Wave Traps

DIFFERENT TYPES

IN MANY cases where difficulty is experienced in
eliminating the signals from a nearby powerful
broadcasting station it will be found advantageous
to use a wave trap in the antenna circuit.

A wave trap is a simple device consisting of a
condenser and a coil, the latter with or without a
primary, depending upon whether the circuit shown
at "A" or "B" is used. In either case the wave trap
should be tuned to the frequency of the interfering
station. It then offers a very high impedance to the
flow of currents of this frequency and in this way
reduces their strength.

The circuit shown in "A" will give most complete elimination of undesired signals but has the
disadvantage that it will also reduce somewhat the
signals from other stations operating on frequencies
adjacent to that of the station causing the interference. In cases of severe interference, however,
the circuit shown at "A" must be used. The
capacity of the variable condenser may be anything
from 0.00025 mfd. to 0.001 mfd., and the coil must
naturally contain sufficient turns so as to tune the
circuit to the frequency of the interfering signals.
There is no reason why a standard condenser and
coil, designed for reception on the broadcast frequencies, cannot be used and it will then be possible

to tune the trap to any frequency in the broadcast
band.

The circuit shown at "B" tunes much sharper
than the circuit shown at "A" but does not give
complete elimination of the undesired signals.
This circuit can be used with satisfaction when the
interference is not very severe. The coil may be any
ordinary tuned radio-frequency transformer.

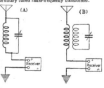

No. 116   RADIO BROADCAST Laboratory Information Sheet   **August, 1927**

Static

POSSIBILITIES OF ELIMINATION

NATURAL electrical disturbances occurring in
the atmosphere are known as "static" or
"strays" and frequently cause serious interference
during the reception of signals. The subject "static"
has been broken up into the following divisions by
DeGroot in an article in *The Proceedings of the
Institute of Radio Engineers.*

(A)—Loud and sudden clicks occurring intermittently. These do not seriously affect reception
and apparently originate in nearby or distant
lightning discharges.

(B)—A constant hissing noise giving the impression of softly falling rain or the noise of running
water. This form usually occurs when there are low-
lying clouds in the neighborhood of the receiving
antenna.

(C)—A third form produces a constant rattling
noise which sounds somewhat like the tumbling
down of a brick wall.

These three forms can be considered as forms of
natural static. The problem of the elimination of
static is a difficult one upon which a great deal of
work has been done and many different schemes
have been devised, most of these schemes making
use of two receiving antennas feeding a common
receiver. The static signals present in the two
antennas are made to balance out each other

whereas the desired signals are not balanced out. In
Morecroft's book, *Principles of Radio Communication,* it is suggested that one of the most promising lines for the development of a static eliminator
has to do with a vacuum-tube detector which can
only produce a limited response and therefore even
with very heavy static the response cannot be more
than the definite peak response of the tube.

Reception is also interfered with to a great extent in many localities by sounds produced by
electrical apparatus, in which category can be
classed the interference caused by various electrical
motors and generators, x-ray apparatus, oil burners,
precipitators, electrical transmission lines, etc.
Their elimination is best accomplished at the source
of the trouble by means of filter circuits such as
those described in Laboratory Sheet No. 77, in the
March, 1927, RADIO BROADCAST.

At the present time it appears that the best
method to overcome natural static is to use a
receiver in conjunction with a loop or a very short
antenna, because with a loop or short antenna a
high signal-to-static ratio can be obtained. Also, to
prevent serious interference with broadcast, programs, high power at the transmitting station is
coming into more common use so that even under
fairly bad conditions of static satisfactory reception
can still be had.

No. 117   RADIO BROADCAST Laboratory Information Sheet   **August, 1927**

Super-Heterodynes

THE OSCILLATOR

IN A super-heterodyne receiver it is necessary to
have one tube acting as an oscillator and functioning to produce radio-frequency energy which, in
combination with the incoming signals, will produce
a third frequency capable of being amplified by the
intermediate frequency amplifier. The amplitude
of the locally generated oscillations, in comparison
with the incoming signals, has a very definite effect
upon the strength of the signal which is finally
detected and some care should therefore be taken in
adjusting the circuit for most efficient operation,
i.e., loudest signals.

In sketch "A" is given the circuit used, probably,
in a majority of super-heterodynes. It has the disadvantage that both sides of the variable condenser
are at high potential and therefore some hand-
capacity effects will be experienced.

In sketch "B" is shown an oscillatory circuit
which is not open to the disadvantage of circuit
"A" and is capable of giving just as good results
in actual practice. In this circuit the rotor plates of
the variable condenser connect to the low-potential
side of the grid coil instead of across the entire coil.
The "low" end of the grid coil connects to the filament and is therefore at ground potential and consequently there is no hand capacity. If a 0.0005-mfd.
variable condenser is used, then coil "L" should
contain 52 turns of No. 24 wire on a 2¼" tube;
for a 0.00025-mfd. condenser the number of turns
should be 65. L₂, the plate coil, should consist, in

either case, of 60 turns of No. 28 wire wound on the
same tube and spaced from the coil "L" by ¼". L₁
is the pick-up coil which should be connected in the
circuit of the first detector tube; it should consist of
10 turns of No. 28 wire preferably wound on a tube
slightly larger than 2¼" so that it can slide over
the other form and the coupling be varied in this
way. Either a 201-A or 199 type tube may be used
in the oscillator without changing the coil constants.

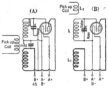

Model A. C. 20
Height, including base, 21⅝ in.

Far ahead of today's accepted standards—the new Amplion Cone with M.A.1 type unit

In quality of reproduction it sets a new high mark for others to strive for

THIS remarkable reproducer has an unusually wide musical range due to the striking developments embodied in the new Cone Unit.

One important feature of this unit is the *felt-lined stylus protection bar* which serves a double purpose. First, it protects the stylus itself against any possible injury. The felt lining, or stylus anchor, neutralizes the harmonics of the stylus itself.

This model employs a new principle in Cone construction—a 14″ cone being mounted on an 18″ sound board which extends toward the center in back of the cone to form a resonating chamber. The unit, cone and sound board are assembled on a rigid bronze base with a handsome bronze bracket.

The sound board is finished in dark walnut, the cone, with the gold-finished cone in the center, gives this instrument both graceful beauty and dignity.

Write to us for price list and literature illustrating and describing the new Amplion line.

THE AMPLION CORPORATION OF AMERICA
280 Madison Ave., New York City

The Amplion Corporation
of Canada Ltd., Toronto

No. 118   RADIO BROADCAST Laboratory Information Sheet   **August, 1927**

Audio Amplifiers

FREQUENCY AND LOAD CHARACTERISTICS

ANY audio amplifying system has two characteristics, equally important, which determine how well it will function. They are generally known as the frequency characteristic and the load characteristic.

The frequency characteristic indicates the relative amplification of the amplifying system of various frequencies between the limits over which the amplifier is to be operated. The frequency characteristic is generally shown in the form of a curve and, of course, a flat curve indicates equal amplification at all frequencies. Slight rises and depressions in the curve in the order of 10 per cent. can be neglected because they are too slight to be noticeable to the ear.

The load characteristic of an amplifier, while not in such common use, is just as important as the frequency characteristic. The load characteristic will show how the total amplification of the system varies with different input voltages at a constant frequency generally of about 1000 cycles. If the amplifier is a good one the amplification will remain constant over the entire range of input voltages at

which the amplifier would normally be worked. If a two-stage amplifier is operated with a 201-A type tube in the output with 90 volts on the plate, it will overload very quickly because a 201-A cannot deliver much power. Consequently, the load characteristic curve of such an amplifier would begin to fall off comparatively quickly, but if a 171 tube with the proper voltages were to be used in place of the 201-A, then the load characteristic would indicate that it was possible to obtain much more power from the amplifier without overloading it.

Both of these characteristics depend upon the type of tubes used and the voltages with which they are supplied, and upon the design of the coupling devices connecting the output of one tube to the input of the next. Frequency and load characteristics can be taken on any part of the complete amplifier but such curves may have very little in common with the characteristics of the complete system. Consequently, although curves on individual units are useful in designing an amplifier, curves on the completed system should always be made to make certain that some factor, such as common coupling in the batteries, is not seriously altering the overall characteristic.

No. 119   RADIO BROADCAST Laboratory Information Sheet   **August, 1927**

Radio-Frequency Choke Coils

THEIR PLACE IN CIRCUIT

IF A very high-gain radio-frequency amplifier is to be constructed, it is essential that radio-frequency choke coils be used in the amplifier to prevent it from oscillating. Neutralization will prevent the production of oscillations due to feed-back through the tube but will not prevent the production of oscillations due to coupling in the battery leads or in a B socket-power device. To prevent instability due to these effects it is necessary that choke coils, L, and bypass condensers, C, be used in the plate circuits of the radio-frequency tubes, as indicated in the diagram on this Sheet. These choke coils offer a very high impedance to the flow of radio-frequency currents and all these currents therefore flow through the bypass condenser connected between the choke coil and the negative filament, instead of through the plate battery.

What size choke and what size condenser should be used? To keep down the cost they should both be as small as possible whereas their effectiveness becomes greater as their size is made larger.

The plate impedance of a 201-A type tube is about 12,000 ohms and it is essential that the condenser which is incorporated to bypass the r.f. currents does not introduce in the plate circuit any great amount of impedance. A 0.003-mfd. condenser will increase the total circuit impedance from 12,000 to about 12,120 ohms, a negligible amount. This value is correct at 500 kilocycles, the lowest frequency used in broadcasting, and at higher broadcast frequencies the effect of the condenser will be even less.

The choke coil's impedance must be large in comparison with that of the condenser so as to cause practically all the current to flow through the condenser and not through the choke. If the choke coil's impedance is made 1000 times greater than that of the condenser only one-tenth of one per cent. of the total radio-frequency current will flow through

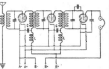

the choke coil and therefore good filtering action will be obtained. If the choke coil's impedance at 500 kilocycles is to be 1000 times greater than the impedance of the condenser then it must be 12,000 ohms. The inductance of a choke coil with an impedance of 12,000 ohms at 500 kilocycles is 38 millihenrys. Most radio-frequency choke coils have an inductance of much more than this.

No. 120   RADIO BROADCAST Laboratory Information Sheet   **August, 1927**

A-Battery Chargers

TRICKLE VERSUS HIGH-RATE CHARGERS

THERE are many different types of A-battery chargers now available; some of them are satisfactory for use as trickle chargers and others only efficient when used to charge the battery at comparatively high rates of charge. The charger employing an electrolytic type of rectifier, for example, is very well adapted for use in trickle charging. It is very efficient, requires little attention, and has long life.

Another very satisfactory type of rectifier for a trickle charger is the so-called dry crystal, which was developed rather recently. A third type of rectifier that can be used for trickle charging is the Tungar but it is not especially efficient as a trickle charger, because of the comparatively large amount of power required to heat its filament.

There are three types of rectifiers that are satisfactory for use in high-rate charging. They are the Tungar, the Vibrator type, and the new cartridge recently developed by Raytheon. All of these chargers are capable of delivering fairly large amounts of rectified current for charging a battery and are fairly efficient when delivering these currents.

There is little to be said regarding the comparative efficiency of the two methods of charging. Trickle charging has the advantage that it requires somewhat less attention than does high-rate charging but it has the disadvantage that it is somewhat difficult to determine just what the best rate of trickle charge should be in order to prevent the battery from being overcharged or undercharged; also slow rates of charge used in trickle chargers are hard on a battery. With a trickle charger, a low-capacity storage battery can be used because it is not called upon to supply any great amount of current for a long period of time.

With high-rate charging, on the other hand, it is usual to charge the battery every one or two weeks and also a fairly large storage battery is necessary in order that it will have sufficient capacity to supply the receiver between charges. It seems to be generally agreed among battery manufacturers, however, that the high charging rate is somewhat better for the battery in that it makes possible longer life. For best results the charging rate should gradually taper off as the battery becomes charged.

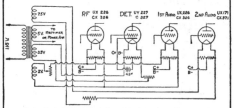

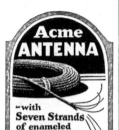

Manufacturers' Booklets

A Varied List of Books Pertaining to Radio and Allied Subjects Obtainable Free With the Accompanying Coupon

READERS may obtain any of the booklets listed below by using the coupon printed on page 255. Order by number only.

1. FILAMENT CONTROL—Problems of filament supply, voltage regulation, and effect on various circuits. RADIALL COMPANY.

2. HARD RUBBER PANELS—Characteristics and properties of hard rubber as used in radio, with suggestions on how to "work" it. B. F. GOODRICH RUBBER COMPANY.

3. TRANSFORMERS—A booklet giving data on input and output transformers. PACENT ELECTRIC COMPANY.

4. RESISTANCE-COUPLED AMPLIFIERS—A general discussion of resistance coupling with curves and circuit diagrams. COLE RADIO MANUFACTURING COMPANY.

5. CARBORUNDUM IN RADIO—A book giving pertinent data on the crystal as used for detection, with hook-ups, and a section giving information on the use of resistors. THE CARBORUNDUM COMPANY.

6. B-ELIMINATOR CONSTRUCTION—Constructional data on how to build. AMERICAN ELECTRIC COMPANY.

7. TRANSFORMER AND CHOKE-COUPLED AMPLIFICATION—Circuit diagrams and discussion. ALL-AMERICAN RADIO CORPORATION.

8. RESISTANCE UNITS—A data sheet of resistance units and their application. WARD-LEONARD ELECTRIC COMPANY.

9. VOLUME CONTROL—A leaflet showing circuits for distortionless control of volume. CENTRAL RADIO LABORATORIES.

10. VARIABLE RESISTANCE—As used in various circuits. CENTRAL RADIO LABORATORIES.

11. RESISTANCE COUPLING—Resistors and their application to audio amplification, with circuit diagrams. DEJUR PRODUCTS COMPANY.

12. DISTORTION AND WHAT CAUSES IT—Hook-ups of resistance-coupled amplifiers with standard circuits. ALLEN-BRADLEY COMPANY.

13. B-ELIMINATOR AND POWER AMPLIFIER—Instructions for assembly and operation using Raytheon tube. GENERAL RADIO COMPANY.

13A. B-ELIMINATOR AND POWER AMPLIFIER—Instructions for assembly and operation using an R. C. A. rectifier. GENERAL RADIO COMPANY.

16. VARIABLE CONDENSERS—A description of the functions and characteristics of variable condensers with curves and specifications for their application to complete receivers. ALLEN D. CARDWELL MANUFACTURING COMPANY.

17. BAKELITE—A description of various uses of bakelite in radio, its manufacture, and its properties. BAKELITE CORPORATION.

19. POWER SUPPLY—A discussion on power supply with particular reference to lamp-socket operation. Theory and constructional data for building power supply devices. ACME APPARATUS COMPANY.

20. AUDIO AMPLIFICATION—A booklet containing data on audio amplification together with hints for the construction. ALL AMERICAN RADIO CORPORATION.

21. HIGH-FREQUENCY DRIVER AND SHORT-WAVE WAVEMETER—Constructional data and application. BURGESS BATTERY COMPANY.

46. AUDIO-FREQUENCY CHOKES—A pamphlet showing positions in the circuit where audio-frequency chokes may be used. SAMSON ELECTRIC COMPANY.

47. RADIO-FREQUENCY CHOKES—Circuit diagrams illustrating the use of chokes to keep out radio-frequency currents from definite points. SAMSON ELECTRIC COMPANY.

48. TRANSFORMER AND IMPEDANCE DATA—Tables giving the mechanical and electrical characteristics of transformers and impedances, together with a short description of their use in the circuit. SAMSON ELECTRIC COMPANY.

49. BYPASS CONDENSERS—A description of the manufacture of bypass and filter condensers. LESLIE F. MUTER COMPANY.

50. AUDIO MANUAL—Fifty questions which are often asked regarding audio amplification, and their answers. AMERTRAN SALES COMPANY, INCORPORATED.

51. SHORT-WAVE RECEIVER—Constructional data on a receiver which, by the substitution of various coils, may be made to tune from a frequency of 16,660 kc. (18 meters) to 1900 kc. (150 meters). SILVER-MARSHALL, INCORPORATED.

52. AUDIO QUALITY—A booklet dealing with audio-frequency amplification of various kinds and the application to well-known circuits. SILVER-MARSHALL, INCORPORATED.

56. VARIABLE CONDENSERS—A bulletin giving an analysis of various condensers together with their characteristics. GENERAL RADIO COMPANY.

57. FILTER DATA—Facts about the filtering of direct current supplied by means of motor-generator outfits used with transmitters. ELECTRIC SPECIALTY COMPANY.

59. RESISTANCE COUPLING—A booklet giving some general information on the subject of radio and the application of resistors to a circuit. DAYEN RADIO CORPORATION.

60. RESISTORS—A pamphlet giving some technical data on resistors which are capable of dissipating considerable energy; also data on the ordinary resistors used in resistance-coupled amplification. THE CRESCENT RADIO SUPPLY COMPANY.

63. RADIO-FREQUENCY AMPLIFICATION—Constructional details of a five-tube receiver using a special design of radio-frequency transformer. CAMPFIELD RADIO MFG. COMPANY.

65. FIVE-TUBE RECEIVER—Constructional data on building a receiver. AERO PRODUCTS, INCORPORATED.

64. AMPLIFICATION WITHOUT DISTORTION—Data and curves illustrating the use of various methods of amplification. AERO APPARATUS COMPANY.

66. SUPER-HETERODYNE—Constructional details of a seven-tube set. G. C. EVANS COMPANY.

70. IMPROVING THE AUDIO AMPLIFIER—Data on the characteristics of audio transformers, with a circuit diagram showing where chokes, resistors, and condensers can be used. AMERICAN TRANSFORMER COMPANY.

72. PLATE SUPPLY SYSTEM—A wiring diagram and layout plan for a plate supply system to be used with a power amplifier. Complete directions for wiring are given. AMERTRAN SALES COMPANY.

80. FIVE-TUBE RECEIVER—Data are given for the construction of a five-tube tuned radio-frequency receiver. Complete instructions, list of parts, circuit diagram, and template are given. ALL-AMERICAN RADIO CORPORATION.

81. BETTER TUNING—A booklet giving much general information on the subject of radio reception with specific illustrations. Primarily for the non-technical home constructor. BREMER-TULLY MANUFACTURING COMPANY.

82. SIX-TUBE RECEIVER—A booklet containing photograph, instructions, and diagrams for building a six-tube shielded receiver. SILVER-MARSHALL, INCORPORATED.

83. SOCKET POWER DEVICE—A list of parts, diagrams, and templates for the construction and assembly of socket power devices. JEFFERSON ELECTRIC MANUFACTURING COMPANY.

84. FIVE-TUBE EQUAMATIC—Panel layout, circuit diagrams, and instructions for building a five-tube receiver, together with data on the operation of tuned radio-frequency transformers of special design. KARAS ELECTRIC COMPANY.

85. FILTER—Data on a high-capacity electrolytic condenser used in filter circuits in connection with a socket power supply units, are given in a pamphlet. THE ABOX COMPANY.

86. SHORT-WAVE RECEIVER—A booklet containing data on a short-wave receiver as constructed for experimental purposes. THE ALLEN D. CARDWELL MANUFACTURING CORPORATION.

88. SUPER-HETERODYNE CONSTRUCTION—A booklet giving full instructions, together with a blue print and necessary data, for building an eight-tube receiver. THE GEORGE W. WALKER COMPANY.

89. SHORT-WAVE TRANSMITTER—Data and blue prints are given on the construction of a short-wave transmitter, together with operating instructions, methods of keying, and other pertinent data. RADIO ENGINEERING LABORATORIES.

90. IMPEDANCE AMPLIFICATION—The theory and practice of a special type of dual-impedance audio amplification are given. ALDEN MANUFACTURING COMPANY.

93. B-SOCKET POWER—A booklet giving constructional details of a socket-power device using either the BH or 313 type rectifier. NATIONAL COMPANY, INCORPORATED.

99. POWER AMPLIFIER—Constructional data and wiring diagrams of a power amplifier combined with a B-supply unit are given. NATIONAL COMPANY, INCORPORATED.

100. A, B, AND C SOCKET-POWER SUPPLY—A booklet giving data on the construction and operation of a socket-power supply using the new high-current rectifier tube. THE Q. R. S. MUSIC COMPANY.

101. USING CHOKES—A booklet with circuit diagrams of the more popular circuits showing where choke coils may be placed to produce better results. SAMSON ELECTRIC COMPANY.

Accessories

22. A PRIMER OF ELECTRICITY—Fundamentals of electricity with special reference to the application of dry cells to radio and other uses. Constructional data on buzzers, automatic switches, alarms, etc. NATIONAL CARBON COMPANY.

23. AUTOMATIC RELAY CONNECTIONS—A data sheet showing how a relay may be used to control A and B circuits. YAXLEY MANUFACTURING COMPANY.

25. ELECTROLYTIC RECTIFIER—Technical data on a new type of rectifier with operating curves. KODEL RADIO CORPORATION.

26. DRY CELLS FOR TRANSMITTERS—Actual tests given, well illustrated with curves showing exactly what may be expected of this type of B power. BURGESS BATTERY COMPANY.

27. DRY-CELL BATTERY CAPACITIES FOR RADIO TRANSMITTERS—Characteristic curves and data on discharge tests. BURGESS BATTERY COMPANY.

28. B BATTERY LIFE—Battery life curves with general curves on tube characteristics. BURGESS BATTERY COMPANY.

29. HOW TO MAKE YOUR SET WORK BETTER—A non-technical discussion of general radio subjects with hints on how reception may be bettered by using the right tubes. UNITED RADIO AND ELECTRIC CORPORATION.

30. TUBE CHARACTERISTICS—A data sheet giving constants of tubes. C. E. MANUFACTURING COMPANY.

31. FUNCTIONS OF THE LOUD SPEAKER—A short, non-technical general article on loud speakers. AMPLION CORPORATION OF AMERICA.

32. METERS FOR RADIO—A catalogue of meters used in radio, with connecting diagrams. BURTON-ROGERS COMPANY.

33. SWITCHBOARD AND PORTABLE METERS—A booklet giving dimensions, specifications, and shunts used with various meters. BURTON-ROGERS COMPANY.

34. COST OF B BATTERIES—An interesting discussion of the relative merits of various sources of B supply. HARTFORD BATTERY MANUFACTURING COMPANY.

35. STORAGE BATTERY OPERATION—An illustrated booklet on the care and operation of the storage battery. GENERAL LEAD BATTERIES COMPANY.

36. CHARGING A AND B BATTERIES—Various ways of connecting up batteries for charging purposes. WESTINGHOUSE UNION BATTERY COMPANY.

37. CHOOSING THE RIGHT RADIO BATTERY—Advice on what dry cell battery to use; their application to radio, with wiring diagrams. NATIONAL CARBON COMPANY.

51. TUBE REACTIVATION—Information on the care of vacuum tubes, with notes on how and when they should be reactivated. THE STERLING MANUFACTURING COMPANY.

54. ARRESTERS—Mechanical details and principles of the vacuum type of arrester. NATIONAL ELECTRIC SPECIALTY COMPANY.

55. CAPACITY CONNECTOR—Description of a new device for connecting up the various parts of a receiving set, and at the same time providing bypass condensers between the leads. KURZ-KASCH COMPANY.

(Continued on page 255)

"RADIO BROADCAST'S" DIRECTORY OF MANUFACTURED RECEIVERS

With this issue of RADIO BROADCAST, we start a comprehensive listing of manufactured receivers. The listing below will appear regularly in this place, corrected to the day we go to press. The Manufactured Set Directory is an additional service to our readers and takes its place with the "Manufacturers' Booklets" section and the "What Kit Shall I Buy?" department, both of which are already very popular. With these three regular features of RADIO BROADCAST, the reader can survey the entire field and, with little trouble, secure the information he wants.

A coupon will be found on page 255. All readers who desire additional information on the receivers listed below need only insert the proper numbers in the coupon, mail it to the Service Department of RADIO BROADCAST, and full details will be sent.

Key Letters

Ant.—Antenna
r—Radio frequency stage
d—Detector
a—Audio frequency stage
t—Transformer coupled
res.—Resistance
i—Impedance coupled
o—Oscillator
O—Output device
C—C-battery connections
Ca—Cable connections
Bp—Binding posts
M—Meter (instrument)
Sh.—Shielded
H—Headphone connection included
pp—Push pull

KEY TO ABBREVIATIONS

m—Meters (wavelength)
TRF—Tuned radio frequency
URF—Untuned radio frequency
Super—Super-heterodyne
Ne—Neutrodyne
bal.—Some means of neutralization
gr—Grid resistances
Br—Bridge neutralization
reg.—Regenerative
int.—Intermediate amplifier
r.f.—Radio frequency
rheo.—Rheostat
mod.—Modulator
pot.—Potentiometer
w—Watts
Wt—Weight in lbs.

Tubes

99—60-mA. filament (dry cell)
01A—Storage battery 0.25 amps. filament
12—Power tube (Storage battery)
71—Power tube (Storage battery)
16B—Half-wave rect. tube
AC—Using a.c. (heater type)
Hmu—High Mu tube for resistance-coupled audio
R—A gaseous rectifier tube
20—Power tube (dry cell)
Tun—Tungar rectifier
MV—Multivalve (several elements in one bulb)
10—Power Tube
00A—Special detector
13—Full-wave rectifier tube
2A—Tungar rectifier

SOCKET-POWER RECEIVERS FOR USE WITH 110-120 VOLTS 60-CYCLE A.C.

No.	Model	Price	Accessories Included	Cabinet Size	No. Dials	Vol. Cont.	No. Tubes	Type Tubes	Cur. Req. Watts	Circuit	Remarks
401.	AMRAD AC9	142	power unit	27 x 9 x 11½	2	res. oh 1st a	3r, d, 2at (6)	5-99, 1-12, 2-16B	50w	TRF Ne.	Series fil. power unit separate. A, B, and C, supplied
402.	AMRAD AC5	125	power unit	27 x 9 x 11½	2	res. on, 1st a	2r, d, 2at (5)	4-99, 1-12, 2-16B	50w	TRF Ne.	Same as above
403.	ARGUS 250B	250	none	35½ x 14½ x 10½	2	res. on 1st a	3r, d, 2at (6)	5-99, 1-10. 2-16B	100w	TRF gr	2TRF stages, and 1 un-tuned stage. Series fil. O. Self contained power unit
404.	ARGUS 375B	375	none	console	2	res. on 1st a	3r, d, 2a. (6)	5-99, 1-10. 2-16B	100w	TRF gr	See above
405.	ARGUS 125B	125	none		2	fil. rheo.	3r, d, 2at (6)	5-99, 1-12. 2-16B	60w	TRF gr	See above
406.	CLEARTONE 110	175 to 375	5 a.c. tubes 1 Rectifier	various sizes	1-2	res.	2r, d, 2at (5)	5-AC. 1-Rectifier	40w	TRF	Built in plate supply. A. C. tubes
407.	COLONIAL 25	250	none	34 x 38 x 18	1-3	ant. sw. and a pot.	2r, d, 2area., lat (6)	2-01A, 3-99, 1-10. 2-16B	100w	TRF bal.	Built in A, B, and C power unit
408.	DAY-FAN DE LUXE	350	complete less tubes	30 x 40 x 20	1	pot. across r.f. tubes	3r, d, 2at (6)	6-01A	300w	TRF	Motor Generator sup-plies d.c. for plate and fil. O
409.	DAYCRAFT 5	170	complete less tubes	34 x 36 x 14	1	pot. in plate of r.f. and a	3r, d, 2at (6)	5-AC	135w	TRF	Reflexed. AC tubes. Built in plate rect.
410.	LARCOFLEX 73	215	none	30 x 42 x 20	1	res. in r.f. plate	4r, d, 2at (7)	6-01A, 1-71		TRF	Sh.
411.	HERBERT LECTRO 120	120	none	32 x 10 x 12	3	rheo in pri. of a.c. trans.	2r, d, 2at (5)	4-99, 1-71. 1-R	45w	TRF	Series fil.
412.	HERBERT LECTRO 200	200	none	20 x 12 x 12	1	rheo. in pri. of a.c. trans	2r. d. lat, pp. (6)	4-99, 2-71. 1-R	60w	TRF	Series fil. Sh. O. Push-pull.
413. MARTI TABLE 414. MARTI CONSOLE 415. MARTI CONSOLE		235 275 325	6-AC tubes and rectifier. Loud speaker with No. 415	7 x 21 panel	2	res. in r.f. plate	2r, d, 3ares. (6)	6-AC. 1-16B	38w	TRF	AC tubes. Built-in plate supply
416.	NASSAU POWER		none	28 x 45 x 18	2	res. in r.f. plate	2r, d, 3a (6)	5-99, 1-10. 1-16B		TRF Br.	Series fil. M, O.
417.	R.C.A. 28	540	all tubes 104 Speaker	26½ x 63 x 17	1	pot.	2d, o, 2at, 3i (8)	7-99, 1-10. 2-16B, 1-874, 1-876		Super	Model 28 may also be obtained without power units. Loop antenna.
418.	RECEPTRAD SUPER-POWER A. C. CONSOLE	390	power unit loud speaker	28 x 50 x 21	2	r.f. res.	(5)	5-01A. 2-2A	90w	Multi-flex	Loop or outside antenna
419.	SUPERPOWER AC CABINET	180	power unit	27 x 10 x 9	2	r.f. res.	(5)	5-01A. 2-2A	90w	Multi-flex, Reg.	See above
420.	SIMPLEX B	250	complete	34 x 36 x 14	1	res. in r.f. plate	(6)	6-AC		TRF	H
421.	SOVERIGN 238	325	7-AC tubes	37 x 52 x 15	2	res. on 2nd. a	(7)	7-AC	45w	TRF bal.	Uses AC tubes run by small trans. Gas rect. for plate
422.	SUPERVOX, JR.	275	complete	28 x 30 x 16	1	ant. coup. and res. in r.f. plate	1r, d, 2at (4)	3-AC, 1-71. 1-16B	40w	TRF	AC tubes, 71 on a.c. O. Sh. d
423.	SUPERVOX, SR.	450	complete	28 x 30 x 16	2	ant. coup. and res. in r.f. plate	2r, d, 2at (5)	4-AC, 1-10. 1-16B	60w	TRF	Sh. O.
469.	FREED-EISEMANN NR 11	225	NR-411 power unit	19½ x 10 x 10½	1	pot.	3r, d, 2at (6)	5-01A, 1-71. 1-R	150w	TRF Ne.	Sh. O.

STANDARD RECEIVERS

No. Model	Price	Accessories Included	Cabinet Size	No. Dials	Vol. Cont.	No. Tubes	Type Tubes	Plate Cur. mA	Circuit	Ant. Length Ft.	Remarks
428. AMERICAN C6 TABLE CONSOLE	30 65	none Spkr.	20 x 8½ x 10 36 x 40 x 17	3	pot.	2r, d, 2at (5)	5-01A	15	TRF semi-bal.	125	Partially Sh. C, Ca.
429. KING COLE VII	80-160	none	varies	2	r. f. pri. shunt	3r, d, lai, la res., lat (7)	7-01A	15-45	TRF	10-100	Steel Sh. O on some consoles. C, Ca, and Bp.
430. KING COLE VIII	100-300	none	varies	1	r. f. pri. shunt	4r, d, lai, la res., lat (8)	8-01A	20-50	TRF	10-100	See above
434. DAY-FAN 6 DAYCRAFT 6 DAY-FAN JR.	110 145 —	none Spkr. —	— 32 x 30 x 34 15 x 7 x 7	1	fil.	3r, d, 2at (6)	5-01A, 1-12 or 71	12-15	TRF	50-120	Ca, C, Sh, O
435. DAY-FAN 7		none		1	fil.	3r, d, la res, 2at (7)	6-01A, 1-12 or 71	15	TRF	50-120	Ca, C, Sh, O
436. FEDERAL	250- 1000	Loop	varies	1	rheo.	2r, d, 2at (5)	4-01A, 1-12 or 71	20.7	TRF bal.	Loop	Ca, C, Sh, Made in 6 models.
437. FERGUSON 10A	150	none	21½ x 12 x 15	1	rheo. 2r.f.	3r, d, 3a (7)	6-01A, 1-12 or 71	18-25	TRF	100	C; Ca, Sh.
438. FERGUSON 14	235	Loop	24 x 12 x 16	2	rheo. 3r.f.	6r, d, 3a (10)	9-01A, 1-12 or 71	30-35	TRF	Loop	C, Ca, Sh. Special bal.
439. FERGUSON 12	85 145	none Spkr.	22½ x 10 x 12	1	rheo. 2 r.f.	2r, d, lat, 2a res., (6)	5-01A, 1-12 or 71	18-25	TRF	100	Ca, C. Partly Sh.
440. FREED-EISEMANN NR8 NR9 NR66	90 100 125	none none none	19¾ x 10 x 10½ 19¾ x 10 x 10½ 20 x 10 x 12	2 1 1	rheo. r.f.	3r, d, 2at (6)	5-01A, 1-71	30	TRF Ne.	100	Ca. NR8 and 9 chassis type Sh. NR66 individual stage Sh.
441. FREED-EISEMANN NR77	175	none	23 x 10½ x 13	1	rheo. r.f.	4r, d, 2at (7)	6-01A, 1-71	35	TRF Ne.	Ant. or loop	Ca, C, Sh.
442. FREED-EISEMANN 800 850		none none	34 x 15½ x 13½ 36 x 65½ x 17½	1	rheo. r.f.	4r, d, 2at (8)	6-01A, 1-71 or 2-01A	35	TRF Ne.	Ant. or loop	Ca, C, Sh. Output stage 2 tubes par. or 1 power tube. O
443. GREBE CR18 (Short-Wave)	100	Set coils	6 x 7¼ x 7	2	rheo.	d, lat (2)	2-01A	8	3 cir. reg.	100	Wavelength range 8-210 m.
444. GREBE MUI	155-320	none	22½ x 9½ x 13	1-2-3	rheo. r.f.	2r, d, 2at (5)	4-01A, 1-12 or 71	30	TRF bal.	125	Bp, C, binocular coils. dials op. singly or together
445. HARMONIC R	75	none	26 x 9 x 9	3	rheo. r.f.	1r, d, 2at (4)	4-01A		TRF reg. d	100	H, C, Bp.
446. HARMONIC S	100	none	26 x 9 x 9	3	rheo. r.f.	1r, d, 3a res. (5)	5-01A		TRF reg. d	100	H, C, Bp.
447. LEUTZ TRANS-OCEANIC	150	none	27 x 8½ x 13¼	1-5	special	4r, d, lat, 3a res. (9)	5-01A, 1-00A, 2-Hmu, 1-71 or 10	20-40	TRF gr	Ant. or Loop	Range 35-3600 m M. Sh. C. Bp. O
448. LEUTZ SILVER GHOST		none	72 x 12 x 20	1-5	special	4r, d, lat, 3a res. (9)	See above	20-40	TRF gr	Ant. or loop	See above
449. NORBERT MIDGET	12	1-MV	12 x 8 x 9	2	rheo.	d, 2at, (1)	1-MV	3	TRF	75-150	C, Bp.
450. NORBERT 2	40.50	1-MV 1-01A	20 x 7 x 5½	2	special	1r, d, 2at (2)	1-MV 1-01A	8	TRF	50-100	C, Bp
451. Norco 66 CONSOLE	130 250	none Speaker	18½ x 8½ x 13¾	1-3	mod. a.f.	2r, d, 3ai (6)	5-01A 1-71	20	TRF	70-90	Sh, C, Ca, O. Drum control
452. ORIOLE 90	85	none	25½ x 11½ x 12½	2	rheo. r.f.	2r, d, 2at (5)	5-01A	18	Trinum	50-100	Ca, C
470. ORIOLE	185	none	25½ x 11½ x 12½	1	rheo. r. f.	(8)	8-01A	25	Trinum	50-100	Ca, C, Sh
453. PARAGON		none	20 x 46 x 17	1	res. r.f. plate	(6)	5-01A, 1-71	40	TRF	100	Double imp. Audio C. Ca, Sh
454. PARAMOUNT V VI	65 75	none none	26 x 7 —			(5) (6)	5-01A 6-01A		TRF	100	Bp, C.
455. PREMIER 6-IN-LINE	60- 150	none	25 x 45 x 16	1-2	rheo. r.f.	3r, d, 2at (6)	4-01A, 1-00A, 1-12	16-18	TRF	100 loop	Ca, C.
456. RADIOLA 20	115	5 tubes	19½ x 11¾ x 16	2	reg.	2r, d, 2at (5)	4-99, 1-20		TRF reg.d	75-150	C, H.
457. RADIOLA 25	165	6 tubes	28 x 37 x 19	1	pot.	o, 2d, 3int. 2at (6)	5-99, 1-20		Super	Loop	Reflexed, C, H.
458. RADIOLA 28	260	8 tubes	26½ x 63 x 17	1	pot.	o, 2d, 3int, 2at (8)	7-99, 1-20		Super	Loop	C. H.
459. STROMBERG CARLSON No. 501 502	180 290	none none	25½ x 13 x 14 28¾ x 50⅞ x 16½	2	rheo. r.f.	2r, d, 2at (5)	3-01A, 1-00A, 1-71	25-35	TRF Ne.	60-100	Ca, C, Sh, O, M, H.
460. STROMBERG CARLSON No. 601 602	225 330	none none	27 x 16¾ x 14⅜ 28¾ x 51½ x 19¾	2	rheo. r.f.	3r,d, 2at (6)	4-01A, 1-00A, 1-71	30-40	TRF Ne.	20-60	Ca, C, Sh, O, M, H.
461. SUN	80	none	23 x 10 x 10	2	res. r.f. plate	2r, d, 2at (5)	5-01A		TRF	100	Bp.
462. CUSTOM BUILT 7	275	com.	7 x 21 panel	2	special	(7)	5-01A, 1-00A, 1-71	40	TRF	100	Bp. C, O. Built in A and B power

(Continued on page 252)

No.	Model	Price	Accessories Included	Cabinet Size	No. Dials	Vol. Cont.	No. Tubes	Type Tubes	Cur. mA.	Circuit.	Ant. Length Ft.	Remarks
463.	Custom Built 9	375	com.	7 x 21 panel	2	Mod.	(9)	8-01A, 1-71	40		Loop	See No. 462
464.	Wright VII	160	none	25 x 15 x 17½	2	res. r.f. plate	3r, d, 3ai (7)	6-99, 1-20	17½	TRF	80	Ca, C, O. Na-Ald Amp.
471.	Voltone XX	50	none	20½ x 8 x 12	2	rheo.	1r, d, 3a res. (5)	1-01A, 1-00A, 2-HMu, 1-71	18	TRF	100	Ca, C, O, Reg
472.	Voltone VIII	140	none	26½ x 8 x 12	3	rheo.	2r, d, 3a res. (6)	4-01A, 1-00A, 1-71	20	TRF	100	Ca, C, O, Reg
473.	Prmco 105	45	none	18 x 7 panel	3	rheo. r. f.	2r. d, 2at, la res. (6)	6-01A		TRF	100	Ca.
474.	Prmco 110	80	none	18 x 7 panel	1	rheo. r.f.	2r. d, 2at, la res. (6)	6-01A		TRF	100	Ca, C. Drum tuning cont.
475.	Penn-C 5	150	none	24 x 70 x 15	3	pot.	(5)	5-01A	15		75	Bp, C, Console
476.	Penn-C 6	95-165	none	24 x 10 x 15	1	pot.	(6)	5-01A, 1-00A	15		75	Bp, C.
477.	Daven Bass Note	150	none	23½ x 12 x 16	2	pot.	2r, d, 3a res. (6)	2-01A, 3-HMu, 1-Power	17	TRF	50-100	Ca, C.
478.	Zimphonic Table / Console	90 / 125	none / none	— / 20 x 40 x 15	1	res.	2rd, d 3a (5)	4-01A, 1-00A. 1-12		TRF	100	H, Ca, C, Sh, reg.
479.	Zimphonic Table / Console	140 / 175	none / none	— / 20 x 40 x 15	1	rheo.	3r, d, 3a (7)	5-01A, 1-00A, 1-12	24	TRF	75	H, Ca, C, Sh, reg
480.	Pfanstiehl Cabinet 30 / Console 302	99.50 / 165	none / Spkr.	17½ x 8½ panel	1	res. plate r.f.	3r, d, 2at (6)	5-01A, 1-12 or 1-71	26-32	TRF	Ant.	Ca, C, Sh.
481.	Pfanstiehl Cabinet 32 / Console 322	135 / 225	none / Spkr.	17½ x 8½ panel	1	res. plate r.f.	3r, d, 3a (7)	6-01A, 1-12 or 1-71	23-32	TRF	Ant.	Ca, C, O. Sh.
482.	Stewart-Warner Table 705 / Console 710	Tentative 115 / 265	none / Spkr.	26½ x 11½ x 13¼ / 29½ x 42 x 17½	2	res. plate r.f.	3r, d, 2at (6)	6-01A	10-25	TRF. bal.	80	Ca, C, Sh.
483.	Stewart-Warner Table 525 / Console 520	Tentative 75 / 117.50	none / Spkr.	19½ x 10 x 11½ / 22½ x 40 x 14½	2	res. plate r.f.	3r, d, 2at (6)	6-01A	24	TRF	80	Ca, C.

What Kit Shall I Buy?

THE list of kits herewith is printed as an extension of the scope of the Service Department of RADIO BROADCAST. It is our purpose to list here the technical data about kits on which information is available. In some cases, the kit can be purchased from your dealer complete; in others, the descriptive booklet is supplied for a small charge and the parts can be purchased as the buyer likes. The Service Department will not undertake to handle cash remittances for parts, but when the coupon on page 255 is filled out, all the information requested will be forwarded.

201. SC FOUR-TUBE RECEIVER—Single control. One stage of tuned radio frequency, regenerative detector, and two stages of transformer-coupled audio amplification. Regeneration control is accomplished by means of a variable resistor across the tickler coil. Standard parts; cost approximately $58.85.

202. SC-II FIVE-TUBE RECEIVER—Two stages of tuned radio frequency, detector, and two stages of transformer-coupled audio. Two tuning controls. Volume control consists of potentiometer grid bias on r.f. tubes. Standard parts cost approximately $66.33.

203. "HI-Q" KIT—A five-tube tuned radio-frequency set having two radio stages; a detector, and two transformer-coupled audio stages. A special method of coupling in the r.f. stages tends to make the amplification more nearly equal over the entire band. Price $69.05 without cabinet.

204. R. G. S. KIT—A four-tube inverse reflex circuit, having the equivalent of two tuned radio-frequency stages, detector, and three audio stages. Two controls. Price $69.70 without cabinet.

205. PIERCE AIRO KIT—A six-tube single-dial receiver; two stages of radio-frequency amplification, detector, and three stages of resistance-coupled audio. Volume control accomplished by variation of filament brilliancy of r.f. tubes, or by adjusting compensating condensers. Complete chassis assembled but not wired costs $42.50.

206. H & H-T. R. F. ASSEMBLY—A five-tube set; three tuning dials, two stage of radio frequency, detector, and 2 transformer-coupled audio stages. Complete except for baseboard, panel, screws, wires, and accessories. Price $35.00.

207. PREMIER FIVE-TUBE ENSEMBLE—Two stages of tuned radio frequency, detector, and two stages of transformer-coupled audio. Three dials. Parts assembled but not wired. Price complete, except for cabinet, $55.00.

208. "QUADRAFORMER VI"—A six-tube set with two tuning controls. Two stages of tuned radio frequency using specially designed shielded coils, a detector, one stage of transformer-coupled audio, and two stages of resistance-coupled audio. Gain control by means of tapped primaries on the r.f. transformers. Essential kit consists of three shielded double-range "Quadraformer" coils, a selectivity control, and an "Amplitrol," price $17.50. Complete parts $70.15.

209. GEN-RAL FIVE-TUBE SET—Two stages of tuned radio frequency, detector, and two transformer-coupled audio stages. Volume is controlled by a resistor in the plate circuit of the r.f. tubes. Uses a special r.f. coil ("Duo-Fotmer") with figure eight winding. Parts mounted but not wired, price $37.50.

210. BREMER-TULLY POWER-SIX—A six-tube, dual-control set; three stages of neutralized tuned radio frequency, detector, and two transformer-coupled audio stages. Resistances in the grid circuit together with a phase shifting arrangement are used to prevent oscillation. Volume control accomplished by variation of B potential on r.f. tube. Essential kit consists of four r.f. transformers, two dual condensers, three small condensers, three choke coils, one 500,000-ohm resistor, three 1000-ohm resistors, and a set of color charts and diagrams. Price $40.50.

211. BRUNO DRUM CONTROL RECEIVERS—How to apply a drum tuning unit to such circuits as the three-tube regenerative receiver, four-tube Browning-Drake, five-tube Diamond-of-the-Air, and the "Grand" 6.

212. INFRADYNE AMPLIFIER—A three-tube intermediate-frequency amplifier for the super-heterodyne and other special receivers, tuned to 3490 kc. (86 meters). Price $25.00.

213. RADIO BROADCAST "LAB" RECEIVER—A four-tube dual-control receiver with one stage of Rice neutralized tuned-radio frequency, regenerative detector (capacity controlled), and two stages of transformer-coupled audio. Approximate price, $78.15.

214. LC-27—A five-tube set with two stages of tuned-radio frequency, detector, and two stages of transformer-coupled audio. Special coils and special means of neutralizing are employed. Output device. Price $65.20 without cabinet.

215. LOFTIN-WHITE—A five-tube set with two stages of radio frequency, especially designed to give equal amplification at all frequencies, a detector, and two stages of transformer-coupled audio. Two controls. Output device. Price $85.10.

216. K H-27—A six-tube receiver with two stages of neutralized tuned radio frequency, a detector, three stages of choke-coupled audio, and an output device. Two controls. Price $66.00 without cabinet.

217. AERO SHORT-WAVE KIT—Three plug-in coils designed to operate with a regenerative detector circuit and having a frequency range of from 19,990 to 2306 kc. (15 to 130 meters). Coils and plug only, price $13.50.

218. DIAMOND-OF-THE-AIR—A five-tube set having one stage of tuned-radio frequency, a regenerative detector, one stage of transformer-coupled audio, and two stages of resistance-coupled audio. Volume control through regeneration. Two tuning dials.

219. NORDEN-HAUCK SUPER 10—Ten tubes; five stages of tuned radio frequency, detector, and four stages of choke, and transformer-coupled audio frequency. Two controls. Price $291.40.

WHAT KIT SHALL I BUY (Continued)

220. BROWNING-DRAKE—Five tubes; one stage tuned radio frequency (Rice neutralization), regenerative detector (tickler control), three stages of audio (special combination of resistance- and impedance-coupled audio). Two controls.

221. LR4 ULTRADYNE—Nine-tube super-heterodyne; one stage of tuned radio frequency, one modulator, one oscillator, three intermediate-frequency stages, detector, and two transformer-coupled audio stages.

222. GREIFF MULTIPLEX—Four tubes (equivalent to six tubes); one stage of tuned radio frequency, one stage of transformer-coupled radio frequency, crystal detector, two stages of transformer-coupled audio, and one stage of impedance-coupled audio. Two controls. Price complete parts, $50.00.

223. PHONOGRAPH AMPLIFIER—A five-tube amplifier device having an oscillator, a detector, one stage of transformer-coupled audio, and two stages of impedance-coupled audio. The phonograph signal is made to modulate the oscillator in much the same manner as an incoming signal from an antenna.

A KEY TO RECENT RADIO ARTICLES

By E. G. SHALKHAUSER

*T*HIS is the twenty-first installment of reference to articles which have appeared recently in various radio periodicals. Each separate reference should be cut out and pasted on 4"x6" cards for filing, or pasted in a scrap book either alphabetically or numerically. An outline of the Dewey Decimal System (employed here) appeared last in the January RADIO BROADCAST.

R344.4. SHORT-WAVE GENERATORS.   TRANSMITTER
QST, May, 1927. Pp. 9-14.   Short-Wave
"A Complete Inexpensive Transmitter," H. P. Westman.
A short-wave transmitter operating with two, UX-171 tubes in a back-to-back Hartley circuit with straight a.c. on the plate, is outlined. Complete construction data, and detailed operating instructions relative to the principle employed in using the full a.c. wave, are given for the benefit of the beginner, who may desire to get into the transmitting game.

R333. THREE-ELECTRODE ELECTRON   Vacuum Tubes
TUBES.   UX-852.
QST, May, 1927. Pp. 20-23.
"The UX-852 Transmitting Tube," R. S. Kruse.
Complete tube data relative to the new UX-852 75-watt transmitting tube are presented. These include the oscillating characteristics, filament curves, amplification factor, plate impedance, mutual conductance, and several static grid and plate voltage and current curves.

R281. 77. QUARTZ.   QUARTZ
QST, May, 1927. Pp. 24-26.   Grinding of
"A Method of Grinding Quartz Plates," P. Mueller.
A method whereby quartz crystals may be ground to parallelism is outlined. The method consists of mounting nine crystal slabs to one plate and grinding; by frequent transpositions, true crystal surfaces are obtained.

R084. MAPS AND CHARTS.   CHART, Tube
QST, May, 1927. Pp. 48-49.   Characteristic
"A Tube Characteristic Chart."
A chart showing the characteristics of plate resistance, amplification constant, and mutual conductance of vacuum tubes, at a glance, has been devised, and is shown. To illustrate the principle, an example is worked out.

R114. STRAYS.   STATIC
Popular Radio. May, 1927. Pp. 427-430.
"Static's New Job as a Life-Saver," Com. S. C. Hooper.
The detection, recording, and the analysis of static signals is said to be of material benefit in locating and studying storm areas, especially at sea, where the observations, here referred to, have been made. Experiences while aboard the U.S.S. Kittery are related.

R382. TRANSMISSION OF PHOTOGRAPHS.   TELEVISION
Popular Radio. May, 1927. Pp. 447-ff.   Baird System
"Television and 'Black Light'," John L. Baird.
The experiments carried on by John L. Baird with his television apparatus are enumerated by the inventor himself. Starting with simple apparatus and making use of many new developments in science, his system now employs rays of light in the infra-red spectrum. Although only in the first stages of development, experiments with the so-called "black light" give promise of great strides being made and the perfection of television, as stated.

R343. ELECTRON-TUBE RECEIVING SETS.   RECEIVER
RADIO BROADCAST. May, 1927. Pp. 24-25.   Short-Wave
"A Balanced Short-Wave Receiver," F. C. Jones.
A non-radiating short-wave receiver, covering the range from 9094 to 5006 kilocycles (30 to 50 meters), is described. A bridge circuit is utilized to prevent the set from radiating. Antenna and ground are connected across the zero potential points of the bridge. Two tubes are found necessary in the described arrangement, the constants of which are discussed in detail.

R343. 7. ALTERNATING-CURRENT   A.C. FOR FILA-
SUPPLY.   MENT
RADIO BROADCAST. May, 1927. Pp. 33-35.
"Filament Lighting from the A. C. Mains," R. F. Beers.
The advantages of series rather than parallel arrangement of filaments are stated as: (1) Total current to be filtered is only that taken by one tube, and (2) higher voltages available for filtering. Circuit diagrams are shown and explained.

R231. AMMETERS.   AMMETER
RADIO BROADCAST. May, 1927. Pp. 38-39. Volume Indi-
"Technical Operation of Broadcasting Stations: cator
No. 15. Volume Indicators," Carl Dreher.
Practical circuit diagram and operating data concerning "volume indicators" as used in broadcasting stations, are given. Use is made of either a d.c. or an a.c. milliammeter connected into the plate circuit of a vacuum tube or coupled to it.

R382. INDUCTORS.   INDUCTOR
RADIO BROADCAST. May, 1927. Pp. 40-42.
"Some Facts About Coil Design," R. Gunn.
The oscillating circuit, consisting of inductance, resistance, and capacity, is found to be a form of voltage amplifier, its amplification being equal to the generated voltage divided by the applied voltage. This factor is spoken of as the "gain." The "gain" depends primarily on the coil rather than the condenser, the latter having comparatively lower losses. The following points are considered of importance:—(1). The gain of the coil should not exceed the limit 250 for good quality reproduction. (2). the exterior field the coil should be as near zero as possible; (3). the distributed capacity should be very low; (4) the coil should be mechanically strong.

A Name
With a Background

Ever since the beginning of radio broadcasting the B-T trademark has identified a long list of successful products.

An ever-increasing army of satisfied customers *know* that that trademark stands for sterling quality.

Personal experience with B-T merchandise has convinced them —it will *convince* you too!

Consider the value of these years of experience—experience that is built into the design and construction of every essential unit used in Counterphase Receivers.

A new Bremer-Tully counterphase six receiver in the one hundred dollar price class is now available in addition to those illustrated.

A wonderful new cone speaker completes the Bremer-Tully line.

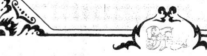

BREMER-TULLY MF(

THE COUNTRY LIFE PRESS, GARDEN CITY, NEW Y

RADIO
BROADCAST

CONTENTS

What Are the Season's Kits

Locating and Remedying Interference

Distinguishing Between Good and Bad Quality

Radio Picture Reception for the Experimenter

A Directory of Manufactured Receivers

A Survey of the A·C·Tubes

35 Cents   Doubleday, Page & Company
Garden City, New York   Sept·1927

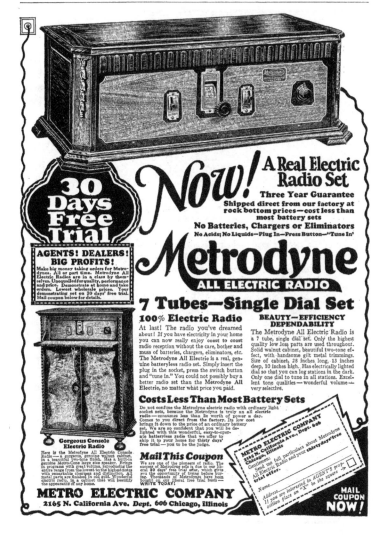

Now! A Real Electric Radio Set

Three Year Guarantee

Shipped direct from our factory at rock bottom prices—cost less than most battery sets

No Batteries, Chargers or Eliminators

No Acids; No Liquids—Plug In—Press Button—"Tune In"

Metrodyne
ALL ELECTRIC RADIO

7 Tubes—Single Dial Set

100% Electric Radio

At last! The radio you've dreamed about! If you have electricity in your home you can now really enjoy coast to coast radio reception without the care, bother and muss of batteries, chargers, eliminators, etc. The Metrodyne All Electric is a real, genuine batteryless radio set. Simply insert the plug in the socket, press the switch button and "tune in." You could not possibly buy a better radio set than the Metrodyne All Electric, no matter what price you paid.

BEAUTY—EFFICIENCY DEPENDABILITY

The Metrodyne All Electric Radio is a 7 tube, single dial set. Only the highest quality low loss parts are used throughout. Solid walnut cabinet, beautiful two-tone effect, with handsome gilt metal trimmings. Size of cabinet, 28 inches long, 13 inches deep, 10 inches high. Has electrically lighted dial so that you can log stations in the dark. Only one dial to tune in all stations. Excellent tone qualities—wonderful volume—very selective.

Costs Less Than Most Battery Sets

Do not confuse the Metrodyne electric radio with ordinary light socket sets, because the Metrodyne is truly an all electric radio—consumes less than 2c worth of power a day. Comes to you direct from the factory. Its low cost brings it down to the price of an ordinary battery set. We are so confident that you will be delighted with this wonderful, easy-to-operate batteryless radio that we offer to ship it to your home for thirty days' free trial—you to be the judge.

Mail This Coupon

We are one of the pioneers of radio. The success of Metrodyne sets is due to our liberal 30 days' free trial offer, which gives you the opportunity of trying before buying. Thousands of Metrodynes have been bought on our liberal free trial basis—WRITE TODAY!

30 Days Free Trial

Gorgeous Console Electric Radio

Here is the Metrodyne All Electric Console Radio—a gorgeous, genuine walnut cabinet in a beautiful two-tone finish. Has a built-in genuine Metro-Cone large size speaker. Brings in programs with great volume, reproducing the entire range from the lowest to the highest notes with remarkable clearness and distinction. All metal parts are finished in old gold. Wonderful electric radio, in a cabinet that will beautify the appearance of any home.

METRO ELECTRIC COMPANY
2165 N. California Ave. Dept. 606 Chicago, Illinois

Still the Standard of Excellence in Audio Transformers

The AmerTran DeLuxe
for 1st and 2nd Stages $10.00 each

MONTH after month, and on into years, speeds the development of new radio units. Yet to-day the AmerTran DeLuxe is still the measure used to judge transformer quality. Time has taken no toll of the AmerTran DeLuxe. Rather, it has broadened its acceptance as *the* faithful amplifier of natural quality.

Research and test laboratories—some of the best known in the country—use the AmerTran DeLuxe as it comes nearest in their opinion to perfection in this type of apparatus. Structural ruggedness has made the AmerTran DeLuxe stand up in the climate of Central America. It has proved itself under all conditions. With the proper tubes and loud speaker, and clear signals from the detector tube, these transformers will produce the truest quality with most volume.

American Transformer Co.

178 Emmet Street   Newark, N. J.

"Transformer Builders for Over 26 Years"

Other AmerTran Radio Products

Type PF52 Power Transformer
Type PF281 Power Transformer for Filament plate supply with new tubes.

Type H-67 Filament Heating Transformer
AmerTran Resistor Type 400
AmerChokes Types 854, 709 and 418

"There's th for Your

$28⁵⁰

with Raytheon B F

IF YOUR SET has five five and a Power tube, now get a Sterling R equipped "B" for it at $28. spend more? Think of it! than $30 you can get ric batteries once and, forevei their place be sure of full-Sterling-filtered "B" curret from the light socket.

Sterling Model RT-81 ha put at 35 mills of 135 volts. and Medium voltages are justable and the high tap ti of the Power tube on most s assures plenty of voltage and exact power regulation for your particu- lar set. RT-81 is no larger than one dry "B" bat- tery, and fits right into the console cabinet or into Radiolas 25 & 28.   RT-

—*Another Popular Sterling M Larger Sets, R-98 "B-C" Pot*
A "Universal" "B-C" model for hig Output at 55 mills, is 180 volts. terminals, two of which are independe All four voltages are variable throt control. Variable High "C" voltage switch. Uses Raytheon B or B H Tub without Tube, $35.50.

"B" POWER UI

THE STERLING MI
2831 Prospect Avenue · Cleve

RADIO BROADCAST. September, 1927. Published monthly. Vol. XI. No. 5. Published at Garden City, N. Y. Subscription price $4.00 a year. Entered at U Garden City, N. Y., as second class mail matter. Doubleday, Page & Company, Garden City, N. Y.

RADIO BROADCAST

WILLIS KINGSLEY WING, Editor

SEPTEMBER, 1927   KEITH HENNEY   EDGAR H. FELIX   Vol. XI, No. 5
Director of the Laboratory   Contributing Editor

CONTENTS

AMONG OTHER THINGS. . .

SOME may look with critical eye at the leading article this month which delves into the subject of ultra-violet, infra-red, and X-rays, for they may think that field a bit removed from radio. But in these days of scientific advance, many fields of experiment which have heretofore been widely separated, are becoming more closely associated, and anything, therefore, having to do with the generation of short electrical waves, is of interest to those who dabble in radio. James Stokley, who contributes this story on ultra short waves, is on the staff of Science Service, that interesting and important Washington organization devoted to telling the public more about science.

OF OUR other authors, M. Thornton Dow, the writer of the article on the remarkable piezo-electric crystal, is at Cruft Laboratory, Harvard University. Homer Davis, whose calculation charts have appeared in RADIO BROADCAST before, is a graduate engineer and a resident of Memphis, Tennessee. Ernest R. Pfaff, who describes the interesting super-heterodyne which may be made with the Jeweller's Time amplifier, is a radio denizen of Chicago and a frequent contributor to the radio press. B. F. Miessner, the author of the Radio Club of America paper on the a.c. tube, is chief engineer of the Garod Corporation and has had an extensive and varied radio experience. In early radio days, he was associated with John Hays Hammond, conducting radio and other experiments. A. T. Lawton, author of the series on eliminating man-made radio interference, is a Canadian, living in Ottawa. He has been associated with the interference-prevention work now being done in the Dominion.

THE short-waves, long the almost exclusive playground and workshop of the amateurs and commercial and military services, are now beginning to harbor broadcast programs. Some of our own American stations are broadcasting their programs on two waves—the standard broadcasting wave and a short wave. This experiment, long exclusively conducted by WGY and KDKA, is now being shared by several others. There are stations abroad, too, providing voice and music on these bands. The article in this issue by Keith Henney tells something of the traffic in these lesser-known bands.

THE subject of transmission of photographs by radio—wrongly termed television by many, for it is in fact, radio-telephotography—is receiving increased attention, and for the readers of RADIO BROADCAST we have arranged to publish an exclusive series of articles. These stories will describe a practical and inexpensive system which can be attached to any good radio receiver. By its use good pictures will result, as experiments, conducted in RADIO BROADCAST Laboratory over a period of more than three years, have shown. An introductory article appears in this number and others will soon follow.

THE Directory of Manufactured Receivers, which appeared in our August number for the first time, appears in a slightly revised form in this number. The present form is much easier to use. We call especial attention to the fact that the Service Department of RADIO BROADCAST will furnish much more detailed information on any or all of the receivers listed if the coupon on page 328 is filled out and sent to us.

—WILLIS KINGSLEY WING.

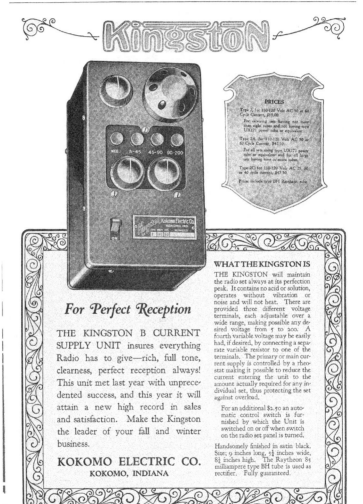

Kingston

PRICES

Type 2, for 110-120 Volt AC 50 or 60
Cycle Current, $35.00.

For receiving sets having not more
than eight tubes and not having type
UX171 power tube or equivalent.

Type 2A, for 110-120 Volt AC 50 or
60 Cycle Current, $42.50.

For all sets using type UX171 power
tube or equivalent and for all large
sets having nine or more tubes.

Type 2C, for 110-120 Volt AC 25, 30,
or 40 cycle current, $47.50.

Prices include type BH Raytheon tube.

For Perfect Reception

THE KINGSTON B CURRENT
SUPPLY UNIT insures everything
Radio has to give—rich, full tone,
clearness, perfect reception always!
This unit met last year with unprecedented success, and this year it will
attain a new high record in sales
and satisfaction. Make the Kingston
the leader of your fall and winter
business.

KOKOMO ELECTRIC CO.
KOKOMO, INDIANA

WHAT THE KINGSTON IS

THE KINGSTON will maintain
the radio set always at its perfection
peak. It contains no acid or solution,
operates without vibration or
noise and will not heat. There are
provided three different voltage
terminals, each adjustable over a
wide range, making possible any desired voltage from 5 to 200. A
fourth variable voltage may be easily
had, if desired, by connecting a separate variable resistor to one of the
terminals. The primary or main current supply is controlled by a rheostat making it possible to reduce the
current entering the unit to the
amount actually required for any individual set, thus protecting the set
against overload.

For an additional $2.50 an auto-
matic control switch is furnished by which the Unit is
switched on or off when switch
on the radio set panel is turned.

Handsomely finished in satin black.
Size; 9 inches long, 5½ inches wide,
8½ inches high. The Raytheon 85
milliampere type BH tube is used as
rectifier. Fully guaranteed.

Sent by Wireless Facsimile Telegraphy

HIGH SPEED WIRELESS TELEGRAPHY
CENTRAL CONTROL ORGANISATION

THE quarter of a century which has witnessed the development of commercial wireless telegraphy from the sending of the first tentative signal to the establishment of high-speed telegraph services to all parts of the world has been a period of incessant progress.

Every year has brought some fresh invention to increase the speed of signalling, or to improve methods of working, but a stage has now been reached when certain basic principles have been established and can be incorporated in standard practice. Two of the most important of these are the ascendancy of continuous wave wireless telegraphy by means of valve transmission, and the distant control of the transmitting and receiving stations from a central office.

These modern methods are to be seen at their highest state of efficiency in the group of Marconi stations comprising Radio House, Ongar, Brentwood, and Carnarvon, from which high-speed commercial services are conducted with France, Switzerland, Spain, Canada, and the United States of America.

The wireless stations at Ongar and Brentwood are situated in Essex, some 20 miles from London, but full control is centred at Radio House, Wilson Street, in the City, the relaying of signals from the land lines to the wireless transmitters at Ongar transmitting station, and from the wireless receivers to the land lines at Brentwood receiving station being entirely automatic. The transmitting plant at Carnarvon used for communication to the United States is also controlled automatically from Radio House, and the signals from the United States are received at Brentwood and relayed automatically to Radio House.

March 12/1927

1

A METHOD OF SPEEDY RADIO TRANSMISSION

Radio telegraph messages, using the present system, when sent must be translated into the code, and at the receiving station, must be decoded and copied. A method has been developed by J. M. Wright of the British Marconi Company for facsimile transmission which eliminates the translation into the code altogether. The sample in the illustration above was transmitted in 100 seconds.

RADIO BROADCAST

VOLUME XI

NUMBER 5

SEPTEMBER, 1927

A Discovery That Newton Missed

Something About Waves of One Trillion Kilocycles Frequency— Ritter's Discovery of Ultra-Violet and Herschel's of Infra-Red Rays —The Uses of Ultra-Violet Rays for Their Health Giving Properties

By JAMES STOKLEY
Science Service

ABOUT two and a half centuries ago Sir Isaac Newton performed an experiment. During his life, this greatest scientist of his day and one of the greatest of all time, performed many experiments, but from this one in particular there came many far-reaching results. From it, more or less directly, came much of our knowledge of radio, of X-rays, the radiations of radium, and of the composition of the most distant stars, obtained by spectrum analysis.

Just what did Newton do? It so happens that we have a very complete account in his own words, left in one of his books, which bore the title of: *Opticks: or, a Treatise of the Reflexions, Refractions, Inflexions, and Colors of Light.* The first edition, of this great work, now very rare, was published in 1704, and on page 18 we read:

THE LIGHT OF THE SUN CONSISTS OF RAYS DIFFERENTLY RE-FRANGIBLE

The Proof by Experiments

In a very dark chamber at a round hole about one third part of an inch broad made in the shut of a window I placed a glass prism, whereby the beam of the sun's light which came in at that hole might be refracted upwards toward the opposite wall of the chamber, and there form a colored image of the sun. The axis of the prism (that is, the line passing through the middle of the prism from one end of it to the other end

parallel to the edge of the refracting angle) was, in this and the following experiments, perpendicular to the incident rays. . . . I let the refracted light fall perpendicularly upon a sheet of white paper at the opposite wall of the chamber, and observed the figure and dimensions of the solar image formed on the paper by that light. This image was oblong and not oval, but terminated with two rectilinear and parallel sides, and two semicircular ends. . . . This image, or spectrum, was coloured, being red at its least refracted end, and violet at its most refracted end, and yellow, green, and blue in the intermediate spaces.

This experiment of Newton's is a very easy one to repeat, for all you need is a prism. If you have one of the kind that used to hang from chandeliers in mid-Victorian homes, it will serve the purpose admirably.

But there was one important thing that

ONE OF THE USES OF ULTRA-VIOLET RAYS

Quartz tube mercury vapor lamps in use as sterilizers for the water supply of a small city. The ultra-violet rays prevent thriving of germs in the water

Newton missed. So far as he could detect, the spectrum began at the violet end and ended at the red, which, together with the colors in between, constituted the complete composition of sunlight. That was because he was human, and had only his sense of sight to guide him. Actually the spectrum that he saw was only a small part of the complete spectrum of sunlight and an infinitesimally small part of the total range of radiation that is now known. The complete spectrum ranges from the longest radio waves and the still longer waves of alternating electric currents, down to the X-rays, and the penetrating radiation recently investigated by Professor Millikan. According to a well-known physicist, Dr. M. Luckiesh, if the visible part of the spectrum were one foot in length, approximately the length that Newton saw it, the spectrum of the total range of radiation would be several million miles long!

When you go to the seashore, and sit on the beach to get tanned by the sunshine, or when you take a snapshot of some member of the family, you demonstrate the presence of one kind of rays that the eye cannot perceive, for many of the chemical effects of sunlight, including that on the skin, and on the photographic film, are due to ultra-violet rays, consisting of waves which are a little

LOOK AT THIS PHOTOGRAPH
A steel surface photographed through a microscope with blue light. Ultra-violet rays would
have given a much clearer picture

too short, and vibrate a little too rapidly,
to be apparent to the eye.

As radio fans who are familiar with the
modern system of designating stations by
frequency rather than wavelength are
aware, the average broadcasting frequency
is in the neighborhood of 1000 kilocycles,
which means that the complete vibration
which produces the wave occurs 1,000,000
times a second. But the longest of the
visible rays, those in the deepest red that
the eye can see, have a frequency of
375,000,000,000 kilocycles. The violet
rays at the other end of the spectrum vibrate about twice as rapidly, and the ultra-
violet rays are between these and those
vibrating with a frequency of 24,000,000,-
000,000 kilocycles!

But Newton, as we have seen, did not
know of these invisible radiations. It was
not many years after the publication of his
Opticks, however, that some of their effects
were noticed, for in 1727 a German physi-
cian, Johann Heinrich Schulze, happened to mix some chalk with nitric acid in
which had been dissolved a little silver. At
the time, he happened to be standing near a

window, and he noticed that where the
light fell on the mixture it became dark.
The silver had dissolved in the nitric acid
forming silver nitrate, and this must have
reacted with some of the impurities in the
chalk, such as ordinary salt, or sodium
chloride, to produce silver chloride, which is
one of the chief constituents of modern
photographic materials.

However, Schulze also missed the discovery of ultra-violet radiation, for he
only tried daylight on his silver chloride
mixture. Half a century later, in 1777, its
discovery was missed by a still narrower
margin. This was by the famous Swedish
chemist, Carl Wilhelm Scheele, who in
1774 had discovered the gas chlorine, and,
three years before that, another gas—
fluorine. Unlike Schulze, who accidentally
came across the effect when he was trying to
do something else, Scheele was studying the
chemical action of light, and he exposed a surface coated with silver chloride to the spectrum obtained from sunlight with a prism.

What must have happened was that the
silver chloride under the red and yellow
part of the spectrum was not darkened at
all, while that under the violet and the
part where no color was visible beyond the
violet, were darkened most. The experiment is one that is easily repeated; a piece
of photographic printing paper may be
used as the silver chloride surface.

But Scheele didn't notice this. Probably
he was working in a room that was not entirely dark, except for the spectrum, and so
the silver chloride was partly darkened all
over due to the white light in the room.
Anyhow, it remained for a German
physicist, named Ritter, to discover that
the invisible extension of the solar spectrum, beyond the violet, produced a greater
effect on the silver chloride than did the
violet itself, and so he discovered the ultra-
violet rays, in 1801. Curiously enough, it
was at almost the same time that the extension of the spectrum in the other direction, beyond the red, was discovered; the
year before, the great English astronomer,

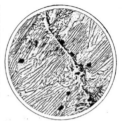

—AND NOW, THIS ONE
The same steel surface photographed with ultra-
violet light, and a magnification of 1600 diameters. Notice the great amount of detail that
can be seen in this picture as compared with the
blue light photograph

Sir William Herschel, discovered the infra-
red rays.

In the century following the discovery
of the invisible rays, there came many discoveries, such as that of methods by which
these rays could be produced artificially.
The most copious source of ultra-violet
rays known is, of course, the sun itself. On
a clear summer day we get great quantities
of them, with possibly even painful results if
we are at the seashore, lying on the beach.
But in winter, even when the sun is shining, it may be low in the sky, and the smoke
and dust in the atmosphere may effectually
keep most of these rays away from us.

As we shall see a little later, ultra-violet
rays are important in keeping people
healthy, and in curing various diseases. For
this reason, as well as for the accommoda-
tion of the physicist who wishes to study
the rays in his laboratory, an artificial
source is important. The ordinary carbon
arc light, particularly if operated at a high
voltage, perhaps 120 (which is high for an
arc light), is rich in these invisible rays;

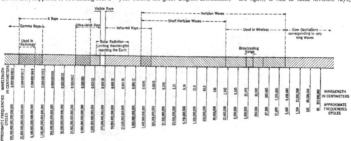

HOW THE FREQUENCY SPECTRUM IS DIVIDED
The spectrum of radiation, from X-rays to alternating-current waves. The penetrating rays, studied by Millikan, would extend from the left end of the
diagram. Naturally, this diagram is not to scale, for if it were, and the short visible part a foot in length, the entire scale would be many millions of miles long

even larger quantities result if we substitute rods of iron for the carbons. When the iron arc is used in laboratories, it is necessary to protect the eyes from its light. Many a physicist making experiments with the iron arc has got as nice a case of sunburn as if he had been at the seashore.

Other metals can be used instead of iron and even more copious quantities of ultra-violet light result, for example, if the electrodes are of silicon. In this case the visible light is reduced nearly to a minimum, and the output of the lamp is almost entirely invisible light. But mercury, or quicksilver, is most commonly used, particularly in the form of the mercury vapor lamp, where an electric discharge is made in an atmosphere of the vapor of mercury instead of air.

As mercury vapor is a gas, it is necessary to completely enclose such a lamp in a transparent tube. Glass, of course, is often used, and the glass-tube mercury vapor lamp is a common light source in photographic and movie studios, as well as in many factories, because it is a very efficient illuminating device. The incandescent gas mantle takes the equivalent of about nine watts of electric power to produce one candle power, the tungsten filament electric lamp about a watt, and the ordinary arc lamp about nine-tenths of a watt; but the glass tube mercury vapor lamp produces the same amount of light with about two-thirds of a watt. Of course the purple color of the light, and the lack of red rays, gives the skin a ghastly pallor, but the light is not harmful to the eyes and is said to be pleasant to work under after one gets used to it.

Such a light doesn't give off much ultra-violet radiation, because these rays are absorbed by glass, but by making the tube of quartz, large amounts of ultra-violet radiation are emitted, and then the lamp gives one candle power for every quarter of a watt. This is the most convenient source of the rays, and is the one now most generally used in laboratories and hospitals. Such lamps are often used for sterilizing water and various food products as the action of ultra-violet light is fatal to many germs. A common use is in swimming pools, for the same water can be filtered and purified by passing it and re-passing it through apparatus which exposes it to ultra-violet rays. The water in a pool so equipped can really be kept purer than if it were emptied and refilled daily.

ULTRA-VIOLET RAYS IN PHOTOGRAPHY

BECAUSE the ultra-violet rays are so short they have an important application in photography of minute objects through the microscope. In modern optical factories it is possible to make lenses for microscopes which will theoretically magnify almost without limit. Actually, such lenses are limited, because a structure must be as large as a wave of light is long in order to reflect it.

If you toss a tennis ball against a wall it bounces back, but if you throw it against a spider web it passes right through. This is because the mass of the wall compared with the ball is very great, but that of the spider web is very small. The case is somewhat analogous to light. The object that reflects it must compare favorably in size with the length of one of the waves.

Since the waves of ultra-violet light are shorter than those of visible light, structures that are too small to be seen under the microscope with ordinary illumination may be seen, or rather photographed, in the ultra-violet ray. It was with a method such as this that the English biologist, Gye, with the assistance of the expert microscopist, J. E. Barnard, was able to make photographs through the microscope of the germs which caused cancer in chickens, and which were beyond the limits of the ordinary microscopic methods. In metallurgical laboratories, ultra-violet microscopic apparatus is used in photographing steel and other metals. As glass stops most of the ultra-violet rays, such a microscope must use lenses made of quartz.

Since ultra-violet light may be used so advantageously in photomicrography, a natural idea is that of using X-rays, the waves of which are much shorter than the ultra-violet. If X rays could be used, structures could be photographed a thousand times smaller than with the visual rays, instead of, perhaps, half the size, as with the ultra-violet. But though X-rays can be deflected through crystals, they can't be focussed, like light, by means of a lens; so X-ray photomicrography is one of the yet unsolved riddles that are so numerous in science.

One of the most important uses of ultra-violet light is its effect on the body. We hear a great deal nowadays about vitamins, those mysterious substances in food about which so little is known, but that are so necessary if we are to keep healthy. One of these vitamins prevents a disease which in some localities is very common among children, namely, rickets. Few people realize just how common this ailment is, but Dr. Alfred Hess, a prominent New York specialist on children's diseases, has estimated that seventy-five per cent. of the children in New York City have at one time or another had at least a mild case of rickets.

However, though there are not many diseases for which the medical profession knows any practically sure-fire remedies, rickets is one of them, for it may be cured with either cod liver oil, which is rich in the anti-rachitic vitamin, or with treatment by ultra-violet light.

The ways of the vitamins, like the "heathen Chinee," are peculiar. But rickets is a disease of the bones, due to lack of salts of calcium, especially calcium phosphate, and the action of the anti-rachitic vitamin seems to be to hasten the deposition of calcium. Cod liver oil normally contains the vitamin, and so it can be used as a cure. Other oils, like cotton-seed oil, or foods such as milk or flour, do not ordinarily contain it, but if they are exposed to ultra-violet light the vitamin seems to be formed, for then the anti-rachitic powers are bestowed on them.

Then also, by exposing the child afflicted to the disease to the beneficial rays of the sun, or of the quartz tube mercury vapor lamp, the calcium deposition is also hastened and the disease relieved. Since egg shells consist largely of calcium, exposure of chickens to ultra-violet light hastens egg laying in the same way that it prevents or cures rickets in children. Such eggs also have anti-rachitic powers, so we may expect to see "eggs from sun-kissed hens" a common article in our markets of the future!

In order that the ultra-violet rays may reach the body, it is necessary that they have an unobstructed path to the skin. Glass stops the rays, so that a sun bath indoors behind ordinary windows is of little value. Windows made of quartz, which can now be fused and made into a variety of shapes, including large windows, will let the rays through. Several kinds of glass have recently appeared on the market which are much less expensive than quartz, but still let the beneficial rays through in large quantity. Ordinary clothing also is opaque to the rays, so that at a sanitarium such as that of Dr. Rollier, in Switzerland, the patients, both children and older people, spend most of their time outdoors with a minimum of clothing. However, the so-called "artificial silk" or rayon, which is so much used, for hosiery and other articles of apparel, is partly transparent to the rays. We can conform with all the standards of modesty, and still get the ultra-violet rays.

Unlike X-rays, ultra-violet rays are not very penetrating. They do not go very deep into the body, but reach only the outer skin, and it has been a puzzle how they have their effect on the bones. Doctor Hess has suggested an explanation which fits in pretty well with the observed facts.

There is a substance known to the chemists as cholesterol, which is found in practically every living animal, especially the skin and brain, a similar substance, phytosterol, being found in plants. Doctor Hess has found that when cholesterol is placed under ultra-violet rays, it achieves the power of preventing rickets, so he believes that this same process takes place in the skin when exposed to ultra-violet light. The radiated cholesterol is then carried by the blood to other parts of the body. To confirm this idea, Doctor Hess took pieces of animals' skins and fed them to rats, as their main article of diet. When the skin had been exposed to ultra-violet rays, the rats remained healthy but another group that were fed with skin, that had not been radiated soon developed rickets.

So another step has been made in understanding how these rays work on the body. Beginning over two centuries ago with New ton, continuing with Ritter's discovery of the rays, our knowledge of them has gradually advanced. Though their medical uses are important, they are not the only applications, and the physicist constantly makes use of them. Just what part these rays play in the daily life of the physicist we shall see in a subsequent article.

Radio Needs No Yearly Models

THE success of the June Chicago Trade Show establishes that function as an annual custom. It gives the manufacturers opportunity to introduce their wares to the trade so that, when the Radio World's Fair in New York and the other fall shows take place, distribution of the new models will be an accomplished fact. The 1927 Trade Show was marked principally by advances in the alternating-current tube sets, deriving their A, B, and C potentials direct from the power mains. Last season might have been styled single-control year; this season marks the beginning of a general market for the a.c. set.

We do feel, however, that the yearly model business cannot long continue to be an advantage to the radio industry. One of the best ways to establish the summer slump as a permanent institution is to promise new and revolutionary changes each fall. Naturally, the public will stop buying about January 1 to await the year's developments. Yearly models are justified if, each year, advances are made of such significance that all the products of the preceding years are justifiably obsolete and the public adopts the custom of scrapping their radio sets every year or two. But the radio ship is approaching a more even keel. Refinement is already replacing revolution in improvement. The purchase of a radio receiver is becoming a four- or five-, and perhaps even a ten-year investment. Consequently, it is of advantage, both to the owner and the manufacturer, to work toward constant refinement and improvement, embodying developments in production as soon as they are perfected rather than annually disrupting the market by "yearly models."

It took the automobile industry twenty years to learn the fallacy of the yearly model plan and we hope that the radio industry will be quick to perceive the hazards and pitfalls of that system. Let us put a stop to the yearly model plan lest it bring us that millstone of the automobile industry—the used model business. A radio receiver does not seriously depreciate in use and hence a used receiver business, the inevitable by-product of the yearly model plan, is of advantage neither to the dealer nor to the set owner.

Will Telephotography Be the Experimenters' Next Field?

A PHOTOGRAPH received by radio in your own home! That is a reward worth working for! We have been watching the efforts of many experimenters in telephotography, hoping to bring a new and alluring field of activity to the home constructor. Essentially, the spirit of the set builder is that of the conqueror who faces and overcomes mechanical and electrical difficulties. When broadcasting first began, the man who built his own receiver was regarded as a sort of wizard among his friends. And indeed, considering the difficulties under which he then labored, he was worthy of that title.

But set building has lost a little of its glamour. It is still, by a wide margin, the most satisfying hobby for one with a flair for electrical experiment. At first, it had, as an additional incentive, the fact that the home-built set was much better and cheaper than anything that could be bought on the market. Competition and technical development, however, have brought such great improvements in the manufactured set, both from the standpoint of price and performance, that the home constructor finds his economic advantage and his pride of achievement both measurably diminished. He must content himself with the fascination of building as his return for the time and money he puts into set making.

And now comes telephotography! Will it be the experimenter's next hobby? It offers a pride of accomplishment which will again make the home constructor a genius among his fellows. And, as for economy, it is impossible to buy telephotographic apparatus; hence, only the home constructor can now possess a telephoto receiver.

In other pages of this issue, we discuss a system of phototelegraphy which seems, from the home constructor's standpoint, to be the most promising development in telephotography. It does not enable the experimenter to secure *perfect* pictures, but it has the great and outstanding advantage of simplicity and low cost. When we are satisfied—and from every indication, it is a matter of weeks

LOOPS THEY USE IN WASHINGTON

Morris S. Strock of the Bureau of Standards radio laboratory with one of the large coil antennas used in experimental work at the Bureau. This coil is employed in receiving long-wave European radio telegraph stations

and months rather than years— that the apparatus will give the experimenter a pleasing and satisfactory field in which to exercise his ingenuity and to secure results commensurate with the effort and expense involved, we will describe the new apparatus with full constructional details. From present estimates, the cost of a complete set of parts will be in the neighborhood of seventy-five or a hundred dollars. No special transmitting apparatus will be required to put photographs on the air because the picture is embodied in an audio-signal which modulates the carrier, as do the speech currents of the studio microphone. Consequently, the only equipment needed by broadcasting stations to put pictures on the air, is a phonograph and a phonograph record of a picture transmission.

Our editorial in June regarding the relation of the set constructor to the set buyer, has aroused a great deal of comment, both from home constructors and from leaders in the industry. Many of our readers have written us in approval of the statement that the parts manufacturer must offer new fields to conquer for the set constructor—new designs and new ideas. We believe the trend is toward new fields and toward improved and economically justifiable combinations of parts for broadcast receiver building. The latter will always interest the home builder, but telephotography, short- and long-wave radio, laboratory experiments and measurements, and ingenious methods of installation, will command ever increasing attention.

A Prosperous Radio Year Forecast

THE fourteen thousand radio manufacturers and dealers who attended the Trade Show were uniformly optimistic. The radio receiver has emerged from a puny complexity to a magnificent simplicity. That large element of the general public which has been awaiting the perfection of radio and the elimination of all maintenance problems has at last

RADIO ABOARD PACIFIC LINERS

The Panama-Pacific Line, has installed long- and short-wave broadcast receivers with power loud speakers located at various points on the ship. The view at the upper left shows one of the 104 loud speakers on deck and the other view shows a public room in which is an electric radio-phonograph — part of the entertainment system

been provided with the receiving set which it desires. High quality of reproduction, tuning reduced to the manipulation of a single calibrated control, and maintenance troubles virtually eliminated, have made radio attractive to millions who, in the past, felt faint at the thought of a hydrometer. That terror of the housewife, three dials which must be turned in the correct ratio to bring in a station, is a bugaboo of the past.

Prosperity to the radio industry and a vastly increased number of listeners mean not only an aggressive and prosperous industry but, much more than that, the establishment of broadcasting in a healthy and wholesome condition. When the ether is cleared so that every station on the air can be received clearly and without interference, throughout its service range, and the number of listeners doubles and triples, the increased income of broadcasting stations will justify tremendously improved programs. Consequently, every owner of a receiving set profits as the art advances. Broadcasting will enjoy a re-awakening of such proportions that any home without a radio will soon be regarded as an object of sympathy and commiseration.

Discovering Ore Deposits By Radio

ALTHOUGH many devices have been announced, alleged to discover the presence of metallic ore under the ground by magnetic or electro-magnetic means, until quite recently very little practical use has been made of them. When ground currents of radio frequency are transmitted from one point to another, the strength of the received impulse is dependent upon the power of the transmitter, the sensitivity of the receiver, the distance, and the conductivity of the soil. The presence of metallic ores exhibits itself by an unusually strong signal for the transmitting power and distance involved. However, no means has yet been developed for identifying the kind of metal involved so that the principal value of such observations has been rather the elimination of unpromising ground than the identification

BEAM SERVICE FROM ENGLAND TO AUSTRALIA IS OPEN

After long tests, the Marconi short-wave beam transmitter has been turned over to the British Post Office. Reliable two-way communication at the rate of 200 words per minute has been established for some time. The antenna system shown is so arranged that, according to the time of day, the beam signals are sent to Australia either by the eastward route (left of illustration), or the antenna at the right is used in which case the signals are sent to Australia in the direction of America. Beam service between England and Canada has been in operation since October, 1926, and announcement of the New York-London beam service is expected shortly. Engineers of the Radio Corporation and the British Marconi Company are understood to be at work with tests and developmental work now

of valuable mineral ores. Recent experiments with portable seismographs, however, have been used with some success in locating salt domes and oil domes in the Southwest. A charge of dynamite is set off and the different sound vibrations are recorded by the seismograph. The relative rate of sound transmission has been used in locating ore bodies. ·

Sulphide ore bodies are good conductors of electricity and hence act like antennas. The radio waves set up in these subterranean antennas create an oscillating current of electricity. These re-radiated waves, however, are so feeble they cannot easily be detected.

In prospecting, a specially designed transmitter is set up in the immediate neighborhood to be prospected. A wavelength is used which will most easily penetrate the earth. The ore body receives strong radio waves from the near-by transmitting station and therefore re-radiates a fairly strong wave. The latter is picked up by a loop receiving outfit stationed close by.

The loop receiver is capable of being rotated around a horizontal and a vertical axis. Telephone receivers are connected with the apparatus and, as the loop is rotated, the position of maximum and minimum sound in the telephone receivers is determined and the instrument readings are recorded. The direction and location of the ore body can then be computed from these readings.

In practical prospecting, when there is an indication of an underground conductor, such as an ore body, and it is desired to make the information as definite as possible, a large number of readings are taken at intervals of twenty-five to fifty feet across the suspected axis of the ore body, and for a distance along the axis as far as it is

SOME OF THE BELIN TELEVISION APPARATUS

The Belin system of television is not radically dissimilar from other systems. In this illustration, a beam of light is reflected from the arc lamp at the left, reflected by a pair of oscillating mirrors, one to send the beam in a horizontal motion, the other for vertical motion. The mirrors (behind the dynamo and the large cogged wheel) throw the beam across the image to be transmitted and on to a photo-electric cell in the large cylinder. These varying light impulses then operate a transmitter

desired to investigate. These readings are then correlated on paper and cross sections made. The location of the ore body shows up quite plainly on such cross sections.

Broadcasters in New York Organize

WE HAD the pleasure of attending one of the early meetings of the New York Broadcast Owners' Association and observing the altruistic attitude of some of the small protesting broadcasters. It was announced in the press that thirty-three New York stations had combined to offer a united protest to the Federal Radio Commission because they had been compelled to split time and to go on the higher frequencies. The meeting was called to organize these thirty-three into a powerful association to present a united protest. Seventeen, as we recall it, attended and, of these, only seven agreed on roll call to protest. Some of these first insisted upon a promise that the constitutionality of the Radio Act would not be questioned but that the committee would confine itself to attempting to seek a better frequency and less division of time for the stations involved. We understand that, after the meeting, one or two additional stations joined.

When we read of the glorious speeches made before the Commission as to the service rendered the public by these stations, it was interesting to recall that the owners had to say regarding their service to the public at their own little meeting. Not once was the matter of serving the public mentioned. The only thing that they whined about was that they could not hope to make as much money out of advertisers if they could not broadcast whenever they pleased. Apparently, the lack of audience has not discouraged these stations, possibly because their sales representatives have succeeded in inducing advertisers to use their facilities in spite of a well nigh non-existent audience. And, when we say advertisers, we do not mean goodwill broadcasters, but broadcasters of the direct advertising which is characteristic of this sort of station.

IN THE CAPTAIN'S SUITE ABOARD THE "LEVIATHAN"

Captain Herbert Hartley uses his own radio receiver in his moments of leisure from his duties. European and American stations are heard with ease at all times during the transatlantic passage. The receiver is displayed among autographed photographs of President Coolidge, General Pershing, Secretary of the Navy Wilbur, and Sir Thomas Lipton

Judge Davis Resigns

JUDGE STEPHEN B. DAVIS announced his resignation from the Department of Commerce after four years of praiseworthy service as administrative supervisor of the Bureau of Navigation, Bureau of Fisheries, Steamboat Inspection Service, Patent Office, and Coast and Geodetic Survey. His service to the Department of Commerce will, however, be most widely remembered by the public for his direction of broadcasting matters. Judge Davis' new book, *The Law of Radio Communication,* recently published, is the only volume on the subject available and discusses the legal phases of radio control with complete competence and thoroughness.

Where Are the Listener Organizations?

ONE of our readers, Mr. R. L. Hitchcock of Detroit, Michigan, suggests that RADIO BROADCAST start an association of listeners among its readers and offers also to pay annual dues of one or two dollars. We had hoped that our recent editorial on this subject would bring out some strength in existing listener organizations but, aside from a few letters from indignant executive secretaries immediately following it, there have been no signs of activity. Considering that, at the moment, the very foundations of broadcasting are threatened by the effort of a few inferior pirate stations to set aside the Radio Act, certainly, if the listener organizations had the slightest voice or influence, there should be some manifestation of activity at this time. While we are pleased at this suggestion as expression of confidence on the part of our readers, we would certainly rather see an independent outside organization formed. Only if the demand were insistent and widespread, would we consider any such task. Why do not some of the existing organizations get busy? We would be glad to help any meritorious listener group to the best of our ability.

The Month In Radio

UNPOPULAR BROADCASTING STATIONS

THE Federal Radio Commission has been haled into court and application for an injunction to set aside its powers has been made. At present the action is pending. We are very doubtful that an injunction will be granted but, if it ever is, the result will be nothing less than a catastrophe to the radio industry. It will interest the protesting stations that 11.2 per cent. of the listeners who responded to our questionnaire, which appeared in the May issue, demanded the elimination of WJAZ because it started the radio chaos last season by tieing up the regulatory power. Seven times as many listeners demanded its elimination as considered it a favorite local or out-of-town station.

"If the report is true," writes one of these in a letter accompanying his questionnaire, "that WJAZ has been assigned a desirable wavelength, then the radio public has been not only insulted,

© Underwood and Underwood
DR. LOUIS COHEN
——— Washington, D. C. ———

Consulting Radio Engineer. A special statement prepared for RADIO BROADCAST:

"*Standardization may be very desirable from an economic and commercial standpoint, but if the standardization process sets in prematurely it may be a serious handicap to the further growth and development of an art. The radio broadcasting art illustrates this. We have come to regard the broadcast frequency band, 500 to 1500 kc., as being more or less fixed, and all broadcasting is confined to channels within this band of frequencies. The extension of the frequency range to, say, 2000 kc., which would greatly increase the number of channels for broadcasting, is being opposed on the ground that the receivers have been standardized to cover only the present limited range, and was done at a time when it was considered very difficult to design a commercial receiver suitable for the higher frequencies. Engineering development, however, has since progressed to a point where it is entirely feasible to build a receiver suitable for the higher frequencies as well as the lower frequencies, but because of the standardization which has set in altogether too early in the art, we are reluctant to take advantage of the new developments which would benefit the entire art.*"

but outraged, as this is the station that brought us the whole mess of trouble and should be made to take the leavings, if any."

This is the tenor of many written complaints which we have received, which indicate the permanent unpopularity which accrues to any station or organization which destroys the public's radio entertainment. If any other station again succeeds with a similar move, it will incur the violent hatred of the radio audience, regardless of any noise about martyrdom which it may make. Station WJAZ was not the most unpopular station in the United States, KFNF being the winner of our unpopularity contest. One hundred and four out of each 500 answering our questionnaire demanded the elimination of that station. This is almost 100 per cent. of those

within its range. Actually one listener in 500 called KFNF a sixth local favorite and three named it as an out of town favorite. The public knows what stations it likes to listen to, even if the broadcasters and their commercial sponsors do not.

HAZELTINE PATENTS SUSTAINED

FEDERAL JUDGE GROVER N. MOSKOWITZ handed down a decision recently in the Hazeltine-Grebe suit, favorable to the former. The defense was based largely on the grounds that prior inventions displaced those of Professor Hazeltine and that his original patent, filed shortly after he had left the employ of the Navy in radio work, contained a clause permitting the general use of his invention. An accounting of the profits received by the defendant during the period of the alleged infringement was ordered.

R. C. A. VS. HAZELTINE PATENTS

A DECISION favorable to the Radio Corporation of America, which sued the Twentieth Century Corporation for selling radio receivers in violation of the Hartley and Rice patents, was handed down in the Circuit Court of Appeals, to which it had been brought from a district court. This decision is of the utmost importance and should be studied closely by every manufacturer in the field. It establishes the relation of the R. C. A. patents to those held by the neutrodyne group. As we understand it, the patent establishes the basic nature of Hartley and Rice's work in preventing the generation of oscillatory currents through neutralization methods and that the Hazeltine inventions are, in effect, a practical and improved method of utilizing the Rice and Hartley inventions. Since the Hazeltine method of neutralization is a simple and effective one, we would not be surprised if many manufacturers find it desirable to secure licenses under both the R. C. A. patents and the Hazeltine patents. License under the R. C. A. patents does not necessarily imply, under this decision, freedom from suit for infringement under the Hazeltine patents.

The patent situation is too complex to be accurately summed up briefly. It is clear, however, that stability in the radio industry would be fostered if it were made convenient for set manufacturers to obtain blanket licenses at a sufficiently low rate to permit them to utilize all the necessary inventions in the field. If Hazeltine licenses are valuable, and it appears under the recent decision that they are, some form of combination license ought to be worked out. Manufacturers ought to be free to develop the radio art, designing, building, and selling better receiving sets, rather than squandering time in conferences with patent lawyers.

NEW R. C. A. LICENSEES

ADDITIONAL licensees under the R. C. A. patents are the powerful Crosley Radio Corporation, the Splitdorf Company, the F. A. D. Andrea Company, the Charles Freshman Company, and the Freed Eisemann Company. By the time this appears in print, there will be many others. The Crosley contract is said to provide for a seven and a half per cent. royalty and the statement is made that more than half a million dollars has already been paid to the R. C. A. by the Crosley Corporation and an additional three quarters of a million by other concerns.

SAN SALVADOR boasts of a broadcasting station, installed in its national theater, with the call letters AQN. It broadcasts three evenings a week on 482 meters, 625 kilocycles

IN HIS address before the Radio Manufacturers' Association, assembled at the Chicago Trade Show, Federal Radio Commissioner Orestes H. Caldwell pointed out that, in six years of broadcasting, the radio industry had succeeded in placing radio equipment in but one fourth of the twenty-two million homes of the United States. Considering that there are eighteen million pleasure automobiles, sixteen million wired homes, sixteen million telephones, and eleven million phonographs, the radio industry has a lot of unfinished business ahead of it. The 1927–1928 season will see a marked change in the ratio of radio to non-radio homes.

THE General Electric Company recently demonstrated its short-wave radio telephone equipment designed for use on long freight trains for communication between engineer and conductor. Although the train was a mile and a quarter long, not the slightest unreliability was observed, the signals being loud and clear under all conditions. Up to this time, it has been necessary for the conductor to use the emergency brake to stop the train or to rely on whistles or flare lights which often fail because of bends in the tracks or poor weather conditions.

WE HAVE before us the souvenir booklet sent by the National Carbon Company to those who write in comment on its Eveready Hour programs. It is a twenty-page booklet, neatly bound, and nicely printed, with an absorbing intimate story of the famous people and the regular artists with Eveready Hour. The character of printed matter sent out by the important commercial broadcasters is of sufficient merit and deserves wider circulation than it now enjoys. Listeners who have not tried the experiment of commenting on the better programs have a pleasant surprise in store for them.

CONGRESSMAN EMMANUEL CELLER, who is very expert in getting large publicity out of small matters, again calls attention to the abuses of radio censorship. Some of the cases cited by him in a story in the New York Sunday *World* are such outrages as the refusal of KOA to broadcast an inflammatory address by De Valera, a pacifist address by Mrs. Mary H. Ford, an address by a Smith College professor, criticising the policy of the Government in its Near East diplomacy, and various similar matters. When one considers what proportion of audiences of millions are interested in the opinions of the persons named, no serious injustices have been done. Since there is a limit to the time which can be devoted by the larger stations to expressions of private opinions, this so-called censorship must be exercised. Stations have been moderately intelligent in their discriminations and the attempt to make this a burning question, considering the insignificance of the cases cited, is either a publicity stunt or a marked failure to appreciate how liberal the broadcasting stations have been with their facilities.

LISTENERS all over the country report a marked improvement in receiving conditions as a result of the allocations made effective June 15. Many individual cases of interference are noted and the Commission is doing its best to adjust them by shifting station assignments. But until there is a radical reduction in the number of stations, or time splitting is more widely practiced, the ether will never be entirely cleared. The Radio Commission has done extraordinarily well with its problem. Had any one

predicted at the Washington hearings last spring that it would eliminate practically no stations, there would have been few ready to say that the Commission could effect any material improvement in broadcasting conditions. That it has done so, under the conditions, shows that the President appointed a committee of ingenious and hardworking men who have taken upon themselves the most difficult method of solving a well nigh impossible problem.

SENATOR EDWARDS of New Jersey has formulated a verbose protest about the Federal Radio Commission's treatment of station WAAT, which a few listeners in its immediate vicinity may be familiar with. The Commission, fortunately, has turned a deaf ear to strictly political pressure and has based its ratings of stations upon expressions of broadcast listeners. Under the circumstances, WAAT has not fared badly. RADIO BROADCAST readers, in their 452 listings of favorite local New York stations, give five votes to WAAT, approximately one per cent.

DR. L. W. AUSTIN, physicist of the Radio Transmission Research Laboratory of the United States Bureau of Standards, was awarded the 1927 medal of honor by the Institute of Radio Engineers. This is the highest recognition which the radio engineering fraternity can offer its fellows.

THE Ninth Radio District, with Chicago as its headquarters, boasts of 233 active broadcasting stations out of a grand total of 694 now on the air. This is approximately 34 per cent. of the stations. Chicago may pride itself on being the noisiest place on earth.

THE French inter-colonial radio service has been extended to equatorial Africa by the opening of a radio telegraph station at Brazzaville. The system consists of a station at Banako, West Africa; Tananarivo, Madagascar; Saigon, Indo-China; Bijbouti, Somaliland; and the new station in western Africa. Owners of long-wave receivers now have some new marks to shoot for.

FAILURE to obtain a receiving license in Greece, as required by the Director of Greek Telegraphs, Telephones and Posts, may result in twelve months' imprisonment and a fine of 100,000 drachmas. Licenses are somewhat less expensive.

THE meeting of the stockholders of the Marconi Wireless Telegraph Company, held in London recently, was a stormy scene, well attended by hundreds of stockholders. The company has undergone a drastic financial reorganization, including such measures as a reduction of nearly 46⅔ per cent. in its total outstanding stock. These courageous steps, coupled with the economy and extensions of short-wave beam telegraphy, may quite conceivably make future stockholders' meetings much happier affairs.

FIELD strength measurements made in the Eighth Radio District by Radio Supervisor S. W. Edwards have revealed the now oft-confirmed fact that location is of as great importance as antenna power in determining the service area of a broadcasting station. The most serious factor which prevents equal radiation in all directions from a radio antenna is the shadow effect of large surrounding masses of both conducting and non-conducting materials. As the cost of transmitting installations goes up, port-

able transmitters will be used to determine the characteristics of any transmitting point before a station is erected. It appears that irregularities in transmission characteristics, due to location, can be analyzed with a portable low-power transmitter and that great increase of power does not change the general nature of shading effects. It is not unusual to find that a broadcasting station of considerable power may serve only fifty or seventy-five miles in one direction, but is easily heard four or five hundred in another.

A TELEPHOTOGRAPHIC radio line (erroneously called television transmission by wireless in the Department of Commerce's Trade Commissioner's report) has been placed in operation between Vienna and Berlin, the time required for transmission being only twenty seconds, considerably faster than any commercial system we know of on this side of the Atlantic.

THE Hydrographic Office of the Navy Department, in collaboration with various other branches of the Navy, has announced further results in their study of weather forecasting methods, based on observation of the strength and direction of static through loop receivers. The study has indicated that the use of new devices which have been developed, to measure static intensity in standard units, will make it possible to plot the highs and lows of static in the same way that we now plot barometric pressures. Static intensity is said to bear a definite relation to barometric pressure, thus offering a valuable aid to reliable weather forecasting.

THE Bakelite Corporation of New York has succeeded in inducing the Tariff Commission to recommend to the President that the importation of certain forms of synthetic, phenolic resin be no longer permitted. Apparently this material made it impossible for the complainants to exercise their rights under certain of their patents, because the material could be used in violation thereof without difficulty.

COMMERCIAL beam wireless service between Great Britain and Australia was recently begun at rates lower than the existing cable rates.

PRESIDENT COOLIDGE has appointed the personnel of the United States delegation to the International Radio Telegraph Conference, to be held in Washington during October. This body, in spite of its name, which was adopted before radio telephony had been invented, will discuss broadcasting problems. The members of the American delegation are Herbert Hoover, Secretary of Commerce; Senator James E. Watson of Indiana; Senator Ellison D. Smith of South Carolina; Representative Wallace H. White, Jr., of Lewiston, Maine; William R. Castle, Jr., Assistant Secretary of State; Alternate, William R. Vallance, Assistant Solicitor of the Department of State; Maj. Gen. Charles M. Saltzman, Chief Signal Officer, U. S. A.; Capt. Thomas T. Craven, Director of Naval Communications, U. S. N.; W. D. Terrell, Chief of Radio Division, Department of Commerce; Owen D. Young, Chairman of the Board of Directors, General Electric Company; Alternate, Samuel Reber, Colonel, U. S. A., retired; John J. Carty, Chief Engineer, American Telephone and Telegraph Company; Stephen Davis, former Solicitor, Department of Commerce; John Beaver White, Electrical Engineer, and John Hays Hammond, Jr.

Piezo-Electric Crystals

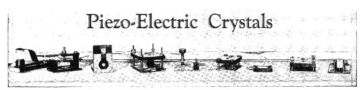

VARIOUS KINDS OF MOUNTINGS FOR PIEZO-ELECTRIC CRYSTALS
The mounting at the extreme left holds a spherical crystal

A Simple Explanation of the Theory Involved in Quartz Crystal Applications—Their Use for Precision Calibration of Wavemeters

By M. THORNTON DOW

CONSIDERING the very large number of radio sets in use to-day, it is not surprising that there exists a national irritation over the uninvited and exasperating whistlings which intrude at many inopportune moments during otherwise enjoyable radio programs. Though the whistling has been dignified by the descriptive term "heterodyning," under any name it must be universally unpopular.

Very often, however, the phenomenon of heterodyning is a most useful one to scientists and in the laboratory. For instance, the information obtainable by this means makes it possible to find with precision the number of times per second piezo-electric crystals oscillate when excited electrically. Among the crystals illustrated in an article entitled "Piezo-Electric Crystals" in RADIO BROADCAST for January, 1927, was one which vibrates at the strikingly high rate of 6,000,000 times per second. It is quite usual for interested persons to ask the question: "How do you know that the crystal oscillates with a frequency of 6,000,000 times per second?" A truthful but uninstructive answer is: "The whistles tell us." How these whistles are made to divulge these secrets of the crystal is somewhat briefly described in what follows.

As pointed out in the previous article, a single quartz crystal, or some other piezo-electric crystal, can be used within a vacuum-tube circuit to act as an oscillator giving any one of two or three (or more) principal frequencies which hold remarkably constant; and when an oscillator, either an electric or crystal-electric one, is operating at some one principal frequency, called the fundamental, there are usually simultaneously available in effect numerous other frequencies which are whole number multiples of the fundamental. These other frequencies are often spoken of as "the harmonics" but in order to specify them clearly it is best to assign to each frequency a number which, if multiplied by the frequency of the fundamental, gives the frequency of the harmonic. On this basis the fundamental is harmonic number one; twice the fundamental frequency is harmonic number two; and so on. If, then, the fundamental, the predominant frequency of the oscillations, is known, the simple multiplication by a whole number gives the frequency of any chosen harmonic. In the case of a crystal oscillator it is easy to see that, since the fundamental frequency is very nearly constant, all the harmonic frequencies are likewise just as constant.

WITHIN A CRYSTAL OSCILLATOR
When compared with the wiring of many radio sets this looks simple indeed

With these facts in mind, we can rightly inquire what they have to do with the whistling notes originating at times in radio circuits. A familiar fact is that whistling notes, called beat notes, are heard in any oscillating tube circuit whenever its frequency (a harmonic or fundamental) differs by an audible amount (that is, within the range of audible frequencies) from some fundamental or harmonic frequency of another oscillating tube circuit which is near enough to the first for the requisite transfer of energy. Although each tube itself may be producing oscillations which are inaudible, yet the interaction between the two circuits results in a beat note which may be adjusted to an audible frequency.

To illustrate the facts noted, we may consider a crystal-controlled circuit giving a fundamental inaudible frequency of 100,000 cycles per second set up, as in the diagrams of Fig. 1, in the vicinity of an oscillator having variable frequency. In a telephone receiver in one of the plate circuits we should hear a beat note of 1000 oscillations per second when the variable oscillator had been adjusted to give 101,000 cycles per second. Thus, the frequency difference between the two inaudible frequencies was made audible, in the form of a beat note.

By further adjusting the variable oscillator toward 100,000 cycles per second the beat note can be decreased in frequency until it is so low as to become inaudible, under which conditions we may be certain that the variable frequency is within, say, 50 cycles or less of 100,000 cycles per second. To decrease the variable frequency further, to 99,000 cycles per second, would again

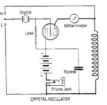

CRYSTAL OSCILLATOR

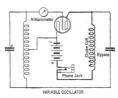

VARIABLE OSCILLATOR

FIG. 1

bring in a beat note of 1000 cycles per second. The width of the region of inaudible (subaudible) beats just passed through is determined by the limitations of the human ear and of the apparatus used. In most cases a beat note below about 50 cycles per second would not be heard. In practice it is possible to approximate closely to *zero beat* (that condition existing when the two notes are of exactly equal frequency and no beat note is heard) by setting the dial at the middle of the silent region which is confined, often narrowly, between the two regions of clearly audible beats whose frequencies decrease as the inaudible band is approached from either side. Examine Fig. 2.

If the occasion demands, it is possible to actually make the *zero beat* frequency audibly evident by employing a third oscillator to give an auxiliary constant audible beat note whose intensity varies except when the variable frequency oscillator gives zero beat with the first oscillator.

All that has been said above regarding frequencies is true of every frequency, whether fundamental or harmonic, having the requisite amount of energy; when we hear a beat note, therefore, we are not at once certain which pair of frequencies from the two harmonic systems of the oscillators is producing the beats. We can be certain, however, that the audible beat frequency is the difference in frequency between some harmonic of one oscillator and some harmonic of the other oscillator, and, by methods described in part farther on, we can determine the respective harmonic numbers in question. The important thing to remember is that the audible beat note is used to indicate a very definite relation between frequencies which in themselves may be far above the range which is audible.

These indicators bridge the gap between such frequencies as fall above a few thousand cycles per second, which by any stretch of the imagination could not be measured by simple mechanical means, and those frequencies in the few hundred cycles per second class which are readily measured by mechanical-photographic means. An adaptation of the familiar telephone receiver, for example, makes it possible to photograph oscillations of a few hundred, or say for instance, a thousand cycles per second. It is a familiar fact that the metal diaphragm just back of the hole in the ear piece of the telephone receiver can be made to vibrate audibly at this frequency. If the vibratory motion of the diaphragm were made to operate a small mirror suitably, a spot of light reflected from the mirror would be deflected with a vibratory motion. A photographic record could be made by the spot of light, simultaneously with a similar record from the motion of the pendulum of an accurately regulated clock, upon a moving photographic film. The resulting print would look like the illustration of Fig. 3. The time represented by the distance between any two lines could be made by the swinging pendulum could be read from the clock. The number

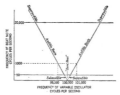

FIG. 2

of oscillations of the mirror within the boundaries set by any two time-lines can be counted, just as the number of box-cars between the locomotive and caboose of a long freight train can be counted. The number of oscillations per second can then be computed from the information so obtained.

CALIBRATING CRYSTALS

TO PUT the facts outlined in the preceding paragraphs to work to tell us the frequency of piezo-electric or other electric oscillating circuits, a series of oscillators are set up as suggested

Records made from pendulum

Record made by excitation from oscillator

FIG. 3

by Fig. 4. Suppose that "A" is the crystal oscillator which we had suspected of giving oscillations of frequency in the neighborhood of 6,000,- 000 cycles per second. "D" is a variable low-frequency oscillator which can be photographically calibrated as above described. "B" and "C" are two other variable oscillators. The fundamental frequency of "B" may be adjusted until its harmonic number 10 gives zero beat with the fundamental of "A"; harmonic 20 of oscillator "C" may be similarly set against the fundamental of "B"; harmonic 30 of oscillator "D" may be set to equal the fundamental of "C." Under these conditions we now know that the fundamental frequency of "A" is 10 x 20 x 30 = 6000 times the fundamental frequency of "D." If a photograph of "D" shows its frequency to be 1019.8 cycles per second it is safe to believe that the frequency of A is 6000 x 1019.8 = 6,118,800 cycles per second, which would certainly justify the nickname "six-million cycle crystal." Thus we can measure extremely high frequencies by first measuring in an electro-mechanical way a much lower frequency which has some simple and definite relation, as indicated by beat notes, to the very high frequency.

For an instructive and very interesting article which describes unusual results of some recent experiments of Professor Wood of Johns Hopkins University with piezo-electric crystals, readers would do well to read "A New Magic," by Frank Thone, in the *Century Magazine*, for February, 1927. A scientific report on methods of using the piezo-electric crystal for precision wavemeter calibration was published in 1923 by Prof. G. W. Pierce of Harvard University in the *Proceedings of the American Academy of Arts and Sciences.* Since that time frequency meters covering the range from more than 150 million cycles per second down to audible frequencies have been built and crystal-calibrated at Cruft Laboratory, Harvard. Another paper by Professor Pierce on the measurement of the velocity of sound at high frequencies, published in October, 1925, by the American Academy of Arts and Sciences, includes a detailed description of some of the methods noted here for calibrating crystals and for using them in precision work.

When once a crystal has in the above manner been calibrated it is a standard against which other instruments for measuring frequency may be checked by methods much simpler in procedure but which in many instances again make use of the whistle of beat notes. Other crystals, by chance or by careful cutting, giving beats with the standard, may be precisely calibrated in short order. The case, however, of precision wavemeter calibrations against a standard crystal is typical of what may be done with, and represents one of the earlier laboratory uses for, the piezo-electric crystal; methods for such work developed in the university laboratory a few years ago have now become common and are regarded as standard practice.

CALIBRATING WAVEMETERS

A WAVEMETER is essentially an instrument for measuring frequency and therefore may better be called a frequency meter. Consisting merely of a coil connected to a condenser, the latter usually variable, a wavemeter gives a marked response at each setting of the condenser to a particular frequency. The response takes the form of a very rapid increase in the current circulating in the wavemeter when the condenser setting is adjusted toward a value at which the wavemeter is in tune to the frequency of the oscillator inducing the current. The evidence of the response may be given by a current meter built into the wavemeter, by a meter in a circuit of the oscillator to which the wavemeter is being tuned, or by other means.

Although for many wavemeters calculation will give the approximate range of frequency, or wavelength, covered by the scale of the instrument, only an experimental comparison against frequencies already known can give a dependable calibration. The frequencies for comparison may be determined by using another wavemeter which has already been standardized, but in any case the frequencies must at one time or another have been measured by some such methods as described for the case of the crystal. Because the frequency of the piezo-electric oscillator depends almost wholly upon the physical dimensions of the crystal, it is especially well adapted for meeting the requirements of a standard of frequency. So, ultimately, if our needs demand frequent measurement of frequency with high accuracy we shall find it necessary to calibrate a wavemeter against a piezo-electric crystal. The wavemeter is the instrument of greater utility in radio work in general, while the crystal is the more highly dependable as a standard of frequency. In

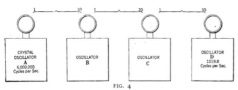

| 1 | 10 | 1 | 20 | 1 | 30 |

CRYSTAL OSCILLATOR A 6,000,000 Cycles per Sec.

OSCILLATOR B

OSCILLATOR C

OSCILLATOR D 1019.8 Cycles per Sec.

FIG. 4

this particular connection then, what is usually needed is not an original determination of unknown frequencies but, rather, the *corrections* applicable to frequencies already known approximately. The fact that corrections to a calibration, which through frequent use has come to be held somewhat in doubt, are about to be carried out, greatly simplifies the experimental procedure and is justified as an illustrative case because it is so common in practice.

The procedure may be most briefly indicated by recollecting two facts before mentioned: A variable electric oscillator can be adjusted to give subaudible, or zero, beat note when in proximity to a piezo-electric crystal oscillator; a wavemeter (having the required range) can be adjusted to respond to the fundamental frequency of the electric oscillator. Thus, through the intermediary of the variable oscillator, the wavemeter calibration can be compared against the crystal frequencies. Suppose that there are 40 harmonics developed in the crystal circuits and 40 harmonics also developed in the variable oscillator circuits, then there are $40 \times 40 = 1600$ different settings of the variable oscillator which will give subaudible, or zero, beat notes, since each harmonic of the latter may in turn be adjusted by beats to equal in frequency each one of the 40 harmonics of the crystal oscillator. There is, therefore, the possibility of obtaining 1600 different settings on the scale of the wavemeter, each setting representing a definite frequency related to the fundamental standardized frequency of the crystal. There may be more than, or less than, 40 harmonics available from each circuit; for any given case there usually are more than enough for a satisfactory calibration.

And now, in deciding upon which pair of harmonics is in each case fixing the setting, we utilize to the fullest extent the old calibration of the wavemeter which we are correcting. We note, furthermore, whether at each setting the beat note obtained is strong, medium, or weak in intensity, for as a general rule the smaller the number specifying the harmonic involved the more intense is the beat note response, and this information aids in identifying the harmonics. To illustrate we shall assume that a standard crystal fundamental frequency is 100,000 cycles per second and that for a setting of a frequency meter, when tuned to an oscillator which had been set against the crystal by the beat method, the older calibration reads 134,670 cycles; this reading is presumably in error by some small amount, since it is taken from the previous calibration now being corrected. The question is, "what is the *correct* reading?" In the course of the experimental work we had noted that by comparison with beats obtained on the average this beat note had been among the louder ones. This leads us to suspect that the harmonic numbers associated with the case were rather small. Proceeding blindly we might calculate that:

$$101 \times 100,000 = 75 \times 134,667 \text{ (Very nearly)}$$

which indicates that if the variable oscillator fundamental were 134,667 cycles per second then its 75th harmonic equalled in frequency the 101st harmonic of the crystal. But when we note:

$$4 \times 100,000 = 3 \times 133,333 \text{ (Very nearly)}$$

we can feel safer in believing that the auxiliary oscillator stood not at 134,670 cycles per second as indicated on the doubted calibration of the frequency meter, but at 133,333 cycles since, in this case, harmonics 3 and 4 would have been used as against harmonics 75 and 101 in the other computation; this recognizes the spirit of our memorandum—which showed that the beat note was among the louder ones observed. The old calibration was, then, at this point, in error by 1340 cycles in a total of 133,333, an error of

A PRECISION CRYSTAL OSCILLATOR
This instrument is being used at Cruft Laboratory, Harvard, and is making possible some very accurate measurements

nearly one per cent., which past experience, we will say, has shown to be not unreasonable.

Having looked at an illustrative example of an isolated calibration point let us get a better view of actual experimental procedure, tie in a few odds and ends of useful information, and thus better understand how the process of calibrating a wavemeter against a piezo-electric crystal may be methodically carried out with a minimum of difficulty from doubtful points which arise for the novice, and even for the expert, at times.

A photograph on this page was taken in a college laboratory. It shows a typical arrangement of apparatus for students' use in the calibration of a wavemeter by means of a crystal. A crystal is chosen which gives as one of its *fundamental* frequencies a frequency not too near the range covered by the wavemeter; the

telephone or amplifier is connected into the plate circuit (indicated as possible in the diagram of Fig. 1.) either of the crystal or of the auxiliary oscillator, usually but not necessarily in the circuit having the lower fundamental.

To begin with, settings of the variable oscillator giving only loud beat notes are chosen throughout the range of the wavemeter; to suppress the weaker responses the distance between the crystal oscillator and the variable oscillator is, for the time being, reasonably increased, or the coils of the two oscillators may be oriented to give the same result. Since these loud responses are surely associated with harmonics having the lower identification numbers, the correct frequency can be determined without ambiguity from the approximate frequency taken from the earlier calibration and from calculation based on the known fundamental frequency of the crystal. By placing the two oscillators closer together or by otherwise making the coupling closer and, possibly, by using an amplifier to the telephones, the weaker and higher-numbered harmonics are made available, and more data are taken to fill in the gaps between those at first recorded. When the frequencies, or wavelengths, are plotted against the scale readings of the wavemeter, some orderly smooth curve, should result. Points much in error become evident at once and the correct assignment of harmonic numbers for these points is now easily made.

A little experience soon dispels the bewilderment which the novice usually feels at the beginning. It soon teaches him the difference between the lively well-behaved whistle of the beat notes which hold the key to all the data of the experiment, and the liquid transitory chirps which speak of the crystal's willingness to oscillate in other ways but which are useful only under other circumstances. He learns that to place the variable oscillator too near the crystal oscillator, or the wavemeter too near the variable, causes undue reactions in the circuits, evidenced by their meters, and is, so far as practicable, to be avoided. A lesson or two makes it an easy matter to pick out many harmonics as individualities, their characteristics of mark being tied up with their intensity and breadth of audible band. A harmonic of small index, in addition to its marked intensity, results in a more leisurely "running of the scale" when the beat note is tuned-in and out by the dial of the auxiliary oscillator.

A TYPICAL SET-UP
Apparatus in a students' laboratory for calibrating a wavemeter by means of a crystal. On the left is the crystal oscillator, on top of the amplifier cabinet. In the center of the photograph is the auxiliary oscillator. The wavemeter to be calibrated stands at the right

Home-Constructing Transformers and Chokes for Power-Supply Devices

How the Number of Turns, Dimension of Core, Gauge of Wire, Etc., May Be Determined Without Resorting to Complicated Mathematical Calculations

By HOMER S. DAVIS

WITH steady improvements being made in the perfection of socket power-supply devices, there is hardly any doubt that they will eventually be even more widely used than at present. The building of these power devices offers an interesting field to the home constructor, resulting in a demand not only for data on the complete assemblies, but also for information about the actual design and construction of the individual units in these devices. In this article, dealing with power transformers and iron-core choke coils, the mathematical complexities of design are reduced to simple arithmetic by means of the calculation charts, and practical information is at the same time offered about the more common methods of construction.

The power transformer is an alternating-current device for changing electric power at one voltage to electric power at another voltage. It consists of two coils of wire wound upon a magnetic core. The winding which receives electric power from the source is called the primary winding, and the winding which delivers electric power to the load is called the secondary winding, the function of the core being to transmit the electric energy from the primary to the secondary, through the medium of magnetic energy. The voltage available at the secondary terminals depends upon the ratio of turns in the two windings. That is to say, if the secondary has, for instance, twice as many turns as the primary, the secondary voltage will be twice as great as that impressed upon the primary; conversely, if the secondary has half as many turns as the primary, its voltage will be only half as great. If the transformer serves to increase the voltage, it is called a step-up transformer; if to decrease the voltage, a step-down transformer.

A small part of the power put into the transformer is used up in overcoming the losses. The copper losses are due to the resistance of the windings, and may be reduced by choosing wire of reasonably ample size. Other losses are due to hysteresis and eddy currents in the core, and may be reduced by the proper choice of flux density, and by using a laminated core of sheets of silicon steel. The power consumed in overcoming these losses is dissipated in the form of heat, so that when the transformer is put under load, the core and windings will slowly warm up to some steady value above room temperature. Although it is possible to design transformers to operate at comparatively cool temperatures, a certain amount of heating is permissible, and transformers that run warmer can be built somewhat cheaper and made more compact. The temperature rise, however, should not be so great as to damage the insulation.

When the transformer is put under full load the voltage across the secondary terminals will drop a small amount. This is what is referred to as voltage regulation, and is expressed by the percentage of ratio of the ratio of this drop in voltage to the original, or no-load, voltage. Good regulation in a transformer is desirable, and depends upon the resistance of the windings and the leakage of magnetic energy between them. The magnetic leakage may be reduced by arranging the transformer as compactly as possible with the winding adjacent and close to the core, by keeping the path around the core as short as possible, and by making good neat joints in the core. As will be shown later, it is possible to use either a large core with relatively few turns of wire, or a small core with many turns. For good regulation, the larger core with fewer turns is preferable.

In general there are two types of transformers, differing chiefly in the arrangement of their cores. One is known as the shell type, the other, the core type; both are shown in Fig. 1. The core type is the simpler of the two to construct, and is the one considered in this article.

Turning now to the design, the fundamental formula for the transformer is:

$$E = 4.44 \frac{BANF}{10^8} \qquad (1)$$

where E is the supply voltage, B the flux density in lines per square inch of core area, A the net area of a cross section through one leg of the core, N the number of turns in the primary winding, and F the frequency of the supply voltage. To avoid the labor of solving formulas such as this, the calculation charts accompanying this article have been prepared, the above formula being represented by Chart III. Flux density is the measure of magnetic energy used in the core to link the primary and secondary coils. Its magnitude depends upon the material used in the core, and for the silicon sheet steel commonly used it is taken as 60,000 lines per square inch. This value was used in the right hand scale, Chart III.

Information as to the supply voltage and frequency may be obtained from the local electric light company. With these values known, and the flux density assumed, it can be seen from formula (1) that it is possible to use either a large core with few turns, or a small core with many turns. To establish a working ratio, the number of turns may be figured from the following formula, corresponding to Chart II:

$$\text{Turns per volt} = \frac{C}{\sqrt{\text{Watts Output}}} \qquad (2)$$

where C is a constant, the value of which may be so chosen as to give any desired proportion between core and copper. The number of turns in any coil is then the product of the turns per volt and the voltage of that coil. In order to obtain better regulation, and also because wire, especially in the smaller sizes, is much more expensive than core material, the transformer with a larger core and fewer turns is preferable. A value of 35 for C will give a good ratio from the home constructor's standpoint and has been used in Chart II. If for any reason the reader wishes to use a different ratio, the new value of turns per volt must be figured from formula (2) and then employed in Chart II to find the total turns in each winding. The larger the value of C used, the greater the ratio of copper to iron, and vice versa.

PROCEDURE IN TRANSFORMER DESIGN

BEFORE taking up the solution of a typical example, it is advisable to have in mind a general idea of the proper procedure to follow. The first step is to ascertain the output rating, or secondary voltage and current, as required by the apparatus to which the transformer is connected. Then the power output in watts is the product of the voltage and the current in amperes, and may for convenience be obtained from Chart I. If there is more than one secondary winding, the total output, of course, is the sum of them all. The number of secondary turns to give the required voltage may be figured by multiplying the turns per volt, from formula (2) by the secondary voltage, or it may be obtained from Chart II. About five per cent. more turns should actually be used on the secondary, to compensate the voltage drop under full load due to losses and regulation. The number of turns required for the primary winding is then found from Chart II in the same manner.

The next thing to be determined is the size of the core. The cross section area may be obtained from formula (1) or from Chart III. The net area is usually assumed to be about 90 per cent. of the gross area, to allow for oxide coating or other insulation between the core laminations. The use of a square core is advisable, to facilitate the winding of the coils as well as the general construction.

The subsequent step is to choose a size of wire for each winding large enough to carry the current without objectionable heating. It is customary in figuring wire sizes to use a unit of area called the circular mil, which is the area of a wire one thousandth of an inch in diameter. For steady loads, (continuous over a period of hours), 1500 circular mils of area should be provided for each ampere of current; if the load is intermittent (in use over short periods of time), this may be reduced to 1000 circular mils per ampere. The current-carrying capacities of the different sizes of wire have been set down

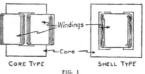

CORE TYPE   SHELL TYPE

FIG. 1

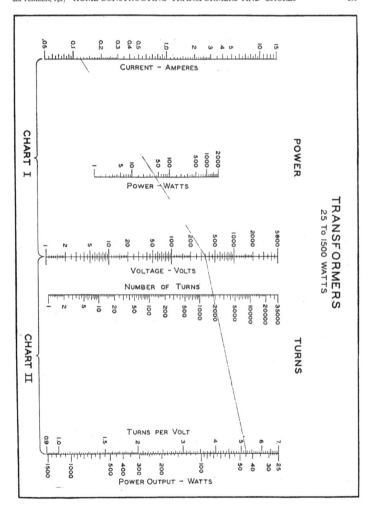

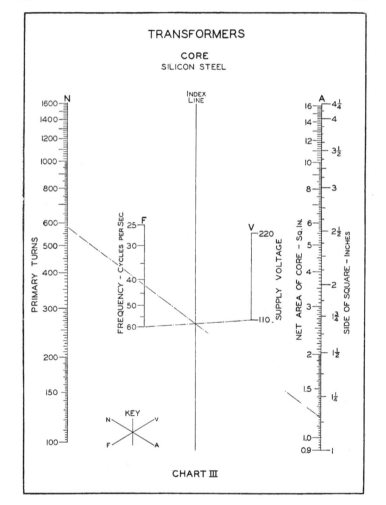

TRANSFORMERS

CORE
SILICON STEEL

CHART III

in the copper wire table accompanying this article. The secondary current having been previously determined, the correct size of wire may be found by referring to the second or third column of this table. The primary current may be obtained from the primary, or input, power, which exceeds the output power by an amount sufficient to overcome the transformer losses. The percentage ratio of output to input is the measure of efficiency of the transformer. Home constructed transformers of about 1500 watts rating may be as high as 95 per cent. efficient, decreasing to about 90 per cent. at 100 watts, and then probably dropping to about 80 per cent. at 50 watts, or even lower for smaller sizes. To find the primary current, therefore, first divide the power output by the probable efficiency, which gives the power input; the primary current is then equal to this input power divided by the supply voltage, or it may be found by means of Chart I. The proper size of wire for the primary may now be obtained from the table.

All that remains is to design a core of required cross section, with an opening or window in it large enough to contain the windings. This is most readily done by making a full size pencil drawing of the transformer, similar to Fig. 2 (on this page). After drawing in one side of the transformer, choose a tentative width for the windings and from the wire table, "turns per linear inch" columns, figure how many turns per layer may be wound in the primary coil. Either double cotton-covered wire or single cotton-covered enameled wire may be used. Except in the smaller sizes, the use of plain enameled wire in transformers is not advisable, and then only is it permissible when a layer of paper is placed between each layer of wire. Enameled wire has the added disadvantage that it cannot be shellaced in place. The number of layers is then found by dividing the total number of turns by the number of turns per layer. From this the depth of the winding may be estimated, allowing for any insulating paper between layers, if used. The primary coil may now be put on the drawing, leaving at least an eighth of an inch for insulation between it and the core. The depth of the secondary winding may be found in the same manner, using the same width as the primary. In placing the secondary on the drawing, allow about a quarter of an inch between it and the primary winding, to take care of insulation and discrepancies in actual construction; or if it is planned to wind it directly over the primary on the same leg of the core, allow for about an eighth of an inch of insulation between them. Then draw in the rest of the core, allowing about a quarter of an inch clearance between it and the ends of the windings.

It can be readily seen from such a drawing whether the transformer is well proportioned from the standpoint of good voltage regulation. If it looks like either A or B in Fig 3, assume a different value for the width of the windings and try another layout. A little care and judgment here will be well repaid in final results.

The solution of a typical example

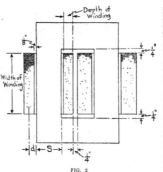

FIG. 2

will better illustrate the use of the calculation charts. Suppose a transformer to operate from a power supply of 110 volts and 60 cycles, is to be designed for use in a plate supply unit employing a Raytheon type BH rectifier. A secondary voltage of 350 is required by this tube; its capacity is 85 milliamperes of direct current, but it is safer to design the transformer to handle about 120 milliamperes (0.12 amperes) of alternating current to allow for that bypassed by the filtering network. If a power amplifier stage is to be included, a filament winding delivering 0.5 amperes at 5 volts will be required for, say, a UX-112 (CX-312) or UX-171 (CX-371) type tube.

The power output of the high-voltage stage is found from Chart I by drawing a straight line from 0.12 on the current scale to 350 on the voltage scale, which will be found to cross the middle scale at 42 watts. The power output of the low-voltage winding is similarly found to be 2.5 watts. The sum of these gives 44.5 watts as the power output of the transformer. The number of turns to put on each winding is found from Chart II. For the 350-volt winding, draw a line from 350 on the voltage scale to 44.5 on the power output scale, intersecting the third scale at 1850 turns. Five per cent. more turns should be added to take care of the voltage drop due to regulation, making 1940 turns for the 350-volt winding under full load. Since the Raytheon tube is a full-wave rectifier, two 350-volt windings, or one 700-volt winding (3880 turns) with a center tap, will be required. Using the chart in the same manner for the filament winding gives 26 turns, and adding about five per cent. makes 27 turns. Similarly, the primary will be found to require 580 turns; the allowance for regulation is not used here of course, as the voltage of the supply mains is fairly steady and independent of the transformer.

With the number of turns for each winding determined, the next step is to find the size of core cross section from Chart III. The four scales of this chart are to be used as shown in the key; that is, the inner or frequency and voltage scales are always connected together, and likewise the outer two scales are always used together. With a power supply of 110 volts and 60 cyles, a line is drawn between these two points. Through the point at which this line intersects the index line, draw a second line from 580 on the primary turns scale until it meets the core area scale. This shows that a core having a net cross section area of 1.18 square inches is required; the nearest size is 1⅛″ square.

In determining the size of wire for the secondary, it is noted that each 350-volt winding delivers current only half the time, and therefore a wire of half the area might be used. However, the resistance of the winding would thereby be dcubled, introducing a detrimental effect upon the voltage regulation, and for this reason it would be better to compromise on a larger size. With a secondary current of 0.12 amperes, choose a size of wire to carry, say, 0.10 amperes, which, at 1500 circular mils per ampere, requires a No. 28 wire as shown by the wire table. For the filament winding delivering 0.5 amperes, No. 21 wire would be satisfactory.

Before the size of wire for the primary can be found, the primary current must be known. Assuming an efficiency of 75 per cent. for this size of transformer, the power input to the primary will be 44.5 divided by 0.75, or about 60 watts. The current is then found from Chart I by drawing a line from 110 on the voltage scale through 60 on the power scale until it meets the current scale, at 0.54 amperes. From the table, No. 21 wire is required for the primary.

The rest of the design consists in laying out the full size drawing of the transformer as described earlier. This layout being completed, it is often desirable before ordering the material, to estimate the weights of wire and core steel required.

	COPPER WIRE TABLE								
SIZE B & S	CURRENT CAPACITY AMPERES		RES. PER 1000 FT. OHMS	TURNS PER LINEAR INCH			WEIGHT—LBS. PER 1000 FT.		
	1000 C. M. PER AMP.	1500 C. M. PER AMP.		D. C. C.	S. C. E.	ENAM.	D. C. C.	S. C. E.	ENAM.
8	16.3	11.0	.628	7.1	7.3	7.7	51.2	51.2	50.6
9	13.1	8.7	.792	7.8	8.1	8.6	40.6	40.6	40.3
10	10.4	6.9	1.00	8.8	9.0	9.6	32.2	32.2	31.8
11	8.2	5.5	1.26	9.8	10	11	25.6	25.6	25.2
12	6.5	4.4	1.59	11	11	12	20.4	20.4	20.0
13	5.2	3.5	2.00	12	12	13	16.2	16.2	15.9
14	4.1	2.7	2.52	14	14	15	12.9	12.9	12.6
15	3.3	2.2	3.18	15	16	17	10.3	10.2	10.0
16	2.6	1.7	4.02	17	18	19	8.21	8.14	7.93
17	2.0	1.4	5.06	18	20	21	6.54	6.47	6.28
18	1.6	1.1	6.38	20	22	24	5.24	5.14	4.98
19	1.3	.86	8.05	22	24	27	4.22	4.09	3.96
20	1.0	.68	10.2	24	27	30	3.37	3.26	3.14
21	.80	.53	12.8	28	30	34	2.68	2.60	2.49
22	.64	.43	16.1	30	33	38	2.17	2.07	1.97
23	.51	.34	20.4	33	36	42	1.73	1.66	1.56
24	.40	.27	25.7	36	40	47	1.40	1.33	1.25
25	.32	.21	32.4	38	44	53	1.13	1.06	.988
26	.25	.17	40.8	42	48	59	.914	.849	.784
27	.20	.13	51.5	45	52	66	.726	.678	.622
28	.16	.11	64.9	48	57	74	.668	.543	.494
29	.13	.084	81.8	52	62	82	.480	.433	.392
30	.10	.067	103	56	67	92	.396	.346	.310
31	.080	.053	130	60	73	103	.326	.281	.246
32	.063	.042	164	63	79	116	.270	.228	.196
33	.050	.033	207	66	86	130	.227	.185	.155
34	.040	.026	261	70	92	145	.193	.150	.123
35	.032	.021	329	73	99	164	.160	.123	.008
36	.025	.017	415	77	103	182	.136	.101	.078
37	.020	.013	523	80	113	206	.120	.084	.062
38	.016	.010	660	84	130	233	.103	.071	.049
39	.012	.008	832	90	128	261	.094	.061	.039
40	.010	.007	1050	95	136	290	.084	.053	.031

The weight of each size of copper wire is approximated by first finding the total length of wire in each winding. The mean or average length of a single turn may be found from Chart IV. Here the left-hand scale represents the width of the square side of the core, shown as "S" in Fig. 2, while the right-hand scale is the distance from the edge of the core out to the center layer of the winding in question, as "d" in Fig. 2. To use the chart it is necessary merely to measure these two distances on the full-size drawing of the transformer, locate each of these values on their respective scales on the chart, and draw a line between these two points; the point at which this line crosses the center scale is the average length of a turn in feet. For instance, if a 1⅛″-core has a 2000-turn winding, the center layer of which is 1¼″ out from the core, a line drawn between these two figures on the chart will intersect the center scale at 0.505 feet, the average length of a turn. Multiplying this value by the total number of turns, 2000, gives 1010 feet as the length of wire in the winding. The weight of the wire may be found from one of the last three columns of the copper wire table; the length of wire in feet divided by 1000 and multiplied by the weight per thousand feet gives the total weight.

The weight of the core is obtained by subtracting the area of the window in square inches from the gross area of the core as it appears on the drawing; multiplying by the depth, and then

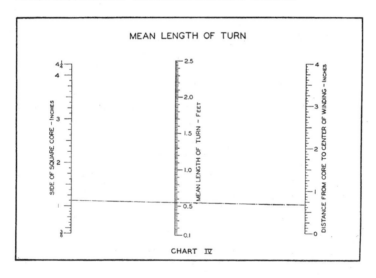

multiplying this result by the factor 0.28, which gives the weight in pounds. For example, suppose a 1″ core measures 3″ by 4″ along its outside edges. The window will then measure 1″ by 2″, with an area of 2 square inches. Subtracting this from the gross area of 12 square inches leaves 10 square inches of net area. Multiplying

this by the depth, 1″, gives 10 cubic inches as the volume of the core. The volume multiplied by 0.28 gives 2.8 pounds as the weight of the core. If the number of laminations is desired, it may be calculated by dividing the depth of the core by the thickness of a single sheet, which is usually 0.014 inches. The shape to cut the laminations will be discussed later.

As a final operation it may be desirable to check the resistance of each winding to make sure it is not excessive. Knowing the length of wire and looking up the resistance per thousand feet in the wire table, the total resistance is a matter of simple arithmetic. The voltage drop under full load due to resistance (exclusive of flux leakage, etc.) is then the product of the resistance and the current.

Summarizing, the general procedure to follow in designing a transformer may be outlined as follows:

(1.) Characteristics of secondary windings, i. e., voltage, current and power.
(2.) Total number of turns in each winding.
(3.) Size of core cross section.
(4.) Wire sizes.
(5.) Physical dimensions from full-size drawing.
(6.) Amounts of wire and core material.

The next and final article will take up the actual construction of the transformers, and the fundamentals of choke coil design will also be discussed in detail.

RADIO PICTURE EQUIPMENT IN THE LABORATORY

These two photographs give some idea of the compactness of the radio photograph equipment which has been in the course of development in the RADIO BROADCAST for some time. Left, the transmitter, above, the receiver

Pictures by Radio

What Radio Broadcast *Laboratory* Is Doing to Aid the Home Constructor to Develop *Phototelegraphy*

DURING the years of 1924 and 1925 two demonstrations of phototelegraphy were given, one by radio and the other by wire. These were milestones in the progress of the art of transmitting pictures to distant points by electricity. The first demonstration, making use of the system developed by R. H. Ranger of the Radio Corporation of America, was given on December 2nd, 1924, when pictures were transmitted from New York to London utilizing a mechanism with which the picture is produced by means of a pen making ink marks on paper. In 1925, a second system, developed in the Bell Telephone Laboratories, was demonstrated, and the system was adapted to use to full advantage the facilities of the Bell system. It made use of photo-electric cells which are sensitive to light and which are capable of controlling electric currents in accordance with the strength of the light impressed on the cell. The Bell system is similar in many ways to the Korn system with which, in 1907, some very good pictures were transmitted from Paris to London.

A great deal of valuable, and not too technical, information on picture transmission is given in a book recently published by the D. Van Nostrand Company, entitled *Wireless Pictures and Television*, and in this book will be found fairly complete descriptions of both the RCA and Bell Laboratory systems. The author is Mr. T. Thorne Baker, himself a scientist who has been actively interested in phototelegraphy, and anyone interested in the subject will find the book well worth reading.

Most of the press accounts describing the Ranger and Bell Laboratory demonstrations have made some reference to the possibility that perhaps, at some later date, means will become available to the general public of receiving, via radio, photographs of important public events, news items, etc., as well as ordinary broadcasting. In order for such a use to become common it is, of course, essential that the system used be simple to operate and fairly cheap in first cost and upkeep. How soon can such a development of picture transmission be expected?

IS AMATEUR PICTURE RECEPTION NEAR?

READERS of RADIO BROADCAST will be interested to know that during the last four years or so experimental work has been going on in the Laboratory on a system of picture transmission which has the essential characteristics of simplicity and low cost that make it especially well adapted for use by amateurs and home-constructors. No technical knowledge is required in order to use the system and it is hardly more difficult to operate than the tricky squealing radio receiver of four years ago. Photographs on this page show the picture transmitting and receiving apparatus in use in the Laboratory.

The system used in RADIO BROADCAST Laboratory has been developed to the point where good pictures can be transmitted by either wire or radio. To transmit a picture 4" x 5" requires about two minutes. The picture is received on a piece of ordinary photographic paper which is finally developed like an ordinary photographic print. There is more "kick" in producing a picture which has been received by radio in your own home than there ever was in "pulling in" the Coast at 2 A. M.

It is quite possible that picture transmission by radio will follow, to a certain extent, the same course that radio followed only a comparatively few years ago when it first became popular. The first picture receivers will probably be home constructed and the home-constructor will, at the beginning, take an important part in stimulating the rapid development of picture transmission. The first users of picture receivers will be the same persons who, when radio first started, constituted the larger part of the radio audience.

The problem in picture transmission with which the Laboratory is concerned at present, is the development of apparatus easily constructed and operated at home. The various parts must be manufactured and made generally available so that their assembly into a complete unit will not be difficult. RADIO BROADCAST is making extensive arrangements to present workable plans to the home constructor, and the latter may expect reasonably soon to see a most interesting series of articles in this magazine which will open this new and fascinating field to him. Complete constructional information will aid the experimenter to start work with no delay. The pictures will be transmitted by broadcasting stations and will be received on a small unit which can be attached to any efficient radio receiver. The number of experiments that can be made with a small picture transmitter costing not more than $100.00 is practically unlimited.

The amateur experimenter has long looked to the transmitting of photographs by radio with envious eyes. Up to the present, experiments of this sort have been confined to luxurious laboratories bristling with engineers and a wealth of apparatus. But now it will shortly be possible to assemble a picture receiver which will not cost much more than $100.00, attach it to a good broadcast receiver, and jump into what will soon be an enthralling hobby. In the meantime we invite correspondence from our readers who have already given thought to this new field of experiment or who have perhaps done some experimenting on their own in picture transmission.

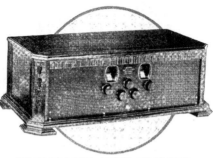

An Eight-Tube
Receiver Which Is
Remarkably Sen-
sitive, Has Knife-
Like Selectivity,
and Is Simple to
Construct

The Intermediate
Frequency Used,
112 Kilocycles, Is
Unusual but Has
Considerable Ad-
vantages

THE LABORATORY "SUPER" AS IT LOOKS AFTER COMPLETION
Compactness is one of the features of this exceptionally efficient eight-tube
super-heterodyne. Many five or six tube receivers require just as large a cabinet

Building the Laboratory "Super"

By ERNEST R. PFAFF

WHEN McMurdo Silver described a super-heterodyne receiver in the January,
1925, RADIO BROADCAST, it represented
about as satisfactory a receiver of that type as
could then be built. In the intervening two and
one-half years, many thousands of these receivers
have been constructed, and while the design
was not blessed by an absolute freedom from the
vicissitudes that often attend the building of
radio sets at home, many builders have reported
truly remarkable reception records, while as far
as is known, practically all have enjoyed quite
satisfactory results. In the meantime, many
experiments have been carried on in an endeavor
to improve the original receiver, and a new model

has finally been developed which, judging from
comparative tests, bids fair to eclipse the very
best performance of the original circuit. In the
design of the latest improved model, described
herewith, cognizance has been taken of the
engineering advances made since the development of the first circuit, as well as of the many
helpful comments and suggestions of the builders
of the original receiver.

The improved Laboratory model super-
heterodyne employs eight tubes in a comparatively conventional arrangement—a first detector, an oscillator, three long-wave (intermediate) amplifiers, a second detector, and two audio
stages. The first detector circuit is very similar

to the conventional short-wave regenerative
circuits now so popular. The circuit is regenerative, a small 75-mmfd. midget condenser controlling regeneration, while a 0.00035-mfd. condenser
does the tuning. The coil system consists of a
conventional Silver-Marshall plug-in coil, so
connected that both regeneration and tuning
condensers are at ground potential, with consequent total absence of hand-capacity effect.
No provision is made to use a loop, as it has been
found that, for extreme selectivity, the use of
an antenna, the coupling to which is variable,
provides for greater flexibility than a loop.

The oscillator circuit is very similar to that of
the first detector, and offers few unusual points

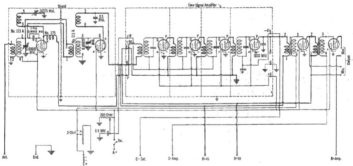

THE CIRCUIT DIAGRAM OF THE EIGHT-TUBE SUPER-HETERODYNE

other than that it is grid-tuned with a 350-mmfd. condenser and there is complete absence of hand-capacity effect. Its output at different wavelengths is sufficiently constant for practical requirements, as is its calibration, while the coupling to the first detector is variable.

The long-wave, or intermediate, amplifier is possibly the most unique and interesting part of the receiver, for it is a completely shielded assembled unit, or catacomb. The copper can 15 inches long, 5 inches wide, and 3 inches deep, contains four individual stage compartments, each holding an r. f. transformer and its attendant tuning capacity, a tube socket, and the necessary wiring and bypass condensers. Three r. f. stages and a detector are employed, and the whole unit is tuned to exactly 112 kilocycles. The reason for the selection of this intermediate frequency is that very satisfactory low-resistance air-core tuned r. f. transformers may be built for operation there.

Another reason for the selection of the odd 112-kilocycle amplifier frequency is because of decreased interference possibilities. Normally, in a super employing, say, a 50-kilocycle intermediate frequency, two stations 50 kilocycles away will heterodyne each other and be received without the use of the local oscillator at all, selectivity being dependent upon the selectivity of the antenna tuner and the local coil pick-up. As the intermediate frequency is increased, this possibility decreases, since it is far easier for an antenna tuner to discriminate between stations 112 kilocycles apart than between powerful locals 30, 50, or even 60 kilocycles apart. Further, powerful stations are generally spaced on even 10-kilocycle separations, so that the odd 112-kilocycle frequency is a greater aid to selectivity.

Coil pick-up is, of course, absent in the shielded amplifier, and wiring pick-up is almost negligible since all wiring is very close to the grounded metal panel or chassis. Complete shielding of the first detector and oscillator sections prevents pick-up of strong local stations on the coil systems themselves, though in receivers to be operated in the country, or in non-congested broadcasting centers, these two shields might be omitted.

The audio amplifier offers no unusual points except one of very great value in an ultra-selective receiver. This is the 5000-cycle cut-off, or fall-off in amplification, which occurs in this amplifier. Normally, frequencies bove 5000

cycles do not contribute to realism of reproduction, according to no less an authority than the Bell Telephone Laboratories, while in the range above 5000 cycles lie practically all parasitic amplifier noises, atmospheric disturbances, and the only too prevalent heterodyne squeals. These the .5000 cycle cut-off tends to decrease very markedly, and almost entirely eliminate.

The amplification possibilities of the receiver are interesting, as compared to, say, a good bridge-balanced or neutralized eight-tube r. f. receiver. Such an r. f. set would employ, perhaps, four r. f. stages, a detector, and three audio stages, and would cost from $250 to $1000. The probable r. f. gain at best would be 10 per stage (including the detector for simplicity), while the assumption of three impedance-coupled audio stages would allow 8 per stage as an average audio gain with a 112 output tube. Thus, the overall voltage amplification would be about $10^4 \times 8^3$, or 51,200,000 times amplification for a weak signal. In the Laboratory Super, it is safe to assume a voltage amplification of, say, 25 times for the first detection and frequency conversion. The overall amplification of the intermediate amplifier is problematical, but the normal computed, and actual, gain of an r. f. stage such as is used in the Laboratory Super is about 20. Thus four tubes (three amplifiers, and a detector, coupled by four similar tuned circuits), might conservatively be counted on for a total gain of 10 to the fourth power. The audio gain is about 20 per stage or 20^3 overall, so the total gain may be assumed to be $25 \times 10^4 \times 20^3$, or 100,000,000.

These figures seem borne out in practice, for the Laboratory Super will bring in on the loud speaker signals inaudible on a good seven-tube shielded and neutralized r. f. receiver. The selectivity appears sufficient to allow reception of out-of-town stations 10 kilocycles away from locals in Chicago, and, in fact, is so great as to allow the reception of a frequency band only three to four kilocycles wide if desired.

The wavelength range of the receiver with standard coils is from 30 to 3000 meters (10,000 to 100 kc.). Regular broadcast range coils cover the range of 200 to 550 meters (1500 to 545 kc.), but it can be seen that the receiver is adapted to any class of broadcast reception by virtue of its wavelength flexibility.

The assembly of the Laboratory Super is very simple if the standard parts recom-

mended for its construction are procured. The completely pierced and decorated steel panel ²nd chassis can be obtained as a standard foundation unit which reduces the whole assembly operation simply to one of mounting parts with a screw driver, machine screws, and nuts. If preferred, a bakelite or wood chassis might be employed with slight alterations in design, but the change is not recommended unless the receiver is to be located well away from powerful broadcasting stations, since the steel chassis and panel contribute materially to the wiring shielding of the receiver. The whole construction is very simple and the one big bugaboo of super construction—uncertainty as to long-wave-transformer matching and operation under varying conditions of home assembly—is totally eliminated. The assembled intermediate amplifier recommended, known as the 440 "Jewelers' Time Signal Amplifier," is a laboratory-calibrated amplifier, the operation of which will not vary appreciably even with the widest variation of standard tube characteristics to be encountered in practice. The original Laboratory Super was constructed with the following parts:

1—Van Doorn Panel and Chassis Unit, with Hardware	$ 8.50
1—Carter 0.00015-Mfd. Grid Condenser, with Clips	.50
1—Carter M-200 Potentiometer . . .	.75
2—Carter 0.5-Mfd. Condensers . . .	1.80
1—Silver-Marshall 275 R. F. Choke . .	.90
1—Silver-Marshall 342 0.000075-Mfd. Regeneration Condenser	1.50
1—Carter 3-Ohm Rheostat	.50
1—Carter Battery Switch	.50
4—Carter No. 10 Tipjacks	.40
1—Polymet 2-Megohm Leak . . .	.25
2—Silver-Marshall 220 Audio Transformers	16.00
4—Silver-Marshall 511 Tube Sockets .	2.00
2—Silver-Marshall 805 Vernier Drum Dials	6.00
1—Silver-Marshall 440 Time-Signal (Intermediate) Amplifier	35.00
2—Silver-Marshall 515 Coil Sockets .	2.00
2—Silver-Marshall 111A Coils . . .	5.00
9—X-L Binding Posts	1.35
2—Silver-Marshall 320 0.00035-Mfd. Variable Condensers	6.50
Total	$89.45

(If shielded oscillator and first detector are desired, add $4.00 for two Silver-Marshall 631 stage shields)

METAL PANEL AND CHASSIS ARE SPECIFIED ALTHOUGH THEY ARE OF BAKELITE IN THIS PHOTOGRAPH

The assembly of the set may be accomplished very easily if the following suggestions are carefully watched.

Upon the chassis should be mounted the detector and oscillator assemblies—inside the stage shield pans if shields are to be used. The end mounting screw of each 511 tube socket is used to join the A minus to the chassis, so a lug should be placed under the screw head, to be soldered to the F minus socket terminal, and the under side of the chassis scraped bright for good contact with the fastening nut. One terminal of the 0.00015-mfd grid condenser should be bent at right angles and fastened directly under the "G" terminal screws. The single long screw holds the 275 choke coil in the detector stage assembly.

The binding posts should be mounted in the nine holes at the rear of the chassis using the insulating washers to positively insulate them from the chassis. The "Ground" post grounds to the metal chassis, and the fastening screw of this post holds one end of the second 0.5-mfd. condenser tightly to the chassis, while the free end must be bent up clear and free of the metal chassis. The four tipjacks mount, using insulating washers, at the right rear of the chassis.

The intermediate amplifier mounts with four 8-32 screws, the chassis being scraped bright for good contact with the screws (the A minus connection is made to the amplifier through the contact between the amplifier shield and chassis). The two audio-amplifier tube sockets mount using their rear fastening screws to connect the F minus posts to the chassis.

All possible wiring should be done on the chassis before proceeding further, leaving free the wire ends that will connect to the instruments on the front panel, and to the two audio transformers, as, if they were put on first, it would be impossible to make the three connections to the right end of the intermediate amplifier.

The potentiometer should be mounted as shown, using insulating washers to thoroughly insulate its frame from the panel. The rheostat and the midget condenser are similarly mounted, except that care is taken to make good contact between them and the panel.

The drive mechanisms of the dials should be dropped into the bracket bearings intended for them, the shafts pushed through the holes in the front panel, and the two brackets bolted to the panel, using the screws provided. One variable condenser fastens to either bracket, using the shaft mounting nut provided. A drum should be slipped over each condenser shaft, with set screw loosened, and pushed up until the drum scale edge is just ready to enter the crack in the

drive mechanism shaft. With a knife blade this crack should be widened to receive the drum scale edge, and the drum pushed well up on the condenser shaft. The scale should then be adjusted to read 100 degrees against the indicator points in the panel windows, when the condenser plates are entirely disengaged, upon which the set screw in the drum dial hub should be tightened on the condenser shaft. With the knobs fastened on the drive shafts, the condenser dials should rotate if the knobs are turned.

The connections to the condensers, rheostat, and potentiometer should be made before fastening the panel to the chassis. After they have been put in, machine screws and nuts serve to hold panel and chassis together. The "On-Off" switch mounts in the one remaining panel hole, with insulating washers to thoroughly insulate it from panel and chassis (it may previously have been connected in circuit, and allowed to hang on the wiring until ready to be mounted).

TESTING AND OPERATING

BEFORE actually hooking up the set for test, or before cabling and lacing the under chassis wiring, the set should be carefully checked. The A battery alone should be connected, a single tube inserted in any socket, and the "On-Off" switch and the rheostat turned on. The tube should light in any and all sockets, with brilliancy slightly varied by rheostat adjustment. The A plus wire should be successively connected to the plus 45, plus 90, plus Amp., minus 4½, and minus Amp. binding posts. If the wiring is correct, the tube will light with the A battery connected only to the proper posts. The remaining batteries should be connected. A B socket-power device may be used, but only a good model employing a glow tube voltage regulator, such as the one described by Howard E. Rhodes in the June RADIO BROADCAST.

To operate the set, all tubes should be inserted, except the first detector tube. A cx-371 (ux-171) goes in the right rear socket; a cx-340 (ux-240) in the adjacent rear socket. The rest of the tubes are cx-301-A's (ux-201-A's). With the "On-Off" switch on, the rheostat should be turned to within ½ inch to ½ inch of the full right position. If the potentiometer ("Gain" knob) is turned to the right, a "plunk" will be heard at some point. This can be detected by varying the oscillator drum, which should cause a number of shrill whistles to be heard. The "Gain" knob should always be operated just to the left of the "plunk" point, i. e., to the right of which squeals were heard when the "Oscillator" dial was varied. The receiver is least sensitive when the "Gain" knob is at the left, and most sensitive

when the "Gain" knob is just to the left of the "plunk" point.

The first detector tube should be inserted and the midget condenser set all out. The antenna coil rotor should be set at 45 degrees, and the oscillator rotor should be all in. A small antenna, 30 to 60 feet long, should be used, or even a larger one if the set is not too close to powerful local stations. Stations may be tuned-in using the two drum adjustments only. Weak stations may be intensified by turning up the "Regeneration" condenser on the front panel. This condenser functions similarly to the "Gain" knob, in that, as it is turned to the right to interleave the plates, signal strength on weak stations will increase up to the point where the first detector oscillates, and the signal turns into a squeal. Adjusting the midget condenser will react slightly on the setting of the "Antenna" drum.

Tuning operations for weak out-of-town stations are as follows: With the rheostat to within ¼ inch to ½ inch of the extreme right position (adjusted to give 5 volts to the filaments of all tubes), the "Gain" knob should be turned to the right until it is just below the "plunk" or oscillating point for the intermediate amplifier. This adjustment is independent of the wavelength of any signal being received. The two drum dials should then be varied to tune-in stations. In order to cover the entire broadcast band, the "Antenna" dial should be varied in steps of one degree at a time, and for each one-degree step the "Oscillator" should be varied over a range of 15 degrees above and below the "Antenna" dial setting. Once a weak station is heard, it may be strengthened by turning the midget condenser farther in and resetting the "Antenna" dial slightly. Each station should be tuned-in at one point on the "Antenna" dial and at either one or two points on the "Oscillator" dial. If stations are heard at two or more points on the "Antenna" dial and "Oscillator" dial, it is not the fault of the receiver, but is probably due to re-radiation of the transmitted signal from local steel structures, etc., which energy, while very weak, may nevertheless be picked up by the Laboratory Super and amplified as it would be on no other receiver. It is just this sensitivity that accounts for the set's remarkable long-distance performance.

The position of the antenna coil rotor should generally be at about 45 degrees. With a small antenna, it may work best all in; with a large antenna, it should be at nearly right angles. The sharpness of tuning of the antenna dial depends upon the setting of this rotor, as well as that of the midget condenser. The oscillator rotor should be adjusted once on a very weak signal at about 300 to 350 meters, and, once set for maximum volume, may be left alone.

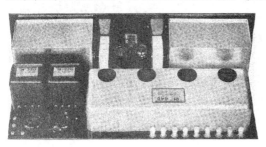

A BEHIND-THE-PANEL VIEW

In this photograph the Laboratory Super has been completely wired and is now ready for connection to the necessary source of A, B, and C power

We Need Better Radio Salesmen

The Sheep and the Goats of the Radio Sales Fraternity—The Goats
Need Better Preparation for their Job Which is Not a Simple One

By CARL DREHER

Drawing by Franklyn F. Stratford

I KNOW little about the mystic process of selling radio equipment to the ultimate user, and disclaim, at the beginning of this discussion, any idea of instructing experts in the merchandising art. My experience has all been in other directions. By these modest words, I hope to avoid harsh looks and nasty expressions from the commercial gentlemen on whom my sustenance depends, although, happily, theirs depends equally on me. With all this ignorance, occasionally I do venture into a radio $s_t o_t e$ in order to purchase some trifle which I am unable to cadge lawfully from the company whose payroll I adorn. On these pilgrimages I have occasionally made observations of possible interest to radio men engaged in trying to earn money.

Some of the most insufferable little jackanapes ever expelled from high school are behind the counters of radio stores, and there are also many decent, sober, competent fellows, the two types being found, often, five feet apart. The latter make money for the stores in which they serve, and, I hope, for themselves. As for the former, I should be sorry to hear that any of them starved, but, abstractly, it would seem to me a highly salutary social phenomenon. That they are not an asset to the businesses which afford them shelter is a foregone conclusion. Furthermore, their lives are in danger. I do not speak of myself, my homicidal impulses being mild outside of family circles, but imagine an ex-president of the Institute of Radio Engineers, for example, entering one of the radio emporiums on Cortlandt Street in the dizzy metropolis, to buy a grid leak, and being patronized by some beardless youth who was playing marbles on the Brooklyn lots two years ago. Such things happen, and is it not conceivable that some worthy pioneer, taking the law into his hands, will reach behind the counter, carry off one of these lightweights, and, declining the polite offers of the porters at the foot of Liberty Street, plunge his struggling burden into the waters of the Hudson?

The development of these radio counter boys and sales clerks is not hard to visualize. The process starts with a young fellow of high school graduation age, no worse and no better than thousands of others, faced with the necessity of earning his living. He has built a few receiving sets, and, as his education has not trained him for anything in particular, and the sets work part of the time, he decides that radio must be his forte. Through an advertisement or otherwise he secures a job in a radio store of the sort which sells everything from rubber overcoats designed to cure vacuum tubes of the notion that they are microphones, to $600 phonograph-radio combinations in Louis XV cabinets. His responsibilities and authority are ill defined, the principal idea being that he is to do whatever he is told and act as a door-mat for the rest of the staff. He unloads thousands of loud speakers from trucks, crates and uncrates sets, hunts through mountains of excelsior for missing parts, and sweeps out the store. In time he learns something about the business and is promoted to an assistant sales clerk in charge of the binding post counter. He now faces the public. And now something happens to him, inwardly.

All day long people come in and ask him questions. Our hero doesn't know much, but it is clear that lots of people know even less, about radio, at any rate. Some of the questions are sensible, and some are idotic, but all must be answered. This is what happens when a horde of people of every grade of intelligence and aptitude invade a highly technical field. The clerk answers the questions fired at him as well as he can. When he does not know the answer, he bluffs. This vice of pretence he shares with much larger carnivora, even up to the presidents of corporations, sometimes, and high officers of government on land and sea. When he bluffs, he is rarely caught, because, as we remarked,

his clients know even less than he does, and no stenographer takes down his words. The more he gets away with it, the better he thinks he is. By the time he gets to be a full fledged retail

"PLUNGE HIS STRUGGLING BURDEN INTO THE WATERS"

salesman, with shiny current supply sets, amplifying transformers containing a pound of iron, and variometers colored like a pair of pajamas, loading down the shelves behind him, he feels like the Oracle of Delphi. And like the priestess who intoned her prophecies above the cavern of noxious vapors, he is mainly a fraud.

For, when you come down to it, this business of selling radio is intricate enough, and as yet there is little systematic preparation for the job. When the job is well done, it is merely the result of a fortunate accident. There are some qualified, resourceful men selling radio parts and sets here and there—the fortuitous combination of opportunity, experience, and aptitude. When a customer enters, not knowing what he wants except that he doesn't wish to pay much for it, they cross-examine him skillfully and give him what he needs, within the limits of his expenditure, carefully explaining what he is not getting before he leaves the store. Encountering an experienced radio man, they secure what he asks for without loss of time, perhaps dropping a deft suggestion which, very likely, it will be profitable for him to listen to.

The job being complicated, and the results, under present conditions, decidedly haphazard, why should not more attention be given to this problem? A well-known chain of barber shops advertises the claim that its tonsorial employees are trained in a school which it maintains, that they may become adept in the superior technique of the establishment. I presume that the course is a brief one, that few of the professors on the faculty have Heidelberg Ph. D.'s, and that the whole advertisement is ninety-per-cent wind. Even so, my observation has been that the barbers afore-mentioned are decidedly better than the average run, that they wash their hands at the proper time, cut hair neatly, and do not take the skin with the beard. If it profits a barber establishment to select and train its operating personnel with some care, should not the same principle apply to a far more technical business like radio? There is money in radio, too, and there might be more if we looked out for these little things. We should discard, also, several freight-train loads of the current bunk about salesmanship. A good salesman is primarily one who knows his goods, senses accurately the needs of his clients, and has the normal energy and social qualifications required in any other business. To the-devil with personal magnetism, hypnotism, the psychological approach, and the whole bag of tricks, if he only knows that a high impedance speaker will work across a low impedance output, but not vice versa, and that a paper grid leak will not carry fifty milliamperes. There are plenty of ex-commercial operators and zealous amateurs who can do good work as radio salesmen, although they know nothing about salesmanship. Some of them, praise be, are already in the retail establishments of the industry, and the proprietors would do well to get in more of them before the next season sends them into the bankruptcy courts

284

The New A. C. Tube

UNCERTAINTY regarding the constants of the new a.c. tubes reigns supreme, as a glance at the table on this page will show. While it has been impossible to secure all makes of tubes designed to operate from raw a.c., those measured in the Laboratory are representative at least, and should give the reader something to think about. Whether or not he is to throw away his batteries and present set of tubes is for him to decide, but if the tubes are good and his battery still possesses its appropriate lead plates and sulphuric acid, he need be in no hurry to make changes.

The a.c. tube has not sprung up overnight. It has been the dream of engineers for a long time and, in the case of the McCullough tube, has been in actual production for several years. It is difficult to understand why there should be any general concern about the situation now that it is possible to choose between a.c. tubes that operate on 1 volt and consume 2 amperes and those that require a voltage of 15 at 0.35 amperes, but there is always something akin to hysteria in the radio business when anyone makes a suggestion that is more or less new.

The a.c. tubes are of two kinds, those that have filaments in the usual sense, and those that have heaters which operate from a.c. and which, in turn, heat a cathode which supplies the electron stream which the radio set actually makes use of. The first a.c. tube in this country was the McCullough tube, which is a heater type requiring a voltage of 3 and a heater current of one ampere. Others tested in the Laboratory in recent weeks are made by Marathon (Northern Manufacturing Company), Sovereign (of Chicago), and of course the Cunningham. It has been impossible up to the time this is written to get R. C. A. a.c. tubes into the Laboratory. The data given here on the C-326 and C-327 (taken from early tubes sent to the Laboratory) will probably not be far from the final design which, according to the Cunningham Company, has not as yet been reached.

The heater type of tube, of which the Mc-Cullough is the forerunner, is not new to the industry. The stout filament type, which has been associated with the name of Mr. B. F. Miessner (although others had worked on the tube prior to Mr. Miessner's papers delivered before the Radio Club of America) is newer and works on different principles. It has a true filament, very heavy and rugged, heated by a lot of current at a low voltage. Readers interested in the theory underlying this tube, which is the result of a nice piece of research on Mr. Miessner's part, will do well to read his articles in RADIO BROADCAST for February and March, 1927, and a paper in this issue.

It does not seem possible to use the latter type of tube as a detector of the familiar grid leak and condenser type. For this reason the R. C. A. and Cunningham recommend that the 227 (C-327) or heater type be used as a detector and the rugged filament type for all other positions. There is no reason why the heater tube

cannot be used throughout the set; indeed, receivers utilizing the McCullough tube have been used this way for some time. The rugged filament type may be used with a C-battery detector.

A cursory perusal of the data given here shows that the average constants of these new tubes are better than the average d.c. tube designed for the same purpose. For example, the table gives the average plate impedance and amplification factor of the familiar 201-A (C 301-A) type. It is possible with either the heater or rugged filament type to build much better tubes than this table indicates is now being done. Such tubes have been designed to have characteristics as near to those of the d.c. tubes now used as possible. It is futile for a tube manufacturer to

build a tube that has half the impedance of present tubes, if only one or two set manufacturers have gumption enough to build a set around them. On the other hand, gradually changing tubes toward those with greater gain and gradually altering the design of radio-frequency transformers to go with them, will ultimately bring about receivers which are more selective and which have greater gain per tube.

Some of these a.c. tubes are noisy while others from the same allotment and with practically the same electrical characteristics are quiet. It seems imperative that a.c. tubes must be highly exhausted. Hum is kept at a minimum by biasing the heater with respect to the cathode, as indicated in Fig. 1. The biasing voltage is not always the same. Some receivers, particularly those with high-quality amplifiers (going down below 60 cycles and using high-quality loud speakers), will hum if a transformer carrying 60-cycle current is anywhere near them. The solution at times is to change the relative positions of the cores of the audio transformers, or chokes, and that of the power transformer. An electrostatic shield, properly grounded, between primary and secondary of a power transformer, is also of considerable aid in stubborn cases of 60-cycle hum.

Some manufacturers of a.c. tubes have been ambitious enough to put out a whole series of a.c. tubes, adding to the general purpose tubes those with high amplification factors, and power tubes. The Van Horne Company lists no less than eight tubes designed for straight a.c. operation or for series filament connection. The Marathon list is made up of general-purpose tubes, high-mu tubes, detector, general purpose and power tubes. The Arcturus line includes detector, general purpose and power tubes.

It is unfortunate that all of these tubes require different values of filament, or heater, current and voltage. This puts a big burden on the manufacturers of transformers, and upon those who like to construct their receivers at home. For example, the Cunningham heater tube utilizes 1.75 amperes at 2.5 volts, while the rugged filament type demands a voltage of 1.5 and a current of 1.05 amperes. The Arcturus type, on the other hand, uses a voltage of 15, which can be conveniently supplied by toy transformers stocked by practically every electrical dealer.

From several engineers who have been responsible for the development of these tubes, comes the suggestion that high values of C bias are desirable from the standpoint of hum, and from others comes the report that at high frequencies, the heater type is subject to some strange vagaries. It seems that the resistance of certain mechanical parts that go into the construction of these tubes decreases at high frequencies, which naturally will cut down the overall gain of a tube. It seems to be true, too, that some trouble is had at times with the vacuum when the oxide coated low-voltage filament is used. Perhaps this explains why some tubes, for no apparent reason, are noisy.

The Arcturus tube is unusual.

A. C. TUBES									
Heater Type									
NAME	E_f	I_f	E_p	E_g	I_p	μ	R_p	G_m
C-327 McCullough	2.5	1.75	90	-4.5	4.2	8.0	9200	940
Sovereign	3.0	1.0	90	-4.5	4.2	8.6	9400	870
Marathon 608	3.0	1.5	90	-4.5	4.6	8.5	9100	935
Marathon 615	5.5	1.0	90	-4.5	4.2	7.3	9500	775
Marathon 630	5.5	1.0	90	-1.0	3.2	12.0	19000	635
Marathon 605	5.5	1.0	90	-1.0	1.1	28.0	40000	680
Arcturus	5.5	1.0	135	-12.5	14.0	4.5	4400	1000
	15.0	0.35	90	-4.5	3.1	10.5	12150	870
Filament Type									
C-326 Armor A. C.	1.5	1.05	90	-4.5	5.0	8.5	8300	1030
100	1.0	2.0	90	-4.5	3.8	7.8	11200	660
Van Horne	1.0	2.0	90	-4.5	4.0	9.5	10000	980
CX-301-A (UX 201-A)	5.0	0.25	90	-4.5	2.0	8.0	12000	675

E_f = Filament Volts
I_f = Filament Current
E_p = Plate Volts
E_g = Grid Volts
I_p = Plate Current
μ = Amplification Constant
R_p = Plate Impedance
G_m = Mutual Conductance

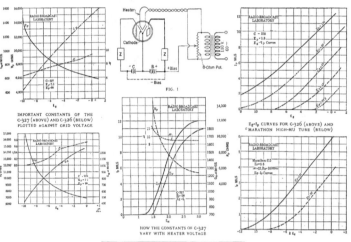

IMPORTANT CONSTANTS OF THE
C-327 (ABOVE) AND C-326 (BELOW)
PLOTTED AGAINST GRID VOLTAGE

FIG. 1

E_g-I_p CURVES FOR C-326 (ABOVE) AND
MARATHON HIGH-MU TUBE (BELOW)

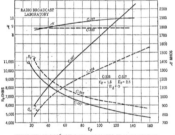

HOW THE CONSTANTS OF C-327
VARY WITH HEATER VOLTAGE

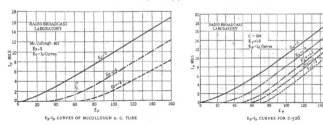

CONSTANTS OF C-326 AND C-327 AS RELATED TO PLATE VOLTAGE

It is a heater type, the heater current being about 0.35 amperes at 15 volts. The heater is a rather heavy carbon filament which at the top is electrically connected to the cathode. This makes it possible to use a standard four-prong base—an obvious advantage which other heater tubes do not possess. The R. C. A. and Cunningham heater tubes require a special five-prong base, while the others, of which the McCullough is the prototype, have two heater terminals at the top of the tube, or at the side, to which are attached the a.c. terminals.

We noted at the Chicago Radio Trade Show that 35 per cent. of the receivers shown were a.c. operated. Not many of these sets used a.c. tubes, but used some other method of eliminating hum when using standard d.c. tubes.

All in all, the a.c. tube is something to watch. It is still in the development stage, and the work of many tube engineers and chemists will be involved before the final answer is ready. The a.c. tube is not destined to supplant the present type of tube we all use in the next six months or in any other period so short. As soon as the Laboratory gets more data on the operation of the new tubes, they will be made available.

E_p-I_p CURVES OF McCULLOUGH A. C. TUBE

E_p-I_p CURVES FOR C-326

Should We Name Our Broadcasting Stations?

THE following bulletin was recently sent to all station owners by the National Association of Broadcasters:

On many occasions during the past few months, during informal discussions among our members, and occasionally during official hearings, the question of continuing the designation of a station by announcement of its call letters over the air has attracted keen interest.

The use of combinations of letters to designate a radio station originated properly in "point to point" communication some years back, and as a natural outgrowth of this, with the advent of broadcasting stations this type of designation has been continued. It was never specifically designed for *radiophone* broadcasting stations.

It is a well-known fact that due to the phonetic similarity of many letters in the alphabet (for example, B may sound like D, E, P or T) the average owner of a radio receiving set makes many mistakes in identifying the station he has tuned, which in turn causes the credit due that station to be entirely misplaced.

At this point, those who favor abolishing call letters immediately advance the argument that the sole reason for a station existing is to create personality for itself, and immediately draw the analogy between the station and a boat or a yacht. For official records, the Government designates all vessels under a license form, as, for instance, KX-109. However, immediately, regardless of whether it is a pleasure or commercial craft, its owner christens it with an appropriate name which lends personality to the ship, and the license designation of letters and numbers is never given further consideration except in connection with the license to operate.

From this analogy, the proponents of the idea ask, "Why is not an announcement, such as 'The Mayflower, Cincinnati' of more value to the station and easier to identify than the announcement 'This is Station WBDT'?" The first is at once suggestive of a personality and entirely distinguishable while the latter is negative and easily confused.

The opponents of such an idea point out that hundreds of thousands of dollars have been spent, in many instances building up the prestige of a certain combination of letters, which in some cases correspond to the trade slogan of the owner of a station. Undoubtedly such stations would be slow to consider favorably the idea of relinquishing their call letters.

However, the discussions have been so frequent and active by both sides, that it is with the thought of determining what the real consensus of opinion is that your vote is asked on the enclosed ballot.

We tremble. We hold our breath. We pace about apprehensively our pockets loaded with horse shoes. Suppose they decide to do it! What manner of mon-

strous monikers may we not have to listen to! We think the suggestion a highly indecent and immoral one—like giving small, and notably reckless boys, shining, loaded revolvers to play with. Needless to say the broadcasters will jump at the idea; tacking names on things is their meat. What grand and glorious, resplendent and superlative station names may we not expect. Give a look what the gents did back in the days when they were coining slogans for their stations:

"Kall For Dependable Magnolene"
"Where The Sun Shines Every Day"
"Stephens College Where Friendliness is Broadcast Daily"
"Kant Fool Us Long Horns."
"The Best Little Station"
"The World's Largest Grease Spot"
"Kum To Hot Springs"
"We Sell Goods Cheaper"
"Where Cheer Awaits U"
"Where Harrisburg Broadcasts Gladness"
"One of Indiana's Most Beautiful Little Cities,
And The Home Of The First Automobile"
"Watch Mercer Attain Zenith"

AT WJAX, JACKSONVILLE

Isaac Wessell of the Jacksonville Little Symphony Orchestra—broadcast by WJAX—and his old faithful bass fiddle. The press agent claims that it is over two hundred years old and has been repaired in Budapest, Leipsig, New York, Boston, and other places, and that it has been cracked so often that there is hardly a sound spot on the instrument

"We Are Never Tired"
"Sunshine Center of America"
"Known For Neighborly Folks"
"Keep Folks Quoting The Bible"

Names like the suggested one, "Mayflower," wouldn't be so bad, but there'll never be enough of them for seven hundred stations. Arrived at the second hundred, there will be a terrible struggle to create some appropriate and unduplicated collection of syllables. Moreover we, the poor listeners, will have to do all the dirty work of memorizing the new names, and it was. only at the cost of much travail that we got the multitude of call letters finally straight in our mind. Also we shall have to junk our receiving set cabinets and install new ones with three-foot dials whereon to inscribe the new and lengthy appellations.

We submit that letters, or even numbers, are quite as easy to memorize in large quantities as names. Imagine what a boon it would have been if Mr. Pullman had only started off in the dim ages, by christening his sleeping cars with call letters. Have you ever returned from the observation platform at a late hour and attempted to crawl into lower twelve in "Grassmere" or "Graymere," only to discover, amidst much feminine shrieking, that you really belonged in lower twelve in "Grassbeer"?

If names must be used, let us, f'revvens sakes, have some system about it. Let all the stations in Georgia be lovely ladies, and those in Mississippi, flowers; those in New York, drinks and those west of the Rockies, kitchen utensils, and so forth. Thus would we have an entertaining and easily catalogued roll call of such stations as: "Gertrude," "Josie," "Maizie," "Clarabelle"—"Hyacinth," "Petunia," "Erysipelas," "Plastodolphus" — "Old Crow," "Tom and Jerry," "Benedictine," "Kummel"—"Rolling Pin," " Potato Masher," "Egg Beater," etc.

But we quite disagree with the main contention that call letters can have no personality. To us they have always seemed very vivid personalities. What matter the fact that the personalities of the letters, as we interpret them, seldom correspond with the personality of the station mentioned.

WEEI, for instance, is a thin, squeaky sort of a name. It has a long skinny nose which is rather red and which it blows at intervals. It is querulous and complaining and takes its tea rather weak.

KDKA, on the other hand, is a robust, high pressure name. It is six feet two; bulging with muscles and wears red, sweaty underwear. It is the dynamic type, bubbling over with energy and talks with explosive loudness. Likes to squeeze little boys' wrists till they squeal.

WEAF is a sanctimonious assembly of letters. It is tall and gaunt and affects loosely flapping black garments. Its cheeks are sunken, its mouth inexpressibly sad, and it wears horn rimmed glasses over a pair of weak watery blue eyes. Delights in eighteen-carat Baptist Camp Meetings.

WJZ is a high sounding and celestial name. It wears purple robes and quite a high crown. Is preceded by two pages blowing trumpets and is followed by a processional of High Priests chanting its praises. Expects the populace to bend its knee before it and occasionally smites one down. Has a large Roman nose.

CNRO is an ugly, snarling name; WOC is bland, honest and open faced, belongs to the Rotary Club; WBBM is a nasty, nagging name, drums on the table with its finger tips while playing bridge; KFOO is·boorish and stupid, given to belching at the dinner table and so on, and so forth, through the list. WHAM is breezy—a calliope-like individual. WCY is a deep-voiced soul with a sombrero and an educated voice.

Where Broadcasting Reigns Supreme

THE making known of great national events, while they are actually taking place is, after all, radio's unique contribution, and the one field in which it reigns supreme without competition from phonographs, theaters, churches, or newspapers.

And it is greatly to radio's credit that it does this job so thoroughly and well. As an example of a national broadcast well done, may we be permitted to hearken back as far as last Memorial Day? The official ceremonies at the Arlington National Cemetery, as broadcast by the N. B. C. chain, we listened to from start to finish and found not only interesting but entertaining. The ceremony began with an overture by the United States Marine Band, then a "Call to Order" by Major General John L. Clem. Following a soprano solo, the "Star-Spangled Banner," accompanied by the Marine Band, the original order establishing Memorial Day was read. Then two more musical interludes: the solo "Out of the Night, a Bugle Blew," accompanied by buglers, an enormously effective and dramatic work for such an occasion, and the "Battle Hymn of the Republic" sung by the Imperial Male Quartet. Then the *piece de resistance* President Coolidge's address, which, agree with him or not, was an expert piece of speech construction. The program was concluded with the reading of an original "poem" by Clagett Proctor, which was bad verse and the only stupid moment of the entire ceremony, and a brief address by Commander-in-Chief Means of the United Spanish War Veterans. Each different feature of this varied program, the voices of the speakers, the distant notes 'of the bugles, the ensemble singing of the quartette, the vocal solos and the massed volume of the Marine Band, was "picked up" in perfect style and broadcast with such clarity that the radio listeners must have heard the program more effectively even than those actually present at Arlington.

The broadcasting of the hullabaloo incidental to Lindbergh's arrival was likewise thoroughly done—perhaps in spots too thoroughly for the

MRS. ANNETTE R. BUSHMAN, PROGRAM DIRECTOR, WEAF

interest of this particular listener. For instance at the banquet tendered Lindbergh in NewYork. We would have been quite content had all the speeches of eulogy been omitted and only that of the flyer broadcast. Never have we heard worse blah sprung at a banquet, and sprung by such eminent leaders, divines and statesmen! The supposedly tongue-tied airplane mechanic gave a valuable, though doubtless unheeded, lesson in public speaking to the much touted, so-called orators who shared his platform. But the nation as a whole was interested in every and any detail of the flyer's reception and credit must be given to the National Broadcasting Company for slipping up on no smallest part. The N. B. C. established a number of new records in covering the event:

Miles of Wire Line Used—14,000.
Number of Engineers Involved—350.

Pick-up Points—Washington, 5; New York, 7.
Number of Stations—50.
Estimated Audience—35,000,000.
Number of "Radio Reporters"—Washington, 4; New York, 6.
Longest Continuous Program Devoted to One Subject—11½ hours.

SUNDAY NIGHTS AT WPG

WPG is offering a series of Sunday night concerts during the summer from the Steel Pier. Operatic arias, duets and concert selections make up the programs which are arranged by Jules Falk. Among the artists scheduled for the series are: Julia Claussen, mezzo-soprano; Marie Sundelius, soprano; John Uppman, baritone; Berta Levina, contralto; Doris Doe, contralto; Arthur Kraft, lyric tenor; Elsa Alsen, soprano; Julian Oliver, tenor; Edwin Swain, baritone; Marie Tiffany, soprano and Paul Althouse, tenor.

NOVELTY THAT IS GENUINE

A COMMENDABLE move on the part of WBAL was the arranging of a program of "first-time" numbers—meaning selections that had never previously been heard over the air. Two of the numbers "Air de Ballet" and "An Irish Tune" were contributed by Gustav Klemm, WBAL's program supervisor who is also a composer. Other selections were "Carioca" and "Batuque," Brazilian tangos by Ernesto Nazareth, "Indian Summer" by Sturkow-Ryder and "Barn Dance" by James Rogers. The recital was given by Sol Sax, pianist, during one of the WBAL Staff Concerts which are on the air Wednesday evenings from 9 to 10 o'clock Eastern Standard Time.

NEW PROGRAM MATERIAL

H. V. KALTENBORN, whose editorial discussions of the outstanding happenings of the week constitute perhaps the best "one man show" of all radio's offerings, is to resume his talks from WOR on October 10. We are informed

THE CRYSTAL STAGE STUDIO OF WOW

Many of the broadcasting stations have made use of this method of accommodating studio visitors without confusion or danger of interference with the program. The stage is insulated from the auditorium by plate glass across the entire stage. Visitors hear the artists via loudspeakers placed behind the grills at the upper left and right

that he is at present in the Far East making a study of the Chinese revolution. He is to visit the important centers in South China and North China and will also visit the Philippines to study problems associated with colonial administration and the independence movement. On his return journey he will travel overland through Manchuria and Korea to get first hand information concerning Japanese colonization and penetration and will spend some weeks in Japan before sailing home.

HURRAH FOR THE A. S. C. A. P.

A MOVEMENT has been started (may it get farther than the ten preceding starts!) by the American Society of Composers, Authors and Publishers to prevent the too frequent broadcasting of popular compositions, according to E. C. Mills, representative of the Association. It is contended that experience has proved beyond a doubt that the excessive broadcasting of a composition quickly destroys its market in published and recorded forms.

' It is not at all unusual to hear a number in popular demand broadcast in any particular area from six to a dozen times in an evening," says Mr. Mills. " Long before the public has had opportunity to purchase the rolls and phonograph records, or the music in sheet form, the composition has been ' blasted to death' and the public is weary of even hearing it.

"No composition should be rendered more than once in an evening in the same form. If played by one orchestra it should not be included in the program of another; sung once during an evening, it should not be sung again. From the broadcaster's viewpoint, as well as for the welfare of the composition and its owner, the public appetite should not be surfeited.

" It looks very much as though it were going to be necessary for program directors to exercise the same sort of jurisdiction over their programs as managers of vaudeville theatres do. In the theatres it is the custom to prohibit more than one rendition of the same composition in any program. The rule is that whatever act first rehearses a certain song at the beginning of the engagement shall have the right to use that composition during the appearance at that theatre.

' If it happens that other acts have the same song in their routine, they are required to substitute something else. If this were not done in vaudeville, patrons would in many cases hear the same song two or three times during the course of a show."

GOOD STUFF IN CHICAGO

IF YOU are anywhere near Chicago the best thing available during the daytime in that neck of the woods is the noon musical period furnished by WGN —almost a full two hours of instrumental music of a caliber quite as good as most of the expensive evening programs. From 12:40 to 1:00 P. M. a luncheon concert by the Drake Concert Ensemble and the Blackstone String Quintet; 1 to 2 P.M. Lyon and Healy artist recital; 2 to 2:30 a luncheon concert. And frequently the succeeding half hour, 2:30 to 3 which is

THE "COOKIES"

Lucile and Nell Cook, who have been heard in harmony over WEAF, New York, and WEBH, WJJD, Chicago

MARIE HILL OF KTCL, SEATTLE

A Cadillac roadster, equipped with a portable broadcasting set is used by Miss Marie Hill, studio hostess at KTCL in conducting what is known as ' Juniata's Shopping Tour," a feature of the daily woman's hour program from this station.

AN ARTISTS' BUREAU AT WLW

WLW, following a practice already in efficient operation at several broadcasting stations, has organized an "Artists' Bureau," an agency for managing the outside bookings of artists appearing regularly on its programs. In the photograph are Emil Heerman, concert master of the Cincinnati Symphony Orchestra, Lydia Dozier, soprano, and Powel Crosley.

turned over to staff soloists, is equally pleasant.

Another Chicago daytime feature that comes as a welcome relief from the unending recipe and household hints lectures is the Overture Hour, WMAQ, 10 to 11 A. M. A varied program, made up according to the expressed wishes of the listeners, is played by the Whitney Trio. We have taken occasion to praise this trio before, while their playing can not yet be called inspired or resplendant with subtle nuances they are at least thoroughly routined and play with a coördination surprisingly good for such a young organization.

THUMB NAIL REVIEWS

WHO—The "Automatic Adjutators," on an advertiser's program, doing some pretty fair harmony singing with fine pianissimo effects. The tenor soloist did as good a job by "Three For Jack" as we have yet heard by radio. It is a surprising fact that songs of the ilk of "Three For Jack" are so seldom heard in radio programs; songs, if you are not familiar with this particular piece, which rely for their effect largely upon the words. Perhaps it is because so few soloists can sing them effectively. They require an intelligence astute enough to fully grasp the meaning of, and properly interpret, the verse without at the same time so neglecting the music as to relapse into mere recitative. Of course "On the Road to Mandalay" and its sister song are given plenty of airings, but there is a host of other "speaking-songs" that are never heard. They are generally light in mood, sometimes quite funny, and mostly designed for male voices. We offer, gratis, the suggestion that a complete program be arranged of these songs and turned over to a capable singer. We venture to guess that it would be popular.

WLBW—Somebody playing classics on the piano in acceptable style. A premature fading out prevents further record.

WJAY—One of those so called "nut" hours. A reading of original poems by the announcers involved, which was not so hot. Some piano work which coaxed an occasional grin. Professor Bumguesser, burlesquing the radio seers, of which we had an epidemic recently, undertook to give answers to the various problems perplexing his listeners, along the lines of: Q. ". . . should I have the family wash in the back yard? Signed, "Puzzled Housewife." A. "No, the neighbors might fall out the windows." and Q. "Should I allow my boss to help me get the mail out or should I stick the stamps on myself?" A. "They will be much more effective on the letters."

WGBS—Morry Leaf, the " Eskimo Ukist" and "Bad-Time Story-Man" burlesquing the gwan-to-bed programs and singing some swell original ditties.

WHT—"Al and Pat" on the "Your Hour" (Announcer Pat Barnes and Organ-

WOR—A good instrumental ensemble and an unusual one: a trio made up of harp, violin and flute. Called the French Trio and under direction of Mme. Savitskaya.

COMMUNICATIONS

WE QUOTE a couple of paragraphs from a long and interesting letter from G. W. Ferens of New Zealand.

The class of programme transmitted is, in a small way, similar to those broadcast by your well-known stations. You are probably aware that the receiving conditions in New Zealand are considered by experts to be the finest in the world due to our geographical position. For instance, a receiver that is not capable of consistent reception of Australian broadcasts every night of the week over a distance of 1400 to 2000 miles is no good whatsoever in this country. A powerful receiver has no trouble at this time of the year in bringing in Pacific coast Stations and even Chicago. In fact, I have many a time listened to KFON at Long Beach on the loud speaker for two hours and heard it as well as a listener in the Middle West would receive that station. In passing, I would like to mention that KFON is the best American broadcasting station heard in this part of the country.

I trust that you will now understand how acceptable the information in your columns is to readers here, when we can, in many cases actually hear the stations and artists of which we read about. In particular the photographs of the artists and the general description of programs are very interesting and it gives one quite a thrill to hear an artist or an orchestra who, through your columns, one considers an old acquaintance.

ASHTABULA, OHIO.

SIR:

Re: Sunday Broadcasting, July issue, you have ably covered a subject that has been too long and too much ignored by anyone actively interested in radio programs . . . Church services have a legitimate place in broadcasting but they are not enjoyable to the exclusion of everything

D. R. P. COATS

Former program director and announcer at CKY, has taken charge of a new station erected by a manufacturer at Moose Jaw, Saskatchewan, CJRM

else, and neither are the heavy, slow musical features and hymns.

A. K.

WASHINGTON, D. C.

SIR:

On the Eveready hour the substitute manager one evening put on the 1812 Overture and then: "After the tumultuous emotions of this great work we will soothe ourselves back to normal by a soprano and tenor duet" and put on the grand seduction scene from Samson et Delilah. What dye mean "normal"?

E. S. S.

LOS ANGELES, CALIFORNIA.

SIR:

The announcer at KFON has just informed his listeners that the orchestra would next play, "Folies Bergères," but, now for the big laugh, he announced it as "Foley's Brassiers," in the approved and generally accepted American pronunciation.

I. W. H.

AT WSM, NASHVILLE

The Andrew Jackson Hotel Trio, Vivian Olson, cellist; Marguerite Shannon, pianist, and Clair Harper, violinist, broadcasting daily through WSM

AT 3 LO, MELBOURNE, AUSTRALIA

This young woman is not giving a lecture on tooth paste, but has just concluded a recital of "Songs at the Piano." She is Lee White and is, we are told, very popular with Australian audiences

ist Al Carney—we had their names backwards last time we mentioned them—beg pardon!) Between the two of them they put on the complete graduating exercises of a "deestrict school," the unctuous trustee, the quavering voiced teacher, and the smart pupils. We recommend them to you. Humorists of the old, ever-reliable, school.

KFAB—A station operated by an automobile dealer. We sat through a long and tedious direct advertising talk just to see how far they would go with it. They went plenty far enough, a full fifteen minutes was devoted to a long winded discussion on why the car they sold was equipped with cast iron pistons. The main issue seemed to be whether or not wear and tear on the piston was more disadvantageous than wear and tear on the cylinder. Long arguments pro and con were duly considered and the proper and predetermined conclusions reached. Aside from the effrontery in offering such a bald piece of advertising at all, the discussion was of such a technical nature that it could not conceivably have interested anybody less than an automotive engineer. What average motor car buyer gives a whoop, we ask you, whether his car has cast iron pistons or not? And how many other listeners besides ourself—who is paid to do so—lasted out the boring dissertation?

WCFL—"The Two Peppers" delivering some not bad close harmony and some very high-powered self accompanying, and solo work, on the guitar and banjo.

WEBH—The Indiana Male Quartet singing a good novelty song having to do with Whispering Sweet Whispers.

WJZ—A first rate concert now being broadcast every Sunday evening at 9:30 Eastern Daylight Saving Time. A studio orchestra consisting of eighteen instrumental soloists in woodwinds, strings and brass, under the direction of Hugo Mariani. The first program, typical of the ones following, was made up of the lighter and most popular compositions of Goldmark, Strauss, Massenet, Wagner, Schubert, Chaminade, Tschaikowsky, Herbert, and Saint-Saens.

WMCA—Art Gillham, singer of sentimental ballads, par excellence, accompanying himself on the piano.

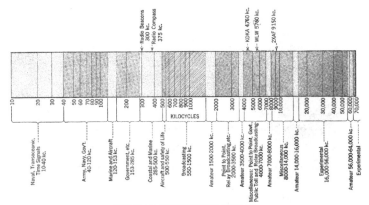

FROM TEN TO SEVENTY-THOUSAND KILOCYCLES

This diagram shows very clearly how the frequency spectrum between these two widely separated points is divided, and a very good idea of the comparative width of the band set aside for broadcasting may be had

Why Not Try the Short Waves?

In Which the Appetite of the Broadcast Listener Is Whetted—A Receiver for Exploring the Nether Regions of the Wavelength Spectrum, Where a Real Taste of Internationalism Is Possible

By KEITH HENNEY

Director of the Laboratory

OUT of a total number of frequencies available for radio communication amounting to some 60 million, the broadcasting band occupies only one million—a very small band indeed. What goes on in the other bands has been, until recently, a closed book to the vast majority of listeners because of the language used there—the International Morse code. Now, however, there are at least three American broadcasting stations using voice and music to modulate waves in these vast open spaces outside of our familiar broadcasting band, so that adventuring into the other 59 million frequencies is not too devoid of interest.

For many months the General Electric Company has put programs into the ether experimentally on a wavelength of 32.79 (9150 kc.) meters, and the 63-meter (4760 kc.) signals of the Westinghouse station KDKA are already too well known to need introduction. Recently WLW at Cincinnati has been using a band at 52 meters (5750 kc.) for rebroadcasting their regular programs and, according to reports, a 30-meter (9990-kc.) station operated in Eindthoven, Holland, can be heard during the early evening hours in the United States. Also, as this is being written (July) a station near Quebec is testing on approximately 32 meters (9380 kc.) and signing cs.

Here are five landmarks for which anyone equipped with a short-wave set can go seeking. Fortunately, it is possible to build a receiver for the high frequencies in a very simple manner

indeed, and with removable coils one can cover the entire range from below about ten meters (30,000 kc.) to something beyond the broadcasting band, thereby opening up a vast expanse of receiving territory unfamiliar to the average listener, and filled with interesting goings-on.

The usual short-wave receiver uses only two tubes since headphone reception is the rule. The first tube is a detector, which may be made to oscillate, or not, as the operator desires, and the second tube is used in a transformer-coupled amplifier. If loud speaker signals are desired it is only necessary to provide an additional stage, preferably equipped with a power tube.

The illustration forming the heading of this article shows the part of the spectrum that broadcasting utilizes and also gives an idea of the many services which take place in the other parts of this spectrum. Above the broadcasting frequencies come the many-tongued bands filled with amateur transmitters. In the United States, for example, there are several sections in which amateurs may operate, that immediately above our broadcasting section (or below if we think in meters), then another at 3750 kilocycles, or 80 meters, one at 7500 kilocycles or 40 meters, and, still others at 15,000 and 60,000 kilocycles, or 20 and 5 meters respectively.

Communications take place on these higher frequency bands that would be considered miraculous on lower frequencies. For example, on February 20th, 1927, the following stations were heard at 2 EJ, one of the stations operated by RADIO BROADCAST Laboratory, on or below the so-called 40-meter (7500-kc.) band:

EST	STATION	LOCATION
3:35 P. M.	NL 4 X	West Indies
	G 5 XY	England
3:50 "	FO A 3 B	South Africa
3:55 "	G 5 XY	England
4:00 "	EK 4 UAH	Germany
4:07 "	OA 5 WH	Australia
4:20 "	F 8 IB	France
6:35 "	SR I AR	Brazil
6:35 "	NQ 5 AZ	Cuba

Thus, in a space of one hour and a half, it was

Coil No.	Wave-length Range Meters	Kilocycle Range	Secondary, L_2 Total Turns	Secondary, L_2 Length of Winding Inches	Secondary, L_2 Size Wire	Secondary, L_2 Insulation	Tickler, L_3 Total Turns	Tickler, L_3 Size Wire	Tickler, L_3 Insulation
1	15-28	19,990-10,710	3	5/32	18	Enamel	2*	26	d.s.c.
2	30-52	9994-5766	8	17/32	18	Enamel	4*	26	d.s.c.
3	57½-111	5260-2701	19	9/16	18	Enamel	6*	26	d.s.c.
4	119-228	2399-1199	40	1½	22	d.s.c.	15*	26	d.s.c.
5	235-550	1276-545	81	1½*	26	d.s.c.	21*	26	d.s.c.

*Close wound.

Note: Tuning condenser is 100 mmfd. Primary, L_1, consists of 10 turns, wound in a space of 9-32nds inch, of No. 22 cotton-covered wire, and is 2¾ inches in diameter. Secondary is 3 inches in diameter.

possible to listen-in on the whole world. At another time BZ 1 IB, a Brazilian Amateur, was heard talking to GMD, the Dyott Expedition. It would be possible to go on indefinitely with stories of long distances covered and interesting experiences, but the preceding paragraphs should give the uninitiated some idea of what can be expected in the short-wave provinces.

Kits of short-wave coils, condensers, etc., are now available from several manufacturers, and with these, receivers may be constructed. All of these receivers will radiate and can disturb near-by listeners, but with loose coupling to the antenna and an additional audio stage if necessary to make up for the decrease in signal strength, this radiation is reduced to a minimum.

In actual practice, the detector is on the point of oscillating when voice or music is being received, and when code signals are copied the detector is actually oscillating, but is near the point where oscillations are liable to cease. In these conditions the detector is most sensitive.

Attention is called to the two articles by Frank C. Jones, 6 AJF, on non-radiating short-wave receivers. Mr. Jones has made a substantial contribution toward the ideal receiver and the set described by him in RADIO BROADCAST in May, 1927, approaches very closely to a truly non-radiating receiver.

Each of the coils designed for the short-wave kits differs from the others, Silver-Marshall using a form similar to broadcast coils, Aero Products using spaced windings on slotted forms, REL supplying the familiar basketweave style, and those of Gross or Hammarlund consisting of spaced windings on a celluloid form.

WINDING THE COILS

FOR the experimenter who likes to make his own, the following description of coils similiar to the Aero inductances will be helpful. The number of turns for the various frequency bands to be covered is given in the table on this page.

An insulating tubing three inches in diameter can be used without introducing much loss. The tickler coil is, in each case, 2¾ inches in diameter and is made self-supporting by using collodion or some other binder. This coil is placed inside the secondary at a position near the grounded end of the coil and need not have a great deal of mechanical strength as it is protected by the secondary. The same primary is used for every coil and should be fastened on a hinge which makes it possible to vary the coupling as desired. This coil is made self supporting, is wound without spacing, and a binder is used to keep it in shape. It is wound with ten turns of No. 22 double silk-covered wire. The primary may be mounted on the sub-panel of the receiver. It will be found, in operation, that the setting of the coupling need not be changed very often and once adjusted properly may be left in a fixed position. Secondary coils Nos. 1, 2, and 3 are wound with No. 18 enameled wire spaced approximately the diameter of the wire. Secondary coil No. 4 is space wound with No. 22 double silk-covered wire, the dimensions of the coil being 1½ inches by three inches. Coil No. 5, which covers the broadcast band, is wound with No. 26 double silk-covered wire and is not spaced. The coils are mounted on flat strips of bakelite, and four connections are brought out from each coil in the form of pins, mounted on the bakelite strips. The pins are designed to fit into corresponding slots in the sub-panel mounting. Three pins are spaced at one end of the bakelite strip with a fourth apart from the others, making it impossible to plug the coils in the wrong way. Almost any type of plug system having four contacts may be used. Suggestions have been made from time to time for using an old base of a vacuum tube and a tube socket for this purpose. If such a device is used care should be exercised to see that too much capacity is not introduced, otherwise the wavelength ranges will be considerably affected.

This covers the mechanical details of the coils. It is not necessary to go into details of the wiring of the receiver shown in Fig. 1 as most of our readers are familiar with this part of set building. It is important, however, especially in short-wave work, to make the grid lead very short and to keep all wires rigidly fixed in their respective positions. The actual order of wiring is not mandatory but it is suggested that the filament wires be completed first.

The circuit used for the several kits on the market is fundamentally the same. That used in the Aero kit is shown in Fig. 1. The Radio Engineering Laboratories kit consists of a complete set of parts, including panel, wire, transformer, coils, etc., and lists at $36. Silver-

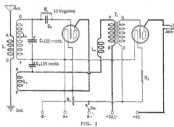

How the Higher Frequencies Are Allocated

AN IDEA of what goes on in the higher frequency bands may be gleaned from the following table, which shows how the various frequencies were divided up by the Fourth National Radio Conference:

Kilocycles	Meters	Service
500-550	600-545	Aircraft and fixed safety of life stations.
550-1500	545-200	Broadcasting only.
1500-2000	200-150	Amateur only.
2000-2250	150-133	Point to point.
2250-2300	133-130	Aircraft only.
2300-2750	130-109	Mobile and Government mobile only.
2750-2850	109-105	Relay Broadcasting only.
2850-3500	105-85.7	Public toll service, Government mobile, and point-to-point communication by electric power supply utilities, and point-to-point and multiple-address message service by press organizations, only.
3500-4000	85.7-75.0	Amateur, Army mobile, naval aircraft, and naval vessels working aircraft, only.
4000-4525	75.0-66.3	Public toll service, mobile, Government point to point, and point-to-point public utilities.
4525-5000	66.3-60.0	Relay broadcasting only.
5000-5500	60.0-54.5	Public toll service only.
5500-5700	54.5-52.6	Relay broadcasting only.
5700-7000	52.6-42.8	Point to point only.
7000-8000	42.8-37.5	Amateur and Army mobile only.
8000-9050	37.5-33.1	Public toll service, mobile, Government point to point, and point-to-point public utilities.
9050-10,000	33.1-30.0	Relay broadcasting only.
10,000-11,000	30.0-27.3	Public toll service only.
11,000-11,400	27.3-26.3	Relay broadcasting only.
11,400-14,000	26.3-21.4	Public service, mobile, and Government point to point.
14,000-16,000	21.4-18.7	Amateur only.
16,000-18,100	18.7-16.6	Public toll service, mobile and Government point to point.
18,100-56,000	16.6-5.35	Experimental.
56,000-64,000	5.35-4.69	Amateur.
64,000-400,000	4.69-0.7496	Experimental.
400,000-401,000	0.7496-0.7477	Amateur.

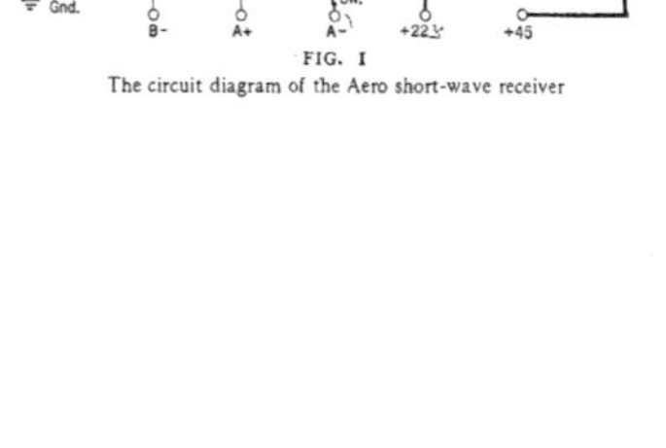

FIG. I

The circuit diagram of the Aero short-wave receiver

RADIO BROADCAST Photograph

THE RECEIVER DESCRIBED IN THIS ARTICLE
The parts list and panel layout for this particular receiver are given
on this page while the circuit diagram appears as Fig. 1, on page 291

RADIO BROADCAST Photograph

MADE IN THE LABORATORY
The short-wave receiver here shown employs REL coils, National condensers, X-L binding posts, etc. The circuit is basically the same as that
shown in Fig. 1

Marshall's kit includes coils, coil socket, and three condensers and lists at $23.00. Other kits of coils are sold by the Aero Products Company as indicated on the list of parts in this article, and that sold by J. Gross which lists at $27.90. Aero also stocks a short-wave drilled and engraved foundation unit listing at $5.75.

THE ANTENNA

THE antenna used with these receivers may be of practically any type, long or short. A long outside antenna will naturally give the better DX results, though the coupling between the primary and secondary coil will have to be reduced or perhaps a series condenser inserted in the lead-in, if it is found that the set refuses to oscillate over a certain band. This is probably due to absorption by the antenna. A little experimenting with the coupling or perhaps changing the antenna length or position will usually remedy this.

Another point which is important in making a set of this character oscillate evenly over the entire band is the grid leak. For best results a high-resistance leak should be used, say of 6-10 megohms. The detector tube should go in and out of oscillation easily when the regeneration condenser is varied. As seen from the circuit diagram the set may be grounded on the negative return (dotted lines) or not, according to which connection gives the better results. A little time spent in becoming familiar with the adjustments of the set will tend to insure better results and at the same time preclude unnecessary disappointments.

The variable condenser used for tuning the set illustrated in Fig. 1 is one of 100 micro-microfarad (0.0001-mfd.) capacity. While this condenser does not quite cover the entire bands it does

spread out the available bands and makes tuning much easier. If it is desired to cover every wavelength over the full range a small shunt condenser may be placed across the main condenser and be operated with a switch. A straight frequency-line condenser is used for tuning. The regeneration control condenser may be of any type as this control is not at all critical.

A list of parts for this Aero receiver, shown

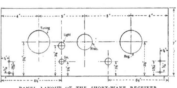

PANEL LAYOUT OF THE SHORT-WAVE RECEIVER

diagramatically in Fig. 1, made in the Laboratory, and used at 2 GY, is given below, and panel and sub-panel layouts are shown in Figs. 2 and 3.

LIST OF PARTS

L₁, L₂, L₃—Short-Wave Kit Consisting of Coils Nos. 1, 2, and 3 with Primary and Plug-In Mounting, Aero Products Company (15-111m.)	$12.50
L₄—Coils Nos. 4 and 5, Aero Products Company (110-550 m.)	8.00
L₅—Silver-Marshall Short-Wave Choke	.60
C₁—Hammarlund Variable Condenser, 100 Mmfd.	4.25
C₂—Precise Variable Condenser, 135 Mmfd.	2.00
C₃—Sangamo 0.0001-Mfd. Grid Condenser with Mounting	.40
R₁—Grid Leak, 10 Megohms	.50
R₂—Frost 20-Ohm Rheostat	.50
R₃—Amperite and Mounting, 1-A	1.10
T₁—Samson 6:1 Audio Transformer	5.00
Two Benjamin Spring Sockets	1.50
Yaxley Single-Circuit Jack	.50
Yaxley Combination Switch and Pilot Light	.75
Two Benjamin Brackets	.70
Five Eby Binding Posts	.75
Two Karas Vernier Dials	7.00
Radion Hard Rubber Panel 7 x 18 x 7/32 Inches	2.50
Radion Hard Rubber Sub-Panel 5 x 16½ x 7/32 Inches	1.65
Wire, Screws, Etc.	.50
TOTAL	$50.70

ACCESSORIES NEEDED

2 CX-301-A Type Tubes
1 45-Volt B Battery
Pair of Phones
Antenna and Ground Connections

RADIO BROADCAST Photograph

ANOTHER SHORT-WAVE RECEIVER
Silver-Marshall parts form the basis
of this efficient and compact receiver

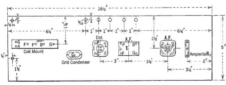

A SUB-PANEL LAYOUT OF THE RECEIVER

Drawings by Franklyn F. Stratford

Design And Operation Of Broadcasting Stations

17. Attenuation Pads

THE instrument known as a "pad" among broadcasters is usually a simple resistive network used to cut down signal level by a definite amount. It may be adjustable, but more frequently various sizes are supplied with the terminals led to jack strips so that a loss of, say, ten, twenty, or forty TU may be introduced at will. A general and sufficiently exhaustive treatment of passive networks in general may be found in Chapters VIII and XI of *Transmission Circuits for Telephonic Communication*, by K. S. Johnson. (D. Van Nostrand Co.) The most thorough work on the subject of artificial lines is no doubt *Artificial Electric Lines*, by A. E. Kennelly, Sc. D. (McGraw-Hill Book Company). Being by Doctor Kennelly, this book leaves little to be said on the subject; the reader must be warned, at the same time, that the mode of approach is through hyperbolic geometry and the calculus, quite in the Heaviside and Steinmetz tradition. Here we shall present only a restricted, elementary, practical treatment of one aspect of the subject, for the benefit of those of the boys who have not gone to M. I. T., but manage to broadcast anyway.

Fig. 1 shows a typical pad of the "H" form. With the resistances shown this network will introduce a loss of 20 transmission units, while presenting an impedance of 500 ohms in either direction. It might be used in the output of an amplifier terminated in a transformer with a 500-ohm secondary, between this secondary and a line, in order to keep the current in the line within allowable cross-talk limits, or for a similar purpose. Fig. 2 illustrates a line which will lose something a little less than 40 TU, but at

"THE READER MUST BE WARNED".

a lower impedance—about 246 ohms, forwards and backwards. It will be noticed that the sum of the two side resistances and the shunt member gives the approximate impedance presented by the network, when the resistance of the shunt member is low compared to the rest of the artificial line on the other side and the apparatus tied thereto. This qualification must not be neglected, for, as in the 10-TU pad shown in Fig. 3, the impedance may be 500 ohms in each direction, while the sum of the three members of the network proper is 125 plus 125 plus 375, equalling 625. However, the 375-ohm shunt member is paralleled by 750 ohms (125 plus 125 plus the 500 ohms of the actual line) which brings the resultant down to 250 ohms, and 125 plus 125 plus 250 equals 500.

In Figs. 1, 2, and 3 we have shown H-shaped artificial lines of various sizes, and with the same impedance looking in either direction. This type of line exhibits complete symmetry, the resistance being the same on each side of the pair, and the impedance the same on each side of the cross-member. Except for the consideration of bilateral symmetry, any such H-network may be replaced by a T-network with three members instead of five. For example, take the 10-TU H circuit of Fig. 3. This may be replaced, if perfect symmetry is not required, by the T circuit of Fig. 4. The only change is in the shifting of one 125-ohm resistance to the same side as the other, leaving no series resistance on one side and collecting 250 ohms on the other. The ohmic relations are unaffected. A different effect is secured when the network is unsymmetrical about the cross member. Then the impedance looking backwards and forwards is changed. To illustrate: take the 40-mile pad of Fig. 2, presenting some 240 ohms at both terminals. Assume that it has become necessary to match a 500-ohm circuit on the right. The network might then be modified as shown in Fig. 5. Toward the left the impedance remains sensibly unchanged, but toward the right the impedance is now approx-

Fig. 1

Fig. 2

Fig. 3

imately 500 ohms. This network would match a 200-ohm output to a 500-ohm input. At the same time, of course, the attenuation of the pad has been changed. The magnitude of the change may readily be calculated, but instead of taking this network as an example we shall outline in a more general way the simple theory involved. All that is required is an acquaintance with Ohm's Law.

Fig. 6 shows a network of the H-type, with series resistances x_1 and x_3 and shunt resistance y. The impedance on either side of this network is Z. In the calculation we shall also use the quantity r, which represents the resistance of the shunt combination formed by y paralleled by $x_3 + Z + x_3$. E_1 is the voltage across the input of the network, E_y the voltage across the shunt member y, E_2 the output voltage of the device. The current in the left-hand rectangle is i_1, that in the right-hand rectangle is i_2. With these

We may now apply the last two formulas to the artificial line of Fig. 3, which was stated to be of 10-TU size. Substituting $x = 125$, $y = 375$, and $Z = 500$ in (14), we find

$$r = \frac{(375)\,(\,250 + 500)}{250 + 500 + 375}$$
$$= \frac{(375)\,(750\,)}{1125} = 250$$

This value ($r = 250$) being substituted in (13) with the other numerical data, we have

$$L = 20 \log_{10} \frac{(500 + 250)\,(250 + 250)}{(500)\,(250)}$$
$$= 20 \log_{10} 3$$
$$= 20\,(0.477)$$
$$= 9.54\ TU$$

This answer is sufficiently close to the actual attenuation of the pad for all practical purposes, and the formulæ derived in the discussion above may be used in calculating the constants of

casting about for the proper etiquette under the circumstances, my two friends, having settled the inevitable dispute as to who should have the privilege of paying for the ride, joined me, and together we entered, through a second door, the main body of the church. It was quite light, in contrast to the gloom of many church auditoriums, and empty, save for us and one fair penitent who knelt before the altar at the far end.

I need not tell my readers that it is the inviolate custom of all good broadcasters in the technical end of the business, on entering a hall of any kind, to clap their hands, whistle, yell, sing, and stamp on the floor in a learned manner, in order to test the acoustics of the enclosed space. The eyes of Mr. Hanson and myself rolled upward toward the groined ceiling of the church, and both of us felt the impulse to clap our hands, at least, but we stood as still as the statues of the saints confronting us on every

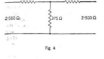

Fig. 4

Fig. 5

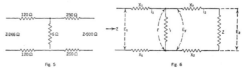

Fig. 6

symbols we may carry out a sufficiently accurate calculation.

First we may write

$$E_1 - 2\,x_1\,i_1 = E_y \qquad (1)$$
And
$$E_y - 2\,x_3\,i_2 = E_2 \qquad (2)$$
Whence
$$E_1 - 2\,x_1\,i_1 - 2\,x_3\,i_2 = E_2 \qquad (3)$$
Or
$$E_1 - 2\,x_1\,i_1 = E_2 + 2\,x_3 i_2 \qquad (4)$$
Also
$$i_1 = \frac{E_1}{2\,x_1 + r} \qquad (5)$$
And
$$i_2 = \frac{E_2}{Z} \qquad (6)$$

Substituting (5) and (6) in (4), we obtain

$$E_1 - \frac{2\,x_1\,E_1}{2\,x_1 + r} = E_2 + \frac{2\,x_3\,E_2}{Z} \qquad (7)$$
Or $E_1 \left(1 - \frac{2\,x_1}{2\,x_1 + r}\right) = E_2 \left(1 + \frac{2\,x_3}{Z}\right) \qquad (8)$
Simplifying
$$\frac{E_1}{E_2} = \frac{(Z + 2x_3)\,(r + 2\,x_1)}{Z\,r} \qquad (9)$$

The value of r is found readily on the principle that the resultant conductivity (the reciprocal of resistance) of paths in parallel equals the sum of the individual conductivities:

$$\frac{1}{r} = \frac{1}{y} + \frac{1}{2\,x_3 + Z} \qquad (10)$$

Clearing of fractions

$$y\,(2\,x_3 + Z) = r\,(2\,x_3 + Z) + r\,(y)$$
Whence $r = \frac{y\,(2\,x_3 + Z)}{2\,x_3 + Z + y} \qquad (11)$

By definition, the TU loss of the network is given by

$$L = 20 \log_{10} \frac{E_1}{E_2}$$

(See this department in RADIO BROADCAST for September and October, 1926, for a discussion of the meaning of the telephonic "Transmission Unit.")

Hence, from (9) above

$$L = 20 \log_{10} \frac{(Z + 2\,x_3)\,(r + 2\,x_1)}{Z\,r} \qquad (12)$$

From (11) and (12) we may calculate the loss introduced by any H-network where the impedance is unchanged from input to output. In the usual practical case, when in addition $x_1 = x_3 = x$, (12) becomes

$$L = 20 \log_{10} \frac{(Z + 2\,x)\,(r + 2\,x)}{Z\,r} \qquad (13)$$
And $r = \frac{y\,(2x + Z)}{2\,x + Z + y} \qquad (14)$

other networks of the same type. The simplest procedure is usually to draw up a network for the required purpose on the basis of previous experience, make a calculation of the TU drop corresponding thereto, modify the constants as indicated by the result of the calculation, and by successive changes and calculations to approach the desired value. With a little experience a very few operations will be found sufficient.

Broadcasters In Church

AWHILE ago my distinguished colleague, O. B. Hanson, who presides over the engineering destinies of the National Broadcasting Company, a prominent architect of New York, and I, were engaged in an argument concerning the merits and demerits of various commercial sound-absorbing materials. Harassed by the fireproofing zealots of the Building Department, who care little about the yearning of a broadcaster for acoustically flat studios, we sat sweating amid sample slabs of expensive special plasters and nondescript compounds, which we glared at through magnifying glasses, scratched with pencils, and chewed with our teeth. Finally, having exhausted our learning without reaching a conclusion, we leaped into a taxicab and sped to a point on First Avenue where, we had been informed, there stood a church whose domed ceiling had been treated with one of the materials in which we were interested, so that we might see and hear the virtues of this substance for ourselves.

Leaving my companions to pay the taxi driver, I approached the door of the edifice, which, I noted, was of the Catholic denomination. The door was open, and, removing my hat, I stepped into a sort of lobby, in which votive lights were burning. It was only at this point that I realized the incongruity of our errand with the primary purpose of the building. We had come neither for prayer, meditation, nor the confession of our sins, but to test the acoustics of an auditorium, a deed in itself neither good nor bad, but certainly not of a religious character. While I was

side. We twitched and opened our mouths, but no sounds came forth. Such an inhibition is astounding in men who have been broadcasting as long as we have. The architect seemed to suffer less. Architects are concerned with the building of houses, rather than with the noises generated in them after completion.

"Now that we are here," I whispered, hunching my shoulders fearfully at the lady kneeling with her back to us, "how shall we proceed? What noise can we make without damning our selves, if it has not been done already? What shall we do?"

"Don't you know any monsignors or high church dignitaries who might furnish us with a dispensation to make a racket in this church?" mumbled Mr. Hanson, wiping the perspiration from his brow with his hat.

"I do not move in those circles," I answered regretfully. "Once, at a reception on Pearl Street, I was introduced to a rabbi, but I left him to talk to a pretty girl. How about you?"

"I am an Episcopalian," confessed Mr. Hanson, in a tragic whisper.

"Cough!" the architectural gentlemen suddenly ejaculated, voicelessly. I flashed him a grateful look, and emitted several genteel but loud reverberations from the region of my diaphragm. Emboldened by the echoes, Mr. Hanson snapped his fingers under his coat, whirled, and had been informed, there stood a whoop. After about half a minute the lady at the altar crossed herself and rose, perhaps to send for an ambulance. If so, she has certainly increased her chances of ascending to heaven, because she neglected her own salvation to aid a fellow being. But it was equally possible that she was going to call the sacristan, so the three of us turned and fled. We slowed down on the next block and my colleague Hanson addressed me.

"I have spent a bad ten minutes," he said. "I'm never going into a church with you again. Knowing your record, I was afraid you would start to yell and howl as you do up at Carnegie Hall and rise to impress the ushers, with the result that we would have been thrown out on our ears into First Avenue and clubbed by all the cops in the parish. Our social reputation would have been ruined if we had had to go to jail with you. Finally, I must say that you cough in a disgusting manner."

But I was unable to answer. I was still coughing.

That SOS Question Once More

THE behavior, actual and ideal, of broadcasters when an sos call is on the air, constitutes a question of perennial interest which has been discussed a number of times in this department, and will no doubt bob up again at intervals. Mr. Ray Newby of San Jose, California, wrote us about it some time ago, and his view is worth noting. Mr. Newby says:

I wish to take exception to the suggested complete shutdown of broadcasting stations in the vicinity of distressed vessels, and to take sides with the critic mentioned, who was in favor of a station coming on the air at intervals during sos shutdowns with its call letters and the reason for its silence.

The article in contradiction to this practice was well taken, but there are two sides to all questions. As an ex-ship operator, broadcast, etc., allow me to present mine.

With present-day commercial receivers there is no possibility of a broadcast station interfering with ship traffic carried on at 600 meters and above, and as to the 300-meter wave for ships, I think this is automatically eliminated.

Now for the main reason: Did it ever occur to you that the ship operators also listen to broadcast programs? Well, that is the case, even though the Chief Operator says that one ear must be on the stand-by wave (600 meters) at all times while the vessel is at sea. Now, if an operator's favorite broadcast station reminded him at intervals that there was an sos on the air, no doubt he would get both ears working where they were supposed to be, and possibly be of some assistance to the sos'er.

I claim from experience that this is entirely possible, and it is also probable that some one-operator oil boat, or perhaps a private yacht, might be within a stone's throw of the ship in trouble.

Mr. Newby takes a practical view of the matter, and what he says is pertinent enough. It is unfortunately true that some marine operators, standing their watches at sea, neglect their immeasurably important duty in order to listen to jazz on the higher kilocycle channels. I can understand why they do it, and I might be tempted myself if I were standing midnight-to-eight watches again—but that stunt is going to put one of the boys into jail yet. Mr. Newby's argument recognizes, and in some slight measure tolerates, this practice. I doubt whether it is wise to condone what amounts to criminal negligence, even to this degree. The great majority of ship radio operators are wholly reliable; they would as soon think of leaving their marine listening channels to hear some entertainment as they would push ahead of women and children into the life boats on a sinking vessel. The relatively few men who neglect their stand-by function on 600 meters are controlled to some extent, also, by the necessity of making a log entry every fifteen minutes.

The point regarding the one-operator boats and private yachts not required to stand a regular radio watch, seems to me well-taken. In the case of yachts carrying only a broadcast receiver it would seem, however, that little good would be accomplished even if it became known, through a broadcast station's announcement, that an sos was abroad. In the absence of a telegraph operator the people on the yachts would have no idea where the ship in distress was

to be found. The chance of succor from this quarter is hardly great enough to justify permitting powerful broadcast stations on the higher wavelengths to make lengthy announcements giving the location of the vessel in trouble, when these communications might interfere with the radio telegraph stations in direct charge of the situation. Furthermore, in actual practice stations below 360 meters, or well inland, generally do not shut down at all for sos calls, leaving only a fraction of the broadcast stations for possible utility in these situations.

My view has been right along that a careful study should be made of the possibility of interference by broadcast stations with distress signals, that those · entertainment transmitters which may conceivably interfere should shut down and not let out another note beyond a bare sos sign-off formula till the clear signal is given by the radio telegraph station in charge, and that where there is no possibility of interference, broadcasting should be permitted to continue without interruption.

With more general high power broadcasting, however, another consideration arises. A powerful radio telegraph coastal station may have about five kilowatts in the antenna. A few of the broadcasters, existing and projected, possess ten times that energy. When an sos is heard, these stentorian voices become silent. Yet it is conceivable that under conditions of heavy static and a distress call far out at sea those fifty kilowatts might come in handy. It might be practicable to arrange some of the very high powered broadcasting stations so that they could be controlled telegraphically, in case of need, from the nearest Naval District Communication Office, the wavelength being shifted, by an automatic wave-changer, to 600 meters. The main item of expense would be the telegraph line. If the handling of sos signals were facilitated in even a small percentage of cases, it would be money well spent.

Blame It on Radio

UNDER this caption we have repeatedly reprinted denunciations of radio broadcasting, which has been blamed for every thing but prohibition and the World War, and would be blamed for those little annoyances as well if they had not antedated it. One of the most persistent of these eruptions of scapegoat psychology is the notion that broadcasting, or wireless communication in general, is responsible for heavy rainfall and floods. Only recently a noted French statesman is reported to have issued a discourse on this subject, suggesting solemnly that the excessive amount of broadcasting in Europe caused the Seine to overflow its banks. A domestic specimen follows below:

SUGGESTS RADIO INTERFERENCE CAUSED SOUTHERN FLOODS

Special to the New York Times

WASHINGTON, May 7.—The Federal Radio Commission has received a letter from a man in Hurricane, W. Va., suggesting that radio interference may be responsible for the present floods. He wrote:

"In view of the excessive downpour of rain and the havoc wrought by floods and the inability of dirt farmers to plow or plant lately, might it not be possible that 'high-wattage' from the many broadcasting stations is so magnetizing the ether, similar to lightning descending to earth, producing magnetic disturbances and rain?

"There surely is some cause. If broadcasting were suspended for ten days or two weeks in America we might find the culprit or nolle the indictment."

Toward the close of the World War there was a summer during which a great deal of rain fell. There resulted great suffering among summer hotel proprietors and the managers of carnivals, fairs, medicine shows, and the like. The

organ of the latter, the *Billboard*, printed an advertisement, or remarked in one of its departments—I don't remember which—that the unusual amount of rainfall must be caused by the firing of the heavy guns in France. The bankrupt concessionaires were advised to go into the umbrella business. Exactly how the detonation of the guns, which could not even be heard a few hundred miles away—and it takes doggone few dynes per square centimeter to make an audible sound, and a dyne, withal, is a pitiably small force, such as might be exerted by a section of a human hair less than an inch long—just how that sort of disturbance was to cause rain to fall . . . but why labor the point? Schiller said all there was to be said when he declared, "Against stupidity the gods themselves fight in vain." And if there is any field in which superstition is rampant, it is the perennial topic of the weather, in its relation to the other phenomena of the universe. After all, radio only gets a little tar from that stick. Ignorant people have always feared the "powers of the air."

As for the suggestion of the gentleman from Hurricane, W. Va., I announce myself in favor of it, although not for the reason he mentions. I could use a two weeks' vacation very nicely almost any time.

Correction On The Gamma Rays

MR. S. WHITTEN of Berkeley, California, justly takes us to task in the following letter:

Please refer to the center column of Page 107 of your June, 1927 issue, in which the wavelength of the gamma rays of radium is given as about 200 Angstroms.

A few years ago some bright gentlemen got a lot of credit for measuring X-rays as long as 15 Angstroms, and Millikan measured ultra-violet rays as short as 400 Angstroms. So I'm afraid someone misinformed you and dropped you in a dark and lonesome part of the spectrum. Therapeutic X-rays are around one Angstrom and gamma rays are one-tenth to one-twentieth as long. (See Page 96 of the *Bell Technical Journal*, January, 1927, for a fine discussion).

I like your department and wouldn't be as rude as this, but we radio bugs are "powerful ignorant" about some things and I hate to see it spread.

I'm glad to see that Mr. Stratford uses log-log slide rules in his pictures. Few of us can use them any other way.

The article in the *Bell System Technical Journal* to which Mr. Whitten refers is by Dr. Karl K. Darrow, on "Contemporary Advances in Physics—XII. Radioactivity," and, curiously enough, it was lying on my desk when I perpetrated the lamentable bull now dragged into the light. Doctor Darrow remarks:

The gamma-rays are spread out into a spectrum, and sometimes lines are discernible in the spectrum; but the line of shortest wavelength thus far measured (so far as I know) is at 0.052 Angstrom units or 52 X-units, and there are certainly many others at much shorter wavelengths which the crystal spectroscope does not diffract far enough outward to be small.

As for Mr. Franklyn Stratford, whose name has adventitiously entered the discussion, let it be known that by vocation he is an engineer, engaged, furthermore, in the active practice of the telephone and telegraph arts. The sketches with which he has amused the readers and the author of "As the Broadcaster Sees It" for now, over two years, are merely one of his relaxations from TU calculations and struggles with quadruplex balances.

The Causes of Poor Tone Quality

How to Systematize the Search for and Remedy the Causes of Distortion in Radio Receivers

By EDGAR H. FELIX

THE radio listener may, with the aid of modern amplifiers and reproducers, secure tonal quality which is little short of perfect. Advances made this last twelvemonth have lifted radio reception from near-mediocrity to such a standard that it is possible to reproduce music difficult to distinguish from the original. The full force of this statement, however, cannot be realized unless one has opportunity, over a period of hours, to compare a last year's receiver with the best kind of modern equipment. For this purpose, switches must be installed so that one may change from an inferior amplifier system to a modern one without the loss of an instant. After such a test, there are few listeners who would be content with anything less than the best reception attainable.

In general, distortion arises either from improper operation and adjustment of the receiving set and its associated equipment, or from imperfect design of parts in the set itself. When improper operation or adjustment is the cause the trouble may lie almost anywhere in the radio receiver. Given a high-grade amplifier system and a loud speaker adequate to handle the output of the receiver, distortion may yet exist because of failure to work within the power or voltage limitations of one or more parts of the receiver. The most frequent cause of curable distortion is overloading of vacuum tubes.

Any judgment formed must be based only upon the reproduction of high-grade broadcasting stations. Not more than ten per cent. of the broadcasting stations on the air exercise sufficient care in the placing of their artists before the microphone, in the adjustment of the input amplifier, and in determining the percentage of modulation, to be classed as being fit to listen to by the owner of a really high-grade receiving set capable of true reproduction.

Overloading may arise in almost any part of the radio receiver. Briefly, it is manifested by blurred, ringing, or harsh effects, particularly noticeable on loud signals. A receiver so overloaded but otherwise satisfactory, gives high-grade reproduction when tuned to nearby stations which deliver a strong signal providing the volume is so adjusted that the music is only quietly and softly heard. Under those conditions, a good receiver gives adequate amplification of low, middle, and high registers, meeting all the requirements constituting good quality. Yet, when volume is increased, displeasing distortion may occur. Oftentimes, this kind of overloading may be corrected by adjustments.

DISTORTION ARISING IN THE R. F. AMPLIFIER

THE first point of distortion which we will consider is that occurring in the radio-frequency amplifier. Distortion in this part of the receiver is not always due to too powerful a signal but frequently to the influence of regeneration. It may be present, therefore, even though the signal itself is weak. Appreciable regeneration in the radio-frequency amplifier is fatal to good quality because it cuts off the high frequencies and causes the lower frequencies to be drummy and ringing.

No matter what the circuit used, good quality cannot be expected if the radio-frequency amplification is mainly regenerative. With the

most modern receivers, sensitiveness and volume may be controlled by a means of varying the gain in the radio-frequency amplifier. With such receivers, it is not difficult to determine if distortion is due to overworking of the radio-frequency amplifier.

Tune to the most powerful nearby station, and then weaken the signal to moderate volume, at the same time observe the quality carefully. Now tune-in a station at a moderate distance which is delivering considerably less energy than the first station. For instance, if the first station is a 500-watter, twenty-five miles distant, pick a second station of equal power forty of fifty miles away. Then, by adjustment of the radio-frequency amplifier control, bring the volume of the second station up to the same point of moderate volume as the first. Provided the quality, of

---

READERS of RADIO BROADCAST have made so many requests for specific information as to A and B power devices, tubes, single-control sets, and the various other subjects treated in this series of practical articles on how to select and judge the quality and performance of manufactured products, that we have arranged to make it more convenient to take advantage of the information which we can make available.

If you are interested in learning of manufactured products, embodying the features of high-grade tone quality, mentioned in this article, and the one which preceded it, address the Service Department, RADIO BROADCAST, Garden City, New York, and the necessary information will be sent to you. Preferably use one of the two forms below in wording your letter or postal:

(1.) I desire to convert my . . . . receiver into one giving the best tone quality. Please send me information as to suitable apparatus which will accomplish this purpose.

(2.) I desire information as to manufactured receiving sets which give the high-grade tone quality described in the series of articles on that subject in the July, August, and September issues of RADIO BROADCAST.—THE EDITOR.

---

transmission of both stations is equal, there will be a marked falling off of quality with the more distant station because of the regenerative influence. An additional stage of radio-frequency amplification, or limiting reception to only nearby stations, is the cure for this condition.

It must not be assumed that absence of a regeneration control implies that no regeneration occurs in the receiver. If stations, particularly distant ones, fall into tune with a hiss and give good quality only when tuned-in exactly, but distort badly if de-tuned slightly, the regenerative effect is likely to be present. The result is that the "highs" are lost and the "lows" are drummy and throaty.

One source of regeneration, both in the audio and radio amplifier, arises from inadequate bypassing and failure to keep the radio and audio signals out of the plate potential leads. This may be caused by magnetic coupling between transformers or by conductive coupling in the power supply leads. The latter difficulty is obviated by the use of choke coils in the plate supply leads and adequate bypass condensers to shunt the radio- and audio-frequency currents.

This precaution of adequate choking and by-passing is usually neglected in home-built receivers.

The high-grade manufactured receiver of 1927 employs three and four stages of fully shielded radio-frequency amplification. Its range is likely to be little more than that of the home constructor's receiver with one stage of radio frequency and a regenerative detector. It may deliver but little more signal to the detector than does the delicate receiver of the home constructor. The principal advantage of the four-stage r. f. receiver is not greater efficiency, but the fact that every element in it has a capacity far in excess of that which it is called upon to exert. Under all conditions it operates far below the overloading point. Similarly, a small four-cylinder automobile covers the ground just as rapidly and, carries just as many passengers as a powerful straight-eight. The latter's reserve power, however, gives a smooth, gliding power, impossible with the small, four-cylinder car. Reserve capacity, in the case of a radio receiver, means a realistic tone with all the necessary volume variations between soft and loud to portray the varying feeling of musical compositions.

OVERCOMING DETECTOR OVERLOADING

THERE is a different form of overloading possible with high-gain radio-frequency amplifiers when too powerful a signal is offered to the detector tube. For example, home-made super-heterodyne receivers, equipped with a volume control only in the audio system (particularly if fed by an outside antenna) may overload the detector because too powerful a voltage is delivered to the grid of this tube. When the radio-frequency system is equipped with a volume control, usually through a filament rheostat in the first stage, no trouble is encountered from this source because it is possible to keep the volume below the point where the detector or audio stages overload. In absence of this well nigh essential control, reduce the signal input by considerably shortening the antenna. If a long antenna is considered necessary for distance work, install both a long and a short one, employing the latter for enjoying nearby stations.

Two stages of radio-frequency amplification, so neutralized that there is no regenerative amplification, require a fair sized antenna to produce a good signal, even from nearby stations. Three tuned stages of radio-frequency amplification work satisfactorily with a small, indoor antenna, while four stages pick up sufficient signal to work from a loop.

The use of a C-battery detector in preference to a grid leak and condenser, somewhat increases the handling capacity of the detector tube and reduces the danger of overloading, due to excessive radio-frequency energy. The best results are attained if the audio amplifier system is sufficiently good to give normal loud speaker volume when the detector signal, as heard in the phones connected in the detector circuit, is so weak as to be practically inaudible. Under those conditions, reception from a good station is characterized not only by absence of noticeable distortion but by great variations in volume following those existing in the original music.

When, however, the signal heard in the detector circuit with a pair of phones is comfortably loud, it is quite likely that at least the second stage of audio-frequency will be overloaded, unless a tube capable of handling considerable power and with correct plate and grid voltages is used.

The recommended C-battery voltages must invariably be used if overloading is not to be experienced even in the first stage, a precaution frequently neglected.

The same principle applies to power tubes. Generally speaking, the higher the amplification factor, the smaller the variation in grid voltage which may be impressed upon it without overloading. For example, the 112 (cx-312) type gives higher amplification than the 171 (cx-371) both of which, incidentally, may employ the same plate voltage. If powerful signals from nearby stations overload a 112 (cx-312) tube, the substitution of a 171, using proper grid biasing voltage, will very probably overcome overloading.

The output stage, to give a comfortably loud signal in a good sized living room, must employ a 371, a 310 tube, or perhaps, two 112 tubes in parallel. The 120 tube is of sufficient power to give satisfactory music only in a small room, unless used in connection with an amplifier which neglects the lower tones. When such semi-power tubes, rather than power tubes, are used, it is simply a matter of operating the set at low volume to attain good quality.

It is not often realized that the improvement of the audio-frequency system of a receiver by the substitution, for example, of three stages of resistance coupling for two poor transformers, or by the use of the highest grade of transformers now available in preference to a pair of low or medium priced transformers, means that the power handling capacity of the audio system, tubes and reproducer, may be easily exceeded. If, by this process of renovation, the lowest frequency amplified is brought from 200 down to 125 cycles, the voltage swings in the audio system may increase fivefold as a result. The hum from a poor-filtered power supply is now easily audible. The output tube and the reproducer, which might have been satisfactory under the old conditions, may overload considerably when better transformers are installed. By substituting, for the 120, a 371 or 310 output tube, satisfactory reproduction of low tones is secured. It must be said here that no receiver employing a 201-A type tube as the second audio amplifier is capable of quality reproduction. It is bound to overload and distort.

Quality of reproduction requires that every link in the chain—radio-frequency amplifier, detector, audio-frequency transformers, tubes in the audio stages, and reproducer—be adequate to do its work. Improvement of any one part of the receiver may simply uncover other weaknesses.

REPRODUCER PROBLEMS

A POOR loud speaker usually fails to reproduce a wide scale of frequencies, missing either "lows" or "highs." When a 171 or a 210 (Cunningham 371 or 310) tube is used, an output transformer, or choke and condenser, should be employed so that the direct-current component of the plate current does not pass through the loud speaker winding. This is absolutely essential. The coupling device must be carefully engineered. Listeners still using the old diaphragm type of speaker with the short narrow necked horn need not expect good reproduction because such a horn cannot release the lower tones properly. Cones of small diameter give good reproduction of the high tones but are incapable of setting up the lows. Again we must realize that distortion in the radio-frequency amplifier, or

that overloading the detector, audio stages, or loud speaker, is fatal to true reproduction and is best cured by strengthening the weak link or maintaining the receiver at gentle volume.

One of the most difficult causes of distortion for the receiving set owner to analyze for himself is the kind which creeps upon him gradually. When the receiving set is first installed, it may give amazingly good reproduction but quality does not attract the attention of the listener. This often occurs because the voltage of the power supply falls or amplifier tubes lose their emission or A batteries run down. With socket B power supply, gradual deterioration of the rectifier tube may bring this trouble. B batteries used too long bring about the same result. Amplifier tubes themselves are often used unconsciously long after their emission has fallen below the point where they carry the low tones without distortion.

RESONANCE PEAKS AND THEIR CONTROL

DISTORTION is sometimes of the resonant variety, that is, confined to a particular frequency rather than spread over a wide range. This is particularly true of horn type loud speakers or poor transformer-coupled audio amplifiers. When the loud speaker is on top of the cabinet, it frequently causes the cabinet to be vibrated mechanically and that, in turn, causes the tube elements themselves to vibrate, producing distortion due to mechanical resonance of tubes, etc. If the loud speaker is moved twenty feet away from the set, the mechanical energy of the sound wave imparted to the tubes is reduced to a minute fraction.

Microphonic tubes usually have a characteristic ringing sound fairly familiar to the listener, and moving the loud speaker some distance from the set often prevents a continuous "sing" produced by mechanical resonance. Sometimes this cure is inconvenient to apply and less troublesome means may be successful. First try changing the tubes about in different sockets, giving particular attention to the detector tube.

Tap each tube lightly and see which one causes the most ringing noise. Frequently the microphonic tube may be identified in this way. Exchange that tube with each of the other stages until the trouble is minimized. With dry cell tubes, it often pays to wrap a cloth around the detector tube or, better yet, to use one of the weighted arrangements or vibration dampers made for the purpose.

A convenient expedient is to hang the reproducer from the moulding so that it is mechanically free and clear of wooden floors and furniture. Plaster walls do not transmit the frequencies to which cabinets resonate nearly as readily as do tables and floors.

Super-sensitive detector tubes often introduce the resonance type of distortion and this effect may frequently be minimized by burning the filament rather brilliantly.

Those who read the article in last month's Radio Broadcast on quality of reproduction should have no difficulty in detecting the absence of either "highs' or "lows." Whatever the kind of amplifier the home constructor likes, transformer, resistance, impedance, or combinations thereof, it is entirely possible to secure relatively perfect reproduction. A loss of a broad band of frequencies, if no regenerative or resonant effects are present, almost invariably points to an incapable reproducer or the use of coupling devices which give a filtering effect.

The experimenter who takes pride in the manipulation of a delicate, efficient, and sensitive receiving set is most likely to neglect tonal reproduction. Experiment is usually concerned with problems of efficiency—getting the greatest distance, the best possible selectivity, or the highest gain with the minimum number of tubes. The attainment of these objectives is in opposition to securing true tonal quality. More than half of those answering a recent questionnaire maintain two or more sets, one for experimental and long-distance purposes and a receiver for local high-grade reception which meets, or attempts to meet, the modern standards of good tone quality.

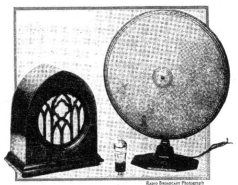

CONCOMITANTS OF GOOD QUALITY

The use of power tubes, together with good reproducers, is essential if good quality signals of adequate volume are desired. The photograph shows a baffleboard "Peerless" cone and a Western Electric cone, together with a 171 type power tube

New Receiver Offerings for the Fall

What the Radio Industry Has in Store for the Set Builder—Advance Information about the Hammarlund-Roberts, LC 28, Silver-Marshall Line, Infradyne, Aero Products Kit, Loftin-White, Strobodyne, and Others

By THE LABORATORY STAFF

DRUM tuning controls, complete shielding, and high-gain radio-frequency amplifiers will·be the predominant trend in kit receivers available to home constructors this fall. All of the new kits also provide for the use of power tubes in the output stage so that good quality can be obtained, and in many cases it is possible to operate the receiver either from batteries or from alternating current, using, in the latter case, a.c. tubes or special rectifier tubes with the filaments of the tubes in the receiver connected in series. Many modern kit receivers are the equal in every respect of much more expensive manufactured sets in the same class, and because of the many kits to be available this coming season, the home constructor will have a wide choice of different types and will doubtless find some one kit which meets his own needs particularly well.

One of the most popular kit receivers of 1927 was the Hammarlund-Roberts "Hi-Q," so it is interesting to note that a new Hi-Q receiver has been developed for the fall. This new "Hi-Q" receiver will contain three stages of tuned radio-frequency amplification, using automatic coupling variation between the primaries and secondaries of the radio-frequency transformers. All of the radio-frequency stages will be carefully neutralized by the Roberts method and the receiver will be completely shielded by the use of a Van Doorn metal sub-panel and heavy aluminum interstage shields. The set has been designed so that most of the wiring may be done beneath the sub-panel, and the final appearance of the receiver is thereby improved. The tuning of this new ' Hi-Q' receiver is accomplished by a unique drum control which has not the slightest amount of backlash and which is illuminated by means of a small flashlight bulb placed above the indicator. The drum is actually in two sections, each section controlling two of the variable condensers. Radio-frequency choke coils and bypass condensers are used in each radio-frequency stage and likewise in the plate circuit of the detector stage, so as to prevent coupling in the power supply. The audio amplifier will consist of a two stage transformer-coupled unit with an output filter in the plate circuit of the power tube.

The LC28, which will be featured by *Popular Radio* during the fall, will contain three completely shielded radio-frequency stages so designed that they cannot oscillate. The receiver will be tuned by two drum controls one controlling the antenna circuit and the other controlling a gang condenser which tunes the other two stages. The radio-frequency amplifier and detector have been designed as a single unit which will be described in *Popular Radio* in a preliminary article. Later articles will then tell how to use this high-gain radio-frequency amplifier with several different audio amplifying systems. The set will be engineered so that it can be operated from batteries or from a.c., using alternating current tubes in the latter case. An unusual feature about the receiver is that it is laid out in such a manner that very short leads are pos-

sible and the complete wiring can be done with slightly less than five feet of wire. Also. the receiver will be so designed ,that it will give satisfactory operation using no outside antenna but merely a very short indoor antenna, or the ground system may also be used as part of the antenna system.

The Silver-Marshall Company will place before home constructors three kits, including an improved design of the shielded six, which proved so popular last year. The receiver will be tuned by means of a new type drum control and will be designed to operate on either a.c. or d.c. Silver-Marshall will also market in kit form a four-tube receiver designed for a.c. or d.c. operation. The third Silver-Marshall unit presents essentials for a super-heterodyne, a description of which appears elsewhere in this issue of RADIO BROADCAST.

Many improvements are evident in the new 1928 model of the Infradyne receiver which was featured by *Radio* during the season of 1927. The new model will be tuned by two large drum dials and there will be two supplementary controls, one for adjusting the volume and the other for adjusting the sensitivity. A special filament switch is used and wired so that in one position all the filaments are turned off and when in the center position the filaments of the radio-frequency, audio-frequency and detector tubes are lighted while the tubes in the special Infradyne amplifier are automatically cut out. When the filament switch is in the third position all the tubes are lighted and the complete receiver is in operation. A single turn of this switch therefore makes available a five-tube single-control tuned radio-frequency receiver for those who desire extreme simplicity, or a ten-tube Infradyne offering high sensitivity and selectivity. The audio amplifier will be a two-stage transformer coupled affair with an output transformer. The receiver is so designed that very little wiring need be done above the sub-panel.

Aero Products, Incorporated, have done considerable work on a new seven-tube receiver to be put out in kit form and which will contain three stages of tuned radio-frequency and three stages of resistance-coupled audio-frequency amplification. This receiver will be somewhat unique in that no shielding will be used, the new Aero coils having been designed so that they are rather long in comparison with their diameter, this causing their magnetic fields to be quite closely confined; as a result, it is possible to make up a high-gain radio-frequency amplifier without the use of shielding. The set will measure about 22 inches long and it is understood that it will be engineered for a.c. operation.

The Loftin-White receiver, after enjoying a popular year in 1927, will be slightly redesigned and placed among the kits available in the fall. Very few details regarding the receiver are available but it is understood that the receiver will very likely be an a.c. model with single control and shielding.

Super-heterodyne fans will find much interest in the new Strobodyne, an eight-tube set using a new type of frequency changer which causes the

circuit to function very much like a supe[r] regenerative receiver. In practical operation will be found possible to receive a station at on one point on the dial, a feature in which th receiver is different from many other supe[r] heterodynes, in which stations can always picked up at two adjacent points. The ne Strobodyne will contain one stage of radi frequency and three stages of intermediat former-coupled audio amplifier with a cho condenser combination in the output of t power tube. The tuning will be controlled by tv main dials but there will also be several suppl mentary controls consisting of one potenti meter, three rheostats, and a volume contr The radio frequency stage and first detect stage will be operated by a single dial and t oscillator will be operated by the other main di oscillator will be carefully shielded with thr individual stage shields.

Incidentally, we might mention here that article is at present being prepared by t laboratory staff describing the various supe heterodyne kits now being sold. This artic will shortly appear and should prove very inte esting because it will be crammed full of valuab dope on many super-heterodyne receivers.

The Arthur H. Lynch Company has design a unique unit which should prove to be u usually popular because of its adaptability many different circuits. Briefly, the unit whi they have developed consists of five tube socke mounted on a bakelite panel measuring abo 6" x 12". Four of the sockets are arranged alo the rear part of this small sub-panel and o socket is placed in the center toward the fro There is on the sub-panel, besides the socke three resistance-coupled amplifier stages whi are wired into the circuit when the assembly purchased. We therefore have a unit consisting a complete resistance-coupled amplifier, a sock for a detector tube, and a fifth socket in t front part of the panel for a radio-frequen tube. The home-constructor may then use a r. f. and detector circuit which he prefers a will find space on either side of the r. f. tu socket for the placement of the necessary c and condensers. The unit may be used as p of a receiver with a 14" or 18" front panel.

Besides the kits which we have mention above, there will be several others on which detailed information was available at the ti this article was written. The Karas Elect Company is working on two kits which w incorporate the "equamatic" system, featuri automatic coupling variation. The Bruno Rad Corporation and the Grimes Radio Engineeri Company are also designing new kit receiv for the fall but no information concerning th characteristics is available. It is also likely th a new Browning-Drake receiver will be design The Radio Receptor Company, Incorporat will continue to sell the kit which they man factured last year, which consists of three sta of tuned radio-frequency amplification, a sta of reflexed audio, and two straight audio stag

Suppressing Radio Interference

Practical Hints on How Radio Interference May be Minimized With
Actual Data on How It Has Been Accomplished in Numerous Cases

By A. T. LAWTON

NOTWITHSTANDING the fact that we have really good broadcasting stations and radio receivers of high quality, a surprising number of broadcast listeners are deprived of normal reception because of local outside interference. This interference may take the form of harsh nondescript noises, so-called "static," crashing, buzzing, clicking noises, etc., which either mutilate the program or blot it out altogether.

Almost any piece of electrical apparatus is a potential source of disturbance to the radio receiver. Wherever an electric spark is formed, waves of high-frequency electrical energy are sent out. They are identical (though untuned and uncontrolled) with those emitted from the broadcasting station. Consequently, they are picked up by the radio installation and amplified in the same manner as radio music or speech. There are two different methods of suppressing the interference due to sparking electrical apparatus. One is to eliminate the spark by discarding the apparatus or by placing it in a condition so the spark will not occur. The other is to confine the electrical waves set up by the apparatus to a very small area and thus prevent interference to reception of radio programs. The following paragraphs, and other articles of this series which will appear in RADIO BROADCAST from time to time, will give definite data on the various causes of interference and, in most cases, the solution of the problem. These data were collected over a period of two and one-half years, during which a 6000-mile patrol with a radio-equipped car was made, taking in the combined interference of more than one hundred and thirty-two towns and cities.

OIL BURNING FURNACES

THE problem of interference from the oil-burning furnace is a most important one to-day and will be discussed first. Certain types of oil-burning furnaces will interfere with reception one block or more distant. At close range they will probably put the set out of business for whatever period the electrical ignition system is functioning.

We say "certain types" of burners because there are many types on the market, and no two seem to act alike, especially when suppressive measures are being considered. Any oil-burning furnace which has its ignition system covered in several enclosures of metal and has all its associated wiring run in conduits, will cause a minimum of interference. This is a condition obtaining only occasionally; the majority of furnaces cause a lot of radio noise.

The fact that the ignition of some furnaces

RADIO BROADCAST published some of the first articles which appeared in the technical press dealing with the reduction of man-made or "artificial" interference. These were written by A. F. Van Dyck of the Radio Corporation of America and appeared in this magazine for April, May, and July, 1924. This article, the first of a pretentious series, includes some of the material covered by Mr. Van Dyck, and also a large amount of additional definite information of immeasurable practical value to the man faced by an interference problem in his own locality. Every conceivable source of interference is noted in these articles. The treatment is intensely practical, for Mr. Lawton has crystallized here the results of his work for two and one half years in more than 132 cities.
—THE EDITOR

operates only for ten seconds each time the furnace starts does not help matters. Eight such installations were recently investigated in less than one block on the same street. Being of the thermostat control type, they started up and stopped at frequent and irregular intervals. The combined interference made reception in the whole district unsatisfactory. Three other furnaces in the same town were fitted with continuous type ignition system, i. e., the spark coil operated for fifteen minutes or more at a time

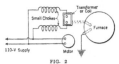

FIG. 2

and did not shut off automatically after lighting the gas. This type is a bad offender.

In considering methods of interference elimination it should be noted that two types of ignition coil are in general use. The first, an ordinary spark coil with vibrating contact, and the second, a straight 100 to 1 ratio transformer giving 11,000 volts on the secondary and operating directly off the 110-volt supply. The former creates much more disturbance than the latter.

While interference from the motor on the oil

burner is possible, in most cases the major disturbance is confined to the high-tension spark. For this reason it is usual, when installing apparatus to eliminate the radio interference, to concentrate on the spark coil, or transformer circuit, rather than on the main 110-volt supply lines.

The first and most promising method is to bridge the transformer primary with two 2-mfd. condensers connected in series, with the mid-point grounded. The interference from five furnaces has been entirely eliminated by this arrangement, shown diagramatically in Fig. 1. Thermostats, control wires, etc., have been omitted in this diagram for the sake of simplicity. The following points should be noted:

(1) Connection of the condensers is made as close to the coil as possible.

(2) Keep the ground wire short; if it is abnormally long, elimination of the interference is not complete.

(3) Two 500-volt condensers are used (costing about $2.00 each). We are dealing with a 110-volt circuit but the surge causing radio noise may be many times this voltage—sufficient to puncture a low voltage condenser.

(4) The capacity of the condensers should be at. least 2 microfarads each. One-half or one-microfarad condensers are not usually effective.

(5) Ground connection to the furnace itself may prove more effective than direct connection to, say, a water pipe.

Another method is shown in Fig. 2, where small choke coils are substituted for condensers. The current in the coil primary circuit varies from one ampere to two and one-quarter amperes according to the type of manufacture, and in winding the choke coils, wire of sufficient size to carry this current without heating must be used. For intermittent service, No. 18 d.c.c. wire may be employed; for the continuous electrical ignition type it might be as well to use No. 16 d.c.c. since, in a narrow coil wound ten or twelve layers deep, the radiating surface is small.

Approximately one hundred and thirty turns are required on a wooden bobbin of the measurements given in Fig. 3. No particular effort need be made to have this winding done neatly in regular layer fashion, as jumble winding is usually just as effective as straight winding.

Taken by itself, while not usually as effective as the condenser method, the choke coil method becomes an important factor in obstinate cases. Several furnaces failed to respond to either of the foregoing arrangements but all interference caused by them was successfully obliterated by a combination of the two, as shown in Fig. 4.

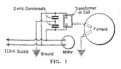

FIG. 1

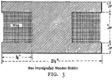

FIG. 3

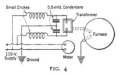

FIG. 4

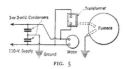

FIG. 5

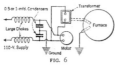

FIG. 6

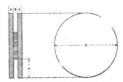

110 Turns No.12 D.C.C. Wire on Wax Impregnated Hardwood Bobbin. Winding Grove ½″ wide. 2¼″deep. A, B, C, ½″each : E, 2¼″, D, 7¼″

FIG. 7

It will be noted that 0.5-mfd. condensers are used instead of the two-microfarad capacity type. Care should be taken to see that the condensers are attached to those leads of the choke coils which are connected to the spark coil or transformer. Attaching them to the opposite, or supply line, end, is of practically no value.

One might imagine that this "surge trap," incorporating as it does, the merits of two methods, would be an all-round more efficient eliminator. As a matter of fact it is not necessarily so. Certain furnaces which respond to the choke coil method again become noisy when condense,s are attached, and the only sure check is to try each method individually.

We now come to those cases where interference is caused by the furnace motor as well as by the ignition spark. In this instance, we apply the above methods to the main supply lines with condenser and choke values altered.

Fig. 5 shows an arrangement which has proved successful on several furnaces fitted with the straight transformer coil. This circuit is electrically the same as that given in Fig. 1, the difference being that the condensers are connected close up to the motor instead of near to the transformer. This condenser method did not work in every case, however, as two furnace installations were found in which the addition of the condensers only increased the trouble.

Choke coils wired as shown in Fig. 6 and of the specifications shown in Fig. 7, entirely eliminated the interference from these two offenders; adding condensers brought all the trouble back again, so these latter were dispensed with.

Fig. 8 shows a method which may be applied to the main supply lines, No. 12 wire chokes being used. This method has been found necessary in certain cases.

Practically all furnace motors operating on 110-volt circuits draw a current of 5 amperes. Allowing 2 amperes for spark coil consumption we have a total current of 7 amperes to handle. On first consideration No. 12 wire may seem unnecessarily large but when the necessary reduction in current-carrying capacity is made for. layer depth in winding and proper heat radiation it will be seen that only normal precautions are taken to ensure a reasonable factor of safety.

Tests carried out with single-layer coils, i. e., 100 turns of wire wound in a single layer along a tube 3½ inches in diameter, were not conclusive. Complications rendered the findings void and the dimensions of the completed surge trap were such as to make its use objectionable.

The data on bank wound coils, as applied to furnaces, are also incomplete though some tests have been carried out with satisfactory results. For intermittent current under 8 amperes, however, little can be gained by substituting banked coils for jumble winding unless it can be determined that, in the former, a smaller quantity of wire can be used with equal results.

PLACING THE COILS CORRECTLY

BEFORE going any further we should stress a point, the importance of which is often overlooked. Let us suppose that you have eliminated the trouble by one of the choke coil methods, say Fig. 6. In this case you have deliberately or unconsciously made the right connections and placed the proper coil faces together. For the sake of compactness it is usual to install these chokes one on top of the other in a standard metal outlet box, and by simply turning one coil over we may bring back the interference sixty per cent. or more although the connections remain unchanged.

Contradictory results in actual practice indicate that so far as furnace installations are concerned no specific method of connection will suit all cases, so if interference elimination is not complete on first trial, then lay the two flat coils face to face and turn the top one over. If this does not produce the desired result reverse the leads of one coil only and repeat the turning test.

You are not so likely to run into this trouble if the inside lead of one coil and the outside lead of the other be connected to the furnace and the two free ends to the 110-volt supply line, but if two inside leads are connected to the furnace and the two outside leads to the line, or vice versa, and both coils are wound in the same direction, then the "turning" test may be necessary.

The reason for this little complication, which, by the way, has caused some investigators to abandon their experiments on the brink of success, is that in one setting the flux of one coil "helps" that of the other, but when the top coil is turned over, both are, as electrical men say, "bucking" or opposing. Just whether they should "help" or "buck" is a question more likely to be settled by the individual furnace than by any other authority. Different installations give different results, but in all cases one or other of the settings proves effective.

Paradoxical as it may seem, all this care and attention to coil positions, etc., proved unnecessary in three instances; the furnace interference in these three cases was reduced to zero when coils were installed in the lines, regardless of the manner of connection or relation of one coil to the other. This is quite in keeping, however, with the whims and vagaries of radio interference.

Perhaps we should condition the statement made a little while ago that "in all cases, one or other of the settings proves effective." This is literally true so far as the ignition systems under discussion are concerned, but in the course of extended patrol one occasionally meets the old belt driven magneto type ignition. Some day it may yield, but so far, all efforts to eliminate its interference have failed. The best method we can suggest now is to take it out and substitute a transformer with surge traps.

High tension leads on modern type furnaces are comparatively short and do not need any particular attention, but on older types, where

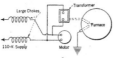

FIG. 8

the lead to the spark plug is two or three feet long, it may be necessary to run a metal pipe, of say, one inch diameter, over this to shield it, first of all running a piece of rubber hose over the high tension lead. The latter precaution is taken not altogether for the sake of insulation but to increase the distance from the wire itself to the metal shield so that there is less probability of foundering the spark energy. In most cases grounding this shield increases its efficiency as an eliminator.

Oil burning furnaces, as a rule, are hooked up to the electric supply lines by an ordinary cable and bayonet plug-receptacle fitting. If for any reason the circuit is opened here in the course of making tests the plug and receptacle should be carefully marked so that the plug is inserted the same way in the receptacle each time it is replaced.

Remembering that one side of this line is grounded and "dead" and the other side very much alive, you will appreciate the desirability of guarding against the introduction of unnecessary complications. As a test, however, there is no objection to reversing this plug during experimental work to find out if, in conjunction with the surge traps, a reversal of the supply leads has any beneficial effect.

The first thing to do when setting out to clear up some determined source of interference is to talk the matter over with some man in the electrical business, preferably one conversant with radio, and get an estimate of the cost of making up the necessary filter coils. If you do not care to stand the whole expense personally, interview the furnace owner and ascertain if he is willing to meet you half way. In a good many instances you will find that the owner, once he is made aware of the disturbance unconsciously created, will gladly shoulder the whole or part of the expense.

A more equable apportionment of the costs, however, is desirable; we must remember that the neighbor is just as much entitled to run his furnace as we are to run our radio receivers. In a recent instance, twelve radio fans in the immediate vicinity of an offending furnace were approached on the question of furnishing the necessary interference elimination apparatus and all but two agreed to a division of costs. The installation came to six dollars; for sixty cents each they rid themselves of an aggravating interference that spoiled winter evening reception for a previous three years.

ELECTRO-MEDICAL THERAPEUTIC APPARATUS

BROADCAST listeners living in the vicinity of a hospital are quite familiar with the class of interference generated by high-tension electromedical treatment machines. Even the humble violet ray for home treatments can spoil reception over a radius which will include a large number of residences.

Larger machines, as used by medical men, can

be distinctly heard a distance of eight city blocks, and in the near vicinity radio interference is extremely heavy. When this is multiplied by the number of machines in every city, it becomes obvious that the total interference from this source alone is serious.

Details and particulars of the various types in common use would serve no useful purpose. Ninety per cent. reduction of interference caused by one of the small machines was obtained by inserting 100-turn banked choke coils of No. 16 d.c.c. wire in each line near the machine, and bridging the circuit with condensers. This large percentage of reduction was not evident in the same building, only forty per cent. reduction being noted here, but in the neighboring residences, only a short distance away, practically all the noise had disappeared. A second machine, closely resembling this one but of greater power, failed to respond sufficiently to this treatment to warrant the expense of installing a surge trap.

So far as the larger apparatus is concerned we have experimented with every known surge trap connected at the machine, where preventative measures usually are most effective, and without exception, all proved useless.

The reason is obvious. If, while walking in a room where one of these machines is being operated, you place your hand near a radiator or lamp fixture or any piece of metal, grounded or ungrounded, sparks will pass from your body to the metal, and vice versa. The whole atmosphere is charged with electricity as also is the water piping, gas piping, metal fixtures, etc. And by the same token the electric lighting wires are charged, all the more because they are directly connected to the offending apparatus, and interfering surges set up are carried out over the distribution system.

It is hardly possible to clear up this interference to the satisfaction of broadcast listeners in the immediate vicinity, say within 100 feet, but it can be stopped from radiating very far beyond, an actual case on record proving this conclusively.

The necessary choke coils in this case are connected, not at the machine, but in the residence main supply lines, preferably at the meter. No condensers are used. Condensers in any combination, in the particular case referred to, brought all the interference back after it had been cleared up with choke coils alone.

Coils for the purpose should be bank wound, three layers deep, and should consist of 150 turns of d.c.c. copper wire on a 3½-inch tube of insulating material. See Fig. 9.

It must be remembered that these coils are required to handle the house load, not merely the current drawn by the therapeutic machine, and the size of wire used will be governed by this. It will probably be No. 8 or 10, B. & S. gauge.

Base-wound Coil No. 8 D.C.C. Wire

FIG. 9

Allowance is not made for electric cooking ranges or fireplace coils as these are usually on a separate circuit. If their associated wiring runs close by the lighting mains, however, it is desirable to separate them or else introduce additional heavy coils.

This interference reaches near-by listeners in two ways; first, by radiation from the surges set up in the city distribution wires, and second, by direct radiation from the source.

Preventive measures already described take care of the first channel; direct radiation can only be reduced by shielding the offending apparatus with a metal screen. More interference comes from the high-tension flexible cables than from the coil or vibrator itself, so that shielding of the vibrator proper has no beneficial effect.

We find too that any amount of metal screening in close proximity to the equipment interferes with free use of its various parts by the operator, and the only satisfactory way to get around this problem of direct radiation is to shield the entire room with a metal screen, an expensive arrangement but one which is more or less common in up-to-date hospitals where X-ray machines are in daily use.

X-RAY EQUIPMENT

THERE is a marked difference between the sound characteristics of X-ray interference and that propagated by the apparatus just discussed. High-frequency treatment machines give rise to an indefinite "mushy" noise—more or less of a blanketing interference—while that from the X-ray is more in the nature of a hard buzzing.

Where a rotary synchronous rectifier is used you can safely count on the installation setting up violent interference. In fact, it constitutes a good, healthy wireless transmitter and interferes, to a greater or lesser extent, all over a small town. In addition to the spark disturbance several cases are on record where the motor driving this rectifier caused enough trouble to bother reception within a radius of 300 feet.

From theory, we should say that choke coils inserted in the main supply at the house meter would reduce the total radiation though it is questionable if complete elimination can be se-

cured. Preventive measures taken on the machine itself gave unsatisfactory results.

So far as our experience goes, we have never yet gotten any trouble from the X-ray tube itself. Also, two complete installations were checked while operating on full power without any interference being noted. These installations used tube rectifiers instead of rotary synchronous rectifiers.

Possibly old or defective tube rectifiers might give rise to some trouble but no definite case appears on record and the substitution of tubes for the rotary rectifier seems to be the best solution of this problem.

Since the change over from rotary synchronous to tube rectification involves considerable expense, it is our policy to place the situation frankly—and courteously—before the medical men using such equipment, and with very few exceptions all have agreed to refrain from using the apparatus during the evening broadcast hours, or at least, limit its operation to urgent cases which cannot be left over.

DENTAL MOTORS

THERE is every probability of your getting fairly loud radio interference if your radio set is being operated in the vicinity of a dental office. The motor commonly used is of the universal type, which may be operated off direct or alternating current. It runs so smoothly and silently that sometimes with the naked ear it is difficult to tell whether the machine is running or not, while at the same time radio sets for four or five houses either side of its location experience a high pitched buzzing interference.

Fig. 10 shows three methods of clearing up this trouble; No. 1 does not always prove effective; No. 3 is only required in obstinate cases; and No. 2 will usually do the trick.

A little removable plate at the base of the engine stand gives easy access to the motor supply leads, and makes a convenient point at which to cut in with surge traps. Water piping runs up the centre of this column also, facilitating good ground connections.

Remember that the coil in (b) Fig. 10, must be inserted in the live side of the line, not the grounded side. The live side is easily found with an ordinary incandescent lamp and socket. One lamp lead is attached to a water pipe or other convenient ground and the free lead touched on the two motor mains in turn; the lamp lights only when touched on the live wire.

Little consideration need be given the second motor usually found in dental offices; we refer to the lathe or "working" motor. Its interference as a rule is negligible and its periods of operation are short.

(a)　　　(b)　　　(c)

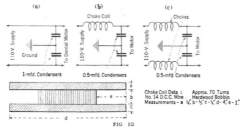

FIG. 10

The Three-Element A. C. Vacuum Tube

A Paper Delivered Before the Radio Club of America Which Discusses the Theory Underlying A. C. Operation of Vacuum Tubes

By BENJAMIN F. MIESSNER

Chief Engineer, Garod Corporation

IN A RECENT paper delivered before the Radio Club of America by the author, and printed in the February and March, 1927, issues of RADIO BROADCAST, it was shown that two principal causes for hum are present within tubes of the ordinary types when their filaments are heated with alternating current. One of these is due to the fluctuations of filament temperature with fluctuations of the energizing current, and the other is due to the effect of the positive leg of the filament acting as a plate electrode of periodically varying potential. It was further pointed out and demonstrated that at least one of the commercially available standard receiving tubes, when subjected to the proper operating conditions, would function normally as an amplifier without the introduction of an objectionable amount of disturbing hum. A commercial broadcast receiver, the Garod Model EA, which has a. c. excitation of all filaments except the detector (this latter has a 199 type tube heated by the combined plate current of all tubes), was demonstrated. Control of the two types of hum resulted in the neutralization of one by the other within the tube itself. Numerous curves were shown, which indicate the relation of the hum amplitude to the operating voltages on most of the standard types of receiving tubes. It was also pointed out in the latter paper that the author's experiments with special tubes had demonstrated the possibility of accentuating the desirable characteristics of standard tubes by a tube of new design, which could be used not only in the amplifier stages, but in the detector stage as well, which requires hum elimination of an extremely high order because of the large amount of amplification behind it in the receiver.

The purpose of the present paper is to amplify some points in the previous discussion, which time did not then permit, and more particularly to describe the a. c. tube, which was discussed briefly and which is now being made available through commercial channels, for use as a detector as well as a radio or audio amplifier.

In the author's researches on a. c. filament excitation, it was found that there are three distinct cases, requiring separate analysis and discussion, depending upon the use of the tube so operated. In audio-frequency amplifiers, with which the hum measurements of the first paper were principally concerned, we have one set of conditions. Another set of conditions obtains when the tube is used as a radio-frequency amplifier; and a third, and considerably more complicated set of conditions, are presented when the tube is used as a detector. Here both the radio and audio effects are present in the same tube, and some very interesting actions have been observed.

HUM IN AUDIO-FREQUENCY AMPLIFIER TUBES

AS BEFORE stated, the previous paper dealt principally with the hum in audio-frequency amplifier circuits. It was shown that the temperature type of hum resulted from the changes in temperature of the filament due to the variable heating current flowing through it. It was shown further, that the degree of the temperature variation is due to what was termed "thermal inertia" of the filament, this "thermal inertia" being dependent upon the ratio of heat storage capacity of the filament, to its heat dissipating ability. The heat storage capacity depends upon the volume or cubical contents of the filament as well as upon its heat absorbing ability, which is called specific heat. This storage capacity may be likened to the electrical storage capacity of a condenser wherein the area of its electrodes compares with the volume of the filament, and its dielectric constant compares with the specific heat of the filament. The capacity in either case is proportional to the product of the two factors. It was shown that the amount of this temperature hum was greatest with thin low-heat capacity filaments, such as those used in the 199 type of tube, and smallest in the heaviest filaments, such as those used in the 112 type of tube. It was shown further, that the amount of temperature variation depended not alone upon the ratio of surface or radiating area to the mass of the filament, but also upon the actual operating temperature of the filament itself. The radiation losses for heat are proportional approximately to the fourth power of the temperature, and inasmuch as these radiation losses are the chief ones to be reckoned with as lowering the filament temper-

ature during periods of small or no current, it is seen that very low temperatures greatly facilitate temperature stability.

The temperature stability of such a filament, with a pulsating heat input occasioned by its alternating heating current, and with a heat output or load consisting chiefly of radiation losses, and partly of conduction losses, depends, as above stated, upon the ratio of its heat storage capacity to its heat dissipating ability. The former has already been defined, and the latter depends upon the operating temperature and the radiating or surface area, neglecting the conduction losses through lead wires. Obviously our problem, in securing high stability, is to obtain a high ratio of storage capacity to dissipating ability.

One method of accomplishing this is to use a filament of round cross-section. This form is to be preferred because the ratio of volume, or storage capacity, to surface, or dissipating ability, is greatest. The strip form of filament so common in the oxide-coated types, decreases this ratio, and is therefore to be avoided, unless increased area is necessary for other reasons.

Another method of increasing this ratio is to use a filament of large diameter. Since the volume is proportional to the second power of the radius, while the surface area is proportional to its first power only, it is seen that the ratio of volume to area of the filament increases rapidly with increasing diameter, and as large a diameter as possible, consistent with other considerations, is to be desired.

A third method, as before noted, consists in reducing the operating temperature to as low a value as possible consistent with other considerations so that the heat dissipation rate of the filament is low. Remembering that the radiation losses are proportional to the fourth power of the temperature, it will be seen that merely cutting the operating temperature in half, as by going from a thoriated tungsten to an oxide platinum filament, may permit of a very great decrease in radiation losses and a correspondingly greater temperature stability.

The temperature stability of an a. c. energized filament may be understood better perhaps by comparing it with the voltage stability across a condenser fed from a rectifier and feeding a load resistance. In both cases we have a pulsating input and a more or less steady output of energy. If the output load is small, the energy level of temperature or voltage remains steady. If the load be large, there will be a large decrease in energy level during periods of little or no input, and consequent great instability.

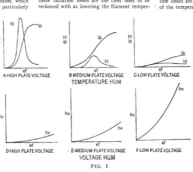

A—HIGH PLATE VOLTAGE   B—MEDIUM PLATE VOLTAGE   C—LOW PLATE VOLTAGE
TEMPERATURE HUM

D—HIGH PLATE VOLTAGE   E—MEDIUM PLATE VOLTAGE   F—LOW PLATE VOLTAGE
VOLTAGE HUM

FIG. 1

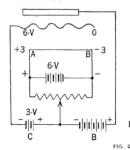

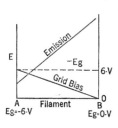

FIG. 2

To summarize then, the method of securing high temperature stability is to use a filament of round cross section, large diameter low temperature, and high specific heat.

The curves presented in the former paper show that the temperature type of hum is at its maximum value at the point of greatest steepness of the filament-voltage plate-current curve of any given tube. That is to say, if a static plate-current curve be plotted with variable filament voltages, the greatest amount of plate current variation will be obtained for a given amount of filament voltage variation at the steepest point in such a curve, and it follows that under dynamic conditions, where the filament current is constantly changing at a high rate, such as by the action of a 60-cycle impressed voltage, the greatest amount of temperature fluctuation would be obtained at a filament voltage corresponding to the steepest point of the plate current curve.

The curves show also that at high plate voltages, where the plate current curve is steepest, the temperature type of hum is greatest, and at low plate voltages, where the plate current curve is comparatively flat, the temperature hum is low or has entirely disappeared. This is shown by the typical curves of Fig. 1.

In curve "A," for a plate voltage that is high compared to the filament voltage, a high temperature peak is produced, due to the steep plate current curve. In "B," where the plate voltage is reduced and the plate current curve is thereby flattened, the temperature peak is much reduced. In "C," where the plate voltage is low and the plate current curve is relatively very flat, the temperature hum has almost completely disappeared. In "D" is shown the low-voltage type of hum produced when the plate voltage is high compared with the filament voltage. In "E," where the plate voltage is reduced, the hum has considerably increased, while in "F," where the plate voltage is very low, the voltage hum has become very high.

These curves represent the general relationship between the two types of hum and the plate voltage with a tube of any given type: With a fixed plate voltage and differing filament voltages the same general results will be obtained, as the ratio of plate to filament voltage is the factor determining the degree of the voltage type hum. This is explained by the fact that there are within the tube two electrodes of positive potential and competing with each other for the emission of the filament. With an a. c. filament voltage of, say, five volts, and a d. c.

plate voltage, say, 135 volts, the plate electrode, by reason of its over-powering attraction for the electrons, attracts to itself most of the emission, while the positive filament end with its small voltage gets only a small part of the emission. When, however, the plate voltage is reduced to a low value, such as twenty-five volts, the five-volt potential of the filament becomes relatively important and it now takes a much greater proportion of the emission than it did under the high-voltage plate condition.

It should be remembered here that the attraction of the plate electrode is considerably modified by the presence of the intervening negatively charged grid, and that no such intervening grid is present between the positive side of the filament and the negative side, and this further contributes to the effectiveness of the positive filament end as a contender with the plate for the electron emission.

As a result of these various studies, the previous paper pointed out that the temperature type of hum could be reduced by increasing the ratio of mass to radiating surface of the filament and by lowering the temperature, and that the voltage hum could be decreased by reducing the filament voltage to a low value and by separating the filament ends as far as possible, that is, by use of a filament of the straight type. Another method is to so construct the grid of the tube, that it surrounds the two filament sides in a V-

type filament and thereby shields the emitting negative side from the attracting positive side, by virtue of its relatively high negative bias.

The latter paper also discussed the variation of emission from the two sides of the filament and explained that the negative leg was considered to be responsible for the chief part of the emission, at least during those portions of the applied a. c. cycle where the voltage was high. In addition to the reasons given for this non-uniform emission, there is another simple explanation which leaves no doubt on this point. This has been proposed by the author's assistant, Mr. Charles T. Jacobs.

Consider a filament excited by a six-volt battery; its associated grid and plate circuits are returned to the mid-point of a potentiometer connected across the battery and filament, as shown in Fig. 2. The grid is biased negatively by a three-volt battery. An examination of this circuit discloses that the negative bias with respect to the negative end of the filament is zero, while the bias with respect to the positive end of the filament is six volts. It is seen that in the case of the negative end of the filament, the C-battery voltage is exactly neutralized by the oppositely poled voltage of one half the A battery. However, for the positive end, one half the A battery voltage is added to the negative bias, producing an overall bias between the grid and the most positive end of the filament of six volts. At the mid-point of the filament, the A voltage neither adds nor subtracts, so that the voltage of the grid with respect to this point is that of the C battery, or negative 3 volts. We can conceive, therefore, that the emission of the positive end of the filament compared to that of the negative end must be very small, because of the much greater negative bias on the grid with respect to the positive end, and that the emission will vary uniformly from one end to the other. The conditions here may be understood by reference to the graphs showing the distribution of emission over the filament as shown in this diagram. If "A" is the positive end of the filament, three volts positive with respect to the mid-point, and "B" is the negative end of the filament, three volts negative with respect to the mid-point, then the potential difference between the grid and "B" is zero volts. The accompanying graph shows in general the distribution of emission over the filament from "A" to "B," being lowest at "A" and highest at "B."

If now the polarity is reversed, as shown in Fig. 3, we see that the distribution of emission is

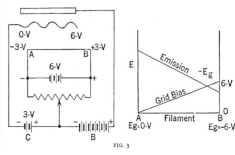

FIG. 3

the same but reversed with respect to the filament ends. In the third case, shown in Fig. 4, wherein the filament supply voltage passes through the zero point, shown by the omission of the battery, the whole filament will be at a uniform potential difference with respect to the grid, and equal to the voltage of the C battery or three volts. The emission of the filament in this case will have an amplitude equal to the mean value of the amplitude over the filament as a whole, as shown in either Fig. 2 or Fig. 3.

If now, we exchange for the six-volt battery a source of alternating current, such as a transformer secondary, having a peak voltage of six and, for sake of simplicity, a sinusoidal waveform of 60 cycles frequency, we have a continuously changing condition wherein at some instances we have one or the opposite set of polarities on the filament with varying voltage, and at other instances, we have no voltage whatsoever, that is, when the applied voltage passes through the zero point in its cycle. The action under these circumstances is considerably more complex than in the steady state given by the battery, and can best be expressed by graphs showing the separate effects, their phase relations, and the net result of all of them.

In Fig. 5, this a.c. filament excitation substituted for the A battery is shown, together with graphs indicating the nature of the emission variation to the plate with respect to the two

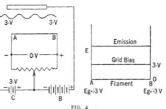

FIG. 4

filament ends. With a peak filament voltage of six, a mid-point grid return, and a C bias of 3 volts, as indicated, both ends of the filament, "A"–"B," will vary from zero to six volts with respect to the grid, during a cycle of the filament voltage. Graph No. I shows the filament voltage; graph No. 2, shows the variation in emission from "A;" and graph No. 3, shows the corresponding variation from "B." Graph No. 4, which is a straight line, represents the sum of "A" and "B," and shows the total plate current under this condition to be constant, neglecting the slight effect of the B return and any slight departure from symmetry in the geometry of the the filament, grid, and plate at the two ends of the filament. Returning now to Fig. 4, we can make this quite clear. If the emission and grid bias lines be given an oscillatory rotating motion

in opposite directions, each about its point of mean amplitude, it i obvious that the total emission i always constant

If, now, the grid and plate return be made at some point off the central point of the potentiometer, i can be shown that the emission wil no longer undulate symmetrically about the center point and therefor remain constant, so that the tw effects will be unbalanced, and a 60-cycle variation of plate curren will result. In Fig. 6, the dynami conditions with a.c. on the filamen and the potentiometer slider dis placed somewhat from the center, ar shown. Here again Graph No. 1 rep resents the exciting voltage across th ends of the filament; No. 2, the emission from on end of the filament; No. 3, the emission from th other end of the filament; and No. 4, the sum o curves Nos. 2 and 3, which is a 60-cycle vari ation. It will be explained later how this 60 cycle grid voltage can be made to counteract i 60-cycle ripple in the plate voltage.

These curves and this analysis are, of course based upon the assumption that the tube i operating as a proper amplifier, with the norma negative grid bias coinciding approximately with the center of the straight line portion of the grid voltage plate-current curve of the tube. Th action in the ease of a detector acting as a plat rectifier is different from that of the amplifier a above explained, and will be discussed at lengt later in this paper, the latter half of which wi appear soon in RADIO BROADCAST.

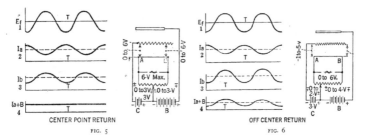

CENTER POINT RETURN

FIG. 5

OFF CENTER RETURN

FIG. 6

A World's Record!

Tokyo, Japan to Hoboken, N. J. with ZETKA *Clear-Glass* Tubes

East Meets West in 7270 Mile Radio Hop

ON April 15, 1927, Arthur Wald of Hoboken, New Jersey, made a new World's Record for DX reception—7270 air miles . . . Station JOAK of Japan—the longest distance bridged in radio history, according to the New York Evening Telegram.

Zetka Clear-Glass Tubes—five Z-201-A's and one Z-112 Power Tube—brought in this monumental achievement, across the Pacific Ocean, across the entire United States—proving once again the unparalleled *distance-volume* characteristics of these finer, clear-glass tubes.

The new ZP-201-A all-socket Power Tube is another great Zetka achievement. A power tube that goes into every socket, without re-wiring—that gives greater volume, greater distance, purer tone—yet consumes less "A" Battery current—that lasts longer and increases in efficiency with service—*that brings radio to its highest point of efficiency.*

Every tube is serial numbered and guaranteed—and the price is as much an achievement as the tube itself—only $2.50 each. Hear them today, at your dealer's.

There is a Zetka Clear-Glass Tube for every radio purpose —each one an outstanding achievement

"I RECOMMEND the use of Zetka Tubes to any one who is deeply enough interested in radio to want the best equipment it is possible to secure.

"Zetka tubes have given me infinitely more power, greater distance, better tone qualities, and a complete freedom from microphonic noises, way beyond any other tubes I have ever used."

Arthur Wald

ZETKA
The *Clear-Glass* Tube

ZETKA LABORATORIES, Inc. 73 Winthrop St., Newark, New Jersey

Superior Tone Quality
plus
Artistic Appearance

IMPROVEMENT in reproduction is the outstanding radio achievement of the past year. Engineering methods developed for radio have given a new lease on life to the phonograph industry.

The audible range of radio receivers and phonographs has been greatly extended by the intelligent application of sound engineering fundamentals gleaned from research in the electrical and acoustical field.

Radio and Phonograph engineers have recognized in Lata Balsa — the marvelous, tropical wood with no inherent acoustic resonance points—the ideal diaphragm, and have begun to apply it to sound reproducers of more than ordinary excellence. Lata Balsa is rapidly gaining favor in both fields.

A good loud speaker must be fed by a good audio amplifier. Radio engineers agree that there is no means of providing tone fidelity to compare with resistance coupling.

The complete Lata Balsa Reproducer Kit which you may assemble in a few minutes into a wonderful reproducer of tone quality and an artistic piece of furniture—
13" x 24", $8.00   21" x 36", $10.00
Lata Balsa Balanced Armature Reproducer Unit designed especially for use with Lata Balsa Reproducer Unit....$8.00

Resistance coupling and its manifold applications have been the work of the Engineering Department of Arthur H. Lynch, Inc., for a long period. In their search for a reproducer which would demonstrate its superiority in a conclusive manner they have made exhaustive tests on the Lata Balsa Reproducer and find the combination to be ideal.

Musicians, Music Lovers and Acoustic Experts agree with them. Artists and artistically inclined housewives recognize in the decorated Lata Balsa Reproducer, which may be made into a firescreen or wall decoration, a relief from the unsightly appearance of the old fashioned "loud speaker."

These scientific wonders are now available to every man and woman at but slight cost. They are easy to apply to any receiver either old or new.

Everything necessary for the building of a modern three-stage resistance coupled amplifier........................$9.00

JOBBERS AND DEALERS
We have a most attractive proposition on these rapidly moving lines. A word from you will bring full details.

RETURN THIS COUPON—

ARTHUR H. LYNCH, Inc.
General Motors Building
1775 Broadway
NEW YORK CITY

Sole Sales Representatives

ARTHUR H. LYNCH, Inc.
General Motors Bldg., 1775 Broadway
At 57th Street, New York City
Gentlemen:
Please send me the items checked, for which I enclose my money order or cash.
Lata Balsa Reproducer Kit, 13 x 24 ..... $8.00
"   "   "   21 x 36 ..... 10.00
"   "   "   Unit 8.00
Lynch Resistance-Coupled Amplifier Kit . 9.00
Name.....................................
Address..................................
City-State...............................

BALSA WOOD
REPRODUCER, Corp.
331 Madison Avenue
NEW YORK CITY

New!

THORDARSON
POWER
TRANSFORMERS

130 M. A. FULL WAVE RECTIFIER

Here is a power unit that will satisfy the ever increasing demand for improved quality of reception. A split secondary 550 volts either side of center, makes possible full wave rectification, using two 216-B or two 281 tubes. Current capacity, 130 milliamperes. The low voltage secondary, 7½ volts, will supply two UX-210 power tubes, enabling the use of push-pull amplification in last audio stage. The Double Choke Unit 2099 is designed for this power unit. Contains two individual chokes of 30 henries, 130 milli-amperes capacity each.

T-2098 Transformer, 4½" x 5¼" x 5¾"
List Price, $20.00
T-2099, Choke Unit
3¼" x 4⅞" x 5⅞" high
List Price $14.00

*R*ealistic tone
quality, that elusive
but much talked of characteristic of radio reception—
can be obtained only through the
use of apparatus of the finest materials
and workmanship. For years Thordarson
transformers have been the choice of many
discriminating manufacturers of quality receiving sets. Follow the lead of the leaders. If you
enjoy good music *specify Thordarson transformers*

THORDARSON ELECTRIC MANUFACTURING CO.
Transformer Specialists Since 1895
WORLD'S OLDEST AND LARGEST EXCLUSIVE TRANSFORMER MAKERS
Huron and Kingsbury Streets — *Chicago, Ill. U.S.A.*

POWER PUSH-PULL TRANSFORMER and CHOKE

Quality reproduction that cannot be obtained with straight audio amplification, is made possible through the Thordarson power push-pull combination. This arrangement is designed for use with power tubes only and has sufficient capacity for all tubes up to and including the UX-210. Makes an ideal power amplifier when used with power supply unit T-2098.

Input transformer couples stage of straight audio to stage of push-pull. Output choke is center-tapped with 30 henries on either side of center tap. Dimensions of both transformer and choke, 2½" x 2½" x 3" high.

Input Transformer T-2408
List Price, $8.00
Output Choke T-2420
List Price $8.00

A. C. TUBE FILAMENT SUPPLY

The new R. C. A. and Cunningham A. C. filament tubes will be very popular with the home constructor this season. The Thordarson Transformer T-2445 is designed especially for these tubes. Three separate filament windings are provided.

Sec. No. 1, 1½ volts, will supply six UX-226 amplifier tubes.

Sec. No. 2, 2½ volts, will supply two UX-227 detector tubes.

Sec. No. 3, 5 volts, will supply two 5 volt power tubes.

In addition to the above, this transformer is equipped with a receptacle for the B-supply input plug. Supplied with six-foot cord and separable plug for attachment to the light circuit. Transformer in compound filled, crackle-finished case. Dimensions — 2¾" x 5¾" x 4¾".

A. C. Tube Supply, T-2445
List Price, $10.00

THORDARSON ELECTRIC MFG. CO.
500 W. Huron St., Chicago, Ill.

Gentlemen:
Please send me your booklets describing your new power supply transformers.

Name
Address
City   State

The *Radio Broadcast*
LABORATORY INFORMATION
SHEETS

THE Radio Broadcast Laboratory Information Sheets are a regular feature of this magazine and have appeared since our June, 1926, issue. They cover a wide range of information of value to the experimenter and to the technical radio man. It is not our purpose always to include new information but to present concise and accurate facts in the most convenient form. The sheets are arranged so that they may be cut from the magazine and preserved for constant reference, and we suggest that each sheet be cut out with a razor blade and pasted on 4" x 6" filing cards, or in a notebook. The cards should be arranged in numerical order. An index appears twice a year dealing with the sheets published during that year. The first index appeared on sheets Nos. 47 and 48, in November, 1926. Last month an index to all sheets appearing since that time was printed.

The June, October, November, and December, 1926, issues are out of print. A complete set of Sheets, Nos. 1 to 88, can be secured from the Circulation Department, Doubleday, Page & Company, Garden City, New York, for $1.00. Some readers have asked what provision is made to rectify possible errors in these Sheets. In the unfortunate event that any such errors do appear, a new Laboratory Sheet with the old number will appear.

—THE EDITOR.

No. 121   Radio Broadcast Laboratory Information Sheet   **September, 1927**

The Hertz Antenna

CHARACTERISTICS

ONE of the commonest antenna systems used by amateurs for transmitting purpose is the Hertz system. This antenna, in its simplest form, consists of two straight wires located diametrically opposite each other as indicated in the drawing on the accompanying curve. The length of the two wires bears a definite relation to the fundamental wavelength at which the antenna system will tune and this relation is indicated by the curve, which is reprinted from *QST* of May, 1926. The relation between the length of the antenna system and the fundamental wavelength is a constant; the length L is equal to the wavelength divided by a constant, 1.3.

It is possible to obtain radiation on any wavelength by using different lengths of antenna and counterpoise. Suppose we wish to transmit on 40 meters (7500 kc.) and the antenna system is to be operated on the fundamental wavelength. Then from the curve the length of the antenna "A" would be 31 feet and the length of the counterpoise "B" would also be 31 feet. It would also be possible to transmit on 40 meters using the third harmonic of the antenna, in which case the antenna would be of such a size as to have a fundamental wavelength equal to 40 times 3 or 120 meters. If supplied with energy at a frequency corresponding to 40 meters,

however, the antenna would radiate energy at this frequency very efficiently even though its natural wavelength is 120. If such a system of transmission were to be used, the length of the antenna and the counterpoise would each be 93 feet.

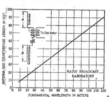

No. 122   Radio Broadcast Laboratory Information Sheet   **September, 1927**

Testing Radio Receivers

FEATURES TO CONSIDER

IT IS obviously of distinct advantage to test radio receivers in accordance with some standardized test procedure so that the results obtained from different receivers can be readily compared. If such a method is used the manufacturer will be able to have before him information which will tell him definitely just how his product compares with those of other manufacturers and also the buyer of a receiver will have certain definite data upon which to base his decision in buying a receiver. Considerable information on methods of testing radio receiving sets is given in the Technologic Paper of the Bureau of Standards, No. 256. In this paper it is suggested that the following tests be made on a receiver:

(A) Frequency range.
(B) Vibration Test, which determines how well the set has been constructed mechanically and whether it will be able to withstand the ordinary shocks obtained in transportation.
(C) Sensitivity.
(D) Selectivity.

These tests are especially effective in indicating how well the set has been engineered from an electrical standpoint. A test should also be made of fidelity, to determine how well the receiver is capable of reproducing voice and music.

From the standpoint of the average user these tests are not conclusive because he is interested in

other things besides the electrical efficiency of the receiver or its fidelity of reproduction. In a laboratory test one receiver might show up much better than another in regard to sensitivity and selectivity but the good results might only be obtained with very accurate adjustments. Obviously, a single-control receiver lacking somewhat in sensitivity and selectivity in comparison with another receiver might actually give somewhat better results when operated by an ordinary buyer with little knowledge of the circuit. As is stated in the paper mentioned above, it is really very difficult to judge the performance of any particular receiving set on the basis of any one trial of its operation, largely due to the widely different types of receivers and conditions under which they are best operated. The skill of the operator very largely determines the degree of satisfaction that will be obtained from any given receiver. In practice it will very likely be found best to just make available to the prospective purchaser some figures of merit indicating the sensitivity, selectivity, and fidelity and then to let him determine for himself whether the receiver in his hands gives satisfactory results.

Many letters are received from readers requesting comparisons between different receivers but to give conclusive information of this sort is impossible. The choice of the receiver which one finally purchases after trying out many sets is governed by many factors on which no laboratory measurements can be made.

S-M IS READY—

For A. C. Tubes or Any New Developments!

THE 440 time signal amplifier is a fully assembled three stage, 112 K. C., long wave amplifier-detector catacomb. It consists of three air-core low-resistance tuned R. F. stages and detector, with all necessary by-pass condensers, etc., mounted in a copper and brass catacomb, providing individual stage and over-all shielding.

Each 440 amplifier is accurately matched in the S-M laboratories to exactly 112 K. C., and every amplifier is guaranteed to within one-half of one percent. The selectivity of the 440 is tremendous—it may be made 10 K. C., or even 5 K. C., at will, while the sensitivity is guaranteed greater than that of any amplifier that might be built of individual parts. Price $35.00.

The fall season finds S-M 220 audio and 221 output transformers still the acknowledged leaders in the high-quality transformer field. And they are accorded the sincerest form of flattery—imitation—for this year other manufacturers, profiting by their phenomenal success copy the 220 characteristics, introduced by S-M a year ago—5000 cycle cut-off, rising low frequency characteristics, and plenty of iron and copper to make a good job.

But S-M audio transformers stand supreme as the finest available—the only types ever backed by a guarantee of BETTER reproduction or your money back. That's the S-M guarantee—and the return average is less than one in every 4000 sold—3999 satisfied customers out of every 4000. 220 Audio Transformer, price $8.00. 221 Output Transformer, $7.50.

The S-M Unipacs, introduced this June, have taken the country by storm. Everywhere, builders are realizing what true distortionless reproduction really is—for the larger Unipacs (power amplifiers and ABC power supplies) can deliver from 5 to 300 times more—and purer signals—than any standard receiver amplifier, which they replace. Prices range from $62.00 to $93.25 for wired and unwired models.

The new 1927 model Improved Shielded six has all the points that made the original the most popular high-priced TRF kit ever offered. The new "Six" has greatly improved selectivity—so great that will allow 10 to 20 K.C. separation of local and distance stations. The volume has been increased. The tone quality of the new 630 is just as fine as last year's model—GUARANTEED UNCONDITIONALLY to be equal or superior to any other set, regardless of price.

The new 630 can be built for battery, eliminator, or complete A. C operation. The A. C. model is absolutely batteryless, and uses the 652A, ABC power supply. (Any 1926 Shielded six can easily be brought up to date, or adapted for complete A. C. operation.) No matter what developments come, the 630 will always be one of the finest of sets, amply satisfying the most discriminating of fans. Price $95.00.

The S-M 652A unit is the famous 652 "Reservoir B" supply, developed into a complete ABC power supply. Thus, with a 652A kit at $36.50, plus as many A. C. tubes as your set needs, you make it entirely A. C. operated, with all batteries eliminated. Or any new set you are building can dispense with batteries if you use the 652A. Or if you want to change over your present B eliminator, new S-M ABC power transformers are available.

We can't tell you the whole story of new S-M developments, so if you'll just fill in the coupon, and mail it with 10c to cover postage, we'll send you free more up-to-the-minute advance radio information than you could buy in a text book.

Silver-Marshall, Inc.
838 West Jackson Blvd.   Chicago, U. S. A.

No. 123　　RADIO BROADCAST Laboratory Information Sheet　**September, 1927**

Characteristics of 171 Type Tube

STATIC AND DYNAMIC

ON LABORATORY Sheet No. 124 are given several curves for the 171 tube. It will be noted that one curve is marked static and another dynamic. The dynamic characteristic is, as its name implies, a curve indicating how the tube will function under actual operating conditions. The static characteristic curve, although valuable in giving an idea of the general characteristics of a tube, gives no indication at all of the tube's actual performance. Under actual operating conditions a tube always operates with a certain load in its plate circuit and consequently a curve taken to indicate the tube's performance should be made with some load in the plate circuit. The curve marked "dynamic" was taken when the tube had 4000 ohms resistance in its plate circuit. The difference between the static and the dynamic curves is considerable.

The curves were taken with 180 volts on the plate and 40.5 volts on the grid. In order that an amplifier may give good quality, its plate-current grid-voltage characteristic must be straight from zero grid vol-

tage to twice the d. c. voltage on the grid. The static characteristic, although straight from 40.5 volts to zero volts, is very curved at voltages greater than 40.5. It might be judged from this curve that the tube's performance would be very poor. However, if a dynamic characteristic is taken, we find that the characteristic remains straight from zero grid voltage down to about 85 volts and consequently the tube would actually give good results.

The other curves on Sheet No. 124 are dynamic characteristics taken with different resistances in the plate circuit. Curve No. 1 was made with 1000 ohms resistance, No. 2 with 2000 ohms, and No. 3 with 8000 ohms. It will be noticed that as the resistances increase the straight portion of the curve becomes greater and greater. The curves all cross at about 40 volts because this grid voltage represents the initial d. c. potential placed on the grid and the curves are made by increasing and decreasing the grid voltage about this average value. It is necessary in taking the curves to adjust the plate voltage each time so that with the different resistances the same plate current is obtained at 40.5 volts on the grid.

No. 124　　RADIO BROADCAST Laboratory Information Sheet　**September, 1927**

Curves of the 171 Type Tube

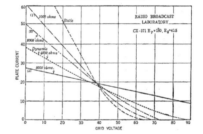

No. 125　　RADIO BROADCAST Laboratory Information Sheet　**September, 1927**

The Morse Code

No. 126   Radio Broadcast Laboratory Information Sheet   Septem

Condenser Reactance

HOW IT IS CALCULATED

IF A condenser is connected in series with an a. c. ammeter to a source of alternating current, a certain amount of current will flow in the circuit, depending upon the size of the condenser and the frequency of the current. If the voltage of the source is divided by the current, the quotient will be the "reactance" of the condenser in ohms. For example, if the frequency of the current being supplied by the source of potential was 60 cycles and the voltage was 110 volts and the size of the condenser was 1 mfd., we would find that, 0.412 amperes of current would flow through the circuit. Then 110 volts divided by 0.412 gives 2666, which is the reactance in ohms at 60 cycles of a 1 mfd. condenser. The reactance of a condenser depends upon its size and upon the frequency of the current. It can be calculated by means of the following formula:

$$\text{Reactance} = \frac{10^4}{6.28\ FC}$$

Where F is the frequency in cycles per second and C is the capacity of the condenser in microfarads.

In many calculations it is necess reactance of a particular condens quency, and for this reason, on La No. 127, is given a table of condense capacities between 0.001 and 10 mfd from 60 to 1,000,000 cycles. From th herewith it is evident that the rea denser is inversely proportional to the condenser and inversely prop frequency. Doubling the size of the fore gives half the reactance, an frequency of the current also halv of the condenser. Remembering the a simple matter to calculate mental of almost any capacity not given Laboratory Sheet No. 127. For ex condenser at 100 cycles has ⅓ of th 1 mfd. condenser at 100 cycles. Sin of the latter size at 100 cycles is reactance of a 3-mfd. condenser m vided by 3, or 533⅓ ohms. A 0.000 at 1,000,000 cycles has a reactance has a reactance of 1600 ohms at 0.0001-mfd. condenser at this freq condenser likewise has a reactance

---

No. 127   Radio Broadcast Laboratory Information Sheet   Septem

Condenser Reactance

CONDENSER CAPACITY IN MFDS.	REACTANCE IN OHMS AT VARIOUS FREQUENCIES						
	60	100	250	500	1000	10,000	100,0
0.001	2666000	1600000	640000	320000	160000	16000	160
0.005	533200	320000	128000	64000	32000	3200	32
0.01	266600	160000	64000	32000	16000	1600	16
0.1	26660	16000	6400	3200	1600	160	1
0.5	5332	3200	1280	640	320	32	3.
1.0	2566	1600	640	320	160	16	1.
2.0	1333	800	320	160	80	8	0.
4.0	666	400	160	80	40	4	0.
8.0	333	200	80	40	20	2	0.
10.0	267	160	64	32	16	1.6	0.1

This table shows how the reactance of various capacities varies with different frequencies. The reactance of a condenser varies inversely with its capacity and with the frequency. See Laboratory Sheet No. 126.

No. 128   Radio Broadcast Laboratory Information Sheet   Septen

B Power Units

DESIRABLE CHARACTERISTICS

A B-POWER unit is essentially a device to supply plate voltage to a radio receiver but such a unit has several characteristics besides the ability to supply proper voltages that are important in determining how satisfactorily it will operate. Modern batteries for the plate supply of a receiver can hardly be improved on. Their voltage is constant, they are perfectly quiet in operation, and leave little to be desired as a source of plate potential. The expense of operating a multi-tube receiver using power tubes from batteries is considerable, however, but a B power unit, properly designed, affords an excellent source of high plate potential at moderate cost.

What are the desirable characteristics of such a unit? It must first of all be capable of supplying the proper voltages to a receiver. Either insufficient or excessive voltage will adversely affect the operation of a receiver in many cases and it is therefore essential that some care be taken to make certain that the voltages being supplied are correct.

The power unit must deliver those voltages with a minimum of a. c. hum. Low hum output is only obtained with a properly designed transformer and filter system. The various filter chokes should be shielded so that magnetic coupling between them will not be possible and it is also necessary that some means be used to electrostatically shield the high-

voltage secondary windings fro winding to prevent any line noise mains getting into the filter system output of the unit noisy. This sh the primary and secondary may by means of a grounded copper sh primary and secondary windings may be accomplished quite effect the filament winding, supplying between the primary and high-v The filament winding, being at g therefore acts as a very effective plate supply unit generally ind of proper magnetic shielding, or in A third desirable characteristic is good regulation, which determ the output voltage will change wit amount of current being supplied particular plate supply device mi 90 volts at the 90-volt tap with a If, however, the regulation was p ceiver only required 4 mA. from the actual voltage at this tap mi as 130 volts; if the unit had good voltage would not be more than 4 mA. A power units with poor regu cause receivers to "motor boat" or in some other way, and good regul variation of output voltage with therefore a very desirable characte

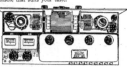

Manufacturers' Booklets

A Varied List of Books Pertaining to Radio and Allied Subjects Obtainable Free With the Accompanying Coupon

READERS may obtain any of the booklets listed below by using the coupon printed on page 322. Order by number only.

1. FILAMENT CONTROL—Problems of filament supply, voltage regulation, and effect on various circuits. RADIALL COMPANY.

2. HARD RUBBER PANELS—Characteristics and properties of hard rubber as used in radio, with suggestions on how to "work" it. B. F. GOODRICH RUBBER COMPANY.

3. TRANSFORMERS—A booklet giving data on input and output transformers. PACENT ELECTRIC COMPANY.

4. RESISTANCE-COUPLED AMPLIFIERS—A general discussion of resistance coupling with curves and circuit diagrams. COLE RADIO MANUFACTURING COMPANY.

5. CARBORUNDUM IN RADIO—A book giving pertinent data on the crystal as used for detection, with hook-ups, and a section giving information on the use of resistors. THE CARBORUNDUM COMPANY.

6. B-ELIMINATOR CONSTRUCTION—Constructional data on how to build. AMERICAN ELECTRIC COMPANY.

7. TRANSFORMER AND CHOKE-COUPLED AMPLIFICATION—Circuit diagrams and discussion. ALL-AMERICAN RADIO CORPORATION.

8. RESISTANCE UNITS—A data sheet of resistance units and their application. WARD-LEONARD ELECTRIC COMPANY.

9. VOLUME CONTROL—A leaflet showing circuits for distortionless control of volume. CENTRAL RADIO LABORATORIES.

10. VARIABLE RESISTANCE—As used in various circuits. CENTRAL RADIO LABORATORIES.

11. RESISTANCE COUPLING—Resistors and their application to audio amplification, with circuit diagrams. DEJUR PRODUCTS COMPANY.

12. DISTORTION AND WHAT CAUSES IT—Hook-ups of resistance-coupled amplifiers with standard circuits. ALLEN-BRADLEY COMPANY.

13. B-ELIMINATOR AND POWER AMPLIFIER—Instructions for assembly and operation using Raytheon tube. GENERAL RADIO COMPANY.

14. B-ELIMINATOR AND POWER AMPLIFIER—Instructions for assembly and operation using an R. C. A. rectifier. GENERAL RADIO COMPANY.

16. VARIABLE CONDENSERS—A description of the functions and characteristics of variable condensers with curves and specifications for their application to complete receivers. ALLEN D. CARDWELL MANUFACTURING COMPANY.

17. BAKELITE—A description of various uses of bakelite in radio, its manufacture, and its properties. BAKELITE CORPORATION.

20. POWER SUPPLY—A discussion on power supply with particular reference to lamp-socket operation. Theory and constructional data for building power supply devices. ACME APPARATUS COMPANY.

20. AUDIO AMPLIFICATION—A booklet containing data on audio amplification together with hints for the constructor. ALL AMERICAN RADIO CORPORATION.

21. SHORT-WAVE RECEIVER—Constructional data on a receiver which, by the substitution of various coils, may be made to tune from a frequency of 16,660 kc. (18 meters) to 1900 kc. (150 meters). SILVER-MARSHALL, INCORPORATED.

21. HIGH-FREQUENCY DRIVER AND SHORT-WAVE WAVEMETER—Constructional data and application. BURGESS BATTERY COMPANY.

46. AUDIO-FREQUENCY CHOKES—A pamphlet showing positions in the circuit where audio-frequency chokes may be used. SAMSON ELECTRIC COMPANY.

47. RADIO-FREQUENCY CHOKES—Circuit diagrams illustrating the use of chokes to keep out radio-frequency currents from definite points. SAMSON ELECTRIC COMPANY.

48. TRANSFORMER AND IMPEDANCE DATA—Tables giving the mechanical and electrical characteristics of transformers and impedances, together with a short description of their use in the circuit. SAMSON ELECTRIC COMPANY.

49. BYPASS CONDENSERS—A description of the manufacture of bypass and filter condensers. LESLIE F. MUTER COMPANY.

50. AUDIO MANUAL—Fifty questions which are often asked regarding audio amplification, and their answers. AMERTRAN SALES COMPANY, INCORPORATED.

51. SHORT-WAVE RECEIVER—Constructional data on a receiver which, by the substitution of various coils, may be made to tune from a frequency of 16,660 kc. (18 meters) to 1900 kc. (150 meters). SILVER-MARSHALL, INCORPORATED.

52. AUDIO QUALITY—A booklet dealing with audio-frequency amplification of various kinds and the application to well-known circuits. SILVER-MARSHALL, INCORPORATED.

56. VARIABLE CONDENSERS—A bulletin giving an analysis of various condensers together with their characteristics. GENERAL RADIO COMPANY.

57. FILTER DATA—Facts about the filtering of direct current supplied by means of motor-generator outfits used with transmitters. ELECTRIC SPECIALTY COMPANY.

59. RESISTANCE COUPLING—A booklet giving some general information on the subject of radio and the application of resistors to a circuit. DAVEN RADIO CORPORATION.

60. RESISTORS—A pamphlet giving some technical data on resistors which are capable of dissipating considerable energy; also data on the ordinary resistors used in resistance-coupled amplification. THE CRESCENT RADIO SUPPLY COMPANY.

63. RADIO-FREQUENCY AMPLIFICATION—Constructional details of a five-tube receiver using a tuned radio-frequency transformer. CANFIELD RADIO MFG. COMPANY.

64. FIVE-TUBE RECEIVER—Constructional data on building a receiver. AERO PRODUCTS, INCORPORATED.

64. AMPLIFICATION WITHOUT DISTORTION—Data and curves illustrating the use of various methods of amplification. ACME APPARATUS COMPANY.

66. SUPER-HETERODYNE—Constructional details of a seven-tube set. G. C. EVANS COMPANY.

70. IMPROVING THE AUDIO AMPLIFIER—Data on the characteristics of audio transformers, with a circuit diagram showing where chokes, resistors, and condensers can be used. AMERICAN TRANSFORMER COMPANY.

72. PLATE SUPPLY SYSTEM—A wiring diagram and layout plan for a plate supply system to be used with a power amplifier. Complete directions for wiring are given. AMERTRAN SALES COMPANY.

80. FIVE-TUBE RECEIVER—Data are given for the construction of a five-tube tuned radio-frequency receiver. Complete instructions; list of parts, circuit diagram, and template are given. ALL-AMERICAN RADIO CORPORATION.

81. BETTER TUNING—A booklet giving much general information on the subject of radio reception with specific illustrations. Primarily for the non-technical home constructor. BREMER-TULLY MANUFACTURING COMPANY.

82. SIX-TUBE RECEIVER—A booklet containing photographs, instructions, and diagrams for building a six-tube shielded receiver. SILVER-MARSHALL, INCORPORATED.

83. SOCKET POWER DEVICE—A list of parts, diagrams, and templates for the construction and assembly of socket power devices. JEFFERSON ELECTRIC MANUFACTURING COMPANY.

84. FIVE-TUBE EQUAMATIC—Panel layout, circuit diagrams, and instructions for building a five-tube receiver, together with data on the operation of tuned radio-frequency transformers of special design. KARAS ELECTRIC COMPANY.

85. FILTER—Data on a high-capacity electrolytic condenser used in filter circuits in connection with A socket power supply units, are given in a pamphlet. THE ABOX COMPANY.

86. SHORT-WAVE RECEIVER—A booklet containing data on a short-wave receiver as constructed for experimental purposes. THE ALLEN D. CARDWELL MANUFACTURING CORPORATION.

88. SUPER-HETERODYNE CONSTRUCTION—A booklet giving full instructions, together with a blue print and necessary data, for building an eight-tube receiver. THE GEORGE W. WALKER COMPANY.

89. SHORT-WAVE TRANSMITTER—Data and blue prints are given on the construction of a short-wave transmitter, together with operating instructions, methods of keying, and other pertinent data. RADIO ENGINEERING LABORATORIES.

90. IMPEDANCE AMPLIFICATION—The theory and practice of a special type of dual-impedance audio amplification are given. ALDEN MANUFACTURING COMPANY.

91. B-SOCKET POWER—A booklet giving constructional details of a socket-power device using either the BH or 313 type rectifier. NATIONAL COMPANY, INCORPORATED.

92. POWER AMPLIFIER—Constructional data and wiring diagrams of a power amplifier combined with a B-supply unit are given. NATIONAL COMPANY, INCORPORATED.

100. A, B, and C SOCKET-POWER SUPPLY—A booklet giving data on the construction and operation of a socket-power supply using the new high-current rectifier tube. THE Q. R. S. MUSIC COMPANY.

101. USING CHOKES—A folder with circuit diagrams of the more popular circuits showing where choke coils may be placed to produce better results. SAMSON ELECTRIC COMPANY.

ACCESSORIES

22. A PRIMER OF ELECTRICITY—Fundamentals of electricity with special reference to the application of dry cells to radio and other uses. Constructional data on buzzers, automatic switches, alarms, etc. NATIONAL CARBON COMPANY.

23. AUTOMATIC RELAY CONNECTIONS—A data sheet showing how a relay may be used to control A and B circuits. YAXLEY MANUFACTURING COMPANY.

25. ELECTROLYTIC RECTIFIER—Technical data on a new type of rectifier with operating curves. KODEL RADIO CORPORATION.

26. DRY CELLS FOR TRANSMITTERS—Actual tests given, well illustrated with curves showing exactly what may be expected of this type of B power. BURGESS BATTERY COMPANY.

27. DRY-CELL BATTERY CAPACITIES FOR RADIO TRANSMITTERS—Characteristic curves and data on discharge tests. BURGESS BATTERY COMPANY.

28. B BATTERY LIFE—Battery life curves with general curves on tube characteristics. BURGESS BATTERY COMPANY.

29. HOW TO MAKE YOUR SET WORK BETTER—A nontechnical discussion of general radio subjects with hints on how reception may be bettered by using the right tubes. UNITED RADIO AND ELECTRIC CORPORATION.

30. TUBE CHARACTERISTICS—A data sheet giving constants of tubes. C. E. MANUFACTURING COMPANY.

31. FUNCTIONS OF THE LOUD SPEAKER—A short, nontechnical general article on loud speakers. AMPLION CORPORATION OF AMERICA.

32. METERS FOR RADIO—A catalogue of meters used in radio, with connecting diagrams. BURTON-ROGERS COMPANY.

33. SWITCHBOARD AND PORTABLE METERS—A booklet giving dimensions, specifications, and shunts used with various meters. BURTON-ROGERS COMPANY.

34. COST OF B BATTERIES—An interesting discussion of the relative merits of various sources of B supply. HART-FORD BATTERY MANUFACTURING COMPANY.

35. STORAGE BATTERY OPERATION—An illustrated booklet on the care and operation of the storage battery. GENERAL LEAD BATTERIES COMPANY.

36. CHARGING A AND B BATTERIES—Various ways of connecting up batteries for charging purposes. WESTINGHOUSE UNION BATTERY COMPANY.

37. CHOOSING THE RIGHT RADIO BATTERY—Advice on what dry cell battery to use; their application to radio, with wiring diagrams. NATIONAL CARBON COMPANY.

53. TUBE REACTIVATOR—Information on the care of vacuum tubes, with notes on how and when they should be reactivated. THE STERLING MANUFACTURING COMPANY.

54. ARRESTERS—Mechanical details and principles of the vacuum type of arrester. NATIONAL ELECTRIC SPECIALTY COMPANY.

55. CAPACITY CONNECTOR—Description of a new device for connecting up the various parts of a receiving set, and at the same time providing bypass condensers between the leads. KURZ-KASCH COMPANY.

(Continued on page 322)

"RADIO BROADCAST'S" DIRECTORY OF MANUFACTURED RECEIVERS

¶ A coupon will be found on page 328. All readers who desire additional information on the receivers listed below need only insert the proper numbers in the coupon, mail it to the Service Department of RADIO BROADCAST, and full details will be sent.

KEY TO TUBE ABBREVIATIONS

99—60-mA. filament (dry cell)
01A—Storage battery 0.25 amps. filament
12—Power tube (Storage battery)
71—Power tube (Storage battery)
16B—Half-wave rectifier tube
Hmu—High-Mu tube for resistance-coupled audio
20—Power tube (dry cell)
10—Power Tube (Storage battery)
00A—Special detector
13—Full-wave rectifier tube
26—Low-voltage high-current a. c. tubes
27—Heater type a. c. tube

DIRECT CURRENT RECEIVERS

NO. 424. COLONIAL 26

Six tubes; 2 t. r. f. (01–A), detector (12), 2 transformer audio (01-A and 71). Balanced t. r. f. One to three dials. Volume control: antenna switch and potentiometer across first audio. Watts required: 120. Console size: 34 x 38 x 18 inches. Headphone connections. The filaments are connected in a series parallel arrangement. Price $250 including power unit.

NO. 425. SUPERPOWER

Five tubes: All 01–A tubes. Multiplex circuit. Two dials. Volume control: resistance in r. f. plate. Watts required: 30. Antenna: loop or outside. Cabinet sizes: table, 27 x 10 x 9 inches; console, 28 x 50 x 21. Prices: table, $135 including power unit; console, $390 including power unit and loud speaker.

A. C. OPERATED RECEIVERS

NO. 508. ALL-AMERICAN 77, 88, AND 99

Six tubes; 3 t. r. f. (26), detector (27), 2 transformer audio (26 and 71). Rice neutralized t. r. f. Single drum tuning. Volume control: potentiometer in r. f. plate. Cabinet sizes: No. 77, 21 x 10 x 8 inches; No. 88 Hiboy, 25 x 38 x 18 inches; No. 99 console, 27½ x 43 x 20 inches. Shielded. Output device. The filaments are supplied by means of three small transformers. The plate supply employs a gas-filled rectifier tube. Volt-meter in a. c. supply line. Prices: No. 77, $150, including power unit; No. 88, $210 including power unit; No. 99, $285 including power unit and loud speaker.

NO. 509. ALL-AMERICAN "DUET"; "SEXTET"

Six tubes; 2 t. r. f. (99), detector (99), 3 transformer audio (99 and 12). Rice neutralized t.r.f. Two dials. Volume control: resistance in r.f. plate. Cabinet sizes: "Duet," 23 x 56 x 16½ inches; "Sextet," 22½ x 13½ x 15½ inches. Shielded. Output device. The 99 filaments are connected in series and supplied with rectified a.c., while 12 is supplied with raw a.c. The plate and filament supply uses gaseous rectifier tubes. Millammeter on power unit. Prices: "Duet," $160 including power unit; "Sextet," $220 including power unit and loud speaker.

NO. 511. ALL-AMERICAN 80, 90, AND 115

Five tubes; 2 t.r.f. (99), detector (99), 2 transformer audio (99 and 12). Rice neutralized t.r.f. Two dials. Volume control: resistance in r.f. plate. Cabinet sizes: No. 80, 23½ x 12½ x 15 inches; No. 90, 37½ x 12 x 12½ inches; No. 115 Hiboy, 24 x 41 x 15 inches. Coils individually shielded. Output device. See No. 509 for power supply. Prices: No. 80, $135 including power unit; No. 90, $145 including power unit and compartment; No. 115, $170 including power unit, compartment, and loud speaker.

NO. 510. ALL-AMERICAN 7

Seven tubes; 3 t.r.f. (26), 1 untuned r.f. (26), detector (27), 2 transformer audio (26 and 71). Rice neutralized t.r.f. One drum. Volume control: resistance in r.f. plate. Cabinet sizes: "Sovereign" console, 30½ x 60½ x 19 inches; "Lorraine" Hiboy, 25½ x 53½ x 17½ inches; "Forte" cabinet, 25½ x 13½ x 17½ inches. For filament and plate supply: See No. 508. Prices: "Sovereign" $460; "Lorraine" $360; "Forte" $270. All prices include power unit. First two include loud speaker.

NO. 403. ARGUS 250B

Six tubes; 2 t.r.f. (99), 1 untuned r.f. (99), detector (99), 2 transformer audio (99 and 12). Stabilized with grid resistances. Two dials. Volume control: resistance across 1st audio. Watts required: 100. Cabinet size: 35½ x 14½ x 10½ inches. Output device. The 99 filaments are connected in series and supplied with rectified a.c., while the 12 is run on raw a.c. The power unit requires two 16-B rectifier tubes. Milliammeter included in d.c. supply. Price $250.00 including self-contained power unit. Other models: No. 125, $125.00; console model, $375.00.

NO. 401. AMRAD AC9

Six tubes; 3 t.r.f. (99), detector (99), 2 transformer (99 and 12). Neutrodyne. Two dials. Volume control: resistance across 1st audio. Watts consumed: 50. Cabinet size: 27 x 9 x 11½ inches. The 99 filaments are connected in series and supplied with rectified a.c., while the 12 is run on raw a.c. The power unit, requiring two 16-B rectifiers, is separate and supplies A, B, and C current. Price $142 including power unit.

NO. 402. AMRAD AC5

Five tubes. Same as No. 401 except one less r.f. stage. Price $125 including power unit.

NO. 484. BOSWORTH, B5

Five tubes; 2 t.r.f. (26), detector (99), 2 transformer audio (special a.c. tubes). T.r.f. circuit. Two dials. Volume control: potentiometer. Cabinet size: 23 x 7 x 8 inches. Output device included. Price $175.

NO. 406. CLEARTONE 110

Five tubes; 2 t.r.f., detector, 2 transformer audio. All tubes a. c. heater type. One or two dials. Volume control: resistance in r. f. plate. Watts consumed: 40. Cabinet size: varies. The plate supply is built in the receiver and requires one rectifier tube. Filament supply through step down transformers. Prices range from $175 to $375 which includes 5 a.c. tubes and one rectifier tube.

NO. 407. COLONIAL 25

Six tubes; 2 t. r. f. (01–A), detector (99), 2 resistance audio (99). 1 transformer audio (10). Balanced t.r.f. circuit. One or three dials. Volume control: Antenna switch and potentiometer on 1st audio. Watts consumed: 100. Console size: 34 x 38 x 18 inches. Output device. All tube filaments are operated on a. c. except the detector which is supplied with rectified a.c. from the plate supply. The rectifier employs two 16-B tubes. Price $250 including built-in plate and filament supply.

NO. 507. CROSLEY 602 BANDBOX

Six tubes; 3 t.r.f. (26), detector (27), 2 transformer audio (26 and 71). Balanced t.r.f. circuit. One dial. Cabinet size: 17½ x 5½ x 7½ inches. The heaters for the a.c. tubes and the 71 filament are supplied by small transformers available to operate either on 50 or 60 cycles. The plate current is supplied by means of rectifier tube. Price $65 for set alone, power unit $60.

NO. 408. DAY-FAN "DE LUXE"

Six tubes; 3 t.r.f., detector, 2 transformer audio. All 01–A tubes. One dial. Volume control: potentiometer across r.f. tubes. Watts consumed: 300. Console size: 30 x 40 x 20 inches. The filaments are connected in series and supplied with d.c. from a motor-generator set which also supplies B and C current. Output device. Price $350 including power unit.

NO. 409. DAYCRAFT 5

Five tubes; 2 t.r.f., detector, 2 transformer audio. All a. c. heater tubes. Reflexed t.r.f. One dial. Volume control: potentiometers in r.f. plate and 1st audio. Watts consumed: 135. Console size: 34 x 36 x 14 inches. Output device. The heaters are supplied by means of a small transformer. A built-in rectifier supplies B and C voltages. Price $170, less tubes. The following have one more r.f. stage and are not reflexed: Daycraft 6, $195; Dayrole 6, $235; Dayfan 6, $110. All prices less tubes.

NO. 469. FREED-EISEMANN NR11

Six tubes; 3 t.r.f. (01–A), detector (01–A), 2 transformer audio (01–A and 71). Neutrodyne. One dial. Volume control: potentiometer. Watts consumed: 150. Cabinet size: 19½ x 10 x 10½ inches. Shielded. Output device. A special power unit is included employing a rectifier tube. Price $225 including NR-411 power unit.

NO. 487. FRESHMAN 7F-AC

Six tubes; 3 t.r.f. (26), detector (27), 2 transformer audio (26 and 71). Equaphase circuit. One dial. Volume control: potentiometer across 1st audio. Console size: 24½ x 41½ x 15 inches. Output device. The filaments and heaters are all supplied by means of small transformers. The plate supply requires one 13 rectifier tube. Price $160 including tubes.

NO. 411. HERBERT LECTRO 120

Five tubes; 2 t.r.f. (99), detector (99), 2 transformer audio (99 and 71). Three dials. Volume control: rheostat in primary of a.c. transformer. Watts required: 45. Cabinet size: 32 x 10 x 12 inches. The 99 filaments are connected in series, supplied with rectified a. c., while the 71 is run on raw a. c. The power unit uses a Q. R. S. rectifier tube. Price $120.

NO. 412. HERBERT LECTRO 200

Six tubes; 2 t.r.f. (99), detector (99), 1 transformer audio (99), 1 push-pull audio (71). One dial. Volume control: rheostat in primary of a. c. transformer. Watts consumed: 60. Cabinet size: 20 x 12 x 12 inches. Filaments connected same as above. Completely shielded. Output device. Price $200.

NO. 410. LARCOFLEX 73

Seven tubes; 4 t.r.f. (01–A), detector (01–A), 2 transformer audio (01–A and 71). T.r.f. circuit. One dial. Volume control: resistance in r.f. plate. Console size 30 x 42 x 20 inches. Completely shielded. Built-in A, B and C supply. Price $215.

NO. 490. MOHAWK

Six tubes; 2 t.r.f., detector, 2 transformer audio. All tubes a.c heater type except 71 in last stage. One dial. Volume control: rheostat on r.f. Watts consumed: 40. Panel size: 12½ x 8½ inches. Output device. The heaters for the a.c tubes and the 71 filament are supplied by small transformers. The plate supply is of the built-in type using a rectifier tube. Prices range from $65 to $245

NO. 413. MARTI

Six tubes: 2 t.r.f., detector, 3 resistance audio. Al tubes a.c. heater type. Two dials. Volume control resistance in r.f. plate. Watts consumed: 38. Panel siz 7 x 21 inches. The built-in plate supply employs on 16-B rectifier. The filaments are supplied by a smal transformer. Prices: table, $235 including tubes an rectifier; console, $275 including tubes and rectifier console, $325 including tubes, rectifier, and loud speake

NO. 416. NASSAU POWER

Six tubes; 2 t. r. f. (99) detector (99), 3 audio (99 an 10). Two dials. Volume control; resistace in r.f. plat circuit. Bridge circuit. Cabinet size: 28 x 45 x 18 inche The 99 filaments are connected in series and supplie with rectified a. c., while the 10 is supplied with ra a. c. The power unit requires one 16-B tube. Output de vice. Milliammeter on the front panel indicates filamen current.

NO. 417. RADIOLA 28

Eight tubes; 2 detectors (99) 1 oscillator (99), transformer audio (99 and 10), 3 intermediate frequenc (99). Super-heterodyne. One dial. Volume contro potentiometer on intermediate stages. Console size 26½ x 63 x 17 inches. The 99 filaments are connected i series and supplied with rectified a.c., while the 10 tub is supplied with raw a.c. The power unit requires tw 216-B's, one 874, and one 876, the latter two being regu lator tubes, Loop operated. Shielded. Output devic Price $540 including all tubes and No. 104 loud speake Model 28 also sold without the power units and lou speaker.

NO. 418. SUPERPOWER A. C.

Five tubes; "Multiplex" circuit using 01–A tube Two dials. Volume control: resistance in r. f. plat Console size: 28 x 50 x 21 inches. Antenna: loop or out side. The filaments of the 01-A tubes are supplied wit rectified a. c. The rectifier employs two tungar tube Price $390 including loud speaker and power uni Cabinet size: 27 x 10 x 9 inches and lists for $180 includ ing power unit.

NO. 420. SIMPLEX B

Six tubes; 3 t.r.f., detector, 2 transformer audi All tubes a.c. heater type. One dial. Volume contr resistance in r.f. plate. Headphone connection. Consol size: 34 x 36 x 14 inches. The heaters are supplied b means of a small a. c. transformer, the B and C volta being obtained from a built-in plate supply unit. Pri $250, complete.

NO. 421. SOVEREIGN 238

Seven tubes of the a.c. heater type. Balanced t.r Two dials. Volume control: resistance across 2nd audi Watts consumed: 45. Console size: 37 x 52 x 15 inch The heaters are supplied by a small a. c. transform while the plate is supplied by means of rectified using a gaseous type rectifier. Price $325, includi power unit and tubes.

NO. 422. SUPERVOX, JR. AND SR.

Four tubes; 1 t.r.f., detector, 2 transformer aud All tubes a. c. heater type except last which is a One dial. Volume controls: antenna coupler and resi ance in r.f. plate. Watts required: 40. Console si 28 x 30 x 16 inches. Shielded detector. Output devi The heaters are supplied by a small transformer wh the 71 filament is supplied with a. c. Price $275 co plete. The Supervox Sr. has one more stage of t. r. has two dials, requires 60 watts, is completely shield and has output device. Price $450, complete.

NO. 488. U-FLEX 5

Four tubes; 2 t.r.f., crystal detector, 2 transfor audio. Reflexed. Three dials. Volume control: resista in r.f. plate. Cabinet size: 28 x 9 x 9½ inches. A C is used as a rectifier for the plate supply. Headph connection. Price, $125.00

BATTERY OPERATED RECEIVERS

NO. 512. ALL-AMERICAN 44, 45, AND 66

Six tubes; 3 t.r.f. (01–A, detector) 01–A, 2 transfor audio (01-A and 71). Rice neutralized t.r.f. Dr control. Volume control: rheostat in r.f. Cabinet si No. 44, 21 x 10 x 8 inches; No. 55, 25 x 38 x 18 inch No. 66, 27½ x 43 x 20 inches. C-battery connectio Battery cable. Antenna: 75 to 125 feet. Prices: No. $70; No. 55, $125 including loud speaker; No. 66, $ including loud speaker.

NO. 428. AMERICAN C6

Five tubes; 2 t.r.f, detector, 2 transformer au All 01–A tubes. Semi balanced t.r.f. Three dials. P current 15 mA. Volume control: potentiometer. Cabi sizes: table, 20 x 8½ x 10 inches; console, 36 x 40 x inches. Partially shielded. Battery cable. C-batt connections. Antenna: 125 feet. Prices: table, console, $65 including loud speaker.

NO 433. ARBORPHONE

Five tubes; 2 t.r.f., detector, 2 transformer au All 01-A tubes. Two dials. Plate current: 16 mA. ume control: rheostat in r.f. and resistance in r.f. p C-battery connections. Binding posts. Antenna: for various lengths. Cabinet size: 24 x 9 x 10½ in Price: $65.

Better Results *from* **POWER CIRCUITS**

Centralab Power Rheostat

HERE are the new Centralab units designed especially for use in socket power circuits to carry continuously and unusually heavy current for their size, providing smooth acting control under all conditions.

Centralab Power Rheostat is warp-proof, heat-proof, permitting continuous operation at temperatures of 482° F. and beyond. Resistance wire is wound on metal core, asbestos insulated. Core expands with wire, insuring smooth action. Narrow resistance strips give small resistance jumps per turn, further insurance of even regulation. Compact 2" diameter, 1" behind panel. Ohms—300, 250, 150, 50, 15, 6, 3, .2, .5—price $1.25.

POWER Centralab Potentiometer

This new unit is identical with the Power Rheostat except for an additional terminal, and is especially suited to obtain variable voltages for detector tube and variable "C" bias in socket power circuits. 15, 150, 250 ohms, $1.50; 2,000, $1.75; 5,000, $2.00.

4-th TERMINAL Centralab Potentiometer

With an added semi-variable contact arm, this new potentiometer is identical to the above units. The 4th terminal is adjustable behind panel to any resistance value. 175 ohm unit gives a variable voltages in ABC power circuits. 250 ohms is used with the new Raytheon ABC. The 2,000 is used for "C" bias in such circuits as AmerTran Power Pack. Two 6,000 ohm units in series across output of a "B" eliminator gives best possible voltage regulation. 175, 250 ohms, $2; 2,000, 3,000, 5,000, $2.25.

At your dealer's, or C.O.D. Send for new ABC power circuits and circuits for improved "B" power control.

Central Radio Laboratories
22 Keefe Ave., Milwaukee, Wis.

Centralab

BETWEEN
80 and 5000 Cycles
is the effective range of
Radio Reproduction

Type 285 Transformers

The type 285 transformers are available in two ratios as follows:

| Type 285-H | 1 to 6 |
| Type 285-D | 1 to 3 |

Price **$6.00** each

Type 387-A

The function of the Speaker Filter is to protect the speaker windings from the direct current while allowing an unimpeded flow of alternating frequency current. This reduces the amount of distortion and prolongs the life of the speaker. Such a device should always be used especially between a power tube in the last audio stage and the speaker.

Price **$6.00**

In buying parts for an audio amplifier it should be borne in mind that no better tone quality can be reproduced than is radiated from the broadcasting station or than can be sustained by the loudspeaker regardless of what method of coupling may be used.

The frequency range of the better broadcasting stations is about 80 to 5000 cycles; the frequency range of the better loudspeaker is about 80 to 7000 cycles; thus making the effective range of radio reproduction between 80 and 5000 cycles. This represents approximately the full orchestral and vocal range.

The logical amplifying device to use in covering the effective range of radio reproduction, then, is the one which will amplify uniformly all frequencies between 80 and 5000 cycles, with the greatest gain per stage and lowest operating cost. Such a device is a properly designed audio frequency amplifying transformer.

While other methods of coupling may have a more uniform frequency curve over a wider range than a transformer, there is much to be said in favor of using good transformers because of the greater gain in amplification per stage and the lower operating cost.

The General Radio Type 285 Transformers are designed to have a high inductance value, but with lower capacity. This combination sustains both the upper and lower ends of the amplification curve to the same degree as the middle portion and is accomplished by using a larger core of a very high quality of selected steel and proper adjustment of coil turns.

When you overhaul your old receiver or build your new one why not assure yourself of good tone quality with plenty of volume at moderate cost by using a pair of General Radio Type 285 transformers?

Have you thought of operating your set entirely from the light socket by using the new A. C. tubes and a plate supply unit?

Write for our circular showing how to rewire the filament circuits of a standard four tube set to use the new A. C. tubes.

GENERAL RADIO COMPANY
Cambridge   Massachusetts

GENERAL RADIO
PARTS AND ACCESSORIES

NO. 431. AUDIOLA 6

Six tubes; 3 t.r.f. (01-A), detector (00-A), 2 transformer audio (01-A and 71). Drum control. Plate current: 20 mA. Volume control: resistance in r.f. plate. Stage shielding. Battery cable. C-battery connection. Antenna: 50 to 100 feet. Cabinet size: 28½ x 11 x 14½ inches. Price not established.

NO. 432. AUDIOLA 8

Eight tubes; 4 t.r.f. (01-A), detector (00-A), 1 transformer audio (01-A), push-pull audio (12 or 71). Bridge balanced t.r.f. Drum control. Volume control: resistance in r.f. plate. Stage shielding. Battery cable. C-battery connections. Antenna: 10 to 100 feet. Cabinet size: 28½ x 11 x 149 inches. Price not established.

NO. 485. BOSWORTH B6

Five tubes; 2 t.r.f. (01-A), detector (01-A), 2 transformer audio (01-A and 71). Two dials. Volume control: variable grid resistances. Battery cable. C-battery connections. Antenna: 25 feet or longer. Cabinet size 15 x 7 x 8 inches. Price $75.

NO. 513. COUNTERPHASE SIX

Six tubes; 3 t.r.f. (01-A), detector (00-A), 2 transformer audio (01-A and 12). Counterphase t.r.f. Two dials. Plate current: 32 mA. Volume control: rheostat on 2nd and 3rd r.f. Coils shielded. Battery cable. C-battery connections. Antenna: 75 to 100 feet. Console size: 18½ x 40½ x 15½ inches. Prices: Model 35, table, $110; Model 37, console, $175.

NO. 514. COUNTERPHASE EIGHT

Eight tubes; 4 t.r.f. (01-A), detector (00-A), 2 transformer audio (01-A and 12). Counterphase t.r.f. One dial. Plate current: 40 mA, Volume control: rheostat in 1st r.f. Copper stage shielding. Battery cable. C-battery connections. Antenna: 75 to 100 feet. Cabinet size: 30 x 12½ x 16 inches. Prices: Model 12, table, $225; Model 16, console, $335; Model 18, console, $365.

NO. 506. CROSLEY 601 BANDBOX

Six tubes; 3 t.r.f., detector, 2 transformer audio. All 01-A tubes. Balanced, t.r.f. One dial. Plate current: 40 mA. Volume control: rheostat in r.f. Shielded. Battery cable. C-battery connections. Antenna: 75 to 150 feet. Cabinet size: 17½ x 5½ x 7½. Price, $55.

NO. 462. CUSTOM BUILT 7 AND 9

Seven tubes; 5-01A, 1-00-A, 1-71. T.r.f. circuit. Two dials. Plate current: 40 mA. Volume control: special. Binding posts. C-battery connections. Output device. Antenna: 100 feet. Built-in A and B supply. Panel: 7 x 21. Price $275 completely equipped. The Custom Built 9 has 9 tubes with built-in speaker and loop. Price $375 completely equipped.

NO. 477. DAVEN BASS NOTE

Six tubes; 2 t.r.f. (01-A), detector (HMu), 3 resistance audio (HMu and power). Two dials. Plate current: 17 mA. Volume control: potentiometer. Battery cable, C-battery connections. Antenna: 50 to 100 feet. Cabinet size: 23½ x 12 x 16 inches. Price $150.

NO. 434. DAY-FAN 6

Six tubes; 3 t.r.f. (01-A), detector (01-A), 2 transformer audio (01-A and 12 or 71). One dial. Plate current: 12 to 15 mA. Volume control: rheostat on r.f. Shielded. Battery cable, C-battery connections. Output device. Antenna: 50 to 120 feet. Cabinet sizes: Daycraft 6, 32 x 30 x 34 inches; Day-Fan Jr., 15 x 7 x 7. Prices: Day-Fan 6, $110; Daycraft 6, $145 including loud speaker; Day-Fan Jr. not available.

NO. 435. DAY-FAN 7

Seven tubes; 3 t.r.f. (01-A), detector (01-A), 1 resistance audio (01-A), 2 transformer audio (01-A and 12 or 71). Plate current: 15 mA. Antenna: outside. Same as No. 434. Price $115.

NO. 497. EXCELL GRAND

Five tubes; 2 t.r.f., detector, 2 transformer audio. All 99 tubes. Na-Ald drum control Plate current: 15 mA. Volume control: rheostat in r.f. Binding posts. C-battery connections. Headphone connection. Antenna; 75 feet. Cabinet size: 21 x 16 x 14 inches. Price $100, including loud speaker.

NO. 503. FADA SPECIAL

Six tubes; 3 t.r.f. (01-A), detector (01-A), 2 transformer audio (01-A and 71). Neutrodyne. Two drum control. Plate current: 20 to 24mA. Volume control: rheostat on r.f. Coils shielded. Battery cable. C-battery connections. Headphone connection. Antenna: outdoor. Cabinet size: 20½ x 13½ x 10½ inches. Price $95.

NO. 504. FADA 7

Seven tubes; 4 t.r.f. (01-A), detector (01-A), 2 transformer audio (01-A and 71). Neutrodyne. Two drum control. Plate current; 43mA. Volume control: rheostat on r.f. Completely shielded. Battery cable. C-battery connections. Headphone connections, Output device. Antenna: outdoor or loop. Cabinet sizes: table, 25½ x 13½ x 11½ inches; console, 29 x 50 x 17 inches. Prices: table, $185; console, $285.

NO. 436. FEDERAL

Five tubes; 2 t.r.f. (01-A), detector (01-A), 2 transformer audio (01-A and 12 or 71). Balanced t.r.f. One dial. Plate current: 20.7 mA. Volume control: rheostat on r.f. Shielded. Battery cable. C-battery connections. Antenna: loop, Made in 6 models. Price varies from $250 to $1000 including loop.

NO. 505. FADA 8

Eight tubes. Same as No. 504 except for one extra stage of audio and different cabinet. Prices: table, $300; console, $400.

NO. 437. FERGUSON 10A

Seven tubes; 3 t.r.f. (01-A), detector (01-A), 3 audio (01-A and 12 or 71). One dial. Plate current: 18 to 25 mA. Volume control: rheostat on two r.f. Shielded. Battery cable. C-battery connections. Antenna: 100 feet. Cabinet size: 21½ x 12 x 15 inches. Price $150.

NO. 438. FERGUSON 14

Ten tubes: 3 untuned r.f., 3 t. r.f. (01-A), detector (01-A), 3 audio (01-A and 12 or 71). Special balanced t.r.f. One dial. Plate current: 30 to 35 mA. Volume control: rheostat in three r.f. Shielded. Battery cable. C-battery connections. Antenna: loop. Cabinet size: 24 x 12 x 16 inches. Price $235, including loop.

NO. 439. FERGUSON 12

Six tubes; 2 t.r.f. (01-A), detector (01-A), 1 transformer audio (01-A), 2 resistance audio (01-A and 12 or 71), Two dials. Plate current: 18 to 25 mA. Volume control: rheostat on two r.f, Partially shielded. Battery cable. C-battery connections. Antenna: 100 feet. Cabinet size: 22½ x 10 x 12 inches. Price $85. Consolette $145 including loud speaker.

NO. 440. FREED-EISEMANN NR-8, NR-9, AND NR-66

Six tubes; 3 t.r.f. (01-A), detector (01-A), 2 transformer audio (01-A and 71). Neutrodyne. NR-8, two dials; others one dial. Plate current: 30 mA. Volume control: rheostat on r.f, NR-8 and 9; chassis type shielding. NR-66, individual stage shielding. Battery cable. C-battery connections. Antenna: 100 feet. Cabinet sizes: NR-8 and 9, 19½ x 10 x 10½ inches; NR-66 20 x 10½ x 12 inches. Prices: NR-8, $90; NR-9, $100; NR-66, $125.

NO. 441. FREED-EISEMANN NR-77

Seven tubes: 4 t.r.f. (01-A), detector (01-A), 2 transformer audio (01-A and 71). Neutrodyne. One dial. Plate current: 35 mA. Volume control: rheostat on r.f. Shielding. Battery cable. C-battery connections. Antenna: outside or loop. Cabinet size: 23 x 10½ x 13 inches. Price $175.

NO. 442. FREED-EISEMANN 800 AND 850

Eight tubes; 4 t.r.f. (01-A), detector (01-A), 1 transformer (01-A), 1 parallel audio (01-A or 71). Neutrodyne. One dial. Plate current: 35 mA. Volume control: rheostat on r.f. Shielded. Battery cable. C-battery connections. Output: two tubes in parallel or one power tube may be used. Antenna; outside or loop. Cabinet sizes: No. 800, 34 x 15½ x 13½ inches; No. 850, 36 x 65½ x 17½. Prices not available.

NO. 443. GREBE CR18 (SHORT-WAVE)

Two tubes; detector, 1 transformer audio. All 01-A tubes. Three-circuit regenerative. Two dials. Plate current: 8 mA. Volume control: rheostat on detector and regeneration. Headphone connection. Binding posts. Wavelength range; 8 to 210 meters. Antenna: 100 feet. Cabinet size: 16 x 7 x 7½ inches. Price $100 including set of coils.

NO. 444. GREBE MU-1

Five tubes; 2 t.r.f. (01-A), detector (01-A), 2 transformer audio (01-A and 12 or 71). Balanced t.r.f. One, two, or three dials (operate singly or together). Plate current: 30mA. Volume control: rheostat on r.f. Binocular coils. Binding posts. C-battery connections. Antenna: 125 feet. Cabinet size: 22½ x 9½ x 13 inches. Prices range from $95 to $320.

NO. 445. HARMONIC R AND S

Four tubes; 1 t.r.f., detector, 2 transformer audio. All 01-A tubes. S type has three resistance audio and 5 tubes. Regenerative detector and t.r.f. Three dials. Volume control; rheostat on r.f. Binding posts. C-battery connections. Headphone connection. Antenna: 100 feet. Cabinet size: 26 x 9 x 9 inches. Prices: R, $75; S. $100.

NO. 426. HOMER

Seven tubes; 4 t.r.f. (01-A); detector (01-A or 00A); 2 audio (01-A and 12 or 71). One knob tuning control. Volume control: rotor control in antenna circuit. Plate current: 22 to 25 mA. "Technidyne" circuit. Completely enclosed in aluminum box. Battery cable. C-battery connections. Cabinet size, 8½ x 19½ x 9½ inches. Chassis size, 6½ x 17 x 8 inches. Prices: Chassis only, $80. Table cabinet, $95.

NO. 502. KENNEDY ROYAL 7. CONSOLETTE

Seven tubes; 4 t.r.f. (01-A), detector (00-A), 2 transformer audio (01-A and 71). One dial. Plate current: 42 mA. Volume control: rheostat on two r.f. Special r.f. coils. Battery cable. C-battery connections. Headphone connection. Antenna: outside or loop. Consolette size: 36½ x 35½ x 19 inches. Price $220.

NO. 498. KING "CRUSADER"

Six tubes; 2 t.r.f. (01-A), detector (00-A), 3 transformer audio (01-A and 71). Balanced t.r.f. One dial. Plate current: 20 mA. Volume control: rheostat on r.f. Coils shielded. Battery cable. C-battery connections. Antenna: outside. Panel: 11 x 7 inches. Price, $115.

NO. 499. KING "COMMANDER"

Six tubes; 3 t.r.f. (01-A), detector (00-A), 2 transformer audio (01-A and 71). Balanced t.r.f. One dial Plate current: 25 mA. Volume control: rheostat on r.f. Completely shielded, Battery cable. C-battery connections. Antenna: loop. Panel size: 12 x 8 inches. Price $220 including loop.

NO. 429. KING COLE VII AND VIII

Seven tubes; 3 t.r.f., detector, 1 resistance audio, 2 transformer audio. All 01-A tubes. Model VIII has one more stage t.r.f. (eight tubes). Model VII, two dials. Model VIII, one dial. Plate current: 15 to 50 mA. Volume control: primary shunt in r.f. Steel shielding. Battery cable and binding posts. C-battery connections. Output devices on some consoles. Antenna: 10 to 100 feet. Cabinet size: varies. Prices: Model VII, $80 to $160; Model VIII, $100 to $300.

NO. 500. KING "BARONET" AND "VIKING"

Six tubes; 2 t.r.f. (01-A), detector (00-A), 3 transformer audio (01-A and 71). Balanced t.r.f. One dial. Plate current: 19 mA. Volume control: rheostat in r.f. Battery cable. C-battery connections. Antenna: outside. Panel size: 18 x 7 inches. Prices: "Baronet," $70; "Viking," $140 including loud speaker.

NO. 501. KING "CHEVALIER"

Six tubes. Same as No. 500. Coils completely shielded. Panel size: 11 x 7 inches. Price, $210 including loud speaker.

NO. 447. LEUTZ "TRANSOCEANIC" AND "SILVER GHOST"

Nine tubes; 4 t.r.f. (01-A), detector (00-A), 1 transformer audio (01-A) 3 resistance audio (HMu and 71 or 10). Grid resistance in t.r.f. Wavelength range: 35 to 3600 meters. One to five dials. Plate current: 20 to 40 mA. Volume control: special. Shielded. Binding posts. C-battery connections. Voltmeter, Output device. Antenna: outside or loop. Cabinet sizes: "Transoceanic," 27 x 8½ x 13½ inches; "Silver Ghost," 72 x 12 x 20 inches. Prices: "Transoceanic," $150; Other not available.

NO. 489. MOHAWK

Six tubes; 2 t.r.f. (01-A), detector (00-A), 3 audio (01-A and 71). One dial. Plate current: 40 mA. Volume control: rheostat on r.f. Battery cable. C-battery connections. Output device. Antenna: 60 feet. Panel size: 12½ x 8½ inches. Prices range from $65 to $245.

NO. 449. NORBERT "MIDGET"

One multivalve tube; detector, 2 transformer audio. Two dials. Plate current: 3 mA. Volume control: rheostat. Binding posts. C-battery connections. Headphone connection. Antenna: 75 to 150 feet. Cabinet size: 12 x 8 x 9 inches. Price $12 including multivalve.

NO. 450. NORBERT 2

Two tubes; 1 t.r.f., detector, 2 transformer audio. One multi-valve tube and one 01-A. Two dials. Plate current: 8 mA. Volume control: special. Battery cable. Headphone connection. C-battery connections. Antenna: 50 to 100 feet. Cabinet size: 20 x 7 x 5½ inches. Price $40.50 including multivalve and 01-A tube.

NO. 451. NORCO 66

Six tubes; 2 t.r.f. (01-A), detector (01-A), 3 impedance audio (01-A and 71). Drum control. Plate current: 20 mA. Volume control: modulator on audio. Shielded. Battery cable. C-battery connections. Output device. Antenna: 70 to 90 feet. Cabinet size: 18½ x 8½ x 13½ inches. Price $130. Price of console, $250 including loud speaker.

NO. 452. ORIOLE 90

Five tubes; 2 t.r.f., detector, 2 transformer audio. All 01-A tubes. "Trinum" circuit. Two dials. Plate current: 18 mA. Volume control: rheostat on r. f. Battery cable. C-battery connections. Antenna: 50 to 100 feet. Cabinet size: 25½ x 11½ x 12½ inches. Price $85. Another model has 8 tubes, one dial, and is shielded. Price $185.

NO. 454. PARAMOUNT V AND VI

Five and six tubes. All 01-A t.r.f. circuit. Binding posts. C-battery connections. Antenna: 100 feet. Panel size: 26 x 7 inches. Prices: V, $65; VI, $75

NO. 453. PARAGON

Six tubes; 2 t.r.f. (01-A), detector (01-A), 3 double impedance audio (01-A and 71). One dial. Plate current: 40 mA. Volume control: resistance in r.f. plate. Shielded. Battery cable. C-battery connections. Output device. Antenna: 100 feet. Console size: 20 x 46 x 17 inches. Price not determined.

NO. 475. PENN C-6

Six tubes; 2 t.r.f. (01-A), detector (00-A), 3 truphonic audio (01-A). Phasatrol. One dial. Plate current: 15 mA. Volume control: potentiometer. Binding posts. C-battery connections. Antenna: 75 feet. Cabinet size: 24 x 10 x 15 inches. Prices range from $95 to $165. A console model having three dials and five tubes sells for $150.

NO. 480. PFANSTIEHL 30 AND 302

Six tubes; 3 t.r.f. (01-A), detector (01- 2A), trans former audio (01-A and 71). One dial. Plate current 26 to 32 mA. Volume control: resistance in r.f. plate Shielded. Battery cable. C-battery connections. An tenna: outside. Panel size: 17½ x 8½ inches. Prices: No. 30 cabinet, $99.50; No. 302 console, $165 including loud speaker.

NO. 455. PREMIER 6-IN-LINE

Six tubes; 3 t.r.f. (01-A), detector (00-A), 2 transformer audio (01-A and 12). One or two dials. Plate current: 16 to 18 mA. Volume control: rheostat in r.f. Battery cable. C-battery connections. Antenna: 100 feet or loop. Cabinet size: 25 x 45 x 16 inches. Prices range from $60 to $150.

NO. 481. PFANSTIEHL 32 AND 322

Seven tubes: 3 t.r.f. (01-A), detector (01-A), 3 audio (01-A and 71). One dial. Plate current: 23 to 32 mA. Volume control: resistance in r. f. plate. Shielded. Battery cable. C-battery connections. Output device. Antenna: outside. Panel: 17½ x 8½ inches. Prices: No. 32 cabinet, $135; No. 322 console, $225 including loud speaker.

NO. 473. PRMCO 105 AND 110

Six tubes; 2 t.r.f., detector, 2 transformer audio, 1 resistance audio. All 01-A tubes. Volume control: rheostat on r.f. Battery cables. Headphone connection. Antenna: 100 feet. Panel size: 18 x 7 inches. Price No. 105, $45. The No. 110 has C-battery connections, shielding, and drum tuning control. Price $80.

NO. 456. RADIOLA 20

Five tubes: 2 t.r.f. (99), detector (99), 2 transformer audio (99 and 20). Balanced t.r.f. and regenerative detector. Two dials. Volume control: regenerative. Shielded. C-battery connections. Headphone connections. Antenna: 75 to 150 feet. Cabinet size: 19½ x 11¾ x 16 inches. Price $115 including all tubes.

NO. 457. RADIOLA 25

Six tubes; oscillator (99), 2 detectors (99), 3 intermediate r.f. (99), 2 transformer audio (99 and 20). Drum control. Super-heterodyne with two reflexed stages. Volume control: potentiometer on intermediate grids. Shielded. C-battery connections. Headphone connection. Antenna: loop. Cabinet size: 28 x 37 x 19 inches. Price $165 with tubes and loop. Can be operated on a. c. with special attachments.

NO. 458. RADIOLA 28

Eight tubes; oscillator (99), 2 detectors (99), 3 intermediate r.f. (99), 2 transformer audio (99 and 20). Super-heterodyne. Drum control. Volume control same as model 25. Shielded. C-battery connections. Headphone connection. Antenna: loop. Console size 26½ x 63 x 17 inches. Price $260 with tubes and loop. Can be operated on a.c. with special attachments.

NO. 493. SONORA F

Seven tubes; 4 t.r.f. (01-A), detector (00-A), 2 transformer audio (01-A and 71). Special balanced t.r.f. Two dials. Plate current: 45 mA. Volume control: rheostat in r.f. Shielded. Battery cable. C-battery connections. Output device. Antenna: loop. Console size: 32 x 45½ x 17 inches. Prices range from $350 to $450 including loop and loud speaker.

NO. 494. SONORA E

Six tubes; 3 t.r.f. (01-A), detector (00-A), 2 transformer audio (01-A and 71). Special balanced t.r.f. Two dials. Plate current: 35 to 40 mA. Volume control: rheostat on r.f. Shielded. Battery cable. C-battery connections. Antenna: outside. Cabinet size: varies. Prices; table, $110; semi-console, $140; console, $240 including loud speaker.

NO. 495. SONORA D

Same as No. 494 except arrangement of tubes; 2 t.r.f., detector, 3 audio. Prices: table, $125; standard console, $185; "DeLuxe" console, $225.

NO. 482. STEWART-WARNER 705 AND 710

Six tubes; 3 t.r.f., detector, 2 transformer audio. All 01-A tubes. Balanced t.r.f. Two dials. Plate current: 10 to 25 mA. Volume control: resistance in r.f. plate. Shielded. Battery cable. C-battery connections. Antenna: 80 feet. Cabinet sizes: No. 705 table, 26½ x 11½ x 13¼ inches; No. 710 console, 29¾ x 42 x 17¾ inches. Tentative prices: No. 705, $115; No. 710 $265 including loud speaker.

NO. 483. STEWART-WARNER 525 AND 520

Same as No. 482 except no shielding. Cabinet sizes: No. 525 table, 19¾ x 10 x 11¾ inches; No. 520 console, 22½ x 40 x 14 ⁷⁄₁₆ inches. Tentative prices: No. 525, $75; No. 520, $117.50 including loud speaker.

NO. 459. STROMBERG-CARLSON 501 AND 502

Five tubes; 2 t.r.f. (01-A), detector (00-A), 2 transformer audio (01-A and 71). Neutrodyne. Two dials. Plate current: 25 to 35 mA. Volume control: rheostat on 1st r.f. Shielded. Battery cable. C-battery connections. Headphone connections. Output device. Panel voltmeter. Antenna: 60 to 100 feet. Cabinet sizes: No. 501, 25½ x 13 x 14 inches; No. 502, 28 ⁵⁄₁₆ x 50 ⅞ x 16¾ inches. Prices: No. 501, $180; No. 502, $290.

NO. 460. STROMBERG-CARLSON 601 AND 602

Six tubes. Same as No. 549 except for extra t.r.f. stage. Cabinet sizes: No. 601, 27⁵⁄₁₆ x 16¾ x 14⁵⁄₁₆ inches; No. 602, 28¾ x 51½ x 19¾ inches. Prices: No. 601, $225; No. 602, $330.

NO. 461. SUN

Five tubes; 2 t.r.f., detector, 2 transformer audio. All 01-A tubes. Two dials. Volume control: resistance in r.f. plate. Binding posts. Antenna: 100 feet. Cabinet size: 23 x 10 x 10 inches. Price, $80.

NO. 491. SUPERFLEX A4

Four tubes; 1 t.r.f., detector, 2 transformer audio. All 01-A tubes. Special circuit. One dial. Plate current 10 to 15 mA. Volume control: capacity. Battery cable. C-battery connections. Headphone connection. Antenna: outside. Cabinet sizes: table, 24 x 10 x 9½ inches; console, 25 x 44 x 14 inches. Prices: table, $80; console, $139.50.

NO. 486. VALLEY 71

Seven tubes; 4 t.r.f. (01-A), detector (01-A), 2 transformer audio (01-A and 71). One dial. Plate current: 35 mA. Volume control: rheostat on r.f. Partially shielded. Battery cable. C-battery connections. Headphone connection. Antenna: 50 to 100 feet. Cabinet size: 27 x 6 x 7 inches. Price $95.

NO. 471. VOLOTONE XX

Five tubes; 1 t.r.f. (01-A), detector (00-A), 3 resistance audio (HMu and 71). Balanced t.r.f. Two dials. Plate current: 18 mA. Volume control: rheostat on r.f. Battery cable. C-battery connections. Output device. Antenna: 100 feet. Cabinet size: 20½ x 8 x 12 inches. Price $50.

NO. 472. VOLOTONE VIII

Six tubes. Same as No. 471 with following exceptions; 2 t.r.f. stages. Three dials. Plate current: 20 mA. Cabinet size: 26½ x 8 x 12 inches. Price $140.

NO. 464. WRIGHT VII

Seven tubes; 3 t.r.f. (99), detector (99), 3 impedance audio (99 and 20). Na-Ald audio amplifier. Two dials. Plate current: 17 mA. Volume control: resistance in r.f. plate. Battery cable. C-battery connections. Output device. Panel voltmeter. Antenna: 80 feet. Cabinet size: 25 x 15 x 17½ inches. Price $160.

NO. 478. ZIMPHONIC 6

Six tubes; 2 t.r.f. (01-A), detector (00-A), 3 audio (01-A and 12). One dial. Regeneration and t.r.f. Plate current: 22 to 24 mA. Volume control: resistance in r.f. plate. Coils shielded. Battery cable. C-battery connections. Headphone connection. Antenna: 75 to 100 feet. Panel size: 21 x 7 inches. Prices: table $90; console, $125.

NO. 479. ZIMPHONIC 7

Seven tubes; Same as No. 478 except for one more stage t.r.f. Completely shielded. Console size: 20 x 40 x 15 inches. Prices: table, $140; console, $175.

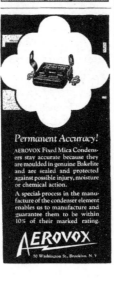

The Best in Radio Programs

The first to review programs, RADIO BROADCAST has maintained its interesting department of news and comment about broadcasters and their programs in "The Listeners' Point of View." A regular and lively department, conducted by John Wallace from his "listening-post" in Illinois.

WHAT KIT SHALL I BUY (*Continued*)

220. BROWNING-DRAKE—Five tubes: one stage tuned radio frequency (Rice neutralization), regenerative detector (tickler control), three stages of audio (special combination of resistance- and impedance-coupled audio). Two controls.

221. LR4 ULTRADYNE—Nine-tube super-heterodyne; one stage of tuned radio frequency, one modulator, one oscillator, three intermediate-frequency stages, detector, and two transformer-coupled audio stages.

222. GREIFF MULTIPLEX—Four tubes (equivalent to six tubes): one stage of tuned radio frequency, one stage of transformer-coupled radio frequency, crystal detector, two stages of transformer-coupled audio, and one stage of impedance-coupled audio. Two controls. Price complete parts, $50.00.

223. PHONOGRAPH AMPLIFIER—A five-tube amplifier device having an oscillator, a detector, one stage of transformer-coupled audio, and two stages of impedance-coupled audio. The phonograph signal is made to modulate the oscillator in much the same manner as an incoming signal from an antenna.

A KEY TO RECENT RADIO ARTICLES
By E. G. SHALKHAUSER

*T*HIS is the twenty-second installment of references to articles which have appeared recently in various radio periodicals. Each separate reference should be cut out and pasted on 4″ x 6″ cards for filing, or pasted in a scrap book either alphabetically or numerically. An outline of the Dewey Decimal System (employed here) appeared last in the January RADIO BROADCAST.

R343.7. ALTERNATING-CURRENT SUPPLY. SOCKET-POWER.   B-Battery.
RADIO BROADCAST. May, 1927. Pp. 43-45.
"Perfecting the B Socket-Power Device," H. E. Rhodes.
The various parts of a good B socket-power device are discussed. The causes of "motor-boating" in these devices, and its possible elimination, are outlined, circuit diagrams and graphs of the operation of a typical B socket power unit being given.

R301.6. MEASUREMENTS WITH HIGH-   Measurements, FREQUENCY BRIDGE.   Vacuum-Tube.
RADIO BROADCAST. May, 1927. Pp. 46-50.
"Methods of Measuring Tube Characteristics," K. Henney.
The writer discusses tube constants, their characteristics, and the bridge circuits used in obtaining these data. The various bridges used to measure amplification factor, plate impedance, mutual conductance, input conductance, and power output, are shown, and experimental data of a variety of tubes on the market are given.

R131. CHARACTERISTIC CURVES; GENERAL   GRID BIAS PROPERTIES.   VOLTAGES.
RADIO BROADCAST. April, 1927. Pp. 36-38.
"A High-Mu Tube At Work," J. E. Anderson.
Here is presented a discussion on the effect of grid bias for various plate voltages when using high-mu tubes and resistance coupling. With the circuit arrangement shown and the curves obtained, the writer presents a detailed analysis of the curves and states why a negative bias is necessary.

R384.1. WAVEMETERS.   WAVEMETERS.
Radio. April, 1927. Pp. 39-ff.   Measurement of.
"How to Calibrate a Wavemeter," C. T. Burke.
Methods used in obtaining fundamental standards of frequency with the aid of quartz crystals and a series of coupled oscillators are outlined. How this standard may be used in calibrating other wavemeter circuits, methods of checking, calibration of frequencies above the fundamental, etc., are other topics discussed by the author.

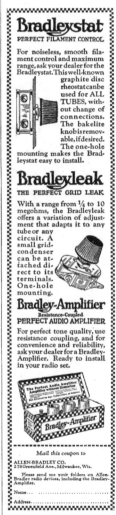

R370. SIGNAL INTENSITY MEASUREMENTS. SIGNAL INTENSITY. *Radio News.* April, 1927. Pp. 1220-ff. "A New Field for Experimentation." J. F. Rider.

This article describes a layout of apparatus which may be used to record the intensity of carrier waves of different stations broadcasting, in order to obtain data for the analysis of transmission problems and fading. The apparatus needed is listed: it consists of a receiver, equipped with meters, and an oscillator for purposes of receiver calibration. The method of plotting signal strength curves, and ,pa "colars on the proper operation of the set, are given.

R112.5. RESISTANCE COUPLING. COUPLING, *Radio News,* April, 1927. Pp. 1250-ff. *Resistance.* "Does Resistance Coupling Give Best Quality?" S. Harris.

The disadvantages, as compared to the advantages, of resistance-coupled amplifier circuits, are given. Two main disadvantages are said to be the presence of the blocking condenser, which reduces amplification at the lower frequencies, and the low amplification per stage as compared to transformer coupling. The effects of these disadvantages are discussed, and remedies are suggested.

R144. HIGH-FREQUENCY RESISTANCE. RESISTANCE. *Physical Review.* Jan., 1927. Pp. 161-171. *High-Frequency.* "The Resistance of Copper Wires at Very High Frequencies." W. M. Roberts.

At frequencies of the order of 107 cycles, the distributed capacity of single loops of wire are said to cause sufficient unequal current distribution in the loop to account for large apparent discrepancies between observed and calculated resistances. For a given frequency, more uniform current distribution is gained by decreasing the size of the loop and simultaneously increasing the capacity of the tuning condenser. Curves are plotted with ratio of observed to calculated resistance as ordinate and condenser setting as abscissa. For all curves taken, the ratio fell well below 1.45 and was still decreasing as far as data were taken. The presence of oxide on copper wire is said to have no appreciable effect on the resistance.

R126.8. LOUD-SPEAKING REPRODUCERS. LOUD-SPEAKERS. RADIO BROADCAST. April, 1927. Pp. 587-600. "A Fundamental Analysis of Loud Speakers." J. F. Nielsen.

The quality of radio broadcast programs when reproduced depends in part on the loud speaker. The nature of the signal which is to be reproduced determines entirely the method of loud-speaker construction. A study concerning facts of speech and musical tones and harmonics is, therefore, presented. The desirable characteristics of loud speakers and the mechanism that is to reproduce these characteristics is taken up in a mathematical discussion under the following: (1) The motor element, which converts the electrical impulses into corresponding mechanical vibrations; (2) the coupling system, which transmits the mechanical vibrations from motor to diaphragm; (3) the diaphragm or loading device, which radiates the mechanical vibrations into the air as waves of sound. Distortion is said to result from saturation of armature and pole faces, and from iron losses.

R113.5. METEOROLOGY. METEOROLOGY. *Popular Radio.* March, 1927. Pp. 327-ff. *Earth Blankets.* "The Three Blankets Around the Earth." C. E. Free.

The writer presents information relative to the nature of space surrounding the earth. Three layers of gases are said to be found varying in height and having different temperatures. The lower layer or blanket, about seven miles high, contains mixed gases and varies considerably in temperature. The middle layer, about 80 degrees below zero Fahrenheit, is about 22 miles in height. The upper layer, about 400 to 600 miles high, is supposed to have a temperature of about 80 degrees above zero Fahrenheit. This latter is said to serve as the protecting blanket, surrounding the earth, against the many meteors which would otherwise destroy everything on the surface. It also serves as the reflecting layer for many radio waves, being commonly called the Heaviside Layer. It is ionized by the ultra-violet rays from the sun. In it the aurora is said to be displayed. Above this layer, space at a temperature of 460 degrees below zero is supposed to exist.

R142. RADIO TELEPHONE SYSTEMS. TELEPHONY. *Radio News.* March 1927. Pp. 1086-ff. *Trans-Atlantic.* "Hello, London! 'Are You There, New York?'", G. C. B. Rowe.

The apparatus and the single side-band transmitter system as used in the new transatlantic radio telephone station are described. The principle of single side-band transmission, with its partial secrecy and saving of "watts per mile," is based on the heterodyne method of frequency amplification.

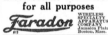

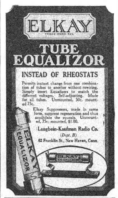

When the **Greatest Show in History thrills the World**

".....You're *there* with a Crosley"

$35

$85

$65

The BANDBOX $55
A 6 tube Receiver

These approved cabinets have been selected by Powel Crosley, Jr., as ideal consoles, acoustically and mechanically, for the installation of the Crosley "BANDBOX." Genuine Musicones built in. Crosley dealers secure them from their jobbers through

H. T. ROBERTS CO.,
914 S. Michigan Ave.,
Chicago, Ill.
Sales Agents for Approved Console Factories:

SHOWERS BROTHERS COMPANY

THE WOLF MFG. INDUSTRIES

A tremendous Crosley radio achievement for 1927-28

RECENT court decisions which clarified the radio patent situation have paved the way for still greater Crosley triumphs.

Now—completely available to Crosley—and amplifying Crosley supremacy in the fullest measure, are the enormous resources, great discoveries and ideas embodied in patents of

1—The Crosley Radio Corporation.
2—The Radio Corp. of America.
3—The Westinghouse Co.
4—The General Electric Co.
5—The American Telephone & Telegraph Co.
6—The Hazeltine Corporation.
7—The Latour Corporation.

under which Crosley is now licensed to manufacture.

Here are the seven big things which represent radio's greatest advancement, brought together by Crosley and combined with the experience, mass production method and leadership of the Crosley organization. No wonder a waiting radio world pronounces the "Bandbox" at the unprecedented price of $55, Crosley's paramount achievement.

The Bandbox is Shielded

Radio coils are surrounded by magnetic fields similar in every respect to the magnetic field around the earth that moves the needle of a compass but around radio coils these fields make nuisances of themselves by feeding back on each other. Heretofore it has been customary to make inefficient coils with inefficient fields to prevent such feeding back. The Crosley Bandbox incorporates copper shields around each coil to prevent such feeding back. The coils consequently can be made and are very much more efficient. The amplification of the receiver is, therefore, much higher—the sensitivity is greatly increased. Condensers are also completely shielded from each other in separate metal compartments. Hitherto, only high priced sets have enjoyed this super radio advantage.

There Is No Oscillation

The Bandbox employs completely balanced or neutralized radio frequency stages to prevent oscillation, instead of the common form of losser method. More costly, to be sure, but extremely necessary in achieving such results as are obtained by this marvel of radio reception.

For Sharpness—The Acuminators

This is another big "Bandbox" feature which permits full brass band power for those who want their dance notes strong and loud. For others, it cuts volume down to a soft and gentle murmur, without distortion.

Volume Control

This is another big "Bandbox" feature which permits full brass band power for those who want their dance notes strong and loud. For others, it cuts volume down to a soft and gentle murmur, without distortion.

Illuminated Dial

A Master Station Selector has an illuminated dial for easy reading in shadowy corners. A single knob permits full tuning for ordinary reception of local, nearby and super-powered stations.

Installation Simplified

A woven cable, containing vari-colored rubber covered leads makes installation and hook-up easy for the veriest novice. No waiting for the radio service man, should the batteries be changed,

Easily Adapted to Consoles

Simply remove screws in escutcheon and in base of set. Lift off metal case. Chassis now stands ready for installation in console cabinet. Opening in console cabinet permits control shafts to protrude. Escutcheon screws in place and—Presto! the console radio is complete.

For A. C. Operation

a special Bandbox is available at $65, wired especially for use with

IMPROVED MUSICONES
Although Musicones improve the reception of any radio set, they are perfect affinities in finish, beauty and reproduction effectiveness for Crosley Radios. A new model built in the form of a Colonial Tilt-Table with brown mahogany finish, stands 4 feet high. Price $27.50.

16-inch Super Musicone (As pictured with Bandbox) $12.75

12-inch Ultra Musicone $9.75

the Crosley Power Converter at $60. This special Bandbox utilizes the new R.C.A. AC tubes which have made the operation of radio receivers direct from house current so simple, efficient and dependable. The first three tubes employed in the AC model are UX 226. These go into the radio frequency sockets. The detector tube is UY 227, with indirectly heated emitter. Another UX 226 is used in the first audio stage. Raw AC current heats the filament of all UX 226 tubes. Power tube UX 171 is in the last audio socket. This makes the "dog houses" rumble sonorously and the bass drums deeply boom.

The Power Converter

The power converter which smooths the alternating current is a marvel of engineering ingenuity. Only half the size of an ordinary "A" storage battery, it supplies the required A, B and C currents, without hum. Finished in brown frosted crystalline.

There are models for 25 and 60 cycle current. A snap switch shuts down the set and power converter completely.

Price of Power Converter—$60

You owe it to yourself to see the "Bandbox" and listen to its remarkable performance. If you cannot easily locate the nearest Crosley dealer, his name and address will be supplied on request. Write Dept. 20.

CROSLEY RADIO

Crosley Radio
is licensed only for Radio amateur, Experimental and Broadcast Reception.

Crosley recommends the use of one UX171 power tube, or which are furnished at stan Bandbox. While Radiotron gives a superior performance

of five 201-A Radiotrons and Cunningham equivalents, dard tube prices, with each UX171 is a 135-volt tube. It for 135-volt "B" batteries.

The Crosley Radio Corporation

Powell Crosley Jr., Pres.

Cincinnati, Ohio

Prices slightly higher west of Rocky Mountains

7 Tube Set
Simple Dial Radio

30 DAYS FREE TRIAL

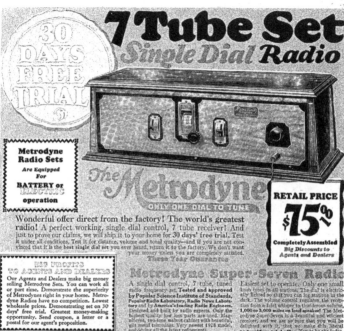

The Metrodyne
ONLY ONE DIAL TO TUNE

Metrodyne Radio Sets Are Equipped For BATTERY or ELECTRIC operation

Wonderful offer direct from the factory! The world's greatest radio! A perfect working, single dial control, 7 tube receiver! And just to prove our claims, we will ship it to your home for 30 days' free trial. Test it under all conditions. Test it for distance, volume and tonal quality—and if you are not convinced that it is the best single dial set you ever heard, return it to the factory. We don't want your money unless you are completely satisfied.

Three Year Guarantee

BIG PROFITS TO AGENTS AND DEALERS

Our Agents and Dealers make big money selling Metrodyne Sets. You can work all or part time. Demonstrate the superiority of Metrodynes right in your home. Metrodyne Radios have no competition. Lowest wholesale prices. Demonstrating set on 30 days' free trial. Greatest money-making opportunity. Send coupon, a letter or a postal for our agent's proposition.

Metrodyne Super-Seven Radio

A single dial control, 7 tube, tuned radio frequency set. Tested and approved by Popular Science Institute of Standards, Popular Radio Laboratory, Radio News Laboratory and by America's leading Radio Engineers. Designed and built by radio experts. Only the highest quality low loss parts are used. Magnificent, two-tone walnut cabinet with beautiful gilt metal trimmings. Very newest 1928 model, embodying all the latest refinements.

Easiest set to operate. Only one small knob tunes in all stations. The dial is electrically lighted so that you can log stations in the dark. The volume control regulates the reception from a faint whisper to thunderous volume, 1,000 to 3,000 miles on loud speaker! The Metrodyne Super-Seven is a beautiful and efficient receiver, and we are so sure that you will be delighted with it, that we make this liberal 30 days' free trial offer. You to be the judge.

MAIL COUPON BELOW

Let us send you proof of Metrodyne quality—our 30 days' free trial offer and 3 year guarantee

Mrs. Wm. Leffingwell, Westfield, N. J., writes: "The Metrodyne Radio I bought of you is a wonder. This is as good as any $325 machine I have ever seen."

N. M. Greene, Maywood, Ill., writes: "My time is up and the Metrodyne works fine. I get Havana, Cuba, Oregon, Florida, Denver, Texas, Boston, Canada, all on the speaker."

J. W. Woods, Leadville, Colo., writes: "Received the 7 tube Metrodyne in fine condition. Had it up and working the same day received. Was soon listening to Los Angeles, San Diego, Oakland and other California points; also St. Louis, Kansas City and other east and south stations—all coming in fine. All more than pleased, tone enjoying full."

We will send you hundreds of similar letters from owners who acclaim the Metrodyne as the greatest radio set in the world. A postal letter or the coupon brings complete information, testimonials, wholesale prices, and our liberal 30 days' free trial offer.

Metrodyne Super-Six

30 Days' Free Trial—3 Year Guarantee

Another triumph in radio. Here's the new 1928 model Metrodyne 6 tube, two dial, long distance tuned radio frequency receiving set. Approved by leading radio engineers of America. Highest grade low loss parts, completely assembled in a beautiful walnut cabinet. Easy to operate. Dials easily logged. Tune in your favorite station on same dial readings every time—no guessing.

Mr. Howell, of Chicago, said: "While five Chicago broadcasting stations were on the air I tuned in seventeen out-of-town stations, including New York and San Francisco, on my loud speaker horn, very loud and clear, as though they were all in Chicago."

We are one of the pioneers of radio. The success of Metrodyne sets is due to our liberal 30 days' free trial offer, which gives you the opportunity of trying before buying. Thousands of Metrodynes have been bought on our liberal free trial basis.

6 Tube Set $48.50 RETAIL PRICE Completely Assembled

MAIL THIS COUPON or send a postal or letter. Get our proposition before buying a radio. Deal direct with manufacturer— SAVE MONEY—WRITE NOW!

METRO ELECTRIC COMPANY
2161-71 N. California Ave. • Dept. 16 • Chicago, Illinois

RADIO
BROADCAST

CONTENTS

Announcing the 1928 Complete Sets
What the Set Manufacturer Offers the Public
A Radio Picture Receiver for Every Experimenter
Should the Small Broadcaster Exist?
What Loud Speaker Shall I Buy?

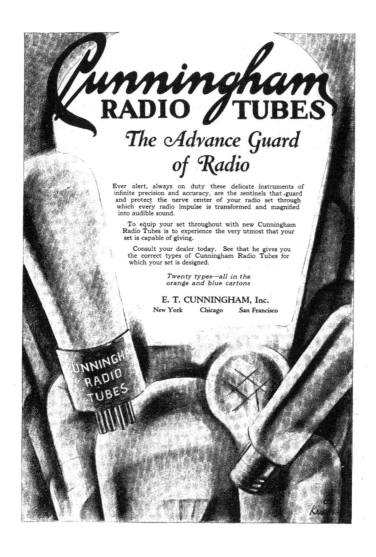

Two *tubes saved*
pay for this voltmeter

Here is a sturdy mantel-clock model voltmeter with moving-vane movement—three times as accurate as other makes of the same type. Acid and jar-proof. Ask your dealer. Price, $4.50.

ADJUST the voltage across your tubes to "right signal strength", say the direction books. But what is right signal strength?

To guess may prove expensive. It doesn't take long to burn out tubes when you give them too high a voltage. Too low gives weak reception.

A Westinghouse voltmeter soon pays for itself by the tubes it saves—you can buy the most accurate moving-vane radio voltmeter made for only $4.50. With a Westinghouse voltmeter you *know* what is right signal strength—you gain battery life and better reception all around. Your dealer can show you several types of Westinghouse instruments.

The new Cabinet Portable Voltmeter made for sets with plug-in jacks. Antique bronze face plate, with gilt dial and pointer. Will fit any arrangement of jacks. Price, $6.50 (west of Rockies, $7.00).

Rectigon and Rectox chargers and Micarta panels and tubing are also Westinghouse products.

WESTINGHOUSE ELECTRIC & MANUFACTURING COMPANY
Offices in All Principal Cities   *Representatives Everywhere*
Tune in with KDKA—KYW—WBZ—WBZA

Westinghouse

The finest money will buy. Approved and used by the U. S. Navy. Jeweled movement. Price, $10.00.

The New Improved Hi-Q Six—the creation of ten foremost American Radio Engineers—a receiving instrument that is far in advance of its time.

Exclusively CUSTOM-BUILT
By Yourself at Home . . . from our Simple
Instructions . . . and at Great Savings!

NO ordinary standards can be applied to this latest improved Hammarlund-Roberts Receiver, for it is the result of a determination to produce America's very finest instrument —*absolutely regardless of cost!*

Every modern constructional feature has been incorporated. Each part is the most efficient known to radio science, and the entire group has been purposely selected for perfect synchronization.

Complete isolation of four tuned circuits plus Automatic Variable Coupling effects maximum and uniform amplification over the entire wave band. Distortion is totally eliminated. Oscillation is utterly absent. Symphonic transformers

ASSOCIATE MANUFACTURERS

Acme Wire Company
Benjamin Electric Mfg. Company
Carter Radio Company
H. H. Eby Mfg. Company
Hammarlund Mfg. Company
International Resistance Company
Radiall Company
Samson Electric Company
Sangamo Electric Company
Yaxley Mfg. Company

Completely drilled panel and sub-panel are foundation for easy building.

and a power tube faithfully reproduce the full musical scale. Selectivity, even in crowded areas is something to marvel at. And tonal quality simply *MUST* be *heard to be appreciated!*

Such a set, factory made, and sold through usual channels, would possibly cost around $300.00, but through following our simple instructions you can purchase all parts for only $95.80 and build this supreme receiver yourself—a CUSTOM-BUILT set which gives you CUSTOM-BUILT results at a saving of $100 to $150.

Get the complete Hi-Q Instruction Book from your dealer—or write us direct. Price 25 cents.

Hammarlund ROBERTS Hi-Q SIX

HAMMARLUND-ROBERTS INC.   1182 Broadway   New York City

RADIO BROADCAST

WILLIS KINGSLEY WING, Editor

OCTOBER, 1927

KEITH HENNEY
Director of the Laboratory

EDGAR H. FELIX
Contributing Editor

Vol. XI, No. 6

CONTENTS

AMONG OTHER THINGS. . .

SO MUCH is now being accomplished in the complete set field beginning with this issue, RADIO BROADCAST will devote much of its space to reflecting the technical and other advances being made. The policy of this magazine remains as before, to present the news of radio surrounded by as much as possible of what the technical radio worker refers to as "dope." And much "dope" there is in this complete set side of radio as many articles which will appear in following issues will strikingly demonstrate. In all the other fields of radio endeavor which RADIO BROADCAST has covered heretofore—the construction of radio-receiving apparatus, laboratory experiments, short-wave communication, broadcasting from the technical and program side, and many others which we have covered to the satisfaction of our readers, RADIO BROADCAST will be as active as before. The expansion in scope of our text pages is directly designed to keep our readers in close touch with all sides of radio.

WHERE is the radio experimenter who has not read of the rapid progress of the transmission of photographs by wire and radio and hoped that he would soon be able to share in the fascination of this new and intensely modern field? For the past few years, RADIO BROADCAST has watched the progress of the art, hoping that devices would be developed within the scope of the amateur laboratory and pocketbook. The leading article on page 341 describes the background of the Cooley "Rayfoto" system which in a very few weeks will be made available through the pages of RADIO BROADCAST to the home experimenter. To be able to construct a radio photograph receiver for less than $100 should appeal very strongly to the experimental fraternity who are crying for something new to do. Here, in a manner of speaking, it is.

MANY pages of this issue are devoted to showing the offerings of the set-makers for the coming season. Later issues of this magazine will describe in greater detail interesting technical features of these many receivers from many manufacturers.

THOSE of a technical turn of mind will read with great interest David Grimes' story on page 367 describing the theoretical features of a radio receiving system which is one of the most interesting that has come to our attention in many moons. Articles to follow by Mr. Grimes will describe the circuit constants and practical information about the system.

THE Federal Radio Commission is making every effort to popularize the use of the expression "frequency in kilocycles" instead of the familiar "wavelength in meters." It is hard going for some, but doing one's thinking in kilocycles does remove many serious complications from calculations. Ever since its August, 1925, issue RADIO BROADCAST has standardized the use of kilocycle designations, always printing at the same time, the equivalent wavelength in meters. We are in sympathy with the wishes of the Radio Commission and invite the expression of our readers on this subject.

—WILLIS KINGSLEY WING.

Doubleday, Page & Co.
MAGAZINES
COUNTRY LIFE
WORLD'S WORK
GARDEN & HOME BUILDER
RADIO BROADCAST
SHORT STORIES
EDUCATIONAL REVIEW
LE PETIT JOURNAL
EL ECO
FRONTIER STORIES
WEST

Doubleday, Page & Co.
BOOK SHOPS
(Books of all Publishers)
LORD & TAYLOR BOOK SHOP
PENNSYLVANIA TERMINAL (2 Shops)
NEW YORK: GRAND CENTRAL TERMINAL
38 WALL ST. and 526 LEXINGTON AVE.
848 MADISON AVE. and 166 WEST 52ND ST.
ST. LOUIS: 223 N. 8TH ST. and 4914 MARYLAND AVE.
KANSAS CITY: 920 GRAND AVE. and 206 W. 47TH ST.
CLEVELAND: HIGBEE CO.
SPRINGFIELD, MASS.: MEEKINS, PACKARD & WHEAT

Doubleday, Page & Co.
OFFICES
GARDEN CITY, N. Y
NEW YORK: 285 MADISON AVENUE
BOSTON: PARK SQUARE BUILDING
CHICAGO: PEOPLES GAS BUILDING
SANTA BARBARA, CAL.
LONDON: WM. HEINEMANN LTD.
TORONTO: OXFORD UNIVERSITY PRESS

Doubleday, Page & Co.
OFFICERS
F. N. DOUBLEDAY, President
NELSON DOUBLEDAY, Vice-President
S. A. EVERITT, Vice-President
RUSSELL DOUBLEDAY, Secretary
JOHN J. HESSIAN, Treasurer
LILLIAN A. COMSTOCK, Asst. Secretary
L. J. McNAUGHTON, Asst. Treasurer

DOUBLEDAY, PAGE & COMPANY, Garden City, New York

Copyright, 1927, in the United States, Newfoundland, Great Britain, Canada, and other countries by Doubleday, Page & Company. All rights reserved.
TERMS: $4.00 a year; single copies 35 cents.

Tone Quality Superb

From Nature's Sounding Board

Lata Balsa Reproducer
Model 150, Price $50

THE strides made in the past few years by the broadcasting stations in bettering the quality of their transmission have been remarkable. The same principles have been applied to the phonograph, resulting in the new records and machines which are miles in advance of the older type.

These improvements are now available to every listener-in. However, there are some radio receivers which will not reproduce the wonderful music being broadcast by the improved stations.

By applying the same principles to radio receivers which have proved so helpful in the broadcasting and phonograph fields it is now possible to make the most out of the broadcasting. By utilizing the scientific aids now available old and new receivers may be made to produce music which in every way resembles the original. There is as much difference between this new form of reproduction and the old as there is between the new phonograph and the old scratchy, squawky cylinder machine of yesteryear.

Resistance coupling simplifies receiver amplifier construction and greatly reduces the cost. It is recognized by leaders in the search for the best tone quality as being the ideal amplifying system. It permits the passage of the very low notes and the very high ones with the same ease. It brings to the loudspeaker a true but greatly amplified picture of what is picked up from the broadcasting station.

There is little use in having this wonderful

Everything necessary for the building of a modern three-stage resistance coupled amplifier which may be applied to any receiver. $9.00

thing available if we do not make use of it. It can now be utilized very simply. Radio artists, editors and many musicians have expressed both pleasure and amazement on hearing the tone portrayal of symphony orchestras, as well as jazz bands which Nature's Sounding Board has made possible.

Nature's Sounding Board is known as the Lata Balsa Reproducer. It is more than a "loud speaker." It is science's latest contribution to the musical art. It will make any good radio receiver sound much more natural and pleasing.

Lata Balsa Reproducers are available in completed, artistically decorated models ranging in price from $30 to $50. Kits are also available in two sizes. The famous Lata Unit, or electrical driving mechanism is now ready for delivery and is ideal for use with the Lata Balsa Reproducer Kits.

The accompanying illustration will give you some idea of the attractiveness of the reproducer— to realize its superior tone quality you must hear it.

The Lata Balsa Kits are complete and may be assembled in a very short time by anyone. No mechanical skill or electrical knowledge is required to follow the simple instructions. Once you have heard this wonderful reproducer no other will satisfy you. The coupon below is for your convenience in ordering. Use it today and you will never regret it.

The number 2 Lata Balsa Kit and Reproducer unit assembled. You can duplicate this assembly in a very short time by using this kit and complete instructions

RETURN THIS COUPON—

THE MOST POWERFUL BROADCASTING STATION IN THE WORLD

A view of the transmitting panel of the 100-kilowatt transmitter at the South Schenectady laboratories of the General Electric Company. Experimental programs are sent out with this set, using the standard wavelength of WGY, from midnight to one a. m., Eastern Time. The transmitter went on the air August 4th, marking the first time that power as great as 100 kilowatts has been used in broadcasting. The three water-cooled 100-kilowatt tubes in the foreground have an attachment for clearing out the gas from the tube while it is on the panel

RADIO BROADCAST

Volume XI   Number 6

OCTOBER, 1927

NOW—*You Can Receive Radio Pictures!*

You Can Build Your Receiver for Less Than One Hundred Dollars—How It Works

By KEITH HENNEY

Director of the Laboratory

THERE is an undeniable thrill in witnessing an extraordinary scientific event, such as receiving one's first broadcast program, or in talking across the Atlantic for the first time, or in seeing photographs that have been sent by radio from England, or in talking via short-wave amateur radio with a fellow "ham" on a steamship a thousand miles up the Amazon river. In fact, as the French astronomer, Camille Flammarion says, "Unless one has a stone instead of a heart and a lump of fat in the place of a brain, it is difficult not to feel some emotion over the achievements of science."

The writer's first contact with Austin G. Cooley, who has been responsible for the picture transmission system to be described in RADIO BROADCAST, produced one of those technical thrills. Cooley was working down in the "shack" at 2 GY, the short-wave station of the RADIO BROADCAST Laboratory, a small place with scarcely room enough to stow his extensive gear. On one bench was a metal drum on which a photograph of a young lady with a large and floppy hat turned over and over. On the other side of the "shack" was another rotating drum covered with photographic paper with a small spot of light playing on it, and filling the shack was an undefinable sound-like that of a high speed motor whirring away with whining and somewhat obnoxious tone. In two or three minutes, Cooley stopped the motor, took the photographic paper from the receiving drum, developed it in rather dim daylight, and there was the young lady again, hat, white plumes and all. Between the transmitter and receiver was an artificial telephone line 300 miles long so that for all practical purposes the young lady's picture had been sent to Garden City from, say, Washington, D. C.

Since that time, Cooley has produced several picture systems, one of which is applicable to present wire telegraph channels so that facsimile copies of original messages may be transmitted at a cost not exceeding present rates. Another system—the one in which we as radio experimenters are interested has been developed to the extent that with a not very expensive attachment to the ordinary radio set, one can receive at home photographs, letters, telegrams, pictures, or text torn from newspapers. This system is known as the Cooley "Rayfoto."

These radio-transmitted pictures are comparatively good reproductions, quite satisfactory when one considers the simplicity of the apparatus and the compromises which were necessary in order to obtain them, it must be remembered that it is about seven years since the first small group of amateurs heard the human voice, and music, coming in by wireless. The early poor quality of reception must not be forgotten; nor must we lose sight of the fact that it has taken since 1835, when S. F. B. Morse performed his epoch-making experiment of transmitting slow and uncertain telegraph signals via wires, to send messages at the rate of several hundred characters per minute.

The Cooley pictures are better than most of the static-laden ones that have come across the Atlantic. Some of them are reproduced here. Improvement in detail and shading of the Cooley Rayfoto pictures will come. The important fact is that here is something the experimenter can have for his own and can have the sport of developing his experience as this apparatus is improved and refined.

To know the Cooley apparatus is to know Cooley, to visualize the young M. I. T. student who preferred his own researches into the then unknown field of picture transmission to the prescribed studies with the result that he was dropped from the college register during his fourth year; his long nights of work, sleeping—when sleep was necessary—beside his apparatus on a laboratory bench. One should know of twenty-four-hour days in the Arctic on the MacMillan Expedition in 1926 when Cooley

> ¶ *Radio pictures at home—that's what the experimenter has been waiting for. Now it is possible. For less than $100, you can build your own picture receiver, connect it to your broadcast set and jump into a new and fascinating field of experiment. Exclusive articles in* RADIO BROADCAST *in the November number and in following issues will tell you how to build and operate the receiver.*

worked under great odds, trying to send picture signals by short waves back to the United States. This background is necessary to understand the years of effort that have gone before the present development. This development has been slow, and full of disappointments but the final result has been that his system is workable, comparatively simple to operate, and what is more important in the eyes of the average experimenter —it is inexpensive.

THE LURE OF HOME EXPERIMENT

WHY should one want to receive pictures by radio at home? As well ask: Why have thousands of experimenters built receivers for the reception of music, or code? In this, there has been always the thrill of accomplishment; making something with one's hands, something that works: creation. Secondly, there is the hope in everyone's heart that television, the art of transmitting scenes from life itself by wire or radio, shall become practical within his time. Before television shall be an accomplished fact, we must conquer the idiosyncrasies of the transmission of still life: pictures, an art known as telephotography. By introducing legions of experimenters to the general problem of sending and receiving pictures by radio or wire, Cooley may have a share in advancing the day when the more difficult task of committing moving pictures to the ether shall be solved. The amateur has been thanked by engineers for his persistence in developing short-wave channels; many have said that to him alone belongs the credit for advancing the art of communication by the use of those very high frequencies. At any rate, he has had his share in making possible communication by short waves across vast distances with inexpensive apparatus. Who knows what his share may be when the ultimate success of telephotography and television has been achieved?

The transmission of a picture by the Cooley system differs from the transmission of music in one important respect: it takes a single audio frequency and performs tricks on it which at the receiving end are translated from sounds to other effects which influence a photographic paper. For example, good broadcasting requires that a band of frequencies 10,000 cycles wide shall be transmitted. These frequencies at the receiver are reconverted to sound waves. The picture system takes a given frequency, say 1000 cycles, and transmits its form of intelligence on it. In a way it is less complicated than broadcasting music; in other ways broadcasting pictures is more complex.

The broadcasting station which may elect to transmit pictures as well as music or speeches will use its present equipment. Since pictures according to the Cooley system are first translated into sound impulses, and since the phonograph record is a means of recording for future use such sound impulses, thereby defeating two of nature's limitations, time and space, the broadcasting station that will invest in a phonograph turn-table needs no other equipment to fill the ether with pictures.

YOUR REGULAR RADIO RECEIVER IS USED

THOSE who will receive pictures will use their present broadcasting receiving set which should be equipped and operated so that overloading and severe distortion of other sorts do not occur. As in transmitting sounds, distortion at the receiving end produces an unintelligible or garbled reproduction. The receiver must be within the "service area" of the transmitter—that is, inside the distance at which fading takes place; the operator must be able to develop and fix ordinary developing-out paper, such as Azo, Velox, or other papers with which the amateur photographer is already well familiar. Here is an opportunity for the radio enthusiast who has dropped his photographic hobby to bring developing trays from the attic, and to renew his acquaintance with hydroquinone, metol, HQ, and hyposulphite of soda!

The present simplified system transmits and receives original photographs full of strong contrast, such as newspaper photographs, or those in which a loss of detail will not mar the recognizable features of the transmitted picture. The examples reproduced here will give an idea of the work that can be done. These limitations are largely due to the simplification of the receiving equipment, and, as experimenters become more familiar with the present apparatus, more refinements will naturally follow, refinements which will make it possible to receive greater detail, and naturally better pictures.

The process of getting the picture into the ether is not too complicated for anyone to understand. The picture is wrapped around an aluminum drum which revolves in front of a very strong light. As the drum revolves it moves along a shaft, so that the beam of light eventually has

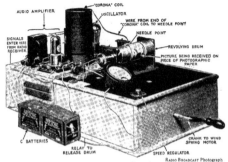

Radio Broadcast Photograph

THE COOLEY "RAYFOTO" RECEIVER

Simplicity is the keynote of the Cooley picture transmission system. Signals in the form of an audio frequency tone enter at the point marked on this photograph, are amplified in a single-stage transformer-coupled amplifier, after which they are placed on the oscillator circuit through a modulation transformer not visible here. The signals in the oscillator circuit drive the corona circuit whose output is conducted to the rotating drum and photographic paper by the wire and needle point shown here. The drum is rotated by the familiar spring motor which usually turns phonographic records. The milliammeter, which is essential, usually reads the oscillator current

covered the entire picture, illuminating a small bit of it at a time. The light which is reflected by the picture passes into a sensitive photo-electric cell, another of modern science's marvels. When the beam of light falls on a black portion of the picture, most of the light will be absorbed and little reflected and consequently what passes into the photo-electric cell will be small—just as the part that comes to the eye is small; which explains what we mean by the word "black."

This sensitive photo-electric cell, whose history dates back many years and touches the lives of many famous scientists, consists of a potassium plate and a second metallic electrode which has a battery attached to it maintaining it at a potential positive with respect to the potassium,

much as the plate of a vacuum tube is maintained positive with respect to the filament by the B batteries. When light shines on this potassium plate, electrons, those omnipresent negative electrical charges, are given off, their passage to the second electrode toward which they are attracted constituting an electric current. This minute current, usually of the order of a few millionths of an ampere is greatly amplified and interrupted by another current of 1000-cycle frequency. The result is that what were originally black and white visual images have been translated into sound impulses which are then impressed on the radio transmitter, or a telephone line, and are sent out like speech or music.

At the receiving end the process must be reversed, i. e. sound impulses must be translated into light impressions, back into those original black and white spots. Assuming that nothing happens in the intervening space, no static is picked up, no fading has lost the signal, at the receiving end the conventional broadcasting set may be used, and if a loud speaker is attached to the output of the receiver in the normal manner we should hear this varying note which is about two octaves above middle C on the piano.

WHY LOCAL AMPLIFICATION IS NEEDED

IN MOST cases it will be necessary to boost this wavering note to sufficient strength to operate the rest of the translating apparatus which consists of an oscillator, similar to that used in superheterodyne receivers except that its wavelength is usually above the broadcasting band, and finally what Cooley calls a "corona" coil. It will remind old-time electrical experimenters of the Tesla coil, a device used to transform a.c. voltages into other voltages high enough to produce electrical discharges across intervening space, in this case about a quarter of an inch.

Under ordinary conditions, the carrier of the transmitter tuned-in but no picture being on the air, the oscillator will be oscillating so feebly that no corona is taking place, but when the signal arrives it supplies power to the oscillator so that a nice fat discharge is produced whose

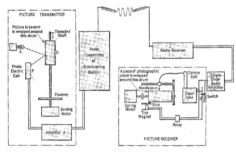

HOW THE COMPLETE SYSTEM WORKS

The owner of receiving apparatus need not worry about how the pictures get on the air, but in case he is interested, here is a layout of the complete system, transmitter, waves in the ether, receiver and all! The switch at the right of the receiving equipment and the trip magnet are part of the start-stop system that insures synchronism between transmitter and receiver

Radio Broadcast Photographs

THE COMPLETE RECEIVING APPARATUS

Here is the entire receiving apparatus. The receiver in this case was a well known set operating from a loop, picking up signals on 208 meters which originated in another part of the Laboratory. They are tuned-in and sent through the picture apparatus which is described in greater detail in the other photographs

total cost of the apparatus, exclusive of batteries, tubes, and broadcast receiver will be less than $100. All of the electric apparatus can be made from material easily available to the experimenter. Complete descriptions by Mr. Cooley of this equipment and how to operate it will follow.

A word about the availability of pictures. Inasmuch as the original equipment for a broadcasting station need involve no more than the expense of a phonograph, there is a certain source of these radio pictures in every broadcast station. Broadcast stations will find in the transmission of the Cooley "Rayfotos" a sensational extension of their activities and by the time this article is in the hands of experimenters, and by the time the receiving equipment is ready, pictures will be on the air. The Cooley system is being exhibited at the New York Radio Show this month.

And there you are: Pictures by radio—and in your own home.

intensity varies with the shading of the transmitted picture.

The counterpart of the rotating drum of the transmitter will be found at the receiver. Here, instead of an electrical motor we use for sake of simplicity a phonograph turn-table which is geared to a small aluminum drum around which the sensitive photographic paper is wrapped. The corona discharge sprays the paper from a fine needle and in some way not well understood affects the emulsion of the photographic paper. Whether this is an electrical or chemical effect or whether there is enough light from the corona to expose the paper is not fully known.

But for Cooley's purpose it does not matter. When developed, the paper shows black spots where the discharge was heavy; light spots where the corona was weak.

Now in all picture systems, some means must be provided to keep the receiver in exact synchronism with the sender; when the latter starts the receiver must start, and not in the middle of the picture which as anyone can see would have certain disadvantages! A simple scheme for hold-

ing the receiver and transmitter together has been employed. It is known as the "start-stop" system, and is very simple and flexible. In operation the receiving drum revolves slightly faster than the transmitting drum and it therefore completes a revolution in a slightly shorter time. At the end of each revolution the receiver drum is held by a trigger until the transmitting drum completes its revolution. A signal is transmitted then that releases the receiving drum so the two start off together. The radio signals are not strong enough in most cases to operate the trip magnet that releases the receiving drum at the beginning of each revolution so this magnet is operated through a more sensitive relay. Both the trip magnet and relay can be seen in the picture of the apparatus shown on this page. The single pole double throw switch S is really part of the trip magnet. When the armature of the trip magnet is against the stop on the drum, terminals 2 and 3 on the switch are pushed together and therefore all of the energy from the audio amplifier passes into the relay. When the synchronizing impulse is received it activates the relay and the trip magnet thus releases the drum and also causes the switch S to make contact between terminal 1 and 2 and then all of the energy passes into the oscillator.

The present apparatus transmits 4 x 5 photographs at a rate of one and one-half inch per minute or a little over 3 minutes for a picture.

The Cooley receiving apparatus consists of first of all one's broadcast receiver, then certain mechanical parts which will be on the market soon; and then certain electrical apparatus which any experimenter can build and operate. The

A CLOSE-UP

The modulation transformer, which was hidden in the view on page 342 by the corona coil is shown here. All of the parts for this apparatus have been especially designed, and will be available soon. A good idea of the gearing mechanism and of the width of the drum on which the picture is received may be had from this photograph. The picture may be as wide as the metal drum. The needle point actually rides on the paper as is shown here

WHAT THE PICTURES LOOK LIKE

These photographs have been transmitted by the Cooley Rayfoto system. Pictures may be five inches wide and about six inches long and a little over three minutes for each picture is required. These pictures have not been retouched

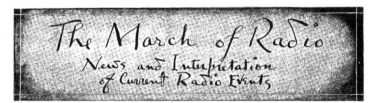

Modern Radio Receivers Are a Good Investment

THE Radio World's Fair opens the radio season for the broadcast listener. Fitting its significance, the Fair is held in mid-September at the huge Madison Square Garden in New York.

No radio season has ever started more auspiciously. Technical progress, represented by simple, easily operated and maintained radio sets, many powered directly from the electric light mains, with broadened tonal range, bring the manufactured receiver to a standard so high that revolutionary change can no longer be expected. Greater broadcasting programs and improved receiving conditions make the possession of these modern receivers all the more desirable.

Every industry goes through the same cycle of growth; first, it has a discouraging struggle for recognition, then a boom period with almost day to day improvement, and finally, stabilization, with slow, continued and healthy progress, keen competition among leaders and the gradual elimination of the less capable who thrived only during the boom period.

In stating that radio has reached a slower level of development and that the purchaser may now select his receiver with the confidence that it will not be hopelessly obsolete within a year or two, or even four or five years, we draw our conclusion from fairly simple and indisputable premises. Improvement of any device is a matter of rendering it so simple that no technical knowledge is required for its operation, so rugged and self-sustaining that but little attention is needed to maintain it in good condition, and so pleasing in appearance that it harmonizes with the most luxurious surroundings.

High standards of simplicity have been attained in radio when we have many receivers in which five circuits are simultaneously controlled by one dial and calibrated so that stations may be promptly logged. The only perishable elements in some receiving sets now are the filaments of the vacuum tubes themselves which need be replaced perhaps only once a year. In tone quality, radio

has reached a standard of reproduction far above that which satisfied the phonograph industry after twenty years of prosperity. This standard can be raised at will with our present knowledge and facilities, although the cost of doing so is prohibitive and the effect hardly noticeable to the average listener. As to appearance, there is still room for some improvement by the attainment of greater compactness in the more powerful receivers, but models are available for which no apologies need be made even in the most exacting surroundings.

Radio has certainly not reached its limit of development. Improvements will continue always and this year's sets will always be better than last. Two or three years ago, a radio receiver was but a one-year investment for those who would possess the best available in the market. This year's standards are making the latest and best models perhaps a five-year investment in approximately the best radio reception attainable.

It is a strange paradox that the substantial developments in manufactured receiving sets are not more widely appreciated by the technically minded radio enthusiast. From extensive contact with

COLONEL LINDBERGH IN SCHENECTADY

"Aviation and radio are twin sisters engaged in the joint enterprise of overcoming time. Distances are no longer measured in miles, but in hours, minutes, and seconds. Colonel Lindbergh reduced the distance from New York to Paris from days to hours. Radio has brought all parts of the world into talking distance of the United States," said Martin P. Rice of the General Electric Company. Above: Martin P. Rice, and at the extreme right Colonel Lindbergh

the better informed element of radio listeners, we have analyzed this lack of familiarity as attributable in part to the experimenter's preoccupation with the intricacies of set building and the failure of the radio set manufacturer to lay before this group, so invaluable to the building of reputations, the real facts about the design and construction of his receiving set. Glittering generalities about performance do not intrigue the radio experimenter who, in past years, has successfully excelled in these proclaimed qualities with his home-made contraptions.

But a new day has dawned. The experimenter cannot deny the superiority of the better manufactured sets. Already he is turning his attention to new fields. While building radio receivers was still an experiment, the successful outcome of which depended upon skill, ingenuity and patience, this hobby had no rival in the hearts of those who considered the soldering iron an instrument of conquest. The element of mystery is disappearing. Set building has become the following of a well established formula. The thrill of accomplishment still exists but the procedure is so well charted that the joy of exploration and discovery is practically gone. These missing elements, which satisfy the experimenter's insatiable desire to overcome obstacles and to conquer the unknown are being rapidly supplied by new lines of endeavor. Telephotography and, some day, television, international short-wave broadcast reception, modern installations with remote control of the radio receiver, laboratory experiments, home motion pictures, aviation, mechanical models and a thousand one avenues of expression beckon the experimentally inclined.

The pages which follow attempt to reflect accurately the trends in the interests of our readers. The manufactured receiver deserves an increasing amount of attention, as do new fields of development like telephotography and short-wave reception. We do not propose, in the least, to neglect the set builder, but rather to keep him in touch with the latest developments of the art. We feel the broadened scope of our pages will appeal to each of the three principal elements of our reader audience, the experimental group; those professionally interested in radio, ranging from dealer to engineer to manufacturer; and the radio enthusiasts who follow these pages to keep abreast of progress in radio.

Listening to World-Wide Broadcasting is Near

PRIOR to the development of broadcasting, experimenters built receiving sets having a wavelength range of from 200 to 25,000 meters, so that they could eavesdrop on every available radio channel. Experience has taught us to build more efficient receivers for much narrower ranges. Indeed the present broadcast band is the widest for which a high grade receiving set with easy control can now be made. Many of the services conducted on non-broadcasting channels are gradually being rendered by radio telephony rather than radio telegraphy. The objection that laborious study of the radio code is necessary to listen to them is thereby eliminated.

Almost every day of the year, new radio services open up. The possessor of a long-wave receiver picks his signals from every section of the globe. A slight movement of the dials may transfer his attention from a station in Java to another in Iceland. There are now nearly one hundred stations which can be heard in any part of the world.

Perhaps the most important development of all, so far as the broadcast listener is concerned, is the increasing interest in short-wave broadcasting. 2 XAF, the 32-meter (9150-kc.) Schenectady transmitter of the General Electric Company, is actually supplying the world with radio programs. Its signals are frequently and regularly heard in England, South Africa, South America, New Zealand and Australia. An increasing number of American broadcasting stations, including WGY, KDKA, WRNY, WLW, WAAM, WRAH and WHK are, or soon will be, broadcasting their programs on short wavelengths. British, German, and Dutch stations are within range of American short-wave sets. Short-wave programs are readily heard at great distances during the summer when the program range of most standard wavelength broadcast receivers is very limited. Several companies are already preparing to meet the demand for short-wave receivers.

The question may arise whether short-wave broadcasting will not replace our present services. A short-wave transmitter is ideal for long distance transmission but, because of fading and the skip distance effect, is of little or no local service. Within two hundred to six hundred miles of short-wave stations, their signals are usually inaudible but even moderate power delivers a strong signal for immense distances beyond the dead area. Hence every high frequency station requires an exclusive channel, placing a definite limit on the number which may be accommodated throughout the world. Receiving sets, working on the high frequencies, are necessarily of the regenerative type and consequently, at the present time at least, no vast number of receiving sets can use these channels without causing destructive interference. Undoubtedly, the radiation problem will be overcome, but fading and transmission irregularities will prevent the use of short waves in the essential local and regional broadcasting services.

Since short-wave stations practically cover the earth with their signals, it is feasible to assign but one powerful short-wave broadcaster to a frequency. Consequently, it is desirable to select a short-wave broadcasting band at once and allot frequencies so as to meet the needs of all the countries of the earth. This matter should be considered in the International Radio Telegraph Conference, convening in Washington in October, lest confusion and congestion arise later. There is no need for a great number of short-wave broadcasters in any one country, and some means should be devised for limiting their increase before the problem becomes as serious as the congestion now obtaining on the conventional wavelengths.

A Profound Study of Radio Law

THE *Law of Radio Communication* by Stephen Davis, former Solicitor of the Department of Commerce, recently published by McGraw-Hill, is a valuable contribution to radio literature. It has been awarded the Linthicum Foundation Prize by the faculty of law of Northwestern University, a distinction which will be recognized by the legal profession as one of no small moment.

The book deals comprehensively with all phases of radio communication and interprets the Radio Act of 1927 in the light of precedents already established by the Courts. The book studies exhaustively the complex problems raised by the existence of broadcasting stations, although the author has been seriously handicapped by lack of established legal precedents and decisions on most vital issues. Clearly, many important legal questions which will harass the Courts during the next few months are yet to be settled.

The preponderance of evidence and precedents which Judge Davis cites leads one to conclude that the question of confiscation of property, involved in cancellation of station licenses, is one which will be decided against the regulatory power of the Commission. On the other hand, established stations, which find their service curtailed by interference from newcomers, seem to have ample grounds for securing restraining injunctions on the grounds of prior rights. Reading these parts of the Judge's book leads us to regret all the more that our suggestion that station priorities be recognized, was not embodied in the law. That was urged in these pages long before the Radio Act was finally drafted.

The author considers federal jurisdiction and its relation to state and local regulation of radio communication, copyright questions, libel and slander, as well as the significance of every phase of the recent Act.

HOW MUCH IT COSTS TO BROADCAST Rate cards are used in selling "time" on the air just as publishers use rate cards in selling "space." The rates in these tables show exactly what it costs to send a program over the entire United States. A large part of the charges are consumed in the high cost of the special telephone wires connecting the stations

For the time being, it is the only authority available to the lawyer handling cases for broadcasting stations. We hope, for the good of radio, that Judge Davis will soon find it necessary to write a revised edition occasioned by court decisions to the unanswered questions which he so ably raises. His book thoroughly establishes Judge Davis as the outstanding American authority on the law of radio communication.

The Danger of Direct Advertising on the Air

THE United States Radio Society has sent us its literature, including a code of regulations and by-laws for local affiliated chapters of the society. By following these regulations, any local radio organization may become affiliated in the national group, and means are provided for representation of each chapter in the national deliberations. In absence of local chapters, a membership can be secured by individuals who may write Paul A. Greene, Managing Director of the Society, at the Temple Bar Building in Cincinnati Ohio.

The Society recently forwarded the returns from a questionnaire to the Federal Radio Commission which emphasized the unpopularity of certain broadcasters permitting blatant advertising, particularly two well known nuisances in Shenandoah, Iowa. The St. Joseph, Missouri *Commercial News*, writing of these "advertising sta-

tions," says "one of the best known of these nuisance stations reported, the first week in February, that it had in a month sold about 45,000 pieces of dress goods, amounting to approximately 175,000 yards. Figuring the mileage, this amounts to 99 $\frac{7}{16}$ miles of dress goods." Continuing, the article comments on the competition which this direct advertising offers to the small local merchant.

One of the claims made to the Federal Radio Commission by one of these stations was that it lowered the price of goods to the farmer. Refuting this claim, the St. Joseph newspaper says:

One of the best known stations , not far from St. Joseph, recently put on an active sale of cans of smoked salt for use in butchering on the farm. It offered two cans of a good quality of smoked salt for $2.50, postage prepaid. Orders were accepted for not less than two cans. One Iowa farmer who fell for this great bargain was shown the identical product by his local merchant for only a dollar a can. Another radio bargain was ten pounds of prunes for $3.50, of a quality purchasable locally in one pound packages for twenty-eight cents.

The Month In Radio

HENRY OBERMEYER of the Consolidated Gas Company of New York informs us that his company received more than 31,000 letters from its customers as a result of a radio course on homemaking which required

A TRANSCONTINENTAL CANADIAN BROADCAST

For the first time in history, twenty-three Canadian stations from Halifax to Vancouver were hooked up in broadcasting the Diamond Jubilee Celebrations of the Dominion. A feature of the broadcast was the program of carillon music from the Peace Tower in the Parliament Buildings at Ottawa (in insert, left). The illustration below shows the control room in the Parliament Buildings with telephones leading to the telegraph and telephone offices. Commander C. P. Edwards, head of Canadian radio, is at the extreme right

four and a half hours of broadcasting time to transmit. One of the important features of the gas company's radio campaign was that an amount equal to fifty per cent. of the broadcasting expense was spent in newspaper advertising to call the attention of radio listeners to the feature. It is an excellent example of successful commercial broadcasting which won its return by the real value of its service. ‡ ‡ ‡ The Bureau of Standards is prepared to calibrate crystal oscillators used in maintaining broadcasting stations on their assigned frequencies, for a nominal fee. Stations contemplating taking advantage of this service should first apply to the Bureau, giving call letters, assigned frequency, type, make and description of the oscillator to be calibrated. This information is required because the Bureau will not accept for calibration instruments which are not so constructed that they will remain in adjustment permanently. ‡ ‡ ‡ WHAM, Rochester, New York, which will soon have a 5000-watt equipment of the latest type, WTMJ, Milwaukee and WJAX, Jacksonville, Fla., have been added to one of the other of the N. B. C. chains, giving still greater coverage to its programs. ‡ ‡ ‡ The National Electric Manufacturers' Association recently completed a study of the question of the number of hours a week the listener uses his radio set. This information has an important bearing on the sale of tubes and maintenance accessories. According to the NEMA figures one listener out of a thousand uses his radio about twenty-two hours out of twenty-four; one out of a hundred, twenty hours out of twenty-four; one-tenth of the total more than seven hours a day and four out of five in excess of thirty hours a week, which means five hours a night, six nights out of seven. The statistics were evidently obtained by making a record of the listening habits of a handful of the most rabid enthusiasts who could be found. We would particularly like to meet the person who uses his radio receiver twenty-two hours out of twenty-four. It is quite apparent that Mark Twain was right about statistics. Editor's Note: Mark Twain said: "There are lies, damn lies and statistics." ‡ ‡ ‡ The Canadian radio industry is developing rapidly. There are now at least a dozen first class Canadian-made broadcast receivers, and several factories engaged in the manufacture of radio tubes and batteries. Broadcasting has prospered with the backing of such powerful concerns as the Canadian National Railways and the Northern Electric Company. A patent pool, similar to that of the Radio Corporation of America, is licensing a number of first class manufacturers. On April 1, licensed broadcast receiving stations numbered 134,486, as compared with 92,000 at the end of the previous year. The fees collected from the listener are applied to the elimination of interference and the proper administration of commercial radio telegraphy. ‡ ‡ ‡ Danish statistics inform us that, on April 1, 1927, there were 130,805 radio receivers in that country, almost equally divided between crystal and vacuum tube sets, and an increase of 50,000 over the previous year. ‡ ‡ ‡ There is a great increase in the demand for radio receiving sets in Brazil and broadcast enthusiasm is spreading throughout the country with great rapidity. Most of the stations are along the seaboard and long range receivers are consequently in special demand. American receivers are making the greatest headway in the market.

LICENSES AND WHAT THEY MEAN

THE most important patent decision rendered during recent months, which is, however, still subject to future appeal, was that of Judge Thacher, favorable to the Radio Corporation against E. J. Edmond & Co., Atwater Kent jobbers. The decision establishes the validity and scope of the Alexanderson patent 1,173,079, generally known as the tuned radio frequency patent. The result of this decision has been to place all concerns making multi-tube sets, not holding R. C. A. licenses in considerable jeopardy. About twenty-four important concerns have already secured Radio Corporation licenses and those outside the pale are making strenuous efforts to secure licenses. So far, no set company licenses have been granted which do not guarantee a minimum royalty to the R. C. A. of $100,000 a year. Just what the future holds for small companies is, at the moment, doubtful. Radio dealers are universally demanding apparatus which is duly licensed and the position of those who cannot, because of small production, guarantee such substantial royalties is most doubtful. The Radio Corporation could not long withstand the adverse criticism which would result were they to force smaller manufacturers out of business.

The object of the patent law is to protect the rights of inventors so that they may derive just compensation for their inventive efforts. The use of patents for restraint of competition, even though patents themselves are an intentional and desirable monopoly, is not supported by public opinion. So long as a concern is willing to recognize inventors' rights and pay royalties it should not be prevented from engaging in competitive business. Dealers are justified in insisting upon licensed apparatus and it appears that the Radio Corporation and its licensees are the only ones who may legally make efficient, high grade radio receivers under present conditions. However, if patents are used in a coercive way to strangle smaller concerns, we doubt that it will result ultimately to the benefit of the patent holders. No one questions that the recent decisions have established definite supremacy of the R. C. A. in the patent situation, but the advantage should not, and probably will not, be used to force independents out of business. ǂ ǂ ǂ The Hazeltine Corporation won a vitally important decision against A. H. Grebe & Co. sustaining the Hazeltine patents 1,489,228 and 1,533,858. The distinction between the Rice and Hartley disclosures, upon which Grebe placed principal reliance, and that of Hazeltine, is clearly set forth in the decision. Although it is granted that both Rice and Hartley disclosed principles of neutralization, the superiority of the Hazeltine method, utilizing a capacity feed back of a voltage to neutralize the regenerative effect, is established as an improvement of great importance to the radio art. Professor J. H. Morecroft appeared as an expert for Grebe. The Court, in its opinion, quoted an item from the "March of Radio," which Professor Morecroft formerly prepared for RADIO BROADCAST, acknowledging the importance of the Hazeltine disclosure. ǂ ǂ ǂ A decision handed down in the Federal District Court at Baltimore sustained Messrs. Willoughby and Lowell over James Harris Rogers in their patent for a submarine reception system. Rogers claimed prior conception, but Willoughby and Lowell clearly anticipated him in reduction to practice. ǂ ǂ ǂ After yielding to a decree *pro confesso*

secured by John A. Victoreen against the Radio Art Company and others; the latter developed a substantially different device from that previously judged to infringe. An action was brought requesting that Radio Art be adjudged guilty of contempt but, upon submission of a brief that the new device was substantially different from that on which the earlier decree had been granted, the Court ordered the plaintiff to file a new suit. The case is cited in the interests of those who have yielded to consent decrees. ǂ ǂ ǂ The Eisler Engineering Company of Newark announces that it successfully defended itself against the General Electric's suit which sought to prevent it from manufacturing tipless tubes. The case was started in October, 1924, and the final decision is of interest to all vacuum tube manufacturers. ǂ ǂ ǂ A suit has been filed by the Balsa Wood Company, Inc. against the Balsa Laboratories Company, Inc. in connection with their patent 1,492,982, describing diaphragms for sound reproducing devices. ǂ ǂ ǂ One claim of V. K. Zworykin's patent 1,634,390, covering a secrecy system of transmission was denied because of an earlier patent, Alexanderson's 1,426,944. ǂ ǂ ǂ The Brandes Products Corporation has petitioned for a writ of certiorari which may bring the dispute regarding the Hopkins patent before the Supreme Court. Originally the Lektophone Corporation of Jersey City sued the Western Electric Company, which

suit resulted in a decision in favor of the latter. The R. C. A. has already paid $200,000 for rights under the Hopkins patents, which will prove an unnecessary investment unless they are sustained before the Supreme Court. ǂ ǂ ǂ The Westinghouse Company secured an injunction against the Kenwood Radio Company in connection with Armstrong's radio patent 1,113,149.

THE COLUMBIA BROADCASTING CHAIN

THE United Independent Broadcasters, of which Major J. Andrew White, pioneer sports announcer, is a leading official, has given out a list of stations which will form its new network. The key station is WOR, which is the third most popular station in New York, exceeded only by WEAF and WJZ in number of listeners. The other stations are WEAM and WNAC in New England; WFBL, WMAK, western New York; WCAU and WJAS, Pennsylvania; WADC, WAIU, WKRC, Ohio; WGHP, Michigan; WMAQ, Chicago; and KMOX, St. Louis. Although, in point of numbers and station standing, the stations comprising the chain are not everywhere in the lead, it would not take long, given good programs, for such a chain to corner a good part of the radio audience.

Nothing will help the broadcasting situation so much as real competition to the N. B. C. chains, so that both organizations will conduct a nip and tuck battle for program supremacy.

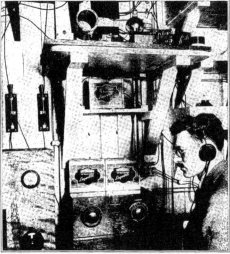

THE RADIO EQUIPMENT ABOARD THE "MORRISSEY" IN THE FAR NORTH

Edward Manley, of Marietta, Ohio, is shown before the radio apparatus aboard the Putnam-Baffin Island expedition. The *Morrissey*, known to the short-wave code world as VOQ has a generator-powered short-wave transmitter, a battery-powered transmitter using a UX-852 tube (on the top shelf), a short- and long-wave receiver, and a portable battery-operated transmitter. Signals from VOQ have been clearly heard at the short-wave laboratory of RADIO BROADCAST and occasional radio dispatches from the expedition have appeared in the New York *Times*

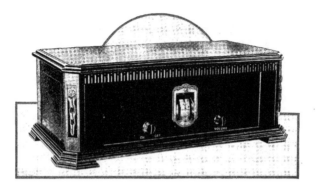

The 1928 "Hi-Q" Has an Extra R. F. Stage

By JOHN B. BRENNAN

Cascaded Stages Result in Sharpness of Tuning but Avoid the Cutting Off of Side-Bands

Variable Interstage Coupling Provides for Equal Amplification Throughout Broadcasting Band

IN ANY discussion centering upon the predominant requirements of a modern receiver, it will generally be admitted that there are two outstanding things to strive for—good quality and a high degree of selectivity. Strangely enough, these two coveted qualities are diametrically opposed to one another—difficult of attainment in the combination.

Nowadays, a receiver is judged by its ability to faithfully reproduce music and speech, and this critical attitude on the part of the listener has resulted in the setting up of a remarkably high standard so far as the audio channel of the modern receiver is concerned.

The other outstanding requirement, that of selectivity, represents a problem which did not exist in the early days of radio. To-day the United States has over six-hundred broadcasting stations whereas there were barely a hundred five years ago.

It needs no stretch of the imagination to realize that there is no receiver which, in a given location, can tune-in all six-hundred of these stations. With an exceptionally good receiver, however, listeners may tune-in more stations—not merely the far-distant stations—than with just a fairly good receiver. Yet, in spite of this, there is no advantage gained in such a case unless the reception is of good enough quality reproduction to warrant listening to it.

To illustrate briefly the difficulties involved, reference to Fig. 1 will prove helpful. Here a suppositional case is presented where the vertical rectangular sections represent the ten-kilocycle bands occupied by several stations whose carrier frequencies are equally separated by a space of twenty kilocycles. The curve b represents the tuning characteristic of a selective circuit which

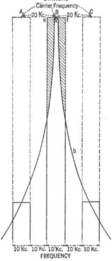

STATIONS

Carrier Frequency

A ← 20 Kc. → B ← 20 Kc. → C

e

b

10 Kc. 10 Kc. 10 Kc. 10 Kc. 10 Kc.
FREQUENCY

FIG. I

is slightly regenerative and which is set for resonance with the incoming signal of station B. It will be observed that, while a section of the curve falls in that part of the space taken up by stations A and C, the amplification of the signals from thest two stations is much below the amplification of the signal to which the circuit is tuned. Maximum amplification of signal A is represented by X, the peak of the curve b. It may be said of this circuit that it is highly selective but it will also be observed that the quality of the signal is impaired due to the fact that all the side-band frequencies, e to f, within the ten-kilocycle band of the station to which the circuit is tuned, are not amplified equally. The shaded portion shows how unequal this amplification is and indicates the rapid cutting off of the side bands. A highly desirable curve would be one where equal amplification of all the frequencies within the band, e to f, is obtained, and where there will be no response obtained to stations in adjacent frequency bands.

Now, going to the other extreme, in circuits wherein there is no inherent regeneration, and where the characteristic curve indicates broadness of tuning, as in Fig. 2, it will be noted that the shaded portion has become greatly reduced and therefore practically an equal amplification of frequencies within the band to which the circuit is tuned, is obtained. It will also be noted, however, that in such a circuit the amplification of signals obtained from adjacent stations, even though the circuit is not tuned to them, is sufficient to cause them being heard as a background to the desired signal. It has been determined that where a number of slightly broadly tuned stages are arranged in cascade, each successive stage filtering the output of the preceding stage, and

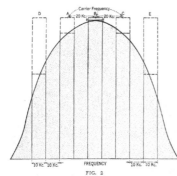

FIG. 2

production into a practical receiver. It is well known that a three-stage radio-frequency amplifier can be built to give a high degree of amplification and smooth operating characteristics; that is, it will have freedom from self oscillation, etc., if it is desired to work at a single wavelength, say, five hundred meters. However, where, under these conditions, the circuits are re-tuned to, say, two hundred meters, they will go into violent oscillation and destroy the stability which they formerly possessed. Explained in another way, this means that, due to the fact that radio-frequency amplifier circuits have the tendency to oscillate, this oscillation becoming more pronounced as the wavelength is decreased, a circuit satisfactory for operation at five hundred meters is wholly unsatisfactory for operation at two hundred meters, due to inherent oscillation, unless, of course, some corrective feature, either mechanical or electrical, is incorporated in the circuit to prevent this oscillation and yet at the same time maintain the circuit at its high degree of amplification.

It has been recognized that some system of securing uniform amplification at all broadcast frequencies is greatly to be desired and various methods for attaining this end have come into prominence during the last year. These systems in the main can be divided into two classes: (1) The mechanical control of coupling, such as is used in the King "Equamatic," "Hi-Q," and Lord systems and, (2) electrical systems, such as those relying on the function of combinations of capacity, inductance, and resistance—the Loftin-White and "Phasatrol" methods, for example.

A little calculation will readily show that these systems can be designed to produce a curve approaching the desired straight line indicating uniform amplification shown in Fig. 4. Since the resultant curve from the use of a fixed coupling between primary and secondary is so far removed from a straight line characteristic, as will be seen by Fig. 4, even remedial systems,

where precautions have been taken to guard against inherent regeneration in the several stages, that the ideal response curve is approximately attained. This is graphically shown in Fig. 3.

From Fig. 2 we have observed that the response curve for a single stage is quite broad. It is reproduced again in Fig. 3 as curve No. 1. However, when a second tuned stage, having the same characteristics, is added, the curve obtained is somewhat altered in form. The top of the curve has practically remained the same but its sides have taken a deeper slope, as shown by curve No. 2. In adding a third stage, the curve obtained is similar to the curve No. 3, wherein there is a further constriction of the sides of the curve. The fourth stage confines the curve within limits which very nearly approach the ideal.

For those who desire a highly technical discussion on the points just outlined, it is recommended that reference be made to Professor L. A. Hazeltine's paper on the subject, which is contained in the June, 1926, issue of the *Proceedings of The Institute of Radio Engineers*.

THE NEW 'HI-Q' RECEIVER

IN THE new Hammarlund" Hi-Q" Six receiver, four tuned stages are employed and the final response curve obtained from their use is similar to No. 4 of Fig. 3. In order to secure this result it was necessary to take precautions in setting up the four tuned stages to prevent the desired signal being received by any other way than through the antenna into the first tuned stage, and so on, up to the detector stage. Therefore, the necessity for completely shielding each of the several tuned stages arose. Also, each of the several tuned stages had to be completely filtered by the use of radio-frequency chokes and bypass condensers so that there were not present any intercoupling effects which would cause unstableness in operation and thereby defeat the purpose of the use of these several stages. Were not the radio-frequency currents confined to their own individual stages by means of chokes and bypass condensers, and also by the individual stage shields, intercoupling effects would undoubtedly be caused either by capacitative coupling from the wiring of the circuits, inductive coupling between the coils of the various tuned circuits, or resistance coupling in the batteries, since one set of batteries is employed to furnish the B potential and its impedance is common to all the tubes.

It was realized that all this must be achieved without sacrificing the sensitivity of the circuits, or in other words, their ability to bring in weak signals. This means that the various interstage transformers employed in the tuned circuits must be efficient enough so that weak antenna signals are built up to a strength sufficient to give a good response in the loud speaker considering that two stages of audio-frequency amplification are employed.

It is this desire for sensitivity that complicates the problem of combining these principal features of a high degree of selectivity and a high degree of re-

FROM BEHIND THE PANEL

The shield walls have been removed for this photograph so that it may be clearly seen how all the parts are disposed

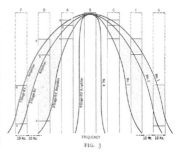

FIG. 3

though not absolutely perfect, are undoubtedly a step in the right direction.

In the Hammarlund "Hi-Q" Six the results obtained by the use of four well designed tuned stages depended not only upon complete isolation of each of the four stages by means of shielding but by a combination of mechanical and electrical systems which, in the end, produced as near as possible the desired straight line of amplification. The mechanical feature consisted of automatically varying the coupling between the several primary and secondary coils by means of a cam located on the shaft of the condensers, so that a variable degree of coupling was obtained to give an equal transfer of energy between primary and secondary regardless of whether the circuits were tuned to two hundred or five hundred meters. By the use of the cam method it is possible to secure the correct degree of coupling at any dial setting due to the fact that the shape of the cam can be predetermined to provide the desired coupling. Since all precautions are taken to prevent feedback by isolating the several tuned stages, through the use of radio-frequency chokes, bypass condensers, and individual stage shields, only the tube capacity of the tubes employed in these stages remains to cause undesired coupling. The electrical means referred to above is obtained by the use of grid "suppressors"—r e s i s-tance units located in the grid circuits of these tubes. These grid "suppressors" will cancel the tendency of the tube capacity to act as a coupling agent and to cause feed-back.

From the detector grid to the loud speaker is a fixed circuit with constant audio amplification. The volume control, situated in the radio-frequency amplifier circuit, permits regulation of its overall amplification so that the correct amount of signal can be fed to the detector grid to produce a volume of signal from the loud speaker which is satisfactory to the listener. The control, a filament rheostat which regulates the voltage applied to the first three radio-frequency amplifier tube filaments, prevents the overloading of the detector, especially during reception from powerful local stations. Two stages of audio

READY FOR THE BATTERIES

This illustration clearly shows the gauged condenser arrangement employed in the new "Hi-Q" Six

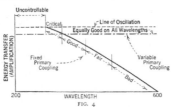

FIG. 4

frequency amplification are employed, good tone of signals being maintained by the use of high-quality audio-frequency transformers. To secure perfect amplification it is necessary that the advantageous results to be obtained by the use of expensive transformers be not offset by the unintelligent use of tubes in the audio amplifier. In the last or power audio stage either a UX-112 (CX-112) or UX-171 (CX-371) type of tube should be employed with the correct B and C batteries.

From the accompanying photographs of the

new "Hi-Q" Six receiver, it will be noted that directly behind the panel there are two metal boxes located at either end, between which is a central space in which is located the drum dials controlling the several tuning condensers. The left-hand box shield is divided into two compartments housing the first and second radio-frequency amplifier stages with their associated chokes and bypass condensers. In the right-hand box shield, also divided by means of a central wall, are located the third radio-frequency amplifier and the detector stages. The two tuning condensers in each of these box shields are located on common shafts which terminate at the central space in the drum dials. These dials are insulated from the shafts by means of insulated coupling units. The following parts are used in building the "Hi-Q" Six:

1—Samson Symphonic Transformer	$10.00
1—Samson Type HW-A3 Transformer (3-1 Ratio)	5.00
4—Hammarlund 0.0005-Mfd. Midline Condensers	22.00
4—Hammarlund "Hi-Q" Six Auto-Couple Coils	12.00
4—Hammarlund Type RFC-85 Radio-Frequency Chokes	8.00
1—Hammarlund Illuminated Drum Dial	6.00
1—Sangamo 0.00025-Mfd. Mica Fixed Condenser	.40
1—Sangamo 0.001-Mfd. Mica Fixed Condenser	.50
1—Pr. of Sangamo Grid Leak Clips	.10
1—Carter IR-6 "Imp" Rheostat, 6 Ohms	1.00
1—Carter "Imp" Battery Switch	.75
1—Durham Metallized Resistor, 2 Megohms	.50
3—Parvolt 0.5-Mfd. Series A Condenser	3.00
6—Benjamin No. 9040 Sockets	4.50
3—Eby Engraved Binding Posts	.45
2—Amperites No. 1-A	2.20
1—Amperite No. 112	1.10
1—Yaxley No. 660 Cable Connector and Cable	3.00
1—Hammarlund-Roberts "Hi-Q" Six Foundation Unit (containing drilled and engraved Westinghouse Bakelite Micarta panel, completely finished Van Doorn steel chassis, four complete heavy aluminum shields extension shafts, screws, cams, rocker arms, wire, nuts and all special hardware, required to complete receiver)	15.90
Total	$96.00

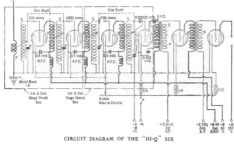

CIRCUIT DIAGRAM OF THE "HI-Q" SIX

¶ The work of the artist and interior decorator—as well as the labors of the radio engineer—are evident in 1927–28 radio receivers. This Splitdorf "Abbey" set, designed by Noel-S. Dunbar after an old-world jewel case is a six-tube receiver, priced at $100. This page and those which follow strikingly demonstrate how the manufacturer is successfully harmonizing technical receiver design with the decorative demands of the home. Modern receivers are not one-year investments because the receiver bought to-day will be up-to-date for some time to come; for this, the engineer is responsible and in later issues of this magazine we shall report much of his fascinating work.

1928 Radio Receivers

FRESHMAN G-4
Here is a new electrically operated Freshman receiver, complete and ready to operate as delivered. The price is $235.00, which includes new RCA a. c. tubes

OFTEN it is said that a radio receiver, to be fully appreciated by the feminine half of the domestic republic, must be encased, in housings which are esthetically as well as technically satisfactory. It is natural and right that, as radio has become more an accepted part of the equipment of every home, women have had an increasing voice in the selection of the radio receiver. What, then, do the radio manufacturers offer to the prospective purchaser? To be brief, the radio receiver of 1927-28 is a thing of beauty as well as of utility. On these two pages, and on others in this issue, are grouped illustrations of the sets which have come from the manufacturers' designing laboratories. Various makers have grappled with the important problem of appearance, and, as these illustrations s h o w, have met it in widely different

THE STEWART-WARNER 520
The tendency towards single-control is again manifest in this six-tube receiver. Two tuned and one untuned stages of r. f., detector, and two transformer-coupled audio stages, comprise the circuit. The cabinet is of dull polished walnut veneer. The price, $125.00

BOSCH 66
Another single-control receiver employing six tubes, this one being priced under a hundred dollars. The beauty of the cabinet is a feature but the efficiency of the circuit itself has in no way been neglected. The illuminated dial is graduated into both kilocycles and arbitrary numbers

GREBE'S "SYNCHROPHASE" SEVEN
The five tuned stages (four r. f. and detector) are tuned by means of a single dial and vernier, the rigidity of the tuning condenser assembly insuring permanency of the accurate factory adjustment. The receiver has been carefully shielded, while the special coils employed enhance the selectivity and permit maximum amplification. These coils are of the binocular type and are wound with Litz wire

"COMMANDER"
Here is a King six-tube completely shielded t. r. f. receiver which makes use of single control tuning. There are three r. f. stages. The loop may be folded into the cabinet. Price $220.00

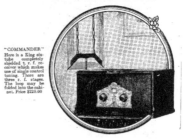

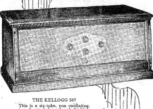

THE KELLOGG 507
This is a six-tube, non oscillating, t. r. f. set, balanced and shielded. "Inductance tuning" is employed, a system which requires no variable condensers and which permits equal amplification throughout the frequency spectrum covered. The price of the receiver is $196.00

# of Beauty and Utility

ways. Some of these models achieve their decorative effect through the dignity of utter simplicity; others are notable for the use of ornament in the treatment of the control elements. The metal tuning and control escutcheon is supplied with some decoration, slightly reminiscent of that simplicity to be found in the grouping of instruments on present day automobile dashboards. Illumination of the control escutcheon is also a feature which has much in common with the instrument panel on the modern motor car. From the representative models shown on these pages, the house-wife can gain an excellent idea of the appearance of moderate priced radio receivers which are offered to suit her taste as well as that of her husband. Her ideas of the necessary limitations of her domestic decorative scheme should blend with her husband's technical opinions.

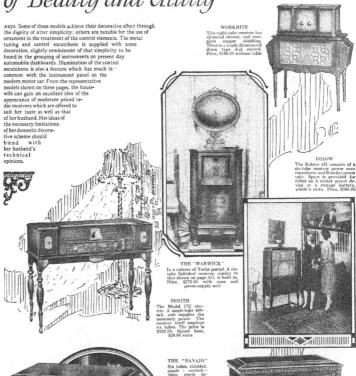

WORKRITE

This eight-tube receiver has all-metal chassis, and complete copper shielding. There is a single illuminated drum type dial control. Price, $160.00 without table

BELOW

The Kolster 6H consists of a six-tube receiver power cone reproducer, and B socket power unit. Space is provided for either an A socket power device or a storage battery, which is extra. Price, $265.00

THE "WARWICK"

In a cabinet of Tudor period. A six-tube Splitdorf receiver, similar to that shown on page 351, is built in. Price, $275.00 with cone and power-supply unit

ZENITH

The Model 17E electric. A single-tube 400-mA. unit supplies the necessary power. The receiver itself employs six tubes. The price is $350.00. Spinet base, $20.00 extra

THE "NAVAJO"

Six tubes, shielded, single - control—these words describe this new Mohawk receiver just as they do other prominent receivers of the season, for this circuit combination predominates in 1928 models. The "Navajo" sells for $67.50. Equipment to adapt it for a. c. operation is obtainable at additional cost

MU-RAD

Two t. r. f. stages, detector, and three audio stages are included in this receiver. The audio stages, all of the double-impedance form, give uniform response between 30 and 5000 cycles, according to the manufacturers. Single-control tuning is featured. Price $98.00. In a console, equipped for complete a. c. operation, the receiver sells for $265

THE "BANDBOX"

This is a recent presentation of the Crosley Corporation. It is a six-tube neutrodyne made in two models, for battery or lamp-socket operation, is shielded, and has one-dial tuning. The prices are $55 and $65, for battery and a. c. operation respectively

The
Neutrodyne
Group

A LOOP NEUTRODYNE

It is a Freed-Eisemann eight-tube single-control receiver, retailing for $395.00 with a cone loud speaker, as illustrated

THE RADIO-
PHONOGRAPH

This Stromberg-Carlson masterpiece consists of a complete a. c. operated seven-tube single tuning control receiver combined with a modern electric phonograph. A concealed loop obviates the use of outdoor antenna. Price is $1245 completely equipped with cone loud speaker

FADA
Three r. f. stages, two tuning controls, one-piece steel chassis, and equalized amplification, are features of the Fada "Special," retailing at $95.00

FOR THE "BANDBOX"
This console has been specially designed for the Crosley "Bandbox," illustrated on the left-hand page, The "Bandbox," the chassis of which is completely shielded, may quickly be removed from its metal cabinet and inserted in the more pretentious console

TO THE lay radio reader the term "Neutrodyne" means a type of receiver only—a well-known type of long standing merit. To the engineer the term signifies much more. It recalls the researches of Professor Louis Hazeltine of Stevens Institute of Technology; it brings the patent office to his mind; it indicates one of the present group of radio set manufacturers who have availed themselves of Professor Hazeltine's inventions.

For four years, the neutrodyne group, now consisting of Amrad, Fada, Freed-Eisemann, Crosley, Murdock, Workrite, Gilfillan, Howard, Garod, and Stromberg-Carlson have built receivers that were representative of the best work of the best engineers. These receivers have become known for their reliability, their sensitivity, simplicity, and generally excellent service. They need no introduction to the American radio public. Of late, those who read the foreign radio press have noted the large number of times the French, English, Spanish, and German papers have described the neutrodyne, indicating that

THE FADA 7
It is optional whether loop or antenna is used with this receiver, which has four stages of r. f. An improvement in the detector circuit reduces the possibilities of overloading. This seven-tube receiver sells for $185.00

Europe is following advances in American radio with great interest.

For the coming radio season, the neutrodyne group has been more than busy, as the accompanying photographs will show. In common with other well organized receiver manufacturers, the group has spent money and time in research for better radio components, for even greater simplicity of operation, and for greater fidelity of reproduction. A glance will show cabinet work that cannot help but amuse those who remember the early days of radio, when a conglomeration of wires and roughly assembled apparatus, frequently boasting no kind of housing whatever, was principally useful in collecting dust.

AMRAD'S "THE WINDSOR"
This artistic cabinet houses a very efficient seven-tube neutrodyne chassis. Tuning is accomplished by a single dial and a further adjustment is provided for volume control. Either an antenna or loop may be used and all parts are copper shielded. A tone filter is incorporated. The price is $195

THE "MAYFLOWER"

Here is a five-tube completely a. c. operated receiver by Cleartone, of Cincinnati. Complete with a. c. tubes, loud speaker, and built-in power unit, the "Mayflower" lists at $250.00

A MARTI RECEIVER

Another electric receiver, this one employing six tubes, three of which constitute the audio channel, resistance coupling being featured. Tuning is accomplished by means of two dials

Servants of Your Light Socket

The Radio Receiver Powered Directly from the Light Socket is like the Automobile with a Self-Starter   Pushing the Button Starts the Machine--Models on This Page are Representative of the Season's A. C. Set Offerings.

TO THE LEFT

The well-known Loftin-White circuit, six tubes, is employed in the Arborphone Model 253 receiver. Tuning is accomplished by means of a single dial calibrated in wavelengths. A Peerless cone loud speaker is included, and the receiver may be used with either batteries or socket-power devices. The cabinet is of matched walnut, curly maple and rosewood veneers, and gumwood. Price, $250.00

A NEW DEPARTURE IN POWER SUPPLY

The Day-Fan Electric Company, of Dayton, Ohio, has placed on the market receivers which are powered by motor generators, all batteries being eliminated. The efficiency of the motor generator principle of transforming alternating current to direct current has long been recognized, but its application to the powering of a radio receiver has hitherto been neglected, so far as receivers commercially available are concerned

AN ECONOMICAL A. C. RECEIVER

The manufacturers of this receiver, the Argus Radio Corporation, of New York City, state that the cost of operating this receiver is no more than one fifth of a cent per hour. The receiver is completely a. c. operated, and employs no batteries or chargers. There are six tubes in all, three of which are r. f. in a combination of tuned and untuned stages. The final audio stage employs a 210 type tube, a plate potential of 400 volts being applied to it. Provision is made so that the audio amplifier may be used in conjunction with a phonograph pick-up device. Price $195.00, less tubes and loud speaker

LEFT

This is a Freed-Eisemann electric-ally operated receiver, listing at $295.00. A neutrodyne circuit is used, and extraordinary selectivity is claimed by the manufacturers

AN ALL-AMERICAN HIBOY

The receiver employs six tubes, and may be obtained either to use the new a. c. tubes or for ordinary tubes, batteries or power units being required in the latter case

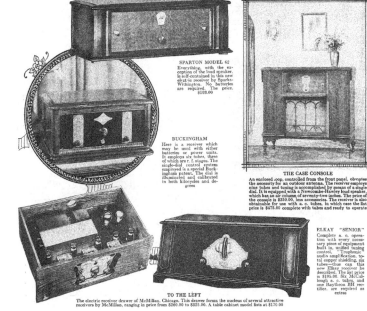

SPARTON MODEL 62

Everything, with the exception of the loud speaker, is self-contained in this new electric receiver by Sparks-Withington. No batteries are required. The price, $108.00

BUCKINGHAM

Here is a receiver which may be used with either batteries or power units. It employs six tubes, three of which are r. f. stages. The single-dial control system employed is a special Buckingham patent. The dial is illuminated and calibrated in both kilocycles and degrees

THE CASE CONSOLE

An enclosed loop, controlled from the front panel, obviates the necessity for an outdoor antenna. The receiver employs nine tubes and tuning is accomplished by means of a single dial. It is equipped with a Newcombe-Hawley loud speaker, which has as air column of seventy-two inches. The price of the console is $350.00, less accessories. The receiver is also obtainable for use with a. c. tubes, in which case the list price is $475.00 complete with tubes and ready to operate

ELKAY "SENIOR"

Complete a. c. operation with every necessary piece of equipment built in, unified tuning control, "Truphonic" audio amplification, total copper shielding, six tubes—thus can this new Elkay receiver be described. The list price is $195.00. Six McCullough a. c. tubes, and one Raytheon BH rectifier, are required as extras

TO THE LEFT

The electric receiver drawer of McMillan, Chicago. This drawer forms the nucleus of several attractive receivers by McMillan, ranging in price from $260.00 to $325.00. A table cabinet model lists at $170.00

THE NEW RADIOLA 17
A receiver which makes use of the new
a. c. tubes in the first five stages and a
171 type tube in the final (second) audio
stage. The receiver may therefore be
supplied with A power from the house
lighting circuit without the use of additional equipment. B and C batteries
are also eliminated by the use of a suitable power device which is built in.
Price $130.00

Radiolas

A NEW CONE
The new 100-A cone
loud speaker priced
at $35.00

A CUSTOM-BUILT RADIO RECEIVER
The Radiola 32 combines the well-known RCA eight-tube super-heterodyne and a
baffleboard loud speaker in a single cabinet. Since a loop, directionally variable, is
built in this receiver, it may be used where facilities for outdoor antenna erection
are not available although binding posts are provided for antenna and ground if
such are used. All current is obtained by plugging into the house lighting supply.
The model is obtainable for either a. c. or d. c. operation. Price $895.00 complete

THE super-heterodyne and the tuned radio-frequency circuit,
with or without regenerative action, continue as the foundations
of the 1927-28 line of receivers offered by the Radio Corporation
of America. A parchment cone, with properly balanced drive for
the moderate power of the battery set or again the high power of the
socket-power set, likewise continues as the basis of the Radiola loud
speaker offerings. Alternating-current operation, with certain refinements particularly by way of rectifier tubes of increased output and
better voltage regulation, makes a bid for favor in the higher priced
combinations. The Radio Corporation has also added the new a. c.
tubes as a means of electrifying even the moderate-priced Radiolas,
the UX-227 heater type for the detector socket and the rugged filament
UX-226 for all other positions.

Starting out with the requirements of the modest home, there is the
new Radiola 16, which fulfills the most rigid requirements of sensitivity, selectivity, ample volume, simplified operation, and excellent
tone quality when used in combination with the new Radiola 100-A
loud speaker. The Radiola 16 is a new uni-control six-tube receiver,
embodying the well-known and perfected tuned radio-frequency
circuit with three stages of radio-frequency amplification, a detector,
and two stages of audio-frequency amplification, and utilizing five
UX-201-A and one UX-112 tubes. The power tube in the last stage
spells ample volume without possibility of distortion due to overloading. The internal construction of this Radiola is extremely rugged.
The operation is reduced to one tuning control which can be logged
for the station call letters. There is also a volume control, and a filament switch which starts and stops the reception of programs. To
simplify the operation still further, the filament rheostats have been
dispensed with. Radiola 16 may be operated from batteries or from a
socket-power device. The cabinet of this receiver is of mahogany
finish, and measures 16½″ long, 8¼″ high, and 7½″ deep. The weight is
14½ pounds, and the price, $69.50 less accessories.

Next comes the Radiola 17, aimed to produce, for a moderate price,
a receiver completely operated from the a. c. house lighting mains.
Simplicity of operation and maintenance are the main features of
Radiola 17. It has three stages of radio-frequency amplification, a
detector, and two stages of audio-frequency amplification. The new
UX-226 a. c. tubes are used in the radio-frequency stages and in the
first audio-frequency stage, while the new UY-227 a. c. tube is used as
a detector. The last audio-frequency stage employs a 171 power tube.
The B and C voltages are obtained from a power supply unit built into
the set and employing the new high-power UX-280 full-wave rectifier.
There are only three controls on this set, one knob for tuning, one
for volume control—to regulate the output of the receiver, and a
power control switch to turn the receiver on or off.

Inside this new Radiola 17 is located a switch whereby adjustment
may be made for any variation in local line voltages between 105 to
125 volts. The size of the mahogany cabinet is 25¼″ long, 7⅞″
deep, and 8½″ high. The receiver weighs 36½ pounds, and retails for
$130.00 without accessories.

The well-known Radiola 20, with its two stages of radio-frequency
amplification, its regenerative detector, and two stages of audio
frequency, employing the economical, UX-199 and UX-120 dry-cell

for 1928

tubes, continues to occupy its place in the RCA line. This Radiola has proved a popular favorite, and for that reason it has been retained among the present offerings. It now lists for $78.00 less accessories. Likewise with Radiola 26, the six-tube portable super-heterodyne with its self-contained loop, loud speaker, and dry batteries, which has provided vacationists and travelers with satisfactory and convenient radio service, and which retails at $225.00 complete. Radiola 25, a six-tube super-heterodyne priced at $165.00 with tubes, is also being retained for 1928 offering.

Turning to the higher-priced offerings, there remain the well-known Radiola 28, or eight-tube super-heterodyne, and the Radiola 104 loud speaker, which, in combination, provide an exceptionally sensitive set which may be completely powered from the a. c. house lighting system. The price of the Radiola 28 is $260.00 with tubes. The 104 loud speaker units cost $275.00 or $310.00, for a.c. or d.c. operation respectively. Radiola 32, a newcomer to the Radio Corporation's line, with its handsome walnut grained cabinet in period design, consists of an eight-tube super-heterodyne, loop, Radiola 104 loud speaker, and is operated from the house lighting system. It provides maximum radio results in the most compact and attractive form. Certain refinements in design have permitted the inclusion of the powerful 104 loud speaker actually in the same cabinet as the super-heterodyne. The new 32 receiver has a self-contained loop which is turned by means of a knurled dial mounted near the loud speaker grille. Another feature is the small electric light bulb which is concealed in the top of the operating compartment to illuminate the dials and also to serve as a warning indicator when the current is turned on. The power is automatically switched off when the set is closed. The cabinet measures 52" high, 72" wide, and 17¾" deep. The list price is $895.00, which is inclusive of all accessories.

A somewhat lower-priced model, but likewise characterized by a distinctive cabinet and entirely self-contained equipment, is the Radiola 30-A, comprising the eight-tube super-heterodyne with the new 100-A loud speaker and an arrangement whereby the set may be powered from the house lighting supply. The Radiola 30-A measures 42½" high, 29" wide, and 17¾" deep, and its list price is $495.00, including all the necessary accessories. Unlike the more expensive Radiola 32, no loop is contained in the cabinet of the 30-A, antenna and ground connections being provided on the rear. A short indoor antenna is recommended for use with this receiver.

Both of the latter, Radiolas, 32 and 30-A, may be had for alternating current or direct-current operation. Furthermore, the 104 loud speaker unit is now available in a direct-current model, as mentioned above.

The new Radiola 100-A loud speaker referred to above is the improved loud speaker for use with all receivers, whether they are operating on batteries or with socket-power equipment. The cone itself is of smaller diameter than the well-known 100 type, which it replaces and, in addition, embodies a newly designed drive that provides increased response with even better tone quality than its predecessor. The cone is enclosed in an attractive metal case, suggestive of a mantelpiece clock, with silk screen bezel, the whole being finished in dull bronze. This new tone loud speaker measures approximately 15" x 11" and retails for $35.

THE RADIOLA 20

This is not a new season's model but it has been carried over on account of its popularity. It is a five-tube t. r. f. receiver with knurled-dial tuning. Its price is $78.00 and it may readily be adapted for a. c. operation

RADIOLA 30-A

The circuit employed in this new receiver is identically the same as that of the "Radiola" 36, i. e., an eight-tube super-heterodyne and it is operated from the light socket. The 30-A retails for considerably less than the 32, however, its list price being $495.00 complete. The differentiating features of the less expensive model are: (1) The 100-A cone is employed instead of the more powerful and more expensive balloonboard loud speaker, and (2) no loop is supplied. Antenna and ground connections are provided, and for average reception an ordinary indoor antenna will be satisfactory

A POWER CONE REPRODUCER

In addition to the cone loud speaker, this Kolster power assembly comprises a B power device to furnish the necessary voltages to operate any commercial receiver. It is operated from the 60-cycle 110-volt a. c. house supply. The price is $150.00 without tubes. Two rectifier tubes (CX-316-B or UX-216-B), a regulator tube, and a power tube (CX-310 or UX-210), are necessary

VALLEYTONE MODEL 52

A five-tube receiver to suit the average pocket, its list price being $85.00, less accessories. Tuning is accomplished by means of two dials. A simple switching arrangement built into the set makes the use of a power tube optional

LEFT

The "Spinet," a seven-tube balanced receiver by Colin B. Kennedy, Incorporated. There are four r. f. stages but just one tuning control. The finish is of two-tone antique mahogany, and there is ample storage space for power accessories. Price, $195.00

FOR B AND C VOLTAGES

Grebe offers an attractive socket power unit for radio receiving sets of from five to seven tubes. There are three B taps, 180, 90, and 22 volts, and two C values, 40 and 4 volts. Special design features have been applied to obviate "motor boating"

Some New Offerings for

Loud Speakers, Cabinets and Accessories offered by Various Makers Show Definite New Trends—

FOR THE SET CONSTRUCTOR

Keeping space with the commercial set manufacturer, the home constructor nowadays builds his receiver into a luxurious piece of furniture. This beautiful walnut cabinet has a sliding frame for a radio panel not larger than 9 1" x 21", and 14" deep behind panel. The list price varies from $80.00 for the straight cabinet to $105.00 to include a Peerless cone. It is a product of the Radio Master Corporation, Bay City, Michigan

THE BOSCH "NOBATTRY"

This attractive unit is designed to deliver three B voltages, the taps being labelled "low," "medium," and "high." The list price of the device is $48.00. Taps on the transformer provide for three separate voltage ranges

the 1928 Radio Season

All, while Serviceable and of Interesting Technical Design Are Made to Harmonize with Their Eventual Setting—the Home.

THE "MINSTREL"

A successful antenna operated receiver is this seven-tube, single-control, Apex set. The circuit employed is of the "Technidyne" variety, one that has become increasingly popular lately, and which makes use of a new form of neutralization. Less accessories, the "Minstrel" lists for $225.00

ABOVE

This is the Pfanstiehl Model 30 six-tube receiver. Its chassis is carefully shielded and equipped with flexible cable connections. Price, $105.00

DESIGNED FOR A RADIO SET

Here is another example of the cabinet makers' efforts to produce a radio cabinet which is in every respect equally as beautiful as the furniture surrounding it. The finish of this particular cabinet is in walnut and maple, and it is equipped with a Farrand cone loud speaker. It is designed to accommodate all standard makes of radio receivers, and is a product of the Musical Products Distributing Company, New York City. Price, $95.00

RIGHT

The "Baby Grand," by Audiola, Chicago, comes with either an eight-tube or six-tube chassis, the former listing at $275.00 and the latter at $225.00. There are two knobs on each model, one for tuning, and the second for volume control. Individual stage shielding adds to the general efficiency of the set. A special long air-column loud speaker is built in

B POWER

The Kellogg B power unit has the usual three taps, for "detector," "medium," and "high," but additional flexibility is obtainable by means of the knobs on the front panel. The detector voltage may be adjusted to any value up to 45, while the medium voltage tap may be set for anything up to 90 volts. The high-voltage tap is for a 171 type tube. Forty dollars is the price of the unit.

AN ELLIPTICAL CONE

The frame and base of this new Splitdorf cone is of rich antique walnut finish, deeply carved, the result being a loud speaker which will enhance almost any setting. The elliptical shape of the cone makes possible deep richly resonant tones, the short diameter of the ellipse favoring the high notes, and the long diameter sustaining the deep tones. A combination of design features enables the loud speaker to handle great power. The list price is $35.00

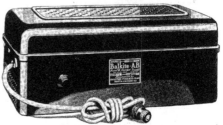

A UNIT THAT SUPPLIES BOTH A AND B CURRENT

Unlike many other units designed to supply both A and B current, this one does not use a storage battery in conjunction with a trickle charger. A rectifier and electrolytic filter condenser, combined in one cell, constitute the A unit. There are no tubes to burn out, the only attention necessary being the addition of water every three or four months. This Balkite "AB" is obtainable in two patterns, both with an output of 6 volts, 2 amperes, for the A supply, but one has a B output of 40 mA., 135 volts, while the second gives 55 mA., 180 volts. The prices are $59.50 and $67.50 respectively

AN AUTOMATIC CHARGER

A charging unit developed by the Westinghouse Company is incorporated in this useful Apco device. Full-wave rectification is accomplished by means of a series of special analysis copper discs in the transformer circuit. An automatic relay is combined with the Westinghouse unit so that, once connected to the set, the charger is spontaneously set in operation when the receiver is not being used. The list price of this Apco unit is $16.50

A AND B POWER FROM ONE UNIT

Acme, of Cleveland, is responsible for this useful power combination. Included in the unit is a B socket-power device, a 40 ampere-hour storage battery, trickle charger, and automatic control switch. The unit may be obtained for 25- or 60-cycle house supply, and for sets with varying amounts of tubes, the prices being between $95.00 and $108.50

THREE LOUD SPEAKERS IN DISGUISE

The group of books, in the first place, is nothing less than a loud speaker in concealment. A forty-inch serpentine tone chamber winds behind the embossed leather book bindings. This "Choral Cabinet" lists at $50.00, or $75.00 if the equestrienne bronzes are included. The lamp too, in addition to serving its obvious purpose, hides an air-column loud speaker. Depending upon the finish, it lists at $35, $50.00, or $60.00. The carved picture frame to the left also conceals a loud speaker, the tone chamber of which is a thirty-inch serpentine winding. The hand carved black walnut model lists at $35.00, but other models are less expensive—even as low as $20.00. Manufactured by Frank R. Porter, Washington, District of Columbia

A New Regulator Tube

Why It Is Desirable to Employ a Regulator Tube in B Devices—Details of a New Glow Tube Which Involves Some Novel Features

INNER CONSTRUCTION
The bulb has been removed to show the electrodes of the "R" tube

By JAMES MILLEN

THE voltage-regulator tube received its rather descriptive nickname, "glow tube," from the performance of some of the early models which, having an open top cathode, permitted an observer to see the ionized gas or "glow" in the space between the anode and cathode.

This tube is a device which, when connected across the 90-volt output of a B power-supply unit, will maintain that voltage at a constant value, regardless of any variations in load, within reasonable limits, applied to the power unit. Without the use of such a regulator, the voltage supplied by the device will fall off rapidly as the current drain is increased.

By using a regulator tube, it is possible to construct a B power unit, the 90-volt tap of which will, for all practical purposes, deliver just 90 volts to any radio receiver whether it has 3 or 10 tubes. In the case of the majority of radio receivers the 90-volt tap supplies the B power for the radio-frequency and first audio stages. In the case of a super-heterodyne, however, the 90-volt tap will also have to supply the B power for the intermediate-frequency amplifier. Incidentally, as the load drawn by the detector tube, in most sets, is substantially the same, a suitable resistor with mid-tap may be connected across the voltage regulator tube terminals to provide a fixed 45-volt tap. Now, it also happens that, when the regulator tube is used, variations in load on the 90- and 45-volt taps will have no effect upon the maximum voltage tap of the power unit. The voltage at this point will, however, vary with the load at the high-voltage tap; but as we know that the usual load to be applied to the high-voltage tap is a UX-171 or CX-371 type power tube, and that at 180 volts this tube will draw a plate current of 20 mA., we can so design our power unit to supply exactly 20 mA. at 180 volts.

The result is a fixed-voltage power device which will supply 45, 90, and 180 volts to almost any standard radio receiver. Fixed-voltage control, however, is but one of the many points of merit of a "glow-tube" equipped power-supply unit.

The action of the tube in holding the voltage of the output circuit constant serves also to eliminate the small ripples which may be present as a result of incomplete filtering, and thus makes possible a reduction in the capacity, and therefore the expense, of the final filter condenser. In fact, the tube, when in operation, has many properties in common with a large fixed condenser. One of these properties is extremely low a.c. impedance which, when combined with its instantaneous response as a voltage regulator, entirely eliminates the annoying "motorboating" effect which generally results when an attempt is made to use one of the ordinary B power units with many forms of amplifiers.

A further advantage accruing from the construction of a B power unit employing a voltage regulator tube is an economical one. With its many electrical advantages, it need cost no more than that of a high-grade power unit of the conventional type. In the first place, the use of the "glow tube" results in a saving of the number of high-voltage filter condensers required—usually an expensive item. In the second place, it permits the use of fewer and lower voltage—and thus less expensive—bypass condensers at the various voltage taps.

The first type of voltage regulator tube to appear on the market was of the two-element type and, like the prototype of most things, had its disadvantages. Different tubes varied considerably in characteristics, some having a working voltage of as low as 70, while others had over 100. Then, the voltage across individual tubes would vary quite a bit between conditions of no load and full load. Another fault was that, should the power unit be heavily loaded for any reason, the tube would cease to glow and, before again becoming operative, the voltage across it would have to rise to a rather high value, which eliminated the possibility of partially providing for the initial cost of the tube by the use of low-voltage bypass condensers.

But perhaps the most serious objection to former voltage regulator tubes was the tendency of many of them to oscillate and introduce noise into the output circuit of the power unit. The use of a power unit equipped with such a tube would often introduce sufficient noise into the loud speaker to interfere with satisfactory reception. But now, after several years of research and experimental work, a new form of voltage regulator has been perfected.

Its small size and low cost permit its ready use in many cases where the older form of tube was out of the question. Its silent non-oscillating operation, its long life, its close control of voltage (a variation of but 3 volts between a condition of no load and full load being allowed), and the use of a third or "keep-alive" element, are all features which make it one of the most interesting developments in the B power-supply field since the introduction of the filamentless full-wave gaseous conduction rectifiers.

The "keep alive" circuit is a most unique arrangement of what really amounts to a tube within a tube. The inner, or secondary tube, operates without load from a high-voltage point of the filter circuit and keeps the gas within the glass tube ionized at all times. Thus, should the voltage across the operating electrodes at any time fall below the value required to maintain the gas in an ionized state, the potential of the third element will be sufficient to maintain ionization. In order to minimize the extra drain on the rectifier and filter, a forty-thousand ohm resistor is employed to limit this parasitic current to approximately three milliamperes.

As a result of the third element and its associated "keep alive" circuit, the high starting voltage required by former types of regulator tubes is eliminated.

The base contacts on the new "R" tube, as it has been named by the manufacturers, the Raytheon Company, are so arranged as to permit its use in B units and combination power amplifier B units designed originally for the UX-874 (CX-374)—the original voltage regulator tube placed on the market. When so used, the third element does not operate.

While the tube may be added, along with suitable resistors, to some of the heavy-duty type B power units now on the market which have not been specially designed for use with a glow tube, it cannot be used with the average run of socket power-supply units due to the extra load it imposes upon the rectifier and filter chokes. Likewise, in home constructing a power unit employing the "R" tube, chokes capable of handling at least 85 milliamperes of direct current without core saturation must be selected. At this time we are familiar with only two chokes on the market that meet this condition—the National and the Amertran. The rectifier tube should be of the 85- and not the 60-mA. type. The power transformer should have a double high-voltage secondary with a voltage of 300 across each side.

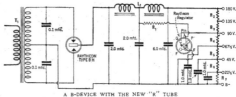

A B-DEVICE WITH THE NEW "R" TUBE
The new Raytheon tube has five leads, the fifth connecting to the metal pin in the tube base

The 1928 "Equamatic" Receiver

A SYMMETRICAL FRONT PANEL REWARDS THE CONSTRUCTOR OF THE 1928 "EQUAMATIC" RECEIVER

Automatic Variation of Coupling between Primaries and Secondaries of Coil Units Provides for Equal Amplification Throughout Broadcasting Band—Simplicity of Control Is Marked Improvement over Earlier Model

By JULIAN KAY

ONE of the most difficult disadvantages to overcome in amplifying high frequencies by means of successive tuned stages is the apparent discrimination of the amplifier against the lower radio frequencies. The longer-wave stations on the average tuned r.f. set come in comparatively poorly, while there is no end of amplification on the shorter-wave stations. Any one who has operated one of these sets will vouch for this fact, and any service man who must answer questions from his clients will agree that his main difficulty is in explaining why such-and-such a set will not get KSD or KYW, longer-wave stations, while it will get less powerful and often more distant shorter wavelength stations.

Now the reason for this lack of "pep on the longer waves is not difficult to explain. Whenever a tube has inductance in its plate and grid circuits, that tube does its best to regenerate, and if a certain balance between these inductances is maintained, this regeneration will break into open oscillation. Squeals, howls, and general instability result. This inductance is, however, necessary to transfer energy from one circuit to another. The tendency to oscillate increases as the frequency increases and as the amount of inductance increases, and more inductance is needed to transfer energy on the lower frequencies than on the high; but with this maximum amount of inductance the set will probably be uncontrollable on shorter waves.

Thus, sets that get the longer-wave stations with maximum r.f. amplification cannot be held down on the shorter waves; those that are especially stable on short waves as quiet as the proverbial grave for DX on the longer waves.

The receiver shown in the photographs in this article does not suffer from the faults enumerated

The Facts About This Receiver

Type of Circuit	Tuned radio-frequency.
Number of tubes.	Five. Two stages of r. f. with 201-A (301-A) tubes; detector, using special detector tube; first audio stage, 201-A (301-A); second audio stage, 112 (312) or other semi-power tube.
Features	Maximum r. f. gain obtained by use of special automatic variable coupling arrangement. Only two tuning controls although there are three tuned stages. Switch for one or two a. f. stages.
Frequency range	1500 kc. to 500 kc. (200–600 meters).

above. It is the outgrowth of a popular receiver of last year, the Karas "Equamatic," and employs a mechanical means of maintaining "constant coupling" between the output of one tuned radio-frequency stage and the input of the next stage. This "constant coupling" means that the plate circuit of the preceding tube will have just enough inductance in it to transfer maximum energy at each frequency to which the circuit is tuned, but not enough inductance to make it oscillate. As the condensers are varied, the coupling is varied simultaneously, and in the proper proportion, to prevent troublesome oscillations, but at the same time to maintain more or less

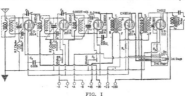

FIG. I

Circuit diagram of the new "Equamatic" receiver

constant amplification over the entire broadcasting frequency band.

The "Equamatic" system is not new to the readers of RADIO BROADCAST, for it was described in several issues of this magazine in 1926 by Zeh Bouck. The receiver itself differs somewhat in various ways from the set of last year; in each of these several respects it is better than last year's set. In the first place, it is a two-dialed receiver, an obvious advantage over the three-dial receiver, previously described. The change is accomplished by means of a lever pantograph arrangement which permanently connects together the variable tuning capacities of the second radio-frequency amplifier and the detector circuits. Additional means have been taken to insure maximum amplification and complete stability, partly by filtering the radio-frequency circuits with chokes and condensers and also by a new means of partially neutralizing the inter-electrode capacity of the tube—the capacity that causes regeneration. In addition, an output filter has been added to the layout so that a power tube can be used without the d.c. plate current of this tube going through the windings of the loud speaker.

The receiver, then, consists of a two-stage radio-frequency amplifier with constant gain over the entire broadcasting band, a grid leak and condenser detector, a conventional two-stage transformer-coupled amplifier, and an output filter. The antenna or first-stage amplifier is tuned with the left-hand dial, while the second dial takes care of the other two tuned circuits, which are made to tune exactly alike. Ease of adjustment is secured by means of the ganged condenser arrangement without apparent loss in either selectivity or sensitivity over the three-dial model described in RADIO BROADCAST for October, 1926.

Left. When the condenser is tuned for maximum wavelength, the coil should be coupled in this manner, allowing the largest transference of energy *Right.* Minimum coupling, for shortest wavelength

RADIO BROADCAST Photograph

FIG. 2

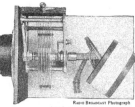

RADIO BROADCAST Photograph

FIG. 3

A special Yaxley interstage and filament switch is employed to cut out the final audio amplifier when desired, automatically shutting off the filament current to the tube in this circuit. Thus the loud speaker can be operated on local stations from the first audio amplifier, and when more volume is desired, the second amplifier may be thrown into the circuit.

A list of parts for the receiver follows:

T₁, T₂—Karas Type 28 Audio Transformers	$16.00
L—Karas Output Filter	8.00
C₁, C₂, C₃—Karas Type 17 0.00037-Mfd. Variable Condensers . .	15.75
L₄, L₅, L₆—Karas "Equamatic" Coils .	12.00
Two Karas Micrometric Dials 0–100 .	7.00
Three Karas Sub-Panel Brackets . .	.70
Karas Control System, Including Complete Hardware	3.00
L₁ L₅—Karas or Samson ¹⁰⁰ Millihenry R. F. Chokes	5.00
Formica 7″ x 24″ Engraved Front Panel	5.68
Formica 9″ x 23″ Drilled Sub-Panel .	6.00
R₁—Carter 10-Ohm Rheostat (Gold Arrow)	1.00
R₂—Carter 20-Ohm Rheostat (Gold Arrow)	1.00
Sw.—Yaxley No. 69-B Interstage Switch (Gold)	1.25
J—Carter No. 10 Tip Jacks . . .	.20
C₇—Sangamo 0.00025-Mfd. Fixed Condenser with Clips	.50
R₄—Durham 2-Meg. Grid Leak . .	.45
R₅—1A Amperite	1.10
R₆—112 Amperite	1.10
Yaxley Cable Plug	3.00

Five Benjamin Cushion Sockets . . 3.75
C₄, C₅—Samson Mica Neutralizing Condensers 3.50
C₇, C₈—Sangamo 0.0001-Mfd. Fixed Condensers 80
C₉—Sangamo 0.006-Mfd. Fixed Condenser 85
C₁₀, C₁₁—Sangamo 0.001-Mfd. Fixed Condensers 1.00

The circuit diagram of the receiver is shown in Fig. 1, and panel and sub-panel layouts are also given in this article. Attention is called to the two neutralizing condensers connected from the plate of each radio-frequency tube to the lower end of each input inductance, providing an additional means of stabilizing the receiver at high frequencies. The photographs show how simply the receiver goes together. The sub-panel, provided by the Formica Company, has three white lines engraved in it to show the position of the three tuning coils, which can be easily adjusted since they are attached to the sub-panel by means of machine screws.

After the receiver is properly wired, the next step is to adjust the variable primary coils, which are rotated when the condensers are turned. The condensers should be turned until the plates completely interleave, and in this position the primary coils should be parallel to the larger secondary inductance, as shown in Fig. 2. They should be fixed in this position. Under these conditions, the loosest coupling will exist between primary and secondary at the highest frequencies, as shown in Fig. 3.

The receiver can now be connected to its A, B, and C batteries. The diagram shows that the r.f. tubes are run with 90 volts on the plate. A negative bias of about one volt is placed on the grid of the r.f. tubes, and is obtained through the drop across the filament rheostat. The C bias battery for the audio tubes is fixed in the receiver, as the photograph of the complete set shows. The final audio tube should be a UX-112 (CX-312), with about 135 to 150 volts on the plate and a grid bias of negative 9. A separate ground wire can be run to the set, or the A battery may be grounded instead. Either method of getting the receiver grounded will be satisfactory.

This receiver is inherently easier to adjust for a non-oscillating condition than a bridge circuit, where an accurate and exact balance is necessary. The three tuning condensers are turned to the longest wave setting, or 100 on dial, and then, as the wavelength is slowly decreased again, with both the dials being turned in synchronism, a point will be found where the set breaks into oscillation. Then the neutralizing condensers should be screwed down slightly until oscillation stops, and the wavelength decreased again by rotating the condenser dial. This process should be repeated until the shortest-wavelength adjustment, zero on the dials, is reached. In actual practice, it will be found that no oscillation will occur if the condensers are adjusted properly, and at the same time it will be discovered that the longer-wave stations will not have lost any of their original strength, a phenomenon that few balanced receivers possess.

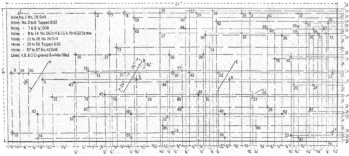

A SUB-PANEL TEMPLATE FOR THE 1928 "EQUAMATIC" RECEIVER

The builder of this new "Equamatic" receiver should have no difficulty in making a one hundred per cent. workable set at the first trial. The parts have been so placed that interaction between circuits, which ordinarily is a difficult problem with which to cope, is minimized, and the progressively increasing coupling between circuits at the longer wavelengths provides the proper interstage energy transfer. The first "Equamatic" coil, which transfers energy from the antenna circuit to the receiver proper, may be arranged to have a slightly different rate of change of coupling if desired—if the user wishes somewhat greater selectivity—for example. In this case, the coupling at the lowest frequency may be decreased by making the primary and secondary not parallel, but turned at a slight angle. In general, however, the greatest coupling should be used at all times, and a somewhat smaller antenna used if greater selectivity is desired.

The condensers which tune this receiver, if those specified are used, have a straight frequency-line characteristic, a decided advantage now that transmitting stations are again operating on ten-kilocycle separations. If it is desired to use a 171 type tube, somewhat different connections must be made to the last two tubes and their plate circuits. As wired according to Fig. 1, the last two tubes get the same B voltage, this being necessary on account of the interstage switch. Additional prongs would be necessary if the two audio tubes needed different B voltages. In the vicinity of broadcasting stations, however, the field strength is sufficient to give all the volume one can desire by using a semi-power tube of the 112 type. At some distance from stations, the 112 tube with its greater amplifi-

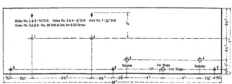

Holes No. 1 & 2 - ¾"Drill   Holes No. 3 & 4 - ⅛"Drill   Hole No. 7 - ¾" Drill
Holes No. 5,6,& 8 - No. 26 Drill & Ctk. for 6-32 Screw

FRONT PANEL DRILLING INSTRUCTIONS

cation factor will have an actual advantage over the 171 type, and louder signals will be obtained. Thus weaker stations will operate a loud speaker.

Now that we have outlined the electrical features of this new "Equamatic" circuit, there are several mechanical notes which should be brought to the attention of the home construc-

BY WAY OF CONTRAST   RADIO BROADCAST Photograph
Here is the 1927 "Equamatic" which is now improved. Note the three-dial tuning

tor. The condensers, for instance, are new, in that they are equipped with removable shafts so that insulated ones may be substituted for the metal ones, and complete insulation from hand capacity thereby realized. In some circuits this is important; in the present receiver, hand capacity will not be apparent owing to the fact that the rotor plates of the condensers are at

ground potential. The output filter, which is new in this year's model, consists of a conventional choke and condenser combination and, as may be seen in an accompanying photograph, is built into a round metal box similar to the audio transformers. The latter, incidentally, have exceptionally high primary inductance, which is necessary for full reproduction of low audio frequencies. Their turns ratio is about 2.5 to 1.

In all receivers employing two or more tuned circuits, tuning is inherently sharp, making necessary some form of fine adjustment. In this receiver the designers have used the familiar Karas dials which, with their 63 to 1 reduction, and their exceptionally smooth action, make tuning a joy. The writer found that the receiver tuned with great ease although it was sharp enough to please a critical fan.

All in all, the home constructor should find this receiver a well designed, simple-to-construct, high-quality set, and in building one, he will be equipping himself with a receiver of which he can be deservedly proud.

LOOKING DOWN ON THE 1928 "EQUAMATIC"
The arrangement of parts is neat and efficient. Note the lever pantograph arrangement
which makes possible simultaneous adjustment of the second and third tuned circuits

RADIO BROADCAST Photograph

A New Principle of R. F. Tuning

The "Octa-Monic" System Resonates the Second Harmonic Instead of the More Usual Fundamental—Increasing the Selectivity

By DAVID GRIMES

UP TO the present time there have been just three fundamental radio circuits, and every type of radio receiver has employed some variation of these three arrangements. First, there was the regenerative system; next radio-frequency amplification was developed; and last, the War and Major Armstrong produced the super-heterodyne.

Regeneration is a principle by which increased signals may be obtained by a properly phased feed-back or reënforcement of the incoming signals. When radio broadcasting first became popular in 1922, the regenerative type of circuit was generally used. The feed-back action made the receiver very sensitive for distant reception and very selective at the particular frequency for which the receiver was tuned. As the number of stations increased, and two or more powerful local stations became established in many communities, the regenerative receiver ceased to give satisfaction because of its broadness of tuning. Near-by locals could not, therefore, be tuned out. The necessity for extra selectivity was met with sets with tuned radio-frequency amplification. But the standard two-stage tuned radio-frequency sets soon became inadequate because inherent resistances in the vacuum tubes and associated circuits limited the ultimate selectivity to be obtained therefrom. Attempts to offset these resistances by feed-back circuits have been somewhat successful in increasing the selectivity but in nearly every case the audio quality has suffered. The RGS Inverse Duplex, described in the January, February, and March, 1927 issues of RADIO BROADCAST, employed a tuned radio-frequency circuit with an automatic negative resistance circuit which greatly increased the selectivity at all wavelengths. The limiting factor was the tone quality, however, as super-selectivity tends to cut the side-bands on a carrier wave, thus sacrificing the high pitch audio tones. The selectivity of the RGS Inverse Duplex was carried to such a point that a special audio circuit was employed to give extra am-

plification to the high-pitched tones in an effort to provide for the reduction in side-bands.

The super-heterodyne came into general use mainly because of its selective properties and its extreme sensitivity, permitting loop operation. But the side-band limitation so characteristic of a really selective tuned radio-frequency set became very noticeable in super-heterodyne circuits employing low-frequency intermediate transformers. Here the plus and minus 5000-cycle variation which must pass through to insure good tone quality is a very large percentage of the intermediate carrier wave of 30,000 or 50,000 cycles. The resonant curve must therefore be quite broad, which somewhat offsets the very

DAVID GRIMES

advantage which the super-heterodyne possesses. Recently, super-heterodynes using higher intermediate carriers than the broadcast frequencies have been introduced to overcome the above detriment.

It is easy to see from this discussion that there is a real need for a new type of super-selective circuit that overcomes the side-band limitation. With this thought in mind, a series of tests has been conducted during the last few months, which has resulted in the development of the "Octa-Monic" principle. This is a new system for obtaining a higher degree of selectivity and follows from the considerations below: It had long been realized that selectivity was a geometric function and that placing one tuning circuit after another created the geometric conditions. Naturally, every geometric function occurring in radio was then investigated with a view to employing that feature for selectivity. One of the most promising was the generation of second harmonics and it is this that forms the basis of the new RGS "Octa-Monic" receiver.

It is well known that a vacuum tube has a curved characteristic unless a very high external

impedance exists in the plate circuit. This curved characteristic leads to distortion which is another name for the generation of harmonics. In the plate circuit of the average r. f. amplifier is only a small impedance, therefore in this circuit exist the second harmonic as well as the fundamental image of what occurs in the grid circuit. It is from this second harmonic that the "Octa-Monic" gets its name. The idea of securing increased selectivity, not through the use of additional circuits, but by the use of an inherent tube characteristic, is believed to be new.

Reference to Fig. 1 shows the familiar grid-voltage plate-current characteristic of a standard vacuum tube. The curve follows a square law, and in doing such, causes some interesting results. It is desirable when using such a tube for an amplifier to operate the grid voltage on a relatively straight part of the curve, so that fairly undistorted amplification will result. For perfect amplification, this curve should be a straight line. The very fact that it is not, however, makes the vacuum tube a valuable device, as the non-linear performance permits its use as a modulator, oscillator, detector, and a harmonic generator. It is its use in this last capacity that is least understood and least employed.

Now, if a special negative C bias is placed on such a vacuum tube so that the grid variations take place about some such point as A in Fig. 1, the conditions will be favorable for the generation of second harmonics. Such harmonics are always the result of unequal amplification in the two halves of a carrier wave. By operating at point A it will be seen that each positive half of the carrier wave is amplified to a greater extent than each negative half. Such an unbalanced carried wave may be resolved into a balanced carrier wave of the same fundamental frequency plus another carrier wave of twice the frequency. This is graphically illustrated in Fig. 2 and is known as Fourier's Theorem. Briefly stated, it is as follows:

If we have any single valued periodic curve, that is, one having only one value of the ordinate to one value of the abscissa, and repeating itself at regular intervals, then, no matter how irregular the curve may be, provided it does not exhibit

FIG. 1

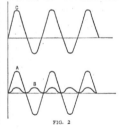

FIG. 2

discontinuities, it is always possible to imitate this curve exactly by adding the ordinates of superimposed simple periodic or sine curves of suitable amplitude and phase difference, having wavelengths which are in the integer relation to each other.

Curve C in Fig. 2 shows the unbalanced carrier wave in the plate circuit of the tube operated at point A on the characteristic curve in Fig. 1. Curve C is the equivalent of Curves A and B in Fig. 2. Curve A is the fundamental carrier wave similar to the one impressed on the grid of the tube. Curve B is the second harmonic wave.

None of the graphic representations pertaining to the second harmonic reveal the secret of selectivity. They merely indicate that such a second harmonic is present. The mathematical equation outlining the performance of the tube, however, does tell the story. The equation for the unbalanced carrier wave shown by Curve C in Fig. 2 is as follows:

$$C = y\sigma\left(\frac{E_p}{\mu} + E_g + s\right) e \sin pt + \frac{\sigma e^2}{2}\cos\left(2\ pt + \pi\right)$$

The first part of the equation represents curve A in Fig. 2—the fundamental carrier. The second part of the equation represents Curve B in Fig. 2 —the second harmonic component.

Now it will be noted that this second harmonic component is proportional to the square of the input voltage, e, and it is this squared function which gives the geometric requirement for selectivity. Reference is here made to Fig. 3, which shows a harmonic generator tube with the grid circuit tuned to some frequency, F, in the broadcast band. The resonant curve of this tuned input circuit is shown at A. Now the second harmonic currents, 2F, in the plate circuit of this tube, are proportional to the square of the input voltages on the grid. As the incoming frequencies are varied, slightly, from the true resonant value of the circuit, the voltages on the grid fall off according to curve A. As the voltage has decreased from its maximum, y, to the half way point at x, due to a change in frequency, a, the second harmonic current decrease is much more rapid and abrupt than in the case of the fundamental current, resulting in a very sharp resonant curve in the plate circuit of the harmonic generating tube. In these figures the fundamental and second harmonic have been plotted so that they have equal maximum amplitudes to show the relative sharpness of resonance.

The next step was to incorporate such a principle into a working receiving circuit. The first

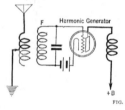

FIG. 3

attempt is shown in Fig. 4. Here the antenna is coupled directly to the tuned input of the harmonic generating tube. The grid is tuned to the fundamental frequency of the broadcast station to be received. The second harmonics are set up in the plate circuit where there is a tuned transformer, whose tuning condenser resonates the secondary at the second harmonic frequency. Signals are changed to audio frequencies in the standard detector tube in the same manner as any other circuit. Selectivity superior to that obtained by a two-stage tuned radio-frequency system was noted, although the sensitivity of the arrangement was much less.

This would obviously be so because no amplification is present in Fig. 4, and because the amplitude of the harmonic is less than that of the fundamental. In all of these diagrams, which are

roughly drawn, no attempt has been made to show relative amplitude— only relative sharpness. In fact, quite a bit of energy is sacrificed in securing the second harmonic component. This sacrifice is more than justified because of the real selectivity obtained, and after all, amplification can always be added to a receiver. The additional amplification is easily added by the arrangements shown in Fig. 5, where one stage of radio frequency is placed ahead of the harmonic generator tube. This additional stage does not need to be particularly selective as the second harmonic principle gives all of this necessary. For this reason, the additional tube has been called a coupling or amplifying stage—the circuits being designed primarily for amplification and not for selectivity. Its input circuit is closely coupled to the antenna to give maximum energy pickup with which to start. In order to gain some idea of the real improvement offered by the "Octa-Monic" as compared with tuned radio frequency, reference is made to Fig. 6. This shows the schematic arrangement of a standard two-stage tuned radio-frequency receiver with the resonant curves of the various parts of the circuit beneath and numbered to correspond. Curve 1 shows the sharpness of tuning in the grid of the first tube. Curve 2 indicates that the fundamental currents in the plate circuit of the first tube follow the same resonant curve. Of course, the energy is greater in the plate circuit but the curves have been drawn to the same scale for the purposes of comparison. Curve 3 shows a gain in selectivity through the addition of the second tuned circuit, while Curve 5 shows the final gain in selectivity through the employment of the 3rd tuned circuit. It can be seen that only

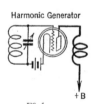

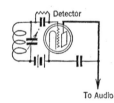

FIG. 4

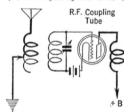

FIG. 5

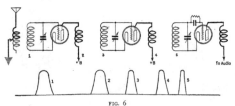

FIG. 6

band of the carrier frequency plus and minus 5000 cycles to- insure good tone quality and 10,000 cycles variation is a much smaller percentage of the second harmonic currents than of the broadcast frequencies. For example, at 1000 kc. a 10,000-band represents 1 per cent. while in the second harmonic circuit this 10 kc. band is only 0.5 per cent. of 2000 kc. In this unique way a high degree of selectivity has been secured with a faithfulness of tone quality that is remarkable.

This RGS "Octa-Monic" arrangement is a fundamentally new development. The existence and generation of second harmonics has long been known, but the application of these currents to the selectivity problem is a new feature. It has created a fourth and distinct type of radio circuit.

The next article in the series will discuss this interesting development and will contain plenty of information in connection with the building of the RGS receiver.

the tuning circuits offer any selectivity whatsoever. Fig. 7 shows the schematic sketch of the "Octa-Monic" system and the associated resonant curves beneath the several circuits involved. Curve 1 indicates the sharpness of tuning in the grid circuit of the harmonic generator. This has the same shape as the first tuning circuit in the system. In the plate circuit the second harmonic currents follow the resonant Curve 2, and the harmonic tuning circuit in the input to the detector still further sharpen this up to resonant Curve 3. This resonant curve is similar in shape to the tuning curve No. 5, in the system. The resonant curves in Fig. 7 are drawn on the same scale for the sake of comparison. It must be remembered that Curve 2 representing the second harmonic selectivity has much less energy in it than Curve 1.

It would seem, offhand, that this excessive selectivity in the "Octa-Monic" would raise havoc with the side-bands, as is the case in tuned radio-frequency circuits. The very harmonic currents themselves, however, offer the solution. At the high frequencies existing in the harmonic

circuit the resonant peak may be much sharper than at broadcast frequencies without harming quality. A tuned circuit must be able to pass a

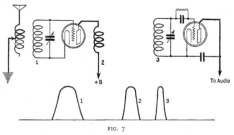

FIG. 7

Selling It by Radio

Using Radio in Sales Promotion: By Edgar H. Felix. Published by McGraw-Hill Book Co., Inc., New York City. 386 pages. Price, $5.00.

IT IS not without a touch of embarrassment that I prepare to review this pioneer work on the subject of broadcasting as an aid to merchandising. The author has written in kindly terms of my own masterpieces on various occasions, and transported me over the highways in his private fleet of Minerva sedans. In finding his work good, therefore, I lay myself open to the charge of log-rolling. To preserve my honor, I should like to pan the author and his treatise, but he has done too effective a job. The book is original, informing, and vigorous. Its blemishes are slight and will probably be visible only to the reviewers.

Using Radio in Sales Promotion is written primarily for advertisers; it aims to instruct the man who contemplates spending a part of his advertising appropriation for time on the air. Throughout it names names, gives examples drawn from experience, and answers the questions which must occur to the man who signs the checks. What interests the client interests the professional broadcaster equally, whether he (or she) owns, manages, operates, sells time, writes publicity, or performs in the studio. Professional advertising men are in a similar position. So there is Felix's potential audience —in these groups and their fringes. He is qualified to speak to all of them, for he has himself

been a broadcaster, advertising man, merchandising consultant, and writer; and incidentally he is, among other things, a Contributing Editor of RADIO BROADCAST.

The author starts with a *résumé* of the seven-year history of the broadcasting art, and in his first chapter he sounds a note which recurs throughout the book; broadcasting is a medium, not for forthright advertising, but for securing the good will of the consuming public. Radio receivers are purchased for entertainment and instruction. A trade name, a brief slogan, a very few words of product description, may be mixed with the entertainment and instruction, but Felix always has his eye fixed on a psychological dividing line which cannot be crossed with either propriety or safety. Broadcasting is primarily for the advertiser who can afford to wait, who has the resources to build up public good will, who is not too short either in his bank account or his patience. As for the little fellow who demands sales tomorrow morning—Felix advises him to take his money to the local newspaper. The Midwestern broadcasting station which I am reliably informed at this writing, is selling over the air by direct description, gasoline, mattresses, gloves, overalls, radio apparatus, cancer and rupture cures, and a 50-cent chicken dinner, is outside of Felix's pale. "He (the broadcaster) does not earn the right to inflict selling propaganda in the midst of a broadcasting entertainment any more than an agreeable week-end guest may suddenly launch into an insurance solicitation at Sunday dinner," is the attitude expressed and implied on every other page of *Using Radio in Sales Promotion.*

The reason for the fundamental similarity of station programs is shrewdly analyzed in the second chapter. Almost all of them are planned to appeal to all tastes, thus including the largest possible audience. Artistic standards, not fundamental program appeal, are the variable quantity. Almost all the stations are shooting at the same target, but with guns of different calibre.

In the chapter on 'Building a Broadcasting Station" and that on "The Broadcasting Station" near the end of the work, Felix discusses technical aspects of the broadcast situation, such as wavelength congestion, station organization, operation, and the like. He does not attempt to tell the reader how to build a station, and in fact advises him against trying it. What he does is to outline the formidable problems which must be faced, and the general economics of the situation. In discussing operation he describes the functions of the various members of the staff and the principles underlying their work. In many other chapters, as in those on "Potential Audiences of Broadcasting Stations" and "Selecting a Commercial Broadcasting Feature," Felix enters the engineering branch, and, dealing with field strengths and microphone characteristics, he exhibits only minor lapses from accuracy.

Using Radio in Sales Promotion is written in a business-like, straightforward style, but it is not devoid of sharp hits and epigrams. I quote a few: "An expert is usually one who has not yet learned enough of a subject to be aware of his ignorance."—"Broadcasting is free and therefore freely criticised."—CARL DREHER.

A DEPARTURE IN LOUD SPEAKERS
The Rola Model 20 combines a baffleboard loud speaker of remarkable efficiency with floor screen measuring three feet high by two feet wide. The price is $85

AN "ACME" OFFERING
This novel loud speaker makes use of a double cone arrangement, good amplification of the low notes being possible on account of the free edge cones. The manufacturers state that an output device is not necessary with this loud speaker. Price $25.00

THE A. K. CONE
The new Atwater Kent Model E employs a novel method of cone suspension, permitting satisfactory response to minute vibrations. Price $36.00

THE CROSLEY MUSICONE
A baffleboard effect is obtained by the use of the tilt-table surface. Price $27.50

1928 *Loud*

TRIMM "CONCERTO"
A $10.00 cone, fourteen inches in diameter. The seventeen-inch cone retails for $16.00

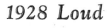

OPERADIO
The "Junior" model has a 30-inch exponential air column while the "Senior" has one of 54-inches. The prices are $15.00 and $25.00

THE BOSCH 'LIBRARY AMBATONE"

A NEW MAGNAVOX
Loud speaker and B supply in one unit, a. c. operated. A 210 type power tube should be used. The loud speaker is of the dynamic type. Price for the unit, $110.00

IN GOTHIC DESIGN
A beautiful and well-proportioned free-edge periphery cone. The reproduction from the "Peerless" is in keeping with its appearance. Price $35.00

FADA "PEDESTAL"
This is a twenty-two inch free floating cone of attractive Grecian design. The overall height is fifty-five inches. Price $50.00

THE "CASTLE"
A two-tone bronze relief pattern enhances the appearance of this Tower seventeen-inch cone. Price $9.50

GREBE "20-20"
The numerals denoting this model refer to the cone diameter and the fact that it is constructed at an angle of twenty degrees. Price $35.00.

ALGONQUIN
A full floating cone of a moisture proof impregnated fabric. The Algonquin Electric Company retails this cone at $15.00

BALDWIN
A special unit is responsible for the good tone of the Baldwin "99." This instrument sells for $28.50

ANOTHER ROLA ART MODEL
A large and effective baffle surface is responsible for excellent tone quality. Rola manufactures less expensive models, from $28.50 up. The one illustrated is priced at $45.00

Speakers

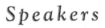

UTAH
This $30.00 instrument is of 5-ply natural finished walnut

AN INNOVATION
An up-to-date post-war map of the world surrounds the Symphonic Globe loud speaker, a $35.00 product of the Symphonic Sales Corporation

A 22-INCH CONE
The new Sparton loud speaker, a product of the Sparks-Withington Company, retails at $30.00

IN DISGUISE
A cone loud speaker in tip-table form—the Teletone Model 70. Price $40.00

THE MOHAWK "PYRAMONIC."
The pyramidal construction in the diaphragm is responsible for very excellent quality from this $25.00 instrument

THE "ADVENTURER"
Another bas-relief pattern of the Tower Manufacturing Corporation. The ship is suitably and vividly colored. The price $9.50

ANOTHER PEDESTAL
The Baldwin loud speaker illustrated elsewhere on this page is also obtainable in the attractive form shown here. Price $39.50

The Listeners' Point of View

SHOULD THE SMALL STATION EXIST?

By JOHN WALLACE

THE case for the small broadcaster is seldom heard, because he is small, possibly and because the public through habit and for other reasons, directs its attention to the larger and better-known stations. Has the small broadcaster a place? Does he serve a local need? How can the small station better serve its public? An interesting letter from a reader in Devils Lake, North Dakota, discussing this topic follows:

I have been reading with a great deal of interest your department in RADIO BROADCAST and obtained very much enjoyment out of it. To believe that you are about as biased an individual as ever wrote.

The particular thing that seems to cause you the greatest amount of worry is the small station. I own and operate a small station, as far as power output goes. Some of the things you say get under my hide. Picture yourself out here in North Dakota, living in a small town or on a farm vainly twisting the dials looking for entertainment. None to be had. Then, " This is KEJM The University of North Dakota" or WDAY or KFYR or KDLR as the case might be. A program of good entertainment for an hour or two comes on the air. You listen, you are pleased, you swear by the small station, you pick up a paper and note the many excellent programs being sent out from a multitude of stations, *none of which are heard here*, except a few months in winter. Would you favor the abolishing of the small station? Would you dismantle your radio set in February and not use it again until November? Would you if you turned on your set night after night and listen for stations that cannot be heard? Would you? Or would you listen to the very excellent programs from our local small station?

The programs sent out here are not the high type of the big Eastern stations but they fill the bill. An example of our type of program, Monday evening at 9:30-10:00 the Oriental Trio, the pianist a graduate of one of the largest schools in the country, a student of a noted teacher in Berlin, the soloist a former member of the Minneapolis Symphony, the violinist a former member of the Detroit Symphony all doing their best. At 10:00 P. M. an hour of dance music played by an 8-piece organization composed of men who have been playing together for three years and who know and understand music. At 11:00 P. M. fifteen minutes of the finest kind of pipe organ music. And on a Wednesday evening a program of band music played by a band of sixty boys, "The Governor's Boy Concert Band of Devils Lake" an organization of six years standing under the leadership of the finest bandmaster in this section of the country and playing overtures and marches of the highest class. The soloist with the band is a lady who has recently returned from that famous music school in Philadelphia where you must have talent to get in. That was our program last week. Daily we run the noon hour program playing the finest available recordings of symphony orchestra and famous singers, played through an electric pick up. Daily we broadcast the weather reports and market information and the station is eagerly listened to.

You may wonder at our quality. We have the finest and newest type microphones, the best amplifying equipment that we can buy, tested and balanced pick up lines, a monitoring operator at each end, every piece of equipment meter-checked at both ends and we pride

ourselves on good quality. KDLR has a studio 18 by 25 built by ourselves following Sabine's formulas. For band concerts we have a band shell, the only in the state of North Dakota, that will seat a 100-piece band, built so that the focal point is 50 feet from the front of the shell and here we pick up every instrument with but one "mike." The shell so built that a speaker in the shell can talk in a whisper and the "mike" will pick it up and put it through an amplifier at any volume desired without its being overridden by hiss.

You, living in a city with the choice of a score of fine stations at your command, find cause to kick and kick loud, long, and lustily at what you hear. Come out here once and hear one station or possibly two, both in the same state, and after you are here awhile I guarantee you will go back home and enjoy what you hear. Then after you are home give a thought to the rest of the world and try to do a little boosting and cease knocking that about which you apparently know so little.

I hope that you have not been bored by the length of this thing but allow me to say another thing, you have a wonderful magazine, but do use a little judgment.

BERT WICK   Operator, KDLR.

Mr. Wick's point is well taken—even if he does take us to task at the same time. In our own defense, let us say that he overestimates our dislike of the small station. We have not said many harsh words about it. More than likely he reaches this conclusion because of the fact that we ignore the small station in these columns. If we do neglect it too much, there are two reasons. First, the small local stations, being low powered, are difficult to "get" often enough to enable fair comment. Secondly, local stations are of especial interest only to the persons within their limited range and hence do not logically demand space in an article that considers radio nationally.

Our correspondent asks us: "Would you favor the abolishing of the small station?" Our answer is yes and no. For small stations situated as are those he mentions in his letter he gives an excellent vindication. There are long periods during which it is impossible for isolated communities to receive the distant stations, even though powerful, with any sort of satisfactory reception. They are entitled to local stations, of low power, to fill in these hiatuses. Moreover the very fact that the program is local must give it a certain interest to those in the hinterlands—who take a keen interest in local things, in contrast to the city dwellers who don't know their alderman's name nor the people in the apartment below. The farmers and other dwellers in the Centerville area are just as likely to prefer the wares of Centerville's 10-watt broadcasting station to those of wcco as they are likely to prefer the Centerville *Bugle* to the Minneapolis *Tribune*. Probably they are personally acquainted with most of the performers who appear before the Centerville station's microphone, or at least they know the announcer's cousin Nellie or their nephew goes to the same District school as the Staff Organist's son. Hence their interest in the local program is natural enough, and since they enjoy it they are entitled to it.

THE SMALL STATION HAS NO PLACE IN CITIES

BUT we are all in favor of abolishing the small station from the metropolitan areas. If the Hyde Park district of Chicago, say, wants to publish a little four-page newspaper to retail the local gossip to the provincial minded of its neighborhood, and to advertise the wares of its local drug stores and meat markets, well and good. Nobody outside that section has to buy the paper, nor need even know that it exists. But when that same Neighborhood Association starts to operate a radio station to afford opportunity to the piano pupils of the neighborhood and to peddle the aforementioned meat market's wares, it becomes, however low its power, somewhat of a public nuisance. Even where it doesn't actually interfere with the first-class stations it needlessly clutters up the dials and is an unpleasant distraction while tuning. This complaint holds also for stations in suburbs and in small towns within a hundred miles or so of some large city.

We have never heard our correspondent's station KDLR; we hope to hear it when DX is again possible. Certainly the program he describes sounds like a very creditable one. Mere magnitude and wealth are not the only requisites for a good broadcasting station. While large financial resources must necessarily allow a station to present more varied and elaborate offerings they do not guarantee artistry in that station's mode of presentation.

It is perfectly possible for some small station, somewhere, to achieve a reputation equal, or even superior, to the large city stations. Whether or not you like the "Little Theater" there is no denying the fact that some Little Theater productions, financed on a shoe string, are superior in artistry and in manifestation of good taste to many of the costly Broadway productions. The same opportunity presents itself in the operating of a small radio station. An individual gifted with good taste and a certain sort of genius which we shall not attempt to label could, conceivably, operate a station *in time* which listeners the country over would labor with flattened ears to pick up. He would have to be an individual of extraordinary personality and that personality would have to manifest itself in every detail of the programs he offered. He would have to have complete supervision over every item he broadcasts and, essentially, do all his own announcing. In contrast to the impersonal, routine, and frequently "high hat" manner affected by the metropolitan stations, his mode of presentation would have to be decidedly personal and intimate. He would have to establish a very real bond between himself and his audience. The musical programs he offered would have to express his own tastes, as would every other sort of program he offered from time to time. And if he were a big enough man, his tastes would be broad, and he would; in consequence, attract a big enough audience. He would not have to have a lot of artists, but the few he had would have to be good, and show, in all their work, his fine Italian hand.

His position as "editor" of all that his station offered would very closely parallel that famous instance in the history of American journalism wherein an editor of unusual genius, though attached to an unimportant newspaper in an unimportant Mid-Western town, gathered to himself such fame that his editorials were quoted the country over and his newspaper even enjoyed a scattered circulation over many remote states.

And finally his reward would be—considered in the form of dross dollars—about as great as is the financial reward of the operator of a Little Theater.

Why Not a Station Operated by a Foreign Language Group?

THERE are several broadcasting stations, mostly among the smaller ones, devoted to some special group of listeners. For instance there are the two labor stations which aim their output especially at the man in overalls; and then the stations operated by some religious group; as the Roman Catholic and Christian Scientist. It has puzzled us, from time to time, that no station has ever come into existence operated by a foreign born, or foreign descended, group for the benefit of its fellows in this country who cling to the mother tongue.

While we have no statistics at hand to lend weight to this proposal, isn't it said that there are as many Italian speaking people in New York as in the city of Naples, and more Swedes in Chicago than in the city of Stockholm? If these statements be exaggerated at least the "foreign-language" populations of these cities do run into large figures. Chicago, besides its vast area around Belmont Avenue where one can examine the store signs for blocks without seeing a name that does not end in -son, has populous settlements of persons of Polish descent, Italian and German. These people particularly the older generation, have recourse among themselves almost exclusively to their former native tongue. Their sons and daughters though they prefer the American idiom, have, nevertheless been acquainted with the second language since childhood. It seems to us they should be interested in hearing programs conducted entirely in their own language, featuring artists and speakers of their race, and in general devoted to their racial interests. The Great War is remote enough now so that no idiot would be likely to call the manoeuvre unAmerican.

There have been foreign language programs from various stations in the United States, but only intermittently. Some French programs were broadcast by a New York station. WIBO in Chicago has for nearly two years broadcast a program in Swedish every Sunday morning from 8:35 to 10:00. Inquiry at WIBO reveals the fact that this program has been enthusiastically received by Swedish people, not only in Chicago but by long distance experts in Wisconsin and Minnesota—both of which states have a large Swedish population. The same station recently inaugurated a program in German on Sunday afternoons from 2:30 to 3:30.

There is no doubt at all that a foreign language station, properly situated, would enjoy a large enough audience to make its efforts worth while. Nor does there seem to be much doubt that it could get artists; suppose an Italian station sent out a call for Italian singers! But there may be some question as to how well it could support itself. Perhaps here is the rub and the reason that none has so far appeared. However the foreign communities in the large cities manage to support flourishing newspapers, which are packed with

advertising. If the bonds of a common fatherland hold these people close enough together to make a newspaper self supporting, it would seem, by analogy, that an unpretentious broadcasting station would not be an impossibility.

The value to those, like ourself, who do their best talking and listening in the American patois, would be, perhaps, not large. But it would be a great thing for students of foreign languages to have a station they could tune-in on, on occasions, to supplement their reading and writing work. Moreover it would give a pleasant cosmopolitan tang to the air. And picture the delight of a listener out in Kansas tuning through to some Italian station in New York and imagining he had got Rome!

Thumb-Nail Reviews

WE WRITE in the middle of the summer, at a time when DX conditions are at their worst, so instead of trying to review some static-mingled distant squawks we shall confine ourself to considering some of the chain features which are received over a wide area.

The summer season, may we say, has been better than any previous summer in radio's

AN ESKIMO LULLABY AT WGY
In the Schenectady studio, Trixie Abkla, Eskimo, sings a good night lullaby for her two sons Miles and Billy. Up North, where the night is long, this domestic duty need be discharged only once a year

history. Many features quite up to winter time standards were heard. The stations supported the annual vociferous protestation that "there is no summer slump in radio!" So many different agencies are engaged in re-iterating this phrase every year, particularly radio manufacturers, that it occurs to us that they do protest too much, and that a slump none the less exists. If it does exist it is because radio listeners have been able to discover other doings more intriguing than sitting in the front parlor of a warm summer's evening, and not because the stations have not made an effort to please.

THE RADIOTRONS (Blue Network) These artists in their program of alternated orchestra and voice strike a mood that nicely ties the whole shebang together. In the case of the program we have in mind the flavor persisting throughout was a decidedly saccharine one; such numbers as Tumble Down Shack in Ath-

lone, When You're in Love, Listening, Honolulu Moon, and Selections from the "Red Mill" following each other in sentimental succession. But Erva Giles has a beautiful soprano voice for these songs and the Radiotron's much featured Vaughn De Leath always delights us with her mellow alto and throaty half-spoken passages.

IPANA TROUBADOURS (Red Network)—continue to be one of the most reliable orchestras that are to be heard weekly. Their plucked instrument section (they have another section—strings) continues to lead in its field, which is probably to be expected considering all the practice it has had over the past two and a half years. ELK'S MALE QUARTET (Blue Network)—The quartet is an unfailing radio standby. Mayhap our interest in the quartet springs from the fact that every one of us is a potential singer in a quartet and stands ready to contribute his services thereat under proper stimulus, such as a shower room, a clam bake, or a couple of stiff snorts. Be that as it may, the Elks' is an excellent quartet.

CONTINENTALS (Blue Network)—The Continentals is a very able organization made up of a chorus and soloists under the direction of Cesare Sodero and specializing in opera selections. The soloists include Astrid Fjelde, soprano—and an excellent one; Elizabeth Lennox, contralto; Julian Oliver, tenor, and Frederic Baer, baritone. The program is well selected and does not stick too exclusively to the threadbare arias. Tuesday evenings at nine Eastern time.

STROMBERG-CARLSON ORCHESTRA (Blue Network)—This is George Olson's band, playing under a trade name, and a first rate band it is, as others than ourself will assure you. Perhaps much of the credit for the smooth and beguiling manner in which it plays is due to its orchestrator, one Eddie Kilfeather, who specially arranges many of its numbers.

STADIUM CONCERT (Blue Network)—The Stadium concerts occupied, as last year, first place on the list of summer offerings. We were afforded mingled emotions in hearing our Mr. Stock as guest conductor of the Philharmonic musicians, what with the current prospects of having no Chicago Symphony Orchestra this coming season, due to wage disputes. (We wonder why the Orchestral Association didn't offer to sell broadcasting privileges to meet the increased demand for funds). To quibble about minor points: we saw no reason why the announcer should descend from his ponderous and dignified perch at intervals to refer to one of his fellow workers as "Jimmy." If they will be high-hat let them be consistent.

CITIES SERVICE CONCERT (Red Network)—An orchestra under the direction of Rosario Bourdon and a chorus called the Cavaliers. But what shall we say about them? we have exhausted our critical imagination. At any rate they are consistently good. They specialize in the lighter composers, Herbert, Friml, Straus, and so forth, with other occasional novelties.

ROYAL HOUR (Blue Network)—After having distributed roses the length of this column it is now our privilege to get mad. The Royal Hour's thirty-minute broadcast is a throw-back to radio's worst days. Advertising is jammed in so thickly that the musical part hardly has a chance to stick its nose out before it is drowned in further advertising. Every number on the program is introduced with a labored reference to the sponsoring company and its product.

Your A. C. Set

*How to Search for Defects
When Your A. C. Oper-
ated Receiver Goes Dead*

EDGAR H. FELIX

CALL THE SERVICE MAN

"Shut off the power, call for a trained professional service expert, and leave the set alone"
—the advice to perplexed owners of a. c. sets in time of trouble by the head of the service
department of one of the largest radio manufacturing concerns

THE head of the service department of one of the largest radio manufacturing concerns in the country was asked at a meeting recently what advice he would give to the owner of an a.c. operated set which has suddenly gone out of order.

"Shut off the power, call for a trained professional service expert, and leave the set alone," was his answer. "The troubles that arise with a.c. receivers are few and easily remedied by experts, but an experimentally inclined novice who doesn't know what he is doing can actually do great damage if he starts to tinker aimlessly. When a person gets sick, he calls on his doctor; when his automobile goes out of order, he brings it to the service station; but when a radio set goes wrong, three out of five listeners proceed boldly with soldering irons to change wires, to short-circuit transformers, to reverse connections, and to apply high voltages here and there, just to find out what happens. A competent man, familiar with the functioning of radio sets and their associated power supply, makes an inspection, analyzes the difficulty, and replaces whatever is necessary in a few minutes at a minimum of expense; a venturesome experimenter usually tinkers until he has done serious damage, then calls in an expert and, when he gets the bill, wonders why radio repairs are expensive. There is nothing mysterious or really troublesome about modern a. c. sets, and servicing them is usually a matter of the utmost simplicity."

Careless tinkering may prove an unhealthy pastime for the would-be radio expert who is accustomed to learn his radio by practical experience. Large condensers of several microfarads capacity, charged to 500 volts, may impress their effectiveness upon him with such force that he will find himself involuntarily, violently, and uncontrollably seated on the floor several feet from the radio set. But, if he temper his examination with knowledge, most of the likely faults of a .c. operated sets can be readily identified by their symptoms and, in some cases, the necessary measures for repair taken.

Servicing the a. c. set has brought a new dignity to the job of repairing radio sets because it has converted the receiver to an electric power device of considerable magnitude. Indeed, so marked is the change involved in the engineering design of a. c. sets that some of the earlier models, designed by radio engineers, failed prin-

cipally because of inadequate knowledge of power engineering.

The excessive service grief which attended some 1925 and 1926 a. c. sets has been largely cured by the application of the principles of power engineering to radio. These pioneer models taught us the stresses and surges to which the filter condensers are subject with the consequence that the latest receivers have condensers of considerably greater voltage-carrying power than those of past seasons. New rectifier tubes of greatly increased capacity have solved voltage regulation and life problems, while such refinements as ballast tubes and regulator tubes give the designer wide latitude in his work. Also we have new types of tubes which avoid the use of a rectifier by using alternating current for the A supply. All of these advances have brought us to the threshold of a new era in radio set design and the elimination of the burden of maintenance attention. Given capable engineering and adequate quality of materials, the modern alternating-current set, whether it be of one type or another, is reasonably reliable and a marked improvement in convenience over its less advanced predecessor.

Certain parts of the alternating-current receiver, however, depreciate with long use. It is no more possible to design a wear-proof radio receiver than it is to design a wear-proof automobile. Vacuum tubes lose their emission and condensers, unless of tremendously greater voltage capacity than their service requirements demand, eventually break down. It is only a matter of time, however, when the service attention required by a.c. sets will be reduced to a matter of annual inspection to replace rundown tubes. Indeed, some of the latest sets have actually reached that stage.

But we cannot expect a radio receiver to require no attention whatever. Filter condensers are really moving parts. They are constantly subjected to considerable voltages, and are charged and discharged rapidly and continuously. The atomic structure of their components is constantly strained so that in time they become less capable of resisting these rapidly changing voltage strains. The experience of the last two years has resulted in the widespread use of bypass and filter condensers which will, under ordinary conditions, serve for a period of years, rather than months, as was the standard heretofore. To men-

tion but one, if you have opportunity to look at the output condenser in the power supply of a Stromberg-Carlson a. c. receiver, you will observe that its size compares with that of condensers used in receivers of a year or two ago almost as does a match box with a steamer trunk! Expensive as these improvements are, they are an economy because they approach the ideal—"plug the set into the light socket and the receiver is ready to operate for a period of years without attention."

TWO TYPES OF A. C. SETS

IN GENERAL, there are two outstanding types of alternating-current sets—first, those in which the alternating current of the line is converted into direct current by rectifiers, smoothed by filters and then used to power filaments of the more familiar type of vacuum tubes and, second, those with tubes utilizing raw alternating current, either of the rugged filament or heater element type. Already considerable experience has been gained with the first type of set and its service and engineering difficulties are quite well known. With the second, there is less experience, and which of the two kinds will prove the ultimate winner is hard to say at this writing.

Tubes using raw a. c. for their A supply nevertheless need d.c. for B and C power, and therefore a rectifier of some kind is necessary unless A and C batteries are employed. The use of these tubes requires the introduction of an appreciable 60-cycle a. c. current right into the radio receiver, in and near the radio-frequency elements, introducing hum difficulties, and the life of the tubes under ordinary service conditions is yet to be determined. Considering that RCA, Cunningham, McCullough, Van Horne, and others are making a.c. tubes, it is certain that there has been much fruitful development work done since the first of this type of tube was introduced several years ago.

As to the first type of receiver, that is, those using conventional tubes, usually of the 199 type, with filaments in series and powered by a rectifier system, there is a wealth of practical experience in servicing them, gained during the last two seasons. The largest sellers in this class have been the Radiola super-heterodyne outfits used in connection with the 104 loud speaker and an a.c. rectifier power unit; the Garod, and the Zenith. By consulting not only manufacturers but the service heads of responsible dealer organizations, it has been possible to learn what the service problems in connection with the sets are. The author is particularly indebted to the R. T. M. Service and to Mr. H. T. Cervantes of Haynes Griffin Radio Service for much of the information which follows, based on their experience with thousands of these sets in the New York area.

The procedure for locating trouble, the significance of symptoms, and the remedies, are described with a view to aiding the set owner in determining whether his set requires the services of an expert service man or whether a simple adjustment will put it back in service. Breakdowns with a.c. sets generally impose a strain on the rest of the outfit and, consequently, a certain amount of knowledge of possible difficulties may avoid more serious damage.

When the set goes dead, either shut off the power until the service man takes charge, or else immediately determine whether the rectifier tubes light and, if the set has them, note also if the ballast and the regulator tubes glow. If these tubes appear to be functioning, reduce the filament voltages by means of the filament rheostat. Sets with filament rectifiers do not, of course, give visual indication of their condition and inspection is of no avail. The usual procedure, with such rectifiers, is to test with a rectifier tube known to be good.

Next check the condition of the radio, detector, and audio tubes of the set. This is easy, if you can plug in phones in the detector and audio stages. A signal in the phones indicates that all the tubes supplying them, including radio-frequency and detector stages, and sometimes the first audio stage, are in good working order, as is the power supply for all but the final output tube.

In absence of this convenient test, examine the tubes in the radio receiver. If necessary, darken the room to do so because tubes of the dry-cell type do not generally light very brightly. With some series filament receivers, all the tubes go out when one filament fails. With others, a resistor is used in parallel with each filament as a protective measure and it will pass enough current to light the remaining tubes in the receiver when one or two filaments are burned out.

If you do not locate the tube out of order by inspection, substitute a good tube for each one in the receiver, one at a time, remembering, of course, the first instruction to keep the filament rheostats low.

In the case of sets without the filament resistors in parallel, an even more serious strain is impressed upon the filter condensers when one radio tube goes out, unless a regulator tube is employed. In some cases, when the filament load is removed from the rectifier system, the voltage builds up greatly, sometimes sufficiently, in fact, to cause the breakdown of filter condensers.

The capable service man who, from a brief inspection of the receiver, decides that one of the radio or audio tubes may be dead, tests the tubes with a tube tester rather than placing his reliance upon the power supply of the set to do so. This avoids undue overload of the rectifier condensers. The manufacturer's instruction book often explains the best course to follow when the receiver goes dead and, understanding the reasons for it, the reader will not leave his power supply turned on if his tube filaments do not light, should the manufacturer advise him not to do so. The best test of radio and audio tubes is to use a tube tester or to replace the entire lot. Otherwise substitute a good tube in each socket, one at a time, leaving the remaining tubes in their sockets.

Assuming that rectifier and ballast tubes light and that radio and audio tubes have been proved good, the next thing to examine is the filament-type rectifier tubes. In most cases, if the rectifier tube or tubes show a blue or purple glow, it means that they have begun to leak. If so, place the set out of service at once, as there is no other repair possible than replacement of the rectifier tubes.

When there are two rectifier tubes, it may happen that the plates of one of them turns bright red while the other remains its normal color. Usually, this is a direct and simple indication that the tube which looks normal has lost its emission and the entire load of the set is being drawn from the other tube, which becomes red as a consequence of this overloading. The remedy is obviously replacement of the rectifier tube which does not overload.

On the other hand, when both rectifier tubes become bright red, it is due to an unusual load upon them, usually a broken-down bypass or filter condenser. Should this occur, the set should be placed out of service at once because this places an extreme overload on the rectifier tubes and the only remedy is replacement of the defective condensers. This is a job for an experienced service technician.

With those sets using regulator tubes, the simplest check which a service man is usually in a position to apply at once is the substitution of a good regulator tube to check the condition of the tube in service. When the regulator tube does not give a colored glow, either violet or pink, it is not receiving its necessary voltage supply, due to the shorting of a bypass condenser, failure of one or both of the rectifier tubes, a failing ballast tube or a radio or audio tube, the elements of which are short-circuited. It is the purpose of the regulator tube to keep the voltage supply constant. A prolonged loud signal or continued strong static may increase the brilliancy of its glow. But, if the antenna is disconnected any flashing of the regulator tube which occurs indicates that it is defective.

This completes the only frequent troubles encountered with a. c. sets. It is needless to go into the usual defects—loose connections, defective plug or flexible cord to power line, lost emission of power tubes, dirty contacts with socket, incorrect adjustment of reproducer, and other difficulties which are not specifically assignable to the a. c. set. They are the kind of difficulties which occur with increasing rareness as the mechanical and electrical design of receiving sets improves.

It may be worth while, however, to mention that a sudden increase of hum in the reproducer,

which has previously been quiet, may be due to the breaking down of a bypass condenser, a change in the load conditions of the power line, or reversed line connections. The latter is sometimes correctable by pulling out the plug, giving it half a turn, and re-inserting it. The power-line noise may require the attention of the power company as, quite frequently, it is the result of using an appliance in the immediate vicinity which sets up power surges in the line. These can be cured by installing an interference preventor at the device.

From the foregoing, it is obvious that there is nothing unusual or startling about the service requirements of a.c. sets. They introduce new problems but, once these problems have become familiar, they will not cause any serious modifications in radio servicing. The radio trade is learning that it must take its service responsibilities seriously while the consumer is recognizing that service is something which he should pay for, just as he does when he has his watch or automobile repaired. Free service has worked as much harm for the consumer as it has for the industry itself, because dealers giving extensive free service ultimately fail in business. And there can be no redress when the dealer has gone out of business. Amateur servicing with power sets, also, may result in more serious damage, just as does inexpert tinkering with an automobile. Just why there is hesitancy in admitting these simple facts is a little hard to understand, since, with a little consideration, anyone will understand that free service, inexpert service, and promises of perfect reliability do not promote satisfaction to the radio user.

It is premature to set up rules for the selection of a. c. sets at this stage because the relative merits of the various a. c. systems have not yet been fully proved by user experience. One of the safest guides is to consider the engineering reputation and stability of the manufacturer and the service reputation of the dealer from whom the set is bought. A manufacturer who has a reputation worth losing will not skimp by using a cheaper filter condenser in the hope that it will not break down prematurely, nor will a dealer of established standing in his community jeopardize his standing by taking on a line without adequate test and engineering examination. The day of the a. c. set is here. It means simplified radio and adequate power, essential to beauty of reproduction. It means a broader market for radio receivers, with the inevitable consequences of larger production and more radio for the money.

MAKING TESTS ON AN A. C. UNIT
The two-microfarad condensers on the RCA 104 loud speaker power unit are here undergoing tests for possible defects

AS THE BROADCASTER SEES IT

By CARL DREHER

Going to Jail for Radio in Germany

IN THE United States, a good many, broadcast listeners are in jail, but not for listening to the programs. They went to prison first and listened to the radio afterward, broadcasting being one of the inducements offered in all up-to-date penitentiaries. In Germany, however, radio has been the direct instrumentality whereby one gentleman went behind the bars. It appears that he was a *Schwarzhoerer*, which means, literally, a black listener, i.e., one who neglects to pay a fee and secure a license for his receiving set, as required by German law. The first time this recalcitrant BCL was caught he was fined and his set confiscated. The second time the authorities, with due process of law, clapped him into the jug for three weeks. If they catch him a third time presumably he will be tried by a court martial of studio managers and shot some Greenwich Mean Time morning.

As a practicing broadcaster, I regard this incident as a bad precedent, and am relieved that it occurred in a foreign country. For, if a broadcast listener may be sent up for a little radio bootlegging in this style, it is only a matter of time when the broadcasters themselves will spend their week-ends behind stone walls with spikes and sentry boxes on top. A program impresario permitting one of his flock of baritones to sing "Rolling Down to Rio" will be sent away for six months. The singer himself will merely receive a black eye at the hands of one of the catchpolls, for delicate psychological tests have proved that all baritones are subject to an irresistible impulse to warble this tune. When one of the water-cooled tubes lies down in the middle of a program, interrupting the festivities for a minute or so, the engineer of the station will be set to breaking rock for one year. When the transmitter deviates from the assigned frequency, the jolly technician will be separated from his wife and children one year for each kilocycle high or low and if another station has been heterodyned the owner thereof will be permitted to extract the offending engineers' teeth. This may seem a cruel, unusual, and hence unconstitutional punishment, but it is a sound maxim of law that a statute may be constitutional in one age and unconstitutional under later circumstances, and some judge may reverse the process and decide that, as broadcasting was not known to the Founding Fathers, new and appropriate punishments may properly be devised for erring broadcasters. On this principle, announcers who read telegrams of appreciation to the radio audience will be forced to eat a pound of coarsely ground Celotex. Announcers who read their own poetry over the air will be thrown to the tigers

in the nearest zoo, thus relieving the tax-payers in two ways.

Some of these penalties, I admit, appeal to me, and I have derived pleasure in contriving them. But, as I said, I disapprove the principle, since I am a broadcaster myself, and might be one of the first to be dragged to the hoosegow. I therefore appeal for moderation in this instance, and shall telegraph to my Congressman requesting that our government make representations to the Reich on behalf of the imprisoned *Schwarzhoerer*.

Radio's "Aristocracy of Brains"

I OFTEN reflect on the melancholy fact that the men who really made radio, the Marriotts, Alexandersons, Fessendens, Hogans and the rest, are not as well known as the more popular announcers of fifty-watt stations. Compared to the coruscating luminaries of the networks, like MacNamee and Cross, the engineers do not shine at all. For this there are, of course, sufficient reasons, psychological and sociological; one might as well lament that Dr. G. W. Crile is not as well known a man as Babe Ruth. I shall not waste my tears in this cause, yet I cannot refrain, coming down to cases, from pointing an accusing finger at our contemporary and friend, *Popular Radio*, which on Page 402 of its April, 1927, number, prints a picture with the caption, "A Group of Radio's 'Aristocracy of Brains.'" The description which follows is reprinted verbatim:

Dr. Ralph Bohn (in front of the desk) is receiving the Liebman Memorial Prize of the Institute of Radio Engineers from Mr. Donald McNicol, former President of the Institute. Dr. Bohn himself is the new president. At Mr. McNicol's left is Mr. John V. N. Hogan, Contributing Editor of *Popular Radio*. Next to him is Mr. R. H. Mariott. Behind Dr. Bohn are Professor Michael I. Pupin, of Columbia University, Mr. L. E. Whittemore, and Mr. E. F. W. Alexanderson, well-known inventor and radio engineer of the General Electric Company.

Children, what is wrong with this description? In the first place, the President of the Institute of Radio Engineers is Dr. Ralph *Bown*. The prize is awarded each year in memory of Colonel Morris N. *Liebmann*. Mr. Hogan's middle initials are *V. L.* Robert Henry *Marriott*, the first President of the Institute, spells his name with two "r's." Finally, the gentleman in the picture just behind Professor Pupin looks suspiciously like Dr. J. H. Dellinger, and not at all like Mr. Whittemore. The total of errors appears to be five. I call loudly for a more flattering vigilance

in such matters on the part of Messrs. Kendall Banning and Laurence M. Cockaday. They run altogether too good a magazine to permit such a high concentration of mistakes to the cubic centimeter of printer's ink to pass unnoticed, especially in reference to the Brahmins of the radio art.

Memoirs of a Radio Engineer. *XX*

MY FIRST assignment at the Aldene factory was in the test shop. This was a good-sized room in the old stone building, which later became merely an annex to the modern factory structure erected to fill the war-time requirements of the Army and Navy. Along one wall there was a switchboard controlling the supply of a.c. and d.c. to various outlets. For the rest the room was crowded with quenched spark transmitters of all sizes from one-quarter to five kilowatts, miscellaneous apparatus in various states of disarray, work tables, meters, and a few men: W. H. Howard, the Chief of Test; Baldwin Guild, now practicing patent law with Pennie, Davis, Marvin, and Edmonds; my present colleague O. B. Hanson; a tall gentleman named Lieb; a more medium-dimensioned gentleman named West, and myself. I may have omitted someone, but I think this comprised the list in May, 1917.

The engineering staff of the Marconi Company had the privilege of using the facilities of the test room for experiments when they could get in. By experiments I do not mean research on electron velocities or anything else at all recondite, but merely such incidental tests and measurements as always accompany the design of apparatus. This work had to be sandwiched in between the routine test functions of the department. Inasmuch as I was known to possess some engineering training, I was assigned temporarily as a sort of assistant to engineers who required work to be done in the test room. My first job was handed to me laconically by Mr. Woodhull, one of the transmitter engineers. A five-kva transformer was trundled into the test shop and dumped off. 'Measure the iron losses," said Mr. Woodhull, and disappeared.

I gazed at the transformer. It contained plenty of iron in the form of a shell core, and no doubt there were losses. I could identify the primary leads, which were to be fed at 110 volts a.c., and the secondary terminals, which were expected to deliver juice at 10,000 volts to the quenched spark gap of the five-kilowatt transmitter. I had never measured the iron losses of a transformer, but I had some vague recollection, from

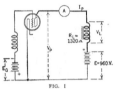

FIG. 1

my studies, that the procedure was to put rated voltage across the primary, leave the secondary open, and measure the power input with a watt-meter under this no-load condition. So I looked around for a wattmeter. The only one I could find had a full scale deflection of 7.5 kw., so I hooked it up and threw the switch. Nothing happened. I disconnected the wattmeter and tested it on another set which was under load. There was nothing obviously wrong with the instrument; it read several kilowatts. I put it back in my circuit and tried again, still without results. After an hour or so Woodhull came back and wanted to know what the iron losses of his transformer were. With much embarrassment I confessed that I did not know, and showed him my predicament. He looked over the circuit, and then glanced at the wattmeter. The scale made him laugh. "The iron losses are only of the order of a hundred watts," he told me. "You can't read that value on this scale. Get a 0.5-kw meter out of the storeroom and try again." Following these directions, I measured the iron losses of the 5-kva transformer at 70 watts, if I remember the figure correctly. Then we went ahead with some other tests involving loads.

This incident illustrates what the engineering student just out of school is up against. He usually has only a vague idea of magnitudes. I had studied under first-rate teachers and my preparation in some directions, was not at all bad. But there were numerous other fields in which I really did not know whether the answer would come out in millimeters or meters, or watts or kilowatts. If that wattmeter had swung up to a reading of a kilowatt or two for the iron losses of a small transformer, I should not have been surprised. A bonehead idea? Certainly, at that

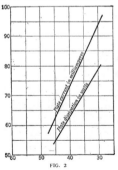

FIG. 2

place and time. But what is a good engineer? One who has, in the course of practice, got rid of a few thousand of such bonehead ideas. He gets paid largely for the relatively accurate judgment of physical magnitudes (always checked by measurement) and economic results (always checked by the balance sheet) which he has been able to put in their place.
(To Be Continued).

Technical Problems for Broadcasters. *IV*

THE plate feed of a fifty-watt (oscillator rating) tube used as a reactance-coupled amplifier in a broadcast transmitter is shown in Fig. 1. Plate potential is supplied by a storage battery with an output voltage E of 960 volts, the internal resistance being negligible. The reactor in the plate circuit has 50 henrys inductance and the d.c. resistance R_L is 1320 ohms. The resistance drop across it is designated by V_L, and this will necessarily depend on the mean current plate current I_P which is indicated by a milliammeter A in the plate circuit. This plate current varies with the negative grid bias E_G according to the upper characteristic curve of Fig. 2. The problem is, Given a safe plate dissipation of 70 watts for the tube in question, what is the minimum allowable negative grid bias under the conditions shown?

Solution

We are told that 960 volts is the steady output potential of the storage battery under all conditions, since its internal resistance may be neglected. Hence, in order to find the actual voltage V_P on the plate of the tube, all that is required is to find the voltage drop along the reactor for various plate currents, and subtract these values from 960. The plate dissipation W_P of an amplifier of the type shown is the product of the mean plate current and voltage, as indicated by d. c. instruments in the circuit. We may therefore draw up the following table:

E_G Volts	I_P Milli- Amperes	V_L Volts	V_P Volts	W_P Watts
30	94	124	836	78.5
35	83	105	855	71.0
40	72.5	95.5	864.5	62.5
45	62	81.5	878.5	54.4

The first two columns are obtained from the plate current-grid voltage characteristic. The V_L column follows when the value of plate current in each case is multiplied into 1320, the d.c. resistance in ohms of the reactor, giving the direct voltage drop across the winding. Subtracted individually from 960 volts, these values yield the actual plate potentials corresponding to the various grid bias figures. The plate dissipation is then calculated in each instance as the product of the V_P and I_P. As it happens, the plate dissipation results are numerically of about the same order as the plate currents in milli-amperes, so that they may be plotted in the form of the lower curve of Fig. 2. (This is an accident, resulting from the fact that the plate potentials are near 1000 volts, but of course an independent characteristic could be drawn in any case.) From this curve we note that 70 watts dissipation corresponds to 35.5 volts negative grid bias. The answer to the problem is, therefore, that 35.5 volts is the minimum allowable bias to be used with the tube in question if the plate is not to be overheated.

Abstract of Technical Article. *VI*

"Making the Most of the Line—A Statement Referring to the Utilization of Frequency Bands in Communication Engineering," by Dr. Frank B. Jewett. Presented before Philadelphia

Section of the American Institute of Electrical Engineers on Oct. 17, 1923. Reprinted May, 1924 by Western Electric Company, Inc. (Bell Telephone Laboratories, Inc.).

ALTHOUGH telephone wires used in connection with broadcast transmission are in general not utilized for other communication services during program periods, technical broadcasters will find it instructive to learn something about the multiplication of channels on expensive long distance circuits by the use of separate frequency bands for different purposes. Doctor Jewett, who is himself responsible for much of this development, has outlined the main features in this paper.

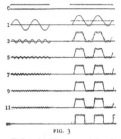

FIG. 3

The first point which concerns us is the analysis of a square-topped wave, reproduced here as Fig. 3. Such a wave is produced in a direct current telegraph system which has neither capacity nor inductance, as the circuit is made and broken by a key. It may be established mathematically and graphically that such a square-topped wave is composed of a continuous direct current and a number of sine wave alternating currents, comprising a fundamental frequency corresponding to the keying frequency, and, theoretically, an infinite number of odd harmonics. Fig. 3 shows this d.c., and the a.c. components up to the eleventh harmonic, with the resultant wave form in each case, and the ideal rectangular shape which is secured when all the harmonics are preserved. Fig. 4 is a diagram showing the same components in their relative amplitudes. If the current at the receiver is to be a strict reproduction of that at the keying end, frequencies considerably higher than the keying frequency must be transmitted. With hand sending it may be necessary to pass over the line frequencies as high as 40 cycles, to preserve the wave shape. With a multiplex printer the range is preferably extended to 100 cycles per second. Even in telegraphy, we note, accurate reproduction (what we call "good quality" in broadcasting) requires widening the frequency band.

Instead of keying a direct current for telegraph

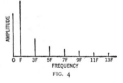

FIG. 4

communication, we may make and break an alternating current carrier. The resulting wave is shown in Fig. 5, the carrier being assumed to obey a sine law, and the line to be without inductance or capacity. This wave may be resolved into a fundamental and harmonics, as shown in Fig. 6. The difference between Fig. 4 and Fig. 6 is that in the former (d.c.) case the harmonics are grouped in ascending order on the positive side of the horizontal axis, starting with the frequency O, which is the direct current component, while with an a.c. carrier the harmonics are grouped on either side of the carrier frequency. In the latter method we employ a process of modulation and remodulation (now more

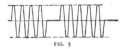

FIG. 5

In low frequency telephony a complex combination of frequencies corresponding to the sounds of speech and music must be transmitted, with fair fidelity in the case of commercial toll circuits and with considerable fidelity for broadcast purposes. In high frequency telephony, whether by radio or line carrier current, the same process of modulation and demodulation is carried out, but the representative chart of current components shown in Fig. 7 indicates the infinite number of harmonic mixtures and proportions. This particular combination corresponds to the sound of long o modulated by the human voice, and is relatively lacking in high overtones.

In transmitting a number of signal waves over a common circuit it is necessary to generate and

separate various kinds of currents. The devices used in the latter process fall generally into two classes:

1. Elements such as transformers, condensers, and choke coils, which discriminate between alternating and direct currents.
2. Filters.

The use of (1) is familiar in urban telephone systems of the central energy type. Low frequency (usually 16-cycle) alternating current is employed to ring subscribers' bells, direct current is supplied from a common battery for the transmitters, and alternating currents of voice fre-

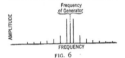

FIG. 6

commonly called demodulation). The process of modulation consists in taking a group of currents which have a certain relation to each other, mixing them with a carrier and thus translating them to a more convenient frequency band, while not disturbing the relation between them, and then, in demodulation, getting back the original group at its original frequencies. This is what we do in radio telephony as well as in the carrier telegraph illustration here used. Fig. 6, it will be noted, shows a carrier with both sidebands. For the transmission of intelligence only one of the sidebands need be retained, a locally generated carrier being substituted for the transmitting carrier, at the receiver, for the purpose of demodulation.

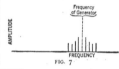

FIG. 7

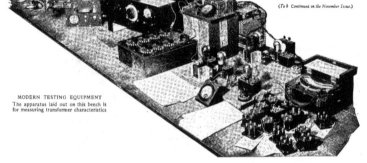

Line

Telephone Set

Telegraph Set

Telegraph Set

FIG 8

quency pass between the subscribers and operators. Direct current is supplied only when the receiver is off the hook, being controlled by the hook switch. The operator's ringing key controls the low frequency a.c. used for signalling. The voice currents pass from one trunk circuit to another through repeating coils, but d.c. is confined to subscribers' loops by the same agency. (Audio amplifiers, and such devices as output transformers for loud speakers, utilize inductances and capacities for similar purposes.) A.c.-d.c. separation is also required in joint use of telephone and direct current telegraph services on one line. Fig. 8 shows one such arrangement, blocking condensers being used to keep the d.c. telegraph currents out of the telephone set, and chokes to keep the voice currents away from the relays and sounders. A series connection of chokes and a parallel connection of condensers (by-pass) is a protection against a.c. while a parallel choke and series condenser arrangement discriminates against d.c.

(To b Continued in the November Issue.)

MODERN TESTING EQUIPMENT
The apparatus laid out on this bench is for measuring transformer characteristics

Home-Constructing Transformers and Chokes for Power-Supply Devices

The Second of Two Articles Intended to Obviate Tricky Mathematical Calculations when Building These Units—The Design of Choke Coils, and the Actual Construction of Both Chokes and Transformers, Is Discussed

By HOMER S. DAVIS

CHOKE coils are used in radio work chiefly to introduce opposition to the flow of alternating current, while at the same time allowing direct current to pass easily. In general, the lower the frequency of the alternating current, the more the inductance required. At radio frequencies, air-core chokes may provide sufficient opposition, but at audio frequencies and for smoothing out the ripple in power-supply units, an iron core is provided to make possible a greater inductance in a more compact unit, since the magnetic flux will flow through iron more easily than through air.

A different situation exists in the core of the choke coil than in the transformer core. In the latter, the magnetic flux due to the alternating current has the core all to itself, while in the case of the choke it has to share the core with the flux due to the direct current. To make certain that the direct current flux does not saturate the core, an air gap must be placed somewhere in the core. This greatly complicates the design; and a reasonable length of gap is difficult to maintain. Too great a gap will reduce the inductance and therefore require a larger choke for the same amount of inductance, while too small a gap will increase the harmonics, thus impairing the filtering action. It is generally agreed that a good value to use is one that uses up about ninety per cent. of the magnetizing force. For an ordinary silicon steel core this means a gap about 0.005 inches long for each inch of core length. The actual value to give best results can be determined only by trial with the other apparatus with which the choke coil is to work. Only the core type of construction will be considered in this article, as was done in the case of the transformer, design data for which were given in last month's article.

The inductance of the choke coil is proportional to the cross section area of the core, to the square of the number of turns of wire, and inversely to the length of the air gap. This is expressed mathematically as:

$$L = 3.2 \frac{AN^2}{G10} \quad \dots \quad (1)$$

where L is the inductance in henries, A the net area of the core cross section in square inches, N the number of turns, and G the equivalent air gap in inches. The equivalent air gap is defined as the total reluctance of the core, including the actual air gap, reduced to an air gap of the same cross section which will replace it. The value of equivalent air gap to use in designing the choke is uncertain. It is best figured from the formula:

$$G = 3.2 \frac{NI}{B} \quad \dots \quad (2)$$

in which N is the number of turns of wire, I the current in amperes, and B the flux density in lines per square inch. The value of I being predetermined, adjust the number of turns so as to give a flux density of not more than about

50,000 lines per square inch, with a reasonable value of G. This is relatively easy with small currents, but with the larger currents it is very difficult to maintain reasonable flux density. The value of G increases with the size of the choke, starting with about 0.05 inches as a minimum.

To avoid the mathematical difficulties of using the above formulas, calculation charts have been devised to replace them, formula No. 1 being represented by Chart II and formula No. 2 by Chart I.

The first things to begin with in designing a choke coil are the inductance and the current capacity. A reasonable length of air gap is then decided upon, and a flux density of about 50,000 lines per square inch assumed to start with. The number of turns can then be found from formula No. 2 or Chart I, after which the size of core may be figured from formula No. 1 or Chart II. Here it may be found that either the amount of copper or the size of the core is excessive from the standpoint of economical and compact design, in which case it may be necessary to try again with a larger air gap or a smaller flux density, or both. Several trials are often necessary. before a reasonable design is arrived at. Then with these vital factors settled, the next step is to choose the size of wire to handle the current, just as was done with the transformer.

Here again it is convenient to lay out a full-size drawing of the choke coil, to see just how its proportions will work out. As with the transformer, a compact arrangement should be striven for. Enameled wire is best for the choke, as the voltage difference per turn is small and thicker insulation would make the choke unnecessarily bulky. Allow for about $\frac{1}{32}''$ of insulation between the winding and the core. Insulating papers between layers are seldom used with choke coils. If the layout looks unwieldy with all the winding on one side of the core, it may be split into two coils on opposite legs. Although this requires less wire per turn, about 10 per cent. more turns should be added to each coil to make up for the effect of leakage of flux between them. With the drawing completed, the length around the center line of the core may be measured and the theoretical length of air gap computed on the basis of 0.005 inches per inch of core length, and compared with the value used in the design, to make sure that the latter was not assumed too small. The amounts of wire and core material may be estimated in the same manner as with the transformer. The resistance of the choke coil is an important factor, and, knowing the total length of wire and looking up its resistance per thousand feet in the wire table, on page 277 of the August article, it may be readily estimated. If the resistance proves greater than desirable, use a larger size of wire.

The solution of a typical example should

clear up any doubtful points remaining. Suppose a 20-henry choke coil is desired, capable of carrying 85 milliamperes of direct current. Where a choke is connected directly to the output of a rectifier without previous filtering, the maximum value of the rectified alternating current may be as high as 1.57 times the value of the direct current. Since both the alternating and direct currents contribute to the flux in the core, and only the direct current component has been used in the formulas, it is customary to allow for the alternating current component by modifying the value of flux density substituted in the formulas. About 35,000 lines per square inch will be satisfactory. Assume an equivalent air gap of about 0.05 inches. The number of turns of wire to wind on the coil may be found from Chart I. The key at the bottom of the chart shows which scales to connect together. Accordingly, draw a straight line from 0.085 on the current scale to 35,000 on the flux density scale. Through the point at which this line crosses the index line, draw a second line from 0.05 on the equivalent gap scale until it meets the turns scale, at 6500 turns in this case. Next, the size of core to use is obtained from Chart II. Draw a line from 20 on the inductance scale to 0.05 on the equivalent gap scale. Through the point at which this line intersects the index line, draw a second line from 6500 on the turns scale until it meets the core area scale. This point indicates a net area of 0.74 square inches, which is approximately equivalent to a core $\frac{7}{8}''$ square.

As mentioned before, the design is not so straightforward for the larger currents. It is difficult to arrive at a compact, economical design, and frequently it is necessary to make several trials. Sometimes a faulty design will not be evident from the plain figures, but will show up when the full-size drawing is made. The only remedy is to try again.

Having settled on the number of turns and area of the core, the next step is the determination of the wire size. This is done in exactly the same manner as with the transformer. Then comes the full-size layout, estimation of materials, and resistance calculation. If the resistance proves excessive for the particular use to which the coil is to be put, choose a larger size of wire and make a new layout. The design is then complete.

CONSTRUCTION

THE preparation of the core of the transformer or choke coil should be undertaken first. Several sources of the material are open to the constructor. Oftentimes a junked pole transformer may be obtained for a small sum from the shops of the local power company or from a junk yard, the core of which may be removed, cleaned with kerosene or alcohol, and cut down to the required size. For use in a B socket power-supply device, the core of a toy

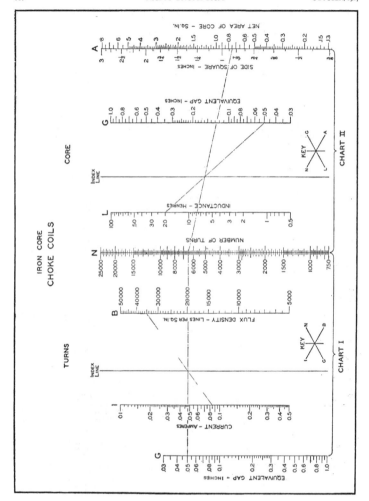

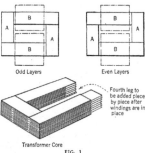

Odd Layers   Even Layers

Fourth leg to
be added piece
by piece after
windings are in
place

Transformer Core
FIG. 1

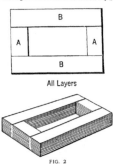

All Layers

FIG. 2

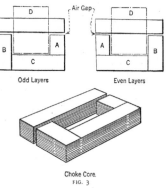

Odd Layers   Even Layers

Choke Core
FIG. 3

transformer will often be found to be about the right size without cutting. Ordinary stovepipe iron may be cut to size and used, but it will increase the losses, resulting in excessive heating and waste of power unless a much larger core is used. The use of regular transformer steel is recommended wherever possible. Motor or armature repair shops sometimes handle it. Supply houses that cater to transmitting amateurs often stock it and will cut it to any desired size. Their advertisements may be found in current amateur periodicals or their addresses obtained from a neighboring "ham." This source lacking, purchase the best available grade of soft sheet iron or steel from a local tin shop. It should be not more than 0.014" thick. It is best to cut the material to size with squaring shears, or to have a tinsmith do it, as accurate work is essential. Unless the pieces are square and of uniform size, poor joints at the corners will be inevitable and the resulting small air gaps will increase the core losses due to reduction of effective area. All rough edges should lie flat.

If an old core is used, the construction of the transformer will of course depend upon whatever conditions are encountered. But assuming that the raw material is available, several types of core arrangement are possible. The one shown in Fig. 1, using interleaving joints, is most convenient for transformers. Two different sizes of pieces are required, shown as A and B, with enough of each to stack twice as thick as the completed core when tightly clamped. Another way of building up the core is indicated in Fig. 2, but this type is not practical unless the pieces of the core are very accurately cut and closely butted together. This construction may be used for choke coils, however, since an air gap is required somewhere in the core, its location being immaterial as long as the requisite total length is provided. Fig. 3 illustrates a modified construction for chokes which has the advantage of being easier to clamp. In this case the pieces are cut to four different sizes, with enough of each to make up the full thickness of the core.

Some sort of insulation between the laminations should be provided, to reduce the core losses due to eddy currents. Commercial transformer steel usually comes with an oxide coating for this purpose. Ordinary rust is often sufficient when the oxide coating is not present, but it is better to apply a thin coat of shellac to one side of each piece, allowing it to dry thoroughly before assembling.

If the interleaving type of construction is used, the building up of the core may be expedited by providing some sort of square corner as a guide, such as a cigar box with two adjacent sides removed. Only three sides of the core should be assembled at first, alternating the layers as shown in Fig. 1; the fourth side is to be put in piece by piece after the coils have been slipped into place. The partially completed core is carefully removed to a vise and clamped, and the legs that receive the coils bound tightly with a layer of tape, or with heavy string, which may be later unwound as the coils are pushed onto the core. The joints should be carefully trued up, making sure that the pieces butt together well. A wooden mallet, or a rawhide or lead hammer may be used for this purpose.

Preparing the coils is not difficult if done properly. They should be evenly wound in layers, not only to make for compactness, but to prevent the possibility of two turns of widely different voltages coming together. Coils of small enameled wire should have a layer of paper between each layer of wire, to keep the winding even and to improve the insulation. If special insulating papers are not available, tracing paper such as used by draftsmen, will serve the purpose nicely; ordinary wrapping paper can also be used. These papers may be omitted with the larger sizes of enameled wire, since they are easier to wind evenly and the voltage difference between layers is less. They may be dispensed with entirely if cotton-covered wire is used.

Windings of cotton-covered wire may be made moisture proof and more rigid by applying a thin coat of shellac to each layer as wound, and baking the finished coil to dry it out. Enameled wire cannot be treated in this way, however, as the shellac may soften the insulation. Small coils of enameled wire may be made rigid by dipping them in melted wax, such as a mixture of beeswax and rosin. Ordinary paraffin is not suitable, as it may soften if the transformer runs at all warm. Large coils may be painted with insulating varnish or black asphaltum paint.

The windings may either be made self-supporting or wound on a fibre spool which is slipped over the core. The former is the more common method. A square wooden mandrel the same size as the core cross section is first prepared, cut to the same length that the winding is to be, and sanded smooth. For the larger core sizes it may have to be built up of several thicknesses of wood glued together. Flanges a quarter of an inch thick are then screwed to each end, as shown in Fig. 4; these are later to be removed in order that the winding may be slipped off.

Some means of rotating the winding form must be provided. A convenient arrangement is to

drive a spike in the center of one end of the mandrel, grind or file off the head, and secure the end of the spike in the chuck of a hand or breast drill which is clamped horizontally in a vise. A lathe may also be used. A small geared emery wheel is excellent for the purpose. Another common method is to fasten the mandrel to a wooden disc which is clamped or wired to the flywheel of a sewing machine. If a geared emery wheel or hand drill is used, the number of turns of wire may be computed by determining the turns ratio and then counting the turns of the crank. Another way is

to insert the spindle of a revolution counter into a hole in the free end of the mandrel, as shown in Fig. 4.

As the coils are to be taped when finished, wind a layer of heavy string around the mandrel and fasten the ends. This is to be unraveled after the coils are wound, enabling them to be slipped off easily and allowing room for the tape. A layer or two of fibre is then cut to size and squarely bent around the string-covered mandrel and glued in place. On high-voltage windings, above about 500 volts, add a few layers of empire cloth, which may be obtained at motor repair shops.

WINDING THE COIL

THE coil is now ready to be wound. Pass the free end of the wire through a small hole drilled in the flange, leaving enough wire to provide a lead to the panel. If fine wire is used, the lead and the first few turns should be of insulated flexible wire, to avoid any possibility of the lead breaking off. The fine wire is then soldered to this, carefully cleaned of any excess flux, and the joint covered with a bit of tape. The rest of this first layer is then wound. If the wire is cotton-covered, this layer, and each succeeding layer as wound, may be given a thin coat of shellac as a binder, or insulating papers may be used as described above, for enameled wire. The wire should be wound as tightly as its strength will permit. It may be necessary here to wear a glove on the guiding hand, or to pass the wire over a piece of cloth held in the hand. Guiding the wire will be made easier if the hand is held as far away from the coil as is convenient. The turns should be kept close together and not allowed to overlap, and extreme care should be exerted to avoid any possibility of shorted turns. A shorted turn acts as an independent short-circuited secondary and will burn out as soon as the transformer is connected to the line. Where taps are brought out, the turns per layer should be so arranged, if possible, that the taps will come at the end of a layer. Be especially watchful here for shorted turns. In finishing the winding, pass the end of the wire through another small hole in the flange and fasten the last few turns with a bit of sealing wax. With fine wire the last few turns and the lead should be of flexible wire just as at the start. If another winding is to be placed over the first, they should be separated by several layers of friction tape or empire cloth.

Where a split secondary is used, the two coils may be wound side by side, each covering half the mandrel, rather than one on top of the other, to insure their having symmetrical characteristics. An end flange may be removed while a fibre separator is slipped in place, and one side filled tightly with cloth strips while the other is being wound, to prevent the separator being crowded over. A split or center-tapped filament winding of only a few turns may be wound by means of a pair of wires, each comprising half of the total number of turns, the end of one wire being connected externally to the beginning of the other so as to put them in series. The connection between the two is used as the center tap.

The finished winding may be removed from the form by taking off one of the flanges and pulling out the layer of string, after which the coil can be slipped off. It may then be taped, as in Fig. 5, using ordinary friction tape and advancing half the width with each turn. High-voltage coils

should be covered with empire cloth tape. The tape should not be allowed to bunch at the inside corners, or trouble will be experienced in inserting the core. The coils should not be taped too heavily, or they will not cool well.

Small coils of enameled wire may be conveniently made by winding them on spools made of pieces of fibre glued together. After being wound the wire may be covered with a layer or two of heavy paper on one layer of friction tape.

The finished coils, after painting or moisture-proofing, may now be placed on the core, unwinding the string on each leg as its coil is pushed on. The fourth leg of the core is then put in, piece by piece, hammering the corners up tightly. If the winding fits too loosely, extra core pieces may be forced in, taking care not to damage the insulation of the coil, or small wooden wedges may be driven between the coil and the core.

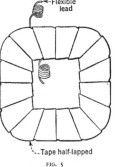

Flexible lead

Tape half-lapped

FIG. 5

The whole assembly should be made rigid, to reduce any audible hum when the transformer is in use.

Methods of mounting the transformer or choke coil are shown in Fig. 6. Angle iron, strap iron, or square lengths of wood may be used for clamping the core. Another method is to drill a hole

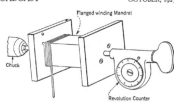

Flanged winding Mandrel

Chuck

Revolution Counter

FIG. 4

in each corner of the core for a small bolt, but the diameter of the hole should not exceed one-fifth of the width of the core leg, otherwise the effective area of the core will be reduced. A panel of bakelite or hard rubber may be provided to hold the terminals, which may be ordinary binding posts or soldering clips. The leads may be brought up to the back of the panel and soldered to the terminals. They should be kept well separated from each other and insulated with "spaghetti" or similar tubing. Flexible insulated hook-up wire makes good leads. The terminals of the transformer should be plainly and permanently labeled with their voltages.

The transformer should be tested before being connected to any other apparatus. First connect the primary to the line without any load on the secondary, to detect any shorted turns, which will show up as excessive heating or as an actual burn-out. If after several hours the transformer is only slightly warm, it is probably all right. The terminal voltages may be checked with an a. c. voltmeter, if desired.

After the choke coil is completed the air gap must be adjusted to the proper value. This can only be done experimentally, by connecting the choke to its associated apparatus and changing the gap until the best filtering action is obtained. The gap should then be filled with cardboard or a cloth pad and the core permanently clamped, to prevent the gap being gradually closed by the magnetic pull between the two parts of the core. The inductance of the choke cannot be measured directly. If the necessary meters are available, it may be found as follows. Connect the choke in series with a battery and measure the voltage across the coil and the current in amperes through it. Its resistance may then be calculated from Ohm's Law:

$$R = \frac{V}{I}$$

Then connect it across a source of a. c. voltage of known frequency, such as the 110-volt line, and again measure the current and voltage. The impedance is then:

$$Z = \frac{V}{I}$$

The inductance, L, may then be calculated from the formula:

$$Z = \sqrt{R^2 + (2\pi f L)^2}$$

Or its equivalent:

$$L = \sqrt{\frac{Z^2 - R^2}{(2\pi f)^2}}$$

The writer wishes to acknowledge his indebtedness to Prof. F. S. Dellenbaugh, Jr., of the Massachusetts Institute of Technology, for the design formulas applying to the choke coil.

  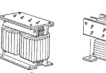

FIG. 6

Not a Super~Heterodyne
Not Tuned Radio Frequency
Not Regenerative
But

All models of the R. G. S. *"OCTA-MONIC"* will be on display at the Radio World's Fair, New Madison Square Garden, between the 19th & 24th of September inclusive at the booth of the R. G. S. Mfg. Co., Inc.

Tireless Performance

These gulls fly and fly until we wonder *how* such stamina can be contained in so frail an object.

Just so with CeCo Tubes. A strong combination of frail materials. Glass for a covering; hair-like wires for filament; fine spun metal for grid.

But so carefully engineered, so cleverly assembled so skillfully exhausted, so thoroughly tested that their durability is astounding to the radio operator and fan who judges CeCo performance by ordinary standards.

You expect *MORE* of CeCo Tubes—and *get* more.

A Type for Every Radio Need

General Purpose Tubes, Special Purpose Tubes, Power Tubes, Filament Type Rectifiers, Gas Filled Rectifiers and A. C. Tubes.

Ask your radio dealer for complete data sheet of CeCo Tubes.

C. E. MANUFACTURING CO., Inc., Providence, R. I.

Write for Data Sheet giving characteristics of all CeCo Tubes

Announcing Our New

Gas Filled Rectifier
(NO FILAMENT)
TYPE D-G

Maximum Volts—300   *Maximum Cur.*—85 *M-A*

Long Life without decrease in output is assured if these values are not exceeded.

Easy Filtration. Less strain on Filter condensers and smoother output with less Hum or Ripple.

These tubes are tested in a Standard rectifying circuit using well designed parts. The unit is connected to a ripple test position, and tube checked both by phones and observed on an oscillograph, insuring a perfect tube which will give excellent results in *well designed and constructed units.*

PRICE $5.00

Makes a Good "B" Eliminator—BETTER

Restored
Enchantment

*This is the Eveready
Layerbilt that gives
you Battery Power for
the longest time and
the least money.*

THERE is no doubt of it—
radio is better with Battery Power.
And never was radio so worthy of the
perfection of reception that batteries,
and batteries alone, make possible.
Today more than ever you need
what batteries give—pure DC. Direct Current, electricity that flows
smoothly, quietly, noiselessly. When
such is the current that operates
your receiver, you are unconscious
of its mechanism, for you do not
hear it humming, buzzing, crackling.
The enchantment of the program is
complete.

Batteries themselves have improved, as has radio. Today they
are so perfect, and so long-lasting,
as to be equal to the demands of the
modern receiver. Power your set
with the Eveready Layerbilt "B" Battery No. 486. This is the battery
whose unique, exclusive construction
makes it last longer than any other
Eveready. Could more be said? In
most homes a set of Layerbilts lasts
an entire season. This is the battery
that brings you Battery Power with
all its advantages, conferring benefits
and enjoyments that are really tremendous when compared with the
small cost and effort involved in replacements at long intervals. For
the best in radio, use the Eveready
Layerbilt.

Radio is better with Battery Power

At a turn of the dial a radio program comes to you. It is clear.
It is true. It is natural. You thank the powers of nature that have
once more brought quiet to the distant reaches of the radio-swept
air. You are grateful to the broadcasters whose programs were
never so enjoyable, so enchanting. You call down blessings upon
the authority that has allotted to each station its proper place. And,
if you are radio-wise, you will be thankful that you bought a new
set of "B" batteries to make the most out of radio's newest and
most glorious season.

NATIONAL CARBON CO., INC. ⬛ New York—San Francisco

Unit of Union Carbide and Carbon Corporation

Radio Batteries
—they last longer

Tuesday night is Eveready Hour Night—8 P. M., Eastern Standard Time

WEAF—*New York*	WGR—*Buffalo*	WGN—*Chicago*	WRC—*Washington*
WJAR—*Providence*	WCAE—*Pittsburgh*	WOC—*Davenport*	WGY—*Schenectady*
WEEI—*Boston*	WSAI—*Cincinnati*	WCCO { *Minneapolis* *St. Paul*	WHAS—*Louisville*
WDAF—*Kansas City*	WTAM—*Cleveland*		WSB—*Atlanta*
WFI—*Philadelphia*	WWJ—*Detroit*	KSD—*St. Louis*	WSM—*Nashville*
		WMC—*Memphis*	

Pacific Coast Stations—9 P. M., Pacific Standard Time

KPO—KGO—*San Francisco*	KFI—*Los Angeles*
KFOA—KOMO—*Seattle*	KGW—*Portland*

*Have you heard the new Victor record by the Eveready Hour Group—orchestra and
singers—in Middleton's Down South Overture and Dvořák's Goin' Home?*

The *Radio Broadcast*
LABORATORY INFORMATION
SHEETS

THE RADIO BROADCAST Laboratory Information Sheets are a regular feature of this magazine and have appeared since our June, 1926, issue. They cover a wide range of information of value to the experimenter and to the technical radio man. It is not our purpose always to include new information but to present concise and accurate facts in the most convenient form. The sheets are arranged so that they may be cut from the magazine and preserved for constant reference, and we suggest that each sheet be cut out with a razor blade and pasted on 4" x 6" filing cards, or in a notebook. The cards should be arranged in numerical order. An index appears twice a year dealing with the sheets published during that year. The first index appeared on sheets Nos. 47 and 48, in November, 1926. In July, an index to all sheets appearing since that time was printed.

The June, October, November, and December, 1926, issues are out of print. A complete set of Sheets, Nos. 1 to 88, can be secured from the Circulation Department, Doubleday, Page & Company, Garden City, New York, for $1.00. Some readers have asked what provision is made to rectify possible errors in these Sheets. In the unfortunate event that any such errors do appear, a new Laboratory Sheet with the old number will appear.

—THE EDITOR.

No. 129   RADIO BROADCAST Laboratory Information Sheet   **October, 1927**

The Type 874 Glow Tube

HOW IT FUNCTIONS

THE type 874 tube is a special voltage regulator designed for use in B power units to maintain the voltages, supplied by the unit, constant. An ordinary B power unit operated without a glow tube has a comparatively poor regulation, *i.e.,* the voltage changes considerably with changes in the amount of current being drawn from the unit. It would obviously be of decided advantage if this voltage could be made to remain practically constant at all loads. The power unit could then be used with any receiver irrespective of the amount of current being drawn by it (within reason) with the knowledge that the actual voltages designated on the binding posts of the B device were being supplied. How the glow tube functions to maintain the voltage constant may be understood by reference to the curve A. This curve is plotted by measuring the voltage across the glow tube with various load currents and it should be noted that the voltage across the tube is practically 90 at all loads up to more than 40 mA. In ordinary operation, when there is no current being drawn from the 90-volt tap, the glow tube current is about 45 milliamperes. Then, if current is drawn for a receiver from the 90-volt tap, which would ordinarily cause the voltage to go down, the current through the glow tube automatically decreases, providing for the current required by the set. The voltage thereby is maintained at exactly 90.

Curve B illustrates the curve of output voltage

that might be obtained from a B power unit not using a glow tube. At no load the voltage is 123, while at a load of 10 mA. the voltage drops to 90. If, however, the receiver requires 20 milliamperes, the actual voltage available would be only 60 volts.

No. 130   RADIO BROADCAST Laboratory Information Sheet   **October, 1927**

Data on Honeycomb Coils

NO. OF TURNS	INDUCTANCE, AT 800 CYCLES, IN MILLIHENRIES	NATURAL WAVELENGTH, METERS	DISTRIBUTED CAPACITY IN MMFD.	WAVELENGTH RANGE, METERS	
				0.0005-MFD. CONDENSER	0.001-MFD. CONDENSER
25	.039	65	30	120 to 245	120 to 355
35	.0717	92	33	160 to 335	160 to 480
50	.149	128	31	220 to 485	220 to 690
75	.325	172	26	340 to 715	340 to 1020
100	.555	218	24	430 to 930	430 to 1330
150	1.30	282	17	680 to 1410	680 to 2050
200	2.31	368	16	900 to 1880	900 to 2700
249	3.67	442	15	1100 to 2370	1000 to 3410
300	5.35	535	17	1400 to 2870	1400 to 4120
400	9.63	666	13	1800 to 3830	1800 to 5500
500	19.5	806	13	2300 to 4870	2300 to 7000
600	21.6	1045	14	2800 to 5700	2800 to 8200
750	34.2	1300	14	3500 to 7200	3500 to 10400
1000	61	1700	13	4700 to 9600	4700 to 13800
1250	102.5	2010	11	6000 to 12500	6000 to 18000
1500	155	2710	13	7500 to 15400	7500 to 22100

Balkite has pioneered—
but not at public expense

Balkite "A" Contains no battery. The same as Balkite "AB" below, but for the "A" circuit only. Not a battery and charger but a perfected light socket "A" power supply. One of the most remarkable developments in the entire radio field. Price $32.50.

Balkite "B" *Has the longest life in radio.* The accepted tried and proved light socket "B" power supply. 300,000 units in use show that it lasts longer than any device in radio. Three models: "B"-W, 67-90 volts, $22.50; "B"-135,* 135 volts, $32.50; "B"-180, 180 volts, $39.50. Balkite now costs no more than the ordinary "B" eliminator.

Balkite Chargers
Standard for "A" batteries. Noiseless. Can be used during reception. Prices drastically reduced. Model "J,"* rates 2.5 and .5 amperes, for both rapid and trickle charging, $17.50. Model "N"* Trickle Charger, rate .5 and .8 amperes, $9.50. Model "K" Trickle Charger, $7.50.

Special models for 25-40 cycles at slightly higher prices.

Prices are slightly higher West of the Rockies and in Canada.

The great improvements in radio power have been made by Balkite.

First noiseless battery charging. Then successful light socket "B" power. Then trickle charging. And today, most important of all, Balkite "AB," a complete unit containing no battery in any form, supplying both "A" and "B" power directly from the light socket, and operating only while set is in use.

This pioneering has been important. Yet alone it would never have made Balkite one of the best known names in radio. Balkite is today the established leader because of Balkite performance in the hands of its owners. Because with 2,000,000* units in the field Balkite has a record of long life and freedom

from trouble seldom equalled in any industry. Because the first Balkite "B," purchased 5 years ago, is still in use and will be for years to come. Because to your radio dealer Balkite is a synonym for quality. Because the electrolytic rectification developed and used by Balkite is so reliable that today it is standard on the signal systems of most American as well as European and Oriental railroads. Because Balkite is permanent equipment. Balkite has pioneered — but not at the expense of the public.

Today, whatever type of set you own, whatever type of power equipment you want, whatever you want to pay for it, Balkite has it. And production is so enormous that prices are astonishingly low.

Your dealer will recommend the Balkite equipment you need for your set.

BALKITE "AB"
Contains no battery. A complete unit, replacing both "A" and "B" batteries and supplying "A" and "B" current directly from the light socket. Contains no battery in any form. Operates only while the set is in use. Two models: "AB" 6-135,* 135 volts "B" current, $59.50; "AB" 6-180, 180 volts $67.50.

FANSTEEL PRODUCTS COMPANY, Inc.
North Chicago, Illinois

FAN STEEL
Balkite
Radio Power Units

No. 131　　　RADIO BROADCAST Laboratory Information Sheet　　　·October, 1927

Resistance-Coupled Amplifier

DATA ON CONSTANTS

ON LABORATORY Sheet No. 132 is given a circuit diagram of a resistance-coupled amplifier using the new type 240 high-mu tube (Cunningham type 340). To obtain satisfactory operation from such an amplifier it is essential that several points be given careful consideration. In the first place it is essential that excessive C bias is not used on any of the high-mu tubes. The following values should be used in combination with the voltage shown in the circuit diagram to prevent overloading: 1 volt on the first stage, 3 volts on the second stage, and 40.5 volts on the 171 power tube. The second consideration of great importance in the construction of such an amplifier is that the coupling condensers, C_2, C_3, and C_4, be of the best quality that can be obtained. Even, a small amount of leakage across the condenser, due to faulty insulation, will permit some of the plate potential to leak through it to the grid of the next tube and this will not only cause distortion but very frequently will make the amplifier absolutely inoperative. Use, only the best of mica condensers.

It is, of course, also essential that the plate and grid resistance be noiseless in operation but it is not necessary that they be exactly of the values given in the circuit. A variation of ten or twenty per cent. in these values is quite unimportant. The plate supply for the amplifier may either be a well con-

structed B power unit or batteries. No trouble whatsoever should be experienced when operating the unit from new batteries, but it is possible that "motor-boating" troubles will develop when the amplifier is used with some B power units. The overall gain is comparatively high and difficulties of this sort become more pronounced as the amplifi-

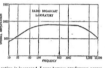

cation is increased. Large bypass condensers across the output of the power unit will frequently be necessary in order to prevent the occurrence of "motor-boating." The frequency characteristic of the complete amplifier is shown by the accompanying curve.

No. 132　　　RADIO BROADCAST Laboratory Information Sheet　　　October, 1927

Resistance-Coupled Amplifier Circuit

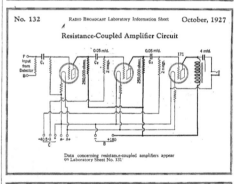

Data concerning resistance-coupled amplifiers appear on Laboratory Sheet No. 131

No. 133　　　RADIO BROADCAST Laboratory Information Sheet　　　October, 1927

Care of Power Supply Units

FREQUENT CHECKING NECESSARY

MANY, modern radio receiver installations employ a B power unit for the plate voltage, and a storage battery in conjunction with a trickle charger for the filament supply, the entire combination being controlled by means of an automatic relay. If well manufactured units are used throughout, such an installation should require practically no attention other than the addition of water to the storage battery and the trickle charger, if the latter is of the electrolytic type.

In order to make certain that the entire power plant is functioning satisfactorily, it is a good idea to make some simple tests every six months or so. Little can go wrong with the B power unit without it becoming noticeable in the operation of the receiver. If the rectifier tube deteriorates the volume produced by the receiver will be lowered and also the quality will be impaired. A total failure of the power unit will, of course, mean that it will be impossible to hear anything at all on the receiver. The simplest check to make on the A power unit in order to make certain that it is functioning satisfactorily is to take a hydrometer reading of the storage battery. If the battery reads "fully charged" it is possible that the trickle charging rate is ex-

cessive and it will be a good idea to somewhat reduce the rate and then make frequent tests with the hydrometer to determine how the battery is standing up. If the total charge in the battery now gradually decreases it will be best to increase the rate of trickle charge again. If, on the other hand, the battery continues to remain in a fully charged condition, we have a good indication that the previous rate of trickle charge was too high and that, very probably the battery was being continually over charged, which is very harmful. If a hydrometer reading of the battery indicates that the battery is very low the trickle charge rate should be increased so that the battery is brought up to practically full charge and then the rate should be adjusted so as to maintain the battery in this condition. The contacts in the relay controlling the installation should be inspected every so often. There is a certain amount of sparking at the contacts which tends to pit them and it might be necessary to smooth them with a piece of emery cloth. Badly pitted contacts in the relay might at times prevent the unit from closing the trickle charger circuit and consequently the battery will not always be charged while the receiver is not being operated.

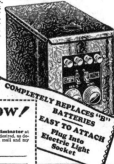

No. 134   RADIO BROADCAST Laboratory Information Sheet   **October, 1927**

Loud Speaker Horns

THE EXPONENTIAL TYPE

A CORRECTLY designed horn makes a very good type of loud speaker. The best horn is one which radiates most uniformly over the required range of frequency and it has been proved mathematically that the exponentially shaped horn conforms closely to this requirement. A horn is of the exponential type when its cross section area doubles at equal intervals along its length. For example, a horn would be of the exponential type if at the orifice it had an area of 1 square inch, and an area of ½ square inch, 1 square inch, and 2 square inches, at distances of 1, 2, and 3 feet respectively from the orifice. The rate of expansion determines the lowest frequency of which the horn will be a good sound producer. A horn which doubles in area every foot will reproduce down to about 64 cycles, and a horn which expands twice as rapidly will only reproduce well down to 128 cycles.

A properly designed horn should be free from noticeable resonance, and to prevent this the mouth of the horn should be made large enough to transmit the sounds coming from it without any great amount of back pressure. In the design of loud speaker horns it has been found that, if the mouth is made comparable to ¼ of the wavelength corresponding to the low frequency cut-off point of the horn, the resonance in the horn will be negligible. The wavelength in feet is determined by dividing the velocity of sound in feet per second, which is 1120, by the frequency. For example, a horn whose cut-off frequency is to be 32 cycles, corresponding to a wavelength of 39 feet, should have a mouth of 39 divided by 4, or 9¾ feet. These facts indicate definitely that a horn, to be a good one, must be large. Small horns, whether they are or are not exponential, cannot radiate the low frequencies.

The horn makes it possible for a comparatively small diaphragm to get a good grip on the air and thereby produce a large volume of sound. The small diaphragm and the large horn may be replaced by a large diaphragm, as is done in a cone type loud speaker.

The material of which the horn is made is important. Although a horn may be well designed, and constructed to the correct size, total length and expansion per unit length, it may still fail to give really good results because of resonant effects in the material used in the construction. The material used should have no marked resonant frequency unless it is very low, where it might help to increase the low note radiation.

No. 135   RADIO BROADCAST Laboratory Information Sheet   **October, 1927**

Closely Coupled Circuits

RESONANCE CURVES

IF TWO circuits are coupled together by a condenser, as shown in the sketch, and they are both adjusted so that they are tuned to slightly different frequencies, we will find that a resonance curve of

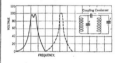

the combination has the form shown by the solid curve. The resonance curve of either separate circuit alone would have the form indicated by the dotted curve. It is evident from the resultant curve that the combination of these two circuits produces a resultant characteristic curve which is quite broad and flat on the top in comparison with the quite sharp peak of either circuit alone. This double peaked effect is a characteristic of closely coupled circuits and has been used to some extent in radio receivers.

An ordinary resonance circuit consisting of a single coil and condenser has a comparatively sharp resonance curve and therefore frequencies slightly above or below the resonant frequency are not amplified as well as the latter and, therefore, the tuned circuit tends to cut down the amplification of the side bands of the incoming wave and this causes some loss of high frequencies. If a receiver is made up, however, with two coupled circuits, such as we have indicated, this cutting of the side bands will not take place because the flat top of the resonance curve can be made sufficiently broad so as to give equal amplification over a band 10,000 cycles wide and therefore practically equal amplification can be obtained at all frequencies 5000 cycles above or below the carrier frequency. The circuit has not been used in actual practice to any great extent because of the difficult tuning required and because of the careful adjustments necessary.

No. 136   RADIO BROADCAST Laboratory Information Sheet   **October, 1927**

Carrier Telephony

THEORY AND USES

THE use of power lines for the dissemination of intelligence is becoming increasingly common throughout the country. Large power companies have in many cases installed radio equipment for inter-communication between various power plants; these radio-frequency signals are transmitted over the power lines rather than through the air, and, in this way less interference is encountered. The system has also been used in some communities in order to make it possible to receive musical programs at home by connecting a suitable device directly to the power socket.

For commercial use, this system has certain advantages, such as lack of interference, which make its use valuable, but it is unlikely that the system will ever replace broadcasting. The number of different stations that can be "tuned-in" by a subscriber using the system is naturally limited, and this is a definite disadvantage.

The system actually differs very little from that of ordinary broadcasting, the major difference being that the power of the transmitter, instead of being radiated into the air by means of an antenna, is coupled directly to the power line. The coupling between the transmitter and the power line is generally made through high-voltage coupling condensers and special filter and protective circuits. At the receiving end an ordinary radio receiver can be used to detect the signals. It also must, of course, be coupled in some way to the transmission line. The system is generally operated in duplex so that transmitting or receiving can be accomplished at any of the various terminals of the system.

In carrier telephony it has generally been found best to use carrier frequencies somewhat above 50,000 cycles. For comparatively low radio frequencies, around 10,000 cycles, there is considerable loss in the various power transformers in the line, and at frequencies intermediate between about 10,000 and 50,000 cycles there will very likely be sharp resonance peaks causing excessive loss at particular frequencies. Above 50,000 cycles an ordinary transmission line is fairly satisfactory as a transmitting medium.

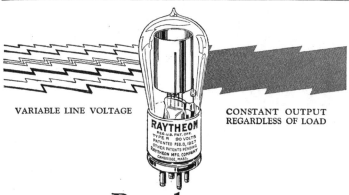

VARIABLE LINE VOLTAGE

CONSTANT OUTPUT
REGARDLESS OF LOAD

Raytheon
Voltage Regulator Tube

EXPERIENCE gained in seven years of exhaustive research, experiments and tests, have enabled the Raytheon Research Laboratories to produce a new and fundamentally improved Voltage Regulator Tube of marked characteristics.

Raytheon—Type R, when incorporated in the proper B Power circuits, maintains constant voltage on the 90 and lower voltage taps and greatly improves the voltage regulation on the 180 and 135 volt taps, *regardless of variations in either the line voltage or load current.* The variable voltage controls can be thus eliminated and the construction of the unit simplified.

Furthermore, this new tube has a very pro-

nounced effect in eliminating the last vestige of ripple from the output, and when connected to an amplifier will completely eliminate "motor-boating".

A new feature — *the starting anode* — incorporated in type R, maintains constantly a state of ionization in the tube to prevent any "going out" regardless of the load fluctuation.

Raytheon — Type R, can be used in any power unit circuit now on the market employing a glow tube with greatly improved results. Diagrams of an approved Type R tube installation, in connection with the heavy duty Raytheon BH tube, can be had from the Raytheon Research Laboratories upon request.

Raytheon BH—a 125 milliampere rectifying tube, because of its constant heavy duty output, is especially designed for use in conjunction with the Raytheon Type R Regulator Tube in light socket power units.

Raytheon, Type R, 90 volts, 60 milliamperes—Price $4.00

Raytheon K—the compact efficient rectifier, new in principle and construction. For A battery chargers and A battery eliminator units.

Raytheon BA—a 350 milliampere rectifying tube for complete battery elimination. Watch for the new ABC power units employing this rectifier.

RAYTHEON MANUFACTURING COMPANY
CAMBRIDGE, MASS.

THE HEART OF RELIABLE RADIO POWER

New Transformers for A. C. Power Supply

The AmerTran Power Transformer Type PF-281, $25.00 each

AS IN audio transformers, AmerTran products stand first in the power transformer field. They are up-to-date in design, well made and dependable.

Type PF-281, illustrated above, becomes virtually an A-B-C eliminator when used with AC tubes and the proper filter circuit for DC voltages of from 425 to 650 volts, plate current 110 milliamperes. This unit is designed for use with the new UX-281 rectifying tube, and has a 750 volt plate winding which enables it to be used with a UX-281 or 216-B rectifying tube. In addition, there are filament heating windings for the new AC tubes. Therefore, this single unit will convert AC house current into filament and plate current, and grid bias potential. Used with types 709 and 854 AmerChokes in the filter circuit, a receiver may be constructed to operate entirely from the house lighting circuit.

Type H-67 Heater Transformer is a new unit recommended for use with the RCA UX-226 raw AC amplifier tubes and the UY-227 detector tube. It also has a third filament winding capable of handling two UX-171 tubes. In connection with the new AC tubes, type H-67 becomes the power source for the filament and is therefore a real "A" battery eliminator. This transformer sells for $12.00.

Write for Booklet, "Improving the Audio Amplifier," and data on Power Supply Units.

AMERICAN TRANSFORMER CO.
178 Emmet Street   Newark, N. J.

We also make Audio Transformers Choke Coils and Resistors.

Transformer Builders for Over Twenty-Six Years

Manufacturers' Booklets

A Varied List of Books Pertaining to Radio and Allied Subjects Obtainable Free With the Accompanying Coupon

READERS may obtain any of the booklets listed below by using the coupon printed on page 396. Order by number only.

1. FILAMENT CONTROL—Problems of filament supply, voltage regulation and effect on various circuits. RADIALL COMPANY.

2. HARD RUBBER PANELS—Characteristics and properties of hard rubber as used in radio, with suggestions on how to "work" it. B. F. GOODRICH RUBBER COMPANY.

3. TRANSFORMERS—A booklet giving data on input and output transformers. PACENT ELECTRIC COMPANY.

4. RESISTANCE-COUPLED AMPLIFIERS—A general discussion of resistance coupling with curves and circuit diagrams. COLE RADIO MANUFACTURING COMPANY.

5. CARBORUNDUM IN RADIO—A book giving pertinent data on the crystal as used for detection, with hook-ups, and a section giving information on the use of resistors. THE CARBORUNDUM COMPANY.

6. B-ELIMINATOR CONSTRUCTION—Constructional data on how to build. AMERICAN ELECTRIC COMPANY.

7. TRANSFORMER AND CHOKE-COUPLED AMPLIFICATION—Circuit diagrams and discussion. ALL-AMERICAN RADIO CORPORATION.

8. RESISTANCE UNITS—A table sheet of resistance units and their application. WARD-LEONARD ELECTRIC COMPANY.

9. VOLUME CONTROL—A leaflet showing circuits for distortionless control of volume. CENTRAL RADIO LABORATORIES.

10. VARIABLE RESISTANCE—As used in various circuits. CENTRAL RADIO LABORATORIES.

11. RESISTANCE COUPLING—Resistors and their application to audio amplification, with data. DEJUR PRODUCTS COMPANY.

12. DISTORTION AND WHAT CAUSES IT—Hook-ups of resistance-coupled amplifiers with standard circuits. ALLEN-BRADLEY COMPANY.

13. B-ELIMINATOR AND POWER AMPLIFIER—Instructions for assembly and operation using Raytheon tube. GENERAL RADIO COMPANY.

15A. B-ELIMINATOR AND POWER AMPLIFIER—Instructions for assembly and operation using an R. C. A. rectifier. GENERAL RADIO COMPANY.

16. VARIABLE CONDENSERS—A description of the functions and characteristics of variable condensers with curves and specifications for their application to complete receivers. ALLEN D. CARDWELL MANUFACTURING COMPANY.

17. BAKELITE—A description of Various uses of bakelite in radio, its manufacture, and its properties. BAKELITE CORPORATION.

19. POWER SUPPLY—A discussion on power supply with particular reference to lamp-socket operation. Theory and constructional data for building power supply devices. ACME APPARATUS COMPANY.

20. AUDIO AMPLIFICATION—A booklet containing data on audio amplification together with hints for the construction. ALL AMERICAN RADIO CORPORATION.

21. HIGH-FREQUENCY DRIVER AND SHORT-WAVE WAVEMETER—Constructional data and application. BURGESS BATTERY COMPANY.

46. AUDIO-FREQUENCY CHOKES—A pamphlet showing positions in the circuit where audio-frequency chokes may be used. SAMSON ELECTRIC COMPANY.

47. AUDIO-FREQUENCY CHOKES—Circuit diagrams illustrating the use of chokes to keep out radio-frequency currents from definite points. SAMSON ELECTRIC COMPANY.

48. TRANSFORMER AND IMPEDANCE DATA—Tables giving the mechanical and electrical characteristics of transformers and impedances, together with a short description of their use in the circuit. SAMSON ELECTRIC COMPANY.

49. BYPASS CONDENSERS—A description of the manufacture of bypass and filter condensers. LESLIE F. MUTER COMPANY.

50. AUDIO MANUAL—Fifty questions which are often asked regarding audio amplification, and their answers. AMERTRAN SALES COMPANY, INCORPORATED.

51. SHORT-WAVE RECEIVER—Constructional data on a receiver which, by the substitution of Various coils, may be made to tune from a frequency of 16,660 kc. (18 meters) to 1999 kc. (150 meters). SILVER-MARSHALL, INCORPORATED.

52. AUDIO QUALITY—A booklet dealing with audio-frequency amplification of Various kinds and the application to well-known circuits. SILVER-MARSHALL, INCORPORATED.

55. VARIABLE CONDENSERS—A bulletin giving an analysis of various condensers together with their characteristics. GENERAL RADIO COMPANY.

57. FILTER DATA—Facts about the filtering of direct current supplied by means of motor-generator outfits used with transmitters. ELECTRIC SPECIALTY COMPANY.

59. RESISTANCE COUPLING—A booklet giving some general information on the subject of radio and the application of resistors to a circuit. DAVEN RADIO CORPORATION.

60. RESISTORS—A pamphlet giving some information on resistors which are capable of dissipating considerable energy; also data on the ordinary resistors used in resistance-coupled amplification. THE CRESCENT RADIO SUPPLY COMPANY.

62. RADIO-FREQUENCY AMPLIFICATION—Constructional details of a five-tube receiver using a special design of radio-frequency transformer. CANFIELD RADIO MFG. COMPANY.

63. FIVE-TUBE RECEIVER—Constructional data on building a receiver. KARO PRODUCTS, INCORPORATED.

64. AMPLIFICATION WITHOUT DISTORTION—Data and curves illustrating the use of various methods of amplification. ACME APPARATUS COMPANY.

66. SUPER-HETRODYNE—Constructional details of a seven-tube set. G. C. EVANS COMPANY.

70. IMPROVING THE AUDIO AMPLIFIER—Data on the characteristics of audio transformers, with a circuit diagram showing where chokes, resistors, and condensers can be used. AMERICAN TRANSFORMER COMPANY.

72. PLATE SUPPLY SYSTEM—A wiring diagram and layout plan for a plate supply system to be used with a power amplifier. Complete directions for wiring are given. AMERTRAN SALES COMPANY.

80. FIVE-TUBE RECEIVER—Data are given for the construction of a five-tube tuned radio-frequency receiver. Complete instructions, list of parts, circuit diagram, and template are given. ALL-AMERICAN RADIO CORPORATION.

81. BETTER TUNING—A booklet giving much general information on the subject of radio reception with specific illustrations. Primarily for the non-technical home constructor. BREMER-TULLY MANUFACTURING COMPANY.

82. SIX-TUBE RECEIVER—A booklet containing photographs, instructions, and diagrams for building a six-tube shielded receiver. SILVER-MARSHALL, INCORPORATED.

83. SOCKET POWER DEVICE—A list of parts, diagrams, and templates for the construction and assembly of socket power devices. JEFFERSON ELECTRIC MANUFACTURING COMPANY.

84. FIVE-TUBE EQUAMATIC—Panel layout, circuit diagrams, and instructions for building a five-tube receiver, together with data on the operation of tuned radio-frequency transformers of special design. KARAS ELECTRIC COMPANY.

85. FILTER—Data on a high-capacity electrolytic condenser used in filter circuits in connection with a socket power supply unit, are given in a pamphlet. THE ABOX COMPANY.

86. SHORT-WAVE RECEIVER—A booklet containing data on a short-wave receiver as constructed for experimental purposes. THE ALLEN D. CARDWELL MANUFACTURING CORPORATION.

88. SUPER-HETERODYNE CONSTRUCTION—A booklet giving full instructions, together with a blue print and necessary data, for building an eight-tube receiver. THE GEORGE W. WALKER COMPANY.

89. SHORT-WAVE TRANSMITTER—Data and blue prints are given on the construction of a short-wave transmitter, together with operating instructions, methods of keying, and other pertinent data. RADIO ENGINEERING LABORATORIES.

90. IMPEDANCE AMPLIFICATION—The theory and practice of a special type of dual-impedance audio amplification are given. ALDEN MANUFACTURING COMPANY.

93. B-SOCKET POWER—A booklet giving constructional details of a socket-power device using either the BH or 313 type rectifier. NATIONAL COMPANY, INCORPORATED.

94. POWER AMPLIFIER—Constructional data and wiring diagrams of a power amplifier combined with a B-supply unit are given. NATIONAL COMPANY, INCORPORATED.

100. A, B, AND C SOCKET-POWER SUPPLY—A booklet giving data on the construction and operation of a socket-power supply using the new high-current rectifier tube. THE Q. R. S.-MUSIC COMPANY.

101. USING CHOKES—A folder with circuit diagrams of the more popular circuits showing where chokes coils may be placed to produce better results. SAMSON ELECTRIC COMPANY.

ACCESSORIES

22. A PRIMER OF ELECTRICITY—Fundamentals of electricity with special reference to the application of dry cells to radio and other uses. Constructional data on buzzers, automatic switches, alarms, etc. NATIONAL CARBON COMPANY.

23. AUTOMATIC RELAY CONNECTIONS—A data sheet showing how a relay may be used to control A and B circuits. YAXLEY MANUFACTURING COMPANY.

25. ELECTROLYTIC RECTIFIER—Technical data on a new type of rectifier with operating curves. KODEL RADIO CORPORATION.

26. DRY CELLS FOR TRANSMITTERS—Actual tests given; well illustrated with curves showing exactly what may be expected of this type of B power. BURGESS BATTERY COMPANY.

27. DRY-CELL BATTERY CAPACITIES FOR RADIO TRANSMITTERS—Characteristic curves and data on discharge tests. BURGESS BATTERY COMPANY.

28. B BATTERY LIFE—Battery life curves with general curves on tube characteristics. BURGESS BATTERY COMPANY.

29. HOW TO MAKE YOUR SET WORK BETTER—A nontechnical discussion of general radio subjects with hints on how reception may be bettered by using the right tubes. UNITED RADIO AND ELECTRIC CORPORATION.

30. TUBE CHARACTERISTICS—A data sheet giving constants of tubes. C. E. MANUFACTURING COMPANY.

31. FUNCTIONS OF THE LOUD SPEAKER—A short, nontechnical general article on loud speakers. AMPLION CORPORATION OF AMERICA.

32. METERS FOR RADIO—A catalogue of meters used in radio, with connecting diagrams. BURTON-ROGERS COMPANY.

33. SWITCHBOARD AND PORTABLE METERS—A booklet giving dimensions, specifications, and shunts used with various meters. BURTON-ROGERS COMPANY.

34. COST OF B BATTERIES—An interesting discussion of the relative merits of various sources of B supply. HARTFORD BATTERY MANUFACTURING COMPANY.

35. STORAGE BATTERY OPERATION—An illustrated booklet on the care and operation of the storage battery. GENERAL LEAD BATTERIES COMPANY.

36. CHARGING A AND B BATTERIES—Various ways of connecting up batteries for charging purposes. WESTINGHOUSE UNION BATTERY COMPANY.

37. CHOOSING THE RIGHT RADIO BATTERY—Advice on what dry cell battery to use; their application to radio, with wiring diagrams. NATIONAL CARBON COMPANY.

53. TUBE REACTIVATOR—Information on the care of vacuum tubes, with notes on how and when they should be reactivated. THE STERLING MANUFACTURING COMPANY.

54. ARRESTERS—Mechanical details and principles of the vacuum type of arrester. NATIONAL ELECTRIC SPECIALTY COMPANY.

55. CAPACITY CONNECTOR—Description of a new device for connecting up the Various parts of a receiving set, and at the same time providing bypass condensers between the leads. KURZ-KASCH COMPANY.

(Continued on page 396)

68. CHEMICAL RECTIFIER—Details of assembly, with wiring diagrams, showing how to use a chemical rectifier for charging batteries. CLEVELAND ENGINEERING LABORATORIES COMPANY.

69. VACUUM TUBES—A booklet giving the characteristics of the Various tube types with a short description of where they may be used in the circuit. RADIO CORPORATION OF AMERICA.

77. TUBES—A booklet for the beginner who is interested in Vacuum Tubes. A non-technical consideration of the Various elements in the tube as well as their position in the receiver. CLEARTRON VACUUM TUBE COMPANY.

87. TUBE TESTER—A complete description of how to build and how to operate a tube tester. BURTON-ROGERS COMPANY.

91. VACUUM TUBES—A booklet giving the characteristics and uses of Various types of tubes. This booklet may be obtained in English, Spanish, or Portuguese. DEFOREST RADIO COMPANY.

94. RESISTORS FOR A. C. OPERATED RECEIVERS—A booklet giving circuit suggestions for building a.c. operated receivers, together with a diagram of the circuit used with the new 400-mil ampere rectifier tube. CARTER RADIO COMPANY.

97. HIGH-RESISTANCE VOLTMETERS—A folder giving information on how to use a high-resistance voltmeter, special consideration being given the Voltage measurement of socket-power devices. WESTINGHOUSE ELECTRIC & MANUFACTURING COMPANY.

102. RADIO POWER BULLETINS—Circuit diagrams, theory constants, and trouble-shooting hints for units employing th BH or B rectifier tubes. RAYTHEON MANUFACTURING COMPANY.

105. A. C. TUBES—The design and operating characteristics of a new a.c. tube. Five circuit diagrams show how to convert well-known circuits. SOVREIGN ELECTRIC & MANUFACTURING COMPANY.

MISCELLANEOUS

38. LOG SHEET—A list of broadcasting stations with columns for marking down dial settings. U. S. L. RADIO, INCORPORATED.

41. BABY RADIO TRANSMITTER OF 9XH-9EK—Description and circuit diagrams of dry-cell operated transmitter. BURGESS BATTERY COMPANY.

42. ARCTIC RADIO EQUIPMENT—Description and circuit details of short-wave receiver and transmitter used in Arctic exploration. BURGESS BATTERY COMPANY.

43. SHORT-WAVE RECEIVER OF 9XH-9EK—Complete directions for assembly and operation of the receiver. BURGESS BATTERY COMPANY.

44. ALUMINUM FOR RADIO—A booklet containing much radio information with hook-ups of basic circuits, with inductance-capacity tables and other pertinent data. ALUMINUM COMPANY OF AMERICA.

45. SHIELDING—A discussion of the application of shielding in radio circuits with special data on aluminum shields. ALUMINUM COMPANY OF AMERICA.

58. HOW TO SELECT A RECEIVER—A commonsense booklet describing what a radio set is, and what you should expect from it, in language that any one can understand. DAY-FAN ELECTRIC COMPANY.

67. WEATHER FOR RADIO—A Very interesting booklet on the relationship between weather and radio reception, with maps and data on forecasting the probable results. TAYLOR INSTRUMENT COMPANIES.

73. RADIO SIMPLIFIED—A non-technical booklet giving pertinent data on Various radio subjects. Of especial interest to the beginner and set owner. CROSLEY RADIO CORPORATION.

74. THE EXPERIMENTER—A monthly publication which gives technical facts, valuable tables, and pertinent information on Various radio subjects. Interesting to the experimenter and to the technical radio man. GENERAL RADIO COMPANY.

75. FOR THE LISTENER—General suggestions for the selecting, and the care of radio receivers. VALLEY ELECTRIC COMPANY.

76. RADIO INSTRUMENTS—A description of Various meters used in radio and electrical circuits together with a short discussion of their uses. JEWELL ELECTRICAL INSTRUMENT COMPANY.

78. ELECTRICAL TROUBLES—A pamphlet describing the use of electrical testing instruments in automotive work combined with a description of the cadmium test for storage batteries. Of interest to the owner of storage batteries. BURTON-ROGERS COMPANY.

95. RESISTANCE DATA—Successive bulletins regarding the use of resistors in Various parts of the radio circuit. INTERNATIONAL RESISTANCE COMPANY.

96. VACUUM TUBE TESTING—A booklet giving pertinent data on how to test Vacuum tubes with special reference to a tube testing unit. JEWELL ELECTRICAL INSTRUMENT COMPANY.

98. COPPER SHIELDING—A booklet giving information on the use of shielding in radio receivers, with notes and diagrams showing how it may be applied practically. Of special interest to the home constructor. THE COPPER AND BRASS RESEARCH ASSOCIATION.

99. RADIO CONVENIENCE OUTLETS—A folder giving diagrams and specifications for installing loud speakers in various locations at some distance from the receiving set. YAXLEY MANUFACTURING COMPANY.

What Kit Shall I Buy?

THE list of kits herewith is printed as an extension of the scope of the Service Department of RADIO BROADCAST. It is our purpose to list here the technical data about kits on which information is available. In some cases, the kit can be purchased from your dealer complete; in others, the descriptive booklet is supplied for a small charge and the parts can be purchased as the buyer likes. The Service Department will not undertake to handle cash remittances for parts, but when the coupon on page 308 is filled out, all the information requested will be forwarded.

201. SC FOUR-TUBE RECEIVER—Single control. One stage of tuned radio frequency, regenerative detector, and two stages of transformer-coupled audio amplification. Regeneration control is accomplished by means of a variable resistor across the tickler coil. Standard parts; cost approximately $18.81.

202. SC-11 FIVE-TUBE RECEIVER—Two stages of tuned radio frequency, detector, and two stages of transformer-coupled audio. Two tuning controls. Volume control consists of potentiometer grid bias on r.f. tubes. Standard parts cost approximately $60.35.

203. "HI-Q" KIT—A five-tube tuned radio-frequency set having two radio stages, a detector, and two transformer-coupled audio stages. A special method of coupling in the r.f. stages tends to make the amplification more nearly equal over the entire band. Price $65.05 without cabinet.

204. R. G. S. KIT—A four-tube inverse reflex circuit, having the equiValent of two tuned radio-frequency stages, detector, and three audio stages. Two controls. Price $69.70 without cabinet.

205. PIERCE AIRO KIT—A six-tube single-dial receiver two stages of radio-frequency amplification, detector, and three stages of resistance-coupled audio. Volume control accomplished by Variation of filament brilliancy of r.f. tubes or by adjusting compensating condensers. Complete chassis assembled but not wired costs $42.50.

206. H & H-T. & F. ASSEMBLY—A five-tube set; three tuning dials, two steps of radio frequency, detector, and a transformer-coupled audio stages. Complete except for baseboard, panel, screws, wires, and accessories. Price $35.00.

207. PREMIER FIVE-TUBE RECEIVER—Two stages of tuned radio frequency, detector, and two steps of transformer-coupled audio. Three dials. Parts assembled but not wired. Price complete, except for cabinet, $35.00.

208. "QUADRAFORMER VI"—A six-tube set with two tuning controls. Two stages of tuned radio frequency using specially designed shielded coils, a detector, one stage of transformer-coupled audio, and two stages of resistance-coupled audio. Gain control by means of tapped primaries on the r.f. transformers. Essential kit consists of four r.f. transformers, two dual condensers, three small condensers, three choke coils, one 500,000-ohm resistor three 100,000-ohm resistors, and a set of color charts and diagrams. Price $41.50.

209. GEN-RAL FIVE-TUBE SET—Two stages of tuned radio frequency, detector, and two transformer-coupled audio stages. Volume is controlled by a resistor in the plate circuit of the r.f. tubes. Uses a special r.f. coil ("Duo-Former") with figure eight winding. Parts mounted but not wired, price $37.50.

210. BREMER-TULLY POWER-SIX—A six-tube, dual-control set; three stages of neutralized tuned radio frequency, detector, and two transformer-coupled audio stages. Resistances in the grid circuit together with a phase shifting arrangement are used to prevent oscillation. Volume control accomplished by variation of B potential on r.f. tube. Essential kit consists of four r.f. transformers, two dual condensers, three small condensers, three choke coils, one 500,000-ohm resistor three 100,000-ohm resistors, and a set of color charts and diagrams. Price $41.50.

211. BRUNO DRUM CONTROL RECEIVERS—How to apply a drum tuning unit to such circuits as the three-tube regenerative receiver, four-tube Browning-Drake, five-tube Diamond-of-the-Air, and the "Grand" 6.

212. INFRADYNE AMPLIFIER—A three-tube intermediate-frequency amplifier for the super-heterodyne and other special receivers, tuned to 3400 kc. (86 meters). Price $25.00.

213. RADIO BROADCAST "LAB" RECEIVER—A four-tube dual-control receiver with one stage of Rice neutralized radio frequency, regenerative detector (capacity controlled), and two stages of transformer-coupled audio. Approximate price, $78.15.

214. LC-27—A five-tube set with two stages of tuned radio frequency, a detector, and two stages of transformer-coupled audio. Special coils and special means of neutralizing are employed. Output device. Price $85.20 without cabinet.

215. LOFTIN-WHITE—A five-tube set with two stages of radio frequency, especially designed to give equal amplification at all frequencies, a detector, and two stages of transformer-coupled audio. Two controls. Output device. Price $85.10.

216. K.H.-27—A six-tube receiver with two stages of neutralized tuned radio frequency, a detector, three stages of choke-coupled audio, and an output device. Two controls. Price $86.00 without cabinet.

217. AERO SHORT-WAVE KIT—Three plug-in coils designed to operate with a regenerative detector circuit and having a frequency range of from 19,900 to 2300 kc. (15 to 130 meters). Coils and plug only, price $12.50.

218. DIAMOND-OF-THE-AIR—A five-tube set having one stage of tuned-radio frequency, a regenerative detector, one stage of transformer-coupled audio, and two stages of resistance-coupled audio. Volume control through regeneration. Two tuning dials.

219. NORDEN-HAUCK SUPER 10—Ten tubes: five stages of tuned radio frequency, detector, and four stages of choke and transformer-coupled audio frequency. Two controls. Price $395.40.

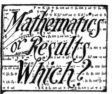

WHAT KIT SHALL I BUY? (Continued)

220. BROWNING-DRAKE—Five tubes: one stage tuned radio frequency (Rice neutralization), regenerative detector (tickler control), three stages of audio (special combination of resistance- and impedance-coupled audio). Two controls

221. LR4 ULTRADYNE—Nine-tube super-heterodyne; one stage of tuned radio frequency, one modulator, one oscillator, three intermediate-frequency stages, detector, and two transformer-coupled audio stages.

222. GREIFF MULTIPLEX—Four tubes (equivalent to six tubes); one stage of tuned radio frequency, one stage of transformer-coupled radio frequency, crystal detector, two stages of transformer-coupled audio, and one stage of impedance-coupled audio. Two controls. Price complete parts. $50.00.

223. PHONOGRAPH AMPLIFIER—A five-tube amplifier device having an oscillator, a detector, one stage of transformer-coupled audio, and two stages of impedance-coupled audio. The phonograph signal is made to modulate the oscillator in much the same manner as an incoming signal from an antenna.

USE THIS COUPON FOR KITS

RADIO BROADCAST SERVICE DEPARTMENT
Garden City, New York.

Please send me information about the following kits indicated by number:

....
....
....
....
....

Name ....

Address ....
    (Number)   (Street)

    ....
    (City)   (State)

ORDER BY NUMBER ONLY. This coupon must accompany each order.
RB1027

A KEY TO RECENT RADIO ARTICLES
By E. G. SHALKHAUSER

THIS is the twenty-third installment of references to articles which have appeared recently in various radio periodicals. Each separate reference should be cut out and pasted on 4" x 6" cards for filing, or pasted in a scrap book either alphabetically or numerically. An outline of the Dewey Decimal System (employed here) appeared last in the January RADIO BROADCAST.

R344.4 SHORT-WAVE GENERATORS.   TRANSMITTER.
QST, April, 1927. Pp. 15-16.   *Short-Wave.*
"A 15-Meter Commercial Station-2 XS."
The General Electric station 2 XS, operating on 15 meters, built for the purpose of transmitting to Buenos Aires 365 days in the year, regardless of weather conditions, is described. The set is rated at 7 kw. and radiates its energy from a horizontal doublet with reflector.

R134. DETECTOR ACTION   DETECTOR ACTION.
QST, April, 1927. Pp. 17-18.
"Which Is the Detector Tube?" L. W. Hatry.
The difference in operation of detectors and amplifier in standard radio circuits is discussed. Whether a tube functions as a detector as an amplifier, or both, depends upon various factors in the circuit, such as grid return, connection of grid leak, etc.

R330. ELECTRON TUBES.   ELECTRON TUBES.
QST, April, 1927. Pp. 26-30.   *UX-240.*
"CX-340—UX-240," R. S. Kruse.
In making a comparison between the UX-201-a and the new UX-240 amplifier tube, the writer presents the details upon which good amplification depends, namely: (1) The amplification constant of the tube; (2) the plate impedance of the tube; (3) the impedance of the transformer. The UX-240 is said to be excellent for c. w. and for resistance-coupled broadcast sets. As a detector, this new tube is said to eliminate considerable interference because of its peak characteristics.

R251.2. THERMO-ELEMENT.   THERMO-COUPLE.
QST, April, 1927. Pp. 31-32.   DEVICE.
"A Sensitive Thermo-Couple." B. J. Chromy.
A very inexpensive and useful thermo-couple, made of tellurium and platinum wire, is described. The method of constructing the device and calibrating its range, together with a wiring diagram and experimental data, are given.

R 371.3. ELECTROLYTIC RECTIFIERS.   RECTIFIERS.
QST, April, 1927. Pp. 34-38.   *Dry Electrolytic.*
"Developments in Dry Electrolytic Rectifiers," R. S. Kruse.
A new type of rectifier, using a combination of copper, magnesium, and composition discs (the last mentioned consisting of zinc selenide and copper selenide, among other things), has been perfected for A battery chargers up to 3 amperes output, and also for A battery substitutes, as stated. Complete details concerning mechanical and electrical characteristics of this new type of dry rectifier are given.

R007.1. UNITED STATES LAWS AND REGULATIONS. LAWS.
QST, April, 1927. Pp. 39-44.   *Radio Act, 1927.*
"The New Radio Law."
The complete text of the new Radio Law, as enacted by Congress in February, 1927, is printed for the benefit of the readers of *QST*.

R113.5. METEOROLOGICAL.   METEOROLOGY.
QST, April, 1927. Pp. 45-46.   *Distance Calculations.*
"How Far Is It?" C. C. Knight.
A method whereby distances on the earth's surface from place to place may be determined with accuracy, is outlined in terms of spherical trigonometric formulas. As stated, the plan is very simple and the average person may obtain very good results using these formulas. Distance on flat maps are very inaccurate ordinarily, since it is difficult to make allowances for the curvature of the earth.

R361. ELECTRON-TUBE VOLTMETERS.   VOLTMETER.
QST, April, 1927. Pp. 47-53.   *Electron-Tube.*
"The Most Useful Meter," R. F. Shea.
A comparison made between ordinary types of meters used in radio measurements and the vacuum-tube voltmeter shows how far superior the latter is when making accurate tests for small current and Voltage Values. Five types of commercial Vacuum-tube Voltmeters, their apparatus, and circuit arrangement, and their operation, are outlined in detail. The following are some of the many measurements that can be made with this handy instrument: (1) Obtaining gain through a radio-frequency or audio-frequency amplifier; (2) Checking the wave form of an oscillator; (3) Measuring high impedance; (4) Employing it as a wavemeter of high precision.

R381. CONDENSERS.   CONDENSERS.
QST, April, 1927. Pp. 53-57.   *Electrolytic.*
"Electrolytic Filter Condensers," L. F. Lenck.
The operation of electrolytic filter condensers, employing aluminum "pie-plates" and ammonium phosphate in their make-up in smoothing out the rectified a. c., is outlined. The method of forming plates and precautions to take in getting good results are discussed in detail.

R344.4. SHORT-WAVE GENERATORS.   TRANSMITTER.
RADIO BROADCAST, April, 1927. Pp. 570-573. *Short-Wave.*
"A 20—40—80—Meter Transmitter," K. Henney.
Supplementing data given in the April, 1926, and Nov. 1926, RADIO BROADCAST, on the construction and operation of a short-wave transmitter, additional information is presented concerning a similar set which operates from the power mains, using two 216-a rectifiers and two UX-210 oscillators. The circuit is of the well-known Hartley pattern. Considerable information on details of construction is presented so that any one interested in transmitters may set up the outfit. Data on DeForest transmitter tubes are also given.

R351. SIMPLE OSCILLATORS.   OSCILLATOR.
QST, March, 1927. Pp. 23-27.   *Calibrator.*
"Quartz Crystal Calibrators," A. Crossley.
Two circuits, in which the quartz crystal may be used, to set up continuous oscillations, are discussed. These employ either a crystal placed between plate and grid or between grid and filament of a vacuum tube, the latter method being preferred. With large inductance and small capacity in the phase adjusting circuit of the plate, many harmonics are said to be obtainable. The crystal calibrator, as described, may be used to measure accurately inductances, capacities, frequency meters, transmitters, etc.

R536. MINING.   MINING AND RADIO.
Radio. Feb., 1927. Pp. 28-ff.
"Experiments in Radio Prospecting." J. G. Alverson.
The author gives some of his findings regarding the propagation of radio waves through the crust of the earth, showing how the wave-front of the electromagnetic wave changes with a change in rock, earth, and material. Data are presented of many observations made with a small transmitter operating on a wavelength of 1000 meters (300 kc.) and using an ordinary type receiver.

R343.5. SUPER-HETERODYNE SETS.   SUPER-HETERODYNE.
Radio. Feb., 1927. Pp. 25-ff.   *Best 1927 Model.*
"The 1927 Model of Best's Super-Heterodyne," G. M. Best.
The improved model of Best's super-heterodyne receiver is outlined for the set builder. The set consists of nine tubes, with two tuning dials, and is completely shielded.

R082. MAPS AND CHARTS.   CHART.
Radio. Feb., 1927. Pp. 140-141.   L. C. / Data.
"Chart for Radio Circuit Calculations," E. L. Hall.
A graphic chart, showing the relation between frequency, inductance, and capacitance for values as used in radio circuits, is drawn for purposes of rapid calculation. The accuracy of results obtainable may be within several per cent. of the mathematical calculation, as stated. The range may be extended beyond those actually shown by a simple process.

R330. GENERATING APPARATUS;   TRANSMITTER.
TRANSMITTING SETS.
QST, March, 1927. Pp. 33-37.   *Flexible.*
"A Flexible Transmitter," F. J. Marco.
A transmitter is described which uses the Armstrong tuned-grid tuned-plate circuit. It is said to be capable of a quick change to either the 20-, the 40-, or the 80-meter band (1500-, 750-, or 3750-kc. band) by means of plug-in coils. Tubes giving an output power up to 7.5 watts may be used with the apparatus. Complete circuit diagrams, methods of tuning, and operation data are given.

The Majestic

Most popular "B" Power Unit for radio sets in the world

The Super "B" illustrated above is only $29.50, complete with Majestic Super~Power B~Rectifier Tube.

GRIGSBY ~ GRUNOW ~ HINDS ~ CO.   4572 ARMITAGE AVE. CHICAGO~ILL

R113.5. METEOROLOGICAL. METEOROLOGY.
Proc. I. R. E. Feb., 1927, Pp. 83–97.
"The Correlation of Radio Reception with Solar Activity and Terrestrial Magnetism," G. W. Pickard.
Atmospheric changes and the resulting magnetic storms affecting radio reception are discussed. Atmospheric changes are said to be due to radiations and emissions from the sun.
The author compares observed radio reception data, obtained over a continuous period, with meteorological data, and finds a close correlation of sunspot disturbances and radio reception on the upper broadcast wave spectra. Graphs drawn from observed data obtained from station wwss show that, when average results are compared with magnetic storms, both variations appear to be more or less in phase.

R201. GENERAL METHODS AND APPARATUS. MEASUREMENTS.
Proc. I. R. E. Feb., 1927, Pp. 99–111. *Receiver.*
"Importance of Laboratory Measurements in the Design of Radio Receivers," W. A. MacDonald.
A series of thirteen fundamental tests are outlined for determining individual and overall characteristics of commercial receivers. A series of graphs gives the results obtained from a typical circuit and depicts to what degree a high precision standard may be used.

R134. DETECTOR ACTION. DETECTOR.
Proc. I. R. E. Feb., 1927. Pp. 113–153. *Action.*
"A Theoretical and Experimental Investigation of Detection for Small Signals," E. L. Chaffee and G. H. Browning.
Data are given on the detecting properties of diode and triode valves as used in radio circuits. The mathematical presentation is expressed in terms of the circuit impedances and the first and second partial differential coefficients of the static characteristic curves of the device, taken at the points on the curves determined by the steady polarizing voltages. It is then assumed that the impressed signal is so small that for any given steady voltages these coefficients can be assumed constant within the range of the variation, due to the signal voltage.

R430. INTERFERENCE ELIMINATION. INTERFERENCE.
QST. March, 1927. Pp. 9–14. *Elimination.*
"Cures for 'Power Leaks,'" R. S. Kruse.
The article states that power leaks and other sources of high-frequency interference may be eliminated by proper filtering. The "Tobe," the "Dayfan," and other interference eliminating devices are described.

R142.5. INDUCTIVE COUPLED CIRCUITS. TRANSFORMERS.
QST. March, 1927. Pp. 20–22. *Tuned R.F.*
"The Theory of a Tuned R. F. Transformer," G. H. Browning and F. H. Drake.
A mathematical presentation, supplemented by experimental curves of tuned radio-frequency transformers, is given. The results show that the secondary inductance should be made as large as possible, the losses made very small, and the primary–secondary capacity kept as low as possible, in order to obtain maximum efficiency.

R140. RADIO CIRCUITS. MASTER-OSC.
QST. March, 1927. Pp. 38-ff. AND POWER-AMP.
"How Our Tube Circuits Work. Part 4." CIRCUITS.
R. S. Kruse.
In this fourth of a series of articles on radio circuits the writer takes up the many phases of oscillator circuits with and without amplifier hook-ups. A variety of subjects dealing with this type of circuit are taken up, as for instance "Wobbulation," its causes and remedy; steadiness of wave with and without the use of a crystal; the characteristics of the Armstrong circuit; single- and multi-stage amplifiers and their use as wave-changers in order to prevent troublesome feedbacks.

R300. MISCELLANEOUS. FUTURE
Popular Radio. March, 1927. Pp. 229-ff. OF RADIO.
"Radio in 1950 A. D.," Dr. Lee DeForest.
The author enumerates some of the future possibilities found in the stored-up energy of waves as utilized for radio transmission and reception. The newly discovered cosmic ray, having a wavelength of 0.0004 angstrom units, and traveling at a much greater speed than light waves, as stated by Millikan, is said to have many possibilities.

R142.7. AUDIO-FREQUENCY AMPLIFIERS. AMPLIFIERS.
Popular Radio. March, 1927. Pp. 253-ff. *Audio-*
"Audio Amplifiers," E. L. Bowles. *Frequency.*
The underlying principles covering audio-frequency reproduction and amplification are outlined in an elementary discussion. Causes and effects of distortion, at either the transmitting or receiving end, may be corrected, it is said, by proper design and construction of the apparatus.

R201.4. SHIELDING AND GROUNDING. SHIELDING.
Radio News. Feb., 1927. Pp. 988-ff.
"Shielding in Radio Receivers," M. L. Hartmann and H. Meagher.
A brief and elementary discussion on the theory of shielding of radio receivers is given. Some of the benefits of shielding are said to be increased amplification, increased selectivity, better neutralization, and decreased body-capacity effects.

R175.3 ELECTROLYTIC RECTIFIERS. RECTIFIERS.
The Transmitter. Dec. 1927., Pp. 10-ff. *Chemical.*
"Practical Chemical Rectifiers," C. R. Stedman.
In constructing a practical chemical rectifier, the writer suggests that pure aluminum should be used. Both the lead and aluminum plates, however, should be of the same size. The solution recommended is either bicarbonate of soda or ammonium phosphate. Constructional details and wiring diagrams are given.

R342.7. AUDIO-FREQUENCY AMPLIFIERS. AMPLIFIERS.
Radio World. Feb., 19, 1927, Pp. 5. *Audio-Frequency.*
"More Volume from Lower Ratio in Audio Transformers, a Frequent Condition," N. B. Humphrey.
In analyzing the importance of proper relation of audio transformer winding ratios to tube resistance, it is stated that consideration should be given the various factors involved from an engineering standpoint when selection is made. Usually, it is said, more volume and less distortion is obtained from properly constructed low-ratio transformers than those of high ratio.

R582. TRANSMISSION OF PHOTOGRAPHS. PHOTOGRAPH
Radio World. Feb., 26, 1927. Pp. 4–5. TRANSMISSION.
"First Television Hookups Elucidate Alexanderson and Baird Plans," H. Wall.
The Alexanderson and the Baird systems of television are shown in schematic diagram. The two systems are compared and the advantages and disadvantages outlined.

R113.4. IONIZATION; HEAVISIDE LAYER. HEAVISIDE
Radio World. Feb., 26, 1927. Pp. 8. LAYER.
"Radio Ceiling Again Verified."
Experiments conducted by the Carnegie Institute regarding the existence and probable nature of the Kennelly-Heaviside layer have established that such a layer actually exists at an altitude varying between 90 and 130 miles. Whether the radio waves are reflected or refracted from this upper layer is not definitely known. Fading may be explained on the assumption that interfering waves neutralize each other.

R030. TEXTBOOKS. TEXTBOOKS.
The New York Sun. Radio Section. Nov. 6, 1926. *Radio.*
"An Expert's 3-Foot Shelf of Radio Books," S. R. Winter.
The following is a list of textbooks and published material selected by Doctor Dellinger, of the Bureau of Standards, for the radio man:
1. "Signalling Through Space Without Wires,"—H. Hertz.
2. "The Principles of Electric Wave Telegraphy and Telephony,"—J. A. Fleming.
3. "Electric Waves,"—H. Hertz.
4. "The Principles Underlying Radio Communication,"—Bureau of Standards.
5. "Radio Instruments and Measurements,"—Circular 74, Bureau of Standards.
6. "Robinson's Manual of Radio Telegraphy,"—United States Navy.
7. "How to Become a Wireless Operator,"—Chas. B. Hayward.
8. "Direction and Position Finding,"—R. Keen, of Sheffield, England.
9. "I. C. S. Handbook for Radio Operators,"—International Correspondence Schools.
10. "Radio Telephony for Amateurs,"—S. Ballantine.
11. "Radio for Everybody,"—A. C. Lescarboura.
12. "Radio for All,"—H. Gernsback.
13. "Elements of the Mathematical Theory of Electricity and Magnetism,"—J. J. Thompson.
14. "Handbook of Technical Instruction for Wireless Telegraphists,"—C. Hawkhead and H. M. Dowsett.
15. "Wireless Telegraphy and Telephony,"—H. M. Dowsett.
16. "The Thermionic Vacuum Tube and Its Application," H. J. Von der Bijl.
17. "The Radio Amateur's Handbook,"—F. E. Handy.
18. "Yearbook of Wireless Telegraphy,"—Wireless Press, Ltd. (Published annually.)
19. "Electromagnetic Oscillations in Oscillators and Radio Telegraphy,"—J. Zenneck.
20. "Wireless Telegraphy and Telephony,"—Chas. R. Gibson.
21. "Radio Laws and Regulations of the United States,"—U. S. Department of Commerce.
22. "The Realities of Modern Science,"—John Mills.
23. "Electric Oscillations and Electric Waves,"—G. W. Pierce.
24. "Principles of Radio Communication,"—J. H. Morecroft.
25. "Principles of Radio Transmission and Reception with Antenna and Coil Aerials,"—Bureau of Standards.
26. "Methods of Measurement of Properties of Electrical Insulating Materials,"—Bureau of Standards.
27. "Handbook of Safety Rules for Radio Installations,"—Bureau of Standards.
28. "Wireless Telegraphy and Telephony,"—W. H. Eccles.
29. "Thermionic Tubes in Radio Telegraphy,"—John Scott-Taggart.
30. "Radio Broadcast's Knockout Receiver,"—Doubleday, Page and Company.
31. "How to Build Your Radio Receiver,"—Popular Radio Publishing Company.
32. "The Radio Service Bulletin,"—United States Department of Commerce. (Monthly).
33. "Introduction to Line Radio Communication,"—Signal Corps Pamphlet.
34. "Modern Radio Operation,"—J. O. Smith.
35. "The C. W. Manual,"—J. B. Dow.
36. "Radio for Amateurs,"—A. H. Verrill.
37. "Circular No. 24,"—Bureau of Standards.
38. "Copper Wire Tables,"—Bureau of Standards.
39. "The Radio Direction Finder and Its Application to Navigation,"—Bureau of Standards.
40. "Formulas and Tables for the Calculation of Mutual and Self-Inductance,"—Bureau of Standards.
41. "The Outline of Radio,"—J. V. L. Hogan.
42. "Qualitative Experiments in Radio Transmission,"—Bureau of Standards.

R134.8. REFLEX ACTION. REFLEX
QST. Feb., 1927. Pp. 21–27. ACTION.
"Developments in Tuned Inverse Duplex," David Grimes. Part 2.
In this second of two articles, the Inverse Duplex System of amplification in the radio and audio circuits, developed for duplexing, are discussed. For the audio stage, a transformer-resistance-transformer arrangement is said to be the most desirable and effective. In order to obtain equal amplification over the entire broadcast band, a radio frequency-audio frequency filter circuit, comprising a choke coil and a large fixed condenser, is connected between stages. Circuit diagrams and panel layouts are shown.

R344.5. ALTERNATING CURRENT SUPPLY. A.C. FOR
Radio Broadcast. Feb., 1927. Pp. 303–306. FILAMENTS.
"A.C. As a Filament-Supply Source," B. F. Miessner.
The writer discusses the problems encountered in lighting the filaments of radio vacuum tubes from the regular direct or alternating current obtained from the light socket. Filter systems, devised to eliminate the hum in the a.c. line, which presents many varying characteristics, as shown in graphs, are given. The 112 type tube, as compared to the 199, the 201-a type, and the wx-12 type, is considered the best tube to use for radio-and audio-frequency amplification in a.c. lighted filament circuits.

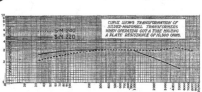

SM 87%
At 30 Cycles!

In this standardized black shielding case are housed the famous 220 audio and 221 output transformers, 222 output, 230 push-pull input, 231 push-pull output, 325 filament, 329, 329A and 330 power transformers, 331 Unichoke and 332 condenser bank. And a new super-power full wave ABC supply transformer is on the way —type 328, at $18.00, for one or two 216B or 281 tubes.

Silver-Marshall now offers two smaller size audio transformers for replacement work in old sets, wherever price and size is a consideration. Type 240 audio transformer is equal or superior to the majority of high-grade audio transformers, but does not reproduce frequencies below 80 cycles to the extent that the famous 220 does. Its single stage amplifier curve is shown above—in two stages, the 240's afford practically the same 5000 cycle cut-off as do 220's, although this is not evident in the single stage curve above. 241 output transformer offers the same low frequency compensation as type 221 and 222. Due to their small size, these transformers will fit in almost any of the older receivers, and once installed, will work wonders in tone quality improvement, for their performance nearly equals that of 220's and 221's. Size 3⅞″ high, 2½″ wide, 2⅜″ deep, weight 2 lbs. 4 ozs. each. Price, 240 audio $6.00, 241 output $5.00.

Laurence Cockaday, for the preferred audio amplifier for the LC-28 receiver uses two 240's and a 241 with an S-M power supply!

At 30 cycles, an S-M 220 audio transformer in a standard amplifier circuit gives 87% of the amplification obtained at 1000 cycles, while its curve is substantially flat from 100 to 1000 cycles. Above 2000 cycles, the curve for a single stage falls off gradually, while in a standard two stage amplifier circuit, the curve is substantially flat up to 5000 cycles above which frequency it falls off rapidly to keep static, heterodyne squeals and "set noise" at a minimum.

The above paragraph sums up at once the desirable characteristics of an audio amplifier for realistic recreation of broadcast programs, and the actual performance of S-M audio transformers. It is just this fact that has made 220's the choice of over half of the designers of the new 1927-1928 circuits, for engineers know that the short cut to the finest of quality is to use S-M audios. Experienced fans know this too, as is proven by the fact that 220's have outsold every other transformer in their class by a wide margin for over a year. And S-M audios are signally favored by being used in more broadcasting stations than any other types. WCAE, WBBM, WEBH, KFCR, WTAQ, KGDJ, WLBF, and many others. WCFL, the "Voice of Labor" checks quality of all programs with them. Nathaniel Baldwin, Inc., famous speaker experts, test with 220's and 221's.

Your guarantee of quality is to use S-M 220's and 221's in every circuit you build, and you'll find that over half the popular 1927 and 1928 circuits will give you just this same guarantee of quality. But S-M promises unconditionally that you can improve any set by using 220's and 221's, and backs the promise by the offer of your money back if 220's and 221's don't give you more satisfactory quality than you've ever heard before.

The 220 audio is the biggest value on the market, and its performance measures up to its 4 pound size. It contains more steel and copper than any other transformer—the measure of transformer merit. Price $8.00.

221 output transformer not only protects loud speakers against power tube plate currents, but compensates low frequencies for all loud speakers. Price $7.50, or with cord and tip jacks, No. 222, $8.00.

230 push-pull input and 231 push-pull output transformers are priced at $10.00 each.

The New Shielded Six Is Ready!

The Improved Shielded Six is ready, the very latest model of this excellent receiver which has over a year of successful and satisfying performance to its credit. The Improved model has vastly increased selectivity, greater distance getting ability, and the same fine tone that has made almost every builder say of the original, "That's the finest set I've ever heard!"

This year the Six offers the additional possibilities of push-pull amplification with 210 tubes for the man who wants the utmost. All in all, the Six deserves the reputation as the finest tuned R. F. kit you can buy, equalled only by $200 to $400 factory-built sets. Yet it's priced at but $95.00 for the complete kit, or $145.00 assembled, in cabinet, and guaranteed to satisfy you. S-M will be glad to tell owners how last year's model can be converted to the Improved Six, or push-pull 210 amplification installed with simple changes.

We can't tell you the whole story of new S-M developments, so if you'll just fill in the coupon, and mail it with 10c to cover postage, we'll send you free more up-to-the-minute advance radio information than you could buy in a text book.

Silver-Marshall Inc.,
838 West Jackson Blvd.   Chicago, U. S. A.

"RADIO BROADCAST'S" DIRECTORY OF MANUFACTURED RECEIVERS

¶ A coupon will be found on page 416. All readers who desire additional information on the receivers listed below need only insert the proper numbers in the coupon, mail it to the Service Department of Radio Broadca and full details will be sent. New sets are listed in this space each mon

KEY TO TUBE ABBREVIATIONS

99—60-mA. filament (dry cell)
01-A—Storage battery 0.25 amps. filament
12—Power tube (Storage battery)
71—Power tube (Storage battery)
16-B—Half-wave rectifier tube
80—Full-wave, high current rectifier
81—Half-wave, high current rectifier
Hmu—High-Mu tube for resistance-coupled audio
20—Power tube (dry cell)
10—Power Tube (Storage battery)
00-A—Special detector
13—Full-wave rectifier tube
26—Low-voltage high-current a. c. tube
27—Heater type a. c. tube

Direct Current Receivers

NO. 424. COLONIAL 26

Six tubes; 2 t. r. f. (01-A), detector (12), 2 transformer audio (01-A and 71). Balanced t. r. f. One to three dials. Volume control: antenna switch and potentiometer across first audio. Watts required: 120. Console size: 34 x 38 x 18 inches. Headphone connections. The filaments are connected in a series parallel arrangement. Price $250 including power unit.

NO. 425. SUPERPOWER

Five tubes: All 01-A tubes. Multiplex circuit. Two dials. Volume control: resistance in r. f. plate. Watts required: 30. Antenna: loop or outside. Cabinet sizes: table, 27 x 10 x 9 inches; console, 28 x 50 x 21. Prices: table, $135 including power unit; console, $390 including power unit and loud speaker.

A. C. Operated Receivers

NO. 508. ALL-AMERICAN 77, 88, AND 99

Six tubes; 3 t. r. f. (26), detector (27), 2 transformer audio (26 and 71). Rice neutralized t. r. f. Single drum tuning. Volume control: potentiometer in r. f. plate. Cabinet sizes: No. 77, 21 x 10 x 8 inches; No. 88 Hiboy, 25 x 38 x 18 inches; No. 99 console, 27½ x 43 x 20 inches. Shielded. Output device. The filaments are supplied by means of three small transformers. The plate supply employs a gas-filled rectifier tube. Voltmeter in a. c. supply line. Prices: No. 77, $150, including power unit; No. 88, $210 including power unit; No. 99, $285 including power unit and loud speaker.

NO. 509. ALL-AMERICAN 'DUET' ;" SEXTET"

Six tubes; 2 t. r. f. (99), detector (99), 3 transformer audio (99 and 12). Rice neutralized t.r.f. Two dials. Volume control: resistance in r.f. plate. Cabinet sizes: "Duet," 23 x 56 x 16½ inches; "Sextet," 22½ x 13½ x 15½ inches. Shielded. Output device. The 99 filaments are connected in series and supplied with rectified a.c., while 12 is supplied with raw a.c. The plate and filament supply uses gaseous rectifier tubes. Millammeter on power unit. Prices: "Duet," $160 including power unit; "Sextet," $220 including power unit and loud speaker.

NO. 511. ALL-AMERICAN 80, 90, AND 115

Five tubes; 2 t.r.f. (99), detector (99), 2 transformer audio (99 and 12). Rice neutralized t.r.f. Two dials. Volume control: resistance in r.f. plate. Cabinet sizes: No. 80, 23½ x 12½ x 15 inches; No. 90, 37½ x 12 x 12½ inches; No. 115 Hiboy, 24 x 41 x 15 inches. Coils individually shielded. Output device. See No. 509 for power supply. Prices: No. 80, $135 including power unit; No. 90, $145 including power unit and compartment; No. 115, $170 including power unit, compartment, and loud speaker.

NO. 510. ALL-AMERICAN 7

Seven tubes; 3 t.r.f. (26), 1 untuned r.f. (26), detector (27), 2 transformer audio (26 and 71). Rice neutralized t.r.f. One drum. Volume control: resistance in r.f. plate. Cabinet sizes: "Sovereign" console, 30½ x 60½ x 19 inches; "Lorraine" Hiboy, 25½ x 53½ x 17½ inches; "Forte" cabinet, 25½ x 13½ x 17½ inches. For filament and plate supply: See No. 508. Prices: "Sovereign" $460; "Lorraine" $360; "Forte" $270. All prices include power unit. First two include loud speaker.

NO. 403. ARGUS 250B

Six tubes; 2 t.r.f. (99), 1 untuned r.f. (99), detector (99), 2 transformer audio (99 and 12). Stabilized with grid resistances. Two dials. Volume control: resistance across 1st audio. Watts required: 100. Cabinet size: 35½ x 14½ x 10½ inches. Output device. The 99 filaments are connected in series and supplied with rectified a.c., while the 12 is run on raw a.c. The power unit requires two 16-B rectifier tubes, Milliammeter included in d.c. supply. Price $250.00 including self-contained power unit. Other models: No. 125, $125.00; console model, $375.00.

NO. 401. AMRAD AC9

Six tubes; 3 t.r.f. (99), detector (99), 2 transformer (99 and 12). Neutrodyne. Two dials. Volume control: resistance across 1st audio. Watts consumed: 50. Cabinet size: 27 x 9 x 11½ inches. The 99 filaments are connected in series and supplied with rectified a.c., while the 12 is run on raw a.c. The power unit, requiring two 16-B rectifiers, is separate and supplies A, B, and C current. Price $142 including power unit.

NO. 402. AMRAD AC5

Five tubes. Same as No. 401 except one less r.f. stage. Price $125 including power unit.

NO. 484. BOSWORTH, B5

Five tubes; 2 t.r.f. (26), detector (99), 2 transformer audio (special a.c. tubes). T.r.f. circuit. Two dials. Volume control: potentiometer. Cabinet size: 23 x 7 x 8 inches. Output device included. Price $175.

NO. 406. CLEARTONE 110

Five tubes; 2 t.r.f., detector, 2 transformer audio. All tubes a. c. heater type. One or two dials. Volume control: resistance in r. f. plate. Watts consumed: 40. Cabinet size: varies. The plate supply is built in the receiver and requires one rectifier tube. Filament supply through step down transformers. Prices range from $175 to $375 which includes 5 a.c. tubes and one rectifier tube.

NO. 407. COLONIAL 25

Six tubes; 2 t. r. f. (01-A), detector (99), 2 resistance audio (99). 1 transformer audio (10). Balanced t.r.f. circuit. One or three dials. Volume control: Antenna switch and potentiometer on 1st audio. Watts consumed: 100. Console size: 34 x 38 x 18 inches. Output device. All tube filaments are operated on a. c. except the detector which is supplied with rectified a.c. from the plate supply. The rectifier employs two 16-B tubes. Price $250 including built-in plate and filament supply.

NO. 507. CROSLEY 602 BANDBOX

Six tubes; 3 t.r.f. (26), detector (27), 2 transformer audio (26 and 71). Neutrodyne circuit. One dial, Cabinet size: 17½ x 5½ x 7½ inches. The heaters for the a.c. tubes and the 71 filament are supplied by windings in B unit transformers available to operate either on 25 or 60 cycles. The plate current is supplied by means of rectifier tube. Price $65 for set alone, power unit $60.

NO. 408. DAY-FAN 'DE LUXE"

Six tubes; 3 t.r.f., detector, 2 transformer audio. All 01-A tubes. One dial. Volume control: potentiometer across r.f. tubes. Watts consumed: 300. Console size: 30 x 40 x 20 inches. The filaments are connected in series and supplied with d.c. from a motor-generator set which also supplies B and C current. Output device. Price $350 including power unit.

NO. 409. DAYCRAFT 5

Five tubes; 2 t.r.f., detector, 2 transformer audio. All a. c. heater tubes. Reflexed t.r.f. One dial. Volume control: potentiometers in r.f. plate and 1st audio. Watts consumed: 135. Console size: 34 x 36 x 14 inches. Output device. The heaters are supplied by means of a small transformer. A built-in rectifier supplies B and C voltages. Price $170, less tubes. The following have one more r.f. stage and are not reflexed: Daycraft 6, $195; Dayrole 6, $235; Dayfan 6, $110. All prices less tubes.

NO. 469. FREED-EISEMANN NR11

Six tubes; 3 t.r.f. (01-A), detector (01-A), 2 transformer audio (01-A and 71). Neutrodyne. One dial. Volume control: potentiometer. Watts consumed: 150. Cabinet size: 19½ x 10 x 10½ inches. Shielded. Output device. A special power unit is included employing a rectifier tube. Price $225 including NR-411 power unit.

NO. 487. FRESHMAN 7F-AC

Six tubes; 3 t.r.f. (26), detector (27), 2 transformer audio (26 and 71). Equaphase circuit. One dial. Volume control: potentiometer across 1st audio. Console size: 24½ x 41½ x 15 inches. Output device. The filaments and heaters and B supply are all supplied by one power unit. The plate supply requires one 80 rectifier tube. Price $175 to $350, complete.

NO. 536. SOUTH BEND

Six tubes. One control. Sub-panel shielding. Binding Posts. Antenna: outdoor. Prices: table, $130, Baby Grand console, $195.

NO. 537. WALBERT 26

Six tubes; five Kellogg a.c. tubes and one 71. Two controls. Volume control: variable plate resistance. Isofarad circuit. Output device. Battery cable. Semi-shielded. Antenna: 50 to 75 feet. Cabinet size: 10½ x 29½ x 16½ inches. Prices: $215; with tubes, $250.

NO. 538. NEUTROWOUND, MASTER ALLECTRIC

Six tubes; 2 t.r.f. (01-A), detector (01-A), 2 audio (01-A and two 71 in push-pull amplifier). The 01-A tubes are in series, and are supplied from a 400-mA. rectifier. Two drum controls. Volume control: variable plate resistance. Output device. Shielded. Antenna: 50 to 100 feet. Price: $360.

NO. 545. NEUTROWOUND, SUPER ALLECTRIC

Five tubes; 2 t.r.f. (99), detector (99), 2 audio (99 and 71). The 99 tubes are in series and are supplied from an 85-mA. rectifier. Two drum controls. Volume control: variable plate resistance. Output device. Antenna: 75 to 100 feet. Cabinet size: 9 x 24 x 11 inches. Price: $150.

NO. 490. MOHAWK

Six tubes; 2 t.r.f., detector, 2 transformer audio. All tubes a.c heater type except 71 in last stage. One dial. Volume control: rheostat on r.f. Watts consumed: 40. Panel size: 12½ x 8½ inches. Output device. The heaters for the a.c tubes and the 71 filament are supplied by small transformers. The plate supply is of the built-in type using a rectifier tube. Prices range from $65 to $245.

NO. 413. MARTI

Six tubes: 2 t.r.f., detector, 3 resistance audio tubes a.c. heater type. Two dials. Volume co resistance in r.f. plate. Watts consumed: 38. Pane 7 x 21 inches. The built-in plate supply employ 16-B rectifier. The filaments are supplied by a transformer. Prices: table, $235 including tubes rectifier; console, $275 including tubes and rec console, $325 including tubes, rectifier, and loud sp

NO. 417 RADIOLA 28

Eight tubes; five type 99 and one type 20. control. Super-heterodyne circuit. C-battery co tions. Battery cable. Headphone connection. Ant loop. Set may be operated from batteries or fro power mains when used in conjunction with the 104 loud speaker. Prices: $260 with tubes, ba operation; $570 with model 104 loud speaker, operation.

NO. 540 RADIOLA 30-A

Receiver characteristics same as No. 417 except type 71 power tube is used. This model is design operate on either a. c. or d. c. from the power m The combination rectifier—power—amplifier unit two type 81 tubes. Model 100-A loud speaker is tained in lower part of cabinet. Either a short in or long outside antenna may be used. Cabinet 42½ x 29 x 17¾ inches. Price: $495.

NO. 541 RADIOLA 32

This model combines receiver No. 417 with the 104 loud speaker. The power unit uses two typ tubes and a type 10 power amplifier. Loop is compl enclosed and is revolved by means of a dial on the p Models for operation from a. c. or d. c. power m Cabinet size: 52 x 72 x 17¾ inches. Price: $895.

NO. 539 RADIOLA 17

Six tubes; 3 t. r. f. (26), detector (27), 2 transfo audio (26 and 27). One control. Illuminated Built-in power supply using type 80 rectifier. Ante 100 feet. Cabinet size: 25 1/16 x 7⅞ x 8⅛. Price: without accessories.

NO. 421. SOVEREIGN 238

Seven tubes of the a.c. heater type. Balanced Two dials. Volume control: resistance across 2nd a Watts consumed: 45. Console size: 37 x 52 x 15 in The heaters are supplied by a small a. c. transfor while the plate is supplied by means of rectified using a gaseous type rectifier. Price $325, inclu power unit and tubes.

NO. 517. KELLOGG 510, 511, AND 512

Seven tubes; 4 t.r.f., detector, 2 transformer a All Kellogg a.c. tubes. One control and special switch. Balanced. Volume control: special. Output vice. Shielded. Cable connection between power su unit and receiver. Antenna: 25 to 100 feet. Panel x 27½ inches. Prices: Model 510 and 512, consoles, complete. Model 511, consolette, $365 without speaker.

NO. 496. SLEEPER ELECTRIC

Five tubes; four 99 tubes and one 71. Two cont Volume control: rheostat on r.f. Neutralized. C Output device. Power supply uses two 16-B tu Antenna: 100 feet. Prices: Type 64, table, $160; T 65, table, with built-in loud speaker, $175; Type table, $175; Type 67, console, $235; Type 78, con $265.

NO. 522. CASE, 62 B AND 62 C

McCullough a.c. tubes. Drum control. Volume trol; variable high resistance in audio system. C-bat connections. Semi-shielded. Cable. Antenna: 100 Panel size: 7 x 21 inches. Prices: Model 62 B, com with a.c. equipment, $185; Model 62 C, complete a.c. equipment, $235.

NO. 523. CASE, 92 A AND 92 C

McCullough a.c. tubes. Drum control. Induc Volume control. Technidyne circuit. Shielded. Ca C-battery connections. Model 92 C contains ou device. Loop operated. Prices: Model 92 A, table, $ Model 92 C, console, $475.

Battery Operated Receivers

NO. 512. ALL-AMERICAN 44, 45, AND 66

Six tubes; 3 t.r.f. (01-A, detector) 01-A, 2 transfor audio (01-A and 71). Rice neutralized t.r.f. D control. Volume control: rheostat in r.f. Cabinet si No. 44, 21 x 10 x 8 inches; No. 55, 25 x 38 x 18 inc No. 66, 27½ x 43 x 20 inches. C-battery connecti Battery cable. Antenna: 75 to 125 feet. Prices: No. $70; No. 55, $125 including loud speaker; No. 66, $ including loud speaker.

NO. 428. AMERICAN C6

Five tubes; 2 t.r.f, detector, 2 transformer au All 01-A tubes. Semi balanced t.r.f. Three dials. P current 15 mA. Volume control: potentiometer. Cab sizes: table, 20 x 8½ x 10 inches; console, 36 x 40 x inches. Partially shielded. Battery cable. C-bat connections. Antenna: 125 feet. Prices: table, console, $65 including loud speaker.

Above is the master Ray-O-Vac 45-volt "B" battery, with the new construction, made especially for sets using four or more tubes.

Improve Reception
and reduce operating expense

EMINENT engineers say that for the best radio reception the "B" power supply should have as little internal resistance as possible.

Otherwise signals are liable to be distorted in amplification, and natural, rounded tones cannot come out of the loud speaker.

The special formula used in making Ray-O-Vacs produces batteries that have only from $\frac{1}{8}$ to $\frac{1}{3}$ the internal resistance of ordinary sources of "B" power supply.

At the same time, this special formula makes batteries that deliver a strong, steady voltage over an unusually long time. It gives them *staying power*.

And the long-life of Ray-O-Vac batteries is now still longer, because a new method of construction prevents internal short circuits and prolongs the life of the battery from 10% to 15%.

There are twice as many radio owners using Ray-O-Vacs today as there were a year ago. They know what low internal resistance and staying power in "B" batteries mean.

Ray-O-Vac batteries are sold by the leading dealers in radio equipment and supplies. If you have any difficulty getting them write us for the name of a nearby dealer who can supply you.

FRENCH BATTERY COMPANY
MADISON, WISCONSIN

Also makers of flashlights and batteries and ignition batteries

NO. 485. BOSWORTH B6

Five tubes; 2 t.r.f. (01-A), detector (01-A), 2 transformer audio (01-A and 71). Two dials. Volume control: variable grid resistances. Battery cable. C-battery connections. Antenna: 25 feet or longer. Cabinet size 15 x 7 x 8 inches. Price $75.

NO. 513. COUNTERPHASE SIX

Six tubes; 3 t.r.f. (01-A), detector (00-A), 2 transformer audio (01-A and 12). Counterphase t.r.f. Two dials. Plate current: 32 mA. Volume control: rheostat on 2nd and 3rd r.f. Coils shielded. Battery cable. C-battery connections. Antenna: 75 to 100 feet. Console size: 18¾ x 40¼ x 15½ inches. Prices: Model 35, table, $110; Model 37, console, $175.

NO. 514. COUNTERPHASE EIGHT

Eight tubes; 4 t.r.f. (01-A), detector (00-A), 2 transformer audio (01-A and 12). Counterphase t.r.f. One dial. Plate current: 40 mA. Volume control: rheostat in 1st r.f. Copper stage shielding. Battery cable. C-battery connections. Antenna: 75 to 100 feet. Cabinet size: 30 x 12¼ x 16 inches. Prices: Model 12, table, $225; Model 16, console, $335; Model 18, console, $365.

NO. 506. CROSLEY 601 BANDBOX

Six tubes; 3 t.r.f., detector, 2 transformer audio. All 01-A tubes. Neutrodyne. One dial. Plate current: 40 mA. Volume control: rheostat in r.f. Shielded. Battery cable. C-battery connections. Antenna: 75 to 150 feet. Cabinet size: 17¾ x 5½ x 7½. Price, $55.

NO. 434. DAY-FAN 6

Six tubes; 3 t.r.f. (01-A), detector (01-A), 2 transformer audio (01-A and 12 or 71). One dial. Plate current: 12 to 15 mA. Volume control: rheostat on r.f. Shielded. Battery cable, C-battery connections. Output device. Antenna: 50 to 120 feet. Cabinet sizes: Daycraft 6, 32 x 30 x 34 inches; Day-Fan Jr., 15 x 7 x 7. Prices: Day-Fan 6, $110; Daycraft 6, $145 including loud speaker; Day-Fan Jr. not available.

NO. 435. DAY-FAN 7

Seven tubes; 3 t.r.f. (01-A), detector (01-A), 1 resistance audio (01-A), 2 transformer audio (01-A and 12 or 71). Plate current: 15 mA. Antenna: outside. Same as No. 434. Price $115.

NO. 503. FADA SPECIAL

Six tubes; 3 t.r.f. (01-A), detector (01-A), 2 transformer audio (01-A and 71). Neutrodyne. Two drum control. Plate current: 20 to 24mA. Volume control: rheostat on r.f. Coils shielded. Battery cable. C-battery connections. Headphone connection. Antenna: outdoor. Cabinet size: 20½ x 13¾ x 10½ inches. Price $95.

NO. 504. FADA 7

Seven tubes; 4 t.r.f. (01-A), detector (01-A), 2 transformer audio (01-A and 71). Neutrodyne. Two drum control. Plate current: 43mA. Volume control: rheostat on r.f. Completely shielded. Battery cable. C-battery connections. Headphone connections. Output device. Antenna: outdoor or loop. Cabinet sizes: table, 25½ x 13½ x 11½ inches; console, 29 x 50 x 17 inches. Prices: table, $185; console, $285.

NO. 436. FEDERAL

Five tubes; 2 t.r.f. (01-A), detector (01-A), 2 transformer audio (01-A and 12 or 71). Balanced t.r.f. One dial. Plate current: 20.7 mA. Volume control: rheostat on r.f. Shielded. Battery cable. C-battery connections. Antenna: loop. Made in 6 models. Price varies from $250 to $1000 including loop.

NO. 505. FADA 8

Eight tubes. Same as No. 504 except for one extra stage of audio and different cabinet. Prices: table, $300; console, $400.

NO. 437. FERGUSON 10A

Seven tubes; 3 t.r.f. (01-A), detector (01-A), 3 audio (01-A and 12 or 71). One dial. Plate current: 18 to 25 mA. Volume control: rheostat on two r.f. Shielded. Battery cable. C-battery connections. Antenna: 100 feet. Cabinet size: 21½ x 12 x 15 inches. Price $150.

NO. 438. FERGUSON 14

Ten tubes; 3 untuned r.f., 3 t. r.f. (01-A), detector (01-A), 3 audio (01-A and 12 or 71). Special balanced t.r.f. One dial. Plate current: 30 to 35 mA. Volume control: rheostat in three r.f. Shielded. Battery cable, C-battery connections. Antenna: loop. Cabinet size: 24 x 12 x 16 inches. Price $235, including loop.

NO. 439. FERGUSON 12

Six tubes; 2 t.r.f. (01-A), detector (01-A), 1 transformer audio (01-A), 2 resistance audio (01-A and 12 or 71). Two dials. Plate current: 18 to 25 mA. Volume control: rheostat on two r.f. Partially shielded. Battery cable. C-battery connections. Antenna: 100 feet. Cabinet size: 22½ x 10 x 12 inches. Price $85. Consolette $145 including loud speaker.

NO. 440. FREED-EISEMANN NR-8, NR-9, AND NR-66

Six tubes; 3 t.r.f. (01-A), detector (01-A), 2 transformer audio (01-A and 71). Neutrodyne. NR-8, two dials; others one dial. Plate current: 30 mA. Volume control: rheostat on r.f. NR-8 and 9; chassis type shielding. NR-66, individual stage shielding. Battery cable. C-battery connections. Antenna: 100 feet. Cabinet sizes: NR-8 and 9, 19½ x 10 x 10½ inches; NR-66 20 x 10½ x 12 inches. Prices: NR-8, $90; NR-9, $100; NR-66, $125.

NO. 501. KING "CHEVALIER"

Six tubes. Same as No. 500. Coils completely shielded. Panel size: 11 x 7 inches. Price, $210 including loud speaker.

NO. 441. FREED-EISEMANN NR-77

Seven tubes: 4 t.r.f. (01-A), detector (01-A), 2 transformer audio (01-A and 71). Neutrodyne. One dial. Plate current: 35 mA. Volume control: rheostat on r.f. Shielding. Battery cable. C-battery connections. Antenna: outside or loop. Cabinet size: 23 x 10½ x 13 inches. Price $175.

NO. 442. FREED-EISEMANN 800 AND 850

Eight tubes; 4 t.r.f. (01-A), detector (01-A), 1 transformer (01-A), 1 parallel audio (01-A or 71). Neutrodyne. One dial. Plate current: 35 mA. Volume control: rheostat on r.f. Shielded. Battery cable. C-battery connections. Output: two tubes in parallel or one power tube may be used. Antenna: outside or loop. Cabinet sizes: No. 800, 34 x 15½ x 13½ inches; No. 850, 36 x 65½ x 17½. Prices not available.

NO. 444. GREBE MU-1

Five tubes; 2 t.r.f. (01-A), detector (01-A), 2 transformer audio (01-A and 12 or 71). Balanced t.r.f. One, two, or three dials (operate singly or together). Plate current: 30mA. Volume control: rheostat on r.f. Binocular coils. Binding posts. C-battery connections. Antenna: 125 feet. Cabinet size: 22½ x 9½ x 13 inches. Prices range from $95 to $320.

NO. 426. HOMER

Seven tubes; 4 t.r.f. (01-A); detector (01-A or 00A); 2 audio (01-A and 12 or 71). One knob tuning control. Volume control: rotor control in antenna circuit. Plate current: 22 to 25 mA. "Technidyne" circuit. Completely enclosed in aluminum box. Battery cable. C-battery connections. Cabinet size, 8¾ x 19½ x 9½ inches. Chassis size, 6¾ x 17 x 8 inches. Prices: Chassis only, $80. Table cabinet, $95.

NO. 502. KENNEDY ROYAL 7. CONSOLETTE

Seven tubes; 4 t.r.f. (01-A), detector (00-A), 2 transformer audio (01-A and 71). One dial. Plate current: 42 mA. Volume control: rheostat on two r.f. Special r.f. coils. Battery cable. C-battery connections. Headphone connection. Antenna: outside or loop. Consolette size: 36½ x 35½ x 19 inches. Price $220.

NO. 498. KING "CRUSADER"

Six tubes; 2 t.r.f. (01-A), detector (00-A), 3 transformer audio (01-A and 71). Balanced t.r.f. One dial. Plate current: 20 mA. Volume control: rheostat on r.f. Coils shielded. Battery cable. C-battery connections. Antenna: outside. Panel: 11 x 7 inches. Price, $115.

NO. 499. KING "COMMANDER"

Six tubes; 3 t.r.f. (01-A), detector (00-A), 2 transformer audio (01-A and 71). Balanced t.r.f. One dial. Plate current: 25 mA. Volume control: rheostat on r.f. Completely shielded. Battery cable. C-battery connections. Antenna: loop. Panel size: 12 x 8 inches. Price $220 including loop.

NO. 429. KING COLE VII AND VIII

Seven tubes; 3 t.r.f., detector, 1 resistance audio, 2 transformer audio. All 01-A tubes. Model VIII has one more stage t.r.f. (eight tubes). Model VII, two dials. Model VIII, one dial. Plate current: 15 to 50 mA. Volume control: primary shunt in r.f. Steel shielding. Battery cable and binding posts. C-battery connections. Output devices on some consoles. Antenna: 10 to 100 feet. Cabinet size: varies. Prices: Model VII, $80 to $160; Model VIII, $100 to $300.

NO. 500. KING "BARONET" AND "VIKING"

Six tubes; 2 t.r.f. (01-A), detector (00-A), 3 transformer audio (01-A and 71). Balanced t.r.f. One dial. Plate current: 19 mA. Volume control: rheostat in r.f. Battery cable. C-battery connections. Antenna: outside. Panel size: 18 x 7 inches. Prices: "Baronet," $70; "Viking," $140 including loud speaker.

NO. 489. MOHAWK

Six tubes; 2 t.r.f. (01-A), detector (00-A), 3 audio (01-A and 71). One dial. Plate current: 40 mA. Volume control: rheostat on r.f. Battery cable. C-battery connections. Output device. Antenna: 60 feet. Panel size: 12½ x 8½ inches. Prices range from $65 to $245.

NO. 449. NORBERT "MIDGET"

One multivalve tube; detector, 2 transformer audio. Two dials. Plate current: 3 mA. Volume control: rheostat. Binding posts. C-battery connections. Headphone connection. Antenna: 75 to 150 feet. Cabinet size: 12 x 8 x 9 inches. Price $12 including multivalve.

NO. 450. NORBERT 2

Two tubes; 1 t.r.f., detector, 2 transformer audio. One multi-valve tube and one 01-A. Two dials. Plate current: 8 mA. Volume control: special. Battery cable. Headphone connection. C-battery connections. Antenna: 50 to 100 feet. Cabinet size: 20 x 7 x 5½ inches. Price $40.90 including multivalve and 01-A tube.

NO. 452. ORIOLE 90

Five tubes; 2 t.r.f., detector, 2 transformer audio. All 01-A tubes. "Trinum" circuit. Two dials. Plate current: 18 mA. Volume control: rheostat on r. f. Battery cable. C-battery connections. Antenna: 50 to 100 feet. Cabinet size: 25½ x 11½ x 12½ inches. Price $85. Another model has 8 tubes, one dial, and is shielded. Price $185.

NO. 453. PARAGON

Six tubes; 2 t.r.f. (01-A), detector (01-A), 3 double impedance audio (01-A and 71). One dial. Plate current: 40 mA. Volume control: resistance in r.f. plate. Shielded. Battery cable. C-battery connections. Output device. Antenna: 100 feet. Console size: 20 x 46 x 17 inches. Price not determined.

NO. 543 RADIOLA 20

Five tubes; 2 t. r. f. (99), detector (99), two transformer audio (99 and 20). Regenerative detector. Two drum controls. C-battery connections. Battery cable. Antenna: 100 feet. Price: $78 without accessories.

NO. 480. PFANSTIEHL 30 AND 302

Six tubes; 3 t.r.f. (01-A), detector (01-2A), transformer audio (01-A and 71). One dial. Plate current: 23 to 32 mA. Volume control: resistance in r.f. plate. Shielded. Battery cable. C-battery connections. Antenna: outside. Panel size: 17¾ x 8½ inches. Prices: No. 30 cabinet, $105; No. 302 console, $185 including loud speaker.

NO. 515. BROWNING-DRAKE 7-A

Seven tubes; 2 t.r.f. (01-A), detector (00-A), 3 audio (Hmu, two 01-A, and 71). Illuminated drum control. Volume control: rheostat on 1st r.f. Shielded. Neutralized. C-battery connections. Battery Cable. Metal panel. Output device. Antenna: 50-75 feet. Cabinet, 30 x 11 x 9 inches. Price, $145.

NO. 516. BROWNING-DRAKE 6-A

Six tubes; 1 t.r.f. (99), detector (00-A), 3 audio (Hmu, two 01-A and 71). Drum control with auxiliary adjustment. Volume control: rheostat on r.f. Regenerative detector. Shielded. Neutralized. C-battery connections. Battery cable. Antenna: 50-100 feet. Cabinet, 25 x 11 x 9. Price $105.

NO. 518. KELLOGG "WAVE MASTER," 504, 505, AND 506.

Five tubes; 2 t.r.f., detector, 2 transformer audio. One control and special zone switch. Volume control: rheostat on r.f. C-battery connections. Binding posts. Plate current: 25 to 35 mA. Antenna: 100 feet. Panel: 7½ x 25½ inches. Prices: Model 504, table, $75, less accessories. Model 505, table, $125 with loud speaker. Model 506, consolette, $135 with loud speaker.

NO. 519. KELLOGG, 507 AND 508.

Six tubes; 3 t.r.f., detector, 2 transformer audio. One control and special zone switch. Volume control: rheostat on r.f. C-battery connections. Balanced. Shielded. Binding posts and battery cable. Antenna: 70 feet. Cabinet size: Model 507, table, 30 x 13½ x 14 inches. Model 508, console, 34 x 18 x 54 inches. Prices: Model 507, $190 less accessories. Model 508, $320 with loud speaker.

NO. 427. MURDOCK 7

Seven tubes; 3 t.r.f. (01-A), detector (01-A), 1 transformer and 2 resistance audio (two 01-A and 12 or 71). One control. Volume control: rheostat on r.f. Coils shielded. Neutralized. Battery cable. C-battery connections. Complete metal case. Antenna: 100 feet. Panel size: 9 x 23 inches. Price, not available.

NO. 520. BOSCH 57

Seven tubes; 4 t.r.f. (01-A), detector (01-A), 2 audio (01-A and 71). One control calibrated in kc. Volume control: rheostat on r.f. Shielded. Battery cable. C-battery connections. Balanced. Output device. Built-in loud speaker. Antenna: built-in loop or outside antenna, 100 feet. Cabinet size: 46 x 16 x 30 inches. Price: $340 including enclosed loop and loud speaker.

NO. 521. BOSCH "CRUISER," 66 AND 76

Six tubes; 3 t.r.f. (01-A), detector (01-A), 2 audio (01-A and 71). One control. Volume control: rheostat on r.f. Shielded. C-battery connections. Balanced. Battery-cable. Antenna: 20 to 100 feet. Prices: Model 66, table, $99.50. Model 76, console, $175; with loud speaker $195.

NO. 524. CASE, 61 A AND 61 C

T.r.f. Semi-shielded. Battery cable. Drum control. Volume control: variable high resistance in audio system. Plate current: 35 mA. Antenna: 100 feet. Prices: Model 61 A, $85; Model 61 C, console, $135.

NO. 525. CASE, 90 A AND 90 C

Drum control. Inductive volume control. Technidyne circuit. C-battery connections. Battery cable. Loop operated. Model 90-C equipped with output device. Prices: Model 90 A, table, $225; Model 90 C, console, $350.

NO. 526. ARBORPHONE 25

Six tubes; 3 t.r.f. (01-A), detector (01-A), 2 audio (01-A and 71). One control. Volume control: rheostat. Shielded. Battery cable. Output device. C-battery connections. Loftin-White circuit. Antenna: 75 feet. Panel: 7½ x 15 inches, metal. Prices: Model 25, table, $125; Model 252, $185; Model 253, $250; Model 255, combination phonograph and radio, $600.

NO. 527. ARBORPHONE 27

Five tubes; 2 t.r.f. (01-A), detector (01-A), 2 audio (01-A). Two controls. Volume control: rheostat. C-battery connections. Binding posts. Antenna: 75 feet. Prices: Model 27, $65; Model 271, $99.50; Model 272, $125.

NO. 528. THE "CHIEF"

Seven tubes; six 01-A tubes and one power tube. One control. Volume control: rheostat. C-battery connection. Partial shielding. Binding posts. Antenna: outside. Cabinet size: 40 x 22 x 16 inches. Prices: Complete with A power supply, $250; without accessories, $150.

NO. 529. DIAMOND SPECIAL, SUPER SPECIAL, AND BABY GRAND CONSOLE

Six tubes; all 01-A type. One control. Partial shielding. C-battery connections. Volume control: rheostat. Binding posts. Antenna: outdoor. Prices: Diamond Special, $75; Super Special, $65; Baby Grand Console, $110.

Universally Used by Amateur Builders

Adopted by Leading Set Mfrs.

First Choice of Professional Builders

Endorsed by Foremost Radio Engineers ~

Sold Everywhere by Leading Dealers

FIRST CHOICE
of the Keenest Minds in Radio!

Durham standard resistors are made in ranges from 500 ohms to 10 megohms. Durham Powerohms for "B" Eliminators and Amplifier circuits are made in 2.5 watt and 5 watt sizes in ranges from 500 to 100,000 ohms.

Adopted by Leading Radio Manufacturers

Philco
Fansteel Products Co.
Kellogg-Switchboard
Western Electric
F. A. D. Andrea
Sterling Mfg. Co.
Kokomo Electric
Garod Radio Corp.
Browning-Drake
Howard Radio
A-C Dayton

FIRST CHOICE—because Durham was the first and original "metallized filament" resistor—because years of heavy production and the confidence of leading radio manufacturers have given us *time* to produce a perfect product.

Durham Resistors and Powerohms are the leaders in their field because their uniform, unfailing accuracy and absolute reliability have been proved time and time again.

This is why they are the first choice of foremost engineers, leading manufacturers, professional set builders and informed radio fans who demand quality results.

Like Durham Resistors and Powerohms, Durham Resistor Mountings are also the leaders in their field. The only upright mountings made; takes minimum space—made of high resistance moulded insulation—best quality tension-spring bronze contacts. Single and double sizes.

RESISTORS & POWEROHMS
INTERNATIONAL RESISTANCE CO., Dept. D, 2½ South 20th Street, Philadelphia, Pa.

NO. 530. KOLSTER, 7A AND 7B

Seven tubes; 4 t.r.f. (01-A), detector (01-A), 2 audio (01-A and 12). One control. Volume control: rheostat on r.f. Shielded. Battery cable. C-battery connections. Antenna: 50 to 75 feet. Prices: Model 7A, $125; Model 7B, with built-in loud speaker, $140.

NO. 531. KOLSTER, 8A, 8B, AND 8C

Eight tubes; 4 t.r.f. (01-A), detector (01-A), 3 audio (two 01-A and one 12). One control. Volume control: rheostat on r.f. Shielded. Battery cable. C-battery connections. Model 8A uses 50 to 75 foot antenna; model 8B contains output device and uses antenna or detachable loop; Model 8C contains output device and uses antenna or built-in loop. Prices: 8A, $185; 8B, $235; 8C, $375.

NO. 532. KOLSTER, 6D, 6G, AND 6H

Six tubes; 3 t.r.f. (01-A), detector (01-A), 2 audio (01-A and 12). One control. Volume control: rheostat on r.f. C-battery connections. Battery cable. Antenna: 50 to 75 feet. Model 6G contains output device and built-in loud speaker; Model 6H contains built-in B power unit and loud speaker. Prices: Model 6D, $80; Model 6G, $165; Model 6H, $265.

NO. 533' SIMPLEX, SR 9 AND SR 10

Five tubes; 2 t.r.f. (01-A), detector (00-A), 2 audio (01-A and 12). SR 9, three controls; SR 10, two controls. Volume control: rheostat. C-battery connections. Battery cable. Headphone connection. Prices: SR 9, table, $65; consolette, $95; console, $145. SR 10, table $70; consolette, $95; console, $145.

NO. 534. SIMPLEX, SR 11

Six tubes; 3 t.r.f. (01-A), detector (00-A), 2 audio (01-A and 12). One control. Volume control: rheostat. C-battery connections. Battery cable. Antenna: 100 feet. Prices: table, $70; consolette, $95; console, $145.

NO. 535. STANDARDYNE, MODEL S 27

Six tubes; 2 t.r.f. (01-A), detector (01-A), 2 audio (power tubes). One control. Volume control: rheostat on r.f. C-battery connections. Binding posts. Antenna: 75 feet. Cabinet size: 9 x 9 x 19½ inches. Prices: S 27, $49.50; S 950, console, with built-in loud speaker, $99.50; S. 600, console with built-in loud speaker, $104.50.

NO. 481. PFANSTIEHL 32 AND 322

Seven tubes: 3 t.r.f. (01–A), detector (01–A), 3 audio (01–A and 71). One dial. Plate current: 23 to 32 mA. Volume control: resistance in r. f. plate. Shielded. Battery cable. C-battery connections. Output device. Antenna: outside. Panel: 17¾ x 8½ inches. Prices: No. 32 cabinet, $145; No. 322 console, $245 including loud speaker.

NO 433. ARBORPHONE

Five tubes; 2 t.r.f., detector, 2 transformer audio. All 01-A tubes. Two dials. Plate current: 16 mA. Volume control: rheostat in r.f. and resistance in r.f. plate. C-battery connections. Binding posts. Antenna: taps for various lengths. Cabinet size: 24 x 9 x 10½ inches. Price: $65.

NO. 431. AUDIOLA 6

Six tubes; 3 t.r.f. (01–A), detector (00–A), 2 transformer audio (01–A and 71). Drum control. Plate current: 20 mA. Volume control: resistance in r.f. plate. Stage shielding. Battery cable. C-battery connection. Antenna: 50 to 100 feet. Cabinet size: 28½ x 11 x 14½ inches. Price not established.

NO. 432. AUDIOLA 8

Eight tubes; 4 t.r.f. (01-A), detector (00-A), 1 transformer audio (01-A), push-pull audio (12 or 71). Bridge balanced t.r.f. Drum control. Volume control: resistance in r.f. plate. Stage shielding. Battery cable. C-battery connections. Antenna: 10 to 100 feet. Cabinet size: 28½ x 11 x 149 inches. Price not established.

NO. 542 RADIOLA 16

Five tubes; 3 t. r. f. (O1-A), detector (O1-A), 2 transformer audio (01-A and 112). One control. C-battery connections. Battery cable. Antenna: outside. Cabinet size: 16½ x 8¼ x 7½ inches. Price: $69.50 without accessories.

NO. 456. RADIOLA 20

Five tubes: 2 t.r.f. (99), detector (99), 2 transformer audio (99 and 20). Balanced t.r.f. and regenerative detector. Two dials. Volume control: regenerative. Shielded. C-battery connections. Headphone connections. Antenna: 75 to 150 feet. Cabinet size: 19½ x 11¾ x 16 inches. Price $115 including all tubes.

NO. 457 RADIOLA 25

Six tubes; five type 99 and one type 20. Drum control. Super-heterodyne circuit. C-battery connections. Battery cable. Headphone connections. Antenna: loop. Set may be operated from batteries or from power mains when used with model 104 loud speaker. Price; $165 with tubes, for battery operation. Apparatus for operation of set from the power mains can be purchased separately.

NO. 493. SONORA F

Seven tubes; 4 t.r.f. (01–A), detector (00–A), 2 transformer audio (01–A and 71). Special balanced t.r.f. Two dials. Plate current: 45 mA. Volume control: rheostat in r.f. Shielded. Battery cable. C-battery connections. Output device. Antenna: loop. Console size: 32 x 45½ x 17 inches. Prices range from $350 to $450 including loop and loud speaker.

NO. 494. SONORA E

Six tubes; 3 t.r.f. (01-A), detector (00–A), 2 transformer audio (01–A and 71). Special balanced t.r.f. Two dials. Plate current: 35 to 40 mA. Volume control: rheostat on r.f. Shielded. Battery cable. C-battery connections. Antenna: outside. Cabinet size: varies. Prices: table, $110; semi-console, $140; console, $240 including loud speaker.

NO. 495. SONORA D

Same as No. 494 except arrangement of tubes; 2 t.r.f., detector, 3 audio. Prices: table, $125; standard console, $185; "DeLuxe" console, $225.

NO. 482. STEWART-WARNER 705 AND 710

Six tubes; 3 t.r.f., detector, 2 transformer audio. All 01–A tubes. Balanced t.r.f. Two dials. Plate current: 10 to 25 mA. Volume control: resistance in r.f. plate. Shielded. Battery cable. C-battery connections. Antenna: 80 feet. Cabinet sizes: No. 705 table, 26½ x 11½ x 13¼ inches; No. 710 console, 29¾ x 42 x 17½ inches. Tentative prices: No. 705, $115; No. 710 $265 including loud speaker.

NO. 483. STEWART-WARNER 525 AND 520

Same as No. 482 except no shielding. Cabinet sizes: No. 525 table, 19¾ x 10 x 11½ inches; No. 520 console, 22½ x 40 x 14¼ inches. Tentative prices: No. 525, $75; No. 520, $117.50 including loud speaker.

NO. 459. STROMBERG-CARLSON 501 AND 502

Five tubes; 2 t.r.f. (01–A), detector (00–A), 2 transformer audio (01–A and 71). Neutrodyne. Two dials. Plate current: 25 to 35 mA. Volume control: rheostat on 1st r.f. Shielded. Battery cable. C-battery connections. Headphone connections. Output device. Panel voltmeter. Antenna: 60 to 100 feet. Cabinet sizes: No. 501, 25½ x 13 x 14 inches; No. 502, 28 ½ x 50 7/16 x 16¾ inches. Prices: No. 501, $180; No. 502, $290.

NO. 460. STROMBERG-CARLSON 601 AND 602

Six tubes. Same as No. 549 except for extra t.r.f. stage. Cabinet sizes: No. 601, 27 5/16 x 16¾ x 14 7/16 inches; No. 602, 28¾ x 51½ x 19¾ inches. Prices: No. 601, $225; No. 602, $330.

NO. 486. VALLEY 71

Seven tubes; 4 t.r.f. (01–A), detector (01–A), 2 transformer audio (01–A and 71). One dial. Plate current: 35 mA. Volume control: rheostat on r.f. Partially shielded. Battery cable. C-battery connections. Headphone connection. Antenna: 50 to 100 feet. Cabinet size: 27 x 6 x 7 inches. Price $95.

NO. 472. VOLOTONE VIII

Six tubes. Same as No. 471 with following exceptions; 2 t.r.f. stages. Three dials. Plate current: 20 mA. Cabinet size: 26½ x 8 x 12 inches. Price $140.

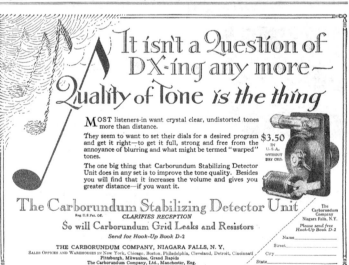

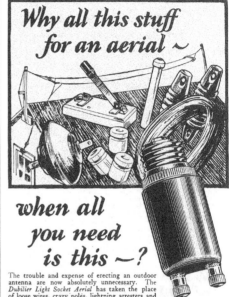

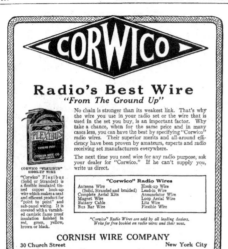

Making Radio Installations Safe

Safety Provisions Specified by the Board of Fire Underwriters in Making Antenna Installations—The Lightning Hazard —Danger From Power Lines

By EDGAR H. FELIX

AN ONCOMING bolt of lightning never stops to argue. It has only one thing in mind—an incredible hurry to reach Mother Earth. Your little, puny pound of copper antenna wire, strung up a few feet above the surface of the earth, has no more influence in attracting a bolt of lightning than a mosquito has to entice you to a symphony concert. If lightning ever strikes your antenna, it is merely a coincidence. The presence or absence of an antenna on a building has nothing to do with either the attraction or repulsion of lightning.

There is in the atmosphere at all times, but more especially during the summer months, a certain amount of static electricity present. Any body of metal or wire tends to collect this atmospheric electricity if it is not connected to ground, and the electrical potentials built up may become of dangerous proportions. A lightning arrester provides an easy path to ground for these charges which tend to accumulate on the antenna, but is so constructed that it does not afford a path to ground for the radio signals. When there is a thunder-shower in the immediate vicinity, the potentials induced in the antenna are greater than at other times, and almost continuous discharges take place through the arrester to ground.

With a lightning arrester properly installed, there seems to be very little danger from this atmospheric electricity, or induced currents from near-by electrical disturbances. To support the contention, let us consider the evidence of experts. Victor H. Tousley, Chief of Electrical Inspectors in the Chicago District, reports that of the 34 cases where the Chicago Fire Department was called out in response to a fire caused by lightning, only 12 of the buildings had antennas; and, in only four of these was there any evidence that lightning had followed the antenna wire to the ground.

William S. Boyd, an electrical engineer of Chicago, has, for some years, been compiling statistics regarding fires caused by lightning striking radio installations. His report lists only 15 damaged by lightning during the years 1922 to 1924—not a very formidable figure. Six of these accidental "hits" damaged only the antenna because the lightning arrester did its work effectively.

In the entire district of Philadelphia, there is a record of only one fire caused by lightning striking a radio antenna, according to an officer of the Fire Underwriters' Association in that city. The antenna and lightning arrester were destroyed, but no damage was done to the receiving set.

Another hazard, which depends purely on the matter of placement of the antenna and good workmanship in putting it up, is the likelihood of the antenna coming into contact with electrical wires, either low tension or high tension. The most important provisions, formulated by

(Continued on page 412)

A "Million-Dollar" Front
For Your Receiver

The New

HAMMARLUND
Illuminated
Drum Dial

An up-to-the-minute tuning improvement every set-builder will want to install.

FRONT VIEW

HAMMARLUND waited to produce a drum dial that would make the single-control of tuning condensers really practicable.

Local stations can now be tuned in over the entire wave band by the simple movement of two fingers. Distant stations, requiring a finer adjustment, are brought in by a slight realignment of the individual halves of the dial.

Viewed from the front, the new Hammarlund Drum Dial gives to any receiver a delightful, professional finish. The bronze escutcheon plate, richly embossed and oxidized, endows the panel with a classic beauty.

BACK VIEW

Mechanical Features

Over-size die-cast frame; Bakelite drums, with knurled edges; translucent celluloid wavelength scales, illuminated by a small electric light, with handy switch, connecting with the "A" Battery circuit. Adaptable to all standard panel proportions.

HAMMARLUND MANUFACTURING COMPANY
424-438 W. 33rd Street, New York

Already many leading radio designers have officially specified Hammarlund Precision Products for their latest circuits.

Hammarlund
For Better Radio
PRECISION
PRODUCTS

Dealer inquiries invited concerning several other new and appealing Hammarlund developments, having a wide sales appeal.

Safe Radio

the Board of Underwriters, are abstracted as follows:

PERTAINING TO THE LIGHTNING HAZARD

1. LIGHTNING ARRESTER. Each lead-in wire shall be equipped with an approved lightning arrester, operative when 500 volts or more are impressed on it, and connected to a ground either inside or outside the building, as near to the arrester as possible. The arrester should not be installed near any inflammable material. If a grounding switch is used, it should shunt the arrester, and should have a capacity of 30 amperes at 250 volts.

2. APPROVED GROUNDS. The ground used is preferably made to a cold-water pipe, where such a pipe is available and is in service and connected to the street mains. Other permissible grounds are: the grounded steel frames of buildings, the grounded metal work in the building, and artificial grounds such as driven pipes, rods, plates, cones, etc. Gas pipe shall not be used for a ground. The ground pipe should be installed so as to be safe from mechanical injury. An approved ground clamp should be used for connecting the wires, and the pipe should be thoroughly cleaned.

3. GROUND WIRE. The protective grounding conductor may be of bare or insulated wire made of copper, bronze, or approved copper-clad steel. The ground wire shall in no case be less in current-carrying capacity than the lead-in wire, and in no case shall it be smaller than No. 14 if copper, nor smaller than No. 17 if of bronze or copper-clad steel. The ground wire should be run in as straight a line as possible from the protective device to the ground connection. Inside wiring should be fastened in a workmanlike manner and should not come closer than two inches to electric conductors, unless porcelain tubing forms a permanent separation from such conductors. This last applies to all inside wires.

7. FUSES IN GROUND LEADS. No fuses shall be used in any lead-in conductors or in the ground wire.

REQUIREMENTS FOR PROTECTION FROM HIGH VOLTAGE

8. INSTALLATION NEAR HIGH VOLTAGE MAINS. The antenna and counterpoise, outside the building, shall be kept well away from any wires carrying a potential of 600 volts or more, including railway, trolley, or feeder wires, in order to avoid accidental contact. These voltages are sufficient to cause shock dangerous to life and to cause fires. Antenna installations near electrical wires of less than 600 volts potential must be installed in a durable manner and shall be provided with suitable clearances so that there is no possible chance of accidental contact, due to sagging or swinging.

9. SPLICES IN ANTENNA WIRE. Splicing in antenna or lead-in wires should be soldered unless made with an approved splicing device.

10. LEAD-IN CONDUCTORS. Lead-in conductors shall be of approved copper-clad steel or other metal which does not corrode excessively, of no smaller gauge than No 14, unless approved copper-clad steel is used when the gauge may be No. 17.

11. LEAD-IN CONDUCTORS OUTSIDE THE BUILDING. The lead-in conductor shall not come closer than four inches to any electric light or power wires unless separated therefrom by a porcelain tube or other firmly fixed nonconductor.

12. LEAD-IN INSULATOR. An approved lead-in insulator should be used.

13. STORAGE BATTERY CIRCUIT. An important and new provision rarely obeyed is the requirement that fuses of not less than 15-amperes capacity shall be installed and located preferably at or near to the battery. An approved circuit breaker of the same minimum capacity may be used. The leads from the battery to the receiver shall consist of conductors having approved rubber insulation.

Clearly, these rules are simple and easily com-

(Continued on page 414)

Vee Dee
All-Metal Cabinet—
For 1927-28 Hook-Ups!

Model 250
For A. C. or
Battery Sets

Using
7x18
7x21
8x18
8x21
Panels

Inside dimensions 25"x14¼"x9½". Hinged top—with stay joint. Rigidly formed for strength and appearance. Felt foot rest—rubber lid stops. A welded job doing away with troubles of swelling, shrinking, cracking, splitting and uncertain fit.

The original beauty of natural wood grains combined with the efficiency of all metal construction! By our photo litho process, we reproduce mahogany and walnut hardwood, and novelty finishes, so gorgeous in their conseption that they excite the admiration of all who see it.

Spacious interior dimensions are demanded for housing all the latest hook-ups! Vee Dee metal cabinets are designed for that purpose. 90% of all the 1927-1928 hook-ups are covered by the dimensions of Vee Dee No. 250 cabinet illustrated above. Beautiful—practical! Low price!

Metal Panel for Citizens Super Eight

Constructed in accordance with
Citizens Radio Call Book and

McMurdo Silver, Remler, Cockaday co-ordinated designing

Metal Panel and Chassis for Silver-Marshall 1927 Model Super-Heterodyne

Complete assembly consisting of panel and chassis, fully drilled, beautiful wood finish with special two-color decoration, all fibre bushings and washers included, also screws, bolts and hardware accessories.

S. C. 2 Assembly Unit

Complete Panel, Chassis, 7x18, fully drilled. Beautiful wood grain finish, handsomely decorated. Kit includes all necessary bushings, washers and hardware accessories.

Unipac Housing

Especially designed and provided for Silver-Marshall Power Hook-ups—including cabinet and chassis, drilled and with all small hardware.

Metal Panels and Chassis in All Standard Sizes

JOBBERS, DEALERS—WRITE FOR PRICES

Set Builders—If your dealer cannot supply you, write direct

The Van Doorn Company
160 North La Salle Street   *Factory, Quincy, Illinois*   Chicago, Illinois

Safe Radio

plied with. Compliance with them gives you a safe installation. Failure to follow them invites unnecessary and totally avoidable risk. Furthermore, failure to comply invalidates your fire insurance policies automatically.

BOOK REVIEW

A Home-Constructor's Handbook

PRACTICAL RADIO CONSTRUCTION AND REPAIR. By James A. Moyer and John F. Wostrel. Published by the McGraw-Hill Book Company, Incorporated; New York. 319 Pages and 157 Illustrations. Price $2.00.

PRACTICAL Radio Construction and Repair" contains a potpourri of useful and practical information. Its authors have compiled their pages out of a large practical knowledge and experience in set building. The title implies a handbook of radio construction and repairing—a reference volume for the dealer's service man. It does not fail in that respect so much from incompleteness as it does from poor arrangement. But it has much to commend it.

The first part of the book deals with, what may be termed, the accessories of the radio receiver—the antenna, the vacuum tube and its power supply, and, tucked away with these subjects, a chapter on the tool equipment needed for radio construction. Then follows a series of chapters on radio-frequency amplifiers, describing several methods of controlling regeneration and including a few words about super-heterodyne amplification. Next follows a brief description of systems of audio-frequency amplification.

The subsequent few chapters deal with constructional details. The first set to be described in detail is the Browning-Drake receiver. The writers then again return to audio-frequency amplifiers, presenting a strong case for a popular resistance-coupled amplifier kit. The superiority of resistance coupling over transformer coupling is emphasized by means of a set of curves. Apparently, an inferior transformer and a superior resistance amplifier were used as the basis of the test. A chapter on impedance amplification, devoted to a description of the Thordarson amplifier, follows that comparing resistance and transformer coupling.

Returning again to receiving sets, the authors next describe the "Universal," the Acme reflex, the Cotton super-heterodyne, and a short-wave receiver. In general, these descriptions follow closely the conventional magazine style of exposition, with lists of parts, and details for assembly and operation. No commercial receivers are described or even considered.

The appeal of this book is confined generally to the home constructor and his problems and not to those of the dealer service man; an outstanding exception is the last chapter, where the service man is suddenly remembered. This last chapter makes interesting reading, containing, as it does, a fairly complete and practical set of troubles likely to be encountered in a radio receiver under various conditions. A few brief samples are quoted to show the type of information in this section:

SET GIVES SQUEALING SOUND CONSTANTLY.— A constant squealing sound in a receiving set is generally due to a defective vacuum tube which in the course of time has become soft or gassy. Continuous squealing may also be due to a very badly run-down B battery or, to a less degree, to a worn-out C battery. A burned-out primary

(Continued on page 416)

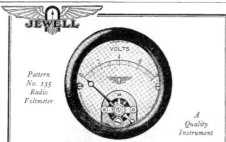

*Pattern
No. 135
Radio
Voltmeter*

*A
Quality
Instrument*

Your Radio Will Be Better

Voltmeter control of filament voltage will make any radio receiving set better. It enables you to retain that nicety of balance in filament emission, which brings in reception clear and exact with maximum volume. It will prevent premature burnout of the radio tubes due to excessive filament voltage, for, with a voltmeter mounted on the panel and connected to the filaments, you will know at all times just what voltage is being applied to the filaments.

The Jewell Pattern No. 135 voltmeter is good looking and rigidly constructed and is the ideal instrument for filament control. The black enamelled case is two inches in diameter and contains a miniature, but very high grade, D'Arsonval moving coil type movement, which is equipped with a zero adjuster. Movement parts are all silvered and the scale is silver etched with black characters.

The addition of this meter to your set will improve its appearance besides being a great help to better and economical reception

Write for descriptive circular No. 776 and ask for a copy of our radio instrument catalog No. 15-c

Jewell Electrical Instrument Co.

1650 Walnut Street, Chicago

"27 Years Making Good Instruments"

Book Review

winding of one of the transformers or a poor connection in the jack connecting the loud speaker and closing the battery circuits will also cause this same effect.

SET OPERATES WHEN PLUG OF TELEPHONE RECEIVER IS PUT IN JACK OF DETECTOR CIRCUIT, BUT DOES NOT OPERATE IN STAGES OF AMPLIFICATION.—This difficulty may be due to reversed connections of the A battery, or the primary winding of the audio-frequency transformer in the detector circuit may be burned out.

No serious attempt is made to correlate the practical information in this last section in a manner lending itself to easy reference.

The set constructor will find much of interest in the book because it is, on the whole, reliable and accurate. We might pick out statements here and there to quibble over, such as that in the chapter on audio-frequency amplifiers. Speaking of audio-frequency transformers, the authors say: "Some manufacturers claim that they can make transformers which amplify consistently over a wide range of wavelengths," a correct, though rather unconventional way of referring to audio-frequencies. Another case of a different kind: "If the sound volume from the loud-speaker is weak, tap the vacuum tubes with a finger nail to determine whether or not the audio amplifiers are operating satisfactorily; if they are not, the tapping by the finger nail will cause a ringing sound in the loud speaker." To include, in a final summary of testing instructions, so broad a generality, having so many conceivable exceptions, is, to say the least, somewhat careless.

A general criticism which may be made of many radio books is the superficial knowledge of their authors. Moyer and Wostrel are certainly exempt from any such accusation because they have the outstanding virtue of knowing what they are talking about. But they have not met the needs of service men working for the radio dealer; the authors' viewpoint is that of the amateur set builder.

The preferred circuits for home construction are well selected and described; there are plenty of diagrams to browse among and, to one who reads the book from cover to cover, many new facts are likely to be discovered. But, if one wants to find out in a hurry such a practical constructional point as whether the jack should be so wired that the sleeve or the tip of the plug goes to the B-plus lead, or the quickest way of finding safely which tube is burned out with a set in which the filaments are wired in series, there is no telling if, how, and where the desired information will be found.

EDGAR H. FELIX.

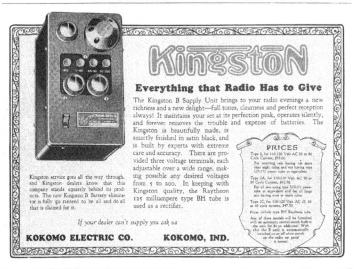

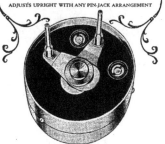

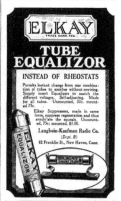

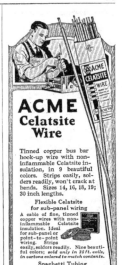

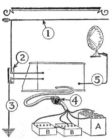

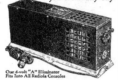

—when the crown stands or falls....

"—You're *there* with a Crosley..."

Shielding is necessary in a modern radio receiver. The more sensitive the set is, the more you need it. Some sets are merely housed in a metal case. This helps to keep strong local

signals from breaking through, but it is even more important to keep them where they belong after you get them, the proper way, from the antenna. A set has tubes, condensers and coils. Here is a coil. The lines around it are the magnetic field. You know the earth's magnetic field will work a compass down in a mine, or up in a plane (it certainly worked for Lindbergh) and the fields around unshielded coils get all mixed up and the set howls and squeals and has to be choked off by turning down the filaments in the tubes.

Now if the coils are housed in copper shields the fields can't mess each other up, and the tubes can do a real job of amplifying. The coils in Crosley sets have these copper shields, and there isn't anything better.

Then there are the condensers, and if it wasn't for the shield around them, the fields would act like those

in the coils, and the results would be just as bad, or worse. It isn't enough to shield the coils and the condensers, because even the wiring of the set has fields around it. This too is shielded, as it is in all really high grade sets.

Of course, it's all in knowing how to do it, but that's why Crosley sets can be as good as the best without costing half as much.

The BANDBOX
A 6 Tube Receiver *of* $55
unmatchable quality at
LIGHT SOCKET MODEL $65

Many features of this set have been found heretofore only in the most expensive radio. Since Crosley is licensed to manufacture under nearly all important radio patents, this combination with Crosley leadership and experience, naturally produced an amazing radio, the remarkable value of which can be judged by the following features incorporated and by seeing it and hearing it at your dealers.

1. Completely shielded coils, condensers and wiring.
2. Acuminators for sharper tuning. 3. Completely balanced genuine neutrodyne. 4. Volume control. 5. Single tuning knob. 6. Illuminated dial. 7. Single cable outside connections. 8. Designed for easy installation in consoles. 9. Beautiful frosted brown crystalline finished cabinet.

AC model using new R. C. A. AC tubes and working directly from electric light socket through Crosley Power Converter is $65. Power Converter $60 extra.

Hear this wonderful new contribution to the enjoyment of radio. If you cannot find one of the 16,000 Crosley dealers near you, write Dept. 20 for his name and literature.

$35

$65

$85

CONDENSERS PROPERLY HOUSED

CROSLEY RADIO

The Crosley Radio Corporation, Powel Crosley, Jr., Pres. Cincinnati, Ohio.

Crosley Radio is licensed only for Radio Amateur, Experimental and Broadcast Reception.

Prices slightly higher west of the Rocky Mountains

RADIO BROADCAST

VOLUME XII

NOVEMBER, 1927, to APRIL, 1928

GARDEN CITY   NEW YORK

DOUBLEDAY, DORAN & COMPANY, INC.

1928

INDEX

(*Illustrated Articles.   Editorials in Italics)

All Electric Radio

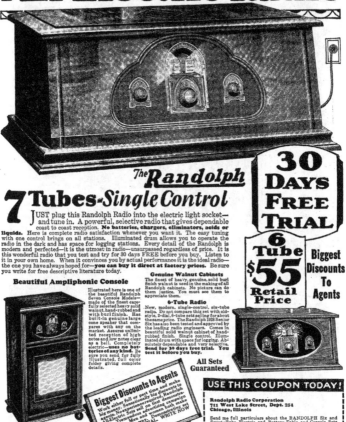

The Randolph

7 Tubes-*Single Control*

JUST plug this Randolph Radio into the electric light socket—and tune in. A powerful, selective radio that gives dependable coast to coast reception. **No batteries, chargers, eliminators, acids or liquids.** Here is complete radio satisfaction whenever you want it. The easy tuning with one control brings on all stations. Illuminated drum allows you to operate the radio in the dark and has space for logging stations. Every detail of the Randolph is modern and perfected—it is the utmost in radio—unsurpassed regardless of price. It is this wonderful radio that you test and try for 30 days FREE before you buy. Listen to it in your own home. When it convinces you by actual performance it is the ideal radio—the one you have always hoped for—**you can buy it direct at factory prices.** Be sure you write for free descriptive literature today.

Beautiful Ampliphonic Console

Illustrated here is one of the beautiful Randolph Seven Console Models—made of the finest carefully selected heavy solid walnut, hand-rubbed and with burl finish. Has built-in genuine large cone speaker that compares with any on the market. Assures unlimited reception of high notes and low notes clear as a bell. Completely electric—**uses no batteries of any kind.** Be sure you send for fully illustrated, full color folder giving complete details.

Genuine Walnut Cabinets

The finest of heavy, genuine, solid burl finish walnut is used in the making of all Randolph cabinets. No picture can do them justice. You must see them to appreciate them.

6-Tube Radio

New, modern, single-control, six-tube radio. Do not compare this set with old-style, 2-dial, 6-tube sets selling for about the same price. The Randolph 1928 Senior Six has also been tested and approved by the leading radio engineers. Comes in beautiful solid walnut cabinet of hand-rubbed finish. Single control. Illuminated drum with space for logging. Absolutely dependable and very selective. **Send for 30 days free trial. You test it before you buy.**

All Sets Guaranteed

30 DAYS FREE TRIAL

6 Tube $55 Retail Price

Biggest Discounts To Agents

RANDOLPH RADIO CORPORATION
711 West Lake Street   Dept. 254   Chicago, Ill.

Many times in the old days, while I trudged home after work to save car-fare, I used to pass enviously at the shining cars gliding by me, the prosperous men and women within. Little did I think that inside of a year, I too, should have my own car, a decent bank account, the good things of life that make it worth living.

I Thought Success Was For Others

Believe It Or Not, Just Twelve Months Ago
I Was Next Thing To "Down-and-Out"

TO-DAY I'm sole owner of the fastest-growing Radio store in town. And I'm on good terms with my banker, too—not like the old days only a year ago, when often I didn't have one dollar to knock against another in my pocket. My wife and I live in the snuggest little home you ever saw, right in one of the best neighborhoods. And to think that a year ago I used to dodge the landlady when she came to collect the rent for the little bedroom I called "home"!

It all seems like a dream now, as I look back over the past twelve short months, and think how discouraged I was then, at the "end of a blind alley." I thought I never had had a good chance in my life, and I thought I never would have one. But it was waiting up that I needed, and here's the story of how I got it.

I WAS a clerk, working at the usual miserable salary such jobs pay. Somehow I'd never found any way to get into a line where I could make good money.

Other fellows seemed to find opportunities. But—much as I wanted the good things that go with success and a decent income—all the really well-paid vacancies I ever heard of seemed to be out of my line, to call for some kind of knowledge I didn't have.

And I wanted to get married. A fine situation, wasn't it? Mary would have agreed to try it—but it wouldn't have been fair to her.

Mary had told me, "You can't get ahead where you are. Why don't you get into another line of work, somewhere that you can advance?"

"That's fine, Mary," I replied, "but what line? I've always got my eyes open for a better job, but I never seem to hear of a really good job that I can handle." Mary didn't seem to be satisfied with the answer but I didn't know what else to tell her.

It was on the way home that night that I stopped off in the neighborhood drug store, where I overheard a scrap of conversation about myself. A few burning words that were the cause of the turning point in my life!

With a hot flush of shame I turned and left the store, and walked rapidly home. So that was what my neighbors—the people who knew me best—really thought of me!

"Bargain counter shiek—look how that suit fits," one fellow had said in a low voice. "Bet he hasn't got a dollar in those pockets." "Oh, it's just 'Useless' Anderson," said another. "He's got a wish-bone where his back-bone ought to be."

As I thought over the words in deep humiliation, a sudden thought made me catch my breath. Why had Mary been so dissatisfied with my answer that "I hadn't had a chance?" *Did Mary secretly think that too?* And after all, wasn't it *true* that I had a "wish-bone" where my back-bone ougt to be? Wasn't that why I never had a "chance" to get ahead? It was true, only too true—and it had taken this cruel blow to my self-esteem to make me see it.

With a new determination I thumbed the pages of a magazine on the table, searching for an advertisement that I'd seen many times but passed up without thinking, an advertisement telling of big opportunities for trained men to succeed in the great new Radio field. With the advertisement was a coupon offering a big free book full of information. I sent the coupon in, and in a few days received a handsome 64-page book, printed in two colors, telling all about the opportunities in the radio field and how a man can prepare quickly and easily at home to take advantage of these opportunities. I read the book carefully, and when I finished it I made my decision.

WHAT'S happened in the twelve months since that day, as I've already told you, seems almost like a dream to me now. For ten of those twelve months, *I've had a Radio business of my own!* At first, of course, I started it as a little proposition on the side, under the guidance of the National Radio Institute, the outfit that gave me my Radio training. It wasn't long before I was getting so much to do in the Radio line that I quit my measly little clerical job, and devoted my full time to my Radio business.

Since that time I've gone right on up, always under the watchful guidance of my friends at the National Radio Institute. They would have given me just as much help, too, if I had wanted to follow some other line of Radio besides building my own retail business—such as broadcasting, manufacturing, experimenting, sea operating, or any one of the score of lines they prepare you for. And to think that until that day

I sent for their eye-opening book, I'd been wailing "I never had a chance!"

NOW I'm making real money. I drive a good-looking car of my own. Mary and I don't own the house in full yet, but I've made a substantial down payment, and I'm not straining myself any to meet the installments.

Here's a real tip. You may not be as bad-off as I was. But, think it over—are you satisfied? Are you _making enough money, at work that you like? Would you sign a contract to stay where you are now for the next ten years, making the same money? If not, you'd better be *doing* something about it instead of drifting.

This new Radio game is a live-wire field of golden rewards. The work, in any of the 20 different lines of Radio, is fascinating, absorbing, well-paid. The National Radio Institute—oldest and largest Radio home-study school in the world—will train you inexpensively in your own home to take Radio from A to Z and to increase your earnings in the Radio field.

Take another tip—No matter what your plans are, no matter how much or how little you know about Radio—clip the coupon below and look their free book over. It is filled with interesting facts, figures, and photos, and the information it will give you is worth a few minutes of anybody's time. You will place yourself under no obligation—the book is free, and is gladly sent to anyone who wants to know about Radio. Just address J. E. Smith, President, National Radio Institute, Dept. O-94, Washington, D. C.

Some folks judge cheese by the holes

Besides Micarta panels and tubing for better insulation and radio testing instruments for better reception. Westinghouse also makes Rectigon and Rectox for better battery charging.

They say, "the more holes the better the cheese." You might make this test for cheese a rule when buying radio panels. At any rate, the more holes you plan to drill, the more certain you'd better be that it's a Micarta panel. Only a strong material can carry a load of condensers, transformers, coils and sockets after the panel has been honeycombed with holes to receive them.

The particular set builder chooses Micarta for both panel and sub-panel — its high insulation qualities allow the most compact arrangements. Micarta is easy to drill, engrave or cut — put a hole close to the edge, or close to another hole, and no crack will appear between.

You get a choice of striking finishes in black, mahogany, walnut grain or walnut burl. And you'll be surprised how inexpensive it is to use the best. Micarta panels and tubing can be obtained from Micarta Fabricators, Inc., whose addresses are given below.

WESTINGHOUSE ELECTRIC & MANUFACTURING COMPANY
Offices in Principal Cities   •   Representatives Everywhere
Tune in sometime with KDKA — KYW — WBZ — WBZA

Westinghouse

© 1927 W. E. & M. Co.

MICARTA FABRICATORS, Inc.
OF NEW YORK
309 Canal Street
New York City

MICARTA

MICARTA FABRICATORS, Inc.
OF ILLINOIS
500 South Peoria Street
Chicago, Ill.

Announcing

THE A. C. RADIO TUBE
FOR YOUR PRESENT SET

ARCTURUS A. C. TUBES
DETECTOR—AMPLIFIER—POWER
Four Prongs—Fits Present Sockets—For all D. C. Sets

BETTER RECEPTION
MORE RELIABLE   MORE CONVENIENT
LESS EXPENSIVE

Now you can have unfailing quality reception, the convenience and reliability
of A. C. Tubes, with but a few simple
changes, in any D. C. Set.

Arcturize Your Present Set

No matter what set you now own it will
pay you to get complete information
about this latest development in radio.
You will find added satisfaction in the
perfect reception you get with Arcturus
A. C. Tubes.

Write today for complete information
mentioning make and model of your
present set.

ARCTURUS RADIO COMPANY, Inc.
251 Sherman Avenue, Newark, N. J.

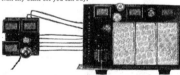

OCT 19 '27 ©Cl B761184C

RADIO BROADCAST

Willis Kingsley Wing, Editor

NOVEMBER, 1927
Keith Henney
Director of the Laboratory
Edgar H. Felix
Contributing Editor
Vol. XII,

CONTENTS

AMONG OTHER THINGS

FOR more reasons than one, the New York Radio generally accepted as the event which crystalliz in all things radio. This is written as the show closed impressive show we have never seen. "Impressive" point of view of the exhibits, certainly, but more imp cause of the tremendous interest in all things radio, des by all sorts and conditions of people who thronge Square Garden. Careful observation of the crowds interest indicated that while the complete sets drew tention, the home-assembled receivers—built from equally interesting. This definite and lively inter home-built sets is especially important in view of the ments of some knowing radio sages who aver that hor is seriously on the decline.

ACTUAL transmission and reception of "still" p radio was demonstrated thousands of times week of the show by Austin G. Cooley who set up Cooley "Rayfoto" transmitter and receiver in a sp provided through the courtesy of G. Clayton Irwin, J of the show. The picture converter or transmitter connected to a small radio transmitter and through broadcast receiver, the pictures were received rapidity and success before the very eyes of eager cr simplicity and speed of the receiver astounded thos the demonstration, and every visitor was eager to l he could build the apparatus and how soon picture sent and where he could get information. Experimen transmissions from various broadcasting stations w even before you read these words; complete info how the system works, how to build and operate exclusively in this and following issues of RADIO l And those who wish to receive printed matter details of the system should at once address a le undersigned who will see that all information is mai The impressive success of the Cooley "Rayfoto" de proves beyond all question that a new era has daw home experimenter, and to be frank, we are as enthu the possibilities opened up as the keenest of experi

A WORD about the authors in this issue: the author of the absorbing leading article is a very figure in aviation and radio. Ralph Langley, who scheme for numbering broadcast channels, is executi to the president, Crosley Radio Corporation. H recently in charge of receiver design for the Gene Company. Howard E. Rhodes who describes wha A-power units is one of the able technical staff of th James Millen, who is a consulting engineer and Long Island, will shortly desert these parts and settle

IN THE next issue we shall have an important artic Nakken on the shielded grid tube indicating what means to American radio. There will be valuable co articles and a description of the technical features of manufactured receivers—information never publish Austin Cooley will tell how to build a Cooley receiver—facts for which many experimenters are w
—Willis Kingsle

Doubleday, Page & Co.
MAGAZINES
Country Life
World's Work
Garden & Home Builder
Radio Broadcast
Short Stories
Educational Review
Le Petit Journal
El Eco
Frontier Stories
West Weekly

Doubleday, Page & Co.
BOOK SHOPS
(Books of all Publishers)
Lord & Taylor Book Shop
Pennsylvania Terminal (2 Shops)
New York: Grand Central Terminal
38 Wall St. and 526 Lexington Ave.
848 Madison Ave. and 166 West 32nd St.
St. Louis: 223 N. 8th St. and 4914 Maryland Ave.
Kansas City: 920 Grand Ave. and 206 W. 47th St.
Cleveland: Higbee Co.
Springfield, Mass.: Meekins, Packard & Wheat

Doubleday, Page & Co.
OFFICES
Garden City, N. Y
New York: 285 Madison Avenue
Boston: Park Square Building
Chicago: Peoples Gas Building
Santa Barbara, Cal.
London: Wm. Heinemann Ltd.
Toronto: Oxford University Press

Doubleday, Page
OFFICER
F. N. Doubleday, Pr
Nelson Doubleday,
S. A. Everitt, Vice-P
Russell Doubleday,
John J. Hessian, Tr
Lillian A. Comstock,
L. J. McNaughton, A

DOUBLEDAY, PAGE & COMPANY, Garden City, New York

Copyright, 1927, in the United States, Newfoundland, Great Britain, Canada, and other countries by Doubleday, Page & Company. All rights reserved.
TERMS: $4.00 a year; single copies 35 cents.

When you have decided to use house current for your receiver—

..what Power Unit will you buy?

A battery eliminator isn't one of those things you can just look at and say it's good or bad. Nor can you determine its qualities with a few moments' trial. How then can you be assured of a reliable unit of lasting satisfaction? Over 700,000 satisfied users of light socket power units will tell you to be sure it is Raytheon *approved* and equipped with the Raytheon long life rectifying tube.

You can identify a Raytheon type power unit by the green *Seal of Approval* on the front. Only units that have been tested and approved can use this symbol or the Raytheon gaseous type tube.

Our Technical Service Department will be glad to answer questions or send the latest Radio Power Bulletin covering in detail any subject on light socket power in which you may be interested.

RAYTHEON MANUFACTURING CO., Cambridge, Mass.

Type BH
125 m.a. 300
volts

Type BA
A-B-C power
350 m.a.

Raytheon

THE HEART OF RELIABLE RADIO POWER

Type A
2½ Amps.

Type R
90 V. 60 m.a.

ONE AEROPLANE ON WHICH PROPER RADIO EQUIPMENT IS USED

THE Dornier-Napier Whale of Captain F. T. Courtney, originally designed for a flight from London to the United States and return, has a 500-watt Marconi i.c.w. transmitter, using a 200-foot trailing antenna, an eight-tube r.f. amplifier ahead of a four-tube super-heterodyne, usable on commercial wavelengths, and a Marconi Bellini-Tosi direction finding antenna. The all-metal construction of the ship introduced special receiving problems which had to be solved. Note in the lower illustration part of the receiving antenna rising over the motor nascelle. The top photograph shows the radio controls. A wind-driven generator supplies power for the transmitter and charges special storage batteries. In case of a forced landing, a 40-foot mast can be erected, and the batteries are made to supply current to the motor generator which runs the main set. Few airships have been so completely equipped.

What's the TROUBLE with Aircraft RADIO?

"Anonymous"

WHO is to blame for the fact that radio communication is not in general use in flying? Is it the radio engineer? Or can it be shown that the fault lies with the airman? Why did not Lindbergh and Chamberlin use radio? Who knows in what different manner the fatal flight of Nungesser and Coli might have ended had there been radio equipment aboard the *White Bird*? What of the *Golden Eagle* and the *Miss Doran*? Had these planes been equipped with radio would they have been lost? Probably not. There is reason to believe that even though forced down, so well would they have been followed by radio watchers on land that they might have been quickly found.

Commander Byrd made good use of radio at times in the flight that ended just short of Paris. But did he, schooled in the Navy and certainly aware of the possibilities for its use, make the most of his radio? One wonders. Why, when approaching the French coast, was he unable to learn the kind of weather awaiting him at Paris?

Hegenberger and Maitland were able to use their radio equipment but a small part of the time on their flight to Hawaii. Receiver trouble developed soon after leaving the Pacific Coast and it was not until they were within eight hundred miles of their goal that they were able to pick up signals again. The preparations for their flight were said to have been most thorough. The radio must surely have been thoroughly tested before the take-off, yet it failed them in time of need. They had

planned to fly the course laid down by the radio beacon. To do this it was necessary to make continuous use of the radio receiver. Fortunately, when it failed, they were prepared to navigate by better-known means. Such was the thoroughness of Army Air Corps methods of preparation.

However, there has been no excuse for the lack of radio equipment of some sort on all of the trans-oceanic flights. The disturbance created by the ignition system which is almost always offered as an argument against it, is not an absolute bar to the use of radio. Ignition disturbance has no effect on the radio transmitter. Even a receiver could have been used to some extent in the presence of ignition noise. This is particularly true for a plane in which the cabin is located some little distance from the engine. Furthermore, the receiver could have been successfully used while passing over vessels at sea. The ship's transmitter under the circumstances of such short range would have pushed signals

through the ignition disturbance at least sufficiently to have given information on weather and course.

RADIO MUST BE USED ON LONG FLIGHTS

IT WOULD be very interesting to know the reasoning which led to a decision to leave radio out of the plans for some of these flights. Undoubtedly the real reasons will not be given to the public. One strongly suspects that the lack of ability to handle radio on the part of the crew aboard each of these planes had a great deal to do with the matter. Of course, Lindbergh flew alone and could have made little use of any kind of radio equipment for that reason. Chamberlin knew little or nothing about radio, and it is likely that Levine, his passenger, knew less; inexperienced as he was in such matters. There is no telling how much Coli or Nungesser or Captain Hamilton, the British pilot, knew about the use of radio equipment.

None of these flights should have been undertaken without radio equipment, and a competent radio operator to handle it. On some of the flights one of the pilots acted as radio operator. This did not prove entirely satisfactory. Hegenberger, who flew with Maitland to Hawaii was fairly familiar with radio apparatus, but when his receivers (there were two aboard) became inoperative, he was unable to locate the source of trouble. It is doubtful if his knowledge of radio was sufficient to have enabled him to diagnose trouble as a trained radio operator could have done.

A NONYMOUS" conceals the identity of an individual who is excellently qualified to write on the closely related problems of the airplane and radio. All we can say is that he is an expert who is well known in both fields. The author knows aviation—not from a swivel-chair vantage point, but from long flying experience and he knows radio from both the practical and distinctly technical angles. Too few radio men know anything about the problems that the aviator has to meet, and too few of the airplane folk know radio. Certainly there is a middle ground on which both may meet and this article is the first of several which will discuss this interesting field. The increasing fatality list of those attempting stupid and pointless trans-oceanic flights has demonstrated to almost the whole world that long-distance flying must be made safer and surer by every means at our command. And through radio will come much of this essential surety.

—THE EDITOR.

The desire to carry passengers on these flights has prevented a good radio operator being present. Miss Doran in the Pacific flight, and the Princess Lowenstein-Waltheim in the Atlantic flight of Captain Hamilton, and Philip Payne in *Old Glory* should have been replaced with radio operators, and, at least, radio receiving equipment. The fact that Brock and Schlee flew successfully to England without radio is no proof that radio was not needed. Redfern carried neither companion nor radio. He should not have been permitted to leave without both. And his companion should have been a good radio man.

Thus it is seen that some of these fliers were unmindful of the value of radio, and that others were unable to make the most of equipment which they had chosen to use.

Who is to blame that the value of radio has been so vastly underestimated in these flights? The question is important. Upon the correctness of the answer depends in a great measure the solution of one of the problems which at present confronts aviation.

WHY AVIATORS DON'T LIKE RADIO

THERE have been many discussions on this subject between flying folk and men interested in radio. These discussions have usually been of a character to which the terms "heated," and sometimes "overheated," could justly be applied. Generally, the debates ended only in disagreement. The pilot and the engineer have not been brought to the same way of thinking. Not only have they disagreed as to who is to blame for the neglect of radio, but the pilot has strenuously objected times without number to the use of radio on his plane.

It is, of course, true that the military and naval flier has on occasion done much with the equipment designed for him by the radio engineer. Very often it was only because that flier was a member of an organization, in which obedience to an order is almost instinctive, that he made real use of his equipment. Often, it is true, he was pleased with the results of his effort and so converts to the cause of radio have gradually been made. They are, however, all too few. As for the commercial fliers, apparently little belief in the need for radio exists. One never hears of radio being used on their planes. Not even the Air Mail companies, or our own Post Office department, have seen fit to equip mail planes with even a receiver with which to receive information on the weather. The Air Mail for a time carried out experiments with radio but no

practical or extensive use has yet been made of it.

What has the radio engineer to say for himself in the face of this obvious disdain on the part of the flier for radio? Were he a psychologist it might occur to him that the feeling of the pilots about the matter might be based on something inherent in the flying profession, or in the flier's training. Could he put himself in the place of the average war-trained flier he would remember that in the exciting days of the war the urge to fly was the strongest thing in his life. It was for that reason that he joined the Air Service instead of going into some other branch of the Service. That was why he worked as he never had worked before during preliminary training days at ground school. His eyes were always lifted to the men in the air. Everything but flying was subordinated. Nothing appealed to him either at ground school or at the flying

AT AN EUROPEAN AIRPORT

Elaborate means of radio communication are required by law on passenger air routes in some European countries. The two radio towers at the Tempelhof, Berlin, airport, are clearly discernable in this aerial picture

school to which he was later ordered, as strongly as the airplane and flying. Motors, navigation, gunnery, photography, radio—all had to be learned; but he learned them, for if they were not learned he would not be taught to fly. But flying was the thing—devil take the rest. Whom did he worship the most, his gunnery instructor or his radio instructor, or any of the other ground instructors? None of these. He worshipped the man who taught him how to fly: Usually his flying instructor was the biggest man on his horizon. His radio instructor was usually a non-flier, "Keewee" being the term contemptuously applied to any ground officer of the Air Service. Usually this man made no impression, or a poor one at the most. He often stood between the cadet and his flying goal. For all who would fly must, in addition to many other things, learn to send and receive radio signals. If he could not pass the speed test he could not fly. That was the regulation, and many

a good man was "shot down" by the elusiv *da-dit-da* before he ever had a chance to learn to fly, and was accordingly sent awa from flying school. At the advanced flyin school came practice with actual transmi ting and receiving of signals while in the a: This was usually even more boring tha the practice in the code room. Generall the radio failed to function. Anyhow, wl wanted to sit in the rear cockpit of a sh which was being flown by someone el and fiddle with knobs and dials and try pick up the faint signals bravely endeavc ing to penetrate the noise and roar of tl motor and the disturbance created by tl ignition system?

Thus was built up all through the flie: training, a genuine dislike for radio. As many of the present fliers were war-traine it is little to be wondered at that radio st has no appeal for them, and that the av age flier has but little faith in it. A man w generally judged by t ability to handle h "ship." If he was clev with radio, providing ! was able to fly, he w forgiven by his fellows.

Experienced fliers a among the most conser ative of men, strange that may seem. Little they relish change or inn vation. They have be flying through all kin of weather and over : kinds of country witho the use of radio. Wl change now? Radio is ju another thing to wor about. It probably wor work anyway, and t receivers in the helm hurt your ears and y can't hear your moto So poor old radio goes consolation to the amate who has been such a go friend all these years. of which the radio eng neer has probably not realized.

TECHNICAL PROBLEMS IN THE PLANE

IN ADDITION to the obstacles form by the fliers' attitude, there have be many technical difficulties to overcon Chief of these is the interference caused the ignition system of the airplane engi This has been a most serious obstacle a has not been completely overcome. It true that, by completely shielding t ignition system, the troublesome noise c be reduced to a point where very satisf; tory reception in an airplane is attained, t such shielding is difficult to install and ev more difficult to maintain.

How is a motor shielded to reduce t interference? How does the ignition syst of a motor produce interfering noises in radio receiver?

The ignition system consists of a hi and low-tension side. The low-tension si consists of everything from the switches the low-voltage side of the magneto in magneto ignition; and everything from the battery, including switches, generator, meters, voltage regulator, and distributors in the battery type of ignition. In the high-tension side we have everything from the high-tension side of the magneto in the first type of ignition, and from the distributors in the second type, down to the spark plugs. In these systems every make and break contact, as in voltage-regulator relay or distributor, produces a disturbance each time the circuit is opened or closed, which should be regular and very frequent, otherwise the pilot has something much more serious to worry about than the QRM from his ignition. The spark plug has not been mentioned in detail yet. Usually there are two of these little short-wave transmitters in each cylinder of the motor. The average airplane engine runs at speeds of from 1400 to 1800 revolutions per minute. This means that in an eight-cylinder, four-stroke-cycle engine, equipped with but one spark plug per cylinder, there will be at 1500 r.p.m., six thousand sparks per minute, or one hundred sparks per second. This produces a noise in a radio receiver which resembles the noise produced by a stream of shot on a loose tin roof. Oscillograms of this QRM indicate that part of the noise is due to induction, just as the "click" heard in a receiver when an electric light switch nearby is opened or closed, is caused by the change in current. The rest of the noise is produced by the oscillating spark in the gap in the spark plug itself. This is a true electro-magnetic disturbance of a definite wavelength. Apparently, then, it should be easy to reduce this interference by means of a short-wave trap; and so it should, but due to the difference in the constants of these small oscillating systems, the use of wave traps has not proved very satisfactory. Up to the present time the most satisfactory method of freeing the receiver of this annoying disturbance is by shielding the whole ignition system.

Completely shielding the ignition system requires that every wire and electric device about the whole plane which carries current be covered with an electric shield. This is usually a braided copper sleeve, slipped over the wire, or a metal container for such devices as regulators, distributors, and switches. This shield must be connected to the ground of the plane. The ground of an airplane consists of all the metallic parts, such as the motor, brace wires, cables and fittings. If you have a few inches of frayed shielding it will cause all the noise to come right back to the receiver. Shielding produces a hazard, the danger of which may be readily realized. If there is faulty insulation anywhere in the system, the vital ignition current will jump through to the ground and out goes part or all of the ignition, depending upon where the break occurs, and, it is needless to say, down comes the plane, to make as safe a landing as the pilot can. It would appear that the solution of this problem is to use nothing but the best of insulation. This is more difficult than it sounds. When a high-tension lead is shielded a corona discharge takes place through the insulation to the grounded shield. The corona produces a chemical change in insulation and it no longer insulates, the engine ceases to "percolate," and the aviator to "aviate."

Now, the pilot knows all this and his feeling for radio has increased in warmth, but not in a direction the radio engineer would like to see. The old feud still exists. The pilot says the engineer loads his plane with hazardous equipment, and the engineer says the pilot is too fussy about what happens to him.

THE FUTURE—WORK FOR ALL

AND so it stands, until the necessity for radio communication between the air and the ground is made apparent to all concerned with flying. That this necessity exists there is no doubt in the minds of many besides the radio engineer, but now the demand for radio is insufficient to induce very much research on these problems.

Such problems can be and are, of course, being worked on in laboratories. However, there is a definite limit to what can be done in a laboratory on the ground. The conditions existing in a plane—the vibration, the noise of wind and motor and ignition, cannot be adequately reproduced in a laboratory; nor can engineers conceive of the conditions except by repeatedly experiencing them in test flights. What I am driving at is this. There should be a laboratory in which the ground and air work is connected and closely related. The engineer should be placed in a position not only to see the problem as the flier sees it, but both flier and engineer should be encouraged to work together. Confidence in the ability and work of the engineer will then come to the flier. Better radio sets will be built, and let us hope that they will be built as a part of the plane and not tacked on—an afterthought. Airplane designers will make provisions for these sets and the power required to operate them. Then, and then only, will pilots want radio, and make good use of what they get.

Before passenger carrying air lines are permitted to operate either in this country or on trans-oceanic flights, this matter of radio should be included in the regulations covering the safety and inspection of the planes. The Department of Commerce should make regulations to fit the needs of the moment. Because commercial aviation is in its growing stage, the regulations should be fairly elastic. But before passengers are permitted to risk their lives, regulations regarding suitable radio equipment and personnel to operate it should be laid down. These should cover all long flights, whether over water or land. By long flights is meant anything over 500 miles.

The radio beacon has had but a very short test outside of the Air Corps experimental tests. But it is apparent even on such short trial that regular flights over long distances of water should not be thought of without contemplating the use of such a beacon. For regular passenger routes over land, the beacon should be depended upon at least for night flying. However, the story of this beacon and its possibilities is too long to include here.

As in so many other things, practice and test are essential to development, and this is no less true of radio on aircraft. The more use made of it the more experience gained. Radio has a very definite and important place in aviation, and it is only to be regretted that use has not been made of it on all transatlantic and transpacific flights. It is likely that the unsuccessful flights would not have had so tragic an ending, had radio played the part that it must come to play in the future of aviation.

WHEN RADIO WAS NO MORE THAN A DREAM

King Edward VII receives a lesson in aeronautics at the hands of Wilbur Wright. Planes of ten years hence, equipped with powerful radio transmitters and receivers, will probably be as much in advance of present day design as are the planes of to-day as compared to this fragile looking craft of Wright's

What is the Matter With Radio Advertising?

FROM time to time, trade associations and better business bureaus formulate codes of ethics for the guidance of writers of radio advertisements. These codes aim to curb exaggerated claims as to long distance reception, quality of tone and other excesses so freely used in radio announcements.

The beautifully worded hyperboles, characterizing modern advertising, have received such spirited attacks recently, that we may look forward to saner and more informative advertising copy. So great is public interest that a book on this subject, *Your Money's Worth*, is threatening to become a best seller. Radio advertising receives its share of scathing criticism from these authors who leave no one unscathed.

Imagination—at least—is lacking when an entire industry depends upon a few standardized general appeals to sell its products to the public. If the advertising is to be believed, all receiving sets possess unbelievable selectivity, marvelous sensitiveness and magnificent tone quality, regardless of price. Rarely does any enlightening information appear in a radio advertisement by which a prospective purchaser may judge the superiority of one receiver over another. Magical phrases are concocted, playing upon the ignorance of the non-technical, to suggest fancied engineering superiority. The uninitiate must be guided by such medicine-man hokum as "utilizing the new intra-paralytic principle of interference submergence," "delightful tone quality obtained with the mastertonic sliding trombone transformers," or "securing magical selectivity by the matched prismatic quartz inductances."

Aside from such senseless and meaningless technical appeals, most radio advertising confines itself to generalized boasts. The same charge may be made not only against the advertising of radio sets, but that of automobiles, iceless refrigerators, and any mechanical or electrical product. The readers of RADIO BROADCAST frequently demand that some comparative technical tests be made to form a basis of judging the relative qualities of sets.

We have given considerable thought to this problem and we would unhesitatingly publish comparative information, could we discover a method of making comparative tests which would not involve the human element and which would be a real test of merit, taking into consideration all of the factors which contribute to the desirability of a radio receiver.

Take, for example, the factor of gain in the radio-frequency amplifier. We may impress a standard modulated signal from an oscillator upon a receiver and measure the resultant fluctuations in plate current of the detector circuit, thus giving an evaluation in the over-all gain of the radio-frequency amplifier. We may also obtain a selectivity curve for each receiver which gives a fair index to that quality. Furthermore, given an adjustable audio frequency oscillator, with which to modulate the incoming test signal, we can determine with a fair degree of accuracy the tonal range and characteristics of the audio frequency amplifier. These three tests would give an index to the three major qualities of a receiving set, namely its sensitiveness, selectivity and fidelity.

Unfortunately, carrying out such tests is far from simple. Most receivers have a gain control in the radio-frequency amplifier system which greatly complicates laboratory tests as a means of comparing receiving sets. Testing a five-tube receiver, the gain might well be adjusted as high as possible, so that it would show maximum amplification per stage. However, when so adjusted, it is likely to show more than normal dis-

SENATORE MARCONI TESTING BEAM TRANSMISSION
The inter-continental beam transmitters of the Marconi Company, now in operation, resulted from a long series of tests. This illustration shows Senatore Marconi testing a short-wave transmitter from a boat on a lake at Livorno, Italy, in 1916

tortion in its audio-frequency amplifier. On the other hand, more conservative adjustment of the radio frequency gain would handicap its sensitivity rating, although it might improve its showing with respect to tone quality. Five engineers could test a number of receivers and secure entirely different results.

If a sufficient number of test conditions are fixed so that the element of adjustment would be minimized, some receiving sets would be unduly handicapped by the test conditions in one respect or another. Consequently, laboratory comparisons, with the test methods we now have available, do not, for the present at least, seem to offer a means of supplanting generalities in radio advertising. But we may look forward to developments in this direction, as our experience with laboratory measurements of sets increases.

Another possible method of making advertising copy more informative is to give a few outstanding facts regarding a receiver, such as number of tubes, number of controls, and other specifications. But the number of tubes in a receiver is hardly a guide to its efficiency. There are ten-tube receivers which give no better results than other six-tube sets. The writer, for instance, has a four-tube receiving set with a 210 tube in the output, which he would confidently enter in any contest for sensitiveness, selectivity and tone quality. But, as a commercial product, it is practically useless. It takes an expert to tune the set and the filters, chokes and by-pass condensers, which are a part of it, would not fit into two set cabinets of normal dimensions. So the listing of specifications is hardly a panacea for indefiniteness in radio advertising.

What remains to assist the honest advertiser in preparing truly informative copy? If we rule out bunk, generalities and specifications, of what may the set manufacturer speak without being frowned upon? Only three general points suggest themselves—outward appearance, price, and reputation, the same factors which the automobile industry has found successful as selling appeals.

Another possibility is to consider some one, simple, technical detail—the thickness of shielding, the strength and rigidity of the chassis, or the accuracy with which tuning circuits are matched—as an indication of the skill and care displayed throughout the whole receiver. Such a policy has advantages, being informative, specific, interesting, and, above all, based on facts instead of on generalities.

Prestige and reputation are the product of years of successful manufacture, and, consequently, production figures and value of sets sold by a manufacturer are a foundation of fact by which an old established manufacturer may distinguish himself from others.

A method, which has been successful in other fields, is to "sell" the engineer who designs the product. Certain companies have engineering and research staffs of acknowledged competence and reputation, whose designs are worthy of great public confidence.

A thorough and detailed study of the radio receiver and those who build it, on the part of the advertising copy writer, is the best preparation for writing advertising which features facts rather than fancy.

Action from the Radio Commission

THE Federal Radio Commission has begun suit against station KWKH, which it charges with the misdeed of using three times the power permitted by its license, for forty successive days. As a result, KWKH is liable to fines aggregating $20,000 at the rate of $500 a violation. If the Commission has a good case and wins out in the courts, it will certainly gain wide respect. The numerous violations of the Commission's regulation as to maintenance of assigned frequencies are likewise subject to fines of five hundred dollars a day. Certain stations frequently wander as much as ten kilocycles from their channels. The former WSOM, for example, was found at different times, within eight days, 24.8, 23.9, 12.5 and 16.1 kc. from its assigned channel.

The Commission, in a public statement, threatened to eliminate about twenty-five of the most flagrant wavelength wobblers but, as usual, grew softhearted in the end and gave them additional grace. Heterodyning is far too widespread to make listening to any but relatively nearby stations any very great pleasure.

The Commission's claim, however, that practically all heterodyning is due to frequency wobbling is not entirely founded on fact. There are altogether too many assignments of stations to the same frequency whose carrier waves are bound to create interference. The clearest broadcasting channels as a matter of fact, are at this time the higher frequencies between 1250 and 1500 kc. On these frequencies, we find mostly low-powered stations which do not interfere with each other.

The numerous hearings held in Washington, upon demand of some of the stations now assigned to these superior channels, are based on the fallacious superstition that the lower frequencies are the most desirable. At one time, when the lower frequencies were reserved for the better stations, while as many as twenty and thirty low- and medium-powered stations were huddled on the lower end of the broadcast band, the ambition to leave the higher frequencies was justified. Although conditions have changed, prejudice against the higher frequencies persists.

Mr. May, seeking a lower frequency for his advertising station, KMA, for example, testified before the Commission that it was a well known fact among radio engineers that the channels below 350 meters were "practically no good for broadcasting purposes," although, as an expert brought out, KDKA, KOA, WBBM, WOK, and numerous other stations, occupying these allegedly unsatisfactory frequencies, have built up nationwide audiences.

The claim that stations do not "get out" on the very high frequencies is made because the public is not accustomed to looking for its programs on these channels. There are too few worthwhile stations using them. Why not assign a few really good stations to the higher frequencies, so as to distribute the public's attention throughout the broadcast band?

Prospects for Patent Pooling

THE Radio Manufacturers' Association is looking into the matter of patent pooling and seeking to inaugurate a system of cross-licensing in the same manner that the automobile industry accomplished this through the National Automobile Chamber of Commerce. There is one great difference, however, in the radio situation and that lies in the fact that a single group has already concentrated most of the patents in its own hands and consequently no one has much to offer it for bargaining purposes.

We learn of the formation of a Radio Protective Association in Chicago with the object of battling against "radio monopoly" which, say the sponsors for the new organization, "will be taken to Congress, to the Department of Justice and to the Courts."

No matter how much outsiders may protest, there is no question about the fact that the Radio Corporation of America has in its hands most of the essential patents to the manufacture of the radio receiver and it is not at the mercy of any outside group. A patent is an entirely legal monopoly created by legislation in accordance with provisions in the Constitution of the United States. Furthermore, the Radio Corporation is extending licenses to competing companies on what appears to be a fair basis. A rather large minimum royalty guarantee is required of the set

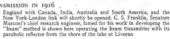

TESTING "BEAM" TRANSMISSION IN 1916

Senatore Marconi's principal assistant in the development of the short-wave "beam" is C. S. Franklin who is here shown on the lake at Livorno, Italy, testing a short-wave receiver with transmissions from the parabolic reflector shown in the accompanying photograph. (*Right*) The "beam" system of short-wave communication has already satisfactorily linked England with Canada, India, Australia and South America, and the New York-London link will shortly be opened. C. S. Franklin, Senatore Marconi's chief research engineer, famed for his work in developing the "beam" method is shown here operating the beam transmitter with its parabolic reflector from the shore of the lake at Livorno

maker; said to be $100,000 a year, which effectively throttles the small producer. Under the patent law, a patent holder has full rights to deny the issuance of licenses to anyone he chooses and, therefore, unless the legal attitude of the patent law is completely reversed, the R.C.A. is entirely within its rights.

The object of the patent law is to assure that inventors are encouraged and properly rewarded. Times have changed and invention is much less a product of individual genius than it is the marshalling of many minds, research facilities and laboratory experience. The reward, instead of going to individual inventors and their backers, now goes to large corporations which make it possible for the complex invention of this day to be made.

The major purpose of the patent is thus fulfilled, both under modern conditions and under those which obtained in the past. We may add a new interpretation in that the patent monopoly shall not be used in restraint of competition and compel patent holders to extend licenses to all those willing to pay just license fees. This plan is followed in Canada. But such a course in this country would be a new situation, a reversal of precedents. It would require new legislation. A possible and, indeed, probable solution of the present radio situation is that the Radio Corporation will extend licenses to smaller concerns on a smaller minimum guarantee, but upon a higher percentage of royalties than it extends to those guaranteeing $100,000 a year.

The radio industry is suffering from the existence of too many incompetent small manufacturers which are bound, in time, to be eliminated by natural economic processes. Hastening their passing by patent pressure is a painful but effective method which, however, reacts unfavorably against those exerting it. But, whatever the considerations animating the policy, the legality of the R. C. A.'s present patent course does not appear to be open to question.

Is Direct Advertising a Service?

A NUMBER of the direct advertising stations have appeared before the Commission, claiming great losses of audience and service range because of their high frequency assignments. Mr. May, speaking for KMA, recently spent three and a half hours on the stand, a record for a single witness before the Commission to date, to prove himself the most popular announcer in the United States and his station the greatest service to humanity of any station, in the corn belt. 450,000 people wrote him during the first seven months of the year, a larger number than practically any but one or two key chain stations can claim.

On the other hand, every questionnaire,

not specially circulated by the stations themselves or by farm papers, indicate the wholehearted public condemnation of direct advertising by radio. RADIO BROADCAST's questionnaire, in which 10,886 expressions of approval and disapproval were made, found KFNF the most popular broadcasting station in the country, 18.8 per cent. of the audience demanding its removal. Considering the fact that those who answered this questionnaire were distributed all over the United States, this seems to represent about 100 per cent. of the listeners within the annoyance range of this station. WJAZ won the disapproval of 15 per cent. of the listeners, most of this vote being a spite vote because WJAZ upset the Radio Act of 1912, rather than because of present day program unpopularity; while KMA came out third with condemnation from 13 per cent. of those answering.

SIGNORA MARCONI

The illustration shows the wife of the noted Italian wi[t] radio receiver fitted up for her use in their palace in [I] Signora Marconi was formerly the Countess Maria C[1] Bezzi Scali

However, 450,000 people do not write a station for nothing. There is no question but that there is a field for the local broadcasting station in the service of the small local merchant. The public, however, resents being-sold harness, glue, tires, and laundry service in the guise of entertainment. The mail order buyer in the rural district is about the only group which responds. Evidently, in spite of the harsh dislike which we have of the direct advertising stations, we must confess that they have an audience and, as such, deserve consideration, but only in proportion to the importance of that audience.

Radio Engineering To-day

R ALPH H. LANGLEY of the Crosley Company writes us at some length in comment on D. A. Johnson's criticism of radio engineers, which we headed, some months ago, "There Are no Radio Engineers." Mr. Langley points out

fendants' device, employing a flexible rubber liaison member, held in place by a rigid frame and covered by an ornamental hood, is an infringing device. ❧ ❧ ❧ The following sets are now licensed under R.C.A. patents: Zenith, Splitdorf, Stromberg-Carlson, Bosch, Crosley, All-American, Freed-Eisemann, Howard, King, Fada, Federal, Murdock, Freshman, Amrad, Steinite, Gilfillan, Day-Fan, Bremer-Tully, Atwater Kent, Federal-Brandes, A. H. Grebe, Pfansteihl and United States Electric (Apex, Case, Slagle, Workrite, and Sentinel).

The Month In Radio

THE evolution of marine radio communication was recently described by T. M. Stevens, General Superintendent of the Marine Department of the R. C. A. Broadcasting considerably hastened the adoption of a continuous wave transmission on a new series of channels, greatly mitigating interference with broadcasting. In 1922, there were twelve spark stations, using principally the waves of 450 and 600 meters, along the coast from Cape May to Bar Harbor. Both on account of congestion and because of the protests of broadcast listeners, seven of these twenty years spark stations are now closed down and the remainder have been replaced by more efficient vacuum tube transmitters. Three hundred ship spark transmitters have also been converted into modified tube transmitters so that they no longer interfere with broadcasting programs.

A few small independent companies are still compelled to use spark transmitters, while many foreign ships with spark transmitters are still working in a manner to interfere with broadcast listening. It is understood that the independent radio companies, operating spark stations, are experiencing difficulty in obtaining properly licensed vacuum tube transmitting equipment. The foreign ship interference will probably be tackled by the International Conference at Washington. Under the circumstances, spark interference with radio programs is likely to be a thing of the past within two or three years, and, possibly sixty to eighty per cent. of the interference is already eliminated. ❧ ❧ ❧ Things have changed for ship operators since the writer pounded the key some twelve years ago. In those days, the emolument was sixteen dollars a month and now it averages a hundred. Considering that the work is generally pleasant and practically all expenses are paid, the radio operator's lot is one to be envied, when compared with that of the clerk with his dull routine and the artisan with his arduous and confining tasks. The radio operator's principal complaint, as we have gathered from interviewing a few, is that once senior operator on a desirable ship, contact with superiors is so limited that the opportunities for advancement are practically nil. Nevertheless, most of the executives of commercial radio companies were once "brass pounders." There is no employment more romantic, responsible and broadening than that of radio operating for the young enthusiast, seeking a career of adventure and promise. ❧ ❧ ❧ The listeners of KFWO, an efficient little 250-watter at Avalon, owned by Lawrence Mott of short-wave fame, have been receiving play-by-play reports of the games played in Chicago by the Cubs. Why this station should go so far afield to present its listeners with this feature is explained by the fact that Mr. William Wrigley, Jr., is so interested in the doings of the Cubs that, while he summered at Catalina, play-by-play reports were sent him by telegraph.

Mr. Mott suggested to Mr. Wrigley that these play-by-play reports be diverted to KFWO and then broadcast. Colonel Green has a rival! ❧ ❧ ❧ The Egyptian government plans to erect a broadcasting station. There are already three thousand sets in operation which, to receive the principal European programs, must be highly sensitive. Eighty-five per cent. of the population of Egypt lives within 150 miles of Cairo and hence a single station can greatly stimulate a market which American manufacturers may do their share in supplying. ❧ ❧ ❧ Any listeners, hearing broadcasting station SOL, have been victims of a slight error which is excusable, due to the distance involved. They are doubtless hearing station XOL, operated by the Tientsin Government in China. Its power is 500 watts and it uses a wavelength of 480 meters. A special license is required from the Chinese government to act as an importer of radio sets and one American Company has taken advantage of this privilege by conforming with the regulation. ❧ ❧ ❧ A beam station, another link in the Marconi worldwide service, has recently been opened for commercial use at Johannesburg, South Africa. ❧ ❧ ❧ WLW, using its short wavelength, supplied an Australasian program recently, enjoyed by listeners of 2 FC, Sydney, Australia, and 1 YZ, Auckland, New Zealand. America is the largest exporter of broadcasting programs in the world. ❧ ❧ ❧ The interference problems of Australia are causing distressing controversy. A new 15-kw. broadcaster is to open at Wellington on 420 meters. What worries the Australians is if Sydney on 440 and Adelaide on 400 meters will not suffer serious interference. Cautious fellows these Australians! ❧ ❧ ❧ JOAK, Tokio, already frequently heard on the Pacific Coast on its thousand watts, is to go on 40,000 watts, which should certainly bring it within range of a good part of the United States during early morning, midwinter hours. It won't be long now before a few American broadcasters will have to close down because of foreign interference. ❧ ❧ ❧ There are 206,334 listeners in Australia, duly licensed and paying license fees. ❧ ❧ ❧ The British Broadcasting Corporation issued a statement recently that it had discovered the advantages of rating stations in terms of kilocycles rather than meters. The advantages of the kilocycle rating have become used in this magazine since August, 1925. In talking to the members of the Federal Radio Commission, we have been pleased to notice that, though at the first the word "wavelength"

was rather frequently in the conversation, it did not take long for the Commission to adopt "frequency" as the only practical term to designate the radiation of a broadcasting station.

WHO REPRESENTS THE LISTENER?

OUR editorial some months ago, entitled "Where Are the Listeners' Organizations?" has brought forward a good deal of correspondence from ambitious would-be executive secretaries, disillusioned leaders who have attempted to form local organizations and readers requesting RADIO BROADCAST to sponsor such an organization. A number have expressed the opinion that listener organizations would be more of a nuisance than an aid to broadcasting. W. W. Waltz, for example, writes that, although in his area WJZ, WEAF, WGY and KDKA are the obvious program leaders, there is a certain advertising station which any Philadelphia listener will recognize, "whose sole idea is to sell every ampere that can be forced off of their antenna. There is no use in trying to describe the junk they broadcast. Everything from near-dirty stories to grand opera selections by the most horrible orchestras in existence. One complaint after another has been made, officially and otherwise, in regard to the manner of operation of this station. Their equipment is modern, but it is adjusted to give a wave like a spark set. And believe it or not—is the most popular station in the city!" The conclusion to be drawn is that no organization can be truly representative of listener tastes.

HOW LONG, OH LORD, HOW LONG?

WE TAKE a special delight in reminding the authors of publicity statements boasting of revolutionary inventions, of the prior discovery and origin of these same inventions, in the hope that more care and conservatism may be displayed, as time goes on, by the publicity romance writers. We note that Mr. C. Francis Jenkins, who has spent many years in research in telephotography, announces the development of radio guiding channels to keep, aviators on a definite course and of a receiving set giving visual indication of deviation from the guiding course. The former has already been widely used experimentally, especially by the Navy Department, and is a well known invention. The visual indicator is not so widely used, although its development in direction-finding apparatus was recorded in these columns several months ago. About twenty-five ships on the Great Lakes are already equipped with the visual direction indicator.

FADING TESTS AT MELBOURNE, AUSTRALIA

Station 3 LO at Melbourne has made a gift to the University of Melbourne for research on the causes of radio fading. R. O. Cherry, working under Professor Laby of the University, is here seen calibrating the portable receiver for measuring signal intensities. The set is carried in an automobile

METERS, KILOCYCLES, OR "CHANNEL NUMBERS"?

By RALPH H. LANGLEY
Crosley Radio Corporation

*O*NE of the most practical and interesting suggestions tending to simplification of radio as far as the non-technical user of radio receivers is concerned is that of Mr. Langley, which he so interestingly discusses in this article. The use of radio receivers will become more and more widespread as the receiver becomes more simple to operate. Great strides in this direction have been made, what with single-control operation and direct light-socket powering of sets. But still, thousands of listeners who don't even know the difference between alternating and direct current, try to solve the dual mysteries of wavelengths and kilocycles which confront them in their local newspaper radio programs and on the dials of their receiving sets. Mr. Langley rightly asks, why should they bother with this wavelength-kilocycle terminology? Frequency calculations in kilocycles—or meters if you belong to that school—are important and necessary for the engineer and the technician, but the listener has no earthly concern for them.

A committee of the National Electrical Manufacturers' Association has been appointed to consider Mr. Langley's suggestion and to take appropriate action. That committee consists of R. H. Langley, chairman; L. W. Chubb, George Lewis, M. C. Rypinski, J. M. Skinner, R. H. Manson, and A. E. Waller. —THE EDITOR.

*P*ERHAPS there is no such thing as "the average broadcast listener." But millions of them come pretty close to the average, and I wonder just what they think when they hear the announcer say that he is broadcasting "on a frequency of twelve hundred and sixty kilocycles." In all but a very few cases, I venture to say that their thoughts have nothing to do with the meaning of these words. Last year it was a "wavelength of two hundred and ninety one and one tenth meters" and that was even worse. Why so many numbers and so many strange words?

Wavelength in meters, and frequency in kilocycles; related to each other by some mathematical law, and yet not related to anything the man in the street has ever heard of. Even the radio engineer must resort to a tabulation or a slide rule to translate one into the other, and yet each and every broadcast listener is expected to use them when he wants to hear his favorite stations. The newspapers print them, and you are expected to know, or somehow to find out, where they all come on the dials of your receiver.

The change from "wavelength" to "frequency" was, of course, a very logical one. It can easily be demonstrated that the current in your receiver or in the distant transmitter has a frequency. The wave out in space is the thing that has a wavelength (as well as a frequency). Primarily we are not interested in the wavelength out in space—but the currents in the receiver—which the listener can hear. Then again, the wavelength listings were irregular and had to be given with at least four figures and a decimal point. The frequencies are given in three or four figures, and the last one is always a zero, because the frequencies are spaced in multiples of ten. But they start at 550 and stop at 1500, and the system is still far from being simple for Mr. Average Listener to understand.

Some manufacturers have tried to put these strange numbers on the dial of the receiver when it was built. Then if you knew and could remember the wavelength or the frequency of the

station you wanted to get, you could set the receiver to that point, and there was the station. There were a lot of mechanical difficulties in doing this, but more than anything else, it was the complexity of the numbers themselves that kept the conventional "zero to one hundred" dial on the sets. Here, of course, is another set of numbers, that must be read from a dial and related to the wavelength or the frequency or the call letters of the stations. It is no secret that the average listener does not know to whom he is listening, or how to find a particular station, except in the case of a very few that are near to him. The others are too hard to find, and many that he could hear and hear well, he does not bother with.

It would be possible to record the locations of our homes and places of business by their latitude and longitude. Your home address, for example might be given as "north 43° 28′ 37.42″, east 76° 18′ 58.13″." That would be just about as easy and just about as logical and just about as technically correct as wavelength in meters or frequency in kilocycles for a broadcasting station. But our houses and our offices are conveniently numbered and so are our telephones. Why not, then, use plain simple numbers for the broadcast frequencies and wavelengths?

"This is station XYZ on Channel 16." When you want station XYZ again, you will turn to the number 16 on the dial. There will be numbers on the dial running from 1 to 96, representing the 96 broadcast channels. You will soon remember the fact that your favorite stations are at 16, 23, 38, 67, and 84. If you notice in the paper that station PQR, on channel 53, is giving an unusually good program, there will be no difficulty about finding it. And the numbers will be the same on all receivers. When you trade in the old set, or when you go over to John's house, you will not be at a loss to know where to find the stations.

A DIAL WITH NINETY-SIX NUMBERS

*I*T WILL be more desirable, of course, to arrange a dial with these simple numbers, than it is to make one that reads in frequencies from 550 to 1500, or in wavelengths from 199.9 to 545.1 with tenths on every one of the 96 of them. And there will not be any unnecessary "meters" or "kilocycles" tied to them. They will just be plain numbers like the one on your front door. You can have a table showing the wavelengths and frequencies corresponding to the channel numbers if you want it; the newspapers and the magazines will print them. But the average listener will not want any such list; he will have no use for it.

Some day the range of frequencies allotted to broadcasting may be increased. When this is done, it is almost certain to be in the direction of the short waves. Then our series of 96 numbers will have to be extended, from 96 up. By starting the number series at the long-wave or low-frequency end, we shall leave room for expansion into the short waves, and we shall also have the smaller numbers for those channels now assigned to the larger and more widely known stations.

RADIO RECEIVERS *for* $175 *or less*

GOOD NEWS *for the* AVERAGE CITIZEN

IT IS an erroneous impression that to possess a modern radio receiver combining both artistic merit and fidelity of reproduction one must spend an inordinate amount of money. The fact is that set makers have produced electrically good receivers and housed them in cabinets that will grace any home—and all at a genuinely moderate price. This and the following pages show attractive moderate-priced receivers ranging in cost from $175 to less than $100. The inspiration for the Bosch 76 receiver shown above is Gothic and it is evident how effectively it may be combined with the furnishings of many a living room. This six-tube RFL circuit receiver uses either a loop or antenna and has an interesting volume control and vernier tuning adjustment. Its price is $175

THE "MILAN"

A cone loud speaker is supplied with this Apex receiver, the price for chassis, cabinet, and loud speaker, being only $135.00. The circuit comprises six tubes, the audio amplifier employing impedance coupling, which is responsible for excellent quality of reproduction. The set is fully shielded and a single dial controls the tuning

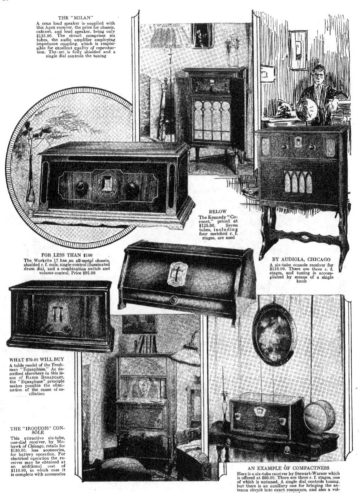

BELOW

The Kennedy "Coronet," priced at $125.00. Seven tubes, including four matched r. f. stages, are used

BY AUDIOLA, CHICAGO

A six-tube console receiver for $110.00. There are three r. f. stages, and tuning is accomplished by means of a single knob

FOR LESS THAN $100

The Workrite 17 has an all-metal chassis, shielded r. f. coils, single-control illuminated drum dial, and a combination switch and volume control. Price $95.00

WHAT $70.00 WILL BUY

A table model of the Freshman "Equaphase." As described elsewhere in this issue of RADIO BROADCAST, the "Equaphase" principle makes possible the elimination of the cause of oscillation

THE "IROQUOIS" CONSOLE

This attractive six-tube, one-dial receiver, by Mohawk of Chicago, retails for $130.00, less accessories, for battery operation. For electrical operation the receiver may be obtained at an additional cost of $110.00, in which case it is complete with accessories

AN EXAMPLE OF COMPACTNESS

Here is a six-tube receiver by Stewart-Warner which is offered at $80.00. There are three r. f. stages, one of which is untuned. A single dial controls tuning, but there is an auxiliary one for bringing the antenna circuit into exact resonance, and also a volume control

OF THE NEUTRODYNE FAMILY
For $100.00 it is now possible to obtain a single-control neutrodyne receiver. The Freed-Eisemann NR-9 is such a receiver and, in addition to the single-control feature, it is completely shielded and has a pilot light on the front panel. Equal amplification throughout the broadcasting frequency spectrum is claimed by the manufacturers

BELOW
The "Warwick," at $138.00, is one of the offerings of Amrad for the 1927-28 season. It is a single-control neutrodyne, completely shielded, and may be used with either loop or antenna

BY SPLITDORF
This is a receiver for the man who still believes in two-dial tuning. There are six tubes in all, three of these being r. f. stages. It is, wired for use of a power tube in the last audio stage. The price is $75.00. The elliptical cone shown is priced at $35.00

McMILLAN'S "RIDGEWAY"
Yet another shielded six-tube, single-control receiver this one being priced at $110.00. A long air column and deep-toned horn are included. There is ample space for batteries or power equipment

ZENITH MODEL 12
Single control is again manifest in this six-tube receiver, four variable condensers being manipulated by a simple movement of the illuminated dial. The chassis is of metal, and the receiver is completely shielded. The cabinet is of mahogany. Price $100.00

What Receiver Shall I Buy?

THE moderate-priced receivers exhibited at radio shows throughout the country this Fall are attracting widespread attention because the offerings in this class more than those in any other price range present greater values than ever before. Radio is now old enough so that those who bought radio sets two and three years ago are now thinking about replacing the old outfit with a more modern and satisfactory one. These pages show a few of these decidedly interesting receivers which can be had at a moderate price and which at the same time guarantee excellent electrical performance. These receivers are simple to control, more than ordinarily compact, and what is of growing importance, are handsome. There are table receivers for those who have but little space for a set, and more pretentious console sets for those who want both a radio receiver and an attractive piece of furniture. Practically all the console models in the medium-price range not only provide space for a loud speaker but also have convenient compartments for A and B socket power units and the convenient relay switch which, through the on-off switch on the receiver panel, controls both A and B units. Many buyers are interested in the console set with these compartments because they can at any time purchase A and B units for their set and, in effect, completely "socket-power" it.

A ROLA CONE

Here is a table model of the well-known Rola cone loud speaker, retailing at $28.50. This unit is equipped with a low-pass electrical filter for the elimination of tube distortion. The heavy turned-wood disk serves as an acoustical baffle surface.

FOR A AND B CURRENT

This Philco power unit will provide 180 volts at 60 mils., and therefore is adequate for a receiver employing six or more tubes, with a power tube in the output stage. The unit, costing $79.50, is supplied in an attractive metal case. A built-in trickle charger keeps the A battery in good condition

THE "UNIPOWER"

A trickle charger and storage battery are combined in this useful device by Gould. It may be obtained in either a four- or six-volt output form

A CONSOLE LOUD SPEAKER

The exponential horn type of loud speaker, of which this is an example, is becoming increasingly popular, due to its excellent reproducing qualities. The one here is by Temple, Chicago, and lists at $65.00

AN INEXPENSIVE CHARGER

A six-volt storage battery may be quickly charged with this device, a product of the Valley Electric Company, St. Louis. The charging rate is 6 amperes and the price, $19.50

FOR A AND B CURRENT

Another device which is capable of converting your a. c. house current into suitable power for your receiver—by Exide. The B output is 180 volts maximum, while the A supply is six volts. A trickle charger keeps the A battery well charged. The approximate cost of operating this device is one cent an hour

AN EXPERIMENTAL SET-UP

RADIO BROADCAST Photograph

This photograph, taken about four years ago in RADIO BROADCAST Laboratory, shows some
of the early photograph transmission and reception apparatus designed by the author

How the Cooley "Rayfoto" System Works

By AUSTIN G. COOLEY

THE articles announcing a system of radio picture reception appearing in the September and October issues of RADIO BROADCAST have attracted widespread attention among radio experimenters. Even without specific data as to the actual operation of the Cooley "Rayfoto" system, experimenters have been fairly besieging the writer since the appearance of these articles and the demonstration of the "Rayfoto" transmitter and receiver at the New York Radio Show.

All the obstacles to making this new field available to the experimenter are being removed, one by one. Engineers are busy designing components and manufacturers are busy getting into production to meet the demand. And, for the broadcasters, an important method of supplying broadcasting stations with 'picture' programs has been evolved. In this article we shall sketch briefly just how the Cooley "Rayfoto" system functions, what each part does, and what its purpose is. These technical details will give the experimenter a clear picture of what the difficulties are and what technical knowledge is needed for him to assemble and operate the apparatus. The Cooley "Rayfoto" recorder is no more difficult to build than a five-tube receiver.

THE SYSTEM IN BRIEF

IN A few words, the cycle of transmitting and receiving a "Rayfoto" picture is as follows: The subject, any ordinary positive or negative print, is placed on the drum of the transmitter or convertor which revolves and feeds it along a shaft before an optical system, which, in turn, focuses the reflected light on to a photo-electric cell.

The amplified currents from this cell are 800-cycle audio-frequency currents varying in amplitude in accordance with the subject. These currents control the radio transmitter output and the signals are received on a conventional broadcast receiver. They are then fed into the "Rayfoto" printer which produces a corona discharge in accordance with the strength of the received signal. The corona discharge

All you need for picture reception is a standard receiver, an oscillator, a stop-start motor mechanism, photographic paper, and enthusiasm. The important part of the receiving mechanism is the motor mechanism. With oscillators and receivers we are all familiar. The motor mechanism and all other necessary components duly approved and labelled with the Cooley Rayfoto label will soon be on the market. Those eager to be the first in their communities to receive pictures by radio may send their names and addresses to RADIO BROADCAST and these will be sent to the manufacturers making the parts. The total cost will not be more than $100.—THE EDITOR

takes place at the point of a corona needle which feeds along a revolving drum as the needle traces over a photographic paper wrapped around the drum. At the end of each revolution of the drum the received signals are diverted from the printer unit to a relay which is actuated when a synchronizing signal is received at the beginning of the revolution of the convertor drum. This relay in turn operates the trip magnet which releases the recorder drum so it may start off at the same time as the convertor drum. After the needle has fed along the entire length of the paper, the latter is removed from the drum, developed, washed, fixed, and washed again. The result is a picture of a prize fighter who has been knocked out a few minutes before; or a picture of a railroad wreck just occurred; or maybe a picture of some sweet young thing who may have won a bathing beauty contest in the afternoon.

Phototelegraphy is not complicated and involves nothing that is really new in physical science, but many of the 'kinks' involved must be well understood if good initial success is to be expected. Most of the difficulties ordinarily involved in picture reception work will be prevented because manufacturers will supply equipment especially designed for the purpose and if the experimenter understands the principles of the system and can handle amplifier and oscillator circuits, he should have no difficulty in setting up his "Rayfoto" recorder and having good picture reception right from the beginning.

All systems of phototelegraphy have one

limitation in common: They can transmit only one shade and one unit area of the picture at a given instant and therefore transmission must be accomplished by dividing up the subject to be transmitted into thousands of small areas.

The "Rayfoto" and many other systems transmit the signals for each unit area in rapid succession and the resultant signal varies in amplitude in accordance with the shading of the picture. The speed at which the impulses are transmitted depends largely upon the ability of the electrical impulses on the recording medium. The corona method of printing used by the Cooley "Rayfoto" system is capable of printing faster than any other system the author knows of, but for simplicity and low first cost we are using a signal frequency of about 800 cycles per second, which does not permit printing as rapidly as is possible with the system when higher frequencies are used. The possibility of operation at higher frequencies has been taken into consideration in designing the present equipment so that the speed of transmission can gradually be increased without necessitating any radical changes in equipment.

As explained in the October issue of RADIO BROADCAST, the picture or subject to be transmitted is placed, at the transmitting station, upon the drum of the picture transmitter, which we will hereafter call the "convertor." A small spot of the picture is illuminated and the reflected light from this spot actuates a photoelectric cell, the signals from which control the radio transmitter after the photo-electric cell currents have been sufficiently amplified. Each time the drum is revolved, the spot of light traverses a different path an eighteenth of an inch wide across the picture. The line is broken up into 480 sections by the optical system so that 480 electrical impulses are transmitted every revolution of the drum and each impulse corresponds in intensity to the reflected light from a small area of the picture. The result is that 480 electrical impulses are transmitted for each

revolution of the drum, or about 800 per second when the drum is making one hundred revolutions per minute. Running at this speed the drum feeds along the shaft T, Fig. 1, at the rate of one and a quarter inches per minute. The drum is two inches in diameter and about five inches long. This will give us an operating speed of four minutes for a five-by-six-inch picture.

The beginning of each revolution is marked by an impulse made up of twenty strong 800-cycle signals in succession. This impulse is used at the receiver to start the recording drum off at exactly the same time as the transmitter drum, for it is necessary that the two drums start off together. To accomplish synchronism in this way, known as the "stop-start" method, the recording drum must start a revolution at the same instant as the transmitter drum. It is necessary that the recorder drum run slightly faster than the convertor drum, then stop at the end of the revolution for an instant until the convertor drum completes its revolution. A trip magnet operated by the strong synchronizing impulse releases the recording drum at the proper time.

This trip magnet is operated through a relay which is connected to the rest of the system only after the revolution of the recording drum has been completed. Between the time the recorder drum stops and the time the synchronizing impulse is received, there must be no strong signals received, so we paste a strip of white paper at the end of the picture being transmitted so the signals will be weak while the recorder drum is stopped. Should a crash of static or some other disturbance be received during this waiting period, or "recorder lap," as it is called, the recorder drum will be released in advance of the synchronizing signal. By making the recorder lap very small, the danger of such a "static slip" will be reduced proportionately. The wider the white strip on the picture being transmitted, the greater will be the chances of a good start after a static slip so that the only marring effect will be one line slightly out of place.

We will consider here a few of the principles involved that affect the characteristics of the received picture. In picture work, we wish to reproduce at the recorder shades of light and dark corresponding exactly to those of the transmitted subject. The light reflected from the subject varies the current through the photoelectric cell in a ratio almost directly with the intensity of the light, and this current, after amplification, is made to control the power input to the radio transmitter modulator which therefore varies directly proportionately to the reflected light, due to the characteristics of the Heising modulator.

The final modulated radio signal sent out over the air will vary as the square root of the reflected light. The received signal is amplified lineally in the radio-frequency stages of the receiver. The detector output varies as the input squared, however, and therefore the current in the plate circuit of the detector will be directly proportional to the reflected light. The signal then can be amplified in the audio amplifier and delivered to the "Rayfoto" printer with an intensity directly proportional to the reflected light at the transmitter.

Limited by the data available at the present time, this is as far as we can go with the signal and know definitely what we are doing in the way of maintaining the proper signal ratio through the various circuits. We have no exact data on the relation of the input to output of the Cooley "Rayfoto" printer. Also we do not know the relation of the power delivered by the "Rayfoto" printer to the effect it has on the receiving paper. This factor is quite flexible and can be controlled considerably by the selection of the printing paper and its time of development in the photographic solutions. The printing paper we recommend today probably will not be the paper you will be using next year. It is therefore necessary to have some control over the system so we may match our amplification characteristics to conform with those of the receiving paper we may choose to use. For example, if the received picture does not show sufficient contrast in the lighter shades but too much in the darker shades, we must adjust our amplification characteristics to correct for this. One way it can be done is to reduce the filament voltage on one of the amplifier tubes so that the strong signals are cut off somewhat by running over the top knee of the characteristic curve while the signals of lower value are on the straight portion of the curve. Additional correction may be obtained by reducing the time of development in the photographic solution.

The most convenient place for signal characteristic control is in the detector circuit, because of its "squared" characteristic. This characteristic may be varied considerably by proper proportioning of the grid condenser and grid leak. If the grid leak can be brought down to a very low resistance, say 500 ohms, and the plate voltage made adjustable over a range of from 4 to 40 volts, additional control of considerable value will be gained. Instead of varying the plate voltage, a variable grid battery may be used.

RADIO BROADCAST Photograph
SOME EARLY "RAYFOTO" EQUIPMENT IN RADIO BROADCAST LABORATORY
The apparatus on the right of this picture is a Cooley photograph transmitter and in the center is an amplifier and "corona" apparatus. The picture receiver at the left has been redesigned in many ways to make its operation as simple as possible. This apparatus was photographed four years ago

EFFICIENT AUDIO STAGES NECESSARY

A GOOD picture must not only represent exact shadings of the subject but it must also show up most of the small details of the original. A poor audio amplifier system will blur up the details in black shades and will not permit any of the details in the light shades to appear. The amplifier must not oscillate at any audio or super-audio frequency or even tend to oscillate. Oscillations in audio amplifiers most generally occur because of feed-back through the B batteries from one stage to another and can be prevented by the use of low-resistance batteries, a very large condenser across batteries of moderately high resistance, or by the use of independent batteries for the audio amplifier. The first pictures transmitted will contain sufficient contrast so that imperfect amplifying characteristics will not appear very noticeable. Nevertheless, the progressive experimenter should try to keep one step ahead of the game.

The plate current drain on the B batteries due to the Cooley "Rayfoto" printer will be about 10 or 15 milliamperes, so that the total current drain of the printer and an ordinary five-tube receiver will be in the neighborhood of 45 milliamperes. However, this additional drain of 15 milliamperes will only be present when the printer is being used and, since it will not be operated for long periods at a time, an ordinary set of B batteries should be good for many months of service. A total voltage of about 200 volts is required.

Naturally an amplifier that can be operated without oscillating is much more efficient than one that tends to oscillate and which therefore requires the introduction of some loss to prevent oscillations. In many cases, however, it is more convenient to use an amplifier we already have and which can be "doctored" up a little to make it serviceable for "Rayfoto" work. A resistance across the secondary of one or more of the transformers will prevent the amplifier from oscillating. The required resistance may vary between 100,000 ohms and 2 megohms.

Many broadcast receivers have sufficient amplification in their own system so that additional audio amplification is not necessary. You may test out your receiver in the following manner to determine whether any additional amplification is required to operate the recorder: Place a milliammeter in the plate current of the last amplifier stage; cut the current down to 0.2 milliamperes by increasing the C battery potential; short-circuit the loud speaker terminals; then tune-in a local broadcasting station. If the milliammeter jumps up over 15 milliamperes, no additional amplifier stage is needed. Even if it only goes to ten mils. it will not be necessary to use the added stage but this amount of current will allow only a very small margin of safety.

If an added stage of amplification is required, a special transformer should be used, one that is capable of operating without saturation and which will not produce oscillations in the audio system. Special transformers for this work will soon be available.

The Cooley "Rayfoto" printer is the device for producing the corona discharge that affects the photographic recording paper. It converts the received audio-signal into a fluctuating source of light corresponding to the transmitted signal. This unit consists of a modulated oscillator feeding a corona coil. The corona discharges are secured from the high-voltage side of the corona coil secondary winding.

Readers may wish to have some explanation of the nature of the corona we refer to here. Visually, the corona discharge at the needle point is similar in appearance to those produced by a violet-ray machine. This discharge occurs when a difference of potential of 13,000 to 26,000 volts per centimeter (which, incidentally, won't hurt you) exist around the needle point. This potential is produced by the radio-frequency amplifying transformer, known as the corona coil. The primary of this coil is part of a vacuum-tube oscillator operating at a frequency of 333 kc. (about 900 meters). The plate circuit is supplied by the signals from the radio receiver. After being amplified to supply enough power to the modulation transformer, these signals are strong enough to produce a strong corona discharge when strong signals are received. For the sake of efficiency and shading, about one hundred volts of direct current is supplied, in series with the modulation transformer to the plate of the oscillator. This boosting voltage must not be sufficient to produce a corona that will print when weak signals are coming through.

The oscillator of the "Rayfoto" printer radiates for some distance if the frequency is high, and to prevent such interference we have chosen the reasonably low frequency of about 333 kc.

We do not recommend an oscillator frequency corresponding to more than this unless careful shielding is used.

The "Rayfoto" recorder is the mechanical unit of the system which consists of the receiving drum driven by a motor and controlled with a "stop-start" system of synchronizing. A screw feed arrangement feeds the corona needle along the drum as it revolves so that the needle moves along at approximately the same speed that the convertor drum as the transmitter moves along its shaft.

The "stop-start" mechanism consists of a slip clutch between the motor drive and the drum, and a trip magnet arrangement that stops the drum at the end of each revolution until the synchronizing impulse is received. This impulse trips the armature of the magnet which operates through a relay. By this system, the transmitter and receiver are synchronized about twice a second, thereby eliminating much delicate and expensive synchronizing apparatus.

The recorder drum is the same size as the one at the transmitter but since the recorder has a slight "lead," that is, runs slightly faster than the convertor drum, the received picture will be stretched out a small amount, depending upon the amount of lead. To compensate this, the gears between the drum and screw feed shaft will be of such a ratio that the needle will feed along a little faster than the transmitter drum so that the proper proportions are restored. As a result the received picture will be slightly larger than the one transmitted.

It is desired to keep the lead as small as possible so as to prevent excessive stretching of the picture. Also, if the lead is too much, the trip magnet may be tripped from a subject signal instead of the synchronizing signal. If the lead is too small, the synchronizing signal may be received before the recorder drum has finished its revolution and has switched the relay in the circuit. Consequently the relay will be operated by the next strong subject signal.

Complete constructional data for a Cooley "Rayfoto" receiver which may easily be made at home is scheduled for next month's RADIO BROADCAST.

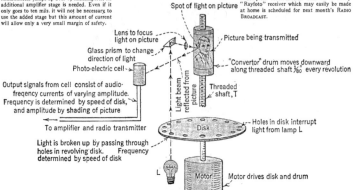

A PICTURE DIAGRAM OF THE COOLEY TRANSMITTER

This drawing shows graphically how the Cooley picture transmitter produces electrical signals varying in strength in accordance with the shading of the picture being transmitted

REFINEMENTS OF THE MODERN RADIO RECEIVER

By EDGAR H. FELIX

THE modern manufactured receiver has become a precision product, built under most exacting conditions prescribed by skilled engineering departments. It is no longer a heterogeneous combination of parts, wired together in conformation to circuits supposedly possessed of magical qualities. Performance is the product of a thousand and one engineering decisions carried out with a care so far above and beyond that which the average buyer can appreciate that engineering refinements are no longer considered suitable as selling arguments by which to sway the buyer's preferences.

Where should the by-pass condenser across the filament leads be placed? Is the improved performance attained by placing it directly underneath the tube sockets sufficient to warrant a special moulding? Does a one per cent. difference in moisture content of the insulating paper of fixed condensers reduce losses sufficiently to justify an additional cost of twelve cents per receiver and does that involve an increasing percentage of condenser breakdowns? Should three more turns be used on the radio-frequency transformer primary to get slightly improved quality or does that involve a sacrifice in selectivity too great to be permitted under present broadcasting conditions? Should the audio-frequency system be designed to cut off at 5500 cycles or at 4800, in the first case giving slightly improved reproduction; in the other, slightly reducing the effect of certain types of interference noises?

It is such highly technical questions as these, clouded in a veil of mystery to all but the experienced radio engineer, that makes one radio set better than another. The placing of a socket half an inch one way or another may make an imperceptible difference in performance, but it is the multiplication of such details, carefully determined after engineering study, that assures the buyer of his money's worth.

In a sense, we have come to a parting of the ways between the factors that make real radio performance and those which make up the buyer's mind between one radio set and another. There is a premium on the little, superficial improvements which the buyer can appreciate because they are the only practical ways of expressing engineering ingenuity to the ultimate consumer.

ONE REFINEMENT OF FADA

THE Fada receivers of this year, for example, employ a new simplified power switch and volume control, an obvious convenience which any prospective purchaser will appreciate. One control takes the place of two. Hidden, in the beautiful cabinet, is a chassis made of $\frac{7}{8}$ inch pressed automobile body steel. It is supported on a three-point suspension with absolute rigidity so that the parts mounted in it cannot get out of alignment. The accurate matching of variable condensers contributes not only to selectivity but to quality of reproduction. In past years, an accuracy of one per cent. in capacity throughout the tuning range has been considered satisfactory. The Fada condensers are matched to an accuracy of $\frac{1}{8}$ of 1 per cent. and the same standard is applied to the tuning inductances coupled with them. These are a few of the hidden values which make for good performance.

The shaft on which the tuning drums and variable condensers are mounted is one-half-inch flash copper plated piston rod steel and is gauged to a tolerance of .0005 of an inch! The day of the curtain rod condenser support is over. The pistons in your automobile are gauged to no closer tolerance.

Advertisements shout uniformly about the most selective receiver with the best tone quality, but give the discriminating buyer no real facts to help him appreciate the performance of a receiver. Generalities may sell the uninformed and help to create name familiarity, but sterling worth, built in by ingenious engineering and painstaking manufacture, is hardly ever conveyed to the reader of advertising.

GOOD THINGS YOU DON'T SEE

ANOTHER instance of superficial selling points which make an obvious appeal to the uninformed buyer and the equally important hidden refinements which contribute even more significantly to good performance is found in the Freed-Eisemann receiver. Several models are equipped with a voltmeter so that the set owner can readily check the A, B, and C voltage applied to every tube of his receiver. Since accurate voltage supply is of vital importance in the performance of the receiver, the selling value of that feature is obvious. But how many buyers know of the two special bonding clips which ground the shielding of the detector stage in order to dissipate more readily the radio-frequency currents generated in that shield? It is a minor point, but an expression of the engineering care which makes the modern radio receiver.

Recently, the writer visited the Stromberg-Carlson factory at Rochester, New York. A complete understanding of the refinement which is concealed in the cabinet of the Stromberg-Carlson receiver hardly ever penetrates beyond the monument to engineering skill and idealistic production standards which that new factory actually is. One could write a thousand words on how the cord by which you tap the a. c. power line is made! Only a detail, but it assures unfailing service for a period of years. It means no frayed cord and no breakdowns, an advantage which passes practically unnoticed in the attention of almost every buyer of that receiver. But to give him that advantage, special engineering standards have been set for every item of material used in the flexible cord. The fine copper wires which, woven together, make an everlasting cable, are ten times as strong as ordinary wire. Special flexible conductor, which does not break if sharply bent, is employed. The individual strands are so fine that they cut the finger like a razor blade. Covering these wires are insulating materials adding a factor of safety far above and beyond that considered necessary. And finally, selected cotton is woven over the insulating material, giving a mechanical strength so that the copper wire itself is relieved of most of its load. Last, but not least, comes an outer covering of silk so woven that there will be no untwisting of the cable and it will hold its lustre for a period of years. Outwardly, there is but little to distinguish this little engineering masterpiece from an ordinary power connecting cord which will fray, untwist, and break in the course of time, particularly if it must be pulled out each time the vacuum cleaner is used. Probably not one salesman in ten thousand selling the Stromberg-Carlson receiver ever considers this refinement—one of a thousand which conscientious engineering has built into that product.

SOMETHING ABOUT CONDENSERS

ONE feature which every radio enthusiast appreciates is the advantage of straight frequency-line tuning condensers over the straight capacity-line type. The desirability of even spacing of stations over the tuning dial throughout the broadcast range is obvious. But, with the almost universal tendency toward multiple condensers, needed to obtain single control, the necessity for straight frequency-line condensers has caused many an engineer gray hairs. It is very difficult to secure uniformity in quantity with condensers having the peculiarly shaped plates necessary in straight frequency-line tuning. With straight capacity-line used in connection with matched inductances, uniformity is easily attained and tuning circuits readily matched. All that need be done to match the stages is to adjust the condensers at any point on the wavelength scale, after the receiver is assembled. Once that is done, absolute accuracy is likely to obtain at all dial settings. But straight capacity-line condensers mean that, at the short wavelength end of the dial, stations are hopelessly crowded, while, at the upper end, they are widely and wastefully separated.

A fine example of engineering refinement in meeting this problem is embodied in the Federal receiver. Condensers with square plates, sliding in rigid grooves, assure absolute uniformity of capacity variations and attain a standard of accuracy almost impossible to secure with condensers having specially formed plates to secure the straight frequency-line effect. But the buyer of a Federal does not sacrifice the advantages of straight frequency-line tuning by the use of these condensers. An ingenious and well designed gang tuning control mechanism gives him all the advantages of straight frequency-line tuning

The mechanism is a masterpiece of mechanical design.

Another example of true engineering beauty is embodied in the antenna tuning compensator which is a part of the mechanism. With multiple tuned circuits, the designer has the choice of several ways to compensate any variations in antenna capacity. He may use a broadly tuned antenna or input stage which gives but little or no amplification. Such a stage contributes its share of tube noise and accentuates nearby station interference. Or else he may employ a sharply tuned stage which has a separately adjusted compensating condenser. Of course, the most efficient and satisfactory method to the user is an antenna stage which tunes sharply and contributes its share of amplification. Those having receivers with a vernier antenna compensating condenser have noticed that, although they have a main tuning dial which gives the set the appearance of one control, they must actually adjust two dials—the main tuning control and the compensator—to tune-in a station properly.

With the Federal receiver, the compensator is geared with the main tuning adjustment and automatically keeps the antenna circuit in step with all the rest throughout the tuning range.

THE R.F. OSCILLATOR USED BY BOSCH TO MATCH COILS

The author would not be surprised to discover that the electrical law determining the correct compensating adjustment needed for all types of antenna systems and working out the mechanical arrangement which assures adherence to that law was a bigger engineering job than designing the entire radio set marketed in 1924.

All this engineering precision applied in design may be nullified by carelessness in production. The electrical and mechanical measurements made in the modern radio plant are of an order of precision unrivaled in any field of quantity production. Atwater Kent, for example, makes 159 precise tests, each requiring engineering knowledge to perform, in producing a single receiver. Some of these tests entail mechanical precision measurements; others electrical measurement of currents of mere millionths of an ampere. But every test contributes to the purchaser's assurance of reliable radio service.

MATCHING INDUCTANCES IN QUANTITY

HOW one manufacturer applies engineering precision to production is best expressed in the words of William F. Cotter, radio engineer for the American Bosch Magneto Corporation:

'The subject of sorting and matching of inductance coils is one to which we have given considerable thought. Its importance is recognized by every manufacturer. With one test

fixture capable of handling the job, it would be comparatively easy to install a check of the manufactured product and be assured of a high degree of accuracy. However, when two or more test fixtures are required to handle the volume of coils manufactured, and where it is desired to check these coils at radio frequency, a more serious problem presents itself.

"We have employed several methods over the last three years The one I describe, however, represents our latest development and is the result of all our past experience.

"A radio-frequency oscillator is built up around the standard 201-A tube. Incorporated in the tuned circuit of the oscillator at nearly ground potential is a resistance of approximately 10 ohms. Across this resistance is shunted in series the coil to be tested and a variable condenser with its separate vernier condenser. When this series circuit is brought to resonance, the total resistance of the shunt circuit comes down somewhat in the neighborhood of the 10-ohm resistance, and one-half of the oscillation current is diverted to the series tuned circuit. Included in this tuned circuit is, of course, the usual thermo-galvanometer.

"By means of a properly chosen vernier condenser, the coils coming through can be checked for accuracy and sorted in as many divisions as experience shows necessary. However, while this is suitable for one test fixture, the problem of keeping two or three fixtures oscillating at the same frequency presented itself. Naturally, if the frequency of the individual oscillators varies, say 10 per cent, it is quite impossible to group coils together just by dial readings.

"We have solved this problem by building a crystal oscillator as one of the test fixtures. Other oscillators of ordinary type are built and each is equipped with a vacuum-tube detector. The oscillators are placed so that it is possible to couple in some of the energy from the crystal oscillator. By means of a headset, the operator adjusts his oscillator to zero beat with the crystal oscillator. Usually, it is not necessary to check this setting more than two or three times a day. By this method, coils may be checked and sorted with an accuracy of a few tenths of one per cent."

That is engineering refinement. It is the kind of "detail" which makes the 1927 radio receiver a precision product.

The broadest appeal to the radio buyer, and one to which most people respond, by and large, is the outward beauty of the product and the name reputation of the manufacturer. Faced, as a prospective buyer is, with numerous products attractive from these standpoints, he is easily swayed from one brand to another by superficial selling points. The more discriminating and intelligent buyer—and in the radio field, because of the host of persons who have a smattering of technical knowledge, this class is predominant and influential—looks to the hidden qualities, the expression of engineering ingenuity and manufacturing skill, as well as the performance qualities, in deciding between one receiver and another. The glittering generalities which have characterized successful advertising and have successfully built up huge quantity production, are at last beginning to suffer a reaction. There are too many "best" automobiles and too many "finest" radio sets. But facts, the refinements, the details, presented to those who can understand them, are indisputable evidence of inherent quality.

A RADIO receiver, as any engineer will tell you, is a complicated and highly organized machine, designed to perform several functions none of which is independent of the others. It is because of these interconnected functions that the engineer will wish he were in Europe if you ask him:

"What is the best radio?"

An automobile, on the contrary, is comparatively simple. It has but one function to perform, it must take energy in some inert form, say unconfined gasoline, and convert it into some other dynamic form which is useful in carrying someone somewhere.

To answer a question regarding the best automobile, then, is simpler, especially since the automobile industry has been established long enough for the most expensive car to be usually the best. Unfortunately, radio has not even this truth to go on, for, all things considered, the most expensive radio does not always pan out to be the best.

A radio must do three things, and therefore there are three problems of design. They deal with:

1. SENSITIVITY. The receiver must be sensitive enough to pick up the signals one wants and amplify them sufficiently to give good loud speaker volume.
2. SELECTIVITY. The receiver must accept signals from the one station the owner wants and reject all others.
3. FIDELITY. The loud speaker signals must be a faithful reproduction of the original in tone and in relative volume.

Thus a perfectly selective and sensitive receiver would pick up any station operating at any distance on any frequency, and would turn a deaf ear to all others, no matter how near the given station in frequency or distance, or how great their power. If the receiver has 100 per cent. fidelity of reproduction, signals as loud as the original would come from the loud speaker, and with all tones exactly as they originate in the studio.

Needless to state, there is no such receiver.

There are several reasons for this. As mentioned above, these varied functions of a receiver depend upon each other, and not always to the same degree. For example, an infinitely selective receiver would be in the present state of the art, highly unsatisfactory from the standpoint of fidelity. Advertising writers to the contrary, one cannot have something for nothing, even in radio.

There are those, however, who desire to know which of two receivers is the better, and among these are the reputable manufacturers themselves. It is for this reason that set testing methods have changed.

Not so long ago, when a receiver had been put together and wired someone took it to a test bench where it was hooked to batteries and an antenna. If signals came out of a horn somewhere the set worked. It was sent to the dealer at once. This was an obvious test.

The obvious, as we all know, is not always the best. In the case of good radio receivers, this older, obvious test has given way to more scientific tests which do not depend upon a smoke covered antenna or the often tired ears of a test man, but upon tireless and unemotional instruments.

In spite of the fact that these tests have been practised—sporadically, we suspect—in the better known laboratories, it has been only within the last half year that descriptions of them began to appear in the technical literature. So new is the industry that standard tests have not been developed, nor have engineers even agreed among themselves regarding even the nature of these tests. Probably while this is being read, committees of the various radio organizations and radio engineering societies will be weighing conditions of test and endeavoring to set definitions of the perfect receiver, definitions that everyone will recognize.

By KEITH HENNEY
Director of the Laboratory

WHAT HAPPENS INSIDE THE BOX?

NOW, before trying to find how much of anything a receiver does, it is well to get an idea of what happens inside the box, to discover if possible what should happen and how much. Then we shall have some idea of what to look for and what the relative order of magnitude will be. For example, harking back to the automobile, almost everyone knows that an automobile has a carburetor. Some people know what it is for; the writer does not. There is also a clutch and some other apparatus which everyone who is about to buy or run or design or fix a car should know all about. He should know what these various pieces of equipment are for, what happens if you leave some of them at home, and when the car runs properly, what the magnitudes of the various operations are. The quantity of gas and oil per mile, or hour, revolutions per minute, miles per hour, ability to climb hills, 'pick-up," etc., are all terms that have both definition and dimensions.

A radio receiver also has several pieces of component apparatus. Each has a different function, and in each piece of apparatus something happens when the set is operating properly. What we must know first is what is each part for, how it does its work, and then how much should a good unit do? This will enable us to define a perfect receiver, and if we have laboratory equipment and patience, or if the manufacturers will supply us with the proper data, we shall be able to decide just where in the scale of goodness our particular receiver stands.

The inner works of the receiver, except superheterodynes, consist of three parts: a set of tubes which serve as radio-frequency amplifiers, boosting in magnitude the incoming signals without change in form and very sharply tuned; a detector more broadly tuned which changes the signals into an audible form; and third, audio amplifiers which bring the detector signals to loud speaker volume. The last are not tuned at all, or at least very broadly.

The radio frequency amplifiers are intimately connected with the transmitting station. The receiving antenna or loop is situated in a reservoir of energy, part made by man, part by nature; part useful, part disturbing. A fair share of this energy is created by broadcasting stations which pour their output into what scientists call the "ether" and what the layman calls the "air."

The energy one wishes to receive speeds from the broadcasting station with the velocity of light, so fast that it may be heard from the loud speaker before it is heard in the rear of an auditorium in which the broadcasting may occur. This is because sound travels through air at 1100 feet a second and radio waves through the ether at 186,000 miles a second.

The receiving antenna is almost as intimately connected with the transmitter as if a wire joined them metallically, although less efficiently to be sure. What comes out of the transmitter sets up a voltage across the receiving antenna. Naturally, the stronger the transmitter, the nearer to the receiver, or the higher the receiving antenna the greater is the received voltage.

Here is where we start on our measuring expedition. How can we measure the relative strength of a transmitter at a given locality?

HOW RECEIVED ENERGY IS MEASURED

IN THEORY the problem is about as follows. The actual voltage across a given antenna is measured by the substitution method. That is, a given deflection on a meter is secured from the distant station. Then a voltage which can be read on another meter is substituted for the distant transmitter and when the proper deflection is secured the voltages are equal. The process may be changed as follows. The given deflection is secured. Then a much greater voltage, which

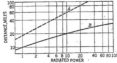

RADIATED POWER

CONDITIONS UNDER WHICH SETS WORK

Estimates agree that a radio set must have delivered to its antenna a signal strength of between one and ten millivolts per meter for "good" service. This means that the received signal will then be strong enough to over-ride ordinary static and local electrical disturbances. This curve shows the increase in power necessary at the broadcasting station to increase the range at which a field strength of ten millivolts per meter is delivered. Curve A shows the power required to lay down this field strength *without* any absorption of the wave; Curve B shows the unit power needed to lay down an equivalent field strength *with* all sources of absorption included. Note that at a distance of about 20 miles from the station, only about 1.1 units power is required for ten millivolts while, with absorption, ten units are required to produce the equivalent signal

can be easily measured, is cut down in known steps until the same deflection is noted. In this manner the field strength of a given station may be definitely measured.

Since the field strength at a given receiver varies with the antenna height, the usual basis of comparison is field strength per meter height. It is merely the actual voltage measured divided by the effective height of the receiving antenna. This is expressed in millivolts per meter and is a factor which is a measure of the effectiveness of a given transmitter at a given locality at a given time of day.

A given number of millivolts per meter will produce a certain loud speaker response with a given receiver. The more sensitive the receiver the greater loud speaker signal will be secured from a given field strength, or conversely, a given

signal may be produced by a weaker field strength the more sensitive the receiver.

Here we are talking dimensions of magnitude. What we want to know is the field strength that will override static and other interference to produce a good lusty loud speaker signal, one that will be good, day or night, rain or shine.

The following table is taken from Dr. Alfred N. Goldsmith's paper in the *I. R. E. Proceedings* for October 1926 and shows what may be expected from various field strengths.

Signal Field Strength	Nature of Service
0.1 millivolt per meter	poor service
1.0 " " "	fair service
10.0 " " "	very good service
100.0 " " "	excellent service
1000.0 " " "	extremely strong

So far so good. Let us see how powerful a station must be to deliver such a field strength over a certain distance. Again quoting from Doctor Goldsmith's paper we have the following data.

Antenna Power	Service Range
5 watts	1 mile
50 "	3 miles
500 "	10 "
5000 "	30 "
50,000 "	100 "

There seems to be some regular progression here between the power and the range of station —in fact a law exists stating that the range of the station varies with the square of the power of the station. That is, to double the range we must quadruple the power. To increase the service range three times we actually increase the power the square of three or approximately ten times.

Now quoting Lloyd Espenschied in the *Bell System Technical Journal* (January 1927), we find:

Fields between 5 and 10 millivolts per meter represent a very desirable operating level, one which is ordinarily free from interference and which may be expected to give reliable year-round reception, except for occasional interference from nearby thunder storms.

From 0.10 to 1 millivolt per meter, the results

may be said to run from good to fair and even poor at times.

Below 0.1 millivolt per meter, reception becomes distinctly unreliable and is generally poor in summer.

Fields as low as 0.1 millivolt per meter appear to be practically out of the picture as far as reliable, high quality entertainment is concerned.

WHAT POWER DO STATIONS DELIVER?

FROM a given station the field strength falls off according to an inverse law, that is, if we double the distance we shall halve the field strength: "a 5 kw. station may be expected to deliver a field of 10 millivolts from 10 to 20 miles away and a 1.0 millivolt field not more than 50 miles away."

In a Bureau of Standards paper, "General Report on Progress of Radio Measurements," (April, 1924) the following data were published. When WEAF was transmitting with 3 kw., its field strength at 10 miles was 32 millivolts per meter. When KDKA has a nominal power of 10 kw. its field at 10 miles was 43 millivolts.

All of these statements may be expressed graphically and the curves on these pages contain much meat for thought. What everyone wants is good lusty signals from a high quality station, day and night, without resorting to regenerative receivers to boost the volume at the cost of fidelity, without being forced to listen to nearby poor quality stations riding in on an adjacent channel, or without having his program more than liberally punctuated with static or extraneous noises. It is up to broadcasting stations to produce a field strength that will insure programs and transmission of this desirable quality. It is up to design engineers to produce receivers that will serve their owners with loud speaker signals from the field strengths laid down by high quality stations. Mathematics alone is not infallible; some experiment and laboratory work must go with it to make certain all the factors have been considered.

Subsequent articles will deal with methods by which engineers check up on the soundness of their designs; methods by which sensitivity, selectivity, and fidelity may be measured.

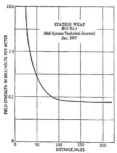

STATION WEAF
(610 Kc.)
(Bell System Technical Journal)
Jan. 1927

HOW FIELD STRENGTH DECREASES

This curve, from the *Bell System Technical Journal* graphically shows how the field strength of WEAF falls off with distance. Any receiver, to be subject to tests which indicate anything, must operate under standard conditions, which are gradually being agreed upon

Fairchild Aerial Camera Corp'n

A RADIO "MAP" OF WEAF'S SIGNALS

Although made some time ago when WEAF was transmitting from Walker Street, New York, this illustration shows that in various parts of New York City WEAF signals were very faint indeed. A receiver, no matter how good, could not successfully "pull in" this station. The answer is greater power and location of broadcasting stations away from areas of great absorption

Do You Own a Battery-Operated Set?

Many Fine Types of A-Power Supply Units Are Now Available to Convert Your Battery-Operated Set to One Which Requires Almost No Attention

By HOWARD E. RHODES

NOT so long ago, the only way to heat the filaments of your tubes was to use a battery. And practically everyone who had a radio set had a storage battery; few indeed, even in the olden days, used dry cells to light their tube filaments unless that was a necessity. Then the storage battery passed through a cycle of development. The crude battery which radio borrowed from the automobile industry was dressed up. The case became polished wood or even glass and special precautions were taken by the makers to keep the acid electrolyte where it should be, for this battery was not to be housed in the interior regions of a motor car but in the parlor of high society. And now, to compete with the steady old battery come socket A-power units. What are they? How much do they cost? How many tubes will they supply? Do they need regular attention? Can a socket A-power unit be installed and be depended upon to light the filaments of the tubes "when, as, and if wanted"? These and goodness knows how many other questions are being asked by technical and non-technical radio folk these days as the offerings of 1927 are more and more widely announced.

The owner of a good radio set of some years back realizes that the tempting new 1927 models, operated directly from the light socket, probably are as superior to his outfit in convenience and performance as Whiskery is to Dobbin. But for one reason or another our loyal owner decides to keep his receiver. Can't he buy gadgets to turn his set into a light-socket outfit? Why not, for there are plenty of good B socket-power units and a goodly number of A socket-power units advertised? Well, so he can. He can buy a reliable A socket-power unit, a good B socket-power unit, a relay switch, and there you are—complete light-socket operation! He has achieved convenience which he seems to be pursuing strenuously. The economics of the change is another matter. Of the convenience and reliability there is no question.

Take the case of a set owner who bought a receiver a year ago. He may not be quite ready to buy a new outfit, but complete light-socket operation tempts him. His receiver must be operated by these A and B power units without any sacrifice in tone quality or volume. If the power unit can

not accomplish this—and for a reasonable length of time, without renewal of parts—they are not worth purchasing. There are, fortunately, many A-power units capable of giving as satisfactory reception as can be obtained from the unadorned storage battery.

The storage battery as a source of filament power is in many ways an almost ideal device. The current it supplies is perfectly steady. Its voltage is practically constant during a greater part of its discharge, and the slight decrease in voltage that does take place as the battery becomes discharged does not affect the operation of the receiver adversely because tubes of present-day design will operate satisfactorily at slightly lower than rated filament voltage. Automatic A-power units have been developed because the public demands convenience. The necessity of

THE BALKITE A-POWER SUPPLY
This device will supply filament current to receivers having up to eight tubes and it requires practically no attention. Inside the case is a transformer and electrolytic rectifier and an electrolytic condenser. The list price is $32.50

charging A batteries is a serious annoyance to many radio users.

TWO GENERAL TYPES OF A UNITS

THE radio set can be operated from the light socket by the use of A and B power units, or by designing the receiver to operate with special a.c. tubes, receiving their filament current from the power mains. The purchaser of a new set finds his problem largely solved, for the makers of light-socket sets have engineered their sets beforehand. It is to the owner of a battery-operated set that this article is addressed.

A power units fall into two classes:

(a) units using a rectifier and filter system connected through a transformer to the a.c. line. (See Table I).
(b) units using a special storage battery in conjunction with a trickle charger. (See Table II).

The various A-power units listed in Table I are all essentially similar in design but they differ in minor ways that are of interest. All A-power units in Table I must contain (1) a step-down transformer (to lower the voltage of the line to the proper value required by the rectifier unit); (2) a rectifying unit (to change the reduced a.c. to a sort of d.c.); and (3) a filter system (to smooth out the product of the transformer-rectifier circuit). The filter must eliminate the "hum" which is always present in unfiltered, rectified a.c.

Enormous capacity in the filter condenser is necessary to remove this troublesome hum. With electrolytic condensers a capacity of 30,000 mfd. can be attained in a reasonable space and at reasonable cost, and such a large capacity as this is necessary for adequate filtration. The electrolytic condensers are shipped dry and when they are put into service, distilled water is gradually added to the condenser container. The contained chemical, which in some cases is potassium hydroxide, dissolves in the water. When it completely dissolves, the unit is ready for use.

These electrolytic condensers require practically no upkeep. Every six months or so a small amount of distilled water must be added. If the user is absent minded and lets the water get too low, the unit is not damaged, but indicates its need of attention by causing the unit to produce an audible hum which is heard in the loud speaker. If, "by a set of curious chances," too much water is put into the condenser unit, it will fail to function properly. Excess liquid can and must be removed with a syringe. When the water is first put into the condenser can some heat will be generated for a short time.

The Balkite and Abox units both use the form of electrolytic condenser discussed above. An interesting feature of each of these two outfits is that the chemical rectifier electrode is immersed in the same electrolyte used for the condenser. The outer plates of the condenser act as the second electrode of the electrolytic rectifier. This

TABLE I

NAME OF UNIT	PRICE	TYPE OF RECTIFIER	MAX. NUMBER OF ¼ AMP. TUBES UNIT CAN SUPPLY	POWER INPUT FROM LINE	SIZE OF CONTAINER L x W x H
Abox	$32.50	Electrolytic	8	—	
Balkite A	32.50	Electrolytic	8	100	11¼ x 6¼¼ x 8¼¼
Electron Electric A					
Regular	45.00	Tube	7	—	12¼ x 7¾ x 9
Giant	49.50	Tube	12	32	
Marco A Socket Power No. 500	60.00	Tube	10	47	11¼ x 7¼ x 9
Sterling A Supply	42.50	Tube	8	—	7 x 11 x 8
Valley Socket A	39.50	Tube	12	36	9¼ x 5¼ x 11
White A Socket Power	43.50	Tube	9	39	11¼ x 7 x 6¼

ingenious scheme achieves very compact construction. Both these units are supplied with external taps which insert resistance in the secondary circuit to control the output voltage. It is always advisable to use the lowest resistance tap that gives satisfactory results. In the Abox unit, there is a film of oil on top of the electrolyte which prevents excessive evaporation of the fluid. Both units require the addition of a small amount of water every six months but do not require any other attention.

Another interesting A-power unit uses the Raytheon "A" cartridge—a new type of dry rectifier. Units of this type are made by Electron, Marco, Valley, and Sterling. Some of these use one Raytheon cartridge, others two. The Marco product, for example, boasts two cartridge rectifiers. A rheostat in the primary circuit allows regulation for different loads. A meter on the front of the panel simplifies proper adjustment. Inside the box is a relay with silver contacts—to avoid sticking—so that the power unit can be controlled by the filament switch in the receiver.

Either one or two Raytheon "A" cartridges may be used in the Valley A-power unit, depending on the number of tubes in the receiver to be supplied. Full-wave rectification obtains with the use of two cartridges. Valley suggests using the full complement of two cartridges for receivers with seven or more tubes. A single cartridge will suffice for more modest receivers. This Valley device also uses an electrolytic condenser to smooth the output of the cartridge rectifiers. Since it is shipped without liquid, the dry chemical may rattle in the condenser can and excite some curiosity on the part of the purchaser. The addition of water, according to directions, makes all things as they should be. Between the rectifier and filter system is the control rheostat which provides sufficient regulation for various loads imposed by the receiver.

The A-power unit from Sterling uses a single Raytheon cartridge, is equipped with an automatic filament circuit relay, a control rheostat, and a meter to facilitate the correct voltage adjustment.

The White unit among those listed in Table I is the only one on which we have information which uses a Tungar or Rectigon tube as a rectifier. The transformer in this device delivers about 8 volts to the plate and about 2 volts to the filament of the rectifier tube. The rectifier tube has a rating of 2 amperes, which, according to its manufacturers, is conservative; the unit, when operating under normal loads, should therefore have a long life. The filter circuit in this device, besides the usual electrolytic condenser, contains a 4-henry choke to assure complete elimination of hum. A calibrated rheostat to control the output is connected between the tube and filter system and the meter scale on the front panel enables the user to adjust the unit accurately and with ease. A six-foot cord with a pendant switch is supplied to control the a.c. input.

YOU WON'T RECOGNIZE THE STORAGE BATTERY

ALL the units listed in Table I are grouped there because they provide a source of A supply by utilizing a rectifier and filter system, while those of Table II combine a storage battery and trickle charger. There are many radio users who are convinced that the ideal A-socket power unit is one that is innocent of liquid of any sort. An unfortunate experience with the old storage battery may have instilled this dislike of an A-supply involving liquids. But as radio has developed and the inevitable and fortunate process of refinement has occurred, ingenious ways have been found to mould the storage battery into a highly desirable product indeed. Another school of manufacturing thought therefore has worked along these lines. They have taken the storage battery, designed it exactly to fit modern radio needs, and in the process have succeeded in producing a unit which has none of the disadvantages always quoted against it. Since any of the two distinct types of socket-power A units listed in Tables I and II supply satisfactory A potential to the receiver, and differ largely in the electrical means used to produce the direct current for the tube filaments, whether one chooses one type or the other is entirely a matter of personal preference.

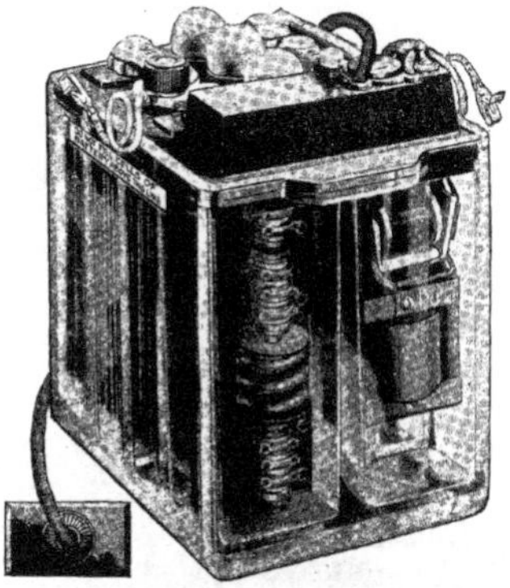

THE VESTA GLASS ENCASED A-POWER UNIT
This is a combination storage battery and trickle charger combined with a relay so that its operation is entirely automatic. A distinctive feature is that the entire unit is enclosed in a moulded glass case so that all the parts are visible. The list price is $47.50

Storage-battery makers, since radio became popular, have sought to reduce the routine attention demanded by the storage battery. Today's battery requires only the occasional addition of distilled water. Keeping the battery "up" is automatically accomplished by a trickle charger. By a study of the demands on storage batteries used by a wide variety of radio owners, sufficient data have been collected to accomplish a storage battery-trickle charger combination which needs only slight attention.

The principles of operation of this type of device have not been changed, but this year many important improvements have been made which insure satisfactory service and almost entire freedom from user attention. Unusually thick plates, especially designed vent caps, built-in "state of charge" indicators, conveniently located controls to vary the charging rate, and special cell construction to insure long life in trickle-charging service—all contribute to make the combination trickle charger and storage battery a convenient and satisfactory A socket-power unit.

Let us discuss some of the points of interest in these devices. The Acme A-power unit, type APU-6, is designed to supply 8 to 10 tubes. The battery unit can be charged at two rates: ½ and 1½ amperes. This wise provision permits adjustment of the unit to take care of the demands of a receiver with many tubes, or the lesser current requirements of a set with fewer tubes.

The Westinghouse "Autopower" has much to commend it. Our friends in East Pittsburgh have combined in a compact unit both a storage battery and an efficient trickle charging device, the latter developed during the last year. This rectifier, which is the heart of the charger, is interesting enough to merit a slight digression. Several years ago, it was found possible to make a solid body of matter conduct electricity more freely in one direction than in the opposite one. This was the origin of Rectox, the trade name of the rectifier used in the "Autopower" and some units by other makers, operating under Westinghouse licenses. The first materials to show this property offered three times as much resistance to the passage of electric current in one direction as in the other. The present Rectox units, developed after considerable research, have increased this resistance ratio of 3 to 1 to as high as 20,000 to 1 in the final units. The life of this rectifier unit is said to be indefinite.

A special clip on the front of the "Autopower" makes it possible to obtain three different rates of trickle charging. In addition, a "booster" rate can be used to revivify the battery if the receiver has been used for an excessive length of time. (One thinks of the 11½-hour continuous Lindbergh broadcast of last June!) The unit contains a relay, which, when the set is turned on, automatically disconnects the a.c. from the trickle charger and connects it instead to two leads terminating in a plug on the side of the "Autopower" unit into which the connecting cord of a B-power unit is connected. When the radio receiver is turned off, the relay automatically closes the trickle-charger circuit and the battery begins to charge. At the same time, the relay opens the

TABLE II

NAME OF UNIT	WATTS INPUT	PRICE	MAX. NUMBER OF TUBES UNIT CAN SUPPLY	TYPE RECTIFIER USED IN CHARGER	SIZE L x W x H
Acme A power	—	$35.00	10	Tube	11¼ x 7 x 9¼
Autopower	22	35.00	10	Copper oxide	11 x 6¾ x 9¾
Basco A power	35	40.00	12	Tube	12½ x 5½ x 10
Compo	26	42.50	8	Tube	10½ x 5½ x 8½
Exide Radio Power	17	31.90	10	Tube	11 x 5 5/16 x 9
Philco A Socket Power (603)	—	32.50	6	Electrolytic	12¾ x 9½ x 7¾
Unipower—AC-6-K	24	39.50	10	Electrolytic	11¾ x 7 5/16 x 10
Universal	—	32.50	8	Dry disk	8 5/16 x 8 x 7½
Vesta A Power	—	37.50	—	Dry rectifier or electrolytic	9½ x 7½ x 9½
Greene-Brown	29	30.00	10	Tube	8¾ x 3¼ x 10¾

circuit to the B-power unit so that this unit is automatically disconnected. The "Autopower" requires no attention except the occasional addition of distilled water to the battery.

WHAT THE UNITS CONTAIN

THE Basco A-power unit contains in a single case the storage battery, the rectifier, an automatic relay (similar to the type just described above), an emergency switch, transformer, fuses, and a terminal board. The battery is an all-glass Exide unit with a capacity of 45 ampere hours. It is equipped with colored indicator balls to show the condition of charge. A thin film of oil on the surface of the electrolyte prevents undue evaporation and also prevents spraying and corroding of battery terminals. This battery has a large water space and the ordinary user will not have to add distilled water oftener than every half year. The Basco A unit is connected to the receiver just as if it were an ordinary battery and when the receiver is turned on, the current from the battery flows to the tubes and at the same time passes through an automatic relay which closes a circuit and makes 110 volts (for your B-power unit) available at a plug on the side. When you are through using the set, turn off the switch. The relay automatically opens the B-power unit circuit and puts the battery on charge. A Raytheon "A" rectifier is used as the charging rectifier. This rectifier has the advantage that its rate of charge automatically decreases as the battery becomes charged. Danger of overcharging is decreased. The "emergency" switch mentioned above is used to recondition the battery after it has stood idle for some time. Turning this switch recharges the battery at a high rate and inconvenience is reduced to a minimum.

A 35-ampere hour battery in a composition jar with a special cellulose moisture-proof pad on top of the plates and a paste electrolyte are features of the Compo A-unit. An eye-dropper full of distilled water in each cell about every four months is all the attention the unit requires.

Three rates of trickle charging are available: 0.2, 0.4, and 0.6 amperes.

The Exide model 3A 6-volt A-power unit is designed to supply constant voltage direct current for the operation of the filaments of the tubes in any standard radio set. It comprises a storage battery and trickle charger with three taps, each affording a different charging rate. This rate depends, of course, on the number of tubes in the receiver and the number of hours the set is used. The battery is a standard Exide unit of excellent design and construction and contains ample space for excess electrolyte over the tops of the plates, thus making necessary the addition of distilled water only once or twice a year. The makers recommend it be used with the extra plug for the a.c. supply for a B-power unit. The unit has a visible charge indicator consisting of two small colored balls so that the condition of the battery can be told at a glance. The entire unit is contained in a nicely finished sheet steel enameled case, fitted with two carrying handles.

The Philco A socket-power unit also affords a dependable source of filament potential. Philco has refined this unit in many ways during the last year, to make it entirely fool-proof and economical in operation. The model 603A A-power unit, listed in Table II, consists of a high-efficiency transformer and rectifier with a battery especially designed for trickle charging service. The battery has unusually thick plates and separators. Spray-proof construction, preventing the leakage of electrolyte from the battery, and the built-in state-of-charge indicator, are two important improvements. These heavy plates and separators insure long life and freedom from the danger of internal short-circuit. Without the built-in Philco indicator there would be no simple means of determining the condition of the battery except through the use of a hydrometer and then it is used there is always the possibility that some acid will be spilled, incurring the righteous wrath of the housewife. Special vent-caps have been

incorporated in the Philco units which make possible the addition of water to the battery without removing them. And water need not be added to these cells oftener than twice a year. In normal operation the vent-caps need never be removed. Philco units employ an "economizer" which permits the user to adjust the charging rate to the lowest current consumption which will, at the same time, keep the battery properly charged as shown by the visual indicators. By using the lowest possible rate, gassing of the electrolyte is prevented, and this reduces the frequency with which water need be added. Three charging rates are available with a ' booster' rate for emergency use. The batteries are in a glass container. Philco units can be had for operation on 25- 30- 40- 50- or 60- cycle a.c. Type A603 is designed to supply up to six tubes and type A-36 is designed to supply up to ten tubes. The latter type contains a dry trickle charger which provides three rates: 0.25, 0.5, 0.26 amperes and a 1.0-ampere rate for booster service.

The Unipower type Ac-6K provides, according to its makers, three unique features. First, a "Kathanode" cell construction which insures long battery life; secondly, an automatic cut-off in the rectifier cell which suspends charging if the user fails to add water when necessary, and third, five charging rates with a high rate of 1½ amperes—meeting the requirements of all grades of receivers.

In the "Kathanode" design, porous glass wool mats are fitted against the positive plates to prevent the shedding of active material which frequently occurs if the battery is overcharged. The glass mats, by capillary action, draw fresh acid to the plates, increasing efficiency. The Unipower, cased in rubber, contains three "Kathanode" constructed battery cells, a rectifier cell, a transformer, as well as the essential switches, terminals, and connections. All these cells are watered at once and the rectifier is designed so that when the level of electrolyte exposes the tops of the cell plates, the charging current is automatically cut off until water is added. The makers feel this safeguard is essential to the proper operation of the battery. On the front of the unit, a dial regulates the charging rate, which ranges from 0.25 to 1.5 amperes in five steps.

A Rectox dry disk rectifier is used in the Universal A-power unit. The 36-ampere-hour battery is assembled in a three-compartment glass jar with mounted hard rubber covers. This A-power unit has a visible state-of-charge indicator, and the whole device is supplied in a steel container.

One of the first battery-trickle charger combinations received in the Laboratory in which glass was the container was the handsome Vesta A-power unit. Vesta now makes two A-power units, one containing an electrolytic trickle charger and the other a dry trickle charger. A visible charge indicator shows the state of the battery; when the three colored balls float at the top of a small compartment, the battery is fully charged and as the charge decreases, one ball after another gradually sinks to the bottom of the compartment. The Vesta unit has a socket into which the a.c. plug for the B socket-power unit may be plugged.

So the A-power devices of 1927 look and perform very differently from the indiscriminate units with which the radio user of some years ago was content. If a variety of A-power devices are offered the purchaser and he does not know what type to use, he should ask his local dealer to install them in his home so he can easily choose the one which best fits his own needs and his local conditions.

A PHILCO UNIT

This Philco unit incorporates several interesting features among which are a visible indicator of the state of charge of the battery and special vent caps on the battery which absolutely prevent any acid from leaking out of the battery. These vent caps do not have to be removed in order to add water to the battery

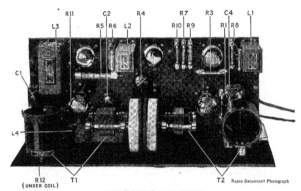

L3   R11   C2   R4   R7   R3   C4   L1
R5 R6   L2   R10 R9   R1 R8
C1
L4
R12   T1   T2   Radio Broadcast Photograph
(UNDER COIL)

A NEW BROWNING-DRAKE RECEIVER

The new Browning-Drake receiver shown above is to be described constructionally next month. It has been designed for complete
a.c. operation although batteries may be used if desired. This first article discusses the various a.c. tubes suitable for the purpose

Electrifying the Browning-Drake

A Discussion of the New A. C. Tubes and How They May
Be Incorporated in a New Design Browning-Drake Receiver

By JAMES MILLEN

WITH the availability of really good a.c. tubes, another and important step toward the ideal radio set is made. With the a.c. tube, no storage battery or A-power unit is required. All that is necessary is merely a compact little transformer for decreasing the line voltage to a suitable operating value. As far as actual performance is concerned, the new a.c. tubes are essentially the same as the well-known 201-a or 301-a type tubes. The person with a set equipped with standard tubes will not improve the performance of his set by changing it over for a.c. tube operation. If his storage battery, charger, and tubes are in good condition, there is nothing to be gained by such a change. If, however, the batteries have about run their useful life, or if the charger has died of old age, the new tubes offer a number of worthwhile attractions to the home constructor. First, they open new fields for experiment; second, they enable him to build a completely lamp-socket operated receiver for less money than a battery operated receiver with its associated storage battery and charger, and at the same time there results a receiver somewhat simpler to maintain.

Once a few of the little tricks of the use of a.c. tubes are acquired one will have no difficulty in constructing any of the popular circuits for a.c. operation or in replacing old tubes in any standard receiver with new a.c. tubes. Perhaps the best way of acquiring this knowledge is to carefully follow the details in connection with the construction of some popular circuit for use with the new tubes. With this in mind, we have selected the Browning-Drake as one of the most popular receivers which has been described in past issues of RADIO BROADCAST, and have redesigned it not only for complete a.c. operation, but also to incorporate the latest ideas on layout, audio amplification, and other slight modifications of the original Browning-Drake circuit. Furthermore, the set has been so designed that it may, if desired, be wired for battery operation where the constructor is not so fortunately situated as to have a.c. on tap. The photograph gives an idea of how the completed receiver looks. Complete construction data on this set will be given in the next article. In this article we will consider some of the general problems involved in the use of a.c. tubes. First of all let us consider the different a.c. tubes available for all but the last audio stage. The last audio, or power tube, be it of the 112, 171, or 210 variety, may be operated on raw a.c. just as well as on batteries. No special a.c. tube is required; therefore, in the last audio stage.

It will be seen from the table on the next page that the a.c. tubes may be divided into two general types, i. e., those using a low-voltage high-current filament, and those having a separate heater element. The heater type tubes are better suited as detectors than the filament type, but either type are about equally well suited as radio and audio amplifiers. Since the heater tubes are, in general, more expensive and have shorter lives,

it is advisable to restrict their use to the detector socket.

The different filament heating transformers available are mostly designed for direct operation with the RCA-Cunningham tubes without the use of rheostats or other resistors. The voltage taps on some of the transformers available at present are:

Manufacturer	Taps (in Volts)
Amertran	1.5, 2.5, 5.0
Dongan	1.5, 2.5, 5.0
General Radio	2, 3.5, 5.0, 7.5
Modern	1.5, 2.5, 5.0
National	1.5, 2.5, 5.0
Silver-Marshall	1.5, 2.5, 5.0
Thordarson	1.5, 2.5, 5.0

When a.c. tubes of other manufacturers are used with transformers having the proper taps for the RCA-Cunningham tubes, special rheostats made by General Radio and Carter should be used in the low-voltage transformer leads. When tubes of the Armour-Van Horne type are used throughout, then two short lengths of resistance wire with a total resistance of about 0.1 ohms should be inserted in the leads to the detector and audio amplifier tubes so that they operate at a slightly lower voltage than the radio-frequency amplifier tube. Several manufacturers make special resistors for just this use.

Where the Kellogg tube is used only as a detector, the 3.5-volt filament transformer winding will be found just right. Where Kellogg tubes are used throughout, then the 1.5- and 2.5-volt windings should be connected in series (that is, so that their voltages add rather than subtract) to give 4 volts which may be dropped down to the desired 3 volts with a suitable rheostat or fixed resistor. The filament voltages required by any of the tubes are far from critical and the tubes will be found to perform excellently with voltages considerably below the rated values. Operating the detector at a lower voltage often results in almost complete elimination of any hum. If the heater voltage of a UY-227 detector is excessive, the set will cease to regenerate, and, in fact, practically stop operating. Generally, about 2.2. volts seems to work best with the 227's when used as detectors and a six-inch length of wire from an old-ten ohm rheostat, in series with one of the 2.5-volt transformer leads, will give this lower voltage.

The 1.5- and 2.5-volt transformer windings should not be center-tapped as potentiometers located close to the tube sockets are necessary for the best results. The 5-volt winding for the 171 or the 7.5-volt winding for the 210, however, may just as well have a center tap and thus eliminate the need for one potentiometer. The detector and the power-tube filament circuits should be wired with No. 18 equivalent rubber covered twisted wire. The proper size wire for the radio and first audio stages, containing high-current tubes, may be determined by estimating the total current drawn by these tubes from the table of characteristics and then selecting a wire that will carry such a current from the table below. In the case of the Browning-Drake receiver using RCA tubes, No. 18 may be used, but if the Van Horne-Armour type tubes are used, then No. 16 will be necessary. The following table gives the current-carrying capacity of rubber covered copper wire:

Wire Size	Current	
14	11	amperes
16	6	"
18	3	"
20	1.5	"

THE R. F. AMPLIFIER

EITHER the heater or the filament type of tube will work well in the radio stages, but because of its longer life, lower cost, and simpler connections, the filament type is generally to be preferred. There is, however, one real advantage that the heater types have over the filament types when used with some circuits, and that is lower inter-electrode capacity, which often facilitates neutralization. The filament type a.c. tube may be employed in the r.f. stage of a Browning-Drake receiver with materially improved results over those obtained with the customary 199 type tube.

While frequently no negative grid bias is employed on the r.f. tubes in a battery operated receiver, the use of this bias is essential with the a.c. filament type tube. This biasing voltage may be obtained from a C battery or

by utilizing the voltage drop across a suitable resistor which can also provide the bias for the first and second audio-frequency stages.

The optimum r.f. tube plate voltage for minimum hum does not seem to be at all critical and the 67½-volt tap on the average B supply unit gives as good results as any, with less tendency for the radio-frequency stage to oscillate than when the 90-volt terminal is used. The C bias on the r.f. tube should be a little more negative, for a given plate voltage, than on the a.f. stages. The use of a somewhat lower plate voltage on the r.f. tube than on the a.f. tubes permits the use of the same C voltage on both the audio- and radio-frequency tubes.

The use of a.c. tubes and a B power unit make two of the forms of volume controls considered more or less standard with battery operated receivers—the r.f. filament rheostat and the variable series resistor in the r.f. plate circuit—unsuited for the electric receiver. There are, however, at least two other systems of volume control which will give satisfactory results. One is a variable antenna coupling coil, and the other is a variable resistor across the primary of the r.f. transformer. By this means it is possible to control the volume by varying the r.f. input to the detector circuit.

A potentiometer across the filament circuits of both the radio and first audio stages must be employed. As the voltage is low, this unit may be a 30-ohm rheostat with a third connection made to the "open" end of the winding. As this potentiometer may, from time to time, require a minute change of adjustment, it is well to locate it in some convenient place on the sub-panel. The potentiometer should not in general be mounted on the front of the panel, as for best results it must be hung directly across the filament leads at about an equal electrical distance from all the tubes. The adjustment of this potentiometer is quite critical, and a very slightly different setting is frequently required at night than during the day in order almost completely to eliminate all the hum—and the hum can certainly be reduced to a very low order if the receiver is carefully constructed and adjusted.

A. C. TUBES IN THE AUDIO STAGES

AS THERE is nothing to be gained by the use of the more expensive heater type tube in the first audio stage, the filament type is to be recommended. As already mentioned, the one potentiometer and grid bias resistor serves both the radio and the audio stages. In the case

of the UX-226 (CX-326) tubes with 90 volts on the plate, the grid bias should be adjusted until the drop across its terminals, as measured with a high-resistance voltmeter, is about 6 volts. In the case of the Browning-Drake receiver to be described in detail next month, a fixed 500-ohm wire-wound resistor is used to obtain C bias and this value of resistance is just right. Any of the several different forms of audio amplification may be employed with excellent results.

Where the grid bias for several stages is obtained by taking the voltage drop across one resistor, as in this case, then the use of a "grid return filter" in each stage is recommended and such filters have been used in the a.c. Browning-Drake receiver. These filters merely consist of a 0.1-megohm resistance and a 1-mfd. condenser connected so as to prevent any of the audio-frequency currents from flowing through the grid bias resistance. In the last or power stage, the 171 is recommended as the tube best suited for home use. A 2000-ohm wire-wound resistor will automatically provide the proper grid bias for this tube regardless of the plate voltage, within reasonable limits. A loud speaker protective device to eliminate the direct current from the loud speaker windings should be employed.

THE DETECTOR

WHILE either form of a.c. tube may be used as a detector, the UY-227 type of heater type tube has several advantages over the filament type. First, the a.c. hum can be, for all practical purposes, entirely eliminated. The hum from a filament type a.c. tube is not what could in any way be termed objectionable, yet it is there. The heater tube may be used with either a grid-leak condenser arrangement or with C bias, whereas, the 226 type of tube, while it will function quite well with a grid-leak condenser, is better suited for plate rectification. Plate rectification, however, is not as sensitive as the grid-leak condenser arrangement and its use in connection with an all a.c. operated receiver also leads to other complications. The Kellogg a.c. tube may be used as a detector with excellent results.

In using the heater type tube as a detector, either a negative or a positive bias of about 40 volts or so should be applied to the heater element by means of a potentiometer. In some instances a positive bias seems best and in others, a negative, and either of these biases are readily obtainable from the 40-volt tap supplying C bias to the power tube or the plus 45-volt tap for the detector. The adjustment of this bias voltage is not at all critical, and once set, will require no further attention. In fact, a fixed resistor with center tap, such as the type 438 General Radio, will serve the purpose excellently. This resistor is so designed as to mount directly on the terminals of the detector tube socket.

In a second article which will appear next month, constructional details and adjustment suggestions on the a.c. Browning-Drake receiver will be given.

A. C. TUBES								
Heater Type								
Name	Ef	If	Ip	Rp	Mu	Gm	Ep	Eg
C-327	2.5	1.75	4.2	8600	7.8	905	90	−4.5
UY-227	2.5	1.65	3.5	10350	8.7	860	90	−4.5
McCullough	3.0	1.0	4.2	9400	8.6	870	90	−4.5
Sovereign	3.0	1.5	4.6	9100	8.5	935	90	−4.5
Marathon	5.5	1.0	4.2	9500	7.3	775	90	−4.5
Arcturus	15.0	0.35	3.1	12150	10.5	870	90	−4.5
Magnatron	2.5	1.50	4.6	8700	9.3	1070	90	−4.5
Filament Type								
CX-326	1.5	1.05	4.6	9000	8.5	935	90	−4.5
UX-226	1.5	1.05	4.4	9150	8.7	950	90	−4.5
Armor	1.0	2.4	3.8	11200	7.8	690	90	−4.5
Van Horne	1.0	2.0	4.4	9000	9.0	1000	90	−4.5
CeCo	1.5	1.05	2.8	14200	9.2	730	90	−4.5
Magnatron	1.5	1.05	4.0	10800	8.8	830	90	−4.5

Ef = Filament Volts   Ip = Plate Current
Ep = Plate Volts   Rp = Plate Resistance
Eg = Grid Volts   Mu = Amplification Factor
If = Filament Current   Gm = Mutual Conductance

"Our Readers Suggest—"

Rewiring an Atwater Kent Receiver for A. C. Tubes

THERE seems little doubt in the minds of engineers that the alternating-current tube will eventually find a place in the majority of radio receivers. It is in anticipation of this eventuality that RADIO BROADCAST has already devoted considerable space to the problems of A battery elimination and the characteristics of a. c. tubes. We are interested in the following description of how a reader, Henry March, of New York, altered a popular type of receiver for a. c. operation, necessitating few and simple circuit changes. He writes:

"It has been my pleasure to discover that the Atwater Kent Model 35 receiver can be easily adapted to a. c. operation through the use of a. c. tubes. I presume that the same simplicity of conversion holds true for many other receivers—a fact that may interest your readers.

"I rewired my receiver for Arcturus tubes (type 28 amplifier, type 26 detector, and type 30 power tube), choosing these tubes because of the fact that they plug into the four-prong socket which is standard equipment on practically all receivers wired for storage battery tubes. Thus no additional filament wiring or special sockets are

required—greatly simplifying the necessary changes, which are illustrated clearly in the accompanying diagrams. Fig. 1 shows the original wiring in the receiver. The parts of the circuit to be changed have been drawn in heavy lines. Fig. 2 shows the circuit with the changes made.

"All grounds have been eliminated from the filament circuit. The lower terminals of all r. f. and a. f. secondaries, excepting that of the power tube, have been grounded. The detector grid return to the potentiometer has been eliminated, and the return is now effected through a 4.5-to 9.0-volt C battery, *positive to the grid*. Detector C minus is connected to the B minus post. The plus terminal of the main C battery also connects to B minus. Minus 1.5 C battery is

grounded (supplying the r.f. tubes), while 22.5 minus runs to the power tube in the usual manner. A Centralab modulator is connected across the secondary of the first audio transformer as a volume control.

"The a. c. filaments are operated from an Ives step-down 'toy' transformer (type 204) at the 14.5 volt tap. All plate voltages remain the same as in the d.c. set, excepting that 180 volts is applied to the output power tube, increasing the possible undistorted power output of the receiver."

for the adaptation of d. c. receivers to a. c. operation. Much of this is covered diagrammatically in the accompanying circuits.

All grounds must be eliminated from the filament circuit. Ground all secondaries (filament side) having the same negative bias. A bias of minus 1.5 to 3 volts is generally applied to all r. f. grids.

Run the two filament wires as close together as possible, lacing or twisting them when convenient. Be sure that all plus filament posts are connected together. Connect minus B to what previously were the positive posts. Connect the r. f. and the a. f. C plus and the detector C minus to B minus.

Eliminate all filament rheostats and potentiometer r. f. controls. It is not practicable to use these forms of volume and sensitivity control with a. c. tubes. With the potentiometer device, sensitivity is governed by varying the bias on the r. f. tubes which, with a. c. tubes, would introduce hum at certain adjustments. A 250,000-ohm variable resistor connected across the r.f. secondary preceding the detector tube is a preferred volume control.

Receivers wired for four-prong base a. c. tubes can be used with d. c. tubes at any time, merely by substituting an A battery for the transformer. No other changes are necessary for d.c. operation of such a receiver.

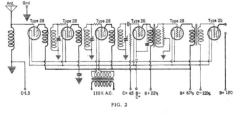

Type 28 Type 28 Type 28 Type 26 Type 28 Type 30

C-1.5   110 V. A.C.   C+ 45 B- B+22½   B+ 67½ C—22½   B+ 180

FIG. 2

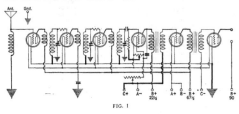

C+ A- B+ 22½   A+ B- B+ 67½ C-   B+ 90

FIG. 1

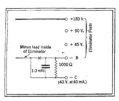

FIG. 3

C Bias from a Mayolian Socket Power Unit

B-SOCKET power units in the future will undoubtedly incorporate extra resistors making it possible to secure C bias for at least the power tube. It is not difficult to incorporate this feature in the average power unit along the lines described by a contributing reader: James J. Corrigan, of Des Moines, Iowa:

"I have a Mayolian B power unit, the utility of which I have doubled by adding an extra resistance and bypass condenser. The drop across the resistance supplies the C voltage to my power tube.

"The lead to the negative binding post is broken at 'X' inside the case (Fig. 3). A 1000-ohm, two-watt resistor is connected in the break and by passed with a 1.0-mfd. condenser. The post marked B minus connects as usual to the receiver, while the C bias voltage is tapped in the eliminator side of the resistor. A forty-volt C battery is supplanted by this means.

STAFF COMMENT

THIS is a simple and practical method of C battery elimination, readily applicable to all eliminators giving voltages, under load, in excess of 180. The C voltage is necessarily subtracted from the B voltage, and the compromise is sometimes undesirable. If your eliminator has a no-load potential of about 250 volts, C elimination is quite worth while. Many B-socket power units fill the bill. Among them are: Kodel, Burns, Greene-Browne, Kellogg, and General Radio.

However, the use of a fixed resistor is not recommended as it is almost impossible to secure the right bias. It is suggested that a variable resistor, connected as shown in Fig. 4, be used instead. Amsco Products manufacture a zero to 2000-ohm variable resistor known as a Duostat, made especially for this purpose. It is equipped with two variable arms, making it possible to secure two C bias potentials, one for the power tube and one for the other a. f. tubes. Each arm of the Duostat must be bypassed with a 1.0-mfd. condenser. Other variable 2000-ohm C bias resistors are made by Carter and Electrad.

A rough adjustment of the bias potentials can be made by ear. However, a much more scientific

job can be done with the aid of a small milliampere meter, reading up to 25 milliamperes. This should be placed in the plate circuit of the tube on which the bias is being adjusted. The variable arm is moved until, on a loud signal, the needle is motionless, or practically so. Any movement of the needle is an indication of distortion. If the needle kicks up, turn down the resistance (lowering the C bias); if the needle kicks down, increase the resistance.

This careful adjustment is generally made only on the output tube. The meter is connected in series with the loud speaker, or the primary winding of the output device if such is used. As the power handled in the preceding tubes is generally small, a rough adjustment by ear is adequate.

Getting High Notes from the Resistance-Coupled Set

I HAVE a Ferguson Model 12 receiver, in which were incorporated three stages of resistance-coupled amplification. I operated this set in conjunction with a Western Electric 540 AW cone loud speaker. While the tone quality of this combination was distinctly superior to that of the average set, there was, at times, a disconcerting rumble on low notes, which quite counteracted my pleasure in the unusual reproduction of these

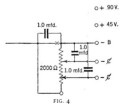

FIG. 4

low frequencies. There seemed to be a resonance point in the output system in the neighborhood of fifty cycles. A friend of mine has an impedance-coupled set, which, while quite free from the particular disturbance I mention, is distinctly partial to higher notes. It occurred to me that a compromise between resistance and impedance coupling might be ideal in my particular case.

Upon the advice of an experienced fan, I removed the coupling resistor from the second audio stage, and ran two wires from the prongs to the primary of an old audio-frequency transformer. I left the grid leak exactly the same as when resistance coupling was used (See Fig. 5). The result is most gratifying. There is no

longer any rumble on the troublesome notes, and it seems to me that the speaking voice is cleared up a bit . . . it is more natural. Also there is a slight improvement on the higher notes such as are occasionally reached by sopranos and violins. A certain vague sense of muffled sound has altogether disappeared.

STAFF COMMENT

THE experimenter writing the above experience, Frank Wendell, of Los Angeles, has accomplished what is being done nightly in the large broadcasting stations, where the process of balancing the scale of frequencies is known as "equalization." With outside or "nemo" pickups, transmitted over landline to the broadcasting station, certain frequencies are transmitted with less fidelity than others, and the boosting up of the delinquent tones is accomplished in much the same manner as our correspondent brought up his high notes.

The average cone loud speaker in comparison with the average horn, is much better on the low notes. The same holds true of the resistance-coupled amplifier as compared with other amplifying systems; but this type of amplifier also has a distinct cut-off on high frequencies. The combination, therefore, is one that favors the low frequencies—often to such an extent that there exists the low-frequency rattle referred to.

In the case under consideration, the high notes have been boosted by substituting a reactance in place of the resistance. It is probable that the response curve of the reproducing system has been leveled out a bit. That is, all frequencies reach the ear with a closer approach to their relative amplitudes or volumes.

Taking out a coupling resistor and substituting a comparatively low inductance choke coil will always increase the amplification of the higher notes more than it increases the amplification of the low notes. The lower the inductance of the choke coil, the more will be the difference. There is no reason why the average broadcast fan should not improve reception by "equalizing" his receiver in this manner. A resistance-coupled amplifier (in any receiver) most easily lends itself to changes of this nature.

VARYING THE AMOUNT OF EQUALIZATION

AN ORDINARY amplifying transformer is probably the most readily available form of inductance or choke coil. The primary, in the case of the average transformer which may be on hand, should be used.

The high notes will be brought up most if only the primary of the transformer is used. There will be less difference from straight resistance coupling if the secondary is used. Different degrees of equalization will be obtained if the primary and secondary are connected in series, with, first, the grid and plate posts strapped (using the B and F posts as terminals) and, secondly, with the grid and B posts strapped.

The grid leak of the tube output putting to the choke coil is not touched but the bias applied through the leak should be increased by about 4.5 volts.

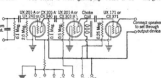

FIG. 5

LISTENERS—GUESTS OR CUSTOMERS?

By John Wallace

A READER at Long Beach, California, addressed us not long ago as follows:

SIR:

May I be permitted to call to your attention the excellent and timely article appearing on page 15 of *Radio News* for July, 1927, entitled "The Fly in the Ointment," by one Nellie Barnard Parker?

A great many listeners hereabouts were struck by the miserably poor taste displayed by the writers of the "can't-the-announcer-be-choked" and the "sprayed-with-petroleum" telegrams to which the writer refers; and one of us, at least, was equally impressed by the sportsmanship displayed by the announcer in reading such telegrams to us at all. I think that you will agree with the author of the article in question that '. . . when a company has spent thousands of dollars to broadcast a program, it has bought the right to let you know who your host is and what it has to sell.'

I, for one, would like to see a similar stand taken by RADIO BROADCAST; and I believe that such a stand, in your columns, would more nearly present the average "Listener's Point of View" than does much that now appears there.

G. I. RHODES.

Here, indeed, was an invitation for your department editor to adopt a policy—and if there is anything an editor, of any variety, keeps an eagle eye out for, it is "policies." Policies are what enable him to get his stuff written. So we swam into the article, a most entertaining one.

The specific fly in the ointment complained of was an incident in connection with the broadcasting of opera by KFI and KPO last season, as the indirect advertising donation of a certain petroleum corporation.

The opera broadcast was unquestionably one of the outstanding musical treats afforded West Coast listeners that season. In the intervals between the acts the announcer read a number of telegrams of commendation and explained, with some precision, just who was financing the broadcast. "And then," says the author of the article referred to, "right out of the sky, came the fly in the ointment! A man wired in: 'We are enjoying the program but can't the announcer be choked off and let us have opera without telegrams and advertising?' And pretty soon another 'guest' wired his objections against being 'sprayed with petroleum' while he listened. Clever, yes, but it struck one listener at least that when a company has spent thousands of dollars to broadcast a program, it has bought the right to let you know who your host is and what it has to sell."

That was the case in question. We are hardly fitted to pass on its merits since we didn't hear the broadcast. It is quite possible that the number of telegrams read and the amount of advertising dished out were entirely within the bounds of reason. In fact this seems probable if there were but two unfavorable comments on it. As the writer points out with some show of logic, the

reading of a telegram of commendation suggests the sending of them to other listeners, and if a large number is received "they are permanently bound and the next time there seems a possibility of interesting some firm in paying the fiddler for an expensive program, this bulky volume is brought forth." Thus the telegram reading may in some cases react finally to the listeners' benefit.

But departing from this particular instance wherein the adverse criticism may not have been entirely warranted, the writer goes on to generalize and takes the stand that adverse and destructive criticism of a better-than-average program is never justifiable. This, it seems to us, is stretching the point to absurdity. She says:

You are free to steer your airship where you please, casting out your line knowing that there are just as good tunes on the air as ever were caught. Such being the case, why send in thoughtless messages to mar the perfect pleasure of your host? Let him sing his little solo without having the anvil chorus crab the act!

It is only fair to say that those who criticise the big programs are in the minority, but there are just enough of them to destroy that fine edge of joy and what-a-good-boy-am-I feeling the sponsors and operators have.

Every graduate operator of a radio is a supercritic of the air. Like an insect of the ether, the true radio bug goes sniffing through the air with his little feeler; when he "contacts" with something he likes, he settles upon it with a pleasant little hum. But if it pleases him not, he is liable to plant a sting, if he is that kind of a bug. How much nicer it would be if he would remember that the sponsors and announcers are just big boys trying to get along! They are not inoculated against praise. It takes on them beautifully and they break out with brighter and better

THE SANKA AFTER DINNER COFFEE HOUR AT WEAF

Heard over this station on Tuesday evenings at 7:30. They should receive some kind of reward for getting the maximum number of words into the title. Anyhow, here are the performers

programs. They invite and welcome constructive criticism and helpful suggestions, but mere "razzing" and discourtesy never fanned a generous impulse into flame. Just be human, kindly and courteous, remembering that the announcer, like the fiddler, is doing the best he can.

And don't be the fly in the ointment!

We quote this writer at such length because hers is a point of view that is all too widely held, namely: that the purveyor of radio programs is your host and that all the rules for polite drawing room conduct should operate in your attitude toward him.

When a man invites you to his home for dinner he does so as a private individual, and however burnt the potatoes may be, it is not common politeness for you to throw them at him. But if the same man sets up a restaurant and you happen in there to eat, you are perfectly justified in calling him all sorts of names if his chef has too highly seasoned the lobster *thermidor*. He has removed himself from the rôle of private individual and become a purveyor to the public. He has become, to use the word loosely, an artist, and by universal assent any and all of the products of the artist are open to criticism and he may not protest. By his very act of setting himself up as an artist he tacitly agrees to submit to any opprobrium that the citizenry feels inclined to hurl at him. This is true of every sort of artist —chef, singer, movie producer, poet, electric refrigerator manufacturer, sculptor, street cleaner, painter or sponsor of broadcast programs.

If a man wants to buy himself a box of paints and surreptitiously records on canvas his impression of the cherry tree in the back yard or sunset on the drainage canal no one has a right to comment on the way he does it. It is entirely his own affair as long as he keeps it his own affair by contenting himself with hanging the finished works on his own wall. But if he starts sending his pictures to the exhibition galleries he, by that gesture, professes himself to be an artist, and his work to be art; and he automatically becomes perfectly legitimate meat for anyone to pounce upon who cares to.

If what he exhibits as art is inexcusably bad, the good name of Art is threatened. And since Art is not his own private possession but is held by common consent to be in the custody of the great unwashed public, it is incumbent upon that public to weed out with vituperatives anything that threatens to cast a smirch upon it. The commentary that the public makes is known as Criticism. Criticism may be of many kinds, favorable or unfavorable, constructive or destructive, gentle or splenetic, competent or incompetent. The writer of the article discussed, and those of the same misguided frame of mind, would object to any criticism that does not fall into the category of favorable or constructive. This is obviously silly and results from a complete misconstruction of the function of criticism. Gentle-spirited sentimentalists get all hot and bothered and are filled with great sympathetic aches when some public performer gets it in the neck from a sharp tongued critic. They decry the critic as mean and lacking in human qualities. But in the case of a genuine artist their sympathy is wasted. A true artist doesn't mind adverse criticism—much; he is his own best judge of whether his work is good or bad. On the contrary he is rather stimulated by it. Splenetic or strongly biased criticism may be far more effective in egging him on to do better work than soporific boquets. The only criticism to which he is likely to object is the incompetent kind—and of this there is, of course, plenty.

The two critics of the KFI-KPO opera broadcast may have been incompetent; they may not have been aware of all the facts, viz.: that a certain amount of advertising was necessary if the broadcast was to pay for itself. As we have said, we did not hear the program and do not know whether this reading of telegrams was carried to excess or not. But not all criticism of radio programs by minority calamity howlers is incompetent. A great deal of it is very much to the point (including, of course, all our own sage pronunciamentos.)

The fact that a majority of the listeners are perfectly satisfied with the way any given radio program is presented does not mean that any criticism on the part of a few of the minority is worthless. The oft-repeated phrase about *giving the public what it wants* is, at best, just a phrase. True, some effort is made in this direction, but the public is not at all sure what it does want, seldom expresses itself on the subject, and finally, finds it the course of least resistance to *take what it gets*.

The masses continue to be satisfied with what they are getting until something better comes along. Then they accept the improvement with the same placid satisfaction—perhaps wonder why they were so easily pleased with the old—but make not the slightest effort to secure further betterment. It is up to the minority kickers and mud slingers to secure for them these improvements.

Your average radio listener was perfectly satisfied with broadcasting as it existed in 1923. His unimaginative and uncritical mind could conceive of nothing better. He was getting programs made up largely of cheap jazz and cheaper talks. To live up to their views, the advocates of "throw-away-your-hammer-and-get-a-horn" would have to argue that things should have been left for him just as they were. He was satisfied; his cup of joy was full; why attempt to overfill it?

But what has happened since then? Programs have improved and his taste has improved with them. He has thrown away the cup and has graduated to the mug, which also is filled to the brim. Having a mug, will he now demand a schooner? He will not.

The conclusion that we have been laboring, somewhat heavily, to reach, is that it is to the mud slingers and knockers, the minority critics—or "Flies In the Ointment"—that most of the credit is due for the rapid strides that radio has made. Back in radio's dark ages at least fifty per cent. of every station's time was devoted to unendurable tin pan jazz. The passive listeners stood for it. The knockers objected. It was eliminated and the passive listeners found themselves with fifty per cent. more entertainment for their money and all through no effort of their own.

Radio has grown up considerably but it still has a few bad habits hanging over from its infancy. It is up to the knockers to knock these out. If radio is to be a Bigger and Better man than it was a boy it is up to the knockers to pummel it into this new shape. The soft soapers and dispensers of ointment can do no more for it than to make it a self satisfied mollycoddle. Let there be more flies in the ointment!

The British Broadcasting Company Gets Razzed

WE ARE unable to give any very valuable dissertation on broadcasting conditions in England, at this distance. But from what we read there seem to be continual rumblings in the tight little isle, and most of them to the effect that the British Broadcasting Company is too highbrow. We have just received a copy of a thirty-eight page pamphlet by one Corbett-Smith flaying the administration of the B.B.C. A decidedly long-winded affair, it gets down to points occasionally:

When one sets out to give a radio entertainment, whether music, poetry, drama, speech, "variety," or anything else, one visualizes (or should visualize) not the few who are already educated in some measure to appreciate the best in these various forms of art, but the vast many of our people to whom beauty has hitherto been a closed book—the great mass of our folk who have never heard good music or noble poetry or any of our incomparable English literature—and so who pretend impatiently to disdain these niceties of civilization, as they would call them. . . . *Every single radio program should be so built and presented as to form a perfect fusion of art, education and popular entertainment.*

The type of mind which is usually associated with scholastic education is hopelessly out of place in radio work. And there is another cause of the B. B. C.'s failure. It is the born showman of a very special quality that is needed. The man with the widest possible range of interests, with "an acute sense of the inter-relationship of every kind of activity." Radio entertainment demands not the depth of the scholar but the breadth of the sensitive man of the world.

Showmanship, in varying degree, is needed for every single feature of radio entertainment. The "superior person" may sometimes scoff; but that person does not interest us. We have to compel and to rivet attention. We need, also, strong and vivid personality. The personality of the leader of men, not of a cold-blooded corporation. And we need absolute sincerity, both of purpose and utterance.

Now the B. B. C. have not begun to appreciate anything of this. Instead of making a Charles Dickens their director of programs they have put in a Matthew Arnold, the apostle of culture. Dickens enjoyed everybody and everything, even Fagin and Mr. Murdstone. That was the secret of his art and of his success. The B. B. C. seem to enjoy nothing, not even themselves.

It is necessary to emphasize this total lack of

sympathy with the people at large, because it strikes at the root of the matter. The B. B. C. are forever vaunting the intensely democratic character of their service when, in fact, it is about the most aristocratic business in the country. The House of Lords is an assembly of plebs beside it.

Wherein, if Mr. Corbett-Smith's words are to be taken as true—and he certainly sounds convincing—we see that a government-controlled monopoly is not one of the best ways of providing satisfactory radio programs. The point that the writer pounds in throughout the length of his diatribe is one which, we think, is well worth making, namely: that radio's principal service is, after all, for the masses. The so-called intellectual class is not interested in radio at all. Its members do not own receiving sets nor would they listen to one if it were given them. This is not due merely to snobbishness; their time is otherwise occupied, and of other means of entertainment they have more at their hand than they can make use of.

But we in the United States have no reason to fear such a state of affairs as Mr. Corbett-Smith complains of. Radio stations in this country are operated essentially for the masses. This is the natural result of a competitive system which depends for its reimbursement on advertising, direct or indirect. A maximum number of listeners must be the aim of every station which is not endowed or privately financed. In fact, here, a condition exactly the opposite of that alleged to exist in England is likely to obtain. A majority of stations, in their devotion to the masses, quite neglect the upper fringe of listeners. This is not true of the two score or so better stations. Careful and intelligent planning has enabled them to present programs appealing to the widest possible range of tastes. Their procedure is, first, to arrange a program that definitely appeals to the great mass of listeners, and secondly, to further manipulate it so as to effect a compromise with the upper crust of listeners.

We, from viewing the subject too closely, are likely to forget how exceedingly well this has been done. Take, as an example, the Atwater Kent Hour. A straight appeal to the masses is made in the making up of these programs. While the selections are limited to the classics, and to the opera, it is almost exclusively the sure-fire hits and tried and true tunes that finally find their way on to the program. But while the highbrow may think some of the tunes are banal and overworked, they've got him on another score: he cannot afford to ignore the importance and artistry of the performers Atwater Kent employs to put them over.

BOB CASON AND REBER BOULT
Artists at station WLAC, pianist and baritone respectively. They call themselves the "Thrift Twins" for some reason not apparent in the photograph

ANITA DEWITTE HALL OF KOIL
She is the versatile program director, organist, pianist, and "Mother Hubbard" of the staff

The R. G. S. "Octamonic" Circuit

How Laboratory Discoveries Were Moulded to Produce the Commercial Design of a Sensitive and Selective Set—Details of a Striking 1928 Development

By DAVID GRIMES

THE first article in this series (RADIO BROAD-CAST for October) described the conception and theory of the fundamental "Octamonic" principle, which obtains a high degree of selectivity by a function of the vacuum tube rather than by any special circuit contraptions. It was shown that the super-selectivity did not impair the tone quality as is the experience in tuned radio-frequency circuits. The high frequency of the second harmonic current permits a very sharp resonance curve without unduly compromising the side band amplification which is absolutely necessary for the proper reproduction of the high-frequency audio tones.

Other points of invention were also discussed, but it is a long road from invention to commercial design. It is one thing to build a laboratory model which proves the principle of an idea and quite another thing to plan the construction of a radio receiver which will meet all of the commercial conditions encountered in the field without a great many operating controls.

The purpose of this article is to reveal the design and constructional information which have been found necessary. These data have been acquired only after a great amount of original investigations, for there appeared to be little or no information available on the subject of second harmonic generation, tuning, amplification, and detection. The subjects will be discussed in the order in which they occur in our laboratory notebook. While the order may appear to be unusual, the facts were accumulated in just that sequence.

The first study was confined to the harmonic generating tube. The proper operation of this tube insures the success of the entire receiver. The first article showed a C potential bias on the grid of the harmonic generator—this bias causing the tube to operate on the lower knee of the grid voltage-plate current characteristics. This point of the characteristic gives the greatest amount of second harmonic energy as the unequal amplification between the two halves of the carrier wave is greatest at this point. However, with the standard commercial types of vacuum tubes operating on standard units of B potential such as 22 volts, 45 volts, or 90 volts, the amounts of negative C bias required for harmonic generation do not correspond with the commercial units of C potential available with the standard C battery. For instance, the maximum amount of second harmonic energy appeared to be generated by a standard cx-301-a tube operating on 45 volts plate potential with about 2 volts minus grid potential: Such a C bias cannot be obtained conveniently from dry batteries.

Of course, the easiest way to obtain a 2-volt negative bias on the grid of the harmonic generating tube is to utilize the principle of an IR drop. By running the filament return of the tube through a fixed 6-ohm resistance, a 2-volt drop may be obtained and if this fixed resistance is placed on the negative filament lead a negative 2-volt bias is available for the grid. Fortunately the operation of the second harmonic tube was not affected by running its filament on 4 volts, the remaining A battery potential available for the filament after 2 volts had been extracted by the fixed resistance for the grid bias. In fact,

it was found that the filament of the harmonic tube could be run much lower than this without in any way impairing its second harmonic generating properties. This is explained graphically by Fig. 1 which shows the grid voltage-plate current characteristic of a vacuum tube which is operated at various filament voltages. The various filament temperatures materially affect the upper portions of the characteristic but have

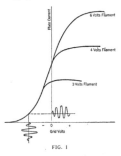

FIG. 1

little or no effect on the lower knee of the curve. The filament voltages only affect the saturation point of the tube.

In the vernacular, this is a fortunate "break" in design work as it affords a very simple arrangement for the harmonic generator which is very stable in its performance and exceedingly inexpensive. As a matter of fact, a series of tests shows that the filament voltage of the tube

could be cut down as low as 2½ volts before the second harmonic currents were affected, and the C bias could vary from 1½ to 2 volts. This more than covers the variation in A battery potential during the period between full charge and discharge of the A battery. The 45 volt B battery on the harmonic generator was also found to be a non-critical factor. Fairly large amounts of second harmonic currents were generated by this tube when the voltage had dropped as low as 34 volts or was raised as high as 50 volts.

PROBLEMS IN THE HARMONIC GENERATOR

EXCESSIVE B potentials on the harmonic generator created an unusual and peculiar difficulty. There existed a tendency toward oscillation on the short wavelengths of the input tuning condenser to this harmonic generator when the plate voltage was boosted too high. The source of this oscillation is not obvious and evaded detection for some little time. One is accustomed to expect oscillation in a tube only when there is a deliberate external feed back circuit or through the internal electrode capacities only when the plate circuit is tuned to the same frequency as the grid circuit—such as occurs in a tuned radio-frequency system. As seen in Fig. 2, the plate circuit of the harmonic generator is tuned an octave higher in frequency than the grid circuit and under these conditions the well known ordinary oscillation cannot occur. As a matter of fact the primary of the second harmonic transformer possesses considerable effective inductance as the result of the tuning to the higher octave. The number of turns in the harmonic transformer primary alone is insufficient to cause oscillation in the harmonic tube unless the secondary is tuned to the higher octave. When this is done, there is an increase in effective inductance over and above the actual inductance which causes the oscillatory difficulties mentioned, with excessive plate voltages on the harmonic tube.

The remedy for the difficulty outlined above

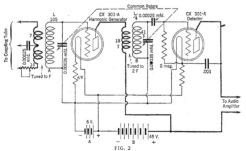

FIG. 2

lies in reducing the number of primary turns in the harmonic transformer to such a point that the effective inductance at the highest commercial voltage will not produce the oscillation described at the short broadcast wavelengths. Fig. 3 shows the design details of the tuned harmonic transformer used for connecting the output of the harmonic generator to the input of the detector tube. It will be noted that the secondary of this transformer has been made unusually small—much smaller than would be expected for merely tuning the double frequency involved. Commercial considerations have controlled the design of this transformer as well. The general tendency in the modern design of the radio receiver is to combine as many of the tuning condensers as possible on one shaft—exercising the proper care in the balancing of the condensers and coils so that they will tune alike for all the broadcast wavelengths. In the R. G. S. "Octamonic" design, it seemed desirable to combine the tuning condenser on the secondary of the harmonic transformer with the tuning condenser on the input to the harmonic generator. The problem is not as simple as the combining of condensers controlling similar circuits. The harmonic condenser must always tune to half the wavelength of the fundamental tuning condenser in the input of the harmonic generator. Furthermore, another limitation is imposed because all available gang condensers have been designed for tuned radio-frequency circuits and have therefore equal capacity in all the individual members of the gang.

This means that the second harmonic transformer must be so designed that a standard 0.0035-mfd. tuning condenser must tune the half wavelength band from 100 to 275 meters at exactly the same settings on the dial as a similar condenser tunes the fundamental coil for the respective corresponding fundamental wave between the 200 and 550 meter broadcast band. A consideration tuning formula shows that this can easily be accomplished if the inductance of the secondary of the harmonic transformer is made exactly equal to $\frac{1}{4}$ of the inductance of the fundamental tuned secondary on the input to the harmonic generator tube.

SOURCES OF SELECTIVITY

ANOTHER fortunate "break" aids in the ganging of these two condensers, as one of them is extremely sharp in tuning while the other is relatively broad. The real selectivity of the receiver is obtained by the tuning condenser on the input to the harmonic generator as the harmonic currents generated in this tube are proportional to the square of the resonant input carrier voltages resulting from this tuning condenser. The tuning condenser across the secondary of the harmonic transformer is no sharper than the ordinary tuning circuit on the input to a detector tube. Slight variations are therefore permissible in the coils and condensers without jeopardizing the performance of the receiver.

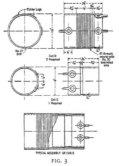

FIG. 3

The two gang condensers may have the same grounded rotor shaft as there is no need for any special insulating between these circuits. It is true that there are different grid biases on the two tubes to be tuned, but the positive bias on the grid of the detector tube may be supplied by the grid leak connecting from the plus filament of the detector tube directly to the grid, while the negative C bias on the harmonic generator is supplied through the common rotor and the secondary of the fundamental transformer in the grid circuit of the harmonic generator. The blocking condenser in the grid circuit of the detector effectively separates the positive potential on the grid of the detector and the negative C bias which exists in the rest of the harmonic secondary by virtue of the common grounded rotor shaft. As the negative potential is obtained by a resistance drop, as previously described, it is obvious that both the high-frequency currents in the detector grid circuit and the broadcast frequency currents in the harmonic grid circuit must return to their respective filaments through this resistance. Feed-back difficulties and oscillation would absolutely occur at this point in a tuned radio-frequency system, but no difficulties are encountered in the R. G. S. "Octamonic" because these two carrier currents are of different frequency and cannot, therefore, interfere with one another.

One thing should be made very clear at this point of the discussion. There is a fundamental difference between detection as such, and the generation of second harmonics. As discussed in the first article of the series the operation of a tube on the knee of its characteristic curve will produce not only second harmonic but

audio currents as well. In this circuit the tube is acting not only as a harmonic generator but, incidentally, as a detector. No method is known at present for the efficient generation of second harmonics without the incidental detector action occurring simultaneously. However, in the detector stage, two possibilities are present. Either the grid leak system of detection or the C battery system of detection may be used. The grid leak system is slightly more sensitive on very weak signals while the C battery system will deliver more audio energy on local reception without distortion. A study of these two types reveals some interesting facts. There is present, along with the detector action, some incidental generation of second harmonics, when the C battery detector is employed. This would be expected from the considerations already discussed in connection with the harmonic generator. However, second harmonic currents are almost wholly absent in a detector tube employing the grid leak system. This means, that detector action cannot be confused with harmonic generation. They may or may not occur simultaneously. The grid leak detector simply cannot be used in the generator stage for the creation of second harmonics. The harmonic generator is not a detector.

WHAT TYPE OF DETECTOR CIRCUIT?

IT NOW remains to determine which type of circuit should be used in the detector stage. Both the C battery and grid leak circuits were subjected to an extensive series of tests. The grid leak system was found to be much more satisfactory and much more stable. The tone quality was not impaired by the grid leak system and the distortion which occurred on local reception when using the C battery system, entirely disappeared when the grid leak system was substituted. The results were so consistently contrary to those anticipated that considerable data was gathered in an effort to explain the cause. Fig. 4 shows graphically just what occurred and why it is desirable to employ the grid leak system in the bona fide detector. It will be noted that the incidental detection occurring in the plate circuit of the harmonic generator is represented by an increase in the plate current—the increase being proportional to the modulation on the incoming carrier waves. Quite the reverse takes place in the detector circuit. Here the plate current decreases upon detection due to the choking action of the grid leak and condenser in the grid circuit. The decrease in plate current is proportional to the modulation on the incoming carrier waves. The rectified or audio currents existing in the plate of the harmonic generator are not utilized but, in turn, flow through the B battery circuit. The detected or audio currents in the plate circuit of the detector go through the primary of the first audio transformer and then flow through the B battery circuit. With these two audio currents opposing each other in the B battery at all times there is no audio voltage drop occurring therein. If a C-battery detector were employed the two audio currents would increase and decrease simultaneously, causing excessive audio voltage drops in the common B battery.

As brought out in the previous article, the R. G. S. "Octamonic" receiver obtains its super-selectivity through the sacrifice of some of the radio-frequency energy. But as radio-frequency energy is very easily obtained by any number of r.f. amplifying circuits and selectivity is not so easily obtained, the sacrifice is well worth while. However, some form of r.f. amplification must be placed ahead of the "selectivity" circuits just discussed. Various r.f. arrangements have been investigated and the one shown in Fig. 5 is

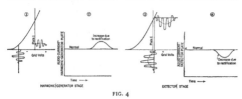

FIG. 4

recommended. This shows one stage of radio-frequency amplification only as one stage has been found to be more than ample for operating the harmonic generator on even the most distant stations. The antenna is very closely coupled to a tuning circuit which serves mainly to bring the antenna circuit to resonance at the frequency desired. The carrier wave is then amplified and applied to the harmonic generator through a special equalizing coupling circuit which is designed to pass all of the broadcast frequencies with approximately undiminished amplitude on to the harmonic generator. The theory of the operation of this unusual coupling is rather simple. The total winding consists of 21 turns which is the proper primary for the longest wavelength of 550 meters. Then a tap is taken off at 7 turns which is approximately the proper primary for the shortest wavelength of 200 meters. A 0.00025-mfd. fixed condenser is connected between the tap and the filament of this amplifying or coupling tube. A variable non-inductive 250-ohm equalizing resistance is placed in series with the total winding. The short waves pass readily through the fixed condenser to the filament while the longer waves tend to pass more and more through the entire winding because of the increasing reactance of the fixed condenser to the lower frequencies of the longer waves.

This first amplifying tube has been designated a coupling tube since its main function is purely a coupling and amplifying action rather than any aid to the tuning. The coupling to the antenna is made as close as possible so as to derive the maximum amount of energy therefrom throughout the broadcast band. Such close coupling makes the tuning very broad and non-critical—the real super-selectivity of the receiver being created by the tuning condenser on the input to the harmonic generator. In actual operation, this antenna condenser appears to be sharper in its tuning than it really is because, after all, it has an indirect effect, though broad, on the amount of energy being transformed into second harmonic currents by the generating tube. This action gives it an apparent sharpness greater than that which is really occurring in the antenna circuit.

CURIOUS MODULATION EFFECTS

ONE very important factor in the design of the coupling tube circuit is *modulation*. Great care must be exercised in the design of this circuit to avoid any possibility of rectification action even on the louder signals. Otherwise, the extreme selectivity of the harmonic generator will be somewhat modified by a cross-talk or

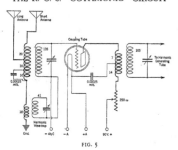

FIG. 5

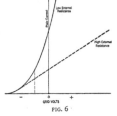

FIG. 6

modulating action between a near-by high-powered local station and a distant station operating on an adjacent carrier channel. The action is as follows: The receiver in New York City is tuned to a weak station such as woo in Philadelphia operating on 508 meters. It is found that several dial degrees of silence are obtained between this Philadelphia station and WEAF, New York City, operating on 492 meters. However, WEAF is coming in with considerable field strength impressing considerable 492 meter energy on the grid of the coupling tube along with the energy from the Philadelphia station. If any rectification occurs on the strong signals from WEAF, audio currents will be set up in the plate circuit of this radio coupling tube which correspond to the program being sent out by WEAF. It must be remembered here that the first circuit is broad—its function being amplification and not tuning. As a result, more WEAF energy may be present than the energy coming from Philadelphia even though the antenna tuning condenser has been set in favor of the Philadelphia station.

Now, the audio currents occurring in the plate circuit of the coupling tube as a result of the rectification of the WEAF carrier wave, will cause a plate voltage variation in this tube which corresponds to the program on the carrier wave of WEAF. This action will, in turn, affect and vary the amplification of any other carrier wave coming through the tube at the time, such as the Philadelphia station which is desired. The audio modulation or WEAF's program will then impress itself on the carrier wave of the Philadelphia station in the same manner that the original audio currents at the studio of WEAF impressed themselves on the original carrier wave being sent out from WEAF on 492 meters. The result is that, while several dial degrees of silence are obtained between WEAF and woo, as soon as the Philadelphia station is tuned-in, the program from WEAF is found also to exist thereon in the form of cross-talk or cross-modulation.

The remedy is to operate the coupling tube on the straightest portion of the grid-voltage plate-current characteristic curve. For the standard cx-301-a tube, this requires 90 volts on the plate and 4½ volts negative bias on the grid. This point is very essential. In addition, it is desirable to have the maximum of coupling to the harmonic generating tube, not only from the standpoint of energy transfer, but also for the purpose of obtaining a fairly high effective resistance in the plate circuit of the coupling tube at the particular frequency for which the harmonic generating tube is tuned. It is a well known fact that the resistance of a primary winding increases considerably at the resonant frequency of the secondary. At the same time the reactance passes through zero, of course. The closer the coupling the greater is the effective resistance of the plate circuit and resistance in the plate circuit tends to flatten out the characteristic curve of the coupling tube. This is shown in Fig. 6. This flattening of the characteristic curve still further reduces any tendency toward rectification in the coupling tube.

A detailed discussion on the audio amplifier as well as a full explanation of the theory and operation of the harmonic wave-trap shown in the antenna circuit will be taken up in the next article of the series. The next article will also describe in further detail the best wiring arrangements.

The "Equaphase"

Better Control of Oscillation Is a Feature of This New Receiver— The Story of Some of the Difficulties Surmounted in Its Design

By J. O. MESA

ANY radio engineer can make a single receiver work in the laboratory but when a factory turns out five hundred a day, each of which must be thoroughly tested, the problem becomes somewhat more complex. It becomes one not only of manufacturing small mechanical parts to a high degree of precision and of simplified assembly so that mistakes are difficult to make, but one of following a circuit that is electrically sound and as foolproof as possible.

Circuits that are highly sensitive are often highly critical in their adjustment, necessitating that they pass through the hands of a well trained tester before they can be released. What every set manufacturer wants is a receiver design such that manufacturing costs and assembly problems are reduced to the bone, that testing methods are neither complicated nor expensive in point of time, and that adjustments are few. Simplicity of design is not the controlling factor, for the simplest circuit must embody the same trouble producing elements as the most complex. For example, every engineer knows that inductance in the plate circuit of a radio-frequency amplifier is necessary to transfer energy from one circuit to the following tube; but he knows too well that including this inductance tends to make the previous tube oscillate. One method of preventing oscillation is to feed back to the input circuit some of the energy that appears in the output circuit in such magnitude and phase that the tube is no longer unstable. Owing to the fact that unity coupling is necessary to obtain complete prevention of oscillation (neutralization) at all frequencies, it is impossible to neutralize the amplifier over more than a narrow band at one time. Often the circuits are balanced at the shortest wavelengths, for the tendency to oscillate is greatest there; but this is apt to result in poor transfer of energy on the longer waves.

If it were possible to neutralize a receiver completely throughout its tuning range and to include sufficient primary inductance with close enough coupling to the secondary to produce proper amplification on the longer waves, the design engineer's problem would be simple. Unfortunately, in spite of the neutralization process, it is often impossible to utilize the utmost desirable amount of primary inductance and coupling, and the longer waves suffer.

A method that has been used to maintain the circuits in a two-or three-stage radio-frequency amplifier free from oscillations is that of including

resistances in the grid circuits of the amplifying tubes. It was found in the Freshman laboratories that, with sufficiently large primary coils, the resistances became rather large, of the order of 500 ohms. This method of "holding down" an amplifier is shown in Fig. I, and in Fig. 2 is the equivalent circuit.

Mathematics will show that for every resistance, another, smaller in actual value, can be substituted, as in Fig. 2, in the tuned circuit itself to accomplish the same end, namely, cessation of troublesome oscillations. In fact, when such external resistances are used, the selectivity of the circuit suffers, and at the same time the amplification falls off, showing conclusively that this external resistance has in effect added considerable damping to the tuned circuit itself.

There is a very serious objection to the use of

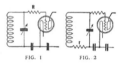

FIG. I   FIG. 2

resistances as in Fig. I. Under the usual conditions obtaining in a high-amplification circuit, the value of the stabilizing resistance is somewhat critical. If the correct value is exceeded, the overall voltage gain of the radio-frequency amplifier is considerably reduced, while if the resistance is too small, the circuit oscillates violently. Furthermore, any slight variations in the component parts which make up the radio-frequency amplifier change the value of resistances needed for best operation.

The solution, naturally, lies in very close manufacturing limits on the values of inductance, capacity, and resistance, and for some time receivers employing the resistance arrangement of Fig. I were built in the Freshman plant. In spite of the fact that the resistances were held accurate to within plus or minus 1.25 per cent, difficulty occurred from oscillation or poor selectivity and lack of amplification too frequently for comfort. It became necessary to develop another stabilizing system.

The new method of overcoming oscillations without the disadvantages previously mentioned was found by Mr. W. L. Dunn, Jr., and was developed by the engineers of the Freshman laboratory. This method is based on a principle which has been used for a long time in telephone equalizing circuits, and has been discussed mathematically by K. S. Johnson in his *Transmission Circuits for Telephonic Communication* and by Morecroft on page 92 of the new edition of his well known *Principles of Radio Communication*.

The circuit which has been utilized is shown in Fig. 3, and it may be seen to consist of a coil and condenser in parallel with a resistance in each branch of the parallel circuit. If the resistances in the two paths are equal to each other and

equal to the square root of the inductance divided by the capacity, or:

$$R_L = R_C = \sqrt{L/C}$$

the impedance of the circuit looked at from the standpoint of the previous tube is a pure resistance at all frequencies.

Therefore, the plate circuit of the previous amplifier has no inductive reactance in it and so the tube cannot oscillate, provided the values of inductance, capacity, and resistance are properly adjusted.

Limiting the magnetic feed back in the circuit by using small diameter coils and placing them at right angles to each other, and by using proper bypass condensers across impedances which are common to more than one stage, naturally aid in keeping the amplifier performing its required tasks.

The method of applying this interesting case of parallel resonance is shown in Fig. 4. Owing to the fact that some resistance is reflected into the circuit when a secondary coil is coupled to it and is tuned to resonance, the actual values of resistance used are different, 380 ohms being used in the condenser, or plate branch, and 350 ohms in the inductance branch. The capacity used is about 100 micro-microfarads and is of the fixed-variable type, that is, a fixed condenser having attached to it a small variable capacity which may be adjusted in the factory so that the receiver does not oscillate.

When it is realized that closer coupling and greater primary inductance may be used when such a circuit is employed, the gain over the grid resistance method of stabilization is evident. Nearly double the coefficient of coupling may be used between primary and secondary in this new system.

THE RECEIVER

THE schematic diagram of a battery operated receiver using this principle is shown in Fig. 5. This receiver is now being manufactured with large quantity production methods. It has been found that it is not subject to the difficulties encountered in the case of the grid resistances method of preventing oscillation. The condenser in the plate circuit of the radio-frequency tubes consists of curved spring plates with mica dielectric which can be flattened by means of a screw which is accessible to the inside of the cabinet. This condenser is adjusted at the factory

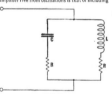

FIG. 3.

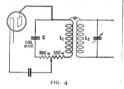

FIG. 4.

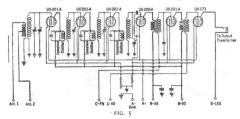

· FIG. 5

so that the amplifier does not oscillate and does not require readjustment thereafter so long as tubes having the same plate impedance are used in the radio-frequency amplifier. When a.c. tubes are used a smaller capacity is needed owing to the high low impedance of these tubes compared with the average d.c. tube.

The radio-frequency transformers have secondaries which are wound on a small bakelite tubing and have individual turns slightly spaced. It has been found that this type of winding can be controlled so that the inductance of the coil may be held to within about 0.5 per cent. of accuracy when manufactured in large quantities. The primary of the radio-frequency transformer is a spiral wound on a wooden form, and is placed at the ground end of the secondary, thus having the advantage of a comparatively large coupling and at the same time small capacity between the primary and secondary.

The audio transformers are mounted on the tube shelf and have a 4 to 1 turn ratio. The secondary windings are of the split or balanced type which have very low distributed capacity and very low capacity between windings, with the result that a quite uniform amplification is obtained between 100 and 5000 cycles per second. The radio-frequency transformers as well as all the other apparatus which is not accessible to the controls on the front panel, are mounted on a spring suspended metal shelf. This shelf carries the tubes and is provided with rubber dampers so that there is no microphonic feed back.

The variable condensers used are selected so that their capacities are equal to within 0.25 per cent. at two points. Since the plates are of heavy construction and have rather wide spacing, the capacities of the condensers in any receiver are practically identical throughout their entire range. The arrangement of the front panel of the "Equaphase" may be seen in the photograph of this receiver which appears on page twenty.

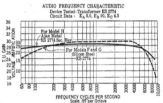

AUDIO FREQUENCY CHARACTERISTIC
Device Tested: Transformer ES 2774
Circuit Data : E_A 5.0, E_B 90, E_C 4.5

INTERESTING CURVES

Showing the characteristics of the audio transformers used in the "Equaphase"

The volume control of the battery operated receivers is obtained by means of a rheostat in series with the filaments of the radio-frequency tubes. This method has been used for several years on Freshman Masterpiece receivers and has left nothing to be desired.

As mentioned before, a slight modification of

the constants of the elements used in the plate circuit of the radio-frequency amplifiers is necessary to adapt the circuit to the use of the alternating current tubes of the UX-226 (CX-326) type. Since the circuit does not oscillate and has high amplification throughout the broadcast range, the volume control can be obtained by means of a potentiometer across the secondary of the first audio-frequency transformer. This potentiometer has the so-called logarithmic scale, that is, its resistance varies so that equal angular increments on the control produce equal increments of volume. A schematic diagram of the house current operated "Equaphase" receiver and its power supply is shown in Fig. 6. The plate supply is of the conventional type using a UX-280 full-wave rectifier tube and having a two-section filter. Various plate voltages are obtained from taps on a resistor connected across the output of the filter. The grid bias for the various tubes is obtained from the drop across the resistance in the plate circuit of the tubes.

Two independent tests are made on the sensitivity of each receiver before it is packed. These tests are made by two men who check each other's work without either knowing the other's. Around the laboratory is fed a continuous 1000-cycle tone of a certain amplitude which is maintained constant. This is available at each test bench, and is used to modulate a small radio-frequency oscillator. A very small part of this energy is picked up on a dummy antenna whence it goes through the receiver just as a radio signal would. The test man listens to it comparing it with the standard input 1000-cycle tone by means of an attenuation box which is placed in the receiver output. In this way he reads the relative amplification of the receivers, and after fixing a certain standard he can reject any which fall below the required limit. Phonograph music is also fed around the test benches so that the test man can listen to music as well as his standard 1000-cycle signal.

Any slight error in a component part is discovered in this way before the receiver leaves the factory, in which case it can be sent back to the repair bench for final adjustment.

It can easily be appreciated that this method of preventing oscillation is one which does not affect the selectivity of the receiver once it is properly adjusted. The selectivity of the "Equaphase" is quite satisfactory enough for congested districts.

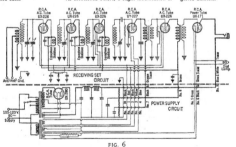

FIG. 6

The Improved "Shielded Six"

By JOHN E. McCLURE

IT IS seldom that a radio receiver design, or kit, outlives a single season's popularity, and when the exception comes along, it gives assurance that it must be an unusually fine set. Such is the "Shielded Six." During this last year certain refinements of design have been developed, and the new improved model is now ready for the 1927-28 season.

Mechanically, the design of the receiver is one of the prettiest of kit jobs, and the "Shielded Six" looks more like a carefully worked-out assembly for quantity production in a modern factory than a kit receiver. The whole set builds up progressively on a die-formed and pierced steel chassis, which is a radical departure from the often makeshift packing-case baseboards to which the home constructor is accustomed. The panel also is of metal, bronze, attractively decorated in the fashion of the new expensive factory-built sets, being utilized for this purpose.

Electrically, the circuit design involves three stages of tuned radio-frequency amplification with controlled regeneration, a grid-biased detector tube, and two stages of transformer-coupled audio-frequency amplification. In these respects the improved model is very like the original, and the only really startling improvements found in the set are circuit changes resulting in greatly increased selectivity, and the addition of vernier tuning dials found necessary because of the greatly increased sharpness of tuning.

The antenna stage, or first radio-frequency amplifier, is left unshielded in the new model to increase the coil pick-up to a point where the receiver may be operated in apartment houses with no antenna at all, and yet give adequate loud speaker volume on powerful local stations. If the second, third, or fourth coils were unshielded, selectivity would be affected, for energy pick-up on these coils would affect the selective tuning action of the tuned circuits. Shielding not only prevents pick-up of external interference, but possibly more important, entirely prevents extraneous interstage coupling.

Losses in the r.f. circuits have been reduced to a minimum through the use of quite low resistance inductances, wound on threaded ribs of bakelite coil forms in such a fashion that they are practically air-supported. These inductances are tuned by means of newly designed, very rugged, condensers providing a semi-straight-frequency, straight-wavelength tuning curve which gives most satisfactory spacing of stations over the tuning dial scales. The set may be adapted for loop reception without a single change except to pull out the antenna coil and clip on two loop leads to coil socket posts 3 and 6. The new set seems amply selective for present broadcasting conditions, for, in Chicago, it will cut through a mass of thirty local stations and bring in out-of-town stations.

While four tuned circuits are employed, only two tuning controls are used, this being made possible through extremely accurate matching of condensers and coils. All coils for a set are matched to within a quarter of one per cent. and the condensers are all checked and held within one per cent. of each other. Since the tuning of the three right-hand circuits is substantially identical, a mechanical link serves to turn all three condensers as the right-hand dial is turned. The antenna stage condenser is tuned separately.

The audio-frequency amplifier has a highly desirable characteristic in that it amplifies practically uniformly all frequencies between 30 and 5000 cycles, above which frequency there is practically no amplification. The highest fundamental note of any musical instrument is 4192 cycles and the 5000-cycle upper limit allows plenty of leeway for the handling of the highest fundamental frequency in music, and it has been proved that frequencies above 5000 cycles do not contribute, in any measure, to fidelity of reproduction: Thus the interference caused by heterodyne squeals developed between transmitting stations, by atmospheric noise and by the tube and battery noises, all of which are generally above 5000 cycles, are almost entirely absent.

A group of people listening to the improved "Shielded Six" receiver operating in conjunction with a good cone loud speaker and receiving an organ program, will actually feel the vibration of the room in which the receiver is located, as

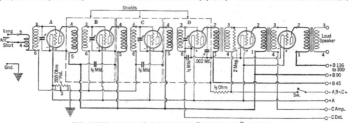

THE CIRCUIT DIAGRAM OF THE IMPROVED "SHIELDED SIX" RECEIVER

PREPARATORY TO PLACING THE STAGE SHIELDS IN POSITION

The arrangement of the Silver-Marshall triple link motion is distinctly shown in this photograph. The transformers are, from left to right, output, second audio, and first audio

The parts needed for the Improved Model are listed below. They total exactly $95.00. While this price may seem a bit high in that fifty-dollar six-tube sets, completely constructed, are available, it must be remembered that the "Shielded Six" has been designed with unusual care and that all the parts have been very carefully matched to insure satisfactory operation.

3—S-M 631 Stage Shields	$ 6.00
2—S-M 316A Matched Condensers .	9.00
2—S-M 316B Matched Condensers .	9.00
4—S-M 515 Coil Sockets	4.00
3—S-M 118A Matched Coils . . .	7.50
1—S-M 118A Coil, Matched with Above	2.50
6—S-M 511 Tube Sockets	3.00
2—S-M 220 Audio Transformers . .	16.00
1—S-M 221 Output Transformer . .	7.50
1—S-M 632 Triple Link Motion . .	2.50
1—Polymet-0.25-Meg. Grid Leak . .	.25
1—Polymet Resistor Mounting . .	.50
1—Carter 0.002-Mfd. Condenser . .	.50
3—Carter 105 or Polymet Condensers, 0.5 Mfd.	2.70
1—Carter M-200 Potentiometer (200 Ohms)	.75
2—Carter No. 10 Tip Jacks . . .	.20
1—Carter H-⅛ Resistor	.25
2—Marco Walnut Vernier Dials . .	5.00
1—636 Terminal Strip, 1½ x 11 inches .	2.00
1—Crowe 633 Drilled and Engraved Metal Panel, Size 7 x 21 Inches .	8.50
1—Crowe 634 Steel Chassis, Size 12 x 19½ x 1½ Inches	6.00
1—Carter No. 12 Antenna Switch .	.60
1—Carter "Imp" Battery Switch . .	.50
1—Coil Kellogg Fabricated Hook-up Wire, Screws, Nuts, Lugs, and Complete Building Instructions .	.25
Total	$95.00

they would were they in the original building in which the organ itself was located.

A. C. OPERATION

A. C. OPERATION is so ridiculously simple with the "Shielded Six" as to require little description and if Sovereign or equivalent heater tubes are used, lighted from a filament transformer, only minor changes need be made in the wiring. The standard Sovereign tubes have the heater leads coming out on top and all tubes should have their heater elements connected in parallel and to the 2½-volt winding of a filament heating transformer. It is also well to ground one side of this winding. The F—terminal of each tube socket should be connected to the receiver chassis while the F+ terminal of each tube socket should be ignored. The 200-ohm potentiometer should be eliminated and in its place a Carter 6000-ohm potentiometer used (the two center shields will have to have their corners clipped away to accommodate the new 6000-ohm potentiometer). Terminals 6 of the three left-hand r.f. transformers, which previously connected to the center arm of the potentiometer, should ground to the chassis. The 0.5-mfd. potentiometer bypass condenser is eliminated. The new potentiometer should be connected with one end to the chassis, the other end to the +90 binding post, and the arm connecting to one end of the B-90 bypass condenser and to terminals 5 of the three right-hand r.f. transformers. Five Sovereign a.c. tubes are used, with a cx-371 (ux-171) type tube in the last audio stage. The two filament leads from this latter power tube socket are run directly to the 5-volt terminals of the filament transformer. The center tap of

a Frost FT 64 resistance shunting the power tube filament connects through a 2000-ohm Carter fixed resistance to the chassis of the receiver or ground. A regular B supply, such as the Silver-Marshall 652A (which will also supply A potential to the tubes) will then furnish B potential to the entire receiver, and A, B, and C potential to the second audio stage, while a 4½-volt C battery will have to be used for the detector and the first audio stage (terminal 4 of the first audio transformer should connect to the C—Det. binding post of the receiver).

THERE IS AN ADVANTAGE IN LEAVING ONE TUNED STAGE UNSHIELDED

The antenna tuning stage, the only one of the four tuned stages not shielded, has purposely been left in this condition. By so doing, the coil pick-up is increased to a point where the receiver may be operated in apartment houses with no antenna at all, and adequate volume obtained on the loud speaker when receiving powerful local stations

The IMPROVED Aristocrat

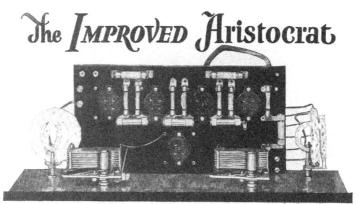

A SUITABLE LAYOUT THAT MAKES FOR SIMPLICITY OF CONSTRUCTION

There are many possible variations of design for the new "Aristocrat," the layout here, in fact, not being exactly similar to that described in the text. The "deck," for example, is mounted away from the front panel in the model described, and one variable rheostat takes the place of the two ballasts shown in the photograph. New Eby sockets have been substituted for the older pattern ones shown, and binding posts are used for Antenna, Ground, and Loud Speaker

By ARTHUR H. LYNCH

EXACTLY two years ago RADIO BROAD-CAST first described the "Aristocrat" receiver. This receiver became unusually popular and many interesting letters were received telling of the good results that were obtained. It was a five-tube affair consisting of a stage of tuned radio frequency followed by a regenerative detector, and the audio circuit comprised a three-stage resistance-coupled amplifier. Correspondence is still received from many readers regarding the receiver and evidently many "Aristocrats" are still giving good service. We do not intend in this article to describe a radically new "Aristocrat" receiver. The original circuit was carefully thought out and even though two years have elapsed since it was first described there are only minor ways in which it can be improved. An "Aristocrat" receiver carefully constructed in accordance with the original description would be found selective, sensitive, and capable of giving good quality reproduction in the majority of cases; there are, however, a few rearrangements that might be made in the mechanical design which will make the construction of this receiver simpler and better looking.

Before going into the details concerning these suggested changes, a very brief description of the circuit in its revised form, with special reference to the ways in which it differs from the original, will be given. The circuit diagram of the new "Aristocrat" is given in Fig. 1. An important difference between the new and the old set is immediately evident to those who are familiar with the original circuit, i. e., that the antenna stage now uses a variocoupler instead of a tapped coil. Antenna tuning in the original receiver was accomplished by means of the taps on the primary of the antenna coil and by proper use of this adjustment it was possible to obtain high efficiency

from the receiver with various lengths of antennas. The new antenna coil of the "Aristocrat" contains a secondary with primary inside it, variable coupling between the two coils accomplishing the same results as did the taps in the original coil; with the new arrangement the adjustment can be made more readily and more accurately. The variable antenna coil is a distinct improvement and should be incorporated in the new "Aristocrat" receiver and might also be used to advantage in receivers constructed according to the original circuit.

The detector stage of the new "Aristocrat" remains the same as the original circuit. The audio amplifier is arranged so that somewhat greater voltage is placed on the plate circuits of the first two stages than was originally used. The new high-mu tubes should be operated with at least 135 volts on their plates.

Simplicity of construction is the keynote of the new receiver. The improvements that have been made in the constructional features of the "Aristocrat" are, first, the use of a metal panel of special design and, secondly, a new and unique type of sub-panel construction. The special metal panel is designed to accommodate two variable condensers of the single-hole mounting type and the panel is also made for use with illuminated dials. There are three additional holes

PLATE VOLTAGE	NEGATIVE C BATTERY VOLTAGE REQUIRED			
	Power	Semi-power	Audio	Radio
45	—	—	—	0
90	16.5	6.0	—	3
135	27	9.0	1.5	—
180	40.5	12.0	3.0	—

in the panel, the one at the left being for the antenna rotor control, the center one for the rheostat knob, and another hole at the right is for the regeneration control. This panel and dial combination can be used in constructing any number of circuit combinations where there are only two tuning controls and its use in the "Aristocrat" is a good example of its utility. The panel measures seven by eighteen inches and is a product of the Wireless Radio Company, of Brooklyn, New York.

The new special sub-panel, or "deck" as it is called, has five Eby DeLuxe sockets mounted on it and audio amplifying equipment for a five-tube receiver. It is made of Westinghouse Micarta, and is built to accommodate ten binding posts. In the model illustrated the six-wire cable obviates the use of six of these binding posts, connection for the batteries being made directly to the wiring of the receiver by means of this six-wire cable. The audio amplifier is a three-stage resistance-coupled affair and both the resistances and condensers are held in place on the "deck" by clips so that constructors may use any values which they may prefer. The person who wishes to experiment can procure additional values of resistance and condensers than those which come with the deck and substitute them when desired.

There are available many different antenna couplers and three-circuit tuners that may be used in the "Aristocrat." In the particular receiver illustrated in the photographs accompanying this article, Sickles "Aristocrat" coils have been used in conjunction with Cogswell condensers. The Cogswell antenna tuning condenser has been made in a very ingenious fashion. Its stator plates form one side of the neutralizing condenser, while the other plate of the latter is mounted on a pair of hinges and is adjusted by turning a small screw which pushes against an eccentric cam. It is all very small, very simple, convenient, and very effective, as well as being cheap. At the end of this article there is given a list of those parts used in the model that is illustrated, and the photographs and the circuit diagram should enable experienced constructors to build the receiver with little difficulty.

NEW TUBES

IT IS possible to procure an improvement in results from either an old or new "Aristocrat" by making proper use of the several new types of tubes that have become available since the "Aristocrat" first made its bow. A special detector tube, for example, may be used in the receiver instead of a 201-A type and it will increase the sensitivity and volume very considerably. In this case the detector grid return should go to negative A instead of to positive as indicated in Fig. 1, unless a special Ceco type H detector tube is used, when no change is necessary. Ceco type G high-mu tubes should be used in the first and second stages of the audio amplifier. High-mu tubes of other manufacturers should not be used unless the condenser and resistor values are changed to comply with the specifications of the individual makers. The output tube should be of the semi-power type with proper C and B voltages. Ceco makes special radio-frequency amplifier tubes which will give slightly increased gain in the r. f. stage. They are known as the type K tubes. These new tubes, without regard to any other improvements that might be made, will give greater distance, sharper tuning, and more volume than can be obtained when ordinary tubes are used.

In the table accompanying this article there are given data on the C and B voltages that should be used on the various tubes. The column headed "Power" gives the voltages when a UX-171, CX-371, or Ceco J-71, is used in the output stage. The column head "Semi-Power" gives the required voltages when a UX-112, CX-112, or Ceco type F is used in the output. The voltages given under the column headed "Audio" refer to the high-mu tubes in the audio amplifier. The values given under the column headed "Radio" apply to either 201-A's or special radio-frequency

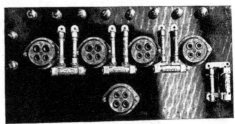

THE LYNCH "DECK"

Its utilization considerably simplifies receiver construction. The list of parts below tells just what the "deck" comprises

amplifier tubes. If a 171, 371, or J-71 power tube is used in the output of the receiver, an output filter or output transformer should be used in the plate circuit of the tube to protect the windings of the loud speaker from the high plate current. As shown, the circuit is wired for a 112 or Ceco F type tube. C bias for the first two audio tubes is obtained by inserting a battery in the common grid lead at the point marked "X," the positive terminal of the grid battery connecting to negative A.

The results obtainable from the improved "Aristocrat" receiver do not suffer at all in comparison with the original set, while the total cost of building the receiver has been materially reduced. The automobile business is not the only one in which the honest claim that production methods enable you to purchase a better product for a lower price. In the case of the improved "Aristocrat," production methods have been applied to radio.

L₁, L₂—Sickles "Aristocrat" Coils . $ 4.50
C₄—Cogswell Variable Condenser,
    Type A, 0.00035 Mfd. . .   2.25
C₄—Cogswell Variable Condenser,
    Type B, 0.00035 Mfd., Neu-

tralizing Condenser (C₅) Attached 2.75
C₆, C₇—Sangamo 0.004 Mfd. Fixed
    Condensers 1.20
Four Eby Binding Posts . .   .60
Six-Wire Cable 60
Wireless Radio Company's
Panel, 7 x 18 Inches, with
Mounting Brackets, Illuminated Dials, and Filament
Rheostat (R₁) 4.50
Two Kurz-Kasch Knobs . .   .50
Lynch Deck, Including the
Following Parts, Mounted and
Ready for Wiring:
Westinghouse Micarta Panel
Seven Resistor-Condenser
Mounts
R₂, R₃, R₄—0.1-Meg. Metallized
Resistors
R₄, R₄, R₇—0.5-Meg. Metallized Resistors   12.50
R—2-Meg. Metallized Resistor
C₃, C₈, C₉—0.006-Mfd. Tubular
Condensers
C₇—0.00025-Mfd. Tubular
Condenser
Five Eby DeLuxe Sockets

              Total $29.40

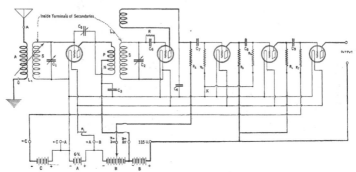

THE CIRCUIT DIAGRAM OF THE "ARISTOCRAT" RECEIVER

Suppressing Radio Interference

Interference from Motion Picture Theatres, Telephone Exchanges, Arc Lamps, Incandescent Street Lamps, Flour Mills, Factory Belts, Electric Warming Pads, Precipitators, Etc., Is Discussed, and Remedies Are Suggested

By A. T. LAWTON

ALTHOUGH this article is really the second of a series, the first of which appeared in the September RADIO BROADCAST, it is nevertheless complete in itself, and the reader who is suffering from interference of any of the forms outlined here will profit considerably from a study of this paper. The data presented here result from a two-and-a-half-year study conducted by the author in more than 132 cities. The forms of interference covered in the September article were those due to oil-burning furnaces, perhaps the most common source of man-made static, X-ray equipment, and dental motors. The first kind of interference to be considered here is that originating at motion picture theatres.

MOTION PICTURE THEATRES

THE radius of interference from this source is ordinarily about 200 yards, occasionally greater, depending on exterior wiring. In the great majority of cases the direct-current generator is responsible for the trouble. Contrary to popular belief, the arc lamps themselves cause practically no interference; in fact, there is often less disturbance with the arcs lighted than before they are "struck," i. e., with the generator running unloaded. The difference in certain cases is decided, absorption of the interference occurring as soon as the arcs are put on.

Fig. 1 shows a method used to eliminate this interference with success in actual practice. Five-ampere fuses are used. If the commutator is badly worn, it should be turned down in a lathe, and we might remark here that the quality of the carbon brushes used have a noticeable effect on the intensity of commutation interference.

Squirrel cage induction motors are in common use for driving these generators and, ordinarily, give no trouble unless some defect is present.

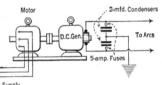

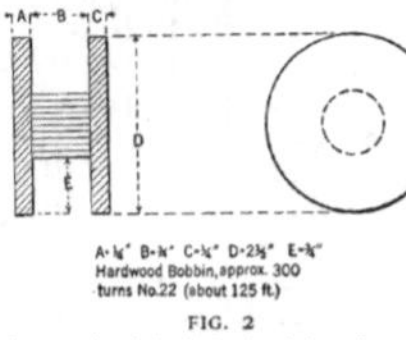

A-⅛" B-¼" C-½" D-2⅛" E-¾"
Hardwood Bobbin, approx. 300 turns No.22 (about 125 ft.)

FIG. 2

Constructional details for the choke coils recommended for elimination of interference originating in telephone exchanges

TELEPHONE EXCHANGES

IT IS probable that in most cities interference from this source has been cleared up. The larger operating companies have been active in this regard, but in smaller towns and rural communities much trouble exists. On the larger type motor ringers, high-tone and low-tone (sometimes referred to as trouble tone and howler) circuits are mainly responsible. Complete elimination is secured by inserting a choke coil in each of the two brush leads, close up to the machine. Details of the coil required are shown in Fig. 2.

Complications arise if connection is made at a distance of more than four inches from the brushes—incomplete elimination resulting.

The greatest offenders in the category of telephone ringing apparatus are pole changers and frequency converters. These constitute standard equipment in thousands of small exchanges; in some larger exchanges they are operated only after 10 P. M. when the rush hours have passed. The interference is of a rapid clicking nature and may carry half a mile or more, depending on the proximity and layout of the city distribution and telephone wiring.

For pole changers, definite and conclusive results are obtained by inserting the coils described in Fig. 2 in the ringing leads and at least 95 per cent. of the trouble disappears. Up to the present we know of only one instance where this method failed—and peculiarly enough, a simpler method cleared up the trouble. A single one-half microfarad condenser bridged across the contacts gave 100 per cent. elimination.

Frequency converters are a different proposition, operating off a. c. instead of d. c. as in the case of pole changers. The surge trap applicable to pole changers gives only 50 or 60 per cent. reduction.

A special surge trap is made up for this source by companies manufacturing the frequency converter, and it will be found more economical in the long run to purchase this complete rather than endeavor to make it up locally. For all practical purposes, complete elimination is obtained through the installation of the special surge trap.

Automatic telephones, now coming into such general use, contribute their little quota of dis-

FIG. 1

turbance. Usually they affect radio receivers only in the same residence where the dialing operation is being carried out, although several cases are on record where radio sets of next-door neighbors were also affected.

A single condenser of one microfarad capacity placed across the dialing circuit will cut out the clicking noise and does not appear to have any detrimental effect on the speech transmission or proper functioning of the line. However, permission of the telephone company should be obtained before making any attachments.

ARC LAMPS

STRANGE as it may seem, flickering arc lamps cause practically no radio interference. If the arc is jumping violently, however, then clicks are recorded on radio receivers in the vicinity; a short distance away interference is not material. This doesn't mean that arc lamps cause no disturbance. On the contrary, during a recent investigation one arc lamp practically killed radio reception for eleven city blocks along one street. It was burning perfectly steadily and showed no sign whatever of defect.

The characteristics were slow clicking in dry weather, fast clicking in moist atmosphere, and rapid clicking during rainy weather. On a five-tube set with loud speaker, the noise was violent, resembling very much the operating of a pneumatic hammer. The source of trouble here was a minute fissure in the composition head ring which, filling up with moisture, caused a high-resistance short across the 4000-volt lines. Evidently the spark dried up the moisture at each crossover, and time was required for the path to reform, otherwise we should have gotten steady buzzing here.

During all the days this case was under observation not once did it come on coincident with the lighting of the arcs, but always twenty minutes or twenty-five minutes afterward. It took time to develop, possibly a slight heating and consequent expansion of the parts being involved.

If arc lamp interference comes on directly the lamps are lighted the source is very likely to be line trouble, such as wires scraping on iron bracket arms, loose splices, etc.

In many localities where it is prevalent arc lamp circuit interference starts up before the lamps are lighted, perhaps twenty-five or thirty minutes previous to lighting, and it is quite regular every evening.

This is caused by the rectifying tubes in the power house or sub-station. These tubes are "warmed up" for service prior to the line being switched in; the operation takes twenty-five minutes or so, and radio-frequency surges pass out on the line despite the open switch, so even before the lamps are actually lighted interference starts. Intensity of the interference is increased when the lights come on and continues all night, until the daylight shut-down. Generally speaking, new rectifying tubes do not give trouble, nor older ones, paralleled. Overloaded tubes, however, or those which have become hard through long usage, are liable to set up interfering surges that will travel long distances over the system.

The obvious thing to do with a defective lamp is to have it repaired; the same thing applies to line defects. Rectifying tube disturbance, by far the worst trouble because of its continuous nature and wide range, can be cleared up by putting a choke coil as described in Fig. 3, in each outgoing d. c. feeder.

Two hundred turns of No. 16 d. c. c. wire are required on a wooden cylinder 3½ inches diameter and about 12 inches long. Longitudinal slots are cut in the cylinder for insertion of fibre strips which keep the wire off the wood and provide adequate heat radiation facilities. This is for 4-ampere arcs; for 6-ampere arcs it may be as well to use No. 14 wire.

It seems that in certain instances of this nature, seventy-turn chokes were large enough to give satisfactory elimination; in the particular case in mind they were ineffective.

INCANDESCENT STREET LAMPS

IT MAY come as a little surprise when we say that, given any city where the street lighting system consists of, say two thousand series arcs and two thousand series incandescent lamps, on various circuits, more radio interference will be caused by the incandescent lamp circuits than by the arc system.

This is due to less careful installation in the case of the incandescent system—not because of any inherent defect. The condition is general.

Of one hundred cases of radio interference due to faults on series incandescent lighting systems:

42 were caused by down-lead wires scraping on the iron brackets.
15 by loose connections of the wires at the lamps.
10 by internal defects in the lamp fixture proper.
10 by partial shorting of the wires in conduit prior to connection at the lamp.
9 by poor line splices.
5 by leakage or spitting at the disc fuses in the lamp head.
5 by lamps loose in their sockets.
4 by defective mercury or other type automatic time switches.

It may be remarked here that sources on a series lighting system giving rise to radio interference are most difficult to locate. What occurs at a defective lamp seems to be duplicated in many lamps either side of the faulty one; intensity values of the interference are misleading and very careful observation is required.

FLOUR MILLS

IN PRACTICALLY every flour mill the chlorine process of bleaching has been superseded by the electrical method. This consists of a twenty-thousand volt spark oscillating directly in the path of a blast of air, the latter becoming ozonated, passes on to the grain in process of crushing.

Direct radiation is confined to a few hundred feet. The source, however, is a vigorous one and distribution wiring carries the disturbance to great distances; it is capable of mutilating radio reception practically all over the average small town.

Methods of elimination are simple and definite. One hundred-and-fifty turn choke coils wound on three or three and a half inch tubing will kill at least 85 per cent. of the noise and will not interfere in the least with normal operation.

Operating current here is 12 to 15 amperes; if the coils are not banked, No. 10 wire will be suitable. To conform to Electrical Inspections

requirements, enclosure of the coils in a standard outlet box is recommended, the knockout holes of which have been opened and covered with fine wire gauze. This latter affords good ventilation while preventing grain dust accumulating on the coils.

Self-contained bleachers of the arc type cause no trouble. They are, however, being rapidly displaced by the more efficient spark type.

FACTORY BELTS

HIGH-SPEED belts are a fruitful source of radio interference, especially noticeable in cold, dry weather. Friction between the pulleys and belting causes a static charge to form on the belt and this, after rising to a high potential, will spark to nearby metal objects, setting up a "crackling" noise in the receiver.

As most heavy rotary machines are well connected to ground one would imagine that this static charge would filter away gradually, and fail to build up to any material potential. The fact is, and it can be demonstrated, that the film of oil in the machine bearings is sufficient to insulate the rotating parts from ground.

A "static collector" is used to get rid of this trouble. It consists of a wiping contact of springy

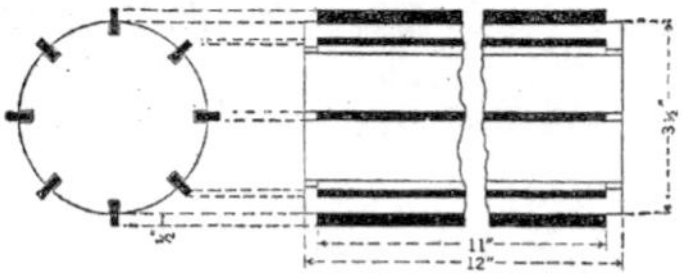

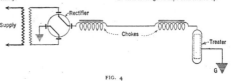

FIG. 3

metal, at all times resting on the belt in motion and permanently connected to earth.

A metal laced belt can set up quite a loud clicking interference. Every time the metal lacing passes over an iron pulley the click is heard. No such effect is noted, of course, where wooden pulleys are used, but in the cases cleared up we simply removed the metal lacing and substituted rawhide.

ELECTRIC WARMING PADS

WHETHER interference from this source is to be regarded as serious or not depends on how far away you live from the offending pad. The radius is about two dwellings either side of the one in which the pad is being used, assuming that the houses are close together.

Little thermostats inside the pad automatically break the supply current when the pad becomes sufficiently heated and switch it on again when the elements cool sufficiently. This alternate opening and closing of the 110-volt supply line sets up clicks which are extremely annoying to broadcast listeners in the immediate vicinity.

Intensity of the trouble varies with the make of pad and also with length of service. It seems that corroded or burned contacts are responsible for most of the trouble and before going to the expense of purchasing condensers, etc., it is usual to open the warming pad and readjust the thermostat contacts after cleaning them up properly.

A different problem is presented in the case of hospitals. Take a literal, specific example: Two hundred patients, two hundred pairs of telephones (no loud speaker allowed) and, incidentally, two hundred warming pads (one for each cot), were in use. Every time a patient gets restless and kicks out his feet he jars the pad thermostats and treats the other hundred and ninety-nine to a series of sharp clicks usually resulting in reciprocation.

Substitution of quiet types for the noisy ones would seem to be the best recommendation.

PRECIPITATORS

THIS apparatus is used in the treatment of various ores as well as for the purpose of smoke and dust precipitation. Its radio interference can be heard ten miles; at five miles it hurts reception and in the vicinity of the plant, normal reception is impossible.

As in the case of the notorious oil-burning furnace, methods of elimination which clear up the trouble in one installation fail to give the desired results when applied to another plant.

Several cases have been cleared up by using high-frequency chokes, i.e., placed in the high-tension circuit. The coils consist of 500 turns of No. 18 or No. 20 bare or covered wire on a tube three and one half inches in diameter. Individual turns should be spaced about one eighth inch apart except at the ends, where one-quarter inch spacing is recommended.

Fig. 4 shows an arrangement which has given satisfactory results. It may be necessary in other cases to split this 500 turn choke, i.e., putting 250 turns near the rectifier and 250 near the treater.

In the types where the high-tension energy for the rectifier is obtained through transformer action direct from a transmission line, interference is naturally heavier than that experienced from the more or less self-contained motor generator type although the actual energy in the former is less than in the case of the latter. Average energy values will be about 70 milliamps. at 30,000 volts and 80 milliamps. at 50,000 volts, respectively.

Considerable experimental work has been carried out by different precipitation plants in this connection and while special treatment was found necessary in several instances practically all cases investigated have been cleared up.

FIG. 4

AS THE BROADCASTER SEES IT

BY CARL DREHER

Drawing by Franklyn F. Stratford

Putting Freak Broadcasting in Its Place

IN ALL human affairs the tendency is toward quiescence and boredom. Among vigorous peoples this drift toward monotony is resisted by a constant seeking for innovations. Now and then some fruitful bit of originality rewards the quest. In the main, however, the innovations are failures. They have no deep roots in human desires or interests, and, after the first glance, they amuse the normal spectator even less than the old shows he is weary of. The spirit is praiseworthy, but the results are dreadful, especially in the arts. Consider the poets, for example. Here and there one of them, distressed by his inability to convince the world that he is another Shelley, decides to be entirely original by printing a magazine of verse with a fake Nujol advertisement on the cover, no capital letters, and half the words upsidedown. His originality is hard on the compositors. To be original, in the sense of doing something that is not commonly done, is very easy. To that degree you can be an exponent of novelty, by entering the nearest drug store and ordering a pineapple soda with chocolate cream. But to be fruitfully original requires more than twenty cents.

We are similarly beset in the art of broadcast entertainment. The innovations are many, but most of them are of the twenty-cent variety. I do not intend to itemize all the varieties of freak broadcasting; it would be impossible, and, besides, a certain portion of this department is consecrated to sensible subjects. A few samples will suffice.

The broadcasting of alleged disembodied spirits seems to me to fall into the category of futilities politely hinted at above. I have no animus against the spirits as such. It is true that I have never seen one, and, since studying psychopathology, have doubted their existence. Nevertheless, as I have not seen everything in the universe, and don't expect to, I acknowledge that such things as ghosts may exist. But why broadcast them? If a ghost wants to see me, let him or her call on me at my office, or in the dark reaches of the night. But as a broadcast listener, I like to be entertained. As a broadcast listener, I also object to the imputation that I am a total ass. No doubt I am an ass in some respects, but not to the extent that I can be kidded, on hearing the tinkling of glass, a noise like a $6 saxophone, and some mumbling through my loud speaker, into believing that authentic goblins are disporting themselves in the studio of the puissant broadcaster who is striving to instruct me. And, when

a committee of spiritualist investigators assure me that everything is aboveboard, I guffaw openly at their discourse. Who are they to tell me so? What do they know about the tricks of broadcast transmission?

The method used in broadcasting the shades is to turn on the microphone and, with the studio doors locked and no one in the room, to listen for mysterious sounds on the station carrier, which is assumed to be quiet. The investigating committee watches the studio doors and snoops around otherwise at their discretion. But can they assert with any semblance of plausibility that they know all the sources of input to the speech amplifiers of the transmitter? Nothing could be simpler than to rig up an additional microphone somewhere in the building and, paralleling it with the transmitter in the empty studio, to broadcast any sounds one cares to. Not even that is necessary. One of the operators can tap a tube in the speech amplifier and make noises which will seem inexplicable to the ghost-chasers. The business of spiritualistic investigation is at best full of complications, and to complicate it further by adding the technical intricacies of a broadcast transmitter is beyond all sense. Whether there is fraud or not—and certainly in connection with struggling stations avidly bent on cadging every possible square inch of newspaper space the possibility of deception is not remote—the pretence of scientific investigation under such conditions is simply silly. The broadcast listeners may not be Newtons, Goethes, and Mommsens, but they are not voodoo worshippers either. The studio manager who first conceived the idea of broadcasting spirits may have been original, but he omitted to mix a few brains with his originality.

Even more infantile is the menagerie broadcasting stunt. The only justification I can see for it is in connection with a children's hour. The noises made by sea lions and rhinoceroses

are neither agreeable nor intellectual, and, being full of steep wave fronts and tones outside the band of transmittable frequencies, they don't get over anyway. If a man roared into the transmitter it would sound as much like a lion when it got through the loud speakers as if a lion did the roaring.

As one listener, I ask to be spared such buffooneries. I respect the urge for originality, but it must take a more convincing shape than in such procedures, which have no other use than to get some station's publicity matter a transient hearing. If there is nothing interesting left to say, and nothing beautiful left to play, my counsel is to shut down the transmitter and economize on electricity at the rate of three cents per kilowatt hour.

Background Noises

MR. A. S. DANA of Seymour, Connecticut, writes us as follows:

The better grade of broadcasters have made such improvement in their quality that there is but one factor which could be improved in so far as my ability to judge quality goes.

Is it not possible for them to reduce or eliminate the noise background which accompanies the transmission? In other words, cannot the equipment be improved so that it is not possible to tell when a station is on the air unless speech or music enters the microphone? According to present standards, it is easy to locate and tune-in a station when they are not broadcasting simply by the racket which occurs when the station is in tune.

Mr. Dana seems to think that the noise in question is all generated at the transmitter. Actually there are three possible sources. In some cases all are in evidence, and in others none. Background noise may have its inception right at the microphone. The transmitter itself has a degree of internal hiss caused by current passing through the carbon in the case of a microphone; and by tube irregularities in the first stages of the amplifier associated with a condenser transmitter. But if the microphone is well designed and the carbon of high quality, and not too old, the sensitivity to external sounds is so great that the hiss is seldom audible. In the case of speech input some slight hiss is usually observable, and occasionally during pianissimo passages of musical performances, but with proper microphone care the internal

"LET THE GHOST CALL AT MY OFFICE"

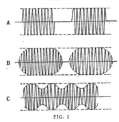

FIG. 1

noise level is negligible. If the tubes in the associated amplifier of a condenser transmitter are carefully picked this instrument is practically noiseless with normal inputs. Of course any type of transmitter will pick up room noises, and broadcasters cannot always secure their material in perfectly quiet places. But in general we may say that most well regulated stations transmit practically noise-free modulation. By that I mean that at a distance of a few feet from the monitoring cone in the station, which is assumed to emit a loud signal during modulation, no sound is audible in the intervals. Of course, by placing one's ear on the cone one can hear plenty of rustle, but that is going out of one's way. It is like saying that the alarm clock in my house, which I can hear ticking 33 feet away if I listen hard enough, is causing a disturbance which should be eliminated.

But after leaving the transmitting antenna, the modulated wave must run the gauntlet of electromagnetic disturbances in the medium between transmitter and receiver. The metaphor I have used here may be an unfortunate one, if it confirms the popular supposition that the carrier, in some mysterious way, picks up noises on its journey through space. People jump to this conclusion because they find, in tuning their receivers, with the sensitivity control well down, that they hear nothing over a certain section of the broadcast frequency range, and then, running across a blank carrier, they get a more or less audible background. The actual sequence here is more along the following lines, however: The receiver has been picking up slight disturbances—static, distant violet-ray machines, transmission lines, bells, and the like, right along, but at a level below audibility, when the receiver sensitivity is low. But the carrier coming in increases the receiver sensitivity through its heterodyne amplification, hence the noises come up when the receiver is tuned to a carrier. Of course neighborhoods vary in the relative strength of external disturbances, and anywhere there is a variation with time; normally the atmosphere may be quiet, but when there is local lightning plenty of crashes and rumbles will be picked up by all receivers. However, Mr. Dana's question probably does not include these relatively rare periods of acute disturbance.

Finally we must take into account the internal noises of the receiver itself. There is a tendency for slight gaseous irregularities in the radio frequency tubes to be amplified through each successive stage until quite a noticeable rustle results in the loud speaker. But if the tubes are properly exhausted there should be no trouble from this source. I have an eight-tube receiver, and, in testing it as I write with the sensitivity control all the way up and the loop removed to eliminate r.f. input, I am unable to hear any

sound whatsoever. With the loop in position I can hear the elevator motors in nearby apartments and a medley of undifferentiated noises, but then in practice I should never think of using the receiver in this state of excessive sensitivity. The receiver may also develop internal noises through regeneration at radio or audio frequency, or through an impure plate or filament supply. When at a low level, many such sources of disturbance manifest themselves as rustling or murmuring sounds which may be ascribed to the broadcast transmitter.

The increase of transmitting power has undoubtedly reduced background noise in radio reception, by permitting the use of less stages of amplification and lower sensitivity at the receiver, for the same signal volume. But the factor of modulation depth comes into the problem forcefully. A weakly modulated carrier simply amplifies static and interference at the expense of its intelligence-bearing side bands. Deep modulation is highly desirable on this account, and limitations on adequate modulation—which means 80-90 per cent. peaks, are as bad as inadequate field strength. Of course the best thing is to get rid of the carrier altogether, but that is a technical step feasible, at this stage, only in radio circuits professionally operated at both ends.

Abstract of Technical Article. VII

Making the Most of the Line—A Statement Referring to the Utilization of Frequency Bands in Communication Engineering, by Dr. Frank B. Jewett. Presented before Philadelphia Section of the A. I. E. E. on October 17, 1923. Reprinted May, 1924 by Bell Telephone Laboratories, Inc.

(*Continued from October* RADIO BROADCAST.)

IN A carrier-current telegraph system free of capacity and inductance, a series of dashes made at the transmitting key will be reproduced accurately as oscillations of the carrier frequency within a rectangular envelope, as

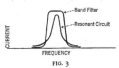

FIG. 3

shown in Fig. 1-A. If, now, inductance and capacity be inserted into the circuit so that it becomes resonant to the carrier frequency, the same keying action will produce, at the receiving end, a trace like that of Fig. 1-B. The circuit now has a certain "stiffness," so that it takes some time for it to reach the full amplitude of oscillation at the carrier frequency, and, again,

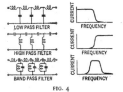

FIG. 4

after the key has been opened, the oscillations continue in a decaying wave train like that of the old spark transmitters. The faster the rate of signalling, the more serious the distortion; if, for example, dots and dashes are made too rapidly, the amplitude will never drop to zero at all, as shown in Fig. 1-C, although complete breaks are made at the key. The reason for this appears in Fig. 2. The top curve (A,) is a typical resonance peak, showing how the circuit efficiency, by virtue of the tuning effect, varies with frequency. This property, as we saw in the first instalment of this abstract, is a valuable means of frequency discrimination. But the effect also involves changing the amplitude of the various components shown in the input currents of Fig. 2-B to the output currents of Fig. 2-C. As we saw in the earlier discussion, 2-B includes the components of a square-topped wave. The suppression of the higher harmonics and the exaggeration of the carrier frequency have destroyed the rectangular wave shape. We could get it back by sacrificing selectivity—by broadening the circuit—but then we sacrifice also the power of discrimination on which we must depend if we are to make the most of the line. Obviously what is needed is a form of frequency discrimination which will pass a certain *band* of frequencies with substantially equal efficiency, and cut off sharply frequencies outside of this band. Reactive networks known as "filters" have been devised by telephone engineers to give this effect. Fig. 3 shows the difference in transmission characteristics between a resonant circuit and a filter designed to pass a band of frequencies in the same neighborhood. Fig. 4 illustrates the principal types of filters and their respective properties. Such circuits are of the utmost importance, not only in the practical communication arts, but also in investigations of the nature of speech and music. (Cf. Jones: "The Nature of Language," abstracted in April, 1927, RADIO BROADCAST.)

Besides the property of selective frequency transmission, the characteristic impedance of such networks is of importance. Fig. 5 illustrates two types of band filter with substantially similar elements, but designed for different connections. *A* is feasible for parallel connections, the impedance being very high for all frequencies save the band the network is designed to pass. But when the terminating elements are as shown in *B* the impedance to frequencies other than those in the transmitted region is low, so that such filters may only be connected in series. By the use of networks suitably designed and connected a number of carrier frequencies may be delivered to a line without mutual absorption,

and then separated for individual demodulation
at the receiving end of the line.

By the methods described above the multiplex-
ing of lines is accomplished. As an example,
Jewett gives the schematic circuits for the multiplication of telephone channels over an open wire
toll line.

Up to about 100 cycles per second the line is
available for d.c. telegraph purposes. Above this
point comes the d.c. telephone band with its
300-2800 cycle range, approximately. From
3000 cycles up to about 35,000 may be used for
carrier-current telephone channels. It is customary in most cases to use the frequencies below
20,000 cycles for transmission from east to west,
and those above this figure for transmission from
west to east. The attenuation suffered by currents in the upper range is naturally greater and
correspondingly higher amplification is required
for equal received energy. A band about 2500
cycles wide must be allowed for each carrier
channel, and a space of 1000 cycles is required
for separation between channels with band filters

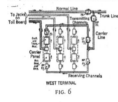

WEST TERMINAL

FIG. 6

possessing the usual discrimination characteristics.

Fig. 6 shows one terminal of a carrier telephone system. At this end we shall follow out the
steps involved in multiplex transmission. As the
line is also used for d.c. telegraphy and telephony,
a low-pass filter is inserted in the metallic circuit
to prevent the carrier frequencies from reaching
subscribers through the toll board. Likewise in
the carrier line, a two-way high pass filter prevents the currents below 3000 cycles from being
absorbed in the carrier apparatus. Three pairs
are shown leading from the toll board to the
carrier equipment. In the case of the channel
which is shown in heavy lines, the voice currents
pass first through a low frequency circuit which
permits the passage of current between the normal line and either the transmitting or receiving side of the carrier equipment, but which
blocks currents between these two halves in a

FIG. 8

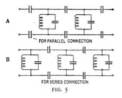

A   FOR PARALLEL CONNECTION

B   FOR SERIES CONNECTION

FIG. 5

vertical line. The importance of this feature will
appear later on. When the subscriber on this end
talks the voice currents generated in his desk
set modulate the output of one of the carrier
oscillators, and, passing through a band filter,
which selects one of the side bands, merge with
other side bands and go to the trunk line. First,
however, it will be noted that they pass through
a low-pass filter, designed to transmit the east-
bound group of frequencies below 20,000 cycles.
This filter prevents the transmitting carrier
circuits from absorbing incoming energy intended for the receiving channels.

At the other terminal of the line, sketched in
Fig. 7, the receiving process may be traced.
Again the common line is connected to a low-
pass and a high-pass filter. At this crossroads
the low-pass filter selects currents associated
with d. c. telegraphy and telephony and admits
them to the composite set. The high pass filter
selects the carrier side bands coming in from the
west terminal and conducts them to the grouping filters, where the incoming currents are led
to the receiving modulators. Again a band filter
selects the appropriate side band for each branch.
The demodulated currents pass to their respective jacks on the toll operators' board, through
the low frequency balancing circuits. If it were
not for these circuits obviously the received
voice currents would be sent back through the
transmitting carrier equipment, instead of being
confined to the subscriber at the east terminal.

Jewett sums up the process physically as
follows:

We have . . . followed through, from one toll
board to the other, a particular signal and have
seen how it is moved about on the frequency scale
to a position which identifies it from other similar
signals, how it is associated with such signals on a
common line, transmitted to the distant terminal,
isolated at the receiving end from these other
signals and finally restored to its original position
upon the frequency scale.

When a circuit is to be multiplexed for telegraph the range between
3000 and 10,000 cycles is
normally devoted to this
purpose. The directional
dividing line is then usually
at 6000 cycles, frequen-
cies below this point being
used for transmission from
west to east and frequen-
cies above 6000 for transmission from east to west.
Various combinations of
carrier telephone and telegraph are possible. One
layout shown in Jewett's
paper comprises the fol-
lowing facilities: 2 full
duplex normal telegraph
channels; 1 normal tele-
phone channel; 10 full
duplex carrier telegraph

ALUMINUM
AT THE RADIO SHOWS

THIS year's Radio Shows demonstrate that Aluminum has been adopted for shielding by more of the leading manufacturers and Radio Engineers than ever before.

The RGS "Octa-Monic" is an outstanding example of the use of Aluminum in prominent sets. The specifications call for Aluminum Box Shields to insure amplification, tone quality, sensitivity and selectivity.

The standard "knock-down" ALUMINUM BOX SHIELDS— 5" x 9" x 6"—are adaptable to many hook-ups.

Write for new booklet, "Aluminum for Radio," telling of the advantages of Aluminum in Radio apparatus.

ALUMINUM COMPANY OF AMERICA
2464 Oliver Building, Pittsburgh, Pa.
Offices in 18 Principal American Cities

ALUMINUM IN EVERY COMMERCIAL FORM

Employing the New "Octa-Monic" Principle

Frequency! Not a Super-Heterodyne!
Fundamentally New
"OCTA-MONIC"*

Automatic Wavetrap for prevention of heterodyning and whistling resulting from stations operating on one-half wave-length or on first octave beat.

Automatic Filament Control.

Employs 135 Volts or 180 Volts. Draws 22 mils.

Each R. G. S. "Octa-Monic" is carefully tested with scientific apparatus and under actual broadcasting conditions before it leaves the laboratories; while every piece of apparatus is just as thoroughly tested before it is built into this receiver.

The R. G. S. "Octa-Monic" is a closely co-ordinated Receiver built of quality apparatus. Careful tests are the basis for the choice of each piece of apparatus, tests that not only determine the merits of each individual part, but more importantly its relation to the whole receiver.

Standard Cunningham tubes (5 CX301-A's and 1 CX371, Power tube in last stage) and Western Electric Cone are recommended for best results.

The R. G. S. "Octa-Monic" is highly attractive in appearance. It is built on five-ply, specially shellaced sub-panel (20" x 9") to which is mounted a beautiful walnut finished, standard size panel (7" x 21") that will fit any good cabinet or fine console 7" x 21". The panel and drum escutcheons are trimmed in dull bronze.

You will find your R. G. S. "Octa-Monic" mighty easy to operate.

There are but two drums with vernier adjustments and two control knobs, one of which is an ordinary volume control and filament switch, the nearest approach to tuning efficiency, **possible.** Stations actually "click" or "tumble-in" as you slowly revolve your drums.

The customary need of wooden screw-drivers or involved balancing devices is entirely removed in the R. G. S. "Octa-Monic." Major or minor adjustments are unnecessary. The R. G. S. "Octa-Monic" is free from ordinary service. Tuning condensers are the only moving parts, and as a consequence, there are no fussy mechanisms, either mechanical or electrical, to get out of order.

The R. G. S. "Octa-Monic" operates satisfactorily on either a good "B" battery eliminator or batteries without "motor-boating" or howling.

Orders cannot be accepted for individual pieces of apparatus or blueprints.

The R. G. S. "Four" employing the Inverse Duplex System (1) R. G. S. "Four" Kit, all parts, complete instructions, $74.40. (2) Chassis, assembled according to latest laboratory methods, $84.40.

All prices slightly higher west of Denver. Canadian and Export prices on request.

Go to your dealer to-day and insist on a demonstration. If he hasn't stocked the R. G. S. "Octa-Monic" yet, tear off and mail to us the attached coupon with the required information. Every effort will be made to arrange a demonstration for you.

Arrange for that demonstration now because you have a real radio thrill waiting for you. In the R. G. S. "Octa-Monic" you will hear radio at its best. And when you hear the R. G. S. "Octa-Monic" you will know why it is: *"The Synonym of Performance."*

All models of the R. G. S. "Octa-Monic" and the R. G. S. "Four" are fully protected by Grimes Patents issued and pending.

Trade Mark Registered.

DEALERS: *Write for Complete Merchandizing Proposals*
R-G-S Manfg. Co., Inc.
Staten Island   **New York**

Battery or "B" Eliminator Models

R. G. S. "Octa-Monic" Kit
of parts including all required apparatus, complete instructions and blue-prints, ready to build, $84.60.

R. G. S. "Octa-Monic" Chassis
completely assembled according to latest laboratory methods, (closely co-ordinated and specially designed apparatus, eight foot De Hery cable, etc., etc.) with complete operational instructions, ready to operate, $89.60.

R. G. S. "Octa-Monic" Receiver
housed in an attractive, well-designed, walnut table cabinet, $104.60.

R. G. S. "Octa-Monic" Tuning Kit
including all necessary apparatus and complete blue-prints and instructions, $63.60.

R. G. S. "Octa-Monic" Tuning Chassis
completely assembled according to latest laboratory methods with complete instructions and ready to wire to your favorite amplifier, $66.60.

Price Note
The apparatus required to build the radically new and fundamental R. G. S. Octa-Monic actually lists at over $100.00.

R. G. S. MANFG. CO., Inc.
Staten Island, New York

Gentlemen:
Please arrange with my dealer, whose address I have printed below, for a demonstration of the new and revolutionary R.G.S. "Octa-Monic." I am much interested in this receiver but this request for a demonstration and literature, you understand, entails no obligation on my part.

My Name ....
Street ....
City or State ....
My Dealer's Name ....
His Address ....

Built For Modern R.G.S. THEM TROLI ONLY *Broadcast Conditions*

The *Radio Broadcast*
LABORATORY INFORMATION
SHEETS

THE RADIO BROADCAST Laboratory Information Sheets are a regular feature of this magazine and have appeared since our June, 1926, issue. They cover a wide range of information of value to the experimenter and to the technical radio man. It is not our purpose always to include new information but to present concise and accurate facts in the most convenient form. The sheets are arranged so that they may be cut from the magazine and preserved for constant reference, and we suggest that each sheet be cut out with a razor blade and pasted on 4" x 6" filing cards, or in a notebook. The cards should be arranged in numerical order. An index appears twice a year dealing with the sheets published during that year. The first index appeared on sheets Nos. 47 and 48, in November, 1926. In July, an index to all sheets appearing since that time was printed.

The June, October, November, and December, 1926, issues are out of print. A complete set of Sheets, Nos. 1 to 88, can be secured from the Circulation Department, Doubleday, Page & Company, Garden City, New York, for $1.00. Some readers have asked what provision is made to rectify possible errors in these Sheets. In the unfortunate event that any such errors do appear, a new Laboratory Sheet with the old number will appear.

—THE EDITOR.

No. 137   RADIO BROADCAST Laboratory Information Sheet   **November, 1927**

Operating Vacuum Tubes in Parallel

METHODS AND RESULTS

IT IS sometimes desirable to operate several tubes in parallel in order to obtain a greater power output, and it is of interest to know how efficiently this may be done.

If two tubes are to be used in parallel in the output of an audio amplifier the two sockets are wired so that the grid of one tube connects to the grid of the other tube and the two plates connect together. The two filaments are also connected together. The result is that from these two tubes we will have only four leads—one from the grids, another from the plates, and two others from the filaments.

The amplification constant of the combination will be equal to the constant of a single tube, provided both of the tubes have the same constant. If one of the tubes had a low amplification constant and the other a high constant the resultant amplification constant of the two would be equal to the arithmetic mean. If the amplification constant of one tube is six and the other four, the resultant amplification constant will be five.

The resultant plate impedance will be equal to one half the impedance of a single tube, and if unlike tubes are used, the total impedance can be calculated by the simple laws governing resistances in parallel. The combined impedance can be stated as follows:

Imped. of one tube × Imped. of other tube
───────────────────────────────────────
Imped. of one tube + Imped. of other tube

The greatest power output is obtained when the two tubes have identical plate impedances and amplification constants. Fortunately, however, a very large fraction of the total power of the two tubes can be obtained even if they differ considerably.

To illustrate, two tubes might be connected in parallel, the amplification constants of which are in a ratio of 2 to 1, and the plate impedances of which are equal, and from the combination we could obtain 90 per cent. as much power as could be obtained if the tubes were operated in separate circuits. If, with equal amplification constants, the plate impedances are in a ratio of 2 to 1, the total power will be about 90 per cent. of the maximum possible value. It is evident, therefore, that the total power will not be decreased by any great amount even if tubes quite widely differing in characteristics are used. From two perfectly matched tubes, feeding into a load resistance equal to their combined plate impedance, we can obtain twice as much power as can be obtained from a single tube feeding into a load resistance equal to its plate impedance.

No. 138   RADIO BROADCAST Laboratory Information Sheet   **November, 1927**

The Unit of Capacity

CALCULATION AND FORMULAS

THE capacity of a condenser is stated in terms of the quantity of electricity it will hold per volt. When a condenser stores a specific quantity of electricity known as a coulomb and there is an electrical pressure of one volt across its terminals then the capacity of the condenser is one "farad." A condenser must be very large to have a capacity of

where
C = capacity of condenser in microfarads
K = dielectric constant
A = total area of dielectric between plates in square inches
d = thickness of dielectric in inches

Example:
What is the capacity in microfarads of a condenser having 2000 plates? The dielectric consists of paraffined paper 0.002 inch thick. The part of

Vaseline	Ebonite	Glass	Mica	Paraffin Wax	Porcelain	Quarts	Resin	Shellac	Castor Oil	Olive Oil	Petroleum Oil
2.0	3.0	7.0	6.0	2.5	4.0	4.5	2.5	3.5	5.0	3.0	2.0

a farad and therefore a millionth part of a farad has been adopted as the practical unit and it is called the "microfarad," meaning one-millionth of a farad. Capacities smaller than one microfarad can be expressed in micro-microfarads, corresponding to a millionth of a microfarad.

The capacity of a condenser may be computed from the general equation:

$$C = \frac{2250 \ AK}{10^9 d}$$

each sheet of dielectric actually between the plates has an area of 6.3" x 8".

From the table in this sheet it will be seen that the constant of the dielectric is 2.5.

The total area, A, of the dielectric is:—
$A = 6.3 \times 8 \times 2000$
= 100,000 square inches, approximately

Therefore
$$C = \frac{2250 \times 100,000 \times 2.5}{10^9 \times 0.002}$$
= 28.1 microfarads

▪▪▪ Modern

Here is the Eveready Layerbilt "B" Battery No. 486, Eveready's longest-lasting provider of Battery Power.

Radio is better with *Battery* Power

NOT because they are new in themselves, but because they make possible modern perfection of radio reception, batteries are the modern source of radio power.

Today's radio sets were produced not merely to make something new, but to give you new enjoyment. That they will do. New pleasures await you; more especially if you use Battery Power. Never were receivers so sensitive, loud-speakers so faithful; never has the need been so imperative for pure DC, Direct Current, that batteries provide. You must operate your set with current that is smooth, uniform, steady. Only such current is noiseless, free from disturbing sounds and false tonal effects. And only from batteries can such current be had.

So batteries are needful if you would bring to your home the best that radio has to offer. Choose the Eveready Layerbilt "B" Battery No. 486, modern in construction, developed exclusively by Eveready to bring new life and vigor to an old principle—actually the best and longest-lasting Eveready Battery ever built. It gives you Battery Power for such a long time that you will find the cost and effort of infrequent replacement small indeed beside the modern perfection of reception that Battery Power makes possible.

NATIONAL CARBON CO., INC.
New York ▯▮▮ San Francisco
Unit of Union Carbide and Carbon Corporation

Tuesday night is Eveready Hour Night
—9 P. M., Eastern Standard Time

WEAF–New York	WOC–Davenport
WJAR–Providence	WCCO–{ Minneapolis
WEEI–Boston	St. Paul
WFI–Philadelphia	KSD–St. Louis
WGR–Buffalo	WDAF–Kansas City
WCAE–Pittsburgh	WRC–Washington
WSAI–Cincinnati	WGY–Schenectady
WTAM–Cleveland	WHAS–Louisville
WWJ–Detroit	WSB–Atlanta
WGN–Chicago	WSM–Nashville
WMC–Memphis	

Pacific Coast Stations—
9 P. M., Pacific Standard Time

KPO–KGO–San Francisco	KFI–Los Angeles
KFOA–KOMO–Seattle	KGW–Portland

EVEREADY
Radio Batteries
—they last longer

The   air   is   full   of   things   you   shouldn't   miss

No. 139

Inductive Reactan

HOW IT IS CALCULATED

IF AN inductance coil is connected in series with an a. c. ammeter to a source of alternating current, a certain amount of current will flow in the circuit, depending upon the size of the coil and the frequency of the current. If the voltage of the source is divided by the current, the quotient will be the "reactance" of the coil in ohms. For example, if the frequency of the current being supplied by the source of potential was 60 cycles and the voltage was 110 volts and the coil had an inductance of 1.0 henry, we would find that 0.292 amperes of current would flow through the circuit. Then 110 volts divided by 0.292 gives 377, which is the reactance in ohms at 60 cycles of a coil with an inductance of 1.0 henry. The reactance of a coil depends upon its inductance and upon the frequency of the current. It can be calculated by means of the following formula:

$$Reactance = 6.28 \, FL$$

where F = the frequency of the current in cycles

per secon henries.

In many reactance and for th given a t tween 0.0 to 100,000 it is evid proportio directly p size of th the react doubled. simple m of any c Sheet No third the cycles. Si 100 cycle of a 30-h 18,840 oh

No. 140

Coil Reactance

COIL INDUCTANCE IN HENRIES	REACTANCE IN OHMS AT			
	60	100	250	500
0.01	3.77	6.28	15.7	3
0.05	18.8	31.4	78.5	15
0.1	37.7	62.8	157	3
0.5	188.5	314	785	1.5
1.0	377	628	1,570	3,1
2.0	754	1,256	3,140	6,2
5.0	.885	3,140	7,850	15,7
10.0	3,770	6,280	15,700	31,4
20.0	7,540	12,360	31,400	62,8
30.0	11,310	18,840	47,200	94,2
40.0	15,060	24,720	61,800	123,6
50.0	18,850	31,400	78,500	157,
100.0	37,700	62,800	157,000	314,

This table shows how the reactance of various inductance coil atory Sheet No. 139 explains what inductive reactance is and u

No. 141

A. C. Tube Da

"HEATER" AND FILAMENT TYPES

ON THIS Laboratory Sheet are given data on the new a. c. tubes, type UY-227 (C-327) and type UX-226 (CX-326). The former tube is of the heater type whereas the latter is of the a. c. filament type. The heater tube requires a special five-prong socket whereas the type 26 may be used with any standard socket. The filament voltage and current of the type 27 are 2.5 volts and 1.75 amperes respectively. The type 26 requires a filament voltage of 1.5 volts and the filament current is 1.05 amperes. The filament current of these tubes is quite large, especially so in a multi-tube receiver, and for this reason it is essential in wiring the filament leads

that hea rent req will tell

(B

Table of these grid volt

TABLE No. 2

TYPE OF TUBE	PLATE VOLTAGE	NEGATIVE GRID VOLTAGE	PLATE CURRENT
UY-227 & C-327	90 135 180	5 9 13.5	3 5 6
UX-226 & CX-326	90 135 180	6 9 13.5	3.7 6 7.5

Balkite has pioneered —
but not at public expense

Balkite "A" *Contains no battery. The same as Balkite "AB" but for the "A" circuit only.* Not a battery and charger but a perfected light socket "A" power supply. One of the most remarkable developments in the entire radio field. Price $32.50.

Balkite "B" *One of the longest lived devices in radio.* The accepted tried and proved light socket "B" power supply. The first Balkite "B," after 5 years, is still rendering satisfactory service. Over 300,000 in use. Three models: "B"-W, 67-90 volts, $22.50; "B"-135," 135 volts, $32.50; "B"-180, 180 volts, $39.50. Balkite now costs no more than the ordinary "B" eliminator.

Balkite Chargers

Standard for "A" batteries. Noiseless. Can be used during reception. Prices drastically reduced. Model "J,"* rates 2.5 and .5 amperes, for both rapid and trickle charging, $17.50. Model "N"* Trickle Charger, rate .5 and .8 amperes, $9.50. Model "K" Trickle Charger, $7.50.

Special models for 25-40 cycles at slightly higher prices

Prices are higher West of the Rockies and in Canada

The great improvements in radio power have been made by Balkite First noiseless battery charging. Then successful light socket "B" power. Then trickle charging. And today, most important of all, Balkite "AB," a complete unit containing no battery in any form, supplying both "A" and "B" power directly from the light socket, operating only while the set is in use.

This pioneering has been important. Yet alone it would never have made Balkite one of the best known names in radio. Balkite is today the established leader because of Balkite performance at the hands of its owners.

Because with 2,000,000 units in the field Balkite has a record of long life and freedom from trouble seldom equalled in any industry.

Because of the first 16 light socket "B" power supplies put on the market, Balkite "B"

Balkite "AB" *Contains no battery.* A complete unit, replacing both "A" and "B" batteries and supplying radio current directly from the light socket. Contains no battery in any form. Operates only while the set is in use. Two models: "AB"-135," 135 volts "B" current, $59.50; "AB"-180, 180 volts, $67.50.

alone remains in its original form; all others have either been radically revised in principle or completely withdrawn.

Because the first Balkite "B," purchased 5 years ago, is still in use and will be for years to come.

Because to your radio dealer Balkite is a synonym for quality.

Because the electrolytic rectification developed and used by Balkite is so reliable that today it is standard on the signal systems of most American as well as European and Oriental railroads.

Because Balkite is permanent equipment. Balkite has pioneered — but not at the expense of the public.

Today, whatever type of set you own, whatever type of power equipment you want (with batteries or without), whatever you want to pay for it, Balkite has it. And production is so enormous that prices are astonishingly low.

Your dealer will recommend the Balkite equipment you need for your set.

FANSTEEL PRODUCTS COMPANY, INC., NORTH CHICAGO, ILLINOIS

Licensees for Germany:
Siemens & Halske, A. G. Wernerwerk M
Siemensstadt, Berlin

Sole Licensees in the United Kingdom:
Messrs. Radio Accessories Ltd., 9-13 Hythe Rd.
Willesden, London, N. W. 10

FANSTEEL
Balkite
Radio Power Units

No. 142   RADIO BROADCAST Laboratory Information Sheet   November, 1

Obtaining Various Voltages from a B-Power Unit

VALUES AND CURRENT-CARRYING CAPACITY

IT IS comparatively simple to calculate the resistance values required in the output circuit of a B-power unit in order to obtain any specific voltages, This Laboratory Sheet will explain how to calculate the values of these resistances.

Consider the fundamental output circuit of a B-power unit as illustrated in the sketch. The diagram of the rectifier and filter has been omitted since they play no important part in the calculation of resistance values. Suppose tap No. 1 is to be 45 volts and is to be used to operate a detector tube. We will assume that the load current through R_1 is 3 milliamperes, or 0.003 amperes. This is an average figure for the load current and can generally be used in this type of calculation. If the voltage at tap No. 1 is to be 45, then the Voltage drop across resistance R_1 must be 45. The resistance of R_1 will be equal to the voltage across it divided by the current through it or, in this case, 45 divided by 0.003, which gives 15,000 ohms as the value of R_1. The voltage at tap No. 2 is to be 90. Since the voltage drop across R_1 is 45, it follows that the voltage drop across R_2 will also be 45 in order to make the total voltage between the negative B and tap No. 2 equal to 90. The current flowing through the resistance R_2 will be equal to the load current at 3 milliamperes plus the current drawn by the detector tube, which is 1 milliampere, Therefore the value of resistance R_2 will be equal to the voltage across it, 45, divided by the current through it, which is 0.003 plus 0.001, or a total of 0.004 amperes. This gives a value of 11,250 ohms for R_2. Suppose that

the total drain from the 90-volt tap is 10 m peres. Then, the total current flowing throu will be equal to 10 plus 1 plus 3, or 14 milliam If the maximum voltage available from the unit is 180 and the voltage at terminal No. : be 90, it follows that, the voltage drop acr must be 90. Ninety volts divided by 0.014 ac gives 6400 ohms as the value of R_1.

Resistance units for B power units are u rated in watts and it is essential that the resis used be capable of carrying the necessary without overheating. The load in watts being dled by a resistance can be determined by mu ing the resistance in ohms by the square current in amperes. In this particular exampl

$$\text{Watts through } R_1 = 15000 \times 0.003^2 = 0.135 \text{ watts}$$
$$\text{Watts through } R_2 = 11250 \times 0.004^2 = 0.18 \text{ watts}$$
$$\text{Watts through } R_1 = 6400 \times 0.014^2 = 1.25 \text{ watts}$$

No. 143   RADIO BROADCAST Laboratory Information Sheet   November, 1

Solenoid Coil Data

UNITS FOR THE BROADCAST BAND

THIS Laboratory Sheet gives the data necessary to wind the secondaries of solenoid type coils for use with 0.0005-mfd., 0.00035-mfd., or 0.00025-

mfd. Variable condensers. The wavelength ra the coil will be approximately 200 to 550 n The coils may be wound on hard rubber or ba tubing, or some type of self-supported windin be used.

DIAMETER OF TUBE in INCHES	SIZE OF WIRE	NUMBER OF TURNS OF D.C.C. WIRE REQUIRED WITH VAR SIZES OF TUNING CONDENSERS		
		0.0005 mfd.	0.00035 mfd.	0.00025 mf
3½	28	28	38	50
	26	31	42	54
	24	34	46	58
	22	38	50	64
	20	42	55	72
3	28	35	48	62
	26	39	53	67
	24	43	55	73
	22	47	61	81
	20	51	67	88
2½	28	42	54	63
	26	45	58	73
	24	48	63	80
	22	51	70	90
	20	53	78	98

No. 144   RADIO BROADCAST Laboratory Information Sheet   November, 1

The Transmission Unit

CORRECTION OF LABORATORY SHEET NO. 114

TWO errors occurred in LABORATORY SHEET No. 114 published in the August, 1927, RADIO BROADCAST. In the last line in the first column, the word "natural" should be changed to read "common," and in the first line in the second column, the same change should be made.

The chart on this sheet makes it possible to determine easily the number of telephone transmission units if the current or voltage ratio is known. For example, from the curve it is evident that if two voltages or two currents are in a ratio of 5, then the TU difference between them is 14. If we are dealing with powers rather than currents or voltages, it is merely necessary to divide the number of TU obtained from the curve by 2 in order to determine the TU difference of two powers. For example, two powers in the ratio of 8 to 1 have a TU difference of 9. To determine this value we look up the number of TU corresponding to a ratio of 8 which gives 18 and then divide by 2.

To illustrate the use of the curve we might take an audio amplifier requiring a tenth of a volt input to produce three volts at the output. If we wanted to know the overall gain in TU we would divide three by 0.1, which gives 30, This ratio on the curve corresponds to a 29.5 TU voltage gain.

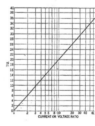

CURRENT OR VOLTAGE RATIO

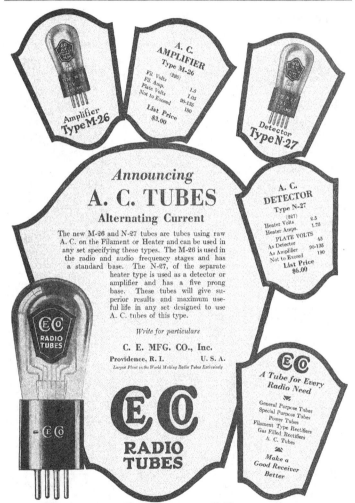

Manufacturers' Booklets

A Varied List of Books Pertaining to Radio and Allied Subjects Obtainable Free With the Accompanying Coupon

READERS may obtain any of the booklets listed below by using the coupon printed on page 64. Order by number only.

1. FILAMENT CONTROL—Problems of filament supply, voltage regulation, and effect on various circuits. RADIALL COMPANY.

2. HARD RUBBER PANELS—Characteristics and properties of hard rubber as used in radio, with suggestions on how to "work" it. B. F. GOODRICH RUBBER COMPANY.

3. TRANSFORMERS—A booklet giving data on input and output transformers. PACENT ELECTRIC COMPANY.

4. RESISTANCE-COUPLED AMPLIFIERS—A general discussion of resistance coupling with curves and circuit diagrams. COLE RADIO MANUFACTURING COMPANY.

5. CARBORUNDUM IN RADIO—A book giving pertinent data on the crystal as used for detection, with hook-ups, and a section giving information on the use of resistors. THE CARBORUNDUM COMPANY.

6. B-ELIMINATOR CONSTRUCTION—Constructional data on how to build. AMERICAN ELECTRIC COMPANY.

7. TRANSFORMER AND CHOKE-COUPLED AMPLIFICATION—Circuit diagram and discussion. ALL-AMERICAN RADIO CORPORATION.

8. RESISTANCE UNITS—A data sheet of resistance units and their application. WARD-LEONARD ELECTRIC COMPANY.

9. VOLUME CONTROL—A leaflet showing circuits for distortionless control of volume. CENTRAL RADIO LABORATORIES.

10. VARIABLE RESISTANCE—As used in various circuits. CENTRAL RADIO LABORATORIES.

11. RESISTANCE COUPLING—Resistors and their application to audio amplification, with circuit diagrams. DEJUR PRODUCTS COMPANY.

12. DISTORTION AND WHAT CAUSES IT—Hook-ups of resistance-coupled amplifiers with standard circuits. ALLEN-BRADLEY COMPANY.

13. B-ELIMINATOR AND POWER AMPLIFIER—Instructions for assembly and operation using Raytheon tube. GENERAL RADIO COMPANY.

13a. B-ELIMINATOR AND POWER AMPLIFIER—Instructions for assembly and operation using an R. C. A. rectifier. GENERAL RADIO COMPANY.

16. VARIABLE CONDENSERS—A description of the functions and characteristics of variable condensers with curves and specifications for their application to complete receivers. ALLEN D. CARDWELL MANUFACTURING COMPANY.

17. BAKELITE—A description of various uses of bakelite in radio, its manufacture, and its properties. BAKELITE CORPORATION.

19. POWER SUPPLY—A discussion on power supply with particular reference to lamp-socket operation. Theory and constructional data for building power supply devices. ACME APPARATUS COMPANY.

20. AUDIO AMPLIFICATION—A booklet containing data on audio amplification together with hints for the constructor. ALL-AMERICAN RADIO CORPORATION.

21. HIGH-FREQUENCY DRIVER AND SHORT-WAVE WAVEMETER—Constructional data and application. BURGESS BATTERY COMPANY.

46. AUDIO-FREQUENCY CHOKES—A pamphlet showing positions in the circuit where audio-frequency chokes may be used. SAMSON ELECTRIC COMPANY.

47. RADIO-FREQUENCY CHOKES—Circuit diagrams illustrating the use of chokes to keep out radio-frequency currents from definite points. SAMSON ELECTRIC COMPANY.

48. TRANSFORMER AND IMPEDANCE DATA—Tables giving the mechanical and electrical characteristics of transformers and impedances, together with a short description of their use in the circuit. SAMSON ELECTRIC COMPANY.

49. BYPASS CONDENSERS—A description of the manufacture of bypass and filter condensers. LESLIE F. MUTER COMPANY.

50. AUDIO MANUAL—Fifty questions which are often asked regarding audio amplification, and their answers. AMERTRAN SALES COMPANY, INCORPORATED.

51. SHORT-WAVE RECEIVER—Constructional data on a receiver which, by the substitution of various coils, may be made to tune from a frequency of 16,660 kc. (18 meters) to 1999 kc. (150 meters). SILVER-MARSHALL, INCORPORATED.

52. AUDIO QUALITY—A booklet dealing with audio-frequency amplification of various kinds and the application to well-known circuits. SILVER-MARSHALL, INCORPORATED.

56. VARIABLE CONDENSERS—A bulletin giving an analysis of various condensers together with their characteristics. GENERAL RADIO COMPANY.

57. FILTER DATA—Facts about the filtering of direct current supplied by means of motor-generator outfits used with transmitters. ELECTRIC SPECIALTY COMPANY.

59. RESISTANCE COUPLING—A booklet giving some general information on the subject of radio and the application of resistors to a circuit. DAVEN RADIO CORPORATION.

60. RESISTORS—A pamphlet giving some technical data on resistors which are capable of dissipating considerable energy; also data on the ordinary resistors used in resistance-coupled amplification. THE CRESCENT RADIO SUPPLY COMPANY.

62. RADIO-FREQUENCY AMPLIFICATION—Constructional details of a five-tube receiver using a special design of radio-frequency transformer. CANFIELD RADIO MFG. COMPANY.

63. FIVE-TUBE RECEIVER—Constructional data on building a receiver. AERO PRODUCTS, INCORPORATED.

64. AMPLIFICATION WITHOUT DISTORTION—Data and curves illustrating the use of various methods of amplification. ACME APPARATUS COMPANY.

66. SUPER-HETERODYNE CONSTRUCTION—Constructional details of a seven-tube set. G. C. EVANS COMPANY.

70. IMPROVING THE AUDIO AMPLIFIER—Data on the characteristics of audio transformers, with a circuit diagram showing where chokes, resistors, and condensers can be used. AMERICAN TRANSFORMER COMPANY.

73. PLATE SUPPLY SYSTEM—A wiring diagram and layout plan for a plate supply system to be used with a power amplifier. Complete directions for wiring are given. AMERTRAN SALES COMPANY.

80. FIVE-TUBE RECEIVER—Data are given for the construction of a five-tube tuned radio-frequency receiver. Complete instructions, list of parts, circuit diagram, and template are given. ALL-AMERICAN RADIO CORPORATION.

81. BETTER TUNING—A booklet giving much general information on the subject of radio reception with specific illustrations. Primarily for the non-technical home constructor. BREMER-TULLY MANUFACTURING COMPANY.

82. SIX-TUBE RECEIVER—A booklet containing photographs, instructions, and diagrams for building a six-tube shielded receiver. SILVER-MARSHALL, INCORPORATED.

83. SOCKET POWER DEVICE—A list of parts, diagrams, and templates for the construction and assembly of socket power devices. JEFFERSON ELECTRIC MANUFACTURING COMPANY.

84. FIVE-TUBE EQUAMATIC—Panel layout, circuit diagrams, and instructions for building a five-tube receiver, together with data on the operation of tuned radio-frequency transformers of special design. KARAS ELECTRIC COMPANY.

85. FILTER—Data on a high-capacity electrolytic condenser used in filter circuits in connection with A socket power supply units, are given in a pamphlet. THE ABOX COMPANY.

86. SHORT-WAVE RECEIVER—A booklet containing data on a short-wave receiver as constructed for experimental purposes. THE ALLEN D. CARDWELL MANUFACTURING CORPORATION.

88. SUPER-HETERODYNE CONSTRUCTION—A booklet giving full instructions, together with a blue print and necessary data, for building an eight-tube receiver. THE GEORGE W. WALKER COMPANY.

89. SHORT-WAVE TRANSMITTER—Data and blue prints are given on the construction of a short-wave transmitter, together with operating instructions, methods of keying, and other pertinent data. RADIO ENGINEERING LABORATORIES.

90. IMPEDANCE AMPLIFICATION—The theory and practice of a special type of dual-impedance audio amplification are given. ALDEN MANUFACTURING COMPANY.

92. SOCKET POWER—A booklet giving constructional details of a socket-power device using either the BH or 313 type rectifier. NATIONAL COMPANY, INCORPORATED.

94. POWER AMPLIFIER—Constructional data and wiring diagrams of a power amplifier combined with a B-supply unit are given. NATIONAL COMPANY, INCORPORATED.

100. A, B, AND C SOCKET-POWER SUPPLY—A booklet giving data on the construction and operation of a socket-power supply using the new high-current rectifier tube. THE Q. R. S. MUSIC COMPANY.

101. USING CHOKES—A folder with circuit diagrams of the more popular circuits showing where choke coils may be placed to produce better results. SAMSON ELECTRIC COMPANY.

ACCESSORIES

22. A PRIMER OF ELECTRICITY—Fundamentals of electricity with special reference to the application of dry cells to radio and other uses. Constructional data on buzzers, automatic switches, alarms, etc. NATIONAL CARBON COMPANY.

23. AUTOMATIC RELAY CONNECTIONS—A data sheet showing how a relay may be used to control A and B circuits. YAXLEY MANUFACTURING COMPANY.

25. ELECTROLYTIC RECTIFIER—Technical data on a new type of rectifier with operating curves. KODEL RADIO CORPORATION.

26. DRY CELLS FOR TRANSMITTERS—Actual tests given, well illustrated with curves showing exactly what may be expected of this type of B power. BURGESS BATTERY COMPANY.

27. DRY-CELL BATTERY CAPACITIES FOR RADIO TRANSMITTERS—Characteristic curves and data on discharge tests. BURGESS BATTERY COMPANY.

28. B BATTERY LIFE—Life curves with general and curves on tube characteristics. BURGESS BATTERY COMPANY.

29. HOW TO MAKE YOUR SET WORK BETTER—A non-technical discussion of general radio subjects with hints on how reception may be bettered by using the right tubes. UNITED RADIO AND ELECTRIC CORPORATION.

30. TUBE CHARACTERISTICS—A data sheet giving constants of tubes. C. E. MANUFACTURING COMPANY.

31. FUNCTIONS OF THE LOUD SPEAKER—A short, non-technical general article on loud speakers. AMPLION CORPORATION OF AMERICA.

32. METERS FOR RADIO—A catalogue of meters used in radio, with connecting diagrams. BURTON-ROGERS COMPANY.

33. SWITCHBOARD AND PORTABLE METERS—A booklet giving dimensions, specifications, and shunts used with various meters. BURTON-ROGERS COMPANY.

34. COST OF B BATTERIES—An interesting discussion of the relative merits of various sources of B supply. HARTFORD BATTERY MANUFACTURING COMPANY.

35. STORAGE BATTERY OPERATION—An illustrated booklet on the care and operation of the storage battery. GENERAL LEAD BATTERIES COMPANY.

36. CHARGING AND B BATTERIES—Various ways of connecting up batteries for charging purposes. WESTINGHOUSE UNION BATTERY COMPANY.

37. CHOOSING THE RIGHT RADIO BATTERY—Advice on what dry cell battery to use; their application to radio, with wiring diagrams. NATIONAL CARBON COMPANY.

38. TUBE REACTIVATOR—Information on the care of vacuum tubes, with notes on how and when they should be reactivated. THE STERLING MANUFACTURING COMPANY.

44. ARRESTERS—Mechanical details and principles of the vacuum type of arrester. NATIONAL ELECTRIC SPECIALTY COMPANY.

55. CAPACITY CONNECTOR—Description of a new device for connecting up the various parts of a receiving set, and at the same time providing bypass condensers between the leads. KURZ-KASCH COMPANY.

(Continued on page 64)

The Celestial Stradivarius
The Advanced Hi-Q* Six
BOTH CUSTOM-BUILT!

The Hi-Q Six—the newest advance in radio—four completely isolated tuned stages—Automatic Variable Coupling—symphonic amplification. A non-oscillating, super-sensitive receiver that assures maximum and uniform amplification on all wave lengths and establishes a totally new standard of tonal quality.

JUST as Antonio Stradivari gave the priceless *Custom-built* violin to musicians of his day, so does Hammarlund-Roberts offer music lovers of our day the *Custom-built* Radio.

The advanced "Hi-Q Six"—designed by ten of America's leading manufacturers—made with America's finest parts—incorporating every modern constructional feature—and built under your own eyes from plans so complete, so exacting and so clear cut that the only outcome can be absolute radio perfection.

In addition to its unprecedented performance, the Hi-Q Six offers equally unprecedented economy, for by building it yourself you can save at least $100.00 over the cost of finest factory-assembled sets. Complete parts including Foundation Unit chassis, panels, with all wire and special hardware cost only $95.80.

The Hi-Q Instruction Book tells the complete story with text, charts, diagrams and photos. Anyone can follow it and build this wonderful instrument. Get a copy from your dealer or write us direct. Price is 25 cents.

HAMMARLUND-ROBERTS, INC.
1182 Broadway   Dept. A   New York City

Associate Manufacturers

NATIONAL

AN ENTIRELY NEW AND UNIQUE
HEAVY-DUTY *BETTER*-B

Supplies

Detector voltages, 22 to 45, adjustable;
R. F. voltages from 50 to 75;
A. F. voltages from 90 to 135;
Power tube voltage 180 fixed.

An Exclusive Feature

Tubes and by-pass condensers are protected against excessive and harmful
voltages.

Designed for lasting service with
liberal factors of safety.

A Strictly Heavy-Duty
Power Unit

Output rating is 70 mils at 180 volts.
Uses R. C. A. UX-280 or Cunningham
CX-380 Rectron.

*Licensed under patents of Radio
Corporation of America and Associated Companies.*

For 105-115 Volts, 50-60 cycles A. C.
List price with cord, switch and plug,
$40. Rectifier tube $5.
*Write National Co., Inc., W. A. Ready,
Pres. Malden, Mass. for new Bulletin B-124*

NATIONAL-B
Type 7180
A "B" That's Built for Service

See our Exhibit, Booth No. 8, Chicago Show, Oct. 10th-16th

NATIONAL TUNING UNITS — THE HEAVENLY TWINS
More National Tuning Units have been used by set builders than all other similar components combined.
Standard since 1923   *Approved By*   The OFFICIAL Design.
BROWNING & DRAKE

TONE FILTER
FOR BETTER TONE AND
SPEAKER PROTECTION

FILAMENT TRANSFORMER
NO. 171 FOR THE
NEW A.C. TUBES

THE ORIGINAL
VELVET VERNIER DIAL
TYPE A

POWER TRANSFORMER
FOR PLATE
SUPPLY UNITS

R-26 BATTERY-CHARGER
2½ TO 5 AMPERES

IMPEDAFORMER
FOR QUALITY AUDIO

ILLUMINATED VELVET VERNIER
DIAL TYPE C

FILTER CHOKES
TYPE 60

What Kit Shall I Buy?

THE list of kits herewith is printed as an extension of the scope of the Service Department of RADIO BROADCAST. It is our purpose to list here the technical data about kits on which information is available. In some cases, the kit can be purchased from your dealer complete; in others, the descriptive booklet is supplied for a small charge and the parts can be purchased as the buyer likes. The Service Department will not undertake to handle cash remittances for parts, but when the coupon on page 68 is filled out, all the information requested will be forwarded.

201. SC FOUR-TUBE RECEIVER—Single control. . One stage of tuned radio frequency, regenerative detector, and two stages of transformer-coupled audio amplification. Regeneration control is accomplished by means of a Variable resistor across the tickler coil. Standard parts: cost approximately $58.85.

202. SC-II FIVE-TUBE RECEIVER—Two stages of tuned radio frequency, detector, and two stages of transformer-coupled audio. Two tuning control. Volume control consists of potentiometer grid bias on r.f. tubes. Standard parts cost approximately $60.35.

203. "HI-Q" KIT—A five-tube tuned radio-frequency set having two radio stages, a detector, and two transformer-coupled audio stages. A special method of coupling in the r.f. stages tends to make the amplification more nearly equal over the entire band. Price $63.95 without cabinet.

204. R. G. S. KIT—A four-tube inverse reflex circuit, having the equivalent of two tuned radio-frequency stages, detector, and three audio stages. Two controls. Price $69.70 without cabinet.

205. PIERCE AERO KIT—A six-tube single-dial receiver; two stages of radio-frequency amplification, detector, and three stages of resistance-coupled audio. Volume control accomplished by variation of filament brilliancy of r.f. tubes or by adjusting compensating condensers. Complete chassis assembled but not wired costs $42.50.

206. H & H-T. R. F. ASSEMBLY—A five-tube set; three tuning dials, two steps of radio frequency, detector, and a transformer-coupled audio stages. Complete except for baseboard, panel, screws, wires, and accessories. Price $50.00.

207. PREMIER FIVE-TUBE ENSEMBLE—Two stages of tuned radio frequency, detector, and two steps of transformer-coupled audio. Three dials. Parts assembled but not wired. Price complete, except for cabinet, $35.00.

208. "QUADRAFORMER VI"—A six-tube set with two tuning controls. Two stages of tuned radio frequency using specially designed shielded coils, a detector, one stage of transformer-coupled audio, and two stages of resistance-coupled audio. Gain control by means of tapped primaries on the r.f. transformers. Essential kit consists of three shielded double-range "Quadraformer" coils, a selectivity control, and an "Amptirol," price $17.50. Complete parts $70.

209. GEN-RAL FIVE-TUBE SET—Two stages of tuned radio frequency, detector, and two transformer-coupled audio stages. Volume is controlled by a resistor in the plate circuit of the r.f. tubes. Uses a special r.f. coil ("Duo-Former") with figure eight winding. Parts mounted but not wired, price $37.50.

210. BREMER-TULLY POWER-SIX—A six-tube, dual-control set; three stages of neutralized tuned radio frequency, detector, and two transformer-coupled audio stages. Resistances in the grid circuit together with a phase shifting arrangement are used to prevent oscillation. Volume control accomplished by variation of B potential on r.f. tube. Essential kit consists of four r.f. transformers, two dual condensers, three small condensers, three choke coils, one 500,000-ohm resistor, three 1500-ohm resistors, and a set of color charts and diagrams. Price $41.50.

211. INFRADYNE AMPLIFIER—A three-tube intermediate-frequency amplifier for the super-heterodyne and other special receivers, tuned to 3400 kc. (86 meters). Price $25.00.

212. RADIO BROADCAST "LAB" RECEIVER—A four-tube dual-control receiver with one stage of Rice neutralized tuned-radio frequency, regenerative detector (capacity controlled), and two stages of transformer-coupled audio. Approximate price; $78.15.

213. LC-27—A five-tube set with two stages of tuned-radio frequency, a detector, and two stages of transformer-coupled audio. Special coils and special means of neutralizing are emp loyed. Output device. Price $85.20 without cabinet.

214. LOFTIN-WHITE—A five-tube set with two stages of radio frequency, especially designed to give equal amplification at all frequencies, a detector, and two stages of transformer-coupled audio. Two controls. Output device. Price $85.10.

215. K.H-27—A six-tube receiver with two stages of neutralized tuned radio frequency, a detector, three stages of choke-coupled audio, and an output device. Two controls. Price $86.00 without cabinet.

216. AERO SHORT-WAVE KIT—Three plug-in coils designed to operate with a regenerative circuit and having a frequency range of from 19,990 to 2306 kc. (15 to 130 meters). Coils and plug only, price $13.50.

217. DIAMOND-OF-THE-AIR—A five-tube set having one stage of tuned-radio frequency, a regenerative detector, one stage of transformer-coupled audio, and two stages of resistance-coupled audio. Volume control through regeneration. Two tuning dials.

218. NORDEN-HAUCK SUPER 10—Ten tubes: five stages of tuned radio frequency, detector, and four stages of choke- and transformer-coupled audio frequency. Two controls. Price $391.40.

220. BROWNING-DRAKE—Five tubes; one stage tuned radio frequency (Rice neutralization), regenerative detector (tickler control), three stages of audio (special combination of resistance- and impedance-coupled audio). Two controls.

WHAT KIT SHALL I BUY? (*Continued*)

221. LR4 ULTRADYNE—Nine-tube super-heterodyne: one stage of tuned radio frequency, one modulator, one oscillator, three intermediate-frequency stages, detector, and two transformer-coupled audio stages.

222. GREIFF MULTIPLEX—Four tubes (equivalent to six tubes); one stage of tuned radio frequency, one stage of transformer-coupled radio frequency, crystal detector, two stages of transformer-coupled audio, and one stage of impedance-coupled audio. Two controls. Price complete parts, $10.00.

223. PHONOGRAPH AMPLIFIER—A five-tube amplifier device having an oscillator, a detector, one stage of transformer-coupled audio, and two stages of impedance-coupled audio. The phonograph signal is made to modulate the oscillator in much the same manner as an incoming signal from an antenna.

224. BROWNING-DRAKE—Five tubes: one stage tuned radio frequency (with special neutralization system), regenerative detector (tickler control), three stages of audio (special combination of resistance- and impedance-coupled audio). Two controls.

225. AERO SHORT-WAVE Transmitting Kit consists of interchangeable coils to be used in tuned-plate tuned grid circuit. Kits of coils, two choke coils, and mountings, can be secured for 20-40 meter band, 40-80 meter band, or 90-180 meter band for $13.00.

USE THIS COUPON FOR KITS

RADIO BROADCAST SERVICE DEPARTMENT
Garden City, New York.

Please send me information about the following kits indicated by number:

....

....

....

....

Name.......................................

Address....................................
    (Number)   (Street)

....
    (City)   (State)

ORDER BY NUMBER ONLY. This coupon must accompany each order.
                RB 11-27

Thumb Nail Reviews

WLS—A skit having to do with various and droll adventures around the lion's cage in a circus and centering about one J. Walter Sapp. The mechanically simulated lions' roars were perfectly swell. As for the spoken lines, they were not at all bad, but suffered from high-schoolish and unconvincing delivery—a frequent enough radio play complaint.

WOR—The Kapellmeister String Quartet, excellent interpreters of chamber music, playing on this occasion the Schubert Quartet in D minor.

WBBM—The station's own string trio performing its routine tasks with great gusto and a splendid attack.

WJZ—The Arion Male Chorus singing "Sleep Kentucky Babe" in mellow fashion and introducing some tricky guitar effects against a background of humming.

JOHN WALLACE.

USE THIS COUPON FOR COMPLETE SETS

RADIO BROADCAST SERVICE DEPARTMENT
RADIO BROADCAST, Garden City, New York.
Please send me information about the following manufactured receivers indicated by number:

....

....

....

....

Name.......................................

Address....................................
    (Number)   (Street)

....
    (City)   (State)

ORDER BY NUMBER ONLY
This coupon must accompany each order.   RB 11-27

"RADIO BROADCAST'S" DIRECTORY OF MANUFACTURED RECEIVERS

¶ A coupon will be found on page 68. All readers who desire additional information on the receivers listed below need only insert the proper numbers in the coupon, mail it to the Service Department of Radio Br and full details will be sent. New sets are listed in this space eac

KEY TO TUBE ABBREVIATIONS

99—60-mA. filament (dry cell)
01-A—Storage battery 0.25 amps. filament
12—Power tube (Storage battery)
71—Power tube (Storage battery)
16-B—Half-wave rectifier tube
80—Full-wave, high current rectifier
81—Half-wave, high current rectifier
Hmu—High-Mu tube for resistance-coupled audio
20—Power tube (dry cell)
10—Power Tube (Storage battery)
00-A—Special detector
13—Full-wave rectifier tube
26—Low-voltage high-current a. c. tube
27—Heater type a. c. tube

Direct Current Receivers

NO. 424. COLONIAL 26

Six tubes; 2 t. r. f. (01-A); detector (12), 2 transformer audio (01-A and 71). Balanced t. r. f. One to three dials. Volume control: antenna switch and potentiometer across first audio. Watts required: 120. Console size: 34 x 38 x 18 inches. Headphone connections. The filaments are connected in a series parallel arrangement. Price $250 including power unit.

NO. 425. SUPERPOWER

Five tubes: All 01-A tubes. Multiplex circuit. Two dials. Volume control: resistance in r. f. plate. Watts required: 30. Antenna: loop or outside. Cabinet sizes: table, 27 x 10 x 9 inches; console, 28 x 50 x 21. Prices: table, $135 including power unit; console, $390 including power unit and loud speaker.

A. C. Operated Receivers

NO. 508. ALL-AMERICAN 77, 88, AND 99

Six tubes; 3 t. r. f. (26), detector (27), 2 transformer audio (26 and 71). Rice neutralized t. r. f. Single drum tuning. Volume control: potentiometer in r. f. plate. Cabinet sizes: No. 77, 21 x 10 x 8 inches; No. 88 Hiboy, 25 x 38 x 18 inches; No. 99 console, 27½ x 43 x 20 inches. Shielded. Output device. The filaments are supplied by means of three small transformers. The plate supply employs a gas-filled rectifier tube. Voltmeter in a. c. supply line. Prices: No. 77, $150, including power unit; No. 88, $210 including power unit; No. 99, $285 including power unit and loud speaker.

NO. 509. ALL-AMERICAN "DUET"; "SEXTET"

Six tubes; 2 t. r. f. (99), detector (99), 3 transformer audio (99 and 12). Rice neutralized t.r.f. Two dials. Volume control: resistance in r.f. plate. Cabinet sizes: "Duet," 23 x 56 x 16¼ inches; "Sextet," 22½ x 13½ x 15½ inches. Shielded. Output device. The 99 filaments are connected in series and supplied with rectified a.c., while 12 is supplied with raw a.c. The plate and filament supply uses gaseous rectifier tubes. Milliammeter on power unit. Prices: "Duet," $160 including power unit; "Sextet," $220 including power unit and loud speaker.

NO. 511. ALL-AMERICAN 80, 90, AND 115

Five tubes; 2 t.r.f. (99), detector (99), 2 transformer audio (99 and 12). Rice neutralized t.r.f. Two dials. Volume control: resistance in r.f. plate. Cabinet sizes: No. 80, 23½ x 12¼ x 15 inches; No. 90, 37½ x 12 x 12½ inches; No. 115 Hiboy, 24 x 41 x 15 inches. Coils individually shielded. Output device. See No. 509 for power supply. Prices: No. 80, $135 including power unit; No. 90, $145 including power unit and compartment; No. 115, $170 including power unit, compartment, and loud speaker.

NO. 510. ALL-AMERICAN 7

Seven tubes; 3 t.r.f. (26), 1 untuned r.f. (26), detector (27), 2 transformer audio (26 and 71). Rice neutralized t.r.f. One drum. Volume control: resistance in r.f. plate. Cabinet sizes: "Sovereign" console, 30½ x 60½ x 19 inches; "Lorraine" Hiboy, 25½ x 53¼ x 17½ inches; "Forte" cabinet, 25½ x 13½ x 17½ inches. For filament and plate supply: See No. 508. Prices: "Sovereign" $460; "Lorraine" $360; "Forte" $270. All prices include power unit. First two include loud speaker.

NO. 401. AMRAD AC9

Six tubes; 3 t.r.f. (99), detector (99), 2 transformer (99 and 12). Neutrodyne. Two dials. Volume control: resistance across 1st audio. Watts consumed: 50. Cabinet size: 27 x 9 x 11¼ inches. The 99 filaments are connected in series and supplied with rectified a.c., while the 12 is run on raw a.c. The power unit, requiring two 16-B rectifiers, is separate and supplies A, B, and C current. Price $142 including power unit.

NO. 402. AMRAD AC5

Five tubes. Same as No. 401 except one less r.f. stage. Price $125 including power unit.

NO. 536. SOUTH BEND

Six tubes. One control. Sub-panel shielding. Binding Posts. Antenna: outdoor. Prices: table, $130, Baby Grand console, $195.

NO. 537. WALBERT 26

Six tubes; five Kellogg a.c. tubes and one 71. Two controls. Volume control: variable plate resistance. Isofarad circuit. Output device. Battery cable. Semi-shielded. Antenna: 50 to 75 feet. Cabinet size: 10½ x 29½ x 16¼ inches. Prices: $215; with tubes, $250.

NO. 484. BOSWORTH, B5

Five tubes: 2 t.r.f. (26), detector (99), 2 transformer audio (special a.c. tubes). T.r.f. circuit. Two dials. Volume control: potentiometer. Cabinet size: 23 x 7 x 8 inches. Output device included. Price $175.

NO. 406. CLEARTONE 110

Five tubes; 2 t.r.f., detector, 2 transformer audio. All tubes a. c. heater type. One or two dials. Volume control: resistance in r. f. plate. Watts consumed: 40. Cabinet size: varies. The plate supply is built in the receiver and requires one rectifier tube. Filament supply through step down transformers. Prices range from $175 to $375 which includes 5 a.c. tubes and one rectifier tube.

NO. 407. COLONIAL 25

Six tubes; 2 t. r. f. (01-A), detector (99), 2 resistance audio (99), 1 transformer audio (10). Balanced t.r.f. circuit. One or three dials. Volume control: Antenna switch and potentiometer on 1st audio. Watts consumed: 100. Console size: 34 x 38 x 18 inches. Output device. All tube filaments are operated on a. c. except the detector which is supplied with rectified a.c. from the plate supply. The rectifier employs two 16-B tubes. Price $250 including built-in plate and filament supply.

NO. 507. CROSLEY 602 BANDBOX

Six tubes; 3 t.r.f. (26), detector (27), 2 transformer audio (26 and 71). Neutrodyne circuit. One dial. Cabinet size: 17½ x 5½ x 7½ inches. The heaters for the a.c. tubes and the 71 filament are supplied by windings in B unit transformers available to operate either on 25 or 60 cycles. The plate current is supplied by means of rectifier tube. Price $65 for set alone, power unit $60.

NO. 408. DAY-FAN "DE LUXE"

Six tubes; 3 t.r.f., detector, 2 transformer audio. All 01-A tubes. One dial. Volume control: potentiometer across r.f. tubes. Watts consumed: 300. Console size: 30 x 40 x 20 inches. The filaments are connected in series and supplied with d.c. from a motor-generator set which also supplies B and C current. Output device. Price $350 including power unit.

NO. 409. DAYCRAFT 5

Five tubes; 2 t.r.f., detector, 2 transformer audio. All a. c. heater tubes. Reflexed t.r.f. One dial. Volume control: potentiometers in r.f. plate and 1st audio. Watts consumed: 135. Console size: 34 x 36 x 14 inches. Output device. The heaters are supplied by means of a small transformer. A built-in rectifier supplies B and C voltages. Price $170, less tubes. The following have one more r.f. stage and are not reflexed: Daycraft 6, $195; Dayrole 6, $235; Dayfan 6, $110. All prices less tubes.

NO. 469. FREED-EISEMANN NR11

Six tubes; 3 t.r.f. (01-A), detector (01-A), 2 transformer audio (01-A and 71). Neutrodyne. One dial. Volume control: potentiometer. Watts consumed: 150. Cabinet size: 19¾ x 10 x 10½ inches. Shielded. Output device. A special power unit is included employing a rectifier tube. Price $225 including NR-411 power unit.

NO. 487. FRESHMAN 7F-AC

Six tubes; 3 t.r.f. (26), detector (27), 2 transformer audio (26 and 71). Equaphase circuit. One dial. Volume control: potentiometer across 1st audio. Console size: 24½ x 41¼ x 15 inches. Output device. The filaments and heaters and B supply are all supplied by one power unit. The plate supply requires one 80 rectifier tube. Price $175 to $350, complete.

NO. 421. SOVEREIGN 238

Seven tubes of the a.c. heater type. Balanced t.r.f. Two dials. Volume control: resistance across 2nd audio. Watts consumed: 45. Console size: 37 x 52 x 15 inches. The heaters are supplied by a small a. c. transformer, while the plate is supplied by means of rectified a.c using a gaseous type rectifier. Price $325, including power unit and tubes.

NO. 517. KELLOGG 510, 511, AND 512

Seven tubes; 4 t.r.f., detector, 2 transformer audio. All Kellogg a.c. tubes. One control and special zone switch. Balanced. Volume control: special. Output device. Shielded. Cable connection between power supply unit and receiver. Antenna: 25 to 100 feet. Panel 7½ x 27½ inches. Prices: Model 510 and 512, consoles, $495 complete. Model 511, consolette, $365 without loud speaker.

NO. 496. SLEEPER ELECTRIC

Five tubes; four 99 tubes and one 71. Two controls. Volume control: rheostat on r.f. Neutralized. Cable. Output device. Power supply uses two 16-B tubes. Antenna: 100 feet. Prices: Type 64, table, $160; Type 65, table, with built-in loud speaker, $175; Type 66, table, $175; Type 67, console, $235; Type 78, console, $265.

NO. 538. NEUTROWOUND, MASTER ALLECTRIC

Six tubes; 2 t.r.f. (01-A), detector (01-A), 2 audio (01-A and two 71 in push-pull amplifier). The 01-A tubes are in series, and are supplied from a 400-mA. rectifier. Two drum controls. Volume control: variable plate resistance. Output device. Shielded. Antenna: 50 to 100 feet. Price: $360.

NO. 413. MARTI

Six tubes: 2 t.r.f., detector, 3 resistan tubes a.c. heater type. Two dials. Vol resistance in r.f. plate. Watts consumed: 7 x 21 inches. The built-in plate supply 16-B rectifier. The filaments are supplie transformer. Prices: table, $235 includin rectifier; console, $275 including tubes console, $325 including tubes, rectifier, and

NO. 417 RADIOLA 28

Eight tubes; five type 99 and one ty control. Super-heterodyne circuit. C-bat tions. Battery cable. Headphone connecti loop. Set may be operated from batteries power mains when used in conjunction wi 104 loud speaker. Prices: $260 with tu operation; $570 with model 104 loud s operation.

NO. 540 RADIOLA 30-A

Receiver characteristics same as No. 41 type 71 power tube is used. This model i operate on either a. c. or d. c. from the The combination rectifier—power—ampli two type 81 tubes. Model 100-A loud sp tained in lower part of cabinet. Either a or long outside antenna may be used. 42½ x 29 x 17¾ inches. Price: $495.

NO. 541 RADIOLA 32

This model combines receiver No. 417 w 104 loud speaker. The power unit uses tubes and a type 10 power amplifier. Loop enclosed and is revolved by means of a dial Models for operation from a. c. or d. c. Cabinet size: 52 x 72 x 17¾ inches. Price:

NO. 539 RADIOLA 17

Six tubes; 3 t. r. f. (26), detector (27), audio (26 and 27). One control. Illu Built-in power supply using type 80 recti 100 feet. Cabinet size: 25 x 7⅞ x 8⅛ without accessories.

NO. 545. NEUTROWOUND, SUPER

Five tubes; 2 t.r.f. (99), detector (99), and 71). The 99 tubes are in series and are an 85-mA. rectifier. Two drum controls. trol: variable plate resistance. Output dev 75 to 100 feet. Cabinet size: 9 x 24 x 11 $150.

NO. 490. MOHAWK

Six tubes; 2 t.r.f., detector, 2 transfor tubes a.c heater type except 71 in last st Volume control: rheostat on r.f. Watts c Panel size: 12½ x 8½ inches. Output device for the a.c tubes and the 71 filament are small transformers. The plate supply is c type using a rectifier tube. Prices range $245

NO. 522. CASE, 62 B AND 6

McCullough a.c. tubes. Drum control. trol: variable high resistance in audio syst connections. Semi-shielded. Cable. Anter Panel size: 7 x 21 inches. Prices: Model 6 with a.c. equipment, $185; Model 62 C, a.c. equipment, $235.

NO. 523. CASE, 92 A AND

McCullough a.c. tubes. Drum contr volume control. Technidyne circuit. Shi C-battery connections. Model 92 C co device. Loop operated. Prices: Model 92 Model 92 C, console, $475.

Battery Operated Recei

NO. 542. PFANSTIEHL JUNIO

Six tubes; 3 t.r.f. (01-A), detector (01 Pfanstiehl circuit. Volume control: variabl r.f. plate circuit. One dial. Shielded. Batt battery connections. Etched bronze panel. door. Cabinet size: 9 x 20 x 8 inches. Price accessories.

NO. 512. ALL-AMERICAN 44, 45,

Six tubes; 3 t.r.f. (01-A, detector) 01-A, audio (01-A and 71). Rice neutralized control. Volume control: rheostat in r.f. No. 44, 21 x 10 x 8 inches; No. 55, 25 x 3 No. 66, 27½ x 43 x 20 inches. C-battery Battery cable. Antenna: 75 to 125 feet. P $70; No. 55, $125 including loud speaker; including loud speaker.

NO. 428. AMERICAN C6

Five tubes; 2 t.r.f. detector, 2 transf All 01-A tubes. Semi balanced t.r.f. Thre current 15 mA. Volume control: potentiom sizes: table, 20 x 8½ x 10 inches; console, inches. Partially shielded. Battery cabl connections. Antenna: 125 feet. Prices: console, $65 including loud speaker.

30 Days FREE TRIAL 5 Year Guarantee

MARWOOD

The 1928 Sensations!
8 Tube-1 Control

ALL ELECTRIC 8 Tube 1 Control $98 RETAIL PRICE

Big Discount to Agents From this Price

Has Complete A-B Power Unit

A REAL ALL ELECTRIC Radio with one of the best A-B power units on the market—no batteries needed—at the world's lowest price. This Marwood can't be excelled at ANY price. If you have electricity in your home, just plug into the light socket and forget batteries. No more battery trouble and expense. Costs less than 2c a day to operate. Always have 100% volume. ALL ELECTRIC Radios are high priced because they are new. We cut profit to the bone and offer a $250.00 outfit for $98.00 retail price. Big discount to Agents. Don't buy any Radio 'till you get details of this sensational new ALL ELECTRIC Marwood.

All Electric
or Battery Operation

AGAIN Marwood is a year ahead—with the Radio sensation of 1928—at a low price that smashes Radio profiteering. Here's the sensation they're all talking about—the marvelous 8 Tube Single Control Marwood for BATTERY or ALL ELECTRIC operation. Direct from the factory for only $69.00 retail price—a price far below that of smaller, less powerful Radios. Big discount to Agents from this price. You can't beat this wonderful new Marwood and you can't touch this low price. Why pay more for less quality? To prove that Marwood can't be beat we let you use it on 30 Days' Free Trial in your own home. Test it in every way. Compare it with any Radio for tone, quality, volume, distance, selectivity, beauty. If you don't say that it is a wonder, return it to us. We take the risk.

New Exclusive Features

Do you want coast to coast with volume enough to fill a theatre? Do you want amazing distance that only super-power Radios like the Marwood 8 can get? Do you want ultra-selectivity to cut out interference? Then you must test this Marwood on 30 Days' Free Trial. An amazing surprise awaits you. A flip of your finger makes it ultra-selective—or broad—just as you want it. Every Marwood is perfectly BALANCED—a real laboratory job. Its simple one drum control gets ALL the stations on the wave band with ease. A beautiful, guaranteed, super-efficient Radio in handsome walnut cabinets and consoles. A radio really worth double our low price.

Buy From Factory—Save Half

Why pay profits to several middlemen? A Marwood in any retail store would cost practically three times our low price direct-from-the-factory price. Our policy is highest quality plus small profit and enormous sales. You get the benefit. Marwood is a pioneer, responsible Radio, with a good reputation to guard. We insist on the best —and we charge the least. If you want next year's improvements NOW—you must get a Marwood—the Radio that's a year ahead.

AGENTS
Make Big Spare Time Money

Get your own Radio at wholesale price. It's easy to get orders for the Marwood from your friends and neighbors. Folks buy quick when they compare Marwood quality and low prices. We want local agents and dealers in each territory to handle the enormous business created by our national advertising. Make $100 a week or more in spare time demonstrating at home. No experience or capital needed. We show you how. This is the biggest season in Radio history. Everybody wants a Radio. Get in now. Rush coupon for 30 days' Free Trial, beautiful catalog. Agents' Confidential Prices and Agents' New Plan.

MARWOOD RADIO CORP.

5315 Ravenswood Avenue
Department A-17 Chicago, Illinois

Only $69 RETAIL PRICE

Big Discount to Agents From This Price

Get Our Discounts
Before You Buy a Radio

Don't buy any Radio 'till you get our big discounts and catalog. Save half and get a Radio that IS a Radio. Try any Marwood on 30 Days' Free Trial at our risk. Tune in coast to coast on loud speaker with enormous volume, clear as a bell. Let your wife and children operate it. Compare it with any Radio regardless of price. If you don't get the surprise of your life, return it. We take the risk. Don't let Marwood low prices lead you to believe Marwood is not the highest quality. We have smashed Radio prices. You save half.

6 Tube—1 Control

This is the Marwood 6 Tube, 1 Control for BATTERY or ALL ELECTRIC operation. Gets coast to coast on loud speaker with great volume. Only $47.00 retail. Big discounts to Agents. Comes in handsome walnut cabinets and consoles. This low price cannot be equaled by any other high grade 6 tube Radio. Has the volume of any 7 tube set. If you want a 6 tube Radio you can't beat a Marwood and you can't touch our low price.

$47 RETAIL PRICE Big Discount to Agents from This Price

Rush for Free Trial

NO. 485. BOSWORTH B6

Five tubes; 2 t.r.f. (01-A), detector (01-A), 2 transformer audio (01-A and 71). Two dials. Volume control: variable grid resistances. Battery cable. C-battery connections. Antenna: 25 feet or longer. Cabinet size 15 x 7 x 8 inches. Price $75.

NO. 513. COUNTERPHASE SIX

Six tubes; 3 t.r.f. (01-A), detector (00-A), 2 transformer audio (01-A and 12). Counterphase t.r.f. Two dials. Plate current: 32 mA. Volume control: rheostat on 2nd and 3rd r.f. Coils shielded. Battery cable. C-battery connections. Antenna: 75 to 100 feet. Console size: 18¼ x 40¼ x 15½ inches. Prices: Model 35, table, $110; Model 37, console, $175.

NO. 514. COUNTERPHASE EIGHT

Eight tubes; 4 t.r.f. (01-A), detector (00-A), 2 transformer audio (01-A and 12). Counterphase t.r.f. One dial. Plate current: 40 mA. Volume control: rheostat in 1st r.f. Copper stage shielding. Battery cable. C-battery connections. Antenna: 75 to 100 feet. Cabinet size: 30 x 12½ x 16 inches. Prices: Model 12, table, $225; Model 16, console, $335; Model 18, console, $365.

NO. 506. CROSLEY 601 BANDBOX

Six tubes; 3 t.r.f., detector, 2 transformer audio. All 01–A tubes. Neutrodyne. One dial. Plate current: 40 mA. Volume control: rheostat in r.f. Shielded. Battery cable. C-battery connections. Antenna: 75 to 150 feet. Cabinet size: 17½ x 5½ x 7½. Price, $55.

NO. 434. DAY–FAN 6

Six tubes; 3 t.r.f. (01-A), detector (01-A), 2 transformer audio (01-A and 12 or 71). One dial. Plate current: 12 to 15 mA. Volume control: rheostat on r.f. Shielded. Battery cable, C-battery connections. Output device. Antenna: 50 to 120 feet. Cabinet sizes: Daycraft 6, 32 x 30 x 34 inches; Day-Fan Jr., 15 x 7 x 7. Prices: Day-Fan 6, $110; Daycraft 6, $145 including loud speaker; Day-Fan Jr. not available.

NO. 435. DAY–FAN 7

Seven tubes; 3 t.r.f. (01-A), detector (01-A), 1 resistance audio (01-A), 2 transformer audio (01-A and 12 or 71). Plate current: 15 mA. Antenna: outside. Same as No. 434. Price $115.

NO. 503. FADA SPECIAL

Six tubes; 3 t.r.f. (01–A), detector (01–A), 2 transformer audio (01–A and 71). Neutrodyne. Two drum control. Plate current: 20 to 24mA. Volume control: rheostat on r.f. Coils shielded. Battery cable. C-battery connections. Headphone connection. Antenna: outdoor. Cabinet size: 20½ x 13½ x 10½ inches. Price $95.

NO. 504. FADA 7

Seven tubes; 4 t.r.f. (01–A), detector (01–A), 2 transformer audio (01–A and 71). Neutrodyne. Two drum control. Plate current: 43mA. Volume control: rheostat on r.f. Completely shielded. Battery cable. C-battery connections. Headphone connections. Output device. Antenna: outdoor or loop. Cabinet sizes: table, 25½ x 13½ x 11½ inches; console, 29 x 50 x 17 inches. Prices: table, $185; console, $285.

NO. 436. FEDERAL

Five tubes; 2 t.r.f. (01–A), detector (01–A), 2 transformer audio (01–A and 12 or 71). Balanced t.r.f. One dial. Plate current: 20.7 mA. Volume control: rheostat on r.f. Shielded. Battery cable. C-battery connections. Antenna: loop. Made in 6 models. Price varies from $250 to $1000 including loop.

NO. 505. FADA 8

Eight tubes. Same as No. 504 except for one extra stage of audio and different cabinet. Prices: table, $300; console, $400.

NO. 437. FERGUSON 10A

Seven tubes; 3 t.r.f. (01–A), detector (01–A), 3 audio (01–A and 12 or 71). One dial. Plate current: 18 to 25 mA. Volume control: rheostat on two r.f. Shielded. Battery cable. C-battery connections. Antenna: 100 feet. Cabinet size: 21½ x 12 x 15 inches. Price $150.

NO. 438. FERGUSON 14

Ten tubes; 3 untuned r.f., 3 t. r.f. (01–A), detector (01–A), 3 audio (01–A and 12 or 71). Special balanced t.r.f. One dial. Plate current: 30 to 35 mA. Volume control: rheostat in three r.f. Shielded. Battery cable, C-battery connections. Antenna: loop. Cabinet size: 24 x 12 x 16 inches. Price $235, including loop.

NO. 439. FERGUSON 12

Six tubes; 2 t.r.f. (01–A), detector (01–A), 1 transformer audio (01–A), 2 resistance audio (01–A and 12 or 71). Two dials. Plate current: 18 to 25 mA. Volume control: rheostat on two r.f. Partially shielded. Battery cable. C-battery connections. Antenna: 100 feet. Cabinet size: 22½ x 10 x 12 inches. Price $85. Consolette $145 including loud speaker.

NO. 440. FREED-EISEMANN NR-8 NR-9, AND NR-66

Six tubes; 3 t.r.f. (01–A), detector (01–A), 2 transformer audio (01–A and 71). Neutrodyne. NR-8, two dials; others one dial. Plate current: 30 mA. Volume control: rheostat on r.f. NR-8 and 9: chassis type shielding. NR-66, individual stage shielding. Battery cable. C-battery connections. Antenna: 100 feet. Cabinet sizes: NR-8 and 9, 19½ x 10 x 10½ inches; NR-66 20 x 10½ x 12 inches. Prices: NR-8, $90; NR-9, $100; NR-66, $125.

NO. 501. KING "CHEVALIER"

Six tubes. Same as No. 500. Coils completely shielded. Panel size: 11 x 7 inches. Price, $210 including loud speaker.

NO. 441. FREED-EISEMANN NR-77

Seven tubes; 4 t.r.f. (01–A), detector (01–A), 2 transformer audio (01–A and 71). Neutrodyne. One dial. Plate current: 35 mA. Volume control: rheostat on r.f. Shielding. Battery cable. C-battery connections. Antenna: outside or loop. Cabinet size: 23 x 10½ x 13 inches. Price $175.

NO. 442. FREED-EISEMANN 800 AND 850

Eight tubes; 4 t.r.f. (01–A), detector (01–A), 1 transformer (01–A), 1 parallel audio (01–A or 71). Neutrodyne. One dial. Plate current: 35 mA. Volume control: rheostat on r.f. Shielded. Battery cable. C-battery connections. Output: two tubes in parallel or one power tube may be used. Antenna: outside or loop. Cabinet sizes: No. 800, 34 x 15½ x 13½ inches; No. 850, 36 x 65½ x 17½. Prices not available.

NO. 444. GREBE MU-1

Five tubes; 2 t.r.f. (01–A), detector (01–A), 2 transformer audio (01–A and 12 or 71). Balanced t.r.f. One, two, or three dials (operate singly or together). Plate current: 30mA. Volume control; rheostat on r.f. Binocular coils. Binding posts. C-battery connections. Antenna: 125 feet. Cabinet size: 22½ x 9½ x 13 inches. Prices range from $95 to $320.

NO. 426. HOMER

Seven tubes; 4 t.r.f. (01-A); detector (01-A or 00A); 2 audio (01-A and 12 or 71). One knob tuning control. Volume control: rotor control in antenna circuit. Plate current: 22 to 25 mA. "Technidyne" circuit. Completely enclosed in aluminum box. Battery cable. C-battery connections. Cabinet size, 8½ x 19½ x 9½ inches. Chassis size, 6½ x 17 x 8 inches. Prices: Chassis only, $80. Table cabinet, $95.

NO. 502. KENNEDY ROYAL 7. CONSOLETTE

Seven tubes; 4 t.r.f. (01–A), detector (00–A), 2 transformer audio (01–A and 71). One dial. Plate current: 42 mA. Volume control: rheostat on two r.f. Special r.f. coils. Battery cable. C-battery connections. Headphone connection. Antenna: outside or loop. Consolette size: 36½ x 35½ x 19 inches. Price $220.

NO. 498. KING "CRUSADER"

Six tubes; 2 t.r.f. (01–A), detector (00–A), 3 transformer audio (01–A and 71). Balanced t.r.f. One dial. Plate current: 20 mA. Volume control: rheostat on r.f. Coils shielded. Battery cable. C-battery connections. Antenna: outside. Panel: 11 x 7 inches. Price, $115.

NO. 499. KING "COMMANDER"

Six tubes; 3 t.r.f. (01–A), detector (00–A), 2 transformer audio (01–A and 71). Balanced t.r.f. One dial. Plate current: 25 mA. Volume control: rheostat on r.f. Completely shielded. Battery cable. C-battery connections. Antenna: loop. Panel size: 12 x 8 inches. Price $220 including loop.

NO. 429. KING COLE VII AND VIII

Seven tubes; 3 t.r.f., detector, 1 resistance audio, 2 transformer audio. All 01–A tubes. Model VIII has one more stage t.r.f. (eight tubes). Model VII, two dials. Model VIII, one dial. Plate current: 15 to 50 mA. Volume control: primary shunt in r.f. Steel shielding. Battery cable and binding posts. C-battery connections. Output devices on some consoles. Antenna: 10 to 100 feet. Cabinet size: varies. Prices: Model VII, $80 to $160; Model VIII, $100 to $300.

NO. 500. KING "BARONET" AND "VIKING"

Six tubes; 2 t.r.f. (01–A), detector (00–A), 3 transformer audio (01–A and 71). Balanced t.r.f. One dial. Plate current: 19 mA. Volume control: rheostat in r.f. Battery cable. C-battery connections. Antenna: outside. Panel size: 18 x 7 inches. Prices: "Baronet," $70; "Viking," $140 including loud speaker.

NO. 489. MOHAWK

Six tubes; 2 t.r.f. (01–A), detector (00–A), 3 audio (01–A and 71). One dial. Plate current: 40 mA. Volume control: rheostat on r.f. Battery cable. C-battery connections. Output device. Antenna: 60 feet. Panel size: 12½ x 8½ inches. Prices range from $65 to $245.

NO. 543. ATWATER KENT, MODEL 33

Six tubes; 3 t.r.f. (01-A), detector (01-A), 2 audio (01-A and 71 or 12). One dial. Volume control: r.f. filament rheostat. C-battery connections. Battery cable. Antenna: 100 feet. Steel panel. Cabinet size: 21½ x 6½ x 6½ inches. Price: $90, without accessories.

NO. 544. ATWATER KENT, MODEL 50

Seven tubes; 4 t.r.f. (01-A), detector (01-A), 2 audio (01-A and 12 or 71). Volume control: r.f. filament rheostat. C-battery connections. Battery cable. Special band-pass filter circuit with an untuned amplifier. Cabinet size: 20½ x 13 x 7½ inches. Price: $150.

NO. 452. ORIOLE 90

Five tubes; 2 t.r.f., detector, 2 transformer audio. All 01–A tubes. "Trinum" circuit. Two dials. Plate current: 18 mA. Volume control: rheostat on r. f. Battery cable. C-battery connections. Antenna: 50 to 100 feet. Cabinet size: 25½ x 11½ x 12½ inches. Price $85. Another model has 8 tubes, one dial, and is shielded. Price $185.

NO. 453. PARAGON

Six tubes; 2 t.r.f. (01–A), detector (01–A), 3 double impedance audio (01–A and 71). One dial. Plate current: 40 mA. Volume control: resistance in r.f. plate. Shielded. Battery cable. C-battery connections. Output device. Antenna: 100 feet. Console size: 20 x 46 x 17 inches. Price not determined.

NO. 543 RADIO

Five tubes; 2 t. r. f. (99), d
former audio (99 and 20). Reg
drum controls. C-battery con
Antenna: 100 feet. Price: $78 w

NO. 480. PFANSTIEI

Six tubes; 3 t.r.f. (01–A), d
former audio (01–A and 71).
23 to 32 mA. Volume control
Shielded. Battery cable. C-b
tenna: outside. Panel size: 17½
30 cabinet, $105; No. 302 c
loud speaker.

NO. 515. BROWNIN

Seven tubes; 2 t.r.f. (01-A),
(Hmu, two 01-A, and 71). Ill
Volume control: rheostat on 1s
ized. C-battery connections.
panel. Output device. Antenn
30 x 11 x 9 inches. Price, $145.

NO. 516. BROWNIN

Six tubes; 1 t.r.f. (99), de
(Hmu, two 01-A and 71). Dru
adjustment. Volume control: rl
tive detector. Shielded. Neutra
tions. Battery cable. Antenna
25 x 11 x 9. Price $105.

NO. 518. KELLOGG "V 504, 505, AN

Five tubes; 2 t.r.f., detecto
One control and special zone s
rheostat on r.f. C-battery con
Plate current: 25 to 35 mA. A
7½ x 25½ inches. Prices: Moc
accessories. Model 505, table,
Model 506, consolette, $135 wi

NO. 519. KELLOGG,

Six tubes, 3 t.r.f., detector, 2
control and special zone switch
stat on r.f. C-battery connecti
Binding posts and battery ca
Cabinet size: Model 507, table
Model 508, console, 34 x 18 x
507, $190 less accessories. Mo
speaker.

NO. 427. MUR

Seven tubes; 3 t.r.f. (01-A),
former and 2 resistance audio (
One control. Volume control:
shielded. Neutralized. Battery
nections. Complete metal ca
Panel size: 9 x 23 inches. Price

NO. 520. BOS

Seven tubes; 4 t.r.f. (01-A),
(01-A and 71). One control ca
control: rheostat on r.f. Shiel
battery connections. Balanced.
loud speaker. Antenna: built-in
100 feet. Cabinet size: 46 x 16
including enclosed loop and lou

NO. 521. BOSCH "CRU

Six tubes; 3 t.r.f. (01-A), d
(01-A and 71). One control. V
on r.f. Shielded. C-battery
Battery cable. Antenna: 20 to
66, table, $99.50. Model 76, c
speaker $195.

NO. 524. CASE, 61

T.r.f. Semi-shielded. Batter
Volume control: variable high
tem. Plate current: 35 mA. Ar
Model 61 A, $85; Model 61 C,

NO. 525. CASE, 90

Drum control. Inductive vol
circuit. C-battery connections
operated. Model 90-C equipp
Prices: Model 90 A, table, $22
$350.

NO. 526. ARBOR

Six tubes; 3 t.r.f. (01-A), d
(01-A and 71). One control. V
Shielded. Battery cable. Outpu
nections. Loftin-White circuit.
7½ x 15 inches, metal. Prices:
Model 252, $185; Model 253, $
ation phonograph and radio, $

NO. 527. ARBOR

Five tubes; 2 t.r.f. (01-A), d
(01-A). Two controls. Volum
battery connections. Binding
Prices: Model 27, $65; Model
$125.

NO. 528. THE

Seven tubes; six 01-A tube
One control. Volume control: r
nection. Partial shielding. Bi
outside. Cabinet size: 40 x
Complete with A power suppl
sories, $150.

NO. 529. DIAMOND SPECI AND BABY GRAN

Six tubes; all 01-A type. One
ing. C-battery connections. Vo
Binding posts. Antenna: out
Special, $75; Super Special, $6
$110.

. 530. KOLSTER, 7A AND 7B

es; 4 t.r.f. (01-A), detector (01-A), 2 audio 2). One control. Volume control: rheostat ied. Battery cable. C-battery connections. to 75 feet. Prices: Model 7A, $125; Model lt-in loud speaker, $140.

31. KOLSTER, 8A, 8B, AND 8C

es; 4 t.r.f. (01-A), detector (01-A), 3 audio nd one 12). One control. Volume control: .f. Shielded. Battery cable. C-battery con- odel 8A uses 50 to 75 foot antenna; model output device and uses antenna or detach- lodel 8C contains output device and uses built-in loop. Prices: 8A, $185; 8B, $235;

32. KOLSTER, 6D, 6G, AND 6H

3 t.r.f. (01-A), detector (01-A), 2 audio 2). One control. Volume control: rheostat tery connections. Battery cable. Antenna: t. Model 6G contains output device and speaker; Model 6H contains built-in B nd loud speaker. Prices: Model 6D, $80; 165; Model 6H, $265.

33. SIMPLEX, SR 9 AND SR 10

; 2 t.r.f. (01-A), detector (00-A), 2 audio 2). SR 9, three controls; SR 10, two con- control: rheostat. C-battery connections. le. Headphone connection. Prices: SR 9, onsolette, $95; console, $145. SR 10, table te, $95; console, $145.

NO. 534. SIMPLEX, SR 11

3 t.r.f. (01-A), detector (00-A), 2 audio 2). One control. Volume control: rheostat. onnections. Battery cable. Antenna: 100 table, $70; consolette, $95; console, $145.

5. STANDARDYNE, MODEL S 27

2 t.r.f. (01-A), detector (01-A), 2 audio). One control. Volume control: rheostat tery connections. Binding posts. Antenna: inet size: 9 x 9 x 19½ inches. Prices: S 27, 50, console, with built-in loud speaker, 00, console with built-in loud speaker,

481. PFANSTIEHL 32 AND 322

es: 3 t.r.f. (01-A), detector (01-A), 3 audio 1). One dial. Plate current: 23 to 32 mA. trol: resistance in r. f. plate. Shielded. e. C-battery connections. Output device. tside. Panel: 17¾ x 8½ inches. Prices: No. $145; No. 322 console, $245 including

NO 433. ARBORPHONE

Five tubes; 2 t.r.f., detector, 2 transformer audio. All 01-A tubes. Two dials. Plate current: 16 mA. Volume control: rheostat in r.f. and resistance in r.f. plate. C-battery connections. Binding posts. Antenna: taps for various lengths. Cabinet size: 24 x 9 x 10½ inches. Price: $65.

NO. 431. AUDIOLA 6

Six tubes; 3 t.r.f. (01-A), detector (00-A), 2 transformer audio (01-A and 71). Drum control. Plate current: 20 mA. Volume control: resistance in r.f. plate. Stage shielding. Battery cable. C-battery connection. Antenna: 50 to 100 feet. Cabinet size; 28½ x 11 x 14½ inches. Price not established.

NO. 432. AUDIOLA 8

Eight tubes; 4 t.r.f. (01-A), detector (00-A), 1 transformer audio (01-A), push-pull audio (12 or 71). Bridge balanced t.r.f. Drum control. Volume control: resistance in r.f. plate. Stage shielding. Battery cable. C-battery connections. Antenna: 10 to 100 feet. Cabinet size: 28½ x 11 x 14½ inches. Price not established.

NO. 542 RADIOLA 16

Six tubes; 3 t. r. f. (01-A), detector (01-A), 2 transformer audio (01-A and 112). One control. C-battery connections. Battery cable. Antenna: outside. Cabinet size: 16½ x 8¼ x 7½ inches. Price: $69.50 without accessories.

NO. 456. RADIOLA 20

Five tubes: 2 t.r.f. (99), detector (99), 2 transformer audio (99 and 20). Balanced t.r.f. and regenerative detector. Two dials. Volume control: regenerative. Shielded. C-battery connections. Headphone connections. Antenna: 75 to 150 feet. Cabinet size: 19½ x 11¾ x 16 inches. Price $115 including all tubes.

NO. 457 RADIOLA 25

Six tubes; five type 99 and one type 20. Drum control. Super-heterodyne circuit. C-battery connections. Battery cable. Headphone connections. Antenna: loop. Set may be operated from batteries or from power mains when used with model 104 loud speaker. Price; $165 with tubes, for battery operation. Apparatus for operation of set from the power mains can be purchased separately.

NO. 493. SONORA F

Seven tubes; 4 t.r.f. (01-A), detector (00-A), 2 transformer audio (01-A and 71). Special balanced t.r.f. Two dials. Plate current: 45 mA. Volume control: rheostat in r.f. Shielded. Battery cable. C-battery connections. Output device. Antenna: loop. Console size: 32 x 45½ x 17 inches. Prices range from $350 to $450 including loop and loud speaker.

NO. 494. SONORA E

Six tubes; 3 t.r.f. (01-A), detector former audio (01-A and 71). Special Two dials. Plate current: 35 to 40 mA. rheostat on r.f. Shielded. Battery c connections. Antenna: outside. Cabir Prices: table, $110; semi-console, $14 including loud speaker.

NO. 495. SONORA D

Same as No. 494 except arrangem t.r.f., detector, 3 audio. Prices: table, console, $185; "DeLuxe" console, $22

NO. 482. STEWART-WARNER 7

Six tubes; 3 t.r.f., detector, 2 tra All 01-A tubes. Balanced t.r.f. Two rent: 10 to 25 mA. Volume control: plate. Shielded. Battery cable. C-batt Antenna: 80 feet. Cabinet sizes: No. x 11½ x 13½ inches; No. 710 console, inches. Tentative prices: No. 705, $265 including loud speaker.

NO. 483. STEWART-WARNER

Same as No. 482 except no shielding No. 525 table, 19¾ x 10 x 11¼ inches: 22½ x 40 x 14 inches. Tentative price No. 520, $117.50 including loud speak

NO. 459. STROMBERG-CARLSON

Five tubes; 2 t.r.f. (01-A), detector former audio (01-A and 71). Neutrod Plate current: 25 to 35 mA. Volume on 1st r.f. Shielded. Battery cable. C- tions. Headphone connections. Outpu voltmeter. Antenna: 60 to 100 feet No. 501, 25½ x 13 x 14 inches; No. x 16¾ inches. Prices: No. 501, $180;

NO. 460. STROMBERG-CARLSON

Six tubes. Same as No. 549 except stage. Cabinet sizes: No. 601, 27 x 1 No. 602, 28¼ x 51½ x 19½ inches. Prices No. 602, $330.

NO. 486. VALLEY 7

Seven tubes; 4 t.r.f. (01-A), detector former audio (01-A and 71). One dia 35 mA. Volume control: rheostat shielded. Battery cable. C-battery cor phone connection. Antenna: 50 to 1 size: 27 x 6 x 7 inches. Price $95.

NO. 472. VOLOTONE V

Six tubes. Same as No. 471 with tions; 2 t.r.f. stages. Three dials. P mA. Cabinet size: 26½ x 8 x 12 inches

A KEY TO RECENT RADIO ARTICLES

By E. G. SHALKHAUSER

THIS is the twenty-fourth installment of references to articles which have appeared recently in various radio periodicals. Each separate reference should be cut out and pasted on 4" x 6" cards for filing, or pasted in a scrap book either alphabetically or numerically. An outline of the Dewey Decimal System (employed here) appeared last in the January RADIO BROADCAST.

R402. SHORT-WAVE SYSTEMS. SHORT-WAVE
QST. July, 1927. Pp. 8-14. TRANSMISSION.
"Short-Wave Radio Transmission and Its Practical Uses," C. W. Rice. Part 1. .
The ionization of the atmosphere through cosmic radiation as well as through propagation of electron streams from the sun determines the nature of electromagnetic waves, as stated. How this ionization affects day and night and also seasonal variations is explained from experimental data obtained to date. Comparison is made between auroral phenomena and the theory of ionization as well as the effect of terrestrial magnetism on the motion of the electron. Skip distance effect is said to be due to the bending of the waves in the upper atmosphere, the degree of bending depending on the wavelength.

R201.6. HIGH-FREQUENCY BRIDGE. BRIDGE,
QST. July, 1927. Pp. 15-20. High-Frequency.
"A Bridge to Measure Capacity, Power Factor, Resistance, and Inductance," J. Katzman.
The purpose of the article is to describe and explain the important factors of the Wien Series Resistance Bridge when used to measure C, L, R and power factor accurately to 1/10 of 1 per cent.

R344.3. TRANSMITTING SETS. TRANSMITTERS,
QST. July, 1927. Pp. 24-28. Tuning.
"Some Light on Transmitter Tuning," R. A. Hull.
The construction of a shielded oscillator and its use in tuning transmitter circuits for good signal output are outlined. Good plate and filament supply regulation is one of the main requirements. The proper method of tuning various circuits to adjust the wavelength of the oscillator and the antenna, and the correct amount of grid excitation to be used are told. Key thumps can be greatly reduced by having proper coupling and antenna tuning.

R402. SHORT-WAVE SYSTEMS. SHORT-WAVES.
QST. July, 1927. Pp. 29-30.
"An Investigation of the 5-Meter Band," E. M. Guyer and O. C. Austin.
Some problems on the construction and the operation of 5-meter transmitters are related, photographs of several sets being shown with a list of material for their construction appended.

R342. AMPLIFIERS. KEYING
QST. July, 1927. Pp. 31-35. AMPLIFIERS.
"Keying the Amplifier," A. G. Shafer.
A keying system, whereby a specially constructed key is placed in the grid circuit of one of the amplifier tubes, is utilized to prevent key thumping. The system consists of changing the capacity of the coupling capacity to such an extent as to prevent proper transfer of energy from the oscillator without actually breaking any part of the circuit.

R344.3. TRANSMITTING SETS. TRANSMITTER,
QST. July, 1927. Pp. 38-40. Short-Wave.
"A Constant Frequency Transmitter," W. H. Hoffman.
A non-crystal oscillator, capable of maintaining a constant frequency output, yet flexible enough that the frequency may be shifted to other amateur wavelengths, is described and illustrated.

R344.5. ALTERNATING-CURRENT SUPPLY. SOCKET POWER,
Radio. July, 1927. Pp. 25-ff. "A-B-C"
"ABC Socket Power For Large Tubes," G. M. Best.
A discussion on the assembly and the operation of several combination ABC socket power units and the results obtained when used with a Browning-Drake receiver are given. The Raytheon 350-mA. tube is used with each combination. All wiring details of the units, including those of the Browning-Drake receiver, are shown.

R160. RECEIVING APPARATUS. RECEIVER,
Radio. July, 1927. Pp. 29-ff. Single-Control.
"Trouble Shooting the Single-Control Set," M. P. Gilliland.
In adjusting single-control receivers for selectivity the following points are said to be of importance: Proper neutralizing of all radio-frequency stages; balancing of tuned circuits. For volume control a shunt resistance across the secondary of the first audio transformer is recommended.

R330. ELECTRON TUBES. ELECTRON
Radio. July, 1927. Pp. 47-ff. TUBES.
"Vacuum Tube Characteristics."
The characteristics of dry cell tubes, power tubes, high-mu tubes, and special detector tubes, also of the new a.c. filament tubes such as the UX-226, the UX-280, the UX-281, the UX-227, are given. The quadratron, the Emerson multivalve, the Sovereign A-C tube, the Van Horne A-C tube, the new A-C Magnatron tubes and the Armor A-C 110 tube are described.

R160. RECEIVING APPARATUS. RECEIVER
Proc. I. R. E. May, 1927. Pp. 287-305. MEASUREMENTS.
"Notes on Radio Receiver Measurements," T. A. Smith and G. Rodwin.
The comparison of radio receivers electrically involves the three main points: sensitivity, selectivity and fidelity, as stated. The method of test and of making and interpreting the curves presented are outlined.

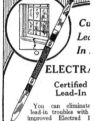

R376.3. LOUD-SPEAKING REPRODUCERS.   LOUD
Proc. I. R. E. May, 1927. Pp. 363-376.   SPEAKERS.
"Loud-Speaker Testing Methods." I. Wolff and A. Ringel.
An electric oscillator method is used in obtaining quantitative measurements in testing loud speakers. The output of an oscillator, having a continuously variable frequency is fed to the loud speaker. The results are recorded on a revolving drum mechanism. The curves, showing variation of sound pressure against frequency, reveal interesting characteristics, as explained.

R1113.3. DIRECTIONAL VARIATIONS.   TRANSMISSION
Proc. I. R. E. May, 1927. Pp. 377-385.   PHENOMENA.
"High Angle Radiation of Short Electric Waves." S. Uda.
The paper describes some accounts of experimental work on the field distribution due to straight vertical unloaded antenna operating at one of its harmonics. Short waves of 2.66 meters were employed, and observations have been made with the grounded and the ungrounded antennas.
The paper also gives the test results on a new wave projector devised by the author with special reference to high angle radiation of short electric waves.

R344.5. TRANSMITTING SETS.   TRANSMITTER.
Proc. I. R. E. May, 1927. Pp. 397-400.
"The Tuned-Grid Tuned-Plate Circuit Using Plate-Grid Capacity for Feed-Back. A Derivation of the Conditions for Oscillation." J. B. Dow.
Mathematical equations are developed showing the conditions required for oscillation in the tuned-grid tuned-plate circuit of a transmitter.

R162. SELECTIVITY OF RECEIVERS.   RECEIVERS,
Proc. I. R. E. May, 1927. Pp. 401-423.   Selective.
"Selectivity of Tuned Radio Receiving Sets." K. W. Jarvis.
The question of modern receiver design, incorporating selectivity, fidelity of reproduction, and adequate sound volume, is discussed. The resonance circuit and its requirements are analyzed mathematically and curves presented showing relation between amplification and electivity of radio-frequency stages. Discussing quality of reproduction the problem of regeneration, the phase shift of the side bands and the transient response of the circuit are mentioned.

R1113.3. DIRECTIONAL VARIATIONS.   TRANSMISSION
Proc. I. R. E. May, 1927. Pp. 425-430.   PHENOMENA.
"Radio Phenomena Recorded by the University of Michigan Greenland Expedition—1926." P. C. Oscanyan, Jr.
The experiences encountered by the writer when using short waves for transmission on Maligiak Fiord, North of the Arctic Circle, are related. It was noted that when attempting to receive signals from stations working on wavelengths of 50 meters or below, complete screening was effected when the receiver was placed at the foot of a hill which is of a height greater than 17 degrees from the horizontal of the station. Photographs of the station are shown.

R500. APPLICATIONS OF RADIO.   APPLICATIONS.
RADIO BROADCAST. Aug. 1927. Pp. 199-202.   Paper
"Saving Paper!" J. Millen.   -weighing.
The device illustrated and described consists of an oscillating circuit coupled to a tuned circuit. a thermal meter recording the deflection when in resonance. The material to be measured acts in the capacity of a dielectric, thus changing the frequency of the resonant circuit, this change being recorded on the thermal meter.

R134.8. REFLEX ACTION.   REFLEX
RADIO BROADCAST. Aug. 1927. Pp. 208-210.   CIRCUIT.
"Have You a Roberts Reflex?" J. B. Brennan.
Improvements which can be made in the Roberts circuit consist in increasing the sensitivity and selectivity, improving the quality of reproduction, making it more stable in operation, and increasing its volume. These changes are discussed in detail.

R376.5. LOUD-SPEAKING REPRODUCERS.   LOUD
RADIO BROADCAST. Aug. 1927. Pp. 211-212.   SPEAKERS.
"The Balsa Wood Loud Speaker."
Data on the assembly and the properties of the new Balsa wood loud speaker are given. The wood is obtainable in kit form, and by careful assembly of the parts a speaker of excellent reproducing qualities is said to result. Suggestions concerning changes and improvements are offered for those who experiment with this type of loud speaker.

R344.3. TRANSMITTING SETS.   TRANSMITTER.
RADIO BROADCAST. Aug. 1927. Pp. 213-217.   Short-Wave.
"A Flexible Short-Wave Transmitter." H. E. Rhodes.
The construction of a portable telegraph-telephone transmitter for short-waves, using tuned-plate tuned-grid circuit, is outlined, many data being given concerning the general characteristics of the circuit employed. The set operates between 7900 kc. and 2650 kc. (38 to 113 meters). A series of tests were carried on, the results of which are shown graphically and discussed in detail. These include: 1, The effect of varying the grid leak resistance; 2, the effect of varying the resistance of either the tuned-grid or tuned-plate circuit; 3, the effect of varying the coupling between the plate and the antenna coils; 4. the effect of varying the plate voltage.

R300. RADIO MEASUREMENTS AND STANDARDIZATION.
RADIO BROADCAST. Aug. 1927. Pp. 224-226.   TONE
"Judging Tone Quality." E. H. Felix.   QUALITY.
The subject of distortion in radio receivers is discussed from the standpoint of the listener when trying to discriminate between good and poor tone quality. What is desired is faithful reproduction throughout, from microphone to loud speaker. Because of the importance of harmonics in distinguishing different instruments it is essential that frequencies up to 6000 cycles be reproduced. Suggestions as to methods which can be used in judging the reproducing qualities of receivers are offered.

R320. CAPACITY.   CAPACITY MEASUREMENTS.
RADIO BROADCAST. Aug. 1927. Pp. 227-228.
"Condenser, Coil, Antenna Measurements." K. Henney.
The measurements of variable and fixed condensers, distributed capacity of inductance coils and of antenna capacity and inductance can readily be made with the aid of a calibrated modulated oscillator. Data for the use of this instrument and typical measurements are presented.

R343. Electron-Tube Receiving Sets. Receiver Neutrodyne, AC.
Radio Broadcast. Aug. 1927. Pp. 232–234.
"Constructing a Five-Tube Neutrodyne," H. E. Rhodes.
A shielded, two radio-frequency, detector and two audio-frequency tube neutrodyne, using the new a. c. tubes, is shown and details for its construction given. Great sensitivity, selectivity and ease of operation are claimed for this circuit arrangement.

R330. Electron Tubes. Electron Tubes.
Radio Broadcast. Aug. 1927. Pp. 238–240. High mu.
"Use of Tubes Having High Amplification," A. V. Loughren.
The amplification characteristics of high-mu tubes are treated. The discussion analyzes the frequency characteristics of each stage in a resistance-coupled amplifier and the choice of the amplification factor. Oscillographs and curves show the results to be expected.

R270. Signal Intensity. Signal Intensity.
Radio News. July, 1927. Pp. 12–13. Broadcast.
"The Service Area of a Broadcast Station," S. R. Winters.
Results of measurements made with a loop test set by S. W. Edwards, radio supervisor of the 8th radio district, on signal strength from several broadcast stations, are given. These show to what extent steel buildings, static, electrical disturbances and other noises affect radio reception at a distance. The working standard of 10,000 micro-volts per meter intensity was used to determine a reliable reception area about the station.

R3.2.7. Audio-Frequency Amplifiers. Transformers.
Radio News. July, 1927. Pp. 25–ff. Coupling.
"Why Loud Speaker Coupling Devices are Necessary," I. F. Jackowski.
An explanation is given of the necessity of coupling the loud speaker to the audio amplifier through some coupling transformer and auxiliary apparatus, in order to bypass the direct-current component of the power tube output energy.

R800. (535.3) Photo-Electric Phenomena. Crystals.
Radio News. July, 1927. Pp. 32–ff. Photo-electric.
"Light-Sensitive Crystals," G. C. B. Rowe.
The construction of simple light-sensitive cells, using ordinary metals such as copper, zinc, etc. or molybdenite and the substance selenium, is described. The numerous applications of such cells are mentioned and diagrams show how such cells may be used by the experimenter.

R330. Electron Tubes. Electron Tubes.
Radio News. July, 1927. Pp. 50–51. High mu.
"A New Electron Tube," S. Harris.
A tube having a fourth element has been developed for use in circuits where objectionable feed-backs are encountered. With the aid of the fourth element, known as the "shielded grid," the effect of plate to grid capacity has been eliminated. It is stated that the amplification obtainable with this tube is as high as 200 per tube at 50 kc.

R387. 1. Shields. Shielding.
Radio News. July, 1927. Pp. 52–ff.
"The Effects of Shielding," H. A. Zahl.
The effect that shielding has on the electrical properties of circuits is discussed in detail, with curves shown, and the method of making the measurements is described.

R201. Frequency, Wavelength Frequency.
Measurements Measurements.
Exp. Wireless (London). July, 1927. Pp. 394–401.
"The Exact and Precise Measurement of Wavelength in Radio Transmitting Stations," R. Braillard. (Concluded).
The description of the wavemeter is continued from the previous article and the method of standardization is outlined. Its accuracy is said to be exceptional, transmitters being adjustable to a variation limit of 17.8% of their wave.

R134.75. Super-Heterodyne. Super-Heterodyne.
Exp. Wireless (London). July, 1927. Pp. 402–411.
"Design and Construction of a Super-heterodyne Receiver," P. K. Turner. (Concluded).
In the last of a series of articles on the super-heterodyne the author discusses the intermediate stages of amplification and the low-frequency stages, and proceeds to give a detailed analysis of the actual construction of the set.

R800 (621.354). Batteries, Secondary. Batteries.
Amateur Transmitter. April, 1927. Pp. 10–ff. Edison.
"Edison Storage B Batteries," H. Rodloff.
The construction of small Edison cells from standard parts is described. Considerable information as to their characteristics and their properties is related.

R382. Inductors. Choke
Amateur Transmitter. May, 1927. P. 11. Coils.
"Radio-Frequency Choke Design," Wm. Zeidlik.
In order to obtain maximum efficiency in the operation of any short-led transmitter, properly designed radio-frequency choke coils are essential, as stated. The method of determining correct coils for such purposes is outlined.

R800(621.313.7). Rectifiers. Rectifiers.
Amateur Transmitter. May, 1927. Pp. 12–15. Electrolytic.
"Electrolytic Rectifiers, Lead-Aluminum Type," J. E. Hall.
The theory and the principle underlying electrolytic rectifiers is given. Information concerning the electrolytes used, the forming process, the heating of the cells and the capacity of the units constructed, is outlined in detail.

R344.3. Transmitter Sets. Transmitter.
Amateur Transmitter. June, 1927. Pp. 7–ff. Short-Wave.
"Master Oscillator Kinks," R. M. Ehret.
The design and construction of a master-oscillator, power-amplifier transmitter, using two UX-210 tubes are outlined in detail. The circuit differs somewhat from the usual, but is considered to give very good results and a sharp signal when adjusted properly. Complete circuit diagram and list of parts are presented.

MAGNAVOX
Magnetic
Cone Speaker

M-7 Unit

Distortion-free on power tube volume

This speaker goes far beyond previous magnetic cone reproducers. By reason of the new type pole piece construction, patented by Magnavox, new beauty of tone and new range of equalized volume are possible.

The M-7 passes low frequencies down to about 100 cycles with substantial volume. It also reproduces unusually high frequencies without distortion provided tubes are not being overloaded. It is extremely sensitive and responds easily and with little energy to weak signals and low notes. Takes volume from biggest sets and power tube.

The unit is only 8½" inches in diameter,—it fits into any radio or phonograph cabinet and is simple to install, only 4 screws to turn. Unit list price $15.00.

Warwick Cabinet Model

Has standard M-7 unit mounted on beautiful burl walnut circle on enameled metal base. List $27.50.

Dynamic Power Cone Speaker

Built under electro-dynamic patents made famous by Magnavox. Operates from A battery. Gives full power volume but at a fraction of the cost of other power speakers. You should hear this speaker and realize the great advance in musical reproduction. R-4, 6 volt unit only $50. In mahogany cabinet $75.

R-S, 110 volt D. C. unit only for Electric phonograph and A. C. circuits, $55.

Send for Speaker Bulletins

They give full information on Magnavox magnetic and dynamic type speakers. We will give you name of nearest dealer.

The Magnavox Co.
Oakland   California

Chicago Sales Office   1315 So. Michigan Ave.

2-ampere Tungar

5-ampere Tungar

G-E Trickle Charger

New Low Prices
(East of the Rockies)

2-ampere Tungar . new $14
5-ampere Tungar . new $24
G-E Trickle Charger now $10
Tungar causes no radio interference. It cannot blow out tubes. An overnight charge costs about a dime.
It is a G-E product developed in the Research Laboratories of General Electric.
The 5- or 5-ampere Tungars charge 2-, 4- and 6-volt "A" batteries, 24- to 96-volt "B" batteries in series; and auto batteries, too. No extra attachments needed.

Merchandise Department
General Electric Company
Bridgeport, Connecticut

Battery Chargers
made and guaranteed
by
General Electric
now $10 - $14 - $24

When you buy a General Electric battery charger for your radio batteries, you get a thoroughly tested and proved product.

The Tungar [the name of the General Electric battery charger] was developed by a staff of technical experts in the experimental and testing laboratories of the world's greatest electrical organization.

More than 1,000,000 Tungars are in use today. For Tungar long ago established its reputation for dependable, trouble-free economical service.

Just turn on a Tungar at night ... in the morning your radio storage batteries are pepped up and ready for active duty.

Your dealer can help you. Ask him to show you the popular 2-ampere Tungar that gives both trickle and boost charging rates. It charges "A" and "B" radio batteries and auto batteries, too.

Tungar—a registered trademark—is found only on the genuine. Look for it on the name plate.

GENERAL ELECTRIC

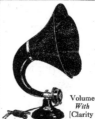

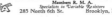

Hoyt MODEL 100 TUBE TESTER

The Hoyt Model 100 Tube Tester is still as popular as ever and in use by many thousands of Radio dealers and jobbers. It will give all the usual tests in determining whether tubes are good or not, and with a minimum of time.

List price, with full instructions, $35.

Send for price list B-11

BURTON-ROGERS CO.
Sole Selling Agents
Boston   Massachusetts

SICKLES
Diamond-Weave Coils

THE IMPROVED
ARISTOCRAT
CIRCUIT

DESCRIBED BY ARTHUR LYNCH

Uses our new coil combination consisting of an improved variable antenna coupler and a three circuit tuner. Both coils are equipped with single hole panel mounting and shaft for panel control.

They are the result of exhaustive laboratory research and years of experience in the manufacture of high grade inductance coils and quality radio apparatus.

LYNCH IMPROVED ARISTOCRAT COIL COMBINATION No. 28, $4.50

There are Sickles Diamond Weave Coils for all Leading Circuits.

The F. W. Sickles Co.
132 Union Street
SPRINGFIELD, MASS.

COIL PRICES

No. 30 Shielded Transformer....	$2.00 each
No. 24 Browning-Drake............	7.50 Set
No. 18A Roberts Circuit............	8.00 "
No. 25 Aristocrat Circuit..........	8.00 "
No. 28 Lynch Aristocrat..........	4.50 "

What did you learn from the
RADIO SHOWS?

YOUR visit to the major radio shows this Fall will convince you of the marked trend towards the use of voltmeter control of radio sets. This is the final safeguard of set performance for both the manufacturer and the owner.

Eventually every set owner will come to the same conclusion if he desires the best performance with continuous radio satisfaction.

Regardless of type or hook-up, all sets are designed for operation at a definite predetermined filament voltage. Only at that voltage will a set give the best results.

Not only is a voltmeter necessary, but anything short of the best injects just that much uncertainty into your radio operation.

Model 528 A. C. Miniature Instruments

Made as double-range Voltmeters, Single-range Ammeters

When you buy a Weston meter your reception problems are solved. Engineers will tell you that Weston instruments have no equal and, in the light of the service they will render, are the least expensive in the long run.

Portable types of testing instruments are always desirable because of their adaptability to all testing conditions. With the new trend towards A. C. operation Weston has developed a companion A. C. line to the Model 489 D. C. instruments. These new instruments known as Model 528 are designed for testing the voltages of A. C. operated sets.

Write for Circular V and learn of their unmatched characteristics before making your final selection of A. C. portable instruments.

WESTON ELECTRICAL INSTRUMENT CORPORATION
179 Weston Ave., Newark, N. J.

Weston
PIONEERS SINCE 1888
INSTRUMENTS

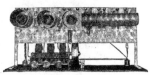

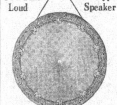

When the ball goes 'round the end for 40 yds.

"--You're there with a Crosley."

The Crosley Radio Corporation:
Can you explain "balancing," so folks can understand it? We technical people know that Hazeltine's neutrodyne principle is a wonderful thing. If you can make it understandable to laymen you're good.
Yours truly,
Van Camp Hdwe. Co., Ind., Ind.

Nature always puts obstacles in our way. When men begin to study a new invention or discovery they find that there are many problems to solve before a successful device can be built. This was the case with the steam engine, the printing press, the automobile, the aeroplane, and every other major invention that you can think of.

The vacuum tube is, perhaps, one of the most remarkable inventions ever made. We found that we could use it to amplify the radio signals. But when we tried to tune these amplifiers, so that they would help us select the desired signal, we found that the vacuum had a tendency to misbehave.

When a tube is used to amplify, the output voltage is much stronger than the input voltage. This is the natural result of the amplification. But there is a path back through the tube through which some of the strong output voltage can get back to the input side of the tube. This voltage is then again amplified and again returns, getting stronger each time, the result being that the tube goes wild. It becomes a miniature broadcasting station on its own hook.

If we can provide a second path from the output circuit to the input circuit, so arranged that the voltage which comes back through this second path is opposed to the voltage through the tube itself we can prevent the trouble. This is called "balancing" because the second path is adjusted so that it exactly balances the path through the tube.

The Hazeltine method of balancing (or neutralizing) this path through the tube has several unique advantages over all the other methods that have been proposed. This is why Crosley radios use the Hazeltine "neutrodyne" method.

THIS new Crosley Bandbox
6 TUBE RECEIVER *de luxe*
is the national radio hit at $55.

The "All American" radio of 1928! With license to participate in the enormous radio resources of The Radio Corporation of America, The General Electric Co., The Westinghouse Co., The American Telephone and Telegraph Co., and The Hazeltine and The Latour Corporations, the Crosley Bandbox of 1928 is an "eleven" of super-efficient features and amazing co-ordinated performance. In it are incorporated:

1—The best idea of balancing.
2—The best ideas of shielding.
3—The best ideas of sharp tuning.
4—The best idea of controlling volume.
5—The best idea of station selection.
6—The best idea of finish and color.
7—The best idea of power tube use.
8—The best idea of console installation.
9—The best idea of power supply connections by enclosing all leads in a cable.
10—The best idea of AC tube operation.
11—The best idea of converting AC current to necessary radio DC.

Operation of the Bandbox receiver from house current is possible with the AC model at $65, which uses the new amazing R.C.A. AC tubes. Power converter costs $60 more.

These new Bandbox receivers are now on display at over 16,000 Authorized Crosley dealers. Their faultless reception of the many wonderful events constantly on the air is proving such a startling demonstration that a national enthusiasm sweeps the country in the natural exclamation—"You're *there* with a Crosley!" If you cannot locate the nearest dealer, write Dept. 20 for his name and literature.

'Approved Con...

$65

$35

Selected by Powel Crosley, J... acoustically and mechanica... installation of the Crosley ... Genuine Musicons built b... dealers secure them from th... through

H. T. ROBERTS C...
1340 S. Michigan A...
Chicago, Ill.

Sales Agents for Appr...
Console Factories

Showers Brothers Con...
The Wolf Mfg. Indus...

IMPROVED MUSICONES

Musicones improve the reception of any radio set. They are perfect affinities in beauty, and reproductive effectiveness for Crosley Radios. A tilt-table model with brown mahogany finish stands 36 inches high, $27.50—16-inch Super-Musicone as pictured above with "Bandbox", $12.75—12-inch Ultra-Musicone, $9.75.

CROSLEY RADIO

THE CROSLEY RADIO CORPORATION
Powel Crosley, Jr., Pres.   Cincinnati, Ohio
Prices slightly higher west of the Rocky Mts.

Crosley is licensed only for Radio Amateur, Experimental and Broadcast Reception.

Every ALL-AMERICAN

TRADE MARK

Audio Unit *Guarantees* Natural Reproduction **!**

Rauland-Lyric Audio Transformer. Type R-500. Price.....$9.00

Rauland-Trio Imped-ance Unit. Type R-300, for Intermediate Stage. Price....$6.00 Type R-310, for Final Stage. Price..........$6.00

Audio Transformer
Type R-14, 3 to 1 Ratio, $4.50
Type R-15, 5 to 1 Ratio, $4.50
Type R-13, 10 to 1 Ratio, $4.50

FINER radio reproduction than you have ever before experienced . . .

All those finer over-tones retained in their true and natural value, without muffle or distortion . . .

Smooth, flawless amplification . . . not a distorted note anywhere over the entire musical range . . . here is just what radio manufacturers and the public have been hoping to see developed!

All this is GUARANTEED by the makers of ALL-AMERICAN Audio Units.

These points of superiority have placed All-American in a position of leadership since 1919 . . . have made ALL-AMERICAN'S the largest selling transformers in the world.

ALL-AMERICAN Audio Units as shown here are the final refinement in radio manufacture. Any dealer, distributor or manufacturer will find it worthwhile to get the latest illustrated descriptive booklets—mailed free on request.

ALL-AMERICAN RADIO CORPORATION

4211 Belmont Avenue
Chicago, Ill.

THE COUNTRY LIFE PRESS, GARDEN CITY, NEW YORK

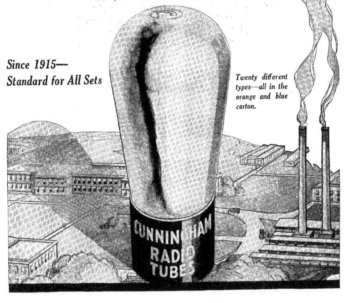

More Money

Be a Trained Radio Expert

If you're earning a penny less than $50 a week, get my free book of info about the radio business. Trained Radio Experts are needed in mo 20 different lines of this new and growing profession (300,000 new o created by the swift growth of Radio in past few years). Why go alon; or $35 or $45 a week all your life? Study Radio and after only a sh land yourself a REAL job with a REAL future! Be a man who has m his pocket and in the bank—don't scrimp and scrape for the rest of yo

Salaries of $50 Up to $250 a Wee

Not Unusual in Radio

The good positions in Radio pay all the way from $50 on up to $15 and even $250 a week. Suppose you don't climb to the very top, but t do advance to a position that pays you $125 a week, year in and y Any chance to make that much where you are now? Then send for book, and learn about a field where there's some *real opportunity*. good men, *if they have the right training*, can work their way into re salaried jobs!

Money Back If You Aren't Fully Satis

I'll give you all the training you need to get into any line of the Radio business. And I back up this training by a signed agreement to refund every penny you pay me if I don't give you exactly the training you need. After you finish my training, you'll be the judge. If you think I've earned my tuition fee, I keep it. If not, ask for it and you'll get it right back.

GET BOO
Find out for about bigger pay for you in Radi $2,000,000 in $304,000,000 in that's the record Radio industry. big Radio jobs a for the man who k John Fetzer sent free book—now h Engineer at WE designs and buil cast stations. T cox sent for the now he's in his o business and repe its as high as $7 day! All informa be sent to

Six Big Practical Outfits Given You to Help You Learn

I teach you both the "why" and the "how" of Radio. You learn to DO a thing, and you learn WHY it's done. I send you, WITH-OUT EXTRA COST with your course, six big practical outfits of material to experiment and work with. These outfits are the real thing—not toys. The parts they contain will build approximately 100 different Radio circuits. With all this material you do practical work from start to finish of your training. You get your hand in, and you get confidence in yourself. Then when you run into a Radio problem later on, on the job, you KNOW you can do it because you've Already done it. With these six outfits of practical material. With me you don't learn to be a "paper Radio Expert"—you learn to be the kind of expert that shows his worth on the payroll. Full details in my big book—sent free.

Send for Free Book of Information

Broadcasting and commercial Radio land station work appeal to a lot of men—it's a big, growing field and fascinating work. My course prepares you thoroughly to get into this field and make good.

Six million Radio receiving sets in use in the United States means millions of dollars are going to Radio Service and Repair men everywhere. For this work you must be trained.

see coupo nex pag

Radio manufacturing has grown faster in the last 6 years than any other big business ever did. It's 500 times as big as it was. That means lots of good chances for the trained Radio Expert.

Radio operators on board ship go every-where—see everything. You sail the world over, all your expenses paid, and draw a good salary besides. It's the life of Reilly.

RADIO BROADCAST. December, 1927. Published monthly. Vol. XII, No. 2. Published at Garden City, N. Y. Subscription price $4.00 a year. Entered at the Garden City, N. Y., as second class mail matter. Doubleday, Page & Company, Garden City, N. Y.

LEARN QUICKLY, EASILY

Train at Home in Spare Minutes

Stay home! Hold your job! I'll bring your Radio training to you, and you can learn in your spare time after work. No need to go to a strange city and live for months on expense when you learn my way. You study in the quiet of your own home. As for this training—it's written just as I would talk—in straightforward, everyday, understandable language. You'll get it, and you'll get it quickly—in a few months' spare time—because I've made it so clear and interesting! No particular education needed—lots of my successful students didn't finish the grades.

Earn $15, $20, $30 Weekly Right Away "On the Side"

Deloss Brown, South St., Foxboro, Mass., made $1,000 from spare-time Radio jobs before he even finished my course. H. W. Colbentz, Washington, averaged $45 a week; Leo Auchampaugh, 6432 Lakewood Ave., Chicago, made $500 before graduation; Frank Toomey, Jr., Piermont, N. Y., made $833 while taking the course. All this done IN SPARE TIME away from the regular job, while these fellows were still studying the course—and they're only a few of hundreds. As soon as you start this training I begin teaching you practical Radio work. Then a few weeks later, I show you how to make use of it in spare time, so you can be making $15, $20, $30 a week "on the side," all the while you're learning.

64-Page Book Sent Free for the Asking

My big book of Radio information won't cost you a penny, and you won't be under any obligation by asking for it. It's put hundreds of fellows on the way to big pay and brighter futures. Sending for it has been the turning-point where many a man has made his start toward real Success. Get it. See what it's going to mean to you. Send coupon TO-DAY!

Address: J. E. SMITH, President
National Radio Institute
Dept. P-94, Washington, D. C.

JOBS WAITING!

In 7 short years — 300,000 new jobs in Radio! Lots of jobs open right now, for those who have the training. The Radio industry has grown by leaps and bounds—so fast it has had to take whatever sort of men it could get. Such men, if they haven't trained themselves in the meantime, are losing out and will keep on losing out. They'll be replaced by men with the KNOW-HOW. But it's trained men ONLY that are needed.

Over 1,000 Openings for Trained Men NOW!

One great Radio manufacturing concern alone has over 1,000 openings to give my graduates this year. These men will be needed all over the United States. Any graduate of mine who stands well in his home town is eligible for this work. The head of the above mentioned concern—one of the biggest Radio organizations in the country—is a graduate of mine. He knows what my training did for him. When he wants new men for his organization he wants men with the same training.

I can't possibly graduate enough men this year to fill these openings. So there will be more openings with this one concern than there will be graduates to accept them.

But there are other openings to choose from, too. My school has trained more Radio Experts than any other school in the world. It's the oldest and largest Radio home-study school in the world. There are N. R. I. trained men in almost every Radio concern of any importance in this country. Many Radio employers are themselves my graduates. That's where you get your "stand-in" as an N. R. I. graduate yourself.

Every graduate of my course is entitled to Life-Time Employment Service, without a penny's charge, from my helpful Employment Department.

Full Information Sent with Free Book

My Free Book contains full information about the Radio employment situation, and the advantages I'm in a position to give you. Also about my Life-Time Employment Service, and Life-Time Consultation Service, too.

Mail Coupon To-day!

RICH REWARDS *in* RADIO

This big 64-page book, printed in two colors, crammed with interesting facts and photos about money-making opportunities in Radio, sent free to everyone who clips the coupon. No obligation by sending for the book—it's absolutely free. One of the most valuable books about Radio ever written.

6 big out fits given

Your "B" Battery Eliminator
will give you better service with

Q · R · S
(Trade Mark Registered)

Gaseous
Rectifier Tubes
ARE BETTER

60 Milliamperes - **$4.50**
85 Milliamperes - 4.50
400 Milliamperes - 7.00

Ask for Catalog of full line of Standard Tubes.

Guaranteed

The standing of the Q·R·S Company, manufacturers of quality merchandise for over a quarter of a century, establishes your safety.

Orders placed by the leading Eliminator Manufacturers for this season's delivery, approximating Four Million Dollars' worth of Q·R·S Rectifier Tubes, establishes the approval of Radio Engineers.   Ask any good dealer.

THE Q·R·S COMPANY
Manufacturers
Executive Offices: 306 S. Wabash Ave., Chicago
Factories: Chicago—New York—San Francisco—Toronto, Canada—Sydney, Australia—Utrecht, Holland
Established 1900. References—Dun, Bradstreet, or any bank anywhere

How do you know that your set is operating at its best?

If you are one of those who believe that the ear is a trust-worthy organ, you will find that it has led you far astray when you adopt the scientific method of toning up your set.

Spend a few minutes with your radio—*and a Weston Volt-meter*—before broadcasting time arrives.

Check up the voltages of your set and note the difference in the power and quality of reception, simply by making the quick little adjustments that your Weston tells you are required.

With this Model 489
Portable Weston Voltmeter

correct radio operation is made simple and also *less costly*. You merely test the tubes and batteries once a week—like winding a clock. You do not make a habit of guessing the time—then why guess at the faithfulness of radio set reproduction?

Just ask your dealer for Model 489, having 1000 ohms per volt. Suitable for B-Eliminator or battery operated sets. Other models of lower resistance for battery testing. The complete line of Weston Radio Instruments is fully de-scribed in Circular J mailed upon request by writing to—

Weston Electrical Instrument Corporation
179 Weston Avenue   Newark, New Jersey

WESTON
RADIO
INSTRUMENTS

They
stand the surges

PARVOLT
WOUND
CONDENSERS

ARE specified most frequently in circuits where service is hard and only a *good* condenser can stand up.

Use the proper Parvolt Wound Condensers in the filter circuits of current supply units and *know* that all condenser problems are completely elim-inated.

Parvolt Wound Condensers are made for 3 maximum serv-ice voltages:

Type A—400 Volts d.c. continuous duty
Type B—800 Volts d.c. continuous duty
Type C—1000 Volts d.c. continuous duty

THEY STAND THE SURGES

Made and guaranteed by

ACME WIRE COMPANY
New Haven, Connecticut

No "A"
Batteries with
these TELEVOCALS

The new Televocal A. C. 226 and 227 Tubes operate direct from A. C. current and require no "A" Battery. Also Televocal T. C. 112-A and 171-A Power Tubes are made with oxide-coated fila-ments taking but half the current—¼ amperes as against ½ amperes here-tofore.

For best results insist on Televocal Quality Tubes for all purposes. All standard types featured by the exclu-sive Televocal Support-all fully guar-anteed.

TELEVOCAL CORP'N
Televocal Building
Dept. S-3, 588 12th St., West New York, N. J.

TELEVOCAL
QUALITY
TUBES

From one radio fan to another

Pins screw in this way for jacks running horizontally.

Pins screw in this way for jacks running vertically.

For battery testing.

How to say "Merry Christmas" is no problem — if the person you wish to remember is "radio-active." Give him what you'd like yourself — a voltmeter.

If his is one of the newer sets, get one of the new Westinghouse Cabinet Portables. Plugs right into the jacks provided — looks and performs like a permanent cabinet fitting, but is easily removable for testing "A", "B" and "C" batteries or for other purposes.

Styles in voltmeters are changed by this new Cabinet Portable. The rich, antique bronze finish — the clever adjustable features — make this a fitting companion piece for the most expensive radio receiver.

An *accurate* voltmeter warns continuously of voltage variations that affect good reception. Its use prevents prematurely burned-out tubes and prolongs battery life.

PRICE COMPLETE	
Range	Price
(0-5 and 150) or (0-5 and 50)	$6.50
West of Rockies to Pacific Coast	7.00
As your dealer's	

Good dealers have other Westinghouse models. For instance, the most accurate moving-vane instrument made, at $5.00. Or a highly sensitive jeweled movement at $10.00. Either, a gift!

WESTINGHOUSE ELECTRIC & MANUFACTURING COMPANY
Offices in all Principal Cities   *Representatives Everywhere*
Tune in with KDKA—KYW—WBZ—KFKX

Westinghouse

NOV 29'27   ©CI B761185

RADIO BROADCAST

WILLIS KINGSLEY WING, Editor

DECEMBER, 1927

KEITH HENNEY
Director of the Laboratory

EDGAR H. FELIX
Contributing Editor

Vol. XII, No. 2

CONTENTS

AMONG OTHER THINGS. . .

PROBABLY the most interesting article in this issue from the point of view of the experimenter is the constructional data and operating and assembly instructions on the Cooley "Rayfoto" radio picture receiver. By the time this magazine is in the hands of its readers, all the essential apparatus will be available on the market and nothing will delay the experimenter in his experience in this new field. RADIO BROADCAST is glad to forward the names of readers who are interested in receiving printed matter and late bulletins to manufacturers who are supplying the various parts for the "Rayfoto" apparatus. After the appearance of Mr. Cooley's November article, a great number of our readers wrote us for this information which has been supplied. A letter should at once be addressed to the undersigned, asking for additional data in case you have not already written.

WASHINGTON is the center of interest these days, what with the International Radio Conference and the changes in the Federal Radio Commission. The death of Commissioner Dillon is a great loss to radio in the United States and it will be next to impossible to fill his place. The resignation of Commissioner Bellows removes one of the ablest members of the Commission, but President Coolidge has filled his place through the appointment of Sam Pickard, former secretary to the radio body. Mr. Pickard is a likeable and able individual and we believe his appointment is a wise one. Carl H. Butman, of Washington, was appointed as Secretary to succeed Mr. Pickard. Mr. Butman has long served RADIO BROADCAST as its Washington news representative and we are indeed pleased that the Commission has so wisely chosen a man who knows radio problems so well.

A WORD about the authors in this issue: William J. Brittain is an English writer on radio and scientific topics who has just returned from a European trip to see what is being done in television. Theodore H. Nakken is a research engineer for the Federal Telegraph Company. He is a pioneer in photo-electric cell work and is unusually familiar with radio progress abroad. Austin Cooley, whose "Rayfoto" picture apparatus has attracted national attention, is a native of the state of Washington, received his technical training at M.I.T., and except for his trip in 1926 with the MacMillan Arctic expedition, has been in New York for the past four years. John F. Rider is a well-known New York technical writer who is at work on an interesting series of "fact" articles about manufactured receivers.

THE next issue will contain another story about the Cooley "Rayfoto" radio picture system and its operation, as well as interesting data about push-pull power amplification. Another of Mr. Rider's articles about manufactured receivers will be featured as well as a wealth of constructional matter.

—WILLIS KINGSLEY WING.

Doubleday, Page & Co. MAGAZINES	Doubleday, Page & Co. BOOK SHOPS	Doubleday, Page & Co. OFFICES	Doubleday, Page & Co. OFFICERS
COUNTRY LIFE	*(Books of all Publishers)*	GARDEN CITY, N. Y	F. N. DOUBLEDAY, *President*
WORLD'S WORK	LORD & TAYLOR BOOK SHOP	NEW YORK: 285 Madison Avenue	NELSON DOUBLEDAY, *Vice-President*
GARDEN & HOME BUILDER	PENNSYLVANIA TERMINAL (2 Shops)	BOSTON: PARK SQUARE BUILDING	S. A. EVERITT, *Vice-President*
RADIO BROADCAST	NEW YORK: GRAND CENTRAL TERMINAL	CHICAGO: PEOPLES GAS BUILDING	RUSSELL DOUBLEDAY, *Secretary*
SHORT STORIES	38 WALL ST. and 526 LEXINGTON AVE.	SANTA BARBARA, CAL.	JOHN J. HESSIAN, *Treasurer*
EDUCATIONAL REVIEW	848 MADISON AVE. and 166 WEST 32ND ST.	LONDON: WM. HEINEMANN LTD.	LILLIAN A. COMSTOCK, *Asst. Secretary*
LE PETIT JOURNAL	ST. LOUIS: 223 N. 8TH ST. and 4914 MARYLAND AVE.	TORONTO: OXFORD UNIVERSITY PRESS	L. J. McNAUGHTON, *Asst. Treasurer*
EL ECO	KANSAS CITY: 920 GRAND AVE. and 206 W. 47TH ST.		
FRONTIER STORIES	CLEVELAND: HIGBEE CO.		
WEST WEEKLY	SPRINGFIELD, MASS.: MEEKINS, PACKARD & WHEAT		
THE AMERICAN SKETCH			

DOUBLEDAY, PAGE & COMPANY, Garden City, New York

Acme E4-B Supply, $35
Acme B Power Supply units were
the first on the market to use a
Raytheon tube as rectifier. The
E-4, above, is the latest B Power
Supply achievement of the Acme
engineers. Price includes
Raytheon BH Tube.

This Year's Programs
Deserve *Such a Speaker!*

PROGRAMS now available to you have become so great that they deserve the finest speaker radio science has been able to construct after years of experience—the Acme K-1A ($25).

With cones on both sides, each 13 inches in diameter, and with two motors instead of one to "feed" sound to these cones—you enjoy the advantage of two perfect speakers working as one.

Look for This Symbol

And ··· what an addition, this Acme K-1A, to home furnishings! Its graceful design blends with furniture background as no other can.

For resonant volume beyond belief... have your dealer show you the Acme PA-1 Power Amplifier ($12.50). It uses socket power whether your set is electrified or not. Makes a power speaker of your present speaker without additional drain on storage batteries.

Acme's two booklets tell (1) how to improve any radio set; (2) the story of lamp socket operation. Fill in the coupon below for the one you want.

ACME
—for amplification

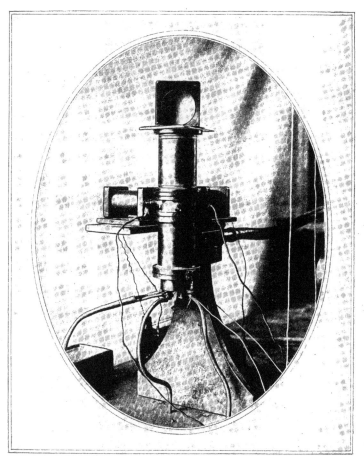

TELEVISION APPARATUS OF A EUROPEAN SCIENTIST

THIS equipment constitutes the television receiver developed by M. Holweck, who is collaborating with Edouard Bélin, in the design of television equipment. M. Holweck is specializing in the receiving side of the installation. The received picture appears on the small circular screen at the top of the receiver shown in this illustration. Numerous other European scientists are devoting their time to the development of tele-vision schemes, and many and promising are the reports emanating from the various laboratories. M. Bélin is, of course, a Parisian and has done most of his work in France. The short story which begins on the suc-ceeding page is from the pen of one who has visited many of the pioneers in the television field in Europe, and the information has, therefore, been obtained at first hand.

VON MIHÁLY'S TRANSMITTING APPARATUS

TELEVISION IN EUROPE
By WILLIAM J. BRITTAIN

WHAT is Europe doing towards the furtherance of television? America already knows quite a lot about the work of Baird, and the public has made his name known in most countries. But aside from this, little is known of the progress of the many experimenters in this fascinating art on the other side of the Atlantic.

Recently the author went from England to find out what the Continental men are doing, what their apparatus is like, and whether they are preparing a surprise for the world, and in Berlin was found the man preparing the surprise. He is Dénes von Mihály, a young Hungarian, and chief consulting engineer to A. E. G. (the General Electric Company of Germany). An engineer brought from America for the purpose is making a simplified version of Von Mihály's apparatus to be shown in Berlin and London as a preliminary in forming television companies there.

The vital feature of Von Mihály's method is an oscillograph which consists of a tiny mirror mounted on twin wires. The mirror vibrates between two electro-magnets at speeds which sometimes reach thousands of times a second. Light reflected from the object—a face, a scene, or whatever it may be—is focussed by a specially constructed set of Zeiss lenses upon the mirror. The mirror, vibrating rapidly, sees each point of the object in turn, in the manner necessary for television, and flashes it to a photo-electric cell.

Von Mihály has made his own cell, and it sends out currents corresponding exactly to the intensity of light or depth of shadow of each tiny point as it is reflected upon it.

In his receiving apparatus Von Mihály again uses vibrating mirrors. An electric lamp, shining brightly or becoming dim as the current from the transmitter is strong or weak, is concentrated by lenses upon mirrors which repeat the action of the mirror at the sender and zig-zag a beam of light over a ground glass screen. The varying light beam, covering the screen eight times a second, makes up the picture.

To ensure that the sending and receiving mechanisms are working exactly in time—so that the mirror at the receiving end is shining light upon the centre of the screen at the same fraction of a second as the mirror at the transmitting end is "seeing" a bright part in the center of the object—Von Mihály uses a tuning fork arrangement on the same principle as those that have been used by experimenters in photo-telegraphy. A tuning fork in the receiver, kept vibrating by an electro-magnet, acts as a switch, regulating current to other magnets which allow a wheel to progress one cog for every impulse, and so regulate the vibrations of the mirror. The apparatus at each end now fills a table, but Von Mihály says he can simplify it to work as a home set in conjunction with a one-tube radio receiver.

Behind this assurance is a secret. The secret is in a small black cylinder, five inches by two and a half inches. The inventor calls it his "little black wonder." He will not tell the world what is inside, but told the author that with it it is possible to do away with the great amplifiers necessary in other systems.

"Television sets for the home," he said, "will be simple and yet give a boxing match or a horse race. They will be sold in a few months for the equivalent of a hundred dollars."

Von Mihály has been working for thirteen years on television. He first became interested when he was twenty, after hearing a lecture on photo-telegraphy by Professor Arthurn Korn. He carried on his work during the war, and on July 7, 1919, gave his first crude demonstration of television. Ministers in the laboratory of the Telephon Fabrik in Budapest then saw on a screen the images of the letters M. D. and REX transmitted from the young engineer's home laboratory in another part of the city.

It was the writer's privilege to be present at a recent demonstration of Mihály's apparatus. The results obtained were considerably better than those of the early demonstrations referred to above, and the images were clearer than those seen by the author on Baird's screen. On the picture of a "televised" boy it was possible to see the collar, the wavy outline of the hair, the shape of the ear, the forehead, the eye, the nose, and the mouth, the latter merging into shadow on the left side of the face.

OTHER EXPERIMENTERS

PROFESSOR Max Dieckmann, whom I met in his station near Munich, Germany, has up to the present no result like this to show. He has achieved results, but has scrapped the transmitter and other apparatus that gave them.

"I used mirrors," he told me, "but I came to the conclusion that no mechanism could ever be made light enough and accurate enough for television. I am therefore trying to make use of electrons. By two electro-magnets, alternated by currents of different frequencies, I make the stream of electrons—or the cathode ray—zig-zag over the object, and I am now experimenting with devices to register the result of this 'exploring.'

"With electrons I think I have the real instrument for television. Electrons are almost weightless and can travel at any speed we need. All mechanism has a weight and inertia that in my opinion will always drag down efforts at perfect television. By perfect television I mean, of course, the reception of images as fine as published photographs. It is possible now to have crude television. You can have a picture on as large a screen as you like, but the larger the screen is the larger must be the patches making up the picture. Distance of transmission, too, offers little difficulty. We must concentrate on producing a finer image, and I believe electrons will enable us to do it."

Professor Dieckmann is retaining his

PROFESSOR MAX DIECKMANN

former receiver which already uses electrons. The receiver is like a bottle. The received currents vary the flow of electrons from a tube fixed to the neck of the "bottle." By magnets similar to those in his new transmitter the varying flow of electrons is made to zig-zag over a screen at the bottom of the "bottle" which glows as the electrons touch it. When a strong current, showing that a light part of the object is being encountered at the transmitter, sets off a heavy flow of electrons, the screen glows strongly at that part, and the glowing patches make up the picture.

With Mr. Rudolf Hell, his chief assistant, Professor Dieckmann is working with enthusiasm at his latest apparatus.

Mirrors form an essential part of the apparatus of M. Edouard Bélin, the scientist famed for his systems for phototelegraphy, who has large stations at La Malmaison, near Paris. M. Bélin inspired cartoons with a television machine thirty years ago. His latest apparatus looks businesslike.

Two rectangular mirrors, about half an inch long, set at right angles, are made to oscillate by cranks and connecting rods driven by an electric motor. A beam of light shines on the mirrors and is reflected zig-zag in the usual way. For his object M. Bélin uses his hand. Light from the hand as the beam passes over it is caught by an eighteen-inch concave mirror at the bottom of a drum which concentrates the light on a photo-electric cell held by an arm half-way down the drum. With this apparatus, M. Bélin told me, he can record fifty thousand flashes of light and shade a second.

M. Holweck, collaborator with M. Bélin, is responsible for the receiver. He has designed a special form of cathode ray oscillograph in which as complete a vacuum as possible is kept by an air pump, also of his own design. M. Holweck is working to perfect the fluorescent screen so that it will vary its glow exactly according to the strength of the stream of electrons. He has also made the apparatus more sensitive so that a difference of potential of five volts between the grid and the filament will absolutely cease the flow of electrons. This means that slight differences of light and shade in the object, and therefore, tiny differences in the current received, are recorded on the screen.

Promising results are being obtained. At present M. Bélin is transmitting only the silhouettes of his hand. On the screen the outline of the hand can be seen clearly. The hand can be seen to move, and the fingers to bend. A silhouette of the profile of a face, and a simple photographic negative, have been transmitted with equal success.

"Our work is progressing gradually" said M. Bélin. "We have found it better to pass over the object a bright spot of light rather than illuminate fiercely the whole object. It we used flood lighting to obtain the same brilliancy as our spot light gives us over a person's face the intensity would be insupportable.

"Earlier in the year we were sending over our object in a thousand points; now we have reached two thousand five hundred. We cover the object eight times a second which means that in our ordinary experiments twenty thousand signals are flashed a second. We are greatly encouraged by our present results. In a few months we should have something to offer the world."

This is the stage European inventors have reached. Each one of them is watching carefully every step forward by other workers and trying to go a step further. Von Mihály is confident that all his system needs now is to be put on the market. Dieckmann and his young assistants are working quietly but eagerly. And all the time I was at the *établissement* Edouard Bélin I was filled with the boyish enthusiasm which permeates the atmosphere there.

Of hopes and plans it would be possible to write pages but in this article an attempt has been made to keep plainly to facts to let America know just what Europe is doing in television.

VON MIHÁLY'S TELEVISION RECEIVING APPARATUS

How the Radio Commission Can Set Radio to Rights

A S THE peak of the radio season approaches, we look upon the situation with considerable satisfaction. Last year, broadcasting was in chaos and the Radio Act had not been passed; this year, progress has been made in the direction of restoring order. Public interest in radio is at a maximum; the Radio Show at New York broke all records for attendance at an industrial exposition. Manufacturers and dealers report brisk sales. Broadcasting now has the stimulus of two competing chains. Everywhere there is activity and progress.

The only sore spot in the radio situation is in the regulation of broadcasting. The Commission went about its task with diligence as soon as it was formed. It cleared the Canadian channels and put the stations back on even ten-kc. channels. Then it spaced the New York and Chicago broadcasters at fifty-kc. intervals, forcing time sharing in some cases to make it possible. After these commendable steps had been taken, the Commission confined its activities to juggling a channel here and switching a station there.

We understood that the assignments of June 15 were merely an experiment, a stopgap measure effective only until a comprehensive plan of allocation could be worked out, which would mean an end to the heterodyne whistle. The persistently optimistic announcements of the Commission that the broadcasting situation is now remedied give the impression that the Commission considers its major task completed.

At the opening of the Radio World's Fair, Admiral Bullard pleaded for more time to give the Commission an opportunity to do its work; at the Radio Industries Banquet, he made numerous proposals to the radio industry, many of them no less than amazing, but nowhere have we had a simple, direct statement of the future plans of the Commission. Does the Commission consider its task virtually completed or will it devote itself to a radical improvement of broadcasting conditions?

The Admiral's speech at the banquet contained some striking indications. Briefly, he stated that broadcasters should find a way to fix the responsibility for statements made in radio advertising; that direct advertising stations should be taxed; that radio ought to be a public utility regulated by public service commissions; that provision should be made to link up broadcasting for national sos. calls, perhaps for such occasions as the loss of the President's racoon; that motors for electric elevators should be re-designed; and most ingenious and amazing, that receiving sets should be equipped with crystals to permit of greater selectivity.

A few words at the very end of this astounding speech were devoted to the Commission's plans. With regard to the high power stations serving the long distance listener, "the Commission is looking forward to a time when the listener, on any night of good reception, can hear broadcasting stations from the Atlantic to the Pacific, from Canada to Mexico, without interference, on channels cleared for them, not by arbitrary rulings of the government, not by fixed and necessarily discriminating classifications, but by the normal, logical process of demonstrated fitness and capacity to render a great public service. Such a development is entirely practicable on the basis of allocations now in force. It requires no sweeping changes, but only a clear picture of the ideal to be attained, and a steady careful improvement of existing conditions. . . ."

Thus the ingenious Commission will by "orderly and natural, rather than by autocratic and arbitrary methods" bring us

A RADIO TOUR OF THE CONTINENT
Capt. L. F. Plugge, an English radio enthusiast, spent the months of July and August on a tour which the accompanying map shows. There were two radio-equipped cars, one of which is illustrated. Each had a loop-operated super-heterodyne and a short-wave transmitter operating on 6660 kc. (45 meters). Intercommunication was attempted and reception conditions along the route noted

these ideal listening conditions. No one, unless it be the broadcast listener, will be imposed upon; only stations which elect by natural processes to eliminate themselves will be taken off the broadcasting lists.

The listener unless he lives within the shadow of a broadcasting station, that is, in that short distance which engineers like to call the service range, must put up with disagreeable heterodyne whistles. Only if we use "arbitrary" methods, which means actually applying the regulatory powers with which the Commission is endowed, can we hope for fewer stations. The natural tendency is toward increasing the number of stations and the power they use. The Commission leans upon a broken reed, if it expects "normal, logical processes" to eliminate stations. Rubber spine methods cannot help the broadcasting situation. There is only one solution, which we repeat, like Cato and his "Carthage must be destroyed," and that is the elimination of at least four hundred broadcasting stations.

What Can the Commission Do?

S ECTION IV of the Radio Act authorizes the Commission to classify broadcasting stations, to prescribe the nature of service rendered by each, to assign bands and powers, and to determine the location of stations. There is no limitation on how far it may go in its work of classification.

Why does not the Commission use these powers? Why does it not classify broadcasting stations as (1) national, (2) regional, (3) local; divide the country into geographical areas and prescribe exactly how many

stations of each class shall be licensed in each of those areas?

Public convenience and necessity clearly establish the point that interference among stations should not be tolerated and certainly the Commission should be competent, if it earns its keep, to determine how many stations of various powers will be accommodated in the present broadcasting band. In fact, all of these points have been analyzed for it by qualified experts in precise and unequivocal terms.

NATIONAL STATIONS, to which exclusive channels should be assigned, might be defined as follows: (1) *Power*, 10,000 watts or over; (2) *Service*, fifty hours a week or more; (3) *Location*, at least ten miles from all centers of 100,000 or more population and at a point more than fifty miles from the nearest national station and not within 200 miles radius of more than five national stations.

REGIONAL STATIONS, sharing channels with other regional stations more than 1000 or 2000 miles distant: (1) *Power*, 2000 to 5000 watts; (2) *Service*, at least twenty-five hours a week, and (3) *Location*, not more than 100,000 population within a five mile radius, nor more than five regional stations within 100 miles.

LOCAL STATIONS: (1) *Power*, between 250 and 500 watts; (2) *Service*, at least twenty-five hours; and (3) *Location*, such that there are not more than five local service stations within a hundred mile radius.

Such a program would, of course, require the elimination of stations in a few of the congested areas, a blessing to the radio audience. The stations so eliminated need not go out of business, but merely consolidate with others serving the same area. Stations of less than 250-watt power should be ruled off the air at once, not because they themselves contribute seriously to congestion but because their channels might better be assigned to national or regional stations.

Concrete suggestions, which are not only logical, but also require the exercise of some of the "arbitrary" powers conferred upon the Commission by law, may be in order. We respectfully suggest the promulgation and actual observance of regulations for the accomplishment of four objectives, the constitutionality of which cannot be questioned:

1. ALL STATIONS should be required to adhere to their frequencies and those failing to do so, after occupying their assigned channels for more than thirty days, should be fined $500 for each violation noted, *without any further consideration of their cases*. The Commission has been buncoed by whining station managements into the belief that staying on a channel requires extraordinary equipment and engineering genius. A station failing to adhere to its channel is not technically competent and not worthy of a franchise on the air. Furthermore, after its fourth offense, a station's license should be cancelled, without further consideration of the case. The ether space thus regained should not be assigned to a new ether nuisance, but

utilized in relieving congestion where it exists. The Commission's leniency with regard to channel wobbling, to which it attributes practically all heterodyning, is a remarkable example of unwarranted bashfulness and consideration the stations don't deserve. The five hundred dollar fine for each violation of the Commission's regulations gives the Radio Act plenty of teeth but, to our knowledge, the Commission has never tried them out.

2. THE COMBINATION and consolidation of broadcasting in congested areas should be encouraged by guaranteeing to the consolidators the combined broadcasting privileges of the stations so consolidated. For example, four, full time, 500-watt stations, combined into one, should be permitted a power increase to 2000 watts, or two, half-time stations, forming one, should receive full time. Furthermore, all local and regional stations, not sharing the same channel, which combine, should be guaranteed privileged consideration on the basis of program merit, should they seek to secure full time on a single channel by challenging another station.

3. POWER INCREASES to local and regional stations shall not be granted where congestion exists, unless other stations, having a power equivalent to the increase, be absorbed. Thus, for example, for a thousand-watt station in New York to jump to 1500 watts, it should be necessary for it to absorb a 500-watt station.

National stations, on the other hand, serving large areas, should be encouraged to increase power, because they require clear channels and failure to employ the maximum power means that they are not making full use of the channel assigned to them.

4. THE COMMISSION, empowered to assign hours of broadcasting to stations, should conserve ether space by limiting licenses only to hours actually used by the stations concerned. It has left problems of time division to the stations themselves, instead of utilizing its power to help in relieving congestions. There are many broadcasting stations which are assigned fifty per cent. of the time on a channel which use only ten per cent. of it, while the other station on the channel is required to remain silent, although it has program material to fill the unused time. In congested areas, the assignment of the time should be based upon the average hours which a station broadcast over the same period in the preceding year. Increase over this time should be granted only upon the basis of program merit and service, or the unused time held to encourage consolidations and to accommodate other stations.

The present assignment of forty channels to New York and Chicago, nearly half the ether space in the eastern part of the United States, is an imposition upon the listener. Yet new stations are being licensed in New York and Chicago, although six stations in each of those cities have corralled ninety per cent. of the audience. This concentration of broadcasting facilities in two centers

A COMPLETE RADIO INSTALLATION ON AN AIRPLANE

Although the *Ville de Paris*, the Sikorsky airplane built for Captain Fonck, the noted French flier, never started toward Paris, plans for the flight were exceptionally complete. Top right shows the small transmitter and a larger set below it. In the center is a regenerative receiver and at the extreme bottom, the antenna reel. The motor generator unit is at the extreme left and supplies plate current for the transmitter. The generator and propeller can be swung out through the fuselage when in use

of population forces the rural listener to contend with heterodyning all over the dials and precludes power increases in rural areas where better and bigger stations are actually needed.

The Federal Radio Commission has worked long and hard with its problem. It has done the best possible job without seriously disturbing or curtailing the privileges of the broadcasting station owners. But, so long as it fails to regard its duty as serving the interests of the listening public, and fails to use the ample powers conferred upon it by the Radio Act to reduce the number of stations on the air, ether congestion will remain the unhealthy disease of the broadcasting situation.

$100,000 to Improve Broadcasting

THE National Association of Broadcasters appropriated the sum of $100,000 to make a scientific study of broadcasting. It plans to employ field engineers and program specialists to visit individual stations throughout the country. The procedure of the Association in the effective utilization of this fund has not yet been established. If it is sensibly administered, very valuable contributions can be made in the technical, economic and program problems of the broadcaster. From the technical standpoint, studio methods, as they affect transmission quality, and the correct operation of the broadcasting stations to help in eliminating ether congestion are fruitful subjects for research. The Association might well help in determining just what the capacity of the broadcasting band is with regard to power, service range and geographical location of stations.

In the field of program technique, critical study of the outstanding features and systematic examination of voice and musical instruments which make good broadcasting could be very helpful. An investigation of the possibilities of building high grade programs by the use of recording methods, as suggested by Edgar H. Felix in a speech before the Association, might also be studied with a view to investigating its practicability. Mr. Felix suggested the recording of "scenes," blending the voices of speakers and pick-up music through mixing panels and the "editing" of programs much as films are cut and assembled, until the ideal feature is assembled. When thus worked over and perfected, it may be presented as often and through as many stations as its popularity warrants, without further cost for talent. This suggestion may result not only in better planned and coördinated programs, but it may help to reduce the mounting wire costs which commercial broadcasters now meet.

In the field of commercial broadcasting, a close study of the methods used to associate the commercial program with the product of its sponsor and to secure the most effective results in a manner pleasing to the listener might help to increase the effectiveness of commercial broadcasting, an end

© Henry Miller
THE LATE COL. JOHN F. DILLON

Colonel Dillon, member of the Federal Radio Commission from the Pacific Coast, died early in October. His loss will be keenly felt by the Commission and the radio world at large. A practical radio man of wide experience, Colonel Dillon had served in various technical capacities in the Signal Corps, and as radio inspector in charge of the Eighth District when headquarters were in Cleveland in 1913 and 1914. He was later transferred to San Francisco as Radio Supervisor for the Sixth District and it was from this duty that his appointment as a Radio Commissioner called him. His wide practical experience with government, amateur, and commercial radio made Colonel Dillon one of the most valuable members of the Radio Commission

which is necessary to aid economic stabilization of broadcasting stations.

The National Association of Broadcasters is to be commended for its foresight in making this substantial expenditure, which is likely to be returned many fold through better broadcasting and larger audiences.

What to Tell the Consumer—And Where

AFFLICTED with the expanded craniums resulting from mushroom growth, the larger manufacturers of the radio industry are often flattered into advertising excesses which ultimately cause financial embarrassment. As typical of this trend, we received a dealer notice, not long ago, describing a new type of A, B, and C power device which was to make its debut to the world principally through three publications having a combined circulation of over three million copies. Although a prophet is not often recognized in his own country, so frequently has the folly of plunging into expensive national mediums been demonstrated to the radio industry, that most manufacturers first make an effort to sell the merits of their products among the more influential radio listeners.

The general public has been too frequently fooled by innovations to become immediate buyers through the medium of an advertising flash in national weeklies. They are inclined to consult the most expert enthusiast whom they can reach before they are willing to risk their money on a

device which may fail. The more successful manufacturers establish their products among the more influential groups of radio buyers before they plunge recklessly into national campaigns in behalf of products which do not have behind them the weight of acknowledged approval of the better informed radio enthusiast. The influence of the radio enthusiast, like halitosis, is often the insidious element which prevents the success of the national advertising campaign which is not supported by the goodwill of well informed broadcast listeners and constructors.

WHAT BROADCASTERS WANT

A LIST of hearings scheduled by the Federal Radio Commission early in October indicates the evils of requiring hearings upon all applications, regardless of their merit. For example, WBAW, Nashville, Tenn., a 100-watt station, operated by a drug concern, seeks to increase its power to 10,000 watts, making it necessary for nineteen stations to defend themselves against this unwarranted incursion of their service range by the drug store carrier. There is no channel available for any new 10,000-watt stations anywhere.

Another hearing is demanded by WJBL of Decatur, Illinois, operated by a dry goods store, calling for a power increase which would damage the service of ten stations, including such widely recognized stations as WBAL and WJAX.

WORD, the Peoples Pulpit Association in Chicago, seeks to occupy the channel of WTAS and WBBM, both well established and serving large groups. There is little question but that the defending stations will be able to show the Commission the presumptuousness of those demanding these hearings, but it is unfortunate that lawyers, witnesses and disorganization of station staffs are required to do so.

"RADIO INDUSTRY" STANDARDS

H. B. RICHMOND, Director of the Engineering Division of the R. M. A., perhaps inspired by our suggestions as to the desirability of one set of radio standards rather than two, in an article in the R. M. A. News, suggests that the R. M. A. and the N. E. M. A. should combine their work of writing radio standards. Although, as Mr. Richmond points out, the R. M. A. has ten times as many members as the Radio Division of the N. E. M. A., the long engineering experience of the older organization and the great importance of the manufacturers comprising it, makes its coöperation in writing standards of vital importance. Mr. Richmond's fair exposition of the situation is a long step toward affecting a consolidation of the standards committees of both organizations, vitally necessary if either of them are to be in the least effective.

WHY THE SOUTH HAS FEW STATIONS

SENATOR Simmons of North Carolina recently launched an attack upon the Federal Radio Commission, declaring that it showed favoritism to stations in the North. Illinois, Nebraska and Missouri, with a population of fourteen million, have more licenses to broadcast than the eleven states of the south with their population of twenty-seven million.

The Senator is correct in his facts, but he disregards the point that the south has not been sufficiently progressive to erect its share of stations with the consequence that the Northerners have already filled their wavelength bands. So long as the Commission disregards future needs by filling the ether bands with New York and

Chicago stations, there is not room enough for better broadcasting service in the more remote areas.

HOW THE RADIO BEACON WORKS

THE radio beacon operated at Hadley Field, New Jersey, the terminus of the New York-San Francisco air mail route, has proved remarkably satisfactory. Two directional antennas are used, set at right angles. By means of a mechanical keying device, the letter "A" (dot dash) is sent from one antenna and the letter ' N" (dash dot) is sent from the second. The transmissions are so timed that the dots and dashes exactly interlock so that, at the points where the signals from both transmitters are received equally, a continuous dash is heard. That point of equal signal strength is exactly midway between the directional signals of the two antennas. The radio listener aboard the plane can determine from the signals he hears whether he is exactly on the course or to the right or to the left of his course, because, in the former case, he will hear the steady dashes, while off his course, he will hear either A or N, depending on whether he is to the left or the right of it. The closer to the landing field he approaches, the more narrow the dashes. A few hundred feet from the beacon station, a deviation of ten or twenty feet from the course is clearly indicated by the signal in the headphones.

THE NEW WEAF TRANSMITTER

IN SPITE of its 50 kw., the initial broadcasts of weaf at Bellmore proved a disappointment to many New York listeners who have depended upon weaf for their principal program service. There are large areas within twenty-five miles of New York which, due to the change of location, now receive a weaker signal than the 50-kw. transmitter than they did from the old 5-kw. at West Street. There have been other instances when the removal of stations, even a short distance from the congested areas to permit increase of power, have actually reduced the number of persons served.

The transmitting apparatus at Bellmore is the last word in perfected control. The operator in charge sits before his desk and manipulates a number of buttons controlling each operation in the station, which has the proportions of a fair sized power house. If one of the water-cooled rectifier, oscillator or modulator tubes burns out, a light indicates the faulty tube. Pressure of a control button takes it out of service and connects a substitute without interruption of broadcasting.

The receiving set used to maintain the sos watch has a range of several thousand miles and will be used to advantage by weaf's operator. weaf's sos watches already have the remarkable record of being the first to hear sos calls in the New York area and notify naval and coast guards in one case out of each three and of hearing the sos simultaneously with naval and coast guard stations in the same proportion. Most broadcasting stations continue blithely on the air through sos calls until the silence of the ether around them impresses them with the fact that there must be something wrong.

NEWS OF THE PATENT FIELD

ELEVEN claims of F. A. Kolster's patent 1,637,615, were declared invalid in a decision by the Second Assistant Commissioner of Patents on the grounds that the applicant's combination claim to a radio compass having a coil form of antenna was not novel and was well known at the

time the applicant entered the field. ⫟ ⫟ ⫟ The Patent Office *Gazette* mentions the following suits over radio patents: Westinghouse vs. Allen Rogers, Armstrong 1,113,149; Radio Frequency Laboratory, Inc. vs. Federal Radio Corporation, Warren patent 1,603,432. ⫟ ⫟ ⫟ The Dubilier Condenser Company has filed against the Radio Corporation of America on various socket power patents. ⫟ ⫟ John V. L. Hogan filed against the American Bosch Magneto Company, Stewart Warner Corporation, Freed-Eisemann, Freshman, and Splitdorf for recognition of his patent 1,014,002, and also against a large department store for its sale of Crosley, Stromberg Carlson, Federal and Fada sets which he alleges infringe his basic patent. ⫟ ⫟ ⫟ A. H. Grebe and Company, Stewart-Warner, and the Consolidated Radio Corporation (Wells Gardner, Chicago and Precision Products Company, Ann Arbor, Michigan) are now R. C. A. licensees.

The Month In Radio

THE Eastman Kodak Company suggests that Radio Broadcast encourage the use of the term "phototelegraphy" rather than "telephotography" in referring to the radio transmission of pictures. "Telephotography" is used among photography experts to denote the taking of pictures over long distances by the use of special lenses, although Webster approves the use of the term to describe the transmission of pictures by radio or wire. Indeed, so extensive has been this latter use in scientific circles that it would require much more than the approval of Radio Broadcast to bring about a change in the accepted terminology. Perhaps a compromise may be suggested which may help to eliminate the confusion. Why not refer to telephotography in the sense of transmitting pictures by wire or radio, as "radio photography" or "wire photography," as the case may be? ⫟ ⫟ ⫟ Radio will perform a new feat in eliminating the isolation of explorers when the Army Signal Corps and the Pathe Company participate in their exploration of the Grand Canyon of the Colorado. The expedition will traverse the entire length of the canyon, taking moving pictures and collecting data of scientific and educational value. Accompanying the explorers will be a radio telephone transmitter which will be used to link them with broadcasting station kgo, from which reports will be broadcast through a chain of stations. The explorers will venture into dangerous and heretofore inaccessible parts of the canyon. ⫟ ⫟ ⫟ We note in the list of changes ordered by the Federal Radio Commission, authorization to move kfkx from Hastings, Nebraska, to Chicago, Illinois. kyw has shared its channel with kfkx. The result of the move is to give Chicago listeners the full use of the channel without increasing station congestion. Nebraska and the great open spaces, however, suffer a curtailment of broadcasting service. ⫟ ⫟ ⫟ koil is now transmitting its regular programs on 4910-kc. (61.06 meters), as well as its regular channel in the broadcasting band.

koil is a member of the Columbia chain. ⫟ ⫟ ⫟ The Mackay Companies purchased the Federal Telegraph Company's communication system, according to a recent announcement. The Federal Company's equipment consists of high powered arc stations installed along the Pacific Coast for point-to-point service in California, Washington and Oregon, and ship-to-shore service on the Pacific. ⫟ ⫟ ⫟ We are opposed to the radiation of the same program by stations covering the same service area. The listener is entitled to as much variety as the congested ether permits and the employment of two channels to do what may be done effectively with one is a waste of ether space. This practice is frequently indulged in by chain stations. ⫟ ⫟ ⫟ A novel use of broadcasting was employed by the United Gas Improvement Company of Philadelphia to warn its customers that gas service had been temporarily discontinued because of damage by an accidental blast. Undoubtedly, this prevented many accidents upon the resumption of service. ⫟ ⫟ ⫟ The *American Agriculturist* should be able to write a volume on the service of radio to the farmer as a result of the contest which it recently announced. It offers not too large prizes to farmers writing the best letters on the service which radio renders them. There have been many instances of thousands of dollars of saving through weather and market information. ⫟ ⫟ ⫟ Broadcast Listeners in Germany now number 1,713,899, according to *Wireless Age*, an increase of 78,171 in a three months' period. ⫟ ⫟ ⫟ Thirty million dollars worth of radio apparatus was involved in international trade in 1926, of which about thirty per cent. consisted of American shipments, twenty-five of German, and twenty per cent. of British. Exports from the United States decreased twelve per cent. in 1926 as compared with 1925, but the figures for the first half of this year show a revival of business. During the first half of 1927, American exports, were $3,705,861, an increase of $450,000 over the same period for the previous year. ⫟ ⫟ ⫟ Our British contemporary, *Popular Wireless*, made some measurements as to the radiation range of a two-tube receiver, consisting of one stage of r. f. and detector. The set was presumably a non-radiating one, but actually its radiations were readily heard at a distance of twelve miles, although but fifty volts of plate battery were used. The radiations were found to blanket an area of nearly two hundred square miles in which some five million people reside.

A TUG CAPTAIN WHO CAN TELEPHONE FROM HIS BOAT

More than forty British-Columbian tug-boats, used in towing lumber on the waterways, are equipped with 50-watt radiophone sets, tuned to 1507 kc. (199 meters). The view above shows a Captain's cabin and the complete receiving and transmitting installation

APPLICATIONS OF THE FOUR ELECTRODE TUBE

AN ENGLISH SHIELDED-GRID TUBE
In England and on the Continent, four-electrode tubes have been available for some time. The original research is credited to Schottky in Germany and the "shielded-grid" tube which has recently appeared in this country is credited to Dr. A. W. Hull

By THEODORE H. NAKKEN

A REVIEW of the progress of receiver design, which is possible by turning over the advertising pages of some early radio magazines, would offer some surprising evidence. We would see that mechanical improvements, refinements, and modern methods have been the cause of radical changes in receiver pattern, and have so simplified operation of tuning as to make the modern receiver seem as far in advance of its forbears as is the present-day automobile ahead of the automobile of fifteen years ago. Yet we would note that there has been no basic change in the type of circuit used. The regenerative receiver of ten years ago still stands unchallenged as a sensitive device for translating signals from a distant broadcasting station.

In searching for the reason of this lack of change in circuit arrangement, it will occur to us that we have reached a limit, and that it is almost impossible to obtain greater amplification than the present-day receiver gives us. And this limit is easily located as lying in the inherent characteristics of the vacuum tube as manufactured to-day. Even with its better filaments and better all round design, the vacuum tube of to-day has exactly the same fundamental characteristics as it had when first conceived and built as an experiment. It follows, then, that if any improvements in receivers are to be expected, such improvements will not be realized before radically improved vacuum tubes are made available.

But if we boldly lay the lack of actual progress at the door of the commercial vacuum tube, we must state why the tube should be responsible and how its inherent faults can be eliminated. The indictment against the present-day vacuum tube covers in the main two points—lack of amplification and the tendency to cause oscillations due to inter-element capacity. Another charge that may be brought forward is inefficiency, but this is almost identical with its lack of amplification. How to improve these conditions seems at the present time more important than all other efforts combined to make better receiver circuits, and so we will try to indicate shortly why the vacuum tube is inefficient, and how we can largely do away with the inter-element capacity, so as to get better all around performance from any circuit.

The ordinary vacuum tube contains three elements—filament, grid, and plate. The filament acts as a source of electrons when heated; the plate, by virtue of its positive potential, causes these electrons to be attracted to itself and thus establishes a plate current; the grid,

interposed between filament and plate, governs the amount of electrons that can reach the plate, acting, therefore, as a controlling element of the plate current. The grid, generally being held at a negative potential, tends to prevent electrons from wandering away from the source (the filament). The plate attracts electrons only by virtue of its high positive potential, and overcomes the repelling effect of the grid.

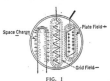

FIG. 1

In the three element tube there are three static fields which govern the tube's functions. From this diagram it is seen that there are two negative fields, both of which impede the flow of electrons to the plate

It is not only the grid that tends to repel the electrons emitted by the filament but this repellent action is also exercised by the electrons themselves. In fact, we may say that the fila-

FOR the last four years, foreign radio periodicals have contained a wealth of articles on the advantages of the double-grid tube. These tubes are chiefly used by our foreign neighbors because of their economy, but it has been inevitable that these tubes should make their appearance in this country. More than a year ago, two manufacturers brought sample double-grid tubes to the Laboratory but the time was not yet ripe for their general introduction. In April, 1926, Dr. A. W. Hull of the General Electric Laboratories described his "shielded-grid" tube in the Physical Review and on October 1st the New York newspapers carried the announcement of the Radio Corporation that a "shielded" grid tube—the UX-222—was in the process of commercial development and would be ready for the general public "some time in the future." Believing that our readers would be interested in a review of important information on double-grid tubes, the following article was prepared at our request by Mr. Nakken who is familiar with the use and operation of multi-grid tubes on the Continent.—THE EDITOR.

ment is surrounded by a cloud of electrons, which therefore constitute a negative charge, trying to drive the electrons back instead of allowing them free passage to the plate. See Fig. 1. Hence the plate must not only overcome the effect of the negative grid, but it also must nullify the effect of this cloud of electrons, which generally is called the space charge, in addition to its duty to attract electrons and thus establish the plate current.

There are two combined factors then which tend to retard the flow of electrons from filament to plate—space charge and the grid, and both are counteracted by the plate potential. It follows that part of the plate potential is utilized only to overcome the repellent action of space charge and grid, and of course, as far as amplification goes, this part of the plate potential is virtually useless. We may then say that the statement to the effect that the tube is inefficient is proved, the more so when it has been established that, in most designs, only from 10 to 15 per cent. of the plate potential is actually available for the establishing of plate current, and the remaining potential serves the purpose indicated.

We know that the space charge is virtually constant and its effect is added to that of the grid effect. The space charge, having its sphere of influence much nearer to the source of electrons than the grid, is much more powerful in its action, and thus a variation in grid potential, while representing a comparatively large change of the grid action on the flow of electrons, is decreased in its effect by the fact that it represents only a comparatively small change in the total sum of the retarding action of grid and space charge combined. Here again we may say that the tube is proved to be highly inefficient, but now in the sense that the presence of the space charge prevents the grid from being fully effective.

It follows from the foregoing remarks that the main reason for the inefficiency of the vacuum tube may be sought in the presence of the space charge. In fact, if the latter were absent, we would need only a small plate potential to obtain the identical results as at present, with the additional advantage that the grid would be fully effective because the grid field would be the only factor governing the magnitude of the electron flow to the plate, instead of only part of the sum of two factors, of which the second one, the space charge, is by far the greater. The truth of the matter is that, if only the space charge were absent, the grid effect would be from three to four times greater than at present, i.e., without any further changes in the tube the amplification factor would jump from, say, 8 to 30, yet the internal impedance of the tube would remain the same

THE FOURTH ELEMENT

WHEN we consider the static fields present in the vacuum tube we will see that we can count three—space charge, grid field, and plate field. The former two are negative while the latter is positive. The space charge, as we have seen, is a constant, or virtually so, and must be nullified by part of the positive plate field. If, then, a second positive field were introduced, nearer the filament, and thus nearer the space charge, a fairly low potential field would easily nullify the latter's effect. Obviously this can easily be done by a fourth element, which would, of necessity, be placed either between filament and grid, or between grid and plate. This element, however, should not obstruct the flow of electrons from filament to plate, hence it should be an open structure, and for this reason logically take the form of a very open grid. In this way the four-element (double-grid) tube was born.

Let us consider for a moment that such a grid is placed between filament and grid, as in Fig. 2. Due to the construction of the tube it is much nearer the filament than the plate, and as the influence of such a field is inversely proportional to the cube of the distance, it becomes apparent that, if this grid is placed at, say, one third of the distance between filament and plate, its field is 27 times more effective than the plate field. Thus, if in the ordinary tube 90 volts is used on the plate, approximately 3 volts would suffice on this fourth element to completely do away with the space charge effect. This, first of all, increases the percentage effectiveness of any

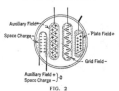

FIG. 2

When an auxiliary positive field is introduced into the tube, the negative field due to the filament is overcome, leaving the total negative field (which is detrimental to the progress of electrons to the plate) in a much reduced condition

potential change on the controlling element, the grid proper, so that we reach automatically a much higher amplification factor, and secondly,

seen in Fig. 3, there is nothing strange in the hookup of a four-element tube, the extra electrode being hooked directly to some part of the B battery.

We will now consider the second possibility in construction, i.e., that of placing the fourth element between grid and plate. Of course it must take the form of an open grid, as its purpose again is only to create a positive field, to be used to nullify the space charge effect.

Let us suppose that the tube is now so constructed that this element is placed halfway between filament and plate, in which case it follows that its effect on the space charge is eight times greater than the same potential on the plate. If, then, normally the plate has a potential of 90 volts, a positive potential of 12 volts will be equally effective when applied to the fourth element, so that once more the plate voltage can be decreased to, say, 22½ volts. Due to its open construction the positive grid offers no obstruction to the flow of electrons, and itself draws only a very small current. Once more we make the grid fully effective in its influence upon the flow of electrons, so that the amplification factor of the tube has been materially increased.

But simultaneously we have attained another effect, which merits close investigation. The positive grid, being held at a constant positive potential by the expedient of connecting it to a point on the B battery, may be stated to be constantly at a certain potential above ground potential. But after all, it is grounded. As it is interposed between plate and grid, it has the effect of splitting the capacity between these two elements into two capacities. And as this structure is very open, its capacity to each of the elements

makes it possible to decrease the plate voltage considerably, say to ten or fifteen volts, and still retain a tube of the same general characteristics as the three-element original.

It should be noted here, that we have assumed that this fourth element is built into an ordinary tube. The result then is that we have not increased the capacity between the plate and grid, and thus have not increased the tendency of the tube to oscillate due to capacitive feedback. This is a very important consideration, because it is easy enough to build an ordinary three-element tube with as high an amplification factor, as is done with modern high-mu tubes. But the latter is accomplished by narrowing the grid, i.e., by increasing the plate-to-grid capacity, and hence such tubes are almost completely unfit for radio-frequency amplification. In such a tube the tendency for capacitive feedback is increased tremendously, and this capacity affords an easy path for the signal potentials to escape via the plate and become ineffective. As will be

be readily seen in Fig. 4, because it acts the same as if a grounded plate were inserted between two condenser plates. And as its structure is very open, its capacity to each of the elements of the tube is very small indeed, smaller in fact than the capacity between plate and grid originally was. As the two capacities are in series, the resultant capacity between plate and grid is smaller than either one of the two, and hence we have, in this particular construction of the double-grid tube, almost completely eliminated the plate-grid capacity, with all its baneful effects on receiver efficiency.

Thus, this type of vacuum tube has even greater advantages than when the positive grid is placed between filament and grid. We have created a tube which is highly efficient as to

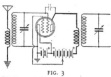

FIG. 3

This diagram is that of a single radio-frequency amplifier using a double-grid tube, the inner, grid being at a positive potential with respect to the filament. The grid-plate capacity remains unchanged

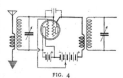

FIG. 4

If the outer grid of a double-grid tube is made positive, the resultant grid-plate capacity of the tube is greatly reduced. At the same time it is possible to build tubes with much greater amplification factor. The plate-grid capacity is reduced owing to the fact that two "condensers" are now in series

THE "INNER WORKS" OF AN ENGLISH SHIELDED GRID TUBE

One really ought to call them "shielded-plate" tubes, for the grid differs but little from that in ordinary tubes, while the plate is housed behind the shield. This illustration and that which heads · · · · · this article are reproduced from *Wireless World* (London)

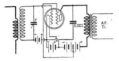

FIG. 5

In this detector circuit the outer grid is positive, the inner grid biased negative to prevent over-loading

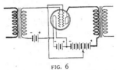

FIG. 6

A transformer-coupled audio amplifier stage using a tube whose outer grid is positive to reduce the space charge and make the plate voltage more effective

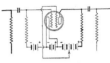

FIG. 7

A resistance-coupled low frequency amplifier with the outer grid of the tube positive

plate potential, its amplification factor has been increased considerably, and the plate-grid capacity has been largely eliminated, so that the tube may be called self stabilizing.

No wonder then that the European amateur uses these double-grid tubes quite extensively, for the upkeep of a small receiver with tubes of this kind is very economical.

Let us for a moment imagine what can be done

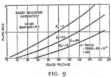

FIG. 9

Measurements made in the Laboratory of RADIO BROADCAST show that when the inner grid is positive, the mutual conductance of the tube may rise to as high as 800 while the plate impedance is 125,000 ohms, indicating a voltage amplification factor of 100. Plate voltage-plate current curves are shown here

with tubes of this kind. In an ordinary receiver employing two r.f. stages we may be glad if we get a voltage amplification of about eight per radio stage, so that the total amplification before detection is only sixty four. With tubes of this new design, and an amplification factor of, say, 25, we get an amplification before detection of 625 under the same circumstances, and with less trouble, because the capacitive feedback is as much more easily controlled.

An ordinary detector gives an additional amplification of about four, so that with the commercial receiver and ordinary tube the detector delivers a signal with a voltage amplification of about 250. The new tube as a detector would give an amplification of about 12, so that its signal would represent an amplification of 7500 times after the two r.f. stages—and this is voltage amplification only.

Due to this enormous amplification, the conventional condenser and grid leak should of course be discarded for a negative potential on the detector grid, as shown in Fig. 5, because otherwise the detector would surely be overloaded.

For audio amplification the type of double-grid tube used is almost immaterial, but as only the one type (with extra element between plate and grid) gives great advantage of decreased inter-element capacity, and thus will be employed in the r.f. stages, we may just as well use it for the audio stages too. With a good three to one transformer one stage will give us an amplification of about 90, so that the total reaches, after the first stage, 675,000, as against

ordinary tubes, with the same transformer, 24 for one audio stage and a total amplification of 6000. One perceives that almost unlimited perspectives in receiver design are opened up, that

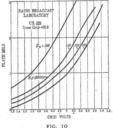

FIG. 10

Grid voltage-plate current curves on the new UX-222 tube with the inner grid positive. Note that the grid voltage lines are only two-tenths volts apart indicating a large amplification factor

enormous volume may be expected, and distance undreamed of may be covered.

A study of the diagrams will reveal that the tubes are hooked up almost in the same way as ordinary tubes, with the exception of the posi-

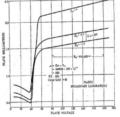

FIG. 11

The most interesting curve of all—the plate voltage-plate current data with positive outer grid. Note the negative resistance at low plate voltages, the rapid rise when the plate voltage equals that of the outer grid, and the very flat straight portion where the tube is ordinarily worked

tive grid connection. Figs. 6 and 7 show different audio stage hookups.

The names "double-grid" tube is, in the author's opinion, a misnomer. Generally we call the controlling element the grid, and as the fourth element in no way serves as a controlling element, it should not be called a grid, but simply the auxiliary element, or fourth element. Others have called the peculiar action of the fourth element between grid and plate a shield-

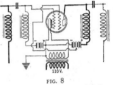

FIG. 8

The circuit of a special a.c.-operated double grid tube in process of development.

ing one, and call a tube thus constructed a shielded grid tube. It will be perceived that there is a good reason for this name, because the grid is actually more or less shielded against the plate effect, causing it to be remarkably stable.

EDITOR'S NOTE

THE curves presented here were made in the Laboratory and show the interesting characteristics of the UX-222—the R. C. A. "shielded grid" tube. Followers of our tube articles should note the extremely flat plate-current plate-voltage curve indicating an impedance—with the outer grid positive—of about 650,000 ohms, the negative resistance or falling characteristic at low plate voltages, and the high amplification factor of 222 secured by multiplying the mutual conductance by the plate impedance. If it is possible to place a load in the plate circuit of this tube, say at broadcast frequencies, of 650,000 ohms, a voltage amplification of 111 will result, compared with the usual amplification of about 10 for a single tube and its accessory apparatus.

With the inner grid positive, the mutual conductance rises, the plate impedance falls, and the amplification factor drops, to about 100. Under these conditions the tube can be used in a resistance—or impedance—coupled low-frequency amplifier.

Experimenters will delight in this tube. Its possibilities are many and diverse. It will not revolutionize the radio industry, newspapers to the contrary, nor will it produce an entirely new era in receiver design. It is just one more step toward the ultimate goal of—what? RADIO BROADCAST will publish additional data as it is available on the use of tubes of this type.

By Way of Introduction

NOT many months ago, Carl Dreher suggested in his department in RADIO BROADCAST, that a radio broadcast program was almost the most ephemeral thing in the world. Thousands of dollars are spent to engage talent, wires covering half a continent are hired, advertising is scheduled in newspapers, several studio rehearsals are held, and finally the elaborate program is put on the air. For an hour it lasts—but it can never be repeated. If you did not hear it, all the king's horses and all the king's men couldn't put it into your loud speaker again.

If it is not possible to reproduce a complete radio program in one's own home, one can at least recreate the equivalent. A very great number of well-known radio artists are regularly recording for each of the important phonograph companies. Their records—electrically cut—are available everywhere.

These pages list a few of the records made by artists who are perhaps better known to the radio listener than to the average purchasers of phonograph records. Here are fine recordings made by the favorites of the Atwater Kent hour, and the famous artists of the Victor, Brunswick and Columbia hours. As for the jazz bands, the comedy duos, and other entertainers with a more local fame, they, too, are forever at your beck and call on the black discs.

One of the most important advances made in recent years—for which we must thank the scientists—is the great progress made in the reproduction of music and speech by electrical means. All radio folk know how audio amplification has been improved, what with new amplifiers of excellent characteristics, better loud speakers, and so forth. An equally important improvement has taken place in the phonograph field. Now the phonograph, with its electrically cut records and its acoustically excellent exponential horns or cone loud speakers, will rival the musical fidelity of the best radio receiver.

NEW RECORDS BY RADIO FAVORITES

Released Since September

WHAT DO WE DO ON A DEW-DEW-DEWY DAY	Shilkret-Victor Orchestra		
IS IT POSSIBLE?	Kentucky Serenaders	20819	*Victor*
THE TAP TAP	Kahn's Orchestra		
IF I HAD A LOVER	Shilkret and Victor Orchestra	20827	"
PRESIDENT COOLIDGE WELCOMES COLONEL LINDBERGH AT WASHINGTON, D. C., JUNE 11, 1927—PARTS 1 AND 2	Hon. Calvin Coolidge	35835	"
PRESIDENT COOLIDGE WELCOMES COLONEL LINDBERGH AT WASHINGTON, D. C., JUNE 11, 1927—PART 3	Hon. Calvin Coolidge	35836	"
COLONEL LINDBERGH REPLIES TO PRES. COOLIDGE	Colonel Charles A. Lindbergh		
COLONEL CHARLES A. LINDBERGH'S ADDRESS BEFORE THE PRESS CLUB OF WASHINGTON, D. C., JUNE 11, 1927			
COLONEL LINDBERGH'S SOUVENIR RECORD—*Concluded*	Colonel Charles A. Lindbergh	35834	"
CIRIBIRIBIN (WALTZ SONG)			
IL BACIO (THE KISS) (ARDITI)	Bori	1262	"
INDIAN LOVE CALL (FROM ROSE-MARIE)	Virginia Rea		
ROSE-MARIE (FROM ROSE-MARIE)	Murphy	4015	"
OLD BLACK JOE (FOSTER)			
UNCLE NED (FOSTER)	Tibbett	1265	"
ACTUAL MOMENTS IN THE RECEPTION TO COLONEL CHARLES A. LINDBERGH AT WASHINGTON D. C.,—PARTS 1 AND 2		20747	"
AT DAWNING			
THE WALTZING DOLL	Victor Concert Orchestra	20668	"
KENTUCKY BABE			
MIGHTY LAK' A ROSE	Vaughn De Leath	20664	"
ANGELS WATCHING OVER ME			
CLIMBIN' UP THE MOUNTAIN	Utica Jubilee Singers	20665	"
SAM'S BIG NIGHT			
THE MORNING AFTER	"Sam 'n' Henry"	20788	"
JUST LIKE A BUTTERFLY	Franklyn Baur		
JUST ANOTHER DAY WASTED AWAY	Marvin-Smalle	20758	"
UNDER THE MOON	Stanley-Marvin		
SING ME A BABY SONG	Vaughn De Leath	20787	"
YOU DON'T LIKE IT—NOT MUCH	The Happiness Boys	20756	"
OH JA JA			

The public has suffered rapid education. They have learned that faithful reproduction of the original is possible in radio sets and in phonographs alike. And if the growth of broadcasting were not enough to sharpen the interest in music of all kinds, the new phonographs and the new records have come along to broaden the domestic entertainment horizon.

The radio receiver has taken its place as a musical instrument—a medium of entertainment—along with the phonograph and the piano. The radio set in the public consciousness is no longer merely a scientific marvel and myste[...] And since they are so closely related because what they can bring to the home, the pho[...] graph and the radio set have been drawn clos[...] together in association of ideas and in act[...] physical form. Those who wish to buy a combi[...] tion radio-phonograph can choose many f[...] models from five or six well-known manufact[...] ers. Those who already have a radio recei[...] which they wouldn't trade for the royal thr[...] of Roumania can make their audio amplifier a[...] loud speaker do double duty in reproduc[...] phonograph records or a radio program—acco[...] ing to the whim of the owner. All one needs [...] side a good loud speaker system is any kind [...] turntable which will twist the record at an ev[...] speed of 78 revolutions per minute, and a g[...] electro-magnetic pick-up. And there are many [...] the latter on the market.

A COMBINATION RADIO-PHONO-GRAPH FROM BRUNSWICK

The Brunswick Panatrope-Radiola, model 138C. This instrument contains a Radiola 28 super-heterodyne receiver with enclosed loop, which can be controlled by the dial in the front of the left-hand panel. On the right is the Panatrope and below it the cone loud speaker working out of a UX-210 amplifier which is also the amplifier for the radio set. The instrument, complete with all tubes for 60-cycle a. c. operation lists at $1150

FAVORITES IN CHICAGO

"Sam 'n' Henry"—Correll and Gosden who nightly disport before the twin microphones of WGN and amuse countless WGN listeners. Their verbal antics are embalmed on Victor records, listed above

the RADIO Set

New Records by Radio Favorites

SOMETHING TO TELL STOP, GO!	Shilkret-Victor Orchestra	20682	*Victor*
I AIN'T GOT NOBODY OODLE	Coon-Sanders Orchestra	20785	"
MY WIFE'S IN EUROPE TO-DAY A LITTLE GIRL—A LITTLE BOY—A LITTLE MOON	Fry's Million Dollar Pier Orch.	20726	"
BABY FEET GO PITTER PATTER SOMETIMES I'M HAPPY	Vaughn de Leath	3608	*Brunswick*
WHEN DAY IS DONE NO WONDER I'M HAPPY	Radio Franks, White and Bessinger	3588	"
AIN'T THAT A GRAND AND GLORIOUS FEELING? MAGNOLIA	Harry Richman	3583	"
JUST A MEMORY (VOCAL CHORUS BY ELLIOT SHAW) JOY BELLS (VOCAL CHORUS BY VAUGHN DE LEATH)	Harold Leonard and His Waldorf- Astoria Orchestra	1104D	*Columbia*
OOH! MAYBE IT'S YOU (VOCAL CHORUS BY FRANKLYN BAUR) SHAKING THE BLUES AWAY (VOCAL CHORUS BY FRANKLYN BAUR)	Harry Reser's Syncopators.	1109D	*Columbia*
JUST A MEMORY MY HEART IS CALLING	Franklyn Baur	3590	*Brunswick*
NO WONDER I'M HAPPY JUST ONCE AGAIN	Ernie Golden and His Hotel McAlpin Orchestra	3604	"
BABY FEET GO PITTER PATTER THERE'S ONE LITTLE GIRL WHO LOVES ME	Abe Lyman's California Orchestra	3605	"
FANTASY ON ST. LOUIS BLUES PARTS 1 AND 2	Don Vorhees and His Earl Car- roll's Vanieties Orchestra	1078D	*Columbia*
AIN'T THAT A GRAND AND GLORIOUS FEELING? I AIN'T THAT KIND OF A BABY	Paul Ash and His Orchestra	1066D	"
LEONORA PAREE!	Leo Reisman and His Orchestra	1083D	"
HERE AM I—BROKEN HEARTED HAVANA (VOCAL CHORUSES BY FRANKLYN BAUR)	Cass Hagan and His Park Central Hotel Orchestra	1085D	"
SWANEE SHORE MEET ME IN THE MOONLIGHT	Harry Reser's Syncopators	1087D	"
DO YOU LOVE ME?—VOCAL CHORUS BY F. BAUR HONEY—VOCAL CHORUS BY VAUGHN DE HEATH	The Columbians	1068D	"
AIN'T THAT A GRAND AND GLORIOUS FEELING? VO-DO-DO-DE-O BLUES	Van and Schenck	1071D	"
THAT SAXOPHONE WALTZ I COULD WALTZ OR FOREVER WITH YOU SWEETHEART	Art Gillham-The Whispering Pianist	1081D	"
GID-AP, GARIBALDI OH! YA! YA!	Billy Jones and Ernest Hare (Happiness Boys)	1074D	"
FOR THEE (POUR TOI) (GORDON) FROM OUT THE LONG AGO (STRATTON AND DICK)	Barbara Maurel	140M	"
JUST ONCE AGAIN (CHORUS BY F. BAUR) LOVE AND KISSES	Paul Ash and His Orch.	1090D	"
ARE YOU HAPPY? GIVE ME A NIGHT IN JUNE	Ipana Troubadours	1098D	"
SONG OF HAWAII HAWAIIAN HULA MEDLEY	South Sea Islanders	1111D	"
TWO BLACK CROWS, PART 3 TWO BLACK CROWS, PART 4	Moran and Mack	1094D	"
MAGNOLIA PASTAFAZOOLA	Van and Schenck	1093D	"
THE VARSITY DRAG (VOCAL CHORUS BY BAUR, JAMES, AND SHAW)	Cass Hagan and His Park Central Orchestra	1114D	*Columbia*
DANCING TAMBOURINE	The Radiolites		
GOOD NEWS (VOCAL CHORUS BY BAUR, SHAW AND LUTHER) LUCKY IN LOVE	Fred Rich and His Hotel Astor Orchestra	1108D	"

THIS RADIO PROGRAM IS RECORDED

Colonel Lindbergh before the Washington microphones which carried his welcome-home ceremonies to the entire nation. Victor has made four excellent records from this event.

FOR the first time, phonograph records of a radio broadcast program are offered to the public. Victor has the distinction of pioneering and they offer three double-face records of the national welcome to Colonel Charles A.. Lindbergh at Washington. On these three records you have the voice of President Coolidge, the interspersed announcements of Graham McNamee, a short address by Colonel Lindbergh, and his longer speech at the National Press Club. It's all there and if you close your eyes, it isn't hard to imagine that the events are just taking place—the cheers of the crowd, the applause which interrupts the speakers, the blare of the bands, and the quiet unruffled voice of Lindbergh.

The Victor Company arranged a direct wire from Washington, culminating in their studios through which their recording apparatus got the same program as each of the broadcasting stations. The ceremonies were recorded on forty-six record surfaces and finally edited down to the six surfaces now available. It is a good job from any point of view and Victor is to be congratulated. It is time that some of the historic events which are being offered to the radio listener with impressive regularity were preserved in permanent form. The next offering will be a championship fight, we suppose.

GUESS WHO

None other than Billy Jones and Ernest Hare—known to Eastern listeners of WEAF on Friday nights as the Happiness Boys. Their songs are recorded by Victor and Columbia

A PHONOGRAPH—RADIO COMBINATION FROM VICTOR

"Automatic Electrola-Radiola No. 955" is what Victor calls this beautiful instrument. Records are changed automatically and groups of 12 can be played without attention.

An 8-tube super-heterodyne with enclosed loop operating entirely from the light socket through the power supply for the vacuum tube amplifier and the cone speaker used alike for phonograph and radio reproduction. The radio receiver panel can be used in three positions. List price, $1550

MAKE YOUR OWN RADIO PICTURE RECEIVER

THIS article is the third in the series explaining the use and operation of the Cooley "Rayfoto" picture receiving system. The Cooley apparatus was demonstrated in actual operation at the New York Radio show and attracted an astounding amount of interest. Governor Alfred E. Smith of New York, who made the opening address of the Show, transmitted a part of a picture of himself, reproduced by the Cooley system over WJZ and other stations of the Blue network. Many of the readers of this article undoubtedly heard that interesting broadcast. The subject of radio photograph reception is so large that it can be discussed only in part on each article. Our readers are advised to preserve carefully each of the articles in RADIO BROADCAST *on this system, beginning with the first story in the October, 1927, issue. Pictures will be sent by broadcasting stations, using their regular assigned wavelength, and no tuning changes in your present receiver are necessary. Readers are urged to write us their experiences with the construction of the recorder. The development of the Cooley "Rayfoto" system opens for the first time to the American experimenter an important "next step" in radio development.*

—THE EDITOR.

By AUSTIN G. COOLEY

THE articles on the Cooley "Rayfoto" system in the October and November issues of RADIO BROADCAST explained how the system works, told something of the results to be expected, and gave some details regarding the operation of a picture receiver. This third article in the series gives some general and particular information about the system, and diagrams necessary for the construction of a Cooley receiver are also presented.

Many of the units for use in the radio picture receiver have been especially designed for the purpose and therefore possess the necessary characteristics for good results. They have been designed to take care of the present requirements of the receiver and are so flexible that they will still be suitable for use as the system may be gradually developed. Considerable care has been taken in this matter so that it will not be necessary to scrap any of the parts as the natural development of increased speed of reproduction and better quality are consummated. The approved parts, which have all been carefully tested, are made under the Cooley "Rayfoto" trade mark.

Arrangements are now under way to supply various broadcasting stations with phonograph records which will enable them to put Cooley pictures on the air. The radio editor of your local newspaper is the best source of information.

The complete set-up for picture reception by means of the Cooley system consists of three distinct units:—the radio receiver proper, the amplifier-oscillator unit, and the printer assembly. The first—the radio receiver—should be capable of quality reproduction of radio programs for if it falls down in this respect it will assuredly do so when called upon to detect and amplify the incoming modulated wave which has super-imposed upon it the audible note representing the picture being transmitted.

Passing from the last audio stage of the receiver proper, the "picture signal" is further amplified in the amplifier-oscillator unit and it then modulates the output of the oscillator in accordance with the modulation produced by the picture. The varying output of the oscillator is made to cause corresponding variations in the output of the corona coil, and thus the intensity of the needle point discharge is made to produce an effect on the proper tallying with the original

picture. The corona coil is included in the second unit although the actual needle at which the discharge occurs is naturally a part of the third—the printer-unit. This third unit consists of the needle, the drum upon which the photographic printing paper is wrapped, and the mechanism which causes the drum to revolve. It is purchasable as a whole, for there are few who possess the mechanical ability and facilities for the construction of such an intricate piece of mechanism.

The construction of the amplifier-oscillator unit from the approved parts is a simple matter. Fig. 1 shows a suggested layout while Fig. 2 is the schematic diagram. The following parts are necessary for this unit:

T_1—"Rayfoto" Amplifying Transformer
R_1—Variable Shunt Resistance for Primary of T_1
R_2—200-Ohm Variable Resistance Capable of Carrying 100 Mils.
R_3—12-Ohm Filament Rheostat, ½-Ampere Capacity
R_4—"Rayfoto" Relay
T_2—"Rayfoto" Modulation Transformer
C_1—0.1-Mfd. Condenser
R_5—0.01-Meg. Grid Leak and Mounting
C_2, C_3—0.0005-Mfd. Fixed Condensers
C_4—0.0005-Mfd. Variable Condensers.

L_1—"Rayfoto" Corona Coil
L_2—Radio-Frequency Choke Coil
S_1—Filament Switch
S_2—Push Button or Special Switch
J_1, J_2, J_3—Double Contact Short-Circuiting Jacks
R_6—Filament Ballast Resistance
One Telephone Plug
Milliameter, 0-25-mil. Scale
Two Sockets
Fourteen Binding Posts
Base-Board
Panel
Brackets
"Rayfoto" Printer Unit

Although wide deviations from the layout shown in Fig. 1 are permissible, it is also quite possible that considerable "grief" will be experienced in many cases where original schemes are attempted. We therefore suggest that the home constructor follow our plan of layout and construction very religiously, at least on his first set. The photographs indicate very clearly the arrangement of the apparatus and the experienced home constructor should have little difficulty in putting the apparatus together.

All but the radio-frequency circuits may be wired up in any convenient manner. We find

PICTURES RECEIVED BY THE COOLEY SYSTEM

These two photographs have not been retouched and were received at the demonstration at the New York Radio Show. Governor Smith, in his radio address opening the Show let radio listeners hear how his picture, above, sounded. The heading above shows Mr. Cooley and a part of his apparatus as set up in operation at the Show. Governor Smith's picture is on the receiving drum

that ordinary No. 18 rubber covered fixture wire is very easy to handle and makes a reasonably neat job. Each circuit should be properly tested out, as will be explained later, before the wires are laced up in bundles. The radio-frequency circuit should be wired up with considerable care, the use of bus bar wiring or rigid wires having a fair amount of spacing between them, being recommended. No particular care need be taken to prevent losses in the radio-frequency oscillator circuit. The secondary of the corona coil, and its lead to the corona needle, however,

require very special attention. This subject will be covered in another paragraph.

When the "Rayfoto" printer has been completed and set up with all connections to batteries, proceed as follows for testing and adjusting:

Place two 201-A type tubes in the sockets and see that the filaments are properly lighted and a good range of brilliancy is controlled by the rheostat, R₄, of the amplifier tube. With the input terminals open, plug a pair of phones in meter jack, J₂, for the amplifier. Turn the filaments on and off a couple of times to see if the

THE PRINTER UNIT COMPLETE

With its spring motor. If the user desires to use the motor in a phonograph which he already has, the illustration on the next page shows a special unit made for that purpose

proper click is obtained in the phones. If it sounds satisfactory, plug in the milliammeter. If you find the milliammeter reading down scale, reverse the connections to it. Adjust the C battery until the plate current is a little less than one milliampere. The plate voltage on the amplifier should be about 180 volts and the C bias to bring the plate current down to 1 mA., will have to be around 22½ volts.

Now connect a piece of wire between the plate terminal of the a. f. amplifier tube socket and the B plus terminal of the modulation transformer T₃, and connect the input of the amplifier to the output of the radio receiver. Tune-in any broadcast signal and watch the milliammeter to see if it varies in accordance with the incoming signal. The phones may be plugged into jack J₂ and the gain control resistance, R₄, varied to determine if the proper control is obtained.

Now connect the lead from the corona coil to

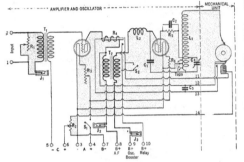

FIGURE 1

The circuit diagram of the amplifier and oscillator is given in this figure

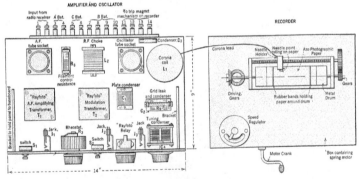

FIGURE 2

The suggested layout of apparatus shown in this drawing may be followed when you construct your amplifier-oscillator unit. The various identifying letters on the parts refer to similar letters on the circuit diagram given in Fig. 1. The lead from the corona coil, L₁ to the needle holder on the recorder should not be over three feet in length and should be supported where necessary by silk thread. The parts for the oscillator-amplifier unit can be purchased and then assembled at home, and no special precautions are necessary in constructing the unit. The wiring may be done in straightforward fashion. The corona lead is a No. 38 wire and should be carefully handled, otherwise it will break. If the lead is broken by accident, unwind about three turns from the coil to make a new lead. The record unit cannot be home constructed and must therefore be purchased as a complete unit. The recorder depicted in this sketch is a complete unit containing a spring motor. R₁ and R₆ not a part of the early model illustrated, may be located at any convenient point in the layout. A fixed condenser was used for C₄ in the first model. It is preferable that it be variable as indicated above.

the corona needle. Give this lead no more support than absolutely necessary. If this important lead needs support, it should only be by suspending it from small threads. The lead should be made as short as possible and in no case should be over three feet long. Place a small piece of Azo No. 4 photograph paper on the drum and let the needle rest on it. The paper may be held around the drum by means of two rubber bands.

Disconnect the input of the amplifier from the radio receiver and connect the booster terminal to about 90 volts of battery. Cautiously plug the meter in the oscillator circuit jack, J₃. If the meter registers over fifteen milliamperes, the circuit is not oscillating properly. Adjust the variable condenser until the current is brought down to less than ten milliamperes.

Now watch the point of the corona needle to see if a very small corona discharge takes place. If not, re-tune condenser C_4. If this does not bring results, try the various taps on the oscillator coil. If you are still unable to obtain any visible corona, increase the booster voltage to about 150 and try the different taps again. It may be found that all the condenser capacity is required to obtain a discharge. The length of the lead wire should then be shortened or a small fixed condenser, of about 0.0005-mfd., may be placed across the variable condenser. Only a good mica condenser should be used. If the discharge is strong enough to burn a hole in the paper, the booster voltage should be reduced. The taps should be tried to determine the best position of maximum discharge.

With the printing circuit operating properly, connect the input to the radio receiver and tune-in any broadcast signal. Take off the short circuiting wire on the trip magnet switch terminals and allow the drum to revolve with the needle riding on the paper. If the trip magnet does not trip from the radio signals, release the drum by operating the switch, S₂ across the relay circuit. Watch the needle point to see if the corona varies with the incoming signal.

GENERAL HINTS

AFTER the "Rayfoto" printer and recorder are set up for operation, trouble may be experienced in many cases due to feed-back from the corona circuit about which considerable information was given in the November issue. This trouble will depend greatly upon the characteristics of the radio receiver. To help avoid it, the experimenter should provide a separate set of small 22½-volt B batteries for the printer. As an additional precaution, the printer and recorder should be placed at a considerable distance from the radio receiver, say eight or ten feet, if convenient. After the apparatus is set up and working properly, attempts may be made to reduce this distance. Also experiments can be made with the battery circuits to determine the feasibility of operating the radio receiver and printer with the same B batteries. The experimenter will be aided greatly by the use of low-resistance B batteries. The filaments of the tubes in the "Rayfoto" printer may be operated from the same storage battery used for the radio receiver. All the above precautions should be taken to prevent feed-back before attempting to tune up the "Rayfoto" printer. If feedback still occurs after tuning up the printer, additional steps may be taken to suppress it.

If there is any feed-back from the printer circuit, the discharge will be strong and continuous. A high reading on the meter when no signals are coming in indicates feed-back.

The last resort to prevent feed-back is to shield the printer circuit. The shielding should not be attempted before the unit is set up and ready

THREE VIEWS OF THE AMPLIFIER-OSCILLATOR UNIT
The upper view taken from the side, shows the home-made relay mounted on the front panel and the corona coil is in the foreground of the picture. The center view shows the front panel layout and a top view is given below

RADIO BROADCAST Photograph

USING A PHONOGRAPH AND ITS MOTOR

The printer unit depicted above is designed for use with any standard phonograph and utilizes the spring motor in the phonograph for the operation of the drum

for operating because in many cases the shielding will do more harm than good if it cannot be first tested out in an experimental way.

Due to the different characteristics of radio receivers, we cannot give information here that will cover every case. In general, the proper results may be obtained by running very small gauge grounded wires parallel to the corona feed wire and separated from it by a number of inches. This may necessitate re-tuning the oscillator circuit. Experiments with shield wires and shield plates will eliminate feed-back in most cases.

In many cases it will help matters to adjust the neutralizing on the radio receiver or slightly de-tune one stage of radio-frequency amplification.

With the printer circuit working properly, we may now adjust the relay and trip magnet. Adjust the relay contacts so that they close before the armature strikes the magnet pole. With the contacts *closed*, you should be able to slide a piece of thin paper (the thickness of this page) between the pole and armature. The gap between the contacts when *open* should be equal to two thickness of the paper used on the cover of this magazine. If sparking is bad, it may be necessary to increase this slightly. A condenser across the contacts will do more harm than good as it will cause the contacts to stick unless the spring tension is excessive.

With the trip magnet contacts open, adjust the C battery so that the plate current of the amplifier is about five mils. Adjust the spring tension of the relay so as to barely hold the contacts open with the five mils. in the circuit.

After this adjustment is made, return the C battery adjustment to its original point.

The voltage for the trip magnet should be as small as practical, consistent with strong operation of the magnet. The power for this magnet may be taken from the batteries operating the radio receiver if small batteries are used for the printer circuits.

With all the foregoing adjustments and tests made, the experimenter is ready to test his set out on picture signals. It is well to have the developer and fixer solutions made up in advance although it will not hurt to let the undeveloped picture stand for a considerable time if the paper is protected from light. Instructions will be found on the developer tubes and fixing powder cartons for making up the solutions so we will not cover that information here. Regular trays for the solutions should be used for the sake of convenience although any enamel or glass tray or dish will serve the purpose. In mixing up the solutions, it will do no harm if they are made a little more concentrated than the manufacturers specify, using about 6 oz. instead of 8 oz. of water.

With the trip magnet properly adjusted, the speed of the drum should be adjusted and checked by counting the number of revolutions of the drum. If a converter drum speed of 100 r.p.m. is used, the recorder drum speed should be about 105 r.p.m. The clutch should be free enough on the first tests so that the turntable will revolve with the drum in its locked position. By loosening up on a set screw on the collar directly below the spring, the clutch friction may be regulated.

To receive a Rayfoto picture, tune-in signals

from the station transmitting the signals so that they are received with maximum intensity. If the meter in the amplifier circuit runs up over fifteen mils. on the strong signals, reduce the intensity by the volume control on the radio set or the gain control on the printer amplifier. In most cases it will be good practice to reduce the radio-frequency input to the detector circuit.

The minimum signal should be between one and four milliamperes. Adjustment of the minimum signal is more important than that of the maximum for if the minimum is too high, it will operate the relay when it should not, and if too low, it will cause irregular relay operation.

While the synchronizing pulse is being transmitted from the station sending pictures, the drum should revolve and trip regularly with only a brief period of rest or lap. By regulating the speed of the driving motor the lap can be regulated. If the lap should be very long, you will find that the speed is too slow and that the drum is tripping only on every second synchronizing impulse. Increase the speed.

After the drum is adjusted to trip regularly, a "range" should be taken; that is, the signal should be increased to the point where the relay trips from the minimum rather than the synchronizing signal, and then the signal should be decreased to the point where the relay does not trip at all. The operating adjustment should be about half way between the two points. The operating range of the relay should be as large as possible. If it is very small, it may be increased by increasing the amplifier tube's C-battery potential so that the plate current is practically zero when no signals are being received.

When the synchronizing adjustments are made and the paper placed on the drum, you are ready to start as soon as the picture signals begin. The paper should be drawn up to fit the drum tightly. If the center of the paper at the lap bulges out very far, a rubber band may be placed around the center of the drum and slid along when the needle approaches it. Care should be taken not to get the hand close enough to the corona feed wire to detune the circuit.

After the picture is received the paper is placed in the developing solution until developed to the proper density. It is then dipped in the washing bath and placed in the fixing solution. After remaining in the fixing solution for ten or fifteen seconds, it may be removed temporarily for observation purposes but should be replaced in this solution for ten or fifteen minutes and washed in running water for five or ten minutes, if it is desired to preserve the picture indefinitely.

In some cases the experimenter will experience excessive blurring, lack of detail, streaks in the pictures, improper contrast, etc. In the next issue of RADIO BROADCAST we will give more complete information that will enable the experimenter to clear out any such troubles should they arise.

The Freedom of the Air

EXCERPT from a news item in the New York *Times* of May 3, 1927, under the caption, "Pacifist Talk Hushed by Radio Station WGL:"

"We are proud that Mrs. Corson is a woman," Mrs. Ford said, "proud that she comes from Denmark, that country which upholds an ideal of peace, that country which said to the enemy, 'If you must cut through our country, even if you must cut through our women and children'"—

At this juncture Mr. Isaacson cut out the microphone through which she was speaking and substituted one in the studio through which music was broadcast as a stopgap.

Mr. Isaacson later explained to the radio audience what had happened. In discussing the incident later he said:

"We believe in free speech and I have always been willing to extend the use of our station to anyone to express his views, but there are certain things which are dictated by good taste. This was not the time nor the occasion for such a speech."

Excerpt from a news item in the New York *Sun* of July 18, 1927, under the caption, "Worm Controversy To Be Aired from WGL:"

Fred B. Shaw, one time international fly fish-

ing champion, will have an opportunity to discuss President Coolidge and his angleworm fishing to his heart's content to-night when WGL will allow him to broadcast, uncensored, his speech which was barred last week by another station.

"Our broadcast policy has always upheld free speech on the air," declared Dr. Charles D. Isaacson, program director of WGL, in extending the invitation to Mr. Shaw yesterday, "and for that reason we are only too happy to extend the privileges of our broadcast station to you."

A vexing question is thus cleared up for all time. What is free speech à la Isaacson? It is freedom to discuss angleworm fishing. But not to discuss war!

Beauty *the* Keynote *of*

AN INSTALLATION that is completely self-contained—the Bosch Model 57. Although there is a very efficient loop built in, provision has been made for the use of an outside antenna where desirable. The receiver employs seven tubes, and is tuned by a single main dial. A cone loud speaker is harbored within the cabinet. Price, $340. For power operation, $100. extra

ANOTHER receiver which may be operated with either loop or outside antenna. This one is the Fada 8. There are, as the model number implies, eight tubes, but there is a switching arrangement whereby either seven or the eight may be used. There are four t.r.f. stages. All the radio and audio stages and tubes are individually shielded. When not in use the loop is folded behind the cabinet. Price, $400.

THE set builder can now obtain a cabinet for his receiver which is every bit as beautiful as those which house the most expensive of factory-built receivers. The Model 80 cabinet of the Musical Products Distributing Company, New York, here illustrated, is in combination walnut and satinwood. The set compartment measures approximately 28 inches long, 10 inches high, and 15 inches deep. Price, $250.

the New Radio Receivers

Making the Most of Surroundings

It is easy enough to discuss the offerings of the current radio season abstractly but one should visualize this year's radio sets in the proper domestic surroundings to appreciate what great strides have been made toward making radio a truly domestic bit of furniture. The illustrations on these two pages show radio equipment in home settings and who will deny their grace?

AN INTERESTING design in loud speakers, the new Amplion "Fireside." This loud speaker has a large cone mounted on a big sound-board, exceptional fidelity of reproduction being possible. The screen stand and long cord render the "Fireside" readily portable, making it easy to place the loud speaker anywhere in the room or outside on the porch. The panelling is of embossed walnut, attractively curved, combining a grille front and back. The height is 36½ inches, and the cone is 16½ inches. Price, $97.50, with 20-foot cord

A TYPICAL example of the trend of the more expensive radio receiver, in which utility and beauty are in combination. We are taken back by this Winthrop secretary to America's younger days, when money was scarce and furniture was made to serve dual purposes. Nowadays a return to this state of affairs is demanded since home space is so scarce. The radio receiver built into the Winthrop is the popular Splitdorf single-control six-tube set. The price, complete with power operation and loud speaker, is $600.

What B DEVICE Shall I BUY?

By HOWARD E. RHODES

TO WRITE of B power units in general terms, with the object of assisting in their wise selection, is not difficult because there are simple rules that can serve to guide the prospective purchaser. In the first place it should be realized that the satisfaction which any power unit gives in service frequently bears a close relation to its cost, for power units can be built to meet almost any price. Cheap units, constructed of inferior materials, are often capable of giving as good results as more expensive devices, during a single demonstration, but whether the cheaper device will stand up for as long a time in service is certainly open to question.

The first rule in the purchase of a B power unit should be insistance upon a well-known make, purchased through a reputable dealer, for only from such a source can you be assured of obtaining satisfactory service. One of the simplest and most satisfactory methods of appraising a B power unit to make certain that it will satisfactorily operate your receiver, is to give it a trial lasting several days in your own home, under actual working conditions. Only a reputable dealer selling a good product can afford to do this for you. A cut price dealer, with little or no interest in his customer once the sale has been made, cannot afford to sell a unit to you on the basis of a trial lasting several days. The unit purchased from a reputable dealer might cost somewhat more, but the higher price is justified because of the better service and greater assurance of satisfaction which you can obtain.

There are many things about a B power unit which must be taken more or less on faith. You can't tell, by looking at the device, what kind of chokes are used or if the condensers in the filter circuit have a sufficiently high voltage rating so as to prevent any possibility of breakdown. Then again, the design of the transformer supplying the rectifier and filter circuit is something that cannot be examined when you buy the unit. It is only by relying upon the reputation of the dealer and of the manufacturer whose product he sells, that you can have any assurance of a properly designed unit. If care is taken in the selection and use of a B power device, it will give satisfactory operation and lasting service for years.

In the list appended, we have given the important characteristics of about twenty-five well-known B power units. Since there are being made at the present time about one hundred and thirty of these units, it will be seen that the list is by no means exhaustive.

The proper selection of a B power unit is a matter of knowing the total plate-current drain of the receiver and of then finding a device that will supply the correct voltages to your receiver at this current drain. You must make certain that the maximum voltage available from the device when supplying a milliampere load equal to that of your receiver is sufficient to supply the power tube you are using. The maximum output voltage of the device should be 180 volts, or slightly more if a 171 type tube, with 40.5 volts

grid bias, is used in the output; if a 112 type tube is used, with 9 volts grid bias, the maximum voltage should be about 135 volts. The information given in the appended list includes the maximum output voltage which the units will supply at various current drains.

Of course, it is also essential that the power unit be capable of supplying the other tubes in the receiver with the voltages they require. These voltages are obtained from voltage taps on the power unit and in most cases the voltage that is obtained from any one of these taps varies with the current being drawn from it and the other voltage taps, and for this reason it is not possible to give any definite figures for the voltage output from these taps. Many power units use adjustable or semi-adjustable resistances so that the desired voltages can be obtained by proper adjustment of the resistance. These resistances should not be adjusted by guess but should be adjusted with the aid of a high-resistance voltmeter. This is a service any reliable dealer can give you.

There are some power units (those using glow tubes) which give practically constant output voltages independent of the load. If such a unit is purchased it may be connected to any receiver with assurance that the actual voltages being supplied to the receiver will be near enough to those marked on the terminals of the power unit for satisfactory operation.

As will be seen in the accompanying list, some of the devices have been approved by the National Board of Fire Underwriters and many of the other devices have been submitted for test but have as yet not been approved. The submission of B power units to the National Board of Fire Underwriters for their approval is a distinct step in the right direction. It gives the prospective purchaser the assurance that the unit conforms to definite standards designed to make certain that the operation of the device will be entirely safe.

The power input to a power unit is a measure of the cost of operating the unit. The input power to the average B power unit is about 25 watts, which is the amount of power required by a small electric light bulb. If we obtain power from the electric light company at the rate of ten cents per kilowatt hour, it will cost just about one quarter of a cent per hour to operate the average plate-supply device. To this cost of operating, we should, of course, add depreciation on the unit and the cost of tube renewals which must be made on an average of once a year.

All B power units for use with an alternating-current supply must use rectifiers of some sort. A majority of units use a tube for this purpose, but there are also quite a large group that use electrolytic rectifiers; thousands of B power units using either type of rectifiers have given satisfactory service. When purchasing a power unit avoid those using unknown makes of rectifiers. Whether you purchase a power unit

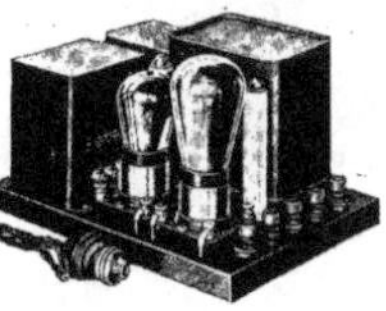

B POWER UNITS FOR YOUR RECEIVER

The power unit at the left is the Erla Steadivolt BC Converter and it utilizes a Raytheon BH rectifier tube and a glow tube to maintain the output constant at all loads. It will supply up to 80 milliamperes at 180 volts. It lists at $40 including tubes. The Exide unit in the center and the Burns power device at the right both contain adjustable resistance units to regulate the voltage at the various terminals and further information regarding these two devices will be found in the listings on these pages

using an electrolytic or tube rectifier is a matter of personal preference.

WHEN YOU USE A B-POWER UNIT

A LOUD hum audible in the output of a receiver operated in conjunction with a B power unit may be due to coupling between the receiver and the power unit itself. If the hum is due to such a cause it can generally be eliminated by placing the power unit in some other position relative to the receiver. Also if hum is to be prevented it is essential that the negative B be grounded directly, and in some cases it is necessary to connect a 1-mfd. 200-volt condenser between the grounded negative B terminal and one side of the input power lead. If these simple remedies do not eliminate the hum it is likely that there is some defect in the unit itself and the dealer should be consulted for, if the power unit is operating properly it should produce practically no audible hum in the output. Every plate-supply unit must contain a transformer designed to step up the input voltage to an amount depending upon what output voltage is required and upon what type of rectifying element is used. The rectifier, whether it be a tube or an electrolytic device, modifies the alternating current obtained from the power supply and changes it to a pulsating direct current. The filter circuit smooths out the pulsation in the rectified current and delivers a steady fluctuating direct current to the output terminals of the device. Some power units use half-wave rectifiers and others use a full-wave rectifier. In the first case, only half of the alternating voltage is used and when full-wave rectification is used both halves of the alternating voltage wave are utilized. The filter system may contain either one or two sections. Whether a half- or full-wave rectifier is used or whether a one- or two-section filter is used is something that need not particularly concern the prospective purchaser. The design of the filter in either case should be such as to eliminate any hum, and so long as the device which you buy does not hum excessively, you may be sure that the filter circuit has been correctly designed for the type of rectifier used. The coils and condensers used in the filter circuit do not become weak with age, and a filter system capable of giving an output voltage free from hum will continue to do so unless some unit in the system completely fails.

Facts About Some B Units

ACME APPARATUS CO., MODELS E-1, E-2, E-3, AND E-4

Maximum output voltage of a.c. models: 205 at 20 mA., 185 at 30 mA., and 160 at 50 mA. Uses Raytheon BH rectifier. Two adjustable voltages. Two-section filter. Approved by National Board of Fire Underwriters. Size: 8¾ x 3½ x 7½ inches. Models E-1, E-3, and E-4 for operation on 110 volts 60 cycles a.c. Model E-2 for operation on 120 or 220 volts d.c. Model E-1, with cable, for one- to twelve-tube sets, $50; Model E-2, with cable, for one- to twelve-tube sets, $25; Model E-3, with cable, for one- to eight-tubes sets, $85; Model E-4, with binding posts, for one- to eight-tube sets, $35.

ACME ELEC. AND MFG. CO., MODEL BE-40

Maximum output voltage: 200 at 20 mA., 185 at 30 mA., and 145 at 50 mA. Uses QRS rectifier. Two adjustable voltages. Full-wave rectifier. Two-section filter. Approved by National Board of Fire Underwriters. For operation on 110 volts 50-60 cycles a.c. Recommended for five- to eight-tube sets with power tubes. Size: 7¾ x 11½ x 3¼ inches. Price, with tube, $34.50.

ALL-AMERICAN, MODELS A-1, A-3, A-4, AND A-8

Maximum output voltage of Model A-8; 200 at 20mA., 180 at 30 mA., 140 at 50mA. (with "high-low" switch in "high" position); 140 at 20 mA., 120 at 30 mA., and 70 at 50 mA. (with "high-low" switch in "low" position). Other models supply about 10 volts less. Approval pending by National Board of Fire Underwriters. Raytheon rectifier is used. Two-section filter. Adjustable output voltages. Model A-1 for operation on 110 volts 50-60 cycles a.c.; Model A-3, 25-40 cycles; Model A-4, 220 volts 50-60 cycles a.c.; Model A-8, 110 volts 50-60 cycles a.c. Size: Models A-1, A-3, A-4, 9 x 7 x 5¼ inches; Model A-8, 3 x 7 x 10 inches. Prices, including rectifier; Model A-1, $37.50; Model A-3, $42.50; Model A-4, $42.50; Model A-8, $27.50.

BURNS, MODELS 750-A, 750-B, AND 800-B

Maximum output voltage of Models 750-A and 750-B; 190 volts at 30 mA., and 180 volts at 50 mA. Model 800-B, 205 volts at 20 mA., and 190 volts at 30 mA. Uses Raytheon rectifier. Amplifier and detector voltages adjustable. Two-section filter. Approximately 20 watts a.c. input with 30 mA. load. Designed to operate 171-type power tubes. Size: Models 750-A and 750-B; 7¼ x 10¼ x 6¾ inches; Model 800-B, 4¾ x 10¾ x 5¾. Prices: Models 750-A and 750-B, $47.50 with tube; Model 800-B, $35 with tube.

AMRAD, MODEL NO. 280

Maximum output voltage; 200 at 20 mA., 180 at 30 mA., and 165 at 50 mA. Uses type 280 or 213 thermionic rectifier. One-section filter. Fixed output voltages. Full-wave rectifier. Twenty watts a.c. input at 30 mA. load. For operation on 100-120 volts 60 cycles a.c. Size: 10¼ x 6¾ x 7¾ inches. Price, without tube, $45.

ARCO B POWER

Maximum output voltage: 180 volts at 50 mA. Uses filamentless rectifier. Two-section filter. Full-wave rectification. Fifteen watts input with 30 mA. load. Size: 9¾ x 3¾ x 8¾ inches. For operation on 110 volts 60 cycles a.c. Price, without tube, $32.50.

BREMER-TULLY B POWER

Maximum output voltage: 216 at 20 mA., 195 at 30 mA., and 150 at 50 mA. Uses Raytheon BH rectifier. Output voltages adjustable by fixed steps. Two-section filter. Full-wave rectifier. Seven watts input with 30 mA. load. Approved by National Board of Fire Underwriters. Will supply receivers having up to ten tubes. For operation on 110-115 volts 60 cycles a.c. Size: 4¾ x 9¾ x 6¾ inches. Price: $37.50, without tube.

BASCO B POWER

Maximum output voltage: 250 at 20 mA., 230 at 30 mA., 190 at 50 mA., and 175 at 60 mA. Uses Raytheon BH rectifier. Two-section filter. Full-wave rectification. Fixed detector voltage. Other voltages variable. A special primary rheostat functions to regulate output to supply different types of power tubes. Twenty watts input with 30 mA. load. Size: 12 x 4½ x 7½ inches. Price, with tube, $35.

KING, TYPES M AND V

Maximum output voltage: 238 at 20 mA., 215 at 30 mA., and 180 at 50 mA. Uses type 213 rectifier. One-section filter. Full-wave rectification. Type M has variable detector and amplifier voltages. Type V has variable detector voltage and fixed 90, 135, and power tube taps. Twenty watts input with 30 mA. load. Can supply up to nine tubes. Size: Type V, 9½ x 4½ x 7 inches; Type M, 11 x 6 x 6 inches. Prices: Type M, $45.00; Type V, $37.50.

EXIDE, MODEL 9-B

Maximum output voltage: 208 at 20 mA.,

MAJESTIC, SUPER-B POWER UNIT

Maximum output voltage: 186 at 20 mA., 156 at 30 mA., and 112 at 50 mA. Adjustable output voltages. Special "high-low" switch gives two voltage ranges. Full-wave rectification. Two-section filter. For operation on 110 volts 60 cycles a.c. Size: 10½ x 5¾ x 9 inches. Price, complete with tube, $29.50.

FRESHMAN, MODEL A

Maximum output voltage: 220 volts at 40 mA. Uses tube rectifier in full-wave circuit. Adjustable output voltages. Two-section filter. Unit supplies following C voltages: -4½ -9, -40. Thirty watts input with 30 mA. load. Designed to supply sets using up to seven tubes. Price complete: $45.

MUTER B POWER UNITS

Two types made, one for Raytheon BH tube, the other for 213 or 280 type rectifier. Maximum output voltage (Raytheon type): 200 at 20 mA., 180 at 30 mA., and 150 at 50 mA. Other type gives output voltages about 10 volts lower. Full-wave rectification. Two-section filter. Fixed output voltages. Twenty-one watts input with 30 mA. load. For operation on 110-120 volts 60 cycles a.c. Prices: Raytheon model, $26.50; other model, $24.50.

THREE MORE LINE SUPPLY DEVICES

The Acme B power unit incorporating two variable resistances to compensate differences in load imp at the right. The All-American Constant B is illustrated at the center and contains a "high-low" swi voltages. The C potential as well as B potential can be obtained from the

194 at 30 mA., and 180 at 40 mA. Uses electrolytic rectifier, arranged in bridge circuit for full-wave rectification. Two intermediate and detector voltages adjustable. Two-section filter. For operation on 105-125 volts 50-60 cycles a.c. Ten watts input with 30 mA. load. Approved by National Board of Fire Underwriters. Designed to supply sets with six or more tubes. A 112 or 171 type power tube may be used. Size, 6½ x 11½ x 9½ inches. Price: $42.50.

GENERAL RADIO, TYPE 445

Maximum output voltage: 200 at 20 mA., 185 at 30 mA., and 160 at 50 mA. Uses type 280 rectifier tube. Two-section filter. Output voltages adjusted by means of sliding taps on wire-wound resistance. C voltage available for power tube. Twenty-five watts input with 30 mA. load. Designed to meet specifications of National Board of Fire Underwriters. Automatic switch breaks the 110-volt a.c. input circuit when cover is removed. Size: 15½ x 7 x 7 inches. Price: $55.00, without tubes.

FREED-EISEMANN, MODEL 16

Maximum output voltage: 135 at 30 mA. Uses type 280 rectifier tube. Also uses type 874 glow tube to maintain output voltages constant independent of the load. Two-section filter. Three C voltages available:-4½, -9, and -27. Twenty-five watts input with 30 mA. load. For operation on 105-120 volts 60 cycles a.c. Size: 7 x 7 x 9½ inches. Price: $35.00, without tubes.

PRESTO-O-LITE "SPEEDWAY" B

Maximum output voltage: 188 at 20 mA., 175 at 30 mA., and 148 at 50 mA. Fixed output voltages. Compensation for variations in line voltage obtained by adjusting three-point switch. One-section filter. Full-wave rectification. Twenty-five watts input with 30 mA. load. Uses Raytheon BH rectifier. Size: 6 x 8 x 8 inches. Price: $37.00, including tube.

SENTINEL, MODEL B-C

Maximum output voltage: 180 at 80 mA. Uses two rectifier tubes: Two variable voltages. Unit supplies two C voltages, -4½ (fixed) and -4½ to -45. Voltage control to compensate variations in line voltage. Beverly model is equipped with special instrument used to read the voltages being supplied by the various taps on the power unit. Prices, including two tubes, regular model, $44.50; Beverly model, $65.00.

SILVER-MARSHALL, TYPE 656

Maximum output voltage: 170 at 20 mA., 160 at 30 mA., and 140 at 50 mA. Unit uses UX-243 (CX-313) rectifier and UX-284 (CX-384) glow tube. Glow tube maintains output voltages practically constant independent of load. Two-section filter. Twenty-five watts input with 30 mA. load. Size: 6½ x 7½ x 5 inches. Price: $38.50.

STEWART, U-80

Maximum output voltage: 190 at 20 mA., 175 at 30 mA., and 140 at 50 mA. Two-section filter. Full-wave rectification. Thirty watts input with

Measuring the "GAIN" of *your* RECEIVER

Stromberg-Carlson engineers testing the audio-frequency characteristics of one of their No. 744 receivers. The apparatus consists of a "beat frequency" oscillator which produces the audio tones, and meters for measuring the extent to which these frequencies are amplified within the receiver

By KEITH HENNEY
Director of the Laboratory

OUR present broadcasting structure is made up of three intimately connected components, the broadcasting station, the receiving equipment, and the intervening medium. The broadcasting station has one *raison d'etre*—to translate sound impulses into electrical waves; the receiver's only purpose is to accomplish the opposite—to translate these electrical waves back into sound impulses. The intervening medium is the connecting link between the transmitter and the receiver, an inefficient link it is true, performing its task with many vagaries, and for reliability's sake it might be very well displaced by a metallic conductor. The radio medium, however, has the advantage that for broadcasting purposes, the communication is radiated in all directions, and is not confined to a direct path between two points.

We have already outlined in the November RADIO BROADCAST what the transmitter does when it lays down a "field strength" about the receiver, how this field strength is measured, how much is necessary for various degrees of service, and how field strength and the attributes of sets, selectivity, sensitivity, and fidelity, are related.

It was pointed out that the greater the field strength, or the more sensitive the receiver, the more powerful the corresponding loud 'speaker signal. We are now faced with the problem of ascertaining how sensitive a receiver must be to deliver a certain signal from a certain field strength, how selective a set must be to shut out unwanted signals in favor of the desired program, and what the degree of fidelity must be to furnish sufficient realism to make a receiving equipment no longer a "radio"-but a musical instrument.

What we must do is to answer the following questions, presupposing a station to deliver a certain field strength at a certain point:

How loud will be the resulting signal from a certain receiver? How can it be measured? What will happen if another station a given number of kilocycles away goes on the air and lays down a certain field strength about the receiver? If the receiver is sufficiently sensitive and selective, how do these factors influence the fidelity?

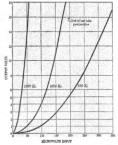

FIG. 1

When one measures the output voltage of a receiver of several years ago as the input voltage is changed, he gets a curve similar to those shown above. Note the steep 1300-kc. curve indicating extreme sensitivity, and the decreasing sensitivity as the radio frequency is decreased

Since radio receivers were first made, qualitative answers to these questions have been the stock in trade of all engineers. It is only within the last year that quantitative answers have been generally available, especially when one considers the receiver as a single unit, and desires answers to his questions not with regard to the component parts that make up that receiver but with respect to the ensemble equipment. Methods for determining the characteristics of coils, condensers, transformers, and other individual units that go to make up a "radio," have been known and used for several years. In England great emphasis has been laid upon and many arguments built around such measurements as contrasted with those which include everything in a receiver between antenna and loud speaker. While it is true that the characteristics of such units can be combined to produce a fair approximation of what the complete receiver will do, an overall measurement carries much more conviction to the engineer.

An interesting experiment was carried out at station WOR some time ago, and more recently by WLW, to enable listeners to determine how low and how high in audio frequencies their receivers were responsive. At WLW, where a Wurlitzer organ forms part of the studio equipment, continuous tones varying from the lowest to the highest organ note were put on the air. First the pure note of the open diapason was transmitted and then it was played with various harmonics added. This enabled the listener to determine not only the acoustic limits of his receiver and loud speaker but the change in quality as the harmonics modulated the original pure note.

In laboratory the business of performing the same experiment or that of investigating the sensitivity and selectivity of the receiver under better controlled conditions consists in moving the transmitter nearer the receiver, and decreasing its power accordingly. This eliminates the vagaries of the intervening medium, and when the transmitter is finally connected metallically to the receiver a set-up results which is sufficiently flexible that everything can be varied and measured at the same time. A miniature broadcasting station is necessary. This must consist of an r.f. oscillator whose output can be regulated and measured, an audio-frequency oscillator variable from the lowest to the highest audio tone ordinarily broadcast and relatively free from harmonics, and some means of combining these two generators of electric and sound waves.

The Hazeltine Laboratories, under the direction of Chief Engineer MacDonald, made such tests on receivers which were under development there, and the results were published in the *Proceedings of the I. R. E.* in February, 1927, the first paper to be published in this country on such laboratory practice. The emphasis here, however, was more on component parts, such as radio-frequency amplifiers, coupling coils, and audio amplifiers than on the receiver as a whole. The data as published were most interesting.

In a discussion of this paper, appearing in the April, 1927, *Proceedings of the I. R. E.*, H. D. Oakley and Norman Snyder of the General Electric Laboratories described the methods used several years ago at the Schenectady Laboratory for measuring receivers.

The equipment was housed in two shielded rooms, one of which contained the radio and audio oscillators as well as a control device for regulating the voltage which was put on the receiver under test in the adjoining room. A standard Radiola 100 loud speaker was placed across the output of the receiver, and the voltage across it measured as the input voltage was varied.

To test the sensitivity of the set, that is, to tell how many output volts could be delivered with a given input radio voltage, the following procedure was carried out. The generator was set going at a certain radio frequency and this was modulated at 1000 cycles to a given degree of modulation. The input to the receiver was varied in small steps until the output tube overloaded. This was considered the upper working limit of the receiver. A specimen curve of output voltage plotted against input voltage is shown in Fig. 1. The receiver was a standard set of several years ago, and is not indicative of the better types of modern sets.

The data show that the receiver at 1300 kc. was roughly 6.5 times as sensitive as at 560 kc. and that to produce an output voltage of 16 at 1300 kc. required an input of only 51 microvolts compared to 335 required at 560 kc. At 1000 kc. the voltage required was roughly 175. The output voltage at 560 kc. divided by the input voltage gives a rough voltage gain of 46,000; at 1000 kc. the ratio is 102,000 and at 1300 kc. the gain is 300,000.

The data showed that the output voltage for each of the three frequencies was proportional to the input voltage squared, for which the detector tube is responsible.

TESTING SELECTIVITY

TO TEST the selectivity of the receiver, it was tuned to, say, 560 kc., and the output voltage read as the transmitting generator in the first of the two shielded rooms was varied in frequency but kept at constant amplitude. The receiver was then set at some other frequency and a similar set of data was taken. Specimen curves shown in Fig. 2, are taken from the *Proceedings of the I. R. E.*

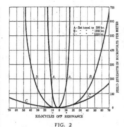

FIG. 2

These queer curves show the selectivity of a rather poor receiver. At 560 kc. the set is selective, at 1300 kc., curve C, it is as broad as the proverbial barn door. They are indicative of a poorly engineered receiver and were made on sets sold several years ago

The curves plotted from data obtained in this manner are given in Fig. 2, and show the field strength required to produce a given signal which differed from the frequency setting of the receiver by a certain number of kilocycles. For example, if the receiver were accurately tuned to 560 kilocycles, a signal 10 kilocycles off resonance, say 570 kc., required a field strength of 150 microvolts per meter to produce an interfering signal. At 1300 kc., a signal 66 kc. away, or 1366 kc., having a field strength of 150 microvolts per meter, would produce the same interference.

In other words the receiver was roughly one seventh as selective at 1300 kc. as it was at 560 kc. which, coupled with the fact that it was nearly seven times as sensitive at the same frequency, may demonstrate why the higher broadcasting frequencies were not so highly regarded by engineers of transmitting stations a year or so ago. There is no reason why a care-

fully designed and engineered receiver cannot be equally sensitive over the broadcasting band of 1000 kc. If the band were to be extended, in the direction of still higher frequencies, the problem placed upon design engineers would be considerable, but would not be insurmountable.

Receivers of the present day are better than these curves show. Methods of maintaining equal gain over rather wide frequency bands are well known, and up-to-date receiver manufacturers make every effort to include in their products the results of all well-known inventions. A receiver which squawks at 1300 kc. and is practically silent at 560 kc. is a poorly designed set, and should not be placed in the same class as others in which care has been taken to avoid just such criticism.

The method of measuring and rating receivers employed by the Radio Corporation of America was described in *The Proceedings of the I. R. E.* in May, 1927, by T. A. Smith and George Rodwin. An arbitrary loud speaker signal is set up and all measurements are made with a view toward determining the field strength required to produce this signal, which is that corresponding to an average audio-frequency (r.m.s.) voltage of 15 across a 5000-ohm resistance when a 400-cycle note modulates the transmitter to a degree of 50 per cent.

Having determined how much output voltage the receiver will deliver when a certain input voltage due to a certain field strength is impressed on it, mathematics will tell how much voltage or power will be delivered at other input levels, up to the overloading point of the amplifier tubes. The following relations express the manner in which transmitter antenna power, input receiver voltage, output voltage and power are interconnected:

Field strength α transmitter antenna current.
Field strength α square root of transmitter power.
Input receiver voltage α field strength.
Output receiver voltage α field strength squared.
Output receiver voltage α transmitter power.
Output receiver power α output voltage squared.
Output receiver power α input voltage to the fourth power.

FIG. 3

The receiver is put through its paces in this shielded room at the General Electric Company. Only signals that are meant for the receiver arrive at its input via shielded wires; all others are excluded by the shielding surrounding the six room surfaces.

Output receiver power α field strength to the fourth power.
Output receiver power α transmitter power squared.

The Greek letter alpha in the above relations means "is proportional to."

Doubling the transmitter antenna current multiplies the transmitted power by four, doubles the field strength, doubles the input receiver voltage, multiplies the receiver output voltage by four, and multiplies the power into the loud speaker by sixteen. These relations may be connected to what happens in one's receiver by the following facts. The average stage of audio amplification has a voltage gain of twenty-five, a two-stage affair having a voltage gain of roughly 300, or 50 TU, if a 171 type tube is used as the output tube. A good radio-frequency amplifier should have a voltage gain of about 50 TU, or 300, so that the overall gain in voltage from a modern well engineered receiver should be in the neighborhood of 100,000, or 100 TU. These figures in power amplification become respectively, for the two-stage amplifier and for the complete receiver, 30,000 and 10,000,000—truly enormous amplification.

In actual practice the R. C. A. engineers do not measure the voltage across the resistance in the output of the receiver while the input voltage is varied. An interesting short cut is used instead, which is possible from the phenomenon accompanying the function of detection.

When the receiver is tuned to a carrier wave, modulated or not, the average d.c. detector current changes, increasing when a C bias detector is employed, decreasing when the conventional grid leak and condenser method is used. Greater changes occur with greater field strengths, or the more nearly the receiver is tuned to the incoming signals. The change in detector plate current, then, is a measure of the effectiveness of the field strength or the sensitivity of the receiver.

To produce the arbitrary 15-volt signal across the 5000-ohm resistance in the receiver output requires a certain change in detector plate current. Once this is determined the audio amplifier can be dispensed with and all measurements may be made by noting the change in detector plate current. This method obviates the necessity of using modulated signals.

In the R. C. A. Laboratory the input voltages

FIG. 4

Signal generating apparatus used at the General Electric Laboratories for testing the characteristics of receivers. This apparatus is housed in a shielded room, and consists of a Heising modulated generator capable of oscillating at any frequency in the present broadcasting band modulated at any audio frequency between 40 and 10,000 cycles

are fed to the receiver through a dummy antenna consisting of an inductance of 28 microhenries which has a resistance of 2 ohms, a capacity of 0.0004 mfd., and a resistance of 23 ohms. The curves obtained in this way show the same general characteristics as those given in the General Electric report, i. e., low gain at low radio frequencies, high gain and poor selectivity at high frequencies. At the same time there is considerable loss of the higher audio frequencies at the longer wavelengths, due to the excessive sharpness of tuning, or selectivity.

While it is true that only a few of the larger

and better-known receivers are engineered with these thoughts and these laboratory measurements in mind, it is a fact that more and more radio manufacturers are becoming aware that good engineering is a priceless asset. The article in this issue of RADIO BROADCAST on the Fada receivers, by John F. Rider, and others to follow on other well-designed receivers, proves this statement. The Laboratory is preparing data on manufactured sets using the methods of measurements mentioned above and as fast as the material is ready, it will be presented to RADIO BROADCAST's readers.

All About Patents

INVENTIONS AND PATENTS, THEIR DEVELOPMENT AND PROMOTION. By Milton Wright, LL.B. Published by the McGraw-Hill Book Company, Incorporated. New York. Price, $2.00. Pages, 225.

A VERY useful contribution to the bewildered inventor, throbbing with the thrill of a discovery, is the sage and practical counsel of Mr. Milton Wright, as embodied in his new book, "Inventions and Patents." There is nothing assuming about the writer's style; the work is not overburdened with technical legal arguments, although the subject is a highly technical one, and there is no obscure language to confuse the uninitiated.

The subject matter of the volume is devoted broadly to all the problems which face the inventor. He is told what is patentable and what is not patentable, what constitutes a practical invention, what steps to take to facilitate securing a patent and how to protect it after it is secured, how to select a good patent attorney, how a patent should be applied for, how to obtain financial support, what the problems of marketing and merchandising are, how to sell patents outright, on a territorial and on a royalty basis, and what steps to take against infringers. There you have it in one long sentence; certainly the scope of the book is broad enough to be a real aid to the floundering inventor.

Valuable cautions and dangerous pitfalls, which are the usual stumbling blocks to the uninitiated patent seeker and inventor, are disclosed. For example, how to keep records which aid in establishing date of conception and reduction to practice, is explained so that the inventor, heeding the advice given, will have no difficulty in later sustaining his invention in the courts. And again, the vital subject of how to select a patent lawyer and how to get the greatest value from his services is presented simply and

clearly. The book abounds in practical illustrations which serve to clarify the force of the writer's arguments.

The reviewer does not hesitate to recommend a thorough reading of this volume to all those who believe they have a patentable idea and those who contemplate obtaining a patent. It is certain either to cause them to abandon the idea because it offers little or no possibility for profit, or else to secure a better and more easily protected patent. Considering that only one patent out of a hundred secured by hopeful inventors proves profitable, the discouragement of the impractical is as valuable a service as the encouragement of the promising. In this respect, Mr. Wright's dispassionate and constructive point of view differs materially from the flamboyant literature and booklets which unscrupulous patent lawyers distribute in the hope of inveigling misguided inventors to obtain patents, whether their ideas show promise or not.

—E. H. F.

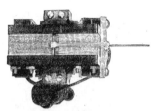

A LOUD SPEAKER ELEMENT
This is the driving unit used in the type 20-20 cone speaker
manufactured by A. H. Grebe. The speaker is priced at $35.00

A PACENT OFFERING
This well made instrument uses a balanced armature construction that insures quality reproduction. Price, $35.00

L O U D   S P E A K E R S

**THE "ALGONQUIN" CONE
SPEAKER**
An artistically designed cone priced at
$15.00, the product of the Algonquin Elec-
tric Company

A POWER CONE WITH B SUPPLY
The perfectly free mounting used in this Magnavox combination is responsible for its excellent reproducing qualities. It contains a 210
power-amplifier and B supply. Price, $242.00

THE "NEUTROWOND" REPRODUCER
An attractive loud speaker finished
in American walnut. Price $35.00

AN INNOVATION
This interesting loud speaker made
by Frank B. Porter, Washington
utilizes the structure above the base
to conceal a tonal chamber. The actual element is with base. Various
models retail for from $50.00 to
$150.00

FADA
This 17-inch table cone sells for $25.00, Model
315 A. Fada manufactures other more expensive
models selling for up to $50.00

FOUR OR SIX-VOLT A-POWER
A combination storage battery and full-wave dry rectifier available in four and six-volt models, made by Triple A Specialty Co., Chicago. Price: either model. $39.50

THE BASCO A-B POWER UNIT
This device contains a B-Power unit and a storage battery-trickle charger combination. A visual indicator shows the state of charge of the battery. Price $75.

A TWO AMPERE "TUNGAR" BATTERY CHARGER
A well known product of the General Electric Company for charging A and B storage batteries. Price, $18.00

POWER RECEIVED

THE "ACME" B POWER UNIT
For use with receivers containing up to twelve tubes, including a 171 type power tube. A Raytheon rectifier is used. Price: $50.00

ANOTHER B POWER UNIT
This unit supplies up to 135 volts at a 60 milliampere load—more than enough for most receivers. Manufactured by the All-American Company and priced at $27.50

A POWER AMPLIFIER AND B SUPPLY
An A-B-C power supply for receivers using a.c. tubes. The unit contains one stage of power amplification using a 210 tube. Manufactured by the Radio Receptor Co., New York, and priced at $60.00

FACTS ABOUT THE FADA "SPECIAL" SET

By John F. Rider

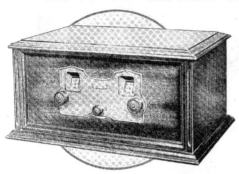

JUST as the research laboratory is the prime mover of every business, whether cheese, steel, or clothing, it is also the heart of the radio industry. The radio public is awakening to the fact that research is a prime mover in the radio industry; that research, and research only, can produce faithful reproduction, ample volume, satisfactory selectivity, ease of control, and perfect stability. The result is recognition of research as the paramount factor, and consistent with this recognition, is the gradual stabilization of the industry—its gradual ascension to an impregnable position.

Research is directly responsible for every good receiver and for every development in radio receiver design since the day KDKA commenced its broadcasting activities. A good radio receiver cannot be produced without a research background.

As an example, let us consider a typical receiver, the design of which may be laid directly at the door of research. This receiver consists of three stages of radio-frequency amplification, a non-regenerative detector and two stages of audio-frequency amplification. It is known as the Fada "Special." But before we can enter into the mechanics of the receiver, we must first ascertain why the electrical design used was selected.

The problem placed before the engineering department was the development of a radio receiver limited to six tubes. The apportioning of the tubes was the first problem. How many stages of radio-frequency amplification should there be; and how many stages of audio-frequency amplification? Having developed audio-frequency transformers with known response characteristics and known gain per stage, and knowing that a two-stage audio amplifying system using their transformers would give the proper amount of amplification, the engineering department decided upon two stages of transformer-coupled audio amplification. Since the detector unit utilizes but one tube, the remaining three tubes can be applied to the radio-frequency system.

The development of the radio-frequency system brings to light many interesting features. Should the stages be tuned or untuned or a combination of both? Should the stages be shielded,

and what material shall be used for the shielding? Since the receiver utilizes an antenna as the pick-up system, the three stages of radio-frequency amplification will give a high degree of sensitivity. The demand for selectivity necessitates the use of tuned stages of radio-frequency amplification. But the development of a three-stage tuned radio-frequency amplifier does not mean a simple decision to use three stages. Consideration must be accorded to the wavelength response of such a system. The average system possesses wavelength characteristics which fall in amplification as the wavelength is increased i.e., the amplifying power of the radio-frequency amplifier is high at the shorter wavelengths and as the tuning dials are manipulated to tune to the longer wavelengths, the amplification decreases, and at 550 meters is a fraction of that at 250 or 300 meters. This situation must be avoided; it is desirable that the receiver should possess equal amplification over the entire broadcast frequency spectrum.

Since the allocation of frequencies to broadcasting stations is such that excellent selectivity can be obtained with two stages of well-designed and shielded tuned radio-frequency amplification, the third stage can be utilized to balance the two tuned stages and give the system the desired wavelength response characteristic. The radio-frequency system would, therefore, consist of two stages of tuned radio-frequency amplification and one stage of untuned radio-frequency amplification.

The decision to shield the individual stages was immediate, since shielding, if properly carried out, is conducive to better radio receiver operation and consequently better radio reception and better stability is thus attained in the radio-frequency stages because coil interaction is eliminated. By the elimination of coil interaction, neutralization is made more effective. Better tone quality is also obtained, because by eliminating coil interaction, the side-band characteristics planned in the design of the tuned stages are actually obtained. Selectivity is augmented, because direct coil pick-up is precluded. The elimination of coil pick-up also means greater amplification in the radio-frequency system.

The selection of the shielding material is

governed by the effect the shield has upon the inductances used in each stage. In order to minimize the electrical effect upon the coils, a material with a high conductivity must be used, since high conductivity means lower losses. Aluminum was decided upon, and the shield takes the form of a can, completely enclosing the radio-frequency transformer. With the shields of proper diameter and properly located with respect to the coils, the losses introduced are so small as to be entirely negligible.

In view of the fact that the radio-frequency coils are shielded, it is possible to make use of the most efficient type of winding—the single-layer solenoid. Without shields, a cascade system employing such coils would be quite difficult to control. The selection of the single-layer solenoid was also based upon the fact that it can be wound with the greatest degree of accuracy, particularly so when the winding form is grooved, and the turns are wound in these grooves.

The receiver is to be dual tuned, requiring two condensers controlled by one drum dial and another single condenser controlled by the other drum dial. Such control is simple because of the precision methods employed in the testing and matching of the coils and tuning condensers. Each tuned transformer consists of three windings, the primary, the secondary, and a neutralizing winding. Each of these windings is matched on a radio-frequency testing instrument, to within one eighth of one per cent. The coil under test is plugged into an oscillator circuit and a resonance point obtained with a standard condenser. This condenser is so graduated that a 10 per cent. variation in resonance is spread out across the 100-division dial. The dial settings for the resonance point for each winding are noted and the coils segregated according to these figures. The result is that each group of three coils consists of coils with windings which never vary more than one eighth of one per cent.

The same precision in testing applies to the tuning condensers, and because of the mechanical design of these condensers, full accuracy is maintained during the operating life of the unit. Each completed tuning condenser is made to within one per cent. plus or minus, of its rated capacity, and each condenser in a group of three is matched to within one eighth of one per cent. of the others. The matching of the variable condensers is carried out by means of a special radio-frequency measuring instrument designed for the purpose.

The construction of variable condensers which will not vary more than one per cent. calls for detailed engineering. The brass used for the plates must be very accurate, the tolerance limit being 0.0005 of a mil in thickness. To assure perfect alignment of the condenser plates, and a smooth rotary action, large bearings are used, these latter being approximately ⅜", in diameter. To further assure perfect alignment of the brass plates, each plate is individually stipled and leveled.

But the precision construction and matching of coils and condensers is not sufficient to assure perfect operation. It is necessary to assure perfect mechanical support for these units—supports which will be identical in every receiver of similar design. It is necessary to select a base for the condensers which will assure easy operation, not for a short while, but for years to come.

Again engineering comes to the fore, with a pressed steel chassis ⅛" thick, punched out on a 100-ton press. One operation punches all the holes necessary and also forms the chassis. The result is uniformity of mounting holes.

R. F. CHARACTERISTICS

WITH the condensers, coils, and shields on hand, we go back to the radio-frequency system. The overall gain curve of the three-stage radio-frequency amplifier, consisting of the two tuned stages and the one untuned stage, is shown in Fig. 1. Here we see a beautiful example of research and engineering. With the exception of the zone between 200 and 212 meters, (1500 and 1410 kc.) the amplification does not vary more than 11 per cent. from 212 to 550 meters. Between 200 and 212 meters, the curve rises with the increase in wavelength, and the difference between the lowest point, 200 meters, and the highest, 250 and 500 meters, is only 17 per cent. With such small variance in amplifying power, the owner can manipulate the dials of his receiver from 200 to 550 meters, and know that the sensitivity of the system is practically uniform over the complete scale.

But the design of a radio-frequency system does not consist solely of the development of an amplifier which will possess the wavelength response curve shown. It is also imperative to accord detailed consideration to the shape of the resonance curve of each individual tuned stage, since the resonance curve manifests a great influence upon the tone quality of the receiver. In this respect, there is a close association between the radio-frequency amplifier and the audio-frequency system. Many owners of radio receivers are unaware of the effect the resonance curve of the radio frequency stages has upon the tone quality obtainable with the receiver employing three stages of radio-frequency amplification. Fans are too prone to overlook the side-band characteristics of the radio-frequency stage. They forget that while the radio-frequency stage is tuned to the frequency of the carrier wave, it is also necessary to consider that this carrier wave is modulated by audio frequencies ranging from 30 to 5000 cycles. Also that the effect of these modulating frequencies is to create a modulated carrier wave whose frequency spectrum is 10,000 cycles wide. In other words, if the carrier wave (unmodulated) is 750,000 cycles (400 meters) when modulated, this wave is broadened to cover from 745,000 to 755,000 cycles. The 5000 cycles above the carrier and the 5000-cycle-band below the carrier constitutes the side-bands. Hence the resonance curve of the radio-frequency stage must be broad enough to cover this band of 10,000 cycles even though the circuit is actually tuned to 750,000 cycles. If the curve is too sharp, some of the higher side-band frequencies will be suppressed. If the curve is too broad, selectivity will be marred. Hence both selectivity and sideband suppression must be considered in the design of the radio frequency amplifier. With a known value of "Q", which is the factor of selectivity, being the ratio of the reactance to the resistance of the circuit at a certain frequency, it is imperative to know the side-band suppression in the radio-frequency amplifier and to give it consideration in the design of the associated audio-frequency amplifying equipment. An example of various degrees of sideband suppression in tuned circuits is

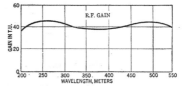

R.F. GAIN

GAIN IN T.U.

WAVELENGTH, METERS

FIG. 1

The curve shows the gain from the antenna to the input to the detector. The sensitivity of the r. f. amplifier is high and the amplification flat

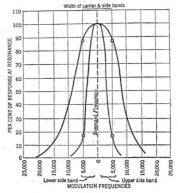

Width of carrier & side bands

PER CENT OF RESPONSE AT RESONANCE

Resonant Frequency

Lower side band ◄—— ——► Upper side band
MODULATION FREQUENCIES

FIG. 2

Tuned circuits that are too selective impair quality. The outer curve shows the response in a well-designed circuit

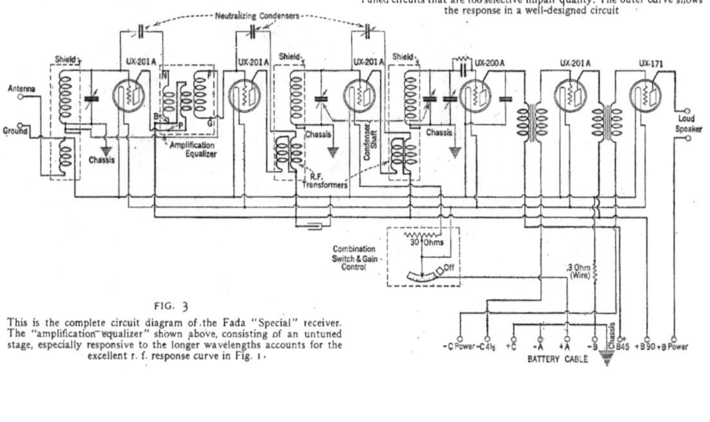

FIG. 3

This is the complete circuit diagram of the Fada "Special" receiver. The "amplification equalizer" shown above, consisting of an untuned stage, especially responsive to the longer wavelengths accounts for the excellent r. f. response curve in Fig. 1.

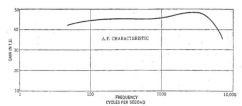

FIG. 4, OVERALL A. F. CHARACTERISTICS

shown in Fig. 2. Curve A shows 80 per cent. suppression on 5000 cycles and curve B shows 15 per cent suppression at 5000 cycles. Curve B is broader than A, but the proper amount of selectivity is obtained by virtue of the cumulative effect of a number of stages.

THE DETECTOR

FROM the radio-frequency system we pass on to the detector circuit. A choice of two systems of detection is available—the grid bias method or the grid leak-condenser arrangement. Because of the increased sensitivity, resulting in greater output, the grid leak-condenser system is used.

From the detector we pass to the audio-frequency amplifying system. We made mention in a previous paragraph that transformers were used, but the design of a transformer-coupled audio amplifier cannot be consummated by simply deciding upon transformers. Sometimes these characteristics of the transformers to be used are exactly what the requirements call for; sometimes they are not. With specific requirements on hand, audio-frequency transformers must be designed to fill the need. The design is a detailed process. First the tubes to be used must be decided upon, and their electrical constants must be taken into consideration. The core material for the transformers must be selected, and the inductance of the primary and secondary windings must be calculated in order that the transformer possess certain predetermined characteristics. The method of winding must be decided upon so that distributed capacity is low and so that satisfactory response on the higher audio frequencies is obtained.

Detailed consideration must be accorded to the regeneration existing in the completed audio-frequency amplifier. This is very important. The overall response curve of a two-stage audio system with regeneration in the amplifier will differ from that of a single unit. If the single unit is designed to match the radio-frequency system, the operation of the completed two-stage amplifier will be entirely different. It is also essential to consider the loud speaker to be used. This unit, too, possesses operating characteristics which must be taken into account. The combined operating characteristics of the radio-frequency amplifying system and the audio-frequency amplifying system must be such as to produce best results with a particular loud speaker or with a group of good loud speakers.

The completed two-stage audio amplifier of the receiver under consideration—the Fada "Special" possesses the overall audio frequency characteristics shown in Fig. 4. The amplification is shown on the ordinate or the left vertical line. The audio frequencies are shown on the abscissa or the horizontal line. The frequencies

are plotted on a logarithmic scale. As is evident, the curve is practically flat from 50 to 1000 cycles, rises from 1000 to 3000 cycles, and then falls gradually to 5000 cycles. The maximum difference in amplification between the lowest and the highest points is only 12.5 per cent., which difference is negligible, since the average ear will not discern intensity variations of such small proportions.

The development of the receiver is completed. Let us now consider the engineering involved in the testing of the receiver. Each receiver must undergo various tests during the process of production. The designing of this testing equipment is also in the hands of the engineering staff. Without testing equipment all the effort placed in the design of the individual parts and systems will have been for naught. Without a testing department the life of a radio plant would be very short.

The first test is a continuity test of the assembled chassis. This makes necessary testing with meters in each and every circuit, showing the voltage across the tube filament, the filament current, the plate current, the plate voltage and, continuity in the grid circuit. The filament current and filament-voltage meters show the operating action of the units incorporated in these circuits. The same is true of the plate-voltage and plate-current meters. Open circuits in the plate circuit will be shown on these me-

ters. By simultaneously testing all the circuits, it is easy to select the faulty circuit if one is present in the receiver. The location of the fault is also noted. By having meters in every circuit it is unnecessary to hunt haphazardly.

The second test is to determine the efficacy of the neutralizing system, and the adjustment of the neutralizing units. In this test the assembled and wired chassis is connected to a series of meters, and the input system is coupled to a dummy antenna which obtains its energy from a local radio-frequency oscillator. The dummy antenna simulates an average outdoor installation. The meters show excessive regeneration in any of the tuned circuits, when these circuits are made resonant to the frequency of the oscillator. The neutralizing system is then adjusted until all signs of excessive regeneration in the radio-frequency amplifying system disappears. Incidentally, this same method of testing is employed to determine the overall gain of the radio-frequency amplifier.

When measuring the amplifying power of the radio-frequency system, from the grid of the first radio-frequency amplifying tube to the grid of the detector tube, a constant predetermined radio-frequency signal is fed into the radio-frequency system. The input voltage is held constant on all wavelengths covered by the tuning system. The voltage across the grid filament circuit of the detector tube is measured with a vacuum-tube voltmeter.

The third test applied to the receiver is the "air" test, i.e., the receiver is connected to an outdoor antenna and outside broadcasting stations are tuned-in. This test is a final check of all the tests applied to the receiver during the process of manufacture. The overall gain of the radio-frequency amplifier and the audio-frequency amplifier is again ascertained. With respect to the measurements of the audio-frequency system and the transformers used, each transformer is individually tested against a standard before being placed into service in the amplifier. Then the completed two-stage unit is again tested under actual operating conditions. With a known constant input, the total gain is finally measured with a tube voltmeter—the last of a series of thorough and efficacious tests.

WHAT THE CHASSIS LOOKS LIKE

"Our Readers Suggest—"

*T*WO pages of RADIO BROADCAST *will regularly be devoted to publishing contributions from readers who have made interesting improvements in the use of ready-made radio products. These suggestions may deal with complete radio receivers, socket-power units, "kinks" in the placing of loud speakers, or slight circuit or mechanical changes in apparatus in general use. Our readers have a wealth of experience along these lines and these pages offer an opportunity for them to share their findings. Typewritten contributions from readers are welcomed which, if published, will be paid for at our regular space rates. In addition, a monthly award of $10 will be paid for the best contribution published each month. Address all contributions to Complete Set Editor, RADIO BROADCAST, Garden City, New York.—THE EDITOR.*

A Short-Wave Converter for any Radio Receiver

THERE is to-day sufficient material being broadcast below 100 meters (3000 kc.) to interest the serious fan and to justify the construction of simple apparatus for its reception. In some cases the use of a short-wave receiver will make possible the reception of important programs beyond the range of the conventional receiver. The construction of a short-wave receiver is often an expensive proposition, and converters heretofore described have been rather complicated affairs. It is the intention of the writer to describe a simple and inexpensive converter which, when attached instantly to any broadcasting set, makes it possible to receive on wavelengths between 15 and 125 meters (20,000 and 2400 kc.) No change is made in the present receiver but by means of the converter the former is alternately available for short- or broadcast-wave reception.

THE SHORT-WAVE CONVERTER

The short-wave converter takes the form of a very simple and incidentally highly efficient short-wave receiver, the output of which is connected to the audio-frequency amplifier of the present broadcast receiver. A simple plug-in arrangement makes the change a matter of a few seconds.

The following is a list of the parts used in the short-wave converter illustrated and described:

L1, L2, L3—Set Aero Short-Wave Coils.
C1—Amsco 0.00025-Mfd. S. F. L. Condenser.
L4—Silver-Marshall Choke, No. 275
R1—Clarostat 0-500,000-Ohm Resistor.
R2—Amsco 3-Megohm "Grid Gate," with Mounting.
R3—Amsco 20-Ohm Rheostat.
C2—Tobe 0.00025-Mfd. Fixed Condenser.
C3—Tobe 0.001-Mfd. Fixed Condenser.
Three Four-Foot Lengths of Flexible Wire.
Sub-Panel Brackets, Hardware, and Old Tube Base.
National Type C Dial.
Amsco Floating Socket.
7 x 12 x 3/16 Inch Celeron Panel.
7 x 11 x 1/2 Inch Wood Baseboard.
Two "XL" Laboratories "Push" Binding Posts.

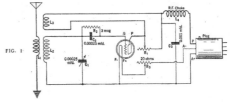

THE FRONT PANEL

The construction of the short-wave converter is best described in the accompanying illustrations. However, a word regarding the connecting plug may be of assistance.

The three wires leading respectively from the radio-frequency choke coil, A-battery minus, and A-battery plus, are led to the base of a discarded tube, as made clear by reference to Fig. 1. The glass of the old vacuum tube is broken off and the base cleaned out. The three wires are soldered to terminals inside the base, one to the A-plus plug, one to the A-minus plug, and one to the plate terminal. These terminals may be identified by holding the tube base, bottom down and the side pin toward you. With the base in this position, the two rear posts are A plus and A minus respectively from left to right, and the

left-hand front post is the plate terminal. The base of the tube is now filled with a wax compound, such as the top of a discarded B battery. This is easily done by placing small pieces of the wax in the tube base and melting them with a hot soldering iron. The receiver may be wired with bus bar, but the author found coded flexible wire more convenient. All leads should be made as short as possible.

The function of the choke coil is important. If the Silver-Marshall one is not available, one may be made by winding 100 turns of 26 d.c.c. wire at random on a wood spool, 1/2 inch in diameter with 3/8-inch wooden core.

To operate the short-wave converter, remove the detector tube from the regular broadcast receiving set and place it in the tube socket of the converter. Next select the plug-in coil from the Aero set covering the wave band in which you wish to receive, and plug it into the coil jacks. Then insert in the detector tube socket of the regular broadcast receiver the plug made from the old tube base. When the antenna and ground have been changed to their respective posts on the converter, you are ready to listen-in. To do so simply leave the loud speaker where it is, or, if phones are used, these may be plugged in as usual in any stage for which a jack is provided on your particular receiving set. Turn the Clarostat until the receiver oscillates. Tune-in a station and clear up the signal by a further adjustment of the Clarostat or rheostat as required.

PERRY S. GRAFFAM.
Boston, Massachusetts.

STAFF COMMENT

WHILE the importance of short-wave reception cannot be overstressed, statements regarding its immediate and direct utility to the fan must be qualified. RADIO BROADCAST does not care to encourage the use of radiating short-wave receivers, and the more simple sets necessarily fall into this category. Serious experimenters, broadcast enthusiasts desiring to take up code work, and fans in isolated districts are, however, undoubtedly justified in conducting experiments along these lines. Mr. Graffam's inexpensive arrangement offers perhaps the simplest introduction into the realm of megacycles.

Short-wave reception is by no means an un-alloyed bliss as some avid publicity men would have us believe. Ninety-eight per cent. of the transmissions carried on below 100 meters is inter-communicative code work and the two per cent. of radio telephonic transmission is often marred by high speed fading.

Antenna Compensation in a Single-Control Receiver

ONE of the major problems in single-control multi-tuned circuit receivers is the elimination of the detuning effect of the antenna on the first radio frequency stage. Loose and variable coupling between the antenna primary and the first r.f. secondary is generally employed to compensate this influence. These arrangements, unfortunately, often lower the signal response of the receiver, and the set still functions best with antennas of definite electrical characteristics.

The arrangement proposed overcomes these difficulties and offers the following advantages:

It eliminates the antenna effect on any receiver.
It can be attached to any receiver without making more than one simple change.
No additional controls are required.
Sensitivity is never reduced. On the contrary it is often increased.
Any length antenna may be used with the receiver without making additional changes.
The device acts as a partial blocking stage in case oscillations are set up in the tuned amplifiers.

In brief, the device causes the radio-frequency impulses to be applied across a resistor to an extra radio-frequency tube, which is outputted to the original antenna primary.

The following is a complete list of parts necessary to make the change:

C_1—0.001-Mfd. Coupling Condenser. L_4—R. F. Choke Coil. Socket. 201-A Type Tube. R_1—$\frac{1}{2}$-Ampere Ballast Resistor. Sw.—Battery Switch. R_2—1000-Ohm Resistor.   Six Binding Posts.

This apparatus is easily wired on a baseboard in accordance with the diagram, Fig. 2.

The antenna is disconnected from the receiver and wired to post number one. The ground remains connected to the receiver and is also wired to post number two of the new stage.

Turn on the filaments to the receiver proper and the switch to the extra stage. Run a wire from the A battery plus post on the set to post number four. The extra tube will probably light. If it does not light, repeat the test with a wire from the negative A post.

If the extra tube lights with one side grounded and the other side connected to the A battery circuit, it is an indication that one side of the A circuit is grounded, as it will be in 90 per cent. of receivers. If this is the case, proceed as follows:

Leave the wire that lights the filament connected to post number four. Connect the antenna post on the receiver to post number five.

If, upon making the tests with the filament wires to post number four, the tube does not light, indicating that the filament circuit is not grounded in the receiver, the filament plus wire should be connected to post four and the filament minus wire to post three. The tube will now light, of course. Connect the antenna post on the receiver to post number five, and terminal six to the plus 90 volts, and your set is ready for operation.

There is no change in the operation of the receiver whatever. If taps are provided on the antenna primary of the original receiver, slightly higher efficiency may be secured by experimenting with them.

If the experimenter is a bit more of an expert,

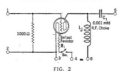

FIG. 2

the filament lead, or leads, to the antenna coupler tube may be led from the tube side of the set switch, making it possible to operate the antenna coupling tube by the same switch that controls the set filaments.

HAROLD BOYD.
Winchester, Virginia.

AN EMERGENCY AN-TENNA

Fifty turns of bell wire are usually satisfactory

STAFF COMMENT

THE arrangement suggested by Mr. Boyd is quite practicable and is already used in several commercial and kit receivers. The tube coupling device is in the nature of an untuned radio-frequency amplifying stage, and, in some instances, may result in increased sensitivity. The r.f. gain, however, is generally small in comparison with the utility of the arrangement.

One thousand ohms is about the optimum value for the antenna coupling resistor. It seems to make little difference whether the resistor is inductive or not. The Lynch, Amsco "Metaloid," Carter, and Electrad are excellent resistors for this purpose. A variable zero to 2000 ohm resistor may be substituted for the fixed resistor recommended by Mr. Boyd and used as a volume control.

In a few cases it may be found that the load imposed by the coupling arrangement on the first r.f. transformer is no more favorable to tandem tuning than that of the average antenna. This can be compensated either by adjusting the compensating condenser across the first tuning section (if the main condenser is so equipped), or by winding a special primary over the first secondary. From six to ten turns of wire may be placed directly over the coil. One end is connected to post number five in Fig. 2 and the other terminal to plus 90 volts.

Increasing Response from a Loop Receiver

ALL receivers are more or less affected by location, but the loop set seems to be particularly susceptible to adverse conditions imposed by position. This is due to the fact that the loop is often surrounded by steel building framework and similar obstructions which the open antenna can rise above. Disadvantages of this nature were impressed upon the writer when moving from the top floor of a New York City apartment house to the second floor of the same building. The receiver, a

Radiola super-heterodyne, worked perfectly in its original position eight floors higher up, but lost perhaps seventy-five per cent. in sensitivity when brought a hundred feet nearer the ground.

Its operation was brought up to normal by a simple device, thrown together in five minutes. A twenty-five foot antenna was strung on the roof of the building. Three turns of wire, harmonizing with the wire on the loop, were wound around the loop frame, twisting slightly about the original loop wire to keep it in place. One end of this extra wire was connected to the short antenna, while the other end ran to ground.

Sufficient energy is transferred to the loop circuit from the open antenna to compensate the losses imposed by an inferior location.

A. J. HOWE.
New York City.

STAFF COMMENT

THE device recommended by our contributor is, of course, a simple antenna coupler, the extra turns of wire functioning as the primary and the loop itself as the secondary. The arrangement is effective. On the ordinary super-heterodyne (that is other than the second harmonic type) the shortest antenna giving satisfactory reception should be used, to reduce the possibility of radiation.

The directional effect of the loop is largely eliminated by coupling to an open antenna in this manner. However, the selectivity of the super-heterodyne is such that this effect can be safely dispensed with.

A coupling device of this type is made commercially by the Jenkins Radio Company, Davenport, Iowa.

A Temporary Antenna for the Traveling Fan

EXTRA tubes and batteries are easily available in an emergency, but the occasion is seldom provided for, so a word as to an excellent makeshift antenna may not be amiss. A first class one, often equal to the average outdoor variety, may be secured by wrapping fifty turns of bell wire around a telephone desk stand. See Fig. 3. One end of the wire is connected to the antenna post on the receiver and the usual ground is used.

The writer, an inveterate radio enthusiast, discovered the possibilities of this arrangement when traveling across the country with a Fada neutrodyne, the operation of the receiver being somewhat limited by the facilities of the average hotel bedroom.

Subsequent experiments show this type of antenna to be equally efficacious with other receivers.

ALFRED A. MARKSON.
New York City.

STAFF COMMENT

THE use of the telephone as a substitute antenna is by no means a novelty, although the exact system of connection outlined here is not that generally advocated, but probably as efficient. The more widely used application of this idea is found in the use of a small metal plate upon which the telephone is stood, and which is connected to the antenna binding post of the receiver. Such metal plates, especially cut for the purpose, are commercially available. As the latter are not always immediately obtainable in an emergency, Mr. Markson's idea is a useful one.

HOW TO IMPROVE YOUR OLD RECEIVER

A radio set of other years can be brought up to date by improving the audio quality through new transformers, tubes and loud speakers or by the purchase of a complete power-supply-amplifier unit. These changes help greatly. R.f. Changes are not suggested

By EDGAR H. FELIX

THE articles appearing in the September and October issues of RADIO BROADCAST, dealing with the judging and attainment of good tone quality, resulted in hundreds of letters from readers, asking specifically how certain makes of receivers could be converted to give the high-grade tone quality described. It was the writer's intention to answer the letters directly in these columns, but their number grew so large that it would require an entire issue of the magazine to meet the demand for information. This article is based upon the questions raised in the letters and will serve as a concentrated answer to these letters.

Hundreds of thousands of radio enthusiasts are owners of receiving sets sufficiently selective to be satisfactory, but falling far short of the latest standards of tonal reproduction. So long as the receiver meets the simple requirement of being sufficiently selective, but not too selective, it can be converted to give good tone quality. The writer does not mean to imply that the radio-frequency end of the modern receiver is not as greatly improved as the audio end and that the most satisfactory measure, after all, is not to discard entirely the obsolete receiver. But not everyone is in a position to employ this remedy; some of us must reconcile ourselves at the high frequencies, where it is not especially needed, and lack of sensitiveness at the low-frequency end, where it is most desired. Simplicity of control, and equal amplification throughout the wavelength scale, are features embodied only in the latest receivers. But, given satisfactory selectivity, an old receiver may be greatly improved so far as tone quality is concerned.

Exceedingly sharp tuning, such that high-power stations within fifty or a hundred miles are heard with considerable volume only when tuned-in precisely and always disappear with a whizz and hiss when detuned but one or two degrees from exact resonance, indicates selectivity too great for the attainment of good tone quality. Oftentimes a receiver behaving in this way can be made to tune more broadly, so that neither the low or high audio frequencies are cut off, by installing a somewhat longer antenna.

Having once determined that the radio-frequency end of the receiver does not tune too sharply, improvement of tonal quality is a matter of re-vamping the audio system. The essential requirements for good tonal quality are: (1) Audio-frequency transformers of sufficiently good quality to pass the entire tonal range; (2) tubes of sufficient power and emission to adequately handle signals of considerable magnitude (3) a power supply assuring correct A, B, and C voltages to every tube under actual operating conditions; and (4) a loud speaker capable of setting up sound waves throughout the audio scale.

Prior to recent developments in transformer design and material, resistance-, and impedance-coupled amplification were the only systems, within reach of the experimenter, capable of handling broad tonal range. These systems under proper conditions are not excelled in quality output by high-grade modern transformers, but require an extra stage so that they are not easily incorporated in a manufactured receiver, unless it is especially designed to accommodate them. During the last year, transformer development has reached such a point that two stages may be used to give the best of tone quality.

Transformers can be manufactured at a cost as low as forty cents, although the actual raw materials which go into the better transformers cost as much as eight times that figure. Expensive iron alloys, which magnetize and demagnetize rapidly, and high-inductance windings, are essential if the entire tonal range is to be amplified: Under no circumstances, can cheap transformers serve as well as well known expensive ones. In replacing transformers, to make the job worth while, confine yourself to the best. Some of the better receiving sets of earlier vintages, are not equipped with suitable transformers and the substitution of such makes as Amertran, Ferranti, Silver Marshall, Thordarson, General Radio, Rauland Lyric, Modern, All American, Pacent, Sangamo, and Samson, to mention some of the better ones, is decidedly worth while.

REPLACING OLD TRANSFORMERS

TO DETERMINE whether such substitution is feasible, open the cabinet and examine the audio-frequency transformers. See if they are easily removed and if the four terminals are so marked that you can put labels on the wires before you remove them, indicating the correct filament, grid, plate, and B+terminals. This will prevent confusion when you put the new transformers in place. Adhesive tape is a convenient form of label. Measure the space available for transformers because cheap transformers are often small. The high grade ones, with which you replace them, are likely to be somewhat larger and hence may not fit in the space provided for the old transformers. Where the problem requires moving of sockets and other parts, your local dealer can replace the transformers for you. His charge should be between two to five dollars, plus the cost of the transformers themselves.

The next link in the chain of audio reproduction concerns the tubes used. You cannot hope to secure good quality, if you do not use a power tube in the last stage. The UX-201-A (CX-301-A) tubes in the output stage are capable of only moderate volume with good quality. If you are attaining fair quality with such tubes now, after replacement of the transformers they may prove unsatisfactory, because the added energy in the low frequencies, impressed upon the output tube by the new transformers, will not be handled satisfactorily.

In the case of the storage-battery receiver, wired with but a single C battery connection, both the first and second stages are usually supplied with the same C battery voltage, generally 4½. Re-wiring of the set, however, is not necessary to put in an UX-171 (CX-371) or a UX-112 or CX-312. Manufacturers such as Na-Ald, have developed special sockets with flexible cable connections, enabling you to add the necessary grid and plate voltages to operate power tubes, without any wiring changes.

There is one exception to the general rule that replacement of the transformers and addition of a power tube will bring you better tone quality. Certain reflex receivers, which enjoyed a fairly wide sale three and four years ago, are not adapted to this simple conversion. The use of a grid bias and plate potential satisfactory for audio purposes may render the radio-frequency amplifier of these reflex sets quite unstable. Many of these receivers are such heavy consumers of plate current that discarding them is an economy. It is not worth while to attempt to improve them. You must treat them as you would an inherited 1902 one cylinder-automobile.

With dry-cell tube receivers, the largest output tube available is the 120 type. This is a great improvement in power handling capacity over the 199 type, but it is still far from sufficient to attain really good tonal quality. The further down the tonal range the reproducing system goes—and that is what gives body and richness to music and naturalness to speech—the greater must be the power available in the output tube.

The owner of such a dry-cell tube receiver need not, however, abandon hope, because he may employ a one-stage power amplifier, receiving its filament, grid and plate potentials, directly from the light socket, and employing the UX-210 (CX-310) in the output. This tube is of even greater power handling capacity than the UX-171 (CX-371) and, hence, capable of magnificent tonal quality, provided good transformers and reproducers are used in connection with them. These light socket units may also be used with storage battery outfits and are recommended to B battery users. The use of a power tube in the output stage considerably increases B battery drain and, as a consequence, the use of a light socket amplifier unit is an economy.

Such power supply devices are manufactured by General Radio, Farrand, Radio Receptor, Pacent, National, Timmons, Amertran, and the Radio Corporation. They furnish A, B, and C power, not only for the 210 or 171 tube, but also B and in some cases C voltages for the receiving set itself. Adding these amplifier-power supply devices to the existing receiver is a simple matter. The tonal reproduction secured is still dependent upon the grade of loud speaker and first stage transformer used but, so far as available power is concerned, the purchase of a good power amplifier and B supply unit settles that question.

SELECTING A REPRODUCER

HAVING now supplied transformers which actually amplify the entire range of tonal frequencies, having installed tubes of adequate power handling capacity, and having supplied them with the correct A, B, and C voltages, it is next necessary to obtain a loud speaker capable of setting up sound waves throughout the entire tonal range. A loud speaker which is seemingly satisfactory with poor transformers and power supply, often fails when the high-grade transformers and tubes are wired into circuit, because of the larger load and greater frequency range thereby impressed upon it.

Remedying an audio system requires that the entire audio system be put in good condition, because any one weak link will destroy the effectiveness of all the other remedial measures. If you have four worn out tires on your car, replacing one, two, or three of them is not sufficient to give you immunity from tire trouble. Many a person, dissatisfied with tone quality, has replaced his loud speaker and then wondered why great improvement did not result. In fact, it often happens that an exceptionally good loud speaker will sound worse than a bad one with a poor set. The good loud speaker sets up sound waves in exact accordance with the electric signal furnished it. A poor loud speaker may be so designed as to exaggerate the low notes, thus providing for their deficiency in a defective au-

dio system. When a good loud speaker is substituted, the absence of low notes, due to unsuitable transformers, becomes more conspicuously apparent.

Every reproducer has an actuating element which sets up the sound waves—a sort of paddle which sets up air vibrations. With good reproducing systems, the loud speaker must be capable of setting up low frequencies as well as high ones, and consequently the actuating element, vibrating diaphragm, or cone surface, must often be large if it is to be successful. The horn, if one is used, must also be of large size, so that it does not impede the radiation of low frequencies. A long, goose-necked horn chokes off the low frequencies, while a large exponential horn can give you much of the true magnificence of the organ.

The writer cannot attempt to list entirely all good cones and horns, but he has personally tested Western Electric, Farrand Sr., Balsa, Rola, and Amplion, and found them capable of handling the output of 171 and 210 type tubes throughout the tonal range attainable by the best of amplifier systems.

Inferior loud speakers fail not only because they are incapable of mechanically setting up waves by reason of small moving surface or confined tubular horn areas, but also because the electromagnetic unit is incapable of handling the large output which is necessary to secure good tone. With the 210 and 171 types of tubes, an output transformer or choke and condenser feed to the loud speaker is absolutely necessary to eliminate the direct-current component from the speaker winding. We desire only to have the audio-frequency fluctuations in the loud speaker winding so that magnetic saturation is avoided. Silver-Marshall, General Radio, Federal, National, Pacent, Samson, Thordarson, Amertran, Muter, Amsco, and others make output devices.

One question which appeared in many of the letters received was the demand for more volume, or specific questions to the same effect, such as whether the use of a 171 tube would increase volume. None of the measures described have for their purpose the increasing of volume output of the receiver. The use of large power tubes simply increases the amount of signal volume which can be handled without distortion due to overloading.

By using the grade of transformers, tubes, and loud speakers mentioned, a signal so weak that it can hardly be distinguished by the phones in the detector output circuit is amplified to comfortable living room volume. If the signal is not sufficiently strong to give such volume, the remedy does not lie in additional audio-frequency amplification but in the use of a more sensitive receiving set. The best results are obtained if the detector tube's output is a signal just strong enough to be discernible in the headphones. Those complaining of weak signals should look to improving antennas and to increasing radio-frequency amplification. The audio system should not be expected to make up for deficiencies in the radio-frequency amplifier.

As a matter of fact, most of the receiving sets, even those of two and three years ago, are adequately sensitive. Many complaints of reduced volume may be attributed to the continued use of depreciated power supply and tubes, whose filaments have lost their emission.

It is an essential requirement of good tone that

A Quality Five-Tube A. C. Receiver

By JAMES MILLEN

FRONT VIEW OF THE RECEIVER

THIS article describes the construction of an a.c. operated receiver, the new type a.c. tubes being used to accomplish the electrification. In the preceding article in this series, published in last month's RADIO BROADCAST, some general information on a.c. tubes was given.

In explaining how to use these a.c. tubes with an actual receiver, we have chosen a circuit which embodies some of the features of the Browning-Drake receiver. Strict adherence to the instructions contained in this article will result in a light socket operated receiver equaling in sensitivity and selectivity a receiver operated on standard storage-battery type tubes.

Before going into details regarding the construction of this a.c. receiver, we will point out how the circuit differs from that of Browning-Drake sets. First, let us consider the r.f. amplifier. Most previous designs of the Browning-Drake receiver have used a 199 type tube as the r.f. amplifier, because the tendency for this tube to oscillate is less than with a 201-A type tube. A. c. tubes, however, have characteristics similar to the latter type and, since an a.c. tube is used

in the r.f. stage of the receiver described here, it becomes necessary to devise some practical method of preventing this r.f. amplifier from oscillating.

In Fig. 4 is shown, at "A," the circuit of the original Browning-Drake radio-frequency amplifier using Hazeltine neutralization; at "B" we see the Browning-Drake circuit using the Rice system of neutralization. At "C" is shown the circuit arrangement for use of a 226 type a.c. tube. In the grid circuit will be noticed a non-inductive resistor, having an ohmic value of approximately 1000 ohms. It is the use of this grid resistor or, as it is more generally called, "suppressor," that makes possible the balancing of the circuit. As this resistor is not placed in the tuned circuit, it has no detrimental effect upon the selectivity of the receiver.

The arrangement shown at "C," Fig. 4, is used in the final model of the receiver illustrated in this article, and it will be noted that the plate voltage is fed to the plate of the tube through a choke coil, L_4, instead of through the primary winding of the r.f. transformer. The former method (feeding the voltage through a choke

coil) tends to somewhat stabilize the operation of the receiver, especially when a socket-power unit is used for the B supply.

The antenna is coupled to the first coil in the usual manner, i.e., through a 50-150 microfarad midget variable condenser, to a center tap on the coil. If the antenna is over 40 feet in length, the connections should be as indicated in the diagram, but if a shorter antenna is used, the lead from the midget condenser may connect directly to the grid end of the coil instead of the center. In congested localities, excellent reception, with greatly improved selectivity, is obtained by using a 3-foot length of bus bar for an antenna. The bus bar should be attached directly to the grid end of the coil.

THE AUDIO AMPLIFIER

THE audio channel employed is capable of producing excellent tone quality and at the same time lending itself equally well for use with either the new a.c. tubes or the standard storage battery tubes. The wiring of the audio channel is shown in the complete circuit diagram, Fig. 2.

In the first stage is employed an impedance

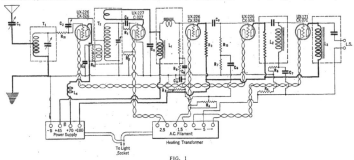

FIG. 1

Complete circuit diagram of the complete a.c. operated receiver

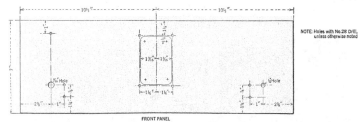

NOTE: Holes with No.28 Drill,
unless otherwise noted

FRONT PANEL

FIG. 2
How to drill the front panel

coupling unit which is incorporated a radio-
frequency choke coil; in the second stage, resistance coupling is used, while the third stage employs a special arrangement of resistance and
impedance with the impedance in the grid circuit-
of the power tube so as to eliminate any tendency
of the amplifier to "motor-boat" when used with
some types of B power units. A tone filter, L_2, is
incorporated in the output circuit as a protective
device for the loud speaker.

As will be noted from the circuit diagram, Fig. 1
resistance-capacity filters (R_6-C_7, R_7-C_6, and
R_9-C_9) are employed in the grid circuits of each
of the audio tubes. These filter circuits prevent
"motor-boating" and make the operation of the
audio amplifier entirely stable under all conditions. The grid bias resistances, R_2 and R_4, as
well as the mid-tapped resistance, R_9, across the
filament of the detector tube, are all of the fixed
variety so as to eliminate needless adjustment
and enable the home constructor to obtain the
same excellent performance from his receiver as
from the original laboratory model.

The various photographs, working drawings,
and circuit diagrams accompanying this article
give all the necessary details regarding the construction of this receiver and only a few ad-

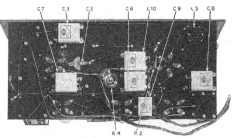

FROM THE UNDER SIDE
The letters correspond with the parts list on page 137. On page 33 of RADIO BROADCAST for
November, 1927, appeared a back panel view of this set to which reference may be made

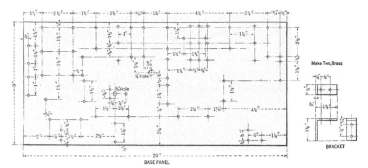

Make Two, Brass

BRACKET

BASE PANEL

FIG. 3
How to drill the base panel

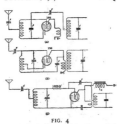

FIG. 4

The circuit of "C" is used in this set. This alteration in the usual Browning-Drake circuit is necessary because of the special problems arising from the use of a.c. tubes

ditional hints need be given in order to make possible the easy construction of the receiver.

As the illustrations show, the a.c. filament heating transformer is not built into the set, it being more convenient in this case to mount it as a separate unit within the usual battery compartment of the console if the latter is employed.

The first construction step is to drill the front and sub-panels in accordance with the details given in Figs. 2 and 3.

The next step is to mount the condensers, coils, sockets, and other parts on the sub-panel, as shown in the illustrations. Then, the sub-panel may be almost completely wired—all before attaching the front panel, which merely carries the dial, volume control resistor, and tickler adjustment.

The small General Radio neutralizing condenser should be disassembled and built right into the sub-panel after discarding the triangular bakelite back.

In the receiver shown in the illustrations, the resistor mountings and sockets were also disassembled and the contacts remounted directly on the sub-panel. Much needless work can be saved however by retaining the bases of the resistor mountings and sockets.

Instead of numerous binding posts the use of two cables is recommended. One cable should consist of the seven low-voltage a.c. leads to the filament transformer while the other cable should have four leads to the B power supply unit.

As the loud speaker cord may be plugged directly into the tip jacks on the front of the tone filter, it is necessary to provide but two binding posts—for the antenna and ground.

The wire to use for all connections as well as the cables should preferably be No. 18 tinned flexible rubber-covered wire. Such wire may be obtained from Acme and Corwin in different colors for making the cable and to facilitate the tracing of the set wiring itself.

With many UY-227 type detector tubes it will be found that the most satisfactory operation is obtained when the heater voltage is slightly below the rated 2.5 volts. For this reason it is recommended that a 6-inch length of resistance wire from an old 6-ohm rheostat be connected in series with one of the 2.5-volt a.c. leads, preferably right at the transformer terminal panel.

ADJUSTING AND OPERATING THE RECEIVER

THE adjustments necessary in order to obtain the best performance from the completed receiver are few and easily made. First, connect the antenna, ground, loud speaker, B-power unit, and filament heating transformer, and then turn on the 110-volt supply and wait for about a minute or so for the detector tube to reach its proper operating temperature. If a high-resistance voltmeter is available, the next step is to set the detector B voltage to approximately 45 and the r.f. B voltage to 70. Should a suitable voltmeter not be available, the r.f. and detector B voltage may be set by guess work and then readjusted for best results later. The next step is to set the potentiometer, R_1, for minimum hum. Generally this adjustment will be obtained when the contact arm is somewhere near the center of its arc. Occasionally, if the receiver should develop a slight hum, a slight readjustment of the potentiometer will remedy the trouble.

The antenna series condenser should be adjusted so that the two tuning condenser scales read somewhat alike.

The neutralizing condenser may now be adjusted so that the receiver does not oscillate at the shortest wavelength when the tickler coil is set for minimum regeneration. Generally the proper setting is with the movable plate of the neutralizing condenser turned in about a third of the way.

When making any of these adjustments, the volume control should be set for maximum volume in which position the receiver has the greatest tendency to oscillate. The following is a list of parts recommended for use in the receiver described in this article:

LIST OF PARTS

T₁—National B-D1E Tuning Unit (Without Dial)	$ 8.25
T₂—National B-D2E Tuning Unit (Without Dial)	11.75
National Drum Tuning Control	6.00
L₃—National Impedaformer, 1st Stage Type	5.50
L₄—National Impedaformer, 3rd Stage Type	5.50
L₅—National Tone Filter	6.50
R₂—General Radio No. 439 Center-Tap Resistor	.60
R₃—Lynch 500-ohm Suppressor	1.15
C₁—Precise No. 940 Midget Condenser, 50-150 Mmfd.	1.75
C₂—General Radio Midget Neutralizing Condenser	1.25
R₆, R₇, R₈—Lynch 0.1-Meg. Standard Metalized-Filament Resistors	2.25
R₇—Lynch 0.1-Meg. Type C Metal-ized-Filament Resistors	1.00
R₁₀—Lynch 0.5-Meg. Standard Metal-ized-Filament Resistors	.50
R₁₁—Lynch 1000-ohm "Suppressor"	.75
R₅—Lynch 2000-Ohm Type P Wire Wound Resistor	1.25
R₁—Lynch 2-Meg. Standard Metal-ized-Filament Resistor	.50
C₃, C₆, C₇, C₈, C₉—Tobe 1-Mfd. By-pass Condensers	4.50
C₇—Tobe 0.1-Mfd. Bypass Condenser	.60
R₉—Electrad Royalty Variable Resistor, Type K	1.50
C₄—Sangamo 0.00025-Mfd. Mica Condenser	.35
C₅—Sangamo 0.001-Mfd. Mica Condenser	.40
L₂—Samson No. 85 R. F. Choke	1.50
R₄—Carter 20-Ohm Midget Potentiometer	.75
Two Eby Binding Posts	.30
Four—General Radio No. 439 UX Sockets	2.00
One General Radio No. 438 UY Sockets	.50
Eight Lynch Single Resistor Mountings	2.80
Bakelite Panel, 7 x 21 Inches	2.75
Bakelite Sub-Panel, 9 x 20 Inches.	2.75
Wire, Etc.	.50
TOTAL	$75.70

ACCESSORIES

One CX-371 (UX-171) or Ceco J-71 Tube	$ 4.50
One CY-327 (UY-227) or Ceco R-27 Tube	6.00
Three CX-326 (UY-226) or Ceco R-26 Tubes	9.00
One UX-280 (CX-380)	5.00
One National No. F226 Filament Heating Transformer	10.00
One B-Power Unit, National Type M	40.00
TOTAL	$74.50

Looking Back

THE STORY OF RADIO. By Orrin E. Dunlap, Jr. Published by the Dial Press, New York. Price $2.50; pages, 226; illustrations, 15.

THIS book, by the Radio Editor of the New York *Times*, is a literary effort to squeeze some more thrills out of radio for the benefit of laymen who desire knowledge but do not want to struggle for it. There is nothing technical in its two hundred or so pages but, in a journalistic and often very interesting fashion, it gives a history of radio progress, and manages to touch on such topics as transmission theories, fading, radio direction finders, and piezo-electric control. The various branches of radio communication, such as aircraft work and transatlantic radio telephone circuits, are described; there are several pages on auditory phenomena; short waves and television are not neglected. The history of radio is told in the

first person, presumably by the spirit of the ether, or some loquacious band of waves. The effort to sustain an appeal to the imagination results in some very silly passages, the worst one appearing in the introduction:

Will the millions and billions of musical scores and countless numbers of spoken words ever return from the Infinite? Will the waves all roll back some day, all intermingled, the music of centuries, the works of all composers a hopeless jumble, a babel of voices, all so powerful electrically that the onslaught of invisible waves will burn up the ether and radio will be no more?

The answer is that this catastrophe will positively not occur, unless at that time God sees fit to suspend the second law of thermodynamics, retroactively.

But, with the exception of some of the chapter headings and captions, which have no conceivable relation to the text, this drivel is not

representative of the contents of Mr. Dunlap's book. He is, in point of fact, an old radio man, and a Member of the Institute of Radio Engineers, and when he remembers that there is only one inimitable Judge Rutherford, he does a good job. What he says is, in the main, accurate, and jazzed up only within the limits permissible in such a book. He has at his fingers' ends, or in his scrap book, about all the interesting things that have ever happened in radio, and in "The Story of Radio" relates them for an audience to which they will be utterly new. The events of the war in the radio field dramatic, SOS episodes, the old Herald station, OHX, silent these fifteen years, all live again in Dunlap's pages, and it is pleasing to see their appearance in a more or less permanent record. Give the book to your son as a birthday present, if you have not already bestowed it on him for Christmas.

—C. D.

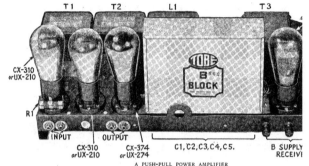

A PUSH-PULL POWER AMPLIFIER

The amplifier illustrated above is designed to give excellent quality reproduction of radio programs. The push-pull amplifier uses cx-310 (ux-210) power tubes which are capable of supplying to a loud speaker large amounts of undistorted power. The amplifier is constructed on a pressed steel sub-base

A NEW "TWO-TEN" POWER AMP

By William Morrison

THE combined push-pull power amplifier and light socket B power unit described on pages 163 and 164 of the July RADIO BROADCAST has recreated considerable interest in push-pull amplification. The device described consisted of a single-stage push-pull power amplifier built into a steel case and chassis assembly which also housed the power supply equipment. The latter furnished A, B, and C power to the push-pull amplifier and B power for the radio receiver as well. While this unit, termed for simplicity a "Unipac," possessed considerable merit, its power output, even with a pair of 171 type power tubes, would appear to be insufficient for really substantially distortionless reproduction, assuming that from 1.5 to 2 watts of power is required for good dance music volume, and that the amplifying system should possess a fairly flat frequency characteristic from 30 to 5000 cycles.

The undistorted power output obtainable from the previously described unit is higher than is generally obtained from receiver output stages, and, in fact, is greater than is often obtained from some of the more popular power packs employing a 210 type tube, the operating voltages of which are often less than they should be.

As a result of the interest that has been displayed in this earlier push-pull "Unipac," a higher-powered model has recently been developed employing a pair of UX-210 (CX-310) type amplifier tubes capable of delivering from 3 to 4 watts of undistorted power to a good loud speaker. It is probably quite safe to say that this is one of the most powerful receiving amplifiers yet developed for the home constructor, and the quality of reproduction it provides is really remarkable. After listening to the push-pull amplifier of the type described here, the significance of the popular phrase "tube overloading," as applied to conventional receiving amplifiers, is really brought home.

This newer combination is illustrated herewith, and closely resembles the push-pull model previously described. The new "Unipac" consists of a full-wave rectifier, which may use either ux-216-B (cx-316-B) or the new UX-281 (cx-381) type tubes, and a push-pull amplifier stage with the two UX-210 (cx-310) power tubes. A good idea of the details of the device may be gained from the detailed circuit diagram, in which the parts are lettered to agree with the list of parts on the next page.

The power supply uses a large, full-wave power transformer supplying 7.5 volts from two separate windings for lighting the rectifier and amplifier tubes. Its primary is designed for 105- to 120-volt, 60-cycle, lighting circuits, while a split high-voltage secondary supplies 550 volts a.c. (r.m.s.) to the plates of the rectifier tubes. Due to good transformer design, the ux-216-B (cx-316-B) rectifier tubes will deliver from 500 to 530 volts of unfiltered d.c. at a 106 mA load. This voltage is about all that may safely be used upon 210 type amplifier tubes

after a 40-volt drop has [...]
filter system. The filter [...]
490 volts d.c., of which [...]
bias on the amplifier tube [...]
450 volts being actual p[...]
to the push-pull amplifie[...]
be quite good since eac[...]
is called upon to furnish [...]
tubes are actually capab[...]
A single UX-281 (cx-381[...]
nearly the same power [...]
type tubes, but the use [...]
rectifier, such as the type[...]
discouraged as increasing [...]
and almost invariably res[...]
high value of hum in the [...]
type rectifier tubes, how[...]
output than two 216-B t[...]
volts, and their use is reco[...]
because of the increased [...]
cause of their probable I[...]
coated filaments and rati[...]

The filter system is [...]
is used in the smaller "U[...]
use of 1000-volt condense[...]
because of the high voltag[...]
selective and "brute-forc[...]
ployed, resulting in very [...]
current values.

The amplifier stage con[...]
input transformer with a [...]
tube, and a split windi[...]
matching the impedance [...]
amplifier tubes to that o[...]
similar loud speaker at 30 [...]
age gain of the amplifier i[...]

CONSTRU[...]

THE construction of t[...]
simple, involving onl[...]
number of standard part[...]
chassis, the wiring up o[...]

Facts About This Amplifier

Circuit: Single stage push-pull power amplifier

Tubes, Two cx-310 (ux-210) tubes in push-pull amplifier, two cx-316-B (ux-216-B) rectifiers, one cx-374 (ux-274) glow tube.

Cost: $83.25, without tubes. (Tubes: $38.50)

This power amplifier is capable of supplying 3 to 4 watts of undistorted power to a loud speaker. Complete A, B, and C power is obtained directly from the light socket. The rectifier and filter system are designed to supply the power amplifier tubes with about 500 volts for the plate and the necessary C bias. The unit is arranged to replace the second audio stage in a receiver. The unit is encased in a nicely finished metal cabinet

THE POWER AMPLIFIER WHEN COMPLETED
IS HOUSED IN A METAL CABINET

Rectifier Tubes (cx-381 or ux-281 Optional)	$15.00
One ux-274 or cx-374 Ballast Tube	5.50
Two ux-210 or cx-310 Power Tubes	18.00
	$38.50

The "Unipac" will furnish ample power to a radio receiver at 45 and 90 volts with voltage held constant by a potential dividing resistance and a glow tube voltage-regulator, preventing high open-circuit voltages to develop, which might damage receiver condensers. The amplifier replaces the conventional second audio stage of a receiver, the input tipjacks connecting to the first audio stage output terminals of the receiver, and the loud speaker connecting to the output jacks of the "Unipac."

In operation, all tubes will get quite hot, as will the larger Ward-Leonard resistor. This is correct, as is a slight warmth noticeable in the power transformer core. It is necessary always to see that the B minus post is grounded, directly or indirectly through a condenser.

final attaching of the cabinet or ventilated housing after the unit has been tested. The parts needed are listed below:

T₂—328 Super Power Transformer	$18.00
L₂—331 "Unichoke"	8.00
T₁—230 Push-Pull Input Transformer	10.00
T₃—231 Push-Pull Output Transformer	10.00
Five 511 Tube Sockets	2.50
C₁, C₂, C₃, C₄, C₅—Type 662 Condenser Block	18.00
R₂—651 Resistor (Ward-Leonard) Set	7.00
Four Frost 253 Tipjacks	.60
R₁—Frost FT64 Balancing Resistance	.50
Van Doorn 661 Steel Chassis and Cabinet, with Hardware	8.00
Three Eby Binding Posts (B—, +45, +90)	.45
Twenty-Five Feet Kellogg Fabricated Hook-Up Wire	.20
	$83.25

Unless otherwise noted, all the parts listed above are manufactured by Silver-Marshall.

The tubes required for the operation of the "Unipac" are as follows:

Two ux-216-b or cx-316-b Half-Wave

A 171 PUSH-PULL AMPLIFIER

The circuit of the 171 type push-pull amplifier described in the July, 1927, RADIO BROADCAST is given above. This amplifier is capable of delivering about 2 watts of power to a loud speaker

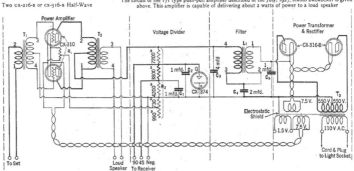

A 210 PUSH-PULL AMPLIFIER

The circuit diagram of the super-power amplifier described in the article given in this illustration. The 550-volt transformer, T₃, at the right, supplies the two rectifier tubes with sufficient voltage so that when it is rectified and filtered each of the power amplifiers will receive about 500 volts each. A glow tube is incorporated in the circuit to maintain the voltages constant, independent of the load. The loud speaker is fed with energy through an output transformer.

THE DX LISTENER FINDS A CHAMPION

By JOHN WALLACE

THE DX hound comes in for a lot of unwarranted disparagement. He is viewed askance by his more enlightened brethren as a benighted soul with a perverse idea of what radio ought to be used for. His greatest delight, as they picture it, is in attacking his receiving set with a screw driver and soldering iron, disemboweling it and putting it together a different way. He is said to prefer the faint whisper of call letters from some mission station in heathen Africa to hearing the Prince of Wales sing Frankie and Johnny from the local station. In his defense it has been iterated from time to time that his experimenting made radio what it is to-day. This vindication seems to us to be still a pretty sound one.

There are two types of DX hounds, the radicals and the conservatives. The extremists care for nothing but distance. They will labor into the small hours of the night to pick up the signal of a station half a continent away. Their standard of reception fidelity is not exacting—all they ask is that the *B*s be distinguishable from the *P*s. Once they have gained their quarry (which Mr. Webster describes as "the entrails of the prey, given to the hounds") their interest is over and they start in pursuit of some other station. We are in no special sympathy with this hunting breed of DX hound, but we will not have it said that he is quite useless. He has forced an increase in the range of receiving sets.

But the conservative hound likes to have a little sport with his catch after he's run it down. Once he has got a distant station he labors patiently at tuning it to shut off all extraneous noises. Not until he has the program coming in with the clarity of a local one is he satisfied. Then, if the program is a good one, he listens to it. With the pleasure he experiences from the program is an added stimulation in the realization that he is eavesdropping on a scene transpiring some hundreds of miles away with no connection between him and that remote city but some great open space and an assortment of ether waves. If a sympathy with his endeavours and an occasional emulation of them makes us a DX hound, then—we are a DX hound. Those who advance the protest that radio is now so much a matter of fact that they cease to marvel at its wonders present not half so good a defense of their attitude as they do a confession of their lack of imagination.

After all, the unique thing about radio is not that it brings music into your home—the phonograph did that years ago—but that it conquers distance. It was radio's ability to conquer distance that gave it its initial impetus, that seized the public imagination and bounded it along to an unprecedentedly swift success. It seems a bit of the basest ingratitude—a sort of biting the hand that feeds you—for radio to turn its back now on the characteristic that gave it birth. And, turning its back on it, it is, what with its two latest developments: chain broadcasting and wired broadcasting. Chain broadcasting, with its extended use of telephone wires has made the listener in large cities content to receive his distant programs from a station perhaps a few blocks away. Wired radio, while it still exists only as a rumor, is likely to come any time soon. Here the program will be circulated on the

already existing electric power lines which enter almost every civilized household, and the program received will have spent no instant of its life bounding on an ether wave. This seems to us a distinct retrogression, technically at least. Of course we do not argue that a trip through the ether makes a program any better; under present conditions the wired program is frequently better in quality. But the one is genuinely *radio*, the other simply glorified telephony. In short, wired radio, and, to a lesser degree, chain broadcasting, summarily renounce the fundamental principle upon which radio was founded—space annihilation.

This renunciation seems to us premature. The possibilities of radio broadcasting have not been completely exhausted. Hardly a score of years of experimentation has been completed. Certainly the idea is worth a score more.

It may be very practically objected that atmospheric conditions, over which man has no control, simply render it impossible to extend further the range and reliability of radio reception. This is practically true. Theoretically it is false and since this is a theoretical article we will press our point further. A demonstrable increase in range has been effected within the past few years by increasing the power of transmitters and the efficiency of receivers. There has been no sign from heaven to indicate that this increase has reached its limit. So, having utterly no knowledge of the mechanical problems of radio, but an unbounded faith in the uncanny powers of its technicians we argue that they can further perfect it if they try. But if interest in long range broadcasting is abandoned it's a cinch they won't try.

Radio's unique contribution to what is drolly referred to as the "progress" of civilization is the conquering of space. We repeat ourself? As a follower of radio programs are we well aware that it has made an enormous contribution to mankind by making music once again part of his daily life. In this rôle it has been of incalculable benefit. However, this rôle, important as it is,

DAVID BUTTOLPH

The gifted young conductor of the National Concert Orchestra, regularly heard through the red network of the National Broadcasting Company

is none the less a secondary one. It is possible for a man in his home to survive the evening without an after dinner concert, but a man at sea on a ship with a hole in it is dependent upon radio for his life. The city dweller can go to a concert hall when he craves music and entertainment, but the dweller at a lonely outpost in Canada is dependent upon radio for any break in the monotony of his existence. A catastrophe such as a cyclone or an earthquake may cut off wire communication with a devastated area but radio may still be able to penetrate and convey important, perhaps life saving, messages. Or, to cite a fanciful but none the less valid instance of the primary importance of radio as a long distance agent: in time of war an invading force could throttle all wired communication within the nation but a few well entrenched transmission stations could still reach the entire populace.

As we have said, further extension of the range of broadcasting, particularly in the face of the apparently unsurmountable difficulties it has already met, is dependent upon a sustained interest in achieving this end. So we think the vast army of DX hounds instead of being reviled should rather be looked upon as a desirable faction and, an important balancing element in radio's development.

While, personally, we are most frequently interested in the musical things radio has to offer, we look forward to the time when it will put us in easy touch with foreign shores. Perhaps some further use will be made of short-wave broadcasting and reception to this end. There would be a kick in that which not even the staunchest deprecator of DX could deny. But such an entertaining, and indeed instructive, state of affairs will not have been reached until after we first overcome the not inconsiderable distances in our own U. S. If we ever do this it will be due to the persistency of the DX hounds.

What We Thought of the First Columbia Broadcasting Program

SUNDAY, the eighteenth of September, witnessed the début of the long heralded Columbia Broadcasting System. The evening of Sunday, the eighteenth of September, witnessed your humble correspondent, tear stained and disillusioned, vowing to abandon for all time radio and all its works and pomps. We have since recovered and will go on with our story. The broadcast divided itself into three successive parts, descending in quality with astounding speed.

PART ONE: THE VAUDEVILLE

This program came on in the afternoon, after a half hour's delay due to mechanical difficulties—a heinous sin in this day of efficient transmission, but excusable, perhaps, in a half-hour-old organization. This opening program, at least, was auspicious. The performers were of superlative excellence. Bits from a light opera were well sung. A quartet gave a stirring rendition of an English hunting song. A symphony orchestra played some Brahms waltzes. A soloist sang "Mon Homme" in so impassioned a fashion that she must have swooned on the last note. Then a dance orchestra concluded the program with some good playing. The offerings were of such high quality that it was doubly disconcerting to have them strung together with a shoddy "continuity"—especially with such stupid and overdone continuity as the "and-now-parting-from-Paris-we-will-journey-to-Germany" type.

Continuity is a device used to bolster up weak programs. It is a bit of psychological trickery designed to keep the listeners listening even while their own good sense tells them that there is nothing being offered worth listening to. A good steak doesn't need to be served with sauce, but there's nothing like some pungent Worcestershire for camouflaging the defects of a bad one. The items offered on this afternoon program were good enough to serve ungarnished, and were cheapened by the introductory blah.

PART TWO: THE UPROAR

"Uproar," let us hasten to explain, is Major J. Andrew White's way of pronouncing Opera. We seek not to poke fun at this announcer; he is one of the best we have. (Though we think both Quin Ryan and McNamee outdid him in the recent fight broadcast): But his habit of tacking Rs on the end of words like Americar and Columbiar doesn't fit into a high-brow broadcast as well as it does in a sports report. The Uproar was "The King's Henchmen" by Deems Taylor. Evidently no effort was spared to make the broadcast notable. A good symphony orchestra was utilized, capable singers were employed, and Deems Taylor himself was intrusted with the duty of unfolding the plot. But after all it was ' just another broadcast." Musical programs into which a lot of talk is injected simply will not work. One or the other has to predominate. Either make it a straight recitation with musical accompaniment—or straight music with only a sparing bit of interpretative comment.

Mr. Taylor's music for this opera is delightful, the singing was admirable, but the tonal effect was disjointed and unsatisfactory. The composer outlined the story, but, enthralling as it may be on the stage, it was impossible to visualize the action with any degree of vividness from his words. We felt continually aware that there was really no action taking place, and the effort at make believe was too strenuous and detracted from an enjoyment of the music. It was less effective, even, than a broadcast from the regular Opera stage. Here the piece is likely to be more familiar and it is possible to conjure up its pantomime from remembrances of performances seen.

It is our humble and inexpert opinion that program designers are barking up a wrong tree and wasting a lot of energy in their unceasing attempts to fit spoken words into musical programs. But if they will persist let us suggest that they are going about the job in a blundering way with no proper realization of its difficulty. All present essays in this line fall into two classes: those which attempt to relate starkly the necessary information in a minimum number of words, and those which attempt to give a spurious arty atmosphere by the meaningless use of a lot of fancy polysyllables.

Neither method works. The first is distracting and effectively breaks up any mood or train of thought that may have been induced by the music. The fancy language system, besides being obviously nauseating, takes up too much time.

Program makers may as well realize soon as later that the simple possession of a fountain pen doesn't qualify a man for writing "script" or other descriptive text. It is a job calling for the very highest type of literary ability and one that can't be discharged by just anybody on the studio staff. The properly qualified writer should be able to state the information tersely, but, with all the vividness of a piece of poetry. Each word he uses must be selected because it is full of meaning, and of just the right shade of meaning. Any word not actively assisting in building up a rapid and forceful picture in the listener's mind must be sloughed off. A further complication: the words can't be selected because they look descriptive in type, but because their actual sound is descriptive. Altogether an exacting job; it would tax the ability of a Washington Irving.

It is highly improbable that a genius at writing this sort of stuff will ever appear; the ether wave is yet too ephemeral a medium to attract great writers. But there is no question that scriveners of some literary pretensions could be secured if the program builders would pay adequately for their services. This they will never do until they realize the obvious fact that the words that interrupt a program are just as conspicuous as the music of the program itself. It is incongruous, almost sacrilegious, to interrupt the superb train of thought of Wagner or Massenet to sandwich in the prose endeavours of Mabel Gazook, studio hostess, trombone player and "script" writer.

PART THREE: THE EFFERVESCENT HOUR
' O dear ! O dear ! Whither are we drifting !

You have all heard the ancient story of the glazier who supplied his small son with a sack of stones every morning to go about breaking windows. Comes now a radio advertiser who deals in stomach settling salts with a program guaranteed to turn and otherwise sour the stomach of the most robust listener. The Effervescent Hour was the first commercial offering of the new chain and far and away the worst thing we ever heard from a loud speaker. We thought we had heard bare faced and ostentatiously direct advertising before, but this made all previous efforts in that line seem like the merest innuendo. The name of the sponsoring company's product had been mentioned ninety-eight times when we quit counting. An oily voiced soul who protested to be a representative of the sponsoring company engaged with announcer White in sundry badinage before each number, extolling the virtues of his wet goods and even going so far as to offer the not unwilling announcer a sip before the microphone. Stuck in here and there amidst this welter of advertising could actually be discovered some bits of program! But such program material it was. First the hackneyed "To Spring" by Grieg. Then "Carry Me Back to Old Virginia." Next some mediocre spirituals followed by a very ordinary jazz band and culminating with a so-called symphony orchestra which actually succeeded in making the exquisite dance of the Fée Dragée from the "Nutcracker Suite" sound clumsy and loutish—no mean achievement.

One long interruption occured while special messages were given to soda jerkers while the country o'er, inviting them to enter a prize contest for the best encomium to the advertiser's wares. But the most aggravating interruptions were the frequently spaced announcements: "This is the voice of Columbia—speaking." This remarkable statement was delivered in hushed and reverential tones, vibrant with suppressed emotion, a sustained sob intervening before the last word. It was positively celestial. We have given a rather complete résumé of this program, but it may be warranted by the fact that probably not a dozen people in the country, beside ourself, heard it. No one not paid to do so, as we are, could have survived it. Perhaps this indictment of Columbia's opening performance is unkind in the light of subsequent offerings. Our stomach is still unsettled. Furthermore we will not make use of any of the Effervescent Hour's salts to settle it!

THIS MONTH'S prize for the ugliest and most cacophonous coined name plastered on any troup of radio performers is hereby awarded by unanimous and enthusiastic vote to wow's popular entertainers the Yousem Tyrwelder Twins!

A FAMILIAR WBZ-WBZA PROGRAM GROUP
The Hotel Statler Ensemble Group, one of the best of the dinner orchestras in the New England territory. From left to right: Helen Clapham, Hazel McNamara, Katherine Stang, leader, and Virginia Birnie

BY CARL DREHER

Drawing by Franklyn F. Stratford

Radio As An Electro-Medical Cure-All

THAT electricity plays a considerable rôle in the physiological functions is a fact well known and already extensively investigated. But outside of the area of verified or verifiable observations there is, as in every other division of science, a penumbra of dubious ideas, and beyond that lies what Theodore Roosevelt called, in one of the most apt of phrases, the lunatic fringe. Roosevelt was concerned with the field of politics, but politics have no monopoly of lunatics—nor knaves. The two are frequently coupled.

I am forced to these melancholy reflections on re-reading a newspaper article which was clipped for me during the summer. Under the caption, "Metal Lingerie As Radio Shield," it related the adventures of an afflicted governess in the radio realm. It seems that for years the lady suffered from "mysterious burns, bruises, blisters, and internal pains," which, of course, the doctors were unable to explain or cure. Thereupon a learned scientist (non-medical) came to her rescue. He subjected the sick woman to extensive tests, including the effect of ultra-violet, infra-red and X-rays, as well as short and long radio waves. She was very sensitive to all these oscillations and the professor decided that they might be responsible for her pains. He designed for her some metal screen lingerie to act as a shield against the nefarious oscillations. The method of keeping a ground on this intimate shield, as the wearer moves about, is not disclosed. Nor, unfortunately, are the results of the treatment reported. The article does state, however, that the afflicted woman, while previously in hospital near a radio station, heard sounds like the wind whistling through the shrouds of a ship, she would awake in the night with pains in her neck and ears, and hear a "dream-like voice." At this time she spent ten weeks in an insane asylum in the hope of being relieved.

There she was probably on the right track, and the fact that she entered the asylum voluntarily indicates some degree of insight, with a favorable prognosis if the patient came under the care of a skilled psychiatrist able to give her the requisite attention. She is almost certainly a mental case. The fact that she was bothered in the tests by electrical oscillations proves precisely nothing. If, as part of her psychosis, she was convinced that electrical waves made her ill, she would exhibit symptoms during any tests in the course of which she knew or suspected such waves were being generated. Even skin maladies may be of hysterical origin; this is the modern explanation of the "stigmata" which, in the Middle Ages, were taken for crosses printed on the bodies of certain persons by divine intervention, just as people might be possessed by devils through the machinations of Satan. Both beliefs are still firmly held in some parts, although their influence as a whole has decreased inversely with the spread of scientific ideas.

If radio waves were capable of exerting physiological effects professional radio workers in transmitting stations would certainly manifest whatever symptoms could result. Spending eight hours or more each day in an atmosphere where the field strength is many volts per meter, some of them, after thirty years, should be lamentably corroded in sensitive regions. But I have never heard of a radio man quitting a transmitting station because the waves were hurting him. I have seen them quit because they did not like the cooking, or the movies in the near-by town, but not one of them ever seemed to realize that Hertzian oscillations were whizzing through him at a velocity of 186,000 miles a second and might cause his vital juices to curdle. This seems to me cogent enough. While many individuals might be resistant, surely among some thousand of exposures a considerable amount of pathology, definitely traceable to the ether waves, would by this time have accumulated. As for the pitiably feeble emanations a few miles from a station, which cannot even be heard until they are amplified on a grandiose scale, doing a man any harm—that chance is about as great as One-Eyed Connelly's hopes of becoming President of the United States. And the possibility of benefiting a patient physiologically, save incidentally through entertainment or education, is equally large.

But some of the apostles of the late Doctor Abrams' medical credo know better. One of them relates in a learned journal of his cult the story of his efforts to benefit the human race by "Improvement in Electronic Diagnostic and Treatment Apparati; Broadcasting Electro-Magnetic Radio Treatment Waves." He has made his diagnostic circuit bigger and better, he feels, by adding "amplifying attachments" as follows:

1. A solenoid with its South attached to the

"HE SUBJECTED THE SICK WO-MAN TO EXTENSIVE TESTS"

dynamizer and its North connected by the head-band electrode of the subject, t through subject and grounded plates, to a stob driven two feet into the earth.

2. The South end of diagnostic set is con to a second metal stob. These stobs are set or ten feet apart *on the magnetic merid* ascertained by the compass.

3. A small high-frequency machine f creasing the electric tension to drive all p of the radiant force of disease through the nostic set and subject; and to stimulate th ject's reflexes so that they will act with highest efficiency. . . .

What the "stob" part refers to I canno The word is not in the dictionary.

But the Doctor continues:

Later I added a one-stage radio ampli the above diagnostic outfit which multipl findings four times. Instead of carrying th leading from the last reflexophone to the band of the subject, it is carried to the po side of the one stage radio amplifier's transf and then a second wire is carried from the tive side of the amplifier's transformer t head-band electrode on the subject.

Now comes the actual radio application idea. A two-stage radio amplifier, accordi the description, is attached to the " oscilloclast." The electronic treatment is let loose on the world, first on a small scale

In the electric store two doors away we se specimens for electronic examination of people and secured specimens from two peo a store in the same building with the treat instruments. We then made electronic exa tions of the six specimens and recorded wh fections those people were carrying. After ning the instruments some twenty days, we secured specimens from the same people made reëlectronic examinations. We found four of the people were negative of their ca infections—the other two not negative but their infections greatly reduced, showing that or few more days of broadcasting v render them negative.

These experiments proving that infections could be destroy a distance of one hundred fifty or more feet through air, brick v glass doors, windows, etc., we de to broadcast the treatment wave for a mile or more. We had ma five-stage radio-transmitting b casting instrument that multiplie eighteen multiplication, of maste chine, by fifteen hundred times ing a total multiplication of m machine twenty-seven thou (27,000) times—we then had er a broadcasting aerial on top of a story building and connected by the instruments, oscilloclast, two-amplifier, one treatment short cir ed unit and the five-stage r transmitter to the aerial. We started up the instrument. The secured a large radio receivin (with loud speaker) from the el company and set it up in a se passenger automobile and three a skilled electrician and radio mechanic, chauffeur and myself, tuned at the curb in front of the electric company and could hear the working of the oscilloclast—we then drove five squares and tuned in and could hear the instrument and at intermediate points for a little more than a mile, showing that the treatment waves were being broadcast a mile or more.

We then secured specimens for electronic examination from a distance of half a block up to three miles and from many intermediate points. The number of specimens secured from beginning of broadcasting, from November 1, 1926, to present date, January 20, 1927, were thirty-three. We have found that out of that number twenty-three have been made negative—the balance, ten, have been greatly reduced, showing that it will only require a week or ten days further broadcasting of the treatment waves to render them negative.

The amplified treatment machines are run by batteries that are fed from the electric light socket and will last from four to six months; otherwise batteries only last a few days. The five-stage radio-transmitter is also fitted up with a large battery supplied by electric socket and it in turn supplies the dry batteries, making the apparatus very efficient and durable.

. . . In the last six weeks or more, with broadcasting outfit we have treated electronically a population of fifty thousand (50,000) on an average of two hours per day, making a total of one hundred thousand (100,000) hours per day. If the broadcasting electronic treatment waves have rendered negative two-thirds of the fifty-thousand (50,000) population and reducing the other one-third, as was proven in the thirty-three test cases (two-thirds made negative, one-third reducing) taken out of the same population we can readily see the great benefit and per day value to the people. Giving the value of one dollar per hour for treatment of each individual (which is a low tariff fee) we have a total of $100,000.00 per day.

If a transmitter fed from batteries can benefit the surrounding population to the extent of $100,000. per day, conceive of the value of a 100-kw. outfit devoted to the same philanthropic purpose! Why not make it 10,000 kw., while we are at it? Assuming that the professor's present equipment has a capacity of 20 watts, the 10,000-kw. set would be worth $50,000,000,000, a day to the citizens of the United States, on the valuation basis assumed in the first place. This is of the general order of the amount of business transacted in the country in a year. It is clear, therefore, that a stupendous wealth producing agency is in the hands of the electronic practitioner, which, I suppose, makes him feel very bad.

For my own part, I have some qualms. By broadcasting these electronic treatments, the learned Doc cures all the people within reach of their ailments. But why only the people? Any such general specific must also be good for animals. Horses will be cured of the blind staggers, tubercular monkeys will rise from their beds, sick cockroaches will report at the office fit for work. The rats, pediculi, and bed-bugs (Cimex lectularius) may benefit even more than human beings, who may thereby be crowded off the earth. I hope that the electronic broadcaster will consider this aspect of the matter and quiet my fears if he can.

Some Catalogues

BULLETIN No. 1 of H. F. Wareing and Associates, on *Modulator Reactors*, will prove of interest to some broadcasting stations. This company, whose address is 401 Pereles Building, Milwaukee, Wis., is in the business of supplying apparatus and service to broadcasting stations. The first bulletin includes a discussion of modulator reactor design, and a price list of types stated to be suitable for trans-

mitters from 50-watt to 10-kilowatt size. The corresponding currents for which the chokes are built vary from 0.25 to 5.00 amperes, at d.c. voltages of 1000-5000. Ten- thirty-, and fifty-henry reactors are available. Bulletins on other broadcast station equipment are to be issued by H. F. Wareing and Associates at intervals, according to the announcement reaching us.

The Samson Electric Company of Canton, Massachusetts, distributes a "Radio Division Price List" including, besides the usual radio parts sold to receiver constructors, such items as microphone input transformers, tube-to-line and line-to-tube transformers, mixer equipment, and other specialized broadcast transmitter and public address material. They have made up a blueprint showing a small public address system assembled with their parts. Provision is made for microphone, radio set, and phonograph pick-up. There are three 0.25 ampere (filament) tubes, and apparently the output stage is push-pull, using 5- or 7.5-watt tubes. On this basis the volume capacity should approximately equal that of the Western Electric 3-A size P. A. system.

Design and Operation of Broadcasting Stations

18. Piezo-Electric Control

THE field of piezo-electric phenomena includes the generation of electrical potentials in various substances by the application of physical pressure, and, conversely, changes in physical dimensions directly correlated with electrical conditions. It is not a new division of physics; the piezo-electric properties of quartz, for example, were investigated by P. and J. Curie in 1889. The effect itself was discovered by the brothers some years earlier.

Prof. W. G. Cady, about eight years ago, began the work which resulted in the application in the radio art of crystal frequency control. He reported his investigation of "The Piezo-Electric Resonator" in the *Proceedings of the Institute of Radio Engineers*, Vol. 10, No. 2, April, 1922. Oscillating crystals are used as frequency standards in wave meters and monitoring units, some practical forms much used by broadcasting stations being those produced commercially by the General Radio Company. RADIO BROADCAST has printed two comprehensive articles by M. T. Dow on crystal circuits and measurement applications (January and September, 1927). A direct form of frequency control of particular interest to broadcasters is that in which the transmitter functions as a radio-frequency amplifier, the master oscillator being a crystal-controlled tube. Some recent engineering publications on this aspect include three in the I. R. E. *Proceedings*: A. Crossley: "Piezo-Electric Crystal-Controlled Transmitters" (Vol. 15, No. 1, Jan., 1927); A. Meissner: "Piezo-Electric Crystals at Radio Frequencies" (Vol. 15, No. 4, April, 1927); and H. E. Hallborg: "Some Practical Aspects of Short-Wave Operation at High Power" (Vol. 15, No. 6, June, 1927). In the present article an attempt will be made to introduce the subject to broadcast operators who have not worked with crystal-controlled transmitters so that when they are called on to operate such equipment they will be in possession of some of the elementary facts.

In itself the use of a crystal is no guarantee of frequency stabilization to any required degree of accuracy. Some broadcasters seem to believe that the use of a crystal in almost any kind of holder, with some sort of radio-frequency amplifier following, will insure constant frequency radiation. Actually a crystal is of little use unless

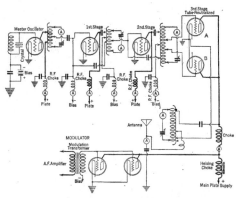

THE CIRCUIT DIAGRAM OF A CRYSTAL-CONTROLLED TRANSMITTER

In actual practice, tubes A and B in the third stage generally consist of two banks of power tubes. The filaments of only one set of tubes are lighted and they deliver power to the antenna circuit while the other bank, with the filaments not lighted, acts as a neutralizing circuit to prevent the active bank from breaking into self oscillation. In operation, if an accident happened to the active bank, the filaments could immediately be turned off and the filaments of what was the inactive bank turned on. The latter bank of power tubes would then deliver power to the antenna while the other bank functioned as a neutralizing condenser

very specific and delicate conditions of operation are maintained for it.

A piezo-electric substance for radio crystal control purposes must have certain internal atomic properties, and it must be hard, durable, and not easily broken down physically or electrically. Quartz is the best commercial product so far offered to fill these requirements. The manufacture of quartz crystals for radio purposes is a specialized subject and, as few broadcasters are likely to attempt grinding their own crystals, need not be discussed here. The crystal should be optically ground to oscillate at only one fundamental frequency. If the frequency is to remain constant, the crystal must be maintained under constant physical conditions as a prerequisite. This includes an unvarying contact pressure and temperature. When the temperature changes the dimensions of the crystal change and the natural frequency varies proportionally. The crystal must be kept clean; a drop of oil or water introduced between the holder and the quartz slab will usually stop oscillations altogether. It follows that in a broadcasting station installation the crystals are usually kept in a dust-proof box whose temperature is thermostatically controlled. Some commercial crystals, in addition, are sealed into small individual containers, provided with lugs designed to slip under binding posts. The actual contact with the crystal is inside the container. If springs or screw clamps are used to make contact with an open crystal care must be taken to secure parallel movement of the metal surfaces, so that the crystal is not subjected to pressure on part of its surface and left untouched elsewhere. A loose contact leads to brushing, heating, and possible cracking of the crystal. The capacity of the crystal holder should be small in order that its piezo-electric variations may exert the maximum governing effect on the circuit of which the crystal is a part. It will be noted that there is some design analogy between the old-style crystal detector stand and the quartz crystal holder of a modern tube transmitter; each has protean forms. Some illustrations of actual crystal holders will be found with the RADIO BROADCAST papers mentioned in the bibliography, and Crossley includes a detailed description of the contact requirements in his paper.

The radio-frequency energy in the initial crystal circuit may amount to a fraction of a watt, while the final stage may deliver many thousand watts to the antenna. It is clear that great care must be taken to prevent feed-back, parasitic oscillations, and unstable circuit conditions along the line. Under some conditions of circuit imbalance the crystal is likely to overheat and be damaged. The transmitter may stop oscillating. In the endeavor to control regeneration and secure circuit stability, designers have frequent recourse to shielding and neutralization of successive amplifier stages, and sometimes push-pull radio frequency amplification is employed, resulting in a series of balanced circuits analogous to those of low frequency telephone practice.

In a crystal-controlled telephone transmitter, modulation may take place in the final stage or at an intermediate point after the crystal but before the final stage. The advantage of modulation at a low power level lies in the possibility of securing ample modulator capacity relative to

the radio frequency energy to be modulated. But if, for example, modulation takes place in one of the earlier stages at a power level of, say, 50 watts, care must be taken not to impair the audio-frequency characteristic of the transmitter by cutting of the side bands in successive tuned stages, and of course the power tubes must have sufficient capacity to handle the peaks of modulation. The Bell Telephone Laboratories engineers seem to incline toward low power modulation, while the Westinghouse and General Electric engineers prefer to wait until the full radio frequency power is developed before impressing the audio frequency on the carrier.

The power of successive stages depends on tube characteristics and the use to which the transmitter is to be put. Crossley shows a 150-600-kc. telegraph transmitter in which the crystal controls a 7.5-watt tube, which is followed by a 50-watt impedance-coupled amplifier, a 1-kw. tuned amplifier stage, and the final 20-kw. stage feeding the antenna. These figures and some others in this paragraph represent the nominal oscillator ratings of the tubes in question; the powers actually developed are generally less. Meissner mentions a telephone transmitter in which, after the crystal tube, 5-watt,75-watt, 500-watt, and 3-kw. stages are found, the last named supplying 3 kw. to the antenna. The National Broadcasting Company's Bound Brook telephone transmitter (made by Westinghouse for R.C.A.) uses a 7.5-watt tube in the crystal stage, swinging a 250-watt tube, followed by two 250's in parallel (500 watts) before the final 40-kw. bank. Bellmore, built by General Electric for the N. B. C., uses more stages; the crystal, likewise governing a 7.5-watt tube, is coupled to another of the same size; then follow a 50-watter, two fifty-watters in parallel (100 watts), a 1000-watt tube, a single 20-kw. tube run at about a quarter of its rating, and the final 50-kw. stage. In both of these American transmitters, plate modulation of the final power stage is used. In the Bellmore transmitter the power of the crystal stage is purposely kept low as one of the design considerations, only about 0.5 watt being generated, while Crossley mentions getting 21 watts from a crystal-governed tube of the same type, in one of the U. S. Navy experiments. This divergence shows how design calculations determine operating conditions.

Fig. 1 is a schematic circuit diagram of a crystal-controlled telephone transmitter without the audio amplifier and the power supply to the modulator and r. f. output stage. The power ratings of the successive stages have been omitted because, as indicated above, various sizes of tubes might be employed in such a chain, according to the final output required, the tube characteristics, and other design factors.

The Small Broadcaster

IN A letter of some length, which we should print in full if the space were available, Mr. Robert A. Fox, formerly owner and manager of Station WLBP of Ashland, Ohio, takes me severely to task for my past animadversions on incompetent broadcast technicians, which he apparently thinks were aimed exclusively at small stations, and then goes on to a penetrating discussion of the small stations' economic problems.

Mr. Fox points out that in arranging high quality programs the small station is at a disadvantage, ' for the simple reason that musicians do not charge you for their time according to the power of your station, but in accordance with the number of hours required of them.'' The advertiser, on the contrary, will pay more or less proportionately according to the power. The members of the station staff are in the same position as the musicians. The result is that one man must sometimes assume the staggering burden of acting as "station monitor operator, announcer (doing a nemo job of it), operator; chief engineer as well as salesman, financier, publicity agent, and program director,'' all this with one assistant. Even then the structure collapses under the fixed expense, and Mr. Fox concludes, "The small broadcaster is economically doomed.'' But, he insists, the failure is economic, not personal—"the fellows who have been operating under 1000 watts have more brains than those above 1000 watts.''

They may not have more brains—nature does not distribute brains according to antenna watts, either in direct or inverse ratio—but they certainly have more courage. And, while economically they may be sick, they may yet survive, on some other basis, to see a better day, No one can say, at this juncture, that the small neighborhood station will not find a place in community life, with some form of cooperative support, in a frequency band wherein it can serve local interests without interfering with the large stations and networks aspiring to national coverage, and be in turn protected from interference by them.

As to the less material matter of my own attitude toward small enterprises, it is a curious commentary on our American attitude toward criticism in general that when a man states baldly, in public, unpleasant facts about institutions or people, he is immediately suspected of being hostile to those institutions and people. That it is his right and duty, once he has set up as a critic, technical or social, to discriminate between what he finds good and bad, is a basic fact not sufficiently recognized among us, in radio and elsewhere. There is still a lot of bad broadcasting and incompetent operation going on. No one with a pair of ears and the tonal discrimination of a tomcat can think otherwise. There is also a large and growing element of good showmanship, efficient operation, and skilled personnel, among both large and small stations, and I have not been backward in giving credit to those responsible for such progress. The standards have been lower among smaller stations, because of the lack of resources and, sometimes, because of the lack of time and skilled personnel. All these factors go together. If a man tries to act as announcer, engineer, operator, program director, and publicity representative of a station he will inevitably turn out a half-baked job in each capacity. He may be a hero, but he is not a broadcaster by 1928 standards. One may admire his courage and still tell him what one thinks of his audio frequency band and the quality of his sopranos. As for constructive contributions, I have tried to do my part by writing technical articles which are of use largely to the smaller broadcasters, because the information contained in them is common property among the operators of the bigger stations. Let that be weighed against my refusal to be a member of a cheering squad.

NATIONAL

NATIONAL · B

Type 7180—A "B" That's Built for Service

A Strictly Heavy-Duty Power-Unit Supplying

Detector Voltages 22 to 45 adjustable
R. F. Voltages - 50 to 75 adjustable
A. F. Voltages - 90 to 135 adjustable
Power Tube Voltage - 180 - fixed

Protects Tubes, Condensers

against excessive and harmful voltages. Designed for lasting service with liberal factors of safety.

Rating

Output Rating is 70 mils at 180 volts. Uses R. C. A. UX 280 or Cunningham CX 380 Rectifying Tube.

Licensed Under Patents of Radio Corporation of America and Associated Companies. For 110-120 volts, 50-60 cycles A. C. only.

List Price with Cord, Switch and Plug $40.00.   Rectifier Tube $5.00

Write National Co. Inc., W. A. Ready, Pres., Malden, Mass.,   for new Bulletin 124

NATIONAL POWER TRANSFORMERS and FILTER CHOKES for AMPLIFIERS
Specified generally. Transformer Type R, center tapped 600 v. secondary.
Price $12.50. Transformer Type U, both 600 and 460 v. secondary $14.50.
NATIONAL FILTER CHOKES - per pair in heavy case - Type 80, $10.00.

NATIONAL TUNING UNITS — THE HEAVENLY TWINS
More National Tuning Units have been used by set builders than all other similar components combined.
Standard since 1923   Approved By   The OFFICIAL Design
BROWNING & DRAKE

The *Radio Broadcast*
LABORATORY INFORMA
SHEETS

THE RADIO BROADCAST Laboratory Information Sheets are a regu magazine and have appeared since our June, 1926, issue. They c of information of value to the experimenter and to the technical radio purpose always to include new information but to present concise and the most convenient form. The sheets are arranged so that they ma magazine and preserved for constant reference, and we suggest that ea with a razor blade and pasted on 4″ x 6″ filing cards, or in a notebook be arranged in numerical order. An index appears twice a year dealii published during that year. The first index appeared on sheets Nos. vember, 1926. In July, an index to all sheets appearing since that tim

All of the 1926 issues of RADIO BROADCAST are out of pi set of Sheets, Nos. 1 to 88, can be secured from the Circulai Doubleday, Page & Company, Garden City, New York, for $1 oo $ asked what provision is made to rectify possible errors in these She tunate event that any such errors do appear, a new Laboratory S number will appear.

—TH

No. 145   RADIO BROADCAST Laboratory Information Sheet   I

Loud Speakers

GENERAL CONSIDERATIONS

IT HAS been realized for some time that a large diaphragm type of loud speaker is capable of giving somewhat better frequency response than can be obtained from a short horn. These large diaphragm loud speakers have generally been called cones because the large diaphragm in most cases takes the form of a right circular cone.

There are certain essential characteristics which must be striven for in designing a loud speaker of this type. What we desire in the diaphragm is to obtain a large surface of great stiffness or rigidity and, at the same time, extreme lightness. If such a material can be obtained, a very satisfactory loud speaker could be made consisting simply of a sheet of the material freely supported at the edge. Such a material having a high ratio of stiffness divided by mass is difficult to obtain, and it has been necessary to devise diaphragm shapes which will give the necessary stiffness and which will still be light. This is the reason why a cone shape has generally been used, for it will give the necessary characteristics. Recently there was described in RADIO BROADCAST the Balsa wood loud speaker, which represents an attempt to obtain a large flat diaphragm using a light material, with the required stiffness obtained through the use of slats radiating from

the center. Because of the ex wood it is possible to obtain ratio of stiffness to mass.

It is, of course, essential if it is to radiate sound effec as possible so as to require energy to move it. It is d diaphragm shall move and tance it encounters in mov to the energy required to diaphragm and set up so Any of the available energ purposes represents a loss.

An excellent book, *Wire* published in England by II ten by N. W. McLachlan speaking of cone type loud |

"There is a wide field fo perimental work regarding phragms of various shape measurement, coupled wit possible to pave the way and to evolve a diaphragm to those now used. Until t main in ignorance of the a various frequencies. The hu diaphragm to be better tha give exact data."

No. 146   RADIO BROADCAST Laboratory Information Sheet   I

B Power Device Characteristics

TYPICAL CURVES

ON THIS Laboratory Sheet are given a group of curves, supplied by the Raytheon Manufacturing Company, which show how the output voltage of a typical B power unit varies with the transformer voltage. The circuit diagram of the rectifier and filter system used in making these tests is given on the curve. The curves apply when a type BH or similar tube is used as the rectifier. These curves indicate the following facts:

(A.) That the slope of all of the curves is the same. This is to be expected because the slope is determined entirely by the resistance of the circuit, which does not vary.

(B.) That an increase of 50 volts in the transformer voltage is effective in producing an average of 75 volts increase in the output voltage.

(C.) That the output voltages of the system at no load have approximately the same value as the transformer voltages.

(D.) That the total resistance of the rectifier-filter system is about 1340 ohms. (This value is determined by dividing the difference of any two voltages on the straight portion of any one curve by the difference of the corresponding load currents.) The resistance of the choke coils used was known to be 600 ohms so that the effective resistance of the rectifier is about 740 ohms.

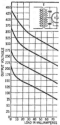

Power

THE modern battleship, typifying power is a true symbol of the engineering genius, the mechanical and electrical precision embodied in Ceco Tubes.

Judge these triumphant Ceco Tubes by RESULTS achieved in your own set—Buy them by the name Ceco, choosing in consultation with your radio dealer the general or special purpose tubes best suited to your set.

You may justly expect more of Ceco Tubes and you'll GET more.

Steadier Performance—Longer Life.

FOR UNDISTORTED POWER IN YOUR SET USE

CeCo Type "E" (120)
A power amplifier tube for the last stage of audio frequency. For use with dry cells.

Fil. Volts............3.
Fil. Amp............. .125
Plate Volts 90-180.
UX Base—Long Prongs.
List Price $2.50

CeCo Type "F" (112)
A power amplifier tube for the last stage of audio frequency. For use with storage battery, or A. C.

Fil. Volts..............5.
Fil. Amp............. .5
Plate Volts 90-180.
UX Base—Long Prongs.
List Price $4.50

CeCo Type "J-71" (171)
Out-put tube will handle the largest loud speaker at full volume. At ninety volts it will handle twelve times the undistorted volume of the ordinary "A" type.

Fil. Volts....................5.
Fil. Amp....................5
Plate Volts 90-180.
UX Base—Long Prongs.
List Price $4.50

There's a CeCo Tube for Every Radio Need

General Purpose Tubes
Special Purpose Tubes
Power Tubes
Filament Type Rectifiers,
A. C. Tubes

Ask your Dealer for Complete Data Sheet

C. E. MFG. CO., Inc.
PROVIDENCE, R. I.

Largest plant in the World devoted exclusively to making of Radio Tubes

No. 147   RADIO BROADCAST Laboratory Information Sheet   **December, 1927**

"Gain"

SIMPLE MATHEMATICAL CALCULATION

THE diagram on this Sheet shows an ordinary tuned circuit with a source of high-frequency voltage, e, in series with it. The voltage e can be considered to be the voltage induced in the tuned circuit from another coil, the primary of a radio-frequency transformer for example. This voltage will cause a current to flow in the tuned circuit and the ratio of the voltage E, developed across the entire circuit, to the voltage e, induced in the circuit, is known as the "gain" of the tuned circuit. The more efficient the tuned circuit is, the greater will be the "gain." We will now derive a mathematical expression for the "gain" of a tuned circuit. The current, I, flowing in a tuned circuit at resonance is:

$$I = \frac{e}{R} \qquad (1.)$$

where e = the voltage induced in the circuit and R = resistance of the circuit. The current flowing through the inductance coil, L, generates a potential across the coil, determined as follows:

$$E = \omega L I \qquad (2.)$$

where ω = 6.28 times the frequency of the current, L = inductance of coil in henries, and I has the same meaning as in equation (1.) Substituting in equation (2.) the value for I given in equation (1.) we have

$$E = \frac{\omega L e}{R} \qquad (3.)$$

and dividing through by e we get:

$$\frac{E}{e} = \frac{\omega L}{R}$$

But, as stated previously, the ratio of E to e is the gain of the circuit. Therefore:

$$Gain = \frac{\omega L}{R} \qquad (4.)$$

This final expression indicates that, to obtain greatest efficiency from a tuned circuit, it is essential that the ratio of the inductance reactance (ωL) to the resistance of the coil should be made as large as possible.

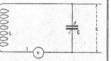

No. 148   RADIO BROADCAST Laboratory Information Sheet   **December, 1927**

An A. C. Audio-Frequency Amplifier

WHAT PARTS TO USE

THE introduction of the new a. c. tubes makes possible the construction of an a. c. audio amplifier with the necessary A, B, and C voltages supplied directly from the light socket. The list of parts necessary to construct such an amplifier is given on this Sheet. The circuit diagram is given on Laboratory Sheet No. 149.

An amplifier of this type is well suited for use with a small receiver consisting of one or more stages of radio-frequency amplification and a detector. The circuit has been so designed that B voltages for the r. f. and detector tubes can be obtained from the radio amplifier device.

The following parts are necessary to construct this amplifier:

A—A. C. Tube, Type UX-171 (CX-371) or Equivalent.
B—CX-171 (CX-371) or Equivalent.
T_1, T_2—Two High-Quality Audio Transformers.
T_3—Filament-Lighting Transformer to Supply Tube A.
T_4—Power-Supply Transformer Designed for use in 171 Type B Power Units.

L_4, L_5—Filter Choke Coils.
L_6—Output Choke Coil.
C_1, C_4—1-Mfd. Bypass Condensers.
C_2, C_3—2 Mfd. Filter Condensers.
C_5—1-Mfd. Filter Condenser.
C_6, C_8, C_9—1-Mfd. Filter Condensers.
C_7—2.4-Mfd. Fixed Condenser.
R_1—30-Ohm Center-Tapped Resistance.
R_2—1500-Ohm Fixed Resistance Capable of Carrying 4 mA.
R_3—2000-Ohm Fixed Resistance Capable of Carrying 20 mA.
R_4—Tapped Resistance for Use in Output of B Power Units.

In wiring this amplifier, be sure to twist the filament leads to the two tubes, to prevent hum. C bias for the first tube, A, is obtained from resistance R_3, and resistance R_2 supplies the output tube with grid bias.

The input terminals of the amplifier should connect to the output of the detector tube, terminal No. 1 connecting to the plate and terminal No. 2 to the detector B plus.

To prevent hum it is essential that the negative B be carefully grounded.

No. 149   RADIO BROADCAST Laboratory Information Sheet   **December, 1927**

Circuit Diagram of an A. C. Audio Amplifier

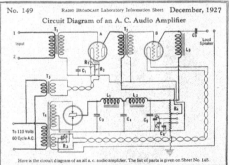

Here is the circuit diagram of an all a. c. audio amplifier. The list of parts is given on Sheet No. 148.

Silent Magic

Here is the Eveready Layerbilt "B" Battery No. 486, Eveready's longest-lasting provider of Battery Power.

TURN your radio dial, and presto! you turn your home into a theater, a concert hall, a lecture room, a cabaret, a church, or whatever you will. Turn the dial and your attentive ear does the rest. That is all there is to this magic of radio.

Or almost all. If a radio set is to work at its very best, attracting no attention to itself, creating for you the illusion that can be so convincing, you must pay a little attention to the kind of power you give it. There is but one direction, a simple one—use Battery Power. Only such power is steady, uniform, silent. It is called by scientists pure Direct Current. Any other kind of current in your

Radio is better with *Battery* Power

radio set may put a hum into the purest note of a flute, a scratch into the song of the greatest singer, a rattle into the voice of any orator.

Don't tamper with tone. Beware of interfering with illusion. Power that reveals its presence by its noise is like a magician's assistant who gives the trick away. Use batteries—use the Eveready Layerbilt "B" Battery No. 486, the remarkable battery whose exclusive, patented construction makes it last longest. It offers you the gift of convenience, a

gift that you will appreciate almost as much as you will cherish the perfection of reception that only Battery Power makes possible.

NATIONAL CARBON CO., INC.

New York [ECC] San Francisco

Unit of Union Carbide and Carbon Corporation

Tuesday night is Eveready Hour Night—9 P. M., Eastern Standard Time

WEAF–*New York*	WOC–*Davenport*
WJAR–*Providence*	KSD–*St. Louis*
WEEI–*Boston*	WCCO–{*Minneapolis St. Paul*
WFI–*Philadelphia*	
WGR–*Buffalo*	WDAF–*Kansas City*
WCAE–*Pittsburgh*	WRC–*Washington*
WSAI–*Cincinnati*	WGY–*Schenectady*
WTAM–*Cleveland*	WHAS–*Louisville*
WWJ–*Detroit*	WSB–*Atlanta*
WGN–*Chicago*	WSM–*Nashville*
	WMC–*Memphis*

Pacific Coast Stations—
9 P. M., Pacific Standard Time
KPO–KGO–*San Francisco*
KFOA–KOMO–*Seattle*
KFI–*Los Angeles*
KGW–*Portland*

EVEREADY
Radio Batteries
—they last longer

The air is full of things you shouldn't miss

No. 150   RADIO BROADCAST Laboratory Information Sheet   December, 1927

Oscillation Control

THE USE OF NEUTRALIZATION

IT HAS been pointed out many times that an ordinary three-element tube has an inherent tendency to oscillate due to the feed-back that occurs from the plate circuit to the grid circuit through the grid-plate capacity, indicated by dotted lines in the accompanying diagram. This diagram represents the circuit of a single-stage of tuned radio-frequency amplification, using the Rice system of neutralization, and the following explanation will make clear why the tube tends to oscillate and why the tendency to oscillate can be overcome by using some system of neutralization.

When a tube acts as an amplifier, the voltage developed in the plate circuit is greater than the voltage originally impressed on the grid circuit and, consequently, if the plate circuit is coupled to the grid circuit in any manner whatsoever, current will tend to flow from the point of high potential, that is the plate, to a point of lower potential, in this case the grid. If this current flowing to the grid circuit has the same phase as the original signal impressed on the grid, then the grid voltage will become somewhat greater and will be equal to the original signal in the grid circuit plus the voltage induced in the grid circuit from the plate. An increase in the grid voltage again produces an increase in plate voltage which in turn reacts back on the grid until the voltage is increased to a point where the losses in the circuit are overcome, and then the tube breaks into continuous oscillation.

It should be evident that if we can place in the circuit some device that will impress a potential

on the grid kind of an equal and opposite to that caused by the coupling between the grid and plate, then the resultant effect will be zero and the tendency for the circuit to build up and break into continuous oscillation will be nullified. The Rice system of neutralization is one way of doing this, the circuit for which is shown in the accompanying diagram. The grid-plate capacity is shown in dotted lines and this is the capacity through which current flows from the plate to the grid circuit and which ordinarily causes the tube to oscillate. This capacity is neutralized in the Rice system by connecting condenser Cn as indicated.

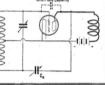

No. 151   RADIO BROADCAST Laboratory Information Sheet   December, 1927

Single-Control

BOOSTING SENSITIVITY

ON LABORATORY Sheet No. 33, October, 1926, some facts were given regarding the tandem tuning of several condensers, to decrease the number of controls. It was pointed out that, to obtain single control, it is necessary to overcome the effect of the antenna circuit in some manner, and that a common method of doing this is as indicated in sketch A on this Sheet. The owner of a receiver of this type may greatly increase its sensitivity by connecting a variometer between the antenna and ground posts as indicated in sketch B. This, of course, adds one more control to the set but in those cases where greater sensitivity is necessary, the additional control is justified.

The increase in sensitivity that results when the variometer is used in the antenna circuit is due to the fact that it brings the antenna into resonance with the signals being received and the resultant gain in amplification is practically equal to that which would be obtained from an additional stage of radio-frequency amplification.

In some cases when this variometer is used, it will be found that the receiver tends to oscillate, or actually does oscillate, when all of the circuits are brought into resonance. Fortunately, however, most single-control receivers have a volume control in the radio-frequency system and it will be found that, by cutting down the volume control, a point will be reached where the set will stop oscillating and usually the actual volume obtained with the antenna circuit tuned will be much greater than

that obtained before with the volume control turned to the "maximum" position. The tendency of the circuit to oscillate can also be lessened by somewhat decreasing the r. f. plate voltage.

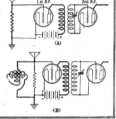

No. 152   RADIO BROADCAST Laboratory Information Sheet   December, 1927

Speech

SOURCES OF INFORMATION

THE nature of speech has been the subject of many scientific inquiries and many of the investigations in connection with speech have been recorded in various scientific journals.

Back in 1873, Alexander Graham Bell, familiar to us as the inventor of the telephone, did considerable work in analyzing speech and in "devising methods of exhibiting the vibrations of sounds optically," and much of the recent research has been done by engineers and physicists associated with the laboratories of the telephone companies.

A bibliography is given below of some of the important articles and books on the subject with which we are familiar. This bibliography is by no means complete in itself but if the references given are studied it will be found that some of them contain many references to other papers on the subject. J. B. Crandall's article, in the October, 1925, Bell System Technical Journal, in particular, contains about twenty-six references to other sources of information on speech and related subjects.

REFERENCE SOURCES

Bell System Technical Journal

C. F. Sacia and C. J. Beck: "The Power of Fundamental Speech Sounds." July, 1926.

I. B. Crandall: "Sounds of Speech." October, 1925.

C. F. Sacia: "Speech Power and Energy." October, 1925.

Irving B. Crandall: "Dynamical Study of the Vowel Sounds." January, 1927.

C. R. Moore and A. S. Curtis: "An Analyser for the Voice Frequency Range." April, 1927.

Journal of the American Institute of Electrical Engineers

Jones: "The Nature of Language." April, 1924.

Martin and Fletcher: "High-Quality Transmission and Reproduction of Speech and Music." March, 1924.

Maxfield and Harrison: "Method of High Quality Recording and Reproducing of Music and Speech Based on Telephone Research." March, 1926.

Books

Miller: *Science of Musical Sounds*. Second Edition. Macmillan.

Sabine: *Collected Papers on Acoustics*. Harvard University Press.

The great improvements in radio power *have been made by* Balkite

Balkite "A" *Contains no battery. The same* as Balkite "AB," but for the "A" circuit only. Enables owners of Balkite "B" to make a complete light socket installation at very low cost. Price $35.00.

Balkite "B" *One of the longest lived devices in radio.* The accepted tried and proved light socket "B" power supply. The first Balkite "B," after 5 years, is still rendering satisfactory service. Over 300,000 in use. Three models: "B"-W, 67-90 volts, $22.50; "B"-135, 135 volts, $35.00; "B"-180, 180 volts, $42.50. Balkite now costs no more than the ordinary "B" eliminator.

Balkite Chargers
Standard for "A" batteries. Noiseless. Can be used during reception. Prices drastically reduced. Model "J,"* rates 2.5 and .5 amperes, for both rapid and trickle charging, $17.50. Model "N"* Trickle Charger, rate .5 and .8 amperes, $9.50. Model "K" Trickle Charger, $7.50.

Special models for 25-40 cycles at slightly higher prices. Prices are higher West of the Rockies and in Canada.

FIRST noiseless battery charging. Then successful light socket "B" power. Then trickle charging. And today, most important of all, Balkite "AB," a complete unit containing no battery in any form, supplying both "A" and "B" power directly from the light socket, and operating only while the set is in use. The great improvements in radio power have been made by Balkite.

The famous Balkite electrolytic principle

This pioneering has been important. Yet alone it would never have made Balkite one of the best known names in radio. Balkite is today the established leader because of Balkite performance in the hands of its owners.

Because with 2,000,000 units in the field Balkite has a record of long life and freedom from trouble seldom equalled in any industry.

Because the first Balkite "B," purchased 5 years ago, is still in use. Because to your

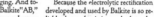

Licensed under Andrews Hammond Patent

Balkite "AB" *Contains no battery.* A complete unit, replacing both "A" and "B" batteries and supplying radio current directly from the light socket. Contains no battery in any form. Operates only while the set is in use. Two models: "AB" 6-135,* 135 volts "B" current, $64.50; "AB" 6-180, 180 volts, $74.50. Special model for Radiola 28, $63.50.

radio dealer Balkite is a synonym for quality.

Because the electrolytic rectification developed and used by Balkite is so reliable that today it is standard on the signal systems of most American as well as European and Oriental railroads. It is this principle that does away with the necessity of using tubes for rectifying current — that makes all Balkite Radio Power Units, including the new Balkite "A" and "AB," permanent equipment with nothing to wear out or replace.

Balkite has pioneered — but not at the expense of the public.

Radio power with batteries or without

Today, whatever type of radio set you own, whatever type of power equipment you want (with batteries or without) Balkite has it. And production is so enormous that prices are as astonishingly low. *Your dealer will recommend the Balkite equipment you need for your set.*

FANSTEEL PRODUCTS COMPANY, INC., NORTH CHICAGO, ILLINOIS

Licensees for Germany:
Siemens & Halske, A. G. Wernerwerk M
Siemensstadt, Berlin

Sole Licensees in the United Kingdom:
Messrs. Radio Accessories Ltd., 9-13 Hythe Rd.
Willesden, London, N. W. 10

FANSTEEL

Balkite

Radio Power Units

Manufacturers' Booklets

A Varied List of Books Pertaining to Radio and Allied Subjects Obtainable Free With the Accompanying Coupon

READERS may obtain any of the booklets listed below by using the coupon printed on page 168. Order by number only.

1. FILAMENT CONTROL—Problems of filament supply, voltage regulation, and effect on various circuits. RADIALL COMPANY.
2. HARD RUBBER PANELS—Characteristics and properties of hard rubber as used in radio, with suggestions on how to "work" it. B. F. GOODRICH RUBBER COMPANY.
3. TRANSFORMERS—A booklet giving data on input and output transformers. PACENT ELECTRIC COMPANY.
4. RESISTANCE-COUPLED AMPLIFIERS—A general discussion of resistance coupling with curves and circuit diagrams. COLE RADIO MANUFACTURING COMPANY.
5. CARBORUNDUM IN RADIO—A book giving pertinent data on the crystal as used for detection, with hook-ups, and a section giving information on the use of resistors. THE CARBORUNDUM COMPANY.
6. B-ELIMINATOR CONSTRUCTION—Constructional data on how to build. AMERICAN ELECTRIC COMPANY.
7. TRANSFORMER AND CHOKE-COUPLED AMPLIFICATION—Circuit diagrams and discussion. ALL-AMERICAN RADIO CORPORATION.
8. RESISTANCE UNITS—A data sheet of resistance units and their application. WARD-LEONARD ELECTRIC COMPANY.
9. VOLUME CONTROL—A leaflet showing circuits for distortionless control of volume. CENTRAL RADIO LABORATORIES.
10. VARIABLE RESISTANCE—As used in various circuits. CENTRAL RADIO LABORATORIES.
11. RESISTANCE COUPLING—Resistors and their application to audio amplification, with circuit diagrams. DEJUR PRODUCTS COMPANY.
12. DISTORTION AND WHAT CAUSES IT—Hook-ups of resistance-coupled amplifiers with standard circuits. ALLEN-BRADLEY COMPANY.
13. B-ELIMINATOR AND POWER AMPLIFIER—Instructions for assembly and operation using Raytheon tube. GENERAL RADIO COMPANY.
13a. B-ELIMINATOR AND POWER AMPLIFIER—Instructions for assembly and operation using an R. C. A. rectifier. GENERAL RADIO COMPANY.
14. VARIABLE CONDENSERS—A description of the functions and characteristics of variable condensers with curves and specifications for their application to complete receivers. ALLEN D. CARDWELL MANUFACTURING COMPANY.
17. BAKELITE—A description of various uses of bakelite in radio, its manufacture, and its properties. BAKELITE CORPORATION.
19. POWER SUPPLY—A discussion on power supply with particular reference to lamp-socket operation. Theory and constructional data for building power supply devices. ACME APPARATUS COMPANY.
20. AUDIO AMPLIFICATION—A booklet containing data on audio amplification together with hints for the constructor. ALL-AMERICAN RADIO CORPORATION.
21. HIGH-FREQUENCY DRIVER AND SHORT-WAVE WAVE-METER—Constructional data and application. BURGESS BATTERY COMPANY.
46. RADIO-FREQUENCY CHOKES—A pamphlet showing positions in the circuit where audio-frequency chokes may be used. SAMSON ELECTRIC COMPANY.
47. RADIO-FREQUENCY CHOKES—Circuit diagrams illustrating the use of radio-frequency chokes to keep out radio-frequency currents from definite points. SAMSON ELECTRIC COMPANY.
48. TRANSFORMER AND IMPEDANCE DATA—Tables giving the mechanical and electrical characteristics of transformers and impedances, together with a short description of their use in the circuit. SAMSON ELECTRIC COMPANY.
49. BYPASS CONDENSERS—A description of the manufacture of bypass and filter condensers. LESLIE F. MUTER COMPANY.
50. AUDIO MANUAL—Fifty questions which are often asked regarding audio amplification, and their answers. AMERTRAN SALES COMPANY, INCORPORATED.
51. SHORT-WAVE RECEIVER—Constructional data on a receiver which, by the substitution of various coils, may be made to tune from a frequency of 16,660 kc. (18 meters) to 1999 kc. (150 meters). SILVER-MARSHALL, INCORPORATED.
52. AUDIO QUALITY—A booklet dealing with audio-frequency amplification of various kinds and the application to well-known circuits. SILVER-MARSHALL, INCORPORATED.
59. VARIABLE CONDENSERS—A bulletin giving an analysis of various condensers together with their characteristics. GENERAL RADIO COMPANY.
57. FILTER DATA—Facts about the filtering of direct current supplied by means of motor-generator outfits used with transmitters. ELECTRIC SPECIALTY COMPANY.
59. RESISTANCE COUPLING—A booklet giving some general information on the subject of radio and the application of resistors to a circuit. DAVEN RADIO CORPORATION.
60. RESISTORS—A pamphlet giving some technical data on resistors which are capable of dissipating considerable energy; also data on the ordinary resistors used in resistance-coupled amplification. THE CRESCENT RADIO SUPPLY COMPANY.
62. RADIO-FREQUENCY AMPLIFICATION—Constructional details of a five-tube receiver using a special design of radio-frequency transformer. CAMFIELD RADIO MFG. COMPANY.
63. FIVE-TUBE RECEIVER—Constructional data on building a receiver. ACME APPARATUS COMPANY.
64. SUPER-HETERODYNE—Constructional details of a seven-tube set. S. G. C. EVANS COMPANY.
70. IMPROVING THE AUDIO AMPLIFIER—Data on the characteristics of audio transformers, with a circuit diagram showing where chokes, resistors, and condensers can be used. AMERICAN TRANSFORMER COMPANY.
71. PLATE SUPPLY SYSTEM—A wiring diagram and layout plan for a plate supply system to be used with a power amplifier. Complete directions for wiring are given. AMERTRAN SALES COMPANY.

80. FIVE-TUBE RECEIVER—Data are given for the construction of a five-tube tuned radio-frequency receiver. Complete instructions, list of parts, circuit diagram, and template are given. ALL-AMERICAN RADIO CORPORATION.
81. BETTER TUNING—A booklet giving much general information on the subject of radio reception with specific illustrations. Primarily for the non-technical home constructor. BREMER-TULLY MANUFACTURING COMPANY.
82. SIX-TUBE RECEIVER—A booklet containing photographs, instructions, and diagrams for building a six-tube shielded receiver. SILVER-MARSHALL, INCORPORATED.
83. SOCKET POWER DEVICE—A list of parts, diagrams, and templates for the construction and assembly of socket power devices. JEFFERSON ELECTRIC MANUFACTURING COMPANY.
84. FIVE-TUBE EQUAMATIC—Panel layout, circuit diagrams, and instructions for building a five-tube receiver, together with data on the operation of tuned radio-frequency transformers of special design. KARAS ELECTRIC COMPANY.
85. FILTER—Data on a high-capacity electrolytic condenser used in filter circuits in connection with A socket power supply units, are given in a pamphlet. THE ABOX COMPANY.
86. SHORT-WAVE RECEIVER—A booklet containing data on a short-wave receiver as constructed for experimental purposes. THE ALLEN D. CARDWELL MANUFACTURING CORPORATION.
88. SUPER-HETERODYNE CONSTRUCTION—A booklet giving full instructions, together with a blue print and necessary data, for building an eight-tube receiver. THE GEORGE W. WALKER COMPANY.
89. SHORT-WAVE TRANSMITTER—Data and blue prints are given on the construction of a short-wave transmitter, together with operating instructions, methods of keying, and other pertinent data. RADIO ENGINEERING LABORATORIES.
90. IMPEDANCE AMPLIFICATION—The theory and practice of a special type of dual-impedance audio amplification are given. ALDEN MANUFACTURING COMPANY.
91. B-SOCKET POWER—A booklet giving constructional details of a socket-power device using either the BH or 313 type rectifier. NATIONAL COMPANY, INCORPORATED.
94. POWER AMPLIFIER—Constructional data and wiring diagrams of a power amplifier combined with a B-supply unit are given. NATIONAL COMPANY, INCORPORATED.
100. A, B, AND C SOCKET-POWER SUPPLY—A booklet giving data on the construction and operation of a socket-power supply using the new high-current rectifier tube. THE Q. R. S. MUSIC COMPANY.
101. USING CHOKES—A folder with circuit diagrams of the more popular circuits showing where choke coils may be placed to produce better results. SAMSON ELECTRIC COMPANY.

Accessories

22. A PRIMER OF ELECTRICITY—Fundamentals of electricity with special reference to the application of dry cells to radio and other uses. Constructional data on buzzers, automatic switches, alarms, etc. NATIONAL CARBON COMPANY.
23. AUTOMATIC RELAY CONNECTIONS—A data sheet showing how a relay may be used to control A and B circuits. YAXLEY MANUFACTURING COMPANY.
25. ELECTROLYTIC RECTIFIER—Technical data on a new type of rectifier with operating curves. KODEL RADIO CORPORATION.
26. DRY CELLS FOR TRANSMITTERS—Actual tests given, well illustrated with curves showing exactly what may be expected of this type of B power. BURGESS BATTERY COMPANY.
27. DRY-CELL BATTERY CAPACITIES FOR RADIO TRANSMITTERS—Characteristic curves and data on discharge tests. BURGESS BATTERY COMPANY.
28. B BATTERY LIFE—Battery life curves with general curves on tube characteristics. BURGESS BATTERY COMPANY.
29. HOW TO MAKE YOUR SET WORK BETTER—A non-technical discussion of general radio subjects with hints on how reception may be bettered by using the right tubes. UNITED RADIO AND ELECTRIC COMPANY.
30. TUBE CHARACTERISTICS—A data sheet giving constants of tubes. C. E. MANUFACTURING COMPANY.
31. FUNCTIONS OF THE LOUD SPEAKER—A short, non-technical general article on loud speakers. AMPLION CORPORATION OF AMERICA.
32. METERS FOR RADIO—A catalogue of meters used in radio, with connecting diagrams. BURTON-ROGERS COMPANY.
33. SWITCHBOARD AND PORTABLE METERS—A booklet giving dimensions, specifications, and shunts used with various meters. BURTON-ROGERS COMPANY.
34. COST OF B BATTERIES—An interesting discussion of the relative merits of various sources of B supply. HARTFORD BATTERY MANUFACTURING COMPANY.
35. STORAGE BATTERY OPERATION—An illustrated booklet on the care and operation of the storage battery. GENERAL LEAD BATTERIES COMPANY.
36. CHARGING A AND B BATTERIES—Various ways of connecting up batteries for charging purposes. WESTINGHOUSE UNION BATTERY COMPANY.
37. CHOOSING THE RIGHT RADIO BATTERY—Advice on what dry cell battery to use; their application to radio, with wiring diagrams. NATIONAL CARBON COMPANY.
53. TUBE REACTIVATION—Information on the care of vacuum tubes, with notes on how and when they should be reactivated. THE STERLING MANUFACTURING COMPANY.
54. ARRESTERS—Mechanical details and principles of the vacuum type of arrester. NATIONAL ELECTRIC SPECIALTY COMPANY.
55. CAPACITY CONNECTOR—Description of a new device for connecting up the various parts of a receiving set, and at the same time providing bypass condensers between the leads. KURZ-KASCH COMPANY.

(Continued on page 168)

Vital Factors

in attaining

High Quality Reproduction

Type 285
Audio Transformer

The Type 285 transformers give high and even amplification of all tones common to speech, instrumental, and vocal music. Available in two ratios.
Type 285-H Audio Transformer.
Price $6.00
Type 285-D Audio Transformer.
Price $6.00

Type 367
Output Transformer

This unit adapts the impedance of an audio amplifier to the input of any coupling type speaker, thus promoting better tone quality and protecting the speaker windings against possible damage from A. C. voltages. Similar in appearance to the Type 285.
Type 367 Output Transformer.
Price $5.00

High quality reproduction depends upon three things: correctly designed coupling units, proper use of amplifier tubes, and an efficient reproducing device.

For over a decade the subject of audio frequency amplification has been extensively studied in the laboratories of the General Radio Company with particular attention given to the design of coupling units.

As a result of this exhaustive research the General Radio Company has been, and is, the pioneer manufacturer of high quality Audio Transformers, Impedance Couplers, and Speaker Filters.

The latest contribution to quality amplification is the type 441 Push-Pull Amplifier, which is mounted on a nickel finished metal base-board and is completely wired.

If the amplifier of your receiver is not bringing out the rich bass notes and the mellow high tones as well as those in the middle register why not rebuild your amplifier for *Quality Reproduction* with *General Radio* coupling units?

Write for our Series A of amplification booklets describing various amplifier circuits and units.

GENERAL RADIO COMPANY
Cambridge, Mass.

Type 373
Double Impedance
Coupler

Many prefer the impedance coupling method of amplification to resistance coupling as lower plate voltages may be used and greater amplification may be obtained. The Type 373 is contained in a metal shell and connected in a circuit in precisely the same manner as a transformer.
Type 373 Double Impedance Coupler.
Price $6.50

Type 387-A
Speaker Filter

The Type 387-A consists of an inductance choke with condenser. It offers a high impedance to audio frequency current and forces these currents to pass through a condenser into the speaker, thereby improving tone quality and protecting the speaker windings.
Type 387-A Speaker Filter.
Price $6.00

Type 441 Push-Pull Amplifier

The Type 441 is completely wired and consists of two high quality push-pull transformers, with necessary sockets and resistances mounted on a nickel finished metal base board. It may be used with any power or semi-power tube to increase the undistorted output of the amplifier with the result that better quality is reproduced from the loudspeaker with more volume than is obtained from other methods of coupling.
Licensed by the R. C. A. and through terms of the license may be sold with tubes only.
Type 441 Push-Pull Amplifier .... **Price $20.00**
Type UX-226 or CX-326 Amplifier Tube .... " 3.00
Type UX-171 or CX-371 Amplifier Tube .... " 4.50

Type 445 Plate Supply and Grid Biasing Unit

The Type 445 meets the demand for a thoroughly dependable light socket plate supply and grid biasing unit that is readily adaptable to the tube requirements of any standard type of receiver. Any combination of voltages from 0 to 200 may be taken from the adjustable "B" voltage taps. A variable grid bias voltage from 0 to 50 is also available. The unit is designed for use on 105 to 125 volt (50 to 60 cycle) A. C. lines and uses the UX-280 or CX-380 rectifier tube.
Licensed by R. C. A. and through terms of the license may be sold with tube only.
Type 445 Plate Supply and Grid Biasing Unit .... **Price $55.00**
Type UX-280 or CX-380 Rectifier Tube .... " 5.00

GENERAL RADIO

LABORATORY EQUIPMENT

Parts *and* Accessories

"RADIO BROADCAST'S" DIRECTORY OF MANUFACTURED RECEIVERS

¶ A coupon will be found on page 172. All readers who desire additional information on the receivers listed below need only insert the proper numbers in the coupon, mail it to the Service Department of RADIO BROADCAST, and full details will be sent. New sets are listed in this space each month.

KEY TO TUBE ABBREVIATIONS

99—60-mA. filament (dry cell)
01-A—Storage battery 0.25 amps. filament
12—Power tube (Storage battery)
71—Power tube (Storage battery)
16-B—Half-wave rectifier tube
80—Full-wave, high current rectifier
81—Half-wave, high current rectifier
Hmu—High-Mu tube for resistance-coupled audio
20—Power tube (dry cell)
10—Power Tube (Storage battery)
00-A—Special detector
13—Full-wave rectifier tube
26—Low-Voltage high-current a. c. tube
27—Heater type a. c. tube

DIRECT CURRENT RECEIVERS

NO. 424. COLONIAL 26

Six tubes; 2 t. r. f. (01–A), detector (12), 2 transformer audio (01-A and 71). Balanced t. r. f. One to three dials. Volume control: antenna switch and potentiometer across first audio. Watts required: 120. Console size: 34 x 38 x 18 inches. Headphone connections. The filaments are connected in a series parallel arrangement. Price $250 including power unit.

NO. 425. SUPERPOWER

Five tubes: All 01–A tubes. Multiplex circuit. Two dials. Volume control: resistance in r. f. plate. Watts required: 30. Antenna: loop or outside. Cabinet sizes: table, 27 x 10 x 9 inches; console, 28 x 50 x 21. Prices: table, $135 including power unit; console, $390 including power unit and loud speaker.

A. C. OPERATED RECEIVERS

NO. 508. ALL-AMERICAN 77, 88, AND 99

Six tubes; 3 t. r. f. (26), detector (27), 2 transformer audio (26 and 71). Rice neutralized t. r. f. Single drum tuning. Volume control: potentiometer in r. f. plate. Cabinet sizes: No. 77, 21 x 10 x 8 inches; No. 88 Hiboy, 25 x 38 x 18 inches; No. 99 console, 27½ x 43 x 20 inches. Shielded. Output device. The filaments are supplied by means of three small transformers. The plate supply employs a gas-filled rectifier tube. Voltmeter in a. c. supply line. Prices: No. 77, $150, including power unit; No. 88, $210 including power unit; No. 99, $285 including power unit and loud speaker.

NO. 509. ALL-AMERICAN "DUET"; "SEXTET"

Six tubes; 2 t. r. f. (99), detector (99), 3 transformer audio (99 and 12). Rice neutralized t.r.f. Two dials. Volume control: resistance in r.f. plate. Cabinet sizes: "Duet," 23 x 56 x 16½ inches; "Sextet," 22¼ x 13½ x 15¼ inches. Shielded. Output device. The 99 filaments are connected in series and supplied with rectified a.c., while 12 is supplied with raw a.c. The plate and filament supply uses gaseous rectifier tubes. Milliammeter on power unit. Prices: "Duet," $160 including power unit; "Sextet," $220 including power unit and loud speaker.

NO. 511. ALL-AMERICAN 80, 90, AND 115

Five tubes; 2 t.r.f. (99), detector (99), 2 transformer audio (99 and 12). Rice neutralized t.r.f. Two dials. Volume control; resistance in r.f. plate. Cabinet sizes: No. 80, 23½ x 12½ x 15 inches; No. 90, 37½ x 12 x 12½ inches; No. 115 Hiboy, 24 x 41 x 15 inches. Coils individually shielded. Output device. See No. 509 for power supply. Prices: No. 80, $135 including power unit; No. 90, $145 including power unit and compartment; No. 115, $170 including power unit, compartment, and loud speaker.

NO. 510. ALL-AMERICAN 7

Seven tubes; 3 t.r.f. (26), 1 untuned r.f. (26), detector (27), 2 transformer audio (26 and 71). Rice neutralized t.r.f. One drum. Volume control: resistance in r.f. plate. Cabinet sizes: "Sovereign" console, 30½ x 60½ x 19 inches; "Lorraine" Hiboy, 25½ x 53½ x 17½ inches; "Forte" cabinet, 25½ x 13½ x 17½ inches. For filament and plate supply: See No. 508. Prices: "Sovereign" $460; "Lorraine" $360; "Forte" $270. All prices include power unit. First two include loud speaker.

NO. 401. AMRAD AC9

Six tubes; 3 t.r.f. (99), detector (99), 2 transformer (99 and 12). Neutrodyne. Two dials. Volume control: resistance across 1st audio. Watts consumed: 50. Cabinet size: 27 x 9 x 11½ inches. The 99 filaments are connected in series and supplied with rectified a.c., while the 12 is run on raw a.c. The power unit, requiring two 16-B rectifiers, is separate and supplies A, B, and C current. Price $142 including power unit.

NO. 402. AMRAD AC5

Five tubes. Same as No. 401 except one less r.f. stage. Price $125 including power unit.

NO. 536. SOUTH BEND

Six tubes. One control. Sub-panel shielding. Binding Posts. Antenna: outdoor. Prices: table, $130, Baby Grand console, $195.

NO. 537. WALBERT 26

Six tubes; five Kellogg a.c. tubes and one 71. Two controls. Volume control: variable plate resistance. Isofarad circuit. Output device. Battery cable. Semi-shielded. Antenna: 50 to 75 feet. Cabinet size: 10½ x 23½ x 16½ inches. Prices: $215; with tubes, $250.

NO. 484. BOSWORTH, B5

Five tubes; 2 t.r.f. (26), detector (99), 2 transformer audio (special a.c. tubes). T.r.f. circuit. Two dials. Volume control: potentiometer. Cabinet size: 23 x 7 x 8 inches. Output device included. Price $175.

NO. 406. CLEARTONE 110

Five tubes; 2 t.r.f., detector, 2 transformer audio. All tubes a. c. heater type. One or two dials. Volume control: resistance in r. f. plate. Watts consumed: 40. Cabinet size: Varies. The plate supply is built in the receiver and requires one rectifier tube. Filament supply through step down transformers. Prices range from $175 to $375 which includes 5 a.c. tubes and one rectifier tube.

NO. 407. COLONIAL 25

Six tubes; 2 t. r. f. (01–A), detector (99), 2 resistance audio (99). 1 transformer audio (10). Balanced t.r.f. circuit. One or three dials. Volume control: Antenna switch and potentiometer on 1st audio. Watts consumed: 100. Console size: 34 x 38 x 18 inches. Output device. All tube filaments are operated on a. c. except the detector which is supplied with rectified a.c. from the plate supply. The rectifier employs two 16-B tubes. Price $250 including built-in plate and filament supply.

NO. 507. CROSLEY 602 BANDBOX

Six tubes; 3 t.r.f. (26), detector (27), 2 transformer audio (26 and 71). Neutrodyne circuit. One dial. Cabinet size: 17½ x 5¼ x 7¾ inches. The heaters for the a.c. tubes and the 71 filament are supplied by windings in B unit transformers available to operate either on 25 or 60 cycles. The plate current is supplied by means of rectifier tube. Price $65 for set alone, power unit $60.

NO. 408. DAY-FAN 'DE LUXE'

Six tubes; 3 t.r.f., detector, 2 transformer audio. All 01–A tubes. One dial. Volume control: potentiometer across r.f. tubes. Watts consumed: 300. Console size: 30 x 40 x 20 inches. The filaments are connected in series and supplied with d.c. from a motor-generator set which also supplies B and C current. Output device. Price $350 including power unit.

NO. 409. DAYCRAFT 5

Five tubes; 2 t.r.f., detector, 2 transformer audio. All a. c. heater tubes. Reflexed t.r.f. One dial. Volume control: potentiometers in r.f. plate and 1st audio. Watts consumed: 135. Console size: 34 x 36 x 14 inches. Output device. The heaters are supplied by means of a small transformer. A built-in rectifier supplies B and C voltages. Price $170, less tubes. The following have one more r.f. stage and are not reflexed: Daycraft 6, $195; Dayrole 6, $235; Dayfan 6, $110. All prices less tubes.

NO. 469. FREED-EISEMANN NR11

Six tubes; 3 t.r.f. (01–A), detector (01–A), 2 transformer audio (01–A and 71). Neutrodyne. One dial. Volume control: potentiometer. Watts consumed: 150. Cabinet size: 19¾ x 10 x 10½ inches. Shielded. Output device. A special power unit is included employing a rectifier tube. Price $225 including NR-411 power unit.

NO. 487. FRESHMAN 7F-AC

Six tubes; 3 t.r.f. (26), detector (27), 2 transformer audio (26 and 71). Equaphase circuit. One dial. Volume control: potentiometer across 1st audio. Console size: 24½ x 41½ x 15 inches. Output device. The filaments and heaters and B supply are all supplied by one power unit. The plate supply requires one 80 rectifier tube. Price $175 to $350, complete.

NO. 421. SOVEREIGN 238

Seven tubes of the a.c. heater type. Balanced t.r.f. Two dials. Volume control: resistance across 2nd audio. Watts consumed: 45. Console size: 37 x 52 x 15 inches. The heaters are supplied by a small a. c. transformer, while the plate is supplied by means of rectified a.c using a gaseous type rectifier. Price $325, including power unit and tubes.

NO. 517. KELLOGG 510, 511, AND 512

Seven tubes; 4 t.r.f., detector, 2 transformer audio. All Kellogg a.c. tubes. One control and special zone switch. Balanced. Volume control: special. Output device. Shielded. Cable connection between power supply unit and receiver. Antenna: 25 to 100 feet. Panel 7½ x 27½ inches. Prices: Model 510 and 512, consoles, $495 complete. Model 511, consolette, $365 without loud speaker.

NO. 496. SLEEPER ELECTRIC

Five tubes; four 99 tubes and one 71. Two controls. Volume control: rheostat on r.f. Neutralized. Cable. Output device. Power supply uses two 16-B tubes. Antenna: 100 feet. Prices: Type 64, table, $160; Type 65, table, with built-in loud speaker, $175; Type 66, table, $175; Type 67, console, $235; Type 78, console, $265.

NO. 538. NEUTROWOUND, MASTER ALLECTRIC

Six tubes; 2 t.r.f. (01-A), detector (01-A), 2 audio (01-A and two 71 in push-pull amplifier). The 01-A tubes are in series, and are supplied from a 400-mA. rectifier. Two drum controls. Volume control: variable plate resistance. Output device. Shielded. Antenna: 50 to 100 feet. Price: $360.

NO. 413. MARTI

Six tubes: 2 t.r.f., detector, 3 resistance audio. All tubes a.c. heater type. Two dials. Volume control resistance in r.f. plate. Watts consumed: 38. Panel size 7 x 21 inches. The built-in plate supply employs one 16-B rectifier. The filaments are supplied by a small transformer. Prices: table, $235 including tubes and rectifier; console, $275 including tubes and rectifier; console, $325 including tubes, rectifier, and loud speaker.

NO. 417 RADIOLA 28

Eight tubes; five type 99 and one type 20. Drum control. Super-heterodyne circuit. C-battery connections. Battery cable. Headphone connection. Antenna: loop. Set may be operated from batteries or from the power mains when used in conjunction with the model 104 loud speaker. Prices: $260 with tubes, battery operation; $570 with model 104 loud speaker, a. c. operation.

NO. 540 RADIOLA 30-A

Receiver characteristics same as No. 417 except that type 71 power tube is used. This model is designed to operate on either a. c. or d. c. from the power mains. The combination rectifier—power—amplifier unit uses two type 81 tubes. Model 100-A loud speaker is contained in lower part of cabinet. Either a short indoor or long outside antenna may be used. Cabinet size: 42½ x 29 x 17¾ inches. Price: $495.

NO. 541 RADIOLA 32

This model combines receiver No. 417 with the model 104 loud speaker. The power unit uses two type 81 tubes and a type 10 power amplifier. Loop is completely enclosed and is revolved by means of a dial on the panel. Models for operation from a. c. or d. c. power mains. Cabinet size: 52 x 72 x 17¾ inches. Price: $895.

NO. 539 RADIOLA 17

Six tubes; 3 t. r. f. (26), detector (27), 2 transformer audio (26 and 27). One control. Illuminated dial. Built-in power supply using type 80 rectifier. Antenna: 100 feet. Cabinet size: 25⅛ x 7⅞ x 8⅛. Price: $130 without accessories.

NO. 545. NEUTROWOUND, SUPER ALLECTRIC

Five tubes; 2 t.r.f. (99), detector (99), 2 audio (99 and 71). The 99 tubes are in series and are supplied from an 85-mA. rectifier. Two drum controls. Volume control: variable plate resistance. Output device. Antenna: 75 to 100 feet. Cabinet size: 9 x 24 x 11 inches. Price: $150.

NO. 490. MOHAWK

Six tubes; 2 t.r.f., detector, 2 transformer audio. All tubes a.c heater type except 71 in last stage. One dial. Volume control: rheostat on r.f. Watts consumed: 40. Panel size: 12½ x 8¼ inches. Output device. The heaters for the a.c tubes and the 71 filament are supplied by small transformers. The plate supply is of the built-in type using a rectifier tube. Prices range from $65 to $245.

NO. 522. CASE, 62 B AND 62 C

McCullough a.c. tubes. Drum control. Volume control: variable high resistance in audio system. C-battery connections. Semi-shielded. Cable. Antenna: 100 feet. Panel size: 7 x 21 inches. Prices: Model 62 B, complete with a.c. equipment, $185; Model 62 C, complete with a.c. equipment, $235.

NO. 523. CASE, 92 A AND 92 C

McCullough a.c. tubes. Drum control. Inductive volume control. Technidyne circuit. Shielded. Cable. C-battery connections. Model 92 C contains output device. Loop operated. Prices: Model 92 A, table, $350; Model 92 C, console, $475.

BATTERY OPERATED RECEIVERS

NO. 542. PFANSTIEHL JUNIOR SIX

Six tubes; 3 t.r.f. (01-A), detector (01-A), 2 audio. Pfanstiehl circuit. Volume control: variable resistance in r.f. plate circuit. One dial. Shielded. Battery cable. C-battery connections. Etched bronze panel. Antenna: outdoor. Cabinet size: 9 x 20 x 8 inches. Price: $80, without accessories.

NO. 512. ALL-AMERICAN 44, 45, AND 66

Six tubes; 3 t.r.f. (01–A, detector) 01–A, 2 transformer audio (01-A and 71). Rice neutralized t.r.f. Drum control. Volume control: rheostat in r.f. Cabinet sizes: No. 44, 21 x 10 x 8 inches; No. 55, 25 x 38 x 18 inches; No. 66, 27½ x 43 x 20 inches. C-battery connections. Battery cable. Antenna: 75 to 125 feet. Prices: No. 44, $70; No. 55, $125 including loud speaker; No. 66, $200 including loud speaker.

NO. 428. AMERICAN C6

Five tubes; 2 t.r.f. detector, 2 transformer audio. All 01–A tubes. Semi balanced t.r.f. Three dials. Plate current 15 mA. Volume control: potentiometer. Cabinet sizes: table, 20 x 8½ x 10 inches; console, 36 x 40 x 17 inches. Partially shielded. Battery cable. C-battery connections. Antenna: 125 feet. Prices: table, $30 console, $65 including loud speaker.

NO. 485. BOSWORTH B6

Five tubes: 2 t.r.f. (01-A), detector (01-A), 2 transformer audio (01-A and 71). Two dials. Volume control: variable grid resistances. Battery cable. C-battery connections. Antenna: 25 feet or longer. Cabinet size 15 x 7 x 8 inches. Price $75.

NO. 513. COUNTERPHASE SIX

Six tubes; 3 t.r.f. (01-A), detector (00-A), 2 transformer audio (01-A and 12). Counterphase t.r.f. Two dials. Plate current: 32 mA. Volume control: rheostat on 2nd and 3rd r.f. Coils shielded. Battery cable. C-battery connections. Antenna: 75 to 100 feet. Console size: 18¾ x 40½ x 15½ inches. Prices: Model 35, table, $110; Model 37, console, $175.

NO. 514. COUNTERPHASE EIGHT

Eight tubes; 4 t.r.f. (01-A), detector (00-A), 2 transformer audio (01-A and 12). Counterphase t.r.f. One dial. Plate current: 40 mA. Volume control: rheostat in 1st r.f. Copper stage shielding. Battery cable. C-battery connections. Antenna: 75 to 100 feet. Cabinet size: 30 x 12½ x 16 inches. Prices: Model 12, table, $225; Model 16, console, $335; Model 18, console, $365.

NO. 506. CROSLEY 601 BANDBOX

Six tubes; 3 t.r.f., detector, 2 transformer audio. All 01–A tubes. Neutrodyne. One dial. Plate current: 40 mA. Volume control: rheostat in r.f. Shielded. Battery cable. C-battery connections. Antenna: 75 to 150 feet. Cabinet size: 17¼ x 5½ x 7½. Price, $55.

NO. 434. DAY–FAN 6

Six tubes; 3 t.r.f. (01-A), detector (01-A), 2 transformer audio (01-A and 12 or 71). One dial. Plate current: 12 to 15 mA. Volume control: rheostat on r.f. Shielded. Battery cable, C-battery connections. Output device. Antenna: 50 to 120 feet. Cabinet sizes: Daycraft 6, 32 x 30 x 34 inches; Day-Fan Jr., 15 x 7 x 7. Prices: Day-Fan 6, $110; Daycraft 6, $145 including loud speaker; Day-Fan Jr. not available.

NO. 435. DAY-FAN 7

Seven tubes; 3 t.r.f. (01-A), detector (01-A), 1 resistance audio (01-A), 2 transformer audio (01-A and 12 or 71). Plate current: 15 mA. Antenna: outside. Same as No. 434. Price $115.

NO. 503. FADA SPECIAL

Six tubes; 3 t.r.f. (01–A), detector (01–A), 2 transformer audio (01–A and 71). Neutrodyne. Two drum control. Plate current: 20 to 24 mA. Volume control: rheostat on r.f. Coils shielded. Battery cable. C-battery connections. Headphone connection. Antenna: outdoor. Cabinet size: 20½ x 13¾ x 10½ inches. Price $95.

NO. 504. FADA 7

Seven tubes; 4 t.r.f. (01–A), detector (01–A), 2 transformer audio (01–A and 71). Neutrodyne. Two drum control. Plate current: 43mA. Volume control: rheostat on r.f. Completely shielded. Battery cable. C-battery connections. Headphone connections. Output device. Antenna: outdoor or loop. Cabinet sizes: table, 25¾ x 13½ x 11½ inches; console, 29 x 50 x 17 inches. Prices: table, $185; console, $285.

NO. 436. FEDERAL

Five tubes; 2 t.r.f. (01–A), detector (01–A), 2 transformer audio (01–A and 12 or 71). Balanced t.r.f. One dial. Plate current: 20.7 mA. Volume control: rheostat on r.f. Shielded. Battery cable. C-battery connections. Antenna: loop. Made in 6 models. Price varies from $250 to $1000 including loop.

NO. 505. FADA 8

Eight tubes. Same as No. 504 except for one extra stage of audio and different cabinet. Prices: table, $300; console, $400.

NO. 437. FERGUSON 10A

Seven tubes; 3 t.r.f. (01–A), detector (01–A), 3 audio (01–A and 12 or 71). One dial. Plate current: 18 to 25 mA. Volume control: rheostat on two r.f. Shielded. Battery cable. C-battery connections. Antenna: 100 feet. Cabinet size: 21½ x 12 x 15 inches. Price $150.

NO. 438. FERGUSON 14

Ten tubes; 3 untuned r.f., 3 t. r.f. (01–A), detector (01–A), 3 audio (01–A and 12 or 71). Special balanced t.r.f. One dial. Plate current: 30 to 35 mA. Volume control: rheostat in three r.f. Shielded. Battery cable. C-battery connections. Antenna: loop. Cabinet size: 24 x 12 x 16 inches. Price $235, including loop.

NO. 439. FERGUSON 12

Six tubes; 2 t.r.f. (01–A), detector (01–A), 1 transformer audio (01–A), 2 resistance audio (01–A and 12 or 71). Two dials. Plate current: 18 to 25 mA. Volume control: rheostat on two r.f. Partially shielded. Battery cable. C-battery connections. Antenna: 100 feet. Cabinet size: 22½ x 10 x 12 inches. Price $85. Consolette $145 including loud speaker.

NO. 440. FREED–EISEMANN NR–8 NR–9, AND NR–66

Six tubes; 3 t.r.f. (01–A), detector (01–A), 2 transformer audio (01–A and 71). Neutrodyne. NR–8, two dials; others one dial. Plate current: 30 mA. Volume control: rheostat on r.f. NR–8 and 9; chassis type shielding. NR–66, individual stage shielding. Battery cable. C-battery connections. Antenna: 100 feet. Cabinet sizes: NR–8 and 9, 19¾ x 10 x 10½ inches; NR–66 20 x 10½ x 12 inches. Prices: NR–8, $90; NR–9, $100; NR–66, $125.

NO. 501. KING "CHEVALIER"

Six tubes. Same as No. 500. Coils completely shielded. Panel size: 11 x 7 inches. Price, $210 including loud speaker.

NO. 441. FREED–EISEMANN NR–77

Seven tubes; 4 t.r.f. (01–A), detector (01–A), 2 transformer audio (01–A and 71). Neutrodyne. One dial. Plate current: 35 mA. Volume control: rheostat on r.f. Shielding. Battery cable. C-battery connections. Antenna: outside or loop. Cabinet size: 23 x 10½ x 13 inches. Price $175.

NO. 442. FREED–EISEMANN 800 AND 850

Eight tubes; 4 t.r.f. (01–A), detector (01–A), 1 transformer (01–A), 1 parallel audio (01–A or 71). Neutrodyne. One dial. Plate current: 35 mA. Volume control: rheostat on r.f. Shielded. Battery cable. C-battery connections. Output: two tubes in parallel or one power tube may be used. Antenna: outside or loop. Cabinet sizes: No. 800, 34 x 15½ x 13½ inches; No. 850, 36 x 65½ x 17½. Prices not available.

NO. 444. GREBE MU–1

Five tubes; 2 t.r.f. (01–A), detector (01–A), 2 transformer audio (01–A and 12 or 71). Balanced t.r.f. One, two, or three dials (operate singly or together). Plate current: 30mA. Volume control: rheostat on r.f. Binocular coils. Binding posts. C-battery connections. Antenna: 125 feet. Cabinet size: 22¼ x 9½ x 13 inches. Prices range from $95 to $320.

NO. 426. HOMER

Seven tubes; 4 t.r.f. (01-A); detector (01-A or 00A); 2 audio (01-A and 12 or 71). One knob tuning control. Volume control: rotor control in antenna circuit. Plate current: 22 to 25 mA. "Technidyne" circuit. Completely enclosed in aluminum box. Battery cable. C-battery connections. Cabinet size, 8¾ x 19½ x 9½ inches. Chassis size, 6¾ x 17 x 8 inches. Prices: Chassis only, $80. Table cabinet, $95.

NO. 502. KENNEDY ROYAL 7. CONSOLETTE

Seven tubes; 4 t.r.f. (01–A), detector (00–A), 2 transformer audio (01–A and 71). One dial. Plate current: 42 mA. Volume control: rheostat on two r.f. Special r.f. coils. Battery cable. C-battery connections. Headphone connection. Antenna: outside or loop. Consolette size: 36½ x 35½ x 19 inches. Price $220.

NO. 498. KING "CRUSADER"

Six tubes; 2 t.r.f. (01–A), detector (00–A), 3 transformer audio (01–A and 71). Balanced t.r.f. One dial. Plate current: 20 mA. Volume control: rheostat on r.f. Coils shielded. Battery cable. C-battery connections. Antenna: outside. Panel: 11 x 7 inches. Price, $115.

NO. 499. KING "COMMANDER"

Six tubes; 3 t.r.f. (01–A), detector (00–A), 2 transformer audio (01–A and 71). Balanced t.r.f. One dial. Plate current: 25 mA. Volume control: rheostat on r.f. Completely shielded. Battery cable. C-battery connections. Antenna: loop. Panel size: 12 x 8 inches. Price $220 including loop.

NO. 429. KING COLE VII AND VIII

Seven tubes; 3 t.r.f., detector, 1 resistance audio, 2 transformer audio. All 01–A tubes. Model VIII hrs one more stage t.r.f. (eight tubes). Model VII, two dials. Model VIII, one dial. Plate current: 15 to 50 mA. Volume control: primary shunt in r.f. Steel shielding. Battery cable and binding posts. C-battery connections. Output devices on some consoles. Antenna: 10 to 100 feet. Cabinet size: varies. Prices: Model VII, $80 to $160; Model VIII, $100 to $300.

NO. 500. KING "BARONET" AND "VIKING"

Six tubes; 2 t.r.f. (01–A), detector (00–A), 3 transformer audio (01–A and 71). Balanced t.r.f. One dial. Plate current: 19 mA. Volume control: rheostat in r.f. Battery cable. C-battery connections. Antenna: outside. Panel size: 18 x 7 inches. Prices: "Baronet," $70; "Viking," $140 including loud speaker.

NO. 489. MOHAWK

Six tubes; 2 t.r.f. (01–A), detector (00–A), 3 audio (01–A and 71). One dial. Plate current: 40 mA. Volume control: rheostat on r.f. Battery cable. C-battery connections. Output device. Antenna: 60 feet. Panel size: 12½ x 8½ inches. Prices range from $65 to $245.

NO. 543. ATWATER KENT, MODEL 33

Six tubes; 3 t.r.f. (01-A), detector (01-A), 2 audio (01-A and 71 or 12). One dial. Volume control: r.f. filament rheostat. C-battery connections. Battery cable. Antenna: 100 feet. Steel panel. Cabinet size: 21¼ x 6¾ x 6¼ inches. Price: $75, without accessories.

NO. 544. ATWATER KENT, MODEL 50

Seven tubes; 4 t.r.f. (01-A), detector (01-A), 2 audio (01-A and 12 or 71). Volume control: r.f. filament rheostat. C-battery connections. Battery cable. Special band-pass filter circuit with an untuned amplifier. Cabinet size: 20¾ x 13 x 7¾ inches. Price: $120.

NO. 452. ORIOLE 90

Five tubes; 2 t.r.f., detector, 2 transformer audio. All 01–A tubes. "Trinum" circuit. Two dials. Plate current: 18 mA. Volume control: rheostat on r. f. Battery cable. C-battery connections. Antenna: 50 to 100 feet. Cabinet size: 25½ x 11½ x 12½ inches. Price $85. Another model has 8 tubes, one dial, and is shielded. Price $185.

NO. 453. PARAGON

Six tubes; 2 t.r.f. (01–A), detector (01–A), 3 double impedance audio (01–A and 71). One dial. Plate current: 40 mA. Volume control: resistance in r.f. plate. Shielded. Battery cable. C-battery connections. Output device. Antenna: 100 feet. Console size: 20 x 46 x 17 inches. Price not determined.

NO. 543 RADIOLA 20

Five tubes; 2 t. r. f. (99), detector (99), two transformer audio (99 and 20). Regenerative detector. Two drum controls. C-battery connections. Battery cable. Antenna: 100 feet. Price: $78 without accessories.

NO. 480. PFANSTIEHL 30 AND 302

Six tubes; 3 t.r.f. (01–A), detector (01– 2A), transformer audio (01–A and 71). One dial. Plate current: 23 to 32 mA. Volume control: resistance in r.f. plate. Shielded. Battery cable. C-battery connections. Antenna: outside. Panel size: 17¾ x 8½ inches. Prices: No. 30 cabinet, $105; No. 302 console, $185 including loud speaker.

NO. 515. BROWNING-DRAKE 7-A

Seven tubes; 2 t.r.f. (01-A), detector (00-A), 3 audio (Hmu, two 01-A, and 71). Illuminated drum control. Volume control: rheostat on 1st r.f. Shielded. Neutralized. C-battery connections. Battery Cable. Metal panel. Output device. Antenna: 50–75 feet. Cabinet, 30 x 11 x 9 inches. Price, $145.

NO. 516. BROWNING-DRAKE 6–A

Six tubes; 1 t.r.f. (99), detector (00-A), 3 audio (Hmu, two 01-A and 71). Drum control with auxiliary adjustment. Volume control: rheostat on r.f. Regenerative detector. Shielded. Neutralized. C-battery connections. Battery cable. Antenna: 50–100 feet. Cabinet, 25 x 11 x 9. Price $105.

NO. 518. KELLOGG "WAVE MASTER," 504, 505, AND 506.

Five tubes; 2 t.r.f., detector, 2 transformer audio. One control and special zone switch. Volume control: rheostat on r.f. C-battery connections. Binding posts. Plate current: 25 to 35 mA. Antenna: 100 feet. Panel: 7½ x 25½ inches. Prices: Model 504, table, $75, less accessories. Model 505, table, $125 with loud speaker. Model 506, consolette, $135 with loud speaker.

NO. 519. KELLOGG, 507 AND 508.

Six tubes, 3 t.r.f., detector, 2 transformer audio. One control and special zone switch. Volume control: rheostat on r.f. C-battery connections. Balanced. Shielded. Binding posts and battery cable. Antenna: 70 feet. Cabinet size: Model 507, table, 30 x 13½ x 14 inches. Model 508, console, 34 x 18 x 54 inches. Prices: Model 507, $190 less accessories. Model 508, $320 with loud speaker.

NO. 427. MURDOCK 7

Seven tubes; 3 t.r.f. (01-A), detector (01-A), 1 transformer and 2 resistance audio (two 01-A and 12 or 71). One control, Volume control: rheostat on r.f. Coils shielded. Neutralized. Battery cable. C-battery connections. Complete metal case. Antenna: 100 feet. Panel size: 9 x 23 inches. Price, not available.

NO. 520. BOSCH 57

Seven tubes; 4 t.r.f. (01-A), detector (01-A), 2 audio (01-A and 71). One control calibrated in kc. Volume control: rheostat on r.f. Shielded. Battery cable. C-battery connections. Balanced. Output device. Built-in loud speaker. Antenna: built-in loop or outside antenna, 100 feet. Cabinet size: 46 x 16 x 30 inches. Price: $340 including enclosed loop and loud speaker.

NO. 521. BOSCH "CRUISER," 66 AND 76

Six tubes; 3 t.r.f. (01-A), detector (01-A); 2 audio (01-A and 71). One control. Volume control: rheostat on r.f. Shielded. C-battery connections. Balanced. Battery cable. Antenna: 20 to 100 feet. Prices: Model 66, table, $99.50. Model 76, console, $175; with loud speaker $195.

NO. 524. CASE, 61 A AND 61 C

T.r.f. Semi-shielded. Battery cable. Drum control. Volume control: variable high resistance in audio system. Plate current: 35 mA. Antenna: 100 feet. Prices: Model 61 A, $85; Model 61 C, console, $135.

NO. 525. CASE, 90 A AND 90 C

Drum control. Inductive volume control. Technidyne circuit. C-battery connections. Battery cable. Loop operated. Model 90-C equipped with output device. Prices: Model 90 A, table, $225; Model 90 C, console, $350.

NO. 526. ARBORPHONE 25

Six tubes; 3 t.r.f. (01-A), detector (01-A), 2 audio (01-A and 71). One control. Volume control: rheostat. Shielded. Battery cable. Output device. C-battery connections. Loftin-White circuit. Antenna: 75 feet. Panel: 7½ x 15 inches, metal. Prices: Model 25, table, $125; Model 252, $185; Model 253, $250; Model 255, combination phonograph and radio, $600.

NO. 527. ARBORPHONE 27

Five tubes; 2 t.r.f. (01-A), detector (01-A), 2 audio (01-A). Two controls. Volume control: rheostat. C-battery connections. Binding posts. Antenna: 75 feet. Prices: Model 27, $65; Model 271, $99.50; Model 272, $125.

NO. 528. THE "CHIEF"

Seven tubes; six 01-A tubes and one power tube. One control. Volume control: rheostat. C-battery connection. Partial shielding. Binding posts. Antenna: outside. Cabinet size: 40 x 22 x 16 inches. Prices: Complete with A power supply, $250; without accessories, $150.

NO. 529. DIAMOND SPECIAL, SUPER SPECIAL, AND BABY GRAND CONSOLE

Six tubes; all 01-A type. One control. Partial shielding. C-battery connections. Volume control: rheostat. Binding posts. Antenna: outdoor. Prices: Diamond Special, $75; Super Special, $65; Baby Grand Console, $110.

NO. 531. KOLSTER, 8A, 8B, AND 8C

Eight tubes; 4 t.r.f. (01-A), detector (01-A), 3 audio (two 01-A and one 12). One control. Volume control: rheostat on r.f. Shielded. Battery cable. C-battery connections. Model 8A uses 50 to 75 foot antenna; model 8B contains output device and uses antenna or detachable loop; Model 8C contains output device and uses antenna or built-in loop. Prices: 8A, $185; 8B, $235; 8C, $375.

NO. 532. KOLSTER, 6D, 6G, AND 6H

Six tubes; 3 t.r.f. (01-A), detector (01-A), 2 audio (01-A and 12). One control. Volume control: rheostat on r.f. C-battery connections. Battery cable. Antenna: 50 to 75 feet. Model 6G contains output device and built-in loud speaker; Model 6H contains built-in B power unit and loud speaker. Prices: Model 6D, $80; Model 6G, $165; Model 6H, $265.

NO. 533. SIMPLEX, SR 9 AND SR 10

Five tubes; 2 t.r.f. (01-A), detector (00-A), 2 audio (01-A and 12). SR 9, three controls; SR 10, two controls. Volume control: rheostat. C-battery connections. Battery cable. Headphone connection. Prices: SR 9, table, $65; consolette, $95; console, $145. SR 10, table $70; consolette, $95; console, $145.

NO. 534. SIMPLEX, SR 11

Six tubes; 3 t.r.f. (01-A), detector (00-A), 2 audio (01-A and 12). One control. Volume control: rheostat. C-battery connections. Battery cable. Antenna: 100 feet. Prices: table, $70; consolette, $95; console, $145.

NO. 535. STANDARDYNE, MODEL S 27

Six tubes; 2 t.r.f. (01-A), detector (01-A), 2 audio (power tubes). One control. Volume control: rheostat on r.f. C-battery connections. Binding posts. Antenna: 75 feet. Cabinet size: 9 x 9 x 19¼ inches. Prices: S 27, $49.50; S 950, console, with built-in loud speaker, $99.50; S 600, console with built-in loud speaker, $104.50.

NO. 481. PFANSTIEHL 32 AND 322

Seven tubes: 3 t.r.f. (01-A), detector (01-A), 3 audio (01-A and 71). One dial. Plate current: 23 to 32 mA. Volume control: resistance in r. f. plate. Shielded Battery cable. C-battery connections. Output device. Antenna: outside. Panel: 17¾ x 8¼ inches. Prices: No. 32 cabinet, $145; No. 322 console, $245 including loud speaker.

NO 433. ARBORPHONE

Five tubes; 2 t.r.f., detector, 2 transformer audio. All 01-A tubes. Two dials. Plate current: 16 mA. Volume control: rheostat in r.f. and resistance in r.f. plate. C-battery connections. Binding posts. Antenna: taps for various lengths. Cabinet size: 24 x 9 x 10½ inches. Price: $65.

NO. 431. AUDIOLA 6

Six tubes; 3 t.r.f. (01-A), detector (00-A), 2 transformer audio (01-A and 71). Drum control. Plate current: 20 mA. Volume control: resistance in r.f. plate. Stage shielding. Battery cable. C-battery connection. Antenna: 50 to 100 feet. Cabinet size: 28¼ x 11 x 14¼ inches. Price not established.

NO. 432. AUDIOLA 8

Eight tubes; 4 t.r.f. (01-A), detector (00-A), 1 transformer audio (01-A), push-pull audio (12 or 71). Bridge balanced t.r.f. Drum control. Volume control: resistance in r.f. plate. Stage shielding. Battery cable. C-battery connections. Antenna: 10 to 100 feet. Cabinet size: 28¼ x 11 x 14¼ inches. Price not established.

NO. 542 RADIOLA 16

Six tubes; 3 t. r. f. (01-A), detector (01 A), 2 transformer audio (01-A and 112). One control. C-battery connections. Battery cable. Antenna: outside. Cabinet size: 16½ x 8¼ x 7½ inches. Price: $69.50 without accessories.

NO. 456. RADIOLA 20

Five tubes: 2 t.r.f. (99), detector (99), 2 transformer audio (99 and 20). Balanced t.r.f. and regenerative detector. Two dials. Volume control: regenerative. Shielded. C-battery connections. Headphone connections. Antenna: 75 to 150 feet. Cabinet size: 19¼ x 11¾ x 16 inches. Price $115 including all tubes.

NO. 457 RADIOLA 25

Six tubes; five type 99 and one type 20. Drum control. Super-heterodyne circuit. C-battery connections. Battery cable. Headphone connections. Antenna: loop. Set may be operated from batteries or from power mains when used with model 104 loud speaker. Price; $165 with tubes, for battery operation. Apparatus for operation of set from the power mains can be purchased separately.

NO. 493. SONORA F

Seven tubes; 4 t.r.f. (01-A), detector (00-A), 2 transformer audio (01-A and 71). Special balanced t.r.f. Two dials. Plate current: 45 mA. Volume control: rheostat in r.f. Shielded. Battery cable. C-battery connections. Output device. Antenna: loop. Console size: 32 x 45¼ x 17 inches. Prices range from $350 to $450 including loop and loud speaker.

NO. 494. SONORA E

Six tubes; 3 t.r.f. (01-A), detector (00-A), 2 transformer audio (01-A and 71). Special balanced t.r.f. Two dials. Plate current: 35 to 40 mA. Volume control: rheostat on r.f. Shielded. Battery cable. C-battery connections. Antenna: outside. Cabinet size: Varies. Prices: table, $110; semi-console, $140; console, $240 including loud speaker.

NO. 530. KOLSTER, 7A AND 7B

Seven tubes; 4 t.r.f. (01-A), detector (01-A), 2 audio (01-A and 12). One control. Volume control: rheostat on r.f. Shielded. Battery cable. C-battery connections. Antenna: 50 to 75 feet. Prices: Model 7A, $125; Model 7B, with built-in loud speaker, $140.

NO. 495. SONORA D

Same as No. 494 except arrangement of tubes; 2 t.r.f., detector, 3 audio. Prices: table, $125; standard console, $185; "DeLuxe" console, $225.

NO. 482. STEWART-WARNER 705 AND 710

Six tubes; 3 t.r.f., detector, 2 transformer audio. All 01-A tubes. Balanced t.r.f. Two dials. Plate current: 10 to 25 mA. Volume control: resistance in r.f. plate. Shielded. Battery cable. C-battery connections. Antenna: 80 feet. Cabinet sizes: No. 705 table, 26¼ x 11½ x 13 13/16 inches; No. 710 console, 29¾ x 42 x 17¼ inches. Tentative prices: No. 705, $115; No. 710, $265 including loud speaker.

NO. 483. STEWART-WARNER 525 AND 520

Same as No. 482 except no shielding. Cabinet sizes: No. 525 table, 19¾ x 10 x 11¾ inches; No. 520 console, 22¼ x 40 x 14 11/16 inches. Tentative prices: No. 525, $75; No. 520, $117.50 including loud speaker.

NO. 459. STROMBERG-CARLSON 501 AND 502

Five tubes; 2 t.r.f. (01-A), detector (00-A), 2 transformer audio (01-A and 71). Neutrodyne. Two dials. Plate current: 25 to 35 mA. Volume control: rheostat on 1st r.f. Shielded. Battery cable. C-battery connections. Headphone connections. Output device. Panel voltmeter. Antenna: 60 to 100 feet. Cabinet sizes: No. 501, 25¼ x 13 x 14 inches; No. 502, 28 11/16 x 50 7/16 x 16¾ inches. Prices: No. 501, $180; No. 502, $290.

NO. 460. STROMBERG-CARLSON 601 AND 602

Six tubes. Same as No. 549 except for extra t.r.f. stage. Cabinet sizes: No. 601, 27 3/16 x 16¼ x 14 3/16 inches; No. 602, 28¾ x 51¼ x 19¾ inches. Prices: No. 601, $225; No. 602, $330.

NO. 472. VOLOTONE VIII

Six tubes. Same as No. 471 with following exceptions; 2 t.r.f. stages. Three dials. Plate current: 2-mA. Cabinet size: 26¼ x 8 x 12 inches. Price $140.

NO. 546. PARAGON "CONGRESS"

Six tubes; 2 t.r.f. (01-A), detector (01-A), 3 impedance-coupled audio (two 01-A and 12 or 71). One main control and three auxiliary adjustments. Volume control: resistance in r. f. plate circuit. Plate current: 40 mA. C battery connections. Tuned double-impedance audio amplifier. Output device. R. F. coils are shielded. Cable or binding posts. Cabinet size: 7 x 18 x 9 inches. Price $90.00; without cabinet, $80.00.

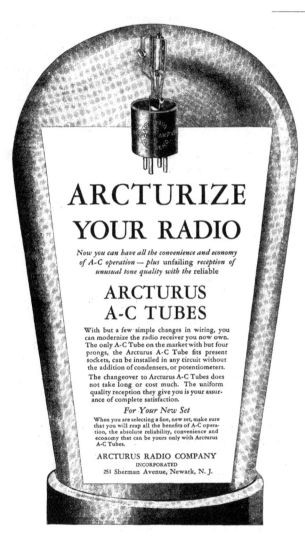

ARCTURIZE
YOUR RADIO

*Now you can have all the convenience and economy
of A-C operation — plus unfailing reception of
unusual tone quality with the reliable*

ARCTURUS
A-C TUBES

With but a few simple changes in wiring, you
can modernize the radio receiver you now own.
The only A-C Tube on the market with but four
prongs, the Arcturus A-C Tube fits present
sockets, can be installed in any circuit without
the addition of condensers, or potentiometers.

The changeover to Arcturus A-C Tubes does
not take long or cost much. The uniform
quality reception they give you is your assurance of complete satisfaction.

For Your New Set

When you are selecting a fine, new set, make sure
that you will reap all the benefits of A-C operation, the absolute reliability, convenience and
economy that can be yours only with Arcturus
A-C Tubes.

ARCTURUS RADIO COMPANY
INCORPORATED
251 Sherman Avenue, Newark, N. J.

Your Set Needs Protection

Jewell
Lightning
Arrester

*About Two-Thirds
Actual Size*

IT is hardly reasonable to put your money into an expensive radio set without providing it with the best possible protection against lightning. Yet people are doing this every day and risking the chance of severe damage, if not complete destruction of the delicate mechanism in their receiving set.

The Jewell Lightning Arrester will provide as good protection as can be obtained. It consists of a glazed brown porcelain case enclosing an accurately calibrated air gap. The case is sealed with a moisture-proof compound, which makes it equally suitable for indoor or outdoor installations.

As a further indication of the reliability of the Jewell Lightning Arrester, it has passed all tests prescribed by the Underwriters' Laboratories and is listed as standard by that organization.

Write for descriptive circular No. 1019.

"27 Years Making Good Instruments"

Jewell Electrical Instrument Co.
1650 Walnut St., Chicago

Improved
Positive
VOLT
CONT
for
"B" Elimina

HEAVY D
Centrala
Potentiom

A new Centralab variable resista all wire-wound and has exceptio current-carrying capacity.   Unit tically heat-proof and will dissipa watts through entire resistance wit of burning out. This new unit h vantages for regulating the out eliminators. A single turn of the full resistance variation with th remaining constant at any knob that the knob or panel can be c volts.
Has sufficient current-carrying permit shunting a low resistance rectly across the "B" power o assures a constant current load high to reduce the output voltage tiller to workable pressure even set is not connected. By prevent tremely high no load voltage, protect the filter condensers from —the most common source of tro eliminators.
Centralab Potentiometer control a much better voltage regulation tha tained with series resistors. Any ch rent at the set will produce a very in the voltage at the taps. This fa with the non-inductive of Centr Duty Potentiometers, insures bes ity with a minimum of by pass c

"B" Power Circuit for Best
Voltage Regulation

This circuit is suitable for any "B an output of 200 volts or less. Th are Centralab Heavy Duty-Potenti 005 and HP-008 and Power Rheo Centralab Heavy Duty Potentiom used as straight variable resistan necting the center terminal and terminal. They may be used to tances on old eliminators as well ing new ones. The diameter is 1 Because of their heavy current-c acity, Centralab Heavy Duty P are adaptable to transmitter work as Variable Grid Leak resistan voltage dividers and various pos ceiver circuits:
Resistances - 2,000 - 3,000 - 5
- 10,000 - 15,000 - 20,000 -
Price $2.00

At your dealer's, or C. O. D for true folder "Resistances a Function in Radio Circuits."

CENTRAL RADIO LABOR
22 Keefe
Milwauke

Central

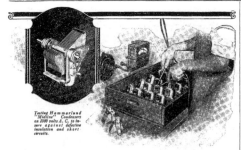

Testing Hammarlund "Midline" Condensers on 1000 volts A. C. to insure against defective insulation and short circuits.

Hammarlund *Precision* Means Your *Protection*

THE word "Precision" as applied to Hammarlund Condensers, Coils, Chokes and Shields means precisely what it says.

It means soundness of design, accuracy of manufacture, fineness of finish and scientific testing. Precision has given Hammarlund Products undisputed leadership in their respective fields.

From the very first, Hammarlund has stood for quality and Hammarlund has never deviated from that principle.

But you pay no premium for Hammarlund reputation. Hammarlund Products are standard-priced. You can pay far more and get far less than Hammarlund Products give you.

Write for the Hammarlund descriptive folder before planning your new receiver

HAMMARLUND MANUFACTURING COMPANY
424-438 W. 33rd Street, New York

More than a score of radio designers officially specify Hammarlund Precision Products for their latest circuits

For Better Radio **Hammarlund** PRECISION PRODUCTS

Dealer inquiries invited concerning several new and appealing Hammarlund developments, having a wide sales demand.

Samson Power Block No. 210. The only block which will supply 500 volts at 80 mils to two 210 tubes

Powerize with Samson Units for Best Results

For new SAMSON Power Units insure the best there is in radio current supply by

1. Doing away with hum, motor boating and poor voltage regulation.
2. Remaining so cool after 24 hours continuous operation under full load that they will be well within the 20° rise of temperature specified by the A. I. E. E.
3. Being designed to more than meet the specifications adopted by the National Board of Fire Underwriters.
4. Insuring safety against shock because of protected input and output terminals.

5. Insuring for all tubes the correct filament voltages specified by their manufacturers.
6. Compensating for lighting circuit voltage variation by the use of a special input plug and terminal block to which is attached a 6 ft. flexible rubber-covered connecting cord and plug.
7. Being more economical in operation than other units of same power rating.
8. Living up to the name plate rating.

Limited space prevents us from listing the fourteen types that we make. Our Power Units bulletin descriptive of these is free for the asking. In addition, our construction bulletin on many different "B" Eliminators and power Amplifiers will be sent upon receipt of 10c in stamps to cover the mailing cost.

MANUFACTURERS SINCE 1882

CANTON   MASS.

ALUMINUM
IN RADIO—A SIGN OF QUALITY

IN the fine sets of many of the leading manufacturers you will find that Aluminum is the prevailing metal for shielding, variable condensers and chasses.

The reason for the widespread preference for Aluminum is expressed in three characteristics which are possessed in combination only by Aluminum:

High Electrical Conductivity, Workability, Lightness

Added to these characteristics, Aluminum is impervious to common corrosion.

The exhaustive experiments of leading manufacturers that have proved the superiority of Aluminum are a guide to the layman in building or in buying.

Information on the subject of Aluminum in Radio will be sent on request. Write for the booklet, "Aluminum for Radio." It is sent without cost.

Aluminum shielding is used in the RGS Octa-Monic.

ALUMINUM COMPANY OF AMERICA
2464 Oliver Building, Pittsburgh, Pa.
Offices in 18 Principal American Cities

ALUMINUM IN EVERY COMMERCIAL FORM

A name worth looking for

Whether buying parts, kit, or complete set, it is a mighty good plan to look for the name "Faradon."

Long life, convenience, and dependability are built into Faradon Condensers. These points of quality are reasons why you will find Faradons in most of the high grade sets.

Assure yourself of freedom from condenser trouble. Look for the name "Faradon."

WIRELESS SPECIALTY APPARATUS CO.
Jamaica Plain
BOSTON, MASS.
U. S. A
Est. 1907

Faradon

Electrostatic condensers for all purposes

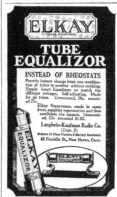

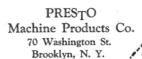

Improved Tone— More Volume

WITH your present socket power supply a stage of AmerTran Push-Pull, using two power tubes can be operated at *no* increased cost over the single power tube. It brings the advantages of greater tone reality and added volume. Tube distortion and harmonics are suppressed— the hum caused by raw A. C. on the filaments of the power tubes is eliminated.

For those wishing to utilize their present high voltage power packs we recommend the 210 combination of Push-Pull. Volume will be greater and only a slight change required in the plate supply.

AmerTran Push-Pull Input and Output transformers use high permeability alloy cores with multiple windings to give high inductive coupling and low capacity coupling. The Input transformer, has approximately the same primary impedance as the second stage AmerTran DeLuxe. It is suitable for use ahead of any pair of standard power tubes.

The plate impedance of two tubes connected push-pull is double the impedance of a single tube. Since various types of power tubes have different values of plate impedance, this company provides output transformers to correspond with the power tubes and the speakers. The impedance ratios provide for the greatest transfer of energy from 60 to 100 cycles, because at these low frequencies more energy is required to drive the loud speaker mechanism.

AmerTran Push-Pull Transformers Are Now Available in These Types:

Push-Pull Input Type 151 . . . . $15.00 each
Push-Pull Output Type 252 (Impedance ratio 4:1,) for two UX-210's or similar power tubes. $15.00 each
Push-Pull Output Type 271 (Impedance ratio 2:1), for two UX-171 tubes connected push-pull . . . . $15.00 each

AMERICAN TRANSFORMER CO.
178 Emmet Street   Newark, N. J.

We also make Audio and Power Transformers, Chokes and Resistors.

Transformer Builders for Over Twenty-Six Years

What Kit Shall I Buy?

THE list of kits herewith is printed as an extension of the scope of the Service Department of RADIO BROADCAST. It is our purpose to list here the technical data about kits on which information is available. In some cases, the kit can be purchased from your dealer complete; in others, the descriptive booklet is supplied for a small charge and the parts can be purchased as the buyer likes. The Service Department will not undertake to handle cash remittances for parts, but when the coupon on page 172 is filled out, all the information requested will be forwarded.

201. SC FOUR-TUBE RECEIVER—Single control. One stage of tuned radio frequency, regenerative detector, and two stages of transformer-coupled audio amplification. Regeneration control is accomplished by means of a Variable resistor across the tickler coil. Standard parts: cost approximately $58.85.

202. SC-II FIVE-TUBE RECEIVER—Two stages of tuned radio frequency, detector, and two stages of transformer-coupled audio. Two tuning controls. Volume control consists of potentiometer grid bias on r.f. tubes. Standard parts cost approximately $60.35.

203. "HI-Q" KIT—A five-tube tuned radio-frequency set having two radio stages, a detector, and two transformer-coupled audio stages. A special method of coupling in the r.f. stages tends to make the amplification more nearly equal over the entire band. Price $63.05 without cabinet.

204. R. G. S. KIT—A four-tube inverse reflex circuit, having the equivalent of two tuned radio-frequency stages, detector, and three audio stages. Two controls. Price $69.70 without cabinet.

205. PIERCE AIRO KIT—A six-tube single-dial receiver; two stages of radio-frequency amplification, detector, and three stages of resistance-coupled audio. Volume control accomplished by variation of filament brilliancy of r.f. tubes or by adjusting compensating condensers. Complete chassis assembled but not wired costs $42.50.

206. H & H-T. R. F. ASSEMBLY—A five-tube set; three tuning dials, two steps of radio frequency, detector, and 2 transformer-coupled audio stages. Complete except for baseboard, panel, screws, wires, and accessories. Price $50.00.

207. PREMIER FIVE-TUBE ENSEMBLE—Two stages of tuned radio frequency, detector, and two steps of transformer-coupled audio. Three dials. Parts assembled but not wired. Price complete, except for cabinet, $35.00.

208. "QUADRAFORMER VI"—A six-tube set with two tuning controls. Two stages of tuned radio frequency using specially designed shielded coils, a detector, one stage of transformer-coupled audio, and two stages of resistance-coupled audio. Gain control by means of tapped primaries on the r.f. transformers. Essential kit consists of three shielded double-range "Quadraformer" coils, a selectivity control, and an "Ampitrol," price $47.50. Complete parts $70.13.

209. GEN-RAL FIVE-TUBE SET—Two stages of tuned radio frequency, detector, and two transformer-coupled audio stages. Volume is controlled by a resistor in the plate circuit of the r.f. tubes. Uses a special r.f. coil ("Duo-Former") with figure eight winding. Parts mounted but not wired, price $37.50.

210. BREMER-TULLY POWER-SIX—A six-tube, dual-control set; three stages of neutralized tuned radio frequency, detector, and two transformer-coupled audio stages. Resistances in the grid circuit together with a phase shifting arrangement are used to prevent oscillation. Volume control accomplished by Variation of B potential on r.f. tube. Essential kit consists of four r.f. transformers, two dual condensers, three small condensers, three choke coils, one 500,000-ohm resistor, three 1500-ohm resistors, and a set of color charts and diagrams. Price $41.40.

212. INFRADYNE AMPLIFIER—A three-tube intermediate-frequency amplifier for the super-heterodyne and other special receivers, tuned to 3400 kc. (86 meters). Price $35.00.

213. RADIO BROADCAST "LAB" RECEIVER—A four-tube dual-control receiver with one stage of Rice neutralized tuned-radio frequency, regenerative detector (capacity controlled), and two stages of transformer-coupled audio. Approximate price, $78.15.

214. LC-27—A five-tube set with two stages of tuned-radio frequency, a detector, and two stages of transformer-coupled audio. Special coils and special means of neutralizing are employed. Output device. Price $85.20 without cabinet.

215. LOFTIN-WHITE—A five-tube set with two stages of radio frequency, especially designed to give equal amplification at all frequencies, a detector, and two stages of transformer-coupled audio. Two controls. Output device. Price $85.10.

216. K.H.-27—A six-tube receiver with two stages of neutralized tuned radio frequency, a detector, three stages of choke-coupled audio, and an output device. Two controls. Price $86.00 without cabinet.

217. AERO SHORT-WAVE KIT—Three plug-in coils designed to operate with a regenerative detector circuit and having a frequency range of from 10,900 to 2306 kc. (15 to 130 meters). Coils and plug only, price $12.50.

218. DIAMOND-OF-THE-AIR—A five-tube set having one stage of tuned-radio frequency, a regenerative detector, one stage of transformer-coupled audio, and two stages of resistance-coupled audio. Volume control through regeneration. Two tuning dials.

219. NORDEN-HAUCK IMPROVED SUPER 10—Ten tubes; five stages of tuned radio frequency, detector, and four stages of choke- and transformer-coupled audio frequency. Two controls. Price $260.00.

220. BROWNING-DRAKE—Five tubes; one stage tuned radio frequency (Rice neutralization), regenerative detector (tickler control), three stages of audio (special combination of resistance- and impedance-coupled audio). Two controls.

What Kit Shall I Buy? (Continued)

221. LR4 Ultradyne—Nine-tube super-heterodyne; one stage of tuned radio frequency, one modulator, one oscillator, three intermediate-frequency stages, detector, and two transformer-coupled audio stages.

222. Greiff Multiplex—Four tubes (equivalent to six tubes); one stage of tuned radio frequency, crystal detector, two stages of transformer-coupled audio, and one stage of impedance-coupled audio. Two controls. Price complete parts, $60.00.

223. Phonograph Amplifier—A five-tube amplifier device having an oscillator, a detector, one stage of transformer-coupled audio, and two stages of impedance-coupled audio. The phonograph signal is made to modulate the oscillator in much the same manner as an incoming signal from an antenna.

224. Browning-Drake—Five tubes; one stage tuned radio frequency (with special neutralization system), regenerative (five detector flicker control), three stages of audio (special combination of resistance- and impedance-coupled audio). Two controls.

225. Aero Short-Wave Transmitting Kit consists of interchangeable coils to be used in tuned-plate tuned grid circuit. Kits of coils, two choke coils, and mountings, can be secured for 20-40 meter band, 40-80 meter band, or 90-180 meter band for $12.00.

Distress Calls From The Air

THE recent distress calls from aircraft in flight have been referred to in the press as the first from aircraft, even by some of our contemporary radio publications which should know better. Seventeen years ago, a dirigible, the *America*, attempted a transatlantic flight, trailing its reserve fuel supply on a 340-foot steel cable below it. After making a flight of several hundred miles, the crew found that the tugging trailer threatened the destruction of the craft. Jack Irwin summoned aid by radio and the steamship *Trent* responded. No lives were lost. This was the first S O S from the air!

A FIVE-KW. broadcasting station is to be erected on Mt. Victoria, overlooking the harbor of Wellington, New Zealand. This will be the most powerful broadcasting equipment in the Pacific.

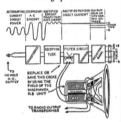

From the Manufacturers

THE paragraphs below represent in part interesting announcements of new products which have been received by RADIO BROADCAST from important manufacturers within recent weeks. Many of our readers will welcome this section where they can find concise information on the interesting new offerings. Space does not permit the more complete discussion and remarks which many of these announcements deserve. Readers can preserve these columns and those which will appear in the future in this section of RADIO BROADCAST. Further information on the apparatus described may be obtained by writing to the Service Department of RADIO BROADCAST. It is hoped that those readers who write the manufacturers directly will refer to this department when writing.—THE EDITOR.

New Offerings by—

AMSCO PRODUCTS CO. INC.
Extra-spaced variable condensers, known as Duospace to be used in power oscillators, small transmitters or in receiving circuits where the $\frac{7}{16}$ inch spacing between rotor and stator plates contribute toward extremely accurate alignment of single-control circuits.

SAMSON ELECTRIC CO.
Power chokes with an inductance of 30 henries under the following conditions: No. 30, 30 milliamperes, d.c. resistance 470 ohms, price $5.00; No. 380, 80 milliamperes, d.c. resistance 2000 ohms, price $11.00; No. 312, 120 milliamperes, d.c. resistance 290 ohms, price $12.00. Power transformers, No. 132 rated at 40 watts, furnishes 200 volts plate supply at 80 milliamperes from 60 cycles, fitted with windings to furnish filament voltage for UX-280 and two 171's. Price, $15.00. No. 162, designed to furnish a 500-volt plate supply to tubes such as the 210, rated at 75 watts, filament supply for UX-281 and 210. Secondary voltage, 725. Price, $19.00. No. 253, same as No. 132 but to be used on 25-cycle supply. Price $19.00. No. 256 same as No. 162 but to be used on 25-cycle supply. Price $27.50 No. 232, designed to supply 200 volts direct current to receiving set. Similar to No. 132 but supplies in addition filament current for five 226 type tubes, two 227, and two 171's. Rated at 100 watts. Price, $16.50. No. 262, similar to No. 232 except supplying power for a 210 amplifier. No. 463 furnishes a.c. voltage for all a.c. tubes. Has windings for 1.5-, 2.5-, and 5-volt tubes. Price, $14.50. No. 259 same as 463 except for 25 cycles. Price, $17.50. No. 217, rated at 20 watts. Supplies filament voltage for two 171's or two 210 tubes. Price, $12.00. Power block No. 713 to furnish necessary energy at 180 volts 50 milliamperes, rated at 30 watts, supplies filament windings for UX-280 and 171 tube, and contains chokes and transformer. Price, $24.50. No. 718, supplies 200 volts at 80 milliamperes. Otherwise same as No. 713. Price, $33.50.

JEFFERSON ELECTRIC MFG. CO.
Output transformer with 30 inches of cord and equipped with telephone tip and pin type jacks. Model 395. price $6.00. Filament supply transformer, furnishing voltage for eight 1.5-volt tubes, one 2.5-volt detector tube, and two 5-volt tubes. Price $7.50. Plate supply transformer to be used with UX-280 rectifier tube, 5-volt center-tapped filament winding, 520 volt center-tapped 125 milliampere plate winding, price $10.00. Combination filament and plate voltage transformer, price $15.00. Plate supply choke passing 125 milliamperes, price $6.00.

BENJAMIN ELECTRIC MFG. CO.
New style of the well known Benjamin floating socket; also Y type socket for heater type tubes. Straight frequency-line condenser single hole mounting, No. 9080 capacity .00025. mfd. price $3.50; No. 9081 capacity .00035 mfd. price $3.75; No. 9082 capacity .0005 mfd. price $4.00.

WESTINGHOUSE ELECTRIC AND MFG. CO.
Rectox A battery charger. Uses copper oxide rectifying elements providing full-wave rectification. Charges at $\frac{3}{4}$ amperes, 6 volts.

Do You Know How to Stop
That Rasping Cackle
In Your Loud Speaker

Protect It From

Paralyzing "B" Current with
The New Muter Clarifier~

[OUTPUT TRANSFORMER]

Haven't you often wondered just why it's so difficult to get clear, natural, enjoyable reception? Haven't you many times felt like throttling that scratching, squawking, inhuman voice? Thousands of set owners are surprised to learn that their speakers *are* being throttled—constantly—by paralyzing high "B" voltage current. That's exactly what ails them.

The Muter Clarifier is especially designed to prevent this; it diligently protects the Speaker and coils, assuring vast improvement in tone, quality and volume. Nothing else serves the same purpose. This compact, attractive little Instrument is easily attached in a few moments

without disturbing set. Try out to your own pleasure on our liberal guarantee—you won't recognize your set.

See Your Dealer—or Send Direct

Dealers can quickly secure Muter Products for you from leading jobbers. However, should you have any difficulty obtaining them from your dealer, mail coupon direct to us. Prompt shipment of Clarifier, complete with phone cords attached, will be made upon receipt of price or C. O. D. if you wish. Give your speaker a chance to show what good reception means. MAIL COUPON TODAY!

LESLIE F. MUTER CO., 76th & Greenwood Av., Dept. 822-T, Chicago, Ill.

MUTER
DEPENDABLE PRODUCTS
The Complete Quality Popular Priced Line. Send for comprehensive catalog.

Use This Coupon!

LESLIE F. MUTER CO.,
76th & Greenwood Ave., Dept. 822-T Chicago, Ill.
☐ Send Muter Clarifier at once, postage prepaid. $5.00 is enclosed.
☐ Send C. O. D.
☐ Send complete Muter catalog.

Name.............................
Address...........................
City...............State..........

WOULD YOU HELP YOURSELF TO OUR CASH BOX—if we said "GO AHEAD"?

ALL RIGHT, THEN—GO AHEAD! . . . .
THE CASH IS THERE. . . . And it's easy to get . . . .
The Holiday Season is approaching with its heavy drain on your income. What are you doing about it?
Will it be just as it was last year? Or do you really want some spare-time cash.
If you are in earnest, let us tell you how you can get it.
There's much to gain and nothing to lose.

Write: Agents' Service Division

DOUBLEDAY, PAGE & CO.   Garden City, N. Y.

STATEMENT OF THE OWNERSHIP, MANAGEMENT, CIRCULATION, ETC., required by the Act of Congress of August 24, 1912, of RADIO BROADCAST, published monthly at Garden City, New York for October 1, 1927. State of New York, County of Nassau.

Before me, a Notary Public in and for the State and County aforesaid, personally appeared John J. Hessian, who, having been duly sworn according to law, deposes and says that he is the treasurer of Doubleday, Page & Company, owners of Radio Broadcast and that the following is, to the best of his knowledge and belief, a true statement of the ownership, management (and if a daily paper, the circulation), etc., of the aforesaid publication for the date shown in the above caption, required by the Act of August 24, 1912, embodied in section 411, Postal Laws and Regulations, printed on the reverse of this form, to wit:

1. That the names and addresses of the publisher, editor, managing editor, and business managers are: *Publisher*, Doubleday, Page & Co., Garden City, N. Y.; *Editor*, Willis Wing, Garden City, N. Y.; *Business Managers*, Doubleday, Page & Co., Garden City, N. Y.

2. That the owner is: (If owned by a corporation, its name and address must be stated and also immediately thereunder the names and addresses of stockholders owning or holding one per cent. or more of total amount of stock. If not owned by a corporation, the names and addresses of the individual owners must be given. If owned by a firm, company, or other unincorporated concern, its name and address, as well as those of each individual member, must be given.) F. N. Doubleday, Garden City, N. Y.; Nelson Doubleday, Garden City, N. Y.; S. A. Everitt, Garden City, N. Y.; Russell Doubleday, Garden City, N. Y.; John J. Hessian, Garden City, N. Y.; Dorothy D. Babcock, Oyster Bay, N. Y.; Alice De Graff, Oyster Bay, N. Y.; Florence Van Wyck Doubleday, Oyster Bay, N. Y.; F. N. Doubleday or Russell Doubleday, Trustee for Florence V. Doubleday, Garden City, N. Y.; Janet Doubleday, Glen Cove, N. Y.; W. Herbert Eaton, Garden City, N. Y.; S. A. Everitt or John J. Hessian, Trustee for Josephine Everitt, Garden City, N. Y.; William J. Neal, Garden City, N. Y.; Daniel W. Nye, Garden City, N. Y.; E. French Strother, Garden City, N. Y.; Henry L. Jones, 285 Madison Ave., N. Y. C.; W. F. Etherington, 50 East 42nd St., N. Y. C.

3. That the known bondholders, mortgagees, and other security holders owning or holding 1 per cent. or more of total amount of bonds, mortgages, or other securities are: (If there are none, so state.) NONE.

4. That the two paragraphs next above, giving the names of the owners, stockholders, and security holders, if any, contain not only the list of stockholders and security holders as they appear upon the books of the company but also, in cases where the stockholder or security holder appears upon the books of the company as trustee or in any other fiduciary relation, the name of the person or corporation for whom such trustee is acting, is given; also that the said two paragraphs contain statements embracing affiant's full knowledge and belief as to the circumstances and conditions under which stockholders and security holders who do not appear upon the books of the company as trustees, hold stock and securities in a capacity other than that of a bona fide owner; and this affiant has no reason to believe that any other person, association, or corporation has any interest direct or indirect in the said stock, bonds, or other securities than as so stated by him.

5. That the average number of copies of each issue of this publication sold or distributed, through the mails or otherwise, to paid subscribers during the six months preceding the date shown above is........ (This information is required from daily publications only.)

(Signed)   DOUBLEDAY, PAGE & COMPANY
       By John J. Hessian, *Treasurer*.
Sworn to and subscribed before me this 16th day of September, 1927.
[SEAL]   (Signed) Frank O'Sullivan.
(My commission expires March 30, 1928.)

KingstoN

MAKE SOMEBODY'S CHRISTMAS happier with the gift of clearer, richer, sweeter music, finer radio reception and a new radio experience. The Kingston B Current Supply Unit is a gift that renews itself every day of the year. With the Kingston you are free from B Batteries, and your set is always at its perfection peak. Beautifully and expertly built, smartly finished in satin black, the Kingston has three different voltage terminals, each adjustable over a wide range, making possible any desired voltage from 5 to 200, while a fourth variable voltage can easily be had by connecting a separate variable resistor to one of the terminals. Size: 9 inches long, 8¼ inches high, 5½ inches wide. If your dealer can't supply you, write us.

Kingston service goes all the way through, and Kingston dealers know that this company stands squarely behind its products. The new Kingston B Battery eliminator is fully guaranteed to be all and do all that is claimed for it.

PRICES

Type I, for 110-120 Volt AC 50 or 60 Cycle Current, $55.00
For receiving sets having not more than eight tubes and not having type UX171 power tube or equivalent.

Type IA, for 110-120 Volt AC 50 or 60 Cycle Current, $42.50
For all sets using type UX171 power tube or equivalent and for all large sets having nine or more tubes.

Type 2C, for 110-120 Volt AC 25, 30 or 40 cycle current, $47.50
Prices include type BH Raytheon tube.
Any of these models will be furnished with an automatic control switch built in the unit for $2.50 additional. With this the B unit is automatically switched on or off when switch on the radio set panel is turned.

Make this a Kingston Christmas!

KOKOMO ELECTRIC CO.   KOKOMO, IND.

THOUSANDS
of our graduates
have made good!

You, too, can make rapid strides of advancement in Radio, the field of fascinating work and attractive salaries. Trained radio men are always in high demand—as ship operators, sailing to foreign lands, and in big jobs on shore.

Study At Home

Radio Institute of America offers the finest radio instruction obtainable—qualifies you to pass the U. S. Government Commercial or Amateur License Examination. The cost of the course is low. Time payments are allowed. And you can study at home! Mail the coupon now!

RADIO INSTITUTE OF AMERICA
328 Broadway, New York City

Without Equal

Do you still fuss with "A" Batteries and chargers? Is your set noisy? Do you get microphonics? How about the tone? Is the music clear, clean, full overtoned?

Why not put Sovereign A-C Tubes into your set? Throw away your "A" Batteries, "A" Eliminators and Chargers. Be free from all battery noises and bother, and get the kind of reception you've dreamed of. Clear, clean, true music with full overtone and without hum or cracking.

Sovereigns use standard sockets. They are easy to install. Anybody can do it in a few minutes. From then on it's simply a matter of throwing a switch, for your power is unfailing and plenty.

Have your set up-to-date. Write for treatise on A-C Tubes and Receivers with typical wiring diagrams included, if your dealer cannot supply you.

Sovereign Electric & Mfg. Co.
127 N. Sangamon St.   Chicago, Illinois

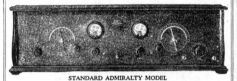
400 CENTS RESISTANCE

is the subscription price of Radio Broadcast for one year. This is less than the cost of a good low loss variable condenser. Let us enter your subscription to begin with the next issue.

Send $4.00 with your name and address to Radio Broadcast, Garden City, N. Y.

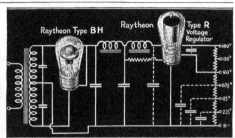
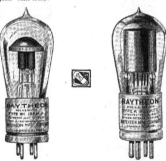

Bremer-Tully Discard Old Standards of Comparison

THREE WINNERS

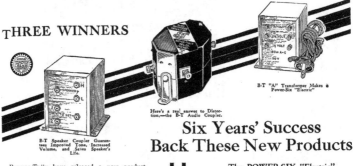

B-T "A" Transformer Makes ö
Power-Six "Electric"

Here's a real answer to Distortion,—the B-T Audio Coupler.

B-T Speaker Coupler Guarantees Improved Tone, Increased Volume, and Saves Speaker's Life.

Six Years' Success
Back These New Products

Bremer-Tully have released a new product which should create world-wide discussion.

The old idea of "Amplification Curves" as a basis for comparing audio transformers has been discarded. *It never was anything more than a very secondary matter.*

B-T have always maintained that the real problem was "Harmonic Distortion,"—and—

Now *they have proved it* in the new B-T AUDIO COUPLER.

It is more than a transformer,—and better,—although no one ever produced a better transformer than BREMER-TULLY.

A constant Impedance Core,—an air-gap,—Tertiary Loading Coil,—and finest laminations, combine to produce Q U A L I T Y U N - EQUALLED anywhere, regardless of size or price.

"Harmonic Distortion" has had practically no attention.—the attempts being to improve transformers by *increased size,—cast cores,* etc. You can readily appreciate that BREMER-TULLY could not release this product if it was not superior.

Type 3-31 is for First Stage; Type 2-22 for second stage or for all stages where three stages are used,—as in replacing Resistance Coupling. Bring your set up to date with a set. Price each, $6.00.

Read more in BETTER : TUNING—See Coupon.

BETTER TUNING is a booklet of 80 pages answering the latest questions in radio. Thousands praise it, and profit by it.
SEND THE COUPON.

The POWER-SIX "Electric"

Everyone knows the record of the B-T Power Six.

You can now run yours from the light-socket, without batteries.

Use the new B-T "A" Transformer, instead of storage battery and charger. Price $7.50.

Complete Diagrams and Instructions for making the change or building a new electric set, $1.00.

This is what you've been waiting for.

The Speaker Coupler

Put a B-T Speaker Coupler between your power tube and speaker and improve reception wonderfully. The difference is amazing.

You will get Better Tone, and particularly with air column speakers or horns, much Greater Volume.

You will protect your speaker from heavy current flow and prolong its life.

You can match your speaker and power tube through various combinations, as shown, without the use of tools. Simply insert cord tips.

Speaker Coupler, suitable for table use, or may be mounted inside cabinet; Price $7.50.

BREMER-TULLY MFG. CO. CHICAGO, ILL.

520 S. CANAL

COUPON

Send 80-page Booklet.

I am interested in Sets.......... .... Parts........................

I am a Dealer..................Consumer...................or...............

Print Name ....

Address ....

"A" "B" "C" &

POWER

AMPLIFIERS
for any Job

HOWARD A. C. RECEIVER

Leads the field

Enormous Power

B B L SPEAKER

Best Broadcast Loudspeaker

SUPERB TONE QUALITY

Ross V. Starkweather
Wholesale Electric Radio
Tools and Supplies
Room 508
19 S. Wells Street, Chicago, Ill.

Nathaniel BALDWIN
LOUD SPEAKERS

A Famous Speaker

Benj. Franklin once said:
"My performance devotes itself entirely to thy service and will serve thee faithfully and if it has the good fortune to please its master, its gratification enough for the labor of—"
Poor Richard

Baldwin "99"
The New !!

is without comparison. To-day, as 15 years ago, Baldwin is still the standard by which others are judged and there is real pride and satisfaction in owning one. The New Baldwin "99" can be used on any set.

Ask your dealer to demonstrate it.

Price $28.50

Write For Booklets
—BALDWIN UNITS—
Loudspeaker units, Phono adapters, and head sets are standard the world over.

J. W. & W. L. WOOLF
Distributors and Exporters For
Nathaniel Baldwin, Inc., 227-229 Fulton St., N. Y.
In Canada Baldwin International Ltd., Toronto

Get Your Unipac Here!

WE carry the complete line of Silver-Marshall Unipacs, the amplifier and power supply recommended by all authorities wherever the finest in power output is demanded. Send your order to us—12 hour service.

All Standard Lines Carried

Our stock is one of the most complete in the country—standard sets, parts and accessories. We do not sell "job lots" or "close-outs." Send for your buyers' guide now. Lowest wholesale prices to community set builders, dealers and agents.

The Barawik Company
132 No. Jefferson Street
Chicago, Ill.

Acme ANTEN

—with
Seven Stran
of enameled
Copper Wire

Best outdoor antenn buy. Seven strands ed copper wire. Pres imum surface for resists corrosion; t improves the signal diameters equal to s 16. (We also offer stranded bare, and tinned antenna.)

Loop Antenna

Sixty strands of No. 38 wire for flexibility, 5 st 36 phosphor bronze stretching. Green or covering; best loop wir make.

Battery Cab

A rayon-covered cable c 5, 6, 7, 8 or 9 vari-colore Flexible Celatsite wire for connecting batteries or eliminator to set. Plainly tabbed; easy to connec an orderly appearance.

Acme Celatsite W

Tinned copper b up wire with mable Celatsite i 9 beautiful col easily, solders re crack at bends. 18, 19; 30 inch le

Flexible Cela for sub-panel v

A cable of fine, tinne copper wires with no inflammable Celatsite i sulation. Ideal for sub-panel or point-to-point wiring. Strips easily, solders readily ful colors; *sold only i in cartons colored to ma*

Spaghetti Tu

Oil, moisture, acid p dielectric — used by I neers. Nine colors, for to 18; 30 inch length make tinned bus bar, square, in 2 and 2½ ft. l

Send for folde
THE ACME WIRE CC
New Haven, Co

ACME
MAKES BETTE

Any Bird Can Twitter
—BUT—
It Takes a Bull to Bellow

I T takes real power output—real voltage—for a power amplifier to reproduce the lowest note on a bass viol or the beat of the kettle drums so that it is really recognizable.

The 660-210 Unipac is a combined last stage amplifier and full 425 volt power supply. Its amplifier consists of a stage of transformer coupled audio, in push-pull arrangement, employing two 210 super power amplifying tubes. Its undistorted power output is three or four times greater than that of any amplifier yet developed. And, this does not necessarily mean that the volume delivered is such that an auditorium is needed to keep tone values within reason—the volume is under control—but it does mean that the tremendous power required to reproduce tones in the lower frequencies—below 100 cycles—is there at all times.

The 660-210 Unipac is entirely self-contained—it connects to any 105-120 volt, 60 cycle alternating current. It furnishes its own A, B and C power—and supplies B current to the entire receiver as well. 1.5 volt filament current is also available for first stage audio A. C. tubes. The 660-210 Unipac in kit form is $83.25 or $93.25 completely wired.

While the 660-210 Unipac stands alone for its undistorted power output and handling capacity, other models are available covering practically all requirements in the audio amplifying and power fields. The 660-171 Unipac with its two 171 tubes in push-pull arrangement, possessing of course less amplification than the 660-210, will furnish ample undistorted output for the average home—more than any single tube 210 amplifier that is on the market. It delivers all ABC power to the amplifier and 45 and 90 volt potential to the receiver as well. The 660-171 Unipac kit is $64., or $74. completely wired.

The 660-171A Unipac, similar to the 660-171 will, in addition, furnish complete A power to the receiver when new A. C. tubes are employed. The 660-171A Unipac kit is $66., or $76. completely wired.

Various models may be built up as phonograph amplifiers—as complete two-stage amplifiers—in fact, the constructor can build up practically any combination of audio amplifier system and power supply that he demands. All are described in the Unipac booklet which we will gladly send. An outstanding example of Unipac flexibility is the LC-28 Unipac, developed for Laurence Cockaday, a two-stage amplifier and power supply for the LC-28 receiver, employing the new S-M 240 and 241 transformers. A similar model has been specified by Glen Browning as the official amplifier for the new Browning-Drake Two Tube Tuner. Complete parts for the LC-28 Unipac, $81.25.

We Could Charge More But— A *Better* Transformer Can't Be Made

That's a cold statement about the S-M 220 audio transformer, but it takes steel and copper to make the best transformer. The size and weight of the S-M core—the countless turns of wire—these are the things that produce real transformer quality. And where will you find a heavier or higher grade silicon steel core—where will you find more copper than in the S-M 220? The 220 has from 25 to 50 per cent more steel and copper in its construction than any other transformer on the market. We could charge from 25 to 50 per cent more for them than we do, but at no price can you get a better transformer. Do you wonder why it's sold on an unconditional guarantee to give you better tone quality than any other audio amplifying unit available? The 220 audio is $8.00, and the 221 output is $7.50. They are priced low, not merely made to sell at a low price.

We can't make the 220's cheaper but if you desire a transformer somewhat lower in price taking up a little less room, and with a little less core and wire, the new 240 audio and 241 output transformers are superior to most other transformers, and are far superior to anything in their price field. They have the same general characteristics as the famous 220's and 221's, but provide slightly less accentuation of frequencies below 80 cycles. They have the same 5000 cycle cut-off for which 220's are famous, eliminating many of the objectionable whistles and heterodyne squeals. The 240 audio sells for $6.00, and the 241 output at $5.00. Hard to beat at any price. Impossible to equal at these prices.

S-M audios have outsold all other transformers. They are specified in more prominent circuits than any other. The answer is obvious—they are better—that's all.

SILVER-MARSHALL, Inc.

838-A West Jackson Blvd.   Chicago, U. S. A.

We can't tell you here the full Unipac story, nor of the new developments in audio amplification and complete A, C. operation, but if you'll send in the coupon we'll send you complete data.

The A. C. Improved Shielded Six is ready—a completely light socket operated batteryless set using the new A. C. tubes. It's illustrated above, with its complete dry ABC power unit—less than 7 inches square.

$55.

At this price the Crosley Bandbox is Radio's most astonishing success, *not* because the price is low, *but* because the set is magic!

*T*HE ability of the new Bandbox is amazing. Its simple operation is easily understood and its wonderful performance is at the command of any hand that can turn the dial.

Millions are making up their minds today to buy a radio. Millions will replace obsolete sets with new, up-to-date receivers this fall.

Experienced radio owners will look first for 3 fundamental points and to every set they consider will address these questions:

1. Is it selective?
2. Is it sensitive?
3. Is it easy to operate?

Satisfied on these points they will look for:

1. Single dial control
2. Illuminated dial
3. Volume control
4. Single cable leads
5. Console installation adaptability.
6. Reasonable price.

Millions will look at the Crosley Bandbox. This amazing little set is now displayed by more than 16,000 dealers.

One dealer alone expects to sell a million dollars worth of Bandboxes this season.

Crosley dealers from Maine to California have this wonderful receiver hooked up for immediate demonstration and will explain its matchless performance in a manner somewhat like this:

The Crosley Bandbox is a 6-tube receiver.

The circuit of this set is of the excellence you would expect from a group of skilled engineers suddenly given the pick of the world's radio patents to work with.

Crosley has always given the radio world its biggest value for its dollar. Contemplate the perfection possible when the doors of the research and development laboratories of The Radio Corporation of America, The General Electric Co., The Westinghouse Co., The American Telephone & Telegraph Co., and the Hazeltine and Latour Corporations were thrown open.

Licensed under their patents!

Tremendous! Wonderful! Significant!

Simply it means that millions will possess the best radio performance possible at the low prices for which Crosley is already famous.

The Crosley Bandbox is totally and completely shielded. Every element is absolutely separated from every other element by solid shielding. Coils are covered with copper. This could have been done cheaper but efficiency would have been sacrificed. Condensers are housed in cadmium-plated steel. All wiring is separated and shielded from all other parts of the receiver. Solid, sturdy, substantial, the entire set is assembled on a heavy metal chassis.

The tuned radio frequency amplification stages have been absolutely balanced through use of the Neutrodyne principle. The set is a genuine Neutrodyne!

To the initiated this means much. To the layman it manifests itself only as a radio receiver that does not squeal or howl when you are trying to get a station.

The foregoing answers for the Crosley Bandbox the three fundamental questions set forth in the fourth paragraph which the experienced radio owner is asking of every set he sees this fall.

The shielding makes the Bandbox highly selective—the circuit, acutely sensitive, and the design, extremely easy to operate.

The Bandbox is operated with a single station selector (one dial.)

In most localities and in most owners' hands the single station selector will find all the programs anyone could possibly wish. But there are some owners who demand greater ability like the possessors of 50 horse-power motor cars who may never step on it but like to be conscious it's there. For such have the Acuminators been designed. Far

MUSICONE
TILT-TABLE
$17.50

away stations of weak power, but perhaps good music are captured by the use of these little auxiliary tuners. Their function is best likened to a pair of field glasses. As the lens bring the distant scene to nearby aspect, so do the acuminators bring the remote station signals up to open filling volume. Ordinary one dial radios can never perform like this. Hair line tracking of the condensers together is difficult—but the Acuminators, little second-hand adjustments exclusive to Crosley give the Bandbox a substantial command of the air and all that is in it.

The dial of the Bandbox is illuminated. A detail? A refinement added but not as an excuse to raise the price. For shadowy corners and dim eyesight it recommends itself.

Volume Control is necessary on good radio today. Nearby and high powered stations send terrific impulses into the receiver. Detuning has been a favorite method of softening this loud reception but with easy, closer and closer together on the dial detuning particularly in large cities proves an overlapping of programs. The ear like the eye is only good for one thing at a time. Under the towers of the heaviest stations the Volume control of the Bandbox cuts the loudest blast down to a veritable whisper. No distortion whatsoever!

A single cable leads all outside and power connections from the Bandbox. In the brown fabric covered cable lies each lead covered with colored rubber for protection, accuracy and easy assembly. Tidy housewives appreciate it.

The adaptability of the Bandbox to installation in all types of cabinets is a feature. The metal case of the Bandbox lifts off

When HISTORY is
in the making—

".. You're *there* with a Crosley"

6 Tube Crosley
BANDBOX $55.

the chassis. This leaves the closely grouped dial, switch and volume control shafts to be stuck through holes in the panel of any sort of cabinet. The escutcheon is quickly screwed over them and the console installation is not only complete but has no earmarks of a makeshift.

Much has influenced the $55 price of the Bandbox.

First, an ideal and an idea.

Then, a working out of the idea.

And now, the constant possession of the ideal.

Back before radio became the entertainment force it is today Powel Crosley, Jr., held an ideal that the things which give people pleasure should be made to sell at low prices so that millions may enjoy them.

When radio was a bundle of hair pins turned with the knobs from typewriter carriages, he had the idea that if he could make radios in sufficient quantities, he could supply millions with a means of enjoying this new source of pleasure at moderate prices.

Every radio year has been a year of mass production experience to Crosley. This year saw an investment of over half a million dollars in equipment that a fine radio might be made at such speed and in such quantities that price of nearly half a hundred dollars could be maintained.

Throughout the country millions examine the Bandbox today. They see it the achievement of an organization who began its development when radio as we know it today began. Its success has been tremendous if clamorous demands from dealers are any indication. Sceptics, the unbiased and the radio wise have pronounced it GREAT. Even at any price it would be a sensation, for its performance ranks with the most expensive and fanciest radio receivers on the market.

Write Dept. 20 for descriptive literature

An AC model Bandbox takes its power from the electric light.

Former power supply with its constant annoyance and expense is entirely eliminated.

POWER CONVERTER
$60.00

The new R.C.A. AC tubes provide clear, smooth and loud reception comparable in every way to the most efficient wet storage battery power.

Alternating current ripple is smoothed out in the compact little power converter which is sold with the AC Bandbox. The device needs no attention—is half the size of an ordinary storage "A" battery and matches the Bandbox in finish and color.

The AC Bandbox is $65

The Power Converter is $60

This gives you a complete, direct AC radio adaptable to any type of installation you may choose—bookcase, console, desk, cabinet, armchair or tuck it away on the corner of the table—for $125.00.

THE CROSLEY MUSICONES ARE AS OUTSTANDING IN THEIR FIELD AS THE BANDBOX IS IN ITS

Back in early 1925 radio's audibility above the single telephone ear unit depended upon a horn. Unnatural and harsh it laid a handicap on a fast developing industry. Only at great cost could its limitation be surmounted. Suddenly Crosley offered the radio world a cone speaker at $15.50. Instantly the demand exceeded the supply. Promptly loud speaker sales were gauged by the leadership of the Musicone.

Today that position is still maintained. The cause is plain. Mechanical refinement and improvements of the Musicone since its inception have been constant and considerable.

Price of the Musicone has shown a steady downward trend from $17.50 in 1925 down to $9.75 at the present time.

Only a national acceptance could make this possible and only a remarkable value could create such a national acceptance.

Today the Musicone is a perfect reproducer through—

1. A new, metallurgical discovery giving many times the vibrating capacity of the actuating unit as formerly.
2. Bakelite coil cores impervious to moisture.
3. Securely coated wire that does not permit deterioration in damp climates.

Crosley owners find a perfect affinity between the Bandbox and the Musicone aside from their physical appearance.

The Musicones are sold in two sizes:

The Ultra Musicone (12-inch)............ $ 9.75
The Super Musicone, as illustrated (16-inch)... 12.75

$65 $85.

$35.

APPROVED CONSOLES

"I want the public to have as great a value in consoles this year as I have given them in the Bandbox," said Powel Crosley, Jr.

Prominent furniture manufacturers through their long experience promised beautiful cabinets at moderate prices. Designs submitted were admired, praised, tested, approved! The Musicones were built in. Crosley dealers now sell them. Purchasers may know they are best suited for Crosley radio by looking for the "approved label" in each one. Crosley dealers get these cabinets only from The H. T. Roberts Co., located at 1340 S. Michigan Ave., Chicago, Sales representative for The Showers Brothers Co., Bloomington, Ind., and The Wolf Manufacturing Industries, Kokomo, Ind.

CROSLEY RADIO

THE CROSLEY RADIO CORP.
Powel Crosley, Jr., Pres., Cincinnati, Ohio

Crosley is licensed only for
Radio Amateur, Experimental and
Broadcast Reception

Montana, Wyoming, Colorado, New Mexico and West prices slightly higher

The Best
in Radio
MUST BE
CUSTOM
BUILT !

TAKE a circuit continuously improved over years of experiment by ten of America's leading manufacturers working as a unit towards radio perfection—add the finest parts in the industry —put them together under your own eyes, following the creators' simple instructions—and the result is the "Hi-Q Six"—*a CUSTOM-BUILT Receiver to meet your exact individual needs.*

"CUSTOM-BUILT" usually implies high cost. With the Hi-Q Six it means economy, for the complete approved parts cost only $95.80 and produce a radio instrument that cannot be duplicated in factory-made sets at more than twice the price.

"CUSTOM-BUILT" always means quality. So with the Hi-Q Six—equal and maximum amplification on *all* wave lengths—knife-like selectivity—super-sensitiveness— absolute freedom from oscillation—and tonal quality as natural as the original.

For complete details send 25 cents for a copy of "Hi-Q Construction Manual."

HAMMARLUND-ROBERTS, INC., 1182-A BROADWAY NEW YORK CITY

The basis of Radio Efficiency and Economy— the "Hi-Q" Book explains simply and completely how to build this remarkable receiver. The Hi-Q Foundation Unit includes ready-drilled steel chassis, decorated panels, shields and all special hardware—a combination which simplifies construction.

Associate Manufacturers

RADIO
BROADCAST

CONTENTS

How to Use the Screen-Grid Tube

A Directory of Manufactured Receivers

Hints on Operating Your Cooley Rayfoto Receiver

The Phonograph Joins the Radio Receiver

A Push-Pull Amplifier and B Supply

Inside the Complete Receiver

35 Cents   Doubleday, Page & Company
                Garden City, New York   JAN · 192

Cunningham
RADIO TUBES

SINCE 1915 STANDARD FOR ALL SETS

Gifts

that bring happiness at
Christmas time and
throughout the year

Radio sets and radio equipment make immensely popular Christmas gifts.

To assure the utmost in Yule-tide Radio enchantment, see that Cunningham Radio Tubes are on duty in every socket. Every owner of a radio will be delighted to receive a set of new tubes.

Buy them in combination of five or more. Your dealer will tell you the correct types of Cunningham Radio Tubes for which any radio set is designed.

Don't use old tubes with new ones. Use new tubes throughout.

E. T. CUNNINGHAM, Inc.

New York   Chicago   San Francisco

Never *a warp when you use* MICARTA

Home set builders may secure standard sizes of Micarta panels, with directions for easiest machining, direct from local radio dealers.

It's tough, too — you'll never find a chip around a drilled hole or a sawed edge on a Micarta Panel. Produced under great heat and pressure, it has fine texture and strong bond — permitting clean machining and engraving, and smooth, accurate finishing.

Micarta is used for both main and sub-panels — its high insulating qualities allow the most compact arrangements. Its strength is amazing — it will not soften, warp or sag under the most severe radio requirement, nor will it crack between holes, be they ever so neighborly.

Micarta finishes — jet black, mahogany, walnut grain and walnut burl — permit striking treatments in fine cabinet design. It will pay you to communicate with Micarta Fabricators, Inc., regarding Micarta in manufacturing quantities, to your specifications.

WESTINGHOUSE ELECTRIC & MANUFACTURING COMPANY
Offices in Principal Cities *Representatives Everywhere*
Tune in sometime with KDKA — KYW — WBZ — WBZA

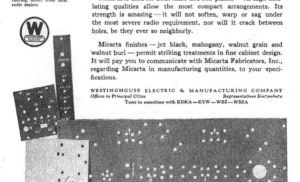

Westinghouse
© 1927 W. E. & M. Co.

MICARTA FABRICATORS, Inc.
OF NEW YORK
309 Canal Street
New York City

MICARTA

MICARTA FABRICATORS, Inc.
OF ILLINOIS
500 South Peoria Street
Chicago, Ill.

More Money

Be a Trained Radio Expert

If you're earning a penny less than $50 a week, get my free book of informat about the radio business. Trained Radio Experts are needed in more th 20 different lines of this new and growing profession (300,000 new openii created by the swift growth of Radio in past few years). Why go along at $ or $35 or $45 a week all your life? Study Radio and after only a short ti land yourself a REAL job with a REAL future! Be a man who has money his pocket and in the bank—don't scrimp and scrape for the rest of your da

Salaries of $50 Up to $250 a Week

Not Unusual in Radio

The good positions in Radio pay all the way from $50 on up to $150, $2 and even $250 a week. Suppose you don't climb to the very top, but that y do advance to a position that pays you $125 a week, year in and year o Any chance to make that much where you are now? Then send for my f book, and learn about a field where there's some *real opportunity*. Wh good men, *if they have the right training*, can work their way into really salaried jobs!

Money Back if You Aren't Fully Satisfie

I'll give you all the training you need to get into any line of the Radio business. And I back up this training by a signed agreement to refund every penny you pay me if I don't give you exactly the training you need. After you finish my training, you'll be the judge. If you think I've earned my tuition fee, I keep it. If not, ask for it and you'll get it right back.

Six Big Practical Outfits
Given You to Help You Learn

I teach you both the "why" and the "how" of Radio. You learn to DO a thing, and you learn WHY it's done. I send you, WITH-OUT EXTRA COST with your course, six big practical outfits of material to experiment and work with. These outfits are the real thing—not toys. The parts they contain will build approximately 100 different Radio circuits. With all this material you do practical work from start to finish of your training. You get your hand in, and you get confidence in yourself. Then when you run into a Radio problem later on, on the job, you KNOW you can do it because you've Already done it. With these six outfits of practical material. With me you don't learn to be a "paper Radio Expert"—you learn to be the kind of expert that shows his worth on the payroll. Full details in my big book—sent free.

Send for Free Book of Information

Broadcasting and commercial Radio land station work appeal to a lot of men—it's a big, growing field and fascinating work. My course prepares you thoroughly to get into this field and make good.

Six million Radio receiving sets in use in the United States means millions of dollars are going to Radio Service and Repair men everywhere. For this work you must be trained.

Radio manufacturing has grown faster in the last 6 years than any other big business ever did. It's ten times as big as it was. That means lots of good chances for the trained Radio Expert.

Radio operators on board ship go everywhere—see everything. You sail the world over, all your expenses paid, and draw a good salary besides. It's the life of Reilly.

GET BOOK!

Find out for yours about bigger pay waiti for you in Radio! Fro $2,000,000 in 1920 $304,000,000 in 1926 that's the record of t Radio industry. Plenty big Radio jobs are waiti for the man who KNOW John Fetrer sent for f free book—now he's Ch Engineer at WEMC a designs and builds bro cast stations. T. M. W cox sent for the book now he's in his own Ra business and reports pr its as high as $70 in o day! All information v be sent to you free, wit out obligation. Just m coupon on opposite pa

see coupon next page

LEARN QUICKLY, EASILY

Train at Home in Spare Minutes

Stay home! Hold your job! I'll bring your Radio training to you, and you can learn in your spare time after work. No need to go to a strange city and live for months on expense when you learn my way. You study in the quiet of your own home. As for this training—it's written just as I would talk—in straightforward, everyday, understandable language. You'll get it, and you'll get it quickly—in a few months' spare time—because I've made it so clear and interesting! No particular education needed—lots of my successful students didn't finish the grades.

Earn $15, $20, $30 Weekly Right Away "In the Bag"

Deloss Brown, South St., Foxboro, Mass., made $1,000 from spare-time Radio jobs before he even finished my course. H. W. Colbentz, Washington, averaged $45 a week; Leo Auchampaugh, 6432 Lakewood Ave., Chicago, made $500 before graduation; Frank Toomey, Jr., Piermont, N. Y., made $833 while taking the course. All this done IN SPARE TIME away from the regular job, while these fellows were still studying the course—and they're only a few of hundreds. As soon as you start this training I begin teaching you practical Radio work. Then a few weeks later, I show you how to make use of it in spare time, so you can be making $15, $20, $30 a week "on the side," all the while you're learning.

64-Page Book Sent Free for the Asking

My big book of Radio information won't cost you a penny, and you won't be under any obligation by asking for it. It's put hundreds of fellows on the way to big pay and brighter futures. Sending for it has been the turning-point where many a man has made his start toward real Success. Get it. See what it's going to mean to you. Send coupon TO-DAY!

Address: J. E. SMITH, President
National Radio Institute
Dept. 1-0, Washington, D. C.

JOBS WAITING!

In 7 short years — 300,000 new jobs in Radio! Lots of jobs open right now, for those who have the training. The Radio industry has grown by leaps and bounds—so fast it has had to take whatever sort of men it could get. Such men, if they haven't trained themselves in the meantime, are losing out and will keep on losing out. They'll be replaced by men with the KNOW-HOW. But it's trained men ONLY that are needed.

Over 1,000 Openings for Trained Men NOW!

One great Radio manufacturing concern alone has over 1,000 openings to give my graduates this year. These men will be needed all over the United States. Any graduate of mine who stands well in his home town is eligible for this work. The head of the above mentioned concern—one of the biggest Radio organizations in the country—is a graduate of mine. He knows what my training did for him. When he wants new men for his organization he wants men with the same training.

I can't possibly graduate enough men this year to fill these openings. So there will be more openings with this one concern than there will be graduates to accept them.

But there are other openings to choose from, too. My school has trained more Radio Experts than any other school in the world. It's the oldest and largest Radio home-study school in the world. There are N. R. I. trained men in almost every Radio concern of any importance in this country. Many Radio employers are themselves my graduates. That's where you get your "stand-in" as an N. R. I. graduate yourself.

Every graduate of my course is entitled to Life-Time Employment Service, without a penny's charge, from my helpful Employment Department.

Full Information Sent with Free Book

My Free Book contains full information about the Radio employment situation, and the advantages I'm in a position to give you. Also about my Life-Time Employment Service, and Life-Time Consultation Service, too.

Mail Coupon To-day!

RICH REWARDS in RADIO

This big 64-page book, printed in two colors, crammed with interesting facts and photos about money-making opportunities in Radio, sent free to everyone who clips the coupon. No obligation by sending for the book—it's absolutely free. One of the most valuable books about Radio ever written.

6 big outfits given

Send No Money

J. E. SMITH, President
National Radio Institute
Dept. 1-0, Washington, D. C.

Dear Mr. Smith: Kindly send me your big free book "Rich Rewards in Radio," giving all information about the big-money opportunities in Radio and how you will train me to take advantage of them. I understand this places me under no obligation, and that no salesmen will call on me.

Name....................................

Address.................................

Town....................... State..............

Acme PA-1
Power Amplifier $12.50

Acme E-4
B-Power Supply
$35

Look for this
ACME head

Acme K-1A
Speaker $25

Don't condemn your present radio! Give it the equipment it deserves!

THERE is one way to become as proud of your radio set as you were the day you bought it. Equip it with one or more of these three units! You'll hear the most amazing difference immediately—and that's absolutely guaranteed.

1 Acme B-Power Supply—eliminates your dry batteries—gives you a constant maximum of power—saves you more than its cost. And the Acme B-Power Supply is the one that is known to stand up indefinitely. Price $35.

2 The most scientifically perfect speaker ever devised—the double free-edge cone Acme K-1A. Shaped to fit the highest and lowest notes—instead of forcing them to fit its shape. And beautiful in its graceful simplicity. Price $25.

3 Acme now offers a single stage-power amplifier to add to any set. All you have to do is connect your set, your B supply and your speaker to the amplifier,' put in a power tube, plug into the lamp socket, and your radio enjoyment increased a hundred-fold. Price $12.50.

ACME
–for amplification

How Does Your Receiver Work?

What Antenna is Best?

How Radio Tubes Function?

These questions and many others are answered in Walter Van B. Roberts' book, "How Radio Receivers Work."
Today—send one dollar, for your copy, to the Radio Broadcast Booklet Dept., Garden City, N. Y.

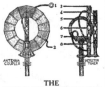

JAN -5 1928   ©CI B763288

RADIO BROADCAST

WILLIS KINGSLEY WING, Editor

JANUARY, 1928   KEITH HENNEY   EDGAR H. FELIX   Vol. X

Director of the Laboratory   Contributing Editor

CONTENTS

AMONG OTHER THIN[

IT IS a sad duty to record the death of the Cl Federal Radio Commission, Admiral W. F which occurred in Washington on Thanksgiving Bullard, who served in the United States Navy years, for a very long time was close to the cen almost all of its branches. His loss will be keenly by those who knew him as a likable and able especially by the Radio Commission itself. WI Commission went to work on March 15, two had a background of technical radio experience. were Admiral Bullard and Colonel Dillon. Deat both. The Commission at this writing now cons Chairman E. O. Sykes, O. H. Caldwell, Sam H. A. Lafount. Not one of these members has a background which would enable them to better the complicated problems which confront them.

THE reports of international conferences, subject, usually make rather dull reading f public and the Washington Radio Conference exception to this rule. The proceedings may not the results are certainly important. There has be of international agreement since the London conf and radio progress has been so rapid since then t of that Convention were hopelessly inadequate t needs. There have been many rocks and shoals the present conference, which, at this writing, up its work, but through good management and desire for general accord, the delegates have drawing up a Convention which well meets the today. Not the least important decision reached was that dealing with the international assignm in the frequency spectrum. In that respect, we a the future needs of short-wave communication commercial, and amateur work were provided fo teurs had a hard fight, but room has been save result of which the broad-minded directors of Radio Relay League may well be proud.

THE issue of RADIO BROADCAST before you extremely interesting articles. The story Rhodes on the problems of push-pull amplificati helpful and should cast much light on a form which is again being revived after several years disuse. . . . Those who are anxious to know screened-grid tube will do will find Keith Henne valuable indeed. As soon as possible, RADIO B give its readers data on receiving circuits whi with the tube; the latter has just been released

THOSE of our readers who would like to ha forwarded to the manufacturers of the sp the undersigned, and printed matter containi formation will be sent them. . . . The next Ra will contain an article describing a new su entirely operated from a. c., which has much to both from the design and appearance point will also be many other articles of interest.

—WILLIS KIN[

How's Your Old Audio Amplifier?

THORDARSON 171 TYPE POWER AMPLIFIER

Built around the Thordarson Power Compact R-171, this power amplifier supplies "A," "B," and "C" current for one UX-171 power tube and B-voltage for the receiver. Employs Raytheon B. H. tectifier.

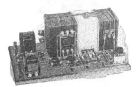

THORDARSON 210 TYPE POWER AMPLIFIER

This amplifier, mounted on a special metal chassis, uses the Thordarson Power Compact R-210. Provides "A," "B," and "C" current for one UX-210 power tube and "B" voltage for the receiver. Employs one 216-B or 281 rectifier.

THORDARSON 210 PUSH-PULL POWER AMPLIFIER

This heavy duty power amplifier operates two 210 power tubes in push-pull and has an ample reserve of power for "B" supply for the heaviest drain receivers. Built with Thordarson Power Transformer T-2098, and Double Choke Unit T-2099.

A Home Assembled Thordarson Power Amplifier Will Make Your Receiver

A Real Musical Instrument

IMPROVEMENTS in the newer model receiving sets are all centered around the audio amplifier. There is no reason, however, why you cannot bring your present receiver up to 1928 standards of tone quality by building your own Thordarson Power Amplifier.

With a screw driver, a pair of pliers and a soldering iron you can build any Thordarson Power Amplifier in an evening's time in your own home. Complete, simple pictorial diagrams are furnished with every power transformer.

The fact that Thordarson power transformers are used by such leading manufacturers as Victor, Brunswick, Federal, Philco and Willard insures you of unquestionable quality and performance.

Give your radio set a chance to reproduce real music. Build a Thordarson Power Amplifier.

Write today for complete constructional booklets sent free on request.

THORDARSON

THORDARSON ELECTRIC MANUFACTURING CO.
Transformer Specialists Since 1895
WORLD'S OLDEST AND LARGEST EXCLUSIVE TRANSFORMER MAKERS
Huron and Kingsbury Streets — Chicago, Ill. U.S.A.

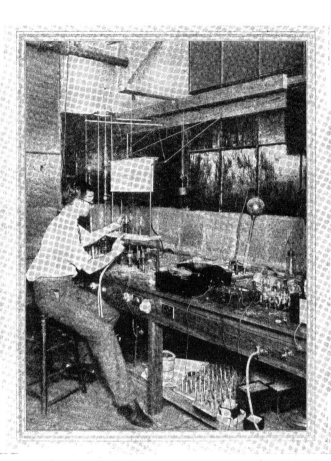

In the Laboratory of a Tube Manufacturer

Where diligent investigation precedes production of any new kind of tube. The particular laboratory here shown is given over to the study of rectifier tubes—the kind you are accustomed to use in your power-supply devices to rectify the a. c. of the house supply. This investigator is shown engaged in the construction of an experimental tube, which will afterwards be tested and, as likely as not, discarded. It is said that about 99 per cent. of these hand-made experimental tubes are discarded without even being put into production. When we consider the expense of such fireless research, more and more do we recognize the logic of the statement, that it is only the larger and more "moneyed" manufacturers who are capable of producing dependable radio equipment

STUDYING GASES IN THE LABORATORY
Looking into the depths of ionized helium atoms with a spectroscope. Charles Grover Smith,
who is shown in this picture, spends considerable of his time studying gases, especially helium

RADIO ENLISTS THE HELIUM ATOM

By VOLNEY G. MATHISON

IT'S all so useless, that's why I'm quitting," complained the young college graduate to the chief of the laboratory in which he had lately gotten a job. "What can ever come out of measuring the ohmic resistances of a cubic millimeter of about a million different substances under a couple of hundred different temperatures? That's the job you've given me—nothing but an endless measuring of ohms. If this is what you call scientific research——"

"It· is," interrupted the electro-chemist. "It's the backbone of it—prolonged patience at tedious work." .

"Seems tedious enough. But I could stand that if I could see some results ahead. I can't see any."

"Well, there may be no directly important outcome of what you're doing. You are simply adding to the stock of scientific knowledge in a prosy way, working on a huge book of temperature-resistance tables. You can't tell in advance what it may lead to. Take, for example, the helium-gas rectifier tube used in radio B-supply devices. It was never invented, it just grew out of a lot of laborious work like this. American factories are turning out about 20,000 of these tubes a day at present and the patent profit is close to a dollar and a half apiece. There's a return to pure research at the rate of something like $30,000 a day. That soon ·pays for a lot of slow laboratory grinding, doesn't it?"

"Yes," reiterated the young graduate, "but that tube wasn't developed by any such work as I'm doing here—measuring the resistances of rust, rocks, roots, cocoanut shell, and the end years away. It's all so discouraging!"

As a matter of fact, the helium-gas tube for B-devices was the result of a great amount of purely scientific work of the tedious and rather unfocused sort that this discouraged young chemist was assigned to, though, perhaps, the experiments involved were a good deal more technical. A few years ago we began studying the actions of electrons in gases. At that time nobody knew much about electrons—maybe we don't yet—and the experiments were entirely general in kind. The aim was to find out something, not to invent something. One young fellow had the job of finding out definitely whether or not electrons emitted from a cathode into a tube of gas passed through the gas without colliding with its atom centers.

An atom of any kind of substance, as nearly every one now is aware, consists of a group of protons and electrons surrounded by planetary electrons. The space in between seems to contain nothing but electric tension, though now late in 1927 we are about to believe that this tension consists of a flurry of particles called etherons that are so small they make an electron look like a balloon in comparison with an apple, and move almost twice as fast as light. Matter,even

solid steel, is nearly all emptiness, and it is the crudeness of our undeveloped physical senses that makes us think ótherwise.

So this young man set about trailing a lot of wandering electrons through a wilderness of gases to find out what they did in there. He had a photographing outfit that would show the paths of the electrons. Many gases and pressures had to be tried. In one series of experiments each photograph showed the flight of 200,000 electrons, and that young man took 100,000 photographs separately, one after the other, before he got one collision of an electron with the middle of an atom of the gas under test. He kept on, and got a total of eight collisions in 400,000 photographs. Even when it is possible to use a fast automatic machine to take such photographs, they all have to be individually and laboriously examined.

The photographer found out something else. He found out that when an electron hit an atom of gas square in the middle, it blew the atom to pieces, producing a shower of unattached electrons and unsmashable groups of electrons and protons that were named alpha particles. Alpha particles were identified with a spectroscope as being the same as a mysterious gas which had been seen by astronomers pouring over the sun, and which they had named helium—a 'sun" word.

The alpha particles or helium atoms can be

195

quite easily robbed temporarily of one planetary electron, but they cannot be further broken up except with the most extreme difficulty. Helium gas was therefore recognized as a valuable electrical conducting medium that would tend to act at all times with unchanging properties and that would refuse to eat, corrode, or combine in any way with the metal tips of the electrodes by which electricity would be fed into the gas. No particular electrical use for this gas had yet been thought of, though it was proposed that metal-ended glass cartridges of helium might be valuable as ultra-stable resistors capable of withstanding enormous currents and pressures.

In photographing the flights of electrons through helium gas, it was observed that, when an electron struck an atom of helium, the electron ran off with one of the planetary electrons of the helium atom, carrying it to the anode.

A sketch of this performance would be almost inconceivably out of proportion since there would have to be represented the trillions times trillions of atoms in the tube, while electrodes about the size of the state of Georgia would be necessary to preserve the proper scale if the atoms were one half inch across. If the helium nucleus were drawn the size of a pea, the planetary electrons would be placed a quarter of a mile away. Such is the ghastly emptiness of matter—of even the "solid" walls of a glass tumbler from which we drink billions of electrons and protons in the peculiarly ordered state we call water; and of the stout iron bars upon which the jail-bird leans his head. But solidity, although physically an illusion, is nevertheless real because it is a manifestation of a powerful electrical condition.

To return now to the story of the doings in the helium tube under electrical pressure, we learn that the unfortunate gas atom that has lost an electron is said to be ' ionized!'' This is all that ionization means, i. e., a breaking away of the ionized atom, is over-positively charged and flies with terrific force toward the negative or cathode element, which it strikes violently and from which it extracts an electron from the inflowing stream in the wire leading to the cathode —the entrance element from the outside supply. The gas atom, now once more in a normal state, rebounds from the cathode and flies about jubilantly until it is again robbed of an electron, when it instantly goes through the performance just described all over again.

One remarkable thing about this action which is not yet understood is the fact that the bombardment of the cathode by ionized gas particles causes a liberal release of free electrons—of extra electrons beside those taken to balance the ionized atoms. In other words, by hammering a negative electrode with positively charged gas particles, we get a discharge of electrons. Perhaps the "yanking" out of the electron needed by the atom for itself is so violent that several extra electrons are hustled out along with it.

Some of these extra electrons in their flight to the positive or exit element collide with other gas molecules, and the action, the pounding of gas particles on the cathode and the emission of electrons, is continuous as long as electrical pressure is applied to the device. A current consequently

flows through the tube. In this manner electricity gets through a cold-electrode rectifier tube.

HOW A TUBE RECTIFIES

THE foregoing does not explain how current gets through the tube in only one direction. It would seem that, when the tube is put in an a. c. circuit, each element would become alternately an anode and a cathode, an exit and an entrance, and that the flow would be equally back and forth through the gas in the tube. But since the current flow depends upon hitting an electrode with gas particles, it becomes apparent that, by making one electrode big and the other one little, you can facilitate the hitting in one direction and minimize it in the other. When the larger electrode is negative, it will get thoroughly hammered with gas particles and will release many electrons; consequently a large

BEHIND THE SCENES IN THE LABORATORY
Preparing a globe of pure helium for research experiments. This kind of work never stops, and in every progressive tube concern, the laboratory is always ahead of the factory—is always working on "something better."

current will flow through the device. On the other hand, when the smaller electrode goes negative under the reversing a. c. current, it will not be battered so much by the flying gas particles as the other electrode was and therefore will not emit many electrons. A small current will, however, flow. The big electrode is a better emitter because it has more surface for the gas particles to "rip" electrons out of. When the big electrode is positive and the small electrode is negative, the ionized gas atoms do fly toward the small electrode but most of them miss it because it is comparatively hard to hit. Some particles strike the small element while it is negative and release electrons which fly across to the large electrode. This produces a "back-current" through the tube. All helium-gas rectifier tubes have this "back-current." The amount of it determines the efficiency of the rectifier and it usually is but a small percentage of the current in the other direction.

The "back-current'' is reduced by making the anode small and hard, while the other large

electrode is treated with radioactive earths to encourage it to release electrons. Besides the difference in size of the bombarding areas of the two elements, the electric field produced by the smaller electrode is less attractive to the ionized gas particles than that of the larger one with the result that it is hit less violently.

Both of the elements are made principally of nickel, which has been purified in a hydrogen furnace to remove all impurities.

The description so far has dealt with a tube in which there are only two elements. The usual helium-gas rectifier tube has three elements—a hat-shaped or tubular cathode and two anodes. This is simply two rectifier tubes in one, using a common cathode, and enables both halves of the a. c. wave to be used. The action then is exactly the same as that already described, except that the emission of electrons is alternately from the cathode to first one anode and then to the other—to each one as it becomes positive in turn. And the cathode is almost continuously under bombardment, because it is always negative with respect to one anode or the other. At the same time that the flight of electrons takes place from the cathode to the positive anode, a "back-current" emission travels from the temporarily negative anode toward the cathode. Some electrons also fly between the two anodes, causing "leak" current. These "leak" and "back-currents," while not wanted, cannot be entirely eliminated, and as long as they are small, cause no serious trouble. In the factories, each tube is tested on a machine that measures the "back-current," and if excessive, the tube is discarded. Too much "back-current" through a tube will cause a B-device to hum.

It is worth pointing out that the popular conception of the movement of electricity through a B-device rectifier and filter system is erroneous. The general idea is that a "positive current" gets across the rectifier tube into the filter where it is tanked and choked into smoothness. But it should be remembered that current is purely electronic flow—it consists only of moving electrons in a conductor—and these electrons flow only from negative to positive. It is very confusing to the novice and entirely unnecessary to say that the current flows one way and the electrons the other. There is no such current flowing against the movement of the electrons. It simply doesn't exist. In the early days of electrical science, long before any kind of a radio vacuum-tube had even been thought of, experimenters misunderstood some of the actions of electricity, and the positive-to-negative idea of current flow was one of the consequences of their lack of knowledge. In the business of science, there is no sensible reason for compromising with mistakes or twisting them around to meet new facts. It is being attempted all the time in religion; they should be simply left out of the story.

The electrons flowing in a B-device circuit enter the negative wires, pass direct to the filaments of the vacuum-tubes in the radio receiver, are emitted from the hot filaments to the plates, pass from the plates through the various circuits to the positive lead of the B device, thence through the chokes to the cathode element of the rectifier tube, and from this point they are sprayed alternately to the anode elements as these become positive in turn under the inductive action of the a. c. current in the primary power-supply windings. The filter chokes and tanks prevent a voltage and current fluctuation in the line during the intervals between the spraying or emission surges through the rectifier tube.

So deeply ingrown is the current-flow conception that in a recent issue of a well-known radio magazine there was a cut of a helium-gas tube with the hat-shaped cathode marked "anode" and the two anodes marked "cathodes." The accompanying text used the same terminology, which was wrong even in the light of the old theory. It should be clear that the positive side of a B-device filter is of negative potential compared to the negative winding of the power transformer to which the rectifier-tube anodes are connected. The large cathode of a helium rectifier is equivalent to the incandescent filament of the filament type tube.

WHY A TUBE DETERIORATES

SINCE the helium-gas rectifier tube contains no heated filament emitter to burn out or become lifeless through deterioration, many users of the device feel that it ought to last almost forever—for years and years at any rate, and wonder why it sometimes has a short life.

As a matter of fact, the developers of the tube themselves thought at first that it would have a life of 10,000 hours or more of continuous use, but soon found that such was not the case. The principal thing that brings the life of a helium-gas rectifier tube to an end is the fact that the helium gas in the bulb disappears. Helium gas is inert, it will not combine with anything, so far as we know at present; it is genuinely strange, therefore, that it should disappear from a hermetically-sealed bulb. It seems that the ionized gas particles pound the cathode with such force that some of them are driven deep into the metal and stay there—become occluded or imprisoned.

After a certain length of time so many gas atoms are bound in the metal that the tube becomes very hard, the vacuum rises, and the bombardment of the cathode becomes meager, owing to the reduced number of molecules of gas to do the battering. The current output of the tube then falls off to such a point that it must be discarded.

The life of many a good helium-gas tube has been quickly brought to an end through the breaking down of condensers in filter circuits. Cheap inferior condensers in both home-made and factory-built power devices usually go to pieces after a few weeks or months of use, with the result that the rectifier tube is placed in a dead short-circuit. The heavy current flow quickly burns off the tips of the anodes in the tube. The helium gas itself cannot be injured by any current. Helium gas will carry currents so great that they will instantly explode copper conductors of the same cross-sectional area; but under such currents the gas particles quickly drive themselves deep into the negative electrodes and are as good as lost.

Some of the cheaper helium-gas tubes now on the market may be short-lived through the presence of impurities in the helium, which would destroy the electrodes. Extremely pure helium must be used. This gas is purified by passing it through copper tubes filled with cocoanut charcoal and maintained at the temperature of liquid air—more than 250° below zero, Fahrenheit; then to a steel reservoir; then to a second battery of tubes of charcoal surrounded with liquid air to remove oxygen or nitrogen which might come away from the walls of the reservoir. The helium is admitted until the charcoal is partly loaded with it; next it is pumped off with vacuum pumps and the impurities remain in the charcoal, which is itself purified and used over again.

It is interesting to consider that absolute purification of anything, liquid, gaseous, or solid, is almost impossible. Imagine for a moment that you could mark molecules in some such way that you could identify them when you saw them again. Assume that you took a glass of water with the molecules of water all thus marked and stirred it into the waters of the oceans of this earth, that you waited a couple of million years for thorough mixture, and that you then walked up to the nearest hydrant in your vicinity and casually drew a glass of water, you would find about 2000 of your marked molecules in it! Of course, we may be a few molecules out on this estimate, but that is roughly the correct mathematical number, because there are 2000 times as many molecules in a glass of water than there are glasses of water on the earth.

Again, molecules of air, if admitted into a

highly-evacuated 25-watt electric light bulb through a hole so small that they had to flow in single file, would take about 100,000,000 years to fill it to atmospheric pressure. A little reflection will show that "purity," like everything else, is probably only a relative condition. "Some of 'em is more pure, and some of 'em is less pure, but none of 'em is *all* pure," said the sour cynic, and while he wasn't speaking of gases, and was not a scientist for that matter, he was uttering profound truth.

THE TIN "HAT"

WHAT is the queer tin "hat" for in the modern gaseous rectifier tube? Nobody outside of the research laboratories seems to know. Even some of the "bootleggers" making these tubes don't know.

In the early experimental forms of the helium-gas tube, great trouble was met with owing to the disruption of the cathode element under the hammering of the ionized gas atoms. The earlier tubes had disk shaped cathodes, and the gas atoms pulled out electrons with such violence that tiny pieces of solid metal were often ripped loose from the electrode. These metal bits were thrown against the glass walls of the tube, blackening it, and resulting in the speedy destruction of the cathode.

For this reason the peculiar tin-hat form of cathode was evolved because, with this arrangement, the bombardment of the cathode and emission of electrons is entirely internal. The action all takes place inside of the electrode. If a bit of metal is torn from the cathode at any point, it is hurled across the inner chamber and thrown back onto the element somewhere else. There is no loss of metal because the cathode is continually built up as fast as it is torn down.

It seems that, if we, as human beings, could be temporarily reduced to the size of, say, an atom, together with a corresponding ability to see small objects, and were then placed upon the cathode of an operating helium-gas rectifier tube, we would have an impression of standing among ranges of heaving mountains of metal in a state of furious convulsion and uproar, bombarded with enormous meteors of helium and full of volcanic upheavals and earthquake-like shocks, while electrons would arise like clouds of steam on all sides, or spurt out like fiery sparks.

A VIEW OF THE LABORATORY IN WHICH THE HELIUM-GAS RECTIFIER TUBE WAS PERFECTED

Can The Serious Problem of Radio Patents Be Settled?

THE recent adjudication of several important patents, such as those of Hazeltine and Alexanderson, has forced upon the radio industry the long deferred day of reckoning with inventive genius. Conscientious and established manufacturers have proceeded promptly to obtain licenses under Radio Corporation patents which make available to them the work of some of the world's greatest laboratories. By assuming an annual royalty guarantee of one hundred thousand dollars a year, charged at the rate of 7½ per cent. of the cost of radio receivers, they become licensed under R. C. A., A. T. & T., Western Electric, General Electric, Westinghouse and Wireless Specialty patents.

Having assumed this substantial burden, the licensees considered their patent difficulties disposed of. But some quickly found that licenses under Hazeltine and Latour patents are also necessary to freedom from patent difficulty and, probably quite reluctantly, signed the Hazeltine licenses with the additional burden of a 2½ per cent. royalty and an annual guarantee of thirty thousand dollars a year. This duty performed, the manufacturer dismissed patent trouble and consecrated himself to the problem of selling newer and better radios. And then came the independent inventor to disturb his peace of mind. Old patents were dug up, demanding recognition. New patents, just issued, added to the swarm. Some of these inventions are as worthy of recognition as those covered by the Radio Corporation license. Others may be worth-

less and which will not withstand the test of adjudication.

The weary manufacturer's answer to those demanding additional royalties is becoming less and less courteous. He is now paying all that the traffic will bear. Unless some remedy is offered, his answer to patent holders soon will be: "A plague upon your patents!"

The Prospects of a Patent Pool

SOME manufacturers have united in defensive groups to protect themselves against the swarm of inventions which now confronts them. They foresee the necessity for so great an increase in the price of radio receivers by reason of patent royalties that the public will no longer be able to afford them. Faced with the alternatives of excessive royalties or occasional injustice to the legitimate inventor, the manufacturers have, quite naturally, tended to the latter course.

No doubt, some of the inventors, whose claims for royalties are being disregarded or opposed, will eventually win adjudications, and triple damages, if they are sufficiently patient and prosperous to afford the protracted legal battle which must precede such a result. It is quite possible that combined resistance to the inventor may, in some cases, prove costly, because it is not reasonable to assume that Radio Corporation, Latour and Hazeltine patents are the only ones which will be favorably adjudicated.

Combined resistance, however, is the only

course open to the manufacturer because there are too many unadjudicated patents demanding attention. It would be suicidal to agree to a license under all of them; the cost of radio sets to the consumer would double and sales resistance would fourfold.

Most of the executives in the radio field wish to concentrate their attention on the design, manufacture and merchandising of radio equipment, but patent problems now require an alarming proportion of their time. Naturally, the leaders of the radio industry are nervous. At every trade convention and meeting, we hear talk of an all-inclusive patent pool.

Unfortunately, no patent pool can be successfully organized unless it has the unanimous support of all radio manufacturers and patent holders. To relieve the patent situation economically and painlessly, there must be a single, powerful, radio trade organization. Nevertheless, the N. E. M. A. and the R. M. A. still—to all outward appearances—indulge in short-sighted rivalry. An unofficial canvass of ninety per cent. of the membership of one of these organizations reveals that all but two were in favor of consolidation. To make a consolidation possible, one or two leaders in the R. M. A. must for a moment forget that their organization has the largest number of members and the youngest blood, and one or two members of the N. E. M. A. must forget some remarks made several years ago over matters long since settled. And both groups must cease suspecting each other.

© Henry Miller

CHANGES IN THE FEDERAL RADIO COMMISSION

Henry A. Bellows, Sam Pickard, and Carl H. Butman. Commissioner Bellows was appointed from the Minneapolis district where he had long ably managed WCCO. His resignation, effective November 1st, left a vacancy which was filled by the President in appointing Sam Pickard who has been Secretary to the Commission since its appointment. Carl Butman is the new Secretary succeeding Mr. Pickard and has many years of experience in Washington as news correspondent, specializing in radio, to aid him in his new post. Mr. Butman has for some time been Washington correspondent for RADIO BROADCAST

The industry is headed for one of the most dangerous shoals in its career. Doubtless, it will weather it successfully. But how long the shoal will impede its progress depends upon how successful it is in placing the ship in the hands of one pilot, instead of two squabbling deck hands, to guide it past the patent whirlpool. It will require leadership of the highest order to establish a patent pool.

In the future, there will be a new and greater industry, much greater than we imagine to-day. The radio receiver is but a nucleus of a home entertainment device which will rival the automobile in usefulness and entertainment value and, in the end, its gross sales figures will be as large as those of the envied motor car industry. The radio receiver, the phonograph, the motion picture machine and the television receiver will, some day, be available in a single, compact, home-entertainment device. The public will pay as much for a versatile means of home entertainment as for an automobile to take them away from home. The more the leaders of the radio industry concentrate upon the development of radio and the establishment of its true market, the sooner they will have a five-billion-dollar industry. At present, the most vital aid in that objective would be turning over radio's patent problems to a patent pool. The alternatives are continued squabbles, continued patent fights, and a radio market still limited to about ten per cent. of the American public.

The Commission Announces a New Policy

THE Federal Radio Commission recently announced a long list of allocation changes which have been made with the purpose of improving the channels of a few of the leading stations of the country. The Commission it is rumored, will hereafter work on the theory that there are a few leading, national stations, which are the favorites of listeners all over the country and therefore deserve clear channels, entirely free of interference to the limit of their range. This is the basis upon which several years ago Secretary of Commerce Hoover worked out the plan of Class A and Class B broadcasting stations and urged on the commission in these columns for more than a year.

Following this plan, WHAZ, which shared time with WGY, was shifted to share time with WMAK, giving powerful and popular WGY a full channel. WJAR, Providence, was shifted from 620 kc. to 800, eliminating widespread heterodyning with WEAF, ten kc. off, experienced throughout southern New England. WEEI, Boston, was shifted from 670 to 650 kc., avoiding a heterodyne imposed upon it by a Chicago station. WTRL, a little station in Tilton, New Hampshire, formerly occupying a channel adjacent to WJZ, was shifted downward in order to eliminate a whistle which it thrust on WJZ's carrier in large parts of New England. WDWM of Asbury Park, New Jersey, was

shifted so as to eliminate conflict with WSAI's carrier. WNAC of Boston and WEAN of Providence, Columbia chain members, were made channel-sharing stations, probably in the interests of better management, and moved quite far into the unpopular higher frequency region. WCAU, a Philadelphia advertising station and now a member of the Columbia network and WKRC of the same chain were demoted from the lower frequency region. WOR now has WOS of Jefferson City, Missouri, as a channel neighbor instead of WSUI, Iowa City. We believe the Missouri station is more powerfully received in Newark and will therefore accentuate slightly the whistle which already mars WOR's programs.

Another station to benefit by the Commission's reallocation is KSD, St. Louis, which is given full time. KSD is one of the pioneers of broadcasting and is deserving of the consideration which the Commission has shown.

The conclusion that N. B. C. stations have fared better than Columbia chain stations is inescapable, but it must not be forgotten that the former do include most of the pioneer stations of the country which have served faithfully and well for years, while most of the latter have not yet won their spurs in public estimation. Clearing the channels of the N. B. C.'s leading stations cannot be criticized, but it might have been better policy to have concentrated less on Columbia stations when the demoting process was begun.

The Commission Suggests Synchronization Schemes

IN A speech before the American Institute of Electrical Engineers, Commissioner O. H. Caldwell made some remarks about the problems of maintaining a broadcasting station on its assigned frequency. He mentioned three methods of accomplishing this purpose, one well known and widely used, one successfully used experimentally but economically prohibitive, and a third which is a rather unfortunate suggestion. It is the Commissioner's idea that, if complete frequency stability could be secured and the heterodyne interference between stations now assigned to the same channels eliminated, more stations could safely occupy the same channel. While this is true, it must be realized that the audio-frequency components of two stations on the same channel also affect each other. When the distant station does not come in with sufficient volume to cause cross talk, it often causes irregular distortion.

Cyrstal control, the first method suggested for synchronizing carriers, is not sufficiently stable to solve the problem. Temperature and humidity changes affect the frequency of the crystal and, consequently, it does not give the absolute regulation necessary for successful occupancy of the same channel by two broadcasting stations whose carrier ranges overlap.

The second method suggested, the use of a wire circuit for the transmission of a con-

A WIRELESS STATION IN 1904

A 35-kw. spark transmitter was erected by the old De Forest Company for the Navy near San Juan, Porto Rico. The illustration shows the receiving installation with Mr. Irodell, operator for the De Forest Company using the receiving equipment which consisted of a "pancake" tuner and an electrolytic detector. It was not until 1906 that a carborundum detector was substituted for the electrolytic one. The call signal of this station, which may be recalled by old-timers, was SA.

trolling frequency, which has been successfully employed by WBZ and WBZA at
Springfield and Boston, has the disadvantage of being prohibitively expensive. For
example, if WOR attempted to eliminate the
heterodyne whistle caused by WOS at
Jefferson City by this method, it would
probably cost some fifty thousand dollars
a year. To stabilize the whole broadcasting
structure would require perhaps five years
to erect sufficient telephone channels for
the purpose and an expenditure of perhaps
twenty million dollars a year in maintenance.

The third suggestion made by the Commissioner was prompted by a suggestion
from WDRC of New Haven, Connecticut,
a 500-watt station. A heterodyne whistle,
originating from the carrier of WAIU, a
5000-watt station in Columbus, Ohio,
about 500 miles distant, had been sufficiently annoying to require drastic measures.
To solve this problem, a receiving station
was installed five miles from New Haven,
connected by wire lines with WDRC's transmitter. By tuning this receiving set carefully so that the heterodyne whistle is
eliminated, WDRC's carrier is adjusted to
coincide with that of WAIU. So long as the
operator is vigilant and skillful, there is no
heterodyne whistle.

But, if the whole broadcasting structure
depended for frequency stability upon
manual control, it would become a sorry
mess. One need but recall the days of the
regenerative receiver, with its heterodyning
carrier of but a tiny fraction of a watt.
Then imagine manually controlled broadcast transmitters with hundreds and thousands of watts power, trying to establish
zero beat with each other. The incident
again emphasizes the fact that the Commission is sadly in need of technical assistance which will help the members to grap-
ple more wisely with their problems. Any
competent engineer could have pointed out
the dangers of this ingenuous panacea.

What Readers Say About Broad-
casting Conditions

THE following are quotations from readers
of these editorials. George Madtes, Radio
Editor of the *Youngstown Vindicator*,
writes: "I have no doubt that the re-allocation
of frequencies has materially helped stations in
New York and Chicago, but it has not attained
the Commission's apparent goal—an arrangement which would permit listeners everywhere to
enjoy the stations nearest them. We are within
fifty miles of stations in Cleveland, Pittsburgh
and Akron and depend upon them for local
service. Our four main stations in these cities,
WTAM, KDKA, WCAE, and WADC are often heterodyned and WADC and WCAE are almost invariably
useless at night."

W. W. Muir of Lockport, New York writes:
"One cannot help but notice the difference between the stations which are operating in the
few wave bands on which there is only one station
and those operating on the frequencies on which
there are more than one station. The stations
which are operating on exclusive channels are
usually free from distortion, the signal being
strong and clear. The stations which are operating on wavelengths on which there are more than
one station show a decided tendency to be mushy
and weak, and have a wide variation in signal
strength from moment to moment....One cannot
help speculating what is apt to take place in the
future. We know that the American public have
had lots of things put over on them without
complaint. It is hard to believe that they will be
willing to stand for the huge joke that it is
possible to successfully operate more than one
powerful broadcasting station on a single frequency without serious interference."

Another correspondent writes from Wyoming
to the effect that KOA, Denver, is the principal
reliance for summer and winter reception of the
entire state. The Federal Radio Commission has
ordered that station to cut its power in half after
seven in the evening. Continuing, he writes: "Practically every strong station near the east coast is
located on the same frequency as some powerful
station on the west coast. While probably they
do not interfere in their home territory, the
heterodyne of the two completely ruins reception
in the Rocky Mountain region. Before the recent
changes, we could usually depend on WOC,

HOW VISITORS SEE THE ATWATER KENT FACTORY

lines. It describes an experimental relay of programs for Sidney as barely recognizable; a parallel attempt to relay Melbourne a few days later as a complete silence. It regrets that so much emphasis has been laid upon the possibilities of international broadcasting and points out that considerable development work is necessary before we can hope for regular and reliable international broadcasting. ꟷ ꟷ ꟷ E. T. SOMERSET writes us from Sussex, England, that he enjoys WGY, WJZ, WLW, WEAF and KDKA on their regular broadcasting channels, but American programs come in with much greater regularity on the high frequencies. 2XAZ, WGY's short-wave twin is the star performer, with KDKA on twenty-six meters and 2XAF following. He has also heard with great clarity, 2XAH, WRNY and WLW on its short wave, and ANH, Radio Malabar, Bendoeng, Java, on 17.4 meters and last, but not least, 2ME, Sidney, Australia. It gave him a particular thrill, he writes, to hear the clock striking four A. M. in Sidney, when it was still seven P. M., British summer time, of the previous night. Mr. Somerset advises American fans to listen for 5 GB, Daventry, England, on a frequency of 610 kc. with 30 kw. output, and Langenberg, Germany, with a 25 kw. output on a frequency of 640 kc. ꟷ ꟷ ꟷ SAM PICKARD, who first gained fame in radio circles as director of the Department of Agriculture Radio Service, has been made Federal Radio Commissioner to succeed Henry A. Bellows, recently resigned to resume the management of WCCO. The Commission loses Mr. Bellows because the gentlemen of the Congress failed to confirm his appointment. He was a useful and hardworking Commissioner. Mr. Pickard is qualified to serve on the Commission because of his familiarity with its problems as its former secretary. Carl H. Butman now becomes Secretary of the Commission. He is well known to the newspaper fraternity and may be helpful to the Commission, not only as an efficient secretary, but in advising it how to handle its relations with the press and the public. ꟷ ꟷ ꟷ "I HAVE come to the conclusion that it is not a practical or even a theoretical advantage to a broadcaster to sponsor a program through any small station. The companies that are marketing national products can use radio advertising to excellent advantage but for local companies to broadcast through a small local station is not good advertising, in my opinion. Their efforts are so mediocre in comparison with the programs sponsored by the big companies and transmitted through the high power of a well equipped, well operated station, that a bad impression is made and no benefit is derived." That is the statement, not of a newspaper publisher, but of Mr. Robert A. Fox, of Ashland, Ohio, who owned and operated station WLPC. Realizing that the small station serves little useful purpose, WLPC requested the Federal Radio Commission to cancel its license and its owner now states that he wishes "about two hundred more stations, now operating, would do the same.' ꟷ ꟷ ꟷ THE DEPARTMENT OF AGRICULTURE Farm Radio Service is being broadcast by eighty-nine radio stations in thirty-four states. Each of these stations will broadcast one or more of the eleven regular farm and household radio services prepared and released by the U. S. Department of Agriculture. Such services as these help to sell radio to the farmer.

NEWS OF THE PATENT FIELD

LEE DEFOREST won a victory over Edwin Armstrong in the United States Circuit Court of Appeals at Philadelphia, which decided that he is the inventor of the regenerative or feed-back circuit and the oscillating audion. Since the right to use both DeForest and Armstrong patents is included in the R. C. A. license, the decision does not affect the R. C. A's licensees particularly. Certain companies, however, operated under licenses granted by Armstrong before his patent was acquired by the Westinghouse Company, appear, through this decision, to be liable for royalties under the DeForest patent. There is a possibility that this case may now reach the Supreme Court, although that body has the power to refuse to consider the matter. ꟷ ꟷ ꟷ THE MACKAY interests announce that the DeForest victory places them on an equal footing with the Radio Corporation of America in the field of wireless communication. They will undertake immediate steps to establish short-wave wireless systems across the Pacific Ocean and throughout the United States. ꟷ ꟷ ꟷ PATENT No.1,639,042, recently issued to Wilford MacFadden of Philadelphia and assigned to the Atwater Kent Manufacturing Company, describes the use of a potentiometer for the stabilization of radio-frequency amplifiers. This system was used extensively before the neutrodyne system of stabilization was developed. ꟷ ꟷ ꟷ THE DUBILIER Condenser Company has notified a number of manufacturers of the scope of patents 1,635,117, 1,606,212, and 1,455,141, describing plate-current supply devices and power amplifiers. Included among prospective defendants under these patents are various Radio Corporation licensees. One of these patents describes a power system comprising rectifiers, filter and choke circuits, using a. c. on the filaments; another, a two-stage power amplifier with alternating current on the filaments and a C battery used to obtain grid bias; plate potential is obtained from a thermionic rectifier.

AMONG THE MANUFACTURERS

THE Sonora Phonograph Company, manufacturers and distributors of phonographs and radio sets, the Bidhamson Company, a patent holding corporation, organized by John Hays Hammond, Jr., Lewis Kausman and others, and the Premier Laboratories, headed by Miller Reese Hutchinson, have recently merged to form a corporation devoted to the manufacture of acoustic devices. ꟷ ꟷ ꟷ ARTHUR D. LORD, receiver in equity of the DeForest Radio Company, has filed a complaint with the Federal Trade Commission on Clause IX of the R. C. A. license contract. This clause specifically forbids R. C. A. licensees to equip and sell licensed radio sets without equipping them with R. C. A. or Cunningham tubes to make them initially operative. In his complaint, Mr. Lord claims that the consumer is penalized because he is forced to take a tube which otherwise might not be his choice. The clause is obviously aimed at independent tube manufacturers. He expresses the belief that this is an attempt at monopoly and restraint of trade, a direct violation of the Federal Trade Commission Act, the Clayton Act and the Sherman Anti Trust Law. ꟷ ꟷ ꟷ IN FULL page newspaper advertisements in the principal newspapers of the country, Mr. A. Atwater Kent announced a price reduction of twenty per cent. in his receiving sets. This reduction, says the announcement, is made possible by tremendous increase in production facilities. Particularly in the lower price classes, we may expect an era of intensive price competition with consequent advantages to the consumer. ꟷ ꟷ ꟷ POWEL CROSLEY, JR., has announced that his Bandbox model will probably not be changed for several years. This is the first time that a manufacturer has ventured such a prediction. ꟷ ꟷ ꟷ THE STEWART WARNER Speedometer Corporation, which has long defied the R. C. A. in patent matters, is the most recent addition to the ranks of those committed to a 7½ per cent. royalty.

A STATEMENT by Dr. J. H. Dellinger, calls attention to a general current misunderstanding regarding short-wave beam communication. The international short-wave beam links confine the radiated energy to a thirty-degree arc which is indeed not concentration in a single narrow path. It represents merely, Doctor Dellinger points out, an economic advantage and not a secrecy system.

Science has been unable to affect a concentration of radiated wave energy, either light, sound, or heat, in a perfect single beam by the aid of any form of reflector, and there seems little ground for hope that we shall soon achieve it with radio telegraphy or telephony. The concept that we may reduce beam transmission to a concentration comparable to that obtainable by wire communication is now untenable.

THE NEW COAST GUARD SHORT-WAVE TRANSMITTER

B. J. Fadden, chief radioman aboard the U. S. C. G. Modoc in ice patrol duty is shown standing beside the 35.5-meter (8500-kc.) transmitter. The transmitter on this wave is used for direct communication between the Modoc while in the North Atlantic ice fields and headquarters in Washington

MAKING FINAL ADJUSTMENTS ON THE PUSH-PULL AMPLIFIER DESCRIBED IN THIS ARTICLE
Measurements of the grid bias voltage are being made. Note the electro-dynamic Magnavox loud speaker in the background. A circular baffle board has been attached to it in the laboratory

PUSH-PULL AMPLIFICATION—WHY?

By HOWARD E. RHODES

THE essential prerequisites for faithful reproduction from a radio set are, first, a properly designed receiver capable of giving reasonably distortionless amplification and, secondly, a good loud speaker fed with power from a source able to supply the necessary energy without overloading. Much of the distortion in receivers is due to tube overloading, which usually occurs to the greatest extent in the last audio tube. The cure for this condition, obviously, is to use a tube, or combination of tubes, in the output circuit that has a high enough power rating so that overloading will not take place. As will be brought out in the following discussion, this requires that "power" tubes be used in the output circuit of the receiver, and at the end of the article some constructional details will be given regarding a push-pull amplifier employing 210 type tubes. Such an amplifier will deliver a large amount of power to a loud speaker without overloading.

Let us first determine approximately what requirements are necessary in the output of a receiver to prevent serious overloading. By the term "overloading," in this discussion, we mean that the input voltage on the grid of the tube is so great as to cause the grid to become positive at times so that current begins to flow regarding the grid circuit. In the operation of any ordinary amplifier, care must be taken that the signal input voltage is never great enough to cause grid current to flow. for, when this does occur, the

input signal will be badly distorted. In determining the characteristics of an amplifier to prevent overloading, we must assume certain values, with the result that the final answer will not be exact, but should nevertheless give a good idea of what conditions must be met. Suppose, to take an average case, that an orchestra is broadcasting and that the ratio of power between the fortissimo and pianissimo passages as played by the orchestra, is a million to one, corresponding to a power ratio of 60 TU. Because of the characteristics of the amplifier

A CLOSEUP
Showing the plug which provides for variations in line voltage

used to pick up this music, it is necessary to cut down this power ratio somewhat so as to keep the pianissimo passages above the noise level and to prevent the fortissimo passages from overloading the amplifier. In practice, this ratio is cut down in the control room at the broadcasting station by an operator in charge of the gain control. The power ratio is, after being cut down, generally about 40 TU into the amplifier system. This corresponds to a ratio of ten thousand to one. Let us assume that this ratio is maintained throughout the entire broadcasting and receiving system, a condition which will be true if there is no overloading at any point. Suppose that the energy in the pianissimo passages as they are reproduced by the loud speaker is 3 microwatts (0.000003 watts).

To get an idea of what this amount of energy represents, it may be compared to the average speech power of a person speaking, which is about 10 microwatts. The energy associated with the fortissimo passages will be 10,000 times as great, or 0.03 watt. It is now necessary to assume a figure for the average efficiency of the loud speaker, but because the efficiency of a loud speaker varies considerably over the range of audio frequencies, it is hardly accurate to assume an average efficiency and have it mean very much. We will do so in this case, however, merely to get some idea of how much power is required. The efficiency of a loud speaker is very low, we will assume it to be 3 per cent., which means that, in order to obtain a given amount of sound energy, we must supply the loud speaker with many times as much electrical energy. The amount of electrical energy required is found by dividing the sound energy output by the efficiency of the loud speaker; in this case we must divide 0.03 watt by 0.03 (3 per cent.) and the quotient, one watt, is the amount of energy the power tube in the receiver must be capable of delivering to the loud speaker during the fortissimo passages. Now let us see what tube or combination of tubes is capable of supplying this power.

The maximum amount of undistorted power that can be obtained from various tubes is given below.

TABLE NO. 1

TUBE TYPE	PLATE VOLTAGE	GRID VOLTAGE	UNDISTORTED OUTPUT WATTS
199	90	− 4.5	0.007
120	135	−22.5	0.110
201-A	90	−9.0	0.055
112	157	−10.5	0.195
171	180	−40.5	0.700
210	450	−48	1.700

When two tubes are used in a push-pull arrangement the maximum power output of the combination is twice that of a single tube.

It is evident from the table that the only tubes delivering, in push-pull arrangement, more than 1 watt of power are the 171 and 210 combinations, and, therefore, these combinations are most satisfactory for supplying a loud speaker with the necessary amount of undistorted power. In practice it will be found that a push-pull amplifier can be overloaded, but this amount of overload is so small as to be negligible.

This treatment of the problem is not exact. It was necessary to assume an average value for the power associated with the pianissimo passages and this first-assumption determines how much power will be required for the fortissimo passages. It is also true that a considerable amount of distortion can be present in the output of a loud speaker without being evident to most of us. The figures do, nevertheless, give an idea of why power tubes must be used, and show that present-day loud speakers cannot be supplied with sufficient undistorted power from tubes other than the 171 or 210 type. Marked improvement in the efficiency of loud speakers will some day make other tubes with a lower power output suitable for use in the last stage of a receiver, but until such an improvement is made, we must make certain that we have plenty of power handling capacity available in the receiver's output.

PUSH-PULL OR PARALLEL TUBES?

AT THIS point there might be some question regarding the relative merits of a push-pull amplifier with two 210 tubes and a parallel arrangement of the same tubes. Let us list the advantages and disadvantages of the two arrangements.

The italics indicate with which arrangement the advantage lies. Although point No. 4 was indicated as an advantage for the parallel arrangement, it is possible, by the use of a special push-pull output transformer, to compensate the higher plate impedance of the push-pull circuit, and the two arrangements will then be equal in this respect. Point No. 5 has not, as yet, been explained, but it is the most important reason for the existence of the push-pull arrangement:

PARALLEL ARRANGEMENT	PUSH-PULL ARRANGEMENT
(1.) *Requires only half as much input voltage from receiver to give same output as push-pull arrangement.*	Requires twice as much input voltage from receiver to give same output power as parallel arrangement.
(2.) Distortion due to overload quite noticeable.	*Slight overload (about 25 per cent.) possible without noticeable distortion.*
(3.) *Voltage gain somewhat higher.*	Voltage gain is somewhat lower.
(4.) Plate Impedance four times smaller than push-pull arrangement.	*Plate impedance four times greater than parallel arrangement.*
(5.) Distortion due to curvature of tube characteristic not eliminated.	*Distortion due to curvature of tube characteristic eliminated.*
(6.) Some hum may result if filaments are operated on a. c.	*Any a. c. hum from filaments eliminated due to push-pull arrangement*

We shall endeavor to explain now how the push-pull amplifier eliminates a certain type of distortion which exists in a simple single-tube amplifier. It is necessary to start the discussion by examining in some detail the characteristics of a cx-310 (ux-210) type tube (or, for that matter, any tube).

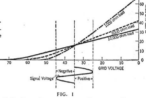

FIG. 1

Grid voltage - plate current curves of a 210 type tube

In Fig. 1 we have drawn several curves for a single tube of the 210 type with a plate voltage of 400 and a grid bias of − 35 volts, and these curves show the relation between the plate current and the grid voltage with various load resistances in the plate circuit. The curve marked 1000 was made with a 1000-ohm resistance in the plate circuit and the curves marked 5000 and 10,000 were made with resistances of 5000 and 10,000 ohms respectively in the plate circuits. These curves are dynamic characteristics in the sense that they indicate how the plate current will vary with different loads in the plate circuit. If a signal having a value of, for example, 10 volts, is impressed on the grid of this tube, it will cause the grid voltage to vary 10 volts either side of its average value of 35 volts. Such a signal voltage is represented in Fig. 1 by the curve, marked "signal voltage," drawn below the grid voltage axis. If the change in plate current due to this voltage is determined on the 10,000-ohm curve by reading the values of plate current at each extremity, we find that, when the voltage is positive, the current rises to 21 milliamperes and that, when the voltage is negative, the current decreases to 11 milliamperes, a drop of 10 milliamperes.

The signal voltage of 10 volts has, therefore, caused the plate current to increase and decrease an equal amount with respect to the average value, the increase and decrease being 5-milliamperes in this case. If the same measurements are made on the 5000-ohm curve we find that the plate current increases 7 milliamperes above the average value but only decreases 6 milliamperes. On the 1000-ohm curve the increase is 14 milliamperes and the decrease is only 11 milliamperes. These values have been arranged in the form of a table:

TABLE NO. 2

OUTPUT RESISTANCE	INCREASE IN CURRENT	DECREASE IN CURRENT
10,000	5	5
5,000	7	6
1,000	14	11

This table indicates clearly that, as the resistance in the output circuit of the tube decreases, the increase and decrease in plate current due to a given signal become unequal. This represents greater change in the plate current than does the negative side.

LOUD SPEAKER CONSIDERATIONS

AND now let us consider the loud speaker. The impedance of a loud speaker is a function of frequency and increases with increase in frequency. At low frequencies, therefore, the loud speaker will have a comparatively low impedance and the tube feeding the loud speaker will then operate on the characteristic corresponding to a low-resistance load in the plate circuit. This characteristic is indicated by the 1000-ohm curve in Fig. 1. At medium frequencies, where the loud speaker's impedance is higher, the tube will operate on a characteristic similar to the 5000-ohm curve, and at high frequencies the tube will operate on a characteristic similar to the 10,000-ohm curve. As indicated by the figures in table No. 2, the 10,000-ohm curve is quite straight and therefore produces little distortion. A small amount of distortion is produced by the 5000-ohm curve, but much greater distortion occurs when the tube operates along the 5000-ohm curve. When a loud speaker is operated from a single 210 tube, this distortion occurs and, if possible, it would evidently be of advantage to arrange the circuit so that no distortion of this type is produced. This leads us to consider push-pull amplification.

The circuit diagram of a push-pull amplifier is given in Fig. 2. In some push-pull arrangements the output choke, L, is replaced by a transformer, but the circuit will function with a simple choke coil as indicated. When a signal is induced in the secondary of the input transformer, T, the voltage relations are as indicated by the plus and minus signs on the diagram. It will be noted that the voltage at one end of the transformer is positive relative to the voltage at the other end, which is negative. The signal voltage impressed on either grid is one half the total voltage across the transformer. Since the two grids are at relatively opposite potential the plate current changes will also be opposite

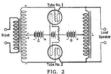

FIG. 2

The circuit connections for a push-pull amplifier

in nature. Referring to Fig. 1, this means that, during the time that the grid of tube No. 2 goes positive, the plate current will increase, and that the plate current of No. 2, as the grid goes negative, will decrease. In Fig. 3, we have represented at A the signal induced in the secondary of the input transformer, T, curve A-1 indicating how the voltage on the grid of tube No. 1 varies and curve A-2 indicating the variation of voltage on the grid of tube No. 2. It should be noted that the voltages are similar, that there is no distortion, and that the voltages are in opposite phase relation to each other (when one is positive the other is negative). Now these voltages cause changes in plate current in accordance with the curves given in Fig. 1, and if the particular signal being amplified is low in frequency, the loud speaker's impedance will be low and the tube's characteristic will have a form similar to the 1000-ohm curve. This curve will produce unequal changes in plate current (see table No. 2) and the curves at B-1 and B-2 in Fig. 3 indicate the change in plate current due to the voltages impressed on the grid. It should be noted that these two curves are distorted (the positive halves are larger than the negative halves) although the distortion of the two curves is similar in nature. These curves at B can be split into two parts, as indicated at C, C-1 and C-2 represent plate current variations exactly similar in form to the grid voltage variations and C-3 and C-4 represent additional plate current variations due to the curvature of the tube characteristic. The point of interest here is that, although the variations in plate current indicated by C-1 and C-2 are out of phase (as they should be) the distotted parts represented by C-3 and C-4 are in phase; that is, they are both positive or negative at the same time. In order to have current flow through the loud speaker, the a. c. voltage at one plate must be opposite in sign relative to the voltage at the other plate. We might consider that the plate whose voltage is negative tries to "pull" some current through the loud speaker while the plate whose voltage is positive tries to "push" some current through the loud speaker, and this gives us an idea of why such an amplifier is termed "push-pull." C-3 and C-4, indicating the distorted part of the plate cur-

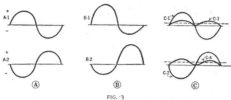

FIG. 3
These curves are used to explain how a push-pull amplifier operates

rent variation produced by the curvature of the tube characteristic, are such that both plates are relatively positive at the same time. These currents, therefore, cannot force any energy through the loud speaker. The only current flowing through the loud speaker is indicated by C-1 and C-2, and it is undistorted. In this way two tubes in a push-pull arrangement eliminate

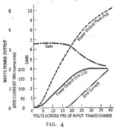

FIG. 4
What happens when an amplifier is overloaded can be determined from these curves. They are explained in the text, col. 3, page 204

a form of distortion present in a simple circuit using a single tube.

In Fig. 4 are given a group of curves obtained from some data on the Samson push-pull amplifier illustrated in this article. The three curves shown in solid lines were made using a single 210 type tube. Note how the gain begins to fall off when the voltage on the grid reaches about 18 volts and this point also corresponds approximately to the point at which grid current begins to flow. The power output also begins to flatten out after more than 18 volts is placed across the input. These three effects, a decrease in the gain, the presence of grid current, and a falling off in power output, are all definite indications of overloading. The dotted curve indicates the power output obtained from two 210's in a push-pull amplifier. This curve also begins to fall off slightly after about 18 volts has been placed on the input, but the change is not as rapid as in the case of a single tube. The power output of the push-pull amplifier at the point where grid current begins to flow is twice as great as that of a single tube.

The Samson push-pull amplifier illustrated in this article is an excellent example of a well-designed unit. The major characteristics of this amplifier are as follows:

(1.) The unit consists of two 210 type tubes in a push-pull arrangement fed from an input push-pull transformer. The unit is designed to connect to the output of the first audio stage in a receiver, and thereby makes possible the attainment of better quality than can be obtained from the smaller tubes ordinarily used in a receiver.

(2.) The power transformer and choke coils have been enclosed in a nicely finished metal case with the various leads brought out through a small terminal box at one end. At the other end of the transformer is a special plug, a Samson feature, which can be turned to different points to compensate differences in line voltage. The condenser bank is also enclosed in a metal case.

(3.) The device will supply B power to a receiver. The circuit incorporates a glow tube which maintains the output voltages from the various terminals practically constant independent of load, and this makes it possible to use the device with almost any receiver with assurance that the voltages marked on the terminals will be equal to the actual voltages delivered by the device. The following voltages are available; 180, 135, 90, 67½, and a variable tap so that accurate adjustment of the detector voltage can be made. The 210 type tubes receive about 500 volts and the C bias is about 40 volts. The device also supplies C potentials as follows: –4.5, –9. and –43.

The circuit diagram of this power amplifier is given in Fig. 5. The following parts were used in the amplifier illustrated in this article.

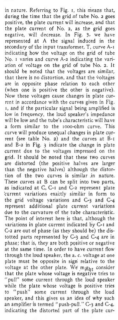

THE LAYOUT OF THE PUSH-PULL AMPLIFIER FROM ABOVE
No d. c. plate current can flow through the loud speaker, and blocking condensers are therefore not necessary in this amplifier for this purpose. However, in the arrangement shown the loud speaker itself is at a potential of 500 volts above ground and a serious shock will be had if the loud speaker and a grounded object are touched at the same time. In order to make the installation entirely safe it is a good idea to connect a 4-mfd. high-voltage condenser in series with each lead to the loud speaker terminals

LISTS OF PARTS

TL—Samson Power Block, Type 210. Containing Power Transformer and Two Filter Choke Coils	$ 37.00
Tobe Condenser Block, for Samson Power Amplifier, Containing the Eleven Necessary Condensers . .	38.00
R₁—Electrad 7200-Ohm Type C "Tru-Volt" Resistance, 50 Watts . . .	2.25
R₂—Electrad 420-Ohm Type C "Tru-Volt" Resistance, 25 Watts . . .	1.50
R₃—Electrad 50,000-Ohm Variable Resistance, Type T-500	3.50
R₄—Tobe 10,000-Ohm Veritas Resistance	1.10
T—Samson Input Push-Pull Transformer, Type Y . . . . }	
L₂—Samson Output Push-Pull Choke, Type Z }	19.50
12—Eby Binding Posts	1.80
4—Benjamin Sockets	3.00
2—RCA ux-210 (Cunningham cx-310) Tubes	18.00

1—RCA ux-874 (Cunningham cx-376) Tube	5.50
1—RCA ux-281 (Cunningham cx-381) Tube	7.50
TOTAL	$138.65

The circuit diagram has been marked with figures corresponding to the terminal markings on the power block and the condenser block. The arrangement of the apparatus on a single large baseboard makes it a simple matter to construct it and the circuit diagrams and photographs given herewith should supply all the necessary information.

All of the transformer cases, and also the case of the condenser block, should be connected to the negative B, as indicated in the schematic diagram, to prevent hum. This grounding can generally be most readily accomplished by running a lead from the negative B to the mounting screws of the various units. The wire can be fastened under these mounting screws. Remember that the voltages delivered by the transformer are very high and therefore care is necessary in making all of the connections. The 50,000-ohm resistance is the only variable control in the entire unit, and it is used to obtain accurate adjustment of the detector voltages.

When the construction has been completed, the special plug on the power input side of the Samson transformer may be inserted in the correct manner and connected to the a.c. light socket. When this is done, the tubes should light and the regulator tube should glow with a pinkish light.

In operation, the transformer block may become somewhat warm, but if it becomes so hot that the hand cannot be comfortably held on it, it indicates some error in the wiring. The input terminals of the device should be connected to the output of the first stage of the receiver and the loud speaker then be connected to the output of the amplifier.

If carefully constructed and properly operated, the unit will be found capable of giving excellent reproduction.

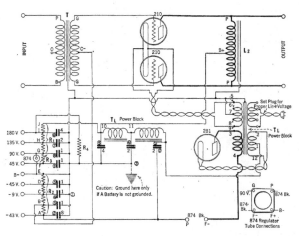

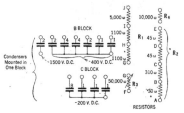

FIG. 5

This is a complete circuit diagram of the Samson push-pull power amplifier. The various numerals on the units correspond with the markings on the terminal blocks on the various parts; numbers in circles indicate terminals on the condenser block. This amplifier employs two 210 type tubes in the push-pull circuit, a type 281 tube as the rectifier, and an 874 regulator tube. The connections to the socket holding the latter tube are indicated in the diagram. The P and F minus terminals on the socket are connected in the filter circuit as shown at the point marked "874 B K." on the diagram. The two corresponding contacts on the tube are short-circuited inside of the tube base during the process of manufacture and, therefore, if the regulator tube is removed from its socket while the power is on, the circuit is automatically opened and damage to the filter condensers thus prevented. The G post on the socket connects to the + 90 voltage tap and the F + terminal connects to B minus

The PHONOGRAPH

TO JUDGE by the sale of radio-phonograph combination instruments and the popularity of the various electromagnetic ·pick-up units, the phonograph and the radio set in combination are climbing to high favor among radio users. If you have a good radio receiver and loud speaker, the purchase of a magnetic pick-up makes your old phonograph up to date and all the fine new electrical recordings then pour out of your loud speaker sounding as well as the best radio program. This magazine and others have contained descriptions on how to bring the old phonograph up to date using the electromagnetic pick-up. These pages on the entertainment that the radio-phonograph offers to the user are a regular feature of RADIO BROADCAST. It is not enough to know'that the combination of the radio and phonograph provides a flexibility of home entertainment that is astounding—we feel our readers would also like to know what the disks actually offer. These columns discuss only the records made by artists well known to broadcasting.
—THE EDITOR.

A New Use for Records

IN THE days when life was simpler, we thought of the phonograph merely as an instrument for the diversion of the multitude of home-loving folk, who were enabled by it to listen to jazz or classical music without budging an inch from their Franklin stoves. Then someone realized that by preserving on its records the speeches, music and other audible accompaniments of events of national importance the phonograph could be made to have a definite historical importance. And now the phonograph has had another burden laid upon it, that of delivering speeches for important individuals at gatherings at which they cannot themselves be present. This was actually done at the opening of the fifth annual convention of the American Institute of Steel Construction in Pinehurst, North Carolina, on October 25. Secretary of Commerce Herbert Hoover was asked to deliver the speech of wel-

A CLOSE-UP OF VICTOR'S NO. 955 "ELECTROLA"

This illustration shows the radio panel of this elaborate combination radio phonograph model. The radio panel can be tipped for convenience in operation as shown here, or it can be tilted upright. Pilot lights make tuning easy. This instrument, completely electrically operated, will play 12 records without stopping, contains built-in loop, power loud speaker (shown behind the grille above) and power supply. Price complete, $1550

come at this convention, but he could not take the time to make the trip there and back. The Institute, determined to have the speech, enlisted the services of the Victor Company, who made a record of a speech which Mr. Hoover delivered in his own office in Washington, and on October 25 this record was reproduced in Pinehurst before the convention. As Mr. Hoover himself remarked, one of the advantages of this method is that it puts a definite time limit on speeches. What a unique device for curtailing long-winded recitations! And what a splendid way to eliminate superfluous oratory, for as the Secretary also noted, and as all who have broadcast well know, a microphone is about as inspiring an audience as a bathroom door knob! We foresee a great future for this branch of phonograph service.

A Review of Recent Records

BUT though the phonograph now has a Mission it still continues to provide amusement and entertainment for those who want it. And, as far as we can see, the recent output of records is much the same as ever; there are many good records, a few poor ones, and a goodly supply of in-betweens. Of the latter the majority seem to be instrumental dance records. The orchestras which play for these are admirable and the recording of their playing is in most cases all it should be, but the selections on which they waste their talent are just about zero in musical worth. The result is like apple pie without the apple. Alas for more Gershwins and Berlins!

On the Victor, Columbia and Brunswick lists are many names familiar to those who sat at home turning the dials of their radio sets through the long winter evenings of 1926-27. They will recognize numerous regular performers and others who have made only a limited number of ethereal bows on such programs as Atwater Kent, Eveready, .and Victor. New names are being .added daily to this register of radio-broadcasters and now that Columbia has its aërial chain, we can expect even more.

Of the recent dance records, *Who Do You Love?* and *I'll Always Remember You* played by Paul Whiteman and His Orchestra (Victor) head the list. This famed outfit have taken two of the best songs now extant and by decorating them with a trick orchestration in the inimitable Whiteman manner have made an unusually good record out of them.

Is It Possible? and *Just Call On Me* played by Leo Reisman and his Orchestra (Columbia) is another grand record. If we hadn't always lived in the belief that Whiteman had no equal, we would say this was as good as the first record on our list. We will say it. Anyway, it is pretty smooth music and we defy you to keep your feet still when you listen to either number.

Gorgeous by Johnny Hamp's Kentucky Serenaders would be an asset to any collection of dance records and its companion on the opposite side, *There's a Trick in Pickin' a Chick-Chick-*

Chicken by Nat Shilkret and His Vic-Vic-Victor Orchestra is just as full of pep. (Victor).

Habitual listeners-in on Harold Leonard and his Waldorf Astoria Orchestra will want his latest Columbia product, *Just A Memory* and *Joy Bells*. You don't have to be told it is good.

Once Again and *No Wonder I'm Happy* as played by Ernie Golden and his Hotel McAlpin Orchestra will have the ring of familiarity to those who tune-in on station WMCA. They are good snappy numbers. (Brunswick.)

Having heard excellent reports of the new Broadway show "Good News" we were disappointed in the three numbers from it which have found their way onto the rubber discs. The title number, *Good News*, and *Lucky in Love* have been recorded for Columbia by Fred Rich and his Hotel Astor Orchestra. and Cass Hagan and his Park Central Hotel Orchestra have done *The Varsity Drag* (also Columbia). If you are one of those who raved over the show you may enjoy the records. You may also like *Dancing Tambourine* by the Radiolites on the reverse side of *The Varsity Drag*, though we can't enthuse over it.

If you stuff cotton in your ears during the seconds devoted to the vocal chorus in *Baby Feet Go Pitter Patter* you may agree with us that this record by Abe Lyman's California Orchestra (Brunswick) is one of the best that have appeared in many moons. In addition to an aversion to vocal choruses in general we detest the words to this particular song. It was an error on someone's part to attach such a·silly lyric to such an excellent tune. However, it's short and the few seconds one sacrifices to get through it are a drop in the bucket and the rest of the record is fine. The other side carries *There's One Little Girl Who Loves Me*, also played by Abe Lyman, and also good.

Another of the better records slightly marred by a vocal chorus is *No Wonder I'm Happy* and *Sing Me a Baby Song* by the George H. Green Trio with Vaughn De Leath doing the vocalizing. (Columbia.)

The Ipana Troubadours, smile vendors under the direction of S. C. Lanin, have done a fine job with *Are You Happy?* and *A Night in June*, of which Frank Harris carols the chorus. (Columbia.)

Inhabitants of Mayor Thompson's Strictly American City will welcome a Columbia disk made by one of Chicago's Municipal Heroes, Paul Ash, and his Orchestra. *Just Once Again* is excellent and its vocal chorus by Franklin Baur makes us eat some of our words just uttered; we must admit it is an addition to the

THE UTICA JUBILEE SINGERS

There are those who are not especially impressed by negro spirituals. But if you like these interesting melodies and want to hear them sung as they should be sung, listen to this group on W2I and associated stations at 9:45 eastern time Sunday evenings. They have made one double-faced recording for Victor, one of the finest recordings of the kind we have heard

Joins the RADIO *Set*

record. In self-justification we insist that Franklin Baur is an exception. The reverse of the record *Love and Kisses*, is not quite up to the Paul Ash standard but that doesn't mean it is bad.

Several other records get only half a vote due to the fact that one face of record is good and the other not. *Stop, Go!* executed by Nat Shilkret and the Victor Orchestra has a unique rhythm, better than any other number on the list but *Something To Tell* is only moderate. *Me and My Shadow* by Phil Ohman and Victor Arden and their Orchestra is good; *Broken Hearted* is not. Even Paul Whiteman could not do much better with that last number but *Collette* on the reverse makes the record decidedly worth buying. (Victor). Don Voorhees tried something tricky with *Soliloquy* (which we are told belongs to the new school of music ushered in by the *Rhapsody in Blue*) and was not very successful, but his more orthodox *My Blue Heaven* is exceedingly good. *Just a Little Cuter* falls rather flat but *Marianette* is excellent dance music. Both come from the orchestra under the baton of Ben Selvin and are recorded by Brunswick.

We don't care much for *Cheerie Beerie Be* or *Waters of the Perkiomen* even though played by Leo Reisman and his Orchestra. *Roodles and I Ain't Got Nobody* are not as good as they ought to be coming as they do from Coon-Sanders Orchestra. And we were very much disappointed in *Who's That Pretty Baby?* and *Barbara* by Paul Specht and his Orchestra. The fault lies in each case not with the orchestras but with the stupid selections they play.

AND NOW FOR COMEDY

OF THE good humorous records recently issued by far the best is *Two Black Crows*, Parts 3 and 4. (Columbia). Of course we don't need to describe it. Everyone knows Moran and Mack and their riotously funny dialogues. What, you dont? Well, go right down to the corner music store and buy this record and their first one, *Two-Black Crows, Parts 1 and 2*. You will remember us in your prayers.

Next in order of importance comes that grand song-perpetrated by the Happiness Boys, Billy Jones and Ernest Hare, *Since Henry Ford Apologized To Me*. (Victor.) It is worth seventy-five cents just to hear this record once, which is fortunate, for we wouldn't give a nickel for the song on the reverse side, *I Walked Back From The Buggy Ride*, by Vaughn De Leath and Frank Harris.

Then the famous Sam and Henry combination (Correll and Gosden of WGN) offer two dialogues called *Sam's Big Night* and *The Morning After*. (Victor.) They are both labelled comic dialogues

but oh! the pathos of Sam's refrain, "Henry, Henry, I'se sick! My head's bout to kill me! . . . I b'lieve I'se gonna die!" It almost makes you want to sign the pledge.

Of the many popular vocal records we nominate for first place, a work of art by the Happiness Boys, *You Dont Like It—Not Much* and *Oh Ja Ja*. (Victor.) This is typical of what they offer to the audience of WEAF every Friday night and it's good! Personally we like them best of all the regular aërial performers. It's personality that does it—plus good voices.

Van and Schenck sing *Magnolia* and *Pastafazoola* for you on a Columbia record. The Radio Franks present *No Wonder I'm Happy* and *When Day Is Done*. (Brunswick). And Johnny Marvin and Ed Smalle do a little duet with *Just Another Day Wasted Away* on a Victor. All of these are good. But on the back of the last mentioned disk is *Just Like a Butterfly* sung by Franklyn Baur. Personally, as little Alice said in *Peggy Ann*, we'd rather have a baked apple. No matter how good the voice, the song is terrible! We feel just as strongly about *Baby Feet Go Pitter Patter* as we have hinted before, and when it is sung by Vaughn De Leath we see spots before our eyes. She could make *Turkey in the Straw* sound like a sentimental ditty and when she has something as mawkish as this to start with . . . words fail us. In case you are still interested in the record the opposite side bears another song by the same lady, *Sometimes I'm Happy* (Brunswick).

The Sunflower Girl of WRAP vocalizes *You Went Away Too Far* and *I Hold The World in The Palm of My Hand* on a Columbia disk. We cannot say she sings them because she has one of those rough and tumble shouts so often heard on the vaudeville stage. It is about as far from being musical as anything could be, but for that sort of thing, she will do.

We don't feel strongly one way or another about *Charmaine!* and *The Far-Away Bells* sung by Franklyn Baur (Columbia), *Ain't That a Grand and Glorious Feeling?* and *Magnolia*, sung by Harry Richman (Brunswick), or *Flutter By, Butterfly* and *I'd Walk a Million Miles* by Art Gillham and his Southland Syncopators (Columbia).

RED SEAL RECORDS AND SUCH

TURNING now to the sublime we find several red seal records made by operatic stars from the Metropolitan firmament, who twinkled before microphones upon occasion last winter, Lucrezia Bori, the beautiful Spanish soprano who is as lovely to look upon as she is to listen to (though you couldn't tell that when you heard her sing in the Victor radio concerts or in the Atwater Kent hour) presents us with the lovely old waltz song by Pestalozza, *Ciribiribin*, and *Il Bacio* by Ardit. (Victor).

Another soprano who was presented to the radio audience by Atwater Kent is Hulda Lashanska In company with Paul Reimers, tenor, she records for Victor two simple but extremely lovely old German folk songs, *Du, Du Liegst Mir Im Herzen* and *Ach, Wie Ist's Möglich Dann*. The harmony of the first selection is particularly notable.

It was a Victor hour that launched Emilio De Gogorza's baritone voice on the ether waves and it is a Victor record which presents his voice again for your permanent enjoyment. He sings two favorites, *O Sole Mio*, and *Santa Lucia*. We think it would be a grand idea for everybody to have this record in his home, if for no other reason than just so that whenever he hears the song murdered by a would-be artist he can play the record on the phonograph and reassure himself that the song is all right after all.

We presume it is rank heresy to say that we prefer to hear the Utica Jubilee Singers sing *Old Black Joe* than to hear Lawrence Tibbett. Tibbett's voice is marvelous, of course, and it is perfectly trained, but he cannot manage this negro melody as expressively as the Utica Jubilee Singers do. If this be treason. . . . The reverse of the record is *Uncle Ned*. We have never heard the Jubileers sing this but we will wager our two cents that they could do it more to our satisfaction than the Metropolitan star has done. Both these songs are so worn out that they need all the expression that can be put into them.

The Utica Jubilee Singers *have* done a record for Victor, *Angels Watching Over Me* and *Climbin' Up the Mountain*. It's perfect! We can say no more. Incidentally it is interesting to note that these singers have just returned from a concert tour of Europe where they were greeted with great acclaim, so their popularity is no longer limited to this continent. They are now heard from WJZ and others on that chain Sunday nights at 9:45, eastern time.

Virginia Rea, staff artist for Eveready, has recorded a popular number which Victor thought good enough for a red seal record, *Indian Love Call* from Rose-Marie; and Lambert Murphy, whom you undoubtedly heard in an Atwater Kent hour, has recorded the title number from the same musical play on the reverse of the record. Though we saw this show at regular intervals through the winter of 1924-25 and heard these numbers on hand organs for the next two winters, we *still* like them.

AN ELECTRIC RADIO-PHONOGRAPH FROM FRESHMAN

This instrument, completely a, c. operated provides the usual Freshman receiver, electric turntable, electric pick-up, record space and loud speaker which serves for radio or phonograph music. Complete with a. c. tubes, $350

ERNIE GOLDEN OF WMCA

He leads the Hotel McAlpin Orchestra, regularly heard through WMCA of New York. The Hotel McAlpin Orchestra has recorded many good dance numbers for Brunswick

THE SCREENED GRID TUBE

By KEITH HENNEY

Director of the Laboratory

STUDENTS of the characteristics and applications of the vacuum tube, and of circuits to do with it, may find themselves somewhat bewildered by the apparent complexity of the screened grid tube when they begin their researches into its idiosyncrasies. They will be impressed at once with the thought that this tube is no ordinary structure, and will appreciate more such names as Schottky of Germany and Hull of the United States—names actively associated in its development. Experimenters here have yet to become familiar with the new tube, which has already taken its share of space in English, French, and German radio periodicals, and it is certain that, with the Radio Corporation's announcement of the UX-222, not a

great length of time will elapse before it will be possible for anyone to obtain these interesting and useful screened grid tubes.

American writers, too, will have considerable to say about the double-grid tube, of which the screened grid tube is a type, as they become more familiar with its operation, and as its possibilities become more apparent.

Imagine a tube with an amplification factor of about 250, and such a small grid-to-plate capacity that it has little tendency to oscillate even though the plate circuit is made highly inductive in reactance. The nearest approach is the standard "high mu" tube with an amplification factor of about 30, but, unfortunately, with sufficient capacity to make oscillation inevitable if sufficient inductance is included in its plate circuit to transfer energy efficiently to a subsequent circuit. The screened grid tube, of which the UX-222 (CX-322) is the precursor, is a most unusual tube. What are its characteristics, its possibilities, its weak points?

Physically it is complicated by having a fourth element within the glass bulb. Picture a cylindrical construction with a 3.3-volt filament in the center, surrounded by a rather coarse grid, then, at some distance, another fine mesh grid, then a cylindrical plate, and the whole surrounded by another very fine grid of about forty spiral turns. The latter two grids, connected in parallel, form one electrode, thus constituting the extra element in the tube. It would be more accurate, for geometrical reasons, to call this tube a screened plate tube, but electrically, as we shall see, it is really the grid which is protected from alternating voltages impressed upon the plate.

The inner coarse grid is connected to a small metal cap which sits on top of the bulb, where the tip used to be, making the overall height about three quarters of an inch higher than standard tipless tubes. The screening grid connects to the usual grid terminal in the base so that the tube fits into the standard UX or UV socket.

As pointed out in the December, 1927, RADIO BROADCAST by Mr. T. H. Nakken, there are two types of double-grid tubes, those in which the inner grid is positive, known as the space-charge tube, and those in which the outer or protective grid is positive. The UX-222 may be used either

way. Let us consider first its action as a shielded-grid device. In this case the outer grid is positive.

We shall place 3.3 volts on the filament, make the inner grid negative, place about 45 volts on the shield, and read the plate current as the plate voltage is changed. The result is shown in Figs. 1 and 2.

When this is done there are unusual results: First, the plate current rises, as is customary with increase in plate voltage. Then the plate current begins to decrease, giving the tube a characteristic like that of the electric arc, i.e., decreasing current with increased voltage; next, after a sharp minimum, the current rises almost per-

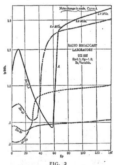

FIG. 1

The plate current of a screened grid tube varies in strange fashion, as these curves show. In these data the effect upon the plate current of changing the plate voltage, with several values of bias voltage, is given. The normal bias is about 1.5 volts, with 45 volts positive on the screen, and about 135 on the plate

FIG. 2

If any reader of RADIO BROADCAST feels that he knows all there is to know about tubes, let him explain the sudden and extensive changes in plate current with change in plate voltage indicated in these graphs. With even greater screen voltages than are shown, here, the plate current may be reduced to zero or even go negative (reverse its direction of flow) at some positive plate voltage

pendicularly and finally flattens out to become practically horizontal. All of this is contrary to what happens in standard tube practice and, to the student of physical phenomena, is extremely interesting.

The slope of this plate voltage-plate current curve represents the plate impedance and, if plotted, an extraordinarily large scale graph would be required owing to the extent to which it changes. For example, in Fig. 1 it is 74,000 ohms near the origin, then it suddenly goes negative to the extent of 100,000 ohms, then positive about 10,000 ohms, and finally becomes about three-quarters of a megohm in value! The tube has a negative resistance, or a dynatron effect, at low plate voltages.

These rapid and extensive changes in internal resistance are due to the varying proportions of current taken by the shield and the plate, both of which attract negative electrons, and to a certain amount of secondary emission which takes place within the tube. At the present moment, however, the detailed explanation of these effects must give way to the more practical information regarding the tube. We are more interested in this article in how the tube works than in "why." It is sufficient to state that the sum of the currents taken by the shield and the plate is constant, the plate current increasing when the shield takes fewer electrons, and vice versa. Under usual operating conditions, i.e., high plate voltages, the shield takes very little current indeed.

Grid voltage-plate current curves appeared on page 111 of RADIO BROADCAST for December, 1927, and will not be repeated here. They conform to what one secures from other tubes of the general-purpose type. They indicate a mutual conductance of about 300 to 400 under average conditions, and an amplification factor of about 250 to 300, values which should be compared to those of standard tubes in Table 1. It is difficult to measure these factors on the ordinary bridge because of the extraordinarily high values of mu and plate impedance involved, and the better plan is to pick them from characteristic curves as we have done here.

MATHEMATICS OF THE TUBE

THE screened grid tube is designed primarily for radio-frequency amplification and, to understand its possibilities in amplifier circuits, we must examine somewhat more critically than usual the processes involved in the ordinary amplifier. Naturally we must have an input and output circuit, and for general analytical purposes we shall consider Fig. 3.

The purpose of the transformer in such circuits is not, as many would have us believe, to increase the voltage step-up from tube to tube, but to obtain a proper impedance for the amplifier plate circuit to look into. Mathematics will show that the maximum voltage amplification will be obtained when the effective primary impedance of the transformer is equal to the internal resistance of the tube, and that under these conditions this amplification is:

$$K_{max} = \frac{L\omega}{2\sqrt{R_p\,R_s}}$$

where R_p = tube impedance, L = secondary inductance, R_s = secondary resistance, and ω = 6.28 x frequency.

When the effective resistance of the tuned circuit at resonance is higher than that of the preceding tube impedance, a step-down transformer must be used. This effective resistance may be found mathematically by substituting the proper values in the following expression:

$$R_0 = \frac{L^2\omega^2}{R_s}$$

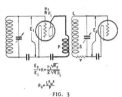

FIG. 3

A diagrammatic representation of the ordinary interstage radio-frequency amplifier, consisting of a transformer, tuned to the frequency desired, connecting two tubes. The voltage gain at resonance is given in the form of two equations

$$\frac{E_2}{E_1} = K = \frac{\mu_1\sqrt{R_2}}{2\sqrt{R_1}R_{s_2}}$$

$$R_2 = \frac{L^2\omega^2}{R}$$

where L = inductance, R_s = high-frequency resistance, W = 6.28 x frequency.

If we use an inductance of 250 microhenries having a resistance of 15 ohms at 1000 kc., this effective resistance will be:

$$R_0 = \frac{(250 \times 10^{-6})^2 \times (6.28 \times 10^6)^2}{15} = 177,000 \text{ Ohms}$$

and if the previous tube is a 201-A with an impedance of 12,000 ohms we shall be compelled to use a step-down transformer to secure maximum amplification and to prevent short circuiting the secondary, to the impairment of the selectivity. Using a tube and coil of these

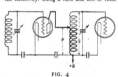

FIG. 4

For analytical purposes the two-winding transformer of Fig. 3 may be replaced by this auto transformer. The same conditions for maximum voltage amplification obtain

characteristics, and with the proper primary, the maximum amplication will be:

$$K_{max} = \frac{8}{2} \times \frac{250 \times 10^{-6} \times 6.28 \times 10^6}{\sqrt{12,000 \times 15}} = 15.0 \text{ (Approx.)}$$

When any receiver designer states that he gets a uniform amplification per stage of much over this, he has neither used his mathematics nor

FIG. 5

In the ordinary tube, some electrostatic lines of force connect the plate and grid because they are at different potential. This means, simply, that some capacity exists between them and it is this capacity that causes trouble in the usual high-frequency amplifier

his vacuum-tube voltmeter to substantiate his statement.

This transformer, with its two windings, can be replaced by an auto transformer for all practical purposes, as shown in Fig. 4. An auto transformer, it will be remembered, is used in the "Universal" receiver previously described in this magazine and in the R. B. "Lab." Circuit, when the plate or primary coil is reversed. The impedance of the circuit, looked at from the preceding tube, must equal the impedance of that tube, and the position of the tap regulates the effective transformation ratio, so that this condition is realized, since the ratio of impedances across secondary and primary is equal to the turns ratio.

If we use a special radio-frequency tube with a higher amplification factor and higher impedance, such as the Ceco Type K, we must move the tap higher toward the grid end, or use more primary turns if we use a transformer, while if a 112 type tube is used with its lower impedance, the tap can be brought further down. Table 1 gives essential data on existing tubes. The approximate $\mu_1\mu_2$ns ratio, in the auto transformer case, can be found by substituting the value of plate impedance in the following equation:

$$\text{Turns Ratio} = \frac{60000}{\sqrt{R_p}}$$

TABLE 1

TUBE	μ	R_p	G_m	TURNS RATIO
199	6.25	16,600	380	1.90
201-A	8.00	12,000	675	2.25
112	8.00	5,000	1,600	3.5
210	7.70	5,000	1,540	3.5
171	3.00	2,000	1,500	5.5
222	250.00	700,000	400	1.0
"K"	13.00	16,800	780	1.9

Now all of this sounds simple to carry out but, practically, there are difficulties ahead—most of them due to the fact that the tube does not act like a one-way street. Some traffic always goes in the opposite direction, because of the grid-plate capacity. As soon as we get the tap on our auto transformer moved high enough toward the grid end to secure maximum amplification, we include sufficient inductance in the plate circuit of the amplifier to make it oscillate, and trouble begins. Therefore we must do one of two things: we must either move the tap down, and lose amplification because our equal impedance condition is no longer satisfied, or we must play neutralization tricks on the amplifier to keep it from oscillating, with perhaps slight loss in amplification as the price of stability.

Here is where the screened grid comes in. Suppose, as in Fig. 5, we have the plate receiving electrons from the filament after passing through the grid in straight lines. Because of the fact that the grid and plate are at different potential there will be electrostatic lines between them, represented by the curved lines. In other words, there is some connection between the plate and the grid, other than that produced by the passage of negative electrons. Now if we surround the plate by a fine grid which is grounded, as shown in Fig. 6, these electrostatic lines do not reach the grid, and the latter is free to function only as a control on the flow of electrons. If, in addition, we make this shield positive with respect to the filament and grid, we neutralize some of the space charge which, in turn, boosts the amplification factor to a very high degree.

If the plate is completely screened, the tube will be a one-way repeater, there will be no tendency to oscillate in the familiar tuned grid-tuned plate circuit, and a little mathematics will show that the amplification is a factor of the mutual conductance of the tube and the external impedance. This external impedance is the effective resistance of the tuned circuit, as already mentioned, and varies as shown below for average coils at usual frequencies. The possible voltage amplification may be easily calculated, with an assumed mutual conductance of 400 micromhos and an infinite plate impedance.

FREQUENCY, KILOCYCLES	$R_0 = \dfrac{L^2\omega^2 \text{ OHMS}}{R_s}$	AMPLIFICATION Gm x Ro
100	400,000	200
1,000	100,000	40
10,000	10,000	4

These values of amplification are considerably greater than is possible with standard tubes such as we all use at the present time. At 1000 kc. (300 meters) the average gain in modern receivers may be as high as 10, and not many sets can do as well without some loss of stability.

Actually, however, these values in the table with the 222 tube will not be attained, since the assumptions on which they were calculated, infinite plate impedance, no grid-plate capacity, and an effective resistance in the plate circuit of 100,000 ohms, are not realized. Since the tube's internal impedance is of the order of a half megohm, or greater, it is considerably more than can be attained by average coils or by coils which will not cut side bands, there is no use in

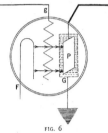

FIG. 6

If a grounded shield is placed around the plate, the lines of force from grid to plate will be interrupted, and fewer changes of plate voltage will affect the grid—in other words, the grid is shielded from the plate

above. If this resistance is equal to that of the tube, approximately one-half the mu of the tube may be realized. With 100,000 ohms in the plate circuit approximately one seventh of the amplification factor, or about 35 may be expected. For maximum voltage gain the effective resist-

amplifier, which had attached to its input a single wire antenna about 35 feet long, and a ground.

With the brass box containing a Rice neutralized amplifier, using a 201-A tube, and with the best position of the plate tap on the detector coil (see Fig. 8), the voltage upon the input to the detector was measured. Then the screened grid tube was used, the whole coil being used in its plate circuit, and the neutralization apparatus was done away with. The voltage was again measured with exactly the same input, and at the same frequency—500 kc. In this case the output voltage was a little over three times the best that could be obtained with the 201-A circuit. Resonance curves showed that the two circuits were about equally selective.

If the 201-A tube gave an amplification of ten, which is reasonable, the new tube had an indicated voltage gain of over 30, which seems to fit in with our calculations explained previously. Two stages would give a gain of 900 compared with 100 for two 201-A amplifiers, or approximately 20 TU, which has about the same effect as adding one stage of audio to existing receivers.

With the antenna described, and with but two tuning circuits, there was no difficulty in separating WEAF and WJZ, 50 kc. apart, when the former was 8 miles away with 50 kilowatts of power in the antenna and the latter was roughly 30 miles distant with somewhat less power. Measurements showed that, with the screened grid tube, WEAF delivered over 4.5 volts to the detector, sadly overloading it.

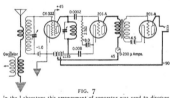

FIG. 7

In the Laboratory this arrangement of apparatus was used to discover the voltage delivered to a detector when the screened grid tube was used as an amplifier. The meter in the detector plate circuit read the change in plate current when a. c. voltages were applied to its grid. It was a calibrated detector, or vacuum-tube, voltmeter

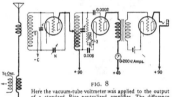

FIG. 8

Here the vacuum-tube voltmeter was applied to the output of a standard Rice neutralized amplifier. The difference between the input voltage in this case and that in Fig. 7 determines the gain in using the screened tube. The meter used in the calibrated detector was a Westinghouse 200-microampere meter, type PX

using a step-down transformer to couple the output of the tube to a succeeding stage, and the whole coil may be used without danger from oscillation.

With the entire coil in the plate circuit, as shown in Fig. 7, the amplification per stage may be figured by somewhat simpler mathematics. In this case we have a simple tuned impedance in the plate circuit of the tube, which at resonance has an effective resistance, as explained

ance must be high, that is, we must use exceptionally good coils of high inductance and low resistance with the probability that fidelity will suffer.

In the Laboratory a simple set-up was employed to examine the tube's behavior. The circuit is shown in Fig. 7 and, as may be seen, consisted of a single amplifier tube followed by a non-regenerative C bias detector which could be calibrated in volts input against change in d. c. plate current.

The shielded tube and its accessory input apparatus was carefully enclosed in a tight brass box which was grounded. The audio amplifier was useful as a kind of monitoring stage to follow what was happening in the preceding circuits. It made it possible to insure against the amplifier oscillating, etc. A 1000-cycle modulated signal was induced into the

The tube may also be used as a space charge grid affair, where the inner grid is positive with respect to the filament and the fine structure which ordinarily screens the plate is used as the signal grid. To test its capabilities the circuit shown in Fig. 9 was used. With an input voltage of 0.1, the gain was about 30, from 60 to 30,000 cycles, and then fell slowly. At the higher frequencies the tube capacities (in the space charge grid tube the capacities are much greater than in the screened grid connection) shunt the output resistance, with consequent loss of amplification.

Here, then, is a tube which must henceforth be carefully considered by all designers of radio equipment, for amplification of the very high frequencies, those of intermediate range such as are used in super-heterodyne receivers and broadcast frequencies, and again at the low audible tones. In all of these ranges it seems probable that greater amplification will be secured than has been possible with ordinary or high-mu tubes. In subsequent articles, we shall discuss the design problems at greater detail and have something to say about the problem of selectivity and fidelity of reproduction.

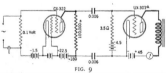

FIG. 9

When the grids of the UX-222 (CX-322) are reversed; i.e., the inner grid is made positive, the amplification factor and plate impedance fall to more usual values, and the tube can be used as an audio- or intermediate-frequency amplifier

WHAT

SET SHALL

I BUY?

By

EDGAR H. FELIX

radio market for a long time to come. Magnetically controlled trickle chargers, which turn on automatically when the set switch is turned off, reduce the maintenance responsibility placed on the user of a storage battery to a minimum. Precautions against overcharging, and occasional filling of the cells with distilled water, are the only attentions required. Indeed, this type of storage battery receiver is often sold to the uninitiated buyer as the last word in socket power operation.

Large storage batteries, used almost exclusively with receiving sets four and five years ago, are likely to require replacement after two to five seasons, the period of service depending upon the care taken of the storage battery by its owner, and the quality of the battery itself. Bear in mind, when such replacement becomes necessary, that the reliability of many an old receiver would be greatly improved by the purchase of a trickle charger outfit, a suggestion which may be of value to many persons desiring to give a Christmas gift costing twenty to thirty dollars.

A device of interest to every storage-battery set owner is the "Abox," which may be substituted for the storage battery or trickle charger-storage battery combination. The "Abox" requires no maintenance attention and operates on an entirely different principle from the storage battery-trickle charger combination. Other manufacturers are marketing devices similar in principle to the "Abox." Balkite might be cited as an example.

A comprehensive article on A-power units, in which twenty or so A units were listed, and their characteristics noted, appeared in the November, 1927, RADIO BROADCAST, beginning on page 30 of that issue.

TUBE POWER UNITS

RECEIVING sets having A, B, and C power supplied from vacuum- or gas-tube rectifier units incorporated in the set, require no maintenance other than renewal of rectifier tubes. There is no periodic adding of distilled water to battery or chemical charger, nor is there any other maintenance problem in connection with

WITHOUT repeating the old arguments which earn for radio the title "the ideal Christmas gift," we are certain that more people select a new radio at this season than at any other. The fundamentals of good selection do not change from year to year, but the improvements which come with each new season always bring with them new factors to consider when purchasing. Indeed, the constant improvement of the radio receiver leads some to await new developments, expecting some sort of radio millennium.

Improvement will always continue in the radio art, and he who awaits perfection will neither purchase nor enjoy radio. Were the same policy followed with respect to the purchase of motor cars, some forty million people would still be walking because the ultimate automobile, after a quarter of a century has been devoted to development in this field, is not yet here.

The radio receiver of to-day is a product which, both from the musical and technical standpoint, is capable of many years of service. It will not be greatly outclassed in the musical quality of its output for a long time. In sensitiveness, selectivity, fidelity of reproduction, simplicity of control, and convenience of maintenance, it has reached high standards. In appearance, efficiency, compactness, simplicity of installation, and automatic operation, considerable progress may still be looked forward to, but none of these factors mean great changes in the fundamental output of the radio receiver, that is, reproduced musical programs. We have passed through the period of revolution and have come to the era of refinement in this radio world of ours. There is no longer any excuse for delay in

purchasing a high-grade manufactured receiver.

Maintenance convenience is the keynote of 1927's advances. The last word in installation instructions now is: ' Plug the cord in the light socket, turn the set switch, and tune." This is indeed a contrast to the requirements of former years. There were antennas to install, batteries to connect with multi-colored cables, chargers to wire, power amplifiers to plug in, and loud speakers to attach. Between these two extremes, there are several stages of convenience, each of which is still represented in this year's products. The term "light socket operation," for example, is applied to power systems varying considerably in convenience, and is a very flexible term. It is advisable, therefore, to ascertain exactly what type of receiver and what degree of convenience you buy.

There are several ways of powering a receiving set "directly from the light mains." Filament power might be supplied from a storage battery-trickle charger unit and plate power from a rectifier unit operating directly from the power mains; or both filament and plate power may come directly from the mains through a rectifier unit; in another case the filament power may be delivered directly from the line to a set using alternating-current tubes, through a step-down transformer, and plate power obtained through a rectifier unit.

The field of usefulness for the storage battery receiver is far from exhausted. Its lower manufacturing costs give it a price advantage, at no sacrifice in attainable performance, over the alternating-current powered receiver. Because of lower cost, it is destined to remain in the

211

these sets. There are a large number of receivers of this type on the market. With skillful engineering and high-grade components, they offer care-free and high-quality reception. If carelessly designed, they may give a marred output because of excessive hum, and unreliability of service due to failures in vital parts. It should not be thought, however, that a set is necessarily good because it does not hum. If poor audio transformers are incorporated they may not be capable of amplifying the low-frequency hum produced by alternating current. The problem of the uninitiated in distinguishing between the inferior and the superior type is perplexing. The name and reputation of the manufacturer, the endorsement of men technically qualified to judge radio products, and the pages of high-grade publications, which censor and check the statements made in their advertising columns, are helpful sources of information and guidance.

Practical tests can be made by the technically untutored buyer, when a set is being demonstrated, which will protect him against the purchase of a power set of inferior design. The principal characteristic of a poorly designed receiver, deriving its A, B, and C potentials directly from the light mains, including both those employing rectifier systems and those using a. c. tubes, is the excessive hum experienced when the set is adjusted to sensitive reception.

A dealer, selling a radio set subject to hum, is likely to concentrate his demonstration upon the reception of strong, near-by stations. Ask him to tune-in a weak station, preferably one fifty or a hundred miles away, during the daytime, or one several hundred miles away at night, requiring that the sensitivity or volume control be turned up all the way to get the station comfortably. Then slightly detune the set. Without the covering effect of the music, you should then get a direct indication of how much the receiver hums under unfavorable conditions provided, of course, that a loud speaker is used which reproduces the very low notes. To be entirely satisfactory, the hum should be so weak that it cannot be heard in a quiet room ten feet from the loud speaker which the prospective purchaser will use.

It is quite possible to attain this standard but it costs money. A great discrepancy in price between two sets having the same sensitiveness, tone quality, and appearance, is often accounted for by the complete absence of hum in the more expensive receiver. The hum test is a simple one and should be made by every purchaser, regardless of his technical qualifications.

MEETING MODERN STANDARDS OF FIDELITY

IF THERE is one quality in radio receiving sets which has been appreciated by manufacturers, it is the ability to produce good tone. The judgment of a receiver's tone quality has already been fully discussed by the author in three recent issues of RADIO BROADCAST and hence it is unnecessary to repeat in great detail the factors applying to this most important requirement. Briefly, to obtain good tone quality requires that the set have: (1) Adequate power supplied the loud speaker by the use of a ux-171 (cx-371) type output tube, or even by the still more powerful ux-210 (cx-310) type tube; (2) an audio amplifying system which covers the tonal scale and (3) a loud speaker adequate to handle the volume and tonal range supplied it. To the non-technical reader, these requirements may seem difficult to appraise. But a simple test reveals a great deal about the tonal capacity of a receiver. Ask the dealer to tune-in a strong, near-by signal and bring it to full volume. Although the music is uncomfortably loud, using

the output tubes mentioned, it should not, even with very strong signals, be scratchy, stringy, or drummy. Music should be simply loud with tonal quality unaffected.

The second test is to listen critically with the set at moderate volume for the instruments producing low tones, like the 'cello, drums, or the organ. If these appear to be in their proper proportion, without being overshadowed by the treble, the receiver is capable of handling low tones.

There is much danger of selecting a set which exaggerates the low tones, a characteristic easily demonstrated in speech. A low, throaty, ringing effect, which makes words difficult to understand, is an indication of over amplification of low notes. A receiver omitting low notes, usually gives harsh, unsympathetic, but clear speech. Speech over the telephone is quite easily understood, but the rich, sympathetic quality of a good voice is lost because of absence of low tones in telephone transmission.

TESTING FOR SELECTIVITY

SELECTIVITY is necessary under modern receiving conditions, particularly in congested areas. Generally speaking, the more tubes a set has, the more likely it is to be selective, because each stage of radio-frequency amplification adds another filter circuit. It does not necessarily follow, however, that a great number of tubes means great selectivity any more than a great number of cylinders means great power in an automobile.

The pick-up system used is a valuable guide in determining the selectivity. Given an equal number of stages of radio-frequency amplification, an antenna set is likely to be less selective than one with a loop. A receiver lacking in selectivity, can often be improved by shortening the antenna.

Selectivity is simply tested but it is hardly possible to set down any definite procedure since local conditions play an important part; while a receiver may be perfectly satisfactory in one district, it might fail badly elsewhere, where the conditions and requirements are different.

There are some receiving sets so lacking in selectivity that the nearest station (we do not mean a so-called "super-power" station) can be heard over one-fourth to one-eighth of the entire dials' scales, while a few, high-grade receivers pick up the same station over only two or three degrees of the dial. The signal should tune-in fairly sharply without long fringes over which it is heard weakly above and below the point where it is heard at full volume. The selectivity with a weak and distant station is no indication of the receiver's performance under ordinary conditions.

EVALUATING SIMPLICITY OF CONTROL

THE third factor, and one of great importance, if the entire family is to use the receiver, is its simplicity. Only three controls are essential to the operation of the receiver: (1.) An "on-off" switch; (2.) a volume control and (3.) station selector. The "on-off" switch should take care of all power supply connections, such as those of the chargers and power supply units, as well as the filaments of the tubes themselves.

The volume control should enable you to bring the loudest, near-by station down to a whisper, without impairing its quality, while the station selector should give you the complete parade of broadcasting stations, up and down the scale, without requiring any other adjustment. Set the volume control to a weak station and turn

the dial from top to bottom and the stations should come in in the order of their frequency, this depending, of course, on their power and distance.

If, at the low end of the dial, the tone quality of the station cannot be cleared up without cutting down the volume control; the receiver is not properly balanced and probably radiates, and thus interferes with other receivers on the short wavelengths. The functioning of volume control and station selector should not be interdependent.

When two station selectors are used, which is frequently the case, one may be calibrated in frequencies, so that the dial may be set to a desired station accurately, and the other arbitrarily calibrated from 1 to 100, suiting it to antennas of different length. It is not inconvenient to operate a receiver with two station selectors, although the ideal arrangement provides but one.

Receivers having a single station selector and designed for antenna (as opposed to loop) operation, generally require an extra stage of radio-frequency amplification generally untuned. This extra stage only contributes little amplification and does not materially affect the selectivity. Because of this consideration, the fallacy of rating a receiver's capabilities by the number of tubes it possesses is obvious. Its sensitiveness and selectivity are dependent upon the number of stages of *tuned* radio-frequency amplification.

The buyer is often perplexed by the great number of receivers, apparently similar, but possessing a wide range of price. The factors of tone quality, selectivity, and volume capacity may be roughly rated upon demonstration, but unreliability develops only in service. Beware of the receiver that is too cheap, particularly one having power supply incorporated in it, because filter condensers may break down and mechanical difficulties may arise in service. It is true that very large quantity production decreases costs but, as with everything else, you do not get something for nothing. The extra cost of purchasing a set having back of it the name of a well-known manufacturer, is a protection against the hidden factor of unreliability. There is no reason for tolerating an unreliable receiver because the instrument which the buyer has at his command nowadays requires virtually no attention other than the periodic renewal of tubes.

The question is often asked, "How much does a good radio cost?" or "What is the best radio set for the money?" There are so many different requirements which must be met, dependent upon receiving conditions where the set is to be installed, and so wide a variation in the skill and temperament of the users, that such general questions cannot be answered.

The qualities every discriminating listener looks for are good tone, reliability, selectivity, convenience, and simplicity.

The prospective purchaser of a radio receiver should be fully cognizant of his requirements before setting forth to the radio store. He may also decide how far he will go in cost to secure the degree of maintenance convenience he desires, as already considered. He will confirm the quality of design of the power supply by listening for hum with the receiver set at maximum sensitiveness with no station tuned-in; its volume capacity and selectivity by tuning in the loudest near-by station to maximum volume; its tonal range by listening for low and high tones; and its simplicity by checking the exact number of operations required to tune-in a desired station. With the aid of these precautions, he will know just what he is buying and be able to compare intelligently one receiver with another.

"Our Readers Suggest—"

OUR Readers Suggest. . ." is a regular feature of RADIO BROADCAST, made up of contributions from our readers dealing with their experiences in the use of manufactured radio apparatus. Little "kinks," the result of experience, which give improved operation, will be described here. Regular space rates will be paid for contributions accepted, and these should be addressed to "The Complete Set Editor," RADIO BROADCAST, Garden City, New York. A special award of $10 will be paid each month for the best contribution published. The award for the December prize contribution "Antenna Compensation in a Single-Control Receiver" goes to Harold Boyd of Winchester, Virginia.
—THE EDITOR.

Eliminating the Loop

A FREQUENTLY suggested method of increasing the sensitivity of loop receivers is that of coupling the loop to a short antenna. This, of course, is a feasible proposition, but when coupling to any sort of an open antenna, why not eliminate the loop altogether?

The system generally advocated when a loop is used in conjunction with some kind of antenna is that of winding several extra turns of wire on the loop-frame and connecting the short antenna span and a ground lead to the two ends of these extra turns of wire. The combination of the loop with an extra winding thus constitutes a simple coupler, but one of rather extravagant dimensions. There is no reason why the size of the coupler cannot be reduced to more conventional dimensions, and an ordinary antenna coupler, such as an Aero or Bruno-Coil, substituted for it. The loop-antenna combination will be no more sensitive than the antenna-small coupler arrangement.

It requires no technical skill to effect the change. The coupler consists of two windings, a primary winding having relatively few turns—from six to twenty—and the secondary, having about fifty turns of wire. The primary should be connected between antenna and ground while the secondary should be substituted for the loop, as shown in Fig. 1.

It is desirable to obtain a coupler of the correct size for the tuning condenser. If the connections from the loop are traced, they will be found to lead to a variable condenser. In the majority of cases this condenser will have a capacity of 0.00035 mfd. The capacity of the condenser can almost always be determined by counting the plates. A condenser having approximately thirteen plates has a capacity of 0.0005 mfd.; one having about 17 plates is probably 0.00035 mfd.; with 23 plates, the capacity will be near 0.0005 mfd. In purchasing a coupler, specify the size condenser it is to be used with. In the few instances where it is difficult to determine the exact

THE CROSLEY "LOWAVE"
Used with a Stromberg Carlson "Treasure Chest." See descriptive matter in the box below

capacity, obtain a coupler for a 0.0005-mfd. condenser. If you find that the receiver no longer responds to the shorter wavelengths, or that the longest desirable wave is attained with a considerable portion of the tuning condenser unused, a few turns can be taken from the secondary.

It is a simple matter to wind your own coupler. The following tables give the correct number of turns of wire:

CONDENSER	TUBING DIAMETER	WIRE D.C.C.	TURNS Pri.	Sec.
13 Plates 0.0005 Mfd.	3-4"	No. 24	14	55
	3"	No. 24	16	64
	2-4"	No. 28	18	70
This Coil Has an Inductance of 283 Microhenries				
17 Plates 0.00035 Mfd.	3-4"	No. 24	12	49
	3"	No. 24	14	58
	2-4"	No. 28	16	64
This Coil Has an Inductance of 245 Microhenries				
23 Plates 0.0005 Mfd.	3-4"	No. 24	11	40
	3"	No. 24	12	40
	2-4"	No. 28	13	50
This Coil Has an Inductance of 175.6 Microhenries				

ARE YOU INTERESTED IN SHORT WAVES?

L AST month in this department we published the description of a simple home-made short-wave converter, an auxiliary attachment which makes it possible to receive wavelengths as short as 18 meters (16,700 kc.) on any receiver. RADIO BROADCAST Laboratory has recently received a commerical instrument, the Crosley "Lowave," of somewhat similar design. Its adaptation to a Stromberg Carlson "Treasure Chest" receiver is shown in an accompanying photograph.

The Crosley "Lowave" is an especially designed circuit, employing three tubes, and so arranged that the radio-frequency amplifier of the broadcast receiver is also utilized. The "Lowave" is connected to batteries according to directions, and one wire is led from it to the antenna post of the broadcast receiver. A small switch on the front of the panel throws the short wave attachment in or out of the circuit.

In tuning over the very high frequencies with such a converter, many broadcasting stations will be heard, but a great majority of them carry a badly distorted signal, which no amount of tuning will clear up. These are the harmonics of standard broadcasting stations, the frequency characteristics of which are invariably badly garbled. True short-wave broadcasting will be received as clearly as long-wave reception.

THE LABORATORY STAFF

I am using a simple electric light plug antenna made by Dubilier. The substitution for the loop was effected in this case not because of poor sensitivity, but merely for esthetic reasons—to do away with the unsightly loop. The successful use of this system is by no means confined to super-heterodynes and may be applied to any loop receiver.

A. J. HOWARD
New York City.

Combining Horn and Cone

CERTAIN of the cone type loud speakers now on the market are designed to compensate the failure of the usual two-stage transformer-coupled amplifier to satisfactorily amplify the very low audio frequencies. The customary audio amplifier feeding into such a loud speaker gives a very satisfactory overall characteristic. If, however, the audio amplifier itself is designed to give a slightly rising low-frequency characteristic, or even "straight-line" amplification, and it be used to feed into a cone of the type described, the bass is emphasized to such an extent that a sense of frequency distortion is introduced.

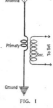

FIG. 1

A three-stage audio amplifier was hooked up in conjunction with a Radio Corporation of America cone loud speaker, model 100 A. The resultant reproduction gave the effect of exaggerated bass. A very great improvement in the quality was accomplished in the following manner: A Radiola horn type loud speaker was obtained and connected in series with the model 100A cone. The horn was shunted by a Centralab Modulator, variable from 0 to 50,000 ohms, as suggested in Fig. 2. Without this shunt, the proportion of the total load taken by the horn would have been excessive, on account of the higher impedance of its windings. With the shunt resistance set at about 25,000 ohms, the load seemed to be about equalized between the two loud speakers. Between 25,000 ohms and 50,000 ohms the effect of the horn was more pronounced, while from 25,000 down to zero, the cone predominated. At a setting of zero the horn was completely shunted from the circuit.

The particular cone used in this circuit gave excellent reproduction of the bass, while the horn brought out the treble clear and distinct. The reproduction resulting from this combination of

213

about 20,000 ohms was the most pleasing the writer has ever listened to. By the turn of a single knob it was possible to bring out the bass or treble to any relative degree to suit the particular ear of the listener.

D. C. Redgrave
Norfolk, Virginia.

Another Cone and Horn Combination

LIKE many other radio fans I recently discarded my horn loud speaker for a cone. When listening attentively, I decided that the cone slighted the higher frequencies somewhat so I dug out the old horn, which rather favors the high notes, with the idea of using the two loud speakers in combination. But the impedance of the loud speakers were so poorly matched that some balancing arrangement had to be devised.

I evolved the scheme illustrated in Fig. 3, employing a half-megohm potentiometer, such as is often used for a volume control. As the slider is moved, one loud speaker is cut in and one eliminated.

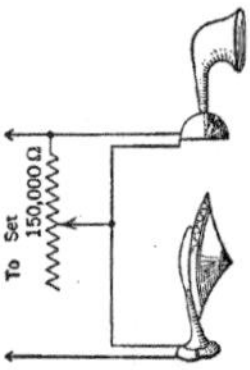

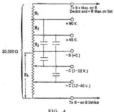

FIG. 2

I soon found, however, that I needed a better criterion than my ears to place the slider where it belonged, so I plugged a pair of good phones in the detector plate circuit and adjusted the potentiometer until the loud speakers duplicated the pitch of the telephone receivers.

H. D. Hatch
Boston, Massachusetts

STAFF COMMENT

THE choice of the systems outlined above by our two readers rests pretty much on the convenience of what you have in the way of a variable resistor. Fig. 3 provides a greater variation in loud speaker selection than Fig. 2 but sufficient variation can be secured with the latter for the purpose of correct balance. In the arrangement shown in Fig. 2 the variable resistor should always be connected across the loud speaker having the higher impedance, i. e., the loud speaker that seems to produce the most sound when the resistor is disconnected.

Mr. Hatch's system for adjusting the relative intensities of the two loud speakers is worthy of special note. While a preference may exist for an emphasis of high or low notes (adjusting to suit individual tastes), this is not necessarily natural reproduction. The phone test method probably provides as correct a balance as can be achieved without elaborate equipment.

C Bias from B Socket Power Units

IN THIS department for November, 1927, an interesting item was published showing how a B socket power device could be changed to supply grid bias voltages as well as plate potentials. In effecting this change, it was necessary to "get inside" the device, break a connection, and make several additions. This method is quite practicable, but there is always a general disinclination to "monkey" with a commercial set-up.

It is possible to obtain the C bias potentials by application of external resistors—no change whatever being necessary in the device itself.

In so doing, several possible faults in the device can be corrected, such as break down of resistors, as may be evident in noisy reception and poor voltage regulation. It is also possible with the external arrangement to secure voltages other than those supplied by the original device for special sets or purposes.

The additional device takes the form of an entirely separate set of resistors, which is connected between the high-voltage and negative posts on the old power unit. The resistors are tapped to supply the desired B and C potentials and are mounted on a base-board.

The number of resistors required is determined by the number of voltage outlets. We shall need one Amsco "Duostat" to supply two variable "C" biases, while one resistor will be required between the negative tap and the lowest positive plate tap, and another resistor for each additional plate tap. The Amsco "Duostat" is equipped with two variable arms, making it possible to secure two variable C bias potentials with the one resistance unit.

The average arrangement is shown in Fig. 4, which requires three fixed resistors in addition to the "Duostat." The fixed resistors may be any satisfactory power resistor, such as those of Electrad, Amsco, Metaloid, Ward Leonard, Durham, Carter, etc.

The method of calculating the values of the resistors takes into consideration the probable plate current drain through each resistor, and is best illustrated by the example of Fig. 4.

An arbitrary value of ten milliamperes is chosen as the loss current—the current through the resistors over and above that drawn by the receiver. The presence of this loss current reduces the variations in voltage with slight differences in load. The higher the loss current the better the voltage regulation.

The output from the average power device receiver combination under load is about 200 volts. By an application of Ohm's Law, which tells us that the resistance is always equal to the voltage drop divided by the current in amperes, or one thousand times the voltage drop divided by the current in milliamperes, it is a simple matter to determine the total resistance necessary. The equation in this case is:

$$R = \frac{200 \times 1000}{10} = 20{,}000 \text{ Ohms}$$

So the total resistance of the "bridge" will be 20,000 ohms, 2000 of which is already apportioned to the "Duostat," R_4. It remains to calculate the values of R_1, R_2, and R_3. To do this, we must consider the probable plate currents through these resistors. A 201-A type tube, properly biased, draws the following currents: r. f. amplifier, 90 volts, 4 milliamperes; detector, 45 volts, 1.5 milliamperes; a. f. amplifier, 90 volts, 2 milliamperes.

To B + Max. on B
Device and + B Max. on Set

B_1

+90 V.

R_2

+45 V.

20,000 Ω

R_3

– B (+C)

– C (1–12 V.)

R_4

– C (12–40 V.)

To B – on B Device

FIG. 4

A COMPLETE COOLEY RAYFOTO RECEIVING INSTALLATION

The incoming radio signals from a standard broadcasting station are received on a standard Freed-Eisemann broadcast receiver the audio output of which is connected to the printer unit and thence to the recorder at the extreme right. Station WOR, on November 5, 1927, transmitted the first series of radio pictures ever to be sent from a broadcasting station which were successfully received by an installation similar to this. This complete Cooley receiver was demonstrated early in November at a special picture show held at the Bamberger store in Newark. There are only two essential units to the Cooley Rayfoto receiving apparatus and they are shown on the table at the right of the loud speaker. Any good broadcast receiver can be used to pick up the Rayfoto signals. The recorder at the extreme right is not the type usually supplied, which is furnished without the spring phonograph motor. The commercial recorder may be attached to the turntable of any standard phonograph, thus decreasing the expense

WHY I INSTALLED A COOLEY PICTURE RECEIVER

By Edgar H. Felix

BEFORE me, as I write, is a photograph of a lady. I treasure it highly. Her face is rather indistinct; in fact, I doubt if I could recognize her from the photograph. Although the circumstances surrounding the acquisition of this photograph are highly romantic and thrilling, I neither hope, seek, nor expect to make her acquaintance. Yet her picture will always remain a treasured memento. It is the beginning of a beautiful friendship, a new addiction which will make my charming wife still more thoroughly a radio widow. My new absorbing interest is the radio reception of photographs and the valued curio is the first photograph ever received on a home-assembled radio picture receiver from any broadcasting station.

Not many months will pass before picture transmission will be a regular feature of broadcasting programs. I do not say this because I have inside information, but because picture reception is an inevitable step in the progress of broadcasting. The only hindrance so far has been the fact that no simple and inexpensive apparatus has been available, a problem which has been solved by the appearance of the Cooley Rayfoto System.

It is easy to speculate on the possibilities of picture reception. Most people think of it as a means of seeing broadcasting artists performing in the studio. Although such a use of picture transmission is quite feasible, it is only a minor application. Imagine a ringside description of a champion prize-fight coming through the loud speaker, and near-by your Rayfoto recorder is grinding out one picture after another of the high lights of each round. Or, imagine the ceremonies at the take-off of the first transatlantic air liner, accompanied by pictures of the immense bird taking on its cargo, taking off with its load, and finally disappearing in the distance.

The new medium has all the diverse possibilities of tonal broadcasting and it will add a new lure to radio reception. Perhaps my imagination runs away with me because I have just

made my first radio picture, but I firmly believe that it is only the first of hundreds, each of improved quality, which I shall pick out of the air.

On November 5, 1927, WOR, the broadcasting station of L. Bamberger & Co., Newark, New Jersey, put the first picture on the air from a public broadcasting station. It was only a preliminary test, to precede a public demonstration. So far as I know, my picture receiver was the only one, not installed at an experimental laboratory, capable of receiving the picture. But so great has been the interest manifested by set builders in picture reception that I am sure thousands of picture receivers will be in operation within a few weeks.

Picture transmission by wire and radio is not a

© Bachrach

AUSTIN G. COOLEY

new art. Pictures flash across the Atlantic by the Ranger system almost daily. The Jenkins system has been available to amateurs for years. The A. T. & T. has been transmitting high-quality pictures over wire lines to all parts of the country for some time. But most of these systems require either expensive receiving apparatus or more than one broadcasting channel, so that they are hardly suited to widespread use without considerable development.

THE COOLEY SYSTEM IS SIMPLE

THE Cooley Rayfoto system is rugged and inexpensive. It uses but a single radio telephone channel for transmission. The parts for a picture recorder cost no more than those required for a good six-tube home built receiver. The quality—and quality in picture reception means detail and accuracy in shading—is not below the standard of tonal reception quality which we had in 1921 and 1922, a quality sufficient to set the world on fire with a broadcasting boom. There is no reason to doubt that picture reception will improve as rapidly as did tone quality in a like period. In fact, all of the experience gained with tone broadcasting is directly applicable to the broadcasting of pictures.

The constructional details of the Cooley system have been given completely in this magazine. The radio element of the Cooley apparatus is no more difficult to assemble than a two-tube radio set. Essentially, it is a stage of audio-frequency amplification, coupled so as to modulate the output of an oscillator. It can be assembled and wired in two hours. The rest of the work is connecting power supply, setting the mechanical unit on the phonograph, and adjusting the relay, corona discharge, and phonograph motor speed. Mr. Cooley, in his articles, has described these matters in some detail; my only purpose in mentioning them here is to give you an idea of how easy it is to get the Cooley receiver in working order.

215

To receive this treasured picture of mine, we adjusted the Rayfoto relay by tuning-in a musical program. A few moments of tinkering with a screwdriver and the relay tripped at each loud modulation peak. Mr. J. L. Whittaker, who did most of the hard work while I looked on, had the relay working in two minutes.

Adjusting the corona offered no special difficulties. By tuning the variable condenser across the oscillator coil, the corona starts as soon as the oscillator starts going. There really isn't anything to it. To get the modulation current right, you plug in the meter to set the input at the correct value. It's only a matter of juggling the input volume control.

Synchronizing the phonograph motor when everything else is working properly is simple when you know how. The phonograph motor should revolve just rapidly enough so that the synchronizing signal releases the drum at each revolution. When it does so, the drum stops with a firm click and then resumes turning with hardly a perceptible stop. When it is working just right, the drum stop makes its click with perfect regularity, about as loud as that of a typewriter key against the platen. If the speed is decreased below this point, there is a drop in the intensity of the clicks and a few may skip entirely. That is the critical speed, when the drum is at "zero beat" with the transmitting drum. At that speed, the synchronizing signal is not utilized in releasing the drum; at a slightly faster speed, when the clicks are regular, it is in proper adjustment.

These three adjustments made, and you are ready to receive the picture. We got the first one quite well, but we did not get the second one. First the grid leak went wrong, dimming the corona. Then the phonograph needle slipped off the bushing. Then the relay stuck. And, finally, the A battery went down. Of course, we didn't get the second picture!

So, you see, there is plenty that may happen when you get your first picture with a Cooley receiver. You have reason to be proud of a successful result. Of course, when there are plenty

of pictures on the air, you won't have to meet all of your troubles in two short minutes. And, more important, is the advantage to be gained if you have a phonograph record of a Cooley picture transmission. I recorded the Cooley transmissions from WOR

The author receiving pictures by means of a Cooley Rayfoto receiver

on a high-quality Dictaphone. By reproducing these records with an electric pick-up, every element of the picture receiver can be put in working order before transmission begins. Such records can probably be used for transmission purposes so that all the broadcasting station needs to put Cooley pictures on the air is a phonograph record and a phonograph.

THE GATEWAY TO TELEVISION

PICTURE transmission is the gateway to television and every red-blooded experimenter longs for the opportunity to become familiar with its problems by actual experience. The Cooley system sends the picture in one minute and to increase this speed of transmission sufficiently to get motion pictures requires a thousandfolding of this speed. Considering the magnitude of the improvement required, it seems impossible that the Cooley system can

ever be the basis for television. A fundamental and startling invention is necessary before so great an increase of speed can be hoped for. There have been demonstrations of television already, but all use electrical machinery so complex and expensive that the tests really proved that television is not yet a practical possibility with our present knowledge.

Someone will some day accomplish the essential invention to make television practical. Very probably the discoverer will be someone who has worked with picture transmission. And the man who does make television possible will go down as one of the world's greatest inventors.

Thousandfolding the speed of any process is a large order. But remember that only twenty-five years ago, when radio telegraphy made its appearance, the detector used was a glass tube filled with iron filings. Electromagnetic waves caused these iron filings to congeal or cohere so that the current from a local battery flowed through them. This operated a sounder. The iron filings were agitated by the clapper of a bell so that they would de-cohere or interrupt the battery current as soon as the radio wave ceased flowing through it. With this crude apparatus, transmission of two, three, four, and five words a minute was the maximum possible. Ranges were matters of a few miles.

Then came the invention of Marconi's magnetic detector. This greatly increased both speed and range. And then the really revolutionary vacuum tube. Now, with high-speed radio telegraphy, we send thousands of characters a minute. From many of the long-wave, transoceanic stations, you can hear the myriad of dots and dashes, so rapid that you cannot distinguish between them. The step from the coherer to radio telephone transmission and reception is no greater than that necessary from telephotography to television. Undoubtedly, someone will make that essential invention to make television possible and I hope it will be an American amateur experimenter, who has enough vision to see in the crude, simple, telephotographic apparatus available to him to-day the possibilities of the great new science and art of the future.

TWO PICTURES RECEIVED ON A RAYFOTO RECEIVER

The picture on the right shows the effect of improper adjustment of the oscillator in the Rayfoto printer unit. The white streaks are produced when the oscillator suddenly stops oscillating. The left-hand picture indicates the improvement in results after the oscillator has been adjusted. For best results the oscillator should not be tuned exactly to the point at which the maximum corona discharge is obtained since, when so adjusted, it is rather critical and easily stops oscillating. The variable condenser controlling the oscillator circuit should always be slightly detuned from that point at which maximum corona discharge is obtained

Suppressing Radio Interference

Every Conceivable Source of Radio Interference Is Considered, Remedial Suggestions Being Offered—Farm Lighting Plants, Railway Signals, Telegraph Lines, Stock Tickers, Street Railways, and Interference Originating in the Receiver Itself, Are Taken up in This Chapter

By A. T. LAWTON

THE studies made by the author in the elimination of interference have been so extensive that it would be hardly possible to combine the various chapters in one issue. An endeavor has been made, however, to eliminate cross references so that each article in the series may be complete in itself. The forms of interference covered in the two previous chapters, printed in the September and November, 1927, issues of RADIO BROADCAST, dealt with interference originating at the following sources: Oil-burning furnaces, X-ray equipment, dental motors, motion-picture theatres, telephone exchanges, arc lamps, incandescent street lamps, flour mills, factory belts, electric warming pads, and precipitators. The information printed results from a two-and-a-half-year study by the author in more than 132 cities. The first form of interference considered in the present chapter, is that originating in farm lighting plants.

FARM LIGHTING PLANTS

INTERFERENCE from this source is confined to rural districts. The characteristic click, click, corresponding to the ignition spark, reveals the source at once; rarely do we get trouble from the commutator.

This continuous clicking is very loud on nearby radio receivers and because of the large number of plants in operation in communities not served by electric power companies, the total interference is very annoying.

Complete elimination of the disturbance created is a relatively simple matter. Nearly all these plants, being bolted down to a wooden platform, are insulated from the ground and two 2-microfarad condensers in series, midpoint connected to the engine frame, and bridged across the outgoing d.c. feeders, should clear up the clicking. Ground connection of the series wire should not be made to a water pipe or earth rod in this case. Even 1-microfarad condensers will be found suitable for small plants.

If the engine bed is of concrete or the plant is grounded in some other way, complete elimination is not obtained by the above method although the reduction (about 80 per cent.) is material. The remaining interference is not likely to be serious. However, where it is desired to

clear this out also, choke coils should prove effective.

It is not uncommon to find the exhaust pipe of such plants carried to a muffler drum in the ground; such a connection to earth offsets the effect of the condensers and more satisfactory results will be obtained if this pipe is cut and a section of asbestos or other suitable piping inserted.

Further experimental work is required in the cases where Electrical Inspections call for the grounding of small lighting plants. Possibly a fine wire choke coil, say, of No. 26 wire, bridged by a standard lightning arrester, would be satisfactory, the coil preventing the accumulation of any static charge and the arrester taking care of any abnormal surge superimposed on the system from outside sources.

As a precaution, all leads around the plant proper, either high- or low-tension, should be cut as short as possible.

Where one or two of the storage cells are used to operate a radio set as well as light the residence, complete elimination of the trouble for that particular set becomes difficult.

Prior to arriving at suitable preventive measures in the various plants investigated, experiments were carried out directly on the ignition system but all methods tried proved of no avail. In the case of "make and break" type ignition, condensers applied directly to the ignition system stalled the engine, apparently neutralizing the effect of the inductance coil, but were quite effective on the outgoing lines and, of course, did not interfere with normal operation.

In all cases attachment of the condensers is made at the switchboard, under the same terminals to which the two lighting mains are connected.

RAILWAY SIGNALS

RAILWAY "wig-wag" signs and crossing bells give rise to heavy clicking radio interference. Fortunately, their periods of operation are limited, but where the trouble is material it can be cleared up by shunting the operating contacts with a resistance of about 350 ohms. This applies to low-voltage d.c. operated bells and signals.

For types operating on 400–600 volts it is necessary to bridge each individual set of contacts with a resistance of the above order.

The vibrating reed type battery charger used in conjunction with these signals is a real offender. Interference from this source has given rise to a large number of complaints but, generally speaking, railway companies are averse to tacking on any surge traps to this equipment and in the dozen or so cases cleared up, the operating company took the chargers out bodily and substituted a type which is silent so far as radio interference is concerned.

LAND LINE TELEGRAPH AND STOCK TICKERS

RAPID clicking interference from the above equipment is a serious matter in towns and smaller cities. Not that there is less telegraph activity in the larger centres, but in the places referred to a comparatively greater number of residences and radio dealers are located in the vicinity of the telegraph offices.

Normal operation of the keys and repeaters set up a vigorous highly damped wave which breaks in on practically any setting of the dials of the average radio set. The clicking from stock tickers is severe even where the wires are enclosed in lead-sheathed cable.

Since it is necessary to apply suppressive measures to each individual line, the factor of cost looms large when we consider the number of lines entering the average city office. Where only a few lines are concerned it is usual to place a 1-microfarad condenser across the key contacts inserting in each condenser lead a resistance of about 20 or 30 ohms.

Strictly speaking, only one resistance unit is required but it makes a very great difference which contact of the key this resistance lead is connected to. If the condenser alone is used, reduction of the interference may be noted but the resultant arcing at the contacts is prohibitive. It is ordinarily supposed that the condenser, in this case should absorb the spark; rather, it turns the spark into an arc and resistance is required to overcome this.

We must remember that many variables enter here; at a given time, one leg of the key may be to ground and the other leg to line, positive or

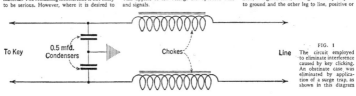

To Key — 0.5 mfd. Condensers — Chokes — Line

FIG. 1
The circuit employed to eliminate interference caused by key clicking. An obstinate case was eliminated by application of a surge trap, as shown in this diagram

Chokes· 130 turns No.26 D.C.C. wire on wooden core 1¼ diam.

negative. Again, the key may be cut in on a through line, positive or negative either way, and these conditions are likely to be reversed many times in the course of a day through jack plugging operations at the switchboard.

In most cases these clicks are unreadable; they occur at the break of the key, not on the closing of the circuit, but in the immediate vicinity of a telegraph office, radio interference may be noted from the local or sounder circuit. This is readable and can usually be eliminated by bridging the relay contacts of the local circuit with a one-half microfarad condenser without any series resistance.

Occasionally, complications enter into the situation, making necessary the use of full surge traps. An obstinate case of key clicking was cleared up only after applying the surge trap described in Fig. 1.

Another difficult case, in connection with a set of repeaters, required choke coils as described in Fig. 2 in each line and 0.5-mfd. condensers, series arrangement midpoint grounded, across the repeater contacts.

Duplex and quad circuits require individual treatment; much experimental work is required before standard eliminators can be recommended for these and the various types of stock tickers in daily operation.

RADIO RECEIVERS

MUCH of the so-called inductive interference has its origin in the broadcast listener's own set. It might be due to any one of the following causes:

(1.) Loose antenna or ground connections will give rise to serious clicking noises when shaken by wind or other agency.

(2.) Loose or corroded connections at the A battery will cause flickering of the tubes and consequent clicking.

(3.) Internal defects in any one cell will cause harsh grating noises. This can be checked with a telephone transformer and pair of headphones, the secondary of the transformer being connected to the battery and primary to the phones.

(4.) Defective B batteries cause a "chirping" or clicking interference. It doesn't follow that because a B battery registers full voltage, 22½ or 45, that it is in good condition. One high-resistance cell of the many incorporated in this battery can render the battery unfit for use although on open circuit it reads up to full strength.

(5.) Oxidized contacts of the tube prongs or spring contacts of the sockets are fruitful sources of trouble. These should be scraped occasionally and the spring strips pulled up gently to insure that good contact is being made.

(6.) Broken lead-in wires cause poor reception and clicking noises. Naturally, a lead-in wire must be insulated where it enters the building but the practice of using covered wire all the way down from the horizontal portion of the antenna is not to be recommended. From swinging in the wind, this wire often breaks inside the insulation but is held up by the fabric so it appears to be continuous whereas the wire itself is separated an eighth of an inch or more.

(7.) Concealed house light wiring in the walls will cause humming noises if the radio set is installed close up to such partitions. This is especially noticeable in the case of single-circuit receivers.

In the case of abnormal humming it is probable that the main supply line (three-wire system) is badly out of balance or the wrong side of the house lighting system is grounded. This latter will not necessarily blow the fuses.

A and B supply devices designed for

sixty-cycle supply will also give some similar trouble when used on twenty-five cycle lines.

Various stray noises will, of course, result from defective tubes, shaky splices, poor construction, connections soldered with acid flux, etc. Before assuming that any given interference comes from the outside, broadcast listeners should disconnect the antenna and ground wires and note if the interference weakens or disappears. If it does, the source of the trouble is outside. If it remains the same and is not responsive to tuning, the trouble, nine times out of ten, is in the set itself.

Unshielded sets may pick up outside interference without the antenna or ground being attached, but a decided weakening will be observed when these are disconnected and this is sufficient proof that the disturbance is not internal.

As a test, the house lighting supply switch should be opened for a moment. If the interference disappears coincident with this, all lights and electrical apparatus in the residence should be checked over for possible faults. A lamp loose in its socket can cause quite a lot of trouble. Partial shorts in the interior wiring will cause trouble over a fairly wide area; such a case

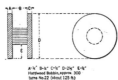

A·¾″ B·¾″ C·½″ D·2½″ E·¾″
Hardwood Bobbin, approx. 300
turns No.22 (about 125 ft.)

FIG. 2

occurred recently where a long nail driven in an attic floor brought a concealed power wire in contact with the grounded plumbing.

ELECTRIC STREET RAILWAYS

INTERFERENCE from this source presents a problem for which no really satisfactory solution has been found. Without question, many radio sets are unnecessarily interfered with by this agency and in such cases relief is possible.

The steady interference under a street-car line is violent; fifty feet either side of this it is almost negligible. Only when cars go by or some very abrupt alteration of the line energy values occurs do we get trouble at this distance.

The following actual test is interesting: On a radio set, the antenna of which ran parallel to a car line, broadcast programs were rendered valueless as mediums of entertainment but when this antenna was changed to run at right angles to the line, good reception became possible. The effect was decided.

If your antenna is at present installed at the front of the building, facing the car line, shift it to the rear as far as possible. If conditions permit, run it at right angles to the line of interference. It is surprising what a difference one or two feet variation in the plane will make here and we suggest when installing, permanently fixing one end of the horizontal portion and temporarily attaching the other end of the horizontal part to a pole which may be carried backward and forward on the roof to test out the exact position of minimum interference.

Relative intensity of the interference is noted at each setting and a permanent pole fixed at the

proper point. We must remember that lighting and power circuits nearer the residence than the trolley feeder are concerned in transferring this disturbance to the antenna, and distortion of direction is pretty sure to occur.

Sparks at the trolley wheel cause clicking interference; large arcs cause none. The bigger the flash, the less the interference. Cars going up grade on full power cause decidedly less interference than cars coasting with practically all power off, only the lights or heaters being in operation. Trolley shoes are no improvement over wheels. Catenary suspension makes no difference.

If the positive lead of the station generator goes direct to the trolley wire, interference will probably result. If this lead goes to the series field winding first, the free end of the series field going to the line, the choking effect of the coils tends to suppress any commutator interference. In addition to checks on the usual 600-volt d. c. systems, observations were carried out on:

(a.) Pantograph System, 6600 volts a.c. copper trolley wire.

(b.) 1500-volt d.c. system using shoes. Copper trolley.

(c.) 2400-volt d.c. pantograph. Copper trolley.

(d.) 1500-volt system d.c. shoe contact. Entire system steel trolley wire.

A great many variable factors entered into the results and for scientific purposes the data obtained are not considered satisfactory. Much time and effort and expense have been devoted to the whole problem of interference from electric railways but, so far, results have not been very encouraging.

It must not be concluded, however, that the proposition is hopeless; greater efforts are being made in every city to keep the rail bonding in good shape and it is probable that more careful attention to car motors and equipment generally will tend to alleviate the situation.

Peculiar effects are often observed on radio sets installed in the vicinity of a car line. It is not uncommon to find that a passing car, even when causing little or no radio interference, will "take away," temporarily, the program being received. As soon as the car moves away a few hundred feet the program comes back, without any alteration of the set tuning whatever.

Again, in two widely separated instances, and some distance away from the car lines, the disturbance on residence radio receivers amounts to a continuous loud roar. Immediately outside both residences no interference can be picked up on a standard six-tube super-heterodyne receiver. Evidently this was a case of electrolysis since an exploring coil on the water piping indicates fluctuating current in the plumbing system. Concealed pipes and wiring in these residences carry the surge also and loop reception is practically no better although the ground wire to the water pipe is cut off.

One of the most curious and perplexing cases of radio interference that we know of had its origin in an electric street railway system. This particular noise affected only three residences, all fairly close together. All efforts to locate the trouble failed but the plumber solved the problem. It so happened that the drain pipe of the centre house became blocked; on digging this up it was found to be full of a fibrous growth and at a crack in the pipe a mass of roots had come through and grown out in the direction of the railway tracks, close up to the rails, in fact.

This offshoot evidently served as a conductor to electrify the plumbing fixtures of the three houses concerned and as the radio set grounds were attached to this, every variation in the railway line voltage set up a disturbance on the radio receivers. When the drain pipe was cleaned and replaced, all interference disappeared.

ARE PROGRAMS GOING IN THE WRONG DIRECTION?

By JOHN WALLACE

SOMETIMES when we sit ourself down to pound out this monthly manifesto concerning matters radio, we attack the job with great gusto and enthusiasm. We are filled with a vast faith in radio's achievements and possibilities and garner a certain amount of personal satisfaction in making our small critical contribution to radio's great ends.

There are other times when we entertain grave misgivings as to whether the subject is actually worth the fifteen clean sheets of typewriter paper we employ to comment on it. Such doubts assail us this very minute. Probably when we are effusing enthusiastic utterances about radio entertainment we are writing romance, and to admit that it is pretty punk is simply to be realistic. Though if you be of the other turn of mind you may claim that it is the former that is realism and that the latter is rank pessimism—which suggests an appropriate variation of Cabell's ingenious epigram: The *optimist* thinks that radio has now reached the greatest heights of achievement. The *pessimist*, alas, is afraid that he's right!

Whatever roseate promises radio may have seemed to have held in the past we are at present thoroughly convinced that things have reached a sorry pass and that radio is standing still—smug, self-satisfied, and inutterably banal. One robin does not make a summer and the indubitably good program to which you may point here and there (only too infrequently!) isn't enough to raise radio to the standard it ought to be maintaining.

If two years ago someone had told us that by the end of 1927 programs would have only attained the level they now actually occupy we would have pooh-poohed him as a person incapable of reading the signs and portents. At the rate things were progressing in 1925 there was every indication that another two years would find us surfeited with high-class program material, of such high standard that we would be loath to absent ourselves from our receiving sets for two nights in a row.

As we look back several years, the greatest hope for improved programs then seemed to lie in the ever increasing number of sponsored programs. Great industries were going to start pouring their gold into the radio stations in payment for indirect advertising—more properly, radio publicity—and great things were going to result. Herein was the solution of all our difficulties: without government operation; without levying of taxes; without philanthropic financing, we were going to lead the world in program standards by utilizing our great American asset—Big Business.

Well the dismal fact of the matter is that none of these things has come about. And the ironical conclusion to which we are forced is that the rise of the sponsored program is responsible for the stand-still that radio has reached at the dawn of this year of grace 1928. In fact, stand-still is putting it mildly; the state of affairs is more exactly a retrogression. All the money, all the ingenuity, and all the labor that is being devoted to the designing of programs is being diligently devoted to efforts *in the wrong direction*—with the result that radio is going to the dogs at a breakneck speed, so rapidly in fact, that to check it will require no little effort.

LET US LOOK AT THE GUIDE POSTS

WHAT is the right direction? It would seem that program makers are too embroiled in their business to glance at the guide posts, too pressed by the strenuous and unceasing job of making programs to take a moment or two off for a little rational reflection on what their job is all about. They persist in refusing to take account of the fact that radio is a new medium, a unique medium and, like any other medium, endowed with its peculiar limitations and peculiar possibilities. Pig-headedly they persist in attempting to reconcile with their duties the traditions of the drama, the opera, the music hall, and the vaudeville stage. This observation has

Radio Times, London

IGOR STRAVINSKY

The British Broadcasting Company, on a recent Sunday afternoon symphony program, presented the modernist composer, Igor Stravinsky, conducting a program of his own works. This is an event in radio annals. True, Stravinsky has been "outmodered" by a group of still younger composers but his compositions are still quite far in advance of the popular taste. It's likely that no American station would have, nor will have for many years, the temerity to present a program of such dubious popular appeal. This does not indicate that the Britisher is any more sophisticated in his tastes than the American; it is our guess that the B. B. C. is enabled to essay such a high-brow program simply because it is non-competitive. On Mr. Stravinsky's program appeared the Overture to Mavra, the Suite from the Fire Bird and a Concerto for Pianoforte with Accompaniment of Wind Instruments. The latter was a premier English performance. The above portrait is from a drawing by the famous French artist, Picasso.

been made often before. We are a trifle abashed at shouting the same tune again. But probably it will have to be stated many times more—and certainly by others than ourselves—before it sinks in.

Without annoying you by enumerating radios'

limitations, which you know as well as we (unless you happen to be a program designer, in which case you probably don't), we will proceed to the proposition that the one species of entertainment that radio is by its intrinsic nature best fitted to put across is instrumental music.

Music is the only existing mode of entertainment that can be assimilated solely by the unaided ear. The radio (together with the phonograph) is the only existing entertainment device which can reach nothing but the ear. Obviously they were made for each other! They should be joined in holy wedlock and wander hand in hand through the new mown hills, happily ever after.

Instrumental music should be the backbone of the radio program; it should predominate every program; it should be the *pièce de résistance* with all the other little absurdities of broadcasting arranged around it like the potatoes around a roast beef. Plays, where you can't see the players, speeches where you can't see the speakers, comedy, where you can't see the comedians—all such like stuff is mere piccalilli and sauce.

So the business of broadcasting is music. And by music we mean music at its fullest realization, which is that of the instrumental ensemble—the symphony, the little symphony, or the chamber group.

In the field of the representative arts are many different mediums—water color, etching, caricature, wood cut, lithograph, miniature, pen drawing, fresco, and countless other specializations. But it is finally in the oil painting that graphic art finds its complete expression. Beside it all other artistic endeavors fall into comparative insignificance. It contains within itself the whole total of their qualities and infinite other ones of its own.

A perfect analogy exists in music. Vocal music is all right in its way, as are also organ recitals, piano gymnastics, jazz bands, and marimbaphones. But they are all dwarfed by the symphony orchestra. The orchestra not only can do all the things that they can do, but can do them better—and with its great musical resources can secure added effects that they can't possibly aspire to.

Of course it is always necessary to get back to the cold fact that, in America at least, the station manager's only business is to give the listeners what they want. But don't they want orchestral music? There has been much talk about how the taste of the listening public has been elevated by radio and much proclaiming that Mr. Average Citizen has reached a stage of enlightenment where he can actually enjoy serious musical compositions. We are inclined to believe that this is true. Moreover, there are thousands of people in the country who needed no "educating" but who already liked such music. It seems fair enough to deduce that genuine music is the "what-they-want" of a sizable section of listeners and potential listeners.

In view of the fact that orchestral music is the best thing that a station can put across, and the best thing that a listener could listen to, it seems fair enough further to deduce that there ought to be a certain amount of it in the air of an evening, available for such persons as wish to seek it out. But is it there?

THE HORRIBLE EVIDENCE

BY WAY of confirming our suspicion that there isn't, we sat ourself down at our receiver the other night and proceeded systematically to get a cross section of what was on the air. At 8:15 P. M. (Central time) we started at the top of the dial and worked our way patiently to the bottom, recording everything that was going on within our receiver's range. The stations encountered extended from Colorado to Texas to New England. At 9:35 P. M. (Central time) the chore was concluded, and if you entertain any delusions that there are a lot of fine things on the air awaiting the turn of a dial, gaze at the cold and cruel statistics we found on our tablet:

1 Jazz Piano, playing "Ain't She Sweet."
2 Jazz Orchestra, dance music.
3 Couple at a Piano, wise cracks, request numbers, ballads.
4 Sentimental Songs, "Sweetheart of Sigma Chi," etc.
5 Soprano, singing the Brindisi from "Lucrezia Borgia."
6 Baritone, singing unidentifiable ballad.
7 Sermon, of the vocal-cord-splitting variety.
8 Tenor, popular ballad, "I'll Forget You."
9 Dance Orchestra, playing "Rio Rita."
10 Old Time Fiddling, with "swing your partners," etc., interpolations.
11 String Trio, playing semi-popular airs.
12 Choral Group, in the finale of a light opera.
13 Soprano, singing the Shadow Song from ' Dinorah."
14 Soprano, singing "Just a Wearyin' for You."
15 Churchill Sisters, singing "Say Au Revoir but Not Good-bye."
16 Soprano, singing some light ditties in French.
17 Dramatization, of Rip Van Winkle.
18 Tenor, solo.
19 String Trio, playing Saint-Saëns' "Swan."
20 Weather Report.
21 Dance Orchestra.
22 Novelty Song, with banjo.
23 Hawaiian Guitar and Mandolin, duet.
24 Travel, talk on Starved Rock, Illinois.
25 Female, reciting poetry.
26 Male Quartet, singing "Bye and Bye."
27 Prize Fight.
28 Brass Band, playing "Moonlight Wonderings."
29 Jazz Orceshtra.
30 Tenor, singing Welsh folk songs.
31 Tenor, singing popular ballad, "Lonesome."
32 Pianologue.
33 Tenor, singing sentimental ballad.
34 Violinist, playing Beethoven—Opus 12 Number 1.
35 Tenor. "A Robin Sings in the Apple Tree."
36 Trio, playing semi-classics.
37 Organ, "Pale Hands I Loved,' etc.
38 Bass, soloist.
39 Dance orchestra, "On a Dew Dew Dewy Day."
40. Female duet, sloppy ballad.
41 Male Chorus, college songs.

And there you are! Nowhere was the orchestral program we had set out in search of. Twenty of the forty-one programs were vocal, practically fifty per cent.! And this in spite of the fact that vocal transmission is one of the lesser effective things broadcasting is capable of. Only three of the programs encountered seemed to hold any promise of suiting our mood of the moment, numbers 5, 13, and 34. But a return to these dial positions found the stations already shifted to something trifling.

But perhaps, we reflected magnanimously, we

had picked out the wrong hour of the evening. So the following night we repeated the procedure, commencing at 7:15 P. M. Central time, and running through to 8:20. Behold our second log:

1 Soprano, singing Nevin's "Rosary."
2 Dance Orchestra, jazz.
3 Baritone, solo "Young Tom o'Devon."
4 Jazz Orchestra.
5 Dance Orchestra.
6 Dance Orchestra.
7 Dance Orchestra—(what, another!).
8 Organ, Mendelssohn's A Major Organ Sonata.
9 Male Quartet, "Back Home Again in Indiana."
10 Banjo, solo with piano accompaniment.
11 Talk, on something or other.
12 Piano, solo.
13 Negro Spiritual, "I Heard From Heaven To-day."
14 Play, Julius Caesar.
15 Small Orchestra, playing "Gypsy Sweetheart."
16 Band, playing march tune.
17 Violin, Schubert's "Ave Maria."
18 Vocal Duet, semi-popular songs.
19 Bible Readings.
20 Tenor, singing sloppy ballad.
21 Trio, playing light stand-bys.
22 Couple, singing novelty songs about sweet mammas.
23 Dance Orchestra.
24 Speech, by some labor leader.
25 String Quintet playing Handel's Water Music Suite—but, alas, even as we listened, this changed into a soprano solo!

Such was our clinic—sixty-six cases examined over a period of two hours and twenty minutes of the most favorable broadcasting time of the evening. It may be objected that we didn't examine all the programs of all the stations during that time period, but we see no argument to show that the cross section we observed was other than representative. Representative, for the most part, of a lot of inanities that only the veriest imbecile, with the meagerest amusement resources conceivable, could dignify with the name of worthwhile entertainment.

Of these sixty-six programs not a single one was orchestral. The few trios we ran across were simply doing their five or ten minute.' turn on a well scrambled variety hour.

WE DON'T NEED SO MUCH VARIETY

THIS frantic search for variety is one of the silliest things in the whole radio broadcasting business. Variety has been set up on a pedestal as one goal to be achieved. The means used to secure it are devious and dull. Variety is necessary, of course, but there are other sorts of variety than that of the vaudeville show. Program directors do not realize that music, real music, contains within itself all the variety that is necessary. If program arrangers only realized it, their job has already been done for them by the great composers.

Practically every hour of program furnished by the broadcasters to-day is a variety program. And where every program is a variety program would it not be a variety. to introduce a program that is not a variety program?

We throw out the foregoing sound suggestion to whomsoever chooses to make use of it (fearing the while that no one will).

To the rising tide of sponsored programs is due the blame for the overwhelming number of variety programs which is rapidly reducing radio to the level of a gigantic and worthless vaudeville

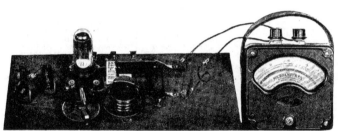

USED IN THE LABORATORY

The photograph of the vacuum-tube voltmeter shown schematically in Fig. 2 indicates a simple breadboard layout. The equipment includes a Yaxley potentiometer, a Pacent rheostat, an Alden socket, Mountford and Crescent resistances, a Sangamo "Parvolt" condenser, and several Fahnestock clips

A VACUUM-TUBE VOLTMETER

By THE LABORATORY STAFF

IN THE study of radio-frequency phenomena, experimenters have always been more or less handicapped by their lack of accurate measuring instruments. It is a simple matter, comparatively, to build a meter that will cover a considerable range when one works at 60 cycles, or at fairly high values of current and voltage, but let the research student attempt to measure currents at a million cycles and of a few microamperes, or voltages of a few microvolts, and his problem has increased in difficulty many fold.

The advent of the vacuum tube has made material progress possible in the realm of radio-frequency measurements, so much progress, in fact, that it is now possible to buy a book that deals with nothing but measurements at frequencies far beyond those used for household purposes, and at values of current and voltage that are prefixed by the word "micro." Such a book, *Radio Frequency Measurements*, by E. B. Moullin, was reviewed in RADIO BROADCAST for October, 1926; no serious experimenter can afford to be without it.

In the present-day radio laboratory there will generally be found, on one hand, a vacuum tube generating currents and voltages of practically any frequency and any amplitude (the modulated oscillator described in the June, 1927, RADIO BROADCAST is an example), while on the other hand is another tube, perhaps exactly similar in type to the generator, the purpose of which is to measure the output of the first tube.

This latter tube, variously known as a "peak" voltmeter, or a vacuum-tube voltmeter, is the laboratory's most versatile and useful instrument. It can be used to measure voltages and currents of any amplitude and frequency, to measure the voltage or power amplification of audio- or radio-frequency amplifiers, the field strength of distant stations, the high-frequency resistance of a coil or condenser, as a level indicator in broadcasting stations or remote control stations—anywhere in fact where a meter is required which takes so little power from the circuit being measured that its presence causes no error in measurement. Briefly, the vacuum-tube voltmeter is a tube acting as a detector, or distorting device, so arranged that alternating potentials on the input may be read as direct current in the output. Unlike other translating mechanisms, the vacuum-tube voltmeter has no moving parts, there is little to wear

out, it has nothing in its construction that is not easily replaceable and, best of all, it can be made to consume almost no power from the source to which it is attached.

Why is such an instrument necessary, and what are its particular qualifications compared to voltmeters with which we are all familiar? Let us suppose, for example, that we wish to measure the voltage delivered by a dynamo which will light a hundred 40-watt lamps, that is, will deliver 4000 watts, or 4.0 kilowatts. Suppose our ordinary moving-coil voltmeter has a resistance of 10,000 ohms and that when placed across the terminals of the dynamo it registers 100 volts, how much current and power does it take from the generator?

Ohm's law, which states that the number of amperes flowing in a circuit is equal to the voltage of the circuit divided by its resistance, tells us that 0.01 amperes will flow, and when we multiply this value of current by the voltage across the meter we find that the product, 1.0 watt, is the actual power required for the meter to give the proper deflection. Now this one watt is a very small part of the power that can be delivered by the dynamo, one four-thousandth as a matter of fact. It will readily be seen that the error of reading of the meter, due to the current it consumes itself, will be so small that in this case as to be negligible.

But suppose our dynamo still had a terminal voltage of 100, but was so small that it could supply only one watt of power? How can we measure its output? It is at once obvious that

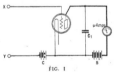

FIG. 1

A C-battery detector and a direct-current meter are the principal parts of a vacuum-tube voltmeter

we cannot use the same voltmeter for it would consume all the power the dynamo is capable of supplying.

The answer lies in making the resistance of the meter so high compared to the resistance of the source of power, the dynamo, that very little current flows through it, and consequently, very little power is drawn from the source, which, in turn, means that applying the meter to the tiny one-watt dynamo will not short-circuit it, as would be the case with a low-resistance voltmeter. This problem has resulted in the design of high-resistance voltmeters useful in measuring the output voltage of plate-voltage supply units.

The problem is even more complicated when we wish to measure alternating currents or voltages of very small magnitude, or at very high frequencies. There is a very definite need for a voltmeter or an ammeter that will measure any values of current, or voltage, at any reasonable frequency, and without taking appreciable power to operate it.

The vacuum-tube voltmeter is such a device. Its input resistance can be made so high that it consumes practically no current from the source being measured, and aside from a small input capacity which may require slight retuning when working with tuned circuits, it has so little effect on the circuit that its presence may be neglected.

C-BATTERY METER

THERE are several types of vacuum-tube voltmeters, the C-battery detector type being the simplest and perhaps the most generally useful. The one described in this article has been designed for reading small values of a.c. voltage although it may be adapted for reading any desired maximum value.

The circuit diagram of a C-battery type of voltmeter is shown in Fig. 1. As can be seen, it is quite simple, consisting only of the tube and its necessary batteries, and a direct-current meter which, for all ordinary measurements, should read not over 500 microamperes.

The question, "how can a direct-current meter in the plate circuit of a detector be made to measure alternating voltages placed on the grid-filament circuit of the tube?" naturally arises

Builders of plate-supply units, and owners of power amplifiers, who have placed a milliammeter in the plate circuit of the power tube, and watched its deflections under strong signals, have possessed the essential part of a vacuum-tube voltmeter—an overloaded tube. Whenever a signal came along which was greater than the C bias could handle, the average plate current changed and caused variations of the milliammeter needle.

But why does the needle wobble?

The normal plate current, as indicated by the milliammeter, is fixed with a fixed value of grid bias and a given plate voltage. When small a.c. voltages (the incoming signals) are placed on the grid circuit, the C bias is changed accordingly and in the plate circuit appears a magnified replica of these input voltages. In other words, the actual voltage on the plate of the tube varies with the input grid-filament voltage and naturally the plate current changes accordingly. If these changes are rapid and symmetrical, so that an increase in current is followed by an equal decrease, the average value of the plate current will remain the same and the milliammeter needle does not move. It is these a.c. plate currents which are amplified and which produce signals; the d.c. current is only a necessary and not directly useful part of the process.

If the changes in plate current are not symmetrical with respect to the value when no a.c. input is applied, the average value of plate current is different, and the plate milliammeter needle jumps about if this average value changes rapidly. The relative amount of plate current change, and whether it decreases or increases from the steady d.c. value with no input, depends upon the fixed C bias, so that we can make the complete unit, tube and meter, into a sensitive indicator of small a.c. voltages if we choose the plate and grid voltages correctly.

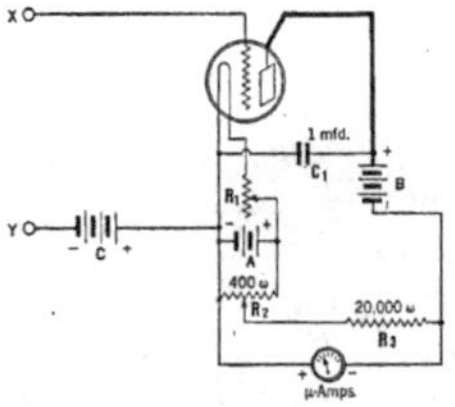

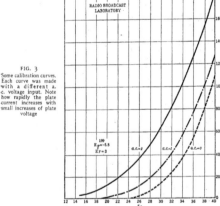

FIG. 2

Here is the circuit diagram of the vacuum-tube voltmeter described in this article. The additional apparatus not shown in Fig. 1 constitutes a bucking-voltage device to keep the normal plate current out of the microammeter and, of course, the filament battery and rheostat

The vacuum-tube voltmeter, then, consists of a tube so biased that input a.c. voltages cause large changes in average d.c. plate current. To make it into a calibrated meter, one merely places known a.c. voltages on the input, reads the plate current change, and plots a curve (typical curves are shown in Fig. 4). The purpose of the bypass condenser, C_1, shown in Fig. 1, is to improve the rectifying or distorting property of the tube.

To prevent the device taking power from the source being measured, the C bias must be great enough that input peak voltages do not make the grid go positive at any time. In practice the bias is such that the normal d.c. plate current is very near zero; this near-zero current is prevented from going through the microammeter by a

"bucking" voltage secured from a battery or a potentiometer across the A battery, as shown in Fig. 2. The high resistance prevents shunting the meter with the potentiometer. The reason for preventing the normal d.c. plate current going through the meter is so that the latter's range will not be limited by having to register both the normal current and the current produced in the process of measurement.

Any type of tube may be used. For convenience, the small 199 type tubes are used in the Laboratory, and should be used with the voltages specified here.

One simple form of vacuum-tube voltmeter used in the Laboratory is shown in the photograph on page 221 and another type, housed in a Ware radio cabinet, on page 223. The microammeter in the first picture (that at the top of page 221) is a Westinghouse instrument Model PX and has a maximum range of 200 microamperes. It is also possible to obtain this model with a full-scale deflection of 500 microamperes. These are particularly good instruments for this work, since the scale length of four inches makes it easy to read accurately, and the little button at the left, when pushed, removes a shunt which protects the meter from overload. In other words, the meter is always protected until one pushes the button, and in a vacuum-tube voltmeter of the type described here, this is a very valuable feature.

The two Westinghouse meters list at about $35. Weston makes a model 301 meter in two ranges which are suitable for tube voltmeters, a 200-microampere deflection at 200 microamperes listing at $33 and another, a 1-milliampere (1000 microamperes) meter at $12. Jewell makes similar instruments at about the same prices. In the photograph on page 223 a Weston 1.5-milliampere (1500 microamperes) meter is shown.

CONSTRUCTION OF THE VOLTMETER

THE photograph on page 221 shows how simple the vacuum-tube voltmeter may be: The small single-pole double-throw switch at the left is used to short the input by connecting the grid directly to the negative post of the C battery when circuits to be measured are being set up. There is no need to place the apparatus in such a small space as shown in this photograph, although the grid lead between voltmeter and external apparatus under test must be as short as possible, and well protected from other leads carrying a.c. voltages.

A list of the apparatus used in the voltmeter shown in Fig. 2 follows. Any of these parts may be substituted by others that are well made, and none of the values of resistance, etc., are critical.

One Tube Socket.
R_1—30-Ohm Rheostat.
R_2—400-Ohm Potentiometer.
R_2—20,000-Ohm Resistance.
C_1—1-Mfd. Bypass Condenser.
One S.P.S.T. Switch.
7—Clips (or Binding Posts).
One Binding Post, Insulated from the Baseboard.
Microammeter.

The experimenter who builds such an a.c. voltmeter for the first time must remember that he has in the circuit a very sensitive and, therefore, expensive meter, namely, the plate current reading device. If the grid circuit of the tube is left open, or if any one of several accidents happen, the meter will be blown up, and all experiments will terminate in an abrupt and disheartening manner. Every step in the construction, calibration, adjustment, and operation must be watched with great caution, and adjustments should be

FIG. 3
Some calibration curves. Each curve was made with a different a.c. voltage input. Note how rapidly the plate current increases with small increases of plate voltage

made only after due deliberation of what will happen once the proposed adjustment is carried out. After the experimenter becomes familiar with his apparatus, he will be able to proceed rapidly from calibration to operation, and to have such a feeling for his instrument that he can predict what each change in voltage, etc., will produce.

CALIBRATION

THE first job after the assembly of the vacuum-tube voltmeter is the calibration process. This is a fairly simple operation and need not be baulked at by the experimenter with average experience in handling measuring instruments. Before starting, bear in mind that too much current in the plate circuit of the tube will ruin the meter, so every step should be proceeded with gradually and with the utmost care.

But before the actual calibration is started, we must decide upon the voltage of the C and B batteries in the voltmeter circuit itself (the A voltage is, of course, that recommended by the manufacturer of the tube used). We are assuming in these experiments that a 199 type tube will be used. The microammeter should not be connected in circuit until the potentiometer switch arm is thrown as far as possible towards the negative end or, preferably, a small piece of paper should be slipped under the movable contact to insulate it from the wire beneath. This prevents any of the voltage across the potentiometer, due to the filament battery, flowing through the microammeter until needed. Since this current is backwards with respect to the meter, it may damage it until some plate current flows. Now adjust the filament rheostat until the voltage is normal.

Before proceeding further, we will test the circuit to see that everything is satisfactory (the actual calibration has not yet begun). We short-circuit the input to the tube by connecting together points X and Y, Fig. 2. Assuming that the microammeter has been connected in circuit and that the B and C batteries are of satisfactory value, a current will flow in the plate circuit and will be indicated by a deflection of the meter needle. The value of this current is dependent upon the voltages of the B and C batteries. Thus it will be seen that the value of these batteries is all important for if they are of such value as to give a larger plate voltage than the microammeter can indicate, it will be at this point that the abrupt and disheartening termination of experiments will occur. Do not, therefore, play about with different values of plate and grid voltages unless you know just what you are doing.

With a 199 type tube, we recommend that the B voltage be 45 and the C voltage be 9 to start with. Under these conditions a reading of 6 microamperes was obtained in experiments in the Laboratory. Fig. 3 shows some typical curves which were obtained in the Laboratory. They were obtained with a 199 type tube using a grid bias (Eg) of 5.8 volts and a filament voltage (Ef) of 3. Of the several curves shown, only that marked A.C. = O interests us at the moment. The term A.C. = O indicates that there is no alternating-current input to the vacuum-tube voltmeter (in other words, X and Y are shorted. See Fig. 2). We note in the curve A.C. = O how rapidly the plate current increases with small increases in plate voltage. At 40 volts plate potential (Ep), for example the plate current (Ip) is 110 microamperes (μ Amps.). If with, say, forty volts plate potential, we *increased* the grid voltage, the plate current would decrease. The other curves in Fig. 3, incidentally, show the effect of placing an a.c. input across the terminals X and Y; in other words, they represent microammeter

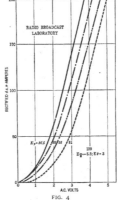

RADIO BROADCAST LABORATORY

$E_p = 36.5$ $E_p = 88$ 81

199
$Eg = 5.8; Ep = 3$

A.C. VOLTS

FIG. 4

Here are some typical calibration curves. The deflections of the microammeter needle caused by various a.c. input voltages are plotted, together with the effect of increasing the plate voltage (Ep) for a given value of C bias (Eg)

deflections due to the sum of the normal plate current and the current caused by input a.c. voltage.

We have digressed a little to go into a brief explanation of the curves in Fig. 3. We had previously arrived at that point where it was recommended that the experimenter use 45 volts B

battery and 9 volts C battery. Under these conditions, it was stated, a plate current of 6 microamperes was obtained in the Laboratory. Do not expect, however, to get a similar reading. Every tube will vary. It is quite possible that you will get no reading whatsoever. Perhaps the reading will be higher than 6 microamperes. Should there be no reading on the microammeter, slightly reduce the C voltage until a reading of a few microamperes is apparent. It is quite possible that a C battery quite a little smaller than a 9-volt one will ultimately be used, but the large value is recommended for safety's sake. The larger the C battery the smaller the plate current and, therefore, the greater the margin of safety.

Suppose that we have adjusted our batteries and have obtained a reading of 10 microamperes. We are satisfied that everything is working as it should, and take the next step, which is to balance out this 10-microampere reading. If we do not balance it out it will always be present when we are measuring a.c. voltages applied across X and Y, and, therefore, the microammeter will not be used to its full advantage. That is to say, if the meter is always indicating 10 microamperes with no input across X and Y, its range will be reduced when an a.c. voltage *is* applied. True indeed, with a 200-microampere meter this 10 microamperes only represents one fiftieth of the whole scale, but possibly the combination of B and C voltages used will give a greater plate reading, perhaps 50 microamperes (such a high value would not be probable with a 199 tube and 45 volts B and 9 volts C), which certainly should be balanced out. The balancing out is accomplished by removing the piece of paper which we previously slipped underneath the potentiometer arm and adjusting the latter slowly until the needle on the meter reads zero. Some experimenters prefer their meters to indicate a few microamperes current when there is no a.c. input across X and Y so that any possible back deflection of the needle will not harm the meter.

The actual calibration of the instrument is our next job, and it is not an extremely difficult

A

VERY COMPLETE VACUUM-TUBE VOLTMETER

The complete instrument may be dressed up to look like this if you do not care for the breadboard layout shown in the photograph on page 221. Voltmeters to read the B and C potentials are included

one. Fig. 4 shows four typical calibration curves made with different values of plate potential. After one has become familiar with the vacuum-tube voltmeter described here, the voltages may be adjusted and all kinds of different curves obtained.

For the process of calibration, some known source of a.c. is necessary.

A fairly accurate calibration may be made by using one of the many step-down transformers that are now being used to supply filament voltage to a.c. tubes. With the primary connected to the 110-volt a.c. house supply the secondary voltage from one of these units will give several points on a perfectly good calibration curve. For example, a transformer which supplies filament power to a receiver using 226, 227, and a power tube, has the following voltage taps available: 5, 2.5, 1.5, 0.75. These voltages can be added to or subtracted from each other by connecting the windings to aid or to buck each other. Fig. 5 shows the complete vacuum-tube voltmeter with the accessory equipment for calibrating it. If, during calibration, but before the maximum known voltage to be applied across X and Y has actually been applied, the microammeter needle is dangerously near the maximum deflection point, it will be necessary to increase the C bias and commence the calibration again.

Vacuum-tube voltmeters can be calibrated by any Laboratory at small cost and it is probable that the constructor can procure such a calibration near-by. The reader should remember that the tube which is to be used should be sent with the instrument and that the whole unit should be very well packed.

The calibration of the instrument is practically independent of frequency, so that the experimenter can compare voltages in circuits operating at audio, intermediate, or broadcast frequencies. He can measure the voltage step-up in an audio transformer or amplifier, or the output of two radio-frequency amplifiers, or plot the resonance curve of an intermediate frequency stage.

It will be noted that the input of the tube looks directly into the device whose terminal voltage is being measured. If d.c. flows through this device, it will be impressed across the input to the voltmeter and ruin either the calibration or the plate meter, or both. Some means, such as that shown in Fig. 6, must be provided for isolating the voltmeter from d.c. potentials existing in the circuit under measurement, when, for example, the output voltage of a resistance-coupled amplifier tube is being measured. With

this arrangement the short-circuiting switch shown in the photograph on page 221 may be omitted, since the grid is always at a safe d.c. potential with respect to the filament. Frequencies in the same range, that is, all audio tones, or all frequencies in the broadcast band, will give similar calibration curves.

If the experimenter has no means of calibrating his instrument, he is not so unfortunate as might be supposed. He may make use of the fact that the deflections of the microammeter are roughly proportional to the input a.c. voltages squared. Thus he may get two deflections representing the gain, say, of two amplifiers. He need not know the actual voltages provided he knows roughly how much greater one is compared to the other. All he needs to do is to divide one

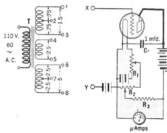

A suitable set-up for calibrating the vacuum-tube voltmeter. T is a transformer which is used with a.c. tubes and, therefore, has several taps on the secondary. Thus several known voltages may be obtained, and these plotted as a graph, similar to Fig. 4

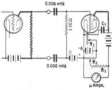

FIG. 6
When the vacuum-tube voltmeter is used to measure an a.c. voltage across the terminals of which is also a direct-current voltage, some means must be employed to prevent this d.c. from getting on the grid of the tube. Condensers as shown here will accomplish this. The diagram shows measurements being made on a resistance-coupled amplifier

deflection by the other and extract the square root. This will be the ratio of the voltages.

The home constructor should be warned again to carefully watch every step when experimenting with the vacuum-tube meter and to be very careful of the microammeter. One mistake and the meter is gone. Always, at the conclusion of an experiment, remove the meter from the circuit first.

Those who wish to read further about the vacuum-tube voltmeter will find a most interesting and instructive series of articles in the English paper, *Wireless Experimenter and Wireless Engineer*, for October and November, 1926. Part of this material was republished in this country in *Lefax* leaflets in February, 1927, and forms the best background for the serious student of this instrument. It is possible to purchase from the Cambridge Instrument Company, of Ossining-on-Hudson, New York, a vacuum-tube voltmeter calibrated and "ready to go." It is a beautiful instrument, although somewhat expensive. The owner, however, may rest assured that he has the best possible apparatus for measuring small a.c. voltages, radio-frequency currents, the gain of his amplifiers, both audio and radio, and for the performance of a hundred valuable experiments.

SOME METERS USED IN THE LABORATORY

The vacuum-tube voltmeter, although an invaluable asset to the RADIO BROADCAST Laboratory, by no means holds a monopoly in usefulness. Here are just a few of the dozens of meters always in use in the laboratory. Included are instruments by Weston, Sterling, Jewell, Dongan, and Hoyt

RADIO FOLK YOU SHOULD KNOW

I. RALPH H. LANGLEY

Drawing by Franklyn F. Stratford

RALPH H. LANGLEY, Assistant to the President of the Crosley Radio Corporation, and, in that capacity, manager of most of the operations carried on by the Crosley interests, was born in New York City on January 5, 1889. As this article goes into print he is, therefore, still two years short of the forty mark. By training, and probably by primary inclination, he is an engineer, but his vision has never been limited to screw-heads and tuning knobs; of late years he has become increasingly an executive figure. An administrator of thirty-eight would be considered very young in most lines of business, but in the radio industry there are a number of them. The fact is that no one can have more than thirty years or so of radio experience, and only a handful of men can point to anything like that. The twenty-year candidates, even, are few. In the case of all such veterans, the early years of experience have a value which is mostly sentimental, for the modern structure of the industry, with its complexities of mass production and public relations sprang up suddenly after the war. Mr. Langley's years of contact with the radio art amount to about nineteen, or half his age, which is quite enough in a field where the race is to the swift rather than the old.

Mr. Langley was born in New York City and lived there until 1916. His boyhood residence was near Morningside Park, under the hill on which the buildings of Columbia University were being erected. Young Langley, gazing at the newly created campus, formed an ambition to drink of knowledge at that fountain, but, as yet, he did not see how the project was to be financed. After finishing his elementary school course, however, he went to De Witt Clinton High School, 1904-1908, and towards the end of his secondary school course succeeded in winning a scholarship which enabled him to enter Columbia.

During the following winter, Mr. Langley's father died, and the son gave up college to take a position with the New York and Queens Electric Light and Power Company. But at Columbia, in the college "Wireless Club," the radio virus had already got into him, and in May, 1910, at the invitation of Emil J. Simon, he turned from electric power to work in Dr. Lee DeForest's laboratory at Park Avenue and 41st Street, where many strange wonders were being performed. Here he met Frederick A. Kolster and other men now prominent in the radio industry of to-day, which had its feeble and often abortive beginnings in just such laboratories. In those early days the courage of the workers made up for the scarcity of good milliammeters. A half year in DeForest's laboratory probably did Langley good, but a school teacher friend, James F. Berry, who had advised him then, now convinced the young man that he would be heavily handicapped in his later career if he did not complete his university course while there was still time. Mr. Langley took the advice and went back to college in the fall of 1910, repeating the sophomore year. But he did not give up radio. The summer-of-1911 he spent working with E. J. Simon once more, this time for the International Wireless Telegraph Company (wasn't it the National Electric Signaling Company then?) at Bush Terminal in Brooklyn. Here he met S. M. Kintner, now in charge of the research activities of the Westinghouse Company. The summer of 1912 found him with the Wireless Improvement Company.

Mr. Langley graduated from Columbia University as an Electrical Engineer in 1913. Edwin H. Armstrong was one of his classmates.

RALPH H. LANGLEY

Michael I. Pupin one of his professors. Professor Arendt, another of Langley's preceptors, had a poor opinion of the wireless game, and advised the young engineer to stay out of it, but Langley promptly joined the Wireless Improvement Company once more.

During the three years Mr. Langley put in with the Wireless Improvement Company, then under the guidance of Colonel John Firth, most of his work was with various types of 500-cycle quenched spark transmitters. The ½ kw. submarine transmitter was one of his early design jobs. Mr. Langley's interests were not, however, confined to commercial matters. He had joined the Institute of Radio Engineers as an Associate in 1912, and in 1914 served as Assistant Secretary. In 1916 he was advanced to the grade of Member of the Institute. In that same year, at the invitation of David Sarnoff, he joined the engineering staff of the Marconi Company, and went to work at the Aldene factory, of which Adam Stein, Jr., was Works Manager. Roy A. Weagant was Chief Engineer of the Marconi Company during the period of Mr. Langley's connection with the firm. The Marconi Company was handling war-time orders, principally for the armed forces of the United States. In 1917 the plant was greatly enlarged, and, running on three shifts twenty-four hours a day, must have employed in the neighborhood of one thousand men. New types of quenched spark transmitters were designed for submarine and aircraft use. In the meantime the manufacture of the standard Marconi marine transmitters and receivers, with auxiliary apparatus, such as Leyden jar condensers, had to be continued, and in the shops one would see standing side by side models of Naval receivers of the SE Types,

with their heavily-varnished bank-wound coils, the older Type 106 tuners in their black cases, and cheap little cargo receivers which looked as if they had just come out of the five-and-ten cent store. But Mr. Langley's concern at this time was with the transmitters, so much so that, in 1918, one of them almost ended his career. This distinction would have gone to the 250-watt aircraft transmitter, equipped with the General Electric pliotrons of the same rating. During one of the tests of the transmitter, Langley, having shut off the filaments of the tubes, reached in and grasped a plate terminal, forgetting that the 1500-volt supply was still on. "That particular set never worked again," states Mr. Langley laconically. "and it was some time before I did." There was evidently nothing wrong with his heart. After the completion of the development work on the sets, he made test flights with them from the air base at Norfolk, Virginia. "But none of these sets was ever used in France," the designer adds, somewhat sadly. The answer to that is that few of the airplanes ever got to France either.

In 1920, the Radio Corporation of America having been formed, the radio engineering and manufacturing activities of the Aldene factory were transferred to the General Electric Company's plant at Schenectady. Adam Stein Jr. became Managing Engineer of the Radio Department there, and Langley was assigned to the Receiver Section, later to become the Engineer-in-Charge thereof. Practically all the broadcast receivers turned out by the General Electric Company have contained one or several of Langley's inventions and design features. Working with Messrs. Carpenter and Carlson, Mr. Langley was responsible for the production of the first Radiola super-heterodyne models, incorporating the sealed "catacomb" construction and the divided cabinet. He spent seven years at Schenectady, leaving for his present executive position with the Crosley Company on February 1, 1927.

During the last three years, Mr. Langley has been much interested in the work of the radio manufacturers' associations. He was vice-chairman of the Radio Section of the Associated Manufacturers of Electrical Supplies, and later, when that body was merged with the National Electrical Manufacturers Association, became Chairman of the Committee on Section Activities in the Radio Division. He also served, in 1926, on the Standardization Committee of the Institute of Radio Engineers.

With his years of experience in radio manufacturing and organization, and his wide acquaintance among radio men, Mr. Langley believes that the next two years will witness remarkable progress in the industry. He points out that many lines of progress have been almost completely blocked until the present season. With patent difficulties largely resolved, notable progress in standardization, adequate Federal control of broadcasting, and the development of exact methods of measurement and quantity production, the economic stability of the industry should approach that of more settled branches of business. Mr. Langley has contributed more than his share in the progress of radio toward that goal.

225

Some Fine Receivers

THE "HIAWATHA"

An attractive six-tube receiver by Mohawk, Chicago. As the chassis picture shows, there are six tubes. The three tuned stages are adjusted by means of a single tuning knob. The chassis depicted is standard in several Mohawk receivers, which range in price from $67.50 to $275.00 for battery operation. For a. c. operation, add $110.00 to the above prices. The retail price of the "Hiawatha" is $165.00. It has a built-in pyramid form loud speaker, for which very excellent reproducing qualities are claimed

FOUR TUNED STAGES

Are incorporated in this six-tube receiver by Bremer-Tully. The "Counterphase" 6-55, to give the receiver its exact name, will delight the advocates of two-dial tuning, for the condensers are arranged in two units of two, and are, therefore, controlled by two knobs on the front panel. There are separate controls for sharp tuning and volume. This new "Counterphase" retails at $110.00. A console model sells for $165.00

"GRANADA" CONSOLE

By Electrical Research Laboratories ("Erla"), of Chicago, The finish is of dark antique walnut combined with birdseye maple. The drawer front is of satinwood with maple overlay. The chassis comprises the equipment for four r.f. stages (three of which are tuned), detector, and two transformer-coupled audio stages. Tuning is accomplished by means of a single dial, and there is a built-in loud speaker. Price $295.00. Furnished with an a.c. converter system, all tubes, and an output filter, price $395.00

and Their CHASSIS

BOSCH, MODEL 57

Here is an interesting receiver typical of the progress of recent years—an example of the modern self-contained installation. The price, $440.90, is not excessive when we consider that the receiver is designed for socket-power operation, has a built-in loop, and also includes a loud speaker of advanced design. This price is reduced to $340.00 if the socket power feature is not desired. The single station selector, which adjusts five separate variable capacities, is graduated in kilocycles

THE FADA 7

Another receiver employing four stages of r.f., but two tuning controls are used with this model. A loop is supplied with the Fada 7 although an outside antenna may be used successfully. The loop fits into a special clamp on the side of the cabinet. The two audio stages are transformer-coupled. A special arrangement in the detector circuit reduces the possibilities of overloading. Price, $185.00

AMRAD'S THE "BERWICK"

A popular six-tube neutrodyne type circuit, employing four tuned stages and single-control tuning. As the chassis illustration shows, complete shielding is featured. A cone loud speaker is built into the "Berwick," which, partly due to the special Amrad tone filter, gives remarkably good quality of reproduction. The price is $195.00

227

How Reliable Are Short Waves?

RECENT EXPERIMENTS of the General Electric Company, of engineers from other companies, and of those interested in private research, have resulted in an explanation of many of the mysteries that have surrounded the short waves. Everyone marvels at the ease with which amateurs communicate with fellow enthusiasts over great distances with small input powers. We have done it ourselves—communicated and marveled both—and great is the "kick" thereof. It is undeniably thrilling to take from one's lamp socket 60-cycle power of less than one fifth that required to heat the average electric iron, and to feed it into a comparatively simple system of apparatus from which it emerges as radio-frequency energy with which we actually ask a man in South Africa how the weather is there, all the time sitting quietly in our den surrounded by unimposing gear. When it is winter in New York it is summer in South Africa, when day here, it is night there, and so on. It is one of the marvels of our time that two people in the security of their homes but separated by 7000 miles can transfer their thoughts instantaneously and economically.

Just how good are these short waves? How reliable is communication? How many hours of the day, how many days of the year, can we send messages via short waves from New York to South Africa?

The results of several investigations point to the following facts which seem fairly well established. Ten meters (30,000 kc.) is probably the shortest useful wavelength. Below about 20 meters (15,000 kc.) the waves prefer to travel in the daylight, and above that wavelength, night time is the best. Below about 45 meters (6660 kc.) curious "skip distances" occur, resulting in regions beyond which signals are heard but within which they are inaudible. For example, on 15 meters (20,000 kc.) during daylight, transmission is not practicable within a distance of 900 miles, which increases to about 1000 miles at night, although it is possible to transmit signals for reception at points more distant than these figures indicate. At 27 meters (11,000 kc.) the daytime skip distance has been reported as 1000 miles and 450 miles at night, these distances being about the same at 33 meters (9086 kc.).

The General Electric experiments show that the 32.79-meter wave is no good at all for short distances. A power output of 500 watts on 65.16 meters (4500 kc.) will, however, transmit commercial daytime signals up to 100 miles.

Short waves seem necessary for extremely long-distance communication. During daytime, waves of the order of 20 meters should be used; waves below about 45 meters are not much good for short-distance work.

Night after night we have heard NAA on 37.5 meters pound away at a terrific rate, we have listened to the Marconi beam stations on 26

meters, AGA at Nauen, Germany, FW in Paris. and others, all bent on getting somewhere in the band between 25 and 45 meters is as busy as the long-wave channels. Amateurs in this country who have a channel 1000 kilocycles wide around 40 meters have been blessed with an excellent assignment which at the time it was doled out was thought to be more or less worthless. More and more amateurs are going to 20 meters and with low powers are accomplishing unheard of records in broad daylight. They have not as yet the feeling for this band and for conditions existing there that they had for the 40-meter band, but it will come when they gain the wealth of experience they have amassed on the longer wavelength band.

Mathematics of the Audio Transformer

IN MOST radio work the mathematics is fairly simple; the difficulty comes when it is necessary to put the mathematical theory into practice. For example, the following mathematics is that underlying the theory of the input transformer for audio-frequency amplifiers. The circuit about which this theory is built up is given in Fig. 1, and its equivalent is shown in Fig. 2. In this mathematics, N is the turn ratio of the transformer.

$$E_s = NE_p = I_g R_g$$
$$E_p = \mu E_g - I_p R_p$$
$$E_s = N (\mu E_g - NI_g R_p)$$
$$I_g R_g = N (\mu E_g - NI_g R_p)$$
$$I_g = \frac{N\mu E_g}{R_g + N^2 R_p}$$
Whence $E_s = \dfrac{R_g \cdot N\mu E_g}{R_g + N^2 R_p}$
$$= \frac{N\mu E_g}{1 + N^2 \frac{R_p}{R_g}}$$

Differentiating this equation with respect to N and solving for a maximum, it is found that:

$$N^2 = R_g/R_p$$

228

and substituting this into the equation above for Es:

$$E_s = \frac{\mu E_g N}{2}$$

All of this assumes that the transformer is perfect, i.e., no d.c. resistance, no magnetic leakage, infinite primary and secondary reactance. It shows that under these conditions the voltage delivered to the input of the tube is one half that delivered to the previous tube multiplied by the turns ratio of the transformer and by the amplification factor of the previous tube.

If the input impedance is one megohm, 1,000,000 ohms, and the plate impedance of the previous tube is 12,000 ohms, the turn ratio will be equal to the square root of the ratio between these two quantities, viz:

$$N = \sqrt{\frac{1,000,000}{12,000}} = 3 \text{ approximately}$$

In order that a large percentage of the a.c. voltage developed in the plate circuit of the previous tube be available across the primary of the input transformer, the impedance of this transformer must be high. If we want to amplify well at 100 cycles, the input impedance should be not less than 30,000 ohms which means that the primary should have no less than 50 henries inductance—which in turn explains why transformers, good ones, cost money, and why a skinny little affair with a few sheets of iron in the core and a little wire on the primary makes radio music sound "something fierce."

Furthermore, if the turn ratio is three, and the inductance of a winding varies as the square of the turn ratio, the secondary inductance must be about 450 henries—and when anyone states that the secondary of an audio transformer makes a good output choke he neglects the fact that one cannot wind up an inductance of 450 henries without adding enough d.c. resistance to prevent the last tube from getting any plate voltage at all.

MATHEMATICS OF THE OUTPUT TRANSFORMER

THE mathematics of an output transformer design is no more difficult than that of the input transformer—and the answer is the same, as the following rigamarole proves. The symbols used in this discussion are the same as for the input transformer. N is the turn ratio of the transformer, and instead of Rg we use Rs.

$$I_1 = \frac{E_s}{R_1} = \frac{NE_p}{R_1}$$
$$E_p = \mu E_g - I_p R_p$$
$$I_p = \frac{\mu E_g}{R_p - \frac{R_1}{N^2}}$$

Whence $I_1 = \dfrac{N}{R_1}\left(\mu E_g - \dfrac{\mu E_g R_p N^2}{N^2 R_p - R_1}\right)$
$$= \frac{N\mu E_g}{N^2 R_p - R_1} = \frac{N\mu E_g}{1} \cdot \frac{1}{(R_p N^2 + R_1)}$$

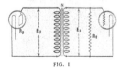

FIG. 1

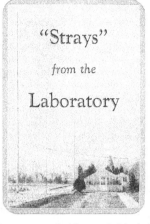

FIG. 2

Differentiating this expression with respect to N and solving for a maximum, ,

$$N^2 = \frac{Rs}{Rp}$$

$$And \ Is = \frac{\mu Eg}{2\sqrt{RpRs}}$$

whence power in Rs $= I_s^2 Rs$

$$= \frac{\mu^2 Eg^2}{8 Rp}$$

If the speaker Rs is in the plate circuit of the power tube without the output transformer, and if the impedance Rs equals Rp the power will be as indicated above, showing that the transformer is indeed an impedance adjusting device and that it is possible to secure maximum power in a load impedance which is not equal to the tube impedance provided the output transformer has the proper turn ratio:

A little consideration will show that if the primary impedance of this transformer is equal to the tube impedance and if the secondary impedance is equal to —"matches," is the usual word—the load impedance one half the voltage existing in the plate circuit will be expended in heating the plate. It is obvious, then, that the primary impedance again should be several times that of the tube impedance, but since the final tube has a low impedance, the UX-171 (CX-371) type for example has an impedance of 2000 ohms, a manufacturer of transformers needs to put nothing like the impedance in his output transformer primary that he puts into an input transformer. Likewise the secondary impedance must be high compared with the load, or speaker. As long as the turns ratio is proper, energy will be transferred from the tube to the speaker with a minimum loss, and for all practical purposes the output transformer may be neglected in calculations.

SOME OF RADIO BROADCAST'S

High-Powered Bunk   readers are technically inclined, and must enjoy the monthly battle of wits among radio advertising writers whose "eulogies" appear in many radio publications. We saw recently a new loud speaker advertised which consisted of a "tone column of new design; by means of a scientifically designed tone distributing chamber a forced crossing of sound waves is accomplished, and a divisional tone chamber of unique design segregates high and low tones, reproducing both with equal facility." Bell Laboratories engineers should take notice of this new scheme for attaining perfect fidelity, and should forget all about loud speakers which will look like a pure resistance to a power tube at all frequencies. All that is necessary is to segregate the low voice from the high one and then re-combine them. Presto!

There is also a new antenna that promises much, for, according to a widely circulated advertisement, it meets the need of radio users, whether for small or large sets. It has been thoroughly tested for over two years under all kinds of conditions and on many kinds of sets. In every set, it has proved its many advantages:

1. Easy to put up or take down.
2. Picks up waves in any direction because it is round.
3. Proven more selective.
4. Greater volume.
5. Greater distance.
6. Helps eliminate static.
7. Small and compact.
8. Neat in appearance.
9. Low in cost—only $7.50.

What more could anyone want?

This kind of bunk is not only confined to the advertising pages of the media in question, sadly to recount. The following passages are quoted from a recent article describing a new and revolutionary receiver that anyone can build for about twice as much as he would have to pay for a well known and thoroughly engineered set.

"Its volume is due to the high amplification in the audio end of the circuit and to the fact that two 71 tubes in push-pull are used in the last stage."

Off hand this seems reasonable until one considers that there are three audio tubes—which furnish most of the amplification—and that there are eight tubes which precede them! It reminds us of the early days of radio when the Laboratory had its hands full weeding out the good apparatus from the poor. A receiver came to Garden City equipped with thirteen tubes, guaranteed to pick up any signal in any part of the world at any time. The total plate current of this super-receiver, including a power tube, was nine milliamperes. It seemed incredible—and was, until we discovered that eight of the thirteen tubes had no plate voltage on them. But let us continue with this totally new receiver.

"With all that amplification and power handling capacity there will be undistorted volume enough to make the welkin ring. But all of this would be merely potential volume were it not for the almost incomprehensible amplification in the intermediate amplifier. It is here where the weak signals from the remote stations are pulled from infinitesimal levels and placed on the plane of the signals from local stations. Both the amplification and selectivity could be expressed in numbers but they would be so large as to be meaningless to the human intelligence."

There you are, and five more columns of it for good measure!

THE NEW SCREENED GRID TUBE

The bulb of a "dead" UX-222 tube was broken so that this photograph of the interior construction might be taken. The RADIO BROADCAST Laboratory is expending considerable time experimenting with these tubes so that the information contained in articles in these pages might be backed up with actual experience. We can promise readers some particularly fine articles along such links in the very near future

DURING recent months, the Laboratory has received for *New Apparatus* test the following apparatus: Power units from Kellogg, Wise McClung, Valley, Universal, Briggs and Stratton, Boutin Electric Co., Sterling Mfg. Co., and Grigsby Grunow; tubes from the following tube plants, Arcturus, R. C. A., Cunningham, Supertron, CeCo, Manhattan Electric Supply Co., Connewey Laboratories, Zetka, Cable Supply Co., Televocal, Van Horne, De-Forest, and Supercraft; audio transformers from Modern, Silver-Marshall, Samson, Tyrman, Sangamo, Amertran, G. W. Walker; the new Eby sockets, a Muter double-impedance amplifier, a Pacent Phonovox electrical pick-up unit, the excellent looking Abbey receiver manufactured by Splitdorf and already illustrated in RADIO BROADCAST, a useful floor cord made by Belden which enables one to place wires under a rug without danger of tripping or of impairing the appearance of the room—the wires may carry house lighting current for a lamp, or loud speaker wires; a complete assortment of Acme Wire Co. Parvolt condensers; rheostats, etc. from Carter; a complete push-pull amplifier and B supply units from Samson, Thordarson, and General Radio; resistances for stabilizing grid circuits, for plate supply circuits, for center taps on a.c. tube circuits, etc., from General Radio, Electrad, Frost, Daven, Amsco, Gardner and Hepburn, Aerovox; and coils from Precision Coil Co.

Many other interesting pieces of apparatus have been received, previous mention of which has been precluded pending test and on account of space limitations. Among the larger items are Silver-Marshall's interesting Time Receiver, a huge and much involved horn loud speaker from Newcomb-Hawley, using a Baldwin Unit, and known as a Console Grand Reproducer, a Peerless loud speaker from the United Radio Corporation of Rochester, and a Holmes Piano loud speaker from the International Radio Corporation, of Los Angeles. The Newcomb-Hawley loud speaker has a very long air column secured by giving the neck several convolutions about the wide opening. The Peerless is an attractive loud speaker with an excellent element. Both loud speakers cover wide frequency ranges and go down particularly well.

The Laboratory was especially interested to receive one of the "tuned" impedance amplifiers about which so much is heard, this one delivered by Mr. Kenneth Harkness in person, giving us the opportunity of meeting him for the first time. Frequency curves are being prepared on this amplifier, and will be ready soon.

Transformers for the new a.c. tubes were received from Amertran, General Radio, Mayolian, and Northern Manufacturing Company. New rheostats came in from Frost, Centralab, and Carter. Some large heavy-duty resistors from the Electro-Motive Engineering Corporation also made their appearance. Among the other items are new Samson apparatus, loud speaker units from the Balsa Wood Corporation, Engineers Service Company, Baldwin, Vitaltone, and Magnaphon Company, a series of new and probably very efficient inductances made by the Precision Coil Company, a "Subantenna" (is there any way of testing this antenna without digging up the garden?), two sheets of beautifully burnished copper for shielding, from C. G. Hussey & Company, of Pittsburgh, condensers from X-L Radio Laboratories, Electrad apparatus and a fine looking White socket power unit from Sioux City, Iowa.

Concomitants

THE AMPLION "GRAND"

AS PLEASING to the eye as to the ear is this magnificent loud speaking device by Amplion. Contrary to what one might justly suppose upon first glance, the "Grand" does not use a horn inconceivably munificent in inches, but employs a combination cone and sound board. The "Grand" lists at $145.00. It is supplied with a 20-foot cord. The walnut cabinet measures 34" x 33" x 18"

BY PACENT

HERE is a seventeen-inch cone loud speaker with several interesting features. Its extreme ruggedness and responsiveness to weak signals are features. The manufacturer claims that this cone has a thirty per cent. greater frequency coverage and sensitivity than most of the popular model cones on the market. Price, $22.50

TO THE LEFT

IS A new power cone loud speaker employing electro-dynamic principles. Due to its unique construction, great volume may be handled without any possibility of rattling. An external source of direct current is necessary for operating this instrument. Either the storage battery or the electric light mains may be employed for this, depending upon which type of loud speaker is used. If the storage battery is employed, the current drain will not exceed that usually drawn by a 171 type tube. The price of the "Aristocrat" is $95.00 for storage battery operation or $90.00 for 110-volt d. c. operation

THE MAGNAVOX "ARISTOCRAT"

DESIGNED FOR HEAVY-DUTY WORK

THERE are four voltage taps on this new B supply device by National, three of which are adjustable. The detector voltage is from 22 to 45; the r.f. voltage, from 50-75; and the a.f. voltage from 90-135. The power-tube voltage tap delivers 180 volts. Price, with cord, switch, and plug, $40.00. Rectifier tube, $5.00

of Good Quality

EXPONENTIAL HORNS

A GOOD exponential horn is capable of very excellent results so far as quality reproduction is concerned, and is exceptionally efficient. The Temple Console to the right employs a 75-inch exponential air column, and is priced at $65.00. The Temple D'um below also has an exponential air column and comes in two models, using 54-inch and 75-inch air columns, priced at $29.00 and $48.50 respectively

A NEW SOCKET-POWER DEVICE

TOGETHER with a relay which automatically switches the house electric light supply to the B device illustrated, or to a trickle charger, depending upon whether the set is in use or not. These devices are by Stromberg-Carlson, of Rochester, and retail at: B device, $65.00 (with tube); relay, $11.00

FOR THE PHONO-RADIO COMBINATION

THE Platter Cabinet Company, North Vernon, Indiana, is responsible for this attractive cabinet of walnut plywood, in which space is provided for a complete radio installation and a phonograph. The price of the cabinet is $72.00. The same manufacturer will supply turntable, motor (electrical or mechanical), radio receiver, electrical pick-up device, and loud speakers of many patterns to fit into the cabinet, if desired

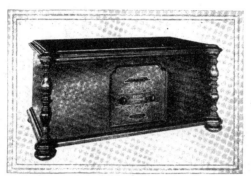

THE GREBE "SYNCHROPHASE" SEVEN
The circuit diagram of this receiver is shown on the next page

How the "Synchrophase" Seven Was Developed

By John F. Rider

WE REQUIRE a seven-tube receiver with single tuning control to be sold at a retail price of $135.00. Produce it." Such, in effect, might have been the wording of a memorandum from the board of directors of A. H. Grebe to the engineers responsible for the design of the "Synchrophase" Seven. Thereupon, a tedious process of laboratory and mathematical experiment sets in, for the "cut and try" methods of "mediæval" radio days are forgotten, since present broadcasting conditions would make such procedure impracticable.

Being familiar with the congestion among stations, the transmitting characteristics of the broadcasting stations, and the power used by the stations, the radio engineering department was able to select the number of tubes to be used as radio-frequency amplifiers, decide upon the detector system, and arrive at a decision concerning the number of stages of audio-frequency amplification necessary to produce a satisfactory receiver output signal. They were aware of the operating channels used by broadcasting stations in various localities, hence they could judge if their tuning system would be capable of separating the stations and of providing radio reception without a background of interference from other stations.

This was, however, not all a matter of guess work. Such perfectly engineered radio receivers are first produced on paper, with the aid of mathematics and the "slip stick" (a nick-name for the slide rule). The amount of time put into the design of radio receivers of modern times is almost inconceivable when compared to that of six or seven years ago.

After due deliberation, the decision of the Grebe engineers was to employ four stages of tuned radio-frequency amplification, a non-regenerative detector, and two stages of transformer-coupled audio-frequency amplification. The question then arose relative to the design of the radio-frequency system. Should it be shielded or unshielded? Development carried out prior to this time resulted in the birth of a new type of fieldless inductance—an inductance which could be used in a multi-stage tuned radio-frequency receiver without fear of coil interaction. This coil, known as the "binocular" inductance, was the brain child of P. D. Lowell, and is the evolutionary development of the toroidal winding. Since this coil is fieldless, interstage shielding was unnecessary; two, three, or four stages of unshielded tuned radio-frequency amplification may, therefore, be used without worrying about excessive regeneration. Eliminate the coil fields and shielding is unnecessary, since coil interaction will not take place. The elimination of shielding means economy without any sacrifice.

The function of shielding is also to preclude direct coil pickup, but since the fieldless coil will not pick up any energy, shielding is unnecessary on this account. The binocular coil, therefore, solved several problems in the design of the "Synchrophase" Seven. The coil itself consists of two solenoidal windings mounted closely together with their axes parallel and with the two windings connected in such a manner that their electromagnetic fields are opposing each other. This means that, if one coil starts to radiate a magnetic field, the other coil simultaneously radiates a magnetic field of equal intensity but of opposite phase or direction, and the two fields combine and counterbalance each other. The result of the combined fields is zero. Hence, no reaction between the tuned transformers is evident.

The same phenomenon applies to the pickup of waves by the transformers. A voltage introduced in one coil is reacted upon by the voltage introduced in the other coil by that passing wave. The two induced voltages react upon each other and the resulting induced voltage is zero.

Hence pickup of energy is not apparent. The fact that these coils are fieldless permits of a layout different from that possible when the coils used have strong fields. Interaction of fields is the greatest objection to the single-layer solenoid. It is a very efficient winding out possesses a very strong field, and total shielding is imperative in a receiver employing a number of single-layer solenoid radio-frequency transformers.

The next problem was the actual design of the binocular inductances. The mere selection of a form of winding does not complete the design problems of the tuned radio-frequency transformers. Solid wire is conventional for the average tuned radio-frequency transformer. Is it best for these coils? Does the form factor of the average coil apply to this type of winding? Can some other wire be used to greater advantage? Here is room for research and experiment.

The first binocular coils were wound with solid wire. Further experiments and calculations by the engineering department proved conclusively that "litz" wire improved the inductance-resistance ratio to a large extent and as far as practical results were concerned, coils wound with "litz" wire could be made to have a more uniform amplification curve over the broadcast frequency spectrum, in addition to improving the selectivity of the tuned circuit. Here again we find a deviation from the conventional path. "Litz" wire has frequently been frowned upon as unsuited for high-frequency work. The designers of the "Synchrophase" have overcome the difficulties and are utilizing "litz" to good advantage.

The occasion next arose for the development of an original system to overcome a frequently occurring fault with conventional tuned radio-frequency amplifiers. It is of paramount importance that the frequency response curve of the radio-frequency system be substantially flat. Furthermore it is desirable that the effects of the

232

grid-filament capacity within the tube be so nullified that the purchaser can use whatever tubes he may choose without fear of upsetting the tuned circuit balance obtained in the factory when the receiver was first tested. With many receivers the neutralizing condenser (or condensers) has to be readjusted whenever any changes in the tubes are made. The frequency response curve of the average tuned radio-frequency amplifier shows maximum response on some short wavelength (high frequency) with a falling characteristic at the wavelength is increased (frequency lowered). The system represented in Fig. 1 by the variable condenser C_1 and the two resistances, R_1 and R_2, was developed and incorporated to attain the two objectives mentioned at the beginning of this paragraph. The action of these units is twofold. First, they eliminate the effect of the grid-filament tube capacity upon the tuned circuit, particularly on the low settings of the tuning condensers, in such manner that tubes may be changed without affecting the original resonance setting. Second, they control the voltage being

give sufficient sensitivity and selectivity. Furthermore, the sideband suppression characteristics of the radio-frequency system are known and the introduction of a variable constant, which would be encountered with a variable regenerative detector, would upset the balance between the radio and audio amplifying systems.

THE AUDIO CHANNEL

THE audio amplifying system is closely associated with the radio-frequency system; in fact, it must be, for the reason that the frequency characteristics of the audio-frequency amplifying transformers are governed by the sideband characteristics of the radio-frequency system. If the sideband suppression of the upper audio register is great in the radio-frequency amplifier, the audio system must possess a certain rising characteristic. The slope of the rise is governed by the degree of suppression in the radio-frequency amplifier. Hence the two systems are closely associated. Being familiar with the sideband suppression in the radio-frequency

in the inclusion of a device which permits tonal flexibility. The problem arose during the process of development when the sales staff mentioned the fact that the aural fancy of the listener-in was apt to vary over a wide range. Could not some device be incorporated which would permit variation of the tone of received speech or music, so as to satisfy the individual tastes of the multitude? Some fans prefer a preponderance of low tones, while others are not so anxious about these low frequencies. The engineers decided that the best location for such a unit would be in the audio amplifying system, but a continuously variable change in the physical structure of the audio-frequency transformers to produce different response was impractical. Hence the "tone color" unit, consisting of a number of fixed capacities which can be shunted across the secondary of the second-stage audio-frequency transformer to change its operating characteristic, was originated. The "tone color" is controlled by a knob on the front of the panel and, by its manipulation, the listener is able to adjust the tone to suit his own taste. The capacities in the "tone color" vary from 0.00008 mfd. downwards.

THE PROCESS OF TESTING

ENGINEERING and originality has made possible the manufacture of all the necessary equipment, exclusive of the cabinets, in the Grebe plant, at Richmond Hill, Long Island.

The manufacture and testing of the tuned radio-frequency inductances is of especial interest. The winding is spaced yet the winding form is not grooved. This is made possible by means of a grooved slider which carries the wire as it is wound on the winding form. The grooves on the slider space the wire, and the turns are kept in place by means of a layer of clear lacquer which is sprayed upon the coil before assembly. Ingenuity in testing now manifests itself. The wire, as mentioned before, is "litz," and it is extremely important that all the turns remain intact. If a single strand is broken it will result in a steep rise in the radio-frequency resistance of the wire, with consequent increase in losses, and lower selectivity. The condition of the finished coils is tested on a d.c. bridge, accurate enough to show one broken strand. The total resistance of all the strands is balanced against a known resistance. One broken strand in the "litz" cable will deflect the meter in the bridge, in which case the coil is rejected. The satisfactory coils' radio-frequency resistances are then measured. The inductance value of a completed coil should be 310 microhenries.

The condensers are matched on a capacity bridge, each one being individually tested against a standard. The condensers are then grouped according to their respective capacities. A control is arranged which shows a variation of 7 micro-microfarads for the complete scale, and extremely small variations are, therefore, detectable. By means of this control it is also possible to determine increased effective resistance of the condenser under test. The bridge is fed from

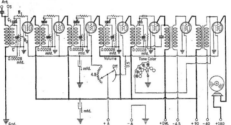

FIG. 1
Circuit diagram of the "Synchrophase" Seven

fed into the grid-filament circuit of the amplifying tube, so that the radio-frequency voltage fed into the tube is practically uniform over the complete tuning scale and the frequency response curve of the tuned system is sufficiently flat. With this arrangement, and the inherent lack of regeneration in the system, a high degree of stability and amplification is afforded.

The four stages provide ample selectivity and sensitivity and are designed to possess sideband characteristics with minimized suppression above 1000 cycles. This consideration is very important, and the presence of excessive regeneration would tend to nullify all the effects of scientific design. But the properties of the fieldless coil guard against this deterimental effect. With sufficient spacing between the inductances, the very small amount of external field, which cannot be eliminated completely, does no harm. Hence the regeneration present is uniform over the scale and is at no time sufficient to cause uncontrollable oscillation.

The detector system is the grid leak-condenser arrangement, affording maximum signal sensitivity and intensity. The compensating system utilized in the radio-frequency stages, is also resorted to in the detector input circuit, thus permitting the use of any detector tube without unbalancing the tuning system. A non-regenerative detector was decided upon because the four stages of tuned radio-frequency amplification

system, two audio stages are decided upon after a study of the amplifying powers of the system; these two stages possess sufficient magnifying power to afford a satisfactory output. The overall response curve of the audio system is shown in Fig. 2. The gain in transmission units is shown on the ordinate and the frequencies are shown on the abscissa. The curve is plotted on a logarithmic scale.

Particular consideration was given to the feminine speaking voice, the frequencies of which are difficult to amplify, and to the overtones of the high-frequency producing musical instruments. The result is that the transformers were designed to function satisfactorily on audio frequencies above 8000 cycles. This is easier said than done. A great deal of work was entailed before suitable transformers were produced. In order to realize satisfactory amplification on the upper audio register it was important to reduce the distributed capacitance of the secondary winding. This was accomplished by the use of three layers of insulation between each layer of winding. The distributed capacity of the secondary winding is approximately 18 microfarads. The importance of a low distributed capacity can be appreciated when one realizes that the higher it is, the more limited will be the frequency range of the amplifying unit.

An example of vision, and a knowledge of the buying public's whims and fancies, is displayed

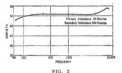

FIG. 2
The overall response curve of the "Synchrophase" Seven audio channel

a 1000-cycle source and the output of the bridge is connected to a single-stage audio amplifier, thus increasing the sensitivity of the system. The operator uses the headphone method of adjusting for capacitance by finding the minimum sound intensity. When the sound is minimum the capacity of the unknown is exactly that of the standard. A condenser with high losses does not permit of complete silence in the phones, and is rejected.

The audio-frequency transformers are also tested in an interesting manner. It is highly important to know the response of the transformers to be placed in the receiver, but it would be a tedious procedure to plot a response curve for each individual unit, hence each transformer is compared with a standard on certain frequencies. A tube oscillator, generating a 500-cycle note and a 6000-cycle note, is connected to a standard amplifier. The output of this amplifier is connected to the transformer to be tested, and the output of the transformer under test is connected to a vacuum-tube voltmeter, the output circuit meter of which gives visual deflections indicative of the response of the transformer under test. The circuit is so arranged that, by means of an anti-capacity switch, the transformer under test can be replaced by the standard and the deflections compared. By means of a switch, the oscillator can be adjusted to generate either the 500- or the 6000-cycle signal. This signal is free of harmonics and is constant at all times. The deflections with the standard transformer are therefore constant.

The complete receiver undergoes several tests, and the method of testing is also original and novel. A buzzer-modulated master oscillator, tuned to 200, 400, and 530 meters (1500, 750, and 566 kc.) feeds a master antenna. The buzzer modulation is accomplished by breaking the plate voltage supply with the interruptions of the buzzer. The operator testing the receiver (several operators are testing at one time) has his own antenna of standard inductance and capacitance value. He first adjusts the receiver at 200 meters by tuning it to resonance with the 200-meter oscillator signal. The receiver output is then noted by means of a tube voltmeter connected to the output circuit of the receiver. After the receiver is adjusted on 200 meters, adjustments are made on 400 and 530 meters, and the second harmonic of 530, which is 265 meters (1130 kc.). In this way each receiver is tested on four wave-

THE GREBE CONE LOUD SPEAKER

lengths. This is indeed a comprehensive test, for it will bring to light any defects in design upon any of the wavelengths within the range employed for broadcasting purposes. If the tube voltmeter does not show standard output on all four waves, the receiver is rejected for a re-examination.

The problem of conductive coupling in the receiver to adjacent leads was overcome by the use of the chassis as the negative filament lead, thus eliminating numerous long leads. The filament wiring in the receiver consists of only the positive polarity wires. The negative lead is formed by the chassis. The condensers are all grounded upon the chassis.

The mechanical alignment is facilitated by punching the complete chassis in one piece. It is made out of aluminum and stamped out on a 60-ton press. The chassis, after the stamping, carries all the mounting holes and brackets, thus assuring correct alignment. The receiver, from start to finish, is carried from one operation to another by means of a conveyer system approximately 1000 feet in length. This conveyer consists of a belt or a roller as the occasion demands,

and the partially assembled receiver moves from one operation to another until it finally reaches the final department.

THE GREBE CONE

THE electrical development of the Grebe loud speaker is also of interest, particularly because this field of endeavor requires highly trained engineering. The development of a loudspeaker does not consist of the mere selection of steel, iron, wire, and a paper cone. Let us consider for a moment an important consideration which the fan generally passes over very lightly. This is the angle of the apex of the cone, and the size of the cone. The information relative to the size of the cone will doubtless be of interest to constructively inclined radio fans. According to the engineer in charge of cone construction, their experiments showed that a 20" cone was the best compromise and that increases above this diameter did not justify the additional space required. Experiments showed that very little is gained by using a cone of larger diameter. Reduction in size, however, showed a material loss.

As to the angle of the apex, 20 degrees is also the best compromise for efficiency and quality. The greater the angle, the greater the efficiency but the poorer the quality of reproduction. The 20-degree angle was considered the best for quality and efficiency. Many articles published in this and other periodicals have stressed the importance of a large value of inductance for loud speaker windings in order to produce satisfactory response on the low frequencies. With this in mind, a value of 1.7 henries was selected as the loud speaker coil inductance. The shape of the armature is also an important consideration, and by using an armature that is wide and short, the lowest moment of inertia is obtained. The selection of the material for the armature also requires care and silicon steel was chosen instead of Swedish iron, because the losses of silicon steel are less on the higher audio register. The difference between Swedish steel, iron, and silicon steel is not appreciable on the lower audio register but it approaches an appreciable value on the higher audio frequencies.

The elimination of harmonics in the average loud speaker is a paramount item because their presence will not permit true reproduction. To attain this result it is necessary to minimize magnetic saturation.

The testing of the loud speaker is carried out by first subjecting it to a series of audio frequencies obtained from a beat note oscillator. This beat note oscillator consists of two radio-frequency oscillators adjusted in a manner which permits the generation of a beat note; this note is passed through several stages of audio-frequency amplification and then into the loud speaker. One of the radio-frequency oscillators, is variable in tuning and the frequency of the beat note is variable between 50 and, approximately, 20,000 cycles. This test will bring to light any defects in the loud speaker mechanism which would result in a rasping sound or a rattle when it is placed into operation. Another test consists of the reproduction of an organ record played upon a talking machine and fed into the loud speaker by means of an electric pick-up and amplifier combination. The organ selection has a wealth of low notes, and these are particularly desired for testing purposes, since the amplitude of these low frequencies is high. This test will bring to light any defects in the placement of the armature. Another test consists of the application of the plate current of a 171 tube through the windings of the loud speaker, first in one direction and then in the other, to test for magnetic saturation when it is in operation.

AN INTERIOR VIEW OF THE "SYNCHROPHASE" SEVEN

AS THE BROADCASTER SEES IT

BY CARL DREHER

Technical Problems for Broadcasters and Others

TECHNICAL problems presented in this section in the past have consisted of numerical problems requiring more or less lengthy solutions, so that only one question could be allowed to an issue. We shall vary these occasionally by a series of questions requiring only brief answers, like those below. The subjects, while not strictly confined to the design and operation of broadcast transmitters and associated apparatus, will have some connection with those considerations of quality in reproduction which are of most interest to the professional broadcast technician.

QUESTION 1. Why does sharp tuning tend to drop out the high audio frequencies associated with a carrier?

Answer. The audio energy of speech or music exists in the side bands accompanying the carrier in question. If, for example, the carrier is of the order of 600 kilocycles, a frequency within the broadcast band, and the audio note being transmitted is 5 kilocycles, the side bands will have a frequency of 595 and 605 kilocycles. A sharply tuned circuit resonant to 600 kilocycles will include perhaps one kilocycle on either side, and will discriminate to a greater and greater degree against the higher audio frequencies, which lie further out from the central or carrier frequency. Hence the 595- and 605-kilocycle currents, in the present example, may be lost altogether, and with them the 5-kilocycle audio note which they would yield on demodulation. By this mechanism, called "side band cutting," the higher musical frequencies are likely to be lost.

QUESTION 2. Why does slight detuning of a sharply resonant circuit, with reference to the carrier frequency, tend to reinforce the higher notes contained in the modulation?

Answer. See Fig. 4, which has reference to the same example used to illustrate the answer to Question 1. Now, however, the circuit has been tuned so that the peak of the resonance curve occurs at 602.5 kilocycles instead of at the carrier frequency of 600.0 kilocycles. As a result, the peak of receptivity, after demodulation, has been shifted from currents of frequencies in the neighborhood of zero cycles to currents in the neighborhood of 2500 cycles per second. If, owing to the steepness of the resonance curve, the cut-off characteristic of the circuit becomes serious one kilocycle to either side of the peak, as assumed in the answer to Question 1 above, with the peak set at 602.5 kilocycles the audio frequencies up to 3.5 kilocycles are nevertheless included in the received band. It will be seen that through this mechanism a sharply tuned radio frequency circuit may be used to some degree as an equalizing or frequency correcting device in reception. This observation may be checked in practice. A similar phenomenon is found in some super-heterodyne receivers.

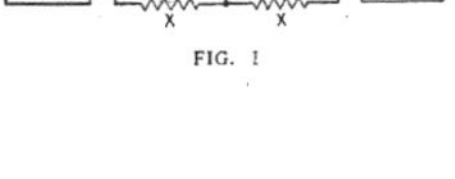

FIG. 1

QUESTION 3. What change in loudness is normally noticeable by the human ear?

Answer. A change of 3 TU is usually observed in speech or music even by listeners ho are not expecting it. This corresponds to a change in energy of 2:1. A practiced listener looking for a change may detect differences of the order of 1 TU in a sustained note, without much difficulty. This is equivalent to an energy change of about 25 per cent.

QUESTION 4. What size steps, in TU, would you allow in a smooth gain control?

Answer. Since, according to the answer to Question 3 above, a change of 3 TU is noticeable, the maximum allowable step in a smooth gain control is 2 TU.

QUESTION 5. What error do non-technical observers invariably make in estimating relative loudness of sounds?

Answer. They underestimate radically. As is well known, the human ear follows a logarithmic response characteristic, which is as much as to say that a large increase or decrease in the stimulating energy results in a slight change in the loudness subjectively perceived. Or, more definitely, *multiplying* the energy by a fixed ratio results only in *adding* to the loudness an increment proportional to the logarithm of the energy ratio. This is expressed mathematically in the formula for the telephonic transmission unit, which is a measure of subjective loudness:

$$TU = 10 \log P_1/P_2$$

where P_1 and P_2 are the powers corresponding to the two telephone currents under comparison. Non-technical listeners, being unaware that hearing is a logarithmic process, usually apply to sounds the standards of measurement and estimating to which they are accustomed in dealing with distances, for example. A broadcast listener, comparing reception from two stations, one of which is a stage of audio amplification above the other, will say that the first station is "fifty per cent. louder," or "one hundred per cent. louder." Actually one a. f. stage may correspond to 20 TU, or a sound energy ratio of 100. In other words, a non-technical listener is apt to speak of an energy ratio of the order of 2:1 where the actual figure is of the order of 100:1.

Commercial Technical Publications

AMONG catalogs and technical publications received we must mention the September, 1927 issue of the General Radio *Experimenter*, a four-page paper sent out monthly to concerns and individuals on the mailing list of the General Radio Company of Cambridge, Massachusetts. [This publication may be secured by our readers by filling out the "Manufacturers' Booklet" coupon appearing in the advertising section in this and other issues. Booklet number 74.—*Editor.*] Thi firm, as is well known, specializes in communication laboratory and measuring apparatus, such as standards of inductance, resistance, and capacitance, oscillators, oscillographs, audibility meters, bridges, artificial telephone lines, etc. Its engineers have played no small part in reducing radio designing to a respectably exact

science. The September, 1927 issue of the *Experimenter* is of special interest to broadcasters in that it contains an article on "Design and Use of Attenuation Networks," by Horatio W. Lamson. The subject was discussed, it may be remembered, in this department for September, 1927 (Pages 293–294). Mr. Lamson's paper covers the same ground, in part, with the addition of a number of formulae and a detailed description of the General Radio Company's variable attenuation networks. Given, as in Fig. 1, a source of impedance Z and an absorbing circuit or "sink" of the same impedance, joined by an H-network as shown, with currents I_0 and 1 leaving the source and entering the sink, respectively, for a definite number of transmission units of attenuation N we may calculate the arms X and Y from

$$X = \frac{Z}{2}\left(\frac{k-1}{k+1}\right) \qquad (1)$$

$$Y = 2Z\left(\frac{k}{k^2-1}\right) \qquad (2)$$

$$\text{where } k = \frac{I_0}{1} \qquad (3)$$

or, in TU
$$k = 10 \text{ N/20} = \text{Antilog N/20} \qquad (4)$$

If, as shown in Fig. 2, we make all five branches of the H-network adjustable by steps, moving five switch arms in unison to the proper switch points through a single control, the characteristic impedance Z of the network may be maintained constant to match the impedances to either side, while the attenuation is varied in steps. This is the principle of the Type 239 Variable Attenuation Network supplied by the General Radio Company. The box is equipped with two multiple switches. In the size which affords a total attenuation of 55 TU, one decade is calibrated in steps of 5 TU, and the other in steps of 0.5 TU. Another size goes up to a total of 22 TU, and in this case the steps are 2 TU and 0.2 TU. Characteristic impedances of 600 and 6000 ohms are carried in stock or built to order. Some types are provided with a center tap for the Y-branch, where a ground may be applied.

A somewhat older type of variable pad supplied by the same manufacturer is adjusted by the addition or subtraction of fixed sections, which are cut in or out by means of four-pole double-throw switches, as shown in Fig. 3. This is known as the Type 249. The manipulation of this form is necessarily somewhat less convenient, but in many cases it serves the same purpose and the cost is about half of the rotary switch form. There are eight-section boxes affording a total attenuation of 110 TU in steps of 1 TU, and six-section boxes with a total of 63

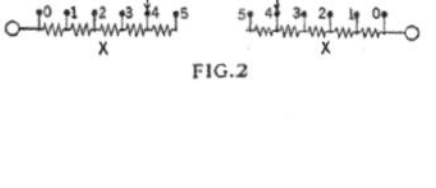

FIG. 2

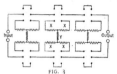

FIG. 3

TU in the same steps. The same characteristic impedances, 600 and 6000 ohms, are available. As explained in the September, 1927, RADIO BROADCAST discussion, the simpler T-networks may be used in place of the H-type where a balanced circuit is not essential. The General Radio Company also supplies the sectional networks in the T-connection, at a somewhat lower price, inasmuch as the number of coil-taps and switch-arms required is less.

The Portrait of the Author

MR. FRANKLYN F. STRATFORD, who embellishes these articles with clever pictures, sometimes sketches my likeness for the customers, so that they may not shoot the wrong man when they take offense at my sooth-sayings. In so doing Mr. Stratford has perpetrated a libel on me. I ask that interested readers turn to the illustration in the November, 1927 issue, over the quotation, "Let the ghost call at my office." The picture includes the ghost, myself, and part of the office. About the ghost I say nothing; Mr. Stratford's idea of a ghost is no doubt as good as anyone's. And my own likeness is not bad. It is true that my nose is depicted as about the size of Cyrano de Bergerac's, and not as well-shaped, but so God made me. My head has a contour not unlike that of a truncated cone, in Mr. Stratford's sketch, but such geometrical outlines are not lacking in esthetic value, and it would be worst if the cone were upside down. What I object to is the office furniture, and particularly the desk. The desk depicted seems about four feet wide. I wish the world to know that my office boy sits

at such a desk. My own desk was made to order in Circassian walnut, and cost $3000. The top is a solid piece of Florentine marble, 150 square feet in area. Nymphs and satyrs are painted on the sides. When next Mr. Stratford draws my desk let him take an extra column and do justice to his subject.

Linguistic Observation

THE vilest French and German I have ever heard—in the sense of execrable accent, not of content, be it said—is on radio broadcast programs, and I refer to many large stations as well as small ones, if you please. The radio announcers, with a trifling number of exceptions, pronounce foreign languages like so many high school freshmen. The better vaudeville circuits are vastly superior in this detail. When a girl like Kitty Doner spills some French before the assembled intelligentsia at the New York Keith's 81st Street, it is Parisian French as you might hear it on the boulevards. But the announcers drive any half-way educated listener to thoughts of homicide whenever they have occasion to pronounce a simple phrase like *Danse Russe* or *Wiener Blut*. What ails the announcers and their bosses? Don't they know any better? Then let them go back to school. Or don't they care? Then let them seek some business in which they don't have to appear before the public.

Memoirs of a Radio Engineer: XXI

INSTEAD of going forward a few more months in the recital of these memories, at this point I wish to regress briefly in order to include an incident which was omitted in its proper place. It returns to my recollection when I hear of some broadcast listener ending his mortal existence by throwing a length of antenna wire across a high tension line, or through a fall from a roof. Such fatalities are too infrequent to deserve mention, perhaps, in a country which takes no account of some 100,000 deaths in four or five years through automobile accidents, but the victims are just as dead, although not killed in a popular manner. I might have joined their number, thus

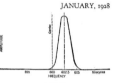

FIG. 4

swelling the total slightly, but for luck. The time was about 1914, when I was feeding a crystal receiving set from a one-wire antenna about four hundred feet long. So long was it, in fact, that it crossed a city street. After some months of fine reception, during which I was able to boast to the other amateurs about the great distances from the phones at which signals might be heard at my house, the antenna yielded to the elements one night, and in the morning it was down. On the street which it crossed there was an electric power line, the cross arms carrying six or nine heavy rubber covered wires. My copper strand now lay across these conductors. I went to the house at the end of the block and looked up at the transmission line. It did not appear dangerous, and I decided that it must be a 110-volt circuit, and that nothing much could happen. However, I decided to be very prudent, so I went below for a bamboo trout rod, made a cast for my antenna wire, hooked it, and lifted it off the transmission line. It fell back, however, and I grasped it impatiently in my bare hands and swung it clear. Nothing happened. What could happen even if a boy radio operator got his antenna tangled up with a 110-volt line covered with an eighth of an inch of rubber? But the other day I passed along that block, and, thirteen years older, looked up at the transmission line, which was unchanged in equipment and fittings, as far as I could judge. It is not a 110-volt line, and probably never was one. It may be carrying its load at a tension of 2200 volts, or perhaps 4400. Either would have been enough. Some people have luck. The insulation on the wires was not frayed.

The Radioman's "Britannica"

DRAKE'S RADIO CYCLOPEDIA. By Harold P. Manly. Published by Frederick J. Drake and Company. Chicago. Price $6.00. Pages, about 800. Illustrations, 950.

MR. MANLY'S 'Cyclopedia" is a comprehensive compilation of all the conceivable information which a set builder and designer needs for ready reference. The reviewer could not, of course, read the entire work in detail because that would require several weeks, but many pages were critically examined. The first impression gained is the completeness with which the author covers his subject.

As an example of the wealth of detail on a particular subject, under the heading of "Receiver, Audio Amplifier for," constructor's circuit diagrams, showing the arrangement of parts, wiring, outstanding performance qualities, and specifications of components, are given for fifteen types of audio amplifier systems, including three-stage choke-coupled, three-stage transformer-coupled, one transformer and two choke-coupled, three-stage double impedance, one transformer and one push-pull stage, two push-pull stages, three resistance-coupled stages, one transformer and three resistance stages with output choke, two transformer stages with output transformer,

two transformer stages with parallel output tubes, two transformer stages with potentiometer control.

In the matter of receiving circuits, the following types are described: Browning-Drake, crystal, four-circuit, long-wave, loop, "N" circuit, neutrodyne, one-tube, regenerative, Rice, Roberts, short-wave, single-circuit, single-control, super-heterodyne (13 pages), super-regenerative, and tuned radio-frequency.

A table, "Coil Design, Advantages and Disadvantages of Types of Coils," is an example of the practical kind of information collected in a manner suited to easy reference. This table lists various elements of coil design, such as type of winding, shape of winding, proportion of winding, wire insulation, wire size, material of winding form, design of winding form, fastening of winding, material of supports and connections to winding, and gives under each of these headings three to six different practices which may be followed. For example, under "Material of Form," paraffined paper or cardboard, fibre and "mud" dielectrics, dry paraffined wood, hard rubber, phenol, bakelite, and glass are listed. Under each of these design possibilities, the characteristics of each type is given in the simplest terms, "poor, fair, good, and best," for the following factors: Durability, most in-

ductance, least resistance, little distributed capacity in small field. Thus, "type of winding, (1) cylindrical, single wire, close wound," is declared the "best" from the standpoint of durability, most inductance, and least resistance, but "poor" from the standpoint of less distributed capacity and small field. Honeycombs, on the other hand, are good from the standpoint of durability, inductance, less distributed capacity, and small field, but only fair from the resistance standpoint. By the aid of the table, the designer may select the type of winding form which best suits his purpose.

In addition to covering constructional information, due space is given to practical operation, general theory, and design. Power supply for A, B, and C potential, design of receivers with alternating-current supply for filament lighting, single-control, shielding, are some of the subjects fully treated, indicating the volume to be up to the minute.

One mysterious feature of the whole work is that hardly a single line of credit is given for sources of information and assistance. Not a page in this book is numbered. The reviewer does not believe, however, that a more satisfactory addition to the experimenter's library, in any one volume, can be made.

—E. H. F.

Now
AC Electric Radio

Licensed under Andrew Hammond patent

To owners of a "B" eliminator:

Balkite "A" is like Balkite "AB" but for the "A" circuit only. It enables you to make an electric installation at very low cost. $35.

Balkite "B"

The accepted, tried and proved light socket "B" supply. One of the longest lived devices in radio. Three models, $22.50, $35, $42.50.

Balkite Chargers

Standard for "A" batteries. Noiseless. Can be used during reception. High rate or trickle. Three models, $17.50, $9.50, $7.50.

There are special models for 25-40 cycle current at slightly higher prices. Prices are higher West of the Rockies and in Canada.

Without the uncertainty of untried apparatus

And without any sacrifice in quality of reception

Of course you want an AC electric receiver. For its convenience. Now you can have it, without the uncertainty of untried apparatus and without sacrificing quality of reception.

Simply by adding Balkite *Electric* "AB" to your present radio set. Balkite *Electric* "AB" replaces both "A" and "B" batteries and supplies radio power from the light socket. It contains no battery in any form. It operates only during reception. It makes any receiver an electric set.

This method makes possible the use in electric reception of standard sets and standard type tubes. Both are tried and proved, and give by far

the clearest and truest reproduction. With this method there is no waiting for tubes to warm up. No difficulty in controlling volume. No noise. No AC hum. No crackling or fading of power. Instead the same high standard of reception to which you are accustomed.

In this method there is nothing experimental, nothing untried. It consists of two of the most dependable products in radio—a standard set and Balkite. And if you should already own a radio set, the cost of equipping it with Balkite is only a fraction of the cost of a new receiver.

By all means go to AC reception. Its convenience is the greatest improvement in radio. But be as critical of an AC receiver as you would of any other. That your AC receiver be a standard set equipped with Balkite *Electric* "AB." Then it will be as clear and faithful in reproduction as any receiver you can buy.

Two models, $64.50 and $74.50. *Ask your dealer.* Fansteel Products Co., Inc., North Chicago, Illinois.

Chicago Civic Opera

on the air Thursday Evenings, 10 o'clock Eastern time. Over stations WJZ, WBZA, WBZ, KDKA, KYW, WGN, WMAQ, WBAL, WHAM, WJR, WLW, WENR, 10:30 Eastern time: WEBH, KSD, WOC, WOW, WCCO, WHO, WDAF.

BALKITE HOUR

Balkite
ELECTRIC AB

{ contains no battery }

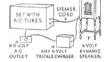
The *Radio Broadcast*
LABORATORY INFORMAT
SHEETS

THE RADIO BROADCAST Laboratory Information Sheets are a regular magazine and have appeared since our June, 1926, issue. They cove of information of value to the experimenter and to the technical radio ma purpose always to include new information but to present concise and a the most convenient form. The sheets are arranged so that they may b magazine and preserved for constant reference, and we suggest that each s with a razor blade and pasted on 4" x 6" filing cards, or in a notebook. Th be arranged in numerical order. In July, 1927, an index to all Sheets a that time was printed.

All of the 1926 issues of RADIO BROADCAST are out of print set of Sheets, Nos. 1 to 88, can be secured from the Circulation Doubleday, Page & Company, Garden City, New York, for $1.00. Som asked what provision is made to rectify possible errors in these Sheets tunate event that any such errors do occur, a new Laboratory Shee number will appear.

—THE E

No. 153   RADIO BROADCAST Laboratory Information Sheet   J

Standard- and Constant-Frequency Stations

BROADCASTERS WITH ACCURATE FREQUENCIES

THE *Radio Service Bulletin*, published monthly by the Radio Division of the Department of Commerce, Washington, District of Columbia, contains a list of standard and constant frequency broadcasting stations as determined by the Bureau of Standards. This bureau makes measurements on an average of about three times a month on the transmissions of a small number of stations and as a result of these tests data are published in the *Bulletin* on those stations which have been found to maintain a sufficiently constant frequency to be useful as standards. These are known as "Standard Frequency Stations." The list of "standard, frequency stations" is supplemented with a list of "constant frequency stations." No regular tests are made on these latter stations but each station in the list employs a special device, such as a crystal, to maintain its frequency accurately so that they can be generally relied upon to maintain their correct frequency.

STANDARD FREQUENCY STATIONS

Call Letters	Frequency Kc.
WEAF	610.00
WRC	640.00
WJZ	660.00
WGY	790.00

WBZ
KDKA
WBAL

CONSTANT FREQUENC

Call Letters

WMAQ
WJAD
WCCO
WTAM
WEAR
WHBM
KGO
KTHS
WCAD
WJJD
WLS
WSM
WKAQ
KOA
KFAB
WBAA
WHK
WMBI
WAHQ
WEBJ
KWUC
KFVS

No. 154   RADIO BROADCAST Laboratory Information Sheet   Ja

The 112-A and 171-A Type Tubes

OPERATING CHARACTERISTICS

TWO new power tubes have recently become available; they are designed especially for use in the output of a receiver. These new tubes employ an improved type of filament which gives high emission at a filament current of 0.25 amperes at 5 volts. They are exactly similar to the older UX-112 and UX-171 type tubes with the exception that the filament consumption is only half that of the older types. The filament of the corresponding 112 and 171 type tubes is 0.5 amperes at 5 volts. The other characteristics of these new tubes remain the same as those of the 0.5 ampere filament tubes. These characteristics are given below.

The UX-112-A (cx-312-A) may be satisfactorily used as a detector, general-purpose tube, or as a power tube in the last stage of a receiver. When used as a detector, the plate voltage should be 45

volts. The UX-171-A (cx-371-A), in the last stage of a receiver, and combination or output transform in the plate circuit to keep the of the loud speaker.

The advantage of these new greater efficiency. Under the sam voltage they produce the sam the corresponding 0.5-ampere half as much filament current.

These tubes must not be subs 112 (cx-112)or UX-171 (cx-171) without changing the values of trol resistances or rheostats if th they take the same filament c type tube, it follows the filament designed for the latter tube ma junction with these new tubes.

TYPE	FILAMENT VOLTS	FILAMENT CURRENT	PLATE VOLTAGE	NEGATIVE BIAS	PLATE IMPEDANCE	AMP. CONSTANT	F Cu
UX-112-A (CX-312-A)	5	0.25	90	6	8800	8	
			135	9	4800		
			157½	10.5	6800		
UX-171-A (CX-371-A)	5	0.25	90	16.5	2500	3	1
			135	27	2700		1
			180	40.5	2000		2

We Could Charge More— But a *Better* Transformer Can't Be Made

IMITATED everywhere—never equalled—the S-M 220 audio transformer stands out to-day as the finest for audio amplification that money can buy just as it did when introduced a year and a half ago. The 220 has been copied in one or more of its characteristics by every high-grade transformer put on the market since then—in the rising low note or in 5000 cycle cut-off, features first offered by S-M. That's proof that the principles the 220 introduced are right—that the market is still trying to catch up with the leader.

Don't be misled by exaggerated claims—for it takes plenty of core and wire to make a good transformer. The 220 has from 25 to 50 per cent more steel and copper in its construction than any other transformer on the market. That alone means the high primary impedance through which real bass note amplification is made possible.

That's why S-M 220's and 221's are specified in more popular receiver designs—why they have outsold every other transformer in their price field. That's why they're sold on an unconditional money-back guarantee to give better quality than any other audio amplifying device available.

We could charge from 25 to 50 per cent more than we do, but at no price can you get a better transformer. The 220 audio is $8.00, and the 221 output is $7.50. They are priced low, but, you can't buy a better audio coupler at any price, for there's none better made.

Just as the 220 transformer, compared with all other makes, gives you the sense of Gibraltar-like sturdiness and dependability truly in a class by itself, so does its performance far outclass that of other transformers.

The Finest Tone You've Ever Heard

—and complete A. C. operation. Complete light socket operation of the Improved Shielded Six using A. C. and power tubes is an accomplished dependable fact. You may build either the battery model with standard tubes or the A. C. model with the new C-327 and CX-326 tubes—and at a cost of less than half what you'd pay for nearly equivalent performance in a factory-built set. Every one of the thousands who built last year's Shielded Six said the same thing—"The Six has the finest tone I ever heard." And now the new and improved 1928 model of this famous receiver, with the same fine tone as the original, and tremendously improved selectivity and distance getting ability is available for light socket, battery, or eliminator operation. Above all the Six is guaranteed to have finer tone than any other set you can build or buy. The 630 Shielded Six kit in the battery or eliminator model is $95.00 and for complete light socket operation $99.00.

The New 240 Audios

We can't make the 220's cheaper but if you desire a transformer somewhat lower in price, taking up a little less room, and with a little less core and wire, the new 240 audio and 241 output transformers are available—superior to most other transformers, and far and away ahead of anything in their price field. They have the same general characteristics as the famous 220's and 221's, but provide slightly less accentuation of frequencies below 80 cycles. They have the same 5000 cycle cut-off for which 220's are famous, eliminating the objectionable whistles and heterodyne squeals of congested broadcasting. The 240 audio sells for $6.00 and the 241 output at $5.00. Hard to beat at any price, they are impossible to equal at these prices. And—you can bring your old set up to the minute using them—they're small enough to fit in most anywhere.

328 Super Power Transformer

A heavy power transformer for either full wave or half wave rectifiers—for UX-281 (or UX-216-B) rectifier tubes. Will furnish 480 volts at 100 milliamperes of thoroughly filtered direct current using two UX-281 tubes, the S-M 331 Unichoke, and only six microfarads of filter condenser. This is power for a 210 push-pull amplifier at full voltage and to furnish receiver B power as well. Consists of two 550 volt secondaries, two 7½ volt, 2½ ampere filament windings and one 1½ volt, 2 ampere filament winding for UX-226 tubes. Price $18.00.

SILVER-MARSHALL, Inc.

838-B West Jackson Blvd.   Chicago, Ill.

Only a few of the real S-M developments are listed here. 10¢ in stamps will bring you more real information on power equipment, A. C. operation, and other pertinent subjects than you can read in a week.

```
SILVER-MARSHALL, INC.
838-B West Jackson Blvd., Chicago.

    Please send me all data on S-M audio transformers, power
equipment, and new A.C. improved Shielded Six.

Name .......................................

Address .....................................
```

The A. C. Improved Shielded Six—a completely light socket operated batteryless set using the new A. C. tubes. It's illustrated above, with its complete dry ABC power unit—less than 7 inches square. Price, 630 A. C. receiver kit, $99.00, and 652A, ABC socket power kit, $34.50.

No. 155   Radio Broadcast Laboratory Information Sheet   **January, 1928**

Wave Traps

THREE CIRCUITS

THE trend of broadcasting, for sometime, has been toward the use of high power, and this has made the problem of selectivity a serious one for many listeners located within a few miles of a high-power broadcasting transmitter. When difficulty is experienced in satisfactorily tuning-out such a station, it will be advisable to incorporate a wave trap in the antenna circuit. Wave traps are very easily constructed and cost little. They consist of any ordinary coil and a condenser, connected in the antenna circuit, and adjusted to absorb a large amount of the energy being received from the interfering station. The traps may be connected in several ways, as indicated on the diagram. The arrangement shown at A will give most complete elimination of the undesired signal but may also cause a considerable decrease in volume of stations operating on adjacent channels. The arrangement shown at B is probably the most flexible manner in which to connect a wave trap. If the coil is arranged with several taps an adjustment can be arrived at which gives most satisfactory results. Arrangement C is only useful in case of mild interference. The circuit tunes very sharply and will effectively eliminate interference provided it is not too great.

In constructing a wave trap, coil L may consist of 47 turns of No. 22 wire on a 3-inch diameter form

if the tuning condenser C₁ has a capacity of 0.0005 mfd.; with a 0.00035 condenser coil L should consist of 60 turns. With either size, coil L₁ may consist of about 15 turns wound at the b end of the secondary coil. With arrangement B taps should be made at about every 10 turns.

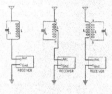

No. 156   Radio Broadcast Laboratory Information Sheet   **January, 1928**

Wavelength-Frequency Conversion

A TABLE FOR THE BROADCASTING BAND

ON LABORATORY Sheet No. 157 is given a wavelength-frequency conversion table covering the broadcasting band. Broadcasting is assigned to channels 10 kc. apart on frequencies that are divisible by 10. It is simple to use the table. If we knew that some station was transmitting on 1000 kc. we can determine from the table the corresponding wavelength, which in this case is approximately 300 meters. The wavelength corresponding to any given frequency can be determined by dividing the frequency in kc. into 300,000.

A 10-kc. separation between broadcasting stations is necessary to prevent bad interference between two stations on adjacent channels. When a broadcasting station is transmitting it actually uses a band of frequencies (side bands) 10,000 cycles wide —5000 cycles either side of the "carrier" frequency. The carrier frequency is the frequency assigned a station by the Federal Radio Commission, but as mentioned above, in the ordinary process of modulation a frequency band 10,000 cycles wide is used.

When a station is transmitting it also radiates a frequency exactly double its carrier frequency. The additional wave is called the second harmonic, being equal in frequency to the carrier frequency multiplied by two. Careful design and operation of the transmitter will keep these harmonics small in amplitude and this is essential if interference is to be prevented. If a station transmits on, say, 600 kc. and also radiates a strong second harmonic with a frequency of 1200 kc., it will interfere with another station transmitting on a carrier frequency of 1200 kc.

Any radio station might be considered to have two ranges; first the broadcasting range, being the distance area over which the program on the station may be received satisfactorily and, secondly, the interference range, being the area over which a station causes interference due to the production of a heterodyne whistle between its carrier and the carrier of another station. The first range is much smaller than the second and a station having a service area of 100 miles will have an interference range of probably about 1000 miles.

No. 157   Radio Broadcast Laboratory Information Sheet   **January, 1928**

Table for Wavelength-Frequency Conversion

Kc.	Meters	Kc.	Meters	Kc.	Meters	Kc.	Meters
550	545.1	800	374.8	1,050	285.5	1,300	230.6
560	535.4	810	370.2	1,060	282.8	1,310	228.9
570	526.0	820	365.6	1,070	280.2	1,320	227.1
580	516.9	830	361.2	1,080	277.6	1,330	225.4
590	508.2	840	356.9	1,090	275.1	1,340	223.7
600	499.7	850	352.7	1,100	272.6	1,350	222.1
610	491.5	860	348.6	1,110	270.1	1,360	220.4
620	483.6	870	344.6	1,120	267.7	1,370	218.8
630	475.9	880	340.7	1,130	265.3	1,380	217.3
640	468.5	890	336.9	1,140	263.0	1,390	215.7
650	461.3	900	333.1	1,150	260.7	1,400	214.2
660	454.3	910	329.5	1,160	258.5	1,410	212.6
670	447.5	920	325.9	1,170	256.3	1,420	211.1
680	440.9	930	322.4	1,180	254.1	1,430	209.7
690	434.5	940	319.0	1,190	252.0	1,440	208.2
700	428.3	950	315.6	1,200	249.9	1,450	206.8
710	422.3	960	312.3	1,210	247.8	1,460	205.4
720	416.4	970	309.1	1,220	245.8	1,470	204.0
730	410.7	980	303.9	1,230	243.8	1,480	202.6
740	405.2	990	302.8	1,240	241.8	1,490	201.2
750	399.8	1,000	299.8	1,250	239.9	1,500	199.9
760	394.5	1,010	296.9	1,260	238.0		
770	389.4	1,020	292.9	1,270	236.1		
780	384.4	1,030	291.1	1,280	234.2		
790	379.5	1,040	288.3	1,290	232.4		

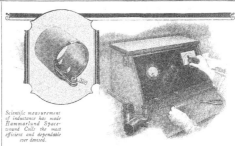

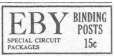

No. 158   Radio Broadcast Laboratory Information Sheet   January, 1928

The Three-Tube Roberts Reflex

CIRCUIT CONSTANTS

THERE have been many requests from readers for further information on the Roberts 3-tube receiver illustrated in the August, 1927 issue of RADIO BROADCAST on page 209. This receiver is a reflex set consisting of a stage of r.f. amplification, a regenerative detector, one stage of reflexed transformer-coupled audio amplification, followed by another straight audio stage. The circuit, which was not given in the article mentioned above, and which many readers have requested, is published on Laboratory Sheet No. 159. The list of parts is given below.

L_1, L_4—R. F. transformer, L_1 may consist of 45 turns of No. 24 wire wound on a 3-inch tube. L_1 should contain 40 turns of No. 24 wire with a tap at each 10 turns. L_1 should be wound alongside the filament end of L_4.

L_2, L_4, L_4—Interstage r.f. transformer. L_3 and L_4 have the same specifications as L_1 and L_4 with the exception that L_4 should be wound with No. 26 or No. 28 wire and should only be tapped at the exact center instead of at every 10 turns. That end of L_4 nearest the grid end of L_1 should connect to the plate of the r.f. tube, the center tap connects to transformer T_5 and the other end of L_4 connects to the neutralizing condenser. L_4 is a movable tickler coil consisting of 20 turns of No. 26 on a 1½ inch tube.

T_1, T_2—Any good audio transformers.
T_F—Any good output transformer.
C_1, C_5—0.0005-mfd. variable condensers.
S_1—Antenna tap switch.
S_2—Filament switch.
J_1—Double-circuit interstage jack.
J_2—Single-circuit jack.
V—Volume control, 50,000-ohm variable resistance.
C_3—Neutralizing condenser, 0.000015 mfd.
C_4—Grid condenser, 0.00025 mfd.
R_1—1-megohm grid leak.
R_2—10-ohm rheostat.
R_3—0.5-ampere fixed filament control resistance.
C_1—0.001-mfd. fixed condenser.
C_2—0.0001-mfd. fixed condenser.
C_6—0.00025-mfd. fixed condenser.
Eleven binding posts
Three sockets
Hook-up wire

For best results a power tube should be used in the last socket. If a 171 type tube is used with 180 volts on the plate, the C bias required is 40.5 volts.

When the receiver has been completed it should be neutralized by tuning-in some station, adjusting the tickler until the detector oscillates and a whistle is heard and then varying the neutralizing condenser until the whistle changes in pitch the least amount (its loudness will change considerably) as C_1 is varied.

No. 159   Radio Broadcast Laboratory Information Sheet   January, 1928

The Three-Tube Roberts Reflex

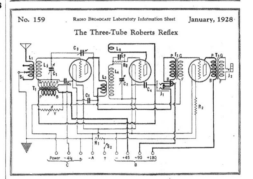

No. 160   Radio Broadcast Laboratory Information Sheet   January, 1928

Fading

HOW IT MAY BE PLOTTED

ON THIS Laboratory Sheet is published a curve showing how the signal strength from station WGY varied over a period of about 10 minutes during a fading test on this station made during the early part of November.

Anyone can make these measurements. To make fading measurements of this sort the only instrument needed is a 1.5-or 2-mA. milliammeter. The meter is connected in the B plus lead to the detector tube; it will read about 1 mA. if the detector is a 201-A type tube using a grid leak and condenser for detection with 45 volts on the plate. When a signal is tuned-in the meter deflection will decrease. The amount of the decrease depending upon the strength of the signal. If the meter deflection with the signal tuned-in is subtracted from the meter deflection when not receiving a signal, the difference will be the amount the meter deflection has changed due to the signal. If the normal plate current is 1 mA. and the signal causes the values to decrease to 0.6 mA, then the deflection due to the signal is 0.4 mA. If this value varies with time it indicates fading and can be plotted as a curve, as shown on this Sheet. An examination of this curve indicates that at the start of the test the meter deflection due to the signal was 0.5 mA. but that after about one minute the signal strength quickly fell to 0.25 mA.

and then increased and decreased several times in rapid succession.

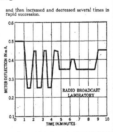

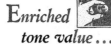

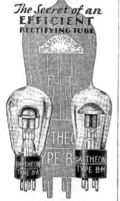

Manufacturers' Booklets

A Varied List of Books Pertaining to Radio and Allied Subjects Obtainable Free With the Accompanying Coupon

READERS may obtain any of the booklets listed below by using the coupon printed on page 254. Order by number only.

1. FILAMENT CONTROL—Problems of filament supply, voltage regulation, and effect on various circuits. RADIALL COMPANY.

2. HARD RUBBER PANELS—Characteristics and properties of hard rubber as used in radio, with suggestions on how to "work" it. B. F. GOODRICH RUBBER COMPANY.

3. TRANSFORMERS—A booklet giving data on input and output transformers. FACENT ELECTRIC COMPANY.

4. RESISTANCE-COUPLED AMPLIFIERS—A general discussion of resistance coupling with curves and circuit diagrams. COLE RADIO MANUFACTURING COMPANY.

5. CARBORUNDUM IN RADIO—A book giving pertinent data on the crystal as used for detection, with hook-ups, and a section giving information on the use of resistors. THE CARBORUNDUM COMPANY.

6. B-ELIMINATOR CONSTRUCTION—Constructional data on how to build. AMERICAN ELECTRIC COMPANY.

7. TRANSFORMER AND CHOKE-COUPLED AMPLIFICATION—Circuit diagrams and discussion. ALL-AMERICAN RADIO CORPORATION.

9. VOLUME CONTROL—A leaflet showing circuits for distortionless control of volume. CENTRAL RADIO LABORATORIES.

10. VARIABLE RESISTANCE—As used in various circuits. CENTRAL RADIO LABORATORIES.

11. RESISTANCE COUPLING—Resistors and their application to audio amplification, with circuit diagrams. DEJUR PRODUCTS COMPANY.

12. DISTORTION AND WHAT CAUSES IT—Hook-ups of resistance-coupled amplifiers with standard circuits. ALLEN-BRADLEY COMPANY.

14. B-ELIMINATOR AND POWER AMPLIFIER—Instructions for assembly and operation using Raytheon tube. GENERAL RADIO COMPANY.

15a. B-ELIMINATOR AND POWER AMPLIFIER—Instructions for assembly and operation using an R. C. A. rectifier. GENERAL RADIO COMPANY.

16. LOUD SPEAKERS—A description of the functions and characteristics of variable condensers with curves and specifications for their application to complete receivers. ALLEN D. CARDWELL MANUFACTURING COMPANY.

17. BAKELITE—A description of various uses of bakelite in radio, its manufacture, and its properties. BAKELITE CORPORATION.

19. POWER SUPPLY—A discussion on power supply with particular reference to lamp-socket operation. Theory and constructional data for building power supply devices. ACME APPARATUS COMPANY.

20. AUDIO AMPLIFICATION—A booklet containing data on audio amplification together with hints for the constructor. ALL AMERICAN RADIO CORPORATION.

21. HIGH-FREQUENCY DRIVER AND SHORT-WAVE WAVE-METER—Constructional data and application. BURGESS BATTERY COMPANY.

26. AUDIO-FREQUENCY CHOKES—A pamphlet showing positions in the circuit where audio-frequency chokes may be used. SAMSON ELECTRIC COMPANY.

27. RADIO-FREQUENCY CHOKES—Circuit diagrams illustrating the use of chokes to keep out radio-frequency currents from definite points. SAMSON ELECTRIC COMPANY.

28. TRANSFORMER AND IMPEDANCE DATA—Tables giving the mechanical and electrical characteristics of transformers and impedances, together with a short description of their use in the circuit. SAMSON ELECTRIC COMPANY.

29. BYPASS CONDENSERS—A description of the manufacture of bypass and filter condensers. LESLIE F. MUTER COMPANY.

30. AUDIO MANUAL—Fifty questions which are often asked regarding audio amplification, and their answers. AMERTRAN SALES COMPANY, INCORPORATED.

51. SHORT-WAVE RECEIVER—Constructional data on a receiver which, by the substitution of various coils, may be made to tune from a frequency of 16,660 kc. (18 meters) to 1990 kc. (150 meters). SILVER-MARSHALL, INCORPORATED.

52. AUDIO QUALITY—A booklet dealing with audio-frequency amplification of various kinds and the application to well-known circuits. SILVER-MARSHALL, INCORPORATED.

56. VARIABLE CONDENSERS—A bulletin giving an analysis of various condensers together with their characteristics. GENERAL RADIO COMPANY.

57. FILTER DATA—Facts about the filtering of direct current supplied by means of motor-generator outfits used with transmitters. ELECTRIC SPECIALTY COMPANY.

59. RESISTANCE COUPLING—A booklet giving some general information on the subject of radio and the application of resistors to audio. DAVEN RADIO CORPORATION.

60. RESISTORS—A pamphlet giving some technical data on resistors which are capable of dissipating considerable energy; also data on the ordinary resistors used in resistance-coupled amplification. THE CRESCENT RADIO COMPANY.

63. AUDIO-FREQUENCY AMPLIFICATION—Constructional details of a five-tube receiver using a special design of radio-frequency transformer. CAMFIELD RADIO MFG. COMPANY.

65. FIVE-TUBE RECEIVER—Constructional data on building a receiver. AERO PRODUCTS, INCORPORATED.

66. AMPLIFICATION WITHOUT DISTORTION—Data and curves illustrating the use of various methods of amplification. ACME APPARATUS COMPANY.

68. SUPER-HETERODYNE—Constructional details of a seven-tube set. G. C. EVANS COMPANY.

70. IMPROVING THE AUDIO AMPLIFIER—Data on the characteristics of audio transformers, with a circuit diagram showing where chokes, resistors, and condensers can be used. AMERICAN TRANSFORMER COMPANY.

72. PLATE SUPPLY SYSTEM—A wiring diagram and layout plan for a plate supply system to be used with a power amplifier. Complete directions for wiring are given. AMERTRAN SALES COMPANY.

80. FIVE-TUBE RECEIVER—Data are given for the construction of a five-tube tuned radio-frequency receiver.

Complete instructions, list of parts, circuit diagram, and template are given. ALL-AMERICAN RADIO CORPORATION.

81. BETTER TUNING—A booklet giving much general information on the subject of radio reception with specific illustrations. Primarily for the non-technical home constructor. BREMER-TULLY MANUFACTURING COMPANY.

82. SIX-TUBE RECEIVER—A booklet containing photographs, instructions, and diagrams for building a six-tube shielded receiver. SILVER-MARSHALL, INCORPORATED.

83. SOCKET POWER DEVICE—A list of parts, diagrams, and templates for the construction and assembly of socket power devices. JEFFERSON ELECTRIC MANUFACTURING COMPANY.

84. FIVE-TUBE EQUAMATIC—Panel layout, circuit diagrams, and instructions for building a five-tube receiver, together with data on the operation of tuned radio-frequency transformers of special design. KARAS ELECTRIC COMPANY.

85. FILTER—Data on a high-capacity electrolytic condenser used in filter circuits in connection with a socket power supply units, are given in a pamphlet. THE ABOX COMPANY.

86. SHORT-WAVE RECEIVER—A booklet containing data on a short-wave receiver as constructed for experimental purposes. THE ALLEN D. CARDWELL MANUFACTURING CORPORATION.

88. SUPER-HETERODYNE CONSTRUCTION—A booklet giving full instructions, together with a blue print and necessary data, for building an eight-tube receiver. THE GEORGE W. WALKER COMPANY.

89. SHORT-WAVE TRANSMITTER—Data and blue prints are given on the construction of a short-wave transmitter, together with operating instructions, methods of keying, and other pertinent data. RADIO ENGINEERING LABORATORIES.

90. IMPEDANCE AMPLIFICATION—The theory and practice of a special type of dual-impedance audio amplification are given. ALDEN MANUFACTURING COMPANY.

93. B-SOCKET POWER—A booklet giving constructional details of a socket-power device using either the BH or 313 type rectifier. NATIONAL COMPANY, INCORPORATED.

94. POWER AMPLIFIER—Constructional data and wiring diagrams of a power amplifier combined with a B-supply unit are given. NATIONAL COMPANY, INCORPORATED.

100. A, B, and C SOCKET-POWER SUPPLY—A booklet giving data on the construction and operation of a socket-power supply using the new high-current rectifier tube. THE Q. R. S. MUSIC COMPANY.

101. USING CHOKES—A folder with circuit diagrams of the more popular circuits showing where choke coils may be placed to produce better results. SAMSON ELECTRIC COMPANY.

22. A PRIMER OF ELECTRICITY—Fundamentals of electricity with special reference to the application of dry cells to radio and other uses. Constructional data on burgers, automatic switches, alarms, etc. NATIONAL CARBON COMPANY.

23. AUTOMATIC RELAY CONNECTIONS—A data sheet showing how a relay may be used to control A and B circuits. YAXLEY MANUFACTURING COMPANY.

25. ELECTROLYTIC RECTIFIER—Technical data on a new type of rectifier with operating curves. KODEL RADIO CORPORATION.

26. DRY CELLS FOR TRANSMITTERS—Actual tests carefully made of dry cells, with curves showing exactly what may be expected of this type of B power. BURGESS BATTERY COMPANY.

27. DRY-CELL BATTERY CAPACITIES FOR RADIO TRANSMITTERS—Characteristic curves and data on discharge tests. BURGESS BATTERY COMPANY.

28. B BATTERY LIFE—Battery life curves with general curves on tube characteristics. BURGESS BATTERY COMPANY.

29. HOW TO MAKE YOUR SET WORK BETTER—A non-technical discussion of general radio subjects with hints on how reception may be bettered by using the right tubes. UNITED RADIO AND ELECTRIC CORPORATION.

30. TUBE CHARACTERISTICS—A data sheet giving constants of tubes. C. E. MANUFACTURING COMPANY.

31. FUNCTIONS OF THE LOUD SPEAKER—A short, non-technical general article on loud speakers. AMPLION CORPORATION OF AMERICA.

32. METERS FOR RADIO—A catalogue of meters used in radio, with connecting diagrams. BURTON-ROGERS COMPANY.

33. SWITCHBOARD AND PORTABLE METERS—A booklet giving dimensions, specifications, and shunts used with various meters. BURTON-ROGERS COMPANY.

34. COST OF B BATTERIES—An interesting discussion of the relative merits of various sources of B supply. HARTFORD BATTERY MANUFACTURING COMPANY.

35. STORAGE BATTERY OPERATION—An illustrated booklet on the care and operation of the storage battery. GENERAL LEAD BATTERIES COMPANY.

36. CHARGING A AND B BATTERIES—Various ways of connecting-up batteries for charging purposes. WESTINGHOUSE UNION BATTERY COMPANY.

37. WHY RADIO IS BETTER WITH BATTERY POWER—Advice on what dry cell battery to use; their application to radio, with wiring diagrams. NATIONAL CARBON COMPANY.

53. TUBE REACTIVATOR—Information on the care of vacuum tubes, with notes on how and when they should be reactivated. THE STERLING MANUFACTURING COMPANY.

54. ARRESTERS—Mechanical details and principles of the vacuum type of arrester. NATIONAL ELECTRIC SPECIALTY COMPANY.

55. CAPACITY CONNECTOR—Description of a new device for connecting up the various parts of a receiving set, and at the same time providing bypass condensers between the leads. KURZ-KASCH COMPANY.

60. VACUUM TUBES—A booklet giving the characteristics of the various tube types with a short description of where they may be used in the circuit. RADIO CORPORATION OF AMERICA.

(Continued on page 254)

"RADIO BROADCAST'S" DIRECTORY OF MANUFACTURED RECEIVERS

¶ A coupon will be found on page 258. All readers who desire additional information on the receivers listed below need only insert the proper numbers in the coupon, mail it to the Service Departi and full details will be sent. New sets are listed

KEY TO TUBE ABBREVIATIONS

99—60-mA. filament (dry cell)
01-A—Storage battery 0.25 amps. filament
12—Power tube (Storage battery)
71—Power tube (Storage battery)
16-B—Half-wave rectifier tube
80—Full-wave, high current rectifier
81—Half-wave, high current rectifier
Hmu—High-Mu tube for resistance-coupled audio
20—Power tube (dry cell)
10—Power Tube (Storage battery)
00-A—Special detector
13—Full-wave rectifier tube
26—Low-voltage high-current a. c. tube
27—Heater type a. c. tube

DIRECT CURRENT RECEIVERS

NO. 424. COLONIAL 26

Six tubes; 2 t. r. f. (01-A), detector (12), 2 transformer audio (01-A and 71). Balanced t. r. f. One to three dials. Volume control: antenna switch and potentiometer across first audio. Watts required: 120. Console size: 34 x 38 inches. Headphone connections. The filaments are connected in a series parallel arrangement. Price $250 including power unit.

NO. 425. SUPERPOWER

Five tubes: All 01-A tubes. Multiplex circuit. Two dials. Volume control: resistance in r. f. plate. Watts required: 30. Antenna: loop or outside. Cabinet sizes: table, 27 x 10 x 9 inches; console, 28 x 50 x 21. Prices: table, $135 including power unit; console, $390 including power unit and loud speaker.

A. C. OPERATED RECEIVERS

NO. 508. ALL-AMERICAN 77, 88, AND 99

Six tubes; 3 t. r. f. (26), detector (27), 2 transformer audio (26 and 71). Rice neutralized t. r. f. Single drum tuning. Volume control: potentiometer in r. f. plate. Cabinet sizes: No. 77, 21 x 10 x 8 inches; No. 88 Hiboy, 25 x 38 x 18 inches; No, 99 console, 27½ x 43 x 20 inches. Shielded. Output device. The filaments are supplied by means of three small transformers. The plate supply employs a gas-filled rectifier tube. Voltmeter in a. c. supply line. Prices: No. 77, $150, including power unit; No. 88, $210 including power unit; No. 99, $285 including power unit and loud speaker.

NO. 509. ALL-AMERICAN "DUET"; "SEXTET"

Six tubes; 2 t. r. f. (99), detector (99), 3 transformer audio (99 and 12). Rice neutralized t. r. f. Two dials. Volume control: resistance in r. f. plate. Cabinet sizes: "Duet," 23 x 56 x 16½ inches; "Sextet," 22½ x 13½ x 15½ inches. Shielded. Output device. The 99 filaments are connected in series and supplied with rectified a. c.; while 12 is supplied with raw a. c. The plate and filament supply uses gaseous rectifier tubes. Milliammeter on power unit. Prices: "Duet," $160 including power unit; "Sextet," $220 including power unit and loud speaker.

NO. 511. ALL-AMERICAN 80, 90, AND 115

Five tubes; 2 t. r. f. (99), detector (99), 2 transformer audio (99 and 12). Rice neutralized t. r. f. Two dials. Volume control: resistance in r. f. plate. Cabinet sizes: No. 80, 23½ x 12½ x 15 inches; No. 90, 37½ x 12 x 12½ inches; No. 115 Hiboy, 24 x 41 x 15 inches. Coils individually shielded. Output device. See No. 509 for power supply. Prices: No. 80, $135 including power unit; No. 90, $145 including power unit and compartment; No. 115, $170 including power unit, compartment, and loud speaker.

NO. 510. ALL-AMERICAN 7

Seven tubes; 3 t. r. f. (26), 1 untuned r. f. (26), detector (27), 2 transformer audio (26 and 71). Rice neutralized t. r. f. One drum. Volume control: resistance in r. f. plate. Cabinet sizes: "Sovereign" console, 30½ x 60½ x 19 inches; "Lorraine" Hiboy, 25½ x 53½ x 17½ inches; "Forte" cabinet, 25½ x 13½ x 17½ inches. For filament and plate supply: See No. 508. Prices: "Sovereign" $460; "Lorraine" $360; "Forte" $270. All prices include power unit. First two include loud speaker.

NO. 401, AMRAD AC9

Six tubes; 3 t. r. f. (99), detector (99), 2 transformer (99 and 12). Neutrodyne. Two dials. Volume control: resistance across 1st audio. Watts consumed: 50. Cabinet size: 27 x 9 x 11½ inches. The 99 filaments are connected in series and supplied with rectified a. c., while the 12 is run on raw a. c. The power unit, requiring two 16-B rectifiers, is separate and supplies A. B. and C current. Price $142 including power unit.

NO. 402. AMRAD AC5

Five tubes. Same as No. 401 except one less r. f. stage. Price $125 including power unit.

NO. 536. SOUTH BEND

Six tubes. One control. Sub-panel shielding. Binding Posts. Antenna: outdoor. Prices: table, $130, Baby Grand Console, $195.

NO. 537. WALBERT 26

Six tubes; five Kellogg a. c. tubes and one 71. Two controls. Volume control: variable plate resistance. Isofarad circuit. Output device. Battery cable. Semi-shielded. Antenna: 50 to 75 feet. Cabinet size: 10½ x 29½ x 16½ inches. Prices: $215; with tubes, $2.50

NO. 484. BOSWORTH, B5

Five tubes; 2 t. r. f. (26), detector (99), 2 transformer audio (special a. c. tubes). T. r. f. circuit. Two dials. Volume control: potentiometer. Cabinet size: 23 x 7 x 8 inches. Output device included. Price $175.

NO. 406. CLEARTONE 110

Five tubes; 2 t. r. f., detector, 2 transformer audio. All tubes a. c. heater type. One or two dials. Volume control: resistance in r. f. plate. Watts consumed: 40. Cabinet size varies. The plate supply is built in the receiver and requires one rectifier tube. Filament supply through step down transformers. Prices range from $175 to $375 which includes 5 a. c. tubes and one rectifier tube.

NO. 407. COLONIAL 25

Six tubes; 2. t. r. f. (01-A), detector (99), 2 resistance audio (99). 1 transformer audio (10). Balanced t. r. f. circuit. One or three dials. Volume control: Antenna switch and potentiometer on 1st audio. Watts consumed: 100. Console size: 34 x 38 x 18 inches. Output device. All tube filaments are operated on a. c. except the detector which is supplied with rectified a. c. from the plate supply. The rectifier employs two 16-b tubes. Price $250 including built-in plate and filament supply.

NO. 507. CROSLEY 602 BANDBOX

Six tubes; 3. t. r. f. (26), detector (27); 2 transformer audio (26 and 71). Neutrodyne circuit. One dial. Cabinet size: 17½ x 5½ x 7½ inches. The heaters for the a. c. tubes and the 71 filament are supplied by windings in B unit transformers available to operate either on 25 or 60 cycles. The plate current is supplied by means of rectifier tube. Price $65 for set alone, power unit $60.

NO. 408. DAY-FAN "DE LUXE"

Six tubes; 3 t. r. f., detector, 2 transformer audio. All 01-A tubes. One dial. Volume control: potentiometer across r. f. tubes. Watts consumed: 300. Console size: 30 x 40 x 20 inches. The filaments are connected in series and supplied with d. c. from a motor-generator set which also supplies B and C current. Output device. Price $350 including power unit.

NO. 409. DAYCRAFT 5

Five tubes; 2 t. r. f., detector, 2 transformer audio. All a. c. heater tubes. Reflexed t. r. f. One dial. Volume control: potentiometers in r. f. plate and 1st audio. Watts consumed: 135. Console size: 34 x 36 x 14 inches. Output device. The heaters are supplied by means of a small transformer. A built-in rectifier supplies B and C voltages. Price $170, less tubes. The following have one more r. f. stage and are not reflexed: Daycraft 6, $195; Dayrole, 6, $235; Dayfan 6, $110. All prices less tubes.

NO. 469. FREED-EISEMANN NRII

Six tubes; 3 t. r. f. (01-A), detector (01-A), 2 transformer audio (01-A and 71). Neutrodyne. One dial. Volume control: potentiometer. Watts consumed: 150. Cabinet size: 19½ x 10 x 10½ inches. Shielded. Output device. A special power unit is included employing a rectifier tube. Price $225 including NR-411 power unit.

NO. 487. FRESHMAN 7F-AC

Six tubes; 3 t. r. f. (26), detector (27), 2 transformer audio (26 and 71). Equaphase circuit. One dial. Volume control: potentiometer across 1st audio. Console size: 24½ x 41½ x 15 inches. Output device. The filaments and heaters and B supply are all supplied by one power unit. The plate supply requires one 80 rectifier tube. Price $175 to $350, complete.

NO. 421. SOVEREIGN 238

Seven tubes of the a. c. heater type. Balanced t. r. f. Two dials. Volume control: resistance across 2nd audio. Watts consumed: 45. Console size: 37 x 52 x 15 inches. The heaters are supplied by a small a. c. transformer, while the plate is supplied by means of rectified a. c. using a gaseous type rectifier. Price $325, including power unit and tubes.

NO. 517. KELLOGG 510, 511, AND 512

Seven tubes; 4 t. r. f., detector, 2 transformer audio. All Kellogg a. c. tubes. One control and special zone switch. Balanced. Volume control: special. Output device. Shielded. Cable connection between power supply unit and receiver. Antenna: 25 to 100 feet. Panel 7½ x 27½ inches. Prices: Model 510 and 512, consoles, $495 complete. Model 511, consolette, $365 without loud speaker.

NO. 496. SLEEPER ELECTRIC

Five tubes; four 99 tubes and one 71. Two controls. Volume control: rheostat on r. f. Neutralized. Cable. Output device. Power supply uses two 16-B tubes. Antenna: 100 feet. Prices: Type 64, table, $160; Type 65; table, with built-in loud speaker, $175; Type 66, table, $175; Type 67, console, $235; Type 78, console, $265.

NO. 538. NEUTROWOUND MASTER ALLECTRIC

Six tubes; 2 t. r. f. (01-A), detector (01-A), 2 audio (01-A and two 71 in push-pull amplifier). The 01-A tubes are in series, and are supplied from a 400-mA. rectifier. Two drum controls. Volume control: variable plate resistance. Output device. Shielded. Antenna: 50 to 100 feet. Price: $360.

NO.

Six tubes: 2 t. r. f., tubes a. c. heater typ resistance in r. f. plate. 7 x 21 inches. The bu 16-B rectifier. The fila transformer. Prices: t rectifier; console, $275 console, $325 includi speaker.

NO. 41

Eight tubes; five ty control. Super-heterod tions. Battery cable. H loop. Set may be oper power mains when use 104 loud speaker. Pri operation; $570 with operation.

NO. 540

Receiver characterist type 71 power tube is operate on either a. c. The combination recti two type 81 tubes. M tained in lower part of or long outside anten 42½ x 29 x 17½ inches. F

NO. 54

This model combines 104 loud speaker. The tubes and a type 10 pov enclosed and is revolve Models for operation f Cabinet size: 52 x 72 x

NO. 539

Six tubes; 3 t. r. f. (audio (26 and 27). Built-in power supply 100 feet. Cabinet size without accessories.

NO. 545. NEUTROW

Five tubes; 2 t. r. f. and 71). The 99 tubes a an 85-mA. rectifier. T trol: variable plate resi 75 to 100 feet. Cabinet $150.

NO.

Six tubes; 2 t. r. f., d tubes a. c. heater type Volume control: rheost Panel size: 12½ x 8½ inc for the a. c. tubes and small transformers. Th type using a rectifier $245.

NO. 522. CA

McCullough a. c. tu trol; variable high resis connections. Semi-shiel Panel size: 7 x 21 inch with a. c. equipment, $ a. c. equipment, $235.

NO. 523. CA

McCullough a. c. t volume control. Techn C-battery connections. device. Loop operated. Model 92 C, console, $

BATTERY OP

NO. 542. PFAN

Six tubes: 3 t. r. f. (Pfanstiehl circuit. Volu r. f. plate circuit. One battery connections. outdoor. Cabinet size: 9 out accessories.

NO. 512. ALL-AM

Six tubes; 3 t. r. f. former audio (01-A an Drum control. Volume sizes: No. 44, 21 x 10 inches; No. 66, 27½ x 4 tions. Battery cable. A No. 44, $70; No. 55, $ 66, $200 including loud

NO. 428.

Five tubes; 2 t. r. f. All 01-A tubes. Semi b current 15mA. Volume sizes: table, 20 x 8½ x inches. Partially shiel connections. Antenna: console, $65 including l

New AERO Circuits For Either Battery or A. C. Operation

The Improved Aero-Dyne 6, and the Aero 7—popular new circuits are built around these marvelous coils

You Should Learn About Them *Now!*

Proper constants for A. C. operation of the improved Aero-Dyne 6 and the Aero Seven have been studied out, and these excellent circuits are now adaptable to either A. C. or battery operation. A. C. blue prints are packed in foundation units. They may also be obtained by sending 25c for each direct to the factory.

AERO Universal Tuned Radio Frequency Kit

Especially designed for the Improved Aero-Dyne 6. Kit consists of 4 twice-matched units. Adaptable to 201-A, 199, 112, and the new 240 and A. C. tubes. Tuning range below 200 to above 550 meters.
This kit will make any circuit better in selectivity, tone and range. Will eliminate losses and give the greatest receiving efficiency.
Code No. U-16 (for .0005 Cond.) ...$15.00
Code No. U-163 (for .00035 Cond.)... 15.00

AERO Seven Tuned Radio Frequency Kit

Especially designed for the Aero 7. Kit consists of 3 twice-matched units. Coils are wound on Bakelite skeleton forms, assuring a 95% air di-electric. Tuning range from below 200 to above 550 meters. Adaptable to 201-A, 199, 112, and the new 240 and A. C. tubes.
Code No. U-12 (for .0005 Cond.) ...$12.00
Code No. U-123 (for .00035 Cond.)... 12.00

NOTE—All AERO Universal Kits for use in tuned radio frequency circuits have packed in each coil with a fixed primary a twice matched calibration slip showing reading of each fixed primary AERO Universal Coil at 250 and 500 meters; all having an accurate and similar calibration. Be sure to keep these slips. They're valuable if you decide to add another R.F. Stage to your set.

A NEW SERVICE

We have arranged to furnish the home set builder with complete Foundation Units for the above named Circuits, drilled and engraved on Westinghouse Micarta. Detailed blue-prints for both battery and A. C. operation and wiring diagram for each circuit included with every foundation unit free. Write for information and prices.

You should be able to get any of the above Aero Coils and parts from your dealer. If he should be out of stock order direct from the factory.

AERO PRODUCTS, Inc.
1772 Wilson Ave.   Dept. 109   Chicago, Ill.

Samson Power Block No. 210—The only block which will supply 500 volts at 80 mils to two 210 tubes.

Powerize with Samson Units for Best Results

For new SAMSON Power Units insure the best there is in radio current supply by

1. Doing away with hum, motor boating and poor voltage regulation.
2. Remaining so cool after 24 hours continuous operation under full load that they will be well within the 20° rise of temperature specified by the A. I. E. E.
3. Being designed to more than meet the specifications adopted by the National Board of Fire Underwriters.
4. Insuring safety against shock because of protected input and output terminals.
5. Insuring for all tubes the correct filament voltages specified by their manufacturers.
6. Compensating for lighting circuit voltage variation by the use of a special input plug and terminal block to which is attached a six ft. flexible rubber-covered connecting cord and plug.

Our Power Units bulletin descriptive of these is free for the asking. In addition, our construction bulletin on many different "B" Eliminators and Power Amplifiers will be sent upon receipt of 10c. in stamps to cover the mailing cost.

Samson Electric Co.

Manufacturing Since 1882   Principal Office, CANTON, MASS.

If TUBES Could Talk!

They would tell you—that only at the precise and definitely prescribed filament current, or temperature, can their tonal qualities, clarity and sensitiveness be brought out to the full. That "A" battery current constantly varies according to the age of the battery and state of charge—and operation with too little or too great current is certain death to efficient tube performance—and too quickly, of the tube itself. That only AMPERITE can automatically supply and control this exact current despite battery variation—as long as sufficient current is to be had. That you should never confuse AMPERITE with fixed filament resistors which do not do the Amperite's job. AMPERITE is sold by dealers everywhere. Price $1.10 mounted (in U. S. A.).

Write for FREE "Amperite Book" of the season's best circuits and latest construction data. Address Dept. R.B.-1

**"AMPERITE-
Watch Dog
of Your
Tubes**

Radiall Company
50 Franklin St., New York

AMPERITE
REG. U. S. PAT. OFF.

The "SELF-ADJUSTING" *Rheostat*

NO. 485. BOSWORTH B6

Five tubes; 2 t. r. f. (01-A), detector (01-A), 2 transformer audio (01-A and 71). Two dials. Volume control: variable grid resistances. Battery cable. C-battery connections. Antenna: 25 feet or longer. Cabinet size 15 x 7 x 8 inches. Price $75.

NO. 513. COUNTERPHASE SIX

Six tubes; 3 t. r. f. (01-A), detector (00-A), 2 transformer audio (01-A) and 12). Counterphase t. r. f. Two dials. Plate current: 32 mA. Volume control: rheostat on 2nd and 3rd r. f. Coils shielded. Battery cable. C-battery connections. Antenna: 75 to 100 feet. Console size: 18½ x 40½ x 15½ inches. Prices: Model 35, table, $110; Model 37, console, $175.

NO. 514. COUNTERPHASE EIGHT

Eight tubes; 4 t. r. f. (01-A) detector (00-A), 2 transformer audio (01-A and 12). Counterphase t. r. f. One dial. Plate current: 40 mA. Volume control: rheostat in 1st r. f. Copper stage shielding. Battery cable. C-battery connections. Antenna: 75 to 100 feet. Cabinet size: 30 x 12½ x 16 inches. Prices: Model 12, table, $225; Model 16, console, $335; Model 18, console, $365.

NO. 506. CROSLEY 601 BANDBOX

Six tubes; 3 t. r. f., detector, 2 transformer audio. All 01-A tubes. Neutrodyne. One dial. Plate current: 40 MA. Volume control: rheostat in r. f. Shielded. Battery cable. C-battery connections. Antenna: 75 to 150 feet. Cabinet size: 17¼ x 5½ x 7½. Price, $55.

NO. 434. DAY-FAN 6

Six tubes; 3 t. r. f. (01-A), detector (01-A), 2 transformer audio (01-A and 12 or 71). One dial. Plate current: 12 to 15 mA. Volume control: rheostat on r. f. Shielded. Battery cable. C-battery connections. Output device. Antenna: 50 to 120 feet. Cabinet sizes: Daycraft 6, 32 x 30 x 34 inches; Day-Fan Jr., 15 x 7 x 7. Prices: Day-Fan 6, $110; Daycraft 6, $145 including loud speaker; Day-Fan Jr. not available.

NO. 435. DAY-FAN 7

Seven tubes; 3 t.r.f. (01-A), detector (01-A), 1 resistance audio (01-A), 2 transformer audio (01-A and 12 or 71). Plate current: 15 mA. Antenna: outside. Same as No. 434. Price $115.

NO. 503. FADA SPECIAL

Six tubes; 3 t. r. f. (01-A), detector (01-A), 2 transformer audio (01-A and 71). Neutrodyne. Two drum control. Plate current: 20 to 24 MA. Volume control: rheostat on r. f. Coils shielded. Battery cable. C.-battery connections. Headphone connection. Antenna: outdoor. Cabinet size: 20 x 13½ x 10½ inches. Price $95.

NO. 504. FADA 7

Seven tubes; 4 t. r. f. (01-A), detector (01-A), 2 transformer audio (01-A and 71). Neutrodyne. Two drum control. Plate current: 43,A. Volume control: rheostat on r. f. Completely shielded. Battery cable. C-battery connections. Headphone connections, Output device. Antenna: outdoor or loop. Cabinet sizes: table, 25½ x 13¼ x 11½ inches; console, 29 - 50 x 17 inches. Prices: table, $185; console, $285.

NO. 436. FEDERAL

Five tubes; 2 t. r. f. (01-A), detector (01-A), 2 transformer audio (01-A and 12 or 71). Balanced t. r. f. One dial. Plate current: 20.7 mA. Volume control: rheostat on r. f. Shielded. Battery cable. C-battery connections. Antenna: loop. Made in 6 models. Price varies from $250 to $1000 including loop.

NO. 505. FADA 8

Eight tubes. Same as No. 504 except for one extra stage of audio and different cabinet. Prices: table, $300; console, $400.

NO. 437. FERGUSON 10A

Seven tubes; 3 t. r. f. (01-A), detector (01-A), 3 audio (01-A and 12 or 71). One dial. Plate current: 18 to 25 mA. Volume control: rheostat on two r. f. Shielded. Battery cable. C-battery connections. Antenna: 100 feet. Cabinet size: 21½ x 12 x 15 inches. Price $150.

NO. 438. FERGUSON 14

Ten tubes; 3 untuned r. f., 3 t. r. f. (01-A), detector (01-A), 3 audio (01-A and 12 or 71). Special balanced t. r. f. One dial. Plate current: 30 to 35mA. Volume control: rheostat in three r. f. Shielded. Battery cable, C-battery connections. Antenna: loop. Cabinet size: 24 x 12 x 16 inches. Price $235, including loop.

NO. 439. FERGUSON 12

Six tubes; 2 t. r. f. (01-A), detector (01-A), 1 transformer audio (01-A), 2 resistance audio (01-A and 12 or 71). Two dials. Plate current: 18 to 25mA. Volume control: rheostat on two r. f. Partially shielded. Battery cable. C-battery connections. Antenna: 100 feet. Cabinet size: 22½ x 10 x 12 inches. Price $85. Consolette $145 including loud speaker.

NO. 440. FREED-EISEMANN NR-8 NR-9, AND NR-66

Six tubes: 3 t. r. f. (01-A), detector (01-A), 2 transformer audio (01-A and 71). Neutrodyne. NR-8, two dials; others one dial. Plate current: 30 mA. Volume control: rheostat on r. f. NR-8 and 9; chassis type shielding. NR-66, individual stage shielding. Battery cable. C-battery connections. Antenna: 100 feet. Cabinet sizes: NR-8 and 9, 19½ x 10 x 10½ inches; NR-66 20 x 10½ x 12 inches. Prices: NR-8, $90; NR-9, $100; NR-66, $125.

NO. 501. KING "CHEVALIER"

Six tubes. Same as No. 500. Coils completely shielded. Panel size: 11 x 7 inches. Price, $210 including loud speaker.

NO. 441. FREED-EISEMANN NR-77

Seven tubes; 4 t. r. f. (01-A), detector (01-A), 2 transformer audio (01-A and 71). Neutrodyne. One dial. Plate current: 35mA. Volume control: rheostat on r. f. Shielding. Battery cable. C-battery connections. Antenna: outside or loop. Cabinet size: 23 x 10½ x 13 inches. Price $175.

NO. 442. FREED-EISEMANN 800 AND 850

Eight tubes; 4 t. r. f. (01-A), detector (01-A), 1 transformer (01-A), 1 parallel audio (01-A or 71). Neutrodyne. One dial. Plate current: 35 mA. Volume control: rheostat on r. f. Shielded. Battery cable. C-battery connections. Output: two tubes in parallel or one power tube may be used. Antenna: outside or loop. Cabinet sizes: No. 800, 34 x 15½ x 13½ inches; No. 850, 36 x 65 x 17½. Prices not available.

NO. 444. GREBE MU-1

Five tubes; 2 t. r. f. (01-A), detector (01-A), 2 transformer audio (01-A and 12 or 71). Balanced t. r. f. One, two, or three dials (operate singly or together). Plate current: 30mA. Volume control: rheostat on r. f. Binocular coils. Binding posts. C-battery connections. Antenna: 125 feet. Cabinet size: 22½ x 9½ x 13 inches. Prices range from $95 to $320.

NO. 426. HOMER

Seven tubes; 4 t. r. f. (01-A); detector (01-A or 00A); 2 audio (01-A and 12 or 71). One knob tuning control. Volume control: rotor control in antenna circuit. Plate current: 22 to 25mA. "Technidyne" circuit. Completely enclosed in aluminum box. Battery cable. C-battery connections. Cabinet size, 8¾ x 19½ x 9½ inches. Chassis size, 6½ x 17 x 8 inches. Prices: Chassis only, $80. Table cabinet, $95.

NO. 502. KENNEDY ROYAL 7. CONSOLETTE

Seven tubes; 4 t. r. f. (01-A), detector (00-A), 2 transformer audio (01-A and 71). One dial. Plate current; 42mA. Volume control: rheostat on two r. f. Special r. f. coils. Battery cable. C-battery connections. Headphone connection. Antenna: outside or loop. Consolette size: 36½ x 35½ x 19 inches. Price $220.

NO. 498. KING "CRUSADER"

Six tubes; 2 t. r. f. (01-A), detector (00-A), 3 transformer audio (01-A and 71). Balanced t. r. f. One dial. Plate current: 20 mA. Volume control: rheostat on r. f. Coils shielded, Battery cable. C-battery connections. Antenna: outside, Panel: 11 x 7 inches. Price, $115.

NO. 499. KING "COMMANDER"

Six tubes; 3 t. r. f. (01-A), detector (00-A), 2 transformer audio (01-A and 71). Balanced t. r. f. One dial. Plate current: 25mA. Volume control: rheostat on r. f. Completely shielded. Battery cable. C-battery connections. Antenna: loop. Panel size: 12 x 8 inches. Price $220 including loop.

NO. 429. KING COLE VII AND VIII

Seven tubes; 3 t. r. f., detector, 1 resistance audio, 2 transformer audio. All 01-A tubes. Model VIII has one more stage t. r. f. (eight tubes). Model VII, two dials. Model VIII, one dial. Plate current: 15 to 50 mA. Volume control: primary shunt in r. f. Steel shielding. Battery cable and binding posts. C-battery connections. Output devices on some consoles. Antenna: 10 to 100 feet. Cabinet size: varies. Prices: Model VII, $80 to $160; Model VIII, $100 to $300

NO. 500. KING "BARONET" AND "VIKING"

Six tubes; 2 t. r. f. (01-A), detector (00-A), 3 transformer audio (01-A and 71). Balanced t. r. f. One dial. Plate current: 19mA. Volume control: rheostat in r. f. Battery cable. C-battery connections. Antenna: outside. Panel size: 18 x 7 inches. Prices: "Baronet," $70; "Viking," $140 including loud speaker.

NO. 489. MOHAWK

Six tubes; 2 t. r. f. (01-A), detector (00-A), 3 audio (01-A and 71). One dial. Plate current: 40mA. Volume control: rheostat on r. f. Battery cable. C-battery connections. Output device. Antenna: 60 feet. Panel size: 12½ x 8½ inches. Prices range from $65 to $245.

NO. 547. ATWATER KENT, MODEL 33

Six tubes; 3 t. r. f. (01-A), detector (01-A), 2 audio (01-A and 71 or 12). One dial. Volume control: r. f. filament rheostat. C-battery connections. Battery cable. Antenna: 100 feet. Steel panel. Cabinet size: 21½ x 6½ 6½ inches. Price: $75, without accessories.

NO. 544. ATWATER KENT, MODEL 50

Seven tubes; 4 t. r. f. (01-A), detector (01-A), 2 audio (01-A and 12 or 71). Volume control: r. f. filament rheostat. C-battery connections. Battery cable. Special bandpass filter circuit with an untuned amplifier. Cabinet size: 20½ x 13 x 7½ inches. Price: $120.

NO. 452. ORIOLE 90

Five tubes; 2 t. r. f., detector, 2 transformer audio. All 01-A tubes. "Trinum" circuit. Two dials. Plate current: 18mA. Volume control; rheostat on r. f. Battery cable. C-battery connections. Antenna: 50 to 100 feet. Cabinet size: 25½ x 11½ x 12½ inches. Price $85. Another model has 8 tubes, one dial, and is shielded. Price, $185.

NO. 453. PARAGON

Six tubes; 2 t. r. f. (01-A), detector (01-A), 3 double impedance audio (01-A and 71). One dial. Plate current; 40 mA. Volume control: resistance in r. f. plate. Shielded, Battery cable. C-battery connections. Output device. Antenna: 100 feet. Console size: 20 x 45 x 17 inches. Price not determined.

NO. 543 RADIOLA 20

Five tubes; 2 t. r. f. (99), detector (99), two transformer audio (99 and 20). Regenerative detector. Two drum controls. C-battery connections. Battery cable. Antenna: 100 feet. Price: $78 without accessories.

NO. 480. PFANSTIEHL 30 AND 302

Six tubes; 3 t. r. f. (01-A), detector (01-2A), transformer audio (01-A and 71). One dial. Plate current: 23 to 32 MA. Volume control: resistance in r. f. plate. Shielded. Battery cable. C-battery connections. Antenna: outside. Panel size: 17½ x 8½ inches. Prices: No. 30 cabinet, $105; No. 302 console, $185 including loud speaker.

NO. 515. BROWNING-DRAKE 7-A

Seven tubes; 2 t. r. f. (01-A), detector (00-A), 3 audio (Hmu, two 01-A, and 71). Illuminated drum control. Volume control: rheostat on 1st r. f. Shielded, Neutralized. C-battery connections. Battery Cable. Metal panel. Output device, Antenna: 50-75 feet. Cabinet, 30 x 11 x 9 inches. Price, $145.

NO. 516. BROWNING-DRAKE 6-A

Six tubes; 1 t. r. f. (99), detector (00-A), 3 audio (Hmu, two 01-A and 71). Drum control with auxiliary adjustment. Volume control: rheostat on r. f. Regenerative detector. Shielded. Neutralized. C-battery connections. Battery cable. Antenna: 50-100 feet. Cabinet, 25 x 11 x 9. Price $105.

NO. 518. KELLOGG "WAVE MASTER," 504, 505, and 506.

Five tubes; 2 t. r. f., detector, 2 transformer audio. One control and special zone switch. Volume control: rheostat on r. f. C-battery connections. Binding posts. Plate current: 25 to 35 mA. Antenna: 100 feet. Panel: 7½ x 25½ inches. Prices: Model 504, table, $75, less accessories. Model 505, table, $125 with loud speaker. Model 506, consolette, $135 with loud speaker.

NO. 519. KELLOGG, 507 AND 508

Six tubes, 3 t. r. f., detector, 2 transformer audio. One control and special zone switch. Volume control; rheostat on r. f. C-battery connections. Balanced. Shielded. Binding posts and battery cable. Antenna: 70 feet. Cabinet size: Model 507, table, 30 x 13½ x 14 inches, Model 508, console. 34 x 18 x 54 inches. Prices: Model 507, $190 less accessories. Model 508, $320 with loud speaker.

NO. 427. MURDOCK 7

Seven tubes; 3 t. r. f. (01-A), detector (01-A), 1 transformer and 2 resistance audio (two 01-A and 12 or 71). One control. Volume control: rheostat on r. f. Coils shielded. Neutralized. Battery cable. C-battery connections. Complete metal case. Antenna; 100 feet. Panel size: 9 x 23 inches. Price, not available.

NO. 520. BOSCH 57

Seven tubes; 4 t. r. f. (01-A), detector (01-A), 2 audio (01-A and 71). One control calibrated in kc. Volume control: rheostat on r. f. Shielded. Battery cable, C-battery connections. Balanced. Output device. Built-in loud speaker. Antenna: built-in loop or outside antenna, 100 feet. Cabinet size: 46 x 16 x 30 inches. Price: $340 including enclosed loop and loud speaker.

NO. 521. BOSCH "CRUISER," 66 AND 76

Six tubes; 3 t. r. f. (01-A), detector (01-A), 2 audio (01-A and 71). One control. Volume control: rheostat on r. f. Shielded. C-battery connections. Balanced, Battery cable. Antenna: 20 to 100 feet. Prices; Model 66, table, $99.50. Model 76, console, $175; with loud speaker $195.

NO. 524. CASE, 61 A AND 61 C

T. r. f. Semi-shielded. Battery cable. Drum control. Volume control: variable high resistance in audio system. Plate current: 85mA. Antenna: 100 feet. Prices: Model 61A, $85; Model 61 C, console, $135.

NO. 525. CASE, 90 A AND 90 C

Drum control. Inductive volume control. Technidyne circuit. C-battery connections. Battery cable. Loop operated. Model 90-C equipped with output device. Prices: Model 90 A, table, $225; Model 90 C, console, $350.

NO. 526. ARBORPHONE 25

Six tubes; 3 t. r. f. (01-A), detector (01-A), 2 audio (01-A and 71). One control. Volume control: rheostat. Shielded. Battery cable. Output device. C-battery connections. Loftin-White circuit. Antenna: 75 feet. Panel: 7½ x 15 inches, metal. Prices: Model 25, table, $125; Model 252, $185; Model 253, $250; Model 255, combination phonograph and radio, $600.

NO. 527. ARBORPHONE 27

Five tubes; 2 t. r. f. (01-A), detector (01-A), 2 audio (01-A). Two controls. Volume control: rheostat. C-battery connections. Binding posts. Antenna: 75 feet. Prices: Model 27, $65; Model 271, $99.50; Model 272, $125.

NO. 528. THE "CHIEF"

Seven tubes; six 01-A tubes and one power tube. One control. Volume control: rheostat. C-battery connection. Partial shielding. Binding posts. Antenna: outside. Cabinet size: 40 x 22 x 16 inches. Prices: Complete with A power supply, $250; without accessories, $150.

NO. 529. DIAMOND SPECIAL, SUPER SPECIAL, AND BABY GRAND CONSOLE

Six tubes; all 01-A type. One control. Partial shielding. C-battery connections. Volume control: rheostat. Binding posts. Antenna: outdoor. Prices: Diamond Special, $75; Super Special, $65; Baby Grand Console. $110.

No. 531. KOLSTER, 8A, 8B, AND 8C

Eight tubes; 4 t. r. f. (01-A), detector (Q1-A), 3 audio (two 01-A and one 12). One control. Volume control: rheostat on r. f. Shielded. Battery cable. C-battery connections. Model 8A uses 50 to 75 foot antenna; model 8B contains output device and uses antenna or detachable loop; Model 8C contains output device and uses antenna or built-in loop. Prices: 8A, $185; 8B, $235; 8C, $375.

No. 532. KOLSTER, 6D, 6G, AND 6H

Six tubes; 3 t. r. f. (01-A), detector (01-A), 2 audio (01-A and 12). One control. Volume control: rheostat on r. f. C-battery connections. Battery cable. Antenna: 50 to 75 feet. Model 6G contains output device and built-in loud speaker; Model 6H contains built-in B power unit and loud speaker. Prices: Model 6D, $80; Model 6G, $165; Model 6H, $265.

No. 533. SIMPLEX, SR 9 AND SR 10

Five tubes; 2 t. r. f. (01-A), detector (00-A), 2 audio (01-A and 12). SR 9, three controls; SR 10, two controls. Volume control; rheostat. C-battery connections. Battery cable. Headphone connection. Prices: SR 9, table, $65; consolette, $95; console, $145. SR 10, table $70; consolette, $95; console, $145.

No. 534. SIMPLEX, SR 11

Six tubes; 3 t. r. f. (01-A), detector (00-A), 2 audio (01-A and 12). One control. Volume control: rheostat. C-battery connections. Battery cable. Antenna: 100 feet. Prices: table, $70; consolette, $95; console, $145.

No. 535. STANDARDYNE, MODEL X 27

Six tubes; 2 t. r. f. (01-A), detector (01-A), 2 audio power tubes. one control. Volume control: rheostat on r. f. C-battery connections. Binding posts. Antenna: 75 feet. Cabinet size: 9 x 9 x 19½ inches. Prices: S 27, $49.50; S 950, console, with built-in loud speaker, $99.50; S 600, console with built-in loud speaker, $104.50.

No. 481. PFANSTIEHL 32 AND 322

Seven tubes; 3 t. r. f. (01-A), detector (01-A), 3 audio (01-A and 71). One dial. Plate current: 23 to 32 mA. Volume control: resistance in r. f. plate. Shielded. Battery cable. C-battery connections. Output device. Antenna: outside. Panel: 17½ x 8½ inches. Prices: No. 32 cabinet, $145; No. 322 console, $245 including loud speaker.

No. 433. ARBORPHONE

Five tubes; 2 t. r. f., detector, 2 transformer audio. All 01-A tubes. Two dials. Plate current: 16mA. Volume control: rheostat in r. f. and resistance in r. f. plate. C-battery connections. Binding posts. Antenna: taps for various lengths. Cabinet size: 24 x 9 x 10½ inches. Price: $65.

No. 431. AUDIOLA 6

Six tubes; 3 t. r. f. (01-A), detector (00-A), 2 transformer audio (01-A and 71). Drum control. Plate current: 20mA. Volume control; resistance in r. f. plate Stage shielding. Battery cable. C-battery connection. Antenna: 50 to 100 feet. Cabinet size: 28½ x 11 x 14½ inches. Price not established.

No. 432. AUDIOLA 8

Eight tubes; 4 t. r. f. (01-A), detector (00-A), 1 transformer audio (01-A), push-pull audio (12 or 71). Bridge balanced t. r. f. Drum control. Volume control: resistance in r. f. plate. Stage shielding. Battery cable. C-battery connections. Antenna: 10 to 100 feet. Cabinet size: 28½ x 11 x 14½ inches. Price not established.

No. 542 RADIOLA 16

Six tubes; 3 t. r. f. (01-A), detector (01 A), 2 transformer audio (01-A and 112). One control. C-battery connections. Battery cable. Antenna: outside. Cabinet size: 16 x 8½ x 7½ inches. Price: $69.50 without accessories.

No. 456. RADIOLA 20

Five tubes; 2 t. r. f. (99), detector (99), 2 transformer audio (99 and 20). Balanced t. r. f. and regenerative detector. Two dials. Volume control: regenerative. Shielded. C-battery connections. Headphone connections. Antenna: 75 to 150 feet. Cabinet size: 19½ x 11½ x 16 inches. Price $115 including all tubes.

No. 457. RADIOLA 25

Six tubes; five type 99 and one type 20. Drum control. Super-heterodyne circuit. C-battery connections. Battery cable. Headphone connections. Antenna: loop. Set may be operated from batteries or from power mains when used with model 104 loud speaker. Price: $165 with tubes, for battery operation. Apparatus for operation of set from the power mains can be purchased separately.

No. 493. SONORA F

Seven tubes; 4 t. r. f. (01-A), detector (00-A), 2 transformer audio (01-A and 71). Special balanced t. r. f. Two dials. Plate current: 45mA. Volume control: rheostat in r. f. Shielded. Battery cable. C-battery connections. Output device. Antenna: loop. Console size: 32 X 45½ X 17 inches. Prices range from $350 to $450 including loop and loud speaker.

No. 494. SONORA E

Six tubes; 3 t. r. f. (01-A), detector (00-A), 2 transformer audio (01-A and 71). Special balanced t. r. f. Two dials. Plate current: 35 to 40mA. Volume control: rheostat on r. f. Shielded. Battery cable. C-battery connections. Antenna: outside. Cabinet size: varies. Prices: table $110; semi-console, $140; console, $240 including loud speaker.

No. 530. KOLST...

Seven tubes; 4 t. r. f. (01-A...) (01-A and 12). One control... on r. f. Shielded. Battery ca... Antenna: 50 to 75 feet. Pric... 7B, with built-in loud speak...

No. 495. S...

Same as No. 494 except t. r. f., detector, 3 audio. Pr... console, $185; "DeLuxe" co...

No. 482. STEWART-W...

Six tubes; 3 t. r. f., dete... All 01-A tubes. Balanced t... rent: 10 to 25mA. Volume... plate. Shielded. Battery cab... Antenna: 80 feet. Cabinet x 11½ x 13½ inches; No. 71... inches. Tentative prices: ... $295 including loud speaker.

No. 483. STEWART-W...

Same as No. 482 except n... No. 525 table, 19½ x 10 x 11... 22½ x 40 x 14½ inches. Tent... No. 520, $117.50 including l...

No. 459. STROMBERG-C...

Five tubes; 2 t. r. f. (01-A... former audio (01-A and 71)... Plate current: 25 to 35mA... on 1st r. f. Shielded Battery... tions. Headphone connectio... voltmeter. Antenna: 60 to... No. 501, 25½ x 13 x 14 inches... inches. Prices: No. 501, $18...

No. 460. STROMBERG-C...

Six tubes. Same as No. 5... stage. Cabinet sizes: No. 60... No. 602, 28¼ x 51½ x 19½ incl... No. 602, $330.

No. 472. VOL...

Six tubes. Same as No. 471... 2 t. r. f. stages. Three di... Cabinet size: 26½ x 8 x 12 in...

No. 546. PARAGO...

Six tubes; 2 t. r. f. (01-A... pedance-coupled audio (two... main control and three aud... control: resistance in r. f. p... 40 mA. C-battery connec... pedance audio amplifier. Ou... shielded. Cable or binding po... inches. Price $90.00; withou...

THIS RADIO CHRISTMAS

by

Zeh Bouck

LIKE all people, I hate to brag. And I sincerely believe that there is no one in this
world who knows more about giving things
radio for Christmas than the writer. I have
had a radio laboratory for some ten years
now, to which manufacturers send samples of
their products. With the exception of an
occasional birthday, or a debt here and there,
this apparatus accumulates until around this
time of the year. And I'd give away a lot more
Christmas presents if radio manufacturers
only supplied green tissue paper and red
ribbons.

So it is most fitting that I should be telling
you how to spend your money on radio Christmas presents—or perhaps it isn't. I find myself in somewhat the reverse situation of
Oscar Wilde's cynic who knew the price of
everything but the value of nothing. At any
rate, our ever-increasing horde of radio
burglars will find what I have to say of un-
equalified use.

From One Dollar Up

WE—or rather you—may start your purchases from a dollar up. (Make sure you
do not start them with a dollar down. This
stretches Christmas out over too long a period
of time with rather fatal results to the Christmas spirit.)

In the case of a radio Christmas present,
not everything within the various price ranges
can be considered. There are many parts
particularly designed for special circuits and
receivers, and unless you are aware that the
recipient is interested in this particular apparatus you will do well to confine yourself, for
the greater part, to accessories.

One Dollar to Three Dollars

This class recommends itself to gifts among
the family—to brother, sister, cousins, aunts,
and all the rest sacred to Gilbert and Sullivan.
The following are really too useful for gifts:
One or two 01A type tubes
A loud speaker extension cord
A battery cable
A light socket antenna (for the gentle-
man with the loop set)
An antenna set
A set of 0.5, 1.0, 1.5, 2.0, 3.0, and 5.0-
megohm metallic grid leaks
At the very beginning we run into the in-
evitable book. The following are on our own
shelves:
FOR THE ENGINEER: *Engineering Mathe-
matics*, D. Van Nostrand and Company
FOR THE AMATEUR: *The Radio Amateur's
Handbook*, The American Radio Relay League,
Hartford, Connecticut, $1.00.
FOR THE AVERAGE FAN: *The Outline of
Radio*, by John V. L. Hogan, Little Brown
and Company, or *How Radio Receivers Work*,
by W. Van B. Roberts, published by RADIO
BROADCAST, Garden City, N. Y., $1.00
A subscription to *The Bell Technical Journal*
(195 Broadway, New York City) at $3.00
will be gratefully received by the engineer in
the family, and a subscription to QST (American Radio Relay League, Hartford, Conn.)
at $2.00 by the amateur or even broadcast fan.

Three Dollars to Five Dollars

As we go up a bit in price, the charity that
should begin at home evidences itself in gifts
outside the family—our purse strings loosen
and we splurge with:
Output filters
A set of five 1.0-mfd. bypass condensers
A telephone headset
A power tube
A filament relay
The following books for engineering folk:
Theory of Vibrating Systems and Sounds by
Crandall, $5.00. (D. Van Nostrand & Co.)

The Thermionic Vacuum Tube by H. J.
Van de Bijl, (McGraw Hill), $5.00
A most acceptable and unique gift may be
effected by obtaining a copy of *Radio Instruments and Measurements* from the Bureau of
Printing and Engraving, Government Printing Office, Washington, for $.75 (no stamps)
and having it bound by Brentano's or some
similar establishment.
And a subscription to RADIO BROADCAST
is never out of place.

Five Dollars to Ten Dollars

A slide rule is always an acceptable present
to the engineering friend who possesses the
inevitable book, if he hasn't a slide rule. Get
a Keuffel and Esser Polyphase Mannheim
ten-inch rule. Do not succumb to the seductive technicalities of various "duplex" and
"log log" designs. The engineer would thank
you, of course, for these last types, and use it,
perhaps, as a straight edge; but that's about
all.
You can always determine whether your
technical friend has the book you want to
give him or not. Just say to him, "I understand that in hyperbolic space the characteristic constant is negative, the degree of
negativity varying directly with the divergence from Euclidean space. I want to check
up on this. Lend me your Crandall, or your
Morecroft or your Van de Bijl, will you?"
If he tells you he hasn't the book, you will
know he has it but doesn't want to loan it to
you.
The following are indispensable books:
*The Manual of Radio Telegraphy and Tele-
phony*, Admiral S. S. Robison, United States
Naval Institute, Annapolis, Md. $5.50
Principles of Radio Communication, J. H.
Morecroft, John Wiley and Sons, New York
City, $7.50 (and worth it!)
An electrical Engineering Handbook
Among parts and accessories, we have:
A QRS or Raytheon Rectifying tube
An output filter
A resistance-coupled amplifying kit
A filament control relay (to control your
power unit and A battery from the set's on-off
switch)
A tube rejuvenator
A Balsa Speaker kit
A loud speaker unit

Ten Dollars to Fifteen Dollars

The customary way to give Christmas gifts
is to establish the understanding that you do
not believe in the exchange of gifts—that you
are giving nothing except cards—and then
give presents anyway to make your friends
uncomfortable. We suggest the following in
the line of this expensive misanthropy:
An "A" supply filter such as the "A-Box"
A *good* trickle-charger
A —— kit.
A Balsa speaker kit

Fifteen Dollars to Twenty Dollars

We are fairly well above the usual engineer-
ing books, but you might try a set of Rabelais.
Boni and Liveright have a new limited edition
selling for $20.00
Then there are:
A complete set of "B" batteries (a really
fine present!)
A cone loud speaker
A set of good audio transformers
A —— kit
A standard high-rate charger
A Balsa speaker kit

Twenty Dollars to Twenty-Five Dollars

We may now leave on the price tags, and
suggest—

A cone loud speaker
A complete set of A-C tubes
An "A" battery and charger combination
A —— kit
A radio table (a good one).

Twenty-Five Dollars to Fifty Dollars

We now leave gifts to the family and friends and consider presents for elevator boys, postmen, janitors, ice-men and others requiring special attention.
A very fine cone or Balsa speaker.
A well-designed "B" and "C" battery eliminator.
A —— kit

From Fifty Dollars up

We started with gifts for the family, and we conclude with the same. If there is a very expensive bit of apparatus for which you have long yearned—some deluxe speaker, perhaps a Norden Hauck Super or a Radiola Borgia model, or maybe a Wheatstone Bridge or a Leeds & Northrup type K potentiometer—why give it to the family for Christmas.
And in closing let me advise you to do your Christmas shopping early—early on the morning of December twenty-fourth.

TO
RADIO
DEALERS!

The R. B. Laboratory Information Sheets have been appearing in Radio Broadcast since June, 1926. They are a regular feature in each issue and they cover a wide range of information of value to the radio experimenter and set builder. We have just reprinted Lab. Sheets Nos. 1-81 from the June, 1926, to April, 1927, issues of Radio Broadcast. They are arranged in numerical order and are bound with a suitable cover. They sell at retail for one dollar a set. Write for dealers' prices. Address your letter to

Circulation Dept.,
RADIO BROADCAST
Garden City,   N. Y.

A HIGH-POWERED 10-TUBE
MODEL FOR 1928

STANDARD ADMIRALTY MODEL

NORDEN-HAUCK
IMPROVED SUPER-10
LICENSED UNDER HOGAN PATENT 1,014,002

"Highest Class Receiver in the World"

This new and advanced design brings to you all the luxury and performance you could possibly desire in a radio receiver. Absolutely nothing has been omitted to provide supreme reception—a new standard previously unknown.

The up-to-the-minute engineering principles found exclusively in the improved Super-10 are comparable to the high compression head and super-charger in an automobile engine.

In the Norden-Hauck Improved Super-10 you have
—*Wonderful Selectivity*
—*Faithful Reproduction with any Degree of Volume*
—*Signals received at only one point*
—*Two major tuning controls for all wave lengths*
—*Wave band 200-550 meters (adaptable down to low waves and as high as 6000 meters with removable coils)*

Material and workmanship conform to U. S. Navy specifications.

The improved Super-10 is sold complete as a manufactured receiver, or we are glad to furnish blue prints and all engineering data for home construction.

Upon request, new attractive illustrated literature will be promptly mailed to you without cost or obligation on your part.

Tear off and mail today

Your correspondence or inquiry for further information is cordially invited

Write direct to
NORDEN-HAUCK, INC.
ENGINEERS
"Builders of the Highest Class Radio Apparatus in the World"

MARINE BLDG., PHILA., PA., U. S. A.
Cable: Norhauck

NORDEN-HAUCK, INC.
MARINE BLDG., PHILA., PA.

Gentlemen:
☐ Without obligation on my part send me complete literature on the new improved Super-10.

☐ I enclose $2.00, for which send me, postpaid, complete Blue Prints and operating instructions for the new improved Super-10.

☐ I enclose $1.00 for which send me Blue Prints of the Model 500 B. C. Power Unit.

Name ....

Address ....

Read Radio Broadcast

every month. You can't afford to miss an issue. Order your copy from your Newsdealer or Radio Store. If you wish to subscribe direct send $4.00 for one year or $6.00 for two years.

DOUBLEDAY, PAGE & COMPANY
Garden City, N. Y.

While your neighbors are cussin' the static

Your set, with its Dubilier Light-Socket Aerial, is bringing the programs in smooth as silk. It's a fact! This little aerial, which you simply attach to the set and plug into the nearest light socket, reduces both static and interference to a marked degree. It uses no current whatever and absolutely eliminates the lightning hazard. Costs you nothing to prove it, for the Dubilier Aerial is sold by all good dealers on a 5-day, money-back basis. If your dealer can't supply you, write direct to us. Price, $1.50.

Dubilier
LIGHT SOCKET AERIAL

If you're planning to build a power-unit make sure that the condenser blocks you intend to use are built to withstand long hours of heavy-duty service. Dubilier blocks have an excessive high factor of safety and a "life" that makes them by far the most economical to buy. Full instructions enclosed with each block unit.

DUBILIER CONDENSER CORP.
4377 Bronx Blvd.   New York, N. Y.

Dubilier
CONDENSERS

---

77. TUBES—A booklet for the beginner who is interested in vacuum tubes. A non-technical consideration of the various elements in the tube as well as their position in the receiver. CLEARTRON VACUUM TUBE COMPANY.

87. TUBE TESTER—A complete description of how to build and how to operate a tube tester. BURTON-ROGERS COMPANY.

91. VACUUM TUBES—A booklet giving the characteristics and uses of various types of tubes. This booklet may be obtained in English, Spanish, or Portuguese. DEFOREST RADIO COMPANY.

92. RESISTORS FOR A. C. OPERATED RECEIVERS—A booklet giving circuit suggestions for building a. c. operated receivers, together with a diagram of the circuit used with the new 400-milliampere rectifier tube. CARTER RADIO COMPANY.

97. HIGH-RESISTANCE VOLTMETERS—A folder giving information on how to use a high-resistance voltmeter, special consideration being given the voltage measurement of socket-power devices. WESTINGHOUSE ELECTRIC & MANUFACTURING COMPANY.

102. RADIO POWER BULLETINS—Circuit diagrams, theory, constants, and trouble-shooting hints for units employing the BH or B rectifier tubes. RAYTHEON MANUFACTURING COMPANY.

103. A. C. TUBES—The design and operating characteristics of a new a. c. tube. Five circuit diagrams show how to convert well-known circuits. SOVEREIGN ELECTRIC & MANUFACTURING COMPANY.

38. LOG SHEET—A list of broadcasting stations with columns for marking down dial settings. U. S. L. RADIO, INCORPORATED.

41. BABY RADIO TRANSMITTER OF 9XN-9XR—Description and circuit diagrams of dry-cell operated transmitter. BURGESS BATTERY COMPANY.

44. ARCTIC RADIO EQUIPMENT—Description and circuit details of short-wave receiver and transmitter used in Arctic exploration. BURGESS BATTERY COMPANY.

43. SHORT-WAVE RECEIVER OF 9XN-9XR—Complete directions for assembly and operation of the receiver. BURGESS BATTERY COMPANY.

58. HOW TO SELECT A RECEIVER—A commonsense booklet describing what a radio set is, and what you should expect from it in language that any one can understand. DAY-FAN ELECTRIC COMPANY.

67. WEATHER FOR RADIO—A very interesting booklet on the relationship between weather and radio reception, with maps and data on forecasting the probable results. TAYLOR INSTRUMENT COMPANIES.

USE THIS BOOKLET COUPON

RADIO BROADCAST SERVICE DEPARTMENT
Radio Broadcast, Garden City, N. Y.

Please send me (at no expense) the following booklets indicated by numbers in the published list above:

....

....

Name...............................

Address............................
        (Name)   (Street)

....
        (City)   (State)

ORDER BY NUMBER ONLY
This coupon must accompany every request. RB 1-28

73. RADIO SIMPLIFIED—A non-technical booklet giving pertinent data on various radio subjects. Of especial interest to the beginner and set owner. CROSLEY RADIO CORPORATION.

74. THE EXPERIMENTER—A monthly publication which gives technical facts, valuable tables, and pertinent information on various radio subjects. Interesting to the experimenter and to the technical radio man. GENERAL RADIO COMPANY.

91. RADIO INSTRUMENTS—A description of various meters used in radio and electrical circuits together with a short discussion of their uses. JEWELL INSTRUMENT COMPANY.

78. ELECTRICAL TROUBLES—A pamphlet describing the use of electrical testing instruments in automotive work combined with a description of the cadmium test for storage batteries. Of interest to the owner of storage batteries. BURTON ROGERS COMPANY.

95. RESISTANCE DATA—Successive bulletins regarding the use of resistors in various parts of the radio circuit. INTERNATIONAL RESISTANCE COMPANY.

96. VACUUM TUBE TESTING—A booklet giving pertinent data on how to test vacuum tubes with special reference to a tube testing unit. JEWELL ELECTRICAL INSTRUMENT COMPANY.

68. COPPER SHIELDING—A booklet giving information on the use of shielding in radio receivers, with notes and diagrams showing how it may be applied practically. Of special interest to the home constructor. THE COPPER AND BRASS RESEARCH ASSOCIATION.

99. RADIO CONVENIENCE OUTLETS—A folder giving diagrams and specifications for installing loud speakers in various locations at some distance from the receiving set. YAXLEY MANUFACTURING COMPANY.

105. COILS—Excellent data on a radio-frequency coil with constructional information on six broadcast receivers, two short-wave receivers, and several transmitting circuits. AERO PRODUCTS COMPANY.

106. AUDIO TRANSFORMER—Data on a high-quality audio transformer with circuits for use. Also useful data on detector and amplifier tubes. SANGAMO ELECTRIC COMPANY.

107. VACUUM TUBES—Data on vacuum tubes with facts about each. KEN-RADIO COMPANY.

108. VACUUM TUBES—Operating characteristics of an a.c. tube with curves and circuit diagram for connection in converting various receivers to a.c. operation with a four-prong a.c. tube. ARCTURUS RADIO COMPANY.

100. RECEIVER CONSTRUCTION—Constructional data on a six-tube receiver using restricted field coils. BODINE ELECTRIC COMPANY.

Now That Everybo[dy]
Demands
Electrically-Opera[ted]
Radios!

Dongan is in Product[ion]
on All Types of
A. C. Tube Transform[ers]

Six months ago Dongan e[n]gineers were preparing f[or] the day when the indust[ry] unanimously accepted com[plete electrical operation [of] receiving sets. For every n[ew] tube brought forth, Donga[n] designed the proper tran[s]former or power unit.

To-day you can secure fro[m] the production line Tran[s]formers and Power Supp[ly] Units for whatever type [of] A C or A B C Tube y[ou] have chosen. For Donga[n] has been in production [on] approved types for ma[ny] months.

Here is the New[est]

No. 6515 Transformer for u[se] with 4 UX 226, 1 UY 227 [A] C Tubes and 1 UX 171 Tub[e]. Together with a B Elimina[tor], tor, this new transform[er] will convert old type set i[nto] to an efficiently operatin[g] A C set.

$4.75 List

This is one of 14 types rang[ing] ing in price from $3.50 [to] $8.00 for use with the ne[w] types of A C Tubes.

Manufacturers
are invited to write for any kind of [infor]mation from our engineering depar[tment].

Fans—order from your dealer or [send] check or money order to factory.

No. 5[...]
A B [...]
POWER [...]
for U[X]
UY 22[...]
171, an[d]
280 T[...]
$22 [...]

DONGAN ELECTRIC MFG.
2991-3001 Franklin St., Detroit, Mic[h.]

TRANSFORMERS of MERIT for FIFTEEN Y[EARS]

110. RECEIVER CONSTRUCTION—Circuit diagram and constructional information for building a five-tube set using restricted field coils. BODINE ELECTRIC COMPANY.

111. STORAGE BATTERY CARE—Booklet describing the care and operation of the storage battery in the home. MARKO STORAGE BATTERY COMPANY.

112. HEAVY-DUTY RESISTORS. Circuit calculations and data on receiving and transmitting resistances for a variety of uses, circuits for popular power supply circuits, d.c. resistors for battery charging use. WARD LEONARD ELECTRIC COMPANY.

113. CONE LOUD SPEAKERS—Technical and practical information on electro-dynamic and permanent magnet type cone loud speakers. THE MAGNAVOX COMPANY.

114. TUBE ADAPTERS—Concise information concerning simplified methods of including various power tubes in existing receivers. ALDEN MANUFACTURING COMPANY.

115. WHAT SET SHALL I BUILD?—Descriptive matter, with illustrations, of fourteen popular re-ceivers for the home constructor. HERBERT H. FROST, INCORPORATED.

104. OSCILLATION CONTROL WITH THE "PHASATROL"—Circuit diagrams, details for connection in circuit, and specific operating suggestions for using the "Phasatrol" as a balancing device to control oscillation. ELECTRAD, INCORPORATED.

A KEY TO RECENT RADIO ARTICLES

By E. G. SHALKHAUSER

THIS is the twenty-fifth installment of references to articles which have appeared recently in various radio periodicals. Each separate reference should be cut out and pasted on 4″ x 6″ cards for filing, or pasted in a scrap book either alphabetically or numerically. An outline of the Dewey Decimal System (employed here) appeared last in the January RADIO BROADCAST.

R402. SHORT-WAVE SYSTEMS. SHORT WAVES, QST. Aug. 1927. Pp. 9–14. ½-Meter
"The ½-Meter Band Officially Opened," B. Phelps and R. S. Kruse.
Detailed information on ½-meter transmitting and receiving sets is presented. It is stated that tubes having the XL filaments have a very short life at these frequencies, best results being obtained from the old UV-202 tubes. The antenna system, and the methods used in determining the length of the wave transmitted, are shown. The field tests indicate the manner in which signals decrease in strength and show the location of dead spots.

R132.1. AMPLIFYING ACTION: INDUCTIVE AMPLIFIERS, COUPLING. Audio.
QST. Aug. 1927. Pp. 15–20.
"Better Audio Amplification for Short-Wave Receivers," L. W. Hatry.
The writer shows the practical use of more than one audio stage of amplification for short-wave receivers. In order to insure more uniform volume from loudspeakers, whether listening to foreign or domestic stations, a switching or a shunt resistance system is described. The type of audio transformer to be used depends greatly upon the type of reception desired, a scheme being shown whereby an amplifier may be made either peaked or flat by using a tuned rejector circuit.

R346. RADIO TELEPHONE SETS TRANSMITTER, (ELECTRON-TUBE). Short-Wave Crystal.
QST. Aug. 1927. Pp. 21–24.
"Cuban 6 XJ," F. H. Jones and H. P. Westman.
The construction and the tuning of a first-class phone station operated on 20 meters, are outlined. A crystal, having a natural period of 159.6 meters, controls the transmitted frequency. The set consists of an oscillator and three amplifiers. Instead of using the Heising constant-current method of modulation, the series method of plate modulation is employed with good results. A circuit diagram and a list of parts are shown.

R113. HARMONIC METHODS. HARMONICS, QST. Aug. 1927. Pp. 34–35. Determination of
"The Identification of Radio-Frequency Harmonics," J. E. Waters.
A method of determining and identifying radio-frequency harmonics when making measurements of radio-frequency oscillations is outlined. Use is made of a standard wave-meter, an oscillator, and a receiver.

R113. TRANSMISSION PHENOMENA. TRANSMISSION, QST. Aug. 1927. Pp. 36–42. Short-Wave.
"Short-Wave Radio Transmission and Its Practical Uses," C. W. Rice. (Continued.)
The variation of signal strength with distance is discussed, taking into consideration the effect of multiple reflection. In order to choose the proper wavelength to use for distant transmission in summer daylight, a theoretical chart is prepared, showing probable performance of different waves. Conclusions drawn point to the following: Below 10 meters distant communication is impossible; the plane of polarization in the sky wave is no determining factor for energy flux density and for ray paths; different waves give best results between two given points; low-angle radiation is best for long-distance work.

R381.71. QUARTZ. QUARTZ. RADIO BROADCAST. Sept. 1927. Pp. 271–273.
"Piezo-Electric Crystals," M. T. Dow.
The writer explains the use of quartz crystal oscillators in the calibration of frequency meters. How to distinguish between the harmonics that are heard when two oscillators are in operation, is fully outlined. Photographs and circuit diagrams illustrate the points in question.

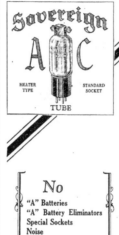

R113. TRANSMISSION PHENOMENA.   TRANSMISSION.
Proc. I. R. E. June, 1927. Pp. 501–517.   *Short-Wave.*
"Some Practical Aspects of Short-Wave Operation at High Power." H. E. Hallborg.
Propagation data on the frequency range of 3000 to 30,000 kilocycles are submitted. A correlation is shown between wave frequency and angle of projection of the wave front. The effect of ionization on the angle of projection is indicated. Some calculations are given of probable values of attenuation constant.
The importance of frequency stabilization is discussed, and three typical circuits for utilizing control crystals are described. Features of the design and adjustment of a 20-kw. power amplifier are also outlined. Antenna and antenna feed systems are discussed, and graphical results of comparisons of various antenna types are given. The relative importance of static at short wavelengths is considered. The author's anticipation of the density of the short wave is summarized.

R612. SHORT-WAVE STATIONS.   STATIONS.
Proc. I. R. E. June, 1927. Pp. 467–499.   *Short-Wave.*
"Short-Wave Commercial Long-Distance Communication." H. E. Hallborg, L. A. Briggs, and C. W. Hansell.
The development of short-wave communication by the Radio Corporation of America is outlined. A summary of short-wave installations, with call letters, wavelengths, and services to which each installation is assigned, is submitted.
Traffic charts showing the diurnal and seasonal characteristic of various wavelength over typical circuits are also shown. An outline of the technical problems inherent to the development of tubes and transmitter circuits is discussed. Methods are described for obtaining proper operation of tubes and transmitters at these very short wavelengths. The paper is illustrated with typical pictures and charts showing transmitter development and traffic performance.

R800 (512.82). COMPLEX VARIABLES.   FUNCTIONS.
Proc. I. R. E. June, 1927. Pp. 519–524.
"Maximization Methods for Functions of a Complex Variable." V. B. Roberts.
The maxima and minima of a function of a real variable are found by equating to zero the derivative of the function. In the case of a function of a complex variable, however, the derivative is a vector quantity, so that conditions may be imposed upon its direction as well as upon its magnitude. These various conditions lead to maxima and minima of the various aspects of the function. Rules are developed for setting up equations giving the various maximizing conditions, and a simple example is given illustrative of the use of the rule.

R240. PHASE DIFFERENCE.   PHASE RELATIONS.
Radio, Aug., 1927. Pp. 21–ff.
"Phase Relations in Radio." J. E. Anderson.
Of importance to radio experimenters is this discussion on the effect of phase relation between current and voltage in amplifier circuits. To illustrate the point, the phase conditions in a 4-tube resistance-coupled amplifier are analyzed in detail. Impedance- and audio-transformer coupled circuits are also discussed and the points to be observed are mentioned.

R334. FOUR-ELECTRODE TUBES   FOUR-ELECTRODE
Radio, Aug. 1927. P. 23–24.   CHARACTERISTICS.
"The Static and Dynamic Characteristics of a Double-Grid Vacuum Tube." H. R. Lubcke.
Static and dynamic characteristics of Van Horne double-grid vacuum tubes are graphed and discussed. It is noted that (1) a change in characteristics accompanies a change in frequency; (2) high plate voltages should be used because of hi-mu characteristics, (3) the inner grid should have a 3-volt C bias and the outer grid a 3-volt C bias; (4) the plate impedance of the tube should match that of the transformer primary in radio-frequency amplification.

R800. (530.) PHYSICS.   ULTRA-VIOLET
    RAYS.
RADIO BROADCAST. Sept., 1927. Pp. 269–265.
"A Discovery That Newton Missed." J. Stokley.
The history of ultra-violet rays, starting with the time of Newton two centuries ago, continuing with their discovery by Ritter, and ending with an explanation of their use made to-day in medical sciences, is given.

RR356. TRANSFORMERS.   TRANSFORMERS.
RADIO BROADCAST. Sept., 1927.   *Construction of.*
Pp. 274–278.
"Home-Constructing Transformers and Chokes for Power-Supply Devices." H. S. Davis.
By means of charts and certain fundamental mathematical formulas, sufficient information may be obtained to design and construct power transformers and choke coils for use in a.c. operated receivers. Of importance is said to be the characteristics of secondary windings, i. e., voltage current and power values, total turns in each winding, size and amount of core and wire, etc.

R330. ELECTRON TUBES.   ELECTRON TUBES.
RADIO BROADCAST. Sept., 1927. Pp. 284–285. *New A. C.*
"The New A. C. Tubes." Radio Broadcast, Laboratory Staff.
Information and data on the operation of various a.c. tubes now on the market show how these may vary in performance. From the writers' viewpoint, a.c. operated tubes still belong in the experimental class.

Erratum

AN ERROR occurred in the circuit diagram of the "Shielded Six" receiver published in the November, 1927, issue of RADIO BROADCAST. The connection indicated between terminal No. 6 of the third coil socket and the negative terminal of the detector tube socket, should be ignored. If this connection is made the C battery will be short-circuited and the detector will operate very inefficiently.

What Kit Shall I Buy?

THE list of kits herewith is printed as an extension of the scope of the Service Department of RADIO BROADCAST. It is our purpose to list here the technical data about kits on which information is available. In some cases, the kit can be purchased from your dealer complete; in others, the descriptive booklet is supplied for a small charge and the parts can be purchased as the buyer likes. The Service Department will not undertake to handle cash remittances for parts, but when the coupon on page 258 is filled out, all the information requested will be forwarded.

201. SC Four-Tube Receiver—Single control. One stage of tuned radio frequency, regenerative detector, and two stages of transformer-coupled audio amplification. Regeneration control is accomplished by means of a variable resistor across the tickler coil. Standard parts: cost approximately $58.85.

202. SC-11 Five-Tube Receiver—Two stages of tuned radio frequency, detector, and two stages of transformer-coupled audio. Two tuning controls. Volume control consists of potentiometer grid bias on r.f. tubes. Standard parts cost approximately $60.35.

203. "HI-Q" Kit—A five-tube tuned radio-frequency set having two radio stages, a detector, and two transformer-coupled audio stages. A special method of coupling in the r.f. stages tends to make the amplification more nearly equal over the entire band. Price $63.05 without cabinet.

204. R. G. S. Kit—A four-tube inverse reflex circuit, having the equivalent of two tuned radio-frequency stages, detector, and three audio stages. Two controls. Price $69.70 without cabinet.

205. Pierce Airo Kit—A six-tube single-dial receiver; two stages of radio-frequency amplification, detector, and three stages of resistance-coupled audio. Volume control accomplished by variation of filament brilliancy of r.f. tubes, or by adjusting compensating condensers. Complete chassis assembled but not wired costs $42.50.

206. H & H-T. r. f. Assembly—A five-tube set; three tuning dials, two steps of radio frequency, detector, and a transformer-coupled audio stages. Complete except for baseboard, panel, screws, wires, and accessories. Price $30.00.

207. Premier Five-Tube Ensemble—Two stages of tuned radio frequency, detector, and two steps of transformer-coupled audio. Three dials. Parts assembled but not wired. Price complete, except for cabinet, $35.00.

208. "Quadraformer VI"—A six-tube set will two tuning controls. Two stages of tuned radio frequency using specially designed shielded coils, a detector, one stage of transformer-coupled audio, and two stages of resistance-coupled audio. Gain control by means of tapped primaries on the r.f. transformers. Essential kit consists of three shielded double-range "Quadraformer" coils, a selectivity control, and an "Ampitrol," price $17.50. Complete parts $70.15.

209. Gen-Ral Five-Tube Set—Two stages of tuned radio frequency, detector, and two transformer-coupled audio stages. Volume is controlled by a resistor in the plate circuit of the r.f. tubes. Uses a special r.f. coil ("Duo-Former") with figure eight winding. Parts mounted but not wired, price $37.20.

210. Bremer-Tully Power-Six—A six-tube, dual-control set: three stages of neutralized tuned radio frequency, detector, and two transformer-coupled audio stage. Resistances in the grid circuit together with a phase shifting arrangement are used to prevent oscillation. Volume control accomplished by variation of B potential on r.f. tube. Essential kit consists of four r.f. transformers, two dual condensers, three small condensers, three choke coils, one 500,000-ohm resistor, three 1,000-ohm resistors, and a set of color charts and diagrams. Price $41.50.

212. Infradyne Amplifier—A three-tube intermediate-frequency amplifier for the super-heterodyne and other special receivers, tuned to 3400 kc. (86 meters). Price $25.00.

213. Radio Broadcast "Lab" Receiver—A four-tube dual-control receiver with one stage of Rice neutralized tuned-radio frequency, regenerative detector (capacity controlled), and two stages of transformer-coupled audio. Approximate price, $76.15.

214. LC-27—A five-tube set with two stages of tuned-radio frequency, a detector, and two stages of transformer-coupled audio. Special coils and special means of neutralizing are employed. Output device. Price $85.30 without cabinet.

215. Loftin-White—A five-tube set with two stages of radio frequency, especially designed to give equal amplification at all frequencies, a detector, and two stages of transformer-coupled audio. Two controls. Output device. Price $85.10.

216. K.H.-27—A six-tube receiver with two stages of neutralized tuned radio frequency, a detector, three stages of choke-coupled audio, and an output device. Two controls. Price $86.00 without cabinet.

217. Aero Short-Wave Kit—Three plug-in coils designed to operate with a regenerative detector circuit and having a frequency range of from 19,000 to 2500 kc. (11 to 130 meters). Coils and plug only. price $12.50.

218. Diamond-of-the-Air—A five-tube set having one stage of tuned-radio frequency, a regenerative detector, one stage of transformer-coupled audio, and two stages of resistance-coupled audio. Volume control through regeneration. Two tuning dials.

219. Norden-Hauck Improved Super 10—Ten tubes; five stages of tuned radio frequency, detector, and four stages of choke- and transformer-coupled audio frequency. Two controls. Price $360.00.

220. Browning-Drake—Five tubes: one stage tuned radio frequency (Rice neutralization), regenerative detector (tickler control), three stages of audio (special combination of resistance- and impedance-coupled audio). Two controls.

A New R. F. System

THE De Forest Company, which from the beginning has been one of the principal opponents of the Radio Corporation in the radio patent field, recently announced that Dr. George A. Somersalo, a Finnish physicist, has developed a new system of radio-frequency amplification which does not conflict with the Alexanderson tuned radio-frequency patent. This system depends upon the use of a filter ahead of the first radio stage, the radio-frequency amplifier being untuned. The Alexanderson patent, however, is not the only one possessed by the Radio Corporation of America and a great battle of wits lies ahead for those who want to do entirely without Radio Corporation licenses. No one who has had extensive contact with the American inventor, however, would venture to predict that it cannot be done.

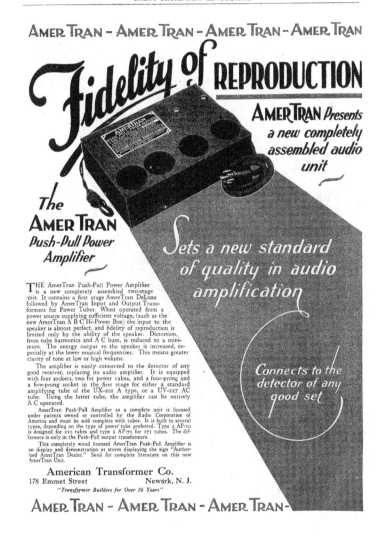

No Falling Off

CC Tubes

Avg. of 27
Other Makes

THE above chart graphically illustrates the longer life and steadier performance of CeCo tubes as shown by an unbiased test of CeCo Tubes in comparison with twenty-seven other makes.

The test was conducted by a nationally known set manufacturer (Name on request).

CeCo Tubes prove to be as efficient after 1,000 hours of use as when new. In fact, hundreds of instances have been called to our attention by users where CeCo Tubes have performed for over 2,000 hours without apparent deterioration.

Experiences like these, together with a noticeable improvement of tone, clarity and volume resulting from the use of CeCo Tubes, have secured the recommendations of well-known radio authorities such as

ARTHUR H. LYNCH	G. M. BEST	HERMAN BERNARD
GLENN H. BROWNING	KENNETH HARKNESS	KEITH HENNEY
LAWRENCE M. COCKADAY	VOLNEY HURD	JAMES MILLEN

There's a CeCo Tube for Every Radio Need

GENERAL PURPOSE TUBES	POWER TUBES
SPECIAL PURPOSE TUBES	RECTIFIERS
A. C. TUBES	

Ask your dealer to help you select the types best suited to your set.

C. E. MFG. CO., INC., Providence, R. I.

Largest plant in the world devoted exclusively to making of Radio Tubes

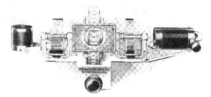

RADIO
BROADCAST

CONTENTS

The Facts About Output Devices

Tested Uses for the New Screen-Grid Tube

A Directory of Manufactured Radio Receivers

An A-C Push-Pull Amplifier and B Supply

How to Build an A-C Super-heterodyne

The Eyes of a Future Air Liner

35 Cents   Doubleday, Doran & Company, Inc.   FEB·1928
Garden City, New York

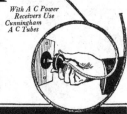

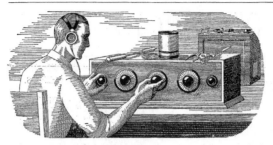

If all the Radio sets I've "fooled" with in my time were piled on top of each other, they'd reach about half-way to Mars. The trouble with me was that I thought I knew so much about Radio that I really didn't know the first thing. I thought Radio was a plaything—that was all I could see in it for me.

I Thought Radio Was a Plaything

But Now My Eyes Are Opened, And
I'm Making Over $100 a Week!

$50 a week! Man 'alive, just one year ago a salary that big would have been the height of my ambition.

Twelve months ago I was scrimping along on starvation wages, just barely making both ends meet. It was the same old story—a little job, a salary just as small as the job—while I myself had been dragging along in the rut so long I couldn't see over the sides.

If you'd told me a year ago that in twelve months' time I would be making $100 and more every week in the Radio business—whew! I know I'd have thought you were crazy. But that's the sort of money I'm pulling down right now—and in the future I expect even more. Why only today—

But I'm getting ahead of my story. I was hard up a year ago because I was kidding myself, that's all—not because I had to be. I could have been holding then the same sort of job I'm holding now, if I'd only been wise to myself. If you've fooled around with Radio, but never thought of it as a serious business, maybe you're in just the same boat I was. If so, you'll want to read how my eyes were opened for me.

When broadcasting first became the rage, several years ago, I first began my dabbing with the new art of Radio. I was "nuts" about the subject, like many thousands of other fellows all over the country. And no wonder! There's a fascination—something that grabs hold of a fellow—about twirling a little knob and suddenly listening to a voice speaking a thousand miles away! Twirling it a little more and listening to the mysterious dots and dashes of steamers far at sea. Even today I get a thrill from this strange force. In those days, many times I stayed up almost the whole night trying for DX. Many times I missed supper because I couldn't be dragged away from the latest-circuit I was trying out.

I never seemed to get very far with it, though. I used to read the Radio magazines and occasionally a Radio book, but I never understood the subject very clearly, and lots of things I didn't see through at all.

So, up to a year ago, I was just a dabbler—I thought Radio was a plaything. I never realized what an enormous, fast growing industry Radio had come to be—employing thousands and thousands of trained men. I usually stayed home in the evenings after work, because I didn't make enough money to go out very much. And generally during the evening I'd tinker a little with Radio—a set of my own or some friend's. I even made a little spare change this way, which helped a lot, but I didn't know enough to go very far with such work.

And as for the idea that a splendid Radio job might be mine, if I made a little effort to prepare for it—such an idea never entered my mind. When a friend suggested it to me one year ago, I laughed at him.

"You're kidding me," I said.

"I'm not," he replied. "Take a look at this ad."

He pointed to a page ad in a magazine, an advertisement I'd seen many times but just passed up without thinking, never dreaming it applied to me. This time I read the ad carefully. It told of many big opportunities for trained men to succeed in the great new Radio field. With the advertisement was a coupon offering a big free book full of information. I sent the coupon in, and in a few days received a handsome 64-page book, printed in two colors, telling all about the opportunities in the Radio field and how a man can prepare quickly and easily at home to take advantage of these opportunities. Well, it was a revelation to me. I read the book carefully, and when I finished it I made my decision.

What's happened in the twelve months since that day, as I've already told you, seems almost like a dream to me now. For ten of those twelve months, I've had a Radio business of my own. At first, of course, I started it as a little proposition on the side, under the guidance of the National Radio Institute, the outfit that gave me my Radio training. It wasn't long before I was getting so much to do in the Radio line that I quit my measly little clerical job, and devoted my full time to my Radio business.

Since that time I've gone right on up, always under the watchful guidance of my friends at the National Radio Institute. They would have given me just as much help, too, if I had wanted to follow some other line of Radio besides building my own retail business—such as broadcasting, manufacturing, experimenting, sea operating, or any one of the score of lines they prepare you for.

And to think that until that day I sent for their eye-opening book, I'd been wailing "I never had a chance!"

Now I'm making, as I told you before, over $100 a week. And I know the future holds even more, for Radio is one of the most progressive, fastest-growing businesses in the world today. And it's work that I like—work a man can get interested in.

Here's a real tip. You may not be as bad off as I was. But think it over—are you satisfied? Are you making enough money, at work that you like? Would you sign a contract to stay where you are now for the next ten years—making the same money? If not, you'd better be doing something about it instead of drifting.

This new Radio game is a live-wire field of golden rewards. The work, in any of the 20 different lines of Radio, is fascinating, absorbing, well paid. The National Radio Institute—oldest and largest Radio home-study school in the world—will train you inexpensively in your own home to know Radio from A to Z and to increase your earnings in the Radio field.

Take another tip—No matter what your plans are, no matter how much or how little you know about Radio—clip the coupon below and look their free book over. It is filled with interesting facts, figures, and photos, and the information it will give you is worth a few minutes of anybody's time. You will place yourself under no obligation—the book is free, and is gladly sent to anyone who wants to know about Radio. Just address J. E. Smith, President, National Radio Institute, Dept. 2O, Washington, D. C.

©CI B765463

JAN 21 1928

RADIO BROADCAST

WILLIS KINGSLEY WING, Editor

FEBRUARY, 1928

KEITH HENNEY
Director of the Laboratory

EDGAR H. FELIX
Contributing Editor

Vol. XII, No. 4

CONTENTS

AMONG OTHER THINGS. . .

WASHINGTON is distinctly the center of radio interest these days. One important conference which means much to radio all over the world is barely concluded with the closing of the International Radio Convention on November 25th when the Federal Radio Commission announces that public hearings will be held in Washington about the middle of January on the question of short-wave allocations. The thoughts of everyone have turned toward the short waves, and the Commission is very wise in holding hearings to enable some sort of ordered development to take place in these many vital channels now and in the years to come. Now all the short-wave channels are government, commercial, military, marine, naval, and amateur services. And, in addition, these channels contain some genuine experimental broadcasting stations and many more stations operating under experimental licenses which are broadcasting without any intelligent reason at all. The Commission is to be praised for its foresight in throwing this question open before it is too late.

THIS issue of RADIO BROADCAST contains articles of undoubted interest. "The Eyes of a Future Air Liner," for example, points out how radio can be applied to the present problems of air navigation. ' Anonymous' cloaks the identity of an authority on radio and aviation—a man who is better qualified to write on these twin subjects than any one we know. There has been much almost hysterical writing about the wonderful possibilities of the screened-grid tube, recently announced, and precious little genuine information about actual experiment with the possibilities of this very interesting tube. The Laboratory Staff, in the article on page 282, presents actual facts about what this tube can do, highly important to every experimenter whose interests lie in this direction. . . . For the first time, too, as far as we know, the facts about output devices are related. From the story by Keith Henney on page 294 you can learn exactly what the different types are, what they will do, and how best to use each type.

REGULAR broadcasting in the New York area of photographs sent by the Cooley Rayfoto system will be established before this issue is in the hands of readers and, accordingly, we publish a story by the inventor, Austin Cooley, presenting some additional technical information about the receiver which is now available in parts form to every interested constructor. Many readers write to request their names be forwarded to the manufacturers of the essential products. Any reader who has not yet done so should address a letter to the undersigned who will forward the request to the companies concerned. Picture broadcasting is here and we prophesy that in the not too distant future great numbers of experimenters will take this field for their own, completely fascinated by it.

WE SHALL soon publish the descriptions of a remarkably inexpensive receiver using the screened-grid tube, a new receiver design by Glenn Browning, a technical description with circuit diagrams and data on the Crosley "Bandbox" set, and an interesting kit for an A-socket power supply which will furnish enough A potential for ten quarter-ampere tubes.

—WILLIS KINGSLEY WING.

Doubleday, Doran & Company, Inc.	Doubleday, Doran & Company, Inc.	Doubleday, Doran & Company, Inc.	Doubleday, Doran & Company, Inc.
MAGAZINES	BOOK SHOPS	OFFICES	OFFICERS
COUNTRY LIFE	(Books of all Publishers)		F. N. DOUBLEDAY, Chairman of the Board
WORLD'S WORK		GARDEN CITY, N. Y.	NELSON DOUBLEDAY, President
GARDEN & HOME BUILDER	LORD & TAYLOR BOOK SHOP	NEW YORK: 244 MADISON AVENUE	GEORGE H. DORAN, Vice-President
RADIO BROADCAST	PENNSYLVANIA TERMINAL (2 Shops)	BOSTON: PARK SQUARE BUILDING	S. A. EVERITT, Vice-President
SHORT STORIES	NEW YORK: GRAND CENTRAL TERMINAL	CHICAGO: PEOPLES GAS BUILDING	RUSSELL DOUBLEDAY, Secretary
EDUCATIONAL REVIEW	38 WALL ST. and 526 LEXINGTON AVE.	SANTA BARBARA, CAL.	
LE PETIT JOURNAL	848 MADISON AVE. and 166 WEST 32ND ST.		JOHN J. HESSIAN, Treasurer
EL ECO	ST. LOUIS: 223 N. 8TH ST. and 4914 MARYLAND AVE.	LONDON: WM. HEINEMANN LTD.	LILLIAN A. COMSTOCK, Asst. Secretary
FRONTIER STORIES	KANSAS CITY: 920 GRAND AVE. and 206 W. 47TH ST.	TORONTO: OXFORD UNIVERSITY PRESS	L. J. McNAUGHTON, Asst. Treasurer
WEST WEEKLY	CLEVELAND: HIGBEE CO.		
THE AMERICAN SKETCH	SPRINGFIELD, MASS.: MEEKINS, PACKARD & WHEAT		

DOUBLEDAY, DORAN & COMPANY, INC., Garden City, New York

Copyright, 1928, in the United States, Newfoundland, Great Britain, Canada, and other countries by Doubleday, Doran & Company, Inc. All rights reserved.
TERMS: $4.00 a year; single copies 35 cents.

DISTANCE
lends
ENCHANTMENT

When you buy or build a radio set make
sure that it has Copper Shielding. Where
long distance reception is desired Copper
Shielding is essential. It is a refinement
to your set that will enable you to hear
the programs of distant stations much
more clearly.

Copper Shielded sets give:

BETTER RECEPTION
FINER SELECTIVITY
IMPROVED TONE QUALITY

By virtue of its easy working qualities and
its high conductivity Copper Shielding
is a decided improvement to any set.

COPPER & BRASS
RESEARCH ASSOCIATION
25 Broadway, New York

*Write for your copy of this
book. There is no cost nor
obligation on your part.*

At the Rugby High-Power Station of the British Post Office

The antenna lead-in as seen from the interior of the transmitter house. This station is one of the transmitters built in England by the British Marconi Company for the British Post Office, for communication with the Dominions. The British transmitter for the transatlantic radio telephone circuit is located at Rugby and the high-power radio telegraph transmitter is used for direct communication with Australia

OUT OF TOUCH WITH THE WORLD
One need not be a meteorologist, versed in the significance of the clouds,
to know what weather is ahead and below, after flying in the upper atmosphere for many hours. A simple application of modern scientific
principles will supply the information to pilots out of sight of the terrain

The Eyes of the Future Air Liner

"Anonymous"

DURING one of the ill-fated attempts to fly to Hawaii this summer the short-wave signals from the radio transmitter aboard the plane were copied in New York. Entire messages were easily received and the progress of the flight followed by newspaper reporters at the receiving stations, although the power of the transmitter wasn't enough to thoroughly warm up a curling iron.

It was a dramatic occurrence. Soon after the pilot took his heavily-loaded, single-engine plane off the runway and headed it westward over an unmarked trail traveled by few, cheerful messages came flashing back telling of the progress of the flight. The confident jaunty messages continued for some time. Then suddenly came warning of disaster. "We're in a spin," was the startling and serious message. The pilot was evidently able to momentarily bring his ship out of the spin, and put it back on an even keel. But not for long for the next message announced that the plane was again in a tailspin. And that was all, the last that was ever heard from that plane. Undoubtedly, the plane, loaded to the limit, was unstable. It could not be held in level flight in a fog or clouds, and stalled, with the resultant crash to the water below. Radio told of the end.

Over 3000 miles away newspaper reporters listening to these messages were impressed with the distance of transmission. In certain New York newspapers much was made over this feat of short-wave transmission. Reading these articles, one was ready to believe that something phenomenal in radio communication had taken place, and that the solution for problems of aircraft radio could be readily found in the use of short waves.

It is not believed that this is exactly so. Every radio amateur knows what can be done with low power on short waves. Bouncing his waves off a reflecting medium which scientists have named the Heaviside Layer, he is able to hop his signals all over the globe. He might well be something of a billiard player to properly angle his da-dit-dit-das so as to drop them on a friend's antenna perhaps on the other side of the world. But the amateur found that there were places, usually not so far away, where his waves could not be received, and other places close to him that reported considerable "fading." This much is known about short-wave transmission. But communication from plane to ground will be useless if not reliable over the entire route of flight. The "skip-effect" and the "fading" characteristic of waves much shorter than 80 meters makes their use in aircraft communication of little value. For what good will be the reception in San Francisco of a message sent by a plane flying from New York to Chicago when the message is intended for someone in New York?

The "skip" distance, or immune zone, is known to vary with the wavelength, the time of day, the time of year, the kind of antenna, and the nature of structures surrounding the antenna. There is a great deal to be learned about short-wave transmission from an airplane before the adoption of a definite near-short wave can be decided upon. The low power required, the simplified apparatus, the light weight, small size, and cost of short-wave equipment tempts one to jump at the proposition of equipping commercial craft with these sets. But research work in this connection must be exceedingly thorough. This means that the period of research must be extended to include all conditions as to season, time of day, and kind of terrain flown over, if work with short waves is to result in anything conclusive.

Before such research is begun it would be well to make a survey of the needs of commercial aviation with regard to radio service. Let us consider this point very briefly. First of all it appears that radio can be put to at least two very good uses on regular express and passenger air lines. Some means of telling the pilot in flight what is happening to the weather along the route ahead of him as well as at his destination would seem to be very desirable. Equally

271

desirable would be a means for telling the pilot when he was off his course and to help him get on it again. These two are the most important requirements at the present time. Nothing need be invented to provide this assistance to aërial travel. It is but a question of application of things now known.

To see just how the two kinds of radio service mentioned might be utilized, let us use our imagination and picture an airway of the near future.

We may start where the most imagination is required and visit New York's great municipal airport. It is just dark and we have arrived in plenty of time for a glance around before we take the early evening express plane for Chicago. There are a number of planes on the "flying line." Busy mechanics are fussing around these while other mechanics and helpers push and wheel planes into and out of the huge squat hangars. Twinkling red, green, and white lights outline the boundary of the airdrome. The red lights indicate obstacles; the green lights show favorable approaches to the landing area, while the white lights show the general outline of the field. A flood light illuminates like day much of the landing area. Smaller lights flood the hangars and the concrete "aprons" between. A huge beacon flashes from a low tower on one of the hangars.

We approach the "flight office" and here we see a busy official marking up on a bulletin board news of the movement of "ships" on the various airways terminating at New York. Another official is marking the latest weather reports on a large weather map of the United States. Here we also see posted forecasts of the weather to be expected along the different airways. We read with no feeling of glee that on the route to Chicago we may expect rain, and low clouds through the mountain region, and somewhat higher clouds with

A SPERRY REVOLVING BEACON
This particular one was installed by the Department of Commerce

showers from Cleveland to Chicago. Anxiously we go to inquire if in view of the weather prospects the Chicago express will leave to-night. We are assured that it will, and are advised to go aboard soon, as the "take-off" will not be delayed.

Aboard the big three-engined air liner we soon find that most of the twenty seats in the cabin are already occupied. We have seats near the front fortunately, and can observe

through a doorway the controls and instruments in the pilot's cabin.

The engines have been warmed up and tested at full throttle and now are idling, the metal propellers turning over lazily, awaiting only the will of the pilot. He is getting into his seat and is talking to his mechanic in a seat at his side, when an official comes aboard to converse a moment with these members of the "crew." We are being "cleared" for Chicago. The official leaves, the doors are secured and we are off. The rumbling muffled roar settles down into a droning beat as the three engines are synchronized in speed.

Below, the twinkling lights of suburban towns glitter against the black of night. Soon, however, we leave the region of more thickly clustered lights and then we are able to pick up some of the rotating lights of marker beacons which show the way westward. They blaze a trail all the way to our destination which will be easy to follow if bad weather is not encountered.

Our attention having been directed so much below we failed to see the mechanic reel out his trailing wire antenna and tune the receiver located in front of him. Now we notice a small panel on the instrument board, just below a compass indicator. In this panel a small white light is blinking slowly—on-off, on-off, without a break. This is evidently a signal. After a time we notice other lights on this panel. There are combinations of reds, greens, and whites, that come on for a few seconds, change, and then go out. More signals. The pilot and mechanic evidence interest but do not seem perturbed.

Looking out of the window at our side we see but a gray black nothingness faintly lighted by the illumination from the lighted cabin. We can see nothing below, and we realize that we are in clouds or fog. A large altimeter in the front wall

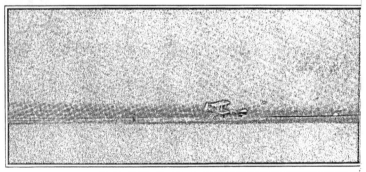

OFF FOR CHICAGO. THE NIGHT AIR MAIL LEAVES HADLEY FIELD, NEW JERSEY, BY THE AID

HIGH-INTENSITY LANDING FIELD FLOOD LIGHTS
One of these Sperry lights will illuminate a 50-acre field

of the cabin shows that we are 3200 feet above sea level. We are in the clouds. If they continue and we cannot fly beneath them and see the trail of lights on the ground, how are we to find our way? How will we know when it is safe to come down again?

We will watch the little panel with the steadily blinking white light and the peculiar combinations of colored lights that come only occasionally. As we watch, the white light blinks unsteadily and then changes to green and flashes regularly once more. We are off our course, to the right, we are told by a passenger just behind us. He has been over this route before and tells us that when the green light flashes we are to the right of the course. If it shows red, we are to the left. When it is white we are exactly on our course. The color combinations which we saw appear for only a short time on the panel were signals concerning the weather ahead. It was in

this way that the pilot knew that it was safe to climb up into the clouds, and so avoid the storm below, and yet steer clear of mountain peaks—reefs in this ocean of air.

For hours we fly through the night, yet there is no sensation of flying. That is because we are, unable to see anything but the dense fog of the cloud bank through which we pass. Our feeling is that our comfortable, lighted cabin is floating in space. The vibration of the engines and the muffled drone of the exhaust nearly lulls us to sleep. We see the needle of the altimeter crawl slowly up. Five thousand feet, six thousand; perhaps the pilot seeks to climb above the cloud into the moonlight which should be above. But no, the altimeter now shows that we are descending a little. Evidently the cloud bank is very thick, and it is not worth while to go above it.

We should be nearing our destination, that is if these red and green and white lights have been

faithful to their trust. And the combinations of colored lights, which now again are gleaming in the dimness of the pilot's compartment. Have they been unerring too? We shall soon know. The engines have been throttled a little. The floor of our cabin slopes forward and the needle of the altimeter is dropping slowly. We are losing altitude, and soon should burst through the "ceiling" of cloud. Suddenly the gray black mass which has engulfed us for hours is swept away. Twinkling lights appear below showing towns and lighted highways. Off to our left we see the rotating beam of a marker beacon, and, some little way ahead, another. We are on our course. Ahead there is a great glow in the sky, and as we draw nearer we realize that it is the light from a large city reflected against the clouds.

Now we pick up a bright flashing red light. It marks the airport of this city. Soon we are circling over it, and can see the line of lighted hangars and a brightly lighted letter "T" on the ground. This is to indicate the direction of the surface wind. We glide smoothly down, circle once more, and come in to land. Leveling off we float along hugging the ground and then with a few gentle bumps and a short roll we come to rest. To taxi up to the line in front of the flight office, and disembark, requires but a moment. We are in Chicago.

Radio has done its work. It told the pilot when he was off his course, and how to get on it again. It told him this and more, by very simple signals. Signals he could see. It told him of storms ahead, of cloud levels, and winds, and he was able to fly over disturbed areas and dangerous areas, and to come down safely at the proper time. No uncomfortable head-phones in a helmet were required, and it was not necessary that he know a telegraph code.

Some, perhaps many, may think the picture drawn is fantastic and only a dream. That remains to be seen. As for the ability to fly in the manner described, through clouds and fog, there are ships and pilots in plenty equal to the task. Let others apply things already known in the radio art, and the radio aids to aërial navigation which have been pictured will be a reality. Short waves may ultimately be the medium whereby this is accomplished, but considerable experiment will be necessary before the vagaries of these high frequencies will be fully understood.

© William E. Arthur & Co., Inc.

POWERFUL FLOOD LIGHTS. THE MAIL WILL BE DELIVERED IN CHICAGO TOMORROW MORNING

OBTAINING SOME DATA ON THE A. C. SUPER-HETERODYNE IN THE "RADIO BROADCAST" LABORATORY

Whenever receivers are described constructionally in RADIO BROADCAST, they are invariably tested out in the magazine's laboratory first. Frequently changes are recommended to the designers, and the technical staff does not give its O. K. unless the receiver comes up to expectations

A 45-Kc. A. C. Super-Heterodyne

By DORMAND S. HILL

IN A season during which a bewildering array of super-heterodyne receiver designs are offered to the radio fan and home builder, it is felt that, in presenting still another "super" to the readers of RADIO BROADCAST, it would be well to define, if possible, the relation of such a receiver to other designs, and to point out just why it was felt necessary that another entry should be made in an already too crowded field.

The forty-five kilocycle socket-powered super-heterodyne described here has as the greatest argument in its favor the fact that, though it is fully as sensitive and selective as the majority of eight, nine, and ten-tube sets of its class, its tuning controls are so simple that peak results are always at the command of even the novice operator, its tone quality is unusually good for super-heterodynes, and, possibly most important, it derives all power from any sixty-cycle alternating-current home light socket.

The set is entirely self-contained, power unit and set being housed in a standard seven by twenty-four inch cabinet, twelve inches deep. The only external equipment required is a good loop and a modern loud speaker. Since there is not a single battery to run down, and as the power unit is entirely dry, the servicing problem becomes one of occasional tube replacements only. The new a.c. tubes, which make possible the socket powered super-heterodyne described here, give promise of greater average life than is experienced with five-volt battery-type tubes.

In view of the foregoing advantages—better performance, excellent tone, and complete self-contained light socket operation—it is felt that the appeal of this super-heterodyne set should be indeed great. There is another factor which tremendously enhances the value of the receiver —its low initial cost. The entire receiver and power unit, constructed of the finest quality parts throughout, will cost only about $145.00, or, with a beautiful walnut cabinet, about $22.00 more. Thus, for less than $170.00 at list prices, the radio fan, or even the novice, can build for him-

self a full socket-powered super-heterodyne and obtain exceptional selectivity and sensitivity, while the 45-kc. super-heterodyne rivals one-dial sets in the simplicity of its tuning. Another factor in the home-built set is the beauty of finish of each individual part—a factor often neglected in all but the most expensive factory-built sets.

The forty-five kilocycle super-heterodyne employs eight tubes in a new a.c. tube circuit, carefully tried and tested. It possesses a great factor of "hum safety" due to the generous use of C-327 (UX-227) heater tubes instead of the greater hum-producing CX-326 (UX-226) type tubes. One C-327 type tube is used as a grid-tuned oscillator, one in a conventional Rice regenerative split loop first detector circuit, and three more follow in a forty-five kilocycle intermediate amplifier, which, in turn, works into a C-327 second detector. Then follows a two-stage low-frequency transformer-coupled amplifier employing a CX-326 (UX-226) first-stage tube, and either a CX-112 (UX-112) or CX-371 (UX-171) output tube with output transformer. The power supply consists of the conventional B device circuit using a CX-380 (UX-280) thermionic rectifier and a special filament heating transformer for all tubes. The power unit delivers 1.5, 2.5, and 5 volts a.c. and 200 to 220 volts pure d.c. to the receiver A, B, and C circuits.

A description of the individual circuit sections will help to provide a clear understanding of the unusual simplicity of the set, and the ease with which it may be built. On the front panel are three knobs. Two of these control the two illuminated tuning drums affecting the 0.00035-mfd. variable condensers, one of the two tuning the loop and the other tuning the oscillator, both with high reduction vernier drives and well-spaced 360-degree scales, numbered 0 to 200. In tuning, both dials tally approximately, stations being heard loudest at one point on the loop drum and at two points on the oscillator drum, although powerful local stations may be

heard at more than two points. The third knob, a volume control, provides adjustment from absolute zero to a full maximum. It is not critical as to setting, but if turned too far to the right, will cause the intermediate amplifier to block.

Looking down into the set from above, at the left front is the plug-in oscillator coil, L_4; behind it is the oscillator tube socket, and to the coil's right is the oscillator condenser, C_9, and drum. At the left rear are the loop and loud speaker connection tip jacks; and at their right, along the rear of the Micarta chassis, is the audio amplifier. The latter consists, from left to right, of an output transformer T_3, the output tube, a 3:1 ratio audio transformer, T_2, the CX-326 (UX-226) first-stage tube, and then the first-stage 3:1 transformer, T_1. The filament balancing resistors, R_4 and R_5, on the audio tube sockets, are clearly visible. The frequency characteristic of the audio amplifier is practically flat from 100 cycles to over 5000 cycles, and then falls off for reasons to be given later. The voltage gain of the audio amplifier, with CX-326 (UX-226) and CX-112 (UX-112) tubes is about 400 overall.

Just in front of the output transformer, and behind the first detector tube socket (second from left), is the knob of the loop regeneration condenser, C_6 in Fig. 1. To the right are the four forty-five kilocycle intermediate transformers, $L_2, L_3, L_4, L_5,$ with the three amplifier and second detector (at right) tube sockets in front of them. At the center of the front panel is the loop tuning drum and condenser, C_1. Between the two tuning drums is the 5000-ohm volume-control potentiometer R_6, and on the sub-base also between the drums is a knob controlling the radio-frequency amplifier C bias by means of the potentiometer, R_4. The first transformer, L_6, of the intermediate amplifier is untuned; the next, L_3, is a tuned stage (an XL 0.0005-mfd. Variodenser, C_4, below the chassis, with adjusting screw projecting up, is used for tuning this filter). Then follows a second untuned transformer, L_4 and, at the right, a second tuned transformer L_5, feeding the de-

POWER UNIT AND RECEIVER PROPER
In the event that the power unit is to be mounted elsewhere than right beside the receiver, the front panel for the latter may be smaller

tector tube. This latter stage is tuned by means of a fixed condenser, C₈.

The power unit, consisting of a power transformer, T₆, a condenser block, C₁₀, filament transformer, T₅, and Clough selective filter choke, L₄, with tube socket, and binding posts (and resistor, R₁₀, below), is on a seven-by-twelve inch chassis at the right of the set chassis, the two being cross-connected by the use of Eby binding posts as shown. An under-chassis view shows the voltage dividing resistor, R₁₀, beneath the power unit chassis, and other obvious parts beneath the receiver chassis. The simplicity of wiring is evident.

SELECTIVITY

THE selectivity of the set is quite good, and allows of ten- to fifteen-kilocycle separation of powerful local and weak out-of-town stations. The frequency band passed is a good ten kilocycles wide in normal operation, thus providing excellent tone quality with the audio channel used. This audio channel, due to its cut-off above 5000 cycles, materially aids the apparent selectivity of the radio-frequency circuits; in fact, the frequency characteristics of radio and audio circuits match very nicely.

An examination of the circuit indicates that all grid returns are brought back to the common B minus lead. First detector and oscillator are operated with approximately forty-five volts plate potential, and with zero grid potential; the grid-circuit returns and heater tube cathodes connect to the B minus lead. The three radio-frequency amplifiers derive an adjustable C bias from the 400-ohm potentiometer, R₁, which is connected between their cathodes and the B minus lead (a voltage is developed by virtue of the plate currents flowing through this resistor). The best bias is one-half to one volt, and, once set, varies automatically with changes in plate

voltage. The C bias resistor and the plate voltage potentiometer are bypassed with 1-mfd. condensers, C₆ and C₇, to prevent radio-frequency coupling. Plate voltage on the radio-frequency amplifier can be varied from zero to about ninety volts, using the 5000-ohm potentiometer (volume control knob), and generally about forty to forty-five volts gives greatest volume and sensitivity without amplifier oscillation. C bias for the detector is obtained from a 5000-ohm resistor, R₄, between the B minus lead and the detector cathode, shunted with a 0.0005-mfd. bypass condenser, C₉.

The C bias for the first audio stage is obtained across a 1500-ohm resistor, R₅, between the B minus lead and the center tap of a 64-ohm resistor, R₁, shunting the filament of the cx-326 (ux-226) audio tube. C bias for the second audio tube is similarly obtained by a 2000-ohm fixed resistor, R₆, between B minus lead and the center tap of a second 64-ohm balancing resistor, R₆, across the last audio tube's filament. This 2000-ohm resistor is bypassed with a 1-mfd. condenser, C₆, to improve low-frequency reproduction. For safety, a 64-ohm resistor, R₆, connected across the 2½-volt heater circuits of the cx-327 (ux-227) tubes, leads to plus 45 volts to prevent hum. The plate voltage of the first audio stage is about 90 volts; the C bias should be 5 to 6 volts—correct for cx-326 (ux-226) tubes. The last audio tube may be interchangeably a cx-112 (ux-112) or a cx-371 (ux-171) type tube, the former for greatest amplification, the latter for best quality on strong signals. With a cx-112 (ux-112) tube, C bias is about 20 volts and the plate voltage about 200—a safe value and one at which the performance of the 112 type tube is quite creditable. The C bias for a cx-371 (ux-171) tube would be about 38 volts, and the plate potential, about 170 volts—entirely satisfactory values, and slightly below the tube's rated maximum.

The construction of the set is quite simple, involving the use of the following parts. If the exact values specified are used, all adjustments for proper a.c. operation are automatic. Substitution may necessitate experiment:

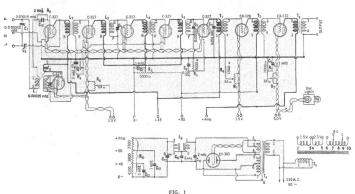

FIG. 1
The circuit diagram of the 45-kc. a.c. super-heterodyne

PARTS LIST

Two Remler 100 Universal Drum Dials	$9.00
L_6, L_4 Remler 600 Interstage Transformers	12.00
L_3, L_4 Remler 610 Tuned Stage Transformers	10.00
C_4, C_5 Remler 658 type 0.00035-Mfd. Condensers, Variable	10.00
T_1, T_2 S-M 240 Audio Transformers	12.00
T_3 S-M 241 Output Transformer	5.00
Six S-M 512 Five-prong Tube Sockets	4.50
Two S-M 511 Tube Sockets	1.00
One S-M 515 Coil Socket	1.00
L_1 S-M 111A Coil	2.50
Two Pairs S-M 540 Mounting Brackets	1.40
C_3 S-M 340 Midget Condenser	1.50
C_{10} Polymet 0.00015-Mfd. Condenser, with Clips	.45
C_9 Polymet 0.00025-Mfd. Condenser	.35
C_{11} Polymet 0.002-Mfd. Condenser	.40
C_4 Polymet 0.005-Mfd. Condenser	.60
R_9 Polymet 2-Megohm Grid Leak	.25
C_6, C_7, C_8 Polymet 1-Mfd. Bypass Condensers	3.00
R_4 Frost F1500 Resistor	.50
R_7 Frost F2000 Resistor	.50
R_1, R_2, R_4 Frost FT64 Resistors	1.50
Five Frost 253 Tip Jacks	.75
R_3 Frost 5000-Ohm De Luxe Potentiometer	2.25
R_8 Polymet 5000-Ohm Resistor	1.00
R_4 Frost 400-Ohm De Luxe Potentiometer	1.25
C_4 XL Model G "Variodenser"	1.50
Sixteen Eby Binding Posts: (8-plain, 2 B—, 2 B+Det., 2 B+Amp., 2 B+Int.)	2.40
Westinghouse Micarta Walnut Panel 7 x 24 x $\frac{7}{16}$ inches	8.00
1 Westinghouse Micarta Sub Panel 12 x 17 x $\frac{7}{16}$ inches	9.00
TOTAL	**$103.60**

POWER SUPPLY PARTS LIST

T_4 S-M 329 Power Transformer	9.00
L_5 S-M 331 "Unichoke"	8.00
T_5 S-M 325 Filament Transformer	8.00
S-M 511 Tube Socket	.50
Pair S-M 540 Mounting Brackets	.70
C_{12} Polymet 14-Mfd. Condenser Block	9.50
R_{10} Ward-Leonard 659 Resistor	2.50
Westinghouse Micarta Power Unit Base, 12 x 7 x $\frac{7}{16}$ Inches	4.00
8 Eby binding posts: (4Plain, 1 B—, 1 B + Det., 1 B + Amp., 1 B + Int.)	1.20
TOTAL	**$43.40**

Panels can either be drilled, following the apparatus layout of the photographs, or procured drilled and engraved from any Micarta dealer. The apparatus should be mounted on them after careful study of the photographs, and in the positions shown. One point to observe is that all cathode leads from the c-327 (UY-227) tubes should be brought out below the chassis by one of the socket mounting screws, as this is good practice to follow wherever practicable.

In wiring the set and power unit, all 1.5-, 2.5-, and 5-volt a.c. leads should be twisted to localize their fields. All grid and plate leads should be of busbar, in spaghetti where necessary, as should the leads to the tuning condensers, while all other low-potential wiring should preferably be of

Kellogg switchboard wire. The leads from the 2.5-volt binding posts of the receiver to F posts of the five-prong tube sockets are each composed of two No. 14 tinned wires in parallel, for high current-carrying capacity. All metal parts of the set and power unit, such as transformer frames or shells and condenser bank case should ground to B minus. The drum dial frames, carrying the dial lights, are taken care of through the lamp wiring to the 5-volt a.c. circuit, which grounds through the F2000 C bias resistor, R_7.

TEST AND OPERATION

AFTER finishing the set and power unit, it would be well to check voltages with a high-resistance voltmeter (1000 ohms per volt). The power unit not connected to the receiver should show about 70 to 80 volts on the 45-volt tap; 120 to 130 on the 90-volt tap; and 220 or more on the high tap. Next, the power unit should be connected to the receiver, no tubes being inserted in the set, and voltages again checked (for short circuits). All voltages should have fallen somewhat. The audio tubes should now be inserted in the set and the loud speaker connected. A strong hum evidences proper operation of the audio channel if B and C voltages (as given previously) check out *approximately*. If posts Nos. 1 and 2 of the first audio transformer are short-circuited, the hum will fall to the actual operating value; it should be so low as not to interfere with reception at average speaking volume. If the hum is stronger, ground the B minus post to a water pipe. If the hum is still too high (and it may be under unusual line or set assembly conditions), the solution is to move the power unit away from the set. If all cx-327 (UX-227)

T4   RECT.   2nd.DET.   L5 3rd.INT.   2nd.INT.   1st.INT.   C2   L1
C1 L4   L3 R3 R4   L2   1st.DET.
T5   OSC.
   C3
L6
C12   C5   R1   T2   R2   T3
T1   1st.AUDIO   C4   2nd.AUDIO
(CX-326)

LOOKING DOWN INTO THE CABINET OF THE A.C. SUPER-HETERODYNE
The various parts are lettered so that they may be easily identified by cross reference to Fig. 1, on the previous page

THE COMPLETE RECEIVER AND POWER UNIT

It is here shown in an attractive looking cabinet, a product of D. H. Fritts and Company, Chicago, Illinois

heater tubes are inserted, the hum pitch will vary as the tubes warm up and, after a minute, will decrease to the average operating value observed with posts Nos. 1 and 2 of the first a.f. transformer shorted (this very low hum is always experienced with CX-326 [UX-226] or other "raw a.c." tubes.)

To operate the set, set the midget condenser all out, connect a Bodine L350 or equivalent good loop to the loop posts, and tune-in stations with the two drums. Set the C bias potentiometer to include ⅓ to ½ of its total resistance in circuit which will give ⅓ to 1 volt negative bias on the r.f. amplifiers. A good starting point would be KDKA, tuned in at about 53 on the loop, and 46 or 58 on the oscillator, or WDAP, at 120 on the loop and 103 or 130 on the oscillator. If the volume knob is set as far as it will go to the left, no signal will get through; advancing the knob

right up to the oscillation point (sometimes called "spill-over," "plunk," or "squeal" point) will increase volume. If the volume knob is turned to the right of this point, only squeals will be heard. Remember, no squeals should be heard on the set, other than station heterodyne squeals of substantially unvarying pitch. On a short-wave station, the midget condenser should be turned in slightly to increase sensitivity; if turned in too far, only squeals will be heard— bear this in mind. Once set, leave it alone, but in the first case, be sure to set it on a 210-or 220-meter (1430 or 1360 kc.) station. The oscillator coil rotor should generally be set at about a 45-degree position. There are four of these points in the full 360-degree arc of rotation. Remember that full 180-degree rotation from any of these points will cause either upper or lower oscillator dial setting to give strongest signals on any

particular station. Therefore, set the rotor on a weak 300- or 325-meter (1000-or 920-kc.) station to produce greatest volume with good selectivity on *either* upper or lower oscillator dial setting, then use whichever point (upper or lower) is loudest, in tuning all other stations. On a weak signal, adjust the XL Variodenser for strongest signals and sharpest tuning of the oscillator dial.

TROUBLE-SHOOTING

REMEMBER, if the set does not give as good results for sensitivity and selectivity or tone as any other average eight- or nine-tube "super," or seven- or eight-tube t.r.f. set, under *identical and simultaneous* operating conditions, it simply is not put together or operated in accordance with the foregoing paragraphs. Therefore, look for your trouble in your own work.

The set should not squeal in operation; if it does, incorrect operation is the cause—look to midget condenser and volume control settings more carefully.

Lack of sensitivity and selectivity should cause careful readjustment of the XL condenser and midget condenser, and a comparison against some other set of known performance.

Hum should be checked as outlined for the a.f. amplifier. If it tests out correctly, and there is hum coming through the r.f. circuits, the trouble is an open grid-circuit and most probably due to poor or incorrect wiring. If loud hum is in the audio amplifier (with first a.f. transformer primary posts, Nos. 1 and 2, shorted) look for defective wiring, open balancing resistor, or power tubes. Move the power unit several feet away from the set when making this test.

In proper operation the receiver should give good super-heterodyne performance, say, average 500- to 2000-mile loud speaker range, on all good stations, hum so low as not to be noticeable through station announcements at average speech volume, and absolutely dependable operation in the hands of a novice, with practically no servicing problem—and all at an operating cost of less than ten cents an hour.

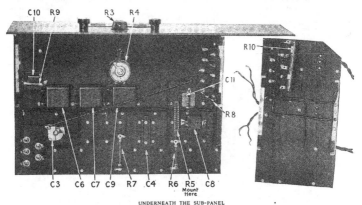

UNDERNEATH THE SUB-PANEL

Here again the parts have been lettered for easy reference. The grid bias resistance for the detector tube was being experimented with at the time this picture was taken, and, therefore, its correct location has been indicated by means of dotted lines

The Commission Improves Broadcasting Conditions

THIS month we chronicle what should be an improvement of a most pleasing nature in the broadcasting situation, the clearing of 25 of the 36 channels between 600 and 1000 kc., accomplished by the Federal Radio Commission's General Order No. 18. Perhaps, by the time these lines are published, the entire 36 channels will be cleared. "Clearing" means that in most cases many stations operating on one channel have been placed elsewhere and coast-to-coast duplication of stations now obtains which "clears" the channel for all practical purposes.

But we thirst for even greater improvement. Having tasted the gratification which comes of the Commission's firm and commendable action, we hope for more. Every move in the direction of clearing channels means larger audiences and, in consequence, still better broadcasting.

With amazing strides in program development and with a Federal Radio Commission working for the best interests of broadcasting, we hope that, sooner or later, ideal reception conditions will be attained. Indeed, the order clearing channels from 600 to 1000 kilocycles, truly a drastic move, promises to bring good reception not only to urban districts but to rural as well, where radio is becoming the most important and essential form of home entertainment and education.

So many listeners who write us are unfair in laying the blame for heterodyne whistles upon the incompetence of the Federal Radio Commission. The Commission is not incompetent; it is impotent. No five men under the sun could solve the present broadcasting tangle without having undisputed power to eliminate stations from the broadcasting band.

In spite of any congressional declarations that the ether may be regulated in the public interest and that none have vested rights to use it, it is quite generally assumed that broadcasting stations have established

what amounts to vested rights in the ether. Having invested capital and legally conducted a public service, station owners contend that they cannot be deprived of the opportunity of continuing that service without compensation. Be that as it may—we contend that stations, which had to wait for broadcasting licenses until the regulatory powers of the Department of Commerce were nullified by unfavorable court decision, have no right on the air. If property confiscation is necessary to secure good broadcasting, let us find a way to confiscate worthless broadcasting stations.

The elimination of two or three hundred of the smaller stations, accomplished by a board with confiscatory powers, need not cost more than two or three million dollars, really a small sum when the social and economic importance of broadcasting and the magnitude of the industry serving it are considered.

We make a serious suggestion for a tax on commercial broadcasting. We realize that our numerous friends among broadcasting station owners will regard us, for a moment at least, as an enemy, seeking to add another to their already numerous burdens. A tax on commercial broadcasting would, at first, certainly penalize the good stations in favor of the bad. But a two per cent. tax, spent solely upon compensating the owners of confiscated stations, could bring radical improvements in broadcasting, such that audiences would

be enormously increased, with inevitably augmented revenues to broadcasting stations and improved program standards.

Furthermore, the amount of tax paid by stations would serve as an indisputable guide to their value. Educational and religious stations, now operating at considerable losses, could charge the organizations sponsoring them, on an hourly basis, an amount sufficient to cover their deficits and thus, having established revenue, suffer no penalty for their non-commercial status. A definite principle of compensation for station condemnation could be established, based upon a year's commercial revenue, plus the physical value of its equipment. Three years of taxation of commercial broadcasting and an honestly administered and efficient condemnation board would leave a strong broadcasting structure, consisting only of the most popular and successful stations, with the weak sisters bought out at minimum cost, and clear channels for the remaining. The stations to continue would be selected not only upon the basis of economic value but upon a definite and scientifically organized plan, taking into account power, service area, and geographical location.

The ultimate effect of such a policy would certainly double, triple, and fourfold the value of commercial broadcasting to program sponsors. Interference-free reception and increased coverage would inevitably result in program improvement. The clear reception of a program on every channel would threefold the usefulness of the receiver and increase program choice, serving as another incentive to add to the listening audience. The radio industry would, therefore, prosper by the sale of more receiving sets. On every hand, for the listener, station owner, and manufacturer, there would be greater prosperity and greater service.

We are indeed aware that, at first sight, to burden a harassed broadcasting industry, suffering from interference and insufficient

THE FEDERAL RADIO COMMISSION

This view was taken a few days before the death of the Chairman, Admiral Bullard, who is third from the left. Harold A. Lafount, the new Commissioner appointed to succeed John F. Dillon, deceased, is shown at the extreme right. Until a Chairman is chosen, Eugene O. Sykes is acting as Vice-Chairman. From left to right: Sam Pickard, E. O. Sykes, W. H. G. Bullard, Carl H. Butman, secretary, O. H. Caldwell, H. A. Lafount

© Henry Miller

revenue, is not likely to be popular among the commercial broadcasters. Too often, when taxation is suggested as a remedy, the cure is worse than the disease.

The Commission, by its recent allocations, has clearly demonstrated that fewer stations per channel means improved reception. But it has made its progress by virtually destroying almost two thirds of the broadcasting frequencies in order to make one third capable of performing maximum service. Is it efficient utilization of valuable ether channels to impair two thirds of those available so that one third may be of maximum service?

The only sensible ideal is to make every channel useful to the utmost of its capacity. Any channel, which serves as a graveyard for ten, twenty, or thirty stations, represents the confiscation of a substantial investment on the part of receiving set owners. If we value the receiving sets in service at $450,000,000—and this is a conservative estimate—broadcast listeners have spent some $5,000,000 a channel. Confiscating two thirds of the broadcast band, by allotting it to chaos, represents a loss of $300,000,000 to listeners.

Is the listener not entitled to free and complete use of his apparatus, considering that his collective investment is much larger than that of the broadcasting stations? Broadcasters who wail about destroyed investments are themselves guilty of destruction a thousandfold greater than that which they suffer.

If confiscation of property is necessary to establish good broadcasting, let us give the Federal Radio Commission full power to condemn broadcasting stations and the funds with which to purchase them. We do not hesitate to condemn private property in order to build improved highways, a process which has made the automobile business one of America's greatest and most serviceable industries. Now let us condemn private property in order to have good highways of the ether. It will make radio a truly great industry, serving every American family and home.

Who Buys This Year's Radio Sets?

THE replacement market will be radio's most profitable sales field this year. The listener who has had experience and contact with radio is readier to appreciate the advantages of socket power operation and the greatly improved tonal quality attained by this year's better receiving sets. The radio trade quite generally overlooks the replacement market, although receivers of two, three, four and five years ago are still widely used. The automobile industry is now subsisting very largely upon replacements which reach a dollar total far in excess of new owner purchases.

To the radio industry, this is a new problem and the approach to the replacement market requires fundamental changes in sales and advertising methods. The industry should seize upon its opportunity en-

thusiastically because three fourths of the selling resistance has already been eliminated by the set owner's previous experience. He is ready to understand and appreciate the significance of modern improvements, with but a simple explanation, laid before him where he is accustomed to absorb his radio knowledge.

The replacement market has the fortunate advantage that those most enthusiastic about radio are efficiently reached through specialized mediums which appeal to the more ardent and, consequently, more responsive radio follower. The publications in the radio field have a combined circulation of nearly one million and, therefore, reach the best one fifth or one sixth of the entire replacement market. Those who purchase radio consumer magazines are naturally more interested in radio than those who do not. They have followed radio for several years at least, and are the most likely to possess obsolete sets. This invaluable group is reached at a lower cost than an equal number of prospects among general magazine circulations. Popular magazines, serving every class of society instead of singling out especially fruitful groups, reach an insignificantly small proportion of already interested buyers and persons sold on radio by actual experience. A general magazine is fortunate if one out of twenty of its readers have the slightest interest in its radio advertising; every reader of a radio magazine is a prospect for a 1928 radio.

Since the general spirit of the better radio magazines is no longer predominantly technical, but covers all phases of the radio science, art, personality, broadcasting, as viewed behind the microphone and

heard before the speaker, progress in manufactured receivers, telephotography and television, and numerous other fields of wide appeal, circulations have changed in character so that they include not only the home constructor and radio engineer, but the dealer and jobber, the wealthy and ardent listener, and the most liberal purchasers of radio equipment. The new situation offers a great opportunity to the manufacturer to sell his wares at minimum cost.

A prophet is without honor in his own country and the radio manufacturer is enthusiast as a sort of demented crank who has no influence upon the purchase of radio sets. However, he readily admits, when pinned down, that it was this very element which built up his industry and that, instead of being cranks, they are pioneers and leaders. Of course, there are a few manufacturers who cannot profitably address themselves to this discriminating and well-informed group, because they do not make a product fit to offer to an expert radio enthusiast. But, in general, the products of the radio industry have reached a standard which makes them highly acceptable to the man with some technical knowledge and appealing to him is an asset which means the sale of not only one but usually a good many radio receiving sets.

Rural Radio Listeners Served by Re-allocations

IN A statement issued by the Commission shortly after its clearing of the 600-to 1000-kc. channels, it points out that millions of listeners in remote communities, presumably a good two thirds

A BRITISH MARCONI SHORT-WAVE RECEIVING STATION
This receiving unit, located at Skegness, England, was built by the Marconi company for the British Post Office and is used as the British end of the India and Australia service. The view above shows a corner of the land line connections and high speed recording instruments. On the left are key and sounder for communication with the transmitting station and Radio Central in London; next are two sets of high-speed recorders with tape pullers and a Wheatstone transmitter for sending to the London Central office

of the radio audience, who have little or no choice of local programs, are especially benefited by the new allocations. It is this group which leans most heavily upon radio's utilitarian and entertainment service. The Commission states that, although the DX listener is served by these reallocations, its principal purpose is to make radio more acceptable to the rural listener.

There is some criticism due to the fact that the thirty or more stations which now have clear channels are affiliated with one or the other of the networks. Therefore, a rural listener, going down the dials on a clear night, would find only three programs available to him, although he may tune-in twenty of the stations. Ultimately, this situation may be alleviated by the assignment of all chain stations to a single channel. This is not feasible at this time because of the great cost entailed in synchronizing stations, a complete and extra wire network being necessary to accomplish this, and, furthermore, because the affiliated stations do not receive their programs exclusively from the chain source. The better stations throughout the country have affiliated themselves with chain organizations and it is therefore inevitable that they should receive favored consideration in the matter of being assigned clear channels. Consequently, the Commission's action is unavoidable and it cannot rightly be accused of favoritism.

Sooner or later, however, it will be necessary to synchronize stations so that the same channel will be used by all stations radiating a program. This will give an added advantage to the rural listener because he will derive signal energy, not only from the nearest of the chain stations to which he tunes, but the combined signal energy from all the stations which energize his set. The Blue Network, for example, combined on a single channel by synchron-

ous broadcasting, would deliver an adequate local service signal almost anywhere in the eastern part of the United States.

It has been suggested that a single synchronizing signal be broadcast from which all stations are to derive the carrier signal. Such a synchronizing signal would have to be of a frequency low enough that its harmonics would include every broadcasting channel. Thus a 10,000-cycle synchronizing signal, radiated by modulating a high-frequency carrier, would be received at every broadcasting station in the country. Those assigned to a 500-kc. frequency, for example, would, by means of a harmonic producer, pick off the fiftieth harmonic of the 10,000-cycle synchronizing signal. This would be sufficient to serve as the carrier for the station. The 51st harmonic would provide the 510-kc. signal and so on throughout the broadcasting band.

Although this process appears to be a simple solution of the frequency stability problem, it is, unfortunately, full of technical pitfalls, sufficient to prevent its immediate general utilization.

We are unable to build harmonic producers of sufficient reliability to obtain the selected harmonic with unfailing certainty. Much investigation is necessary before reliable harmonic producers to cover the entire broadcast band would be available so that they could be employed by the usual broadcasting station technical staff. The harmonic producer is still a laboratory product, requiring the highest kind of engineering talent to secure satisfactory results.

Technical aids will certainly come to the help of the broadcasting situation, but they cannot be relied upon to give prompt and substantial relief, in our present knowledge of the radio art. The mills of the gods and the laboratories of scientists

grind slowly, but, fortunately, they never cease grinding.

Frequency Allocation Outstanding Achievement of International Conference

IT REQUIRES a careful study of the allocations adopted by the International Radio Telegraphic Conference to appraise the labors of that body briefly and it is premature to report on more than its outstanding and obvious achievement, the establishment of a complete schedule of international wavelength allocations. In examining this schedule, it must be borne in mind that nations may permit any kind of emission to any radio station under their jurisdiction at any frequency, under the sole condition that such emission does not interfere with any other country. To stations, which, by their nature, are known to be capable of causing material international interference, the contracting parties agree to assign frequencies in accordance with the schedule which appears below.

The assignments made to amateurs represent a curtailment of their present wide range of channels, although they have not lost completely the right to any of their accustomed bands. Those bands, designated as "amateur" without restriction or division, are exclusive amateur bands and therefore available for international communication. As to the broadcasting bands, the American standard has been adopted, although a few channels in the 150-kc. region have been made available for European broadcasting. A number of small but well distributed short-wave bands for international links are also provided, which take into account the possibilities of both daylight and night transoceanic rebroadcasting. The Conference, evidently, listened appreciatively to the advice of experts in selecting these short-wave bands.

How Radio Channels Are Internationally Assigned

10 to 100 kilocycles (30,000 to 3000 meters)—Point to point service.	550 to 1300 kilocycles (545 to 230 meters)—Broadcasting. See Note 3.	12,825 to 12,350 kilocycles (23.4 to 22.4 meters)—Mobile and fixed.
100 to 110 kilocycles (3000 to 2725 meters)—Point to point and mobile service.	1300 to 1500 kilocycles (230 to 200 meters)—(a) Broadcasting, (b) mobile (on the frequency 1364 kilocycles only), wave length 200 meters).	12,350 to 14,000 kilocycles (22.4 to 21.4 meters)—Fixed.
110 to 125 kilocycles (2725 to 2400 meters)—Mobile.	1500 to 1715 kilocycles (230 to 175 meters)—Mobile.	14,000 to 14,400 kilocycles (21.4 to 20.8 meters)—Amateur.
125 to 150 kilocycles (2400 to 2000 meters)—Mobile, maritime service, general public correspondence only.	1715 to 2000 kilocycles (175 to 150 meters)—Mobile, fixed and amateurs.	14,400 to 15,100 kilocycles (20.8 to 19.85 meters)—Fixed.
150 to 160 kilocycles (2000 to 1875 meters)—(a) Broadcasting, (b) point to point. (c) mobile. Subject to agreement as follows: All regions where broadcasting stations now exist working below 300 kilocycles (above 1000 meters)—Broadcasting; other regions, (b) point to point, (c) mobile. Regional agreements will respect the rights of one another in this band.	2000 to 2250 kilocycles (150 to 133 meters)—Mobile and fixed.	15,100 to 15,350 kilocycles (19.85 to 19.55 meters)—Broadcasting.
	2250 to 2750 kilocycles (133 to 109 meters)—Mobile.	15,350 to 16,400 kilocycles (19.55 to 18.3 meters)—Fixed.
	2750 to 3850 kilocycles (109 to 105 meters)—Fixed stations.	16,400 to 17,100 kilocycles (18.3 to 17.5 meters)—Mobile.
	2850 to 3500 kilocycles (105 to 85 meters)—Mobile and fixed.	17,100 to 17,750 kilocycles (17.5 to 16.9 meters)—Mobile and fixed.
104 to 285 kilocycles (1550 to 1050 meters)—(a) Mobile, (b) point to point, (c) broadcasting. Subject to regional agreement as follows: Europe (a) Mobile (aircraft only). (b) point to point (air services only). (c) point to point (ncr) from 250 to 285 kilocycles (1200 to 1050 meters); (a) broadcasting from 194 and 224 kilocycles, (1550 to 1340 meters); other regions; (a) Mobile, except commercial ships, (b) point to point (aircraft only), (c) point to point (ncr).	3500 to 4000 kilocycles (85 to 75 meters)—Mobile, fixed and amateurs.	17,750 to 17,800 kilocycles (16.9 to 16.85 meters)—Broadcasting.
	4000 to 5500 kilocycles (75 to 54 meters)—Mobile and fixed.	21,450 to 21,550 kilocycles (14 to 13.9 meters)—Broadcasting.
	5500 to 5700 kilocycles (54 to 52 meters)—Mobile.	21,550 to 22,300 kilocycles (13.9 to 13.45 meters)—Mobile.
	5700 to 6000 kilocycles (52.7 to 50 meters)—Fixed.	22,300 to 23,000 kilocycles (13.45 to 13.1 meters)—Mobile and fixed.
	6000 to 6150 kilocycles (50 to 48.8 meters)—Broadcasting.	23,000 to 28,000 kilocycles (13.1 to 10.7 meters)—Not reserved.
285 to 315 kilocycles (1050 to 950 meters)—Special (radio beacons).	6150 to 6675 kilocycles (48.8 to 45 meters)—Mobile.	28,000 to 30,000 kilocycles (10.7 to 10 meters)—Not reserved.
	6675 to 7000 kilocycles (45 to 42.8 meters)—Fixed.	30,000 to 56,000 kilocycles (10 to 5.35 meters)—Not reserved.
315 to 350 kilocycles (950 to 850 meters)—Mobile (aircraft service only). See Note 1.	7000 to 7300 kilocycles (42.8 to 41 meters)—Amateurs.	56,000 to 60,000 kilocycles (5.35 to 5 meters)—Amateurs and experiments.
350 to 360 kilocycles (850 to 830 meters)—Mobile (ncr).	7300 to 8200 kilocycles (41 to 36.6 meters)—Fixed.	60 kilocycles (5 to 0 meters)—Not reserved. Note 1—3331 kilocycles (000 meters) is the international aircraft calling and listening frequency.
360 to 390 kilocycles (830 to 770 meters)—(a) Special (direction finding); (b) mobile, where it does not interfere with direction finding.	8200 to 8550 kilocycles (36.6 to 35.1 meters)—Mobile.	
	8550 to 8900 kilocycles (35.1 to 33.7 meters)—Mobile and fixed.	
390 to 460 kilocycles (770 to 650 meters)—Mobile.	8900 to 9600 kilocycles (33.7 to 31.6 meters)—Fixed.	Note 2—500 kilocycles (600 meters) is the international calling and distress frequency. It may be used for other purposes when it will not interfere with calling.
460 to 485 kilocycles (650 to 620 meters)—Mobile, except damped and radio telephone waves.	9500 to 9600 kilocycles (31.6 to 31.2 meters)—Broadcasting.	
485 to 515 kilocycles (620 to 580 meters)—Mobile (distress calling, &c). See Note 2.	9600 to 11,000 kilocycles (31.2 to 27.3 meters)—Fixed.	Note 3—Mobile services use the band 550 to 1300 kilocycles (545 to 230 meters) on the condition that they do not interfere with the services of any nation using this band exclusively for radio telephone broadcasting.
515 to 550 kilocycles (580 to 545 meters)—Mobile (not open to general public correspondence), except damped and radio telephone waves.	11,000 to 11,400 kilocycles (27.3 to 26.3 meters)—Mobile.	
	11,400 to 11,700 kilocycles (26.3 to 25.6 meters)—Fixed.	
	11,700 to 11,900 kilocycles (25.6 to 25.2 meters)—Broadcasting.	Note—ncr means: Not for general public correspondence.
	11,900 to 12,300 kilocycles (25.2 to 24.4 meters)—Fixed.	
	12,300 to 12,825 kilocycles (24.4 to 23.4 meters)—Mobile.	

New Commissioner Appointed

ON NOVEMBER 14, President Coolidge announced the appointment of Harold A. Lafount of Utah as a member of the Federal Radio Commission to fill the vacancy created by the death of John F. Dillon. Mr. Lafount has used a radio receiving set and therefore approaches the problems of broadcast regulation with a confidence equal to that with which we would assume the command of a battle fleet after a visit to a navy yard. The new commissioner may prove to be the dark horse who brings an ultimate victory over confusion and we wish him every success. But certainly, it has been amply demonstrated that the Commission's problems are highly technical and that the Commissioners are handicapped in their work until they have, at great cost of time and effort, learned the intricacies of broadcasting.

Inside the Radio Industry

A CLASSIFICATION of questionnaires returned by 3546 dealers, made by the Electrical Equipment Division of the Department of Commerce, indicates that 26 per cent. are electrical supply dealers, 20 per cent. radio dealers, 13½ per cent. hardware dealers, 8½ per cent. dealers in musical instruments, and 6 per cent. automobile dealers, 5½ per cent. battery and ignition supplies, 4½ per cent. tires and tire repair shops, and smaller percentages to other classes. A total of 68 different varieties of retail outlets is represented in the classification. It is rather surprising to find the hardware retailer so prominently represented and the music dealer outlet so small in proportion to the whole.

Another comprehensive statistical survey, to be a quarterly investigation conducted by the Department of Commerce, is an enumeration of the stock of radio sets and their principal accessories in dealers' and jobbers' hands. The survey is national in scope and the first returns are based upon 7718 filled-in questionnaires. This number of dealers had 147,548 battery receiving sets on hand and 9548 socket powered sets; ordinary speakers, 153,091; amplifier-speaker units, 5018; B and C batteries, rated in 45 volt units, 525,441; storage A batteries, 77,148. This makes an average of twenty battery-operated receiving sets and a little over one socket power set per dealer, a healthy situation, considering that both dealers and jobbers are represented. Moving less than 150,000 battery-operated sets at this season is not an abnormal demand upon the public.

RADIO INDUSTRY STANDARDS

OUR persistent campaign for the adoption of a single set of standards for the radio industry has now come to an end because both trade organizations have, at last, agreed to work together in this one respect at least. This move, however, is only the first step in our program, which calls for complete unification of the two trade associations. The present step, respecting standards, brings this objective nearer.

The Radio Manufacturers' Association and the Radio Division of the N. E. M. A. have agreed to review all manufacturing standards pertaining to radio and will publish a single industry standard. In cases of dispute, the American Engineering Standards Committee will serve as an arbitration board. The entire industry is to be congratulated upon this outcome of the persistent campaign which a few individuals have waged quietly and persistently for many months.

THE Radio Manufacturers' Association has appointed a Patent Interchange Committee, with A. J. Carter as its chairman, which is to work out a patent pooling plan. It has obtained the consulting services of Mr. C. C. Hanch, who worked out the patent cross license system for the automobile industry. We would suggest that this worthy effort be combined with that along similar lines being made by the Policies Division of N. E. M. A. Without unification of effort, both associations are wasting time and cannot hope to accomplish anything of permanent value.

THE Radio Corporation's quarterly statement for the September quarter announces a net profit of $4,141,355, its largest net operating profit for any quarter. Its earnings per share were $2.80 and, for the same quarter a year ago, $1.53.

NEWS FROM ABROAD

THE writer has had the privilege of talking to Mr. J. M. Bingham, the Chief Engineer of the New Zealand Broadcasting System, who is visiting the United States in order to learn the latest in broadcasting practice from American engineers. As a matter of fact, the entire broadcasting industry of the United States might profitably visit New Zealand in order to learn how to run radio broadcasting successfully. No country has more efficient regulation.

Broadcasting has been placed in the hands of a single company by authority of the New Zealand Parliament. This monopoly is supported by an annual tax upon broadcast listeners. The purchase of every receiving set and every part that goes into a receiving set, down to the last binding post, is recorded by dealers. Government agents have access to their books at all times. No listener can escape the vigilant eyes of government inspectors. The revenue thus gained is divided between the broadcasting stations and the Government, but not in the unsatisfactory ratio which obtains in the British Isles. Eighty per cent. or more of the broadcast tax collected in New Zealand is actually spent in program talent or the erection and maintenance of broadcasting stations. No radio advertising or commercial goodwill broadcasting is allowed or necessary.

One would conclude that, under such an efficient system, with its magic wand to overcome any serious problems and vicissitudes, broadcasting would thrive and grow. Its growth has been steady but not startling. The number of licensees, in a population of one and a quarter million, is about twenty thousand. Four high power broadcasting stations cover the four-hundred-mile length of the country. There are some dead areas, but prospective changes in transmitter location will soon remedy these. Interconnecting wire circuits are being developed but, as yet, there is virtually no chain broadcasting. Naturally, with but four stations, there are no frequency allocation difficulties. There is no economic problem in meeting station maintenance cost. There is little or no evasion of listener tax. Altogether, it is the most efficient broadcasting system in the world.

A DEPARTMENT of Commerce statement advises that there are 685 broadcasters in the United States and its territories; the total in all nations, other than the United States, is 431. Of these, 196 are in Europe, 128 in North America, outside of the United States, 52 in South America, 18 in Asia, in Oceania 28, and in Africa 9. The four most powerful broadcasting stations in this country are WGY, 100,000 watts with 50,000 watts as its usual power; WEAF, 50,000 watts; KDKA, 50,000 watts; and WJZ, 35,000 watts. In Russia, there are two 40,000 watt stations and one of 20,000 watts. The total number of broadcasting stations in numerical order are: Canada 59, Cuba 47, Russia 38, Sweden 30, Australia and Germany 24 each, Argentina 22, United Kingdom 20, France and Mexico 18, Spain 15, Brazil 12, Chile 9, Finland 7, England 6. There are four each in Belgium, Czecho-Slovakia, Uruguay, India, Netherlands, East Indies and New Zealand; three in Italy, Poland, China, Japan and South Africa, and smaller numbers in other countries.

STATION 2 YA AT WELLINGTON, NEW ZEALAND

The New Zealanders have elegant scenery on which to mount their radio stations if this can be taken as typical. 2 YA has 5000 watts power and is one of four stations serving the country

Radio Broadcast Photograph

USING 222 TYPE TUBES IN AN AUDIO AMPLIFIER

This is the conventional resistance-coupled amplifier using two screened-grid tubes and one 171 power tube. The little cap on top of the tube is the signal grid connection. The values of resistance, etc.,' are given on Fig. 5

THE SCREENED-GRID TUBE

By The Laboratory Staff

THE new screened-grid tube, the ux-222 or cx-322, is designed to appeal to the more experimentally inclined of the set constructing fraternity, and while it is comparatively simple to build receivers to utilize some of the advantage of these tubes 'to outdo some of the great task comes after the last wire is in place.

Contemporaries of Radio Broadcast have devoted considerable space to this new tube and, perhaps, are so keenly alive to its potentialities that they have somewhat over-exaggerated its possibilities so far as the average non-technical set builder is concerned. The screened-grid tube has been in the Radio Broadcast Laboratory for several months now, and for a considerable longer period of years in the research laboratories of the R.C.A. and the General Electric Company. As a result of our experiments at Garden City, we must still say that this is an experimental tube, and that hard and fast rules for quickly throwing together coils and con-

TABLE 1

AMPLIFICATION FACTOR	PLATE IMPEDANCE
3/8	2000
8	12,000
15	25,000
30	150,000
35	250,000

densers with screened-grid tubes to connect them, are not yet possible.

Tubes available in the United States differ considerably from those used in England, and in our opinion suffer somewhat in comparison. This is not due to their electrical characteristics but, rather, due to the mechanical arrangement which, in English tubes, seems to have been worked out with more thought toward ease of use.

Readers interested in the tube may find Captain H. J. Round's book, The Shielded Four-Electrode Valve, very helpful. It is published by Bernard Jones Publications, Ltd., 58 Fetter Lane, London, E.C.4, and the price is two shillings and sixpence (a little over sixty cents).

Considerable time has been devoted in Radio Broadcast Laboratory to determine how to use these tubes in radio-frequency amplifiers. Before discussing these experiments in detail, we will indicate the advantages obtained through the use of a 222 screened-grid tube instead of a 201-a type tube in a radio-frequency amplifier. This involves delving somewhat into r.f. amplifier theory to get an idea of how they work and upon what their amplification depends.

When a 201-a type tube is used in a radio-frequency amplifier circuit (see Fig. 1) in conjunction with an ordinary r.f. transformer, the amplification obtained will be a function of the number of turns in the primary coil and its coupling to the secondary coil. With either a very few or a great many primary turns the amplification will be low, but with some definite intermediate number of turns the amplification will become a maximum. Under such conditions the amplification using a 201-a tube will be about 10 at 500 meters (600 kc.) and about 15 at 200 meters (1500 kc.). In order to get even this amplification from a 201-a type tube, a carefully designed circuit is necessary, incorporating some method of neutralization to overcome the effect of the capacity between the grid and plate of the tube, which would otherwise cause the circuit to break into self oscillation.

Somewhat greater amplification can be obtained with an ordinary three-electrode tube if the electrodes are so constructed and arranged as to give the tube a higher amplification constant. Such a tube is called a "high-mu" tube and although such tubes have been used most frequently in resistance-coupled audio amplifiers, they may also be used in an r.f. amplifier. There are, however, several reasons why their use in the latter capacity is not advisable.

The amplification factor of a tube is entirely under the control of the tube engineers, and tubes can be built with little difficulty having constants anywhere from 0.5 to 500. When the amplifica-

tion constant of a tube is increased in this manner, however, the plate impedance also increases and, in fact, increases approximately as the square of the amplification constant, so that the plate impedance increases much more rapidly than the amplification constant. The table on this page will give some idea of how the plate impedance varies as the amplification factor is changed.

Now, in order to obtain considerable gain from a high-mu tube when it is used as a voltage multiplier in an r.f. stage, it is necessary to have a very high impedance load in the plate circuit, so that transformer coupling with a step-up ratio may not be practical. An ordinary high resistance might be used were it not for the fact that it would be shunted by the load resistance' capacities of those tubes preceding and following it. These capacities across the load resistance' would limit the effective impedance of the load in the plate circuit to a low figure, so that the amplification will fall to a comparatively low value. An effective way in which to obtain high gain from a high-mu tube is through the use of a tuned-plate amplifier (see Fig. 2 and compare it with Fig. 1). In such a circuit the output and

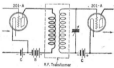

FIG. 1

A transformer is used to connect two tubes working as radio-frequency amplifiers. The number of turns on the primary for maximum amplification is controlled by the input and output impedance of the tubes used. For a given set of tubes and at a given frequency there is a number of turns which will give maximum amplification

input capacities of the tube are in parallel with the condenser which tunes the circuit and have no effect other than to decrease slightly the amount of capacity necessary in the variable condenser in order to tune to any given wavelength. At resonance the effective resistance as measured across the two ends of a tuned circuit is very high and the tuned plate type of amplifier is therefore capable of producing an effective high impedance in the plate circuit—a necessity if high gain is to be obtained, as stated before. The effective resistance at resonance of even a good coil-condenser combination, when it is connected into an amplifier, is, however, probably not more than 100,000 to 200,000 ohms. When the effective resistance in the plate circuit of a tuned plate r.f. amplifier is equal to the plate impedance of the tube, the amplification obtained is one half the mu of the tube. Thus, a tuned circuit giving an effective resistance of 150,000 ohms, and used in conjunction with a tube having an amplification constant of 30 and a plate impedance of 150,000 ohms would produce an effective amplification of one half mu, or 15. It should be evident that the limiting factor preventing the attainment of much larger values of amplification, is the large increase in plate impedance that occurs when the amplification constant is increased. What we want, obviously, is a tube with a very high amplification factor and as low a plate impedance as possible. Such characteristics are not obtainable using an ordinary three-electrode tube because, as indicated previously, the plate impedance increases much more rapidly than the amplification constant. To obtain a high mu with a comparatively low plate impedance, it is necessary to introduce a fourth electrode into the tube. The effect of the fourth electrode was explained in articles on the four-electrode tube in the December, 1927, and January, 1928, issues of RADIO BROADCAST, and will not be repeated here except to point out that this fourth electrode does give a very large increase in the amplification constant without as large an increase in plate impedance as would be obtained with a three-electrode tube. Measurements in the Laboratory indicate that the amplification constant of the 222 tube is about 200, under certain conditions, and the plate impedance is about 800,000 ohms. If an ordinary three-electrode tube, such as a type 240 high-mu tube, were to be constructed to have a mu of 200, the plate impedance would be about five to seven million ohms! A high value of amplification constant with a comparatively low plate impedance is one of the major characteristics and advantages of the 222 screened-grid tube.

The amplification obtained from this tube is a function of the mutual conductance of the circuit and the impedance of the load in the plate circuit. Owing to the fact that the impedance of the tube is so high, higher than its load impedance, the mutual conductance of the circuit is practically equal to that of the tube. If the mutual conductance of the tube is assumed to be 350 micromhos the amplification with various load impedances can be obtained by multiplying the mutual conductance in mhos by the load impedance. The values have been calculated and appear in Table 2. For certain reasons, however, it is likely that we can't get more than about 200,000 ohms in a tuned circuit under operating conditions. If an effective resistance of 200,000 ohms is connected in the plate circuit of a 222 tube, then the amplification will be about 70 (see Table 2). Compare this with the amplification of 10 to 15 obtained with a 201-A type tube.

The 222 tube has another distinct advantage, also important in the construction of high-gain radio-frequency amplifiers. This second advantage is that the effective grid to plate capacity

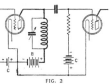

FIG. 2

If the plate impedance of the previous tube is very high the entire input coil of the following tube may be used to approach the maximum amplification possible. This is true when the screened-grid tube is used, for its plate impedance is very high indeed, of the order of one half to one megohm

of the tube is very small. It is this capacity which causes an ordinary tube to oscillate when it is used in an r.f. circuit. Theoretically, it is possible, therefore, to use this new tube in an r.f. amplifier without resorting to any scheme of neutralization to prevent the tube from oscillating.

Some experiments have been made in the Laboratory in an attempt to use these tubes in an r.f. amplifier designed for operation over the broadcast band. The circuit diagram is given here in Fig. 3. The experiments indicated more than anything else the difficulty of operating these tubes in a circuit designed to give the maximum amplification of which the tube is capable. The experiments also indicate that the tube can only be satisfactorily used in circuits containing very complete shielding and very complete filtering in the battery supply leads. This is to be expected, for such shielding and filtering is necessary in a good r.f. amplifier, using 201-A type tubes and, obviously, will be much more essential in an amplifier using these new tubes.

If the amplification of a circuit consisting of a UX-222 tube followed by a tuned circuit is measured, it be will found, with a good coil, to average about 30. The amount of gain obtained depends upon the efficiency of the coils used and increases as the coil is made better.

The first experiments were made with just one stage of r.f. amplification using the 222 tube followed by a detector, and even with very complete filtering and with both of the circuits shielded in aluminum cans the circuit would oscillate on practically all wavelengths. The coils used in the tuned circuits in this receiver were those made by Silver-Marshall, Incorporated, and it is probably true that, if poorer coils had been used, the circuit would have been found stable in operation. In fact, one ohm in series with the tuned circuit at 500 meters (600 kc.) and four ohms in series with the circuit at 200 meters (1500 kc.) was all that was necessary to make the circuit stable. This additional resistance necessary to prevent oscillation is not very great and many coils would have even higher resistances than that of the combined Silver-Marshall coils and resistances.

The same conditions of instability were, of course, also experienced when the receiver was changed over to use 222 tubes in two stages of r.f. amplification. Again, even with good filtering of the battery circuits, it was impossible to

TABLE 2
Effect of Coil Resistance

SERIES RESISTANCE OF COIL	EFFECTIVE RESISTANCE AT RESONANCE	FIGURE OF MERIT	AMPLIFICATION $Gm = 350$
50	50,000	31.5	17.5
25	100,000	63.0	35
17.5	150,000	126	42.5
12.5	200,000	252	70
6.2	300,000	504	105

Conditions: Coil inductance 250 microhenries
Frequency, 1000 kc.

control the circuit, and it persisted in oscillating over the entire broadcast band. The circuit could be stabilized by any of the "losser" methods. In this particular receiver it is found that a resistance of about 1000 ohms connected between the grid and tuned input circuit would prevent the tubes from oscillating. Also stabilization of the amplifier is possible by tapping the plate on to a portion of the output coil rather than connecting it to the top of the tuned circuit, but this method of connection also results in a decrease of amplification. The photograph on page 284 shows.

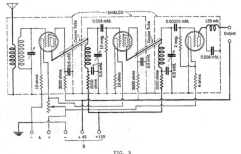

FIG. 3

Here is a two-stage screened-grid broadcast frequency amplifier hook-up. The stage shielding, and the filtering in each plate- and screen-voltage lead must be very good, and leads to the control or signal grid on top the tube must be short and often shielded. The leads connecting one stage to the next should be run in copper tubing which is grounded. Even then it may be found difficult to "hold down" the circuit

how the coil was tapped. Further experiments will be conducted in the Laboratory to indicate the comparative effect of these various methods of stabilizing the circuit where they are necessary, but the use of such methods of preventing oscillation probably decrease the amplification that can be obtained and it will be preferable to design the receiver in such a manner as to give stable operation without any stabilizing devices. It may seem peculiar to the reader that so much difficulty was experienced with oscillation in the r.f. amplifier when one of the major purposes of the tube was to give freedom from instability in r.f. circuits. It should be realized, however, that the tube gives a very high amplification and it is necessary to develop new design features in receivers before these tubes can be utilized to best advantage. When it is realized that very careful design is required to produce stable high-gain radio-frequency amplifiers using 201-A's it should be clear why even greater care is necessary when a 222 tube is used, and perhaps four times the amplification per stage is obtained. It is probable that high-gain amplifiers that are perfectly stable at high radio frequencies will be constructed, but as yet we have not been able to do it in the Laboratory. This should not be construed as a statement to the effect that the possible gain is not greater than that obtainable with 201-A type tubes. for it certainly is—considerably so, but what we wish to infer is that circuits designed to operate with the tube to its best advantage are not available.

There are other problems than those connected with gain and stability. That of securing selectivity is probably even greater than that of maintaining stability over a wide frequency band. Owing to the greater amplification of all signals, the apparent selectivity is poor. When one has a two-stage broadcast frequency amplifier using screened-grid tubes giving three times, to be conservative, the gain per stage of 201-A tubes, the problem of selectivity becomes serious. The solution may be to decrease the gain per stage or to use some type of filter circuit at the input of the receiver capable of increasing the selectivity.

USE IN AUDIO AMPLIFIERS

IN AUDIO-frequency amplifiers the difficulties are not so great, due to the fact that what grid-plate capacity exists in the tube (it is of the order of 0.025 mmfd.) cannot feed back such a large voltage into the input as it can at higher frequencies.

The two uses of this tube have been pointed out in previous articles. It may be used as a screened-grid tube in which the grid-plate capacity has been almost eliminated, or as a tube in which the grids are reversed and one is used mainly to reduce the space charge of the tube so that we have a tube whose values of amplification factor and impedance are much greater than the best of our high-mu tubes, the plate-grid capacity still being appreciable.

Experiments in the Laboratory indicate that the space-charge application of the tube is a good talking point but not much good in practice. If one wants to use this new tube in a low-frequency amplifier, which is where the space-charge tube would be important, all that is necessary is to use the screened-grid tube with proper values of resistances, capacities, and voltages. A photograph on page 282 shows a conventional three-stage resistance amplifier used in the Laboratory

AN AUTO-TRANSFORMER MAY BE USED
In place of the conventional two-winding transformer between tubes, an auto-transformer may be used. This photograph shows how the proper place to tap the coil is determined in the Laboratory. The amplification at each tap is measured and the best number of primary turns found

FIG. 4
When the total a.c. voltage appearing in the plate circuit of the last tube is divided by the input voltage to the first tube, this kind of curve results. In each case the final tube was a 171, and various combinations of high-mu tubes, screened grid tubes, and standard 201-A tubes were measured and the results plotted as shown

to obtain the data given in Fig. 4. Using screened-grid tubes with 180 volts to the 0.25-megohm resistors and 45 volts on the screen, a flat characteristic was obtained from 60 to 10,000 cycles. The total voltage appearing in the plate circuit, roughly three times that across the 1000-ohm resistance in its output, divided by the input voltage, gave a ratio of 2200 which is almost three times that obtainable when UX-240 high-mu tubes were used. Six volts were obtained across the output resistance when the input voltage was roughly five millivolts.

There is no catch in this amplifier, and for anyone who wants a voltage gain of 2200, and can use it, we recommend converting his old resistance amplifier into one using the new tubes.

While the curve in Fig. 4 shows that some amplification may be obtained at low radio frequencies with resistance coupling, much greater voltage step-up will be secured if tuned plate circuits are used, just as we use them at broadcast frequencies. Here the voltage gain possible is a function of but two factors, the mutual conductance of the circuit and the load impedance. The mutual conductance of the circuit is practically constant and equal to the static mutual conductance of the tube. Neither of these statements is true of ordinary tubes in ordinary circuits.

Multiplying the two factors together gives the maximum voltage gain to be expected, which is approached but not reached in practice. With a mutual conductance of 300 and a tuned circuit with an effective impedance of 250,000 ohms, a voltage step-up of 75 should be secured, and with air-core coils, amplification as high as 200 in the intermediate-frequency band has been recorded. When resistance coupling is used, the familiar drop in amplification at high frequencies is noted, due to the stray shunting capacities as well as to the output and input capacities which tend to lower the effective impedance in the plate circuit.

In the amplifier of Fig. 4 a voltage gain of 2200 was indicated when 222 type tubes were used in the first and second stages. This is the total a.c. voltage in the plate circuit of the last tube divided by the input voltage. Figuring that the power tube, a 171, contributes a voltage amplification of 3, the two screened-grid tubes

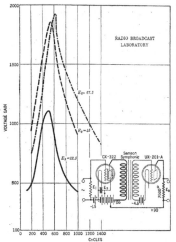

FIG. 5

There is often some advantage in a highly peaked amplifier. These curves were made with a Samson "Symphonic" transformer when used with a screened-grid tube

contribute about 27 each. In the Laboratory a single CX-322 (UX-222) fed into a UX-171 gave a voltage stepup of 120, corresponding to what may be expected from a three-stage resistance amplifier when two 201-A tubes with a 171 are used.

Off hand a voltage amplification of over 3000 with three tubes seems tremendous, but let us consider two transformer-coupled stages plus a final 171 amplifier. If the transformers have a 3:1 turn ratio, and the tubes have a mu of 8, the total voltage amplification from plate circuit of the last tube to input will be $(3 \times 8)^2 \times 3$, which gives 1730.

When transformers are used with the screened-grid tube, the effect of a low primary inductance transformer used with a low-impedance tube is obtained. In other words, a good transformer with a low-impedance tube gives a good characteristic; a high-ratio transformer (low primary inductance) with a high-impedance tube gives a peaked characteristic; a good transformer with a high-impedance tube also gives more amplification about the middle of the audio band than it does at the two ends. Curves of a Samson "Symphonic" transformer, which gives flat amplification with a 12,000-ohm tube, are shown in Fig. 5 and may indicate to code listeners how they can confine the amplification to a rather narrow band of audio frequencies. This transformer naturally resonates at a rather low frequency, and if the amateur or code listener desires his amplifier to peak at a higher frequency he should use a transformer with lower primary inductance, say a 5-1 transformer.

These curves show a discrimination of two to one in voltage amplification between 200 and 500 cycles, and a total amplification of nearly 2000 under certain conditions. In other words, across 7500 ohms in the output of a 201-A type tube, 8.85 volts were obtained with an input of 3.5 millivolts. This is vastly greater amplification than amateurs use to-day, and probably far greater than they need. The average short-wave receiver with one stage of audio, and fair coupling to the antenna, goes down to the noise level, if the audio gain is about 30 to 50. The advantage of using more audio amplification lies in the fact that looser coupling to the antenna may be utilized, with consequent decreased radiation—with the usual type of receiver—and the fact that in good locations, where the noise is far down compared to weak signals, this amplifier will enable the listener to hear practically any radio-frequency disturbance in the ether, whether it is caused by signals or otherwise. The diagram of connections for this amplifier are given in Fig. 5. If still greater discrimination in favor of a certain small band of audio frequencies is desired the screened-grid tube might be connected to the detector output by means of a low-inductance choke, say of two or three henries, as shown in Fig. 6. This will be practically a short-circuit on the low frequencies, and will lower the voltage gain at all frequencies,

but there will be plenty left. The average telephone receiver peaks very sharply between 700 and 1200 cycles and, combined with such a circuit as has been used in the Laboratory, should enable the listener to work through low-frequency static noises. There is some question among amateurs whether it is desirable to tune their audio amplifiers, since so many "hams" use low-frequency sources of plate supply for their power tubes, many using raw a.c. and others using self-rectifying circuits in which the

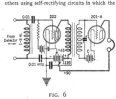

FIG. 6

If still greater discrimination against unwanted audio signals is desired, this manner of connecting the detector to the audio stage might be useful. The low-impedance choke in the input short circuits the low frequencies. The bypass condenser across the audio transformer secondary shunts out the higher audio tones and leaves only a narrow band around 1000 cycles

tone is rather low. It is probable that it is the higher audio tones which should be eliminated from a code amplifier by proper bypassing, for it is these tones due to tube noise and other sources that supply nothing to the amateur but, on the other hand, tax his nervous energy.

The screened-grid tubes which have been tested in the Laboratory are microphonic, and in the transformer-coupled system shown in Fig. 6, some trouble was encountered in keeping the amplifier from singing at the frequency where it amplifies greatest, which was near the mechanical resonant frequency of the elements of the tube. In the resistance amplifier less trouble was encountered, but it must be remembered that we are dealing with a large voltage amplification, which means that the likelihood of trouble from every source is increased.

RADIO BROADCAST Photograph

AN EXPERIMENTAL SET-UP

This is the Laboratory's "brass box" receiver—one r.f. stage using a screened-grid tube, with accessory apparatus, a detector, and a single stage of transformer-coupled audio amplification, are included. Arrangements are made for using a 201-A type tube in place of the 222 type so that the increase in amplification due to the new tube may be directly measured. The detector is calibrated and used as a vacuum-tube voltmeter while the audio stage enables the experimenter to monitor what is going on in the system

RADIO FOLK YOU SHOULD KNOW

WALTER VAN B. ROBERTS

Drawing by Franklyn F. Stratford

M R. ROBERTS is well known to readers of RADIO BROADCAST as a frequent contributor to the magazine, and as the author of the comprehensive and lucid "How Radio Receivers Work" (Doubleday, Page & Co.). He is the inventor of the Roberts receivers, combining cascade radio-frequency amplification and regeneration. Since 1924 he has been on the engineering staff of the Radio Corporation of America, as a receiver and patent specialist. Mr. Roberts was an instructor in Physics and Electrical Engineering at Princeton before assuming his present position.

With some exceptions, among which the Roberts receiver is noteworthy, new receiving circuits christened with the name of the inventor have been lacking both in originality and technical merit. The reason has generally been too much publicity urge and not enough engineering qualifications on the part of the begetter of the great idea. Mr. Roberts is utterly lacking in publicity itch, and very strong in engineering background. From Princeton he has the following degrees: B. S., A. M., E. E., and Ph. D., but he is so modest about them that few people know he has them. The same applies to his scholastic honors in mathematics and physics—membership in Phi Beta Kappa, a medal for electrical research, and several fellowships and scholarships. A characteristic touch is the footnote appended to the bibliography at the end of his excellent little book, "How Radio Receivers Work": "This Bibliography is recommended to radio experimenters who really desire to increase their technical knowledge." The book itself is a very good summary, even under the handicap of some degree of popularization, of technical knowledge in the radio receiver field, but this fact the author apparently declines to recognize.

In May, 1917, Mr. Roberts enlisted in the Signal Corps as a Master Signal Electrician, "a high-sounding title ranking along with some of the higher grades of sergeants, if I remember correctly," as he puts it. He soon returned to Princeton to organize and run the Department of Signalling and Wireless in the School of Military Aviation, an army ground school for aviators. The object of the course was to inculcate the several thousand men who took it with an elementary knowledge of radio theory and practice on the spark-set-and-crystal-detector level, and to teach them to receive and send Morse at low speed. The code students were taught to print each letter separately as received, and cautioned not to look at letters already recorded for fear that they would be influenced in copying what was to follow. "To this day," testifies Dr. Roberts, "I copy code by writing letters separately and get all 'balled up' if I try to see what the message is about before it is all finished." Evidently he found learning the code a very painful process at first, as it is for most people. A monograph could be written about the agonized thoughts, amounting almost to hallucinations, the disturbances of breathing, the involuntary lapses of attention during which the code characters are heard while, in a sort of paralysis,

THERE are many and important figures in the radio world who are not especially well known to readers of this magazine. We have read in great detail of the lives and work of such men as Marconi, Alexanderson, DeForest, Sarnoff, Crosley, Armstrong—to choose names at random—but there are many others who are worthy of note and who are behind the scenes. RADIO BROADCAST will, from time to time, carry stories which will take the reader behind the scenes a bit. A short article in our January issue told something about Ralph H. Langley of the Crosley Company. This article, devoted to Walter Van B. Roberts, describes a man whose name is known all over the world because of his circuit, which was first described in this magazine for April, 1924. Mr. Roberts' activities are not especially well known to our readers and the accompanying article attempts to sketch some of his work.—THE EDITOR.

the mind refuses to decipher them, which afflict the learner, particularly when he is pushed ahead too fast. As far as the intellectual factors were involved, Dr. Roberts naturally had little difficulty, and one of his feats, which proved

WALTER VAN B. ROBERTS

useful in convincing the men that their memory of the code characters would improve with practice, was an ability to receive at ten words a minute any new code made up of four-element combinations of dots and dashes, aft: minutes of study.

Early in 1918, Mr. Roberts obtained mission as First Lieutenant in the Signal and soon after he was transferred to th Engineers Sound-Ranging Service. In branch of artillery practice microphor placed at about half-mile intervals in con or possible places, depending on the terra the nature of the gun fire as well as on th nical requirements of the method. A pair c runs from each transmitter to a central s When an enemy gun fires, the sound, tr at only about 1200 feet per second, is re first by the microphone nearest to it, and an appreciable interval, by the next micro and so on. These times are automatica corded at the central station, and the le of the gun follows after a few minutes of lation. The big job is keeping the lines Mr. Roberts states that half the personn sound-ranging group is required for this p over a hundred breaks in the lines du night of heavy shelling being not unco He thinks that a small radio transmitte ciated with each microphone might be than the wire links.

Mr. Roberts was a technical officer sound-ranging dugouts from June to O 1918, when he was retired to hospital "w excuse for getting a wound stripe," as pleased to phrase it. However, valuable te officers were not sent back for bruised nails, and we may surmise that Mr. Robe not find his way to the rear alone, or un own means of locomotion. He did go to th alone, however, and by almost as painful a It seems that he sailed for France as a officer in charge of one hundred and crates of sound-ranging equipment, inc storage batteries, carboys of acid, win special microphones responsive only t frequency sounds. His first job was to k one hundred and twenty crates together unloading process. He succeeded, presu by exerting the forcefulness of a brigadie eral, which is difficult without the silve With all the stuff in a freight car, Roberts his baggage and a can of corned beef, and himself. He thus traveled in state to the yards outside of Paris, where he remaine days. He did not leave his freight car, beca one knew when it would pull out, and he urgently to be with it at that time. After in the freight car, he finally got close eno the front to permit the equipment to be ferred to three trucks, which moved it in trenches under cover of darkness.

Before the war Dr. Roberts had one oth —he worked on acoustic problems for the ern Electric Company, at $12 per week. I says that this sum was more than he was

Mr. Roberts is married, has two childre lives in Princeton, New Jersey.

286

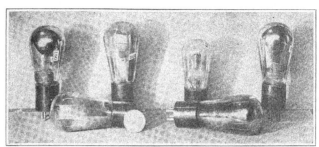

HERE ARE THE NECESSARY TUBES FOR THE UNIT DESCRIBED IN THIS ARTICLE
The metal disc is the size of a half dollar, and is shown to give some idea of the size of the tubes

An A. C. Push-Pull Amplifier and B Supply

By J. E. Coombes

THE introduction of a.c. tubes has greatly simplified the construction of complete receivers which derive all of their power directly from the electric light mains. For those of us who now have a tuner unit or a complete receiver with an out-of-date audio amplifier, the a.c. tube is a logical means whereby we can construct a completely a.c. operated audio amplifier that can be hitched on to the output of the detector tube or to which we can connect an electrical pick-up when we want to play phonograph records. And, incidentally, we can design our amplifier also to supply plate voltage to the detector and r.f. tubes of the radio receiver proper. Such a combination amplifier and B supply device is illustrated in this article.

The device illustrated herewith includes a two-stage transformer-coupled amplifier using a type

227 a.c. tube in the first stage, followed by two type 210 tubes in a push-pull arrangement. Plate potential for these tubes is obtained from two 216-b or 281 type tubes operating in a full-wave rectifier circuit. Filament current for the two 210's is obtained from a filament winding on the power transformer (T_1 in Fig. 1) and filament current for the 227 a.c. tube is obtained from a separate filament transformer, T_2. The latter transformer also contains a winding which can be utilized to supply one or more r.f. tubes in the radio receiver proper, thereby making the entire set operate directly from the power mains. We are assuming that 226 type a.c. tubes will be used in the r.f. amplifier, and a 227 type tube for the detector. The latter will be supplied by the same transformer winding as the first audio stage. The B power part of the amplifier contains a

glow tube to maintain the output voltages constant independent of the load drawn by the radio receiver.

Some readers may merely want to construct the push-pull circuit and B supply rather than a complete two-stage amplifier. For this reason the circuit diagram has been divided into two parts by a dotted line, the push-pull amplifier and B supply being located to the left and the first stage of audio amplification to the right. Neglect the right-hand side of the diagram if only the push-pull amplifier and B supply are to be constructed; and connect the output of the receiver proper to the terminals marked X-Y, the new input posts. This latter unit (the push-pull amplifier) will operate without any appreciable hum although a slight amount of hum is noticeable during the silent periods in a pro-

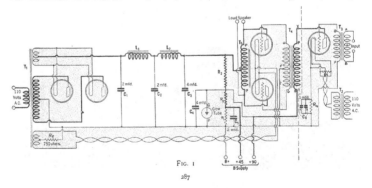

FIG. I

287

gram when using the complete two-stage amplifier.

Naturally this amplifier can handle considerable volume. Volume, however, is not the main objective any more than an 80-mile-per-hour automobile is expected to travel at top speed through traffic. Like the automobile, however, this heavy-duty power amplifier has the liberal fund of reserve power essential to prevent overloading and thereby reproduce a radio program with the least possible distortion.

The power transformer, T_1, contains three separate windings. One winding provides 550 volts either side of the center tap to supply a full-wave rectifier system using two 216-b or 281 type tubes. The latter tube, because of its greater rating, will generally have somewhat longer life than a 216-b tube and will also give a slightly greater output voltage. In addition to the center-tapped high-voltage secondary, the transformer also contains two 7½-volt windings to supply filament current to the rectifiers and the 210 power amplifier tubes. The output of the rectifier is fed into a filter system containing two filter chokes, L_1 and L_2, and three condenser banks, C_1, C_2, C_3. The high-voltage output, (about 450 volts under load) of the filter is fed directly to the plates of the UX-210 (cx-310) tubes.

All of the resistances necessary in the construction are contained in the Ward-Leonard Vitrohm resistor kit, type No. 507-47. R_1 and R_2 in the diagram are two 5000-ohm resistances contained in the kit. R_3 is composed of a combination of resistors. By using different values of resistance at this point the amount of current drawn by the glow tube is varied. Generally, it will be satisfactory to have R_3 consist of the 3500-ohm and 1500-ohm resistances in series, to give a total resistance of 5000 ohms. However, when the tuner unit consists of only a couple of tubes, say a detector and one stage of r.f., this resistance may be increased by connecting the 3000-ohm resistance in series with the two mentioned above. This combination will give a total resistance of 8000 ohms and this additional resistance will decrease, somewhat, the current drain on the filter system and the maximum possible voltage is thereby delivered to the plate of the power tube.

The C bias for the 227 type tube is obtained from the voltage drop across the 1500-ohm resistance, R_4. This resistance is bypassed with a 2-mfd. condenser. C bias for the 210 tubes is obtained from the 750-ohm resistance, R_5, con-

nected to the center tap of the filament winding supplying these tubes. A bypass across this resistance is not necessary because there are no audio-frequency currents flowing through it.

The following table shows how the total output voltage of the rectifier-filter system varies with current drawn from it:

D. C. CURRENT, mA.	D. C. VOLTS
60	537.5
70	515.0
80	487.5
90	462.5
100	440.0
110	427.5
120	417.5

In normal operation the 210's take about 20 mA. each, the glow tube a maximum of 40 mA., the first audio stage requires a maximum of 5 mA., and if the detector and r.f. tubes take a total of 10 mA., then the total load on the B supply unit will be 95 mA. At these current drains the output is about 450 volts and subtracting about 30 or 35 volts which is lost in the C bias resistance, there is left an effective voltage of about 415 volts on the plates of the 210 tubes.

The complete amplifier, as the illustration shows, was constructed on a baseboard, measuring 9 x 24 inches, and most of the wiring can be done under the baseboard by drilling holes through the board directly beneath the terminals on the various units and threading the leads through. The filament leads to the 210's and 227 should be twisted. No special kinks are necessary in the construction and the circuit diagram and illustration in this article should supply all the necessary constructional hints.

To assemble the complete amplifier and B supply exactly as illustrated, the following parts are necessary:

T_1—Thordarson Power Transformer, T-2098
L_1, L_2—Thordarson Double Filter Choke, T-2099
T_2—Thordarson Filament Transformer, T-2445
T_3—Thordarson R-200 Audio Transformer
T_4—Thordarson Push-Pull Input Transformer, T-2408
T_5—Thordarson Push-Pull Output Choke, T-2420
C_1, C_2—Acme Parvolt Filter Condensers, Series B, 2-Mfd.

C_3—Acme Parvolt Filter Condenser, Series B, 4-Mfd.
C_4—Acme Parvolt Filter Condenser, Series A, 4-Mfd.
C_5—Two Acme 1-Mfd. Parvolt Filter Condensers, Series A
C_6—Acme Parvolt Filter Condensers, Series A, 1-Mfd.

R_1—5000 Ohms	All these resistances
R_2—5000 Ohms	are contained in the
R_3—5000 to 8000 Ohms (See Text)	Ward Leonard Vitrohm Resistor Kit,
R_4—1500 Ohms	No. 507-47
R_5—750 Ohms	

R_6—General Radio Center Tapped Resistance
Five Benjamin Sockets, 4-Prong, No. 9040
One Benjamin Socket, 5-Prong, No. 9036
Seven Eby Binding Posts

THE REQUISITE TUBES

Two UX-210 Power Tubes
One UX-874 Glow Tube
Two UX-216-B Rectifier Tubes
One UX-227 A.C. Heater Type Tube

The power lead from T_1 can be plugged into a receptacle which is in the end of the filament transformer, leaving one lead to be plugged into the regular light socket. In operation, the resistances will get quite warm and the regulator tube should glow with a bluish or pinkish glow. The input terminal connecting to the P post of the input transformer should connect to the plate of the detector tube and the other input lead should connect to the B + 45-volt terminal on the B supply. If one side of the A battery is not grounded, it should be.

It will be noted that R.C.A. terminology has been resorted to throughout this article when referring to tube types. Tubes exactly similar to those bearing the R.C.A. stamp are supplied by Cunningham. The following table shows the parallel types:

R. C. A.	Cunningham
UX-226	cx-326
UX-227	cx-327
UX-210	cx-310
UX-216-B	cx-316-B
UX-281	cx-381
UX-874	cx-374

Many independent manufacturers are also supplying tubes the characteristics of which are similar to or approximate very closely those of the tubes specified in this article.

THE COMPLETE PUSH-PULL AMPLIFIER AND B SUPPLY

New Recordings by
Radio Favorites

FAMOUS SYMPHONY CONDUCTORS

THE insert at the top, shows Nikolai Sokoloff, conductor of the Cleveland Symphony, shown in the large oval. Many records have been made for Brunswick by the Cleveland orchestra. Henri Verbrugghen is shown in the left-hand insert of the lower group of three conductors, the center one being Willem Mengelberg, and the right-hand one Walter Damrosch. Verbrugghen is conductor of the Minneapolis Symphony, heard over WCCO; Mengelberg now conducts the New York Philharmonic, heard over WOR; Walter Damrosch, conducting a large group of New York Symphony men, is heard over WJZ and chain each Saturday night. The Cleveland orchestra broadcasts through WEAR.

Recent Symphony Orchestra Recordings

A FEW records which have been recently issued deserve special attention, particularly because they are noteworthy achievements in recording. If you are not aware of the vast improvement that has been made in this art by the development of electrical recording, play a record of the old method and compare it with one of the new method. You will observe the extended tone range of the latter, in which you will hear high and low notes which were lost by the old system of recording. Of course that is taking for granted that your own reproducing apparatus is adequate. If you are still making use of one of the pioneer phonographs, or an antediluvian horn as your loud speaker, all records will sound equally bad. But with good equipment it won't require much imagination to believe that you are in the hall with the orchestra itself instead of in your own home listening to canned music.

One of the new records which would be particularly successful for this demonstration which we suggest is the "1812" *Overture* of Tschaikowsky (Brunswick), in which the composer has plumbed the musical depths with the resounding tones of the tympani, then in the next breath climbed to ecstatic heights with the strings. The full orchestra is brought into play and not an iota of color is lost by the recording of this magnificent overture, which has been played in masterly fashion by the Cleveland Orchestra under the direction of Nikolai Sokoloff.

Nor are any of the delicate shadings lost in the exquisite *Dernier Sommeil de la Vierge* of Massenet, as played for Brunswick by the Minneapolis Symphony under the baton of Henri Verbrugghen. This lovely ecclesiastical music contrasts strangely with the robust *Coppelia Ballet* which appears on the reverse side of the record. This delightful composition by Delibes has as much swing to it as any present day dance number you can think of and a melody as simple as any jazz strain—and furthermore it is many times as soul-satisfying.

Two more classical dance numbers are Johann Strauss's *Artist's Life* and *Tales From the Vienna Woods* (Brunswick). The interpretation of these oft-heard waltzes by Willem Mengelberg, conducting the New York Philharmonic Orchestra, is to their usual rendition as a production of Midsummer Night's Dream by Max Reinhardt is to a high-school performance of the same comedy.

Perhaps the most famous musical selection in the world is the *Wedding March* from Lohengrin. Certain it is that a great many people, be they married or single, know it and love to hear it played. Certain it also is that most of them have never heard it more beautifully interpreted, if as beautifully, as it has been by the Cleveland Orchestra under the direction of Nikolai Sokoloff, for Brunswick. This makes the usual rendition, cloaked in romance and sentiment though it be, seem stale by comparison. And the same may also be said about the Prelude to Act 3 of *Lohengrin*, which selection is on the opposite side of the *Wedding March* (Brunswick).

Another familiar number is the *Song of India* by Rimsky-Korsakow. We once thought that we never wanted to hear this again. That was because a neighbor of ours—oh, years ago—had just invested in a phonograph and a few records, very few! One of them was the *Song of India* popularly recorded. The thrill of having a phonograph was evidently great, for the machine was never silent. It was Spring and all windows were open. Need we say more? But we find that this record by the Cleveland Orchestra is something else again. We can hear it with as much delight as if it were entirely new to us. This same orchestra's version of Tschaikowsky's *Sleeping Beauty* waltz is full of color and feeling and contrasts vividly with the usual performance of this favorite by hotel orchestras and bands, whose treatment of it is seldom more than an adequate reading of the notes (Brunswick).

On the next page we give a list of recent electrically recorded symphony records.

Recent Popular Records

HAVING on various occasions thrown stones in no uncertain fashion at popular music it behooves us to explain ourselves before we proceed to review some of the popular records. While we still contend that it wearies us in bulk, we admit that we thoroughly enjoy it in small doses. A little of it over the radio is an excellent thing; and the same applies to the phonograph.

But the dose must be moderate and we want it palatable.

Though we can't seem to get away from the stumbling block of a limited number of current songs which we are forced to hear over and over again, the better orchestras do their best to vary the music as much as possible by the addition of frills in studied orchestration. Those who orchestrate well can make good music out of mediocre. Take for instance *Good News* and *Lucky in Love* as played for Brunswick by Ben Selvin and His Orchestra. Yes, those are the same selections which we frowned on so severely last month. But you would never know they were the same! That is what Selvin orchestration does for a piece. A little trimming here and there by a steel guitar—and the trick is done. The result: a simply grand dance record!

This orchestra is versatile, too. In *I Could Waltz On Forever* it makes the most of the strings, the sax, and the piano. Even that moss-grown favorite, *Cheerie-Beerie-Bee*, on the reverse, takes on new life under Selvin treatment (Brunswick).

Play-ground in the Sky from " Sidewalks of New York," being a particularly good number to start with, doesn't need much doctoring and it has very wisely been simply treated by this same orchestra. Incidentally, why haven't we heard this selection oftener? It is swell! *Whatever You Are* from the same show isn't as good but the most has been made of it (Columbia).

The fourth record by this outfit is *I Call You Sugar* and *Yes She Do* (Brunswick). Again trick orchestration. Selvin does this instrumental ornamentation extremely well, making it fit into the general scheme of things instead of letting it stand out like the ball on the Paramount Building, as a lesser light would be apt to do. The result is that the records are not ruined for dancing but are improved.

Our old favorite, Ernie Golden, is a past master at this art of orchestration. He rings the bell again with *All By My Ownsome* and *A Night In June*, in which he introduces a steam caliope effect which is grand! We suggest that he get it patented and use it as a musical trade mark (Brunswick).

Don Voorhees has made four recordings on three different discs for Columbia. These four are: *Rain, Baby's Blue, Highways Are Happy*

Ways, and *When the Morning Glories Wake Up In The Morning*. There is a sort of smooth placidity about this orchestra by which you can always identify it. It never gets excited, it is uniformly good and yet it never seems to climb quite to the topmost heights. However, it has personality and that is a lot in these days. We have named the records in the order in which we rate them, the first being by far the best. Listen to the saxophone.

On the opposite side of *Baby's Blue* is *The Calinda* by the Radiolites. It is a very good number with an irresistible swing. Not the least attractive feature of the record is the vocal chorus by Scrappy Lambert, of cough drop fame. (Columbia).

The Radiolites are responsible for another good dance record, *There's A Cradle in Caroline* and *Everybody Loves My Girl*. Neither selection is inspired but each moves along with a smooth rhythm (Columbia).

A record that stands out from the rest is *Charmaine* and *Did You Mean It?* by Abe Lyman's California Orchestra. Both numbers are played with a restraint not often displayed by a dance orchestra. Soft pedals and plaintively insinuating rhythms are a relief after robust, vigorous jazz (Brunswick).

If you have once heard Phil Ohman and Victor Arden stroke, jingle, bang, and otherwise urge on the ivories, you will look forward eagerly to hearing them again. We have and did, and were disappointed by their record for Brunswick, *There's Everything Nice About You and Mine*. Oh, yes, they are good numbers, well played, but there is too much of the orchestra and too little piano. You can always hear good orchestras but there is only one Ohman and Arden.

Two more disappointments were records by Ben Bernie and His Hotel Roosevelt Orchestra and by Vincent Lopez and His Casa · Lopez Orchestra. By rights one can expect the best from these two bands. But they both play as if pay day were at least a month away. The records are *Miss Annabelle Lee* and *Swanee Shore* by the first outfit and *Someday You'll Say "O. K."* and *Just a Memory* by the second. Both are Brunswick.

Only moderately good are the rest: *A Night in June* and *Are You Happy* by the Ipana Troubadours (Columbia); *Feelin' No Pain* and *Ida*

Sweet as Apple Cider by Red Nichols and His Five Pennies (Brunswick); *Manhattan Mary* and *Broadway*, both numbers from "Manhattan Mary," by Cass Hagan and His Park Central Hotel Orchestra (Columbia); *Like the Wandering Minstrel* and *Molly Malone*, from "The Merry Malones," played by The Cavaliers (Columbia); and *No Wonder I'm Happy* and *Sing Me a Baby Song* by the George H. Green Trio (Columbia).

Taken all in all all these dance records that we have reviewed form a good collection. Not one of them is really poor.

If you are a devotee of Roxy you will welcome three records played in the Roxy Theatre by Lew White, the organist, *Broken Hearted* and *Just Like a Butterfly, When· Day is Done* and *Forgive Me, Underneath The Weeping Willow* and *At Sundown* (Brunswick all). It is all typical movie organ music. Many people object to that sort of thing but we like it when it is well done, as this is. Our preference is for *Underneath the Weeping Willow* and second choice is *Broken Hearted*. Neither of these has a vocal chorus and the rest have. Does a vocal chorus go with organ music?

The male counterpart of Vaughn De Leath seems to be Vernon Dalhart. He isn't as gushy, for which let us be truly thankful, but the idea is the same. He offers *My Mother's Old Red Shawl* and *Down On The Farm* on a Brunswick record.

Billy Jones and Ernie Hare, the Happiness Boys, again present us with a little vulgar singing of a nice sort. Well, you know they aren't exactly refined but they are *good*. This time they have recorded *You Can't Walk Back From An Aeroplane* and *Who's That Pretty Baby?* for Columbia.

Art Gillham, the Whispering Pianist, hands out the typical vaudeville sob stuff, piano and recitative, in *Just Before You Broke My Heart* (Columbia). On the other side is *I Love You But I Don't Know Why* which is moderate.

About the only thing to say about *Roam On My Little Gypsy Sweetheart* and *There's A Cradle in Caroline* as sung by the Goodrich Silyertown Quartet is that they have been in better, shows than this (Columbia).

The same might be said of the Anglo-Persians who play *Call of the Desert* on a Brunswick record. But they redeem themselves by the selection on the reverse side, *Down South*. We end on a note of praise for the carpet riders.

New Electrical Symphony Orchestra Recordings

New York Philharmonic Orchestra	Victory Ball—Fantasy. Parts 1 and 2	(Schelling)	1127	Victor
	Victory Ball—Fantasy. Parts 3 and 4	(Schelling)	1128	Victor
	Artist's Life	(Strauss)	50096	Brunswick
	Tales from the Vienna Woods	(Strauss)		
	Marche Slave, Parts 1 and 2	(Tschaikowsky)	50072	Brunswick
	Midsummer Night's Dream—Scherzo	(Mendelssohn)	50074	Brunswick
	Midsummer Night's Dream—Nocturne	(Mendelssohn)		
Chicago Symphony Orchestra	Wine, Woman and Song	(Strauss)	6647	Victor
	Southern Roses	(Strauss)		
	Carnival Overture, Parts 1 and 2 (Op. 92)	(Dvorak)	6560	Victor
	In Springtime—Overture, Parts 1 and 2, Op. 36	(Goldmark)	6576	Victor
	(1.) Serenade (Volkmann, Op. 63.) (2.) Flight of Bee	(Rimsky-Korsakow)	6579	Victor ·
	Valse Triste	(Sibelius)		
	To a Water Lily	(MacDowell)	1152	Victor
	To a Wild Rose	(MacDowell)		
St. Louis Symphony Orchestra	Country Dance, No. 1	(German)	9009	Victor
	Pastoral Dance, No. 3—Merrymaker's Dance, No. 3	(German)		
	Fingal's Cave—Overture, Parts 1 and 2	(Mendelssohn)	9013	Victor
Cleveland Symphony Orchestra	Blue Danube Waltz	(Strauss)	50052	Brunswick
	Tales From Vienna Woods	(Strauss)		
	Danse Macabre	(Saint-Saëns)	50089	Brunswick ·
	Merry Wives of Windsor Overture	(Nicolai)		
	Finlandia	(Sibelius)	50053	Brunswick
	Symphony No. 2	(Brahms)		
	Hungarian Dance, No. 5, G Minor·	(Brahms)	15592	Brunswick
	Valse Triste (Op. 44)	(Sibelius)		
	Slavonic Dance No. 3	(Dvorak)	15591	Brunswick·
	Traümerei	(Schumann)		
	1812 Overture—Parts 1 and 2	(Tschaikowsky)	50090	Brunswick
	Lohengrin: Prelude to Act 3	(Wagner)	15121	Brunswick
	Lohengrin: Wedding Music	(Wagner)		
	Sleeping Beauty Waltz	(Tschaikowsky)	15120	Brunswick
	Song Of India	(Rimsky-Korsakow)		
Minneapolis Symphony Orchestra	Coppelia Ballet—Prelude and Mazurka	(Delibes)	50087	Brunswick
	Dernier Sommeil de la Vierge	(Massenet)		
	Der Freischutz, Overture, Parts 1 and 2	(Weber)	50088	Brunswick
	Melodrama from "Piccilino"	(Guiraud)	15117	Brunswick ,
	Waiata Poi	(Hill)		

"Our Readers Suggest—"

OUR Readers Suggest. . ." is a regular feature of RADIO BROADCAST, *made up of contributions from our readers dealing with their experiences in the use of manufactured radio apparatus. Little "kinks," the result of experience, which give improved operation, will be described here. Regular space rates will be paid for contributions accepted, and these should be addressed to* "The Complete Set Editor," RADIO BROADCAST, *Garden City, New York. A special award of $10 will be paid each month for the best contribution published. The prize this month goes to Conrad Ederle, New York City for his suggestion entitled "Extending Loud Speaker Leads.*

—THE EDITOR.

Extending Loud Speaker Leads

IT IS often very desirable to place a loud speaker some distance from the set in conjunction with which it is operating. The long leads, however, are somewhat bulky, and are especially conspicuous when run along the wood molding.

It was recently the writer's job to install a loud speaker diametrically across the living room from the receiver in such a manner as to merit the approval of several discriminating family critics. The method used to accomplish this task may be of general assistance.

In the first place one of the two long leads generally necessary was dispensed with. There are two available grounds in our living room, a radiator by the receiver, which is used for the set ground, and gas logs in a fireplace, on the mantel over which the loud speaker is mounted. A connection from the loud speaker to the logs was used as one wire, the other loud speaker terminal being connected to the receiver as shown by the dotted lines in Fig. 1. As a matter of experiment, the resistance of the ground circuit (*i. e.*, the matter between the set ground and the loud speaker ground) was determined, and its value found to be 28 ohms. This is not, indeed, very low, but, on the other hand, not high enough to occasion an appreciable loss in signal strength.

A No. 26 single white cotton-covered wire was used as the conductor from the loud speaker to the set. This wire was run inconspicuously along the cream colored molding. However, the last six feet, from the molding across the wall to the loud speaker cord, was necessarily exposed. The visibility of the exposed portion was reduced materially by running the single wire down a corner and gluing the wire in place rather than using tacks. A drop of LePage's glue was placed at one foot intervals along the wire. This was allowed to become tacky before being pressed against the wall. Talcum powder was blown from the palm of the hand, while the glue was still moist, effectively hiding the brown color, and blending successfully with the buff tone of the walls.

CONRAD EDERLE,
New York City.

STAFF COMMENT

AS MR. EDERLE observes, it is often desirable to operate a loud speaker at some distance from the receiver. The desirability of this arrangement may be dictated by esthetic or technical considerations. Some receivers, notably those operating from loop antennas, function best in a definite section of a room.

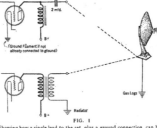

FIG. I

Showing how a single lead to the set, plus a ground connection, can be used with either transformer or choke coil output device to operate a loud speaker at some distance from the receiver

The same holds true with loud speakers when careful consideration is given to acoustic effects.

Belden has produced a flat two-wire extension cord which may be laid under a rug, providing a very convenient form of inconspicuous wiring. While the Belden arrangement was designed ordinarily with the requirements of the 110-volt power line in mind, there is no reason why it should not be adapted to a loud speaker output circuit.

In using the ground return suggested by Mr. Ederle, it is quite essential that an output device be employed—either a choke coil arrangement or a transformer. Unless one of these is used, the high voltage to the plate of the output tube will, in many cases, be shorted over to ground. Both circuit arrangements are shown in Fig. 1.

The Cartridge Type Charger

MANY fans have the connections from the storage A battery to the charger arranged as shown in Fig. 2A, merely closing the 110 volts a. c. circuit whenever the battery needs charging. This scheme is entirely satisfactory when a bulb type charger is used, for the path between the plate and the filament of the tube becomes non-conducting when the filament heating current is cut off and, therefore, no discharge path through the charger is presented to the A battery current.

When the cartridge type charger which has recently appeared on the market is used, however, this scheme is no longer satisfactory. An investigation made to learn why a battery connected to the charger as shown in Fig. 2B seemed to lose its charge in a few days disclosed the reason. An ammeter introduced into the circuit to indicate the rate of charge showed a slight negative reading when the a. c. power was cut off. Further investigation disclosed a reverse current of approximately 180 milliamperes passing continuously when the transformer was disconnected from the power line.

A double-pole double-throw switch was included in the circuit as shown in Fig. 2C. The battery now retains its charge as long as it did before the type of charger was changed.

HERBERT J. HARRIES,
Pittsburgh, Pennsylvania

STAFF COMMENT

MEASUREMENTS in the Laboratory on the new type National full-wave charger, which employs two cartridge recti-

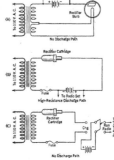

FIG. 2

Various methods of A battery charger connections. C shows a system of connections for a cartridge rectifier

fiers, indicate that practically *no* current will flow if the arrangement in Fig. 2B is employed, suggesting that the switching is unnecessary. We believe that the current Mr. Harries noted was due to a defective cartridge. In such a case the switching arrangement might very well be resorted to since a cartridge that is only slightly defective will still give many hours service. The switching arrangement shown in Fig. 2C may be slightly improved by using a three-blade switch, the extra contacts being used to control the 110-volt line to the charger. When the switch is in the right-hand or charge position, the charger will automatically be turned on.

The Glow Tube

THE use of the glow or regulator tube, such as the UX-874 (CX-374) or the Raytheon type R, effects a general improvement in socket power circuits. The stability of the r. f. circuit is sometimes improved, while the tendency to "motorboat," and oscillation in the audio-frequency amplifier is considerably reduced by its use. The glow tube is connected between the ninety-volt tap and the negative binding post on the power supply unit and is so designed that a practically constant potential is maintained across these posts regardless of the current load.

The regulator tube was specified or furnished

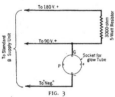

FIG. 3
How to add a glow tube to any B supply device. The addition of the glow tube will increase the stability and all-around efficiency of the circuit.

with very few B supply devices, homebuilt or ready made, previous to the present radio season. The benefits of the glow tube can be obtained with such power sources by connecting the tube externally. The addition of the tube is a matter of a few simple connections. Briefly, the glow tube is connected between the ninety-volt tap and negative, and an additional resistor, having a value of three thousand ohms, is wired between the high-voltage tap and the ninety-volt tap. The regulator tube plugs into a standard UX socket, the grid post on the socket being the anode terminal of the tube and the negative filament terminal the cathode. The anode should always be wired to the ninety-volt post. The diagram of connections is shown in Fig. 3.

HENRY LANDON,
Chicago, Illinois.

STAFF COMMENT

A REGULATOR tube should never be added to any socket power set until it is determined whether the device will supply an additional drain of 30 milliamperes without an excessive voltage drop.

Output Transformer Connections

WHEN connecting up an output transformer it may be noticed that the instrument is usable for either the straight transformer form of output or as a choke with condenser bypass type (Fig. 4). The idea, when I tried it out, proved to be quite successful and showed some possible advantages for the arrangement.

The tone quality, from the two jacks, was different. The transformer coupling was more mellow and softer and the impedance coupling more brilliant in tone with the particular in-

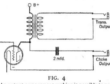

FIG. 4
An output arrangement making it possible to use the transformer as a choke if desired

struments that I used. General tone fidelity was good in both cases so one could select the arrangement most pleasing at the time, or match the reproduction from different loud speakers to some extent.

The double output suggested the simultaneous use of two reproducers and was tried using Sparton and Baldwin loud speakers. Both worked perfectly alone or together in either position and neither reproducer seemed in any case to reduce the volume obtained from the other one alone.

J. B. HOFFMAN,
Kewanee, Illinois.

STAFF COMMENT

MR. HOFFMAN suggests a simple and practical method of connecting two loud speakers with different characteristics to secure a tonal combination more pleasing to the ear than the more usual series connections—always providing, of course, that the loud speakers used have impedances adapted to this particular arrangement. The choke connection will, in some cases, provide more pleasing reproduction than the transformer system. Mr. Hoffman's arrangement may be used in conjunction with the resistor controls suggested for tandem speaker operation in this department last month.

The volume from the individual loud speakers is necessarily reduced when two or more of them are used but the total volume from the receiver may be increased.

A Simple Vernier Condenser

WITH the present-day congestion of radio traffic, any device that may facilitate tuning is worthy of consideration. The writer found that the construction of a simple vernier condenser along the lines to be described improved reception on his tuned r. f. receiver. The vernier is connected across the main tuning condensers in the receiver, and is employed in securing that delicate adjustment so often essential to quality reception.

The mechanical details of the device are illustrated in the drawing, Fig. 5.

The base consists of a piece of spare panel material, 2½ inches long by 2 inches wide. The plates are cut from stiff copper sheets with a 1½-inch radius. The peak of the stator plate is cut away for the shaft of the rotor. The fixed plate is fastened to the base by small screws threaded to the panel, or with nuts.

A long brass screw (See Fig. 5) provides the shaft to the rotor, which is clamped between a nut and the screw head. A shim or several washers between the rotor plate and the panel, space the two plates of the condenser from ⅛ to ¼ of an inch, depending upon the amount of tuning desired.

The closer the plates are arranged together the greater will be the variation. A lock-nut and washer on the other side of the small panel holds the shaft at the desired tension. A hard rubber top of a binding post is used as a knob. In the author's receiver the vernier condenser is mounted on the main tuning panel, as shown in the sketch.

L. B. ROBBINS,
Harwich, Massachusetts

STAFF COMMENT

THERE are many uses for a small variable condenser of this type. Its possibilities as a short-wave vernier, as a compensating condenser across the main tuning capacities in a gang condenser, and as a neutralizing condenser in capacity stabilized circuits, are immediately suggested. With the exception of such instances where the vernier condenser is actually employed for tuning, in which case constant adjustment is necessary, it will be desirable to mount this capacity adjuster on the sub-panel of the receiver.

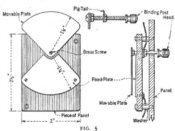

FIG. 5
Constructional details of a useful vernier condenser

Some Reliable Radio
Power-Supply Accessories

AN AUTOMATIC CONTROL

The purpose of this switch is to automatically connect the a. c. power source to either the A battery charger or the B socket-power device, depending upon whether the set is switched off or on. It is a product of U-S-L Radio, Incorporated, Niagara Falls, New York. Price, $3.75.

A AND B POWER

THIS Vesta power unit combines a source of A current with one of B current. Features are: A Westinghouse rectifier is used; built-in hydrometer; side windows enable the state of charge and liquid level to be viewed; there is an automatic relay incorporated. The B supply is 180 volts at either 40 or 60 mils. Price, 40 mils, $72.00; 60 mils, $77.00. The lower picture shows the unit with case removed

AN "ALL AMERICAN" B DEVICE

THERE are four B taps on this unit—45 volts, 67 volts, 90 volts, and power tap. In addition, the two knobs which may be seen on the front panel are for adjusting the 45-volt and 67-volt outputs to any desired amount for most satisfactory operation. There is also a "Hi-Lo" switch for increasing or decreasing the voltage output in accordance with the milliampere drain. The price of this unit, in the 110 v., 50-60 cycle model, is $31.50, less tube

TOWER OF BOSTON

IS responsible for this ruggedly constructed B device. The output is rated at 65 milliamperes, 180 volts. There are five positive B taps—detector, 67 volts, 90 volts, 135 volts, and 180 volts. An automatic power control switch is built in the unit. If a trickle charger is used, this latter control switch automatically switches it in or out of circuit. Price, $32.50

A NEW A-SUPPLY DEVICE, BY GRIGSBY-GRUNOW-HINDS

IT IS here shown with the popular "Majestic" Super-B, in combination with which it is particularly adaptable. This new "Majestic" A unit is completely dry in construction, and uses no acids or liquids whatever. The manufacturers claim that there is absolutely no hum in operation. The maximum output of the "Majestic" A unit is 2¼ amperes at 60 volts, and it lists at $39.50. The Super-B unit, incidentally, lists at $29.50 with tube. The complete "Majestic" A-B supply therefore sells for $69.00

AN INEXPENSIVE B UNIT

BY Modern, of Toledo. The maximum voltage output at 25 mils. is 185 volts, and at 30 mils, 175 volts. The Modern B Compact lists at $26.50 without Raytheon tube. The voltage at the power tube tap may be reduced by inserting a fixed resistor in the fuse clips provided within the case

Why the Output Device?

By Keith Henney

Director of the Laboratory

IN SPITE of its apparent simplicity (it consists of a core of iron, windings of copper wire, and perhaps a paper condenser), an "output device" performs several useful purposes, aside from being the connecting link between one's loud speaker and a power amplifier. In the first place it keeps the d.c. plate current of the last amplifier tube from circulating through the windings of the loud speaker and, secondly, it may adjust matters when a loud speaker and a tube are used whose respective impedances do not "jibe."

From the standpoint of keeping d.c. from the loud speaker, the output device is only necessary when power tubes are used—when the plate current is greater than ten milliamperes. From the standpoint of fidelity the device is necessary when the loud speaker impedance differs considerably from that of the tube out of which it works. In the latter case, only one kind of output device, the output transformer, does any good, the choke-condenser type serving no useful purpose from the standpoint of fidelity unless the choke has taps on it, when it becomes an auto transformer, and not, strictly speaking, a choke.

What does it matter if the d. c. plate current of the last tube in one's amplifier goes through the loud speaker?

There are several effects. The power which the loud speaker winding must dissipate as heat may be found by multiplying the resistance of the winding by the current squared through it. If the loud speaker has a resistance of 1500 ohms and the final power tube is a 171 with a plate current of 20 milliamperes (0.02 amps.), the power lost is 0.6 watts, which may or may not be too great for the winding to dissipate in heat, depending upon whether the winding was made with this thought in view. If the winding will not satisfactorily dissipate this heat (such will eventually become evident by a burnt-out winding) the use of an output device, of suitable electrical dimensions, may be resorted to as a safety measure.

One fiftieth of an ampere (20 mils.) flowing through the loud speaker which has a resistance

of 1500 ohms represents a voltage drop, obtained by multiplying these two values together, of 30 volts, which must be subtracted from the terminal voltage of the plate supply source to calculate what voltage is actually on the plate of the final tube. If the B battery consists of four standard blocks we should have 180 volts and accordingly we place a C bias on the 171 type tube of 40.5, but since there are 30 volts drop in the loud speaker winding, the actual plate voltage is only 150, and 40.5 volts C bias is, therefore, too great. The use of an output device in which the voltage drop is lower than through the loud speaker is to be recommended in such a case.

When d.c. flows through the winding of the loud speaker, the armature, the little lever that moves in accordance with voice-frequency currents and imparts its message to the loud speaker diaphragm, is pulled from its neutral position with respect to the winding, and is much more liable to strike the pole pieces of the magnet under strong signals. In other words, the armature is working under a permanent bias which is neither necessary nor desirable; it forces the armature to work under a hardship that is easily removed by means of the output device.

Some types of output devices have the additional advantage that they remove d.c. plate voltages from the loud speaker tips with the result that these tips or any part of the loud speaker mechanism or the output jack or terminals may be touched while the "juice" is on without danger of shock. No output device, however, will protect one from being shocked

by the loud speaker voltages produced by strong signals. What one hears as kettle drums—or static—is the result of very strong sudden voltages across the loud speaker after they have been translated into sound by the loud speaker diaphragm.

Output devices, then, are used to:

(1.) Keep d.c. from the loud speaker winding.
(2.) Prevent serious loss in plate voltage.
(3.) Prevent heating of the loud speaker winding due to power loss there.
(4.) Prevent placing a mechanical bias on the loud speaker armature.
(5.) Adjust serious impedance differences and, therefore, improve fidelity.

AND SOME OUTPUT DEVICES:

(6.) Keep the loud speaker terminals at low d.c. potentials preventing shock or burn.

As mentioned previously, there are two types of output devices, or loud speaker filters, as they are sometimes called. There is the true transformer, with two windings, of copper wire insulated from each other and wound on an iron core which is insulated from the windings; and there is the choke and condenser combination. The transformer, by its very nature, keeps the d.c. from the loud speaker and d.c. voltages from the terminals. It has the added advantage that, by proper design, differences in impedance which may exist between the loud speaker and the power tube may be adjusted.

The choke-condenser combination consists of a high-inductance copper coil of many turns on an iron core—and the condenser. If the choke is tapped, impedance differences may be adjusted, but for purposes of discussion, it then becomes a transformer, although the two windings are not insulated from each other but possess a certain part of the copper wire in common. This combination of a coil and a condenser may be connected into the power tube circuit in two ways, one of which is better than the other. The transformer may be connected in only one way, which corresponds to the poorer of the condenser-choke connections.

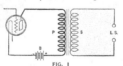

FIG. 1

The output transformer is illustrated in Fig. 1. The d.c. resistance of the primary should be low so that little plate voltage is lost; the a.c. impedances of both windings should be high so that most of the a.c. voltage developed by incoming signals will appear across the loud speaker winding and not be lost on the tube's impedance. The d.c. voltage lost is again the product of the current and the resistance, so that every time one milliampere flows through 100 ohms, a tenth of a volt is prevented from reaching the plate. If the tube requires 20 milliamperes and the resistance is 500 ohms (a typical case) we have lost 10 volts.

The transformer has one disadvantage in that all of the voice-frequency currents must return to the filament of the last tube through the common impedance of the plate supply apparatus, as shown in Fig. 1. Unless this impedance is small, either inherently or unless it be reduced by proper bypassing, the amplifier is liable to "sing," or to cause poor reproduction from otherwise good apparatus, due to audio-frequency regeneration.

Two condenser-choke combinations are illustrated in Fig. 2. The resistance of the choke must be small, for the same reason that it must be small in any case—to prevent plate-voltage loss; the inductance must be high at rather large values of d.c. current, and the condenser must

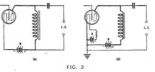

FIG. 2

shows the result. 32 per cent. of the current which should flow through the loud speaker being lost. The condenser which is in series with the loud speaker must take its share of the voice currents which are circulating in this circuit, and for that reason its impedance, compared to that of the loud speaker, must be small.

METHODS OF CONNECTION

IF THE condenser-choke arrangement is connected as in B Fig. 2, all of the a.c. currents in the plate circuit of the last tube return to the filament directly. This connection has the added advantage that neither of the loud speaker terminals are at high d.c. potential with respect to ground. One side connects directly to A minus, which is at ground potential, while the other connects to the condenser and is insulated from the high potential. If the condenser breaks down

used without impedance adjustment with a 4000-ohm, or even higher impedance, loud speaker. But if two 5000-ohm tubes are used in a push-pull circuit so that the resultant impedance is doubled to 10,000 ohms, and one considers a loud speaker whose impedance at 100 cycles is 1000 ohms, trouble will occur.

Here we need a transformer. If the primary and secondary individual impedances are high compared with the impedances of the tube and loud speaker, and provided a good core with good coupling is used, there will be little voltage loss, and practically the entire a.c. voltage on the plate of the last tube will be transferred to the loud speaker, the magnitude depending upon the turn ratio of the transformer, of course.

The trouble mentioned above will not be loss of power due to improper impedance matching so much as distortion due to another cause. All tubes have a plate current plate voltage characteristic that is somewhat curved. This curve produces additional frequencies when incoming signals produce large values of a.c. plate current. If, however, a large impedance is placed in the plate circuit, changes in grid voltage, due to large signals, produce a smaller proportionate change in a.c. plate current, and the characteristic is straightened out. Mathematics shows that when the impedance of the load is twice that of

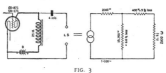

FIG. 3

FIG. 4

be large. The equivalent circuit is shown in Fig. 3. Here is a source of voltage, a series impedance representing the plate resistance of the tube, a shunt impedance representing the choke, and a series impedance representing the condenser, and another impedance representing the loud speaker. The shunt impedance is a bypass for the audio frequencies and naturally must be large compared to the remainder of the circuit, i.e., the series impedance of the loud speaker and the condenser.

Let us take a definite example. We shall consider a 171 type tube with an internal impedance of 2000 ohms, a shunt choke of 25 henries, a condenser of 4 mfd., and a 2000-ohm loud speaker. This latter figure represents the impedance of the loud speaker at some audio frequency, say 100 cycles, and has little to do with its d.c. resistance. At this frequency the choke has an impedance of 15,700 ohms, representing a loss of 6.0 per cent., and the condenser has an impedance of 400 ohms representing a loss of 9.0 per cent., with the result that all but 15 per cent. of the voice-frequency currents flow through the loud speaker, compared to the case in which neither condenser or choke were used when there is no loss. The 15 per cent. loss will be hardly detectable to the human ear. The impedances of the condenser and choke are most important at low frequencies.

Let us consider another case, this time choosing a loud speaker which has an impedance of 1000 ohms at 100 cycles, a tube with an output impedance of 5000 ohms, a choke of 10 henries, and a 1.0-mfd. condenser series impedance. Fig. 4

the B-battery voltage will be put across the loud speaker, therefore a good condenser must be used—one capable of withstanding not only the voltage of the plate supply without puncturing but the added voltages produced across it by the audio frequencies, which, however, will be small if its capacity is large. In A of Fig. 2 one of the loud speaker terminals is "hot" since it is connected to the positive side of the plate supply, which is high above ground potential. In this case nothing happens if the condenser breaks down except that the plate current will divide between the loud speaker and the choke, most of it still going through the latter because of its lower d.c. resistance. The voltages across the condenser in this case are lower than in B of Fig. 2 since there is no steady voltage due to the plate supply but only the audio-frequency voltages, and these will be small owing to the low impedance of the condenser.

The absolute values of the inductance and the capacity may be varied within certain limits depending upon the tubes and the loud speaker used. With a 2000-ohm tube and a 4000-ohm loud speaker, 20 henries is plenty and a capacity of 4.0 mfd. is correct.

Since the effect on audio frequencies of both the choke and the condenser may be neglected, if proper values of each are used, the loud speaker looks directly into the plate circuit of the power tube, and as far as a.c. currents go, might just as well be connected there. This is perfectly proper provided the impedances of the tube and loud speaker do not differ too widely. For example, a 2000-ohm or a 5000-ohm tube may be

the tube, the greatest amount of undistorted power will be secured from the tube. At low frequencies the impedance of most loud speakers is very low, so that the tube's characteristic is curved, and loud low-frequency signals produce a rattle or rumble that is objectionable. For this reason, the loud speaker should not match the tube in impedance, but should have a greater impedance, and if this is not possible, a transformer must be used which makes the loud speaker look like a higher impedance to the tube. This may be done by using a step-up transformer, looking from the loud speaker to the tube, so that the impedance of the former is stepped-up or increased as far as the latter is concerned.

Such is the story of loud speaker filters, or output devices. With power tubes in the plate circuit, when there are over ten milliamperes flowing, they are useful in protecting the loud speaker; when low-impedance speakers are used with high-impedance tube circuits, they are desirable. Unless the plate current is such that the loud-speaker armature is sadly biassed and rattling against the pole pieces of the magnets, it is doubtful if the average listener can tell the actual difference in fidelity whether the device is used or not. If a condenser-choke combination is used the loud speaker should be connected to the negative filament lead or the center tap of an a.c. operated power tube. Connected thus, the owner of the loud speaker will be protected from d.c. shocks, although he can get a severe jolt by holding to the speaker terminals when a kettle drum operator gives his instrument a good "whack."

THE RAYFOTO RECORDER

RADIO BROADCAST Photograph

This photograph illustrates a final model of the recorder. The pictures are recorded on a piece of photographic paper wrapped around the drum. This unit is designed to fit over the turntable of any standard phonograph

Operating Your Rayfoto Picture Receiver

By Austin G. Cooley

THERE are several details concerning the Rayfoto printer and Rayfoto recorder which were not discussed in the article in the December RADIO BROADCAST; a knowledge of these is not necessary in order to construct the printer although necessary to obtain most satisfactory results. In this article we will use several terms that have not been used in preceding articles but which will serve to differentiate between the various units of the Rayfoto apparatus. A description of these terms, with a brief explanation of the function of the various parts which they define, will be found in the table of definitions on this page.

Readers of this series of articles will recall the description given in the November article of the Rayfoto relay and its function. This relay is operated by the plate-current increase produced in the amplifier tube of the printer by the synchronizing signal. There is a natural tendency, however, for the relay contacts to vibrate for a very short time after they first close. This causes irregular operation of the trip magnet which the relay controls, and the irregular operation in turn produces uneven synchronizing which causes jagged effects in the received picture.

To prevent this irregular action of tne relay it was necessary to arrange the circuit so that when the relay contacts closed they would lock tightly together. The most important part of this locking circuit is a resistance, R_3 in the circuit diagram published on page 297.

When the relay, R_4, closes, due to the synchronizing impulse, it causes current to flow through the trip magnet coil and hence through resistance R_3 to minus B. The current through R_3 produces a voltage drop across the resistance of such polarity as to decrease the negative bias on the grid of the amplifier tube. This causes

the plate current to increase a comparatively large amount, and this plate current, flowing through the coils of the relay, R_4, makes it lock fast and prevents the contacts from vibrating. As soon as the armature on the trip magnet releases the drum, the contacts on the trip magnet close and thereby short-circuit the relay so that it is out of the circuit while the drum is making a revolution.

The locking resistance should be adjusted so that the plate current of the amplifier tube, as read at jack J_1, is about 15 milliamperes when the relay and trip magnet contacts are closed.

There is always a certain amount of sparking at the contacts but this causes no harm unless the spark is sufficiently intense to cause an arc after the contacts have opened. This is remedied

by increasing the gap between contacts when they are open.

The contacts should be kept clean although there is nothing to be gained by excessive filing and cleaning. Cleaning the contacts about once a month with a piece of cloth is all that is necessary.

Reliable operation of the synchronizing system is impossible if there are excessive set noises. Such noises can often be traced to noisy batteries or poor connections. Make no attempt to receive Rayfoto pictures until such noises are cleared out. This does not imply that perfect receptive conditions are necessary for picture reception but considerable extraneous noise is not conducive to success.

The photographic paper that has been found most satisfactory is Azo No. 2 semi-matt or semi-gloss, singleweight, size 5 x 7 inches. The room in which the pictures are received must be somewhat darkened and a test can be made to determine if the room is dark enough by placing a piece of paper on the drum, allowing it to remain there about five minutes, and then developing it. If, in the developer solution, it turns gray or black in about 30 seconds there is too much light in the room, and the room will have to be darkened in some way. Of course, in the evening, no difficulties will be experienced and it will generally be found safe to operate the Rayfoto receiver without any shading at a distance of about ten feet or more from a forty-watt electric light.

In wrapping the piece of photographic paper around the drum for this test be sure that the emulsion side is on the outside. The side of the paper with the emulsion on can be determined by biting a piece of the paper with your front teeth. That side of the paper with the

Table of Definitions

Printer: A two-tube unit consisting of a one-stage audio-frequency amplifier and an oscillator. The Rayfoto signals from the radio receiver are amplified in the audio amplifier the output of which is impressed on the Rayfoto modulation transformer which, in turn, modulates the oscillator. The output of the oscillator produces the corona discharge which prints the picture.

Recorder: The mechanical unit for attachment to a phonograph turntable, and consisting of a drum on which a piece of photographic paper is wrapped, a clutch system, and a trip magnet for use in obtaining synchronism.

Relay: This relay is operated by the synchronizing impulse and when the contacts on the relay close, the trip magnet operates and releases the drum on the Rayfoto recorder.

Lap: The interval between the time that the drum on the Rayfoto recorder completes a revolution and the time that the synchronizing impulse is received. The recorder lap was explained fully in the November, 1927, RADIO BROADCAST.

Static Slip: The operation of the Rayfoto relay by a static impulse instead of by the regular synchronizing impulse.

emulsion will take the impression of the teeth or will stick to the teeth.

After experience has been gained in the operation of the system, No. 4 Azo paper can be used and it will be found that this paper can stand more light without being affected. When No. 4 paper is used a greater discharge of corona is necessary than with No. 2. If the signal strength is low so that not much corona is available, it will be best to use No. 1 paper. which is more sensitive.

In making preliminary tests it is a good idea to let the printer operate on broadcast signals and, after a run has been made, to develop the paper for about thirty seconds and determine if the corona discharge from the strong signals is sufficient to print out black.

Be sure to wrap the paper around the drum in the right direction, which is opposite to the direction of rotation of the drum. If the paper is wrapped in the wrong direction the corona needle will catch under the edge of the paper and pull it up.

The speed of the drum at the transmitter has been standardized at 100 revolutions per minute and, therefore, the speed of the drum on the Rayfoto recorder should be near 106 revolutions per minute in order to obtain the correct lap. Adjust the speed of the drum to this value by letting the drum revolve for a minute with the trip magnet armature down, counting the number of revolutions.

If the Rayfoto recorder unit is examined it will be found that the arm which carries the corona needle will slide along the machined shaft located at the rear of the device. If the arm does not move along the shaft very easily, it should be oiled. At no time should the shaft be touched with emery paper or sandpaper. The shaft may be cleaned if necessary by mixing a little Gold Dust powder with some lubricating oil and wiping off the shaft.

To adjust the Rayfoto receiver to operate through static, it is necessary to first adjust the relay and corona discharge. Tune exactly to the station transmitting the pictures. Reduce the signal in the printer by the gain control to a point where the relay just ceases to operate on the synchronizing impulse. Determine whether there is sufficient corona by examining the discharge from the needle point. There should be quite a noticeable discharge. A piece of Azo paper should, of course, be on the drum when any test of corona discharge is being made and if there is not sufficient corona for making a fairly black mark from the strong signals, increase the spring tension on the relay, then boost up the gain until the relay just starts to operate. Check again to see if the corona discharge is sufficient. If not, repeat until a fair discharge is obtained. When working through static, it may be necessary to work with a discharge much lower than normal. Consequently, the print will be weaker, unless a more sensitive paper, such as Azo No. 1 is substituted for that generally used, to make up for the reduced corona.

In making the above adjustment, the drum may be held in such a position that the stop-shoe does not strike the push rod by revolving the drum half a turn. The discharge made by the synchronizing signal may then be observed.

If it is difficult to obtain sufficient control of the relay by the spring adjustment, the gap between the armature and pole tip may be increased by adjusting the contacts.

After these adjustments have been made, tune the radio receiver to a frequency about fifteen kc. above or below the frequency of the picture transmitting station. If the relay operates more than once a second from the static noises, the experimenter might as well turn in

THE RAYFOTO RELAY

A close-up of the relay. It is this relay, actuated by the synchronizing signal, that causes the drum of the recorder to be released, every revolution, at exactly the correct moment

and give up the idea of picture reception for the rest of that night. If the relay does not operate from static noises, the gain should be increased to a point where it just commences to operate; then reduce the gain a small amount. Thus you find the critical point of maximum permissible gain without static tripping the synchronizing relay. After this has been done, tune back to the picture signal and check to see that the gain has not been increased too much.

The next thing to do is to adjust the speed of the recorder. The speed should be such that the recorder lap is as small as is consistent with regular operation of the trip magnet.

THE November RADIO BROADCAST gives some of the essential information on the subject of blurring and detail. It may be well to mention here one or two things the experimenter may check if he experiences trouble. In many cases, the following simple checks will solve the difficulty.

First, be certain of the connections to the modulation transformer, making sure that the proper terminals are used, because this has an important bearing on the operation of the entire system. The primary terminals are No. 1 and No. 2. No. 1 should go to the plate of the input amplifier tube and No. 2 to the meter jack and then to the battery supply. No. 3 of the secondary should be connected to the plate of the oscillator through the r. f. choke and No. 4 goes to the meter jack and then to the booster voltage supply. Also the connections to the primary of the audio transformer in the printer should be reversed to determine which arrangement gives most satisfactory operation.

If there is a large recorder lap, that is, if the receiving drum has sufficient lead to arrive at the end of the revolution considerably ahead of the convertor drum, the relay will be connected in the circuit for a longer period than necessary. Should some static of strong intensity be received during the lap, the drum will be released ahead of the synchronizing impulse. Such jumps, due to static operation of the relay, are known as static slips.

The possibility of these static slips decreases

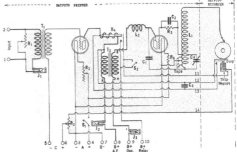

FIG. 1

The circuit diagram of the Rayfoto printer unit. The connections between this unit and the Rayfoto recorder are indicated. Constructional information on the printer was given in the December issue of RADIO BROADCAST. The Recorder unit cannot be home constructed but may be purchased as a complete unit

as the lap is reduced. Careful adjustment of phonograph motor speed is suggested to maintain a minimum lap. If, however, the speed is matched too closely, there is no lap. The trip magnet then operates as the stop shoe strikes the push rod. Such operation is very unstable and produces very jagged pictures.

If the recorder speed is reduced so that it is slightly slower than that of the convertor, the recorder drum will not reach the end of the revolution in time to receive the synchronizing impulse. The relay will then operate on the impulse of the next revolution, or, if there are strong picture signals, during the revolution, and the result will be no picture at all or one that is very badly distorted.

After a little experience, the experimenter will be able to tell from the sound of the recorder whether the lap is correct. Sometimes streaks occur in the Rayfoto pictures because the oscillator tube ceases to function for a moment. Such a streak appears in the Rayfoto picture printed on page 216 of the January RADIO BROADCAST. This frequently occurs when the oscillator is tuned to give the maximum amount of corona. To avoid these streaks, detune slightly with the variable condenser across the oscillator coil. A higher resistance in the grid circuit also tends to reduce the possibilities of streaks.

Streaks are also caused by poor connections and by detuning of the radio receiver by hand or body capacity. For this reason, it is desirable to have a well-shielded receiver.

At present, pictures with fairly strong contrast are being transmitted so that good Rayfoto prints may be made with the average haphazard adjustment of the printer. Some may find that they are getting a little too much corona on the minimum signals and that the intermediate shades are produced as blacks. Correction for the minimum signal can be best accomplished by reducing the booster voltage. The difference between the intermediate shades and blacks can easily be increased by reducing the gain and then allowing a little more time for development of the picture. When reducing the gain, it may be necessary to reduce the spring tension on the relay.

From the foregoing it should be obvious that some skill is required to secure the best possible results. Poor pictures are a matter of poor ad-

RADIO BROADCAST Photograph

THE RECORDER IN PLACE
The photograph illustrates how the Rayfoto recorder is mounted over the turntable of a phonograph

justment of motor speed, amplifier gain, stop-start mechanism. Good pictures are a credit to the experimenter who receives them. In picture reception, the amateur has one advantage over the fan who contents himself with mere audio reception. He obtains permanent and irrefutable evidence of his success—a catalogue of his progress throughout the development of what will some day be a widely practised science and art—the reception of high-grade pictures in the home.

PARTS FOR A RAYFOTO PICTURE RECEIVER

CERTAIN parts for use in a Rayfoto receiver have been especially designed for the purpose and therefore possess the essential char-

acteristics for good results. They have been designed to take care of the present requirements of the system and will also be satisfactory for use as the system may be gradually developed. These special parts, made under the Rayfoto trade mark, are listed below:

L₅—Rayfoto Corona Coil
T₁—Rayfoto Amplifying Transformer
T₂—Rayfoto Modulation Transformer
R₄—Rayfoto Relay
      Rayfoto Printer Unit

The remainder of the parts necessary to construct a picture receiver are given below. Any standard parts conforming with the specifications given below may be satisfactorily used.

R₁—Variable Resistance for Gain Control
R₂—200-Ohm Variable Resistance Capable of Carrying 100 Mils.
R₅—12-Ohm Filament Rheostat, 0.5-Ampere Capacity
C₂—0.1-Mfd. Fixed Condenser
R₃—0.01-Megohm Grid Leak and Mounting
C₁, C₃—0.0005-Mfd. Fixed Condensers
C₄—0.0005-Mfd. Variable Condenser
L₄—Radio-Frequency Choke Coil
S₁—Filament Switch
S₂—Push Button Switch
J₁, J₂, J₃—Double-Contact Short-Circuiting Jacks
R₆—4-Ohm Filament Ballast Resistance
      Telephone Plug
      Milliammeter, 0-25 Milliampere Scale
      Two Sockets
      Fourteen Binding Posts
      Panel
      Panel Brackets

The designating letters preceding the parts given in the above list refer to the lettering on the circuit diagram given on page 297 of this article. Readers interested in constructing a Rayfoto picture receiver and desiring further information regarding the necessary parts may obtain this data by writing to RADIO BROADCAST magazine.

RADIO BROADCAST Photograph

THE RAYFOTO PRINTER UNIT
The parts for the unit, containing the amplifier and oscillator circuits, can be purchased and then wired together as indicated in the photograph. The circuit diagram is given in Fig. 1. At the left is the Rayfoto corona coil. The transformer at the extreme right is the Rayfoto amplifying transformer and at its left is the Rayfoto modulation transformer. The Rayfoto relay can be seen mounted on the panel

THE FREED-EISEMANN "NR-60" ELECTRIC RECEIVER

How the "NR-60" was Engineered

By John F. Rider

GENTLEMEN, we need a new electric receiver, designed for a.c. tubes and equipped with B supply. The receiver must be a single-control set to retail at $150.

The above is the sum and substance of a message delivered to the engineering department of the Freed Eisemann Corporation and is the reason for the birth of the "NR-60 receiver."

The electrical development of this receiver will prove of intense interest to the radio fraternity at large, because it covers a subject very much in the public eye at present.

One can readily see that the problem placed before the engineering department of this organization differs greatly from that usually confronting the average research staff. The request was not for a d.c. receiver, but one designed for use with tubes utilizing raw a.c. upon the filaments and also a B supply unit. Hence we have four requirements. First is the receiver circuit itself, which constitutes a problem replete with many obstacles, particularly so in this day of competition. Secondly, the use of a.c. tubes means the provision in the design of the receiver for the elimination of the 60-cycle hum in the filament supply. Thirdly, the design of the power transformer which will supply all the filament voltages and the a.c. for the plate voltage, later to be rectified by the rectifying tube, is a problem not to be scoffed at. Fourth, is the coördination of all

the parts to produce a satisfactory all-electric receiver. The latter is more easily said than done, as can be attested by many fans.

Let us follow in the order mentioned above and watch the progress of the electrical development. Six tubes are to be used in the receiver. The receiver is to be operated with an outdoor antenna. A certain amount of selectivity is therefore necessary. Considering the status of broadcasting at the present, and the possibility of increased power at the transmitters, provision for satisfactory selectivity must be incorporated in the event that more stations go on the air, station wavelengths are reallocated, or transmitter power is increased. This necessitates that at least three tuned stages be used. Furthermore, the manufacturer specifies a certain antenna length, which, however, is not always obtainable. To assure satisfactory sensitivity in the event that a short antenna is used, three stages of tuned radio-frequency amplification are decided upon. Such decision, however, can be made only after the gain or amplification per stage has been determined mathematically and checked empirically.

In order to permit this determination, a large number of different types of tuned radio-frequency transformers must be first designed on paper and then constructed for experimental purposes. This experimental work is of paramount importance during the development of a radio receiver. The variance in types of the different experimental tuned radio-frequency transformers is found in the type of wire used, the diameters of the winding form, the spacing of the individual turns, the placement of the primary with respect to the secondary, the ratio of inductance to capacity on different wavelengths, the length of the winding form, and many other intricate details. Comparison in a mathematical manner is not sufficient, because phenomena encountered in practice can not be included in the calculations. Hence both mathematical and experimental determinations are necessary.

The actual experimental work of measuring the gain per stage for various radio-frequency amplifying systems is a tedious, detailed procedure. The tuned radio-frequency transformer to be measured is coupled to a tube. A known radio-frequency voltage is fed into the input circuit of this tube. The output of the tuned radio-frequency transformer is then measured at various broadcast wavelengths by means of a vacuum-tube voltmeter. A comparison of the output voltage values of the various radio-frequency transformers is a direct indication of their merit.

After a large number of tests of various forms of inductances, and with the realization that a special system of neutralization was available in addition to the fact that individual shielding was permissible, the staff decided to utilize what is considered to be the most efficient form of inductance, the single-layer solenoid with spaced winding. With this decision the design of the three tuned stages was complete, but what about the antenna system? Should this be untuned, and of the conventional system? The consensus of opinion was to deviate from the conventional and to utilize a tuned antenna circuit. The sensitivity of the complete system would be greater, the receiver output would be greater and, in addition, the frequency response characteristic of the complete r.f. system would be better. Last but not least, the tuned antenna system could be designed so that single-control would be satisfactory, regardless of the length of the antenna employed.

This started another series of investigations, and after a period of time, the variometer tuned type of input was selected. This arrangement possessed several salient features. First, it permits unicontrol. Second it provides greater gain or amplification as the wavelength is increased. This property is inherent in inductively tuned circuits and is diametrically opposite to the phenomena in capacitatively tuned circuits. This is due to the increase of impedance of inductively tuned circuits with increase in wavelength. The action of the variometer-tuned antenna-input circuit by increasing the gain on the upper wavelengths of the normal broadcast spectrum would tend to compensate the falling characteristic of the other stages. This type of tuned antenna input circuit affords a much greater signal voltage to the grid filament circuit of the first r.f. tube. As a matter of fact the difference between such a tuned system, Fig. 1A, and the conventional untuned antenna, Fig. 1B, is of the order of 3 to 1 in favor of the former arrangement (system A).

By the use of a small series capacity, connected between the antenna and the grid input terminal of the first tube, the variance in antenna capacity when different lengths of antenna are used, is practically nullified, and the tuned system is to all intents and purposes, isolated. This means that the setting of the variometer dial will remain constant regardless of the length or type of antenna used.

Stability of the radio-frequency system was the next point of interest. Being in possession of a Hazeltine license, neutralization of this form for the radio-frequency stages was an immediate decision. Having designed the inductances and knowing the operating characteristics, the conventional Hazeltine system was selected. This system utilizes a voltage transfer of reversed phase from one grid circuit to the preceding grid circuit. The value of the feed-back voltage is governed by the design of the secondary of the tuned radio-frequency transformer and is the voltage across a certain portion of this

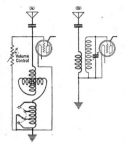

FIG. 1

winding. The voltage is obtained by tapping the secondary winding at a predetermined point.

With respect to shielding, the decision to shield individual stages by enclosing the tube, coil, and condenser in a can, was immediate. The choice, however, of the shielding material, required some consideration. Electrical conductivity and economy are the two important factors. After considering these two points the selection was aluminum.

The experimental work carried out upon the tuned radio-frequency transformers influenced the selection of the type of winding. Now arose the problem of producing "matched" inductances. Accuracy is very important in all single-control units. To overcome slight discrepancies, such as would be occasioned by one or two $\frac{1}{10}$ ths more or less on the coils, each tuned radio-frequency transformer is equipped with a copper vane located at one end of the main inductance. Manipulation of these vanes permits accurate variation of the inductance of the windings, thus facilitating "matching" of the condenser-inductance combinations, and the tuned circuits.

Summarizing the radio-frequency amplifying system we have the following: A tuned antenna input, complete individual shielding, three stages of tuned radio-frequency amplification, complete neutralization, and single tuning control. The maximum capacity values of each tuning condenser is 0.00032 mfd. The shape of the tuning condenser plates affords a modified straight frequency-line variation.

DETECTOR AND AUDIO SYSTEMS

NOW for the detector and audio systems. Utmost sensitivity is desired, hence the grid leak-condenser system of detection is employed.

As to the audio system, the choice must be

made consistent with three factors, economy, results, and knowledge. Only two tubes are available for audio amplification. The best way of obtaining sufficient volume is by means of transformer coupling, and since extensive research work has been carried on to design and produce an excellent audio-frequency transformer, the decision to use transformer coupling was natural. That the research work along this line was of high calibre is shown by reference to Fig. 2. This curve shows the operating characteristic of the audio transformer. It is a 3 to 1 coupling unit, without a pronounced peak on the higher audio frequencies, a characteristic seldom found with the average audio-frequency transformer. The elimination of a sharp peak at some frequency between 3000 and 10,000 cycles is due to scientific coil design and minimization of leakage reactance.

The next problem was the selection of a means of coupling the output tube to the loud speaker. Some coupling medium is necessary because of the heavy output plate current occasioned by the necessity of using a power tube in the output stage. Passing the heavy plate current through the loud speaker windings would injure them, in addition to the possibility of reducing the magnetic strength of the magnets in the event that the polarity of the loud speaker is reversed with respect to the polarity of the plate battery. The design of a coupling unit was imperative. A transformer of very good design was developed, and its frequency operating characteristic when used with a 171 type tube is shown in the accompanying curve, Fig. 3, at the bottom of this page.

Now arose the problem of volume control. Much work was done along this line; as a matter of fact, this work could not be avoided, since the method of controlling receiver signal output with a d.c. receiver is not wholly satisfactory when applied to a.c. receivers. After an extended period the system shown in the wiring diagram on page 301 was selected as being most effective. This is a variable high resistance shunting the tuned input circuit

This arrangement provides a means of adjusting the receiver output by controlling the signal voltage passing into the first radio-frequency amplifier. This arrangement proved satisfactory because it does not display an effect upon the sideband characteristics of the radio-frequency amplifier, nor does it manifest any variation in the degree of neutralization. Yet the control of volume is perfect.

THE TUBES USED

UNDER the existing circumstances, the tubes selected were the RCA 226 and 227, with a 171 in the output, or the equivalent Cunningham 326 and 327 with a 371 in the output. The operating characteristics of these tubes are practically identical to that of the regular d.c. filament tubes, hence the design of the associ-

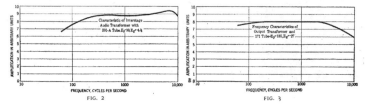

FIG. 2

FIG. 3

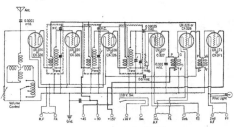

CIRCUIT DIAGRAM OF THE "NR-60"

ated apparatus did not require special pre-cautions. The 226 type of tube was selected for the three stages of radio-frequency amplification and for the first audio stage. The 227 heater type of tube was selected for the non-regenerative detector and the 171 was used as the output tube. A precaution required in the receiver was that twisted cables be used and that they be isolated so that transfer of the 60-cycle hum was mini-mized. That the arrangement was effective is demonstrated by the satisfactory operation of the receiver, despite the fact that the audio-frequency transformers have satisfactory re-sponse on 60 cycles. Were the filament wiring in a position to cause hum due to induction, this hum would be heard with regularity in the loud speaker.

The use of a.c. tubes necessitates a mid-tap for each amplifying system. Experience proved that a variable potentiometer shunting each tube filament circuit was a better means of obtaining an accurate electrical balance than the use of a mid-tapped transformer winding. The use of such a mid-tap requires a separate grid bias for each amplifying system. This grid bias is obtained from the B supply.

The design of the B supply unit and the A supply unit was an interesting problem. Let us consider the power unit as a whole. The power pack supplies all the voltages required for fila-ments, grids, and plates. It is complete in itself encompassing the B device, the A supply, and the output transformer. A study of the wiring dia-gram of the power supply on this page shows an arrangement which can be followed to excellent advantage by many manufacturers of power transformers and by others interested in a.c. tubes of the type mentioned herein. The 226 tubes are rated at 1.5 volts and 1.05 amperes and the 227 is rated at 2.5 volts and 1.75 amperes. The voltage is low and the current is high. If a large number of these tubes are fed through one cable and from one source of supply, that is, from one winding, the total amount of current will reach a fairly high value and any small resistance in the circuit will cause an appreciable voltage drop. To minimize this effect each system of amplification is equipped with a separate filament winding, supplying energy to the tubes in that system. This arrangement also elimi-nates the necessity of inserting various values of resistances to supply the correct voltages, were one single winding used. For example, the maximum filament winding voltage is 5 volts, for the 171 tube. The 227 requires 2.5 volts and the 226 requires 1.5 volts. All of these voltages could

be supplied from one 5-volt winding of proper capacity. But the insertion of the necessary resistances for reducing the 5 volts to the correct value for the other tubes would increase the cost of manufacture and would be less efficient. The method used was found much more efficient and, therefore, adopted. An accurate mid-tap is made possible for the radio-frequency and the first audio-frequency tubes by the use of a po-tentiometer. Furthermore, each filament circuit consists of a pair of twisted cables. The 227 de-tector tube receives a plate voltage of 45 volts from the B unit. By employing filament windings which provide the required filament voltages, all filament controls are eliminated.

The design of the B device was cause for considerable thought. Should it be a half-wave rectifier or a full-wave rectifier? Each possessed certain advantages. The half-wave rectifier is simpler, but the full-wave rectifier affords cer-tain technical and practical advantages. In the first place the frequency of the charging voltage applied to the condensers of the filter system is 120 cycles with a full-wave rectifier and as such the action of the condenser is better; the reactance of the condensers is lower. The filter-ing action improves as the frequency of the hum to be eliminated increases. In addition, the cur-rent capacity of the unit is doubled. Where a

half-wave rectifier would permit 65 mils., a full-wave rectifier would permit 130 mils. at the same voltage. Hence, by using the full-wave rectifier, sufficient current is provided at high voltages.

To eliminate any possible hum due to induc-tion from the power device to the receiver, the complete unit is housed in a shield. The con-densers used in the B device are, in turn, housed in an individual can within the main can. Both cans are grounded to the common ground termi-nal and the main can is at ground potential.

One side of the main line is connected to ground through a 0.1-mfd. condenser, thus by-passing to ground any radio-frequency energy in the power line.

The design of a receiver is not limited solely to the choice and pattern of the individual com-ponents of the receiver. Complete test of the combined parts may show that they are not satisfactory when used together.

The method of testing the radio-frequency transformers applies to the method of testing the gain or amplifying power of the complete radio-frequency amplifier.

The audio-frequency transformers are also tested with vacuum-tube voltmeters. Each trans-former is tested on three frequencies and a definite output must be obtained before the transformer is passed.

A vacuum-tube voltmeter is employed when balancing out the hum in the receiver and by means of the meter reading the value of the existing ripple is ascertained.

After the receiver is assembled ready for the first complete test, it is placed into a test rack. The source of energy supply for this test comes from a crystal-controlled master oscillator cir-cuit tuned to a fixed frequency of 600 kilocycles. This type of oscillator also supplies testing frequencies which are multiples of the fixed frequency. These multiple frequencies are the harmonics generated by the crystal-controlled tube. The tuned circuits in the receiver are adjusted to 600 kilocycles and are brought to resonance by means of the copper vanes associ-ated with the inductances. Then the receiver is tuned to the second harmonic of 600 kilo-cycles, which is 1200 kilocycles, approximately 250 meters. It is then adjusted to perfect reso-nance by means of balancing condensers on the shortest wavelength within the broadcast band.

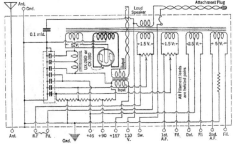

THE POWER SUPPLY OF THE "NR-60"

RADIO

RECEIVERS

REPRESENTING

A WIDE PRICE

RANGE

SPLITDORF'S "BUCKINGHAM"
Designed to blend in with the surroundings of a home decorated in the English style, the period of this cabinet work dates back some four hundred years. The "Buckingham" is a six-tube receiver employing three r.f. stages, tuning being accomplished by an attractive rose-shaped knob which may be seen in the center of the receiver. A complete A, B, C power unit is housed within the cabinet. Behind the cane center panel in the upper part of the chest is the "Maestro Cone Tone" reproducer. The list price of the "Buckingham" is $800.00 complete

THE FERGUSON "HOMER"
Compactness is expressed in this new receiver by J. B. Ferguson. Housed within an exceptionally attractive little cabinet is this seven-tube receiver, employing four r.f. stages, two only of which are tuned, and two audio stages. A single illuminated dial accomplishes the tuning, and there is also a volume-control handle on the panel. Control of volume is obtained by means of variable plate coupling. The audio channel makes use of General Radio transformers. The "Homer" lists at $95.00 as illustrated. Chassis only, $80.00

THE advent of the more expensive, more luxurious, receiver, is not to be heralded as something new, although developments along this line have been very marked of late. Ever since its swaddling clothes days radio has been represented in a surprisingly wide range of prices. That a certain manufacturer charges ten times as much for one receiver as for another using the same circuit and number of tubes, need not deter the man of modest means from investing in the least expensive model. Frequently the chassis in both models is identical, the extra cost being due to refinements of cabinet work, the inclusion of a complete power-supply unit and also, perhaps, a built-in loud speaker

A LOOP RECEIVER

This is the "Ortho-sonic" F40 receiver, by Federal of Buffalo. The circuit makes use of seven tubes in a carefully balanced and shielded circuit. The loud speaker, which is built in and concealed by a silk screen and hand-carved grille, is said to be capable of beautiful tone. The loop may be seen mounted upon the inside panel of the left-hand door. There is ample space for installing the necessary power equipment. The cabinet is of walnut with vermilion inlay and hand carving. Price, without tubes or accessories, $450.00

THE "CRUSADER"

By King, also of Buffalo. Many interesting features distinguish this receiver. Special attention, for example, has been paid to its design that the use of a B-supply device will not complicate matters. There are two r.f. stages and three audio stages, the latter employing a combination of transformer and double-impedance coupling. The three variable, tuning condensers are adjusted by means of a single knob, but there is an auxiliary knob for the first stage so that exact resonance may be obtained. The "Crusader" is completely shielded. The cabinet will fit in with either mahogany or walnut home furnishings particularly well. Price, without accessories, $115.00

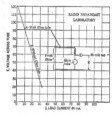
A "NUT-SHELL" EXPLANATION OF GLOW TUBE OPERATION

The functions of a well-known type of regulator tube were fully described on Laboratory Sheet No. 129, which is reprinted above. The principles underlying the performance of the Raytheon R tube, illustrated below, are somewhat different to those of the 874 type tube

Constant B Device Output

By G. F. Lampkin

THE most obvious problem in the construction of a B device is that of filtering —of reducing the hum to a negligible value. In some power units, especially those made during the last few years, so much attention was paid to this problem that another problem, not so apparent, but fully as important, was somewhat neglected. This latter problem is one of voltage regulation—the changing of the B device output voltage when the output current is varied. A cut and dried definition states that voltage regulation of an electrical device is the rise in voltage when full load is thrown off the device, expressed as a percentage of the full-load voltage. The regulation of ordinary electrical apparatus is seldom greater than ten per cent. The regulation of some B power units runs above 100 per cent.

If a 90-volt set of B batteries be connected to a receiver, it may be known with reasonable certainty that the voltage applied to the receiver is 90, whether there are one or six tubes being supplied. If a B device is connected to a receiver and the device has poor regulation, the voltage

GLOW TUBE CONSTRUCTION

The use of a glow tube in B power units is becoming increasingly popular. The Raytheon one here shown is very efficient. It was described at length in the October, 1927, RADIO BROADCAST

may be 150 when supplying one tube but less than half that figure when supplying six. If the experimenter has available a high-resistance voltmeter suitable for the measurement of the output voltages of a B power unit, the voltages may be checked while the power unit is connected to the receiver to make certain that the voltages supplied to the set are correct. Some power units are equipped with variable resistance units so that the correct voltage may be obtained

by the proper adjustment of them. Without a high-resistance voltmeter it is difficult to adjust accurately the voltages of a B supply unit, containing variable resistances to control the voltage. However, if the variable resistances are loosened so that the resistance is as high as possible and are then gradually tightened until the receiver operates satisfactorily, it is possible to adjust the voltages with fair accuracy. There is a danger of shortening the lives of the tubes in the receiver if the resistances are tightened beyond that point which gives satisfactory reception.

Under some circumstances the variation of the d.c. voltage with the load current introduces another disadvantage in that it tends to cause audio distortion. When the loud speaker is connected as indicated in Fig. 1, the plate current delivered by the power unit supplying the entire receiver may have a swing of ten milliamperes when a loud low note is being amplified. A receiver that would give such a signal might draw a total load of about thirty milliamperes and the plate current swing will, therefore, form an ap-

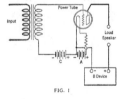

FIG. 1

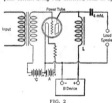

FIG. 2

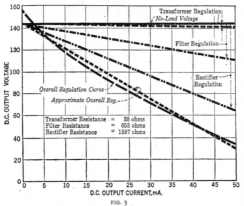

FIG. 3

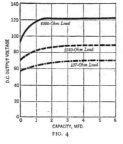

FIG. 4

preciable part of the total load so that the B device voltage will also vary. When the plate current increases, due to the signal, the B device voltage drops and tends to nullify the change; similarly, when the plate current decreases the voltage goes up and again tends to nullify the change. The regulation of the device would thus tend to cut off both the positive and negative peaks of the signal. The filter condenser at the output of the B device can take care of the plate current swings to some extent by charging and discharging on the negative and positive swings respectively. The effect of the condenser in this function is dependent on the frequency, however, and at low frequencies it does not exercise control to any great extent over the output voltage.

When the loud speaker is connected as in Fig. 2, distortion due to a variable load on the power unit is prevented because all the a.c. currents must flow around through the loud speaker and back to the filament and hence do not go through the power supply. As a result, the current drawn through the choke L from the power supply is practically constant and the load on the power

unit does not vary appreciably with the signal.

The regulation of a B device is caused by the internal impedances of the three units that make up the device—the transformer, the rectifier, and the filter. As the current drawn from the device increases, the internal voltage drops also increase and the result is that the output voltage becomes less. A graph of the output voltage of a typical B device for different values of current is shown in Fig. 3. Such a graph is called a regulation curve. It is nearly a straight line and for purposes of simplification it may be replaced by one such as the lowest straight line of the group. The vertical distance between this line and the upper horizontal line, which represents the no-load voltage, is the internal voltage drop for any given current value. The drop at 50 milliamperes load is 144–30, or 114 volts. Dividing voltage by current, $\frac{144}{.05}$, gives the value of total internal resistance as 2280 ohms. This resistance is partly hypothetical, for it represents the overall value

from a.c. input to d.c. output. It might be called the "equivalent" internal resistance.

That part of the regulation which is due to the transformer is comparatively negligible. The secondary voltage of the transformer dropped from 116 to 114.5 when the output current was changed from 0 to 50 milliamperes; this corresponds to a regulation of 1.3 per cent., and an equivalent internal resistance of 38 ohms. The regulation due to the filter is dependent chiefly on the resistance of the filter chokes, and on the capacity of the first filter condenser, i.e., the one immediately after the rectifier. The curves of Fig. 4 show what effect the size of the first condenser has on the output voltage. At light loads, high-resistance, the voltage goes up sharply, then flattens out, as the capacity is increased. For any load, the voltage curve is flat at a value of 4 microfarads. This is a more or less standard value, so the contribution of the first condenser to the filter regulation may be neglected, and only the resistance of the chokes considered. This resistance can be measured with d.c., and was in this case 655 ohms. The equivalent resistance of the rectifier, is, by subtraction, 2280 — (655 + 38) = 1587 ohms. Thus, for this particular B device, the rectifier constituted the major cause of regulation. The individual curves of transformer, rectifier, and filter regulation shown give a good idea as to how the voltage drops are distributed. The total of the distances from the no-load-voltage line to these three curves gives the approximate overall regulation curve.

The problem of producing power units with good regulation is important. As a result several methods are at present in vogue whereby power units can be constructed with comparatively good regulation. The use of a glow tube in the output circuit of a power unit will cause the regulation of the unit to be excellent over the entire range of useful load. A group of curves taken on a power unit utilizing a glow tube are given in Fig. 5. Also it has been found that by decreasing the total resistance across the output of the power unit the regulation can be improved. When the total resistance R, Fig. 6, is reduced, the currents circulating through it, which represent a loss, are increased, but if the power unit has available sufficient capacity, this loss of current is advisable because it improves the regulation of the entire unit.

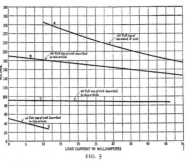

FIG. 5

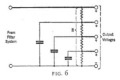

FIG. 6

A FANTASY ON SPONSORED PROGRAMS

By JOHN WALLACE

THIS is station KWOK broadcasting on a wavelength of four yards and six inches by authority of the Federal Boxing Commission. Station KWOK is loaned and saturated by the McSwif Stomach Pump Company and the Quebec Liquor Commission, Chicago Branch, and is situated on top of the Division Street gas holder, holder of high-grade gas. Every Wednesday evening at this time a program of detrimental music is brought to you by buggy from New York through the courtesy of Hamstein, Hoffstein, Snickelby, and Snootch, manufacturers of taxi cabs, doll cabs, bottle caps, and cogs, KWOK, Where Everybody's Sappy. Please stand by for the Eastern pronouncer.

This is WOOP, New York, broadcasting by special license of the city pound with a frequency that is positively dismaying. Every Monday evening at this time our facilities are engaged by the Hoffstein, Hammerstein, Snigelby and Scrooch Company, manufacturers of high grade taxi caps, doll cats, bottle cogs and coops, presenting the "Hof-hac-snack-co Hour" through WOOP together with 43 other stations, four precincts, three wards, and the juice of one lime. Stand by while your local station idemnifies itself.

This is KWOK, Chica-
This is WOOP New Yo- (just to give the boys in the switch yard a chance to do their stuff.)
This is KWOK
'Taint, it's WOOP
'Tis
'Taint
'Tis (O dear, this could go on forever, but at this point both stations join hands in a circle and the girls choose partners.)

You're on your back in the studios of WOOP the key station of the Duodecimal System. I will now tip the microphone over on Mr. Gregory Swallop advertising manager for the Hatstein, Hemstein, Hockelby and Pooch Company who will renounce the program. Mr. Swallop.

Good evening folks. Well once again the "Hic-haec-hoc-co Hour" is with you and I know how happy you are to hear from us and I

just know you are going to enjoy our program. I also know the color of your eyes—yes I do, take those handies down!—and how high up is, and the date of the second coming, and oh lots of other things. Seven days is an awfully long time to wait, isn't it, my ducks? Why of the 17,586,864,203 letters we received during the week 117 were from listeners who died because they couldn't wait. Ha, ha, they must have had jobs in a cafeteria! Ha, ha. Well, I will have my little joke. Well I know you are anxious to hear the program so I will not delay you any longer except to divulge to you that this hour of entertainment is donated to you gratis as a Christmas present, free, through the munificence of Messers —shhh! Come over closer—HINKLESTEIN, HOCKELSTEIN, SNOOPELBY and SNATCH. Don't tell! This high grade firm was foundered in 1898 and has been engaged for forty years in the manufacture of high grade taxi backs, doll craps, bottled cats, and bogs. Our taxi backs are equipped within and without with equillibrated non-actinic colloidal stradilators and sometimes with metaspheroid double-trussed oscillators—a reassuring thing to know, my doves, in case of eggs. Hinc-hoc-sno-co Brand Bottled Cats may be obtained at any grocery store, your money back if the color frays. Our bogs are shipped direct to you from the boggery, untouched by human hand.

Were you Mr. Listener, and you Mrs. Listener too, ever caught in the rain without a taxi cab? You got soaked didn't you? And if you'd of hired a cab you'd of got soaked too, wouldn't you? You should own your own. So I will now present to you the Old Soak who with his White Mule will spirit you away this evening to the Never-never Lands, there to hear lovely strains of music of foreign crimes, rendered for you by the Lard-Werks orchestra. Here take it you bum!

Orchestra plays Chopin's Funeral March

. . . and as the last strains softly strain o'er the distant strains we vault lightly from the saddle of good old Black Beauty while the carriage boy informs us that we bring to you tonight a pogrom of Hebrew heirs collected for you by our research department in the picturesque pathways of Palestine and Palisades Park. And now if you will extinguish your lights, turn off that damned radio, relax, returning to the squatting position at the count of five, we will enter the tonneau of the Brox Bros. Travelling Crane and —wissht!— We find ourselves on a sunny slope in Spain whilst from the half open door of a little tepee in the very shadow of Popocatapetl we hear the gentle warble of the Swiss Marine Band crooning plaintively to its young that lovely lullaby "On the Road to Mandalay-hay."

Orchestra plays the Prelude in C Sharp Minor.

Hello folksch, thisich me. An afore we gawan wisha program thorchestra sgonna play for you our musical trade mark the Funeral March so that whenevah you heah it, wherevah you may be, you may remembah the lovely, "Hing-hangsnig-co Hours" and recall that one of our bogs in your back yard means immediate death for your neighbors' chickens.

Boom! Boom! Boom! Three crashing chords on the piccolos followed by some thin tweedling sounds, four policemen and a cat.

This is WOOP—We Own Our Pants—broadcasting the Himpstein, Hinchbein, Snogerty and Snike program. Stand by please for your local anaesthetic.

This is KWOK broadcasting on a wa— (WOOP's switchman wins this chukker by seventeen words, receiving in award two kewpie dolls and a ham.)

This is WOOP my dears, Old Soak speaking. Our sturdy steed seems to have gone back to pasture and rubbing our eyes we find ourselves on a Kansas farm! Well I swan! And so now the boys are going to play for you a new derangement of that lovely piece by Saahnt Saahns called "The Swan." This is the first time to our knowledge that a female child under the age of fourteen has ever been choked to death over the radio and is presented to you by special arrangement with the composer. Saahnt Saahns was born in Moravia in 1792. He was the son of a sea cook and a bridge keeper's daughter. His great grandfather on the maternal side invented the poodle by interbreeding dachshunds and cotton batting so it is not surprising that the age of three found the gifted young Saahnt conducting his own works in the Ruhr district. There were no bond houses in that day so he decided to become a musician. He never got around to this, however, and in 1693 his unclad body was found in the Thames. Morticians discovered three measures of rye in his stomach and nineteen measures of music tatooed on the elevation nearest London Bridge. Thus was this immoral melody given to the world, which will now be played for you by Gilles de Rais on the flute. Mr. de Rais is the greatest living flautist. He was the first man to flaut across the English Channel. For your interest, his flute is a genuine Strad and is valued at $17.98 an inch except in leap year one cent more. Mr. de Rais will be accompanied by Miss Elva Orkney on the night boat. The Swan. . . .

LESTER PALMER
Chief announcer and program director of station wow, at Omaha

BOB HALL OF KOIL
Mr. Hall is chief announcer, studio director, baritone soloist, 'Uncle Josh' and director "Little White Church on the Hill" at this middle western station

Orchestra plays Tschaikowsky's Nutcracker Suite.

. . . and as this lovely Air for the G String concludes we discover that our old faithful White Mule has stumbled over the last bar and must be shot. (Three sobs followed by a cannon shot and screams of protest from the loud speaker). But hopping blithely on the bullet we find ourselves suddenly conveyed to Alsatia. " I'll Say She Does" will be the next number by the dance orchestra:

Orchestra plays Tschaikowsky's Nutcracker Suite.

And now folks, before this "Squawk-squack-co Hour" comes to a close we want to tell you about the elegant booklet we have prepared just for you. It is printed on décolleté vellum in an edition strictly limited to four million copies, each one signed in six places by the authors, Hipstein, Hopstein, Spiggoty and Speck, manufacturers of waxey jacks, boot jacks, model gats and fogs. It contains not only a copy of the Constitution, multiplication tables up to ten, a list of contributors to the Yale alumni fund of 1876, the population of the principal cities of Denmark and the Lives of the Saints, but also such valuable information as "Eighty-nine Appetizing Ways to Serve Coddled Bats," "How to Re-cog Your Baby," and "How To Make Attractive Lamp Shades, Bird Baths, and Pen Wipers Out of Old Taxi Backs." Write us and tell us how much you enjoyed our program and how it has brought cheer into your life —or leave out all that but anyway give us your name and address and the addresses of twenty of your friends who you think would also be interested in high-grade blueing. And now we conclude our program with the "Squawk-squack-co March."

Orchestra plays Tschaikowsky's Nutcracker Suite

This is WOOP. The program you have just heard was handed to you on a silver platter as the joint offering of Santa Claus, the Celestial Powers, and Hamstein, Hoffstein, Snickelby and Snootch, manufacturers of . . . hey, Joe, what the hell do those buzzards make anyhow?

The Possibilities of the Radio "Talk"

WE WERE pretty nearly won back to radio "talks" the other night after having at one time sworn off them for life. We happened to tune-in WJZ at 6 P.M. Central Time on a Tuesday and heard a Mr. Frank Dole holding forth on the Airedale. Mr. Dole's weekly dog talk, it seems, has been a regular WJZ feature for many years. He is known as an expert on dog life and is kennel editor of the New York *Herald-Tribune.*

However, we didn't know all this when we accidentally stumbled across his program and it had to survive or be tuned-out on its own merits. Our speaker was violating all the rules in the little hand book on elocution. He was dropping the "g's" on words ending in "ing" and mispronouncing some others. His speech could certainly not be described as fluent and was decidedly lacking in that first requisite of elocution —polish. Furthermore, he occasionally got tangled up in his words and would have to start a sentence over again. Some times, even, he paused—an infamous procedure in radio delivery, wasting the station's good time like that! And along toward the end of his dissertation he waxed sentimental—an oratorical device we

heartily abhor. But that's not all: to cap the climax he terminated his talk with some personal messages to friends of his in the old home town in Maine!

Certainly a sufficiently lengthy catalogue of faults to damn any speaker—in the face of which we stoutly maintain that the talk was one of the best we have ever heard on the radio.

Mr. Dole succeeded in that oft talked of, but seldom demonstrated, stunt, to wit: putting across personality. He spoke exactly as he would have if you had cornered him on the street and asked him to tell you something about Airedales. Each one of the "faults" enumerated above contributed to this impression of informality and the net effect was convincing—he sounded as if he *did* know something about Airedales. Your average radio speaker, given the same material, would have made it sound as if he had just looked up the subject in the Encyclopedia Britannica.

Which last remark sums up the general run of radio talks. Almost invariably they give the impression of having been dug out of some book for the occasion. The listener's reaction to such a

FRANK DOLE
Kennel Editor of the New York *Herald-Tribune* who is regularly heard through WJZ speaking on dogs, a subject which he handles remarkably well

delivery is invariably this—"why not look it up in a book myself? it's easier to read than to listen to," click!, and the would-be talker is switched off. The speaker has to add something to the words if they're to mean any more than the same stuff printed. Mr. Dole's addition to his mere subject matter was his indubitable love for dogs, which stuck out all through his dissertation. His was no speech performed for the sole purpose of filling out a fifteen minute radio program. He likes dogs; likes to talk about them and wants other people to like them. His interest in his subject was contagious and we suspect that he may even number non-dog owners in his audience.

Given other speakers with similar qualities of delivery and we don't know but that we would alter our opinions concerning the merits of the radio talk. If we could be assured that enough untutored speakers could be obtained, who would talk naturally and not dress their material up for the microphone, we would suggest an admirable series of radio talks—let some metropolitan station schedule a series of weekly ten minute talks, promising a new speaker, an authority on his subject, each week. Then as speakers get head waiters, bakers, subway guards,

mounted policemen, customs inspectors, window demonstrators, flag pole painters, information clerks, pan handlers, bell hops, ribbon clerks, bootleggers—or any others of the numerous people whose work gives them a unique slant on humanity, and with whom the average man doesn't ordinarily have an opportunity to be clubby. Let them tell the inside dope on their business in their own words.

Of course it would first be necessary for the director of that program to interview the prospective speakers to ascertain whether or no their garrulities would be interesting. He could question them in the course of an hour's interview about entertaining and intimate sidelights on their trade. During the course of the conversation he could keep a topical record of the interesting things that came up. Then the speaker could be furnished with this brief list of reminders, perhaps only five or ten sentences long, and told to go ahead and talk until his ten minutes were up.

The series would succeed or fail according to its convincingness. Absolutely no editing, other than by the method suggested above, could be indulged in, and the speaker would have to be encouraged to talk in his every day language. Any faking of material, or permission of exaggerations for the purpose of putting "punch" in the talk, would defeat its own end. The listeners —a suspicious lot—would immediately decide that the "confessions" were faked in their entirety and delivered by some actor-announcer.

But if the thing were honestly done it would be honestly convincing. Suppose a window washer, inexperienced at formulating his ideas, got "microphone fright" before his speech was two minutes under way, and the program had to be filled out with piano music—the catastrophe would only serve to build up the prestige of the whole series.

INTERESTING news from France, translated by ourself at the cost of much labor and consulting of dictionaries, for your delectation:

Acting upon the Radio Broadcasting decree issued on December 31, 1926, the constitution of a new organization, the "Radio-Diffusion Française," is being prepared by the qualified representatives of literary and artistic groups, of radio manufacturers and dealers, of the press, and of various associations interested in the development of radio.

The "Radio-Diffusion Française" proposes to act upon the suggestions and orders of the statute established by the government by bringing about broadcasts of such quality and interest as will be worthy of French thought, technique, and art.

Under the auspices of the "Radio-Diffusion Française," a complete broadcast of the performance of "La Traviata" at the National Opera House was transmitted by station Radio-Paris, with full power, on January 26, 1927. The opera was offered to the listeners by the "Grands Magasins du Printemps." (The which, as you know, is a department store in Paris).

A NEW day time schedule has been inaugurated by WJR. It is called the Musical Matinee program and lasts from 12:45 to 2 o'clock every day except Sunday. Dance music is alternated with concert music, and occasionally a solist is added. Dance music, is by Charles Fitzgerald and his Rhythm Kings, and light concert numbers by Jean Goldkette's Petite Symphony Orchestra.

MATCHING
R. F. COILS

By

F. J. FOX and R. F. SHEA

Engineering Dept., American Bosch Magneto Corporation

THE CRYSTAL-CONTROLLED MASTER OSCILLATOR
See Fig. 2 for the circuit diagram

THE trend toward simplicity of operation as manifest in the design of radio receivers during the last few years has rendered necessary a rapid development in the technique of production testing. During the time when single-control receivers were unknown, there was less necessity for precision test methods. It was desirable that the coils and tuning condensers be identical in order that the dials might read alike, but considerable tolerance could be allowed without affecting the efficiency of the receiver. It is quite evident that this cannot be permitted with modern receivers, where as many as five or six tuned circuits are coupled together and controlled from one central drive. In this case any appreciable departure from uniformity will reduce the efficiency of the receiver and, consequently, it becomes very necessary to develop and maintain elaborate inspection equipment in order to insure absolute uniformity of all the component parts of the tuned circuits. In this article it is our purpose to analyze the various means of testing and matching radio-frequency inductances.

Most of the methods of testing inductances at radio frequencies involve some variation of the well-known resonant circuit, wherein the coil to be tested is used in combination with a calibrated condenser as a wave trap. This wave trap may be attached to some sort of detector circuit or else it may be incorporated in an oscillator circuit. If the former is used, an oscillator must be used to supply power, so, as a rule, it is more economical to make the tuned circuit part of the oscillator circuit.

The type of apparatus to be designed for this purpose is determined by several considerations. If the test set-up is to be used for matching coils,

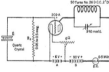

FIG. 2
The master oscillator circuit
with quartz crystal control

its frequency must be fixed and the tuning of the wave trap condenser must not change the power or the frequency of the oscillator. The apparatus must be designed for all possible accuracy consistent with quantity production, and, above all, it must be reliable and fool-proof. The accuracy required depends on whether the coils are to be sorted and then matched, or whether they are to be passed or rejected. If a small receiver using not more than three coils is being built, it is advisable to match coils. If, however, a large set is being manufactured, it is too expensive an operation to match five or six coils. In this case

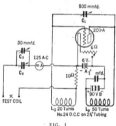

FIG. I
A setup developed for matching coils

it will be found more advisable to hold the coils within a certain percentage variation from a standard, this allowable variation being as much as can be tolerated without serious loss in receiver performance.

In Fig. 1 is shown a coil test set-up which was developed for the purpose of matching coils. This set-up utilizes a Hartley oscillator in connection with a wave trap circuit. L_g and L_p are the grid and plate coils respectively and C_1 is the main tuning condenser. The oscillator may be set at a frequency of about 1000 kc. A resistance of ten ohms is placed in series with the grid coil and the wave trap circuit is shunted across this resistance. The wave trap consists of a thermo-galvanometer in series with the coil to be tested, and a variable condenser, C_5, of such a size that

it will tune the coil to 1000 kc. Across the condenser is a vernier, C_4, of approximately 30 mmfd. capacity, and this vernier will indicate the amount that the test coil deviates from standard. The variable condenser, C_5, is so adjusted that the galvanometer will give a maximum reading when the vernier is set at 15 mmfd., or, let us say, a dial reading of 50, the standard coil being connected at X. If, now, a coil to be tested is placed across X it will usually necessitate a readjustment of the vernier in order to obtain a maximum galvanometer deflection. The amount of this deviation is an indication of the deviation of the coil's inductance from standard. Coils which give vernier readings between 40 and 60 on the dial may be passed since the deviation is, roughly, only plus or minus 1 per cent. By passing all coils coming within the limits and rejecting any which fall outside we can hold coils to as great a degree of uniformity as desired. Or if the coils are to be matched, they are labeled with the reading of the vernier and sorted, all coils bearing the same number being identical.

MATCHING THE TESTING DEVICE

WHILE the above system is extremely useful when small quantities of coils are being tested, it has several disadvantages when applied to quantity production. Chief among these is the difficulty of exactly duplicating two such testing devices. If coil matching is desired, we cannot match coils tested on one set-up with those tested on another set-up.

To overcome this disadvantage, a coil-testing method has been worked out with which it is possible to duplicate measuring set-ups and in this way handle a large volume with a minimum of expense and a maximum of efficiency.

This improved system employs a constant-frequency master oscillator and a number of auxiliary oscillators. The last-mentioned oscillators incorporate the test circuits. The master oscillator maintains the test frequency of 1000 kc. and all the other oscillators are adjusted to this frequency by means of a detector loosely coupled to each oscillator.

Fig. 2 shows a wiring diagram of the master oscillator using a quartz crystal for frequency control. The crystal is connected between the grid and filament of a vacuum tube and is shunted by a grid leak. The plate circuit is tuned by means of a coil and condenser in parallel, and the circuit will oscillate when the natural period

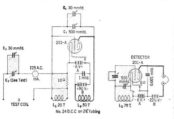

FIG. 3

The circuit of the auxiliary oscillator. A detector circuit is loosely coupled to it

of the crystal is found. A 5-milliampere d. c. meter is connected in the plate supply circuit and this serves to indicate the condition for maximum oscillation. When oscillation begins, the plate current decreases, and maximum oscillation corresponds to minimum plate current. It will be found best to operate with the tuning condenser set for a slightly lower capacity than that for best oscillation in order that oscillation will start when the switch is turned on. Crystals having an accuracy of $\frac{1}{20}$ of 1 per cent. may be obtained from the General Radio Company. Since only constant frequency is required, the crystals need not be ground accurately to any particular frequency, and hence may be obtained at a lower price.

The wiring diagram for the auxiliary oscillator is shown in Fig. 3. The oscillator is the same as that shown in Fig. 1 with the addition of a 30-mfd. vernier condenser, C_4, across the main tuning condenser, C_3, so that the latter may be set and locked and C_4 used to adjust this oscillator to the same frequency as that of the master. Loosely coupled to the auxiliary oscillator is a detector circuit consisting of a coil and condenser and a vacuum-tube detector. A pair of headphones is connected in the plate circuit. The auxiliary oscillator is tuned to the frequency of the master by the heterodyne or beat note method. In other words, the vernier is turned until the beat note heard in the headphones is lowered to "zero beat." The test circuit is placed across the ten-ohm resistance in the grid circuit in the same manner as shown in Fig. 1. This test circuit has the main condenser, C_2, and the vernier condenser, C_3, as before, the large main condenser being used to set the standard coil reading for maximum meter deflection when the vernier is set at mid-scale. The vernier condenser, C_4, is the only piece of equipment that has to be matched in each oscillator set-up. To facilitate this, the condenser is made of heavy plates with extra large spacing, and each condenser is measured on a capacity bridge before being used. This is necessary in order that coils tested on one set-up can be used with coils giving the same reading on another set-up.

A saving of one test set-up may be accomplished by adding to the crystal oscillator a coil test circuit. This is done by inserting a resistance in the plate or tuned circuit and shunting the wave trap circuit across it, as shown in Fig. 4. It is seen that the crystal oscillator is unchanged except for the introduction of the measuring circuit. This test circuit is identical with those used in the previously described oscillators with the exception that it is in the plate circuit instead of the grid circuit of the oscillator.

IN USING this arrangement the following procedure is employed. The crystal oscillator is set to a stable operating condition as described previously. A standard coil is now connected at X and the tuning condenser, C_2, is adjusted until maximum current flows in the test circuit when the vernier condenser, C_3, is set at mid-scale. Limits are set on either side of the vernier mid-scale by inserting coils which are known to be plus or minus the desired amount in inductance. There are a number of ways by which a standard may be obtained. A number of coils may be made in experimental production, and these may be measured on an inductance bridge. One of these is used which is representative of the average. Another method is to run a large number of coils through the coil tester and pick out a coil which is equal to the average. The choosing of limits is a difficult problem. It is possible to set up a radio-frequency amplifier and try coils which are plus or minus standard inductance and vary these limits until the radio-frequency gain is affected appreciably. These coils can then be placed in the tester and the maximum allowable limits determined. It is also possible to compute the limits which may be allowed and from this obtain the capacity variation which will tune these coils to resonance and thus obtain the allowable limits in terms of the vernier condenser capacity. Another and less accurate method is to choose limits from a study of the tests of a large number of coils. In this case the smallest limits possible, consistent with quantity production, may be selected.

All auxiliary oscillators are brought to the same frequency as that of the standard by the "beat frequency" method as described above. The oscillators are checked as frequently as possible, say every few hours.

The size of the tuning condenser (C_2 in the diagrams) necessary to bring the test coil to resonance will of course depend on the test frequency and the inductance of the coil to be tested. The proper value of C_2 can be easily computed from the following equation:

$$C_2 = \frac{1,000,000,000}{0.0395\ f^2\ L}$$

where C is in mmfd., L in microhenrys, and f in kilocycles. Thus, if the test coil has an inductance of 200 microhenrys and if the test frequency is 1000 kc., the capacity required is:

$$C_2 = \frac{1,000,000,000}{0.0395 \times 1,000,000 \times 200} = 127\ \text{Mmfd.}$$

A 150 mmfd. condenser will, therefore, easily serve the purpose.

A photograph of the crystal-controlled oscillator described above accompanies this article. The test coil may be seen in the test position. The eyelets on each end of the coil serve as coil terminals and these are also used to make contact with the tips provided on the test setup. The adjusting condensers are placed inside in order to prevent the operators from tampering with them. The dial on the panel controls the vernier, C_3. The limits are either marked on the dial or they may be posted near the operator.

A photograph of an auxiliary oscillator and detector combination is also shown, in this case the main adjusting condensers are inside the case. The vernier, C_4, used for frequency setting, may be adjusted by means of a screw driver from the outside. The detector coil is on the right-hand side and is placed about eight inches from the oscillator coil so that the coup-

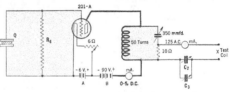

FIG. 4

A saving of one test setup is possible by combining the latter with the crystal oscillator

ling is small. The dial on the panel is for the vernier, C_4.

The system described in this article has been in operation in a well-known laboratory for some time and has given very excellent results.

THE AUXILIARY OSCILLATOR AND DETECTOR

The circuit diagram of this unit is shown in Fig. 3

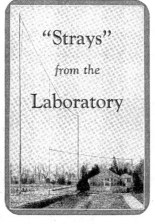

"Strays" from the Laboratory

New Tubes Mean Greater Economy

THE ARRIVAL on the open market of power tubes of the 112 and 171 types (the cx-312-A and cx-371-A, UX-112-A, and UX-171-A) which require but one-quarter ampere, is another step in the evolution of the vacuum tube. Now we hear rumors of new general purpose tubes of the 201-A type which will consume but one-eighth ampere, half the present tube's requirements, again reducing the power required by filaments to the low value of 5.25 watts for the average six-tube receiver. Years ago we used tubes which ate up our batteries to the tune of more than one ampere each—but in those days we did not have six-tube sets—and the new tubes, present and promised, mark one more step in a continual advance toward economy.

The 112-A type of tube, which is a quarter-ampere semi-power tube, has an amplification factor of 8 and an impedance of 5000 ohms, similar to the 210 power tube except that one cannot use plate voltages as high. This semi-power tube, in our opinion, is a tube which should be used more generally than is the 171 or 171-A tube. Two of them in parallel will deliver as much *undistorted* power as a 171 type on one-third the input a.c. voltage and with slightly less plate current drain from one's batteries. This business of requiring adequate volume with smaller input voltages is important to one who dwells over 100 miles from powerful stations. Here more voltage amplification is needed to bring signals to the volume level desired and, since the 171 type of tube with its low mu is essentially a local station power tube, that is, it requires large input signals, its advantage over the 112 type is not always apparent.

Two 112's in parallel, which is a better combination than push-pull when choke-coupled to the average low-impedance loud speaker, can be operated from B batteries economically. A single 171 tube with only 135 volts on the plate draws as much from the B batteries as two 112's in parallel with 157 volts on the plate and requires more than twice the input voltage to deliver less power output. In other words, weaker signals will deliver the volume required when 112 type tubes are employed.

The screened-grid tube is another step toward economy. Its filament operates from 3.3 volts and requires 0.132 amperes—the same filament as is used in the old and, we must say, rather poor, 120 type tubes—and delivers roughly three times the voltage amplification at high frequencies as the 210-A type of tube. Probably as many stages of amplification as we are accustomed to in r.f. circuits will be required with the screened grid tube to furnish sufficient selectivity, but the resultant input to the detector will be much greater than when our present general-purpose tubes are used.

Don't Overwork the A Battery

ONE OF THE most frequent complaints made to radio service men sounds like this: "My radio seems to be all right when I first turn it on, but then fades out. What is wrong?" This is invariably prefaced with the statement that all batteries are in good condition, i.e., that new B batteries are in use and that the A battery has been newly charged.

What is wrong? The answer in nearly every case is that the storage battery is run down, and

that if it is thoroughly charged trouble will disappear like dew in the morning. People using trickle chargers apparently fail to understand or appreciate the fallibility of all things mechanical or electrical. Over the week end, they work the radio for hours at a time giving it little chance to recharge from the trickle charger, until the little booster cannot hold up its end of the bargain longer. The battery is down and nearly out, the signals fade after a few minutes of the program, and the only thing left to do is to cart the battery to the nearest service station.

Better Loud Speakers Are Here

WE TAKE considerable pleasure in announcing that the loud speaker situation seems to be improving rapidly. For years we knew of no loud speaker that could be compared to the W.E. 540-AW, or the larger W.E. cones, either in sensitivity or in tonal range or in what we might call "intelligibility." This latter term is taken from telephone practice to mean the ability to distinguish the various instruments of an orchestra—when listening to it from a given loud speaker—or to recognize various consonants such as d, b, t, s, when spoken singly and without context, or to recognize the identity of an announcer. Many loud speakers convey sounds to us in a comparatively decent fashion, it seems, but when we try to distinguish the piano from a banjo, or to pick out the wood winds from the violins, we realize that something is wrong. All of the instruments seem "blended" together; we cannot distinguish one from another. We wonder how many loud speakers have been sold because the salesman, or the advertising told how carefully the loud speaker had been designed so that a proper "blending" had resulted?

We have listened to two cone speakers recently which will give the 540-AW a close race—the Fada 415-B and the B.B.L. Both of these are rather large cones, of about 24 inches diameter, and both are remarkably good. As a matter of fact they are so good that we have decided to

call upon all of the loud speaker manufacturers to send their instruments into the Laboratory, and to get to the bottom of the present loud speaker situation.

And speaking of loud speakers, it is our opinion that improving the loud speaker will spell trouble for the designers, manufacturers, and owners of a.c. sets. With a Balsa loud speaker which we operate with a Western Electric 540-AW unit out of a single 171 tube with about 160 volts on the plate, the average a.c. set is too noisy for pleasure although on other loud speakers the hum is inaudible. In other words, the a.c. tube either marks the limit of loud speaker development, or the newer and better loud speakers will force a.c. tubes to deliver signals unruffled by a.c. hum. We hope the latter, for the average loud speaker of to-day is less than five per cent. efficient. considering the entire audio band to be passed. Which prompts us to ask Mr. Burke, writing in the *General Radio Experimenter*, what he means when he states that most of the power delivered to a loud speaker by a power tube is transformed into sound waves and radiated. The average power radiated by man talking is about ten microwatts, and yet to get the same degree of loudness, we shove into the loud speaker 500 milliwatts of so-called undistorted power.

New Amateur Regulations

NEWSPAPERS credit the Radio Commission with following the advice of the A.R.R.L. in removing from the so-called 80-meter band the many amateur 'phone stations now existing, and opening up the 20-meter band for radiophone communication. In addition, the amateurs have from 1580 to 2000 kc. (150 to 190 meters) and a nice fat band between 56,000 and 64,000 kc. for communication by 'phone. This sounds fine until one realizes that the lower-frequency band has always been open for amateur 'phones—except when they caused disturbance to broadcast listeners, when they could not be operated until late at night—and the higher frequencies are in the so-called 5-meter band which is probably no good for code and worse for voice. The 80-meter band disturbed no B.C.L.'s, and was high enough up the wavelength scale that fairly good quality of speech could be transmitted.

But let us hasten to admit that we have had radiophone apparatus working in the Laboratory on as low as 3 meters and that, with 20 watts input we were able to hear the ticking of a watch held near the microphone as far away as one mile. We didn't experiment over greater distances.

The new international regulations, probably in effect by 1929, narrow the so-called 40-meter band, but leave the other bands about as before. In this 40-meter band all of the amateurs of the world who once could be found from 30 to 50 meters will be shoved, and instead of listening for South America and South Africa on 27.5 to 33 meters and England and France above our American band, we may find them mixed up with 8's and 9's which are so numerous in this country. We shall be thankful for one thing when these new regulations go into effect—they will automatically remove from the amateur bands the many high-power commercial stations, such as PW in France and several in Germany, which clutter up amateur communications that seem futile to commercial interests, and yet which paved the way into a band that will in ten years carry the bulk of the world's communication.

AS THE BROADCASTER SEES IT

By CARL DREHER

Technical Radio Problems for Broadcasters and Others

QUESTION 1: What is usually wrong with a loud speaker of which it is said: "It sounds all right on music, but not as good on speech?" How may the characteristics of the monitoring loud speaker affect the output of a broadcasting station?

ANSWER: The fault in such a case is usually loss of the high frequencies, causing speech to sound muffled, and even unintelligible if the cut-off point is low enough. In reality the defect is equally present in musical reproduction, but most people are more sensitive to this particular defect in speech. A loud speaker which cuts off on the high end at about 3000 cycles will sound manifestly "tubby" on speech to almost any observer, but many listeners tolerate it readily for music, some even praising the result as "mellow."

The ideal condition for judging the quality of a broadcasting station would naturally be to have the apparatus, including the monitoring equipment, flat over the whole audio band, and to have the same condition hold for the receiving sets. The broadcast operators could then be assured that the listeners would hear everything just as it was heard at the station. Under existing conditions, with many types of receivers and loud speakers in use, the best course for the broadcasting station to follow is the same as if all the receivers were good. Its own monitoring equipment, including the loud speaker, should respond impartially over a band between, say, 100 and 5000 cycles. If the loud speaker at the station is "drummy," or "down" in high frequencies, the operators may tend to over-emphasize this portion of the band in microphone placing, equalization of lines, etc. If, on the contrary, the station loud speaker is "tinny," or relatively lacking in low notes, there may be a preponderance of bass in the station's output without the operators being aware of the fact. While such effects may benefit some listeners whose receiving apparatus requires acoustic correction, it will result in distorted reproduction in both the good receivers and those which have the opposite fault relative to the monitoring circuits at the transmitter. Thus the characteristics of the station loud speakers are often an important element in the fidelity of reproduction attained.

QUESTION 2: What amount of energy is required for effective loud speaker operation under the usual listening conditions?

ANSWER: A well-designed cone loud speaker requires a telephonic level of plus 10 TU. This is roughly the maximum distortionless output of a tube of the 120 power type. Such a tube will de-

liver 110 milliwatts of undistorted output. On the basis of zero level equalling 10 milliwatts, and calculating from the formula for conversion from energy units into TU, which has been cited so often in this department that it need not be repeated, this corresponds to plus 10.4 TU. For broadcast transmitter monitoring it is preferable to use a tube affording more margin, say up to plus 15 TU (the 171 type or equivalent). On the other hand, if the loud speaker is improved in sensitivity 100 milliwatts of audio energy may be ample for its operation.

QUESTION 3: Why does a loud speaker sound "tinny" (lacking in bass tones) when the volume is decreased below the comfortable listening level?

ANSWER: The same effect occurs with natural sources of sound when the volume reaching the ear is decreased. The cause is found in the ear rather than in the source of sound. The ear is much more sensitive to relatively high-pitched sounds (say between 1000 and 4000 cycles) than to bass notes. For example, while the threshold of hearing in the favorable middle portion of the audio range, say at 1000 cycles, may be around minus 60 TU, at 100 cycles it may be no lower than minus 20 TU. When the volume is decreased the bass notes drop below the auditory threshold, so that the ear no longer responds to them, long before the treble disappears and all audition ceases. This phenomenon explains the effect observed. In some cases there may be contributing causes in the loud speaker mechanism.

Death Among the Broadcasters

TOO often we have had occasion, in this department, to discuss recent electrical fatalities among broadcast technicians and ways of avoiding such accidents in the future. It is admittedly impossible to render work with high-tension currents perfectly safe, and there will always remain an irreducible minimum of unavoidable mishaps. A transformer may break down without warning, admitting high voltage to a circuit where no one expects it, lightning surges may take place, psychological lapses sometimes upset the most carefully planned precautions. A fatal casualty in the last class occurred recently at the Daventry station of the British Broadcasting Company. The man killed was W. E. Miller, a Maintenance Engineer. The B. B. C. gives the following account in its announcement:

"Mr. Miller threw in a high-tension switch in

connection with the apparatus at 5 GB and a few minutes later was observed to lean over a guard rail apparently with the object of making an adjustment which should not have been undertaken with the switch on."

The death of Mr. Miller is stated to be the first in the five years of operation of the B. B. C.

The question in such a case is whether mechanical precautions can be elaborated to a point where a man will be protected against his own temporary unawareness of danger. In other words, must a sharp distinction be made between accidents due to circuit breakdowns and the like, and psychological failures resulting in death or serious injury? Knowledge of circuits appears in some cases to be no protection at all. As I have pointed out in previous articles on the subjects, some of the men, or rather boys, working on broadcast transmitters are altogether too young for dangerous jobs, but older men, it must be admitted, are sometimes no more fortunate. What, then, can be done to establish safety by machinery?

One device is to enclose all the high-voltage apparatus in a grounded cage with special doors. The doors may be opened by turning a wheel which cuts off the high tension inside the cage and grounds the plate circuits. The wheel cannot be turned back while the door is open. This comes close to being an absolute safeguard when only one man is involved. He cannot get into the dangerous portion of the station without cutting off the current. If another operator is involved, however, it may happen that one man is working on the apparatus while another, unaware of the fact, closes the door, locks it with the wheel, and burns up the operator inside the cage. A red tag system will obviate this, if it is faithfully followed—but where there is an "if," someone will get killed some day.

A similar device consists of circuit interrupters on doors in panels giving access to tubes from the front of the panel. When the door is opened, for replacement of a tube, or observation, the plate supply to that tube is cut off. But it is sometimes almost imperative to allow a man to observe the operation of the set, from a point behind the panel, while the circuits are energized. If that man starts to touch the equipment, he is on his way to the undertaker. Or, if he leans over a guard rail, as Mr. Miller did, he invites immediate death in the same way. Such a guard rail, it may be pointed out, has scarcely more than a symbolical value.

There are, in sum, two schools of design, working on opposite fundamental assumptions. One group contends that the best thing is to

311

leave the high-tension circuits open and accustom the operators to keeping away from them. If they realize fully, it is argued, that only their own care stands between them and death, they will be careful. The other group carries out the design on the assumption that human beings are irremediably fallible and must be safeguarded by forethought operating through mechanical devices. Probably one system fits some men best while others are safer under the opposite scheme. But unfortunately we do not select personnel on the basis of temperament in such details, nor do we possess adequate psychological data which would enable us to control accidents by such means.

It is in this general direction that I would suggest study and thought. Personally I am not convinced of the superiority of either of the design systems outlined above. There is no doubt that much can be done with mechanical safety devices. One has only to consider the low ratio of accidents in a well conducted vacuum-tube factory to realize that. With highly dangerous potentials distributed through such a plant, and most of the workers on the forty-cents-an-hour level, it is possible practically to eliminate accidents by careful safety engineering and attention to orderly procedure and cleanliness. Nevertheless, with highly-trained technical personnel the margin of safety may be as great at the end of a hooked stick, even with the live parts of the set accessible, if a special effort is made to control the psychological factors. At army aviation fields the pilots are frequently under the scrutiny of surgeons who are expected to detect anomalies of sight and hearing, chronic fatigue, and emotional disorders in time to prevent avoidable accidents. The engineer in charge of a broadcast transmitter should watch his men in somewhat the same way. Usually he is in a position to know when anyone on the staff is in some personal difficulty which might interfere with his ability to keep out of danger. Psychiatrists can cite cases where men working in factories have injured themselves as a result of lack of normal coördination clearly traceable to domestic troubles and other emotional disturbances. In a broadcasting station the part of wisdom would be to detach a man in such a condition from transmitter duty and place him temporarily in the control room or wherever he would be safe.

But, aside from such acute cases, preventive measures of a psychological nature may also be applied at intervals. After the monthly resuscitation drill for members of the staff a discussion may be started on accidents which have occurred in the experience of the men, and how they might have been prevented. This will direct attention to the problem and may in a measure preclude the contemptuous familiarity with the apparatus into which technicians are apt to fall

after nothing has happened for a period. There is no danger of making radio men morbidly conscious of such danger as exists. Their faults are in the opposite direction; they are not afraid enough. Yearly medical examinations, with special attention to the condition of the heart, are very desirable, in that a man with a sound heart who is caught on a high-tension circuit usually has a good chance for his life if he is released quickly, whereas a cardiac case is snuffed out in the first few seconds. To compel or allow men to work excessively long hours is plain criminality on the part of the employer.

HOW ULTRA-VIOLET RAYS AID PHOTOGRAPHY

Ultra-violet photomicrographic apparatus in use at the Bell Telephone Laboratories, New York. The lenses of the apparatus are of quartz and the ultra-violet rays are furnished from an arc light using rods of cadmium or magnesium instead of carbons. A system of quartz prisms, in front of the lamp house, permits the operator to "tune-in" on one particular frequency of the invisible rays

Antennas from 1913 to 1927

BACK in 1913, when I was just getting out of preparatory school, the Marconi Company was engaged in erecting a series of transmitting and receiving stations for transoceanic communication. The receiving stations on the Atlantic seaboard were located at Chatham, Massachusetts, and Belmar, New Jersey. At each of these points the placid skyline of the countryside was broken by a string of four-hundred foot iron masts, mounted on cement emplacements, and guyed to anchorages in the surrounding fields. Much money went into those hollow masts—and never came out again. Within a year the development of the vacuum tube had reached a point where an amplifier, fed from an antenna of moderate height, would produce as readable a signal from over the ocean as the

mile-long antenna strung on four-hundred foot masts. The four-hundred footers were taken down and stored away. When, four years later, I viewed the cement blocks still set in the earth, I reflected with amusement that the engineers of the Marconi Company had been bad prophets, and, with the superficial confidence of a young man, I probably thought I could have done better had I stood in their shoes when the decision was to be made.

In 1923, ten years after that die was cast, I left the Riverhead transatlantic receiving station of the Radio Corporation of America, having served my time as a receiving engineer. The wave antenna at Riverhead, a type developed by Harold H. Beverage, stretched to the southwest over miles of Long Island sand, through forests of scrub pine, and oak, almost to the sea. It did its stretching on thirty-foot telephone poles. It was a good antenna, one which was kind to European signals and not at all kind to the static which came from the opposite direction. Many more such antennas have been built since that time, on the same sort of poles. When I left Riverhead, had anyone told me that transoceanic receiving antennas would ever be built otherwise, I should have expressed polite doubts. Had my informant added that steel towers would rise in Riverhead almost as high as the Marconi masts at Belmar, before I returned to the town, I should have expressed doubts not nearly as polite. Had he stated finally that there would be five of them, with ninety-foot cross arms at the top, I should have taken him, I am afraid, for a harmless lunatic whose aberration led him to imagine towers instead of the more usual pink elephants or snakes.

Another four and one-half years flit past. The short-wave explorations of the ether (if you care to assume one) begin, and soon reach formidable portions. A short-wave receiving system is developed in which vertical antenna wires hang from steel crossarms which, curiously, are supported by steel towers three hundred feet high, while at the other ends of the crossarms reflector wires sweep gracefully to the earth. The wave antennas are still at Riverhead, combined and phased for even greater efficiency, and even more inconspicuous, for five of the three-hundred foot towers have come to keep them company. The fact that none of us expected them did not keep them from coming. They obeyed a different sort of logic than that which ground out conclusions in our brains. They followed the innate logic of invisible oscillations propagated through space and the laws of the materials which men use in communicating over the distances separating one continent from another. And so, for the moment, they stand proudly over the scrub pine woods.

LEARN RADIO and find

Good Pay from the Start Rapid Advancement, Glorious Adventure and Phenomenal Success in A Life Profession of Fascinating Brain-work.

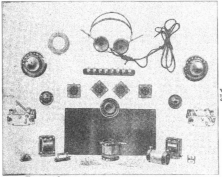

With these parts you can build 12 different circuits.

The Peerless Signagraph for code instruction.

Free with course, all this first-quality equipment for experimental work!

You can learn at home!

R. L. DUNCAN, Director, Radio Institute of America. Author of several volumes on radio

Here is your big opportunity! Radio pays hundreds of millions in salaries each year. In a few years the industry has progressed from almost nothing to one of the most important in the world. And the big demand for trained men continues in all the branches of radio. *Are you going to plod along at a thirty-five dollar a week job when REAL MONEY is waiting for you in radio?*

Our graduates are earning big money as radio designers, as executives with large radio organizations, in broadcasting work, as skilled mechanics, assemblers, servicemen and radio dealers. We have trained thousands of men to become successful radio operators on ships traveling to far corners of the globe where they meet excitement and adventure—to become radio operators in shore stations, sending and receiving radio traffic with countries across the two oceans. And now Opportunity is knocking at your door.

A Brand-new Course Offered by the World's Oldest Radio School

After years of experience the Radio Institute of America has evolved a new radio course—the most up-to-date of any offered today. It starts with the fundamentals of radio and carries you

RADIO INSTITUTE OF AMERICA
Dept. E-2   326 Broadway, New York City

through the most advanced knowledge available. The work has been prepared in simplified form by men who have written many volumes on radio.

Radio Institute of America backed by RCA, G-E and Westinghouse

Conducted by the Radio Corporation of America and enjoying the advantages of RCA's associates, General Electric and Westinghouse, the Radio Institute of America is equipped to give—and does give—the finest radio instruction obtainable anywhere in the world.

Home Study Course

Moreover you need not sacrifice your present employment for you can STUDY AT HOME during your evenings and other spare time. Thousands have successfully completed RIA training and have advanced to important radio positions. So can you with this new course.

Just Off the Press

This new catalog describing the course is just coming off the press. If you want to learn more about the lucrative and fascinating profession of radio send the coupon now for your copy.

The *Radio Broadcast*
LABORATORY INFORMATION
SHEETS

THE RADIO BROADCAST Laboratory Information Sheets are a regular feature of this magazine and have appeared since our June, 1926, issue. They cover a wide range of information of value to the experimenter and to the technical radio man. It is not our purpose always to include new information but to present concise and accurate facts in the most convenient form. The sheets are arranged so that they may be cut from the magazine and preserved for constant reference, and we suggest that each sheet be cut out with a razor blade and pasted on 4″ x 6″ filing cards, or in a notebook. The cards should be arranged in numerical order. In July, 1927, an index to all Sheets appearing up to that time was printed.

All of the 1926 issues of RADIO BROADCAST are out of print. A complete set of Sheets, Nos. 1 to 88, can be secured from the Circulation Department, Doubleday, Doran & Company, Inc., Garden City, New York, for $1.00. Some readers have asked what provision is made to rectify possible errors in these Sheets. In the unfortunate event that any such errors do occur, a new Laboratory Sheet with the old number will appear.

—THE EDITOR.

No. 161   RADIO BROADCAST Laboratory Information Sheet   **February, 1928**

Comparing the 112, 171, and 210 Type Tubes

THEIR RESPECTIVE OUTPUTS

ON LABORATORY SHEET No. 162 are shown three curves that indicate an interesting relation between the three most common types of power tubes, i.e., the 112, 171, and 210 types. The curves indicate the relation between the power output of the tubes and the value of the signal voltage impressed on the grid. The plate impedance and amplification constants of the 112 and 210 type tubes are practically identical and, therefore, the curves for these two tubes coincide from zero up to that point corresponding to the maximum output power of the 112, which is approximately 195 milliwatts, or 0.195 watts.

If a vertical line is drawn at any point on the curve, for example, at A, the points at which this line crosses the various curves will indicate the power output obtained from the particular tube associated with the curve being examined. In this particular case, line A, drawn at the point corresponding to a signal voltage on the grid of 15 volts indicates that, with this value of signal voltage, the power output of a 210 tube with 425 volts on the plate is approximately 0.34 watts. The power output of a 171 at the same point is approximately 0.1 watts. The maximum grid voltage that can be impressed on a 112 without resultant output distortion is about 10.5 volts and, therefore, a 112 tube cannot be used if the signal input voltage is greater than this value. At B we have drawn

another line corresponding to a signal on the grid of 8 volts. Here we find that the power output of a 112 is approximately 0.1 watts and the power output of a 171 about 0.04 watts. It is therefore evident that at low values of input voltage a 112 tube is capable of putting more power into the loud speaker than is a 171. If the signal voltage, however, is in excess of 10½ volts, the 112 cannot be used and the choice then lies between the 210 and the 171. The curves indicate that the 210 will give much more power output than a 171 but it should be realized that much greater plate voltages are necessary on the 210 than on the 171. With 180 volts on the plate the 171 can deliver approximately 740 milliwatts of power, but 250 volts on the plate of the 210 will only permit this tube to handle signal voltages up to 18 volts and the maximum output power will be only 460 milliwatts. From these data the following conclusions can be arrived at:

(1.) For input signals on the grid of the power tube of 10 volts or less the 112 tube will deliver the most power to the loud speaker.

(2.) When more power output is required and only moderate plate voltages are available (not in excess of 200 volts) a 171 is capable of giving greater output than can be obtained from a 210 under similar conditions of plate voltage.

(3.) Where high plate voltages around 400 volts are available the 210 should be used and under the same input signal it will give approximately 2½ times as much power as can be obtained from a 171.

No. 162   RADIO BROADCAST Laboratory Information Sheet   **February, 1928**

112, 171, and 210 Tube Curves

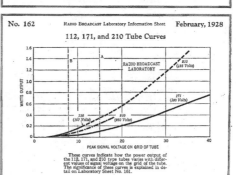

These curves indicate how the power output of the 112, 171, and 210 type tubes varies with different values of signal voltage on the grid of the tube. The significance of these curves is explained in detail on Laboratory Sheet No. 161.

Not What We Say— But what they DO!

That's the Secret of

REMLER

Parts Popularity

REMLER *Twin-Rotor* CONDENSER

Both sets of plates are entirely insulated from the dial shaft and both sets of plates rotate. Body capacity effects are entirely absent. Insulation is of genuine bakelite. Because of the high ratio of maximum to minimum capacity, the condenser will cover an unusually great wavelength range. S. L. Frequency and S. L. Wavelength Types. Capacities .0003S; .0005 and .0001.

Price. . . . $5.00

PERHAPS they are too Good. . . . It would be so easy to cheapen them here and there, to rush them out without the final certainty of maximum performance which is built into every Remler part.

REMLER *Drum* DIAL

The Remler Drum Dial permits quiet velvet smooth, vernier control of any type of condenser. The drum is 15 inches in circumference and is divided into 200 divisions—two for each broadcast channel. Six-volt lamp and bracket are supplied for illumination. It will drive either a single or gang condenser and is adaptable to either right or left-hand drive.

Price. . . . $4.50

But we know we are on the right track when we make every Remler part just as if it were a precision instrument. For 10 years, Remler parts have set the pace as quality radio products. No part has even been cheapened to meet a competitive price. Precision standards govern every operation from the purchase of raw material to final testing of each part. These standards are your guarantee of satisfaction.

REMLER
Division of
GRAY & DANIELSON MFG. CO.
260 FIRST STREET, SAN FRANCISCO

CHICAGO   Eastern Warehouse: Elkhart, Indiana   **NEW YORK**

No. 163 — Radio Broadcast Laboratory Information Sheet — February, 1928

Testing Receivers

USING THE MODULATED OSCILLATOR

THE accurate determination of the characteristics of a radio receiver requires a careful laboratory test, but it is possible to construct comparatively simple apparatus of much practical value for the testing and repairing of receivers. The instrument that will enable us to make such tests is the modulated-oscillator. From a modulated oscillator we can obtain an audio-frequency tone which can be fed into the input of the audio amplifier in a radio receiver and the functioning of the audio amplifier thus checked, or by turning on both r.f. and a.f. oscillators we can obtain a modulated wave which can be used to test both the r.f. and a.f. circuits.

The circuit diagram of a modulated-oscillator will be found on Laboratory Sheet No. 164. The following paragraphs will explain how to use the instrument for testing receivers.

(1.) Audio Amplifiers

Place all the tubes in and connect all the batteries to the amplifier. Do not place the detector tube in its socket. Connect the plate terminal of the detector tube socket to audio output terminal No. 1 on the modulated oscillator. Connect both the B +

detector lead on this receiver and terminal No. 2 on the modulated oscillator to B — on the receiver. Turn on the receiver and audio circuit of the modulated oscillator and adjust potentiometer P to give an output of medium intensity from a loud speaker connected to the output of the audio amplifier. A defect in the amplifier is indicated if the output is low or distorted or both.

(2.) Radio-Frequency Amplifiers.

A test of the r.f. amplifier of a receiver is accomplished by first placing all the tubes in the receiver and connecting all the batteries, and then winding about two turns of insulated wire around the coil on the oscillator, connecting the other end of this wire to the antenna post on the receiver. The oscillator should be located about ten feet away from the receiver. If the a.f. and r.f. tubes in the modulated oscillator are turned on and the receiver's tuning circuits adjusted to resonance, an audiofrequency tone should be audible in the output. Since the a.f. amplifier in the receiver was tested previously, any defect in the operation of the receiver must be located in the r.f. amplifier or detector circuit.

No. 164 — Radio Broadcast Laboratory Information Sheet — February, 1928

A Modulated-Oscillator

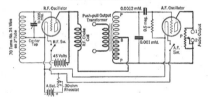

All the constants of the apparatus used in the instrument are given on the diagram. Some information on the use of this instrument will be found on Laboratory Sheet No. 163. The frequency of the audio-frequency oscillations can be varied by using various values of capacity across the push-pull transformer.

No. 165 — Radio Broadcast Laboratory Information Sheet — February, 1928

Audio Amplification

GENERAL CONSIDERATIONS

AN AUDIO system can be considered satisfactory if it amplifies the signals impressed on its input sufficiently to operate adequately a loud speaker and does so without distorting the signals to an extent sufficient to become apparent in the output of the loud speaker. Such performance can only be realised when the amplifier has been correctly designed and is operated properly.

The overall frequency characteristic of an amplifier is frequently quite dissimilar to the characteristic of a single stage. This is especially true of transformer- or impedance-coupled amplifiers and is probably due, in most cases, to coupling in the plate supply. Regenerative effects are thereby introduced into the circuit, which may produce considerable changes in the frequency characteristic of the audio system. Such effects are also present, at times, in resistance-coupled amplifiers and generally cause such an amplifier to "motor-boat."

The solution of such difficulties is either to design the amplifier so that with the regenerative effect present the system has a flat characteristic or to

design two units to have a flat characteristic and then arrange the circuit so carefully that regenerative effects will not be present. This necessitates feeding all the grid and plate circuits through resistances or choke coils and bypassing all the circuits with condensers.

Some recent audio transformers are designed to have a fairly sharp cut-off at about 5000 cycles to reduce the effect of various extraneous sounds, such as tube noise, high-frequency heterodyne whistles, etc., which are composed mostly of frequencies above 5000 cycles. Frequencies above this value add little to the quality of the speech or music and can therefore be eliminated without introducing noticeable distortion. It is doubtful whether the majority of broadcasting stations themselves transmit notes of more than 5000 cycles in frequency.

Also many amplifiers have a tendency to oscillate at very high audio frequencies and sometimes at supersonic frequencies. If the amplifier is designed, however, to give little or no amplification to frequencies much above 5000 cycles, this tendency of the amplifier to oscillate will be nullified.

No. 166   Radio Broadcast Laboratory Information Sheet   **February, 1928**

Acoustics

DAMPING AND REVERBERATION

THE quality of reproduction from any loud speaker depends to a considerable extent upon the room in which it is used and upon the room's furnishings. The reason why the room and its furnishings influence the output of the sound generator, whether it be a piano, phonograph, or a loud speaker, is not difficult to understand, and will be explained briefly here.

In an average room the sounds from a piano, for example, are somewhat damped by the hangings, carpets, furniture, etc., so that they decrease to inaudibility quite rapidly. When the furniture, rugs, etc., are removed and the piano is permitted to stand on the bare boards, the sounds from it will be prolonged and the music will become jumbled, especially when playing forte. This effect is due to the absence of the furniture, which normally acts as a damping agent, and also due to the fact that the piano is resting directly on the floor so that the latter acts to increase the effective area of the sound board. The sounds produced by the piano when it is in direct contact with the floor will be somewhat louder than usual, indicating increased efficiency.

Under any given room conditions the rate at which a sound dies away is the same whether the sound at its beginning is loud or soft. However, the time taken for the sound to become inaudible depends upon the loudness of the original sound and, of course, the louder the sound, the longer it will take to decrease in volume to a point where it is inaudible. In a room containing furnishings that cause considerable damping, we may, therefore, play much louder than in an unfurnished room without causing any excessive blurring.

A room can be too completely damped, when the playing will sound "dead." A certain amount of blurring or intermingling of succeeding chords is considered good, for it adds coloration to the music.

The importance of these matters in relation to the design of the studios in broadcasting stations is evident. The correct amount of damping must be obtained to prevent deadening the music (too much damping) or to obviate difficulties due to reverberation (too little damping.)

No. 167   Radio Broadcast Laboratory Information Sheet   **February, 1928**

Resonant Circuits

GAIN AND SELECTIVITY

THE current at resonance in a tuned circuit is equal to the voltage induced in the circuit divided by the total resistance of the circuit. The actual capacity of the condenser or the inductance of the coil used in the circuit do not enter into the calculation once the induced voltage and the resistance are known. The voltage across the coil in the circuit is equal to a constant times the current times the inductance of the coil and, the voltage is, therefore, larger the greater the inductance of the coil. Since a vacuum tube is a voltage rather than a current operated device, it might appear that best results, *i.e.*, greatest amplification, would be obtained by making the coil very large. When we increase the inductance of a coil by adding to the number of turns, however, we also increase the resistance and the increase in resistance nullifies to some extent the advantage gained through the use of a larger coil.

The selectivity of a tuned stage in a receiver depends upon the series resistance of the circuit; with low-resistance circuits the selectivity is good while with high-resistance circuits the selectivity is poor. The curves on this Sheet indicate the effect of resistance in the tuned circuit. Curve 1 shows the characteristic of a very low-resistance tuned circuit and curve 2 a comparatively high-resistance circuit. Since practically all of the resistance in a tuned circuit is in the coil, it follows that carefully constructed, fairly "low-loss" coils should be used in a radio-frequency circuit. A coil can be made so good as to cut "side-bands," however, and thereby distort the received signal head-band suppression results in the loss of the high audio frequencies in the modulated wave.

If the ratio of the inductive reactance of the coil (6.28 times the frequency times the inductance of the coil) to the radio-frequency resistance of the coil at the same frequency is made much more than 250, distortion of the "side-bands" results.

No. 168   Radio Broadcast Laboratory Information Sheet   **February, 1928**

The Ear

ITS CHARACTERISTICS

THE characteristics of the human ear have been determined and investigated by many different scientists, and some of these characteristics are given below:

(a.) There is a minimum sound intensity below which the ear cannot detect any sounds. A curve was published on Laboratory Sheet No. 109, indicating how this minimum audible intensity varied with frequency.

(b.) There is a maximum intensity of sound above which the auditory sensation is one of pain rather than sound. The intensity and its variation with frequency was also explained on Lab Sheet No. 109.

(c.) There is a lower limit of the pitch of a sound below which the ear will not respond. This lower limit is about 20 cycles.

(d.) There is an upper limit to the pitch of a sound above which the ear will not respond. The upper limit is about 20,000 cycles.

(e.) The ear can distinguish between about 300,000 separate sensations of sound.

(f.) The ear can respond to pressure changes between the pressure required to produce a minimum audible sound and a pressure 100 million times greater. These two pressures correspond to an energy ratio of 10,000 million.

(g.) The ear can distinguish between the loudness of various sounds. At low levels of sound intensity a change of about 25 per cent. is necessary to be distinguishable. At greater intensities a change of 10 per cent. in loudness is detectable by the ear.

(h.) The ear can distinguish between the pitch of various sounds. At medium frequencies a change in frequency of about 0.3 per cent. can be detected; at low frequencies a change of about 1 per cent. is necessary.

A knowledge of these characteristics is useful to the student interested in problems of sound reproduction.

The Advanced "Hi-Q SIX"—the greatest of all Hammarlund-Roberts Receivers—the culmination of years of concerted effort to produce radio's finest instrument, regardless of cost. As beautiful as it is efficient, yet costs only $95.80 for complete approved parts

This Perfect Receiver
CUSTOM-BUILT *for less than* $100

IMAGINE A CUSTOM-BUILT Receiver—designed by ten of America's leading Manufacturers—incorporating latest modern constructional features and America's very finest parts—*and costing you only $95.80!*

The new advanced "Hi-Q SIX" is more than a radio receiver—it is a marvelous musical instrument—a set that produces maximum and uniform amplification over the entire tuning range and that completely eliminates oscillation. These exclusive features — plus four isolated tuned stages — plus symphonic audio amplification and a power tube—result in

the faithful reproduction of all musical frequencies with the full, natural tone quality that radio engineers have sought for years.

You can build the "Hi-Q SIX" yourself and save at least $100.00. Simply get our complete Constructional Manual; buy the approved parts and our Foundation Unit, which contains chassis, shields, panels, all special hardware, etc. Manual contains 48 pages of construction data. Complete description, charts, diagrams and photos. Anyone can follow it and build the de luxe "Hi-Q SIX". 25c from your dealer or direct from us.

Hammarlund ROBERTS Hi-Q SIX

HAMMARLUND-ROBERTS Inc., Dept. A, 1182 Broadway, New York

Associate Manufacturers

Manufacturers' Booklets

A Varied List of Books Pertaining to Radio and Allied Subjects Obtainable Free With the Accompanying Coupon

READERS may obtain any of the booklets listed below by using the coupon printed on page 328. Order by number only.

1. FILAMENT CONTROL—Problems of filament supply, voltage regulation, and effect on various circuits. RADIALL COMPANY.
2. HARD RUBBER PANELS—Characteristics and properties of hard rubber as used in radio, with suggestions on how to "work" it. B. F. GOODRICH RUBBER COMPANY.
3. TRANSFORMERS—A booklet giving data on input and output transformers. PACENT ELECTRIC COMPANY.
4. CARBORUNDUM IN RADIO—A book giving pertinent data on the crystal as used for detection, with hook-ups, and a section giving information on the use of resistors. THE CARBORUNDUM COMPANY.
7. TRANSFORMERS AND CHOKE-COUPLED AMPLIFICATION—Circuit diagrams and discussion. ALL-AMERICAN RADIO CORPORATION.
9. VOLUME CONTROL—A leaflet showing circuits for distortionless control of volume. CENTRAL RADIO LABORATORIES.
10. VARIABLE RESISTANCE—As used in various circuits. CENTRAL RADIO LABORATORIES.
11. RESISTANCE COUPLING—Resistors and their application to audio amplification, with circuit diagrams. DEJUR PRODUCTS COMPANY.
12. DISTORTION AND WHAT CAUSES IT—Hook-ups of resistance-coupled amplifiers with standard circuits. ALLEN-BRADLEY COMPANY.
15. B-ELIMINATOR AND POWER AMPLIFIER—Instructions for assembly and operation using Raytheon tube. GENERAL RADIO COMPANY.
15A. B-ELIMINATOR AND POWER AMPLIFIER—Instructions for assembly and operation using an A. C. A. rectifier. GENERAL RADIO COMPANY.
16. VARIABLE CONDENSERS—A description of the functions and characteristics of variable condensers with curves and specifications for their application to complete receivers. ALLEN D. CARDWELL MANUFACTURING COMPANY.
17. BAKELITE—A description of various uses of bakelite in radio, its manufacture, and its properties. BAKELITE CORPORATION.
19. POWER SUPPLY—A discussion on power supply with particular reference to lamp-socket operation. Theory and constructional data for building power supply devices. ACME APPARATUS COMPANY.
20. AUDIO AMPLIFICATION—A booklet containing data on audio amplification together with hints for the constructor. ALL AMERICAN RADIO CORPORATION.
21. HIGH-FREQUENCY DRIVES AND SHORT-WAVE WAVE-METERS—Constructional data and application. BURGESS BATTERY COMPANY.
25. AUDIO-FREQUENCY CHOKES—A pamphlet showing positions in the circuit where audio-frequency chokes may be used. SAMSON ELECTRIC COMPANY.
27. RADIO-FREQUENCY CHOKES—Circuit diagrams illustrating the use of chokes to keep out radio-frequency currents from definite points. SAMSON ELECTRIC COMPANY.
28. TRANSFORMER AND IMPEDANCE DATA—Tables giving the mechanical and electrical characteristics of transformers and impedances, together with a short description of their use in the circuit. SAMSON ELECTRIC COMPANY.
29. BYPASS CONDENSERS—A description of the manufacture of bypass and filter condensers. LESLIE F. MUTER COMPANY.
50. AUDIO MANUAL—Fifty questions which are often asked regarding audio amplification, and their answers. AMERITRAN SALES COMPANY.
51. SHORT-WAVE RECEIVER—Constructional data on a receiver which, by the substitution of various coils, may be made to tune from a frequency of 16,660 kc. (18 meters) to 1999 kc. (150 meters). SILVER-MARSHALL, INCORPORATED.
52. AUDIO QUALITY—A booklet dealing with audio-frequency amplification of various kinds and the application to well-known circuits. SILVER-MARSHALL, INCORPORATED.
56. VARIABLE CONDENSERS—A bulletin giving an analysis of various condensers together with their characteristics. GENERAL RADIO COMPANY.
57. FILTER DATA—Facts about the filtering of direct current supplied by means of motor-generator outfits used with transmitters. ELECTRAD SPECIALTY COMPANY.
59. RESISTANCE COUPLING—A booklet giving some general information on the subject of radio and the application of resistors to a circuit. DAVEN RADIO CORPORATION.
61. RADIO-FREQUENCY AMPLIFICATION—Constructional details of a five-tube-receiver using a special design of radio-frequency transformer. CAMPFIELD RADIO MFG. COMPANY.
63. FIVE-TUBE RECEIVER—Constructional data on building a receiver. AERO PRODUCTS, INCORPORATED.
64. AMPLIFICATION WITHOUT DISTORTION—Data and curves illustrating the use of various methods of amplification. ACME APPARATUS COMPANY.
66. SUPER-HETERODYNE—Constructional details of a seven-tube set. C. G. EVANS COMPANY.
70. IMPROVING THE AUDIO AMPLIFIER—Data on the characteristics of audio transformers, with a circuit diagram showing where chokes, resistors, and condensers can be used. AMERICAN TRANSFORMER COMPANY.
72. PLATE SUPPLY SYSTEM—A wiring diagram and layout plan for a plate supply system to be used with a power amplifier. Complete directions for wiring are given. AMERTRAN SALES COMPANY.
80. FIVE-TUBE RECEIVER—Data are given for the construction of a five-tube tuned radio-frequency receiver. Complete instructions, list of parts, circuit diagram, and template are given. ALL-AMERICAN RADIO CORPORATION.
81. BETTER TUNING—A booklet giving much general information on the subject of radio reception with specific illustrations. Primarily for the non-technical home constructor. BREMER-TULLY MANUFACTURING COMPANY.
82. SIX-TUBE RECEIVER—A booklet containing photographs, instructions, and diagrams for building a six-tube shielded receiver. SILVER-MARSHALL, INCORPORATED.
83. SOCKET POWER DEVICE—A list of parts, diagrams,

and templates for the construction and assembly of socket power devices. JEFFERSON ELECTRIC MANUFACTURING COMPANY.
84. FIVE-TUBE EQUAMATIC—Panel layout, circuit diagrams, and instructions for building a five-tube receiver, together with data on the operation of tuned radio-frequency transformers of special design. KARAS ELECTRIC COMPANY.
85. FILTER—Data on a high-capacity electrolytic condenser used in filter circuits in connection with a socket power supply units, are given in a pamphlet. THE ABOX COMPANY.
86. SHORT-WAVE RECEIVER—A booklet containing data on a short-wave receiver as constructed for experimental purposes. THE ALLEN D. CARDWELL MANUFACTURING CORPORATION.
88. SUPER-HETERODYNE CONSTRUCTION—A booklet giving full instructions, together with a blue print and necessary data, for building an eight-tube receiver. THE GEORGE W. WALKER COMPANY.
89. SHORT-WAVE TRANSMITTER—Data and blue prints are given on the construction of a short-wave transmitter, together with operating instructions, methods of keying, and other pertinent data. RADIO ENGINEERING LABORATORIES.
90. IMPEDANCE AMPLIFICATION—The theory and practice of a special type of dual-impedance audio amplification are given. ALDEN MANUFACTURING COMPANY.
93. B-SOCKET POWER—A booklet giving constructional details of a socket-power device using either the BH or 313 type rectifier. NATIONAL COMPANY, INCORPORATED.
94. POWER AMPLIFIER—Constructional data and wiring diagrams of a power amplifier combined with a B-supply unit are given. NATIONAL COMPANY, INCORPORATED.
100. A, B, AND C SOCKET-POWER SUPPLY—A booklet giving data on the construction and operation of a socket power supply using the new high-current rectifier tube. THE Q. R. S. MUSIC COMPANY.
101. USING CHOKES—A booklet giving circuit diagrams of the more popular circuits showing where choke coils may be placed to produce better results. SAMSON ELECTRIC COMPANY.
21. A PRIMER OF ELECTRICITY—Fundamentals of electricity with special reference to the application of dry cells to radio and other uses. Constructional data on buzzers, automatic switches, alarms, etc. NATIONAL CARBON COMPANY.
23. AUTOMATIC RELAY CONNECTIONS—A data sheet showing how a relay may be used to control A and B circuits. YAXLEY MANUFACTURING COMPANY.
25. ELECTROLYTIC RECTIFIER—Technical data on a new type of rectifier, with operating curves. KODEL RADIO CORPORATION.
26. DRY CELLS FOR TRANSMITTERS—Actual tests given, well illustrated with curves showing exactly what may be expected of this type of B power. BURGESS BATTERY COMPANY.
27. DRY-CELL BATTERY CAPACITIES FOR RADIO TRANSMITTERS—Characteristic curves and data on discharge tests. BURGESS BATTERY COMPANY.
28. B BATTERY LIFE—Battery life curves with general curves on tube characteristics. BURGESS BATTERY COMPANY.
30. TUBE CHARACTERISTICS—A data sheet giving constants of tubes. C. E. MANUFACTURING COMPANY.
31. FUNCTIONS OF THE LOUD SPEAKER—A short, non-technical general article on loud speakers. AMPLION CORPORATION OF AMERICA.
32. METERS FOR RADIO—A catalogue of meters used in radio, with connecting diagrams. BURTON-ROGERS COMPANY.
33. SWITCHBOARD AND PORTABLE METERS—A booklet giving dimensions, specifications, and shunts used with various meters. BURTON-ROGERS COMPANY.
35. STORAGE BATTERY OPERATION—An illustrated booklet on the care and operation of the storage battery. GENERAL LEAD BATTERIES COMPANY.
36. CHARGING A AND B BATTERIES—Various ways of connecting up batteries for charging purposes. WESTINGHOUSE UNION BATTERY COMPANY.
37. WHY RADIO IS BETTER WITH BATTERY POWER—Advice on what dry cell battery to use; their application to radio, with wiring diagrams. NATIONAL CARBON COMPANY.
53. TUBE REACTIVATION—Information on the care of vacuum tubes, with notes on how and when they should be reactivated. THE STERLING MANUFACTURING COMPANY.
69. VACUUM TUBES—A booklet giving the characteristics of the various tube types with a short description of where they may be used in the circuit. RADIO CORPORATION OF AMERICA.
77. TUBES—A booklet for the beginner who is interested in vacuum tubes. A non-technical consideration of the various elements in the tube as well as their position in the receiver. CLEARTRON VACUUM TUBE COMPANY.
87. TUBE TESTER—A complete description of how to build and how to operate a tube tester. BURTON-ROGERS COMPANY.
91. VACUUM TUBES—A booklet giving the characteristics and uses of various types of tubes. This booklet may be obtained in English, Spanish, or Portuguese. DEFOREST RADIO COMPANY.
92. RESISTORS FOR A C OPERATED RECEIVERS—A booklet giving circuit suggestions for building a, c. operated receivers, together with a diagram of the circuit used with the new 400-milliampere rectifier tube. CARTER RADIO COMPANY.
97. HIGH-RESISTANCE VOLTMETERS—A folder giving information on how to use a high-resistance voltmeter, special consideration being given the voltage measurement of socket-power devices. WESTINGHOUSE ELECTRIC & MANUFACTURING COMPANY.
102. RADIO POWER BULLETINS—Circuit diagrams, theory constants, and trouble-shooting hints for units employing the BH or B rectifier tube. RAYTHEON MANUFACTURING COMPANY.

(Continued on page 328)

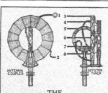

"RADIO BROADCAST'S" DIRECTORY OF MANUFACTURED RECEIVERS

¶ A coupon will be found on page 331. All readers who desire additional information on the receivers listed below need only insert the proper numbers in the coupon, mail it to the Service Depart and full details will be sent. New sets are listec

KEY TO TUBE ABBREVIATIONS

99—60-mA. filament (dry cell)
01-A—Storage battery 0.25 amps. filament
12—Power tube (Storage battery)
71—Power tube (Storage battery)
16-B—Half-wave rectifier tube
80—Full-wave, high current rectifier
81—Half-wave, high current rectifier
Hmu—High-Mu tube for resistance-coupled audio
20—Power tube (dry cell)
10—Power Tube (Storage battery)
00-A—Special detector
13—Full-wave rectifier tube
26—Low-voltage high-current a. c. tube
27—Heater type a. c. tube

DIRECT CURRENT RECEIVERS

NO. 424. COLONIAL 26

Six tubes; 2 t. r. f. (01-A), detector (12), 2 transformer audio (01-A and 71). Balanced t. r. f. One to three dials. Volume control: antenna switch and potentiometer across first audio. Watts required: 120. Console size: 34 x 38 inches. Headphone connections. The filaments are connected in a series parallel arrangement. Price $250 including power unit.

NO. 425. SUPERPOWER

Five tubes: All 01-A tubes. Multiplex circuit. Two dials. Volume control: resistance in r. f. plate. Watts required: 30. Antenna: loop or outside. Cabinet sizes: table, 27 x 10 x 9 inches; console, 28 x 50 x 21. Prices: table, $135 including power unit; console, $390 including power unit and loud speaker.

A. C. OPERATED RECEIVERS

NO. 508. ALL-AMERICAN 77, 88, AND 99

Six tubes; 3 t. r. f. (26), detector (27), 2 transformer audio (26 and 71). Rice neutralized t. r. f. Single drum tuning. Volume control: potentiometer in r. f. plate. Cabinet sizes: No. 77, 21 x 10 x 8 inches; No. 88 Hiboy, 25 x 38 x 18 inches; No. 99 console, 27½ x 43 x 20 inches. Shielded. Output device. The filaments are supplied by means of three small transformers. The plate supply employs a gas-filled rectifier tube. Voltmeter in a. c. supply line. Prices: No. 77, $150, including power unit; No. 88, $210 including power unit; No. 99, $285 including power unit and loud speaker.

NO. 509. ALL-AMERICAN "DUET"; "SEXTET"

Six tubes; 2 t. r. f. (99), detector (99), 3 transformer audio (99 and 12). Rice neutralized t. r. f. Two dials. Volume control: resistance in r. f. plate. Cabinet sizes: "Duet," 23 x 56 x 16½ inches; "Sextet," 22½ x 13½ x 15½ inches. Shielded. Output device. The 99 filaments are connected in series and supplied with rectified a. c.; while 12 is supplied with raw a. c. The plate and filament supply uses gaseous rectifier tubes. Milliammeter on power unit. Prices: "Duet," $160 including power unit; "Sextet," $220 including power unit and loud speaker.

NO. 511. ALL-AMERICAN 80, 90, AND 115

Five tubes; 2 t. r. f. (99), detector (99), 2 transformer audio (99 and 12). Rice neutralized t. r. f. Two dials. Volume control: resistance in r. f. plate. Cabinet sizes: No. 80, 23¾ x 12½ x 15 inches; No. 90, 37½ x 12 x 12½ inches; No. 115 Hiboy, 24 x 41 x 15 inches. Coils individually shielded. Output device. See No. 509 for power supply. Prices: No. 80, $135 including power unit; No. 90, $145 including power unit and compartment; No. 115, $170 including power unit, compartment, and loud speaker.

NO. 510. ALL-AMERICAN 7

Seven tubes; 3 t. r. f. (26), 1 untuned r. f. (26), detector (27), 2 transformer audio (26 and 71). Rice neutralized t. r. f. One drum. Volume control: resistance in r. f. plate. Cabinet sizes: "Sovereign" console, 30½ x 60¼ x 19 inches; "Lorraine" Hiboy, 25½ x 53¾ x 17½ inches; "Forte" cabinet, 25½ x 13½ x 17½ inches. For filament and plate supply: See No. 508. Prices: "Sovereign" $460; "Lorraine" $360; "Forte" $270. All prices include power unit. First two include loud speaker.

NO. 401, AMRAD AC9

Six tubes; 3 t. r. f. (99), detector (99), 2 transformer (99 and 12). Neutrodyne. Two dials. Volume control: resistance across 1st audio. Watts consumed: 50. Cabinet size: 27 x 9 x 11¾ inches. The 99 filaments are connected in series and supplied with rectified a. c., while the 12 is run on raw a. c. The power unit, requiring two 16-B rectifiers, is separate and supplies A. B. and C current. Price $142 including power unit.

NO. 402. AMRAD AC5

Five tubes. Same as No. 401 except one less r. f. stage. Price $125 including power unit.

NO. 536. SOUTH BEND

Six tubes. One control. Sub-panel shielding. Binding Posts. Antenna: outdoor. Prices: table, $130, Baby Grand Console, $195.

NO. 537. WALBERT 26

Six tubes; five Kellogg a. c. tubes and one 71. Two controls. Volume control: variable plate resistance. Isofarad circuit. Output device. Battery cable, Semi-shielded. Antenna: 50 to 75 feet. Cabinet size: 10¾ x 29½ x 16½ inches. Prices: $215; with tubes, $2.50

NO. 484. BOSWORTH, B5

Five tubes; 2 t. r. f. (26), detector (99), 2 transformer audio (special a. c. tubes). T. r. f. circuit. Two dials. Volume control: potentiometer. Cabinet size: 23 x 7 x 8 inches. Output device included. Price $175.

NO. 406. CLEARTONE 110

Five tubes; 2, t. r. f., detector, 2 transformer audio. All tubes a. c. heater type. One or two dials. Volume control: resistance in r. f. plate. Watts consumed: 40. Cabinet size varies. The plate supply is built in the receiver and requires one rectifier tube. Filament supply through step down transformers. Prices range from $175 to $375 which includes 5 a. c. tubes and one rectifier tube.

NO. 407. COLONIAL 25

Six tubes; 2 t. r. f. (01-A), detector (99), 2 resistance audio (99). 1 transformer audio (10). Balanced t. r. f. circuit. One or three dials. Volume control: Antenna switch and potentiometer on 1st audio. Watts consumed: 100. Console size: 34 x 38 x 18 inches. Output device. All tube filaments are operated on a. c. except the detector which is supplied with rectified a. c. from the plate supply. The rectifier employs two 16-b tubes. Price $250 including built-in plate and filament supply.

NO. 507. CROSLEY 602 BANDBOX

Six tubes; 3. t. r. f. (26), detector (27), 2 transformer audio (26 and 71). Neutrodyne circuit. One dial. Cabinet size: 17¼ x 5½ x 7¾ inches. The heaters for the a. c. tubes and the 71 filament are supplied by windings in B unit transformers available to operate either on 25 or 60 cycles. The plate current is supplied by means of rectifier tube. Price $65 for set alone, power unit $60.

NO. 408. DAY-FAN "DE LUXE"

Six tubes; 3 t. r. f., detector, 2 transformer audio. All 01-A tubes. One dial. Volume control: potentiometer across r. f. tubes. Watts consumed: 300. Console size: 30 x 40 x 20 inches. The filaments are connected in series and supplied with d. c. from a motor-generator set which also supplies B and C current. Output device. Price $350 including power unit.

NO. 409. DAYCRAFT 5

Five tubes; 2 t. r. f., detector, 2 transformer audio. All a. c. heater tubes. Reflexed t. r. f. One dial. Volume control: potentiometers in r. f. plate and 1st audio. Watts consumed: 135. Console size: 34 x 36 x 14 inches. Output device. The heaters are supplied by means of a small transformer. A built-in rectifier supplies B and C voltages. Price $170, less tubes. The following have one more r. f. stage and are not reflexed; Daycraft 6, $195; Dayrole, 6, $235; Dayfan 6, $110. All prices less tubes.

NO. 469. FREED-EISEMANN NRII

Six tubes; 3 t. r. f. (01-A), detector (01-A), 2 transformer audio (01-A and 71). Neutrodyne. One dial. Volume control; potentiometer. Watts consumed: 150. Cabinet size: 19¾ x 10 x 10½ inches. Shielded. Output device. A special power unit is included employing a rectifier tube. Price $225 including NR-411 power unit.

NO. 487. FRESHMAN 7F-AC

Six tubes; 3 t. r. f. (26), detector (27), 2 transformer audio (26 and 71). Equaphase circuit. One dial. Volume control: potentiometer across 1st audio. Console size: 24½ x 41½ x 15 inches. Output device. The filaments and heaters and B supply are all supplied by one power unit. The plate supply requires one 80 rectifier tube. Price $175 to $350, complete.

NO. 421. SOVEREIGN 238

Seven tubes of the a. c. heater type. Balanced t. r. f. Two dials. Volume control: resistance across 2nd audio. Watts consumed: 45. Console size: 37 x 52 x 15 inches. The heaters are supplied by a small a. c. transformer, while the plate is supplied by means of rectified a. c. using a gaseous type rectifier. Price $325, including power unit and tubes.

NO. 517. KELLOGG 510, 511, AND 512

Seven tubes; 4 t. r. f., detector, 2 transformer audio. All Kellogg a. c. tubes. One control and special zone switch. Balanced. Volume control: special. Output device. Shielded. Cable connection between power supply unit and receiver. Antenna: 25 to 100 feet. Panel 7¼ x 27¾ inches. Prices: Model 510 and 512, consoles, $495 complete. Model 511, consolette, $365 without loud speaker.

NO. 496. SLEEPER ELECTRIC

Five tubes; four 99 tubes and one 71. Two controls. Volume control: rheostat on r. f. Neutralized. Cable. Output device. Power supply uses two 16-B tubes. Antenna: 100 feet. Prices: Type 64, table, $160; Type 65, table, with built-in loud speaker, $175; Type 66, table, $175; Type 67, console, $235; Type 78, console, $265.

NO. 538. NEUTROWOUND, MASTER ALLECTRIC

Six tubes; 2 t. r. f. (01-A), detector (01-A), 2 audio (01-A and two 71 in push-pull amplifier). The 01-A tubes are in series, and are supplied from a 400-mA. rectifier. Two drum controls. Volume control: variable plate resistance. Output device. Shielded. Antenna: 50 to 100 feet. Price: $360.

NO.

Six tubes: 2 t. r. f.,
tubes a. c. heater ty
resistance in r. f. plate.
7 x 21 inches. The bu
16-B rectifier. The fil
transformer. Prices: t
rectifier; console, $27
console, $325 includi
speaker.

NO. 41

Eight tubes; five ty
control. Super-heterod
tions. Battery cable. I
loop. Set may be oper
power mains when use
104 loud speaker. Pr
operation; $570 with
operation.

NO. 540

Receiver characteris
type 71 power tube is
operate on either a. c.
The combination recti
two type 81 tubes. M
tained in lower part o
or long outside anten
42½ x 29 x 17¾ inches. I

NO. 54

This model combines
104 loud speaker. Th
tubes and a type 10 po
enclosed and is revolve
Models for operation I
Cabinet size: 52 x 72 x

NO. 539

Six tubes; 3 t. r. f. (
audio (26 and 27).
Built-in power supply
100 feet. Cabinet size
without accessories.

NO. 545. NEUTROW

Five tubes; 2 t. r. f.
and 71). The 99 tubes a
an 85-mA. rectifier. T
trol: variable plate resi
75 to 100 feet. Cabinet
$150.

NO.

Six tubes; 2 t. r. f., d
tubes a. c. heater type
Volume control: rheost
Panel size: 12½ x 8¾ inc
for the a. c. tubes and
small transformers. Th
type using a rectifier
$245.

NO. 522. CA

McCullough a. c. tul
trol; variable high resist
connections. Semi-shiel
Panel size: 7 x 21 inche
with a. c. equipment, $
a. c. equipment, $235.

NO. 523. CA

McCullough a. c. tu
volume control. Techn
C-battery connections.
device. Loop operated.
Model 92 C, console, $

BATTERY OP

NO. 542. PFAN

Six tubes: 3 t. r. f. (
Pfanstiehl circuit. Volur
r. f. plate circuit. One
battery connections. E
outdoor. Cabinet size: 9
out accessories.

NO. 512. ALL-AM

Six tubes; 3 t. r. f.
former audio (01-A an
Drum control. Volume
sizes: No. 44, 21 x 10 x
inches; No. 66, 27½ x 4
tions. Battery cable. A
No. 44, $70; No. 55, $1
66, $200 including loud

NO. 428.

Five tubes; 2 t. r. f.
All 01-A tubes. Semi ba
current 15mA. Volume
sizes: table, 20 x 8½ x 1
inches. Partially shiel
connections. Antenna:
console, $65 including l

Durham *Leadership*

The Original Metallized Resistor—Adopted by leading Set Manufacturers—Endorsed by Foremost Engineers—Are Sold by Reputable Dealers!

DURHAM Resistors and Powerohms have won their position of leadership because they have never failed to deserve the recognition of radio leaders who appreciate flawless accuracy and utmost dependability.

Ever since their first appearance, years ago, manufacturers, engineers, professional builders and radio fans have adopted Durhams in steady progression. Today Durhams are in use wherever fine results are a foremost consideration.

If you are not using Durhams now, an experiment with replacements may prove why they are the accepted leaders in their field.

Durham standard resistors are made in ranges from 500 ohms to 10 megohms. Durham Powerohms for "B" Eliminators and Amplifier circuits are made in 2.5 watt and 5 watt sizes in ranges from 500 to 100,000 ohms.

DURHAM
METALLIZED
R E S I S T O R S &
P O W E R O H M S
International Resistance Company
Dept. D, 2½ So. 20th St., Phila.

Rhythm, Symphony or Soprano—

Whatever it is that Radio brings to your set you can get it crystal clear —undistorted with the

CARBORUNDUM
REG. U. S. PAT. OFF.
Stabilizing Detector Unit

IT gives you pure, natural tones with volume. Can be used on practically any set.

The Complete Unit for	*The Detector Alone is*
$3.50	**$1.50**

Your reception will also be improved with Carborundum Grid Leaks

[FROM YOUR DEALER OR SENT DIRECT
Write for Our Hook-Up Book D-2]

THE CARBORUNDUM COMPANY, NIAGARA FALLS, N. Y.
CANADIAN CARBORUNDUM CO., LTD. NIAGARA FALLS, ONT.

The Clearest and Truest Electric Radio

Is a standard Radio Set equipped with

BalkiteElectric"AB"
$64.50 and $74.50. *Ask your dealer*
Fansteel Products Co., Inc., North Chicago, Ill.

Balkite
Radio Power Units

COMPLETE PARTS FOR
REMLER
45 Kc.
AC SUPERHET.

GET prompt, efficient service and lowest prices from Western. Our catalog lists a complete line of kits, parts, sets and accessories.

Dealers—Write for Catalog
WESTERN RADIO MFG. CO.
134 W. Lake St. Dept 42, Chicago

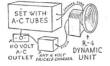

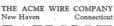

How's Your Old Audio Amplifier?

THORDARSON 171 TYPE POWER AMPLIFIER

Built around the Thordarson Power Compact R-171, this power amplifier supplies "A," "B," and "C" current for one UX-171 power tube and B-voltage for the receiver. Employs Raytheon B. H. rectifier.

THORDARSON 210 TYPE POWER AMPLIFIER

This amplifier, mounted on a special metal chassis, uses the Thordarson Power Compact R-210. Provides "A," "B," and "C" current for one UX-210 power tube and "B" voltage for the receiver. Employs one 216-B or 281 rectifier.

THORDARSON 210 PUSH-PULL POWER AMPLIFIER

This heavy duty power amplifier operates two 210 power tubes in push-pull and has an ample reserve of power for "B" supply for the heaviest drain receivers. Built with Thordarson Power Transformer T-2098, and Double Choke Unit T-2099.

A Home Assembled Thordarson Power Amplifier Will Make Your Receiver

A Real Musical Instrument

IMPROVEMENTS in the newer model receiving sets are all centered around the audio amplifier. There is no reason, however, why you cannot bring your present receiver up to 1928 standards of tone quality by building your own Thordarson Power Amplifier.

With a screw driver, a pair of pliers and a soldering iron you can build any Thordarson Power Amplifier in an evening's time in your own home. Complete, simple pictorial diagrams are furnished with every power transformer.

The fact that Thordarson power transformers are used by such leading manufacturers as Victor, Brunswick, Federal, Philco and Willard insures you of unquestionable quality and performance.

Give your radio set a chance to reproduce real music. Build a Thordarson Power Amplifier.

Write today for complete constructional booklets sent free on request.

THORDARSON

THORDARSON ELECTRIC MANUFACTURING CO.
Transformer Specialists Since 1895
WORLD'S OLDEST AND LARGEST EXCLUSIVE TRANSFORMER MAKERS
Huron and Kingsbury Streets — *Chicago, Ill. U.S.A.*

Samson Symphonic
PUSH-PULL UNITS
Clarify radio reception as a Chemist filters liquids

TRY this experiment. It probably will assist you to obtain better quality of reproduction.

Listen to your set from three or four rooms distant while people talk and the usual house noises are present. Try to understand speech. Do you actually hear the s's and k's or does your imagination supply them? Can you recognize the high notes of a Stradivarius violin, or do all violins sound alike?

Now replace the audio transformers with the Samson Symphonic and Samson Symphonic Push Pull Units. Listen again under the same conditions. Words are now crystal clear and music has a background and a brilliance entirely new to you.

Send for authoritative information from leading engineers which will be sent for about cost. Audio Amplification 25 cents; B Eliminator and Power Amplifier Construction 10 cts.; Make-Em-Better Sheet 5 cts.; and Inductance Units Bulletin 10 cts. All four sent for 50 cts.

Symphonic Push Pull Units

Samson Electric Co.

Manufacturers since 1882   Principal Office, Canton, Mass.

Electrostatic condensers for all purposes

Faradon

WIRELESS SPECIALTY APPARATUS COMPANY
Jamaica Plain
Boston, Mass.

RADIO PANELS
BAKELITE—HARD RUBBER

Cut, drilled and engraved to order. Send rough sketch for estimate. Our complete Catalog on Panels, Tubes and Rods—all of genuine Bakelite or Hard Rubber—mailed on request.

STARRETT MFG. CO.
521 S. Green Street   Chicago, Ill.

Why not subscribe to *Radio Broadcast?* By the year only $4.00; or two years $6.00, saving $2.00. Send direct to Doubleday, Doran & Co., Inc., Garden City, New York.

HAMMARLUND
Midline
CONDENSER

Soldered brass plates with tie-bars; warpless aluminum alloy frame; ball bearings; bronze clock-spring pigtail; full-floating, removable rotor shaft permits direct tandem coupling to other condensers. Made in all standard capacities.

Write for Folder

HAMMARLUND MFG. CO.
424-438 W. 33rd St.   New York

Hammarlund Accuracy
Assured

Michael Angelo, the great artist, once said, "Genius consists in the capacity for taking infinite pains."

While we are too modest to claim' that Hammarlund Products are the result of genius, we *do* insist that they have attained leadership because we *take the pains to make them RIGHT*.

Above is shown how accurate alignment of Hammarlund condenser plates is assured—firmly anchored in a metal holder, each plate in its own slot, while the tie-bars and solder are applied.

For Better Radio
Hammarlund
PRECISION
PRODUCTS

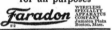

CARTER
New A. C. Adapters

for CX 326
(type tubes)
90c each

for CX 327
(type tubes)
$1.10 each

Use these A. C. Bakelite Adapters in converting your set. Again Carter leads and is just a little ahead. They are sturdy and make positive contacts. Furnished with long lead wires for your convenience in connecting.

Any dealer can supply.

In Canada: Carter Radio Co., Ltd., Toronto
Offices in principal cities of the world

Carter Radio Co.
CHICAGO

EBY
SPECIAL CIRCUIT PACKAGES

BINDING POSTS
15c

LYNCH
RESISTORS

ARTHUR H. LYNCH, Inc.
1775 Broadway   General Motors Building   New York City

Learn the Code at Home With the Omnigraph

Morse and Wireless—taught at home in half usual time and at trifling cost. Omnigraph Automatic Transmitter will send, on Sounder or Buzzer, unlimited messages, any speed, just as expert operator would. Adopted by U. S. Govt. and used by leading Universities, Colleges, Technical and Telegraph Schools throughout U. S. Send 6c for Catalog.

OMNIGRAPH MFG. CO., 13K, Hudson St., New York

NOW A-C OPERATED

Using standard
Cunningham or Radiotron Tubes

The new 2-Dial Karas A-C-Equamatic uses standard Cunningham or Radiotron tubes. It is a powerful, long-range, sweet-toned receiver with full A-C operation that will delight you with its superior performance. Send for com lete data and full size wiring diagrams.

Affiliated manufacturers: Carter, Hammarlund, Sampson, Yaxley, Benjamin, Electrad, International Resistor Co.

KARAS ELECTRIC CO.
4032-B North Rockwell St., Chicago

2 DIAL KARAS USE
KARAS
PARTS
A-C-EQUAMATIC

103. A. C. Tubes—The design and operating characteristics of a new a. c. tube. Five circuit diagrams show how to convert well-known circuits. Sovereign Electric & Manufacturing Company.

41. Baby Radio Transmitter of oxh-oer—Description and circuit diagrams of dry-cell operated transmitter. Burgess Battery Company.

42. Arctic Radio Equipment—Description and circuit details of short-wave receiver and transmitter used in Arctic exploration. Burgess Battery Company.

58. How to Select a Receiver—A commonsense booklet describing what a radio set is, and what you should expect from it, in language that any one can understand. Day-Fan Electric Company.

67. Weather for Radio—A very interesting booklet on the relationship between weather and radio reception, with maps and data on forecasting the probable results. Taylor Instrument Companies.

73. Radio Simplified—A non-technical booklet giving pertinent data on various radio subjects. Of especial interest to the beginner and set owner. Crosley Radio Corporation.

74. The Experimenter—A monthly publication which gives technical facts, valuable tables, and pertinent information on various radio subjects. Interesting to the experimenter and to the technical radio man. General Radio Company.

76. Radio Instruments—A description of various meters used in radio and electrical circuits together with a short discussion of their uses. Jewell Electrical Instrument Company.

78. Electrical Troubles—A pamphlet describing the use of electrical testing instruments in automotive work combined with a description of the cadmium test for storage batteries. Of interest to the owner of storage batteries. Burton Rogers Company.

95. Resistance Data—Successive bulletins regarding the use of resistors in various parts of the radio circuit. International Resistance Company.

96. Vacuum Tube Testing—A booklet giving pertinent data on how to test vacuum tubes with special reference to a tube testing unit. Jewell Electrical Instrument Company.

98. Copper Shielding—A booklet giving information on the use of shielding in radio receivers, with notes and diagrams showing how it may be applied practically. Of special interest to the home constructor. The Copper and Brass Research Association.

99. Radio Convenience Outlets—A folder giving diagram and specifications for installing loud speakers in various locations at some distance from the receiving set. Yaxley Manufacturing Company.

105. Coils—Excellent data on a radio-frequency coil with constructional information on six broadcast receivers, two short-wave receivers, and several transmitting circuits. Aero Products Company.

106. Audio Transformer—Data on a high-quality audio transformer with circuits for use. Also useful data on detector and amplifier tubes. Sangamo Electric Company.

107. Vacuum Tubes—Data on vacuum tubes with facts about each. Ken-Radio Company.

108. Vacuum Tubes—Operating characteristics of an a.c. tube with curves and circuit diagram for connection in converting various receivers to a.c. operation with a four-prong a.c. tube. Arcturus Radio Company.

109. Receiver Construction—Constructional data on a six-tube receiver using restricted field coils. Bodine Electric Company.

110. Receiver Construction—Circuit diagram and constructional information for building a five-tube set using restricted field coils. Bodine Electric Company.

111. Storage Battery Care—Booklet describing the care and operation of the storage battery in the home. Marko Storage Battery Company.

112. Heavy-Duty Resistors: Circuit calculations and data on receiving and transmitting resistances for a variety of uses, circuits for popular power supply circuits, d.c. resistors for battery charging use. Ward Leonard Electric Company.

113. Cone Loud Speakers—Technical and practical information on electro-dynamic and permanent magnet type cone loud speakers. The Magnavox Company.

114. Tube Adapters—Concise information concerning simplified methods of including various power tubes in existing receivers. Alden Manufacturing Company.

115. What Set Shall I Build?—Descriptive matter, with illustrations, of fourteen popular receivers for the home constructor. Herbert H. Frost, Incorporated.

109. Oscillation Control with the "Phasatrol"—Circuit diagrams, details for connection in circuit, and specific operating suggestions for using the "Phasatrol" as a balancing device to control oscillation. Electrad, Incorporated.

116. Using a B Power Unit—A comprehensive booklet detailing the use of a B power unit. Tables of voltages—both B and C—are shown. There is a chapter on trouble-shooting. Modern Electric Mfg. Co.

117. Best Results from Radio Tubes—The chapters are entitled "Radio Tubes," "Power Tubes," "Super Detector Tubes," "A. C. Tubes," "Rectifier Tubes," and "Installation." Gold Seal Electrical Co.

WHAT KIT SHALL I BUY? (Continued)

221. LR4 ULTRADYNE—Nine-tube super-heterodyne; one
stage of tuned radio frequency, one modulator, one oscillator,
three intermediate-frequency stages, detector, and two
transformer-coupled audio stages.
222. GREBE MULTIPLEX—Four tubes (equivalent to six
tubes); one stage of tuned radio frequency, one stage of
transformer-coupled radio frequency, crystal detector, two
stages of transformer-coupled audio, and one stage of
impedance-coupled audio. Two controls. Price complete
parts, $60.60.
223. PHONOGRAPH AMPLIFIER—A five-tube amplifier device having an oscillator, a detector, one stage of transformer-coupled audio, and two stages of impedance-coupled
audio. The phonograph signal is made to modulate the
oscillator in much the same manner as an incoming signal
from an antenna.
224. BROWNING-DRAKE—Five tubes: one stage tuned
radio frequency (with special neutralization system), regenerative detector (tickler control), three stages of audio
(special combination of resistance and impedance-coupled
audio). Two controls.
225. A3no SHORT-WAVE Transmitting Kit consists of interchangeable coils to be used in tuned-plate tuned grid circuit.
Kits of coils, two choke coils, and mountings, can be secured
for 20-40 meter band, 40-80 meter band, or 90-180 meter
band for $12.00.

Short-Wave Notes

JUST as we were about to go to press two
items of utmost interest (insofar as those who
listen for DX on the short waves are concerned)
came to our desk. The first was to the effect
that the short-wave station of the British Broadcasting Corporation now has a regular schedule.
As of December 12th, 1927, this station (5 sw.)
of Chelmsford, England) will transmit daily,
except on Saturdays and Sundays, from 7.30 to
8.30 a.m. E. S. T., and from 2.00 to 7.00 p.m.
E. S. T. This station is on 24 meters (12,500 kc.)
The other item relates to the short-wave
channel of 3LO, Melbourne, Australia. This
latter station will broadcast special programs of
two-hour duration every Sunday afternoon from
1.30 p.m. onwards, E. S. T. The Melbourne
station broadcasts on 36 meters (8325 kc.).

ᏁheCrosley ᏘᎧBandbox is the leading radio of today–*because*

At last! The radio tube that needs no batteries! Here it is functioning quietly, smoothly, powerfully in this new Crosley 6 tube receiver.—the A C Bandbox.

Now, the Crosley A C Bandbox needs no more attention than you pay the electric lamp that lights your home.

Combined with the Crosley facilities for economical manufacture is the patent situation of which Crosley has full advantage. Licensed to manufacture under the patents controlled by the electrical and radio industries, the Crosley Bandbox is a NEW receiver incorporating latest radio developments, the most advanced ideas of radio reception as well as sound reproduction. This outstanding engineering job is best understood when you consider its features are such as are found in radio twice and more its price.

1. Complete shielding of all elements.
2. Absolute balance (genuine Neutrodyne).
3. Volume Control.
4. Acuminators for sharpest tuning.
5. Single cable connection.
6. Single station selector.
7. Illuminated dial.
8. Adaptability to ANY type installation.

The set is solidly mounted on a stout steel chassis. As all controls are assembled together in the front, cabinet panels are easily cut to allow their protrusion. The metal escutcheon is screwed on over the shafts and the installation has all the appearance of being built to order.

Two large furniture manufacturers have designed console cabinets in which the Bandbox can be superbly installed (Showers Bros. Co., of Bloomington, Ind., and the Wolf Mfg. Industries of Kokomo, Ind.). Powel Crosley, Jr., has approved them mechanically and acoustically and has seen to it that the famous Crosley Musicones are built in them so that the best type of loud speaker reproduction may be insured.

The Bandbox is housed in a brown frosted crystaline finished metal case which is easily removed for console installation.

See the new Crosley A C Bandbox at your dealer's NOW! Hear first hand its delightful performance! Enjoy the best in radio at the least cost! Write Dept 20 if you can't locate a dealer!

Crosley Musicones are famous for their value. The new type D Musicone is an extraordinary as its companions and promises great satisfaction in its tone, volume and reproduction.

ULTRA MUSICONE $9.75

SUPER MUSICONE $12.75

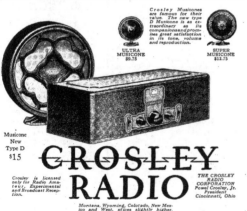

Musicone
New
Type D
$15

CROSLEY RADIO

Crosley is licensed only for Radio Amateur, Experimental and Broadcast Reception.

THE CROSLEY RADIO CORPORATION
Powel Crosley, Jr.
President
Cincinnati, Ohio

Montana, Wyoming, Colorado, New Mexico and West, prices slightly higher.

of these wonderful tubes

The amazing new RCA alternating current tubes—the UX 226 and UY 227—utilize for their filaments and their heating regular house lighting current. Current is stepped down through transformers. Rectifiers are NOT used.

the radio patents of these industries

The research and development work of these great industries — The Radio Corporation of America, The General Electric Co., The Westinghouse Co., The American Telephone & Telegraph Co., and The Hazeltine and Latour Corporations—are available to Crosley engineers in the constant advancement of Crosley radio design.

and the amazing capacity of this MERSHON *Electrolytic* CONDENSER

This is one of Crosley's great features. It is an exclusive Crosley device. It is self-healing—will last indefinitely—never needs attention and eliminates the danger of blown out paper condensers which are causing so much trouble in electrically operated sets.

ALUMINUM

The mark of Quality in Radio

Aluminum Box Shields

Essential in many popular amateur sets

FOR greater selectivity, finer appearance, longer life, lighter weight, use Aluminum Box Shields in the set you build—and look for Aluminum Shielding in the set you buy.

Designers agree on the superiority of Aluminum for shielding. It has become an established factor in radio design—recognized alike by advanced amateur set builders and engineers responsible for commercial production.

Aluminum Company of America's standard box shields, designed especially for amateur sets, are made of heavy Alcoa Aluminum with satin-dip finish, size 5 in. x 9 in. x 6 in. high. They are easily adapted to smaller sizes. They require no soldering.

If your dealer cannot supply you with Aluminum Box Shields send us his name and we will see that he is put in position to service you promptly. Be sure to send, also, for a copy of the new edition of "Aluminum for Radio." It is free.

THE ALUMINUM COMPANY OF AMERICA

2464 Oliver Building   Pittsburgh, Pa.

Aluminum Shields and Other Parts
Zenith Radio Corporation

RADIO Engineers are using more and more Alcoa Aluminum because of the efficiency with which this *one* metal meets the widely differing conditions encountered in radio design.

In many modern sets you find Alcoa Aluminum Shielding used because of its great effectiveness at radio frequencies. You find condenser blades made of Alcoa special Aluminum sheet because of its narrow limits of tolerance, its uniformity and its lightness. You find Alcoa Aluminum foil preferred for fixed condensers. Countless screw machine radio parts of Alcoa Aluminum are being used.

Sand castings and die castings of Alcoa Aluminum make the finest chasses, sub-panels and loud speaker frames, both because of the combined lightness and strength of Aluminum and because of its permanence and finish. Aluminum, decorated in a variety of rare wood grain effects, is ideal for front panels.

Look for Aluminum in the receiving set you buy. It has come to be recognized as the mark of quality in radio. Your copy of our new booklet, "Aluminum for Radio," will be sent free on request.

ALUMINUM COMPANY OF AMERICA

2464 Oliver Building   Pittsburgh, Pa.

ALUMINUM

The mark of Quality in Radio

Now!
It Is Easy To Build A Real Good Radio

THE NEW OFFICIAL BROWNING-DRAKE KIT

THIS new Official Browning-Drake Kit is an advance in radio design and engineering. An exclusive product of the Browning-Drake Corporation, it incorporates electrical and mechanical refinements which simplify construction of receiving apparatus and assure efficient operation.

ONE knob controls the single drum illuminated dial, giving a new smoothness of tuning with absolutely no trace of backlash. Coils and condensers are "precision-placed" in the laboratory.

With this new Kit as a basis, it is easy to build either the new Official Browning-Drake five tube Kit-Set, or the new Official Browning-Drake Two Tube Tuner which may be used with any one of the power amplifiers tested and specified by the Browning-Drake Laboratories.

Attractive cabinets are supplied for these new Kit receivers.

Constructional booklets may be obtained either from your dealer or direct, for 25 cents.

BROWNING-DRAKE CORPORATION

CAMBRIDGE MASS.

BROWNING-DRAKE RADIO

CARBORUNDUM

GRID LEAKS   RESISTORS

for NOISELESS RECEPTION

Just slip a Carborundum Grid Leak into your set and you won't be long in noticing the difference— the improved reception.

For Carborundum Grid Leaks are quiet. They are solid rods of Carborundum that provide an uninterrupted flow of current. There is no noise from arcing. These resistors are dense and they can't disintegrate. Carborundum Grid Leaks are tested at the maximum operating grid voltage, namely 5 volts, and Resistors at 90 volts.

BUY FROM YOUR DEALER OR DIRECT ∽ *Write for Booklet D-2, "Carborundum in Radio"*

THE CARBORUNDUM COMPANY, NIAGARA FALLS, N. Y.
Reg. U. S. Pat. Off.
CANADIAN CARBORUNDUM CO., LTD., NIAGARA FALLS, ONT.
Sales Offices and Warehouses in New York . Chicago . Boston . Philadelphia . Cleveland . Detroit . Cincinnati . Pittsburgh . Milwaukee . Grand Rapids
The Carborundum Co., Ltd., Manchester, England   Deutsche Carborundum Werke, Dusseldorf, Germany
Carborundum is the Registered Trade Name used by The Carborundum Company for Silicon Carbide. This Trade Mark is the exclusive property of The Carborundum Company.

Your Radio Cabinet

should receive *some* attention if you wish to *preserve* its original beauty.

Atmospheric conditions cause the finish to check and lose its lustre, and soon a once beautiful instrument becomes faded and ordinary.

VARNITE RADIO POLISH

is a scientific compound, designed especially for Radio Cabinets, Phonographs and wherever a superior polish and restorative is required.

It restores the grain of the wood to its original beauty, preserves the finish and prevents checking.

Use VARNITE to clean the panel on your Radio, it will prevent "blooming" and that "foggy" appearance.

It is easily applied and leaves no greasy surface to collect dust. Sold by leading radio dealers everywhere or by mail postpaid.

Insist on the genuine, there is no substitute for VARNITE.

Price 50 cents

Manufactured and guaranteed by
DAVIS CHEMICAL COMPANY
93 Massachusetts Ave.   BOSTON, MASS.

Now—A Single Drum Dial by National Co.

—with 360 degree Rotation—easy attachment,—the Velvet Vernier Quality,—just ask your dealer for

NATIONAL

TYPE F ILLUMINATED VELVET VERNIER DIAL
NATIONAL COMPANY, Inc., Malden, Mass.

COMPLETE PARTS FOR

REMLER

45 Kc.
AC SUPERHET.

GET prompt, efficient service and lowest prices from Western. Our catalog lists a complete line of kits, parts, sets and accessories.

Dealers—Write for Catalog

WESTERN RADIO MFG. CO.
134 W. Lake St. Dept 42, Chicago

RADIO PANELS
BAKELITE—HARD RUBBER
Cut, drilled and engraved to order. Send rough sketch for estimate. Our complete Catalog on Panels, Tubes and Rods—all of genuine Bakelite or Hard Rubber—mailed on request.

STARRETT MFG. CO.
521 S. Green Street   Chicago, Ill.

Heavy duty resistors for all standard power packs

This is a good time to subscribe for **RADIO BROADCAST**
Through your dealer or direct, by the year only $4.00
DOUBLEDAY, DORAN & COMPANY, Inc., Garden City, N. Y.

AERO Corona Coil
Used in New
Cooley "Rayfoto"

Of course the new Cooley "Rayfoto" uses an AERO Inductance Coil. This special coil is designed to meet the exact specifications of A. G. Cooley, whose "Rayfoto" receiver so many experimenters will build. For every inductance requirement AERO Coils are proved best—by experts and amateurs as well. Always specify AERO Coils if you want the finest in radio performance.

AERO PRODUCTS, Inc.
1722 Wilson Avenue   Chicago, Ill.

FEB 24 1928

©Cl B766522

RADIO BROADCAST

WILLIS KINGSLEY WING, Editor

MARCH, 1928   KEITH HENNEY   EDGAR H. FELIX   Vol. XII, No.
      Director of the Laboratory   Contributing Editor

CONTENTS

AMONG OTHER THINGS. .

THE many radio experimenters who are looking forw:
to a college course to lead them farther in their chosen fi
will find this month's leading article, by Carl Dreher, of c
siderable value. Mr. Dreher's story on radio instruction
colleges and universities is not exhaustive and there are ma
other colleges which have courses of value to the student w
plans eventually to go into radio work. The article does, ho
ever, answer many of the general questions most frequen
asked us in correspondence.

THIS issue contains a great deal of interesting material
the experimenter and home constructor as well as so
pages of especial importance to the many who are connect
with the business of selling radio apparatus. The description
the Crosley "Bandbox" receiver is one example; Sylvan Har:
article describing the Stewart Warner receiver is another. T
interesting illustrations in the full pages of radio set and
cessory pictures form a useful guide to interesting new pr
ucts. The description of the a.c. Browning-Drake set shows t
popular tuning unit combined with an excellent amplifier u
working on the push-pull principle. Thus far, we have descri
the Samson push-pull amplifier and B-supply, the Thordar
push-pull amplifier and B-supply, and in this issue reference
made to the AmerTran push-pull amplifier. Each of these devi
will give the user excellent audio quality when combined wit
good loud speaker. The Knapp A-Power unit, descri
on page 350, should appeal to many home-constructors
cause of its performance and price. And on page 355 a
following, a remarkably, inexpensive screened-grid receiver
discussed.

FOR those who like to discuss the design and performan
of amplifiers, the leading article in our technical editori
section, "Strays from the Laboratory," will provide plenty
material for discussion. Keith Henney, who writes this depa
ment, will be pleased to hear from those of our readers who ha
ideas on the matter. On page 352, the average characteristics
the 226 and 227 type tubes are presented.,

READERS who are interested in receiving additional i
formation from the makers of the Cooley Rayfoto app
ratus may send their names to the undersigned who will forwa
them. The April RADIO BROADCAST will contain a long-awaite
article on the RADIO-BROADCAST "Lab" circuit, a descriptic
of an interesting short-wave phone and code transmitter, and
wealth of other interesting constructional material.

THE Laboratory Data Sheets, which have been one of tl
most popular features of RADIO BROADCAST since th
first appeared in our June, 1926, issue, are the work of Howa
E. Rhodes of the technical staff of RADIO BROADCAST Labor
tory. Of necessity they cannot be signed, but so many ha'
written us in complimentary terms of them that we feel th
the many readers who have expressed interest in this regul
feature should know to whom they are due. The first eigh
eight Data Sheets, by the way, are available in bound for
from the Circulation Department of Doubleday, Doran ar
Company, Inc., and sell for one dollar.

—WILLIS KINGSLEY WING.

Doubleday, Doran & Company, Inc.	Doubleday, Doran & Company, Inc.	Doubleday, Doran & Company, Inc.	Doubleday, Doran & Company, Inc.
MAGAZINES	BOOK SHOPS	OFFICES	OFFICERS
COUNTRY LIFE	(Books of all Publishers)	GARDEN CITY, N. Y.	F. N. DOUBLEDAY, Chairman of the Boa
WORLD'S WORK		NEW YORK: 244 MADISON AVENUE	NELSON DOUBLEDAY, President
GARDEN & HOME BUILDER	LORD & TAYLOR BOOK SHOP	BOSTON: PARK SQUARE BUILDING	GEORGE H. DORAN, Vice-President
RADIO BROADCAST	PENNSYLVANIA TERMINAL (2 Shops)	CHICAGO: PEOPLES GAS BUILDING	S. A. EVERITT, Vice-President
SHORT STORIES	NEW YORK: GRAND CENTRAL TERMINAL	SANTA BARBARA, CAL.	RUSSELL DOUBLEDAY, Secretary
EDUCATIONAL REVIEW	38 WALL ST. and 526 LEXINGTON AVE.	LONDON: WM. HEINEMANN LTD.	JOHN J. HESSIAN, Treasurer
LE PETIT JOURNAL	848 MADISON AVE. and 166 WEST 32ND ST.	TORONTO: OXFORD UNIVERSITY PRESS	LILLIAN A COMSTOCK, Asst. Secretary
EL ECO	ST. LOUIS: 223 N. 8TH ST. and 4914 MARYLAND AVE.		L. J. McNAUGHTON, Asst. Treasurer
FRONTIER STORIES	KANSAS CITY: 920 GRAND AVE. and 206 W. 47TH ST.		
WEST WEEKLY	CLEVELAND: HIGBEE CO.		
THE AMERICAN SKETCH	SPRINGFIELD, MASS.: MEEKINS, PACKARD & WHEAT		

DOUBLEDAY, DORAN & COMPANY, INC., Garden City, New York

Copyright, 1928, in the United States, Newfoundland, Great Britain, Canada, and other countries by Doubleday, Doran & Company, Inc. All rights reserved.
TERMS: $4.00 a year; single copies 35 cents.

A Short-Wave Transmitter for Imperial Broadcasting

ANIMATED by the success of other European countries in this field, the British Broadcasting Corporation finally permitted itself to fall victim to short-wave broadcasting. The purpose of 5 SW is to unite the radio audiences of the far-flung corners of the British Empire, and reports so far received would indicate that quite a large degree of success is being met with in this effort. The wavelength of 5 SW is 24 meters (12,500 kc.) and it may be heard well in this country. With the exception of Saturdays and Sundays (when the schedule is irregular), 5 SW broadcasts daily from 7:30 to 8:30 A. M. E. S. T., and from 2:00 to 7:00 P. M. E. S. T. The power is 20 kw. The transmitter is at present erected in the experimental laboratories of the Marconi Company at Chelmsford, some thirty miles from London. The apparatus consists of two panels of a Marconi short-wave beam transmitter (shown in the foreground) with the addition of three modulating panels, which may be discerned faintly in the background

WHERE ELECTRIC COMMUNICATION ENGINEERING MAY BE STUDIED
Students at work in the Cruft Memorial Laboratory, Harvard University.
Professor G. W. Pierce is Rumford professor of physics there. In the article below he tells what courses are available at Harvard for the radio man

University Offerings in Radio Education

By Carl Dreher

NOT infrequently readers address this magazine with questions regarding university training in radio. They are interested in such matters as entrance requirements at various institutions, courses offered, prerequisites, and the practical value of the training available. As a rule the questions come, not from the usual group from which college students are drawn, but from young men who feel the need for adding to their knowledge but lack formal preparation. Secondary school graduates, and those who have been fortunate in securing college preparatory education at the usual age, generally are familiar with the educational system beyond this point, but many others have only the vaguest ideas about the procedure of technical study in institutions. They imagine, in some cases, that all universities admit students to certain courses of interest to them, regardless of aptitude or previous training on the part of the student. Actually there is a wide range of flexibility among technical colleges, some offering a wealth of extension courses, with little scrutiny of the applicant's credentials, while others have rigid entrance qualifications, no extension courses, and, in a number of instances, no electives and no variations in the curriculum. The relation of radio engineering to the more fundamental divisions of technology is also frequently misapprehended, the importance of radio being naturally exaggerated in the minds of some of its devotees. All this has made it appear worth while to conduct an inquiry into the subject of university training in radio communication, with a view to supplying information regarding conditions in representative institutions, and also to make clear the relation of university training in radio to other forms of study of the subject.

Letters were written to the professors of Electrical Engineering at ten institutions, with requests for answers to the following questions:

(1.) What communication and radio engineering courses are given as required or elective studies in the electrical engineering division?

(2.) Does your institution offer any extension courses in communication or radio engineering, and, if so, what are the entrance requirements for such specialized work?

Nine out of the ten institutions addressed replied, giving the information sought. Of course a choice of ten universities and technical schools, among several hundred, is not sufficient for com-

© Bachrach
J. H. MORECROFT
Professor of Electrical Engineering, Columbia University, New York City: "I do not recommend that a man specialize too much while he is at college"

339

prehensive conclusions, but it does serve to show the general trend of higher education in radio. Colleges known to be interested in the radio field were rather favored, although not exclusively, and there was some attempt at territorial distribution.

The answer of Prof. J. H. Morecroft at Columbia University is well worth quoting, inasmuch as it represents one reasoned policy in higher technical education:

"Both Professor Slichter and I have always felt that it is extremely foolish for a young man to specialize on a specific branch of engineering work before he is well aware of his aptitude and of the opportunities awaiting him in any special field, and naturally the courses at Columbia are laid out in accordance with this idea.

"The very large companies which are always ready to take any men whom we regard as well fitted for research work have this same idea with regard to specialization. Our largest communication company, for example, does not desire to have men trained in specialized communication theory and practice. Their representatives have always expressed to me their desire for men who are trained rather in the general fields of science and engineering than those who have attempted to specialize in some certain phase of the work.

"Whereas the fundamental principles of communication, including radio, are given to all of our electrical engineering students, there is no special arrangement of courses for those who might desire to specialize in radio. In the senior year the electrical engineering students at Columbia are allowed to take more than half of their work from a list of elective subjects and at this time of course a man desiring to do so can pick out most of his work in the communication field.

"I myself do not recognize radio engineering as being apart from telephone engineering or other similar types and do not recommend that a man specialize too much while he is at college. When he starts to work he will have to specialize to earn his living and will get but little opportunity to carry on studies outside of those required by his daily tasks, so that it seems foolish for him to give up the opportunity for broad training while at college."

These views of Professor Morecroft's are an authoritative statement of what might be called the broad-training policy in engineering education, which in my own opinion, is the best thing for those who fortunately have the time and money to take advantage of it. Columbia does, however, offer an extension course, "Electrical Engineering-to-Radio Communication," which "is not intended for those who are already familiar with radio theory and practice but rather for those who have had no training in this particular field, and is so designed that anyone who has had the equivalent of an ordinary high school course will be able to do the work satisfactorily." The class meets twice a week for an aggregate of three hours. The admission requirements for Columbia University Extension students are very elastic. For "mature students whose chief interest lies outside the University and who have leisure to pursue only a few courses in the late afternoon or at night . . . the sole condition is that they show their ability to pursue the work with profit." In practice this amounts to satisfying the instructor or a supervising professor that one has some background in the subject and an earnest desire to learn more of it. In past years Columbia has given some excellent advanced extension courses in vacuum-tube theory and other branches.

Professor Morecroft is one of the most prominent of American radio engineers, and a past president of the Institute of Radio Engineers. Dr. G. W. Pierce likewise ranks among the highest radio technicians, and is similarly a past president of the Institute. Nevertheless Harvard University, where Professor Pierce is Rumford Professor of Physics and Director of the Cruft Memorial Laboratory, pursues a policy differing from that of Columbia, in that Harvard offers a "Programme of Study in Electric Communication Engineering," and confers the degree of "Bachelor of Science in Communication Engi-

AT HARVARD UNIVERSITY
The Cruft Memorial Laboratory

neering" upon candidates who complete the four-year course satisfactorily. The course is substantially one in electrical engineering, with specialization beginning in the third year. Of the eight courses listed in that year four are radio, or, broadly, communication subjects ("Electric Oscillations and Their Application to Radiotelegraphy and Radiotelephony; Electric Oscillations, Electric Waves, and Radio-Frequency Measurements; Electron Tubes—Amplifiers, Detectors, and Oscillators; Electron

A WELL-EQUIPPED LABORATORY
A view of the electrical-engineering laboratory of the Rennsselaer Polytechnic Institute, Troy, New York. R. P. I. offers special and graduate work in communication to qualified students

Tubes, Advanced Course"). The other courses in the third year are in electrical engineering and mathematics. In the fourth year there are five communication subjects and four courses in electrical engineering.

Harvard University also offers work in communication engineering to graduate students and on occasion confers the degrees of "Master of Science" and "Doctor of Science in Communication Engineering." The admission requirements in all cases are high. In the case of the four-year course leading to the bachelor's degree the entrance requirements are the same as for admittance to the Freshman Class of Harvard College, the academic elements being substantially those provided by a first-class high school or preparatory training.

The Massachusetts Institute of Technology answered the questionnaire through Prof. Edward L. Bowles. In the regular electrical engineering course intended for students who seek training in electrical power engineering, two optional one-term courses, "Principles of Radio Communication" and "Principles of Wire Communication," are available. Students who desire to specialize in electrical communication, after taking the regular E. E. course for two years, may register for the "Electrical Communication Option" at the beginning of the junior year. This option "embraces work covering wire telephony, carrier telephony and radio telephony, also wire telegraphy, carrier telegraphy, and radio telegraphy. The properties and engineering applications of electron tubes are also included." During the third year only one specific communication course, "Electrical Communications, Principles," appears, but there are other modifications in the work, such as the addition of a course in vector analysis, the omission of a course in heat engineering which appears in the regular E. E. outline, etc. In the fourth year there is marked specialization in communication and electro-magnetic theory. Incidentally, a one-term Senior course in "Sound, Speech, and Audition" is included. The degree is that of "Bachelor of Science." M. I. T. also offers a number of limited-attendance five-year "co-operative courses in electrical engineering" which, for the first two years, are the same as the regular E. E. course, while for the last three years, the student's time is equally divided between instruction at the Institute and work in industrial plants. "Option 3," in Communications, is arranged in co-operation with the Bell Telephone Laboratories in New York City. These co-operative courses lead to the degree of "Bachelor of Science" after four years, and "Master of Science" after five years. Admission to the Institute is by examination. There are no extension courses.

At Stevens Institute, Hoboken, N. J., where the Department of Electrical Engineering is headed by Prof. Frank C. Stockwell, succeeding no less a radio man than L. A. Hazeltine, no special attention is paid to communication training. Stevens offers only a single course, leading to the degree of "Mechanical Engineer," and has no electives in any engineering subject. Extension courses are unknown, nor does Stevens accommodate special students. However, as part of the regular work in electrical engineering, a certain amount of radio engineering is prescribed for all students, including laboratory exercises and class room practice.

Undergraduate students at Rensselaer Polytechnic Institute, working for their E. E. degrees, after the usual grounding in mathematics and physics, followed by general electrical engineering courses, receive a communication engineering course in the last seven weeks of the senior year. The following subjects are presented in the order named: "Telephone transmitters; telephone receivers; transformation of medium-frequency alternating currents and electromotive forces; resistance, inductance, and capacitance in medium-frequency alternating-current circuits; distribution of current and electromotive force over telephone lines; electrical filters, transmission line impedance, and equivalent networks; fundamental telephone and telegraph circuits; telephone transmission and its measurement; telephone and telegraph systems and telephone service; vacuum tubes and their application; telephone repeaters and public address systems; multiplex or carrier current telephony; radio telephony and telegraphy; interference and cross-talk."

Rensselaer men seeking the B. S. in Physics are required to take a general course in electrical engineering covering the production, transmission, and utilization of electrical energy for light, power, and communication purposes. In the junior year these students have a seven-week course in radio communication. The topics are as follows: "Underlying electrical theory; properties of oscillatory circuits; antenna systems and radiation; damped and undamped wave radio telegraphy; general properties of the three-electrode vacuum tube; the three-electrode vacuum tube as detector, amplifier, oscillator, and modulator; radio telephony."

R. P. I. offers special and graduate work in communication engineering to qualified students. Applicants who have acquired the physics and mathematics covered in the first and second years at Rensselaer, or in equivalent courses elsewhere, may be admitted to a special course in radio communication whereby practically the entire third year is devoted to laboratory and theoretical study of radio. "We do not advise students to take this special course unless they know positively that they are going to enter some branch of the radio industry at the completion of their course," writes Prof. W. J. Williams, who supplied all the information regarding communication training at R. P. I. It would appear that this special course is in the line of definitely vocational training in radio engineering for men who have the requisite grounding in the general engineering field, including, necessarily, considerable mathematics and physics. Two one-year graduate communication courses are also given in alternate years. One covers wire communication and general electric circuit theory. The other is in radio-communication and advanced electromagnetic theory. About twelve hours of work a week are required. The content of these courses is variable, the effort being to keep the subject matter up to date. This work is necessarily limited to students who have the E. E. or B. S. in Physics from Rensselaer, or equivalent degrees from some other institution.

At the Polytechnic Institute of Brooklyn, on the basis of Prof. Erich Hausmann's reply, four subjects are given as required studies in the curriculum leading to the degree of Electrical Engineer. These are "Telegraphy" (two hours per week of class work in the second semester of the Junior year); "Telephony" (two hours per week of class work in the first semester of the Senior year); "Radio Communication" (two hours per week of class work in the second semester of the Senior year); "Communication Laboratory" (three hours per week of laboratory work in the second semester of the Senior year). The course in radio communication is described as "A practical and theoretical course on the generation of radio oscillations of the sustained and decadent types, damping of wave trains, resonance of single and coupled circuits, plotting of reactance diagrams and resonance curves

reception of electric waves, the use of vacuum tubes as amplifiers, detectors and oscillators, the forms of antennas, and the design of commercial forms of radio telegraphic and telephonic apparatus. Prerequisite, Alternating Current Circuits." The "Communications Laboratory" work covers operation of telephone, telegraph, and fire alarm installations, tests of characteristics of spark, arc, and tube oscillation generators, measurements of such quantities as coupling coefficient and decrement, etc.

The four courses outlined above are also given in the Evening Session of the Polytechnic, the first three during one year and the fourth during the following year, alternating. The prerequisites are "Electricity and Magnetism," "Direct-Current Machines," and "Calculus," For "Telephony," "Alternating Current Circuits" is an additional prerequisite, and the three class room courses must precede the "Communication Laboratory" course. "A few years ago," writes Professor Hausmann, "an elementary evening course in radio was given which did not require a knowledge of calculus: while this course has not been offered for several years, there is no reason why it should not be given

again. What do you think of the demand for such a brief course, say of 15 lectures?" The best way to gauge the demand is to propound the question and to invite any readers who are interested to communicate with Professor Hausmann. The writer believes that there is a sufficient demand for such a course to warrant offering it again.

Stanford University in California, while offering no extension work of any kind, lists four communication engineering courses described by Prof. Frederick Emmons Terman as follows:

(1.) *Lecture course in principles of radio communication.* Three lectures a week for ten weeks, open to seniors who have taken the regular electrical course, and to graduate students.

(2.) *Laboratory course in radio measurements.* Two lectures and one afternoon in laboratory per week for ten weeks. Reports are required. The experiments consist of bridge measurement of vacuum-tube amplification factors, dynamic plate resistance, etc.; radio-frequency resistance; resonance curves; detectors; audio-frequency transformer characteristics; adjustment of vacuum-tube oscillators, etc. This course is open to gradu-

ate students who have completed the fourth year electrical courses, as well as to other students with similar qualifications.

(3.) *A combination lecture and laboratory course on the theory of transmission circuits,* intended to occupy about one third of the student's working time for ten weeks. The subject matter includes the theory of long-distance power and telephone circuits, waves on transmission lines, theory and design of simple and composite wave filters. Probably half of this course applies directly to telephone communication problems. It is open to electrical engineering graduate students in full standing.

(4.) *Another graduate course with the same prerequisites is about fifty per cent. concerned with communication problems,* such as skin effect, transient oscillations, theory of telephone receivers and loud speakers, and harmonics, the treatment being of an advanced mathematical nature. There are two lectures a week for ten weeks.

All these courses are given by Professor Terman, who is in charge of communication and analytical work at Stanford University.

The University of Wisconsin, with its College of Electrical Engineering located at Madison,

ELECTRICAL ENGINEERING BUILDING, UNIVERSITY OF MINNESOTA
The communication laboratories of this fine building occupy the third floor. The University has an experimental radio station and, like R. P. I., operates a broadcasting station

cannot be said to neglect the field of communication engineering. Four resident elective courses in "Radio Telegraphy"; "Electric Amplifiers and Oscillators"; "Radio Circuit Analysis and Design"; and "Telephony and Telegraphy," are provided. The first is substantially the "application of alternating-current theory to radio problems and measurements," accompanied by laboratory work. The second includes "analytical study of the properties of amplifier and oscillator circuits and of the characteristics of tri-electrode thermionic amplifiers," likewise with laboratory sessions. "Radio Circuit Analysis and Design" is described as "a continuation of the above courses, treating such topics as amplifiers and their design, operation of oscillators in parallel, design of oscillators, modulation and demodulation, and analysis and design of radio transmitting and receiving sets." The theory of instruments and lines is treated under "Telephony and Telegraphy," laboratory work being included. The above are all one-semester courses. Other courses bearing on communication problems are also available for graduate students, such as "Advanced Theory of Electric

Circuits" and "Seminar in Electric
Circuit Theory." The content of
both of these courses is largely in
the field of transient and high-
frequency phenomena, behavior of
networks, etc.

A number of correspondence
courses in communication work are
included in the bulletin "Courses in
Electrical Engineering" issued by
the University Extension Division
of the University of Wisconsin.
Course 318, "Principles of the Telephone" is given in three parts:
"Subscriber's Apparatus," "Central Office Equipment," "Aerial
and Underground Construction."
The instruction fee for the first part
is $6, with ten assignments in this
portion. "This course," according
to the catalogue, "includes study
of the laws underlying speech, transmission, of the instruments, switch-
boards, and other apparatus in an
exchange, and of the laying, testing,
and maintaining in good condition of
the circuits outside the exchange.'

Course 329, "Principles of Radiotelegraphy"
treats the standard topics in twenty assignments, for an instruction fee of $12. An under-
standing of trigonometry is stated to be essential. An evening course on "Theory of Radio
Circuits"' is also being given in Milwaukee
under the auspices of the Extension Division.
This is an engineering course, open to "graduates
of scientific courses of college grade or men of
equivalent training," and may carry credit for
degrees. It provides "a quantitative treatment
of radio circuit theory" and aims to demonstrate
"the dependence of radio circuit theory upon
fundamental electric theory." The fee is $10
for the course of eighteen sittings. Four other
radio courses are given by the Extension Division in Milwaukee, these being designed for
amateurs, so that the treatment is more popular
and elementary.

The Professor of Electrical Engineering at
the University of Wisconsin, Edward Bennett,
is a well-known radio engineer and has con-

© Harvard Crimson

IN THE CRUFT MEMORIAL LABORATORY, HARVARD UNIVERSITY
The picture shows a group of students making
measurements during one of the courses

tributed extensively to the literature of the
subject.

For the University of Minnesota, C. M.
Jansky, Jr., Associate Professor, Radio Engi-
neering, writes as follows, after referring to his
paper, "Collegiate Training for the Radio
Engineering Field" in *The Proceedings of the
Institute of Radio Engineers* for August, 1926:

"Collegiate work in the field of radio engineering is given in the Department of Electrical
Engineering of the University of Minnesota.
Students desiring to specialize in communication
engineering in general, or radio engineering in
particular, take their work in this department.
Upon completing the four years' work they
receive the degree of 'Bachelor of Science in
Electrical Engineering,' and upon completing
an additional year of graduate work they receive
the degree of 'Master of Science in Electrical
Engineering.' The student gets his first course
in the communication field in his junior year.
He gets a year's course in radio engineering

in the senior year. Where men are
specializing in radio, I select such
elective courses as will be of particular value to them, such as 'Transient
Electrical Phenomena' and 'Differential Equations.'

"This year the Electrical Department has approximately 80
senior students, about 40 of whom are
registered in radio engineering. Not
all of these 40 will, however, become radio engineers, as many of
them will go into other fields of
electrical engineering.

"We are obtaining an increasing
number of post-graduate students
who are specializing in radio communication. These students, in
addition to advanced courses in electrical engineering, take advanced
courses in physics and mathematics. Where possible, I recommend
the fifth year.

"I believe our building and equipment excel that of the majority of
institutions. We have approximately
7000 square feet of floor space
devoted to communication laboratories and
research rooms. In addition, we operate an
experimental radio station and a broadcasting
station."

The above summaries of communication engineering training which is offered at nine centers
of higher learning afford a view of some of the
main trends in the field. At some institutions
there is little interest in communication engineering except as a part of general electrical
engineering. The philosophy underlying this
attitude is that industry is so highly specialized
that it is hopeless to give a man more than the
broad fundamentals at school, and that if this is
accomplished, he will make his way after he gets
into business by the addition of his common
sense and the training provided by the job
itself. Some of the large communication concerns,
in fact, set up schools of their own for the
engineers on the staff, evidently with the
thought that the technological content of their
particular jobs cannot be secured in any outside
school. Many colleges, on the other hand, teach
the theory and some of the practice of telegraphy
and telephony, by wire and radio, to qualified
students of junior and senior grade. Still others
go further and, in addition to such studies,
offer popular extension courses, correspondence
study, and the like. If there is a demand for
vocational study, for example, in a neighborhood, the local university is probably better
qualified to fill the need than some primarily commercial institution. Study, even if it is
not of the most scholarly sort, is not apt to do
anyone harm, and it may do good, as long as it
is not allowed to interfere with the rigorous and
inexorably thorough training which makes a real
engineer. Again, it may be argued that communication engineering courses deal with fundamentals just as much as the older power
engineering subjects, that nowadays a vacuum
tube is as important a machine as a dynamo.
Different courses are for different people, we
may conclude. If a man has the intellectual
equipment, the money, and the time required for
a thorough study of the subject he intends to
make his lifework, by all means let him spend
four or six years preparing himself. Let him
specialize only after he has mastered the fundamentals. But if at some time a man wants to
learn something special in a superficial way, no
harm is done, provided he knows what he is
getting, and does not take it for more than what
it is.

A CORNER OF ONE OF THE COLUMBIA UNIVERSITY LABORATORIES
Making measurements on a resistance-coupled audio amplifier in the Hartley Laboratory

HOW THE COMPLETED TWO-TUBE TUNER LOOKS
The five knobs, from left to right, are: "On-Off" switch, trimmer condenser (C_5), main tuning control, volume control, regeneration control

An A. C. Browning-Drake Receiver

By Glenn H. Browning

THE Browning-Drake circuit has been popular for a number of years, due, probably, to the fact that the set is very easy to build and operate, sufficiently sensitive to receive most distant stations which are above the noise level, and selective enough to cope with present broadcasting conditions, except when located in extremely congested regions. The sensitivity of the set is primarily due to the tuned r.f. transformer which Dr. Drake and the writer developed a number of years ago at Cruft Laboratory, Harvard University. Also, the antenna tuning system, which is a conductively coupled one, gives more signal strength than any the writer has tested so far. This is especially true when operating the set with a short antenna.

The electrical engineering in the Browning-Drake receiver has been changed very little during the last three years, though minor improvements have been incorporated from time to time. With the introduction of a.c. tubes and the popularity of so-called single control receivers, however, it seems particularly advisable at the present time to change the mechanical layout of the receiver and add any refinements in the electrical design which constant work on the circuit have indicated as worthy of recommendation.

As most radio constructors know, the Browning Drake circuit consists of one stage of tuned r.f. amplification with some type of neutralization, coupled to a regenerative detector. Any form of audio amplification may be used with this tuning arrangement, but, of course, the audio amplifier determines to a large extent the quality of the received signals.

This article deals essentially with the new two-tube receiver unit, employing a.c. tubes and making use of single control, and we will not go into detail about the audio channel. In passing, it might, however, be mentioned that the Browning-Drake Corporation supplies a foundation kit for a five-tube receiver, which includes the necessary sub-panel for the mounting of whatever audio equipment is used, and a larger

base panel than that used in the two-tube tuner unit described here.

After experimenting for some time with a.c. tubes two CX-327 (UX-227) tubes were chosen for the tuner unit. With a tuner constructed according to the instructions in this article, and used with a three-stage amplifier also wired for a.c. tubes, there is very little audible 60-cycle hum in the loud speaker. Sometimes it is necessary to experiment with the voltages used and also with the tubes in order to get the best combination, otherwise there may be some hum in the loud speaker, and the quality would be impaired

The nucleus of the Browning-Drake set described here is what might be called a single drum control unit; it consists of a single illuminated drum type dial driving two variable condensers. The necessary coils for the two-tube

tuner are also an integral part of the drum unit. As antennas differ a great deal, it is advisable to put a small condenser in parallel with the first tuning condenser in such fashion that it may be controlled from the front panel, and variations in antenna length thus compensated. In actual tuning, this condenser is used as a minor control. In Fig. 1, the diagram of the complete tuner unit, this condenser is indicated as C_5.

As will be noted, a slightly different system of neutralization than has heretofore been used is employed; also parallel plate-voltage feed for the r.f. tube is featured. The parallel feed, which consists of an r.f. choke coil in series with the r.f. B-battery lead, together with a 0.5-mfd. condenser, keeps all r.f. current from entering the B supply, and forces this current to go through

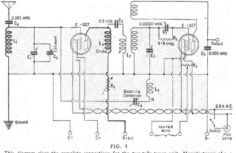

FIG. 1

This diagram gives the complete connections for the two-tube tuner unit. Manufacturers of a. c. tubes of the heater type state that less hum will be evidenced if the heater is biased with respect to the cathode. This bias may be plus or minus, depending upon which gives the better results

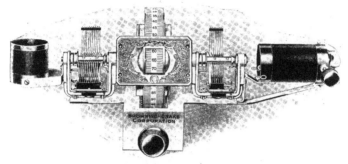

THE NUCLEUS OF THE BROWNING-DRAKE A. C. TUNER

It is known as the single drum control, and comprises (refer to fig. 1, page
343) C_1, C_2, L_1, L_2, L, regeneration coil, and the receptacle for the pilot light ·

the 0.5 mfd. condenser and the primary of the r.f. transformer. Neutralization is then accomplished by means of a few extra turns, L_2, on the r.f. transformer, the end of which is connected through a neutralizing condenser, C_4, to the grid of the tube. In practice, it is necessary to have the stator plates of the neutralizing condenser connected to the grid of the first tube rather than the rotor plates. It is found that a tube with a much larger capacity between grid and plate can be tolerated by the use of the above-mentioned system of neutralization.

When the set builder is located in a very congested region and there are a number of local broadcasting stations within a radius of three or four miles from the receiver, shielding is to be recommended, and a complete shield for the Browning-Drake receiver will probably be commercially available soon. Such shielding eliminates all pick-up by coils and the wiring of

the set, and adds materially to the selectivity of the receiver when it is used under the above conditions. Where the set builder is located in the country or suburbs, shielding is unnecessary.

The parts necessary for the construction of the single-control a.c. Browning-Drake tuner unit are listed below:

LIST OF PARTS

Official Browning-Drake Single Drum Control Kit comprising C_1, C_2, L_1, L_2, L_3, the regeneration coil and pilot light socket	$26.00
Official Browning-Drake Type T-2 Foundation Unit consisting of Westinghouse Micarta drilled and engraved front panel, base panel complete with mounting hardware. Also miscellaneous machine screws, nuts, and wire	13.50
L_4 Browning-Drake Radio-Frequency Choke	2.00
C_2 Browning-Drake 135-Mmfd. Trimmer Condenser	2.50
C_4 Browning-Drake 30-Mmfd. Neutralizing Condenser	1.75
Yaxley Filament Switch No. 10 B-D	.60
C_5 Tobe 0.5-Mfd. Moulded Condenser	.90
R_1 Tobe 8-Meg. Grid Leak . .	.75
C_6, C_7 "Tinytobe" Condensers (0.001 and 0.00007 Mfd.)	.80
C_8 Mica Fixed Condenser (0.0001 Mfd.)	.75
Five Eby Binding Posts (Ant., Gnd., B + Amp., B + Det., B—) . . .	.75
One Set of Shields (Optional) . .	8.00
R. Clarostat	2.25
R_2 Browning-Drake Center-Tapped Resistor	.60
Two Benjamin Five Contact A-C Sockets	2.40
One Flashlight Lamp (2.5 Volt) . .	.10
TOTAL	$63.65

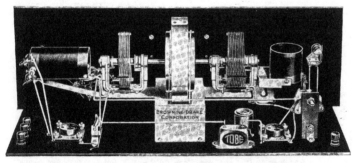

A REAR VIEW OF THE COMPLETE TWO-TUBE UNIT FOR D. C. OPERATION·

After the parts are mounted on the drilled and engraved panel it is a simple matter to complete the wiring. The use
of the single drum control unit greatly helps matters. This layout is similar to that employed in the a. c. model

CONSTRUCTION OF THE RECEIVER

THE receiver is very easy to build and for most set builders detailed constructional information is unnecessary. The leads running to the B supply and filament supply for the a.c. tubes are placed underneath the sub-panel, while the high-potential leads, which carry r.f. current, are placed above the sub-panel. These high-potential leads comprise:

The connection from the stator plate of the first tuning condenser to the grid of the first tube; the plate of the first tube through the 0.5-mfd. condenser, and from the 0.5-mfd. condenser to the primary of the r.f. transformer; the neutralizing lead, which comes from one end of the secondary of the r.f. transformer to the rotor plates of the neutralizing condenser. Of course, the connections between the grid leak, grid condenser, and the grid itself, should be as short as possible. These are the most important connections in the set, and they should be run as directly as possible and kept away from other connections

In connecting the leads' of the tubes, twisted pairs must be run from the power supply to their filaments. This is extremely important for if the wires are not carefully twisted together, a.c. hum may be noticeable. Also, the power apparatus, which should consist of a good B supply device and the necessary transformer for lighting the filaments of the tubes (the latter being either incorporated as a part of the B supply or as a separate transformer), must be kept several feet away from the receiver; otherwise there might be some 60-cycle voltage induced in the audio circuit of the radio set.

As will be noted in Fig. 1, a Clarostat inserted in the B plus lead of the r.f. tube is used for volume control. This the writer has found quite satisfactory as long as the parallel feed shown in the diagram is used. Some experimenting will be necessary on the C battery bias for the r.f. tube, as the voltage may vary between 1½ and 4½, according to the individual characteristics of the tube used.

BALANCING AND OPERATION OF THE RECEIVER

WHEN the set has been connected up 'according to the instructions given, and attached to the power supply (note: the filaments of the C-327 [UY-227] tubes take about 45 seconds to a minute to heat up sufficiently for satisfactory operation), it is ready to balance. Turning the tickler coil in one direction or the other should throw the second circuit into oscillation. This condition may be determined by placing the finger on the stator plates of the second tuning condenser, whereupon a distinct "click" or "plop" will be heard in the loud speaker. Then turn back the ticker coil until the circuit just goes out of oscillation. If the set is improperly balanced, turning the trimmer condenser slightly will throw the circuit into oscillation again, which state can be determined as before. The neutralizing condenser then can be varied by means of a wooden or bakelite screwdriver until turning the antenna trimmer has no effect whatever on oscillations produced in the second circuit.

There is another method which will give almost the exact point of neutralization. Tune-in a local station, remove one of the leads going to the C battery of the r.f. tube and, by careful retuning, the local station can in all probabilities be heard. Set the neutralizing condenser so that a minimum amount of signal is heard. This point

AN AMERTRAN PUSH-PULL AMPLIFIER

on the neutralizing condenser is almost the correct one, and the test given above can then be applied for the exact point. It will be found that neutralization is quite critical. Of course, when neutralizing, the Clarostat should be turned up as far as it will easily go (in a clock-wise direction.)

In operation, the a.c. Browning-Drake is exactly like previous models. The tickler coil may be turned up to a point where the set is just oscillating and the dial rotated, whereupon a station will be indicated by a whistle. Set the dial where the whistle is lowest in pitch and turn back the tickler coil and adjust the antenna trimmer condenser until the signal is plainly heard. Readjustments of the drum dial will then probably be advantageous.

AUDIO-AMPLIFIER SUGGESTIONS

AS HAS already been mentioned, the choice of an audio amplifier is left entirely to the constructor. There are several amplifier units and plate supply units now on the market which have very good characteristics, and it is believed that the combination of such apparatus with the two-tube tuner unit described here will pro-

As mentioned on this page, it is a two-stage push-pull affair, provision being made for the use of either a 201-A type tube or an a. c. type tube in the first audio stage

vide the user with as simple and as effective a receiver as he may need.

As an example of the type of audio-frequency amplifier and power supply that can be constructed and which will work very satisfactorily with the two-tube tuner, there is shown in Fig. 2 the circuit diagram of a combined push-pull amplifier and power unit which may be made from Amertran parts.

The amplifier is a two-stage transformer-coupled affair, designed to use in the first stage a C-327 (UY-227) type a.c. tube. The other two sockets in the amplifier are for the power tubes. Tubes of the CX-371 (UX-171) or CX-310 (UX-210) may be used in conjunction with either a type 271 (for 171 tubes) or a type 152 (for 210 tubes) Amertran push-pull output transformer. With a push-pull amplifier similar to that shown here, the comparatively large amount of undistorted power necessary for high-quality reproduction, can be obtained.

Completely assembled power amplifiers and A, B, C supply units can also be obtained from the Amertran Company. The amplifiers, when combined with the power units, result in a circuit differing slightly from that given in Fig. 2, in that they are designed to use either a CX-301-A (UX-201-A) or a C-327 (UY-227) type tube in the first stage, and CX-371 (UX-171) or CX-310 (UX-210) type tubes in the push-pull stage.

The type 2AP-10 amplifier is designed for CX-310 (UX-210) type output tubes while the type 2AP-71 is for CX-371 (UX-171) type tubes. Both of these amplifiers are the same in external appearance. The necessary Amertran power unit will furnish complete A, B, and C power to the push-pull amplifier and to all the other a.c. tubes used in the receiver proper.

The complete type 2AP amplifier (either type) lists at $60, without tubes. A set of Amertran push-pull transformers, which might be used in the home-construction of a power amplifier, can be purchased for $30. The Amertran power supply (the A, B, C Hi-Power Box) lists at $95.00.

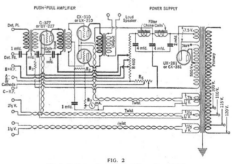

FIG. 2

The circuit diagram of an amplifier which may be used with the two-tube tuner unit. It is made with Amertran parts

THE MARCH OF RADIO

What Is the True Broadcasting Situation?

WHENEVER we see the name of Senator Howell of Nebraska in print, our eyes blaze, because we consider him to be the man who, single handed, did more to obstruct the proper administration of radio during the last nine months than any other Senator. During the closing days of the last session of Congress, his was the only dissenting vote raised to a unanimous consent agreement to vote an appropriation for the maintenance of the Federal Radio Commission.

Thereby deprived of an appropriation, the Federal Radio Commission has been most seriously embarrassed in its labors. Its effectiveness has been crippled because a petulant Senator chose to throw a monkey wrench at the radio listeners of America. The Commission can issue orders to broadcasting stations, but has no means whatever of finding out whether those orders are complied with. When its members travel, they advance the money out of their own pockets.

Instead of a staff of expert observers with high-grade measuring instruments, the Commission has had nothing to guide it but the reports of the public, the complaints of station owners and the protests of politicians, advancing the fatuous claims of every broadcasting station represented in their particular constituencies, regardless of the public interest. The Commission has not been able to employ consulting experts in order to evaluate the numerous panaceas offered it in the direction of frequency stabilization, synchronization of stations, and to determine quantitatively the exact minimum spacing which may prevail among stations of various power combinations without heterodyning within their service areas.

Consequently, when we saw Senator Howell's name in a dispatch from Washington, we expected bad news for radio. The Senator, we find, is now engaged in a war with the ghost of monopoly, which he finds skulking behind every microphone radiating chain programs. He wants a Congressional investigation, perhaps as a vehicle of publicity, perhaps for some other purpose. He may suspect that Mr. Aylesworth, president of the N. B. C., once bought a lunch for a Federal Radio Commissioner or that George McClelland, vice-president of the same company, sent a basket of apples from his Queens fruit ranch to Sam Pickard. Whatever he finds will be sensational and well phrased for newspaper publication. To save time, we offer the proposed Committee's findings even before the hearings are arranged:

Does Monopoly Rule the Commission?

THE Federal Radio Commission, at this writing, has cleared twenty-five channels and picked the best stations in the country to utilize them. Naturally, these stations, since they have been well selected, are the ones which offer the greatest variety and the best type of program. Such programs are not available in every hamlet from which a Congressman comes, but are centered at the musical and artistic centers of the United States. From such centers, a company, the odious and monopolistic National Broadcasting Company, has offered a wire service to which independently owned broadcasting stations have subscribed, making superior programs available to them. This service has been conducted at a tremendous loss, met out of the treasuries of receiving set and accessory manufacturers so that their customers would derive pleasing entertainment as a result of their purchases and therefore continue to patronize them.

Utilizing the chain programs has made the subscribing stations superior in program value to those which must rely for their programs on the Squeedunk church choir and the piano-playing professor at a roadside speakeasy. And so, by selecting good stations for good channels, the Federal Radio Commission has become guilty of inflicting the Red, Blue, and Columbia chain programs upon practically all the

AT THE INSTITUTE OF RADIO ENGINEERS
The old president and the new. On the left, Ralph Bown, president for 1927, greets Dr. Alfred N. Goldsmith who has been chosen to lead the Institute for 1928. Doctor Bown is a radio engineer for the American Telephone & Telegraph Company and Doctor Goldsmith is chief broadcasting engineer of the Radio Corporation of America as well as one of the board of consulting engineers of the National Broadcasting Company

channels in the cleared band. This is the situation which Senator Howell will uncover, perhaps by spending a few hundred thousand dollars of the Government's money to do so.

We recognize, without the aid of a Congressional investigation, that this situation is not an ideal one. It would be much more desirable if only four or five leading stations on each chain, widely separated at such points as New York, Chicago, Denver, and New Orleans, for example, should be given clear channels so that long-distance listeners who are not within the service range of other stations on the chain can secure them via a cleared route.

The rest of the chain stations should also have good channels so that they can be heard throughout their service area without an annoying heterodyne. This requirement, however, does not call for a nationally clear channel, but merely one which is free of excessive congestion. The claimed service range of chain stations, as expressed in their prospectuses, on the basis of which they sell radio advertising, clearly indicates that nationally clear channels are not contemplated by them nor is national service attempted by all or any of their stations.

But certainly such stations as WJZ, WEAF, WOR, WGY, and WGN are worthy of a nationally clear channel. Allowing five clear channels to each of the three chains would account for only fifteen of the twenty-five clear channels. The remainder should be reserved for some of the better stations, offering high grade programs from local sources, for example, WPG, Atlantic City, WBBM, Chicago, WCCO, Minneapolis, and WHAM, Rochester. These latter two do use chain programs but the excellence of their local programs ranks them high indeed.

We understand that WCCO, recognizing that chain programs are available from so many points and are therefore needed only locally, has decided to distribute its chain programs through a smaller Minneapolis station, which covers only the local Minneapolis-St. Paul area. The programs of WCCO will be almost entirely from local program sources, such as its famous symphony orchestra and other individual and distinctive programs, thus contributing to the variety of programs offered listeners near and far. This policy should be encouraged by offering such stations clear channels, thereby increasing program variety.

The long-distance listener, remote from the local service area of any broadcasting station, and that includes sixty or seventy per cent. of the radio audience, is not served if, with twenty-five cleared channels, he can hear but three different programs. Under those circumstances, there is no necessity for clearing such a large number of channels. Each clear channel means that three or four higher frequency channels must be more seriously congested. If greater program variety for the rural listener is not gained by additional channel clearing, it will soon cease. But, if there are sufficient independent stations, originating their own high-grade programs, or more chains, then the clearing process can continue even to the point where we have only clear channels with either independent or chain stations serving them and local channels for low- and medium-power stations.

At the present time, there is a monopoly of good program service, attained by subscription to the three chain programs—the Columbia and the Red and Blue Networks of the National Broadcasting system. This monopoly has been won by good service and by public tolerance. It is hard to name ten or fifteen independent stations truly worthy of cleared channels; the writer confesses he knows of but four, WPG, WCCO, WHAM, and WBBM, the latter more because of its good carrying qualities than because of any especially good programs. Of these, two use chain programs, but to a limited extent. The Commission would not for a moment hesitate to assign clear channels to good independent stations, but most broadcasting stations are hopelessly mediocre. The broadcasting band is much larger than is required to accommodate the good stations of the country. Southern politicians cry discrimination, but they cannot name five stations having program standards sufficiently high to attract consistent audiences outside their local service areas.

The Growing Political Pressure on the Commission

IN CONGRESS, the evaluation of a station is measured by the amount of pressure which the station owner can bring upon his particular Senator and Congressman. A Congressman can readily be induced to raise a large howl about the most puny, insignificant, puerile, useless, annoying, broadcasting station on earth. He can wax eloquent before the Federal Radio Commission and praise alleged services which he has never heard of except from the owners of the stations claiming to render them, an eloquence usually entirely unwarranted. Never would such an alleged statesman ask a local station owner to modify his demands for increased power or a clearer channel in the slightest iota in deference to the national situation and to aid in securing uniformly good broadcasting conditions all over the country.

We visited one of the most annoying and obnoxious stations in New Jersey which spreads a blanket of odoriferous advertising throughout the northern part of the

state. This station owner was able to show us a pile of letters which he had obtained by solicitation from his Senators and Congressmen, commending its service, although they had probably never listened to it voluntarily. These legislators urge the Federal Radio Commission to give that station special consideration over other stations in the metropolitan area of New York, simply because it means political support next time elections come around.

Our prophecy, made before the Radio Act was written, that placing broadcasting regulation in the hands of a Commission at the mercy of Congress would simply make political lobbying the criterion by which broadcasting rights are distributed, may, unfortunately, be fulfilled. The Commission has done its best to educate or to disregard the politicians, but both courses have their difficulties. If the Commission does not heed the politicians and their local needs and impossible requests, it is bound to suffer chastisement by failure to secure confirmation of its members, by curtailing of its appropriations or by special legislation requiring recognition of principles incompatible with good allocation but favoring station owners above the listening public.

Limitations to the Use of High-Frequency Channels

WE HAVE already mentioned the problem of selective fading on short-wave channels, which causes varying audio-frequency distortion, but there are a number of other points, equally potent, which must be considered when appraising the value of the short-wave region for broadcasting purposes. If the number of broadcasting stations on high-frequency

channels continues to increase, the short-wave DX audience will show continued growth. Any great increase in the number of radio receivers working on the short waves will render these waves useless for radio telephone reception because only a radiating regenerative receiver can be made to work on these waves. No sets have yet been built successfully which do not oscillate at the high frequencies, although the new UX-222 tube promises to make short-wave neutrodynes and super-heterodynes practical. With the receivers now available, however, tuning is so critical that the most skilful operator cannot avoid radiating whistles when tuning-in a station. Consequently, wide-spread listening will make the short waves even more disturbing than the broadcasting channels were during the oscillating receiver days of 1922 and 1923.

A channel utilized on short waves is international in scope, almost regardless of the amount of power used. Even a fifty-watt station, broadcasting on short waves, requires an internationally exclusive channel. Therefore, there are a maximum of 2700 channels in the entire world between 10 and 200 meters (30,000 and 1500 kc.), of which the United States might justly claim perhaps 250 for all purposes. Already the band between 25 and 50 meters (12,000 and 6000 kc.), is completely filled with stations and there are few blank spots in the remaining channels. Hence, instead of a vast unused ether territory, as those suggesting a broadcasting band seem to consider it, the high-frequency territory will soon have a congestion problem of its own.

The needs of broadcasting and the needs of the radio listener can be adequately met by a well allocated layout of broadcasting stations confined to the present broadcast-

RADIO ON CANADIAN TRAINS

Many of the crack trains of the Canadian National Railways are equipped with radio receivers to entertain passengers on long daylight runs. The sets were especially designed for this service. Loop operation is not used, every set receiving its energy from a low antenna on the car

ing band. Such an allocation can provide ample program choice, equitable service for all parts of the country, and sufficient long-distance reception to meet the needs of the public for years to come. The present band is the only one ideally suited to broadcast transmission and reception. Stable, non-radiating receivers which can select stations ten kc. apart are easily made. Neither of these features can now be embodied in sets working on the high-frequency band. Below 25 meters (12,000 kc.) channels for radio telephony must be separated by 75 to 100 kc. because we do not yet know how to avoid cross talk with closer spacing.

A further difficulty is introduced by varying hour to hour transmission qualities of the high frequencies, which requires changing of the frequency of a station every few hours if it is to give reliable service between any two points. Short-wave, transatlantic telegraph communication stations need three and four channels in order to maintain satisfactory twenty-four hour service between two points on opposite sides of the Atlantic.

More important than any of these objections to the use of high frequencies for general broadcasting, however, are the needs of the world for short-wave, radio telegraph communication networks. The high frequencies are extremely prolific in telegraph channels. For speech and music, a frequency space ten kc. wide is required but, in a band of this width, fifty short wave transmitters could be accommodated.

This is assuming, of course, that means of establishing perfect frequency stability has been discovered so that a high-frequency station might stay exactly on its channel. While there would be a maximum of 2700 broadcasting channels in the ether space from 30,000,000 to 1,500,000 kc. or 200 down to 10 meters, there are actually about 275,000 potential telegraph channels.

The needs for these channels are literally enormous. Already they are extensively used in high-frequency beam transmission across oceans and even to the opposite side of the earth. They are needed for transmitting news to large numbers of newspapers by syndicate news services. They will be needed for the tens of thousands of aircraft transmitters which will fill the skies within a decade or so and which will depend upon radio for weather information and for the safety of the lives of the passengers. Navigation companies, railroads, bus systems, police and fire systems, and forest and water patrols, will require these channels in increasing numbers. Then there are a host of private communication systems in contemplation where the use of high-frequency radio service will effect tremendous economies and savings which will be reflected in the cost of goods purchased by the public. Department store chains, businesses having factory branches at widely distributed points, ranch owners who have to communicate with employees fifty and

a hundred miles away, lumber operators, packers and shippers, and numerous other services can use short waves and use them effectively.

There is need for ship beacons, fire and signal systems, aircraft service, aircraft beacons, landing field radio beacons, and a host of other services which dwarf radio broadcasting into insignificance so far as social and economic importance is concerned.

As radio listeners and as those participating in a large and prosperous radio industry, we are inclined to exaggerate the importance of broadcasting. The most important feature of the service which radio renders is in promoting the safety of the lives of those at sea. Radio has saved thousands of lives annually by bringing aid to ships in distress. The ships of the air are not as staunch as ships at sea, and radio is much more vital to their safety. Ether congestion will add material problems to the expansion of aerial navigation.

The clamor for short-wave communication in order to soothe the feelings of broadcasting station owners must cease. We have learned the lesson of conservation of forests and now go to great effort to replenish the trees which have been so ruthlessly cut. Once they are assigned to a service, there is no way of planting new ether channels. The mistakes of to-day are not easily rectified to-morrow.

The broadcasting stations, which go on short waves for any reason other than serious research or to effect the rebroadcasting of special international events, are doing so in order that they may claim to advertisers that they cover the civilized world. Without doubt, going down on short waves will bring responses from foreign countries. If all the letters received by all the short-wave broadcasters were put in a heap, it would be found that there are three men in South Africa, about ten in Argentina, and about thirty in Australia and New Zealand who are the sources of seventy-five per cent. of all of these letters. These short-wave transmitting outfits, for which so much publicity value is claimed, are serving audiences of very small numbers.

Broadcasting on the regular channels, it may be claimed, started in the same modest way, serving only small numbers. But it had the opportunity to grow because the medium was suited to good transmission and reception. That consideration does not apply to high-frequency broadcasting. It has DX fascination but not esthetic value. As a means of distributing goodwill programs, the short waves are useless because the musical reproduction is not sufficiently good to please the listener. Those who listen to radio for the entertainment will prefer the stabler low-frequency broadcasting band.

May the publicity-seeking gentlemen who solve the broadcasting problem by newspaper interviews cease misleading the public about the possibilities of short-wave broadcasting!

The First Rayfoto Transmission

THE backers of Austin G. Cooley are exhibiting pleasing conservatism in going forward with their Cooley Rayfoto transmissions. Instead of accepting the numerous offers to broadcast through every station which invites them to the microphone, they are still quietly conducting tests with the aid of qualified experimenters. Some twenty picture recorders are in the hands of experienced set builders and they are making radio pictures, sent occasionally during the early morning hours through WOR, of L. Bamberger & Co. Newark, New Jersey. If these twenty experimenters are successful in securing reliable and satisfactory pictures without special instruction or experience with the Cooley Rayfoto apparatus, more recorders will be distributed until the reliability of the apparatus is fully established. The equipment will not be offered to the general public until it has been fully tested by typical users to their entire satisfaction.

Mr. Cooley's representatives are being flooded with inquiries both from broadcasting stations and from experimenters who desire to purchase the apparatus. Phonograph records for transmission and testing, and the complete equipment, will be distributed as soon as existing receivers have been fully tested. The recorder at RADIO BROADCAST Laboratory has been very satisfactory.

Two Stations Cannot Occupy the Same Ether Space

A NEW YORK *Times* headline proclaims that Dr. Lee DeForest has discovered room for 500,000 broadcasting stations. It is most irritating to read such misleading statements because they afford an excuse for not dealing with broadcasting conditions as they exist in the present. The remedies of to-morrow do not alleviate the problem of to-day.

Scrutiny of Doctor DeForest's remarks shows that the usually accurate *Times* misinterpreted what he said. He merely stated that there are some 6000 channels, with 10-kc. separation, between 200 and 10 meters (1500 and 30,000 kc.) and that perhaps 500,000 radio-telegraph stations could be disposed upon these frequencies if the double-heterodyne method of transmission, requiring the use of low-frequency super-audible, master oscillators, were employed.

Doctor DeForest also suggests that the same system might be utilized in the broadcasting band. Such a course would necessitate the complete scrapping of all transmitting and receiving equipment and add a delicate manipulation somewhat beyond the skill of the average listener, to the tuning process. The remedy would cost a billion dollars—or what you will—and would not even accommodate all our present stations.

Between 10 and 200 meters—considered a radio utopia by experts in publicity and politics, there are 27,500,000 cycles, or 27,500 kilocycles, of frequency space. At 10-kc. separation, which is necessary for ordinary radio telephony, there are 2750 broadcasting channels for distribution among all the nations of the world. Some twenty of these are already used by stations broadcasting on this band as well as their regular longer-wave channels some of which boast that they are broadcasting "to the civilized world" through their short-wave stations.

This is a misconception because the only persons who listen to broadcasting on the short waves now are DX cranks. No civilized person would seek musical entertainment on these short waves because of the existing audio-frequency distortion for which no one has found a cure. Programs are quite distinguishable at enormous distances but they appeal to musical tastes about as much as a five-legged calf appeals to a stock breeder. To suggest relief from broadcasting congestion by pointing to the numerous high frequencies is as practical as offering the vast Arctic spaces to the overcrowded populations of India.

The problems of the broadcasting situation are extremely complicated. Experts have no easy remedies to offer. With the unwearying persistence of Cato, we repeat that excess broadcasting stations must be eliminated. The rights of the listener and the property rights of the broadcaster are in conflict. No one, not even the broadcasting station owner, denies that the listeners' rights are paramount. Therefore, the question centers on how great are the rights of the broadcasting stations involved because the broadcasting station owner must make sacrifices so that the listener may hear undisturbed programs.

The National Association of Broadcasters estimates that the total investment in broadcasting stations, as computed on March 1, 1927, is $19,283,000. These figures were probably obtained in answer to a questionnaire and are therefore the broadcasters' own valuation of themselves. We are inclined to regard the figure as a good guess. The one hundred leading stations are probably the larger part of this investment. That leaves out about ten million dollars of disputed rights involved, of which only half need be confiscated to bring good broadcasting. If that problem is too great for the master minds in Congress to solve, we had better employ some foreign broadcasting experts.

Will the Radio Industry Do It Again?

WITH great secrecy, a score of manufacturers are designing new radio sets to utilize the experimental UX-222 (CX-322) tube. This tube is virtually a laboratory product, the much heralded double-grid type tube, having four elements. Everything points to a repetition of the radio industry's annual suicide because undoubtedly one manufacturer or another will soon startle the world with the statement that he has a set using this great tube which will make all rivals and predecessors obsolete. Then will follow the usual race and a score of manufacturers will announce new UX-222 sets, and all their previous models, some of them hardly perfected by experience, will go into the discard. In this way, the radio industry forces itself into a seasonal state and adopts new methods and types before the old ones are hardly introduced. The way in which the a.c. set was heralded completely ruined the business in battery sets. Everything points to a repetition of precedents, that the a.c. sets will be rendered obsolescent by the new UX-222 tube.

There is no sense whatever in this procedure. There is no reason for stopping the manufacture of touring cars and sedans because the roadster is popular. On the contrary, development should be pursued on all types because all have their particular and distinctive fields of service.

The annual destruction of last season's stocks by the vainglorious announcements of next season's improvements has forced the radio industry to concentrate its production during only four months of the year and to maintain its factories in virtual idleness during the remaining eight months of the year. Manufacturers do not dare to accumulate stock in excess of the immediate demands and the result of hand to mouth manufacture has been to re-

duce the quality and grade of workmen who can be attracted to the industry. Look at the advertising now current, and observe that the a.c. set has in no way reduced the simplicity, the tonal quality, or the usefulness of the battery-powered set.

The first radio set manufacturer who promises the public a revolution in radio next year by his UX-222 tube set should be squelched by the radio trade. He is paving the way to wreck the values of existing stocks of a.c. sets. The industry should determine, from now on, to make its announcements of new styles and types modest and in true perspective with the facts, and leave revolutions to Mexican bandits.

The Month In Radio

THE Secretary of the Navy's annual report states that 119,337 vessels were furnished 267,486 bearings by the Navy Radio Compass Service. The savings to other government departments by the use of the naval radio communications service amounted to about a million dollars.

IN A report on beam and directional short-wave radio telegraphy, issued by the Transportation Division of the Department of Commerce, V. Stanley Shute points out the advantage of the American beam system over the British. The American beam antenna requires only ordinary telegraph poles while the Marconi system uses expensive and complex steel masts. Another feature of the American antenna system is that provision is made for the melting of ice, overcoming serious difficulties in winter transmission. This feature is impossible with the Marconi beam. Other advantages, common to both systems of beam transmission, is the possibility of high-speed transmission, economy of power, and stability of the received signal.

Mr. Merlin Aylesworth, in his New Year statement, informs us that the Red, Blue, and

Pacific Coast networks, during 1927, spent approximately six million dollars in program presentation, of which over two million was for talent on sponsored programs, presented by some fifty American concerns. The Company itself spent a half a million dollars for its own sustaining programs. Wire cost, involving 10,270 miles of regularly used circuits, was in excess of $1,350,000. ‡ ‡ ‡ Our attention is frequently called to the verbose claims made by Norman Baker, operating direct-advertising station KTNT. He writes listeners that "his is the only station in America that is dedicated to the farm, labor, and general public's problems, which we will always fight for against powerful interests. Do you realize that advertising by radio, cuts overhead expenses so low, that merchandise can be sold at prices meaning from twenty-five to fifty per cent. saving to you?"

We have examined some of these prices, for example, the claim that Mr. Baker is selling "tires, far below others, in fact we sell 12,000 miles guaranteed tires as low as others ask for 8000 miles tires." Specifically, the 30 x 3½ oversize cord, clincher tire, guaranteed for 12,000 miles, sells for $7.70 through KTNT's radio emporium. The corresponding Sears, Roebuck, tire sells for $7.75, making the possible saving less than one per cent. Can this be called a substantial saving?

There is no law protecting the listener against untruthful statements in radio advertising, although firms, subjected to unfair competition, may invoke the aid of legal protection on that ground. ‡ ‡ ‡ Station 3 LO of Melbourne, Australia, has begun to transmit on 36 meters, and has been reported in many parts of the world. The station's organization has made application to install a number of relay stations with a view to making its programs available to all parts of the Australian continent. ‡ ‡ ‡ The Mackay interests will erect a short-wave transatlantic radio station on Long Island to supplement their Commercial Cable Company's submarine cable service. The progress which has been made in short-wave radio telegraph communication makes it unlikely that any more cables will be laid in the ocean.

AT THE BUREAU OF STANDARDS

The illustration shows a field set-up for comparing "coil antennas," more familiarly known as loops

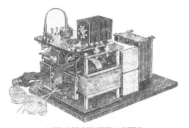

A HOME-CONSTRUCTED A SUPPLY
A view of the A power unit described in this article.
It is here shown with the metal cover removed

A New A Power Unit

By Ralph Barclay

THE convenience and desirability of a radio receiver that requires no attention other than to turn it on and off, is obvious, and to produce such a receiver has been the aim of set designers for the last few years. The problem has now been more less surmounted and it is possible to-day to purchase many receivers which are completely a.c. operated.

The design of apparatus for use in conjunction with existing battery-operated receivers to obtain socket power operation has also been satisfactorily solved, and preceding articles in RADIO BROADCAST have described many B power units and A power units which will give socket power operation of a receiver designed originally for use with batteries. These preceding articles have, however, described completely manufactured units, while it is the purpose of the present article to give the characteristics of the Knapp A power unit, parts for which can be obtained for *home* assembly.

The characteristics of the Knapp A power unit are as follows:

(a.) It can be used to supply A power directly from the light socket, to any receiver using up to about ten 201-A type tubes or combination of tubes drawing an equivalent amount of filament current. No changes at all are necessary in the receiver, the two leads from the unit merely being connected to the A plus and A minus terminals on the set.

(b.) The hum audible in the loud speaker when the tube filaments in the receiver are supplied from the power unit is so low as not to interfere at all with reception.

(c.) The cost of operation is very low—about 25 cents per month if the receiver is an ordinary five-tube set, in use an average of four hours a day.

(d.) The cost of a complete kit of parts from which the A power unit may be constructed is $22.50.

(e.) The device requires no maintenance attention other than to turn it on and off.

The circuit diagram of the device is given in Fig. 1. The transformer, T, steps down the line voltage to the correct value; the various taps on the secondary being used to adjust the output voltage. The taps are all numbered and the movable contact should be clipped onto a low-numbered tap if the receiver is a small one; if the receiver contains a large number of tubes, the clip should be placed on a high-numbered tap. The two condensers, C_1 and C_4 designed especially for use in this A power unit, and the filter choke coils, L_1 and L_2, combine to make the final output of the unit free from hum. C_1 has a capacity of about 1000 to 1500 mfd. and C_2 has a capacity of between 1500 and 2000 mfd. It is the design of these condensers, in which compactness combined with high capacity is embodied, that makes possible the A device described here. The choke coils each have an inductance of about 0.1 henry, a resistance of 3 ohms, and a current-carrying capacity of approximately 3 amps. The rectifier, R, is a dry metallic one and the particular unit used in this power unit contains sixteen pairs of electrodes arranged in a bridge circuit to give full-wave rectification. At times, when the power is first turned on, the rectifier will spark over but this does not injure it in any way for the device is self-healing, a characteristic not found generally in rectifiers of this type. The rectifier has a life, according to its manufacturers, generally in excess of 1000 hours. It is held in place by several clamping screws and can easily be re-placed when necessary. A new rectifier can be obtained for $6.00.

A complete kit from which the power unit can be constructed should contain the following parts:

T—Transformer
Rectifier Unit
C_1, C_2—Special High-Capacity Condensers
L_1, L_2—Choke Coils
Drilled Base Plate (Copper-Plated Steel)
Drilled Top Plate (Brown Bakelite with Studs in Place)
Contact Plate (With Mounting Bracket)
H & H Toggle Switch
Drilled Baseboard
Metal Cover
A. C. Line Attachment Cord with Plugs
Output Cord for Connecting to Set with Polarized Plug
Complete Instructions for Assembly and Operation
Necessary Nuts, Screws, Clamps, etc.

The constructional data supplied with the kit of parts are very complete and with their aid, the building of the unit may easily be accomplished in an hour or two. The baseboard is all drilled and the first job is to fasten all the parts to it. The unit may next be wired with any ordinary kind of rubber-covered wire.

In order that it may operate most satisfactorily, it is essential that either the A plus or A minus be grounded. In most receivers one side of the filament circuit is grounded but in those cases where this is not true, it will be necessary to connect a ground between the A minus, preferably, and the regular ground post on the receiver.

Light socket operation of a radio receiver is, therefore, a simple matter using a Knapp A power unit and good B power unit. The Knapp A power unit is supplied with an extra plug into which the supply lead for the B power circuit may be connected, and the entire installation is then controlled from the single power lead on the A power unit, and turning on your receiver becomes a matter of merely pushing a plug in a light socket receptacle.

FIG. 1
A circuit diagram of the home-constructed A supply

110 Volts
A.C. 60 Cycles

Switch

Rectifier

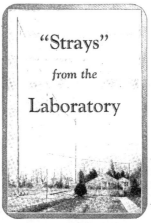

"Strays" from the Laboratory

Poorly-Designed Amplifiers

IN THE Laboratory at the present moment is an a.c. receiver with a well-known name which uses a three-stage audio amplifier, two stages being resistance-coupled and one transformer-coupled. Although the power tube, a 171 type, draws no measurable grid current, and although the plate current of this tube is constant as measured by a d.c. meter, the quality is bad; something seems to overload. What is wrong?

The transformer couples the detector to the amplifier while resistance-capacity units connect together the remaining amplifier stages. There is no C bias on either of the resistance-coupled tubes (this seems to be standard practice, although one has difficulty in understanding why). To load up the final tube, 40 volts are required on its grid. Since the next-to-the-last tube is coupled to this power tube by a resistance-capacity unit, there is no voltage gain except that due to the tube. In other words, the next-to-the-last tube must have 40 volts a.c. in its plate circuit, necessitating about 5 volts a.c. on its grid, if the amplification factor of the tube is 8, which is correct for a 226 type tube. Five volts a.c. on the grid of a tube which has no C bias will cause severe overloading, and when a voltmeter is placed from the plate of this tube to the negative filament, the needle jumps violently. The voltage between negative filament and the plate battery side of the resistor shows a voltage which is found constant at 175 volts.

When the next-to-the-last tube overloads, considerable d.c. plate current is generated. This plate current generated in the tube, due to overloading, is opposite in direction to the no-signal plate current and the voltage drop across the coupling resistor due to this decrease in plate current reduces the voltage drop across the resistor due to the no-signal current. This allows more plate voltage to be impressed on the tube, so that the plate varies from a steady no-signal value of 25 to 50 on the modulation peaks.

Now let us suppose that the transformer is used to couple the second stage of the amplifier to the power tube, which still requires 40 volts a.c. on its grid. The turn ratio is 3 so that across its primary are required 40 ÷ 3, or 13.3 volts which, divided by 8, the mu of the tube, gives 1.67, which is the grid a.c. voltage which the next-to-the-last tube requires. If no C bias is used, much less severe overloading will occur than if 5 volts appear here, and in a fair loud speaker it is true to say that no loss in quality will be noticeable.

The answer is naturally to use C bias on all tubes, but if this cannot be done, the transformer should be used last in the chain. This is a better plan for another reason—that of using as high an impedance as possible in the plate circuit of the detector.

The plate resistors are each of 100,000 ohms value, and for a transformer to have that impedance at 60 cycles—which is as low a frequency as anyone need worry about—its primary must have 160 henries inductance!

This amplifier is clearly a case of poor engineering, and a small amount of pencil and slide rule work would have shown the designers of this receiver what could have been expected before they had even bult one, up and put it on the laboratory bench and tested it with expensive instruments.

Quartz Transmitting Tubes

FROM time to time we read, and are mildly amused, about the new glass substitutes which admit ultra violet into our homes—even in large cities where there is practically no ultra violet because of smoke. During the early development of the General Electric's fused quartz, which transmits 92 per cent. of all radiation, from ultra violet to heat,

we had the pleasure of working with Dr. Berry of the General Electric Company and the Harvard Cancer Commission, on some uses of this beautifully transparent glass.

Fused quartz has another important quality in addition to its transparency to ultra violet. This is its coefficient of expansion which is so low that it is possible to grind telescope lenses of quartz without the many hours of slow work necessary when our best lens glass is used. Under the heat produced by the grinding process the quartz expands so little that there is no danger of cracking, and it is possible to grind the lens to the desired dimensions with less regard to internal mechanical stress.

One of the great difficulties with high-power transmitting tubes lies in the heat generated, which must be dissipated. Without danger to the glassware. Water-cooled tubes mark one step in the development of safe high-power tubes; another may be the use of fused quartz, as is being done experimentally in England.

We remember a demonstration in a motion picture projection room where glass condensing lenses are used which get very hot and which occasionally crack because of the severe mechanical stress in the glass. A pair of lenses had been made—large pieces of convex quartz carefully ground and polished—and were taken to the motion picture house for test. They worked beautifully, of course, and after the show it was necessary to remove the lenses and take them back to the laboratory.

This would have meant a delay of appreciable time for them to cool if they had been made from glass, but despite the operator's disclaiming any responsibility (the quartz lenses were worth their weight in gold) the lenses were removed, thrown into a pail of water, and within a very few moments they were sufficiently cool to be removed to the laboratory.

While there is no prospect that there will be any need to cool a 100-kw. tube as quickly as this, or that anyone would want to carry such a tube around, hot or cold, a tube that is impervious to temperature changes will relieve engineers from one worry at least.

Fused quartz is wonderfully transparent; one can read a paper through a yard of it; it expands but little under application of heat and, as a matter of fact, is difficult to heat because the radiation passes through it without being absorbed. It has another characteristic—it is frightfully expensive.

Quiet A. C. Sets

SOME interesting tests have been made by the manufacturers of Kuprox, one of the dry rectifiers, on using a combination of rectifier and a.c. tubes. In every case the audibility of hum emanating from the receiver was decreased by the use of partially filtered a.c. on the filaments.

One receiver was a standard five-tube set using McCullough a.c. tubes. With d.c. throughout, the audibility of hum—the origin of which was unknown—was 2; with a.c. on the heaters of these tubes, the hum rose to 250 audibility units and dropped to 46 when the Kuprox rectifier and a single choke coil was used to rectify and partially filter the a.c. Another receiver was a well-known product requiring no batteries. With d.c. on the filaments and a socket power device for supplying the plate current, the hum was from 120 to 150 units as measured on a General

General Elec. Co.

A SLAB OF FUSED QUARTZ
The writing would not be legible through a piece of ordinary glass of equal thickness

351

Radio audibility meter. When the special a.c. converter, supplied by the manufacturer furnished power for the filaments, the hum increased to 1000 units, but dropped to from 90 to 120 when the Kuprox was added.

While we do not agree with the manufacturers that Kuprox is "the most important discovery in radio since the first three-element vacuum tube was successfully used", and cannot agree with advertising writers of a.c. sets who state that there is absolutely no hum from their sets, if a Kuprox unit, or other rectifier, and a choke coil, will enable one to use a.c. tubes without appreciable hum, we are willing to advise our friends to go in for a.c. receivers.

At the present moment it is difficult to subscribe whole-heartedly to the apparent craze for a.c. sets. Without a doubt the ultimate receiver will require no attention, will be foolproof, and will operate from any lamp socket, but at the present moment we are not convinced that we should junk our storage battery and charger outfit, which is perfectly quiet and which requires an expenditure of about eight minutes a week to put on and take off the charger, for a newfangled receiver that can be neglected as soon as it is plugged into a socket.

While on the question of a.c. operation, we might give the following data on Radio Corporation of America a.c. tubes, which come from the Technical and Test Department of that concern, and are average data of a great many measurements:

MEASUREMENTS MADE WITH D.C. FILAMENT SUPPLY

UX-226

Ef = 1.5 v.; Ec = –9.0 v.; Ep = 135 v.

Plate Impedance	8000 Ohms
Amplification Factor	8.2
Mutual Conductance	1050 Micromhos
Filament Current	1.05 Amperes
Plate Current	5.2 mA.

UY-227

Ef = 2.5 v.; Ec = –6.0 v.; Ep = 90 v.

Plate Impedance	9810 Ohms
Amplification Factor	8.9
Mutual Conductance	907 Micromhos
Filament Current	1.75 Amperes
Plate Current	3.1 mA.

Into What Impedance Should the Tube Work? CONSIDERABLE uncertainty seems to exist in the minds of popular writers as to whether the impedance into which a tube works should be equal to that of the tube or several times greater. The answer is that it all depends.

If any source supplies power to a load, the

maximum output of power will be absorbed when the effective resistances of the source and the load are equal and when the reactances are equal but opposite in sign. Under these conditions the power absorbed will be the voltage squared divided by four times the effective resistance of the source.

In other words, to supply power the numerical value of the impedances must be equal—to use the usual semi-technical language.

If a tube is to be used as a voltage multiplier, say in a resistance-coupled amplifier, greater amplification will result if the load resistance is several times that of the tube resistance. In a transformer-coupled amplifier, the impedance which the tube looks into—the effective imped-

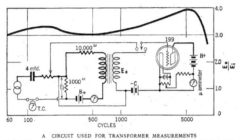

A CIRCUIT USED FOR TRANSFORMER MEASUREMENTS

In the Laboratory the voltage appearing across the secondary of a new type A Sangamo audio transformer was measured with this set up with a constant d.c. flowing through the primary and a constant a.c. voltage impressed upon the primary. The resultant curve is also shown here

ance of the primary of the transformer,—should be several times the impedance of the tube at the lowest audio frequency which it is desirable to transmit. At the higher frequencies a combination of effects takes place to maintain the amplification more or less flat. With a good transformer, that is, one which has a very high primary impedance, and a high impedance tube, the amplification will still be peaked. This is the price one pays for amplification—loss in fidelity. If it were possible to make a tube whose impedance would be low and still have a high amplification factor, it would be possible to use good transformers and have high amplification and a high degree of fidelity. Or, if it were possible to build 10:1 ratio audio transformers which would not "go resonant" and otherwise cause trouble at audio frequencies, we should have high gain and high fidelity—but here again we are in difficult water.

This impedance matching problem prompts one reader to ask if it is wise to add resistance to a tube plate circuit in case we desire the tube to work into a resistance several times the internal impedance of the tube but if the load impedance is not this great. A concrete example will illustrate the question. Suppose we have a 10,000-ohm

tube and a 10,000-ohm load. Since greatest amplification will result if the load has, say 30,000 ohms impedance, shall we add 20,000 ohms to the plate circuit? The answer is no, unless the voltage across the entire 30,000 ohms is made use of, and not that across the 10,000-ohm load alone. If the additional 20,000 ohms can be included in the load, somewhat greater voltage amplification will result.

Audio Transformer Measurements THE LABORATORY has recently had the pleasure of measuring, according to the N. E. M. A. standard method already described in these columns, the voltage appearing across the secondary of the new Sangamo Type A audio transformer when a constant d.c. current flowed through the primary, and when a constant a.c. voltage was impressed upon the primary in series with 11,000 ohms. The circuit is given on this page as well as a curve representing the result of the measurement, which should be interesting to all pursuers of fidelity in reproduction.

As mentioned before, we do not believe measurements, such as this, on single transformers without accessory tubes and common impedance which always exist in a standard two-stage amplifier, mean a great deal, because we have found in practice that the curve obtained by measuring the voltage across an output resistance of a two-stage amplifier may differ altogether from what one obtains by measuring a single stage only. This should not be taken to indicate that a two-stage amplifier using Sangamo Type A transformers would not be as good as is indicated in our curve—it might be better. It all depends upon the care taken in the construction and the amount and kind of feedback existing in the circuits.

A single Silver-Marshall type 220 audio transformer falls off badly above 3000 cycles when measured singly, while a two-stage amplifier using these transformers is perfectly good up to 5000 cycles, the additional amplification at these higher audio tones being due to regeneration. Many two-stage amplifiers will howl or sing at the higher audio frequencies if sufficient impedance exists in the common negative plate-battery lead. This difficulty is easy to remedy, usually, and necessitates the use of a 2-mfd. condenser across the B batteries, or socket power device. The transformers used in such amplifiers usually have a rising characteristic when measured singly, as was done in the Laboratory to measure the Sangamo transformer.

AN EARLY PICTURE OF MR. HOGAN

He is here shown using the first audion ever made, at which time he was DeForest's Laboratory assistant

An Interview With J. V. L. Hogan

By Edgar H. Felix

NO INDUSTRY is more clearly the product of inventive ingenuity than radio. If necessity is the mother of invention, radio must take after its paternal parent—creative imagination. It has grown to its imposing importance, not by virtue of necessity, but by discovering for itself a new field—the bringing of entertainment into the home.

"It is my belief" says John V. L. Hogan, pioneer radio engineer and technical authority, "that a vital and fundamental change has taken place in the spirit of the industry. Instead of a competition of ingenuity, the majority of manufacturers are now content to copy the designs of a few leaders who shape the trend from season to season."

Mr. Hogan is qualified to speak of the spirit of the radio industry because he has been intimately associated with the progress of that industry for more than twenty years. Shortly after Marconi had sent his famous first signal across the Atlantic in 1901, Hogan, then a boy in his teens, witnessed a demonstration of wireless telegraphy at the University of Wisconsin. Two years later, he spent a summer at San Juan, Porto Rico, most of it within the four walls of the radio station there. As it has with so many after him, radio made short work of its victim, imbuing him with its incurable fascination. From then on, radio has been his principal interest in life. He did not lose time in associating himself with the best minds in the art.

During 1906 and 1907, Hogan worked as DeForest's laboratory assistant and, in that capacity, used the first audion ever made. After his association with DeForest, he continued his studies at Yale University, specializing in mathematics and physics. But, before his work was completed, Reginald Fessenden, recognizing his natural ability for conducting experimental work, employed him to assist with his experiments at Brant Rock, Massachusetts. Here, in 1909, the Arlington Naval Station transmitter was designed, assembled, and tested. Mr. Hogan's association with Fessenden extended over a period of many years.

After completion of the Arlington transmitter and its acceptance by the United States Government, Fessenden and his National Electric Signalling Company transferred their activities to Bush Terminal in Brooklyn. Hogan became chief research engineer and later manager of the company.

In the laboratory, operated under his direction, the basic principles for continuous-wave transmission were formulated by Fessenden. The first successful, high-frequency alternator was built for Fessenden at the General Electric Laboratories. E. F. W. Alexanderson, already distin-

J. V. L. HOGAN

guished by his achievements in electric railway locomotive design, was assigned to the problem and successfully built and later perfected the high-frequency alternator. These were fruitful periods of research, both in transmission and reception. Hogan is credited with many inventions, perhaps the most important of which is the detector heterodyne which he described in the Proceedings of the Institute of Radio Engineers in 1913. He was a founder and later became president of that body.

Fessenden is the inventor of the heterodyne principle. In its original reduction to practice, two windings in the telephone receiver were used, one to carry the incoming detector signal and the other to superimpose the local high-frequency oscillations. Hogan's invention is the method of combining the incoming signal with the local heterodyne signal, before detection by an electrical rectifier. Owing to the square law action of the detector, enormous increases in sensitiveness are attained, accounting largely for the effectiveness of the regenerative and super-heterodyne systems. He was also the first to disclose the advantages of single control, so widely applied in broadcast reception, even before the various tuning circuits of receiving sets followed any law of regular and equalized progression.

Identified so closely with the early growth of radio, Hogan is now recognized as a leader in the patent and engineering fields. His views on the patent situation of to-day and the causes underlying its complexity are founded on the best possible authority.

"Contrasting with the early tendency toward extraordinary ingenuity," said Mr. Hogan, in answer to the writers' request that he amplify the statement which opened the interview, "the radio industry has largely adopted the habit and practice of copying the designs developed by

353

the leaders of the industry. For example, the most widely used modern radio receiver, almost a standard design for the entire radio industry, is the alternating-current, tuned radio-frequency set, employing no batteries, self-contained with all the necessary power supply, and using a single-control tuning system.

"Following so closely the same design principles, practically all makers use the same inventions, covered by the same patents. With such a formidable array of patents, some adjudicated and some not, the decisions of the courts are bound to have salutary results upon the economic situation in the radio industry. As an example of the effect of only a single patent, the recent adjudication of the Alexanderson tuned radio-frequency patent was sufficient incentive to more than twenty of the leading manufacturers of the industry to obtain licenses from the Radio Corporation of America, which holds the right to issue licenses under this patent. If future adjudications continue favorable to the Radio Corporation, these licensees will not suffer any great changes in their status. But suppose that the Latour, the later Hazeltine, the Lowell and Dunmore patents, and some of the other patents whose scope and validity are not yet tested in the courts, should be adjudicated in favor of their holders, there may be another set of substantial royalties to pay. The holders of patents applying to vacuum tubes, such as the thoriated filament, pure electron discharge, and the magnesium keeper, may collect large penalties from independent tube manufacturers who have so far disregarded them.

"The radio industry has problems still ahead of it," continued Mr. Hogan, "although there is no doubt that they do not mean its destruction or its paralysis. Their existence is due to a very fundamental weakness in the present conduct of most manufacturers in the field, namely, the tendency to stereotype and imitate.

"The radio industry has not always been prone to follow the designs of its pioneers, but showed independent inventiveness on the part of many individuals. Marconi, in the early days, held a patent on the insulated, grounded antenna. Instead of waiting for the adjudication of that patent, or entirely disregarding it, rival inventors worked out means so that they would not have to use the Marconi antenna system, by devising the loop type of aerial for instance.

"In the field of detectors, after the coherer and the magnetic detectors had been invented, the manufacturers in the field simply developed other forms of detector. The electrolytic of Fessenden, the crystal detector, and the three-element vacuum tube of DeForest, each a valuable contribution to the industry, were invented to improve service and, at the same time, their manufacturers and users were practically cleared of infringement of existing patents. Thus patents, instead of serving as a constricting and restraining influence, were the stimulus to making some of the most important inventions in the radio art.

"Only three sensible courses are open to those who find patents apparently covering devices which they desire to make or actually do make," said Mr. Hogan. "They are either pay, or fight, or don't use. The second of these is often hazardous. Many members of the radio industry elect a still more hazardous course—to use without paying until patents are adjudicated. Naturally, these are likely to have to pay dearly in the end. Many of those who fight do so in the spirit that patents are a danger and a menace, to be fought and destroyed, though this is obviously not a sound position. When forced to do so, they pay reluctantly because there is no other course open to them, rather than encouraging inventions and utilizing them legitimately and without restriction. The alternative of developing new means so as to avoid infringement is the one least used, although it is the course that would contribute most to the progress of radio."

Mr. Hogan cited specific instances of unadjudicated patents which are almost universally used by the industry and which manufacturers are seemingly making no effort to circumvent by developing substitute or improved means. He even showed how, in the field of broadcast transmission, the tendency is to endeavor to make the best of conditions as they are instead of attempting seriously to develop and put into use new means and methods.

Mr. Hogan has had exceptional opportunities in studying transmission phenomena. Fessenden was one of the first to build a successful radio telephone transmitter, having established a record of several hundred miles' range as early as 1906. When the Arlington transmitter was still installed at Brant Rock, prior to its acceptance by the Navy, a most comprehensive series of tests, the first extensive study of radio transmission phenomena, were made with its aid. Data as to attainable range, with various powers, at all hours, were collected aboard the U. S. S. *Birmingham*. Out of this mass of data, the historic and useful Austin-Cohen formula was worked out. Hogan was intimately concerned with these tests, being in entire charge of the Brant Rock end. With such pioneer study of transmission phenomena as a foundation and the many subsequent years of research and contact, his suggestions as to the most effective way to attack the present broadcasting problem are worthy of most profound consideration.

"The limitations of our ether channels are almost as definite and specific as familiar laws of physics," Mr. Hogan stated to the writer. "The number of solid bodies of a certain size that can be fitted into a room of certain dimensions is readily calculated and no one attempts to deny the operation of the law which determines it. In the available ether, likewise, there are just so many broadcasting channels, each of which can accommodate just so many stations of a certain power. There is no evading this law. If we squeeze more programs than fit in our ether space, the programs are certain to be damaged.

"The number of stations which can be accommodated in the broadcasting band is directly a factor of their power. Either we maintain many stations comfortably in the band by reducing their power, or we increase their power and reduce the number of stations. If we exceed the capacity of the ether, as we are doing at the moment, we have confusion. There is no successful evading of technical laws upon which the capacity of the ether is founded, unless we establish new laws by making new discoveries. As in patent tangles, so in ether tangles, inventions may cut the Gordian knot.

"To continue to increase the number and the power of stations on the air, we must attain such objectives as perfect synchronization of carriers and their modulation, limitation of carrier range to the area actually served by programs, or modulation of the carrier in a new way which narrows the band occupied by a fully modulated carrier. None of these things is impossible, but how much more energy is spent by broadcasters in clamoring to stay on the air than in developing the means which will make room for them on the air.

"Founded, as the radio industry is, upon ingenuity and invention, the work of the research laboratory is still its most valuable asset and is still the only really effective gateway to the solution of its problems. Mere imitation is fatal to its growth and to its economic future. To-day, for every dollar spent on research, hundreds of dollars are spent on imitation. The crying need of the industry is research and originality, the employment of engineering genius and continuous technical growth. With proper concentration on these factors, patent difficulties will be mitigated and the fullest potentialities of the industry and its field of service will be developed."

THE FIRST PRACTICAL HIGH-FREQUENCY ALTERNATOR
It was built for Fessenden at the General Electric Laboratories, by E. F. W. Alexanderson. The latter is now in the public eye on account of his television experiments

RADIO BROADCAST Photograph

THIS IS HOW THE COMPLETED RECEIVER LOOKS
It fits into any standard cabinet that will take a 7 x 18 inch panel

A Four-Tube Screened-Grid Receiver

By McMurdo Silver

THE four-tube all-wave receiver described herewith represents what is probably the most profitable application to a radio receiver of the new screened-grid tube, from a performance per dollar per tube standpoint.

The new screened-grid tubes, offered, as they have been, by the tube makers with little or no actual practical operating data, present most attractive possibilities to the experimentally inclined by virtue of the increased amplification that they theoretically make possible. The application of one of these tubes in the four-tube all-wave receiver described here provides a receiver of high sensitivity, selectivity, and general worth, and the methods of utilizing the UX-222 (CX-322) tube should be of considerable interest to the experimentally inclined in view of the dearth of practical operating information; a number of misconceptions concerning the UX-222 tubes have already arisen in the minds of many radio fans.

The receiver pictured in the accompanying illustrations employs one UX-222 tube in a tuned r.f. amplifier stage with conventional transformer coupling between this tube and the detector, despite popular belief that the UX-222 tube should work with tuned impedance coupling. The detector is made regenerative by the use of a fixed tickler winding, with regeneration controlled by a midget variable condenser. The two-stage audio amplifier, employing two large-core heavy 3:1 ratio transformers, provides a practically flat curve from below 100 cycles to over 5000 cycles. The measure of the receiver's

true worth is its performance compared against other sets. Operated in a steel building in Chicago with a forty-foot wire hanging out of a window for an antenna, a model set brought in stations within a range of 1000 to 1500 miles on the loud speaker, while KFI, in Los Angeles, was faintly heard on the loud speaker through local interference. A couple of popular one-dial factory sets on the same table refused to "step out" at all. The four-tube receiver was moved to a

THIS article describes a four-tube receiver which is notable chiefly because of its economy—all the parts listing for a total of $46.75. This feature alone should attract many a home constructor and professional set builder. As the author states in the article, the use of the screened-grid tube increases the voltage to the detector, from a distant station, about twice compared to that delivered by a 201-A type tube. This gain is not all that the new shielded-grid tube is capable of producing, but is about all that is possible in a single stage and with the requisite degree of selectivity. This is a distinct gain, and is desirable, and is due to the new tube alone, but the home constructor should not expect a doubling of voltage before the detector to work miraculous results on extreme DX; it means, simply, that louder signals will be received without the bother of neutralizing apparatus or tricky adjustments. The receiver delivers excellent tone quality.

—THE EDITOR

Chicago suburb, where it brought in stations on the East and West coasts with ample loud speaker volume, pulling in some fifty stations in one evening on a fifty-foot antenna. Yet the parts for the whole set cost less than $50.00.

In developing the set, experiments were started on the basis of individually shielded stages for the r.f. amplifier and detector, using tuned impedance coupling as recommended in the UX-222 data sheets. It was immediately found that, while quite high amplification could be obtained, varying from 20 per stage at 550 meters (545 kc.) to about 55 per stage at 200 meters (1500 kc.), the amplifier was far too broad for practical use. Using the optimum value of coupling between r.f. amplifier and detector, the selectivity was even worse, though the amplification increased. All this was previously predicted by mathematical analyses of the system, which experiments simply served to confirm. Tuned impedance coupling was then abandoned, and a standard tuned r.f. transformer employed, having a secondary coil equivalent to the coil previously employed in the tuned impedance amplifier circuit. This transformer was provided with adjustable values of primary coupling, and a series of amplification measurements were made at different wavelengths, the different primary sizes providing varying degrees of selectivity. Some representative amplification curves are reproduced in Fig. 1. The final transformer selected employs ninety turns of No. 20 plain enamelled wire wound on a threaded moulded bakelite form, with turns spaced to provide low r.f. resistance, this coil being tuned by a 0.00035-mfd. variable condenser to cover the wavelength band from 200 to 550 meters (1500 to 545 kc.). The primary consists of 55 turns of No. 32 d.s.c. wire wound on a 1¼-inch tube slipped inside the secondary form, the primary winding being at the filament end of the secondary. The actual amplification obtained with this coil and a UX-222 tube is shown on the accompanying curves, varying from 11 at 550 meters to 35 at 200 meters for the stage. This value is very low as compared to the mu of the UX-222 tube (about 250) but it is about all that can be realized without adversely affecting the selectivity. Actually, the figure of merit for the UX-222 tube is only about twice that of a UX-201-A, despite its misleadingly high amplification factor, so that it is entirely proper that the actual amplification obtained in practice from the UX-222 tube should only be about double that of a UX-201-A tube, *at substantially equal values of selectivity.* As stated, throwing selectivity to the winds, the gain could be almost redoubled—an impossible condition in practice.

At this point, conclusions arrived at on the laboratory bench were checked experimentally, and it was found that shielding was not necessary or even helpful in the four-tube circuit, providing the coils were spaced about twelve inches apart, as has been done in the final receiver layout. Precautions were taken to keep coupling between the r.f. amplifier and the detector circuits at a minimum by bypassing, but even with shielding it was found that "motor-boating" was experienced when the detector was adjusted for critical regeneration. An analysis of the UX-222 tube's action indicated that this was due to changes in screen grid current reacting on the detector, and vice versa, through the coupling of the B battery. This was eliminated by an r.f. choke in the screen grid lead. The operation then became entirely stable, and the results obtained in tests were quite gratifying.

The physical aspects of the set are well illustrated in the photographs. On the attractively decorated walnut-finished metal front panel are mounted the vernier drum dials controlling the two 0.00035-mfd. tuning condensers, which are of the modified straight frequency-line—straight wavelength-line—type. These drums read nearly alike in operation, log definitely for any station

heard, and are provided with small lamp brackets for illumination. Between them is the 0.00075-mfd. midget condenser controlling regeneration, and below it the 50-ohm rheostat controlling the filament voltage of the UX-222 tube to regulate volume. Attached to the rheostat is an automatic "On-Off" switch turning the set on or off at will.

The a.f. amplifier is unusual in that it employs UX-112-A (CX-312-A) type tubes in both the first and second stages. The first stage operates at 135 volts plate potential, with 4½ volts C bias. These values insure good handling capacity and a low plate impedance, favoring good bass-note reproduction. The second stage should have from

135 to 180 volts of B battery with 9 to 12 volts of C bias. At the higher value, the undistorted power output is nearly 300 milliwatts, a respectable value. Of course, a UX-171 (CX-371) tube could be used with greater undistorted power output, but in this case an output device should be used. The total current consumption of the whole set is about 15 milliamperes, a very low value even for battery operation.

The simplicity of the front panel is in keeping with the general design of the receiver, which is simple in the extreme. All parts not on the panel are mounted on the wood sub-base. The antenna coil and coil socket are at the left end, and the r.f. transformer and its socket at the right end

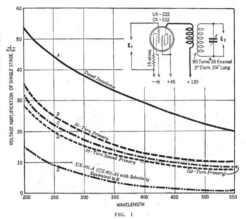

FIG. 1

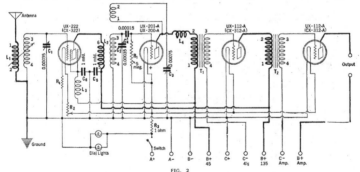

FIG. 2

The schematic diagram of the four-tube receiver which uses a screened-grid r.f. tube

A CLOSE-UP OF THE FRONT PANEL

turned out slowly, the squeal will disappear and the station program be heard. Volume may be controlled by the knob, adjusting R_8. In tuning for local stations, the regeneration condenser can be left set about one quarter in (far enough out so that no squeals are heard), but in tuning for distant stations it should always be turned in far enough to make the detector oscillate, and stations first picked up as a squeal, and then cleared up by turning the regeneration condenser out slowly to cut out the squeal and get the program. The set is most sensitive, the operator will find, with the regeneration condenser just barely out of the squealing condition.

of the base. On the baseboard beneath the variable condensers are fastened the 1.0-mfd. bypass condensers. The positions of all other parts is clearly illustrated in the photographs.

The actual parts used in the set are listed below, and due to the possible complexities that might arise on account of the UX-222 tube, it would be well not to substitute other parts for those specified, particularly as the wavelength ranges might also be affected were parts of the oscillating circuits changed in any way.

LIST OF PARTS

L_1—S-M 111A 190-550 Meter Antenna Coil	$ 2.50
L_2—S-M 114SG 190-550 Meter R.F. Transformer	2.50
Two S-M 515 Universal Interchangeable Coil Sockets	2.00
L_3, L_4—S-M 275 Chokes	1.80
T_1, T_2—S-M 240 Audio Transformers .	12.00
C_3, C_2—S-M 320 0.00075-Mfd. Variable Condensers	6.50
C_3—S-M 342 0.000075-Mfd. Midget Condenser	1.50
C_5—Sangamo 0.00015-Mfd. Condenser	.40
C_6, C_7—Fast 1-Mfd. Condensers . .	1.80
R_1—Polymet 5-Megohm Grid Leak .	.25
R_2—Carter IR50-S 50-Ohm Switch Rheostat	1.50
R_3—Carter H1 1-Ohm Resistance .	.25
R_4—Carter H15 15-Ohm Resistance .	.25
Polymet Grid Leak Mount	.50
Ten Fahnestock Connection Clips . .	.50
Four S-M 511 Tube Sockets . . .	2.00
Two S-M 805 Illuminated Vernier Drum Dials	6.00
7 x 17 x ½ Inch Wood Baseboard with Hardware	1.50
Van Doorn 7 x 18 Inch Decorated Metal Panel	3.00
TOTAL	$46.75

The assembly of the set is very simple. The condensers are first mounted on the dial brackets, the drums attached, and the brackets fastened on behind the front panel, which also carries the dial windows and the small dial lamp brackets. The rear of the panel should be scraped to insure good contact between the panel and dial brackets and the same precaution should be observed in mounting the midget condenser so that its shaft bushing makes good contact with the panel. The switch rheostat should be thoroughly insulated from the metal panel by means of two extruded fibre washers.

All parts mounted on the base are screwed down as shown, taking care that terminal 3 of the antenna coil socket is to the right, and post 3 of the detector coil socket (r.f. transformer) is to the left. The positions of these sockets, audio transformers, etc., should be exactly as

shown, all being screwed down using roundhead No. 6 wood screws from ⅜ inch to 1⅛ inches (for the r.f. chokes) long.

The wiring is simple, and clearly illustrated in the schematic diagram, Fig. 2 in which all instruments represented by symbols, carry exactly the same numbers and markings in the diagram as they do physically. All grid and plate leads are of bus-bar, in spaghetti where necessary, while all low-potential battery wiring is grouped along the center of the base, and is of flexible hook-up wire. After all wiring is done, the central group is cabled, or laced, using waxed shoemaker's thread. The Fahnestock clips are used for battery connections.

Testing and operating the set is very easy, and involves the use of standard accessories as listed below:

Two UX-112-A (CX-312-A) Tubes
One UX-201-A (CX-301-A) Tube
One UX-222 (CX-322) Screened-Grid Tube
One Western Electric Cone Loud Speaker
One 6-Volt A Battery
Three (or Four) 45-Volt Heavy-Duty B Batteries
Two 4½-Volt G Batteries

With the batteries connected and tubes in place, the rheostat should be turned full on, which will give 3.3 volts to the UX-222 tube under average conditions. With the midget condenser all in, the right-hand tuning dial should be rotated until a squeal is heard (every squeal is a station). The squeal should be tuned-in loudest by proper adjustment of the left-hand and right-hand drums. If the midget condenser is then

The rotor in the antenna coil may be adjusted with the fingers to give greatest sharpness of tuning on the left-hand tuning dial. When set at right angles to the coil form, greatest selectivity and least volume will be obtained.

By using a 111D antenna coil and a 114D r.f. transformer in the coil sockets, the set will tune from 500 to 1500 meters (600-200 kc.). A 111E and a 114E coil will go from 1400 to 3000 meters (215-100 kc.). To operate below 200 meters (1500 kc.), the screened-grid r.f. amplifier is cut out entirely and the antenna connected to post 3 of the detector coil socket through a 0.0000-25-mfd. variable midget coupling condenser, such as the S-M 340. With this connection, using the right-hand condenser dial only to tune the regenerative detector, the three-tube set will tune from 70 to 210 meters (4285-1430 kc.), with a 114B coil, from 30 to 75 meters (10,000 to 4000 kc.), with a 114C coil; and down to about 18 meters (16,660 kc.) if the stator winding of another S-M 114C coil is cut down to four turns. It is not desirable to use the UX-222 r.f. amplifier stage below 200 meters, the three-tube portion of the set being amply sensitive for all short-wave reception.

It will be necessary to shunt the regeneration condenser with a fixed capacity of about 0.0001 mfd. to cause oscillation at the frequencies below the broadcast band.

Of course, suitable A or B power devices may be used with the set, glow-tube equipped B units being most satisfactory.

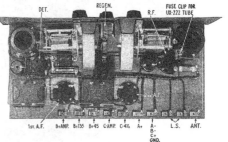

SHOWING THE ARRANGEMENT OF APPARATUS ON THE BASEBOARD

Some

New Loud

THE richness of its selected woods instantly distinguishes this beautiful Bosch receiver (shown to the left), from the commonplace. The model 87 is a seven-tube table type receiver having four stages of balanced radio-frequency amplification, a detector, and two transformer-coupled audio stages. Exceptional selectivity is possible, while provision is made for very exact and gradual control of volume. Needless to say, there is but a single dial for station selecting, and this is graduated in kilocycles. Engineering features make the receiver adaptable for all forms of power supply, while there is a switch on the front panel which not only controls the power for the tubes, but automatically turns on or shuts off the battery charger, B supply device, and all other power equipment. The model 87 is priced at $195.00

ANOTHER unique loud speaker by Amplion—the "Shield." The unusual lines and artistic appeal of the "Shield" lend charm to whatever surroundings it may be required to form a part of. The cabinet construction is entirely new in radio reproducer construction. The new-process embossed walnut panelling is attractively curved, and combines a grille front and back. The cone has a diameter of 16½ inches. Height, 22 inches. The Amplion "Shield" retails at $67.50

A NEWCOMER to the ranks of the electric set is depicted above—in the form of the Kolster 6J table type receiver. The new a. c. tubes make possible the 6J, which is a six-tube receiver employing three r. f. stages, and having built-in power equipment. All that is necessary to start the set operating is to insert the tubes, connect a loud speaker, antenna, and ground (or loop), and plug in to a light socket. Tuning is accomplished by means of a single control. The 6J retails at $250.00

Speakers
and Sets

A BEAUTIFUL example of what Federal, Buffalo, is offering—the E45–60 "Ortho-sonic." While the price ($460.00) of the model shown is a little high for the average man's purse, it is worth while to remember that Federal has an equally attractive five-tube table model for only $100.00. The picture shows a completely a. c. operated six-tube receiver with a built-in loud speaker. Tuning is accomplished by means of a single knob, and the graduated scale is illuminated to facilitate tuning. Its remarkable selectivity is a feature claimed for the E45–60, but this is not obtained at the expense of tone quality. The height is 54½ inches, the depth is 17½ inches, and the width is 30½ inches. The cabinet is an original design in figured walnut with overlay of fiddle-back mahogany. The knobs and pendants are burnished.

THE new Atwater Kent a. c. receiver is shown in the illustration above. It is indeed a remarkable piece of engineering—a complete radio installation, with the exception of the loud speaker—in a metal cabinet 7¼ inches high and 17¼ inches long. The apportioning of the six tubes is as follows: Three r. f. stages, detector, two transformer-coupled audio stages. Single control tuning is featured, and the built-in power unit is shielded from the rest of the receiver. Absolutely no batteries are necessary for operation. The price of the Model 37 Atwater Kent a. c. receiver is $88.00

A WELCOME newcomer to the loud speaker field is pictured to the left. It is the "Air Chrome," a product of the Air-Chrome Studios, Irvington, New Jersey. Tests in RADIO BROADCAST Laboratory have shown this loud speaker to be capable of exceptional tone reproduction, while its efficiency is claimed by the manufacturers to exceed that of the best cones on the market. The "Air-Chrome" is not a cone loud speaker although in appearance it somewhat resembles one. The console model shown retails at $55.00, there being a choice of design in so far as the tapestry front covering is concerned. The loud speaker without cabinet is priced at $25.00

The
Armchair Engineer

Keith Henney
Director of the Laboratory

TO THE RIGHT

This illustration shows a corner of the RADIO BROAD-
CAST Laboratory. As the accompanying article explains,
it is not necessary to have a laboratory or costly instruments at one's disposal to conduct interesting radio
experiments. A piece of paper, a pencil, and a slide rule
are sufficient to enable one to learn a lot about the design
and operation of radio circuits, etc.

THERE are several interesting and perhaps
instructive investigations of receiver design that one may explore without a
laboratory full of expensive apparatus. A slide
rule, a pencil, and some paper, are all that are
necessary.

For example, let us consider a conventional
two-stage transformer-coupled audio amplifier
with an output device to protect the loud speaker
from the d.c. plate current of the last tube. There
are two methods of connecting the loud speaker
to this amplifier, as shown in Figs. 1 and 2. The
turn ratio of the first audio transformer let us
suppose to be T_1, and that of the second to be T_2;
μ_1 and μ_2 are the amplification factors of the
first and second tubes respectively. The detector
of this set-up also secures its plate current from
a common source, represented by the box at the
left. In the first case the loud speaker is connected
to the negative lead of the final amplifier tube.
Now let us suppose that a 1000-cycle note comes
into this system producing an a.c. voltage of 50
across the output choke, which, if it is very good
indeed, will have an inductance of 40 henries,
or an impedance at 1000 cycles of about 250,000
ohms. If the loud speaker has an impedance of
only 4000 ohms at this frequency, most of the a.c.
current will flow through the loud speaker, as is
desired. This 50 volts across the choke is arbitrarily chosen, and the absolute value does not matter for our present discussion. Fifty volts is
probably higher than is encountered in practice.

Fifty volts across the 250,000 ohms impedance
of the choke will send through it a current of
about 0.2 milliamperes which, to return to the
filament of the last tube, must go through the
B battery leads, through the 2-mfd. condenser across the plate supply unit, and
thence to the filament. The 2-mfd. bypass condenser has an impedance of about 80 ohms at
1000 cycles (this may be found by reference to
laboratory sheet No. 127, in the September,
1927, RADIO BROADCAST) and the voltage across
it, obtained by multiplying the impedance by
the current, is roughly 0.016 volts, most of which
is impressed across the primary winding of the
first audio transformer because of its high impedance compared to the plate impedance of the
tube. This tone will go through the amplifier,
and will be amplified accordingly, and if there
are an odd number of stages securing plate voltage from this common source, say a detector and
two audio stages, and if the transformer primaries are "poled" correctly, the final voltage appearing across the output choke will not only be
amplified but will be in phase with the original
voltage.

The problem is to find out how strong this
voltage becomes by going through the amplifier.
If the transformers are 3:1 each and the first
tube has an amplification factor of 8, and the
final tube (an UX-171 or CX-371) an amplification
factor of 3, the maximum amplification will be
the product of these factors, or 3 x 3 x 8 x 3, or
roughly 200. In the plate circuit this voltage will
divide, part being lost on the 2000-ohm plate
impedance of the output tube and part appearing
across the 4000-ohm output impedance (choke
and loud speaker). As a matter of fact two-thirds
of the voltage will appear across the choke so

that the voltage finally appearing there will be
0.016 x 200 x ⅔, or about 2.1. In other words, the
original 50 volts which appeared across the output choke have returned in phase but have been
decreased to 2.1 volts.

Now let us consider Fig. 2. Here all a.c.
components in the plate circuit of the last tube
must go through the plate supply before they can
return to the filament of the power tube. The
impedance of the choke is lowered by the shunt
impedance of the loud speaker, say to 4000 ohms,
and if the same 50 volts appears across this load
impedance the a.c. current through it will be
50/4000 or 12.5 milliamperes, which will produce a voltage of 1.0 across the 2-mfd. condenser,
and this finally appears as 1.0 x 200 x ⅔, or 135,
volts. Since this "feedback" voltage is greater
than the original voltage, and may be in phase
with it, an endless chain results and the amplifier
turns itself into an excellent oscillator, singing
at some frequency determined by the constants
of the circuit, usually at the point where the
maximum amplification of the audio transformers takes place, say about 5000 cycles.

This is exactly what happens in an oscillator:
part of the output is fed back to the input so that
it is amplified through the tube and again impressed across the output.

In the amplifier under discussion the feed-back
is caused by the impedance of the plate supply
unit which is common to all of the amplifier
stages. If the bypass condenser is increased
(which decreases the total impedance) the tendency to oscillate becomes less, since the feed-
back voltage impressed upon the input to the
amplifier decreases with decrease in common impedance. In the Laboratory, a high-quality two-
stage transformer-coupled amplifier with detector getting its plate supply from the same source
as the amplifier sang terrifically when a resistance
of 37 ohms was inserted in the negative B battery
lead when the loud speaker was connected directly across the output choke-condenser combination. With one side of the loud speaker connected to the filament, a total of 670 ohms in the
negative B lead could be tolerated before the
amplifier sang.

Bypassing the common resistance with a
2-mfd. capacity, the tolerance in common impedance was increased, so that with the loud
speaker across the output choke, 200 ohms did

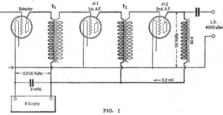

FIG. I

not cause singing, and in the other case 2000 ohms could be tolerated.

This investigation indicates that, when a condenser-choke output device is used, the loud speaker should connect directly to the negative A, or to the filament center tap if a.c. operated, of the last tube. When an output transformer is used, corresponding to the case where the loud speaker is connected across the output device, and all of the a.c. components of the plate circuit must go through the plate supply, a large bypass capacity should be used. This is necessary when B batteries or plate-supply units are used, for the latter have a rather large a.c. impedance at all times, and the former have considerable a.c. impedance when they have been used for some time.

Now continuing our investigation, let us look at this two-stage amplifier again. Let us suppose that at some frequency the loud speaker represents a load to the last tube which takes as much power as a pure resistance equal to twice the impedance of the tube. This is an assumption easily satisfied, and since tube experts in England, and of the General Electric Company in America, have shown that the maximum undistorted output will be attained under these conditions, we shall start off on the right foot at least.

If the loud speaker impedance is twice that of the tube, $\frac{2}{3}$ of the total a.c. voltage appearing in the plate cicuit will be impressed across it.

Now the maximum voltage amplification of the amplifier, which we shall call A, is as follows, when T represents the turn ratio of the transformers and μ the amplification factor of the tubes:

$$A = T_1 \times \mu_1 \times T_2 \times \mu_2$$

and since two-thirds of this appears across the loud speaker, the voltage across the latter is:

$$E_{l.s.} = \tfrac{2}{3} \times A \times E_d$$

where E_d is the a.c. voltage available across the primary of the first audio transformer.

Actually the loud speaker voltage will be less than this figure, since the full transformer ratio, and amplification factor of the tube, cannot be realized, but good design will make it possible to approach this maximum voltage amplification.

Now, looking at a booklet on tubes, we note the maximum undistorted output of a 171 type tube is 700 milliwatts, that the amplification factor is 3, and that the plate impedance is 2000 ohms. Then at the frequency chosen, the loud speaker impedance will look like 4000 ohms resistance to the tube, and since the power, W_0, into it is:

$$W_0 = I^2 Z = \frac{(E_{l.s.})^2}{Z}$$

or $E_{l.s.} = \sqrt{W_0 \times Z} = \sqrt{0.7 \times 4000} = 53$ volts r.m.s.

where Z = loud speaker impedance, $E_{l.s.}$ = voltage across loud speaker, and I = current through loud speaker, we see that there must be 53 volts r.m.s. across the loud speaker to put 0.7 watts into it, and since this is but two-thirds of the total a.c. voltage in the plate circuit of the last tube, the total is $\frac{3}{2} \times 53$, or 79.5 volts.

Since the amplification factor of this tube is 3, 79.5 ÷ 3 = 26.5 r.m.s. or 37.4 peak volts (since peak volts equals 1.41 times r.m.s. volts) which must appear on the grid of the power tube, so that when the booklet states that the grid bias should be 40.5 it shows that our calculations are not far wrong.

Now let us use a UX-201-A (CX-301-A) type tube as the first amplifier and two 3:1 transformers and calculate:

$$A = 3 \times 8 \times 3 \times 3 = 216$$

which shows that the maximum voltage am-

plification of the amplifier, from input to the first transformer to the plate circuit of the final tube, will be 216, and since we need 79.5 volts in the output tube plate circuit the voltage necessary across the primary of the first audio transformer will be:

$$79.5 \div 216 = 0.368 = E_d$$

This, then, is the voltage which the detector must supply to the input of the audio amplifier.

Now in place of the low-mu low-impedance tube in the output or power stage let us use a UX-210 (CX-310), which has a mu of 8 and an impedance of 5000 ohms, or a UX-112 (CX-312) which has the same characteristics. In this case we must use a step-down output transformer of 1.58 turns ratio ($\sqrt{10,000 \div 4000}$) so that our 4000-ohm loud speaker will look like 10,000 ohms to the power tube. Of course the same voltage will be necessary across the loud speaker to deliver 700 milliwatts to it, i.e., 79.5, which, multiplied by 1.58, the turn ratio of the output transformer, gives 125 as the voltage which must appear in the plate circuit of the final tube.

In this case, however, the voltage amplification of the amplifier is increased to:

$$3 \times 8 \times 3 \times 8 = 575$$

and the detector output must be:

$$125 \div 575, \text{ or } 0.218$$

and the input peak volts to the last tube must be:

$$125 \div 8 = 22, \text{ approximately}$$

This calculation shows that the detector must deliver only 0.218 volts compared to 0.368 when

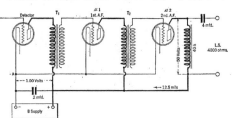

FIG. 2

a 171 type tube is used to furnish the amount of power, 700 milliwatts, to the loud speaker. If, however, the detector can deliver to the input of the amplifier 0.368 volts without distortion, due to detector overloading, the power put into the loud speaker will be:

$$W_0 = \frac{(E_d \times A \times \tfrac{2}{3})^2}{10,000} = 1.4 \text{ watts}$$

In other words the 210 type output tube will deliver exactly twice the power to the loud speaker as is obtained when a low-mu low-impedance tube is used. This difference is entirely due to the higher amplification factor of the tube.

The results of this simple comparison between output tubes are several. In the first place it means that the detector tube can be worked with lower input voltages, the r.f. amplifier need not be "geared up" so high, weaker stations may "load up" the amplifier, and it shows in some measure why more volume results when a UX-210

(CX-310) tube is substituted for a UX-171 (CX-371).

There is one more point which our computation may bring out—the reason why such high voltages are necessary for the plates of UX-210 (CX-310) type tubes. If 40 volts C bias is used on a UX-171 (CX-371) type tube, 180 plate volts are needed. This figure may be calculated by multiplying 40 by 3, the mu of the tube, and by what we may call a factor of safety of 1.5, or 40 x 3 x 1.5 = 180. Now, when we use a UX-210 (CX-310) tube with a C bias of 22 the calculations say 22 x 8 x 1.5 = 263 volts. If, however, the detector delivers 0.368 volts the C bias on the last tube must be 37.5 volts which, in turn, demands a plate voltage of 8 x 37.5 x 1.5 = 450, and there you are.

Our little investigation into amplifier problems has not taken us from our armchair, and all we have needed to determine several interesting facts is a pencil, some paper, and if we possess one, a slide rule. We have learned that output devices, if choke-condenser affairs, should be so connected that the a.c. plate currents return directly to the filament of the last tube, and that if we use an output transformer, or if the a.c. currents do not return directly to the filament, we must bypass as heavily as we can afford the common impedance of the plate supply.

We have learned that we can estimate the maximum voltage amplification of an amplifier by multiplying the turn ratios of the transformers and the amplification constants of the tubes together, and that for distortionless amplification the loud speaker should have roughly twice the impedance of the tube. Under these conditions the voltage across the loud speaker is two-thirds of the total a.c. plate voltage in the plate circuit of the last tube, and the detector output required may be calculated by dividing the a.c. voltage in the last tube's plate circuit by the amplification of the amplifier.

We have learned that a high-mu tube in the final stage of an audio amplifier delivers considerably more power, and requires considerably smaller input voltages to deliver the same power, compared to a low-mu tube. We have learned why such high plate voltages are required on power amplifiers using tubes with high values of amplification factor. Throughout this armchair investigation we have had to make certain assumptions that will make a "hard-boiled" engineer smile. We know but little of how a loud speaker looks to a tube, we have not bothered with vector voltages, we have not tried to be mathematically exact. But we have shown that with a slide rule, a pencil, and some paper we can delve into the subject of receiver and amplifier design.

CJRM, Pioneer Picture Broadcasting Station

By Edgar H. Felix

THE series of articles in RADIO BROADCAST describing the Cooley Rayfoto system, have brought to its editors a surprising number of letters from experimenters who have already worked with telephotography of one kind or another. These, together with the thousands of letters, from those who plan to build Cooley recorders, make it appear quite certain that there will soon be a definite new experimental field—the making of radio pictures in the home. Indeed, we know of quite a number of Cooley outfits already in operation, even though, as we write, broadcasting is only spasmodic.

Perhaps the most interesting broadcasting experiment which has come to light is that of CJRM, of Moose Jaw, Saskatchewan, of which Mr. D. R. P. Coates is manager. An active Radio Picture Club, formed under his leadership, is already in existence, and a group of ardent workers are busy improving their existing picture reception apparatus. The Radio Picture Club is installing a Cooley picture receiving system but, in the meanwhile, is working with a somewhat more crude, picture system.

The transmitting apparatus, built by Mr. Coates, is a remarkable example of resourcefulness, for the facilities available are very limited. The transmitter is made with a cylindrical record phonograph as its basis. A copper plate is mounted on the drum, taking the place of the conventional record. On the copper plate, the picture to be sent is pasted in silhouette or outline form. A stylus passes over the copper drum and completes the circuit for an audio-frequency "howler," the output of which is used to modulate the transmitting carrier. A synchronizing signal, also used in the Cooley receiver, serves to re-check the synchronization at the beginning of each revolution.

Reception is simply a reverse of the transmitting process. A stylus makes a continuous black line by pressing on carbon paper laid over a sheet of white paper, on a revolving cylinder. When the "howler" signal comes in, the stylus is lifted, thus

giving a positive silhouette. Nothing simpler than this system can be imagined, but Mr. Coates tells us it was widely used by the Germans during the War for transmitting military information.

Owing to the fluctuations in drum speed, the synchronization is not good. The difficulty lies in the fact that the phonograph motor is brought to a full stop by the synchronizing system and it does not always resume speed at the same rate when released by the synchronizing control.

Naturally, the Radio Picture Club looks forward to the receipt of its Cooley recorder equipment because that will overcome these difficulties. An ingenious clutch and locking arrangement makes good synchronization easy with the Cooley system, and it is independent of the operation of the phonograph motor. The Cooley stopstart system, and the mechanical unit which is a part of it, is so built that the load on the phonograph motor does not change when the light aluminum drum is stopped at the end of each revolution by the stopstart mechanism. A clutch allows the motor and the turntable to continue revolving, so that there is little or no speed variation.

A PREDICTION

IT IS pleasing to see such enterprise in far-away western Canada. A rather interesting point in Mr. Coates' letter is his point of view with regard to the broadcasting of pictures:

"The present position of the radio picture art reminds me very much of the position of radio seven or eight years ago. I believe that we shall find amateurs dabbling in it for a while and that this will gradually evolve until receiving apparatus is improved and popularized. If we are to follow the line that was taken by broadcasting, we must go ahead and put pictures on the air, no matter how simple and elementary, so that people will be encouraged to build apparatus for the purpose of reproducing the pictures. The fact that there is no one around here yet able to reproduce the pic-

tures does not deter me at all, because I remember broadcasting years ago when our audience was very limited indeed. If we had waited for a big audience before going ahead with broadcasting, we would not be very far advanced at the present time."

Not only do we find that American stations have a similar point of view, but inquiries have come from broadcasting systems, even as far distant as New Zealand, desiring to put Cooley pictures on the air so as to give experimenters an opportunity to develop the picture receiving art.

Mr. Coates is the originator of a system of broadcasting "pictures" of the constellations as an aid to the study of astronomy. He terms these "stellagraphs." The stellagraphs were broadcast by coöperation with a local newspaper which printed a graph paper suitably marked up so that any position on the paper could be given by two reference numbers. A third number in the code indicated the intensity of the star so that a lecturer on astronomy could enable his radio listener to illustrate and draw out the principal constellations of which he was speaking.

The precedent set by Mr. Coates, in suggesting the formation of a radio picture club, is one which should be encouraged. An individual experimenter may hesitate to spend a hundred or a hundred and fifty dollars to go into picture experimentation, but, if five or six club together, the individual cost is small and the experimenters have the benefit of their combined facilities and ingenuity. The Cooley system is sufficiently well developed that it can hardly be termed a hazardous experiment and the amount of special equipment required is not particularly great. We have been advised that a number of substantial prizes will soon be offered for the best Cooley picture reception, through the courtesy of a large radio manufacturer who believes in the future of picture transmission and reception. Naturally, the leaders and pioneers who have the greatest experience, are the most likely to be successful in such a contest.

"Our Readers Suggest—"

OUR Readers Suggest. . ." is a regular feature of Radio·Broadcast, *made up of contributions from our readers dealing with their experiences in the use of manufactured radio apparatus. Little "kinks," the result of experience, which give improved operation, will be described here. Regular space rates will be paid for contributions accepted, and these should be addressed to "The Complete Set Editor,"* Radio Broadcast, *Garden City, New York. A special award of $10 will be paid each month for the best contribution published. The prize this month goes to Fred Madsen, Chicago, Illinois, for his suggestion entitled "A.C. Tube Operation Without Rewiring"*

—The Editor.

A Distortion Indicator

THE average loud speaker is blamed for a multitude of sins which really should be laid at the door of other parts of the receiving circuit. It is not generally understood that a rattle almost identical with the mechanical rattle of an overloaded loud speaker can be caused by a distorting amplifier. The sound of the strain, depending upon the extent of distortion, reproduces in a manner quite deceiving to other than experts, all the various forms of blasting experienced in faulty loud speakers.

If you are experiencing difficulties of this nature, it would be well to investigate the characteristics of your amplifier before discarding your loud speaker as defective. The procedure is simple, as the following illustrations indicate.

A milliammeter capable of carrying the current to the power tube (a zero-to-25 milliampere meter is about right) can be inserted in the plate circuit of any tube, providing a fairly reliable indication as to whether or not that particular tube is introducing distortion. The circuit is shown in Fig. 1.

Any deflection on the meter scale with incoming signals is evidence that the tube in the plate circuit of which it is included is rectifying, i. e., distorting. This distortion can be eliminated by proper biasing. If the needle kicks down, more C battery should be used. If the needle kicks up, the C potential should be lowered. If the C bias is correctly adjusted no movement whatever should be noticed in the needle when a signal of moderate strength is being received. Any movement, up or down, indicates distortion, and the C battery should be adjusted in an endeavor to stabilize the needle.

Quite naturally, every tube has a maximum distortionless output with a given plate voltage, and when this limit is exceeded by applying too powerful a signal to the grid, rectification and distortion will result, regardless of the grid bias. It will be at a minimum, however, if the grid battery is correctly adjusted. If more than a slight flicker of the needle is indicated at the desired volume, a higher plate voltage or a power tube of greater·handling capacity should be employed.

The correct biasing of every tube in the

audio-frequency amplifier can be effected in this manner. As the grid swing applied to other than the last or power amplifier stage is relatively small, however, a rough adjustment (that is the application of approximately the bias recommended by the tube manufacturer for the plate voltage used) will be sufficient in these preceding stages.

Philip Riley
Cincinnati, Ohio.

STAFF COMMENT

MR. RILEY'S distortion indicator is of particular utility when the plate voltage is obtained from a B power-supply device. The voltage supplied by such an arrangement varies with the load, and it is always more or less indeterminant. The correct bias changes with variations in plate potential, so an arbitrarily designated bias voltage rarely affords the maximum distortionless output.

Still more undistorted power can often be obtained from a given amplifier-loud speaker combination by the use of an output device

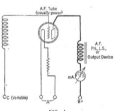

FIG. 1
A simple hook-up to indicate distortion

connected between the loud speaker and the power tube, because the relatively high d. c. resistance of the loud speaker lowers the actual voltage on the plate of the tube, and the use of an output device therefore results in a greater potential being applied to the plate. The loud speaker should never be placed directly in the plate circuit when there is a current of more than 10 mils. flowing. Aside from the relatively powerful current surges, which may damage the delicate windings, the steady direct current tends to draw the armature toward one of the pole pieces, causing the mechanism to hit and rattle on loud signals. There are several commercial output devices which may be instantly connected between loud speaker and power tube. Such filters are manufactured by the National Company, Silver-Marshall, Muter, Ferranti, General Radio. Samson, and Federal, etc.

"Motor-boating"

I HAVE a seven-tube set which employs one stage of resistance-coupled audio amplification and I have been trying for some time to find a B supply device that would not "motor-boat," but without success. It was quite costly to operate the set with a 171 power tube with B batteries so I conceived the idea of using a combination of B battery and B device. It is well known that the r. f. stages and the power tube take most of the B current, so I use the B device on those taps and the battery on the resistance coupled tap. This arrangement is very satisfactory and is giving fine results. The drain on the B·battery is small, hardly noticeable in fact, while the B device operates the tubes drawing the heavy current. Fig. 2 tells the story. The negative of both B supplies are common.

C. R. Yarger
Shenandoah, Iowa.

STAFF COMMENT

THE method suggested by Mr. Yarger·hits directly at the very source of "motor-boating," i. e., a common B supply circuit through which coupling may ·be effected. The arrangement suggested by our correspondent will probably be effective in the most violent cases of ;"motor-boating."

Unfortunately, the purpose of B battery elimination is more or less defeated in this scheme. The current drain on the B battery is, as Mr. Yarger points out, nevertheless very slight.

The connection of an Amrad Mershon electrolytic condenser across the ·high-voltage and negative terminals of the power supply of a motor-boating receiver, will, in the majority of cases, be equally effective in eliminating this disturbance.

A glow tube, connected according to the directions given in "Our Readers Suggest" for February 1928, is also effective.

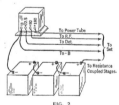

FIG. 2
A simple way of eliminating "motor-boating" in all circuits. The troublesome circuit is fed from a separate plate source

A. C. Tube Operation Without Rewiring

IN THE November, 1927, issue of RADIO BROADCAST, "Our Readers Suggest" department gave directions for the rewiring of a standard radio-frequency receiver and an Atwater Kent 35 for the use of Arcturus alternating-current tubes.

This rewiring necessitated changes in the filament and radio-frequency circuits of such nature that many comparatively experienced fans would hesitate to make. It is extremely difficult to gain access to the wiring of many commercial receivers, and some enthusiasts, adept with the soldering iron and pliers, are reluctant to tamper with a ready-made job.

I have an Atwater Kent 35 receiver which I desired to adapt for a. c. operation without touching the actual wiring of the receiver. I succeeded in doing this along the lines suggested in Fig. 3, and in the accompanying photograph. Arcturus a. c. tubes, which are of special construction, being used for the purpose. The arrangement, briefly, requires the sawing off of the two heater prongs from the a. c. tube bases and soldering lugs in their places, to which the heater connections are made.

The mechanical changes are clearly illustrated in the photograph. Though the prongs are sawed off at the bottom of the base, sufficient metal remains for firm soldering of the lugs. The prongs should be carefully tinned before the lugs are soldered to them in order to insure perfect connection with the heater leads. See that the leads from the elements within the tube are not sawn off when the prongs are removed. They should be firmly soldered to the remaining prong stubs.

The original sockets in the receiver hold the tubes with sufficient rigidity by means of the plate and grid prongs, which remain unchanged. Flexible Braidite is recommended for use in the rewiring of the heater circuits. This can be obtained in different colors facilitating consistent heater connections. The same side of every heater (as determined by their relationship to the pin on the tube base) should be connected together throughout the circuit.

Looking at the bottom of the base with the pin toward you, the left-hand prong is filament plus on a d. c. tube. On the Arcturus a. c. type tube it is a combined heater and cathode connection. It is good practice to wire these prongs with red Braidite, using black for the anode heater leads. Both leads should be twisted, as shown in the photograph. A separate wire is brought out from the cathode of the detector tube to which B minus, C plus, and D minus

WITH HEATER PRONGS REMOVED

An accompaning contribution explains how the Atwater Kent model 35 receiver may be arranged for a.c. tube operation. The main mechanical change requires that the two heater prongs be sawed off and replaced with soldering lugs

are connected. The expression "D" refers to the special biasing battery necessary for the detector tube, the grid of which is made positive. The fundamental diagram is shown in Fig. 3.

An Electrad Royalty 200,000-ohm resistor, shown in Fig. 3, in dotted lines, must be

connected across the secondary of the first audio-frequency transformer where it functions as a volume control. This volume control is mounted externally. The only change actually made in the receiver itself was a slight alteration in the detector grid circuit. The grid leak was removed and a piece of heavy wrapping paper inserted in the prong toward the back of the receiver. A small piece of copper foil, to which a flexible lead was soldered, was laid on top of the wrapping paper so that no connection was made to the prong. The copper foil, however, makes contact with the cap of the grid leak when inserted. This lead was brought out along with the other wires and marked 'D" plus 4.5 volts. This operation automatically opens the connection between the grid leak and the center of a resistance which exists across the filament of the detector in this model of the Atwater Kent receiver.

The tubes were placed in the sockets and the set was ready for operation.

Minus 1.5 volts of C battery were connected to the ground post and minus 22.5 volts of C battery were connected to the power tube. If it is possible to tap off between 4.5 to 9 volts on your first B battery or from the B supply device, the D plus lead from the detector grid leak may be lead directly to such a tap, otherwise it will be necessary to insert a 4.5-volt battery, plus to the grid leak and negative to the B minus.

A plate potential of 90 volts is applied to the plates of the radio-frequency and first audio tubes and 180 volts to the plate of the power tube. Arcturus type A-C 28 tubes are used in the first radio-frequency amplifier and in the first audio amplifier. An Arcturus detector type A-C 26 tube is employed in the detector socket and a power tube type A-C 30 in the power stage. The A plus and A minus leads from the original Atwater Kent receiver are short circuited. Fifteen volts a. c. is applied across the twisted heater wires. This is obtained from an Ives type 225 step-down transformer.

The arrangement as applied here can be adapted to practically any circuit with very few changes.

FRED MADSEN
Chicago, Illinois.

STAFF COMMENT

THE arrangements suggested by Mr. Madsen are quite practical. As he suggests, there are doubtless many instances where the "harness" system is preferable to actual set changes. Commercial "harnesses," which are very similar to the arrangement described, are being placed upon the market to adapt various receivers to different types of a. c. tubes.

In some receivers the prongs of the tube are inserted in metal eyelet holes which protrude a fraction of an inch or more above the surface of the sub-panel. The presence of these eyelets, which are connected electrically to some part of the circuit, will make the receiver inoperative unless care is taken to see that the sawed off prongs or the soldering lugs do not touch them. This is easily accomplished by the use of a pasteboard disk cut to size and punched with two holes to pass the grid and plate prongs.

In some instances it will be more desirable to control volume by means of a zero to 500,000 ohm Electrad Royalty or Centralab resistor connected across the radio-frequency secondary preceding the detector tube. Not all variable resistors of the required ohmic value will work in this arrangement due to the relatively high capacity of several makes.

FIG. 3

Adapting the Atwater Kent Model 35 receiver for a.c. tube operation without making any major changes to the wiring of the receiver itself. The detector grid leak return is broken at X, and brought down, according to the dotted line, to the special "D" plus lead. Dotted lines also show the position of the volume control. The wiring below shows the external wiring to the heaters of the tubes. Posts Nos. 1, 2, and 3 are short-circuited, and no connections whatever are made to them

IF YOU LIKE GOOD MUSIC LISTEN THIS WAY

By JOHN WALLACE

WE HAVE utilized this department before, in fact as recently as last January, to make tearful entreaties to the radio lords to furnish more straight instrumental music. If we may be said to have any platform, that is its principal plank.

In one of these articles we advanced the point that listening to symphonic music via the radio has at least one distinct advantage over listening to the same in a symphony hall, namely: the orchestra cannot be seen. With none of the modesty proper to the father of an idea we aver that the point is an excellent one and well worth dragging out again.

The universal custom of lighting up concert halls with a dazzling effulgence of electric light has about as much to recommend it as would the equipping of art galleries with a fire siren and a ship's bell beside each picture. It is our occasional custom to attend the Friday afternoon concerts of the Chicago Symphony Orchestra, frequented presumably, by the very cream of Chicago's music lovers. Here, as elsewhere, the blaze of light system prevails. In the course of the concert, a surprisingly large proportion of the audience may be observed to be taking advantage of this lavish pyrotechnical display in any one of three ways: (a) reading their programs, (b) embroidering, (c) sitting on the edge of their chairs better to watch the conductor. Anyone who can thus disport himself before Brahms or Beethoven, or even Tschaikowsky, and then later claim that he "heard" the music is either a physiological phenomenon or a prevaricator. To follow honestly the development of a piece of music requires so high a degree of concentration that it is absurd to imagine that an attention divided between the eye and the ear is sufficient. We have essayed the more or less successful subterfuge of slumping down in the seat with one or the other hand before the eyes, but this lays us open to the suspicion of our neighbors that we are (a) a silly fellow feigning intense absorption, (b) asleep. Extinguishing the lights would solve the problem, but this the persons who run concert halls will not do.

To attack the reading of program during a concert is a ticklish proposition. It is "done" by our very best people. Tell them they are demonstrating their ignorance of music by so doing and they will look at you aghast.

It is perfectly possible to look at, and enjoy, a picture without knowing a thing about the painter's life, in fact without even knowing who painted it. This is done constantly, even by connoisseurs. Music, a purer art than painting, is even more independent of its composer and he may be even more easily ignored.

Or it may be protested with still more vigor that "you have to watch the conductor really to 'feel' the music unfolding!" This likewise is pishposh. The gesticulations of the conductor, however graceful or dramatic they may be, have little to do with the music as music. They exist for the purely technical purpose of evoking the proper sounds at the proper time, and the technique of putting the thing across is no more your business than would be the trade name of the pigments your artist used in his painting.

Perhaps we seem to wax too wroth and to be making a mountain out of a mole hill in our vehemence against optical assimilation of music.

But our fury is aroused by the fact that the people who so insult good music in the concert halls are supposedly the very topmost strata of music lovers—the highbrows, no less! If the highbrows of the nation don't know how to listen to serious music what about the masses?

It would seem, perhaps, that the case we are making out for proper concert listening is no very happy one, and that we would place it on the same tedious and exacting plane as listening to a class-room lecture on philosophy or astronomy. This is not entirely true. Certainly we have to concentrate just as much—a symphony by Brahms has just as much meat in it as any four chapters from Kant's *Critique*—but the concentration can be entirely effortless.

Herein, as we have said before, lies the advantage of radio. Given a good receiver and a symphony orchestra properly "picked up" and transmitted and you are all set for a concert

EUNICE WYNN OF KFWB

Miss Wynn is a regular artist of the Hollywood station of Warner Brothers, KFWB. Cute songs, she sings, according to the station

which may nine times out of ten be more enjoyable than one in a mazda-equipped concert hall. You can don your slippers, turn off the lights and park in an easy chair—three separate counts wherein "second hand" radio has it over the first hand thing! With your own private stage thus set you are in an ideal position really to 'hear" the music with a fullest possible realization of what it really has in it.

A popular delusion exists that music should caress the listener and lull him into a pleasant state of lethargy, and that he need do nothing but just" set" and let the vague tides of sound wash over him soothingly. The answer to this is age old: the artist can go only half way.

The reward for going half way is the surprising discovery that there are tasty bon-bons waiting at the half-way point whose existence was never even suspected.

Now our point is that it is easier for the neophyte to cultivate an understanding of serious music by listening to it on the radio than by going to concert halls—simply because he can do it with less distraction.

Devious and many are the ways suggested for learning "how to appreciate music." Most of

them involve too much work (they are the best ones). Others go in too much for the technical and intellectual, which after all is only half the content of music at most. We are going to suggest a short-cut method—one which requires no preparatory work at all.

HOW TO LISTEN TO GOOD MUSIC

DOUBTLESS this system of learning how to follow music has been suggested before—though we have not run across it. It may be objected to as an unscientific method, as an unintellectual method—in short as a too strictly emotional method. But it is, a practical method.

Sit yourself before the receiving set in the afore-mentioned slippered condition when there is some first rate symphonic orchestra program going on (we hope you can find one!). Turn off the lights and the oil burner and otherwise exclude all conceivable distractions and then concentrate on the sound issuing from the loud speaker as though you were entombed in a mine waiting for the faint ring of a distant pick ax. Or strain your ears as though you were trying at three in the morning for 210. Things will immediately begin to happen. Surprising things. A host of sounds will begin to emerge that were formerly just lost in the shuffle. Pick out one of the thinnest and feeblest of these sounds and follow it through the maze like a bloodhound pursuing little Eliza through the forest. Keep on its trail and see what it does—and what some of the big bullying noises do to it. Then for a change pick out some little transitory tune—perhaps only five notes long—wait for it like a cat before a mousehole. Presently it will appear again, perhaps in a different key or on a different instrument, or even disguised with false whiskers. But you will recognize it, and with a glee quite equal to that of the cat when the mouse finally emerges from its burrow.

Next try listening to two tunes—or two instruments—at once. Watch how the two tunes sneak along side by side, some times drawing together and shaking hands; other times running off on by-paths and making faces at each other. Watch them intertwine and overlap and disappear and emerge again with a new suit and their hair combed on the other side. Search out some little insignificant orchestral effect that seems to be buried obscurely away at the bottom of the heap of noises. Watch it while it pussy-foots around the corner and gets itself a drink. Watch it start to swell and swagger and toss its hat around. Presently the sniffer does its stuff and it is strutting around bombastically. Before you know it, it is running the whole works, the other noises fleeing, terrified, to shelter.

In such wise, listening to the symphony orchestra becomes a grand game with yourself just as much a participant as the orchestra. It has much in common with football: there are long end-runs and fake plays and intercepted passes and trick formations; there is team work and tripping, signalling, and shifts, even "time out" where a rest occurs in the music. And both the football game and the symphony are divided into quarters.

The analogy to football is not quite complete. What happens in a football game is largely subject to chance (as we found out just before we swore off betting last season). But what happens in a symphony, far from resulting from chance, results from perfect organization.

Herein our proposed method of learning how to listen fails. It will not reveal to you the organization of the music. That would be asking too much. However, the method we suggest will at least demonstrate to a listener that music is not simply a blur of sound, and will enable him to recognize the elements out of which music is organized.

This will be a first step. Furthermore it will be fun—which is the only excuse music has for existence anyway.

For Program Directors Only

WRITES Paul Hale Bruske, of Detroit: My friend Jimmie is blessed with one of those urbane baritone voices that pleasantly vibrate so many loud speakers in the sun rooms and front parlors of our broad land.

Yes! He's a radio announcer. More than that, he's also a program director. On his shoulders rests the responsibility of seeking, interviewing, choosing, auditioning, hiring, scheduling and rejecting various features of alleged entertainment and enlightenment. His station is a good one.

Jimmie does little if any seeking for talent. He does a microscopical amount of hiring. At interviewing he shines. His waiting room is usually full of folks who believe they have a Mission and a Message. Male and female—blondes and brunettes—old and young—artists and artistes—they are all grist for Jimmie's mill. Auditions occupy a good share of his day. He always takes the name, address and telephone number, and gives them a sweet promise to let them know.

One might think that, with all this wealth of willing workers on call, the programs from Jimmie's station would be replete with variety, and fertile in surprise, as the theatrical notices say. Such is not the result. Far from it. Tune-in on him any evening and here is what you are pretty sure to get:—

6:00—7:00—Dinner concert from the Herculaneum Room of a good hotel. Gwan to Bed dope for the kiddies.
7:00—8:00—News Bulletins, Hot Stuff for Farmers, Organ Recital from a Cinema Palace.
8:00—9:00 (Commercially Tainted Hour)—Somebody's Antiseptic Carolers in solos, duets and quartet numbers. Somebody's Realestate Minstrels in comic songs and dialog.
9:00—10:00 P. M.—(Big Station Feature);—Baby Grand Philharmonic Orchestra in classical and semi-classical numbers, with guest artist soprano or tenor.
10:00 P. M. and on:—Jazz Bedlam from some cabaret or night club.

On Sundays, Jimmie broadcasts church services. In midsummers, he gives us band concerts from the parks. He occasionally hands us a sporting event. He has a weekly silent night. More so than seems usually the case, Jimmie enjoys quite a free hand. His boss is rich, a radio bug and has no personal propaganda to get over—not enough, at any rate, to make it obnoxious.

Some months ago I began to razz Jimmie a bit on the striking lack of originality in his programs. He insisted that he was every bit as good as his competition, and I had to admit it. With him, it was purely a matter of beating competition on one common ground. Jimmie's rut was too deep. He couldn't see over the top.

But one day I got a barb under his hide. Perhaps the boss had just asked about the applause letters. For Jimmie turned on me defiantly.

"What would you do if you were in my place?" he challenged.

That was surely a quick pass of the celebrated

buck but, after all my raillery, I couldn't dodge.

"Well," I stalled, "I'd first try to analyze a bit. Here you are, competing for public attention with from four to forty other stations, depending on reception conditions. You want folks to be tuning you in and then letting the dials alone for a while. You want them to think of you when they think of radio—to talk about your station and your programs. You want more applause letters. You'd even prefer knocks to the present silence. You crave personal glory. You could endure a bigger check in your pay envelope."

"Yes! And how?"

"Shut up; I'm analyzing.

"The main offering of all radio stations is music. On that you've gone about the logical limit. Vocally, instrumentally and in combination, you fellows have probably tried about all the tricks there are. I doubt if there is any new musical dodge which would create more than a ripple of interest. And what we want is a tidal wave.

"But there are at least two channels of approach to your dear invisible audience. The one that isn't music is speech. Let us admit, therefore, that the method you will use in getting folks to talking about your station is the spoken word."

"No chance!" yelled Jimmie. "The dullest thing that comes over the air is a speech. The minute one starts here, I can just feel the people tuning-out. There are only three kinds of radio speeches. There's politics. There's platitudes. And there's propaganda. Each is worse than the other. I'd like to pass a rule that would absolutely prohibit all speeches from this station.

' Why you've no comprehension of the speechifiers we turn down right now," Jimmie continued. "There isn't a public official in town who doesn't think he'd be the hit of the season, if he could only get on the air. Every convention that comes here tries to get time for its Grand High Cockalorum. The ladies with pet charities can't understand why we don't put them on oftener. The boy scouts litter up the place with officers that have a message. Speeches? Take a swift jump into the lake!"

"Wait a minute Jimmie," I begged. "Don't get me as any friend of your three P's. They're even more terrible than you say. But there must be such a thing as interesting talk. The newspapers get it."

Jimmie picked a fresh edition from his desk. Clear across the front, in glaring 72-point, screamed the legend, "MRS. BANCROFT WEEPS ON STAND."

It was just the current divorce case and not much of a case at that—no shooting, no violence of any kind, hardly any real scandal. But both parties were socially prominent. The gentleman was rich. The lady alleged he was cruel and neglectful. The children also socially prominent, took sides. The case was good food for tea-table gossip, so the papers were playing it strong.

Jimmie and I looked at the headline. It seemed suggestive.

"She weeps, Jimmie," I said. "And the people read about it. Wouldn't it be better if they could really hear her weep?

"If you could schedule her to weep into your mike to-night, would they tune-in?"

"You're whoopin' they would," admitted Jimmie.

"Then go get her," I insisted. "Tell her she's got a Mission. Explain that what's happened to her is only a sample of what's happening to thousands of other women. Get her to tell these others what to do. Let her put her case to the whole world, with her own voice and freed from any cross-examination or other rules of any kind. I'll bet she'd jump at the chance. Give those listeners of yours something worth listening to."

"Gosh!" commented Jimmie.

"Then next night, give the same privilege to Mr. Bancroft. Let him talk to the husbands. Have him tell what a real life partner should do. Let him say anything within reason about these wives who spend their days in clubs, and don't have time to cook a dinner for the family. Play him as the outraged American husband, and let him counsel others who feel bad with him. Then let the household arguments rage. Every time the case is mentioned, folks will think of your station and wonder what next."

"Well, that's a good question. What, next?"

"Next will come the poor boob who's just been sentenced to life in the hoosegow for murder of his sweetie's friend husband. Give him a last chance to say farewell to the world. What a cinch it will be for you to write his speech! To-night, he's a man and has a name. To-morrow night, and until he dies, he's nothing but a number. And who's to blame? The woman, of course! And then the moral. All this in his own voice, with the handcuffs rattling every time he turns a page.

"Gruesome? Sure, but will they listen?

"As for me, Jimmie, I've always had a longing to know just what happens at a hospital during a good, major operation on the human torso. I'd rather hear you describe such a thing than get your fresh-from-the-ringside word picture of a good prize fight. That's only a sample of what you can do when you once get your mind on really interesting topics. But let that pass.

"Watch the front pages. They're the best bet. Grab those features hot, give 'em a good, moral, uplifting line of talk; rehearse 'em, and turn 'em loose with their own voices to give us listeners honest-to-goodness heart throbs in the raw. If you're too busy to add this department's duties to those you carry now, hire Pat Montgomery, our old city editor, and put it up to him. Yellow up—and that only means make your stuff interesting."

Jimmie had been getting more pop-eyed with every word. Temporarily, at least, I had him sold. But he cooled off just as quickly. Eventually he admitted that there was something in this idea and promised he'd think it over.

"But no murderers! And no gory operations, either," he declared. "Something controversial, maybe. That's perhaps the secret of making speech interesting. And I may be able to do something with decent celebrities of the day."

A few days later, Jimmie's station announced the first controversial event. It was a debate. On one side was the mayor. Opposed was one of his appointees who had the courage to think for himself. The debate itself wasn't so much. But before it started, right in Jimmie's studio, the mayor fired the appointee, and the latter announced the fact in his address. Jimmie's station got a lot of good publicity, and he was greatly elated.

For several weeks thereafter I was out of touch with Jimmie and with radio. Back in town again I bumped into Jimmie.

"I'll bet I've been missing some hot stuff," I remarked. Jimmie looked blank. Actually I had to remind him of the big idea.

"Oh! We've got something a lot better than that," he boasted. "We've joined a big chain and get our programs right from New York." And then he went on to tell how much better his chain was, on every count, as compared to the others.

However, comma, there are other stations and other cities. Every city has a newspaper, and that newspaper must have a front page. Somewhere, I doggedly insist, there will bob up a program director with the necessary nerve to watch that front page and do what every successful newspaper does—get circulation by being interesting.

AS THE BROADCASTER SEES IT

BY CARL DREHER

How a Famous Artist Broadcasts

THE scene is in the large "B" studio of the National Broadcasting Company at the building on Fifth Avenue in New York City. The room, thirty-six by fifty-two feet, and two stories in height, holds an orchestra of sixty-five men recruited from the New York Philharmonic, with Fritz Busch conducting. The musicians are all in evening dress; their white shirt-fronts gleam in the light streaming down from six giant electroliers. It is a General Motors hour on a Monday night, a somewhat formal occasion, as broadcast features go. About a hundred spectators are grouped around the orchestra, most of the ladies seated, the men standing. I lean against the wall, an idle spectator for the moment, interested, even so, in the effect of the presence of this good-sized crowd on the acoustic characteristics of the room. A concert manager stands on one side of me, and a very famous baritone on the other. We are all looking toward Fritz Busch, who stands, with his baton upraised, in an attitude of commanding tenseness, and at another man who faces the conductor. This man is of a notably handsome and virile aspect; his body is that of an athlete; his features might be those of an intelligent and sensitive business man, but at the same time something of the actor and artist is easily discernible. His head is about three feet from one of the microphones of a double set-up. He stands with his legs well apart and his arms folded across a broad chest. This is John Charles Thomas, the tenor, who is about to sing an aria from Verdi's "The Masked Ball." Interested in the outward manifestations of his technique, I watch the artist carefully.

Fritz Busch now swings his arm downward with an emphatic gesture, and the orchestra begins to play the introductory bars. As the moment for his first note approaches, Thomas raises his left hand and cups it behind his ear. He does this in order better to hear the tones as they reach the microphone, for a man hears his own voice both through the air and through the bones of his head, while others hear him only through the air. His other arm the singer holds across his chest in a rather cramped position, which does not concern him, for he is singing with only moderate volume and does not, for the present, require the full capacity of his lungs. Nevertheless, he is ready for exertion, he has his coat off, and incidentally he is not, like most of the others in the room, in evening dress, but wears a business suit and a blue shirt with soft collar.

He rounds his lips carefully for the notes and sways a little in time with the music. His attitude is one of mingled nonchalance and the greatest circumspection. On the one hand you see a man with every natural advantage for his part, with a reputation made early and securely based on experience; and yet this same man realizes that to sing beautifully is never easy, and that it is only too easy to deviate from the pitch, if only a little, or to falter in the time, if only for a fraction of a second, or to mar a phrase with a breath that is only a little awkwardly drawn. So, seen in one aspect, John Charles Thomas sings effortlessly, and from another angle, he is working as hard as any man in any trade cares to work. The expression of his face changes, sometimes in consonance with the music, but at times, because he is singing more for the audience which cannot see him than for the hundred

---

Interesting Highlights this Month

—How John Charles Thomas Performs Behind the Microphone.

—Matters Which Make Life Hard for the Broadcaster.

—Where to Get New Information on Human Speech Constants.

—Sources of Helpful Printed Information from Manufacturers.

---

spectators, he seems frankly to use his facial muscles to aid his larynx. And again, when a note does not suit him precisely, he frowns critically, engrossed in his own private world of tonal creation. As Thomas maneuvers through a difficult succession of transitions, the concert manager nods approvingly, while the baritone on my right, when I glance at him, is watching the singer with critical professional admiration.

The aria is nearing its end. As he approaches the forte passage at the close, Mr. Thomas moves his right arm down from his chest and lets out something near his full volume. He is much too good a microphone performer to go all the way; it is interesting to see how nicely, without knowing a transmission unit from a kilocycle, he unconsciously compresses his volume range within approximately the 40-TU-width allowable in radio telephone transmission, leaving the control operator little to do. Having had considerable experience singing for the radio and phonograph, and, presumably, having heard the performances of others in the same mediums, this artist has adjusted his

technique, when he sings for the air, to the requirements of that particular machinery. Therein you see one of the reasons for his success. The difference between the artist who goes well over the air and the one who is a flop in the broadcast field is as often one of adaptive intelligence as a matter of voice characteristics. That adaptive intelligence may be intuitive rather than quantitative, but the results are the same, whether reached through the vocal feeling of the artist or the calculations of the engineer—better, in fact, when the artist does the editing himself. So, at his climax, Mr. Thomas roars formidably, but not so recklessly as to cause consternation at Bellmore, where the transmitter technicians are keeping an eye on the modulation peaks. Nor does he walk into the microphone; he gestures somewhat with his free right hand, but his feet remain rooted to the spot where he originally took up his position.

The last fine notes ring out, and for a second the artist stands, still in his part, looking somberly at the microphone, as if, behind its unrevealing diaphragm, he saw, across hill and plain and river, on farms and in cities, the several hundred thousand of his countrymen whom he has held entranced for those few minutes. Then, turning, he smiles broadly at the conductor, and, as the announcer speaks his lines, Thomas walks off-stage, or what amounts to off-stage in a broadcast studio, puts on his coat with a gesture curiously reminiscent of a football player struggling into his blanket as he retires to the sidelines, and looks over his next song before he is once more called before the transmitters.

Note for Lexicographers

IN THE December, 1927 issue, under the heading "Radio As An Electro-Medical Cure-All" we recounted the claims of a quack imbued with the conviction that he is curing the assorted ailments of the populace by saturating them with his own brand of radio waves. The word "stob" occurred several times in the healer's description of his paraphernalia, and, failing to find it in my Webster's Collegiate, a very good dictionary for its size, I was at a loss as to its meaning, except that the context indicated that it was some sort of metal ground stake. Mr. H. G. Reading of Franklin, Pennsylvania, clarifies the subject in the following communication:

"Thinking it queer that there could possibly be a word 'not in the (New Standard) Dictionary,' I looked for 'stob'

367

and find it having several meanings, one of which is 'a long steel wedge' used in coal mines. Doubtless the Doctor wielded the pick for a living at one time."

Mr. Reading's conjecture may possibly have hit the mark. If so, the electronic magic worker in question abandoned a socially laudable profession. We may hope that his adventures in the healing arts, having led him to the conclusion that he can remedy the lesions of a man a mile away by modulating the output of his battery-operated radio transmitter with the stutterings of his "oscilloclast," will end by putting him to work with a pick again, although perhaps not in a coal mine.

Speech Constants

ON PAGES 753–754 of Morecroft's *Principles of Radio Communication* (Second Edition), published by John Wiley & Sons, Inc., there is a useful summary of the results of the research work on speech carried on in telephonic laboratories during the past decade. Most of the figures have already been given in this department in the past, but Morecroft adds one or two new ones gleaned from his reading.

(1) The frequencies encountered in human speech are within the range of 100 to 6000 complete vibrations per second.

(2) The energy contained in speech is carried almost completely by frequencies below 500 that the quality and intelligibility of speech is determined very largely by the frequencies higher than 500.

(3) The average power output of the average normal voice is about 75 ergs per second or 7.5 microwatts.

(4) The average male voice exerts a pressure of about 10 dynes per square centimeter at a distance 3 centimeters from the mouth of the speaker.

(5) The human ear can detect sounds, at a frequency of about 1000 cycles, if the sound pressure is as low as 0.001 dyne per square centimeter. If the pressure exceeds about 1000 dynes per square centimeter at this frequency, the ear is practically paralyzed in so far as sound is concerned and the sensation is one of feeling rather than hearing.

(6) The ratio of peak power in the voice (accented syllable) to average may be 200 to 1. Thus an average voice of 10 microwatts shows peaks of 2000 microwatts.

Professor Morecroft gives references to two articles in the *Bell System Technical Journal* which have not been specifically mentioned in this department. One is the article on "Speech, Power and Energy" in the October, 1925 issue of the *Journal*; the other is a paper on speech analysis by Harvey Fletcher in the July, 1925 issue of the same publication. Both discussions are of direct interest to broadcast technicians.

Catalogs and Commercial Publications

THE number of commercial publications of interest to broadcasters is increasing. Three may be mentioned this month. The Sales Department of the Radio Corporation of America (233 Broadway, New York City, with district offices in

Chicago and San Francisco) is distributing to its dealers a leaflet entitled *Average Characteristics of Receiving Radiotrons*. [This leaflet, listed as No. 69 in our "Manufacturers' Booklets Available List, appearing in the back pages of this magazine, may be secured by our readers by using the coupon indicated.—*Editor*.] Some of the tubes described, such as the UX-210 (7.5 watts oscillator rating) are used in the lower power stages of broadcast speech amplifiers, while others are commonly employed in field equipment. The tubes are classified as "Detectors and Amplifiers"; "Power Amplifiers"; "Rectifiers" and "Miscellaneous." The data given is, first, general physical and electrical information, such as the type of base, outside dimensions, special circuit requirements, possible filament supplies, voltages, and currents. For detection purposes the proper grid return, grid leak, "B" voltage, and plate current are specified. Under "Amplification," for various plate and corresponding bias voltages, the plate current, a.c. plate resistance, mutual conductance, voltage amplification factor, and maximum undistorted output in milliwatts, are included. The alternating current filament and a.c. heater type radiotrons are listed, and the whole list is a useful page to be added to the broadcast engineer's notebook.

J. E. Jenkins & S. E. Adair of 1500 North Dearborn Parkway, Chicago, have issued their *Bulletin No. 4*, dealing with a "Complete Input System" installed at the 1927 Chicago Radio Show. This equipment was used during the entire week of the show, feeding a network of some thirteen of the local broadcasting stations, and a public address system the output of which supplied 22 eighteen-inch cone speakers in the Coliseum. The "Broadcast Amplifier" portion of the apparatus was mounted on a frame 72 inches high by 25 inches wide, which supported the following units: meter panel; a monitor panel, employing a W. E. 205-E or R. C. A. UX-210 tube to feed a loud speaker; an output level indicator; a three-stage broadcast amplifier; a microphone mixing panel; and a telephone jack panel for the broadcast pairs and order wires. The amplifier was the "Type A" described in Jenkins & Adair's *Bulletin 1A*, comprising two stages of amplification using W. E. 102-E or R. C. A. UX-240 tubes, and an output stage for a 205-E or UX-210. A 350,000-ohm wire-wound gain control afforded 11 values of amplification, with equal TU-increments. The 500-ohm output of the amplifier was connected to the level indicator, the monitor amplifier, and the jacks for the outgoing broadcast pairs. The maximum undistorted output available is stated to be 0.8 watt, on a plate voltage of 135 for the small tubes and 350 for the output stage, secured from heavy-duty dry batteries.

The power amplifier portion of the equipment was mounted on a frame only 19 inches wide, but the same height as the broadcast frame (72"). Current was supplied from a generator capable of giving

150 milliamperes at 1200 volts and 10 amperes at 12 volts for the plates and filaments, respectively, of the power tubes, which consisted of two fifty-watters in a push-pull circuit. The actual plate potential was 1000 volts, and separate grid bias batteries enabled each tube to be adjusted to draw 60 milliamperes. Each tube also had a 0–100 milliammeter in the plate circuit, and a quarter-ampere fuse. The push-pull amplifier was fed from a 5-watt stage, which in turn received part of the output of the three-stage broadcast amplifier previously described. A level indicator panel for the P. A. system was also included in this frame, the input being directly connected to the output of the large amplifier with a high resistance in series to reduce the voltage. This instrument gave a check on the volume of the loud speakers, incidentally showing up irregularities in the 3000 feet of line connecting up the loud speakers. The broadcast and P. A. frames, set up with a desk between them, made a neat lay-out. All the parts used in building the units with the exception of such items as tube sockets, etc., were manufactured by J. E. Jenkins & S. E. Adair.

A more elaborate publication is *Samson Broadcast Amplifier Units*, issued by the Samson Electric Company, of Canton, Mass. This is a pamphlet of 24 eight-and-one-half-by-ten pages describing Samson parts for broadcast amplifiers and associated equipment, and incidentally going quite deeply into design considerations. It is not intended for general distribution, but may be obtained free of charge by broadcasters writing for it on their letterheads. Microphone-to-tube transformers are first described, with some advice regarding connections and care of carbon microphones. There appears to be an error on page 5, where the d.c. of a 200-ohm microphone is given as 200 milliamperes per button, instead of 20. The difference is serious! Following there is a discussion of multiple microphone (mixer) operation, but the pamphlet argues that "this practice (using more than one transmitter for pick-up) should be avoided unless it is imperative." The point is highly debatable, but at any rate, as the Samson people sell mixer transformers, it is refreshing to see them state what they believe to be true, regardless of a little economic advantage. Tube-to-line and line-to-tube transformers, as well as other matching devices, interstage impedances and transformers are described in following pages. The discussion of impedance relations in audio circuits, direct current design considerations, and the use of center-tap connections, including devices for obtaining a center tap electrically where the actual winding midpoint is unavailable, is quite thorough. The last seven pages are devoted to "General Considerations"—Interference Level, Gain, Volume Control, Attenuation Networks or Pads, and Pad Design. The Samson Electric Company's engineers have turned out a valuable publication.

The A. C. "Bandbox"

By JOHN F. RIDER

A SHORT time ago, the writer had occasion to visit a friend who maintains a hunting lodge in the Berkshires—in the northwestern part of Connecticut. We spent the day in a virgin forest, chopping trees and hauling timber, for the stock of wood fuel required to heat the lodge had to be replenished. There were five of us at the camp and, at the day's end, after our evening meal, we would gather around the fire and listen to the radio.

Say what you will of the radio in the city, the welcome cheer which it imparted to us up in the woods created a deep impression in our minds. To the woodsman habitually isolated many miles from the city it must indeed be a blessing. Imagine the interior of a rough shack, snowflakes beating against the window panes which are almost hidden by sweeping drifts of snow, a merciless wind screaming through the tree tops, and a temperature of "three above" outside. The stillness within the shack is broken by the crackling of the blazing log fire, and by the deep breathing of the woodsman inhabitant, who sits deep in thought—a few well worn books his only companions. His nearest neighbor may be miles away; it would be useless to try and reach him during the several days the storm rages. And what a different state of affairs exists at this neighbor's. Outwardly there is little to choose between the two shacks. They are both lost in the depths of the snow. Each has its blazing open fire of logs, and in each lives a solitary woodsman. But there is also a radio in the second one, and what a transformation it has wrought. Were you to knock at the door for shelter, you might enter to the strains of a Strauss waltz, a Berlin melody, or perhaps even,

some politician may be monopolizing the attention of the blustering fellow who admits you. He'll discuss the Waldorf Astoria orchestra as if he were accustomed to dine there and listen to it almost every day, and will talk with you of Mr. Coolidge's choices almost before the President even chooses. Or he'll demonstrate how Gene got in the winning punch, compare Damrosch's rendition of Beethoven's "Fifth" with Mengelberg's, talk of the merits of the four-wheel brakes on Dodge Brothers "Fastest Four," and match the "Silver Slipper" orchestra with that of the "Twin Oaks." No, he's never been to New York. He was born up in the woods. Rarely sees an automobile. Yes, lots of the men now have radios up here. Don't need them in the summer. Too much to do outside, and plenty of people around, anyhow. No, neighbor Joe hasn't got one. He's an iconoclast (a word he picked up on the radio). Never knows what's going on in the big cities. He himself wouldn't be without a radio.

THE "BANDBOX"

THE receiver that we used at the lodge in the Berkshires was a Crosley "Bandbox." Reception was good, volume adequate, and the quality of reproduction, excellent. After listening to a concert from Springfield, we tuned-in woy, with the usual result—periodical fading. This phenomenon provoked a discourse by one of the party—Jones, a medical man. He knew all about receivers, although medicine was his forte. His was an analytical mind, and he had grasped the fundamentals of radio. It was simple. He was an old timer. His interest dated back since the day WOR commenced operations.

The discussion finally centred upon the receiver at hand. He was the radio authority. The receiver seemed to work well, but——. Yes, it was selective, but——. The volume was good, but——. Well, he had a ten-tube receiver at home, which he had constructed; it comprised five stages of tuned radio-frequency amplification, push-pull audio amplification, a variable grid leak, etc., etc. In sum and substance, it was a "wow." He hoped that some day we would have the opportunity to listen to his pet receiver!

At this point the writer timidly explained his connection with the radio industry. Fine—just the man he was looking for. Could I explain to him wherein the Crosley differed from any other receiver? Would I point out to him the engineering features of the "Bandbox?" How could this receiver be a scientific receiver, and still sell at its low price? As far as he could see, all manufactured receivers were alike; he would never buy a finished product anyhow. Would I tell him something about the new a. c. operated "Bandbox," which is fundamentally similar, so far as the circuit constants are concerned, to the battery-operated "Bandbox" we were using up in the Berkshires?

The fact that highly scientific design and highly scientific production were absolutely necessary to produce a successful inexpensive receiver never entered his mind. He overlooked the fact completely that an accurate study must be made before even the apportioning of the tubes is made, or before the engineering staff can decide upon how many stages of radio-frequency amplification should be employed. The fact that the receiver is sold at a low price cannot influence the staff to make haphazard decisions. In this particular instance, I told him, the selection of three stages of radio-frequency amplification was a compromise of the sales division and the engineering staff to produce and sell a radio receiver at a reasonable cost for a given amount of radio-frequency amplification and stability. That is to say, calculations were made to determine how much radio-frequency amplification was necessary to provide a certain amount of sensitivity, with a reasonable stability factor. After a study of the sensitivity factor, selectivity was the next consideration. Would the three stages which afford sufficient sensitivity be selective enough? This could be determined mathematically, but it was also determined experimentally. This determination is extremely important in the design of a radio receiver. A low sales price and a high degree of efficiency will immediately "make" a receiver, but low sales price will not compensate deficiency in design.

In view of the demand for single-control receivers, there was much discussion as to what arrangement could be used in the "Bandbox" to permit efficient single tuning control. The receiver has but one major tuning control, and in order to accomplish this, it was necessary to incorporate into the design of the receiver, a circuit arrangement which provided for the different electrical characteristics of various antennas. This was essential in order that true single control would be obtained, otherwise a vernier control would be necessary for the input circuit. Under normal conditions, different antennas, with different electrical constants of capacity and inductance, would alter the setting of the first tuning condenser. In the a. c. "Bandbox" the antenna stage has been left untuned. The remaining stages are tuned, and are isolated from the antenna stage. Hence the settings of the tuning condensers remain uniform regardless of the height or length of the antenna used. A radio-frequency choke is connected across the grid-filament circuit of the first tube. The antenna and ground are connected across the choke, and the action of this latter is to cause the radio-frequency signals induced in the antenna circuit to be impressed across the grid-filament circuit of the tube. The choke must be one with very

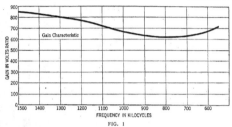

FIG. 1

little distributed capacity, since inherent capacity would afford an easy path for the radio-frequency signals to ground. By reducing the effect of the choke, the magnitude of the signal applied across the grid-filament circuit of the first radio-frequency tube can be reduced, hence the Crosley engineering department immediately visualized an excellent volume control. This is found in the variable resistance which shunts the radio-frequency choke in the antenna circuit. This resistance is non-inductive, free of capacity and, when varied, reduces the impedance of the choke. By reducing the impedance of the choke, its effect is reduced, and the signal input is decreased. Thus we have a volume control, which does not display any effect upon the selectivity of the receiver.

Receivers which oscillate and at the same time perhaps radiate an interfering signal, are not very popular, hence the radio-frequency stages are neutralized—by means of the Hazeltine system. The manner of neutralization gives the system an excellent radio-frequency amplifying characteristic, and helps overcome the usual falling characteristic on the longer wavelengths. Fig. 1 shows the radio-frequency response curve of the system. This graph is graduated from 200 meters (1500 kilocycles) to 545 meters (650 kilocycles). The maximum difference is about 25

per cent., and the minimum point of sensitivity is around 320 meters. The amplification rises on both sides of this point. The maximum gain is found at 200 meters, but even at this frequency, it is not a sharp peak. At no place on the curve is there a sharp peak or depression. Contrary to usual design, the amplification increases as the wavelength setting is increased above 320 meters (940 kc.), the low amplification point of the system. At 350 meters (545 kc.) the difference between its level and the maximum is approximately 16 per cent. These differences are small, and are not noticeable in actual operation. The radio-frequency transformers are single-layer solenoids with bifilar primary windings; that is to say, they have two primary windings. One winding is the regular primary and the other small winding is the inductance utilized in the neutralizing system. The inductance value of the secondary of these transformers is 190 microhenries and the capacity of the tuning condensers is approximately 430 micro-microfarads (0.00043 mfd.). The excellent radio-frequency response characteristic is attributable to a very large extent to the design of the primary winding of each radio-frequency transformer, its position with respect to the secondary winding, and to the coil utilized in the neutralizing system. A close study of the wiring diagram,

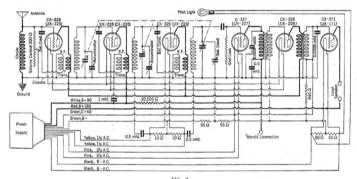

FIG. 2
The circuit diagram of the a.c. "Bandbox" showing the power unit connected for operation

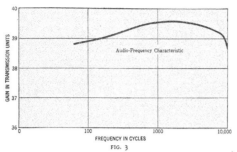

FIG. 3

Fig. 2, will bring to light the fact that this system is different from the conventional split-primary neutralizing arrangements. It is just these little differences which make for the variances in design, and necessitate extensive research, in order that the final result be meritorious and worthy of recognition.

An examination of the receiver shows thorough and complete shielding of all the units, including condensers, coils, and transformers. This is a sound piece of engineering, since it practically isolates the receiver from all external influences. Very often, a. c. operation with the power supply unit adjacent to a receiver utilizing unshielded transformers results in the induction of an interfering hum from the power equipment into the receiver audio-frequency transformers. Shielding the units eliminates all possibility of external interfering influences. Sheet iron, cadmium plated, is used as the shield for the condensers, of which shield the chassis also forms a part. Copper cans are used for the shielding of the radio-frequency transformers. These shields facilitate elimination of coil interaction and external influences, and also tend to increase stability. By selection of copper shields, the design of the coils, and placement of these latter with respect to the shields, the effect of the shield upon the radio-frequency transformer is kept at a very low value, so much so, that its detrimental effects are negligible so far as sensitivity and selectivity are concerned.

THE AUDIO SYSTEM

WITH respect to the audio system, two stages of transformer coupling are utilized and the operating curve is shown in Fig. 3. A study of the curve shows a variation of less than one half of a transmission unit through a frequency spectrum of from 60 to 10,000 cycles. At first glance, one is apt to imagine an operating characteristic very much akin to the majority of two-stage audio units. Upon closer observation, however, the small variation becomes apparent. The gain of the audio channel is low, but with the amplifying powers of the radio-frequency system, the combination affords a very satisfactory overall response. The absence of a definite peak on some audio frequency shows the result of careful study of the leakage reactance factor in transformer design.

A study of the curve shows an overall frequency range of from 60 to 10,000 cycles, with about equal amplification on 60 and 10,000 cycles. The curve rises between 60 and 900 cycles, is fairly flat between 900 and 3000 cycles,

falls gradually to 8000 cycles, and then drops abruptly between 8000 and 10,000 cycles.

OTHER FEATURES

BEING, at the time of my sojourn in the Berkshires, conversant with the features of the a. c. operated "Bandbox," the writer was able to communicate all the above information to

THE POWER UNIT
Its compactness is evidenced by
the photograph on page 369

Jones, whose mind, strangely enough, was gradually allowing itself to be changed to look upon the manufactured receiver in more favorable light.

By this time the other members of the group had evinced interest. Some had read the advertisements describing the receiver, and commenced hurling questions: "What were the accuminators?" "Were there any other interesting features?" The appearance of the accuminator controls deceived the "expert." He was certain that they were resistances or potentiometers, by means of which the grid bias voltages were altered. He was much surprised to hear that they were small vernier condensers, one in shunt with the third radio-frequency stage input tuning condenser and the other in shunt with the detector tuning condenser. These are visible in the wiring diagram and their function is to permit more accurate tuning.

The chassis is of $\frac{1}{16}$" iron and is grounded so that all ground connections of the receiver may be attached to it. The radio-frequency transformers used in the receiver are tested on radio-frequency bridges to within 2 per cent. and are matched with condensers which are also tested at radio frequencies. No coil or condenser combination of a set of three varies more than 1 per cent. from the rated values. In other words, the coils and condensers are sorted into groups and are matched so that the resonance points are identical in the various tuned stages. The grid returns for the untuned and tuned stages are at ground potential and obtain the various bias voltages through resistances, which carry the plate current. The voltage drop across these resistances is utilized as the grid bias.

The tubes in the three stages of radio-frequency amplification of the a. c. "Bandbox" are of the cx-326 (ux-226) type, and this tube is also used for the first audio stage. The detector is a c-327 (uy-227) and the output audio tube is a cx-371 (ux-171).

The electrical balance in the various filament circuits is obtained by means of mid-tapped resistances connected across the filaments, instead of resorting to the use of mid-tapped transformers. The plate voltage applied to the radio-frequency tubes and also the first audio tube is 90 volts. The 90-volt lead also connects to a 20,000-ohm resistance in the set, and is thus dropped to supply the required detector plate voltage. In other words, only two positive plate-voltage leads are available from the unit supplying the receiver. These are the 180-volt lead for the power tube and the 90-volt lead for the

A REAR VIEW OF THE "BANDBOX" CHASSIS

other tubes. A separate 45-volt lead is not provided in the manner usually employed. The power unit also has a negative grid bias 40-volt lead, and the regular B minus lead. The resistances which supply the radio-frequency and detector tubes with grid bias are in series with the B minus.

All of the cx-326 type tubes are connected in parallel and are fed from individual low-voltage filament windings. The midpoint in the filament circuit is obtained by means of a 20-ohm mid-tapped resistance in shunt with this filament circuit. A mid-point is not required in the filament circuit of the c-327 tube since it is of the heater type. The mid-point for the 171 tube filament circuit is obtained by means of a 100-ohm mid-tapped resistance. The filament heating transformer in the power unit supplies three distinct voltages from three distinct windings. These are for the 1.5-volt tube, the 2.5-volt tube, and the 5-volt tube. The cx-326 tubes require 1.5 volts at 1.05 amperes for normal filament operation. The c-327 requires 2.5 volts at 1.05 amperes, and the cx-371 requires 5 volts at 0.5 ampere.

The "Merola" connection indicated on Fig. 2 is the contact point for a phonograph pick-up unit, should it be used. "Merola" is the trade name for a pick-up made by Crosley.

THE POWER UNIT

THE power unit differs in several respects from the conventional. It is obtainable in two forms, the first being suitable for a 60-cycle supply and the second for a 25-cycle supply. The latter may also be used for 60 cycles.

A ux-280 type full-wave rectifying tube is employed, as can be seen in Fig. 4. A well-designed two-section filter is employed and the chokes are of 10 henries inductance each, and have a high current rating. This is important, since filtration is improved when the choke is operating below the maximum current rating.

The condensers in the filter network are of 10-mfd. each, three being used in the form of a single Mershon condenser. The voltage reducing resistance has a total resistance of 5000 ohms, tapped at 1525 ohms. With 280 volts input across the two anodes of the rectifying tube, the total voltage across the output is 220 volts. The rectified voltage is, therefore, sufficient to supply the required negative bias of 40 volts. The transformer contained in the power unit has six individual windings. The primary is tapped for low and high line voltage. The rectifier tube filament voltage is obtained from one secondary

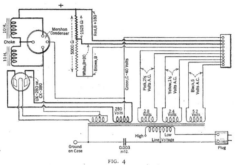

FIG. 4
The circuit diagram of the power supply unit of the a.c. "Bandbox"

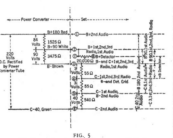

FIG. 5
This diagram shows how the grid voltages are obtained

winding, the rectifier plate voltage from another, while a winding of 2.8 volts is used to supply the 2.5-volt filament of the detector tube.

resistance in the feed line will cause an appreciable voltage drop.

The voltage output of the winding of the power transformer which supplies the 1.5-volt tubes is rated at 2.4 volts, and the voltage output of the winding which supplies the 5-volt power tube is 5.2 volts.

An idea of how the various grid bias voltages are obtained can be gleaned by a study of the wiring diagram of the receiver Fig. 2, and Fig. 5. The three resistances which produce the C bias voltages for the various tubes are shown in both drawings. The main wiring diagram shows the actual position of these resistances, as located in the receiver. The small wiring diagram, on the other hand, shows a much simpler arrangement, such as would be found were these resistances located in the power unit. The first 55-ohm resistance results in a voltage drop of 3 volts. This is the bias applied to the three radio-frequency amplifiers and to the detector grid. The next 55-ohm resistance, in series with the first, produces a voltage drop of 6 volts, which is applied to the grid of the first audio-frequency tube. The 540-ohm resistance produces the voltage drop of 40 volts required for the second audio tube, which, in this case, is a cx-371 (ux-171).

The excess voltage is an excellent feature because it provides for a reasonable voltage drop in the receiver filament circuit. This phase of a. c. tube operation has been the cause of the greatest amount of annoyance. With the high current consumption of low-voltage tubes, the voltage drop in the average receiver circuit, when five or six tubes are used, is sufficient to reduce the available filament voltage to a value below the requirements of the vacuum tubes. The current drain of six or eight cx-326 or c-327 tubes is quite high, and even a small amount of

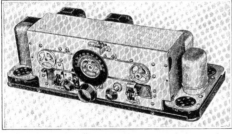

WHAT THE "BANDBOX" LOOKS LIKE WITH THE CABINET REMOVED

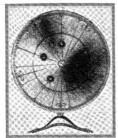

Interesting Loud Speakers and Power Equipment

THE FADA 22″ CONE

ALTHOUGH this picture illustrates only the table model, the 22″ cone is obtainable also in pedestal form at $50.00, and for hanging on the wall at $25.00. The table model shown lists at $35.00. Fada also has a 17″ cone which retails (the table pattern) for $25.00

THE BOSCH A UNIT

WHICH has been designed to supply A current to a receiver employing from four to ten tubes. The unit makes use of a Tungar rectifier tube. The price of this A supply is $58.00, and it is produced by the American Bosch Magneto Corporation, Springfield, Massachusetts

A NEW DYNAMIC LOUD SPEAKER

THE movable coil principle is featured in this attractive new loud speaker, which retails at only $65.00. Two of its connections go to the A battery, but a minimum of current is consumed. The loud speaker (with step-down transformer) without cabinet retails at $47.50. Jensen Radio Manufacturing Company, Oakland, California

IF YOU HAVE AN ATWATER KENT RECEIVER

YOU will find this Atwater Kent B supply especially suitable for use in conjunction with it. The B supply may, however, be used with other receivers. It delivers up to 135 volts, and is for sets consuming not more than 40 mils. An automatic relay is incorporated within the unit, and there is a receptacle in the front for trickle charger plug. Price $39.00

TWO MAGNAVOX LOUD SPEAKERS

THE popular R-4 unit is shown to the left. It is a moving coil electrodynamic loud speaker retailing for $50.00. It should be used in conjunction with a baffleboard. The model to the right lists at $120.00. In addition to being a loud speaker, it also combines a B supply unit and power amplifier. A 210 type tube is used for the latter, while a 216 type tube is employed as the rectifier

A NEW WESTERN ELECTRIC CONE

THE Western Electric 540 AW cone loud speaker has been the standard of comparison for so long that considerable interest is bound to follow the announcement of a new cone by this company. The 560 AW, as this new instrument is known, is a 25″ cone, and it lists at $35.00. The decoration in red and dark brown, is more striking than that on the 540 AW

$\mathcal{N}$ew $\mathcal{R}$ecords

Short Reviews of Recent Releases by
Victor, Brunswick, and Columbia—
A List of Some New Record Albums
—Rimsky-Korsakov's Scheherazade
Suite Obtainable in Complete Form

WE ARE celebrating the four-month birthday of this department by climbing out of our rompers and taking a big step. Henceforth we shall aim to be a guide to all the best phonograph records. You cannot read all the books published each year in order to choose the best volumes, so you consult book reviews and guides. Neither can you while away the hours in your favorite music store listening to the new—and old—offerings of the very active phonograph companies. When you buy records the chances are you act on the advice of the music salesman, or take the word of a friend, or if you happen to be on the mailing list of a music store, you check over the catalogue and select the most likely sounding titles. Any one of these methods is precarious and the element of chance is large in each.

When this department was mapped out in the editorial mind, the word went forth that these two pages were to be devoted to a general review of phonograph records which had been produced by artists who were familiar to listeners-in as broadcast performers. Now we will extend the field to include the records of all artists whether or not they play the dual rôle. Some of the recordings will be briefly reviewed, others will merely be mentioned, and each month there will be a list of records which we consider well worth hearing, and buying, if the spirit moveth.

WHAT HAVE WE HERE?

AFTER inspecting the current supply of records we find that it contains the following ingredients: The usual popular vocal numbers by the usual popular vocal artists; an array of good dance numbers with one outstanding success, *Dream Kisses*; several old favorites rendered superbly by such distinguished artists as Sophie Braslau, John Charles Thomas, Charles Hackett,

and Maria Kurenko; seven minutes of very beautiful choral singing by the Metropolitan Opera Chorus; and an album of Rimsky-Korsakov music, of which, more anon. These ingredients have been highly seasoned with the sentiment which, we are led to believe by song writers' song singers, and phonograph companies, the public cries for, and they have been expertly mixed, and sifted, spread on the discs by the new electrical recording method, and served hot to the public, for prices ranging from seventy-five cents to ten dollars. Taken as a whole there can be no question of the general excellence of the output. Our chief complaint is that it is too sweet for our taste. Is our taste

<hr/>

Don't Miss These

Scheherazade Suite (Rimsky-Korsakov) played by the Philadelphia Orchestra under Leopold Stokowski (Victor). *Cavalleria Rusticana–Gli Aranci Olezzano* and *Immeggiamo Il Signor.* (Mascagni) sung by the Metropolitan Opera Chorus, with Orchestra (Victor).
Rigoletto: La Donna E Mobile (Verdi) and *Cavalleria Rusticana: Siciliana* (Mascagni), sung by Charles Hackett (Columbia).
Liebestraum (Liszt) and *Sheep and Goat Walkin' To Pasture* and *Gigue* (Bach) played by Percy Grainger (Columbia).
Among My Souvenirs and *Washboard Blues* played by Paul Whiteman and His Concert Orchestra (Victor).
Dream Kisses and *Among My Souvenirs* played by the Ipana Troubadours and Ben Selvin respectively (Columbia).
My Lady and *Two Loving Arms* played by Cass Hagan and His Park Central Orchestra and The Cavaliers respectively (Columbia).
A Shady Tree and *There Ain't No Land Like Dixieland To Me* played by Ernie Golden and His Hotel McAlpin Orchestra (Brunswick).
Back Where the Daisies Grow and *Lonely in a Crowd* played by the Park Lane Orchestra (Brunswick).
Liliue and *Honohano Hawalei* by the South Sea Islanders (Columbia).

<hr/>

peculiar or are there others who do not clamor for sentiment as the pervading flavor in their musical diet? Would they, too, like a little humor in their daily slice of song?

More or Less Classic

Cavalleria Rusticana—Gli Aranci Olezzano and *Cavalleria Rusticana—Inneggiamo Il Signor.* By Metropolitan Opera Chorus with Orchestra. (Victor). An expert recording of two of the

most melodious of the choruses of Mascagni's opera, sung with great beauty and restraint.

(a) *Sheep and Goat Walkin' to the Pasture* (Guion), (b) *Gigue* from *First Partita* (Bach), and *Liebestraum* (Liszt). By Percy Grainger (Columbia). Here is variety itself: a humorous tale, a lively jig, and a romance all on the same record, and each feelingly interpreted by this master pianist.

Hungarian Dance No. 1 (Brahms-Joachim) and *Slavonic Dance No.* 2, in E minor, (Dvorak-Kreisler). By Toscha Seidel (Columbia). Two lusty dances played with too mechanical vehemence to suit us.

Lucrezia Borgia: Brindisi (Donizetti) and *Come to Me O Beloved* (Bassani-Malipiero). By Sophie Braslau (Columbia). We prefer the rollicking joyousness of the drinking song to the heavy solemnity of the cantata but that is a matter of opinion. The rich contralto voice of this artist handles both expertly.

Love's Old Sweet Song (Molloy) and *The Sweetest Story Ever Told* (Stults). By Sophie Braslau (Columbia). Miss Braslau digs way down in the bag and brings up some of the old tricks. But she sings beautifully.

Rigoletto: La Donna e Mobile (Verdi) and *Cavalleria Rusticana: Siciliana* (Mascagni) by Charles Hackett (Columbia). So convincingly does this glorious tenor sing Verdi's surprise number that one is almost ready to agree that woman *is* fickle! Well, were it necessary, we would agree to anything for the privilege of listening to Hackett's singing.

Coq D'Or, Hymn to the Sun (Rimsky-Korsakov) and *Song of India* (Rimsky-Korsakov) By Maria Kurenko (Columbia). We would like the first selection better were it minus a few coloratura frills. As for the S. of I. we said what we had to say about that years ago. However, it is beautifully sung.

Smiling Eyes and *Roses of Picardy.* By John Charles Thomas (Brunswick). Why turn this fine baritone voice loose on such shop-worn ballads as these?

"Popular"

Among My Souvenirs and *Washboard Blues* by Paul Whiteman and his Concert Orchestra (Victor). Whiteman at his unique best. You can't dance to these but who wants to? It's music worth *hearing.*

Dream Kisses by the Ipana Troubadours (Columbia). At last we have something different in dance numbers! A soothing, insinuating rhythm built for dancing and played for dancing

374

Interesting Record Albums

Beethoven: *Symphony No. 9, in D minor (Choral)*	ALBERT COATES AND SYMPHONY ORCHESTRA	*Victor*
Beethoven: *Concerto in D major, Violin*	FRITZ KREISLER AND STATE OPERA ORCHESTRA, BERLIN	*Victor*
Brahms: *Symphony No. 1, in C minor*	LEOPOLD STOKOWSKI AND PHILADELPHIA ORCHESTRA	*Victor*
Schubert: *Symphony No. 8, in B minor (Unfinished)*	LEOPOLD STOKOWSKI AND PHILADELPHIA ORCHESTRA	*Victor*
Tschaikowsky: *Casse Noisette (Nutcracker Suite)*	LEOPOLD STOKOWSKI AND PHILADELPHIA ORCHESTRA	*Victor*
Chopin: *Sonata in B minor, for Pianoforte, Opus 58*	PERCY GRAINGER	*Columbia*
Brahms: *Sonata in A major, Opus 100, Violin and Piano*	TOSCHA SEIDEL AND ARTHUR LOESSER	*Columbia*
Ravel: *Ma Mère l'Oye (Mother Goose) Suite for Orchestra*	WALTER DAMROSCH AND NEW YORK SYMPHONY ORCHESTRA	*Columbia*
Dvorak: *Symphony No. 5, "From the New World"*	SIR HAMILTON HARTY AND HALLÉ ORCHESTRA	*Columbia*
Berlioz: *Symphonie Fantastique, Opus 14*	FELIX WEINGARTNER AND LONDON SYMPHONY ORCHESTRA	*Columbia*
Beethoven: *Symphony No. 5, in C minor*	WILHELM FURTWAENGLER AND PHILHARMONIC ORCHESTRA, BERLIN	*Brunswick*
Beethoven: *Symphony No. 7, in A major*	RICHARD STRAUSS AND THE ORCHESTRA OF THE STATE OPERA, BERLIN	*Brunswick*
Handel: *Concerto for Organ and Orchestra No. 4 (Op. 4)*	WALTER FISCHER OF THE BERLIN CATHEDRAL WITH ORCHESTRA	*Brunswick*
Mozart: *Jupiter Symphony, No. 41 Opus 551*	RICHARD STRAUSS AND ORCHESTRA OF THE STATE OPERA, BERLIN	*Brunswick*
Richard Strauss: *Ein Heldenleben*	RICHARD STRAUSS AND THE ORCHESTRA OF THE STATE OPERA, BERLIN	*Brunswick*

The records in the above groups are to be had only in album form. The list is by no means complete but serves to indicate a few of the most interesting complete recordings which are available. The *Rimsky-Korsakov Scheherazade Suite* (Victor) is reviewed elsewhere on this page.

by S. C. Lanin's Toothpaste Boys. On the reverse is *Among My Souvenirs*. Ben Selvin makes as good a dance record out of this as Whiteman did a set piece.

A Shady Tree and *There Ain't No Land Like Dixieland To Me* by Ernie Golden and His Hotel McAlpin Orchestra (Brunswick). Two more hot numbers from the orchestra under the direction of the gent who has musical "it."

Yep! 'Long About June and *Blue Baby* by Ray Miller and His Hotel Gibson Orchestra (Brunswick). The first is a lively down-east-barn-dance sort of number that will make your feet very restless; the second, only another dance tune. Both smoothly played by this excellent Cincinnati orchestra.

Back Where the Daisies Grow and *Lonely in a Crowd* by the Park Lane Orchestra (Brunswick). After hearing these two numbers you will add the P. L. to your list of best orchestras.

Ooh! Maybe It's You and *Shaking the Blues Away* by Ben Selvin (Brunswick). Not quite up to the Selvin mark but you won't want to sit still to either number.

Barbara and *There's a Cradle in Caroline* by Ben Bernie and H. R. Orchestra (Brunswick). Just another disappointment.

Together We Two and *What'll You Do* by Isham Jones Orchestra (Brunswick). Isham has certainly been in seclusion long enough to have dug up better numbers than these for his return engagement.

My Lady by Cass Hagan and His Park Central Orchestra. Simply swell! *Two Loving Arms* by the Cavaliers. A grand waltz. (Columbia).

Are You Happy? and *Kiss and Make Up* by Vincent Lopez and His Casa Lopez Orchestra (Brunswick). Answering the question: No more so than if we'd never heard this record.

Together We Two by Fred Rich and His Hotel Astor Orchestra. (Columbia). Only moderate. *Baby Feet Go Pitter Patter* by Harry Reser's Syncopators. Won't you give just a little something for a decent funeral?

Where Is My Meyer? by Eddie Thomas' Collegians (Columbia). A good nonsense song from the *Chauve Souris* sung by Frank Harris with an orchestral background and a little yodeling for good measure. *Clementine* by Don Voorhees and His Orchestra is only fair.

A Lane in Spain and *There Must Be Somebody Else* by Van and Schenck (Columbia). A very good vocal duet aided by guitar and piano.

Watching the World Go By by Ford and Glenn (Columbia). Why shouldn't it? This isn't enough to stop for. *Are You Thinking of Me To-night?* by Elliott Shaw. Insomnia must be prevalent among song writers.

There's a Cradle in Caroline and *I'll Be Lonely* by Frank Bessinger and Ed Smalle (Brunswick). Good sentimental singing.

I'm Coming, Virginia and *Just a Memory* by the Singing Sophomores (Columbia). Leaves us cold.

Twiddlin' My Thumbs and *The Pal You Left At Home* by the Whispering Pianist (Columbia). Good rubber wasted on tripe.

Liliuae and *Hanohano Hanalei* by the South Sea Islanders (Columbia). Good Hawaiian music magnificently played.

My Blue Heaven and *The Song is Ended* by Jesse Crawford (Victor). If you know anyone who wields a better movie organ than the organist at the Paramount Palace we would like to hear of him. Jesse Crawford is at his very best on these records.

Are You Lonesome To-night? and *Under the Moon* by Lew White (Brunswick). But the Roxy organist isn't far behind.

Estrellita and *Mi Viejo Amor* by Godfrey Ludlow (Brunswick). Good violin solos by the well-known staff artist of the National Broadcasting Company.

A COMPLETE SYMPHONY

Scheherazade—Symphonic Suite (Rimsky-Korsakow). By Leopold Stokowski and the Philadelphia Orchestra. Complete on five doublefaced Victor records.

When the Russian composer Rimsky-Korsakov called this music the *Scheherazade Suite* he did not mean to imply that he was telling, musical word for spoken word, the story of the Sultan Schahriar, who so distrusted women that he vowed to put each of his wives to death after the first nuptial night; and of Scheherazade, the Sultana, who caused him to forsake his vow by entertaining him with fascinating tales for one thousand and one nights, at the end of which time it is to be presumed that he had regained his faith in women. Rimsky-Korsakov merely used the title as a hint to his listeners that the suite was "an Oriental narrative of some numer-

ous and varied fairy tale wonders," told by some one person to her stern husband.

Take the hint or leave it. If you take it you can easily pick out the voice of the stern husband with which the first movement, *The Sea and the Vessel of Sinbad*, opens. It is a bold phrase played in unison by the trombone, the tuba, horns, woodwinds, and strings in their lower range. Then the sweet, timid voice of Scheherazade, in the high trembling notes of the violin, with the harp in the background. Then we hear the long roll of the sea translated into music by the violins. Now and then the lapping of the waves on the vessel. In the second movement, *The Tale of the Prince Kalender*, you can certainly identify the figure of the fakir prince in the now sad, now comic, notes of the bassoon; and there is no mistaking the wild violent dance of the Orient in which brasses, woodwinds, tuba, trombone, bassoon, and strings combine. The third movement tells a love story of *The Young Prince and the Young Princess*. The romance is plainly indicated in the simple tender melodies. And lastly there is *The Festival at Bagdad*, music full of the color and sinuous rhythm of the East. Queer minglings of sounds, seductive strains, call to mind beautiful veiled maidens, snakecharmers, swaying camels, spicy odors, perfumes of the Orient. Then suddenly we swing back again to the sea, this time not the calm, rolling sea of the first movement but a turbulent, treacherous sea, in which the vessel of Sinbad finally sinks, after a mighty crash on the rocks. And as the waters close over the ship the voice of the Sultan, now subdued, tells us that the tale is over and he is pleased, and the Sultana goes back to the opening theme of the strings for her finale.

Beautiful music, rich, colorful, varied—dealing with weird and wonderful events but always essentially human. Played by a master orchestra directed by a master hand—a feast to suit the palate of the most discriminating of music lovers.

RADIO BROADCAST Photograph

A RECEIVER EMBODYING NEW R. F. PRINCIPLES

This six-tube Stewart-Warner receiver employs the features of r.f. design which are described in this article

Designing an R. F. Amplifier

By Sylvan Harris

PERHAPS the most important problem that remains in connection with the design of radio receivers is that of controlling regeneration and the tendency toward self-oscillation. The patent situation in the radio industry to-day accentuates the importance of this matter, but aside from this, and considering only the technical aspect of the problem, there are certain important points concerning which the layman's ignorance is appalling and on which many engineers are doubtful. The most important of these is, perhaps, the magnitude of the amplification which a radio-frequency amplifier can furnish under the most favorable conditions.

This problem will not be discussed here; it will suffice for the present to state that the greatest amplification that a radio-frequency amplifier, having regeneration, will furnish, is determined by the electrical characteristics of the circuits, and takes place when the decrease in amplitude of the feed-back current from stage to stage becomes less than equal to the natural amplification in the opposite direction. That is, a voltage established in the plate circuit of the last amplifier (r.f.) tube, is decreased in each stage looking in a direction toward the input of the amplifier.

In the opposite direction we have the signal voltage passing from stage to stage, but being amplified at each step. When the decrease in amplitude in the direction of feed-back becomes small enough, oscillations are established, and it is at the instant that these oscillations are established that the greatest amplification is obtained.

The maximum possible amplification that can

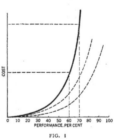

FIG. I

be obtained from the system under given conditions can be computed. A method of making this computation was shown some time ago by Dr. A. W. Hull, in his paper on the shielded-grid tube in the *Physical Review*.

It can be shown that, using the usual tubes, and under the usual average circuit conditions, it is not possible to obtain a gain per stage of more than about 10 or thereabouts, in the broadcasting spectrum, assuming reasonable values for the circuit elements and the 201-A type tube characteristics. Furthermore, the amplification naturally drops off as the frequency increases because of several complicated tube and circuit factors. The amplification we are referring to is the *maximum* amplification obtainable, keeping the adjustments such that the system is always on the verge of oscillation.

This brings to our attention a particular phase of the problem, *viz.*, that of keeping the overall gain of the amplifier constant over the broadcasting frequency range. Since the amplification is naturally greater at lower frequencies than at higher, the only way in which to make it uniform is to operate very close to the point of oscillation at high frequencies, and not so close at low frequencies. On account of difficulties encountered due to the stabilizing elements (grid resistors, etc.), we have many sets which operate well at

376

300 meters but are 'dead" at 500 meters, and many of which are rather "wild" below 300 meters.

On the other hand, due to absorption in metal panels, additional attenuation introduced by neutralizing condensers, etc., we often encounter the condition where the set is "dead" at the higher frequencies also. Of course, these terms are merely relative; it is clear that under any circumstances the design of the receiver, especially when it is intended for production in large quantities, must provide for sufficient tolerance in this matter of operating close to the oscillation point.

Another point which is of paramount importance, especially as regards the production of radio receivers in large quantities, is that of the relation between the performance of a receiver and its cost. Let us call the "performance" of our ideal receiver 100 per cent. and the performance of the receiver which will not work at all, 0 per cent. As regards the cost, let us say we can build a receiver whose performance is zero per cent. for practically nothing. On the other hand, as we improve the performance of our receiver in uniform steps, the cost mounts up and up at an ever-increasing rate, until it would cost an infinite amount to produce the ideal receiver operating at a performance of 100 per cent. The relation between the performance and the cost of a receiver is probably an exponential one, and may be something like the curve of Fig. 1. Thus, we can design and build a receiver having a "performance" of 60 per cent. at a reasonable cost, but when we attempt to improve the performance by 10 per cent. more, we drive the cost way up. Of course, the curve of Fig. 1 is purely qualitative; it is intended merely to illustrate the idea.

The receiver described in these articles has been designed with these ideas in mind; at the same time it must be remembered that there are certain circuit arrangements which are more flexible than others, making it possible to build a better receiver at the same cost, so that actually Fig. 1 should be a family of curves, each curve applying to a different type of receiver. As to the merits of the receiver to be described here, it will be well to leave these to the judgment of the reader rather than run the risk of laying oneself open to criticism.

It is well known that the presence of inductive reactance in the plate circuit of a radio-frequency amplifier tube results in regeneration. Furthermore, if that reactance exceeds a certain critical value, oscillations will be established in that stage, and further, the critical value varies with the frequency. Many different methods have been tried for keeping this reactance below the

critical value, such as varying the coupling in the resonance transformers simultaneously with the tuning condenser. The method used in this circuit is unusual as applied to r.f. amplifiers, although a similar arrangement has been used for controlling regenerative detectors.

Fig. 2 illustrates the fundamental idea of the

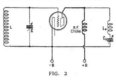

FIG. 2

circuit. The coil L and the condenser C constitute the tuned input circuit. In the plate circuit of the tube there is the usual primary coil of a resonance transformer, L_0. This inductance L_0 is of such value that the circuits would ordinarily oscillate; in other words, L_0 is greater than the critical positive value.

In order to stop the oscillations the condenser C_0 is introduced in series with L_0; the condensive reactance furnished by C_0 is negative in sign, so that the net positive reactance in the plate circuit of the tube is reduced, depending upon the value of the capacity of C_0.

It is clear that the net reactance can be made almost anything we desire by this means. We may so proportion L_0 and C_0 that X_0 may be positive and greater than the critical value, equal to or slightly less than the critical value, or, we may even make X_0 equal to zero, or make it negative. When X_0 is positive and greater than the critical value, the circuit will oscillate; under any other condition it will not oscillate. The special case where X_0 equals zero is known as the "zero reactance plate circuit," and will be discussed later on. The case where X_0 is negative (or capacitive) is not of interest here, for under such conditions the circuit becomes very inefficient, due to the absorption of power at the input of the tube. These conditions have been discussed by J. M. Miller in *Circular No. 351 Bureau of Standards*.

It would evidently be very cumbersome to operate two condensers in each stage of the radio-frequency amplifier, a three-stage amplifier requiring seven condensers, three "twins" and one "single." The obvious thing to do, therefore, is to operate them all together. The problem re-

mains then, to so proportion the inductance and capacity in the plate circuits so that when C_0 is varied at the same rate as C (the tuning condenser) the net plate circuit reactance will be slightly below the critical value at all frequencies.

The first problem in the design was, therefore, to determine whether C and C_0 could be varied at the same rate; this amounts to the same thing as determining whether the plates in the two condensers could have the same shape. In order to determine this, as well as to determine other things that were to follow, a system of measuring the amplification or "gain" per stage was set up, a description of which was presented by the writer in *The Proceedings of the I.R.E.*, July, 1927.

The complete circuit of the receiver is shown in Fig. 3. Each stage is completely shielded, originally in copper cans, but later in aluminum cans. A laboratory set-up was made, in conjunction with the measuring system mentioned above, "single" condensers being used in the receiver, with leads coming out of the cans to which separate plate condensers were connected. Each condenser was individually controlled.

With a constant-frequency signal impressed on the input of the receiver, the condensers in the plate circuits of the r.f. tubes were gradually and simultaneously increased from zero upwards, in steps, their capacities being always kept the same, and measurements of the gain were taken at each step.

At a high frequency, say 1500 kilocycles, a curve similar to that marked f_1 in Fig. 4 was obtained. At the upper limit of the curve the circuits broke into oscillation. Having completed this curve, a similar curve was obtained for a slightly lower frequency, say 1750 kilocycles, illustrated by curve f_2 of Fig. 4. So, a family of curves was obtained, each curve for a different frequency; sufficient curves were obtained so that the operation over the entire broadcasting range of frequency could be studied.

It will be noted that there is an inflection in each of these curves near the upper limit where oscillations begin. Above the point where the inflection begins the system is very critical, so it was evident that the greatest gain per stage that could be utilized with safety at any given frequency is at this point. In Fig. 4, therefore, the points a, b, c, d, represent the greatest amplification that can be obtained with the particular tubes, coils, etc. Through these points a curve can be drawn, marked AB, from which can be taken values to plot a curve of C_0 against the frequency, or against C, which is the information which was required. It was found that on plotting C_0 against C, the curve obtained was very nearly, linear, indicating that it was per-

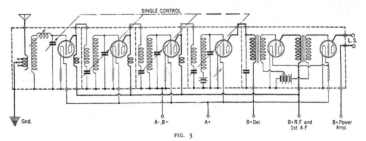

FIG. 3

fectly feasible to vary C_6 at the same rate as C, and that the two condensers may have plates of the same shape and may be driven by the same dial.

It is evidently desirable to have the same shape plates in the two condensers, in the interests of economy as well as simplicity. It was also found that with the given coils, the capacity required in the plate circuit (represented by the points a, b, c, d, of Fig. 4) at various frequencies was very

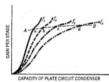

CAPACITY OF PLATE CIRCUIT CONDENSER

FIG. 4

close to that required in the tuning condenser C at each frequency. This relation can be adjusted at will, by simply changing the design of the resonance transformer; it was thought advisable, however, to keep both sections of the "twin" condenser the same, as this would facilitate factory inspection, permitting the inspection of both sections, by the same operator and on the same testing "jigs."

Upon setting up a model of the receiver, it was found, that, whereas the amplification was considerable at lower frequencies, it dropped off very rapidly as the higher frequencies were approached. This was very noticeable when the receiver was tried "on the air;" the ability of the receiver to "perform" at wavelengths shorter than about 350 meters was practically nil. The problem then remained to increase the sensitivity of the receiver at the shorter wavelengths, leaving the sensitivity at the longer wavelengths the same.

The means used for doing this is, as far as the writer is aware, a new one; it is new, not in the sense that it has never been used before, but in the sense that it has been done intentionally and with a definite purpose. There is plenty of coil capacity in plenty of receivers, but as far as the writer has been able to find out, no one has heretofore attempted to put this coil capacity to a good use.

· Up to this point in the work the primary coils of the resonance transformer were wound in a single layer at one end of the secondary, separated from the secondary by about one-quarter of an inch. The idea occurred that by introducing a slight amount of capacity coupling in the resonance transformers, in addition to the magnetic coupling, the regeneration at the higher frequencies would be slightly increased, and the desired increase in amplification would be thereby attained.

Consequently, the primary coil of one of the resonance transformers (that preceding the detector) was wound on a short tube and slipped

inside the secondary. When tuned to a short wavelength (or high frequency), the response of the receiver would gradually increase as this primary is moved up into the secondary further and further, until oscillations begin. This point is located, and the primary coil is then fixed in place slightly below it. Besides the advantage gained by increasing the amplification at the higher frequencies and thus solving the greatest problem which was encountered, there is an additional and important advantage gained by making the primary adjustable at the factory. It is possible by this means to insure uniform production of receivers with regard to sensitivity.

The effect can be explained qualitatively by means of Fig. 5. The curve AB is supposed to represent the amplification curve of a stage of the r.f. amplifier before the coil capacity has been introduced. On introducing the coil capacity the curve rises at the right-hand side, resulting in the curve AC. On the first trial a considerable depression was found in the curve at a frequency of about 850 kilocycles. It was found, however, that by properly adjusting the self-inductance of the primary as well as its location within the secondary, this depression could be made to disappear almost, and the amplification was thus made practically uniform over the entire broadcasting spectrum.

The resulting receiver proved to be very sensitive and selective. No difficulty was experienced in separating the local broadcasting stations in either New York or Chicago, and it was possible in many cases to even tune-in stations between the locals.

There are several other additional features of this receiver which are worthy of mention. One of these is the location of a separate filament resistor in the negative lead of each tube filament of the r.f. amplifier.

By this means a negative bias is placed on each grid, with the result that the receiver is very economical with respect to the B supply. The maximum plate current is about 27 milliamperes, and on receiving local concerts, sufficient volume for ordinary sized homes is obtained with the volume control "just on," and the set drawing only about 10 milliamperes. This latter statement holds also when receiving some powerful stations at fair distances.

Another feature of the receiver is the C battery detector. This type of detector was used instead of the grid leak-grid condenser type on account of its ability to rectify more powerful signals before overloading occurs. It also cuts down the costs by eliminating the grid condenser and leak, and necessary inspection were they used. The difference in sensitivity between the two types of detectors is of secondary importance in this receiver on account of the great sensitivity of the r.f. amplifier.

The input stage of the amplifier is tuned by a "single" condenser, and, in addition to this, has a portion of the secondary winding of the antenna transformer variable. There are also three taps on the primary or antenna coil, so that by means of the taps and variable portion of the secondary, the input stage is made very efficient. The constants of this stage, however, are so chosen that no matter on which of the three points the switch

may be set, it is not possible to "pass over" any except the very weakest signals when varying the condensers.

Another feature of the receiver is the absence of bypass condensers in the r.f. amplifier. The only place where a bypass condenser is used is in shunting the primary of the first audio transformer. The general wiring of the receiver makes it unnecessary to use them, the only avenue of exit for the r.f. currents outside of the grid and

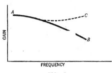

FREQUENCY

FIG. 5

plate leads being the B supply lead, which is blocked by an effective r.f. choke coil.

The greater part of the wiring is in the metal of the receiver—the aluminum cans and base. The ground accommodates the A minus, B minus, and C plus, making the entire wiring job a very simple one indeed. A "matched unit" construction is used, the four cans and their contents being identical in many respects. The first stage differs from the rest only in the coil and single condenser, and the fourth can · (the detector stage) differs only in having an individual filament control. This was deemed necessary, or at least desirable, on account of the varying characteristics of electron tubes. But when once this rheostat is set for a particular tube, there is no further necessity of adjusting it.

The mechanical arrangement of the receiver is also novel. The separate cans are easily removable, for servicing purposes, by simply loosening the screws on the backs of the cans which hold the terminal lugs fastened on to the four or five leads connecting to the can. All the battery supply leads are cabled along the rear of the set outside the cans. Likewise the audio amplifier, of the transformer type, is removable as a unit, in the same manner. There are four "twin" condensers and one "single" condenser in the receiver, all of which are driven by the same mechanism. To facilitate the "matching" of condensers while the lids of the cans are in place, the front steel panel on which the condensers are mounted is slotted, so that the stators of the condensers can be rotated through a small angle and then clamped in place.

The great sensitivity of the receiver described here is due to several things; first the primary self-inductance is a little greater and the coupling is slightly closer than in the usual resonance transformer employed in r.f. amplifiers. Furthermore, it is possible to adjust, and to maintain the adjustment closer to the oscillation point than is usually the case. Next, there is the reduction of grid losses due to the grid bias on each r.f. amplifier tube, and finally there is the tuned input stage.

Suppressing Radio Interference

Some Recently Investigated Sources of Interference—Trouble from High-Tension Lines and Distribution Systems—General Hints on the Location of Sources, the Arrangement of a Patrol Car, and the Receiver to Use—Procedure in Patrol Work and the Many Misleading Clues

By A. T. Lawton

WHEN we consider that a little spark, say one-sixty-fourth of an inch long, can cause severe radio interference, it becomes obvious that almost any piece of electrical apparatus in ordinary commercial use is a potential source of trouble. This does not mean that all electrical accessories do actually give rise to disturbances. On the contrary; considering the expansion of the electrical industry and the universal use of electrical equipment, we may well marvel that the noise level of the average city is so low.

When looking around for the possible cause of interference, however, nothing in the electrical line should be overlooked. It may help a little if we detail some of the sources located recently in following up interference complaints:—

(1.) Printing office linotype motor; normal operation, no fault. Cleared up by method (a) Fig. 1.
(2.) Clippers in barber shop. Normal. Cleared up by method (a) Fig. 1
(3.) Same source as above. Abnormal—commutator grooved. Used method (c) Fig. 1
(4.) Cream tester in dairy plant. Normal. Used (a) Fig. 1
(5.) Battery charger, vibrating reed type. A 1-mfd. condenser across contacts eliminated trouble.
(6.) Same source as above—different manufacture. Required method (a) Fig. 1
(7.) Refrigerator, home type artificial refrigeration. Used method (a) Fig. 1
(8.) Battery Charger, rotary rectifier. Obstinate case. Used method (c) Fig. 1
(9.) Woolen mill, static neutralizer. Defective plug connection. Repaired.
(10.) Brass foundry, lighting wires crossed and sparking, concealed. Confusing, as corresponded with vibration of large motor which was shaking floor. Motor supply cut abruptly but radio interference died gradually with slowing down of motor giving clue to the trouble, which was located and fixed.
(11.) Voltage regulator in a power station. Case still pending.
(12.) Chattering circuit breaker in power plant operated by thermocouple. Breaker readjusted and interference disappeared.
(13.) Washing machine motor, defective. Repaired.
(14.) Thermostat, mounted on wall subject to abnormal vibration. Bad clicking radio interference. Changed location of thermostat to solid wall.
(15.) Constant-current transformer in power station. Case pending.
(16.) Noisy grid leak in complainant's radio set. Referred to service man.
(17.) Elevator in apartment house. Under investigation.
(18.) Bakery oven. Thermostats to regulate heat and operate gas lighting jump sparks. Case pending.

Many other cases of interference reported in the same period had their origin in apparatus already discussed in previous sections while an additional number arose from defects on power lines, all of which were located and rectified.

HIGH-TENSION LINES

IN THIS category we include primary distribution systems operating on from thirty thousand to one hundred and ten thousand volts a.c.

We must frankly admit that the data on interference from this source are far from complete. Difficulties of a practical nature are responsible. Thirty or more towns may be served by say, a

---

UNDER *the title "Suppressing Radio Interference" the author has printed three previous articles in this series, all of which deal, in an enlightening and comprehensive fashion, with the different forms of interference with which the radio listener may have to cope. Each article in the series is complete in itself, and should be read by all radio men whether they are troubled by man-made static or not. Mr. Lawton's articles cover a period of two and a half years' research made in more than 130 different cities. The first article appeared in the September, 1927,* RADIO BROADCAST, *and dealt with interference caused by oil-burning furnaces, electro-medical therapeutic apparatus, X-Ray equipment, and dental motors. The November article dealt with interference from motion picture theatres, telephone exchanges, arc lamps, incandescent street lamps, flour mills, factory belts, electric warming pads, and precipitators. In the* January, 1928, RADIO BROADCAST, *the following sources of interference were dealt with: Farm lighting plants, railway signals, land line telegraph and stock tickers, radio receivers, and electric street railways. The present article is of especial interest to radio clubs and organizations contemplating the construction and operation of interference locating equipment.*

—THE EDITOR.

110,000-volt three-phase system and to cut this line the required number of times for test purposes is out of the question since all towns depending on it for electric power would be interfered with to a prohibitive extent.

Also, when suspicion to any piece of apparatus in connection with such a line arises and arrange-

---

ments can be made to have it investigated at a close range, the line must, of course, be cut prior to detail inspection. Complications enter; foreign matter, deposited at such points as to be a possible cause of the trouble, is burned up by the surge produced on cutting the line. The investigator, not noting any apparent defect, and no spitting of current being possible for an audio check, often removes the source of the trouble unconsciously.

Where several possible sources are observed all must be rectified at once in order to get the line back in operation as quickly as possible, so even if elimination of interference is secured the exact source is not definitely established. Piece-meal elimination is a theoretical idea—desirable from a scientific standpoint but impractical.

It is probable that a slightly leaky insulator on a 110,000-volt line will cause radio interference though certain observations point to the contrary. For instance, a six-petticoat suspension insulator which was spitting vigorously over three petticoats caused absolutely no radio interference on a three-tube regenerative receiver located thirty feet distant. Possibly this caused a disturbance farther out on the line but other disturbances at the same points complicated the determinations.

If a high-tension conductor on a pin insulator is not fastened down securely, audible hissing is usually the result, but it gives rise to no radio interference. Two defective wall bushings on a 12,000-volt line showed considerable spitting, which gave an audible noise, but there was no indication of radio interference on a "patrol" receiver in the vicinity. Out on the line half a mile or so it was impossible to tell whether these constituted material sources or not; if anything, the "mushy" noise usually associated with high-tension lines was slightly augmented.

We must not assume that because a line, d.c. or a.c., produces audible noise, it necessarily gives a corresponding radio interference. Take the case of a d.c. line operated at 700 volts and carrying thirty thousand amperes. The "lines"

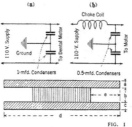

FIG. 1

in this case were composed of several hundreds of wires or, rather, rods, bound together at intervals and were mechanically noisy but variations of this heavy current caused no radio disturbance.

When two separate conductors are used as one on a high-tension line, running close together and attached to the same insulators, there will likely be audible hissing although the voltage at any given point is equal on both wires. Prevailing interference could not be definitely associated with lines of this nature.

It is worth noting that the smelting arcs fed by the thirty-thousand ampere d.c. line referred to above do not set up any serious disturbance beyond about one hundred yards. Even within this distance we are inclined to think that some of the interference noted on test had its origin in associated apparatus because in the near vicinity of three a.c. arcs, operating on 22,000 amperes, reception from station wqy, 300 miles distant, was possible with the receiver within 60 feet of the arcs.

The charging of h.t. lightning arresters sets up heavy interference. Charging is done twice daily as a rule, each operation requiring only a few seconds. Where there are many sub-stations in a restricted area it is usual to arrange to have the arresters charged at 2 A.M. and 2 P.M. instead of during the more significant broadcasting hours.

Thick snow falling between the horn gaps is pretty sure to cause considerable snapping and the interference, being vigorous, is propagated for miles.

Surprising as it may seem, the normal interference field of a 110,000-volt a.c. line has much in common with a 600-volt d.c. trolley feeder. Within fifty or sixty feet of either, reception is doubtful, but an abrupt cut occurs here, and a few feet farther away normal reception is possible, a condition which would seem to indicate magnetic coupling rather than a pick-up through direct radiation.

A.C. lines of fifty-thousand volts or so can, of course, set up currents of very high potential in parallel systems. We know that enough energy is transferred by induction to the overhead ground wire of a 100,000-volt line to supply lighting power to small communities, and advantage is taken of this in many localities for the purpose mentioned. It is obvious then, that where h.t. lines and part of the city distribution system run parallel at close range, any irregularity of the high-tension line will transfer a vigorous surge to the city system—a surge of sufficient initial amplitude to carry it all over the distribution and create widespread interference. The source in a case of this nature was recently located seventeen and a half miles from the town affected.

It has been noted in a few cases, that h.t. systems having telephone lines on the same poles or towers appear to be the greater offenders. Further research is required here and must be carried out with the full coöperation of the line engineers and their qualified assistants. All h.t. switches are not set with a cycle counter although they really should be, and any investigator who happens to be touching bare metal parts of the parallel line telephone installation when the h.t. line is cut will probably have good reason for remembering the incident.

In the case of noisy h.t. lines we do not know of one exception to the following: In sub-zero weather, interference is very bad; in mild weather, it is slight; in rainy weather, when all the insulators and cross arms are dripping wet, it is practically nil. Offhand, we are inclined to believe that this notorious trouble originates, not at any one specific point, but is the result of thousands of small brush discharges all over the

system. Incidentally, we hope that we are wrong; the outlook for elimination is indeed poor if our diagnosis happens to be correct.

JUST what percentage of a city's total interference originates on the local distribution system is a difficult question. One thing is certain; any interference occurring here is serious because of its proximity to the broadcast listener and its wide range. One loose primary cutout on a 2200-volt line can propagate severe interference for two miles in one direction and every wire running parallel picks up the disturbance by induction and spreads it all over the district. Now there are two primary cutouts used in connection with every transformer, except in the case of banked units, and if there are three hundred transformers in the city we have roughly six hundred potential sources of trouble from this agency alone.

Primary cutout interference represents about seventy-five or eighty per cent. of the total trouble caused by power lines in any city at any time. We should remember that this means no material loss of energy to the power company; such faults will not necessarily affect power consumption or even cause the house lights to flicker. For the sake of the good will of the public, however, it is in the interest of the operating company to give this particular accessory careful attention.

Interference from this source takes the form of a hard buzzing. It may be steady or intermittent; gusts of wind shaking the pole may start or stop it and heavy trucks passing along the road will do the same thing. Several cutouts buzzing intermittently will give the effect of wireless telegraph transmission and is easily confused by persons unfamiliar with the code.

A systematic check with a radio receiver should be made of such sources at least twice yearly. In one city, 146 transformers were checked and 41 loose cutouts giving rise to radio interference, were found. In another city the relative figures were higher.

The next fault, in the order of frequency of occurrence, is that of primary leads (2200-volt) spitting to the transformer case, guy wires, or crossarm braces. Severe buzzing results in the first case—practically the same characteristics as primary cutout interference—and it doesn't matter whether the transformer is grounded or not. Spitting to an insulated guy wire gives rise to a heavy clicking interference rather than buzzing. Where the primary lead is scraping on a brace the disturbance is not unlike that produced by the home type violet-ray machine but may, under varying conditions, take a different form.

Tree grounds are next in order. The intermittent "zip" from this source is very annoying. Even a little green twig waving on a bare 2200-volt feeder can set up enough "chirping" to bother reception. Tree trimming is the obvious remedy here but property rights, etc., come up, and much difficulty is often experienced in getting action. It might be as well for tree owners to reflect on the fact that copper has a bad effect on tree growth and if a copper spike be driven in the base of a healthy tree the tree will perish. Chemical analysis of branches, grooved and burned by 2200-volt lines, showed a deposit of copper, not only at the point of contact, but also a short distance either way, just as though the sap were absorbing this fine deposit and spreading it through the tree system.

Interference from the next source, loose splices, is not common. When such faults are present they give rise to heavy clicking. Generally speaking, poor splices or connections are the result of

winter jobs, done in zero weather when the lineman's lot is no sinecure.

It is not usual to find cracked or leaky bushings in transformers. There are instances, of course—a few bad ones—but they are rare. Also, in a thousand odd transformers checked, only one had loose inside connections, and there were three of them in the one transformer. This particular source wiped out reception in two adjoining towns.

WHILE access to a properly equipped radio patrol car is desirable, much good work in the location of sources of radio interference has been done with an ordinary loop receiver and standard automobile.

In order to get an idea of the area covered by any given interference, it is usual to telephone various radio fans throughout the city and find out if interference coinciding in time and character with your own case prevails at the distant points. The distant observer should bring his loud speaker near the house telephone so that you may be able to check the noise from his set and your own at the same time. Another method is for the various observers to keep a log of the characteristics and periods of activity of the trouble under investigation. Synchronized watches should be used here.

Knowing the general location, use is then made of a loop receiver carried around in an automobile. Any good super-heterodyne or radio-frequency receiver with a volume control is suitable.

Once in a great while some use may be made of the directional properties of the loop; for the present, disregard this feature for it is misleading.

Prop the loop in a fixed position, not too close to the side of the car, and drive around the affected district. At some point where the interference is strong, cut down the volume control so that the noise is just audible. Then repeat the patrol. It is probable that another point will be found where the noise comes up slightly; cut down the volume control further and keep lowering it at every increase in the intensity of the disturbance until there are only one or two points on the patrol route where the noise is audible.

An inspection is then made, watching out for garage battery chargers, transformer cutouts, chiropractors' offices, etc., etc. If the trouble seems to be in any way connected with the lighting or distribution system it should be referred to the power company for action.

The foregoing gives a general idea of the method followed in running down radio disturbances. In view, however, of the growing interest and activity along this line, it might be of help to those organizing patrols on a large scale if we go into a little more detail.

THE car should be a six-cylinder one; complicated interference requires that the receiver volume control be cut down until the disturbance is a mere whisper and any mechanical rumbling noise is a detriment to the patrol.

The car body, above the waist line anyway, should be of wood, facilitating inside loop reception. Metal bodies shield an inside loop and weaken the signals in addition to distorting the wave direction.

Measures are taken to prevent the ignition interference unduly affecting reception. No standard method, applicable to all cars, can be recommended, but one of the following methods, or a combination of them, should be effective:

(1.) A one-microfarad condenser across the battery circuit; attach one side to the ammeter terminal and the other under any convenient nut in contact with the chassis. Paint or enamel causing imperfect contact should be scraped away.

(2.) Duplication of method (1.) except that connection is made at the coil primary instead of at the ammeter.

(3.) Seventy-five turn choke coil, enclosed in a metal box, inserted in the lead from the dash switch to the coil primary.

(4.) Complete box screen of fine mesh wire gauze over cylinder head clamped around engine block and covering spark plugs, distributor, and ignition cables. Removable side gate is fitted giving access to the enclosed parts.

Where trouble is experienced from the generator a one-half or one-microfarad condenser from the positive brush to the chassis will help, and if the tail light cable is carrying a surge another condenser will be required here.

All low-tension wiring should be of armored cable but copper braid over the high-tension leads, while it clears up the interference, is objectionable since it tends to founder the spark energy.

It is not desirable to wipe out the ignition click completely; leave a slight trace of it to act as a pilot signal for the receiver.

The layout of interior fittings, i.e., desk, cupboards, etc., will depend on the size of the car and the scope of the patrol. Convenient racks for record books, etc., should be fitted; it is very necessary to have a place for everything and equally necessary that the interior of such a car be kept, at all times, in first-class order.

THE RECEIVER

FOR official patrols, a good stable superheterodyne receiver is desirable. We specify stable because so many "supers" burst into oscillation under the conditions of patrol—especially near high-tension lines. In this case the "mushy" note of oscillation and the normal h.t. interference are almost identical, and accurate determinations are not possible.

Any make of tube known to be microphonic should be ruled out without further consideration. It is not necessary to take the precaution of mounting the sockets on rubber. A volume control of some nature is essential but throwing the set out of tune to accomplish this is not recommended.

The loop is mounted on a swivel base, allowing at least ninety-degree rotation and the three leads to the set should be sewn in a flat strap leaving a distance of three-eighths or half an inch between each wire. A storage A battery and dry cell B batteries are connected to a receptacle which is fixed permanently at some convenient point on the desk front. The corresponding plug and cable is, of course, attached to the "super." A neat box containing spare A and B batteries should also be carried, and should be fitted with a receptacle similar to the desk one.

Where metal body cars are used the loop is mounted outside, on the car roof, the handle for rotation coming down through a special weatherproof fitting.

It is obvious that in the course of extended patrol the receiver will get some rough usage so it is important that very careful attention be paid to all connections and steps taken to make the equipment as rugged as possible. Nothing can be more exasperating than to have this gear fall at a critical moment when tests involving much prearrangement are being carried out.

Condenser bearings must be set up fairly tight so that vibration of the car does not alter the tuning. Mechanically unbalanced condensers, mounted in a vertical plane, which move ever so little through jolting of the car, can throw a patrol into complete chaos. Frequent observation of the filament ammeter is important; a slight drop in filament current after the first intensity observations have been made can cause considerable confusion.

Inclusion of a loud speaker in the equipment is not recommended since this instrument tends to suppress the finer characteristics, harmonics and overtones, of specific interferences, rendering determination of their origin more difficult. Headphones, exclusively, are used on standard patrols.

PATROL WORK

SUCCESSFUL patrol work is largely a matter of experience. Given a suitable car and the necessary receiving equipment, the beginner is apt to become discouraged at his failure to secure immediate results—a condition which will probably last until he learns some of the wiles of the elusive interference and knows the pitfalls to avoid.

In the first place, the directional properties of the loop are practically valueless in city work; its plane for maximum sensitivity will, in every case, be parallel to the adjacent street wiring and instead of pointing toward the source, may point in any direction away from it. There is just one circumstance in which a rotating loop can be of value and it is simply to cover this one condition that we recommend the swivel base.

If the patrol is parked at the corner of intersecting streets where distribution wiring runs, say, north and south on one street and east and west on the other, it is sometimes an advantage to know on which set of wires the interference is stronger. The loop may be turned through ninety degrees while the car is parked directly under the intersection of the wires and a determination arrived at. The system showing the greater disturbance is then followed up.

It is popularly supposed that, at its source, interference is very loud, and that its intensity tapers gradually to zero as we move away. This is only partly correct. As a matter of fact, one often finds that actually at the source, interference is much weaker than, say, fifty feet away. Suppose, however, we start patrol in the vicinity of a source where the noise is very loud. On moving away, say, 250 feet, the disturbance falls to zero; at 500 feet it is up again very strong. At 750 feet it has died again; at 1000 feet it comes up, and so on for a mile or more of straight patrol. We can see that this trouble is a case of forced oscillations creating standing waves on the distribution system and the varying intensities plotted out take the general form of a sine curve.

In contrast to the sine curve, however, successive "bumps" or peaks of intensity will be slightly weaker than the preceding one, that is, when we are moving away from the source. The reverse is true when we approach the source from a distance.

By cutting down the volume control to bare audibility in the vicinity of the heavy bumps it is possible to narrow down the trouble to a small area, since nothing will be heard where only a slight bump prevails. The audibility control should have a knob and pointer rather than a dial; it is desirable to be able to switch on to full volume at intervals but we must know, accurately, the original setting to which to revert. Failure to note this renders comparisons with previous observations inaccurate.

Now, in the course of this patrol, certain complications will probably ehter. We are assuming that the disturbance is strongest near the source —if not directly at the source—and, naturally, are watchful for the strongest "bump." In ordinary reception, say, from some broadcasting station, a difference of half a mile or so in distance between the station and receiver makes very little difference in signal strength. A few feet, however, can make a big difference in the intensity of radio interference being guided by city wiring; simply moving the patrol set over to the other side of the street may make as much

WHERE RADIO COMPETES WITH THE WASHING

The owner of these tenements in Long Island City, New York, does not permit construction of antennas on the roof, with the result that a forest of back yard poles at crazy angles support the necessary antennas. Interference from improperly operated receivers is frequently acute in such congested areas

as 40 per cent. reduction or increase as the case may be. For this reason it is desirable always to keep the same distance from the curb when patrolling, so far as this is possible, and occasional observations should be made to see that the line is still on the same side of the street being patrolled. If it has crossed over at an unobserved point and continued on, on the other side of the street, misleading checks may result.

Every wire, directly connected or not, in the vicinity of a vigorous source of radio interference, will pick up the disturbance and radiate it in different directions; trolley wires, being much nearer the loop than the power wires, will give a heavy indication even when the trouble is not in any way connected with the car system.

Service wires running low across the street will also produce a false peak and, incidentally, give a bogus direction if directional properties of the loop are being counted upon.

Potheads and pole ground wires are notorious misleaders of the unwary. At these points the overhead wiring is brought right down to us and we will get a heavy bump of the noise when passing the pothead or ground wire although the source may be half a mile or more away. The bump here may even be stronger, so far as effect on the receiver is concerned, than at its loudest point near the source.

In the affected area, nearly every street corner or intersection registers a number of confusing bumps. This is due to abrupt physical changes in the circuits, free radiation probably, since high-frequency surges dislike going around sharp corners.

A rise in the disturbance intensity will be noted at each transformer passed during patrol but when the car is running over an iron bridge with side supporting girders, interference will drop practically to zero although the source may be nearby. A peak will be registered at dead ends and circuit stops also, whether the wires are alive or not. Dead wiring is just as effective in propagating radio interference as live wiring and under certain circumstances the actual source may be on a system which is not energized except by induction from some other line. If a dead circuit having on it a partial ground, parallels a high-tension line, the voltage induced in the former is sufficient to cause a spit-over at the imperfect contact, resulting in severe disturbance.

Where complicated interference is being investigated on streets on which electric cars are run, much time will be saved by carrying out the work between the hours of 1 A.M. and 5 A.M. In fact, these are the best hours for radio patrol work up to the point of location to a given pole or residence. Daylight inspection is then necessary. During the early morning hours street car activity is at a minimum and traffic conditions are ideal for concentrated patrol.

One thing to be religiously avoided is any tendency to jump at hasty conclusions; an investigator cannot be too careful on this point. Literally hundreds of reasons, each with a story behind it, can be cited to show how very necessary it is for the patrol man to attack his problem with a perfectly open mind. One example selected at random, may be of interest. Reception over a fairly wide area in a certain town was being spoiled by a strong buzzing interference. The "buzzes" came at one-second intervals and kept up for days at a time, stopping for a few hours or a day, as the case might be, and then starting off again. It was narrowed down to two buildings—a large factory and an electrical power plant. Generator interference close to the plant swamped the disturbance being investigated, but it was learned here that the ' static ground detector" on the factory supply lines (2200 v.) showed an intermittent ground on one phase. The factory electrician had checked everything in detail and insisted there were no grounds on his end of the business; the power company did likewise. Service was not impaired in the least but still the radio interference corresponded exactly with the swinging of the ground indicator, and such a ground on the 2200-volt system would account for our trouble.

We might say that there was absolutely no reason whatever for suspecting this meter; it was a standard instrument made by a very reliable firm. Nevertheless one of the condensers in it had broken down and constituted the source from which our radio interference originated. The trouble was eliminated with despatch.

It must be obvious from what has been said that the location of the source of any given disturbance is not always an easy matter. The work is doubly difficult in the case of intermittent interference or that which periodically alters its intensity. Where several sources are active at the same time, the resultant confusion demands close concentration on the part of the investigator.

On all standard patrols a five- or six-pound sledge hammer is carried. If the characteristics of the interference indicate power line trouble, patrol is carried out until the source is confined to half a block or less on one street. Then the suspected poles are tapped with the sledge and any loose connection of any nature whatever will immediately show up on the patrol receiver, usually as a violent buzzing. Loose primary cutouts, defective lightning arresters, partial grounds, bad splices—all show up definitely and at once.

Care is taken not to hit the pole too hard; an inexperienced investigator can cause damage here. If looking for loose splices, the pole should be tapped in the plane of the overhead wiring since right-angle tapping does not always send the vibration along the wires. A moderate tap in the plane of the wiring will show up defective splices five poles distant.

Literally, thousands of transformer poles have been checked in this way and we know of only three instances in which the method failed. In these instances the cutouts were jammed mechanically solid and could not vibrate but at the same time they were making poor electrical connection. Such cases are extremely rare, the point is worth noting however, in view of a po recurrence.

Poles giving an indication of severe int ence on the first tap are not jarred a second since the loose cutouts are liable to fall an rupt the service.

When a lineman is setting up loose cuto is desirable to check with the sledge hamme radio set, tapping the pole while the linen still aloft. A cutout in proper order for light power purposes may, at the same time, source of radio interference because of som parently insignificant internal sparking.

We might say, in passing, that a good faults have been found in perfectly new accessories and constructions; new installa are checked just as carefully as old con tions.

COMPLAINT RECORDS

THE method of handling complaints, ad by public utility companies and organizations engaged in radio interference ination work, varies according to the service formed and the area covered. The question system finds favor in certain quarters. It plication, however, is limited, and where is being carried out on a large scale the tionnaire is worse than useless.

Details of the recording system for takin of nationwide patrols with which the aut associated would serve no useful purpose it may be remarked, however, that the c authority diagnoses every complaint, an the information given, endeavors to fi source. If this can be done, appropriate 'ren measures are recommended; if not, the n patrol car is ordered to the location and tr experts take over the case, reporting to the sion headquarters on completion of the wc

Here, the information is classified an corded together with other data relative t particular location, i.e., previous cases cl up, specific sources, general noise level, patrol facilities, radio clubs, etc., etc. In way a fairly accurate check is kept on the g interference situation, which facilitates th ing out of future patrol work for the staf gaged.

In reviewing the radio situation generall from direct contact with thousands of broa listeners, we are forced to the conclusion tha greatest need in the radio game to-d a concentrated and determined effort t every town and city of all preventable int ence.

The suppression of every source is a pra impossibility, but it is obvious that the level of any given centre can be conside reduced. Much work along this line has done by different private corporations and ous governments, and we hope that the c not far distant when every city will ha specially equipped radio patrol car to run interfering disturbances and give radi chance that it deserves.

The *Radio Broadcast*
LABORATORY INFORMATION
SHEETS

THE RADIO BROADCAST Laboratory Information Sheets are a regular feature of this magazine and have appeared since our June, 1926, issue. They cover a wide range of information of value to the experimenter and to the technical radio man. It is not our purpose always to include new information, but to present concise and accurate facts in the most convenient form. The sheets are arranged so that they may be cut from the magazine and preserved for constant reference, and we suggest that each sheet be cut out with a razor blade and pasted on 4" x 6" filing cards, or in a notebook. The cards should be arranged in numerical order. In July, 1927, an index to all Sheets appearing up to that time was printed.

All of the 1926 issues of RADIO BROADCAST are out of print. A complete set of Sheets, Nos. 1 to 88, can be secured from the Circulation Department, Doubleday, Doran & Company, Inc., Garden City, New York, for $1.00. Some readers have asked what provision is made to rectify possible errors in these Sheets. In the unfortunate event that any such errors do occur, a new Laboratory Sheet with the old number will appear.

—THE EDITOR.

No. 169   RADIO BROADCAST Laboratory Information Sheet   **March, 1928**

Data on the UX-222 (CX-322)

CONSTRUCTION

THE new UX-222 (CX-322) screened-grid tube is designed especially for use as a radio-frequency amplifier and when used as such it is capable of giving greater amplification than can be obtained from other tubes. A receiver using these tubes does not have to be neutralized. This Laboratory Sheet gives details regarding its construction.

The arrangement of the elements as we look down on a UX-222 (CX-322) tube is indicated in the drawing on this Sheet. At the center is the filament, A, consisting of a single straight wire. Surrounding the filament is the control or signal grid, B. The plate, D, is located between C and E, which are two comparatively coarse "screen grids." The filament terminals, the plate terminal, and the extra grid terminal (grids C and E are connected together inside the tube), are brought down to a standard four-prong base. The signal grid, B, is connected to a small brass cap on top of the tube.

The amplification constant of this new tube is of the order of 200 to 300, the mutual conductance is about 300 micromhos, and its plate impedance is around one megohm. These values will vary widely, depending upon the voltages. The amplification of the tube in an r. f. circuit may average about three times that possible with a 201-A type tube. Three times as much amplification in the r.f. stage is equivalent to 81 times as much power in the loud speaker.

When a 201-A type tube is used as an r. f. amplifier there is a strong tendency for it to oscillate, due to feed-back through the grid-plate capacity. The plate of the UX-222 (CX-322) is shielded from the signal grid by the screen grid C-E, and the tendency toward oscillation due to feed-back through the tube is practically nullified.

The general characteristics of the tube, and the correct voltages to employ when it is used as an r. f. amplifier, are given below:

Filament Volts 3.3
Plate Voltage ....90 to 135
Screen-Grid Voltage .....+45
Signal Grid Bias−1 to−1.5 volts

No. 170   RADIO BROADCAST Laboratory Information Sheet   **March, 1928**

Selectivity and Sensitivity

DESIRABLE CHARACTERISTICS

THE ideal receiver should be as selective as is possible; that is, it should receive a channel of frequencies 10,000 cycles wide (or only 5000 cycles wide in the case of single side-band transmission) equally well, but should not receive other frequencies at all. A receiver for reception of broadcast programs cannot be made any more selective than this without impairing the quality of reproduction. When a receiver is this selective, it will offer a barrier to all frequencies except those lying in the channel to which it is tuned.

The ideal receiver should not need to be any more sensitive than is necessary to amplify interfering noises to more than tolerable loudness under conditions of least interference. When the interference is greater, the sensitivity should be cut down to keep these noises from becoming objectionably loud. In summertime the interfering radio waves manufactured by nature are generally the strongest.

Assuming that an ideal radio receiver is available, there is only one way left (other than the invention of a static eliminator or reducer) to reduce interference to any further extent and thereby increase the distance over which satisfactory reception is possible. This second method of reducing interference is through the use of increased power at the transmitting station. If the signal strength at any given location is increased, the ratio between the signal and the static is thereby increased and reception in this way made freer of interfering noises. Just as in the case of land wire telephony, however, we will probably never be able to put enough power into the ether to give good transmission across the continent in spite of bad interference.

In so far as sensitivity and selectivity are concerned, the super-heterodyne type of receiver is probably the most desirable. These characteristics in a receiver of this type depend, however, in large measure on the design of the intermediate-frequency amplifier. This amplifier can be designed only to amplify a very narrow band of frequencies (a good design for reception of code signals), or, by the use of band-pass filters, the equal amplification of a band of frequencies can be accomplished (a satisfactory design for the reception of ordinary broadcast signals).

No. 171   Radio Broadcast Laboratory Information Sheet   **March,**

The CX-312 (UX-112) and CX-371 (UX-171)

FURTHER COMPARISONS

THE two types of power tubes best adapted for medium B voltages are the CX-312 (UX-112) and the CX-371 (UX-171). The former tube, introduced first, came into immediate favor, and for a time was more popular than the CX-371 (UX-171). This initial preference was due to several factors, the most important ones being, first, the fact that the voltages required by this type were identical with those required by type CX-301-A (UX-201-A) tubes, and therefore, the tube could be substituted without battery changes. Secondly the horn type loud speaker, generally more sensitive than the cone loud speaker, was still popular, and there was less necessity for the greater power output given by the CX-371 (UX-171). A third factor was the misapprehension about battery voltages, man'y not realizing that although the CX-371 could be used to best advantage at the maximum voltage of 180 volts, the quality of reproduction was equally good at 135 volts, and the volume ample for average home service.

During the current season the standing of the two tubes is rapidly being reversed, the CX-371 (UX-

171) assuming the leadership, partly becau large number of new receivers for which th specified, and partly because of better for using the tube to its best advantages provements in audio amplification and speaker design are made, the advantage of u type of tube becomes increasingly appare higher frequencies are usually reproduc factorily by any type of output tube, but t full undistorted reproduction of low freq or the bass notes, a tube having low inter tance, such as the CX-371 (UX-171), is requ

In installing the CX-371 (UX-171), the f caution with which the user has to become is the use of a high grid-biasing, or C, batt age—from 16½ to 40½ volts, depending u B voltage used. With general-purpose tub the power tubes replaced, the use of a C was to a large extent optional with the though the fact that better quality was with this battery was generally recognized Laboratory Sheets Nos. 161 and 162 gave teresting data and curves on the type 112 tubes. The 210 type tube was also covered latter sheets.

No. 172   Radio Broadcast Laboratory Information Sheet   **March,**

A Simple Wavemeter

CONSTRUCTION AND CALIBRATION

A WAVEMETER is a very useful asset to the laboratory of any radio experimenter. A coil of wire and a condenser, connected together properly, are all that is required to make this instrument.

The circuit diagram of a wavemeter is given on the curve published herewith. It is evident from the diagram that the coil is simply connected across the condenser. The coil should, preferably be a solenoid wound on a piece of tubing so that it will be able to withstand some abuse without any alteration in its inductance, and should have sufficient number of turns to cover the frequency band it is to be used on.

The construction of the wavemeter presents no problems. The method of calibrating it and plotting a calibration curve may, however, require some explanation. The procedure is as follows:

(1.) Set the wavemeter at a distance of about two feet from your radio receiver.

(2.) Take the lead from the antenna and wrap one turn around the coil of the wavemeter and then run the antenna lead over to the regular antenna post on your receiver.

(3.) Turn on the receiver and tune-in the signals of some station. Now slowly revolve the dial on the wavemeter and at some point on the dial the signal output of the receiver will decrease. Note the reading on the wavemeter condenser dial which cuts out the signal most completely. Make the same test on some other stations.

(4.) Now draw the curve, using the tained, in a manner similar to that indi this Laboratory Sheet. The wavelength quencies, on which the various stations a emitting can, of course, be obtained from of broadcasting stations.

Such a wavemeter aids materially in th fication of stations heard on a receiver whi calibrated.

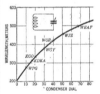

No. 173   Radio Broadcast Laboratory Information Sheet   **March,**

The Regulator Tube

WHY IT IS USED

THE voltage regulator, or glow tube, as it is sometimes called, has found rather wide use in the design of B power units, making them capable of delivering a voltage output that is practically constant over a wide range of load. The output of a power unit not using a glow tube will, of course, vary with the load, although the magnitude of this variation may be held to comparatively low values by good design. A power unit supplying an output voltage that does not depend upon the load may be used with practically any receiver with a knowledge that the voltage actually delivered to the receiver will be correct. Constant voltage output is, however, only one of the advantages accruing from the use of a regulator tube.

The action of the tube in holding the voltage of the output circuit constant serves also to eliminate the small ripples which may be present as a result of incomplete filtering, and this action makes possible a reduction in the capacity, and therefore the expense, of the final filter condenser. In fact, the

tube, when in operation, has many prop common with a large fixed condenser. One properties is extremely low a. c. impedanc when combined with its instantaneous resp voltage regulator, entirely eliminates the a "motor-boating" effect which generally when an attempt is made to use one of the c B power units with many forms of amplifi

The fact that the regulator tube keeps th voltage constant also permits the safe use densers of a lower voltage rating than w permissible if the tube were not used. The c the condensers used in an ordinary powe is fixed by the maximum values of volt they must handle. The voltage output of sor at no load, rises to comparatively high va the condensers must therefore have a rati cient to withstand these voltages. The outpu of a power unit with a regulator tube is even at no load, is values only slightly abo voltage and, therefore, the condensers called upon to withstand voltages greater t rated output of the unit.

No. 174 RADIO BROADCAST Laboratory Information Sheet March, 1928

Grid Bias

WHY IT IS USED

THERE are apparently many, as indicated by letters to the Laboratory, who still feel that the major reason for using C bias on a tube is to reduce the plate current. Although negative bias on the grid of a tube does decrease the plate current, this is not really the most important reason for its use.

C bias is used primarily to reduce distortion and make the tube operate more efficiently. In an ordinary receiver, C bias is most important in the audio-frequency amplifier and we will, therefore, discuss on this Laboratory Sheet the effect of using various values of C bias on the grid of a tube.

The curves A, B, C, and D indicate the distortion of signals which results when too little or too much bias is used. Curve A represents the voltage on the grid of the tube. Curve B shows how the plate current of the tube varies if the bias on the tube is correct. It should be noted that the form of this curve is the same as curve A, indicating that there is no distortion being produced by the tube. If too little bias is used, the positive halves of the input voltage wave will cause the grid to become positive when the grid draws current, and the positive peaks are then cut off as indicated at C. If the bias is too great the tube operates on the lower bend of its characteristic

and this causes the negative half of the signal to be flattened out, as shown in curve D. To prevent distortion, therefore, the proper C bias must be used.

It is especially important that the bias on the last tube be correct, for this tube must handle the greatest amount of signal current and will, therefore, overload and distort most easily. As a matter of information, the correct bias for a 112 or 171 type tube is given below:

TUBE	PLATE VOLTS	C BIAS
112	90	6.0
	135	9.0
	157	10.5
171	90	16.5
	135	27.0
	180	40.5

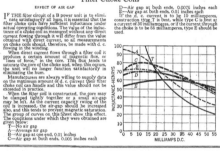

No. 175 RADIO BROADCAST Laboratory Information Sheet March, 1928

Filter Choke Coils

EFFECT OF AIR GAP

IF THE filter circuit of a B power unit is to eliminate satisfactorily all hum, it is essential that the filter choke coils have sufficient inductance under actual operating conditions. The value of the inductance of a choke coil as measured without any direct current flowing through it will differ from the value obtained with direct current, so all measurements on choke coils should, therefore, be made with d. c. flowing in the winding.

When direct current flows through a filter coil it produces a certain amount of magnetic flux, or "lines of force," in the core. This flux tends to saturate the core of the choke and, when this occurs, the unit will no longer function satisfactorily in eliminating the hum.

Manufacturers are always willing to supply data on the maximum amount of d. c. current their filter choke coil can handle and this value should not be exceeded in practice.

When the filter coil is constructed, the core may be clamped tightly together or a small air-gap may be left. As the current capacity rating of the coil is increased, the air-gap should be increased also, and this tends to prevent magnetic saturation. The group of curves on this Sheet show this effect. The conditions under which they were obtained are given below:

T—No air gap
A—Average air gap
B—Air gap at one end, 0.01 inches
C—Air gap at both ends, 0.005 inches each

D—Air gap at both ends, 0.0075 inches each
E—Air gap at both ends, 0.01 inches each
If the d. c. current is to be 10 milliamperes, construction type T is best, while type C is best at a current of 30 milliamperes, or if the current through the choke is to be 55 milliamperes, type E should be used.

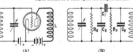

No. 176 RADIO BROADCAST Laboratory Information Sheet March, 1928

How the Plate Circuit Affects the Grid Circuit

REVERSE ACTION

IN WORKING with tubes, we normally consider that the plate circuit is controlled by the grid and that there is no reverse action. This, however, is not strictly true, for the plate circuit does affect the grid circuit in two ways.

In the first place the plate acts as a grid with respect to the regular grid in the tube and large variations in plate voltage have the same effect with respect to the grid as has slightly varying the grid voltage. The reverse effect is generally not appreciable so long as the grid is held negative, as is the case in an amplifier. The reverse effect is important in oscillator circuits, however, where the grid is not always negative. In making an accurate analysis of the action of an oscillator, it would be necessary to consider this effect.

The second manner in which the grid is controlled by the plate is through the grid-plate capacity of the tube. At (A) in the diagram on this Sheet, we have indicated the circuit of an ordinary r. f. amplifier and at (B) is shown the equivalent circuit with the inter-electrode resistances and capacities indicated. R_g is the grid-filament resistance of the tube, C_g the grid-to-filament capacity, C_1 the inter-electrode

capacity between the grid and the plate, R_p the plate filament resistance, C_p the plate-filament capacity, and L the load impedance. Probably the most important of the capacities shown is the grid-plate capacity, C_1, for it is this capacity which permits the grid circuit to be affected by what goes on in the plate circuit. In radio-frequency amplifiers it is this capacity which causes the tube to oscillate.

The diagram at (B) should give some idea of the complexity of the network represented by a tube, and the action of this network of resistances, condensers, and inductances must be understood if the action of a tube in any particular circuit is to be accurately foretold. J. M. Miller, in Scientific Paper of the Bureau of Standards No. 351, carefully and completely determined the dependence of the input circuit of a tube upon the output circuit.

"RADIO BROADCAST'S" DIRECTORY OF MANUFACTURED RECEIVERS

¶ A coupon will be found on page 395. All readers who desire additional information on the receivers listed below need only insert the proper numbers in the coupon, mail it to the Service Department of RADIO BROADCAST, and full details will be sent. New sets are listed in this space each month.

KEY TO TUBE ABBREVIATIONS

99—60-mA. filament (dry cell)
01-A—Storage battery 0.25 amps. filament
12—Power tube (Storage battery)
71—Power tube (Storage battery)
16-B—Half-wave rectifier tube
80—Full-wave, high current rectifier
81—Half-wave, high current rectifier
Hmu—High-Mu tube for resistance-coupled audio
20—Power tube (dry cell)
10—Power Tube (Storage battery)
00-A—Special detector
13—Full-wave rectifier tube
26—Low-voltage high-current a. c. tube
27—Heater type a. c. tube

DIRECT CURRENT RECEIVERS

NO. 424. COLONIAL 26

Six tubes; 2 t. r. f. (01-A), detector (12), 2 transformer audio (01-A and 71). Balanced t. r. f. One to three dials. Volume control: antenna switch and potentiometer across first audio. Watts required: 120. Console size: 34 x 38 inches. Headphone connections. The filaments are connected in a series parallel arrangement. Price $250 including power unit.

NO. 425. SUPERPOWER

Five tubes: All 01-A tubes. Multiplex circuit. Two dials, Volume control: resistance in r. f. plate. Watts required: 30. Antenna: loop or outside. Cabinet sizes: table, 27 x 10 x 9 inches; console, 28 x 50 x 21. Prices: table, $135 including power unit; console, $390 including power unit and loud speaker.

A. C. OPERATED RECEIVERS

NO. 508. ALL-AMERICAN 77, 88, AND 99

Six tubes; 3 t. r. f. (26), detector (27), 2 transformer audio (26 and 71). Rice neutralized t. r. f. Single drum tuning. Volume control: potentiometer in r. f. plate. Cabinet sizes: No. 77, 21 x 10 x 8 inches; No. 88 Hiboy, 25 x 38 x 18 inches; No. 99 console, 27½ x 43 x 20 inches. Shielded. Output device. The filaments are supplied by means of three small transformers. The plate supply employs a gas-filled rectifier tube. Voltmeter in a. c. supply line. Prices: No. 77, $150, including power unit; No. 88, $210 including power unit; No. 99, $285 including power unit and loud speaker.

NO. 509. ALL-AMERICAN "DUET"; "SEXTET"

Six tubes; 2 t. r. f. (99), detector (99), 3 transformer audio (99 and 12). Rice neutralized t. r. f. Two dials. Volume control: resistance in r. f. plate. Cabinet sizes: "Duet," 23 x 56 x 16½ inches; "Sextet," 22½ x 13¾ x 15½ inches. Shielded. Output device. The 99 filaments are connected in series and supplied with rectified a. c.; while 12 is supplied with raw a. c. The plate and filament supply uses gaseous rectifier tubes. Milliammeter on power unit. Prices: "Duet," $160 including power unit; "Sextet," $220 including power unit and loud speaker.

NO. 511. ALL-AMERICAN 80, 90, AND 115

Five tubes; 2 t. r. f. (99), detector (99), 2 transformer audio (99 and 12). Rice neutralized t. r. f. Two dials. Volume control: resistance in r. f. plate. Cabinet sizes: No. 80, 23½ x 12½ x 15 inches; No. 90, 37½ x 12 x 12½ inches; No. 115 Hiboy, 24 x 41 x 15 inches. Coils individually shielded. Output device. See No. 509 for power supply. Prices: No. 80, $135 including power unit; No. 90, $145 including power unit and compartment; No. 115, $170 including power unit, compartment, and loud speaker.

NO. 510. ALL-AMERICAN 7

Seven tubes; 3 t. r. f. (26), 1 untuned r. f. (26), detector (27), 2 transformer audio (26 and 71). Rice neutralized t. r. f. One drum. Volume control: resistance in r. f. plate. Cabinet sizes: "Sovereign" console, 30½ x 60¼ x 19 inches; "Lorraine" Hiboy, 25½ x 53¾ x 17½ inches; "Forte" cabinet, 25½ x 13¾ x 17½ inches. For filament and plate supply: See No. 508. Prices: "Sovereign" $460; "Lorraine" $360; "Forte" $270. All prices include power unit. First two include loud speaker.

NO. 536. SOUTH BEND

Six tubes. One control. Sub-panel shielding. Binding Posts. Antenna: outdoor. Prices: table, $130, Baby Grand Console, $195.

NO. 537. WALBERT 26

Six tubes; five Kellogg a. c. tubes and one 71. Two controls. Volume control: variable plate resistance. Isofarad circuit. Output device. Battery cable, Semi-shielded. Antenna: 50 to 75 feet. Cabinet size: 10¾ x 29½ x 16½ inches. Prices: $215; with tubes, $250.

NO. 522. CASE, 62B AND 62C

McCullough a. c. tubes. Drum control. Volume control; variable high resistance in audio system. C-battery connections. Semi-shielded. Cable. Antenna: 100 feet. Panel size: 7 x 21 inches. Prices: Model 62B, complete with a. c, equipment, $185; Model 62 C, complete with a. c. equipment, $235.

NO. 523. CASE, 92 A AND 92 C

McCullough a. c. tubes. Drum control. Inductive volume control. Technidyne circuit. Shielded. Cable. C-battery connections. Model 92 C contains output device. Loop operated. Prices: Model 92 A, table, $350; Model 92 C, console, $475.

NO. 484. BOSWORTH, B5

Five tubes; 2. t. r. f. (26), detector (99), 2 transformer audio (special a. c. tubes). T. r. f. circuit. Two dials. Volume control: potentiometer. Cabinet size: 23 x 7 x 8 inches. Output device included. Price $175.

NO. 406. CLEARTONE 110

Five tubes; 2, t. r. f., detector, 2 transformer audio. All tubes a. c. heater type. One or two dials. Volume control: resistance in r. f. plate. Watts consumed: 40. Cabinet size varies. The plate supply is built in the receiver and requires one rectifier tube. Filament supply through step down transformers. Prices range from $175 to $375 which includes 5 a. c. tubes and one rectifier tube.

NO. 407. COLONIAL 25

Six tubes; 2. t. r. f. (01-A), detector (99), 2 resistance audio (99). 1 transformer audio (10). Balanced t. r. f. circuit. One or three dials. Volume control: Antenna switch and potentiometer on 1st audio. Watts consumed: 100. Console size: 34 x 38 x 18 inches. Output device. All tube filaments are operated on a. c. except the detector which is supplied with rectified a. c. from the plate supply. The rectifier employs two 16-b tubes. Price $250 including built-in plate and filament supply.

NO. 507. CROSLEY 602 BANDBOX

Six tubes; 3. t. r. f. (26), detector (27), 2 transformer audio (26 and 71). Neutrodyne circuit. One dial, Cabinet size: 17¼ x 5¼ x 7¾ inches. The heaters for the a. c. tubes and the 71 filament are supplied by windings in B unit transformers available to operate either on 25 or 60 cycles. The plate current is supplied by means of rectifier tube. Price $65 for set alone, power unit $60.

NO. 408. DAY-FAN "DE LUXE"

Six tubes; 3 t. r. f., detector, 2 transformer audio. All 01-A tubes. One dial. Volume control: potentiometer across r. f. tubes. Watts consumed: 300. Console size: 30 x 40 x 20 inches. The filaments are connected in series and supplied with d. c. from a motor-generator set which also supplies B and C current. Output device. Price $350 including power unit.

NO. 409. DAYCRAFT 5

Five tubes; 2 t. r. f., detector, 2 transformer audio. All a. c. heater tubes. Reflexed t. r. f. One dial. Volume control: potentiometers in r. f. plate and 1st audio. Watts consumed: 135. Console size: 34 x 36 x 14 inches. Output device. The heaters are supplied by means of a small transformer. A built-in rectifier supplies B and C voltages. Price $170, less tubes. The following have one more r. f. stage and are not reflexed; Daycraft 6, $195; Dayrole, 6, $235; Dayfan 6, $110. All prices less tubes.

NO. 469. FREED-EISEMANN NRII

Six tubes; 3 t. r. f. (01-A), detector (01-A), 2 transformer audio (01-A and 71). Neutrodyne. One dial. Volume control: potentiometer. Watts consumed: 150. Cabinet size: 19¾ x 10 x 10½ inches. Shielded. Output device. A special power unit is included employing a rectifier tube. Price $225 including NR-411 power unit.

NO. 487. FRESHMAN 7F-AC

Six tubes; 3 t. r. f. (26), detector (27), 2 transformer audio (26 and 71). Equaphase circuit. One dial. Volume control: potentiometer across 1st audio. Console size: 24½ x 41½ x 15 inches. Output device. The filaments and heaters and B supply are all supplied by one power unit. The plate supply requires one 80 rectifier tube. Price $175 to $350, complete.

NO. 421. SOVEREIGN 238

Seven tubes of the a. c. heater type. Balanced t. r. f. Two dials. Volume control: resistance across 2nd audio. Watts consumed: 45. Console size: 37 x 52 x 15 inches. The heaters are supplied by a small a. c. transformer, while the plate is supplied by means of rectified a. c. using a gaseous type rectifier. Price $325, including power unit and tubes.

NO. 517. KELLOGG 510, 511, AND 512

Seven tubes; 4 t. r. f., detector, 2 transformer audio. All Kellogg a. c. tubes. One control and special zone switch, Balanced. Volume control: special. Output device. Shielded. Cable connection between power supply unit and receiver. Antenna: 25 to 100 feet. Panel 7¾ x 27½ inches. Prices: Model 510 and 512, consoles, $495 complete. Model 511, consolette, $365 without loud speaker.

NO. 496. SLEEPER ELECTRIC

Five tubes; four 99 tubes and one 71. Two controls. Volume control: rheostat on r. f. Neutralized. Cable. Output device. Power supply uses two 16-B tubes. Antenna: 100 feet. Prices: Type 64, table, $160; Type 65, table, with built-in loud speaker, $175; Type 66, table, $175; Type 67, console, $235; Type 78, console, $265.

NO. 538. NEUTROWOUND, MASTER ALLECTRIC

Six tubes; 2 t. r. f. (01-A), detector (01-A), 2 audio (01-A and two 71 in push-pull amplifier). The 01-A tubes are in series, and are supplied from a 400-mA. rectifier. Two drum controls. Volume control: variable plate resistance. Output device. Shielded. Antenna: 50 to 100 feet. Price: $360.

NO. 413. MARTI

Six tubes: 2 t. r. f., detector, 3 resistance audio. All tubes a. c. heater type. Two dials. Volume control-resistance in r. f. plate. Watts consumed: 38. Panel size 7 x 21 inches. The built-in plate supply employs one 16-B rectifier. The filaments are supplied by a small transformer. Prices: table, $235 including tubes and rectifier; console, $275 including tubes and rectifier; console, $325 including tubes, rectifier, and loud speaker.

NO. 417. RADIOLA 28

Eight tubes; five type 99 and one type 20. Drum control. Super-heterodyne circuit. C-battery connections. Battery cable. Headphone connection. Antenna: loop. Set may be operated from batteries or from the power mains when used in conjunction with the model 104 loud speaker. Prices: $260 with tubes, battery operation; $570 with model 104 loud speaker, a. c. operation.

NO. 540. RADIOLA 30-A

Receiver characteristics same as No. 417 except that type 71 power tube is used. This model is designed to operate on either a. c. or d. c. from the power mains. The combination rectifier—power—amplifier unit uses two type 81 tubes. Model 100-A loud speaker is contained in lower part of cabinet. Either a short indoor or long outside antenna may be used. Cabinet size: 42½ x 29 x 17¾ inches. Price: $495.

NO. 541. RADIOLA 32

This model combines receiver No. 417 with the model 104 loud speaker. The power unit uses two type 81 tubes and a type 10 power amplifier. Loop is completely enclosed and is revolved by means of a dial on the panel. Models for operation from a. c. or d. c. power mains. Cabinet size: 52 x 72 x 17¾ inches. Price: $895.

NO. 539. RADIOLA 17

Six tubes; 3 t. r. f. (26), detector (27), 2 transformer audio (26 and 27). One control. Illuminated dial. Built-in power supply using type 80 rectifier. Antenna: 100 feet. Cabinet size: 25 5/16 x 7¾ x 8¼. Price: $130 without accessories.

NO. 545. NEUTROWOUND, SUPER ALLECTRIC

Five tubes; 2 t. r. f. (99), detector (99), 2 audio (99 and 71). The 99 tubes are in series and are supplied from an 85-mA. rectifier. Two drum controls. Volume control: variable plate resistance. Output device. Antenna: 75 to 100 feet. Cabinet size: 9 x 24 x 11 inches. Price: $150.

NO. 490. MOHAWK

Six tubes; 2 t. r. f., detector, 2 transformer audio. All tubes a. c. heater type except 71 in last stage. One dial. Volume control: rheostat on r. f. Watts consumed: 40. Panel size: 12½ x 8¾ inches. Output device. The heaters for the a. c. tubes and the 71 filament are supplied by small transformers. The plate supply is of the built-in type using a rectifier tube. Prices range from $65 to $245.

NO. 411. HERBERT LECTRO 120

Five tubes; 2 t.r.f. (99), detector (99), 2 transformer audio (99 and 71). Three dials. Volume control: rheostat in primary of a.c. transformer. Watts required: 45. Cabinet size: 32 x 10 x 12 inches. The 99 filaments are connected in series, supplied with rectified a. c., while the 71 is run on raw a. c. The power unit uses a Q. R. S. rectifier tube. Price $120.

NO. 412. HERBERT LECTRO 200

Six tubes; 2 t.r.f. (99), detector (99), 1 transformer audio (99), 1 push-pull audio (71). One dial. Volume control: rheostat in primary of a. c. transformer. Watts consumed: 60. Cabinet size: 20 x 12 x 12 inches. Filaments connected same as above. Completely shielded. Output device. Price $200.

BATTERY OPERATED RECEIVERS

NO. 542. PFANSTIEHL JUNIOR SIX

Six tubes; 3 t. r. f. (01-A), detector (01-A), 2 audio. Pfanstiehl circuit. Volume control: variable resistance in r. f. plate circuit. One dial. Shielded: Battery cable. C-battery connections. Etched bronze panel. Antenna: outdoor. Cabinet size: 9 x 20 x 8 inches. Price: $80, without accessories.

NO. 512. ALL-AMERICAN 44, 45, AND 66

Six tubes; 3 t. r. f. (01-A, detector) 01-A, 2 transformer audio (01-A and 71). Rice neutralized t. r. f. Drum control. Volume control: rheostat in r. f. Cabinet sizes: No. 44, 21 x 10 x 8 inches; No. 55, 25 x 38 x 18 inches; No. 66, 27½ x 43 x 20 inches. C-battery connections. Battery cable. Antenna: 75 to 125 feet. Prices: No. 44, $70; No. 55, $125 including loud speaker; No. 66, $200 including loud speaker.

NO. 410. LARCOFLEX 73

Seven tubes; 4 t.r.f. (O1-A), detector (01-A), 2 transformer audio (01-A and 71). T.r.f. circuit. One dial. Volume control: resistance in r.f. plate. Console size 30 x 42 x 20 inches. Completely shielded. Built-in A, B and C supply. Price $215.

NO. 485. BOSWORTH B6

Five tubes; 2 t. r. f. (01-A), detector (01-A), 2 transformer audio (01-A and 71). Two dials. Volume control: variable grid resistances. Battery cable. C-battery connections. Antenna: 25 feet or longer. Cabinet size 15 x 7 x 8 inches. Price $75.

NO. 513. COUNTERPHASE SIX

Six tubes; 3 t. r. f. (01-A), detector (00-A), 2 transformer audio (01-A) and 12). Counterphase t. r. f. Two dials. Plate current: 32 mA. Volume control: rheostat on 2nd and 3rd r. f. Coils shielded. Battery cable. C-battery connections. Antenna: 75 to 100 feet. Console size: 18½ x 40½ x 15½ inches. Prices: Model 35, table, $110; Model 37, console, $175.

NO. 514. COUNTERPHASE EIGHT

Eight tubes; 4 t. r. f. (01-A) detector (00-A), 2 transformer audio (01-A and 12). Counterphase t. r. f. One dial. Plate current: 40 mA. Volume control: rheostat in 1st r. f. Copper stage shielding. Battery cable. C-battery connections. Antenna; 75 to 100 feet. Cabinet size: 30 x 12½ x 16 inches. Prices: Model 12, table, $225; Model 16, console, $335; Model 18, console, $365.

NO. 506. CROSLEY 601 BANDBOX

Six tubes; 3 t. r. f., detector, 2 transformer audio. All 01-A tubes. Neutrodyne. One dial. Plate current: 40 mA. Volume control: rheostat in r. f. Shielded. Battery cable. C-battery connections. Antenna: 75 to 150 feet. Cabinet size: 17½ x 5½ x 7½. Price, $55.

NO. 434. DAY-FAN 6

Six tubes; 3 t. r. f. (01-A), detector (01-A), 2 transformer audio (01-A and 12 or 71). One dial. Plate current: 12 to 15 mA. Volume control: rheostat on r. f. Shielded. Battery cable. C-battery connections. Output device. Antenna: 50 to 120 feet. Cabinet sizes: Daycraft 6, 32 x 30 x 34 inches; Day-Fan Jr., 15 x 7 x 7. Prices: Day-Fan 6, $110; Daycraft 6, $145 including loud speaker; Day-Fan Jr. not available.

NO. 435. DAY-FAN 7

Seven tubes: 3 t.r.f. (01-A), detector (01-A), I resistance audio (01-A), 2 transformer audio (01-A and 12 or 71). Plate current: 15 mA. Antenna: outside. Same as No. 434. Price $115.

NO. 503. FADA SPECIAL

Six tubes; 3 t. r. f. (01-A), detector (01-A), 2 transformer audio (01-A and 71). Neutrodyne. Two drum control. Plate current: 20 to 24 mA. Volume control: rheostat on r. f. Coils shielded. Battery cable. C.-battery connections. Headphone connection. Antenna: outdoor. Cabinet size: 20 x 13½ x 10½ inches. Price $95.

NO. 504. FADA 7

Seven tubes; 4 t. r. f. (01-A), detector (01-A), 2 transformer audio (01-A and 71). Neutrodyne. Two drum control. Plate current: 43.A. Volume control: rheostat on r. f. Completely shielded. Battery cable. C-battery connections. Headphone connections. Output device. Antenna: outdoor or loop. Cabinet sizes: table, 25½ x 13½ x 11½ inches; console, 29 x 50 x 17 inches. Prices: table, $185; console, $285.

NO. 436. FEDERAL

Five tubes; 2 t. r. f. (01-A), detector (01-A), 2 transformer audio (01-A and 12 or 71). Balanced t. r. f. One dial. Plate current: 20.7 mA. Volume control: rheostat on r. f. Shielded. Battery cable. C-battery connections. Antenna: loop. Made in 6 models. Price varies from $250 to $1000 including loop.

NO. 505. FADA 8

Eight tubes. Same as No. 504 except for one extra stage of audio and different cabinet. Prices: table, $300; console, $400.

NO. 437. FERGUSON 10A

Seven tubes; 3 t. r. f. (01-A), detector (01-A), 3 audio (01-A and 12 or 71). One dial. Plate current: 18 to 25 mA. Volume control: rheostat on two r. f. Shielded. Battery cable. C-battery connections. Antenna: 100 feet. Cabinet size: 21½ x 12 x 15 inches. Price $150.

NO. 438. FERGUSON 14

Ten tubes; 3 untuned r. f., 3 t. r. f. (01-A), detector (01-A), 3 audio (01-A and 12 or 71). Special balanced t. r. f. One dial. Plate current: 30 to 35mA. Volume control: rheostat in three r. f. Shielded. Battery cable. C-battery connections. Antenna: loop. Cabinet size: 24 x 12 x 16 inches. Price $235, including loop.

NO. 439. FERGUSON 12

Six tubes; 2 t. r. f. (01-A), detector (01-A), 1 transformer audio (01-A), 2 resistance audio (01-A and 12 or 71). Two dials. Plate current: 18 to 25mA. Volume control; rheostat on two r. f. Partially shielded. Battery cable. C-battery connections. Antenna: 100 feet. Cabinet size: 22½ x 10 x 12 inches. Price $85. Consolette $145 including loud speaker.

NO. 440. FREED-EISEMANN NR-8 NR-9, AND NR-66

Six tubes: 3 t. r. f. (01-A), detector (01-A), 2 transformer audio (01-A and 71). Neutrodyne. NR-8, two dials; others one dial. Plate current: 30 mA. Volume control: rheostat on r. f. NR-8 and 9; chassis type shielding. NR-66, individual stage shielding. Battery cable. C-battery connections. Antenna: 100 feet. Cabinet sizes: NR-8 and 9, 19½ x 10 x 10½ inches; NR-66 20 x 10½ x 12 inches. Prices: NR-8, $90; NR-9, $100; NR-66, $125.

NO. 501. KING "CHEVALIER"

Six tubes. Same as No. 500. Coils completely shielded. Panel size: 11 x 7 inches. Price, $210 including loud speaker.

NO. 441. FREED-EISEMANN NR-77

Seven tubes; 4 t. r. f. (01-A), detector (01-A), 2 transformer audio (01-A and 71). Neutrodyne. One dial. Plate current: 35mA. Volume control: rheostat on r. f. Shielding. Battery cable. C-battery connections. Antenna: outside or loop. Cabinet size: 23 x 10½ x 13 inches. Price $175.

NO. 442. FREED-EISEMANN 800 AND 850

Eight tubes; 4 t. r. f. (01-A), detector (01-A), 1 transformer (01-A), 1 parallel audio (01-A or 71). Neutrodyne. One dial. Plate current: 35 mA. Volume control: rheostat on r. f. Shielded. Battery cable. C-battery connections. Output: two tubes in parallel or one power tube may be used. Antenna: outside or loop. Cabinet sizes: No. 800, 34 x 15½ x 13½ inches; No. 850, 36 x 65 x 17½. Prices not available.

NO. 444. GREBE MU-1

Five tubes; 2 t. r. f. (01-A), detector (01-A), 2 transformer audio (01-A and 12 or 71). Balanced t. r. f. One, two, or three dials (operate singly or together). Plate current: 30mA. Volume control: rheostat on r. f. Binocular coils. Binding posts. C-battery connections. Antenna: 125 feet. Cabinet size: 22½ x 9½ x 13 inches. Prices range from $95 to $320.

NO. 426. HOMER

Seven tubes; 4 t. r. f. (01-A); detector (01-A or 00A); 2 audio (01-A and 12 or 71). One knob tuning control. Volume control: rotor control in antenna circuit. Plate current: 22 to 25mA. "Technidyne" circuit. Completely enclosed in aluminum box. Battery cable; C-battery connections. Cabinet size, 8½ x 19½ x 9½ inches. Chassis size, 6½ x 17 x 8 inches. Prices: Chassis only, $80. Table cabinet, $95.

NO. 502. KENNEDY ROYAL 7. CONSOLETTE

Seven tubes; 4 t. r. f. (01-A), detector (00-A), 2 transformer audio (01-A and 71). One dial. Plate current; 42mA. Volume control: rheostat on two r. f. Special r. f. coils. Battery cable. C-battery connections. Headphone connection. Antenna: outside or loop. Consolette size: 36½ x 35½ x 19 inches. Price $220.

NO. 498. KING "CRUSADER"

Six tubes; 2 t. r. f. (01-A), detector (00-A), 3 transformer audio (01-A and 71). Balanced t. r. f. One dial. Plate current: 20 mA. Volume control: rheostat on r. f. Coils shielded. Battery cable. C-battery connections. Antenna: outside, Panel: 11 x 7 inches. Price, $115.

NO. 499. KING "COMMANDER"

Six tubes; 3 t. r. f. (01-A), detector (00-A), 2 transformer audio (01-A and 71). Balanced t. r. f. One dial. Plate current: 25mA. Volume control: rheostat on r. f. Completely shielded. Battery cable. C-battery connections. Antenna: loop. Panel size: 12 x 8 inches. Price $220 including loop.

NO. 429. KING COLE VII AND VIII

Seven tubes; 3 t. r. f., detector, 1 resistance audio, 2 transformer audio. All 01-A tubes. Model VIII has one more stage t. r. f. (eight tubes). Model VII, two dials. Model VIII, one dial. Plate current: 15 to 50 mA. Volume control: primary shunt in r. f. Steel shielding. Battery cable and binding posts. C-battery connections. Output devices on some consoles. Antenna: 10 to 100 feet. Cabinet size: varies. Prices: Model VII, $80 to $160; Model VIII, $100 to $300.

NO. 500. KING "BARONET" AND "VIKING"

Six tubes; 2 t. r. f. (01-A), detector (00-A), 3 transformer audio (01-A and 71). Balanced t. r. f. One dial. Plate current: 19mA. Volume control: rheostat in r. f. Battery cable. C-battery connections. Antenna: outside. Panel size: 18 x 7 inches. Prices: "Baronet," $70; "Viking," $140 including loud speaker.

NO. 489. MOHAWK

Six tubes; 2 t. r. f. (01-A), detector (00-A), 3 audio (01-A and 71). One dial. Plate current: 40mA. Volume control: rheostat on r. f. Battery cable. C-battery connections. Output device. Antenna: 60 feet. Panel size: 12½ x 8½ inches. Prices range from $65 to $245.

NO. 547. ATWATER KENT, MODEL 33

Six tubes; 3 t. r. f. (01-A), detector (01-A), 2 audio (01-A and 71 or 12). One dial. Volume control: r. f. filament rheostat. C-battery connections. Battery cable. Antenna: 100 feet. Steel panel. Cabinet size: 21½ x 6½ 6½ inches. Price: $75, without accessories.

NO. 544. ATWATER KENT, MODEL 50

Seven tubes; 4 t. r. f. (01-A), detector (01-A), 2 audio (01-A and 12 or 71). Volume control: r. f. filament rheostat. C-battery connections. Battery cable. Special bandpass filter circuit with an untuned amplifier. Cabinet size: 20½ x 13 x 7½ inches. Price: $120.

NO. 452. ORIOLE 90

Five tubes; 2 t. r. f., detector, 2 transformer audio. All 01-A tubes. "Trinum" circuit. Two dials. Plate current: 18mA. Volume control; rheostat on r. f. Battery cable. C-battery connections. Antenna; 50 to 100 feet. Cabinet size: 25½ x 11½ x 12½ inches. Price $85. Another model has 8 tubes, one dial, and is shielded. Price, $185.

NO. 453. PARAGON

Six tubes; 2 t. r. f. (01-A), detector (01-A), 3 double impedance audio (01-A and 71). One dial. Plate current: 40 mA. Volume control: resistance in r. f. plate. Shielded, Battery cable. C-battery connections. Output device. Antenna: 100 feet. Console size: 20 x 45 x 17 inches. Price not determined.

NO. 543 RADIOLA 20

Five tubes; 2 t. r. f. (99), detector (99), t[...] former audio (99 and 20). Regenerative detec[...] drum controls. C-battery connections. Batte[...] Antenna: 100 feet. Price: $78 without accesso[...]

NO. 480. PFANSTIEHL 30 AND 3[...]

Six tubes; 3 t. r. f. (01-A), detector (01-2[...] former audio (01-A and 71). One dial. Plate 23 to 32 mA. Volume control: resistance in r[...] Shielded. Battery cable. C-battery connecti[...] tenna: outside. Panel size: 17½ x 8½ inches. P[...] 30 cabinet, $105; No. 302 console, $185 inclu[...] speaker.

NO. 515. BROWNING-DRAKE 7-[...]

Seven tubes: 2 t. r. f. (01-A), detector (00-A[...] (Hmu, two 01-A, and 71). Illuminated drum[...] Volume control: rheostat on 1st r. f. Shielded. [...]ized. C-battery connections. Battery Cabl[...] panel. Output device, Antenna: 50-75 feet. [...] 30 x 11 x 9 inches. Price, $145.

NO. 516. BROWNING-DRAKE 6-[...]

Six tubes; 1 t. r. f. (99), detector (00-A) (Hmu, two 01-A and 71). Drum control with adjustment. Volume control: rheostat on r. f. [...] tive detector. Shielded. Neutralized. C-batter[...] tions. Battery cable. Antenna: 50–100 feet. [...] 25 x 11 x 9. Price $105.

NO. 518. KELLOGG "WAVE MASTE[...] 504, 505, and 506.

Five tubes; 2 t. r. f., detector, 2 transform[...] One control and special zone switch. Volume rheostat on r. f. C-battery connections. Bindi[...] Plate current: 25 to 35 mA. Antenna: 100 fe[...] 7½ x 25½ inches. Prices: Model 504, table, accessories. Model 505, table, $125 with loud Model 506, consolette, $135 with loud speake[...]

NO. 519. KELLOGG, 507 AND 50[...]

Six tubes, 3 t. r. f., detector, 2 transformer a[...] control and special zone switch. Volume cont[...] stat on r. f. C-battery connections. Balanced. Binding posts and battery cable. Antenna Cabinet size: Model 507, table, 30 x 13½ x 1[...] Model 508, console, 34 x 18 x 54 inches. Price 507, $190 less accessories. Model 508, $320 [...] speaker.

NO. 427, MURDOCK 7

Seven tubes; 3 t. r. f. (01-A), detector (01-A) former and 2 resistance audio (two 01-A and [...] One control. Volume control: rheostat on r. shielded. Neutralized. Battery cable. C-batt[...] nections. Complete metal case. Antenna: [...] Panel size: 9 x 23 inches. Price, not available.

NO. 520. BOSCH 57

Seven tubes; 4 t. r. f. (01-A), detector (01-A[...] (01-A and 71). One control calibrated in kc. control: rheostat on r. f. Shielded. Battery [...] battery connections. Balanced. Output device. loud speaker. Antenna: built-in loop or outside 100 feet. Cabinet size: 46 x 16 x 30 inches. Pr[...] including enclosed loop and loud speaker.

NO. 521. BOSCH "CRUISER," 66 AN[...]

Six tubes; 3 t. r. f. (01-A), detector (01-A) (01-A and 71). One control. Volume control: on r. f. Shielded. C-battery connections. [...] Battery cable. Antenna: 20 to 100 feet. Price 66, table, $99.50. Model 76, console, $175; [...] speaker $195.

NO. 524. CASE, 61 A AND 61 C

T. r. f. Semi-shielded. Battery cable. Drum[...] Volume control: variable high resistance in a[...] tem. Plate current: 35mA. Antenna: 100 feet Model 61A, $85; Model 61 C, console, $135.

NO. 525. CASE, 90 A AND 90 [...]

Drum control. Inductive volume control. Te[...] circuit. C-battery connections. Battery cab[...] operated. Model 90-C equipped with outpu[...] Prices: Model 90 A, table, $225; Model 90 C, $350.

NO. 526. ARBORPHONE 25

Six tubes; 3 t. r. f. (01-A), detector (01-A), (01-A and 71). One control. Volume control: Shielded. Battery cable. Output device. C-bat[...] nections. Loftin-White circuit. Antenna: 75 fe[...] 7½ x 15 inches, metal. Prices: Model 25, ta[...] Model 252, $185; Model 253, $250; Model 255, [...]ation phonograph and radio, $600.

NO. 527. ARBORPHONE 27

Five tubes: 2 t. r. f. (01-A), detector (01-A) (01-A). Two controls. Volume control: rhe[...] battery connections. Binding posts. Antenna[...] Prices: Model 27, $65; Model 271, $99.50; M[...] $125.

NO. 528. THE "CHIEF"

Seven tubes; six 01-A tubes and one pow[...] One control. Volume control: rheostat. C-bat[...] nection. Partial shielding. Binding posts. [...] outside. Cabinet size: 40 x 22 x 16 inches Complete with A power supply, $250; witho[...] sories, $150.

NO. 529. DIAMOND SPECIAL, SUPER S[...] AND BABY GRAND CONSOLE

Six tubes; all 01-A type. One control. Parti[...]ing. C-battery connections. Volume control: [...] Binding posts. Antenna: outdoor. Prices: [...] Special, $75; Super Special, $65; Baby Grand $110.

Why not subscribe to *Radio Broadcast*? By the year only $4.00; or two years, $6.00, saving $2.40. Send direct to Doubleday, Doran & Company, Inc., Garden City, New York.

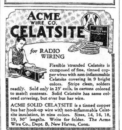

Manufacturers' Booklets

A Varied List of Books Pertaining to Radio and Allied Subjects Obtainable Free With the Accompanying Coupon

READERS may obtain any of the booklets listed below by using the coupon printed on page 393. Order by number only.

1. FILAMENT CONTROL—Problems of filament supply, voltage regulation, and effect on various circuits. RADALL COMPANY.

2. HARD RUBBER PANELS—Characteristics and properties of hard rubber as used in radio, with suggestions on how to "work" it. B. F. GOODRICH RUBBER COMPANY.

3. TRANSFORMERS—A booklet giving data on input and output transformers. PACENT ELECTRIC COMPANY.

4. CARBORUNDUM IN RADIO—A book giving pertinent data on the crystal as used for detection, with hook-ups, and a section giving information on the use of resistors. THE CARBORUNDUM COMPANY.

7. TRANSFORMER AND CHOKE-COUPLED AMPLIFICATION—Circuit diagrams and discussion. ALL-AMERICAN RADIO CORPORATION.

9. VOLUME CONTROL—A leaflet showing circuits for distortionless control of volume. CENTRAL RADIO LABORATORIES.

10. VARIABLE RESISTANCE—As used in various circuits. CENTRAL RADIO LABORATORIES.

11. RESISTANCE COUPLING—Resistors and their application to audio amplification, with circuit diagrams. DEJUR PRODUCTS COMPANY.

12. DISTORTION AND WHAT CAUSES IT—Hook-ups of resistance-coupled amplifiers with standard circuits. ALLEN-BRADLEY COMPANY.

15. B-ELIMINATOR AND POWER AMPLIFIER—Instructions for assembly and operation using Raytheon tube. GENERAL RADIO COMPANY.

15a. B-ELIMINATOR AND POWER AMPLIFIER—Instructions for assembly and operation using an R. C. A. rectifier. GENERAL RADIO COMPANY.

16. VARIABLE CONDENSERS—A description of the functions and characteristics of variable condensers with curves and specifications for their application to complete receivers. ALLEN D. CARDWELL MANUFACTURING COMPANY.

17. BAKELITE—A description of various uses of bakelite in radio, its manufacture, and its properties. BAKELITE CORPORATION.

19. POWER SUPPLY—A discussion on power supply with particular reference to lamp-socket operation. Theory and constructional data for building power supply devices. ACME APPARATUS COMPANY.

20. AUDIO AMPLIFICATION—A booklet containing data on audio amplification together with hints for the constructor. ALL AMERICAN RADIO CORPORATION.

21. HIGH-FREQUENCY DRIVER AND SHORT-WAVE WAVEMETER—Constructional data and application. BURGESS BATTERY COMPANY.

26. AUDIO-FREQUENCY CHOKES—A pamphlet showing positions in the circuit where audio-frequency chokes may be used. SAMSON ELECTRIC COMPANY.

27. RADIO-FREQUENCY CHOKES—Circuit diagrams illustrating the use of chokes to keep out radio-frequency currents from definite points. SAMSON ELECTRIC COMPANY.

28. TRANSFORMER AND IMPEDANCE DATA—Tables giving the mechanical and electrical characteristics of transformers and impedances, together with a short description of their use in the circuit. SAMSON ELECTRIC COMPANY.

29. BYPASS CONDENSERS—A description of the manufacture of bypass and filter condensers. LESLIE F. MUTER COMPANY.

30. AUDIO MANUAL—Fifty questions which are often asked regarding audio amplification, and their answers. AMERTRAN SALES COMPANY, INCORPORATED.

31. SHORT-WAVE RECEIVER—Constructional data on a short-wave receiver. The set is designed to be made to tune from a frequency of 16,660 kc. (18 meters) to 1900 kc. (150 meters). SILVER-MARSHALL, INCORPORATED.

32. AUDIO QUALITY—A booklet dealing with audio-frequency amplification of various kinds and the application to well-known circuits. SILVER-MARSHALL, INCORPORATED.

36. VARIABLE CONDENSERS—A bulletin giving an analysis of various condensers together with their characteristics. GENERAL RADIO COMPANY.

37. FILTER DATA—Facts about the filtering of direct current supplied by means of motor-generator outfits used with transmitters. ELECTRIC SPECIALTY COMPANY.

39. RESISTANCE COUPLING—A booklet giving some general information on the subject of radio and the application of resistors to a circuit. DAVEN RADIO CORPORATION.

62. RADIO-FREQUENCY AMPLIFICATION—Constructional details of a five-tube receiver using a special design of radio-frequency transformer. CAMFIELD RADIO MFG. COMPANY.

63. FIVE-TUBE RECEIVER—Constructional data on building a receiver. AERO PRODUCTS, INCORPORATED.

64. AMPLIFICATION WITHOUT DISTORTION—Data and curves illustrating the use of various methods of amplification. ACME APPARATUS COMPANY.

66. SUPER-HETERODYNE—Constructional details of a seven-tube set. C. E. EVANS COMPANY.

70. IMPROVING THE AUDIO-AMPLIFIER—Data on the characteristics of audio transformers with a circuit diagram showing where chokes, resistors, and condensers can be used. AMERICAN TRANSFORMER COMPANY.

72. PLATE SUPPLY SYSTEM—A wiring diagram and lay-out plan for a plate supply system to be used with a power amplifier. Complete directions for wiring are given. AMERTRAN SALES COMPANY.

80. FIVE-TUBE RECEIVER—Data are given for the construction of a five-tube tuned radio-frequency receiver. Complete instructions, list of parts, circuit diagram, and template are given. ALL-AMERICAN RADIO CORPORATION.

81. BETTER TUNING—A booklet giving much general information on the subject of radio reception with specific illustrations. Primarily for the non-technical home constructor. BREMER-TULLY MANUFACTURING COMPANY.

82. SIX-TUBE RECEIVER—A booklet, containing photographs, instructions, and diagrams for building a six-tube shielded receiver. SILVER-MARSHALL, INCORPORATED.

83. SOCKET POWER DEVICE—A list of parts, diagrams, and templates for the construction and assembly of socket power devices. JEFFERSON ELECTRIC MANUFACTURING COMPANY.

84. FIVE-TUBE EQUAMATIC—Panel layout, circuit diagrams, and instructions for building a five-tube receiver, together with data on the operation of tuned radio-frequency transformers of special design. KARAS ELECTRIC COMPANY.

85. FILTER—Data on a high-capacity electrolytic condenser used in filter circuits in connection with A socket power supply units, are given in a pamphlet. THE ABOX COMPANY.

86. SHORT-WAVE RECEIVER—A booklet containing data on a short-wave receiver as constructed for experimental purposes. THE ALLEN D. CARDWELL MANUFACTURING CORPORATION.

88. SUPER-HETERODYNE CONSTRUCTION—A booklet giving full instructions, together with a blue print and necessary data, for building an eight-tube receiver. THE GEORGE W. WALKER COMPANY.

89. SHORT-WAVE TRANSMITTER—Data and blue prints are given on the construction of a short-wave transmitter, together with operating instructions, methods of keying, and other pertinent data. RADIO ENGINEERING LABORATORIES.

90. IMPEDANCE AMPLIFICATION—The theory and practice of a special type of dual-impedance audio amplification are given. ALDEN MANUFACTURING COMPANY.

93. B-SOCKET POWER—A booklet giving constructional details of a socket-power device using either the BH or 313 type rectifier. NATIONAL COMPANY, INCORPORATED.

94. POWER AMPLIFIER—Constructional data and wiring diagrams of a power amplifier combined with a B-supply unit are given. NATIONAL COMPANY, INCORPORATED.

100. A, B, AND C SOCKET-POWER SUPPLY—A booklet giving data on the construction and operation of a socket power supply using the new high-current rectifier tube. THE Q. R. S. MUSIC COMPANY.

101. USING CHOKES—A folder with circuit diagrams of the more popular circuits showing where choke coils may be placed to produce better results. SAMSON ELECTRIC COMPANY.

22. A PRIMER OF ELECTRICITY—Fundamentals of electricity with special reference to the application of dry cells to radio and other uses. Constructional data on buzzers, automatic switches, alarms, etc. NATIONAL CARBON COMPANY.

23. AUTOMATIC RELAY CONNECTIONS—A data sheet showing how a relay may be used to control A and B circuits. YAXLEY MANUFACTURING COMPANY.

25. ELECTROLYTIC RECTIFIER—Technical data on a new type of rectifier, with operating curves. KODEL RADIO CORPORATION.

26. DRY CELLS FOR TRANSMITTERS—Actual tests given, well illustrated with curves showing exactly what may be expected of this type of B power. BURGESS BATTERY COMPANY.

27. DRY-CELL BATTERY CAPACITIES FOR RADIO TRANSMITTERS—Characteristic curves and data on discharge tests. BURGESS BATTERY COMPANY.

28. B BATTERY LIFE—Battery life curves with general curves on tube characteristics. BURGESS BATTERY COMPANY.

30. TUBE CHARACTERISTICS—A data sheet giving constants of tubes. C. E. MANUFACTURING COMPANY.

32. METERS FOR RADIO—A catalogue of meters used in radio, with diagrams. BURTON-ROGERS COMPANY.

33. SWITCHBOARD AND PORTABLE METERS—A booklet giving dimensions, specifications, and shunts used with various meters. BURTON-ROGERS COMPANY.

34. STORAGE BATTERY OPERATION—An illustrated booklet on the care and operation of the storage battery. GENERAL LEAD BATTERIES COMPANY.

36. CHARGING A AND B BATTERIES—Various ways of connecting up batteries for charging purposes. WESTINGHOUSE UNION BATTERY COMPANY.

37. WHY RADIO IS BETTER WITH BATTERY POWER—Advice on what dry cell battery to use; their application to radio of various condensers together with their characteristics. GENERAL RADIO COMPANY.

53. TUBE REACTIVATOR—Information on the case of vacuum tubes, with notes on how and when they should be reactivated. THE STERLING MANUFACTURING COMPANY.

69. VACUUM TUBES—A booklet giving the characteristics of the various tube types with a short description of where they may be used in the circuit. RADIO CORPORATION OF AMERICA.

77. TUBES—A booklet for the beginner who is interested in vacuum tubes. A non-technical consideration of the various elements in the tube as well as their position in the receiver. CLEARTRON VACUUM TUBE COMPANY.

87. TUBE TESTER—A complete description of how to build and how to operate a tube tester. BURTON-ROGERS COMPANY.

91. VACUUM TUBES—A booklet giving the characteristics and uses of various types of tubes. This booklet may be obtained in English, Spanish, or Portuguese. DEFOREST RADIO COMPANY.

92. RESISTORS FOR A. C. OPERATED RECEIVERS—A booklet giving circuit suggestions for building a, c operated receivers, together with a diagram of the circuit used with the new 400-milliampere rectifier tube. CARTER RADIO COMPANY.

97. HIGH-RESISTANCE VOLTMETERS—A folder giving information on how to measure the voltage measurement of socket-power devices. WESTINGHOUSE ELECTRIC & MANUFACTURING COMPANY.

102. RADIO POWER BULLETINS—Circuit diagrams, theory constants, and trouble-shooting hints for units employing the BH or-B rectifier tubes. RAYTHEON MANUFACTURING COMPANY.

103. A. C. TUBES—The design and operating characteristics of a new a. c. tube. Five circuit diagrams show how to convert well-known circuits. SOVEREIGN ELECTRIC & MANUFACTURING COMPANY.

(Continued on page 393)

What Kit Shall I Buy?

THE list of kits herewith is printed as an extension of the scope of the Service Department of Radio Broadcast. *It is our purpose to list here the technical data about kits on which information is available. In some cases, the kit can be purchased from your dealer complete; in others, the descriptive booklet is supplied for a small charge and the parts can be purchased as the buyer likes. The Service Department will not undertake to handle cash remittances for parts, but when the coupon on page 395 is filled out, all the information requested will be forwarded.*

Book Review

PRINCIPLES OF RADIO COMMUNICATION. By J. H. Morecroft. Second Edition. Published by John Wiley & Sons, Incorporated, New York. Pages, 1001. Illustrations, 831. Price, $7.50.

MORECROFT has gone into a second edition. The revised volume calls for a supplementary review, for after the lapse of six years (the first edition was issued in 1921) much that was important is slipping into desuetude, and technological fancies have become engineering realities. The ordinary engineer, called on to review Morecroft's monumental work, is placed somewhat in the position of a parish priest ordered to review the Bible. It is safe, however, to quote Professor Morecroft himself in his outline of the changes made in the new edition:

The new material incorporated in this edition so increased the size that it was thought advisable to delete much of the first edition. A considerable part of the chapter on Spark Telegraphy has been taken out, therefore, and two chapters of the earlier edition have been deleted. The chapter on radio measurements, and that on experiments, have been omitted.

Notable additions to the older edition occur in Chapters II, IV, VIII, and X. In Chapter II many new data on coils and condensers at radio frequencies are given. In Chapter IV, dealing with the general features of radio transmission, new material on field strength measurements, reflection and absorption, fading, short-wave propagation, etc., has been introduced. In Chapter VIII (radio telephony) a great deal of material on voice analysis has been added; the performance of loud-speaking telephones, frequency control by crystals, etc., has been discussed. In Chapter X, dealing with amplifiers, the question of distortionless amplification has been thoroughly dealt with, some of the material being given for the first time. The questions of radio-frequency amplification, balanced circuits, push-pull arrangements, etc., have been explained.

Principles of Radio Communication is a comprehensive textbook of radio engineering. The author, a Professor of Electrical Engineering at Columbia University, and a Past President of the Institute of Radio Engineers, is one of the outstanding opponents of guesswork in radio technology. When the publishers in their circular describing the book refer to it as the "most complete, accurate, and authoritative book on radio available" they are simply telling the truth. But in justice to the author, who has put a considerable number of years into this job, it should be stated that when the publishers continue: "—for Designers, Engineers, Service Men, Distributors, Dealers, Salesmen, Teachers, Students, Operators, Set Owners," they talk like hasheesh addicts. I hope that Wiley sells as many copies of Morecroft as the publisher of Durant's *Story of Philosophy* has managed to dispose of, to his own and his client's enrichment, but I feel bound to warn Distributors, Dealers, Salesmen, and Set Owners that, with negligible exceptions in their ranks, the only portions of *Principles of Radio Communication* which they can hope to understand are the articles and prepositions. Not that it is an excessively abstruse work; any student of mathematics through the calculus can follow the demonstrations, and any student of radio engineering can read the whole thousand pages with vast profit. But it is a work in radio *engineering*. Its precise virtue is that dealers, salesmen, and the generality of set owners will not understand it.

How radio has grown! Here is a book of a thousand six-by-nine pages, and yet it is largely an outline of principles. If you consulted it for the actual design of a line equalizer or a 10-TU

pad, you would not find either device even mentioned. The author makes no bones about this. "As in the previous edition," he writes in the new preface, "no pretense is made that the book is a treatise on radio practice; in general, only the principles involved in the operation of radio apparatus have received attention. Whatever radio apparatus is discussed is dealt with only to illustrate those principles the text is intended to elucidate." But those principles, by diagram, mathematical and physical analysis, oscillogram, and every other resource of technical instruction, it does clarify. The chapter on vacuum tubes alone runs to 240 pages, a book in itself, and after you have mastered it you know something about tubes. You may burn out the next one you put into a socket anyway, but at least you will not concoct any idiotic theories about it. If you believe that there is a dividing line between principles and practice, and that life is too short to include both, you will, of course, find no use for such a course as Morecroft offers. You had best go out and sell bonds, in that case. But if you are a radio engineer, or want to become one in the only genuine sense of the word, then Morecroft's textbook will be worth more than $7.50 to you.

In a work of this size there are inevitably sections over which other engineers may disagree with the author. The treatment of "Elimination of Strays," pages 340–343, may be cited. In Fig. 16, "one of the early attempts to eliminate 'strays,'" ascribed to De Groot, is shown. This scheme is a neutralization system—one antenna tuned to the desired signal, the other aperiodic, etc. It will not work, which Professor Morecroft knows as well as anyone, but he does not tell the reader that the scheme is worthless, and why. The ingenious and useful wave antenna of Beverage, Rice, and Kellogg, which deserves mention in this section if anything does, is omitted without a word. Roy A. Weagant's name is twice misspelled. But such defects, which might be serious in a lesser work, are overshadowed by the high virtues of Morecroft's imposing contribution to radio science. CARL DREHER.

The radio leadership of 1928

180 volts on the output tube plate!
Gigantic *UNDISTORTED* volume from the Bandbox!

Power! Power! POWER! A feature of the Crosley AC Bandbox that lifts it head and shoulders above competition!

170 to 185 volts on the plate of the power output tube!

Comparative checkings of competitive radios show interesting figures. Under identical testing conditions the Bandbox shows a full 170 to 185 volts on the plate of the 171 power output tube. Other radios show from 100 to 110 and 130 to 140 volts on the plate of output tube. The 171 power tube should have around 180 volts. This better than 40% superiority in one case and 25%, in the other is the difference between *today's* radio and yesterday's.

MUSICONE
Type D
$15

Crosley Musicones are famous for their value. This new style is no exception. Its low price of $15 is in keeping with Crosley traditions. It instantly demonstrated its soundness by immediate and enormous sales.

602 Double Unit
*Ac*BANDBOX
Single Unit 704

The Bandboxes are genuine Neutrodyne receivers. Totally and completely shielded, their acute sensitivity and sharp selectivity is amazing.

They have a single illuminated dial.

Contributing much to the success of this 1928 wonder radio is the Mershon Condenser in the power element of the set. Not being paper, the danger of its blowing out is entirely removed so that the desired *heavy voltage* can be used to produce the acoustic and volume results so greatly desired. IT IS SELF HEALING. It does not have to be replaced as is the case with paper condensers.

The capacity of smoothing condensers in Crosley power units is 30 mf. Other sets use only a fraction of that condenser capacity. Undersize condensers, transformers, etc., are used in order to build down to a price. Crosley builds up to a standard.

The AC Bandbox is purposely made in two models—the 602 in a double unit—the 704 self contained. This is to provide maximum adaptability in all sorts of surroundings and uses.

The 602 double unit provides console cabinet installation in ALL kinds of consoles.

The 704 is for those who want the entire set in one cabinet. The two sets are identical in element, design and performance. The physical difference is solely to meet the human differences of taste, necessity and price! The size of the 704 is 17½ inches long by 12½ inches wide and 6¼ inches high.

Battery Type Bandbox $55

This celebrated model needs no picture, for in appearance it is identical to the 602 receiver pictured above. Its amazing performance has won the radio world this season and its value is as outstanding NOW as the day it was first presented.

SELF CONTAINED
$95

New
401 Dry Cell Type
BANDBOX JUNIOR
$35

A new dry cell receiver with all the features of the Bandbox—selectivity, sensitivity, volume and appearance. For places where AC current or storage battery service is not available or desired.

Approved Console Cabinets manufactured by Showers Brothers Co., of Bloomington, Ind., and Wolf Mfg. Industries, Kokomo, Ind., are sold to Crosley dealers by H. T. Roberts Co., 1340 S. Michigan Ave., Chicago, Sales Representatives.

NEUTRODYNE

Crosley is licensed only for Radio Amateur, Experimental and Broadcast Reception.

THE CROSLEY RADIO CORPORATION
Powel Crosley, Jr., Pres.   Cincinnati, Ohio

Montana, Wyoming, Colorado, New Mexico, and West, prices slightly higher

Write Dept. 20 for descriptive literature

"You're *there* with a Crosley"
~~CROSLEY~~
RADIO

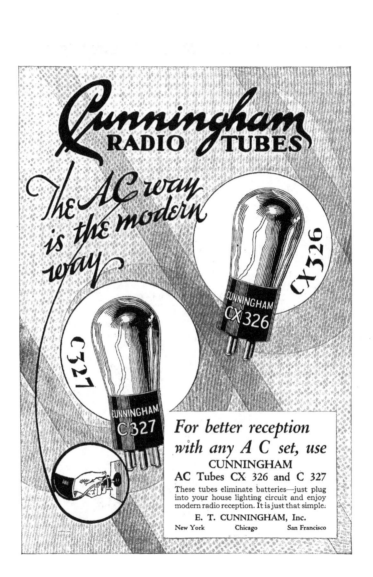

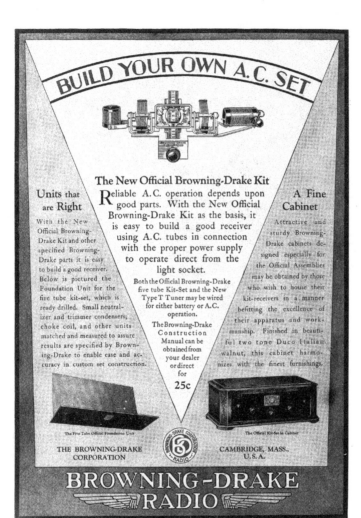

RADIO
BROADCAST

CONTENTS

"I have used ACME PARVOLT Condensers for many years and have never had one break down."

Showing the well-known L. C. 28 wired for A. C. power supply. Only the finest parts are specified in this high class set, including PARVOLT By-Pass Condensers. ACME PARVOLT Filter Condensers are also recommended for use in the power end.

Whether You Buy or Build

a Power Supply Unit for Your Radio

PLAY SAFE WITH PARVOLTS!

WHEN you buy an electrified radio or power supply unit for your receiver, look for ACME PARVOLT Condensers; they are your guide to quality in all other parts. They cost the manufacturer a trifle more, but they are both his and your guarantee against costly condenser break-down.

Should you build your own power supply, be sure to use ACME PARVOLT Condensers and be safeguarded against the possibility of break down. Remember that poor filter condensers have caused untold thousands of dollars worth of loss in the past year or two, for blown out condensers mean blown tubes, burned out transformers and frequently the ruination of speaker units.

Just as PARVOLT By-Pass Condensers have been used for years in high grade

ACME PARVOLT FILTER CONDENSERS
In all required mfd. capacities for 200, 400, 600, 800, 1000, and 1500 Volt D C requirements. Supplied singly or in complete wired blocks for the important power supply units.

ACME PARVOLT BY-PASS CONDENSERS
Supplied in all required mfd. capacities and for all standard working voltages.

receivers, so are PARVOLT Filter Condensers rapidly replacing ordinary condensers in electrified radio. These condensers are wound with the very finest insulating papers combined with highest grade foils. Every detail produced in one of America's most modern plants and under the supervision of experts in condenser design and manufacture.

Uniformity of capacity and uniformity of sizes are two big features. Accuracy of all ratings, based upon the R.M.A. standards, is another guarantee of uninterrupted service. Play safe with PARVOLTS!

Made by THE ACME WIRE CO., New Haven, Conn., manufacturers of magnet and enameled wire, varnished insulations, coil windings, insulated tubing and radio cables.

ACME PARVOLT CONDENSERS
Made by the Manufacturers of
ACME CELATSITE WIRE

ENAMELED AERIAL WIRE	CELATSITE FLEXIBLE and SOLID	ACME SPAGHETTI
Enameled copper wire in both stranded and solid types. Also Acme Leadins, Battery Cables, Indoor and Loop Aerial Wire	*For all types of radio wiring. High insulation values; non-inflammable. 10 colors.*	*A superior cambric tubing for all practical radio and other electrical requirements. Supplied in 10 colors.*

I Will Train You at Home to Fill a Big-Pay Radio Job

IF you are earning a penny less than $50 a week, send for my book of information on the opportunities in Radio. It's FREE. Clip the coupon NOW. A flood of gold is pouring into this new business, creating hundreds of big pay jobs. Why go along at $25, $30 or $45 a week when the good jobs in Radio-pay $50, $75, and up to $250 a week. My book "Rich Rewards in Radio," gives full information on these big jobs and explains how you can quickly become a Radio Expert through my easy, practical, home-study training.

You can build 100 circuits with the six big outfits of Radio parts I give you

Here's the PROOF

Made $185 in Three Weeks' Spare Time

I have met with continued success. For instance, recently I realized a profit of $185 in three weeks for spare time work. I charge $5 an hour. Right now I am making more money in my spare time than I am making in my regular job. I have been making good money almost from the time I enrolled. I am going to give up my present position and open a Radio shop. The N. R. I. has put me on the solid road to success.—Peter J. Dunn, 912 W. Monroe St., Baltimore, Md.

Made $588 in One Month

The training I received from you has done me a world of good. Some time ago during one of our busy months, I made $588. I am servicing all makes of Radio receiving sets. I haven't found anything so far that I could not handle alone. My boss is highly pleased with my work since I have been able to handle our entire output of sets here alone.—Herbert Reese, 2015 South E Street, Elwood, Indiana.

Earns Price of Course in One Week's Spare Time

I have been so busy with Radio work that I have not had time to study. The other week, in spare time, I earned enough to pay for my course. I have more work than I can do. Recently I made enough money in one month's spare time to pay for a $375 beautiful console all-electric Radio. When I enrolled I did not know the difference between a resistor and a coil. Now I am making all kinds of money.—Earle Cummings, 18 Webster Street, Haverhill, Mass.

SALARIES OF $50 TO $250 A WEEK NOT UNUSUAL

Get into this live-wire profession of quick success. Radio needs trained men. The amazing growth of the Radio business has astounded the world. In a few short years three hundred thousand jobs have been created. And the biggest growth of Radio is still to come. That's why salaries of $50 to $250 a week are not unusual. Radio simply hasn't got nearly the number of thoroughly trained men it needs. Study Radio and after only a short time land yourself a REAL job with a REAL future.

YOU CAN LEARN QUICKLY AND EASILY IN SPARE TIME

Hundreds of N. R. I. trained men are today making big money—holding down big jobs—in the Radio field. Men just like you—their only advantage is training. You, too, can become a Radio Expert just as they did by our new practical methods. Our tested, clear training, makes it easy for you to learn. You can stay home, hold your job, and learn quickly in your spare time. Lack of education or experience are no drawbacks. You can read and write. That's enough.

MANY EARN $15, $20, $30 WEEKLY ON THE SIDE WHILE LEARNING

My Radio course is the famous course "that pays for itself." I teach you to begin making money almost the day you enroll. My new practical method makes this possible. I give you SIX BIG OUTFITS of Radio parts with my course. You are taught to build practically every type of receiving set known. M. E. Sullivan, 412 73rd Street, Brooklyn, N. Y., writes, "I made $720 while studying." Earle Cummings, 18 Webster Street, Haverhill, Mass., "I made $375 in one month." G. W. Page, 1807 21st Ave., Nashville, Tenn., "I picked up $935 in my spare time while studying."

YOUR MONEY BACK IF NOT SATISFIED

I'll give you just the training you need to get into the Radio business. My course fits you for all lines—manufacturing, selling, servicing sets, in business for yourself, operating on board ship or in a broadcasting station—and many others. I back up my training with a signed agreement to refund every penny of your money if, after completion, you are not satisfied with the course I give you.

ACT NOW— 64-Page Book is FREE

Send for this big book of Radio information. It won't cost you a penny. It has put hundreds of fellows on the road to bigger pay and success. Get it. Investigate. See what Radio has to offer you, and how my Employment Department helps you get into Radio after you graduate. Clip or tear out the coupon and mail it RIGHT NOW.

J. R. SMITH, President
Dept. 4-O
National Radio Institute
Washington, D. C.

3 of the 100 you can build

Find out quick about this practical way to big pay

RADIO NEEDS TRAINED MEN!

Mail This FREE COUPON Today

J. E. SMITH, President
Dept. 4-O, National Radio Institute
Washington, D. C.

Dear Mr. Smith: Kindly send me your big book "Rich Rewards in Radio," giving information on the big-money opportunities in Radio and your practical method of teaching with six big Outfits. I understand this book is free, and that this places me under no obligation whatever.

Name ....

Address ....

City .... State ....

Employment Service to all Graduates
Originators of Radio Home Study Training

MAR 17 '28   © CI B783718

RADIO BROADCAST

WILLIS KINGSLEY WING, Editor

APRIL, 1928   KEITH HENNEY   EDGAR H. FELIX   Vol. XII, No. 6
             Director of the Laboratory   Contributing Editor

CONTENTS

AMONG OTHER THINGS. . .

THE issue before you contains a variety of articles appealing to all tastes. For the short-wave enthusiast there is the constructional article on the code and telephone transmitter on page 410. For experiments, we offer many articles, such as those on a two-tube, screen-grid receiver, measurements on the "Lab" circuit receiver, the circuit and description of a non-motor-boating resistance amplifier, the technical editorials, "Strays from the Laboratory," descriptions of tests on B-power units for hum characteristics, how to operate the Hammarlund-Roberts "Hi-Q" set on a.c., and the story on how to convert many standard receivers for a.c. operation. Of more general interest, there are the reviews of new phonograph records, "The Listeners' Point of View," the editorial section, "The March of Radio," the invaluable "As the Broadcaster Sees It," the article by Sylvan Harris describing modern methods of phonograph record making, and many others.

OWING to causes which were beyond our control, we are unable at the last moment to present the article on a new tube for B-supply units, promised in the announcement on our cover. This article will appear as soon as it is finally released by the manufacturer.

THE May RADIO BROADCAST will be full of features which will make it one of the most important issues we have had in many months. Lloyd T. Goldsmith of M. I. T. has written a description of a short-wave receiver, and an intermediate-frequency amplifier using the screen-grid tube. This receiver, especially when used for code reception, provides efficiency heretofore impossible before the advent of the screen-grid tube. There will also appear in the May issue a most accurate list of international short-wave stations. Many readers have asked for a description of a power supply circuit for use in direct-current districts and an article describing a practical circuit for this purpose will appear in May.

THE problems of synchronizing broadcasting stations and, in general, of accurate frequency control for radio stations, has assumed great importance in the past year. Edgar H. Felix has prepared an accurate report of what has been accomplished to date and an analysis of the immediate possibilities in an interesting article scheduled for the May number. For radio constructors, Hugh S. Knowles is writing a description of an a.c. operated "Lab" circuit receiver which has many interesting features, besides its efficiency and flexibility of use, to commend it. This story is also scheduled for May.

SEVERAL additional regular features are being planned for the coming numbers of RADIO BROADCAST and it is hoped to start the first of them with the May issue. Each of these features will appeal in a very practical way to a large number of the radio fraternity. . . . Those interested in information about the Cooley Rayfoto system of picture reception who desire to be placed in touch with the manufacturers may address their letters to the undersigned who will forward them.

—WILLIS KINGSLEY WING.

DOUBLEDAY, DORAN & COMPANY, INC., Garden City, New York

MAGAZINES	BOOK SHOPS (Books of all Publishers)	OFFICES	OFFICERS
COUNTRY LIFE	LORD & TAYLOR; JAMES McCREERY & COMPANY	GARDEN CITY, N. Y.	F. N. DOUBLEDAY, Chairman of the Board
WORLD'S WORK	PENNSYLVANIA TERMINAL and 166 WEST 32ND ST.		NELSON DOUBLEDAY, President
GARDEN & HOME BUILDER	848 MADISON AVE. AND 51 EAST 44TH STREET	NEW YORK: 244 MADISON AVENUE	GEORGE H. DORAN, Vice-President
RADIO BROADCAST	NEW YORK: 420 AND 526 AND 819 LEXINGTON AVENUE	BOSTON: PARK SQUARE BUILDING	S. A. EVERITT, Vice-President
SHORT STORIES	GRAND CENTRAL TERMINAL and 38 WALL STREET		
EDUCATIONAL REVIEW	CHICAGO: 75 EAST ADAMS STREET	CHICAGO: PEOPLES GAS BUILDING	RUSSELL DOUBLEDAY, Secretary
LE PETIT JOURNAL	ST. LOUIS: 223 N. 8TH ST. and 4914 MARYLAND AVE.	SANTA BARBARA, CAL.	JOHN J. HESSIAN, Treasurer
EL ECO	KANSAS CITY: 920 GRAND AVE. and 206 WEST 47TH ST.	LONDON: WM. HEINEMANN, LTD.	LILLIAN A. COMSTOCK, Ass't Secretary
FRONTIER STORIES	CLEVELAND: HIGBEE COMPANY		L. J. McNAUGHTON, Ass't Treasurer
WEST	SPRINGFIELD, MASS: MEEKINS, PACKARD & WHEAT	TORONTO: OXFORD UNIVERSITY PRESS	
THE AMERICAN SKETCH			

Copyright, 1928, in the United States, Newfoundland, Great Britain, Canada, and other countries by Doubleday, Doran & Company, Inc. All rights reserved.
TERMS: $4.00 a year; single copies 35 cents.

Francis Gow Smith

THE INTREPID explorer whose recent expedition to the jungles of the Brazilian hinterland was facilitated by the use of short-wave radio equipment carried by the party. Twice weekly time signals were sent out on one of WGY's short-wave channels to aid Mr. Gow Smith in his map making. The special microphone used for the purpose is still known up at Schenectady as the Gow Smith "mike." The story beginning on the next page relates the amusing adventures of the explorer in the smaller towns on his route leading to the heart of the dark continent. Listening-in was a pastime never before indulged in by the inhabitants of many of these towns, and the explorer, who otherwise might have been received coldly, was acclaimed wherever he went, and was embarrassed by the great number of invitations to social functions which he received

MOUNTED BRAVADOS IN A MATTO GROSSO TOWN
The author frequently encountered mounted groups of nomadic ex-convicts whose thirst for amusement seemed to be satisfied only by plundering. Sometimes they will lay waste a whole settlement, murdering and robbing the inhabitants

Thanks to WGY—!

By Francis Gow Smith

THIS is WGY calling Francis Gow Smith, on the Upper Paraguay River, Matto Grosso: The discovery made in the use of your set without antenna or ground while you were on the steamship *Pan America* this last March is expected to have important results in the further development of radio."

This message, broadcast for me by the General Electric Company from Schenectady, and reaching me one spring night in the tiny frontier town of São Luiz de Caceres, forty-five hundred miles south of New York, finally convinced my Brazilian friends that I was not a magician or a faker or a spy but a *bona fide* explorer.

And whatever value there may have been to radio in the experimental short-wave broadcasting that I received during my latest and most adventurous trip into the wilderness of Brazil, there is no doubt in my own mind that these experiments helped greatly to build goodwill for the United States in a backwoods region where North Americans have hitherto been looked upon with suspicion.

Incidentally, I owe my safe return from that expedition very largely to the radio set I carried. It was a neat, portable two-tube affair, specially built for me by RADIO BROADCAST. When I sailed with it for Rio, aboard the *Pan America*, I appreciated it rather as a possible source of recreation during the long months I would be isolated from civilization. I had no inkling of its future utility in making my expedition a success.

Indeed, I was very much disappointed, the first few nights out, when I strung the antenna around my cabin expecting to get news reports and actually getting nothing in the phones but dead silence. Finally one night I turned it over to the radio operators to experiment with.

"What's the matter with the darn thing?" I asked.

"Give it a chance," said they. "Bring it up on deck."

And then one of them, with the headphones rumpling his hair and a delighted grin on his face, remarked:

"What's the matter with the blamed thing? Listen to this! I'm getting New York without antenna or ground. We're close to seventeen hundred miles out."

Every night thereafter we used the set on deck and got code signals from Germany, England, and Japan. The sensitivity of the set

SATURDAY NIGHT!
A community bathing "crevice" makes unnecessary the "hot and cold water" clause in the South American Indians' lease

amazed the radio boys, and was the occasion for that later message from Schenectady, which I received at São Luiz de Caceres.

In Brazil to-day, radio is barely emerging from its infancy. There are two broadcasting stations in Rio and one in São Paulo, while the General Electric Company is erecting another. Entertainment programs and lessons in English are on the air constantly; but outside of the cities and a few prosperous ranches, there is nobody in Brazil equipped to hear the programs.

Here's a nation bigger than the continental United States, with a population of more than thirty million, and increasingly prosperous economic conditions. Some day soon it is bound to be a profitable market for American radio products. And when American radio programs are being received by American radio sets throughout Brazil, a greater influence will have come into being for Pan American goodwill and commercial development than all the spectacular goodwill flights and conferences that we can organize.

For radio will break down among the common people that suspicion which at present is fed by local propagandists to the detriment both of political amity and friendly trade.

I'm positive of this, because I've seen how it works out. When I reached Corumba, on the upper Paraguay River, I was just another of those Yankees, regarded with vague suspicion. I had a few acquaintances there, from previous trips, but no friends to speak of and no entrée into the intimate social life of the town. I put up at the hotel, and asked the proprietor's permission to install my radio.

He looked at me blankly. Corumba has water supply, electric lights, and telephones. But radio? Huh! There was no use trying to persuade the hotel man that the little wooden box I carried could perform any of the radio miracles he had indefinitely heard about. Still, he skeptically authorized me to put up an antenna on his roof.

403

My opening act was unfortunate. One of the poles slipped while I was erecting it, fell through the red tile roof of the house next door and just missed hitting on the head one of the society leaders of the town. She came storming outdoors and gave me frankly a piece of her mind. Soon it was all over town that this crazy Yankee was destroying property with something he called a radio.

But that night I invited the hotel proprietor to bring a few friends into my room. They stood around rather abashed. What was the sense of all this fuss over a box with some wires and two oddly shaped electric light bulbs in it?

And then, singing sweetly in the phones, came from Schenectady a musical program given experimentally on short waves for a man in Johannesburg!

My guests were thrilled as they had never been in their lives. They were almost incoherent in their enthusiasm. The news spread like wildfire through the town. The leading citizens flocked into my room, uninvited; the crowd jammed the corridor outside. Men waited hours for their few minutes' turn with the phones; many went away disappointed and came back night after night until they had heard for themselves, this miracle. The lady next door was placated—she wouldn't even let me pay for the damage to her roof. While the hotel proprietor was in the Seventh Heaven of bliss. His bar had never done such a flourishing business.

SOME FRIENDLY INDIANS OF MATTO GROSSO
While these were only too willing to pose for their photographs, there are others who could not be approached with safety

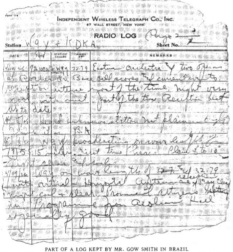

PART OF A LOG KEPT BY MR. GOW SMITH IN BRAZIL
The expedition came to an untimely end owing to the fact that the party was robbed of all its belongings by a gang of bandits. This fragment was saved because Mr. Gow Smith covered it with his foot while he was being searched

For eighteen days, while waiting for the little stern wheeler that would take me up to São Luiz de Cacerés, I had the set working. There was considerable fading at times, but it worked, better after nine at night. Soon I had become the personal friend of the most important business men and politicians in Corumba. They heard dance music from the Waldorf and music from Aeolian Hall; they heard Chauncey Depew speak; they listened to plays in Rochester; they got the news of Amundsen's flight over the North Pole, and of one of the attacks on Mussolini.

I translated the news every night for the local newspaper. Corumba had never before been so intimately in touch with the outside world. The community became pro-American. Nobody could do enough for me; they wanted me to settle down and stay with them; they all said they were going to learn English; and I could have sold the set a hundred times over at any price I asked.

When I left, the leading citizen of the town pressed on me a letter of introduction to an influential friend of his in Cuyaba, the frontier capital of the state of Matto Grosso. I wasn't going anywhere near Cuyaba, but he wanted me to make a special trip there anyhow. His letter read:

"This is to introduce to you my good friend Mr. Francis Gow Smith, who is coming to you with a most marvelous machine called the *radio telephonia*, which has proved in experiments here to be a very amazing thing indeed and I want you to have the privilege of listening to this truly magical instrument."

SÃO LUIZ

WARNED by my experience in Corumba, I rented a big twelve-room house when I reached São Luiz de Caceres. I knew that no hotel room would hold the crowds that would come. And besides, São Luiz boasts no hotel worthy of the name. It's the jumping-off place on the Brazilian frontier—a town of five thousand population, at the edge of the jungle. To the northward beyond it are scattered ranches, a few hamlets of rubber and ipecac gatherers, and then unexplored wilderness, several hundred thousand square miles in extent, peopled sparsely with naked savages.

In São Luiz I was received with greater suspicion than in Corumba. Many of the common people thought I was some sort of gringo spy. Why did I have a camera, if it wasn't to take photographs of sites for future forts, when the dreaded "Colossus of the North" should begin the process of gobbling up Brazil? The people misunderstand the United States completely. They have been filled up with such fantastic bugaboo stories about us that they extend toward us a hazy mixture of dread and dislike such as a child feels toward imaginary giants and dragons.

The inhabitants are practically cut off from the world. They have a weekly four-page newspaper, but it publishes only items concerning local society events—marriages and birthday festivals, and perhaps a sprinkling of political news. Besides, most of the people can't read.

The streets are unpaved; there is no electric light, telephone, or water supply. Water from the river is peddled in barrels about the streets. On the edge of the barren plaza, fronting the river, there stands an unfinished stone church, and the piles of building stones beside it are alive with snakes, lizards, and rats.

Naturally, in this isolated community, I was looked upon with disfavor by the uneducated. My radio set in its mysterious box with the iron handles was set down in their minds with my camera as evidence that I had some mysteriously unfriendly intentions. They couldn't believe I was using the town merely as a jumping-off place for exploration. Why should anybody in his senses penetrate the jungles beyond and get himself shot at with poisoned arrows? No; obviously I had come for no good purpose, doubtless as a secret agent of the United States, either to survey gun emplacements or to detect hidden mineral wealth.

What a change in their attitude my radio worked!

There wasn't a stick of furniture in the house I had rented, except an old table. I set up the radio on that table, and slung my hammock in one of the bare rooms. I hired a boy to help me, went into the jungle and cut down two fifty-foot bamboos, which we erected as antenna masts in the *quintal*, or back garden, of my house.

The villagers watched these proceedings with growing suspicion. Many of them had never even heard of radio. Then I invited the mayor and a few prominent citizens to come and hear the set work. To give a touch of festivity to the occasion, I had hung Eveready electric flashlights around the walls, and softened them with bits of colored paper. The evening was a phenomenal success. Three sets of phones were working all the time. Sometimes a listener, after hearing ecstatically a few bars of music, would drop the phones and scurry out of the house, to round up his wife and children and friends.

Soon I was mobbed, as I had been in Corumba. Every citizen prosperous enough to wear shoes felt himself privileged to come in; the barefoot families humbly congregated about the doors and windows and stared in, imagining some queer sort of magic was going on.

So, within a few nights I was welcomed into the center of the town's social life, invited to all the most elite weddings, birthdays, and funerals, and offered banquets in every home. The bare rooms of my rented house began to fill up with furniture—chairs, tables, wine glasses, and even that rare and valued treasure, a bed! All contributed by the citizens to their distinguished guest—who had done nothing to distinguish himself but bring a radio set into their midst.

But suspicion still smouldered among the poor and illiterate, and I was advised to allay it by having a barefoot soirée, when the unshod portion of the populace could hear the radio work for themselves. Many of them heard it, but were still unconvinced. They'd go out into the back yard and stare up at the aerial; they poked about in the rose bushes, and surreptitiously investigated the empty rooms of the house. Somewhere, they were sure, I had a confederate hidden, to play on a musical instrument.

Then came the three special programs, broadcast by the General Electric Company for my benefit from Schenectady. The dentist in São Luiz understood English, and when the rest of the guests heard him translate the messages addressed to me, the last bit of skepticism vanished. But my reputation was enormously enhanced. Everybody called me affectionately "Mister," or "Mister Yank," and it was said that I was a great and influential millionaire.

Certainly nobody but a millionaire could have a wonderful radio set and receive special news and musical programs sent through the air, forty-five hundred miles over sea and jungle, just so that his evenings on the frontier might not be lonely!

AT CORUMBA

Listening-in with the short-wave receiver built for the expedition by Radio Broadcast

The girls of the town could never quite get over the notion, however, that I was some sort of magician as well. Some of them asked me to tell their fortunes with cards, and when I did so, invariably predicting a forthcoming marriage, the town's regular and previously prosperous fortune-teller was deserted. She couldn't compete with the authentic forecasts that I had to offer—for didn't I get them straight out of the air, wearing my headset and listening wisely while I read the cards?

The guests came to these radio soirées clad in the strangest mixture of costumes. Some of the men wore evening dress; others came in homespun. There were cowboys wearing sombreros, gaudy neckerchiefs, and sidearms; there were ranchers in khaki shirts and big boots. But many of the women, with their bobbed hair, low cut evening gowns and sparkling diamonds, might have just stepped out of a Broadway night club.

I understand that this experimental short-wave broadcasting was of some interest to the General Electric Company. However that may be, it certainly put a backwoods community of Brazil into a turmoil of excitement. The old suspicions of the United States vanished, and the town became a focal point of boosting for Uncle Sam. Every inhabitant wanted a radio, and they all insisted that nowhere else in the world could it be so great a blessing as it will soon be in these backwoods regions where there are only the most meager entertainments and no contact at all with outside events.

When finally my batteries gave out, it was considered a disaster. I left the set in São Luiz, and carried out my expedition into the wilderness, the entire services of the town being put at my disposal while I was making ready. Months later, having been waylaid by bandits, robbed of all my equipment and left fever-stricken and starving in the jungle, I was rescued by a boat coming up the Sipotuba River from São Luiz. They brought me down to the village too weak to walk, put me up in a private home, nursed me back to health, and lavished attentions upon me.

All this in the town that I had entered first amid such an atmosphere of suspicion. And when I was strong enough to leave I was escorted down to the little river steamer by practically the entire populace. Knowing that I had been robbed and left penniless, they thrust handfuls of currency at me; and I even discovered that some of them had secretly stuffed money into my pockets.

Believe me, a radio receiving set will be an essential part of my equipment on my next exploration trip. I don't mind the isolation in the jungles myself, and I dislike to increase the weight of my equipment; but a radio works wonders in building goodwill for the stranger along the frontier. I might not have survived the illness of my last trip but for the friendly attentions which Radio Broadcast's portable set had won for me. And I think that, when it comes to the touchy job of dealing with hostile Indians in the wilderness, a radio would be even more valuable in winning friendship and esteem.

THE CROSS MARKS THE HOTEL "GALILEO," CORUMBA

It was here that Mr. Gow Smith unfortunately permitted one of his antenna masts to crash through the roof of a local society leader's house

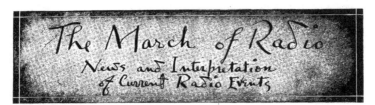

Picture Broadcasting Becomes a Practical Reality

HARDLY a day passes without some news of progress in radio picture transmission. On November 6, WOR broadcast its first complete radio picture, using the Cooley system. On January 13, Dr. E. F. W. Alexanderson publicly demonstrated his television device in Schenectady. On January 26, WEAF broadcast its first picture, utilizing equipment also developed by Doctor Alexanderson. On January 31, WOR began the regular broadcasting of Cooley radio pictures three times a week. Each event is significant and brings nearer the day when the radio reception of pictures becomes an integral part of the broadcasting art.

The transmissions from WOR have been continued since the initial experiment in November, at first occasionally and then on a regular basis. A group of twenty or thirty enthusiasts have installed Cooley Rayfoto receivers and the number increases as rapidly as the equipment is manufactured. These experimenters are typical set builders rather than specially trained professional engineers. Phonograph recordings of actual pictures have also been made successfully which make possible experiments with Cooley apparatus at any time, regardless of the availability of broadcasting. The sponsors of the Cooley system advise that, within two or three months, their equipment will be available in quantity and a rapid growth of the picture reception audience may be expected.

We witnessed a confidential demonstration of Doctor Alexanderson's apparatus at Schenectady some months ago. Several radio channels are required to transmit television images and consequently the system is now restricted to short-wave transmission. The conventional broadcast receiver cannot be utilized, therefore, in the reception of television by this system. The Alexanderson television outfit, which

uses Daniel Moore's improved neon tube, is infinitely simpler than the elephantine apparatus demonstrated by the Bell Laboratories a year ago. The picture, with the Alexanderson system, is scanned eighteen times a second. The received picture is in the form of a pinkish glow, covering an area of three by three inches. It is not rich in detail. Presumably three to five years of development are necessary before this apparatus is capable of reproducing sufficiently good moving pictures at a reasonable cost to the average user.

The picture receiver, demonstrated at WEAF, employs the neon tube and requires about ninety seconds for the reception of a picture. It utilizes high-grade synchronous motors to keep transmitting and receiving

ANOTHER RADIO PICTURE RECEIVER

Readers of this magazine are familiar with the Cooley Rayfoto picture system demonstrated at the New York radio show last September and which was first heard over WOR on November 6, 1927. On January 26, 1928, Dr. E. F. W. Alexanderson of the General Electric Company demonstrated one of his systems of radio picture transmission and reception through WEAF. The illustration shows Doctor Alexanderson and the receiver. The rectifier and amplifier unit is in the box near the wall and the mechanical element with paper on the receiving drum is in the foreground. Pictures are received in 90 seconds and are excellent in quality

drums in step. The transmitted signal was a high-pitched audio note, similar to that used in sending Cooley pictures. No statement has been made as to when the equipment will be on the market. Since the synchronizing equipment is somewhat more expensive than the "stop-start" system used with the Cooley apparatus, its cost may be fairly high. It is capable of excellent detail and possesses considerable reliability.

The adoption of a regular picture broadcasting schedule by WOR, the fourth significant event, is an indication that that station has already established a picture receiving audience and is preparing to extend it. That this audience will grow rapidly as soon as the apparatus is available in quantity is foreshadowed by the fact that the Cooley equipment is no more expensive than the parts for a good broadcast receiver.

At the same time that these various news events occurred in the field of radio vision, a representative of the Baird system, en route to the United States, announced that television between London and New York had been definitely established. One demonstration for the press was held some weeks ago and the first transatlantic television transmission is claimed by Baird's representative in New York.

Possibilities of Still Picture Broadcasting

TELEVISION, or radio vision, naturally has a much greater appeal to the public imagination than the transmission of still pictures. Because of the tremendous number of images which must be broadcast at an extremely high rate of speed, the perfection of radio vision is a problem a thousandfold more complex than radio photography. It is unlikely that television will ever be possible in the present broadcasting band, since it requires a number of channels used simultaneously to transmit all the necessary images. In its present development, the subject of television transmission must stand within a few inches of the scanning device and, as a consequence, only the bust of a single individual can be broadcast. Any rapid motion is blurred. Under the circumstances, present-day television has few, if any, advantages over the transmission and reception of still photographs.

All of the defects in television will, of course, be remedied gradually and the present state of the art is sufficiently advanced to make clear and entertaining radio moving pictures a prospect of the next few years. In the meanwhile, does radio picture reception offer sufficient fascination to promise rapid extension in the homes of broadcast listeners?

The existing systems of radio picture transmission and reception have many practical advantages. The pictures may be radiated from an ordinary broadcasting station and received with the aid of any good broadcast receiver. High-grade reproduction of pictures in the home is quite possible with apparatus now available. Broadcasting of pictures by remote control, with the aid of wire lines, is feasible with all existing systems. A news photographer can take a picture at a broadcasting studio, or at a remote point connected with the studio by wire, and put it on the air within two or three minutes. Radio picture broadcasting serves the same useful purpose that illustrations do in a book, magazine, or newspaper and enhances a radio program to an equal degree. The broadcasting of sporting and news pictures, photographs of prominent artists, and technical diagrams and data accompanying educational lectures is entirely possible and adequately useful or entertaining. Whatever the camera records, so long as the subject is not entirely too fine in detail or lacking in contrast, is suitable material for radio picture transmission.

It is not unreasonable to expect, in radio picture broadcasting, a new field destined to rapid growth. Those who have the foresight to participate in early experiments will reap their reward in the manufacture, marketing, and servicing of picture sets.

Foreseeing these possibilities, RADIO BROADCAST has been diligent in presenting all available information. So far, constructional information has been limited to the Cooley system but, as rapidly as information regarding other systems is available, we will present it in these pages, so that our readers may keep abreast of progress in the art.

The Shrinking Short-Wave Spectrum

FORESEEING the international complications which would result from the indiscriminate assignment of the higher frequencies to the numerous applicants therefor, we have persistently urged conservation of these frequencies.

Not until the hearings in Washington, however, did we realize that the Federal Radio Commission's problem in the short-wave spectrum is already more complex than that existing in the broadcasting field. Competent engineering authorities, with a long background of experience in short-wave transmission, pointed out that a short-wave radio telegraph station requires a channel having a width of 0.2 of one per cent. of the assigned frequency. Thus, immediately below the broadcasting band, a radio telegraph channel must be 3 kc. wide, while, at 30,000 kc., or ten meters, frequency variation is so great that the channel must be 60 kc. wide.

Assuming this separation, or channel width, to be necessary, there are only 1316 channels between 1500 and 30,000 kc. A part of these has been assigned to mobile, amateur, and broadcasting purposes, leaving only 666 channels available to the entire world for point to point short-wave communication. Inasmuch as, even with the most insignificant powers, a radio telegraph transmitter, assigned to these wavelengths, is likely to cause interference in all parts of the world, duplication of station assignments, for operating continuously is quite impossible.

Furthermore, to maintain continuous service over long distances, the varying transmission qualities of these frequencies at different hours of the day and night make it necessary to use two, three, and four channels. Consequently, instead of a host of channels, sufficient to meet the needs of all communication interests, the Army and the Navy, and the considerable

number of private concerns which desire channels, there is only a very limited number of channels to be divided among the numerous applicants. Assuming that the United States is entitled to twenty per cent. of the channels available to all the nations of the world, there are only approximately 126 channels to be considered after discounting those assigned for broadcasting, amateur, and experimental purposes.

The Radio Corporation of America has already established a number of short-wave, transoceanic services. It desires to extend these services greatly and, were all its prospective requirements considered exclusively, there would be no room on the air for any but Radio Corporation stations and the stations of the Army and the Navy. The Mackay interests propose to enter the radio telegraph field and have made demands for channels in numbers sufficient to absorb any reasonable allotment to the needs of the United States.

How the demands of brokerage houses, department stores, newspapers, oil companies, railroads, bus transportation services, and the thousand-and-one other interests can be met with this meager allotment of frequencies is not apparent until considerable technical progress is made in maintaining frequency stability. As soon as we learn how to hold stations within a few hundred cycles of their assigned frequencies, the capacity of the short-wave bands will be increased a hundredfold.

The Commission has announced that it will require two months of study before it can make any decisions. It is hoped that it

THE PRESENT AND PROPOSED MACKAY RADIO SYSTEM

Recently the Mackay Company entered the radio field and took over the Pacific Coast stations of the Federal Telegraph Company. These stations are at, Hillsboro, Oregon, and Clearwater, California. The points shown on the outline map are the basis of a short-wave system to furnish communication which will supplement the present Postal Telegraph wire system. The proposed channels will connect points most subject to storms and other interruptions to wire service. In addition to this land radio system, the old Sayville station on Long Island has been purchased and will shortly be opened for marine communication with ships on the Atlantic. The Mackay system now holds 42 wavelengths in the short-wave spectrum, of which 34 are for a chain of stations for transpacific communication

will establish a definite policy and stick to it. In the broadcasting spectrum, expediency, rather than an established set of principles, has ruled. In view of the complexity of the short-wave problem, it is most urgent that definite principles be enunciated, lest the Commission be later charged with favoritism or discrimination.

The first obvious principle is that there is no short-wave channel available for any service which can be conducted by wire. The Mackay interests, as one of their plans desire to establish an emergency network so that, should wires break down, they may use short-wave radio transmission. There is every reason why a duplicate emergency radio service should be available in national emergencies but, should such a system be established, permission to utilize the radio network should be restricted to grave emergencies.

The requirement that wire services be used where available is complicated by certain economic factors. A chain of department stores, for example, can erect short-wave transmitters and maintain communication at a much lower cost than entailed in using public service wire telegraph channels. The application of this principle requires, therefore, discrimination against private interests in favor of the wire companies. In view of the needs of transoceanic services and the value of an independent American communication service, this discrimination against private interests appears entirely necessary. Furthermore, it is utterly impossible to serve all private interests with the limited facilities available. Only a selected few could be accommodated, were short-wave radio telegraph channels assigned to private interests.

This suggests a second principle which may be established as a policy of the Commission. No short-wave channels shall be assigned for any service unless such service be opened to the general public upon an equitable basis of charges without discrimination.

It is unfortunate that the wire interests, which are so well entrenched and established in the United States, should seek to enter the radio field and compete with existing radio communication companies and that radio communication interests, on the other hand, should seek to compete with the wire interests. For example, application was made for the right to conduct a New York to Montreal service by a radio communications company. These cities are already well linked by wire telegraphy and telephony. Possibly there might be some simplification of the situation if the radio companies decide not to establish radio communication networks where wire services exist and the wire companies, in turn, stay out of the radio field and stick to their well established and profitable wire business.

The stifling of competition is clearly un-American in principle and carrying out this suggestion would obviously tend to stifle competition. But what alternative exists?

We do not grant street railway franchises for routes along the same avenue to two or three rival companies in order to establish competition. If radio channels are limited, they must be regarded as a franchise and be distributed in such a manner as to assure the greatest possible public service.

A further embarrassment in the situation exists in the fact that long-distance radio communication is dominated by a single interest. Whatever the means used to gain this acknowledged ascendancy, the fact remains that it exists. In almost every important line of industry there is a dominant company which has established its acknowledged leadership in the field. This company is always the object of vilification by politicians but, because of sheer strength and competence, merrily goes on occupying its position. Under the circumstances, the Federal Radio Commission is bound to be criticized as the tool of monopoly and its activities will always offer subject matter for spellbinding politicians because the dominating company, if fairly treated, will have a preponderance of assigned channels. For this situation, there is no practical and fair remedy other than the uneconomic proposal of government ownership.

The demands of the press for short-wave channels, although clothed in high-sounding phrases regarding public service and freedom of the press, are really an effort to reduce its wire costs. Signals travel through the ether no faster than they do over wire circuits. The demands of the press for special radio channels are in the same category as all other requests of private interests, excepting in those few instances that wire services are not available. The press already receives preferred consideration from the wire services which carry most of its communications at or below cost.

The Commission Retreats

THE Federal Radio Commission has issued new application blanks, requiring the submission of considerable information on the part of broadcasting stations concerning their technical equipment, their program hours, the proportion of time devoted to commercial broadcasting and facts regarding chain affiliations. The Commission should have gathered this information immediately on its accession, to authority last March.

Congressman Wallace H. White has expressed again and in greater detail his disappointment in the functioning of the Federal Radio Commission. He has urged, as we have in these columns for many months, that the future appointees be possessed of sufficient technical qualifications so that they may perform their duties efficiently.

The members of the Commission have been testifying before the House Merchant Marine Committee and they have hardly had a pleasant time of it. Judge Sykes admitted that a much larger proportion of channels had been assigned to New York and Chicago than those cities deserved, an abuse which we have frequently stressed in these columns for more than a year. The favorable position of chain stations which, the testimony brought out, occupy twenty-one

out of the twenty-five cleared channels, does not please members of the Congressional committee.

The saddest news which we have heard from the Commission so far is the announcement by Acting Chairman Sykes that the 300 stations which were to be scheduled for elimination on March 1 will have their licenses extended. This move, he says, is made because three of the four members of the Commission are not confirmed by the Senate. Naturally, each of these three hundred stations is the pet of some congressman or senator and the unconfirmed members of the Commission could not hope for a confirmation if they took the necessary and drastic action of eliminating such stations. Politics now rule radio, confirming our predictions made when a commission control of radio was first proposed before the Radio Act of 1927 was passed.

Representative White has presented a bill extending the powers of the Commission for another year and including, in addition, some provisions aimed at the Radio Corporation of America and the National Broadcasting Company. An amendment to Section 10 of the Radio Act, which he proposes, is to permit the Commission to refuse a license to a station intended for international commercial communication, if the company operating that station has entered, or intends to enter, into exclusive rights with a foreign country. We understand that the Radio Corporation has made several such agreements. The Commission, however, may grant such a license if the company will maintain just and reasonable rates and will secure and maintain equality of right and opportunity for other American citizens to engage in competitive services.

Other proposed changes in the Act are the strengthening of the powers of the Commission to revoke licenses for false statements in applications, or for failure to observe any of the terms, restrictions, and conditions of the Act or regulations issued by the licensing authority. To the powers of the Commission, Representative White proposes to add that it may fix the hours during which chain broadcasting may be carried on, designate the stations and limit the numbers participating therein, and that it may prohibit commercial broadcasting through chain stations and, in fact, may make any rules and regulations in the public interest, applying to chain broadcasting.

The crowning touch of Representative White's bill is a proposed provision to be included in the Radio Act that it shall be unlawful for any firm to import or ship in interstate commerce any vacuum tube, whether patented or unpatented, which shall have any restrictions of the use to which such tube may be put or which shall have the effect of fixing the price at which the tube may be sold. It is doubtful whether this unusual curtailment of patent rights of a single group is constitutional. This proposal is really an amendment to the patent law and is not properly a part of the act regulating radio communication.

If the fundamental theory of our patent law must be changed, why does not Congress undertake its thorough study and pass a new patent law instead of singling out vacuum-tube manufacture as a special case? High-grade vacuum tubes, suited to all purposes, are available to the public and sold without exorbitant profit. If the R. C. A. is indulging in unfair practices, there are adequate measures which can be taken without depriving it of the benefits of normal patent protection. If the patent monopoly, established by patent law, confers too great powers on the patent holder, then the patent law should be modified. A possibility worth considering is the compulsory issuance of licenses upon an equitable basis to all who apply, thereby assuring patent holders of adequate reward, but preventing the use of patents to establish monopolies or embarrass competition. The principle of singling out a particular patent situation for special legislation is contrary to the principle of equal rights to all.

The Blue Laws of Radio

FAIRFIELD, IOWA, has passed an ordinance prohibiting the use of electrical equipment which causes interference with radio reception between noon and midnight. Interpreted literally, not only do vacuum cleaners, washing machines, electric toasters and flatirons fall within the ban, but also electrical incinerators, elevators, refrigerators, cash registers, fire alarm signals, and railroad block signals. Violations are punishable by a fine of one hundred dollars or thirty days imprisonment or both.

There is nothing reasonable about this ordinance. It is doubtful whether this practical confiscation of property is legal. Electrical interference problems should be solved by aiming at the cause rather than the effect. A modern cash register, for example, should be designed so that it cannot cause electrical interference, although even the best of them do so. Any electrical device can be equipped with suitable interference preventers which will eliminate the possibility of interference with radio reception.

In Providence, Rhode Island, they have invoked an old blue law to embarrass a radio dealer. The ordinance prohibits any person to ring a bell or to use any other instrument or means for the purpose of giving notice of any public sale or auction of any article. This has been interpreted by the police to embrace the use of radio loud speakers in stores, although it is doubtful whether the writers of the blue laws had radio in mind. To be really fair in the matter, the police ought to be prohibited from using whistles.

Here and There

AN APPEAL has been filed by the Westinghouse Electric and Manufacturing Company for a review of the decision of the Circuit Court of Appeals which upheld DeForest against Armstrong in the invention of the feedback circuit. ⁂ THE HAZELTINE Corporation has filed against the Charles Freshman Company and the Stewart-Warner Speedometer Corporation in the Southern District, citing U. S. Patent No. 1,648,808. ⁂ THE ORIGINAL decision of the examiner, rejecting claims 1 to 6 inclusive, 8, 9, 11, 12, 15, 19, 21, and 22 as unpatentable over prior art, was affirmed in the case of re-issue patent 16834, issued to Lloyd N. Knoll on December 27, 1927. The references cited were Bellini, Kolster, and a scientific paper issued by the Bureau of Standards. ⁂ A SUCCESSFUL appeal from the decision of the primary examiner, denying the patentability of several claims of patent 1,654,285, issued to Charles Fortescu, describing a modulation system for quickly absorbing residual energy stored in the antenna system after signal impress has been completed, has been announced by the Board of Appeals. ⁂ CORNELIUS D. EHRET, counsel for the applicant, Frederick A. Kolster, in Patent 1,697,615, called our attention to the fact that our item in the December issue, citing the substance of the opinion of Second Assistant Commissioner of Patents M. J. Moore, implies that the claims referred to were rejected after the patent was issued. While we fail to see how

this conclusion was reached from our item, we are pleased to state, in deference to Mr. Ehret's wishes, that no claims were declared invalid after the patent was issued.

THE FIELD FOR GOOD RADIO SETS

ARTHUR SMITH, radio dealer of Tampa, Florida, sends us a most lucid letter explaining the position of the radio dealer in locations where high-grade local broadcasting is not available. He complains that radio manufacturers concentrate their advertising almost exclusively upon the cheaper models and fail to point out the advantages to the user of the high-grade, super-power, radio receivers, really necessary in such areas. As a consequence, the dealer is compelled to demonstrate the cheaper type of receivers which do not give satisfactory results when remote from good broadcasting. He states as his opinion that less than one per cent. of the listeners in his area are utilizing receiving sets with 210 power output tubes and that most of them are still using the cheapest type of set which has been so forcibly heralded in the advertising columns. He urges that set manufacturers devote more advertising space to high-quality radio sets because the public, once appreciating their capabilities, is quite willing to spend the necessary money for better models.

E. T. CUNNINGHAM announces the introduction of the CX-371-A tube which has the characteristic of the 371 and 171 type except that it has an oxide-coated filament. This reduces the filament current required to a quarter of an ampere, effecting an economy of filament current. The oxide coated filament also gives uniform emission throughout its life instead of gradually falling emission which is characteristic of the thoriated filament.

JUDGE HUGH BOYCE in the Federal Court at Wilmington, Delaware, dismissed the suit of the General Electric Company, charging the DeForest Radio Company with infringement of Langmuir's high-vacuum tube patent. If this decision is sustained in higher courts, to which it will undoubtedly be appealed, one of the most important reliances of the R. C. A. in its hold on the tube situation is destroyed.

THE Federal Trade Commission proposes to broaden the scope of its investigation of the radio combination, amending its formal complaint against the radio group by adding the following charge:

The defendants have: "8. Substantially lessened competition and tended to create a monopoly in the sale in commerce of unpatented parts of chassis, and of unpatented consoles and cabinets, and of other unpatented parts of radio devices," etc. etc.

THE Mackay interests have acquired the famous transatlantic station at Sayville, Long Island, which they will use in ship-to-shore service. Another station for the same purpose is planned somewhere near Norfolk. The system is already operating stations in San Francisco, Los Angeles, and Portland on the Pacific Coast and expects by next summer to have transpacific service to Honolulu and the far east.

THE National Better Business Bureau collected advertising literature at the Radio Show in New York last September and analyzed the inaccuracies and violations of their standard code of radio ethics and advertising practice. They found 232 inaccuracies, of which 39 per cent. were violations of their Rule 4 which calls for the accurate naming of cabinet woods. Twenty-six per cent. violated Rule 2-B, which provides that price quotations state clearly whether the offer includes accessories or not. Sixteen per cent. disregarded Rule 8 which holds that superlative claims lack selling force. The fact that only 58 complaints were investigated during the year 1927, as against 123 in 1925, is taken as an indication of the coöperation which has been extended by the radio industry in the work of the Bureau.

THE International Radio Telegraph Conference has adopted a new schedule of "Q" signals and, in addition, has recognized a number of one-, two- and three-letter combinations, some of which have been widely used but did not heretofore have the stamp of official approval. Prominent among these is the adoption of CQ as the general calling signal, replacing the three-letter combination QST. Some changes were made in the assignments of alphabetical groupings to the various nations from which each assigns call letters to the radio stations under its jurisdiction. The United States, hereafter, will have the entire letter K combinations at its disposal, instead of only a part, as well as the entire range of N and W calls.

AN EARLY FESSENDEN RADIO RECEIVER

—*Courtesy I. V. L. Hogan*

It was "wireless" in those days, however. The antenna and ground circuits enter on the left, through the variable air condensers; the four drums, wound with wire, and each with its handle, illustrate the ingenious method of continuously varying the inductance of the closed circuit; the crystal detector and headset are on the extreme right

THE TRANSMITTER IN ITS FINAL FORM

By the mere throwing of a switch, it may be used for either c.w. or phone signals. It has covered a thousand miles with phone signals during its tests, although this figure is somewhat high to expect for regular work

A Short-Wave Phone and C. W. Transmitter
By Kendall Clough

WITH appetites whetted by the remarkable expanse of their eavesdropping, consummated with the simplest of equipment, the broadcast listener who became interested in short-wave reception, satisfied his curiosity, and listened half way round the globe, is now ready for new fields to conquer. The logical outlet for his enthusiasm lies in the construction of a transmitter, for it is provoking to hear a fellow a thousand miles away pounding out a crystal-clear message, terminating such with a remark to the effect that "I'm using a single 201–A, OM," and not to be able to answer him back, and report a better "watts-per-mile" record.

To supply the demand created by this growing enthusiasm, several well-known parts manufacturers have coöperated in the design of a short-wave radio telephone transmitter which can be built for about the same cost and with the same ease as a good receiver constructed for the reception of broadcasting. While this design is of a low-power type transmitter, it has been repeatedly demonstrated that the power is sufficient to carry on conversations over surprising distances. It will be noted from the photographs that the manner of construction permits the isolation of all the parts carrying radio-frequency currents on the upper "deck," or shelf, while all the circuits associated with the power or voice currents are on the lower "deck." This form of construction insures that the masses of metal contained in power devices, such as transformers, condensers, etc., will not be in the fields of any of the radio-frequency coils, since this would introduce losses therein. Corresponding to the arrangement of the circuits in decks, the front controls are also grouped. Thus, on the upper panel we have the controls for the tuning condensers, the antenna-current meter, and the plate-current meter for

the oscillator tube. On the lower panel we have the plate-current meter for the modulator tubes, the modulator C bias control, the switch for changing from telegraphy to telephony, and the necessary binding posts for the key, microphone, and battery. The circuit diagram of the transmitter is shown in Fig. 1. That portion of the circuit shown on the upper "deck" is the justly famous tuned-grid, tuned-plate circuit, which has been in use for several years. This circuit employs a series feed for the plate voltage to the oscillator rather than the shunt feed ordinarily used. In this way an already efficient oscillator circuit has been improved by eliminating those losses in the choke coil that are bound to occur when using shunt feed. Naturally enough any losses eliminated in the choke coil result in additional energy being available for actual transmitting purposes.

Considerable voltage is developed between the plate coil and the ground by this method, so that it is necessary to use two condensers in series (C_2 in Fig. 1) as an r.f. bypass. An r.f. choke, L_4, serves to keep the radio-frequency currents from finding their way down to the lower "deck."

The power supply consists of a Silver-Marshall 328 transformer, T_1, which supplies the plate current as well as lights the filaments of the oscillator and modulator tubes. In order to secure direct current, which is necessary for phone operation, the high voltage, of T_1 is rectified by means of a gas tube, and filtered with a Silver-Marshall 331 Unichoke, L_5, and Tobe condensers, C_6, and C_7. It will be noted that two modulator tubes, V_2 and V_3, are used in conjunction with one oscillator, V_1, of the same type. This is in accordance with the best practice and while the set is perfectly operable with only one modulator, it is recommended that two be used where the

best quality of transmission is desired. The modulator tubes and the oscillator tube are all of the cx-310 (ux-210) type.

In order to operate the modulator tubes at maximum capacity, it is necessary to amplify the output of the microphone transformer to bring the speech to the proper volume level. This amplification is accomplished by means of a cx-312 (ux-112) tube, V_4, and a Silver-Marshall 240 transformer, T_2. The proper C and B voltages for the cx-312 are secured from the power supply by means of the resistors R_4, R_5, and R_6, the latter supplying the C voltage for this tube. During the preliminary experimental work a cx-326 (ux-226) was used in place of the cx-312 and was lighted from the power transformer. In view of the fact that a 6-volt battery was necessary for the operation of the microphone, however, the same battery was finally used to light a cx-312 instead of the a. c. tube, since a quieter signal from the transmitter resulted.

DETAILS OF ASSEMBLY

THE construction of the transmitter may well commence with the lower "deck" as this unit must be assembled and wired before the work is started on the upper structure, otherwise the latter will hinder the wiring. The lower deck is shown separately in an accompanying illustration. The board for this assembly is screwed to the cleats below, and the screw holes for the equipment are located with the aid of the full-size template supplied with the foundation unit specified in the list of parts.

After screwing down the parts for the lower "deck" and wiring them in accordance with the circuit diagram, the lower front panel and the equipment on it should be screwed in place as shown, after which the wiring of the lower "deck" may be completed. This unit may be tested separately before proceeding with the work. In order to do this the unit is connected, as it would be in operation, with the microphone, storage battery, etc., and the switch on the panel is thrown to the "Phone" position. Most of the resistance, R_2, should be in circuit. The tubes should all light properly and the needle of the modulation meter, M_2, should jump up when the microphone is spoken into. Now, to check the quality, a loud speaker should be connected across the modulation choke, L_6, by means of a long cord leading into another room. It may be necessary to shut the door between the rooms in order to keep the loud speaker from transmitting acoustical energy to the microphone and setting up a continuous howling noise. When the equipment on the lower deck is operating properly, the microphone speech input as heard by another observer at the loud speaker, should be very clear and distinct. The resistance, R_4, should be adjusted during the test until speech is at its clearest point, at which time the modulation meter will indicate from 20 to 30 milliamperes.

The equipment on the upper "deck" should now be assembled from the template and diagram in the same manner as the lower "deck" was, after which the w_3 ove frame (supplied with the foundation kit) may be put together with wood screws. The upper panel, with its equipment, should be attached last. The wiring is next completed in accordance with Fig. 1.

THE LOWER DECK ARRANGEMENT

The circuits associated with the power or voice currents are located on this deck. The two-deck arrangement prevents losses which would otherwise occur

LIST OF PARTS

M_1 Weston Model 425 Thermoammeter, 0-1 Amp.	$13.50
M_2 Weston Model 301 Milliammeter, 0-100 Mils.	8.00
M_3 Weston Model 301 Milliammeter, 0-50 Mils.	8.00
L_1, L_2, L_3 Aero Short-Wave Transmitting Coil Kit (3020K, 4080K, or 9018K)	12.00
L_4 Aero 24R Radio-Frequency Choke (Included with Above)	—
L_5, L_6 S-M 331 Unichokes	16.00
C_1 Cardwell 0.0002-Mfd. Condensers	13.00
C_2 Polymet 0.002-Mfd. Moulded Condensers	.80
C_3 Polymet 0.0005-Mfd. Moulded Condensers	.70
C_4 Polymet 0.00025-Mfd. Moulded Condenser	.55
C_5 Tobe 2-Mfd. 300-Volt Condenser	2.50
C_6 Tobe 2-Mfd. 1000-Volt Condenser	2.50
C_7 Tobe 4-Mfd. 1000-Volt Condenser	6.00
C_8 Tobe 1-Mfd. Condenser	.80
R_1 Aero 810-Type 10-Ohm Resistors	.60
R_2 Polymet 10,000-Ohm 10-Watt Resistor	.75
R_3 Yaxley 2000-Ohm Potentiometer, No. 2000	1.75
R_4 Polymet 25,000-Ohm 10-Watt Resistor	1.25
R_5 Polymet 15,000-Ohm 10-Watt Resistor	1.00
R_6 Polymet 750-Ohm 10-Watt Resistor	.75
R_7 Yaxley 1.5-Ohm Resistor, No. 2L	.15
R_8 Yaxley 100-Ohm Resistance, No. 8100	.25
T_1 S-M No. 328 Transformer	18.00
T_2 S-M No. 240 Transformer	6.00
T_3 S-M No. 242 Transformer	7.00
S_1, S_2, S_3 Yaxley 3-Pole Switch, No. 65	1.60
Five S-M No. 511 Tube Sockets	2.50
Aero Transmitter Foundation Unit	27.00
(Consists of drilled and engraved Westinghouse Micarta upper front panel, 7 x 18 x 4 inches, drilled and engraved lower front panel, 5 x 18 x 4 inches, seasoned walnut lacquered frame kit cut to size for making a stand 16 x 18 x 10¼ inches, with all screw holes drilled that are necessary to put framework together, wiring diagram, and layout sheet.)	
No. 159 Frost Desk Microphone	8.75
	TOTAL $164.50

ACCESSORIES

V_1, V_2, V_3 CX-310 (UX-210) Tubes	$ 27.00
V_4 CX-312 (UX-112) Tube	3.50
V_5 Manhattan No. 2721 Gas Rectifier	7.00
Transmitting Key	1.50
Three Four-Inch Bakelite Dials	1.50
Six Binding Posts	.90
	TOTAL $41.40

Since the accompanying photographs were taken, a "key click" filter, consisting of a 1-mfd. Tobe condenser, C_8, in series with a 100-ohm Yaxley resistor, R_6, has been connected across the key terminals as shown in the circuit diagram. Space is available for these items on the baseboard just behind the key binding posts. Its use will be appreciated by near-by broadcast listeners, who otherwise would hear the key clicks in their receivers.

The set should be tested to insure that it oscillates properly. The plug is inserted in the 110-volt 60-cycle light socket and the switch is thrown to the c.w. side. This should leave only the oscillator tube lighted and, on shorting the "key" binding posts, current will be indicated in the plate meter. Probably this current will cause almost a full-scale deflection but by varying the plate or grid condensers it will snap back to 20

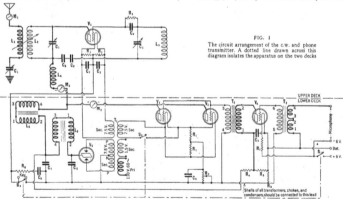

FIG. 1

The circuit arrangement of the c.w. and phone transmitter. A dotted line drawn across this diagram isolates the apparatus on the two decks

or 30 milliamperes at a certain point, indicating
that the tube is oscillating. On connecting an
antenna and ground, and tuning the antenna
condenser, this reading will be increased when
resonance is obtained, and at the same time
some antenna current will be noted. If the coupling is too close the tuning of the antenna will
tend to throw the tube out of oscillation and it
will be necessary to loosen somewhat the coupling
between the hinged primary coil and the plate
coil (L_3 and L_4).

Final tuning should always be done with a
wavemeter in order that transmission may be
within one of the bands licensed by the Government. One of the features of the transmitter is,
however, that due to the interchangeable coil
feature, the set may be tuned to any wave between 18 and 180 meters so that it is not rendered
unserviceable by any slight changes in wavelengths granted by the Government. The transmitter may not be used, of course, unless the
operator has a license which permits him to do so.

Space does not permit us to go into the antenna construction, operating methods, etc., at
this time, and the reader is referred to *The Radio
Amateurs' Handbook*, published by the American
Radio Relay League, Hartford, Connecticut, for
excellent information along this line. This transmitter is now in operation at 2 GY, the RADIO
BROADCAST station at Garden City, New York,
and there is also a similar one now working at
the Aero Products station located at Chicago
and the results that have been obtained in a
limited time are very gratifying. With c.w., on
the 40-meter band, all U. S. districts have been
worked from Chicago as well as NC 5zz in Vancouver, British Columbia. Twenty-meter phone
work has been unusually successful. The following stations have been worked on 20-meter
phone with reports varying from R–5 to R–7:
1 BBM, Harwich, Massachusetts; 1 ASF, Medford,
Massachusetts.; 1 SW, Andover, Massachusetts;
2 BSC. Glen Head, New York; 3 AKS, Phila-

A PHOTOGRAPH OF THE UPPER DECK

All the equipment carrying radio-frequency currents is mounted on this
deck. Fig. 1 will clearly show just what equipment is placed on this deck

delphia; 4 MI, Asheville, North Carolina; and
8 CVJ, Auburn, New York. In all cases where
the transmission has been on phone, the quality
of the speech has been reported to be very fine.
Even greater distances have been worked on
code with the transmitter located at Garden
City.

FOR CODE WORK ONLY

MANY amateurs are interested in c.
transmission to the exclusion of phone.
such cases the transmitter may be construct
for that purpose only at a substantial saving
parts. The conversion requires simply the om
sion of the parts that are necessary for pho
operation since the transmitter described be
is an ideal c.w. transmitter in itself.

The circuit diagram of the outfit wired f
c.w. only is shown in Fig. 2.

The values of the parts shown in Fig. 2 a
exactly the same as those of the parts in Fig.
The only addition is the inclusion of R_9 in t
second diagram. This is a 50,000-ohm Polym
resistor of 10 watts carrying capacity. It li
at $1.50, and is used to prevent the voltage
the final filter condenser from rising to an uns
value when the key is up.

As it will be noted in the above list of par
there are three distinct sets of Aero coils ava
able for transmitting purposes. The most pop
lar kit is the 4080K, which covers a wavelen
range of 36 to 90 meters (8330 to 3330 kc
The 2040K kit covers the band between 18 a
52 meters (18750 and 5770 kc.). The No. 9018
kit is suitable for the band between 90 and 1
meters (3330 and 1670 kc.).

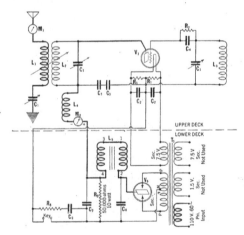

FIG. 2

Here is the transmitter circuit diagram for t
experimenter who wishes only to transm
c.w. signals. The upper and lower deck featu
it will be seen, is still maintained

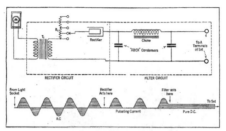

FIG. 1: THE CIRCUIT OF THE "ABOX" UNIT

A SECTIONAL VIEW OF THE "ABOX"
Its circuit arrangement is given in Fig. 1

Electrification Without A. C. Tubes

By Lewis B. Hagerman

"JUST push the plug into the light socket," is the answer most radio set users would like to give to the query: "How do you turn on your radio receiver?" In this classification, so many believe, are included only those receivers using a.c. tubes, and so when they go out to buy an "electric" receiver, they examine it to be sure that it uses a.c. tubes. Also, when they consider converting their battery-operated sets for a.c. operation the problem to most of them becomes one of adapting the set to use a c. tubes. It is possible, however, to electrify a receiver in another way which is frequently much easier to accomplish and generally just as satisfactory. We refer to the use of accessories in conjunction with a receiver originally intended for battery operation so that the equivalent of light socket operation is obtained without the substitution of new tubes.

Electrically there is practically no difference in the operation of a receiver from a.c. tubes or from storage battery type tubes in conjunction with an external A power unit connected to an a.c. source. With a.c. tubes we supply a.c. power either directly or indirectly to the electron emitting surface which then emits electrons. With d.c. tubes we supply a.c. power first of all to a rectifier which in turn supplies power to the filaments, and these become hot and their surfaces then emit electrons. In neither case does the current in the filament enter directly into the operation of the tube; it is merely the agent which causes the electron emitting surface to become hot. Socket power operation is a means of eliminating the problems associated with the *storing* of electric power for the operation of the receiver, such as by means of a storage battery, and any method which enables us to do this implies direct operation of the receiver from the power mains. If you want to electrify your receiver, you can do it by using a good B power unit and a reliable A power unit, such as the "Abox."

Many of RADIO BROADCAST's readers are at present obtaining plate voltage for the operation of their receivers from a B power unit and, therefore, a socket power A unit will complete the electrification of the receiver. When the plus and minus terminals on the "Abox" unit are connected to the corresponding A terminals on the receiver and the power lead is plugged into the light socket, there will be available, from the "Abox" unit, a source of filament current, and from the B power unit, a source of plate

voltage, both obtained directly from the light socket.

Electrically, the problems associated with the design of a satisfactory A power unit are similar to those connected with the design of a B power unit. In both cases the problem is to take alternating current power from the light socket and rectify and filter it so that it will be satisfactory for the operation of the receiver. The problem in the design of an A power unit is that it must deliver large amounts of current, which necessitates a great difference in the values of the constants incorporated in a proper rectifier and filter unit, as opposed to those of a B power device.

An A supply unit must deliver at least two amperes to be universally adaptable to most receivers and this value of current is approximately one hundred times the output in amperes of a low power B device. As the current to be handled increases, the capacity of the input condensers must also be increased in direct proportion, which will be about 100 times, and it is only recently that large capacity condensers of reasonably small physical dimensions have been commercially available at a low price.

Then we have the voltage factor. A given condenser stores more power the greater the voltage; at one hundred and fifty volts, therefore, it will store much more energy than at six volts. To compensate this, the capacity of the filter condenser must be increased in proportion to the difference in voltage, or another twenty-five times. Since the A device delivers current to the filament circuit, any hum will tend to effect the grid bias and be amplified by the tube. The capacity of the filter condensers must be increased about seven times to offset this effect.

Thus it will readily be understood that the filter condenser of an A power-supply device must be one hundred times twenty-five times seven times, or 17,500 times, as great as that used in a B device. The capacity of a B filter condenser is about 4 microfarads; the capacity required for an A power unit is, therefore, in the neighborhood of 70,000 microfarads.

To obtain this the "Abox" Company developed a condenser consisting of a number of nickel and iron plates immersed in a caustic potassium solution, which is not an acid. This solution causes thin films of oxygen and hydrogen

to form on the surface of the plates, and these films constitute the dielectric of the condenser. The caustic solution is one side of the condenser while the plates form the other.

Since the capacity of a condenser increases as the thickness and amount of dielectric decreases, this infinitesimally thin gas film is responsible in part for the tremendous capacity obtained.

This film has several advantageous features. Should an excess voltage be impressed on the condenser, the film immediately breaks down and bypasses the excess energy. When the output returns to normal, the film forms again, and the condenser is as good as new. The bugaboo of burnt-out condensers is thereby done away with.

The capacity of the condenser is far in excess of that required; it has been estimated that its capacity is in excess of 200,000 microfarads. When used with the "Abox" rectifier, it reduces the alternating component of the input pulsating d.c. to less than $\frac{1}{100}$th of its original value.

Both the rectifier and condenser work in the same solution. The addition of distilled water every six months or so is the only maintenance needed. The condenser plates are never affected by use or disuse, and the rectifier electrode has a life of several years and can be replaced in a few seconds at a very low cost. The tapped resistance, R_1, which compensates for the number of tubes used is adjustable from the front of the unit.

Fig. 1 shows the circuit of the complete eliminator. The alternating current from the house lighting circuit is stepped-down from 110 volts to the proper low voltage by the transformer, T_1. The current then flows through the rectifying valve which will pass current in one direction only, thereby eliminating one phase of the alternating current and creating a pulsating direct current. It is next passed through the filter circuit, where it is smoothed, and all variations and pulsations in current are removed. The drawings at the lower part of this diagram show the effect of the rectifier and filter on the alternating current. The rectifier changes the alternating current to pulsating direct current by eliminating one phase of the a.c. wave, and the filter then smooths out the pulsating current, producing practically pure direct current. When this A power unit is used in connection with the average receiver, it will not cause any hum.

413

A "HUDDLE" IN ONE OF THE VICTOR TALKING MACHINE COMPANYS STUDIOS FOR RECORDING, OLD STYLE

How Radio Developments Have Improved

THE electromagnetic phonograph reproducer, also often simply called the "unit" or "pick-up," is acquiring great popularity nowadays on account of the tie-up it creates between the radio receiver and the phonograph. When static is bad, or when radio programs are not to one's taste, it becomes a simple matter to change over to the phonograph and enjoy the best or the worst in musical art, according to the choice of the person who purchases the records.

On the other hand, since most of us are able to afford only a limited number of phonograph records, and must play them many more times than once in order to realize on our investment, such repetition occasionally palls, as it were, and we turn to the "air" to supply us with programs new to our ears.

Although there are a great many "pick-ups" already on the market, and more are coming on every day, their commercial exploitation is relatively new. The development of a new device requires the simultaneous development of a technic particularly suitable to it: At this early stage of the development, it is not to be expected that all those who design reproducers know everything about them, and it must also be remembered that many of those who are working on the problem are radio engineers, and are not versed in the phonographic art.

On the other hand, although the electric reproducer is new to the radio public, it is by no means new to engineers. The writer remembers a demonstration of a piezo-electric reproducer which he witnessed in New York as far back as 1921, and the engineer who developed this reproducer had been working on it for a period of several years before. Electromagnetic pick-ups are likewise fairly old in the art, as also is the capacity type of pick-up, but the advent of these devices for practical and commercial application had to await the development of suitable amplifiers and loud speakers.

There are quite a few phases of the art to consider. These may be listed as follows:

Recording
  Method of sound pick-up (horn, microphones of various types).
  Method of actuating cutting head.
  Amplification
Reproducing
  Type of pick-up (capacity, piezo-electric, carbon, electromagnetic).
  Amplification.
  Conversion into sound (type of loudspeaker)

This list outlines the complete process from beginning to end, which we will describe briefly in the next few paragraphs.

At the recording studio we have a band, orchestra, singers, or other artists furnishing the original music. The sound waves of this music are collected by a horn, in the old "air-line" method of recording, or by a microphone in the newer system of "electrical" recording.

In the "air-line" method, the sound waves, entering the collector horn, were concentrated in it, so that sufficient energy could be obtained for actuating a diaphragm, to which was rigidly fastened a "cutting-head." Under this cutting-head traveled the wax disk known as the "matrix," on which the cutter engraved waves corresponding to the sounds entering the horn. Naturally, the power available for driving or actuating the diaphragm which carried the cutter was limited to that which could be collected from the original sounds in the studio.

In order to obtain sufficient power for cutting the record, it was necessary to use a resonant diaphragm, so that at the outset we have two inherent difficulties in the air-line system of recording; in the first place the horn which collected or concentrated the sound waves was a cause of distortion, due to its "resonance" at various frequencies, and, secondly, the same was true of the diaphragm, which was made resonant in order to operate the cutter satisfactorily.

These difficulties are avoided in the electrical system of recording, in which the sound waves operate directly on the diaphragm of a micro-

phone. The energy pick-up of this microphone is, of course, exceedingly small—much smaller than that picked up by the collector horn in the old system—but the advantage lies in the fact that since the microphone converts the energy of the sound waves into corresponding waves of electric current, it is possible to amplify them to any degree we might desire. On this account distortion need not be permitted at the outset, i. e., as in a horn or resonant diaphragm. On the other hand, we run into the difficulty of distortion in the amplifier or in the microphone.

This is what we referred to previously when we stated that the development of electrical recording and reproducing had to await the development of the amplifier. To-day we can build amplifiers having negligible distortion, and the microphone, collecting such a small amount of power and having no extended surfaces, is inherently far superior to the collector horn of the old system. The main advantage of the "air-line" system of recording and the old method of reproducing, is simplicity. The new electrical systems show to greatest advantage in the recording, for it must be understood that very good quality is obtainable in reproducing by the old system when slowly expanding exponential horns are used in which the resonances have been reduced and the range of response has been extended to include the lower tones. But good reproduction by the old method requires that the recording be done properly, so it is here that the electrical system is especially valuable.

VOLUME CONTROL

ANOTHER feature of the electrical system which is of great importance is the ability to control the volume of reproduction. The phonograph record is a form of mechanical power amplifier, deriving its power to amplify from the motion of the turntable which carries the record. We can understand how this is by considering the old "air-line" system. The acoustic power con-

414

HOW A VICTOR RECORD IS MADE NOWADAYS. FOOTBALL TACTICS HAVE BEEN FORGOTTEN, AND BETTER RECORDS RESULT

Recording and Reproducing—By Sylvan Harris

centrated in the collecting horn and which actuates the diaphragm to which the cutter is attached is very small. After the record is made and is being run on the turntable under the needle of the pick-up, it is the motion of this needle, caused by the rotation of the disk, which furnishes the sounds which come out of the horn. In other words, the power which drives the disk causes the needle to vibrate in the grooves of the record. The waves themselves, in the grooves of the record, furnish no power. It is only the motion of the record which furnishes the power.

We have a very analogous situation in an amplifier; a voltage is impressed on the grid of the amplifier tube, but this voltage is not power. It is only due to the influence of this voltage on the power furnished by the B supply that an amplified reproduction of it occurs in the plate circuit. The alternating grid voltage is similar to the waves on the record; the power of the B supply is analogous to the power in the mechanism which drives the record.

In spite of the inherent amplification in the phonograph record, this amplification is not always sufficient when there are wide ranges of volume in the original music. It is also difficult to obtain all the volume that one might desire for ordinary purposes without introducing considerable distortion, unless electrical amplification supplements the mechanical amplification of the record. First we must amplify the weak output of the microphone, because the microphone pick-up is so small. Yet we must not amplify too much, for we run into other difficulties of recording, as, for instance, where the cutter cuts through from one groove to the next, or where distortion arises in the cutting apparatus.

This brings us to the next phase of the subject —the cutter. This is a specially ground amethyst set into the end of a rigid rod or bar, which, in the old system, was attached to the middle of the diaphragm at the throat of the collector horn. When the diaphragm was set into vibration by the sound waves in the studio, this bar and the

jewel at its end were likewise set into vibration. The jewel, cutting a groove into a heavy wax disk (the matrix), at the same time cuts waves in either the walls of this groove or at its bottom, depending upon the particular construction of the cutting-head, as it is called.

There are two methods of cutting—the "hill-and-dale" cut and the "lateral" cut. In the hill-and-dale method, the cutter vibrates up and down in the groove, so that there are "hills" and "dales" at the bottom of the groove. This was the original method in the phonographic art, and is hardly ever used nowadays. It has been superseded by "lateral" cutting, in which the cutter is made to vibrate "laterally," or from side to side. The result is that the groove cut in the wax, while circular around the center of the disk, is at the same time wavy. We will not discuss the advantages and disadvantages of the two methods here; they are mentioned merely to acquaint the reader with the general ideas involved.

In either case the vibrating bar of the cutting-head is connected (generally through a system of levers) to an armature (in the electrical system) which is actuated by an electromagnet. This electromagnet obtains its power from the amplifier, the input of which is connected to the microphone. The principle of the cutter is the same as that of a loud speaker, excepting that instead of having a cone for the load on the armature, we have the jewel cutting into the wax.

So, in the electrical system of recording we have first the microphone, which may be any one of several types, next, the amplifier, which may also be any one of several types, and finally, the cutting-head, which includes an electromagnet for actuating the armature to which is attached the cutter. There is no standard design of cutting-head, the type used often being the arbitrary choice of the man who does the recording, and the design often being his own. The design of the cutting-head is, however, extremely important, the success of the whole system depending to a very great extent upon it, not so much perhaps, upon the electrical part of it as upon the mechanical arrangement of the levers, the damping of the movable parts, the shape of the cutting point, the depth of the cut, etc.

We will now skip over the actual making of the record; for this is a mechanical process; in this article we are considering only the electrical features. Suffice it to say that from the large, thick wax disk (the matrix) upon which the cutting is done, a "master" is made, and from this master any number of impressions can be made, resulting in the records as they reach the music store.

In the old system of reproduction the vibration of the needle in the grooves of the record actuated a diaphragm of mica or other material, which directly communicated the energy of vibration to the air column of a horn of one type or another. In the newer electrical method power is communicated to the needle by the rotation of the turntable in the usual manner, but now, instead of driving a diaphragm, the needle drives an armature located in the magnetic circuit of a permanent magnet. A coil is also connected in the circuit so that the variations of the magnetic flux caused by the vibration of the armature induce fluctuating voltages in this coil, and these can be impressed on the input of an audio amplifier, the output of which is connected to a high-grade loud speaker.

So we have a means of amplifying the "pick-up" from the record, and of controlling the volume, neither of which could be done by the older method of reproduction.

The main feature of the electrical system of recording and reproducing is the fine quality that can be obtained. Music obtained from old-style records by old-style methods of reproducing is greatly lacking in the bass notes, and sounds thin and hollow. Very fine quality can, on the other hand, be obtained by the new methods. The tie-up between the radio and the phonograph has turned out to be very successful indeed.

415

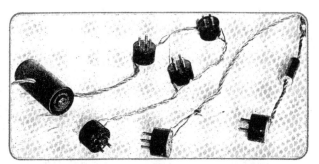

SIMPLE EQUIPMENT FOR UTILIZING A. C. TUBES WITH AN EXISTING RECEIVER
The photograph shows the various components—adaptors, C bias resistors, and cables—of the Carter a. c. harness

Electrifying Your Present Set

By Zeh Bouck

THE introduction of the alternating-current tube has stimulated something in the nature of a mild radio revolution. The advantages of a.c. operation—reliability and economy—in the majority of possible installations, are immediately obvious. This presents the problem of what is to be done with several-hundred thousand receivers of general efficiency, the only deficiency of which is their inability to be operated directly from an alternating-current source of a hundred and ten volts.

From an engineering standpoint this, of course, is merely a mechanical problem. Its solution was a simultaneous by-product of the a.c. tube itself. Any receiver in the world can be rewired for the use of a.c. tubes. In the majority of cases the changes are relatively few and simple. But the actual alteration of a radio receiver, particularly a commercial job, is repugnant to the average fan, and it was up to the manufacturers to provide

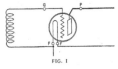

FIG. I

suitable devices for the conversion of battery receivers with little or no alteration of the receiver itself. Almost simultaneously with the production of a.c. tubes—the Cunningham and R.C.A. 226 and 227 types, the Arcturus line, and a host of others—these desired devices appeared upon the market in numerous quantities.

By referring to Figs. 1 and 2 it is obvious that the mechanical and electrical requirements of the new tubes, in reference to receivers originally designed for battery use, can be satisfied by means of a simple adaptor, which will insulate the tube from the original filament terminals

on the socket and at the same time provide two new filament or heater leads, and it was such adaptors that appeared on the market concurrently with the production of a.c. tubes. Fig. 1 indicates the familiar battery arrangement, while Fig. 2 suggests the electrical change effected by a simple adaptor. New filament leads have been provided for, while the former negative A post, to which the lower side of the grid coil or secondary is returned, remains open for biasing purposes.

The manner in which a typical adaptor fits between tube and socket is shown in the photograph, Fig. 3. The use of adaptors necessarily

FOR the convenience of our readers we have collected together here the names of the manufacturers of apparatus for use in converting a receiver for a.c. operation. Many of these names are also mentioned in the text of the article although those of the manufacturers of filament-lighting transformers are an exception for they are not given specific mention in the article. Readers should realize that, to light the filaments of a. c. tubes, a step-down filament transformer is necessary, besides the adaptors and harnesses mentioned in the article. A source of plate potential, and the necessary grid voltage are of course, also, required.

Manufacturers of A. C. Tubes:
Cunningham, R. C. A., Ceco, Arcturus, Sovereign, Televocal.

Manufacturers of Harnesses:
Arcturus, Carter, Eby, Naald, Cornish Wire, Harold Power, and Radio Receptor. The latter two companies sell combined A, B, and C power units and harnesses, as explained in the text.

Manufacturers of Filament Transformers:
Amertran, Dongan, General Radio, Karas, National, Samson, Silver-Marshall, Thordarson, and Ites.

increases the effective height of the tubes which in some cases makes it necessary to slightly gouge out the covers of the receivers, before they can be closed, or with suspended tubes as in the Atwater-Kent Model 35, the receiver must be raised on short legs. In consideration of this occasional inconvenience, the Arcturus a.c. tube, designed for cable or "harness" use, obviates the necessity for the use of adaptors by means of two small screws, one on each side of the base, to which the heater leads are connected. The manner in which the Arcturus a.c. tubes and a.c. cable are mounted in the average receiver is shown in Fig. 4.

"HARNESS" OUTFITS

THE use of adaptors alone solves only half the rewiring problem. The adaptors themselves have to be wired, so the majority of manufacturers producing adaptors are also providing

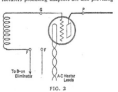

FIG. 2

connecting cables, usually referred to as "harnesses," in the form of braid-covered leads permanently connected to the adaptors. Other manufacturers sell the "harness" complete with an A, B, C power supply unit. Such devices are made in special and general types, prominent among which are those of the Radio Receptor company and Harold J. Power, Inc.

Radio Receptor makes three types of "Powerirers" (combining "harness," adaptors, and complete A, B, and C power supply outfit) designed especially for the Radiolas 20 and 28 and the Atwater-Kent models, which, however, are readily applied to an inclusive list of receivers.

Harold J. Power, Incorporated, manufactures an "A. C. Electrifier" which can supply A, B, and C potentials to any ordinary a.c. receiver. The filament potentials available are 1½, 2½, and 5 volts. Harnesses are available for electrifying the following receivers: Atwater Kent Model 35, Crosley "Bandbox" Model 601, Kolster 6, and universal harnesses for standard 5-, 6-, and 7-tube tuned radio-frequency receivers.

So far as the plain "harness" is concerned, we find that the Cornish Wire Company produces four types of "harnesses," all general designs, for five-and six-tube receivers with Arcturus a.c. tubes and for five-and six-tube receivers with the 226 and 227 type tubes.

The Arcturus Radio Company manufactures one general "harness" and six special "harnesses" or "a.c. cables," for use with the following receivers: The Freshman six-tube "Masterpiece," the Crosley "Bandbox," the Atwater-Kent models 30, 33, and 35, the Stewart Warner model 525, and the Kolster 6-D.

Other manufacturers of "harnesses," generally of a universal type and designed for R.C.A. and Cunningham type tubes, are given in the list on page 416.

FIG. 3
A close-up showing how the adapter plugs into the existing socket and the a. c. tube into the adapter

equivalent substitute. The B and C potentials must still be supplied by either batteries, a power supply device, or a combination of the two. It is only by the use of a B and C socket power unit, in conjunction with a.c. tubes, that the

receiver becomes completely electrified. For operation from a house lighting socket, the following apparatus may be used in combination with a battery receiver:

A.C. Tubes, Plus:
(1.) Filament-lighting transformer, an efficient B power device, and the necessary resistors to supply two C potentials.
Or (2.) a combination power-supply outfit, combining the necessary a.c. and d.c. potentials in a single unit.

THE RECEPTRAD "POWERIZER"

THE Receptrad "Powerizer," mentioned before, is a fine example of a complete power unit. The description given here of the installation of a "harness" and "Powerizer" in an Atwater-Kent receiver is indicative of the general procedure. Details regarding the conversion of other receivers are contained in the direction sheets with the different harnesses and cables.

Fig. 5 illustrates the circuit arrangement of the "Powerizer." Rectification is effected by a ux-280 (cx-380). A 210 type power amplifying tube is an integral part of the "Powerizer," the input of which is fed from the secondary of the second audio transformer in the receiver itself. There is no audio transformer in the "Powerizer," as the diagram shows.

The resistors, R_4 and R_5, are particularly interesting, as the voltage drop across them supplies the C bias potentials to the power tube and

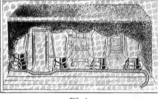

FIG. 4
The Arcturus tube, and a.c. cable installed in a hypothetical receiver. No adapters are required with Arcturus tubes

FIG. 6
An Atwater Kent model 35 receiver with a.c. tubes. The "Powerizer," shown on the next page, is used for power supply

VOLUME CONTROL AND ACCESSORIES

VARIOUS accessories are furnished with the "harness" outfits in accordance with the manufactuer's ideas of his obligations. In almost every instance some form of volume control applicable to a.c. circuits is included in the cable equipment, with the exception of such cases designed for particular d.c. arrangements already provided with an adequate control. The volume control generally consists of a specially tapered 0-to-25,000-ohm potentiometer, the element of which is connected across antenna and ground and the variable arm to the grid of the first tube. This type of volume control is easily attached.

Several manufacturers include C biasing resistors, center tap resistors, and bypass condensers, while others consider this auxiliary equipment a part of the power supply unit rather than a component of the adapting system.

COMPLETE ELECTRIFICATION

THE installation of a.c. tubes does not necessarily mean that a receiver is capable of being operated from the house lighting source without additional assistance. The use of a.c. tubes merely eliminates the A battery, or its

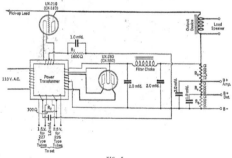

FIG. 5
The circuit diagram of the "Powerizer"

the r.f. amplifying tubes respectively. A resistor connected between the center filament tap of any a.c. tube and B negative, will bias the grid of the tube negatively, providing the grid of the tube is returned (through a secondary or leak) to B negative.

Fig. 6 shows the adaptors and tubes mounted in the Atwater-Kent 35 receiver, and Fig. 7 shows the "Powerizer" itself. The rheostat panel on the Atwater-Kent was removed and a special volume control, supplied by Receptrad, was mounted in its place. An external form of volume control could have been employed.

The following are the steps taken in installing the "Powerizer," along with the time consumed for each operation:

Demounting the Receiver	. .	10 Minutes
Installing Volume Control		30 Minutes
Installation of Adaptors and Tubes		5 Minutes
Reassembly of Receiver	. .	12 Minutes
Connection of "Powerizer"	. .	2 Minutes
TOTAL TIME		59 Minutes

This time would be reduced to 34 minutes by using the external form of volume control.

The result is a thoroughly up-to-date receiver, capable of delivering remarkable volume with fine quality, with a reliability of operation achieved only by complete ' electrification."

Fig. 8 shows a Radiola 28 superheterodyne in which a somewhat similar installation has been made with the Receptrad type 28 "Powerizer."

SPECIAL STABILIZATION

WITH some receivers, special devices (generally resistors in the grid circuits) must be employed, aside from the volume control, to achieve a satisfactory degree of stabilization. The following list considers various receivers that have been successfully adapted for a.c. operation, employing the Receptrad type of "Powerizer" and harness, with notes on special requirements:

FIG. 7
A neat unit—the "Powerizer"

ATWATER-KENT, Models 20, 30, 32, 33, 35: For external control no change is needed.

BOSCH "CRUISER:" Regular harness. Requires re-neutralization. Uses external volume control.

BOSCH, Model 46: Simply plug in adaptors. No volume control necessary.

BOSCH, Models 66 or 76: Requires 400 ohms in grid circuits of r.f. tubes. Also requires re-neutralization. No volume control necessary.

BREMER-TULLY "COUNTERPHASE" 6-37: Requires standard harness. May need re-neutralization or grid resistors. External type volume control.

BREMER-TULLY "COUNTERPHASE" 8: Requires standard type of harness with special distances between adapters. May require re-neutralization or the use of grid resistors. External volume

control. A high resistance of 50,600 or 100,000 ohms should be shunted across the secondary of the 2nd audio transformer.

CROSLEY "BANDBOX:" Simply plug in standard harness. Control may be had by "accuminators" or external. In some cases re-neutralization is necessary.

DAY-FAN 6 JR: Plug in standard harness. External volume control.

FADA SPECIAL 6, 265 A. R. P. 65: Standard harness. External Volume control.

FADA 7 — 475 A-S.F. 45/75: Remove tube housing. This is held in place by 6 or 8 screws. Put in standard harness for seven-tube set with distances between adapters slightly longer. External control.

FADA 8—480 B-S.F. 50/80 B: Requires the use of 1600-ohm resistors in the grids of all of the r.f. tubes. Standard harness for eight-tube set with special distances between the adapters. A high resistance of the order of 50,000 or 100,000 ohms should be shunted across the secondary of the 2nd audio transformer. External type of volume control.

FRESHMAN, Three Dial Type: 1600 ohms in r.f. grid circuits. Special uv harness. External control.

GREBE Mu-1 (5): 700 ohms in grids. Standard harness. Requires re-neutralization. External type control.

GREBE 7: Special distance harness. External type control.

KOLSTER, Model 6D: 700 ohms in grids of r.f. tubes. Standard type harness. No control needed. Sensitivity control on set O. K. Must have very good ground.

PFANSTIEHL 32: Standard harness. External type control which may replace old switch volume control. Connections to switch joined together and leads to old volume control soldered together separately from switch connections. Accomplished by connecting 50,000 ohms between grid and old filament wiring.

RADIOLA 16: Plug in standard harness. External control.

RADIOLA 20: For external control no change is needed.

SILVER-COCKADAY: Requires 3200 ohms in grid of r.f. stage. Standard harness.

SPLITDORF 6: 400-ohm resistors in grids of r.f. tubes. Special distances between adapters. External control.

STEWART-WARNER 525: Standard Harness. No changes. No volume control other than one in set needed. . . .

STEWART WARNER 705: Special distances. No control needed.

STROMBERG-CARLSON 501-A: Special distances. Re-neutralization required in some cases. External control.

THERMIODYNE T.F. 5: Special uv adapters. External type control.

ZENITH, Models 7, 8, 9: Special uv adapters, 400-ohm grid resistors. No volume control needed. All C batteries must be removed and gaps shorted.

ZENITH 11 or 14: Special length harness. External control.

FIG. 8
A "Radiola" Model 28 super-heterodyne with a "Powerizer" mounted in the battery compartment. This installation uses a 210 type power tube, doubly modernizing the receiver. It takes about forty-five minutes to make the revision

The Listeners' Point of View

RADIO MUST BE MADE A NECESSITY
By JOHN WALLACE

MR. H. A. BELLOWS, manager of wcco and former member of the Federal Radio Commission, can generally be counted upon to say something sentient when he speaks about radio. In a talk before the National Electrical Manufacturers' Association last winter he made several suggestions. One of the most important was that if radio is to be a stable institution it must become a necessary institution. This idea seems to us basic. He says:

"Improvement in broadcasting must follow two distinct lines, one of them being better presentation of musical programs. The other is, I think, still more important. Look for a moment at the history of the automobile. For years the automobile was a luxury, and its market was limited to those who felt they could afford luxuries of a rather costly nature. The thing that has made the automobile business what it is to-day is the conversion of a luxury into a necessity. People who would never own a single car as a luxury now own two or three because they regard them as absolute necessities, since our entire mode of living has readjusted itself to this new type of transportation.

"The real future of radio reception lies in a similar conversion of broadcasting service from a luxury into a necessity. This can never be done by programs of musical entertainment alone. You can never persuade the mass of the people that they *must* be able to hear the New York Symphony Orchestra once a week. You can come nearer it with reports of baseball and football games and of boxing matches. The stations in the farm belt of the Middle West have done it admirably, for a part of their audience, with their market reports. For the people in the cities this service feature of radio—this creation of a new necessity—is still largely in the embryonic stage.

"Of course I like to get letters praising our musical programs—we all do—but what I really value is the letter which tells me that thanks to our market service some farmer up in North Dakota has saved two hundred dollars on a shipment of wheat, or the letter saying that some family in Minneapolis has come to find our morning comment on the day's news just as essential a part of the day's routine as the morning newspaper.

"Can we develop some form of centralized teaching so that radio will establish itself in all our schools? Can we coöperate with the great news agencies so that, without competing with the newspapers, radio can be made a dependable and instantaneous means for sending out news flashes? Can we make broadcasting a legitimate agency for communicating between our governing bodies—national, state, and municipal—and the people who pay the taxes? What, in a word, can we do to make radio a necessity?

"Once again, let me cite an illustration. You all know the horrible tedium of civic association annual banquets—the indigestible food, the entertainment half drowned out by clattering waiters, and the dreary reports. We are trying the experiment of holding such an annual meeting by radio. The members of the civic association in question are being cordially invited to dine at home, to listen to reports carefully boiled down, and to do the necessary voting at the close of the meeting by mail.

"'Radio must be a necessity, not a luxury.' This I believe, is the solution of the future of

OSKAR SHUMSKY AT WBAL

Here is your chance to say you heard him "way back when"—provided he turns out to have the kind of future his press agent promises him. At any rate, Oskar Shumsky, whose playing and composition ability have already been praised by none other than Fritz Kreisler, will be heard from wbal Sunday evening, March 18, at 8, eastern time

service which carries radio out of the field of mere entertainment and into that of household necessity."

Mr. Bellows remarks were aimed primarily at the radio manufacturer. He said earlier in his speech: "I do not need to remind you gentlemen that you are all engaged in manufacturing a commodity which of itself is entirely worthless. The finest receiving set in the world is no better than the broadcasting which comes within its range. You do not have to be told what would happen to the radio manufacturing industry if the public should ever become really bored with broadcast programs."

This must have sent the cold chills coursing through the veins of his hearers, all of them with their entire capital and future prosperity inextricably tied up in radio. Particularly since the situation, if not probable, is at least conceivable. The public is notoriously fickle, and if it decided to become bored with broadcasting, all the king's horses and all the king's men couldn't change its whim. Where are the petticoat manufacturers and ostrich plume vendors of yesteryear? How can the man who invests in radio stock to-day be certain that by 1938 the public will not have capriciously shifted over to some other mode of entertainment, that the broadcasting stations will not be abandoned crumbling ruins even as—alas—are the breweries to-day!

Certainly if the radio manufacturers are to have any feeling of security they must see to it that radio becomes a utilitarian device as well as the entertainment device that it now is. If it can be made an indispensable utility like the telephone its longevity is practically assured. But while it remains in the luxury class, like the phonograph, its future is a gamble. The talking machine business, as is well known, almost went on the rocks a little while years back; the new fad which threatened its existence was radio itself.

Mr. Bellows mentions four ways in which radio can be practically utilized: Service to farmers; dissemination of news; agency of communication between governing bodies and the public; and centralized teaching for the schools.

A REGULAR FEATURE AT WGR, BUFFALO

The Hotel Statler Concert Ensemble, which is heard daily from wgr in a program of luncheon music

419

The first mentioned, service to farmers, is already a tried and proved function of radio. A survey recently conducted by wLs shows that the value of radio as a means of entertainment has been equalled and exceeded by its economic utility as a daily aid in the production and marketing operations of the far . It has taken its certain place as an instrument of farm education and has proved its dollars and cents value many times over in the transmission of market news, weather forecasts, and other items of immediate importance to the farmer. "If you had to give up either music or talks on the radio which would you prefer to retain?" was a question put up to a number of farmers by the United States Department of Agriculture. "We will keep the talks," answered 2358, while 1538 answered "music." The most recent government estimates place the number of radios on farms at 1,250,000. The number is now doubtless more than a million and a half. The radio farm and market service has already done much in solving the venerable problem of market gluts, with their resulting demoralization of prices and wrecking of values. Tens of thousands of farmers or their wives hear and tabulate the market returns every day on the particular product in which they are interested, and plan their marketing accordingly.

Anent the second mentioned method, Mr. Bellows asks: "Can we coöperate with the great news agencies so that, without competing with the newspapers, radio can be made a dependable and instantaneous means for sending out news flashes?" The answer is probably no, not without competing with the newspapers. But why should that consideration enter into it? If there is some sort of lively instantaneous news which radio can put over better than the newspapers and in which dissemination it is fore-ordained to supplant the newspaper, let the supplanting start at once. Conservative maneuvers on the part of the news agencies, and on the part of the newspapers, can only succeed in postponing the inevitable, not in suppressing it.

Concerning the third suggested method of giving radio a permanent utilitarian function we can say very little. Our knowledge of the workings of municipal, state, and national administrative bodies is too sketchy to be revealed here in its nakedness. But we see no reason to believe that radio will not some day be an official mouthpiece for governing bodies. And by official we mean Official—that is, there will be certain prescribed governmental broadcasting hours during which time certain prescribed individuals or bodies of individuals will be expected, by duty of their office, to listen-in and make prescribed findings. (Which is about as near as we can get to the jargon of law making.) Anyway we venture to prophesy that before ten years have passed radio will be in some way, if only in a minor way, tied up with governmental administration.

It is in the fourth suggested means of stabilizing radio that we are most interested. The propounder of the idea asks: "Can we develop some form of centralized teaching so that radio will establish itself in all our schools?" The answer is yes; the opportunity is now upon us.

We have made disparaging remarks about radio education in the past and lest we should seem to contradict ourself (not that we object to contradictions; our opinion happens to be still the same) let us reassert that as an "educating" medium radio is the bunk. It is next to useless as far as teaching such stuff as economics, horse-shoeing, sociology, or play-writing goes. But we have always believed that radio is peculiarly well adapted to teaching music appreciation and still think so.

Its possibilities in this field have long been recognized and from time to time during the past couple of years rumors have become current that something was going to be done about it. Mr. Walter Damrosch has been particularly active in keeping the idea before the public. It is one of his pet hobbies. And now, as the result of much scheming and labor on his part, he has carried the idea through to the point where it becomes a live issue.

His plan, as you doubtless already know, is to broadcast a series of his music appreciation lectures next winter over a network of stations during some daytime hour, the talks to be illustrated by his own piano playing and by an orchestra, probably the New York Symphony. So much, in brief, for his end of the arrangement. The other end is where the difficulties come in. Each one of the many thousands of schools throughout the country would have a first-rate receiving set, in perfect operating condition, available in its assembly hall or in one of the larger class rooms. Into this room, at the scheduled hour, would be herded all those pupils who had "Music Appreciation" on their program of courses. They would be provided with advance

MR. WALTER DAMROSCH
Who for very many years was leader of the New York Symphony Orchestra and now, frequently conducts that organization in the capacity of guest conductor. His experiments in broadcasting musical appreciation courses to schools has caused considerable favorable comment

notes on the lecture, sent to their teachers through the mail by Mr. Damrosch, and each lecture-concert would be followed by a written "quiz," also furnished from the central headquarters of the "course."

There is no especial practical difficulty in the way of carrying out this end of the scheme. The rearranging of the schedules to embrace this new hour of school work once every two weeks, or perhaps once a week, could be done with little trouble. Someone on the premises with sufficient intelligence to supervise the upkeep and proper operation of the receiving set could easily be found. Furthermore, receivers and loud speakers of adequate quality to reproduce the lectures satisfactorily are available.

The obstacle to be met with is the difficulty in arousing nation-wide interest in the idea to the point where boards of education and school trustees will undertake the red tape involved in officially adopting the course and appropriating funds for the necessary equipment.

Of course the obvious way to "sell" the idea

to the school masters throughout the country would be by a campaign of publicity and propaganda. But this would involve an enormous expenditure of money. And where is it to come from? All Mr. Damrosch has to offer is the idea and his own time and effort. Who is going to sponsor it?

By way of giving the idea some publicity Mr. Damrosch, with the coöperation of the Radio Corporation of America and the National Broadcasting Company, has already broadcast experimental programs. We think they proved conclusively that Mr. Damrosch will be able to carry off satisfactorily his part of the arrangement if the plan is ever put through. The first program was as follows:

PART I

For Children of the Grammar Schools

(1.) Allegretto from Symphony
No. 8...................... Beethoven
(2.) Entrance of the Little Fauns....Pierne
(3.) Scherzo from Symphony in B
flat,...................... Glazounow

PART II

For Students in the High Schools and Colleges

(1.) Overture to "A Midsummer
Night's Dream".......... Mendelssohn
(2.) Andante from Symphony No.
5...................... Beethoven

Before each of the numbers Mr. Damrosch gave an introductory talk and called attention to certain things to be watched for in the music, much in his familiar manner, except that he modified his material in accordance with the age of the prospective youthful listeners.

You may not be thoroughly in accord with Mr. Damrosch's method of explaining music (we, for instance, think he lays a misleading emphasis on the "story" content of non-program music) but nevertheless he has had some forty years' experience in giving such lectures and ought to know what he's about. Besides we can think off-hand of no one better fitted for the job; no one who combines, as he does, the qualities of authority on the subject, wide and popular renown, distinctive and intriguing personality. We opine that it will be radio's distinct loss if it finds itself unable to take advantage of these rare qualifications while they are available. Pedantry over the radio simply will not work. The pedant needs the help of bodily presence, and often too the help of school regulations, such as the one forbidding sleeping in class, to keep his audience attentive. A radio course in music will succeed or fail according to the personality of its spokesman. If Mr. Damrosch's eminently suitable personality is not made use of it may be a long time before another such personality turns up.

We have said nothing here about the desirability, in theory, of such a course, mostly because, we are confident that its desirability is already granted. Heaven knows, the young hopeful is having enough stuff drummed into him during his school years to "fit him for later life." In this age of educational progress, he is given opportunity to take most anything in "extra" courses from carpentry and cooking to tire repairing and dentistry. These vocational extras are useful. They make for remunerative work hours after he leaves school. But lamentably little attention is paid to equipping him to enjoy his non-working hours in later life. Out of every ten children exposed to eight years or so of such musical training, at least one would discover that he actually liked the stuff. And that, it seems to us, would justify the whole business.

"Our Readers Suggest—"

"OUR Readers Suggest. . ." is a regular feature of RADIO BROADCAST, *made up of contributions from our readers dealing with their experiences in the use of manufactured radio apparatus. Little "kinks," the result of experience, which give improved operation, will be described here. Regular space rates will be paid for contributions accepted, and these should be addressed to "The Complete Set Editor,"* RADIO BROADCAST, *Garden City, New York. A special award of $10 will be paid each month for the best contribution published. The prize this month goes to William C. Duer, Denver, Colorado, for his suggestion entitled "Eliminating A.C. Hum."*

—THE EDITOR.

Testing an Audio Amplifier

I HAVE used the following method for localizing trouble in audio-frequency amplifiers with considerable success.

The average manufactured receiver, as well as most home-built sets, make no provision for outputting the detector circuit to telephone receivers. Thus it is difficult to locate definitely a fault in either the radio- or audio-frequency channels. The system I am recommending will indicate immediately in just which section of the receiver the trouble exists.

Secure a hand microphone. If one is not available, a transmitter button may be used with equally good results if it is backed to a diaphram so that vibration will actuate the button moderately well. Connect this microphone, through a push-button, in series with the primary of a 2.5 to 1 induction coil, such as is available from any junked telephone set, and a six-volt battery. The regular filament battery of the set may be used. The secondary of the induction coil has two leads, with clips, soldered to it for convenience in setting up. The following is the method of operation:

Disconnect the detector plate-voltage lead from the B battery or power-supply device, and attach one of the clips mentioned to the detector plus post on the set. Connect the other clip to the point on the battery or power-supply device on which the lead to the set was formerly connected (22.5 or 45 volts), as shown in Fig. 1. Now push the button in the microphone circuit and speak into the microphone. Speaking into the microphone causes a varying current to pass

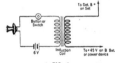

FIG. 1
An interesting circuit arrangement which may be used for testing an audio amplifier

through the primary of the induction coil. This induces current variations in the induction coil secondary, which, being in series with the detector plate lead through the primary of the first audio transformer, in turn affects the battery current in this winding, producing a variation in the balance of the audio-frequency circuit. Speech will be reproduced in the loud speaker as the microphone is spoken into. If the audio circuit is not functioning properly, no speech, or at most, a distorted voice, will issue from the loud speaker, thus indicating the trouble to be in this part of the set.

W. J. MORROW,
Macon, Georgia.

STAFF COMMENT

THE method suggested by Mr. Morrow provides a convenient means of testing the audio-frequency channel of a receiver. For experimental and entertainment purposes a greater response can generally be secured to voice variations by inputting the secondary of the modulation transformer or induction coil to the grid circuit of the detector tube. This is easily accomplished by connecting the secondary terminals across the grid leak, which is generally an accessible portion of the receiver.

Sources of Power-supply Hum

THE output of many receivers taking some of their power from the mains is marred by an excessive 60-cycle hum.

Investigation in many instances reveals the trouble to be due to sources often unsuspected. In the case the writer has in mind, a non-technically trained friend bought a socket-power kit about a year ago, and assembled and wired the device himself. It was less satisfactory than battery power because of the intensity of the hum, which defied all efforts made to eliminate it. He asked the writer to look the outfit over.

Upon opening the output circuit it was found that only part of the hum came from the loud speaker. The laminations in the power transformer were so loose that the resulting vibrations could be heard ten or twelve feet away in a quiet room.

Further examination disclosed that the voltage-divider section consisted of a number of adjustable resistors of the carbon-pile type. Operating the equipment with the power transformer removed from the socket-power baseboard confirmed the belief that a portion of the hum was due to the microphonic action of the carbon-pile resistors brought about by the vibration transmitted from the transformer through the baseboard.

The transformer case was opened and, after the transformer had been warmed up cautiously to about 100° C, was filled with molten battery compound. The primary was energized while the compound was still liquid so that the latter could penetrate between the vibrating laminations.

The hum soon fell to a slight murmur, inaudible six inches from the transformer.

As an additional precaution, the carbon-pile resistors were replaced by the wire-wound type which are now generally available.

HERBERT J. HARRIES.
Pittsburgh, Pennsylvania

STAFF COMMENT

THE microphonic effect caused by loose carbon-pile resistors is unusually interesting. The department editor has run across several such cases in his own experience. Microphonic hum will generally be eliminated when the adjustment of the carbon-pile resistors is tightened.

An R. F. Volume Control

OWNERS of three-dial five-tube radio receivers who live near one or more powerful local stations often find that they have such strong réception that the last audio tube, or even the detector, is overloaded. Even where there is no overloading there is often too much volume for ordinary use, or control, by the usual methods.

This difficulty may be remedied and at the same time more economical operation secured by the addition of a switch so connected in the circuit that the first radio-frequency stage of the set may be laid aside, so to speak, at the desire of the operator. When the first stage is not being used, one tube will be automatically turned out and it will become unnecessary to adjust the first tuning dial. The switch can be installed in any commercial or home-made receiver, and its use will result in no loss of stability or efficiency.

The author recommends a Yaxley radio jack switch No. 60 for this purpose, but any compact double-pole double-throw switch may be used with complete success. The switch may be mounted securely at any convenient place on

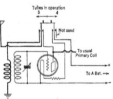

FIG. 2
A simple method of cutting out one radio-frequency stage for local reception. In addition to its economy this method of volume control is conducive to high-quality reproduction from near-by stations

the panel, preferably near the first tuning dial. If a separate rheostat or ballast resistance is not used in connection with the first tube of the receiver, it will be well to obtain a resistance of the proper value for the type of tube in the first stage.

Connect the switch as shown in the diagram, Fig. 2. In one position, all five tubes should light and the first tuning dial should be in operation. In the other position, the first tube should not be lighted, and the first tuning dial should be out of operation. For local work, and when extreme selectivity is not desired, the latter arrangement will be found ideal, since it simplifies operation and prevents the unnecessary use of one tube with resultant extra load on the batteries. This switch will be found to be a very effective, though rough, volume control. Instead of tending to cause distortion, as some volume controls do, it helps to prevent distortion, by preventing tube overloading.

ALBERT R. HODGES,
Clinton, N. Y.

STAFF COMMENT

IN SOME cases a similar degree of volume control can be obtained merely by turning off the filament of the first tube, a switch being provided for this purpose. In some receivers, merely removing this tube from the socket will effect the desired control, sufficient energy being fed through various inductive and capacitative channels to supply an adequate signal on local stations.

Eliminating A.C. Hum

IN SPITE of the many improvements incorporated in the modern socket power device, a.c. hum has not been eliminated in many receiver-power combinations in use to-day.

After trying various methods of getting rid of this nuisance and achieving no real results, I hit upon the hook-up shown in Fig. 3, which really eliminated that a.c. hum.

The filter used was an output device designed to keep the high d.c. current out of the loud speaker windings. It was rated at 30 henries.

By hooking up the 2-mfd. condenser as a shunt across one side and using three of the four posts of the filter, the desired result was obtained. Many of these "tone filters" already contain condensers integral with the choke, which further simplifies the hook-up. Fig. 3 shows the condenser as part of the filter, and is self-explanatory.

WILLIAM C. DUER,
Denver, Colorado.

STAFF COMMENT

THERE seems to be no end of the possible uses of the output device. Here we have an interesting and logical use distinct from the original purpose of the unit. Mr. Duer's sugges-

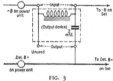

FIG. 3
An arrangement of apparatus which has been successfully applied in the elimination of a. c. hum

tion recommends the use of a filter circuit in addition to that already provided as a part of the power-supply device, which, as our correspondent suggests, is occasionally inadequate.

Phonograph Pick-up Switch

I AM using an electrical pick-up in conjunction with my phonograph and radio receiver. The pick-up is of usual design calling for the removal of the detector tube from its socket and the substitution of a four-prong plug. This arrangement is rather clumsy mechanically so I devised the simple switching arrangement shown in Fig. 4, using a Yaxley No. 60 switch, which is mounted on the panel of the receiver. An extra socket is placed in the receiver into which the pick-up plug is inserted. The switch throws the input to the amplifier from the detector circuit to the pick-up circuit.

FRANKLYN F. STRATFORD,
Jersey City, New Jersey

STAFF COMMENT

ASIDE from the convenience of Mr. Stratford's arrangement, the switch short-circuits the output of the detector when the phonograph is being used, and vice versa, of course, so that there is no danger of "cross talk"

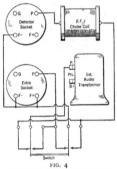

FIG. 4
A convenient switching arrangement for controlling phonograph and radio pick-ups

between two audio circuits. This is particularly desirable when the receiver is powered from the lighting socket, since leaving the detector output open, or suddenly throwing the amplifier to it, may cause oscillations.

Stabilizing With High-Mu Tubes

I HAVE been building and experimenting with a Browning-Drake receiver, and after trying a "Phasatrol" and different neutralizing condensers, I find that a CX-340 (UX-240) tube in the r.f. stage, with ninety volts on the plate, will neutralize easily. A small balancing condenser should be used. Considerable volume, with no sacrifice in efficiency of the set, will result.

MILTON HICKS,
Rockville Centre, Long Island.

STAFF COMMENT

IT IS possible to stabilize many radio-frequency circuits in this manner. The fact that the high-mu tubes have a higher plate impedance than those generally employed in radio-frequency circuits in many cases more than counteracts the increased regenerative effect due to higher amplification constant of the tube, with the result that the circuit into which they are plugged is relatively stable. Also, in the cases of two or more tuned radio-frequency stages, the use of a single high-mu tube, preferably in the radio-frequency stage preceding the detector tube, will have a slight detuning effect upon the tandem circuits due to the introduction of a capacity discrepancy. In other words, one circuit will be slightly off tune, with the usual stabilizing effect.

It is always a good idea to switch the tubes around in an oscillating circuit in an endeavor to arrive at a combination giving the best results.

Another Output Arrangement

SEEING an article on output devices in a recent issue of RADIO BROADCAST, I have decided to send you a diagram of one which I have used for three years. I tumbled on it in fooling around with a choke output and find it is particularly satisfactory as it provides for a head-set without changing the loud speaker circuit.

It requires an amplifying transformer, two condensers (of 2-mfd. and 0.5-mfd.), and two open-circuit jacks.

It will be noted, from the wiring diagram, Fig. 5, that the output of the set is fed through the secondary of the transformer and that the 2-mfd. condenser and loud speaker (in series) are shunted in the usual manner. The variation from the conventional circuit lies in the primary side where you will note that a "jumper" is attached to only one side of the B circuit. This gives plenty of power for the headset without depriving the loud-speaker circuit of enough energy to decrease volume.

Another advantage peculiar to this output device is the fact that the total load on a power-supply device remains constant when the loud-speaker is removed from the circuit, thereby leaving all voltages on tubes the same, regardless of whether phones or loud speaker is employed.

I have made four of these in the past three years using different audio transformers and find that results are equally good; a relatively cheap transformer I found to be as satisfactory as an expensive one.

F. W. WOOLWAY,
Newton Centre, Massachusetts.

STAFF COMMENT

THE average audio transformer secondary has a rather high d.c. resistance so that the voltage actually on the last tube plate will be quite a little below the terminal voltage of the B voltage supply. Experimenters can try reversing the transformer windings, i.e., keeping the primary in the plate circuit and the phones across the secondary.

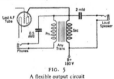

FIG. 5
A flexible output circuit

A SET-UP OF APPARATUS IN ONE OF THE "RADIO BROADCAST" LABORATORIES

The measurements which are outlined in the article were made with this equipment. Note the radio frequency oscillator at the extreme left. It is housed in a large tin wash boiler. The large loud speaker is the new Western Electric 560AW cone

The Four-Tube "Lab" Receiver

By Keith Henney

Director of the Laboratory

ALTHOUGH many receivers have come and gone since June, 1926, when the R. B. "Lab" circuit was first described in RADIO BROADCAST, the latter still represents the criterion by which all other receivers are compared in the Radio Broadcast Laboratory, and the measure by which all four-tube receivers are judged. There are several reasons for the continued popularity of the "Lab" receiver; the most important lies in the fact that, with good component parts and a good layout of apparatus, four tubes in this circuit seem to require a minimum amount of energy from the ether to deliver any required loud speaker power.

The "Lab" circuit receiver is a member of that famous family which includes the Roberts and the Browning-Drake, to mention but two, and it employs, therefore, a single stage of neutralized radio-frequency amplification followed by a regenerative detector and the usual audio amplifier. Since June, 1926, few startling inventions or discoveries to do with circuits have appeared in the field of radio, so the circuit itself has changed but little. Experience has taught, however, just where the inductances and their related parts, condensers and tubes, should be placed behind the panel for most efficient operation, and this alone might be considered quite a big step forward so far as efficient design is concerned. A bad layout of apparatus can mean all the difference between poor and good reception.

The following article, which is preliminary to the publication of a constructional article on the "Lab" receiver, gives some useful information on how receivers are designed, how the component parts are measured, or what the voltage amplification of the detector or the amplifier may be. The information is available as a result of exhaustive tests conducted in the RADIO BROADCAST Laboratory, and although the measurements were made with the prototype of the "Lab" receiver scheduled for description next month, they should appeal to anybody interested in radio receiver design and measurements generally.

A certain amount of engineering data can be collected at home; to gather other data a laboratory equipped with instruments is necessary. A typical "armchair" investigation was described in the March RADIO BROADCAST. It told how one could calculate the voltages appearing at various points in an audio amplifier if one knew the electrical dimensions of his apparatus and the amount of power he desired from his output tube. The data presented below result from a definite series of laboratory experiments designed to learn what is going on between the detector output circuit and the antenna of a receiver; in other words, their purpose is to take up where the armchair engineer left off. As the investigation tells a great deal about the "Lab" circuit it should be interesting to those who already own such receivers, or those who are being introduced to it for the first time.

In the "Armchair Engineer," which was the title of the March article, we showed what voltages were necessary in an audio amplifier when a maximum of 700 milliwatts of undistorted audio power was to be delivered to a loud speaker of 4000 ohms impedance, which conditions obtain when a 171 type tube is in use, delivering its maximum quota of undistorted power. We have chosen 4000 ohms for the impedance of the loud speaker as representing the best value (twice that of the tube) for maximum undistorted power output. Seven hundred milliwatts of power is small compared to that taken by a 40-watt incandescent lamp with which we are all familiar, smaller compared to the 500 watts required by that common home apparatus, the electric iron, but is greatly in excess of the power that can be collected from one's antenna from a distant broadcasting station. In other words, between the antenna and the loud speaker, there must be considerable power amplification. How great is this amplification, how can it be measured? These are questions that a laboratory investigation can answer, and these figures enable a laboratory to form some kind of an opinion on the overall efficiency of a receiver.

In Fig. 1 is a two-stage audio amplifier which includes two transformers whose turn ratios, secondary to primary, are about 3 to 1 at 1000 cycles, a tube the amplification factor of which is 8 (about 7 of which can be realized), and a power tube with an amplification factor of 3. The voltages at several points in this system are given on the diagram, showing that about 0.42 volts must appear across the primary winding of the first audio transformer if a value of 700 milliwatts is to be delivered to 4000 ohms.

It will be noted that the voltage across any point is found by multiplying together the voltages appearing on the grids of tubes in the circuit, the amplification constants of these tubes, and the turn ratios of the audio transformers. Thus the voltage across the output of the second audio tube (note that its effective mu is 7 and not 8) will be approximately:

$$0.42 \times 3 \times 1.26 \times 7 = 8.8$$

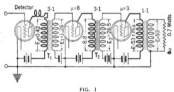

Detector

$\mu=8$ $\mu=3$

3-1   3-1   1-1

0.7 Watts

FIG. 1

In Fig. 1 it will be seen that the voltage appearing across the output of the final tube is not 26.5 x 3 (the grid voltage multiplied by the mu), since only two thirds of the a.c. voltage in this circuit is usefully employed across the output load. The output, then, is about 53 volts.

We resort to Ohm's law to prove that this value of 53 volts is adequate to give the requisite 0.7 watts of power, and in doing so we find that the current in amperes (output voltage divided by loud speaker impedance) is:

$$\frac{53}{4000} \text{ Amperes}$$

And that the power in watts (voltage multiplied by current in amperes) is:

$$\frac{53 \times 53}{4000} = 0.7 \text{ Watts}$$

Thus we have proved that a voltage of 0.42 across the primary winding of the first audio transformer is adequate for our purposes. Given the constants of the transformers, tubes, and loud speaker in an audio circuit, it is a simple matter, therefore, to calculate the requisite voltage for any given power output, across any point of the circuit.

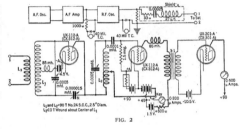

FIG. 2

L₁ and L₂=90 T No.24 S.C.C., 2.5" Diam.
L₃=10 T Wound about Center of L₁

change when a signal is tuned-in. For example, WEAF, 8 miles from the Laboratory and with 50 kw. of power in the antenna, reduces the average plate current of this type of detector to the low value of 300 microamperes, in other words, produces a *change* of 1000 microamperes. Signals from WJZ, 30 miles distant with somewhat

less power, cause a change of 100 microamperes. These changes in detector plate current are then, a measure of the strength of incoming signals.

We must know the voltage across the first audio transformer primary, produced by various r. f. voltage inputs to the receiver and what effect modulation at the transmitter has. We learn from Carl Dreher, Staff Engineer of the N. B. C., that the two stations mentioned above are modulated, on the peaks of the audio signals, to about 60 per cent., so that if we modulate a local oscillator, or generator of radio-frequency waves, to about 60 per cent. and impress known radio-frequency voltages upon our receiver, we can note the change in plate current and the voltage across the primary of the first audio transformer.

Instead of measuring the voltages across the primary, in the Laboratory, we measured those appearing across the secondary by using the first audio tube as a vacuum-tube voltmeter and then calculated that across the primary. This was done by biasing the grid of this tube to about 10.5 volts with 90 volts on the plate, when the plate current was about 30 microamperes, and increased up to about 100 microamperes when there was a reading of 1.5 volts across the transformer secondary. Known audio-frequency voltages impressed across the input of this tube were plotted against plate current to calibrate it as a voltmeter.

The steady detector plate current of 1.3 milliamperes was balanced out by placing a battery and resistance across the microammeter, the battery being poled so that it opposed the flow of steady plate current. A key was placed in the microammeter circuit so that the latter did not read until the key was closed.

We then noted the vacuum-tube voltmeter

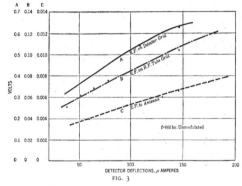

FIG. 3

THE DETECTOR CIRCUIT

HAVING decided that we require 0.42 volts across the primary of the first transformer, we must now set about producing this voltage. Up to the present time our calculations have been purely mathematical but from now on we shall require the help of laboratory instruments before we can conclude our calculations.

Because we desire maximum sensitivity in the new "Lab" circuit we shall use a grid leak and condenser detector the average plate current of which is about 1.3 milliamperes if a 112-A type of tube is used with conventional values of grid leak and plate voltage. We use this type of tube because it is much less microphonic than a 201-A tube, because of its lower plate impedance which gives us better low audio-frequency response, and because it is capable of somewhat greater amplification than its smaller brother tube.

This 1.3 milliamperes plate current will

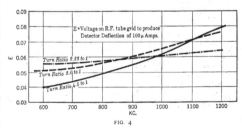

E = Voltage on R.F. tube grid to produce Detector Deflection of 100 μ Amps.

Turn Ratio 2.25 to 1
Turn Ratio 3.0 to 1
Turn Ratio 4.5 to 1

FIG. 4

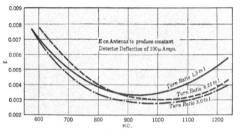

E on Antenna to produce constant Detector Deflection of 100μ Amps.

Turn Ratio 4.5 to 1

Turn Ratio 2.25 to 1

Turn Ratio 3.0 to 1

FIG. 5

deflections against input r. f. voltages and percentage modulation, and the change in detector plate current—which we shall call detector deflections—against r. f. voltages.

The set-up used is shown in the photograph at the head of this article, and in Fig. 2. It comprised an audio-frequency oscillator, a General Radio push-pull amplifier, and a radio-frequency oscillator, the latter housed in a tin wash boiler with a tight fitting lid. Across the output of the audio amplifier is a 40-milliampere thermocouple in series with 1000 ohms to act as a voltmeter for the 1000-cycle modulating energy. Across the output of the radio-frequency generator is a variable resistance in series with a thermocouple which measures the current through this resistance. The voltage drop, at radio frequencies, across this resistance, is utilized to drive our receiver under test. This voltage can be varied by changing the value of resistance used, or by changing the current through it. Large changes in voltage are produced by the former method, small changes by the latter.

These are crude methods of measuring voltage gain compared to the ultra refinements employed in the Crosley Laboratory for example, where an accuracy of better than 2 per cent. is possible, or in the General Electric Laboratory, but they are effective, as the following data will show.

Measuring the plate voltage of the oscillator we find that it is 69 volts, and noting that the current in its tuned circuit varies directly with the plate voltage we assume that the ratio between the fixed plate voltage and the low-frequency voltages from the amplifier fed into the plate circuit of the r. f. generator is a measure of the percentage modulation. The 1000-cycle voltages obtainable are from about 18 to about 35 r. m. s., which give us modulation percentages of from 36 to 71 approximately.

If the modulation is changed while a constant r. f. voltage is applied to the receiver in series with its artificial, or "dummy," antenna, and, secondly, if the modulation is held constant while the r. f. voltage is changed, the following facts are noted: the detector deflections are independent of percentage modulation, the a. f. voltages appearing across the secondary of the first audio transformer vary almost directly as the modulation and as the r. f. voltage applied, and the detector deflections vary approximately as the square of the input r. f. voltages.

We learn that a 660-kc. wave modulated 54 per cent. and producing a change in detector plate current of 100 microamperes produces an audio voltage of 1.67 across the secondary of the transformer. If this is a 3 x 1 turn ratio unit, the

voltage across its primary will be 0.56, which will furnish ample power to the loud speaker from a two-stage transformer-coupled amplifier of the type shown in the photograph of the receiver upon which these measurements were taken.

So far we have established several important points. We have determined the relation between detector deflections and detector output audio voltages; we have found out the relation between audio voltages and radio input voltages; and we have decided that to get about 0.5 volts across the input to the audio amplifier we need a radio signal, 54 per cent. modulated, which will cause a detector deflection of about 100 microamperes. We may now neglect the audio amplifier, and deal only with the part of the circuit between the antenna and the detector plate circuit. What we now desire is to get this 100-microampere deflection with the least input r. f. voltage to the antenna, the greatest amount of selectivity, and the best characteristic, that is, the most even amplification over the broadcasting band.

There are several variables that need in-

vestigating; for example, the number of turns included in the r. f. tube plate circuit must be determined. Let us start by placing various voltages at 660 kc. on the detector input directly, i. e., without going through the tuned circuits or the first tube. Then we shall apply voltages to the grid of the r. f. amplifier, and finally to the artificial antenna. The result is shown in Fig. 3, and was obtained by applying the voltage drop across 20 ohms to the detector, that across 3.3 ohms to the grid of the r. f. tube, and that across 0.191 ohms to the antenna.

To get our 100-microampere detector deflection required about 0.5 volts when applied to the detector input, about 0.085 volts when applied to the r. f. grid, and only 0.005-volts when impressed across the artificial antenna. It is difficult to translate these differences in voltages into "gain," but it is interesting to note that a ratio of about 6 to 1 in voltage appeared by removing the driving r. f. voltage from the detector grid and placing it across the r. f. grid, and that an input voltage of five millivolts produced 100 microamperes detector deflection, and, therefore, if modulated about 54 per cent. produced our required 700 milliwatts of audio power.

Now let us vary the frequency of the impressed voltages and see what happens. The detector input has essentially a flat characteristic over the broadcasting band. Applying voltages to the r. f. grid gives the result shown in Fig. 4, in which various turn ratios between detector input and r. f. tube output are given. At a turn ratio of 2.25 to 1, where nearly one half of the detector coil is in the plate circuit of the previous tube, the curve is flat, and as the turns ratio is increased more input voltage is necessary at high frequencies to deliver the same detector deflection.

When voltages are impressed upon the "dummy" antenna, the characteristic changes, and becomes peaked, with maximum amplification near the middle of the broadcasting band, the ratio between the lowest and highest point in the curve being about two to one. This curve, Fig. 5, gives the input voltage required to give a

A FOUR-TUBE "LAB" RECEIVER

An experimental breadboard layout of the "Lab" receiver which will be described constructionally next month. The measurements outlined in this article were made with the layout shown here

constant deflection of 100 micro-amperes, and when turned upside-down, gives an idea of the overall amplification characteristic.

Investigation of the artificial antenna and coupling coil, consisting of ten turns wound about the center of the r. f. input coil, shows that it resonates at about 900 kc. which accounts for part of the humped characteristic. Part of it is accounted for by the rising characteristic toward the higher frequencies noted in Fig. 5 and part by the looser coupling between antenna and r. f input coil at the lower frequencies.

Placing a coil and variable condenser in series with the antenna, so that the entire system may be resonated to the frequency desired, will bring up the overall amplification of the receiver as well as improve the characteristic. It necessitates an additional tuning control, and has the unfortunate habit of making necessary readjustment of the r. f. tuning condenser whenever any change is made in the setting of the antenna series condenser. For these reasons, and because a 2 to 1 characteristic such as shown in Fig. 5 is not bad enough to worry about, the antenna series resonant system has not been added to the "Lab" circuit. A further study of antenna coupling systems is under way and it is hoped that the result can be given to interested experimenters soon. The turn ratio of any good coil should be 3 to 1 since greatest amplification results thereby.

SELECTIVITY

ONE thing remains to be investigated—the selectivity of the two tuned circuits and that of the receiver as a whole. A small condenser of 0.00005-mfd. capacity bridged across the main tuning condensers and varied through the point which resonates the circuit will give an idea of the selectivity. Fig. 6 shows both the selectivity curve on the r. f. circuit and the overall curve.

A word about the practical result of all this investigation may be interesting. In Garden City, 8 miles from WEAF, it is possible to receive WJZ 50 kc. away without interference. It has been easily possible to receive WJR when WJZ was on the air, a difference of 20 kc., and on the higher frequencies many stations outside of Manhattan, 15 miles distant with perhaps 30 local stations in operation, have been received regularly. In an Ohio town, 100 miles from

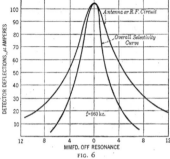

FIG. 6

a broadcasting station, a similar receiver has a choice of over 40 stations without appreciable interference.

In none of these experiments has regeneration been added to the detector, which, in fact, has been operating at poor efficiency, since it has a high-impedance plate circuit at high frequencies. As soon as the regeneration condenser is connected, partially bypassing some of the r. f. load, the detecting efficiency goes up, and adding regeneration increases the overall gain and selectivity at all frequencies, and improves the characteristic. Figs. 2 and 7 give all electrical constants for the two-tube "Lab" circuit. Owing to the superior gain produced by the 112 type of tube, we recommend it in both the r.f. amplifier and detector sockets.

As in all similar tuning systems, the efficiency of the circuit as a whole depends solely upon the quality of the apparatus (especially the coil) put into it, and their relative location behind the

panel. Coil resistance is especially to be avoided in the antenna stage for here the losses in amplification and selectivity are directly proportional to the resistance in the tuned circuit.

Coils of small diameter may be placed as close together as shown in the photograph on page 425, although it will not be possible to neutralize the amplifier at one extreme of the broadcasting band and have it remain neutralized at the other. If neutralization is carried out in the middle of the band, it will be fairly complete at other frequencies and will not oscillate at any frequency. The parts used in the four-tube "Lab" receiver on which the above tests were made, are as follows:

LIST OF PARTS

2—0.0005-Mfd. Remler Condensers
1—0.000055-Mfd. Precise Condenser
1—Type N X-L Condenser
1—0.0001 Mfd. Averovox Condenser
1—0.001-Mfd. Aerovox Condenser
2—Aero Inductances
2—Samson 85-Mh.Chokes
4—Benjamin Sockets
2—Sangamo Type A Transformers
1—Ferranti Output Transformer
2—Amperites, 0.5-Ampere
1—Lynch Resister Mount
1—1.5-Meg. Aerovox Resistor
1—Yaxley 7-Wire Cable
2—CX-312-A (UX-112-A) Tubes
1—CX-371-A (UX-371-A) Tube
1—CX-301-A (UX-201-A) Tube

If the receiver is to be operated in a location where there are many stations, it will be well to shield at least the r. f. stage. In the layout shown in the photograph care has been taken to keep all amplifier apparatus to one side of a middle line, and all detector equipment on the other side of that line. This facilitates the introduction of a shield between the two circuits.

If the approximate layout shown in the photograph is followed, constructors should have no difficulty in making the receiver operate. Any good apparatus may be used; coil dimensions are included in Fig. 2 and commercial inductances from General Radio, Aero Products, Silver-Marshall, and Hammarlund have been used with complete success. An article scheduled for the next issue of RADIO BROADCAST will describe the construction of a four-tube "Lab" receiver in detail.

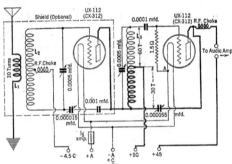

FIG. 7

This is the circuit diagram of the two-tube "Lab" receiver. As will we noted, the 112 type tube is used in both r.f. and detector stages and the gain will be greater than that possible were 201-A type tubes employed

It is with considerable difficulty and some reluctance that we report the first public demonstration in America of Professor Léon Théremin's "Music from the Ether" on January 31st, 1927. We sat in the passing glory of the Metropolitan Opera House surrounded by a typical opera audience, all in evening dress, and watched and heard Professor Théremin and his associate, Dr. J. Goldberg, extract violin-like notes from two beating radio-frequency oscillators, and our emotions, we must admit, were mixed. Musically the demonstration was not good, but one cannot criticize an amateur who makes no effort to set himself up as a musician; as a scientific demonstration before a S. R. O. audience in no less a place than the Metropolitan, it was grand. The audience was enthusiastic, it applauded, and roared "bravo" and "heroique" and begged for more all because they had watched two men without any visible connection with the instrument they were playing, wave their hands before a slender rod standing vertically in the air, and heard the tremolo notes of "Ave Maria," "Song of India," and other old favorites.

We must admit that we were more interested in the audience than in the music or the musicians; we wondered how they would take it—those trembling notes, often off key; we wondered who they were, whether music critics, scientists, or merely curious people, amazed in any sleight-of-hand performances, or demonstrations of forces unseen. We were interested, too, in the electrical mechanism, which is so well known to all workers in radio frequencies, in the loud speakers, great triangular green boxes standing high above the stage and showing off very well against a black velvet drop.

Electrically, Professor Théremin's apparatus is simple. It consists, according to *La T. S. F. Pour Tous*, for December, 1927, of a crystal-controlled generator oscillating at a frequency, let's say, of 100,000 cycles per second, and another oscillator whose frequency may be varied continuously so that the beat note between these two generators, when fed into a detector and amplifier, will cover the entire audio range of tones. Because there is always an attempt on the part of two such generators to "pull together" when their frequency difference is small, some difficulty is noted in securing low or bass notes. An ingenious trick, also well known to radio experimenters, is used here in an attempt to avoid such trouble; the second harmonic of one oscillator is made to beat with the third of the other. In this way Professor Théremin got down pretty low in the frequency scale, but because his loud speakers began to rattle at these frequencies or because he was still bothered by his oscillators "pulling together," these notes were not good.

The loss in audio-frequency output occasioned by using harmonics instead of the fundamentals is not serious; the volume can always be brought up by proper amplification. In fact the audience seemed astonished at the volume produced by these beating generators of radio-frequency waves, for at times it was "sufficient to fill the Metropolitan." Of course the audience did not know, as our readers do, that volume much in excess of this could have been produced if needed; all that was necessary was additional amplification. ——

In Professor Théremin's cabinets, then, were two generators, one fixed in frequency, the other

"Strays" from the Laboratory

variable. Tunes are played by bringing the hand near a metal rod which is attached to the tuned circuit of the variable oscillator, the hand, which is connected to ground through the performer's body, serving to change the capacity of the tuning condenser —"body · capacity," nothing less, the bane of all radio experimenters of a few years ago.

Volume is controlled by another conductor at right angles to the tone control and placed so that it can be approached by the other hand of the performer. It is a single loop of wire and constitutes part of the tuned circuit of the crystal-controlled oscillator. As is well known, when the tuned circuit of such an oscillator is properly adjusted, maximum power is produced, and at any adjustment different from this condition, less power is generated. Thus, bringing the hand near this loop detunes the closed circuit and reduces the power generated in the fixed oscillator, but because of the crystal, its frequency does not change appreciably.

The device is extremely simple to "play," there is no complicated fingering or bowing, as of a violin, and there is no labor on the part of the musician to get fortissimo passages, but in this very simplicity

there are disadvantages. The violin gets its timbre because of the complex manner in which its strings vibrate; its notes are not pure, i. e., good sine waves, but they are more or less complex at the will or skill of the musician. The violin, the tone of which the "Théreminvox" most nearly approaches, can be bowed on more than one string at a time, thereby producing a still more complex wave form, while the electrical instrument can hit but one note at a time, and that note ordinarily is quite pure, like that of a child's voice. By throwing a switch or two, the operator, or player, can bias grids differently and put into the output of one or both oscillators various harmonics, which results in a change of timbre.

The "Théreminvox" glides smoothly from one frequency to another; it cannot, in its present form, jump from one note to another an octave, more or less, distant. It cannot produce staccato effects, or those of plucked strings, nor can it produce any of the effects of percussion instruments.

Let us not be too harsh on Professor Théremin of the Institute Physico-Technique of Leningrad. He is not a musician. He is a physicist and has several inventions to his credit, inventions on such things as methods of measuring weak field strengths, and as early as 1926 made contributions to what has come to be called television. According to a Paris interview published in *La T. S. F. Pour Tous* he stated of his latest development:

This is not at all a plaything for me. It is much more a concrete proof, an incontestable demonstration, of my conception of the arts and the sciences. The two are but one to me. I have always been disturbed by the disdain with which followers of art treat questions of science and of engineering. And conversely, how many times have these savants claimed that art was a word

PROFESSOR LÉON THÉREMIN "PLAYS" HIS "THÉREMINVOX"

devoid of all meaning. To prove to the one that science can render the greatest services in the development of the arts, to demonstrate to the other the fertility of an intimate collaboration of the arts and sciences, is my aim.

Perhaps the orchestra of the future will consist of many of these oscillators and many men before them waving their hands. Certainly there will be plenty of engineers who can put all the harmonics into the tones the musicians desire, and perhaps then the timbre will be more pleasant to musicians.

It is reported that Rachmaninoff turned to a member of the private audience which first witnessed the demonstration of the apparatus, and said, when she shouted "Bravo." "Madame, you exaggerate!"

**a.f. C.
Tube
Troubles**

FROM various sources comes the rumor that set engineers are more pleased with the heater type of tube, the 227 type, from the standpoint of lack of hum and of general operating characteristics, than they are with the straight a. c. filament tube. It is said that set manufacturers would like to use this type only and to forget the low-voltage, high-current tube. How we run in circles, we radio engineers. Three years ago a heater type of tube, the McCullough, appeared on the market after many demands from those (notably Professor Morecroft in RADIO BROADCAST) who knew the possibilities of a unipotential cathode tube. In Canada this tube became very popular; in this country the manufacturers sold a comfortable lot of tubes. But in general the tube people, independent and "corporate," have taken a "high hat" attitude about it. And now, perhaps, it will become the preferred tube after all.

Several readers have noted a strange periodic increase and decrease in filament current in a. c. tubes. These fluctuations are reflected, naturally, in signal strength. On one occasion a reader remarked about the rhythm he noted on his new a. c. set while his neighbors were not so afflicted. What is the reason? Perhaps it is the following:

If an a. c. tube has an open filament or heater, it frequently happens that the broken ends actually make contact so that when the tube is turned on current passes through the filament or heater, causing it to expand in such fashion that the ends are pulled apart. Thereupon the signals disappear and a cooling of the filament or heater takes place. After the cooling process is completed the ends may again come together and signals will be received until expansion causes the ends to part once more. It may happen, too, in some localities, that a severe overvoltage is applied to the heater or filament terminals, due to line fluctuations. The heater tube may get so hot that it refuses to operate. When the voltage goes down the tube cools off and it returns to normal functioning.

The a. c. tube presents a more severe problem than does that of combined B and C voltage supply units. Line fluctuations may cause such changes in the filament, or heater, terminal voltages that the life of the tube is decreased, while in a power outfit the increase in plate voltage is balanced by a corresponding increase in C bias and thus the tubes will not be impaired.

Incidentally, the Samson Electric Company take great pride in the production of a power transformer which has a regulation of 5 per cent. That is, the voltage across the secondary at no load is only five per cent. higher than at full load. We understand some transformers have regulations as poor as 50 per cent., which may

explain why filter condensers in power supply units are prone to end their useful purpose at odd intervals.

Another peculiar case of "fading," which has nothing to do with a. c. tubes but is mentioned here as an interesting case of possible incorrect diagnosis, has been reported by A. H. Grebe and Company. An owner of a Grebe "Synchrophase" in Atlanta, Georgia, reported severe fading from WSB about two miles distant. A. Gilette Clark, the Grebe engineer who investigated the phenomenon, found that the fading took place in regular intervals, that the maximum signal did not increase beyond a certain value which was normal for a station of WSB's power and distance, but abnormal for a case of ordinary fading in which the peak values of signal strength might be somewhat high, and finally, after a little time, it was discovered that the fading coincided with the passing of a trolley car, starting as the car approached and continuing for some time after the car had passed. This gave a clue upon which to work.

"The most logical explanation that occurred to us," says Mr. Clark, "was that the trolley line, coming so close to the transmitter and receiver, acted as a conductor for 'carrier,' or wired-wireless, propagation of the WSB signals. Instead of coming in a direct line, the signals received by this set probably followed the wires. Now, in technical parlance, we find that if 'standing' waves were present on the trolley wire and track, in the fashion of Lecher wires, the trolley cars might act as sliders along the two parallel wires, reducing the volume delivered to the set as they passed through points of high potential and not affecting it when at the nodes.

"Further investigation with regard to the trolley car clue revealed that the lines of this car service ran directly parallel with the WSB antenna system in Atlanta and almost parallel to the set owner's antenna system, although the course altered slightly at about half way between the station and the point of reception. Since the nodes are one half a wavelength apart and the station's wavelength is 476 meters, the action should take place every time the car moves 238 meters, or about 785 feet."

**"Gross
Exaggerations"**

ANOTHER wonderful new invention has come to the attention of the Laboratory. It costs but $4 and anyone wanting a good job can make from $150 to $300 a week selling the gadget to gullible radio listeners! It will:

(1) Eliminate 50 to 90 per cent. of static
(2) Increase volume
(3) Save 30 to 40 per cent. on batteries
(4) Separate short-wavelength stations
(5) Bring in distant stations
(6) Tune-out powerful local stations
(7) Add a stage of amplification to your set

We suspect it is a wave trap, consisting of less than a dime's worth of wire, a cheap variable condenser, a box, and two binding posts.

**Interesting
Technical
Literature**

THREE booklets have arrived in the Laboratory recently. One is a handsomely bound treatise on Cunningham tubes. It contains, in eighty-four large-size pages, data on all tubes made by this well-known tube manufacturer; characteristic curves, as well as a considerable amount of information that will appeal to all who have read the tube articles appearing from time to time in RADIO BROADCAST, are also given. The book sells for $2.50.

$\mathcal{B}$ CARL DREHER

R. F. In Vaudeville

SOME twenty years ago, when most of the great figures in the radio world of to-day were still in the pants business, or sending comments to each other in Morse across the barren acres of West Farms in New York regarding the excellence or otherwise of their "sparks," a popular type of vaudeville act was evolved from the work of Nikola Tesla on high-frequency currents. The Tesla coil consisted of two windings, with non-ferric coupling; the primary was part of a closed oscillating circuit, with a condenser and spark gap. The condenser was charged by a spark coil or high-tension transformer. The secondary of the Tesla coil usually led to a spark gap. The arrangement was no more or less than a wireless transmitter with the open, radiating circuit replaced by another spark gap. In the larger sizes the Tesla coil proper had to be immersed in oil, to control the corona. Very high voltages could be secured at the surface of conductors, including the human body, a man could allow potentials of a few million volts to leap to a metal ball held in his hand, without hurting himself, but scaring the wits out of anyone who did not know what he was up to. Hence the vaudeville application. I did it myself on a small scale, with a Tesla coil fed from a quarter-inch spark coil. I proudly published an article on this marvel which, I asserted without fear of contradiction, was the smallest Tesla coil in the world. The secondary was wound on a test tube, and protected by several layers of empire cloth from the primary, which consisted of a few turns of heavy rubber-covered wire. In its full glory the apparatus threw a half-inch spark which stung slightly when taken on the bare skin, but produced no sensation at all if allowed to jump to metal in contact with the body. I astounded my family and the neighborhood with it, at the age of twelve, while various "professors" were profiting by the same stunt at the sublimated nickelodeons, using much larger coils, of course. Incidentally, Doctor Tesla, at that time an electrical investigator of the highest renown, and in his productive period, was something of an amateur vaudeville actor himself, and once sent Sarah Bernhardt into hysterics by sticking his head into a ball of high-frequency fire generated in his laboratory. Bernhardt was harmed more than he was.

But to return to the vaudeville aspect. It sank into partial oblivion for a decade or two, apparently. It is now being revived. In *Variety*, the theatrical magazine, a review appeared a while ago under the following caption: "Bernays Johnson (4), Electrical Novelty, 18 Mins.; One and Full Stage, Hippodrome (V-P)." The review was by the talented "Abel." He put the seal of his approval on the act, devoting a full column to the task, and ending with this sentence: " nson is a natural for thrill exploitation." The grounds for this verdic are worth recounting.

After speaking of the stunt as a "dignified ballyhoo act," Abel pays tribute to Johnson as a "corking showman without being 'professional' or show-wise in his manner of speech or presentation," which begins with a "scientific" exposition before the curtain. Then follow some high-frequency demonstrations, such as the flipping of a lamp without wires, the frying of an egg ditto, and the transmission of the mysterious energy through a bowl of goldfish and the human body, "employing a comely woman for this demonstration." A beautiful girl will help sell anything, as the hosiery manufacturers know. There is also a "defiance-of-gravity" experiment, the idea of which, as explained by Mr. Johnson, is to improve transportation by allowing the traveler to get up into the air quite a way, whereupon the earth obligingly revolves beneath him, until he lets himself down where he wants to go. Jules Verne stuff.

But this is only the introduction. "The big punch in Johnson's act," Abel tells us, "is his billing as 'the man who defies the electric chair.'" The chair is stated to cost $6000. "Johnson announces that he will receive 350 amperes of current through his body, thrice the quantity necessary to electrocute a human being. Johnson states that it is fear which paralyzes an electrocuted person and not the actual juice that kills him, inferring that possibly the ensuing autopsy has something to do with the physical destruction of a condemned murderer. Johnson contends that by keeping cool and dry (perhaps a trying condition for the average death-house victim), one can withstand the shock. On that theory, Johnson presents its demonstration. To further prove that the juice is actually passing through his body, a bar of metal is burnt to white heat from a wire on his person, the dismembered piece is caught in a pan by a 'nurse.' Two male attendants in regulation prisoner's uniform complete the cast.'

Of course if even an ampere of honest-to-goodness d.c. or low frequency a.c. passed through Bernays' bones his soul would fly to heaven with the speed of light (300,000 kilometers a second), and his body would fall, a veracious carcass, on the Hippodrome stage. But as he is working well up in the kilo- or megacycles he gets away with his fustian, and no doubt provides good entertainment for the customers. He doesn't let it get under his skin, to use the phrase literally. I am told, however, that he is unable to follow his own advice to the death-house inmates to keep cool and dry. At least, one close-up witness to Brother Johnson's demonstration told me that he (Johnson) was sweating copiously and looked uncomfortably scared while working his magic at the last radio show in New York. He may have been thinking of what would happen to him if his Tesla transformer broke down. Such misgivings would not be irrational. I wish Bernays luck, but, in my capacity as a consulting engineer, bestow on him gratis the advice that he had better filter and dry the oil in the said transformer frequently.

Rosa Ponselle Before the Microphone

I CAN see where singing before the microphone must be simpler in some ways for, say, an opera singer, than performing before an audience. Not as much voice need be used, and acting becomes unimportant. But if any one thinks that, on the whole, singing for the radio is less complicated than doing it on the stage, he should watch Miss Rosa Ponselle modulate her way through a Victor hour, as I did recently. Lately I have become interested in how great artists get their effects, as far as one can judge outwardly, and wherein they differ from the amiable but depressing amateurs who helped us out in 1923.

Miss Ponselle is a handsome and robustly built young woman, who looks the prima donna and would get a seat in the subway, even in this unchivalrous age, if she ever rode in it. She does not appear at all nervous as she faces the microphone, although there is some tension in her attitude—mainly energy and determination. She stands about three feet from the microphone and not much farther from the conductor, who is one of the regular maestri from the Metropolitan. The orchestra of about thirty-five is behind her, and a mixed chorus of about the same number is back and to one side, fully thirty feet from the transmitters; but they come through very well. Miss Ponselle, when she is in action, does five things; she sings, acts the rôle in a modified way, varies her distance from the microphone, fopows the conductor, and watches her own monitor or coach who signals to her from the control booth. Talk about coördination! The last stunt is accomplished as follows: The singer faces a side of the studio in which there is a large window through which she can see the group in the monitoring booth, consisting of a few engineers and program people, maybe a vice-president or two, and Miss Ponselle's assistant. The window is double glass, with an air space intervening, and those on the booth side hear only through a loud speaker. Miss Ponselle weaves back and forth as she sings, varying her distance from the microphone according to the loudness of the passages. During piano portions she advances to a point where her mouth is about eighteen inches from the transmitters, while when she wants to hit a note hard she may get as far away as four feet. By this device she sings the aria so that it sounds natural to her, and at the same time she compresses her volume range, as far as input to the microphones is concerned, for the best results on the air. The control operators have little to worry about, but Miss Ponselle has enough, because, after all, she does not hear the results of her divagations in a loud speaker. Her assistant supplies the loud speaker ear for her. When he considers that she is getting too close he moves his hands apart with the familiar "So big!" gesture of the fisherman; when she is

an ordinary artist is one of character as much as physical equipment, which is no doubt true, especially as no one can say where the one begins and the other ends, nor precisely how they interact. · · · ·

Miss Ponselle does five things when she broadcasts. But those are only the main heads. Each one of them includes a multitude of minor coördinations. The singing itself, for example—breathing and voice production and diction and all the rest of it. The acting, as I remarked, is somewhat mechanical in a broadcast rendition, but the fact that the artist retains it even when her audience cannot see her shows how intricately such habit-formations are organized. She evidently feels that, if she dropped it altogether, her vocal expression might suffer. As one watches the headliners one realizes, if it was not plain before, that this business of singing and playing is a hard game, if one wants to excel in it. So many things must be done at once, and all timed to a split second. For myself, I return thanks that my feet were never set on that path. I should as soon think of walking a tight-rope stretched over Niagara Falls and at the same time demonstrating Green's theorem in mathematical physics to the newly-weds on the shore.

leading to the station, either through a transformer or by means of a loud speaker placed before a microphone. I should imagine that the transformer coupling would be preferable. A number of outdoor antennas are available. The procedure of listening and re-broadcasting is quite intricately systematized, there being a scouting operator who catches stations for Mr. Godley, the latter exercising his fine hand only on valuable prospects, such as West Coast broadcasting stations. Mr. Godley listens to the signals as they sound on the air after re-broadcasting, by means of a monitoring receiver tuned to WAAM. Special attention was paid to eliminating electrical noise in the neighborhood of the listening post, so that man-made static has been reduced to a minimum. The DX re-broadcast is said to be a popular feature on WAAM's programs.

WOC

THE well and favorably known Davenport, Iowa station stays on its assigned frequency by means of a beat frequency indicator developed by the Washington Radio Laboratories, of Washington, D. C. This device mixes the output of a crystal oscillator, calibrated by the

TWO OF THE 3 LO STUDIOS AT THE MELBOURNE STATION
The illustration on the right shows the main studio of the station. The studio is only partially acoustically treated and the presence of spectators in the wicker benches helps to deaden the room. In both these studios the use of an illuminated device for communicating with the performers while they are before the microphone is shown. This signal system has gradually been abandoned in the United States with the increase in the number of professional microphone performers

too far from the pick-up he advances the parallel palms of his hands to within a few inches of each other, and when she is hitting it right the coach signals with a short vertical gesture, "palms down, quite like an umpire saying "Safe!" to a runner who has just slid into a base. He also nods his head, to reassure the diva still more. Almost all the time she is right, but no doubt she gets considerable comfort from the signalled affirmations to that effect. It is very comical, as she goes more or less mechanically through the actions natural to the aria—clasping her hands on her chest and so on—to see her glancing constantly, with the hundred-thousand candle power eyes which are part of her operatic equipment, now filled with mute appeal quite apart from the beauty of the notes, to see whether she is treating the microphone properly. It is also impressive. Genius, it has been said, is an infinite capacity for taking pains. It is that, partially, and in a broadcast studio as much as anywhere. Maybe one of the obscure girls in the chorus started with a voice as good as Ponselle's, but was too easy-going to get very far with it. All that amounts to saying, of course, is that the difference between a successful opera singer and

Among the Broadcasters

WAAM

WAAM of Newark, around this time of the year, is just getting through with its winter DX re-transmissions. The transmitter, situated in Newark, New Jersey, is fed for this purpose from a battery of receiving sets in a more rural location at Cedar Grove, presided over by no less a distance shark than Paul F. Godley. It was Paul Godley, as no good radio man has forgotten, who first received American amateur signals in Europe, something like six years ago. He sat in a tent on a beach in Scotland, and picked up the signals of U. S. hams on a Beverage wave antenna. The phones kept his ears from freezing. Now he listens in a sort of dug-out underneath a house—what might be called a radio cellar—and probably keeps more comfortable. The DX re-broadcasting starts at about 1 A. M. Several receivers are used, the main one being an eight-tube loop outfit using a.c. tubes, affording four stages of radio amplification, a detector, and three stages of double-impedance audio amplification. The receiver is coupled to the land line

Bureau of Standards, with radio-frequency from the broadcast transmitter. The latter voltage is picked up by means of a small antenna. The two radio-frequency voltages are mixed in a compartment which is shielded from the standard oscillation generator. A beat frequency is produced in the output of a rectifier tube. This beat frequency indicates its presence in two ways. When it is sufficiently high it produces an audible note through a loud speaker. Below about forty cycles per second the indication is visual. A relay capable of following fluctuations up to forty per second lights a green light when the broadcast transmitter is radiating the same frequency as that generated by the standard oscillator. When the transmitter frequency deviates a red light glows, or the red and green lights flicker in alternation. If this frequency is above fifteen per second both lights appear to be lit continuously, owing to the well-known phenomenon of persistence of vision. The operator of course endeavors to keep the transmitter frequency so close to his standard that the green light glows alone, or the two lights alternate very slowly, corresponding to a beat note of only a few cycles per second.

AT PENNANT HILLS, SYDNEY, AUSTRALIA
The transmitting room whence broadcasting programs received in England, Canada, and the United States have issued. In addition to the broadcasting transmitters, there is a marine transmitter at the extreme left, while the other panels form the short-wave set

The device operates on batteries, the voltage of which must naturally be kept constant. Meters are provided for checking. The highest voltage required is 100. Storage batteries are recommended as a source. Like other piezo-electric controls, this indicator must be kept in a constant temperature box if its indications are to be relied upon.

3 LO, Melbourne

WE PRESENT photographs of the studios and control equipment at the Australian station, 3LO. The main studio, it will be noted, contains plenty of wicker settees, for visitors. There is room for two hundred in the studio, which is therefore as much an auditorium as a broadcasting studio, and spectators are not merely welcome, but necessary. Every one adds a definite amount to the total absorption of the room, and when the place is filled the period of reverberation is markedly reduced. The platform for the performers is partly covered with carpet, as the picture shows, and the wall nearest the camera is draped. The absorption of this end of the studio is fixed, therefore, and with an audience in the body of the room the period is about right for broadcasting, there being enough reverberation to produce a lively effect, and not enough to set up standing waves and disturbing rattles.

The announcer, it would appear from the picture of the main broadcasting hall, sits at the table on the right and talks into his own microphone. The concert-microphone is stage center. Each microphone box is surmounted with a signal lamp. There is also an illuminated device for communicating with the performers during a number; it may be seen hanging from one of the rafters, over the center aisle. This scheme was formerly used at some United States stations—WGY, for example—but with the advent of professional performers it is no longer very serviceable. At most of the large stations many of the artists are on the air several times a week, and they know precisely where to stand and what to do for given effects.

The photograph of the control room shows a good-sized PBX, and an amplifier panel not unlike the American layouts. The boxes on the table may be portable amplifiers.

Sydney, Australia

ANOTHER accompanying photograph gives one a good idea of the transmitting room at the Pennant Hills station whence broadcast programs received in England, Canada, and the United States, have issued. The control table with a monitoring receiver and apparently a small telephone switchboard is in the foreground, with a telegraph table to the right. The transmitter panels stand along the rear wall. The marine transmitter is at the extreme left; the other panels comprise the short-wave set. The design is a combination of panel and pipe-rack construction probably all right if the place doesn't catch fire.

Studio Scandal

MOST tenors start rehearsals with nicely starched collars and cute neckties, only to tear off both before they get halfway through.

Few male singers appear in evening dress. The headliners, with occasional exceptions, dress very informally. Soft collars, in blue or some other fairly bright color, are the rule. When WJZ had its studio on Forty-Second Street in New York City one Italian tenor got down to his undershirt on a warm night. He wouldn't sing in his shirt, so what could the studio staff do? But ultra-modern studios are artificially ventilated, so that the air remains at sixty-eight degrees Fahrenheit, winter and summer, while the humidity is kept at the optimum figure of fifty. Thus the artists remain comfortable, although dressed. They are right, however, in taking off whatever incommodes them. The late Victor Herbert, in the year when he conducted the Stadium concerts during the summer in New York, used to rehearse the men clad in his white flannels from the waist down and only B.V.D.'s from the waist up, on hot mornings. He was a fine old Berserker, and looked better with his chest bared to the zephers than when he appeared, starched perforce, before the *haute monde* in the evening.

Studio Slang

THE operators used to yell, "You're on the air!" when handing the program over to studio or to the field forces. This phrase has changed. "Take it away!" is now *de rigueur*. "It" refers to the program, of course. This phrase seems very picturesque to me and I get a certain kick out of hearing it. The NBC slang for "field" or "outside" annoys me, on the contrary. It is "Nemo" and was originally adopted for purposes of camouflage in some obscure and forgotten contractual tangle. But now that it has become a habit it cannot be eradicated and I expect to hear it for the next forty years, unless broadcasting kills me earlier.

When Orchestra Leaders Sing

THE best conductors seem to have the worst voices, and when they sing a bar to show the musicians what they want even the page boys around the studio snicker behind their handkerchiefs (when they have any). Mr. Setti, the brilliant maestro of the Metropolitan Opera Company who frequently broadcasts incognito, should be excepted. He has a good voice and sometimes sings right through a selection, accompanying the orchestra. He may have been

CONTROL ROOM OF 3 LO, MELBOURNE

an opera singer himself once, before he started conducting for them. Now and then he goes flat, but it does not bother him, for after all his singing is supererogatory. He annoys the control operators, who maintain that his singing interferes with their getting a balance during the rehearsal. But none of them has ever had the nerve to go out and tell him so. They are right; but if a man wants to sing it is generally dangerous to interfere with him. He will let you criticize his wife, his children, and the shape of his head before he will let you stop him from singing when the spirit moves him.

An Engineer's Embarrassment

IT IS true that the more a man is educated, soundly, the less he is surprised at anything. And conversely, if he knows little, he is frequently astonished at perfectly rational behavior which happens to be beyond his comprehension. In the last few months I have had occasion, quite often, to venture into unfinished studios where painters were at work, there to warble various notes or to clap my hands to get some idea of the reverberation. The results have been almost as dreadful as when Mr. Hanson and I invaded a church for a similar purpose, as recounted in these columns a while ago. Some of the painters have nearly fallen off their scaffolds, causing me no small amount of anxiety lest I should give rise to a damage suit against my company. Others have apprehended that I was mocking them, and the looks they directed toward me said as much as that they were ready to paste me in the eye or to go on strike. In most cases I have slunk out after banging my hands together much fewer times than I had intended. Then the painters would glance at each other, and I knew as soon as the sound-proof door closed they would say, "The poor nut!" Well, they may be right, but not on the grounds they thought. What I was doing was just as rational as painting.

Radio Inspectors—Fine Fellows

WITHOUT doubt there is something magical about radio. Have you ever considered the remarkable infrequency of bureaucrats among the government employees in the business? The fact that where I have my office now no transmitting apparatus exists, so that the radio inspectors do not visit the place, has made me think of them. The men in the U. S. Supervisors' offices are a fine lot of fellows. It is rarely that an operator is high-hatted by one of them. On the contrary, the inspectors and their assistants frequently go out of their way in order to help some boy in difficulty about his ticket. They form a marked contrast to the poisonous snobs in the naturalization offices, for instance. Maybe part of the difference is due to radio; I like to think so, anyway.

Commercial Publications

THE Daven Radio Corporation, of 158 Summit Street, Newark, New Jersey announces "Super-Daven" resistance units, wire wound, available in values from 10,000 to 3,000,000 ohms, and accurate to within one per cent. The inductance is given as "practically negligible,"

with the distributed capacity likewise minimized. The temperature coefficient is 0.0001. At a slight additional cost, resistors are furnished with a closer tolerance than one per cent. and with zero-temperature coefficient. These units mount in clip holders. The line is offered for use as laboratory standard resistors, voltmeter multipliers, plate and grid resistors, high-voltage regulators, and in telephone and telephoto work. No doubt broadcasters will find uses for them.

J. E. JENKINS & S. E. Adair, of 1500 North Dearborn Parkway, Chicago, describe in their Bulletin No. 5 a wire-wound. gain control, Type GL 35. This instrument contains 11 separate resistance units, each wound non-inductively with enameled Nichrome wire. The total resistance is 350,000 ohms, arranged in logarithmic steps, resulting in a straight-line TU variation. The resistance values are good to less than one per cent. The housing is an aluminum shield, the end pieces being three inches square; a depth of 4¾ inches is required behind the panel.

IN MARCH we started a review of the pamphlet entitled Samson Broadcast Amplifier Units, issued for limited distribution by the Samson Electric Company of Canton, Massachu

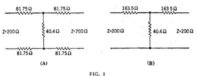

FIG. 1

setts. This detailed review will now be continued. On Page 18 we find a discussion of the design of a volume control arranged in 2-TU steps, with a maximum resistance of 100,000 ohms. The 2-TU drops may be secured by tapping off 79 per cent. of the voltage each time, or 79 per cent. of the resistance, as long as the voltage is proportional to the resistance. The first tap would therefore be at 79,000 ohms. The description is not entirely clear in the pamphlet, and there is an error in that the first tap is given as 59,000 ohms. On Page 19 there is further discussion of gain controls, and the design of one covering a range of 10 TU in 1-TU steps is given. This starts with a minimum, or first tap value, of 161,000 ohms, and continues in progressively increasing steps until the whole 10 TU have been covered. Another potentiometer affords a range of 50 TU in 10 steps of 5 TU each. The total resistance of each device is 500,000 ohms. One of these may be used across the grid and filament of one tube of an amplifier and the other across the input of a succeeding tube, thus providing fine and coarse regulation of gain as required.

The material presented under "Pad Design" will prove extremely useful to many broadcasters although there is no attempt to derive the formulas given nor to adhere to a conspicuously logical sequence in the discussion. The relationship between the impedance on either side of a conventional resistive T- or H-network, the impedances of the legs, and the TU loss introduced, are given at length and worked out to a practical conclusion in the form of a table (Table III). From this tabulation, knowing the TU drop

desired, one may read the value of Z_1 and Z_3 for Z equals 200 ohms and Z equals 600 ohms. Z is the impedance on either side of the pad (only bilaterally symmetrical pads are discussed); Z_1 is the total value of the series or X-legs; Z_3 is the value of the shunt or Y-member. For example, if one requires a 20 TU pad presenting 200 ohms each way, one finds from the table that Z_1 must be 327 ohms, while Z_3 is 40.4 ohms. The result, for an H-network, would be the pad of Fig. 1-A, while if a T-network with the same electrical characteristics is chosen the arrangement will be that of Fig. 1-B. Numerically the difference is simply one of splitting Z_1 into four parts or into two parts, for the corresponding number of legs, depending on whether a balanced network is required or not.

On Page 21 there is apparently an error, L_1 and L_4 having been printed, for Z_4 and Z_b, respectively. The article as a whole seems to have been written by an engineer who knew what he was about but had no unusual skill in presenting the subject for the education of others. This defect is not extremely important, inasmuch as most of the people who consult the pamphlet will be more interested in the results than in the procedure required to reach them. In some ways the treatment is more extensive than that given in the article devoted to the same subject in this department for September, 1927, particularly in the working out of the table, but in other respects it is less thorough than the latter, and readers who are much interested in pad design might go back to the September, 1927, RADIO BROADCAST article after reading the Samson pamphlet, and also to the discussion of the General Radio write-up on pads in the January, 1928, RADIO BROADCAST.

The limits within which all these pad design hold good are well stated in the Samson booklet which broadcasters may secure by writing for i on their letter-heads; this portion is well wort quoting:

"It is assumed that the transformer is ideal, tha is, that it has a negligible resistance in its wind ings, and an infinite input impedance with th output windings open-circuited, and vice-versa It is also assumed that the leg or other circui element into which the transformer is workin has pure resistance of the impedance value give as the impedance which the transformer is de signed to work. It is also assumed that th transformer has no leakage reactance, that i that all magnetic flux which links one windin links the other. In order that all these differen things may hold, it is necessary that the con sideration be based on a certain range over whic the transformer practically meets all these con ditions, and where those not depending on th transformer hold. Of course, as the frequency i reduced more and more, or as it is increase more and more, below and above this rang respectively, these assumptions must fall dowr Therefore, it must not be thought that a 'pa is a pad' over all ranges of frequencies. Howeve over the range that the circuit element is de signed to work, these ideas hold very closely and the simple addition of the attenuation c individual pads to obtain the total attenuatio gives the desired working result, and that is th justification for its wide use in communicatio circles."

No "Motor-Boating"

A Quality Audio Amplifier Which Will Not "Motor-Boat"

By H. O. Ward

RADIO BROADCAST Photograph

A HIGH-QUALITY RESISTANCE-COUPLED AMPLIFIER
The two condensers and the resistance in the lower left-hand corner prevent the amplifier from "motor-boating." The loud speaker connects to the two Fahnestock clips alongside of the output condenser; the two similar clips in the lower right-hand corner are used to read the plate current of the last tube

SEVERAL years ago the resistance-coupled amplifier was much in vogue, and justly so, for it is an excellent type of audio amplifier and is capable of giving practically equal response over the entire audio-frequency range. Its popularity at that time was comparatively short-lived—just why, it is hard to say. Since the introduction of the 240 type tube some time ago, however, there have been indications that the resistance amplifier will stage a comeback.

The amplifier described in this article has a flat frequency response curve from 60 cycles up to about 6000 cycles, it will not "motor-boat" when operated from any ordinary B power unit, and it has a voltage amplification of about 400 from the input of the amplifier to the grid of the power tube. A curve showing the frequency response curve of the amplifier is given in Fig. 1. In Fig. 2 is given the circuit diagram of the amplifier. There is nothing unusual about this amplifier with the exception of the anti "motor-boating" circuit which we have indicated by enclosing it in dotted lines. This anti "motor boating" circuit originated in the engineering department of the E. T. Cunningham Company a short while ago, and the circuit, tested in RADIO BROADCAST Laboratory, has proved very satisfactory. It should be noted that the circuit is arranged so that the plate current of the detector and first audio tubes must pass through resistance R_1.

To test the circuit the amplifier was connected to a B power unit and with the resistance R_1 short-circuited the amplifier immediately began to "motor-boat." As soon as the short-circuit was removed the "motor-boating" stopped. It was found that the value of R_1, necessary to produce stable operation of the amplifier, varies with different B power units. With some power units a resistance of 20,000 ohms is sufficient, while with other power units

a resistance of 100,000 ohms is necessary. Since the latter size resistance seems to be effective in all cases, it is suggested that those who construct the amplifier use this value of resistance. As the plate current of the first tubes in the

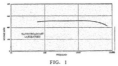

FIG. 1

amplifier and the detector tube must flow through this resistance, it might be thought that there would be an excessive loss in voltage across it. Such is not the case, however, as the following data will show.

For this test the amplifier was set up with a detector circuit connected ahead of it. The detector tube used was also a 240 type tube and the same voltage was applied to the detector tube as was applied to the first and second audio tubes in the amplifier.

This table indicates the voltage actually at the plates of the detector and first audio tubes for two different values of applied voltage. A difference of 10 or 20 volts at the plate of a tube will not change the characteristics of the amplifier and since the loss in voltage due to the resistance R_1 is of this order, it is evident that the amplifier will not be affected.

The following data were obtained:

APPLIED VOLTAGE	R_1 OHMS	VOLTAGE AT PLATE: DETECTOR	FIRST AUDIO
135	0	43	83
135	100,000	40	93
190	0	55	115
190	100,000	43	93

The reason that the detector plate voltage is lower than that of the first audio stage for the same values of resistance is because the current drawn by the latter is less than that drawn by the detector.

The recommended values of plate and grid resistances and coupling condensers are as indicated in Fig. 2 and the frequency response curve shows that the values specified give the desired flat characteristic. If larger values of resistances are used the gain of the amplifier will fall off at the high-frequency end. Larger values of coupling condensers tend to better the response at low frequencies but since the response is satisfactory with the values given, it is a waste of money to use any larger condensers unless you have them on hand.

The correct voltages to use on the grids and plates of the three tubes of the amplifier are given below. The letters B_1, C_1 and B_2, C_2 in the table below refer to similar notation in Fig. 2.

1ST AND 2ND AUDIO TUBES	
Grid Voltage C_1	Plate Voltage B_1
1.0 to 1.5	135
1.5	157
3.0	180

POWER TUBE			
112 TYPE		171 TYPE	
Grid Voltage C_2	Plate Voltage B_2	Grid Voltage C_2	Plate Voltage B_2
6	90	16.5	90
9	135	27	157
10.5	157	33	157
		40.5	180

The following parts were used in the amplifier:

1—Durham 0.1-Megohm Resistor
3—Durham 0.25-Megohm Resistors
3—Durham 2.0-Megohm Resistors
3—Sangamo 0.006-Mfd. Fixed Condensers
6—X-L Binding Posts
3—Benjamin Sockets
1—4-mfd. Tobe Condenser
1—Amertran Choke Coil, Type 854
5—Fahnestock Clips
1—6-Ohm Pacent Rheostat
4—1 Mfd. Dubilier Bypass Condensers
3—Lynch Double Resistor Mounts
1—Lynch Single Resistor Mount

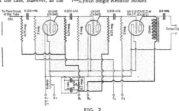

FIG. 2

433

A TWO-TUBE SCREEN-GRID RECEIVER

Maximum efficiency is obtained by using metal front and sub-panels while there is also a metal plate
between the r.f. and detector tube circuits. The active parts used in this receiver are listed on
page 435, and the circuit diagram is also given there. A copper cylinder is used to shield the tube

An Experimental Screen-Grid Receiver

Charles Thomas

THE introduction of the screen-grid tube is most opportune for that section of the radio world which is constantly in search of a new plaything. The UX-222 (CX-322), or screen-grid tube, offers the advantage of an extremely high gain per stage when used with the inside grid as control grid and with a steady polarizing voltage impressed on the screen. As is to be expected, the high amplification factor is accompanied by a very high plate impedance, necessitating a high impedance in the coupling unit if the advantage of the high amplification factor is to be realized to its fullest extent. While the screened-grid tube requires no neutralization, careful shielding is necessary, particularly if several stages are used.

Elaborate shielding is not required if only a single stage is used. For this reason, a single-stage amplifier adapts itself to preliminary experiments with the new tube. The set described here is purely experimental and no claims are made as to the results obtainable. Tests have shown that it is stable, and results in operation approach those obtained with three tuned circuits and fixed input in a three-stage amplifier using standard tubes. The selectivity without regeneration is comparable to that obtained in a normal two-circuit system but the figure of merit of both circuits is somewhat reduced because the first circuit is coupled to the antenna and the second feeds the detector. Two-circuit selectivity is insufficient for broadcasting reception in this country, hence the detector is made regenerative. Regeneration more than compensates the losses

WITHOUT a doubt the screen-grid tube is attracting the attention of every serious experimenter and engineer in the radio field. Readers of RADIO BROADCAST are by this time familiar with the theory of this tube, and something of its operation and application has already been included in the contents of this magazine. The following brief description of a two-tube tuning unit, to which may be added any audio amplifier, is the forerunner of a construction article telling exactly how to build the receiver. This article, and the completed receiver, which is being thoroughly engineered, are products of a well-known engineer, whose real name, unfortunately must be hidden with a pseudonym. Articles on the new screen-grid tube appeared in RADIO BROADCAST as follows: "Applications of the Four-Electrode Tube" December, 1927; "The Screened-grid Tube" January, 1928; "The Screened-grid Tube" February, 1928; "A Four-Tube Screened-grid Receiver" March, 1928 —THE EDITOR.

in the tuning unit feeding the detector and improves both selectivity and sensitivity.

The shielding required, as shown in the photograph, is not very elaborate. Brass panel and baseboard are used, and a brass partition between the two stages is also advisable. The screen-grid tube is enclosed in a copper cylinder which fits closely around the tube and extends about half an inch above it. It is also necessary to shield the short lead between the plate of the screen-grid tube and the detector. In general, shielding is required between all parts of the plate circuit and all parts of the control grid circuit.

The circuit of the experimental set, which includes only detector and radio-frequency amplifier, follows conventional lines. The antenna is coupled through a tapped coil to the grid of the screen-grid tube. The position of the tap controls to a certain extent the gain and selectivity of the set. Its position must be determined experimentally. The plate of the radio-frequency amplifier is coupled to a tuned impedance. Parallel coupling is used, the d. c. plate circuit going through a radio-frequency choke to the plate supply.

The method of regeneration control is somewhat unusual. The regeneration coil L_4 is not appreciably coupled to the grid circuit. Re-

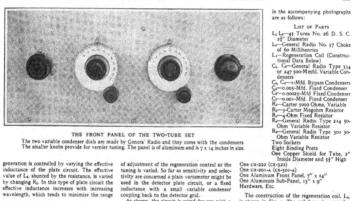

THE FRONT PANEL OF THE TWO-TUBE SET

The two variable condenser dials are made by General Radio and they come with the condensers. The smaller knobs provide for vernier tuning. The panel is of aluminum and is 7 x 14 inches in size.

in the accompanying photographs are as follows:

LIST OF PARTS

L₁ L₂—45 Turns No. 26 D. S. C. 2¾″ Diameter
L₃—General Radio No. 37 Choke of 60 Millihenries
L₄—Regeneration Coil (Constructional Data Below)
C₁, C₂—General Radio Type 334 or 247 500-Mmfd. Variable Condensers
C₃, C₄—1-Mfd. Bypass Condensers
C₅—0.005-Mfd. Fixed Condenser
C₆—0.00025-Mfd Fixed Condenser
C₇—0.001-Mfd. Fixed Condenser
R₁—Carter 5000 Ohms, Variable
R₂—3-Carter Megohm Resistor
R₃—4-Ohm Fixed Resistor
R₄—General Radio Type 214 50-Ohm Variable Resistor
R₅—General Radio Type 301 30-Ohm Variable Resistor
Two Sockets
Eight Binding Posts
One Copper Shield for Tube, 2″ Inside Diameter and 5½″ High
One UX-222 (CX-322)
One UX-201-A (CX-301-A)
One Aluminum Front Panel, 7″ x 14″
One Aluminum Sub-Panel, 13″ x 9″
Hardware, Etc.

generation is controlled by varying the effective inductance of the plate circuit. The effective value of L₄ shunted by the resistance, is varied by changing R₁. In this type of plate circuit the effective inductance increases with increasing wavelength, which tends to minimize the range

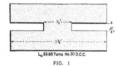

L₄, 93-95 Turns No.30 D.C.C.

FIG. 1

of adjustment of the regeneration control as the tuning is varied. So far as sensitivity and selectivity are concerned a plain variometer might be used in the detector plate circuit, or a fixed inductance with a small variable condenser coupling back to the detector grid.

As shown, the circuit is wired for use with a 6-volt battery and 201-A. type detector tube. Satisfactory operation using the method of obtaining regeneration described was not obtained with a 199 type tube as detector. The UY-227 (C-327) a. c. type tube may be used as a detector if a transformer of proper voltage is used to feed the heater. This tuning and detector unit should be satisfactory for use with alternating-current screened-grid tubes when, as, available. The parts used in the receiver shown

The construction of the regeneration coil, L₄, is shown in Fig. 1. The coil form is a wooden spool with an inside winding diameter of ⅜″ and a groove ⅜″ wide. An outside diameter of 1¾″ provides sufficient winding space. A coil wound with 93-95 turns of No. 30 d. c. c. will be found sufficient for the general run of 201-A tubes. Coils L₁ and L₄ may be General Radio coils type 277c with 10 turns removed.

If loud speaker operation is desired, any standard d. c. audio amplifier or a. c. power audio amplifier may be used with this unit.

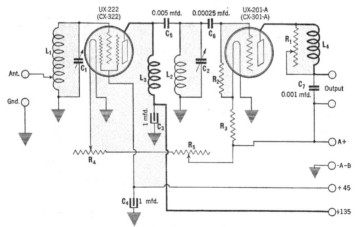

A SCHEMATIC DIAGRAM OF THE TWO-TUBE SCREEN-GRID RECEIVER

THE Roberts receiver, of which the Hammarlund "Hi-Q" is a semi-commercial model, was first introduced to the radio public by Radio Broadcast so many years ago that the author has neither the ambition nor time to go through his files to determine just when Dr. Roberts presented his first article. This momentary reminiscence perhaps has little in common with the point to be discussed in the present writing, but there is significance somewhere in the thought that this is the only circuit, of the many hundreds introduced in broadcasting's nebulous days, that has remained standard and popular to the present time. Simple efficiency is responsible for this consistent popularity.

The 1927–1928 Hammarlund Roberts "Hi-Q," described in Radio Broadcast for October, 1927, departed somewhat from previous models in mechanical and electrical design, though the ultimate effects are consistently in line with previous designs. The last two models of the "Hi-Q" receiver have incorporated variable coupling between the radio-frequency primary and secondary circuits. The possibilities of such an arrangement were pointed out by Zeh Bouck in an article appearing in the September, 1926, issue of this magazine, entitled "Higher Efficiencies in R. F. Amplifiers." The argument, in brief, is as follows:

At every frequency or wavelength there exists an optimum value of coupling between primary and secondary circuits—a value of coupling which provides the maximum signal intensity compatible with quality and stability. This optimum degree of coupling varies, however, with the frequency. To maintain optimum conditions over the entire tuning range, therefore, it is desirable that the coupling be varied with the wavelength. This is accomplished automatically in the Hammarlund-Roberts receiver.

The general characteristics of the Hammarlund "Hi-Q" receiver remain unaltered in the adaptation of this receiver for the use of a. c. tubes, as comparison of the circuits shown in Figs. 1 and 2 with the direct-current arrangement illustrated in the October, 1927, Radio Broadcast, will indicate.

The changes effected have merely been in the nature of the substitution of heater type a. c. tubes for the d. c. ones, accompanied by slight alterations in the constants of the circuit to compensate changes in tube characteristics. The receiver has been redesigned for the use of two different makes of a. c. tubes, the R. C. A. 227 (Cunningham 327) type and the Arcturus a. c. amplifier, detector, and power tubes. The selection of two types of tubes has been suggested by motives of general convenience.

Electrifying the "Hi-Q"

By F. N. Brock

THE "HI-Q" WIRED FOR CUNNINGHAM OR R.C.A. A.C. TUBES

The use of the R. C. A. tube in the "Hi-Q" receiver will be first considered.

The following is a list of the essential parts employed in the construction of the receivers:

1 Samson "Symphonic" Transformer
1 Samson Type HW-A3 Transformer (3–1 Ratio)
4 Hammarlund 0.0005-mfd. Midline Condensers
4 Hammarlund "Hi-Q" Six Auto-Coupled Coils
4 Hammarlund type RFC-85 R. F. Chokes
1 Hammarlund Illuminated Drum Dial
1 Sangamo 0.00025-Mfd. Mica Fixed Condenser
1 Sangamo 0.0001-Mfd. Mica Fixed Condenser
1 Pair Sangamo Grid Leak Clips
1 Durham Metalized Resistor, 2 Megohms
3 Parvolt 0.5-Mfd. Series A Condensers
6 Benjamin No. 9040 Sockets
3 Eby Engraved Binding Posts
1 Yaxley No. 660 Cable Connector and Cable
1 Hammarlund Roberts "Hi-Q" Six Foundation Unit

(Containing drilled and engraved Westinghouse Bakelite Micarta panel, completely finished Van Doorn steel chassis, four complete heavy aluminum shields, extension shafts, screws, cams, rocker arms, wire, nuts, and all special hardware required to complete receiver.)

For the construction of or adaptation of an existing "Hi-Q" receiver to one employing the 227 type tube the following additional parts were used in the adaptation:

5 Benjamin Green Top A. C. 5-Prong Sockets
1 Thordarson Type 2504 Filament Transformer (or Karas AC Former)

1 T200 Electrad Variable Resistance to Permit Temperature Regulation
1 0.5-Mfd. "Parvolt" Series A Condenser
1 200-Ohm "Truvolt" Grid Resistance (Electrad)
1 Samson 30-Henry Choke or a Samson Type O Output Impedance
1 2- or 4-Mfd. Series A "Parvolt" Condenser.
1 Electrad Type J Resistance for Volume Control
1 R. C. A. UY-227 or Cunningham C-327 Tubes
1 UX-171-A or CX-371-A, power tube.

CONSTRUCTIONAL DETAILS

THE construction of the receiver remains practically identical with that of the direct-current models. The general layout of the parts and the mechanical mountings have been described in detail in articles on the d. c. set and in the Hammarlund Roberts "Hi-Q" Six Manual.

The five-prong sockets are mounted in the same places and with the same screws as the old sockets. An extra hole for the cathode lead must, however, be drilled just under the K or cathode terminal. In the a. c. models of the Hammarlund "Hi-Q" the right-hand control (the rheostat in the battery-type receiver) may be used to control a 110-volt line switch, such as the Carter "Imp" type 115.

The similarity of the a. c. and the d. c. mechanical layouts is evidenced by comparing the accompanying photographs with those of the d. c. models which have frequently appeared.

The circuit of the Hammarlund Roberts "Hi-Q" Six receiver employing type 227 tubes is shown in Fig 1, in reference to which the following points are worthy of mention:

All filament or heater wiring should be made with a twisted conductor. It is desirable that consistency be observed in the socket connections with these power leads. In other words, it is preferable that the same heater terminal on each socket be connected to the same heater lead. This is most readily accomplished by employing a twisted pair of two colors. Corwico flexible red and black Braidite is a convenient recommendation.

The red and green leads on the Yaxley cable are not used.

Heater tubes are employed throughout the circuit (with the exception of the output amplifying stage) due to the simplicity and consist-

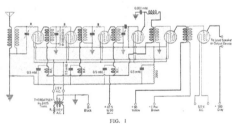

FIG. 1

ency of the circuit arrangement and the low hum characteristics of these tubes.

The Electrad T200 variable resistor is wired in series with the primary of the filament lighting transformer to provide a desirable amount of regulation of the secondary potential. The heaters of the 227 tubes should be operated at as low a temperature as will insure satisfactory reception. With the proper adjustment of the primary resistor it will take about 55 seconds for the tubes to reach an efficient operating temperature after the current is turned on. The life of the tubes will be considerably abbreviated if more than the rated operating filament potential is applied.

The bias to the radio-frequency and first audio transformer tubes is supplied through the drop across the Electrad 200-ohm grid bias resistor.

The pilot light is connected in parallel with the 171-a tube filament.

One side of the 0.5-mfd. bypass condenser across the grid biasing 200-ohm resistor is grounded to the chassis.

Sensitivity and selectivity may be controlled, in the usual manner, by varying the mechanical adjustment controlling the height of the primaries, particularly in the case of the last r. f. stage and the detector stage. Selectivity will also be considerably affected by the tightness of the antenna coupling. In order to attain satisfactory sensitivity and selectivity on the higher frequencies it will occasionally be desirable to use lower values of grid suppressor resistors than those recommended in the d. c. circuit, due to the alteration in the radio-frequency characteristics occasioned by the lower input impedance of the heater cathode type tube. The sensitivity of the receiver may also be increased by employing a higher resistance grid leak. The value of this resistor should, however, be increased cautiously with the possibility of overload on local stations in mind. In the case of a rewired d. c. receiver, originally operating with a radio-frequency plate potential of 67.5 volts from a B supply device, it is desirable to raise the voltage to about 80 to compensate the increased drain. The type J 200,000-ohm resistor is used for a volume control. This is mounted in the left-hand panel hole in the place of the filament switch employed in the battery set. The last three or four turns on the volume control (on the clockwise end, that is) should be clipped in order to give an "open" or maximum volume position.

The potentials, other than the a. c. voltage for the heaters of the tubes, indicated in Fig. 1, may be supplied either from B batteries or from an adequate B supply device, such as the Hammarlund Roberts "Hi-Q" Six power plant described in the Hammarlund Manual. The Thordarson 2504 filament transformer and this power unit will take care of all A, B, and C potentials.

USING ARCTURUS TUBES

FROM an electrical point of view the "Hi-Q" receiver rewired for the use of Arcturus a. c. tubes is practically identical with the 227 type tube design. Mechanically, the Arcturus system offers certain advantages which particularly recommend it for the adaptation of existing battery receivers. Arcturus a. c. tubes are of the four-prong heater type and they plug into the standard UX sockets without the use of adaptors and which, therefore, necessitate neither the use of special sockets nor a comparatively elaborate mechanical rearrangement.

In addition to the essential "Hi-Q" apparatus listed earlier in this article, the following extra

components will be required in the adaption or construction of the Arcturus model:

1 Electrad Royalty Type J Variable Resistor
4 Arcturus Type A. C. 28 Amplifier Tubes
1 Arcturus Type A. C. 26 Detector Tube
1 Arcturus Type A. C. 30 Power Tube
1 Step-Down Transformer, Having a 15-Volt Secondary, Such as the Ives Type 204, or the Thordarson TY-121.

A receiver employing Arcturus tubes is illustrated diagramatically in Fig. 2 and in the accompanying photograph. Referring to Fig. 2, it will be noted that the following alterations have been made on the original d. c. circuit:

The three fixed and one variable filament resistors are eliminated. Similarly all connections between grid returns and filament circuits are broken. The connection between ground and A minus is likewise removed. These changes are best made by completely rewiring the filament or heater circuits with flexible Braidite—red and black wires—twisted into a single pair. Connect the red wire consistently to the positive filament terminals on the sockets. These two leads are wired to the filament lugs on the Yaxley cable post, the red wire being soldered to the plus terminal (polarity, however, being meaningless at this point). Another pair can be led to the switch on the "Hi-Q" which later is connected in series with the *primary* (or 110-volt lead) of the filament lighting transformer for turning on and off the tubes. The switch must not be wired in the conventional manner, i. e., in series with the tubes themselves.

A fifteen-volt pilot light bulb can be secured from any store dealing in electric trains, and should be screwed into the socket provided for this purpose, and wired parallel to the tube circuit.

The grid returns from the radio-frequency amplifier, detector, and first audio-frequency secondaries are brought down to a common lead connected to ground, and this post should also be designated as "C Minus 1.5 volts." The 0.5-mfd. bypass condensers connected from the lower side of the radio-frequency primaries to the filament circuit in the original arrangement should be returned to the plus filament or cathode posts of the respective sockets.

The detector grid leak is disconnected from the A plus terminal of the socket and is brought down to a separate lead or post to be designated

as "D Plus 4.5 Volts." The detector r. f. grid return, i. e., the low-potential end of the secondary coil, is wired, as already indicated, to the common radio-frequency grid return

The grid return from the first audio-frequency amplifier is rewired as described, to the post marked "C Minus 1.5 Volts," which is grounded on the receiver. No change is made in the power tube socket.

A separate wire is led to the plus filament or cathode terminal of the detector tube, designated as "B Minus, C Plus, and D Minus."

The zero to 200,000 ohms Electrad Royalty or any other satisfactory variable resistor is connected across the secondary inputting to radio-frequency tube number two, and mounted in place of the rheostat.

Arcturus type a. c. 28 tubes are used in the first, second, and third r. f. stages and in the first a. f. stage. A detector tube, type a. c. 26, is plugged in the detector socket, and a power tube, type a. c. 30, into the power stage, which feeds the loud speaker.

OPERATION

THE operation of the a. c. "Hi-Q" receiver is practically identical with that of the d. c. model. The indicated connections to batteries and transformer should be made.

A. C. heater type tubes do not function efficiently as soon as the heater current is turned on. With the correct voltage (15 volts) applied to the heater terminals of the Arcturus tubes, it requires just 30 seconds for the tubes to heat to the proper operating point. The filament potential should be adjusted by means of the taps on the transformer until satisfactory operation is obtained. It is needless to say that the heaters must be given three to four minutes to cool before making additional adjustments of this nature. It is desirable, wherever possible, to utilize an a. c. voltmeter for the adjustment of the heater potential.

Any efficient B and C socket power device may be substituted for the indicated battery potentials. Briggs and Stratton are marketing an efficient A, B, and C power unit supplying all the necessary potentials for the operation of Arcturus tubes.

The various points mentioned in the recommended adjustments effecting selectivity and sensitivity in the 227 type tube receiver apply equally as well to the Arcturus arrangement.

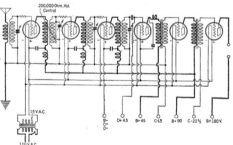

FIG. 2

A COMMERCIALLY AVAILABLE LOFTIN-WHITE RECEIVER
The Arborphone 37-AC set employs 3 r.f. stages, detector, and
two audio stages. The output audio stage is a push-pull one

An A. C. Loftin-White Receiver

By John F. Rider

PROMPTED by the great interest which hinged on the announcement of the Loftin-White circuit several months ago, the writer has endeavored in this article to describe in brief the outstanding features of a commercial receiver which makes use of this interesting circuit. The receiver in question has much to commend it, not the least important of its features being the fact that it is designed to use the new a.c. tubes.

The Arborphone 37-AC set, for that is its name, comprises three stages of tuned radio-frequency amplification, a non-regenerative detector, and two audio-frequency stages. The last audio stage, as will be seen by reference to the accompanying circuit diagram, is a push-pull one, making use of two parallel 171 type power tubes.

The radio-frequency stages use tuned transformers and a stabilizing system developed as a result of the combined efforts of Messrs. Edward H. Loftin, former Lieutenant-Commander, United States Navy, and S. Young White, a well-known radio engineer. The arrangement used in this receiver is really a modified version of the original, but in its function, is very similar.

This radio-frequency amplifying system accomplishes two things. In the first case it stabilizes the circuit, or the individual stages, whichever way we wish to view it, and secondly, it affords a certain uniformity of response over the tuning frequency spectrum.

A glance at the wiring diagram of the receiver shows the plate supply of the r.f. tubes being fed to the tube through a choke, and the plate coupling coil coupled to the plate through a variable condenser, C. This condenser, because of

its function, is greatly responsible for the stabilization of the stage. Its purpose is to change the phase of the alternating potential in the plate circuit due to the a.c. signal impressed upon the grid of that tube, so that it will not combine with the a.c. signal in the grid circuit. The maximum capacity of the phase-shifting condenser employed in such systems is approximately 0.0005 mfd., and it is usually adjusted to a point where the phase shift is such that the stage operates with a definite amount of feedback, or regeneration, which amount, however, is less than that required to cause a continued state of oscillation. The inductance value of the plate feed choke is of such proportion that, when resonated by its distributed capacity, its fundamental is above the longest wavelength which can be tuned-in with the receiver.

Referring to the method of obtaining what is called "constant coupling" between the plate and grid circuits of subsequent tubes, we find the system used differing somewhat from the original Loftin-White arrangement, An idea of the operating principle of the system can be gleaned from a study of the wiring diagram.

The plate and grid coils of the r.f. stages being inductively coupled, a certain amount of mutual inductance exists between the two coils. This mutual inductance is the path of energy transfer between the two coils, but the magnitude of energy transfer varies with the frequency of the signal. The higher the frequency (the shorter the wavelength) the greater the amount of energy transferred. The lower the frequency (the longer the wavelength) the less the energy transferred. But the coupling between the plate and grid circuits is not obtained solely through the mutual

inductance between the two coils. The fixe[d] densers in series with the grid circuits also [ ] tion as coupling capacities, but their co[ ] value is governed by the ratio between thei[ ] capacity and the capacity of the variabl[e] ing condensers in the grid circuits. Now m[ ] capacity behaves in a manner opposite to t[ ] inductance, being more effective on the l[ ] wavelengths and less effective on the s[ ] wavelengths. As the capacity of the tunin[g] denser is increased when tuning for the l[ ] wavelengths, the effect of the fixed con[ ] is increased. The converse is true when t[ ] ceiver is tuned to the shorter wavelength[ ] A graphical representation of the energy [ ] fer by means of mutual inductance and m[ ] capacity for a single stage is shown in Fig.[ ] idea of the overall energy transfer as a re[ ] the combined coupling mediums is also [ ] in Fig 1. The response curve of one stage wi[ ] combined coupling mediums is not a pe[ ] straight line, but has a depression aroun[ ] lower end of the broadcast spectrum than [ ] higher end. These data were obtained [ ] several measurements of different install[ ] which employed the Loftin-White system[ ] greater amplification on the shorter wavel[ ] is probably due to inherent regeneration[ ] complete response curve, however, is of ver[ ] formation. The usual capacity of the va[ ] condensers is 0.0005 mfd. and that of the [ ] coupling condensers is 0.004 mfd.

The use of a push-pull audio output [ ] affords certain advantages not obtained [ ] only one tube is employed in the output.[ ]

MUTUAL INDUCTANCE

MUTUAL CAPACITY

FIG. 1

COMBINED COUPLING

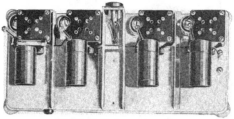

THE SHIELDED STAGES OF THE RECEIVER

it affords a greater signal output with much less distortion. Secondly, the increase in signal output is greater than the proportion of 1 to 2 tubes because somewhat greater input voltage can be tolerated without overloading or distortion. Thirdly, a push-pull output stage minimizes "hum" due to the use of a.c. on the filaments.

The receiver utilizes 226 type tubes for three radio-frequency amplifiers and the first stage of audio. The detector tube is a 227 and the push-pull output tubes are 171's.

Physically, the receiver is an interesting unit. In the first place the radio-frequency and detector systems are completely isolated from the audio and B power supply units. Each tuning stage is individually shielded, the whole forming one large can. The audio-and power supply systems combine to form another shielded unit, thus precluding reaction between the two amplifying systems and minimizing radio-frequency reaction between the power unit and the radio-frequency amplifier. The shielding material is $\frac{1}{32}''$ aluminum of high conductivity and low mass resistance; there is a double thickness between stages. The chassis of the radio-frequency system forms one part of the can and is grounded.

Each can contains the chokes, plate and grid coils, the necessary phase shifting and coupling condenser, and the tube socket. Single-layer solenoids are used for the radio-frequency transformers and are placed parallel to each other. Reaction between these inductances is eliminated by means of the shielding. The phase shifting condenser and its associated radio-frequency choke are located adjacent to the socket connected thereto. The adjustment of the capacity of the phase shifting condenser is accomplished by means of a protruding screw head.

The inductances are wound on bakelite tubes and the turns are spaced 0.002" by means of a machine, as the coil is wound. A layer of collodion sprayed upon the turns keeps them in place. The inductance value of these coils is such that, with the condensers used, the wavelength range is from 200 to 550 meters (1500 to 545 kc.)

The plug located between the two inner sockets and in the groove reserved for the drum dial, carries the connections for the plate and filaments of the tubes in the radio-frequency and detector portion of this receiver. The female portion of this plug is located in the container housing the audio amplifier and the power supply, and the power is fed to the r.f. system by means of this plug.

Rigid sockets are utilized for all tubes, thus showing that very little concern is placed upon the necessity of cushion sockets in the modern receiver. It seems as if we have very little to worry about microphonic tubes. A rigid socket appears satisfactory for the detector tube.

Four tuning condensers are used, one for each stage of radio-frequency amplification. All four are simultaneously controlled from one point and are actuated by means of a small knob attached to a drum dial. The four condensers are divided into two groups of two each, the rotors of each group being on one shaft. The two groups are then coupled together by means of a steel coupling unit. The condenser which tunes the input stage is so arranged that its rotor operates in conjunction with the other rotors, but its stator is located on a rocker arm, which can be actuated by means of a small knob located on the receiver panel. In this way it is possible to make easy adjustment to compensate antenna variations.

The condensers are of the straight wavelength-

line type and are very accurately and rigidly made. The bearings are of fabricated bakelite on steel. The condenser plates are wedged into grooves in the side spacers. All grooves are simultaneously milled with a gang cutter, hence the spaces are uniform. Supplementing this design, in the effort to obtain accurancy, short, stubby plates are used in place of long, thin ones. Bakelite end plates are used on the condensers and this material is used for insulating the rotor from the stator.

The method of testing the variable condensers and the inductances is novel and precise. Two radio-frequency oscillators are adjusted for beat note resonance. One unit is maintained as the standard. The condenser to be tested is applied to the other. Perfect uniformity would mean a zero beat. The tolerance value is a 200-cycle beat note.

The B supply comprises a full-wave rectifier employing a 280 tube. One transformer carries the windings necessary to supply the filament voltages for all the tubes used in the receiver. One winding supplies the 1.5 volts required for the 226's; another winding supplies the 2.5 volts for the 227; a third supplies the 5 volts necessary for the 171's; a fourth supplies the filament voltage for the 280 tube; a fifth supplies the plate voltage for the rectifier tube.

The primary winding of the power transformer is tapped for 110-, 120-, and 130-volt supply. A short-circuiting plug shorts a portion of the winding when the line supply is 110 volts. The entire winding is used for 130-volt systems. All filament windings, with the exception of the 226 winding, are mid-tapped right in the transformer, thus eliminating necessity for mid-tap resistances and adjustments. The 226 winding is equipped with a potentiometer shunt, whereby the correct electrical center can be obtained.

The B supply utilizes a two-section filter, incorporating two chokes and three reservoir condensers. The plate current for the 171's is caused to flow through only one choke, the filtering action of this one section being sufficient for the push-pull tube plates. The voltage reducing resistance is, therefore, a single mid-tapped unit arranged as a potentiometer across the power-unit output. The high end of this resistance supplies the 90 volts required for the plates of the first audio stage and the three radio-frequency amplifiers. This tap is fixed in the process of manufacture. The mid-tap of this resistance supplies the 45 volts for the detector tube plate.

All C bias voltages are obtained directly from the B unit. Because of the heavy plate current drain of the two 171's (approximately 38 to 40 mils.), it is necessary to isolate the loud-speaker winding. This is accomplished by means of an output transformer.

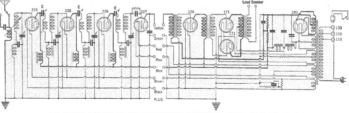

THE CIRCUIT DIAGRAM OF THE LOFTIN-WHITE RECEIVER DESCRIBED IN THIS ARTICLE

Radio Folk You Should Know

3. Lester L. Jones

AS PRESIDENT of the Technidyne Corporation in New York, Lester L. Jones plays an important rôle in radio research and laboratory investigation. His academic training as an engineer was received at the College of the City of New York, which, during the years when Alfred N. Goldsmith was a professor there, probably ranked as the foremost scholastic source of radio engineering personnel in the country. Mr. Jones graduated in 1913 with the degree of Bachelor of Science, *cum laude*. During the summer following his graduation he pursued special work at the College laboratories in various problems of radio engineering, including determination of the action of underground antenna systems, studies of the heterodyne system, using a Poulsen arc, and investigation of the characteristics of the then modern German quenched spark transmitters. The heterodyne tests were conducted in part with the Bush Terminal station of the National Electric Signaling Company, and presumably Mr. John V. L. Hogan was present at the Brooklyn end of the circuit.

In the winter of 1913 Mr. Jones was engaged as a civilian inspector of electrical and radio materials at the Brooklyn Navy Yard. He was responsible for the testing of all the radio transmitting and receiving equipment purchased by the Navy Department and delivered to the New York yard. In about a year this work led to Mr. Jones's promotion to the position of Expert Radio Aide, which included not only the former inspection responsibilities, but also the planning and testing of complete radio installations on battleships, destroyers, and submarines. While he was engaged in these specialized tasks Mr. Jones did not neglect the other branches of radio engineering, and the early issues of the "Proceedings of the Institute of Radio Engineers" frequently contain his name as a participant in the discussions, which were recorded at that time by the devastatingly charming Miss Nan Malkind, the only skilled and accurate radio stenographer who has ever appeared in the

art. In the 1914 "Proceedings," for example, there appeared a learned discussion by Mr. Jones on the subject of why the audion bulb causes a click in the receiving telephones when the filament current is shut off. Dr. Lee De Forest, the author of the paper that evening, observed laconically, "This is probably the correct explanation," a remark which must have been· pleasing to the younger engineer, and which has been preserved for posterity in the "Proceedings," together with many words of contrasting asperity.

As an Expert Radio Aide Mr. Jones was not confined to the New York Navy Yard. At various times his duties carried him to outlying land stations of the Navy Department, such as the post at Guantanamo, Cuba, to suggest improvements and to supervise installations of new apparatus. The position also included design of transmitting equipment for the special conditions of naval radio communication, supervision of manufacture, installation, and testing of models, and the preparation of specifications under which contracts were let for the furnishing of sets in quantity by commercial manufacturers.

The New York yard was primarily a transmitter-developing base. In 1917 Mr. Jones was transferred to the Washington yard, which specialized in naval receiver design and investigation. During the year and a quarter Mr. Jones spent at Washington, he was the civilian in charge of development of naval receiving equipment for use on battleships, submarines, and airplanes, including the well-known two-stage audio amplifier with non-ferric transformer coupling, which became the despair of many a graduate of the Harvard Radio School, although it was probably the best thing in its line at that period. The Washington Yard, incidentally, has some claim to rank with Brant Rock, the Aldene factory of the Marconi Company, and the G. E. test shop at Schenectady, as a nursery of famous radio men. Besides Mr. Jones, at various times during the war period Professor Hazeltine, William H. Priess, Joseph D. R. Freed, and

others worked there on the SE line of naval receivers and auxiliary apparatus.

In addition to these more or less orthodox radio duties, Mr. Jones was charged with investigation of war-time devices offered to the government by inventors confident that the offspring of their brains was required to beat the Germans. Machines for detecting submarines and killing the magnetos of aeroplanes were among them. Some of the ideas were insane and others offered practical possibilities. Only scientific analysis could separate the chaff from the wheat. But Mr. Jones did not spend all his time ·on radio development and related investigations at the Washington yard. At intervals he made observation trips in naval craft, in connection with submarine signaling, search-light communication, and other special problems.

Mr. Jones left the service of the Navy Department in the spring of 1918 and spent a short time in the employ of commercial radio companies which were supplying apparatus to the Army and Navy. In 1919 he established himself as a consulting engineer specializing in radio. Among his clients (in 1920-21) were the Mackay interests, then contemplating establishing their own transatlantic radio circuits on behalf of the Postal Telegraph-Commercial Cable system. The developments considered at that time have only recently been projected anew in the announcement of the Postal Company that long- and short-wave radio channels are to be operated as adjuncts to the cable ·circuits of the company.

Mr. Jones has patented numerous radio inventions at home and abroad. In December, 1925, he was elected a Fellow of the Institute of Radio Engineers. His career is an illustration of the value of broad technical training and experience in the radio engineering field. For every prominent radio man in the technical end who entered the business when broadcasting began to agitate the ether in 1920, there are ten who spent years in developing radio telegraph communication, while wireless telephony ·was still a poet's dream.

The Month's New Phonograph Records

Pagliacci—No Pagliacci 'Non Son! (Leoncavallo). By Giovanni Martinelli (Victor). Martinelli's powerful tenor voice combined with his dramatic ability fit him eminently to sing the emotional Leoncavallo music. He handles these two glorious selections magnificently.

Andrea Chenier Improvviso—Come un bel dì, Parts 1 and 2 (Giordano). By Arnoldo Lindi (Columbia). An imported recording of a fine Italian tenor who just misses being better than that.

Mazurka in B Minor (Chopin) and La Campanella (Liszt-Busoni). By Ignaz Friedman (Columbia). We would like to enthuse over this record because the Columbia Company · has done an excellent job of recording Mr. Friedman's fine display of piano technique · in La Campanella, but how can one enthuse over passionless music?

Dubinuschka and (a) Old Forgotten Waltz and (b) Bouran by the A. & P. Gypsies (Brunswick). If there is aught of the spirit of Terpsichore in you these gypsyish rhythms will make you yearn to express yourself in dance. Meaning: our grading of this offering—50 per cent.

Traumerei (Schuman) and Mazurka in A Minor (Chopin-Kreisler). By Max Rosen (Brunswick). Adequate violin solos unemotionally delivered.

Don't Miss These New Records

Andante Cantabile · (Tschaikowsky) and Theme and Variations (Haydn) played by the Elman String Quartet (Victor).

Pagliacci—Vesti la Giubba and Pagliacci—No Pagliacci Non Son! (Leoncavallo) sung by Giovanni Martinelli (Victor).

Cradle Song (Brahms-Grainger) and Molly on the Shore (Grainger) played by Percy Grainger (Columbia).

Voices of Spring and Enjoy Your Life (Strauss) played by Johann Strauss and Symphony Orchestra (Columbia).

Dubinuschka and (a) Old Forgotten Waltz and (b) Bouran by the A. & P. Gypsies (Brunswick).

'S Wonderful and My One and Only (Gershwin) by the Ipana Troubadours and Clicquot Club Eskimos respectively. (Columbia).

My Heart Stood Still and I Feel at Home With You by George Olsen (Victor).

I Live, I Die For You and Eyes That Love by the Troubadours (Victor).

Beautiful Ohio and Missouri Waltz by Paul Whiteman and His Orchestra (Victor).

A Shady Tree and Dancing Tambourine by Paul Whiteman and His Orchestra (Victor).

WE ARE beginning to comprehend vaguely the extent of the phonograph industry. That we had not done so before is due to the fact that we never could visualize figures. Units, tens, and hundreds we can manage very well but when the thousand mark has been passed our brain reels, and the very numbers jump before our eyes. And so, although we knew that some sixteen hundred recordings were made annually by the Victor, Brunswick, and Columbia companies, we were not impressed because the figure was meaningless. Now we have a dim idea. An average of thirty-four records a month have been reviewed in this department for the last four months. Our statistical department reports that this totals one hundred and thirty-six records. These records occupy a considerable portion of our apartment, to be exact, a couch, a large mahogany office desk, one stool, and three chairs, not to mention the overflow on the floor. Walking has become dangerous and sitting is well-nigh impossible. In another four months the records will have reached the kitchen and we will be forced to take our meals out. If we ever review all the records each month we will move into Carnegie Hall. Nice little industry! .

Many of these we could lose without a tear. Then again there are those we will cherish forever. Already we have formed a permanent attachment for some of this month's supply: two selections from Il Pagliacci sung by Giovanni Martinelli, a Percy Grainger record, two delightful numbers by the Elman String Quartet, two old and one new waltz from the Whiteman organization, and several better-than-usual dance numbers by the usual dance orchestras. These have gone into our library. Into the ash can we would like to put a Ted Lewis record and an Al Jolson song. The rest are chiefly dance records which will provide good entertainment for the moment.

We welcome the appearance of eight waltz numbers. We hope that means that the waltz is coming back but there have been so many false alarms already that we refuse to send out searching parties for our old waltz partners, yet. In the meantime we waltz alone in the privacy of our home.

More or Less Classic

Andante Cantabile (from String Quartet, Op. 11, by Tschaikowsky) and Theme and Variations (from The Emperor Quartet by Haydn). By the Elman String Quartet (Victor). Delicate chamber music exquisitely played by Mischa Elman, Edward Bachmann, William Schubert and Horace Britt. Both performances are richly colored by the beautiful tone of the Elman violin.

Pagliacci—Vesti la Giubba (Leoncavallo) and

441

Cradle Song (Brahms--Grainger) and *Molly on the Shore* (Grainger). By Percy Grainger (Columbia). To realize how thrillingly alive piano music can be one should hear the vibrant beauty of Grainger's rendition of *The Cradle Song. Molly on the Shore* is the familiar Irish reel, jovially played by its composer.

La Boheme: Musetta's Waltz Song (Puccini) and *Mignon: Connais-tu le Pays?* (Thomas). By Maria Kurenko (Columbia). One moment we like this soprano voice exceedingly and the next it develops a harsh pinched nasal quality which is most unpleasant. In spite of this shortcoming we liked *Mignon*.

Voices of Spring and *Enjoy Your Life* (Strauss). By Johann Strauss and Symphony Orchestra (Columbia). Strauss waltzes beautifully played. Need we say more?

Do You Call That Religion and *Honey* by the Utica Institute Jubilee Singers (Victor). Two of the best songs in the repertory of this Negro quartet, sung with the subtle harmony which only Negro voices can achieve.

"Popular" and Such

'S Wonderful by the Ipana Troubadours and *My One and Only* by the Clicquot Club Eskimos (Columbia). *'S Wonderful* now holds first place in our own personal Best Number of the Year Contest. It is a swell Gershwin song and the Troubadours have done it full justice. The Eskimos were not quite as successful with the other Gershwin number but it is worth honorable mention.

My Heart Stood Still and *I Feel at Home With You* by George Olsen (Victor). The A side is runner-up in our contest but the B is a come-down.

(Note: If you want to be a social success you can't afford to be without both the above-mentioned records.)

Together, We Two and *Give Me a Night in June* by Johnny Johnson and His Statler Pennsylvanians (Victor). Despite their age these two numbers remain vigorous, due to the excellent Johnson rejuvenation.

A Shady Tree and *Dancing Tambourine* by Paul Whiteman and His Orchestra (Victor). Your neighbors will cry for, not at, this record. The waltz with its haunting melody is our favorite.

Beautiful Ohio and *Missouri Waltz* by Paul Whiteman and His Orchestra (Victor). Beautiful revivals of the fittest.

I Live, I Die For You and *Eyes That Love* by the Troubadours (Victor). Both these numbers from "The Love Call" have good tunes as backgrounds. Vocal refrains by Lewis James help put them across.

There's One Little Girl Who Loves Me by the Ipana Troubadours and *What'll You Do?* by Leo Reisman and His Orchestra (Columbia). Two melodious dance numbers with a good chorus by Scrappy Lambert in the first.

'S Wonderful and *Funny Face* by Bernie Cummins and His Orchestra (Brunswick). This orchestra unfortunately misses most of the Gershwin subtlety and messes up the Gershwin time, but they can't completely ruin either of the songs.

I'm in Heaven When I See You Smile—Diane and *Worryin'* by the Regent Club Orchestra (Brunswick). Two good languorous waltzes with old fashioned whistling effects.

The Hours I Spent With You and *An Old Guitar* and *An Old Refrain* by Roger Wolfe Kahn and His Orchestra with vocal refrains by Franklyn Baur (Victor). The first is a fair waltz. The second is called a fox trot but it cries out to be tangoed to!

Up in the Clouds and *Thinking of You* by Nat Shilkret and the Victor Orchestra (Victor). Hot and snappy in the usual Shilkret manner.

There's a Cradle in Caroline by Nat Shilkret (Victor). Why didn't Mrs. Victor let Shilkret show the rest of them how to do it at the beginning? *The Song is Ended* by George Olsen and His Music. A good interpretation of a good waltz with a vocal chorus that's terrible!

Down the Old Church Aisle and *Is Everybody Happy Now?* by Ted Lewis and His Band (Columbia). The first number stirs unpleasant memories. Has Ted Lewis been robbing the song cemetery? If so, he'd better replace the corpse. And, oh, Mister Lewis! lay the second number beside the first, while you're at it.

From Saturday Night Till Monday Morning and *She'll Never Find a Fellow Like Me* by Ted Weems and His Orchestra (Victor). At last, a new idea in lyrics! *And* a catchy tune well played. We refer, of course, to the first number; the second is just a really good song on the old, old idea.

Dear, On a Night Like This by Cass Hagan and His Park Central Hotel Orchestra and *I'll Think of You* by Al Lentz and His Orchestra (Columbia). Two smooth, gliding fox trots, if you know what we mean.

Thinking of You and *Up in the Clouds* by Harry Archer and His Orchestra (Brunswick). Direction without enthusiasm.

Where Is My Meyer? by Nat Shilkret and the Victor Orchestra and *Blue Baby* by George Olsen and His Music (Victor). Fast-moving numbers handled by experts.

Make My Cot Where The Cot-Cot-Cotton Grows and *Sugar* by Red Nichols' Stompers (Victor). Why, this orchestra must have been up all night! Or, how do you explain the monotony?

Wherever You Are and *Headin' For Harlem* by Nat Shilkret and the Victor Orchestra (Victor). Franklyn Baur helps the orchestra make the best of two fair numbers.

Worryin' by Don Voorhees and His Orchestra and *Where in the World* by The Cavaliers (Columbia). If they got rid of their worries they might play better, or perhaps it's the song. The other number is not much better.

The Song Is Ended by the Columbians (Columbia). "But the melody lingers on." And why not? It's a good one and well treated by the Columbians. *There Must Be Somebody Else* by the Radiolites. Nice orchestration and a good vocal chorus by Scrappy Lambert, formerly one half of the Trade and Mark combination.

Mother of Mine, I Still Have You and *Blue River* by Al Jolson and William F. Wirges and His Orchestra (Brunswick). Just your mother's boy, aren't you, Al?

Two Black Crows, Parts 5 and 6, by Moran and Mack (Columbia). More an' more Moran and Mack.

Good Records of Operas You Have Heard

DURING the current radio season, parts of many great and popular operas have been heard in the Balkite Hour, relayed from Chicago with the Chicago Civic Opera Company. And on the N. B. C. Networks, many well-liked operas have been done in tabloid form by the National Grand Opera Company. New electrical recordings of some of the most popular operas are offered by the leading phonograph companies. Some of these listed below are new, some not so new, but all are excellent and worth adding to one's collection.

Aida (Verdi)

Celeste Aida	Giovanni Martinelli	Victor
Celeste Aida	Ulysses Lappas	Columbia
Ritorna vincitor }	Elisabeth Rethberg	Brunswick
O patria mia }		
La fatal pietra }	Ponselle-Martinelli	Victor
Morir! si pura e bella! }		
Grand March	Columbia Symphony Orchestra	Columbia
Nel fiero anelito }	G. Arangi-Lombardi and	Columbia
O terra addio }	Francesco Merli	

Faust (Gounod)

Air des Bijoux	Edith Mason	Brunswick
Le Roi de Thule	Florence Easton	Brunswick
Parlate d'amore	Margarete Matzenauer	Victor
Ballet music (four parts on two records)	Sir H. J. Wood and the New Queen's Hall Orchestra	Columbia
Soldiers' Chorus	Victor Male Chorus	Victor
Serenade Mephistopheles	Marcel Journet	Victor
Duet from Garden Scene	Vessella's Italian Band	Brunswick

Il Pagliacci (Leoncavallo)

Prologo, Si puo }	Lawrence Tibbett	Victor
Prologo, Un nido di memorie }		
Selections	Creatore's Band	Victor
Ballatella—"Che volo d'augelli"	Florence Easton	Brunswick
No Pagliacci non son! }	Giovanni Martinelli	Victor
Vesti la Giubba }		

La Traviata (Verdi)

Di Provenza il mar	Giuseppe Danise	Brunswick
Prelude	Capitol Grand Orchestra (Mendoza conducting)	Brunswick

RADIO BROADCAST Photograph

A SET-UP OF APPARATUS FOR MEASURING CHARACTERISTICS OF A AND B UNITS
It is not a difficult matter to measure the voltage output of A and B devices at different loads—This article
explains. More complicated equipment is necessary, however, to determine the amount of hum in the output

Testing A and B Power Units

By Howard E. Rhodes

Laboratory Staff

THE testing of radio power-supply devices sent to RADIO BROADCAST by manufacturers has, for some time, been an important part of the Laboratory's work. What these tests are, how they are conducted, and what apparatus is used to make them, should be of general interest to our readers, and it is the purpose of this article to explain the procedure adopted for these tests. The information given here will also be helpful to manufacturers who, perhaps, contemplate sending power units to our Laboratory and are therefore interested to know to what tests their devices will be subjected. The tests described are applicable to either A power or B power units and the apparatus used in the tests is illustrated in the photograph at the head of this article.

It was the desire of the Laboratory staff to make the tests such that the data obtained would be most useful from the standpoint of the *user* of the device. With this point in mind the following tests were decided upon:

(a.) Determination of the maximum output of the device at various current drains.
(b.) Determination of the amount of hum in the output at various current drains.
(c.) Determination of the cost of operating the unit.

With this information available we can determine whether a device is capable of supplying sufficient current at the correct voltage for the operation of any particular receiver, whether or not the device has a good filter system in it (determined by the amount of hum in the output), and how much it will cost to operate any receiver from a particular power unit. In the following paragraphs we will explain how these tests are made. Although they will be explained separately, all the tests are made at the same time in the Laboratory.

DETERMINING THE OUTPUT

THE circuit used in determining the output of a unit at various current drains is given in Fig. 1. If the unit under test is an A unit, then the resistance R consists of a heavy-duty Carter rheostat with a maximum resistance of 6

ohms so that with the resistance all in the load of the unit will be about 1.0 ampere, which will be indicated in the ammeter A. The voltmeter, V, used to measure the output voltage, may be a Weston Model 301 meter with a maximum reading of 10 volts. By moving the arm on the rheostat the load may be varied so that the A unit is placed under actual working conditions, the load (read on the ammeter) corresponding

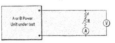

FIG. 1

to what would be drawn by the tube filaments were the A unit to be actually connected to the A posts of a set. At each setting of the resistance R, used to vary the load on the A unit under test should be so adjusted that the voltage on V reads six, which is the value that the unit will be called upon to deliver under actual conditions of operation. At a certain reading of the ammeter it will be found that the voltage shown by the voltmeter is not as high as six, indicating that the A device is being overloaded. The maximum current output of an A device at rated voltage, which is six, can therefore be determined by setting the voltage control knob on the device at maximum (which will boost up the voltage to a figure above six at low values of current) and adjusting the resistance R until the voltmeter reads just six. The ammeter reading then represents the maximum permissible drain of the unit. When the A device is being used in con-

junction with a receiver consuming less than this maximum current output, the voltage control on the device is of course turned to a lower tap, otherwise the voltage output will be excessively high. Data of this kind obtained on three A power units recently tested in the Laboratory are given in the second and third columns of Table 1.

From these data we are able to determine whether an A power unit is capable of supplying filament current to any particular receiver, provided we know the filament current drain of the receiver. Since this merely depends upon the number and type of tubes used in the set, it is easily determined.

If the unit being tested is a B power device, the same circuit is used but instead of the rheostat there is used a variable high resistance—a power Clarostat. The meter M becomes a 0-100 milliammeter and the voltmeter is generally a Westinghouse high-resistance meter with a maximum reading of 250 volts and a resistance of 1000 ohms per volt. Some sample data on four B power units are given in the columns of Table 2.

If we know the total plate-current drain of a receiver we can easily determine from the figures given in Table 2 the maximum voltage the various units will supply at this load. For example if a receiver uses a 171 type tube in the output, on the plate of which we desire to place 180 volts

UNIT No.	LOAD IN AMPERES	VOLTS OUTPUT	WATTS INPUT	HUM VOLTAGE	PER CENT. HUM
1	0	7.1	20	Too small to Measure	
	0.5	6.0	25		
	1.0	6.0	26		
	2.0	4.3	40		
	3.0	4.2	54		
2	0	8	18	0.015	0.187
	1	6	28	0.015	0.25
	2	6	44	0.015	0.25
	3	6	62	0.015	0.25
3	.6	60	0.007	0.12	
	2	.6	71	0.007	0.12
	2.7	.6	101	0.017	0.17
	3	5.2	100	0.005	1.1
	3.5	2.8	100	0.45	7.5

TABLE 1

443

and the total plate current drawn by the receiver is 50 milliamperes, then unit No. 3, supplying only 120 volts at this current drain, would not be satisfactory. Units Nos. 1 and 2 would be more satisfactory as they deliver considerably higher voltage at the current drain specified. Although they do not supply quite as much as 180 volts, a matter of 20 or 30 volts less than 180 on the plate of the power tube does not make a difference sufficient to be noticeable in the output of the loud speaker.

These data also give us the "regulation" of the unit, which is generally specified as the voltage drop per milliampere of load. For example, taking the following data from unit No. 1, Table 2:

Load mA.	Voltage
10	216
40	182
Difference 30	34

and dividing the difference in the voltages by the difference in the loads, we obtain a value of 1.13, which is the voltage drop per milliampere load. This is quite a good value for the regulation. Compare it with the value obtained from unit No. 3, which figures out to be 4.3. Power units with good regulation have the advantage that the voltage they deliver will be more nearly constant at all loads.

Hum

IT IS the function of the filter system in an A or B power unit to filter the output of the rectifier so that the output at the end of the filter system will be as free as possible of any hum or "ripple." Even from a comparatively poorly designed power unit the hum is too small to measure directly. Consequently, it was necessary to construct an amplifier for this test so that the hum voltage could be amplified sufficiently so as to be readily measured. A three-stage resistance-coupled amplifier is being used in the Laboratory for this purpose. Two 240 type tubes are used in the first and second stages and a 201-A type tube is used in the last stage. The circuit diagram is given in Fig. 2.

When an A or B power unit is to be tested for hum the input of the amplifier is connected to the power unit under test, and switch S_1 is thrown to point A. This causes the hum voltage from the power unit to be impressed across the input of the amplifier (note that the d.c. voltage is blocked by the 0.01-mfd. condenser). The amplified hum causes the plate current of the last tube in the amplifier to increase and this increase is indicated by the meter M in the plate circuit. The gain control (a 0.5-megohm po-

tentiometer across the input of the amplifier) is then adjusted so that the meter M gives a deflection that is easy to read. The switch is then thrown to the B position which connects the input of the amplifier to a source of known 60-cycle a.c. voltage the value of which is variable, as will be explained below. The 60-cycle voltage is so adjusted that the reading of the meter in the plate circuit of the output tube is the same as it was when the amplifier was connected to the power unit, and in which case this value of 60-cycle voltage is then equal to the hum voltage impressed upon the input of the amplifier.

Using this method (of connecting to the input of the amplifier a known voltage equal in value to the unknown voltage) makes unnecessary the calibration of the amplifier. It is necessary, however, to have available a source of 60-cycle voltage from which voltages can be obtained comparable in value to the hum voltages ordinarily obtained from radio power units. These voltages, which are around 0.01 volt in the case of a poor unit, can be obtained using the circuit indicated in Fig. 2 as: "Source of known voltage."

The transformer T in this circuit is an ordinary one designed to supply voltages to a.c. tubes. The 1.5-volt winding is used and across its terminals is connected a 6-ohm rheostat with an additional connection soldered to the free end of the resistance wire so that the rheostat might be used as a potentiometer. The voltage across the voltmeter can be adjusted to any value, between 0 and 1.5 volts, by means of the sliding contact on the rheostat. Across the voltmeter are connected, in series, a 4-ohm and two 2-ohm resistances, these resistances constituting a voltage divider the effect of which is to extend the voltmeter range downwards. Connections from these resistances are brought out to four pin jacks marked 1, $\frac{1}{4}$, $\frac{1}{2}$, and 0, indicating the portion of the voltage associated with the particular pin jack. Thus, if P is connected to the jack marked $\frac{1}{2}$, the actual voltage impressed across X-Y is only one half of that indicated by the meter, etc. To the pin P is connected one end of a 400-ohm calibrated potentiometer, the purpose of which is to subdivide the voltmeter readings to even smaller fractions than is possible with the other resistances. By means of these adjustable units it is possible to impress across the input of the amplifier any voltage from 1.5 volts down to about 0.005 volts with an accuracy of not less than about 90 per cent., which is sufficiently accurate for measurements of this type.

Some examples of measurements of this sort are given in columns 5 and 6 of Tables 1 and 2. Column 5 gives the value of 60-cycle voltage that is equal to the hum voltage. Column 6 gives the per-

Unit No.	Load in Milliamperes	Volts Output	Watts Input	Hum Voltage	Per Cent. Hum
1	0	243	9	0.01	0.004
	10	216	11	0.03	0.014
	20	205	14	0.02	0.01
	30	193	16	0.02	0.01
	40	182	19	0.05	0.027
	50	170	22	0.05	0.03
2	0	215	34	0.075	0.035
	10	204	35	0.067	0.033
	20	193	37	0.067	0.034
	40	170	40	0.067	0.039
	50	164	44	0.067	0.041
3	0	270	88	0.27	0.1
	10	250	91	0.26	0.105
	30	210	93	0.25	0.12
	40	145	96	0.315	0.31
	50	120	98	0.165	0.14
4	0	240	17	0.14	0.058
	10	230	18	0.15	0.068
	20	195	20	0.17	0.087
	40	150	24	0.23	0.153
	50	135	26	0.28	0.210

TABLE 2

centage hum in terms of the d.c. voltage output of the device. Power unit No. 1 (Table 1) shows the smallest amount of hum in the output but it could not deliver more than one ampere at 6 volts. Unit No. 3 has more hum voltage than unit No. 1, but is capable of supplying up to 2.7 amperes at 6 volts.

The hum voltage measurements given in Table 2 give some idea of the magnitude of hum voltage obtained from present-day B power units. It is possible to make some interesting mathematical calculations regarding the hum in B power units to indicate the effect in the loud speaker of various values of hum voltage, and this subject will be discussed in an early issue of RADIO BROADCAST. There is room here only to point out briefly the salient points regarding the matter. It can be shown that with a given audio amplifier, capable of amplifying down to 60 cycles, the permissible amount of hum in the output of a B power unit decreases:

(a.) as the amount of audio amplification in the receiver is increased and (b.) as the voltages on the various tubes, especially the detector tube, is increased.

Just how much hum is permissible is, of course, a function of the amount of amplification the audio amplifier gives at the hum frequency and how well the loud speaker will respond to these frequencies and their harmonics.

COST OF OPERATION

THE cost of operation per hour of an A or B power unit is found by first determining the amount of power the device consumes in watts when supplying the receiver, multiplying this power by the cost of power per kilowatt hour, and dividing by 1000. The cost of operation per month can, of course, be found by multiplying the cost per hour by the number of hours the set is in use per month. An example will make the whole calculation clearer. Suppose that a receiver drawing 40 milliamperes of plate current is operated from B power unit No. 4, Table 2, and that the set is in operation on an average of three hours a day. What will be the cost of operation per month, if the cost of power is $0.10 per kilowatt hour? From Table 2 we know that this particular B power unit draws 24 watts of power when supplying 40 milliamperes. Following the information given at the beginning of this paragraph, we multiply 24 by 0.10 and then divide by 1000. This gives $0.0024 as the cost of operation per hour. Since the set is in use 90 hours per month, then the cost per month is 0.21\frac{1}{2}$. Other examples can be worked out in the same manner.

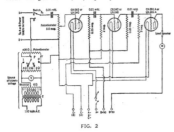

FIG. 2

The *Radio Broadcast*
LABORATORY INFORMATION SHEETS

THE Radio Broadcast Laboratory Information Sheets are a regular feature of this magazine and have appeared since our June, 1926, issue. They cover a wide range of information of value to the experimenter and to the technical radio man. It is not our purpose always to include new information but to present concise and accurate facts in the most convenient form. The sheets are arranged so that they may be cut from the magazine and preserved for constant reference, and we suggest that each sheet be cut out with a razor blade and pasted on 4" x 6" filing cards, or in a notebook. The cards should be arranged in numerical order. In July, 1927, an index to all Sheets appearing up to that time was printed.

All of the 1926 issues of Radio Broadcast are out of print. A complete set of Sheets, Nos. 1 to 88, can be secured from the Circulation Department, Doubleday, Doran & Company, Inc., Garden City, New York, for $1.00. Some readers have asked what provision is made to rectify possible errors in these Sheets. In the unfortunate event that any serious errors do occur, a new Laboratory Sheet with the old number will appear.

—THE EDITOR.

| No. 177 | Radio Broadcast Laboratory Information Sheet | April, 1928 |

Characteristics of Speech

ARTICULATION

CLEAR speech is only possible when the person speaking uses careful articulation. Articulation is especially important in radio for if we do not understand something, we cannot have it repeated. In analyzing speech sounds a clear understanding of how the various sounds are produced is essential.

The human voice consists of sustained and transient notes and noises. The sounds which are ordinarily difficult to recognize (and which therefore require careful articulation, are the transients such as are associated with the sounds "t" and "d" or "p" and "b." These sounds are hard to reproduce accurately for they contain many of the highest frequencies found in sounds of speech.

If we examine the manner in which the sounds "p" and "b," for example, are produced, they will be found to have much in common. They are both produced by first compressing the lips together and then rapidly opening them. To pronounce the word "pa," we first produce the "p" sound by suddenly opening the lips and permitting the air to rush through them and then the vocal chords are set in motion to produce the vowel sound "a." The sylla-

ble "ba," is produced with a very similar motion of the lips but the vocal chords are set into motion and the lips open at the same instant and also there is only a slight rush of air from between the lips. The "pa" sound is characterized by an initial sound of high intensity; the "ba" sound does not have this feature. If the radio loud speaker cannot reproduce accurately the strong portion of the former sound, "pa," it will sound very much like "ba."

Some of the sounds most difficult to reproduce accurately are noises such as the dropping of a book on a table, for these sounds contain frequency components extending throughout the entire range of audible sounds.

The study of how words are formed is very interesting and can best be done with the aid of an oscillograph, which is an instrument with which we can obtain photographic records of the wave form associated with any sound. An analysis of these records, which are sometimes termed "audiograms," is helpful in determining the range of frequencies which must be handled by a radio broadcasting system if the reproduction is to sound natural.

| No. 178 | Radio Broadcast Laboratory Information Sheet | April, 1928 |

The Exponential Horn

THE CUT-OFF FREQUENCY

THE LOWEST frequency transmitted by a horn of the exponential type is determined by the rate of expansion of the cross sectional area of the horn, and to eliminate reflection the diameter of the mouth of the horn (if it is round) must be made equal to one-quarter of the wavelength corresponding to the lowest frequency to be transmitted.

The velocity of sound in air is 1120 feet per second and, therefore, the wavelength (in feet) corresponding to any particular frequency may be found by dividing 1120 by the frequency. The diameter of the mouth of the horn in feet must then be equal to this wavelength divided by 4.

The accompanying curve shows graphically the relation between the diameter of the mouth of a round horn and the cut-off frequency. It should be realized that the diameter of the mouth is not the only factor determining the lowest frequency that the horn will satisfactorily transmit and that the size of the mouth is an indicator of this frequency only if the remainder of the horn has been correctly designed. As shown by the curve, to transmit frequencies down to 64 cycles, for example, it is necessary that the horn's mouth have a diameter of about 4.5 feet.

If the horn is square rather than round, it will be satisfactory to make the area of the mouth equal to that of the equivalent round horn.

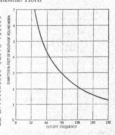

No. 179　　RADIO BROADCAST Laboratory Information Sheet　　**April, 1928**

A Problem in Audio Amplification

THE EFFECT OF TRANSFORMER RATIO

PROBLEM:—The audio amplifier in a receiver
comprises a 3:1 transformer in the detector
circuit, followed by a 201-a type tube in the first
audio stage, a 4:1 audio transformer for the second
stage, and a power output tube. What will be the
effect on the amount of signal voltage supplied to
the grid of the power tube of substituting a 6:1
transformer for the 4:1 transformer?

ANSWER:—Let us first calculate the gain of the
original amplifier. The total amplification to the
grid of the power tube will be equal to the turns
ratio of the first transformer multiplied by the effective amplification of the tube times the turns ratio
of the second transformer. The effective amplification of a tube in a properly designed transformer-
coupled audio amplifier can be taken as about 80
per cent. of the amplification constant of the tube;
for a 201-a type tube, therefore, we take 80 per cent.
of 8, which is 6.4. The total gain of the amplifier is,
therefore:

$$3 \times 6.4 \times 4 = 76.8$$

Similarly the amplification with the 6:1 transformer
substituted for the 4:1 will be:

$$3 \times 6.4 \times 6 = 115.2$$

The substitution of the 6:1 transformer, therefore,
has increased the voltage gain by 50 per cent.; this
represents a gain of 3.6 TU.

Now, the power into the loud speaker is proportional to the square of the signal voltage on the
grid of the power tube feeding the loud speaker.
When the voltage gain is increased 50 per cent.,
therefore, the power into the loud speaker is increased 125 per cent. This corresponds to a power
gain of 3.5 TU which, while not very great, is appreciable. (The minimum gain audible to the ear is
1 TU.)

If the power tube is a 171 type with 40 volts on
the grid, then using the original amplifier, approximately 0.5 volts (40 divided by 76.8) are required
out of the detector tube in order to place 40 volts
signal voltage on the grid of the 171. When the
6:1 transformer is used, only 0.3 volts (40 divided
by 115.2) are required from the detector in order to
"load up" the power tube.

No. 180　　RADIO BROADCAST Laboratory Information Sheet　　**April, 1928**

B Power Unit Characteristics

EFFECT OF TRANSFORMER VOLTAGE

THE CURVES published herewith were made
by the Raytheon Manufacturing Company
using one of their BH type tubes with an ordinary
filter, as indicated in the accompanying circuit
diagram. The curves show the relation between
the voltage, E$_t$, across the secondary of the transformer and the input voltage, E$_b$. The output load
in milliamperes as measured by the meter I is
plotted along the horizontal axis and along the
vertical axis has been plotted the difference between the effective value of the transformer voltage
E$_t$ and the average value of the voltage E$_b$ into
the filter system. The line marked +20, for example, indicates the voltage E$_t$ to be 20 volts
greater than E$_b$; the line marked +20 indicates the
converse.

These curves show that (to take an example)
with a transformer voltage of 300 volts per anode,
the average value of the voltage into the filter is 27
volts higher than the transformer voltage when the
load is 10 milliamperes. At a load of 28 milliamperes
the voltages are equal and at a load of 60 milliamperes the input voltage to the filter has dropped to
a value 25 volts below the transformer voltage.
During these tests the transformer voltage, E$_t$, was
held constant.

Other data showing the effect of various trans-

former voltages, obtained with the same circuit
used here, were given on Laboratory Sheet No. 146,
published in the December, 1927, RADIO BROADCAST.

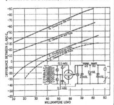

No. 181　　RADIO BROADCAST Laboratory Information Sheet　　**April, 1928**

R. F. vs. A. F. Amplification

A COMPARISON

THE SIGNAL output from a radio receiver may
be increased by augmenting either the audio-
frequency or radio-frequency amplification or by
boosting the detecting efficiency. On this Laboratory Sheet we give briefly the comparative merits
of audio-frequency and radio-frequency amplification. In the accompanying table is shown the
effect on the power into the loud speaker of increasing the a.f. or r.f. amplification. The first column
gives the increase in amplification and the second
column the increase in power into the loud speaker
if this extra amplification is introduced in the audio
amplifier. The third column shows the increase in
power into the loud speaker if the extra amplification is placed in the r.f. amplifier.

This table is based on the fact, first, that the
power into the loud speaker is proportional to the
square of the voltage on the grid of the power tube

and, secondly, that the output of the detector is
proportional to the square of the voltage on its
grid. When the audio-frequency amplification is
multiplied by 10, for example, the power into the
loud speaker is 100 times greater. When the radio-
frequency amplification is multiplied by 10, however, the output of the detector is 100 times greater,
and the power into the loud speaker is 10,000 times
greater. It is evident from these figures, therefore,
that increases in r.f. gain are much more effective in
producing greater signal than increases in audio-
frequency gain.

| Added | Increase in Power into Loud Speaker | |
Amplification	A. F.	R. F.
2	4	16
5	25	625
10	100	10,000
20	400	160,000
50	2500	6,250,000

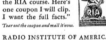
No. 182   RADIO BROADCAST Laboratory I

Filter Conder

HOW TO CONNECT THEM IN SERIES

IF WE desire to place a filter condenser across, for example, a 1000-volt source of direct current, and we have available two large-capacity 500-volt condensers, it is ordinarily not possible to compact them in series across the 1000-volt leads with safety. Why this is so will be explained on this Laboratory Sheet.

At A in the diagram is shown the connection of two condensers, C_1 and C_2, in series across the 1000-volt source. Now, a condenser has a definite d.c. resistance, which is generally very high but nevertheless finite, and this resistance is represented as R_1 and R_2 in B as external resistances across each condenser. A small amount of current will flow through these resistances and the voltage drop across the two resistances will be in direct proportion to the resistances. The resistances of condensers vary widely and therefore it is extremely unlikely that we would have two condensers with the same d.c. resistance. For example, condenser C_1 might have a d.c. resistance of 100 megohms while the d.c. resistance of condenser C_2 might be 900 megohms. The d.c. voltage drops across the two condensers being proportional to the resistances there would then be 100 volts across C_1 and 900 volts across C_2. If the two condensers were both rated at 500 volts, the obvious result would be that condenser C_1 would

No. 183   RADIO BROADCAST Laboratory I

The Type 280 and

THEIR CHARACTERISTICS

THE characteristics of the type 280 and 281 rectifier tubes are given below. These tubes are for use as rectifiers in B power units, the 280 in circuits designed for full-wave rectification and the 281 in half-wave circuits. Two 281 tubes may be used, if desired, to give full-wave rectification:

TYPE 280 FULL-WAVE RECTIFIER

Filament Voltage	5 Volts
Filament Current	2 Amperes
A.C. Plate Voltage (Max. Per Plate)	300 Volts
Max. D.C. Output Current	125 Milliamperes
Max. D.C. Output Voltage	260 Volts
Height of Tube	5⅛ Inches
Diameter of Tube	2⁵⁄₁₆ Inches

TYPE 281 HALF-WAVE RECTIFIER

Filament Voltage	7.5 Volts
Filament Current	1.25 Amperes
A.C. Plate Voltage (Max.)	750 Volts

No. 184   RADIO BROADCAST Laboratory I

Tuning

THE EFFECT OF DISTRIBUTED CAPACITY

A RADIO receiver to cover the broadcasting band must be able to tune-in signals from 550 kc. to 1500 kc., a ratio of 2.73 to 1 in frequency. It can be shown mathematically that, in order to obtain this range, the ratio of the maximum to minimum of the capacity across a tuning coil must be 8.6 to 1 approximately. If we use a tuning condenser with a maximum capacity of 0.0005 mfd. then the minimum capacity across the coil must theoretically be (if the desired tuning range is to be obtained) 0.0005 (divided by 8.6, or 0.000058 mfd. An ordinary condenser might have a minimum capacity of about 0.000025 mfd. and, therefore, it

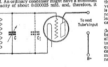

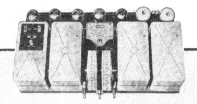

CPSIA information can be obtained
at www.ICGtesting.com
Printed in the USA
BVHW04s2156300918
528774BV00055B/844/P